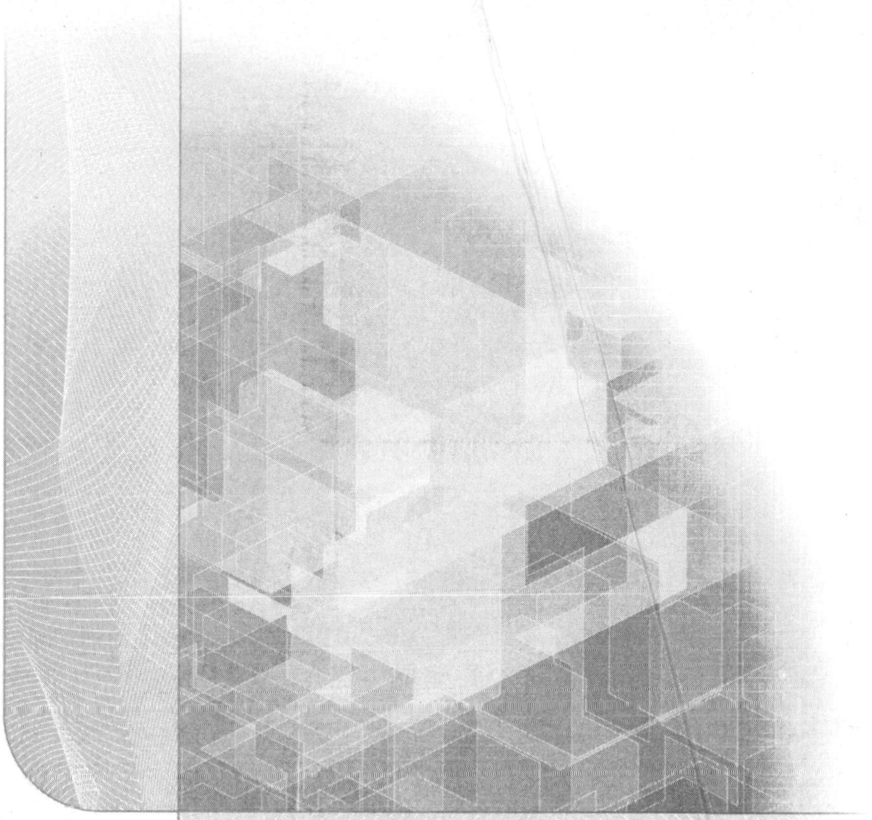

AutoCAD Tutor for Engineering Graphics: 2013 and beyond

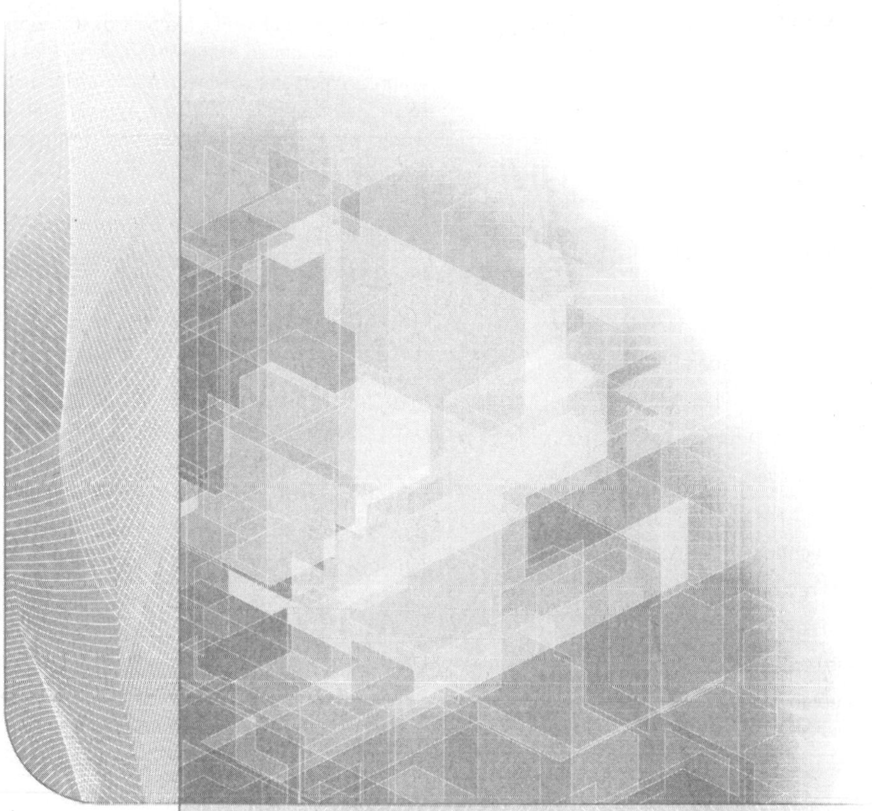

AutoCAD Tutor for Engineering Graphics: 2013 and beyond

KEVIN LANG

Dedication Alan Kalameja
1954–2010
Lifelong Learner, Teacher, and our Friend

DELMAR
CENGAGE Learning®

Australia • Brazil • Japan • Korea • Mexico • Singapore • Spain • United Kingdom • United States

DELMAR
CENGAGE Learning

AutoCAD Tutor for Engineering Graphics: 2013 and beyond

Kevin Lang

Vice President, Editorial: Dave Garza
Director of Learning Solutions: Sandy Clark
Acquisitions Editor: Stacy Masucci
Managing Editor: Larry Main
Senior Product Manager: John Fisher
Editorial Assistant: Kaitlin Murphy
Vice President, Marketing: Jennifer Baker
Marketing Director: Deborah Yarnell
Marketing Manager: Jillian Borden
Senior Production Director: Wendy Troeger
Production Manager: Mark Bernard
Content Project Manager: James Zayicek
Production Technology Assistant: Emily Gross
Senior Art Director: David Arsenault
Technology Project Manager: Joe Pliss
Cover image: Kosarev Alexander/Shutterstock.com
Interior Graphics: Dave Gink/Shadowland Studios

Library of Congress Control Number: 2012940580
ISBN-13: 978-1-1339-6039-3
ISBN-10: 1-1339-6039-1

Delmar
5 Maxwell Drive
Clifton Park, NY 12065-2919
USA

Cengage Learning is a leading provider of customized learning solutions with office locations around the globe, including Singapore, the United Kingdom, Australia, Mexico, Brazil and Japan. Locate your local office at: **international.cengage.com/region**

Cengage Learning products are represented in Canada by Nelson Education, Ltd.

To learn more about Delmar, visit **www.cengage.com/delmar**
Purchase any of our products at your local college store or at our preferred online store **www.cengagebrain.com**

Printed in the United States of America
1 2 3 4 5 6 7 16 15 14 13 12

CONTENTS

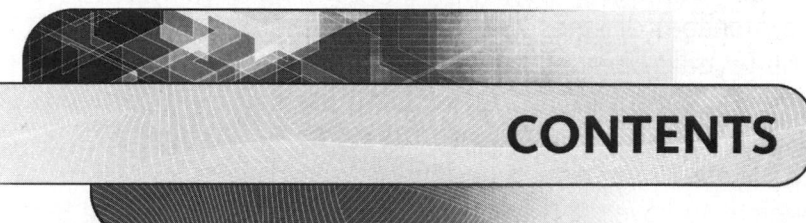

CHAPTER 8 MULTIVIEW AND AUXILIARY VIEW PROJECTIONS 387

CHAPTER 9 CREATING SECTION VIEWS 443

CHAPTER 10 ADDING DIMENSIONS TO YOUR DRAWING 476

CHAPTER 11 MANAGING DIMENSION STYLES 528

CHAPTER 12 ANALYZING 2D DRAWINGS 590

CHAPTER 13 CREATING PARAMETRIC DRAWINGS 632

CHAPTER 14 WORKING WITH DRAWING LAYOUTS 665

CHAPTER 15 PLOTTING YOUR DRAWINGS 708

CHAPTER 16 WORKING WITH BLOCKS 728

CHAPTER 17 WORKING WITH ATTRIBUTES 780

CHAPTER 18 WORKING WITH FILE REFERENCES 813

CHAPTER 19 ADVANCED LAYOUT TECHNIQUES 850

CHAPTER 20 SOLID MODELING FUNDAMENTALS 884

CHAPTER 21 CONCEPT MODELING, EDITING SOLIDS, AND SURFACE MODELING 965

INTRODUCTION

Engineering graphics is the process of defining an object graphically before it is constructed and used by consumers. Previously, this process for producing a drawing involved the use of drawing aids, such as pencils, ink pens, triangles, T-squares, and so forth, to place an idea on paper before making changes and producing blue-line prints for distribution. The basic principles and concepts of producing engineering drawings have not changed, even when the computer is used as a tool.

This text uses the basics of engineering graphics to produce 2D drawings and 3D computer models using AutoCAD and a series of tutorial exercises that follow each chapter. Following the tutorials in most chapters, problems are provided to enhance your skills in producing engineering drawings. A brief description of each chapter follows:

CHAPTER 1 — GETTING STARTED WITH AUTOCAD

This first chapter introduces you to the following fundamental AutoCAD concepts: Screen elements and workspaces; use of function keys; opening an existing drawing file; using Dynamic Input for feedback when accessing AutoCAD commands; basic drawing techniques using the LINE, CIRCLE, and PLINE commands; understanding absolute, relative, and polar coordinates; using the Direct Distance mode for drawing lines; using the Object Snap modes, and Polar and Object Snap Tracking techniques; using the ERASE command; and saving a drawing. Drawing tutorials follow at the end of this chapter.

CHAPTER 2 — DRAWING SETUP AND ORGANIZATION

This chapter introduces the concept of drawing in real-world units through the setting of drawing units and limits. The importance of organizing a drawing through layers is also discussed through the use of the Layer Properties Manager palette. Color, linetype, and lineweight are assigned to layers and applied to drawing objects. Advanced Layer tools such as isolating, filtering, and states and how to create template files are also discussed in this chapter.

CHAPTER 3 — AUTOCAD DISPLAY AND BASIC SELECTION OPERATIONS

This chapter discusses the ability to magnify a drawing image using numerous options of the ZOOM command. The PAN command is also discussed as a means of staying in a zoomed view and moving the image to a new location on the screen. Productive uses of real-time zooms and pans along with the effects a wheel mouse has on ZOOM and PAN are included. Object selection tools are discussed, such as Implied Windowing, Noun/Verb Selection, Selection Cycling, and the Quick Select command, to name a few. Finally, this chapter discusses the ability to save the image of your display and retrieve the saved image later through the View Manager dialog box.

CHAPTER 4 — MODIFYING YOUR DRAWINGS

This chapter is organized into two parts. The first part covers basic modification commands: MOVE, COPY, SCALE, ROTATE, OFFSET, FILLET, CHAMFER, TRIM, EXTEND, and BREAK. The second part covers advanced methods of modifying drawings: ARRAY, MIRROR, STRETCH, PEDIT, EXPLODE, LENGTHEN, JOIN, UNDO, and REDO. Tutorial exercises follow at the end of this chapter as a means of reinforcing these important tools used in AutoCAD.

CHAPTER 5 — PERFORMING GEOMETRIC CONSTRUCTIONS

This chapter discusses how AutoCAD commands are used for constructing geometric shapes. The following drawing-related commands are included in this chapter: ARC, DONUT, ELLIPSE, POINT, POLYGON, RAY, RECTANG, SPLINE, and XLINE. Tutorial exercises are provided at the end of this chapter.

CHAPTER 6 — WORKING WITH TEXT, FIELDS, AND TABLES

Use this chapter for placing text in your drawing. Various techniques for accomplishing this task include the use of the MTEXT and TEXT commands. The creation of text styles and the ability to edit text once it is placed in a drawing are also included. A method of creating intelligent text, called Fields, is discussed in this chapter. Creating tables, table styles, and performing summations on tables are also covered here. Tutorial exercises are included at the end of this chapter.

CHAPTER 7 — OBJECT GRIPS AND CHANGING THE PROPERTIES OF OBJECTS

The topic of grips and how they are used to enhance the modification process for drawings is presented. Grip modes and options are covered, as well as the use of Multi-functional grips. The ability to modify objects through Quick Properties and the Properties Palette are also discussed in detail. A tutorial exercise is included at the end of this chapter to reinforce the importance of changing the properties of objects.

CHAPTER 8 — MULTIVIEW AND AUXILIARY VIEW PROJECTIONS

Describing shapes and producing multiview drawings using AutoCAD are the focus of this chapter. The basics of shape description are discussed, along with demonstrating commands and techniques used for generating orthographic views. Tutorial exercises on creating multiview drawings are available at the end of this chapter segment. This chapter continues by showing how to produce auxiliary views. Items discussed include using the OFFSET and XLINE commands to create true shape projections of inclined surfaces. A tutorial exercise on creating auxiliary views is provided in this chapter segment.

CHAPTER 9 — CREATING SECTION VIEWS

Hatching and hatch editing techniques through the use of the Ribbon's Hatch Creation tab are discussed in this chapter. The ability to apply a gradient hatch pattern is also discussed. Tutorial exercises that deal with the topic of section views follow at the end of the chapter.

CHAPTER 10 — ADDING DIMENSIONS TO YOUR DRAWING

This chapter utilizes various Try It! exercises on how to utilize basic and specialized dimensioning commands to place linear, diameter, radius, and angular dimensions. The powerful QDIM command is also discussed, which allows you to place baseline, continuous, and other dimension groups in a single operation. A tutorial exercise is provided at the end of this chapter.

CHAPTER 11 — MANAGING DIMENSION STYLES

A thorough discussion of the use of the Dimension Styles Manager dialog box is included in this chapter. The ability to create, modify, manage, and override dimension styles is discussed. A detailed tutorial exercise is provided at the end of this chapter.

CHAPTER 12 — ANALYZING 2D DRAWINGS

This chapter provides information on analyzing a drawing for accuracy purposes. The MEASUREGEOM command is discussed in detail, along with the Distance, Radius, Angle, Area, and Volume options. Also discussed are the LIST, ID, and FIELD commands and how they are used to determine the accuracy of various objects in a drawing. A tutorial exercise follows that allows users to test their drawing accuracy.

CHAPTER 13 — CREATING PARAMETRIC DRAWINGS

This chapter introduces the concept of using geometric constraints to create geometric relationships between selected objects. In this chapter, you will learn the constraint types and how to apply them to drawing objects. You will also be shown the power of controlling the objects in a design through the use of parameters. A number of Try It! exercises are available to practice with the various methods of constraining objects. Two tutorials are also available at the end of the chapter to guide you in assigning constraints to objects.

CHAPTER 14 — WORKING WITH DRAWING LAYOUTS

This chapter deals with the creation of layouts, which are utilized for the generation of drawing plots. A layout takes the form of a sheet of paper and is often referred to as Paper Space. Tools for arranging, scaling, and locking viewports are discussed. The creation of multiple layouts for the same drawing is also introduced, including a means of freezing layers only in certain layout viewports. The use of Quick View Drawings and Layouts is also discussed to manage drawing views and layouts. Various exercises are provided throughout this chapter to reinforce the importance of layouts.

CHAPTER 15 — PLOTTING YOUR DRAWINGS

Printing or plotting your drawings is demonstrated through a series of tutorial exercises. One tutorial explains the use of the Add-A-Plotter wizard to configure a new plotter. Plotting from a layout is also demonstrated through a tutorial. This includes the assignment of a sheet size. A tutorial exercise to create a color-dependent plot style is provided. Plot styles allow you to control the appearance of your plot through a series of settings. A final tutorial exercise demonstrates how to plot multiple drawing layouts using the PUBLISH command.

CHAPTER 16 — WORKING WITH BLOCKS

This chapter covers the topic of creating blocks in AutoCAD. Creating local and global blocks for objects such as doors, windows, and electrical symbols will be demonstrated. The Insert dialog box is discussed as a means of inserting blocks into drawings. The chapter continues by explaining the many uses of the DesignCenter. This feature allows the user to display a browser containing blocks, layers, and other named objects that can be dragged into the current drawing file. The use of tool palettes is also discussed as a means of dragging and dropping blocks and hatch patterns into your drawing. This chapter also discusses the ability to open numerous drawings through the Multiple Document Environment and transfer objects and properties between drawings. The creation of dynamic blocks, an advanced form of manipulating blocks, is also discussed, with numerous examples to try out. A tutorial exercise can be found at the end of this chapter.

CHAPTER 17 — WORKING WITH ATTRIBUTES

This chapter introduces the use of attributes in a drawing. Commands such as ATTDEF, ATTDISP, EATTEDIT, and DATAEXTRACTION will be explored to gain a better understanding of attributes. The ATTDEF command is used to define attributes. The ATTDISP command is used to control the display of attributes in a drawing. Once attributes are created and assigned to a block, they can be edited through the EATTEDIT command. Finally, attribute information can be extracted using the DATAEXTRACTION command or Attribute Extraction wizard. Extracted attributes can then be imported into such applications as Microsoft Excel and Access. Various tutorial exercises are provided throughout this chapter to help the user become better acquainted with this powerful feature of AutoCAD.

CHAPTER 18 — WORKING WITH FILE REFERENCES

The chapter begins by discussing the use of File References in drawings. Generally, a file reference is a drawing that is attached to another drawing file. Once the referenced drawing file is edited or changed, these changes are automatically seen once the drawing containing the external file reference is opened again. Performing in-place editing of external references is demonstrated. Importing image files is also discussed and demonstrated in this chapter. A tutorial exercise follows at the end of this chapter to let the user practice using external references.

CHAPTER 19 — ADVANCED LAYOUT TECHNIQUES

This very important chapter demonstrates advanced techniques used in laying out a drawing before it is plotted. The ability to lay out a drawing consisting of various images at different scales is discussed. The ability to create user-defined rectangular and non-rectangular viewports is also demonstrated. Another important topic discussed is the application of Annotation Scales and how they affect the drawing scale of text, dimensions, linetypes, and crosshatch patterns.

CHAPTER 20 — SOLID MODELING FUNDAMENTALS

The chapter begins with a discussion on the use of the 3D Modeling workspace. Creating User Coordinate Systems and how they are positioned to construct objects in 3D is a key concept to master in this chapter. Creating User Coordinate Systems dynamically is also shown. The display of 3D images through View Cube, Steering Wheel, and the `3DORBIT` command are discussed along with the creation of visual styles. Creating various solid primitives such as boxes, cones, and cylinders is discussed in addition to the ability to construct complex solid objects through the use of the Boolean operations of union, subtraction, and intersection. The chapter continues by discussing extruding, rotating, sweeping, and lofting operations for creating solid models from profiles. Tutorial exercises follow at the end of this chapter.

CHAPTER 21 — CONCEPT MODELING, EDITING SOLIDS, AND SURFACE MODELING

This chapter begins with a detailed study on how concept models can easily be created by dragging on grips located at key locations of a solid primitive. The ability to pick and drag subobjects of a solid model and easily change its shape is also discussed. The `FILLETEDGE`, `CHAMFEREDGE`, `3DMOVE`, `3DALIGN`, `3DROTATE`, `MIRROR3D`, `3DSCALE`, `ARRAYRECT`, `ARRAYPOLAR`, `ARRAYPATH`, and `SLICE` commands are discussed as a means of introducing the editing capabilities of AutoCAD on 3D models. Modifications can also be made to a solid model through the use of the `SOLIDEDIT` command. This command provides the ability to extrude existing faces, imprint objects, and create thin walls with the Shell option. The topic of creating and editing procedural and mesh surface models will also be discussed. The editing of faces and edges will be demonstrated as a means of creating a conceptual surface model that can then be converted into a solid. Tutorial exercises can be found at the end of this chapter.

CHAPTER 22 — CREATING 2D DRAWINGS FROM A 3D SOLID MODEL

Once the solid model is created, three methods are available to generate 2D drawing layouts: the `SOLVIEW` and `SOLDRAW` process, the `FLATSHOT` command, and the Model Documentation tools. In the first method the `SOLVIEW` command is used to lay out 2D views of the model, and the `SOLDRAW` command is used to draw the 2D views. Layers are automatically created to assist in the display of 2D drawing objects and with the annotation of the drawing. The `FLATSHOT` command provides another means of projecting 2D geometry from a 3D model. While simple and efficient to use, this command doesn't offer as many view creation options as those available in the other methods. The third method utilizes the model documentation tools: `VIEWBASE`, `VIEWPROJ`, `VIEWSECTION`, and `VIEWDETAIL` commands. This new process is simple and powerful and allows you to create orthographic, isometric, section, and detail views with a pick and place technique. Tutorial exercises are available to assist you in becoming proficient at generating 2D drawings from 3D objects.

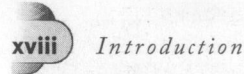

CHAPTER 23 — PRODUCING RENDERINGS AND MOTION STUDIES

This chapter introduces you to the uses and techniques of producing renderings from 3D models in AutoCAD. A brief overview of the rendering process is covered, along with detailed information about placing lights in your model, loading materials through the materials library supplied in AutoCAD, attaching materials to your 3D models, creating your own custom materials, applying a background to your rendered image, and experimenting with the use of motion path animations for creating walk-throughs of 3D models.

CAD CONNECT SITE FROM CENGAGEBRAIN

Extra information is supplied through the CAD Connect web site associated with this book. Drawing files for the book's tutorials and Try It! exercises as well as supplemental information are located at this site. It will also contain future content addressing features of subsequent releases of the software. The access code included with this text enables you to access this web site. Instructions for redeeming your access code and accessing the web site are given below:

Redeeming an Access Code:

1. Go to http://www.CengageBrain.com
2. Enter the Access code in the Prepaid Code or Access Key field, - Register
3. Register as a new user or log in as an existing user if you already have an account with Cengage Learning or CengageBrain.com
4. Open the product from the My Account page

Accessing CAD Connect from CengageBrain - My Account:

1. Sign in to your account at: http://www.CengageBrain.com
2. Go to My Account to view purchases
3. Locate the desired product
4. Click on the Open button next to the CAD Connect site entry, or Premium Website for [core title]

HOW THIS BOOK WAS PRODUCED

The following hardware and software tools were used to create this version of the AutoCAD Tutor Book:

Hardware: Precision Workstation by Dell Computer Corporation

CAD Software: AutoCAD 2013 by Autodesk, Inc.

Word Processing: Microsoft Word by Microsoft Corporation

Image Manipulation Software: Paint Shop Pro by Jasc Software, Inc.

Page Proof Review Software: Adobe Reader 9 by Adobe Corporation

ACKNOWLEDGMENTS

I wish to thank the staff at Cengage Learning for their assistance with this document, especially John Fisher, Glenn Castle, and Stacy Masucci. I would also like to recognize Heidi Hewett, Guillermo Melantoni, and Sean Wagstaff of Autodesk, Inc., for sharing their technical knowledge on the topic of Mesh Modeling that can be found in Chapter 21.

The publisher and author would like to thank and acknowledge the many professionals who reviewed the manuscript to help them publish this AutoCAD text. A special acknowledgment to Jack Johnson, Elizabethtown Community and Technical College, Elizabethtown, KY, and to Divya Tyagi, Project Manager at PreMediaGlobal, Inc.

ABOUT THE AUTHOR

Kevin J. Lang is the Program Coordinator for Engineering Design Graphics at Trident Technical College at Charleston, South Carolina. He has a B.S. in Mechanical Engineering from the University of South Carolina and more than 30 years experience in the engineering field. In addition, he has been teaching CAD-related subjects at the College since 1997. He owes many thanks to Alan, having learned so much from him over the years, including CAD.

DEDICATION

This book is dedicated in memory of Alan J. Kalameja. Alan was a long-time friend and colleague and is greatly missed. Alan authored the AutoCAD Tutor for Engineering Graphics in Release 10, 12, 14, and AutoCAD 2000 through 2010.

CONVENTIONS

All tutorials in this publication use the following conventions in the instructions: Whenever you are told to enter text, the text appears in boldface type. This may take the form of entering an AutoCAD command or entering such information as absolute, relative, or polar coordinates. You must follow these and all text inputs by pressing the ENTER key to execute the input. An icon for most commands is also present to assist in activating a command. For example, to draw a line using the LINE command from point 3,1 to 8,2, the sequence would look like the following:

```
Command: L (For LINE)
Specify first point: 3,1
Specify next point or [Undo]: 8,2
Specify next point or [Undo]: (Press ENTER to exit this
command)
```

Instructions for selecting objects are in italic type. When instructed to select an object, move the pickbox on the object to be selected and press the pick button on the mouse.

If you enter the wrong command for a particular step, you may cancel the command by pressing the ESC key. This key is located in the upper left-hand corner of any standard keyboard.

Instructions in some tutorials are designed to enter all commands, options, coordinates, and so forth, from the keyboard. You may use the same commands by selecting them from the ribbon, pull-down menu area, or from one of the floating toolbars.

Other tutorial exercises are provided with minimal instructions to test your ability to complete the exercise.

NOTES TO THE STUDENT AND INSTRUCTOR CONCERNING THE USE OF TUTORIAL EXERCISES

Various tutorial exercises have been designed throughout this book and can be found at the end of each chapter. The main purpose of each tutorial is to follow a series of steps toward the completion of a particular problem or object. Performing the tutorial will also prepare you to undertake the numerous drawing problems also found at the end of the chapters.

As you work on the tutorials, you should follow the steps very closely, taking the time to process the information being presented. It is highly recommended to both student and instructor that tutorial exercises be performed several times. Completing the tutorial the first time will give you the confidence that it can be done. Completing the tutorial a second or third time will allow you to focus on where certain operations are performed and why things behave the way they do. This will allow you to anticipate each step and have a better idea what operation to perform in each step. Only then will you be comfortable and confident to complete other drawing problems.

The CAD Connect web site (http://www.cengagebrain.com) contains AutoCAD drawing files for the Try It! exercises. To use drawing files, copy files located in the /Drawing Files/ directory to your hard drive. Files cannot be used without AutoCAD.

INSTRUCTOR SITE

An Instruction Companion Website containing supplementary material is available. This site contains an Instructor Guide, testbank, image gallery of text figures, and chapter presentations done in PowerPoint. Contact Delmar Cengage Learning or your local sales representative to obtain an instructor account.

Accessing an Instructor Companion Website site from SSO Front Door

1. Go to: HYPERLINK "http://login.cengage.com/" and login using the Instructor email address and password.
2. Enter author, title or ISBN in the Add a title to your bookshelf search box, click on Search button.
3. Click Add to My Bookshelf to add Instructor Resources.
4. At the Product page click on the Instructor Companion site link.

New Users

If you're new to Cengage.com and do not have a password, contact your sales representative.

CHAPTER

1

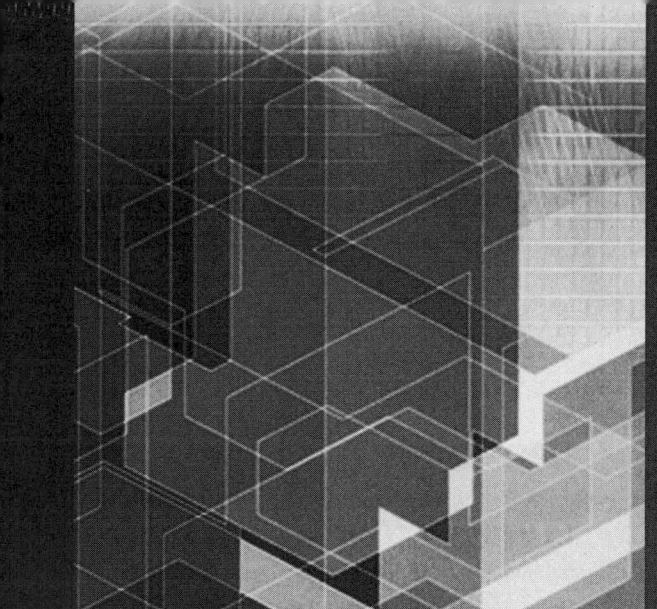

Getting Started with AutoCAD

This chapter will introduce topics necessary to complete a simple AutoCAD drawing. It begins with an explanation of the components that make up a typical AutoCAD display screen. You will learn various methods of selecting commands from this screen, such as through the Ribbon, Application Menu, Quick Access Toolbar, or Command Line. You will be introduced to some essential File commands: QNEW, OPEN, QSAVE, SAVEAS, and CLOSE. These commands will allow you to start, open, save, and close drawings. Once in a drawing, you will utilize some of the Draw and Modify commands: LINE, CIRCLE, PLINE, and ERASE. Technical drawing requires that precise distances and angles be constructed, therefore, you will also be shown methods and tools, which will allow the creation of accurate drawings: Direct Distance mode, Cartesian Coordinates (absolute, relative, and polar), Object Snaps, Object Snap Tracking, and Polar Tracking.

THE DRAFTING & ANNOTATION WORKSPACE

The initial load of AutoCAD displays in a workspace. Workspaces are considered task-oriented environments that use a default drawing template and even launch such items as toolbars and palettes, depending on the workspace. By default, AutoCAD loads the Drafting & Annotation Workspace, as shown in Figure 1.1. This workspace displays the Ribbon, which is used for accessing most essential AutoCAD commands. This workspace contains other items such as the Application Menu, the Quick Access Toolbar, the graphic cursor, the InfoCenter, the View Cube, and the Navigation Bar, as shown in Figure 1.1. Other workspaces supplied with AutoCAD are AutoCAD Classic, 3D Basics, and 3D Modeling. The 3D workspaces will be discussed in greater detail in Chapter 20.

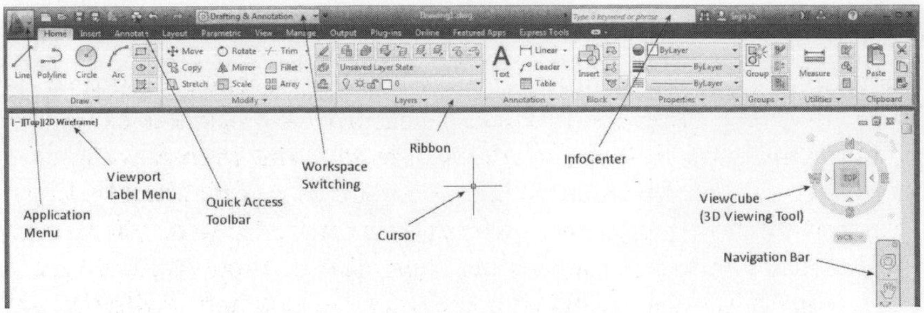

FIGURE 1.1

THE AUTOCAD CLASSIC WORKSPACE

The AutoCAD Classic Workspace, shown in Figure 1.2, provides a layout similar to those found in older versions of AutoCAD. Instead of a Ribbon, most commands are accessed through the Menu Bar or through Toolbars docked around the screen. A Tool Palette is also displayed in this workspace. Commands can be accessed through the palette using a drag and drop method. Although this workspace does not display the Ribbon, the RIBBON command can be used to turn it on if desired.

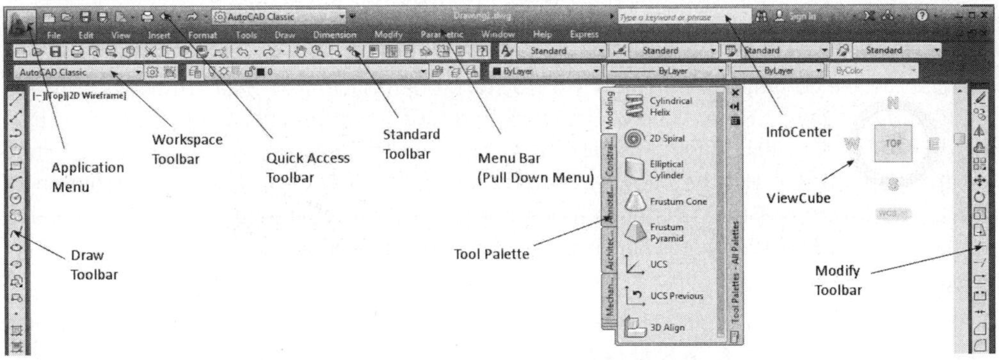

FIGURE 1.2

While major differences occur at the top of the display screen when you are activating different workspaces, most of the tools available at the bottom of the screen are common to all workspaces. Study the various screen components, as shown in Figure 1.3.

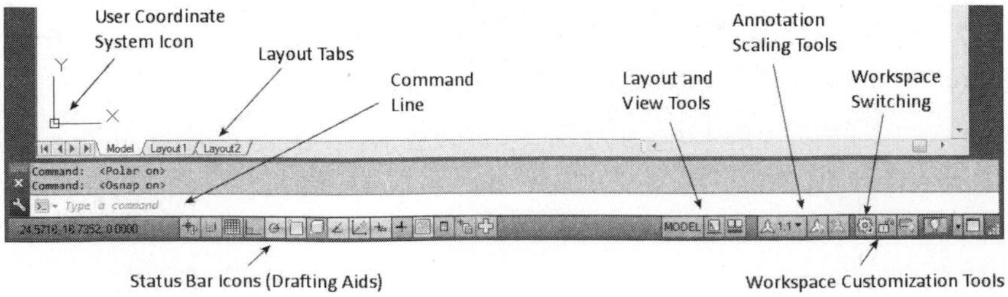

FIGURE 1.3

WELCOME SCREEN

When you start AutoCAD, a Welcome Screen will be automatically displayed, as shown in Figure 1.4. Work, Learn, and Extend panels provide quick access to drawing, learning, and online tools. Use the Work panel to start and open drawing files. For your convenience a list of your most recent working files is available. The Learn panel provides access to learning resources such as getting started videos. The Extend panel provides an online resource for training, tips, advice, products, downloads (apps), and help. Access to social media sites such as Facebook and Twitter is also available.

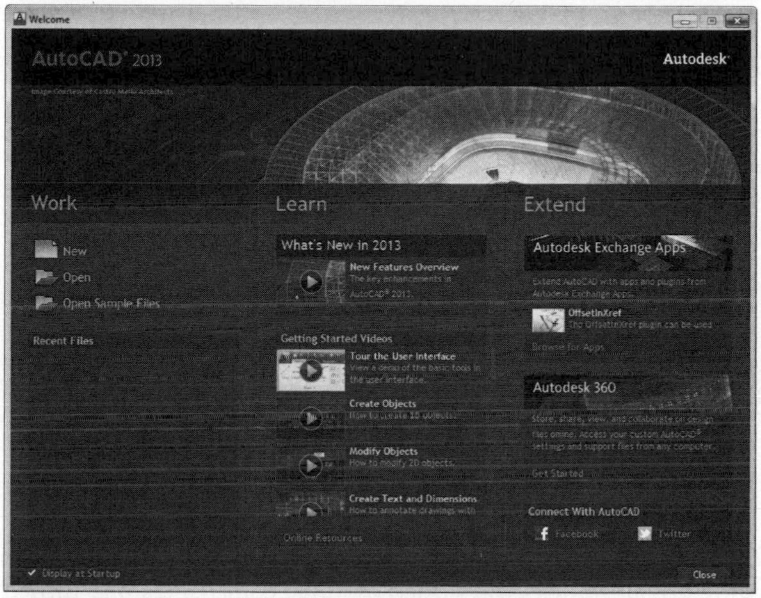

FIGURE 1.4

Remove the check from the check box in the lower left corner of the Welcome window, if you want to disable the automatic display of this window at start-up. The window can be redisplayed at any time by selecting "Welcome Screen …" from the Help drop-down menu on the InfoCenter Toolbar, as shown in Figure 1.5.

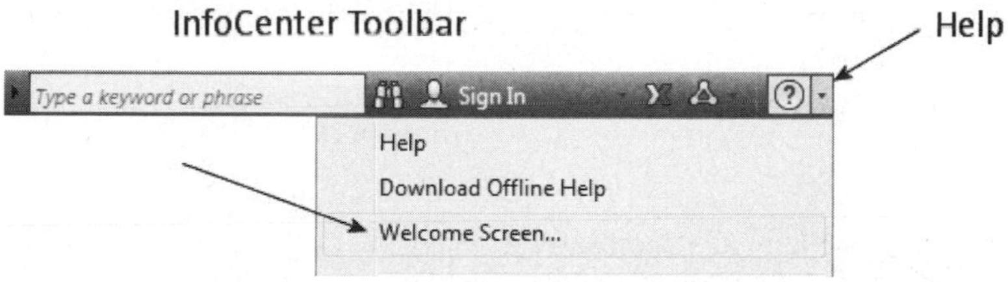

FIGURE 1.5

ACCESSING WORKSPACES

Switching workspaces is easily accomplished by expanding the drop-down list next to the Quick Access Toolbar and clicking on the desired workspace, as shown in Figure 1.6 (Left). A second method is to click the Workspace button on the Status Bar as shown in Figure 1.6 (Right). In addition to selecting one of these pre-existing workspaces, you can also arrange your screen to your liking and save the screen changes as your own custom workspace.

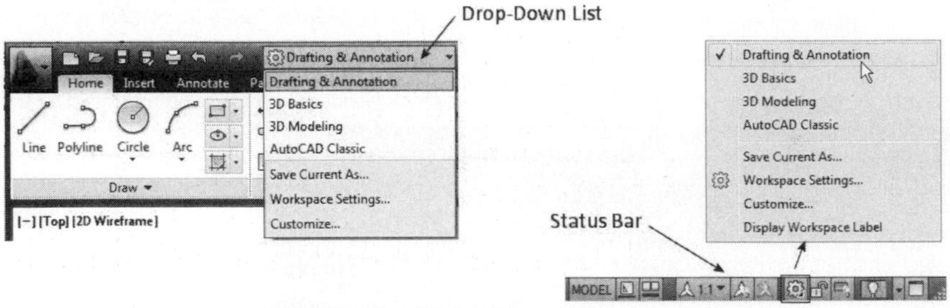

FIGURE 1.6

THE STATUS BAR

The Status Bar, illustrated in Figure 1.7, is used to toggle ON or OFF the following modes (drafting aids): Coordinate Display, Infer Constraints (INFER), Snap, Grid, Ortho, Polar Tracking, Object Snap (OSNAP), 3D Object Snap (3DOSNAP), Object Snap Tracking (OTRACK), Dynamic User Coordinate System (DUCS), Dynamic Input (DYN), Line Weight (LWT), Transparency (TPY), Quick Properties (QP), Selection Cycling (SC), and Annotation Monitor (AM). Click the button once to turn the mode on or off. A button with a blue color indicates that the mode is on. For example, Figure 1.7 illustrates Grid turned on (blue color) and Polar turned off (gray color). Right-clicking on any button in the Status Bar activates the menu shown in Figure 1.7. Clicking on Use Icons will change the graphic icons to text mode icons.

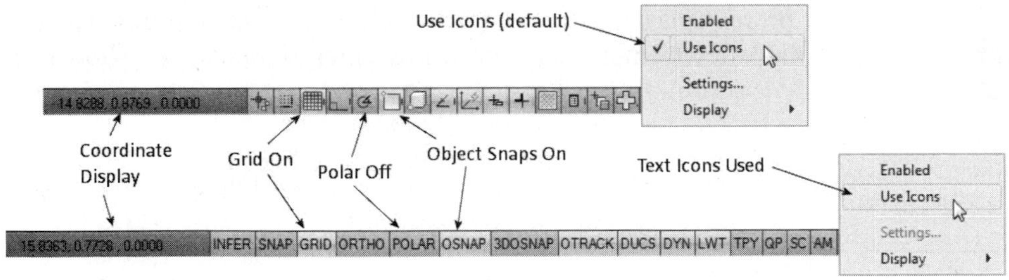

FIGURE 1.7

The following table gives a brief description of each component located in the Status Bar:

Button	Tool	Description
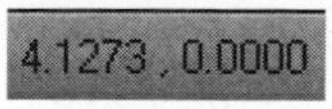	Coordinate Display	Toggles the coordinate display, located in the lower-left corner of the Status Bar, ON or OFF. When the coordinate display is off, the coordinates are updated when you pick an area of the screen with the cursor. When the coordinate display is on, the coordinates dynamically change with the current position of the cursor.
	INFER	Toggles Infer Constraints ON or OFF. When ON, selected constraints are applied while creating or editing geometry (see Chapter 13).
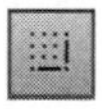	SNAP	Toggles Snap mode ON or OFF. The `SNAP` command forces the cursor to align with grid points. The current snap value can be modified and can be related to the spacing of the grid.
	GRID	Toggles the display of the grid ON or OFF. The actual grid spacing is set by the `DSETTINGS` OR `GRID` command and not by this function key.
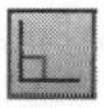	ORTHO	Toggles Ortho mode ON or OFF. Use this key to force objects, such as lines to be drawn horizontally or vertically.
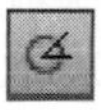	POLAR	Toggles the Polar Tracking ON or OFF. Polar Tracking can force lines to be drawn at any angle, making it more versatile than Ortho mode. The Polar Tracking angles are set through a dialog box. Also, if you turn Polar Tracking on, Ortho mode is disabled, and vice versa.
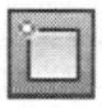	OSNAP	Toggles the current Object Snap settings ON or OFF. This will be discussed later in this chapter.
	3DOSNAP	Toggles the current 3D Object Snap settings ON or OFF. These running object snaps are used for modeling in 3D.
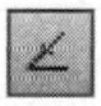	OTRACK	Toggles Object Snap Tracking ON or OFF. This feature will also be discussed later in this chapter.
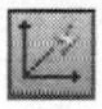	DUCS	Toggles the Dynamic User Coordinate System ON or OFF. This feature is used mainly for modeling in 3D.
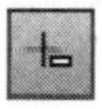	DYN	Toggles Dynamic Input ON or OFF. When turned on, your attention is directed to your cursor position as commands and options are executed. When turned off, all commands and options are accessed through the Command prompt at the bottom of the display screen.
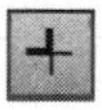	LWT	Toggles Lineweight ON or OFF. When turned off, no lineweights are displayed. When turned on, lineweights that have been assigned to layers are displayed in the drawing.
	TPY	Toggles Transparency ON or OFF to allow the transparency percentage assigned to specific objects to be displayed or not.

(Continued)

Button	Tool	Description
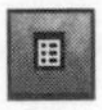	QP	When Quick Properties is turned on, this tool will list the most popular properties of a selected object.
	SC	Toggles Selection Cycling ON or OFF. When selecting overlapped objects, normally the last object created is selected. If selection cycling is on, a list of overlapped objects is provided to choose from.
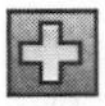	AM	When the Annotation Monitor is turned on, this tool indicates any non-associative annotations.

Right-clicking one of the status buttons displays a shortcut menu. Choose Settings to access various dialog boxes that control certain features associated with the button. These controls will be discussed later in this chapter and also in Chapter 2. The image below shows the shortcut menus displayed by right-clicking on SNAP, POLAR, or OSNAP.

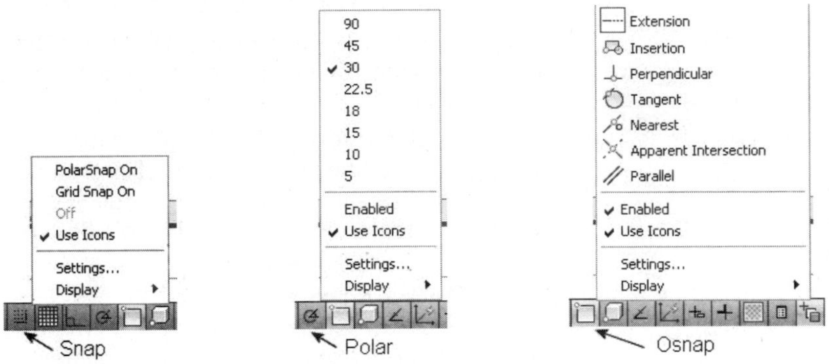

FIGURE 1.8

You can also access most tools located in the Status Bar through the function keys located at the top of any standard computer keyboard. The following table describes each function key.

Function Key	Definitions
F1	Displays AutoCAD Help Topics
F2	Toggle Text/Graphics Screen
F3	Object Snap settings ON/OFF
F4	3D Object Snap settings ON/OFF
F5	Toggle Isoplane Modes
F6	Toggle Dynamic UCS ON/OFF
F7	Toggle Grid Mode ON/OFF
F8	Toggle Ortho Mode ON/OFF
F9	Toggle Snap Mode ON/OFF
F10	Toggle Polar Mode ON/OFF
F11	Toggle Object Snap Tracking ON/OFF
F12	Toggle Dynamic Input (DYN) ON/OFF

NOTE

Most of the function keys are similar in operation to the modes found in the Status Bar except for the following:

When you press F1, the AutoCAD Help window is displayed.

Pressing F2 takes you to the text screen consisting of a series of previous prompt sequences. This may be helpful for viewing the previous command sequence in text form.

Pressing F5 scrolls you through the three supported Isoplane modes used to construct isometric drawings (Right, Left, and Top).

Pressing CTRL+SHIFT+P toggles Quick Properties mode ON or OFF.

Pressing CTRL+SHIFT+I toggles Infer Constraints ON or OFF.

Pressing CTRL+W toggles Selection Cycling ON or OFF.

Additional Status Bar Controls

Located at the far right end of the Status Bar are additional buttons separated into three distinct groups: Layout/View tools, Annotation Scale tools, and Workspace Customization tools. These items include Quick View Layouts, Quick View Drawings, Annotation Scale tools, the Workspace Switching tool, the Toolbar Unlocking tool, the Status Bar Menu tool, and the Clean Screen tool. When annotative objects such as text and dimensions are created, they are scaled based on the current annotation scale and automatically displayed at the correct size. This feature will be discussed in greater detail in Chapter 19. The following table gives a brief description of the remaining buttons found in this area.

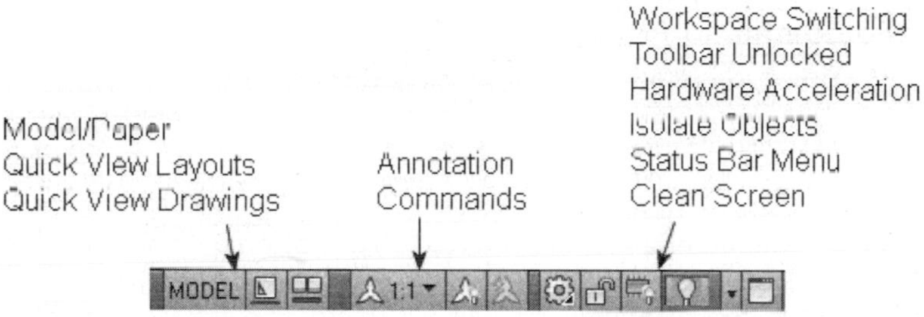

FIGURE 1.9

Button	Tool	Description
	Workspace Switching	Allows you to switch between the workspaces already defined in the drawing.
	Toolbar/Window Positions Toggle	Locks the position of all toolbars on the display screen.
	Isolate Objects	Indicates (turns red) if objects in the view have been isolated or hidden.
	Status Bar Menu Controls	Activates a menu used for turning on or off certain Status Bar buttons.
	Clean Screen	Removes all toolbars from the screen, giving your display an enlarged appearance. Click this button again to return the toolbars to the screen.

COMMUNICATING WITH AUTOCAD

The Command Line

How productive the user becomes in using AutoCAD may depend on the degree of understanding of the command execution process within AutoCAD. One of the means of command execution is through the Command prompt that is located at the bottom of the display screen. As a command is initiated, AutoCAD prompts the user with a series of steps needed to complete this command. In Figure 1.10 on the left, the `CIRCLE` command is chosen as the command. The next series of lines in the command line prompts the user to first specify or locate a center point for the circle. After this is accomplished, you are then prompted to specify the radius of the circle. Shown in Figure 1.10 on the right is the AutoComplete feature. As you start to type in a command, a suggestion list is provided that allows you to select the command before you finish entering the command name.

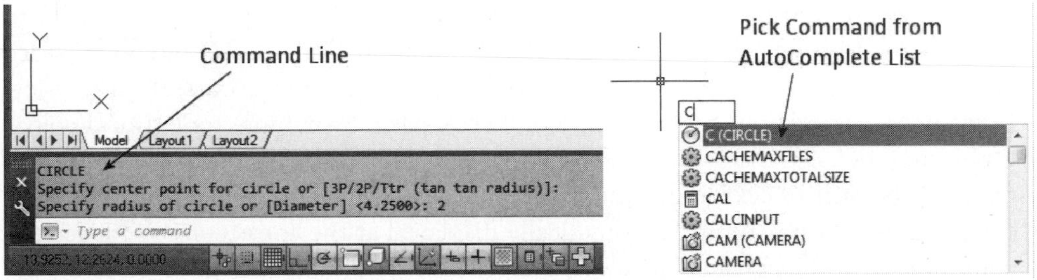

FIGURE 1.10

Understanding the Command Prompt

In the previous image of the command line, notice the string of `CIRCLE` command options displayed as the following:

```
[3P/2P/Ttr (tan tan radius)]
```

Items identified inside the square brackets are referred to as options. Typing in this option from the keyboard activates it. You only need to type in the letters that are capitalized (T—for the Ttr option). It is also possible to select an option from the command line by clicking the option with the left mouse button.

```
Specify radius of circle or [Diameter] <4.2500>:
```

If a number is provided inside angle brackets "<4.2500>," simply press ENTER to accept this default value or type in a new value and then press ENTER. Typically, this value represents the one last used in the command.

Command Line Interface

The command line interface can be modified by changing its color, transparency, size, and location. Change its location by picking and dragging the title area of the interface to a new area, as shown in the following image on the left. The interface can be docked at the top or bottom of the AutoCAD window or simply allowed to float in the drawing screen area. When undocked the interface will appear semi-transparent, allowing you to see objects in the entire drawing area. The number of command lines displayed can be changed by dragging the edge of the interface box or you can temporarily expand the prompt area by pressing the flyout arrow at the right end of the interface. To change the level of transparency or colors used for the interface, press the Customize button, as shown in the following image on the right. A shortcut menu is provided to modify the interface settings. If you close the command line, it can be redisplayed with the COMMANDLINE (CTRL+9) command.

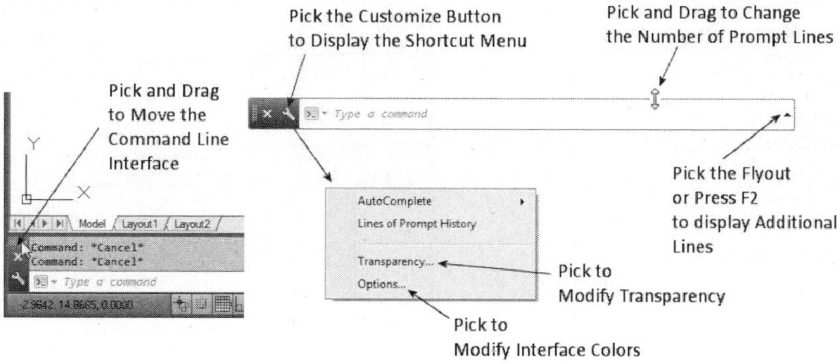

FIGURE 1.11

Dynamic Input

Yet another more efficient means of command execution within AutoCAD is through the Dynamic Input feature, which is activated by clicking the DYN button located in the Status Bar at the bottom of your display screen, as shown in Figure 1.12 (Left). Whether a command is selected from the Ribbon or Menu Bar or entered from the keyboard, you see immediate feedback at your cursor location. Figure 1.12 (Right) illustrates how the CIRCLE Command prompts display at the cursor location. As the cursor is moved around, the Specify center point for Circle prompt also moves. Also notice that the current screen position is displayed. If the down directional arrow is typed on the keyboard, options of the CIRCLE command display. Typing the DOWN ARROW cycles through the available options in executing the CIRCLE command.

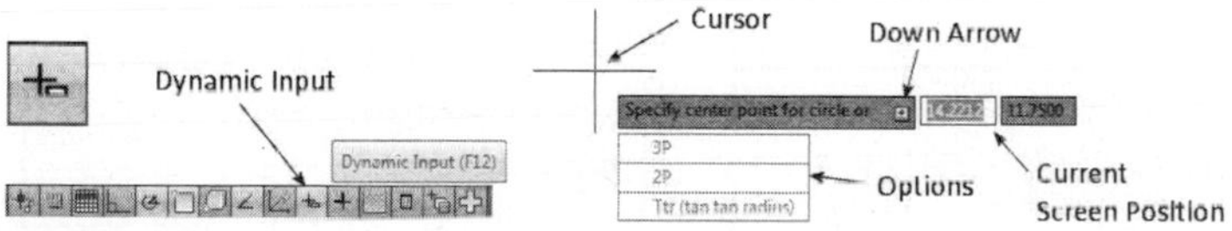

FIGURE 1.12

THE APPLICATION MENU

The Application Menu provides you with the ability to access commonly used AutoCAD tools. Clicking on the Icon in the upper-left corner will display commands that allow you to create, open, save, print, and publish AutoCAD files as shown in Figure 1.13 (Left). A list of recently opened documents is also provided. Simply select a drawing file from the list to open it. Clicking on a command with a cascade arrow next to it, such as the Save As command shown in Figure 1.13 (Right), displays additional related command choices. At the very bottom right of the Application Menu are two buttons, one called Options for launching the Options dialog box and the other called Exit AutoCAD used for exiting the AutoCAD environment. At the top of the Application Menu, a text box is provided to search for commands or other AutoCAD topics.

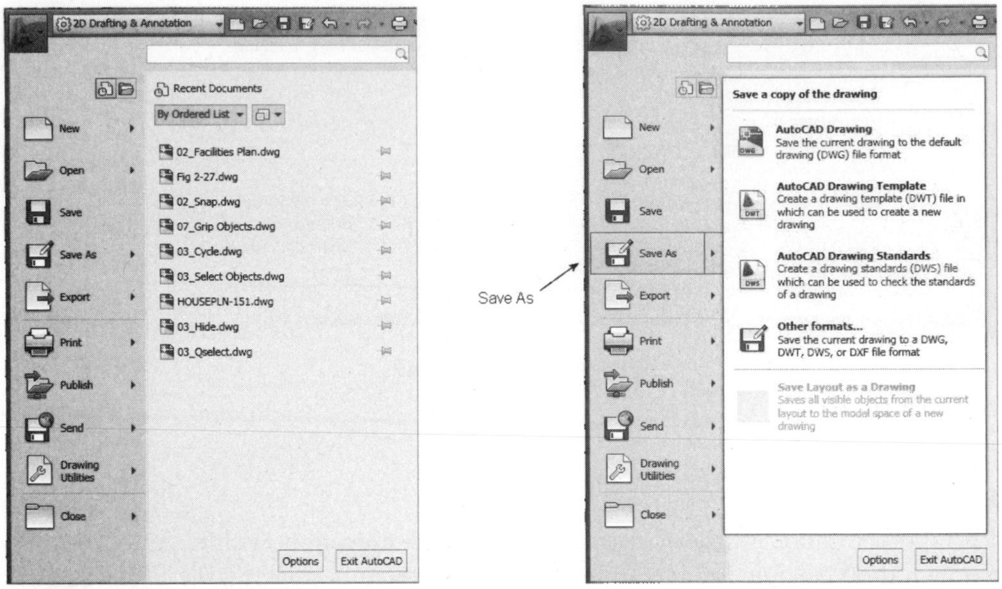

FIGURE 1.13

Document Controls

Clicking on the two document control items in the upper right portion of the Application Menu displays a series of panels used for viewing recent or open documents. The three panels illustrated in Figure 1.14 show the Recent Documents Mode with drop-down lists exposed. You can display existing files in an ordered list or group them by date, size, or file type. When you move your cursor over one of these files, a preview image automatically appears in addition to information about the document.

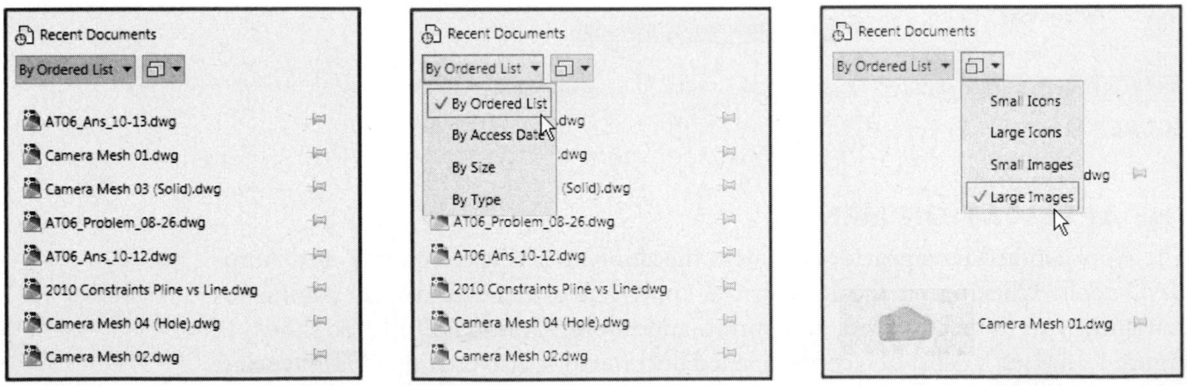

FIGURE 1.14

THE MENU BAR

The Menu Bar is generally associated with the AutoCAD Classic Workspace while the Ribbon with the Drafting & Annotation Workspace. Either display method provides an easy way to access most AutoCAD commands. In the Menu

Bar, various categories exist such as File, Edit, View, Insert, Format, Tools, Draw, Dimensions, Modify, and so on. Clicking one of these category headings pulls down a menu consisting of commands related to this heading.

In the Drafting & Annotation Workspace, you can activate the Menu Bar in the upper part of the display screen by clicking on the arrow located at the end of the Quick Access Toolbar and choosing Show Menu Bar from the menu, as shown in Figure 1.15 (Left). This will display the Menu Bar at the top of the screen, as shown in Figure 1.15 (Right).

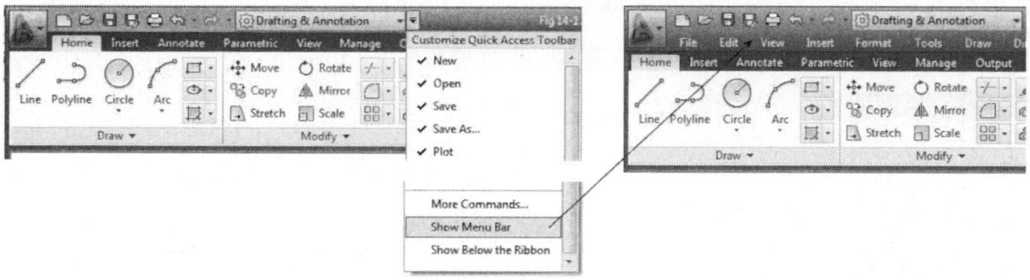

FIGURE 1.15

TOOLBARS FROM THE AUTOCAD CLASSIC WORKSPACE

Besides the Menu Bar, the AutoCAD Classic Workspace utilizes numerous toolbars for command selection. Activating the AutoCAD Classic Workspace will automatically display toolbars, such as Draw, Modify, Standard, Layers, Styles, and Properties. Figure 1.16 shows the AutoCAD Classic Workspace layout.

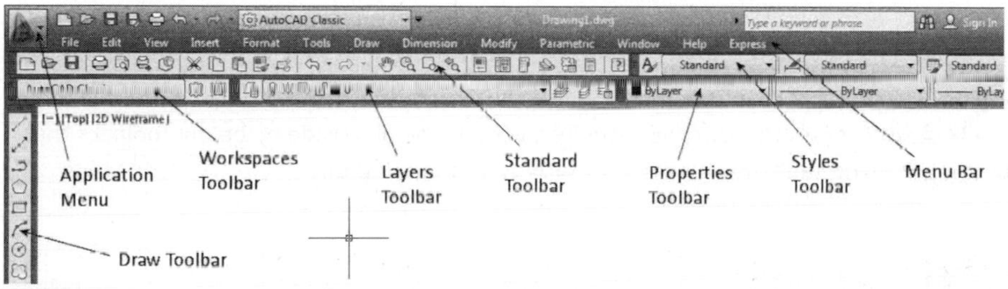

FIGURE 1.16

An example of a toolbar is shown in the figure below. The Zoom Toolbar shown allows you access to most ZOOM command options. When the cursor rolls over a tool, a 3D border is displayed, along with a tooltip that explains the purpose of the command, as shown in Figure 1.17 (Right).

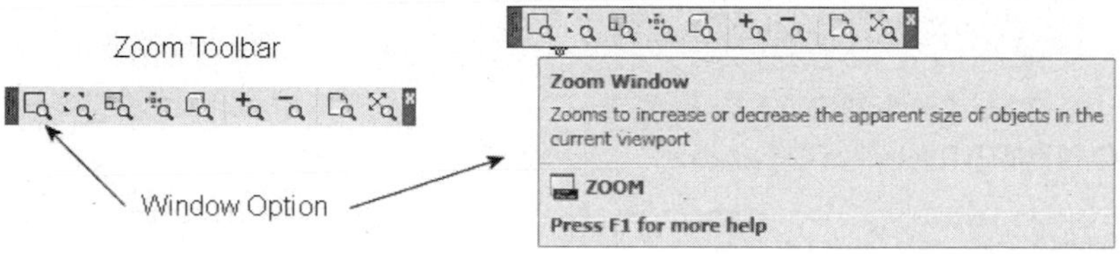

FIGURE 1.17

ACTIVATING TOOLBARS

Many toolbars are available to assist the user in executing other types of commands. When working in the AutoCAD Classic Workspace, six toolbars are already active or displayed: Draw, Layers, Modify, Properties, Standard, and Styles. To activate a different toolbar, move the cursor over the top of any toolbar icon and press the right mouse button. A shortcut menu appears that displays all toolbars, as shown in Figure 1.18. In this example, placing a check beside Text displays this toolbar.

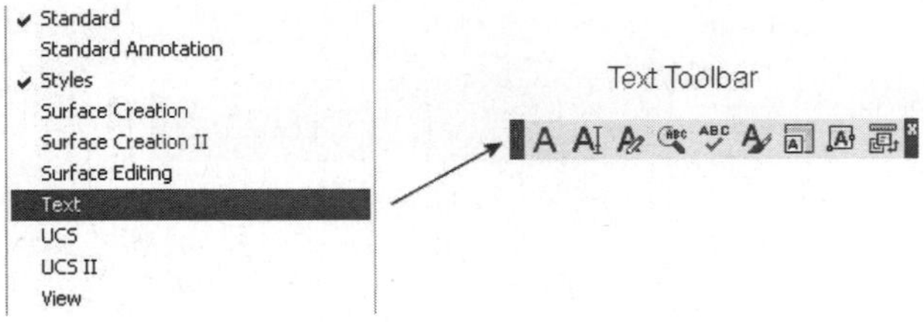

FIGURE 1.18

DOCKING TOOLBARS

In order to maximize drawing screen area, it is considered good practice to line the top or side edges of the display screen with toolbars. The method of moving toolbars to the sides of your screen is called docking. Press down on the toolbar title strip and slowly drag the toolbar to the top of the screen until the toolbar appears to jump. Letting go of the mouse button docks the toolbar to the top of the screen, as shown in Figure 1.19. Practice this by docking various toolbars to your screen.

TIP

To prevent docking, press the CTRL key as you drag the toolbar. This allows you to move the toolbar into the upper or lower portions of the display screen without toolbar docking. Also, if a toolbar appears to disappear, it might actually be alongside or below toolbars that already exist. Closing toolbars will assist in finding the missing one.

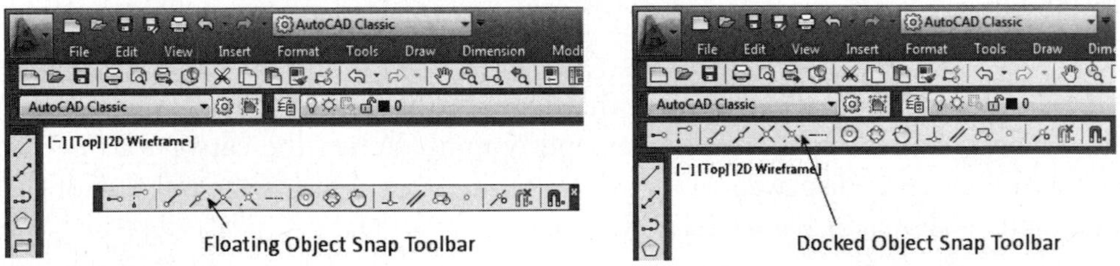

FIGURE 1.19

TOOLBARS FROM THE DRAFTING & ANNOTATION WORKSPACE

While inside of the Drafting & Annotation Workspace, it is possible to display toolbars as in the AutoCAD Classic Workspace. To accomplish this, select the Ribbon's View tab and expand the Toolbars button in the User Interface panel, as shown in Figure 1.20. When the shortcut menu displays, click on AutoCAD to display all of the available toolbars. Select one of the names from the list to show the toolbar on the display screen.

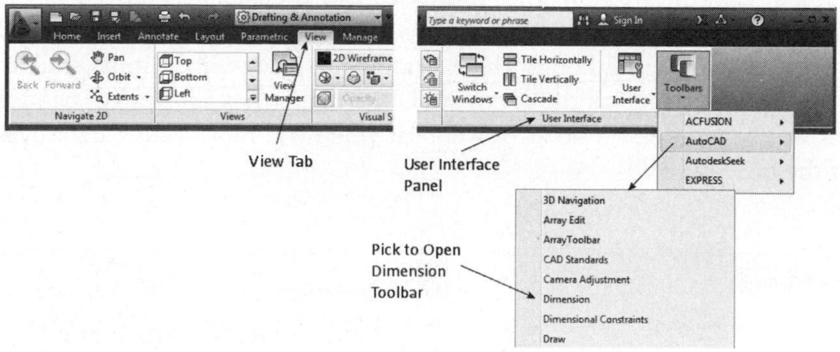

FIGURE 1.20

RIBBON DISPLAY MODES

A small button with an arrow is displayed at the end of the Ribbon tabs. This button allows you to minimize the Ribbon and display more of your screen. Three modes are available, as shown in Figure 1.21: Minimize to Tabs, Minimize to Panel Titles, and Minimize to Panel Buttons.

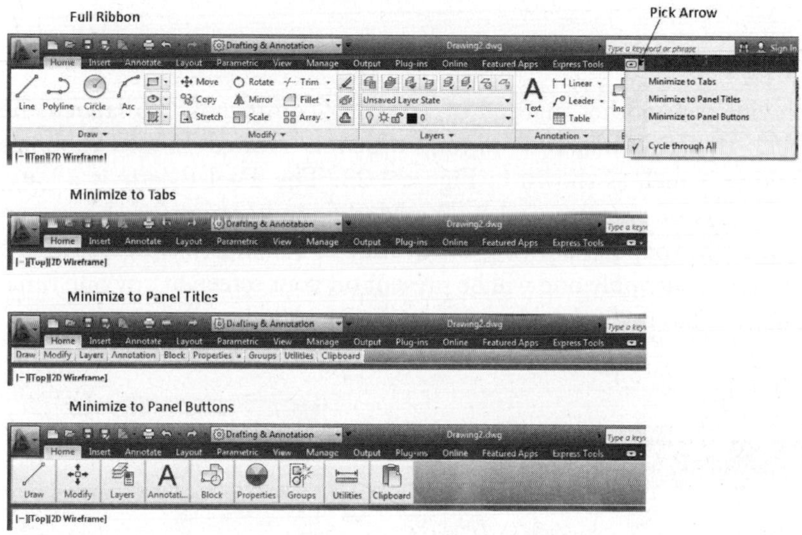

FIGURE 1.21

DIALOG BOXES AND PALETTES

Settings and other controls can be changed through dialog boxes and palettes. Illustrated in Figure 1.21 (Left) is the Drawing Units dialog box and on the right is the Properties Palette. Dialog boxes typically contain text boxes, drop-down lists, and check boxes that allow you to change settings, while palettes are more varied in their appearance but often include small images (icons) that can be selected. The title bar for a dialog box is located at the top of the window but is usually found on the side of a palette. Palettes and dialog boxes can be increased in size by moving your cursor over their borders. When two arrows appear, hold down the pick button of the mouse (the left button) and stretch the window in that direction. If the cursor is moved to the corner, the window is stretched in two directions. These methods can also be used to make the window smaller, although there is a default size, which limits smaller sizes. Some dialog boxes, such as the Drawing Units dialog box illustrated, cannot be sized. This will be indicated if no arrows appear when you move your cursor over the border of the box.

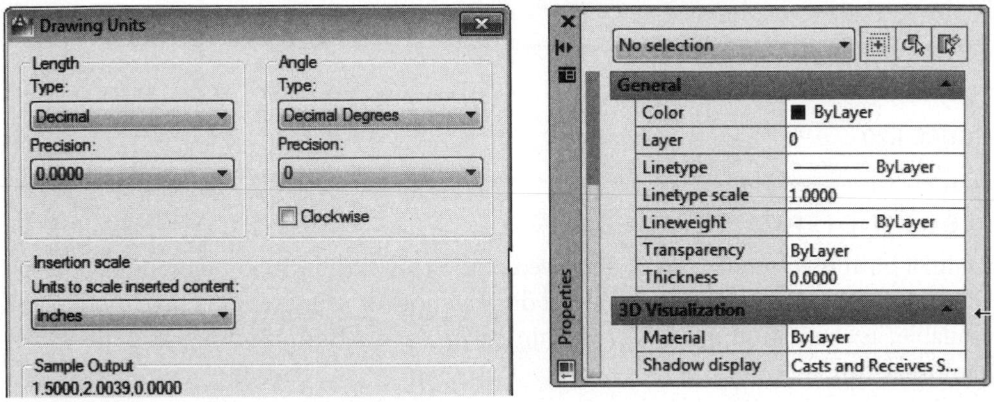

FIGURE 1.22

TOOL PALETTES

Tool palettes provide yet another method of accessing commands. To launch the Tool Palette, click the Tool Palettes button, which is located in the Ribbon's View tab and Palettes panel, as shown in Figure 1.23. The Tool Palette is a long, narrow bar that consists of numerous tabs. The Modify and Draw tabs shown can be used to access the more popular drawing and modify commands. While this image shows two palettes, in reality only one will be present on your screen at any one time. Simply click a different tab to display the commands associated with the tab.

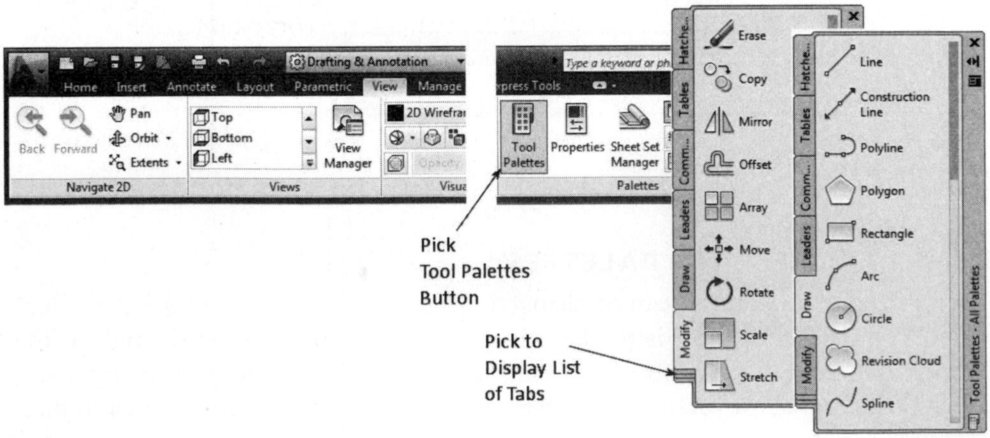

FIGURE 1.23

RIGHT-CLICK SHORTCUT MENUS

Many shortcut or cursor menus have been developed to assist with the rapid access to commands. Clicking the right mouse button activates a shortcut menu that provides access to these commands. The Default shortcut menu is illustrated in Figure 1.24 (Left). It is displayed whenever you right-click in the drawing area and no command or selection set is in progress.

Illustrated in Figure 1.24 (Right) is an example of the Edit shortcut menu. This shortcut consists of numerous editing and selection commands. This menu activates whenever you right-click in the display screen with an object or group of objects selected but no command is in progress.

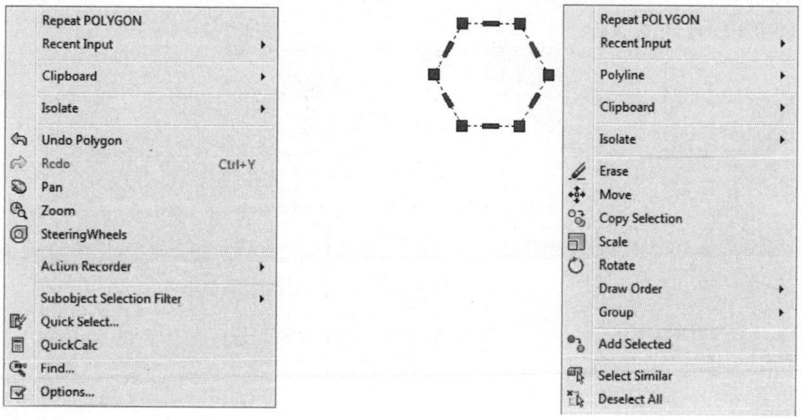

FIGURE 1.24

Right-clicking in the Command prompt area of the display screen activates the shortcut menu, as shown in Figure 1.25 (Left). This menu provides a record of the six most recently used commands and allows the user to repeat the command by simply selecting it from the list. This shortcut menu also provides quick access to settings available for the AutoComplete feature mentioned earlier in this chapter.

Illustrated in Figure 1.25 (Right) is an example of a Command-Mode shortcut menu. When you enter a command and right-click, this menu displays options of the command. This menu supports a number of commands. In Figure 1.25, the 3P, 2P, and Ttr (tan tan radius) listings are all options of the `CIRCLE` command.

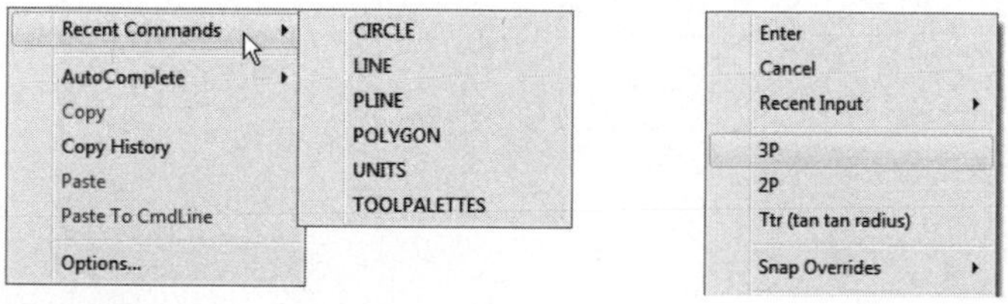

FIGURE 1.25

COMMAND ALIASES

As shown earlier, commands can be executed directly through keyboard entry. This practice is popular for users who are already familiar with the commands. However, users must know the full command name or find it in the suggestion list provided through the AutoComplete feature. To assist with the entry of AutoCAD commands from the keyboard, numerous commands are available in shortened form, referred to as aliases. For example, instead of typing in LINE, all that is required is L. The letter E can be used for the ERASE command, and so on. These command aliases are listed throughout this book. The complete list of all command aliases can be found in the AutoCAD Alias Editor. This dialog box can be accessed through the Ribbon (Express Tools tab > Tools panel > Command Aliases button), as shown in Figure 1.26. The Editor displays all of the commands that have their names shortened. The Add and Remove buttons of the Editor allow you to create new or remove existing aliases, if desired. Keyboard command aliases provides a fast and efficient method of activating AutoCAD commands.

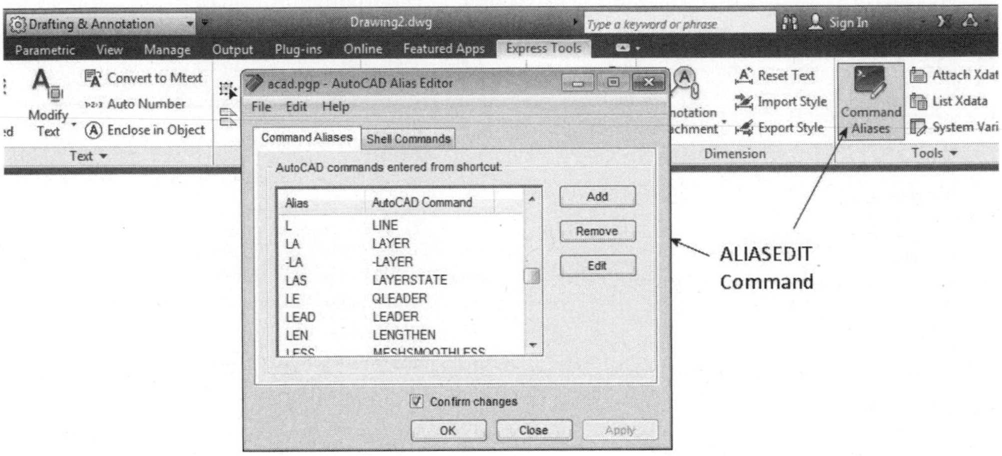

FIGURE 1.26

STARTING A NEW DRAWING

To begin a new drawing file, select the QNEW command using one of the following methods:

- From the Quick Access Toolbar
- From the Application Menu (New)
- From the Standard Toolbar of the AutoCAD Classic Workspace
- From the keyboard (QNEW)

  Command: QNEW

Entering the QNEW command displays the dialog box illustrated in Figure 1.27. This dialog box provides a list of templates to use for starting a new drawing.

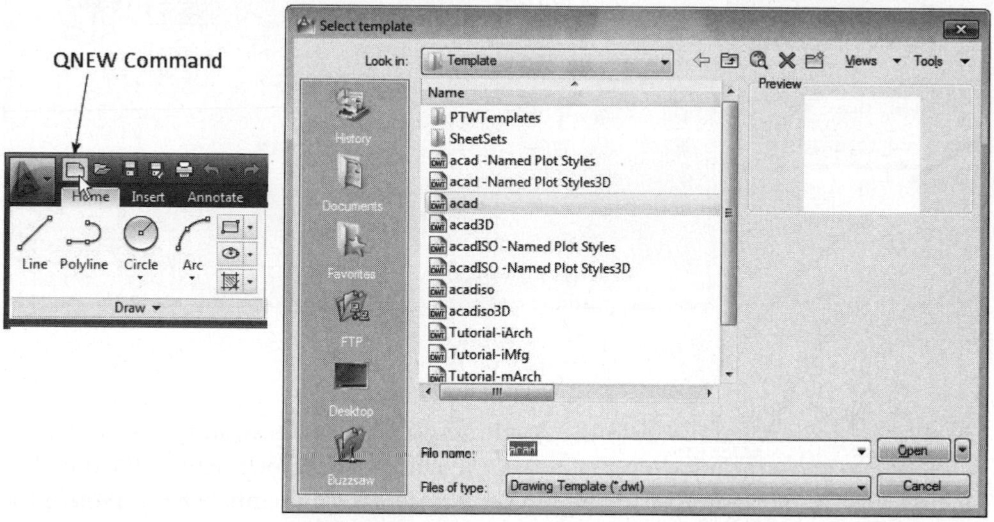

FIGURE 1.27

> **NOTE**
>
> **The QNEW command is similar to the NEW command but provides the option of starting with a preselected template. The template is assigned through the OPTIONS command.**

OPENING AN EXISTING DRAWING

The OPEN command is used to edit a drawing that has already been created. Select this command from one of the following:

- From the Quick Access Toolbar
- From the Application Menu (Open)
- From the Standard Toolbar of the AutoCAD Classic Workspace
- From the keyboard (OPEN)

When you select this command, a dialog box similar to Figure 1.28 appears. Listed in the field area are all files that match the type shown at the bottom of the dialog box. Because the file type is .DWG, all drawing files supported by AutoCAD are listed. To choose a different folder, use standard Windows file management techniques by clicking in the Look in field. This displays all folders associated with the drive. Clicking the folder displays any drawing files contained in it.

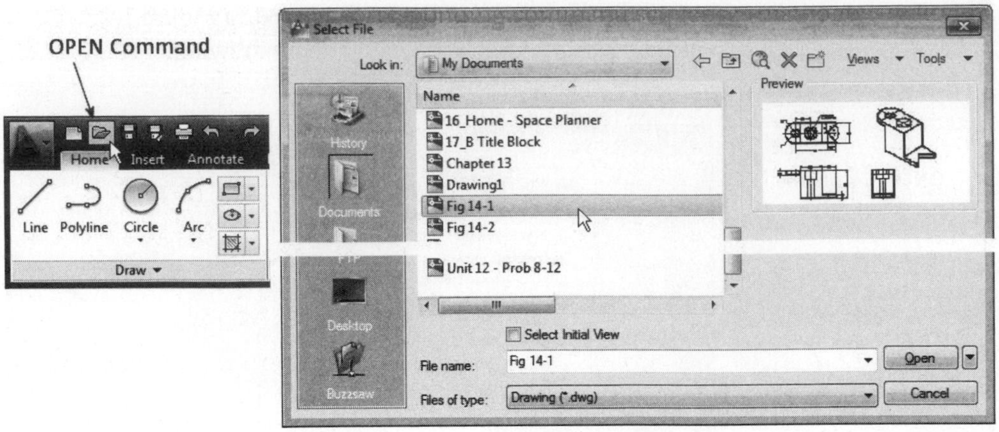

FIGURE 1.28

Additional tools are available in the Application Menu to assist in locating drawing files. These tools include Recent Documents and Open Documents. Illustrated in Figure 1.29 is an example of clicking on the Recent Documents button. Notice the ordered list of all drawings that were recently opened, enabling you to select these more efficiently.

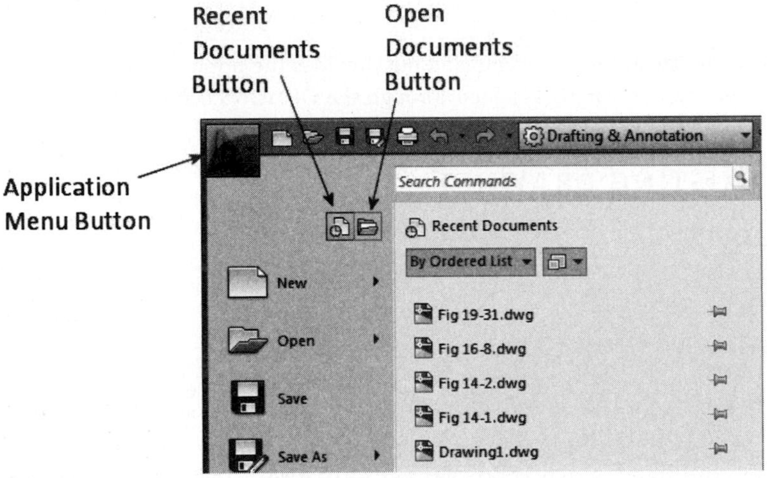

FIGURE 1.29

Clicking on the By Ordered List icon will expand the menu to include a number of options to sort files: By Access Date, By Size, and By Type. You can also change the way drawings display in the listing: Small Icons, Large Icons, Small Images, and Large Images. In Figure 1.30 (Left) the files are sorted By Ordered List. In the image on the right, the display is set to Large Images.

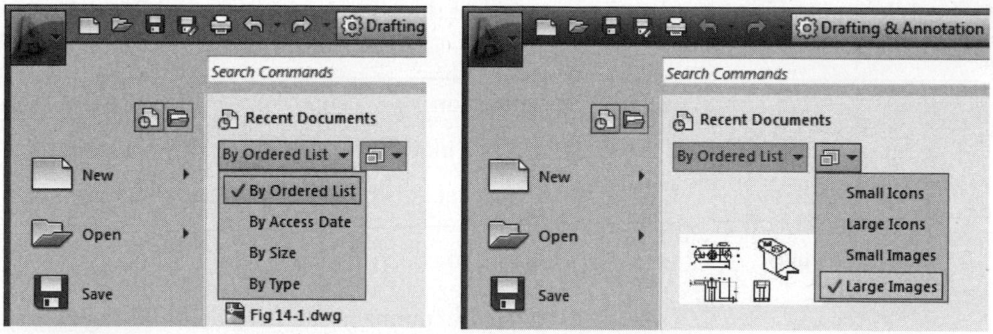

FIGURE 1.30

Because the list of recent documents can get large, certain drawing names drop off, which means you need to look for the drawing again. For drawings that are used most frequently, you can click on the pin icon to change the orientation of the pin, as shown in Figure 1.31. The presence of this pin means that this drawing will always be displayed in the Recent Documents List.

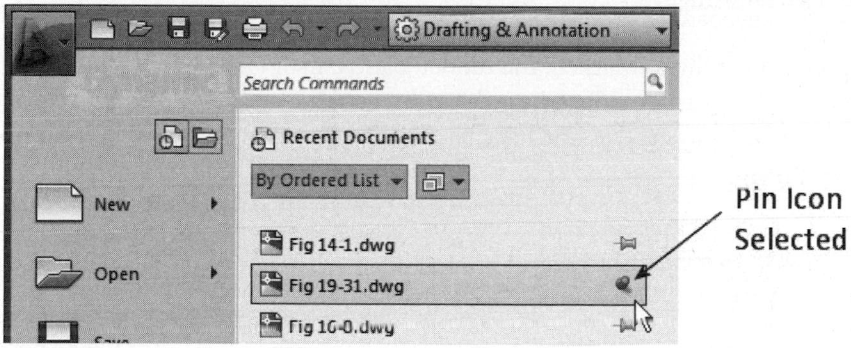

FIGURE 1.31

BASIC DRAWING COMMANDS

The following sections discuss some basic techniques used in creating drawings. These include drawing lines, circles, and polylines; using Object Snap modes and tracking; and erasing objects. Many of the basic drawing tools can be easily accessed through the Ribbon's Draw and Modify panels, as shown in Figure 1.32 (Left). Clicking the down arrow in a panel will display additional commands, as shown for the Draw panel in Figure 1.32 (Right).

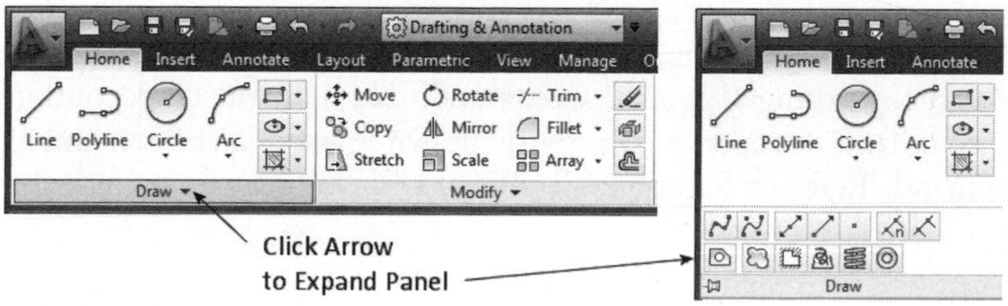

FIGURE 1.32

The following table gives a brief description of the LINE, CIRCLE, and PLINE draw commands:

Button	Tool	Key-In	Function
	Line	L	Draws individual or multiple line segments
	Circle	C	Constructs circles of specified radius or diameter
	Pline	PL	Used to construct a polyline, which is similar to a line except that all segments made with the PLINE command are considered a single object

CONSTRUCTING LINES

Use the LINE command to construct a line from one endpoint to the other. Choose this command from one of the following:

- From the Ribbon > Home Tab > Draw Panel
- From the Menu Bar (Draw > Line)
- From the Draw Toolbar of the AutoCAD Classic Workspace
- From the keyboard (L or LINE)

As the first point of the line is marked, the rubber-band cursor is displayed along with the normal crosshairs to assist in locating where the next line segment will be drawn. The LINE command stays active until the user either executes the Close option or issues a null response by pressing ENTER at the prompt "To point."

TRY IT!

Create a new drawing from scratch. Study Figure 1.33 (Left) and follow the command sequence for using the LINE command.

```
Command: L (For LINE)
Specify first point: (Pick a point at "A")
Specify next point or [Undo]: (Pick a point at "B")
Specify next point or [Undo]: (Pick a point at "C")
Specify next point or [Close/Undo]: (Pick a point at "D")
Specify next point or [Close/Undo]: (Pick a point at "E")
Specify next point or [Close/Undo]: (Pick a point at "F")
Specify next point or [Close/Undo]: C (To close the shape and
exit the command)
```

If a mistake is made in drawing a segment, as illustrated in Figure 1.33, the user can correct the error without exiting the LINE command. The built-in Undo option within the LINE command removes the previously drawn line while still remaining in the LINE command. Refer to Figure 1.33 (Right) and the prompts to use the Undo option of the LINE command.

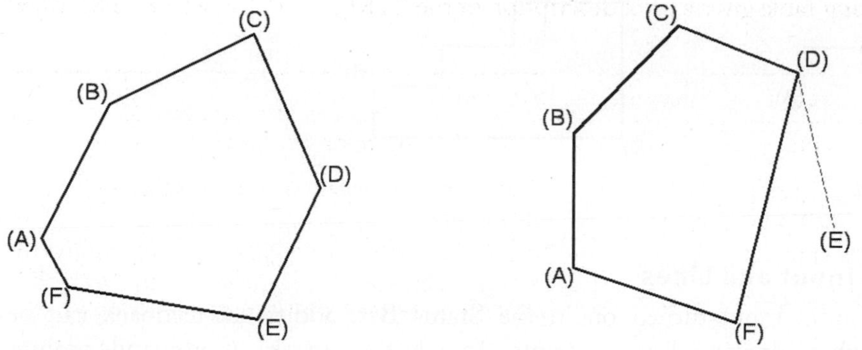

FIGURE 1.33

Command: L *(For LINE)*

Specify first point: *(Pick a point at "A")*

Specify next point or [Undo]: *(Pick a point at "B")*

Specify next point or [Undo]: *(Pick a point at "C")*

Specify next point or [Close/Undo]: *(Pick a point at "D")*

Specify next point or [Close/Undo]: *(Pick a point at "E")*

Specify next point or [Close/Undo]: U *(To undo or remove the segment from "D" to "E" and still remain in the* LINE *command)*

Specify next point or [Close/Undo]: *(Pick a point at "F")*

Specify next point or [Close/Undo]: **End** *(For Endpoint mode)*

of *(Select the endpoint of the line segment at "A")*

Specify next point or [Close/Undo]: *(Press* ENTER *to exit this command)*

Continuing Lines

Another option of the LINE command is the Continue option. The dashed line segment in Figure 1.34 was the last segment drawn before the LINE command was exited. To pick up at the last point of a previously drawn line segment, restart the LINE command and press ENTER. This activates the Continue option of the LINE command.

Command: L *(For LINE)*

Specify first point: *(Press* ENTER *to activate Continue Mode)*

Specify next point or [Undo]: *(Pick a point at "B")*

Specify next point or [Undo]: *(Pick a point at "C")*

Specify next point or [Close/Undo]: **End** *(For Endpoint mode)*

of *(Select the endpoint of the vertical line segment at "D")*

Specify next point or [Close/Undo]: *(Press* ENTER *to exit this command)*

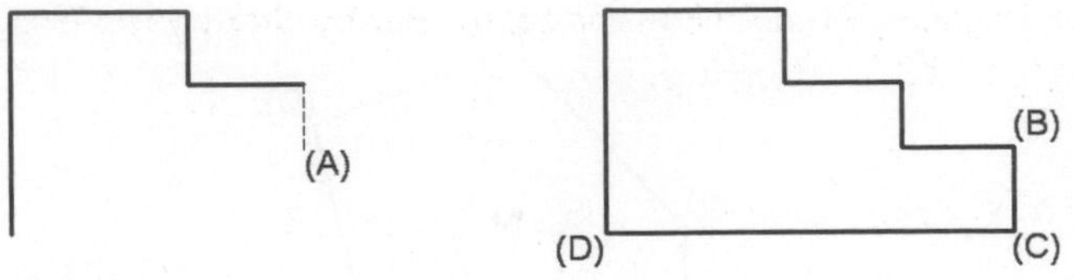

FIGURE 1.34

Dynamic Input and Lines

With Dynamic Input turned on in the Status Bar, additional feedback can be obtained when drawing line segments. In addition to the Command prompt and down arrow being displayed at your cursor location, a dynamic distance and angle are displayed to assist you in the construction of the line segment, as shown in Figure 1.35.

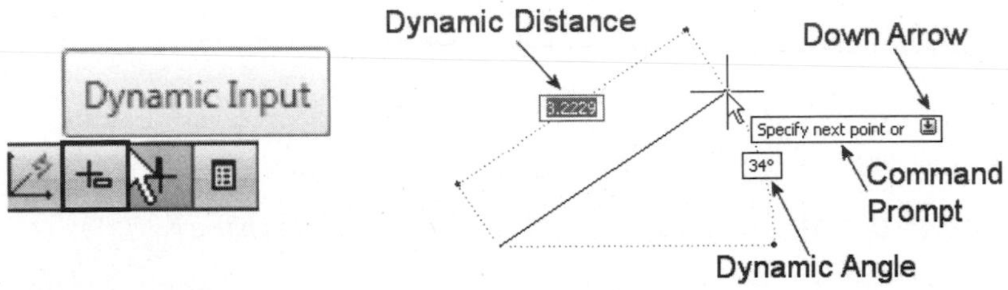

FIGURE 1.35

Command prompts for using Dynamic Input now appear in the drawing window next to the familiar AutoCAD cursor.

- When constructing line segments, dynamic dimensions in the form of a distance and an angle appear on the line. If the distance dimension is highlighted, entering a new value from your keyboard will change its value.
- Pressing the TAB key allows you to switch from the distance dimension to the angle dimension, where you can change its value. The dynamic angle displayed, by default, is only accurate to the nearest degree—you should type in the precise value.
- By default in Dynamic Input, coordinates for the second point of a line are considered relative. In other words, you do not need to type the @ symbol in front of the coordinate. The @ symbol means "last point" and will be discussed in greater detail later in this chapter. When drawing using absolute coordinates, you will probably want to turn Dynamic Input off. A "#" symbol can be used to cancel the automatic use of a relative coordinate.
- You can still enter relative and polar coordinates as normal using the @ symbol if you desire. These older methods of coordinate entry override the default dynamic input setting.
- The appearance of an arrow symbol in the Dynamic Input prompt area indicates that this command has options associated with it. To view these command options, press the DOWN ARROW key on your keyboard. These options will display on your screen. Continue pressing the DOWN ARROW until you reach the desired command option and then press the ENTER key to select it.

- Dynamic Input can be toggled ON or OFF in the Status Bar by clicking the `DYN` button or by pressing the `F12` function key.

THE DIRECT DISTANCE MODE FOR DRAWING LINES

Another method is available for constructing accurate lines, and it is called drawing by Direct Distance mode. In this method, the direction a line will be drawn in is guided by the location of the cursor. You enter a value, and the line is drawn at the specified distance at the angle specified by the cursor. This mode works especially well for drawing horizontal and vertical lines. Figure 1.36 illustrates an example of how the Direct Distance mode is used.

TRY IT!

Create a new drawing from scratch. Turn Dyn Input mode off and Ortho mode on in the Status Bar. Then use the following command sequence to construct the line segments using the Direct Distance mode of entry. Direct Distance mode ensures the line's length is accurate and Ortho mode ensures the angle is accurate.

Command: L *(For LINE)*

Specify first point: 2.00,2.00

Specify next point or [Undo]: *(Move the cursor to the right and enter a value of* 7.00 *units)*

Specify next point or [Undo]: *(Move the cursor up and enter a value of* 3.00 *units)*

Specify next point or [Close/Undo]: *(Move the cursor to the left and enter a value of* 4.00 *units)*

Specify next point or [Close/Undo]: *(Move the cursor down and enter a value of* 1.00 *units)*

Specify next point or [Close/Undo]: *(Move the cursor to the left and enter a value of* 2.00 *units)*

Specify next point or [Close/Undo]: C *(To close the shape and exit the command)*

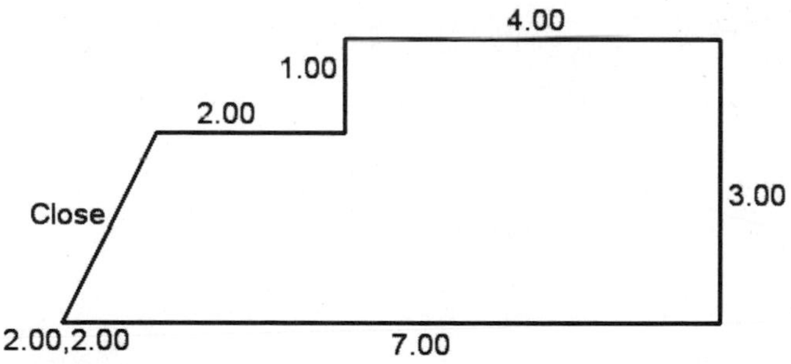

FIGURE 1.36

TIP

If Ortho mode is currently turned on, you can temporarily turn Ortho off while in the `LINE` command by pressing the `SHIFT` key as you drag your cursor to draw the next line.

Figure 1.37 shows another example of an object drawn with Direct Distance mode. Each angle was constructed from the location of the cursor. In this example, Ortho mode is turned off.

TRY IT!

Create a new drawing from scratch. Be sure Ortho mode is turned off. The angles in this drawing are not accurate. Ortho or Polar Tracking mode is normally used with Direct Distance mode to create accurate technical drawings.

Then use the following command sequence to construct the line segments using the Direct Distance mode of entry.

`Command:` `L` *(For LINE)*

`Specify first point:` *(Pick a point at "A")*

`Specify next point or [Undo]:` *(Move the cursor and enter* `3.00`*)*

`Specify next point or [Undo]:` *(Move the cursor and enter* `2.00`*)*

`Specify next point or [Close/Undo]:` *(Move the cursor and enter* `1.00`*)*

`Specify next point or [Close/Undo]:` *(Move the cursor and enter* `4.00`*)*

`Specify next point or [Close/Undo]:` *(Move the cursor and enter* `2.00`*)*

`Specify next point or [Close/Undo]:` *(Move the cursor and enter* `1.00`*)*

`Specify next point or [Close/Undo]:` *(Move the cursor and enter* `1.00`*)*

`Specify next point or [Close/Undo]:` `C` *(To close the shape and exit the command)*

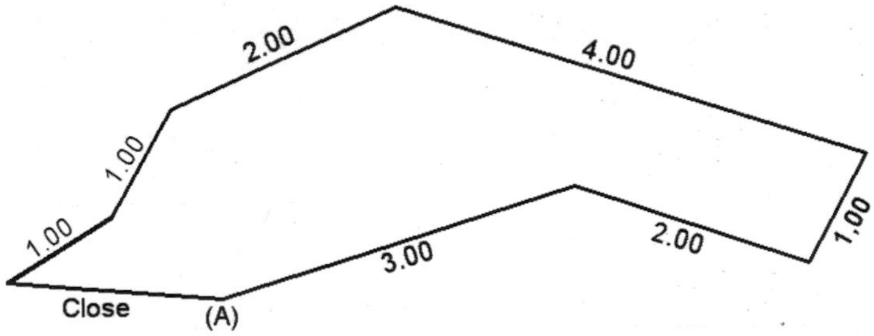

FIGURE 1.37

USING OBJECT SNAP FOR GREATER PRECISION

A major productivity tool that allows locking onto key locations of objects is Object Snap (OSNAP). Figure 1.38 is an example of the construction of a vertical line connecting the endpoint of the arc at "A" with the endpoint of the line at "B." The `LINE` command is entered and the Endpoint mode activated. When the cursor moves over a valid endpoint, an Object Snap symbol appears along with a tooltip indicating which OSNAP mode is currently being used.

> **TRY IT!**
>
> **Open the drawing file 01_Endpoint. Use the illustration in Figure 1.38 and the command sequence below to draw a line segment from the endpoint of the arc to the endpoint of the line.**

```
Command: L (For LINE)
Specify first point: End (For Endpoint mode)
of (Pick the endpoint of the arc at "A" illustrated in the following image)
Specify next point or [Undo]: End (For Endpoint mode)
of (Pick the endpoint of the line at "B")
Specify next point or [Undo]: (Press ENTER to exit this command)
```

Perform the same operation to the other side of this object using the Endpoint mode of OSNAP. The results are shown in Figure 1.38 (Right).

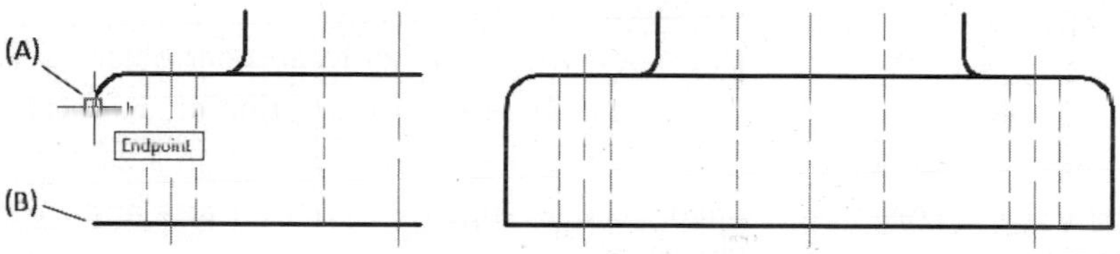

FIGURE 1.38

Object Snap modes can be selected in a number of different ways. Illustrated in Figure 1.39 is the Status Bar. Right-clicking on the Object Snap icon displays the menu containing most Object Snap tools. The following table gives a brief description of each Object Snap mode. In this table, notice the Key-In column. When the Object Snap modes are executed from keyboard input, only the first three letters are required.

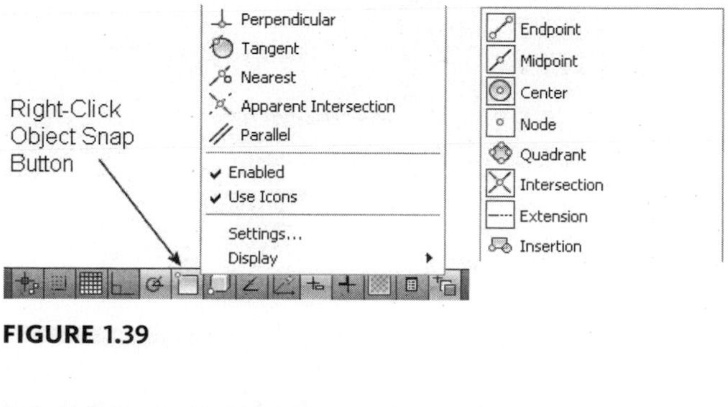

FIGURE 1.39

The following table gives a brief description of each Object Snap mode:

Button	Tool	Key-In	Function
	Center	CEN	Snaps to the centers of circles and arcs
	Endpoint	END	Snaps to the endpoints of lines and arcs
	Extension	EXT	Creates a temporary extension line or arc when your cursor passes over the endpoint of objects; you can specify new points along the temporary line
	From	FRO	Snaps to a point at a specified distance and direction from a selected reference point
	Insert	INS	Snaps to the insertion point of blocks and text
	Intersection	INT	Snaps to the intersections of objects
	Apparent Intersection	INT	Mainly used in creating 3D wireframe models; finds the intersection of points not located in the same plane
	Midpoint	MID	Snaps to the midpoint of lines and arcs
	Midpoint Between 2 Points	M2P	Snaps to the middle of two selected points
	Nearest	NEA	Snaps to the nearest point found along any object
	Node	NOD	Snaps to point objects (including dimension definition points) and text objects (including multiline text and dimension text)
	None	NON	Disables Object Snap
	Osnap Settings	OSNAP	Launches the Drafting Settings dialog box and activates the Object Snap tab
	Parallel	PAR	Draws an object parallel to another object
	Perpendicular	PER	Snaps to a perpendicular location on an object
	Quadrant	QUA	Snaps to four key points located on a circle
	Tangent	TAN	Snaps to the tangent location of arcs and circles

Figure 1.40 shows the Object Snap modes that can be activated when you hold down SHIFT or CTRL and press the right mouse button while within a command such as LINE. This shortcut menu will appear wherever the cursor is positioned in the drawing area.

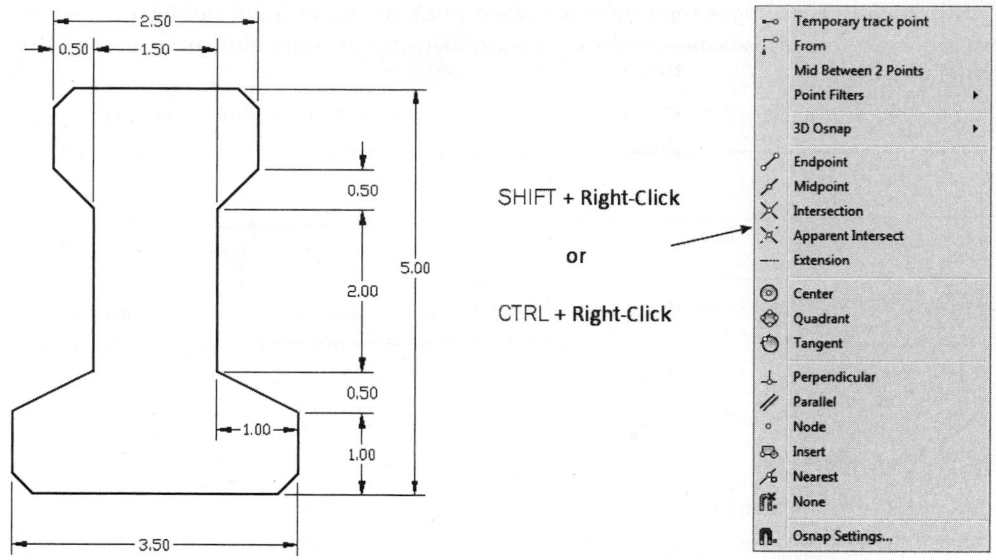

FIGURE 1.40

OBJECT SNAP MODES

Center (Cen)

Use the Center mode to snap to the center of a circle or arc. To accomplish this, activate the mode by clicking the Center button and moving the cursor along the edge of the circle or arc, as shown in Figure 1.41. Notice the AutoSnap symbol appearing at the center of the circle or arc.

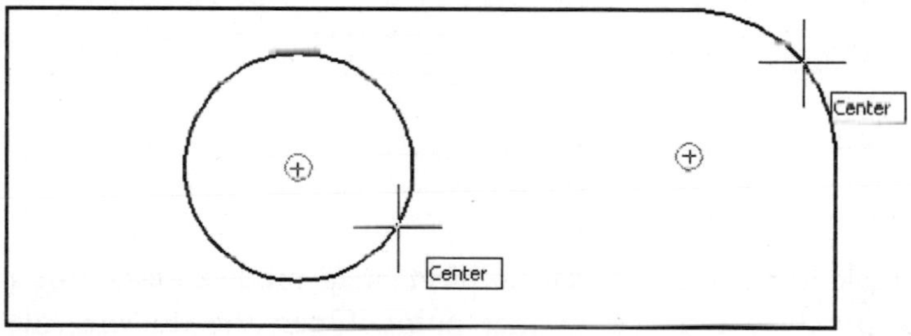

FIGURE 1.41

Endpoint (End)

The Endpoint mode is one of the more popular Object Snap modes; it is helpful in snapping to the endpoints of lines or arcs, as shown in Figure 1.42. One application of Endpoint is during the dimensioning process, where exact distances are needed to produce the desired dimension. Activate this mode by clicking the Endpoint button, and then move the cursor along the edge of the object to snap to the endpoint. In the case of the line or arc shown in Figure 1.42, the cursor does not actually have to be positioned at the endpoint; favoring one end automatically snaps to the closest endpoint.

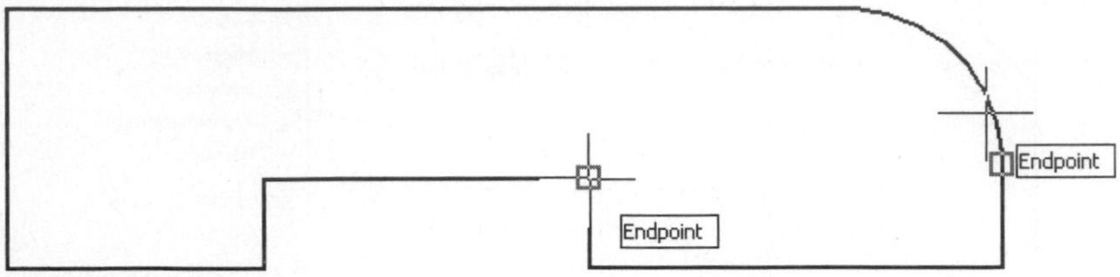

FIGURE 1.42

Extension (Ext)

When you acquire a line or an arc, the Extension mode creates a temporary path that extends from the object. Once the Extension Object Snap is selected, move your cursor over the end of the line at "A," as shown in Figure 1.43, to acquire it. Moving your cursor away provides an extension at the same angle as the line. To un-acquire an extension, simply move your cursor over the end of the line again. A tooltip displays the current extension distance and angle. Acquiring the end of the arc at "B" provides the radius of the arc and displays the current length in the tooltip.

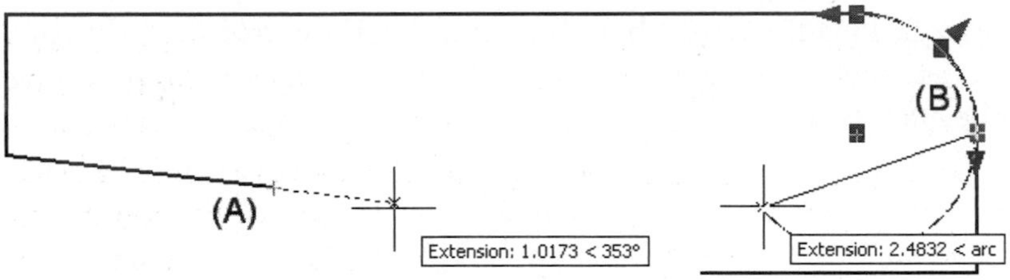

FIGURE 1.43

From (Fro)

Use the From mode along with a secondary Object Snap mode to establish a reference point and construct an offset from that point. Open the drawing file 01_Osnap From. In Figure 1.44, the circle needs to be drawn 1.50 units in the X and Y directions from point "A." The `CIRCLE` command is activated and the Object Snap From mode is used in combination with the Object Snap Intersection mode. The From option requires a base point. Identify the base point at the intersection of corner "A." The next prompt asks for an offset value; enter the relative coordinate value of @1.50,1.50 (this identifies a point 1.50 units in the positive X direction and 1.50 units in the positive Y direction). This completes the use of the From option and identifies the center of the circle at "B." Study the following command sequence to accomplish this operation:

TRY IT!

Open the drawing file 01_Osnap From. Use the illustration and prompt sequence below for constructing a circle inside the shape with the aid of the Object Snap From mode.

```
Command: C (For CIRCLE)
Specify center point for circle or [3P/2P/Ttr (tan tan radius)]: From
Base point: Int (For Intersection Mode)
of (Select the intersection at "A" in the following image)
<Offset>: @1.50,1.50
Specify radius of circle or [Diameter]: D (For Diameter)
Specify diameter of circle: 1.25
```

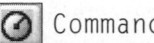

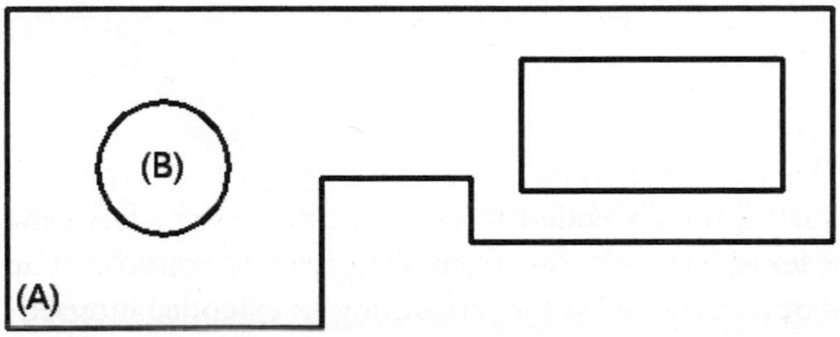

FIGURE 1.44

Insert (Ins)

The Insert mode snaps to the insertion point of an object. In the case of the text object in Figure 1.45 (Left), activating the Insert mode and positioning the cursor anywhere on the text snaps to its insertion point, in this case at the lower-left corner of the text at "A." The other object illustrated in Figure 1.45 (Right) is called a block. It appears to be constructed with numerous line objects; however, all objects that make up the block are considered to be a single object. Blocks can be inserted in a drawing. Typical types of blocks are symbols such as doors, windows, bolts, and so on—anything that is used several times in a drawing. In order for a block to be brought into a drawing, it needs an insertion point, or a point of reference. The Insert mode, when you position the cursor on a block, will snap to the insertion point at "B" of that block.

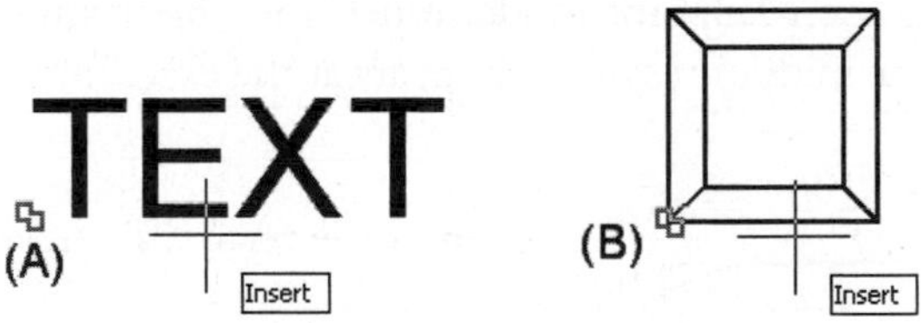

FIGURE 1.45

Intersection (Int)

Another popular Object Snap mode is Intersection. Use this mode to snap to the intersection of two objects. Position the cursor anywhere near the intersection of two objects and the intersection symbol appears. See Figure 1.46.

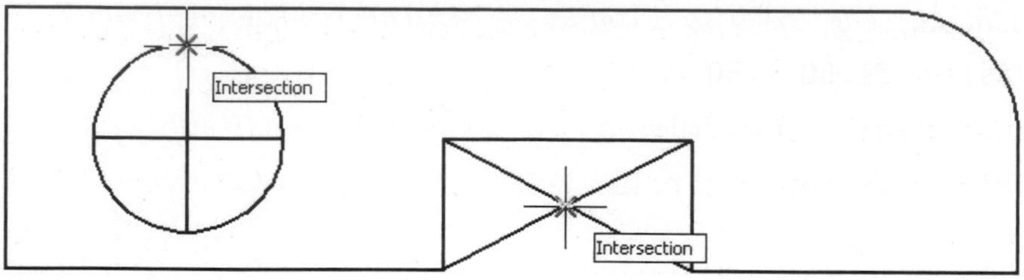

FIGURE 1.46

Extended Intersection (Int)

Another type of intersection snap is the Extended Intersection mode, which is used to snap to an intersection not considered obvious from the previous example. The same Object Snap Intersection button is utilized for performing an extended intersection operation. Figure 1.47 shows two lines that do not intersect. Activate the Extended Intersection mode and pick both lines. Notice the intersection symbol present where the two lines, if extended, would intersect.

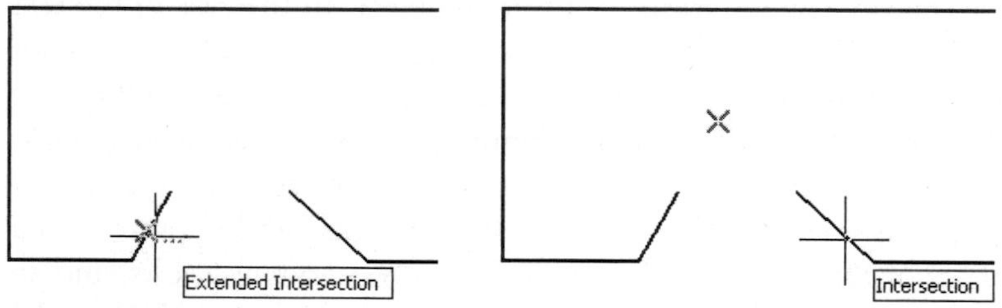

FIGURE 1.47

Midpoint (Mid)

The Midpoint mode snaps to the midpoint of objects. Line and arc examples are shown in Figure 1.48. When activating the Midpoint mode, touch the object anywhere with some portion of the cursor; the midpoint symbol appears at the exact midpoint of the object.

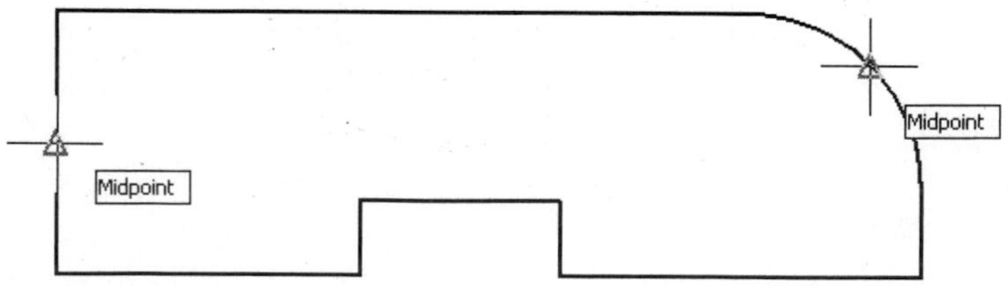

FIGURE 1.48

Midpoint of Two Selected Points (M2P)

This Object Snap mode snaps to the midpoint of two selected points. To access this mode, type either `M2P` or `MTP` at the Command prompt. It can also be found by pressing `SHIFT` + Right Mouse Button to display the Object Snap menu, as shown in Figure 1.49 (Left).

The following command sequence and illustration in Figure 1.49 show the construction of a circle at the midpoint of two selected points.

```
Command: C (For CIRCLE)
Specify center point for circle or [3P/2P/Ttr (tan tan radius)]: M2P
First point of mid: End
of (Pick the endpoint at "A")
Second point of mid: End
of (Pick the endpoint at "B")
Specify radius of circle or [Diameter]: 0.50
```

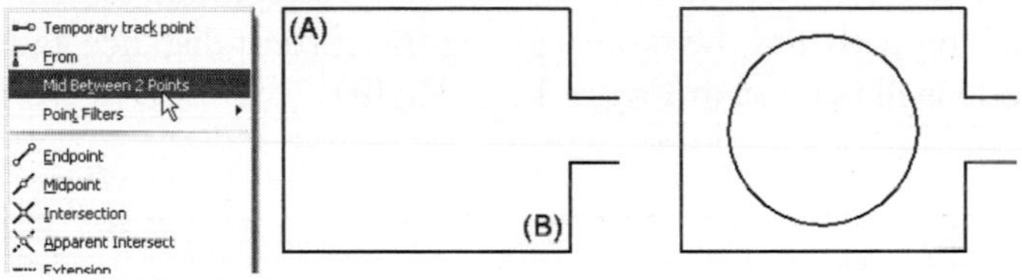

FIGURE 1.49

Nearest (Nea)

The Nearest mode snaps to the nearest point it finds on an object. Use this mode when a point on an object needs to be selected and an approximate location on the object is sufficient. The nearest point is calculated based on the closest distance from the intersection of the crosshairs perpendicular to the object or the shortest distance from the crosshairs to the object. In Figure 1.50, the appearance of the Nearest symbol helps to show where the point identified by this mode is actually located.

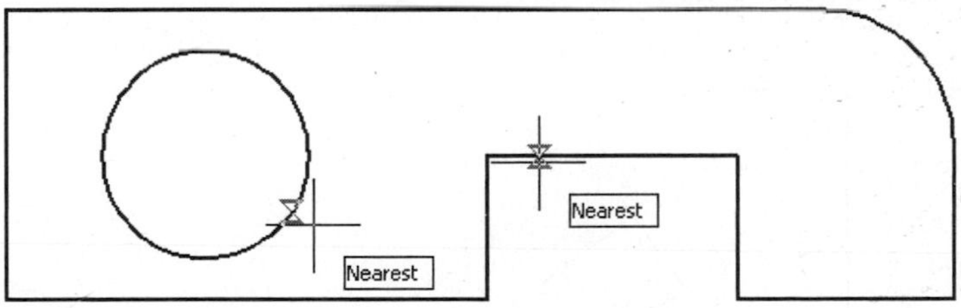

FIGURE 1.50

Node (Nod)

The Node mode snaps to a node or point. Picking the point in Figure 1.51 snaps to its center.

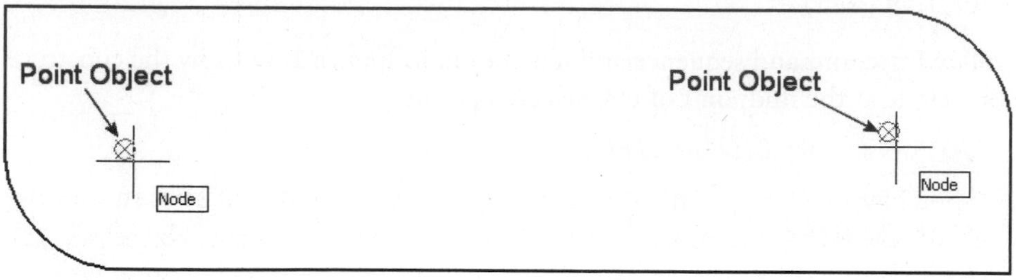

FIGURE 1.51

Parallel (Par)

Use the Parallel mode to construct a line parallel to another line. In Figure 1.52, the `LINE` command is started and a beginning point of the line is picked. With the Parallel Object Snap activated, hover the cursor over the existing at "A" and the Parallel symbol appears. Finally, moving the cursor to the approximate position that makes the new line parallel to the one just acquired allows the Parallel mode to construct a parallel line, the tracking path and the tooltip giving the current distance and angle. The result of this mode is illustrated in Figure 1.52 (Right).

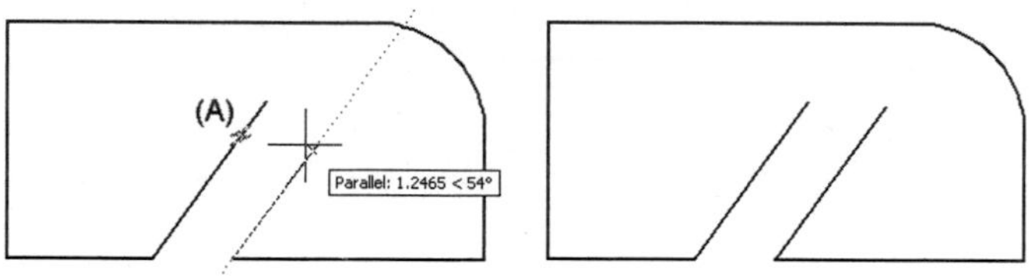

FIGURE 1.52

Perpendicular (Per)

The Perpendicular mode is helpful for snapping to an object normal (or perpendicular) from a previously identified point. Figure 1.53 shows a line segment drawn perpendicular from the point at "A" to the inclined line "B." A 90° angle is formed with the perpendicular line segment and the inclined line "B." With this mode, the user can also construct lines perpendicular to circles.

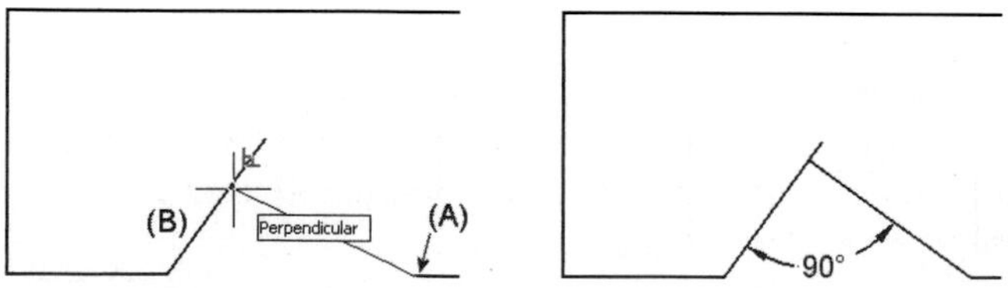

FIGURE 1.53

Quadrant (Qua)

Circle quadrants are defined as points located at the 0°, 90°, 180°, and 270° positions of a circle, as in Figure 1.54. Using the Quadrant mode will snap to one of these four positions as the edge of a circle or arc is selected. In the example of the circle in Figure 1.54, the edge of the circle is selected by the cursor location. The closest quadrant to the cursor is selected.

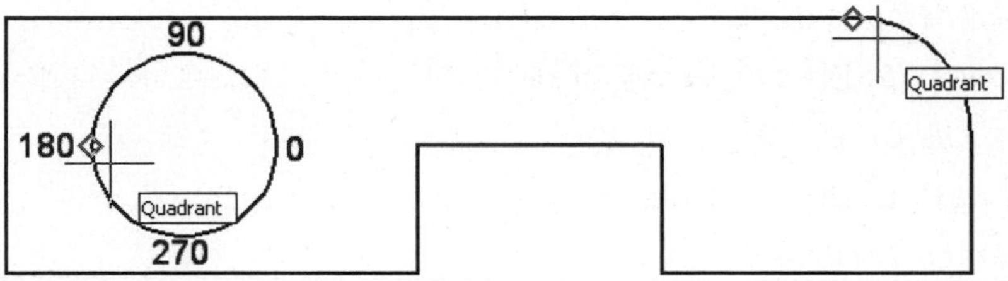

FIGURE 1.54

Tangent (Tan)

The Tangent mode is helpful in constructing lines tangent to other objects such as the circles in Figure 1.55. In this case, the Deferred Tangent mode is being used in conjunction with the LINE command. The point at "A" is first picked at the bottom of the circle using the Tangent mode. When dragged to the next location, the line will be tangent at point "A." Then, with Tangent mode activated and the location at "B" picked, the line will be tangent to the large circle near "B." The results are illustrated in Figure 1.55 (Right).

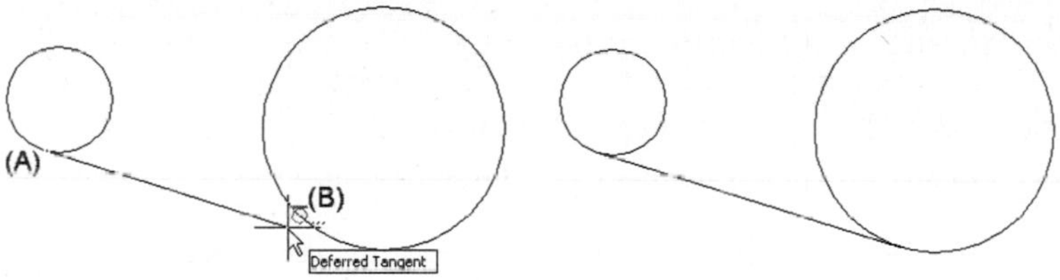

FIGURE 1.55

TRY IT!

Open the drawing file 01_Tangent. Use the previous illustration and the command sequence for constructing a line segment tangent to two circles:

```
Command: L (For LINE)
Specify first point: Tan (For Tangent mode)
to (Select the circle near "A")
Specify next point or [Undo]: Tan (For Tangent mode)
to (Select the circle near "B")
```

TRY IT!

Open the drawing file 01_Osnap. Various objects consisting of lines, circles, arcs, points, and blocks need to be connected with line segments at their key locations. Use the prompt sequence and Figure 1.56 for performing this operation.

```
Command: L (For LINE)
Specify first point: End
of (Pick the endpoint at "A")
Specify next point or [Undo]: Nod
```

```
of (Pick the node at "B")
Specify next point or [Undo]: Tan
to (Pick the circle at "C")
Specify next point or [Close/Undo]: Int
of (Pick the intersection at "D")
Specify next point or [Close/Undo]: Int
of (Pick the line at "E")
and (Pick the horizontal line at "F")
Specify next point or [Close/Undo]: Qua
of (Pick the circle at "G")
Specify next point or [Close/Undo]: Cen
of (Pick the arc at "H")
Specify next point or [Close/Undo]: Mid
of (Pick the line at "J")
Specify next point or [Close/Undo]: Per
to (Pick the line at "K")
Specify next point or [Close/Undo]: Ins
of (Pick on the I-Beam symbol near "L")
Specify next point or [Close/Undo]: Nea
to (Pick the circle at "M")
Specify next point or [Close/Undo]: (Press ENTER to exit this
command)
```

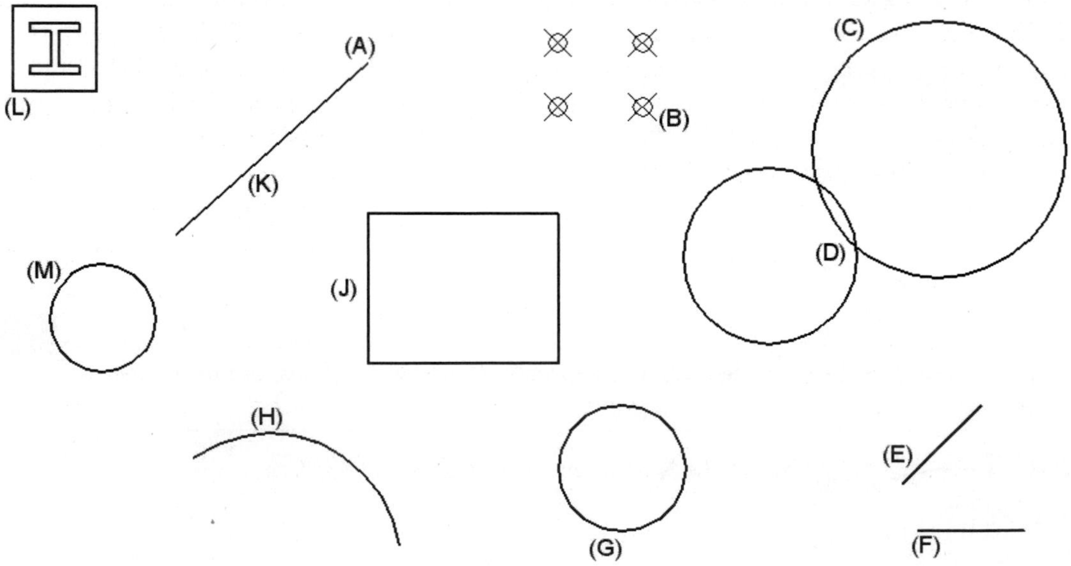

FIGURE 1.56

CHOOSING RUNNING OBJECT SNAP

Object Snap modes selected from the shortcut menu or entered at the keyboard are referred to as Override Object Snaps. While you are ensured of the type of snap selected, you have to select the Object Snap mode every time you use it. It is possible to make the Object Snap mode or modes continuously present through Running Object Snaps. Right-click the OSNAP button located in the Status Bar at the bottom of the drawing area, as shown in Figure 1.57. Pick the desired Running Object Snaps from the shortcut menu displayed or click Settings to activate the Drafting Settings dialog box illustrated in Figure 1.55 (Right). By default, the Endpoint, Center, Intersection, and Extension modes are automatically selected. Whenever the cursor lands over an object supported by one of these four modes, a yellow symbol appears to alert the user to the mode. Other Object Snap modes can be selected by checking their appropriate boxes in the dialog box; removing the check disables the mode.

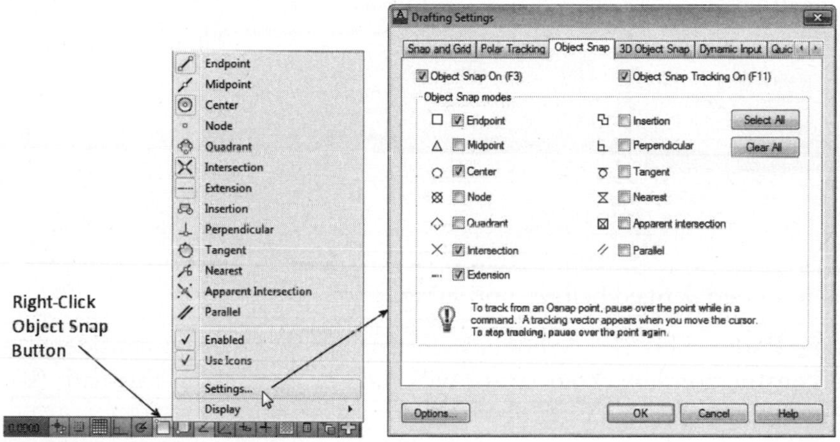

FIGURE 1.57

These Object Snap modes remain in effect during drawing until you click the OSNAP button illustrated in the Status Bar in Figure 1.58; this turns off the current running Object Snap modes. To reactivate the running Object Snap modes, click the OSNAP button again and the previous set of Object Snap modes will be back in effect. Object Snap can also be activated and deactivated by pressing the F3 function key. It is also important to note that anytime you select an Object Snap from the toolbar, cursor menu, or by typing it in, it overrides the Running Osnap for that single operation. This ensures that you only snap to that specific mode and not accidentally to one of the set Running Osnaps.

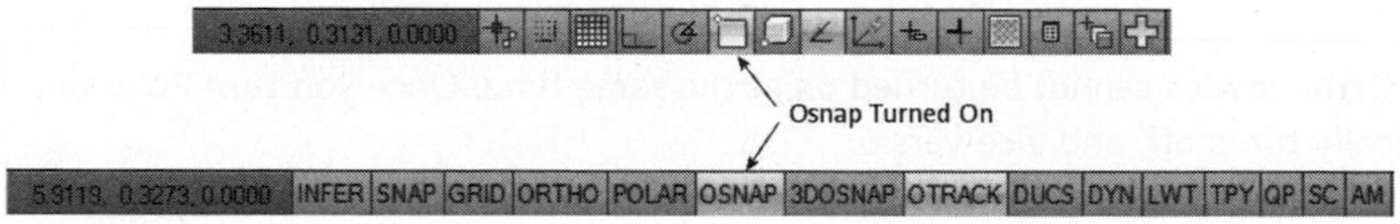

FIGURE 1.58

POLAR TRACKING

Earlier in this chapter, the Direct Distance mode was highlighted as an efficient means of constructing precise length lines. To ensure that the line direction is also accurate, you can use a tool such as Polar Tracking. This mode allows the cursor to follow a tracking path that is controlled by a preset angular increment. The POLAR button located at the bottom of the display in the Status area turns this mode on or off. Right-clicking POLAR in the Status Bar at the bottom of the screen provides a shortcut menu where you may select the tracking angle. You can also select Settings from the menu to display the Drafting Settings dialog box shown in Figure 1.59 (Right). Notice the Polar Tracking tab is already selected.

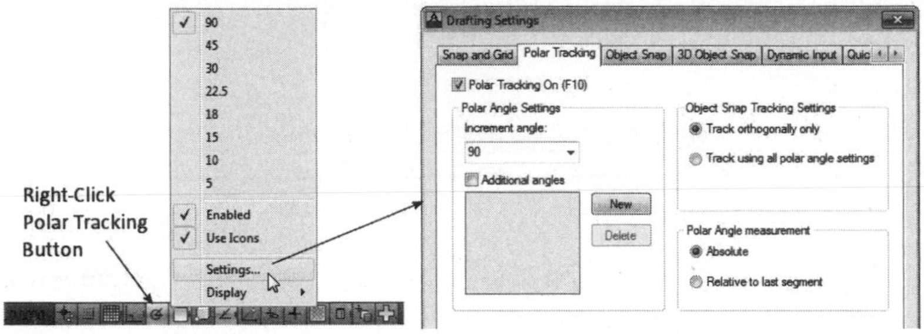

FIGURE 1.59

A few general terms are defined before continuing:

Tracking Path—This is a temporary dotted line that can be considered a type of construction line. Your cursor will glide or track along this path (see Figure 1.60).

Tooltip—This displays the current cursor distance and angle away from the tracking point (see Figure 1.60).

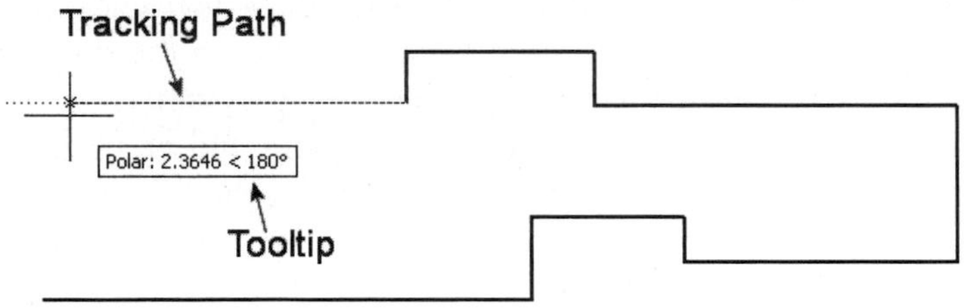

FIGURE 1.60

TIP

Both Polar and Ortho modes cannot be turned on at the same time. Once you turn Polar on, Ortho automatically turns off, and vice versa.

TRY IT!

To see how the Polar Tracking mode functions, construct an object that consists of line segments at 10° angular increments. Create a new drawing starting from scratch. Then, set the angle increment through the Polar Tracking tab of the Drafting Settings dialog box to 10°, as shown in Figure 1.61.

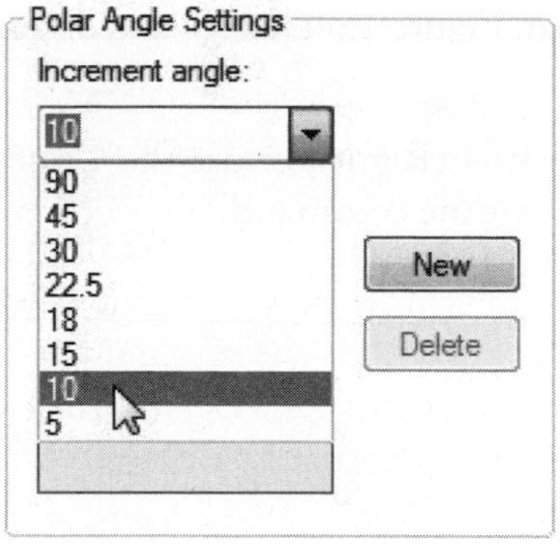

FIGURE 1.61

Start the LINE command, anchor a starting point at "A," and move the cursor to the upper right until the tooltip reads 20° as shown in Figure 1.62 (Left). Enter a value of 2 units for the length of the line segment.

Move the cursor up and to the left until the tooltip reads 110°, as shown in Figure 1.62 (Right), and enter a value of 2 units. (This will form a 90° angle with the first line.)

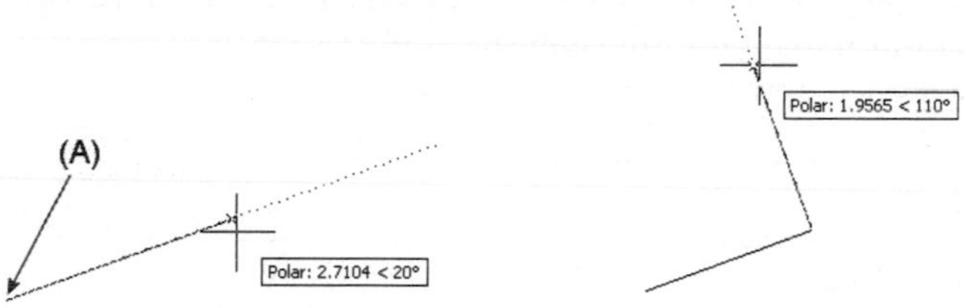

FIGURE 1.62

Move the cursor until the tooltip reads 20°, as shown in Figure 1.63 (Left), and enter a value of 1 unit.

Move the cursor until the tooltip reads 110°, as shown in Figure 1.63 (Right), and enter a value of 1 unit.

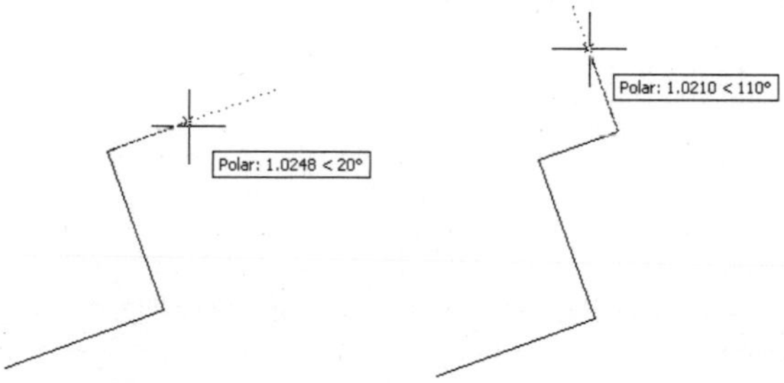

FIGURE 1.63

Move the cursor until the tooltip reads 200°, as shown in Figure 1.64 (Left), and enter a value of 3 units.

Move the cursor to the endpoint, as shown in Figure 1.64 (Right), or use the Close option of the `LINE` command to close the shape and exit the command.

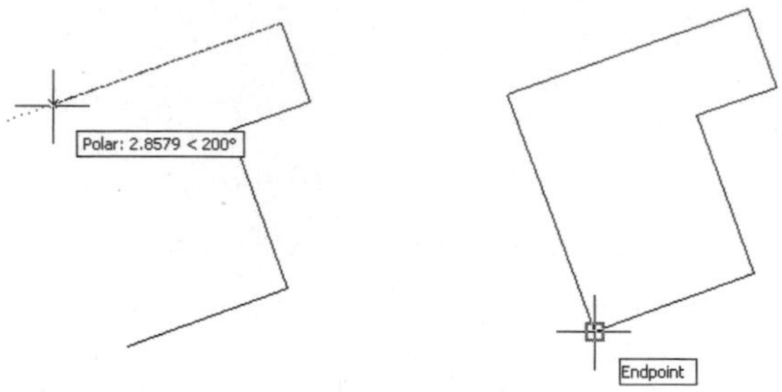

FIGURE 1.64

SETTING A POLAR SNAP VALUE

An additional feature of using polar snap is illustrated in Figure 1.65. Clicking the Snap and Grid tab of the Drafting Settings dialog box displays this option. Clicking the Polar Snap option found along the lower-left corner of the dialog box allows the user to enter a polar distance. When SNAP and POLAR are both turned on and the cursor is moved to draw a line, not only will the angle be set but the cursor will also jump to the next increment set by the polar snap value.

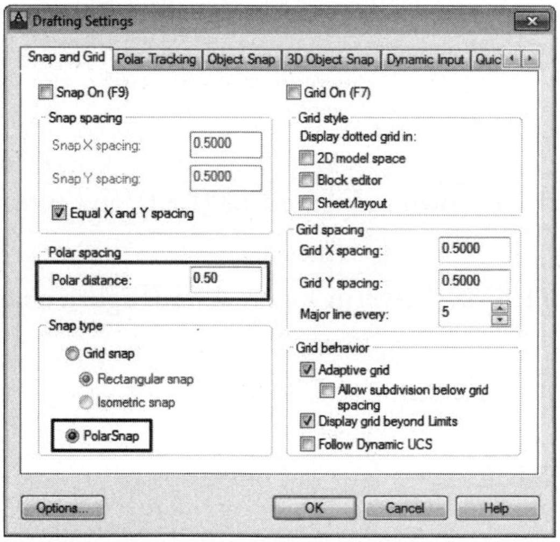

FIGURE 1.65

TRY IT!

Open the drawing file 01_Polar. Set the polar angle to 30° and a polar snap distance to 0.50 unit increments. Be sure POLAR and SNAP are both turned on in your Status Bar and that all other modes are turned off. Begin constructing the object in Figure 1.66 using the Command prompt sequence below as a guide.

```
Command: L (For LINE)
Specify first point: 7.00,4.00
Specify next point or [Undo]: (Move your cursor down until the
tooltip reads Polar: 2.5000<270 and pick a point)
Specify next point or [Undo]: (Move your cursor right until
the tooltip reads Polar: 1.5000<0 and pick a point)
Specify next point or [Close/Undo]: (Polar: 2.0000<30)
Specify next point or [Close/Undo]: (Polar: 2.0000<60)
Specify next point or [Close/Undo]: (Polar: 2.5000<90)
Specify next point or [Close/Undo]: (Polar: 3.0000<150)
Specify next point or [Close/Undo]: (Polar: 1.5000<180)
Specify next point or [Close/Undo]: (Polar: 2.5000<240)
Specify next point or [Close/Undo]: (Polar: 2.5000<120)
Specify next point or [Close/Undo]: (Polar: 1.5000<180)
Specify next point or [Close/Undo]: (Polar: 3.0000<210)
Specify next point or [Close/Undo]: (Polar: 2.5000<270)
Specify next point or [Close/Undo]: (Polar: 2.0000<300)
Specify next point or [Close/Undo]: (Polar: 2.0000<330)
Specify next point or [Close/Undo]: (Polar: 1.5000<0)
Specify next point or [Close/Undo]: (Polar: 2.5000<90)
Specify next point or [Close/Undo]: C (To close the shape)
```

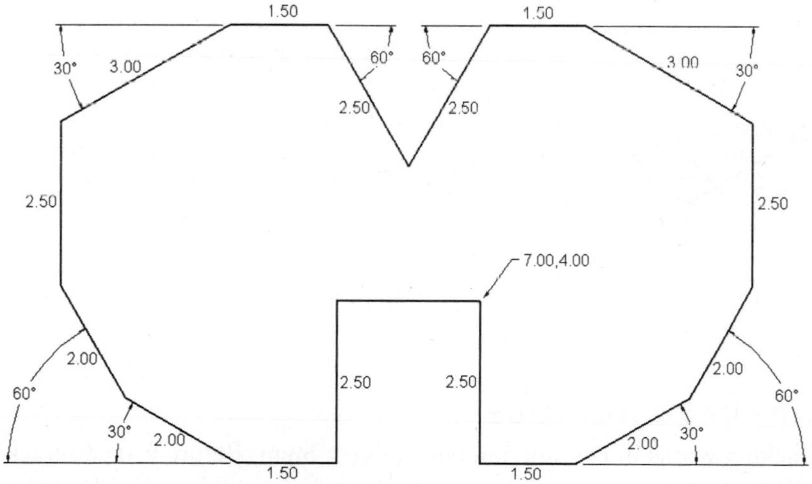

FIGURE 1.66

SETTING A RELATIVE POLAR ANGLE

An additional feature of using polar snap is illustrated in Figure 1.67. When activating the Polar Tracking tab of the Drafting Settings dialog box, located in the lower-right corner are two settings that deal with the Polar Angle measurement; they are Absolute and Relative to last segment.

Absolute—This is the default setting when dealing with Polar Angle measurement. This setting controls all angle measurements based on the position of the current user coordinate system, the icon located in the lower-left corner of all AutoCAD drawing screens.

Relative to last segment—When changing the Polar Angle measurement to Relative to last segment, the Polar Tracking angle is based on the last line segment drawn.

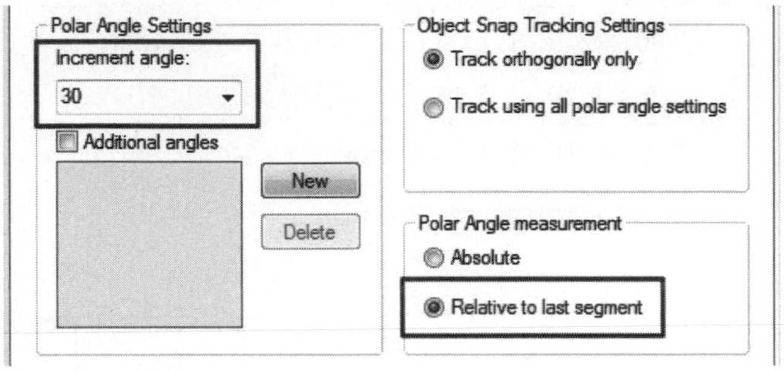

FIGURE 1.67

To get a better understanding of the two different Polar Angle measurement settings, study Figure 1.68. The illustration on the left is an example of the Absolute setting. The 150° angle was drawn from point (A) to point (B). This angle is derived from the absolute position of angle 0° (zero) set by default to the 3 o'clock position as defined in the Drawing Units dialog box. In the illustration on the right, the same line segment is drawn. However, this time the Relative to last segment setting is used. Notice how the 120° angle is calculated. The angle is based on the last line segment, not on an angle calculated in relation to 0° (zero). This is the reason for the parallel line segment at (C).

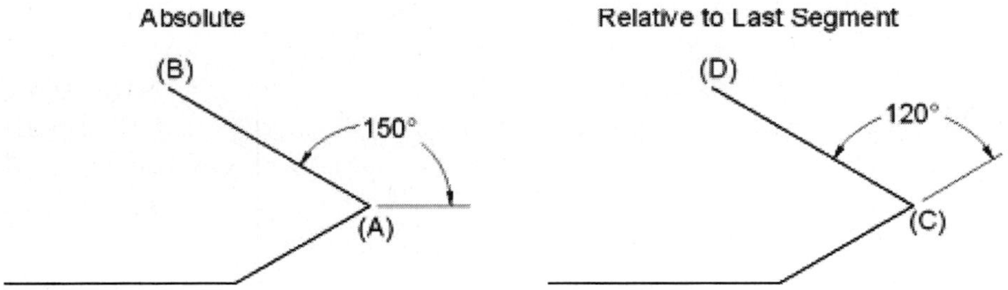

FIGURE 1.68

OBJECT SNAP TRACKING MODE

Object Snap Tracking works in conjunction with Object Snap. Before you can track from an Object Snap point, you must first set an Object Snap mode or modes from the Object Snap tab of the Drafting Settings dialog box. Object Snap Tracking can be toggled ON or OFF with the OTRACK button, which is located in the Status Bar shown in Figure 1.69.

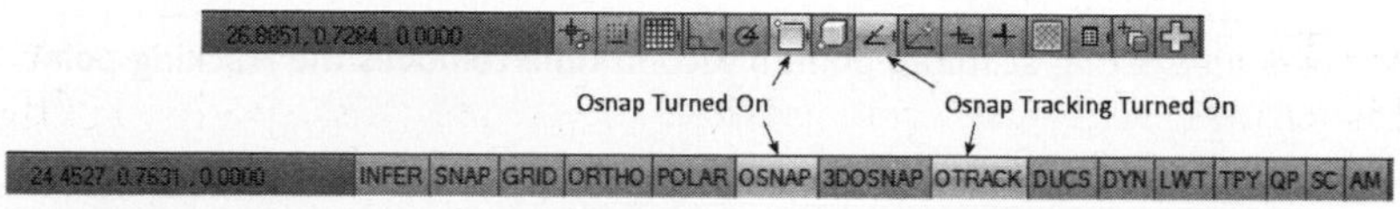

FIGURE 1.69

The advantage of using Object Snap Tracking is in the ability to choose or acquire points to be used for construction purposes. Acquired points are temporarily selected by hovering the cursor over the point versus selecting with the mouse. Care must be taken when acquiring points that the points are in fact not picked. They are used only for construction purposes. For example, two line segments need to be added to the object, as shown in Figure 1.70 (Left), to form a rectangle. Here is how you perform this operation using Polar Tracking.

TRY IT!

Open the drawing file 01_Otrack Lines. Verify in the Status Bar that POLAR, OSNAP, and OTRACK are all turned on. Be sure that Running Osnap is set to Endpoint mode. Enter the `LINE` command and pick a starting point for the line at "A." Then, move the cursor directly to the left until the tooltip reads 180°, as shown in Figure 1.70 (Left). The starting point for the next line segment will be acquired.

Rather than enter the length of this line segment, move your cursor over the top of the corner at "B" to acquire this point (be careful not to pick the point here). Then move your cursor up until the tooltip reads 90°, as shown in Figure 1.70 (Right).

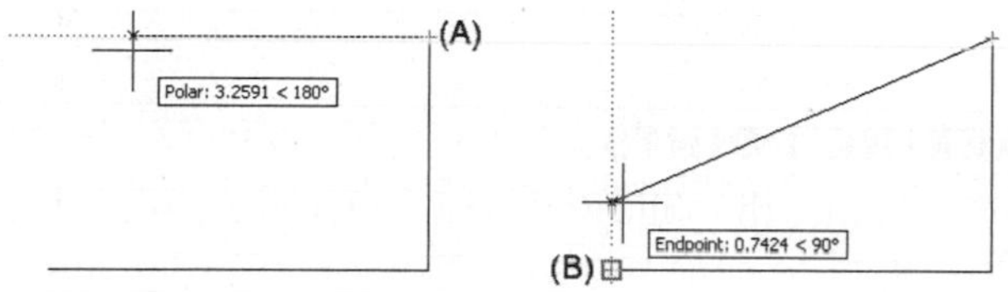

FIGURE 1.70

Move your cursor up until the tooltip now reads angles of 90° and 180°. Also notice the two tracking paths intersecting at the point of the two acquired points. Picking this point at "C" will construct the horizontal line segment, as shown in Figure 1.71 (Left). Finally, slide your cursor to the endpoint at "D" to complete the rectangle, as shown in Figure 1.71 (Right).

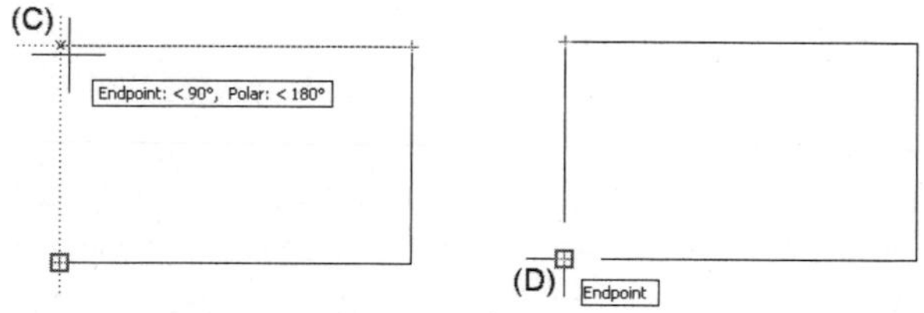

FIGURE 1.71

TIP

Pausing the cursor over an existing acquired point a second time removes the tracking point from the object.

TRY IT!

Open the drawing file 01_Otrack Pipes. Set Running Osnap mode to Midpoint. Set the polar angle to 90°. Be sure POLAR, OSNAP, and OTRACK are all turned on. Enter the `LINE` command and connect all fittings with lines illustrated in Figure 1.72.

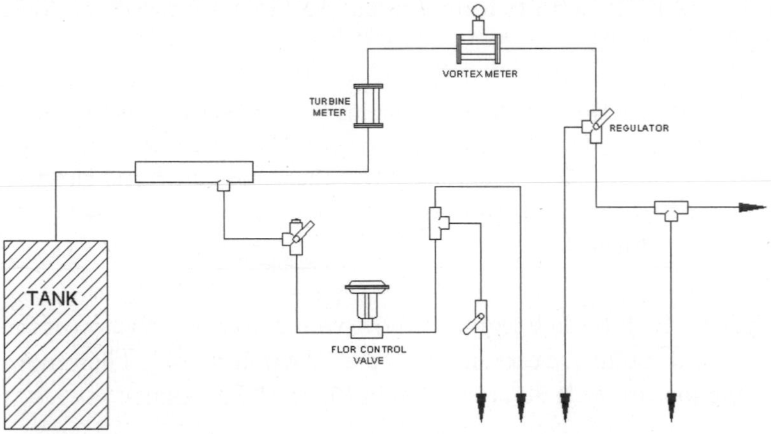

FIGURE 1.72

USING TEMPORARY TRACKING POINTS

Another powerful construction tool includes the ability to use an extension path along with the Temporary Tracking Point tool to construct objects under difficult situations, as illustrated in Figure 1.73.

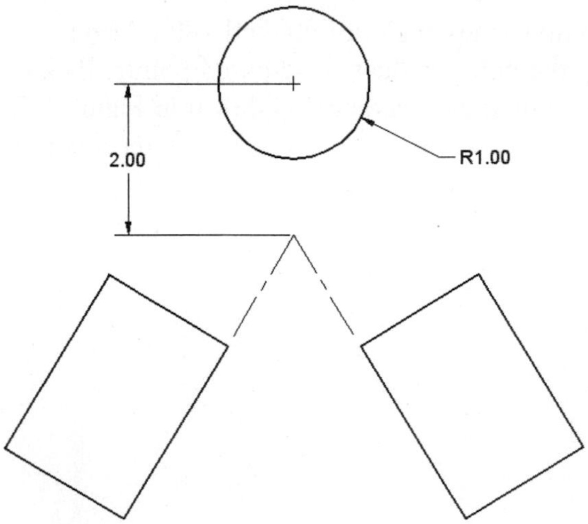

FIGURE 1.73

An extension path is similar to a tracking path except that it is present when the Object Snap Extension mode is activated. To construct the circle in relation to the two inclined rectangles, follow the next series of steps.

TRY IT!

Open the drawing file 01_Temporary Point. Set Running Osnap to Endpoint, Intersection, Center, and Extension. Check to see that OSNAP and OTRACK are turned on and all other modes are turned off. Activate the `CIRCLE` command; this prompts you to specify the center point for the circle. Move the cursor over the corner of the right rectangle at "A" to acquire this point, as shown in Figure 1.74 (Left). Move the cursor up and to the left, making sure the tooltip lists the Extension mode.

With the point acquired at "A," move the cursor over the corner of the left rectangle at "B" and acquire this point. Move the cursor up and to the right, making sure the tooltip lists the Extension mode, as shown in Figure 1.74 (Right).

Move the cursor until both acquired points intersect, as shown in Figure 1.74 (Right). The center of the circle is located 2 units above this intersection. Click the Object Snap Temporary Tracking button from the shortcut menu (`SHIFT` or `CTRL` and press the right mouse button) and pick this intersection, as shown in Figure 1.74 (Right).

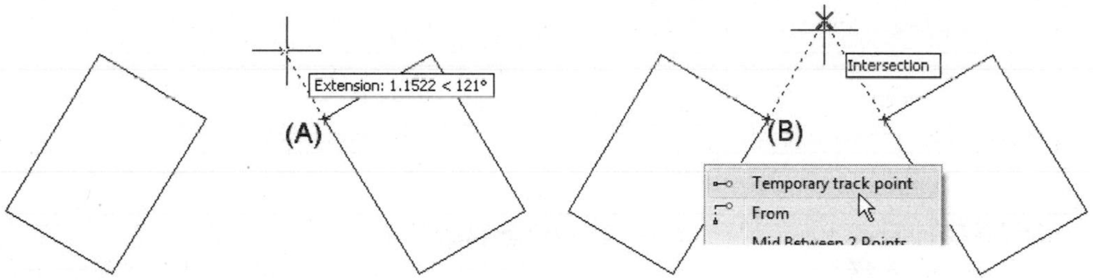

FIGURE 1.74

Next, move the cursor directly above the temporary tracking point, as shown in Figure 1.75 (Left). The tooltip should read 90°. Entering a value of 2 units identifies the center of the circle. Finally, enter the radius for the circle as 1 unit.

The completed construction operation is illustrated in Figure 1.75 (Right).

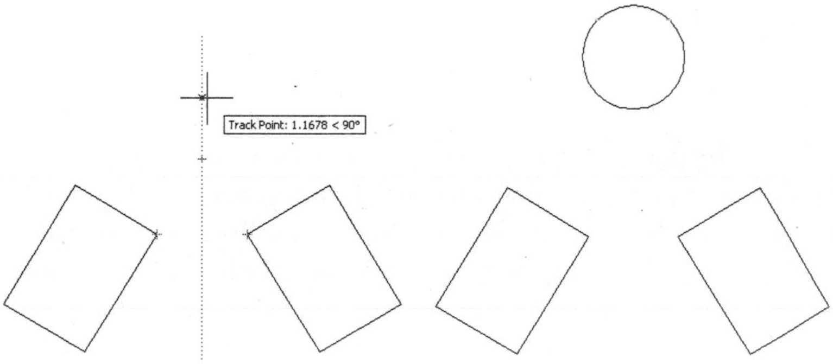

FIGURE 1.75

ALTERNATE METHODS USED FOR PRECISION DRAWING: CARTESIAN COORDINATES

Before drawing precision geometry such as lines and circles, it is essential to have an understanding of coordinate systems. The Cartesian or rectangular coordinate system is constructed of an orthogonal axis intersecting at an origin that creates four quadrants, allowing location of any point by specifying the coordinates. A coordinate is made up of a horizontal and vertical pair of numbers identified as X and Y. The coordinates are then plotted on a type of graph or chart. An example of a rectangular coordinate system is shown in Figure 1.76. The coordinates of the origin are 0,0. From the origin, all positive directions move up and to the right. All negative directions move down and to the left.

The coordinate axes are divided into four quadrants that are labeled I, II, III, and IV, as shown in Figure 1.76. In Quadrant I, all X and Y values are positive. Quadrant II has a negative X value and positive Y value. Quadrant III has negative values for X and Y. Quadrant IV has positive X values and negative Y values.

NOTE

When you begin a drawing in AutoCAD, the screen display reflects Quadrant I of the Cartesian coordinate system, as shown in Figure 1.76. The origin 0,0 is located in the lower-left corner of the drawing screen.

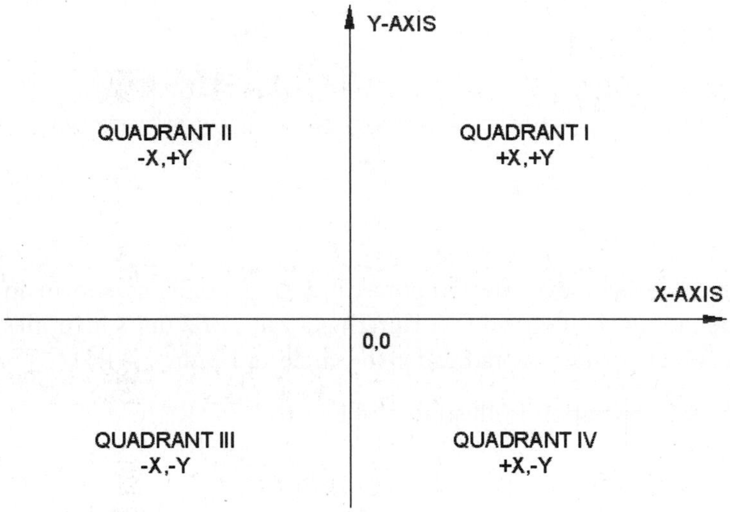

FIGURE 1.76

For each set of (X,Y) coordinates, X values represent distances from the origin horizontally to the right if positive and horizontally to the left if negative. Y values represent distances from the origin vertically up if positive and vertically down if negative. Figure 1.77 shows a series of coordinates plotted on the number lines. One coordinate is identified in each quadrant to show the positive and negative values. As an example, coordinate 3,2 in Quadrant I represents a point 3 units to the right and 2 units vertically up from the origin. The coordinate −5,3 in Quadrant II represents a point 5 units to the left and 3 units vertically up from the origin. Coordinate −2,−2 in Quadrant III represents a point 2 units to the left and 2 units vertically down from the origin. Last, coordinate 2,–4 in Quadrant IV represents a point 2 units to the right and 4 units vertically down from the origin.

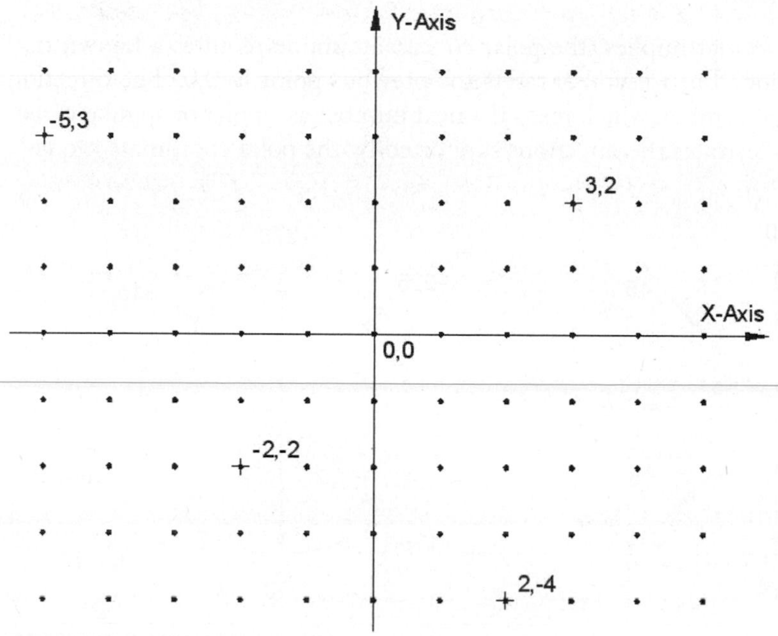

FIGURE 1.77

ABSOLUTE COORDINATE MODE FOR DRAWING LINES

When drawing geometry such as lines, the user must use a method of entering precise distances, especially when accuracy is important. This is the main purpose of using coordinates. The simplest and most elementary form of coordinate values is absolute coordinates. Absolute coordinates conform to the following format:

X,Y

One difficulty with using absolute coordinates is that all coordinate values must refer back to the origin 0,0. This origin on the AutoCAD screen is usually located in the lower-left corner when a new drawing is created.

RELATIVE COORDINATE MODE FOR DRAWING LINES

With absolute coordinates, the horizontal and vertical distance from the origin at 0,0 must be kept track of at all times in order for the correct coordinate to be entered. With complicated objects, this is difficult to accomplish, and as a result, the wrong coordinate may be entered. It is possible to reset the last coordinate to become a new origin or 0,0 point. The new point would be relative to the previous point, and for this reason, this point is called a relative coordinate. The format is as follows:

@X,Y

In this format, we use the same X and Y values with one exception: the At symbol or @ resets the previous point to 0,0 and makes entering coordinates less confusing.

POLAR COORDINATE MODE FOR DRAWING LINES

Another popular method of entering coordinates is the polar coordinate mode. The format is as follows:

@Distance <Direction

As the preceding format implies, the polar coordinate mode requires a known distance and a direction. The @ symbol resets the previous point to 0,0. The direction is preceded by the < symbol, which reads the next number as a polar or angular direction. Figure 1.78 illustrates the directions supported by the polar coordinate mode.

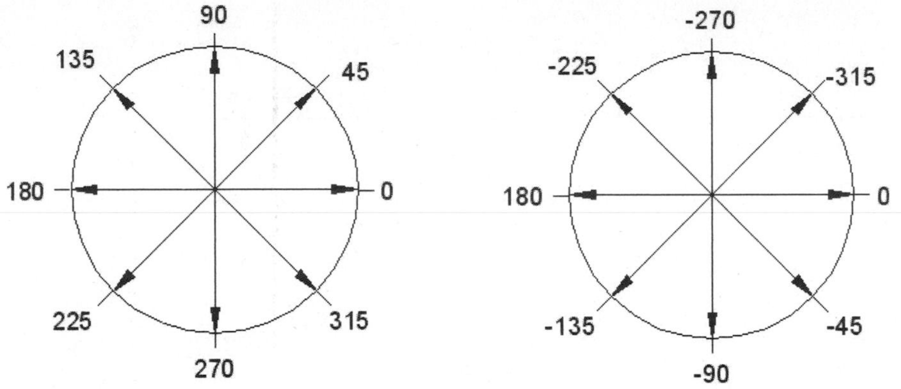

FIGURE 1.78

COMBINING COORDINATE MODES FOR DRAWING LINES

So far, the preceding pages concentrated on using each example of coordinate modes (absolute, relative, and polar) separately to create geometry. While the examples focused on each individual mode, it is important to note that maximum productivity is usually obtained through use of a combination of modes during a drawing session. It is fairly common to use one, two, or three coordinate modes in combination with one another. In Figure 1.79, the drawing starts with an absolute coordinate, changes to a polar coordinate, and changes again to a relative coordinate. The user should develop proficiency in each mode in order to be most productive.

TRY IT!

Create a new drawing file starting from scratch. Turn Dynamic Input off for this Try It! exercise. Use the LINE Command prompts below and Figure 1.79 to construct the shape. When finished, you can turn Dynamic Input back on.

```
Command: L (For LINE)
Specify first point: 2,2 (at "A")Absolute
Specify next point or [Undo]: @3<90 (to "B")Polar
Specify next point or [Undo]: @2,2 (to "C")Relative
Specify next point or [Close/Undo]: @6<0 (to "D")Polar
Specify next point or [Close/Undo]: @5<270(to "E")Polar
Specify next point or [Close/Undo]: @3<180(to "F")Polar
Specify next point or [Close/Undo]: @3<90 (to "G")Polar
Specify next point or [Close/Undo]: @2<180 (to "H")Polar
```

Specify next point or [Close/Undo]: @-3,-3 *(back to "A") Relative*

Specify next point or [Close/Undo]: *(Press* ENTER *to exit this command)*

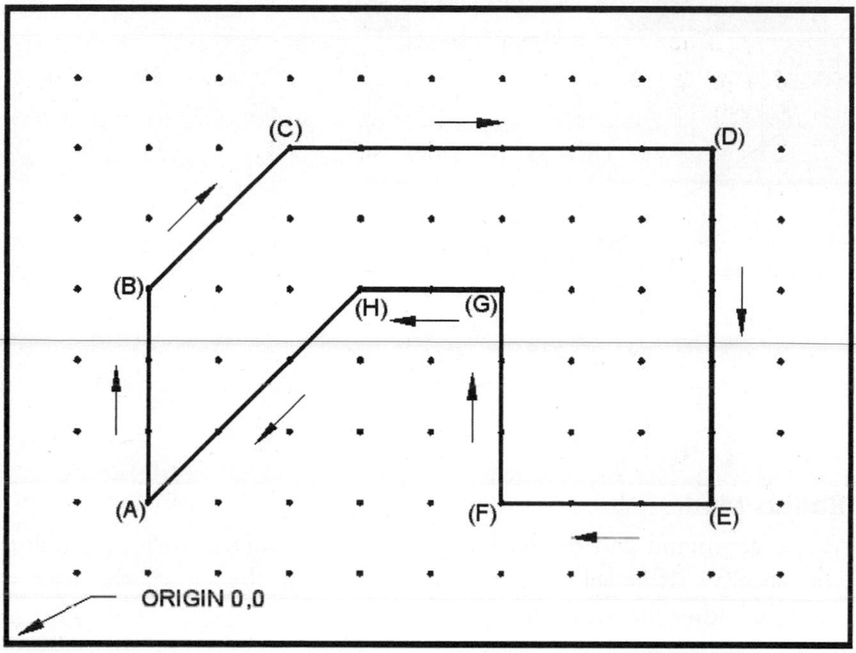

FIGURE 1.79

NOTE

In the previous Try It! exercise it is not necessary to turn Dynamic Input off. In fact, with Dynamic Input on it is not necessary to enter the "@" symbol for relative or polar coordinates. Try repeating the exercise with Dynamic Input on and enter the coordinates without the "@" symbol. Dynamic Input allows you to start with an absolute coordinate but after that it assumes all values are relative. A "#" symbol is required if you want to designate a coordinate as absolute. In the exercise you could finish the line sequence with an absolute coordinate (2,2 instead of @-3,-3). With Dynamic Input on, you would need to enter #2,2.

CONSTRUCTING CIRCLES

The CIRCLE command constructs circles of various radii or diameter. This command can be selected from any of the following:

- From the Ribbon > Home Tab > Draw Panel
- From the Menu Bar (Draw > Circle)
- From the Draw Toolbar of the AutoCAD Classic Workspace
- From the keyboard (C or CIRCLE)

This command can be activated by clicking the Circle icon in the Draw panel of the Ribbon. Selecting the arrow next to the icon displays the cascading menu shown in Figure 1.80. All supported methods of constructing circles are displayed in the list. Circles may be constructed by providing either a radius or diameter. This command also supports circles defined by two or three points and construction of a circle tangent to other objects in the drawing. These last three modes will be discussed in Chapter 5, "Performing Geometric Constructions." Once a method is selected from the menu, that method's icon is then displayed as the default in the Draw panel.

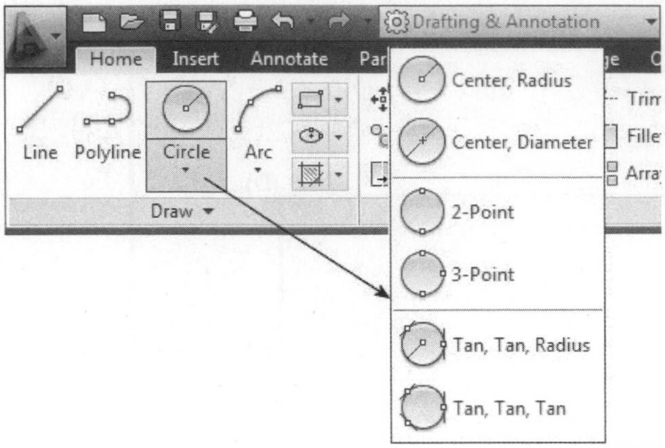

FIGURE 1.80

Circle by Radius Mode

Use the `CIRCLE` command and the Radius mode to construct a circle by a radius value that you specify. After selecting a center point for the circle, the user is prompted to enter a radius for the desired circle.

TRY IT!

Create a new drawing file starting from scratch. Use the `CIRCLE` Command prompts below and the illustration in Figure 1.81 (Left) to construct a circle by radius.

Command: C *(For CIRCLE)*

Specify center point for circle or [3P/2P/Ttr (tan tan radius)]: *(Mark the center at "A")*

Specify radius of circle or [Diameter]: **1.50**

Circle by Diameter Mode

Use the `CIRCLE` command and the Diameter mode to construct a circle by a diameter value that you specify. After selecting a center point for the circle, you are prompted to enter a diameter for the desired circle.

TRY IT!

Create a new drawing file starting from scratch. Use the `CIRCLE` Command prompts below and the illustration in Figure 1.81 (Right) to construct a circle by diameter.

Command: C *(For CIRCLE)*

Specify center point for circle or [3P/2P/Ttr (tan tan radius)]: *(Mark the center at "A")*

Specify radius of circle or [Diameter]: D *(For Diameter)*

Specify diameter of circle: **3.00**

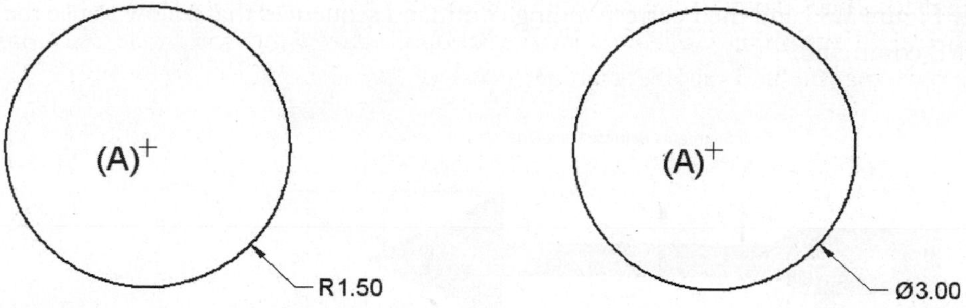

FIGURE 1.81

NOTE If you select the "Center, Diameter" icon in the Ribbon's Draw panel to construct a circle, the Diameter option will automatically be entered for you.

Dynamic Input and Circles

When using Dynamic Input mode for constructing circles, notice the appearance of a Dynamic Radius readout when you drag your cursor, as shown in Figure 1.82. Simply enter the desired radius or press the DOWN ARROW key twice on your keyboard to select the Diameter option, press ENTER, and provide a diameter value.

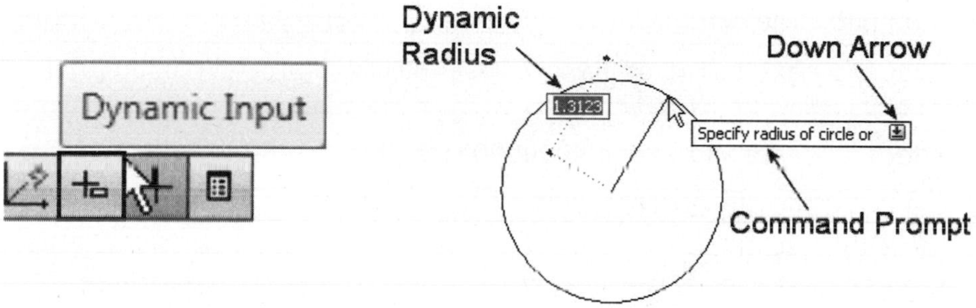

FIGURE 1.82

CONSTRUCTING POLYLINES

Polylines are similar to individual line segments except that a polyline can consist of numerous segments and still be considered a single object. Width can also be assigned to a polyline, unlike regular line segments; this makes polylines perfect for drawing borders, title blocks, and arrow symbols. Polylines can be constructed by selecting any of the following:

- From the Ribbon > Home Tab > Draw Panel
- From the Menu Bar (Draw > Polyline)
- From the Draw Toolbar of the AutoCAD Classic Workspace
- From the keyboard (PL or PLINE)

Study Figure 1.83 and their corresponding command sequences that follow to use the PLINE command.

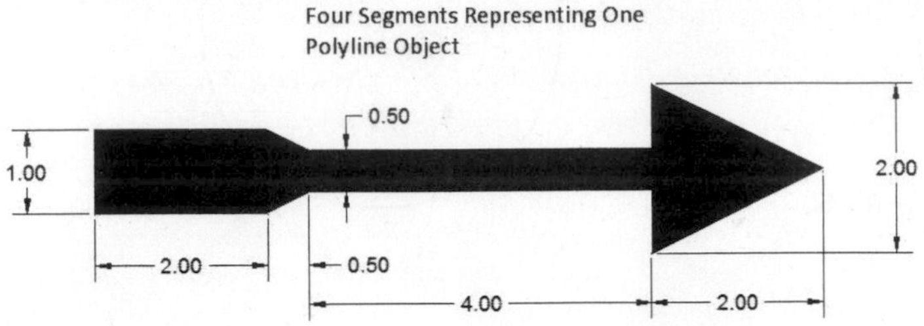

FIGURE 1.83

TRY IT!

Create a new drawing file starting from scratch. Follow the Command prompt sequence and illustration above to construct the polyline. Turn on either the Polar Tracking or Ortho mode and then utilize the Direct Distance mode of entry to construct this object. Try creating the same object using absolute, relative, and polar coordinates.

Command: **PL** *(For PLINE)*

Specify start point: *(Pick a convenient point on the screen)*

Current line-width is 0.0000

Specify next point or [Arc/Close/Halfwidth/Length/Undo/Width]: **W** *(For Width)*

Specify starting width <0.0000>: **1.00**

Specify ending width <1.000>: *(Press ENTER to accept the default)*

Specify next point or [Arc/Close/Halfwidth/Length/Undo/Width]: *(Move the cursor to the right and enter 2.00)*

Specify next point or [Arc/Close/Halfwidth/Length/Undo/Width]: **W** *(For Width)*

Specify starting width <1.0000>: *(Press ENTER to accept the default)*

Specify ending width <1.0000>: 0.50

Specify next point or [Arc/Close/Halfwidth/Length/Undo/Width]: *(Move the cursor to the right and enter 0.50)*

Specify next point or [Arc/Close/Halfwidth/Length/Undo/Width]: **W** *(For Width)*

Specify starting width <0.5000>: *(Press ENTER to accept the default)*

Specify ending width <0.5000>: *(Press ENTER to accept the default)*

Specify next point or [Arc/Close/Halfwidth/Length/Undo/Width]: *(Move the cursor to the right and enter 4.00)*

```
Specify next point or [Arc/Close/Halfwidth/Length/Undo/
Width]: W (For Width)
Specify starting width <0.5000>: 2.00
Specify ending width <2.0000>: 0.00
Specify next point or [Arc/Close/Halfwidth/Length/Undo/
Width]: (Move the cursor to the right and enter 2.00)
Specify next point or [Arc/Close/Halfwidth/Length/Undo/
Width]: (Press ENTER to exit this command)
```

Dynamic Input and Plines

Using Dynamic Input for polylines is similar to lines. As you move your cursor while in the PLINE command, you can observe the appearance of your pline through the Dynamic Distance and Dynamic Angle features, as shown in Figure 1.84. As with all dynamic input modes, pressing the DOWN ARROW on your keyboard displays options for the PLINE command that you can cycle through.

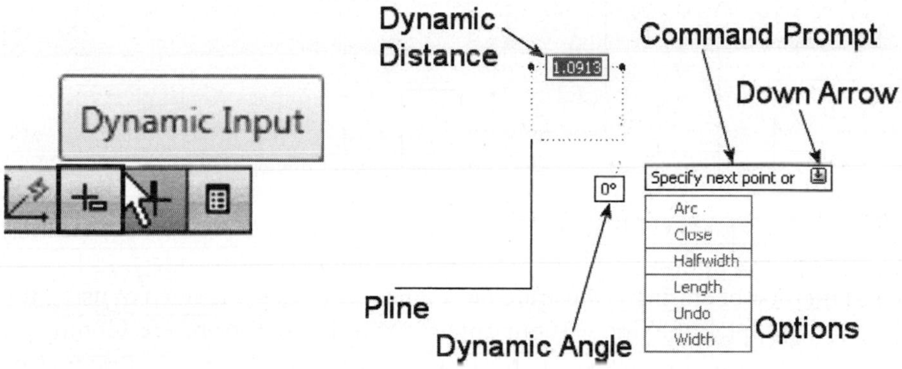

FIGURE 1.84

ERASING OBJECTS

Throughout the design process, as objects such as lines and circles are placed in a drawing, changes in the design will require the removal of objects. The ERASE command deletes objects from the database. The ERASE command is selected from any of the following:

- From the Ribbon > Home Tab > Modify Panel
- From the Menu Bar (Modify > Erase)
- From the Modify Toolbar of the AutoCAD Classic Workspace
- From the keyboard (E or ERASE)

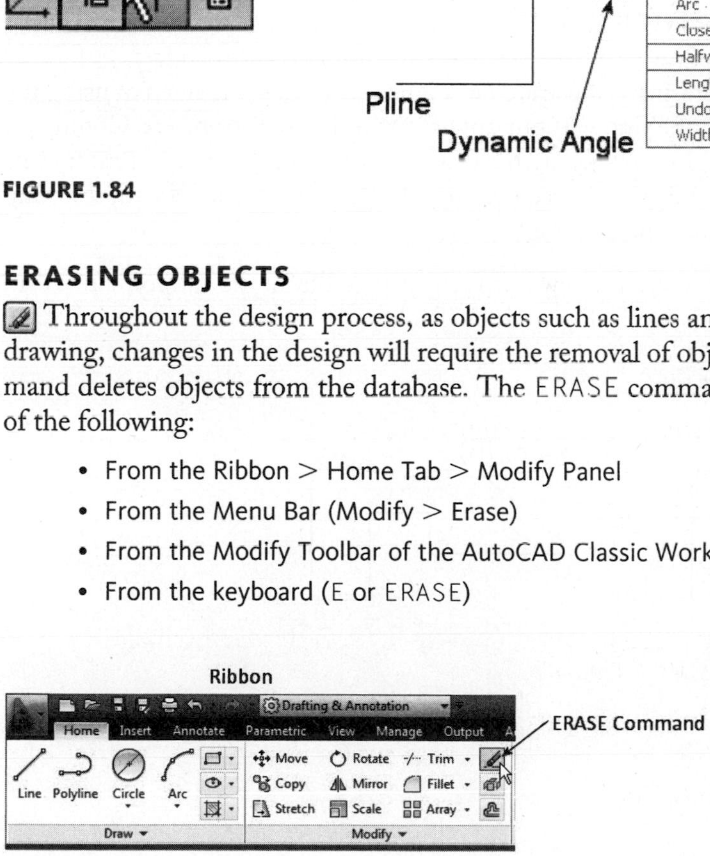

FIGURE 1.85

In Figure 1.86 (Left), line segments "A" and "B" need to be removed in order for a new line to be constructed, closing the shape. Two ways of erasing these lines will be introduced here.

When first entering the ERASE command, you are prompted to Select objects to erase. Notice that your cursor changes in appearance from crosshairs to a pickbox, as shown in Figure 1.86 (Right). Move the pickbox over the object to be selected and select this item. Notice that it will be highlighted as a dashed object to signify it is now selected. At this point, the Select objects prompt appears again. Additional objects may be selected at this point. Once all the objects are selected, pressing enter performs the erase operation.

Command: E *(for ERASE)*

Select objects: *(Pick line "A," as shown on the right in the following image)*

Select objects: *(Press* ENTER *to perform the erase operation)*

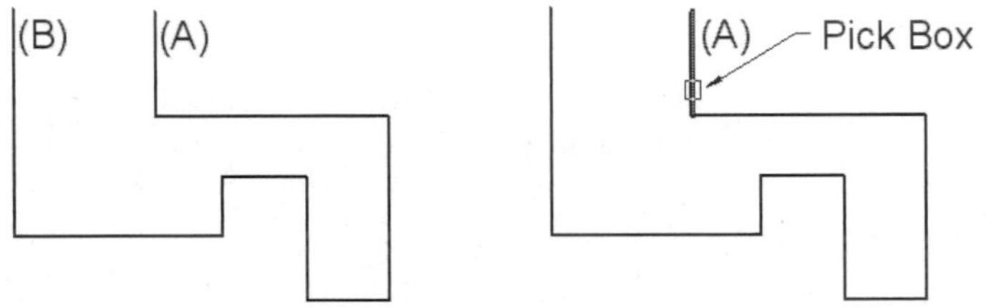

FIGURE 1.86

The second method of erasing is illustrated in Figure 1.87 (Left). Instead of using the ERASE command, pick the line without any command issued from the Command prompt. The line highlights and square boxes appear at the endpoints and midpoints of the line. With this line segment selected, press DELETE on the keyboard, resulting in removal of the line from the drawing.

With both line segments erased, a new line is constructed from the endpoint at "A" to the endpoint at "B," as shown in Figure 1.87 (Right).

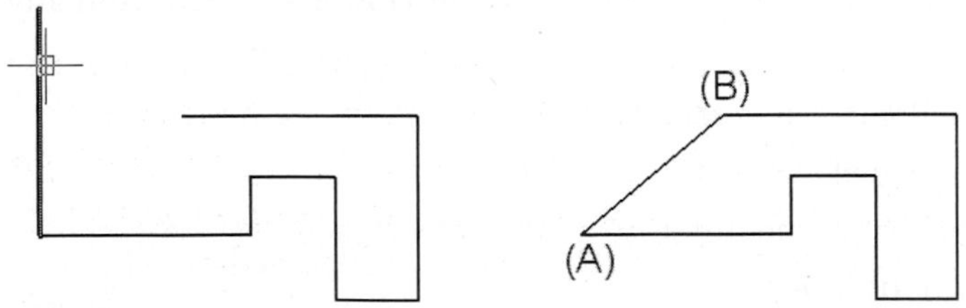

FIGURE 1.87

SAVING A DRAWING FILE

You can save drawings using the QSAVE and SAVEAS commands. The QSAVE command can be selected from the following:

- From the Quick Access Toolbar
- From the Application Menu (Save)
- From the Menu Bar (File > Save)
- From the Standard Toolbar of the AutoCAD Classic Workspace
- From the keyboard (QSAVE)

The SAVEAS command can be selected from any of the following:

- From the Quick Access Toolbar
- From the Application Menu (Save As)
- From the Menu Bar (File > Save As)
- From the keyboard (SAVEAS)

These commands are found on the Quick Access Toolbar and the Application Menu, as shown in Figure 1.88 (Left).

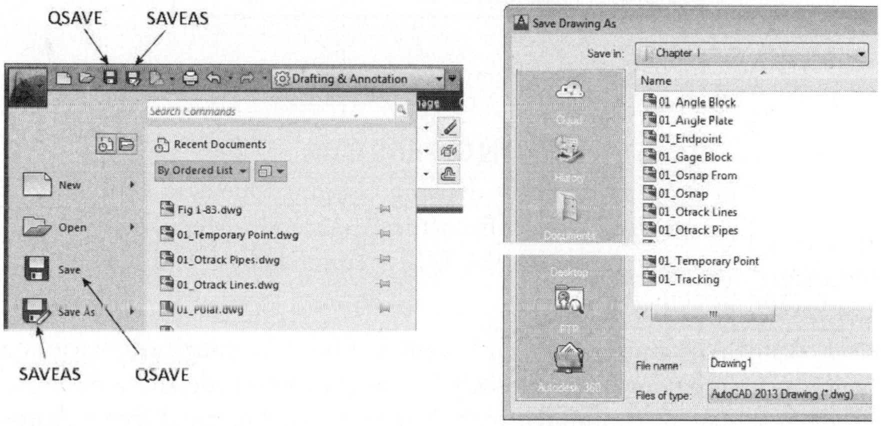

FIGURE 1.88

Save

Selecting Save from the Application Menu, as shown in Figure 1.88, activates the QSAVE command, which stands for Quick Save. If a drawing file has never been saved and this command is selected, the dialog box shown in Figure 1.88 (Right) is displayed. Once a drawing file has been initially saved, selecting this command causes an automatic save and the Save Drawing As dialog box is not displayed.

Save As

Using the SAVEAS command always displays the dialog box shown in Figure 1.88 (Right). Simply click the Save button or press ENTER to save the drawing under the current name, which is displayed in the field. This command is more popular for saving the current drawing in a new location or with an entirely different name. Once a drawing is given a new location or name through this command, it also becomes the new current drawing file.

The ability to exchange drawings with past releases of AutoCAD is still important to many industry users. When the Files of type field is selected in Figure 1.89, a drop-down list appears. Use this list to save a drawing file in AutoCAD 2013, 2010, 2007, 2004, 2000, and even R14 formats. The user can also save a drawing file as Drawing Standard (.dws), a Drawing Template (.dwt), and a Drawing Interchange Format (.dxf). The Drawing Interchange Format is especially useful with opening up an AutoCAD drawing in a competitive CAD system.

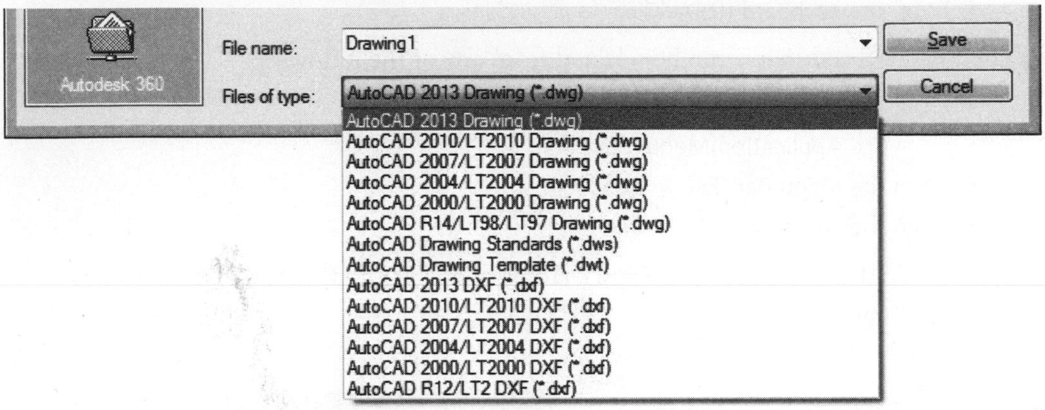

FIGURE 1.89

EXITING AN AUTOCAD DRAWING SESSION

It is good practice to properly exit any drawing session. One way of exiting is by choosing the Exit AutoCAD option from the Application Menu, as shown in Figure 1.90 (Left). You can also use the `QUIT` command to end the current AutoCAD drawing session.

Whenever an AutoCAD drawing session is exited, a built-in safeguard provides a second chance to save the drawing, especially if changes were made and a Save was not performed. You may be confronted with three options, illustrated in the Auto-CAD alert dialog box shown in Figure 1.90 (Right). By default, the Yes button is highlighted.

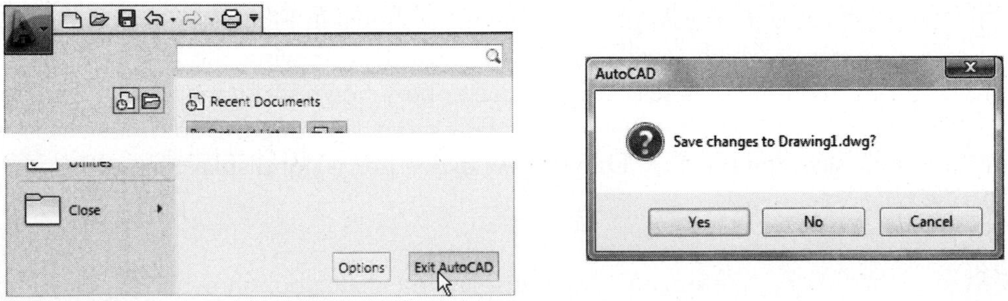

FIGURE 1.90

If changes were made to a drawing but no Save was performed, the user can now save the changes by clicking the Yes button before exiting the drawing. Changes to the drawing will be saved and the software exits back to the operating system.

If changes were made but the user does not want to save them, clicking the No button is appropriate. Changes to the drawing will not be saved and the software exits back to the operating system.

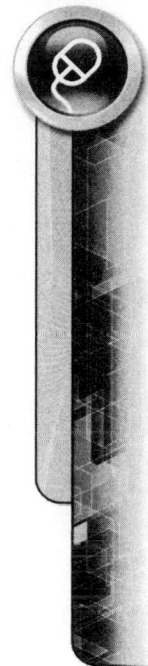

If changes are made to the drawing and the Exit option is chosen mistakenly, clicking the Cancel button cancels the Exit option and returns the user to the current drawing.

TUTORIAL EXERCISE: 01_GAGE BLOCK.DWG

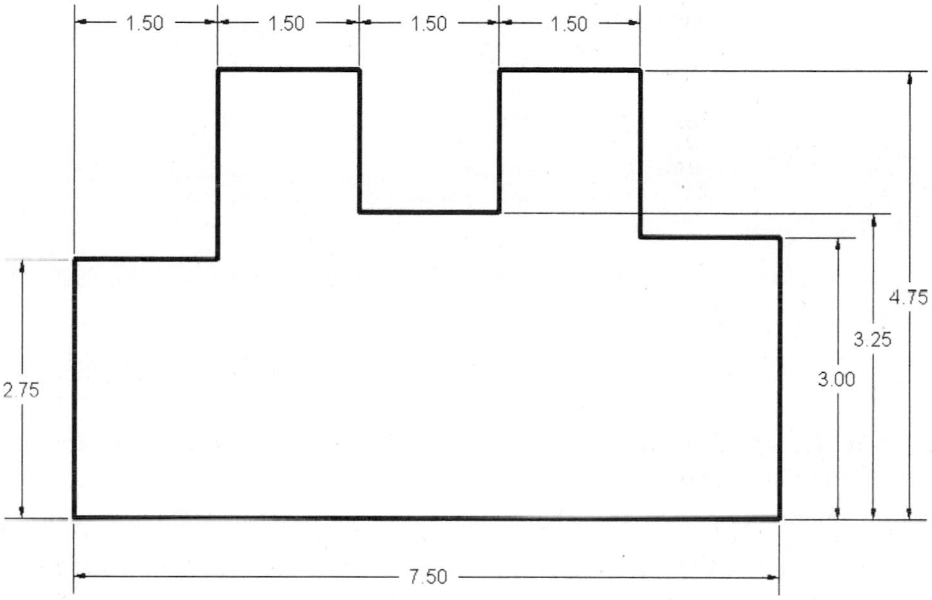

FIGURE 1.91

Purpose

This tutorial is designed to allow you to construct a one-view drawing of the Gage Block using Polar Tracking and Direct Distance mode.

System Settings

Use the current default settings for the limits of this drawing, (0,0) for the lower-left corner and (12,9) for the upper-right corner.

Suggested Commands

Open the drawing file called 01_Gage Block. The `LINE` command will be used entirely for this tutorial in addition to the Polar Tracking and Direct Distance modes. Running Object Snap should already be set to the following modes: Endpoint, Center, Intersection, and Extension.

STEP 1

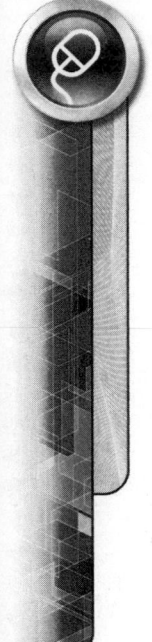

Begin this tutorial by first turning Polar Tracking on. This can be accomplished by clicking Polar in the Status Bar located at the bottom of your display screen. Then activate the `LINE` command and follow the next series of prompt sequences to complete this object.

Command: **L** *(For LINE)*

Specify first point: **1,1**

Specify next point or [Undo]: Move cursor to the right and type **7.5**

Specify next point or [Undo]: Move cursor up and type **3**

Specify next point or [Close/Undo]: Move cursor to the left and type **1.5**

Specify next point or [Close/Undo]: Move cursor up and type **1.75**

Specify next point or [Close/Undo]: Move cursor to the left and type **1.5**

Specify next point or [Close/Undo]: Move cursor down and type **1.5**

Specify next point or [Close/Undo]: Move cursor to the left and type **1.5**

Specify next point or [Close/Undo]: Move cursor up and type **1.5**

Specify next point or [Close/Undo]: Move cursor to the left and type **1.5**

Specify next point or [Close/Undo]: Move cursor down and type **2**

Specify next point or [Close/Undo]: Move cursor to the left and type **1.5**

Specify next point or [Close/Undo]: **C** *(To close the shape)*

TUTORIAL EXERCISE: 01_ANGLE BLOCK.DWG

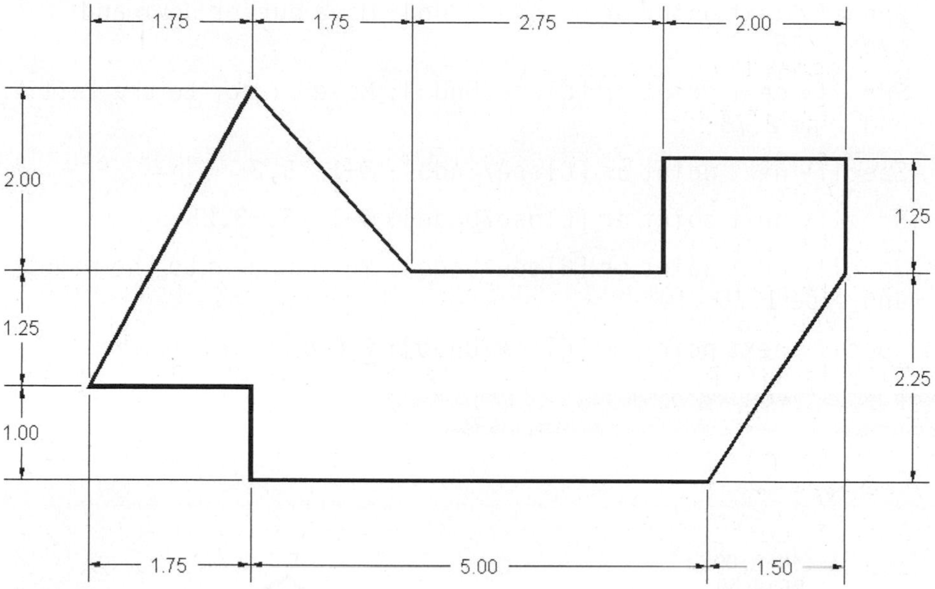

FIGURE 1.92

This tutorial is designed to allow you to construct a one-view drawing of the Angle Block using a combination of Polar Tracking, Direct Distance mode, and relative coordinates.

System Settings

Use the current default settings for the limits of this drawing, (0,0) for the lower-left corner and (12,9) for the upper-right corner.

Suggested Commands

Open the drawing file called 01_Angle Block. The LINE command will be used entirely for this tutorial in addition to the Polar Tracking and Direct Distance modes. Running Object Snap should already be set to the following modes: Endpoint, Center, Intersection, and Extension.

STEP 1

Begin this tutorial by first checking that Polar Tracking is turned on. Then activate the LINE command and follow the next series of prompt sequences to complete this object.

Command: L *(For LINE)*

Specify first point: **3,1**

Specify next point or [Undo]: Move cursor to the right and type **5**

Specify next point or [Undo]: **@1.5,2.25**

Specify next point or [Close/Undo]: Move cursor up and type **1.25**

Specify next point or [Close/Undo]: Move cursor to the left and type **2**

Specify next point or [Close/Undo]: Move cursor down and type **1.25**

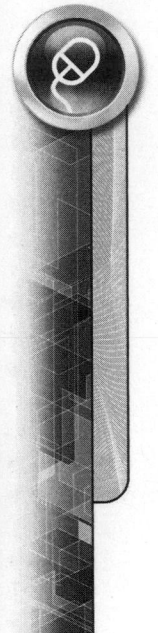

Specify next point or [Close/Undo]: Move cursor to the left and type **2.75**

Specify next point or [Close/Undo]: **@-1.75,2**

Specify next point or [Close/Undo]: **@-1.75,-3.25**

Specify next point or [Close/Undo]: Move cursor to the right and type **1.75**

Specify next point or [Close/Undo]: **C** *(To close the shape)*

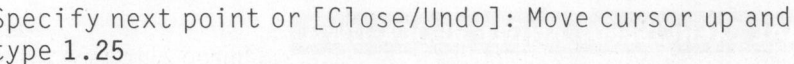

TUTORIAL EXERCISE: 01_ANGLE PLATE.DWG

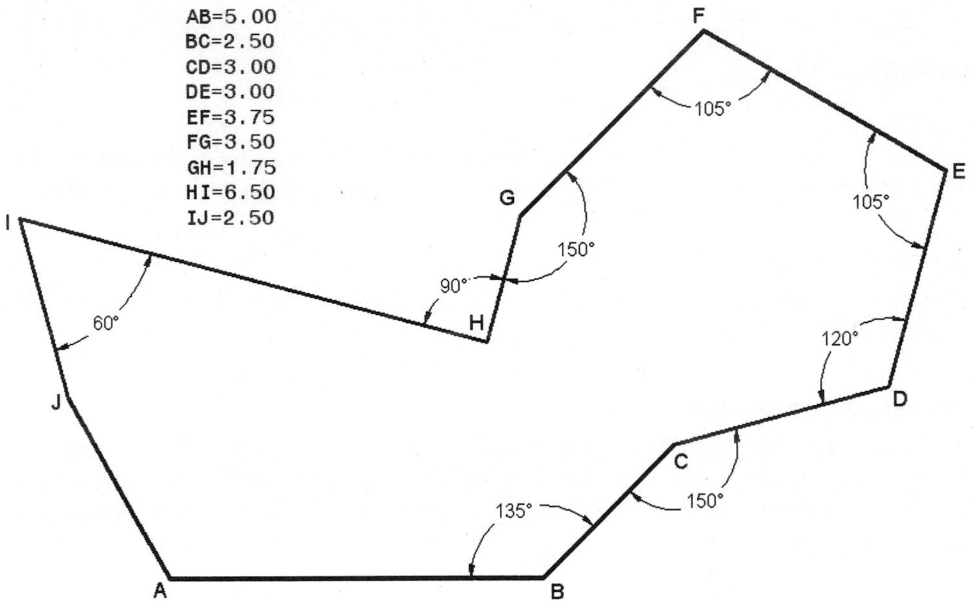

FIGURE 1.93

This tutorial is designed to allow you to construct a one-view drawing of the Angle Plate using Polar Tracking set to relative mode and Direct Distance mode.

System Settings

Use the current default settings for the limits of this drawing, (0,0) for the lower-left corner and (12,9) for the upper-right corner.

Suggested Commands

Open the drawing file called 01_Angle Plate. The LINE command will be used entirely for this tutorial in addition to the Polar Tracking and Direct Distance modes. Polar Tracking will need to be set to a new incremental angle of 15°. Also, Polar Tracking will need to be set to relative mode. Running Object Snap should already be set to the following modes: Endpoint, Center, Intersection, and Extension.

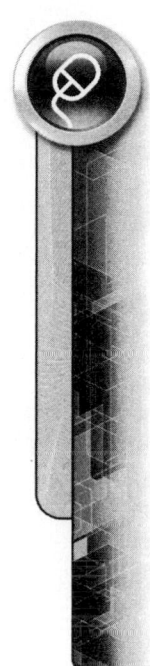

STEP 1

Right-click Polar in the Status Bar at the bottom of the display screen and select Settings from the menu, as shown in Figure 1.94 (Left). When the Drafting Settings dialog box appears, verify the Polar Tracking tab is selected. While in this tab, set the Incremental Angle to 15° under Polar Angle Settings area. Then set the Polar Angle measurement to Relative to last segment, as shown in Figure 1.94 (Right). Click the OK button to save the settings and exit the Drafting Settings dialog box.

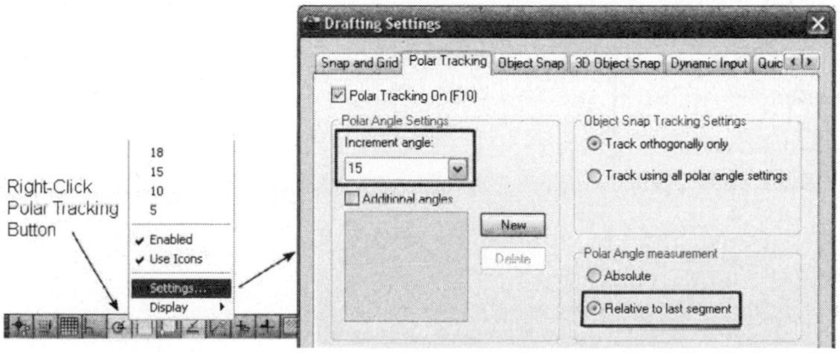

FIGURE 1.94

STEP 2

After setting the angle increment and relative mode under the Polar Tracking tab of the Drafting Settings dialog box, activate the LINE command and follow the next series of prompt sequences to complete this object.

Command: L *(For LINE)*

Specify first point: 3,1

Specify next point or [Undo]: *(Move your cursor to the right until the polar tooltip reads 0° and enter* 5*)*

Specify next point or [Undo]: *(Move your cursor until the polar tooltip reads 45° and enter* 2.5*)*

Specify next point or [Close/Undo]: *(Move your cursor until the polar tooltip reads 330° and enter* 3*)*

Specify next point or [Close/Undo]: *(Move your cursor until the polar tooltip reads 60° and enter* 3*)*

Specify next point or [Close/Undo]: *(Move your cursor until the polar tooltip reads 75° and enter* 3.75*)*

Specify next point or [Close/Undo]: *(Move your cursor until the polar tooltip reads 75° and enter* 3.5*)*

Specify next point or [Close/Undo]: *(Move your cursor until the polar tooltip reads 30° and enter* 1.75*)*

Specify next point or [Close/Undo]: *(Move your cursor until the polar tooltip reads 270° and enter* 6.5*)*

Specify next point or [Close/Undo]: *(Move your cursor until the polar tooltip reads 120° and enter* 2.5*)*

Specify next point or [Close/Undo]: C *(To close the shape)*

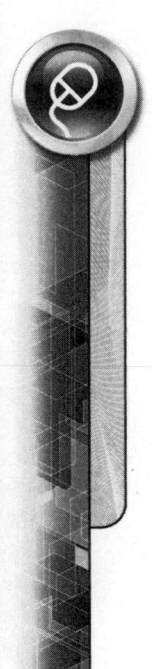

STEP 3

When finished with this problem, change the Increment angle under Polar Angle Settings back to 90° and the Polar Angle measurement back to Absolute, as shown in Figure 1.95.

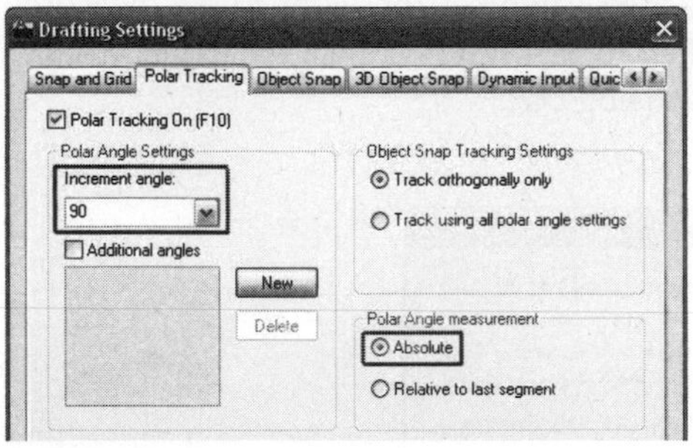

FIGURE 1.95

TUTORIAL EXERCISE: 01_PATTERN.DWG

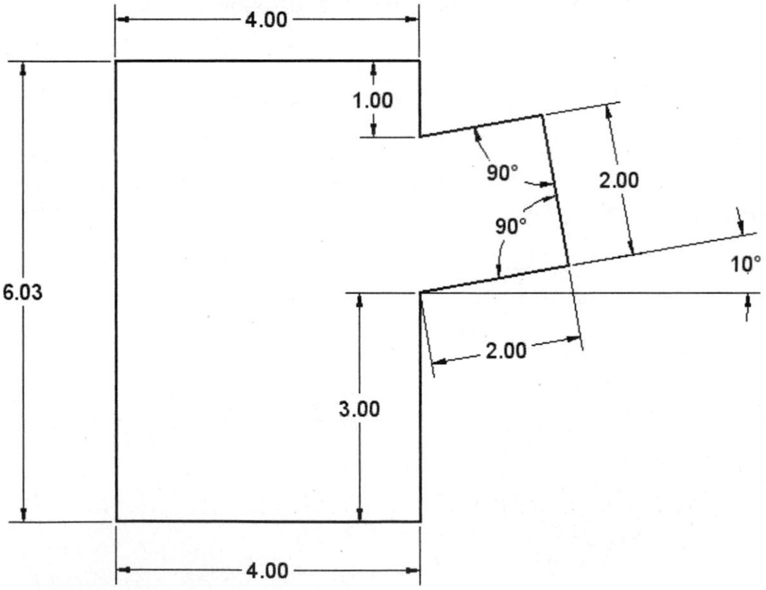

FIGURE 1.96

Purpose

This tutorial is designed to allow you to construct a one-view drawing of the Pattern using Polar Tracking techniques.

System Settings

Use the current default settings for the limits of this drawing, (0,0) for the lower-left corner and (12,9) for the upper-right corner.

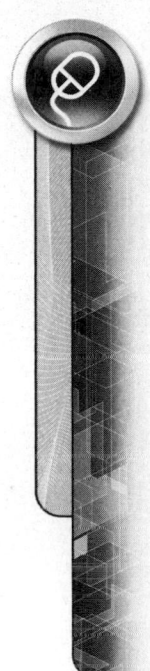

Layers

The following layer has already been created:

Name	Color	Linetype
Object	Green	Continuous

Suggested Commands

Open the drawing file called 01_Pattern. The `LINE` command will be used entirely for this tutorial in addition to the Polar Tracking mode. Running Object Snap should already be set to the following modes: Endpoint, Center, Intersection, and Extension. Dynamic Input has been turned off for this exercise.

STEP 1

Open the drawing file 01_Pattern. Right-click the Polar Tracking icon on the Status Bar and select the Increment angle setting of 10°, as shown in Figure 1.97 (Left). Verify that POLAR, OSNAP, and OTRACK are all turned on.

STEP 2

Activate the `LINE` command, select a starting point, move your cursor to the right, and enter a value of 4 units, as shown in Figure 1.97 (Right). Notice that your line is green because the current layer, Object, has been assigned the green color.

`Command:` L *(For LINE)*

`Specify first point:` 3,1

`Specify next point or [Undo]:` *(Move your cursor to the right and enter* 4*)*

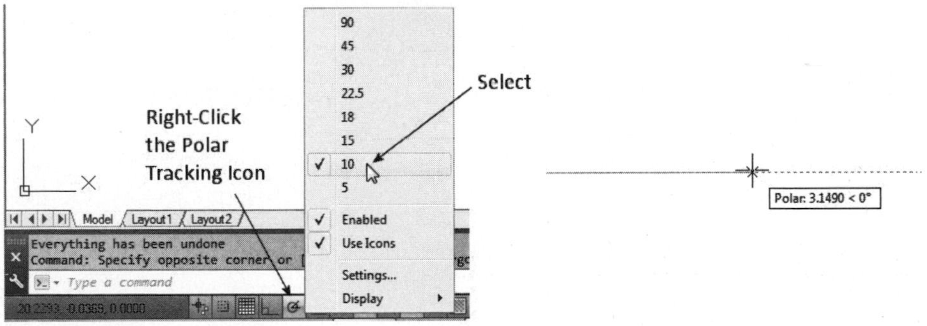

FIGURE 1.97

STEP 3

While still in the `LINE` command, move your cursor directly up, and enter a value of 3 units, as shown in Figure 1.98 (Left).

> `Specify next point or [Undo]:` *(Move your cursor up and enter* `3`*)*

STEP 4

While still in the `LINE` command, move your cursor up and to the right until the tooltip reads 10°, and enter a value of 2 units, as shown in Figure 1.98 (Right).

> `Specify next point or [Close/Undo]:` *(Move your cursor up and to the right at a 10° angle and enter* `2`*)*

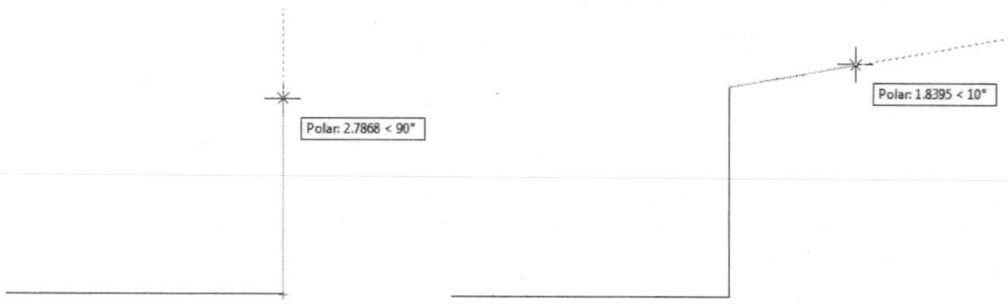

FIGURE 1.98

STEP 5

While still in the `LINE` command, move your cursor up and to the left until the tooltip reads 100°, and enter a value of 2 units, as shown in Figure 1.99 (Left).

> `Specify next point or [Close/Undo]:` *(Move your cursor up and to the left at a 100° angle and enter* `2`*)*

STEP 6

While still in the `LINE` command, first acquire the point at "A." Then move your cursor below and to the left until the Polar value in the tooltip reads 190°, and pick the point at "B", as shown in Figure 1.99 (Right).

> `Specify next point or [Close/Undo]:` *(Acquire the point at "A" and pick the new point at "B")*

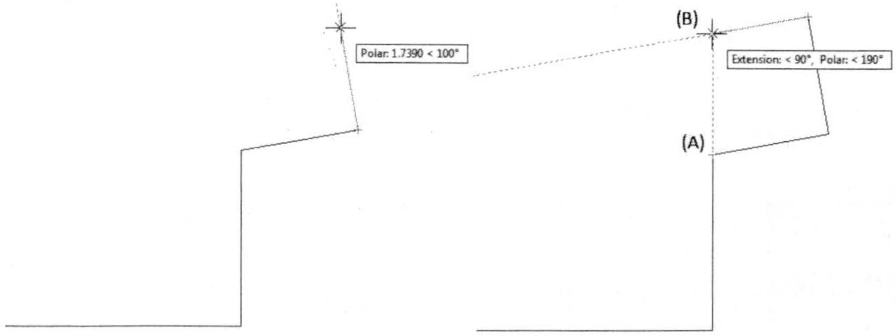

FIGURE 1.99

STEP 7

While still in the LINE command, move your cursor directly up, and enter a value of 1 unit, as shown on the left in Figure 1.100.

> Specify next point or [Close/Undo]: *(Move your cursor up and enter* 1*)*

STEP 8

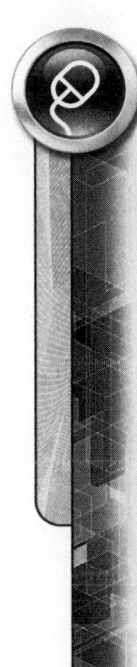

While still in the LINE command, first acquire the point at "C." Then move your cursor to the left until the tooltip reads Polar: < 180°, and pick the point at "D", as shown on the right in Figure 1.100.

> Specify next point or [Close/Undo]: *(Acquire the point at "C" and pick the new point at "D")*

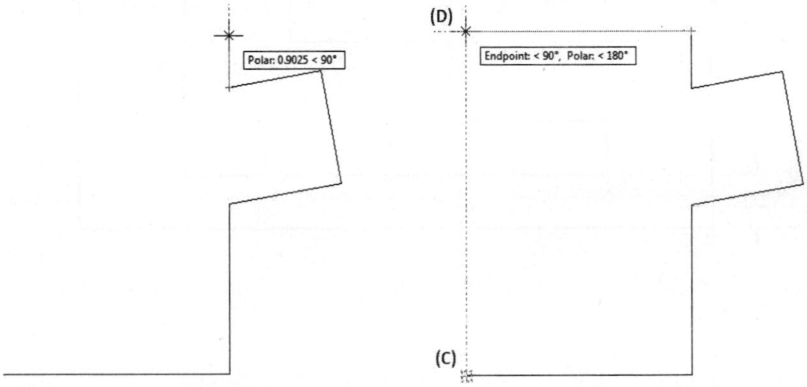

FIGURE 1.100

STEP 9

While still in the LINE command, complete the object by closing the shape, as shown in Figure 1.101.

> Specify next point or [Close/Undo]: C *(To close the shape and exit the* LINE *command)*

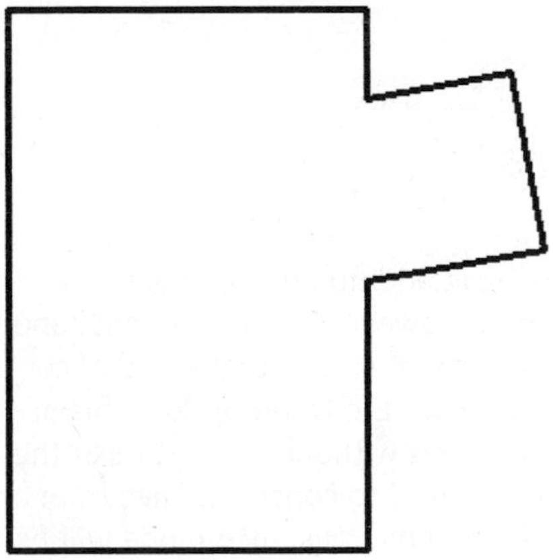

FIGURE 1.101

TUTORIAL EXERCISE: 01_TEMPLATE.DWG

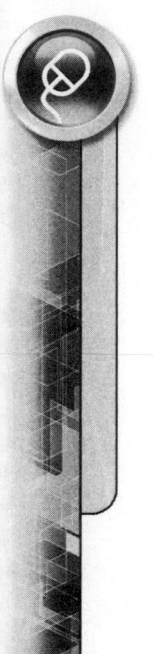

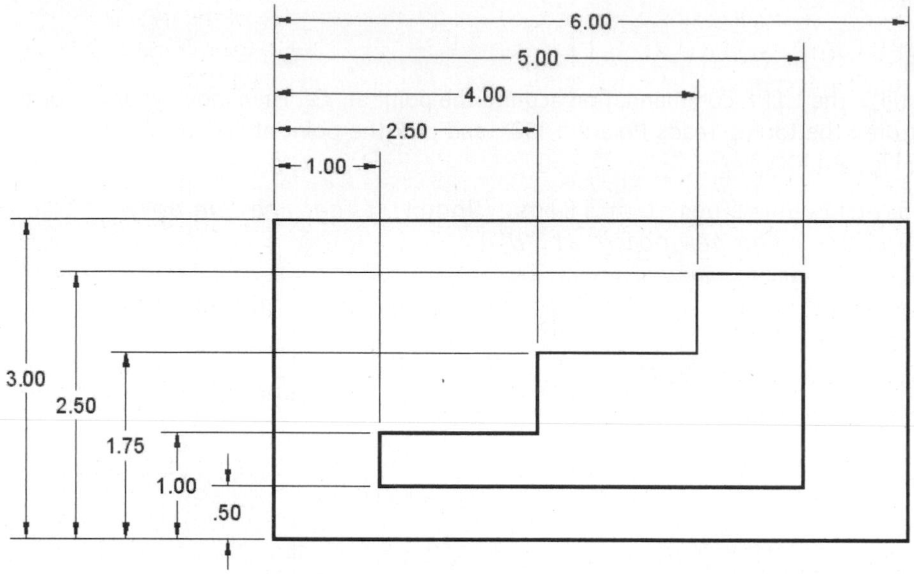

FIGURE 1.102

Purpose

This tutorial is designed to allow you to construct a one-view drawing of the template using Relative Coordinate mode in combination with the Direct Distance mode.

System Settings

Use the current default settings for the limits of this drawing, (0,0) for the lower-left corner and (12,9) for the upper-right corner.

Layers

The following layer has already been created:

Name	Color	Linetype
Object	Green	Continuous

Suggested Commands

The LINE command will be used entirely for this tutorial, in addition to a combination of coordinate systems. The ERASE command could be used (however, using this command will force the user to exit the LINE command), although a more elaborate method of correcting mistakes while using the LINE command is to execute the Undo option. This option allows the user to delete (or undo) previously drawn lines without having to exit the LINE command. The Object Snap From mode will also be used to construct lines from a point of reference. The coordinate mode of entry and the Direct Distance mode will be used throughout this tutorial exercise.

STEP 1

Open the drawing file 01_Template.dwg. Then use the `LINE` command to draw the outer perimeter of the box using the Direct Distance mode. Because the box consists of horizontal and vertical lines, Ortho mode is first turned on; this forces all movements to be in the horizontal or vertical direction. To construct a line segment, move the cursor in the direction in which the line is to be drawn and enter the exact value of the line. The line is drawn at the designated distance in the current direction of the cursor. Repeat this procedure for the other lines that make up the box, as shown in Figure 1.103.

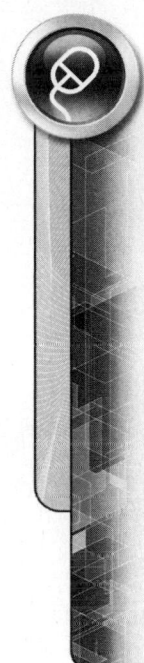

`Command:` `L` *(For LINE)*

`Specify first point:` `2,2`

`Specify next point or [Undo]:` *(Move the cursor to the right and enter a value of* `6.00` *units)*

`Specify next point or [Undo]:` *(Move the cursor up and enter a value of* `3.00` *units)*

`Specify next point or [Close/Undo]:` *(Move the cursor to the left and enter a value of* `6.00` *units)*

`Specify next point or [Close/Undo]:` `C` *(To close the shape)*

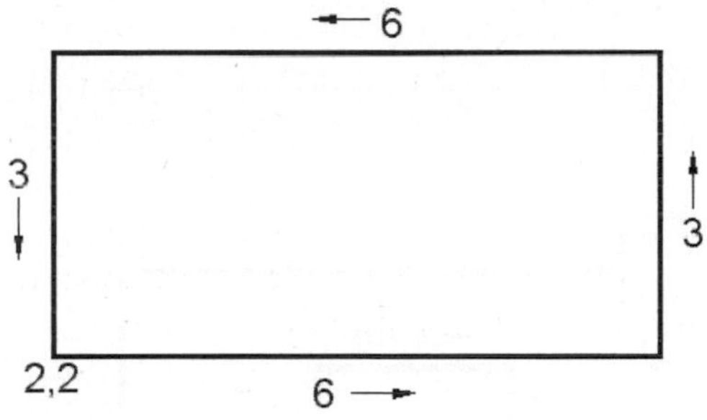

FIGURE 1.103

STEP 2

The next step is to draw the stair step outline of the template using the `LINE` command again. However, we first need to identify the starting point of the template.

Absolute coordinates could be calculated, but in more complex objects this would be difficult. A more efficient method is to use the Object Snap From mode along with the Object Snap Intersection mode to start the line relative to another point. Both Object Snap selections can be activated by holding down SHIFT or CTRL and pressing the right mouse button. Use the following command sequence and image as guides for performing this operation.

`Command:` `L` *(For LINE)*

`Specify first point:` `From`

`Base point:` `Int`

`of` *(Pick the intersection at "A" as shown in the following image)*

`<Offset>:` `@1.00,0.50`

The relative coordinate offset value begins a new line at a distance of 1.00 units in the X direction and 0.50 units in the Y direction.

Continue with the LINE command to construct the stair step outline shown in Figure 1.104. Use the Direct Distance mode to accomplish this task. In this example, Direct Distance mode is a good choice to use, especially since all lines are either horizontal or vertical. Use the following command sequence to construct the object with this alternate method.

> Specify next point or [Undo]: *(Move the cursor to the right and enter a value of* 4.00 *units)*
>
> Specify next point or [Undo]: *(Move the cursor up and enter a value of* 2.00 *units)*
>
> Specify next point or [Close/Undo]: *(Move the cursor to the left and enter a value of* 1.00 *units)*
>
> Specify next point or [Close/Undo]: *(Move the cursor down and enter a value of* 0.75 *units)*
>
> Specify next point or [Close/Undo]: *(Move the cursor to the left and enter a value of* 1.50 *units)*
>
> Specify next point or [Close/Undo]: *(Move the cursor down and enter a value of* 0.75 *units)*
>
> Specify next point or [Close/Undo]: *(Move the cursor to the left and enter a value of* 1.50 *units)*
>
> Specify next point or [Close/Undo]: C *(To close the shape)*

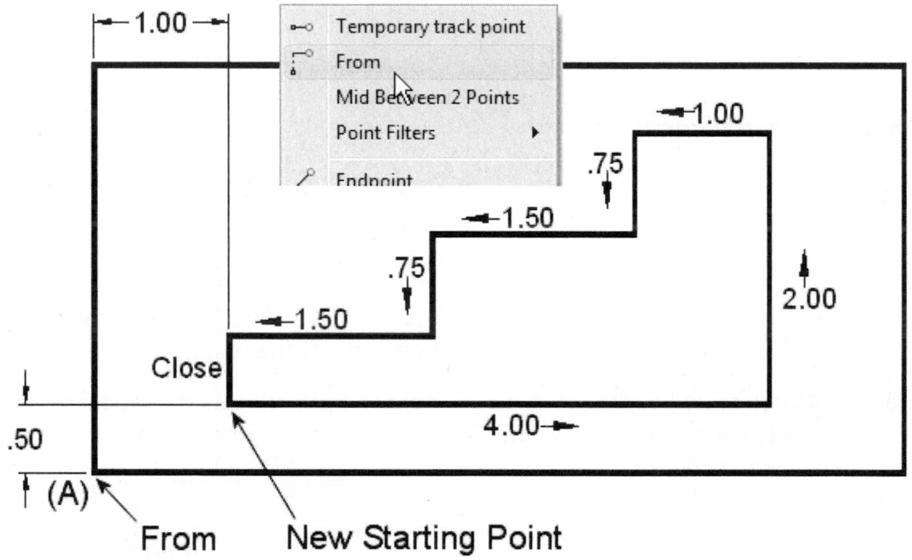

FIGURE 1.104

PROBLEMS FOR CHAPTER 1

Directions for Problems 1–2 through 1–8

Construct one-view drawings from figures (Problems 1-2 to 1-8) using the LINE command along with Direct Distance mode, relative coordinates, and/or polar coordinates.

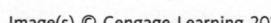

PROBLEM 1-1

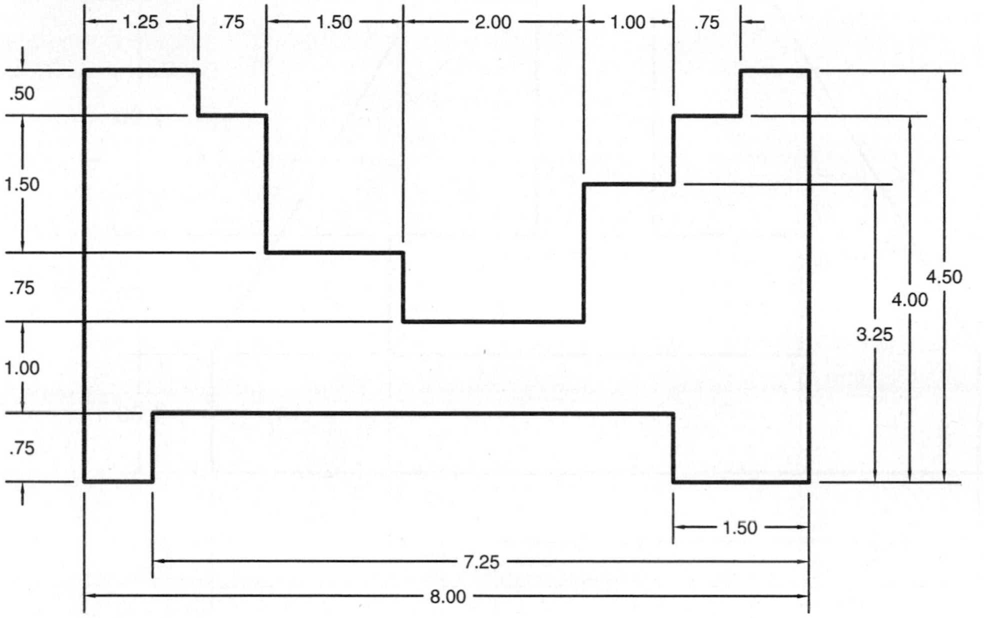

PROBLEM 1-2

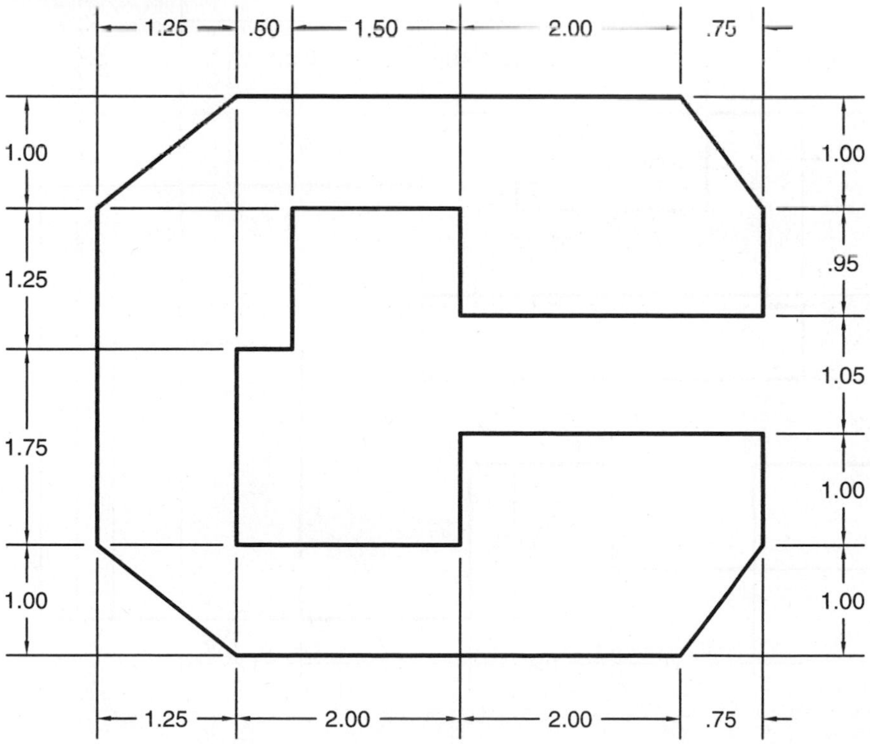

PROBLEM 1-3

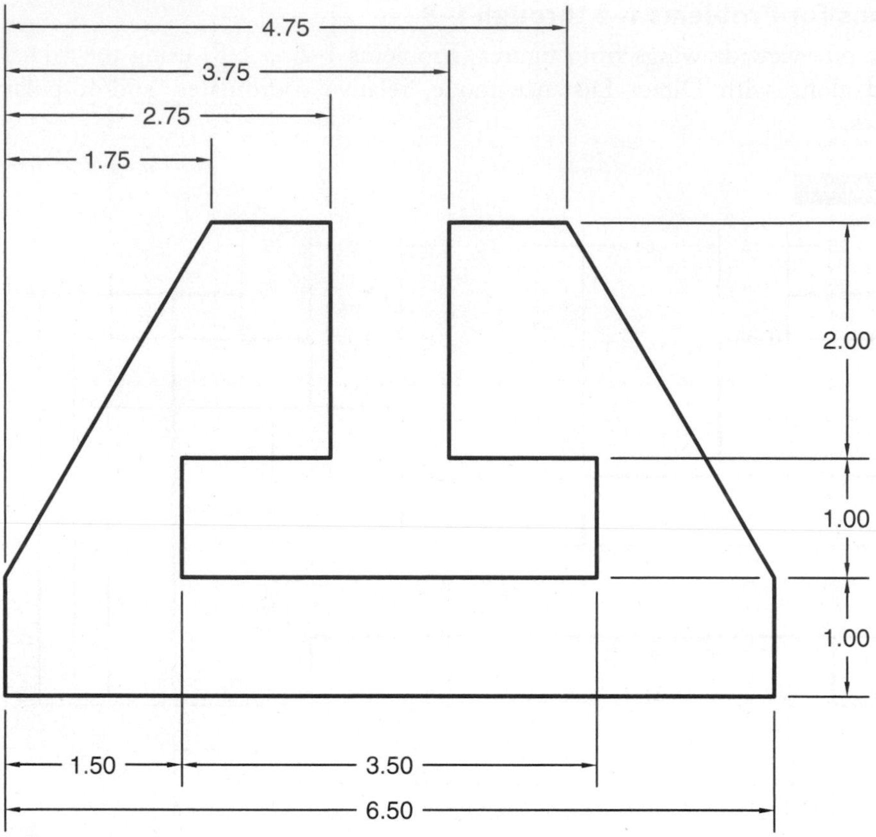

PROBLEM 1-4

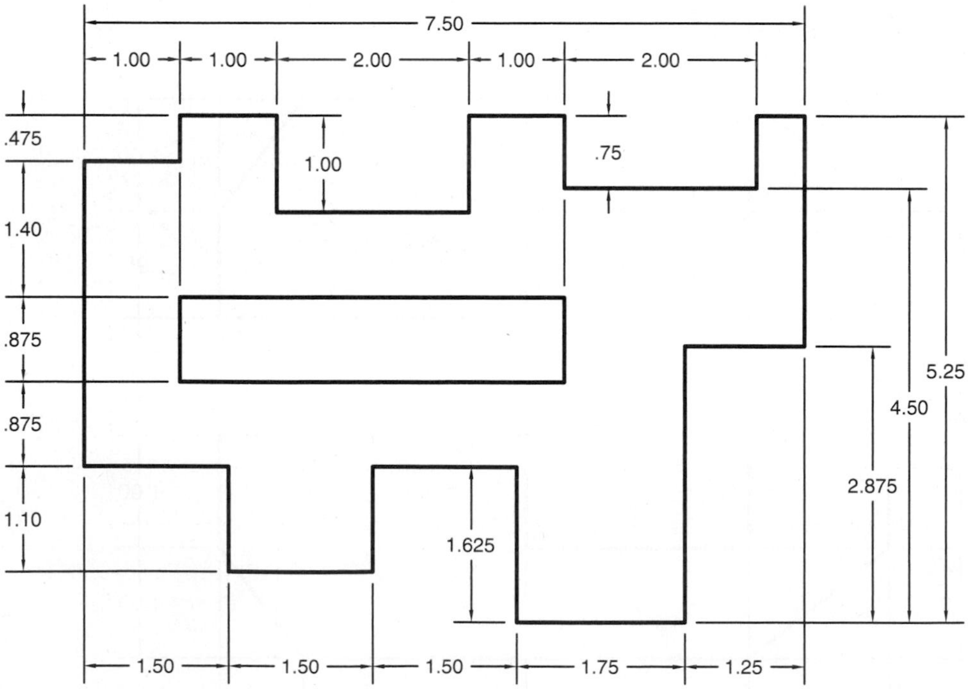

PROBLEM 1-5

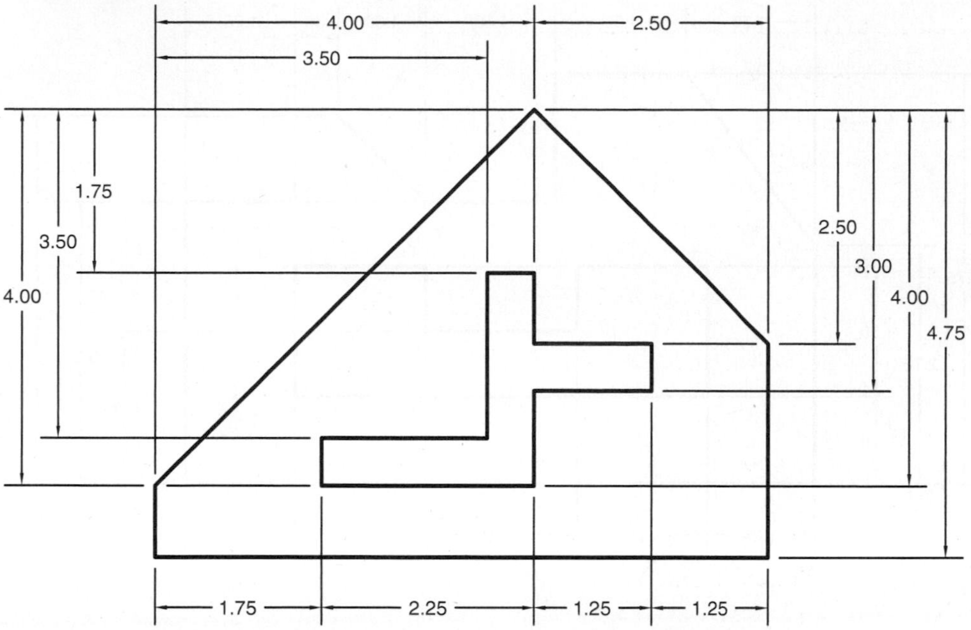

PROBLEM 1-6

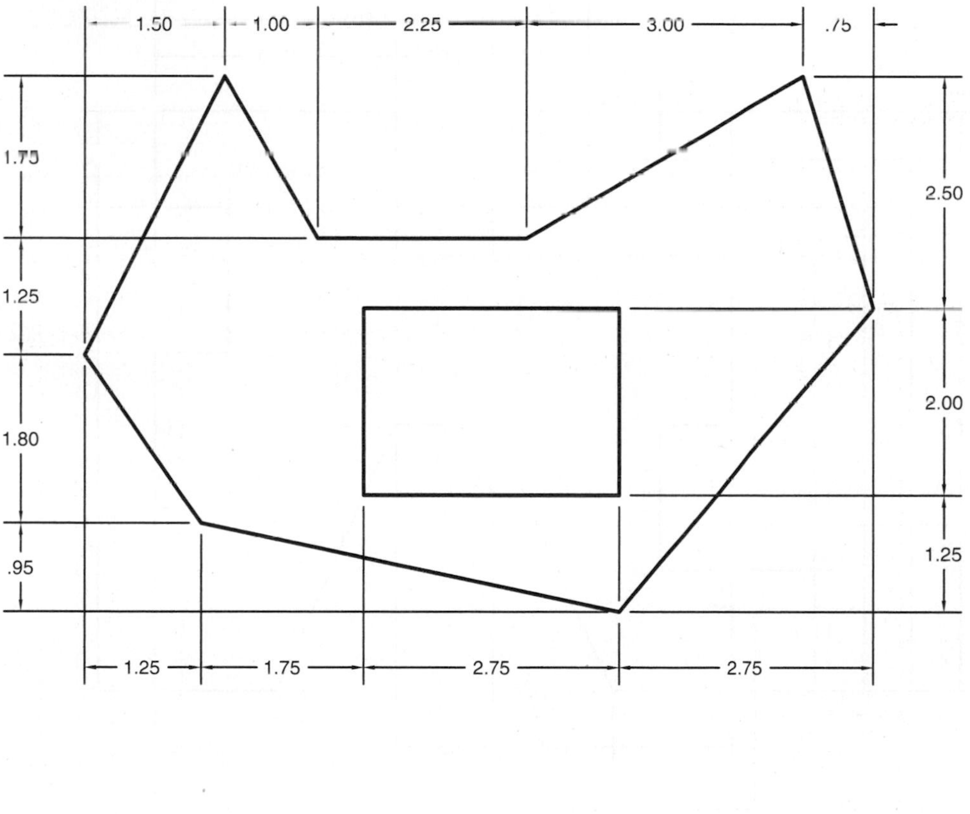

PROBLEM 1-7

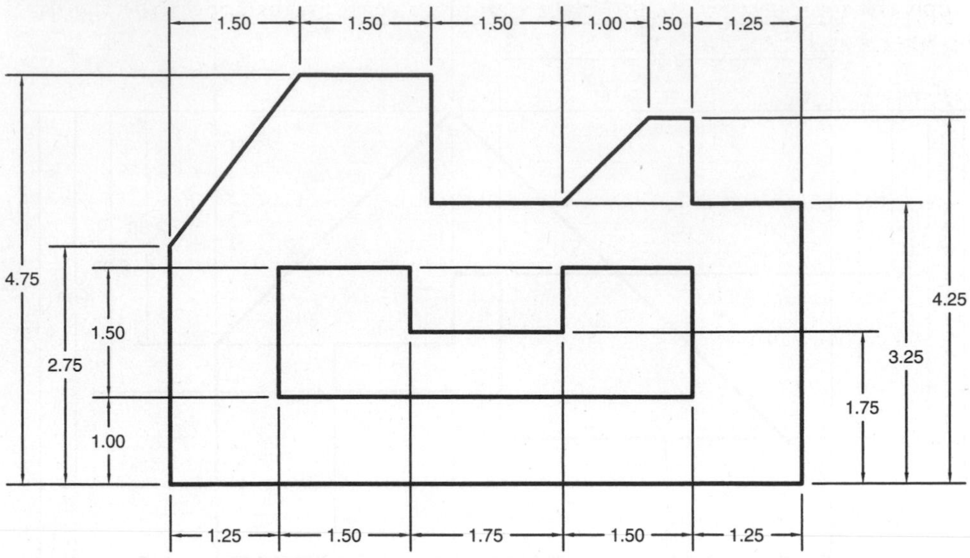

PROBLEM 1-8

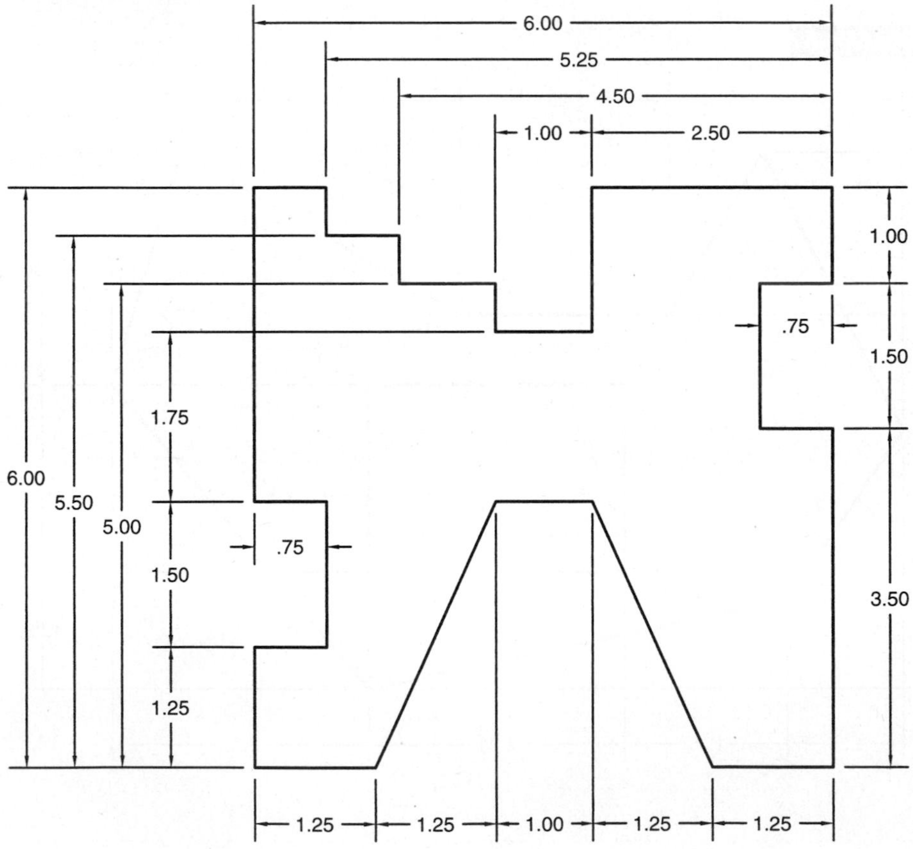

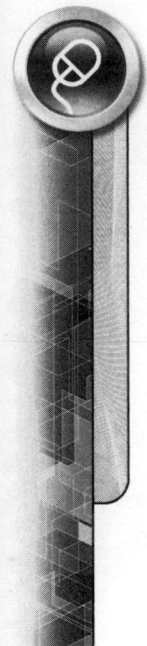

Directions for Problems 1–9

Supply the appropriate absolute, relative, and/or polar coordinates for these figures in the table that follows each object.

PROBLEM 1-9

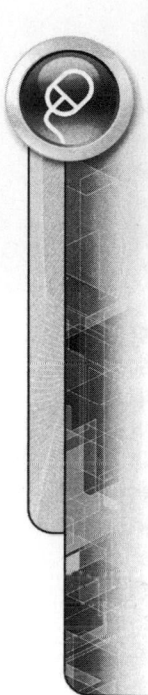

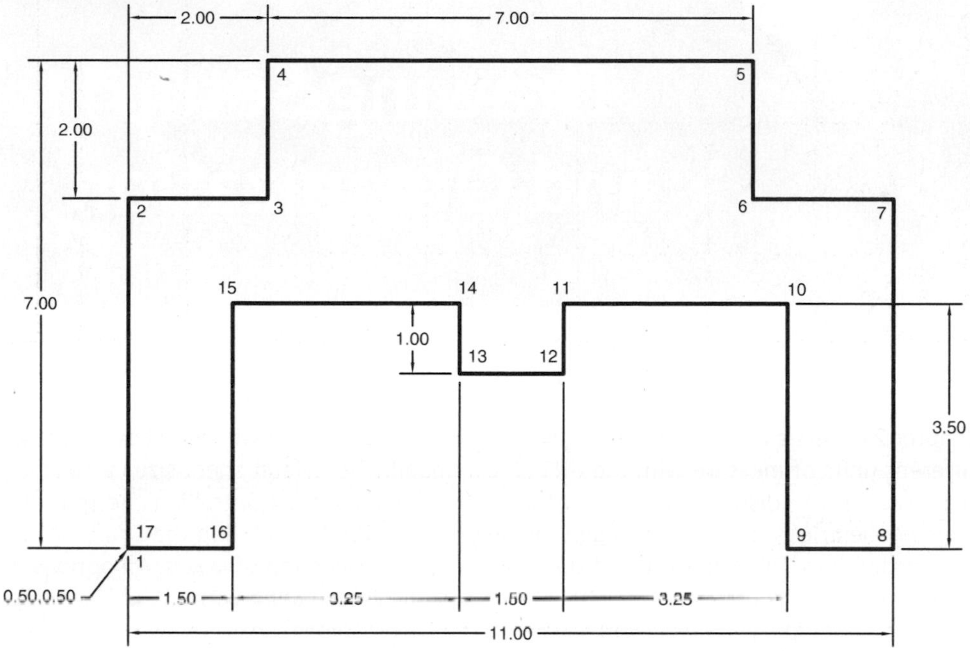

	Absolute	Relative	Polar
From Pt (1)	.50,.50	.50,.50	.50,.50
To Pt (2)	____ , ____	____ , ____	____ , ____
To Pt (3)	____ , ____	____ , ____	____ , ____
To Pt (4)	____ , ____	____ , ____	____ , ____
To Pt (5)	____ , ____	____ , ____	____ , ____
To Pt (6)	____ , ____	____ , ____	____ , ____
To Pt (7)	____ , ____	____ , ____	____ , ____
To Pt (8)	____ , ____	____ , ____	____ , ____
To Pt (9)	____ , ____	____ , ____	____ , ____
To Pt (10)	____ , ____	____ , ____	____ , ____
To Pt (11)	____ , ____	____ , ____	____ , ____
To Pt (12)	____ , ____	____ , ____	____ , ____
To Pt (13)	____ , ____	____ , ____	____ , ____
To Pt (14)	____ , ____	____ , ____	____ , ____
To Pt (15)	____ , ____	____ , ____	____ , ____
To Pt (16)	____ , ____	____ , ____	____ , ____
To Pt (17)	____ , ____	____ , ____	____ , ____
To Pt	Enter	Enter	Enter

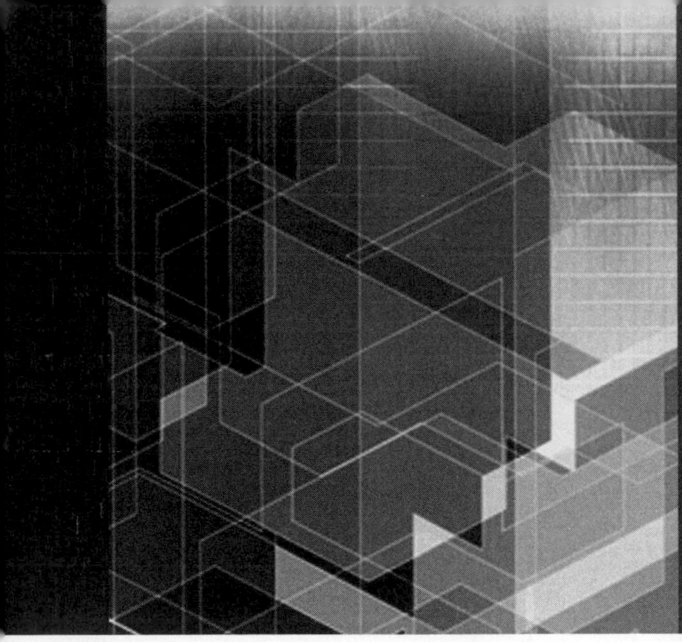

CHAPTER 2

Drawing Setup and Organization

Chapter 2 covers a number of drawing setup commands. The user will learn how to assign different units of measure with the `UNITS` command. The default sheet size can also be increased on the display screen with the `LIMITS` command. Controlling the grid and snap will be briefly discussed through the Snap and Grid tab located in the Drafting Settings dialog box. The major topic of this chapter is the discussion of layers. All options of the Layer Properties Manager palette will be demonstrated, along with the ability to assign color, linetype, and lineweight to layers. The Layer Control box provides easy access to all layers, colors, linetypes, and lineweights used in a drawing. Controlling the scale of linetypes through the `LTSCALE` command will also be discussed. Advanced layer tools such as Filtering Layers and creating Layer States will be introduced. This chapter concludes with a section on creating template files.

SETTING DRAWING UNITS

The Drawing Units dialog box is available for interactively setting the units of a drawing. Choosing Units from the Drawing Utilities heading of the Application Menu activates the dialog box illustrated in Figure 2.1.

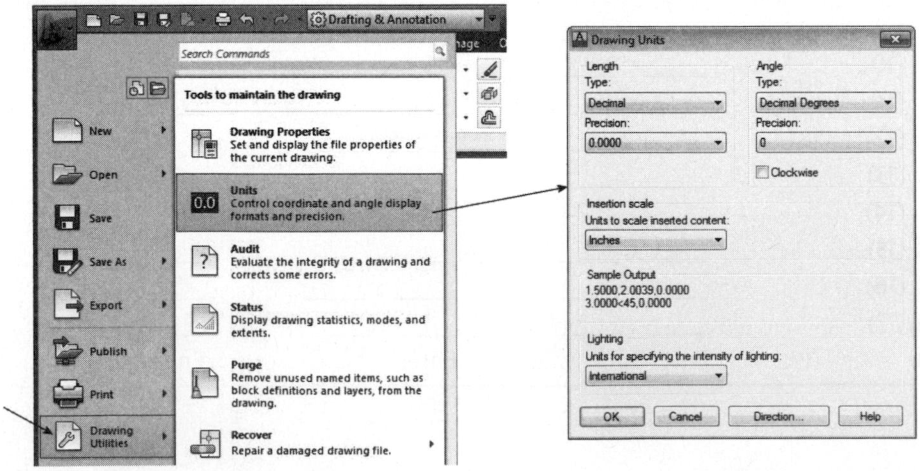

FIGURE 2.1

By default, decimal units are set along with four-decimal-place precision. The following systems of units are available: Architectural, Decimal, Engineering, Fractional, and Scientific (see Figure 2.2 (Left)). Architectural units are displayed in feet and fractional inches. Engineering units are displayed in feet and decimal inches. Fractional units are displayed in fractional inches. Scientific units are displayed in exponential format.

Methods of measuring angles supported in the Drawing Units dialog box include Decimal Degrees, Degrees/Minutes/Seconds, Grads, Radians, and Surveyor's Units (see Figure 2.2 (Center)). Accuracy of decimal degree for angles may be set between zero and eight places.

Selecting the Direction button in the main Drawing Units dialog box displays the Direction Control dialog box shown in Figure 2.2 (Right). This dialog box is used to control the direction of angle zero in addition to changing whether angles are measured in the counterclockwise or clockwise direction. By default, angles are measured from zero degrees in the east and in the counterclockwise direction.

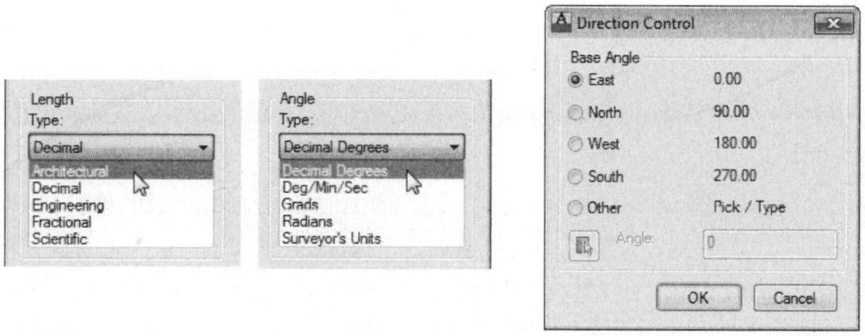

FIGURE 2.2

ENTERING ARCHITECTURAL VALUES FOR DRAWING LINES

The method of entering architectural values in feet and inches is a little different from the method for entering them as decimal units. To designate feet, you must enter the apostrophe symbol (') from the keyboard after the number. For example, "ten feet" would be entered as (10'), as shown in Figure 2.3. When feet and inches are necessary, you cannot use the Spacebar to separate the inch value from the foot value. For example, thirteen feet seven inches would be entered as (13'7), as shown in Figure 2.3. If you do use the Spacebar after the (13') value, this is interpreted as the enter key and your value is accepted as (13'). If you have to enter feet, inches, and fractions of an inch, use the hyphen (-) to separate the inch value from the fractional value. For example, to draw a line seventeen feet eleven and one-quarter inches, you would enter the following value in at the keyboard: (17'11-1/4). See Figure 2.3. Placing the inches symbol (") is not required since all numbers entered without the foot symbol are interpreted as inches.

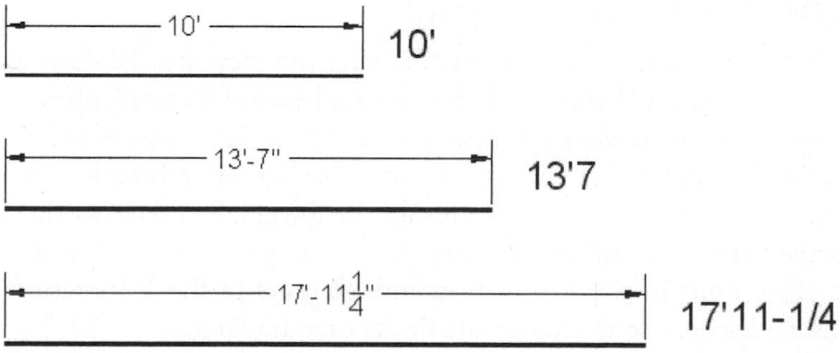

FIGURE 2.3

TRY IT!

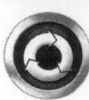

Open the drawing file 02_Architectural. Verify that the units setting is in architectural units by activating the Drawing Units dialog box. With Figure 2.4 as a guide, use the Direct Distance mode of entry with Polar Tracking to construct the shape using architectural values.

Command: L *(For LINE)*

Specify first point: 4',2' *(Enter absolute coordinate)*

Specify next point or [Undo]: *(Move your cursor to the right and enter a value of* 10'*)*

Specify next point or [Undo]: *(Move your cursor up and enter a value of* 4'6*)*

Specify next point or [Close/Undo]: *(Move your cursor to the right and enter a value of* 13'7-1/2*)*

Specify next point or [Close/Undo]: *(Move your cursor up and enter a value of* 4'9-1/2*)*

Specify next point or [Close/Undo]: *(Move your cursor to the left and enter a value of* 7'*)*

Specify next point or [Close/Undo]: *(Move your cursor up and enter a value of* 3'*)*

Specify next point or [Close/Undo]: *(Move your cursor to the left and enter a value of* 16'7-1/2*)*

Specify next point or [Close/Undo]: C *(To close the shape and exit the* LINE *command)*

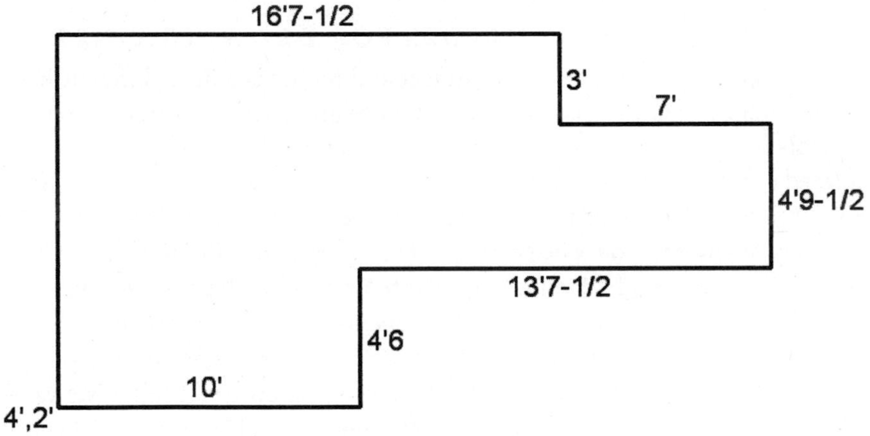

FIGURE 2.4

SETTING THE LIMITS OF THE DRAWING

By default, the size of the drawing screen in a new drawing file measures 12 units in the X direction and 9 units in the Y direction. This size may be ideal for small objects, but larger drawings require more drawing screen area. Use the LIMITS command for increasing the size of the designated drawing area. Enter this command directly at the Command prompt and provide absolute coordinates for the lower left and upper right corners of the new screen area. Illustrated in Figure 2.5 is a single-view drawing that fits on a screen size of 34 units in the X direction and 22 units in the Y direction. Follow the next command sequence to change the limits of a drawing.

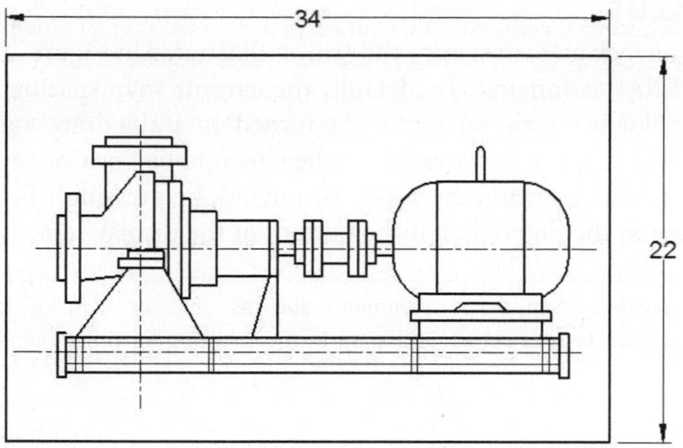

FIGURE 2.5

Command: **LIMITS**

Reset Model space limits:

Specify lower-left corner or [ON/OFF] <0.0000,0.0000>: *(Press* ENTER *to accept this value)*

Specify upper-right corner <12.0000,9.0000>: **34,22**

Changing the limits does not change the current viewing area in the display screen. Before continuing, perform a ZOOM-All to change the size of the display screen to reflect the changes in the limits of the drawing. ZOOM-All can be accessed from the Zoom icon on the Navigation Bar or through the Ribbon from the View tab and Navigate panel.

Command: Z *(For ZOOM)*

All/Center/Dynamic/Extents/Previous/Scale/Window/Object <real time>: A *(For all)*

USING GRID IN A DRAWING

Use grid to get a relative idea as to the size of objects. Grid can also be used to define the size of the display screen originally set by the LIMITS command. The lines or dots that make up the grid will never plot out on paper even if they are visible on the display screen. You can turn the grid on or off by using the GRID command or by pressing F7, or by single-clicking the GRID icon located in the Status Bar at the bottom of the display screen. By default, the grid is displayed in 0.50-unit intervals similar to Figure 2.6 (Left). Illustrated in Figure 2.6 (Right) is a grid that has been set to a value of 0.25, or half its original size.

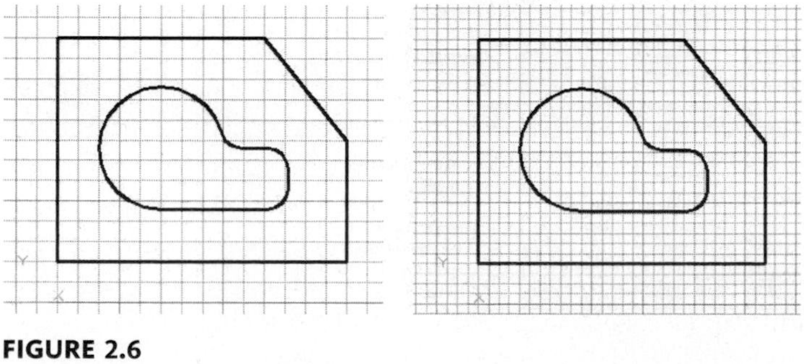

FIGURE 2.6

SETTING A SNAP VALUE

It is possible to have the cursor lock on to or snap to the grid, as illustrated in Figure 2.7; this is the purpose of the SNAP command. By default, the current snap spacing is 0.50 units. Even though a value is set, the snap must be turned on and a draw command, such as LINE or CIRCLE, must be activated for the cursor to be positioned to the grid. Activate the snap feature by using the SNAP command, by pressing F9, or by single-clicking SNAP icon in the Status Bar at the bottom of the display screen.

NOTE

The snap affect will only be present when a draw command, such as LINE or CIRCLE, is activated. Also verify that the Snap type is set to Grid snap in the Drafting Settings dialog box.

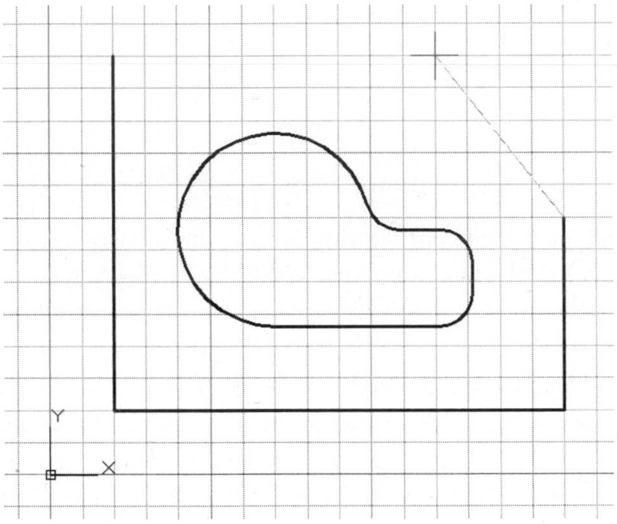

FIGURE 2.7

CONTROLLING SNAP AND GRID THROUGH THE DRAFTING SETTINGS DIALOG BOX

Right-clicking the Snap or Grid icons in the Status Bar displays the menu, as shown in Figure 2.8 (Left). Clicking Settings displays the Drafting Settings dialog box shown in the image on the right. Use this dialog box for making changes to the grid and snap settings.

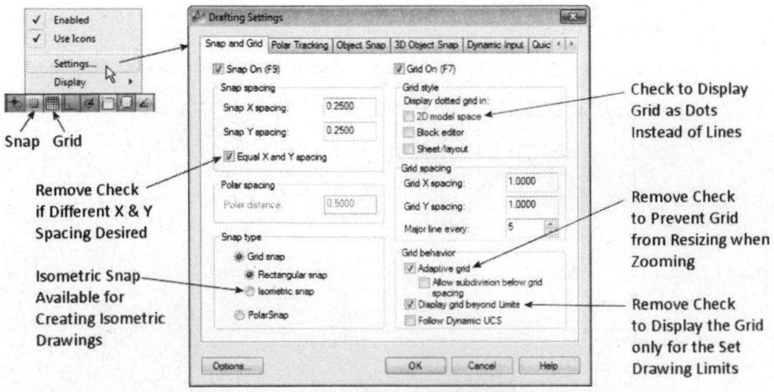

FIGURE 2.8

TRY IT!

Open the drawing file 02_Grid. Activate the Drafting Settings dialog box and experiment with the settings illustrated in Figure 2.8.

CONTROLLING DYNAMIC INPUT

Right-clicking the Dynamic Input icon, located in the Status Bar, and picking Settings launches the Dynamic Input tab of the Drafting Settings dialog box, as shown in Figure 2.9. Various checkboxes are available to turn on or off the Pointer Input (the absolute coordinate display of your cursor position when you are inside a command) and the Dimension Input (the display of distance and angle information for commands that support this type of input). You can also control the appearance of the Dynamic prompts. This can take the form of changing the background color or even assigning a level of transparency to the display of the Dynamic Input.

Clicking the Settings button under Pointer Input and Dimension Input launches the dialog boxes that allow you to change settings to further control Pointer and Dimension Inputs.

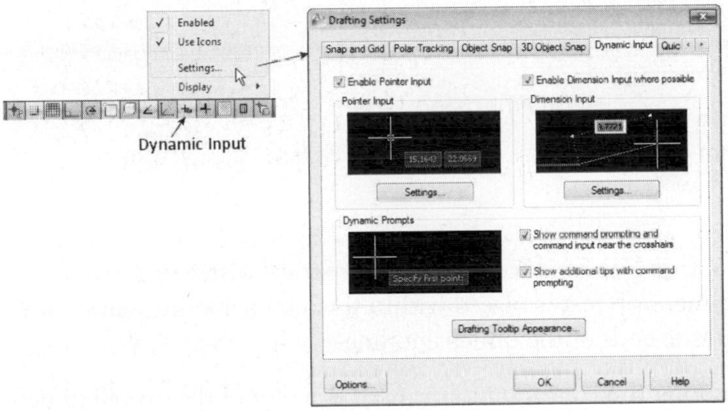

FIGURE 2.9

THE ALPHABET OF LINES

Engineering drawings communicate information through the use of lines and text, which, if used appropriately, accurately convey a project from design to construction. Before you construct engineering drawings, the quality of the lines that make up the drawing must first be discussed. Some lines of a drawing should be made thick; others need to be made thin. This is to emphasize certain parts of the drawing and is controlled through a line quality system. Illustrated in Figure 2.10 is a two-view drawing of an object complete with various lines that will be explained further.

The most important line of a drawing is the object line, which outlines the basic shape of the object. Because of their importance, object lines are made thick and continuous so they stand out among the other lines in the drawing. It does not mean that the other lines are considered unimportant; rather, the object line takes precedence over all other lines.

The cutting plane line is another thick line; it is used to show where a part would be sliced to expose interior details. It stands out by being drawn as a series of long dashes separated by spaces. Arrowheads determine the viewing direction for the adjacent view. This line will be discussed in greater detail in Chapter 9, "Creating Section Views."

The hidden line is a thin weight line used to identify edges that although can't be actually seen in the view, help describe the part shape. It consists of a series of dashes separated by spaces. Whether an edge is visible or invisible, it still must be shown with a line.

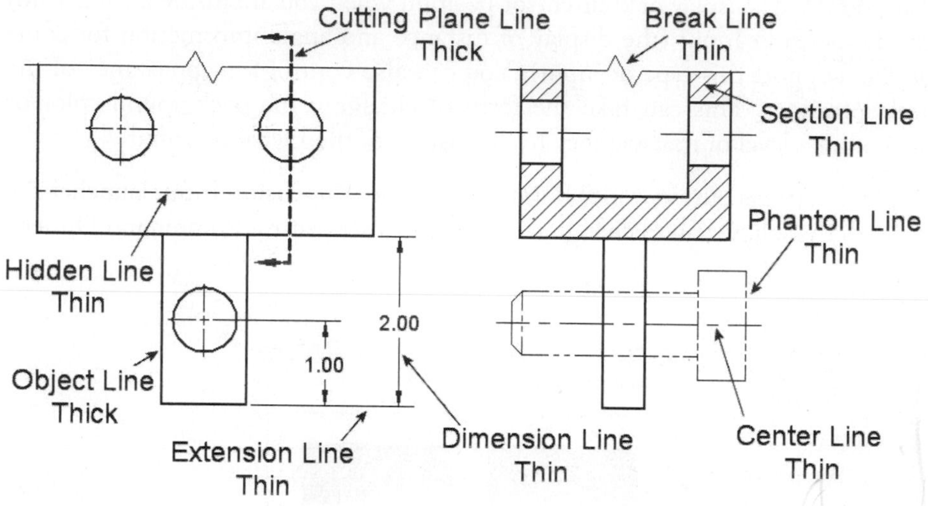

FIGURE 2.10

The dimension line is a thin line used to show the numerical distance between two points. Typically, the dimension text is placed within the dimension line, and arrowheads are placed at opposite ends of the dimension line.

The extension line is another thin continuous line used as a part of the overall dimension. Extension lines provide a means to move dimension lines away from the object into a clear area where they can be easily seen and interpreted.

When you use the cutting plane line to create an area to cut or slice, the surfaces in the adjacent view are section lined using the section line, a thin continuous line.

Another important line used to identify the centers of symmetrical objects such as cylinders and holes is the centerline. It is a thin line consisting of a series of long and short dashes. Centerlines are often used with dimensions to help locate features on a part.

The phantom line consists of a thin line made with a series of two short dashes and one long dash. It is used to simulate the placement or movement of a part or component without actually detailing the component.

The long break line is a thin line with a "zigzag" symbol used to establish where an object is broken to simulate a continuation of the object.

ORGANIZING A DRAWING THROUGH LAYERS

As a means of organizing objects, a series of layers should be devised for every drawing. You can think of layers as a group of transparent sheets that combine to form the completed drawing. The illustration Figure 2.11 (Left) displays a drawing consisting of object lines, dimension lines, and border. An example of organizing these three drawing components by layers is illustrated in Figure 2.11 (Right). Only the drawing border occupies a layer called "Border." The object lines occupy a layer called "Object," and the dimension lines are drawn on a layer called "Dimension." At times, it may be necessary to turn off the dimension lines for a clearer view of the object. Creating all dimensions on a specific layer allows you to turn off the dimensions while viewing all other objects on layers that are still turned on.

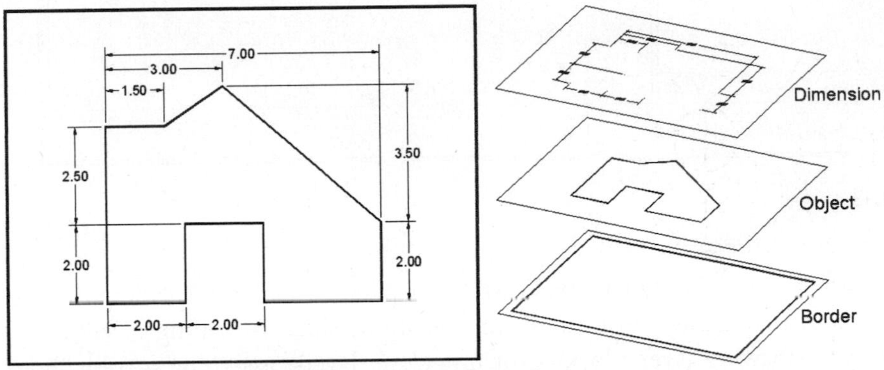

FIGURE 2.11

THE LAYER PROPERTIES MANAGER PALETTE

The Layer Properties Manager palette is the tool used to create and manage layers. This palette is easily activated by clicking the Layer Properties Manager button on the Ribbon, as shown in Figure 2.12.

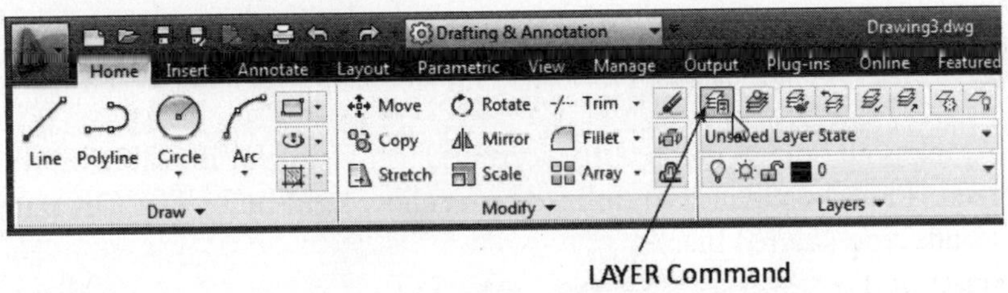

FIGURE 2.12

The Layer Properties Manager palette, illustrated in Figure 2.13, is divided into two separate panes. The first pane on the left is the Tree View pane used for displaying layer filter, group, or state information. The main body of the Layer Properties Manager is the List View pane on the right. This area lists the individual layers that currently exist in the drawing.

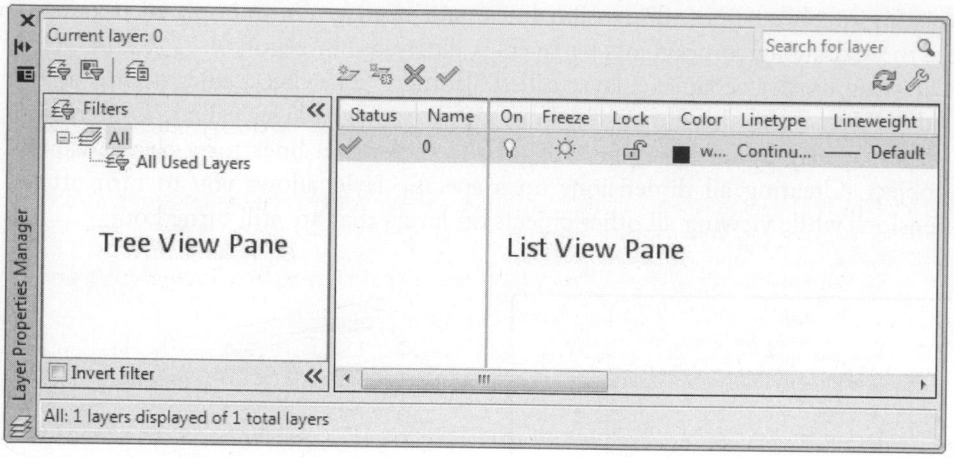

FIGURE 2.13

The layer information located in the List View pane is sometimes referred to as layer states and they allow you to perform operations such as turning layers on or off; freezing or thawing layers; locking or unlocking layers; assigning a color, linetype, lineweight, transparency, and plot style to a layer or group of layer, as shown in Figure 2.14. A brief explanation of each layer state is provided below:

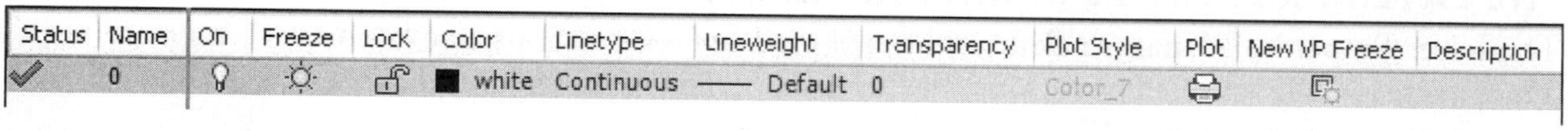

FIGURE 2.14

Status—When a green checkmark is displayed, this layer is considered current.

Name—Displays the name of the layer.

On/Off—Makes all objects created on a certain layer visible or invisible on the display screen. The On state is symbolized by a yellow light bulb. The Off state has a light bulb icon shaded black.

Freeze—This state is similar to the Off mode; objects frozen appear invisible on the display screen. Freeze, however, is considered a major productivity tool used to speed up the performance of a drawing. This is accomplished by not calculating any frozen layers during drawing regenerations. A snowflake icon symbolizes this layer state.

Thaw—This state is similar to the On mode; objects on frozen layers reappear on the display screen when they are thawed. The sun icon symbolizes this layer state.

Lock—This state allows objects on a certain layer to be visible on the display screen while protecting them from accidentally being modified through an editing command. A closed padlock icon symbolizes this layer state.

Unlock—This state unlocks a previously locked layer and is symbolized by an open padlock icon.

Color—This state displays a color that is assigned to a layer and is symbolized by a square color swatch along with the name of the color. By default, the color white is assigned to a layer.

Linetype—This state displays the name of a linetype that is assigned to a layer. By default, the Continuous linetype is assigned to a layer.

Lineweight—This state sets a lineweight to a layer. An image of this lineweight value is visible in this layer state column.

Transparency—This state sets an object's visibility. Applies a transparency level between 0 and 90 for all objects on that layer.

Plot Style—A plot style allows you to override the color, linetype, and lineweight settings made in the Layer Properties Manager dialog box. Notice how this area is grayed out. When working with a plot style that is color dependent, you cannot change the plot style. Plot styles will be discussed in greater detail later in this book.

Plot—This layer state controls which layers will be plotted. The presence of the printer icon symbolizes a layer that will be plotted. A printer icon with a red circle and diagonal slash signifies a layer that will not be plotted.

New VP Freeze—Creates a new layer and automatically freezes this layer in any new viewport.

Description—This state allows you to enter a detailed description for a layer.

CREATING NEW LAYERS

Four buttons are available at the top of the Layer Properties Manager palette, as shown in Figure 2.15. Use these buttons to create new layers, have new layers frozen in all viewports (Chapter 19), delete layers, and make a layer current.

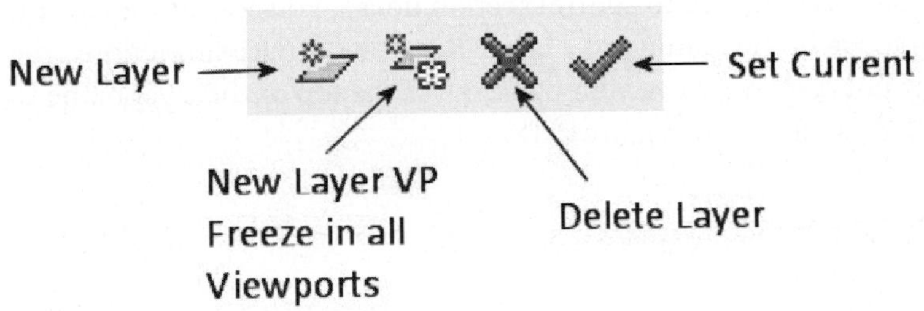

FIGURE 2.15

Clicking the New Layer button of the Layer Properties Manager palette creates a new layer called Layer1, which displays in the layer list box, as shown in Figure 2.16. The layer name is automatically highlighted, so to change it to something more meaningful, you simply type in the new name.

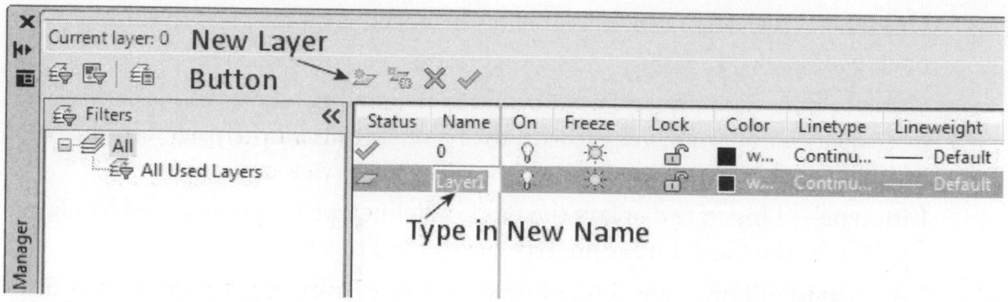

FIGURE 2.16

Illustrated in Figure 2.17 is the result of changing the name of the layer from Layer1 to Object.

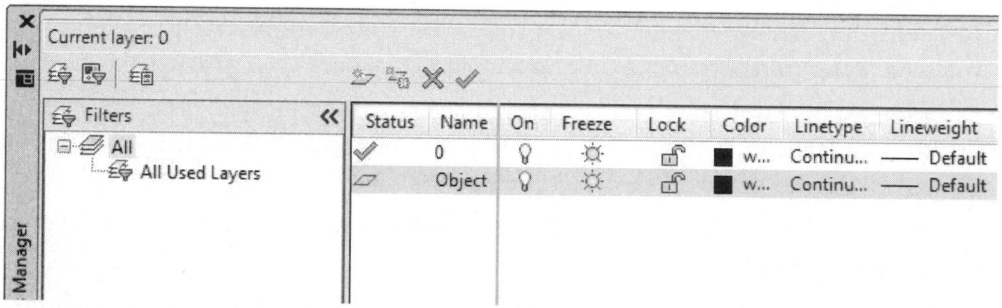

FIGURE 2.17

TIP

Before picking the New Layer button, select a layer that has properties similar to the one you want to create because they will be repeated in the new layer.

You can also be descriptive with layer names. In Figure 2.18, a layer has been created called "Section" (this layer is designed to control section lines). You are allowed to add spaces and other characters in the naming of a layer. Because of space limitations, the entire layer name may not display. Move your cursor over the top of the layer name to view the full description, as shown in Figure 2.18.

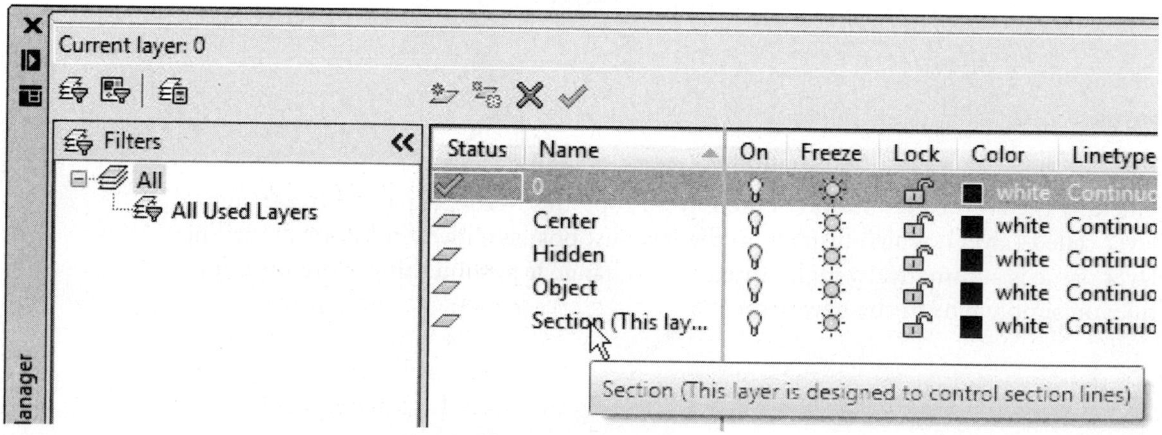

FIGURE 2.18

If more than one layer needs to be created, it is not necessary to continually click the New Layer button. Once a new layer is created and you type in the new name, instead of hitting ENTER, simply type in a comma (,) to display another new layer. Continue creating new layer names followed by a comma until all the layers are created. Instead of using a comma, you could press ENTER twice and achieve the same results.

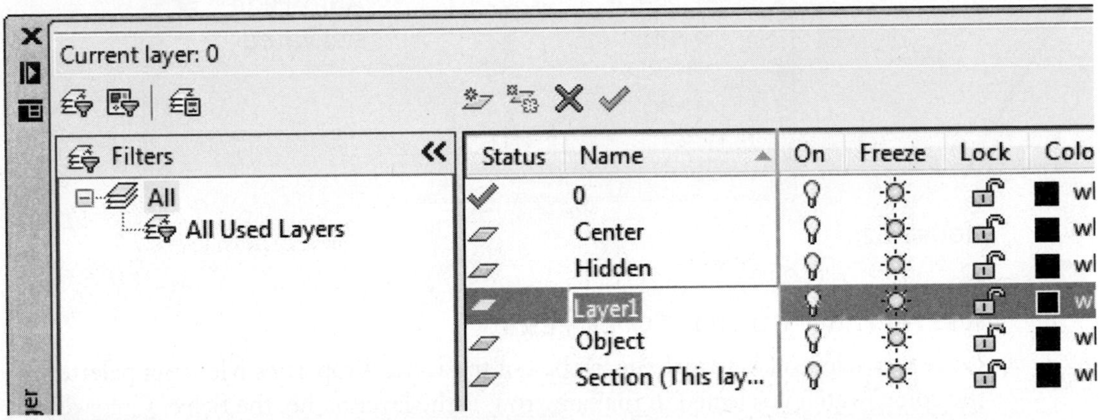

FIGURE 2.19

DELETING LAYERS

To delete a layer or group of layers, highlight the layers for deletion, as shown in the Figure 2.20 (Left). Click on the delete button. The results are displayed in Figure 2.20 (Right) with the selected layers being deleted from the palette.

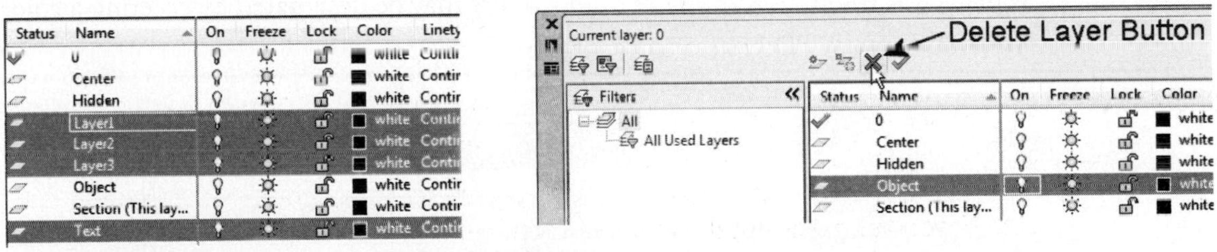

FIGURE 2.20

Only layers that do not contain any drawing geometry or objects can be deleted. Use the CTRL key to select multiple layers.	NOTE

AUTO-HIDING THE LAYER PROPERTIES MANAGER PALETTE

While the Layer Properties Manager palette can display on the screen preventing you from working on detail segments of your drawing, it is possible to collapse or Auto-hide the palette. Right-clicking on the title strip will display a menu, as shown in Figure 2.21 (Left). Clicking on Auto-hide will turn this feature on. This feature can also be set with a toggle button on the title strip. The results are displayed in Figure 2.21 (Right). When you move your cursor away from the palette, it collapses, allowing your drawing to fill the entire screen. Moving your cursor over the title strip of the palette will display it in its entirety.

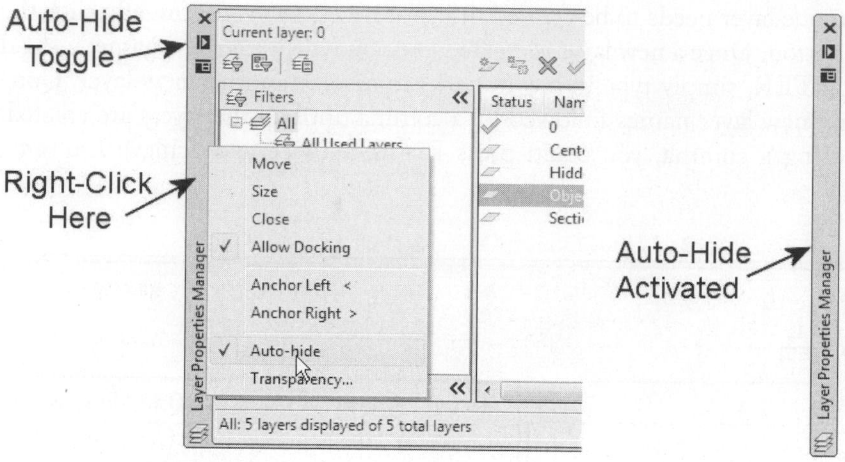

FIGURE 2.21

ASSIGNING COLOR TO LAYERS

Once you select a layer from the list box of the Layer Properties Manager palette and the color swatch is selected in the same row as the layer name, the Select Color dialog box shown in Figure 2.22 is displayed. Three tabs allow you to select three different color groupings. The three groupings (Index Color, True Color, and Color Books) are described as follows.

Index Color Tab

This tab, Figure 2.22, allows you to make color settings based on 255 AutoCAD Color Index (ACI) colors. Standard, Gray Shades, and Full Color Palette areas are available for you to choose colors from. Colors may be designated by entering a color name or index number.

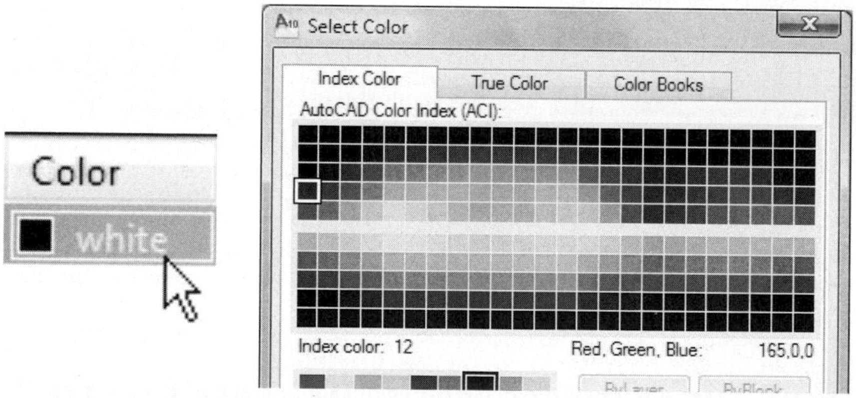

FIGURE 2.22

True Color Tab

Use this tab to make color settings using true colors, also known as 24-bit color. Two color models are available for you to choose from, namely Hue, Saturation, and Luminance (HSL) and Red, Green, and Blue (RGB). Through this tab, you can choose from over 16 million colors, as shown in Figure 2.23 (Left).

Color Books Tab

Use the Color Books tab to select colors that use third-party color books (such as Pantone) or user-defined color books. You can think of a color book as similar to those available in hardware stores when selecting household interior paints. When you select a color book, the name of the selected color book will be identified in this tab, as shown in Figure 2.23 (Right).

NOTE

The Index tab will be used throughout this text. However, you are encouraged to experiment with the True Color and Color Books tabs.

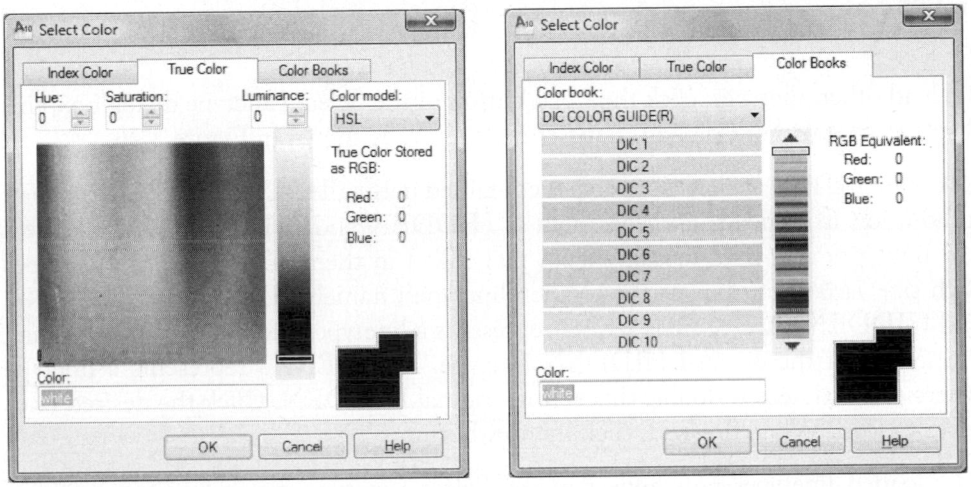

FIGURE 2.23

ASSIGNING TRANSPARENCY TO LAYERS

Assigning transparency to objects through layers is a way to change visual emphasis on those objects on the screen. Once you select a layer from the list box of the Layer Properties Manager palette, select the "0" from the Transparency heading. This action will activate the Layer Transparency dialog box, as shown in Figure 2.24 (Left). Use this dialog box to increase the transparency of objects assigned to this layer up to a value of 90. The transparency effect can be toggled on and off with the Transparency button on the Status Bar, as shown in Figure 2.24 (Right). Transparency only displays on the screen and will not appear on plotted drawings.

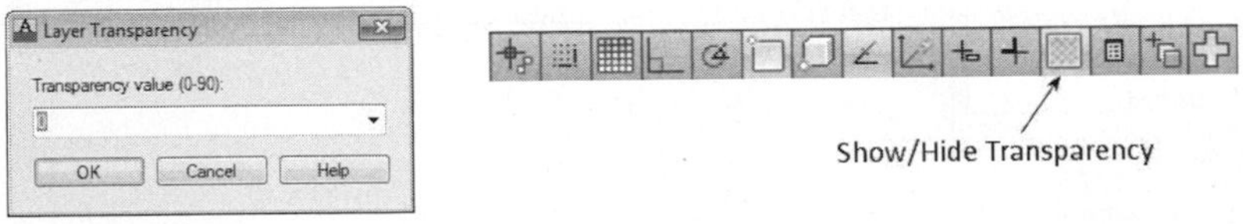

FIGURE 2.24

ASSIGNING LINETYPES TO LAYERS

Selecting the name Continuous next to the highlighted layer activates the Select Linetype dialog box, as shown in Figure 2.25. Use this dialog box to dynamically select preloaded linetypes to be assigned to various layers. By default, the Continuous linetype is loaded for all new drawings.

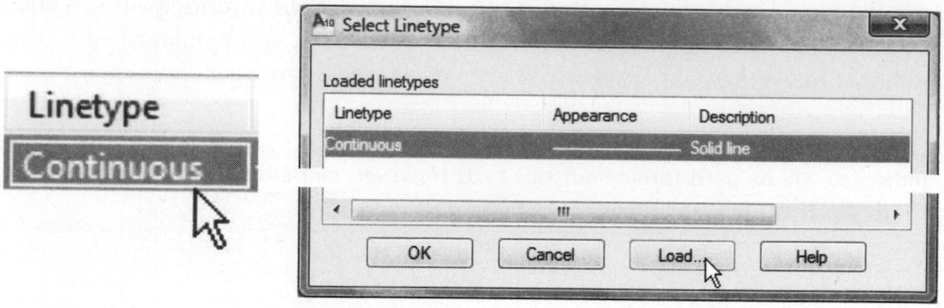

FIGURE 2.25

To load other linetypes, click the Load button of the Select Linetype dialog box; this displays the Load or Reload Linetypes dialog box, as shown in Figure 2.26.

Use the scroll bars to view all linetypes contained in the file ACAD.LIN. Notice that, in addition to standard linetypes such as HIDDEN and PHANTOM, there are a few linetypes that have text or shapes embedded in them. There are also linetypes with size variations, such as the Hidden linetypes; namely HIDDEN, HIDDEN2, and HIDDENX2. The HIDDEN2 represents a linetype where the dash size is half the length of the original HIDDEN linetype. HIDDENX2 represents a linetype where the dash size is double that of the original HIDDEN. Click the desired linetypes to load. When finished, click the OK button.

The loaded linetypes now appear in the Select Linetype dialog box, as shown in Figure 2.26. It must be pointed out that the linetypes in this list are only loaded into the drawing and are not assigned to a particular layer. Clicking the linetype in this dialog box assigns this linetype to the layer currently highlighted in the Layer Properties Manager palette.

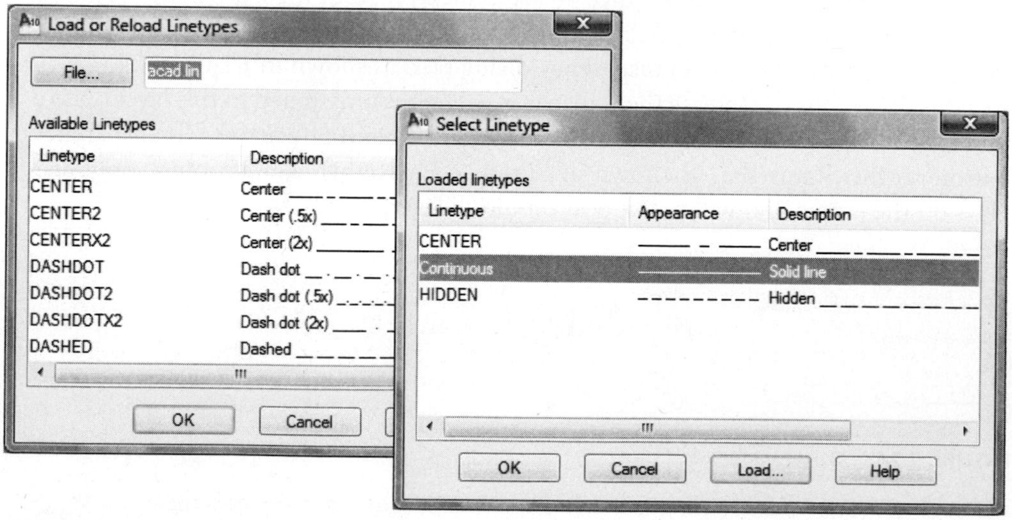

FIGURE 2.26

ASSIGNING LINEWEIGHT TO LAYERS

Selecting the name Default under the Lineweight heading of the Layer Properties Manager dialog box activates the Lineweight dialog box, as shown in Figure 2.27. Use this to attach a lineweight to a layer. Lineweights are very important to a drawing file—they give contrast to the drawing. As stated earlier, the object lines should stand out over all other lines in the drawing. A thick lineweight would then be assigned to the object line layer.

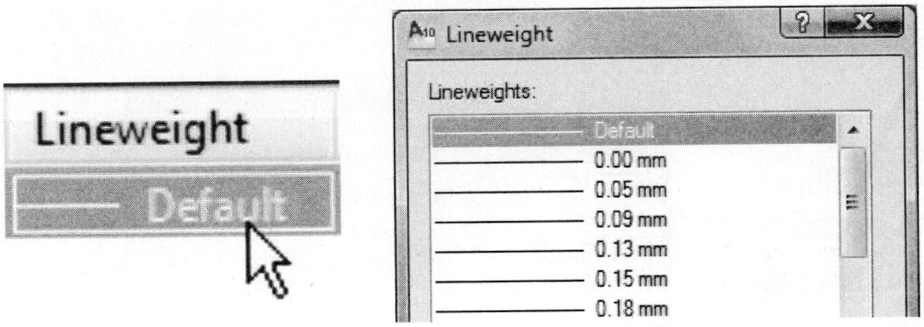

FIGURE 2.27

In Figure 2.28 (Left), a lineweight of 0.50 mm has been assigned to all object lines and a default lineweight (0.25 mm) has been assigned to all other lines. However, all lines in this figure appear to consist of the same lineweight. To display differing lineweights on the screen, utilize the Lineweight feature. This feature is toggled ON or OFF by clicking the Lineweight icon on the Status Bar. Clicking the Lineweight icon, as shown in Figure 2.28 (Right), turns the lineweight function on (the icon will turn a blue color). It should be noted that the layer assigned lineweight will be plotted whether or not the LWT button is activated.

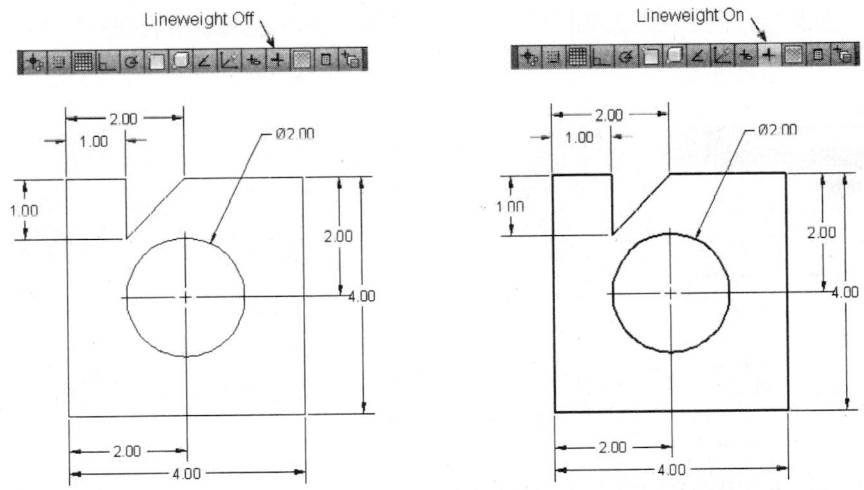

FIGURE 2.28

THE LINEWEIGHT SETTINGS DIALOG BOX

When working with complicated drawings, you might want to turn the Lineweight feature off to better view an area where numerous objects converge. If you prefer to display lineweights, you may find it better to simply change the lineweights' displayed thickness. To give your lineweights a more pleasing appearance, a dialog box is available to control the display of your lineweights. Right-click the LWT button on the

Status Bar and pick Settings, as shown in Figure 2.29 (Left), to display the Lineweight Settings dialog box. This dialog box can also be displayed from the Ribbon (Home Tab > Properties Panel). Notice the position of the slider bar in the Adjust Display Scale area of the dialog box. Sliding the bar to the left near the Min setting reduces the width of all lineweights. If you think your lineweights appear too thin, slide the bar to the right near the Max setting and observe the results. Continue to adjust your lineweights until they have a pleasing appearance in your drawing. This adjustment only affects the appearance of lineweights on the screen, the layer assigned lineweights will be utilized for all plotted drawings.

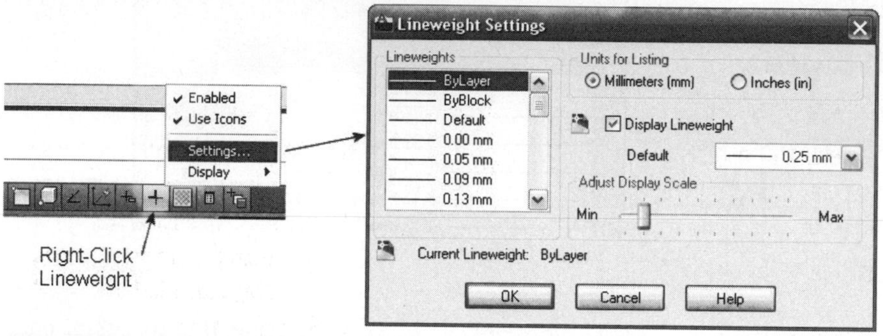

FIGURE 2.29

THE LINETYPE MANAGER DIALOG BOX

The Linetype Manager dialog box can be activated through the Ribbon (Home Tab > Properties Panel), as shown in Figure 2.30 (Left). This dialog box is designed mainly to preload linetypes.

Clicking the Load button activates the Load or Reload Linetypes dialog box illustrated earlier in this chapter. You can select individual linetypes in this dialog box or load numerous linetypes at once by pressing CTRL and clicking each linetype. Clicking the OK button loads the selected linetypes into the dialog box.

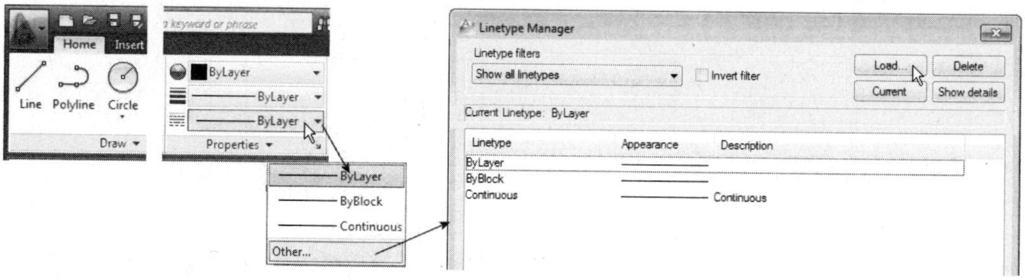

FIGURE 2.30

THE RIBBON'S PROPERTIES PANEL

Typically, you will want to keep the Color, Lineweight, and Linetype properties set to Bylayer, as shown in Figure 2.31 (Left). With these settings, new objects that you create will be assigned the property Bylayer. Their color, lineweight, and linetype are determined by the settings in the Layer Properties Manager palette. Selecting specific properties, as shown in Figure 2.31 (Right), will assign those properties to any new objects created. If you create a new line with the settings shown, its color will be red, its lineweight 0.50 mm (thick), and its linetype hidden. No matter what layer the object is placed on, it will display those properties. It should also be noted that if changes are made in the Layer Manager palette, only those objects that were assigned their properties as Bylayer will update in your drawing.

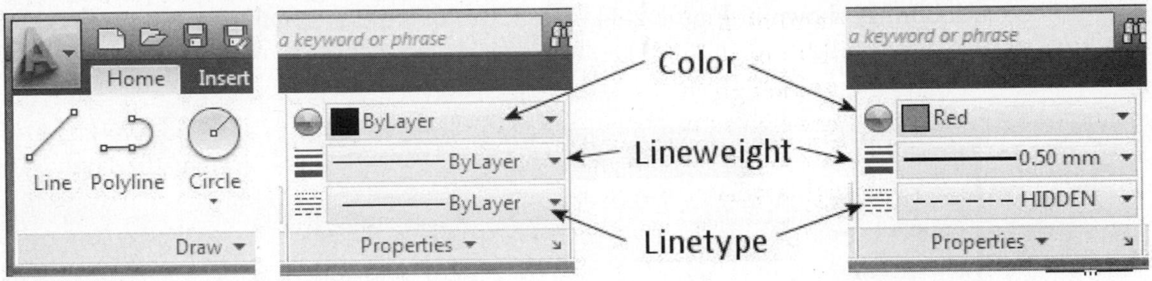

FIGURE 2.31

LOCKED LAYERS

Objects on locked layers remain visible on a drawing; they cannot, however, be selected when performing editing operations. Quite often, it can be difficult to distinguish objects on layers that are locked from normal layers. To assist with the identification of locked objects, a lock icon appears when you move your cursor over an object considered locked, as shown in Figure 2.32 (Left). You can also set a fade factor when viewing locked layers. This is another way of distinguishing regular layers from those that are locked. Expanding the Layers pane in the Ribbon will display the Locked Layer Fading button, as shown in Figure 2.32. Next to this button is a slider bar. When turned on, you can move the slider bar to the left or right depending on the desired amount of fading.

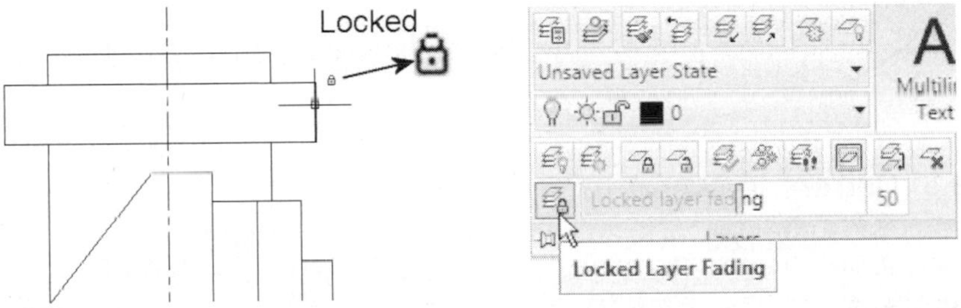

FIGURE 2.32

Notice that Figure 2.33 (Left) has objects on a locked layer faded to 50% while the image on the right in Figure 2.33 is faded 80%. Notice how the objects on the left appear darker than those on the right due to the larger fade factor.

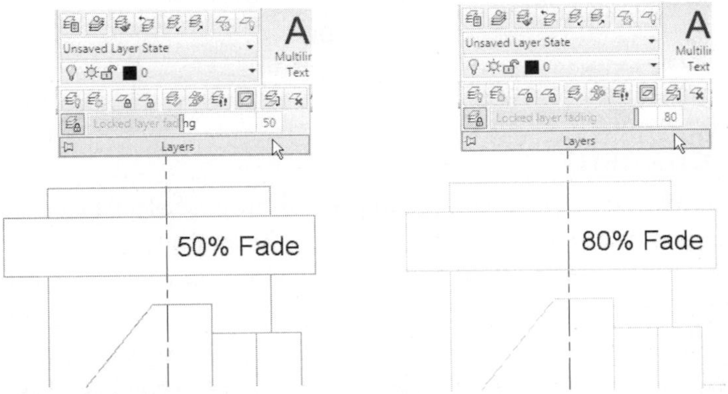

FIGURE 2.33

THE LAYERS CONTROL BOX

The Layers Control box provides a drop-down area to better control the layer properties or states. You can access this feature through the Layer panel of the Ribbon, as shown in Figure 2.34. The Layer Control area will now be discussed in greater detail.

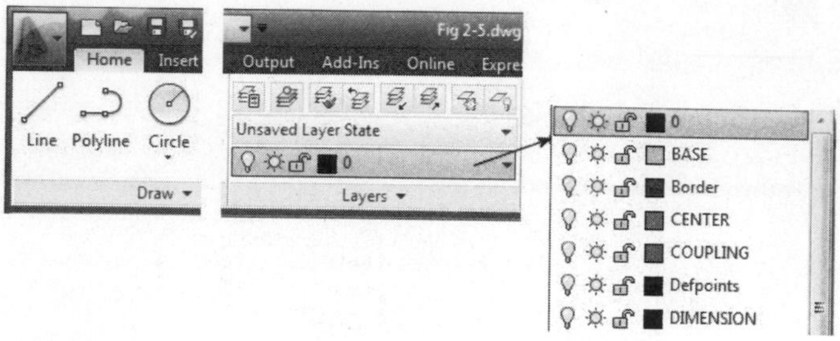

FIGURE 2.34

CONTROL OF LAYER PROPERTIES

Expanding the drop-down list cascades all layers defined in the drawing in addition to their properties, identified by symbols (see Figure 2.35). The presence of the light bulb signifies that the layer is turned on. Clicking the light bulb symbol turns the layer off. The sun symbol signifies that the layer is thawed. Clicking the sun turns it into a snowflake symbol, signifying that the layer is now frozen. The padlock symbol controls whether a layer is locked or unlocked. By default, all layers are unlocked. Clicking the padlock changes the symbol to display the image of a locked symbol, signifying that the layer is locked.

Study Figure 2.35 for a better idea of how the symbols affect the state of certain layers.

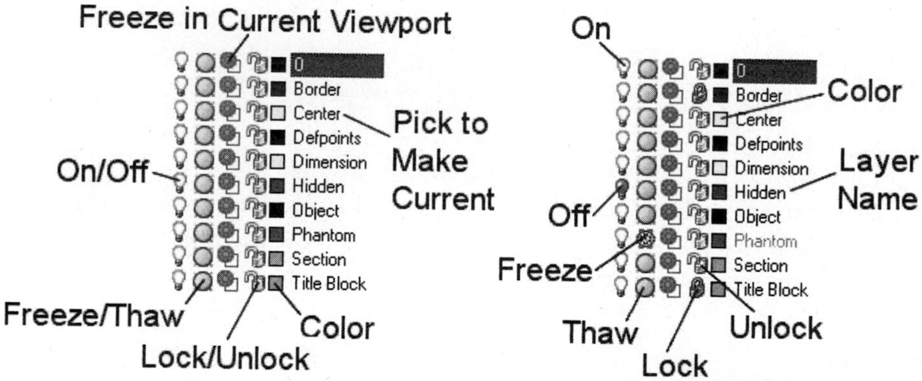

FIGURE 2.35

MAKING A LAYER CURRENT

Various methods can be employed to make a layer current to draw on. Select a layer in the Layer Properties Manager palette and then click the Set Current button to make the layer current, as shown in Figure 2.36 (Left). Double-clicking a layer name in the palette will also set the layer current.

Picking a layer name from the Layer Control box, as shown in Figure 2.36 (Right), provides a quick and simple method to make a layer current. If you forget to set the current layer and objects are accidently placed on the wrong layer, select those items and then pick the correct layer in the Layer Control box. This will transfer the objects to the correct layer.

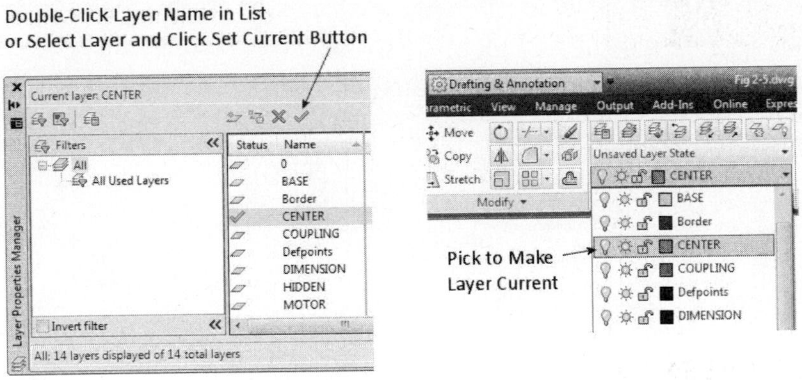

FIGURE 2.36

RIGHT-CLICK SUPPORT FOR LAYERS

While inside the Layer Properties Manager, right-clicking inside the List View pane (layer information area) displays the shortcut menu shown in Figure 2.37. Use this menu to make the selected layer current, to make a new layer based on the selected layer, to select all layers in the dialog box, or to clear all layers. You can even select all layers except for the current layer. This shortcut menu provides you with easier access to commonly used layer manipulation tools.

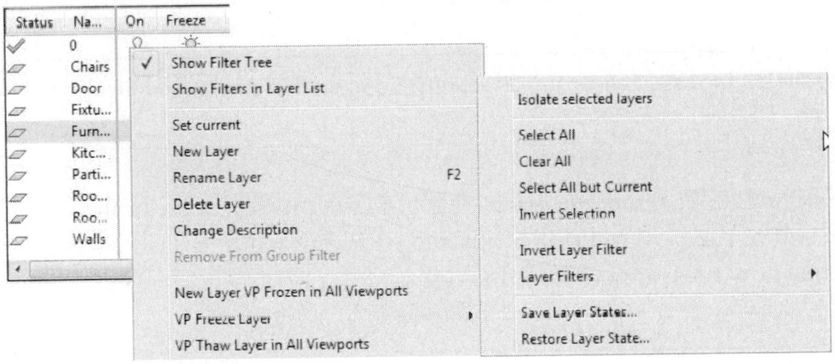

FIGURE 2.37

OTHER RIGHT-CLICK LAYER CONTROLS

If you right-click one of the layer header names (Name, On, Freeze, etc.), you get the menu illustrated in Figure 2.38. This menu allows you to turn off header names as a means of condensing the list of headers and making it easier to interpret the layers that you are using. Other areas of this menu allow you to maximize all columns in order to view all information in full regarding layers. By default, the Name column is frozen. This means when you scroll to the left or right to view the other layer properties, the Name column does not scroll and allows you to view information at the end of the Layer Properties Manager palette.

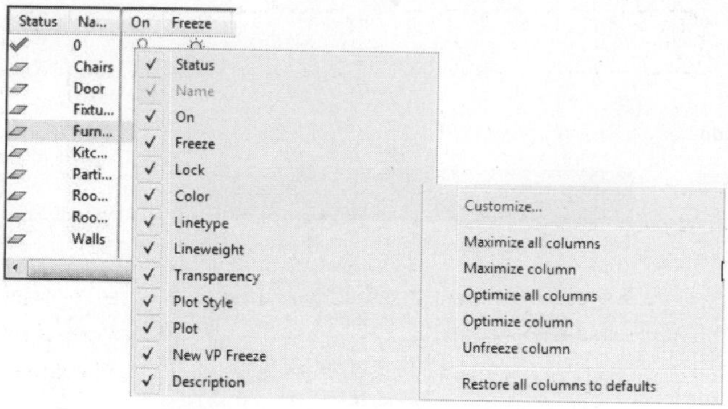

FIGURE 2.38

CONTROLLING THE LINETYPE SCALE

Once linetypes are associated with layers and placed in a drawing, an `LTSCALE` command is available to control their scale. On the left side Figure 2.39, the default linetype scale value of 1.00 is in effect. This scale value acts as a multiplier for all linetype distances. The Hidden linetype, for example, has a dash length of 0.25 units long: if the linetype scale value were set to something larger than 1.00 the dash length would increase and if the scale was set to something less than 1.00 it would shorten. The `LTSCALE` command displays the following command sequence:

```
Command: LTS (For LTSCALE)

Enter new linetype scale factor <1.0000>: (Press ENTER to
accept the default or enter another value)
```

TRY IT!

Open the drawing file 02_LTScale. Follow the directions, Command prompts, and Figure 2.39 below for using the `LTSCALE` command.

In the middle of Figure 2.39, a linetype scale value of 0.50 units has been applied to all linetypes. As a result of the 0.50 multiplier, instead of all hidden line dashes measuring 0.25 units, they now measure 0.125 units.

```
Command: LTS (For LTSCALE)

Enter new linetype scale factor <1.0000>: 0.50
```

On the right side of Figure 2.39, a linetype scale value of 2.00 units has been applied to all linetypes. As a result of the 2.00 multiplier, instead of all hidden line dashes measuring 0.25 units, they now measure 0.50 units. Notice how a large multiplier caused some of the centerlines to appear as continuous lines. The lines are not long enough to display the size of dashes specified.

```
Command: LTS (For LTSCALE)

Enter new linetype scale factor <0.5000>: 2.00
```

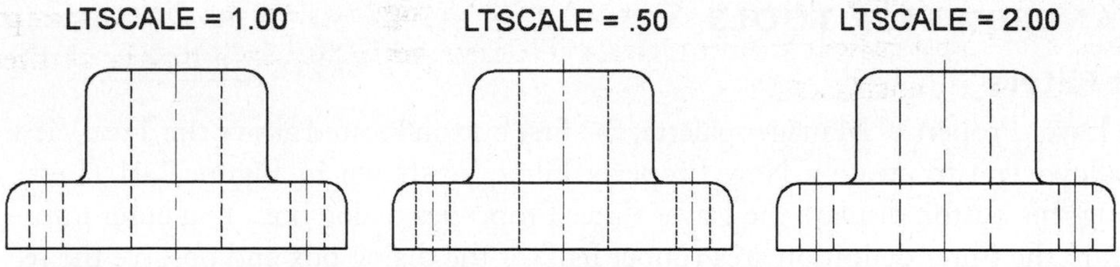

FIGURE 2.39

TRY IT!

Open the drawing file 02_Floor Plan illustrated in Figure 2.40 (Left). This floor plan is designed to plot out at a scale of 1/8" = 1'0". This creates a scale factor of 96 (found by dividing 1' by 1/8"). A layer called "Dividers" is created and assigned the Hidden linetype to show all red hidden lines as potential rooms in the plan. However, the hidden lines do not display. For all linetypes to show as hidden, this multiplier should be applied to the drawing through the `LTSCALE` **command.**

Since the drawing was constructed in real-world units or full size, the linetypes are converted to these units beginning with the multiplier of 96, as shown in Figure 2.40 (Right), through the `LTSCALE` command.

```
Command: LTS (For LTSCALE)
Enter new linetype scale factor <1.0000>: 96
```

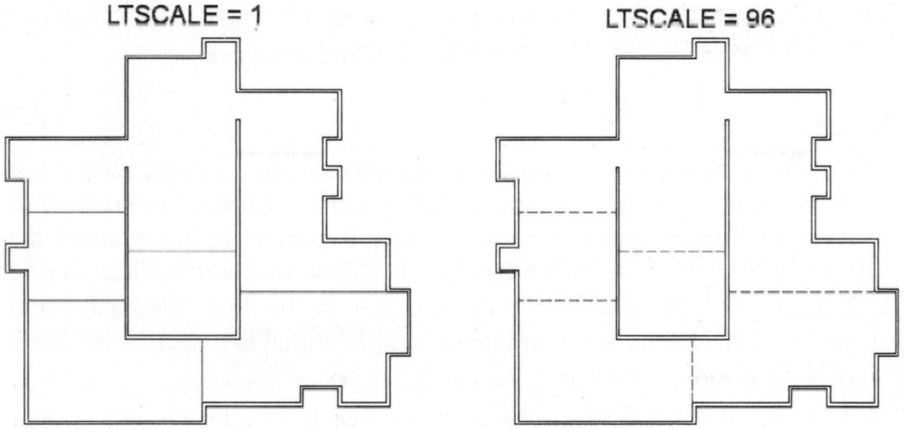

FIGURE 2.40

Although changing the linetype scale did correctly display the linetypes, they will not appear correct in any drawing layouts (discussed in Chapter 14). A better solution is to set the annotation scale (discussed in greater detail in Chapter 19). In the 02_Floor Plan drawing change the LTSCALE back to 1. On the Status Bar change the annotation scale to 1/8" = 1'-0". Perform a REGEN to see the results.

NOTE

If changing the annotation scale did not scale your linetypes correctly, verify that MSLTSCALE is set to "1."

ADVANCED LAYER TOOLS

Layer Filters

In the Layer Properties Manager palette, the first button located above the Tree View pane allows you to create a New Property Filter, as shown in Figure 2.41 (Left). Clicking this button displays the Layer Filter Properties dialog box. You enter information in the Filter definition area (upper half) of the dialog box and observe the results of the filter in the Filter preview area (lower half). In this example, a layer filter has been created in the definition area to identify all layers that begin with Grid*. The results show a number of layers in the preview area. Notice at the top of the dialog box that this filter has been given the name Grid. You can build various named filters as a means of further organizing your layers by function, color, name, and linetype, or even through a combination of these states.

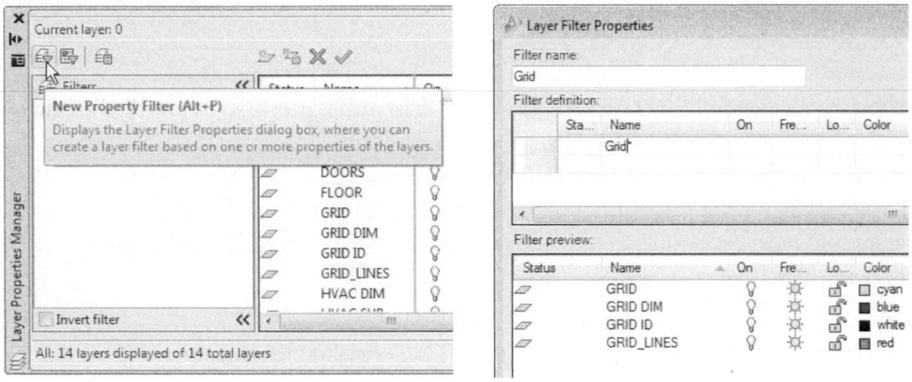

FIGURE 2.41

Layer Groups

Layer groups allow you to collect a number of layers under a unique name. For instance, you may want to group a number of layers under the name Foundation; or in a mechanical application, you may want to group a number of layers under the name Fasteners. Choose this command by clicking the second button illustrated in Figure 2.42 (Left). You could also move your cursor in the Tree View pane and right-click to display the menu shown in Figure 2.42 (Right). Then click New Group Filter.

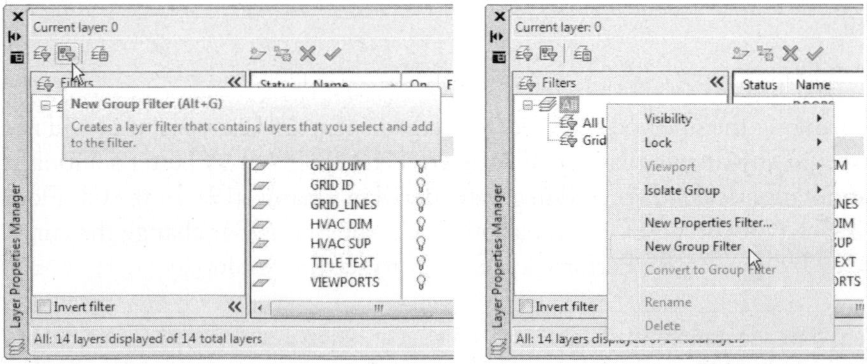

FIGURE 2.42

As shown in Figure 2.43 (Left), a new layer group called HVAC Plan is created and located in the Tree View pane. To associate layers with this group, select the layers in the List View pane and drag the layers to the layer group. You are not moving these layers; rather, you are grouping them under a unique name.

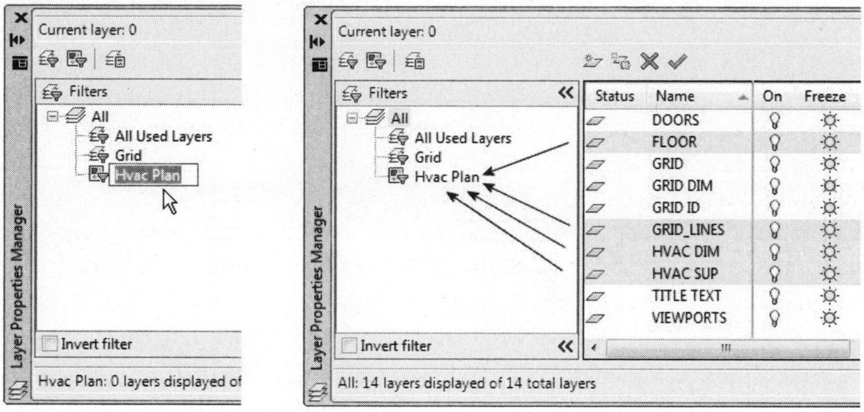

FIGURE 2.43

Clicking the HVAC Plan layer group displays only those layers common to this group in the List View pane of the Layer Properties Manager dialog box, as shown in Figure 2.44 (Left). Notice in the figure on the right that additional groups were created: Electrical Plan, Floor Plan, and Foundation Plan.

> **NOTE**
> **Clicking All in the Tree View pane lists all layers defined in the drawing.**

Additional controls allow you to further manipulate layers. With the HVAC Plan group selected, it is possible to invert this filter. This means to select all layers not part of the HVAC Plan group, place a check in the box next to Invert filter, located in the lower-left corner of the Layer Properties Manager palette. The result of this inverted list is shown in Figure 2.44 (Right).

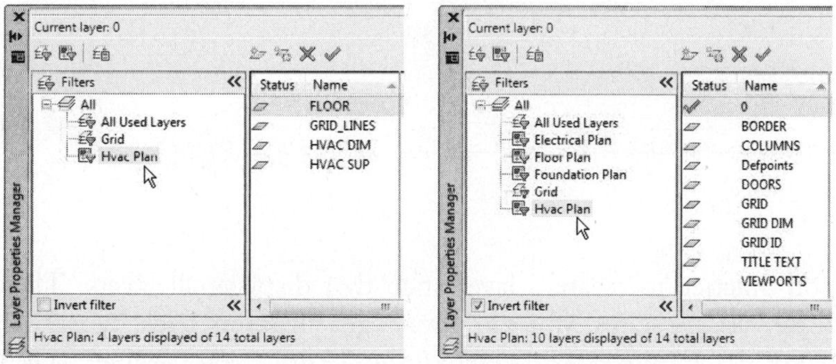

FIGURE 2.44

Layer States Manager

The Layer States Manager allows you to group a number of layer settings under a unique name and then retrieve this name later to affect the display of a drawing. Once you arrange your drawing in various states, such as turning off certain layers, select the

Layer States Manager button, as shown in Figure 2.45 (Left). This launches the Layer States Manager dialog box. Click the New button to assign a unique name in the Layer States Manager dialog box.

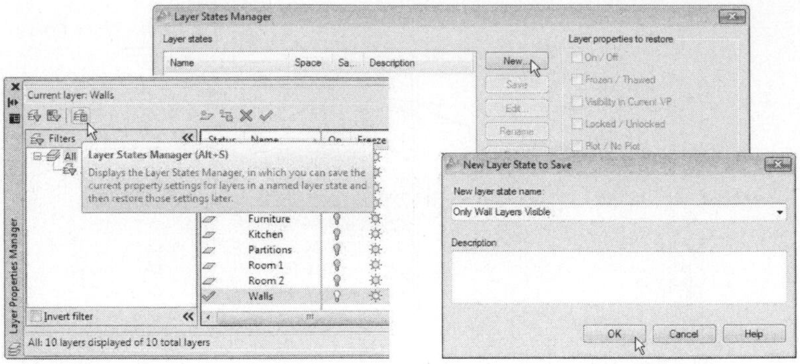

FIGURE 2.45

Once you have created numerous states, you test them out by selecting a name and picking the Restore button or by double-clicking on a name such as "Only Wall Layers Visible," as shown in Figure 2.46 (Left). Your drawing should update to only display the walls, as shown in Figure 2.46 (Right).

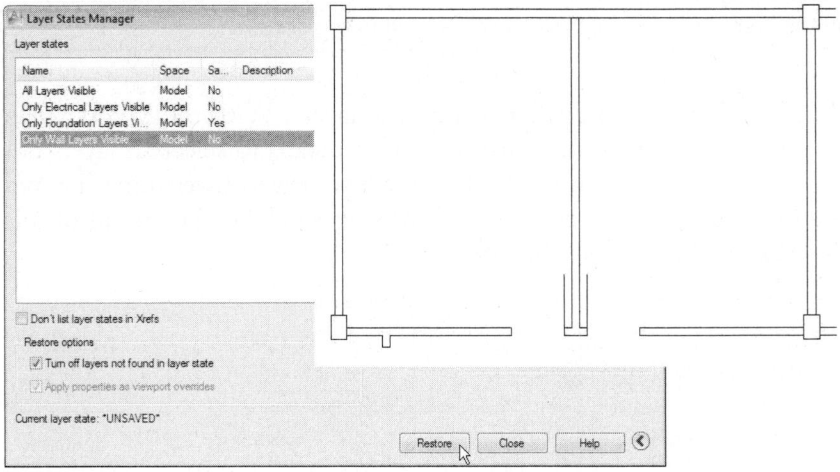

FIGURE 2.46

It is considered good practice to create a layer state that displays all layers. That way, when you call up various states, you can always get all layers back by calling up this all-layers state. In Figure 2.47 (Left) is the "All Layers Visible" state. Double-clicking on this state will display all layers in the drawing, as shown in the Figure 2.47 (Right).

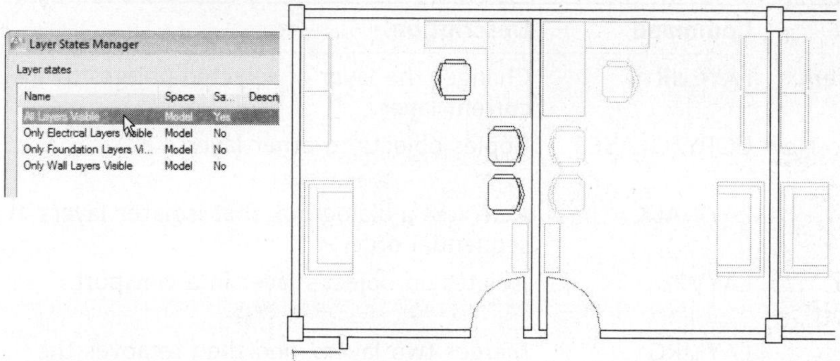

FIGURE 2.47

ADDITIONAL LAYER TOOLS

There are a number of additional layer commands that can be used to further manage layers. The display of the Ribbon's panel and all of these tools is shown in Figure 2.48.

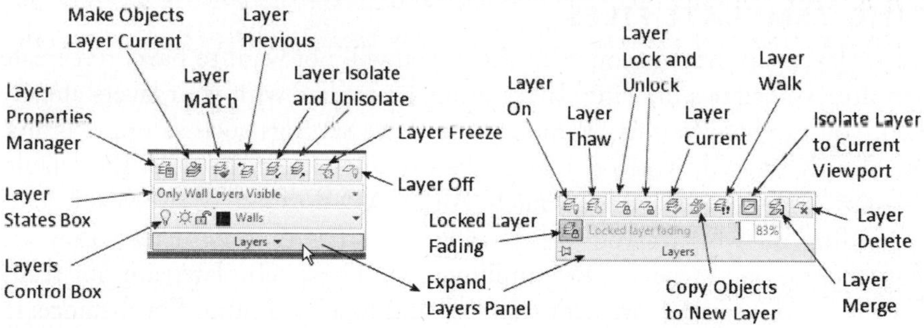

FIGURE 2.48

Menu Bar Title	Command	Description
Make Objects Layer Current	LAYMCUR	Makes the layer of a selected object current
Layer Match	LAYMCH	Changes the layer of selected objects to that of a selected destination object
Layer Previous	LAYERP	Undoes any changes made to the settings of layers such as color, linetype, or lineweight
Layer Isolate	LAYISO	Isolates layers of selected objects by turning all other layers off
Layer Unisolate	LAYUNISO	Turns on layers that were turned off with the last `LAYISO` command
Layer Freeze	LAYFRZ	Freezes the layers of selected objects
Layer Off	LAYOFF	Turns off the layers of selected objects
Turn All Layers On	LAYON	Turns all layers on
Thaw All Layers	LAYTHW	Thaws all layers
Layer Lock	LAYLCK	Locks the layers of selected objects
Layer Unlock	LAYULK	Unlocks the layers of selected objects

Continued

Menu Bar Title	Command	Description
Change to Current Layer	LAYCUR	Changes the layer of selected objects to the current layer
Copy Objects to New Layer	COPYTOLAYER	Copies objects to other layers
Layer Walk	LAYWALK	Activates a dialog box that isolates layers in sequential order
Isolate Layer to Current Viewport	LAYVPI	Isolates an object's layer in a viewport
Layer Merge	LAYMRG	Merges two layers, and then removes the first layer from the drawing
Layer Delete	LAYDEL	Permanently deletes layers from drawings

NOTE

Please refer to the Student Companion site from CengageBrain for additional information on the additional layer tools. The file 02_Facilities Plan.dwg is available in the Try It! folder for you to use while reviewing this information.

CREATING TEMPLATE FILES

Once you set layers up the way you want them, you will not want to have to recreate them each time you start a drawing. By creating a template with your layers already established, you can achieve this. Template files have settings such as units, limits, and layers already created, which enhances the productivity of the user. By default, when you start a new drawing from scratch, AutoCAD uses one of two blank templates depending on whether you are drawing in inches or millimeters; they are Acad.Dwt (inches) and Acadiso.Dwt (millimeters). These templates are not really blank, but the various settings are kept standard and to a minimum. For instance, in the Acad.Dwt template file, the drawing units are set to decimal units with four-decimal-place precision. The drawing limits are set to 12,9 units for the upper-right corner. Only one layer exists, namely layer 0. The grid in this template is set to a spacing of 0.50 units.

If you have to create several drawings that have the same limits and units settings and use the same layers, rather than start with the Acad.Dwt file and make changes to the settings at the start of each new drawing file, a better technique would be to create your own template. Templates often include a company border or they may contain drawing details already completed. Many templates appear empty, when in fact, many drawing tasks are already performed, such as:

- Drawing units and precision
- Drawing limits
- Grid and Snap Settings
- Layers, Linetypes, and Lineweights
- Annotation scale depending on the final plot scale of the drawing

TRY IT!

Create a new drawing file starting from scratch. You will be creating a template designed to draw an architectural floor plan. Make changes to the following settings:

- In the Drawing Units dialog box, change the units type from Decimal to Architectural.

- Set the limits of the drawing to 17',11' for the upper-right corner and perform a ZOOM-All. Change the grid and snap to 3" for the X and Y values.
- Create the following layer names: Floor Plan, Hidden, Center, Dimension, Doors, Windows.
- Make your own color assignments to these layers.
- Load the Hidden2 and Center2 linetypes and assign these to the appropriate layers.
- Assign a lineweight of 0.50 mm to the Floor Plan layer.
- Finally, set the Annotation Scale in the Status Bar to 1"=1'-0".

Once the changes are made, it is time to save the settings in a template format. Follow the usual steps for saving your drawing. However, when the Save Drawing As dialog box appears, first click the arrow on the other side of "Files of type, as in Figure 2.49," to expose the file types.

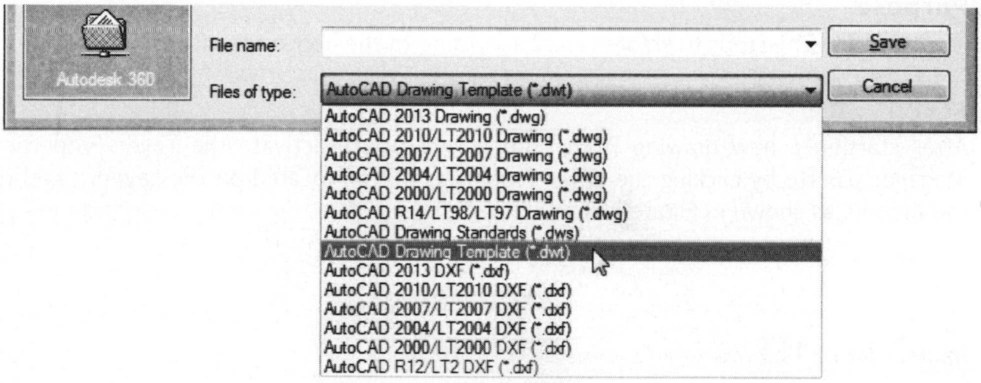

FIGURE 2.49

Click the field that states, "AutoCAD Drawing Template (*.dwt)." AutoCAD will automatically take you to the folder, as shown in Figure 2.50, that holds all template information. Enter the name for your template, such as AEC_B_1=12. The name signifies an architectural drawing for the B-size drawing sheet at a scale of 1" =1'-0".

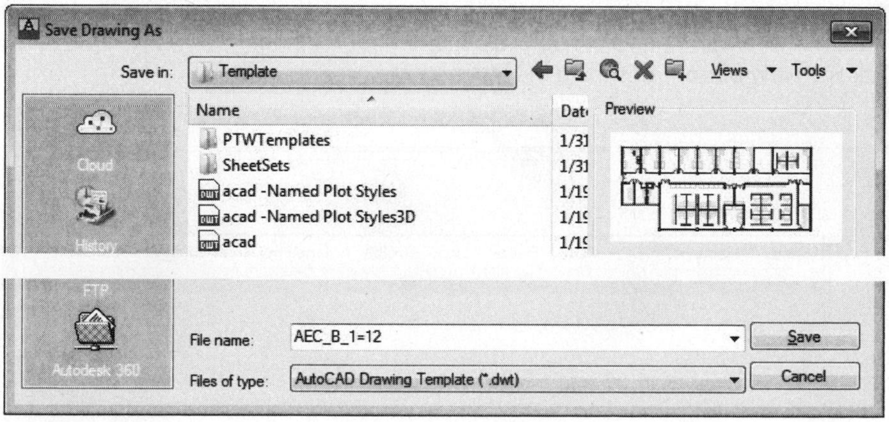

FIGURE 2.50

Before the template file is saved, you have the opportunity to document the purpose of the template. It is always good practice to create this documentation, especially if others will be using your template.

TUTORIAL EXERCISE: 02_CREATING LAYERS.DWG

Layer Name	Color	Linetype	Lineweight
Object	White	Continuous	0.70
Hidden	Red	Hidden	0.30
Center	Yellow	Center	Default
Dimension	Yellow	Continuous	Default
Section	Blue	Continuous	Default

Purpose

Use the following steps to create layers according to the above specifications.

STEP 1

After starting a new drawing from scratch (Acad.dwt), activate the Layer Properties Manager palette by clicking the Layer Properties button located on the Layers panel of the Ribbon, as shown in Figure 2.51.

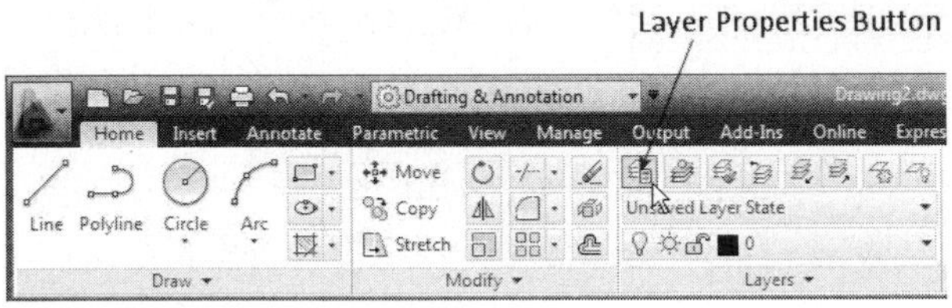

FIGURE 2.51

STEP 2

Once the Layer Properties Manager palette displays, notice on your screen that only one layer is currently listed, namely Layer 0. This happens to be the default layer, which is the layer that is automatically available to all new drawings. Since it is poor practice to construct any objects on Layer 0, new layers will be created not only for object lines but for hidden lines, centerlines, dimension objects, and section lines as well. To create these layers, click the New button, as shown in Figure 2.52.

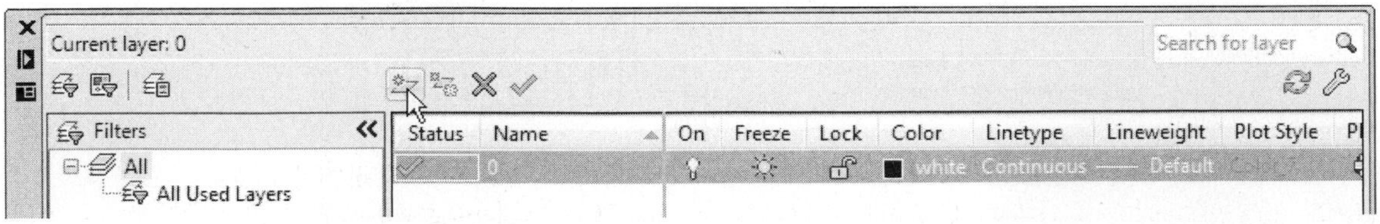

FIGURE 2.52

Notice that a layer is automatically added to the list of layers. This layer is called Layer1, as shown in Figure 2.53. While this layer is highlighted, enter the first layer name, "Object."

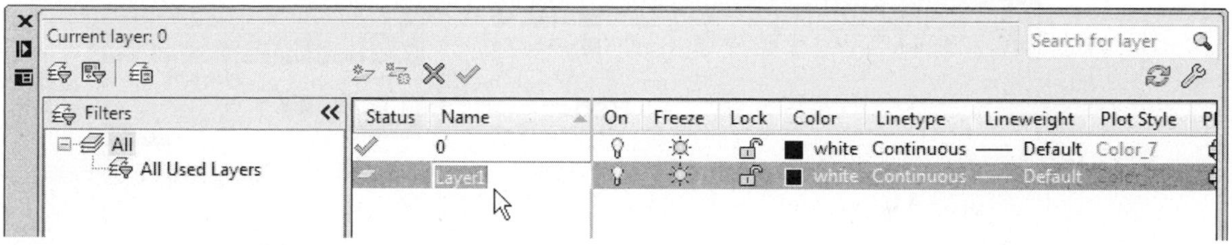

FIGURE 2.53

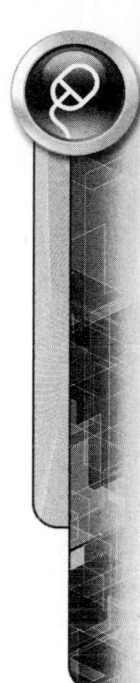

Entering a comma after the name of the layer allows more layers to be added to the listing of layers without having to click the New button. Once you have entered the comma after the layer "Object" and the new layer appears, enter the new name of the layer as "Hidden." Repeat this procedure of using the comma to create multiple layers for "Center," "Dimension," and "Section." Press ENTER after typing in "Section." The complete list of all layers created should be similar to the illustration shown in Figure 2.54.

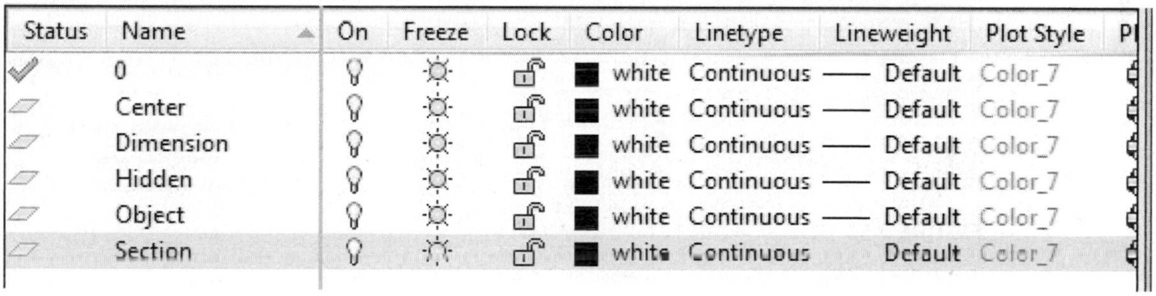

FIGURE 2.54

STEP 3

As all new layers are displayed, the names may be different, but they all have the same color and linetype assignments, as shown in Figure 2.55 (Left). At this point, the dialog box comes in handy in assigning colors and linetypes to layers in a quick and easy manner. First, select the desired layer that will be changed by picking the layer name. A horizontal bar displays, signifying that this is the selected layer. Click the color swatch identified by the box, as shown in Figure 2.55 (Left).

Clicking the color swatch in the previous step displays the Select Color dialog box, as shown in Figure 2.55 (Right). Select the desired color from one of the following areas: Standard Colors, Gray Shades, or Full Color Palette. The standard colors represent colors 1 (Red) through 9 (Gray). On display terminals with a high-resolution graphics card, the Full Color Palette displays different shades of the standard colors. This gives you a greater variety of colors to choose from. For the purpose of this tutorial, the color Red will be assigned to the Hidden layer. Select the box displaying the color red; a box outlines the color and echoes the color in the bottom portion of the dialog box. Click the OK button to complete the color assignment. Continue with this step by assigning the color Yellow to the Center and Dimension layers; assign the color Blue to the Section layer.

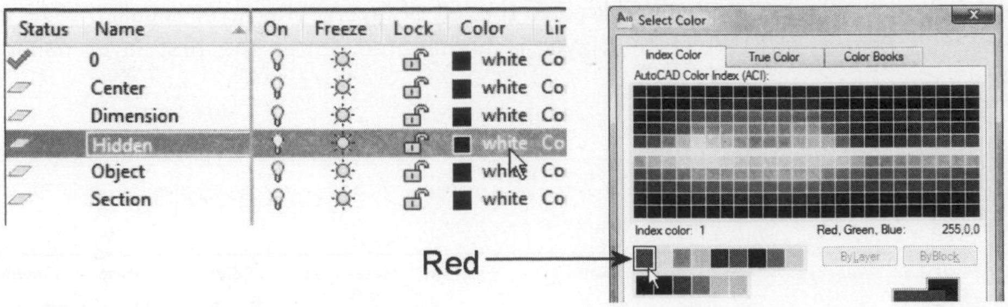

FIGURE 2.55

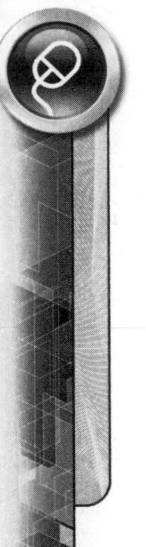

STEP 4

Once the color has been assigned to a layer, the next step is to assign a linetype to the layer. The "Hidden" layer requires a linetype called "HIDDEN." Click the "Continuous" linetype, as shown in Figure 2.56 (Left) to display the Select Linetype dialog box illustrated in Figure 2.56 (Right). By default, and to save space in each drawing, AutoCAD initially only loaded the Continuous linetype. Click the Load button to load more linetypes.

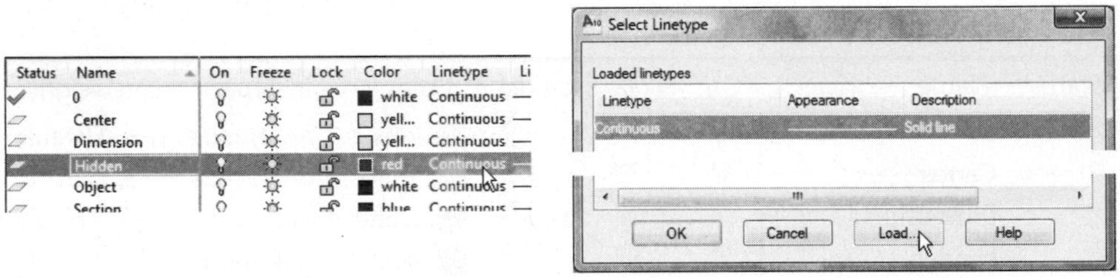

FIGURE 2.56

Choose the desired linetype to load from the Load or Reload the Linetype dialog box, as shown in Figure 2.57 (Left). Scroll through the linetypes until the "HIDDEN" linetype is found. Select the linetype and click the OK button to return to the Select Linetype dialog box.

Once you are back in the Select Linetype dialog box, as shown in Figure 2.57 (Right), notice the "HIDDEN" linetype listed along with "Continuous." Because this linetype has just been loaded, it still has not been assigned to the "Hidden" layer. Click the Hidden linetype listed in the Select Linetype dialog box, and click the OK button. Once the Layer Properties Manager palette reappears, notice that the "HIDDEN" linetype has been assigned to the "Hidden" layer. Repeat this procedure to assign the "CENTER" linetype to the "Center" layer.

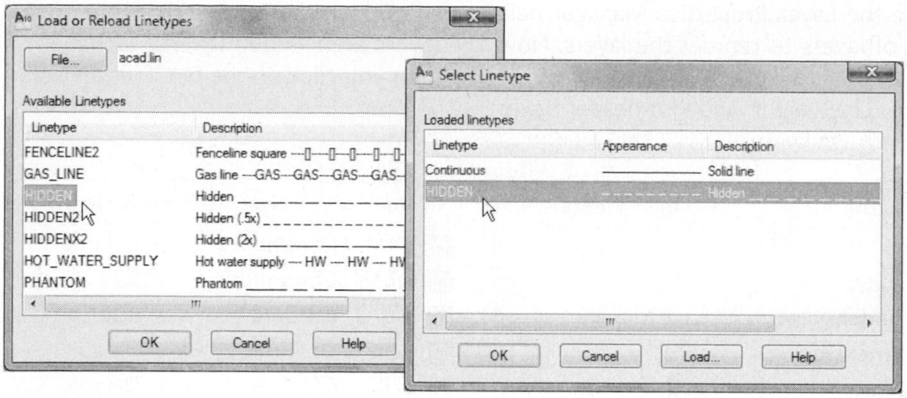

FIGURE 2.57

STEP 5

Another requirement of the "Hidden" layer is that it uses a lineweight of 0.30 mm to have the hidden lines stand out when compared with the object in other layers. Clicking the highlighted default lineweight, as shown in Figure 2.58 (Left), displays the Lineweight dialog box shown in Figure 2.58 (Right). Click the 0.30 mm lineweight followed by the OK button to assign this lineweight to the "Hidden" line. Use the same procedure to assign a lineweight of 0.70 mm to the Object layer.

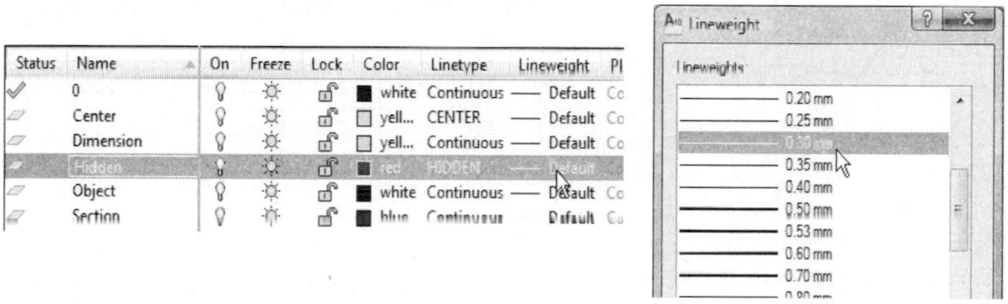

FIGURE 2.58

STEP 6

Once you have completed all color and linetype assignments, the Layer Properties Manager palette should appear similar to Figure 2.59. Clicking the X in the corner of the palette will dismiss the palette, save all layer assignments, and return you to the drawing editor.

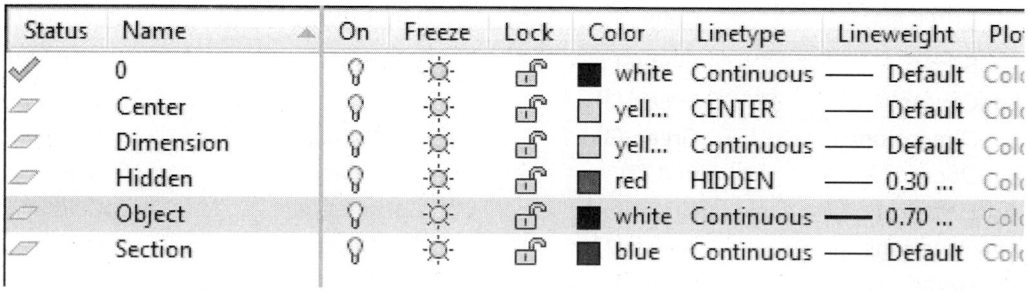

FIGURE 2.59

Activate the Layer Properties Manager palette and click on the Name box at the top of the list of layers to reorder the layers. Now, the layers are listed in reverse alphabetical order (see Figure 2.60). This same effect occurs when you click on the On, Freeze, Color, Linetype, Lineweight, and Plot header boxes.

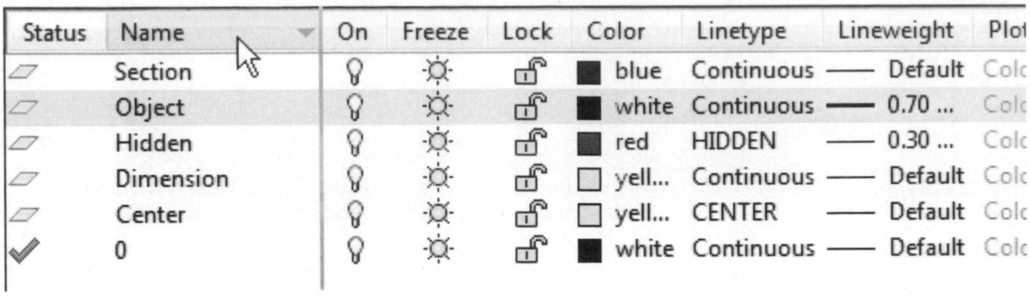

Status	Name	On	Freeze	Lock	Color	Linetype	Lineweight	Plot
	Section				■ blue	Continuous	—— Default	Colc
	Object				■ white	Continuous	—— 0.70 ...	Colc
	Hidden				■ red	HIDDEN	—— 0.30 ...	Colc
	Dimension				□ yell...	Continuous	—— Default	Colc
	Center				□ yell...	CENTER	—— Default	Colc
✓	0				■ white	Continuous	—— Default	Colc

FIGURE 2.60

Figure 2.61 displays the results of clicking the Color header: All colors are reordered, starting with Red (1) and followed by Yellow (2), Blue (5), and White (7).

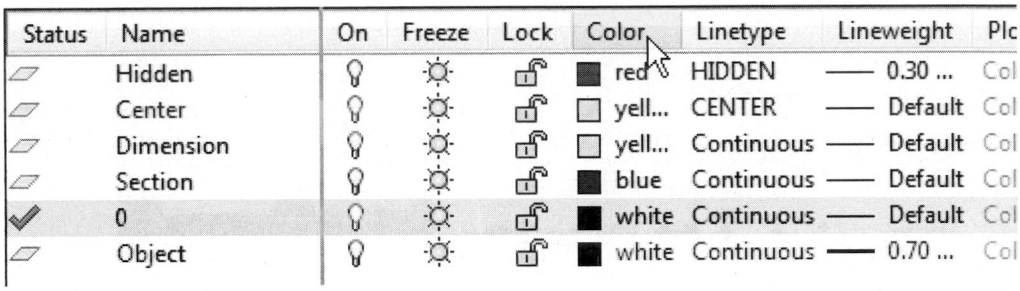

Status	Name	On	Freeze	Lock	Color	Linetype	Lineweight	Plc
	Hidden				■ red	HIDDEN	—— 0.30 ...	Col
	Center				□ yell...	CENTER	—— Default	Col
	Dimension				□ yell...	Continuous	—— Default	Col
	Section				■ blue	Continuous	—— Default	Col
✓	0				■ white	Continuous	—— Default	Col
	Object				■ white	Continuous	—— 0.70 ...	Col

FIGURE 2.61

TEMPLATE CREATION EXERCISES FOR CHAPTER 2

ARCHITECTURAL APPLICATION

Create an Architectural template (dwt) that contains the following setup information.

Use the Drawing Units dialog box to set the units of the drawing to Architectural. Keep the remaining default values for all other unit settings.

Use the Layer Properties Manager palette to create the layers contained in the table below:

Layer Name	Color	Linetype	Lineweight
Border	Blue (5)	Continuous	1.00
Center	Green (3)	Center	0.25
Dimension	Green (3)	Continuous	0.25
Doors	Red (2)	Continuous	0.50
Electrical	Green (3)	Continuous	0.50
Elevations	White (7)	Continuous	0.70
Foundation	Cyan (4)	Continuous	0.50
Furniture	Red (1)	Continuous	0.50

Continued

Layer Name	Color	Linetype	Lineweight
Hatching	Magenta (6)	Continuous	0.50
Hidden	Red (1)	Hidden	0.50
HVAC	Cyan (4)	Continuous	0.50
Misc	Green (3)	Continuous	0.50
Plumbing	Cyan (4)	Continuous	0.50
Text	Green (3)	Continuous	0.50
Viewports	Gray (9)	Continuous	0.25
Walls	White (7)	Continuous	0.70
Windows	Yellow (2)	Continuous	0.50

When finished making the changes to the drawing units and layer assignments, save this file as a new AutoCAD template called ARCH-Imperial.DWT (imperial units are in feet and inches).

MECHANICAL APPLICATION

Create a Mechanical template (dwt) that contains the following setup information.

Use the Drawing Units dialog box to set the units of the drawing to Mechanical. Also change the precision to two decimal places. Keep the remaining default values for all other unit settings.

Use the Layer Properties Manager palette to create the layers contained in the table below:

Layer Name	Color	Linetype	Lineweight
Border	Blue (5)	Continuous	1.00
Center	Green (3)	Center2	0.25
Construction	Gray (8)	Continuous	0.25
Dimension	Green (3)	Continuous	0.25
Hatching	Magenta (6)	Continuous	0.25
Hidden	Red (1)	Hidden2	0.25
Misc	Green (3)	Continuous	0.25
Object	White (7)	Continuous	0.50
Phantom	Cyan (4)	Phantom2	0.50
Text	Green (3)	Continuous	0.25
Viewports	Gray (9)	Continuous	0.25

When you are finished making the changes to the drawing units and layer assignments, save this file as a new AutoCAD template called MECH-Imperial.DWT (imperial units are in inches).

CIVIL APPLICATION

Create a Civil template (dwt) that contains the following setup information.

Use the Drawing Units dialog box to set the units of the drawing to Decimal. Also, change the precision to two decimal places. Keep the remaining default values for all other unit settings.

Use the Layer Properties Manager palette to create the layers contained in the table below:

Layer Name	Color	Linetype	Lineweight
Border	Blue (5)	Continuous	1.00
Building Outline	White (7)	Continuous	0.70
Center	Green (3)	Center	0.25
Contour Lines—Existing	White (7)	Dashed	0.50
Contour Lines—New	Magenta (6)	Continuous	0.50
Dimension	Green (3)	Continuous	0.25
Drainage	Cyan (4)	Divide	0.50
Fire Line	Red (1)	Continuous	0.50
Gas Line	Red (1)	Gas_Line	0.50
Hatching	Magenta (6)	Continuous	0.50
Hidden	Red (1)	Hidden	0.50
Misc	Green (3)	Continuous	0.50
Parking	Yellow (2)	Continuous	0.50
Property Line	Green (3)	Phantom	0.80
Sewer	White (7)	Continuous	0.50
Text	Green (3)	Continuous	0.50
Viewports	Gray (9)	Continuous	0.25
Water Line	Cyan (4)	Continuous	0.50
Wetlands	Green (3)	Dashdot	0.50

When finished making the changes to the drawing units and layer assignments, save this file as a new AutoCAD template called CIVIL-Imperial.DWT (imperial units are in feet and inches).

PROBLEMS FOR CHAPTER 2

PROBLEM 2-1

Open the existing drawing called 02-Storage. Use the Layer Properties Manager along with the Layer Filter Properties dialog box to answer the following layer-related questions:

1. The total number of layers frozen in the database of this drawing is ______.
2. The total number of layers locked in the database of this drawing is ______.
3. The total number of layers turned off in the database of this drawing is ______.
4. The total number of layers assigned the color red and frozen in the database of this drawing is ______.
5. The total number of layers that begin with the letter "L," are assigned the color yellow, and are turned off in the database of this drawing is ______.

PROBLEM 2-2

Open the existing drawing called 02_I_Pattern.Dwg. Use the Layer Properties Manager along with the Layer Filter Properties dialog box to answer the following layer-related questions:

1. The total number of layers assigned the color red and turned off in the database of this drawing is ______.
2. The total number of layers assigned the color red and locked in the database of this drawing is ______.
3. The total number of layers assigned the color white and frozen in the database of this drawing is ______.
4. The total number of layers assigned the color yellow and assigned the hidden linetype in the database of this drawing is ______.
5. The total number of layers assigned the color red and assigned the phantom linetype in the database of this drawing is ______.
6. The total number of layers assigned the center linetype in the database of this drawing is ______.

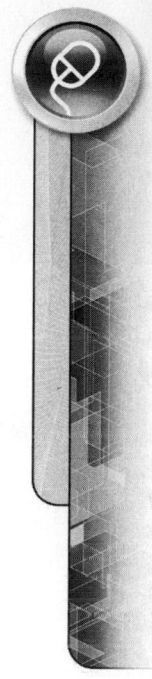

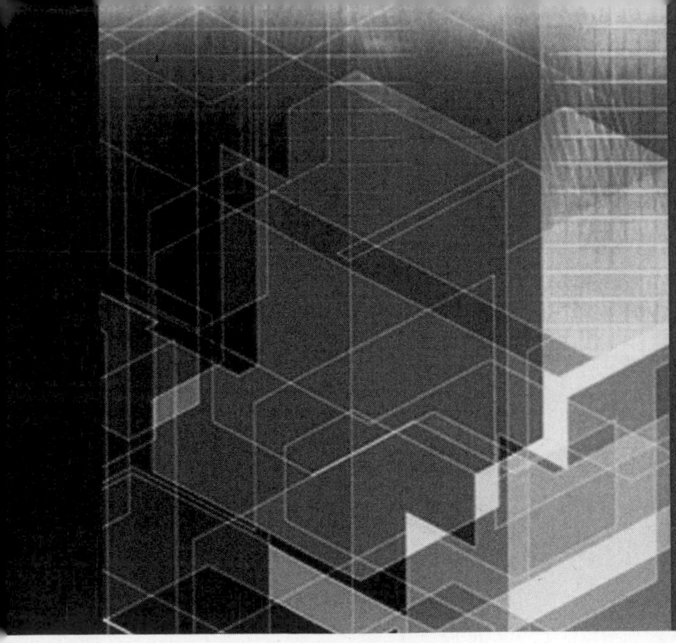

CHAPTER 3

AutoCAD Display and Basic Selection Operations

Chapter 3 introduces you to two important topics: the AutoCAD VIEW commands and object selection methods. A number of options of the ZOOM command will be explained first, followed by a review of panning and understanding how a wheel mouse is used with AutoCAD. User-defined portions of a drawing screen called "named views" will also be introduced in this chapter. The second part of this chapter introduces the creation of selection sets using various tools such as selection options, cycling, noun/verb selection, implied windowing, quick select, and isolate/hiding objects.

VIEWING YOUR DRAWING WITH ZOOM

The ability to magnify details in a drawing or reduce the size of the drawing to see it in its entirety is a function of the ZOOM command. It does not take much for a drawing to become very busy, complicated, or dense when displayed in the drawing editor. Therefore, use the ZOOM command to work on details or view specific parts of the drawing.

When using the Drafting & Annotation Workspace, you can select the ZOOM command and its options through the Ribbon's View tab and Navigate 2D Panel or through the Navigation Bar, as shown in Figure 3.1.

You can also activate the ZOOM command from the keyboard by entering either ZOOM or the letter Z, which is its command alias. Zoom options include zooming in real time, zooming to the previous display, using a window to define a boxed area to zoom to, dynamic zooming, zooming to a user-defined scale factor, zooming based on a center point and a scale factor, zooming in or out, zooming all - to the set drawing limits, or zooming to the extents of the drawing. All these modes will be discussed in the pages that follow.

The following table illustrates each ZOOM command function and the icon or button that it relates to.

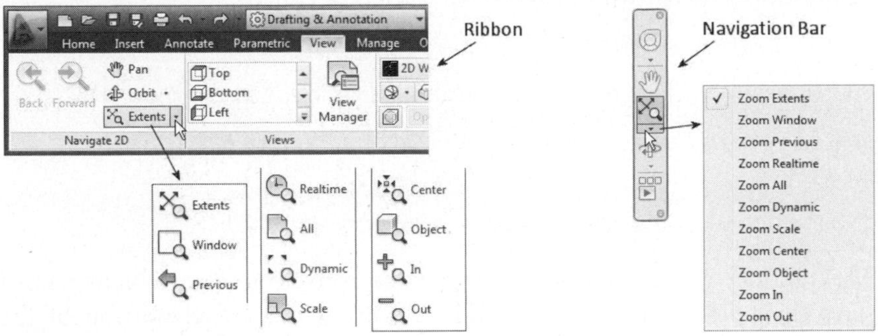

FIGURE 3.1

Button	Tool	Function
	ZOOM-Extents	Zooms to the largest possible magnification based on all objects in the drawing
	ZOOM-Window	Zooms to a rectangular area by specifying two diagonal points
	ZOOM-Previous	Zooms to the previous view
	ZOOM-Realtime	Zooms dynamically when the user moves the cursor up or down
	ZOOM-All	Zooms to the largest possible magnification based on the drawing objects or the set drawing limits, whichever is larger
	ZOOM-Dynamic	Uses a viewing box to zoom in to a portion of a drawing
	ZOOM-Scale	Zooms the display based on a specific scale factor
	ZOOM-Center	Zooms to an area of the drawing based on a center point and magnification value
	ZOOM-Object	Zooms to the largest possible magnification based on the objects selected
	ZOOM-In	Magnifies the display screen by a scale factor of 2
	ZOOM-Out	De-magnifies the display screen by a scale factor of 0.5

The following image illustrates a floor plan. To work on details of this and other drawings, use the ZOOM command to magnify or reduce the display screen. The following command sequence shows the options of the ZOOM command:

```
Command: Z (For ZOOM)
Specify corner of window, enter a scale factor (nX or nXP), or
[All/Center/Dynamic/Extents/Previous/Scale/Window/
Object] <real time>: (enter one of the listed options)
```

Executing the ZOOM command and picking a blank part of the screen places you in automatic ZOOM-Window mode. Selecting another point zooms in to the specified area. Refer to the following command sequence to use this mode of the ZOOM command on the floor plan, as shown in Figure 3.2 (Left).

Command: Z *(For ZOOM)*

Specify corner of window, enter a scale factor *(nX or nXP)*, or

[All/Center/Dynamic/Extents/Previous/Scale/Window/
Object] <real time>: *(Mark a point at "A")*

Specify opposite corner: *(Mark a point at "B")*

The ZOOM-Window option is automatically invoked once you select a blank part of the screen and then pick a second point. The resulting magnified portion of the screen appears, as shown in Figure 3.2 (Right).

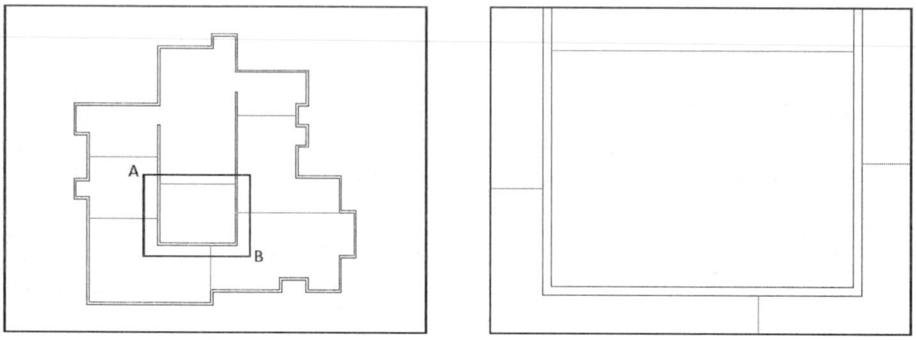

FIGURE 3.2

ZOOMING IN REAL TIME

A powerful option of the ZOOM command is performing screen magnifications or reductions in real time. This is the default option of the command. Issuing the Real-time option of the ZOOM command displays a magnifying glass icon with a positive sign and a negative sign near the magnifier icon. Identify a blank part of the drawing editor, press down the Pick button of the mouse (the left mouse button), and move in an upward direction to zoom into the drawing in real time. Identify a blank part of the drawing editor, press down the Pick button of the mouse, and move in a downward direction to zoom out of the drawing in real time. Use the following command sequence and illustration in the following image on the left for performing this task.

Command: Z *(For ZOOM)*

Specify corner of window, enter a scale factor *(nX or nXP)*, or

[All/Center/Dynamic/Extents/Previous/Scale/Window/
Object] <real time>: *(Press* ENTER *to accept Realtime as the default)*

Identify the lower portion of the drawing editor, press and hold down the Pick button of the mouse, and move the Realtime cursor up; notice the image zooming in.

Once you are in the Realtime mode of the ZOOM command, press the right mouse button to activate the shortcut menu, as shown in Figure 3.3 (Right). Use this menu to switch between Realtime ZOOM and Realtime PAN, which gives you the ability to pan across the screen in real time. The ZOOM-Window, Original (Previous), and Extents options are also available in the cursor menu.

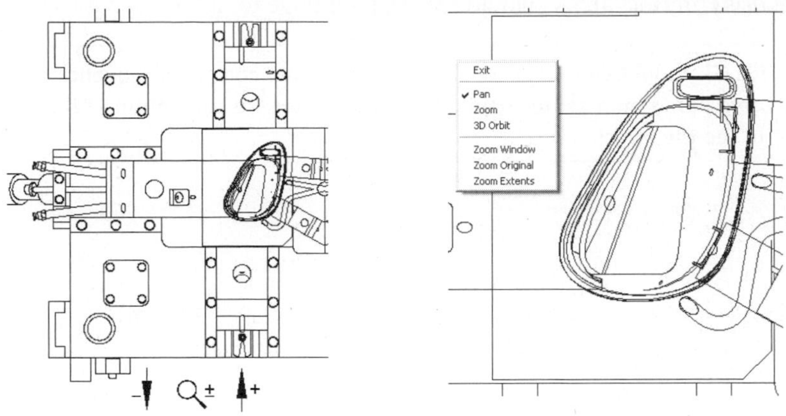

FIGURE 3.3

USING ZOOM-ALL

Another option of the `ZOOM` command is All. Use this option to zoom to the current limits of the drawing as set by the `LIMITS` command. In fact, right after the limits of a drawing have been changed, issuing a ZOOM-All updates the drawing view to reflect the latest screen size. If a drawing's size exceeds the set limits, this option will zoom past the limits to the extents of your current drawing. To use the ZOOM-All option, refer to the following command sequence.

```
Command: Z (For ZOOM)
Specify corner of window, enter a scale factor (nX or nXP), or
[All/Center/Dynamic/Extents/Previous/Scale/Window/
Object] <real time>: A (For All)
```

The illustration in Figure 3.4 on the left shows a zoomed-in portion of a part. Use the ZOOM-All option to zoom to the drawing's current limits, as shown in Figure 3.4 on the right.

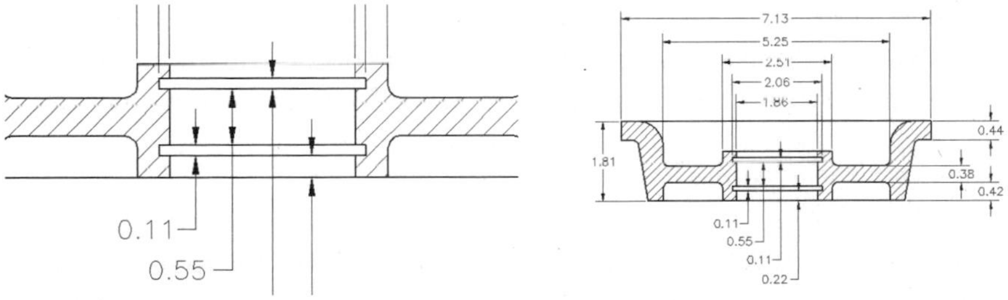

FIGURE 3.4

USING ZOOM-CENTER

The ZOOM-Center option allows you to specify a new display based on a selected center point, as shown in Figure 3.5 (Left). A window height or magnification factor controls whether the image on the display screen is magnified or reduced. If a smaller value is specified for the height, the size of the image is increased (you zoom into the object). If a larger value is specified for the height, the image gets smaller (you zoom out from the object). Instead of specifying the height you can also provide a magnification value. Entering "2X" for example would double the size of the currently displayed object, whereas entering "0.5X" would reduce the image to half its previous size.

TRY IT!

Open the drawing file 03_Zoom Center. Follow the illustrations and command sequence below to perform a zoom based on a center point. Repeat this exercise substituting "2X" or ".5X" for the value to see the effect.

Command: Z *(For ZOOM)*

Specify corner of window, enter a scale factor *(nX or nXP)*, or

[All/Center/Dynamic/Extents/Previous/Scale/Window/Object] <real time>: C *(For Center)*

Specify center point: *(Mark a point at the center of circle "A" as shown on the left)*

Enter magnification or Height <7.776>: 2

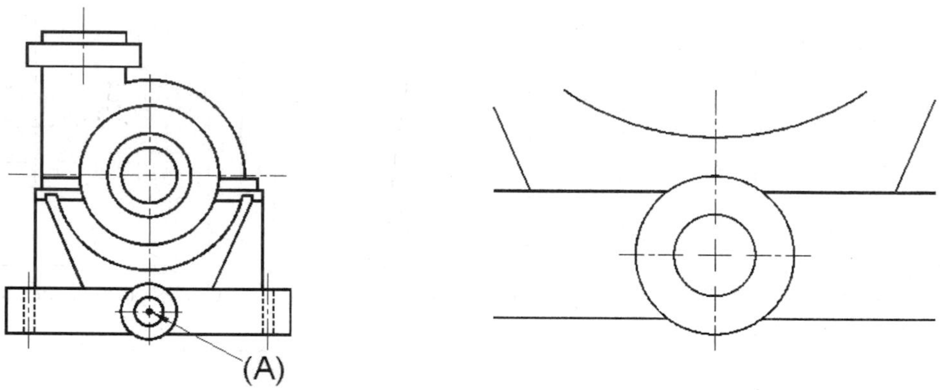

FIGURE 3.5

USING ZOOM-EXTENTS

The view of the pump in Figure 3.6 (Left) reflects a ZOOM-All operation. This option displays the entire drawing area based on the drawing limits. The objects created in this drawing don't take up much of the designated drawing area. Instead of performing a zoom based on the drawing limits, ZOOM-Extents uses the extents of the image to perform the zoom. Figure 3.6 (Right) shows the view displayed as a result of using the ZOOM command and the Extents option.

TRY IT!

Open the drawing file 03_Zoom Extents. Follow the illustrations and command sequence below to perform a zoom based on the drawing limits (All) and the objects in the drawing (Extents).

Command: Z *(For ZOOM)*

Specify corner of window, enter a scale factor *(nX or nXP)*, or

[All/Center/Dynamic/Extents/Previous/Scale/Window/Object] <real time>: A *(For All)*

```
Command: Z (For ZOOM)
Specify corner of window, enter a scale factor (nX or nXP), or
[All/Center/Dynamic/Extents/Previous/Scale/Window/
Object] <real time>: E (For Extents)
```

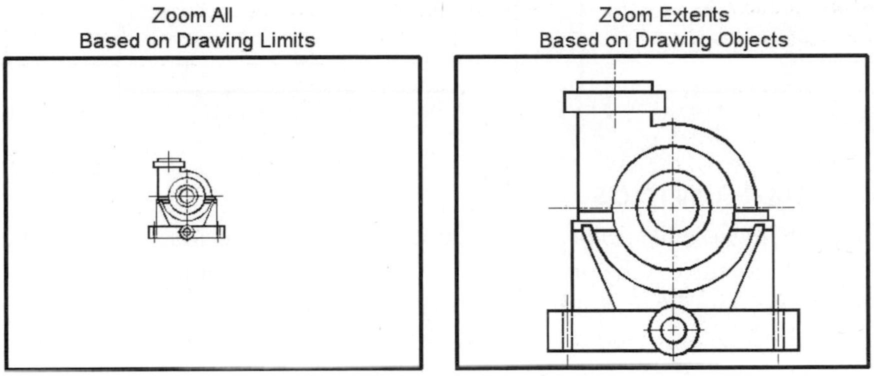

FIGURE 3.6

USING ZOOM-WINDOW

The ZOOM-Window option allows you to specify the area to be magnified by marking two points representing a rectangle, as shown in Figure 3.7 (Left). The center of the rectangle becomes the center of the new image display; the image inside the rectangle is either enlarged, as shown in Figure 3.7 (Right), or reduced. For best results, your window should resemble the rectangular shape of your screen.

TRY IT!

Open the drawing file 03_Zoom Window. Follow the illustrations and command sequence below to perform a zoom based on a window.

```
Command: Z (For ZOOM)
Specify corner of window, enter a scale factor (nX or nXP), or
[All/Center/Dynamic/Extents/Previous/Scale/Window/
Object] <real time>: W (For Window)
Specify first corner: (Mark a point at "A")
Specify other corner: (Mark a point at "B")
```

As demonstrated earlier, the Window option of zoom is automatic; in other words, without entering the Window option, the first point you pick identifies the first corner of the window box. The prompt "Specify other corner:" completes ZOOM-Window, as indicated in the following prompts:

```
Command: Z (For ZOOM)
Specify corner of window, enter a scale factor (nX or nXP), or
[All/Center/Dynamic/Extents/Previous/Scale/Window/
Object] <real time>: (Mark a point at "A")
Specify other corner: (Mark a point at "B")
```

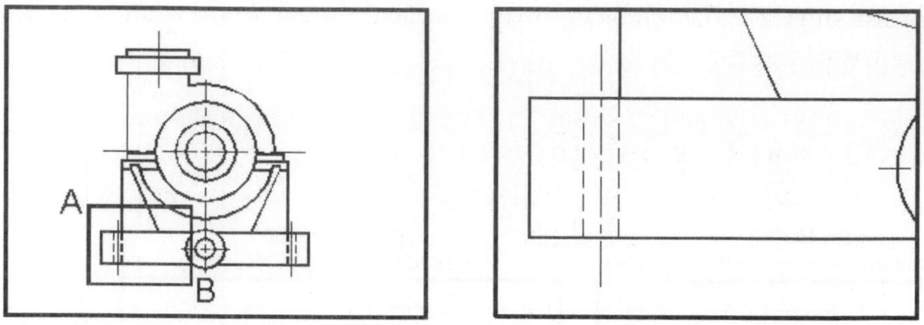

FIGURE 3.7

USING ZOOM-PREVIOUS

After magnifying a small area of the display screen, use the Previous option of the ZOOM command to return to the previous display. The system automatically saves up to ten views when zooming or panning. For example, you can begin with an overall display, perform two zooms, and then use the ZOOM-Previous command twice to return to the original display, as shown in Figure 3.8.

```
Command: Z (For ZOOM)
Specify corner of window, enter a scale factor (nX or nXP), or
[All/Center/Dynamic/Extents/Previous/Scale/Window/
Object] <real time>: P (For Previous)
```

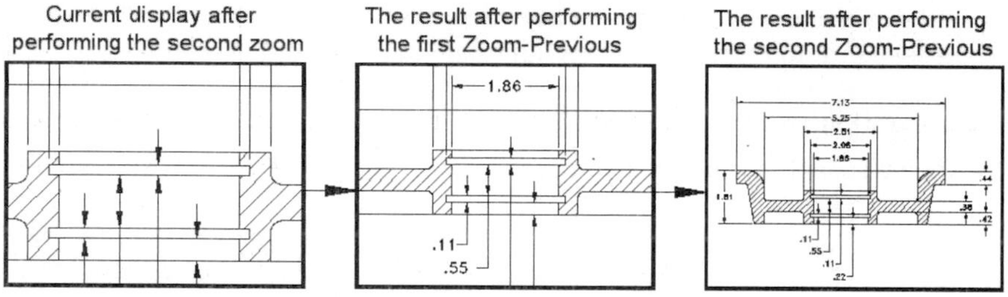

FIGURE 3.8

TIP

Forward and Back buttons are available in the Ribbon (View tab > Navigate 2D panel), which allow you to easily step through previous pan and zoom operations.

USING ZOOM-OBJECT

A zooming operation can also be performed based on an object or group of objects that you select in the drawing. The illustration in Figure 3.9 (Left) displays a facilities plan. The ZOOM-Object option is used to magnify one of the chairs using the following command sequence:

```
Command: Z (For ZOOM)
Specify corner of window, enter a scale factor (nX or nXP), or
[All/Center/Dynamic/Extents/Previous/Scale/Window/
Object] <real time>: O (For Object)
Select objects: (Pick the chair)
Select objects: (Press ENTER to perform a ZOOM-Extents based
on the chair)
```

The results are shown in the following image on the right where the chair is magnified to fit in the display screen.

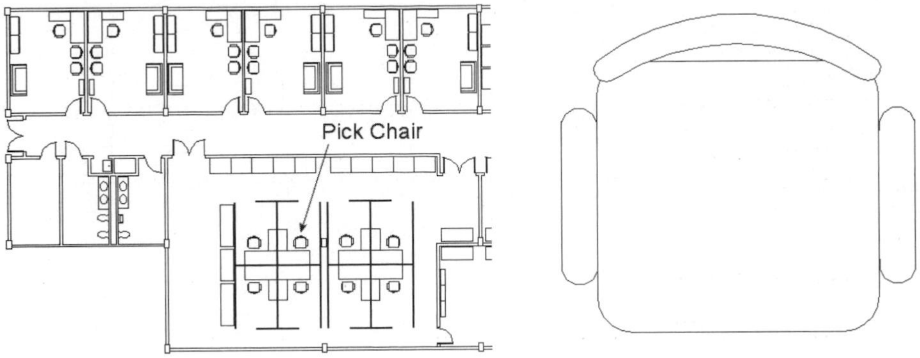

FIGURE 3.9

USING ZOOM-SCALE

In addition to performing zoom operations by picking points on your display screen for such options as zooming to a window or center point, you can also fine-tune your zooming by entering zoom scale factors directly from your keyboard. The zoom scale is based on either the current view size or the drawing limits depending on whether an "X" is or isn't placed after the scale factor number.

TRY IT!

Open the drawing file 03_Zoom Scale. Use the command sequences and illustration in Figure 3.10 for performing a zoom based on a scale factor.

If a scale factor of 0.50 is used, the zoom is performed in the drawing at a factor of 0.50, based on the original limits of the drawing. Notice that the image gets smaller.

```
Command: Z (For ZOOM)
Specify corner of window, enter a scale factor (nX or nXP), or
[All/Center/Dynamic/Extents/Previous/Scale/Window/
Object] <real time>: 0.50
```

If a scale factor of 0.50X is used, the zoom is performed in the drawing again at a factor of 0.50; however, the zoom is based on the current display screen. The image gets even smaller. The objects now appear half the size they did previously.

```
Command: Z (For ZOOM)
Specify corner of window, enter a scale factor (nX or nXP), or
[All/Center/Dynamic/Extents/Previous/Scale/Window/
Object] <real time>: 0.50X
```

Enter a scale factor of 0.90. The zoom is again based on the original limits of the drawing. As a result, the image displays larger.

```
Command: Z (For ZOOM)
Specify corner of window, enter a scale factor (nX or nXP), or
[All/Center/Dynamic/Extents/Previous/Scale/Window/
Object] <real time>: 0.90
```

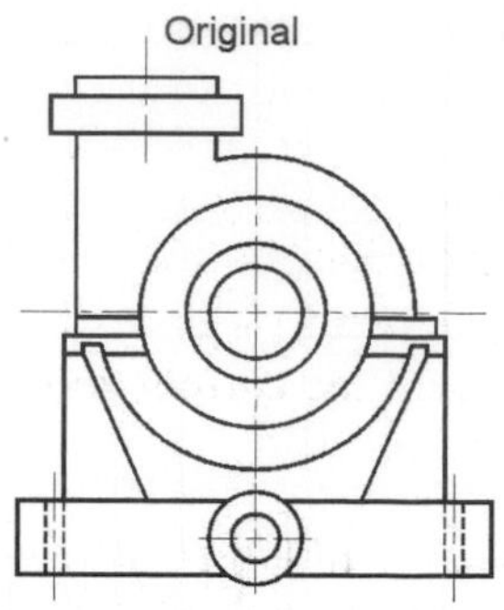

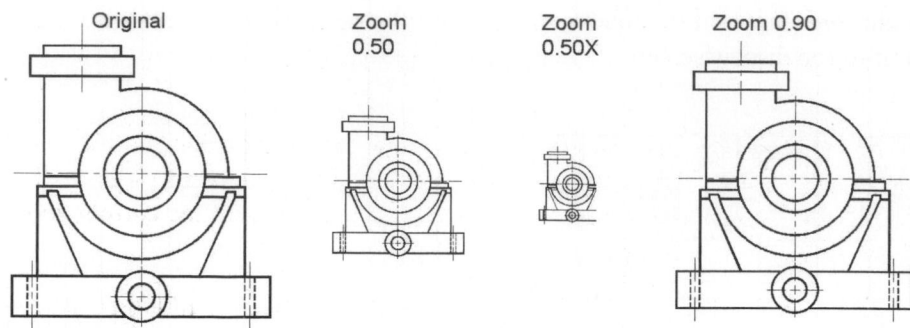

FIGURE 3.10

USING ZOOM-IN

Clicking on this button automatically performs a Zoom-In operation at a scale factor of 2X; the "X" uses the current screen to perform the Zoom-In operation.

USING ZOOM-OUT

Clicking on this button automatically performs a Zoom-Out operation at a scale factor of 0.5X; the "X" uses the current screen to perform the Zoom-Out operation.

PANNING A DRAWING

As you perform numerous ZOOM-Window and ZOOM-Previous operations, it becomes apparent that it would be nice to zoom in to a detail of a drawing and simply slide the drawing to a new area without changing the magnification; this is the purpose of the PAN command. In Figure 3.11, the ZOOM-Window option is used to construct a rectangle around the Top view to magnify it.

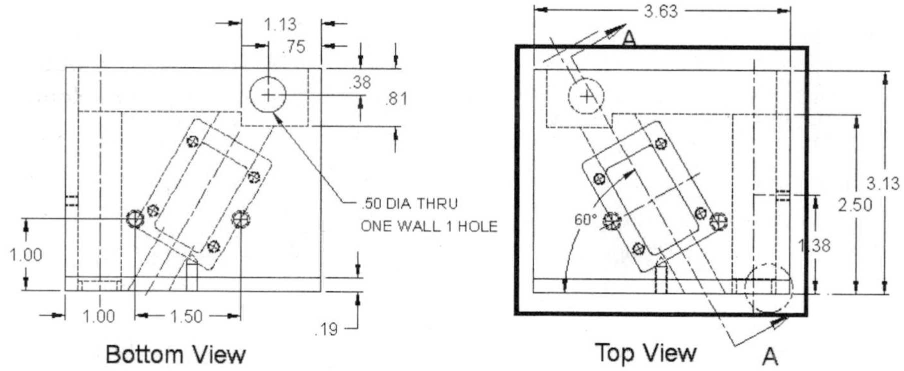

FIGURE 3.11

The result is shown in Figure 3.12 (Right). Now the Bottom view needs to be magnified to view certain dimensions. Rather than use ZOOM-Previous and then ZOOM-Window again to magnify the Bottom view, use the PAN command.

`Command: P (For PAN)`

`Press ESC or ENTER to exit, or right-click to display shortcut menu.`

Issuing the PAN command displays the Hand symbol. Pressing the Pick button down at "A" and moving the Hand symbol to the right at "B," as shown in Figure 3.12 (Right), pans the screen and displays a new area of the drawing in the current zoom magnification.

The Bottom view is now visible after the drawing is panned from the Top view to the Bottom view, with the same display screen magnification, as shown in Figure 3.12 (Left). Pan can also be used transparently; that is, while in a current command, you can select the PAN command, which temporarily interrupts the current command, performs the pan, and then restores the current command.

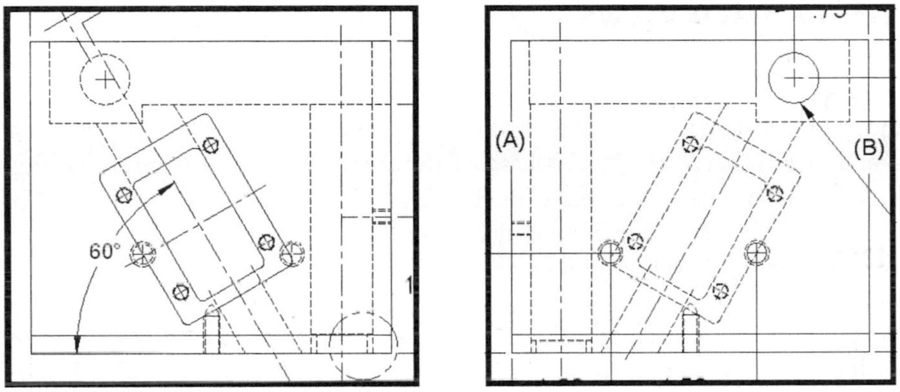

FIGURE 3.12

ZOOMING AND PANNING WITH A WHEEL MOUSE

One of the easiest ways for performing zooming and panning operations is through the mouse. Most computers are equipped with a standard Microsoft two-button mouse with the addition of a wheel, as illustrated in the following image. Rolling the wheel forward zooms in to, or magnifies, the drawing. Rolling the wheel backward zooms out of, or reduces, the drawing.

Pressing and holding the wheel down, as shown in Figure 3.13 (Right), places you in Realtime Pan mode. The familiar hand icon on the display screen identifies this mode.

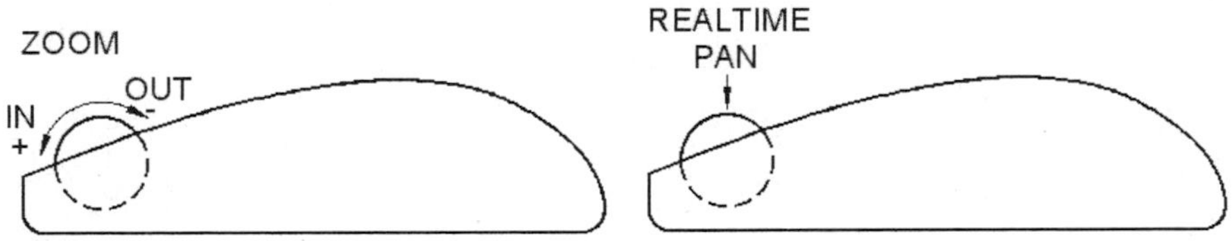

FIGURE 3.13

The wheel can also function like a mouse button. Double-clicking on the wheel, as in Figure 3.14, performs a ZOOM-Extents and is extremely popular for viewing the entire drawing on the display screen.

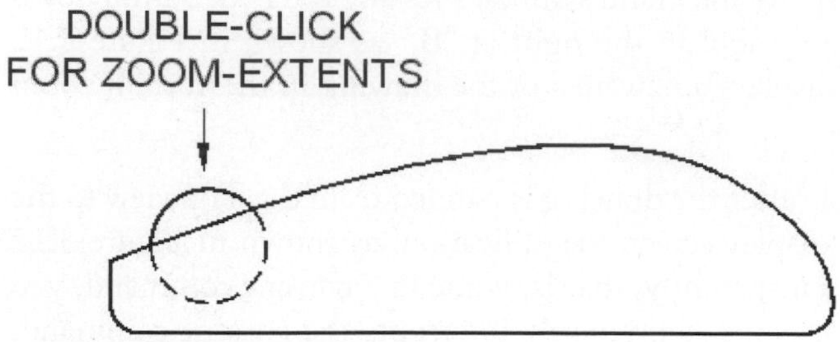

FIGURE 3.14

CREATING NAMED VIEWS

Instead of performing numerous zoom and pan operations to view key parts of a drawing, you can save these views with a name. Then, instead of using the `ZOOM` command, restore the named view with the View Manager to perform detail work. This named view is saved in the database of the drawing for use in future editing sessions. View Manager can be found in the View tab of the Ribbon, as shown in Figure 3.15. You can activate this same dialog box through the keyboard by entering the following at the Command prompt:

`Command: V` *(For VIEW)*

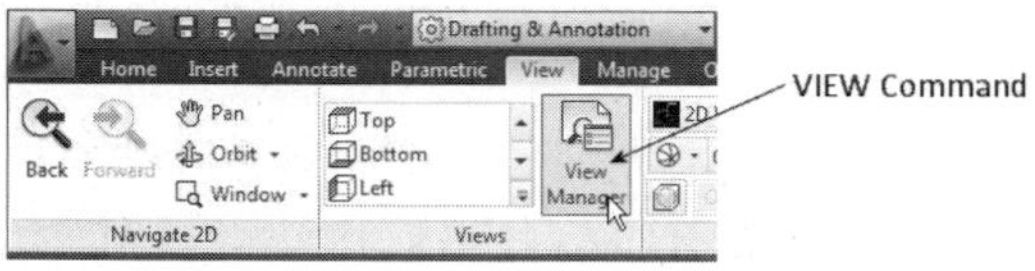

FIGURE 3.15

Initiating the VIEW command will activate the View Manager dialog box, as shown in Figure 3.16.

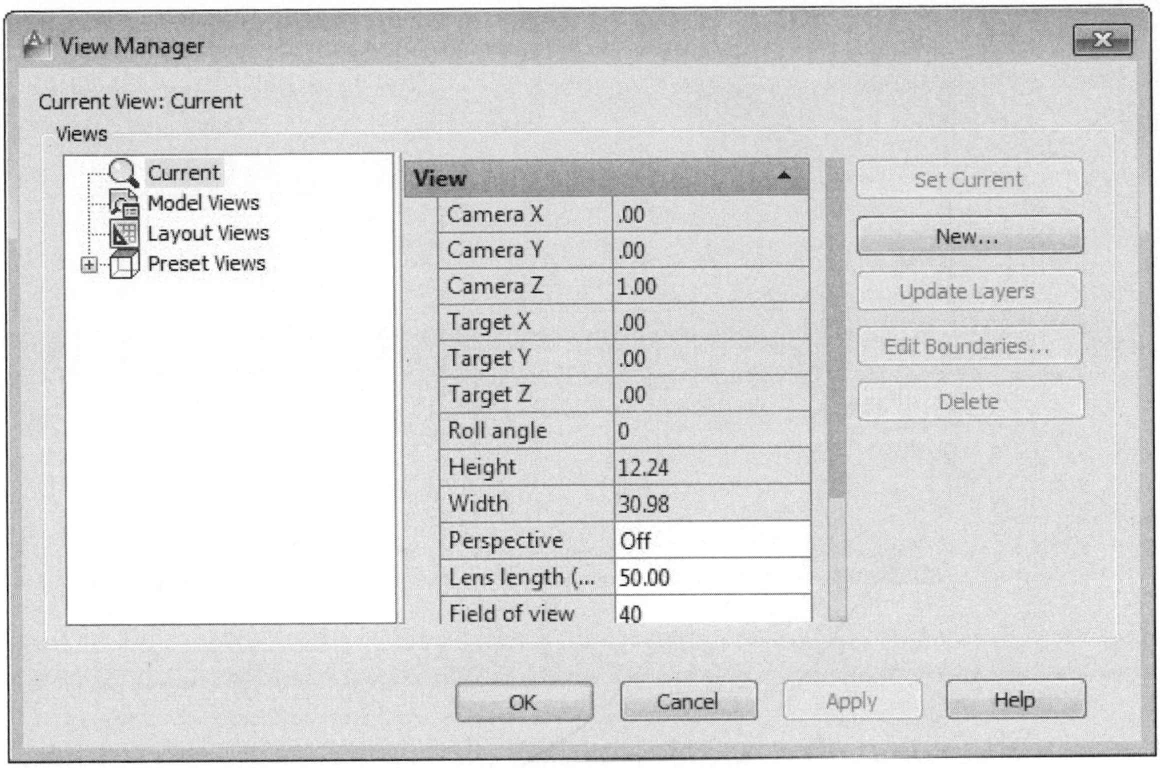

FIGURE 3.16

The following table describes each view type:

View Type	Description
Current	Displays the current view, view information, and clipping properties.
Model Views	Displays a list of all views created along with general, camera, and clipping properties.
Layout Views	Displays a list of views created in a layout in addition to general and view information (height and width).
Preset Views	Displays a list of orthogonal and isometric views, and lists the general properties for each preset view.

TRY IT!

Open the drawing file 03_Views. Follow the next series of steps and illustrations used to create a view called "FRONT."

Clicking the New button in the View Manager dialog box activates the New View/Shot Properties dialog box, as shown in Figure 3.17 (Left). Use this dialog box to guide you in creating a new view. By definition, a view is created from the current display screen. This is the purpose of the Current Display radio button. Views may also be created through the Define Window radio button, which creates a view based on the contents of a window that you define. Choosing the Define View Window button returns you to the display screen where the drawing appears grayed out. You are prompted for the two corners required to create a new view by window.

As illustrated in Figure 3.17 (Right), a rectangular window is defined around the Front view using points "A" and "B" as the corners. This turns the image captured inside of the view white and leaves the other images grayed out. Press ENTER to accept this as the new view. A name must be provided in the dialog box—"FRONT" in this case.

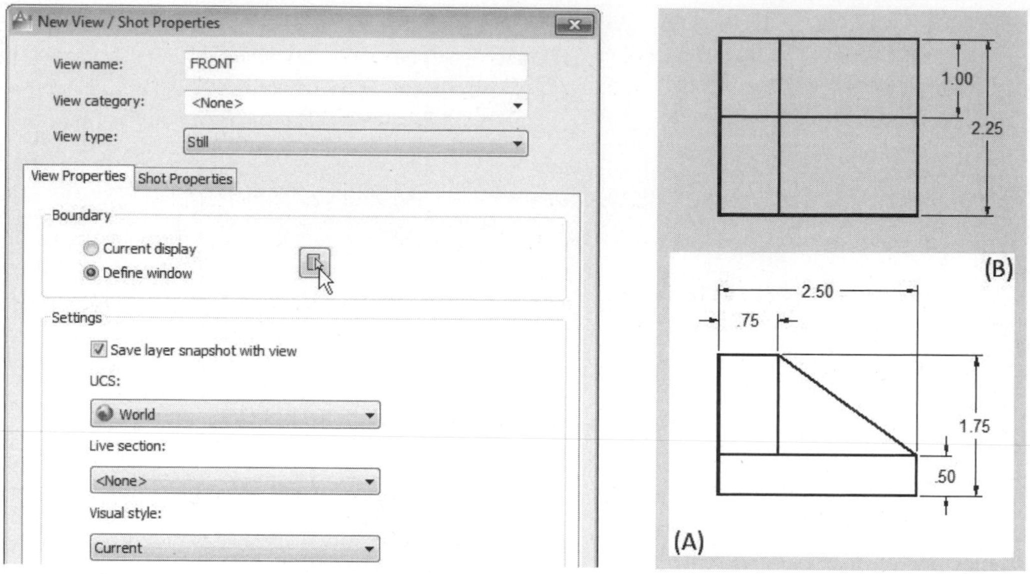

FIGURE 3.17

Accepting the window created in the previous image redisplays the New View/Shot Properties dialog. Clicking OK saves the view name FRONT under the Model Views heading in the View Manager dialog box, as shown in Figure 3.18.

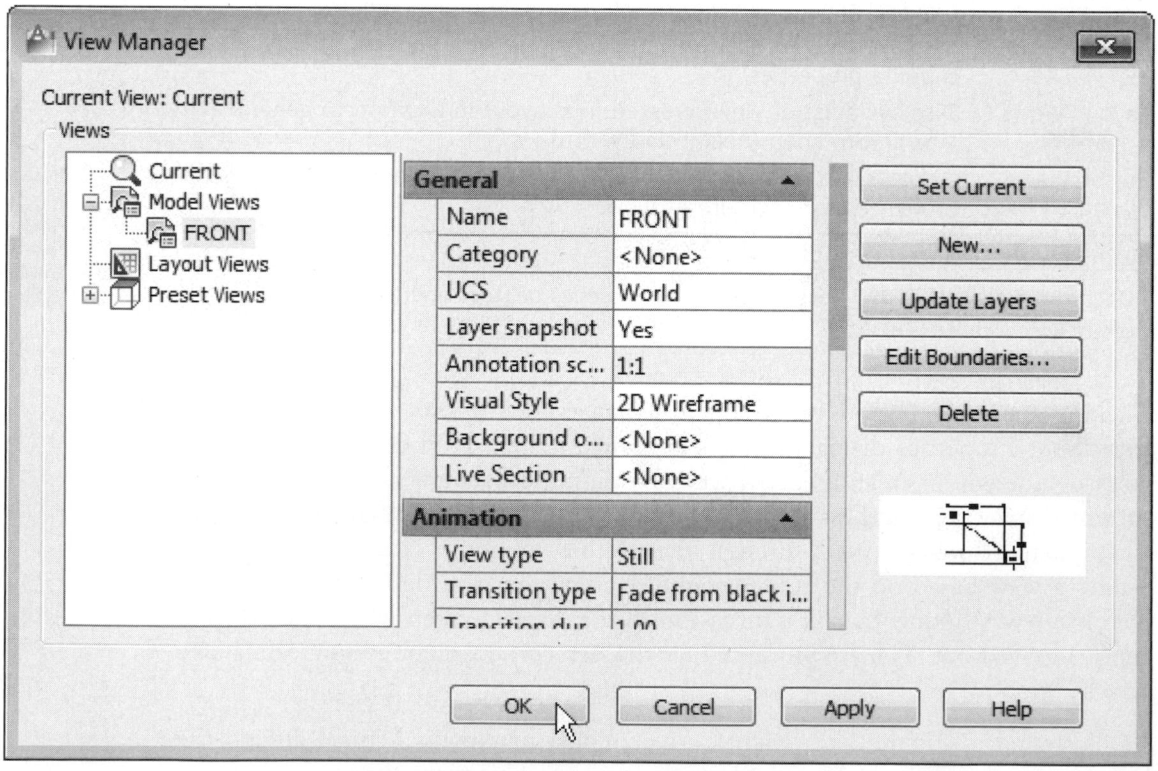

FIGURE 3.18

TIP

By default, layer snapshots are saved with views. Specific layers that are frozen or turned off will be remembered in the newly defined view. This feature can be disabled in the Settings area of the New View/Shot Properties dialog box, if desired.

TRY IT!

Open the drawing file 03_Views Complete. A series of views have already been created inside this drawing. Activate the View Manager dialog box and experiment with restoring a number of these views.

The following image illustrates numerous views of a drawing already created, namely FRONT, ISO, OVERALL, SIDE, and TOP.

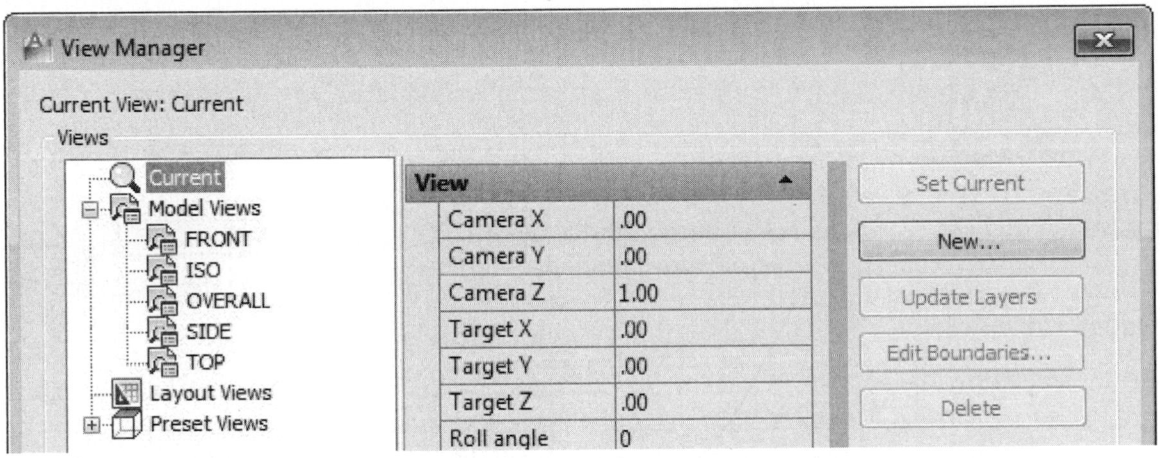

FIGURE 3.19

The quickest way to activate a view is to pick on a defined view name (FRONT) from the Views panel of the Ribbon, as shown in Figure 3.20 (Left). To activate a named view from the View Manager dialog box, you set the defined view name current. Selecting a view name and right-clicking the mouse, displays the shortcut menu used to set the view current, as shown in Figure 3.20 (Right). You can also create

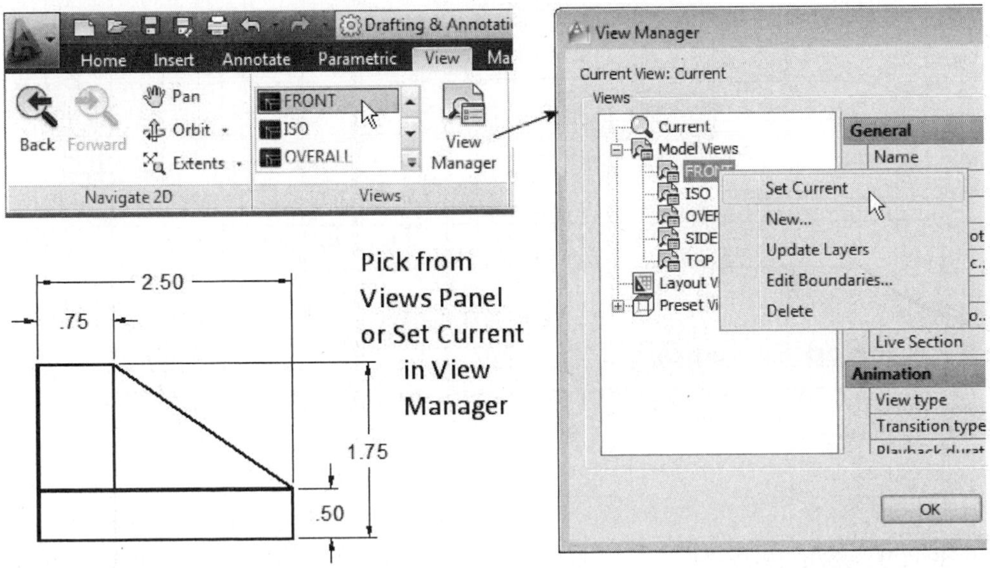

FIGURE 3.20

a new view, update layers, edit the boundaries of the view, and delete the view through this shortcut menu. Instead of using the shortcut menu, you can click the Set Current button, then OK to exit the View Manager dialog box and display the Front view.

Changing the Boundary of a Named View

The View Manager dialog box has an Edit Boundaries button, as shown in Figure 3.21. Use it to change the boundary size of an existing view.

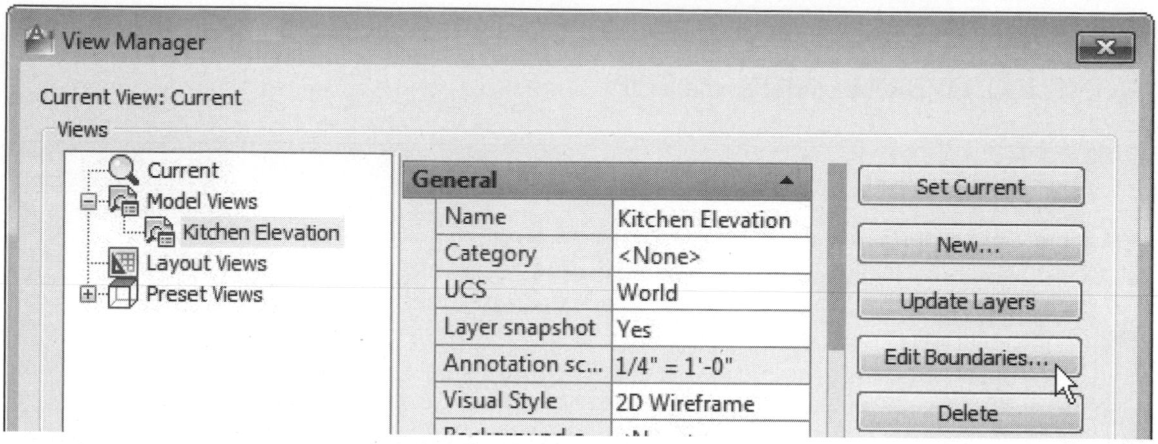

FIGURE 3.21

In Figure 3.22, the white area of the drawing illustrates the current view called Kitchen Elevation. However, this view should also include the cabinet details. Clicking on the view name and then clicking on the Edit Boundaries button allows you to redefine the view boundary box by picking two new corner points.

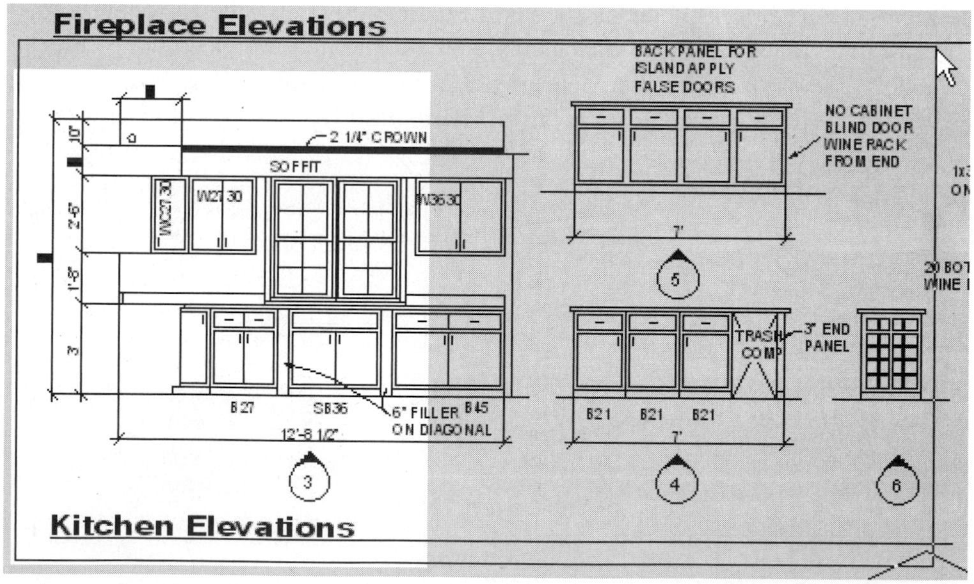

FIGURE 3.22

After you construct this new boundary, the results should be similar to Figure 3.23. The new view is displayed in white. Non-view components have the color gray in the background.

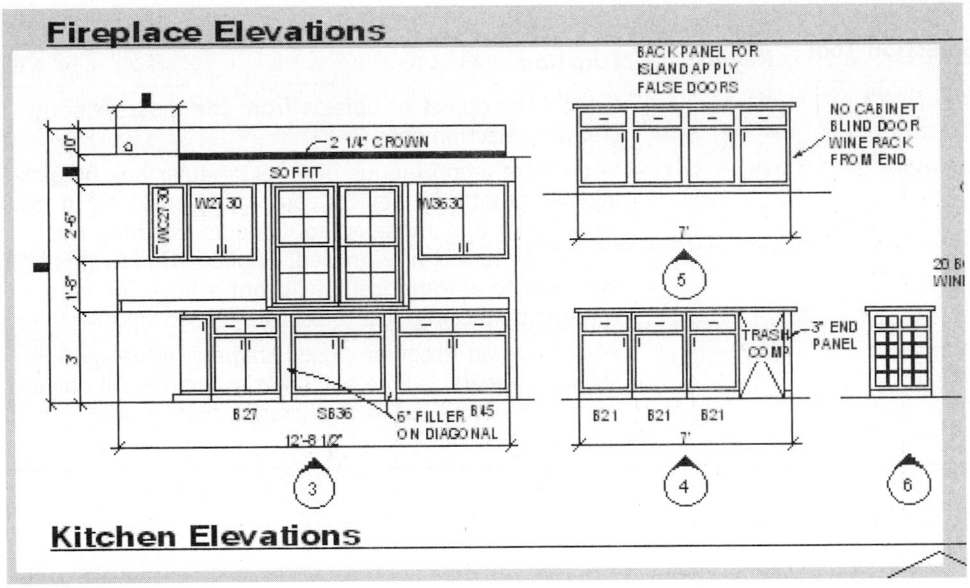

FIGURE 3.23

CREATING OBJECT SELECTION SETS

Selection sets are used to group a number of objects together for the purpose of modifying. Applications of selection sets are covered in the following pages. Once a selection set has been created, the group of objects may be modified by moving, copying, mirroring, and so on. The operations supported by selection sets will be covered in Chapter 4, "Modify Commands." An object manipulation command supports the creation of selection sets if it prompts you to "Select objects."

The commonly used selection set options (how a selection set is made) are briefly described in the following table.

Selection Tool	Selection Key-In	Function
All	ALL	Selects all objects in the current space environment (model or paper) of the drawing. This mode even selects objects on layers that are turned off. Objects that reside on frozen or locked layers are not selected.
Crossing	C	You create a dotted green rectangular box by picking two points. All objects completely enclosed and touching this rectangular box are selected. Implied windowing allows you to automatically (without a key in) create a crossing by simply picking your box from right to left.
Crossing Polygon	CP	You create an irregular closed shape. The polygon formed appears as dotted green line segments. All objects completely enclosed and touching this irregular shape are selected.
Fence	F	You create a series of line segments. These line segments appear dotted. All objects touching these fence lines are selected. You cannot close the fence shape; it must remain open.
Last	L	Selects the last object you created.

Continued

Selection Tool	Selection Key-In	Function
Previous	P	Selects the object or objects from the most recently created selection set.
Window	W	You create a continuous blue rectangular box by picking two points. All objects completely enclosed in this rectangular box are selected. Objects touching the edges of the box are ignored. Implied windowing allows you to automatically (without a key in) create a window by picking your box from left to right.
Window Polygon	WP	You create an irregular closed shape. The polygon formed appears as blue solid line segments. All objects completely enclosed in this irregular shape are selected.

Object Pre-Highlight

Whenever you move your cursor over an object, the object pre-highlights. This helps you determine the correct object before you select it, especially handy in a busy drawing. In Figure 3.24, the cursor is moved over the rectangle that surrounds the 128 text object. This action pre-highlights this rectangle until you move your cursor off this object.

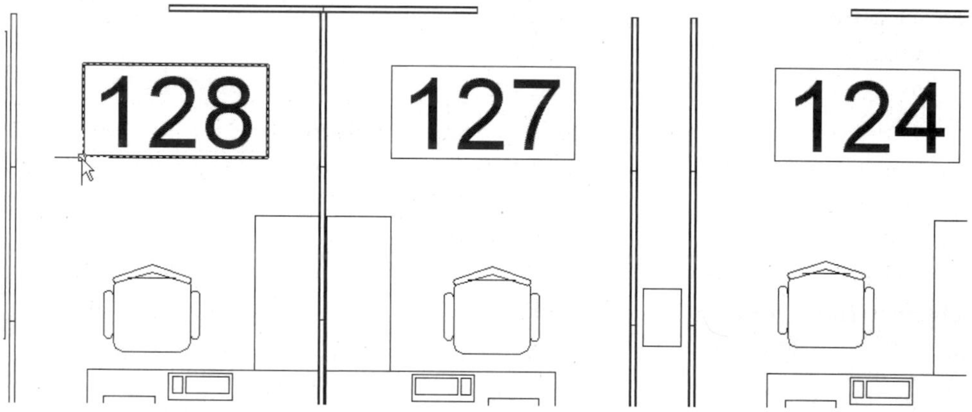

FIGURE 3.24

Selecting Objects by Individual Picks

When AutoCAD prompts you with "Select objects," a pickbox appears as the cursor is moved across the display screen, as shown in Figure 3.25 (Left). As you move your cursor over an object, the object highlights to signify that it is selectable, as shown in Figure 3.25 (Right). An object remains highlighted once it is selected.

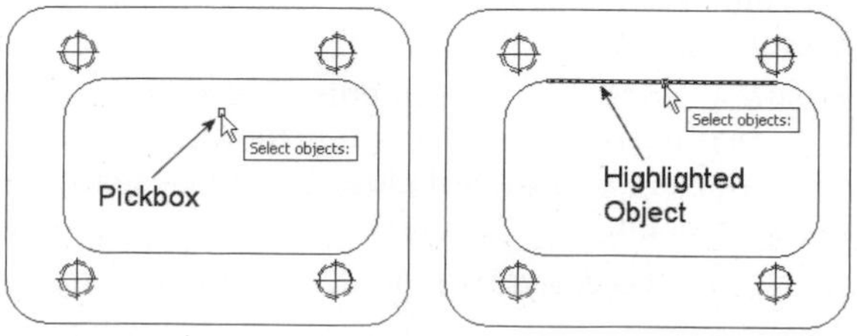

FIGURE 3.25

TRY IT!

Open the drawing file 03_Select. Enter the ERASE **command and, at the Select objects prompt, pick the polyline segment labeled "A," as shown in Figure 3.26 (Left). To signify that the object is selected the rectangle highlights.**

```
Command: E (For Erase)
Select objects: (Pick the object at "A")
Select objects: (Press ENTER to execute the ERASE command.
The rectangle disappears as shown in the following image on
the right)
```

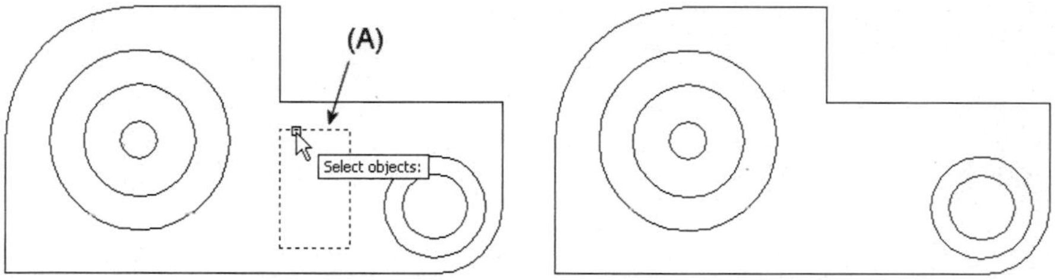

FIGURE 3.26

Selecting Objects by Window

The individual pick method previously mentioned works fine for small numbers of objects. However, when numerous objects need to be edited, selecting each individual object could prove time consuming. Instead, you can select all objects that you want to become part of a selection set by using the Window selection mode. This mode requires you to create a rectangular box by picking two diagonal points. In Figure 3.27, a selection window has been created with point "A" as the first corner and "B" as an other corner. As an additional aid in selecting, the rectangular box you construct is displayed as a transparent blue box by default. When you use this selection mode, only those objects completely enclosed by the blue window box are selected. Even though the window touches other objects, they are not completely enclosed by the window and therefore are not selected.

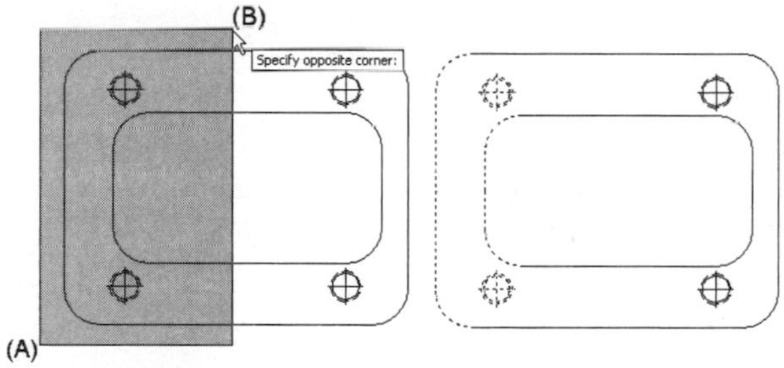

FIGURE 3.27

Selecting Objects by Crossing Window

In the previous example of producing a selection set by a window, the window selected only those objects completely enclosed by it. Figure 3.28 is an example of selecting objects by a crossing window. The Crossing Window option requires two points to define a rectangle, as does the Window selection option. In Figure 3.28, a transparent green rectangle with dashed edges is used to select objects, using "A" and "B" as corners for the rectangle; however, this time the crossing window is used. The highlighted objects illustrate the results. All objects that are touched by or enclosed by the crossing rectangle are selected. Because the transparent green crossing rectangle passes through the four horizontal lines they are also included in this Object selection mode.

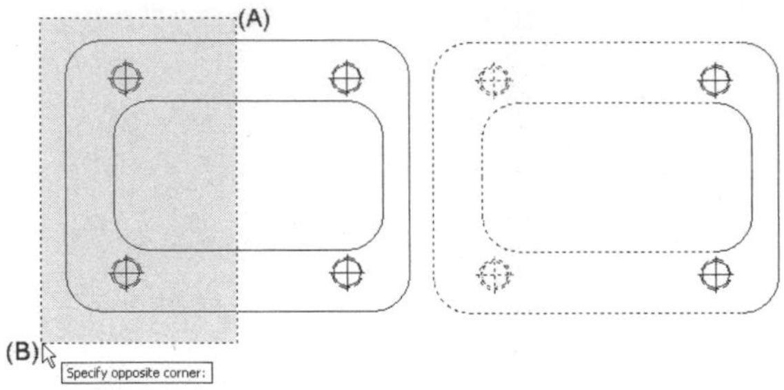

FIGURE 3.28

NOTE

The Crossing and Window methods of selection discussed above do not require any option activation. Implied Windowing is a method (on by default—discussed later in this chapter) that allows you to simply pick left to right for a window and right to left for a crossing.

Selecting Objects by a Fence

Use this mode to create a selection set by drawing a line or group of line segments called a fence. Any object touched by the fence is selected. The fence does not have to end exactly where it was started. In Figure 3.29, all objects touched by the fence are selected, as represented by the dashed lines.

TRY IT!

Open the drawing file 03_Fence. Follow the command sequence below and the illustration to select a group of objects using a fence.

```
Command: E (For Erase)
Select objects: F (For Fence)
Specify first fence point: (Pick a first fence point)
Specify next fence point or [Undo]: (Pick a second fence point)
Specify next fence point or [Undo]: (Pick a third fence point)
Specify next fence point or [Undo]: (Press ENTER to exit fence mode)
Select objects: (Press ENTER to execute the ERASE command)
```

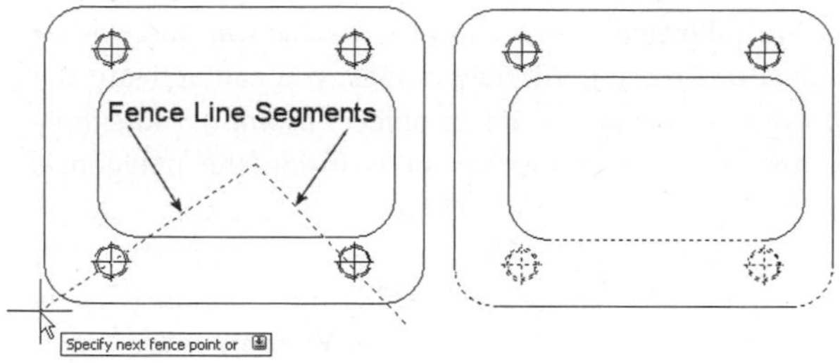

FIGURE 3.29

Removing Objects from a Selection Set

All the previous examples of creating selection sets have shown you how to create new selection sets. What if you select the wrong object or objects? Instead of canceling out of the command and trying to select the correct objects, you can simply remove objects from an existing selection set. As illustrated in Figure 3.30 on the left, a selection set has been created and is made up of all the highlighted objects. However, the outer lines and arcs are mistakenly selected as part of the selection set. To remove highlighted objects from a selection set, press SHIFT and pick the object or objects you want removed. When the highlighted objects are removed from the selection set, as shown in Figure 3.30 (Right), they regain their original display intensity.

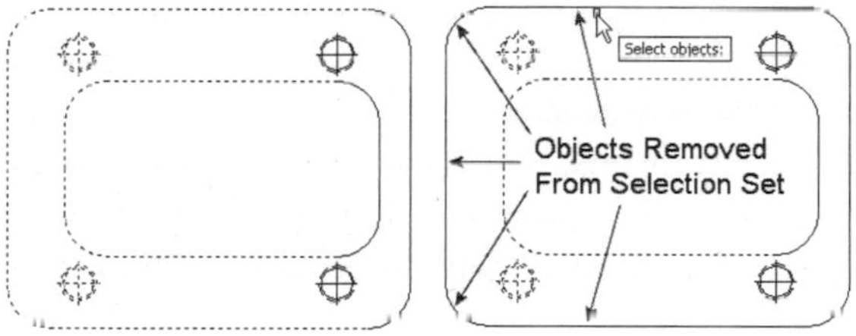

FIGURE 3.30

TIP
Sometimes it is more productive to select all objects and remove just a few objects than it is to try and select around the few.

Selecting the Previous Selection Set

When you create a selection set of objects, this grouping is remembered until another selection set is made. The new selection set replaces the original set of objects. Let's say you moved a group of objects to a new location on the display screen. Now you want to rotate these same objects at a certain angle. Rather than re-selecting the same set of objects to rotate, you could instead use the Previous option or type P at the Select objects prompt. This selects the previous selection set.

Selecting Objects by a Window or Crossing Polygon

When you use the Window or Crossing Window mode to create selection sets, two points specify a rectangular box for selecting objects. At times, you may find that a polygon, as opposed to a rectangle, provides a more convenient shape for selecting the desired objects.

TRY IT!

Open the drawing file 03_Select CP. The following image shows a mechanical part with a "C"-shaped slot. Rather than use Window or Crossing Window modes, you can activate the Crossing Polygon mode by typing Cpolygon or CP at the Select objects prompt. You simply pick points representing a polygon. Any object that touches or is inside the polygon is added to a selection set of objects. As illustrated in Figure 3.31 (Left), the crossing polygon is constructed using points "1" through "5." A similar but different selection set mode is the Window Polygon (WPolygon or WP). Objects are selected using this mode when they lie completely inside the Window Polygon, which is similar to the regular Window mode.

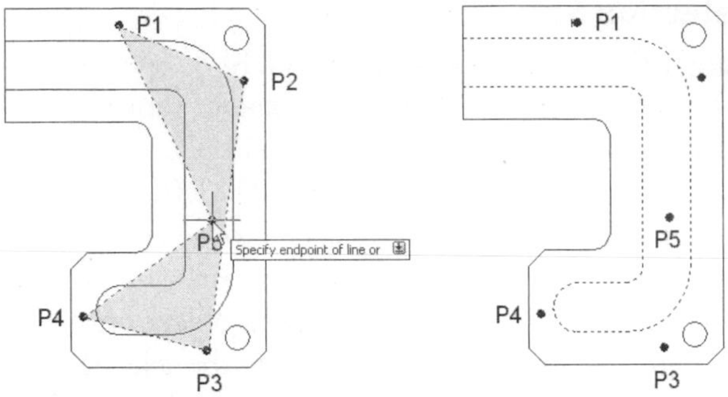

FIGURE 3.31

CREATING OBJECT GROUPS

Saving a group of objects together can make selection for modification fast and simple. Groups can be provided names and descriptions, which can be displayed through the Object Grouping dialog box. Objects can be added and removed from a group and if the group is no longer needed, it can be removed with an "ungroup" tool.

TRY IT!

Open the drawing file 03_Group. Follow the command sequence below and the illustration to create a group of objects.

Command: G *(For GROUP or select the command from the Ribbon as shown in the following image on the left)*

Select objects or [Name/Description]: N *(For Name)*

Enter a group name or [?]: BOLT

Select objects or [Name/Description]: D *(For Description)*

Enter a group description: .75-10UNC-2A

Select objects or [Name/Description]: *(Pick a point at "A")*

Specify opposite corner: *(Pick a point at "B")*

Select objects or [Name/Description]: *(Press* ENTER *to create the group)*

Group "BOLT" has been created.

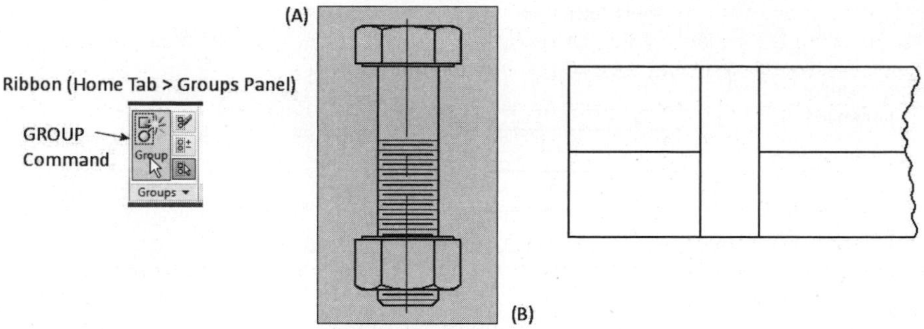

FIGURE 3.32

Move your cursor over any portion of the bolt and notice how the entire group of objects now highlights, as shown in Figure 3.33 (Left). Expand the Groups panel on the Ribbon and select the Group Manager button as shown in Figure 3.33 (Center). This opens the Object Grouping dialog box as shown in Figure 3.33 (Right). In the Group Name area of the dialog box, select Bolt to display the name and description in the Group Identification area of the dialog box. From this dialog box you can manage all of your object groups. You can create new groups or modify existing ones. The Add and Remove buttons allow you to add or remove objects from a group. Click OK to close the dialog box.

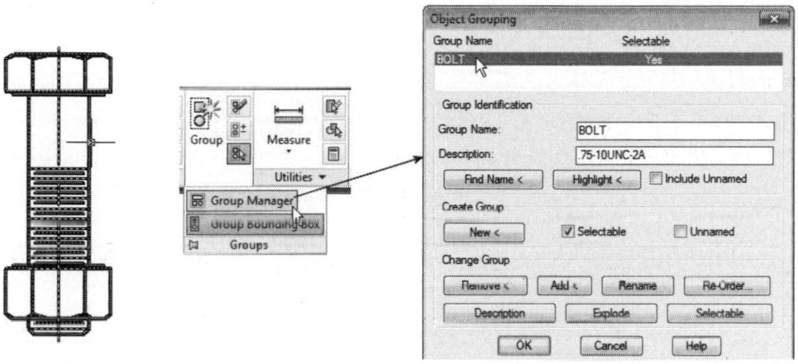

FIGURE 3.33

The Group Edit button on the Ribbon, as shown in Figure 3.34 (Left), can also be used to add or remove objects from a group. Pick the button and follow the command prompt sequence and the illustration to remove 3 threads from the bolt.

`Command: GROUPEDIT` *(or select the command from the Ribbon as shown in the following image on the left)*

`Select group or [Name]:` *(Pick the bolt)*

`Enter an option [Add objects/Remove objects/REName]: R` *(For Remove objects)*

`Select objects to remove from group...`

`Remove objects:` *(Pick the 6 lines as shown in the following image on the right)*

`Remove objects:` *(Press ENTER to remove the lines from the group)*

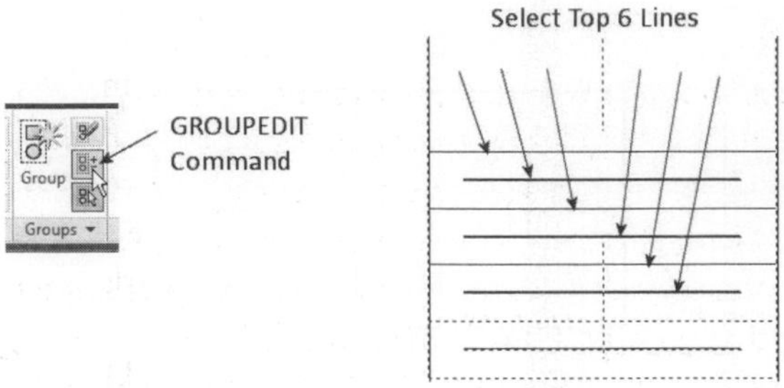

FIGURE 3.34

Finally, move the bolt into position, erase the removed lines, and explode the group using the following command prompt sequences and illustration.

Command: M *(For the MOVE command)*

Select objects: G *(Instead of just picking the bolt, we will use the Group option here)*

Enter group name: BOLT

Select objects: *(Press* ENTER *to Continue)*

Specify base point or [Displacement] <Displacement>: 4.00, 0.00 *(to move 4 units to the right)*

Specify second point or <use first point as displacement>: *(Press* ENTER - the results are shown in the following image on the left*)*

Command: E *(For ERASE)*

Select objects: *(Pick the 6 lines removed from the group earlier)*

Select objects: *(Press* ENTER*)*

Command: UNGROUP *(or select the command from the Ribbon as shown in the following image on the left)*

Select group or [Name]: *(Pick the bolt)*

Group BOLT exploded.

Move your cursor over any portion of the bolt and notice that it is no longer grouped together. Expand the Groups panel on the Ribbon and select the Group Manager button. Notice that in the Group Name area of the Object Grouping dialog box, Bolt is no longer displayed, as shown in Figure 3.35 (Right).

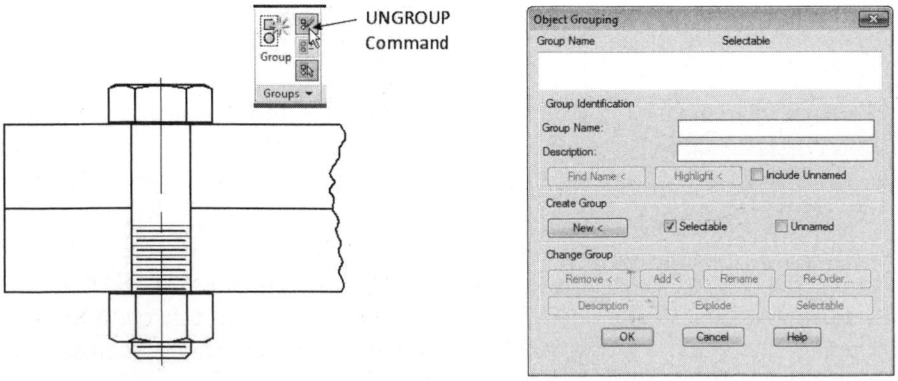

FIGURE 3.35

CYCLING THROUGH OBJECTS

At times, the process of selecting a specific object within a busy area of a drawing can become difficult. Such is the case when objects lie directly on top of each other. As you attempt to select the object to delete, the wrong object gets selected instead. To remedy this, utilize selection cycling. Verify that the Selection Cycling Button on the Status Bar is activated. At the Select objects prompt move your cursor over the overlapped objects. An icon will display indicating that your pickbox is over multiple items. Once you pick, a list of objects is displayed, as shown in Figure 3.36. Continue picking to cycle through the objects or move your cursor over any item listed and pick to select it.

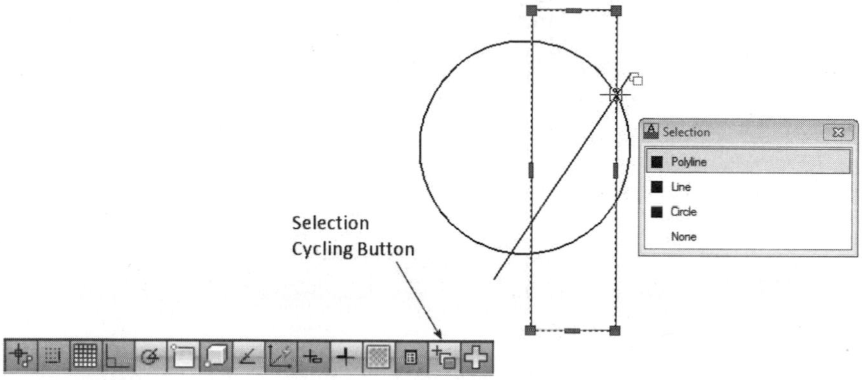

FIGURE 3.36

TRY IT!

Open the drawing file 03_Cycle and experiment using this feature on the object illustrated in Figure 3.36.

NOUN/VERB SELECTION

Instead of entering a MODIFY command and waiting for the Select objects prompt to make a selection set, you can pre-select objects before issuing a command. This is referred to as Noun/verb selection. Objects selected in this manner will not only highlight, but a series of blue square boxes will also appear. These blue objects are called grips and are utilized for making modifications to objects. They will be discussed in detail in Chapter 7. To cancel or deselect the object, press the ESC key and notice that even the grips disappear.

TIP

Pressing CTRL + A at the Command prompt selects all objects in the current space environment (model or paper) of the drawing and displays the blue grip boxes. Objects that reside on a layer that is frozen or locked are not selected.

IMPLIED WINDOWING

Without using any options and with or without issuing any commands, you can select objects by picking or using implied windowing. To utilize implied windowing simply pick a blank part of your screen and move your cursor such that a box is created over the intended objects and then pick again. Picking points left to right has the same effect as using the Window option of Select objects. Only objects completely inside the window are selected. If you pick a blank part of your screen at the Command prompt and move your cursor to the left, this has the same effect as using the Crossing option of Select objects in which any items touched by or completely enclosed by the box are selected.

NOTE

If implied windowing and noun/verb selection methods are not working, verify they are turned on in the Options dialog box (OPTIONS command > Selection tab and Selection modes area). If shift to remove is not working, verify "Use shift to add to selection" is turned off.

THE QSELECT COMMAND

Yet another way of creating a selection set is by matching the object type and property with objects currently in use in a drawing. This is the purpose of the QSELECT (Quick Select) command. This command can be chosen from the Ribbon (Home tab > Utilities panel), as shown in Figure 3.37 (Left), or from a shortcut menu activated by right-clicking in the screen area, as shown in Figure 3.37 in the middle. Initiating this command launches the dialog box shown in Figure 3.37 (Right).

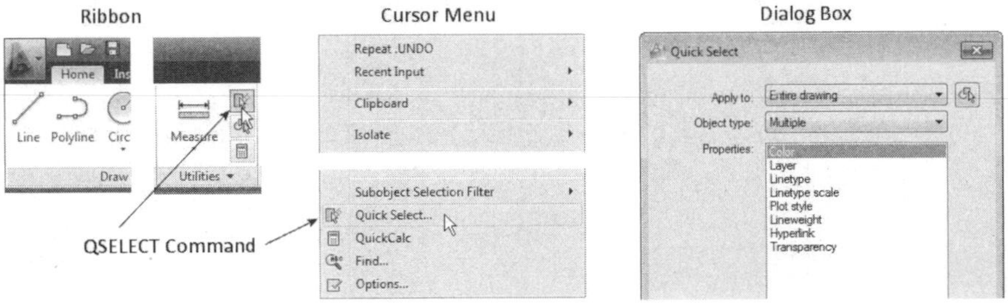

FIGURE 3.37

TRY IT!

Open the drawing file 03_Qselect and activate the Quick Select dialog box. This command works only if objects are defined in a drawing; the Quick Select dialog box does not display in a drawing file without any objects drawn. Clicking in the Object type field displays all object types currently used in the drawing. This enables you to create a selection set by the object type. For instance, to select all line segments whose Color is Bylayer in the drawing file, click on Line in the Object type field, as illustrated in Figure 3.38 (Left).

Clicking the OK button at the bottom of the dialog box returns you to the drawing and highlights all the line segments in the drawing, as shown in Figure 3.38 (Right).

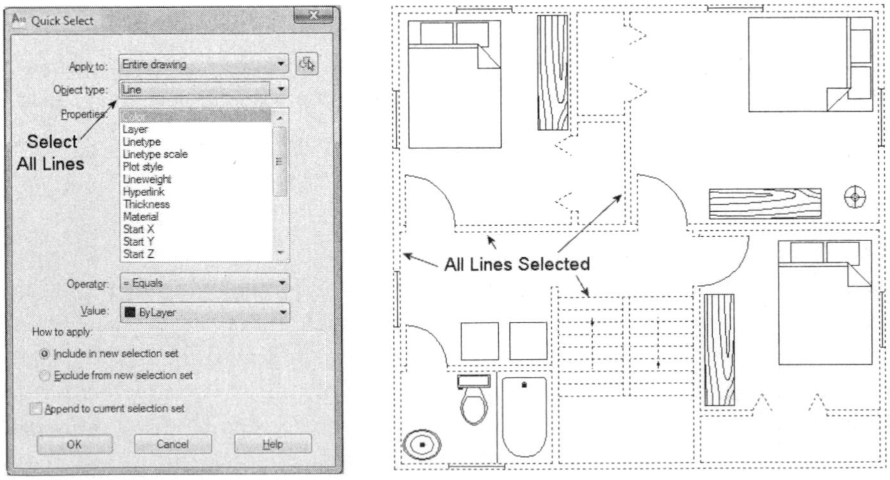

FIGURE 3.38

Other controls of Quick Select include the ability to select the object type from the entire drawing or from just a segment of the drawing. You can narrow the selection criteria by adding various properties to the selection mode, such as Color, Layer, and Linetype, to name a few. You can also create a reverse selection set. The Quick Select dialog box lives up to its name—it enables you to create a quick selection set.

ISOLATE AND HIDE OBJECTS

Earlier in this chapter, we used the VIEW (Named Views) command to save a view so that you could return to it at any time. At times when working with a large and complicated drawing, it might also be helpful to temporarily change the drawing view by simply hiding objects. The ISOLATEOBJECTS, HIDEOBJECTS, and UNISOLATEOBJECTS are commands that can accomplish this temporary simplification of a view. This same type of operation could be accomplished by manipulating layers, but the isolate and hide operations are efficient and require no prior knowledge of layer organization.

AutoCAD CERTIFIED ASSOCIATE EXAM 2013

AutoCAD CERTIFIED PROFESSIONAL EXAM 2013

You can, of course, access the ISOLATEOBJECTS, HIDEOBJECTS, and UNISOLATEOBJECTS commands by typing them in the Command prompt, but the easiest way to perform these operations is through a shortcut menu or a Status Bar button. Right-click in the screen area to display the shortcut menu, as shown Figure 3.39 (Left), or click the Isolate Objects button on the Status Bar, as shown on the right in Figure 3.39. You can also use Noun/verb selection to pre-select objects for these commands.

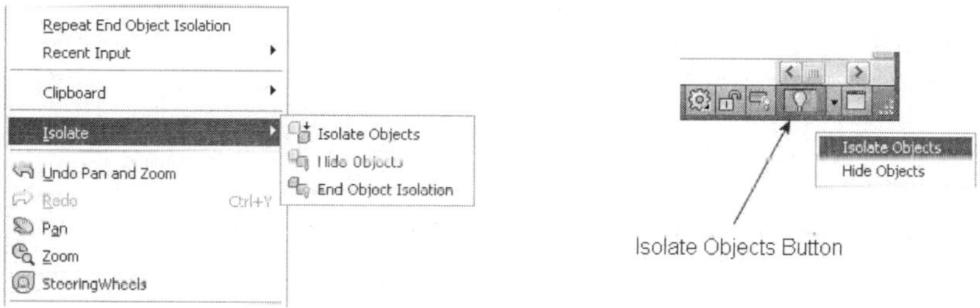

FIGURE 3.39

TRY IT!
Open the drawing file 03_Isolate, right-click in the screen area to display the shortcut menu displayed on the left in Figure 3.40. Select "Isolate Objects" from the menu and at the Select objects prompt pick from point (A) to (B). The result is that everything in the drawing is hidden except for the items selected, as shown on the right in Figure 3.40. To return to the unisolated view, activate the shortcut menu and select "End Object Isolation." Try repeating the procedure utilizing "Hide Objects" and then restore the view with "End Object Isolation."

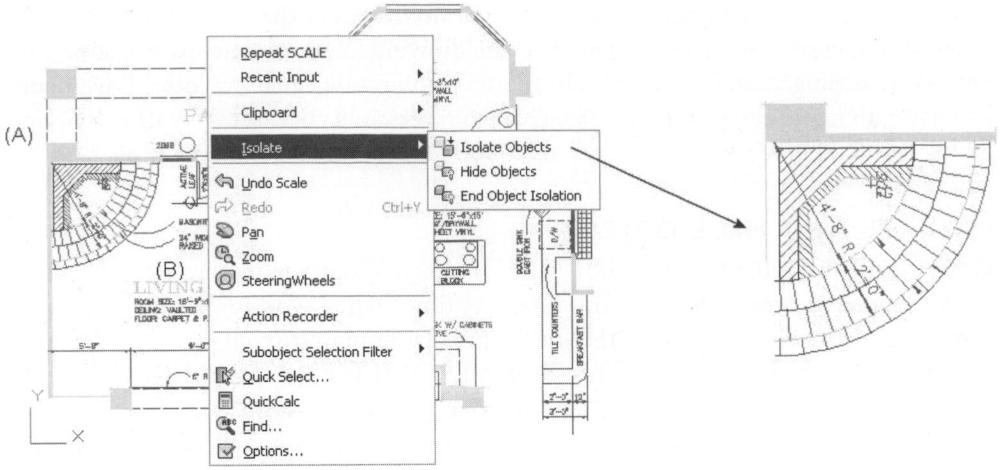

FIGURE 3.40

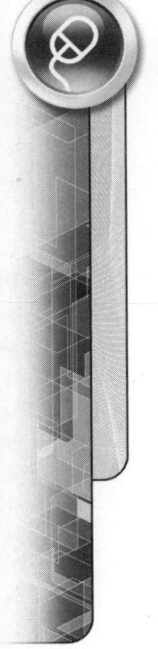

TUTORIAL EXERCISE: 03_SELECT OBJECTS.DWG

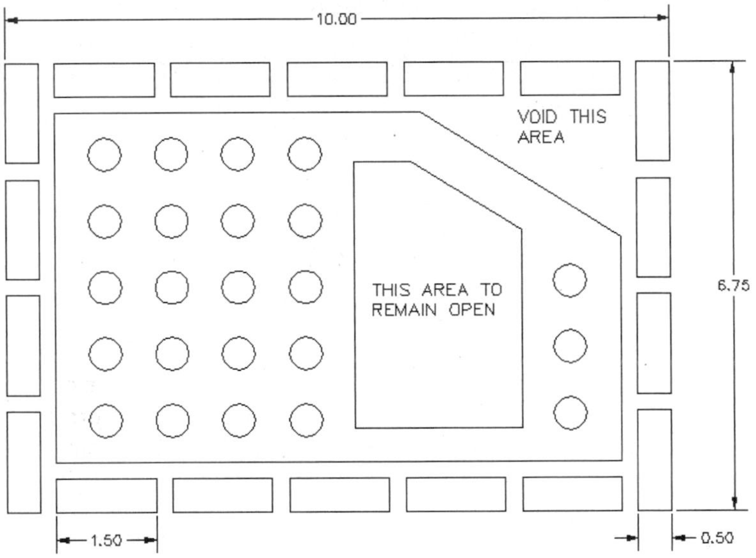

FIGURE 3.41

Purpose

The purpose of this tutorial is to experiment with the Objects selection modes on the drawing shown in Figure 3.41.

System Settings

Keep the current limits settings of (0,0) for the lower-left corner and (16,10) for the upper-right corner. Keep all remaining system settings.

Suggested Commands

This tutorial utilizes the ERASE command as a means of learning the basic Object selection modes. The following selection modes will be highlighted for this tutorial: Window, Crossing, Window Polygon, Crossing Polygon, Fence, and All. The effects of locking layers, noun/verb selection, and quick selection will also be demonstrated.

STEP 1

Open the drawing 03_Select Objects and activate ERASE at the Command prompt. At the Select objects prompt, pick a point on the blank part of your screen at "A." At the next Select objects prompt, pick a point on your screen at "B." Notice that a solid blue transparent box is formed, as shown in Figure 3.42. The presence of this box signifies the Window option of selecting objects.

Command: E *(For Erase)*

Select objects: *(Pick a point at "A")*

Specify opposite corner: *(Pick a point at "B")*

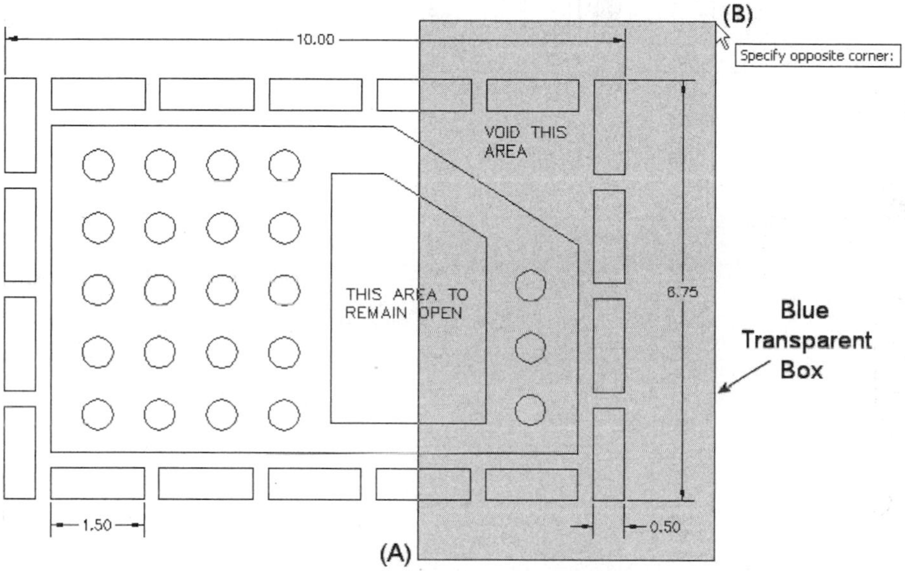

FIGURE 3.42

The result of selecting objects by a window is illustrated in Figure 3.43. Only those objects completely surrounded by the window are highlighted. Pressing ENTER performs the erase operation. Before continuing on to the next step, issue the U (UNDO) command, which negates the previous erase operation.

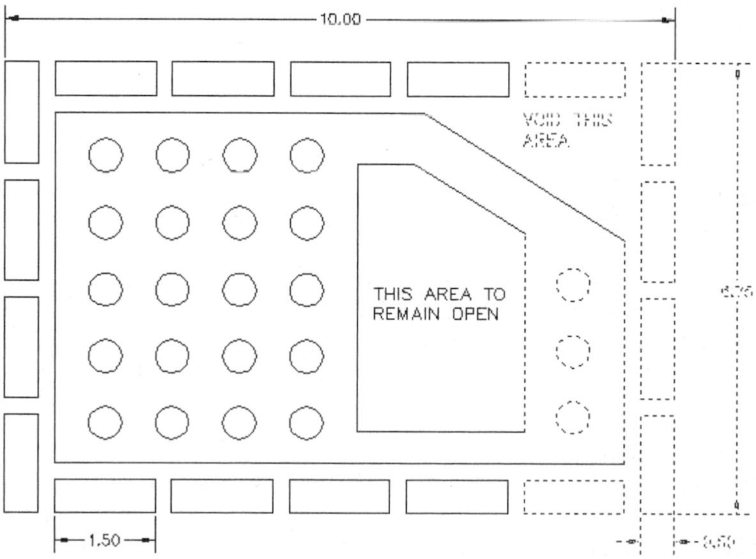

FIGURE 3.43

STEP 2

With the entire object displayed on the screen, activate the ERASE command again. At the Select objects prompt, pick a point on the blank part of your screen at "A." At the Specify opposite corner prompt, pick a point on your screen at "B." Notice that the transparent box is now dashed and green in appearance, as shown in Figure 3.44. The presence of this box signifies the Crossing option of selecting objects.

```
Command: E (For ERASE)
Select objects: (Pick a point at "A")
Specify opposite corner: (Pick a point at "B")
```

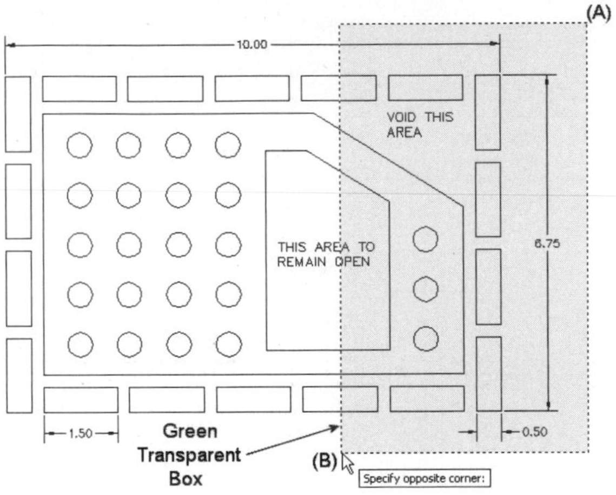

FIGURE 3.44

The result of selecting objects by a crossing window is illustrated in Figure 3.45. Notice that objects touched by the crossing box, as well as those objects completely surrounded by the crossing box, are highlighted. Pressing ENTER performs the erase operation. Before continuing on to the next step, issue the U (UNDO) command, which negates the previous erase operation.

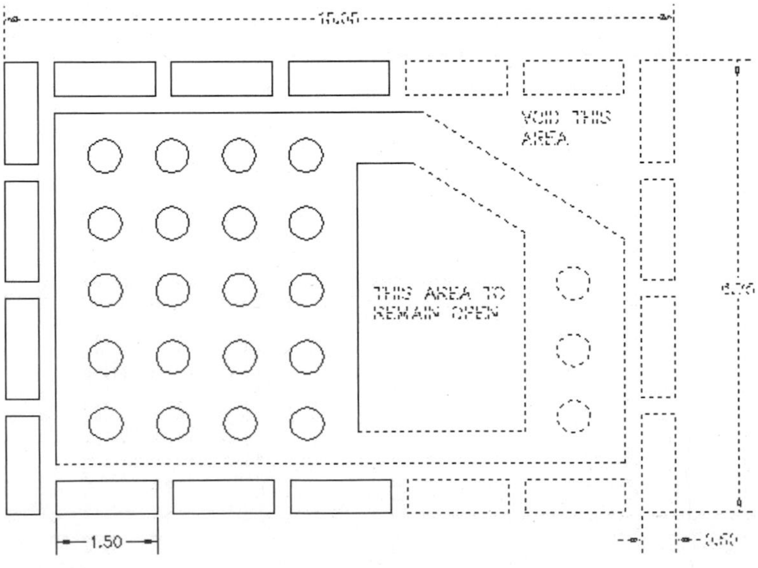

FIGURE 3.45

STEP 3

Activate the ERASE command again. At the Select objects prompt, pick a point on the blank part of your screen at "A." At the Specify opposite corner prompt, pick a point on your screen at "B," as shown in Figure 3.46. The presence of this transparent blue box signifies the Window option of selecting objects.

Command: E *(For Erase)*

Select objects: *(Pick a point at "A")*

Specify opposite corner: *(Pick a point at "B")*

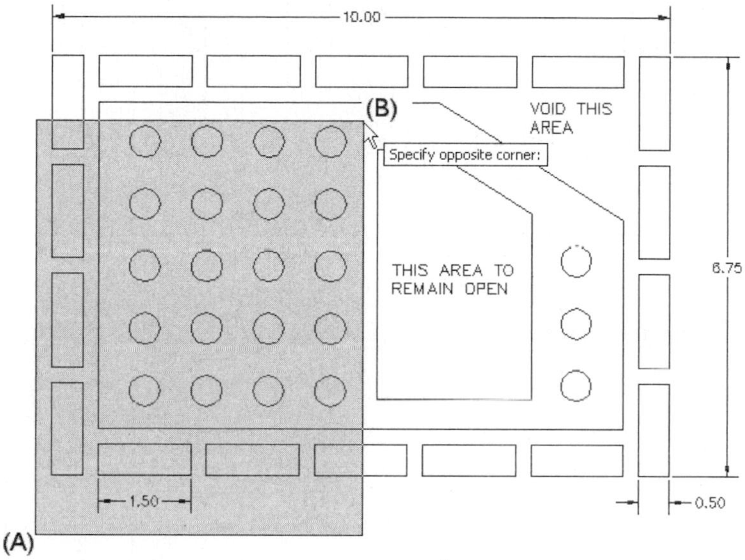

FIGURE 3.46

The result of selecting objects by a window is illustrated in Figure 3.47 (Left). Notice that a number of circles and rectangles have been selected along with a single dimension. Unfortunately, the rectangles and the dimension were selected by mistake. Rather than cancel the ERASE command and select the objects again, press and hold down the SHIFT key and click on the highlighted rectangles and dimension. Notice how this operation deselects the objects (see Figure 3.47 on the right). Pressing ENTER performs the erase operation. Before continuing on to the next step, issue the U (UNDO) command, which negates the previous erase operation.

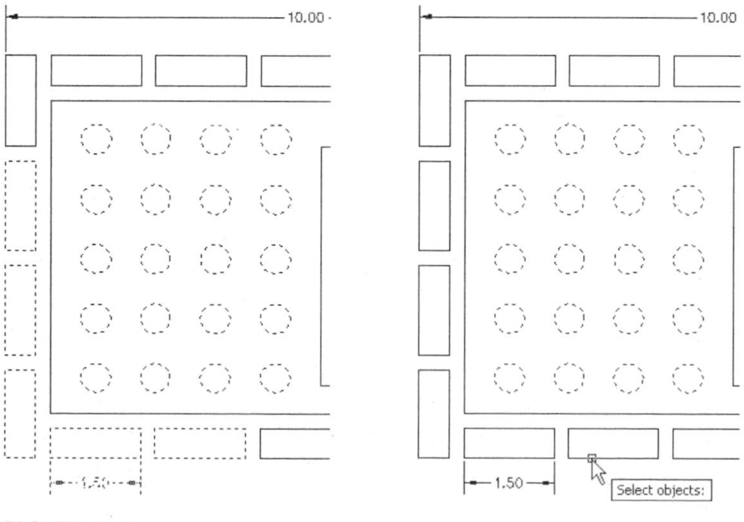

FIGURE 3.47

STEP 4

Yet another way to select objects is by using a Window Polygon. This method allows you to construct a closed irregular shape for selecting objects. Again, activate the ERASE command. At the Select objects prompt, enter WP for Window Polygon. At the First polygon point prompt, pick a point on your screen at "A." Continue picking other points until the desired polygon, in transparent blue, is formed, as shown in Figure 3.48.

NOTE

The entire polygon must form a single closed shape; edges of the polygon cannot cross each other.

Command: E *(For Erase)*
Select objects: WP *(For Window Polygon)*
First polygon point: *(Pick a point at "A")*
Specify endpoint of line or [Undo]: *(Pick at "B")*
Specify endpoint of line or [Undo]: *(Pick at "C")*
Specify endpoint of line or [Undo]: *(Pick at "D")*
Specify endpoint of line or [Undo]: *(Pick at "E")*
Specify endpoint of line or [Undo]: *(Pick at "F")*
Specify endpoint of line or [Undo]: *(Pick at "G")*
Specify endpoint of line or [Undo]: *(Press ENTER)*
11 found

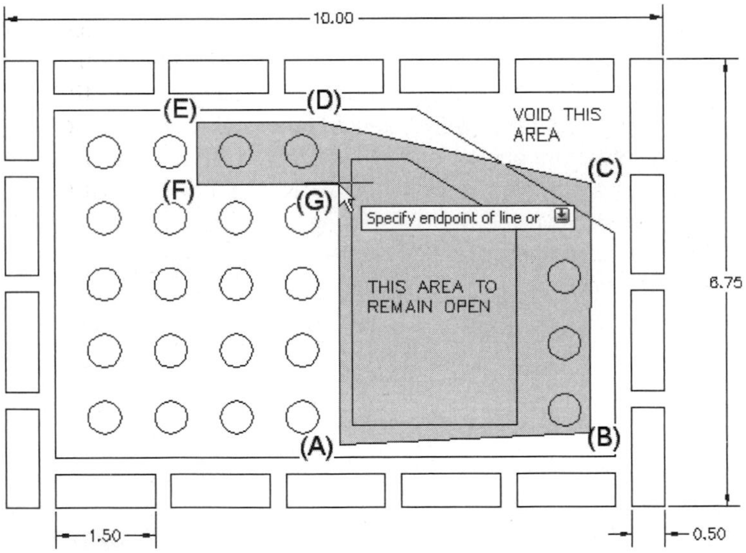

FIGURE 3.48

The result of selecting objects by a window polygon is illustrated in Figure 3.49. As with the Window selection mode, objects must lie entirely inside the window polygon in order for them to be selected. Pressing ENTER performs the erase operation. Before continuing on to the next step, issue the U (UNDO) command, which negates the previous erase operation.

NOTE

A Crossing Polygon (CP) mode is also available to select objects that touch the crossing polygon or are completely surrounded by the polygon. A green polygonal shape displays as it is being created.

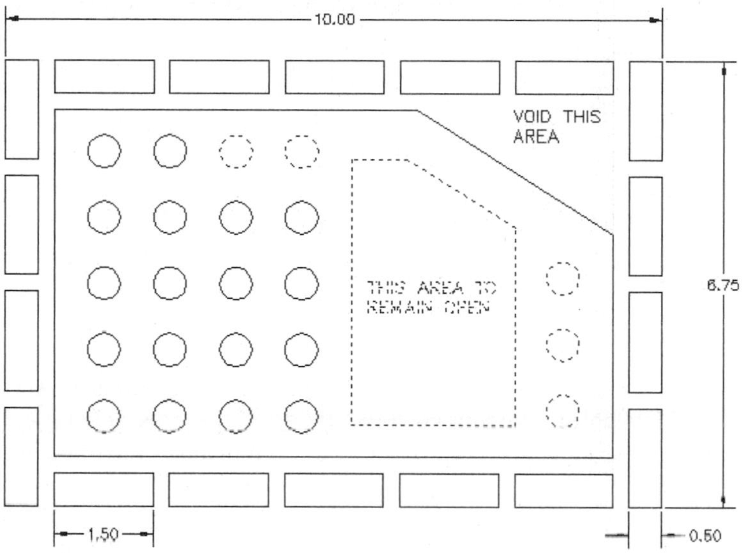

FIGURE 3.49

STEP 5

Objects can also be selected by a fence. This is represented by a crossing line segment. Any object that touches the crossing line is selected. You can construct numerous crossing line segments. You cannot use the Fence mode to surround objects as with the Window or Crossing modes. Activate the ERASE command again. At the Select objects prompt, enter F, for FENCE. At the First fence point prompt, pick a point on your screen at "A." Continue constructing crossing line segments until you are satisfied with the objects being selected (see Figure 3.50).

Command: E *(For Erase)*

Select objects: F *(For Fence)*

Specify first fence point: *(Pick at "A")*

Specify next fence point or [Undo]: *(Pick at "B")*

Specify next fence point or [Undo]: *(Pick at "C")*

Specify next fence point or [Undo]: *(Pick at "D")*

Specify next fence point or [Undo]: *(Pick at "E")*

Specify next fence point or [Undo]: *(Press* ENTER *to create the selection set)*

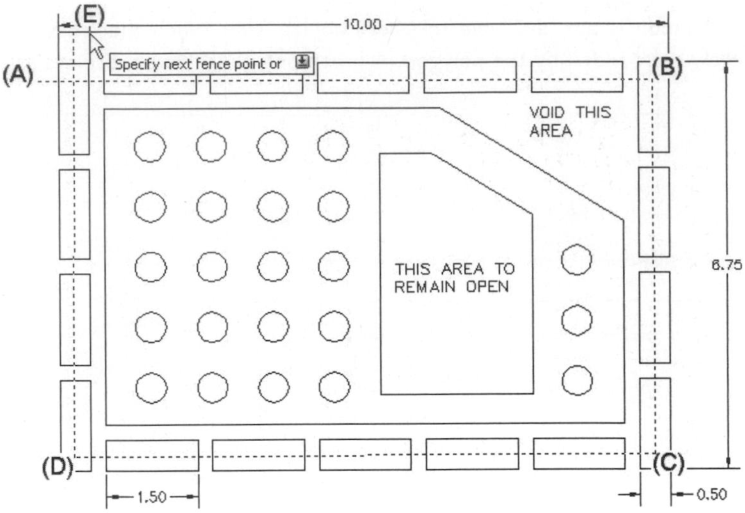

FIGURE 3.50

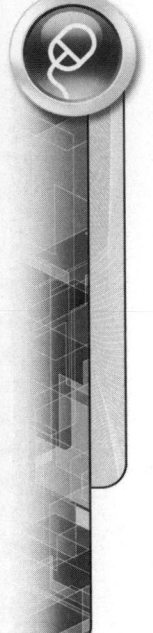

The result of selecting objects by a fence is illustrated in Figure 3.51. Notice that objects touched by the fence are highlighted. Pressing ENTER performs the erase operation. Before continuing on to the next step, issue the U (UNDO) command, which negates the previous erase operation.

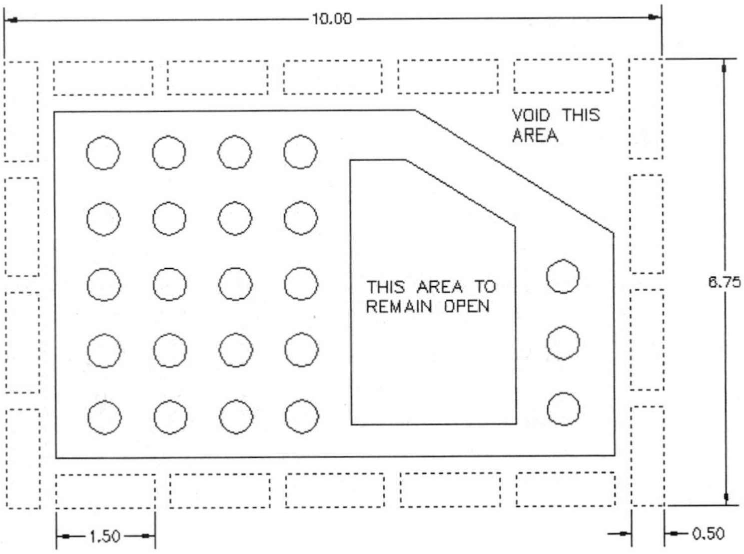

FIGURE 3.51

STEP 6

When situations require you to select all objects in the entire database of a drawing, the All option is very efficient. Activate the ERASE command again. At the Select objects prompt, enter ALL. Notice that in Figure 3.52 that all objects are selected by this option.

Command: E *(For Erase)*

Select objects: All

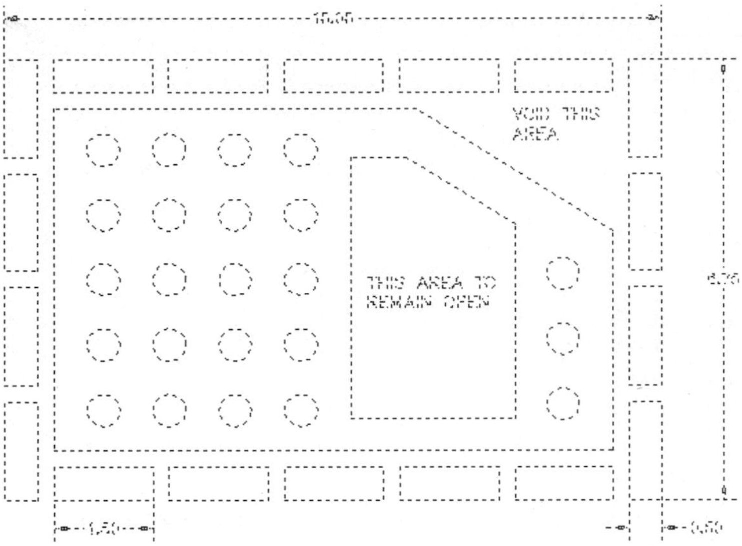

FIGURE 3.52

Pressing ENTER performs the erase operation. Before continuing on to the next step, issue the U (UNDO) command, which negates the previous erase operation.

> **NOTE**
>
> **Care should be exercised when using the All option, because even objects on layers that are turned off will be selected. The All option does not select objects on layers that are frozen or locked.**

STEP 7

Before performing another erase operation on a number of objects, activate the Layer Properties Manager palette, select the "Circles" layer, and click the Lock icon, as shown in Figure 3.53. This operation locks the "Circles" layer.

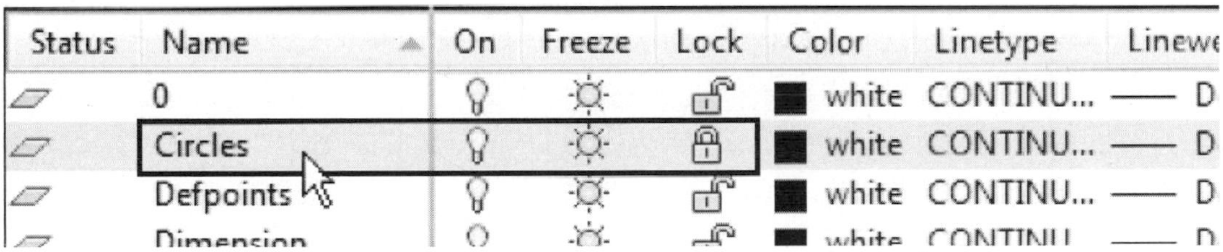
Status
Name
On
Freeze
Lock
Color
Linetype
0
white
CONTINU...
Circles
white
CONTINU...
Defpoints
white
CONTINU...

FIGURE 3.53

To see what effect this has on selecting objects, activate the ERASE command again. At the Select objects prompt, pick a point on the blank part of your screen at "A." At the Specify opposite corner prompt, pick a point on your screen at "B." You have once again selected a number of objects by a green crossing box (see Figure 3.54).

```
Command: E (For Erase)
Select objects: (Pick a point at "A")
Specify opposite corner: (Pick a point at "B")
20 were on a locked layer.
```

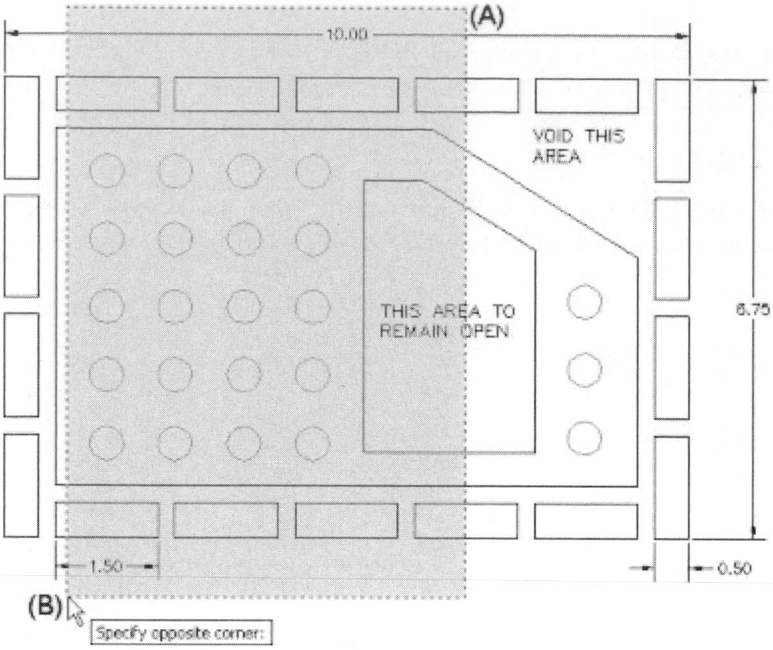

FIGURE 3.54

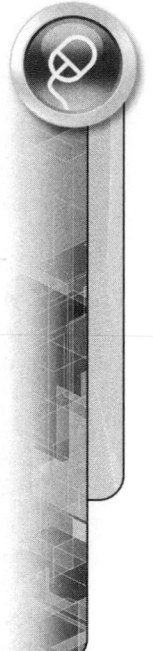

The result of selecting objects by a crossing window is illustrated in Figure 3.55. Notice that even though a group of circles was completely surrounded by the crossing window, the circles do not highlight because they belong to a locked layer. Pressing `ENTER` performs the erase operation. Before continuing on to the next step, issue the `U (UNDO)` command, which negates the previous erase operation.

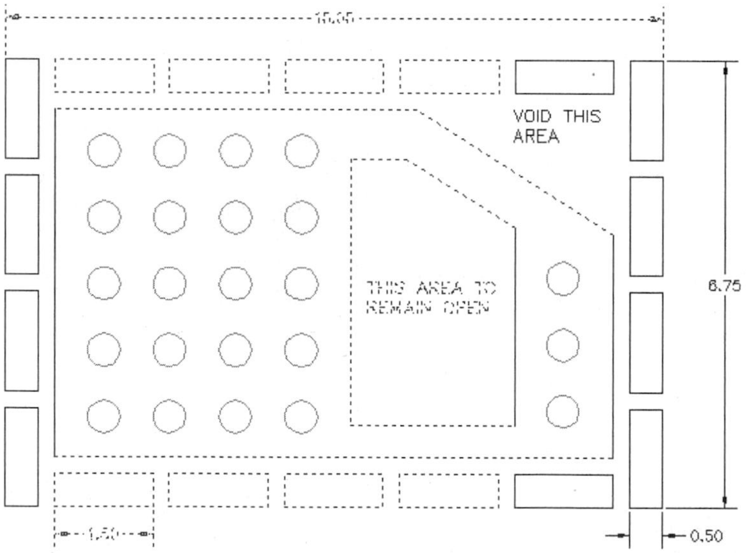

FIGURE 3.55

STEP 8

This final step illustrates the use of the Noun/verb selection and the QSELECT command. First activate the QSELECT command and the dialog box will display, as shown in Figure 3.56 (Left).

Command: QSELECT *(To display the dialog box)*

In the dialog box, for the Object type select Rotated Dimension. Leave the Properties, Operator, and Value as shown. Click OK to select the drawings dimensions, as shown in Figure 3.56 (Right).

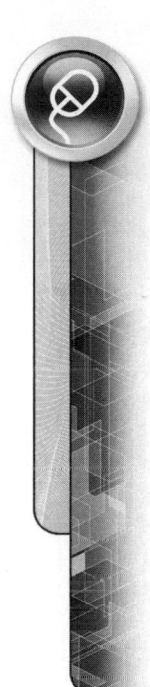

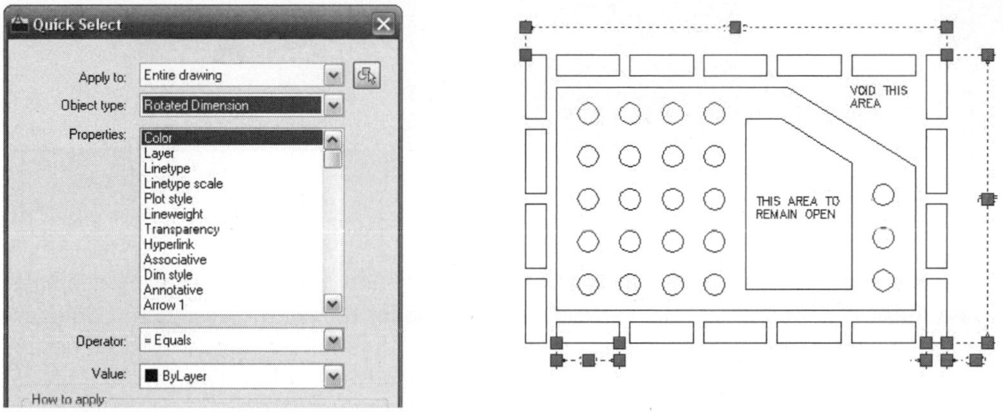

FIGURE 3.56

The dimensions have been pre-selected (noun/verb selection). Now activate the ERASE command to complete the operation.

Command: E *(For ERASE)*

The result is illustrated in the following image.

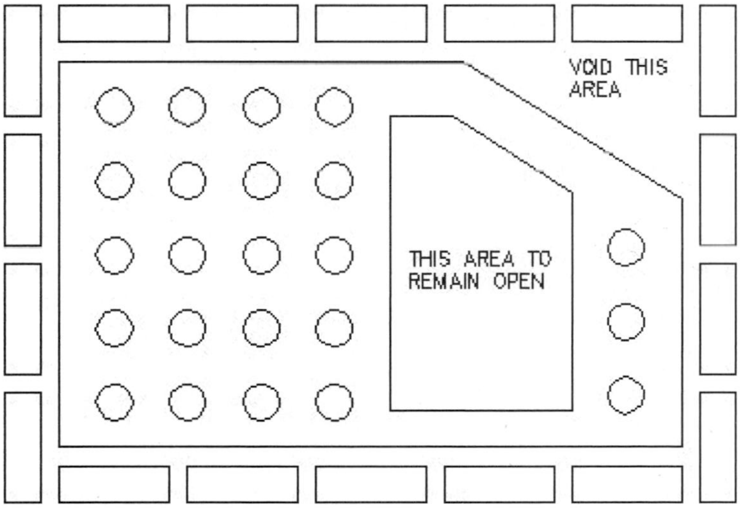

FIGURE 3.57

CHAPTER

4

Modifying Your Drawings

The heart of any CAD system is its ability to modify and manipulate existing geometry and AutoCAD is no exception. Many modify commands relieve the designer of drudgery and mundane tasks, and this allows more productive time for conceptualizing the design. This chapter will break the AutoCAD modify commands down into two separate groupings. The first grouping is called Level I and will cover the MOVE, COPY, SCALE, ROTATE, FILLET, CHAMFER, OFFSET, TRIM, EXTEND, and BREAK commands. The second grouping is called Level II and will cover the ARRAY (ARRAYRECT, ARRAYPOLAR, and ARRAYPATH), MIRROR, STRETCH, PEDIT, EXPLODE, LENGTHEN, JOIN, BLEND, UNDO, and REDO commands. A number of small exercises accompany each command in order to reinforce the importance of its use.

METHODS OF SELECTING MODIFY COMMANDS

As with all commands, you can find the main body of modify commands on the Ribbon (Drafting & Annotation Workspace). Select the Home tab and expand the Modify panel to view all the commands available, as shown in Figure 4.1.

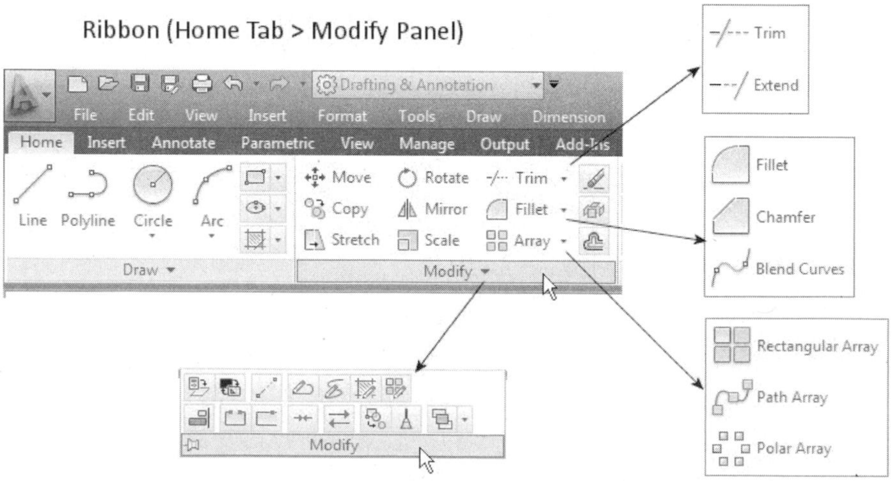

FIGURE 4.1

Image(s) © Cengage Learning 2013

LEVEL I MODIFY COMMANDS

With all the modify commands available in AutoCAD, the following represent beginning, or Level I commands, which you will find yourself using numerous times as you make changes to your drawing. These commands are briefly described in the following table:

Button	Tool	Shortcut	Function
	Move	M	Used for moving objects from one location to another
	Copy	CP or CO	Used for copying objects from one location to another
	Scale	SC	Used for increasing or reducing the size of objects
	Rotate	RO	Used for rotating objects to a different angle
	Fillet	F	Used for rounding off the corners of objects at a specified radius
	Chamfer	CHA	Used to connect two objects with an angled line forming a bevel
	Offset	O	Used for copying objects parallel to one another at a specified distance
	Trim	TR	Used for partially deleting objects based on a cutting edge
	Extend	EX	Used for extending objects based on a boundary edge
	Break	BR	Creates a gap in an object between two specified points
	Break at Point	BR	Breaks an object into two objects at a specified point without a gap present

MOVING OBJECTS

The `MOVE` command repositions an object or group of objects at a new location.

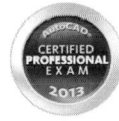

Choose this command from one of the following:

- From the Ribbon > Home Tab > Modify Panel
- The Menu Bar (Modify > Move)
- The Modify Toolbar of the AutoCAD Classic Workspace
- The keyboard (`M` or `MOVE`)
- By right-clicking the mouse after selecting object

Once the objects to move are selected, AutoCAD prompts the user to select a base point of displacement (where the object is to move from). Next, AutoCAD prompts the user to select a second point of displacement (where the object is to move to), as shown in Figure 4.2.

```
Command: M (For MOVE)
Select objects: (Select the bed, as shown in the following image)
Select objects: (Press ENTER to continue)
Specify base point or [Displacement] <Displacement>:
(Select a point at "A")
Specify second point or <use first point as displacement>:
(Mark a point at "B")
```

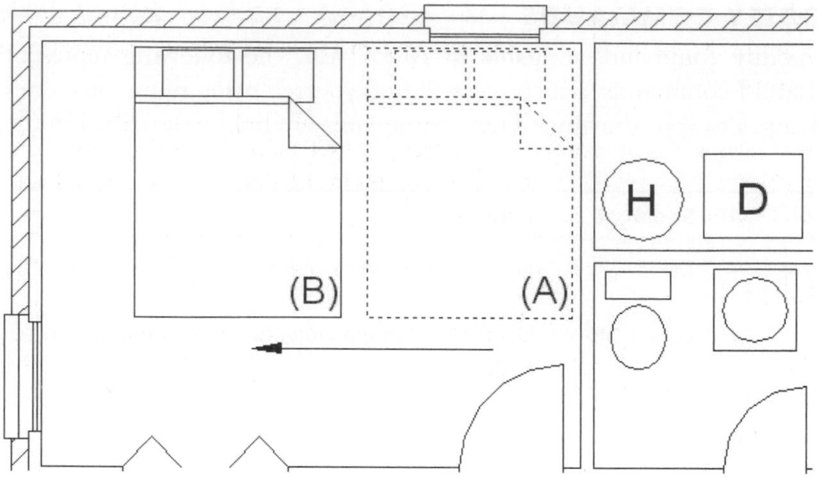

FIGURE 4.2

TRY IT!

Open the drawing file 04_Move. The slot, as shown in Figure 4.3 (Left), is incorrectly positioned; it needs to be placed 1.00 unit away from the left edge of the object. You can use the `MOVE` command in combination with a polar coordinate or Direct Distance mode to perform this operation. Use this illustration and the following command sequence for performing this operation.

`Command: M` *(For MOVE)*

`Select objects:` *(Select the slot and all centerlines)*

`Select objects:` *(Press* `ENTER` *to continue)*

`Specify base point or [Displacement] <Displacement>:` *(Select any convenient point such as "A")*

`Specify second point or <use first point as displacement>:` *(Turn Polar on, move your cursor to the right, and type* `0.50` *at the Command prompt)*

As the slot is moved to a new position with the `MOVE` command, the horizontal dimension will reflect the correct distance from the edge of the object to the centerline of the arc, as shown in Figure 4.3.

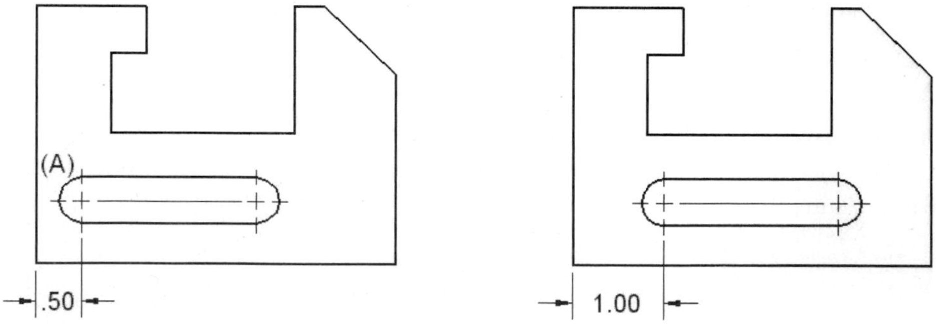

FIGURE 4.3

Press and Drag Move

If accuracy is not important and you simply need to move an object or group of objects to a new approximate location, you can use a press and drag technique. First, select the objects at the Command prompt. Then, press and hold down the left mouse button on one of the highlighted objects (not one of the blue grips), and drag the objects to the new location.

TIP

A Nudge feature is available to move objects slightly in horizontal or vertical direction. After selecting objects, hold the `CTRL` key down and press the desired arrow key on the keyboard to perform this operation.

COPYING OBJECTS

The `COPY` command is used to duplicate an object or group of objects.

Choose this command from one of the following:

- From the Ribbon > Home Tab > Modify Panel
- The Menu Bar (Modify > Copy)
- The Modify Toolbar of the AutoCAD Classic Workspace
- The keyboard (`CP`, `CO`, or `COPY`)
- By right-clicking the mouse after selecting object

Once the `COPY` command is executed, the Multiple Copy mode is on. To duplicate numerous objects while staying inside the `COPY` command, simply keep picking new second points of displacement and the objects will copy to these new locations, as shown in Figure 4.4. Once the copy is completed, press `ENTER` or `ESC` to `EXIT`.

```
Command: CP (For COPY)
Select objects: (Select the chair to copy)
Select objects: (Press ENTER to continue)
Specify base point or [Displacement/mode] <Displacement>:
(Pick a reference point for the copy operation)
Specify second point or [Array] <use first point as
displacement>: (Pick a location for the first copy)
Specify second point or [Array/Exit/Undo] <Exit>: (Pick a
location for the second copy)
Specify second point or [Array/Exit/Undo] <Exit>: (Pick a
location for the third copy)
Specify second point or [Array/Exit/Undo] <Exit>: (Press
ENTER or ESC to exit this command)
```

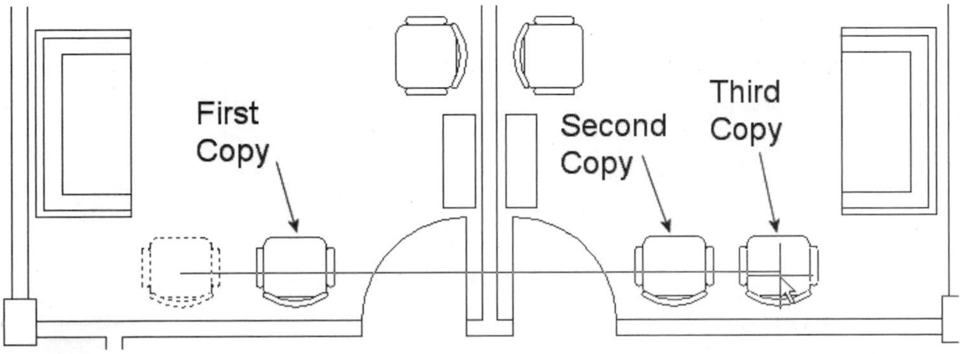

FIGURE 4.4

TRY IT!

Open the drawing file 04_Copy Multiple. Follow the command sequence in the previous example to copy the three holes multiple times. Use the intersection of "A" as the base point for the copy. Then copy the three holes to the intersections located at "B," "C," "D," "E," "F," "G," "H," and "J," as shown in Figure 4.5 (Left). The results are illustrated in Figure 4.5 (Right).

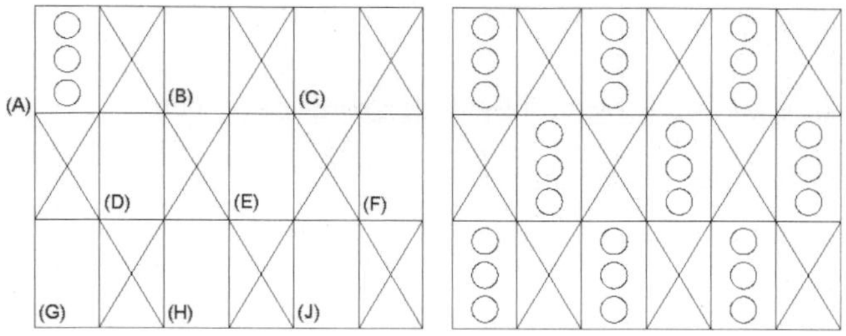

FIGURE 4.5

COPY—Array

If the objects that are to be copied are spaced evenly and are linear (in a single row or column), the Array option can be utilized. After selecting the objects to be copied, specify the number of objects desired and then the spacing. Instead of giving the spacing, a Fit option allows you to pick the beginning and the end points of the array.

TRY IT!

Open the drawing file 04_Copy Array. Study Figure 4.6 and the following prompts for performing this operation. The results of the copy operation are shown in Figure 4.6 (Right).

Command: CP *(For COPY)*

Select objects: *(Pick a point at "A")*

Specify opposite corner: *(Pick a point at "B")*

Select objects: *(Press* ENTER *to continue)*

Specify base point or [Displacement/mOde] <Displacement>: *(Select any convenient point)*

Specify second point or [Array] <use first point as displacement>: A *(For Array)*

Enter number of items to array: 5

Specify second point or [Fit]: *(Move the cursor straight out to the right, verify Polar Tracking is On and enter)* 2.00

Specify second point or [Array/Exit/Undo] <Exit>: *(Press* ENTER *to complete the operation)*

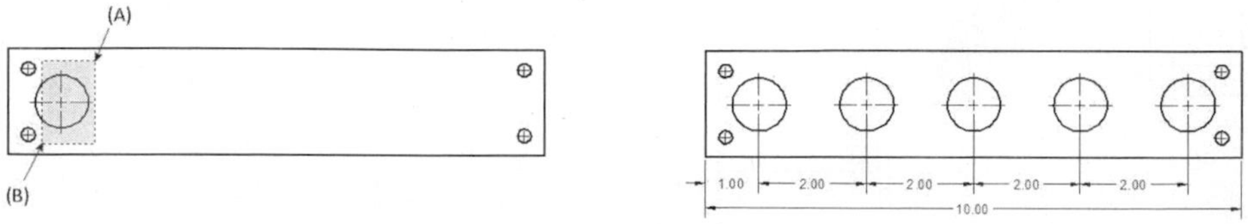

FIGURE 4.6

Press and Drag Copy

As with the MOVE command, you can also use the press and drag technique to copy objects to a new approximate location. First, select the objects at the Command prompt. Then, press and hold down the right mouse button on one of the highlighted objects (not one of the blue grips) and drag the objects to the new location. When you release the mouse button a menu displays, as shown in Figure 4.7. Select the Copy Here item to copy the item or group of items.

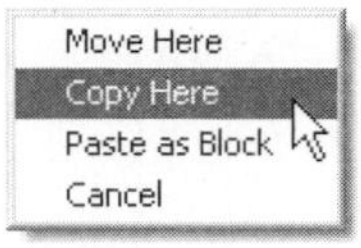

FIGURE 4.7

SCALING OBJECTS

Use the SCALE command to change the overall size of an object. The size may be larger or smaller in relation to the original object or group of objects. The SCALE command requires a base point and scale factor to complete the command. Choose this command from one of the following:

- From the Ribbon > Home Tab > Modify Panel
- The Menu Bar (Modify > Scale)
- The Modify Toolbar of the AutoCAD Classic Workspace
- The keyboard (SC or SCALE)
- By right-clicking the mouse after selecting object

TRY IT!

Open the drawing file 04_Scale1. With a base point at "A" and a scale factor of 0.50, the results of using the SCALE command on a group of objects are shown in Figure 4.8.

Command: **SC** *(For SCALE)*
Select objects: **All**
Select objects: *(Press ENTER to continue)*
Specify base point: *(Select the endpoint of the line at "A")*
Specify scale factor or [Copy/Reference] <1.0000>: **0.50**

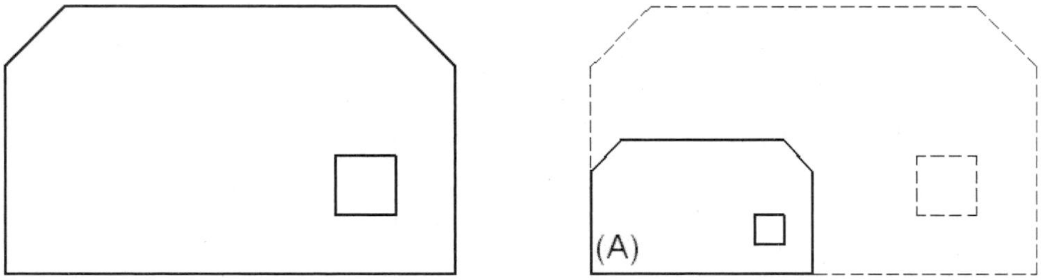

FIGURE 4.8

TRY IT!

Open the drawing file 04_Scale2. The example in Figure 4.9 shows the effects of identifying a new base point in the center of the object.

```
Command: SC (For SCALE)
Select objects: All
Select objects: (Press ENTER to continue)
Specify base point: (Pick a point near "A")
Specify scale factor or [Copy/Reference] <1.0000>: 0.40
```

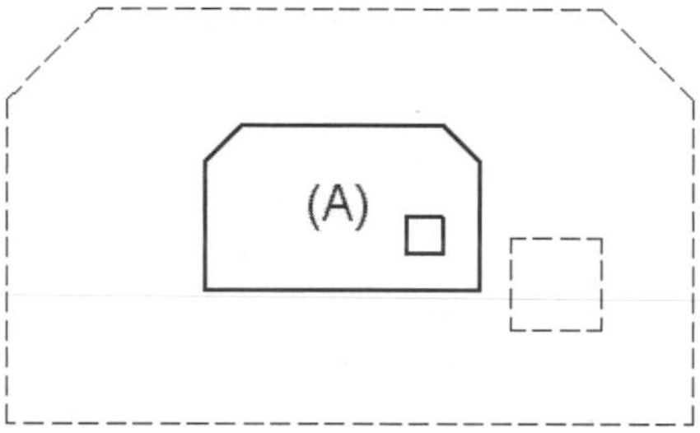

FIGURE 4.9

NOTE

After identifying the base point for the scaling operation, you can use the Copy option to create a scaled copy of the objects you are scaling.

SCALE—Reference

Suppose you are given a drawing that has been scaled down in size. However, no one knows what scale factor was used. You do know what one of the distances should be. In this special case, you can use the Reference option of the SCALE command to identify endpoints of a line segment that act as a reference length. Entering a new length value could increase or decrease the entire object proportionally.

TRY IT!

Open the drawing file 04_Scale Reference. Study Figure 4.10 and the following prompts for performing this operation.

```
Command: SC (For SCALE)
Select objects: (Pick a point at "A")
Specify opposite corner: (Pick a point at "B")
Select objects: (Press ENTER to continue)
Specify base point: (Select the edge of the circle to identify its center using a Center running OSNAP)
Specify scale factor or [Copy/Reference] <1.0000>: R (For Reference)
Specify reference length <1.0000>: (Select the endpoint of the line at "C")
Specify second point: (Select the endpoint of the line at "D")
Specify new length or [Points] <1.0000>: 2.00
```

Because the length of line "CD" is not known, the endpoints are picked after the Reference option is entered. This provides the length of the line to AutoCAD. The final step to perform is to make the line 2.00 units, which increases the size of the object while also keeping its proportions.

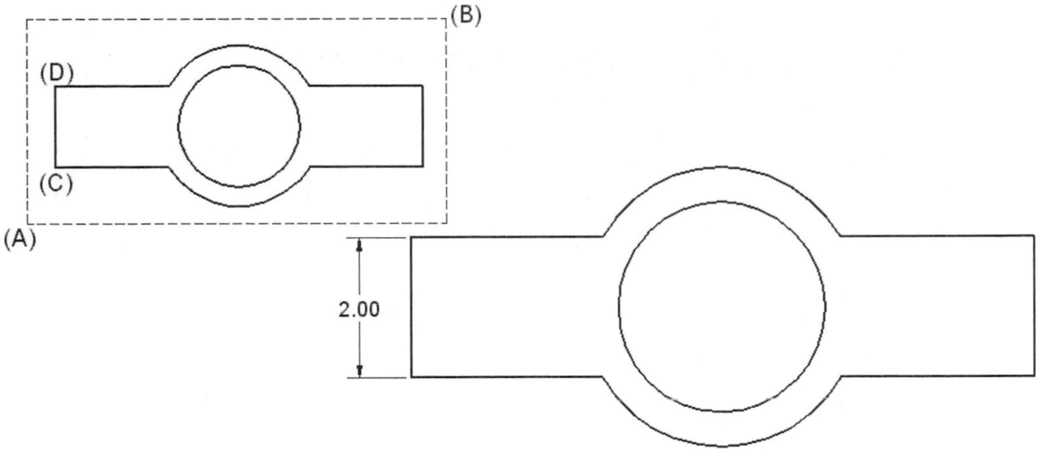

FIGURE 4.10

ROTATING OBJECTS

The ROTATE command changes the orientation of an object or group of objects by identifying a base point and a rotation angle that completes the new orientation. Choose this command from one of the following:

- From the Ribbon > Home Tab > Modify Panel
- The Menu Bar (Modify > Rotate)
- The Modify Toolbar of the AutoCAD Classic Workspace
- The keyboard (RO or ROTATE)
- By right-clicking the mouse after selecting object

Figure 4.11 shows an object, complete with crosshatch pattern, that needs to be rotated to a 30° angle using point "A" as the base point.

TRY IT!

Open the drawing file 04_Rotate. Use the following prompts and Figure 4.11 to perform the rotation.

```
Command: RO (For ROTATE)
Current positive angle in UCS: ANGDIR=counterclockwise
ANGBASE=0
Select objects: All
Select objects: (Press ENTER to continue)
Specify base point: (Select the endpoint of the line at "A")
Specify rotation angle or [Copy/Reference] <0>: 30
```

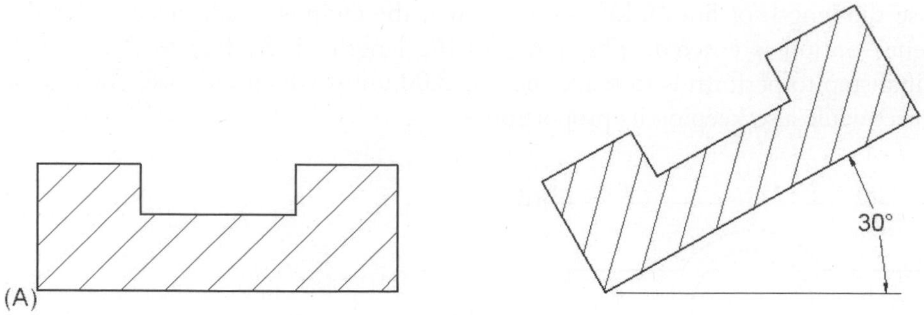

FIGURE 4.11

NOTE

After identifying the base point for the rotating operation, you can use the Copy option to create a rotated copy of the objects you are rotating.

ROTATE—Reference

At times it is necessary to rotate an object to a desired angular position. However, this must be accomplished even if the current angle of the object is unknown. To maintain the accuracy of the rotation operation, use the Reference option of the ROTATE command. Figure 4.12 shows an object that needs to be rotated to the 30°-angle position. Unfortunately, we do not know the angle in which the object currently lies. Entering the Reference angle option and identifying two points create a known angle of reference. Entering a new angle of 30° rotates the object to the 30° position from the reference angle.

TRY IT!

Open the drawing file 04_Rotate Reference. Use the following prompts and Figure 4.12 to accomplish this.

Command: RO *(For ROTATE)*

Current positive angle in UCS: ANGDIR=counterclockwise ANGBASE=0

Select objects: *(Select the object in the following image)*

Select objects: *(Press ENTER to continue)*

Specify base point: *(With a Center OSNAP pick the edge of the circle to locate the center)*

Specify rotation angle or [Copy/Reference] <0>: R *(For Reference)*

Specify the reference angle <0>: *(With a Center OSNAP pick the edge of the circle to locate the center)*

Specify second point: Mid

of *(Select the line at "B" to establish the reference angle)*

Specify the new angle or [Points] <0>: 30

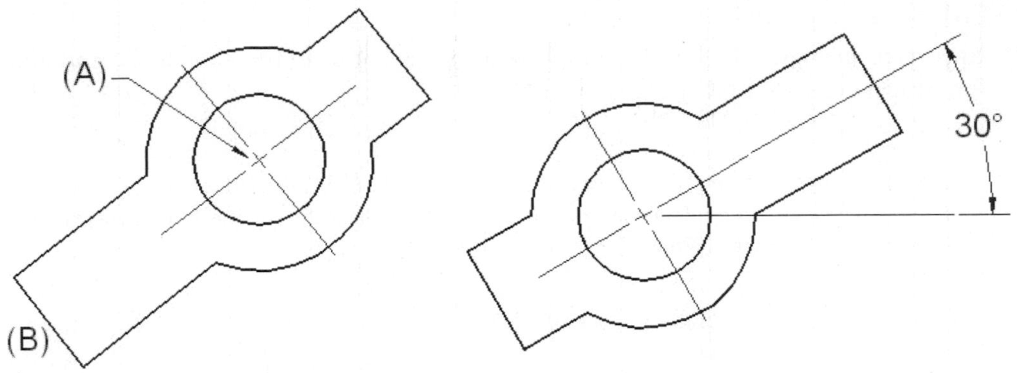

FIGURE 4.12

CREATING FILLETS AND ROUNDS

During some machining processes parts can be left with extremely sharp corners. These corners are then often filleted or rounded for either ornamental purposes or as required by design. Generally a fillet consists of a rounded edge formed at an inside corner of an object, as illustrated in Figure 4.13. A round is formed at an outside corner. Fillets and rounds are also often found on cast parts; the metal forms more easily around a shape that has rounded corners instead of sharp corners. Some drawings have so many fillets and rounds that a note is used to convey the size of them all, similar to "All Fillets and Rounds .125 Radius."

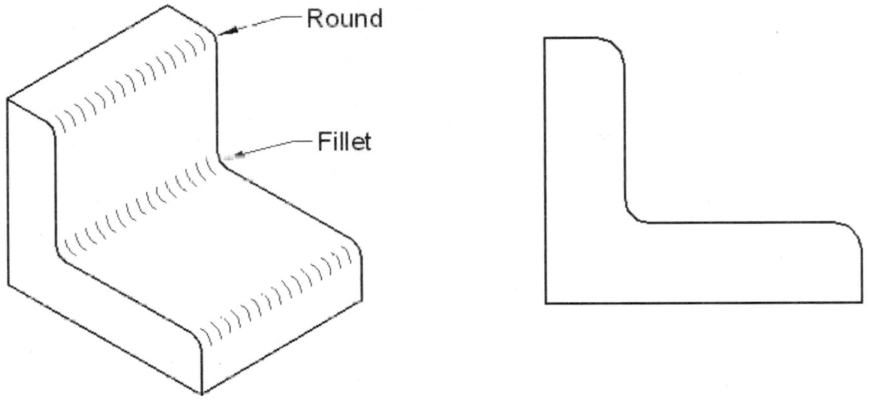

FIGURE 4.13

AutoCAD provides the FILLET command to create fillets or rounds. In the command you provide a radius and select two objects whose intersection should be curved. The result is a fillet of the specified radius at the intersection. The two objects are also automatically trimmed or extended, as necessary. Choose this command from one of the following:

- From the Ribbon > Home Tab > Modify Panel
- The Menu Bar (Modify > Fillet) (Expand icon: Fillet, Chamfer, and Blend)
- The Modify Toolbar of the AutoCAD Classic Workspace
- The keyboard (F or FILLET)

CERTIFIED ASSOCIATE EXAM 2013 | CERTIFIED PROFESSIONAL EXAM 2013

Filleting by Radius

Illustrated in Figure 4.14 is an example of setting a radius in the `FILLET` command for creating rounded-off corners.

TRY IT!

Open the drawing file 04_Fillet. Follow the illustration in Figure 4.14 (Left) and command sequence below to place fillets at the three corner locations. Turn off the Infer Constraints button on the Status Bar, if on.

Command: F *(For FILLET)*

Current settings: Mode = TRIM, Radius = 0.0000

Select first object or [Undo/Polyline/Radius/Trim/Multiple]: R *(For Radius)*

Specify fillet radius <0.0000>: **0.25**

Select first object or [Undo/Polyline/Radius/Trim/Multiple]: *(Select at "A")*

Select second object or shift-select to apply corner: *(Select at "B")*

Command:

(Press ENTER *to re-execute this command)*

Repeat this procedure for creating additional fillets using lines "BC" and "CD." When finished, your display should appear similar to the illustration in Figure 4.14 (Right).

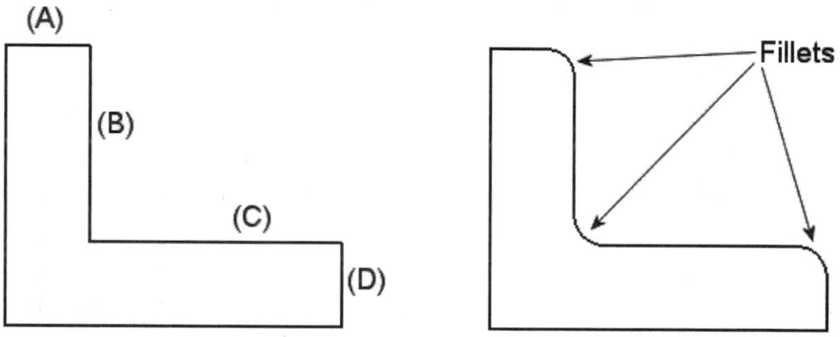

FIGURE 4.14

Fillet as a Cornering Tool

A very productive feature of the `FILLET` command is its use as a cornering tool. To accomplish this, set the fillet radius to a value of 0. This produces a corner out of two non-intersecting objects.

TRY IT!

Open the drawing file 04_Fillet Corner1. Follow the illustration in Figure 4.15 (Left) and the command sequence below for performing this task.

Command: F *(For FILLET)*

Current settings: Mode = TRIM, Radius = 0.5000

Select first object or [Undo/Polyline/Radius/Trim/Multiple]: R *(For Radius)*

```
Specify fillet radius <0.5000>: 0
Select first object or [Undo/Polyline/Radius/Trim/
Multiple]: (Select line "A")
Select second object or shift-select to apply corner: (Select
line "B")
```

Repeat the procedure for the remaining two corners using lines "BC" and "CD." When finished, your display should appear similar to the illustration in Figure 4.15 (Right).

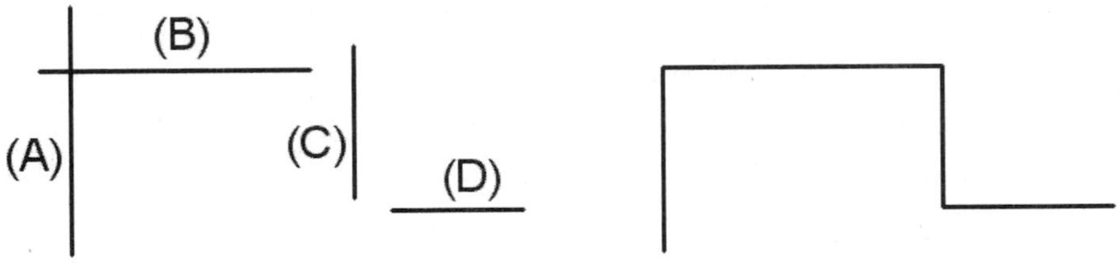

FIGURE 4.15

TIP

Even with a fillet radius set to a positive value, such as 0.50, you can easily apply a corner to two lines by holding down the SHIFT key when picking the second object. The use of the SHIFT key in this example temporarily sets the fillet radius to zero. Releasing the SHIFT key sets the fillet radius back to the current value, in this case 0.50.

Performing Multiple Fillets

Since the filleting of lines is performed numerous times, a Multiple option is available that automatically repeats the FILLET command. This option can be found by picking Multiple from the Cursor menu or by typing M for MULTIPLE at the Command prompt.

TRY IT!

Open the drawing file 04_Fillet Multiple. Activate the FILLET command and verify that the radius is set to 0. Use the Multiple option to make the command repeat. Then click on the corners until your object appears similar to the illustration in Figure 4.16 (Right). If you make a mistake by picking two lines incorrectly, type U to undo this operation and still remain in the FILLET command.

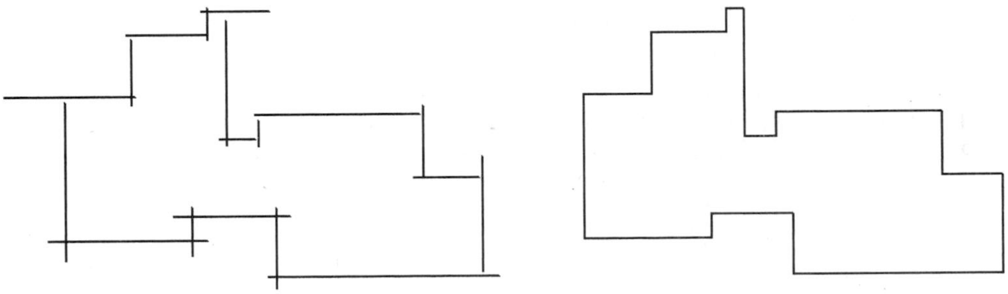

FIGURE 4.16

Filleting Polylines

In the previous examples of using the FILLET command, you had to pick individual line segments in order to produce one rounded corner. You also had to repeat these picks for additional rounded corners. If the object you are filleting is a polyline, you can have the FILLET command round off all corners of this polyline in a single pick.

TRY IT!

Open the drawing file 04_Fillet Pline. Using the FILLET command with the Polyline option produces rounded edges at all corners of the polyline in a single operation. Follow the illustration in Figure 4.17 and the command sequence below for performing this task.

```
Command: F (For FILLET)
Current settings: Mode = TRIM, Radius = 0.0000
Select first object or [Undo/Polyline/Radius/Trim/
Multiple]: R (For Radius)
Specify fillet radius <0.0000>: 0.25
Select first object or [Undo/Polyline/Radius/Trim/
Multiple]: P (For Polyline)
Select 2D polyline: (Select the polyline at "A")
```

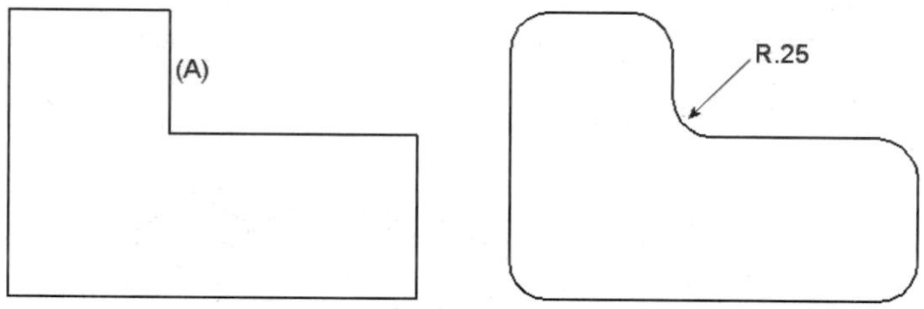

FIGURE 4.17

Filleting Parallel Lines

Filleting two parallel lines, as shown in Figure 4.18, automatically constructs a semicircular arc object connecting both lines at their endpoints. When performing this operation, it does not matter what the radius value is set to.

TRY IT!

Open the drawing file 04_Fillet Parallel. Use the illustration in Figure 4.18 (Left) and the command sequence below for performing this task.

```
Command: F (For FILLET)
Current settings: Mode = TRIM, Radius = 1.0000
Select first object or [Undo/Polyline/Radius/Trim/
Multiple]: (Select line "A")
Select second object or shift-select to apply corner:
(Select line "B")
```

Continue filleting the remaining parallel lines to complete all slots. Quicken the process by using the Multiple option when prompted to "Select first object." Also, notice that the bottom slot can be constructed differently depending on which line is selected first.

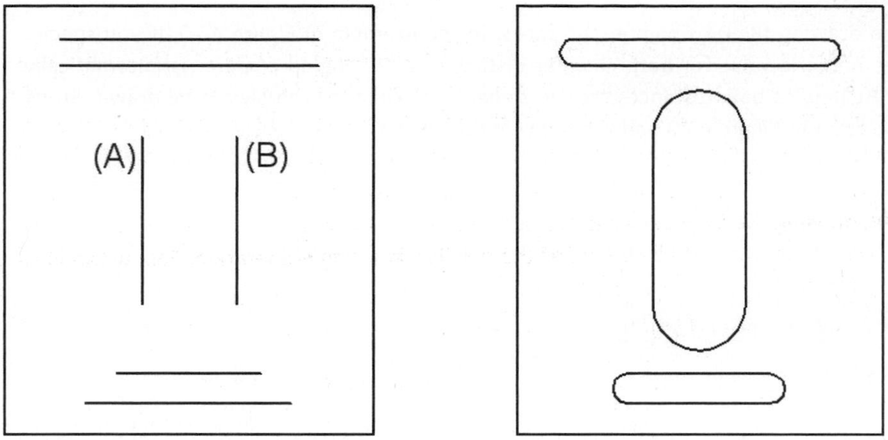

FIGURE 4.18

CREATING CHAMFERS

Besides Fillets, chamfers represent another way to finish the sharp corners of objects. The CHAMFER command produces an inclined surface at an edge of two intersecting line segments. Distances determine how far from the corner the chamfer is made. Figure 4.19 illustrates two examples of chamfered edges; one edge is created from unequal distances, while the other uses equal distances.

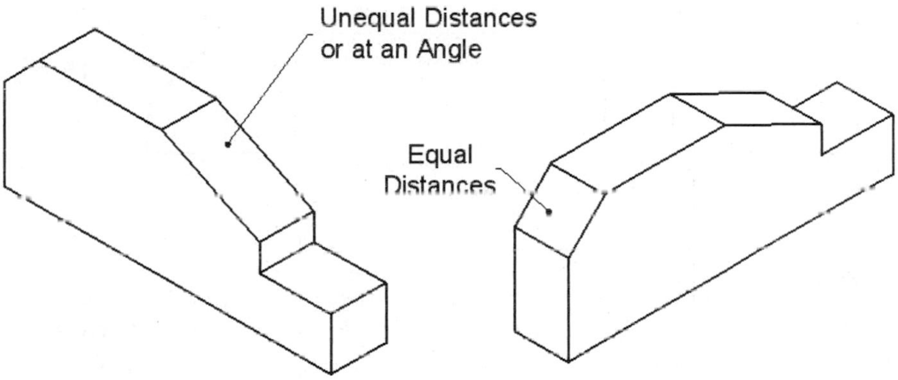

FIGURE 4.19

The CHAMFER command is designed to draw an angle across a sharp corner given two chamfer distances. Choose this command from one of the following:

- From the Ribbon > Home Tab > Modify Panel (Expand icon: Fillet, Chamfer and Blend)
- The Menu Bar (Modify > Chamfer)
- The Modify Toolbar of the AutoCAD Classic Workspace
- The keyboard (CHA or CHAMFER)

Chamfer by Equal Distances

The most popular chamfer involves a 45° angle, which is illustrated in Figure 4.20. You can control this angle by entering two equal distances.

TRY IT!

Open the drawing file 04_Chamfer Distances. In the example in Figure 4.20, if you specify the same numeric value for both chamfer distances, a 45°-angled chamfer is automatically formed. As long as both distances are the same, a 45° chamfer will always be drawn. Study the illustration in Figure 4.20 and the following prompts:

Command: CHA *(For CHAMFER)*

(TRIM mode) Current chamfer Dist1 = 0.5000, Dist2 = 0.5000

Select first line or [Undo/Polyline/Distance/Angle/Trim/mEthod/Multiple]: D *(For Distance)*

Specify first chamfer distance <0.5000>: **0.15**

Specify second chamfer distance <0.1500>: *(Press* ENTER *to accept the default)*

Select first line or [Undo/Polyline/Distance/Angle/Trim/mEthod/Multiple]: *(Select the line at "A")*

Select second line or shift-select to apply corner: *(Select the line at "B")*

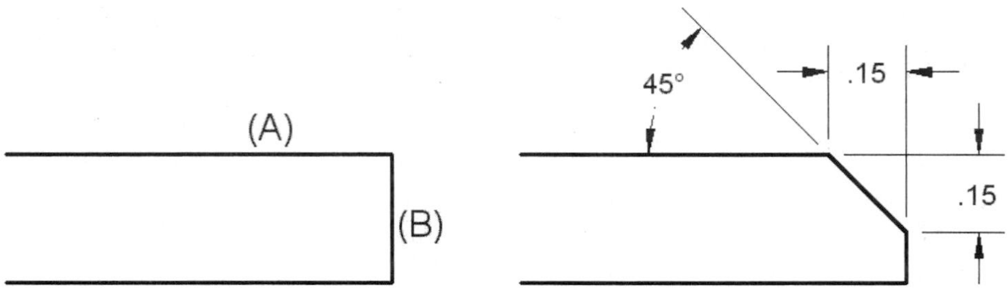

FIGURE 4.20

NOTE

When both chamfer distances are set to 0 (zero) and two edges are selected, the effects are identical to setting the fillet radius to 0 (zero); the CHAMFER **command is used here as a cornering tool.**

Chamfer by Angle

Another technique of constructing a chamfer is when one distance and the angle are given. When the chamfer is made up of an angle other than 45°, it is commonly referred to as a beveled edge.

TRY IT!

Open the drawing file 04_Chamfer Angle. Figure 4.21 illustrates the use of the CHAMFER **command by setting one distance and identifying an angle.**

Command: CHA *(For CHAMFER)*

(TRIM mode) Current chamfer Dist1 = 0.5000, Dist2 = 0.5000

Select first line or [Undo/Polyline/Distance/Angle/Trim/mEthod/Multiple]: A *(For Angle)*

Specify chamfer length on the first line <0.1500>: **0.15**

Specify chamfer angle from the first line <60>: 60

Select first line or [Undo/Polyline/Distance/Angle/Trim/mEthod/Multiple]: *(Select the line at "A"—the angle is measured from the first line selected)*

Select second line or shift-select to apply corner: *(Select the line at "B")*

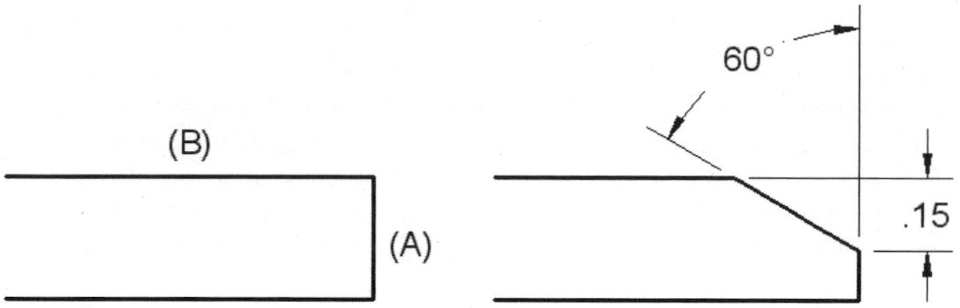

FIGURE 4.21

Chamfering a Polyline

When working with polyline objects, you have the opportunity to select only one of the edges of the polyline. All edges of the polyline will be chamfered to the specified distances or angle.

TRY IT!

Open the drawing file 04_Chamfer Pline. Because a polyline consists of numerous segments representing a single object, using the CHAMFER **command with the Polyline option produces corners throughout the entire polyline, as shown in Figure 4.22.**

Command: CHA *(For CHAMFER)*

(TRIM mode) Current chamfer Dist1 = 0.00, Dist2 = 0.00

Select first line or [Undo/Polyline/Distance/Angle/Trim/mEthod/Multiple]: D *(For Distance)*

Specify first chamfer distance <0.00>: 0.50

Specify second chamfer distance <0.50>: *(Press* ENTER *to accept the default)*

Select first line or [Undo/Polyline/Distance/Angle/Trim/mEthod/Multiple]: P *(For Polyline)*

Select 2D Polyline: *(Select the Polyline at "A")*

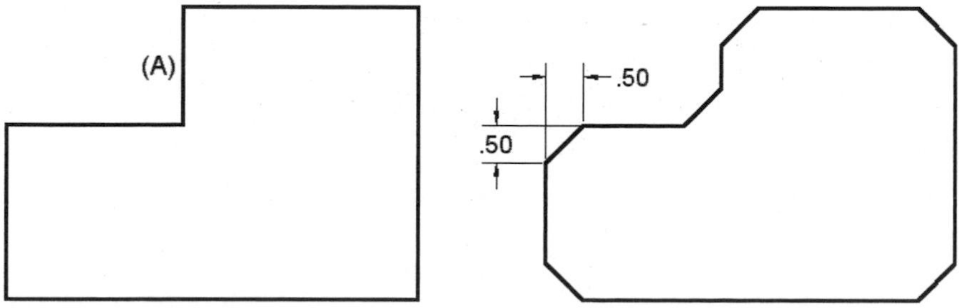

FIGURE 4.22

NOTE

A Multiple option of the `CHAMFER` command allows you to chamfer edges that share the same chamfer distances without exiting and reentering the command.

Chamfer Project—Beam

The following Try It! exercise involves the chamfering of various corners at different distance settings.

TRY IT!

Open the drawing file 04_Chamfer Beam. Using the illustration provided in Figure 4.23, follow these directions: Apply equal chamfer distances of 0.25 units to corners "AB," "BC," "DE," and "EF." Set new equal chamfer distances to 0.50 units and apply these distances to corners "GH" and "JK." Set a new first chamfer distance to 1.00; set a second chamfer distance to 0.50 units. Apply the first chamfer distance to line "L" and the second chamfer distance to line "H." Complete this object by applying the first chamfer distance to line "M" and the second chamfer distance to line "K."

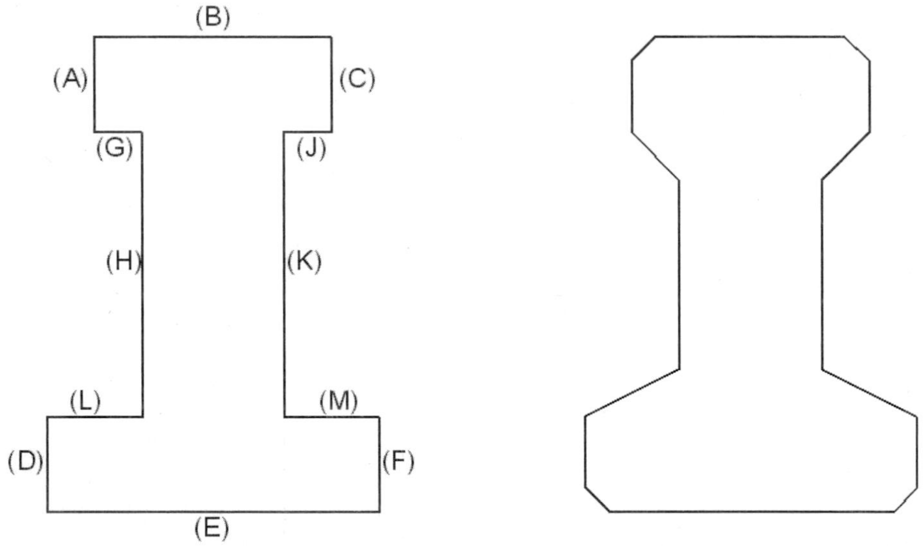

FIGURE 4.23

OFFSETTING OBJECTS

The `OFFSET` command is commonly used for creating a copy of one object that is parallel to another. Choose this command from one of the following:

- From the Ribbon > Home Tab > Modify Panel
- The Menu Bar (Modify > Offset)
- The Modify Toolbar of the AutoCAD Classic Workspace
- The keyboard (`O` or `OFFSET`)

Offsetting Using a Through Point

One method of offsetting is to identify a point to offset through, called a through point. Once an object is selected to offset, a through point is identified. The selected object offsets to that point, as shown in Figure 4.24.

Open the drawing file 04_Offset Through. Refer to Figure 4.24 and command sequence to use this method of the `OFFSET` command.

TRY IT!

Command: **O** *(For OFFSET)*

Current settings: Erase source=No Layer=Source OFFSETGAPTYPE=0

Specify offset distance or [Through/Erase/Layer] <Through>: **T** *(For Through)*

Select object to offset or [Exit/Undo] <Exit>: *(Select the line at "A")*

Specify through point or [Exit/Multiple/Undo] <Exit>: **Nod** *(For OSNAP Node)*

of *(Select the point at "B")*

Select object to offset or [Exit/Undo] <Exit>: *(Press* ENTER *to exit this command)*

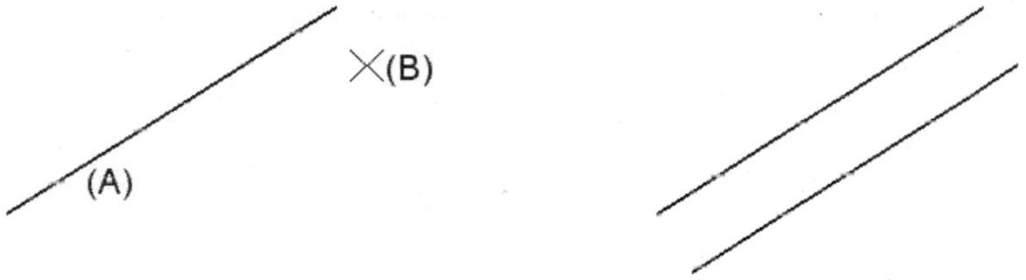

FIGURE 4.24

Offsetting by a Distance

Another method of offsetting is by a specified offset distance, as shown in Figure 4.25, where the objects need to be duplicated at a set distance from existing geometry. Although the `COPY` command could be used for this operation, the OFFSET command is more efficient. This command allows you to specify a distance and a side for the offset to occur. The result is an object parallel to the original object at a specified distance. All objects in Figure 4.25 need to be offset 0.50 units toward the inside of the original object.

Open the drawing file 04_Offset Shape. See the command sequence and Figure 4.25 to perform this operation.

TRY IT!

Command: **O** *(For OFFSET)*

Current settings: Erase source=No Layer=Source OFFSETGAPTYPE=0

Specify offset distance or [Through/Erase/Layer] <Through>: **0.50**

Select object to offset or [Exit/Undo] <Exit>: *(Select the horizontal line at "A"—left image)*

Specify point on side to offset or [Exit/Multiple/Undo] <Exit>: *(Pick a point anywhere on the inside near "B"—left image)*

Repeat the preceding procedure for the remaining lines by offsetting them inside the shape.

Notice that when all lines were offset, the original lengths of all line segments were maintained. Because all offsetting occurs inside, the segments overlap at their intersection points, as shown in the middle of Figure 4.25. In one case, at "A" and "B," the lines did not meet at all. The FILLET command is used to edit all lines to form a sharp corner. You can accomplish this by setting the fillet radius to 0.

```
Command: F (For FILLET)
Current settings: Mode = TRIM, Radius = 0.5000
Select first object or [Undo/Polyline/Radius/Trim/
Multiple]: R (For Radius)
Specify fillet radius <0.5000>: 0
Select first object or [Undo/Polyline/Radius/Trim/
Multiple]: (Select line "A"—middle image)
Select second object: (Select line "B"—middle image)
```

Repeat the above procedure for the remaining lines. Instead of having to restart the command each time, try using the Multiple option.

Using the OFFSET command along with the FILLET command produces the result shown in Figure 4.25 (Right). The fillet radius must be set to a value of 0 for this special effect.

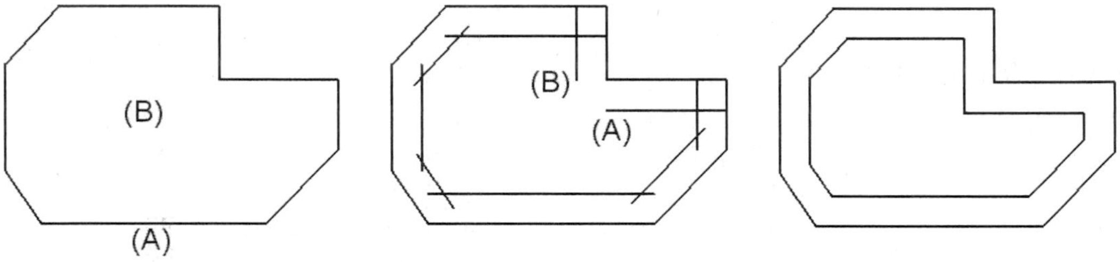

FIGURE 4.25

Performing Multiple Offsets

If you know ahead of time that you will be offsetting the same object the same preset distance, you can use the Multiple option of the OFFSET command to work more efficiently.

TRY IT!

Open the drawing file 04_Offset Multiple. See the command sequence and Figure 4.26 to perform this operation.

```
Command: O (For OFFSET)
Current settings: Erase source=No Layer=Source
OFFSETGAPTYPE=0
Specify offset distance or [Through/Erase/Layer] <Through>:
0.40
```

Select object to offset or [Exit/Undo] <Exit>: *(Pick vertical line "A")*

Specify point on side to offset or [Exit/Multiple/Undo] <Exit>: M *(For Multiple)*

Specify point on side to offset or [Exit/Undo] <next object>: *(Pick at "B")*

Specify point on side to offset or [Exit/Undo] <next object>: *(Pick at "B")*

Specify point on side to offset or [Exit/Undo] <next object>: *(Pick at "B")*

Specify point on side to offset or [Exit/Undo] <next object>: *(Press* ENTER*)*

Select object to offset or [Exit/Undo] <Exit>: *(Press* ENTER *to exit)*

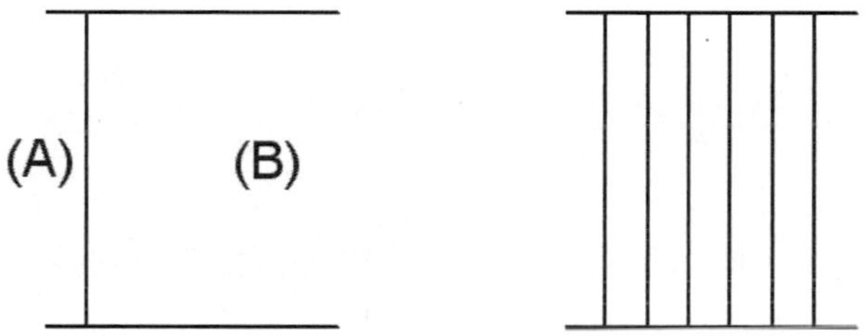

FIGURE 4.26

Other Offset Options

Other options of the OFFSET command include Erase and Layer. When the Erase option is used, the original object you select to offset is erased after the offset copy is made. When using the Layer option, the object being offset can take on the layer properties of the source object or can be based on the current layer. When using the source (the default), the offset copy takes on the same layer as the source object you pick with offsetting. Using the Current Layer mode when offsetting changes all offset copies to the current layer. These extra offset modes allow you more flexibility with using the OFFSET command.

TRIMMING OBJECTS

Use the TRIM command to partially delete an object or a group of objects based on a cutting edge. Choose this command from one of the following:

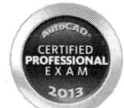

- From the Ribbon > Home Tab > Modify Panel (Expand Icon: Trim and Extend)
- The Menu Bar (Modify > Trim)
- The Modify Toolbar of the AutoCAD Classic Workspace
- The keyboard (TR or TRIM)

Selecting Individual Cutting Edges

As illustrated in Figure 4.27 (Left), the four dashed lines are selected as cutting edges. Next, segments of the circles are selected to be trimmed between the cutting edges.

TRY IT!

Open the drawing file 04_Trim Basics. Use Figure 4.27 and the command sequence below to perform this task.

Command: TR *(For TRIM)*

Current settings: Projection=UCS Edge=None

Select cutting edges ...

Select objects or <select all>: *(Select the four dashed lines in the following image on the left)*

Select objects: *(Press* ENTER *to continue)*

Select object to trim or shift-select to extend or [Fence/Crossing/Project/Edge/eRase/Undo]: *(Select the circle areas at "A" through "D")*

Select object to trim or shift-select to extend or [Fence/Crossing/Project/Edge/eRase/Undo]: *(Press* ENTER *to exit this command)*

The results of performing trim on this object are illustrated in Figure 4.27 (Right).

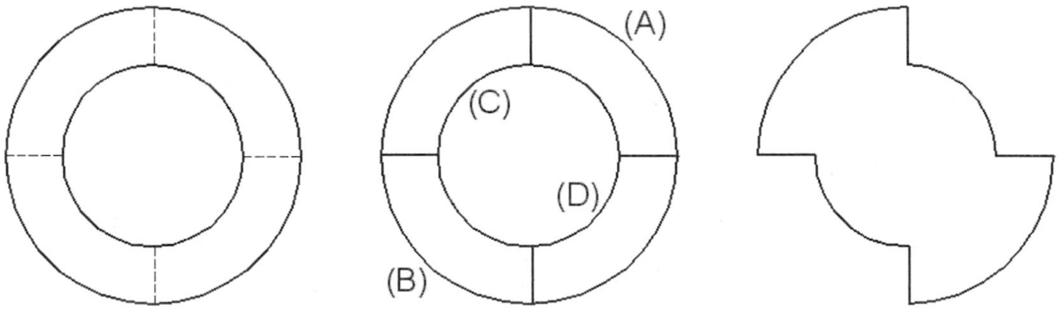

FIGURE 4.27

Selecting All Objects as Cutting Edges

An alternate method of selecting cutting edges is to press ENTER in response to the prompt "Select objects." This automatically creates cutting edges out of all objects in the drawing. When you use this method, the cutting edges do not highlight. This can be a very efficient means of trimming out unnecessary objects. You should, however, examine what you are trimming before using this method; a particularly busy drawing could result in numerous small trim operations.

TRY IT!

Open the drawing file 04_Trim All. Enter the TRIM command; press ENTER at the Select Objects prompt to select all objects as cutting edges. In Figure 4.28, pick the lines at "A," "B," and "E," and the arc segments at "C" and "D" as the objects to trim.

Command: TR *(For TRIM)*

Current settings: Projection=UCS Edge=None

Select cutting edges ...

Select objects or <select all>: *(Press* ENTER *to select all objects as cutting edges)*

Select object to trim or shift-select to extend or [Fence/Crossing/Project/Edge/eRase/Undo]: *(Select segments "A" through "E")*

Select object to trim or shift-select to extend or [Fence/Crossing/Project/Edge/eRase/Undo]: *(Press* ENTER *to exit this command)*

The results of performing trim on this object are illustrated in Figure 4.28 (Right).

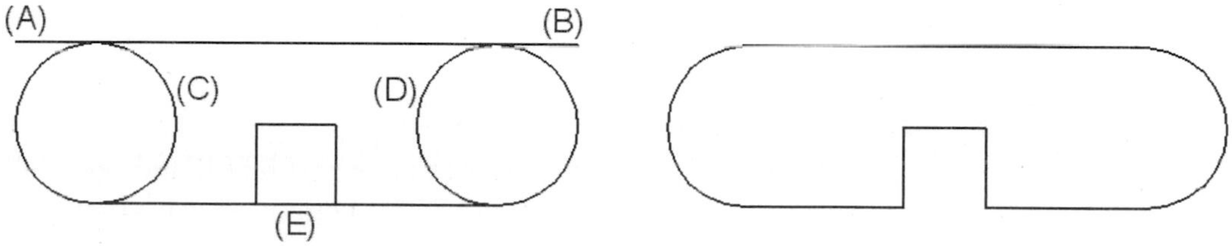

FIGURE 4.28

Trimming by a Crossing Box

When trimming objects, you do not have to pick objects individually. You can erect a crossing box and trim objects out more efficiently.

TRY IT!

Open the drawing file 04_Trim Crossing. Yet another application of the TRIM **command uses the Crossing option of "Select objects." First, invoke the** TRIM **command and select the small circle as the cutting edge. Begin the response to the prompt of "Select object to trim" by clicking a blank part of your screen; this automatically activates Crossing mode. See the illustration in the middle of Figure 4.29.**

Command: TR *(For TRIM)*

Current settings: Projection=UCS Edge=None

Select cutting edges ...

Select objects or <select all>: *(Select the small circle, as shown in the following image on the left)*

Select objects: *(Press* ENTER *to continue)*

Select object to trim or shift-select to extend or [Fence/Crossing/Project/Edge/eRase/Undo]: *(Pick a corner to establish a crossing box, as shown in the middle image)*

Specify opposite corner: *(Pick an opposite corner to complete the crossing box, as shown in the middle image)*

Select object to trim or shift-select to extend or [Fence/Crossing/Project/Edge/eRase/Undo]: *(Press* ENTER *to exit this command)*

The power of the Crossing option of "Select objects" is shown in the following image on the right. Eliminating the need to select each individual line segment inside the small circle to trim, the Crossing mode trims all objects it touches in relation to the cutting edge.

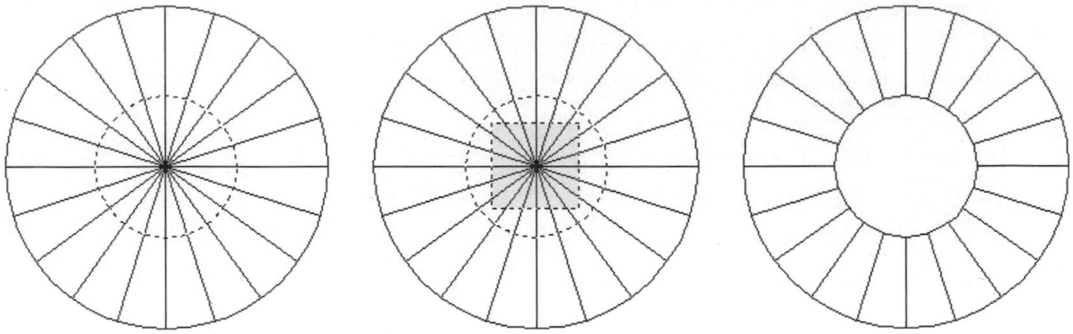

FIGURE 4.29

More about Individual Cutting Edges

Care must be taken to decide when it is appropriate to press ENTER and select all objects in your drawing as cutting edges using the TRIM command. To see this in effect, try the next exercise.

TRY IT!

Open the drawing file 04_Trim Cut. You need to remove the six vertical lines from the inside of the object. However, if you press ENTER to select all cutting edges, each individual segment would need to be trimmed, which would be unproductive. Select lines "A" and "B" as cutting edges in Figure 4.30 and select the inner vertical lines as the objects to trim.

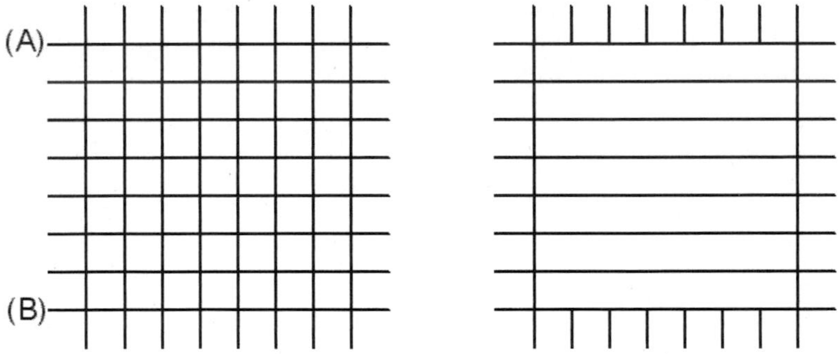

FIGURE 4.30

Trimming Exercise—Floor Plan

Use any technique that you have learned to trim the walls of the floor plan in order for the object to appear similar to the illustration in Figure 4.31.

TRY IT!

Open the drawing file 04_Trim Walls. Using Figure 4.31 as a guide, use the TRIM command to trim away the extra overshoots and complete the floor plan illustrated in the figure.

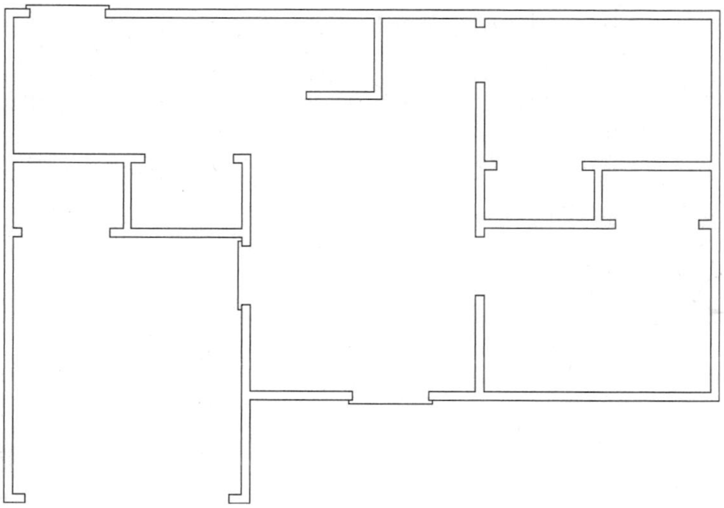

FIGURE 4.31

TIP

While inside the TRIM command, you can easily toggle to the EXTEND command by holding down the SHIFT key at the following Command prompt:

```
Select object to trim or shift-select to extend or [Fence/
Crossing/Project/Edge/eRase/Undo]:
```

EXTENDING OBJECTS

The EXTEND command is used to extend objects to a specified boundary edge. Choose this command from one of the following:

- From the Ribbon > Home Tab > Modify Panel (Expand Icon: Trim and Extend)
- The Menu Bar (Modify > Extend)
- The Modify Toolbar of the AutoCAD Classic Workspace
- The keyboard (EX or EXTEND)

Selecting Individual Boundary Edges

In Figure 4.32, select all dashed objects as the boundary edges. After pressing ENTER to continue with the command, select the lines at "A," "B," "C," and "D" to extend these objects to the boundary edges. If you select the wrong end of an object, use the Undo option to undo the change and repeat the procedure at the correct end of the object.

TRY IT!

Open the drawing file 04_Extend Basics. Use Figure 4.32 and command sequence for accomplishing this task.

Command: EX *(For EXTEND)*

Current settings: Projection=UCS Edge=None

Select boundary edges ...

Select objects or <select all>: *(Select the objects represented by dashes)*

Select objects: *(Press* ENTER *to continue)*

Select object to extend or shift-select to trim or [Fence/Crossing/Project/Edge/Undo]: *(Select the ends of the lines at "A" through "D")*

Select object to extend or shift-select to trim or [Fence/Crossing/Project/Edge//Undo]: *(Press* ENTER *to exit this command)*

TIP

In a similar fashion to the TRIM command, pressing ENTER instead of selecting individual boundary edges automatically creates boundary edges out of all objects in the drawing. When you use this method, the boundary edges do not highlight.

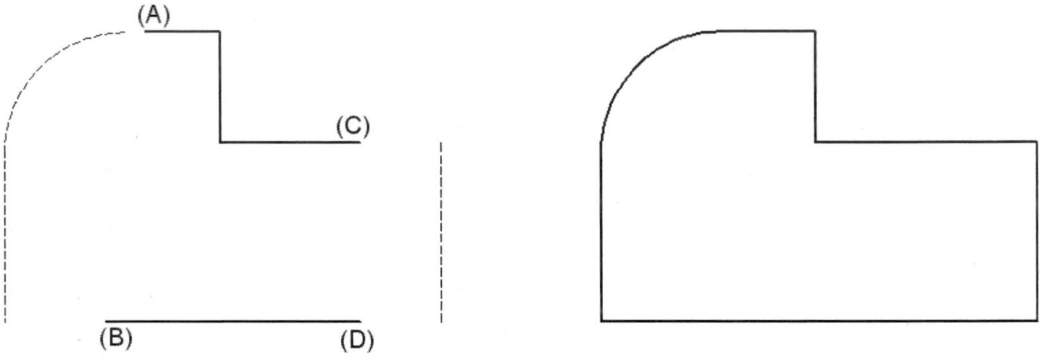

FIGURE 4.32

Extending by a Crossing Box

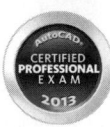

The EXTEND command can be used for extending multiple objects. After identifying the boundary edges, you can use crossing boxes to identify numerous items to extend. This is a very productive method of using this command.

TRY IT!

Open the drawing file 04_Extend Crossing. To extend multiple objects such as the line segments shown in Figure 4.33, you will select the lines at "A" and "B" as the boundary edge and use the Crossing mode to create two crossing boxes, represented by the dashed rectangles. This extends all line segments to intersect with the boundaries.

Command: EX *(For EXTEND)*

Current settings: Projection=UCS Edge=None

Select boundary edges ...

Select objects or <select all>: *(Select the lines at "A" and "B")*

Select objects: *(Press* ENTER *to continue)*

Select object to extend or shift-select to trim or [Fence/Crossing/Project/Edge/Undo]: *(Click a blank area to establish a crossing box, as you did earlier with the TRIM command)*

Specify opposite corner: *(Pick a second point to complete the first crossing box)*

Select object to extend or shift-select to trim or [Fence/Crossing/Project/Edge/Undo]: *(Click a blank area to start the second crossing box)*

Specify opposite corner: *(Pick a second point to complete the second crossing box)*

Select object to extend or shift-select to trim or [Fence/Crossing/Project/Edge/Undo]: *(Press* ENTER *to exit this command)*

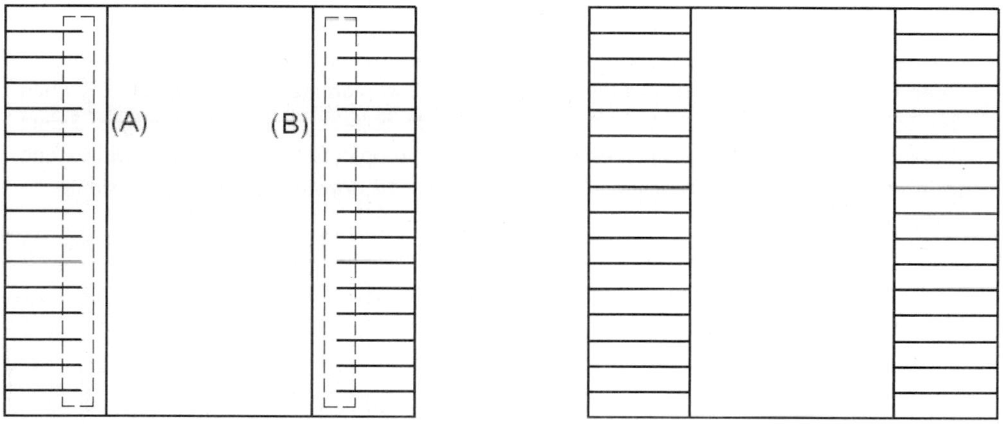

FIGURE 4.33

Toggling from Extend to Trim

While inside the EXTEND command, you can easily toggle to the TRIM command by holding down the SHIFT key at the following Command prompt:

Select object to extend or shift-select to trim or [Fence/Crossing/Project/Edge/Undo]: *(Pressing* SHIFT *while picking objects activates the TRIM command.)*

The next Try It! exercise illustrates this technique.

TRY IT!

Open the drawing file 04_Extend and Trim. First, activate the EXTEND command and press ENTER, which selects all edges of the object in Figure 4.34 as boundary edges. When picking the edges to extend, click on the ends of the lines from "A" through "J," as shown in Figure 4.34 (Left). At this point, do not exit the command. Press and hold down the SHIFT key; this activates the TRIM command. Now pick all of the ends of the lines until your shape appears like the illustration in Figure 4.34 (Right). The use of the SHIFT key when trimming or extending provides a quick means of switching between commands.

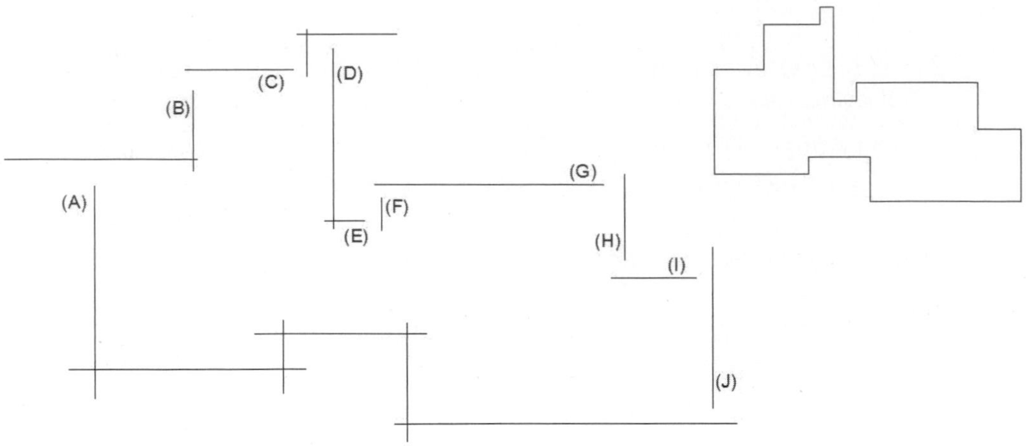

FIGURE 4.34

Extend Exercise—Piping Diagram

TRY IT!

Open the drawing file 04_Extend Pipe. Enter the EXTEND command and press ENTER when the Select objects prompt appears. This selects all objects as boundary edges. Select the ends of all magenta lines representing pipes as the objects to extend. They will extend to intersect with the adjacent pipe fitting. Your finished drawing should appear similar to Figure 4.35.

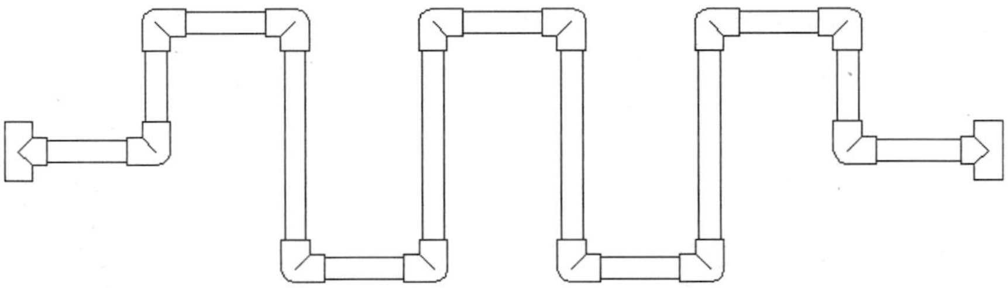

FIGURE 4.35

BREAKING OBJECTS

AutoCAD CERTIFIED ASSOCIATE EXAM 2013 · AutoCAD CERTIFIED PROFESSIONAL EXAM 2013

The BREAK command is often used to remove a segment from an object, but may in some cases be used to split an object without leaving any gap. Choose this command from one of the following:

- From the Ribbon > Home Tab > Modify Panel (Expanded)
- The Menu Bar (Modify > Break)
- The Modify Toolbar of the AutoCAD Classic Workspace
- The keyboard (BR or BREAK)

Breaking an Object

The following command sequence and Figure 4.36 show how the BREAK command is used.

Open the drawing file 04_Break Gap. Turn off Running OSNAP prior to conducting this exercise. Use the following prompts and illustrations to break the line segment.

TRY IT!

Command: BR *(For BREAK)*

Select objects: *(Select the line at "A")*

Specify second break point or [First point]: *(Pick the line at "B")*

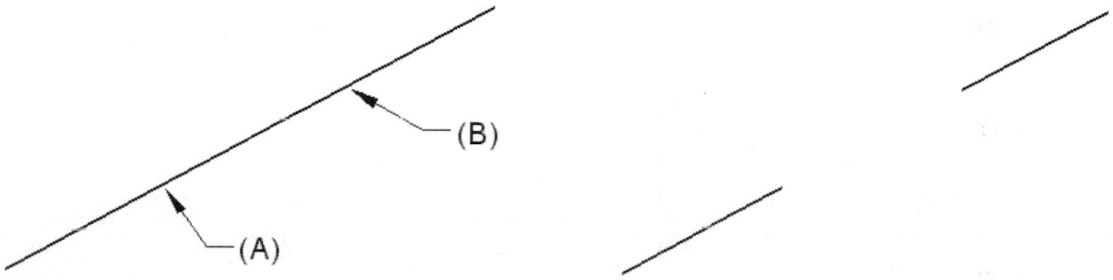

FIGURE 4.36

Identifying a New First Break Point

In the previous example of using the BREAK command, the location where the object was selected became the first break point. You can select the object and then be prompted to pick a new first break point. This option resets the command and allows you to select an object to break followed by two different points that identify the break. The following exercise illustrates this technique.

Open the drawing file 04_Break First. Utilize the First option of the BREAK command along with OSNAP options to select key objects to break. The following command sequence and Figure 4.37 demonstrate using the First option of the BREAK command:

TRY IT!

Command: BR *(For BREAK)*

Select object: *(Select the line)*

Specify second break point or [First point]: F *(For First)*

Specify first break point: Int

of *(Pick the intersection of the two lines at "A")*

Specify second break point: End

of *(Pick the endpoint of the line at "B")*

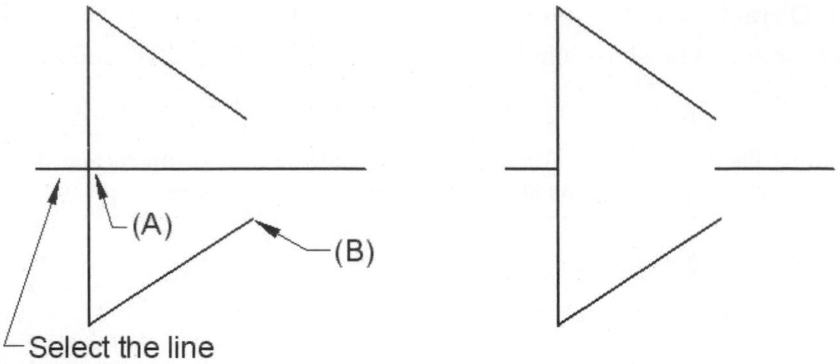

FIGURE 4.37

Breaking Circles

The BREAK command can also be used to break circles into arc segments. There is only one rule to follow when breaking circles: You must pick the two break points in a counterclockwise direction when identifying the endpoints of the segment to be removed.

TRY IT!

Open the drawing file 04_Break Circle. Study the following command sequence and Figure 4.38 for breaking circles. You might want to use a Node OSNAP, although it is not necessary to pick on the object if accuracy is not required.

```
Command: BR (For BREAK)
Select objects: (Select the circle)
Specify second break point or [First point]: F (For First)
Specify first break point: (Pick at "First Point")
Specify second break point: (Pick at "Second Point")
```

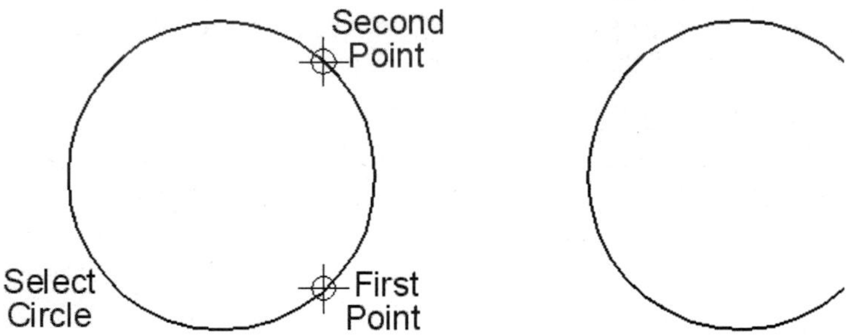

FIGURE 4.38

Break at Point

You can also break an object at a selected point using the Break tool. This breaks an object into two separate objects without leaving a gap. As illustrated on the left in Figure 4.39, the line is highlighted to prove that it consists of one continuous object.

Clicking on the Break tool in the Modify Toolbar activates the following command sequence:

```
Command: BR (For BREAK)
Select object: (Select the line anywhere)
Specify second break point or [First point]: F (For First
point)
Specify first break point: Mid
of (Pick the midpoint of the line)
Specify second break point: @ (For previous point)
```

The results are illustrated in Figure 4.39 (Right). Here the line is again selected. Notice that only half of the line selects because the line is broken at its midpoint.

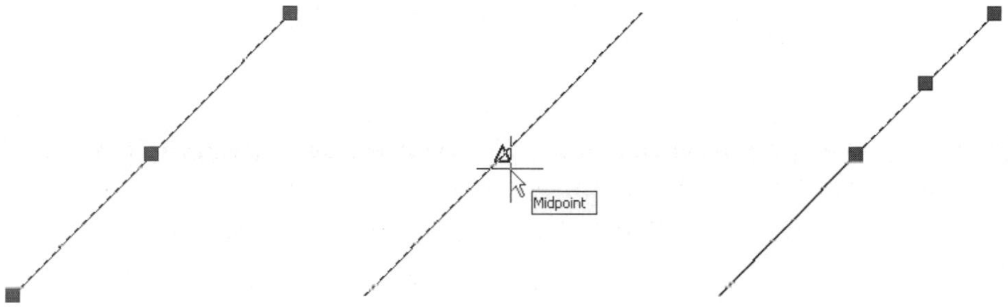

FIGURE 4.39

> **NOTE**
>
> The Break At Point tool automates the command sequence listed above. For this tool to operate correctly you may have to turn off running OSNAP.

LEVEL II MODIFY COMMANDS

This second grouping of modify commands is designed to perform more advanced editing operations compared with the Level I modify commands already discussed in this chapter. These commands are briefly described in the following table:

Button	Tool	Key-In	Function
	Array Rect	AR	Creates multiple copies of objects in a rectangular pattern
	Array Polar	AR	Creates multiple copies of objects in a circular (polar) pattern
	Array Path	AR	Creates multiple copies of objects along a specified path
	Mirror	MI	Creates a mirror image of objects based on an axis of symmetry
	Stretch	S	Used for moving or stretching the shape of an object
	Pedit	PE	Used for editing polylines
	Explode	X	Breaks a compound object such as a polyline, block, or dimension into individual objects
	Lengthen	LEN	Changes the length of lines and arcs

(Continued)

Button	Tool	Key-In	Function
	Join	JO	Joins collinear objects to form a single unbroken object
	Blend	BL	Joins lines or curves together with a connecting spline
	Undo	U	Used for backtracking or reversing the action of the previously used command
	Redo	REDO	Reverses the effects of the previously used UNDO command operation

CREATING ARRAYS

If you need to create copies of objects that form patterns that are either rectangular, circular, or along some specified path, the ARRAY command is available to help with this task. This is a very powerful command that creates associative patterns of objects that can be easily modified. If performing a rectangular array, you will need to supply the number of rows and columns for the pattern in addition to the spacing between these rows and columns. When performing a circular or polar array, you need to supply the center point of the array, the number of items to copy, and the angle to fill. For performing a path array, you select the path curve, the number of items, and the distance between objects. The next series of pages documents methods of performing arrays. Choose this command from one of the following:

- From the Ribbon > Home Tab > Modify Panel
- The Menu Bar (Modify > Array)
- The Modify Toolbar of the AutoCAD Classic Workspace
- The keyboard (AR or ARRAY) (ARRAYRECT, ARRAYPOLAR, and ARRAYPATH)

Once the ARRAY command is executed, you will be immediately prompted to select the type of pattern to perform, as shown in the following command sequence. When selecting an array operation from the Ribbon, Menu bar, or Modify toolbar, you will activate one of the following commands: ARRAYRECT, ARRAYPOLAR, or ARRAYPATH. Clicking one of the three available icons will automatically start the type of array specified. Figure 4.40 shows the icons in the Ribbon for initiating one of the three array types available.

```
Command: AR (For ARRAY)
Select objects: (Select the objects to copy)
Select objects: (Press ENTER to continue)
Enter array type [Rectangular/PAth/POlar] <Rectangular>:
```

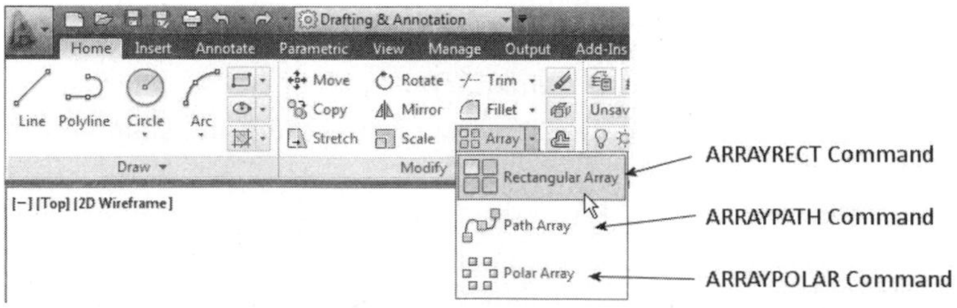

FIGURE 4.40

CREATING RECTANGULAR ARRAYS

Creating Rectangular Patterns with Positive Offset

The ARRAY or ARRAYRECT command allows you to arrange multiple copies of an object or group of objects in a rectangular pattern. When creating a rectangular array, you are prompted to enter the number of rows and columns for the array. A row is a group of objects that are copied vertically in the positive or negative direction. A column is a group of objects that are copied horizontally, also in the positive or negative direction.

TRY IT!

Open the drawing file 04_Array Rectangular Positive. Suppose the object illustrated in Figure 4.41 (Left) needs to be copied in a rectangular pattern consisting of two rows and three columns. The spacing required for this command is from edge to same edge or center to center, not the actual spacing between the objects. The row spacing is 1.00 not 0.50 units and the column spacing would be 2.00 not 1.25. The result is illustrated in Figure 4.41.

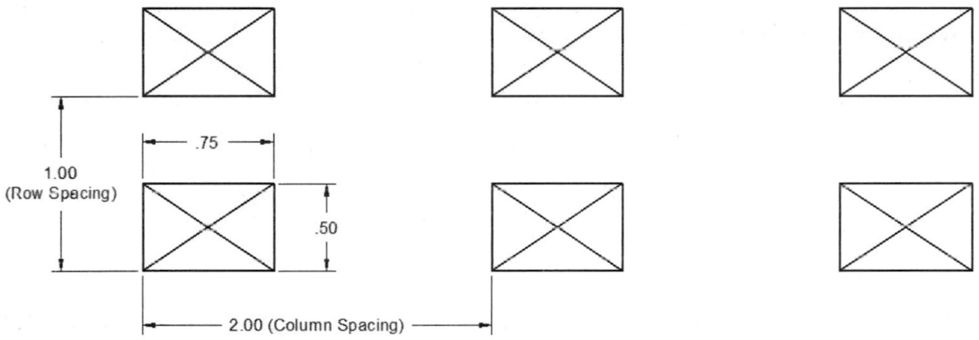

FIGURE 4.41

Clicking Rectangular Array in the Modify panel of the Ribbon displays the following command sequence.

Command: ARRAYRECT

Select objects: *(Select the rectangle and 2 crossing lines)*

Select objects: *(Press* ENTER *to continue, 4 columns and 3 rows are displayed)*

Type = Rectangular Associative = Yes

Select grip to edit array or [ASsociative/Base point/COUnt/Spacing/COLumns/Rows/Levels/eXit] <eXit>: COU *(For COUnt)*

Enter the number of columns or [Expression] <4>: 3

Enter the number of rows or [Expression] <3>: 2

Select grip to edit array or [ASsociative/Base point/COUnt/Spacing/COLumns/Rows/Levels/eXit] <eXit>: S *(For Spacing)*

Specify the distance between columns or [Unit cell] <1.13>: **2.00**

Specify the distance between rows <0.75>: 1.00

Select grip to edit array or [ASsociative/Base point/COUnt/Spacing/Columns/Rows/Levels/eXit]<eXit>: *(Press* ENTER *to exit)*

The rectangular array should now appear as shown in Figure 4.42.

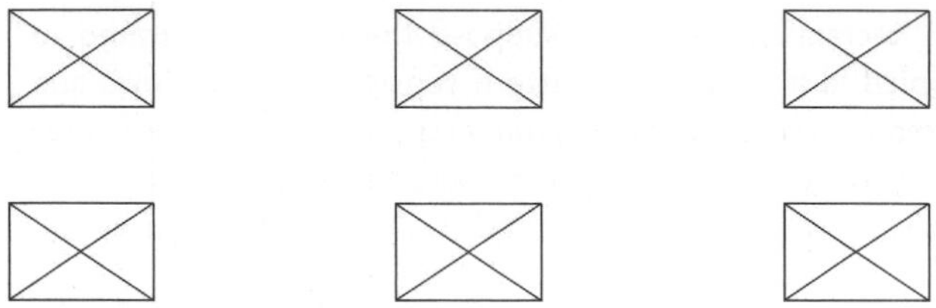

FIGURE 4.42

Creating Rectangular Patterns with Negative Offsets

In the previous array example, the rectangular array illustrates a pattern that runs to the right of and above the original figure. At times these directions change to the left of and below the original object. Direction of the array can be controlled by specifying positive or negative spacing values. A negative row spacing will force the array downward while a negative column spacing will force the array to the left.

TRY IT!

Open the drawing file 04_Array Rectangular Negative. Follow the following Command prompt sequence for performing this operation. The results are shown in Figure 4.43.

Command: ARRAYRECT

Select objects: *(Select the rectangle and 2 crossing lines)*

Select objects: *(Press* ENTER *to continue, four columns and three rows are displayed)*

Type = Rectangular Associative = Yes

Select grip to edit array or [ASsociative/Base point/COUnt/Spacing/COLumns/Rows/Levels/eXit] <eXit>: COU *(For Count)*

Enter the number of columns or [Expression] <4>: 3

Enter the number or rows or [Expression] <3>: 2

Select grip to edit array or [ASsociative/Base point/COUnt/Spacing/COLumns/Rows/Levels/eXit] <eXit>: S *(For Spacing)*

Specify the distance between columns or [Unit cell] <1.13>: −2.00

Specify the distance between rows <0.75>: −1.00

Select grip to edit array or [ASsociative/Base point/COUnt/Spacing/COLumns/Rows/Levels/eXit] <eXit>: *(Press* ENTER *to exit)*

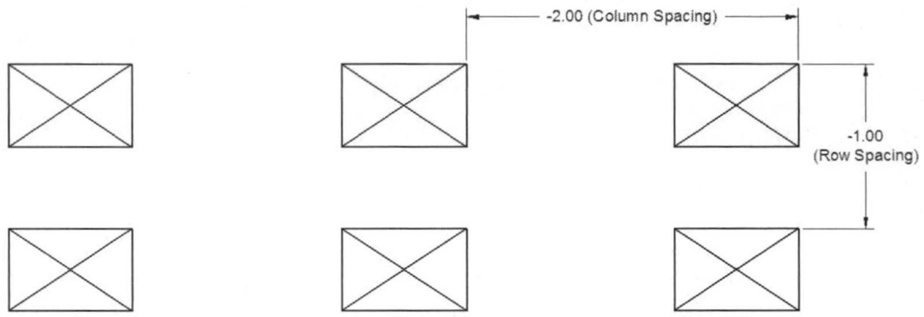

FIGURE 4.43

Modifying Associative Rectangular Patterns

Arrays are now associative, which makes modifying a pattern of objects simpler and more efficient. When selecting an associative array a series of object grips (blue boxes and arrows) are provided. These grips, for example, can be utilized to change the location, count (number of rows or columns), and spacing of your array.

TRY IT!

Open the drawing 04_Array I-Beam and create a rectangular pattern consisting of four rows and five columns. Follow the illustration in Figure 4.44 and the following Command prompt sequence for performing this operation.

Command: ARRAYRECT

Select objects: *(Select the rectangle and I-beam)*

Select objects: *(Press* ENTER *to continue, four columns and three rows are displayed, as shown in the following image on the left)*

Type = Rectangular Associative = Yes

Select grip to edit array or [ASsociative/Base point/COUnt/Spacing/COLumns/Rows/Levels/eXit] <eXit>: *(Pick the Row and Column Count grip)*

Specify number of rows and columns: (Drag the cursor up and to the right until four rows and five columns appear, as shown in the following image on the right - Pick)

Select grip to edit array or [ASsociative/Base point/COUnt/Spacing/COLumns/Rows/Levels/eXit] <eXit>: *(Pick the Column Spacing grip)*

Specify distance between columns: 15'

Select grip to edit array or [ASsociative/Base point/COUnt/Spacing/COLumns/Rows/Levels/eXit] <eXit>: *(Pick the Row Spacing grip)*

Specify distance between rows: 15'

Select grip to edit array or [ASsociative/Base point/COUnt/Spacing/COLumns/Rows/Levels/eXit] <eXit>: *(Press* ENTER *to exit)*

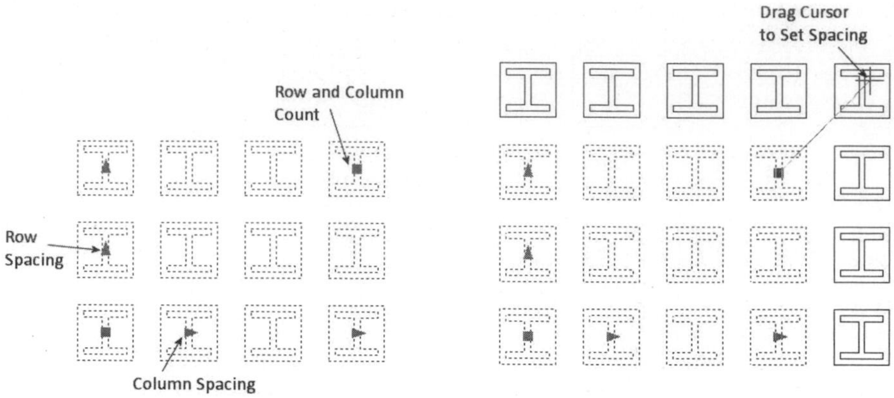

FIGURE 4.44

To modify the array just created, pick the array to display it's grips as shown in Figure 4.45. To add a sixth column, select the Column Count grip and move the cursor to the right until a new column appears - pick. Add a fifth row by using the Row Count grip. The Row and Column grip could have been used to change both counts at the same time.

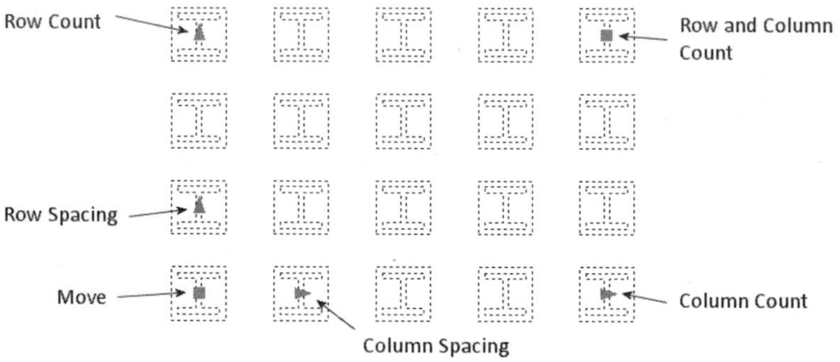

FIGURE 4.45

The final modifications to the array will be made through the Ribbon. When an associative array is selected not only will the grips appear but an Array tab displays on the Ribbon, as shown in Figure 4.46. The results of the previous modification are displayed in the Columns and Rows panels of the ribbon.

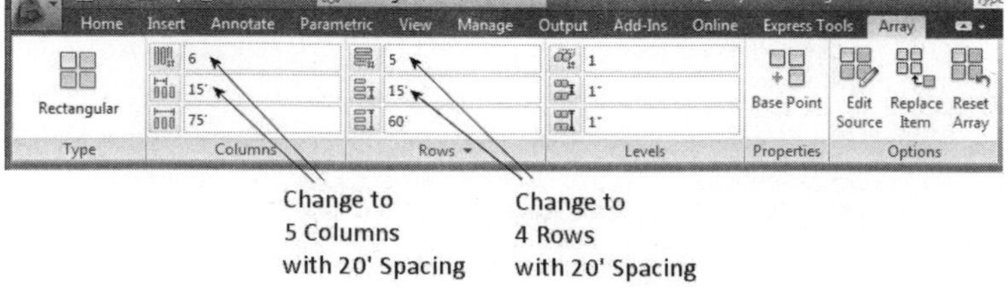

FIGURE 4.46

In the Columns panel of the Ribbon, change the count to 5 and the spacing to 20'. In the Rows panel, change the count to 4 and the spacing to 20'. Pick the Close Array button on the Ribbon. The final results are shown in Figure 4.47.

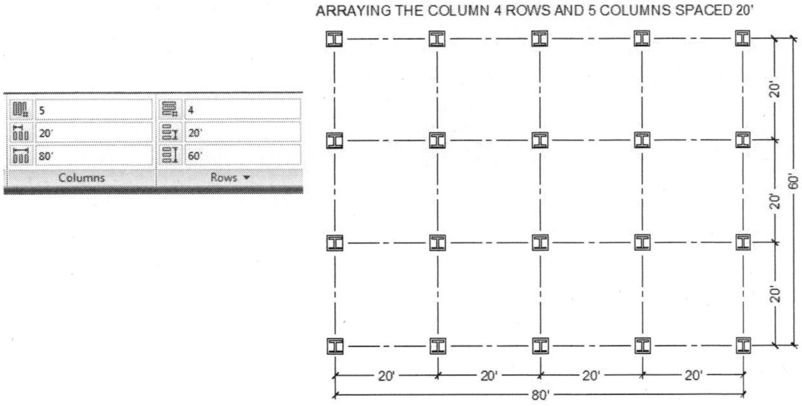

FIGURE 4.47

CREATING POLAR ARRAYS

Polar arrays allow you to create multiple copies of objects in a circular or polar pattern. After selecting the objects to array, you pick a center point for the array in addition to the number of items to copy and the angle to fill.

TRY IT!

> **Open the drawing file 04_Array Polar and activate the** `ARRAYPOLAR` **command, as shown in Figure 4.48 (Left). Follow the illustration and the following Command prompt sequence for creating the array.**

```
Command: ARRAYPOLAR
Select objects: (Select the circle "A" and the center line "B")
Select objects: (Press ENTER to continue)
Type = Polar Associative = Yes
Specify center point of array or [Base point/Axis of rotation]: (Pick the intersection of the center mark at "C", use an Osnap for accuracy - an array of six is displayed)
Select grip to edit array or [ASsociative/Base point/Items/Angle between/Fill angle/ROWs/Levels/ROTate items/eXit] <eXit>: I (For Items)
Enter number of items in array or [Expression] <6>: 4
Select grip to edit array or [ASsociative/Base point/Items/Angle between/Fill angle/ROWs/Levels/ROTate items/eXit] <eXit>: (Press ENTER to exit)
```

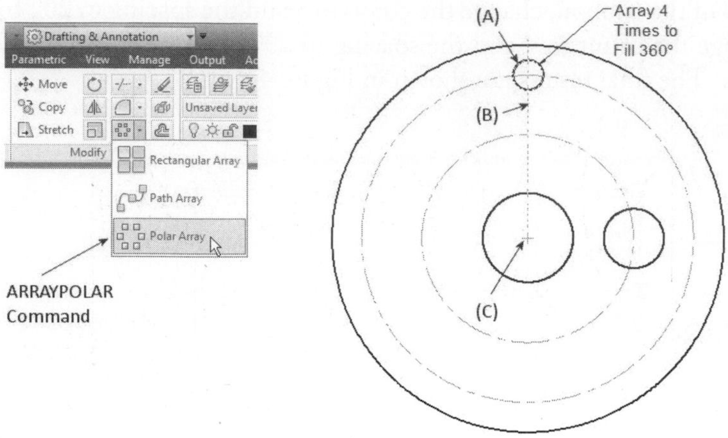

FIGURE 4.48

To complete the part a second polar array needs to be constructed, as shown in Figure 4.49. This image illustrates the creation of four holes that fill an angle of 180° in the clockwise direction. In the following Command prompt sequence, notice that an angle of –180° is given. This negative value drives the array in the clockwise direction. Remember that positive angles drive polar arrays in the counterclockwise direction.

Follow the illustration in Figure 4.49 and the following command sequence for performing this operation. The results are shown in Figure 4.49 (Right).

Command: **ARRAYPOLAR**

Select objects: *(Select the circle "C")*

Select objects: *(Press* ENTER *to continue)*

Type = Polar Associative = Yes

Specify center point of array or [Base point/Axis of rotation]: *(Pick the intersection of the center mark at "D," use an Osnap for accuracy - an array of six is displayed)*

Select grip to edit array or [ASsociative/Base point/Items/Angle between/Fill angle/ROWs/Levels/ROTate items/eXit] <eXit>: **I** (*For Items*)

Enter number of items in array or [Expression] <4>: **4**

Select grip to edit array or [ASsociative/Base point/Items/Angle between/Fill angle/ROWs/Levels/ROTate items/eXit] <eXit>: **F** (*For Fill angle*)

Specify the angle to fill (+=ccw, -=cw) or [EXpression] <360>: **-180**

Select grip to edit array or [ASsociative/Base point/Items/Angle between/Fill angle/ROWs/Levels/ROTate items/eXit] <eXit>: *(Press* ENTER *to exit)*

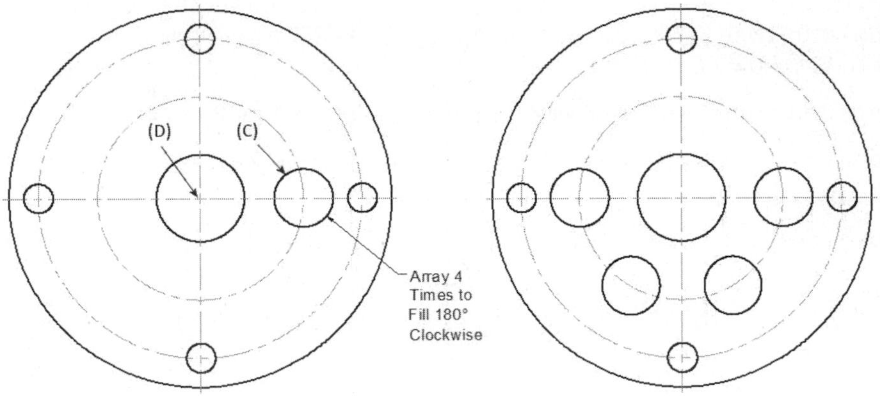

FIGURE 4.49

TRY IT!

Open the drawing file 04_Array Polar Rotate. Illustrated in Figure 4.49 are three different results for arraying non-circular objects. The image on the left illustrates a polar array formed by rotating the square object as it is being copied. In the middle image, the square object is not being rotated as it is being copied. To prevent objects from being rotated as copied you will need to set the ROTate items option to "No". The image on the right was created by setting the ROWs option to "2" and the distance between the rows to "0.36." The incremental elevation setting remained set to "0." The elevation changes the 3D spacing and will be demonstrated in Chapter 21. Follow the Command prompt sequences provided to create the arrays illustrated.

Command: ARRAYPOLAR

Select objects: *(Select the rectangle)*

Select objects: *(Press* ENTER *to continue)*

Type – Polar Associative = Yes

Specify center point of array or [Base point/Axis of rotation]: *(Pick the intersection of the center mark - an array of six is displayed)*

- - - For the image on the left - - -

Select grip to edit array or [ASsociative/Base point/Items/Angle between/Fill angle/ROWs/Levels/ROTate items/eXit] <eXit>: *(Press* ENTER *to exit)*

- - - For the image in the center - - -

Select grip to edit array or [ASsociative/Base point/Items/Angle between/Fill angle/ROWs/Levels/ROTate items/eXit] <eXit>: ROT *(For ROTate items)*

Rotate arrayed items? [Yes/No] <Yes>: N

Select grip to edit array or [ASsociative/Base point/Items/Angle between/Fill angle/ROWs/Levels/ROTate items/eXit] <eXit>: *(Press* ENTER *to exit)*

- - - For the image on the right - - -

```
Select grip to edit array or [ASsociative/Base point/Items/
Angle between/Fill angle/ROWs/Levels/ROTate items/eXit]
<eXit>: ROW (For ROWs)

Enter the number of rows or [Expression] <1>: 2

Specify the distance between rows or [Total/Expression]
<0.36>: 0.36

Specify the incrementing elevation between rows or
[Expresson] <0.00>: (Press ENTER to accept 0.00)

Select grip to edit array or [ASsociative/Base point/Items/
Angle between/Fill angle/ROWs/Levels/ROTate items/eXit]
<eXit>: (Press ENTER to exit)
```

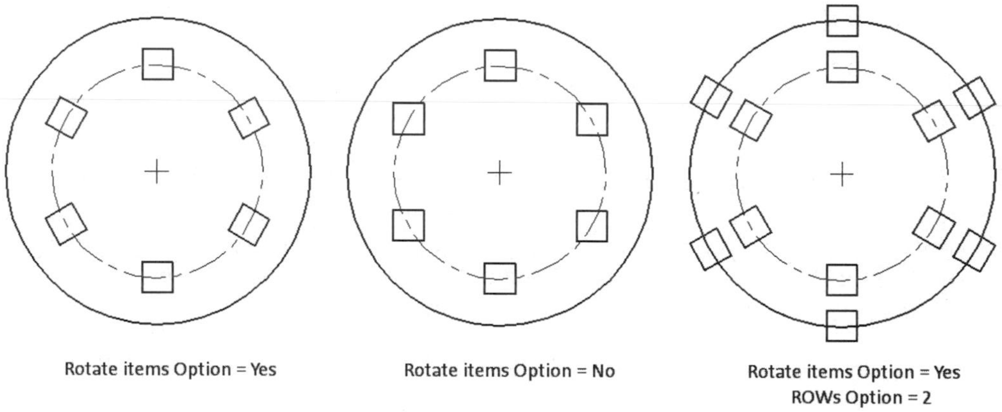

FIGURE 4.50

TRY IT!

Open the drawing file 04_Array Polar Modify. Set Polar tracking to 45° and verify an Intersection Osnap is active. In this exercise, modify the existing associative polar array by using grips and the Array tab of the Ribbon. Select the array to display the grips as shown in Figure 4.51 (Left). The center of the array needs to be relocated. Pick the Move grip and relocate it to the intersection of the center mark at "A." The result is shown in the middle image. Next, modify the radius by selecting the Stretch Radius grip and relocating it to the intersection at "B." The result is shown in the image on the right.

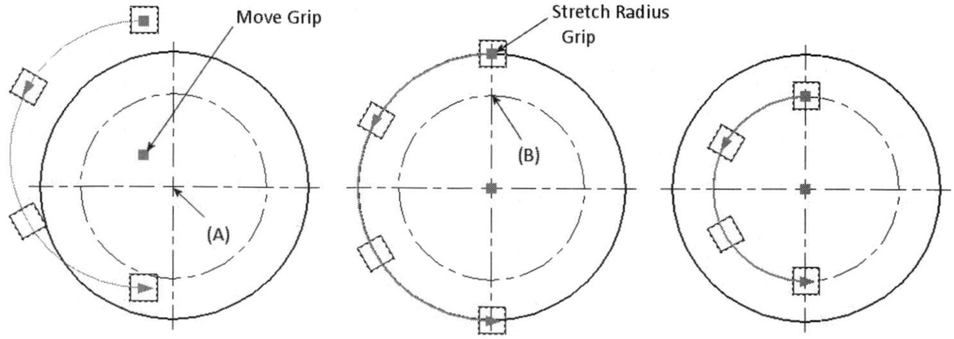

FIGURE 4.51

Pick the Item Count grip as shown in Figure 4.52 (Left) and reposition the cursor counterclockwise until two additional objects are displayed and then pick at "C." This same grip can be used to change the fill angle. Hover the cursor over the grip until a shortcut menu appears. Select "Fill Angle" as shown in Figure 4.51 (Right).

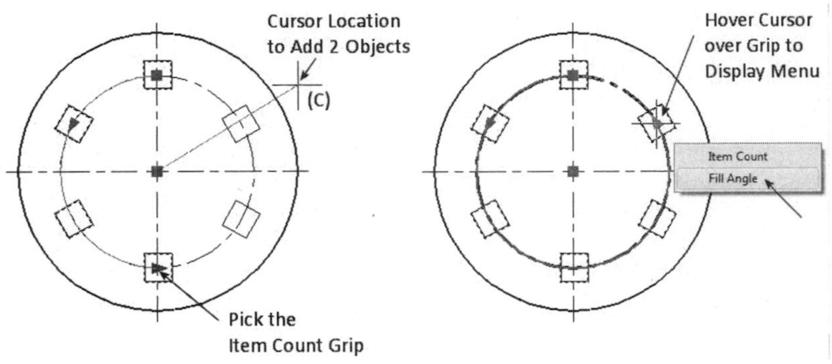

FIGURE 4.52

To change the fill angle to 180°, move the cursor straight down in the 270° direction, as shown in Figure 4.53 (Left) and then pick at "D." The result is shown in the center image. The Ribbon is shown on the right and confirms that there are now six items placed within a 180° fill angle.

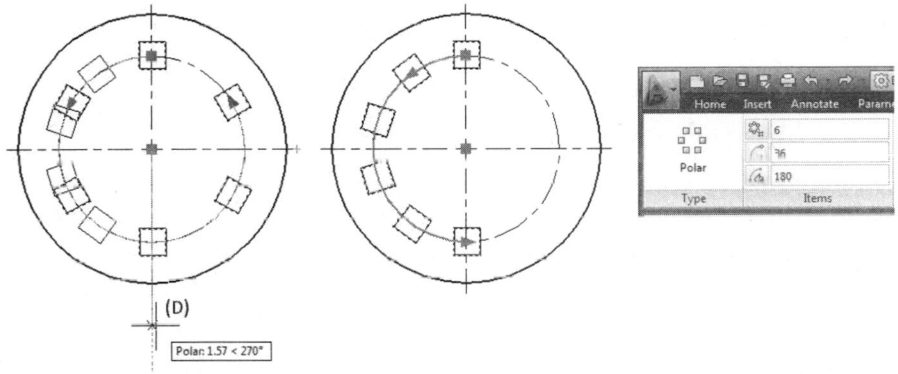

FIGURE 4.53

The final change is to modify the angle between items. Pick the grip as shown in Figure 4.54 (Left). Relocate the cursor counterclockwise until the cursor direction is 135° at "E" and then pick. This will establish a 45° angle between the arrayed items as shown in the center image. Confirm the modification by verifying the setting in the Ribbon, as shown in the image on the right. Of course, you could also have made many of the changes in this exercise by simply changing the values in the Items panel of the Ribbon.

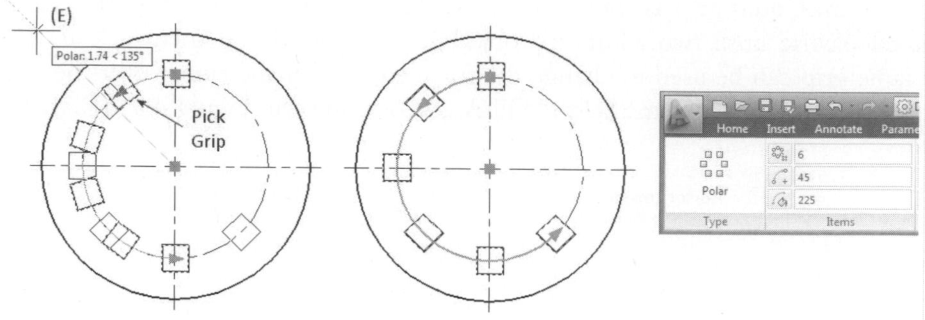

FIGURE 4.54

CREATING ARRAYS ALONG A PATH

Path arrays allow you to create multiple copies of objects along a path curve. The path curve can be a line, a polyline, a 3D polyline, a helix, an arc, a circle, or an ellipse. The objects can be measured (spaced apart by a specified length) or they can be divided (the spacing is determined by the length of the path and the number of items). You can also specify that only a portion of the path will be utilized.

TRY IT!

Open the drawing file 04_Array Path1 and activate the ARRAYPATH command, as shown in Figure 4.55 (Left). The array that will be created in this exercise will consist of 15 items that are divided evenly over the path. Follow the illustration and the following Command prompt sequence for creating the array.

Command: ARRAYPATH

Select objects: *(Select the Rectangle "A")*

Select objects: *(Press* ENTER *to continue)*

Type = Path Associative = Yes

Select path curve: *(Select the Polyline "B")*

Select grip to edit array or [ASsociative/Method/Base point/
Tangent direction/Items/Rows/Levels/Align items/
Z direction/eXit] <eXit>: M *(For Method)*

Enter path method [Divide/Measure] <Measure>: D *(For Divide)*

Select grip to edit array or [ASsociative/Method/Base point/
Tangent direction/Items/Rows/Levels/Align items/
Z direction/eXit] <eXit>: I *(For Items)*

Enter number of items along path or [Expression] <26>: 15

Select grip to edit array or [ASsociative/Method/Base point/
Tangent direction/Items/Rows/Levels/Align items/
Z direction/eXit] <eXit>: *(Press* ENTER *to exit)*

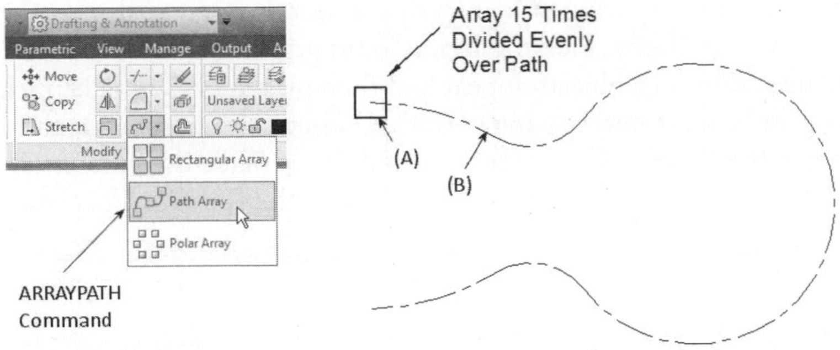

FIGURE 4.55

TRY IT!

Open the drawing file 04_Array Path2. In the previous exercise the Divide method was demonstrated. In this exercise the Measure method will be utilized because a specific spacing of objects is provided. It will, however, not be necessary to use the Method option in the Command prompt sequence since the Measure method is the default. Once the array is created, the array count and spacing will be modified utilizing grips. Follow the illustration in Figure 4.56 and the following Command prompt sequence for creating the array.

Command: ARRAYPATH

Select objects: *(Select the Rectangle "C")*

Select objects: *(Press* ENTER *to continue)*

Type = Path Associative = Yes

Select path curve: *(Select the Polyline "D")*

Select grip to edit array or [ASsociative/Method/Base point/
Tangent direction/Items/Rows/Levels/Align items/
Z direction/eXit] <eXit>: I *(For Items)*

Specify the distance between items along path or [Expression]

<0.36>: 0.75

Specify number of items or [Fill entire path/Expression]
<13>: 10

Select grip to edit array or [ASsociative/Method/Base point/
Tangent direction/Items/Rows/Levels/Align items/
Z direction/eXit] <eXit>: *(Press* ENTER *to exit - the results are shown in the following image on the right)*

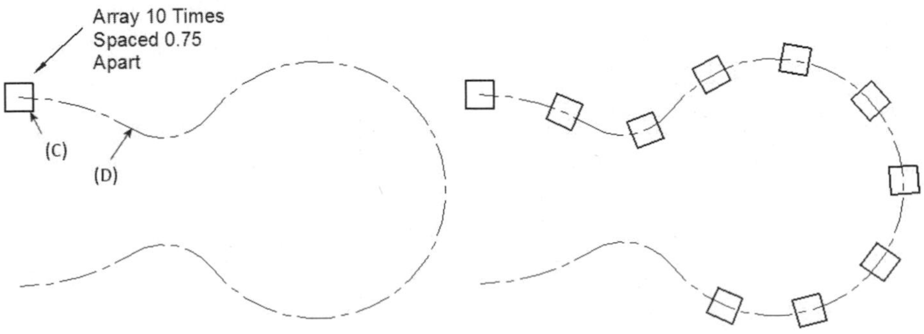

FIGURE 4.56

Next, modify the item count and spacing utilizing grips. Select the array to display the grips as shown in Figure 4.57 (Left). Pick the Item Count grip and move the cursor down the path until two additional objects appear and then pick at "E." Finally, pick the Item Spacing grip and enter a new spacing of 0.83. Confirm the modifications by checking the settings in the Ribbon. Of course, you could have made the same modifications by simply changing the values in the Items panel of the Ribbon. This completes the exercise.

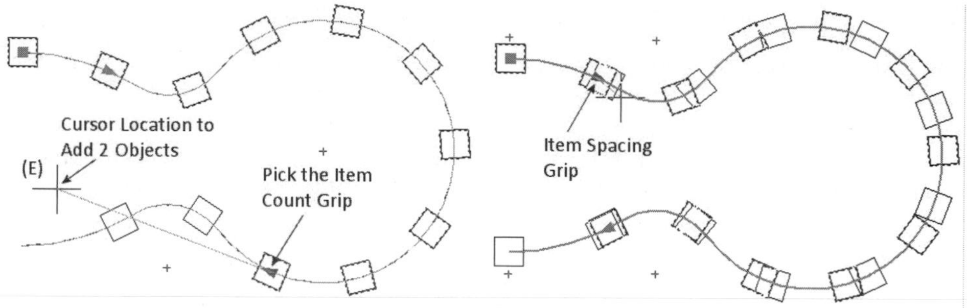

FIGURE 4.57

MODIFYING ASSOCIATIVE ARRAYS

As already demonstrated, grips and the Array tab of the Ribbon can be utilized to modify the count and spacing for associative arrays. Because an associative array maintains a relationship between objects, it is simple to update an array to these new settings. It is also possible to make changes to or even replace individual items in the array. Although associative arrays provide many advantages, for some modifications you may find that you do not want the array to be associative. You can select an option to make the array non-associative as it is created or the array can be exploded after it is created. In the following Try It! exercise we will explore some of these modification techniques.

TRY IT!

Open the drawing file 04_Array Modify. Follow the illustration Figure 4.58 (Left) and the following command sequence for creating the arrays that will be modified.

```
Command: ARRAYPOLAR
Select objects: (Select the circle "A")
Select objects: (Press ENTER to continue)
Type = Polar Associative = Yes
Specify center point of array or [Base point/Axis of rotation]: (Pick the intersection of the center mark at "B," use an Osnap for accuracy)
Select grip to edit array or [ASsociative/Base point/Items/Angle between/Fill angle/ROWs/Levels/ROTate items/eXit] <eXit>: I (For Items)
Enter number of items in array or [Expression] <6>: 15
```

Select grip to edit array or [ASsociative/Base point/Items/
Angle between/Fill angle/ROWs/Levels/ROTate items/eXit]
<eXit>: *(Press* ENTER *to exit)*

Command: **ARRAYPATH**

Select objects: *(Select the Circle "C")*

Select objects: *(Press* ENTER *to continue)*

Type = Path Associative = Yes

Select path curve: *(Select the Polyline "D")*

Select grip to edit array or [ASsociative/Method/
Base point/Tangent direction/Items/Rows/Levels/
Align items/Z direction/eXit] <eXit>: M (*For Method*)

Enter path method [Divide/Method] <Measure>: D (*For Divide*)

Select grip to edit array or [ASsociative/Method/Base point/
Tangent direction/Items/Rows/Levels/Align items/
Z direction/eXit] <eXit>: I (*For Items*)

Enter number of items along path or [Expression] <12>: 7

Select grip to edit array or [ASsociative/Method/Base point/
Tangent direction/Items/Rows/Levels/Align items/
Z direction/eXit]<eXit>: *(Press* ENTER *to exit)*

Command: **ARRAYRECT**

Select objects: *(Select the rectangle "E")*

Select objects: *(Press* ENTER *to continue)*

Type = Rectangular Associative = Yes

Select grip to edit array or [ASsociative/Base point/COUnt/
Spacing/COLumns/Rows/Levels/eXit] <eXit>: COU (*For Count*)

Enter the number of columns or [Expression] <4>: 4

Enter the number or rows or [Expression] <3>: 4

Select grip to edit array or [ASsociative/Base point/COUnt/
Spacing/COLumns/Rows/Levels/eXit] <eXit>: S (*For Spacing*)

Specify the distance between columns or [Unit cell] <0.23>: 0.3

Specify the distance between rows <0.23>: 0.3

Select grip to edit array or [ASsociative/Base point/COUnt/
Spacing/COLumns/Rows/Levels/eXit] <eXit>: (*Press* ENTER *to exit*)

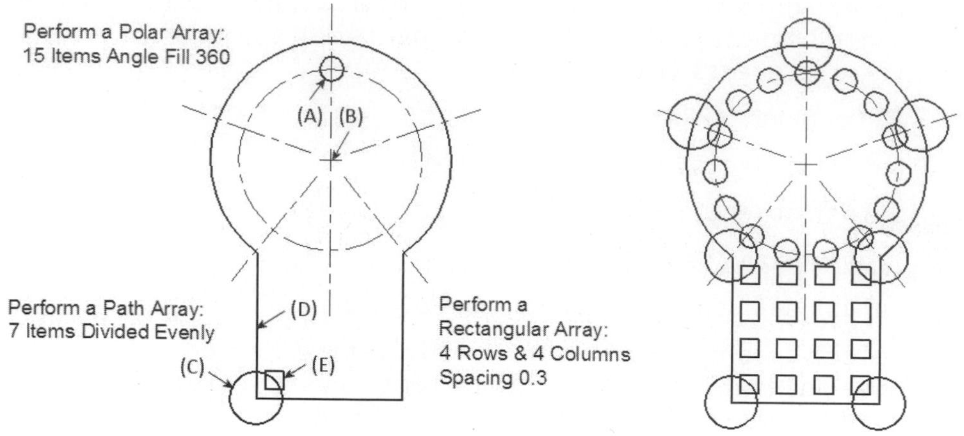

FIGURE 4.58

The first modification will be made to the rectangular array. Four of the rectangles in the array need to be erased. To select subobjects of an array, hold down the `CTRL` key as you pick the items. The selected items can then be erased, moved, rotated, or scaled, if desired. Select the four rectangles indicated in Figure 4.59 (Left). Press the `DEL` key to erase the items. The results are shown in the illustration on the right.

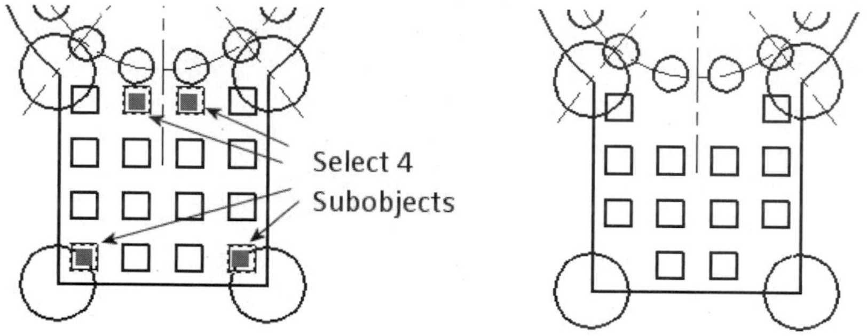

FIGURE 4.59

Next we will modify the polar array. Use the Layer Properties Manager (`LAYER` Command) to turn off the Text layer and turn on the Object-Replace layer. In this modification we will replace the circles in the array with tapered slots. The results are shown in Figure 4.59 (Right). Pick a circle in the polar array to display the grips and the Array tab of the Ribbon. Pick the Replace Item button in the Options panel of the Ribbon, as shown in Figure 4.59 (Left). The `ARRAYEDIT` command will be started with the REPlace option selected for you.

`Command: _arrayedit`

`Enter an option [Source/REPlace/Base point/Items/Angle between/Fill angle/Rows/Levels/ROTate items/RESet/eXit] <eXit>: _rep`

`Select replacement objects:` *(Select the tapered slot "F")*

`Select replacement objects:` *(Press ENTER to continue)*

`Select base point of replacement objects or [Key Point] <centroid>:` *(Pick center of arc at "G," use Osnap for accuracy)*

`Select an item in the array to replace or [Source objects]: S` *(For Source objects)*

Enter an option [Source/REPlace/Base point/Items/Angle between/Fill angle/Rows/Levels/ROTate items/RESet/eXit] <eXit>: *(Press* ENTER *to exit)*

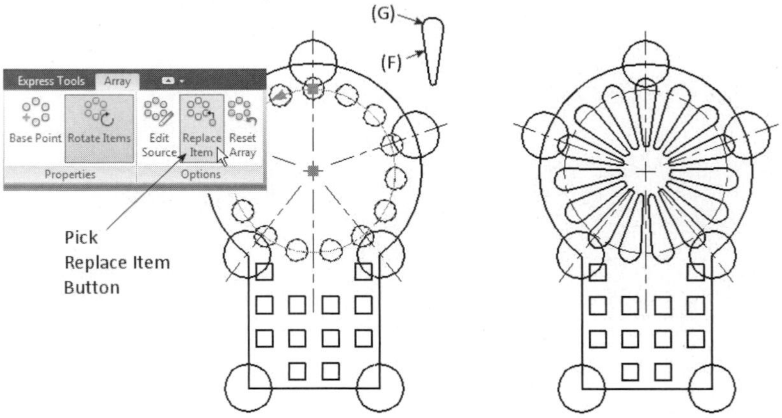

FIGURE 4.60

The final modifications will be made to the path array. First we will edit the array's source item and then we will explode the array to perform some trim operations. Pick a circle in the path array to display the grips and the Array tab of the Ribbon. Pick the Edit Source button in the Options panel of the Ribbon, as shown in Figure 4.61 (Left). The ARRAYEDIT command will be started with the Source option selected for you. After you select the source object to edit, an alert will be shown letting you know you are about to enter the Array editing state. The Alert dialog box is shown in Figure 4.61 (Right). Click OK.

Command: _arrayedit

Enter an option [Source/REPlace/Method/Base point/Items/Rows/Levels/Alignitems/Z direction/RESet/eXit] <eXit>: _s

Select item in array: *(Pick the circle at "H")*

Enter ARRAYCLOSE to exit array editing state.

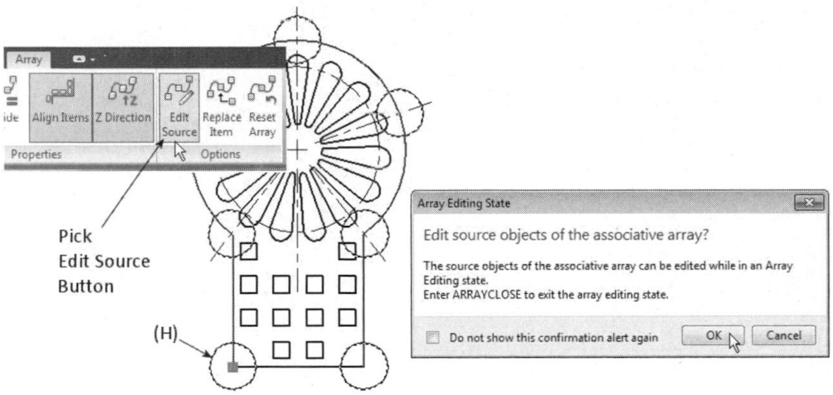

FIGURE 4.61

Once in the array editing state, create a circle using Figure 4.62 and the following command prompt sequence.

Command: C *(For CIRCLE)*

Specify center point for circle or [3P/2P/Ttr (tan tan radius)]: *(Pick the center of the circle at "I," use an Osnap for accuracy)*

Specify radius of circle or [Diameter] <0.08>: **D** *(For Diameter option)*

Specify diameter of circle <0.16>: **0.125**

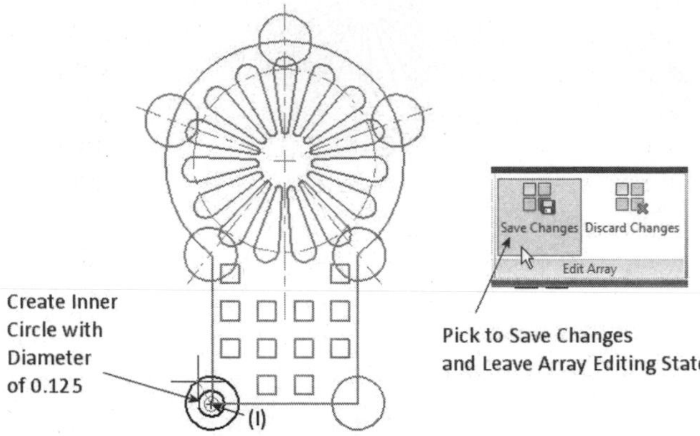
Save Changes
Discard Changes
Edit Array
Create Inner Circle with Diameter of 0.125
(I)
Pick to Save Changes and Leave Array Editing State

FIGURE 4.62

Once the source has been modified, click the Save Changes button in the Edit Array panel of the Ribbon, as shown in Figure 4.62 (Right). Clicking this button records the changes and leaves the array editing state. The results are shown in Figure 4.63 (Left).

For the final modification, use the EXPLODE command to make the path array non-associative. This will allow you to trim the arrayed objects. The results are shown in Figure 4.63 (Right).

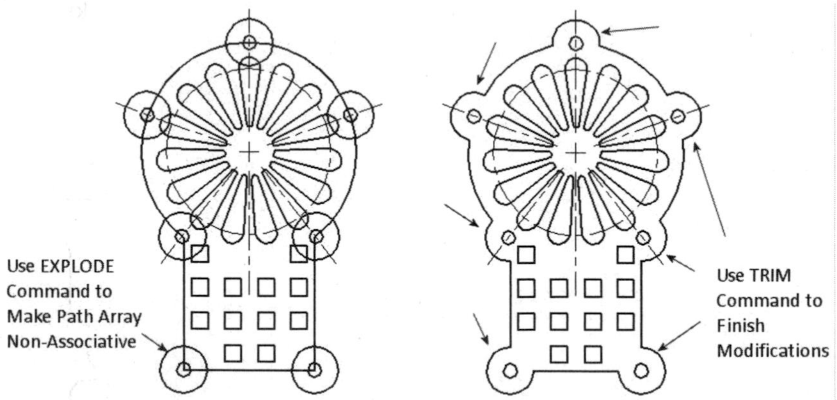
Use EXPLODE Command to Make Path Array Non-Associative
Use TRIM Command to Finish Modifications

FIGURE 4.63

MIRRORED OBJECTS

The MIRROR command is used to create a mirrored copy of an object or group of objects. When performing a mirror operation, you have the option of deleting the original object, which would be the same as flipping the object, or keeping the original object along with the mirror image, which would be the same as flipping and copying. Choose this command from one of the following:

- From the Ribbon > Home Tab > Modify Panel
- The Menu Bar (Modify > Mirror)
- The Modify Toolbar of the AutoCAD Classic Workspace
- The keyboard (MI or MIRROR)

Mirroring and Copying

The default action of the MIRROR command is to copy and flip the set of objects you are mirroring. After selecting the objects to mirror, you identify the first and second points of a mirror line. You then decide to keep or delete the source objects. The mirror operation is performed in relation to the mirror line. Usually, most mirroring is performed in relation to horizontal or vertical lines. As a result, it is usually recommended to turn on Polar or Ortho mode to force orthogonal mirror lines (horizontal or vertical).

TRY IT!

Open the drawing file 04_Mirror Copy. Refer to the following prompts and Figure 4.64 for using the MIRROR command:

```
Command: MI (For MIRROR)
Select objects: (Select a point near "X")
Specify opposite corner: (Select a point near "Y")
Select objects: (Press ENTER to continue)
Specify first point of mirror line: (Select the endpoint of
the centerline at "A")
Specify second point of mirror line: (Select the endpoint of
the centerline at "B")
Erase source objects? [Yes/No] <N>: (Press ENTER for default)
```

Because the original object needed to be retained by the MIRROR operation, the image result is shown in Figure 4.64 (Right). The MIRROR command works well when symmetry is required.

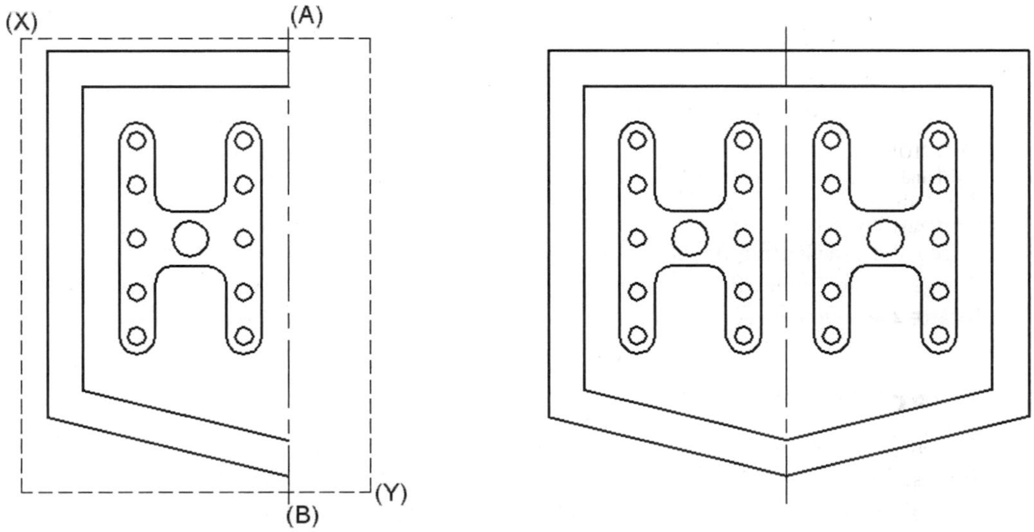

FIGURE 4.64

Mirroring by Flipping

The illustration in Figure 4.65 is a different application of the MIRROR command. It is required to have all items that make up the bathroom plan flip but not copy to the other side. This is a typical process involving "what if" scenarios.

TRY IT!

Open the drawing file 04_Mirror Flip. Use the following Command prompts to perform this type of mirror operation. The results are displayed in Figure 4.65 (Right).

Command: MI *(For MIRROR)*

Select objects: All *(This selects all objects)*

Select objects: *(Press* ENTER *to continue)*

Specify first point of mirror line: Mid

of *(Select the midpoint of the line at "A")*

Specify second point of mirror line: Per
to *(Select line "B," perpendicular to point "A")*

Erase source objects? [Yes/No] <N>: Y *(For Yes)*

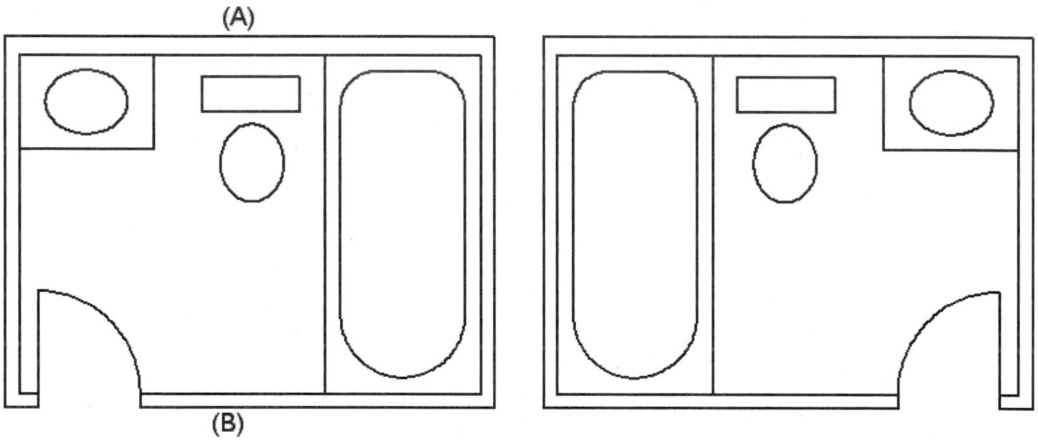

FIGURE 4.65

Mirroring Text

In addition to mirroring other object types, text is an object that we typically do not want flipped, as in Figure 4.66, an image of a duplex complex. Rather than copying and moving the text into position for the matching duplex half, you can still include the text in the mirroring operation without it flipping.

TRY IT!

Open the drawing file 04_Mirror Duplex. Use the MIRROR command and create a mirror image of the Duplex floor plan using line "AB" as the points for the mirror line. Do not delete the source objects. Your finished results should be similar to Figure 4.66.

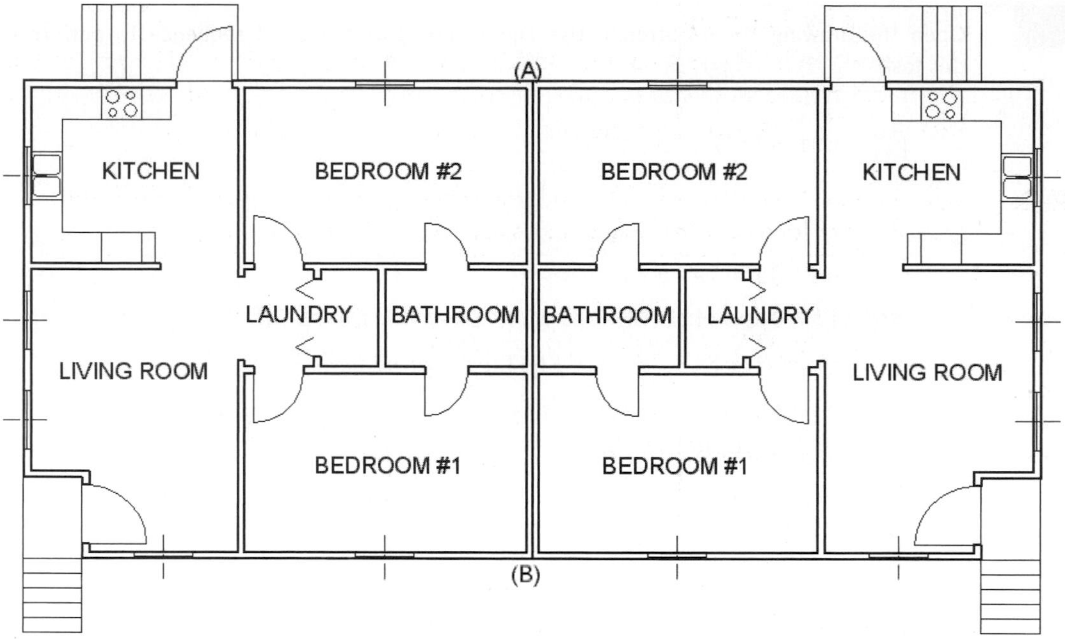

FIGURE 4.66

More Information on Mirroring Text

If text in a drawing does actually flip (the text is backwards and unreadable) during a mirror operation, this is due to the setting of the MIRRTEXT system variable. This variable must be entered at the Command prompt. If this variable is set to a value of 1 (or on), change the value to 0 (Zero, or off). Using the `MIRROR` command now will display the text right-reading.

```
Command: MIRRTEXT
Enter new value for MIRRTEXT <1>: 0(To prevent text from being mirrored)
```

STRETCHING OBJECTS

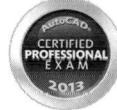

Use the `STRETCH` command to move a portion of a drawing while still preserving the connections to parts of the drawing remaining in place. Choose this command from one of the following:

- From the Ribbon > Home Tab > Modify Panel
- The Menu Bar (Modify > Stretch)
- The Modify Toolbar of the AutoCAD Classic Workspace
- The keyboard (`S` or `STRETCH`)

The Basics of Stretching

To ensure success with stretching operations, you should always select objects using a Crossing option. Items inside the crossing box will move, while those touching the box will change length. In Figure 4.67, a group of objects is selected with the crossing box. Next, a base point is identified at the approximate location of "C." Finally, a second point of displacement is identified directly to the left of the base point (try Polar Tracking and Direct Distance mode). Once the objects selected in the crossing box are stretched, the objects are shifted without any need to extend or trim lines to mend the drawing.

TRY IT!

Open the drawing file 04_Stretch. Use Figure 4.67 and command sequence to perform this task.

```
Command: S (For STRETCH)
Select objects to stretch by crossing-window or crossing-polygon ...
Select objects: (Pick a point at "A")
Specify opposite corner: (Pick a point at "B")
Select objects: (Press ENTER to continue)
Specify base point or [Displacement] <Displacement>:
(Select a point at "C")
Specify second point or <use point place as displacement>:
(With Polar turned on, move your cursor to the left and enter a
value of .75)
```

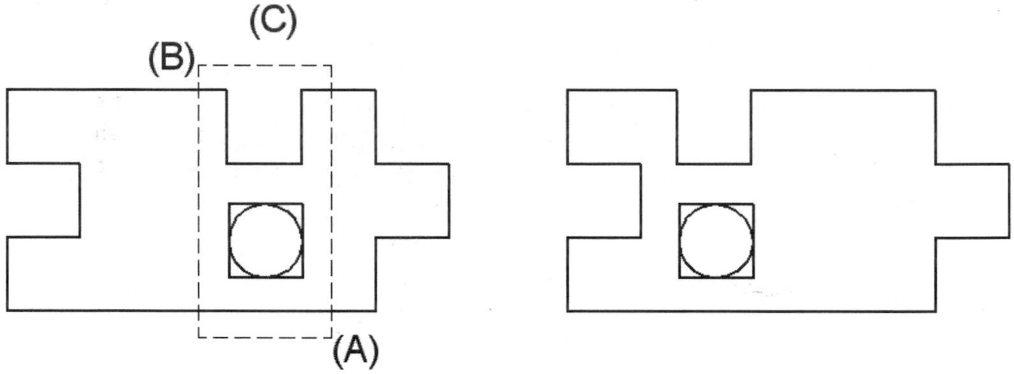

FIGURE 4.67

The results of performing this stretching operation are illustrated in Figure 4.67 (Right).

How Stretching Affects Dimensions

Applications of the STRETCH command are illustrated in Figure 4.68, in which a number of architectural features need to be positioned at a new location. The whole success of using the STRETCH command on these features is in the selection of the objects to stretch through the crossing box. Associative dimensions will automatically move with the objects.

TRY IT!

Open the drawing file 04_Stretch Arch. Use Figure 4.67 to stretch the window, wall, and door to the designated distances using Polar or Ortho and Direct Distance mode. Since each of the stretch distances is a different value, the STRETCH command must be used three separate times.

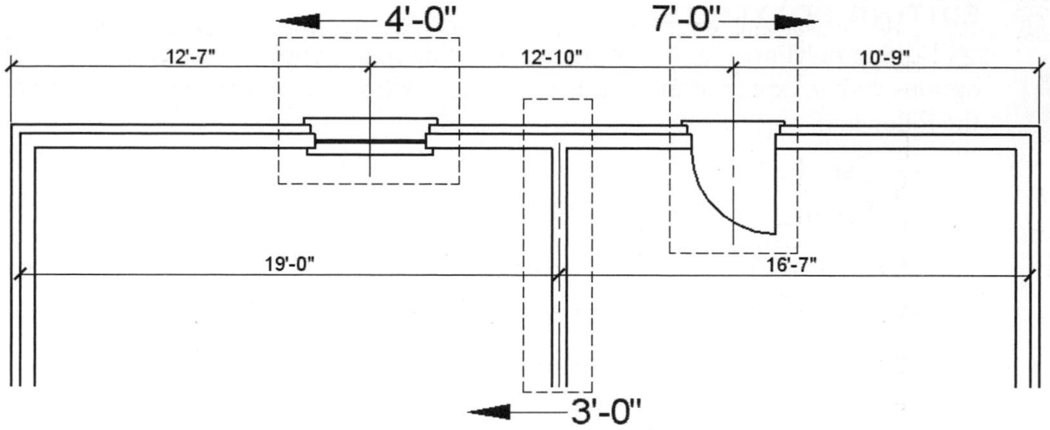

FIGURE 4.68

Stretching Using Multiple Crossing Windows

When identifying items to stretch by crossing box, you are not limited to a single crossing box. You can surround groups of objects with multiple crossing boxes. All items selected in this manner will be affected by the stretching operation.

TRY IT!

Open the drawing file 04_Stretch Fence. Enter the `STRETCH` **command and construct three separate crossing boxes to select the top of the fence boards at "A," "B," and "C," as shown in Figure 4.69. When prompted for a base point or displacement, pick a point on a blank part of your screen. Stretch these boards up at a distance of 12" or 1'-0".**

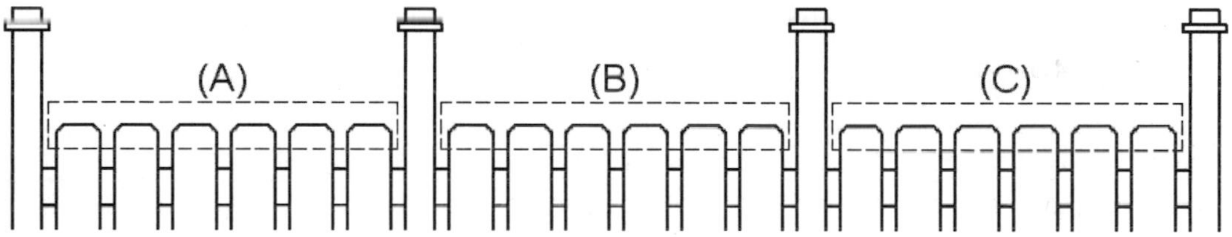

FIGURE 4.69

The results of performing the stretch operation on the fence are illustrated in Figure 4.70. When you have to stretch various groups of objects in a single operation, you can create numerous crossing boxes to better perform this task.

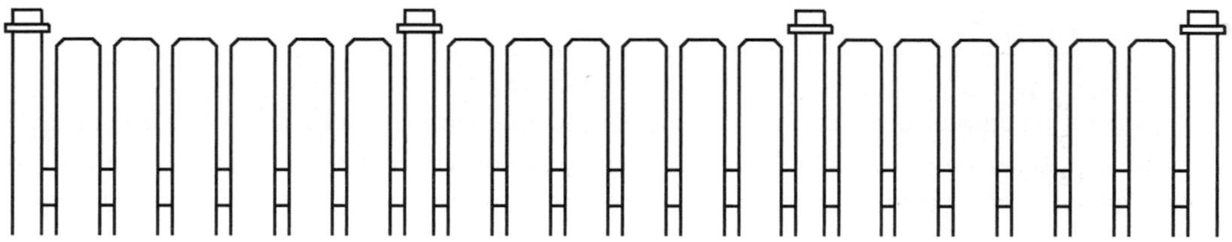

FIGURE 4.70

EDITING POLYLINES

Editing polylines is a productivity tool that has many applications. A few of these options will be explained in the following pages. Choose this command from one of the following:

- From the Ribbon > Home Tab > Modify Panel (Expanded)
- The Menu Bar (Modify > Object > Polyline)
- The Modify II Toolbar of the AutoCAD Classic Workspace
- The keyboard (PE or PEDIT)

Changing the Width of a Polyline

Illustrated in Figure 4.71 on the left is a polyline of width 0.00. The PEDIT command is used to change the width of the polyline to 0.10 units, as shown in Figure 4.71 on the right.

TRY IT!

Open the drawing file 04_Pedit Width. Refer to the following command sequence to use the PEDIT command with the Width option.

Command: **PE** *(For PEDIT)*

Select polyline or [Multiple]: *(Select the polyline)*

Enter an option [Open/Join/Width/Edit vertex/Fit/Spline/Decurve/Ltype gen/Reverse/Undo]: **W** *(For Width)*

Specify new width for all segments: **0.10**

Enter an option [Open/Join/Width/Edit vertex/Fit/Spline/Decurve/Ltype gen/Reverse/Undo]: *(Press* ENTER *to exit this command)*

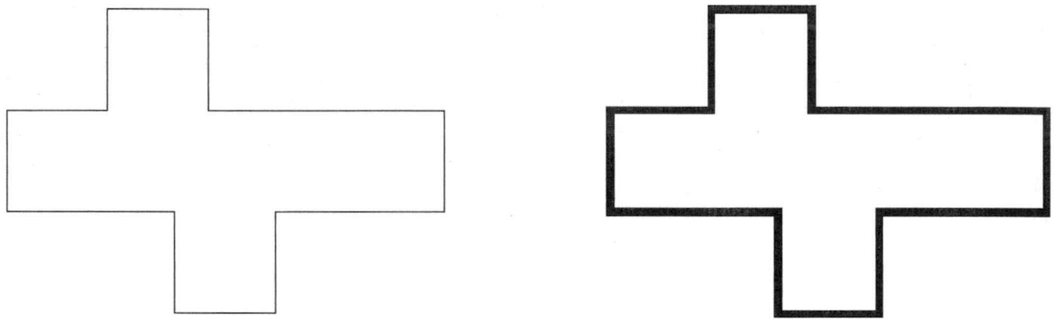

FIGURE 4.71

Joining Objects into a Single Polyline

It is very easy to convert regular objects such as lines and arcs into polylines (circles cannot be converted). As long as the polyline endpoints match with other object endpoints, they can be easily joined into one single polyline object using the Join option of the PEDIT command.

TRY IT!

Open the drawing file 04_Pedit Join. Refer to the following command sequence and Figure 4.72 use this command.

Command: PE *(For PEDIT)*

Select polyline or [Multiple]: *(Select the line at "A")*

Object selected is not a polyline

Do you want to turn it into one? <Y> *(Press ENTER)*

Enter an option [Close/Join/Width/Edit vertex/Fit/Spline/Decurve/Ltype gen/Reverse/Undo]: J *(For Join)*

Select objects: *(Pick a point at "B")*

Specify opposite corner: *(Pick a point at "C")*

Select objects: *(Press ENTER to join the lines)*

56 segments added to polyline

Enter an option [Open/Join/Width/Edit vertex/Fit/Spline/Decurve/Ltype gen/Reverse/Undo]: *(Press ENTER to exit this command)*

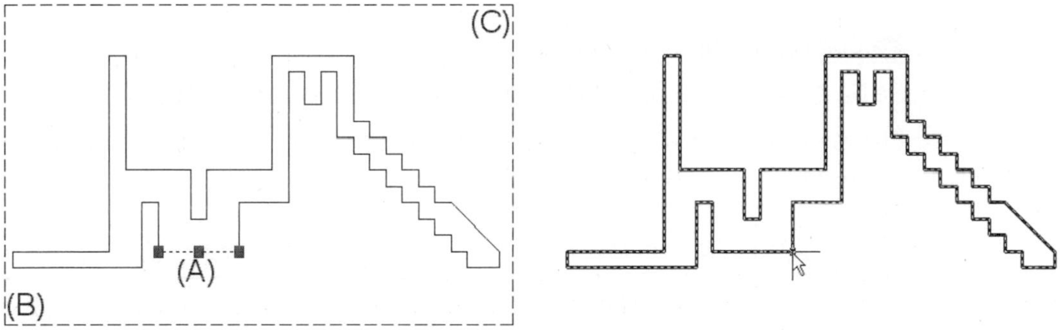

FIGURE 4.72

Curve Generation

Polylines can be edited to form various curve-fitting shapes. Two curve-fitting modes are available, namely, Splines and Fit Curves.

Generating Splines

The Spline option produces a smooth-fitting curve based on control points in the form of the vertices of the polyline.

TRY IT!

Open the drawing file 04_Pedit Spline Curve. Refer to the following command sequence and Figure 4.73 to use this command.

Command: PE *(For PEDIT)*

Select polyline or [Multiple]: *(Select the polyline)*

Enter an option [Close/Join/Width/Edit vertex/Fit/Spline/Decurve/Ltype gen/Reverse/Undo]: S *(For Spline)*

Enter an option [Close/Join/Width/Edit vertex/Fit/Spline/Decurve/Ltype gen/Reverse/Undo]: *(Press ENTER to exit this command)*

The results of creating a spline curve from a polyline are shown in Figure 4.73 (Right).

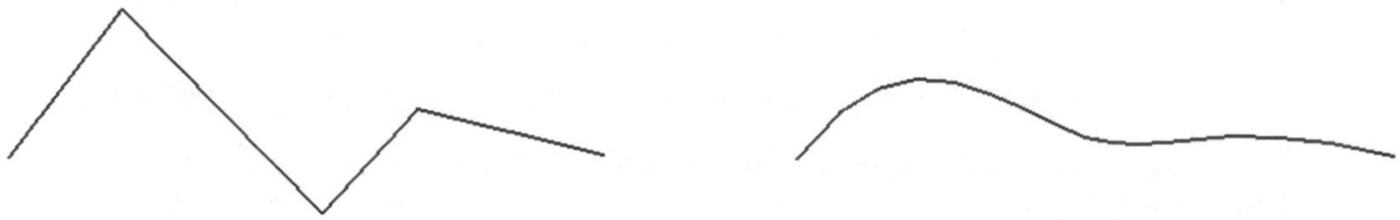

FIGURE 4.73

Generating Fit Curves

The Fit Curve option passes entirely through the control points, producing a more exaggerated curve.

TRY IT!

Open the drawing file 04_Pedit Fit Curve. Refer to the following command sequence and Figure 4.74 to use this command.

```
Command: PE (For PEDIT)
Select polyline or [Multiple]: (Select the polyline)
Enter an option [Close/Join/Width/Edit vertex/Fit/Spline/
Decurve/Ltype gen/Reverse/Undo]: F (For Fit)
Enter an option [Close/Join/Width/Edit vertex/Fit/Spline/
Decurve/Ltype gen/Reverse/Undo]: (Press ENTER to exit this
command)
```

The results of creating a fit curve from a polyline are shown in Figure 4.74 (Right).

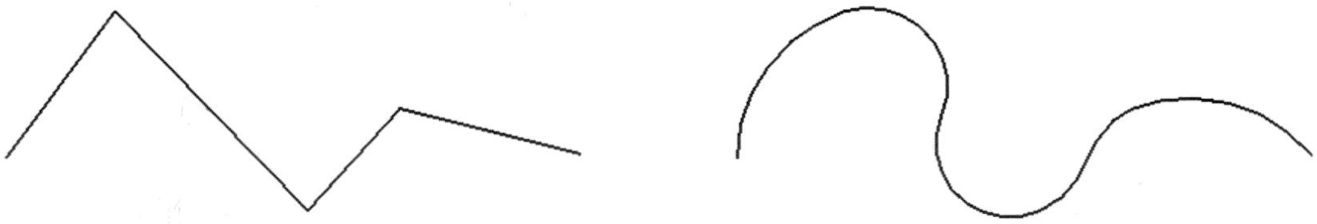

FIGURE 4.74

Linetype Generation of Polylines

The Linetype Generation option of the `PEDIT` command controls the pattern of the linetype from polyline vertex to vertex. In the polyline illustrated in Figure 4.75 (Left), the hidden linetype is generated from the first vertex to the second vertex. An entirely different pattern is formed from the second vertex to the third vertex, and so on. Notice there are no gaps at the vertices. The polyline illustrated in Figure 4.75 (Right) has the linetype generated throughout the entire polyline. In this way, the hidden linetype is smoothed throughout the polyline.

TRY IT!

Open the drawing file 04_Pedit Ltype Gen. Refer to the following command sequence and Figure 4.75 to use this command.

Command: PE *(For PEDIT)*

Select polyline or [Multiple]: *(Select the polyline)*

Enter an option [Close/Join/Width/Edit vertex/Fit/Spline/Decurve/Ltype gen/Reverse/Undo]: L *(For Ltype gen)*

Enter polyline linetype generation option [ON/OFF] <Off>: On

Enter an option [Close/Join/Width/Edit vertex/Fit/Spline/Decurve/Ltype gen/Reverse/Undo]: *(Press* ENTER *to exit this command)*

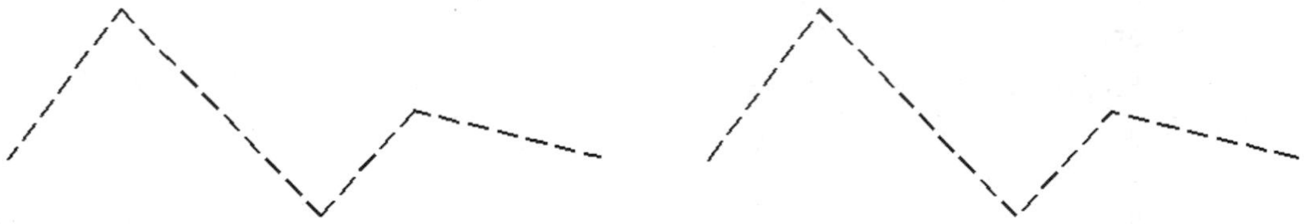

FIGURE 4.75

Offsetting Polyline Objects

Once a group of objects has been converted to and joined into a single polyline object, the entire polyline can be copied at a parallel distance using the OFFSET command.

TRY IT!

Open the drawing file 04_Pedit Offset. Use the OFFSET **command to copy the shape in Figure 4.76 at a distance of 0.50 units to the inside.**

Command: O *(For OFFSET)*

Specify offset distance or [Through/Erase/Layer] <Through>: 0.50

Select object to offset or [Exit/Undo] <Exit>: *(Select the polyline at "A")*

Specify point on side to offset or [Exit/Multiple/Undo] <Exit>: *(Select a point inside the polyline, near "B")*

Select object to offset or [Exit/Undo] <Exit>: *(Press* ENTER *to exit this command)*

Because the object was converted to a polyline, all objects are offset at the same time, as shown in Figure 4.76 (Right).

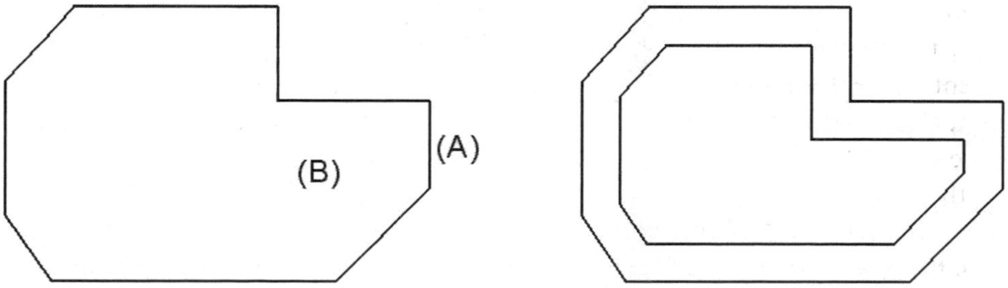

FIGURE 4.76

Multiple Polyline Editing

Multiple editing of polylines allows for multiple objects to be converted to polylines. This is accomplished with the PEDIT command and the Multiple option. Illustrated in Figure 4.77 (Left) is a rectangle and four slots, all considered individual objects. When you run the Multiple option of the PEDIT command, not only can you convert all objects at once into individual polylines, but you can join the endpoints of common shapes as well. The result of editing multiple polylines and then joining them is illustrated in Figure 4.77 (Right).

TRY IT!

Open the drawing file 04_Pedit Multiple1. Use the prompt sequence below and Figure 4.77 to illustrate how the PEDIT command is used.

Command: **PE** *(For PEDIT)*

Select polyline or [Multiple]: **M** *(For Multiple)*

Select objects: **All**

Select objects: *(Press* ENTER *to continue)*

Convert Lines, Arcs and Splines to polylines [Yes/No]? <Y> *(Press* ENTER*)*

Enter an option [Close/Open/Join/Width/Fit/Spline/Decurve/Ltype gen/Reverse/Undo]: **J** *(For Join)*

Join Type = Extend

Enter fuzz distance or [Jointype] <0.0000>: *(Press* ENTER *to accept this default value)*

15 segments added to 5 polylines

Enter an option [Close/Open/Join/Width/Fit/Spline/Decurve/Ltype gen/Reverse/Undo]: *(Press* ENTER *to exit)*

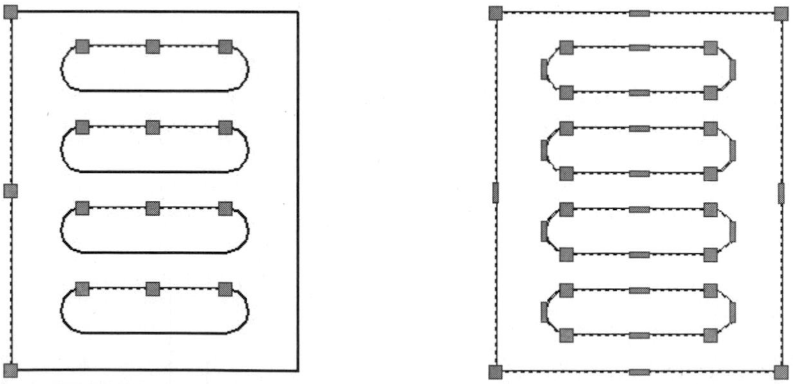

FIGURE 4.77

As a general rule when joining polylines, you cannot have gaps present or overlapping occurring when performing this operation. This is another feature of using the Multiple option of the PEDIT command. This option works best when joining objects that have a small gap or overlap. After selecting the objects to join, you will be asked to enter a fuzz factor. This is the distance used by this command to bridge a gap or trim overlapping lines. You could measure the distance between two objects to determine this value. Study the following example for automatically creating corners in objects using a fuzz factor.

TRY IT!

Open the drawing file 04_Pedit Multiple2. In this example, one of the larger gaps was measured to be 0.12 units in length. As a result, a fuzz factor slightly larger than this calculated value is used (0.13 units). The completed object is illustrated in Figure 4.78 (Right). You may have to experiment with various fuzz factors before you arrive at the desired results.

```
Command: PE (For PEDIT)
Select polyline or [Multiple]: M (For Multiple)
Select objects: All
Select objects: (Press ENTER to continue)
Convert Lines, Arcs and Splines to polylines [Yes/No]? <Y>
(Press ENTER)
Enter an option [Close/Open/Join/Width/Fit/Spline/Decurve/
Ltype gen/Reverse/Undo]: J (For Join)
Join Type = Extend
Enter fuzz distance or [Jointype] <0.25>: 0.13
11 segments added to polyline
Enter an option [Close/Open/Join/Width/Fit/Spline/Decurve/
Ltype gen/Reverse/Undo]: (Press ENTER to exit)
```

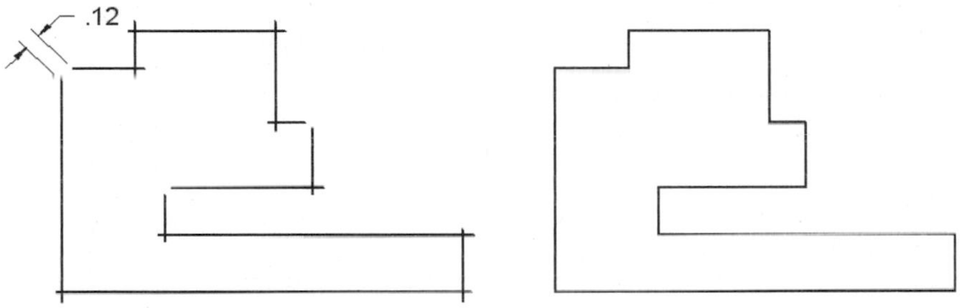

FIGURE 4.78

EXPLODING OBJECTS

Using the EXPLODE command on a polyline, dimension, array, or block separates the single object into its individual parts. Choose this command from one of the following:

- From the Ribbon > Home Tab > Modify Panel
- The Menu Bar (Modify > Explode)
- The Modify Toolbar of the AutoCAD Classic Workspace
- The keyboard (X or EXPLODE)

Illustrated in Figure 4.79 (Left) is a polyline that is considered one object. Using the EXPLODE command and selecting the polyline breaks the polyline into numerous individual objects, as shown in Figure 4.79 (Right).

Command: X *(For EXPLODE)*

Select objects: *(Select the polyline)*

Select objects: *(Press ENTER to perform the explode operation)*

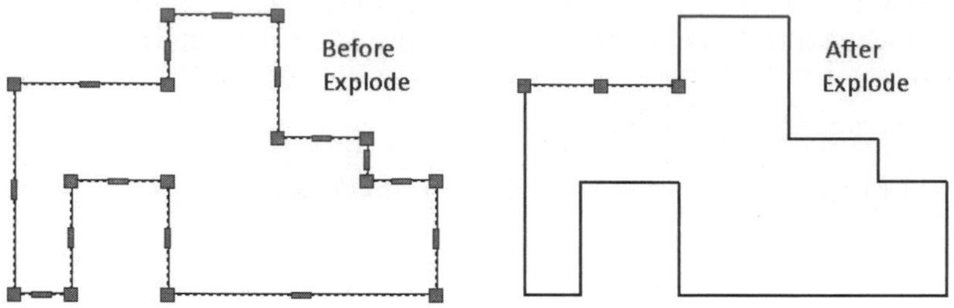

FIGURE 4.79

NOTE

Exploding objects may be necessary in some cases, but in general, it should be avoided; selecting and modifying objects becomes more difficult after breaking them down into individual entities. Avoid exploding dimensions and hatching in particular, because they lose their associative properties. They can no longer be automatically updated as they are modified. (Hatching will be discussed in Chapter 9 and Dimensions in Chapters 10 and 11.)

LENGTHENING OBJECTS

The LENGTHEN command is used to change the length of a selected object without disturbing other object qualities such as angles of lines or radii of arcs. Choose this command from one of the following:

- From the Ribbon > Home Tab > Modify Panel (Expanded)
- The Menu Bar (Modify > Lengthen)
- The keyboard (LEN or LENGTHEN)

TRY IT!

Open the drawing file 04_Lengthen. Use the illustration in Figure 4.80 and the command sequence below for modifying the drawing.

Command: **LEN** *(For LENGTHEN)*

Select an object or [DElta/Percent/Total/DYnamic]: **DE** *(For DElta)*

Enter delta length or [Angle] <0.000>: **0.125** *(Amount to lengthen object by)*

Select an object to change or [Undo]: *(Select the line at "A")*

Select an object to change or [Undo]: *(Select the line at "B")*

Select an object to change or [Undo]: *(Press ENTER to exit)*

Command: **LEN** *(For LENGTHEN)*

Select an object or [DElta/Percent/Total/DYnamic]: **P** *(For Percent)*

Enter percentage length <100.0000>: **200** *(Doubles the length of the object)*

Select an object to change or [Undo]: *(Select the line at "C")*

Select an object to change or [Undo]: *(Press* ENTER *to exit)*

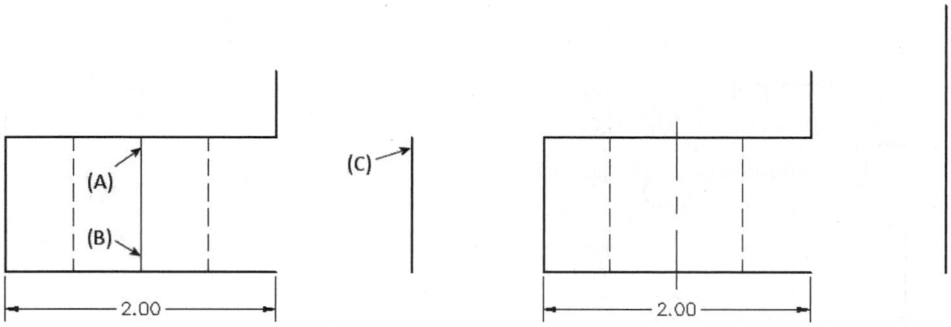

FIGURE 4.80

Command: LEN *(For LENGTHEN)*

Select an object or [DElta/Percent/Total/DYnamic]: T *(For Total)*

Specify total length or [Angle] <1.0000)>: 3 *(Sets new length of the object)*

Select an object to change or [Undo]: *(Select the line at "D")*

Select an object to change or [Undo]: *(Press* ENTER *to exit)*

Command: LEN *(For LENGTHEN)*

Select an object or [DElta/Percent/Total/DYnamic]: DY *(For DYnamic)*

Select an object to change or [Undo]: *(Select the line at "F")*

Specify new end point: END

of *(Select the Endpoint of line "F")*

Select an object to change or [Undo]: *(Press* ENTER *to exit)*

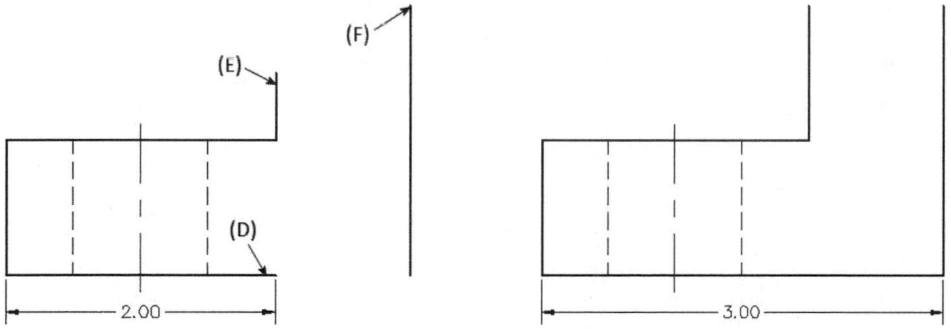

FIGURE 4.81

TIP

The LENGTHEN command can be used to lengthen or shorten objects. To shorten an object, supply a negative delta value, a percent value less than 100, or a total value less than the actual length. Be sure to select the object on the end you want to modify.

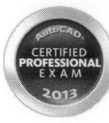

JOINING OBJECTS

For special cases in which individual line or arc segments need to be merged together as a single segment, the JOIN command can be used to accomplish this task. Usually this occurs when gaps occur in line segments and all segments lie in the same line of sight. This condition is sometimes referred to as collinear. Rather than connect the gaps with additional individual line segments, adding unnecessary complexity to the drawing, use the JOIN command to connect all segments as one. Choose this command from one of the following:

- From the Ribbon > Home Tab > Modify Panel (Expanded)
- The Menu Bar (Modify > Join)
- The Modify Toolbar of the AutoCAD Classic Workspace
- The keyboard (J or JOIN)

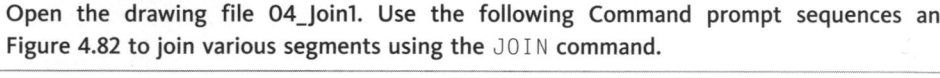

TRY IT!

Open the drawing file 04_Join1. Use the following Command prompt sequences and Figure 4.82 to join various segments using the JOIN command.

Command: J *(For Join)*

Select source object or multiple objects to join at once: *(Select source line "A")*

Select objects to join: *(Select line "B")*

Select objects to join: *(Press* ENTER *to join the segments; notice the created line uses the source object's layer)*

2 lines joined into 1 line

Command: J *(For Join)*

Select source object or multiple objects to join at once: *(Pick at "C")*

Specify opposite corner: *(Pick at "D")*

3 found

Select objects to join: *(Press* ENTER *to join the segments)*

3 lines joined into 1 line

Continue using the JOIN command on the arc segments. Arcs join in a counterclockwise direction, so be careful to select arc "E" before arc "F" in the example. The Close option will be used to change arc "G" into a circle.

Command: J *(For Join)*

Select source object or multiple objects to join at once: *(Select arc "E")*

Select objects to join: *(Select arc "F")*

Select objects to join: *(Press* ENTER *to join the segments)*

2 arcs joined into 1 arc

Command: J *(For Join)*

Select source object or multiple objects to join at once: *(Select arc "G")*

Select objects to join: *(Press* ENTER *to continue)*

Select arcs to join to source or [cLose]: L *(For cLose)*

Arc converted to a circle.

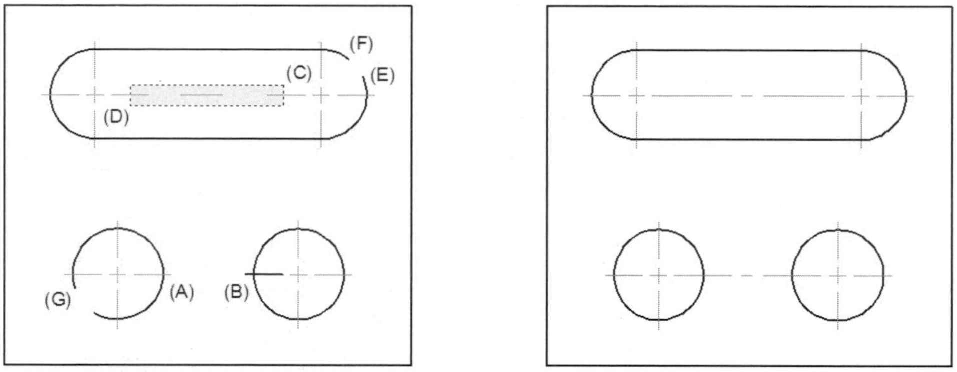

FIGURE 4.82

BLENDING OBJECTS

The BLEND command creates a spline between selected objects. The selected objects can be lines, arcs, helixes, splines, or polylines. Choose this command from one of the following:

- From the Ribbon > Home Tab > Modify Panel (Expand Icon: Fillet, Chamfer, and Blend Curves)
- The Menu Bar (Modify > Blend Curves)
- The Modify Toolbar of the AutoCAD Classic Workspace
- The keyboard (BL or BLEND)

The following command sequence and Figure 4.83 show how the BLEND command is used.

Open the drawing file 04_Blend. Use the following prompts and Figure 4.83 to blend two lines, two arcs, and a helix and spline together.

TRY IT!

Command: BL (For BLEND)

Continuity=Tangent

Select first object or [CONtinuity]: *(Select the line at "A")*

Select second object: *(Select the line at "B")*

Command: BL *(For BLEND)*

Continuity=Tangent

Select first object or [CONtinuity]: *(Select the arc at "C")*

Select second object: *(Select the arc at "D")*

Command: BL *(For BLEND)*

Continuity=Tangent

Select first object or [CONtinuity]: *(Select the helix at "E")*

Select second object: *(Select the spline at "F")*

The results are shown in Figure 4.83 (Right). Try changing the results by picking different ends of the objects and changing the CONtinuity option from Tangent to Smooth.

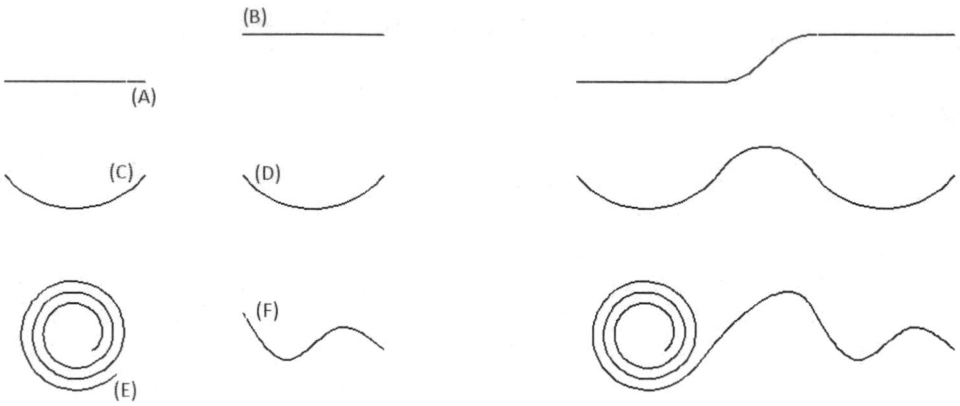

FIGURE 4.83

UNDOING AND REDOING OPERATIONS

The UNDO command can be used to undo the previous task or command action while the REDO command reverses the effects of any previous undo. Choose these commands from one of the following:

- The Quick Access Toolbar
- The Menu Bar (Edit > Undo) or (Edit > Redo)
- The Standard Toolbar of the AutoCAD Classic Workspace
- The keyboard (U or UNDO) or (REDO)
- Select anywhere in the drawing and right-click

For example, if you draw an arc followed by a line followed by a circle, issuing the UNDO command will undo the action caused by the most recent command; in this case, the circle would be removed from the drawing database. This represents one of the easiest ways to remove data or backtrack the drawing process.

Expanding the Undo list found in the Quick Access Toolbar, shown in Figure 4.84 (Left), allows you to undo several actions at once. From this example, notice that the Rectangle, Line, and Circle actions are highlighted for removal.

You can also reverse the effect of the UNDO command by using REDO immediately after the undo operation.

Clicking the REDO command button from the Quick Access Toolbar negates one undo operation. You can click on this button to cancel the effects of numerous undo operations.

As with undo, you can also redo several actions at once through the Redo list shown in Figure 4.84 (Right).

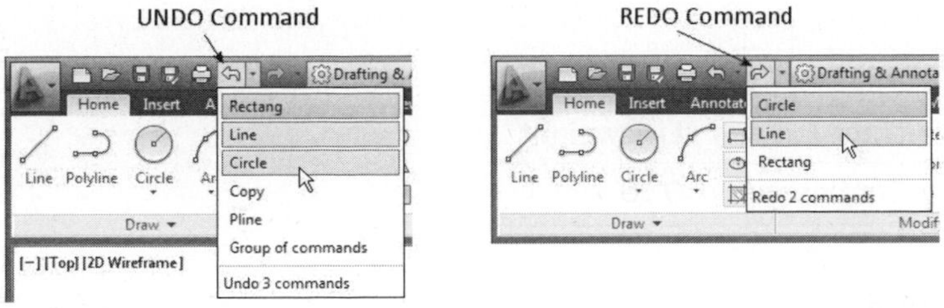

FIGURE 4.84

NOTE

When grouping actions to be undone, you cannot, in Figure 4.84, for example, highlight Rectangle, skip Line, and highlight Circle to be removed. The groupings to undo must be strung together in this dialog box. Redo only works if you have undone a previous operation. Otherwise, redo remains inactive.

TUTORIAL EXERCISE: 04_ANGLE.DWG

Purpose

This tutorial is designed to allow you to construct a one-view drawing of the Angle drawing as shown in Figure 4.85 using the `ROTATE` and `LENGTHEN` commands.

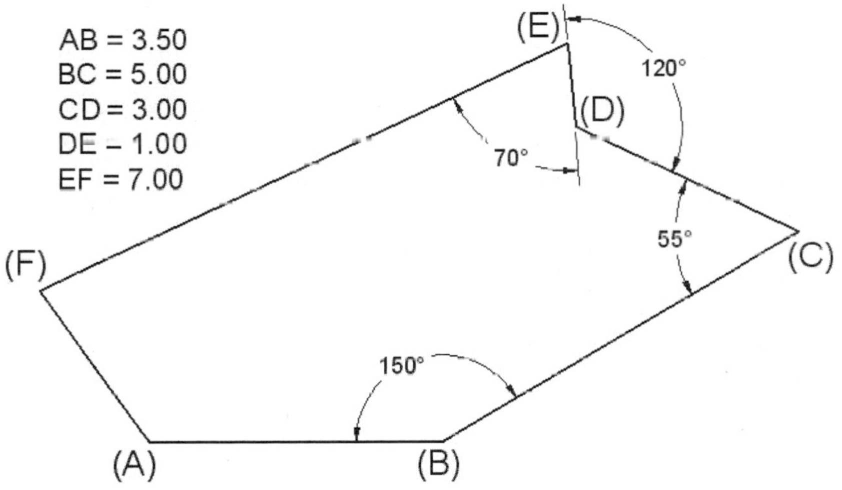

FIGURE 4.85

System Settings

Start a new drawing from scratch using the Acad.dwt template. Use the Drawing Units dialog box and change the precision of decimal units from four to two places. Use the current default settings for the limits of this drawing, (0,0) for the lower-left corner, and (12,9) for the upper-right corner. Check to see that the following Object Snap modes are already set: Endpoint, Extension, Intersection, and Center.

Layers

Create the following layer with the format:

Name	Color	Linetype	Lineweight
Object	White	Continuous	0.50 mm

Suggested Commands

Make the Object layer current. Follow the steps provided to complete the drawing. You will begin this drawing by constructing line "AB," which is horizontal. Use the `ROTATE` command to copy and rotate line "AB" at an angle of 150° in the clockwise direction. Once the line is copied and rotated, use the `LENGTHEN` command and modify the new line to the designated length. Repeat this procedure for lines "CD," "DE," and "EF." Complete the drawing by constructing a line segment from the endpoint at vertex "F" to the endpoint at vertex "A."

Step 1

Draw line "AB" using Polar Tracking and Direct Distance mode, as shown in the following image. (Line "AB" should be a horizontal line.)

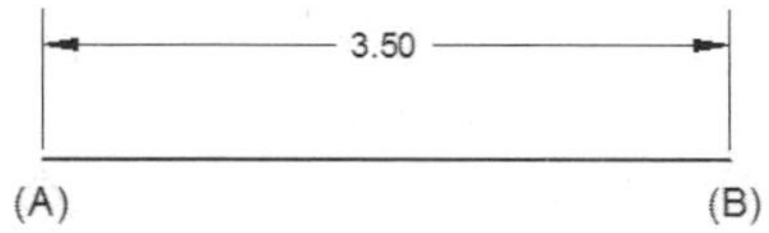

FIGURE 4.86

Step 2

One technique of constructing the adjacent line at 150° from line "AB" is to use the ROTATE command with the Copy option. Select line "AB" as the object to rotate, pick the endpoint at "B" as the base point for the rotation, and enter a value of -150° for the angle. Entering a negative angle revolves the line in the clockwise direction. The Copy option allows you to create a new line without losing the original line.

```
Command: RO (For ROTATE)
Current positive angle in UCS: ANGDIR=counterclockwise ANGBASE=0
Select objects: (Select the line "AB")
Select objects: (Press ENTER to continue)
Specify base point: (With an Endpoint OSNAP pick at "B")
Specify rotation angle or [Copy/Reference] <0>: C (For Copy)
Rotating a copy of the selected objects.
Specify rotation angle or [Copy/Reference] <0>: -150
```

The result is shown in Figure 4.87.

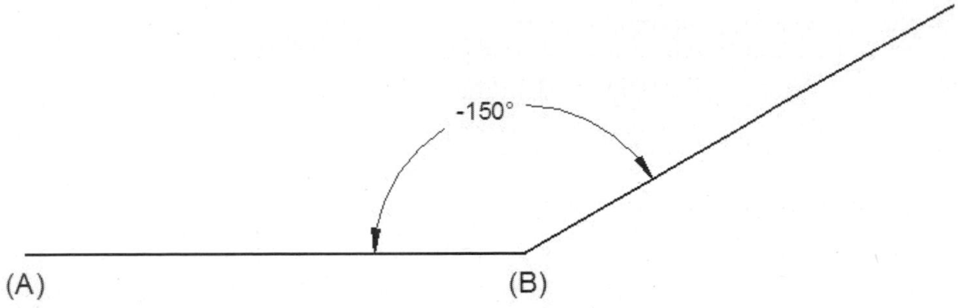

FIGURE 4.87

Step 3

The rotate operation allowed line "AB" to be rotated and copied at the correct angle, namely -150°. However, the new line is the same length as line "AB." Use the LENGTHEN command to increase the length of the new line to a distance of 5.00 units. Use the Total option, specify the new total length of 5.00, and select the end of the line at "1" as the object to change, as shown in Figure 4.88 (Left).

> Command: LEN *(For LENGTHEN)*
>
> Select an object or [DElta/Percent/Total/DYnamic]: T *(For Total)*
>
> Specify total length or [Angle] <1.00)>: 5.00
>
> Select an object to change or [Undo]: *(Pick the end of the line at "1")*
>
> Select an object to change or [Undo]: *(Press* ENTER *to exit this command)*

The result is shown in Figure 4.88 (Right).

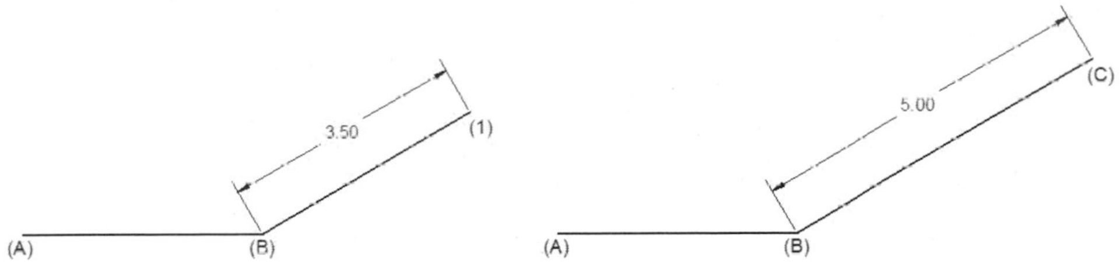

FIGURE 4.88

Step 4

Use the illustration in Figure 4.89 and the command sequence to rotate and copy the next line segment. Select line "BC" as the object to rotate, pick the endpoint at "C" as the base point of the rotation, and enter a value of -55° for the angle. Entering a negative angle copies the line in the clockwise direction.

> Command: RO *(For ROTATE)*
>
> Current positive angle in UCS: ANGDIR=counterclockwise ANGBASE=0
>
> Select objects: *(Select the line "BC")*
>
> Select objects: *(Press* ENTER *to continue)*

```
Specify base point: (With an Endpoint OSNAP pick at "C")
Specify rotation angle or [Copy/Reference] <0>: C (For Copy)
Rotating a copy of the selected objects.
Specify rotation angle or [Copy/Reference] <0>: -55
```

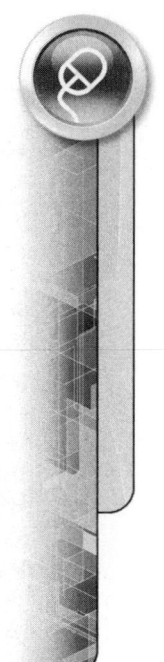

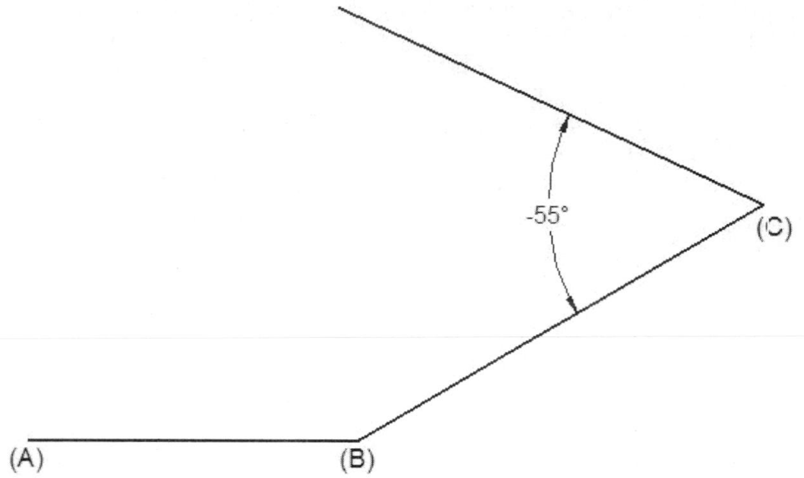

FIGURE 4.89

Step 5

Then use the LENGTHEN command to reduce the length of the new line from 5.00 units to 3.00 units. Use the Total option, specify the new total length of 3.00, and select the end of the line at "1" as the object to change, as shown in the following image on the left.

```
Command: LEN (For LENGTHEN)
Select an object or [DElta/Percent/Total/DYnamic]: T (For Total)
Specify total length or [Angle] <5.00)>: 3.00
Select an object to change or [Undo]: (Pick the end of the line at "1")
Select an object to change or [Undo]: (Press ENTER to exit this command)
```

The result is shown in the following image on the right.

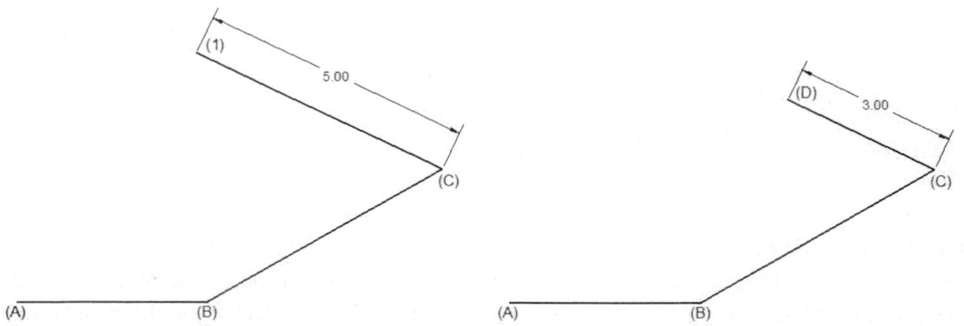

FIGURE 4.90

Step 6

Use the illustration in the following image on the left and the command sequence to rotate and copy the next line segment. Select line "CD" as the object to rotate, pick the endpoint at "D" as the base point, and enter a value of 120° for the angle. Entering a positive angle copies the line in the counterclockwise direction.

Command: RO *(For ROTATE)*

Current positive angle in UCS: ANGDIR=counterclockwise ANGBASE=0

Select objects: *(Select the line "CD")*

Select objects: *(Press* ENTER *to continue)*

Specify base point: *(With an Endpoint OSNAP pick at "D")*

Specify rotation angle or [Copy/Reference] <0>: C *(For Copy)*

Rotating a copy of the selected objects.

Specify rotation angle or [Copy/Reference] <0>: 120

Then use the LENGTHEN command to reduce the length of the new line from 3.00 units to 1.00 unit, as shown in the following image on the right.

Command: LEN *(For LENGTHEN)*

Select an object or [DElta/Percent/Total/DYnamic]: T *(For Total)*

Specify total length or [Angle] <3.00)>: 1.00

Select an object to change or [Undo]: *(Pick the end of the line at "1")*

Select an object to change or [Undo]: *(Press* ENTER *to exit this command)*

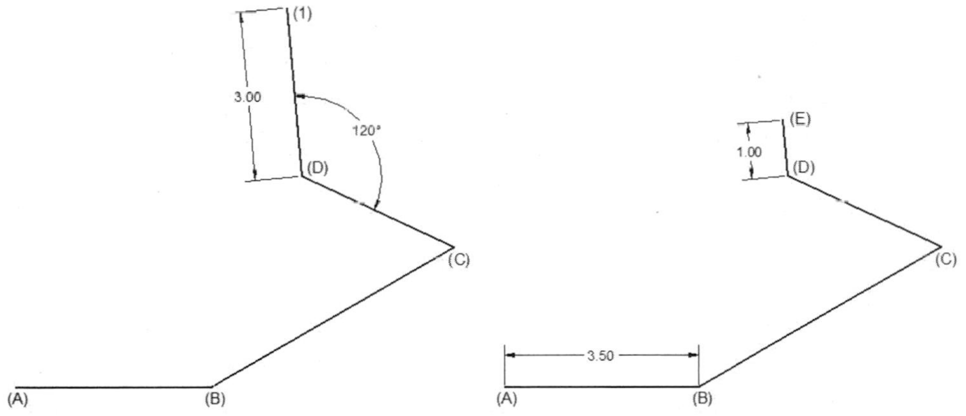

FIGURE 4.91

Step 7

Use the illustration in the following image on the left and the command sequence to rotate and copy the final line segment. Select line "DE" as the object to rotate, pick the endpoint at "E" as the base point, and enter a value of -70° for the angle. Entering a negative angle copies the line in the clockwise direction.

Command: RO *(For ROTATE)*

Current positive angle in UCS: ANGDIR=counterclockwise ANGBASE=0

Select objects: *(Select the line "DE")*

Select objects: *(Press* ENTER *to continue)*

Specify base point: *(With an Endpoint OSNAP pick at "E")*

Specify rotation angle or [Copy/Reference] <0>: C *(For Copy)*

Rotating a copy of the selected objects.

Specify rotation angle or [Copy/Reference] <0>: -70

Then use the LENGTHEN command to increase the length of the new line from 1.00 unit to 7.00 units, as shown in the following image on the right.

Command: LEN *(For LENGTHEN)*

Select an object or [DElta/Percent/Total/DYnamic]: T *(For Total)*

Specify total length or [Angle] <1.00)>: 7.00

Select an object to change or [Undo]: *(Pick the end of the line at "1")*

Select an object to change or [Undo]: *(Press* ENTER *to exit this command)*

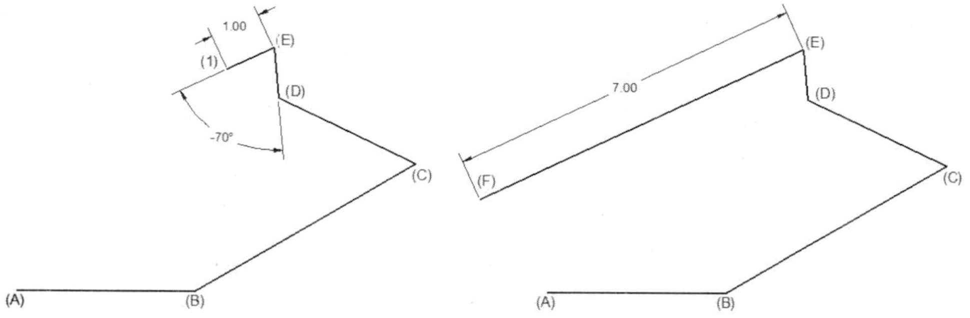

FIGURE 4.92

Step 8

Connect endpoints "F" and "A" with a line as shown in the following image on the left.

Command: L (For LINE)

Specify first point: (Pick the endpoint of the line at "F")

Specify next point or [Undo]: (Pick the endpoint of the line at "A")

Specify next point or [Undo]: (Press ENTER to exit this command)

The completed drawing is illustrated in the following image on the right. You may add dimensions at a later date.

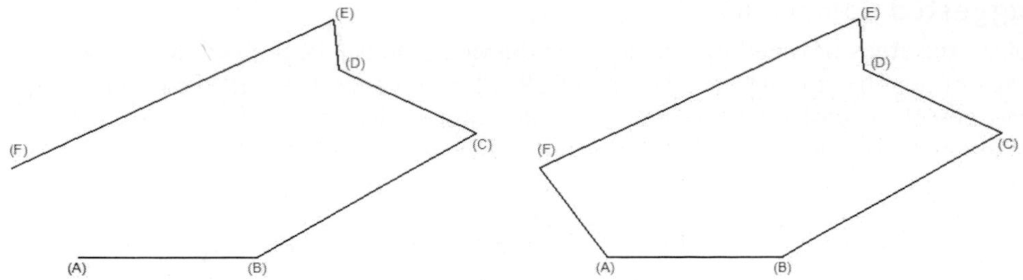

FIGURE 4.93

TUTORIAL EXERCISE: 04_GASKET.DWG

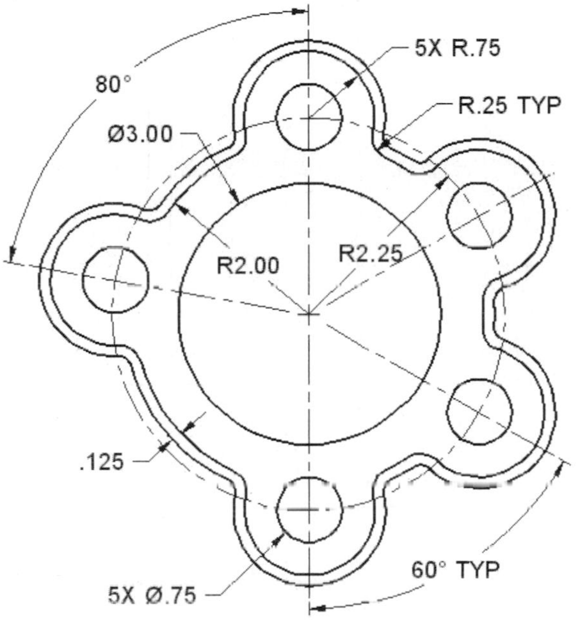

FIGURE 4.94

Purpose

This tutorial is designed to allow you to construct a one-view drawing of the gasket using the Array dialog box.

System Settings

Start a new drawing from scratch using the Acad.dwt template. Use the current default settings for the units and limits of this drawing, (0,0) for the lower-left corner, and (12,9) for the upper-right corner. Check to see that the following Object Snap modes are already set: Endpoint, Extension, Intersection, Center, and Quadrant.

Layers

Create the following layers with the format:

Name	Color	Linetype	Lineweight
Object	White	Continuous	0.50 mm
Center	Yellow	Center2	Default

Suggested Commands

Follow the steps provided to complete the drawing. You will begin by drawing the basic shape of the object using the LINE and CIRCLE commands. Lay out a centerline circle; draw one of the gasket tabs at the top of the center circle. Use polar array operations to create four copies in the −180° direction and two copies in the 80° direction. Trim out the excess arc segments to form the gasket. Convert the outer profile of the gasket into one continuous polyline object and offset this object 0.125 units to the outside of the gasket.

Step 1

After starting a drawing from scratch and creating layers, make the Object layer current and construct a circle of 2.00 radius with its center at absolute coordinate 4.00,4.00, as shown in the following image on the left.

Make the Center layer current and construct a circle of 2.25 radius using the previous center point, as shown in the following image on the right.

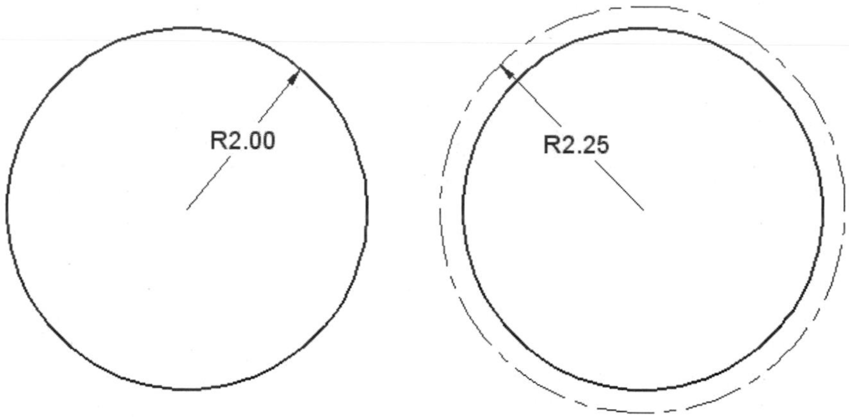

FIGURE 4.95

Step 2

Make the Object layer current again and construct a circle of radius 0.75 from the quadrant at the top of the centerline circle, as shown in the following image on the left. Also construct a circle of 0.75 diameter from the same quadrant at the top of the centerline circle, as shown in the following image on the right.

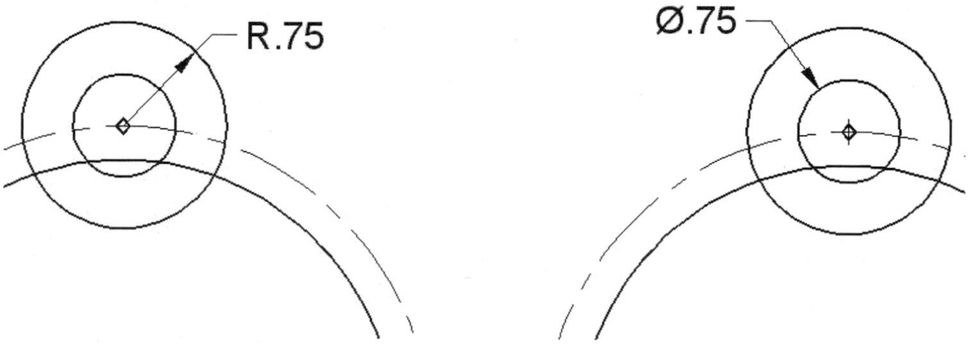

FIGURE 4.96

Step 3

Copy the two top circles just created in a polar (circular) pattern using the ARRAYPOLAR command. Using the illustration in the following image on the left and the command sequence, set the center of the 2.00-radius circle as the center point of the array, change the total number of items to 4, and the angle to fill to -180°. The negative angle drives the array in the clockwise direction, as shown in the following image on the right.

Command: ARRAYPOLAR

Select objects: *(Select the circles "A" and "B")*

Select objects: *(Press* ENTER *to continue)*

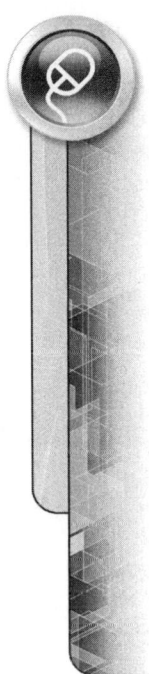

Type = Polar Associative = Yes

Specify center point of array or [Base point/Axis of rotation]: *(Pick the center of the 2.00-radius circle at "C")*

Select grip to edit array or [ASsociative/Base point/Items/Angle between/Fill angle/ROWs/Levels/ROTate items/eXit] <eXit>: I *(For Items)*

Enter number of items in array or [Expression] <6>: 4

Select grip to edit array or [ASsociative/Base point/Items/Angle between/Fill angle/ROWs/Levels/ROTate items/eXit] <eXit>: F *(For Fill angle)*

Specify the angle to fill (+=ccw, -=cw) or [EXpression] <360>: -180

Select grip to edit array or [ASsociative/Base point/Items/Angle between/Fill angle/ROWs/Levels/ROTate items/eXit] <eXit>: *(Press* ENTER *to exit)*

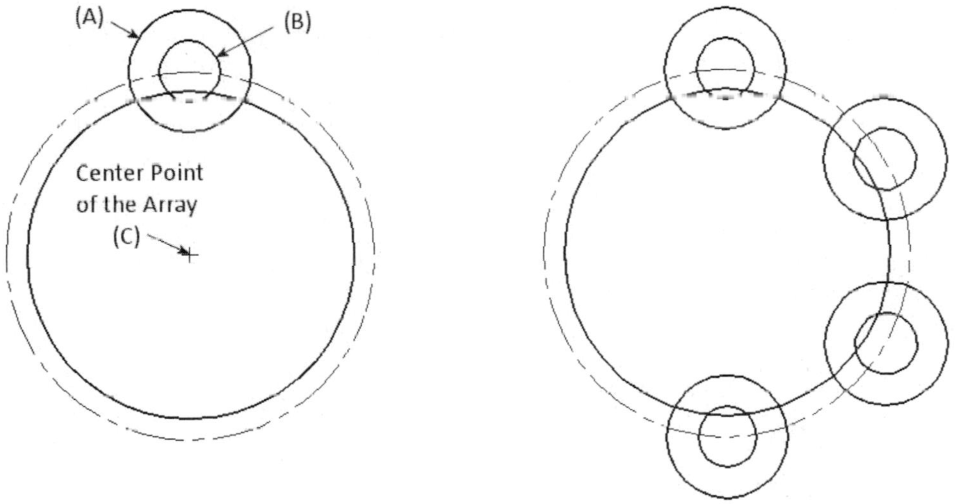

FIGURE 4.97

Step 4

To perform another array operation on the top two circles you will need to break apart the associative array just created. Activate the EXPLODE command and select the array. Now, using the illustration in the following image on the left and the command sequence, set the center of the 2.00-radius circle as the center point of the array, change the total number of items to 2 and the angle to fill to 80°. The positive angle drives the array in the counterclockwise direction, as shown in the following image on the right.

Command: **ARRAYPOLAR**

Select objects: *(Select the circles "A" and "B")*

Select objects: *(Press* ENTER *to continue)*

Type = Polar Associative = Yes

Specify center point of array or [Base point/Axis of rotation]: *(Pick the center of the 2.00-radius circle at "C")*

Select grip to edit array or [ASsociative/Base point/Items/Angle between/Fill angle/ROWs/Levels/ROTate items/eXit] <eXit>: **I** (*For Items*)

Enter number of items in array or [Expression] <6>: **2**

Select grip to edit array or [ASsociative/Base point/Items/Angle between/Fill angle/ROWs/Levels/ROTate items/eXit] <eXit>: **F** (*For Fill angle*)

Specify the angle to fill (+=ccw, -=cw) or [EXpression] <360>: **80**

Select grip to edit array or [ASsociative/Base point/Items/Angle between/Fill angle/ROWs/Levels/ROTate items/eXit] <eXit>: *(Press* ENTER *to exit)*

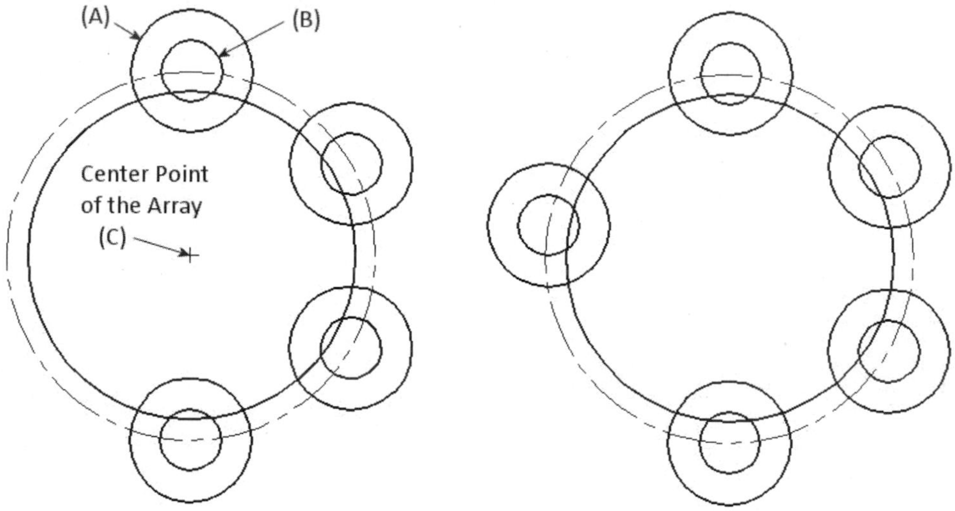

FIGURE 4.98

Step 5

Before you can perform the trim and filleting operations in the next steps, you will need to use the EXPLODE command to break apart the associative array just created. Once the array is non-associative, you can trim out the inside edges of the five circles labeled "A" through "E" using the dashed circle as the cutting edge, as shown in the following image on the left. The results are displayed on the right. It should be noted, that instead of exploding the arrays created in the previous steps, the ASsociative option in the ARRAYPOLAR command could have been used to create non-associative arrays.

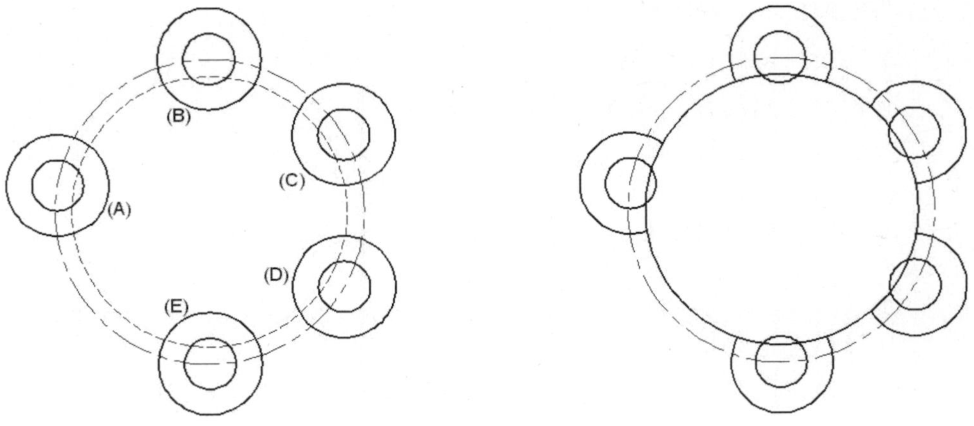

FIGURE 4.99

Step 6

Use the TRIM command again. Select the five dashed arc segments as cutting edges and trim away the portions labeled "A" through "E," as shown in the following image on the left. The results are displayed on the right.

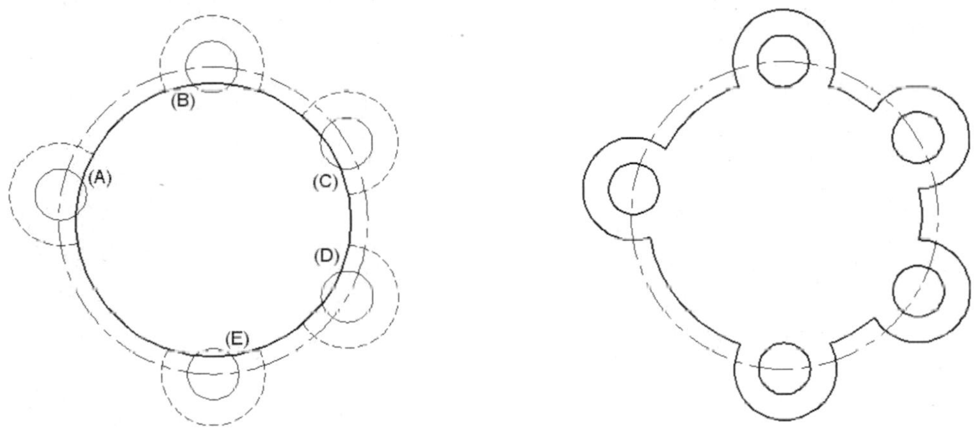

FIGURE 4.100

Step 7

Fillet the inside corners of the 0.75 radius arcs with the 2.00 radius arc using the FILLET command and a radius set to 0.25. These corners are marked by a series of points, shown in the following image on the left. Try using the Multiple option of the FILLET command to make this go faster. The results are displayed on the right.

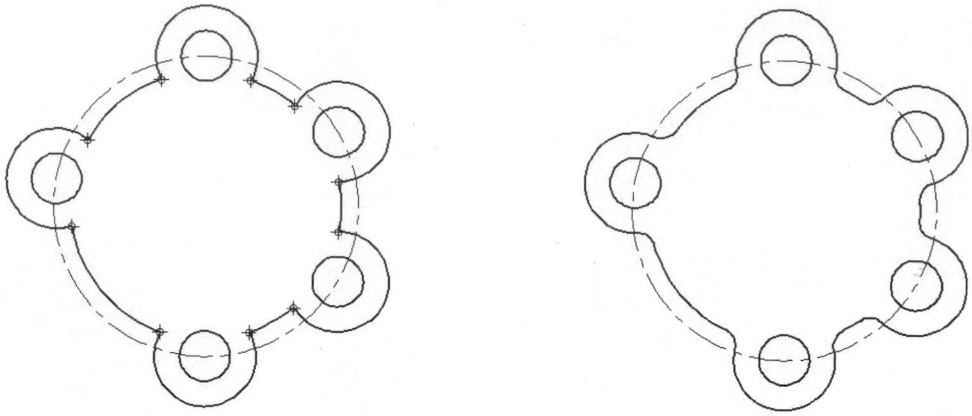

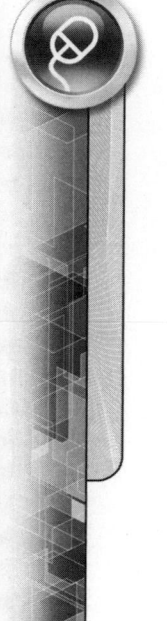

FIGURE 4.101

Step 8

Create the large 1.50-radius hole in the center of the gasket using the CIRCLE command. Change the outer perimeter of the gasket into one continuous polyline object using the PEDIT command. Use the Join option of this command to accomplish this task. Finally, create a copy of the outer profile of the gasket a distance of 0.125 using the OFFSET command. Offset the profile to the outside at "A" as shown in the following image on the left. The completed gasket is displayed on the right.

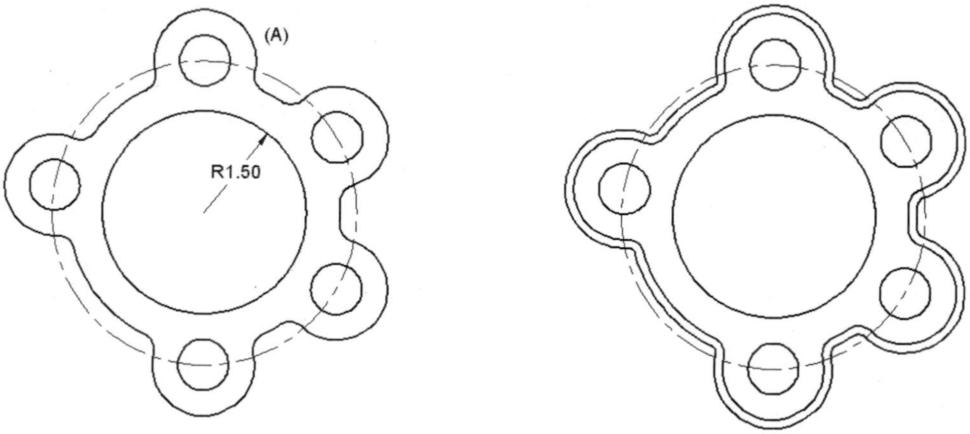

FIGURE 4.102

TUTORIAL EXERCISE: 04_TILE.DWG

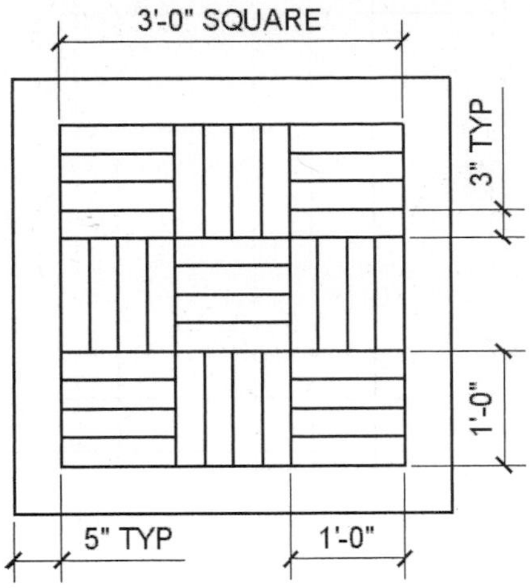

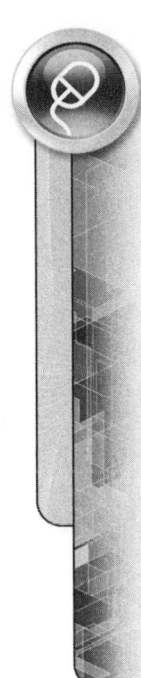

FIGURE 4.103

Purpose

This tutorial is designed to use the OFFSET, COPY, and TRIM commands to complete the drawing of the floor tile shown in the previous image.

System Settings

Start a new drawing from scratch using the Acad.dwt template. Use the Drawing Units dialog box and change the units of measure from decimal to architectural units. Keep the remaining default settings. Use the LIMITS command and change the limits of the drawing to (0,0) for the lower-left corner and (10',8') for the upper-right corner. Use the ZOOM command and the All option to fit the new drawing limits to the display screen.

Check to see that the following Object Snap modes are currently set: Endpoint, Extension, Intersection, and Center.

Layers

Create the following layer with the format:

Name	Color	Linetype	Lineweight
Object	White	Continuous	0.50 mm

Suggested Commands

Make the Object layer current. Follow the steps provided to complete the drawing. You will use the RECTANGLE command to begin the inside square of the tile. The OFFSET command is used to copy the inner square a distance of 5' to form the outer square. The OFFSET and COPY commands are used to copy selected line segments in a rectangular pattern at a specified distance. The TRIM command is then used to form the inside tile patterns.

Step 1

Verify that the current units are set to architectural and that the drawing limits set to 10',8' for the upper-right corner. Be sure to perform a ZOOM-All on your screen. Draw the inner 3'-0" square using the RECTANGLE command, as shown in the following image on the left. Then offset the square 5" to the outside using the OFFSET command, as shown in the following image on the right. Because the Rectangle is actually a polyline, the entire shape offsets to the outside.

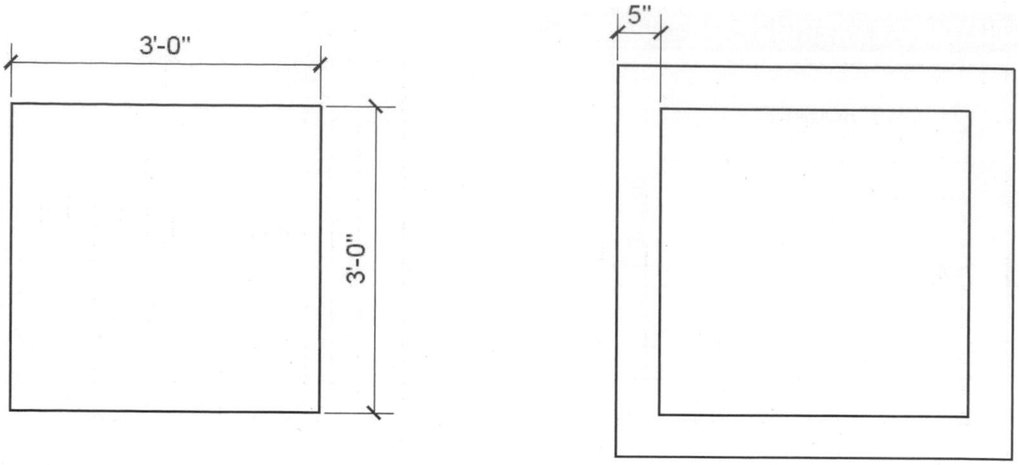

FIGURE 4.104

TIP

When entering inches, it is not necessary to enter the quote (") symbol. Entering 5 is more efficient than entering 5". Entering feet does, however, require the apostrophe (') symbol.

Step 2

Notice that when you click on the inner square in the following image on the left, the entire object highlights because it consists of a single polyline object. Use the EXPLODE command to break up the inner square into individual line segments. Now when you click on a line that is part of the inner square, only that line highlights, as shown in the following image on the right. This procedure is required in order to perform the next step.

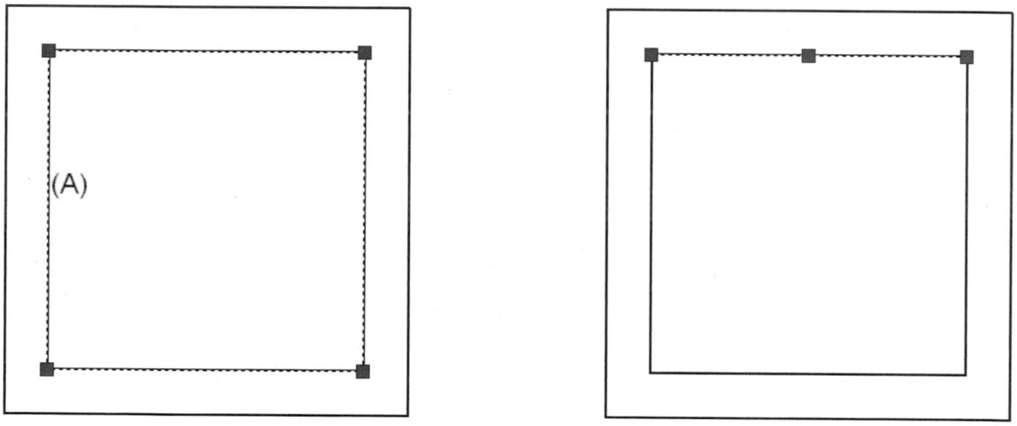

FIGURE 4.105

Step 3

You will now begin laying out the individual tiles with a spacing of 3" between each. There are several methods that could be used to accomplish this. For this step, use the OFFSET command with the Multiple option. Activate the OFFSET command and set the offset distance to 3". Next, pick the line at "A" as the object to offset, as shown in the following image on the left. Select the Multiple option and then after moving your cursor to the right, click the left mouse button 11 times to create 11 new offset lines. Your display should appear similar to the illustration in the following image on the right.

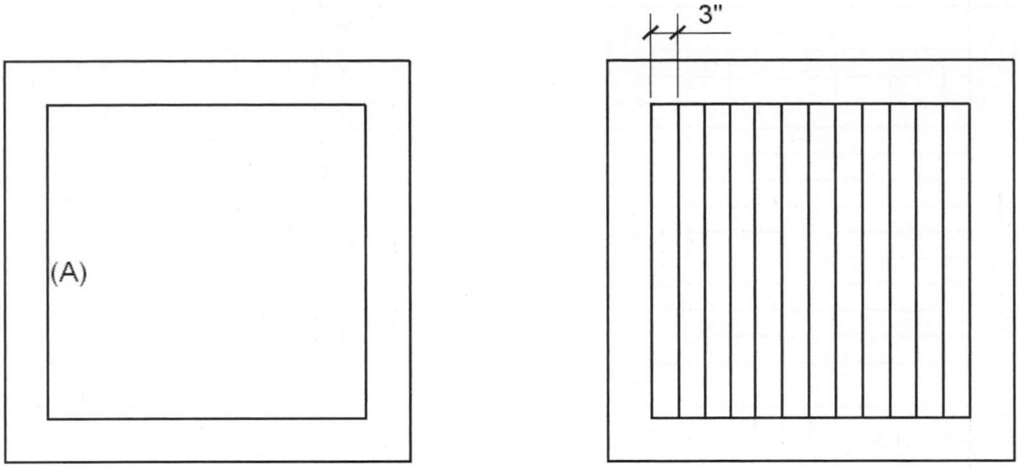

FIGURE 4.106

Step 4

The bottom horizontal line needs to be copied multiple times vertically. You could use the OFFSET command and Multiple option as in the previous step, but instead, try using the COPY command and Array option. Activate the COPY command and pick the line at "A" as the object to copy, as shown in the following image on the left. Select any convenient base point and then select the Array option. Enter 12 as the number of items to array; when prompted to specify the second point, track your cursor straight up and enter 3 for the spacing. Your display should appear similar to the illustration in the following image on the right. The ARRAYRECT command could also have been used for this operation, but the COPY-Array method is often more efficient for creating a single row or single column of objects.

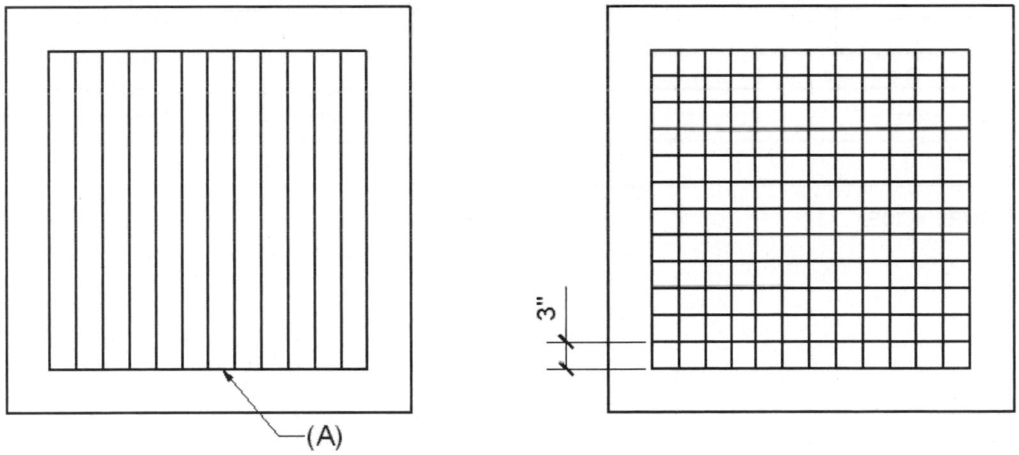

FIGURE 4.107

Step 5

The TRIM command will now be used to clean up the inner lines and form the 3' tiles. When using TRIM, do not press ENTER and select all cutting edges. This would be counterproductive. Instead, select the two vertical dashed lines, as shown in the following image on the left. Then trim away the horizontal segments in zones "A" through "D," as shown in the following image on the right. Remember that if you make a mistake and trim the wrong line, you can enter U in the command line to restore the previous trimmed line and pick the correct line.

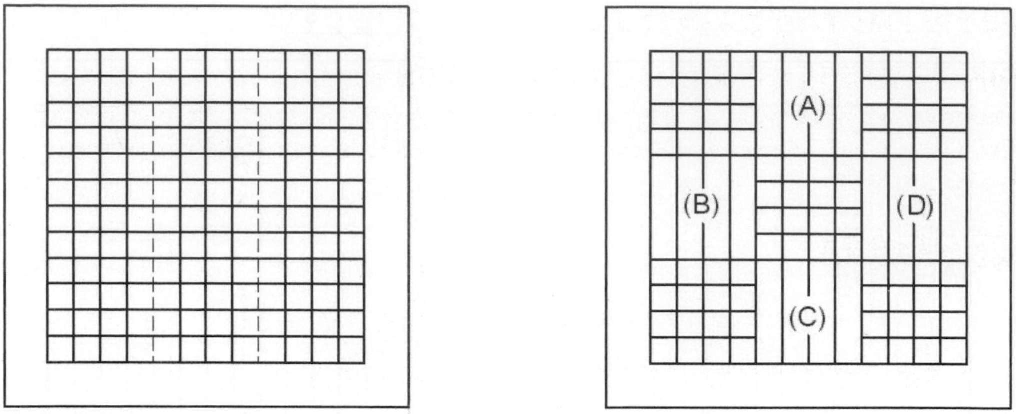

FIGURE 4.108

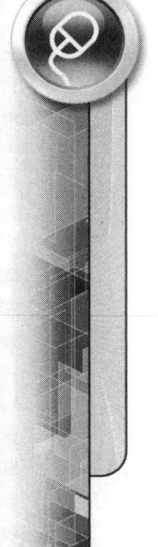

Step 6

Use the TRIM command again to finish cleaning up the object. Select the two horizontal dashed lines as cutting edges, as shown in the following image on the left. Then trim away the vertical segments in zones "A" through "E," as shown in the following image on the right. This completes this exercise on creating the tile.

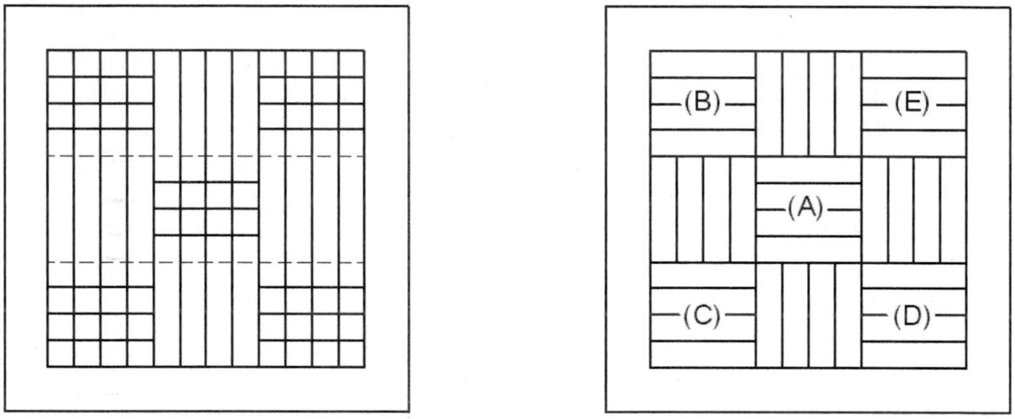

FIGURE 4.109

PROBLEMS FOR CHAPTER 4

DIRECTIONS FOR PROBLEMS 4–1 THROUGH 4–14:

Construct each one-view drawing using the appropriate coordinate mode or Direct Distance mode. Utilize advanced commands such as ARRAY and MIRROR whenever possible.

PROBLEM 4-1

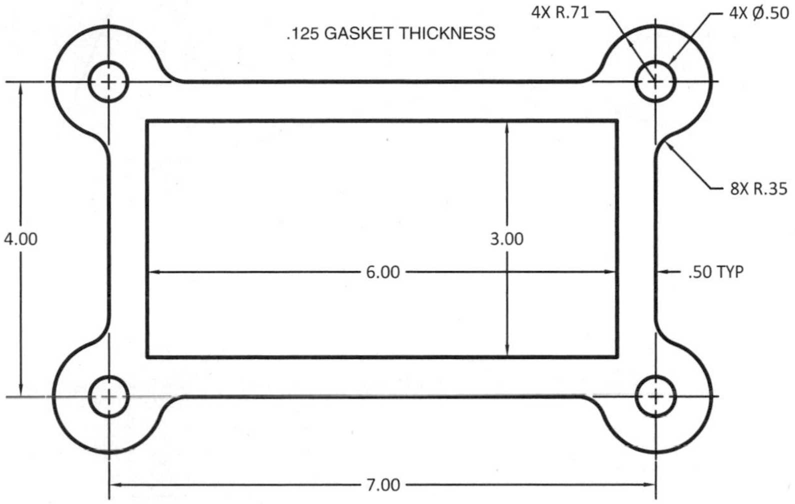

PROBLEM 4-2

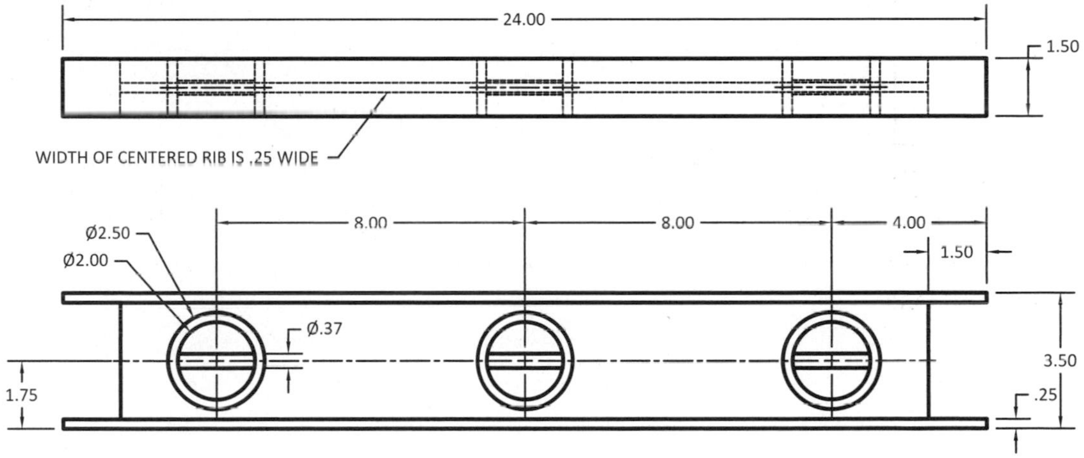

PROBLEM 4-3

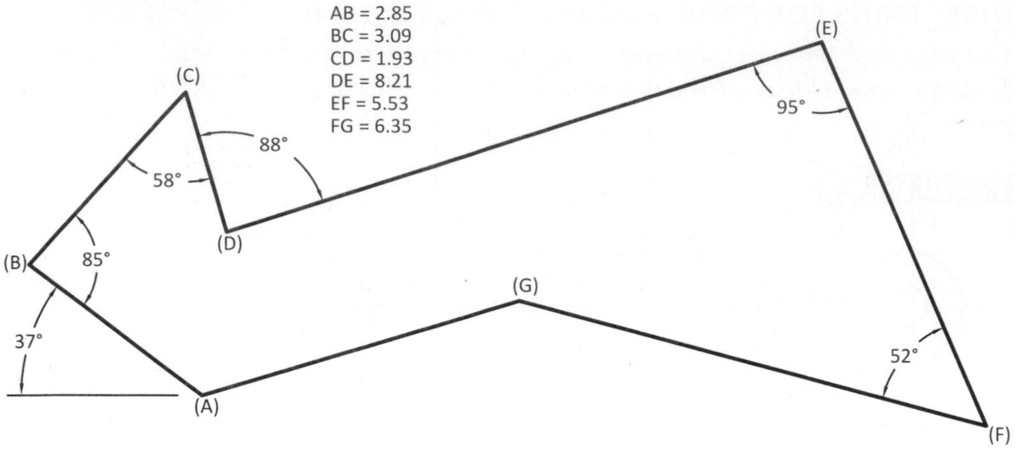

PROBLEM 4-4

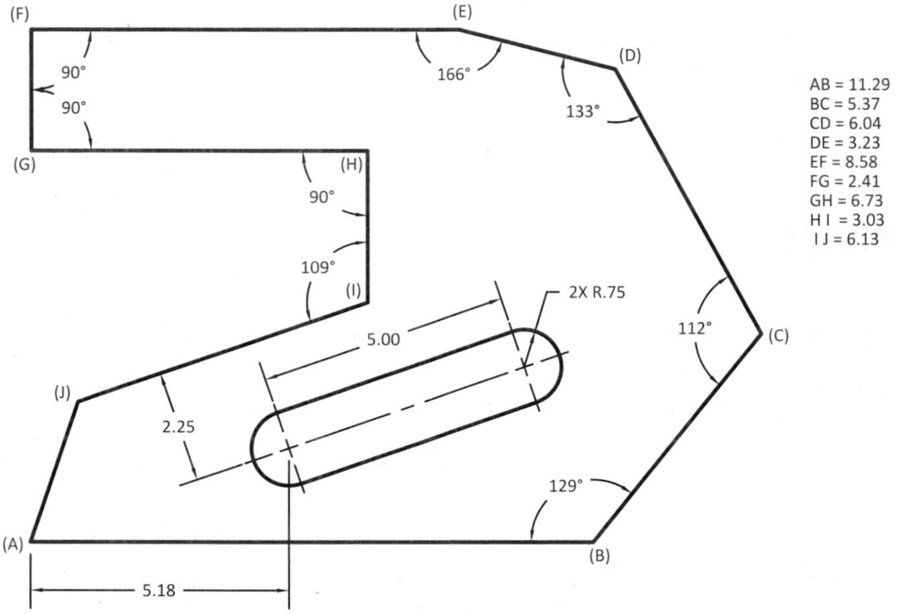

PROBLEM 4-5

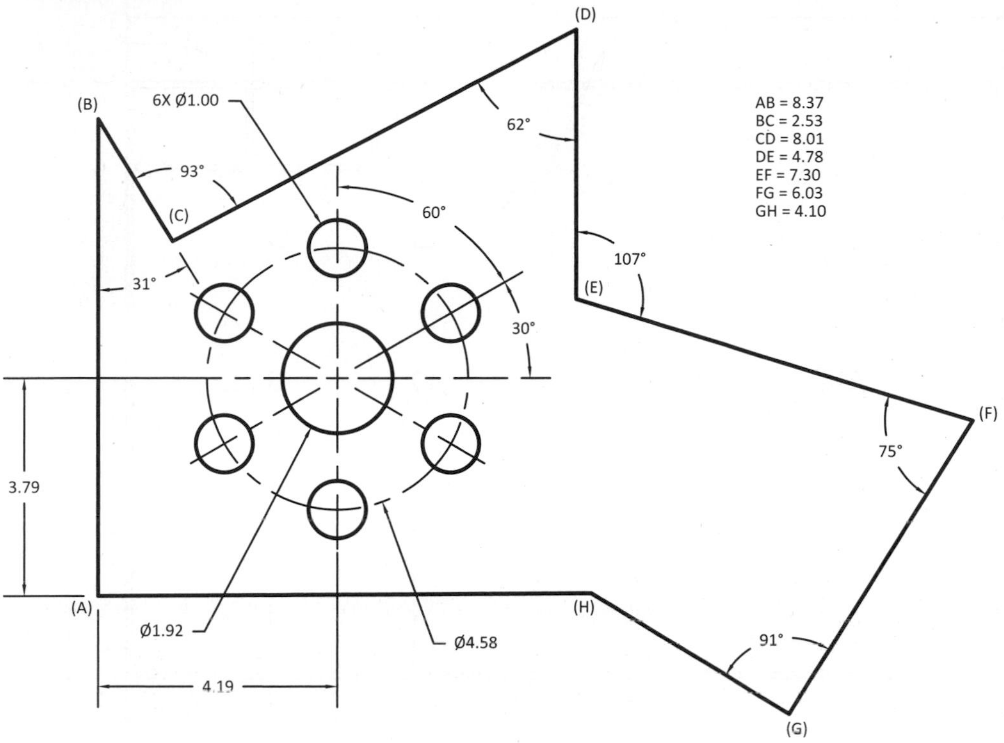

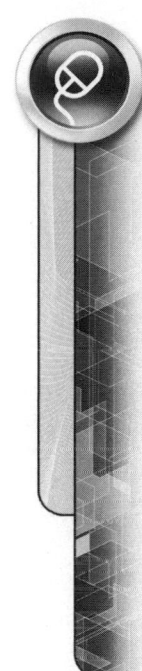

PROBLEM 4-6

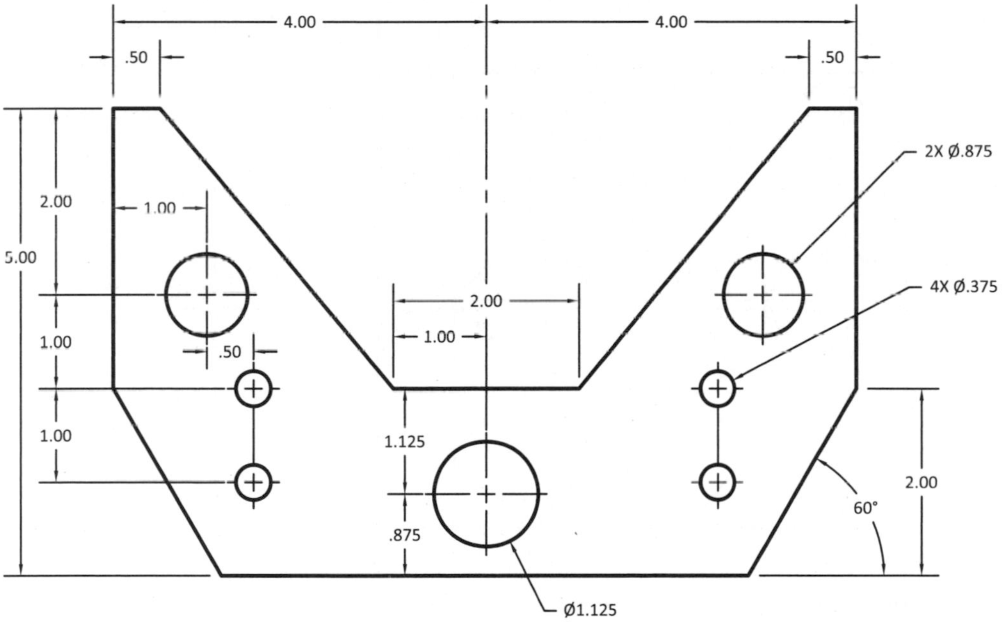

PROBLEM 4-7

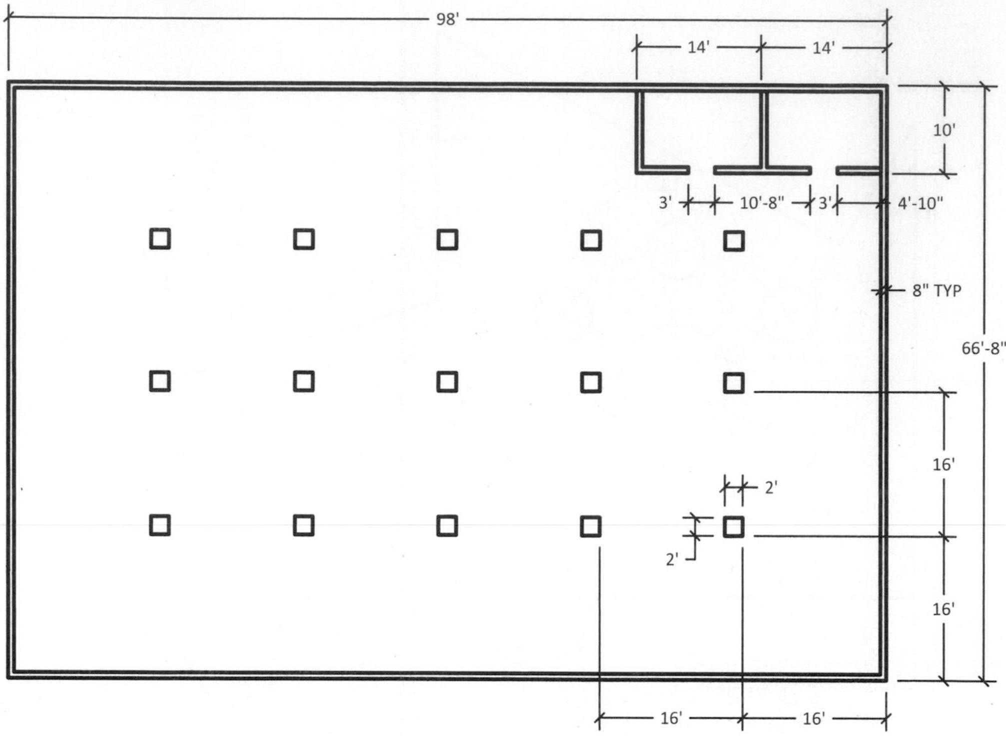

PROBLEM 4-8

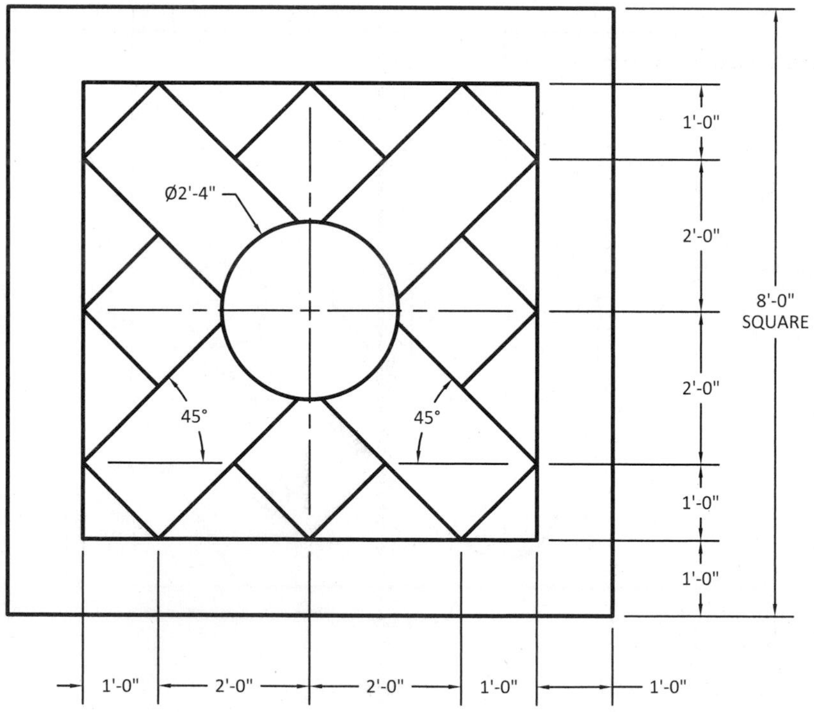

PROBLEM 4-9

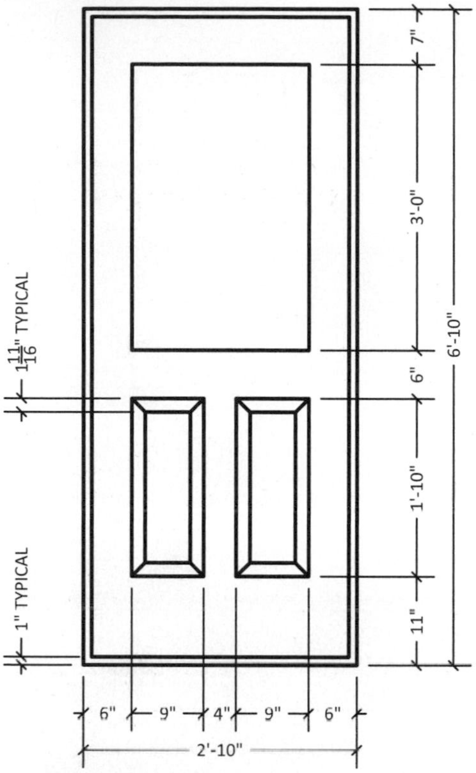

PROBLEM 4-10

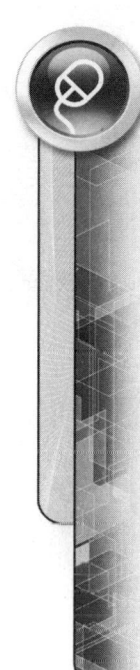

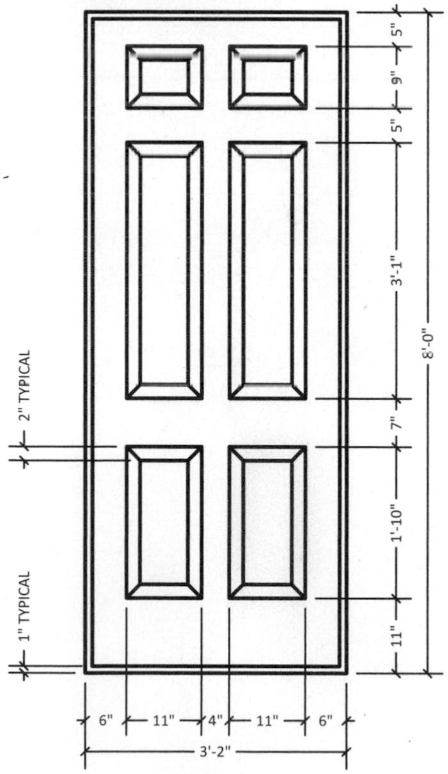

PROBLEM 4-11

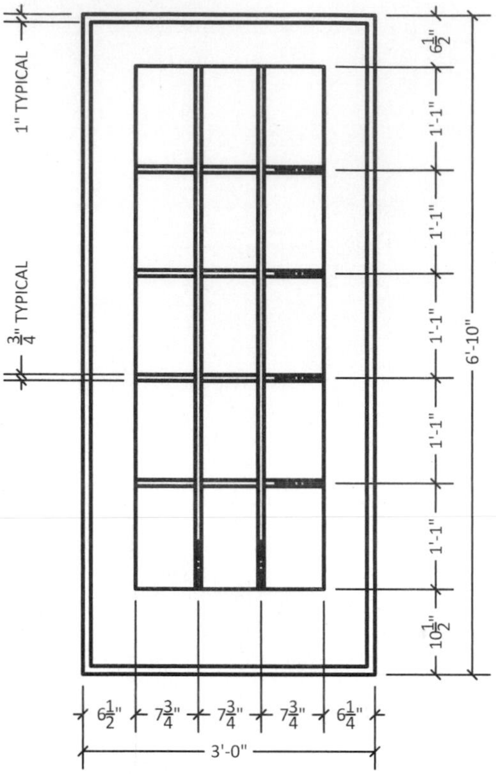

PROBLEM 4-12

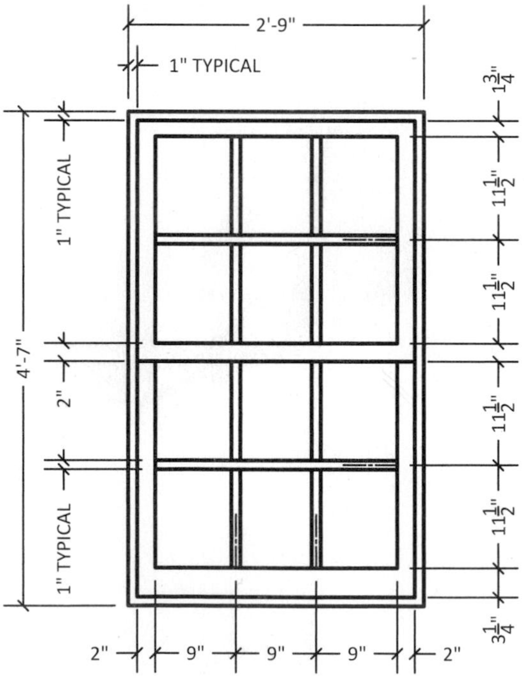

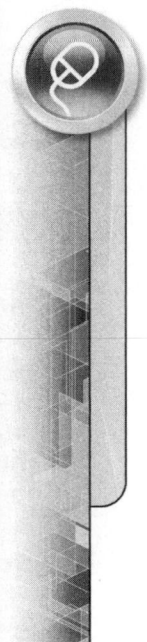

PROBLEM 4-13

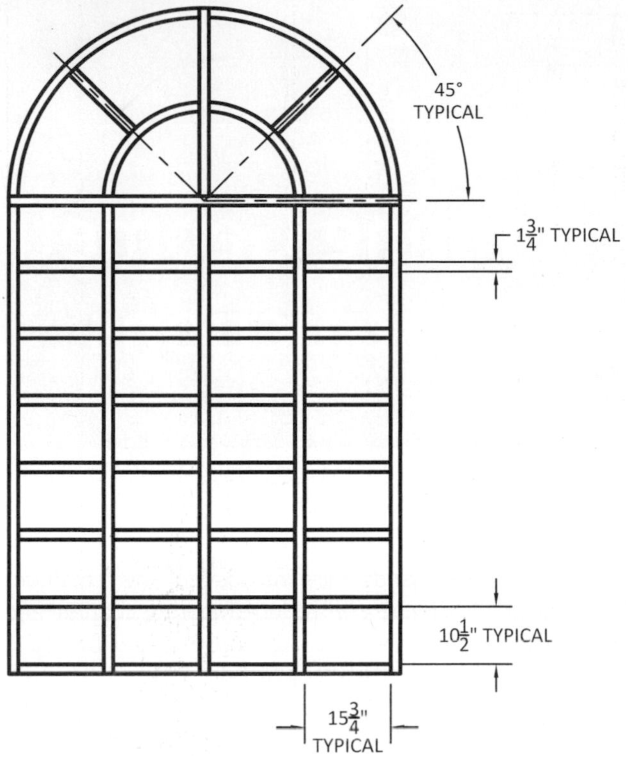

PROBLEM 4-14

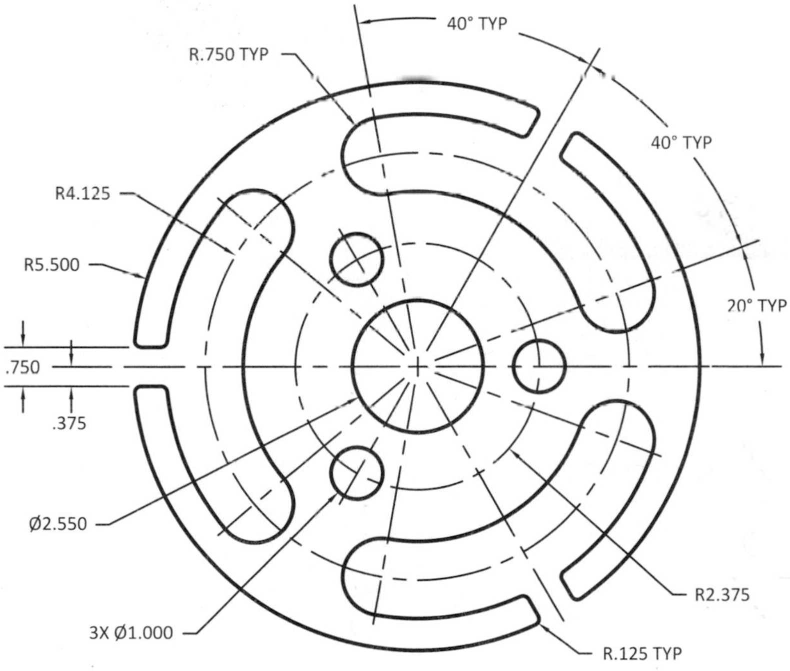

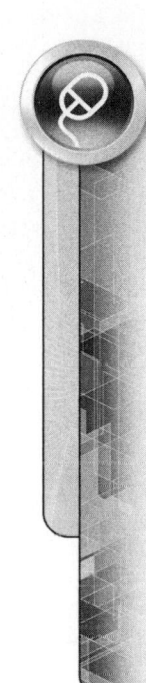

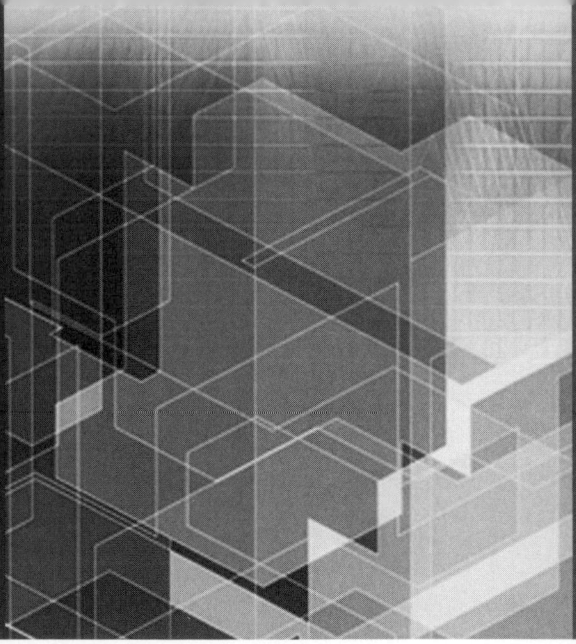

CHAPTER 5

Performing Geometric Constructions

In Chapter 1, the LINE, CIRCLE, and PLINE commands were introduced. The remainder of the drawing commands used for object creation are introduced in this chapter and include the following: ARC, BOUNDARY, CIRCLE-2P, CIRCLE-3P, CIRCLE-TTR, CIRCLE-TTT, DONUT, ELLIPSE, POINT, DIVIDE, MEASURE, POLYGON, RAY, RECTANG, REVCLOUD, SPLINE, WIPEOUT, XLINE, and ADDSELECTED. We will explore various techniques along with the command options that will be utilized to create geometric constructions.

METHODS OF SELECTING DRAW COMMANDS

You can find the main body of draw commands on the Ribbon and Tool Palette as shown in Figure 5.1.

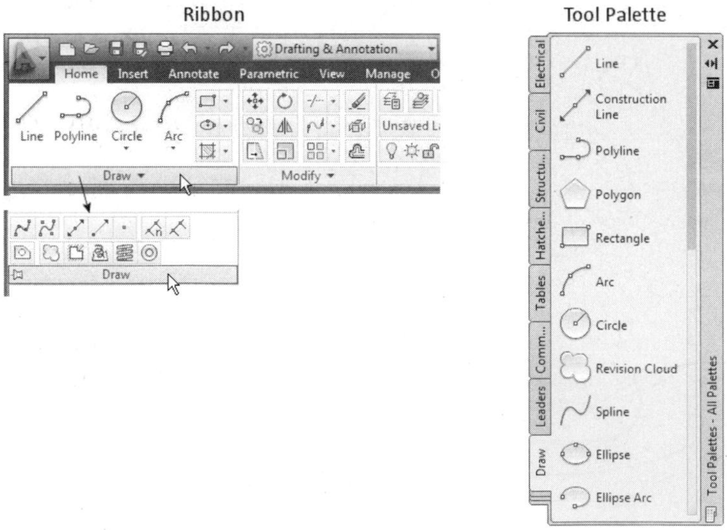

FIGURE 5.1

Refer to the following table for a complete listing of all buttons, tools, shortcuts, and functions of most drawing commands.

Button	Tool	Shortcut	Function
	Arc	A	Constructs an arc object using a number of command options
	Boundary	BO	Traces a polyline boundary over the top of an existing closed group of objects
	Circle	C	Constructs a circle by radius, diameter, 2 points, 3 points, 2 tangent points and a radius, and 3 tangent points
	Donut	DO	Creates a filled-in circular polyline object
	Ellipse	EL	Creates an elliptical object
	Mline	ML	Creates multiline objects that consist of multiple parallel lines
	Point	PO	Creates a point object
	Divide	DIV	Inserts evenly spaced points or blocks along an object's length or perimeter
	Measure	ME	Inserts points or blocks along an object's length or perimeter at designated increments
	Polygon	POL	Creates an equilateral polygon shape consisting of various side combinations
	Ray	RAY	Creates a semi-infinite line used for construction purposes
	Rectang	REC	Creates a rectangular polyline object
	Revcloud	REVCLOUD	Creates a revision cloud consisting of a polyline made up of sequential arc segments
	Spline	SPL	Creates a smooth curve from a sequence of points
	Wipeout	WIPEOUT	Covers or hides objects with a blank area
	Xline	XL	Creates an infinite line used for construction purposes
	Addselected	None	Creates a new object based on the object type selected

CONSTRUCTING ARCS

Use the ARC command to construct portions of circular shapes by radius, diameter, arc length, included angle, and direction. Choose this command from one of the following:

- From the Ribbon > Home Tab > Draw Panel
- The Menu Bar (Draw > Arc)
- The Draw Toolbar of the AutoCAD Classic Workspace
- The keyboard (A or ARC)

Expanding the Arc icon from the Ribbon displays the numerous methods available for creating arcs, as shown in Figure 5.2. By default, the 3 Points Arc mode supports arc constructions in the clockwise as well as the counterclockwise direction. Providing a negative or positive angle in the Angle option also allows a clockwise or counterclockwise direction. Most other arc modes support the ability to construct arcs only in the counterclockwise direction. Three arc examples will be demonstrated in the next series of pages, namely, how to construct an arc by 3 Points, by a Starting point, Center point, and Ending point, and how to Continue arcs.

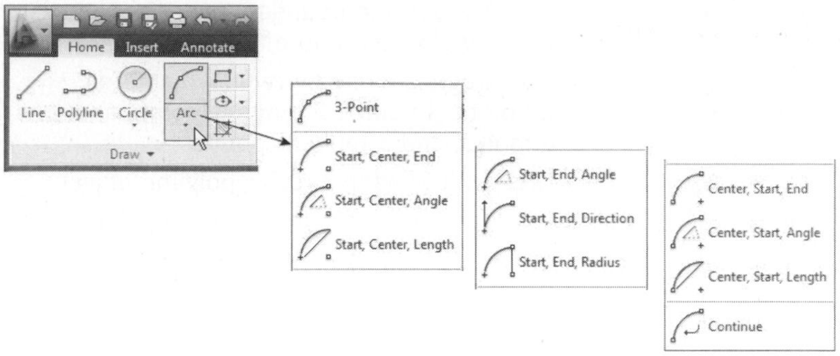

FIGURE 5.2

3 Points Arc Mode

By default, arcs are drawn with the 3 Points method. The first and third points identify the endpoints of the arc. This arc may be drawn in either the clockwise or counterclockwise direction.

TRY IT!

Create a new drawing file starting from scratch. Use the following command sequence and Figure 5.3 to construct a 3-point arc.

```
Command: A (For ARC)
Specify start point of arc or [Center]: (Pick a point at "A")
Specify second point of arc or [Center/End]: (Pick a point
at "B")
Specify end point of arc: (Pick a point at "C")
```

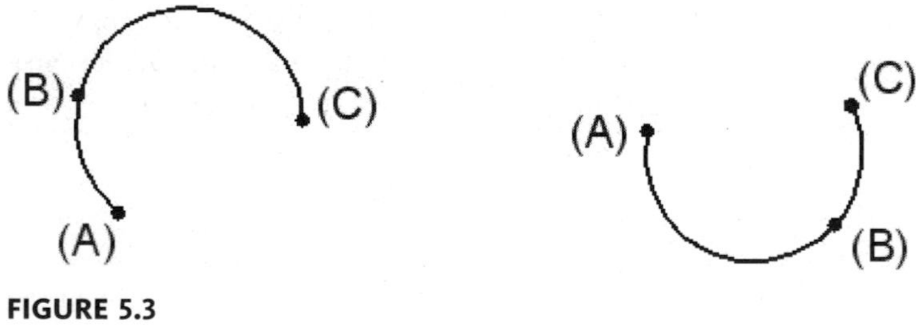

FIGURE 5.3

Start, Center, End Mode

Use this ARC mode to construct an arc by defining its start point, center point, and endpoint. This arc will always be constructed in a counterclockwise direction.

Open the drawing file 05_Door Swing. Use the following command sequence and Figure 5.3 for constructing an arc representing the door swing by start, center, and end points.

TRY IT!

```
Command: A (For ARC)
Specify start point of arc or [Center]: (Pick the endpoint at "A")
Specify second point of arc or [Center/End]: C (For Center)
Specify center point of arc: (Pick the endpoint at "B")
Specify end point of arc or [Angle/chord Length]: (Pick a point at "C")
```

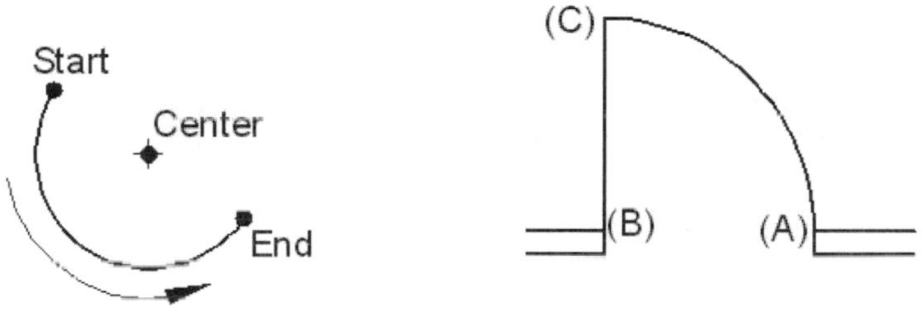

FIGURE 5.4

Because of the numerous options in the ARC command, you will often find that selecting the command from a menu is more efficient. Command options are preselected from the menu and automatically provided in the command sequence.

TIP

Continue Mode

Use this mode to create an arc from the end of a previously drawn arc or line. All arcs drawn through Continue mode are automatically constructed tangent to the previous arc or line. Activate this mode from the Ribbon, as shown in Figure 5.5 (Left). In Figure 5.5 (Right) a new arc begins at the last endpoint of the previous arc.

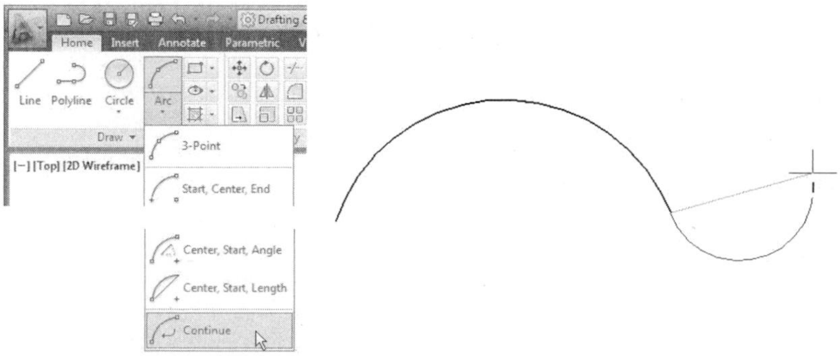

FIGURE 5.5

CREATING A BOUNDARY

The BOUNDARY command is used to create a polyline boundary around any closed shape. Choose this command from one of the following:

- From the Ribbon > Home Tab > Draw Panel (Expand icon: Boundary, Hatch, and Gradient)
- The Menu Bar (Draw > Boundary)
- The keyboard (BO or BOUNDARY)

It has already been demonstrated that the Join option of the PEDIT command is used to join object segments into one continuous polyline. The BOUNDARY command automates this process even more. Start this command by choosing Boundary from the Ribbon, as shown in Figure 5.6 (Left). This activates the Boundary Creation dialog box shown in Figure 5.6 (Right).

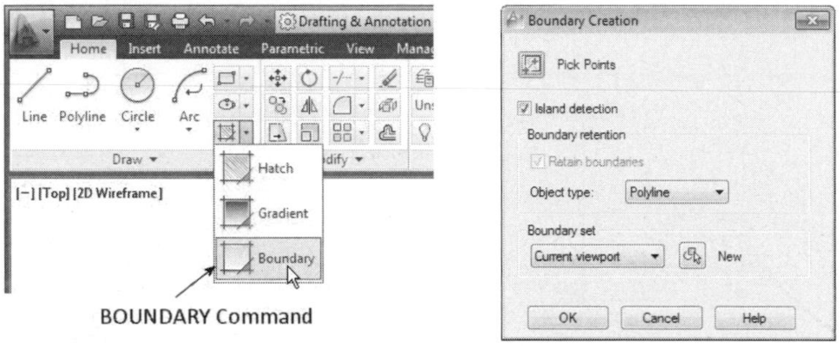

FIGURE 5.6

Before you use this command, it is good practice to create a separate layer to hold the polyline object; this layer could be called "Boundary" or "BP," for Boundary Polyline. Unlike the Join option of the PEDIT command, which converts individual objects to polyline objects, the BOUNDARY command traces a polyline in the current layer on top of individual objects.

TRY IT!

Open the drawing file 05_Boundary Extrusion. Activate the Boundary dialog box and click the Pick Points button. Then pick a point inside the object illustrated in Figure 5.7. Notice how the entire object is highlighted. To complete the command, press ENTER when prompted to select another internal point, and the polyline will be traced over the top of the existing objects.

Command: BO *(For BOUNDARY)*

(The Boundary Creation dialog box appears. Click the Pick Points button.)

Pick internal point: *(Pick a point at "A" in the following image)*

Selecting everything...

Selecting everything visible...

Analyzing the selected data...

Analyzing internal islands...

Pick internal point: *(Press ENTER to construct the boundary)*

boundary created 1 polyline

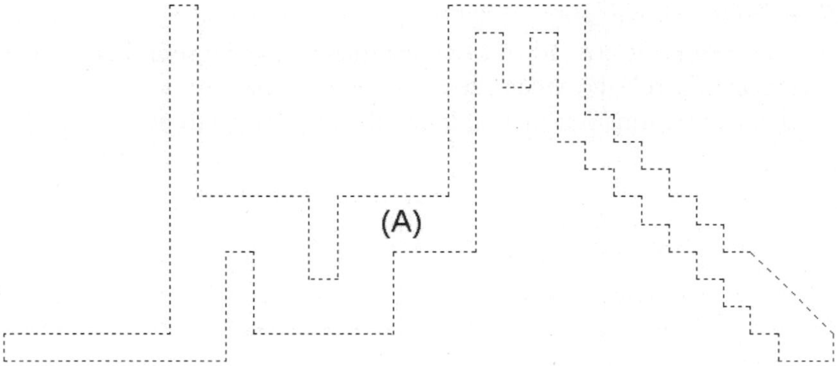

FIGURE 5.7

Once the boundary polyline is created, the boundary may be relocated to a different position on the screen with the MOVE command. The results are illustrated in Figure 5.8. The object on the left consists of the original individual objects. When the object on the right is selected, all objects highlight, signifying that the object is made up of a polyline object made through the use of the BOUNDARY command.

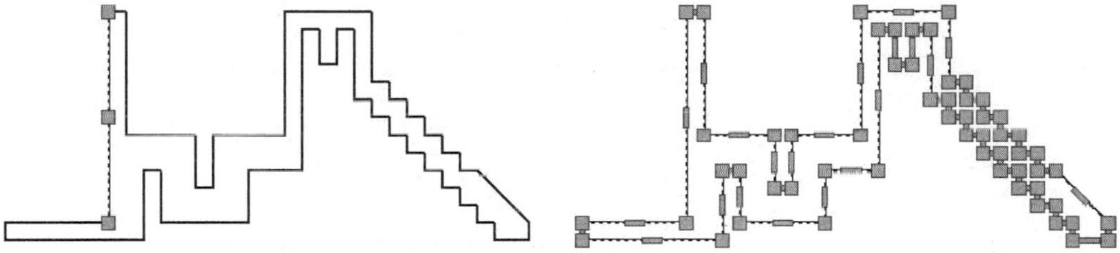

FIGURE 5.8

When the BOUNDARY command is used on an object consisting of an outline and internal islands similar to the drawing in Figure 5.9, a polyline object is also traced over these internal islands.

TRY IT!

Open the drawing file 05_Boundary Cover in Figure 5.9. Notice that the current layer is Boundary and the color is Magenta. Issue the BOUNDARY command and pick a point inside the middle of the object at "A" without OSNAP turned on. Notice that the polyline is constructed on the top of all existing objects. Turn the Object layer off to display just the Boundary layer.

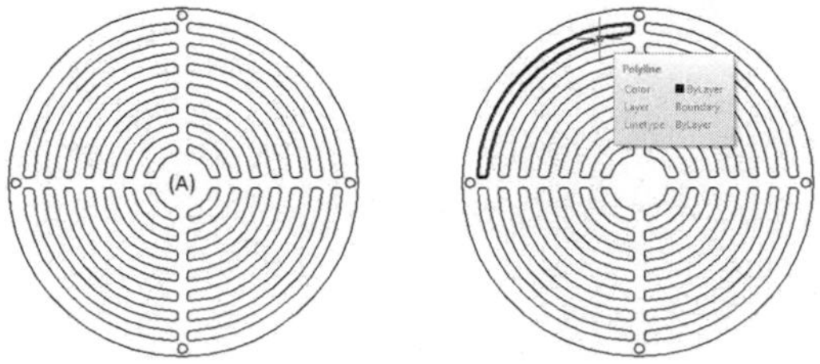

FIGURE 5.9

In addition to creating a special layer to hold the boundary polyline, another important rule to follow when using the BOUNDARY command is to be sure there are no gaps in the object. In Figure 5.10, when the BOUNDARY command encounters the gap at "A," a dialog box informs you that no internal boundary could be found. This is likely because the object is not completely closed. In this case, you must exit the command, close the object, and activate the BOUNDARY command again. It is acceptable, however, to have lines cross at the intersection at "B." While these "overshoots," as they are called, work well when using the Boundary Creation dialog box, this may not be considered a good drawing practice.

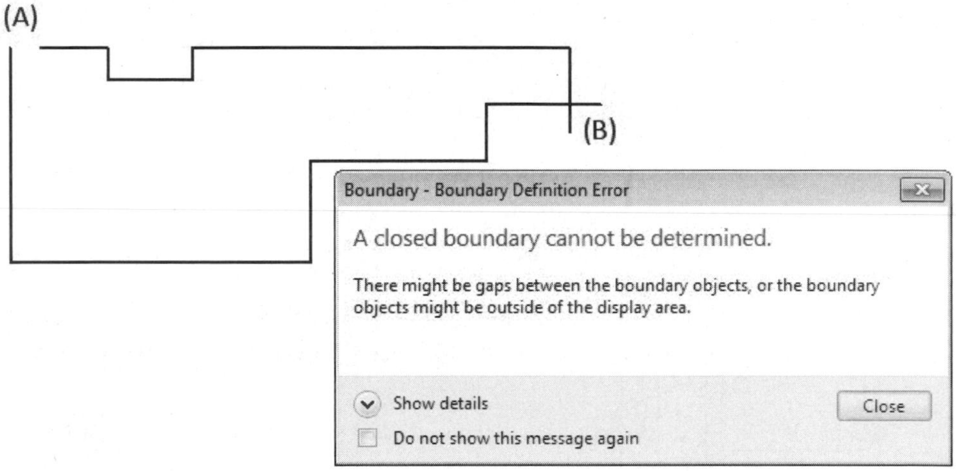

FIGURE 5.10

ADDITIONAL OPTIONS FOR CREATING CIRCLES

The CIRCLE command has several options to assist you with circle creation. These options may be selected from the Circle icon that is found under the Home tab of the Ribbon, as shown in Figure 5.11. The 2 Points, 3 Points, Tan Tan Radius, and Tan Tan Tan modes will be explained in the next series of examples.

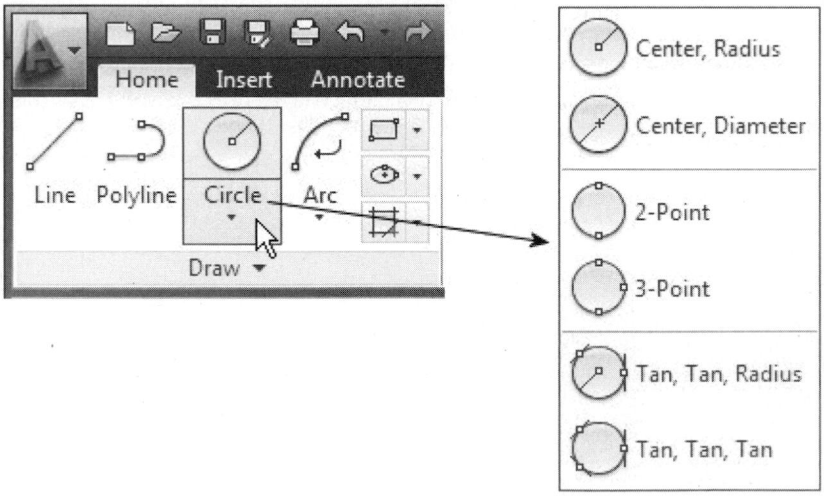

FIGURE 5.11

3 Points Circle Mode

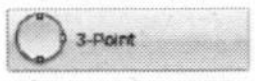
Use the CIRCLE command and the 3 Points mode to construct a circle by three points that you identify. No center point is utilized when you enter the 3 Points mode. Simply select three points and the circle is drawn. Choose this command from the Ribbon as shown in Figure 5.12.

TRY IT!

Open the drawing file 05_Circle_3P. Study the following prompts and Figure 5.12 for constructing a circle using the 3 Points mode.

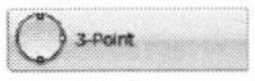

```
Command: C (For CIRCLE)
Specify center point for circle or [3P/2P/Ttr (tan tan
radius)]: 3P
Specify first point on circle: (Pick a midpoint at "A")
Specify second point on circle: (Pick a midpoint at "B")
Specify third point on circle: (Pick a midpoint at "C")
```

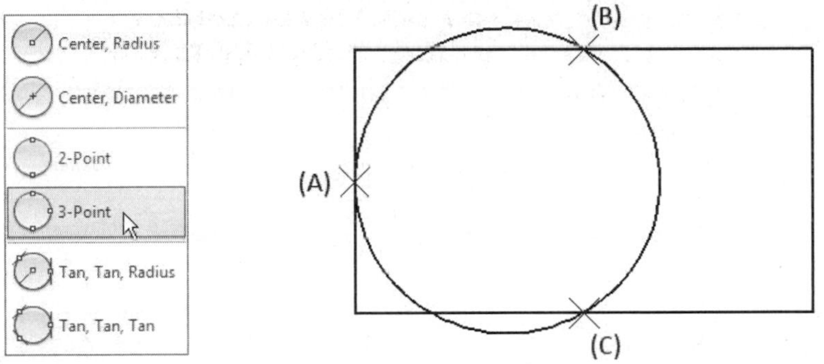

FIGURE 5.12

2 Points Circle Mode

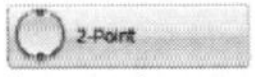
Use the CIRCLE command and the 2 Points mode to construct a circle by selecting two points. Choose this command from the Ribbon, as shown in Figure 5.13. These points form the diameter of the circle. No center point is utilized when you use the 2 Points mode.

TRY IT!

Open the drawing file 05_Circle_2P. Study the following prompts and Figure 5.13 for constructing a circle using the 2 Points mode.

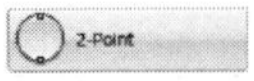

```
Command: C (For CIRCLE)
Specify center point for circle or [3P/2P/Ttr (tan tan
radius)]: 2P
Specify first end point of circle's diameter: (Pick a
midpoint at "A")
Specify second end point of circle's diameter: (Pick a
midpoint at "B")
```

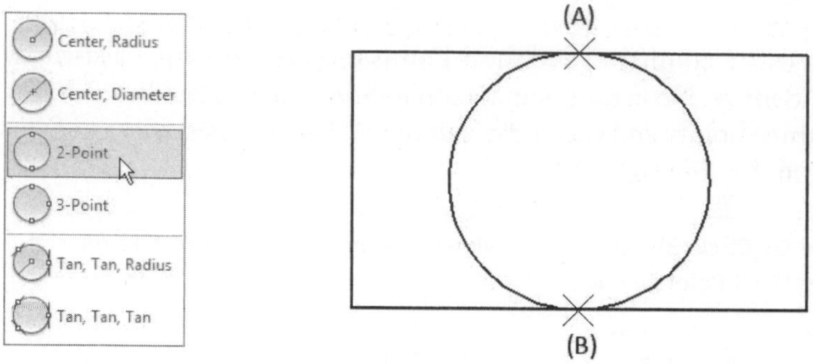

FIGURE 5.13

Constructing an Arc Tangent to Two Lines Using CIRCLE-TTR

Tan, Tan, Radius Illustrated in Figure 5.14 (Left) are two inclined lines. The purpose of this example is to connect an arc tangent to the two lines at a specified radius. The `CIRCLE-TTR` (Tangent-Tangent-Radius) command will be used here along with the `TRIM` command to clean up the excess geometry. To assist during this operation, the OSNAP-Tangent mode is automatically activated when you use the TTR option of the `CIRCLE` command.

First, use the `CIRCLE-TTR` command to construct an arc tangent to both lines, as shown in Figure 5.14 in the middle.

Tan, Tan, Radius `Command: C` *(For CIRCLE)*

`Specify center point for circle or [3P/2P/Ttr (tan tan radius)]: T` *(For TTR)*

`Specify point on object for first tangent of circle:` *(Select the line at "A")*

`Specify point on object for second tangent of circle:` *(Select the line at "B")*

`Specify radius of circle:` *(Enter a radius value)*

Use the `TRIM` command to clean up the lines and arc. The completed result is illustrated in Figure 5.14 (Right). It should be noted that instead of a CIRCLE-TTR and TRIM operation, the `FILLET` command could have more efficiently accomplished this task.

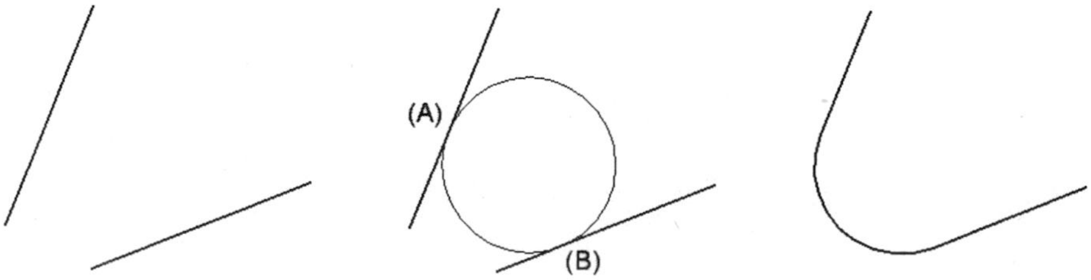

FIGURE 5.14

TRY IT!

Open the drawing file 05_TTR1. Use the `CIRCLE` command and the TTR option to construct a circle tangent to lines "A" and "B" in Figure 5.15. Use a circle radius of 0.50 units. With the circle constructed, use the `TRIM` command and select the circle as well as lines "C" and "D" as cutting edges. Trim the circle at "E" and the lines at "C" and "D." Observe the final results of this operation shown in Figure 5.15.

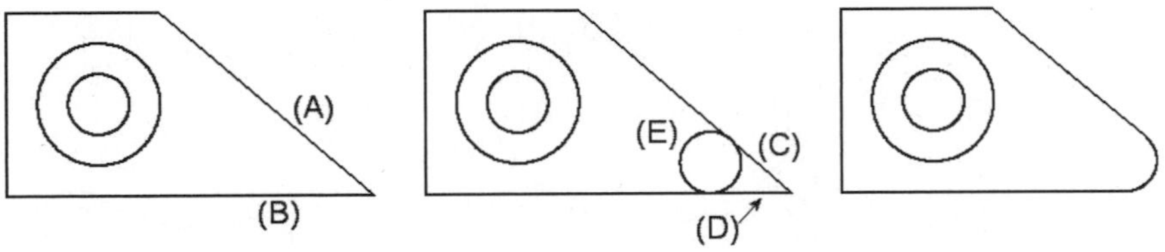

FIGURE 5.15

Constructing an Arc Tangent to a Line and Arc Using CIRCLE-TTR

Illustrated in Figure 5.16 (Left) is an arc and an inclined line. The purpose of this example is to connect an additional arc tangent to the original arc and line at a specified radius. The `CIRCLE-TTR` command will be used here, along with the `TRIM` command to clean up the excess geometry.

First, use the `CIRCLE-TTR` command to construct an arc tangent to the arc and inclined line, as shown in Figure 5.16 in the middle.

Tan, Tan, Radius `Command: C` *(For CIRCLE)*

`Specify center point for circle or [3P/2P/Ttr (tan tan radius)]: T` *(For TTR)*

`Specify point on object for first tangent of circle:` *(Select the arc at "A")*

`Specify point on object for second tangent of circle:` *(Select the line at "B")*

`Specify radius of circle:` *(Enter a radius value)*

Take note that the radius must be greater than or equal to half the distance between the circle and the line; otherwise, the second circle cannot be constructed. Use the `TRIM` command to clean up the arc and line. The completed result is illustrated in Figure 5.16 (Right).

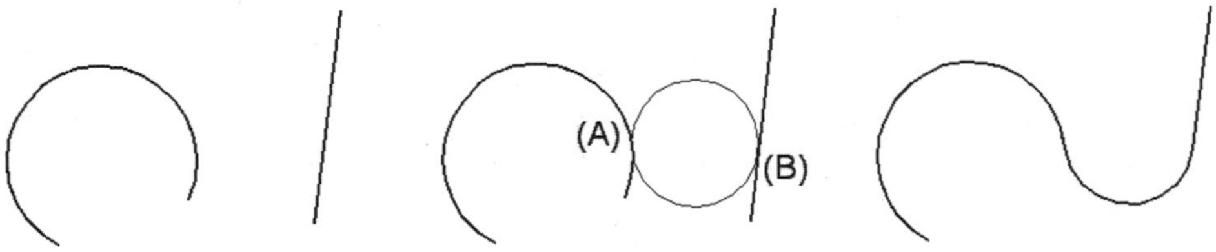

FIGURE 5.16

TRY IT!

Open the drawing file 05_TTR2. Using Figure 5.17 as a guide, use the CIRCLE command and the TTR option to construct a circle tangent to the line at "A" and arc at "B." Use a circle radius value of 0.50 units. Construct a second circle tangent to the arc at "C" and line at "D" using the default circle radius value of 0.50 units. With the circles constructed, use the TRIM command, press ENTER to select all cutting edges, and trim the lines at "E" and "J" and the arc at "F" and "H," in addition to the circles at "G" and "K." Observe the final results of this operation shown in Figure 5.17.

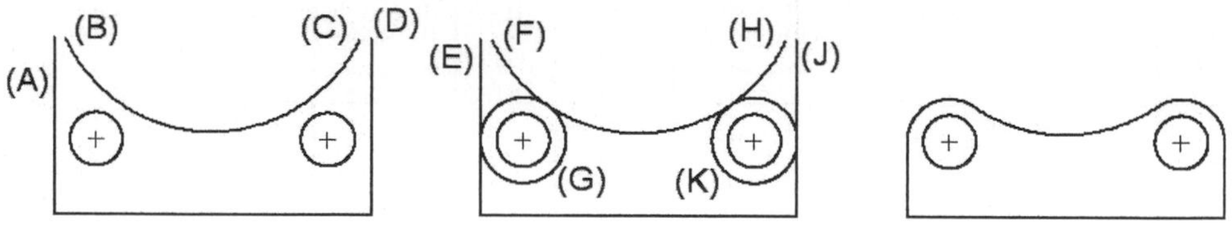

FIGURE 5.17

Constructing an Arc Tangent to Two Arcs Using CIRCLE-TTR

Method #1

Illustrated in Figure 5.18 (Left) are two arcs. The purpose of this example is to connect a third arc tangent to the original two at a specified radius. The CIRCLE-TTR command will be used here along with the TRIM command to clean up the excess geometry.

Use the CIRCLE-TTR command to construct an arc tangent to the two original arcs, as shown in Figure 5.18 in the middle.

Tan, Tan, Radius Command: C *(For CIRCLE)*

Specify center point for circle or [3P/2P/Ttr (tan tan radius)]: T *(For TTR)*

Specify point on object for first tangent of circle: *(Select the arc at "A")*

Specify point on object for second tangent of circle: *(Select the arc at "B")*

Specify radius of circle: *(Enter a radius value)*

Use the TRIM command to clean up the three arcs. The completed result is illustrated in Figure 5.18 (Right).

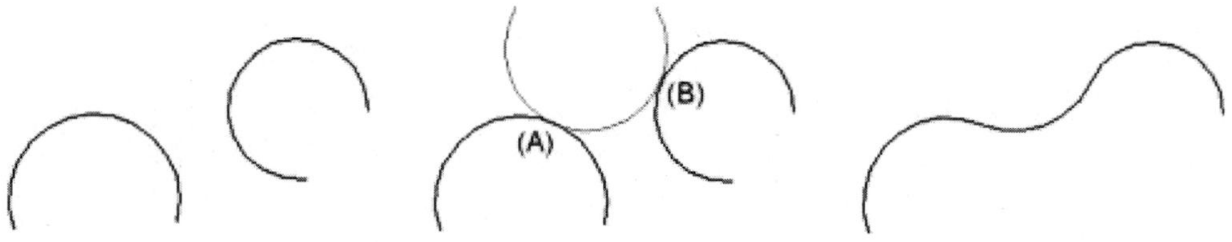

FIGURE 5.18

TRY IT!

Open the drawing file 05_TTR3. Use the `CIRCLE` command and the TTR option to construct a circle tangent to the circles at "A" and "B." Construct a second circle tangent to the circles at "C" and "D" in Figure 5.19. Use a circle radius of 4.50 units for both circles. With the circles constructed, use the `TRIM` command and select both small circles as cutting edges. Trim the circles at "E" and "F" and the opposite end. Observe the final results of this operation shown in Figure 5.19.

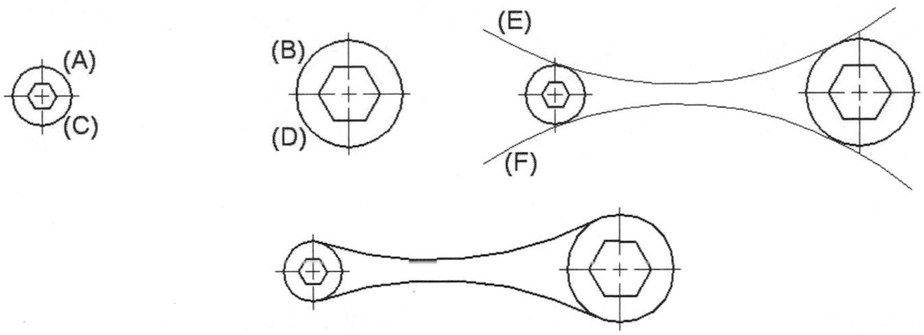

FIGURE 5.19

Constructing an Arc Tangent to Two Arcs Using CIRCLE-TTR

Method #2

Illustrated in Figure 5.20 (Left) are two arcs. The purpose of this example is to connect an additional arc tangent to and enclosing both arcs at a specified radius. The `CIRCLE-TTR` command will be used here along with the `TRIM` command.

First, use the `CIRCLE-TTR` command to construct an arc tangent to and enclosing both arcs, as shown in Figure 5.20 in the middle.

Tan, Tan, Radius `Command: C` *(For CIRCLE)*

`Specify center point for circle or [3P/2P/Ttr (tan tan radius)]: T` *(For TTR)*

`Specify point on object for first tangent of circle:` *(Select the arc at "A")*

`Specify point on object for second tangent of circle:` *(Select the arc at "B")*

`Specify radius of circle:` *(Enter a radius value)*

Use the `TRIM` command to clean up all arcs. The completed result is illustrated in Figure 5.20 (Right).

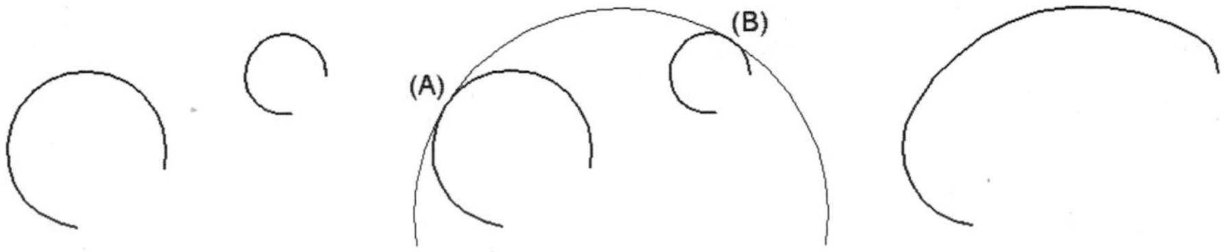

FIGURE 5.20

TRY IT!

Open the drawing file 05_TTR4. Use the `CIRCLE` command and the TTR option to construct a circle tangent to the circles at "A" and "B." Construct a second circle tangent to the circles at "C" and "D" in Figure 5.21. Use a circle radius of 1.50 units for both circles. With the circles constructed, use the `TRIM` command and trim all circles until you achieve the final results of this operation shown in Figure 5.21.

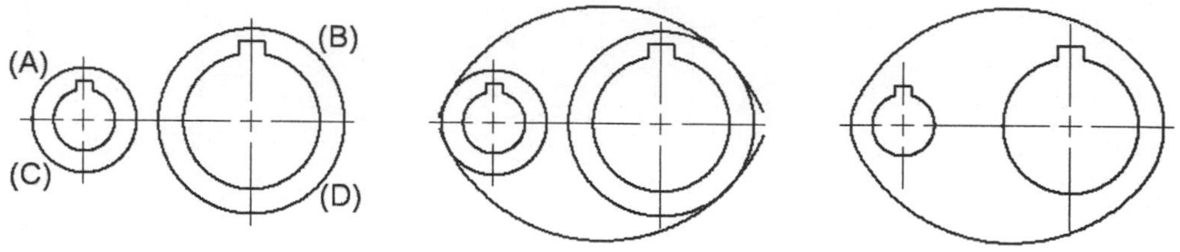

FIGURE 5.21

Constructing an Arc Tangent to Two Arcs Using CIRCLE-TTR

Method #3

Illustrated in Figure 5.22 are two arcs. The purpose of this example is to connect an additional arc tangent to one arc and enclosing the other. The `CIRCLE-TTR` command will be used here along with the `TRIM` command to clean up unnecessary geometry.

First, use the `CIRCLE-TTR` command to construct an arc tangent to the two arcs. Study the illustration in Figure 5.22 in the middle and the following prompts to understand the proper pick points for this operation. It is important to select the tangent points in the approximate location of the tangent to each circle in order to obtain the desired arc tangent orientation.

Tan, Tan, Radius `Command: C` *(For CIRCLE)*

`Specify center point for circle or [3P/2P/Ttr (tan tan radius)]: T` *(For TTR)*

`Specify point on object for first tangent of circle:` *(Select the arc at "A")*

`Specify point on object for second tangent of circle:` *(Select the arc at "B")*

`Specify radius of circle:` *(Enter a radius value)*

Use the `TRIM` command to clean up the arcs. The completed result is illustrated in Figure 5.22 (Right).

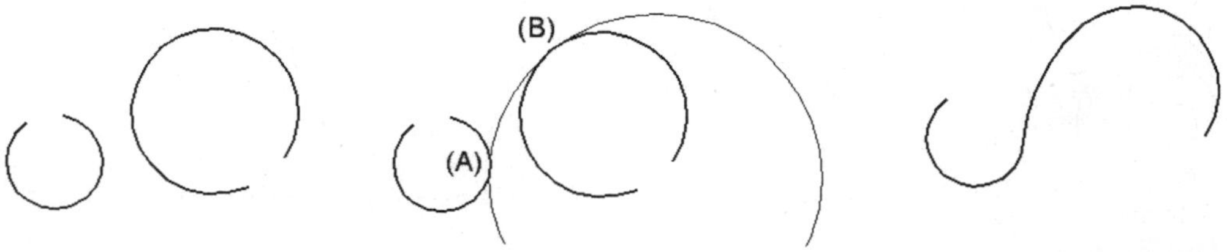

FIGURE 5.22

TRY IT!

Open the drawing file 05_TTR5. Use the `CIRCLE` command and the TTR option to construct a circle tangent to the circles at "A" and "B." Construct a second circle tangent to the circles at "C" and "D" in Figure 5.23. Use a circle radius of 3.00 units for both circles. With the circles constructed, use the `TRIM` command and trim all circles until you achieve the final results of this operation shown in Figure 5.23.

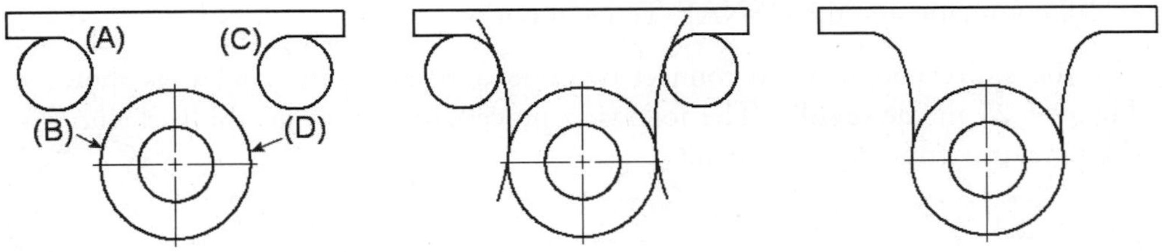

FIGURE 5.23

CIRCLE—Tan Tan Tan

Yet another mode of the `CIRCLE` command allows you to construct a circle based on three tangent points. This mode is actually a variation of the 3 Points mode used together with the OSNAP-Tangent mode three times. If you choose this command from the Ribbon as shown in Figure 5.24 (Left), the Tangent OSNAP is automatically provided. This mode requires you to select three objects, as in the example of the three line segments shown in Figure 5.24 (Right).

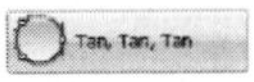


```
Command: C (For CIRCLE)
Specify center point for circle or [3P/2P/Ttr (tan tan
radius)]: 3P
Specify first point on circle: Tan
to (Select the line at "A")
Specify second point on circle: Tan
to (Select the line at "B")
Specify third point on circle: Tan
to (Select the line at "C")
```

The result is illustrated in Figure 5.24 (Right), with the circle being constructed tangent to the edges of all three line segments.

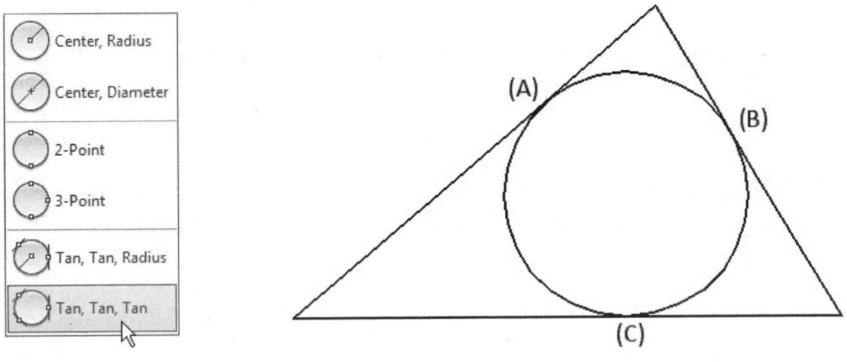

FIGURE 5.24

TRY IT!

Open the drawing file 05_TTT. Use the illustration in Figure 5.24 and the CIRCLE-Tan Tan Tan command sequence for constructing a circle tangent to three lines.

CONSTRUCTING A LINE TANGENT TO TWO ARCS OR CIRCLES

Illustrated in Figure 5.25 (Left) are two circles. The purpose of this example is to connect the two circles with two tangent lines. This can be accomplished with the LINE command and the OSNAP-Tangent option.

Use the LINE command to connect two lines tangent to the circles, as shown in Figure 5.25 in the middle. The following procedure is used for the first line. Use the same procedure for the second.

Command: L *(For LINE)*

Specify first point: **Tan**

to *(Select the circle near "A")*

Specify next point or [Undo]: **Tan**

to *(Select the circle near "B")*

Specify next point or [Undo]: *(Press* ENTER *to exit this command)*

Use the TRIM command to clean up the circles so that the appearance of the object is similar to the illustration in Figure 5.25 (Right).

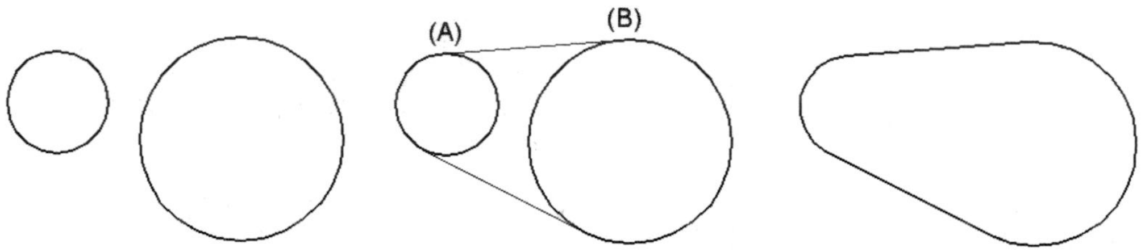

FIGURE 5.25

TRY IT!

Open the drawing file 05_Tangent Lines. Construct a line tangent to the two circles at "A" and "B." Repeat this procedure for the circles at "C" and "D" in Figure 5.26. Use the TRIM command and trim all circles until you achieve the final results of this operation shown in Figure 5.26.

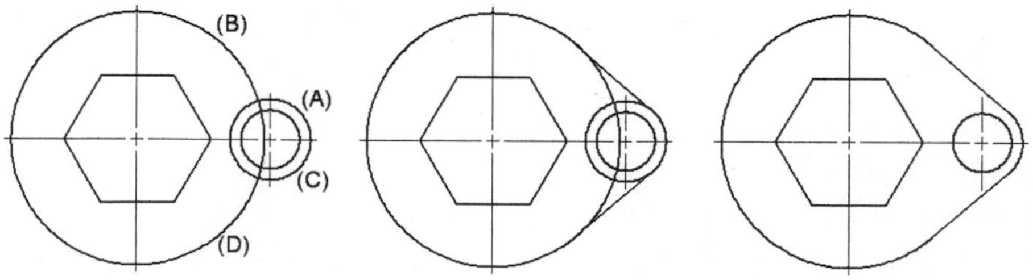

FIGURE 5.26

Quadrant Versus Tangent OSNAP Option

In the previous example, it was shown how a line can be drawn tangent to two circles. The object in Figure 5.27 has lines that are created between the arcs using the same method—the LINE command with Tangent OSNAPs.

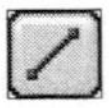 Command: L *(For LINE)*

Specify first point: Tan

to *(Select the arc near "A")*

Specify next point or [Undo]: Tan

to *(Select the arc near "B")*

Specify next point or [Undo]: *(Press* ENTER *to exit this command)*

Note that the angle of the line formed by points "A" and "B" is neither horizontal nor vertical. This is a typical example of the capabilities of the OSNAP-Tangent option.

Because the line between "C" and "D" is horizontal, the tangent points are located at the quadrants of the arcs. In this case, either the OSNAP-Tangent or OSNAP-Quadrant option can be utilized to create a correct line. The vertical line from "E" and "F" can also be created using either OSNAP. The Quadrant option can only be used when the lines to be drawn are perfectly horizontal or vertical; if you are unsure then the OSNAP-Tangent option should be used.

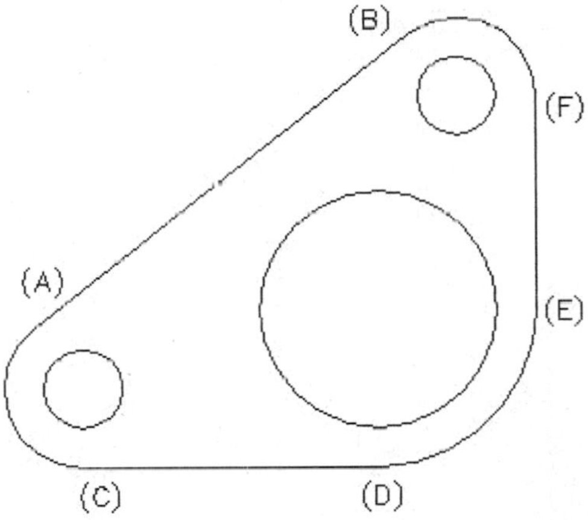

FIGURE 5.27

CREATING FILLED-IN DOTS (DONUTS)

Use the DONUT command to construct a filled-in circle. Choose this command from one of the following:

- From the Ribbon > Home Tab > Draw Panel (Expanded)
- The Menu Bar (Draw > Donut)
- The keyboard (DO or DONUT)

This object belongs to the polyline family. The command used for constructing a donut can be activated from the Ribbon, as illustrated in Figure 5.28 (Left). The first donut on the right has an inside diameter of 0.50 units and an outside diameter of 1.00 unit. When you place donuts in a drawing, the Multiple option is automatically invoked. This means you can place as many donuts as you like until you exit the command.

TRY IT!

Create a new drawing file starting from scratch. Use the following command sequence for constructing this type of donut.

Command: DO *(For DONUT)*

Specify inside diameter of donut <0.50>: *(Press* ENTER *to accept the default)*

Specify outside diameter of donut <1.00>: *(Press* ENTER *to accept the default)*

Specify center of donut or <exit>: *(Pick a point to place the donut)*

Specify center of donut or <exit>: *(Pick a point to place another donut or press* ENTER *to exit this command)*

Setting the inside diameter of a donut to a value of zero (0) and an outside diameter to any other value constructs a donut representing a dot, as shown in Figure 5.28 (Right).

TRY IT!

Create a new drawing file starting from scratch. Use the following command sequence for constructing this type of donut.

Command: DO *(For DONUT)*

Specify inside diameter of donut <0.50>: 0

Specify outside diameter of donut <1.00>: 0.25

Specify center of donut or <exit>: *(Pick a point to place the donut)*

Specify center of donut or <exit>: *(Pick a point to place another donut or press* ENTER *to exit this command)*

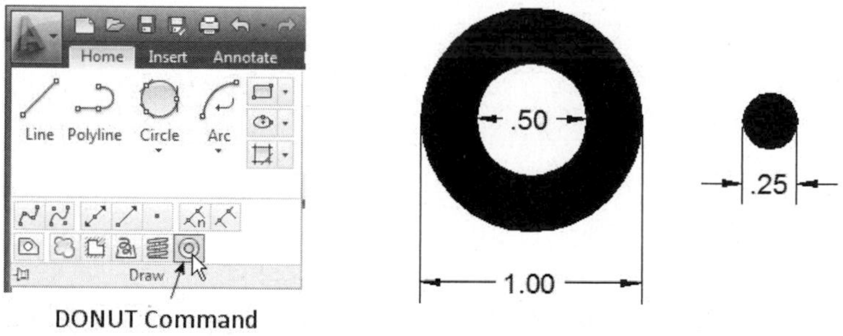

FIGURE 5.28

TRY IT!

Open the drawing file 05_Donut. Activate the DONUT **command and set the inside diameter to 0 and the outside diameter to 0.05. Place four donuts at the intersections of the electrical circuit, as shown in Figure 5.29.**

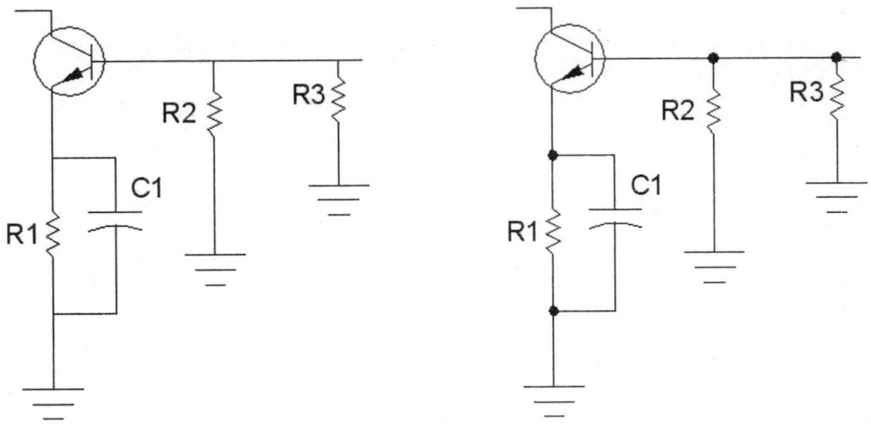

FIGURE 5.29

CONSTRUCTING ELLIPTICAL SHAPES

Use the ELLIPSE command to construct a true elliptical shape. Choose this command from one of the following:

- From the Ribbon > Home Tab > Draw Panel
- The Menu Bar (Draw > Ellipse)
- The Draw Toolbar of the AutoCAD Classic Workspace
- The keyboard (EL or ELLIPSE)

The command used for creating elliptical shapes can be selected from the Ribbon, as shown in Figure 5.30 (Left). Before studying the three examples for ellipse construction, see the illustration in Figure 5.30 (Right) to view two important parts of any ellipse, namely its major and minor diameters.

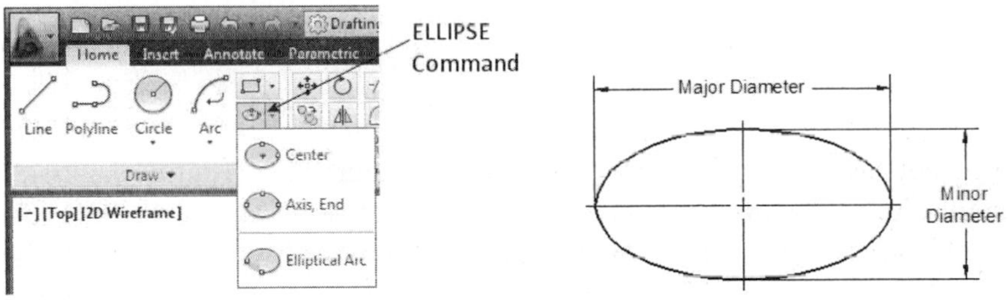

FIGURE 5.30

You can construct an ellipse by marking two points, which specify one of its axes, as shown in Figure 5.31. These first two points also identify the angle with which the ellipse will be drawn. Responding to the prompt "Specify distance to other axis or [Rotation]" with another point identifies half of the other axis. The rubber-banded line is added to assist you in this ellipse construction method.

TRY IT!

Create a new drawing file starting from scratch. Use the illustration in Figure 5.31 and the command sequence to construct an ellipse by locating three points.

Command: EL *(For ELLIPSE)*

Specify axis endpoint of ellipse or [Arc/Center]: *(Pick a point at "A")*

`Specify other endpoint of axis:` *(Pick a point at "B")*

`Specify distance to other axis or [Rotation]:` *(Pick a point at "C")*

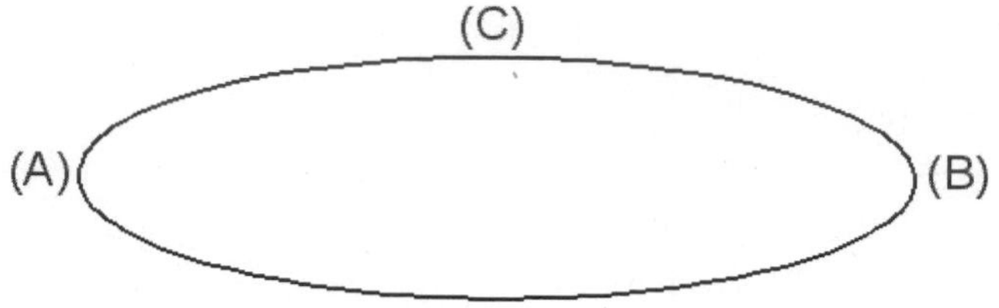

FIGURE 5.31

You can also construct an ellipse by first identifying its center. You can pick points to identify its axes or use polar coordinates to accurately define the major and minor diameters of the ellipse, as shown in Figure 5.32 (Left).

TRY IT!

Create a new drawing file starting from scratch. Use the illustration in Figure 5.32 (Left) and the following command sequence for constructing an ellipse based on a center. Use a polar coordinate or the Direct Distance mode for locating the two axis endpoints of the ellipse. For this example turn Polar or Ortho modes on.

`Command: EL` *(For ELLIPSE)*

`Specify axis endpoint of ellipse or [Arc/Center]: C` *(For Center)*

`Specify center of ellipse:` *(Pick a point at "A")*

`Specify endpoint of axis:` *(Move your cursor to the right and enter a value of* `2.50`*)*

`Specify distance to other axis or [Rotation]:` *(Move your cursor straight up and enter a value of* `1.50`*)*

The last method of constructing an ellipse, as shown in Figure 5.32 (Right), illustrates constructing an ellipse by way of rotation. Identify the first two points for the first axis. Reply to the prompt "Specify distance to other axis or [Rotation]" with Rotation. The rotation angle represents the viewing angle of a circle. At 0° you would see a full circle and at 90° the circle would appear as a line (on edge); 90° is not a valid entry.

TRY IT!

Create a new drawing file starting from scratch. Use the illustration in Figure 5.32 (Right) and the following command sequence for constructing an ellipse based on a rotation around the major axis.

`Command: EL` *(For ELLIPSE)*

`Specify axis endpoint of ellipse or [Arc/Center]:` *(Pick a point at "A")*

`Specify other endpoint of axis:` *(Pick a point at "B")*

`Specify distance to other axis or [Rotation]: R` *(For Rotation)*

`Specify rotation around major axis: 80`

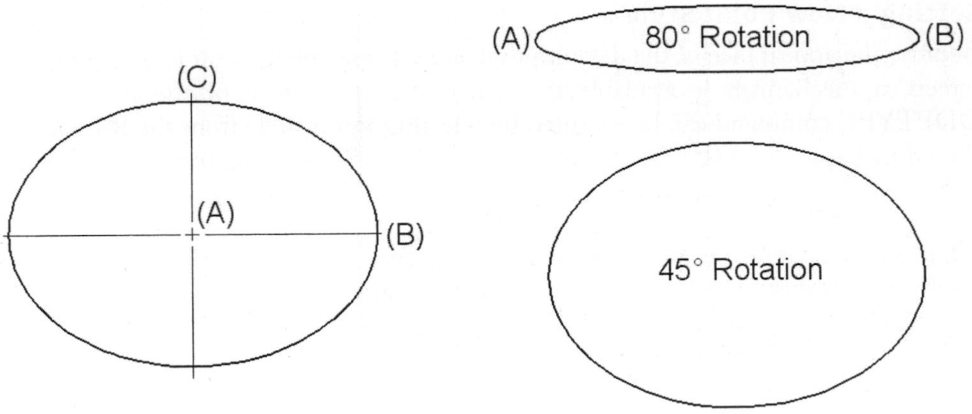

FIGURE 5.32

CREATING POINT OBJECTS

Use the POINT command to identify the location of a point on a drawing, which may be used for reference purposes. The Node or Nearest OSNAP mode can be used to snap to the points. Choose this command from one of the following:

- From the Ribbon > Home Tab > Draw Panel (Expanded)
- The Menu Bar (Draw > Point)
- The Draw Toolbar of the AutoCAD Classic Workspace
- The keyboard (PO or POINT)

When selecting this command from the Ribbon, as shown in Figure 5.33 (Left), you will be automatically placed in a multiple points mode. This allows you to create numerous points in the same operation. By default, a point is displayed as a dot on the screen. Illustrated in Figure 5.33 (Right) is an object constructed using points that have been changed in appearance to resemble Xs.

Command: PO *(For point)*

Current point modes: PDMODE=3 PDSIZE=0.0000

Specify a point: *(Pick the new position of a point)*

Specify a point: *(Either pick another point location or press ESC to exit this command)*

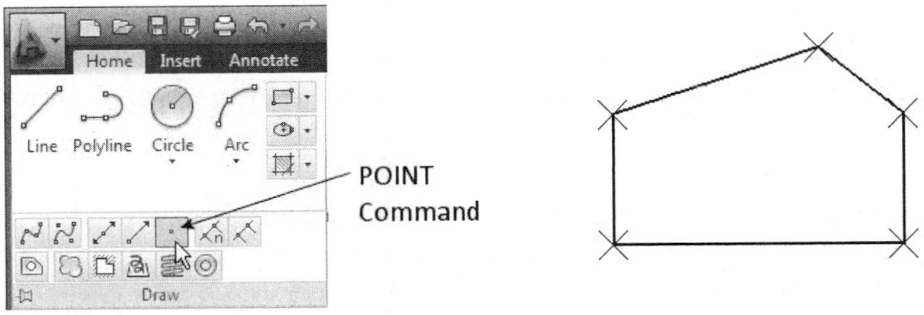

FIGURE 5.33

Setting a New Point Style

Because the appearance of the default point as a dot may be difficult to locate on the screen, a mechanism is available to change the appearance of the point. The DDPTYPE command can be executed by selecting Point Style from the Ribbon, as shown in Figure 5.34 (Left). This displays the Point Style dialog box shown on the right. Use this icon menu to set a different point mode and point size.

TIP

Only one point style may be current in a drawing. Once a point is changed to a current style, the next drawing regeneration updates all points to this style.

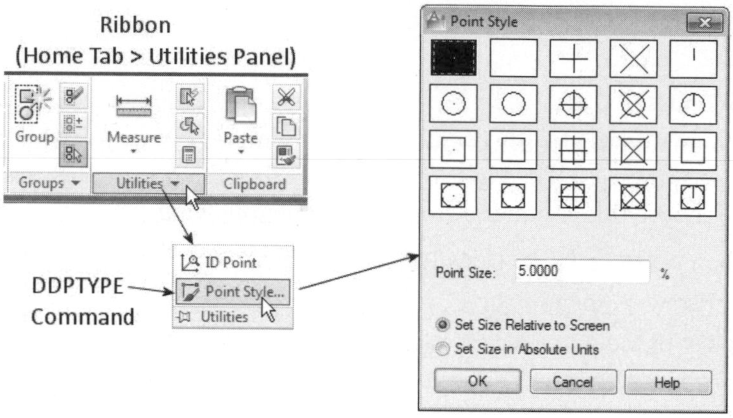

FIGURE 5.34

DIVIDING OBJECTS INTO EQUAL SPACES

Illustrated in Figure 5.35 is an arc constructed inside a rectangular object. In this example, we want to divide the arc into an equal number of parts. This can be a tedious task with manual drafting methods, but thanks to the `DIVIDE` command, this operation is much easier to perform. The `DIVIDE` command instructs you to supply the number of divisions and then performs the division by placing points along the object to be divided. Choose this command from one of the following:

- From the Ribbon > Home Tab > Draw Panel (Expanded)
- The Menu Bar (Draw > Point > Divide)
- The keyboard (`DIV` or `DIVIDE`)

The Point Style dialog box controls the point size and shape. Be sure the point style appearance is set to produce a visible point. Otherwise, the results of the `DIVIDE` command will not be obvious.

TRY IT!

Open the drawing file 05_Divide. Use the `DIVIDE` command and select the arc as the object to divide, as illustrated in Figure 5.35 (Left), and enter a value of 6 for the number of segments. The command divides the object by a series of points, as shown in the middle in Figure 5.35. To finish this drawing, the arc is erased and circles with a diameter of 0.25 are placed at each point, as shown in Figure 5.35 (Right). The OSNAP-Node mode works well with point objects.

Command: DIV *(For DIVIDE)*

Select object to divide: *(Select the arc)*

Enter the number of segments or [Block]: 6

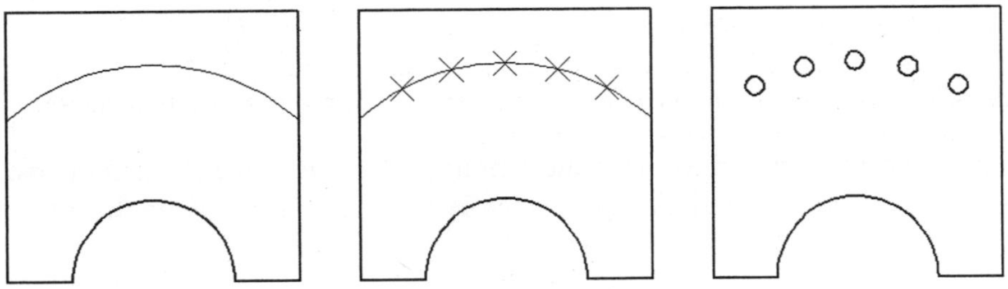

FIGURE 5.35

For polyline objects, the DIVIDE command divides the entire polyline object into an equal number of segments. This occurs even if the polyline consists of a series of line and arc segments, as shown in Figure 5.36.

TRY IT!

Open the drawing file 05_Divide Pline. Use the DIVIDE **command and select the outer polyline as the object to divide, as shown in Figure 5.36 (Left), and enter a value of 9 for the number of segments. The command divides the entire polyline by a series of points, as shown on the right.**

Command: DIV *(For DIVIDE)*

Select object to divide: *(Select the outer polyline shape)*

Enter the number of segments or [Block]: 9

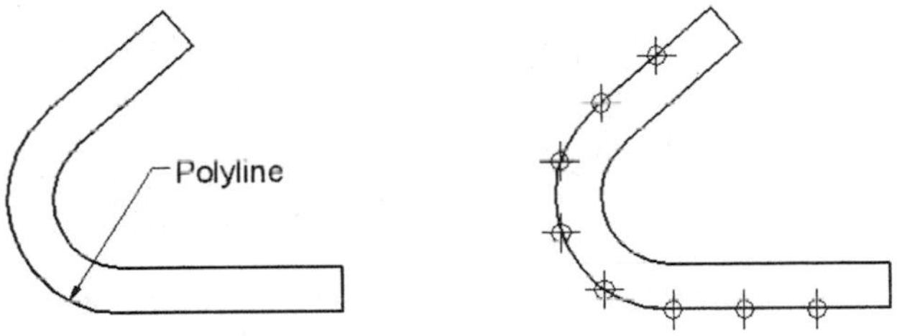

FIGURE 5.36

MEASURING OBJECTS

The MEASURE command takes an object such as a line or an arc and creates points along it depending on the length of segment specified. Similar to the DIVIDE command, it places points along the object selected and does not physically break the object. Choose this command from one of the following:

- From the Ribbon > Home Tab > Draw Panel (Expanded)
- The Menu Bar (Draw > Point > Measure)
- The keyboard (ME or MEASURE)

It is important to note that as points are placed along an object, the measuring starts at the endpoint closest to the point you used to select the object.

TRY IT!

Open the drawing file 05_Measure. Use the illustration in Figure 5.37 and the command sequence below to perform this operation.

Command: ME *(For MEASURE)*

Select object to measure: *(Select the left end of the diagonal line, as shown in the following image on the left)*

Specify length of segment or [Block]: 1.25

The results, shown in Figure 5.37, illustrate various points placed at 1.25 increments. As with the DIVIDE command, the appearance of the points placed along the line is controlled through the Point Style dialog box.

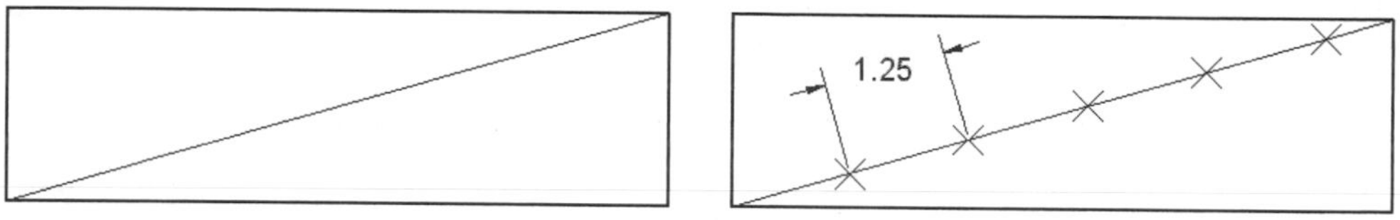

FIGURE 5.37

CREATING POLYGONS

The POLYGON command is used to construct a regular polygon (equal length sides). Choose this command from one of the following:

- From the Ribbon > Home Tab > Draw Panel (Expand icon: Rectangle and Polygon)
- The Menu Bar (Draw > Polygon)
- The Draw Toolbar of the AutoCAD Classic Workspace
- The keyboard (POL or POLYGON)

You create polygons by identifying the number of sides for the polygon, locating a point on the screen as the center of the polygon, specifying whether the polygon is inscribed or circumscribed, and specifying a circle radius for the size of the polygon. Polygons consist of a closed polyline object with width set to zero.

TRY IT!

Create a new drawing file starting from scratch. Use the following command sequence to construct the inscribed polygon, as shown in Figure 5.38 (Left).

Command: POL *(For POLYGON)*

Enter number of sides <4>: 6

Specify center of polygon or [Edge]: *(Mark a point at "A," as shown in the following image on the left)*

Enter an option [Inscribed in circle/Circumscribed about circle] <I>: I *(For Inscribed)*

Specify radius of circle: 1.00

TRY IT!

Create a new drawing file starting from scratch. Use the following command sequence to construct the circumscribed polygon, as shown in Figure 5.38 (Right).

Command: POL *(For POLYGON)*

Enter number of sides <4>: 6

Specify center of polygon or [Edge]: *(Pick a point at "A," as shown in the following image on the right)*

Enter an option [Inscribed in circle/Circumscribed about circle] <I>: C *(For Circumscribed)*

Specify radius of circle: 1.00

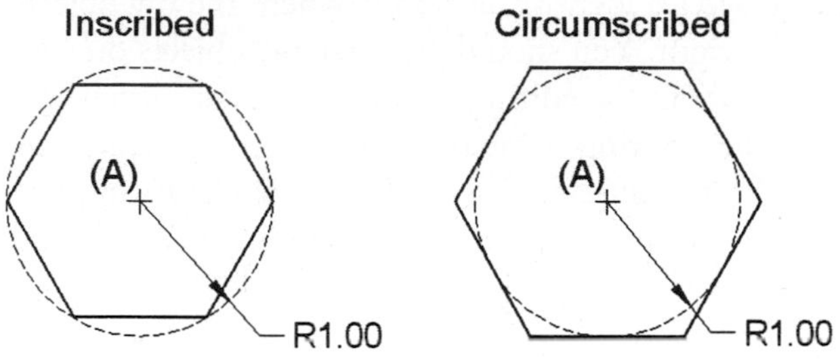

FIGURE 5.38

In many mechanical drawings, the distance between flats is known for regular polygons and the Circumscribed option is the logical choice.	TIP

If you know the length of the polygon side, an additional method for creating polygons is available. This is accomplished by locating the endpoints of one of its edges. The polygon is then drawn in a counterclockwise direction.

Create a new drawing file starting from scratch. Study Figure 5.39 and the command sequence to construct a polygon by edge.	TRY IT!

Command: POL *(For POLYGON)*

Enter number of sides <4>: 5

Specify center of polygon or [Edge]: E *(For Edge)*

Specify first endpoint of edge: *(Mark a point at "A")*

Specify second endpoint of edge: *(Mark a point at "B")*

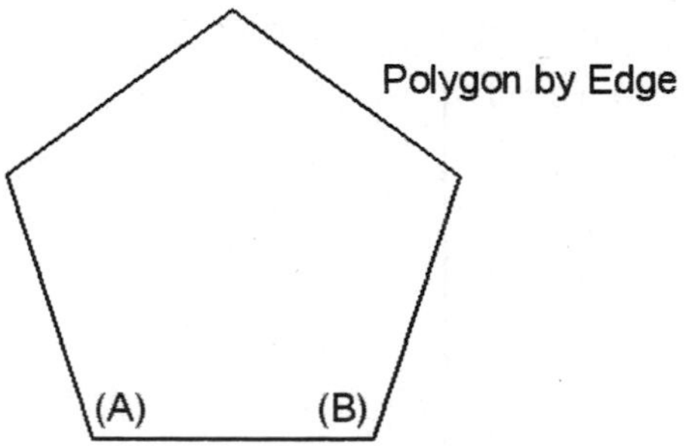

FIGURE 5.39

CREATING A RAY

A ray is a type of construction line object that begins at a user-defined point and extends to infinity in only one direction. This command along with the XLINE command (discussed later in the chapter) is often used for creating projected views (orthographic and auxiliary). Choose this command from one of the following:

- From the Ribbon > Home Tab > Draw Panel (Expanded)
- The Menu Bar (Draw > Ray)
- The keyboard (RAY)

In Figure 5.40, the quadrants of the circles identify all points where the ray objects begin and are drawn to infinity to the right. You should organize ray objects on specific layers. You should also exercise care in the editing of rays, and take special care not to leave segments of objects in the drawing database as a result of breaking ray objects. Breaking the ray object at "A" in Figure 5.40 converts one object to an individual line segment; the other object remains a ray. Study the following command sequence for constructing a ray.

`Command: RAY`

`Specify start point:` *(Pick a point on an object)*

`Specify through point:` *(Pick an additional point to construct the ray object)*

`Specify through point:` *(Pick another point to construct the ray object or press ENTER to exit this command)*

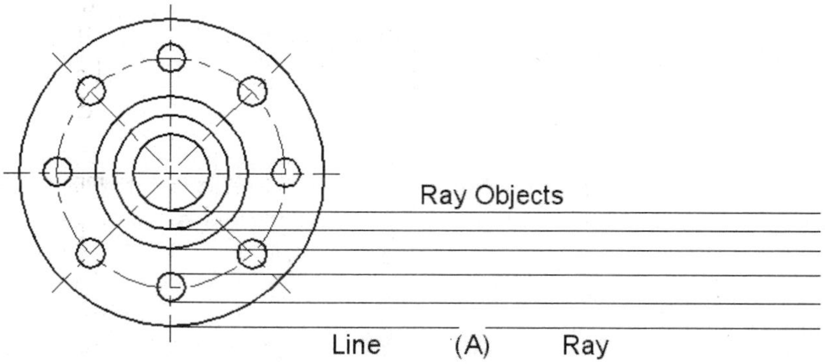

FIGURE 5.40

CREATING RECTANGLE OBJECTS

Use the RECTANG command to construct a rectangle by defining two points, by dimensions, or by area. Choose this command from one of the following:

- From the Ribbon > Home Tab > Draw Panel (Expand icon: Rectangle & Polygon)
- The Menu Bar (Draw > Rectangle)
- The Draw Toolbar of the AutoCAD Classic Workspace
- The keyboard (REC or RECTANG or RECTANGLE)

Rectangle by Picking Two Diagonal Points

As illustrated in Figure 5.41, two diagonal points are picked to define the rectangle. The rectangle is drawn as a single polyline object.

Command: REC *(For RECTANG)*

Specify first corner point or [Chamfer/Elevation/Fillet/Thickness/Width]: *(Pick a point at "A")*

Specify other corner point or [Area/Dimensions/Rotation]: *(Pick a point at "B")*

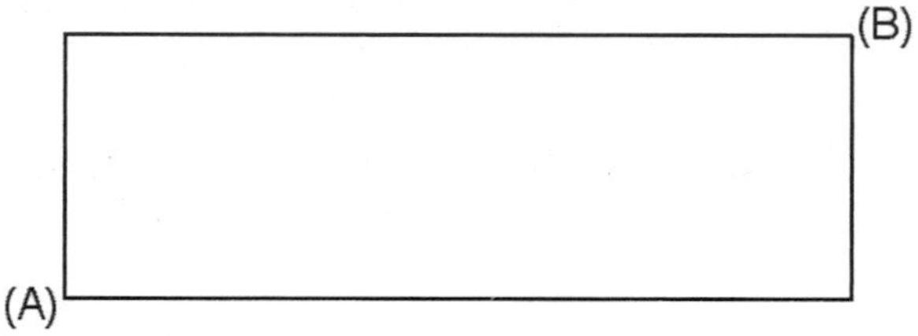

FIGURE 5.41

Changing Rectangle Properties

Options of the RECTANG command enable you to construct a chamfer or fillet at all corners of the rectangle, to assign a width to the rectangle, and to have the rectangle drawn at a specific elevation and at a thickness for 3D purposes. Illustrated in Figure 5.42, a rectangle is constructed with a chamfer distance of 0.20 units; the width of the rectangle is also set at 0.05 units. A relative coordinate value of @4.00, 1.00 is used to construct the rectangle 4 units in the X direction and 1 unit in the Y direction. The @ symbol resets the previous point at "A" to zero. If Dynamic Input is activated on the Status Bar, it is not necessary to type the @ symbol to provide the relative coordinate.

TRY IT!

Create a new AutoCAD drawing starting from scratch. Use the following command sequence image for constructing a wide rectangle with its corners chamfered.

Command: REC *(For RECTANGLE)*

Specify first corner point or [Chamfer/Elevation/Fillet/Thickness/Width]: C *(For Chamfer)*

Specify first chamfer distance for rectangles <0.0000>: **0.20**

Specify second chamfer distance for rectangles <0.2000>: *(Press* ENTER *to accept this default value)*

Specify first corner point or [Chamfer/Elevation/Fillet/Thickness/Width]: W *(For Width)*

Specify line width for rectangles <0.0000>: **0.05**

Specify first corner point or [Chamfer/Elevation/Fillet/Thickness/Width]: *(Pick a point at "A")*

Specify other corner point or [Area/Dimensions/Rotation]: **@4.00,1.00** *(To identify the other corner at "B")*

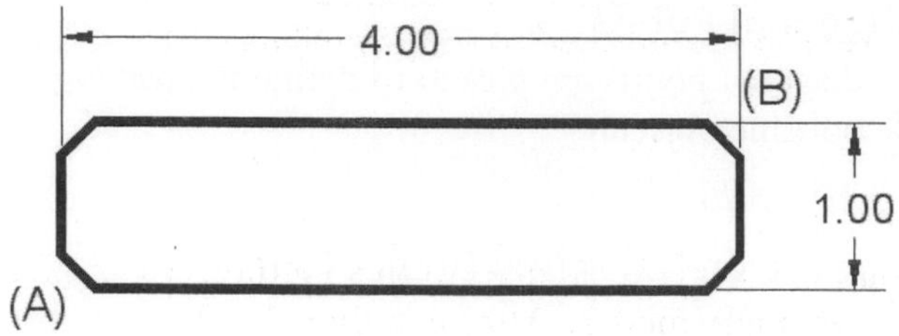

FIGURE 5.42

NOTE

It should be noted that the values supplied in the options are remembered in system variables and the next rectangle created will have those same properties.

Rectangle by Dimensions

Instead of specifying a relative coordinate (@4,1 as in the previous example), the Dimensions option provides another method in which you can supply the length and width dimensions for the rectangle. Issue the option and you will be prompted to provide the length and width. Next, you pick a point on the screen to indicate where the opposite corner of the rectangle will be positioned. There are four possible positions or quadrants the rectangle can be drawn in using this method: upperright, upperleft, lowerright, and lowerleft.

TRY IT!

Create a new AutoCAD drawing starting from scratch. Specify the dimensions (length and width) of the rectangle by using the following command sequence and Figure 5.43.

```
Command: REC (For RECTANGLE)
Specify first corner point or [Chamfer/Elevation/Fillet/
Thickness/Width]: (Pick a point at "A" on the screen)
Specify other corner point or [Area/Dimensions/Rotation]: D
(For Dimensions)
Specify length for rectangles <0.0000>: 3.00
Specify width for rectangles <0.0000>: 1.00
Specify other corner point or [Area/Dimensions/Rotation]:
(Moving your cursor around positions the rectangle in four
possible positions. Click the upper-right corner of your
screen at "B" to anchor the upper-right corner of the
rectangle)
```

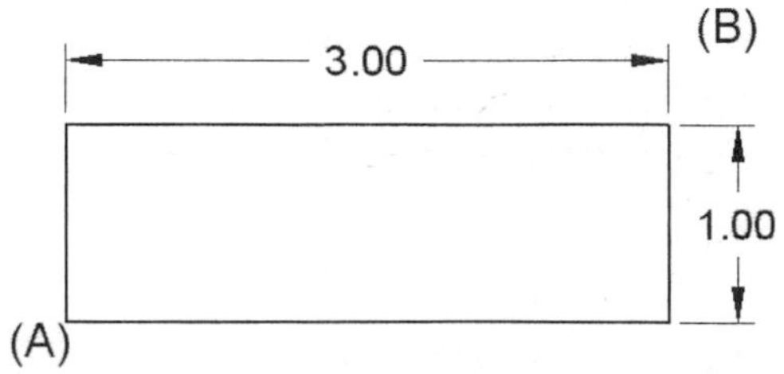

FIGURE 5.43

NOTE

Rectangles can also be constructed based on a user-specified angle by utilizing the Rotation option.

Rectangle by Area

The final method of constructing a rectangle is by area. In this method, you pick a first corner point, as in previous rectangle modes. After entering the Area option, you enter the area of the rectangle based on the current drawing units. You then enter the length or width of the rectangle. If, for instance, you enter the length, the width of the rectangle will automatically be calculated based on the area. In the following command sequence, an area of 200 is entered, along with a length of 20 units. A width of 10 units is automatically calculated based on the other two numbers.

Command: REC *(For RECTANGLE)*

Specify first corner point or [Chamfer/Elevation/Fillet/Thickness/Width]: *(Pick a point to start the rectangle)*

Specify other corner point or [Area/Dimensions/Rotation]: A *(For Area)*

Enter area of rectangle in current units <100.0000>: 200

Calculate rectangle dimensions based on [Length/Width] <Length>: L *(For Length)*

Enter rectangle length <10.0000>: 20

CREATING A REVISION CLOUD

The REVCLOUD command creates a polyline object consisting of arc segments in a sequence. This feature is commonly used in drawings to identify areas where the drawing is to be changed or revised. Choose this command from one of the following:

- From the Ribbon > Home Tab > Draw Panel (Expanded)
- The Menu Bar (Draw > Revision Cloud)
- The Draw Toolbar of the AutoCAD Classic Workspace
- The keyboard (REVCLOUD)

In Figure 5.44, an area of the house is highlighted with the revision cloud. By default, each arc segment has a minimum and maximum arc length of 0.50 units. In a drawing such as a floor plan, the arc segments would be too small to view. In cases in which you wish to construct a revision cloud and the drawing is large, the arc length should be increased. Through a little experimentation an appropriate size can be determined. Once the arc segment is set, pick a point on your drawing to begin the revision cloud. Then move your cursor slowly in either a clockwise or counterclockwise direction. You will notice the arc segments of the revision cloud being constructed. Continue surrounding the object, while at the same time heading back to the origin of the revision cloud. Once you hover near the original start, the end of the revision cloud snaps to the start and the command ends, leaving the revision cloud as the polyline object. Study the Command prompt sequence below and the illustration in Figure 5.44 for the construction of a revision cloud.

Command: REVCLOUD

Minimum arc length: 1/2" Maximum arc length: 1/2" Style: Normal

Specify start point or [Arc length/Object/Style] <Object>: A *(For Arc length)*

Specify minimum length of arc <1/2">: 24

Specify maximum length of arc <2'>: *(Press* ENTER*)*

Specify start point or [Arc length/Object/Style] <Object>: *(Pick a point on your screen to start the revision cloud. Surround the item with the revision cloud by moving your cursor. When you approach the start point, the revision cloud closes and the command exits.)*

Guide crosshairs along cloud path...

Revision cloud finished.

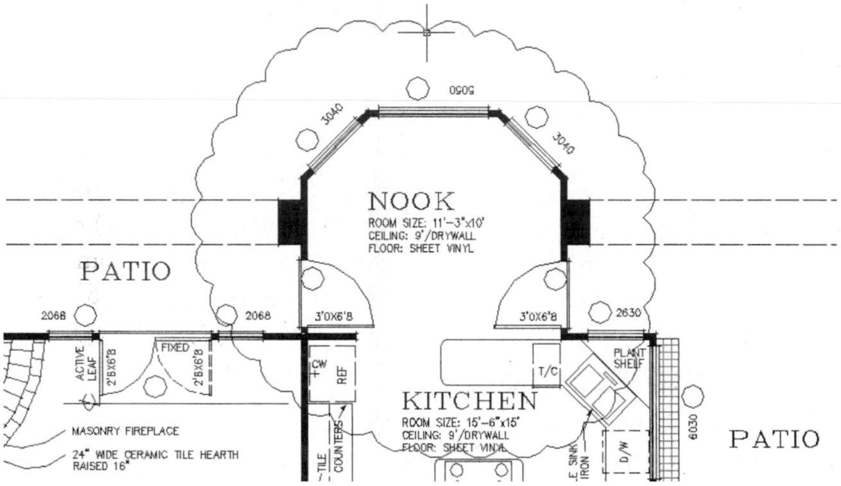

FIGURE 5.44

Calligraphy Revision Clouds

The Calligraphy option of creating revision clouds allows you to enhance the appearance of the cloud and make your drawings more dramatic. Use the following command sequence and Figure 5.45, which illustrate this feature.

Command: REVCLOUD

Minimum arc length: 24.0000 Maximum arc length: 24.0000 Style: Normal

Specify start point or [Arc length/Object/Style] <Object>: S *(For Style)*

Select arc style [Normal/Calligraphy] <Normal>: C *(For Calligraphy)*

Arc style = Calligraphy

Specify start point or [Arc length/Object/Style] <Object>: *(Pick a starting point)*

Guide crosshairs along cloud path...

Revision cloud finished.

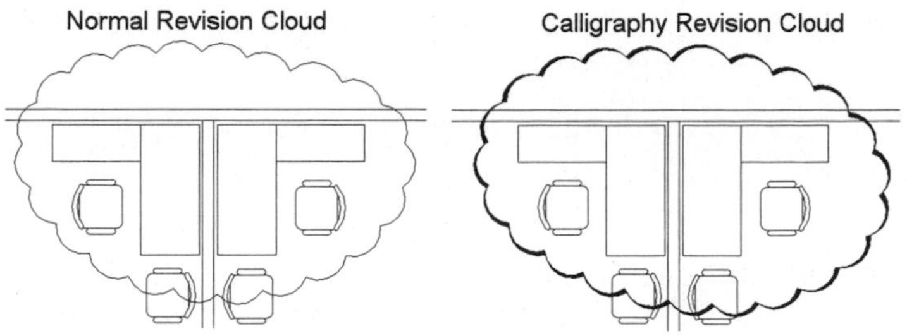

FIGURE 5.45

CREATING SPLINES

Use the SPLINE command to construct a smooth curve given a sequence of points. Choose this command from one of the following:

- From the Ribbon > Home Tab > Draw Panel (Expanded)
- The Menu Bar (Draw > Spline)
- The Draw Toolbar of the AutoCAD Classic Workspace
- The keyboard (SPL or SPLINE)

You have the option of changing the method of creating splines. A Fit option forces the spline to pass through the specified points, while a CV (control vertices) option shapes the curve by an attraction (weight) to the specified points. The basic command sequence follows, which constructs a fit spline segment shown in Figure 5.46 in the center. Entering a different tangent point at the end of the command changes the shape of the curve connecting the beginning and end of the spline.

Command: **SPL** *(For SPLINE)*

Current settings: Method = Fit Knot = Chord)

Specify first point or [Method/Knots/Object]: *(Pick a first point)*

Enter next point or [start Tangency/toLerance]: *(Continue picking points in sequence)*

Enter next point or [end Tangency/toLerance/Undo/Close]: *(Pick another point)*

The results of creating a spline with control vertices is shown in Figure 5.46 (Right).

Command: **SPL** *(For SPLINE)*

Current settings: Method = Fit Knots = Chord

Specify first point or [Method/Knots/Object]:

M *(For Method)*

Enter spline creation method [Fit/CV] <Fit>: **CV** *(For Control Vertices)*

Current settings: Method=CV Degree=3

Specify first point or [Method/Degree/Object]:*(Pick a first point)*

Enter next point:*(Pick another point)*

Enter next point or [Undo]: *(Continue picking points in sequence)*

Enter next point or [Close/Undo]: *(Press* ENTER *to accept)*

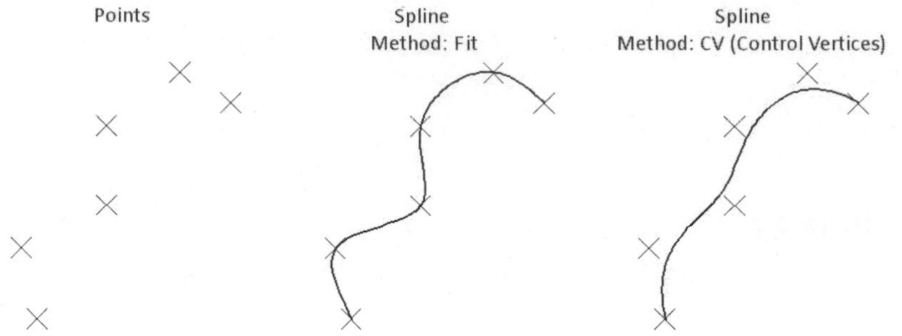

FIGURE 5.46

TRY IT!

Open the drawing file 05_ Spline Rasp Handle, as shown in Figure 5.47 (Left). Turn OSNAP on and set Running OSNAP to Node. Construct a fit spline by connecting all points between "A" and "B." Construct an arc with its center at "C," the start point at "D," and the endpoint at "B." Connect points "D" and "E" with a line segment. Construct another fit spline by connecting all points between "E" and "F." Construct another arc with its center at "G," the start point at "H," and the endpoint at "F." Connect points "H" and "A" with a line segment.

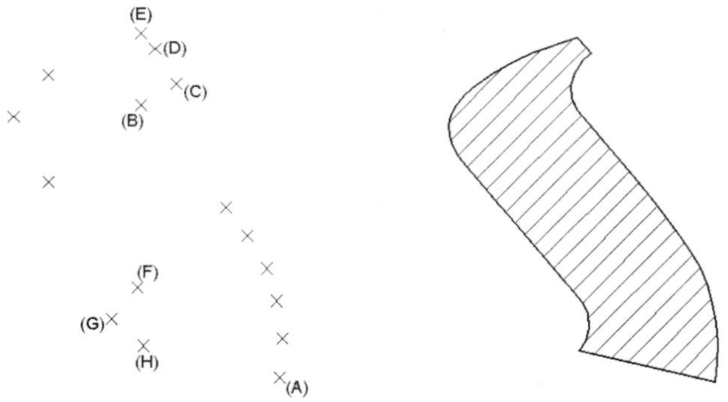

FIGURE 5.47

MASKING TECHNIQUES WITH THE WIPEOUT COMMAND

To mask or hide objects in a drawing without deleting them or turning off layers, the WIPEOUT command could be used. Choose this command from one of the following:

- From the Ribbon > Home Tab > Draw Panel (Expanded)
- The Menu Bar (Draw > Wipeout)
- The keyboard (WIPEOUT)

This command reads the current drawing background color and creates a mask over anything defined by a frame. In the illustration in Figure 5.48 (Left), a series of text objects needs to be masked over. The middle image shows a four-sided frame that was created over the text using the following Command prompt sequence:

```
Command: WIPEOUT
Specify first point or [Frames/Polyline] <Polyline>: (Pick a first point)
Specify next point: (Pick a second point)
Specify next point or [Undo]: (Pick a third point)
Specify next point or [Close/Undo]: (Pick a fourth point)
Specify next point or [Close/Undo]: (Press ENTER to exit the command and create the wipeout)
```

As the text seems to disappear, the wipeout frame is still visible. A visible frame is important if you would like to unmask or delete the wipeout. If you want to hide all wipeout frames, use the following command sequence:

```
Command: WIPEOUT
Specify first point or [Frames/Polyline] <Polyline>: F (For Frames)
Enter mode [ON/OFF/Display but not plot] <ON>: Off
```

Now all wipeout frames in the current drawing are turned off, as shown in the illustration in Figure 5.48 (Right). Instead of turning frames on and off, you can select the "Display but not plot" option, which allows you to see the frames for editing purposes and then hides them on the plot.

NOTE

You could also create a predefined polyline object and then convert it into a wipeout using the Polyline option of the `WIPEOUT` command.

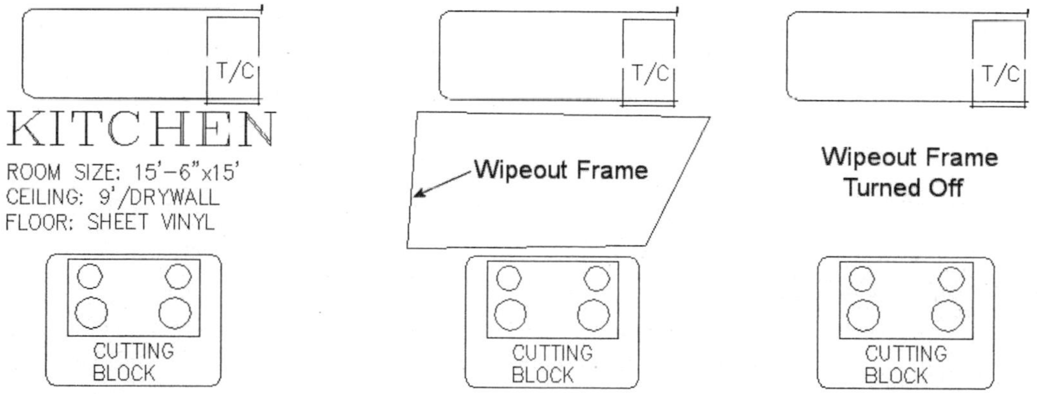

FIGURE 5.48

CREATING CONSTRUCTION LINES WITH THE XLINE COMMAND

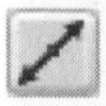
Xlines are construction lines drawn from a user-defined point. Choose this command from one of the following:

- From the Ribbon > Home Tab > Draw Panel (Expanded)
- The Menu Bar (Draw > Construction Line)
- The Draw Toolbar of the AutoCAD Classic Workspace
- The keyboard (`XL` or `XLINE`)

You are not prompted for any length information because the Xline extends to an unlimited length, beginning at the user-defined point and going off to infinity in opposite directions from the point. Xlines can be drawn horizontal, vertical, and angular. You can also bisect an angle or offset an object using the Xline. Xlines are particularly useful in constructing orthogonal drawings by establishing reference lines between views. As illustrated in Figure 5.49 (Left), the circular view represents the Front view of a flange. To begin the creation of the Side views, lines are usually projected from key features on the adjacent view. In the case of the Front view, the key features are the top of the plate in addition to the other circular features. In this case, the Xlines were drawn with the Horizontal mode from the Quadrant of all circles. The following prompts outline the `XLINE` command sequence:

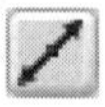
`Command:` **`XL`** *`(For XLINE)`*

`Specify a point or [Hor/Ver/Ang/Bisect/Offset]:` **`H`** *`(For Horizontal)`*

`Specify through point:` *`(Pick a point on the display screen to place the first Xline)`*

`Specify through point:` *`(Pick a point on the display screen to place the second Xline)`*

Since the Xlines continue to be drawn in both directions, care must be taken to manage these objects. Construction management techniques of Xlines could take the form of placing all Xlines on a specific layer to be turned off or frozen when not needed. When editing Xlines (especially with the `BREAK` or `TRIM` commands), you need to take special care to remove all excess objects that remain on the drawing screen. As illustrated in Figure 5.49 (Right), breaking the Xline converts the object to a ray object. Use the `ERASE` command to remove any excess Xlines.

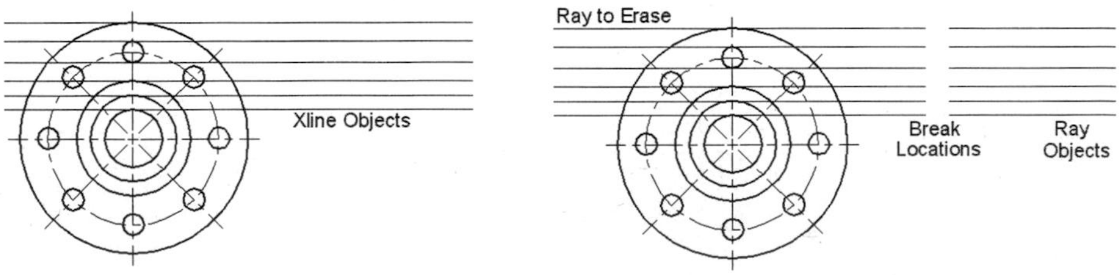

FIGURE 5.49

TRY IT!

Open the drawing file 05_Xline. As shown in Figure 5.50 (Left), three horizontal and vertical Xlines are already constructed. You will corner the Xlines to create the object illustrated in Figure 5.50 (Right). Use the `FILLET` **command and set the radius to zero. Pick the lines at "A" and "B" to create the first corner. Repeat this sequence on the remaining lines to form the other corners.**

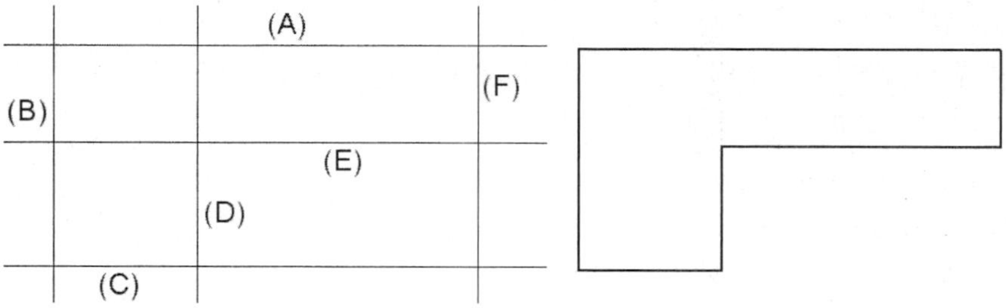

FIGURE 5.50

CREATING OBJECTS WITH THE ADDSELECTED COMMAND

Creates a new object based on the object type selected and with the same general properties of the selected object. Choose this command from one of the following:

- Shortcut Menu
- The Draw Toolbar of the AutoCAD Classic Workspace
- The keyboard (`ADDSELECTED`)

After selecting an object with this command, you will activate the command that created the selected object. If you pick a circle, the `CIRCLE` command is started. If you pick a text object, either the `MEXT` or `TEXT` command is started depending on what created the text. The real advantage, to this command, however, is the transfer of general properties to the new object. Your new object will be created on the same layer as the selected object. There are some special property transfers, such as the height and text style of text objects and the dimension style of a dimension object.

To activate the `ADDSELECTED` command from the shotcut menu, pick the object first and then right-click to display the menu, as shown in Figure 5.51.

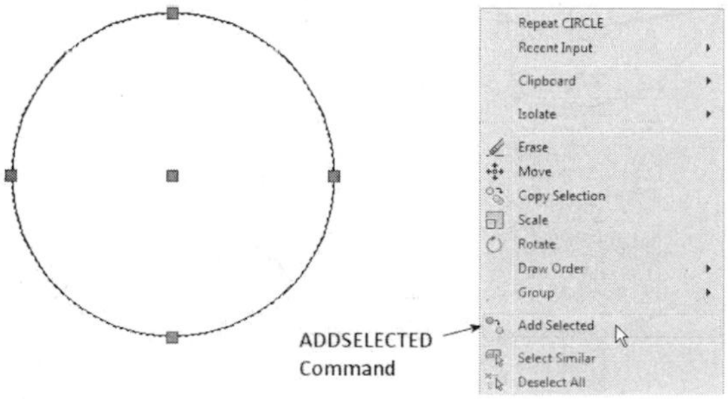

FIGURE 5.51

TRY IT!

Open the drawing file 05_Addselected. As shown in Figure 5.52 (Left), a circle is missing in the Front View and a text object is missing in the Right View. The corrected drawing is shown in the image on the right. The following prompts outline the ADDSELECTED **command sequence:**

Command: **ADDSELECTED** *(or utilize the Shortcut Menu)*

Select object: *(Pick the hidden circle at "A")*

_circle

Specify center point for circle or [3P/2P/Ttr (tan tan radius)]: **CEN** *(For Osnap Center)*

of *(Select the edge of circle "A")*

Specify radius of circle or [Diameter] <0.5000>: **D** *(For Diameter)*

Specify diameter of circle <1.0000>: **1.00**

Command: **ADDSELECTED** *(or utilize the Shortcut Menu)*

Select object: *(Pick the text object at "B")*

_text

Current text style: "SIMPLEX" Text height: 0.1000 Annotative: No

Specify start point of text or [Justify/Style]: *(Pick a point near "C")*

Specify height <0.1000>: *(Press ENTER to accept)*

Specify rotation angle of text <0>: *(Press ENTER to accept)*

Enter text: **RIGHT VIEW**

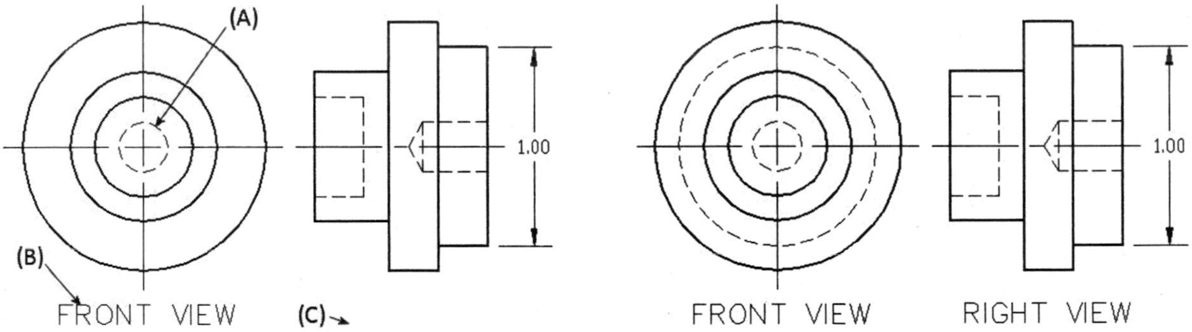

FIGURE 5.52

OGEE OR REVERSE CURVE CONSTRUCTION

An ogee curve connects two parallel lines with a smooth, flowing curve that reverses itself in a symmetrical form.

TRY IT!

Open the drawing file 05_Ogee. To begin constructing an ogee curve to line segments "AB" and "CD," a line is drawn from "B" to "C," which connects both parallel line segments, as shown in Figure 5.53 (Left).

Use the DIVIDE command to divide line segment "BC" into four equal parts. Be sure to set a new point mode from the Point Style dialog box. Construct vertical lines from "B" and "C." Complete this step by constructing line segment "XY," which is perpendicular to line "BC," as shown in Figure 5.53 (Right). Do not worry about where line "XY" is located at this time.

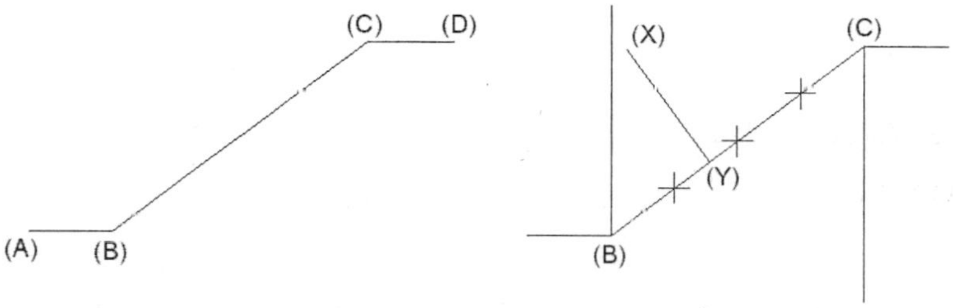

FIGURE 5.53

Move line "XY" to the location identified by the point, as shown in Figure 5.54 (Left). Complete this step by copying line "XY" to the location identified by point "Z," as shown in Figure 5.54 (Left).

Construct two circles with centers located at points "X" and "Y," as shown in Figure 5.54 (Right). Use the OSNAP-Intersection mode to accurately locate the centers. If an intersection is not found from the previous step, use the EXTEND command to find the intersection and continue with this step. The radii of both circles are equivalent to distances "XB" and "YC."

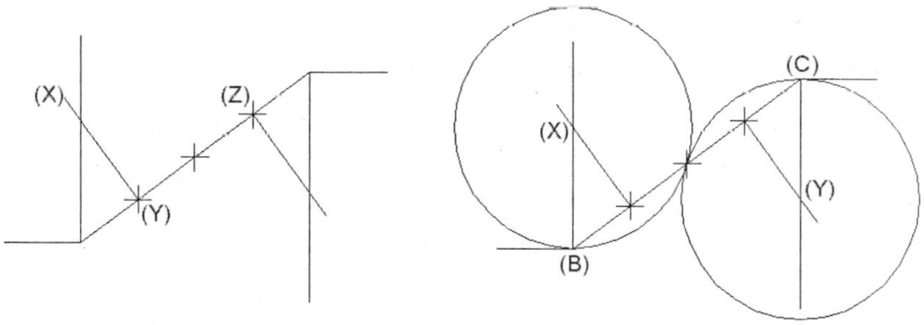

FIGURE 5.54

Use the TRIM command to trim away any excess arc segments to form the ogee curve, as shown in the Figure 5.55 (Left).

This forms the frame of the ogee for the construction of objects, such as the wrench illustrated in Figure 5.55 (Right).

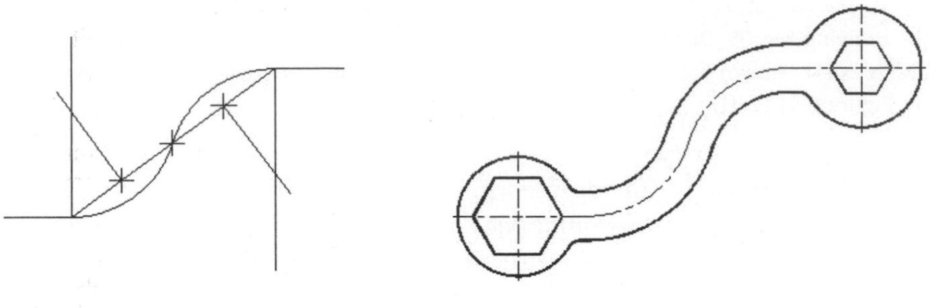

FIGURE 5.55

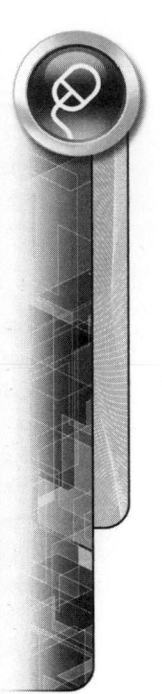

TUTORIAL EXERCISE: 05_GEAR-ARM.DWG

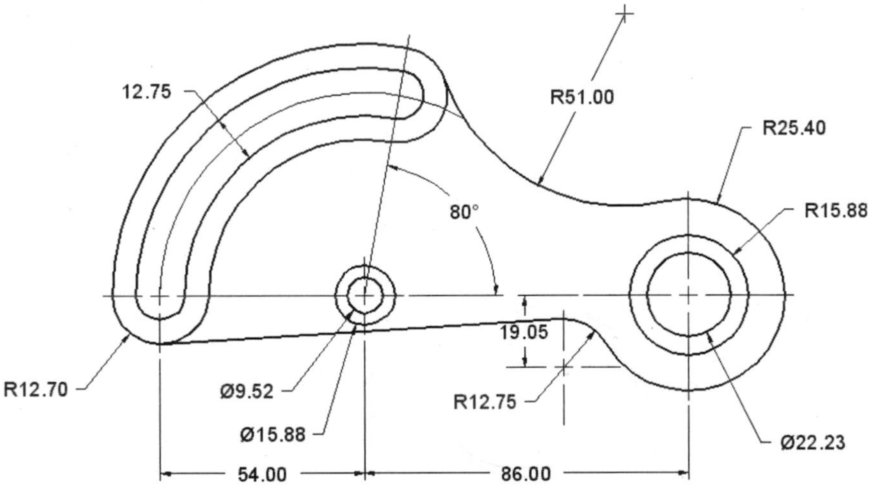

FIGURE 5.56

Purpose

This tutorial is designed to use geometric construction commands to create a one-view drawing of the gear-arm in metric format, as illustrated in Figure 5.56.

System Settings

Start a new drawing from scratch utilizing the Acadiso.dwt template. Use the Drawing Units dialog box to change the number of decimal places past the zero from four to two. Keep the remaining default unit values. Using the LIMITS command, keep (0,0) for the lower-left corner and change the upper-right corner from (420,297) to (265.00,200.00). Perform a ZOOM-All after changing the drawing limits. Check to see that the following Object Snap modes are already set: Endpoint, Extension, Intersection, and Center.

Layers

Create the following layers with the format:

Name	Color	Linetype	Lineweight
Center	Yellow	Center2	Default
Construction	Gray	Center2	Default
Dimension	Yellow	Continuous	Default
Object	White	Continuous	0.50 mm

Suggested Commands

Follow the steps provided to create a new drawing called 05_Gear-Arm. The object consists of a combination of circles and arcs along with tangent lines and arcs. Construction lines will be used as a layout tool to mark the centers of key circles and arcs. The CIRCLE-TTR command will be used for constructing tangent arcs to existing geometry. Use the ARC command to construct a series of arcs for the left side of the gear-arm. The TRIM command will be used to trim circles, lines, and arcs to form the basic shape.

STEP 1

Begin constructing the gear-arm by creating construction geometry that will be used to create all circles. First make the Construction layer current and use the LINE command to lay out the long horizontal line (longer than 140 units). Create a vertical line on the left end of the horizontal line. Then use the OFFSET command to copy the vertical line 54 units to the right and then the same line 86 units to the right. Also, create a line 70 units long at an 80° angle at the intersection, as shown in Figure 5.57.

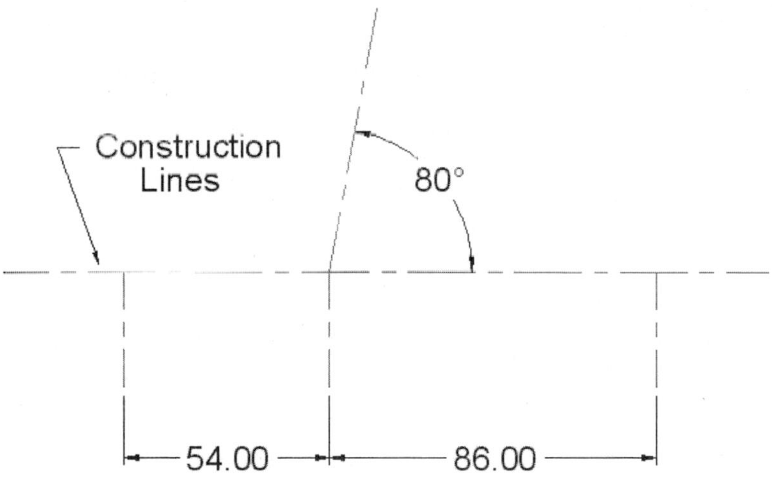

FIGURE 5.57

STEP 2

Continue using the Construction layer to create an arc that will be used to locate the slot detail. This arc is to be constructed using the Center, Start, Angle mode of the ARC command, as shown in Figure 5.58 (Left). Pick "A" as the center of the arc, "B" as the start of the arc, as shown in Figure 5.58 (Right), and for angle, enter a value of –110°. The negative degree value is needed to construct the arc in the clockwise direction.

Create another construction object; this time, use the OFFSET command to offset the long horizontal construction line 19.05 units down. The line shown in Figure 5.58 (Right) was shortened on each end using the LENGTHEN command with the Dynamic option.

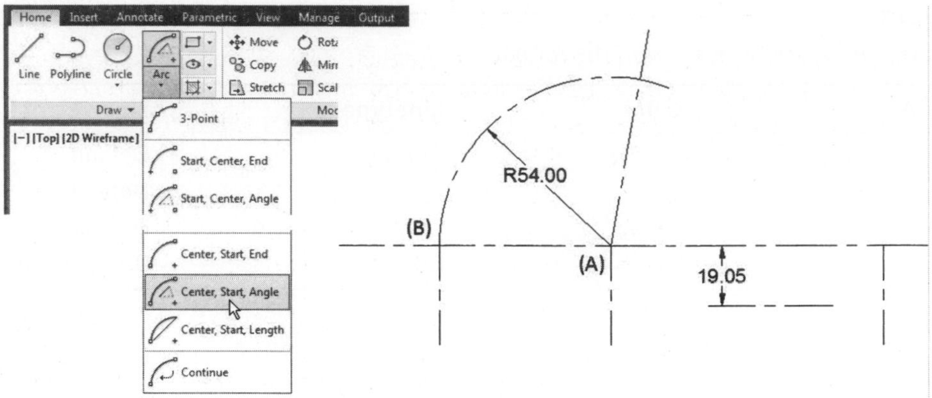

FIGURE 5.58

STEP 3

One final item of construction geometry needs to be created, namely, the circle at a diameter of 76.30. Construct this circle from the intersection of the construction lines at "A," as shown in Figure 5.59 (Left). Convert the circle to an arc segment by using the BREAK command and picking the first break point at "B" and the second break point at "C." This converts the circle of diameter 76.30 to an arc of radius 38.15, as shown in Figure 5.59 (Right).

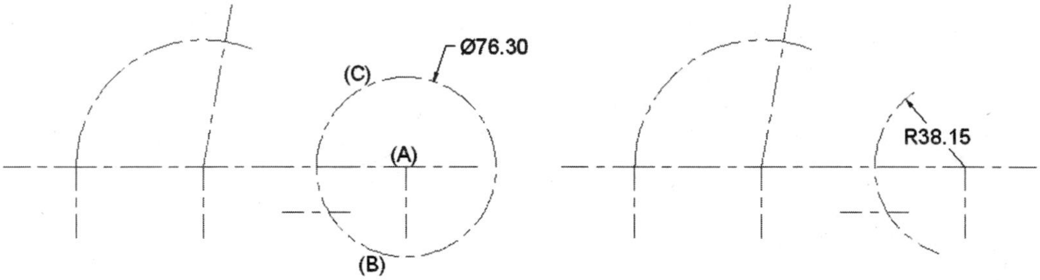

FIGURE 5.59

STEP 4

Set the Object layer current. Using Figure 5.60 as a guide, create all circles using the intersection of the construction line geometry as the centers for the circles. All circles are given with diameter dimensions.

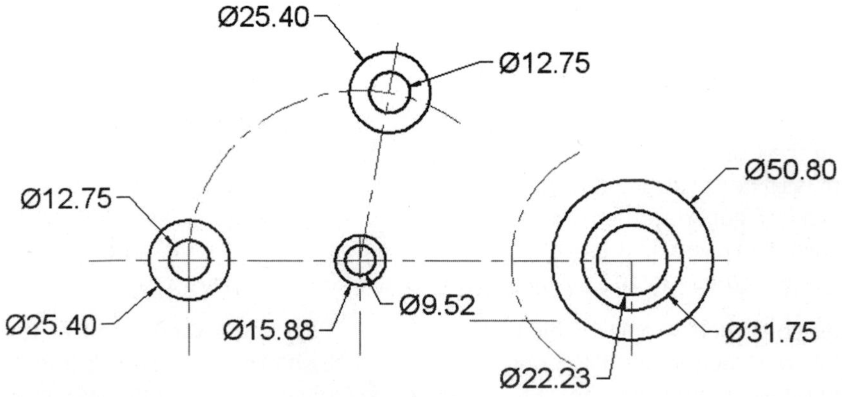

FIGURE 5.60

STEP 5

Use the ARC—CENTER, START, END command to create the arc shown in Figure 5.61 (Left). Pick "A" (Cen OSNAP) as the center of the arc, "B" (Int or Tan OSNAP) as the start, and "C" (Int or Tan OSNAP) as the end of the arc. Follow these same steps to construct the other arc shown in the figure on the left.

When finished creating the outer arc segments, create the two inner arc segments, as shown in Figure 5.61 (Right), using the same ARC—CENTER, START, END command. Continue to use Object Snaps to pick valid intersections in order to create accurate arcs.

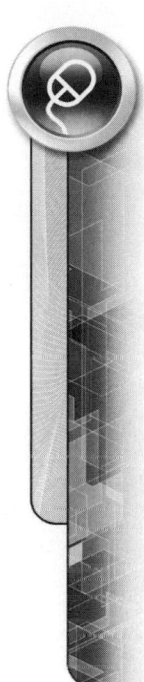

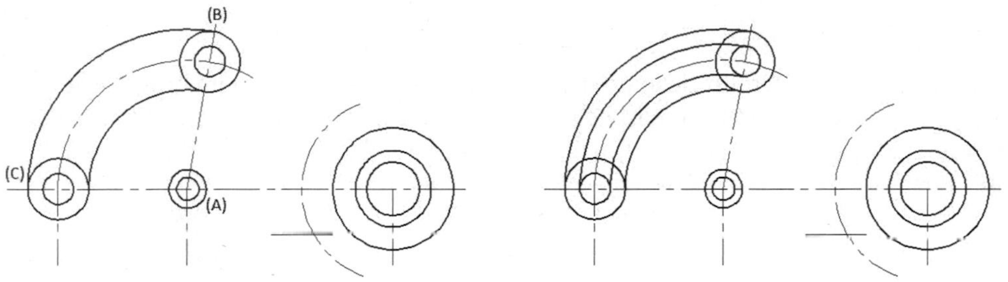

FIGURE 5.61

STEP 6

Use the CIRCLE-TTR command to construct a circle tangent to two existing circles. Pick the two dashed circles as the objects to be made tangent, as shown in Figure 5.62 (Left), and enter a radius value of 51.

After the circle is created, enter the TRIM command, pick the two dashed circles as cutting edges, and pick the upper edge of the large 51-radius circle. The results of this operation are illustrated in Figure 5.62 (Right).

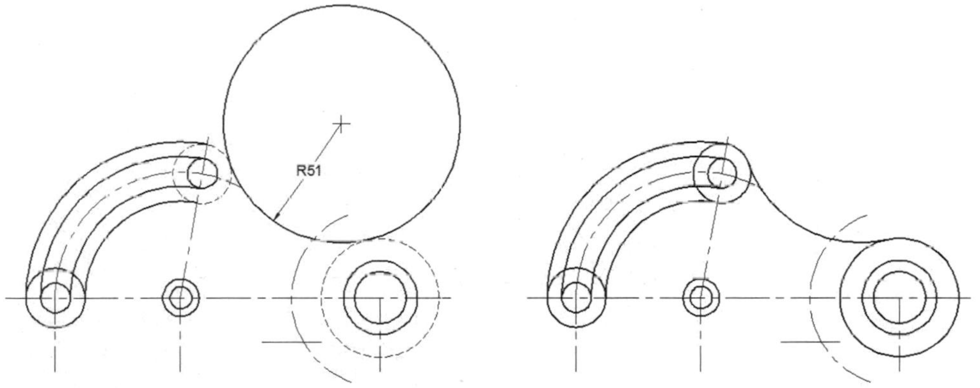

FIGURE 5.62

STEP 7

Enter the TRIM command again, pick the two dashed arcs as cutting edges, and pick the two inside edges of the 25.4-diameter circles ("A" and "B"), as shown in Figure 5.63 (Left). The results of this operation are illustrated in Figure 5.63 (Right).

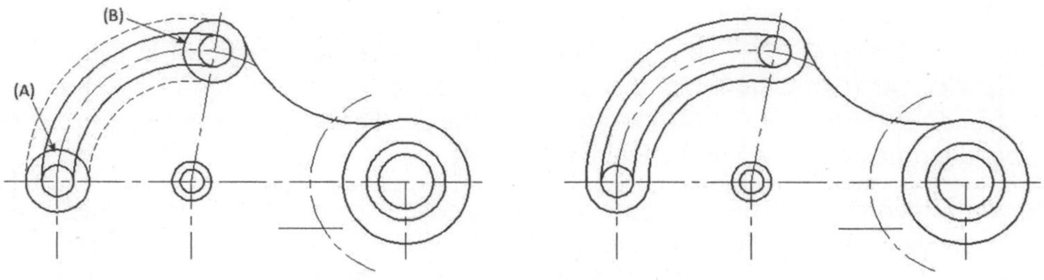

FIGURE 5.63

STEP 8

Do the same procedure to clean up the geometry that makes up the inner slot. Enter the TRIM command, pick the two dashed arcs as cutting edges, and pick the two inside edges of the 12.75-diameter circles ("A" and "B"), as shown in Figure 5.64 (Left). The results of this operation are illustrated in Figure 5.64 (Right).

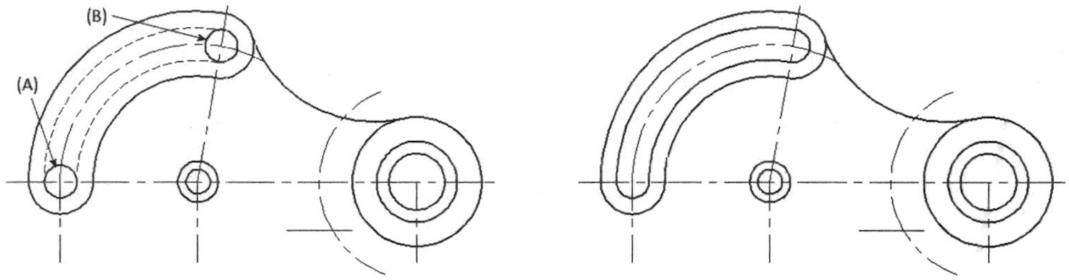

FIGURE 5.64

STEP 9

Construct a new circle of diameter 25.50 from the intersection of the construction line and arc, as shown in Figure 5.65 (Left). Then draw a line segment that is tangent with the bottom of the arc (A) and top of the circle (B) just constructed, as shown in Figure 5.65 (Right).

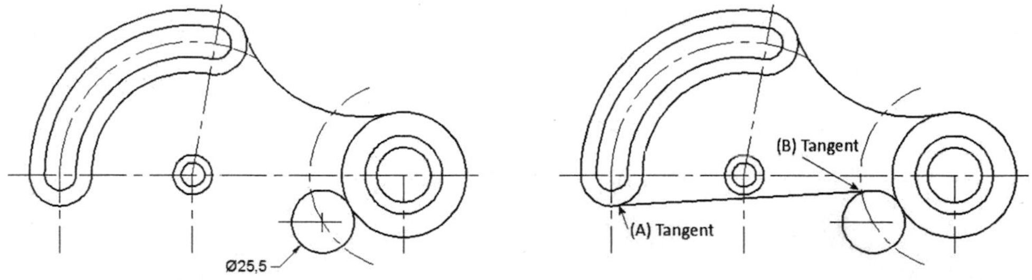

FIGURE 5.65

STEP 10

Use the TRIM command to clean up all unnecessary segments of geometry. Select all dashed objects as cutting edges, as shown in Figure 5.66 (Left). Trim objects until your display appear similar to the illustration in Figure 5.66 (Right).

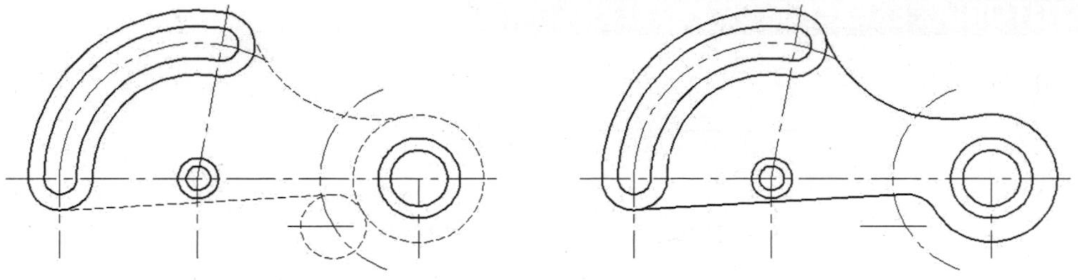

FIGURE 5.66

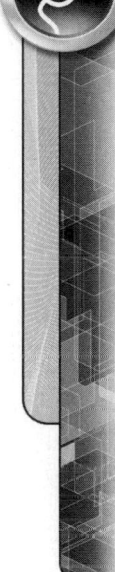

STEP 11

Erase all unnecessary construction geometry, as shown in Figure 5.67 (Left). Notice that the arc and 70° angle line both remain and will be used later to define the centers of the gear-arm.

Next create centerlines to mark the centers of all circles. Switch to the Center layer. A system variable that controls centerlines and marks called DIMCEN must first be entered from the keyboard and changed to a value of −2.5. The negative value constructs the small dash and long segment, all representing the centerline. Entering a value of 2.5 would only mark the center of circles by constructing two short intersecting line segments. Now use the `DIMCENTER` (or `DCE` for short) command and pick the edge of a circle or arc segment to place the centermarks. The results should appear similar to the illustration in Figure 5.67 (Right).

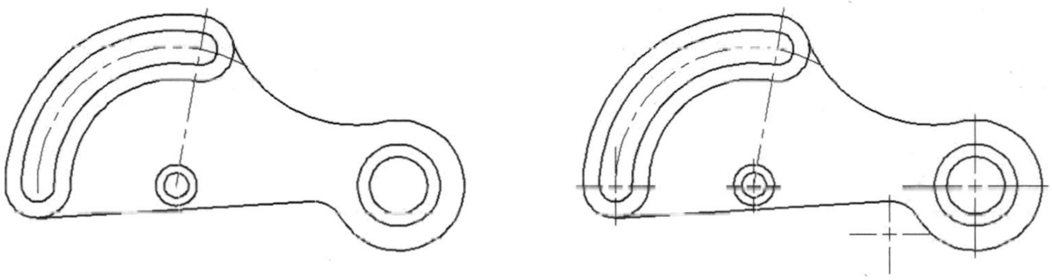

FIGURE 5.67

STEP 12

The completed gear-arm drawing is illustrated, as shown in Figure 5.68 (Left). The finished object may be dimensioned as an optional step, as shown in Figure 5.68 (Right).

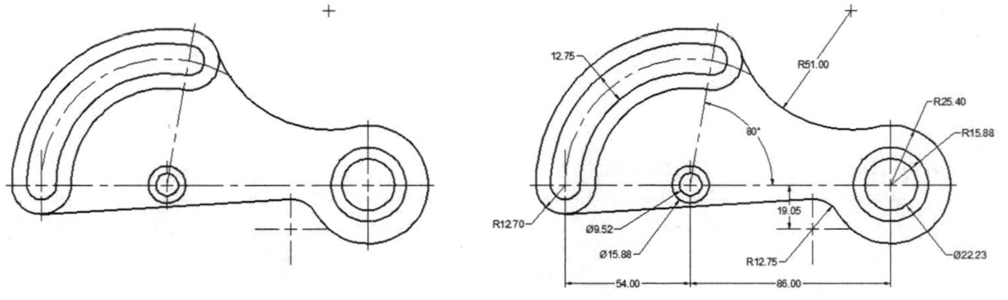

FIGURE 5.68

TUTORIAL EXERCISE: 05_PATTERN1.DWG

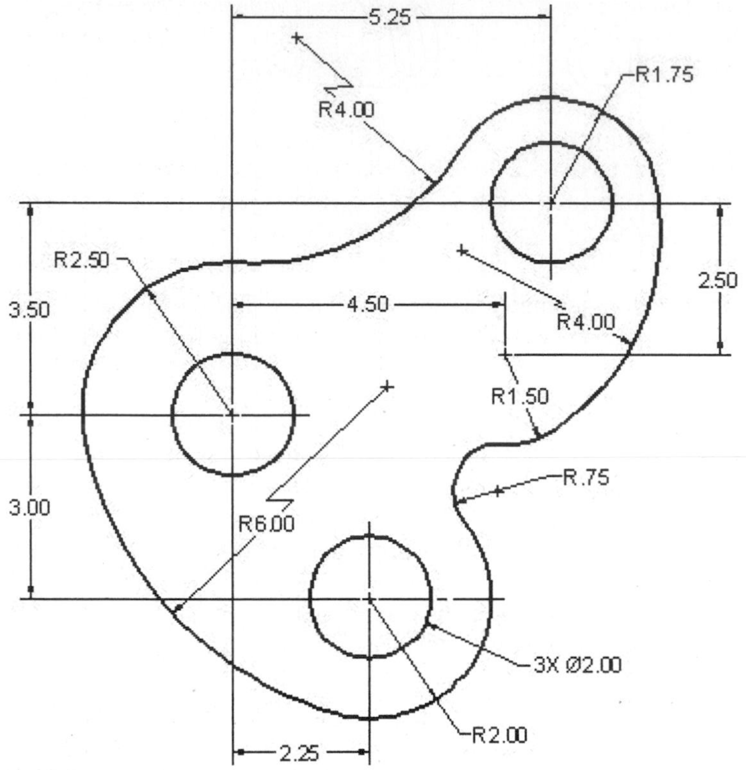

FIGURE 5.69

Purpose

This tutorial is designed to use various Draw commands to construct a one-view drawing of Pattern1, as shown in Figure 5.69. Refer to the following special system settings and suggested command sequences.

System Settings

Start a new drawing from scratch utilizing the Acad.dwt template. Use the Drawing Units dialog box to change the number of decimal places past the zero from four to two. Keep the remaining default unit values. Using the `LIMITS` command, keep (0,0) for the lower-left corner and change the upper-right corner from (12,9) to (21.00,16.00). Perform a ZOOM-All after changing the drawing limits. Check to see that the following Object Snap modes are already set: Endpoint, Extension, Intersection, and Center.

Layers

Create the following layers with the format:

Name	Color	Linetype	Lineweight
Object	White	Continuous	0.50 mm
Center	Yellow	Center2	Default
Dimension	Yellow	Continuous	Default

Suggested Commands

Follow the steps provided. You will begin constructing Pattern1 by first creating four circles using relative coordinates. Then use the `CIRCLE-TTR` command/option to draw tangent arcs to the circles already drawn. Use the `TRIM` command to clean up and partially delete circles to obtain the outline of the pattern. Then add the 2.00-diameter holes followed by the centermarks, using the `DIMCENTER` (`DCE`) command.

STEP 1

Check that the current layer is set to Object. See Figure 5.70 for the dimensions and table for locating the centers of all circles. Use the `CIRCLE` command to create the circle at "A" whose center point is located at coordinate 9.50, 4.50 and which has a diameter of 4.00 units. Locate circle "B" at relative coordinate @−2.25,3.00 and assign a diameter of 5.00 units. The relative coordinate is based on the previous circle. Continue by drawing circle "C" using the center point at relative coordinate @5.25,3.50 and assign a diameter of 3.50 units. Construct the last circle, "D," with center point @−0.75,−2.50 and a diameter of 3.00 units. Notice that when using relative coordinate mode, you do not need to draw construction geometry and then erase it later.

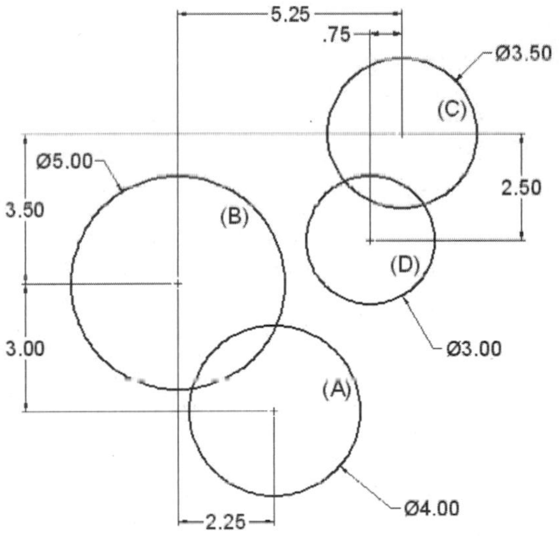

CIRCLE	Center Point Coordinate
A	9.50,4.50
B	@−2.25,3.00
C	@5.25,3.50
D	@−.75,−2.50

FIGURE 5.70

STEP 2

Use the `CIRCLE-TTR` command/option to construct a 4.00-radius circle tangent to the two dashed circles, as shown in Figure 5.71 (Left). Then use the `TRIM` command to trim away part of the circle so your display appears similar to the illustration in Figure 5.71 (Right).

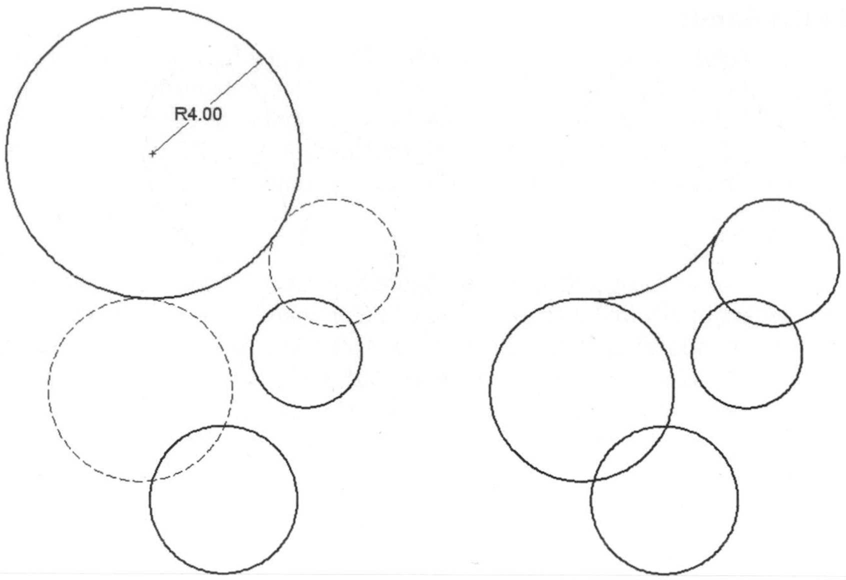

FIGURE 5.71

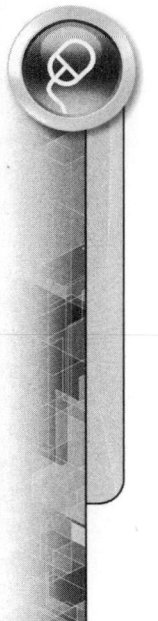

STEP 3

Use the CIRCLE-TTR command/option to construct another 4.00-radius circle tangent to the two dashed circles, as shown in Figure 5.72 (Left). Then use the TRIM command to trim away part of the circle so your display appears similar to the illustration in Figure 5.72 (Right).

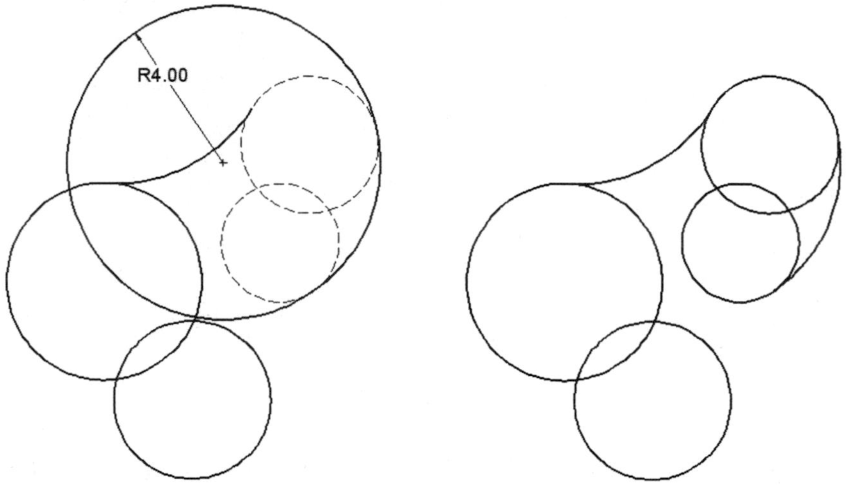

FIGURE 5.72

STEP 4

Use the CIRCLE-TTR command/option to construct a 6.00-radius circle tangent to the two dashed circles, as shown in Figure 5.73 (Left). Then use the TRIM command to trim away part of the circle so your display appears similar to the illustration in Figure 5.73 (Right).

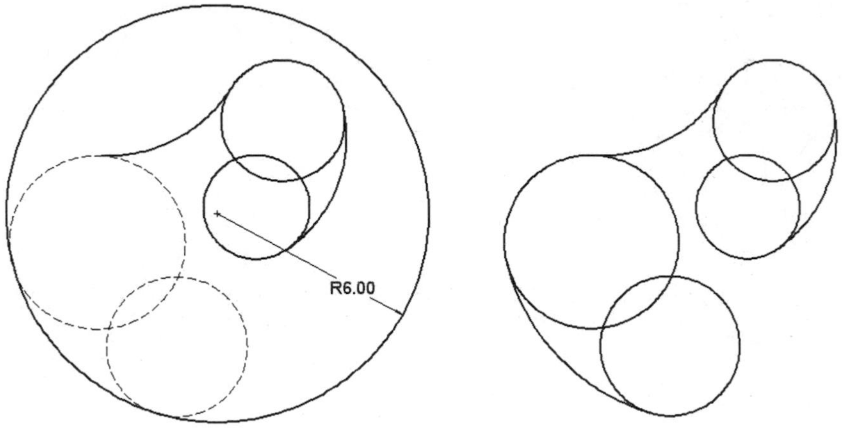

FIGURE 5.73

STEP 5

Use the CIRCLE-TTR command/option to construct a 0.75-radius circle tangent to the two dashed circles, as shown in Figure 5.74 (Left). Then use the TRIM command to trim away part of the circle so your display appears similar to the illustration in Figure 5.74 (Right).

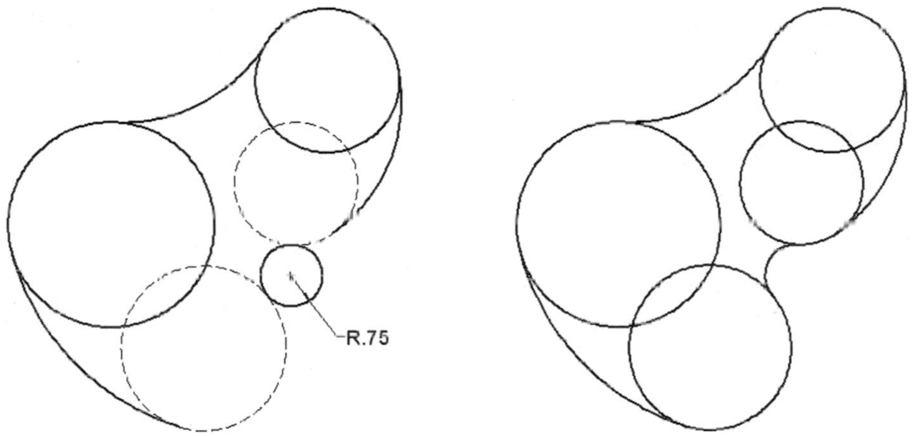

FIGURE 5.74

STEP 6

Use the TRIM command, select all dashed arcs, shown in Figure 5.75 (Left), as cutting edges, and trim away the circular segments to form the outline of the Pattern1 drawing. When finished, your display should appear similar to the illustration in Figure 5.75 (Right).

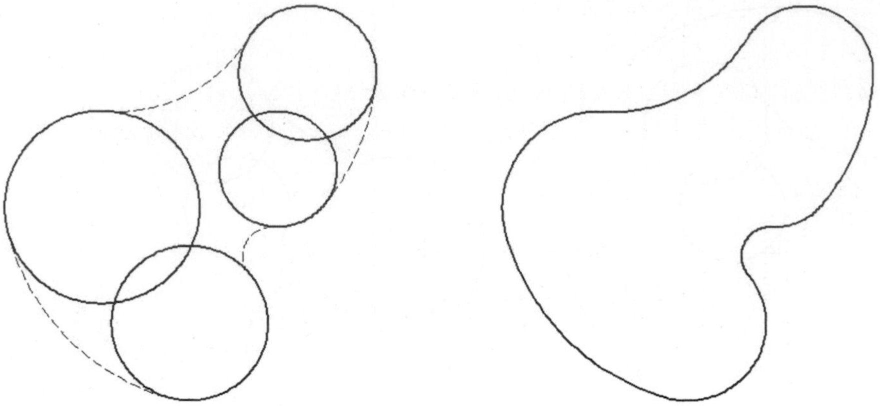

FIGURE 5.75

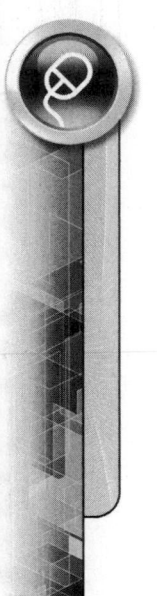

STEP 7

Use the CIRCLE command to construct a total of three circles of 2.00 diameter. Use the centers of arcs "A," "B," and "C," as shown in Figure 5.76 (Left), as the centers for the circles. Switch to the Center layer, change the DIMCEN system variable to a value of –0.09, and use the DIMCENTER (or DCE) command to place center marks identifying the centers of all circles and arcs, as shown in Figure 5.76 (Right).

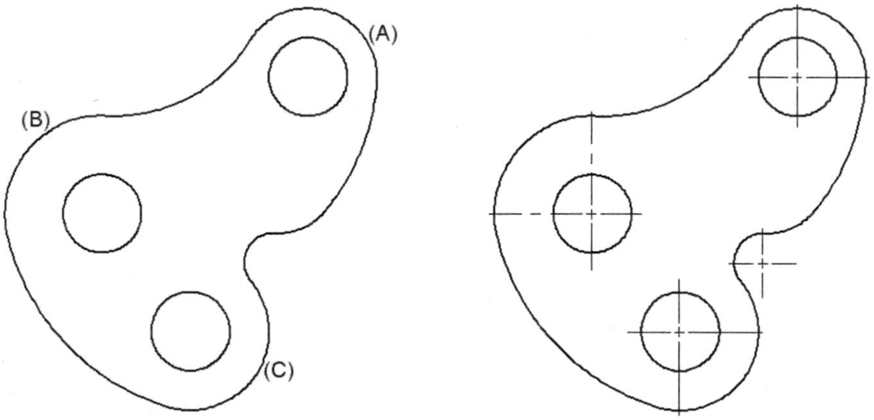

FIGURE 5.76

PROBLEMS FOR CHAPTER 5

DIRECTIONS FOR PROBLEMS 5–1 THROUGH 5–14:

Construct these geometric construction figures using existing AutoCAD commands.

PROBLEM 5-1

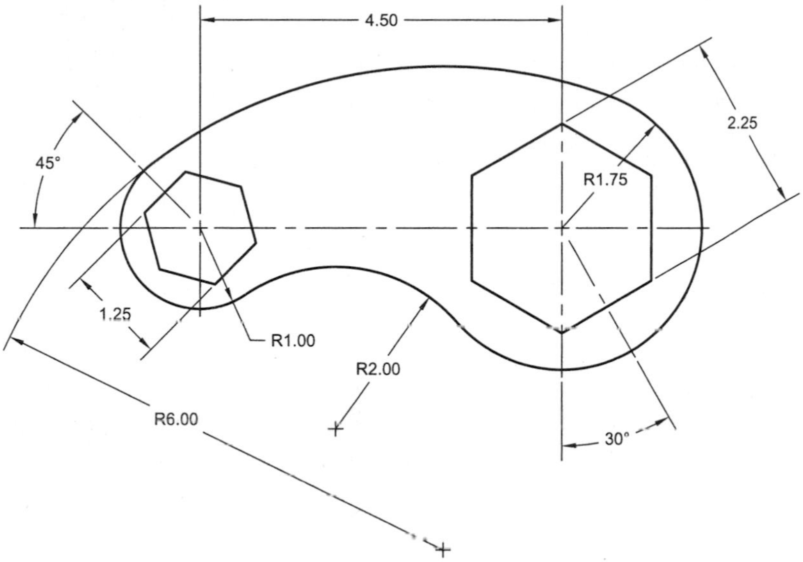

PROBLEM 5-2

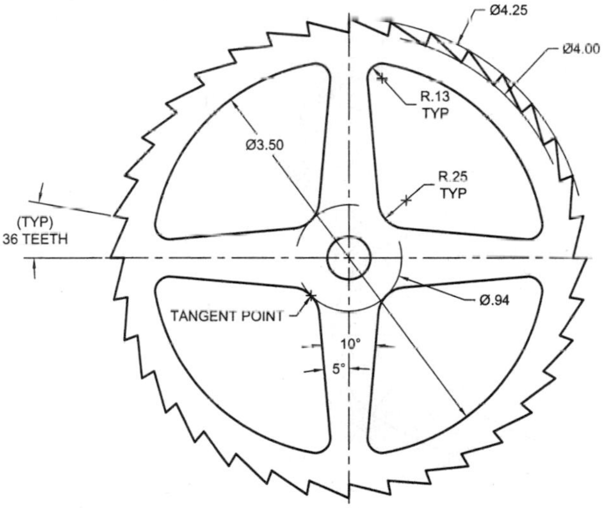

PROBLEM 5-3

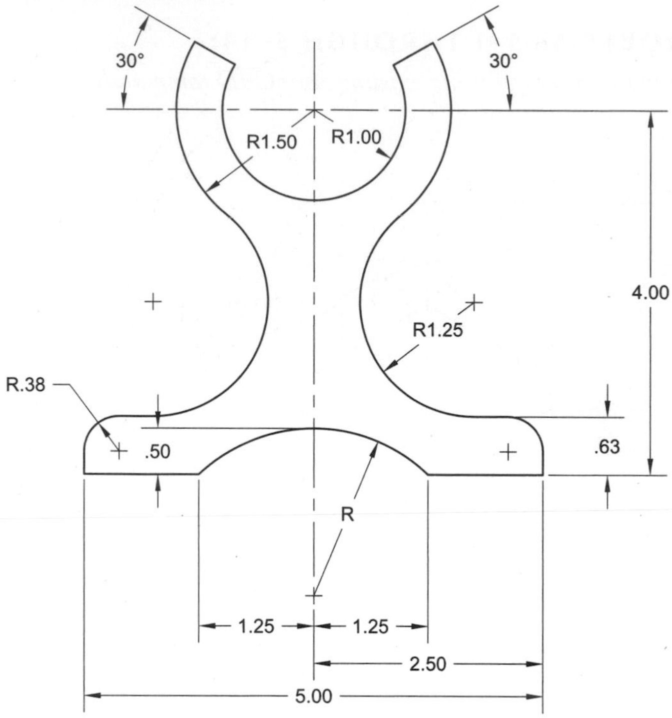

PROBLEM 5-4

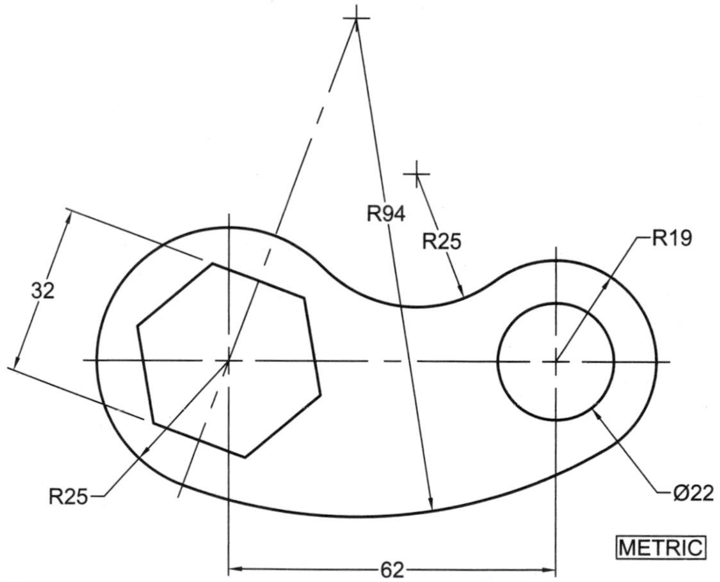

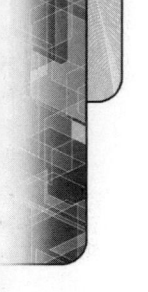

PROBLEM 5-5

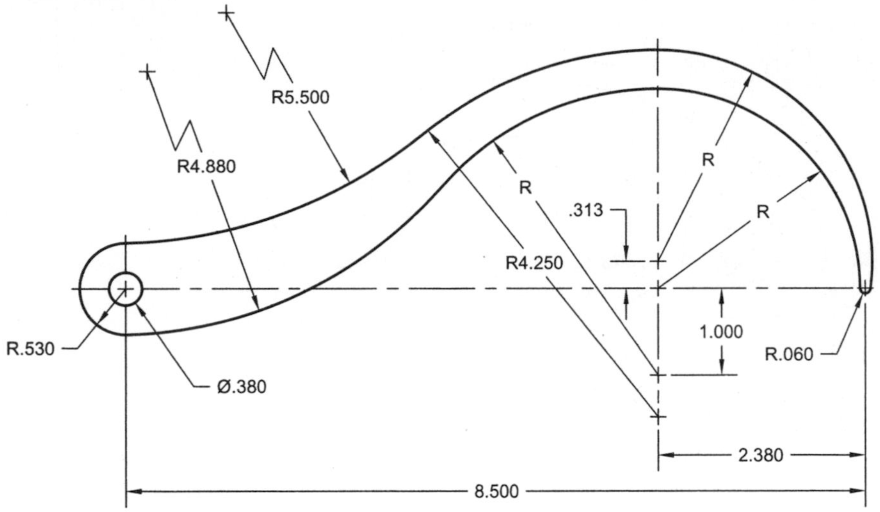

PROBLEM 5-6

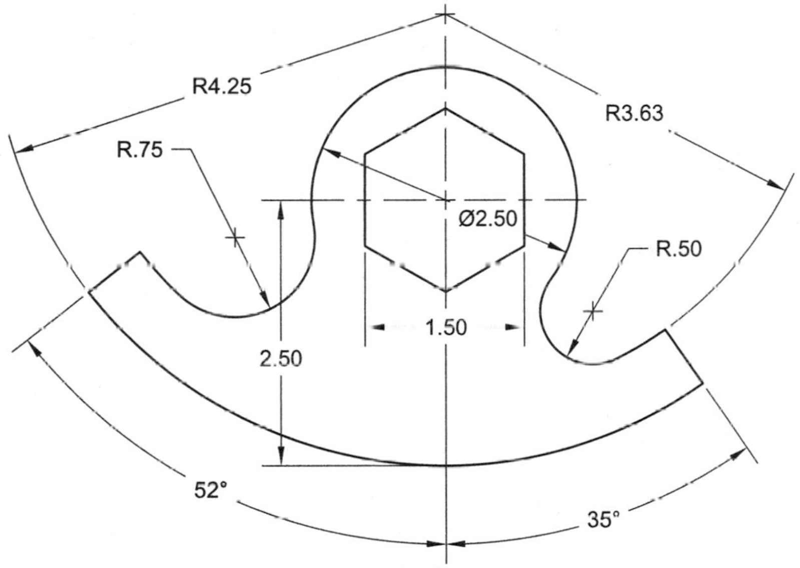

PROBLEM 5-7

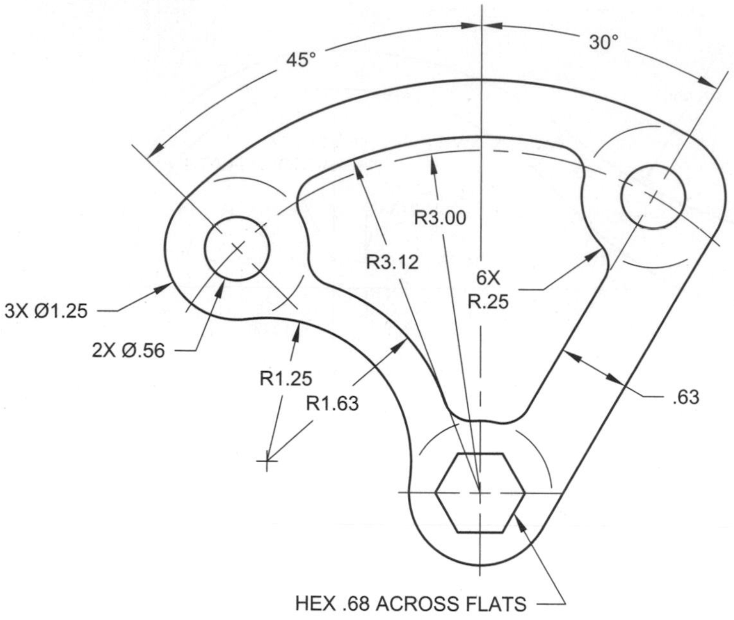

PROBLEM 5-8

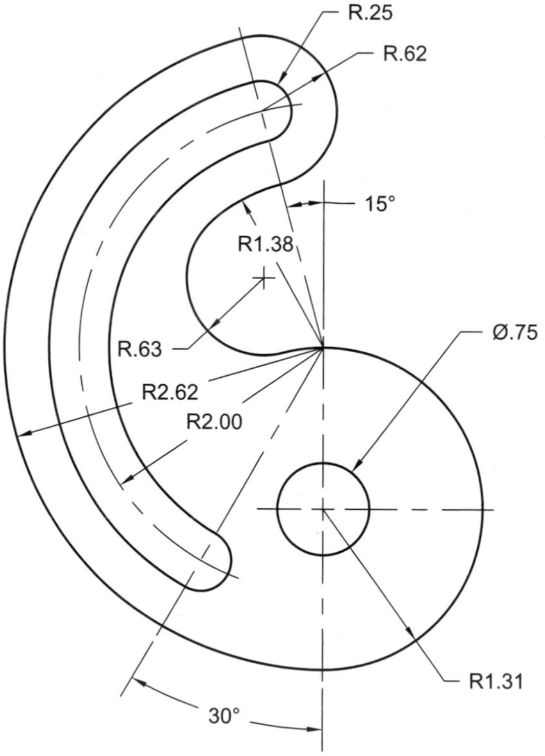

PROBLEM 5-9

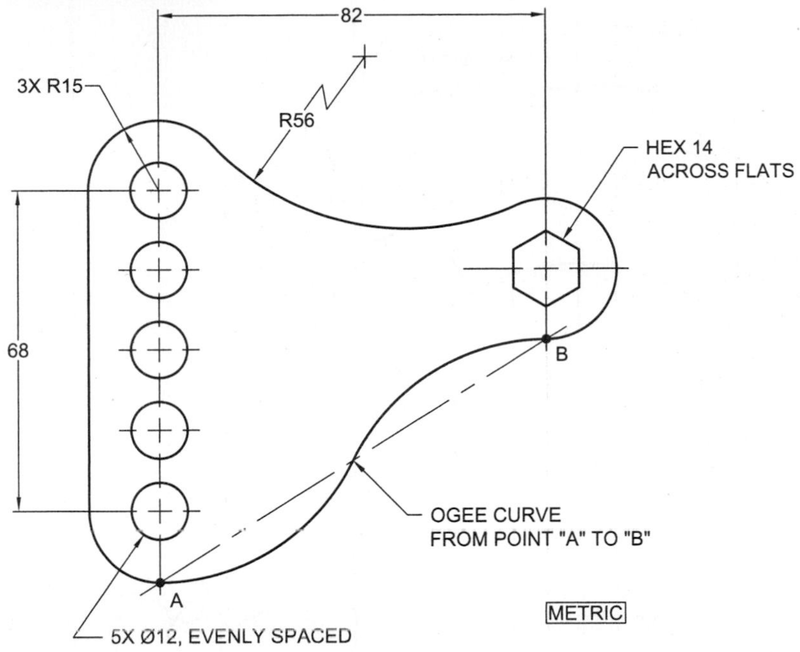
82
3X R15
R56
HEX 14
ACROSS FLATS
68
B
OGEE CURVE
FROM POINT "A" TO "B"
A
METRIC
5X Ø12, EVENLY SPACED

PROBLEM 5-10

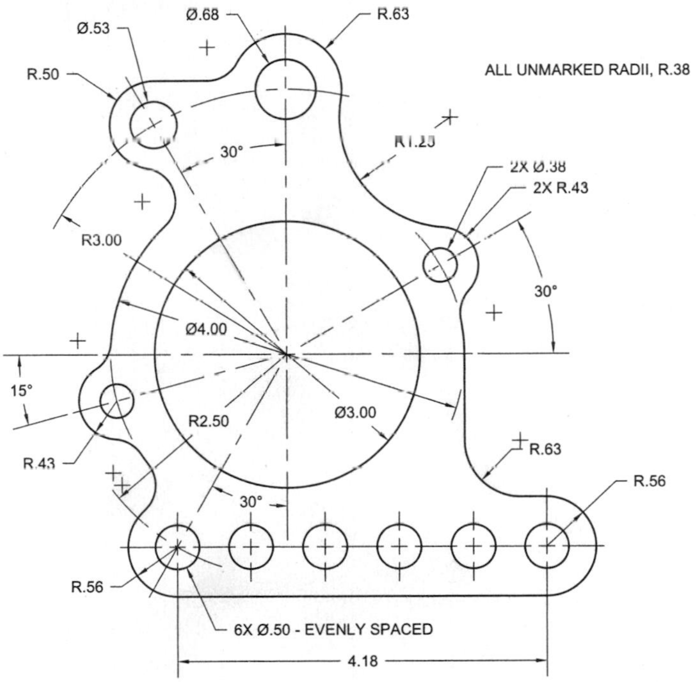
Ø.53
Ø.68
R.63
R.50
ALL UNMARKED RADII, R.38
30°
R1.25
2X Ø.38
2X R.43
R3.00
30°
Ø4.00
15°
R2.50
Ø3.00
R.43
R.63
R.56
30°
R.56
6X Ø.50 - EVENLY SPACED
4.18

PROBLEM 5-11

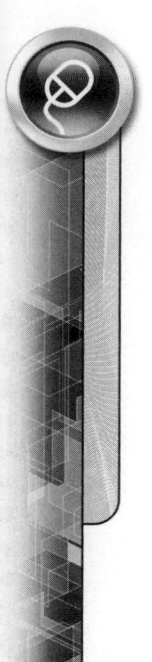

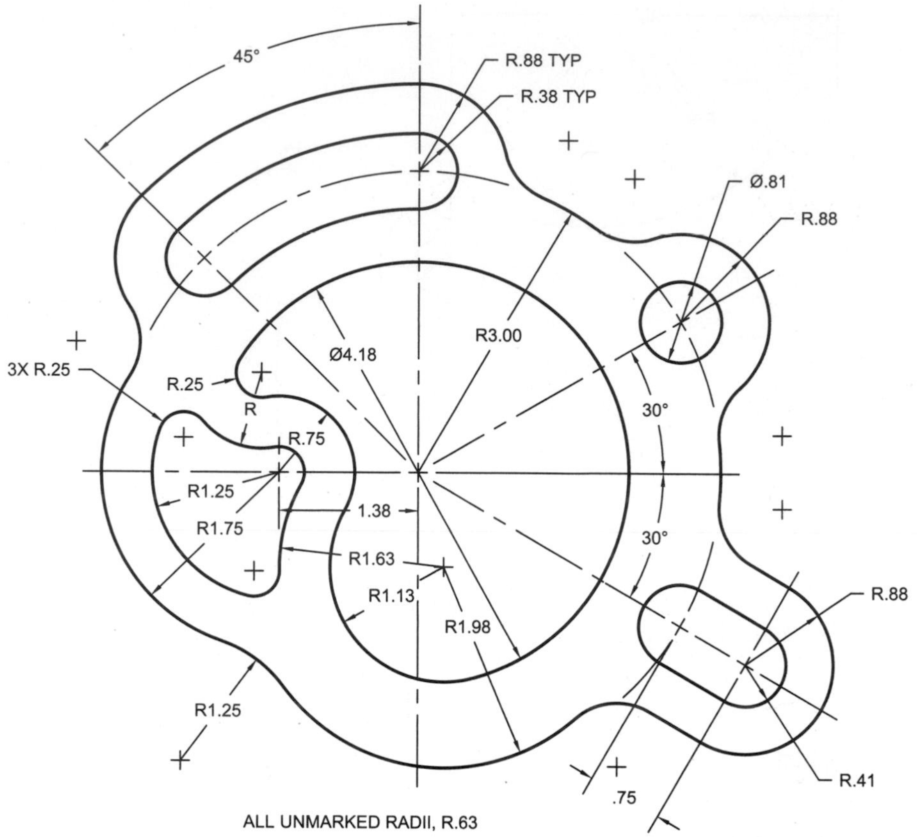
45°
R.88 TYP
R.38 TYP
Ø.81
R.88
Ø4.18
R3.00
3X R.25
R.25
R
R.75
30°
R1.25
R1.75
1.38
30°
R1.63
R1.13
R1.98
R.88
R1.25
.75
R.41
ALL UNMARKED RADII, R.63

PROBLEM 5-12

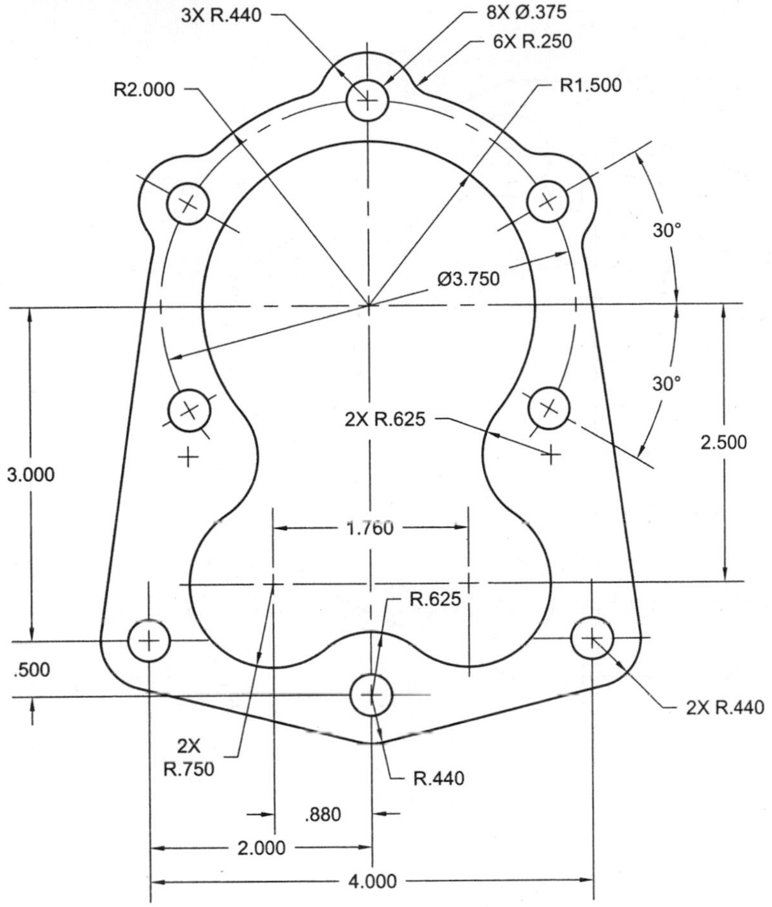

PROBLEM 5-13

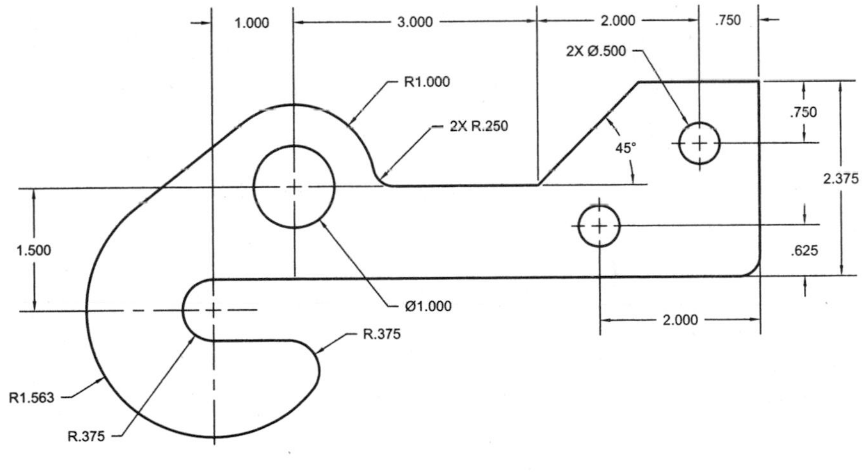

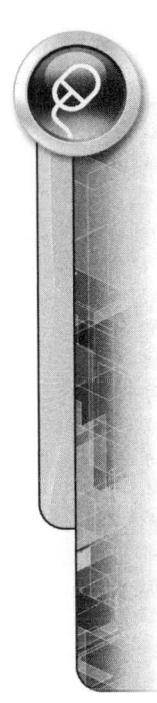

PROBLEM 5-14

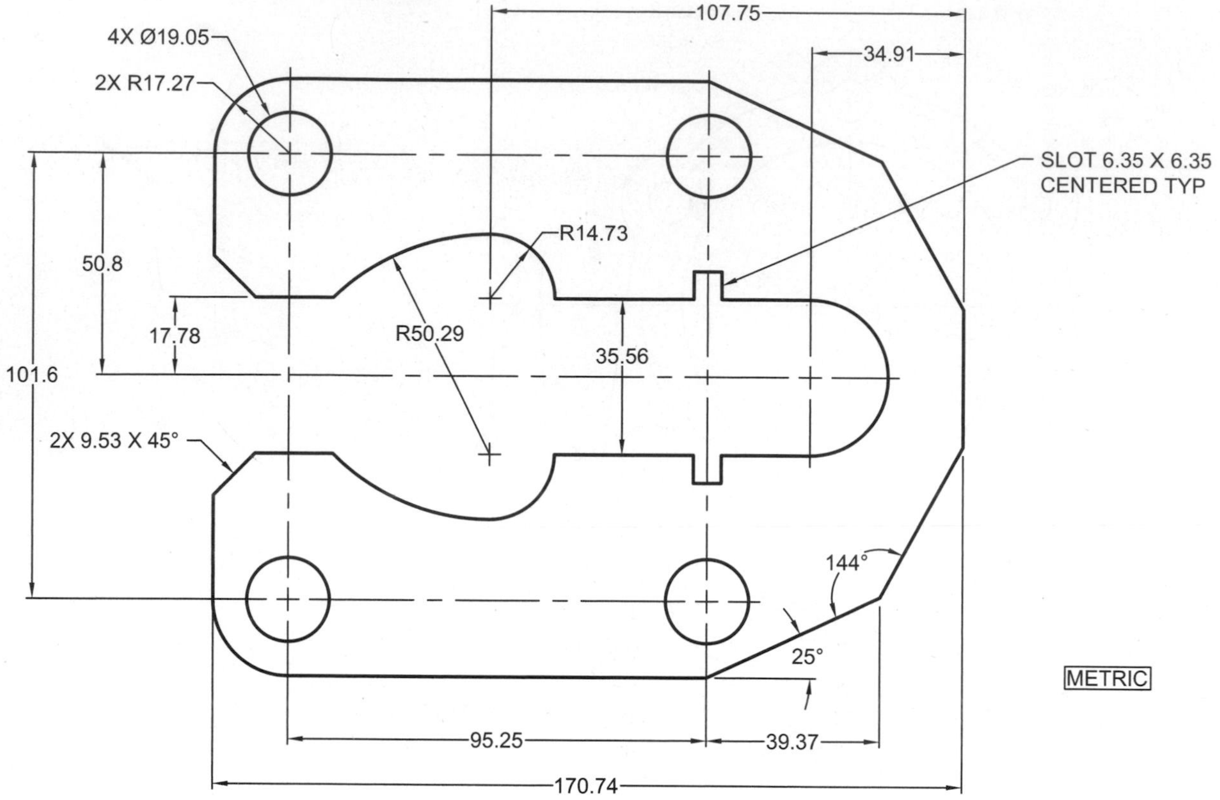

107.75
34.91
4X Ø19.05
2X R17.27
SLOT 6.35 X 6.35
CENTERED TYP
R14.73
50.8
17.78
R50.29
35.56
101.6
2X 9.53 X 45°
144°
25°
METRIC
95.25
39.37
170.74

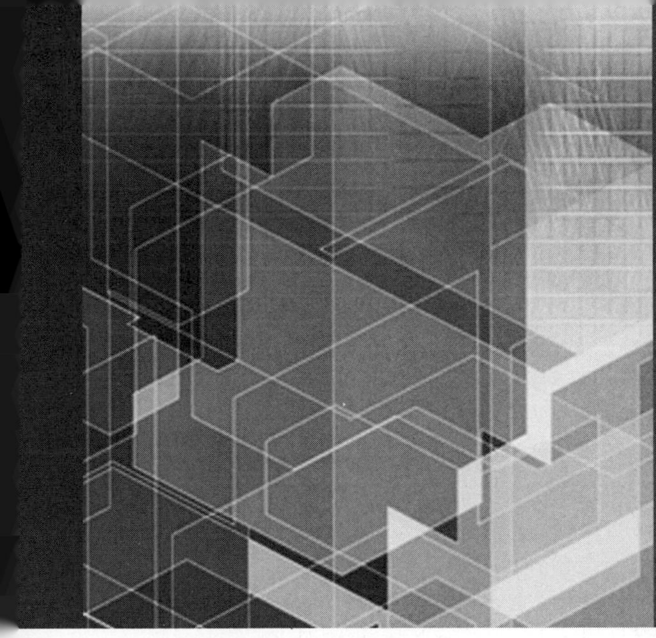

CHAPTER 6

Working with Text, Fields, and Tables

AutoCAD provides a robust set of text commands that allow you to place different text objects and edit those text objects. The heart of placing text in a drawing is through the MTEXT command. This Multiline Text command is basically a word processor inside of AutoCAD. This chapter also discusses the simpler TEXT command, or Single Line Text. This command is most often used for labels. In this chapter, you will also be given information on how to edit text created with the MTEXT and TEXT commands. The SCALETEXT and JUSTIFYTEXT commands will be explained in detail, as will the ability to create custom text styles and assign different text fonts to these styles. Intelligent text in the form of fields will also be discussed in this chapter, as will the creation of tables.

AUTOCAD TEXT COMMANDS

Text commands can be located in the Drafting & Annotation Workspace's Home tab, as shown in Figure 6.1 (Left), or the Annotate tab, as shown in Figure 6.1 (Right). The two main modes for entering text are Multiline Text (the MTEXT command) and Single Line Text (the TEXT command).

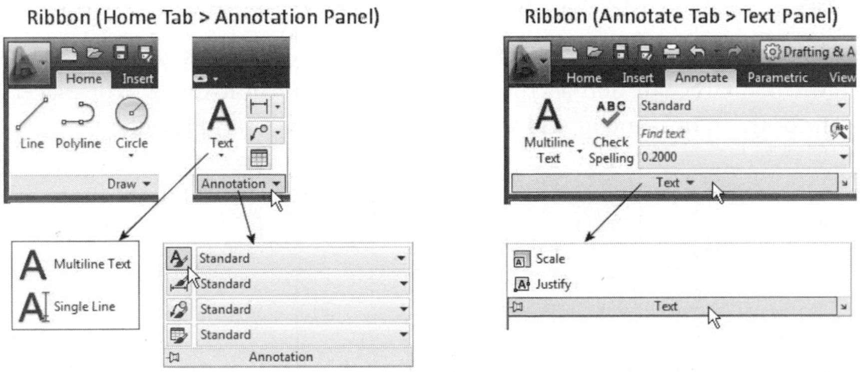

FIGURE 6.1

Study the information in the following table for a brief description of each text command.

Button	Tool	Shortcut	Function
	Multiline Text	MT	Creates paragraphs of text as a single object
	Single Line Text	DT	Creates one or more lines of text; each line of text is considered a separate object
	Edit Text	ED	Used for editing multiline and single line text in addition to dimension and attribute text
	Find and Replace	FIND	Used for finding, replacing, selecting, or zooming in to a particular text object
	Spell Check	SP	Performs a spell check on a drawing
	Text Style	ST	Launches the Text Style dialog box used for creating different text styles
	SCALETEXT	None	Used for scaling selected text objects without affecting their locations
	JUSTIFYTEXT	None	Used for justifying selected text objects without affecting their locations
	SPACETRANS	None	Converts text heights between Model Space and Paper Space (Layout mode)

ADDING MULTILINE TEXT

The `MTEXT` command allows for the placement of text in multiple lines. Entering MT (MTEXT) at the Command prompt displays the following prompts:

`Command: MT` *(For MTEXT)*

`Current text style: "GENERAL NOTES" Text height: 0.2000 Annotative: No`

`Specify first corner:` *(Pick a point to identify one corner of the Mtext box)*

`Specify opposite corner or [Height/Justify/Line spacing/Rotation/Style/Width/Columns]:` *(Pick another corner forming a box)*

Picking a first corner displays a user-defined box with an arrow at the bottom, as shown in Figure 6.2. This box defines the area where the multiline text will be placed. If the text cannot fit on one line, it will wrap to the next line automatically.

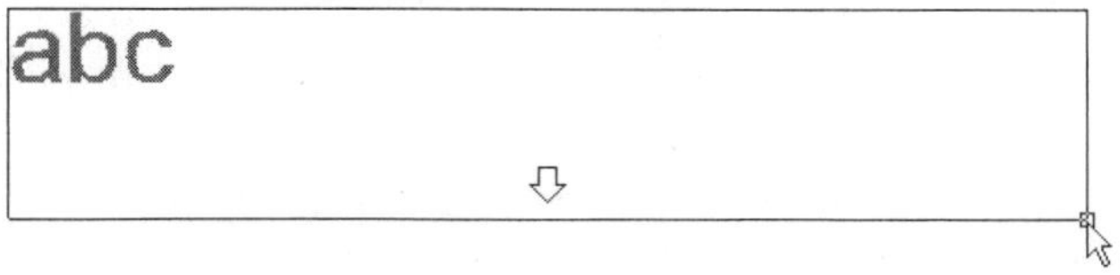

FIGURE 6.2

After you click a second point marking the other corner of the insertion box, the Text Editor Ribbon appears, along with the Multiline Text Editor. The parts of these items are shown in Figure 6.3. One of the advantages of this tool is the transparency associated with the Multiline Text Editor. As you begin entering text, you can still see your drawing in the background.

Notice in the following image other important areas of the Text Editor Ribbon. The current text style and font are displayed, in addition to the text height. You can make the text bold, italicized, or underlined by using the B, I, and U buttons present in the toolbar. When you have finished entering the text, click the Close Text Editor button to dismiss the Text Editor Ribbon and text editor and place the text in the drawing.

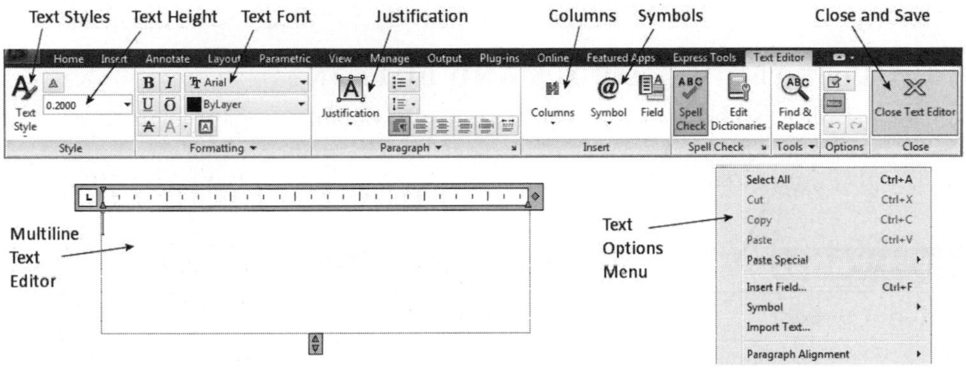

FIGURE 6.3

As you begin to type, your text appears in the Text Editor box, as shown in Figure 6.4. As the multiline text is entered, it automatically wraps to the next line depending on the size of the initial bounding box. Tabs and indents can be utilized as necessary, and even the size of the box can be modified, if required.

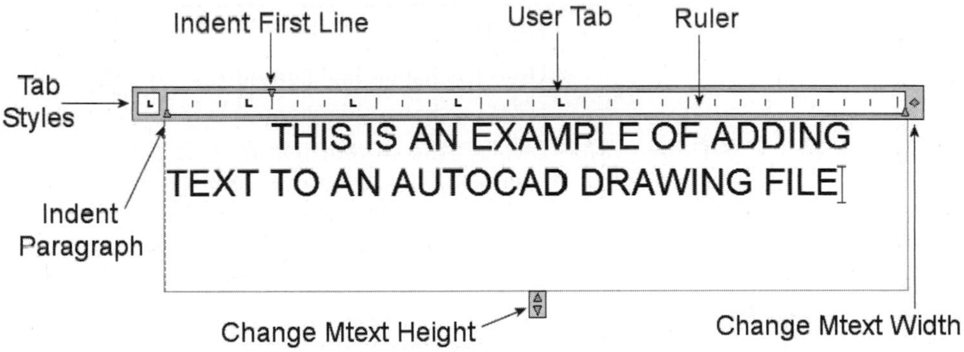

FIGURE 6.4

NOTE

Clicking a blank part of your screen after you have finished entering text in the previous example will also exit the Text Editor Ribbon.

Another method of displaying additional options of the Multiline Text Editor is to right-click inside the text box. This displays the menu shown in Figure 6.5.

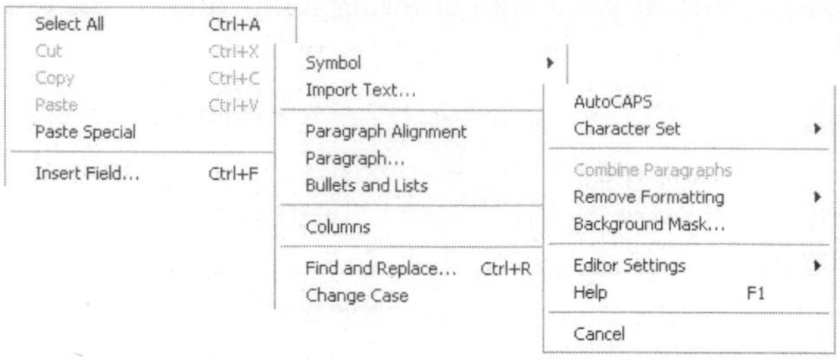

FIGURE 6.5

MULTILINE TEXT CONTROLS AND BUTTONS

The Text Editor Ribbon has a wealth of controls available to manipulate text items. The Text Editor tab is automatically displayed on the Ribbon and the controls are divided into panels. The following tables will explain these controls and buttons.

STYLE PANEL

Button	Tool	Function
Text Style	Text Style Drop-Down	Click to display text styles available in drawing. Pick style to make it current
A	Annotative	Turns annotative scale on or off
0.2000	Text Height	Displays the current text height. Height of selected text can be changed

FORMATTING PANEL

Button	Tool	Function
B	Bold	Used to change highlighted text to bold
I	Italics	Used to change highlighted text to italics
Arial	Text Font Drop-Down	Displays the current text font. Select new font from list. Font of selected text can be modified
U	Underline	Used to underline highlighted text
O	Overline	Used to place a line over the top of highlighted text
ByLayer	Color Drop-Down	Displays current color (bylayer). Select new color from list. Color of selected text can be changed
A	Strike-Through	Used to place a line through highlighted text
aA	Uppercase	Used for changing all highlighted text to uppercase
Aa	Lowercase	Used for changing all highlighted text to lowercase
	Background Mask	Displays a dialog box which allows you to place an opaque background behind text

PARAGRAPH PANEL

Button	Tool	Function
Justification	Mtext Justification	Displays a menu consisting of nine text justification modes
	Bullets & Numbering	Displays a menu used for applying bullets and numbers to text
	Line Spacing	Displays a menu used for changing the spacing in between lines of text
	Default	Returns the paragraph alignment to its default setting
	Left	Justifies the left edge of text with the left edge of the margin
	Center	Justifies Mtext objects to the center
	Right	Justifies the right edge of text with the right edge of the margin
	Justify	Adjusts the horizontal spacing of text so the text is aligned evenly along the right and left margins
	Distribute	Distributes Mtext across the width of the Mtext box and adjusts the spacing in between individual letters
	Fig 6-Paragraph	Displays the Paragraph dialog box used for changing such items as indentations that affect paragraph text

INSERT PANEL

Button	Tool	Function
Columns	Columns	Displays a menu used to control the type of text column generated
Symbol	Symbol	Displays a menu for selecting special text characters and symbols
Field	Field	Used for inserting a field

SPELL CHECK AND TOOLS PANELS

Button	Tool	Function
Spell Check	Spell Check	Turns on or off the As-You-Type spell checker
Edit Dictionaries	Edit Dictionaries	Allows you to load or unload custom dictionaries for spell checking
Find & Replace	Find & Replace	Displays the Find and Replace dialog box

OPTIONS AND CLOSE PANELS

Button	Tool	Function
	More	Used to display additional text options
	Ruler	Displays a ruler used for setting tabs and indents
	Undo	Used to undo the previous text option and remain in the Text Editor Ribbon
	Redo	Used to redo the previous undo
Close Text Editor	Close Text Editor	Used to accept the current Mtext and exit the Text Editor Ribbon

JUSTIFYING MULTILINE TEXT

Various justification modes are available to change the appearance of the text in your drawing. Justification modes can be retrieved by clicking the Justification button found in the Text Editor Ribbon as shown in Figure 6.6. Clicking this button displays the nine modes. The Middle Center justification shown highlighted in Figure 6.6 centers the text in the user-defined box that you specified when the MTEXT command was initiated.

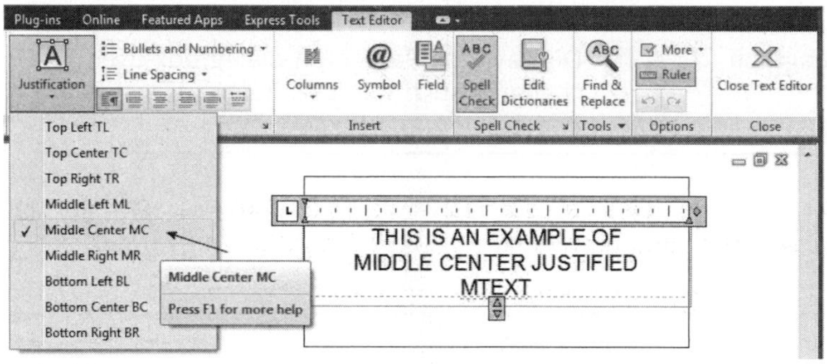

FIGURE 6.6

Additionally, Paragraph Alignment modes are also available in the Paragraph panel of the Text Editor Ribbon, as shown in Figure 6.7. These justification modes are similar to those found in word processors and allow you to align individual paragraphs in the text editor. Paragraphs are separated by hard returns (pressing the ENTER key).

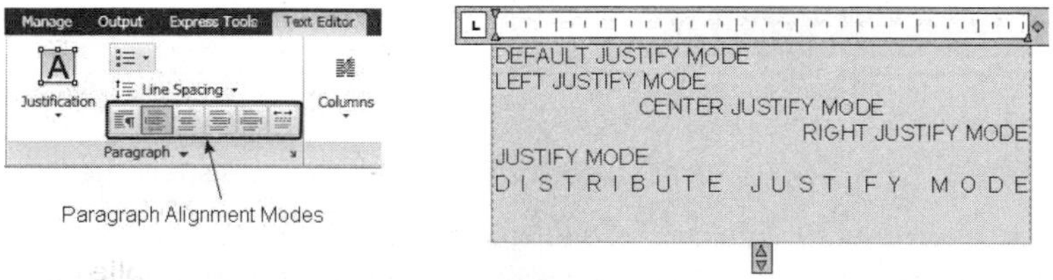

FIGURE 6.7

INDENTING TEXT

Multiline text can be indented in order to align text objects to form a list or table. A ruler, illustrated in Figure 6.8, consisting of short and longer tick marks displays the current paragraph settings.

The tabs and indents that you set before you start to enter text apply to the whole multiline text object. If you want to set tabs and indents to individual paragraphs, click on a paragraph or select multiple paragraphs to apply the indentations.

Two sliders are available on the left side of the ruler to show the amount of indentation applied to the various multiline text parts. In Figure 6.8, the top slider is used to indent the first line of a paragraph. A tooltip is available to remind you of this function.

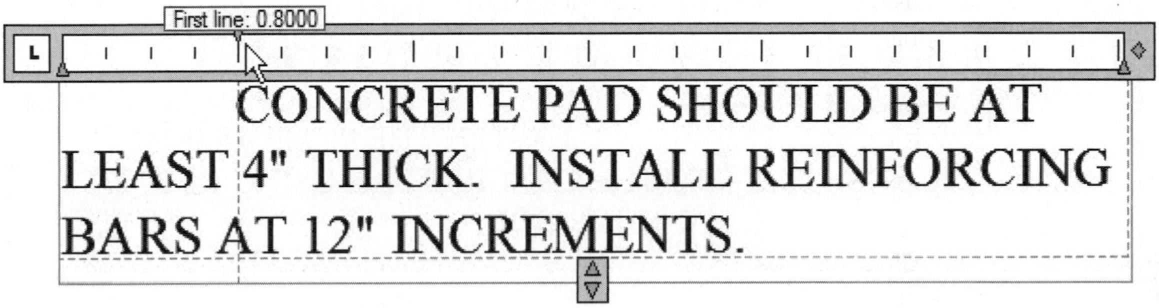
First line: 0.8000
L
CONCRETE PAD SHOULD BE AT
LEAST 4" THICK. INSTALL REINFORCING
BARS AT 12" INCREMENTS.

FIGURE 6.8

In Figure 6.9, the bottom slider is used to control the amount of indentation applied to the other lines of the paragraph. You can see the results of using this slider in Figure 6.9.

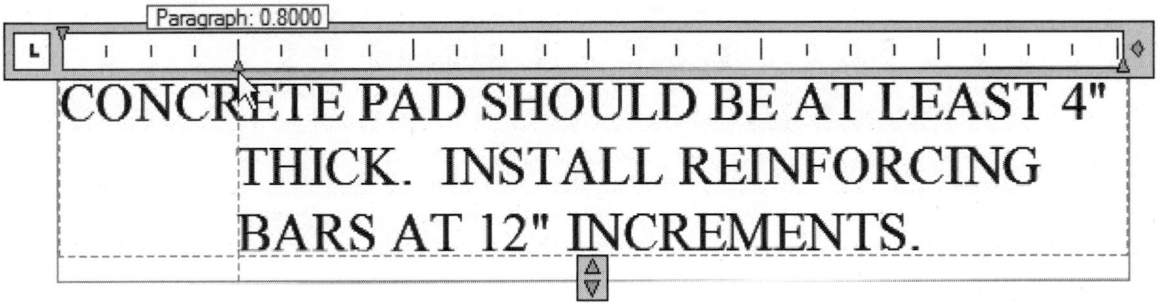
Paragraph: 0.8000
L
CONCRETE PAD SHOULD BE AT LEAST 4"
THICK. INSTALL REINFORCING
BARS AT 12" INCREMENTS.

FIGURE 6.9

> **NOTE**
>
> If you set tabs and indents prior to starting to enter multiline text, the tabs are applied to the whole multiline object.

FORMATTING WITH TABS

You can further control your text by adding tabs. The long tick marks on the ruler show the default tab stops. Clicking the Tab Style button toggles to four different tab modes, as shown in Figure 6.10 (Left). When you click the ruler to set your own tabs, the ruler displays a small tab marker at each custom tab stop, as shown in Figure 6.10. When you anchor your cursor in front of a sentence and press the TAB key on your keyboard, the text aligns to the nearest tab marker. To remove a custom tab stop, drag the marker off the ruler.

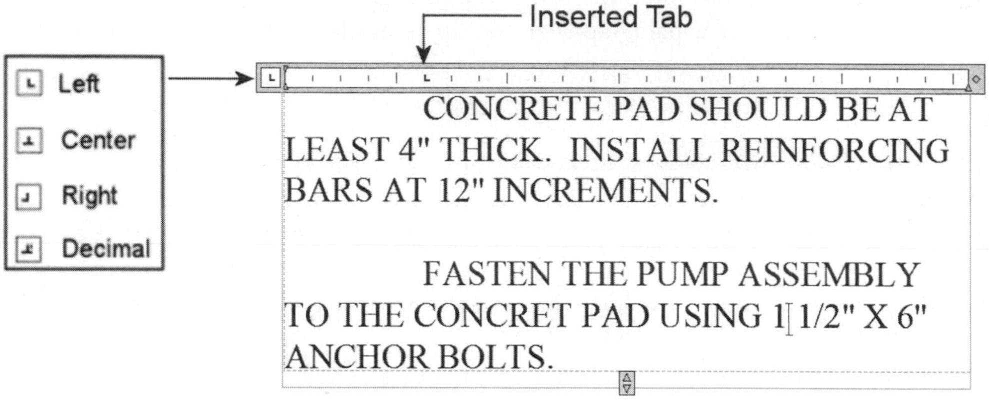

FIGURE 6.10

BULLETING AND NUMBERING TEXT

Multiline text can also be formatted in a numbered, bulleted, or alphabetic character list, either uppercase or lowercase. The four types of lists are illustrated in Figure 6.11.

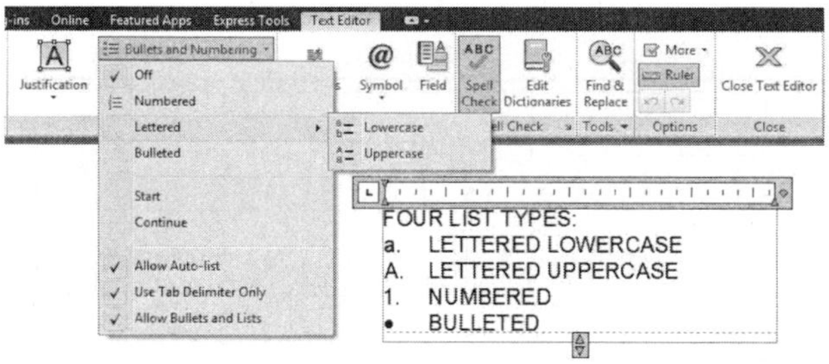

FIGURE 6.11

As lists are created, times occur when you need to change the format for new lists generated in the same multiline text command. In Figure 6.12 (Left), an alphanumeric list was interrupted to generate a numbered list. The menu shown on the right can assist with this operation. When going from the lettered to the numbered list, the Restart mode was utilized. Notice also in the illustration on the left that a paragraph was added with more numbers continuing beneath it. This demonstrates the ability to continue a list.

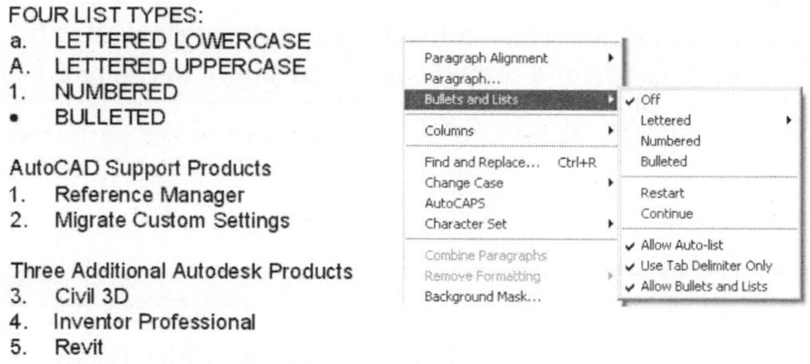

FIGURE 6.12

FORMATTING FRACTIONAL TEXT

Fractional text can be reformatted through the Multiline Text Editor. Entering a space after the fraction value in the following image displays the AutoStack Properties dialog box, also displayed in Figure 6.13. Use this dialog box to enable the auto-stacking of fractions. You can also remove the leading space in between the whole number and the fraction. In Figure 6.13, the fraction on the left was converted to the fraction on the right using the AutoStack Properties dialog box. A Fraction Stack button, which can be used to stack or un-stack a fraction, can be found by expanding the Formatting panel of the Text Editor Ribbon.

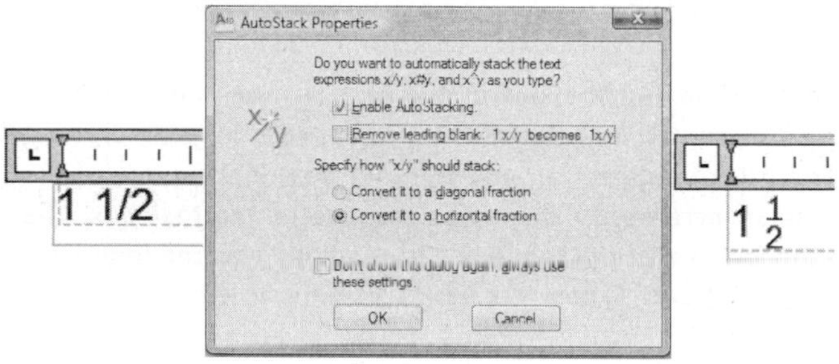

FIGURE 6.13

CHANGING THE MTEXT WIDTH AND HEIGHT

When you first construct a rectangle that defines the boundary of the mtext object, you are not locked into this boundary. You can use the arrow located in the upper right corner to change the column width or the arrows located below the text in the middle to change the column height, as shown in Figure 6.14.

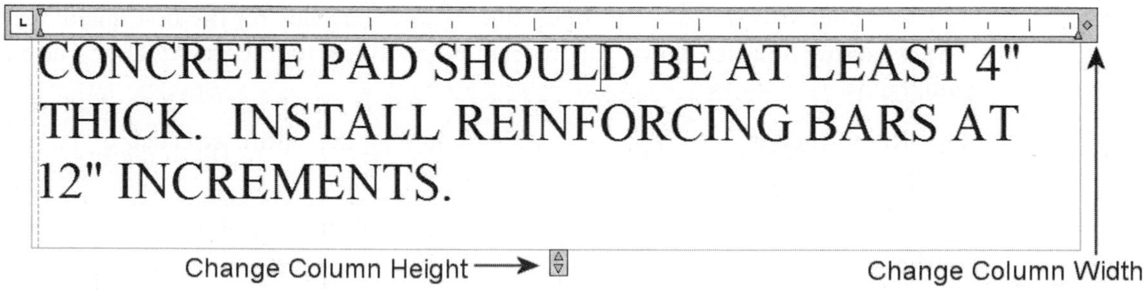

FIGURE 6.14

NOTE

When dragging the center arrows to change the column height, you can form columns by dragging the arrows up and notice the creation of a second column, as shown in Figure 6.15.

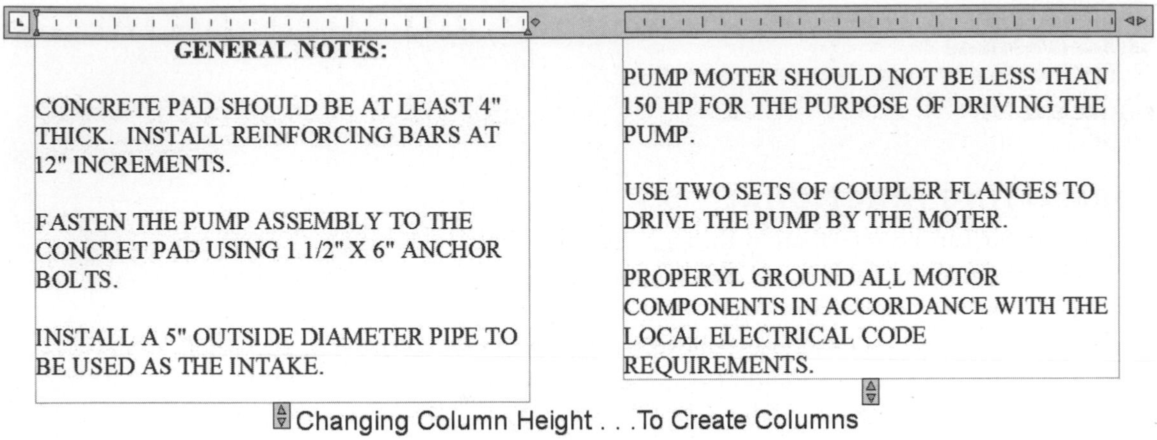
GENERAL NOTES:
CONCRETE PAD SHOULD BE AT LEAST 4" THICK. INSTALL REINFORCING BARS AT 12" INCREMENTS.
FASTEN THE PUMP ASSEMBLY TO THE CONCRET PAD USING 1 1/2" X 6" ANCHOR BOLTS.
INSTALL A 5" OUTSIDE DIAMETER PIPE TO BE USED AS THE INTAKE.
PUMP MOTER SHOULD NOT BE LESS THAN 150 HP FOR THE PURPOSE OF DRIVING THE PUMP.
USE TWO SETS OF COUPLER FLANGES TO DRIVE THE PUMP BY THE MOTER.
PROPERYL GROUND ALL MOTOR COMPONENTS IN ACCORDANCE WITH THE LOCAL ELECTRICAL CODE REQUIREMENTS.
Changing Column Height . . .To Create Columns

FIGURE 6.15

NOTE

Grips provide another method for changing the size of the text box once the text has already been created. Clicking the multiline text object as it appears in your drawing causes blue boxes/arrows called grips to appear around the text box. Picking one of the arrow grips and moving the cursor increases or decreases the size of the text box. The box grip can be used to move the location of the text box. Grips will be covered in greater detail in Chapter 7.

CREATING PARAGRAPHS OF TEXT

Clicking the Paragraph button of the Text Editor Ribbon, as shown in Figure 6.16, opens the Paragraph dialog box. This feature allows you to set indentations for the first lines of paragraphs and for the entire paragraph. In addition to indents, you can also specify the type of tab stop, paragraph spacing, and the spacing of lines in a paragraph.

The Tab area of the dialog box allows you to set left, center, right, and decimal tabs. You can also add or remove tabs using the appropriate control buttons.

The Left Indent area allows you to set the indentation value for the first line of text. The Hanging indent allows you to set indentations to the selected or current paragraphs.

The Right Indent area applies an indentation to the entire selected or current paragraph.

Placing a check in the Paragraph Alignment box activates five alignment properties for the selected or current paragraph.

Placing a check in the Paragraph Spacing box activates controls for setting the spacing before or after the selected or current paragraph. The spacing between two paragraphs is calculated from the total of the After value and the Before value.

Placing a check in the Paragraph Line Spacing box allows you to set the spacing between individual lines in the selected or current paragraph.

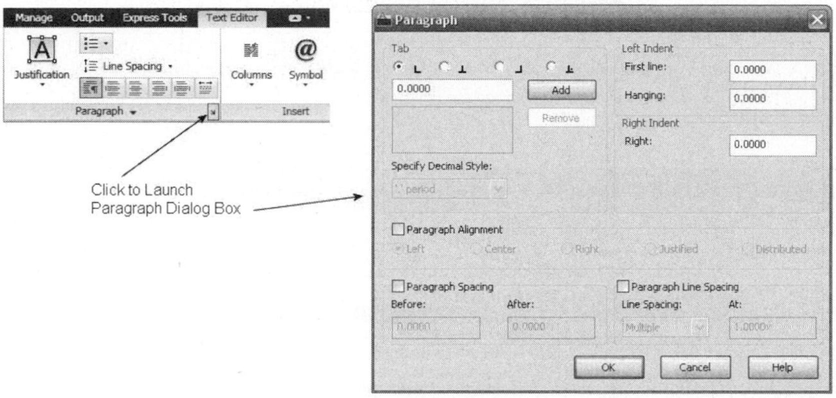

FIGURE 6.16

ORGANIZING TEXT BY COLUMNS

This feature in the Insert panel of the Text Editor Ribbon allows you to create columns of an mtext object. You begin by first creating an mtext object consisting of a single column. You then choose one of the two different column methods (Dynamic Columns or Static Columns) from the Columns menu, as shown in Figure 6.17 (Left).

Dynamic Columns allow columns of text to flow, causing columns to be added or removed. You can control this flow through the Auto height or Manual height modes.

Static Columns allow you to specify the width, height, and number of columns. In this way, all columns share the same height.

Other column options include the following:

No Columns specifies no columns for the current mtext object.

Insert Column Break ALT + ENTER inserts a manual column break.

Column Settings displays the Column Settings dialog box shown in the following image on the right. Through this dialog box, you specify the column and gutter width, height, and number of columns. Grips are used to edit the width and height of the column.

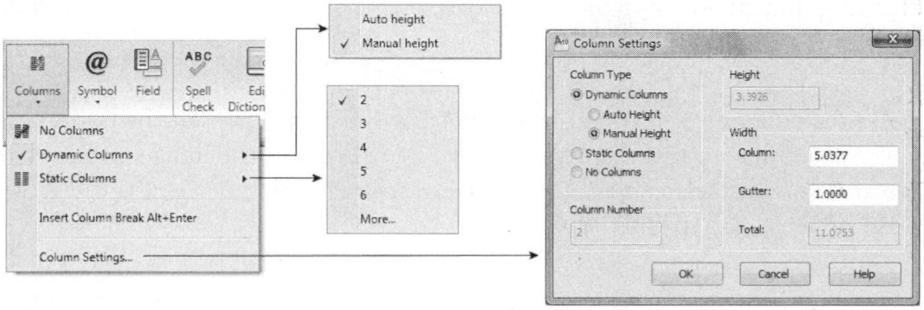

FIGURE 6.17

TRY IT!

Open the drawing file 06_Specifications. The text in this drawing consists of one long column. You will arrange this text in three columns, which will allow all text to be visible on the drawing sheet. Activate the Mutliline Text Editor by entering the `DDEDIT (ED)` command and picking the text object. Click the Columns button located in the Text Editor Ribbon. Click the Dynamic Columns from the menu followed by Manual Height, as shown in Figure 6.18. Next, stretch the column width to approximately 7 inches. Click the Close Text Editor button to dismiss the Text Editor Ribbon. Press `ENTER` to complete the `DDEDIT` command.

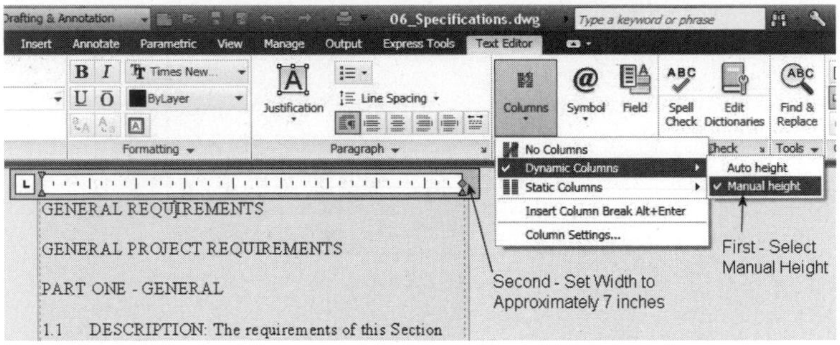

FIGURE 6.18

When you return to the drawing, click the multiline text object and notice an arrow grip present at the very bottom of the text paragraph, as shown in Figure 6.19 (Left). Pick this grip and move the bottom of the paragraph up and pick at a location where you want to break the column. Repeat, to break the second column into a third, as shown in Figure 6.19 (Right).

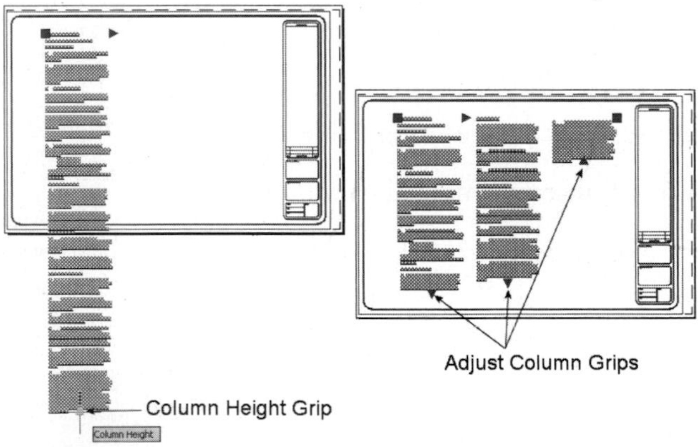

FIGURE 6.19

Zoom in to the columns of text and keep adjusting the arrow grips at the bottom of each paragraph until your text display matches Figure 6.20. In the image, notice that Part Two and Part Three are at the top of their respective columns.

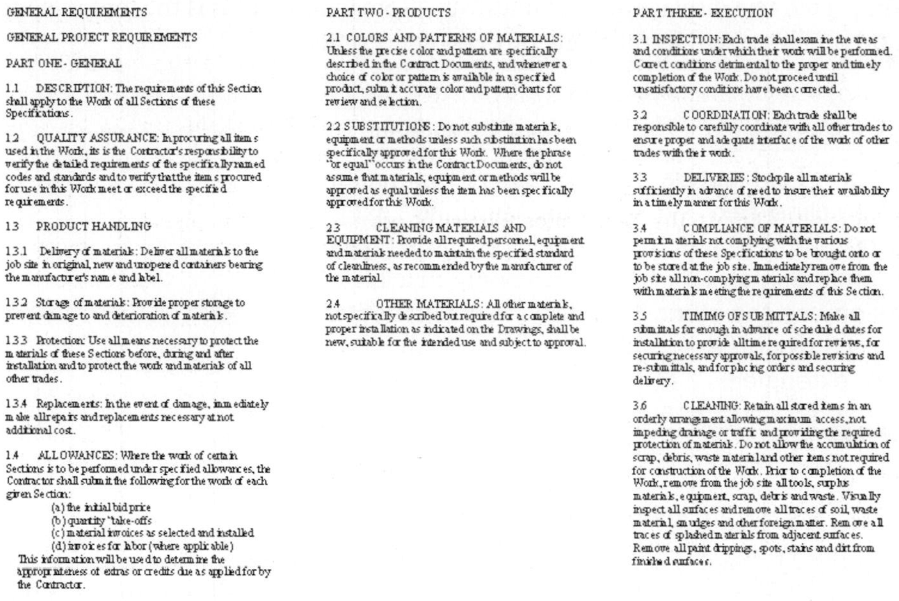

GENERAL REQUIREMENTS

GENERAL PROJECT REQUIREMENTS

PART ONE - GENERAL

1.1 DESCRIPTION: The requirements of this Section shall apply to the Work of all Sections of these Specifications.

1.2 QUALITY ASSURANCE: In procuring all items used in the Work, its is the Contractor's responsibility to verify the detailed requirements of the specifically named codes and standards and to verify that the items procured for use in this Work meet or exceed the specified requirements.

1.3 PRODUCT HANDLING

1.3.1 Delivery of materials: Deliver all materials to the job site in original, new and unopened containers bearing the manufacturer's name and label.

1.3.2 Storage of materials: Provide proper storage to prevent damage to and deterioration of materials.

1.3.3 Protection: Use all means necessary to protect the materials of these Sections before, during and after installation and to protect the work and materials of all other trades.

1.3.4 Replacements: In the event of damage, immediately make all repairs and replacements necessary at not additional cost.

1.4 ALLOWANCES: Where the work of certain Sections is to be performed under specified allowances, the Contractor shall submit the following for the work of each given Section:

(a) the initial bid price
(b) quantity "take-offs
(c) material invoices as selected and installed
(d) invoices for labor (where applicable)

This information will be used to determine the appropriateness of extras or credits due as applied for by the Contractor.

PART TWO - PRODUCTS

2.1 COLORS AND PATTERNS OF MATERIALS: Unless the precise color and pattern are specifically described in the Contract Documents, and whenever a choice of color or pattern is available in a specified product, submit accurate color and pattern charts for review and selection.

2.2 SUBSTITUTIONS: Do not substitute materials, equipment or methods unless such substitution has been specifically approved for this Work. Where the phrase "or equal" occurs in the Contract Documents, do not assume that materials, equipment or methods will be approved as equal unless the item has been specifically approved for this Work.

2.3 CLEANING MATERIALS AND EQUIPMENT: Provide all required personnel, equipment and materials needed to maintain the specified standard of cleanliness, as recommended by the manufacturer of the material.

2.4 OTHER MATERIALS: All other materials, not specifically described but required for a complete and proper installation as indicated on the Drawings, shall be new, suitable for the intended use and subject to approval.

PART THREE - EXECUTION

3.1 INSPECTION: Each trade shall examine the areas and conditions under which their work will be performed. Correct conditions detrimental to the proper and timely completion of the Work. Do not proceed until unsatisfactory conditions have been corrected.

3.2 COORDINATION: Each trade shall be responsible to carefully coordinate with all other trades to ensure proper and adequate interface of the work of other trades with their work.

3.3 DELIVERIES: Stockpile all materials sufficiently in advance of need to insure their availability in a timely manner for this Work.

3.4 COMPLIANCE OF MATERIALS: Do not permit materials not complying with the various provisions of these Specifications to be brought onto or to be stored at the job site. Immediately remove from the job site all non-complying materials and replace them with materials meeting the requirements of this Section.

3.5 TIMIMG OF SUBMITTALS: Make all submittals far enough in advance of scheduled dates for installation to provide all time required for reviews, for securing necessary approvals, for possible revisions and re-submittals, and for placing orders and securing delivery.

3.6 CLEANING: Retain all stored items in an orderly arrangement allowing maximum access, not impeding drainage or traffic and providing the required protection of materials. Do not allow the accumulation of scrap, debris, waste material and other items not required for construction of the Work. Prior to completion of the Work, remove from the job site all tools, surplus materials, equipment, scrap, debris and waste. Visually inspect all surfaces and remove all traces of soil, waste material, smudges and other foreign matter. Remove all traces of splashed materials from adjacent surfaces. Remove all paint drippings, spots, stains and dirt from finished surfaces.

FIGURE 6.20

IMPORTING TEXT INTO YOUR DRAWING

If you have existing text files already created in a word processor, you can import these files directly into AutoCAD and have them converted into mtext objects. As illustrated in Figure 6.21, once the Multiline Text Editor box is displayed, right-click on the drawing screen to display a shortcut menu; and then pick Import Text. An open file dialog box displays, allowing you to pick the desired file to import. The Import Text feature of the Multiline Text Editor supports two file types: those ending in a TXT extension and those ending in the RTF (rich text format) extension. Text with the TXT extension conforms to the current AutoCAD text style regarding font type. Text with the RTF extension remembers formatting and font settings from the word-processing document that it was originally created in.

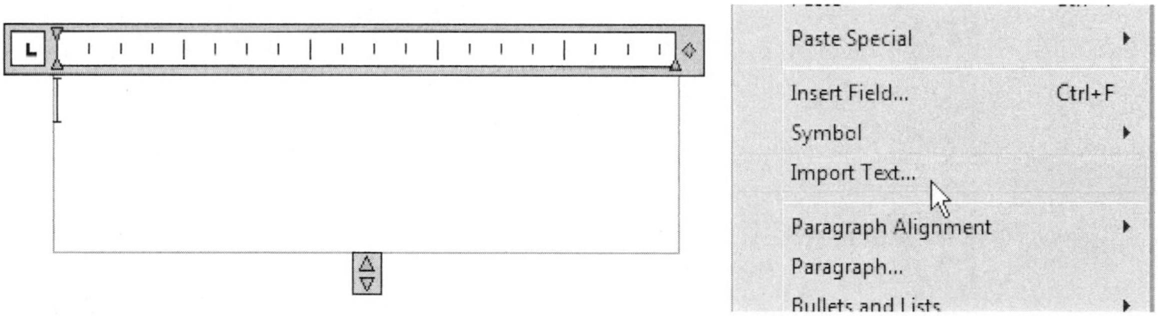

FIGURE 6.21

The finished result of this text-importing operation is illustrated in Figure 6.22. The text on the left, imported with the TXT extension, conformed to the present text style, which has a font of Arial assigned to it. The text on the right was imported with the RTF extension. Even with the text style being set to Arial, the

RTF-imported text displays in the Times New Roman font. This is the font in which it was originally created in the word-processing document.

TXT Extension	RTF Extension
The Import Text feature of the Multiline Text supports two file types; those ending in a TXT extension and the other in the RTF (Rich Text Format) extension.	The Import Text feature of the Multiline Text supports two file types; those ending in a TXT extension and the other in the RTF (Rich Text Format) extension.

FIGURE 6.22

MULTILINE TEXT SYMBOLS

While in the Multiline Text Editor, right-clicking on the text window and selecting Symbol from the shortcut menu, as shown in Figure 6.23, displays various symbols that can be incorporated into your drawing.

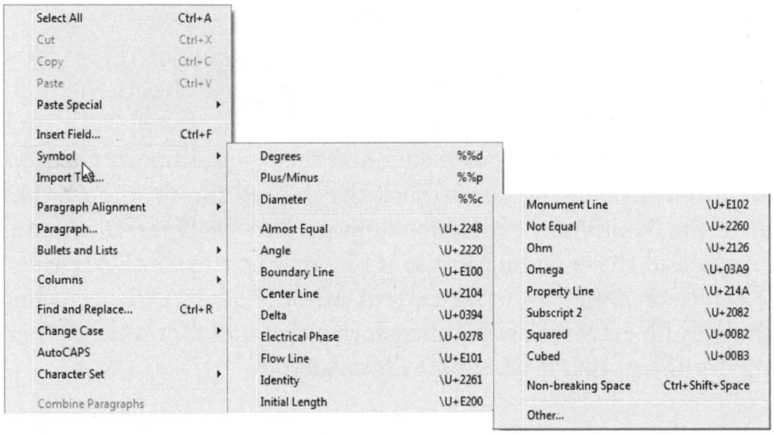

FIGURE 6.23

> **NOTE**
> The same list of symbols shown in Figure 6.23 can be accessed by picking the Symbol button on the Insert panel of the Text Editor Ribbon.

The table illustrated in Figure 6.24 shows the available multiline text symbols along with their meanings.

MTEXT SYMBOLS			
SYMBOL	DESCRIPTION	SYMBOL	DESCRIPTION
°	Degrees	≡	Identity
±	Plus/Minus	OL	Initial Length
Ø	Diameter	ML	Monument Line
≈	Almost Equal	≠	Not Equal
∠	Angle	Ω	Ohm
BL	Boundary Line	Ω	Omega
CL	Center Line	PL	Property Line
Δ	Delta	H_2O	Subscript 2
Φ	Electrical Phase	4^2	Squared
FL	Flow Line	4^3	Cubed

FIGURE 6.24

CREATING SINGLE LINE TEXT

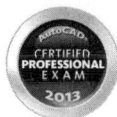

Single line text is another method of adding text to your drawing. The actual command used to perform this operation is TEXT. This command allows you to create single lines of text in a drawing and view the text as you type it. This command can be found on the Ribbon in either the Home or Annotate tabs, as shown in Figure 6.25.

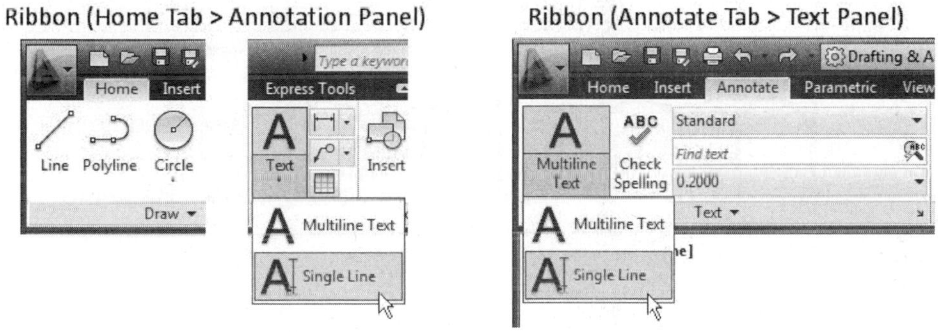

FIGURE 6.25

When using the TEXT command, you are prompted to specify a start point, height, and rotation angle. You are then prompted to enter the actual text. As you do this, each letter displays on the screen. When you are finished with one line of text, pressing ENTER drops the Insert bar to the next line, where you can enter more text. Pressing ENTER again drops the Insert bar down to yet another line of text, as shown in Figure 6.26 (Left). Pressing enter at the "Enter text" prompt exits the TEXT command and permanently adds the text to the database of the drawing, as shown in Figure 6.26 (Right).

Command: DT *(For TEXT)*

Current text style: "Standard" Text height: 0.2000 Annotative: No

Specify start point of text or [Justify/Style]: *(Pick a point at "A")*

Specify height <0.2000>: 0.50

`Specify rotation angle of text <0>:` *(Press* `ENTER` *to accept this default value)*

(Enter text:) **ENGINEERING** *(After this text is entered, press* `ENTER` *to drop to the next line of text)*

(Enter text:) **GRAPHICS** *(After this text is entered, press* `ENTER` *to drop to the next line of text)*

(Enter text:) (Either add more text or press `ENTER` *to exit this command and place the text)*

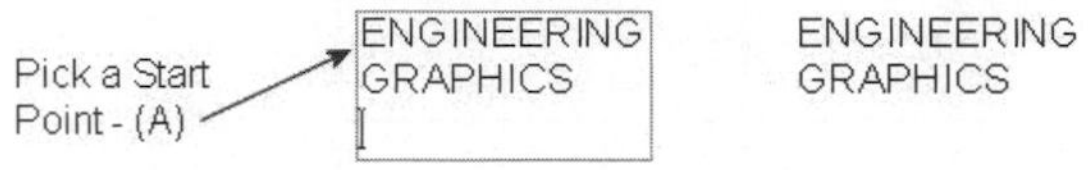

FIGURE 6.26

TEXT JUSTIFICATION MODES

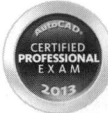

Figure 6.27 illustrates a sample text item and the various locations by which the text can be justified. By default, when you place text, it is left justified. These justification modes are designed to work in combination with one another. For example, one common justification is to have the text middle centered. Use the TEXT command's Justify option and select MC for Middle Center. The same type of combination is available for the other justification modes.

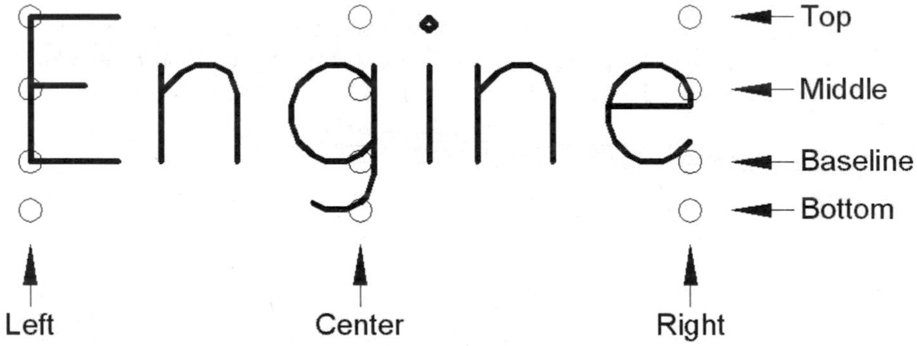

FIGURE 6.27

ADDITIONAL SINGLE LINE TEXT APPLICATIONS

When you place a line of text using the `TEXT` command and press `ENTER`, the Insert bar drops down to the next line of text. This sequence continues until you press `ENTER` at the "Enter text" prompt, which exits the command and places the text permanently in the drawing. You can also control the placement of the Insert bar by clicking a new location in response to the "Enter text" prompt. In Figure 6.28, various labels need to be placed in the pulley assembly. The first label, "BASE PLATE," is placed with the `TEXT` command. Without pressing `ENTER`, immediately pick a new location at "B" and place the text "SUPPORT." Continue this process with the other labels. The Insert bar at "E" denotes the last label that needs to be placed. When you perform this operation, pressing `ENTER` one last time at the "Enter text": Prompt places the text and exits the command. Follow the command sequence below for a better idea of how to perform this operation.

Open the drawing file 06_Pulley Text. Use Figure 6.28 (Left) and command sequence to perform this operation. The results are shown in Figure 6.28 (Right).

TRY IT!

Command: DT *(For TEXT)*

Current text style: "Standard" Text height: 0.2500 Annotative: No

Specify start point of text or [Justify/Style]: *(Pick a point at "A")*

Specify height <0.2500>: *(Press* ENTER *to accept this default value)*

Specify rotation angle of text <0>: *(Press* ENTER *to accept this default value)*

Enter text: BASE PLATE *(After this text is entered, pick approximately at "B")*

Enter text: SUPPORT *(After this text is entered, pick approximately at "C")*

Enter text: SHAFT *(After this text is entered, pick approximately at "D")*

Enter text: PULLEY *(After this text is entered, pick approximately at "E")*

Enter text: *(Enter **BUSHING** for this part and then press* ENTER*)*

Enter text: *(Press* ENTER *to exit this command)*

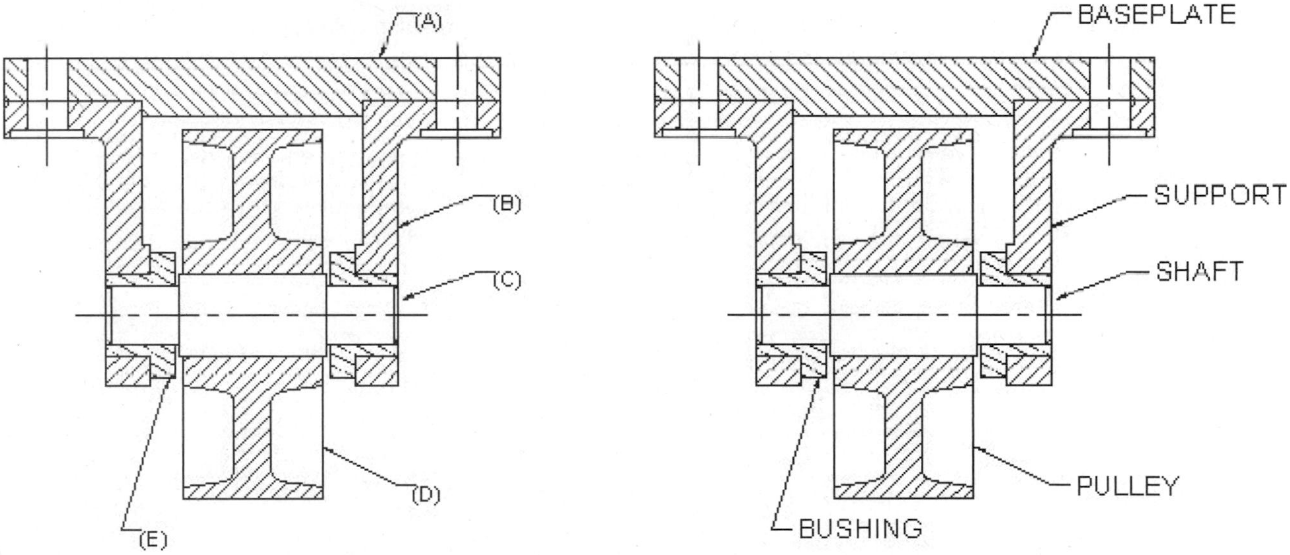

FIGURE 6.28

EDITING TEXT

Text constructed with the MTEXT and TEXT commands is easily modified by double-clicking a text object or by selecting the text object, right-clicking in the drawing area, and picking Edit or Mtext Edit from the shortcut menu. This procedure will issue the DDEDIT command for single line text or the MTEDIT command for multiline text. Double-clicking a multiline text object displays the Multiline Text Editor, as shown in Figure 6.29. This is the same text editor that was used to initially create text through the MTEXT command. Use this editor to change the text height, font, color, and justification.

TRY IT!

Open the drawing file 06_Edit Text2. Activate the Multiline Text Editor by double-clicking the text object. In this dialog box, the text is currently drawn in the RomanD font and needs to be changed to Arial. To accomplish this, first highlight all the text. Next, change to the desired font, as shown in Figure 6.29.

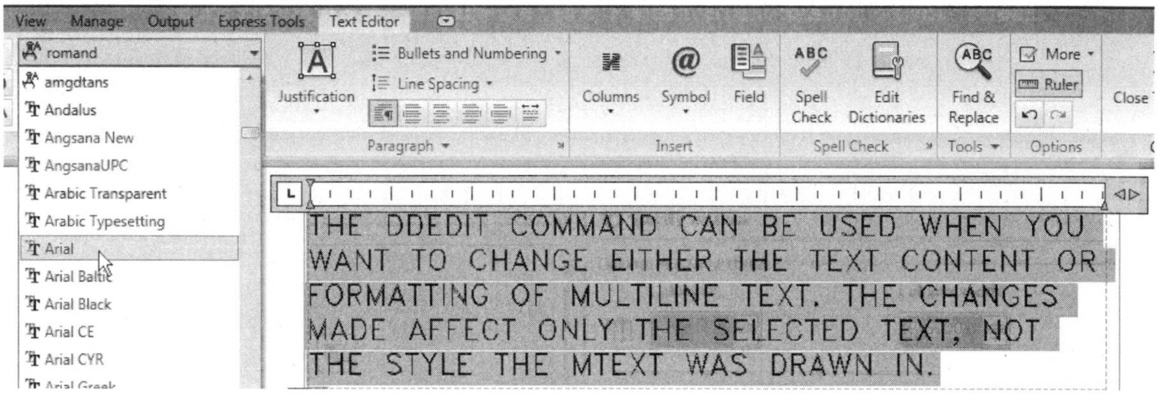

FIGURE 6.29

Since all text was highlighted, changing the font to Arial updates all text, as shown in Figure 6.30. Clicking the Close Text Editor button dismisses the editor and saves the changes to the text font in the drawing.

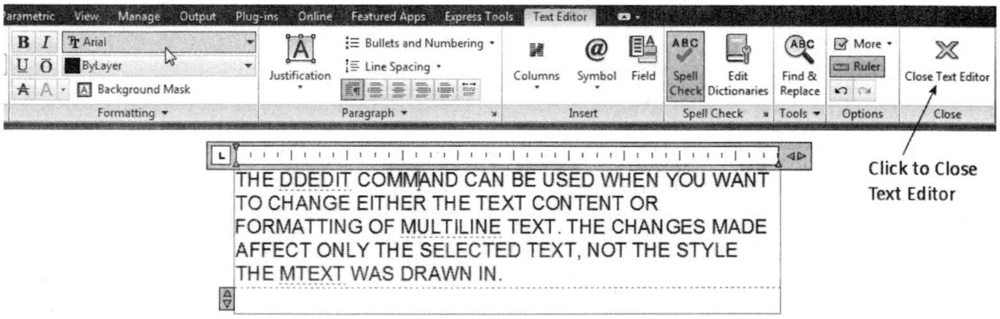

FIGURE 6.30

When editing mtext objects, you can selectively edit only certain words that are part of a multiline text string. To perform this task, highlight the text to change and apply the new formatting style, such as font, underscore, or height, to name a few.

TRY IT!

Open the drawing file 06_Edit Text3. Activate the Multiline Text Editor by double-clicking the text object. In Figure 6.31, the text "PLACING" needs to be underscored, the text "AutoCAD" needs to be changed to a Swiss font, and the text "MTEXT" needs to be increased to a text height of 0.30 units. When performing this type of operation, you need to highlight only the text object you want to change. Clicking the Close Text Editor button dismisses the editor and saves the changes to the text objects in the drawing.

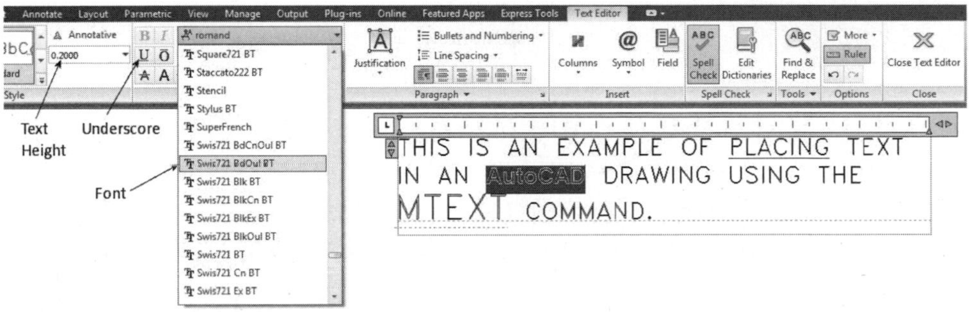

FIGURE 6.31

Using the DDEDIT (ED) command or double-clicking a text object created with the TEXT command, as shown in Figure 6.32 (Left), displays a field, as shown in Figure 6.32 (Right). Use this to change text in the field provided. Pick in the field to add text, highlight text and type over it to replace it, or highlight text and press the DEL key to remove it. The PROPERTIES command (discussed in Chapter 7) can be used if you need to change the font, justification or height of a text object.

Command: ED *(For DDEDIT)*

Select an annotation object or [Undo]: *(Select the text object and an edit field appears. Perform any text editing task and press ENTER)*

Select an annotation object or [Undo]: *(Pick another text object to edit or press ENTER to exit this command)*

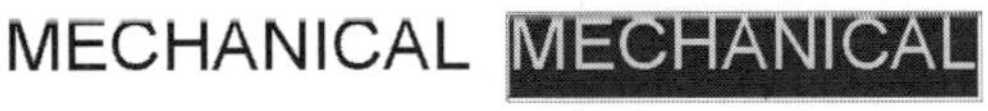

FIGURE 6.32

The TEXT command does support some codes that you might find helpful while typing in or editing text strings, as shown in Figure 6.33.

Code	Type In	Result	Description
%%C	%%C5.00	Ø5.00	Creates a diameter symbol
%%D	30%%D	30°	Creates a degree symbol
%%P	3.000%%P.005	3.000±.005	Creates a plus and minus symbol
%%U	Auto%%UCAD	Auto<u>CAD</u>	Toggles on and off underlining

GLOBALLY MODIFYING THE HEIGHT OF TEXT

The height of a text object can be easily changed through the SCALETEXT command, which can be found in the Annotate tab of the Ribbon (Annotate Tab > Text Panel – Expanded > Scale Button). The important feature of this command is that the actual scaling process has no effect on the justification of the text. This means that after scaling the text, you should not have to move each individual text item to a new location.

TRY IT!

Open the drawing file 06_Scale Text, illustrated in Figure 6.33. All offices in this facilities plan have been assigned room numbers. One of the room numbers (ROOM 114) has a text height of 12", while all other room numbers are 8" in height. You could edit each individual room number until all match the height of ROOM 114, or you could use the `SCALETEXT` **command by using the following command sequence:**

Command: **SCALETEXT**

Select objects: *(Pick all 17 mtext objects, although it is not necessary to pick ROOM 114)*

Select objects: *(Press* ENTER *to continue)*

Enter a base point option for scaling

[Existing/Left/Center/Middle/Right/TL/TC/TR/ML/MC/MR/BL/BC/BR] <Existing>: *(Press* ENTER *to accept this value)*

Specify new model height or [Paper height/Match object/Scale factor] <8.0000>: **M** *(For Match object)*

Select a text object with the desired height: *(Pick the text identified by ROOM 114)*

Height = 12.0000

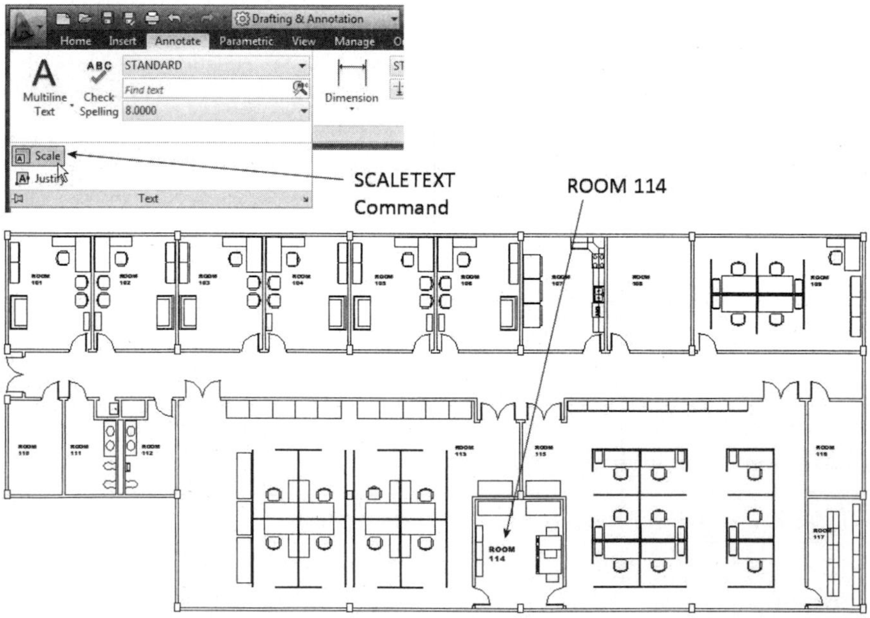

FIGURE 6.33

The results are displayed in Figure 6.34. All the text was properly scaled without affecting the justification locations.

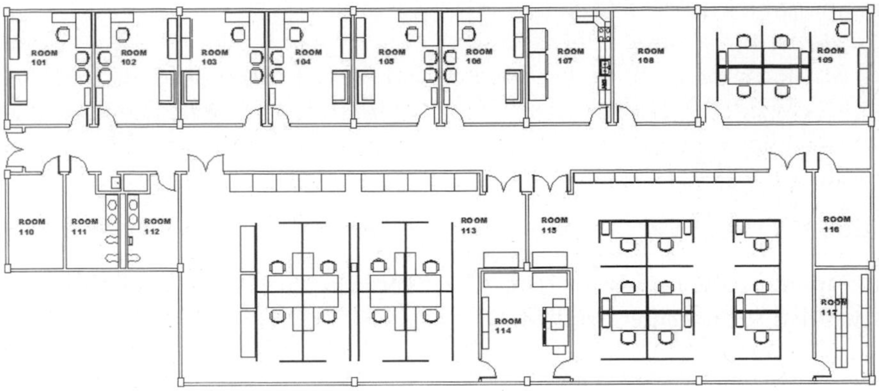

FIGURE 6.34

NOTE

The Quick Select dialog box could be used to select all mtext objects. You could also use a Window to select the `MTEXT` since the `SCALETEXT` command will only select text objects.

GLOBALLY MODIFYING THE JUSTIFICATION OF TEXT

The JUSTIFYTEXT command can be found in the Annotate tab of the Ribbon (Annotate Tab > Text Panel - Expanded > Justify Button). This command allows you to pick multiple text and mtext objects and change their current justification.

TRY IT!

Open the drawing file 06_Justify Text, illustrated in Figure 6.35. All the text justification points in this facilities drawing need to be changed from left justified to top center justified. Use the following command prompt and illustration in Figure 6.35 to accomplish this task.

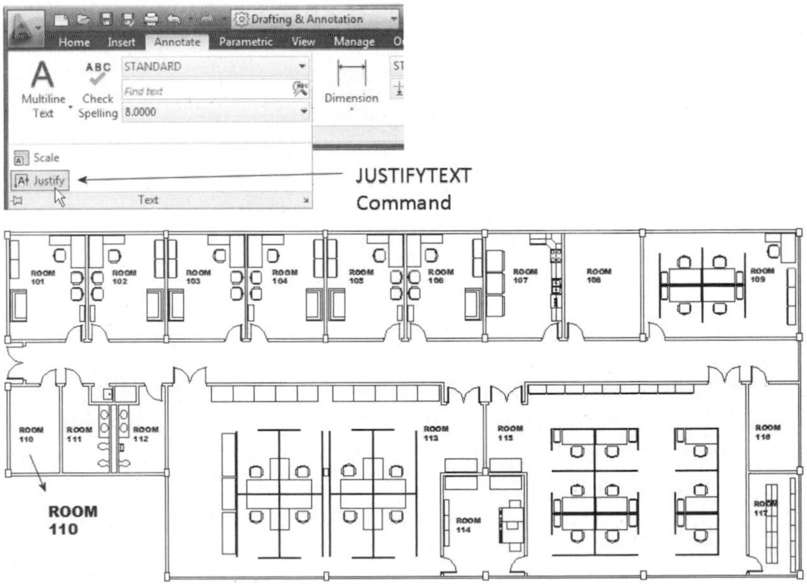

FIGURE 6.35

```
Command: JUSTIFYTEXT
Select objects: (Select all 17 text objects)
Select objects: (Press ENTER to continue)
Enter a justification option
[Left/Align/Fit/Center/Middle/Right/TL/TC/TR/ML/MC/MR/
BL/BC/BR] <Left>: TC (For Top Center)
```

The results are illustrated in Figure 6.36. All text objects have been globally changed from left justified to top center justified.

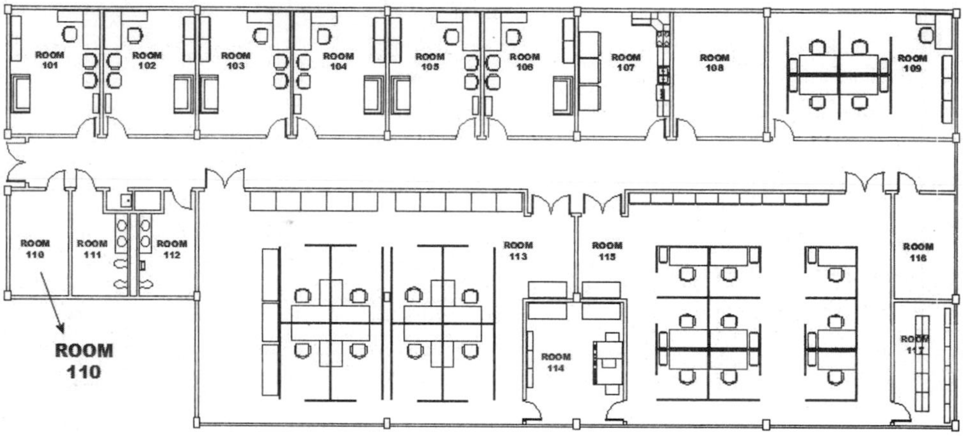

FIGURE 6.36

SPELL-CHECKING TEXT

Spell-checking is an important part of the drawing documentation process. Words, notes, or engineering terms that are spelled correctly elevate the drawing to a higher level of professionalism. Spell-checking a drawing can be activated through the Ribbon, as shown in Figure 6.37.

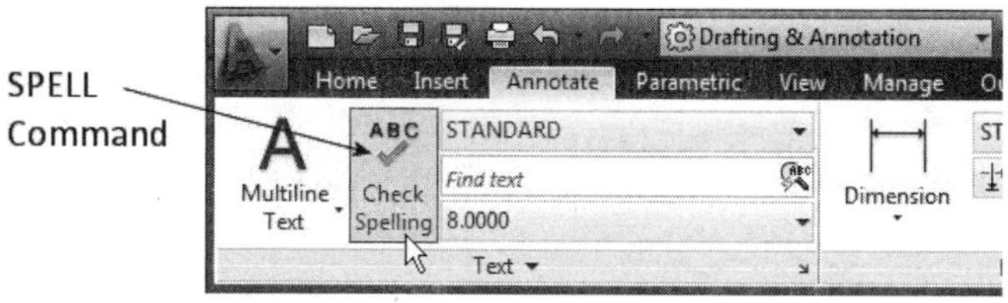

FIGURE 6.37

Issue the SPELL command and click the Start button in the provided dialog box to begin the spell-check. The "Where to Check" drop-down list provides the option of checking the entire drawing or limiting the check to the current space/layout or to only selected text items. The "Not in Dictionary" box displays the first word that is suspected of being spelled wrong. It wasn't found in the current dictionary. A drop-down list and Dictionaries button is provided to select other dictionaries. The Suggestions area displays all possible alternatives to the word identified as being misspelled. The Ignore button allows you to skip the current word; this would be applicable especially in the case of acronyms such as CAD and GDT. Clicking Ignore All skips all remaining words that match the current word. The Change button replaces the word "Not in the dictionary" with the word in the Suggestions box. The "Add to Dictionary" button adds the current word to the current dictionary. In Figure 6.38, the word "DIRECION" was identified as being misspelled. Notice that the correct word, "DIRECTION," is listed as a suggested correction. Clicking the Change button replaces the misspelled word and continues with the spell-checking operation until completed.

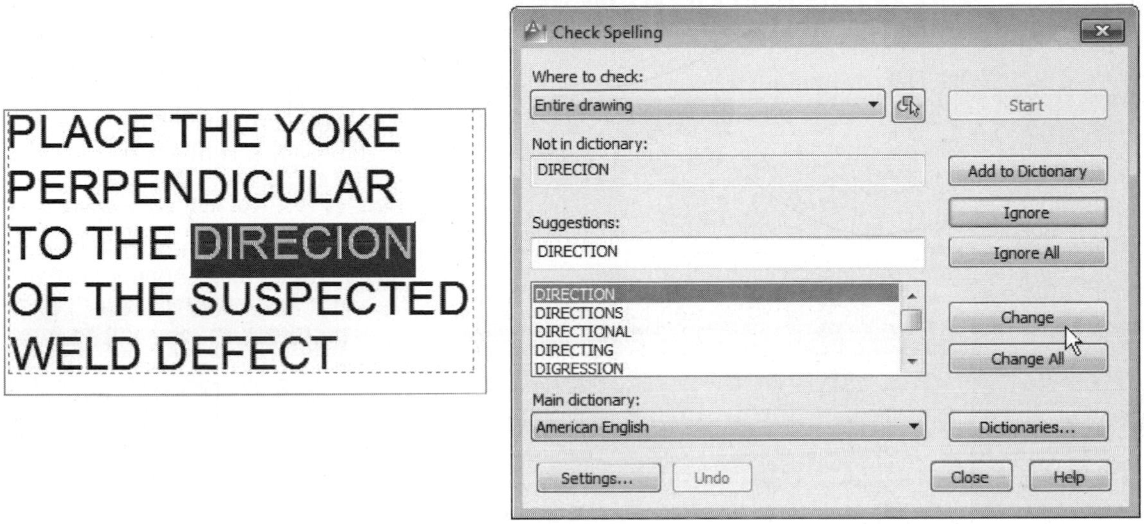

FIGURE 6.38

After completing the spell-checking operation, the mtext object is displayed in Figure 6.39.

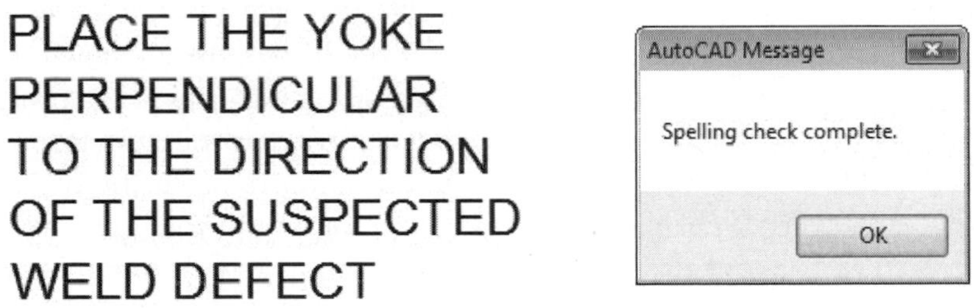

FIGURE 6.39

Open the drawing 06_Spell Check1 and perform a spell-check operation on this mtext object.

TRY IT!

Before	After
THIS IS AN EXAMPL OF PLACING TEXT IN AN AUTOCAD DRAWNG USING THE MTEXT COMMAND.	THIS IS AN EXAMPLE OF PLACING TEXT IN AN AUTOCAD DRAWING USING THE MTEXT COMMAND.

FIGURE 6.40

CREATING DIFFERENT TEXT STYLES

A text style is a collection of settings that are applied to text placed with the TEXT or MTEXT command. These settings could include presetting the text height and font, in addition to providing special effects, such as an oblique angle for inclined text. Choose the STYLE command from the Annotate tab of the Ribbon, as shown in Figure 6.41 (Left), or from the Home tab, as shown in Figure 6.41 (Right).

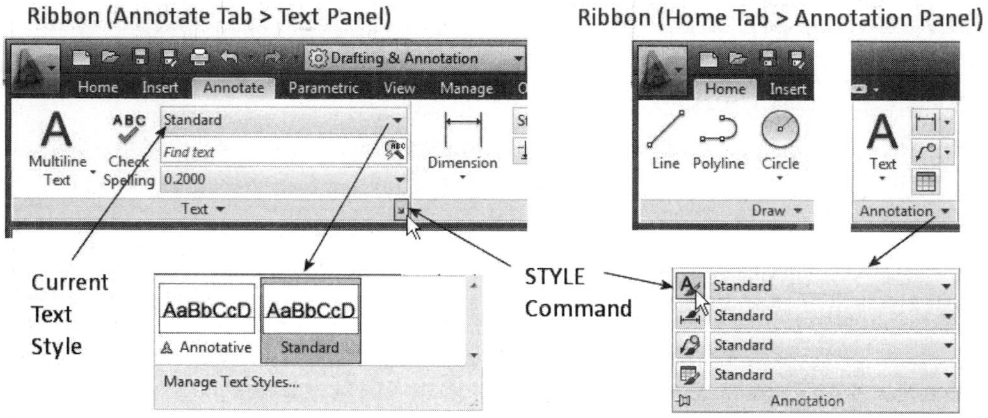

FIGURE 6.41

Initiating the STYLE command will launch the Text Style dialog box, as shown in Figure 6.42, which is used to create new text styles. This dialog box is used to create styles and set the current style. By default, when you first begin a drawing, the current Style Name is STANDARD. Creating multiple styles can improve productivity but be careful not to overuse text style changes, which can be distracting in your drawings. Once a new style is created, a font name is matched with the style. Clicking on the field for Font Name displays a list of all text fonts supported by the operating system. These fonts have different extensions, such as SHX and TTF. TTF or TrueType Fonts are especially helpful in AutoCAD because these fonts display in the drawing in their true form. If the font is bold and filled-in, the font in the drawing file displays as bold and filled-in.

When a Font Name is selected, it displays in the Preview area located in the lower-left corner of the dialog box. The Effects area allows you to display the text upside down, backwards, or vertically. Other effects include a width factor, explained later, and the oblique angle for displaying text at a slant.

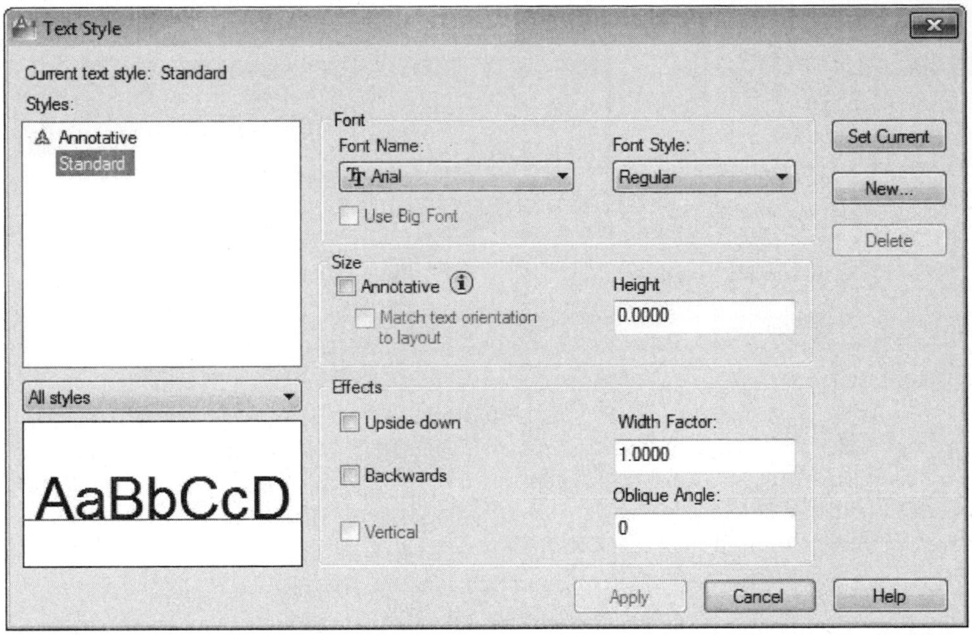

FIGURE 6.42

Clicking the New ... button of the Text Style dialog box displays the New Text Style dialog box, as shown in Figure 6.43. In this example, a new style is created called General Notes. Clicking the OK button returns you to the Text Style dialog box. Clicking on the Font Name field displays all supported fonts. Clicking "romans.shx" assigns the font to the style name General Notes. Clicking the Apply button saves the font to the database of the current drawing file.

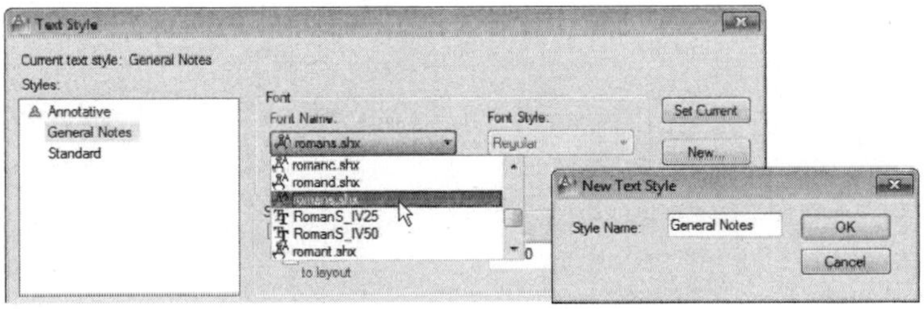

FIGURE 6.43

The Text Style Control Box

Although the Set Current button in the Text Style dialog box can be used to set a current style, the Text Style Control box provides a more convenient method for changing styles. This control box, which can be found in both the Home and Annotate tabs of the Ribbon, as shown in Figure 6.44, provides an easy-to-access drop-down list for selecting the current text style in a user-friendly graphical form.

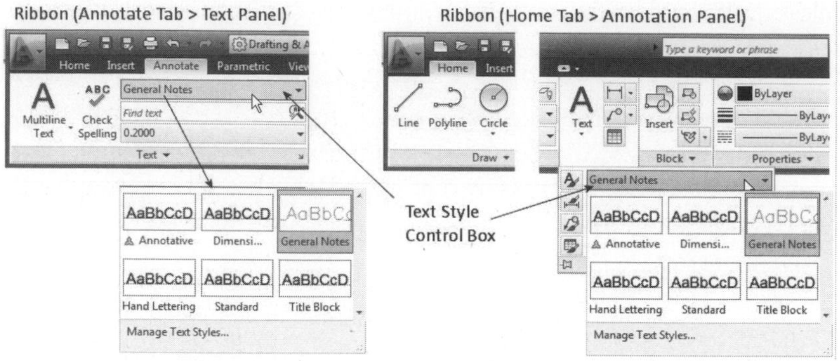

FIGURE 6.44

> **NOTE**
>
> **You can also highlight a text object and use the Text Style Control box to change the selected text to a different text style. This action is similar to changing an object from one layer to another through the Layer Control box.**

FIELDS

Fields are a type of intelligent text that you can add to your drawings. In Figure 6.45, fields are identified by a distinctive gray background, which is not present on regular text. Fields are intended for items that tend to change during the life cycle of a drawing, such as the designer's name, checker, creation date of the drawing, and drawing number, as shown in Figure 6.45.

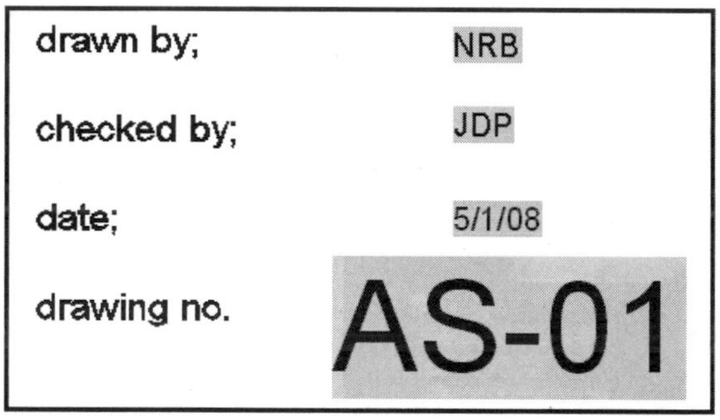

FIGURE 6.45

The FIELD command can be selected from the Insert tab of the Ribbon, as shown in Figure 6.46. This activates the Field dialog box, which consists of Field categories, Field names, Author, and Format information.

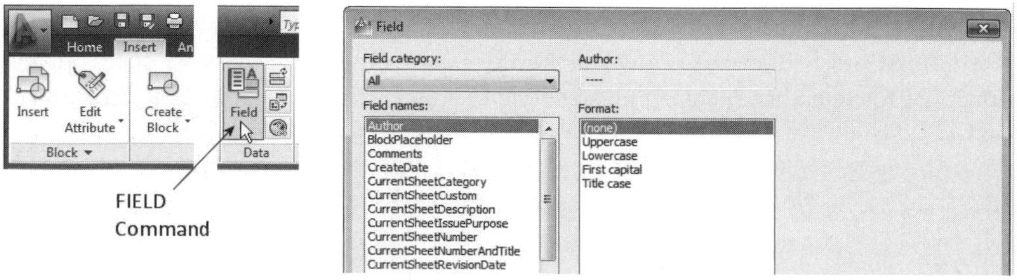

FIGURE 6.46

By default, all fields are available under the Field names area. You can specify a certain Field category to narrow down the list of field names. Illustrated in Figure 6.47 (Left) is a list of all field categories. Selecting the Date & Time category activates specific field names only related to this category, as shown in Figure 6.47 (Right). With the Field name SaveDate highlighted, notice all the examples displayed that pertain to when the drawing was last saved. There is also a Hints area that gives examples of various month, day, and year notations.

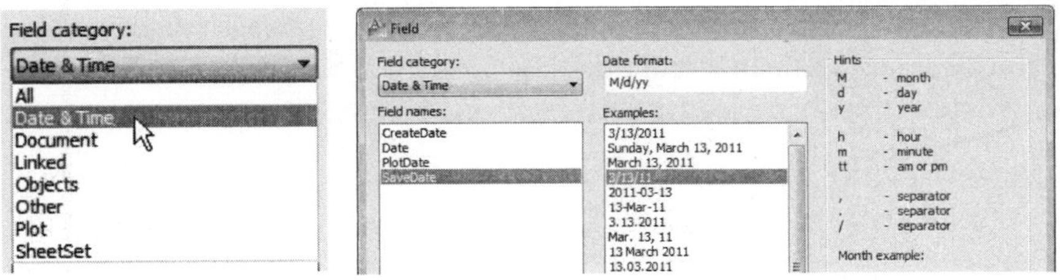

FIGURE 6.47

Fields and Multiline Text

While placing multiline text, you can create a field through the Text Editor Ribbon by selecting the Field button on the Insert panel, as shown in Figure 6.48. This allows you to place numerous fields at one time. Instead of picking the Ribbon button, you can instead activate the Field dialog box by right-clicking on the drawing screen and selecting Insert Field from the provided shortcut menu.

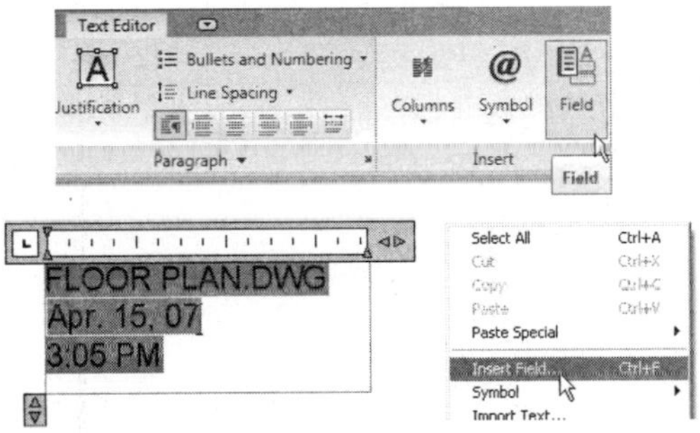

FIGURE 6.48

The results of creating a field are illustrated in Figure 6.49 (Right). In this example, the following field categories were used: Filename and Date. The time entry is a subset of the Date field.

Whenever field information changes, such as the date and time, you may need to update the field in order to view the latest information. Updating fields is accomplished by clicking Update Fields from the Ribbon, as shown in Figure 6.49. When picking this command, you are prompted to select the field. In Figure 6.49 (Right), the date changed automatically. The Filename field (FLOOR PLAN) would automatically update if the drawing were saved with a different name. Many fields will automatically update to reflect the most up-to-date information.

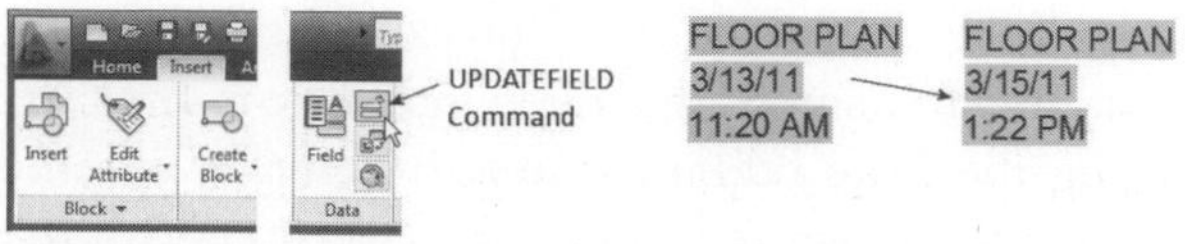

FIGURE 6.49

Fields and Drawing Properties

When creating some fields, information is displayed as a series of four dashed lines. This is to signify that the information contained in the field name has either not been entered or not been updated. In Figure 6.50, the Author, Subject, and Title fields need to be completed before the correct information displays.

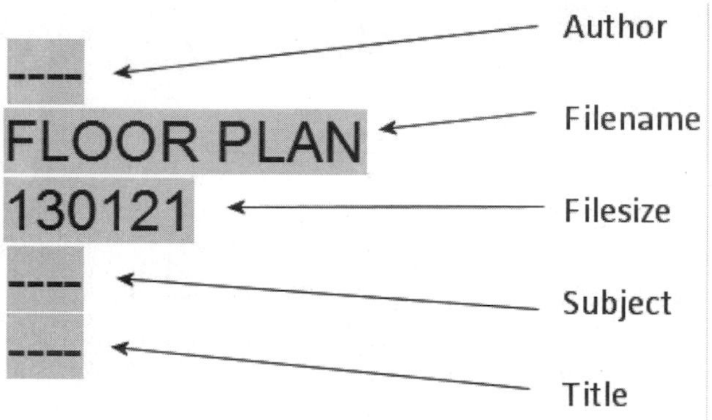

FIGURE 6.50

In the previous illustration of the Author, Subject, and Title field names, activating the Drawing Properties dialog box (pick Drawing Properties from the Application Menu), as shown in Figure 6.51 (Left), allows you to fill in this information. When returning to the drawing, activate the UPDATEFIELD command, and pick the field that needs updating. The information listed in the Drawing Properties dialog box will be transferred to the specific fields, as shown in Figure 6.51 (Right).

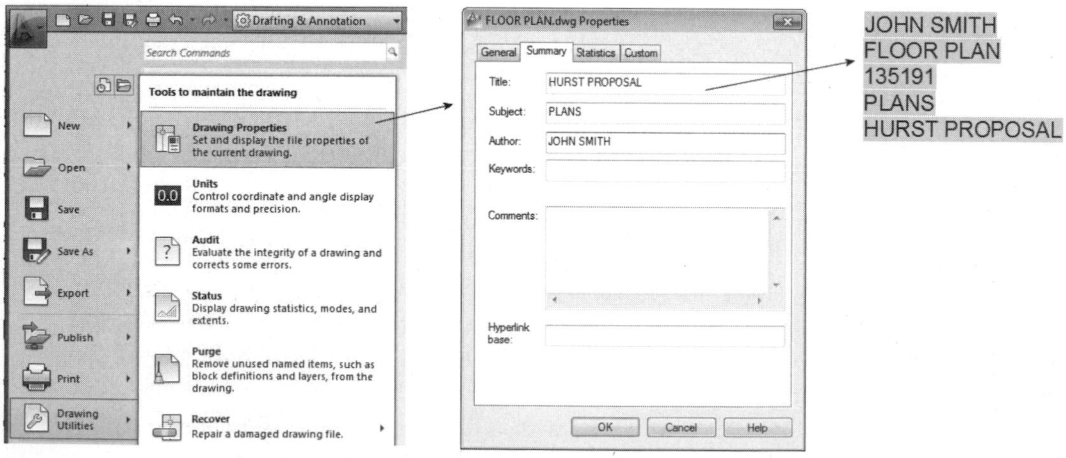

FIGURE 6.51

CREATING TABLES

As a means of further organizing your work, especially text, you can create cells organized in rectangular patterns that consist of rows and columns. The TABLE command is used to accomplish this and can be selected from either the Home or the Annotate tab of the Ribbon, as shown in Figure 6.52.

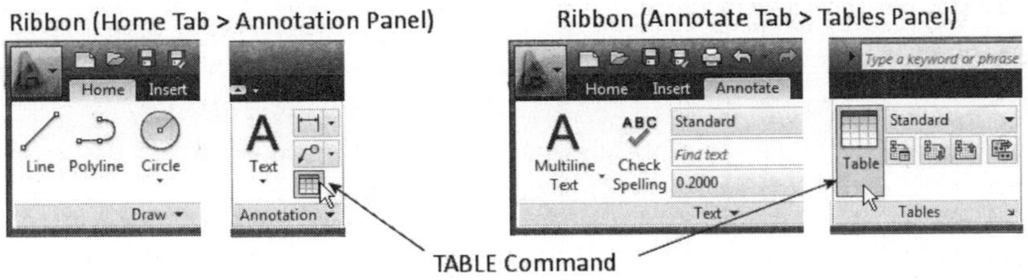

FIGURE 6.52

Using one of the methods shown in Figure 6.52 will activate the Insert Table dialog box, as shown in Figure 6.53. It is here that you specify the number of rows and columns that make up the table, in addition to the column width and row height. Selecting a table style allows you to apply fonts and heights to text objects in the table. Different properties can be displayed for Titles, Headers, and Data cells through the style. This will be discussed later in the chapter. A preview of the table is shown in the lower left corner of the dialog box.

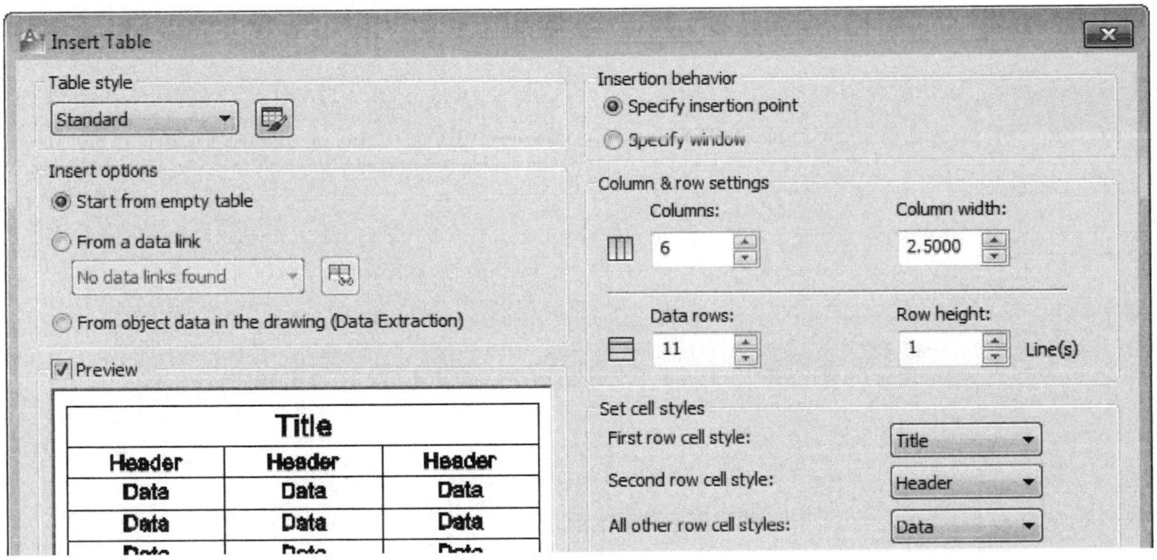

FIGURE 6.53

Once you decide on the table style and the number of rows and columns for the table, clicking the OK button returns you to the drawing editor. Locate the position of the table by picking an appropriate point on the drawing screen. Once the table is inserted, the table appears in the drawing editor and the Text Editor Ribbon displays above the table, as shown in Figure 6.54. In this example, the text "WINDOW SCHEDULE" was added as the title of the table. Notice also that the text height automatically adjusts to the current table style.

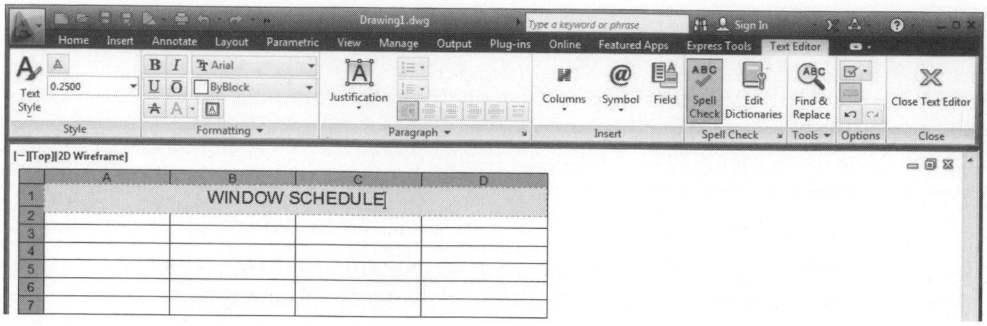

FIGURE 6.54

After you have entered text into the title cell, pressing the TAB key moves you from one cell to another, as shown in Figure 6.55. In this example, cell headers are added to identify the various categories of the table.

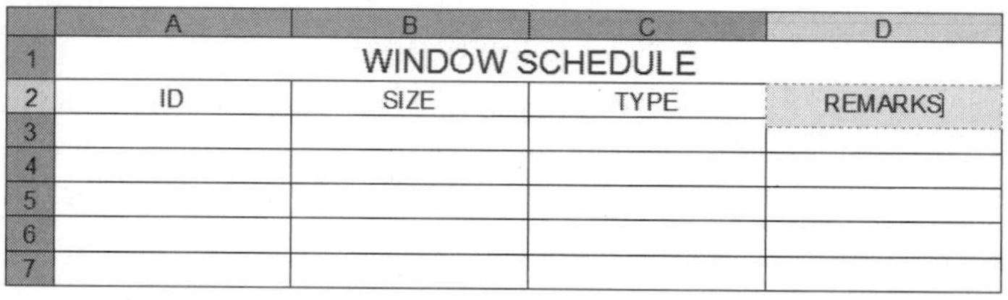

FIGURE 6.55

NOTE

Pressing the TAB key moves you from one cell to another; press SHIFT + TAB to reverse the direction.

Clicking a table entry highlights the cell and displays the Table Cell tab on the Ribbon, as shown in Figure 6.56.

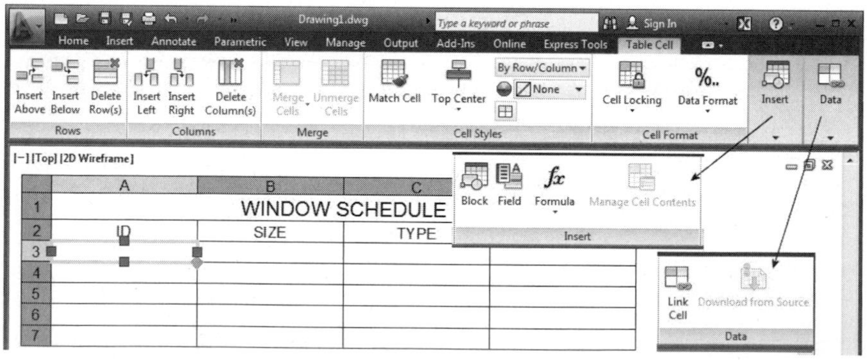

FIGURE 6.56

The Table Cell Ribbon has numerous controls for manipulating tables. The following chart will explain these buttons.

ROWS, COLUMNS, AND MERGE PANELS

Button	Tool	Function
	Insert Row Above	Inserts a row above a selected cell
	Insert Row Below	Inserts a row below a selected cell
	Delete Row	Deletes an entire row
	Insert Column Left	Inserts a column to the left of a selected cell
	Insert Column Right	Inserts a column to the right of a selected cell
	Delete Column	Deletes an entire column
	Merge Cells	Merges all selected cells or cells by row or column
	Unmerge Cells	Unmerges selected cells

CELL STYLES AND CELL FORMAT PANELS

Button	Tool	Function
	Match Cell	Matches the contents of a source cell with destination cell
Top Center	Top Center Align	Displays nine text alignment modes that deal with tables
By Row/Column	Cell Style	Displays a drop-down list that allows you to change to a different cell style
None	Background Color	Changes the cell background color
	Cell Borders	Launches the Cell Border Properties dialog box designed to control the display of cell borders
Cell Locking	Locking	Displays a menu used for locking the cell content, cell format, or both. Also used for unlocking cells
%.. Data Format	Data Format	Launches the Table Cell Format dialog box used for defining the data type of a cell or group of cells

INSERT AND DATA PANELS

Button	Tool	Function
Block	Insert Block	Launches the Insert a Block in a Table dialog box used for placing and fitting blocks in table cells
Field	Insert Field	Launches the Field dialog box designed for creating a field in a table cell
fx Formula	Insert Formula	Displays a menu used for performing summations and other calculations on a group of cells
	Manage Cell Content	Displays the Manage Cell Content dialog box used for changing the order of cell content as well as changing the direction in which the cell content will display
Link Cell	Link Cell	Displays the Select Data Link dialog box that lists the current links with Excel
	Download Changes	Used for downloading changes from a source file to the table

Merging Cells in a Table

There are times when you need to merge cells in the rows or columns of a table. To accomplish this, first pick inside a cell, then hold down the SHIFT key as you pick any additional cells you want to merge with it. To merge the two cells that are shown selected in Figure 6.57, pick the Merge Cells button followed by Merge By Row from the menu.

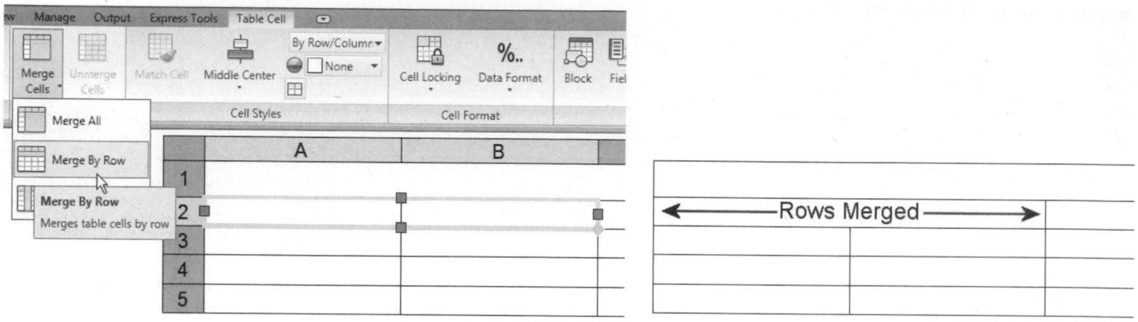

FIGURE 6.57

Modifying a Table

The modification of table cells, table rows, and even the overall size of the table is greatly enhanced through the use of grips. You can achieve different results depending on the table grip that is selected. For instance, first select the table to display all grips. Then, select the grip in the upper-left corner of the table and move your cursor. This action moves the table to a new location. Selecting the grip in the upper-right corner of the table and then dragging your cursor left or right increases or decreases the spacing of all table columns. Selecting the grip in the lower-left corner of the table increases the spacing between rows. Study the following image to see how other grips affect the sizing of a table. Grips will be covered in greater detail in Chapter 7.

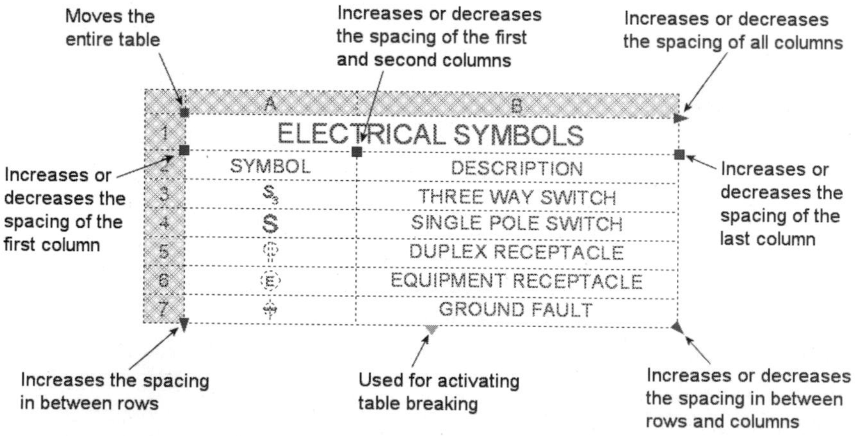

FIGURE 6.58

Autofilling Cells in a Table

Tables can, at times, require a lot of repetitive data entry. Although, you could easily copy and paste identical information from one cell to the next, a powerful tool exists in tables that allows text to be duplicated or even created incrementally, such as the room numbers in Figure 6.59. This is accomplished through the Autofill feature of tables.

ROOM	NAME OF SPACE	FLR	BASE	MATL	FIN	COLOR	MATL	FIN
100	CLASSROOM	CPT	V	GBW	PT		GBW	PT
101	CLASSROOM	CPT	V	GBW	PT		GBW	PT
102	CLASSROOM	CPT	V	GBW	PT		GBW	PT
103	CLASSROOM	CPT	V	GBW	PT		GBW	PT
104	HALLWAY	CPT	V	GBW	PT		GBW	PT
105	PASTOR	CPT	V	GBW	PT		GBW	PT
106	RESTROOM	CPT	V	GBW	PT		GBW	PT
107	OFFICE	CPT	V	GBW	PT		GBW	PT
108	OFFICE	CPT	V	GBW	PT		GBW	PT

FIGURE 6.59

Cell 3A illustrated in Figure on the left is populated with the number "1." If you know the next number below the "1" will be "2" and so on, first click in cell 3A. Notice the light blue diamond shaped grip in the lower-right corner of the cell. Click on this grip and move your cursor in a downward direction as shown in Figure 6.60 in the middle. You will notice your cursor changing to different numbers the more you move the cursor. To automatically fill the numbers "1" through "8," pick in cell 10A. The results are shown in Figure 6.60 on the right with the Autofill feature creating the proper incremental data.

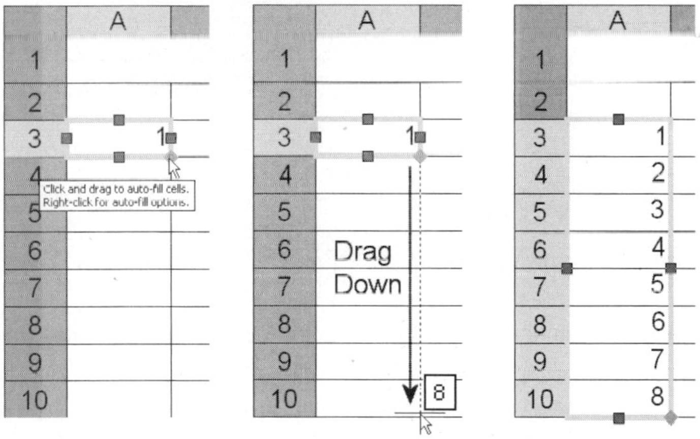

FIGURE 6.60

Autofill also works with regular text entries. If you want to reproduce the word "VINYL" in the remaining cells below, click on the cell with the word. When the light blue diamond grip appears, drag your cursor to the cells that will contain the same text. The results are shown in Figure 6.61 (Right) with the Autofill feature being used to automatically place identical text entries into table cells.

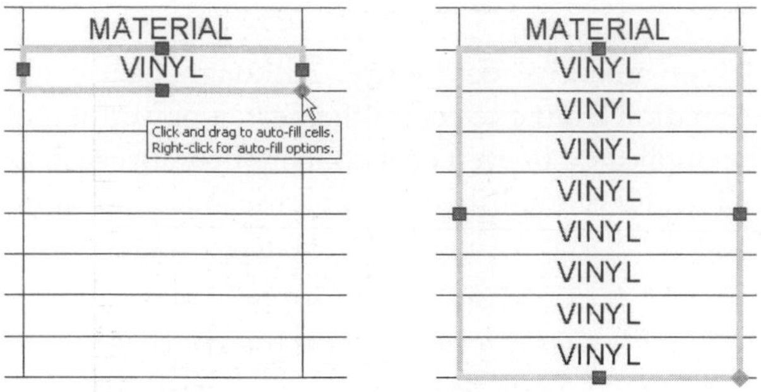

FIGURE 6.61

TABLES AND MICROSOFT EXCEL

Microsoft Excel spreadsheets can easily be imported into AutoCAD drawings as tables. The illustration in Figure 6.62 (Left) represents information organized in an Excel spreadsheet. With the spreadsheet open, copy the appropriate data to the Windows Clipboard. Switch to AutoCAD and while in the drawing editor, pick Paste Special from the Clipboard panel of the Ribbon's Home tab, as shown in Figure 6.62 (Center). This displays the Paste Special dialog box, as shown in Figure 6.62 (Right). Click AutoCAD Entities to merge the spreadsheet into an AutoCAD drawing as a table.

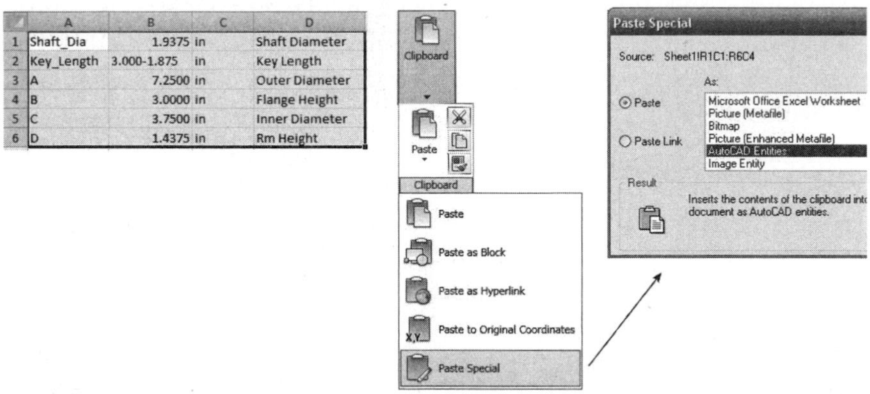

FIGURE 6.62

The initial results of importing an Excel spreadsheet into an AutoCAD drawing are illustrated in Figure 6.63 (Left). The Excel spreadsheet was converted into an AutoCAD drawing object. However, a few columns need to be lengthened and the rows shortened in order to organize the data in a single line. Also, headings need to be added to the top of the table by inserting a row above the top line of data. The completed table is illustrated in Figure 6.63 (Right).

Shaft_Dia	1.9375	in	Shaft Diameter
Key_Length	3.000-.1875	in	Key Length
A	7.2500	in	Outer Diameter
B	3.0000	in	Flange Height
C	3.7500	in	Inner Diameter
D	1.4375	in	Rim Height
BT	0.7500	in	Base Thickness
F	5.3750	in	Bolt Circle Diameter
G	4.5000	in	Base Relief Diameter
H	6.6250	in	Male Connect Diameter
FCD	6.6260	in	Female Connect Diameter
K	0.2500	in	Wall Thickness
L		in	Fillet Radius for

PARAMETER	VALUE	UNITS	DESCRIPTION
Shaft_Dia	1.9375	in	Shaft Diameter
Key_Length	3.000-.1875	in	Key Length
A	7.2500	in	Outer Diameter
B	3.0000	in	Flange Height
C	3.7500	in	Inner Diameter
D	1.4375	in	Rim Height
BT	0.7500	in	Base Thickness
F	5.3750	in	Bolt Circle Diameter
G	4.5000	in	Base Relief Diameter
H	6.6250	in	Male Connect Diameter
FCD	6.6260	in	Female Connect Diameter
K	0.2500	in	Wall Thickness
L	0.1000	in	Fillet Radius for Wall
M	0.1875	in	Other Fillet Radius Values
NBH	5.0000	ul	Number of Bolt Holes
Angle	72.0000	deg	Polar Angle
O	2.0000	in	Bolt Length
P	0.5000	in	Keyway Width
Q	0.2500	in	Half the Keyway Height

FIGURE 6.63

CREATING TABLE STYLES

Table styles, as with text styles, can be used to organize different properties that make up a table. Table styles can be accessed through the Home and Annotate tabs of the Ribbon, as shown in Figure 6.64.

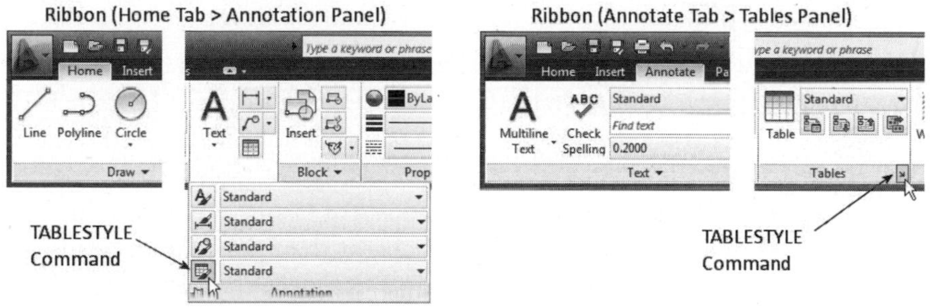

FIGURE 6.64

Clicking on either one of the icons shown in Figure 6.64 will display the Table Style dialog box, as shown in Figure 6.65. Clicking the New button displays the Create New Table Style dialog box, in which you enter the name of the table style.

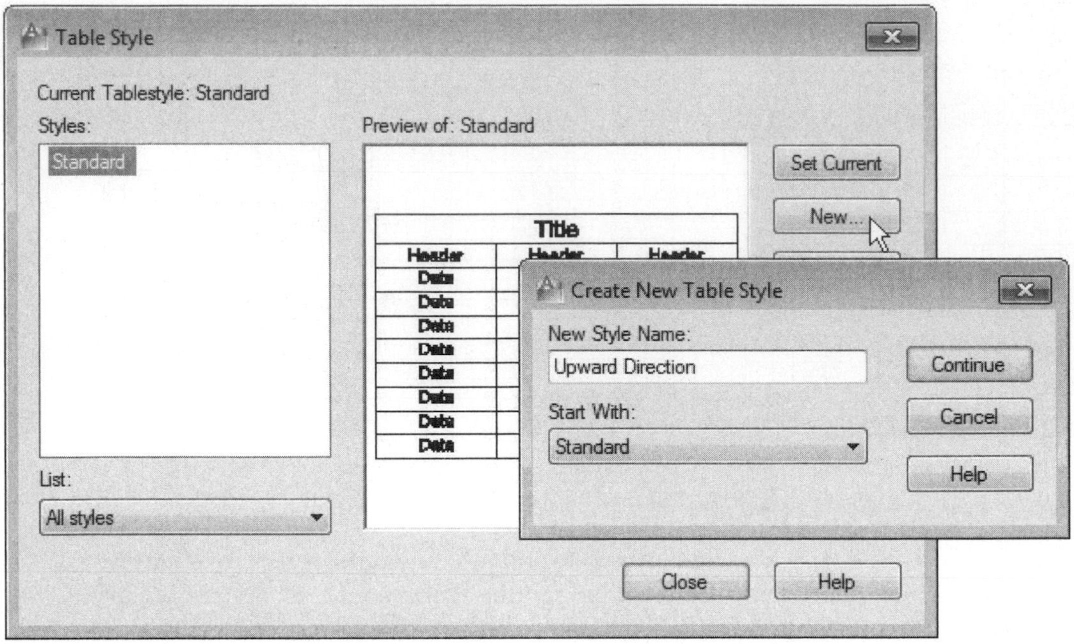

FIGURE 6.65

Clicking Continue allows you to make various changes to a table based on the type of cell. For example, in Figure 6.66 three tabs (General, Text, and Borders) can be used to make changes to the text that occupies a cell in a table based on whether the text is cell data, the column heads of a cell, or even the title cell of the table. These settings can be used to emphasize the information in the cells. For example, you would want the title of the table to stand out with a larger text height than the column headings. You might also want the column headings to stand out over the data in the cells below. Also, as in Figure 6.66, you can change the direction of the table. By default, a table is created in the downward direction. You may want to change this direction to upward.

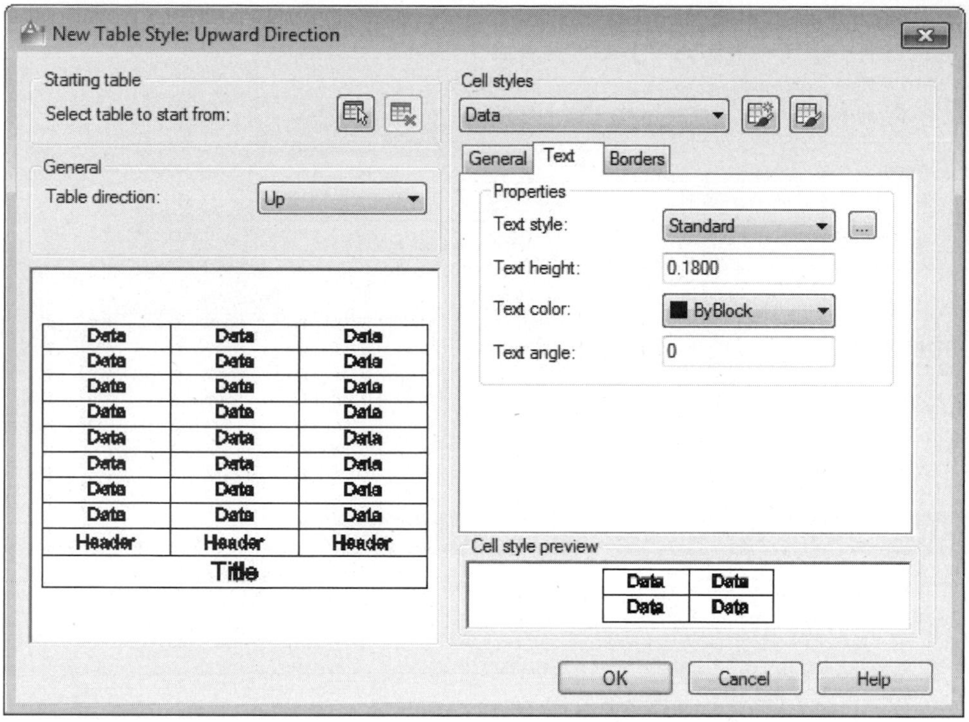

FIGURE 6.66

Figure 6.67 illustrates three tabs containing information that can be modified for the cell type. In this image, the Data cell type is selected.

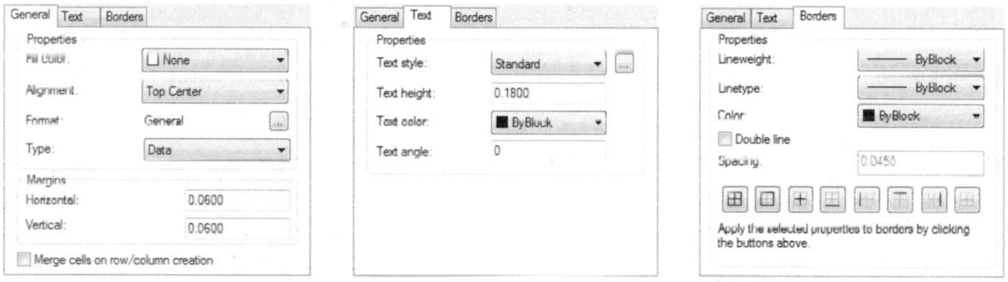

FIGURE 6.67

To make changes to the Header or Title cells, select the cell type in the Cell Styles field, as shown in Figure 6.68, and modify the information in the tabs as desired.

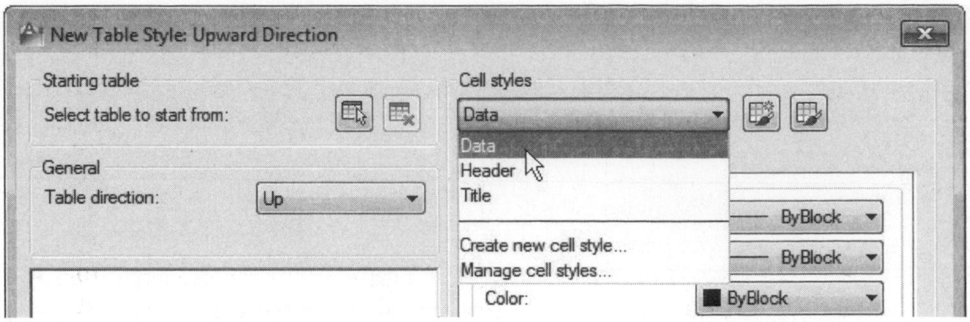

FIGURE 6.68

TUTORIAL EXERCISE: 06_PUMP_ASSEMBLY.DWG

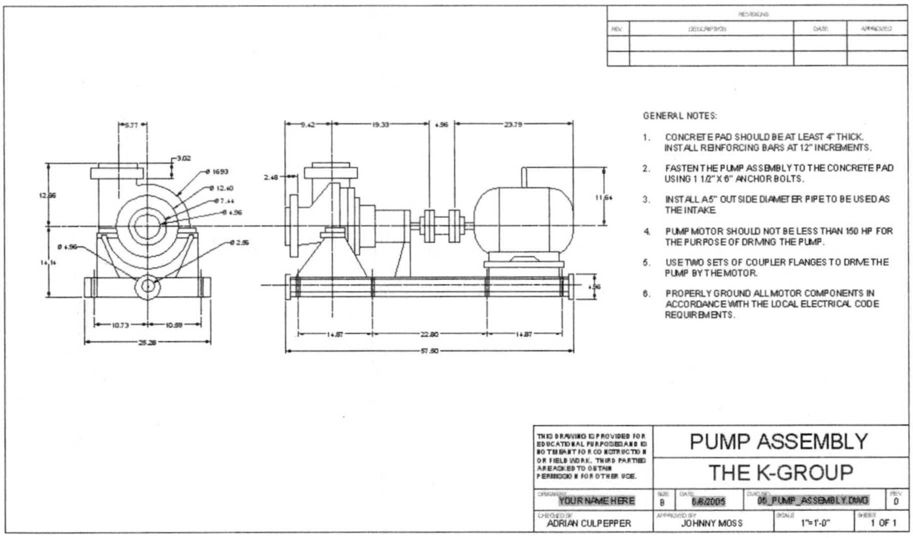

FIGURE 6.69

Purpose

This tutorial is designed to create numerous text styles and add different types of text objects to the title block illustrated in Figure 6.69.

System Settings

In this tutorial you will use an existing drawing file called 06_Pump_Assembly. Follow the steps in this tutorial for creating a number of text styles and then placing text in the title block area. Set the following Object Snap modes: Endpoint and Node.

Layers

Various layers have already been created for this drawing. Since this tutorial covers the topic of text, the current layer is Text.

Suggested Commands

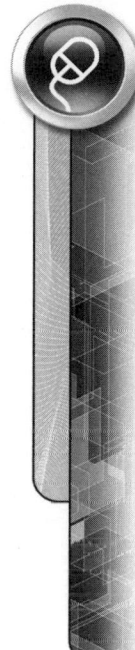

Open the drawing called 06_Pump_Assembly. The purpose of this tutorial is to fill in the title block area with a series of text, multiline text, and field objects. Also, a series of general notes will be placed to the right of the pump assembly. Before you place the text, four text styles will be created to assist with the text creation. Three of the four text styles will be used in the title block, as identified in Figure 6.70. The text style Company Info will be used to place the name of the company and drawing title. Information in the form of scale, date, and who performed the drawing will be handled by the text style Title Block Text. A disclaimer will be imported from an existing .TXT file available from the same folder as the drawing file; this will be accomplished in the Disclaimer text style. Finally, a listing of six general notes, originally created in Microsoft Word in .RTF format, will be imported into the drawing and placed in the General Notes text style. A spell-check operation will be performed on the notes.

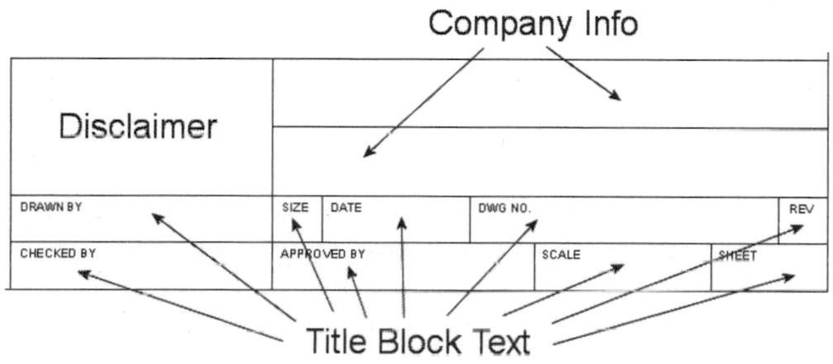

FIGURE 6.70

STEP 1

Before creating the first text style, first see what styles are already defined in the drawing by choosing the STYLE command from the Annotate tab of the Ribbon, as shown in Figure 6.71 (Left). This displays the Text Style dialog box, as shown in Figure 6.71 (Right). The Styles text box displays the current text styles defined in the drawing. Every AutoCAD drawing contains the STANDARD text style. This is created by default and cannot be deleted. Also, the Title Block Headings text style was already created. This text style was used to create the headings in each title block box, for example, "Scale" and "Date."

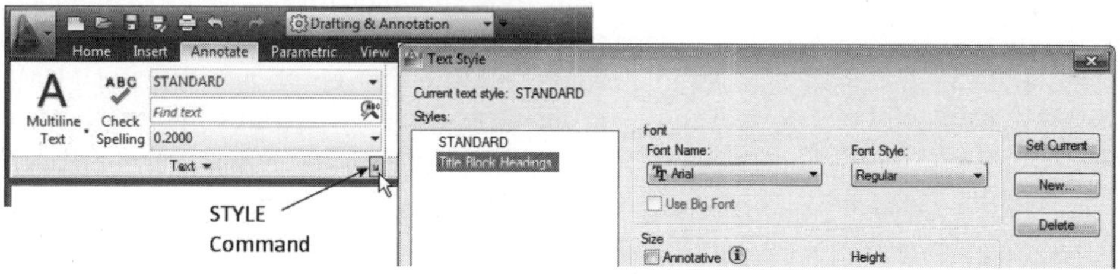

FIGURE 6.71

STEP 2

Create the first text style by clicking the New... button, as shown in Figure 6.72. This activates the New Text Style dialog box, also shown in Figure 6.72. For Style Name, enter Company Info. When finished, click the OK button. This takes you back to the Text Style dialog box, as shown in Figure 6.72 (Right). In the Font area, verify that the name of the font is Arial. Notice the font appearing in the Preview area. If necessary, click the Apply button to complete the text style creation process.

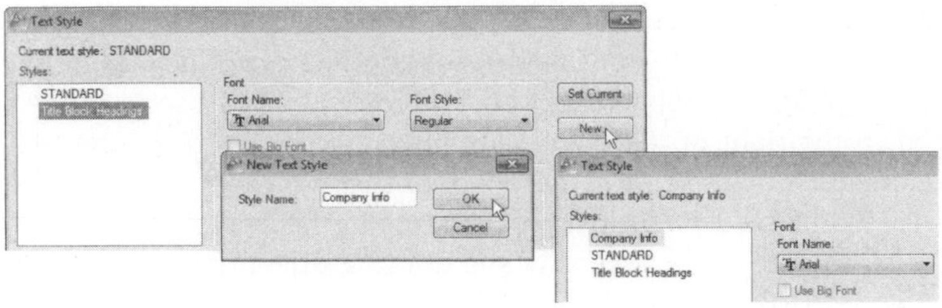

FIGURE 6.72

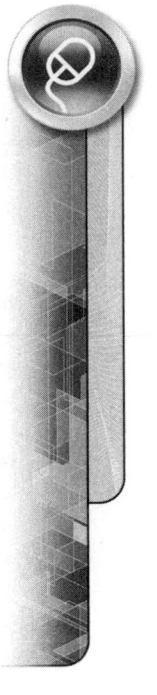

STEP 3

Using the same procedure outlined in Step 2, create the following new text styles: Disclaimer, General Notes, and Title Block Text. Verify that the Arial font is assigned to all of these style names. Keep all other remaining default text settings. When finished, your display should appear similar to the illustration in Figure 6.73.

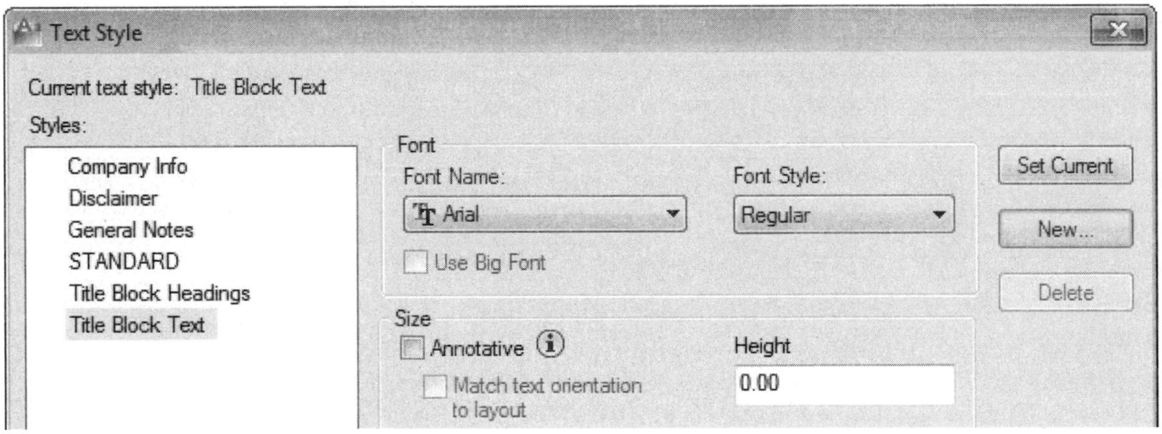

FIGURE 6.73

STEP 4

Make the Company Info style current by selecting it in the Style text box and clicking the Set Current button. Close the Text Style dialog box and activate the MTEXT command. Add two mtext objects under the Company Info text style, changing the text height to .25 units and the justification to Middle-Center. For the first entry, add the text PUMP ASSEMBLY inside the text formatting box. Click the Close Text Editor button to place the text. Repeat the MTEXT command for placing the second text entry, THE K-GROUP in the next title block space.

```
Command: MT (For MTEXT)

MTEXT Current text style: "Company Info" Text height: 0.20
Annotative: No

Specify first corner: (Pick the endpoint at "A")

Specify opposite corner or [Height/Justify/Line spacing/
Rotation/Style/Width/Columns]: H (For Height)

Specify height <0.20>: .25

Specify opposite corner or [Height/Justify/Line spacing/
Rotation/Style/Width/Columns]: J (For Justify)
```

Enter justification [TL/TC/TR/ML/MC/MR/BL/BC/BR] <TL>: **MC** *(For Middle Center)*

Specify opposite corner or [Height/Justify/Line spacing/Rotation/Style/Width/Columns]: *(Pick the endpoint at "B")*

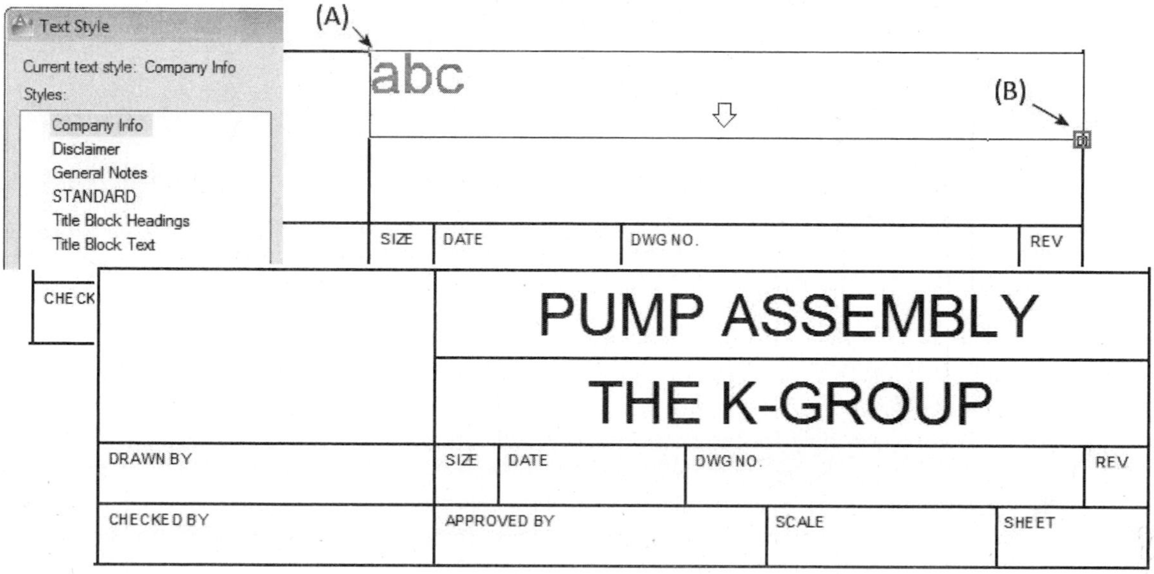

FIGURE 6.74

STEP 5

Change the current text style to Title Block Text, as shown in Figure 6.75 (Left). Begin placing fields into the title block. Click Field under the Insert tab of the Ribbon, as shown in Figure 6.75 (Right). When the Field dialog box appears, pick Author under the Field names and click the OK button to dismiss the dialog box. Before placing the field, set the field height to .10 using the Height option. Utilizing a Node object snap, place the field in the space identified as Drawn By, as shown in Figure 6.75 in the lower left. Notice the appearance of four dashes inside a gray rectangle. This will be filled in when you enter your name in the appropriate area of the Properties dialog box later on in this exercise. Use the following command sequence to assist in the placing of this first field.

Command: **FIELD**

MTEXT Current text style: "Title Block Text" Text height: 0.20

Specify start point or [Height/Justify]: **H** *(For Height)*

Specify height <0.20>: **.10**

Specify start point or [Height/Justify]: *(Pick the point inside of the area identified as Drawn By)*

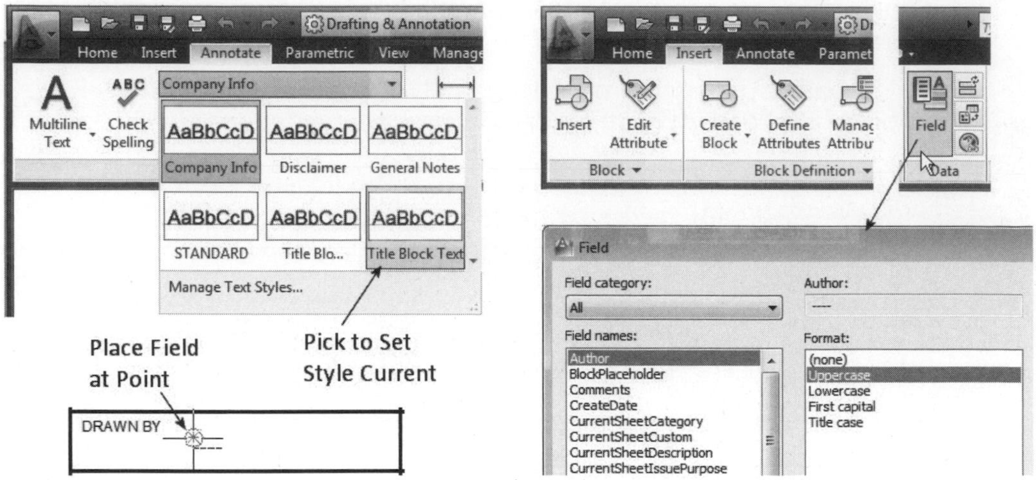

FIGURE 6.75

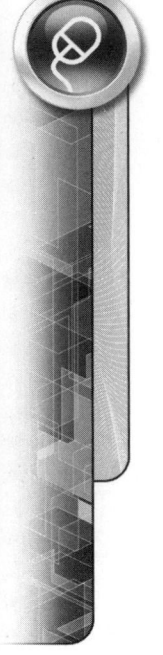

STEP 6

Place a new field for the Date category. Activate the Field dialog box, as shown in Figure 6.76, and pick the Date Field name. Also select the format of the date from the list of examples located in this dialog box. Place this field under the DATE heading, as shown in Figure 6.76. Notice that the date is automatically calculated based on the current date setting on your computer.

```
Command: FIELD
MTEXT Current text style: "Title Block Text" Text height:
0.10
Specify start point or [Height/Justify]: (Pick the point
inside of the area identified as Date)
```

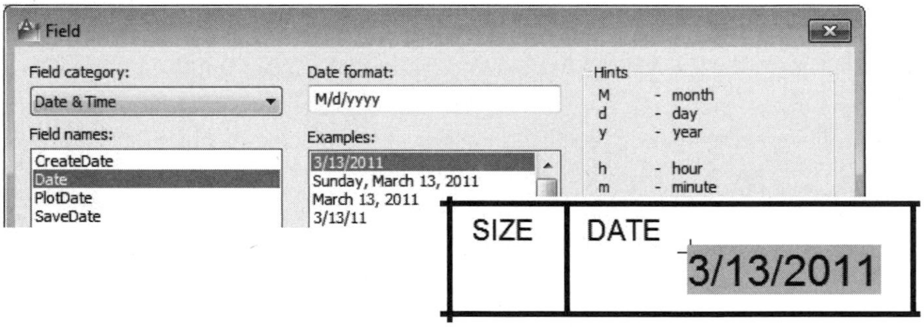

FIGURE 6.76

STEP 7

Place a new field for the Filename category. Activate the Field dialog box, as shown in Figure 6.77, and pick the Filename Field name. Pick Uppercase for the Format and click the radio button next to Filename only. Also remove the check from Display file extension. Locate this field under the DWG NO. heading name, as shown in Figure 6.77.

```
Command: FIELD
MTEXT Current text style: "Title Block Text" Text height:
0.10
```

Specify start point or [Height/Justify]: *(Pick the point inside of the area identified as DWG NO.)*

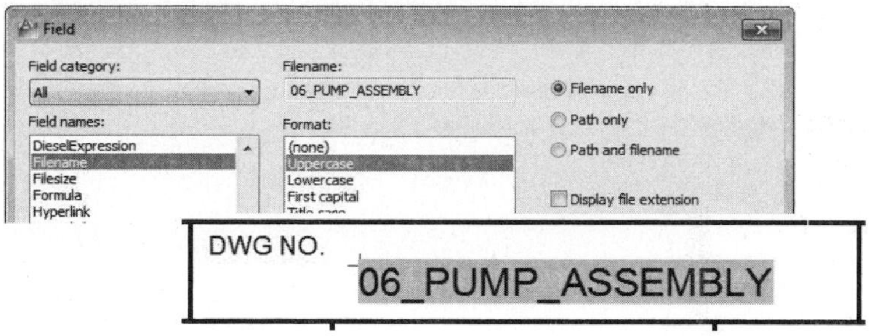

FIGURE 6.77

STEP 8

While still in the Title Block Text style, use the Single Line text command and add the following entries to the title block using the following prompts and Figure 6.78 as guides.

Command: DT *(For TEXT)*

Current text style: "Title Block Text" Text height: 0.10 Annotative: No

Specify start point of text or [Justify/Style]: *(Pick the point at "A")*

Specify height <0.10>: *(Press ENTER to accept this default value)*

Specify rotation angle of text <0d0'>: *(Press ENTER to accept this default value)*

Enter text: ADRIAN CULPEPPER *(After this text is entered, press ENTER twice to exit this command; use the table and figure below to enter the remaining text)*

Text	Title Block Area
B	SIZE (at "B")
JOHNNY MOSS	APPROVED BY (at "C")
1"=1'-0"	SCALE (at "D")
0	REV (at "E")
1 OF 1	SHEET (at "F")

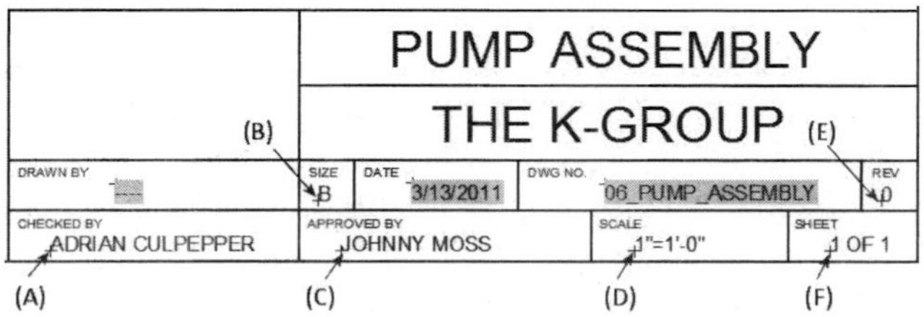

FIGURE 6.78

STEP 9

Activate the MTEXT command, pick a first corner at "A," and pick a second corner at "B," as shown in Figure 6.79. While in the Multiline Text Editor, change the text style to Disclaimer and the text height to 0.08; press the ENTER key to accept the height change. Right-click in the text editor, and pick Import Text from the menu, as shown in Figure 6.79.

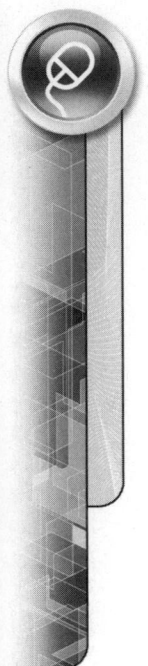

```
[A] Command: MT (For MTEXT)

MTEXT Current text style: "Disclaimer" Text height: 0.10
Annotative: No

Specify first corner: (Pick the point at "A")

Specify opposite corner or [Height/Justify/Line spacing/
Rotation/Style/Width/Columns]: (Pick the point at "B")
```

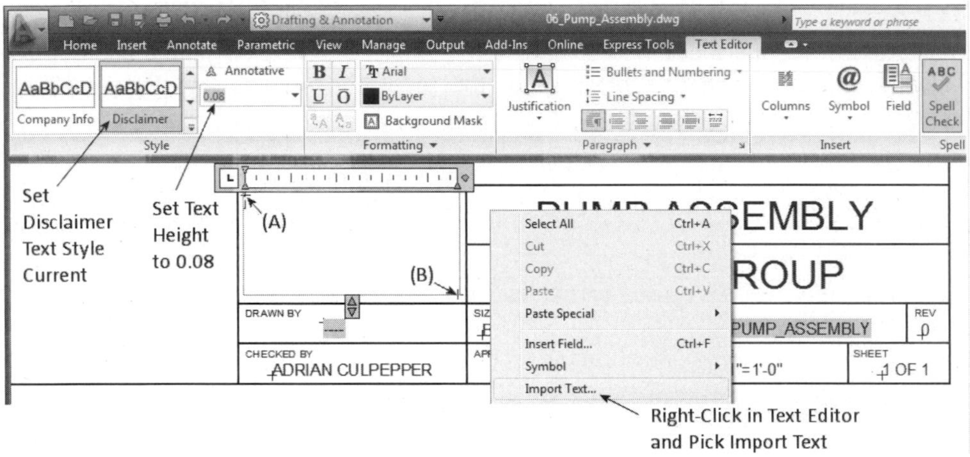

FIGURE 6.79

When the Select File dialog box appears, click the location of your exercise files and pick the file 06_Diaclaimer.txt, as shown in Figure 6.80 (Left). Since this information was already created in an application outside AutoCAD, it will be imported into the Multiline Text Editor. Once the TXT filed is opened, click the Close Text Editor button to place the text in the title block, as shown in Figure 6.80 (Right).

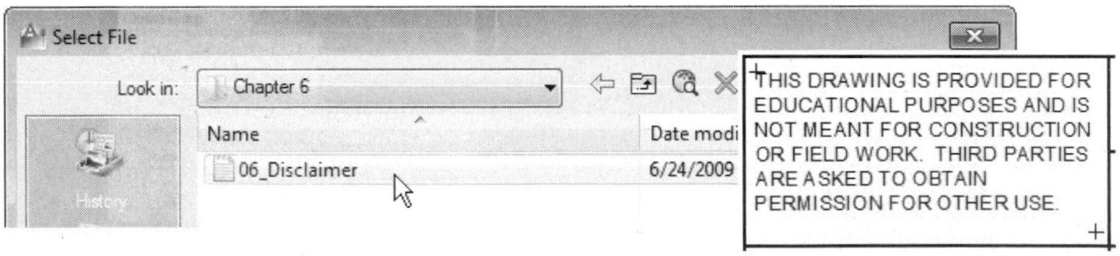

FIGURE 6.80

STEP 10

Before placing the last text object, zoom to the extents of the drawing. Then, click the General Notes text style located in the Annotate tab of the Ribbon, as shown in Figure 6.81 (Left). Use the MTEXT command to create a text box to the right of the Pump Assembly, as shown in Figure 6.81 (Right). This will be used to hold a series of general notes in multiline text format.

A Command: MT *(For MTEXT)*

Current text style: "General Notes" Text height: 0.20 Annotative: No

Specify first corner: *(Pick a point at "A")*

Specify opposite corner or [Height/Justify/Line spacing/Rotation/Style/Width/Columns]: *(Pick a point at "B")*

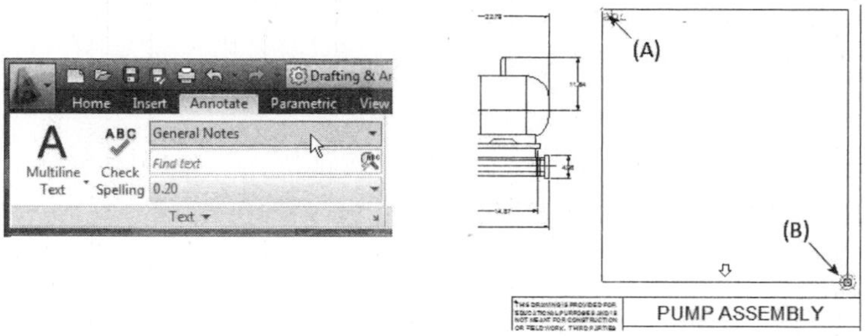

FIGURE 6.81

When the Multiline Text Editor appears, right-click and select the Import Text ... button, as shown in Figure 6.82 (Left). Then click the location of your exercise files and select the file 06_GENERAL_NOTES.rtf, as shown in Figure 6.82 (Right). This file will not be available until you set the "Files of type" to RTF at the bottom of the Select File dialog box. This RTF (rich text format) file was created outside AutoCAD in Microsoft Word.

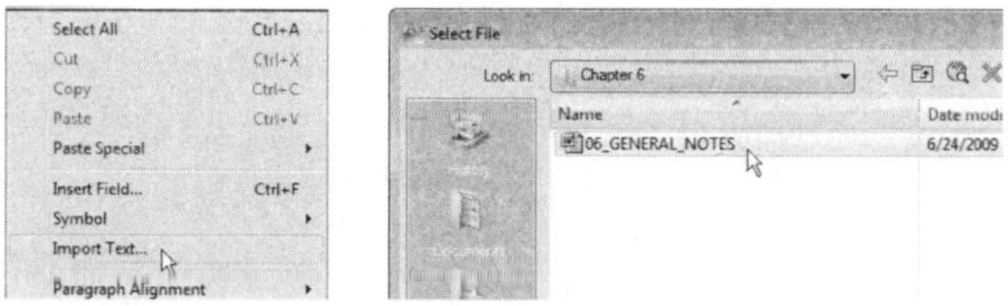

FIGURE 6.82

STEP 11

The results of this operation are displayed in Figure 6.83 (Left). However, the text size and font are not as expected. This is due to the fact that when files are imported in .RTF format, the original format of the text is kept.

While still inside the Text Editor Ribbon, highlight all the general notes and change the size to 0.10 and the font to Arial, as shown in Figure 6.83 (Right).

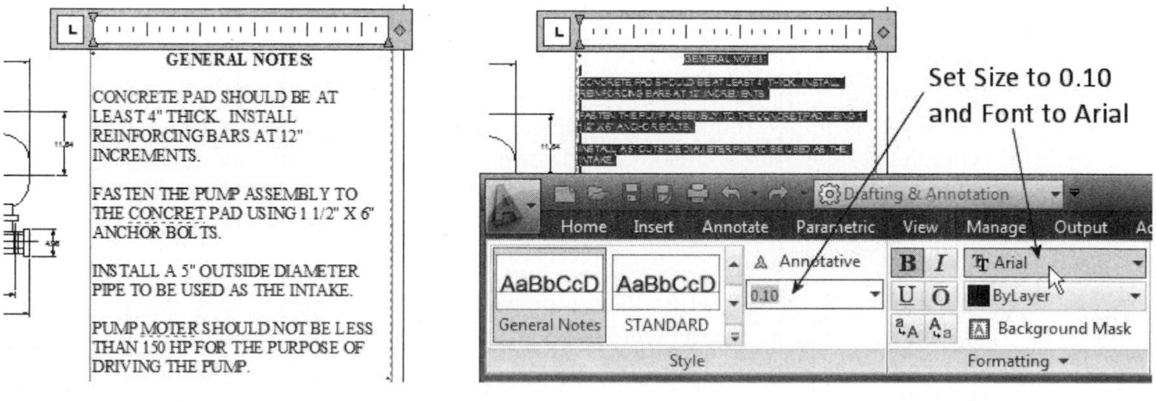

FIGURE 6.83

One more item needs to be taken care of before you leave the Text Editor Ribbon. Highlight all the text directly under the heading of GENERAL NOTES, as shown in Figure 6.84. Click the Numbered button. This creates a numbered list based on all highlighted lines of text, as shown in Figure 6.84. When finished, click the Close Text Editor button to return to the drawing.

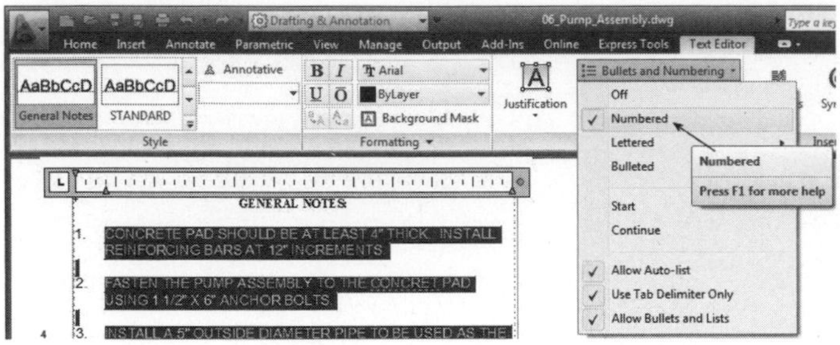

FIGURE 6.84

STEP 12

In this step you will check the spelling for the general notes that were just imported into the drawing. Activate the SPELL command and pick "Selected objects" from the Where to check field. Pick the Select objects button and click the multiline text object that holds the six general notes. When the Check Spelling dialog box reappears, click the Start button as shown in Figure 6.85; use the Change button to make the corrections to the words CONCRETE, MOTOR, and PROPERLY.

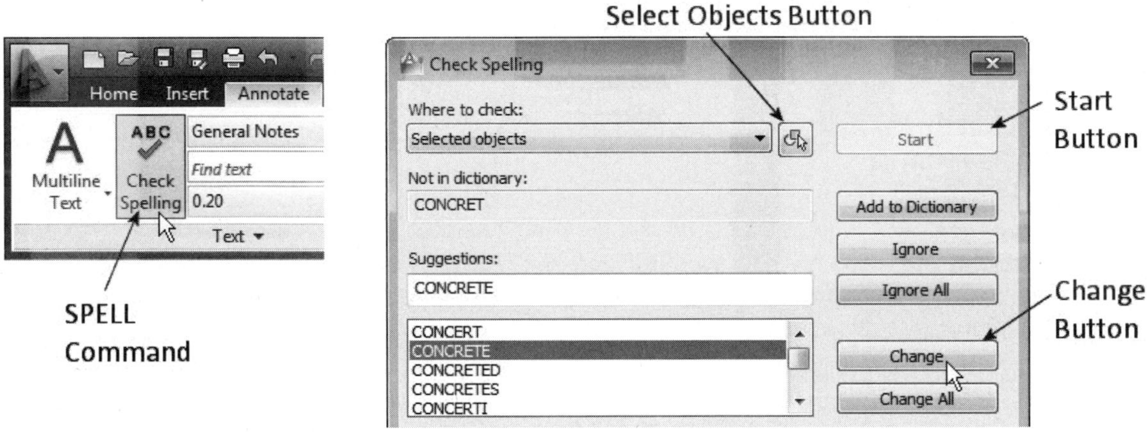

FIGURE 6.85

STEP 13

In a previous step, you created a field for the author of the drawing. However, this field displayed as empty. You will now complete the title block by filling in the information and completing the Author field. Click Drawing Properties, found under the Application Menu, as shown in Figure 6.86 (Left). When the 06_Pump Assembly.dwg Properties dialog box appears, enter your name in the Author text box, as shown in Figure 6.86 (Center).

Once you click OK and return to your drawing, the Author field located in the title block is still not reflecting your name. Click Update Fields, found under the Insert Tab of the Ribbon, as shown in Figure 6.86 (Right), and select the field to place your name in the title block.

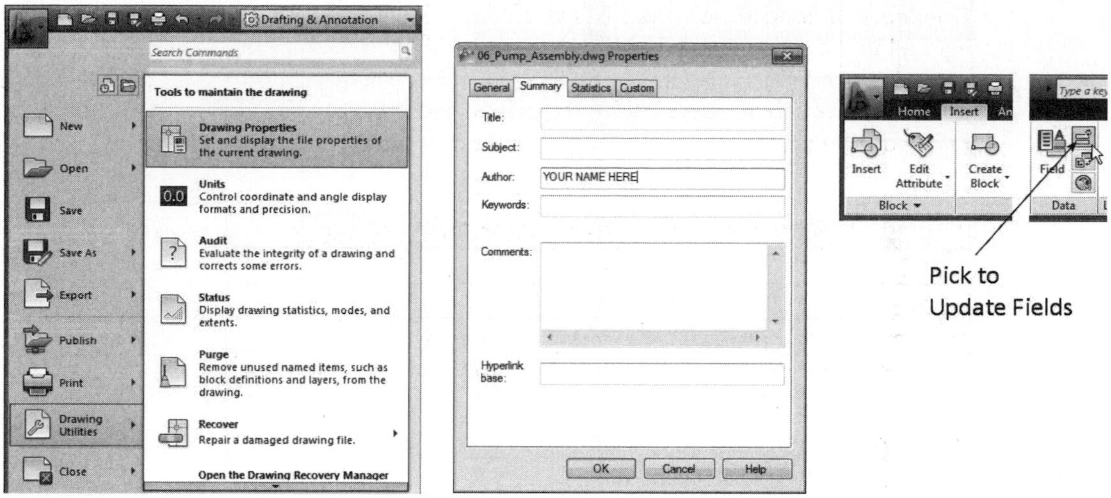

FIGURE 6.86

STEP 14

The completed title block and general notes area are displayed in Figure 6.87. Use the Layer Properties Manger palette to turn off or freeze the Points layer to remove the points from the display.

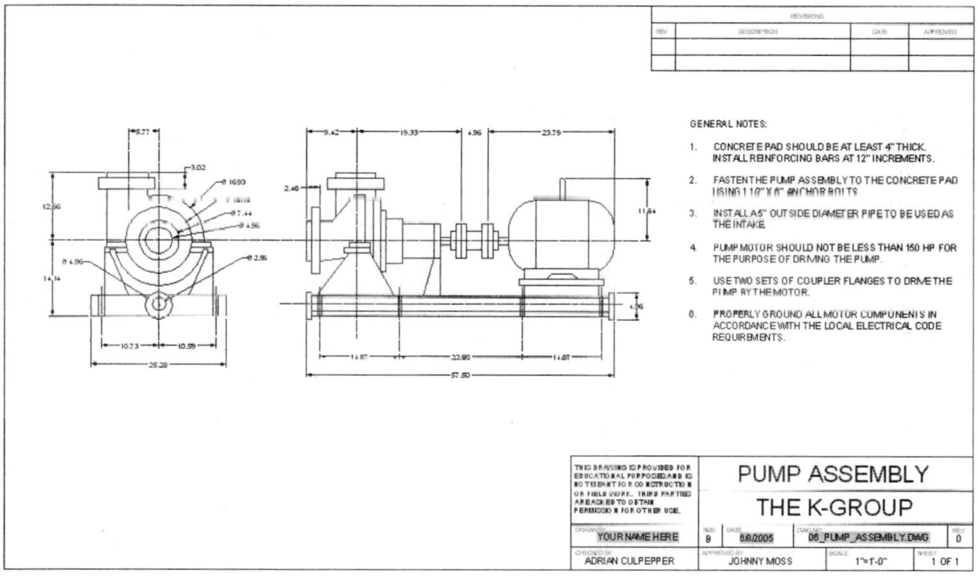

FIGURE 6.87

TUTORIAL EXERCISE: 06_TABLE.DWG

DOOR SCHEDULE					
MARK	WIDTH	HEIGHT	THICKNESS	MATERIAL	NOTES
1	9'-0"	8'-0"	1 3/4"	VINYL	~ ~
2	9'-0"	8'-0"	1 3/4"	VINYL	~ ~
3	9'-0"	8'-0"	1 3/4"	VINYL	~ ~
4	3'-0"	6-8"	1 3/8"	HCW	~ ~
5	2'-4"	6-8"	1 3/8"	HCW	~ ~
6	1'-6"	6-8"	1 3/8"	HCW	~ ~
7	5-0"	6-8"	1 3/4"	METAL CLAD	FRENCH DOOR
8	5'-0"	6-8"	1 3/4"	METAL CLAD	FRENCH DOOR
9	2-8"	6-8"	1 3/4"	METAL	SCREENED DOOR

FIGURE 6.88

Purpose

This tutorial is designed to create a new table style and then a new table object, as shown in Figure 6.88.

System Settings

In this tutorial you will use an existing drawing file called 06_Table. Two text styles are already created for this exercise; namely Architectural and Title.

Layers

A Table layer is already created for this drawing.

Suggested Commands

You will open a drawing called 06_Table and create a new table style called Door Schedule. Changes will be made to the Data, Heading, and Data fields through the Table Style dialog box. Next, you will create a new table consisting of six columns and eleven rows. Information related to the door schedule will be entered in at the Title, Heading, and Data fields.

STEP 1

Open the existing drawing file 06_Table. Activate the Text Style dialog box (STYLE command) and notice the two text styles already created, as shown in Figure 6.89 (Left), namely, Title and Architectural. The Title text style uses the Arial font while Architectural uses the CityBlueprint font. Close the dialog box.

Next, activate the Table Style dialog box (TABLESTYLE Command) and click the New button to create a new table style called Door Schedule, as shown in Figure 6.89 (Right).

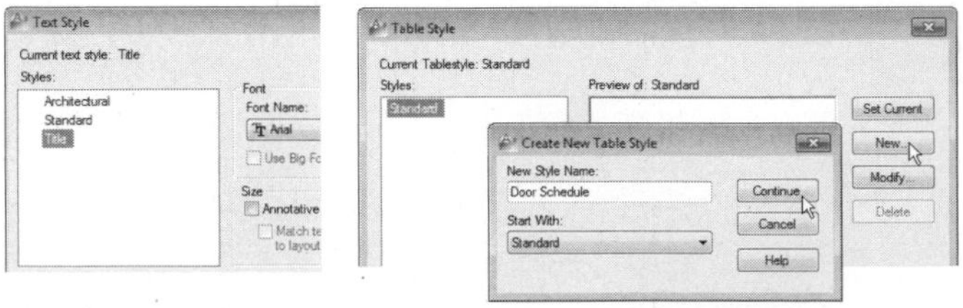

FIGURE 6.89

STEP 2

Clicking the Continue button located in Figure 6.89 takes you to the main New Table Style dialog box. Use the information in Figure 6.90 to change the properties with the Cell styles set to Data.

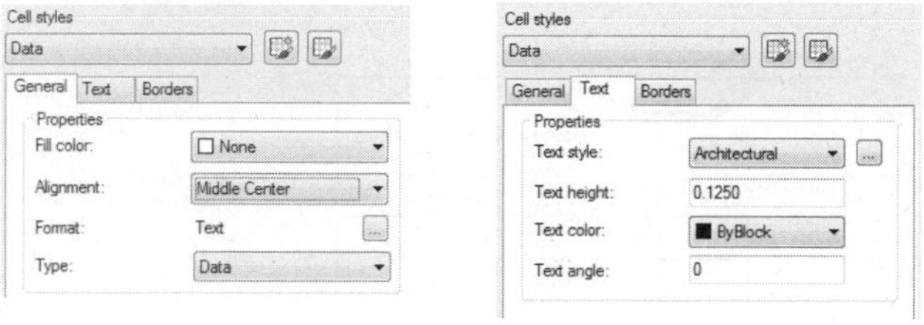

FIGURE 6.90

Use the information in Figure 6.91 to change the properties under the Header cell style.

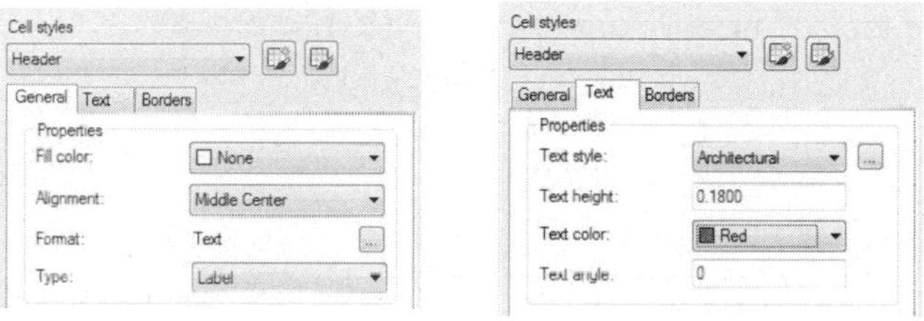

FIGURE 6.91

Use the information in Figure 6.92 to change the properties under the Title cell style.

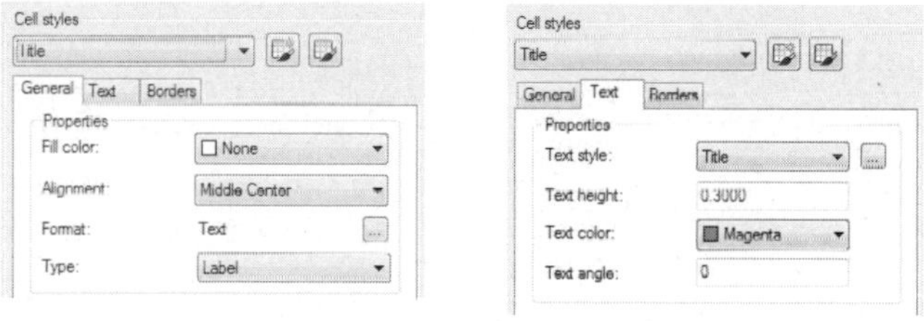

FIGURE 6.92

STEP 3

Click the OK button to return to the Table Style dialog box. Verify the Door Schedule style is the current table style, as shown in Figure 6.93 (you can double-click any table style name from the list to make it current). When finished, close the Table Style dialog box.

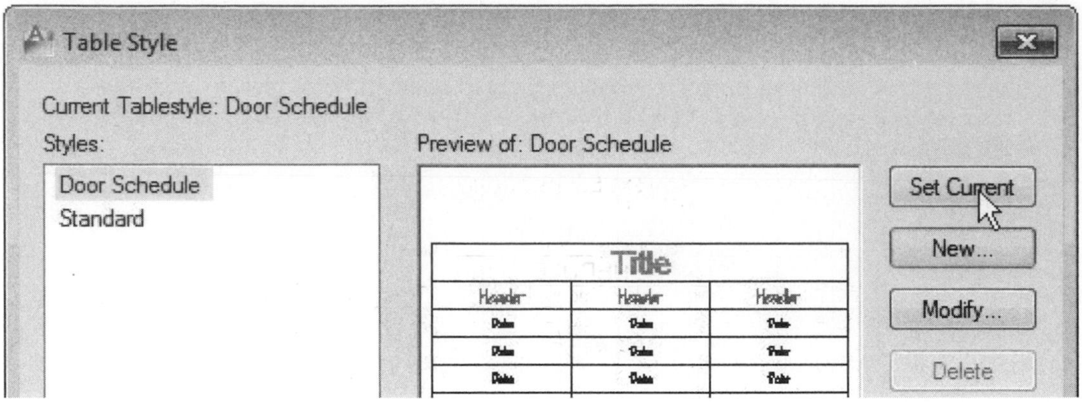

FIGURE 6.93

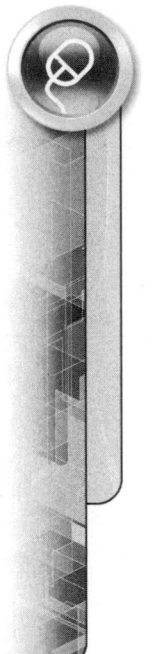

STEP 4

Activate the TABLE command and while in the Insert Table dialog box, create a table consisting of six columns and eleven rows using the new Door Schedule table style. Click the OK button to place the table in a convenient location on your screen.

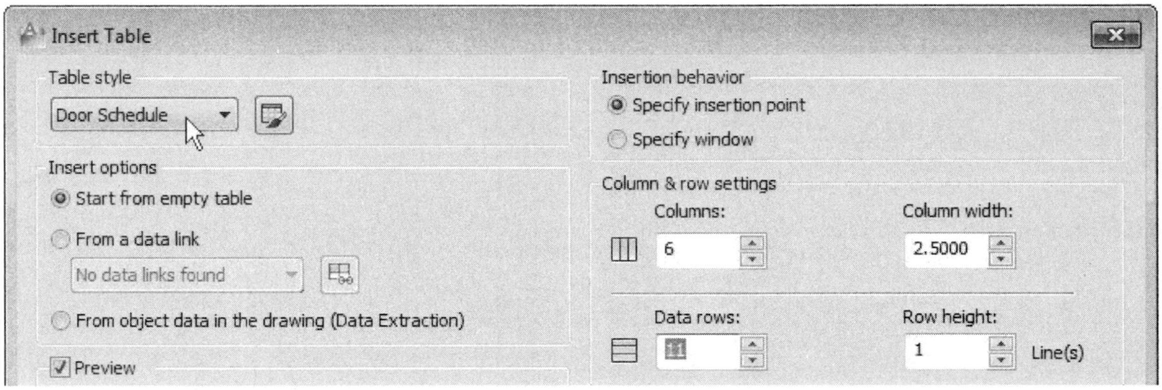

FIGURE 6.94

STEP 5

When the Text Editor Ribbon and table appear, fill out the title of the table as DOOR SCHEDULE, as shown in Figure 6.95. Although we could use the TAB key to move to additional cells to enter data, go ahead and click the Close Text Editor button to exit and place the table.

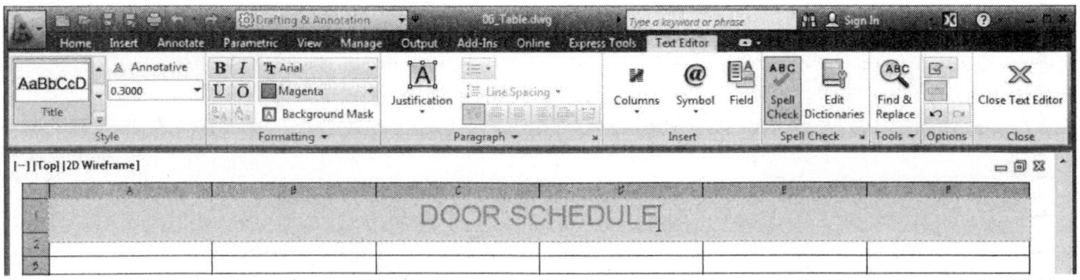

FIGURE 6.95

STEP 6

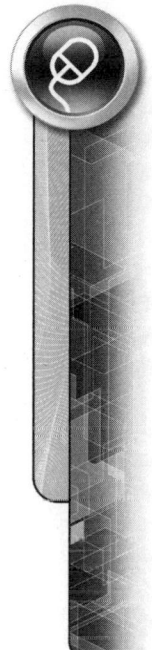

To reenter the table for adding data, double-click inside the first column on the next row, and when the Text Editor Ribbon appears, label this cell MARK. Then press the TAB key to advance to the next column header and label this WIDTH. Repeat this procedure for the next series of column headers, namely HEIGHT, THICKNESS, MATERIAL, and NOTES, as shown in Figure 6.96. Click the Close Text Editor button.

DOOR SCHEDULE					
MARK	WIDTH	HEIGHT	THICKNESS	MATERIAL	NOTES

FIGURE 6.96

STEP 7

As the table appears to be stretched too long, click the table and pick the arrow grip in the upper-right corner. Then stretch the table to the left and pick an appropriate point, which shortens the width of all columns, as shown in Figure 6.97.

DOOR SCHEDULE					
MARK	WIDTH	HEIGHT	THICKNESS	MATERIAL	NOTES

FIGURE 6.97

STEP 8

Complete the table by filling in all rows and columns that deal with the data. Utilize the Autofill technique for incrementing the numbers under the Mark column. To save time, you can also utilize Autofill to copy the "like data" values. The final results are shown in Figure 6.98.

DOOR SCHEDULE					
MARK	WIDTH	HEIGHT	THICKNESS	MATERIAL	NOTES
1	9'-0"	8'-0"	1 3/4"	VINYL	~ ~
2	9'-0"	8'-0"	1 3/4"	VINYL	~ ~
3	9'-0"	8'-0"	1 3/4"	VINYL	~ ~
4	3'-0"	6-8"	1 3/8"	HCW	~ ~
5	2'-4"	6-8"	1 3/8"	HCW	~ ~
6	1'-6"	6-8"	1 3/8"	HCW	~ ~
7	5-0"	6-8"	1 3/4"	METAL CLAD	FRENCH DOOR
8	5'-0"	6-8"	1 3/4"	METAL CLAD	FRENCH DOOR
9	2-8"	6-8"	1 3/4"	METAL	SCREENED DOOR

FIGURE 6.98

TUTORIAL EXERCISE: 06_TABLE_SUMMATIONS.DWG

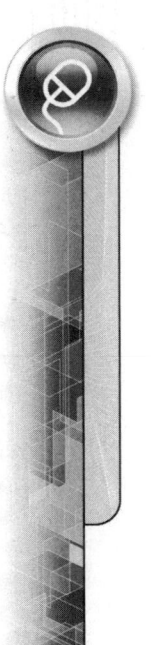

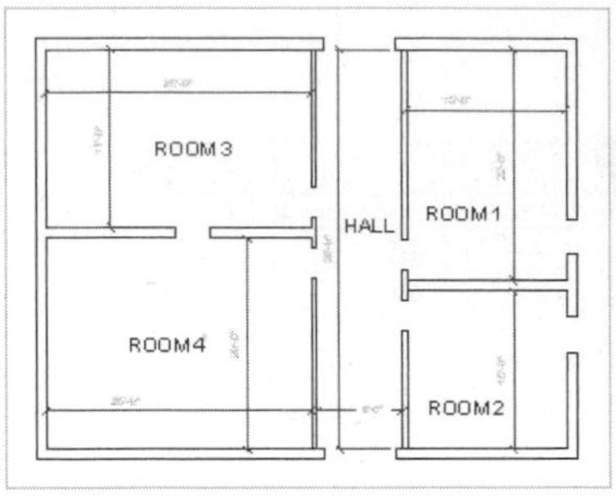

ROOM AREA CHART			
ROOM NUMBER	LENGTH	WIDTH	ROOM AREA (SQ FT)
HALLWAY	38	8	
ROOM 1	22	15	
ROOM 2	15	15	
ROOM 3	17	25	
ROOM 4	20	25	
		TOTAL AREA	

FIGURE 6.99

Purpose

This tutorial is designed to perform mathematical calculations and summations on a table.

System Settings

No special system settings need to be made for this drawing.

Layers

Layers are already created for this drawing.

Suggested Commands

In this exercise you will utilize the drawing file 06_Table_Summations, as shown in Figure 6.99. Opening the drawing displays a Floor Plan layout, which includes an image and a table. Basic mathematical calculations such as addition, subtraction, multiplication, and division can be made on information in a table. Order of operations can also be used on the basic math calculations. You can also perform sum, average, and counting operations. You will insert formulas into this table that will calculate the area of each room. You will then perform a summation on all room areas.

STEP 1

Open the drawing file 06_Table_Summations and zoom in to the table. You will first assign a formula that calculates the area of the Hallway. Click the cell D3 that will hold the area of the Hallway; the Table Cell tab of the Ribbon displays as shown in Figure 6.100. Click the Formula button, and when the menu displays, as shown in Figure 6.100, click Equation.

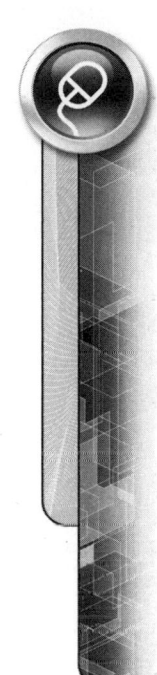

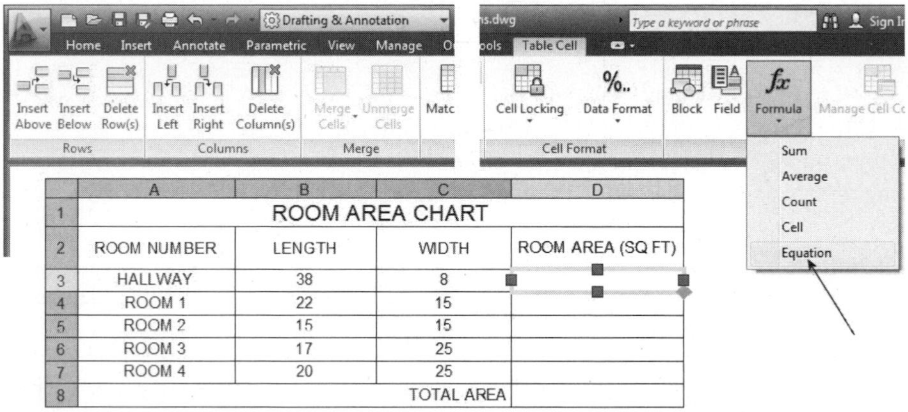

FIGURE 6.100

STEP 2

Clicking Equation in the previous step displays the Text Editor Ribbon, as shown in Figure 6.101. In the highlighted cell, you will need to multiply the Hallway length (38) by the width (8). You should not enter these numbers directly into the equation cell. Rather, enter the column and row number of each value. For example, the value 38 is identified by B3 and the value 8 is identified by C3. Enter these two cell identifiers separated by the multiplication symbol (*). Your display should appear similar to Figure 6.101.

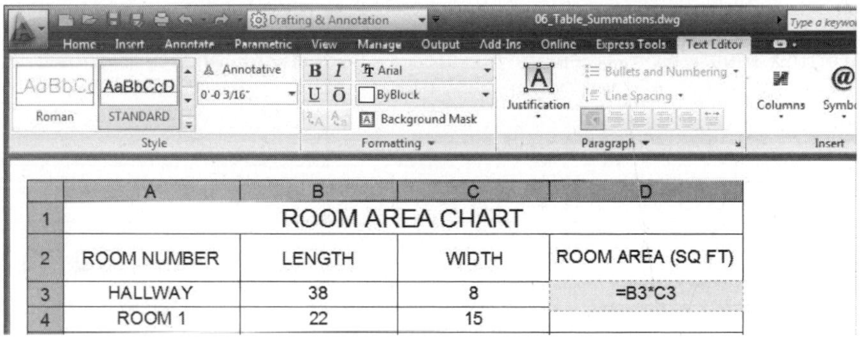

FIGURE 6.101

STEP 3

Closing the text editor creates a field in the cell that calculates the Hallway area, as shown in Figure 6.102. This field is easily recognized by the typical gray text background. Placing information as a field provides flexibility. If the information in the length and width cells changes, the field information will automatically recalculate itself.

ROOM AREA CHART			
ROOM NUMBER	LENGTH	WIDTH	ROOM AREA (SQ FT)
HALLWAY	38	8	304
ROOM 1	22	15	
ROOM 2	15	15	

FIGURE 6.102

STEP 4

Follow Steps 1 and 2 to calculate the area of Room 1. Be careful to use the correct cell-identifying letters and numbers for this area calculation, as shown in Figure 6.103.

	A	B	C	D
1	ROOM AREA CHART			
2	ROOM NUMBER	LENGTH	WIDTH	ROOM AREA (SQ FT)
3	HALLWAY	38	8	304
4	ROOM 1	22	15	=B4*C4
5	ROOM 2	15	15	

FIGURE 6.103

NOTE An equation can be started by simply clicking on a cell and typing the equal "=" sign.

STEP 5

Continue inserting equations to calculate the area of Rooms 2, 3, and 4, as shown in Figure 6.104.

ROOM AREA CHART			
ROOM NUMBER	LENGTH	WIDTH	ROOM AREA (SQ FT)
HALLWAY	38	8	304
ROOM 1	22	15	330
ROOM 2	15	15	225
ROOM 3	17	25	425
ROOM 4	20	25	500
		TOTAL AREA	

FIGURE 6.104

STEP 6

With the individual room areas calculated, you will now perform a summation that will add all current area fields. First, click on cell D8 in which the summation will be performed. When the Table Cell tab appears, click the Formula button followed by Sum, as shown in Figure 6.105.

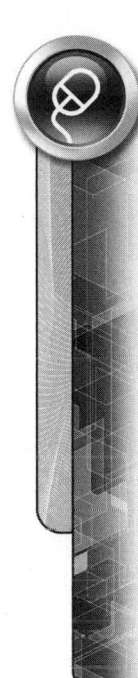

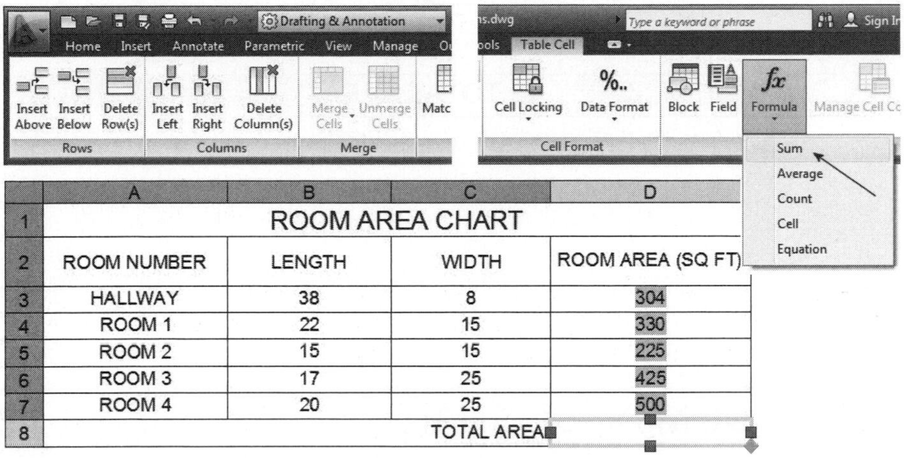

	A	B	C	D
1	ROOM AREA CHART			
2	ROOM NUMBER	LENGTH	WIDTH	ROOM AREA (SQ FT)
3	HALLWAY	38	8	304
4	ROOM 1	22	15	330
5	ROOM 2	15	15	225
6	ROOM 3	17	25	425
7	ROOM 4	20	25	500
8			TOTAL AREA	

FIGURE 6.105

STEP 7

After selecting Sum from the menu in the previous step, you will be prompted for the following:

- Select first corner of table cell range.
- Answer this prompt by picking a point inside the upper-left corner of the cell that displays an area of 304. You will be prompted a second time.
- Select second corner of table cell range:
- Answer this prompt by picking a point inside the lower-right corner of the cell that displays an area of 500.
- This action identifies the cells to perform the summations.

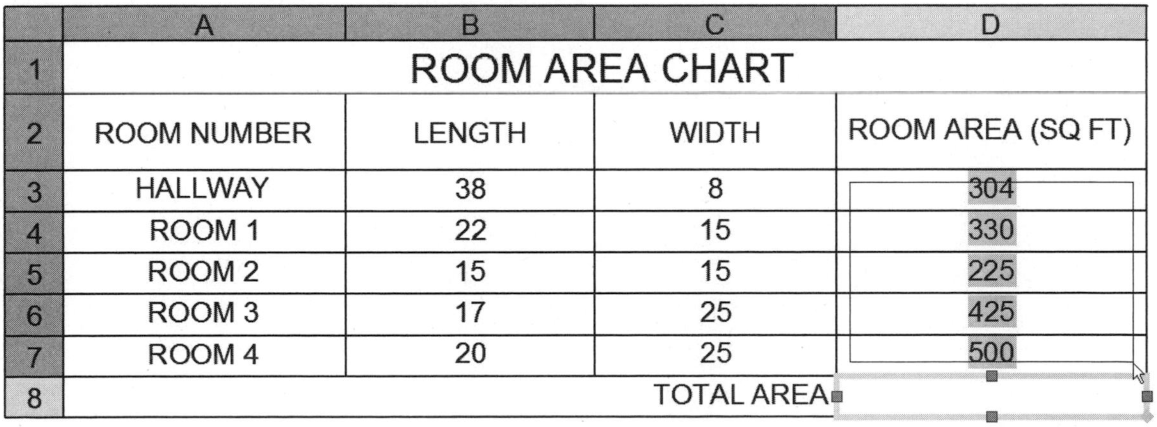

	A	B	C	D
1	ROOM AREA CHART			
2	ROOM NUMBER	LENGTH	WIDTH	ROOM AREA (SQ FT)
3	HALLWAY	38	8	304
4	ROOM 1	22	15	330
5	ROOM 2	15	15	225
6	ROOM 3	17	25	425
7	ROOM 4	20	25	500
8			TOTAL AREA	

FIGURE 6.106

STEP 8

After you select the second corner of the table cell range in the previous step, the equation is automatically created to add the values of cells D3 through D7, as shown in Figure 6.107.

5	ROOM 2	15	15	225
6	ROOM 3	17	25	425
7	ROOM 4	20	25	500
8			TOTAL AREA	=Sum(D3:D7)

FIGURE 6.107

STEP 9

Closing the text editor displays the final results, as shown in Figure 6.108. This completes this tutorial exercise.

ROOM AREA CHART			
ROOM NUMBER	LENGTH	WIDTH	ROOM AREA (SQ FT)
HALLWAY	38	8	304
ROOM 1	22	15	330
ROOM 2	15	15	225
ROOM 3	17	25	425
ROOM 4	20	25	500
		TOTAL AREA	1784

FIGURE 6.108

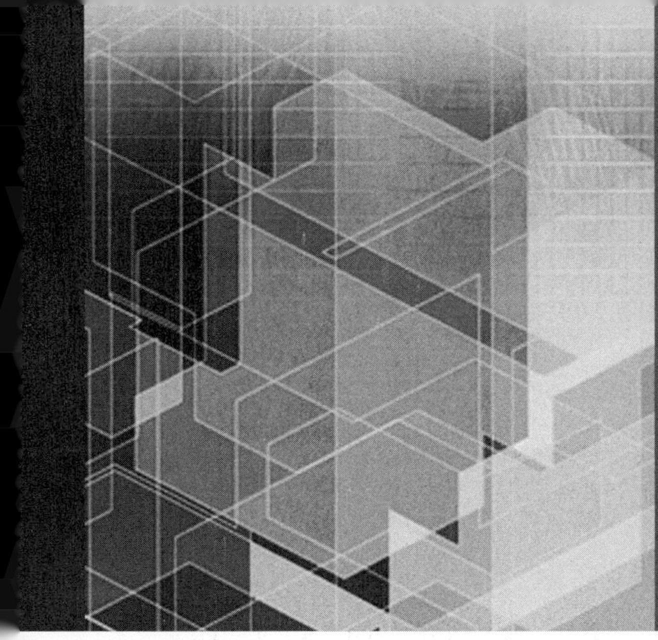

CHAPTER 7

Object Grips and Changing the Properties of Objects

This chapter begins with a discussion of what grips are and how they are used to edit portions of your drawing. Various Try It! exercises are available to practice using grips. This chapter examines a number of methods used to modify objects. These methods are different from the editing commands learned in Chapter 4. Sometimes you will want to change the properties of objects such as layer, color, and even linetype. This is easily accomplished through the Properties Palette. The Match Properties command will also be introduced in this chapter. This powerful command allows you to select a source object and have the properties of the source transferred to other objects that you pick.

USING OBJECT GRIPS

An alternate method of editing is to use object grips. A grip is a small box appearing at key object locations such as the endpoints and midpoints of lines and arcs or the center and quadrants of circles. Once grips are selected, the object may be stretched, moved, rotated, scaled, or mirrored. Grips are at times referred to as visual Object Snaps because your cursor automatically snaps to all grips displayed along an object.

TRY IT!

Open the drawing file called 07_Grip Objects. While in the Command: prompt, click on each object type displayed in Figure 7.1 to activate the grips. Examine the grip locations on each object.

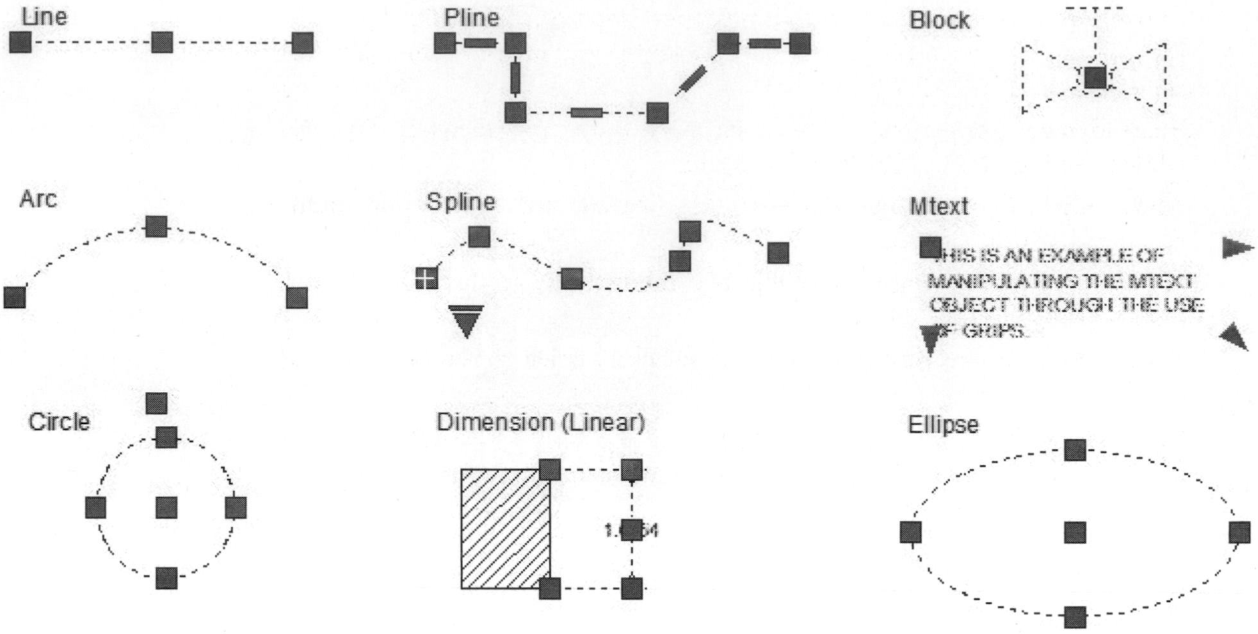

FIGURE 7.1

Changing the settings, color, and size of grips is accomplished under the Selection tab of the Options dialog box. Right-clicking on a blank part of your screen and choosing Options… from the menu, as shown in Figure 7.1 (Left), displays the main Options dialog box. Click on the Selection tab to display the grip settings, as shown in Figure 7.2 (Center).

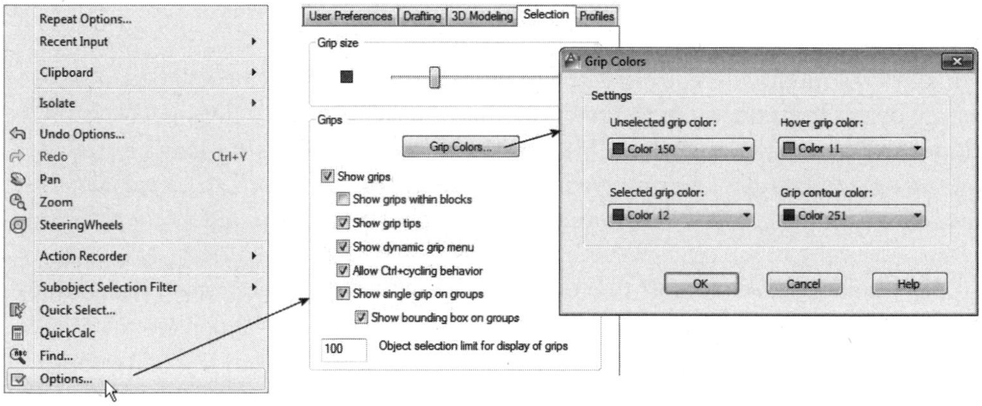

FIGURE 7.2

The grip settings are explained as follows:

Grip size—Use the Grip Size area to move a slider bar to the left to make the grip smaller or to the right to make the grip larger.

Show grips—By default, grips are enabled; a check in the Show grips box means that grips will display when you select an object.

Show grips within blocks—Normally a single grip is placed at the insertion point when a block is selected. Check the Show grips within blocks box if you want grips to be displayed along with all individual objects that make up the block.

Show grip tips—Some custom objects support grip tips. This controls the display of these tips when your cursor hovers over the grip on a custom object that supports grip tips.

Show dynamic grip menu—When checked, a dynamic menu is displayed while hovering over a multi-functional grip.

Allow Ctrl+Cycling behavior—When checked, ctrl-cycling for multi-functional grips is allowed.

Show single grips on groups—Objects combined in a group (GROUP command) display a single grip

Show bounding box on groups—When checked, object groups are enclosed by a bounding box

Object selection limit for display of grips—By default, this value is set to 100. This means that when you select less than 100 objects, grips will automatically appear on the highlighted objects. When you select more than 100 objects, the display of grips is suppressed.

Grip Colors button—Displays the Grip Colors dialog box.

Unselected grip color—Controls the color for an unselected grip. It is good practice to change the Unselected grip color to a light color if you are using a black screen background.

Selected grip color—Controls the color for a grip that is selected.

Hover grip color—Controls the color of an unselected grip when your cursor pauses over it.

Grip contour color—Controls the color for a grip contour.

OBJECT GRIP MODES

Figure 7.3 (Left) shows a typical display of grips. When an object is first selected, the object highlights and the square grips are displayed in the default color of blue. In this example of a line, the grips appear at the endpoints and midpoint. The entire object is subject to the many grip edit modes. As you hover (pause your cursor) over a grip, the grip turns a light reddish color. If over one of the endpoint grips, you will also see a dynamic menu appear, as shown in Figure 7.3 (Center). This menu provides various editing options, depending on the type of object and grip that you hover over. When one of the grips is selected, it becomes hot (active) and it turns a dark red color by default, as shown in Figure 7.3 (Right). With an endpoint grip hot, moving your cursor will stretch the line as the grip is moved toward a new location. This is the result of the stretch editing mode being active. If dynamic input is turned on, distance and angle fields are provided to assist in making changes to this line. Once a grip is selected, the following prompts appear in the Command prompt area:

```
** STRETCH **

Specify stretch point or [Base point/Copy/Undo/eXit]:
(Press the SPACEBAR)

** MOVE **

Specify move point or [Base point/Copy/Undo/eXit]: (Press
the SPACEBAR)
```

** ROTATE **

Specify rotation angle or [Base point/Copy/Undo/Reference/eXit]: *(Press* Spacebar*)*

** SCALE **

Specify scale factor or [Base point/Copy/Undo/Reference/eXit]: *(Press* Spacebar*)*

** MIRROR **

Specify second point or [Base point/Copy/Undo/eXit]: *(Press* Spacebar *to begin STRETCH mode again or enter* **X** *to exit grip mode)*

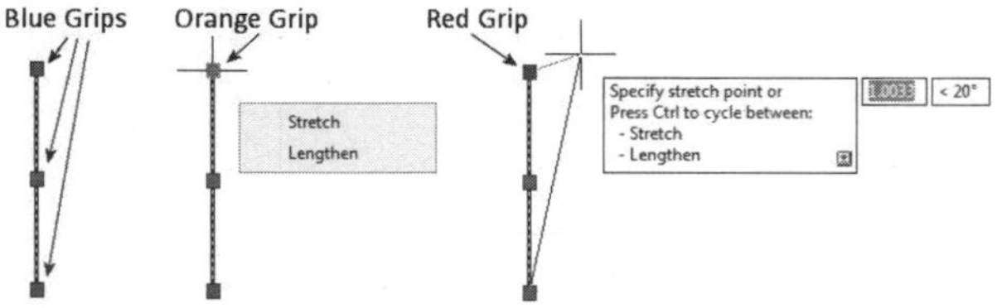

FIGURE 7.3

To move from one edit command mode to another, press the SPACEBAR. Once an editing operation is completed, pressing ESC removes the highlight and removes the grips from the object. Figure 7.4 shows various examples of each editing mode.

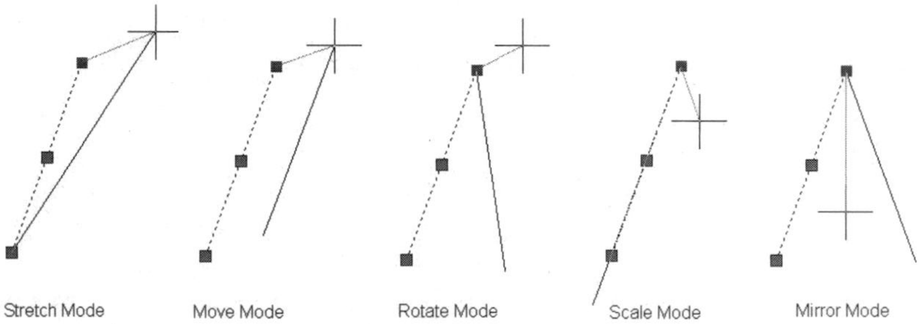

FIGURE 7.4

In the case of using grips with elliptical arcs and multiline text, directional arrows display when grips appear. Use these arrows to guide you through the various directions in which the grips can be stretched, as shown in Figure 7.5.

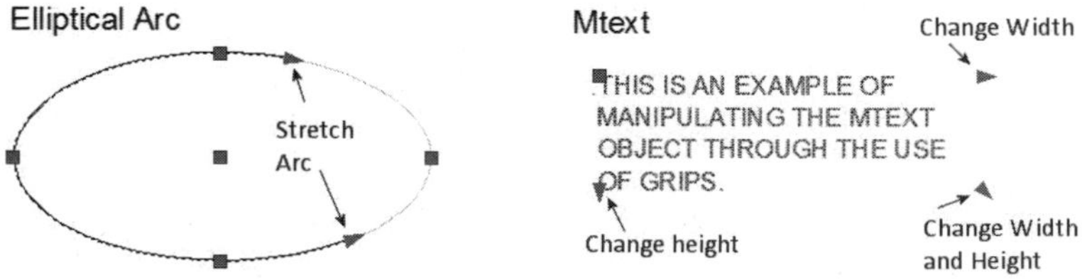

FIGURE 7.5

ACTIVATING THE GRIP SHORTCUT MENU

In Figure 7.6, a horizontal line has been selected and grips appear. The rightmost endpoint grip has been selected. Rather than using the SPACEBAR to scroll through the various grip modes, click the right mouse button. Notice that a shortcut menu on grips appears. This provides an easier way of navigating from one grip mode to another.

TRY IT!

Open the drawing file 07_Grip Shortcut. Click on any object, pick on a grip, and then press the right mouse button to activate the grip shortcut menu.

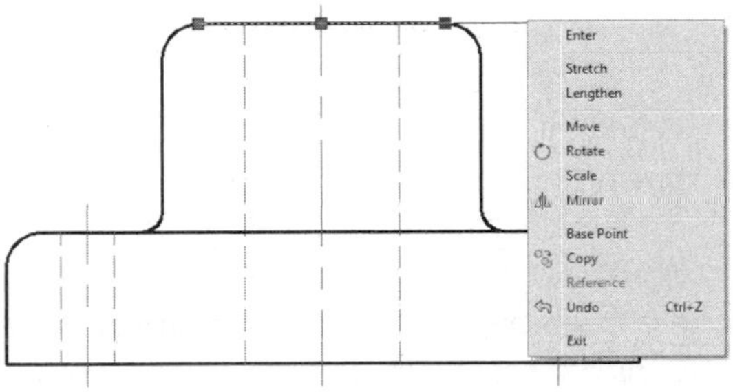

FIGURE 7.6

Using the Grip—STRETCH Mode

The STRETCH mode of grips operates similar to the normal STRETCH command. Use STRETCH mode to move an object or group of objects and have the results mend themselves similar to Figure 7.7. The line segments "A" and "B" are both too long by two units. To decrease these line segments by two units, use the STRETCH mode by selecting lines "A," "B," and "C" at the Command prompt. Next, while holding down SHIFT, select the grips "D," "E," and "F." This selects multiple grips and turns the grips red and ready for the stretch operation. Release SHIFT and pick the grip at "E" again. The stretch mode appears in the Command prompt area. The last selected grip is considered the base point. Moving your cursor to the left and entering a value of 2.00 stretches the three highlighted grip objects to the left a distance of two units. To remove the object highlight and grips, press ESC at the Command prompt.

TRY IT!

Open the drawing file 07_Grip Stretch. Use the illustration in Figure 7.7 and the prompt sequence below to perform this task.

Command: *(Select the three dashed lines labeled "A," "B," and "C" as shown in Figure 7.7. Then, while holding down* SHIFT, *select the grips at "D," "E," and "F." Release the* SHIFT *key and pick the grip at "E" again)*

STRETCH

Specify stretch point or [Base point/Copy/Undo/eXit]: *(Turn on Ortho or Polar mode, move your cursor to the left, and type* 2*)*

Command: *(Press* ESC *to remove the object highlight and grips)*

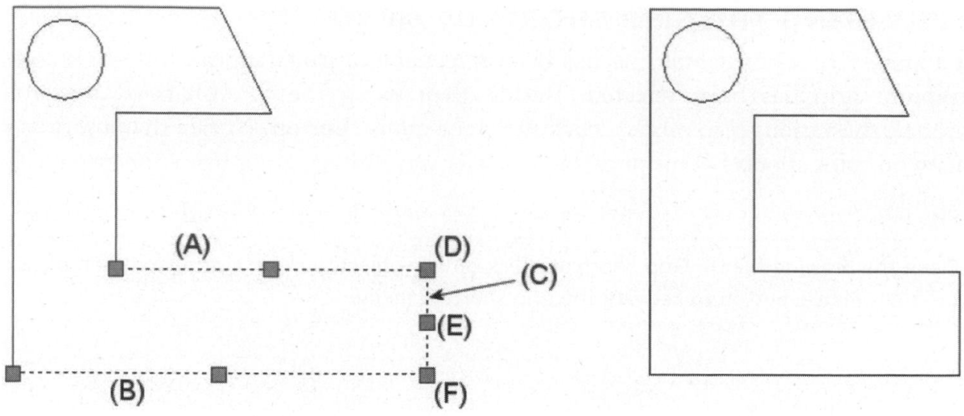

FIGURE 7.7

Using the Grip—SCALE Mode

Using the scale mode of object grips allows an object to be uniformly scaled in both the X and Y directions. This means that a circle, such as the one shown in Figure 7.8, cannot be converted to an ellipse through different X and Y values. As the grip is selected, any cursor movement drags the scale factor until a point is marked where the object will be scaled to that factor. The hot grip will serve as the base point for the scaling operation. Figure 7.8 and the following prompt illustrate the use of an absolute value to perform the scaling operation of half the circle's normal size.

TRY IT!

Open the drawing file 07_Grip Scale or continue modifying the previous Try It! exercise. Use the illustration in Figure 7.8 and the prompt sequence below to perform this task.

Command: *(Select the circle to display grips, and then select the grip at the center of the circle at "A." Press the SPACEBAR until the SCALE mode appears at the bottom of the prompt line or press the right mouse button to activate the grip shortcut menu to choose Scale)*

SCALE

Specify scale factor or [Base point/Copy/Undo/Reference/eXit]: **0.50**

Command: *(Press ESC to remove the object highlight and grips)*

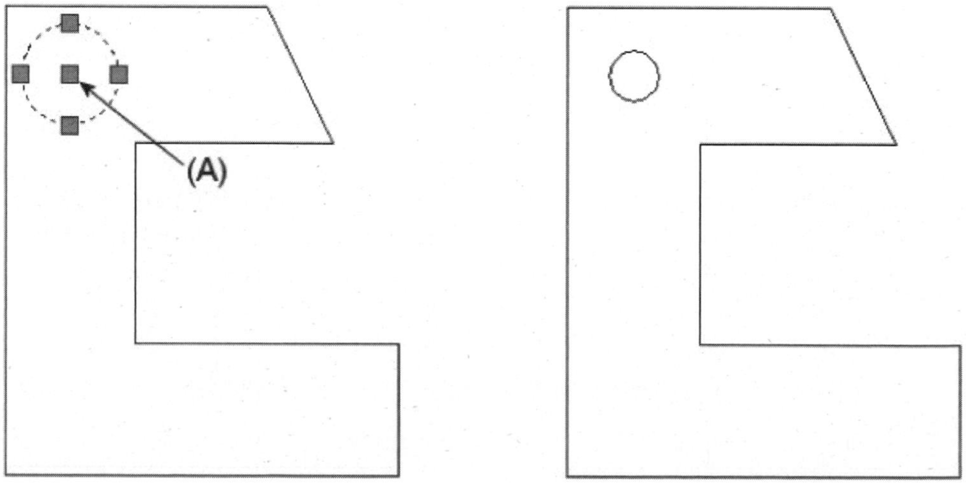

FIGURE 7.8

Using the Grip—MOVE Mode/Copy Option

The Multiple Copy option of the MOVE mode is demonstrated with the circle shown in Figure 7.9 (Left). The circle is copied using Direct Distance mode at distances 2.50 and 5.00. This Multiple Copy option is actually disguised under the command options of object grips.

First, select the circle to display the grips at the circle center and quadrants, and then select the center grip. Use the SPACEBAR or shortcut menu to select the MOVE mode. Issue the Copy option within the MOVE mode to be placed in Multiple MOVE mode, as shown in Figure 7.9.

TRY IT!

Open the drawing file 07_Grip Copy or continue modifying the previous Try It! exercise. Use the illustration in the middle of Figure 7.9 and the prompt sequence below to perform this task.

Command: *(Select the circle to activate the grips at the center and quadrants; select the grip at the center of the circle at "A." Then press the* SPACEBAR *until the MOVE mode appears at the bottom of the prompt line or press the right mouse button to activate the grip shortcut menu to choose Move)*

MOVE

Specify move point or [Base point/Copy/Undo/eXit]: C *(For Copy)*

MOVE *(multiple)*

Specify move point or [Base point/Copy/Undo/eXit]: *(Move your cursor straight down and enter a value of* 2.50*)*

MOVE *(multiple)*

Specify move point or [Base point/Copy/Undo/eXit]: *(Move your cursor straight down and enter a value of* 5.00*)*

MOVE *(multiple)*

Specify move point or [Base point/Copy/Undo/eXit]: X *(For exit)*

Command: *(Press* ESC *to remove the object highlight and grips)*

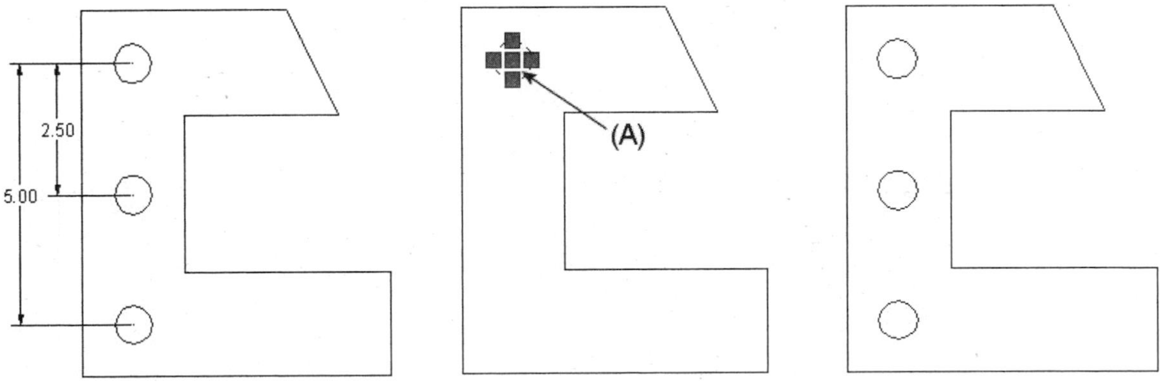

FIGURE 7.9

Using the Grip—MIRROR Mode

Use the grip—MIRROR mode to flip an object along a mirror line similar to the one used in the regular MIRROR command. Following the default prompts for the MIRROR mode performs the mirror operation but does not produce a copy of the original. If the original object needs to be saved during the mirror operation, use the Copy option. Once the mode and options have been selected, locate a base point and a second point to perform the mirror operation. By default the hot grip serves as the base point. The Base point option allows you to place your mirror line at any convenient location. See Figure 7.10.

TRY IT!

Open the drawing file 07_Grip Mirror or continue modifying the previous Try It! exercise. Use the illustration in Figure 7.10 and the prompt sequence below to perform this task.

Command: *(Select the circle in Figure 7.10 to enable grips, and then select the grip at the center of the circle. Press the* SPACEBAR *until the MIRROR mode appears at the bottom of the prompt line or press the right mouse button to activate the grip shortcut menu to choose Mirror)*

MIRROR

Specify second point or [Base point/Copy/Undo/eXit]: C *(For Copy)*

MIRROR *(multiple)*

Specify second point or [Base point/Copy/Undo/eXit]: B *(For Base Point)*

Base point: **Mid**

of *(Pick the midpoint at "A")*

MIRROR *(multiple)*

Specify second point or [Base point/Copy/Undo/eXit]: *(Move your cursor straight up and pick a point)*

MIRROR *(multiple)*

Specify second point or [Base point/Copy/Undo/eXit]: X *(For Exit)*

Command: *(Press* ESC *to remove the object highlight and grips)*

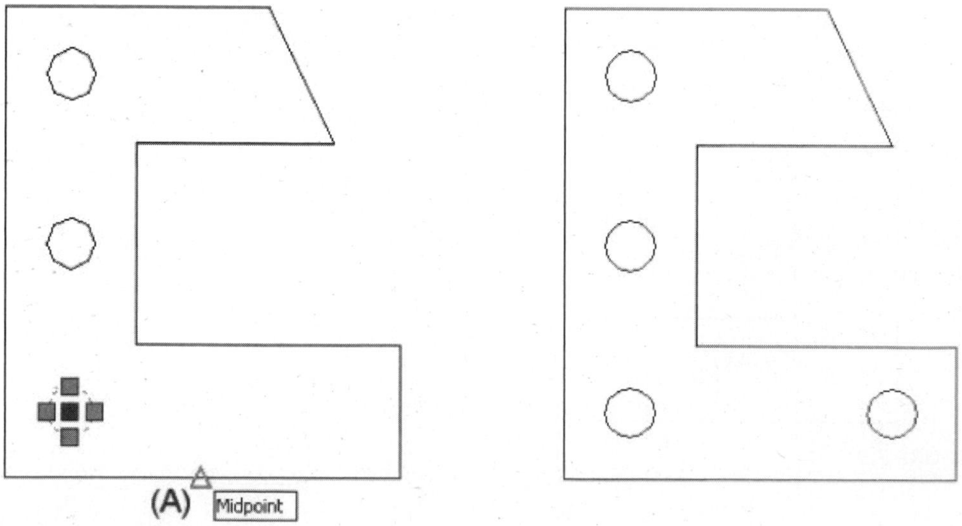

FIGURE 7.10

Using the Grip—ROTATE Mode

Numerous grips may be selected with window or crossing boxes. At the Command prompt, pick a blank part of the screen; this should activate implied windowing. Picking up or below and to the right of the previous point places you in Window selection mode; picking up or below and to the left of the previous point places you in Crossing selection mode. Using either method, select all of the objects shown in Figure 7.11. Selecting the lower-left grip and using the SPACEBAR or shortcut grip menu to advance to the ROTATE mode allows all objects to be rotated at a defined angle in relation to the previously selected grip.

TRY IT!

Open the drawing file 07_Grip Rotate or continue modifying the previous Try It! exercise. Use the illustration in Figure 7.11 and the prompt sequence below to perform this task.

Command: *(Pick near "X," then near "Y" to create a window selection set and enable all grips in all objects. Select the grip at the lower-left corner of the object. Then press the* SPACEBAR *until the ROTATE mode appears at the bottom of the prompt line or click the right mouse button to activate the grip shortcut menu to choose Rotate)*

ROTATE

Specify rotation angle or [Base point/Copy/Undo/Reference/eXit]: 30

Command: *(Press ESC to remove the object highlight and grips)*

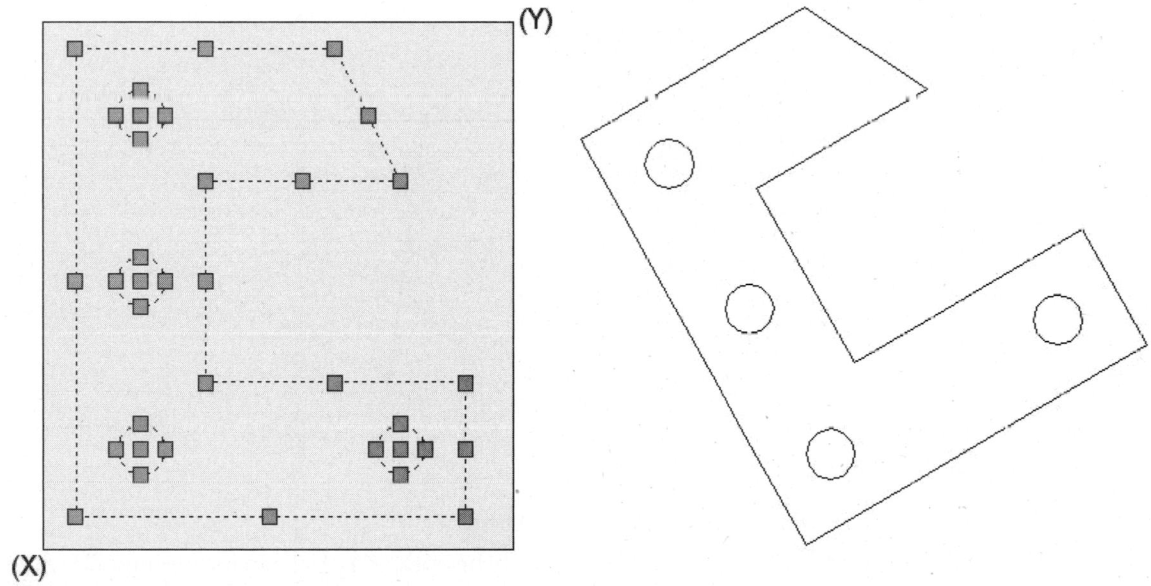

FIGURE 7.11

Using the Grip—Multiple ROTATE Mode

With object grips, you may use rotate mode and copy option to rotate and copy an object. Figure 7.12 illustrates a line that needs to be rotated and copied at a 40° angle. With a positive angle, the direction of the rotation is counterclockwise.

Selecting the line in Figure 7.12 (Left) enables grips located at the endpoints and midpoint of the line. Select the grip at "B" in Figure 7.12 (Center). This hot grip serves as the base point and locates the vertex of the required angle. Distance and angle feedback will appear if dynamic input is turned on. Press the SPACEBAR or activate the grip shortcut menu to enter the ROTATE mode. Enter Multiple ROTATE mode by entering C for Copy when you are prompted in the following command sequence. Finally, enter a rotation angle of 40 to produce a copy of the original line segment at a 40° angle in the counterclockwise direction, as shown in Figure 7.12 (Right).

TRY IT!

Open the drawing file 07_Grip Multiple Rotate. Use the illustration in Figure 7.12 and the prompt sequence below to perform this task.

Command: *(Select line segment "A"; then select the grip at "B." Press the* SPACEBAR *until the ROTATE mode appears at the bottom of the prompt line or click the right mouse button to activate the grip shortcut menu to choose Rotate)*

ROTATE

Specify rotation angle or [Base point/Copy/Undo/Reference/eXit]: C *(For Copy)*

ROTATE *(multiple)*

Specify rotation angle or [Base point/Copy/Undo/Reference/eXit]: 40

Specify rotation angle or [Base point/Copy/Undo/Reference/eXit]: X *(For Exit)*

Command: *(Press* ESC *to remove the object highlight and grips)*

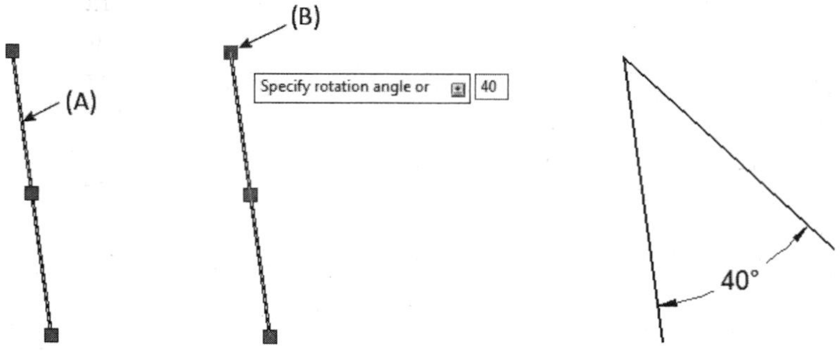

FIGURE 7.12

Using Grip Offset Snap for Rotations

All Multiple Copy modes within grips may be operated in a snap location mode while you hold down the CTRL key. Here's how it works. In Figure 7.13 (Left), the vertical centerline and circle are selected. The objects highlight and the grips appear. A copy of the selected objects is made at an angle of 45° using the appropriate mode and options. Additional copies can now be made, but by holding down the CTRL key they can be placed (snapped) at precise 45° increments.

TRY IT!

Open the drawing file 07_Grip Offset Snap Rotate. Use the illustration in Figure 7.13 and the prompt sequence below to perform this task.

Command: *(Select centerline segment "A" and circle "B"; and then select any one of the grips to activate the grip modes. Press the* SPACEBAR or use the shortcut menu to enter *the ROTATE mode)*

ROTATE

Specify rotation angle or [Base point/Copy/Undo/Reference/eXit]: **C** *(For Copy)*

ROTATE *(multiple)*

Specify rotation angle or [Base point/Copy/Undo/Reference/eXit]: **B** *(For Base Point)*

Base point: **Cen**

of *(Select the circle at "C" to snap to the center of the circle)*

ROTATE *(multiple)*

Specify rotation angle or [Base point/Copy/Undo/Reference/eXit]: **45**

Rather than enter another angle to rotate and copy the same objects, hold down the CTRL key, which places you in offset snap location mode. Moving the cursor snaps the selected objects to any increment of the angle just entered, namely 45°, as shown in Figure 7.13 (Center).

ROTATE *(multiple)*

Specify rotation angle or [Base point/Copy/Undo/Reference/eXit]: *(Hold down the* CTRL *key and move the circle and centerline until it snaps to the next 45° position shown in Figure 7.13 (center) - Pick)*

The Rotate Copy Snap Location mode allows you to create the circles illustrated in Figure 7.13 (Right) without the aid of the ARRAYPOLAR command. Since all angle values are 45°, continue holding down CTRL to snap to the next 45° location, and mark a point to place the next group of selected objects.

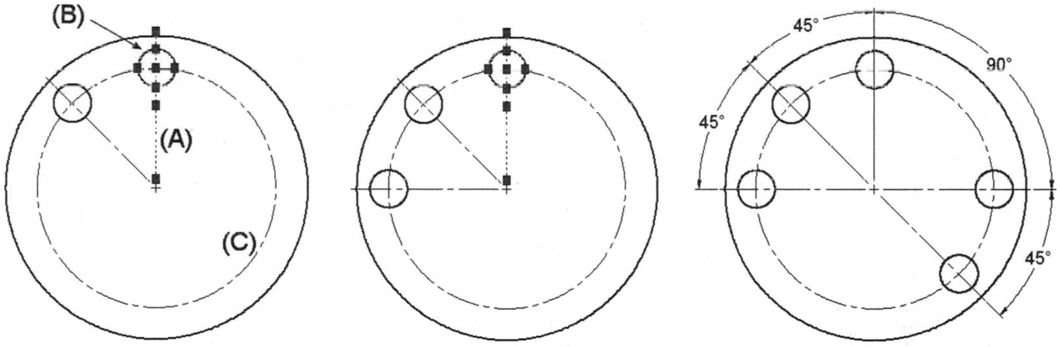

FIGURE 7.13

ROTATE *(multiple)*

Specify rotation angle or [Base point/Copy/Undo/Reference/eXit]: **X** *(For Exit)*

Command: *(Press* ESC *to remove the object highlight and grips)*

Using Grip Offset Snap for Moving

As with the previous example of using Offset Snap Locations for rotate mode, these same snap locations apply to MOVE mode. The illustration in Figure 7.14 (Left) shows two circles along with a common centerline. The circles and centerline are first selected, which highlights these three objects and activates the grips. The intent is to move and copy the selected objects at 2-unit increments.

TRY IT!

Open the drawing file 07_Grip Offset Snap Move. Use the illustration in Figure 7.14 and the prompt sequence below to perform this task.

Command: *(Select the two circles and centerline to activate the grips; select the grip at the midpoint of the centerline. Then press the* SPACEBAR *until the MOVE mode appears at the bottom of the prompt line or click the right mouse button to activate the grip shortcut menu to choose Move)*

MOVE

Specify move point or [Base point/Copy/Undo/eXit]: C *(For Copy)*

MOVE *(multiple)*

Specify move point or [Base point/Copy/Undo/eXit]: *(Move your cursor straight down and enter a value of* 2.00*)*

In the middle of Figure 7.14, instead of entering a new distance to create another copy of the circles and centerline, hold down CTRL and move the cursor down again to see the selected objects snap to the distance already specified.

MOVE *(multiple)*

Specify move point or [Base point/Copy/Undo/eXit]: *(Hold down the* CTRL *key and move the cursor down to have the selected objects snap to another 2.00-unit distance - Pick)*

MOVE *(multiple)*

Specify move point or [Base point/Copy/Undo/eXit]: X *(To exit)*

Command: *(Press* ESC *to remove the object highlight and grips)*

The illustration on Figure 7.14 (Right) shows the completed hole layout, the result of using the Offset Snap Location method of object grips.

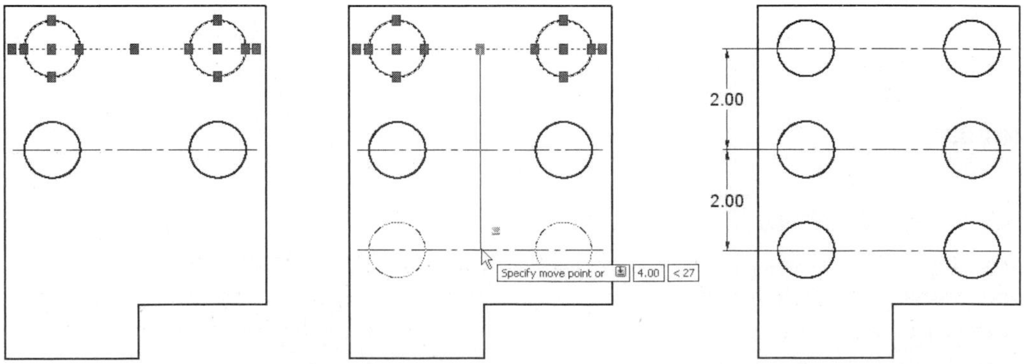

FIGURE 7.14

Grip Project—Lug.Dwg

Open the drawing file 07_Lug. Use the illustration in Figure 7.15 and the prompt sequence below to perform this task.

TRY IT!

Using the illustration in Figure 7.15 as a guide, make the following changes to the lug:

1. Use the Grip-Stretch mode to join the endpoint of the inclined line at (A) with the endpoint of the horizontal line also at (A).
2. Use the Grip-Stretch mode to join the endpoint of the inclined line at (B) with the endpoint of the horizontal line also at (B).
3. Use the Grip-Stretch mode, pick the 1.50 diameter dimension, make the grip located on the dimension text hot, and position the dimension text in the lower-right corner of the object.
4. Use the Grip-Scale mode, pick the 1.50 diameter dimension and the circle, make the grip located at the center of the circle hot, and use a scale factor of .50 to reduce the size of the circle by half. Notice that the dimension recalculates to reflect this change.
5. Use the Grip-Mirror mode, and pick the circle, 1.50 diameter dimension, two vertical lines, both fillets, and the horizontal line at (E). Pick one of the grips to make it hot. Use the midpoint of the long horizontal line of the object as the new base point. Turn Ortho or Polar on and move your cursor down to establish the mirror line.
6. Use the Grip-Stretch mode to increase the size of the 0.50 throat at (F) to a new distance of 1.25. Pick both fillets and the three line segments attached to the fillets. Hold down the SHIFT key and select the square grips on both fillets and the middle of the horizontal line. Release the SHIFT key and pick the grip at the midpoint of the horizontal line again. Move your cursor up (either Ortho or Polar mode needs to be turned on) and type a value of **0.75**.
7. The completed exercise is illustrated in Figure 7.15 (Right).

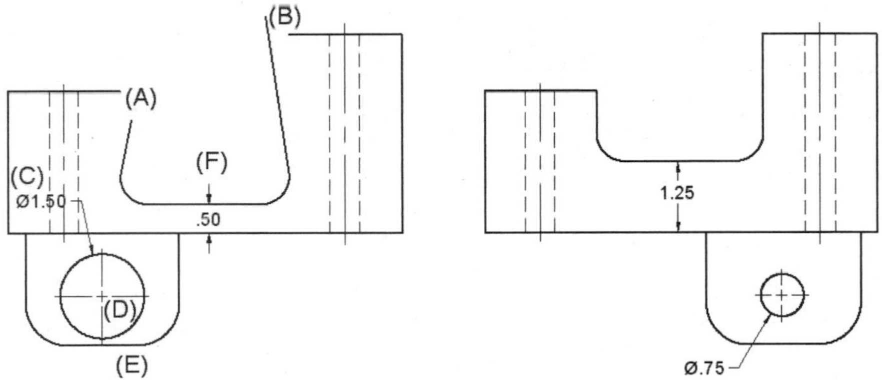

FIGURE 7.15

MULTI-FUNCTIONAL GRIPS

Additional modification capabilities are available for objects through multi-functional grips. Select an object and hover your cursor over one of the grips to display a dynamic menu with various options available. The options displayed will depend on the object type and specific grip selected.

Verify that Dynamic Input is activated on the Status Bar, then hover your cursor over the end grip of a line, as shown in Figure 7.16 (Left). You can select either Stretch or Lengthen option from the menu provided. Select the Stretch option and move your cursor to perform the standard stretch operation. With dynamic input activated you can specify the length and angle of the stretch. If you select the Lengthen option from the dynamic menu, the angle of the line will be fixed and only the length can be edited. Instead of selecting an option from the menu, you can simply pick the end grip of the line to make it active (hot). This automatically places you in the default stretch mode, as shown in Figure 7.16 (Right). Pressing the CTRL key while the grip is active allows you to cycle between the stretch and lengthen options.

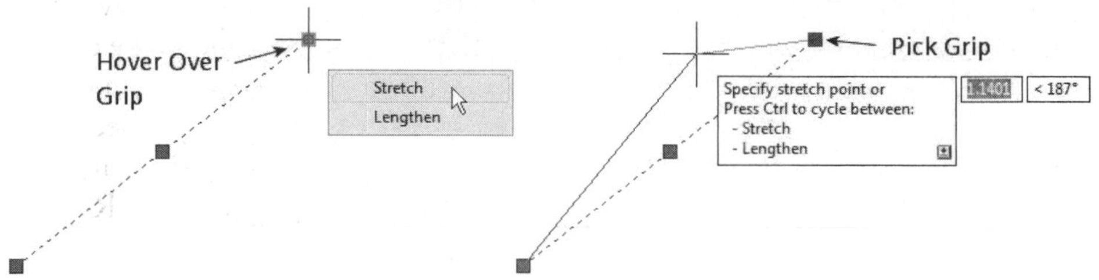

FIGURE 7.16

TRY IT!

Open the drawing file 07_Grip Multi-Functional. Select the sloped line to display its grips and hover your cursor over the upper grip, as shown in Figure 7.17 (Left). When the shortcut menu appears, select Lengthen. Move your cursor to the right, as shown in Figure 7.17 (Right). Enter a value of 1.00 and the line lengthens 1.00 unit. Notice that it is not necessary to specify an angle. However, had your cursor been positioned to the left, the line would have shortened by 1.00 unit.

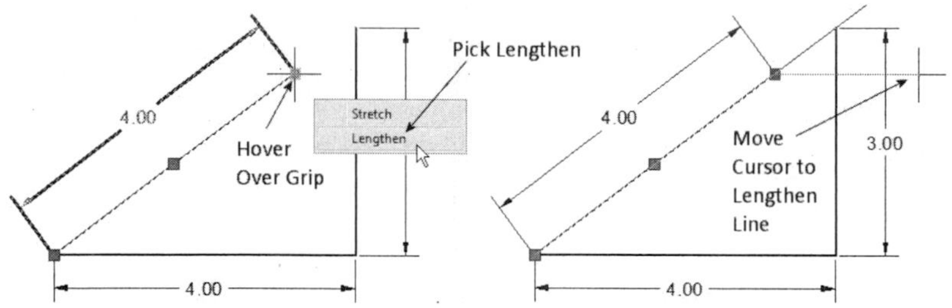

FIGURE 7.17

For the second part of this exercise, we will stretch a grip 4.00 units to the left. If necessary re-select the sloped line and then pick the upper grip. You are automatically placed in the default Stretch mode. Enter the following polar coordinate: @4 < 180. Press ENTER to stretch the grip 4.00 units to the left. The results are shown in Figure 7.18 (Right).

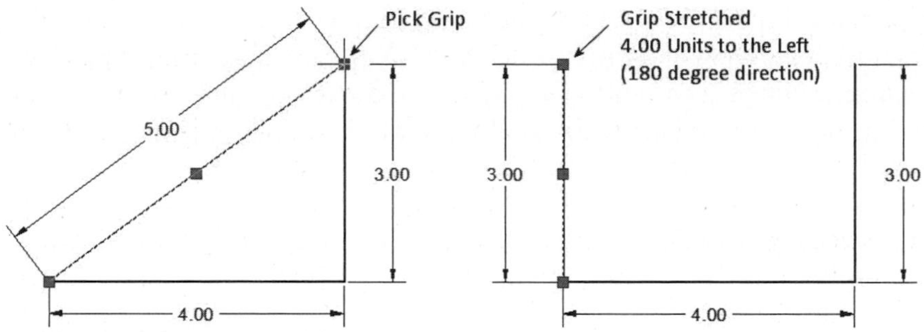

FIGURE 7.18

As mentioned earlier, various multi-functional grip editing options are available depending on the object and grip selected. Shown in Figure 7.19 (Left) is the option menu for the midpoint grip of a polyline segment. Selecting the Stretch option would allow you to stretch the polyline segment, as shown in Figure 7.19 (Left). Select Add Vertex to create a new polyline vertex, as shown in the middle image. You can even select Convert to Arc to round a polyline segment, as shown in Figure 7.19 (Right).

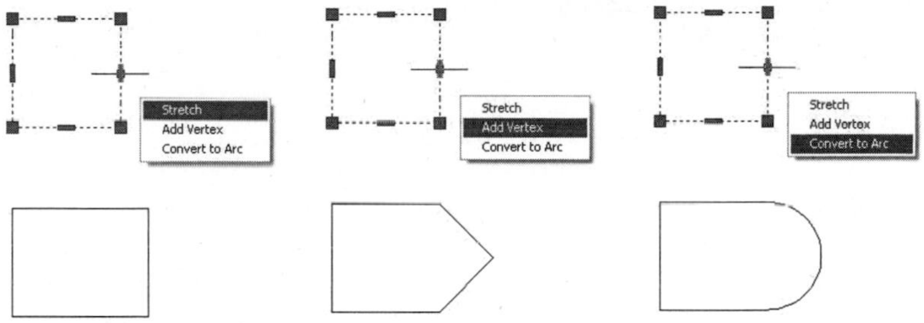

FIGURE 7.19

In Figure 7.20 (Left), a Remove Vertex option was selected. Multi-functional grip options can also be selected from a shortcut menu by picking an object, picking a grip to make it hot, and then right-clicking, as shown in Figure 7.20 (Right).

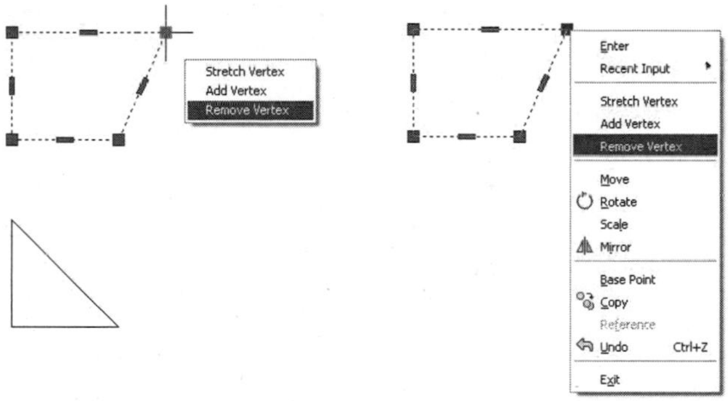

FIGURE 7.20

NOTE

Once a grip is hot, you can also cycle through multi-functional grip editing options by pressing the `CTRL` **key.**

MODIFYING THE PROPERTIES OF OBJECTS

At times, objects are drawn on the wrong layer, with the wrong color, or even in the wrong linetype. The lengths of line segments or the radius values of circles and arcs may be incorrect. Eliminating the need to erase these objects and reconstruct them to their correct specifications, a series of tools are available to modify the properties of these objects. One such tool is the PROPERTIES command which can be selected from the Ribbon, as shown in Figure 7.21 (Left). This will display the Properties Palette as shown in Figure 7.21 (Right).

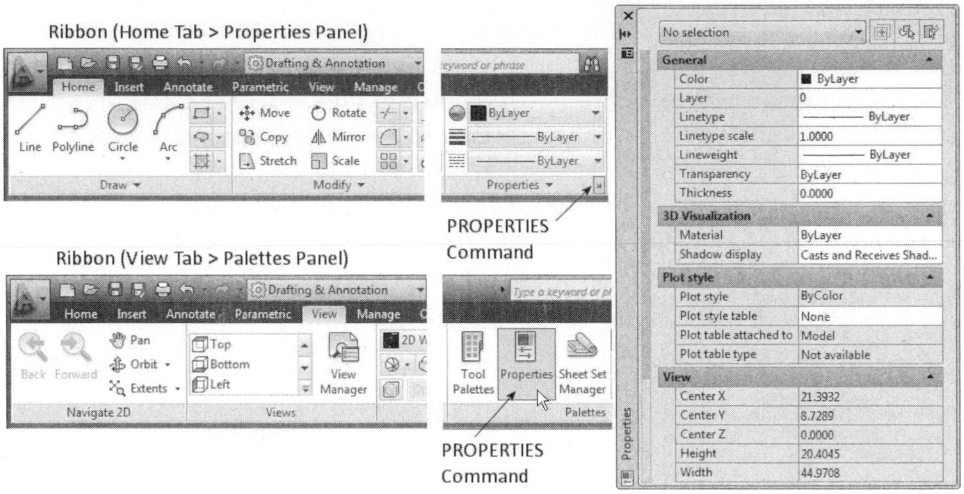

FIGURE 7.21

Illustrated in Figure 7.22 (Left) is a line segment that has been pre-selected, as shown by the highlighted appearance and the presence of grips in the figure.

Right-clicking and selecting Properties from the shortcut menu provides another method of displaying the Properties Palette, as shown in Figure 7.22 (Right). This palette displays information about the object already selected; in this case the information is about the line segment, which is identified at the top of the palette.

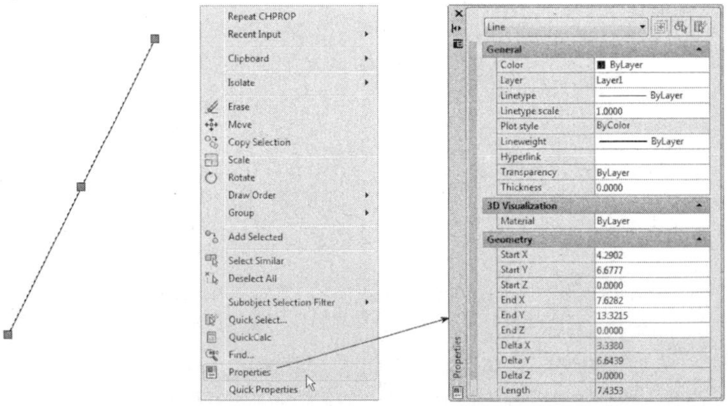

FIGURE 7.22

Certain edit boxes in the Properties Palette will provide a drop-down list of possible changes. In Figure 7.23 (Left), notice that the line was placed on the "0" layer. Clicking in the Layer edit box displays a down arrow, as shown in the middle image.

Clicking the down arrow displays all layers defined in the drawing as shown in the image on the right. Clicking on Object moves the line from the "0" to the "Object" layer.

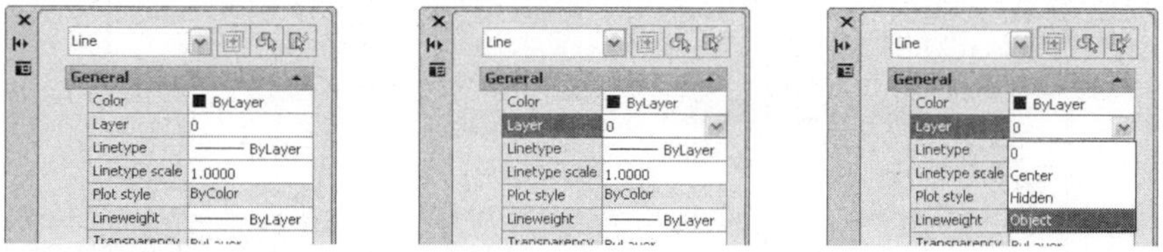

FIGURE 7.23

A number of options that control the Properties Palette are illustrated in Figure 7.24. Right-click on the Properties Palette title bar to display a shortcut menu. You can elect to move, size, or close the Properties Palette from this menu. You can also allow or prevent the Properties Palette from docking (shown docked on the left in Figure 7.24). Auto-hiding displays the Properties Palette when your cursor lies anywhere inside the palette. When your cursor moves off of the Properties Palette, the window collapses to display only the blue side strip as shown in the middle image. In the image on the right the transparency was set to 50%.

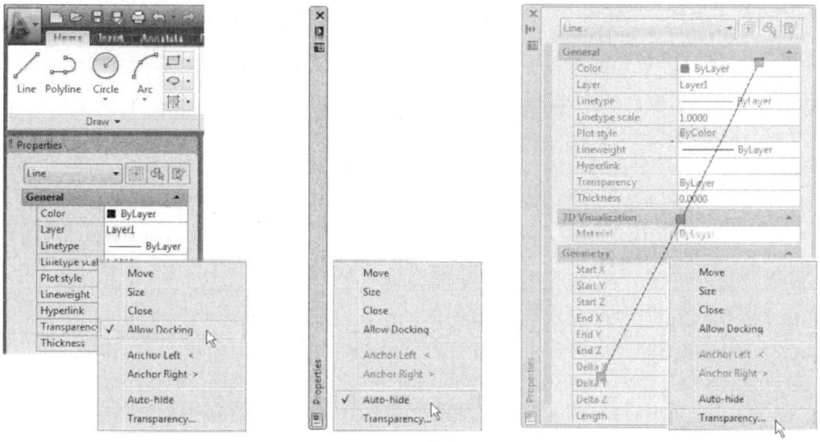

FIGURE 7.24

Three line segments are illustrated in Figure 7.25 and all three segments have been pre-selected, as shown by their highlighted appearance and the presence of grips.

Activating the PROPERTIES command through one of the methods previously discussed displays the Properties Palette, as shown in Figure 7.25. At the top of the palette, it identifies that three lines are selected. You can change the color, layer, linetype, and other general properties of all three lines. However, you are unable to enter the Delta X, Delta Y, and Delta Z length or angle information (these boxes are grayed out). It would prove to be impractical to change any of the geometry data. Whenever you select more than one of the same object, you can change the general properties of the object but not individual values that deal with the object's geometry.

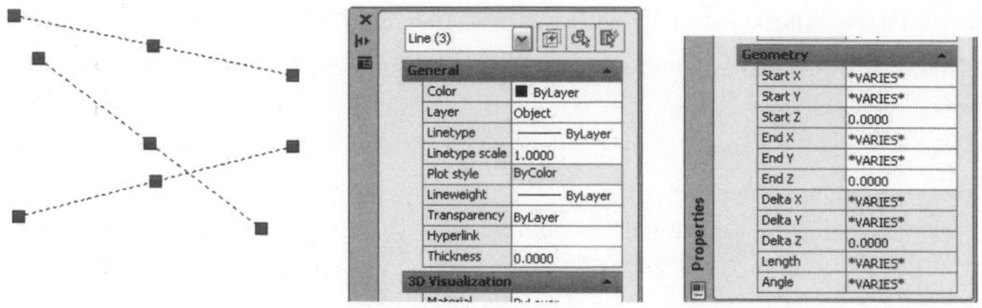

FIGURE 7.25

In Figure 7.26 (Left), an arc, circle, and line are pre-selected, as identified by the appearance of grips. Activating the Properties Palette shown in Figure 7.26 (Right) displays a number of object types at the top of the dialog box. You can click which object or group of objects to modify. With "All (3)" highlighted, you can change the general properties, such as layer and linetype, but not any geometry settings.

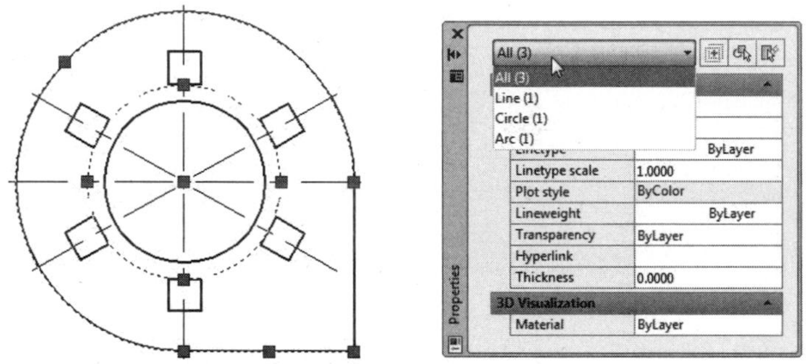

FIGURE 7.26

TRY IT!

Open the drawing file 07_Clutch Properties. What if you need to increase the radius of the circle to 1.25 units? Click on the inner circle and then activate the `PROPERTIES` command. Since only one object was selected, the full complement of general and geometry settings is present for you to modify. Click on the Radius field and change the current value to 1.25 units. Pressing `ENTER` automatically updates the other geometry settings in addition to the actual object in the drawing, as shown in Figure 7.27. When finished, dismiss this dialog box to return to the drawing.

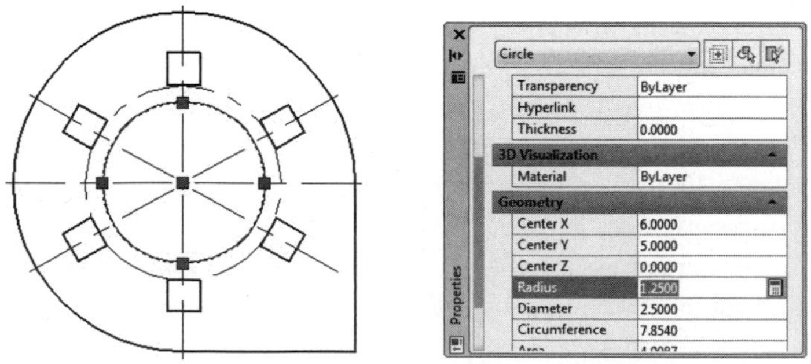

FIGURE 7.27

USING SELECTION TOOLS OF THE PROPERTIES PALETTE

Three buttons at the top of the Properties Palette are available to assist you in efficiently selecting items for modification through the palette. The first is the PICKADD button located in Figure 7.28. To see how it operates, follow the next exercise.

TRY IT!

Open the drawing file 07_PickAdd. A number of object types ranging from lines to text to multiline text and polylines with dimensions are displayed. Activate the Properties Palette and pick on one object. Notice that as it highlights, the grips appear, and information about the object is displayed in the Properties Palette. Suppose, however, that you want to list information about another object. You must first press ESC to deselect the original object. Now select a different object and this information is displayed in the Properties Palette. This time, select the PICKADD button in Figure 7.28 (Left). Notice that the button changes in appearance. Instead of the "plus" sign, a "1" is displayed in Figure 7.28 (Right). Click on any object and notice the information displayed in the Properties Palette. Without pressing ESC, click on another object. The original object deselects and the new object highlights with its information displayed in the Properties Palette. Very simply, the PICKADD button eliminates the need for the ESC key when displaying information about individual objects.

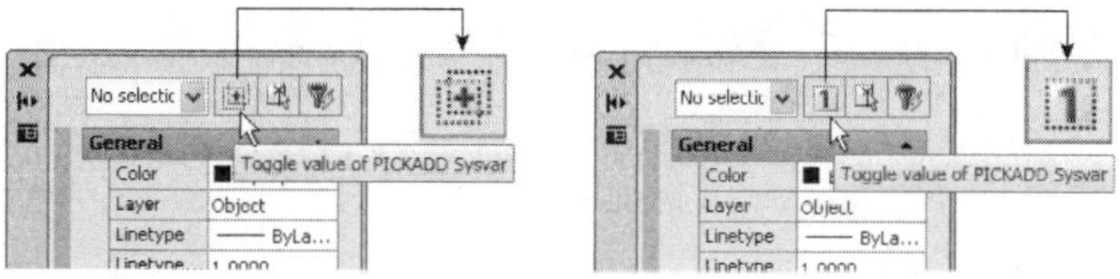

FIGURE 7.28

NOTE

When the PICKADD button is displaying a "1," you can still select multiple objects by holding down the SHIFT key as you pick. This setting remains in effect for all drawings until you change it back. To return to normal selection operations, click the PICKADD button to set it back to the "plus" sign before continuing.

The second button at the top of the Properties Palette is the Select Objects button, as shown in Figure 7.29 (Left). Like the PICKADD button, this button allows you to select a new object without having to use the ESC key to unselect the previous item. You will, however, have to select the button each time you make a new selection set. The third button is the Quick Select button, as shown in Figure 7.29. This button opens the Quick Select dialog box, as shown in Figure 7.29 (Right). This dialog box allows you to build a selection set based on specified properties. For example, you could select all the circles on the Center layer. This dialog box will be discussed in detail later in this chapter.

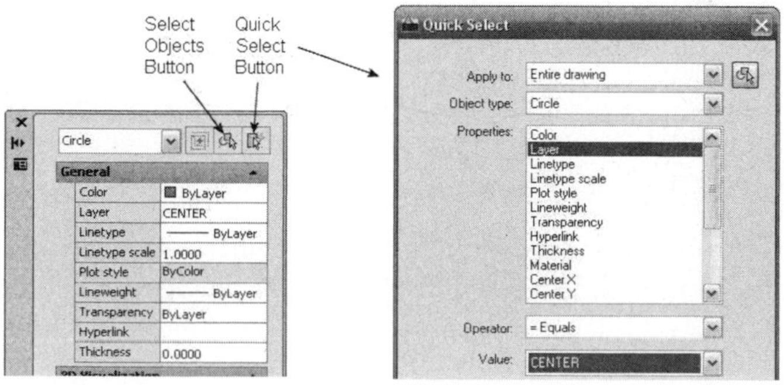

FIGURE 7.29

ROLLOVER TOOLTIPS AND THE QUICK PROPERTIES TOOL

Information concerning drawn objects can be automatically displayed through Roll-over ToolTips and by Quick Properties. Tooltips are displayed by hovering your cursor over an object, such as the line or block, as shown in Figure 7.30 (Left). By default this feature is turned on and displays information on color, layer, and linetype.

The Quick Properties tool not only displays information on selected objects but allows you to modify their properties. Located in the Status Bar at the bottom of the display screen is a Quick Properties button, as shown in Figure 7.30 (Right). When this tool is turned on, you can select an object, such as the text object shown in Figure 7.30 (Right), and a panel appears that allows you to make changes to items such as layer, contents, style, justification, and height. Each object type selected has a default set of properties displayed. This makes Quick Properties a very efficient tool for editing drawing objects.

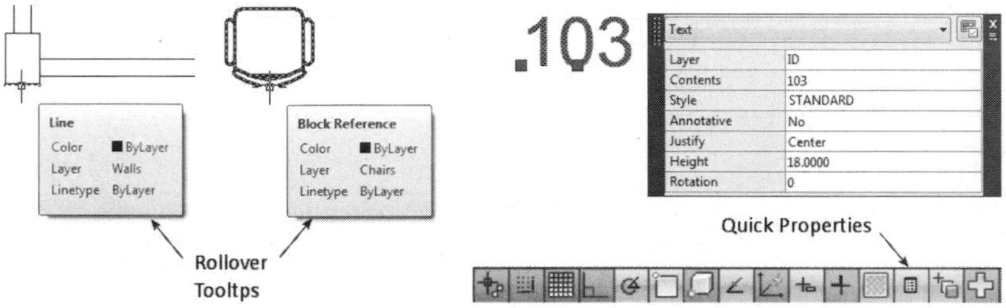

FIGURE 7.30

By default, when you click on an object with Quick Properties turned on, the Quick Properties panel locates itself at a set distance away from the cursor. Whenever you pick a new location on an object, the panel follows along with the cursor location at this set distance. Right-clicking on the Quick Properties palette displays the menu, as shown in Figure 7.31. If you change the Location Mode of the panel from Cursor to Static, the panel will remain in its position no matter where you select the object.

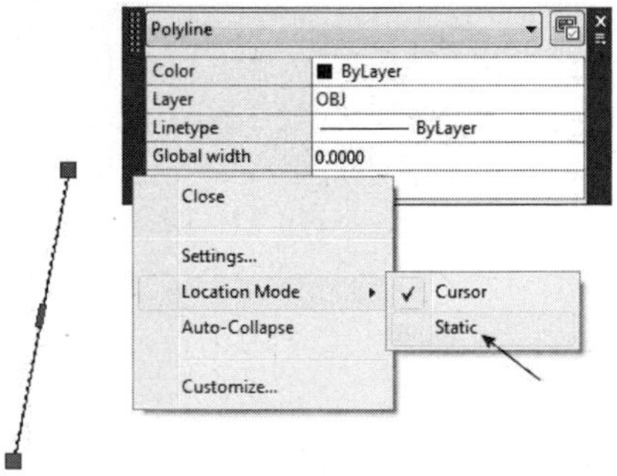

FIGURE 7.31

Additional settings for Quick Properties can be found under the Quick Properties tab of the Drafting Settings dialog box, as shown in Figure 7.32. This dialog box can be displayed by right-clicking on the Quick Properties icon on the Status Bar and selecting Settings… from the shortcut menu provided. Experiment with various settings under this dialog box to see how it affects the display of the Quick Properties panel.

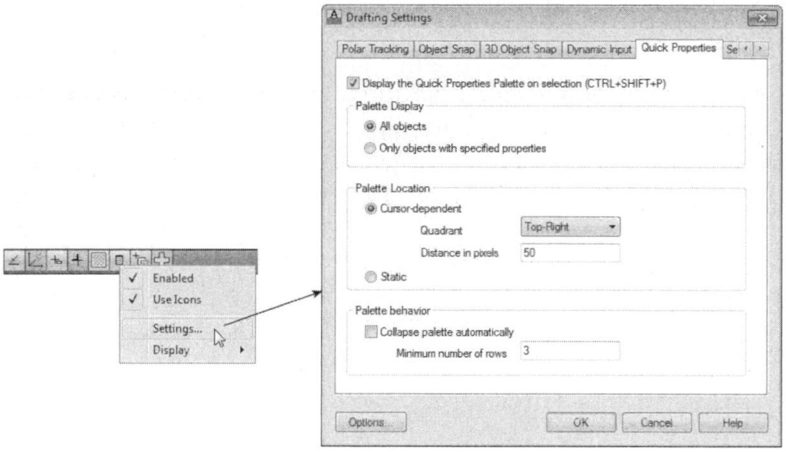

FIGURE 7.32

USING THE QUICK SELECT DIALOG BOX

At various times in a drawing, you need to build a selection set and modify the properties of these selected objects. However, objects in a drawing may not be conveniently grouped together and if you attempt to select objects through the conventional methods, such as window or crossing, you may accidentally ignore a few objects. A tool is available in which you enter certain parameters and AutoCAD selects the objects based on these parameters; this tool is called Quick Select. This tool can be activated by right-clicking on a blank part of your screen to display the menu shown in Figure 7.33 (Left) where you can select Quick Select. You can also choose Quick Select from the Ribbon, as shown in Figure 7.33 (Right).

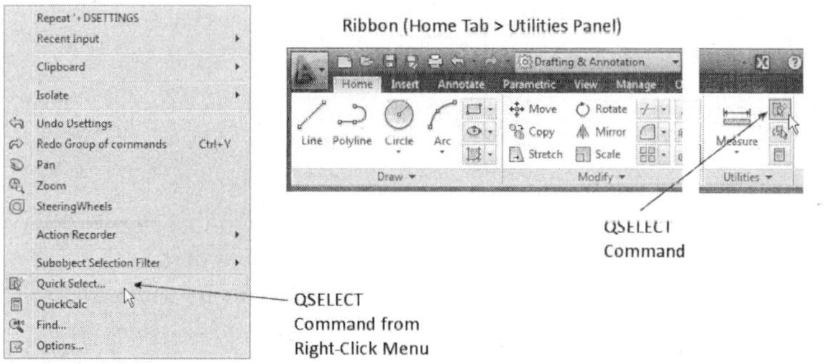

FIGURE 7.33

Clicking on Quick Select from any one of the menus launches the Quick Select dialog box, as shown in Figure 7.34. To have AutoCAD create a selection set by object type and property, first choose the appropriate object type in the dialog box. Notice in the following that two examples are shown, one with Block Reference and the other with Line selected. Once you determine the object type, the property area of this dialog box changes depending on the object. In Figure 7.34, study the two different sets of properties displayed. Notice that the properties for the Block Reference object type greatly differ from those for the Line object type. This, however, is what makes Quick Select so powerful. Not only can you create a selection set based on object type, but you can also narrow your search down based on properties such as layer and color for most objects or even by block name or line length.

By default, whenever you apply the object type and property information, this information becomes the basis for the new selection set. You can also create the inverse of this information; this occurs when you click on the radio button next to Exclude from new selection set entry. If this button is picked and you build a selection set based on a certain object and property, all other objects will be selected except this object and property.

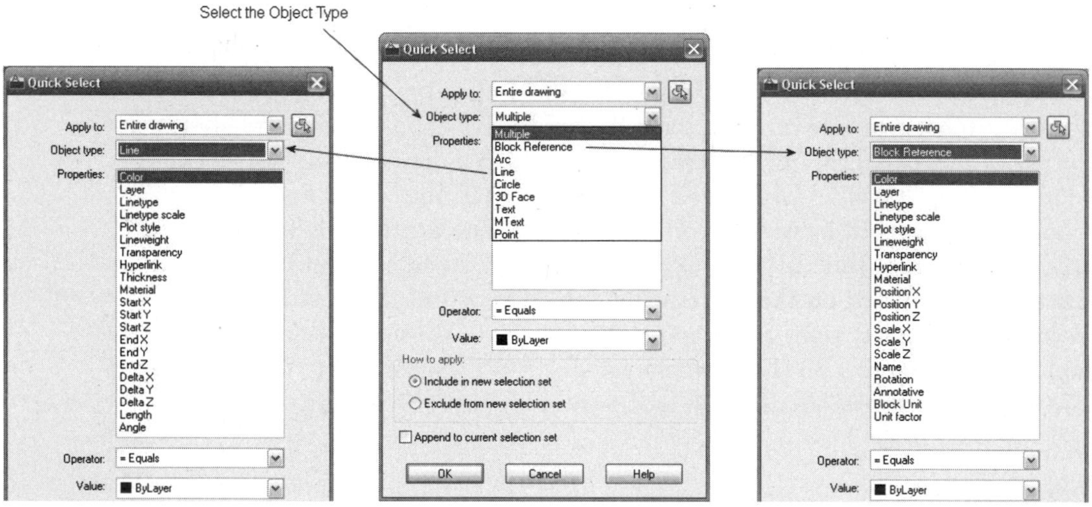

FIGURE 7.34

TRY IT!

Open the drawing file 07_Change Text Height. In Figure 7.35, the room numbers in the rectangular boxes are currently set to a height of 18". All text items need to be changed to a new height of 12". Rather than individually changing each text item, use the Quick Select dialog box to assist with this operation.

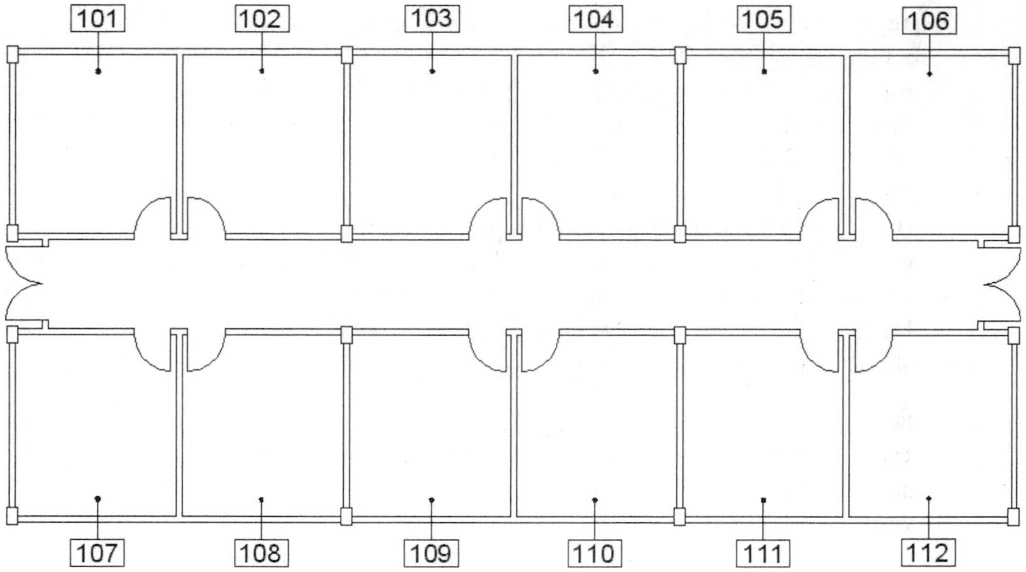

FIGURE 7.35

First, activate the Properties Palette and click on the Quick Select button, as shown in Figure 7.36 (Left). This represents another method of activating Quick Select. Once inside the Quick Select dialog box, click on the Object type window and select "Text," as shown in Figure 7.36 (Center). The Properties selected is Color, the Operator is Equal, and the Value is ByLayer, as shown in Figure 7.36 (Right). You will be selecting all text that is assigned the color ByLayer—all text in the drawing.

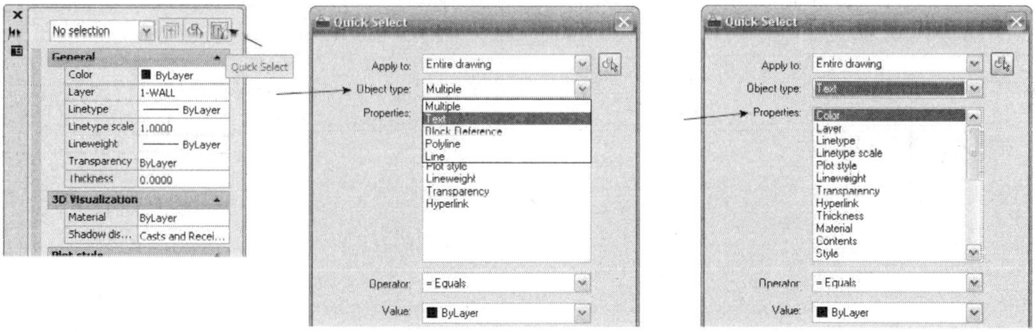

FIGURE 7.36

Clicking the OK button at the bottom of the Quick Select dialog box returns you to your drawing. Notice that all text is highlighted and a text height of 18.0000 is listed in the Properties Palette. Change this value to 12, as shown in Figure 7.37.

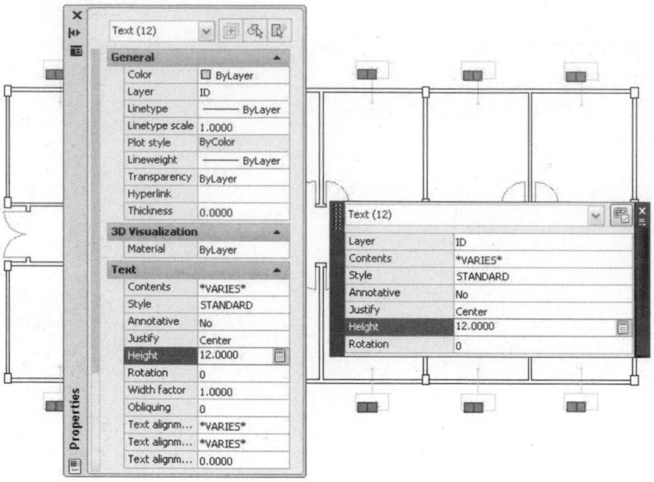

FIGURE 7.37

When finished changing the text height in the Properties Palette, close the palette and notice that all text, as shown in Figure 7.38, has been changed to the new height of 12".

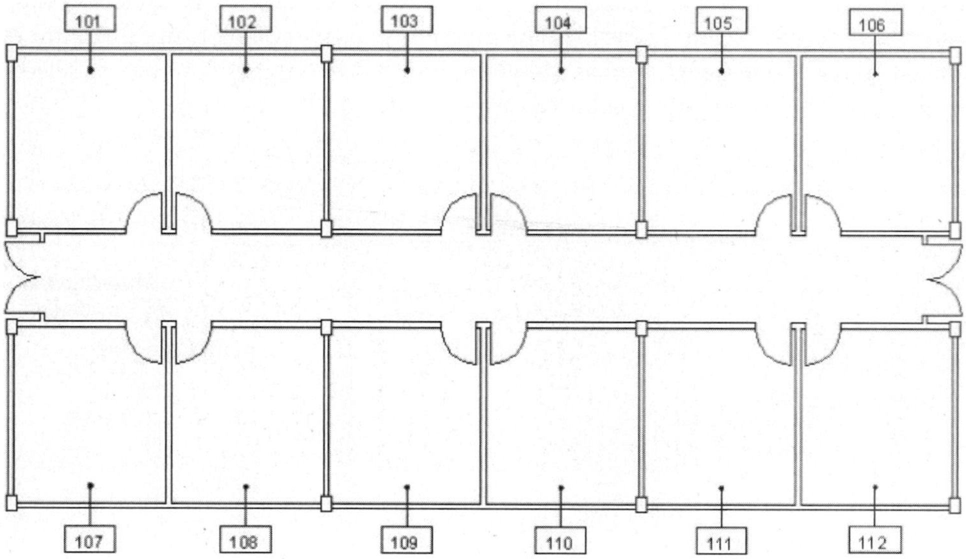

FIGURE 7.38

USING THE QUICKCALC PALETTE

A mathematical calculator is available to assist in making a variety of calculations while still remaining in an AutoCAD drawing. The calculator is activated by clicking on QuickCalc, found by right-clicking on a blank part of your screen to display a shortcut menu or in the Ribbon, as shown in Figure 7.39. This launches the calculator, as shown in Figure 7.39 (Right). When you perform calculations throughout the design cycle, these calculations are stored in a History area. Calculations in this area can be retrieved for later use. As you enter numbers and mathematical functions such as addition and subtraction from the supplied keypads, they are displayed in the Input area of the calculator. Four tabbed areas control the methods by which calculations are made. They are Number Pad, Scientific, Units Conversion, and Variables.

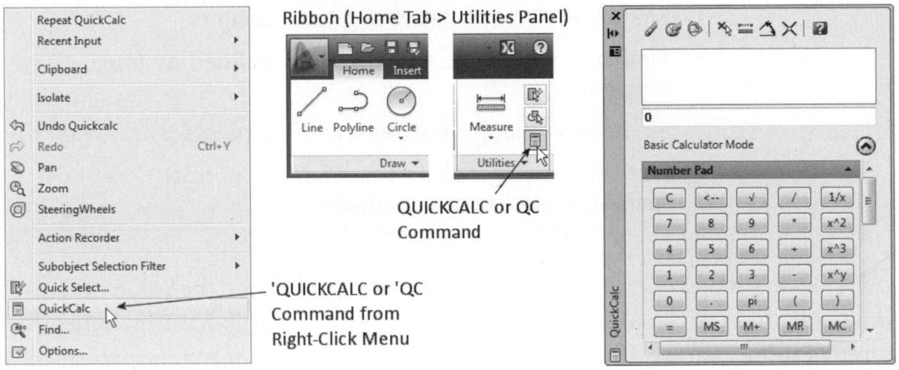

FIGURE 7.39

> The QUICKCALC command can be run transparently. A transparent command is one that can be run inside of another command. If you are running a command that requires entry of a value, you can start the calculator transparently, calculate the value, apply it, and then continue with the original command. AutoCAD knows a command is to be run transparently when an apostrophe is placed in front of the command name.

TIP

Each of the four tabbed areas of the calculator are illustrated in Figure 7.40. A brief description of each follows:

Numeric Pad: This area of the calculator is used to perform the most basic of mathematical functions. Clicking on a number or function places this information in the Input area of the calculator.

Scientific: Use this tabbed area to enter trigonometric and other advanced functions.

Units Conversion: This area is used to convert units from one system to another. You enter the unit type, the unit type to convert from, and the units to convert to. Entering a numeric value to convert displays the converted value.

Variables: This area contains a series of shortcut commands that can be used for performing specialized operations.

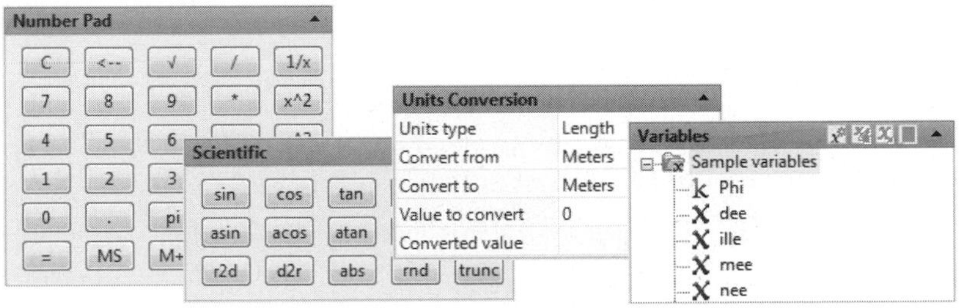

FIGURE 7.40

Variable	Function
PHI	Golden Ratio (1.61803399)
DEE	Finds the distance between two endpoints
ILLE	Finds the intersection of two lines defined by four endpoints
MEE	Finds the midpoint between two endpoints
NEE	Finds the unit vector in the XY plane normal (perpendicular) to two endpoints

When you are using the Properties Palette and you need to change the value of one of the Geometry listings, a calculator icon is present. Clicking this icon launches the QuickCalc palette, as shown in Figure 7.41 (Right). Entering a mathematical formula in the Input area and clicking the Apply button sends the calculated value back to the Properties Palette and changes the highlighted object.

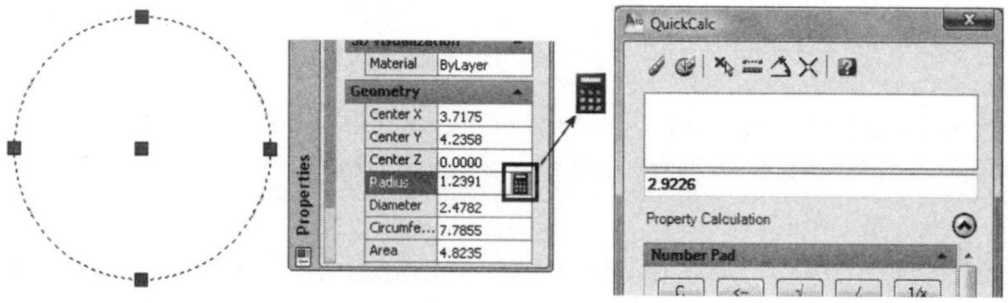

FIGURE 7.41

USING THE LAYER CONTROL BOX TO MODIFY OBJECT PROPERTIES

If all you need to do is to change an object or a group of objects from one layer to another, the Layer Control box can easily perform this operation.

TRY IT!

Open the drawing file 07_Change Layer. In Figure 7.42, select the arc and two line segments. Notice that the current layer is 0 in the Layer Control box. These objects need to be on the OBJECT layer.

Click on the Layer Control box to display all layers defined in the drawing. Then click on the desired layer for all highlighted objects (in this case, the OBJECT layer in Figure 7.42).

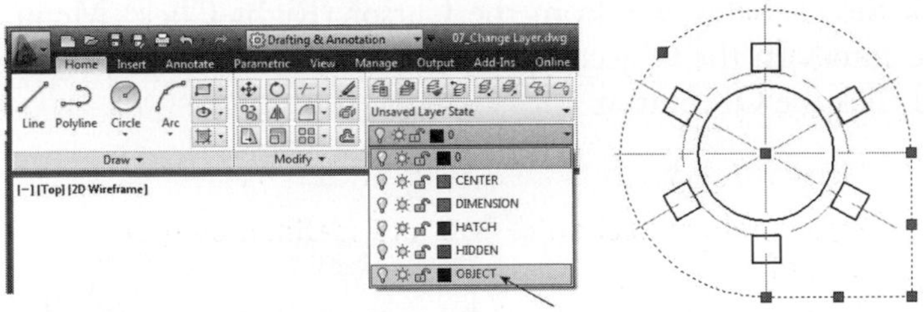

FIGURE 7.42

Notice that in Figure 7.43, with the objects still highlighted, the layer listed is OBJECT. This is one of the quickest and most productive ways of changing an object from one layer to another.

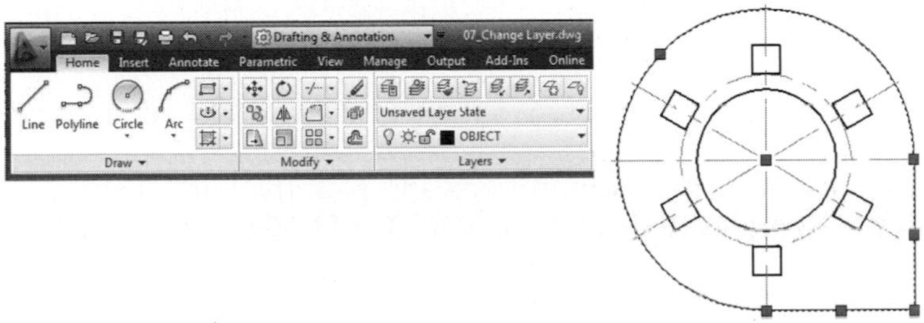

FIGURE 7.43

TRY IT!

Open the drawing file 07_Mosaic. In Figure 7.44, this drawing consists of five different types of objects all drawn on Layer 0: circles, squares drawn as closed polylines, lines, text (the letter X), and text (the letter Y). The Quick Select dialog box will be used to select these object types individually.

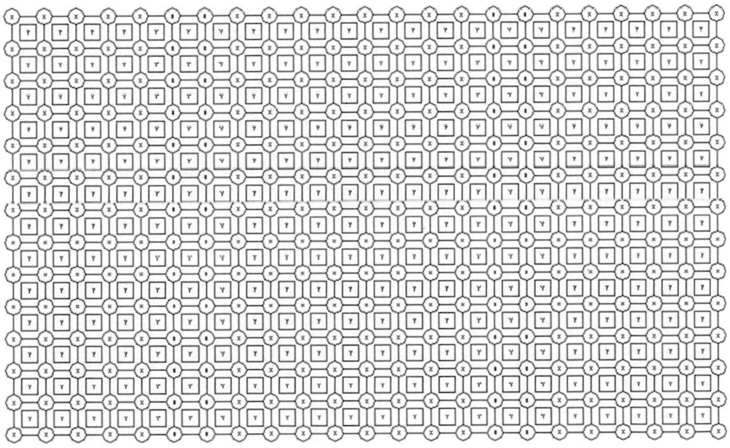

FIGURE 7.44

Once you have selected them, change the objects to the correct layers, which are also supplied with this drawing and identified in Figure 7.45 (Left).

First, activate the Quick Select dialog box from the Cursor (Right-Click) Menu, shown in Figure 7.45 (Center). In the Object type box, select Circle, as you see in Figure 7.45 (Right), and click the OK button.

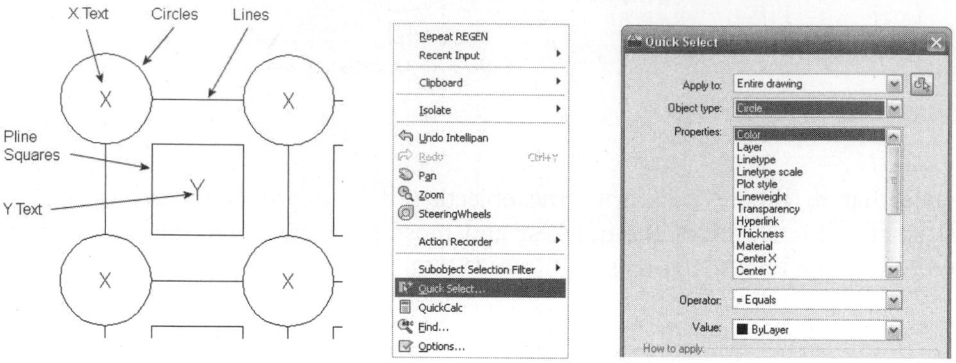

FIGURE 7.45

With all circles selected, click the Layer Control box and pick the Circles layer, as shown in Figure 7.46. All circles in the mosaic pattern should turn red. Press ESC to remove the object highlight.

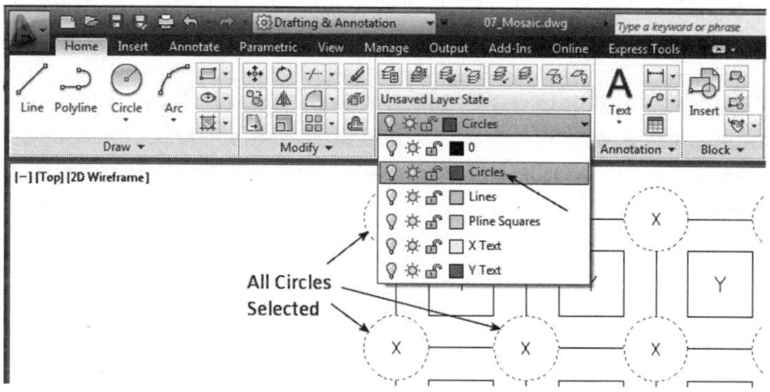

FIGURE 7.46

NOTE

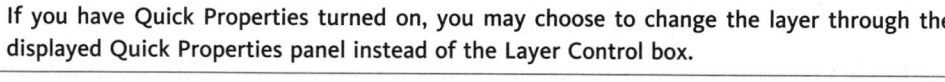

If you have Quick Properties turned on, you may choose to change the layer through the displayed Quick Properties panel instead of the Layer Control box.

Next, you need to select all the text with the letter "X" and change this text to the layer called X Text. Activate the Quick Select dialog box and make the following changes: Change the Object type to Text, set the Properties to Contents, and enter "X" as the value. Your display should be similar to the dialog box in Figure 7.47 (Left). Click the OK button to dismiss the Quick Select dialog box and select all text with the letter "X."

Change all letters to the X Text layer in the Layer Control box, as shown in Figure 7.47 (Right). Press ESC to remove the object highlight and grips. Follow the same procedures for changing all line segments to the Lines layer, all polylines to the Pline Squares layer, and all "Y" text to the Y Text layer.

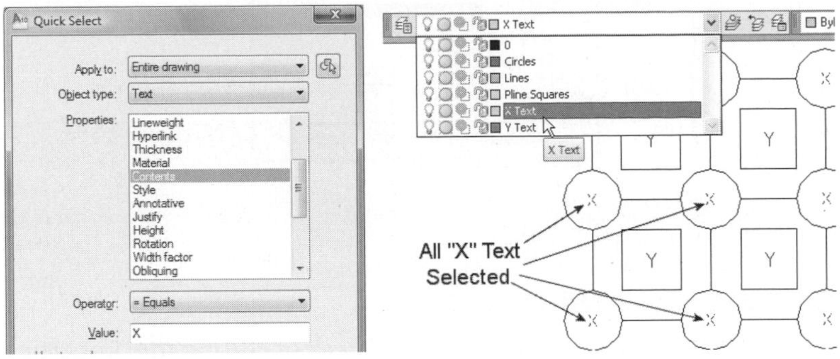

FIGURE 7.47

TRY IT!

Open the drawing file 07_Qselect Duplex. Use the illustration in Figure 7.48 and the information below to change the items to their proper layers.

```
Change all lines to the Walls layer
Change all text to the Room Labels layer
Change the blocks "Door" and "Louver" to the Door layer
Change the block "Window" to the Window layer
Change the blocks "Countertop," "Range," "Sink," and
"Refrigerator" to the Countertop layer
```

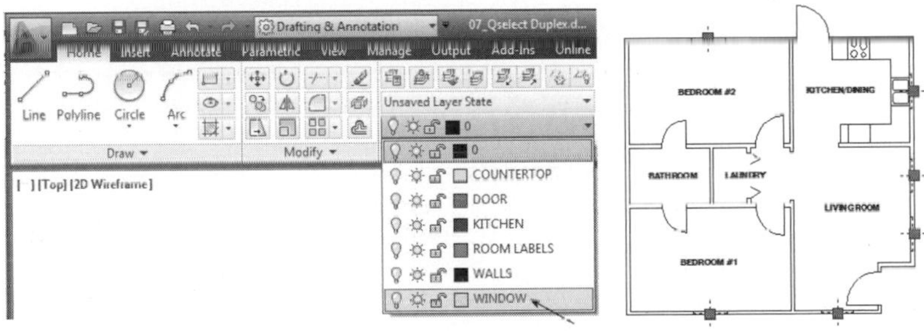

FIGURE 7.48

TIP

For the blocks, see the illustration in Figure 7.49 for supplying the correct information in the Quick Select dialog box; change the Object type to Block Reference. Change the Properties to Name. Click the down arrow next to the Value heading and select the desired block from the list provided. If you select the "Append to current selection set" check box before clicking the OK button, you can restart Quick Select and add other blocks to the selection before changing the highlighted blocks to the correct layer.

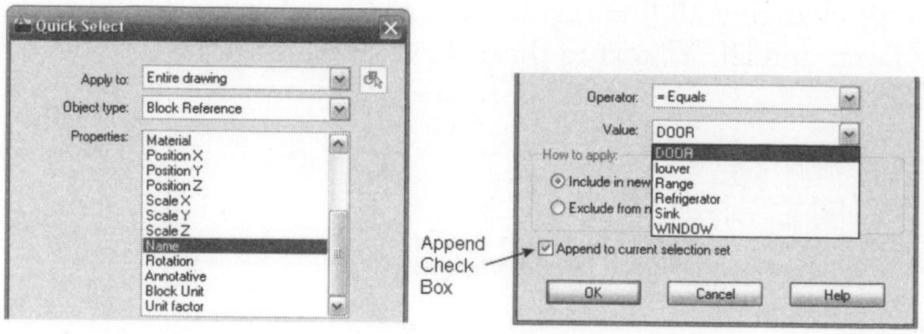

FIGURE 7.49

DOUBLE-CLICK EDIT ON ANY OBJECT

Double-clicking on any object provides you with a quick way of editing the properties of that object. For instance, double-clicking on a line segment launches the Quick Properties Palette, which displays information about the line. Double-clicking on a multiline text object displays the Ribbon's text editor, allowing you to enter, delete, or format words.

TRY IT!

Open the drawing file 07_Double Click Edit. In Figure 7.50, double-click on the magenta centerline and the Quick Properties Palette launches with information about the line. Press ESC to close the Palette. Double-click on the word "BLOCK" to launch the Edit Text field. Press ENTER twice to exit the edit function when finished. Double-click on the sentence "THE OBJECT SHOWN ABOVE IS A WINDOW SYMBOL" to launch the Multiline Text Editor. Click on the Close Text Editor button when finished. Continue by double-clicking on the hatch pattern, circle, dimension, and rectangle and observe the type of editor launched through this method.

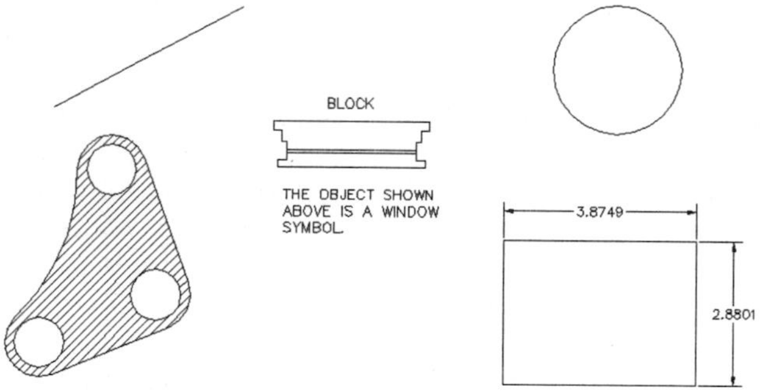

FIGURE 7.50

MATCHING THE PROPERTIES OF OBJECTS

Yet another tool is available for changing the properties of objects—the MATCHPROP command. The advantage of this tool is that you don't have to know the property value to update it, you can simply match the properties to those of an existing object. This command can be selected from the Ribbon, as shown in Figure 7.51.

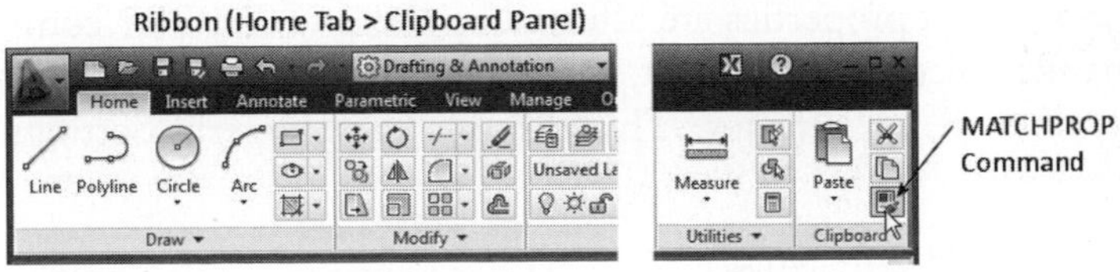

FIGURE 7.51

TRY IT!

Open the drawing file 07_Matchprop Flange. Use the illustration in Figure 7.52 and the command sequence below for performing this task.

When you start the command, a source object is required. This source object transfers all its current properties to other objects designated as "Destination Objects." As shown in Figure 7.52 (Left), the flange requires the object lines located at "B," "C," "D," and "E" to be converted to hidden lines. Using the MATCHPROP command, select the existing hidden line "A" as the source object. Notice the appearance of the Match Properties icon. Select lines "B" through "E" as the destination objects using this icon.

Command: MA *(For MATCHPROP)*

Select source object: *(Select the hidden line at "A")*

Current active settings: Color Layer Ltype Ltscale Lineweight Transparency Thickness PlotStyle Dim Text Hatch Polyline Viewport Table Material Shadow display Multileader

Select destination object(s) or [Settings]: *(Select line "B")*

Select destination object(s) or [Settings]: *(Select line "C")*

Select destination object(s) or [Settings]: *(Select line "D")*

Select destination object(s) or [Settings]: *(Select line "E")*

Select destination object(s) or [Settings]: *(Press* ENTER *to exit this command)*

The results appear in the flange illustrated in the Figure 7.52 (Right), where the continuous object lines were converted to hidden lines. Not only did the linetype property get transferred, but the color, layer, lineweight, and linetype scale information did as well.

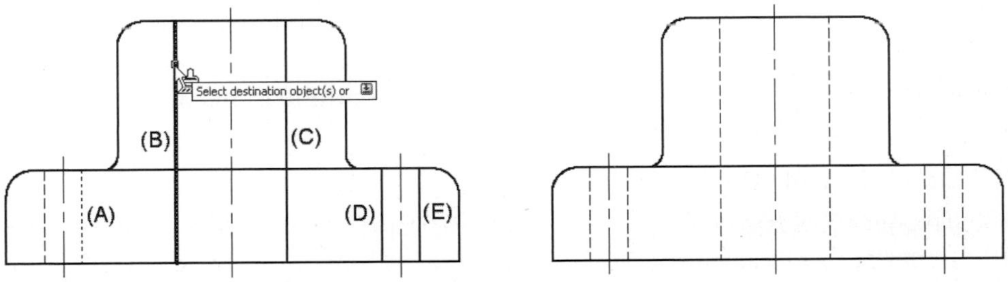

FIGURE 7.52

To get a better idea of what object properties are affected by the `MATCHPROP` command, reenter the command, select a source object, and instead of selecting a destination object immediately, enter S for Settings. This displays the Property Settings dialog box in Figure 7.53.

```
Command: MA (For MATCHPROP)
Select source object: (Select the hidden line at "A" in the previous image)
Current active settings: Color Layer Ltype Ltscale Lineweight Transparency Thickness PlotStyle Dim Text Hatch Polyline Viewport Table Material Shadow display Multileader
Select destination object(s) or [Settings]: S (For Settings; this displays the Property Settings dialog box in Figure 7.53)
```

Any box with a check displayed in it transfers that property from the source object to all destination objects. If you need to transfer only the layer information and not the color and linetype properties of the source object, remove the checks from the Color and Linetype properties before you select the destination objects. This prevents these properties from being transferred to any destination objects.

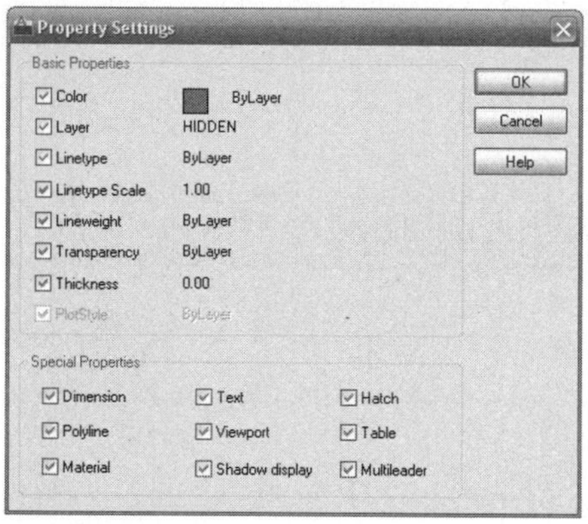

FIGURE 7.53

Matching Dimension Properties

The `MATCHPROP` command can control special properties of dimensions, text, hatch patterns, polylines, viewports, and tables. The Dimension property will be featured next, as shown in Figure 7.54. (Even though the topic of dimensions will not be covered until Chapter 10, this concept is introduced here.)

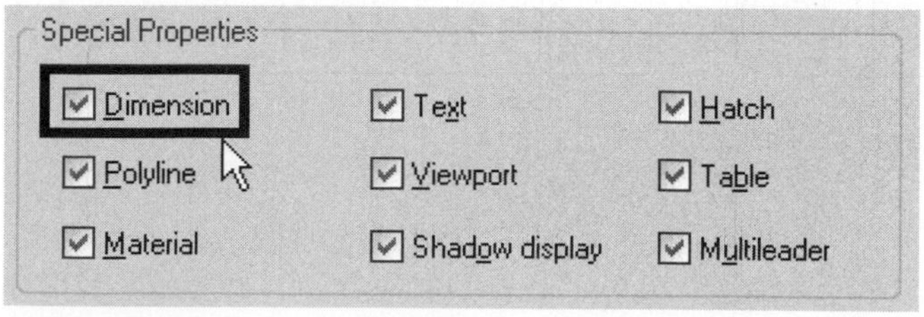

FIGURE 7.54

TRY IT!

Open the drawing file 07_Matchprop Dim. Figure 7.55 shows two blocks: the block assigned a dimension value of 46.6084 was dimensioned with the METRIC dimension style with the Arial font applied. The block assigned a dimension value of 2.3872 was dimensioned with the STANDARD dimension style with the TXT font applied. Both blocks need to be dimensioned with the METRIC dimension style. Issue the `MATCHPROP` **command and select the 46.6084 dimension as the source object and then select the 2.3872 dimension as the destination object.**

```
Command: MA (For MATCHPROP)
Select source object: (Select the dimension at "A" in Figure 7.55)
Current active settings: Color Layer Ltype Ltscale Lineweight Transparency Thickness PlotStyle Dim Text Hatch Polyline Viewport Table Material Shadow display Multileader
Select destination object(s) or [Settings]: (Select the dimension at "B")
Select destination object(s) or [Settings]: (Press ENTER to exit this command)
```

The results are shown in Figure 7.55 (Right), with the METRIC dimension style applied to the dimension with the STANDARD style through the use of the `MATCHPROP` command. Because the text font was associated with the dimension style, it also changed in the destination object.

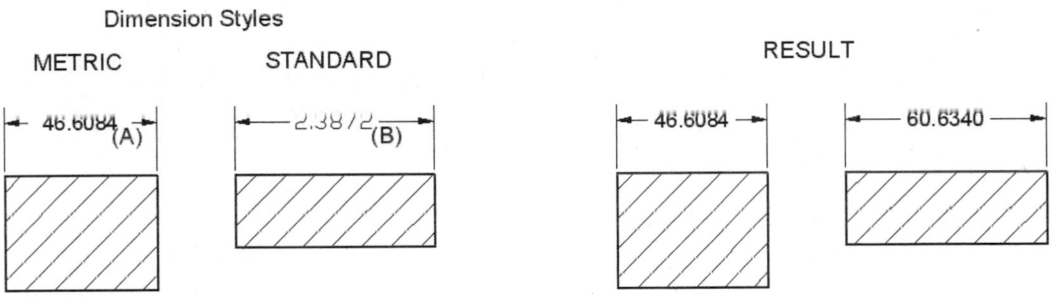

FIGURE 7.55

Matching Text Properties

Figure 7.56 is an example of how the `MATCHPROP` command affects a text object with the Text property.

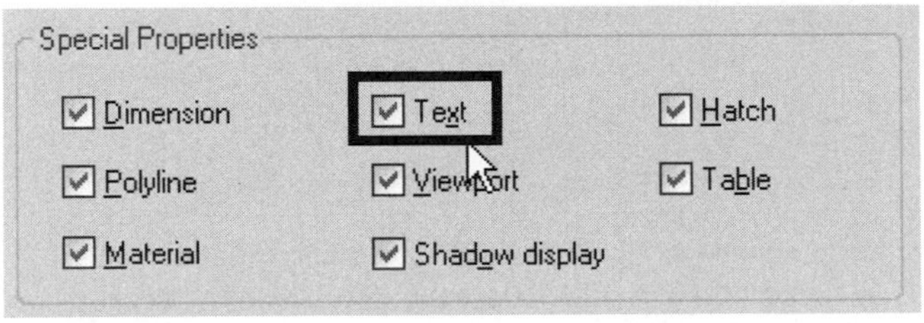

FIGURE 7.56

TRY IT!

Open the drawing file 07_Matchprop Text. Figure 7.57 shows two text items displayed in different fonts. The text "Coarse Knurl" at "A" was constructed with a text style called Arial. The text "Medium Knurl" at "B" was constructed with the default text style called STANDARD. Use the following command sequence to match the STANDARD text style with the Arial text style using the `MATCHPROP` command.

```
Command: MA (For MATCHPROP)
Select source object: (Select the text at "A")
Current active settings: Color Layer Ltype Ltscale Line-
weight Transparency Thickness PlotStyle Dim Text Hatch Poly-
line Viewport Table Material Shadow display Multileader
Select destination object(s) or [Settings]: (Select the text
at "B")
Select destination object(s) or [Settings]: (Press ENTER to
exit this command)
```

The result is shown in Figure 7.57 (Right). Both text items now share the same text style. Notice that the text string stays intact when text properties are matched. Only the text style of the source object is applied to the destination object.

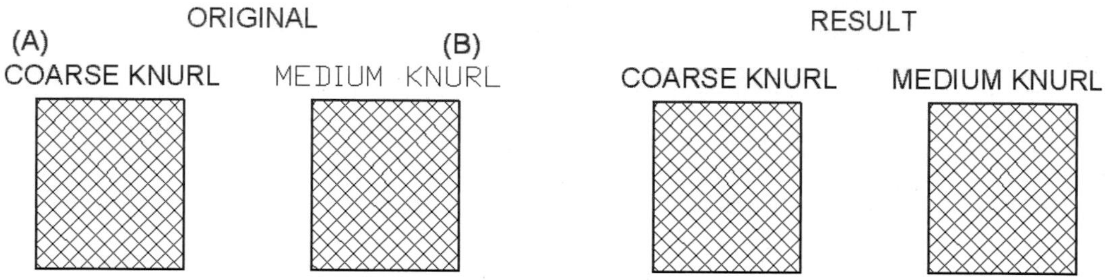

FIGURE 7.57

Matching Hatch Properties

A source hatch object can also be matched to a destination pattern with the `MATCHPROP` command and the Hatch property, as shown in Figure 7.58. (Even though the topic of hatching will not be covered until Chapter 9, this concept is introduced here.)

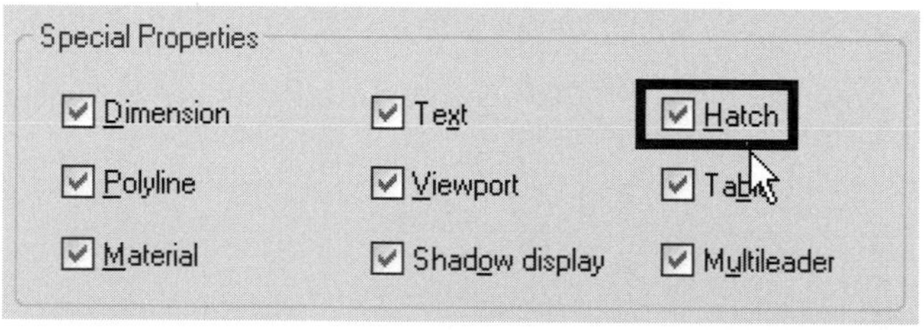

FIGURE 7.58

TRY IT!

Open the drawing file 07_Matchprop Hatch. In Figure 7.59, the crosshatch patterns at "B" and "C" are at the wrong angle and scale. They should reflect the pattern at "A" because it is the same part. Use the `MATCHPROP` command, select the hatch pattern at "A" as the source object, and select the patterns at "B" and "C" as the destination objects.

Command: MA *(For MATCHPROP)*

Select source object: *(Select the hatch pattern at "A")*

Current active settings: Color Layer Ltype Ltscale Lineweight Transparency Thickness PlotStyle Dim Text Hatch Polyline Viewport Table Material Shadow display Multileader

Select destination object*(s)* or [Settings]: *(Select the hatch pattern at "B")*

Select destination object*(s)* or [Settings]: *(Press* ENTER *to exit this command)*

The results appear in Figure 7.59 (Right), where the source hatch pattern property was applied to all destination hatch patterns.

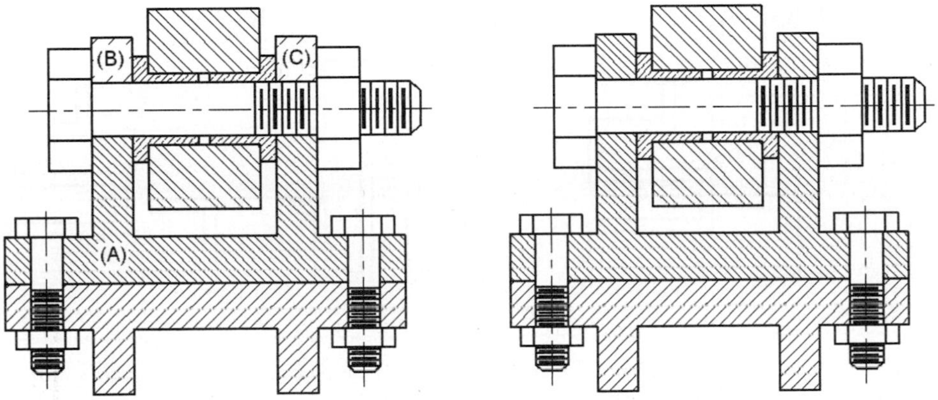

FIGURE 7.59

Matching Polyline Properties

A source polyline object can also be matched to a destination pattern with the MATCHPROP command and the Polyline property, as shown in Figure 7.60.

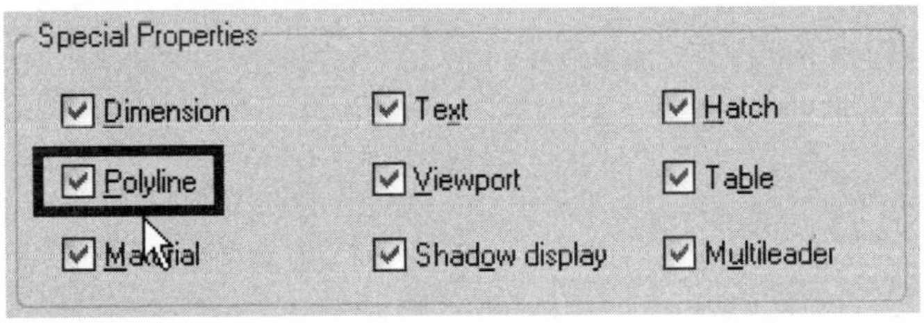

FIGURE 7.60

TRY IT!

Open the drawing file 07_Matchprop Pline. In Figure 7.61, the polyline at "B" is at the wrong width. It should match the width of the polyline at "A." Using the MATCHPROP command, select the polyline at "A" as the source object, and select the polyline at "B" as the destination object.

Command: MA *(For MATCHPROP)*

Select source object: *(Select the polyline at "A")*

Current active settings: Color Layer Ltype Ltscale Lineweight Transparency Thickness PlotStyle Dim Text Hatch Polyline Viewport Table Material Shadow display Multileader

Select destination object*(s)* or [Settings]: *(Select the polyline at "B")*

Select destination object*(s)* or [Settings]: *(Press* ENTER *to exit this command)*

The results appear in Figure 7.61 (Right), where the source polyline property (width, in this example) was applied to the destination polyline.

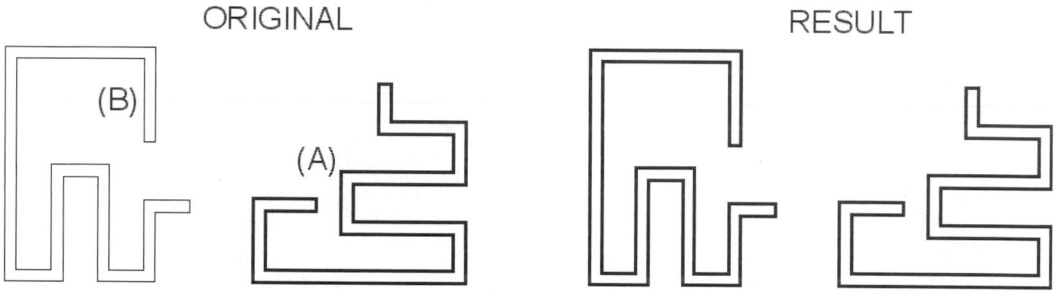

FIGURE 7.61

Matching Viewport Properties

A source viewport object can also be matched to a destination viewport with the MATCHPROP command and the Viewport property, as shown in Figure 7.62. Even though the topic of viewports will not be covered until Chapter 19, this concept is introduced here. Viewports are used to arrange images of a drawing in layout mode before they are plotted out. Matching the properties of a viewport can transfer the scale of one viewport to another. The following Try It! exercise illustrates this.

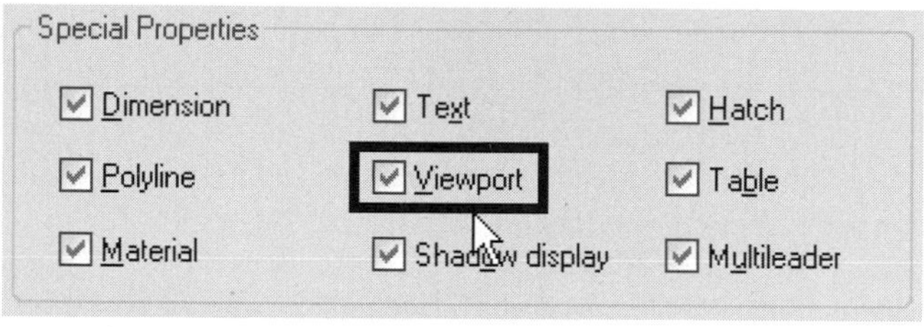

FIGURE 7.62

TRY IT!

Open the drawing file 07_Matchprop Viewport. In Figure 7.63, the viewport image at "A" is at the correct scale. The viewport at "B" should have the same scale as that of viewport "A." Using the MATCHPROP command, select the viewport at "A" as the source object, and select the viewport at "B" as the destination object.

Command: MA *(For MATCHPROP)*

Select source object: *(Select the edge of the viewport at "A")*

Current active settings: Color Layer Ltype Ltscale Lineweight Transparency Thickness PlotStyle Dim Text Hatch Polyline Viewport Table Material Shadow display Multileader

Select destination object*(s)* or [Settings]: *(Select the viewport at "B")*

Select destination object*(s)* or [Settings]: *(Press* ENTER *to exit this command)*

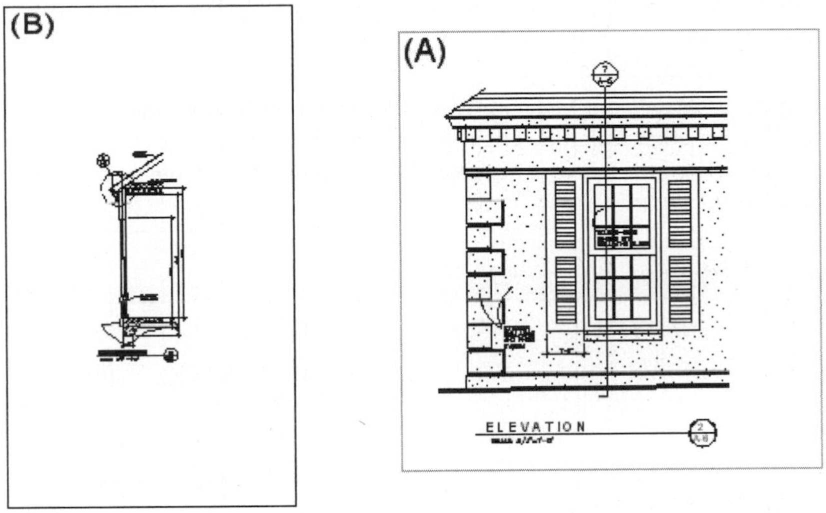

FIGURE 7.63

The results appear in Figure 7.64, where the source viewport property (scale of the image) was applied to the destination viewport. Notice also that the layer of the viewport "B" was changed from 0 to Vports.

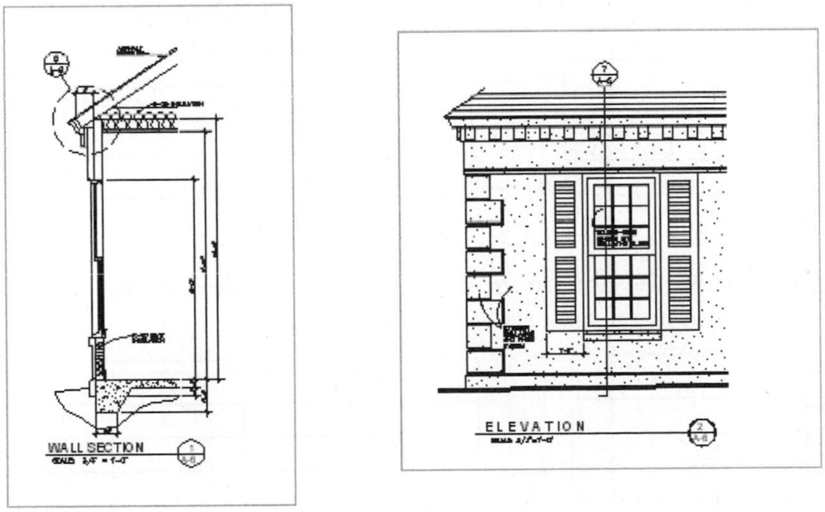

FIGURE 7.64

Match Properties and Tables

A source table object can be matched to a destination table using the Table property, as shown in Figure 7.65. When one table is matched with another, the table style of the source table updates the table style of the destination table.

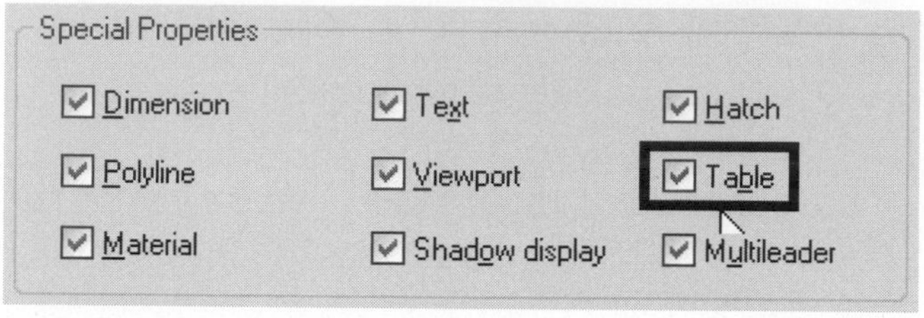

FIGURE 7.65

TRY IT!

Open the drawing file 07_Matchprop Table. In Figure 7.66, one table has the title strip originating at the top and the other table title originates at the bottom. Using the `MATCHPROP` **command, select the table on the left as the source object, and select the table on the right as the destination object.**

```
Command: MA (For MATCHPROP)
Select source object: (Select the table in Figure 7.66 (Left))
Current active settings: Color Layer Ltype Ltscale Line-
weight Transparency Thickness PlotStyle Dim Text Hatch Poly-
line Viewport Table Material Shadow display Multileader
Select destination object(s) or [Settings]: (Select the
table on the right in Figure 7.66)
Select destination object(s) or [Settings]: (Press ENTER to
exit this command)
```

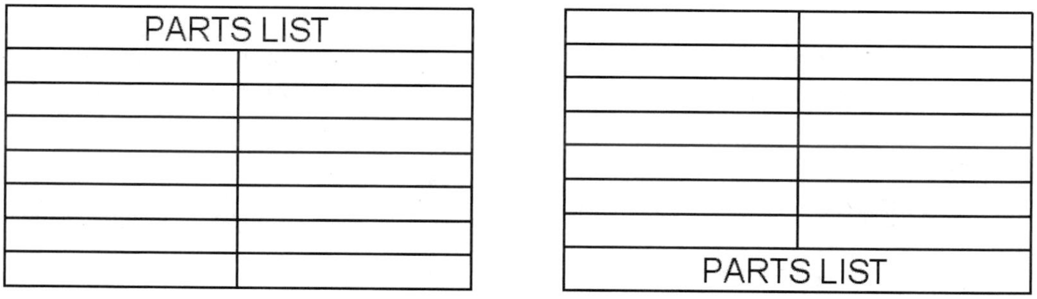

FIGURE 7.66

In Figure 7.67, notice how the table on the right changes to the downward direction arrangement as controlled by the table style.

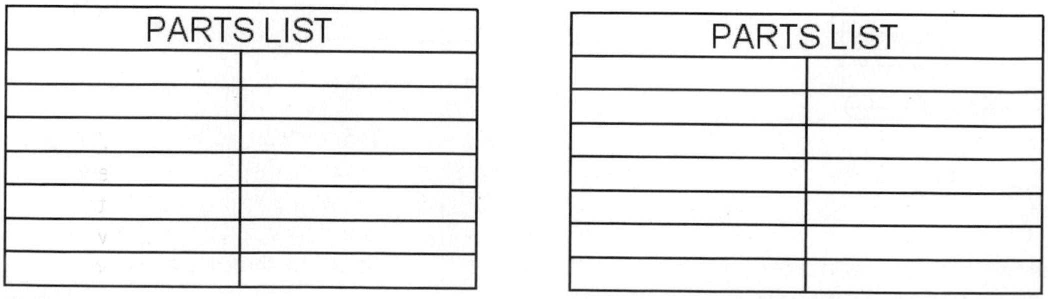

FIGURE 7.67

NOTE

You may have to move the table into position.

TUTORIAL EXERCISE: 07_MODIFY-EX.DWG

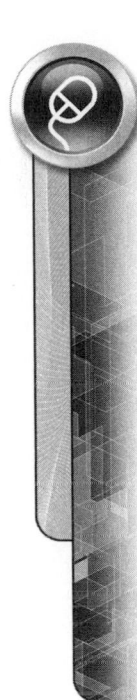

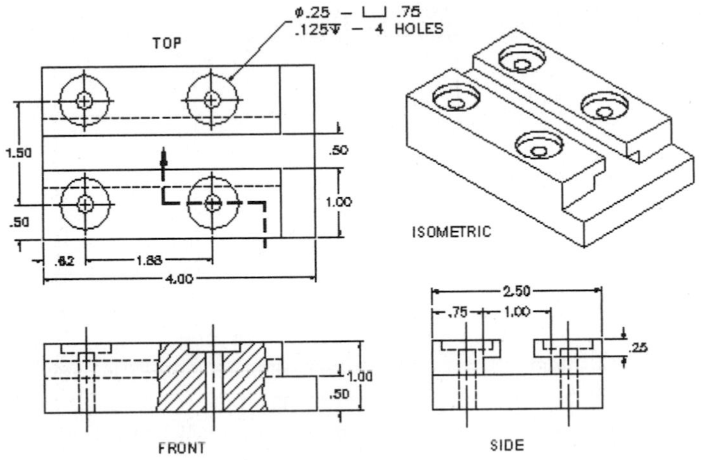

FIGURE 7.68

Purpose

This tutorial exercise is designed to change the properties of existing objects displayed in Figure 7.68.

System Settings

Open an existing drawing file called "07_Modify-EX." Follow the steps in this tutorial for changing various objects to the correct layer, text style, and dimension style.

Layers

Layers have already been created in this drawing.

Suggested Commands

You will begin this tutorial by using the Properties Palette to change the isometric object to a different layer. Continue by changing the text height and layer of the view identifiers (FRONT, TOP, SIDE, ISOMETRIC). The MATCHPROP command will be used to transfer the properties from one dimension to another, one text style to another, and one hatch pattern to another. The Layer Control box will be used to change the layer of various objects located in the Front and Top views.

STEP 1

Loading this drawing displays the objects in a page layout called "Orthographic Views." A Page layout is where the completed drawing is displayed. The drawing is scaled in a viewport with a title block and is ready for plotting. The layout name is present next to the Model tab in the bottom portion of the drawing screen. Since a majority of changes will be made in Model mode, click on the Model tab near the bottom of the display screen. Your drawing will appear similar to Figure 7.69.

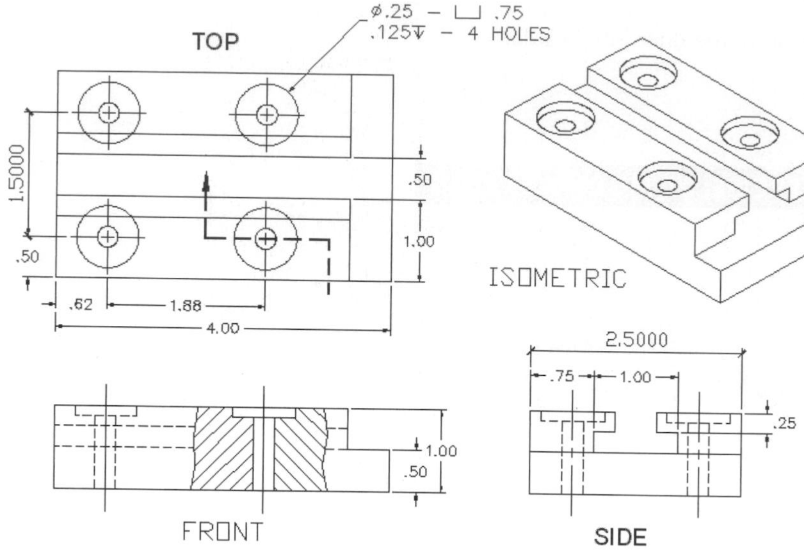

FIGURE 7.69

STEP 2

While in the Model environment, select all lines that make up the isometric view in Figure 7.70. You can accomplish this by using implied windowing - pick from left to right to activate the Window mode. If you accidentally select the word ISOMETRIC, de-select this word (hold the SHIFT key down and select it). Right-click on the drawing screen and activate the Properties Palette from the shortcut menu. Click on the Layer field. This displays the current layer the objects are drawn on (DIM) in Figure 7.70. Click the down arrow to display the other layers and pick the OBJECT layer. This changes all selected objects that make up the isometric view to the OBJECT layer. Press ESC to remove the object highlight and the grips from the drawing.

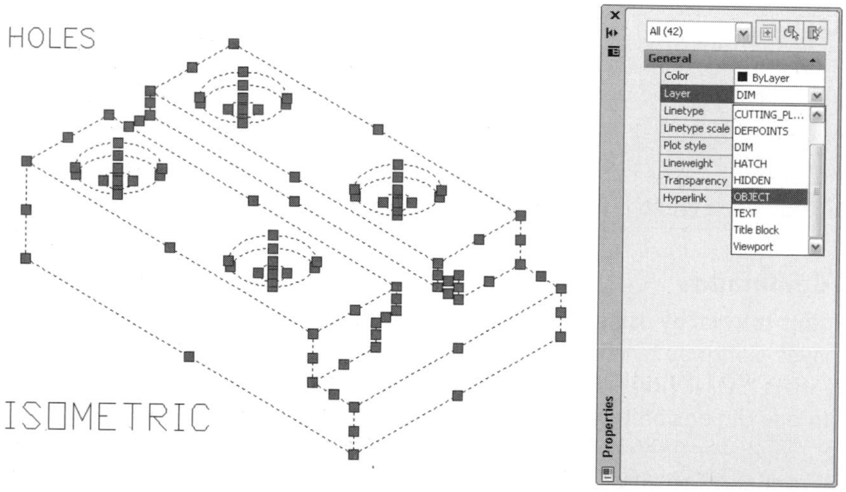

FIGURE 7.70

STEP 3

Select the view titles (FRONT, TOP, SIDE, ISOMETRIC). These text items need to be changed to a height of 0.15 and placed on the TEXT layer. With all four text objects highlighted, and the Properties Palette already active, click on the Layer field, and click the down arrow to place the selected objects to the TEXT layer in Figure 7.71.

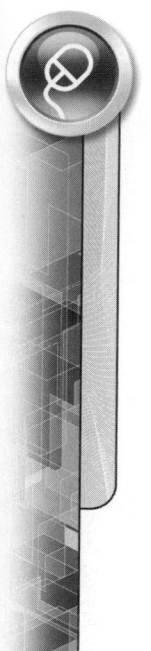

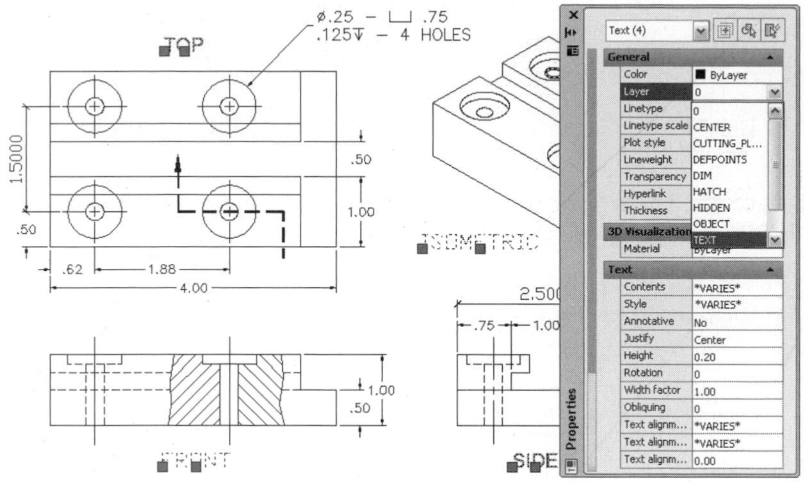

FIGURE 7.71

With the text still highlighted, change the text height in the Properties Palette from 0.20 to a new value of 0.15. Pressing ENTER after making this change automatically updates all selected text objects to this new height in Figure 7.72. Dismiss this palette when finished. Press ESC to remove the object highlight and the grips from the drawing.

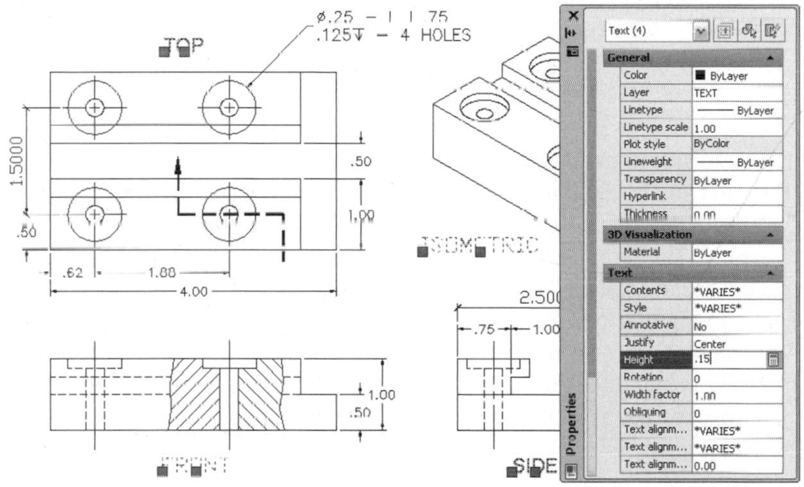

FIGURE 7.72

STEP 4

When you examine the view titles, TOP and SIDE are in one text style while FRONT and ISOMETRIC are in another. To remain consistent in the design process, you should make sure all text identifying the view titles has the same text style as TOP. Activate the MATCHPROP command. Click the text object TOP as the source object. When the Match Property icon appears in Figure 7.73, select ISOMETRIC and FRONT as the destination objects. All properties associated with the TOP text object are transferred to ISOMETRIC and FRONT, including the text style.

Command: MA *(for MATCHPROP)*

Select source object: *(Select the text object "TOP," which should highlight in Figure 7.73)*

Current active settings: Color Layer Ltype Ltscale Lineweight Transparency Thickness PlotStyle Dim Text Hatch Polyline Viewport Table Material Shadow display Multileader

Select destination object*(s)* or [Settings]: *(Select the text object "ISOMETRIC")*

Select destination object*(s)* or [Settings]: *(Select the text object "FRONT")*

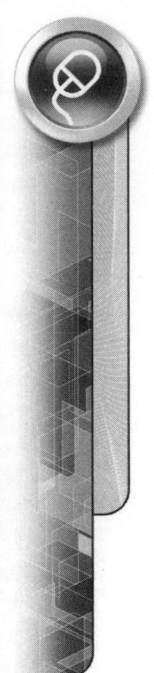

Select destination object*(s)* or [Settings]: *(Press* ENTER *to exit this command)*

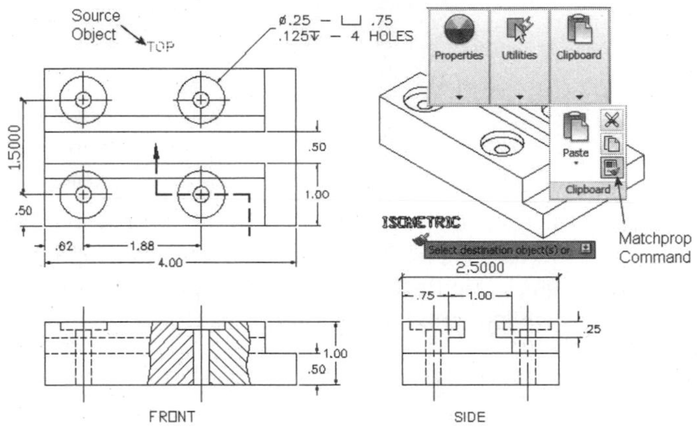

FIGURE 7.73

STEP 5

Notice the dimensions in this drawing. Two dimensions stand out above the rest (the 1.5000 vertical dimension in the Top view and the 2.5000 horizontal dimension in the Side view). Again, to remain consistent in the design process, you should make sure all dimensions have the same appearance (dimension text height, text orientation, whether they are broken inside instead of placed above the dimension line). Activate the MATCHPROP command again by clicking on the button in the Ribbon. Click the 4.00 horizontal dimension in the Top view as the source object. When the Match Property icon appears, as shown in Figure 7.74, select the 1.5000 and 2.5000 dimensions as the destination objects. All dimension properties associated with the 4.00 dimension are transferred to the 1.5000 and 2.5000 dimensions.

Command: **MA** *(for MATCHPROP)*

Select source object: *(Select the 4.00 dimension, which should highlight)*

Current active settings: Color Layer Ltype Ltscale Lineweight Transparency Thickness PlotStyle Dim Text Hatch Polyline Viewport Table Material Shadow display Multileader

Select destination object*(s)* or [Settings]: *(Select the 2.5000 dimension at "A")*

Select destination object*(s)* or [Settings]: *(Select the 1.5000 dimension at "B")*

Select destination object*(s)* or [Settings]: *(Press* ENTER *to exit this command)*

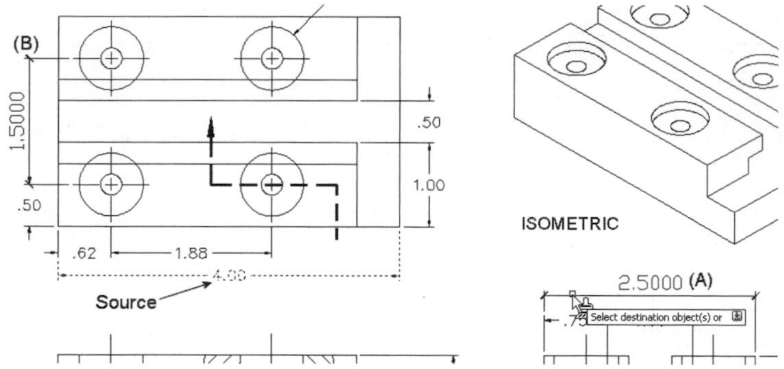

FIGURE 7.74

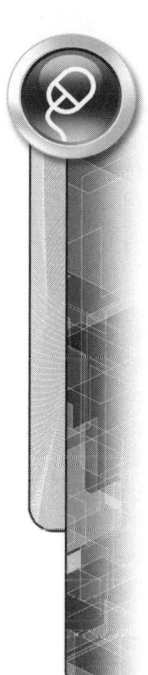

STEP 6

In the Front view, an area is crosshatched. However, both sets of crosshatching lines need to be drawn in the same direction, rather than opposing each other. Activate the MATCHPROP command again by clicking on the button in the Ribbon. Click the left hatch pattern as the source object. When the Match Property icon appears, as in Figure 7.75, select the right hatch pattern as the destination object. All hatch properties associated with the left hatch pattern are transferred to the right hatch pattern.

Command: MA *(for MATCHPROP)*

Select source object: *(Select the left hatch pattern, which should highlight)*

Current active settings: Color Layer Ltype Ltscale Lineweight Transparency Thickness PlotStyle Dim Text Hatch Polyline Viewport Table Material Shadow display Multileader

Select destination object*(s)* or [Settings]: *(Pick the right hatch pattern.)*

Select destination object*(s)* or [Settings]: *(Press* ENTER *to exit this command.)*

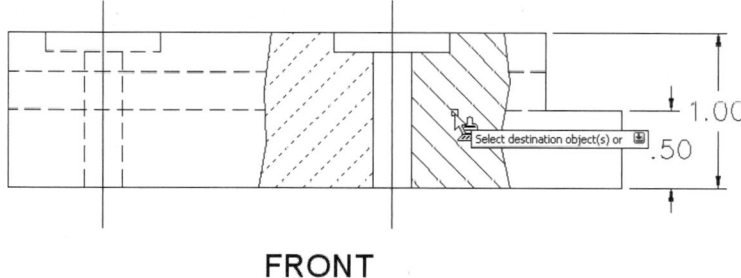

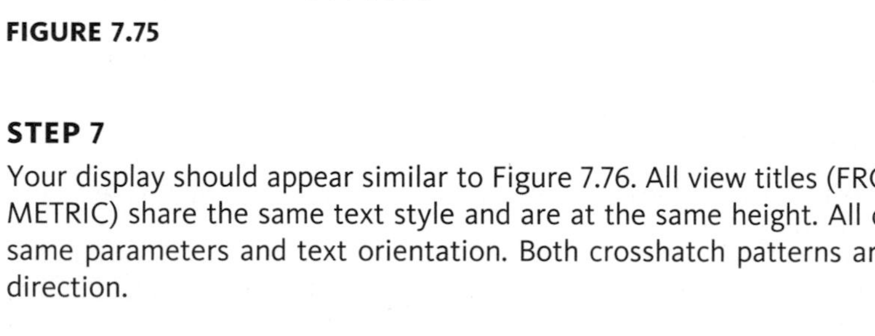

FIGURE 7.75

STEP 7

Your display should appear similar to Figure 7.76. All view titles (FRONT, TOP, SIDE, ISOMETRIC) share the same text style and are at the same height. All dimensions share the same parameters and text orientation. Both crosshatch patterns are drawn in the same direction.

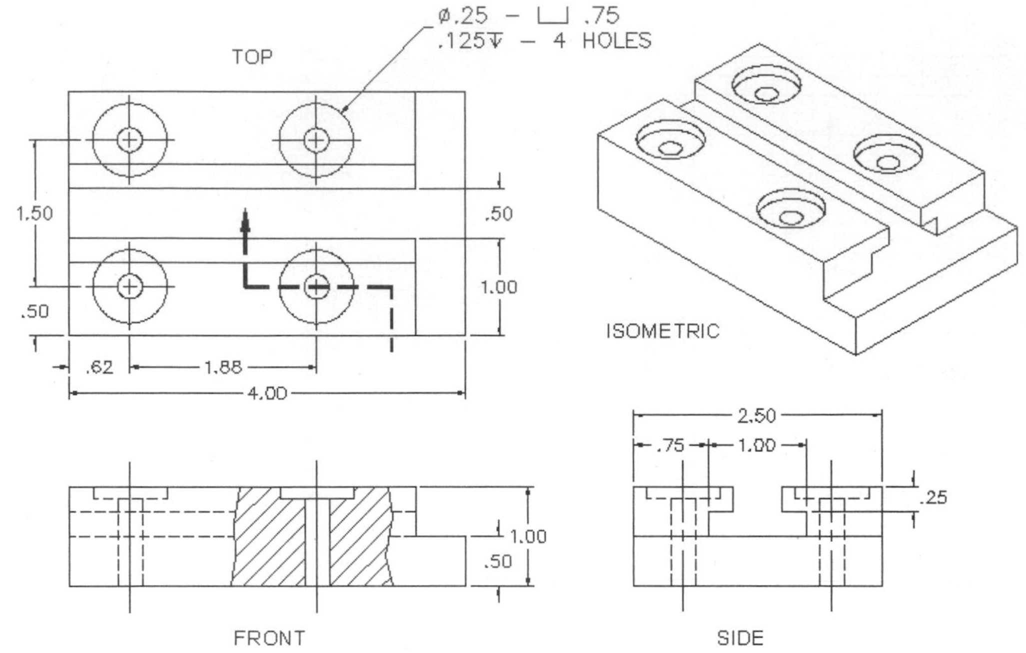

FIGURE 7.76

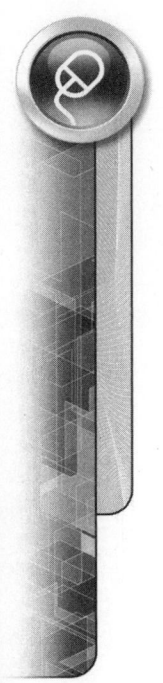

STEP 8

A different method will now be used to change the specific layer properties of objects. The two highlighted lines in the Top view in Figure 7.77 were accidentally drawn on Layer OBJECT and need to be transferred to the HIDDEN layer. With the lines highlighted, click on the Layer Control box to display all layers. Click on the HIDDEN layer to change the highlighted lines to the HIDDEN layer. Press ESC to remove the object highlight and the grips from the drawing.

NOTE

You may need to Regen the drawing in order to get the dashes to appear in the hidden lines.

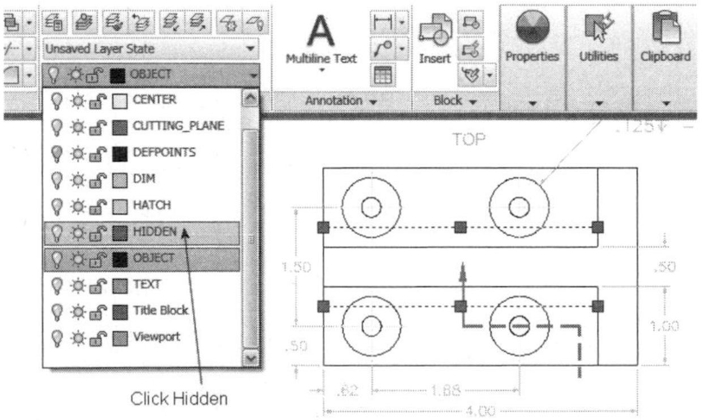

FIGURE 7.77

STEP 9

The two highlighted lines in the Front view in Figure 7.78 were accidentally drawn on the TEXT Layer and need to be transferred to the CENTER layer. With the lines highlighted, click on the Layer Control box to display all layers. Click on the CENTER layer to change the highlighted lines to the CENTER layer. Press ESC to remove the object highlight and the grips from the drawing.

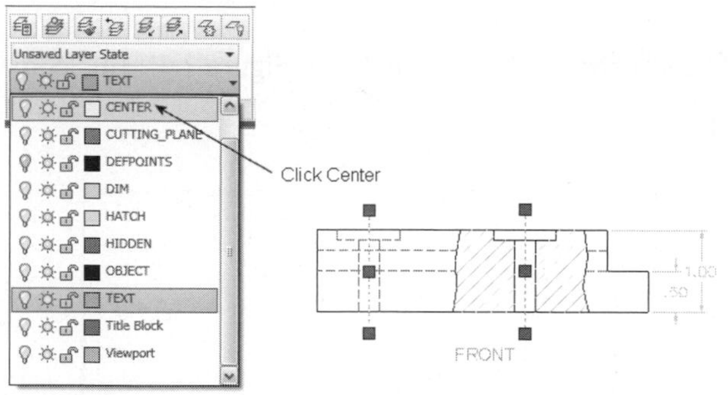

FIGURE 7.78

STEP 10

Your display should appear similar to Figure 7.79. Notice the hidden lines in the Top view and the centerlines in the Front view.

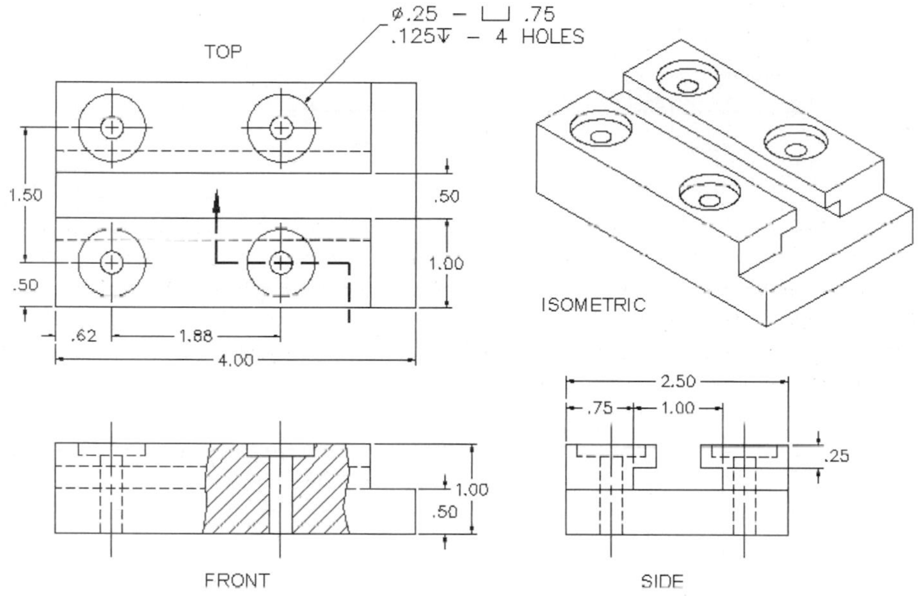

FIGURE 7.79

STEP 11

Click on the Orthographic Views tab. Select the rectangular viewport and change this object's layer to "Viewport," as shown in Figure 7.80.

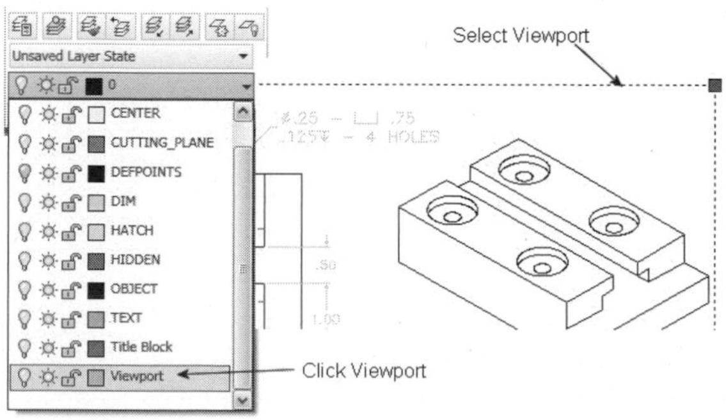

FIGURE 7.80

STEP 12

Turn off the Viewport layer. Your display should appear similar to Figure 7.81. This completes this tutorial exercise.

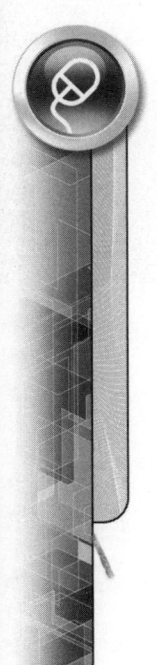

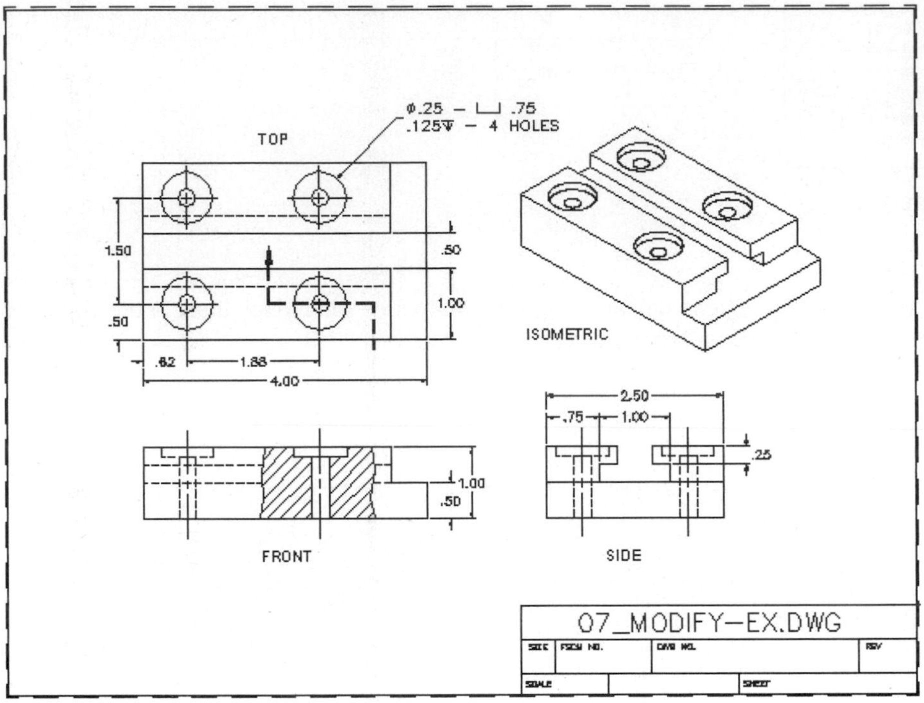

FIGURE 7.81

CHAPTER

8

Multiview and Auxiliary View Projections

Before any object can be manufactured, some type of drawing needs to be created. Not just any drawing, but an engineering drawing consisting of various views showing the object's surfaces in true shape and size so they can be clearly dimensioned. This is typically accomplished by laying out 2D orthographic (front, top, right, etc.) projections of the object organized on the computer screen. The first portion of this chapter introduces the topic of creating multiview projections and includes methods of constructing one-view, two-view, and three-view drawings using AutoCAD commands.

The second portion of this chapter introduces the topic of creating auxiliary views. Sometimes orthographic views cannot clearly describe an object, especially if features are located on an inclined or oblique surface. To show these types of surfaces in true shape and size, an auxiliary view is drawn.

ONE-VIEW DRAWINGS

An important rule to remember concerning multiview drawings is to draw only enough views to accurately describe the object. In the drawing of the gasket in Figure 8.1, Front and Side views are shown. However, the Side view is so narrow that it provides no real significant information. A better approach would be to leave out the Side view and construct a one-view drawing consisting of just the Front view. The material thickness could be given in a note on the drawing.

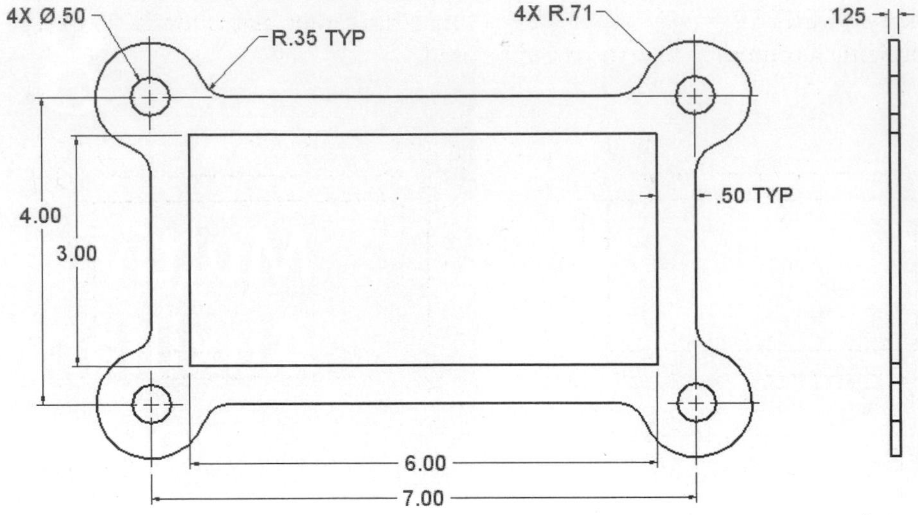

FIGURE 8.1

NOTE

Parts with a constant material thickness are often shown as one-view drawings. Cylindrical objects are also often good candidates for one-view drawings.

Begin the one-view drawing of the gasket by first using the RECTANG command to lay out a 7.00 by 4.00 rectangle marking the centers of all circles and arcs, as shown in Figure 8.2 (Left). A layer containing continuous object lines could be used for the rectangle.

Use the CIRCLE command to lay out all circles representing the bolt holes of the gasket, as shown in Figure 8.2 (Right).

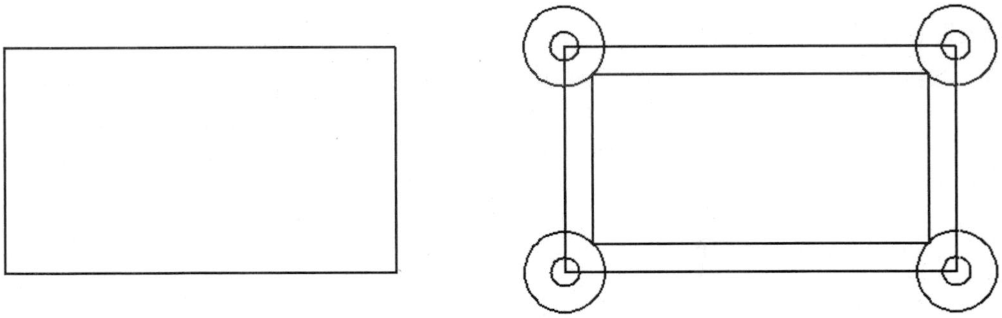

FIGURE 8.2

Use the TRIM command to begin forming the outside shape of the gasket, as shown in Figure 8.3 (Left).

Use the FILLET command set to the desired radius to form a smooth transition from the arcs to the outer rectangle, as shown in Figure 8.2 (Right). It will be necessary to use the EXPLODE command to convert the polylines, created from the trimmed rectangle, into lines before you can create the fillets.

Use the DIMCENTER or DCE command to create the center marks for the circles. A layer utilizing a centerline linetype could be used.

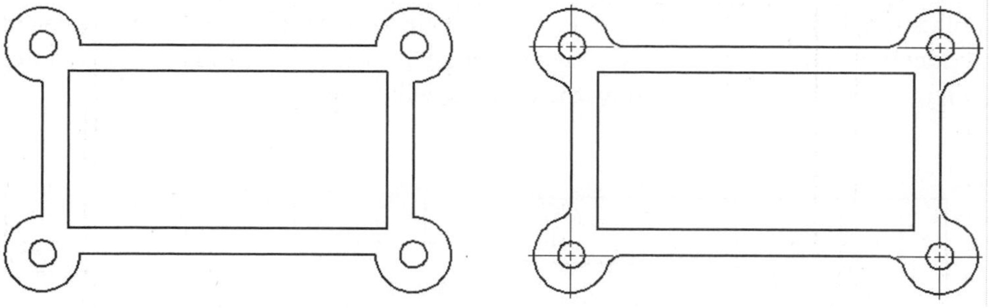

FIGURE 8.3

TWO-VIEW DRAWINGS

Before attempting any drawing, determine how many views need to be drawn. Show only the minimum number of views that are needed to describe an object. Drawing extra views is not only time consuming, but it adds to the complexity of a drawing, making it more difficult to interpret. You must determine which is the correct set of views. The illustration in Figure 8.4 shows the six possible orthographic projections for an object. Several of the views are mirror images: the Right and Left, the Top and Bottom, and the Front and Rear views. Unless one of the views shows a feature better (less hidden lines perhaps), the three standard views are usually selected—Front, Top, and Right Side. This eliminates the need for a Rear, Bottom, or Left Side view. The profiles of the object are best shown in the Top and Front views. The Right Side view does not add any clarification to the object's shape or size. This leaves the Front and Top as the only two necessary views.

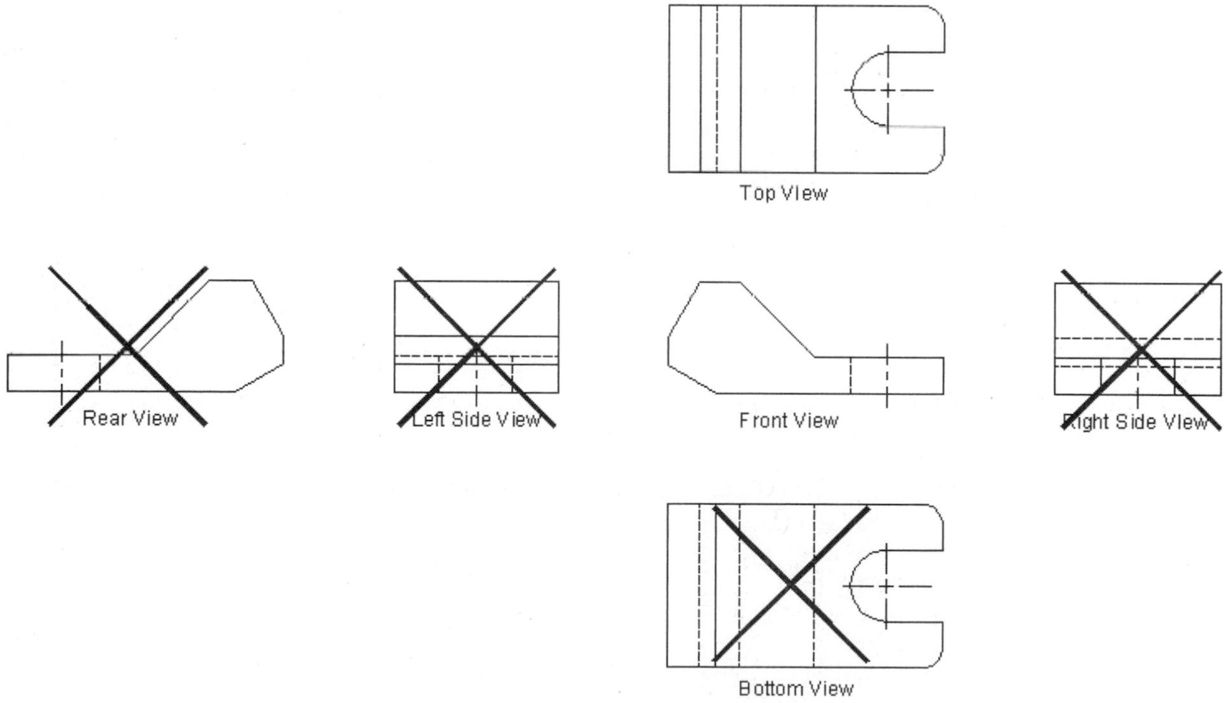

FIGURE 8.4

TRY IT!

Open the drawing file 08_R-Guide. Use Figure 8.5 and descriptions for constructing a two-view drawing of this object.

To illustrate how AutoCAD is used as the vehicle for creating a two-view engineering drawing, study the two-view drawing along with the pictorial drawing illustrated in Figure 8.5 to get an idea of how the drawing will appear.

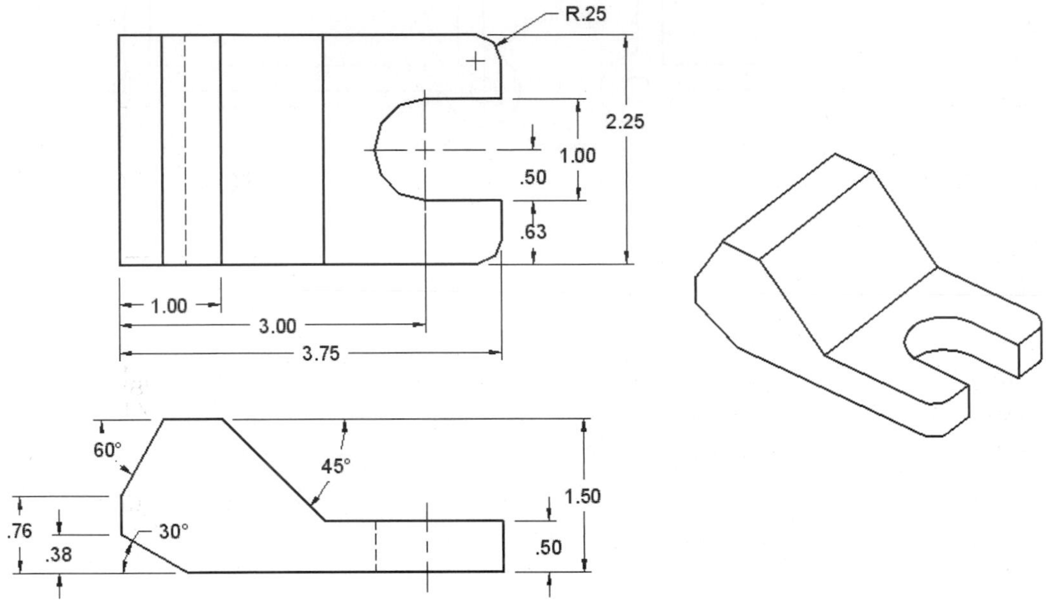

FIGURE 8.5

Begin the two-view drawing by using the `LINE` command to lay out the Front and Top views. You can find the width of the Top view by projecting lines up from the front because both views share the same width, as shown in Figure 8.6 (Left). Provide a space of 1.50 units between views to act as a separator and allow for dimensions at a later time. With the two views laid out, use the `TRIM` command to trim unnecessary lines in order for your drawing to appear similar to the illustration in Figure 8.6 (Right).

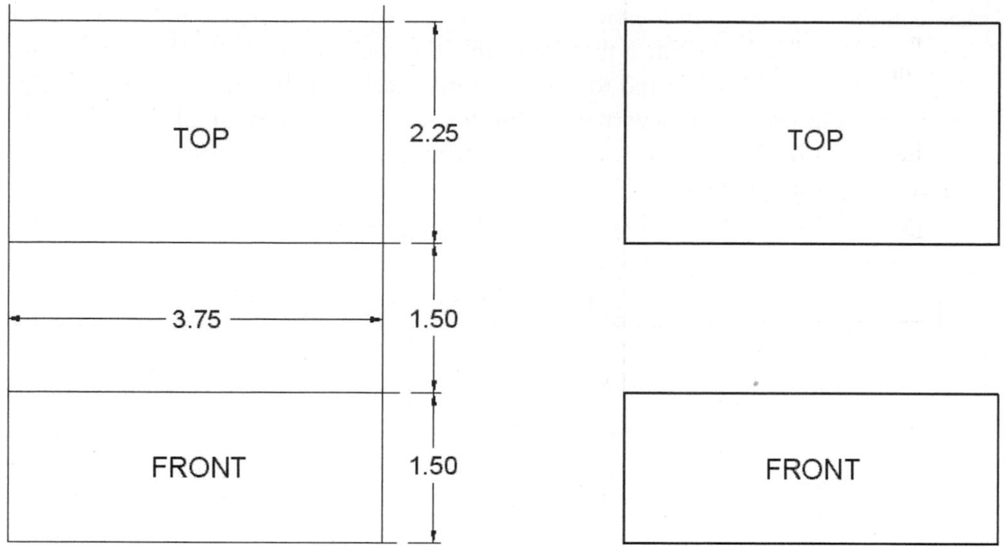

FIGURE 8.6

Next, add visible details to the views, such as arcs, filleted corners, and angles, as shown in Figure 8.7 (Left). Use various editing commands such as TRIM, EXTEND, and OFFSET to clean up unnecessary geometry.

From the Front view, project corners up to the Top view, as shown in Figure 8.7 (Right). These corners form visible edges in the Top view.

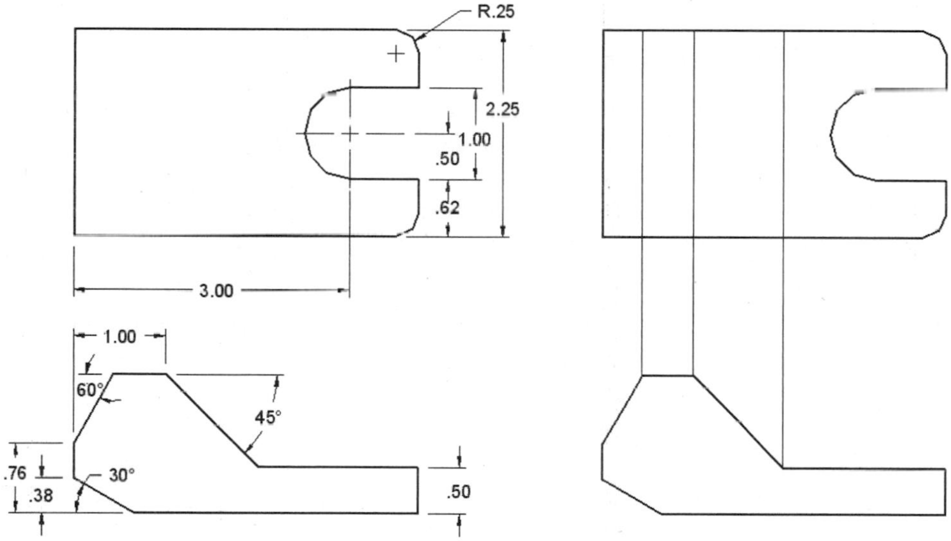

FIGURE 8.7

Use the same projection technique to project features from the Top view to the Front view, as shown in Figure 8.8 (Left). Then use the TRIM command to delete any geometry that appears in the 1.50 dimension space. The views now must conform to engineering standards by showing which lines are visible and which are hidden, as shown in Figure 8.8 (Right). Use the Layer Control box to change the line in the Top view from the Object layer to the Hidden layer. In the same manner, the slot visible in the Top view is hidden in the Front view. Again change the line in the Front view to the Hidden layer. Since the slot in the Top view represents a circular feature, use the DIMCENTER command to place a center marker at the center of the semicircle. To show in the Front view that the hidden line represents a circular feature, add one centerline consisting of one short dash and two longer dashes. Be sure this line is drawn on the Center layer.

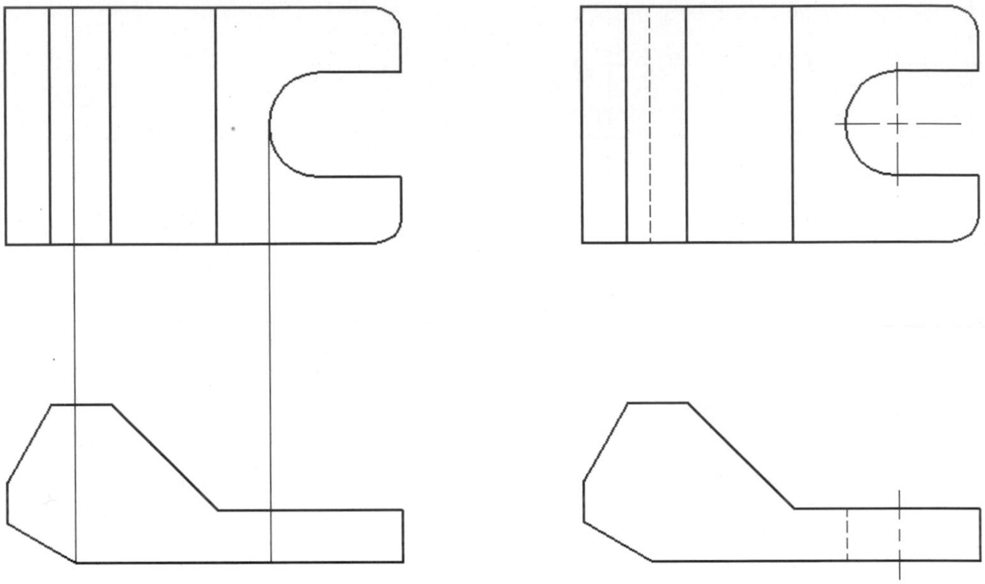

FIGURE 8.8

THREE-VIEW DRAWINGS

If two views are not enough to describe an object, three views may be required: in this case the three standard views—Front, Top, and Right Side views.

TRY IT!

Open the drawing file 08_Guide Block. Use Figure 8.9 and descriptions for constructing a three-view drawing of this object.

A three-view drawing of the guide block, as illustrated in orthographic and pictorial formats in Figure 8.9, will be the focus of this segment. Notice the broken section exposing the Spotface operation above a drill hole. The first line of the Spotface callout indicates that there are four holes with a diameter of 0.25 inches that goes all the way through the part. The second line indicates that the larger diameter hole (Spotface) has a depth of 0.125 inches.

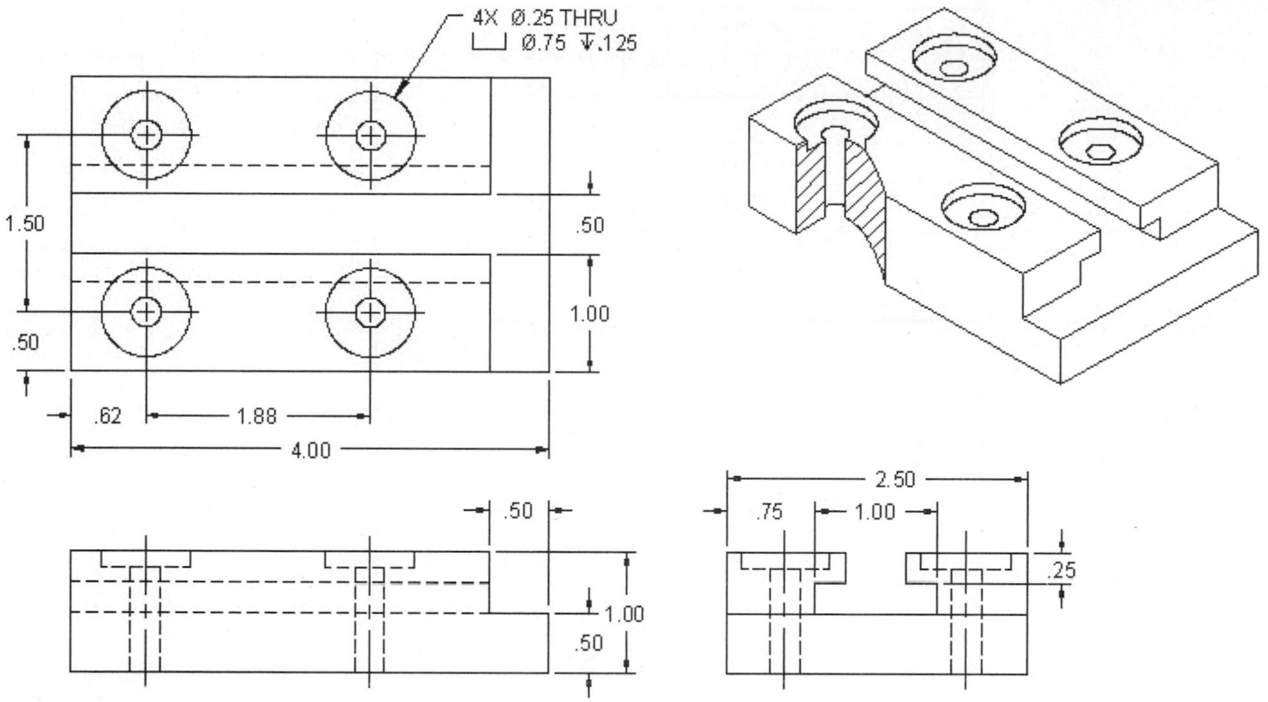

FIGURE 8.9

Begin this drawing by laying out all views using overall dimensions of width, depth, and height, as shown in Figure 8.10. The LINE and OFFSET commands are popular commands used to accomplish this. Provide a space between views to accommodate dimensions at a later time.

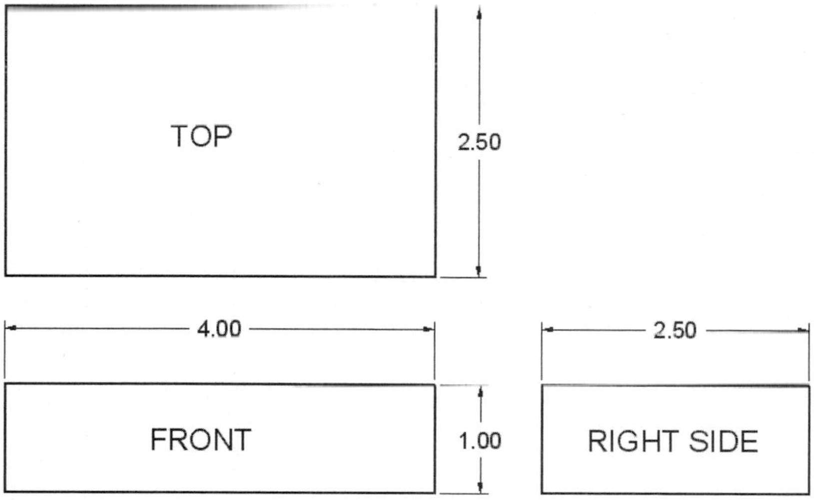

FIGURE 8.10

Next, draw features in the views where they are visible, as shown in Figure 8.11. Since the Spotface holes appear above, draw these in the Top view. The notch appears in the Front view; draw it there. A slot is visible in the Right Side view and is drawn there.

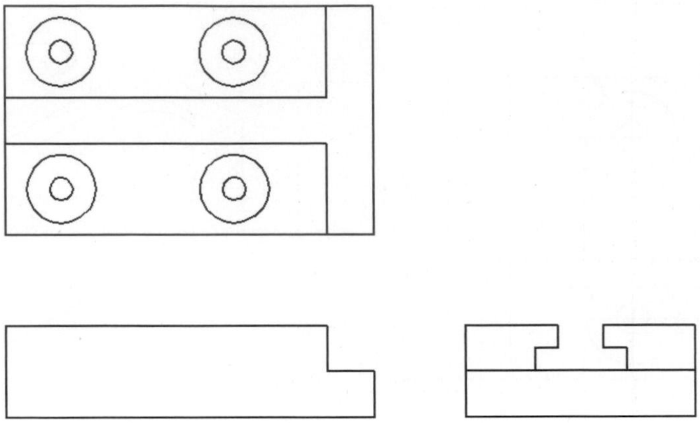

FIGURE 8.11

As in two-view drawings, all features are projected down from the Top to the Front view. To project depth measurements from the Top to the Right Side view, construct a 45° line, as shown in Figure 8.12. Project the holes from the Top view to the Front and Right Side views.

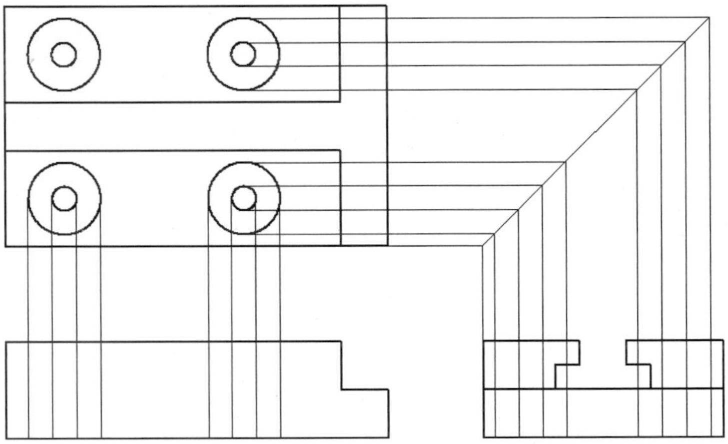

FIGURE 8.12

Use the 45° line to project the slot from the Right Side view to the Top view, as shown in Figure 8.13. Project the height of the slot from the Right Side view to the Front view.

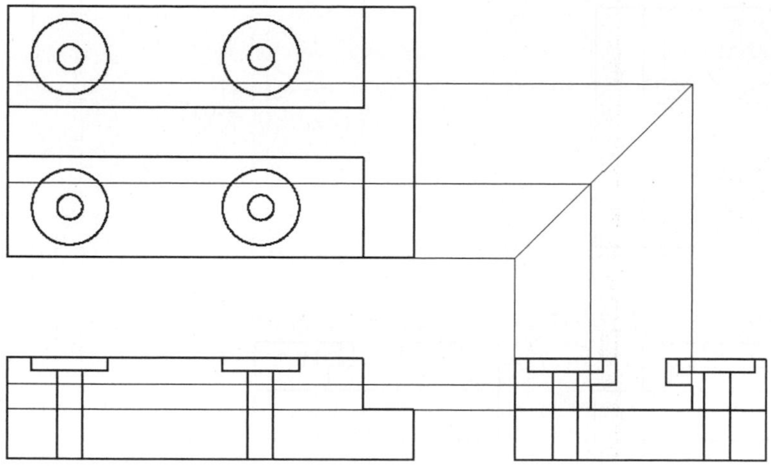

FIGURE 8.13

Change the continuous lines to the Hidden layer where features appear hidden, such as the holes in the Front and Right Side views, as shown in Figure 8.14. Change the lines for the slot in the Top and Front views to the Hidden layer. Erase any construction lines, including the 45°-projection line.

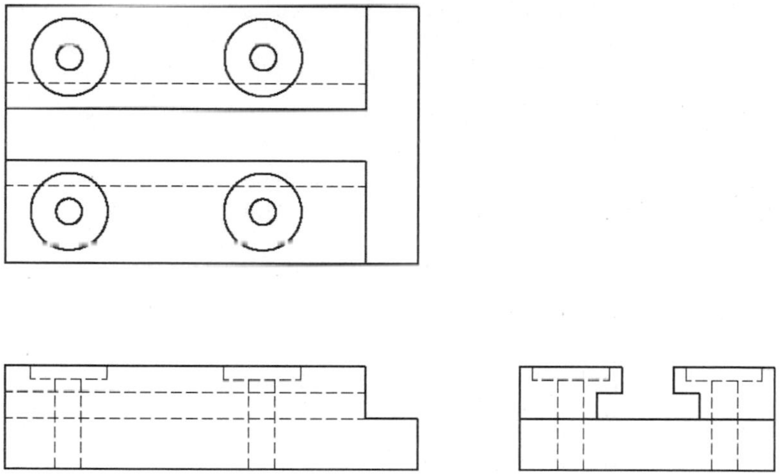

FIGURE 8.14

Begin adding centerlines to label circular features, as shown in Figure 8.15. Either construct these centerlines on the Center layer or change them later on to the Center layer. The DIMCENTER (type DCE at the Command prompt) command is used where the circles are visible. When features are hidden but represent circular features, the single centerline consisting of one short dash and two long dashes is used.

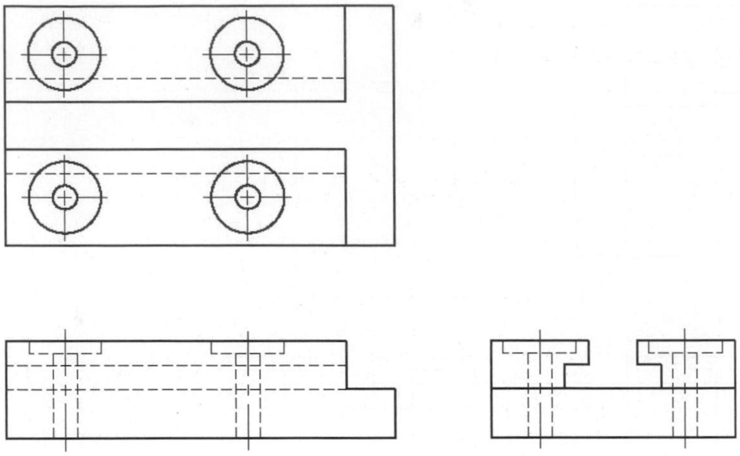

FIGURE 8.15

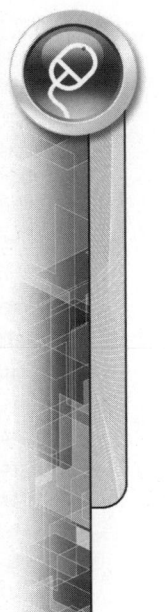

TUTORIAL EXERCISE: 08_ORTHOGRAPHIC BLOCK.DWG

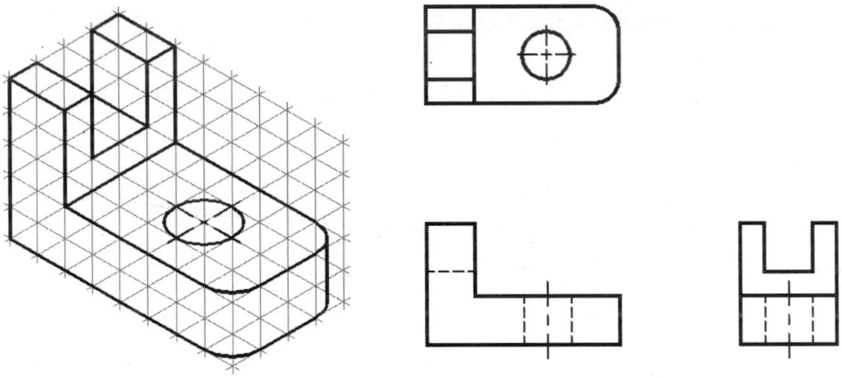

FIGURE 8.16

Purpose

Utililizing the isometric grid sketch provided, construct the Front, Top, and Right Side views of the Orthographic Block, as shown in Figure 8.16.

System Settings

Start a new drawing from scratch using the Acad.dwt template and save it as Orthographic Block.dwg. Use the Drawing Units dialog box to change the precision from four to two decimal places. Keep the remaining default unit values.

Using the LIMITS command, keep (0,0) for the lower-left corner and change the upper-right corner from (12,9) to (15.50,9.50). Perform a ZOOM-All operation to utilize the new limits setting.

Verify that the following Object Snap modes are already set: Endpoint, Center, Quadrant, Intersection, and Extension.

Layers

Create the following layers with the format:

Name	Color	Linetype	Lineweight
Object	Green	Continuous	0.50 mm
Hidden	Red	Hidden2	Default
Center	Yellow	Center2	Default

Suggested Commands

You will begin this tutorial by laying out the three primary views using the `LINE` and `OFFSET` commands. Use a grid spacing of 0.25 units for all distance calculations.

For this exercise, it is not necessary to actually create projection lines; use the grid lines to help you transfer the width, depth, and height dimensions between the views.

STEP 1

Begin constructing the Orthographic Block by laying out the Front, Top, and Right Side views using only the overall dimensions. Do not be concerned about details such as holes or slots; these will be added to the views in a later step. The width of the object shown in Figure 8.17 (Left) is 8 grid units (2 inches). The height of the object is 5 grid units (1.25 inches); the depth of the object is 4 grid units (1 inch). The distance between views is 5 grid units (1.25 inches).

STEP 2

Once the overall dimensions have been used to lay out the Front, Top, and Right Side views, begin adding visible details to the views. The "L" shape is added to the Front view; the hole and corner fillets are added to the Top view; the rectangular slot is added to the Right Side view, as shown in the following image on the right. Refer to the isometric view of this object at the beginning of this tutorial for the dimensions of the "L" shape, the hole, corner fillets, and the rectangular slot.

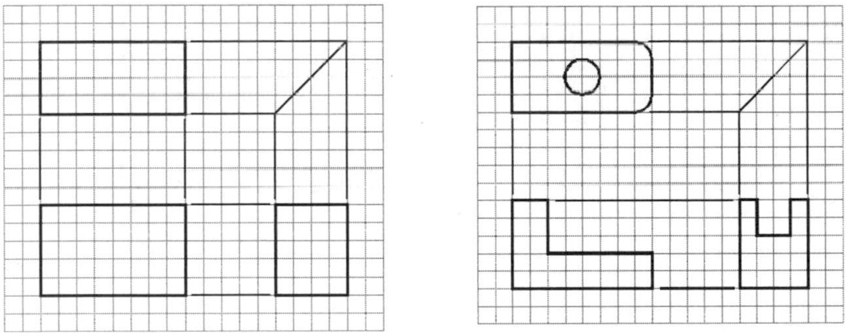

FIGURE 8.17

STEP 3

Begin projecting the visible edges from hole and slot features onto other views. Slot information is added to the Top view and height information is projected onto the Right Side view from the Front view. At this point, add only visible information to other views where required, as shown in Figure 8.18 (Left).

STEP 4

Now project all hidden features to the other views. The hole projection is hidden in the Front view along with the slot visible in the Right Side view. The hole is also hidden in the Right Side view, as shown in Figure 8.18 (Right). Notice how the 45° angle is used to project the hole from the Top view to the Right Side view.

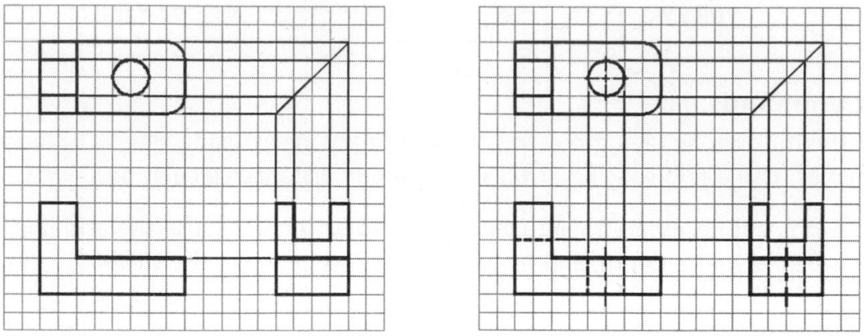

FIGURE 8.18

STEP 5

The completed multiview drawing solution is illustrated in Figure 8.19 (Left). Dimensions are added to document the exact size of the object, as shown in Figure 8.19 (Right). Proper placement of dimensions will be discussed in Chapter 10.

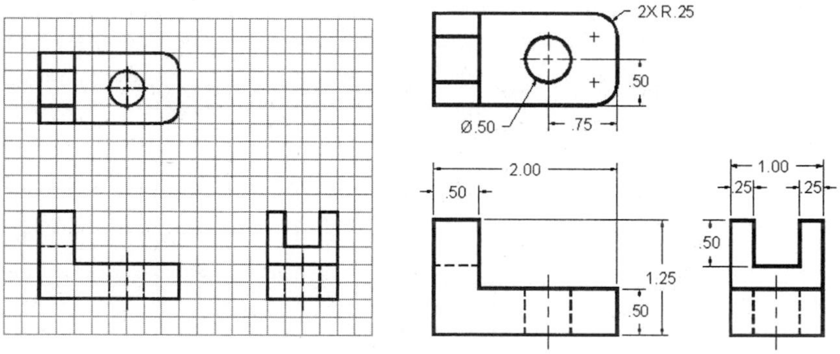

FIGURE 8.19

TUTORIAL EXERCISE: 08_SHIFTER.DWG

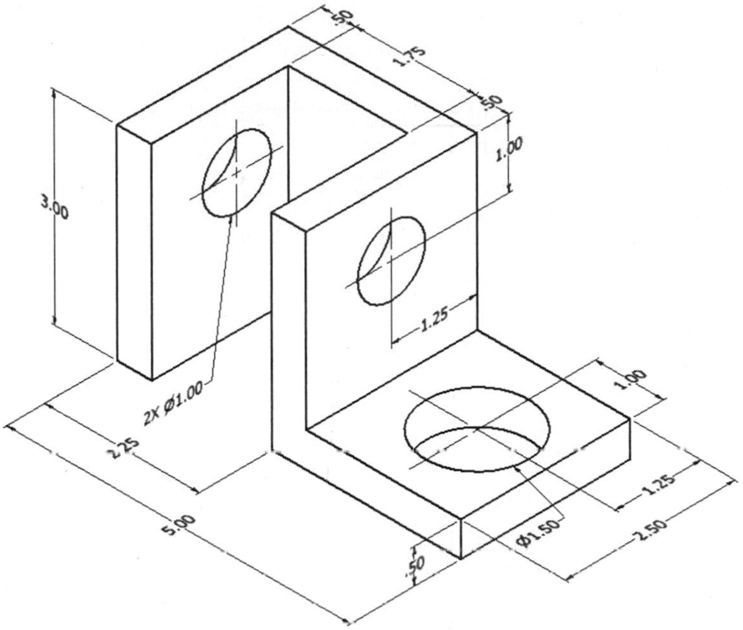

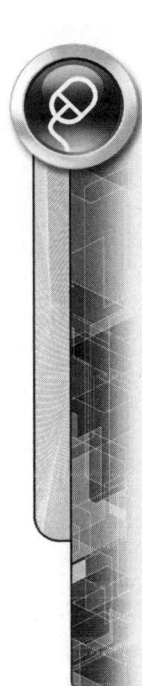

FIGURE 8.20

Purpose

This tutorial is designed to allow the user to construct a three-view drawing of the 08_Shifter, as shown in Figure 8.20.

System Settings

Since this drawing has been started for you and provided (08_Shifter.dwg), no settings are required. Verify that the following Object Snap modes are already set: Endpoint, Center, Quadrant, Intersection, and Extension.

Layers

The following layers have already been created with the following format:

Name	Color	Linetype	Lineweight
Center	Yellow	Center	Default
Dimension	Yellow	Continuous	Default
Hidden	Red	Hidden	Default
Object	Green	Continuous	0.50 mm
Projection	Cyan	Continuous	Default

Suggested Commands

The primary commands used during this tutorial are OFFSET and TRIM. The OFFSET command is used for laying out all views before the TRIM command is used to clean up excess lines. Since different linetypes represent certain features of a drawing, the Layer Control box is used to convert to the desired linetype needed as set in the Layer Properties Manager dialog box. Once all visible details are identified in the primary views, project the visible features to the other views using the LINE command. A 45° inclined line is constructed to project lines from the Top view to the Right Side view and vice versa.

STEP 1

Open the drawing file 08_Shifter. Make the Object layer current. Then use the LINE and OFFSET commands and create the three views displayed in Figure 8.21. The dimensions in this view represent the overall width, height, and depth of the Shifter. Space the views a distance of 1.50 units away from each other.

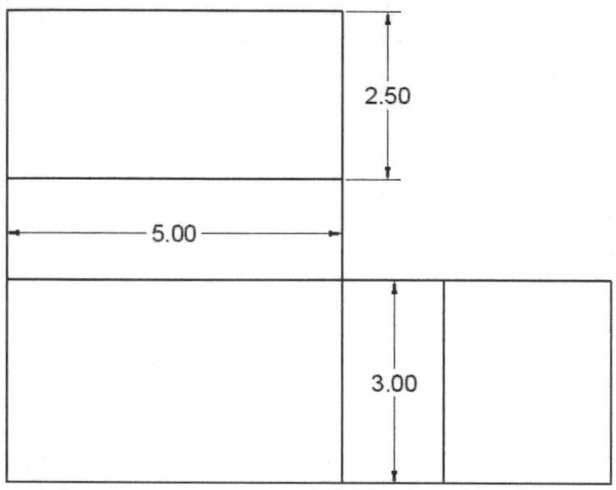

FIGURE 8.21

STEP 2

Continue using the OFFSET command to add the various features to the Front, Top, and Right Side views using the dimensions provided in Figure 8.22. Use the TRIM command to clean up all intersections in order for your display to appear similar to Figure 8.22.

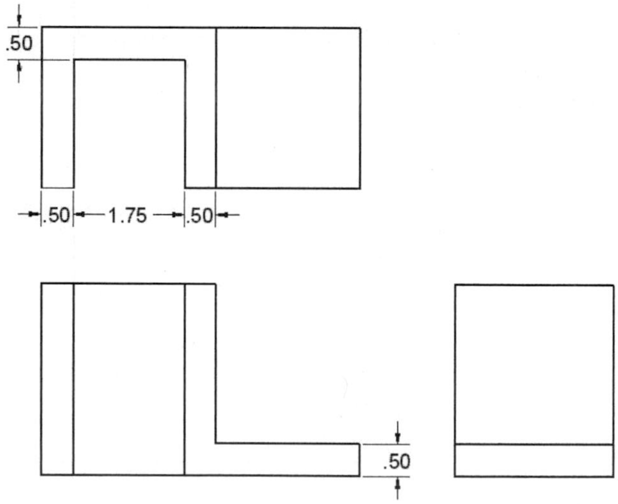

FIGURE 8.22

STEP 3

Add both circles to their respective views, as shown in Figure 8.23. One method of finding the circle center is to use the OFFSET command. Another method would be to use the OSNAP-From mode and the dimensions displayed in Figure 8.23.

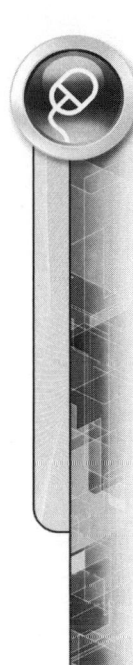

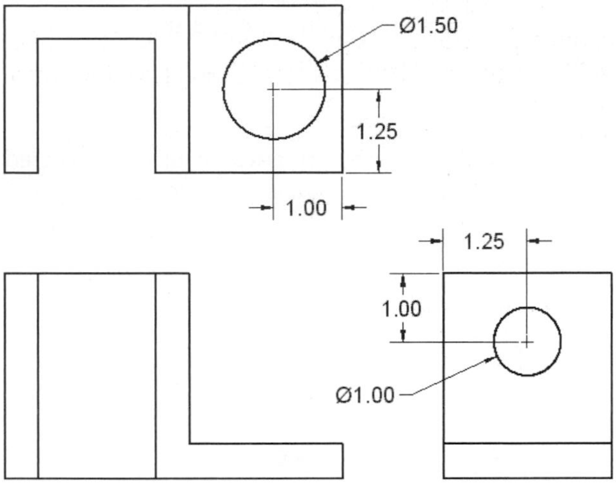

FIGURE 8.23

STEP 4

Project lines from the quadrants of the circles from the Top and Side views into the Front view, as shown in Figure 8.24. These projected lines in the Front view represent the hidden edges of the circles.

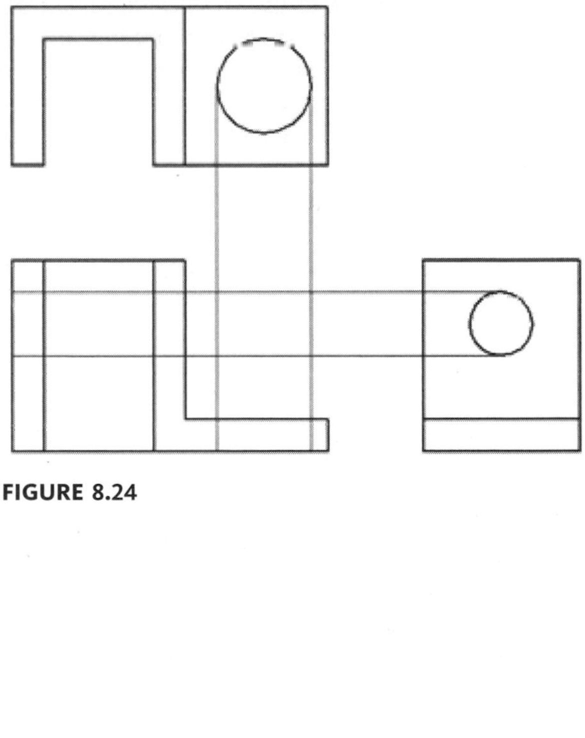

FIGURE 8.24

TIP

Instead of drawing projection lines that have to be trimmed, try using object snap tracking to create the hidden edges of the circles. Make the Hidden layer current and verify that Object Snap, Object Snap Tracking, and Polar Tracking are activated on the Status Bar. Start the `LINE` command and then hover your cursor over a Quadrant object snap to acquire it. Track from the object snap to the adjacent view and pick the start point of the line once you intersect the adjacent view; polar track to the next intersection and end the `LINE` command. Repeat for the other hidden edges of the circles.

STEP 5

Use Figure 8.25 to guide you when trimming the unnecessary portions of the projected lines representing the circle edges in the Front view. Then change these lines, highlighted in Figure 8.35 (Right), to the Hidden layer.

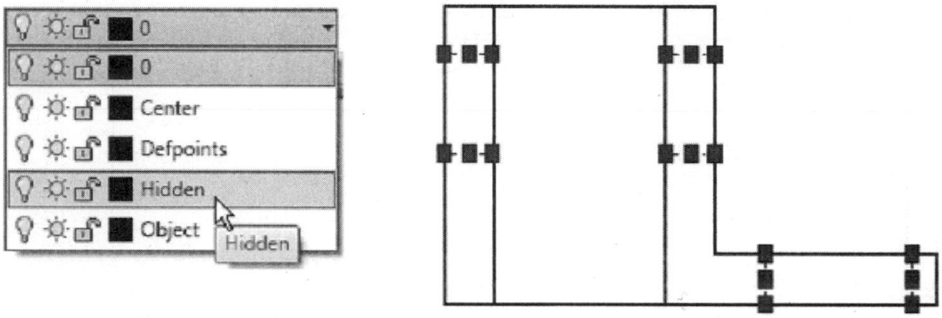

FIGURE 8.25

STEP 6

With the Center layer current, add center marks to both circles using the Dimension Center (`DCE`) command by touching the edge of each circle to place the center mark, as shown in Figure 8.26.

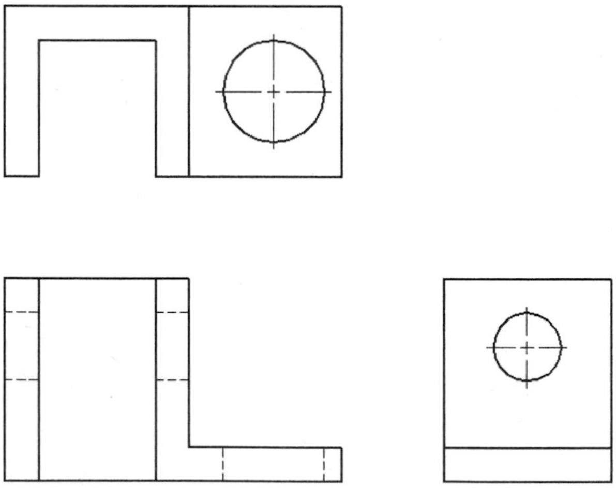

FIGURE 8.26

STEP 7

Project the endpoints of the center marks into the Front view, as shown in Figure 8.27.

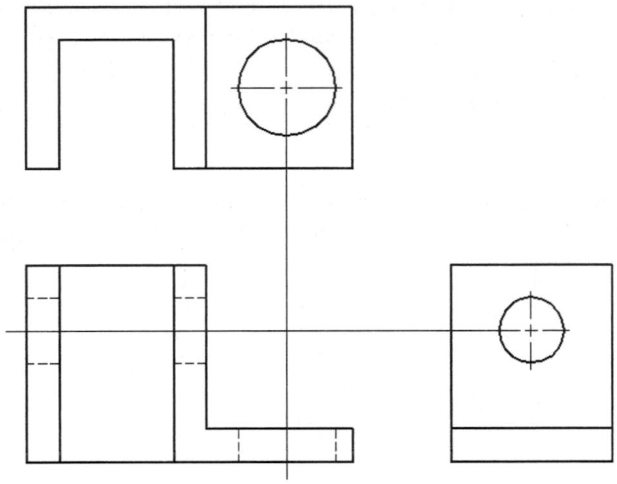

FIGURE 8.27

STEP 8

Use the TRIM command to remove the unwanted part of the projection lines. Use the LENGTHEN command with the DElta option to form the centerlines in the Front view. Set the DElta length to 0.25 and pick each end of the lines to lengthen them. Then change these lines to the Center layer, as shown in Figure 8.28.

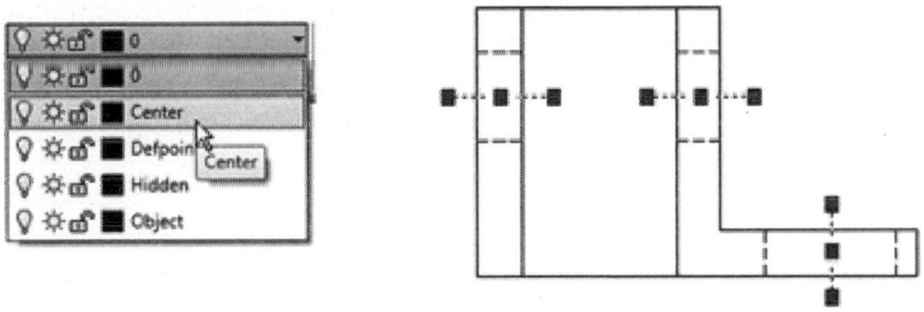

FIGURE 8.28

STEP 9

Use the FILLET command, with the radius set to "0", to create a corner between the Top and Side views, as shown in Figure 8.29. Then create a 45° line from the corner at "A" at a distance of 4 units. This distance is an approximation and will be used to project edges to the Top and Side views.

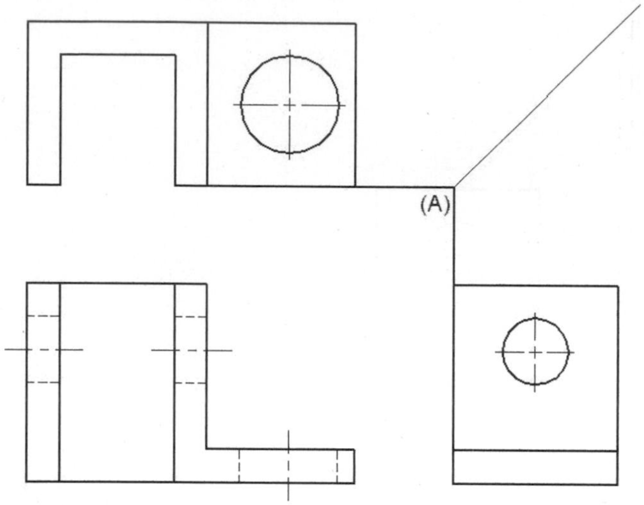

FIGURE 8.29

STEP 10

Project lines from the edges of the circles in the Top view to intersect with the 45° angle. Then continue constructing lines into the Side view, as shown in Figure 8.30.

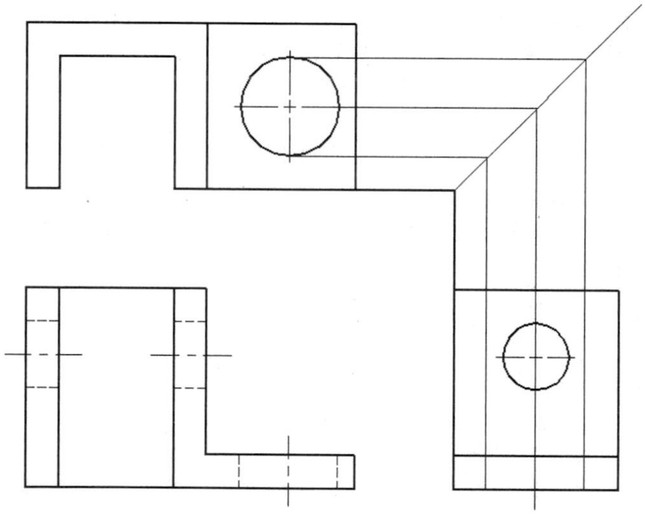

FIGURE 8.30

STEP 11

Use the TRIM and LENGTHEN commands to clean up lines in the Side view, as shown in Figure 8.31. Then change these lines to the Hidden and Center layers.

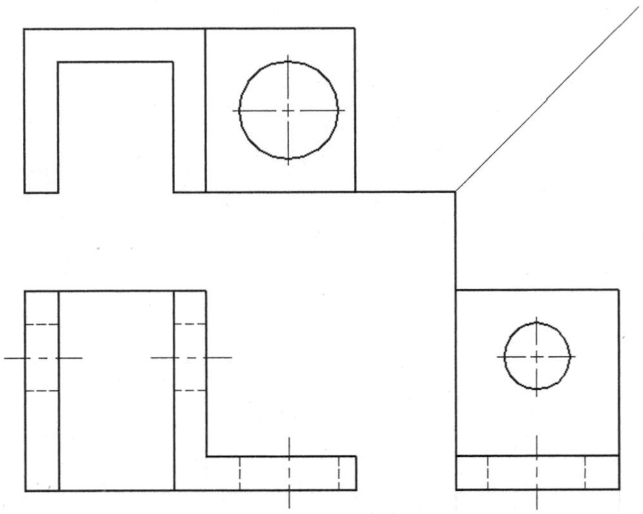

FIGURE 8.31

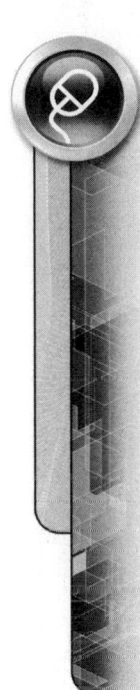

STEP 12

Project lines from the edges of the circles in the Side view to intersect with the 45° angle. Then continue constructing lines into the Top view, as shown in Figure 8.32.

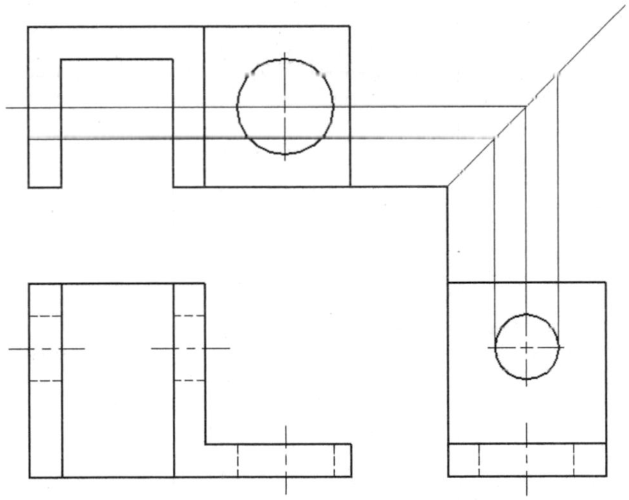

FIGURE 8.32

STEP 13

Use the TRIM and LENGTHEN commands to clean up lines in the Top view, as shown in Figure 8.33. Then change these lines to the Hidden and Center layers.

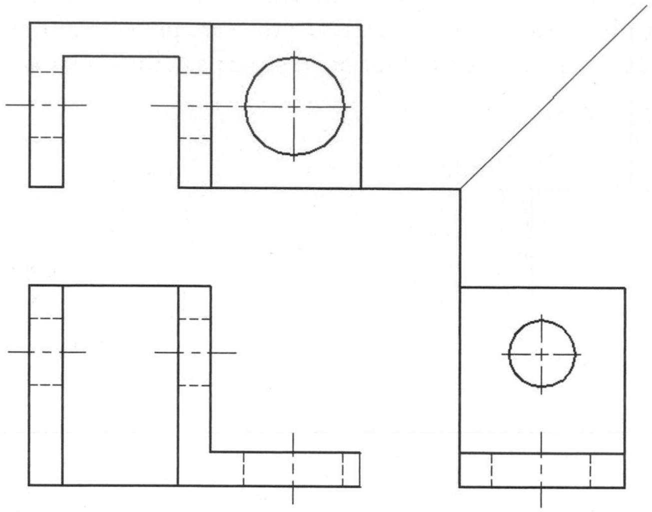

FIGURE 8.33

STEP 14

Erase the 45°-angle line and trim the corner connecting the Top and Side views. Finally, use the EXTEND command to extend the hidden line in the Side view to the top of the line at "A." This line represents a hidden edge. The completed orthographic drawing of the Shifter is shown in Figure 8.34.

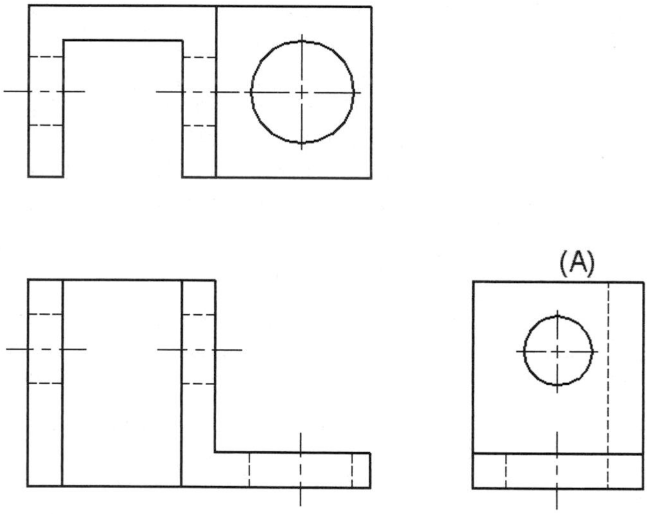

FIGURE 8.34

CREATING AUXILIARY VIEWS

During the discussion of multiview drawings, we discovered that you need to draw only those views of an object that are necessary to accurately describe it. In many cases, this only requires a Front, Top, and Right Side view. Sometimes additional views, such as the Left Side, Bottom, and Back views, are required to show features not visible in the three standard views. Other special views, such as sections, may be required to expose interior details for better clarity. Sections will be discussed in Chapter 9. If an object has unique features that are located on inclined or oblique surfaces, special views called auxiliary views are required. An auxiliary view is projected perpendicular to the sloped surface as shown in Figure 8.35. This projection then shows that surface and its features in true shape and size. This portion of the chapter will describe where auxiliary views are used and how they are projected from one view to another. A tutorial exercise is presented to show the steps in the construction of an auxiliary view.

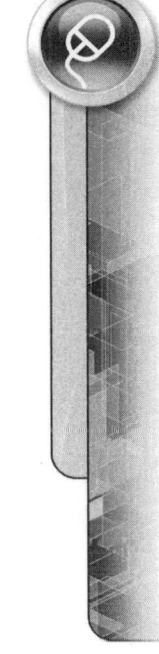

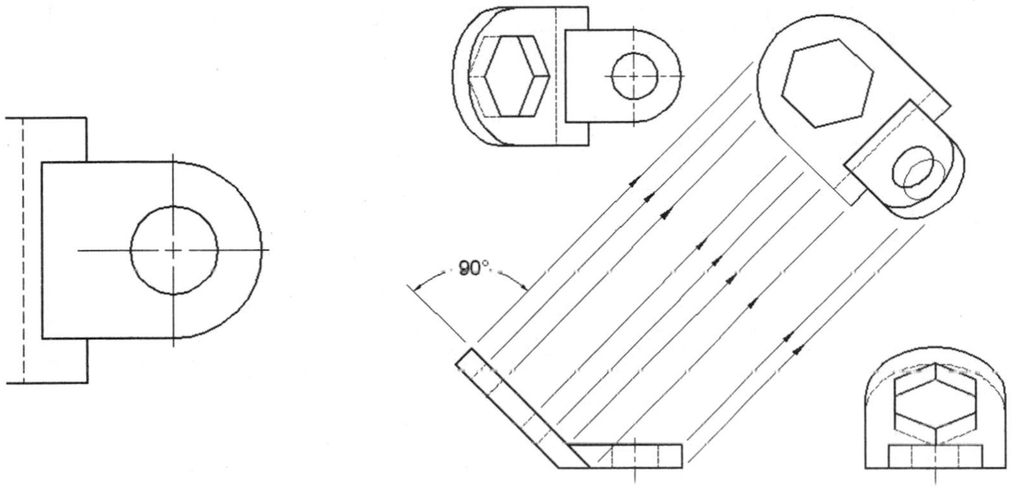

FIGURE 8.35

CONSTRUCTING AN AUXILIARY VIEW

TRY IT!

Open the drawing file 08_Aux Basics. Illustrated in Figure 8.36 (Left) is a basic multiview drawing consisting of Front, Top, and Right Side views. The inclined surface in the Front view is displayed in the Top and Right Side views; however, the surface appears foreshortened in both adjacent views. An auxiliary view of the incline will be made to show its true size and shape. Follow the next series of images that illustrate one suggested method for projecting an auxiliary view. Notice that the current layer is Construction. You will create all lines in this layer. Later you will change the lines to their correct layer to indicate their purpose, such as Object and Hidden.

Begin by using the OFFSET command to create a line parallel to the inclined surface at a distance of 3.25 units. The choice of 3.25 is only significant in that it will be the distance between the Front view and the Auxiliary view. Offset the line just created an additional 1 unit. This establishes the depth of the object as is shown in the Right Side view.

Command: O *(For OFFSET)*

Current settings: Erase source=No Layer=Source OFFSETGAPTYPE=0

Specify offset distance or [Through/Erase/Layer] <Through>: 3.25

Select object to offset or [Exit/Undo] <Exit>: *(Pick inclined surface line "A")*

Specify point on side to offset or [Exit/Multiple/Undo] <Exit>: *(Pick a point on your screen at "B")*

Select object to offset or [Exit/Undo] <Exit>: *(Press* ENTER *to exit this command and perform the operation.)*

Command: O *(For OFFSET)*

Current settings: Erase source=No Layer=Source OFFSETGAPTYPE=0

Specify offset distance or [Through/Erase/Layer] <Through>: 1.00

Select object to offset or [Exit/Undo] <Exit>: *(Pick the line at "B")*

Specify point on side to offset or [Exit/Multiple/Undo] <Exit>: *(Pick a point on your screen at "C")*

Select object to offset or [Exit/Undo] <Exit>: *(Press* ENTER *to exit this command and perform the operation.)*

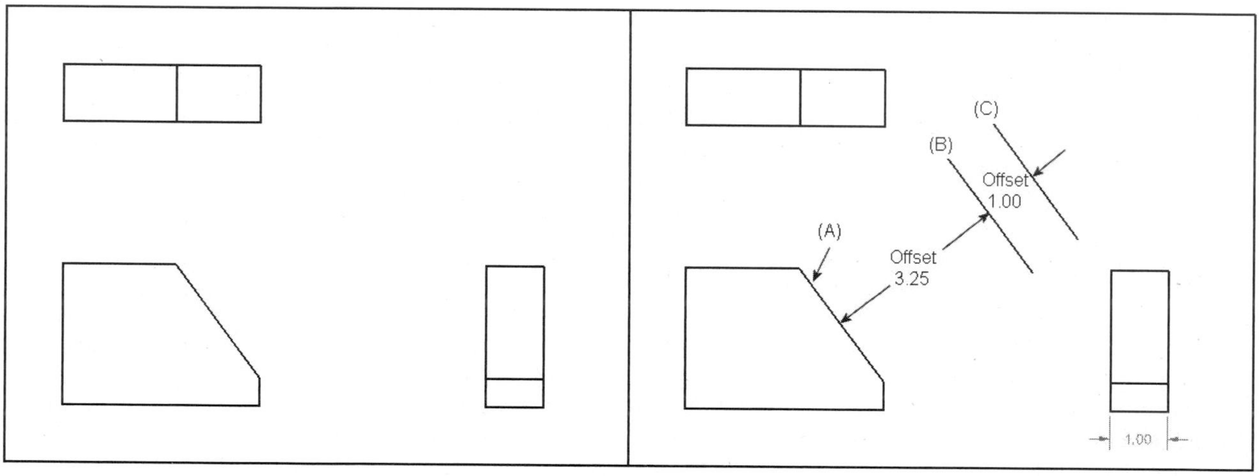

FIGURE 8.36

Construct the perpendicular projection lines that make up the auxiliary view. Use the LINE command and Perpendicular OSNAP to create lines perpendicular to the incline line in the front view at the five endpoint locations, as shown in Figure 8.37 (Right).

Command: L *(For LINE)*

Specify first point: END *(For Endpoint OSNAP)*

of *(Pick the point at "A")*

Specify next point or [Undo]: PER *(For Perpendicular OSNAP)*

of *(Pick the line at "F")*

Specify next point or [Undo]: *(Press* ENTER *to exit this command and perform the operation.)*

Repeat the LINE command sequence listed above for points "B" through "E."

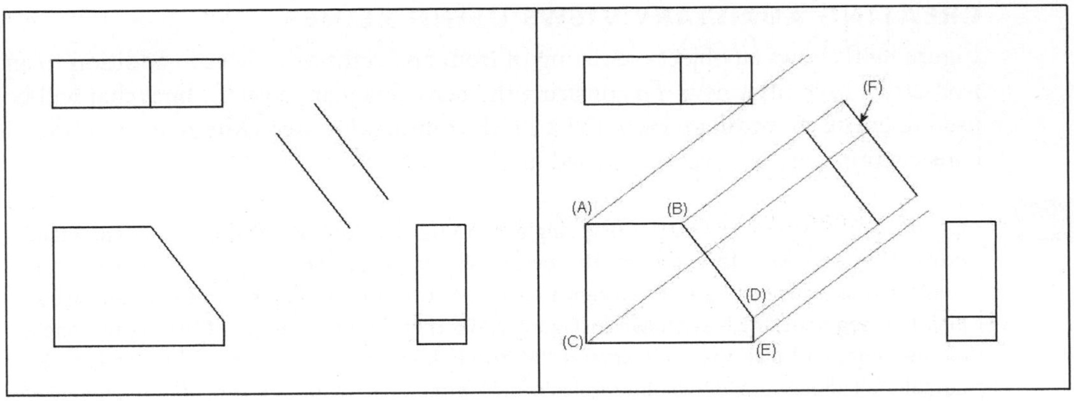

FIGURE 8.37

As shown in Figure 8.38 (Left), use the EXTEND command to extend the two object lines ("A" and "B") to the boundary lines shown as "C" and "D." In Figure 8.38 (Right), use the TRIM command to remove the unneeded portion of the projection lines at the cutting edge "E."

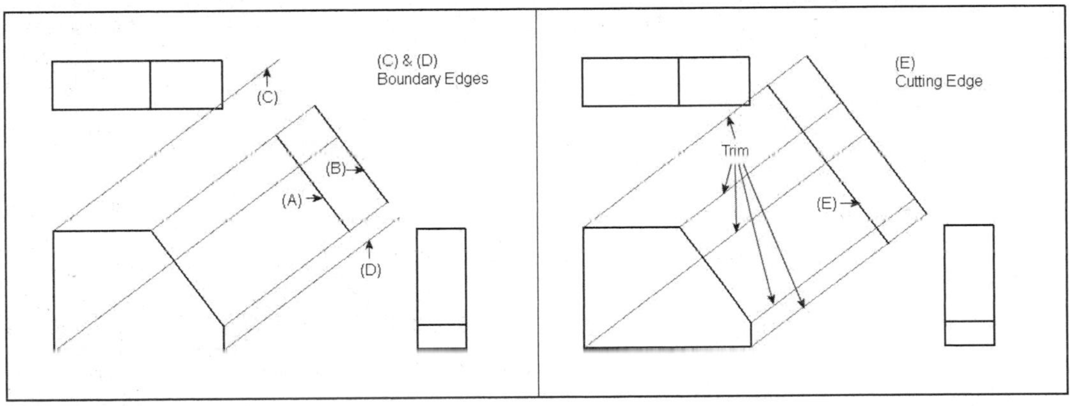

FIGURE 8.38

Change lines to the Object and Hidden layers, as shown in Figure 8.39 (Left). The result is a multiview drawing complete with auxiliary view displaying the true size and shape of the inclined surface, as shown in Figure 8.39 (Right). For dimensioning purposes, aligned dimensions could be used to annotate the distances located in the auxiliary view.

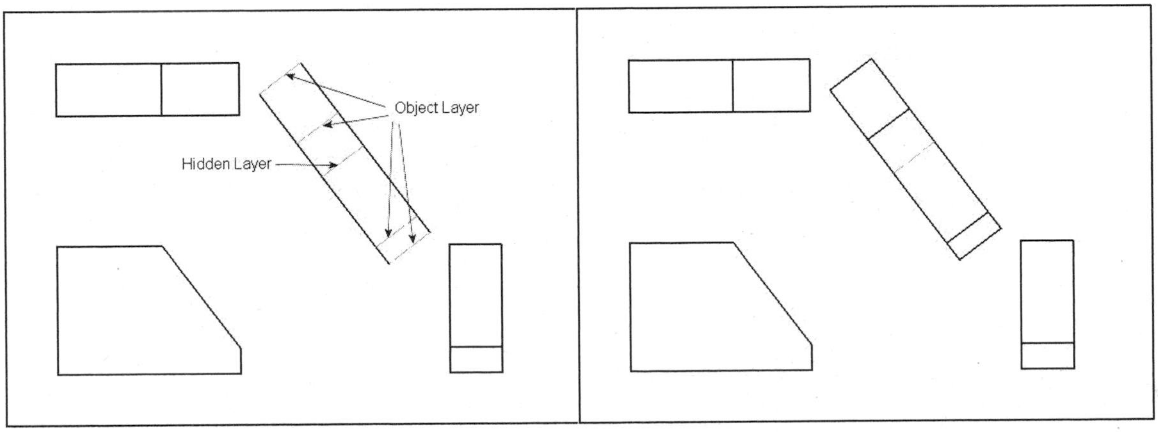

FIGURE 8.39

CREATING AUXILIARY VIEWS USING XLINES

Figure 8.40 shows an object consisting of front and right side views in addition to an isometric or pictorial view. To construct the perpendicular projector lines that will be used to create the auxiliary view, the XLINE command is used. Xlines are considered construction lines and were discussed in Chapter 5.

TRY IT!

Open the file 08_Aux Xlines.dwg. The Angle and Reference options of the XLINE command provide an easy way to construct the perpendicular projection lines. You first enter the Angle option followed by the Reference option. You will be prompted to select an object; pick line segment "AE," as shown in Figure 8.40. Then enter an angle of 90° as the angle of the xline, and pick the endpoints at "A" through "E" to construct the xline objects. By default, all xlines are drawn infinitely in two directions. You can trim, fillet, and even break xlines.

Command: **XL** *(For XLINE)*

XLINE Specify a point or [Hor/Ver/Ang/Bisect/Offset]:

A *(For Ang)*

Enter angle of xline *(0)* or [Reference]: **R** *(For Reference)*

Select a line object: *(Select line segment "AE")*

Enter angle of xline <0>: **90**

Specify through point: *(Pick the endpoints of points "A" through "E")*

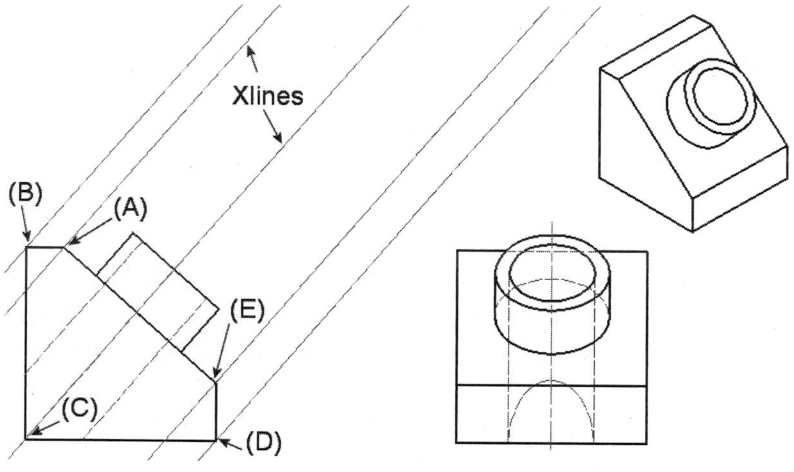

FIGURE 8.40

With the perpendicular projection lines constructed, next construct a base edge that is perpendicular to the projection lines. This base edge normally forms one of the edges of the auxiliary view. In Figure 8.41, line segment "1-2" located in the right side view coincides with xline segment "1-2" found in the auxiliary view.

Command: XL *(For XLINE)*

XLINE Specify a point or [Hor/Ver/Ang/Bisect/Offset]: A *(For Ang)*

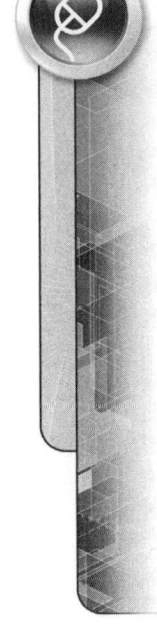

Enter angle of xline *(0)* or [Reference]: R *(For Reference)*

Select a line object: *(Pick one of the existing projection xlines)*

Enter angle of xline <0>: 90

Specify through point: *(Pick a point near "1" in the drawing to locate the base edge of the auxiliary view.)*

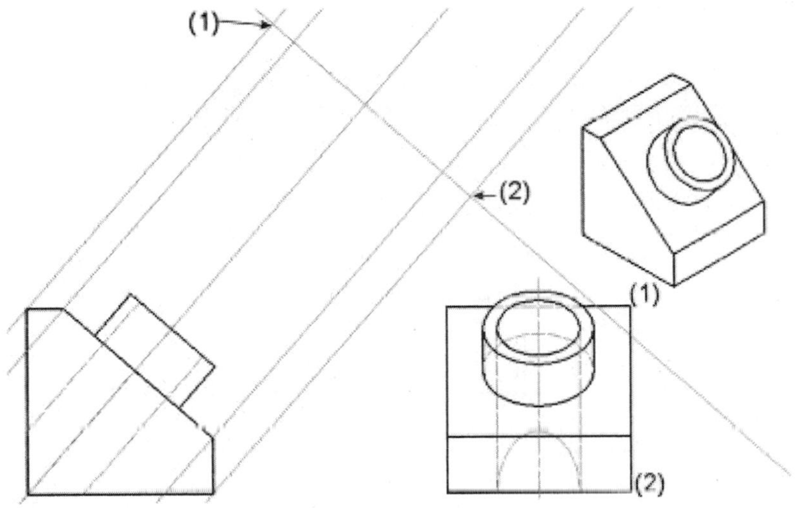

FIGURE 8.41

TRANSFERRING DISTANCES WITH THE OFFSET COMMAND

With the perpendicular projectors constructed along with the base edge using xlines, the next step is to transfer distances from one view and create the auxiliary view. The auxiliary view in our example is a depth type. All views projected from the front view show depth. This means that the distances along the projectors can be transferred from the depth dimensions in Top or Right Side view.

TRY IT!

Open the file 08_Aux Offset.dwg or continue with the previous Try It! exercise. In this example the depth will be transferred from the Right Side view using the OFFSET command. To accomplish this, activate the command and for the offset distance, pick the endpoint at "A" in the Right Side view, as shown in Figure 8.42. For the second point, pick the endpoint at "B." These two endpoints form the depth of the Right Side view. To transfer this distance to the auxiliary view, pick line "C" as the object to offset and then pick a location at "D" as the side to perform the offset shown in Figure 8.42.

Command: O *(For OFFSET)*

Current settings: Erase source=No Layer=Source OFFSETGAPTYPE=0

Specify offset distance or [Through/Erase/Layer] <Through>: *(Pick the endpoint at "A")*

Specify second point: *(Pick the endpoint at "B")*

Select object to offset or [Exit/Undo] <Exit>: *(Pick line "C")*

Specify point on side to offset or [Exit/Multiple/Undo] <Exit>: *(Pick the location at "D")*

Select object to offset or [Exit/Undo] <Exit>: *(Press* ENTER *to exit this command)*

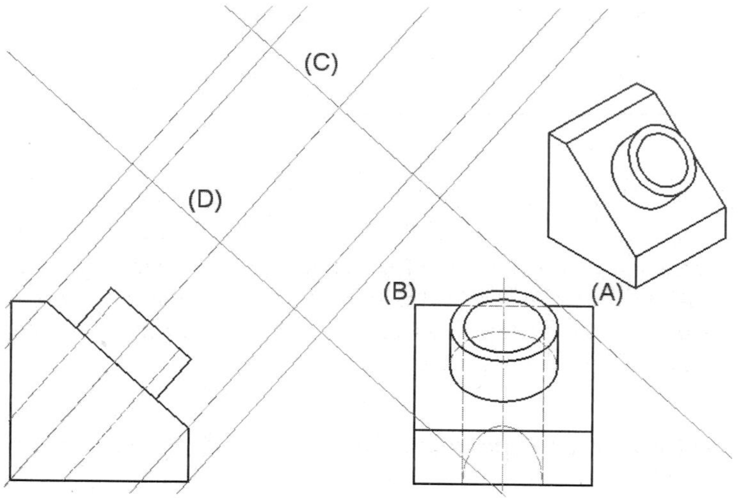

FIGURE 8.42

Continue using the OFFSET command to transfer more distances, such as the hole location, from the Right Side view to the auxiliary view, as shown in Figure 8.42. To complete the auxiliary view use the TRIM and/or FILLET commands to remove excess lines. You may want to add additional projection xlines to help lay out the circles. Place lines and circles on the appropriate layers.

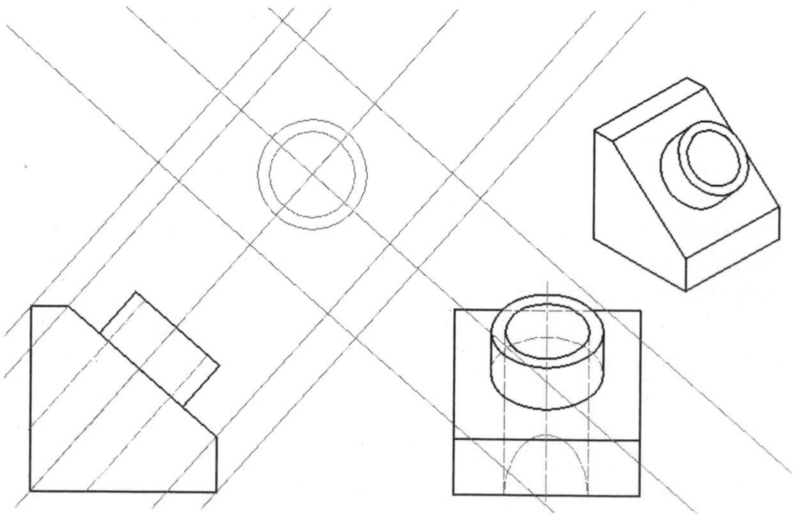

FIGURE 8.43

The completed auxiliary view is illustrated in Figure 8.44.

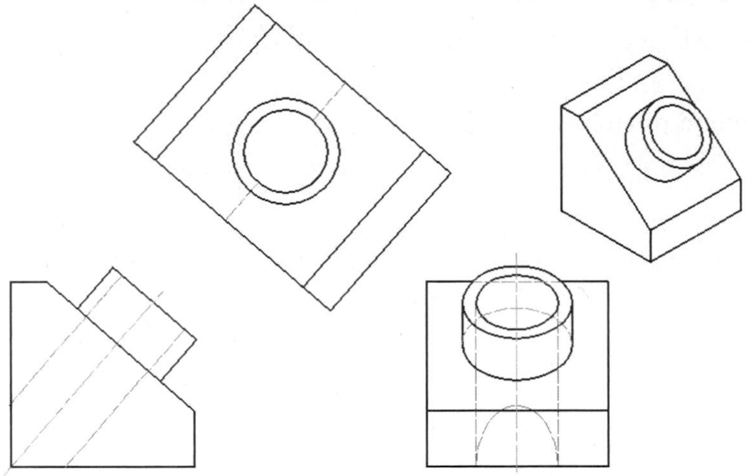

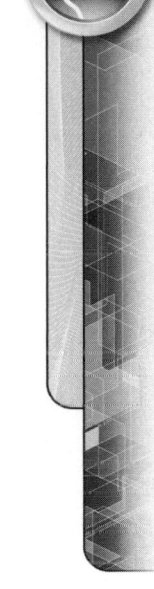

FIGURE 8.44

CONSTRUCTING THE TRUE SIZE OF A CURVED SURFACE

TRY IT!

Open the file 08_Aux Curve.dwg as shown in Figure 8.45 (Left). From Front and Right Side views, an auxiliary view will be created to display the true size and shape of the inclined surface. Notice that a layer called Construction has already been created and is current. This layer will be used throughout this tutorial exercise.

Construct the perpendicular projection lines that make up the auxiliary view. Use the `XLINE` command to create three infinite lines perpendicular to the incline in the Front view at the three locations shown in Figure 8.45 (Right). These form the projection lines used for beginning the auxiliary view.

```
Command: XL (For XLINE)
Specify a point or [Hor/Ver/Ang/Bisect/Offset]: A (For Ang)
Enter angle of xline (0) or [Reference]: R (For Reference)
Select a line object: (Select the inclined line in the Front
view)
Enter angle of xline <0>: 90
Specify through point: (Pick the three locations)
```

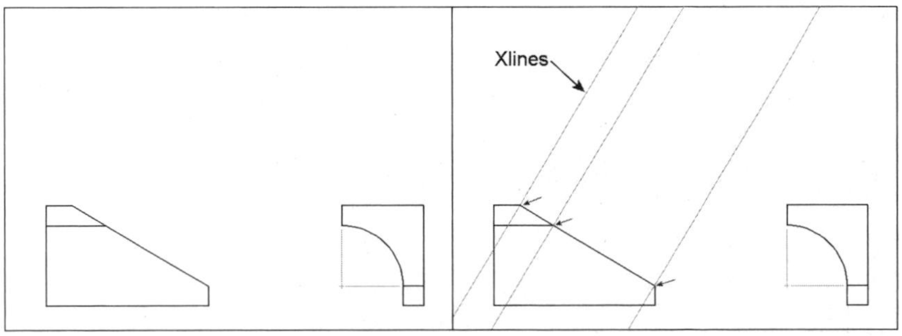

FIGURE 8.45

Create another xline; however, this time the infinite line will be constructed parallel to the inclined line in the Front view. This line will form the back edge of the finished auxiliary view identified by points 1 and 2 shown in Figure 8.46 (Left).

Command: XL *(For XLINE)*

Specify a point or [Hor/Ver/Ang/Bisect/Offset]: A *(For Ang)*

Enter angle of xline *(0)* or [Reference]: R *(For Reference)*

Select a line object: *(Select the inclined line in the Front view)*

Enter angle of xline <0>: *(Press* ENTER *to accept the default value – 0.)*

Specify through point: *(Pick a location)*

Trim the back corners of the projection lines. Your display should appear as shown in Figure 8.46 (Right).

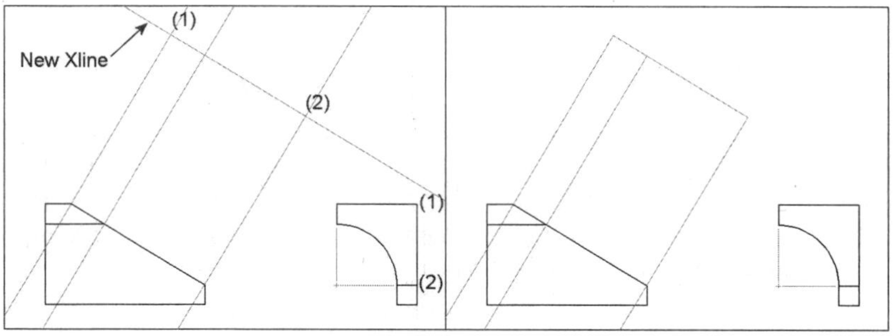

FIGURE 8.46

Set an offset distance from "A" to "B" in the Side view. This distance represents the depth of the object and will be transferred to the auxiliary view. Offset the line located in the auxiliary view this distance in the direction indicated in Figure 8.47 (Left).

Command: O *(For OFFSET)*

Current settings: Erase source=No Layer=Source OFFSETGAPTYPE=0

Specify offset distance or [Through/Erase/Layer] <Through>: *(Pick the endpoint at "A")*

Specify second point: *(Pick the endpoint at "B")*

Select object to offset or [Exit/Undo] <Exit>: *(Pick line "C")*

Specify point on side to offset or [Exit/Multiple/Undo] <Exit>: *(Pick a point on your screen at "D")*

Select object to offset or [Exit/Undo] <Exit>: *(Press* ENTER *to exit this command and perform the operation.)*

Set another offset distance from "A" to "B" in the Side view; then offset back line "C" located in the auxiliary view this distance to "D," as shown in Figure 8.47 (Right).

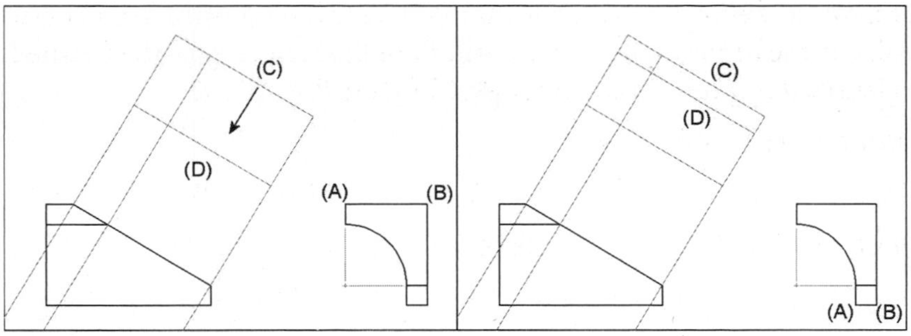

FIGURE 8.47

Trim and erase lines until your display appears as shown in the following image on the left. Construct a horizontal xline from the lower point of the incline. Then offset this xline at 0.25 increments. Your display should appear as shown in Figure 8.48 (Right).

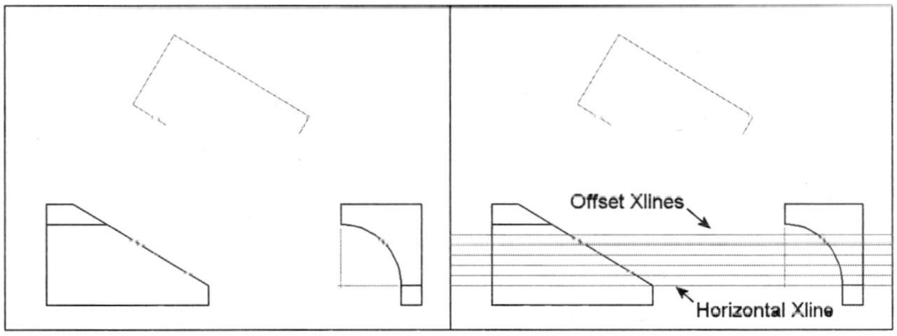

FIGURE 8.48

Construct additional projection xlines at every intersection along the incline. You should have five xlines constructed from these points. Your display should appear as shown in the following image on the left. Create an offset distance from "A" to "B" along line segment 1 in the Side view. Then offset the back line "C" the same distance to "D" in the auxiliary view, as shown in Figure 8.49 (Right).

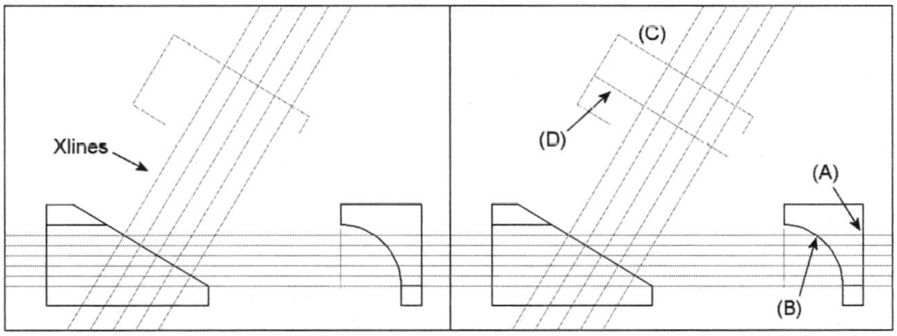

FIGURE 8.49

Place points at their respective intersections, as shown in Figure 8.50 (Left). The purpose of the points is to identify a segment of the curve to be constructed in the auxiliary view. Perform the same series of steps to locate the remaining points that make up the curve in the auxiliary view, as shown in Figure 8.50 (Right).

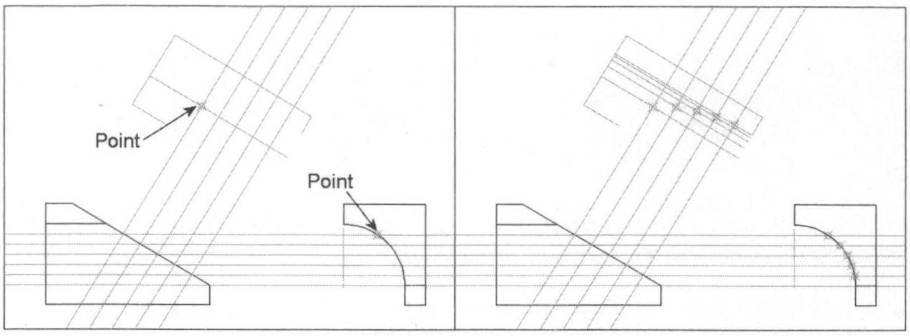

FIGURE 8.50

Make the Object layer current. Use the `SPLINE` command to create a spline connecting all points along the auxiliary view, as shown in Figure 8.51 (Left). Use the OSNAP-Node mode for locking onto each point. To finish, change the visible lines of the auxiliary view to the Object layer. Turn off the Construction layer. Your display should appear similar to the example in Figure 8.51 (Right).

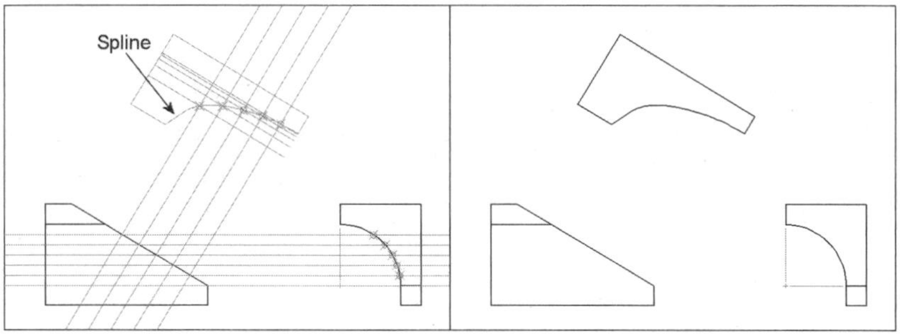

FIGURE 8.51

TUTORIAL EXERCISE: 08_BRACKET.DWG

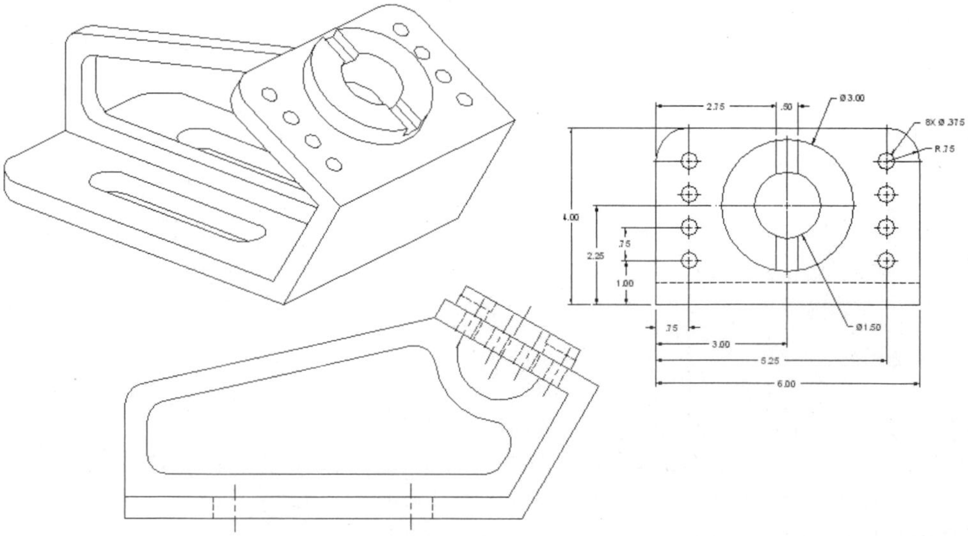

FIGURE 8.52

Purpose

This tutorial is designed to allow you to construct an auxiliary view of the inclined surface for the bracket shown in Figure 8.52.

System Settings

Since this drawing is provided, edit an existing drawing called 08_Bracket. Follow the steps in this tutorial for the creation of an auxiliary view.

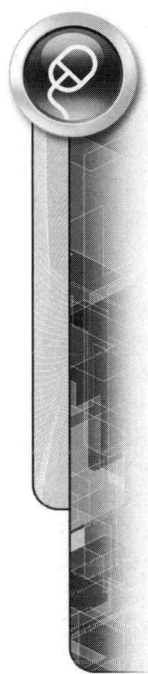

Layers

The following layers have already been created with the following format:

Name	Color	Linetype	Lineweight
CEN	Yellow	Center	Default
DIM	Yellow	Continuous	Default
HID	Red	Hidden	Default
OBJ	Cyan	Continuous	0.50 mm

Suggested Commands

You will begin this tutorial by opening the drawing—08_Bracket. Use the OFFSET command to construct a series of lines parallel to the inclined surface in the Front view. Next construct xlines perpendicular to the inclined surface. Use the CIRCLE command to begin laying out features that lie in the auxiliary view. Use ARRAYRECT to copy the circle in a rectangular pattern. Add centerlines using the DIMCENTER command. Finally, insert a predefined view called Top to complete a three-view drawing of the bracket.

STEP 1

Begin the construction of the auxiliary view by using the OFFSET command to copy a line parallel to the inclined surface edge located in the Front view, as shown in Figure 8.53 (Left). Use an offset distance of 8.50, select line "A" as the object to offset, and pick near "B" as the side to perform the offset. Then use the OFFSET command again to offset line "B" to the side at "C" at a distance of 6.00 units.

The previous two lines formed using the OFFSET command define the depth of the auxiliary view. To determine the width of the auxiliary view, use the XLINE command to construct the two perpendicular projection lines, as shown in Figure 8.53. Project the two xlines from the endpoints of the Front view at "D" and "E" at an angle perpendicular to line "A," as shown in Figure 8.53 (Right).

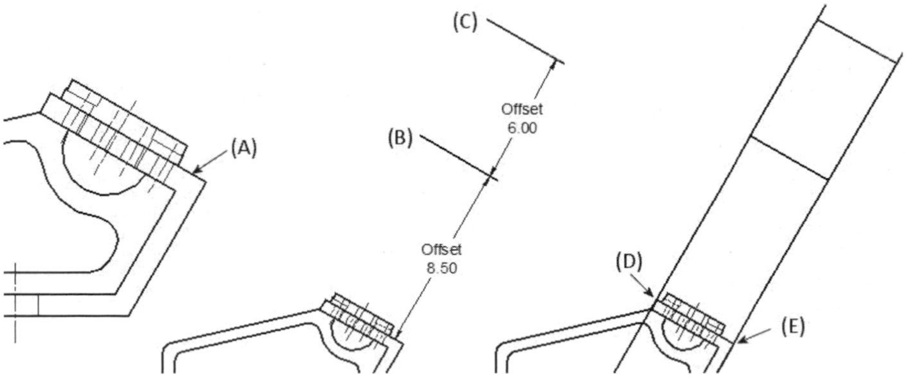

FIGURE 8.53

STEP 2

Use the ZOOM-Window option to magnify the display of the auxiliary view similar to the illustration in Figure 8.54 (Left). Then use the `FILLET` command to create four corners using the lines labeled "A" through "D." The fillet radius should be set to 0 in order to form the corners. The results are displayed in Figure 8.54 (Right).

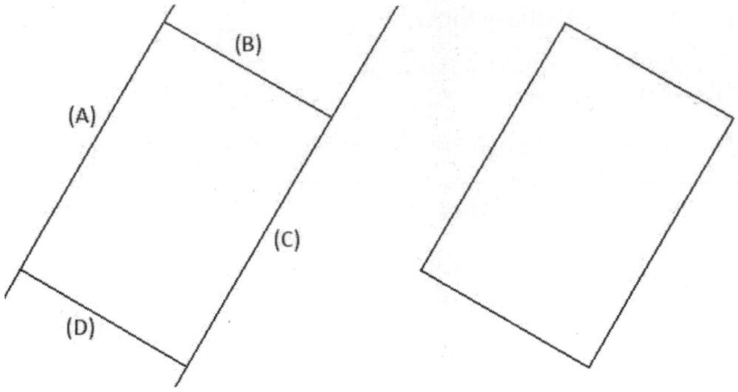

FIGURE 8.54

STEP 3

When you are finished with the filleting operation, activate the View Manager dialog box (VIEW command), as shown in Figure 8.55 (Left), and click the New button. When the New View/Shot Properties dialog box appears, as shown in Figure 8.55 (Right), verify that the view will be defined based on the Current display; a radio button should be active for this mode. Then save the display to a new name, AUX.

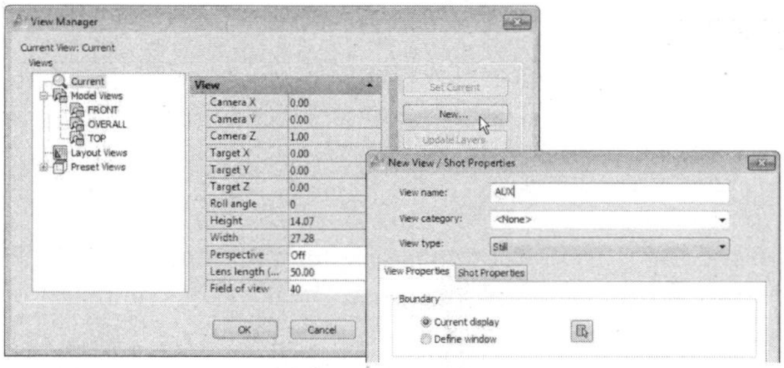

FIGURE 8.55

STEP 4

Use the ZOOM-Previous option or other ZOOM mode to demagnify the screen back to the original display. Use `XLINE` to create the perpendicular projection line "A" from the endpoint of the centerline, as shown in Figure 8.56 (Left). Use the Reference option and pick the centerline. Keep the default angle of 0.

This is one of the construction lines that will be used for finding the centers of the circular features located in the auxiliary view.

Then use the OFFSET command to create the other construction line, as shown in Figure 8.56 (Right). Use an offset distance of 3.00, pick line "B" as the object to offset, and pick a point near "C," as shown in Figure 8.56 (Right).

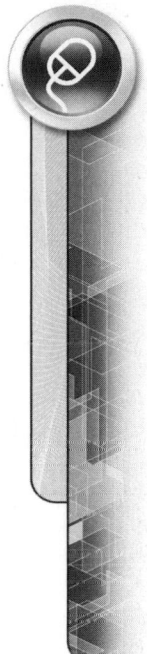

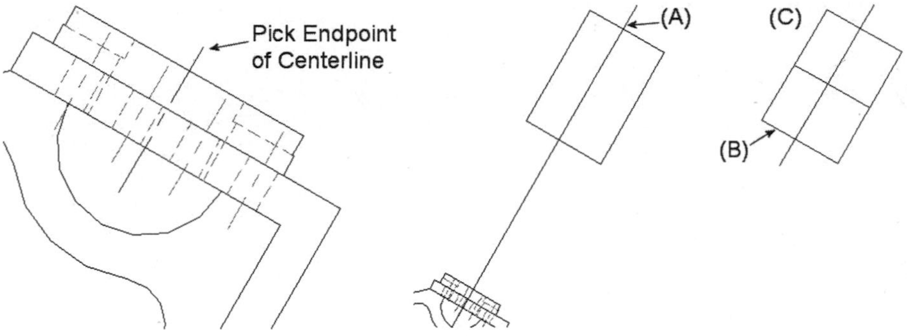

FIGURE 8.56

STEP 5

Activate the View Manager dialog box and set the view AUX current, as shown in Figure 8.57 (Left). Then draw two circles of diameters 3.00 and 1.50 from the center at "A," as shown in Figure 8.57 (Right), using the CIRCLE command. For the center of the second circle, you can use the @ option to pick up the previous point that was the center of the first circle.

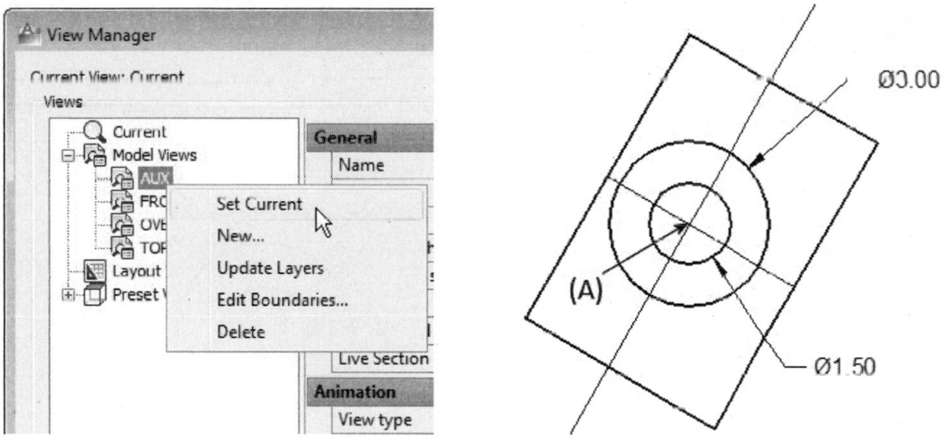

FIGURE 8.57

STEP 6

Use the OFFSET command to offset the centerline the distance of 0.25 units, as shown in Figure 8.58 (Left). Perform this operation on both sides of the centerline. Both offset lines form the width of the 0.50 slot.

Then use the TRIM command to trim away portions of the offset lines, using the two circles as cutting edges. When finished, your display should resemble the illustration in Figure 8.58 (Right).

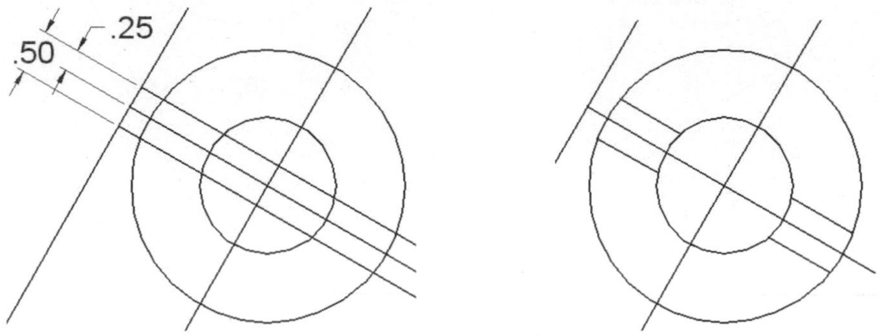

FIGURE 8.58

STEP 7

Use the ERASE command to delete the two lines at "A" and "B," as shown in Figure 8.59 (Left). Standard centerlines will be placed here later, marking the center of both circles.

Two more construction lines need to be made. These lines will identify the center of the small 0.37-diameter circle. Use the OFFSET command to create offsets for the lines illustrated in Figure 8.59 (Right). Offset distances of 0.75 and 1.00 are used to perform this task. Be careful to offset the correct line at the specified distance.

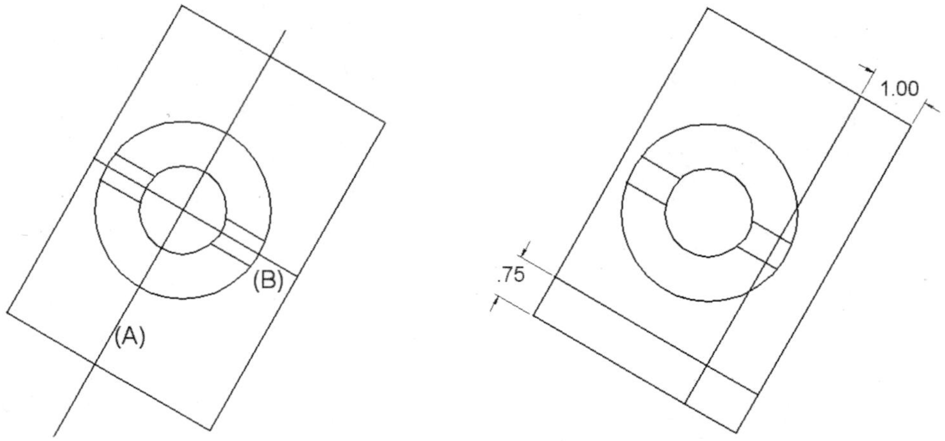

FIGURE 8.59

STEP 8

Draw a circle of 0.37 units in diameter at the intersection of the two lines created in the last OFFSET command, as shown in Figure 8.60 (Left). Then use the ERASE command to delete the two lines labeled "A" and "B."

For the next phase of this step, make the CEN layer current through the Layer Control Box. The DIMCENTER command (or DCE for short) will be used to create the center mark for the 0.37-diameter circle. Type DCE at the Command prompt and pick the edge of the circle at "C," as shown on the right in the following image, to place this center mark.

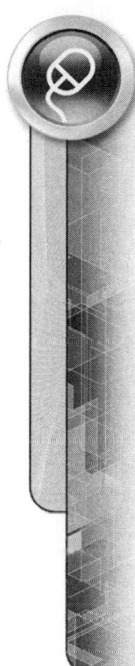

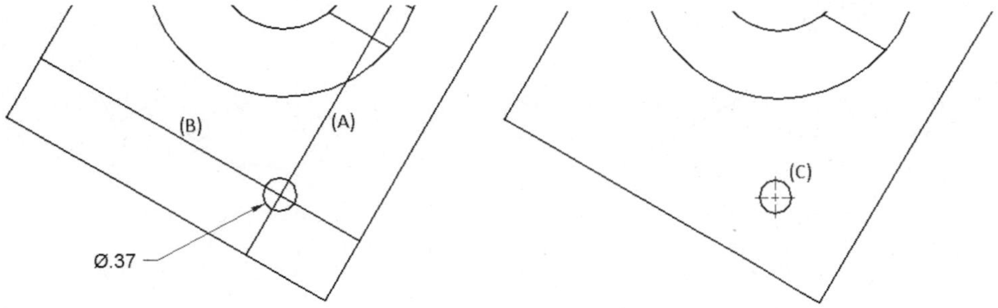

FIGURE 8.60

STEP 9

Since the remaining seven holes form a rectangular pattern, use the ARRAYRECT command and perform a rectangular array based on the 0.37-diameter circle and center marks. The number of rows is four and the number of columns is two. The distance between rows is 0.75 units and between columns is 4.50 units. Use the illustration in Figure 8.61 (Left) and the command sequence to create the array.

Command: ARRAYRECT

Select objects: *(Select the 0.37-diameter circle and center marks)*

Select objects: *(Press ENTER to continue)*

Type = Rectangular Associative = Yes

Select grip to edit array or [ASsociative/Base point/COUnt/Spacing/COLumns/Rows/Levels/eXit] <eXit>: COU *(For COUnt option)*

Enter the number of columns or [Expression] <4>: 2

Enter the number of rows or [Expression] <3>: 4

Select grip to edit array or [ASsociative/Base point/COUnt/Spacing/COLumns/Rows/Levels/eXit] <eXit>: S (For Spacing option)

Specify the distance between columns or [Unit cell] <0.58>: 4.50

Specify the distance between rows <0.58>: 0.75

Select grip to edit array or [ASsociative/Base point/COUnt/Spacing/COLumns/Rows/Levels/eXit]<eXit>: *(Press* ENTER *to exit)*

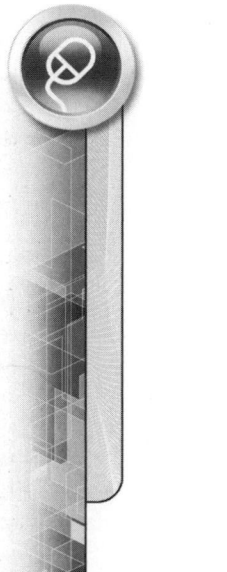

Next, use the ROTATE command to rotate the array such that it is parallel to the edges of the auxiliary view. Activate the ROTATE command and pick the array. For the base point of the rotation, pick a point at the center of the circle at "A" using OSNAP Center or Intersection mode. For the rotation angle, use OSNAP Perpendicular and pick the line near "B." The array should rotate parallel to the edge of the auxiliary, as shown in the following image on the right.

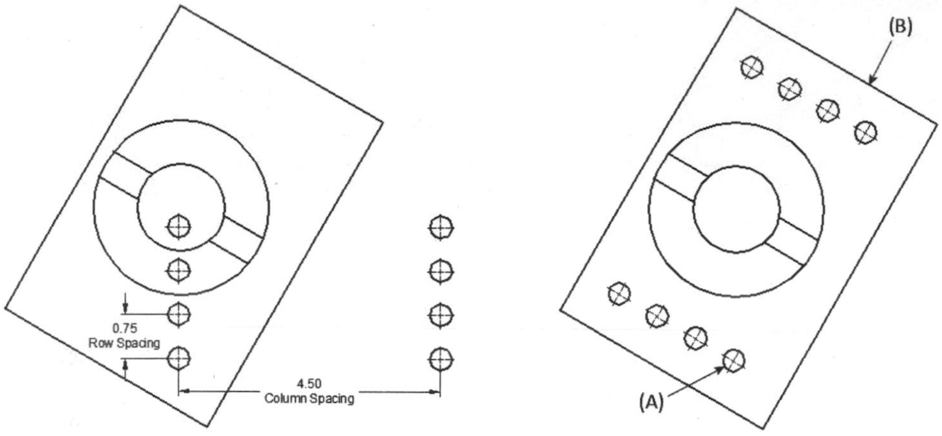

FIGURE 8.61

STEP 10

Use the `FILLET` command set to a radius of 0.75 to place a radius along the two corners of the auxiliary, as shown in Figure 8.62 (Left).

Place a center mark in the center of the two large circles using the `DIMCEN` (or `DCE` for short) command. When prompted to select the arc or circle, pick the edge of the large circle. The results are displayed in Figure 8.62 (Center).

Then rotate the center mark using the `ROTATE` command. Use OSNAP Center to set the base point and OSNAP Perpendicular to set the rotation angle. The results are displayed in Figure 8.62 (Right).

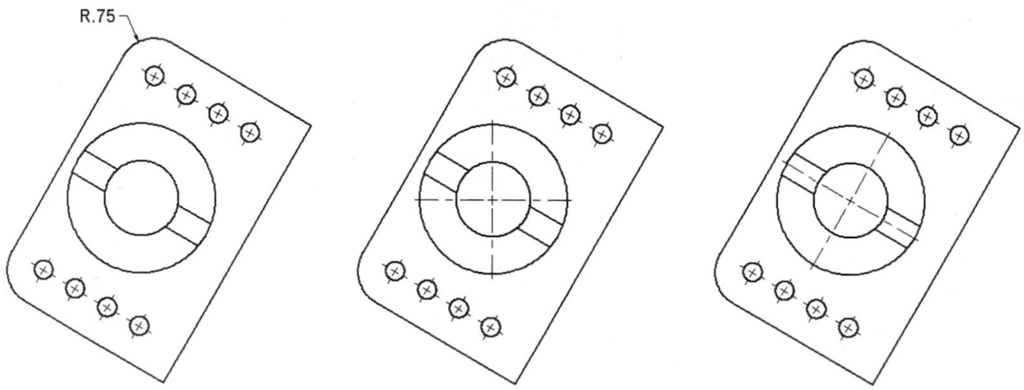

FIGURE 8.62

STEP 11

Activate the View tab on the Ribbon, as shown in Figure 8.63, and set the view named OVERALL current by selecting it from the Views panel. This should return your display to view all views of the Bracket.

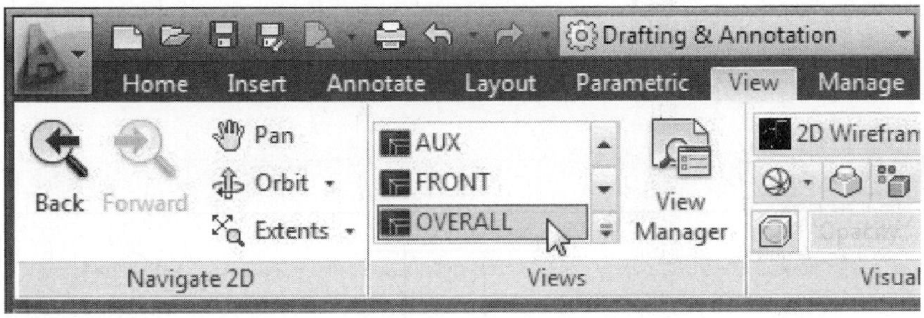

FIGURE 8.63

STEP 12

Complete the drawing of the bracket by activating the Insert dialog box (INSERT command), as shown in Figure 8.64, and inserting an existing block called TOP into the drawing. This block represents the complete Top view of the drawing. Use an insertion point of 0,0 for placing this view in the drawing. Your display should appear similar to Figure 8.64.

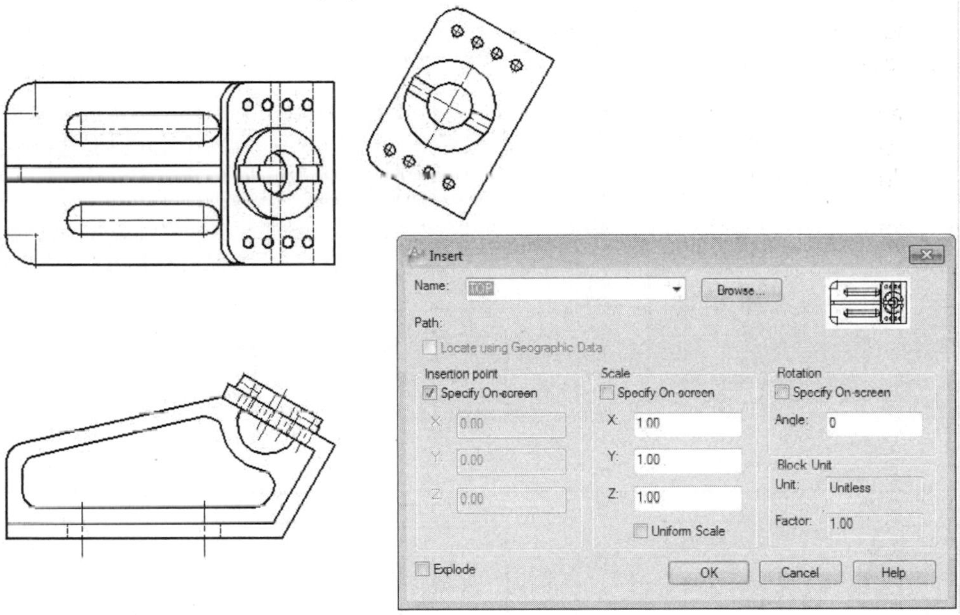

FIGURE 8.64

PROBLEMS FOR CHAPTER 8

DIRECTIONS FOR PROBLEMS 8-1 AND 8-2:

Open each drawing file individually (Pr08-01.dwg and Pr08-02.dwg). Create two new layers called Missing Visible Lines and Missing Hidden Lines. Assign the color red to both of these layers. Assign the Hidden linetype to the Missing Hidden Lines layer. Using projection techniques, find the missing lines for each object and draw in the correct solution on the proper layer.

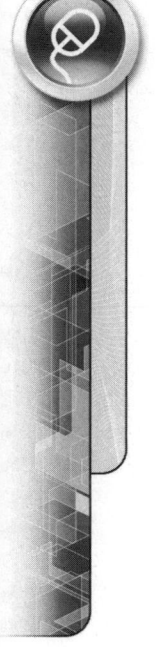

PROBLEM 8-1

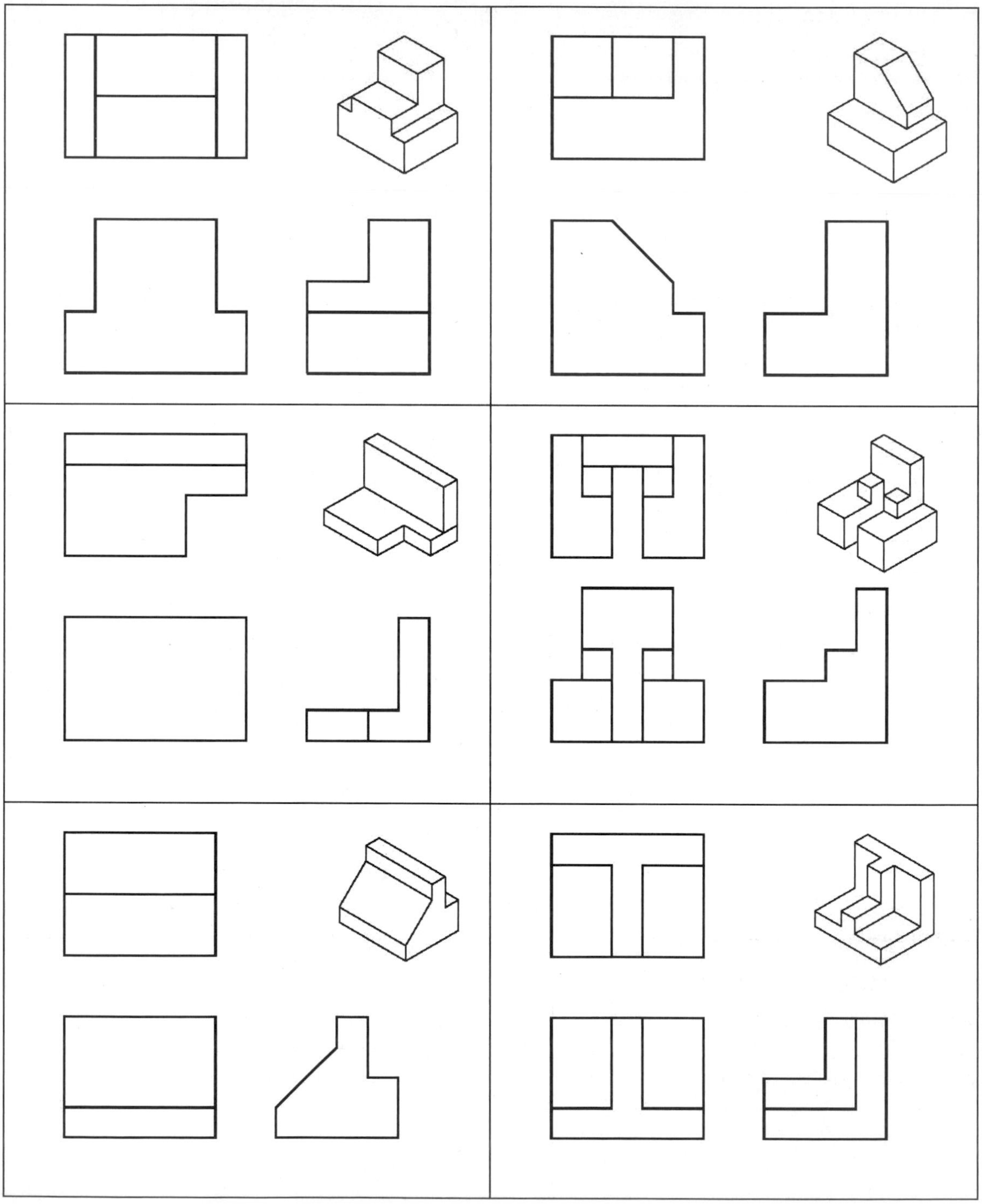

PROBLEM 8-2

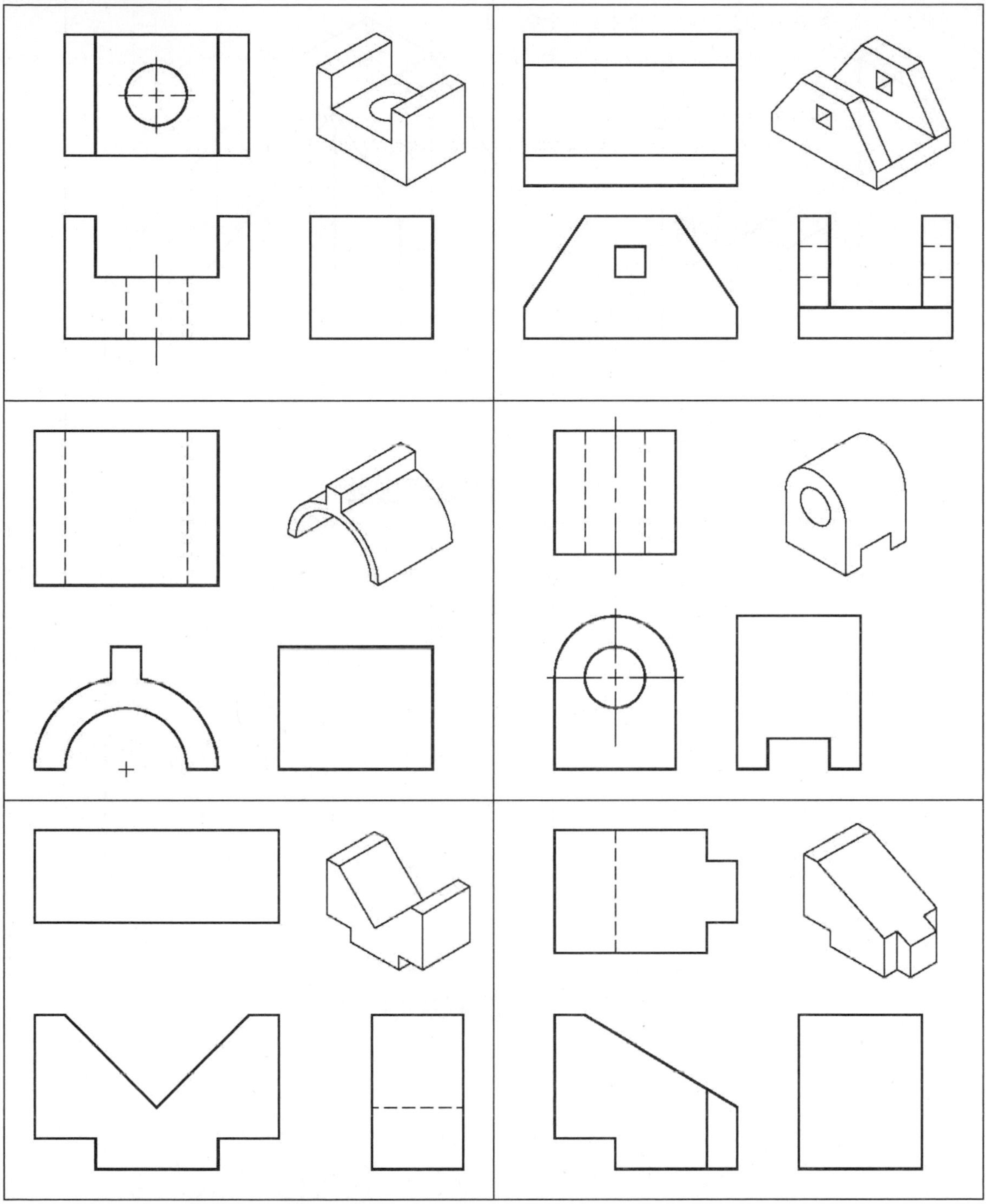

DIRECTIONS FOR PROBLEMS 8-3 THROUGH 8-20:

Construct a multiview drawing by sketching the Front, Top, and Right Side views. Use a grid of 0.25 units to assist in the construction of the sketches. These sketching problems can also be constructed directly in AutoCAD using typical commands and projection techniques.

PROBLEM 8-3

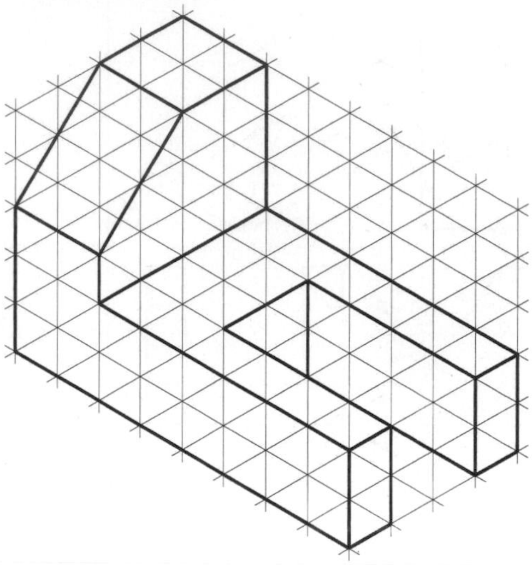

PROBLEM 8-6

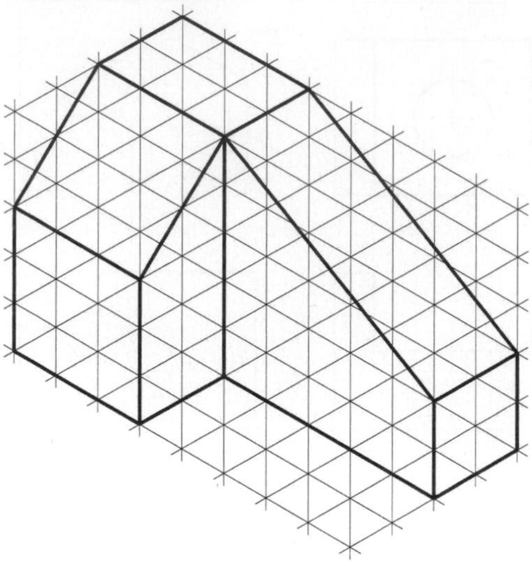

PROBLEM 8-4

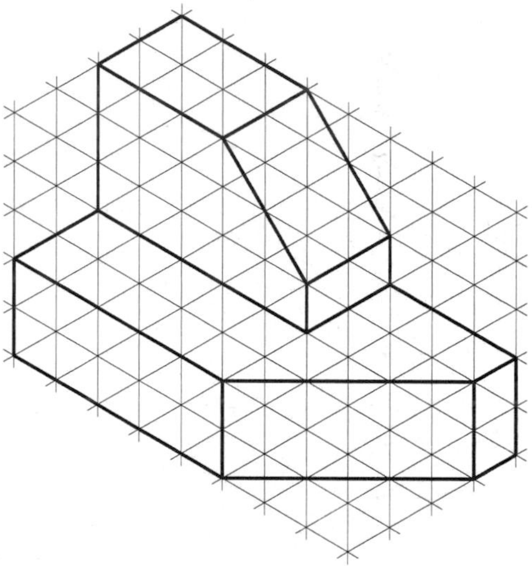

PROBLEM 8-7

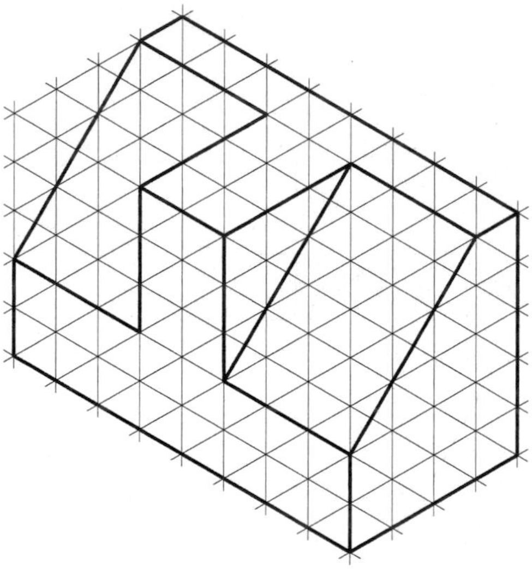

PROBLEM 8-5

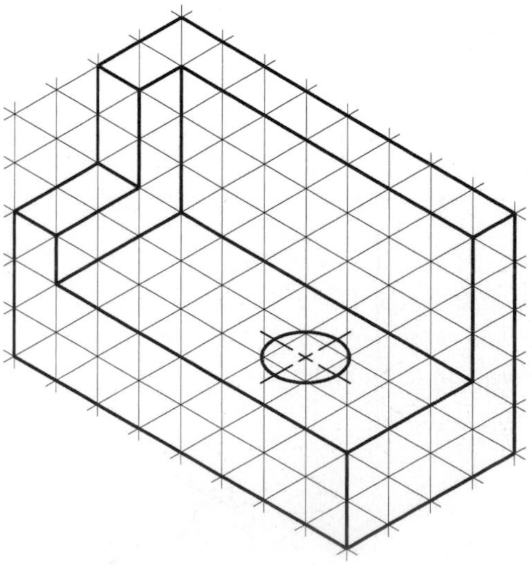

PROBLEM 8-8

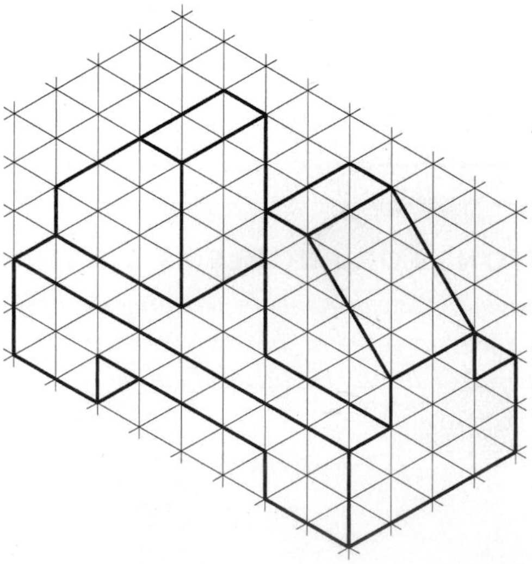

PROBLEM 8-9

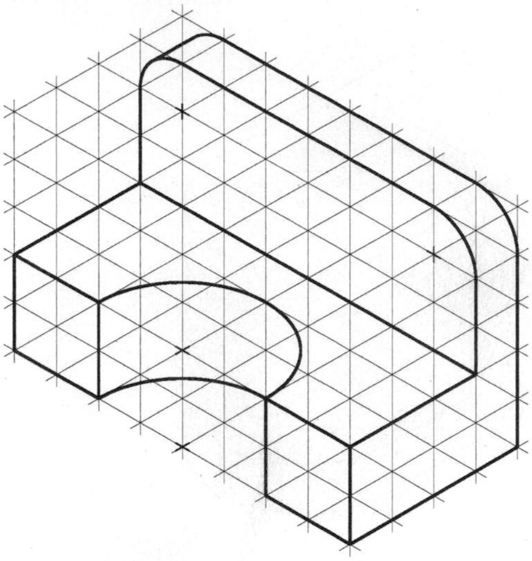

PROBLEM 8-12

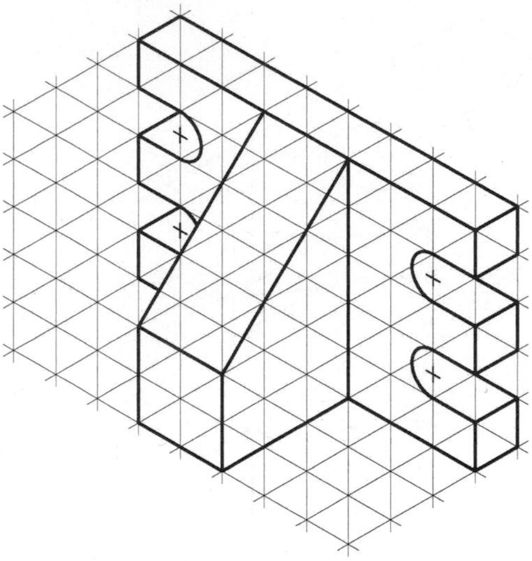

PROBLEM 8-10

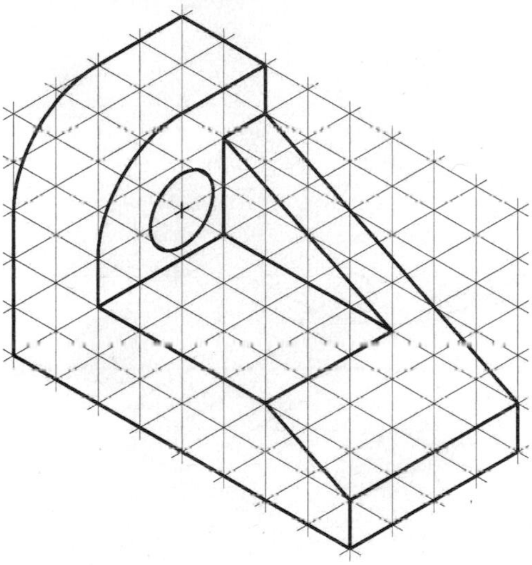

PROBLEM 8-13

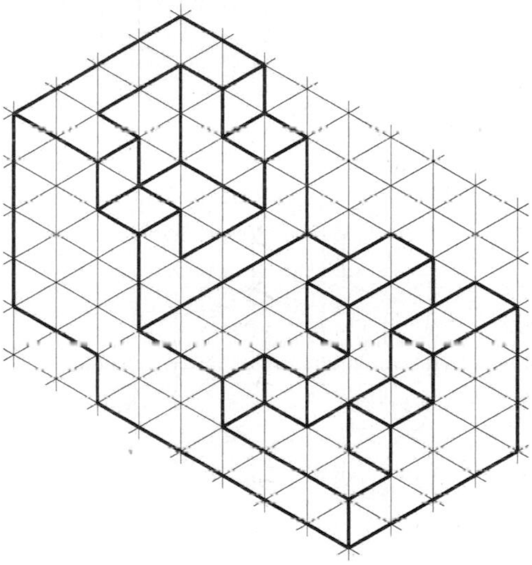

PROBLEM 8-11

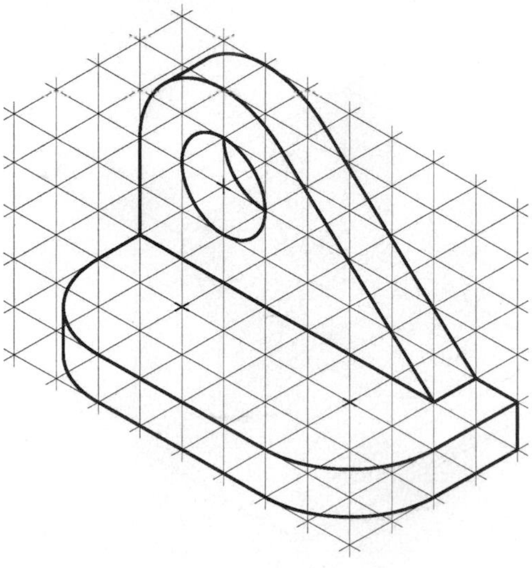

PROBLEM 8-14

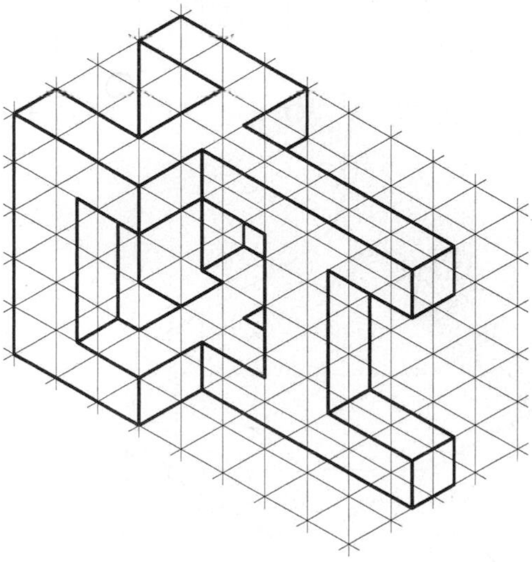

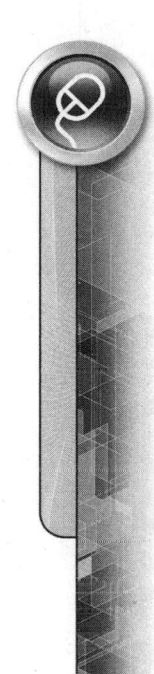

PROBLEM 8-15

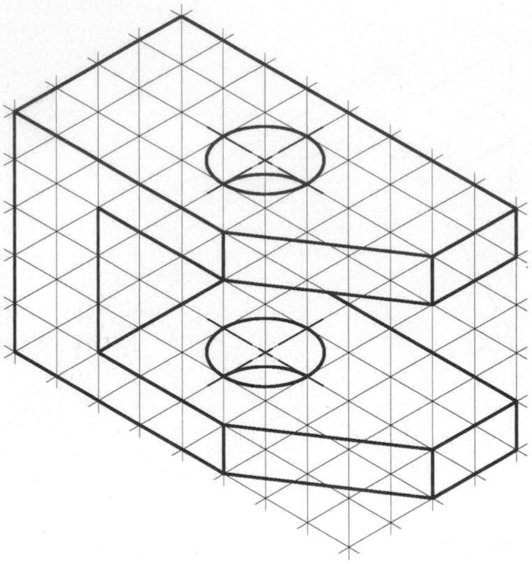

PROBLEM 8-18

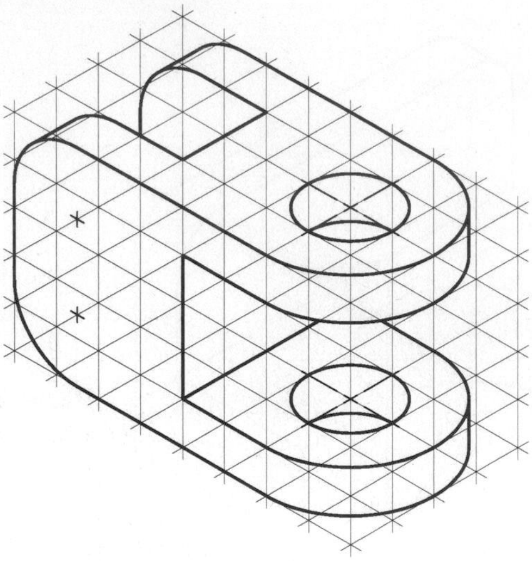

PROBLEM 8-16

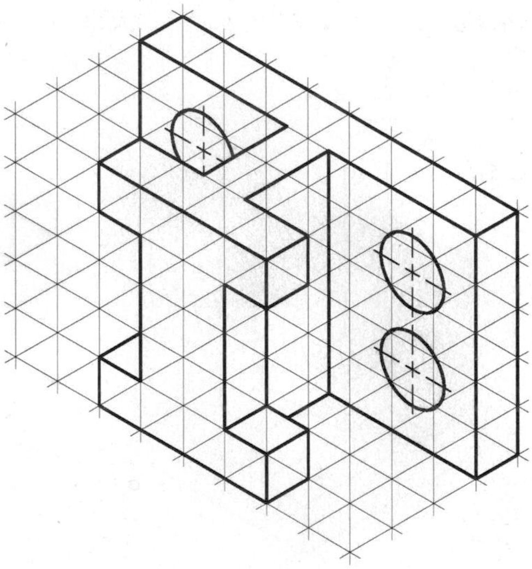

PROBLEM 8-19

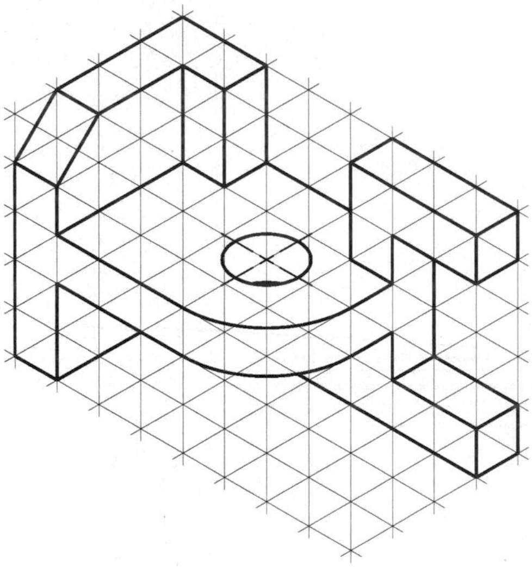

PROBLEM 8-17

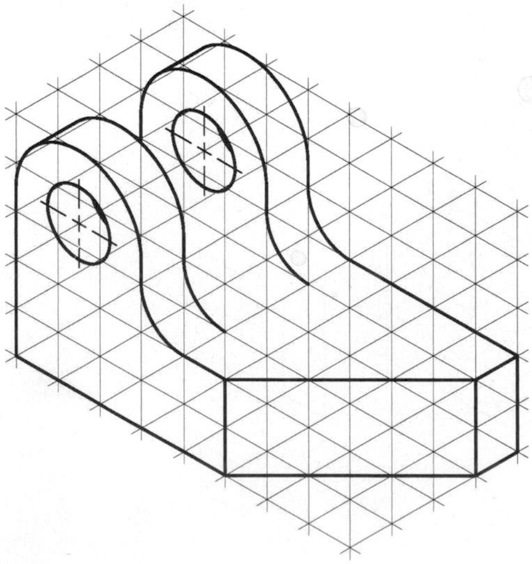

PROBLEM 8-20

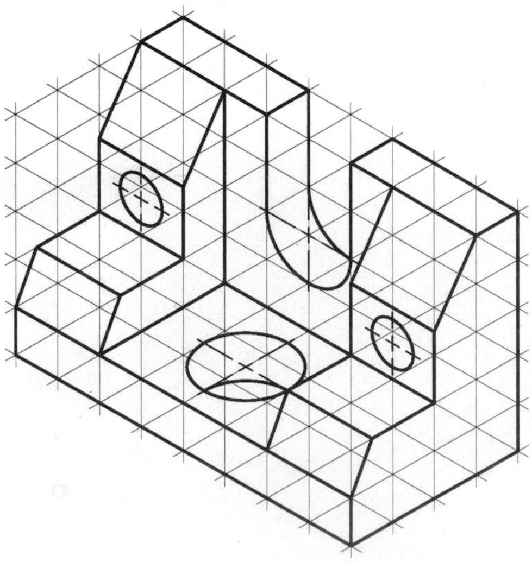

DIRECTIONS FOR PROBLEMS 8-21 THROUGH 8-35:

Construct a multiview drawing of each object. Be sure to construct only those views that accurately describe the object.

PROBLEM 8-21

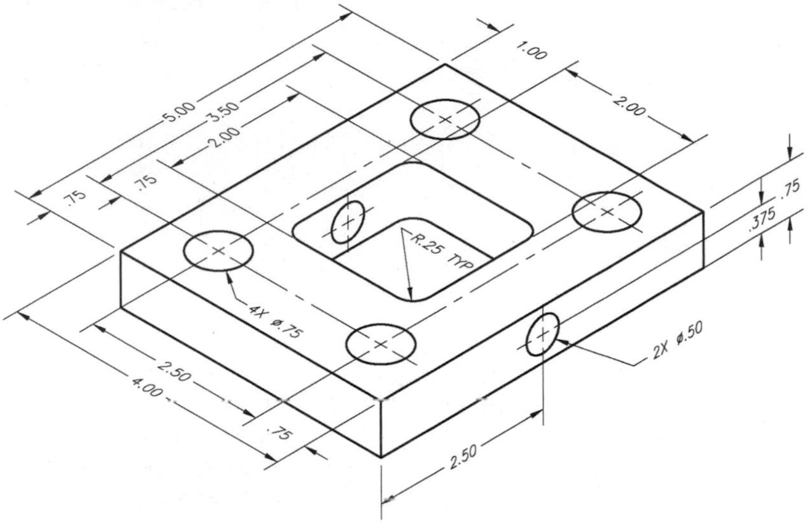

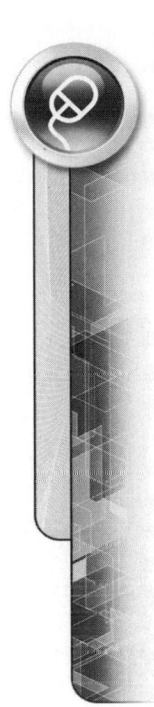

PROBLEM 8-22

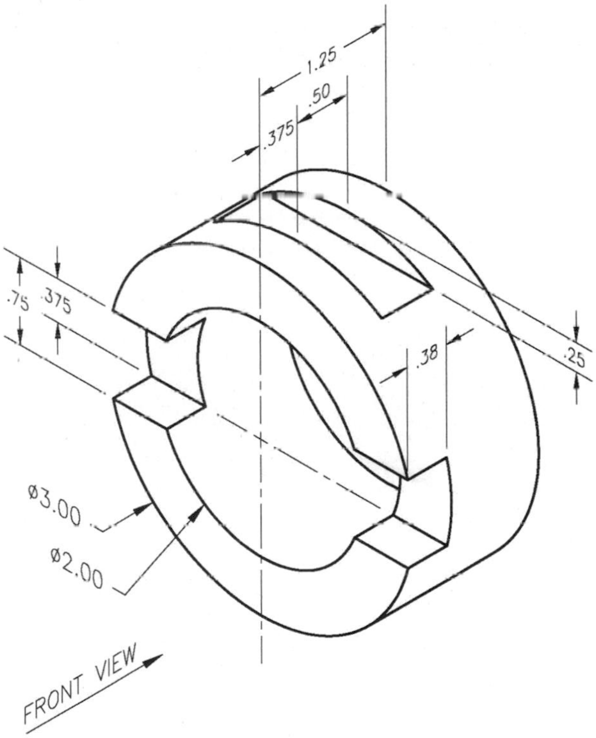

PROBLEM 8-23

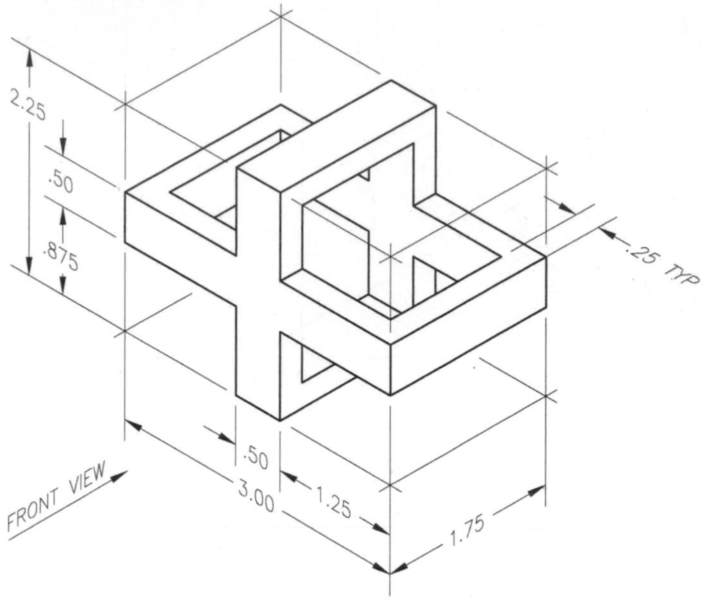

PROBLEM 8-24

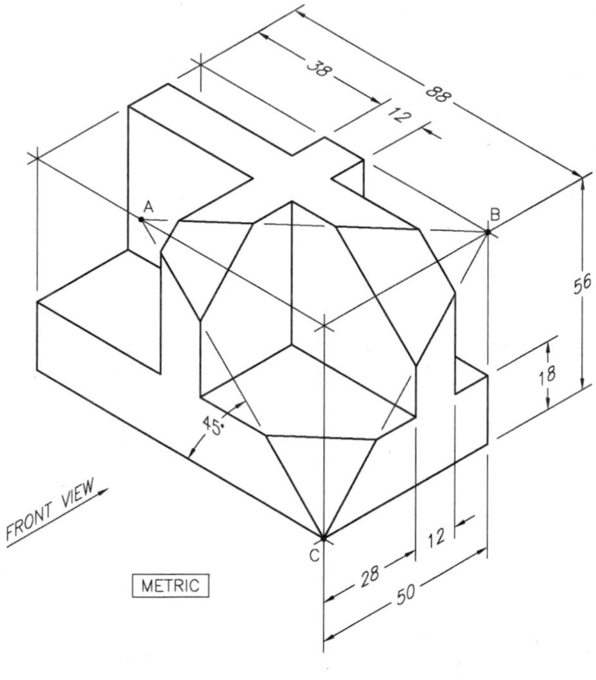

PROBLEM 8-25

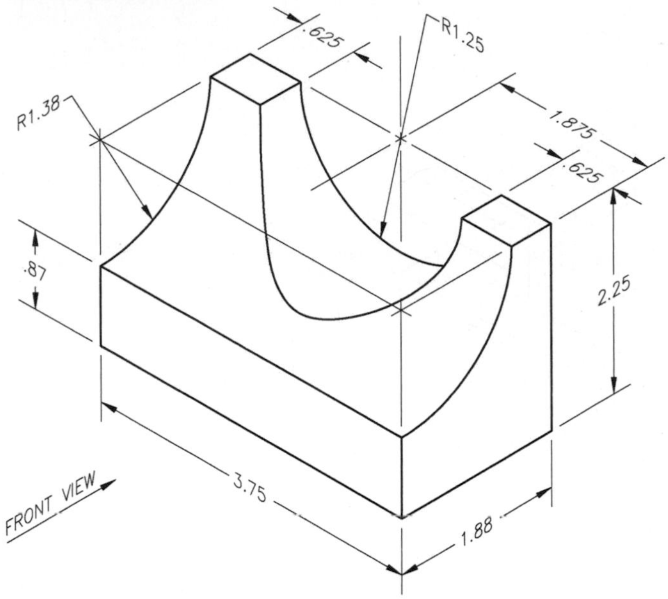

PROBLEM 8-26

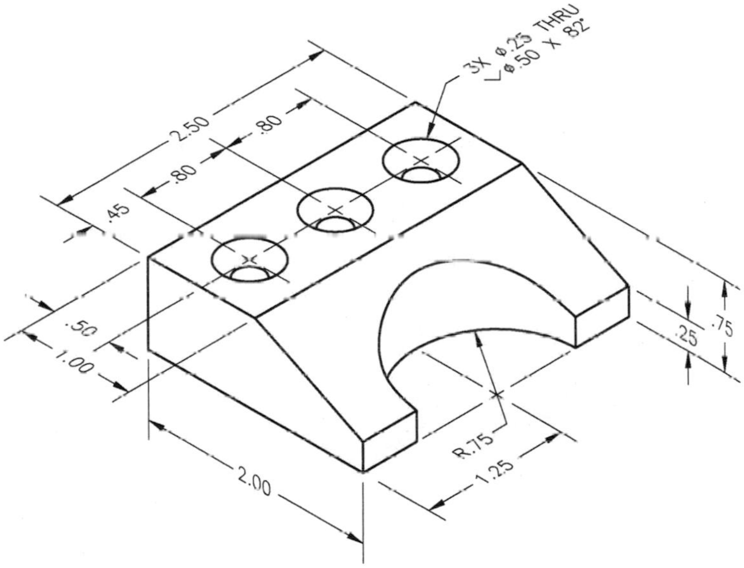

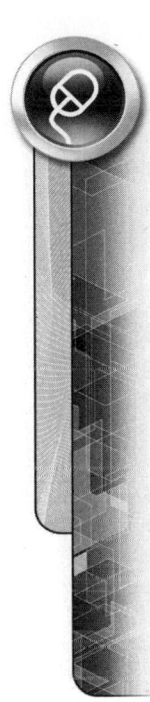

PROBLEM 8-27

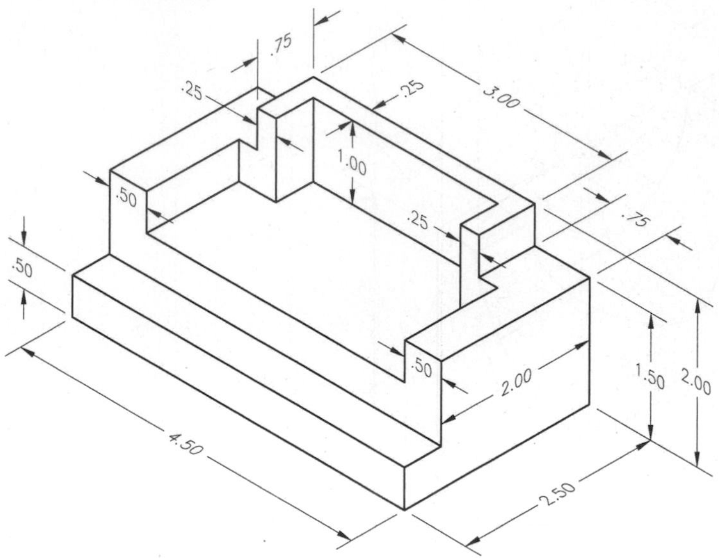

PROBLEM 8-28

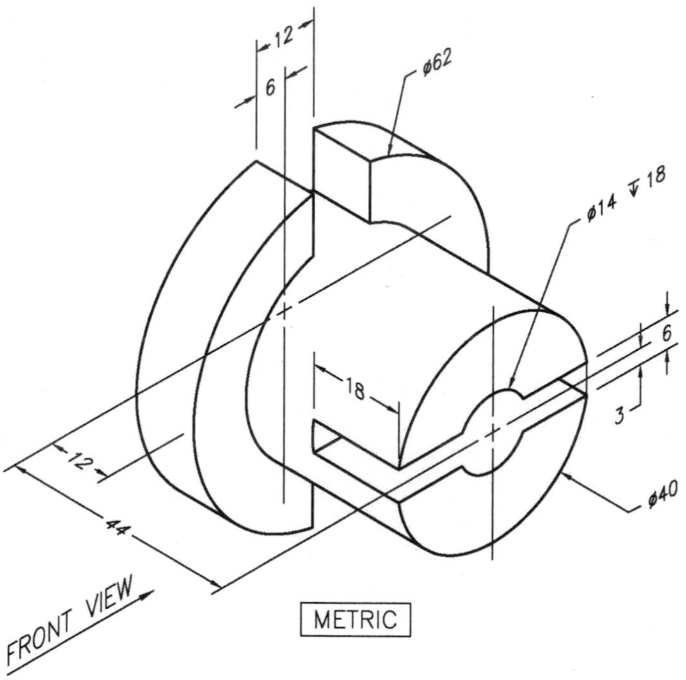

PROBLEM 8-29

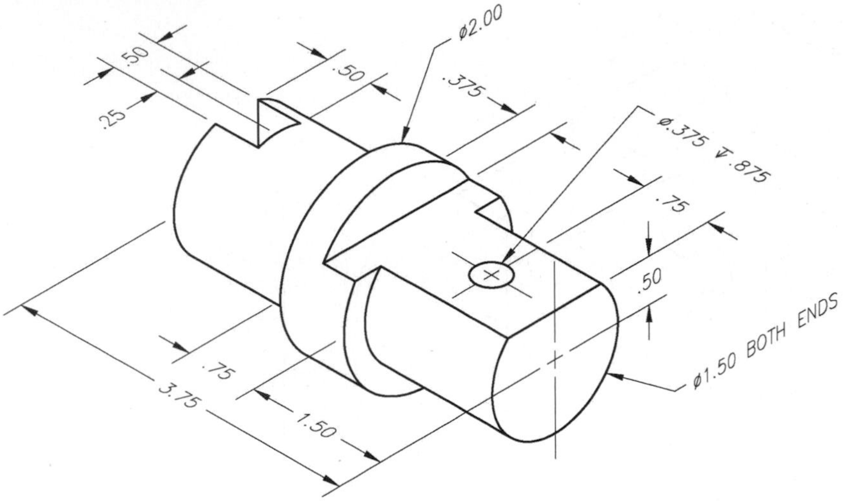

PROBLEM 8-30

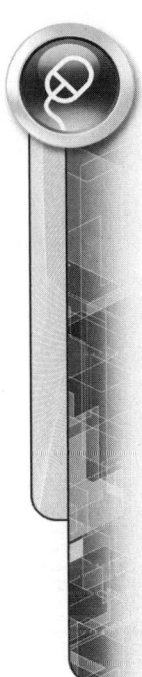

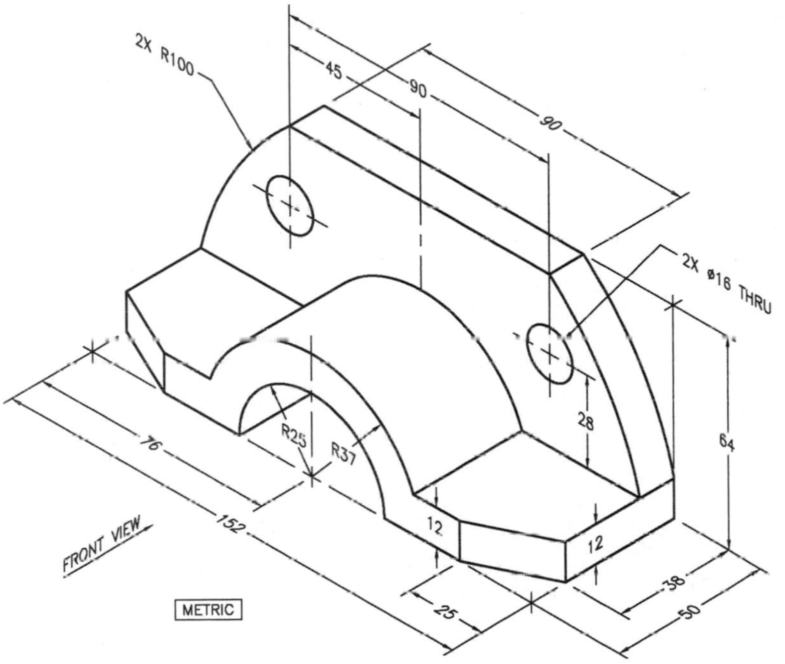

PROBLEM 8-31

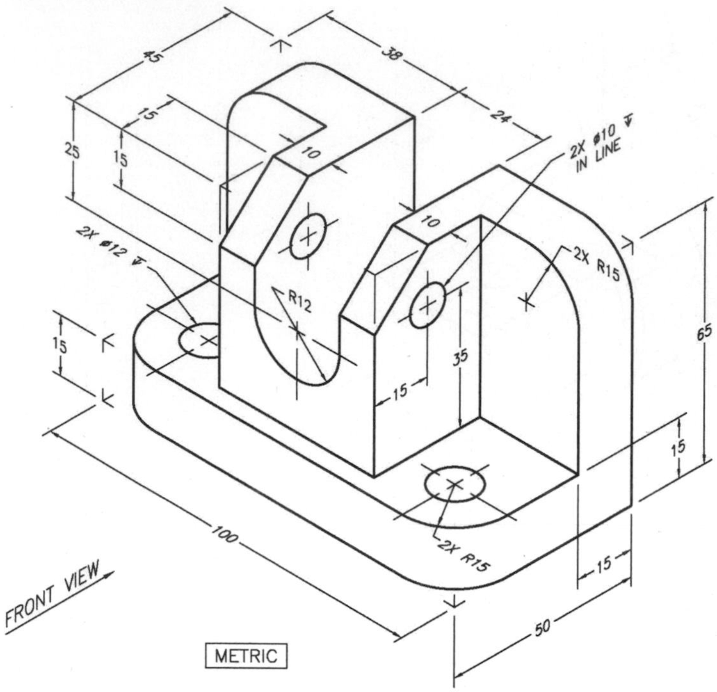

PROBLEM 8-32

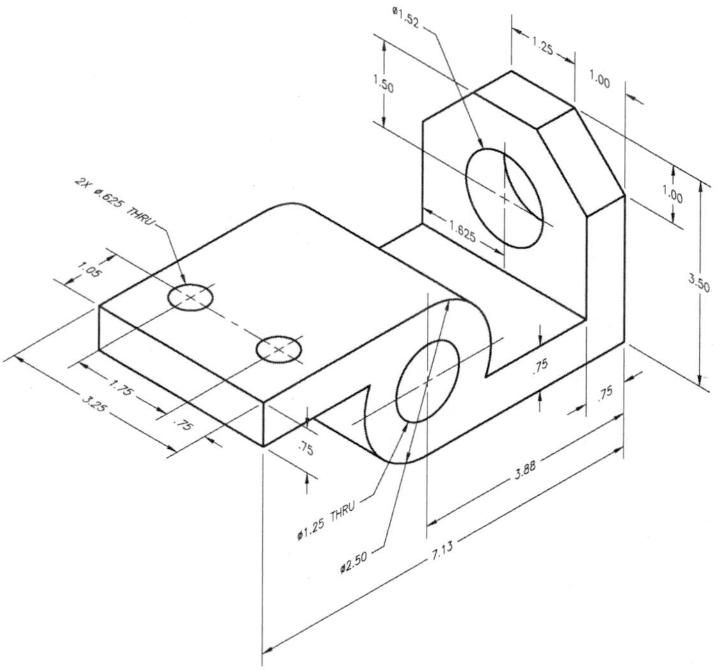

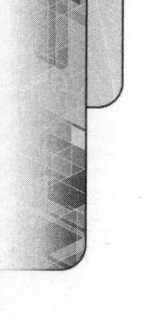

PROBLEM 8-33

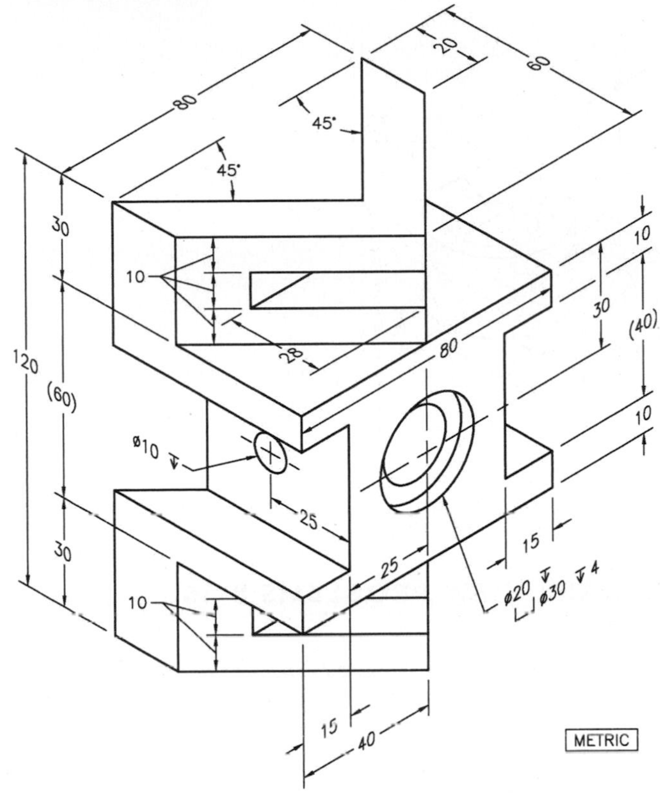

PROBLEM 8-34

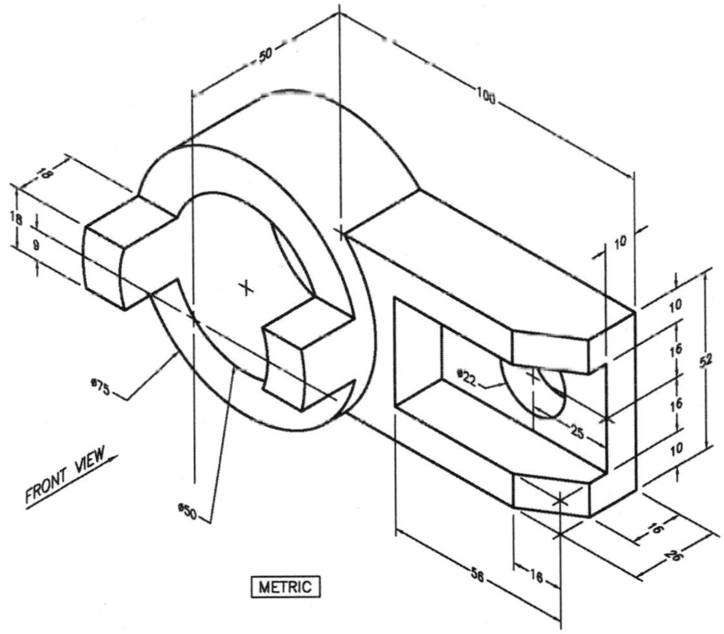

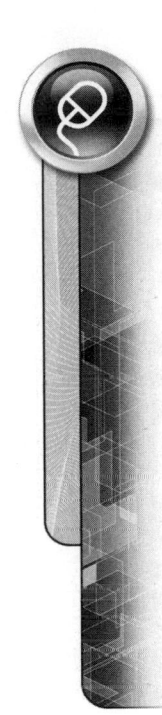

PROBLEM 8-35

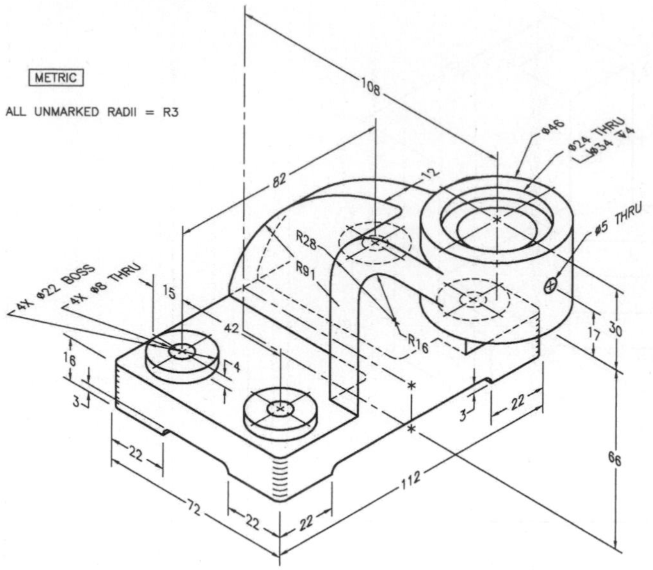

DIRECTIONS FOR AUXILIARY VIEW PROBLEMS 8-36 THROUGH 8-48:

Draw the require views to fully illustrate each object. Be sure to include an auxiliary view.

PROBLEM 8-36

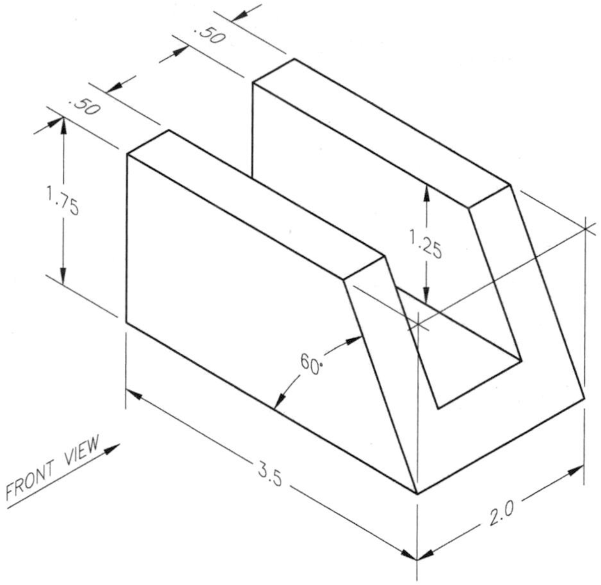

PROBLEM 8-37

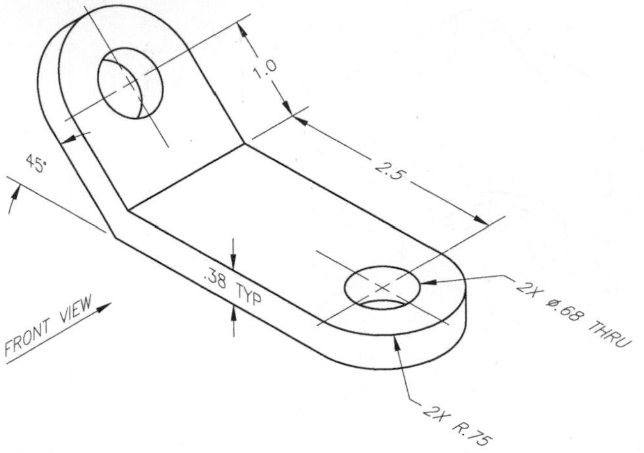

PROBLEM 8-38

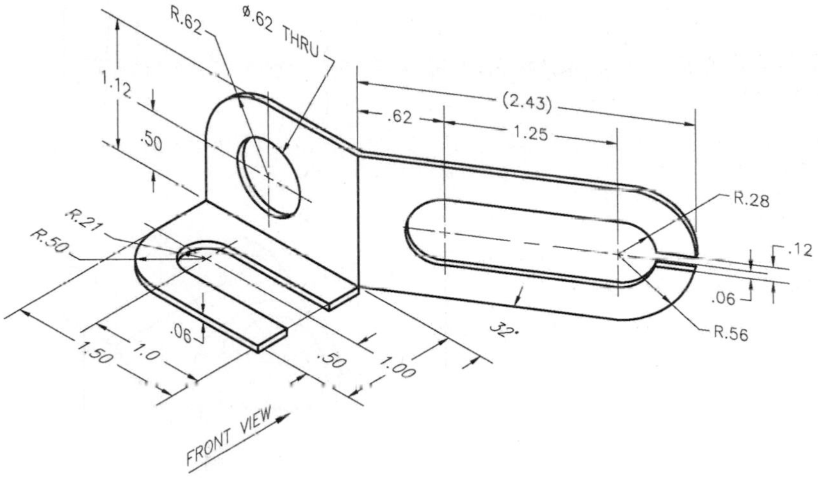

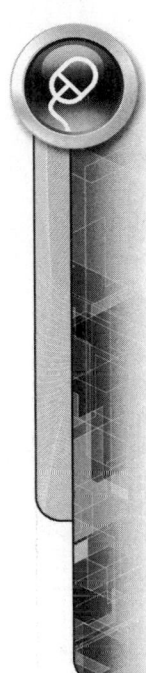

PROBLEM 8-39

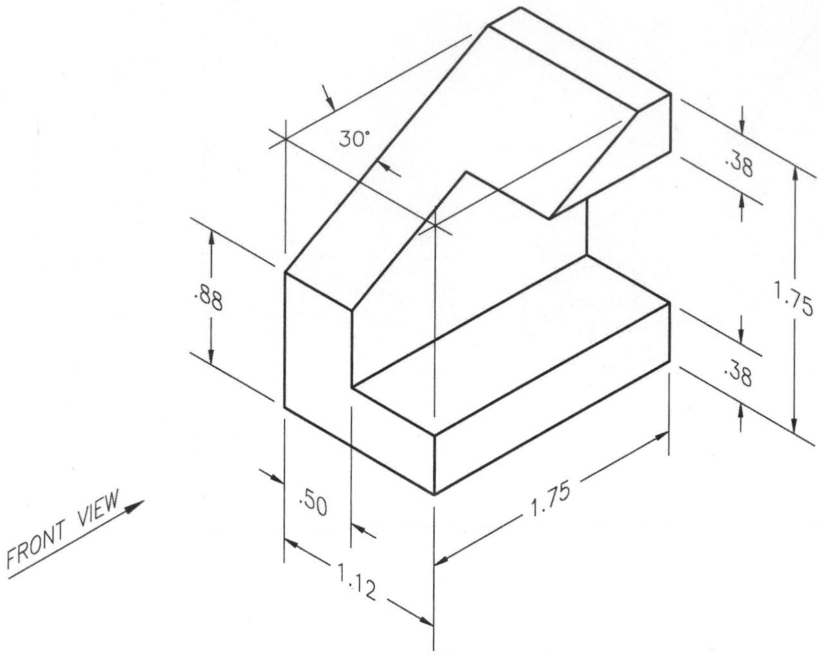

PROBLEM 8-40

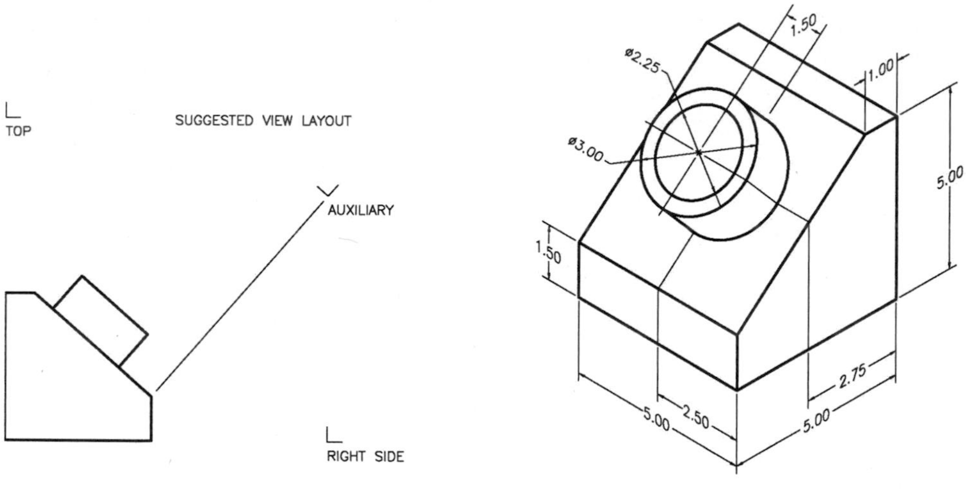

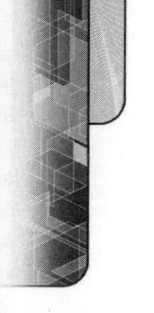

PROBLEM 8-41

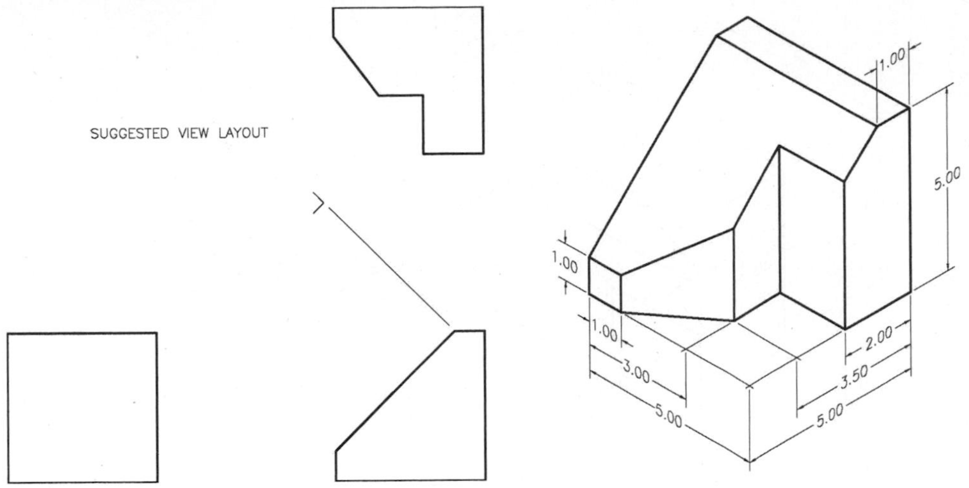

PROBLEM 8-42

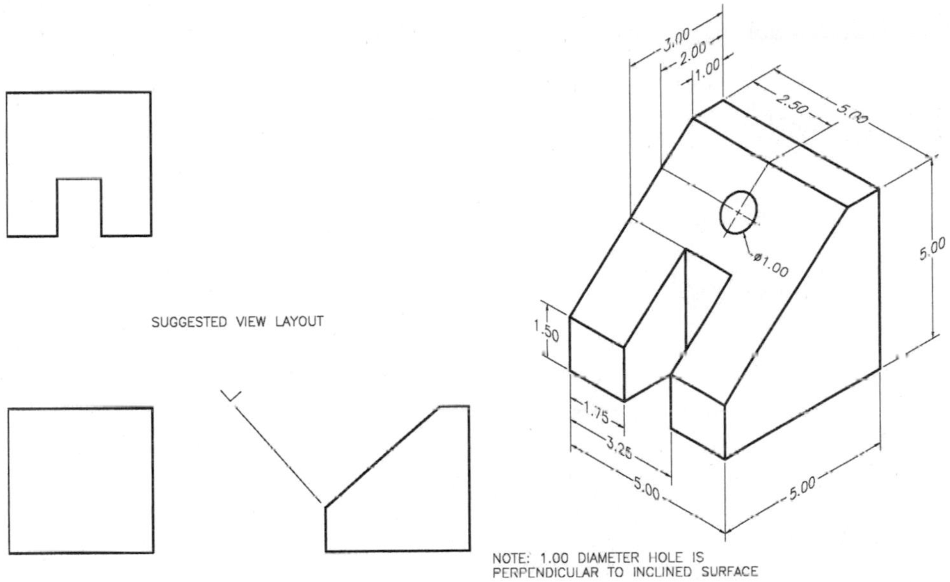

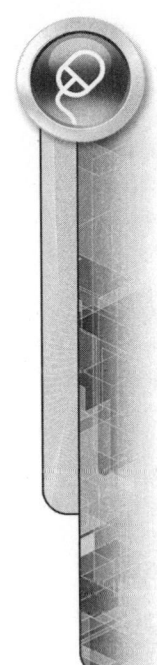

PROBLEM 8-43

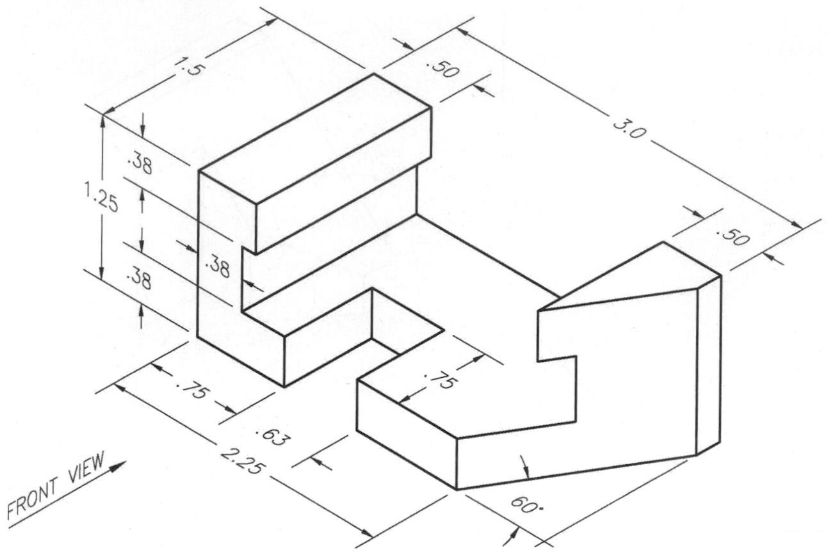

PROBLEM 8-44

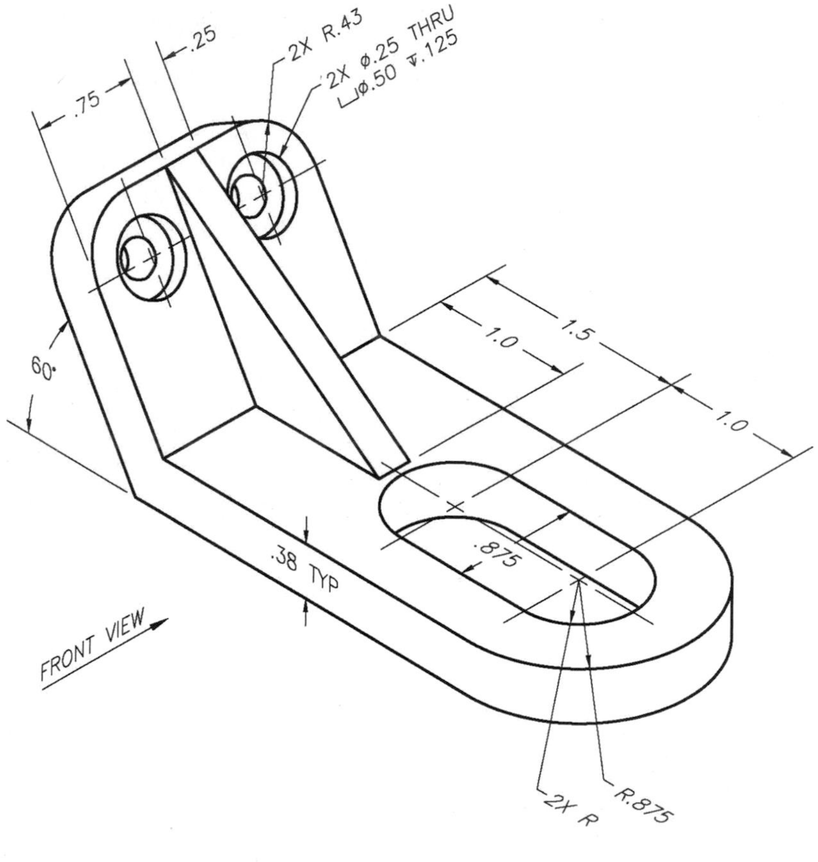

PROBLEM 8-45

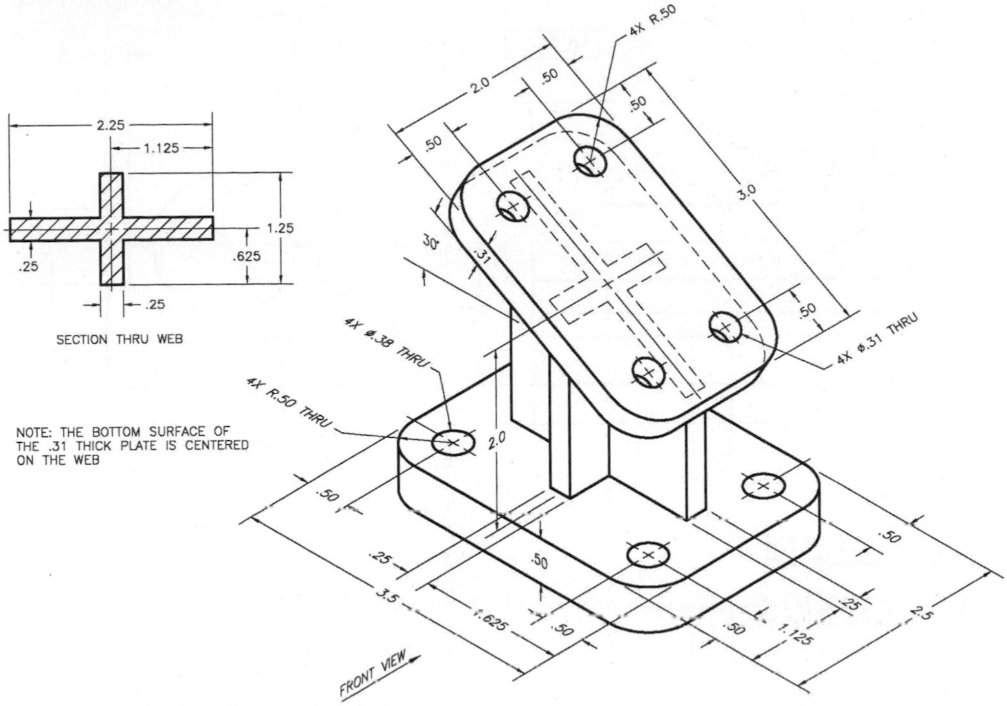

PROBLEM 8-46

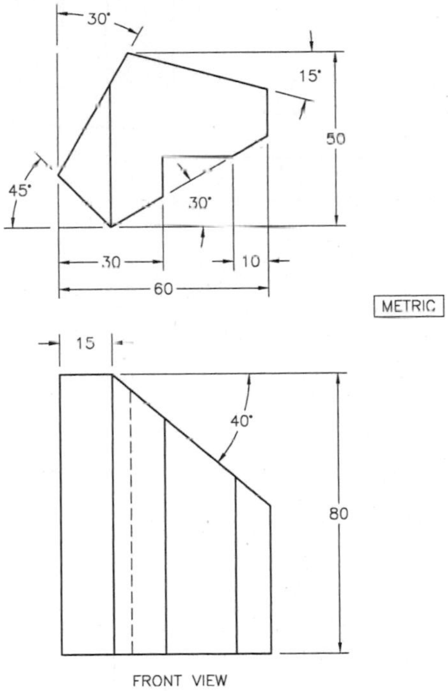

PROBLEM 8-47

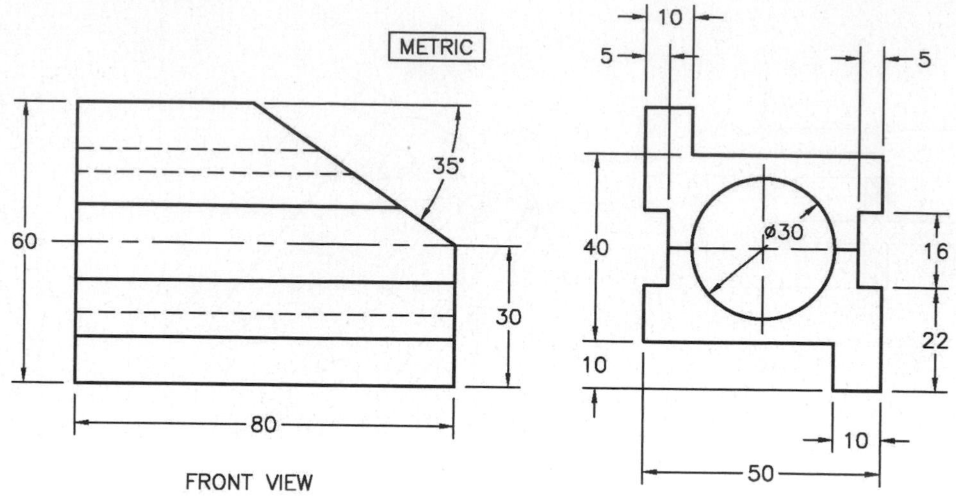

PROBLEM 8-48

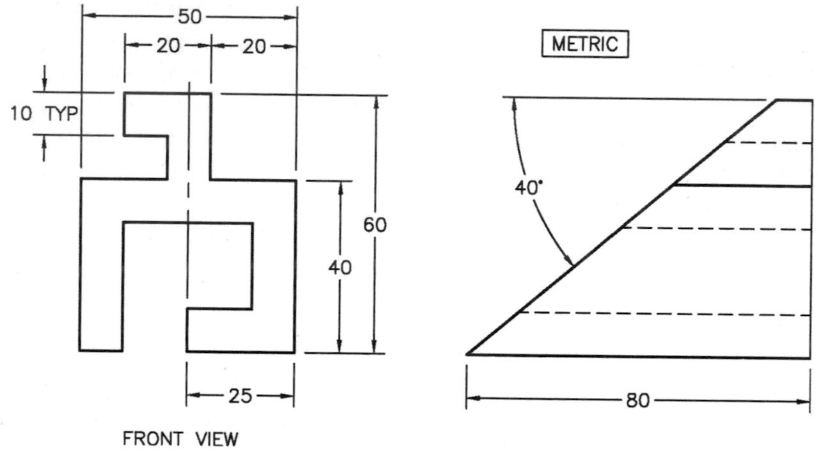

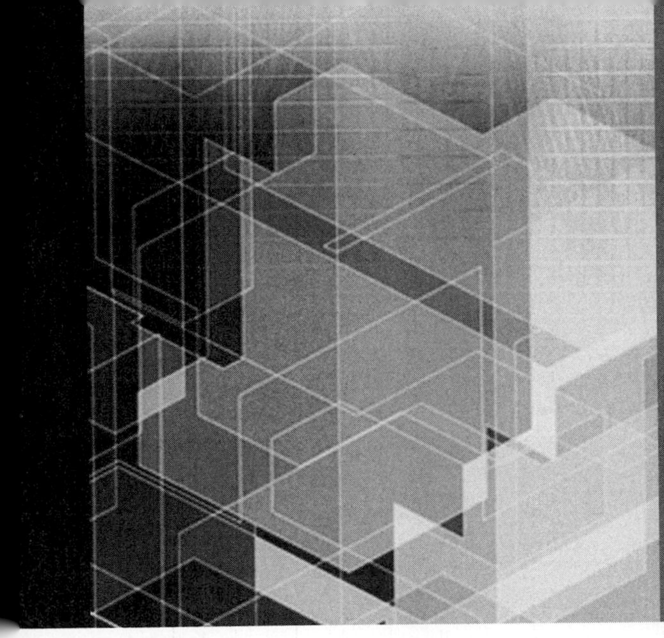

CHAPTER 9

Creating Section Views

This chapter will cover section views, which are created when it becomes necessary to view the interior features of an object. A cutting plane indicates where the object will be sliced to expose its interior. Crosshatching is used to indicate which surfaces were sliced. This chapter will discuss how AutoCAD crosshatches objects through the Ribbon's Hatch Creation Tab (HATCH command). You will be able to select from a collection of many hatch patterns, including gradient patterns, for special effects. Hatch scaling and angle considerations will also be covered. Once hatching is created in a drawing, it can be easily modified through the Ribbon's Hatch Editor, and because hatching is associative, most modifications are automatically performed without the need to recreate the hatching.

THE HATCH COMMAND

In a section view, a uniform pattern of section lines (crosshatching) is placed on each surface that is intersected by the cutting plane. AutoCAD's HATCH command generates these section lines. This command can be chosen from the Drafting & Annotation Workspace's Ribbon, as shown in Figure 9.1.

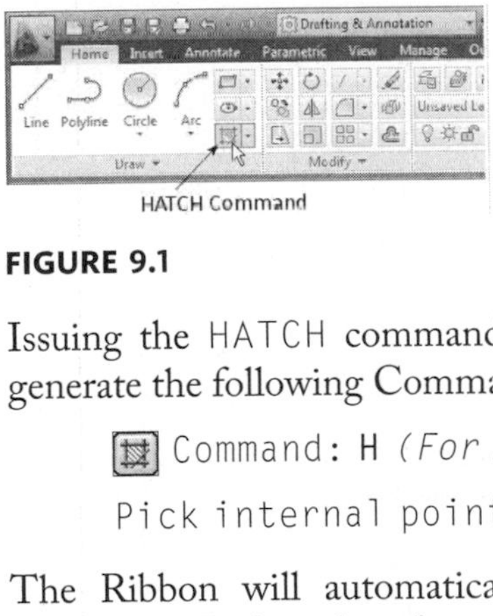

FIGURE 9.1

Issuing the HATCH command from the Drafting & Annotation Workspace will generate the following Command prompt:

```
Command: H (For HATCH)
Pick internal point or [Select objects/seTtings]:
```

The Ribbon will automatically display the Hatch Creation tab, as shown in Figure 9.2. Before identifying the areas to be hatched, you should make any setting changes in the Ribbon's panels. You can select one of the numerous hatch patterns available or you may decide to change the pattern's angle or scale. Hatch pattern settings can also be made through the Hatch and Gradient dialog box by using the settings option of the HATCH command.

```
Command: H (For HATCH)
Pick internal point or [Select objects/seTtings]: T (For seTtings)
```

As shown in Figure 9.2, the Hatch tab of this dialog box displays the same setting options available through the Ribbon's Hatch Creation tab.

NOTE

When using the AutoCAD Classic Workspace (anytime the Ribbon is not displayed) the Hatch and Gradient dialog box will be automatically launched with the HATCH command. This chapter will demonstrate the use of the Ribbon's Hatch creation tab. Utilization of the Hatch and Gradient dialog box is similar.

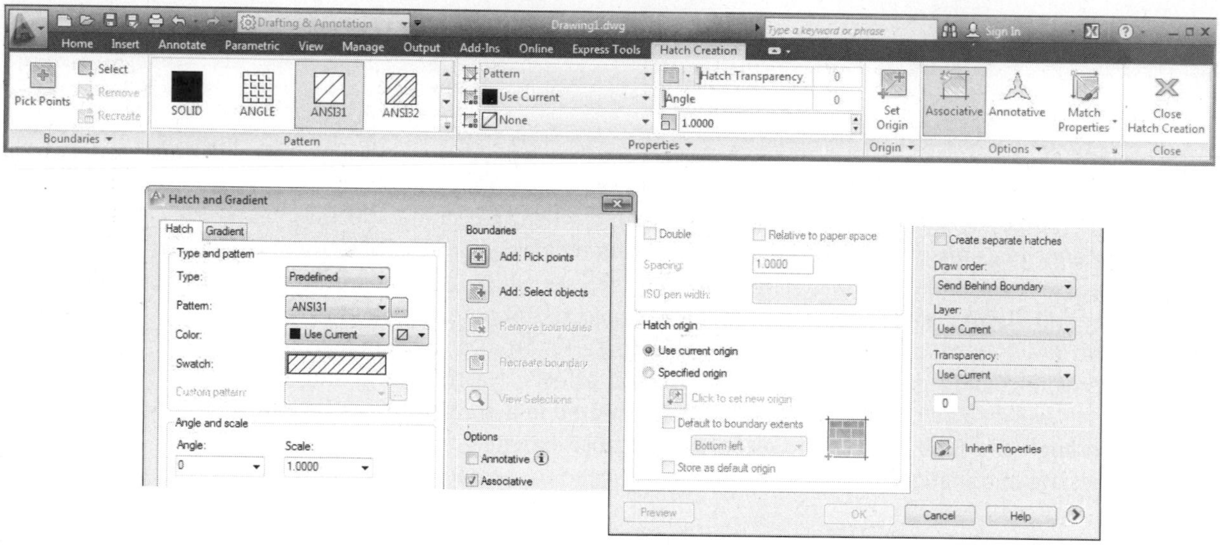

FIGURE 9.2

TRY IT!

Open the drawing file 09_Hatch Basics. The object shown in Figure 9.3 (Left) will be used to demonstrate the boundary crosshatching method. The object needs to have areas "A," "B," and "C" crosshatched.

The HATCH command will identify the boundaries of each closed area to be cross-hatched, as shown in Figure 9.3 (Right).

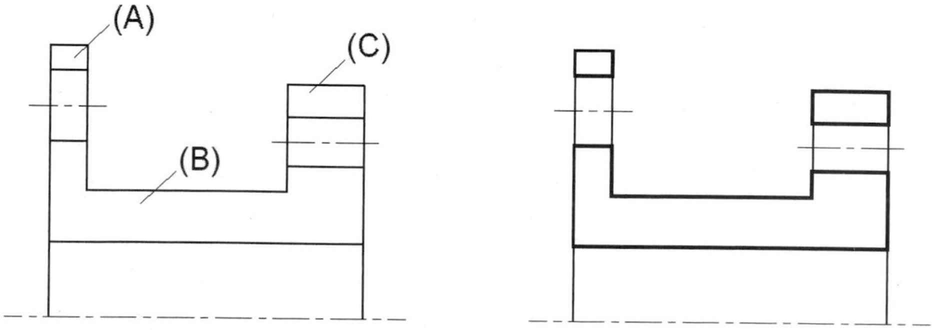

FIGURE 9.3

At the Command prompt, you are ready to pick internal points to identify the areas to be hatched. Moving your cursor over any enclosed area will preview the hatching, as shown in Figure 9.4 (Left).

Click in the three areas required to be hatched, as shown in Figure 9.4 (Center). Any changes to Ribbon settings will update immediately. Press ENTER or click the Close Hatch Creation button to complete the hatching operation. The results are shown in Figure 9.4 (Right).

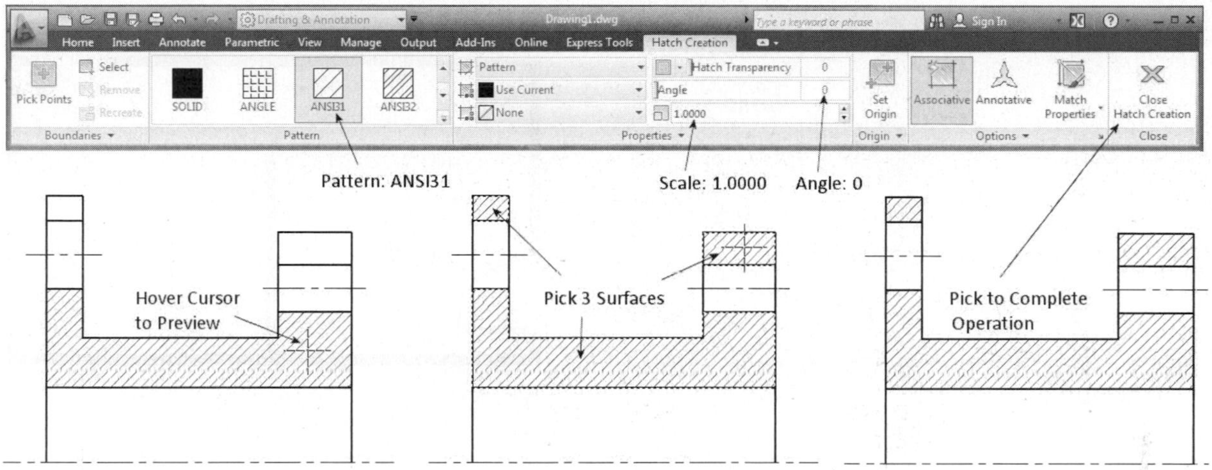

FIGURE 9.4

AVAILABLE HATCH PATTERNS

The Pattern panel of the Hatch Creation tab provides a wide selection of hatch patterns to choose from, as shown in Figure 9.5. Clicking on a pattern in this listing makes the pattern current. If the wrong pattern was selected, simply choose the correct one or expand the rows to view other patterns for use.

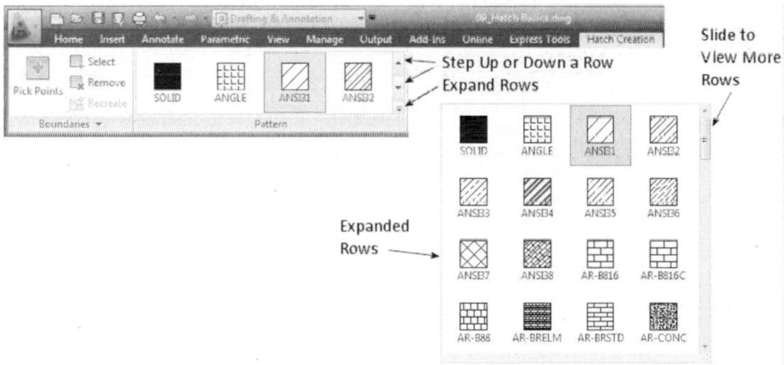

FIGURE 9.5

ARCHITECTURAL AND SOLID FILL HATCH PATTERNS

Hatch patterns typically create section lines, but they can be utilized in other ways. A brick pattern or roof pattern may be placed on an architectural drawing to create a realistic image. A solid fill hatch pattern is sometimes used in a section view on thin cross-sections, but it might also be used to add color to objects in a drawing. A solid pattern is used to fill in an enclosed area with a solid color based on the current layer.

The illustration in Figure 9.6 (Right) shows a typical application of filling in the walls of a floor plan with a solid color.

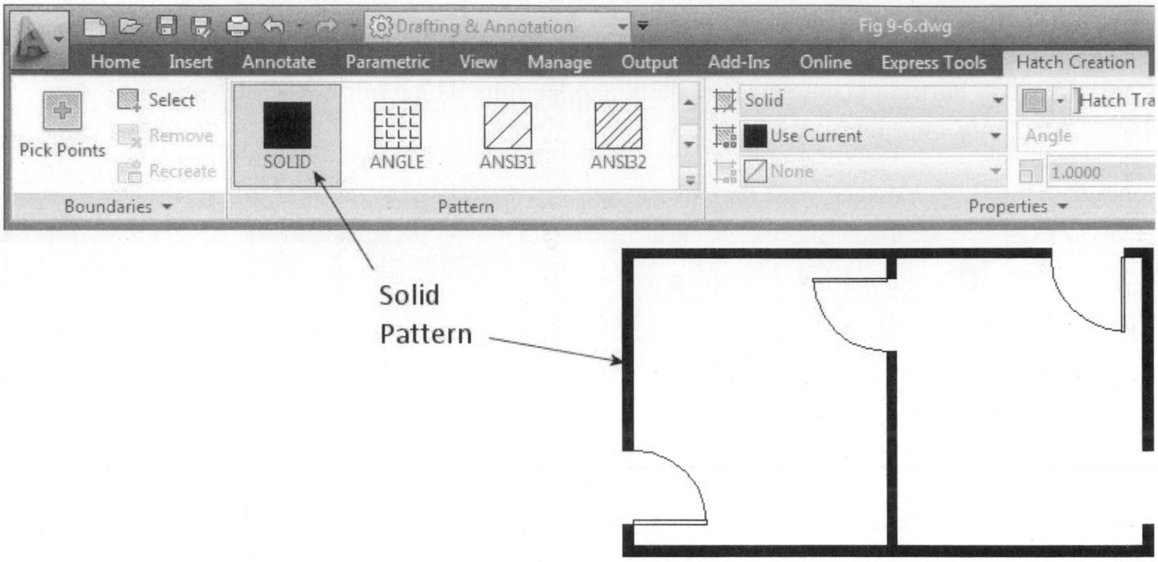

FIGURE 9.6

TRY IT!

Open the drawing file 09_Hatch Arch. The drawing shown in Figure 9.7 (Left) is an elevation view of a one car garage. The `HATCH` **command used with the AR-BRSTD and AR-RSHKE hatch patterns are used to create the image on the right.**

Two hatching operations are required: one with the AR-BRSTD pattern and one with the AR-RSHKE pattern. Pick the internal pick points indicated in Figure 9.7.

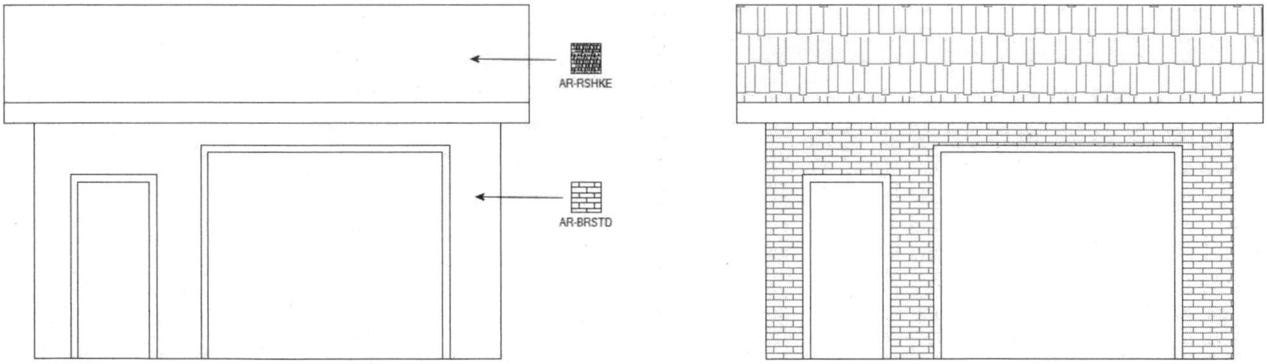

FIGURE 9.7

TRY IT!

Open the drawing file 09_Hatch Solid1. The object shown in Figure 9.7 (Left) illustrates a thin wall that needs to be filled in. The `HATCH` **command used with the SOLID hatch pattern can perform this task very efficiently. When you pick an internal point, the entire closed area is filled with a solid pattern and placed on the current layer with the same color and other properties of the current layer.**

The solid hatch pattern shown in Figure 9.8 (Right) was placed with one internal point pick.

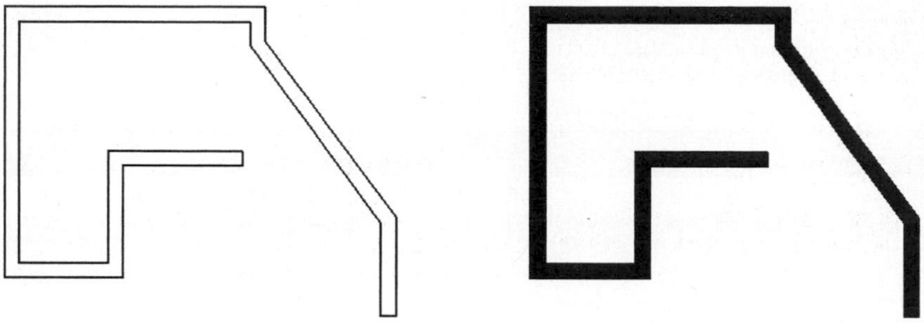

FIGURE 9.8

GRADIENT PATTERNS

A gradient hatch pattern is a solid hatch fill that makes a smooth transition from a lighter shade to a darker shade. Predefined patterns such as linear, spherical, and radial sweep are available to provide different effects. As with the vector hatch patterns that have always been supplied with AutoCAD, the angle of the gradient patterns can also be controlled. Activate the `GRADIENT` command through the Ribbon, as shown in Figure 9.9. As with hatch patterns, they are generated through the Hatch Creation tab of the Ribbon, as shown in Figure 9.9. A few of the controls are explained as follows:

- One Color—When this option is selected, a single color swatch box with a Select Colors button and a Shade and Tint percentage box are made available. The One Color option designates a fill that uses a smooth transition between darker shades and lighter tints of a single color.
- Two Color—When this option is selected, a color swatch box and Select Colors button are available for colors 1 and 2. This option allows the fill to transition smoothly between two colors.
- Color Swatch—This is the default color displayed as set by the current color in the drawing. Expand the box to view the available color options. A Select Colors... button is available to display the Select Color dialog box, similar to the dialog box used for assigning colors to layers. Use this dialog box to select color based on the AutoCAD Index Color, True Color, or Color Book.
- Tint and Shade Percentage—This option designates the amount of tint and shade applied to the gradient fill of one color. Tint is defined as the selected color mixed with white. Shade is defined as the selected color mixed with black.
- Centered Option—Use this option for creating special effects with gradient patterns. When this option is checked, the gradient fill appears symmetrical. If the option is not selected, AutoCAD shifts the gradient fill up to the left. This position creates the illusion that a light source is located to the left of the object.
- Angle—Set an angle that affects the gradient pattern fill. This angle setting is independent of the angle used for regular hatch patterns.
- Gradient Patterns—Nine gradient patterns are available for you to apply. These include linear sweep, spherical, and parabolic.

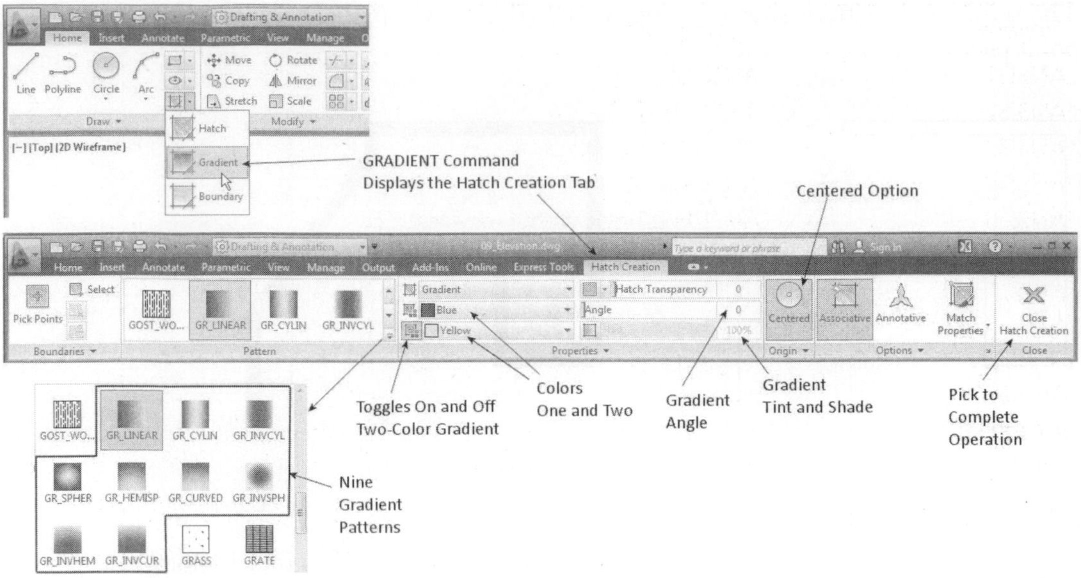

FIGURE 9.9

The example of the house elevation in Figure 9.10 illustrates parabolic linear sweep being applied above and below the house.

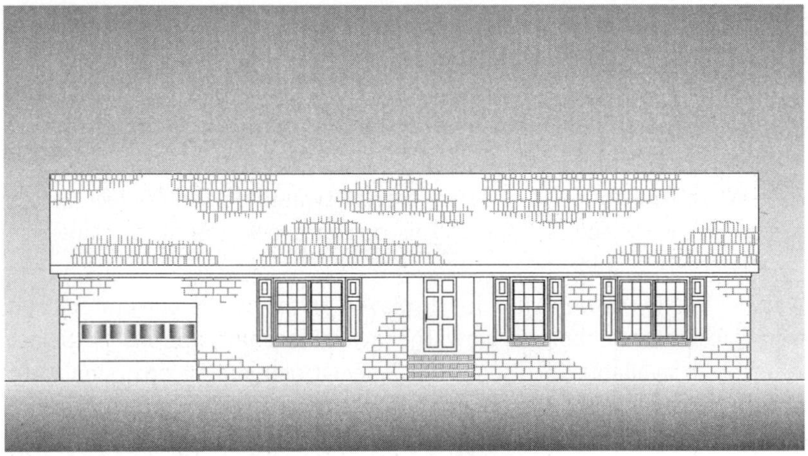

FIGURE 9.10

HATCH PATTERN SYMBOL MEANINGS

Listed below are a number of hatch patterns and their purposes. For example, if you are constructing a mechanical assembly in which you want to distinguish plastic material from steel, use the patterns ANSI34 and ANSI32, respectively. Patterns that begin with AR- have architectural applications. Refer to the following list for the purposes of other materials and their associated hatch patterns.

Pattern	Description
ANSI31	ANSI Cast Iron
ANSI32	ANSI Steel
ANSI33	ANSI Bronze, Brass, Copper
ANSI34	ANSI Plastic, Rubber
ANSI35	ANSI Fire brick, Refractory material
ANSI36	ANSI Marble, Slate, Glass
ANSI37	ANSI Lead, Zinc, Magnesium, Sound/Heat/Electric Insulation
ANSI38	ANSI Aluminum
AR-B816	8x16 Block elevation stretcher bond
AR-B816C	8x16 Block elevation stretcher bond with mortar joints
AR-B88	8x8 Block elevation stretcher bond
AR-BRELM	Standard brick elevation English bond with mortar joints
AR-BRSTD	Standard brick elevation stretcher bond
AR-CONC	Random dot and stone pattern
AR-HBONE	Standard brick herringbone pattern at 45°
AR-PARQ1	2x12 Parquet flooring: pattern of 12x12

ISLAND DETECTION

TRY IT!

Open the drawing file 09_Hatch Islands. Activate the `HATCH` **command and pick a point at "A" in Figure 9.11 (Left), which will define not only the outer perimeter of the object but the inner shapes as well. This ability to detect internal areas is referred to as Island Detection.**

Picking the internal point at "A" will result in the hatch pattern filling the outer area, as shown in Figure 9.11 (Right). If changes need to be made, such as a change in the hatch scale or angle, the preview allows these changes to be made.

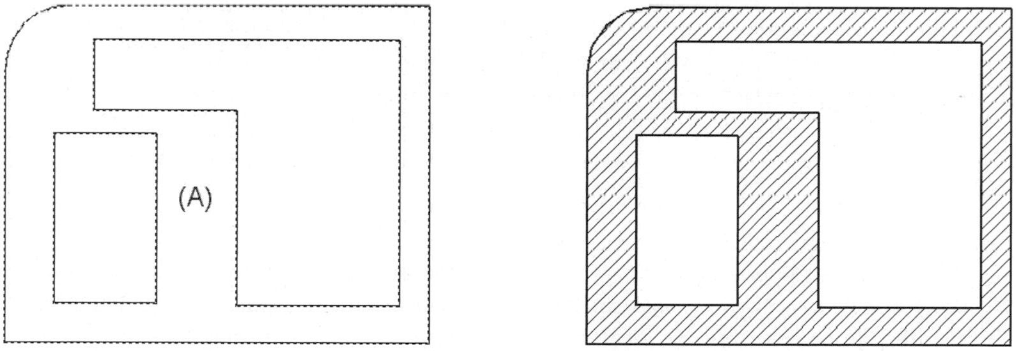

FIGURE 9.11

NOTE

Text objects are also treated as islands and will not be hatched over.

HATCH PATTERN SCALING

Hatching patterns are predefined in size and angle. When you use the HATCH command, the pattern used is assigned a scale value of 1.00, which will draw the pattern exactly the way it was originally created. For the ANSI31 hatch pattern the line spacing is 1/8 inch (0.125).

TRY IT!

Open the drawing file 09_Hatch Scale and activate the HATCH command. Verify that the ANSI31 hatch pattern is selected and pick internal points to hatch the object, as shown in Figure 9.12 (Left).

Entering a different scale value for the pattern either increases or decreases the spacing between crosshatch lines. The illustration in Figure 9.12 (Right) is an example of the ANSI31 pattern with a new scale value of 0.50. The line spacing is now 1/16 inch (0.0625).

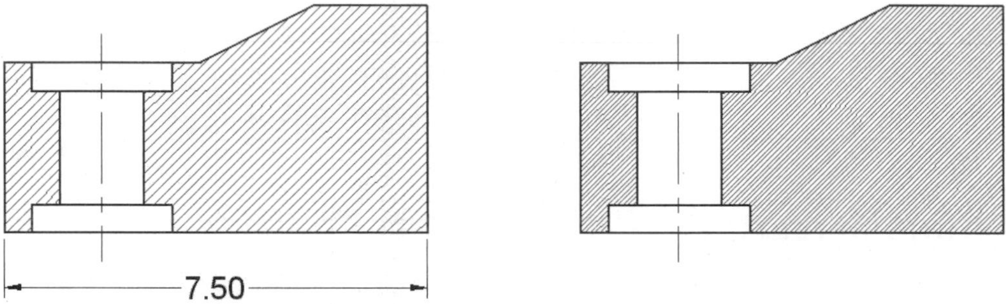

FIGURE 9.12

As you can decrease the scale of a pattern to hatch small areas, you can also scale the pattern up for large areas. The illustration in Figure 9.13 has a hatch scale of 4.00, which quadruples all distances between hatch lines (the new spacing is 1/2 inch or 0.5).

When hatching large areas, a larger scale factor is almost always required. Typically the hatch scale should be set to the drawing scale factor (the reciprocal of the drawing scale). The drawing below is four times larger than the previous one shown. To fit it on a drawing sheet, a scale of 1:4 could be used. Its plotted size would be 7.50. If you did not change the hatch scale, the line spacing would appear 1/32 inch (.03125). Change the hatch scale to 4 (the scale factor—the reciprocal of 1/4) and it will appear on a plot as designed—1/8 inch (0.125).

NOTE

When crosshatching small areas, you will also need to change the hatch scale setting. If using a scale of 4:1, you should set the hatch scale to 0.25. It should also be noted that architectural scales (AR-) are created at sizes utilized for these drawing and typically do not require scale changes.

TRY IT!

Open the drawing file 09_Hatch Scale Large. Activate the HATCH command and change the pattern scale to 4. Pick internal points to crosshatch the object, as illustrated in Figure 9.13.

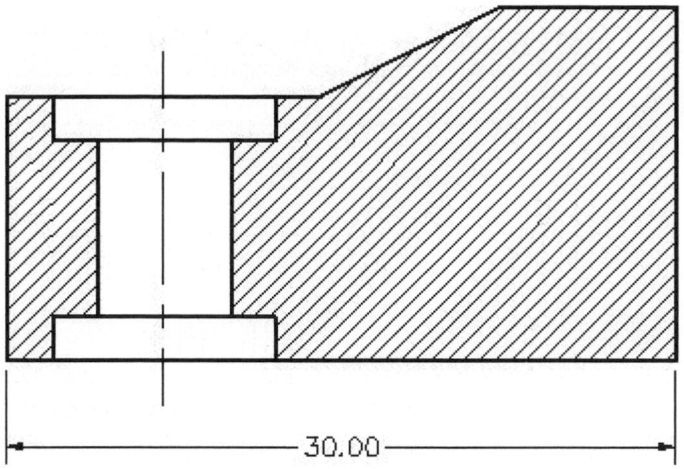

FIGURE 9.13

NOTE

Different hatch patterns are created and utilized on metric and imperial drawings. A 0.125 spacing on a metric drawing would be inadequate. Be sure to utilize the correct template type for your drawings (Acadiso.dwt for metric and Acad.dwt for imperial).

HATCH PATTERN ANGLE MANIPULATION

As with the scale of the hatch pattern, depending on the effect, you can control the angle for the hatch pattern within the area being hatched. By default, the HATCH command displays a 0° angle for all patterns.

TRY IT!

Open the drawing file 09_Hatch Angle. Activate the HATCH command and hatch the object, as shown in Figure 9.14 (Left), keeping all default values. The angle for ANSI31 is drawn at 45°—the angle in which the pattern was originally created.

Experiment with the angle setting by entering any angle different from the default value of 0° to rotate the hatch pattern by that value. This means that if a pattern were originally designed at a 45° angle, like ANSI31, entering a new angle for the pattern would begin rotating the pattern starting at the 45° position.

Entering an angle other than the default value rotates the pattern from the original angle to a new angle. In the illustration in Figure 9.14 (Right), an angle of 90° has been applied to the ANSI31 pattern. Providing different angles for patterns is required if the section lines appear parallel or perpendicular to significant object lines in your drawing. It is also often used in section assemblies for different parts that are in contact with each other. The change in hatch angle helps distinguish the different parts in the assembly.

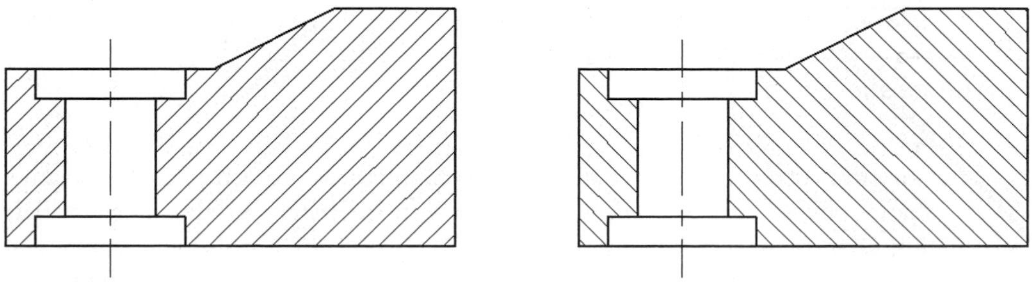

FIGURE 9.14

TRY IT!

Open the drawing file 09_Hatch Assembly. With Figure 9.15 (Left) as a guide, hatch the section assembly using the pattern, scale, and angle settings designated for each area. The results are shown in Figure 9.15 (Right).

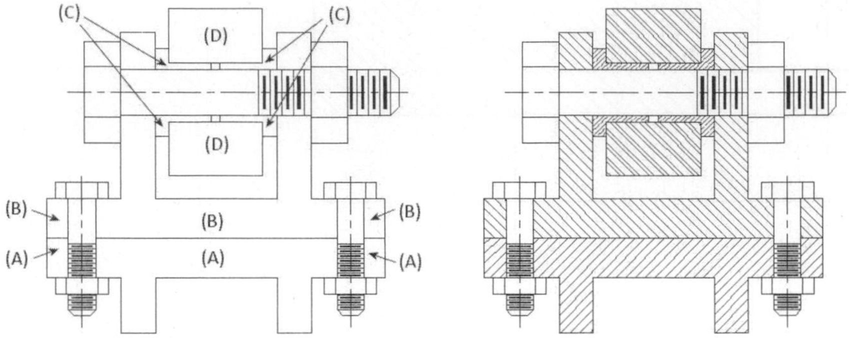

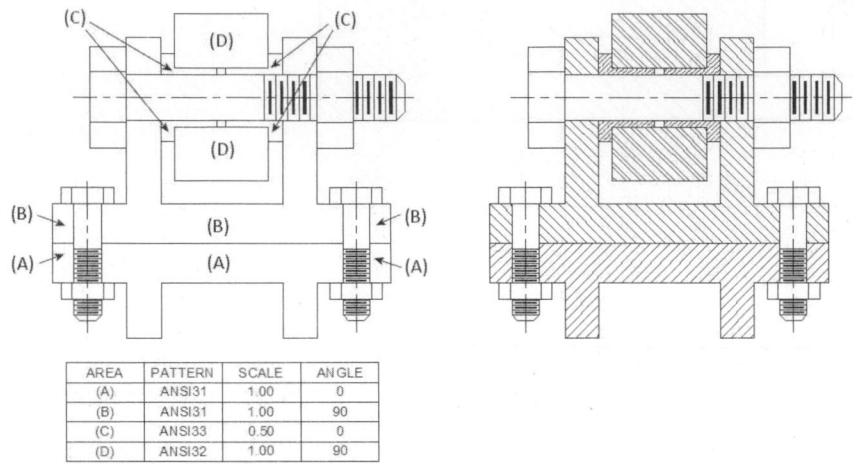

AREA	PATTERN	SCALE	ANGLE
(A)	ANSI31	1.00	0
(B)	ANSI31	1.00	90
(C)	ANSI33	0.50	0
(D)	ANSI32	1.00	90

FIGURE 9.15

MODIFYING ASSOCIATIVE HATCHES

TRY IT!

Open the drawing file 09_Hatch Assoc. In Figure 9.16, the plate needs to be hatched in the ANSI31 pattern at a scale of one unit and an angle of 0°, and the two slots and three holes are to be considered islands. Enter the HATCH command, verify the hatch settings (including making sure Associative is selected in the Options panel of the Ribbon), and pick an internal point somewhere inside the object, such as at "A," as shown in Figure 9.16.

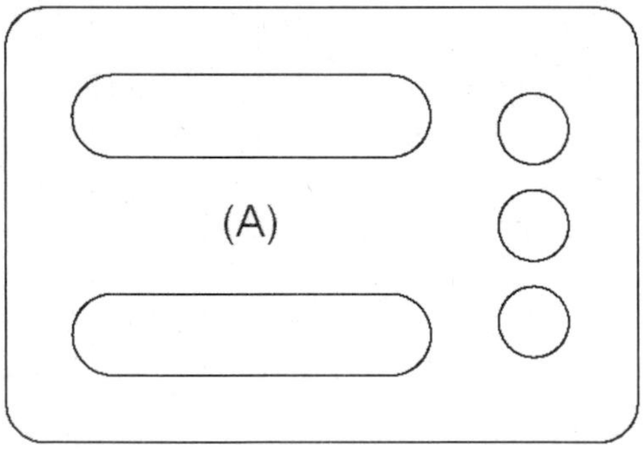

FIGURE 9.16

It is very important to realize that all objects highlighted are tied to the associative crosshatch object. Each shape works directly with the hatch pattern to ensure that the outline of the object is being read by the hatch pattern and that the hatching is performed outside the outline. After completing the hatching operation, the results should appear similar to the object in Figure 9.17 (Left).

Associative hatch objects may be edited, and the results of this editing have an immediate impact on the appearance of the hatch pattern. For example, the two outer holes need to be increased in size by a factor of 1.5; also, the middle hole needs to be repositioned to the other side of the object, as shown in Figure 9.17 (Right). Not only does using the SCALE and MOVE command allow you to resize and reposition the holes, but when the modify operations are completed, the hatch pattern mends itself automatically.

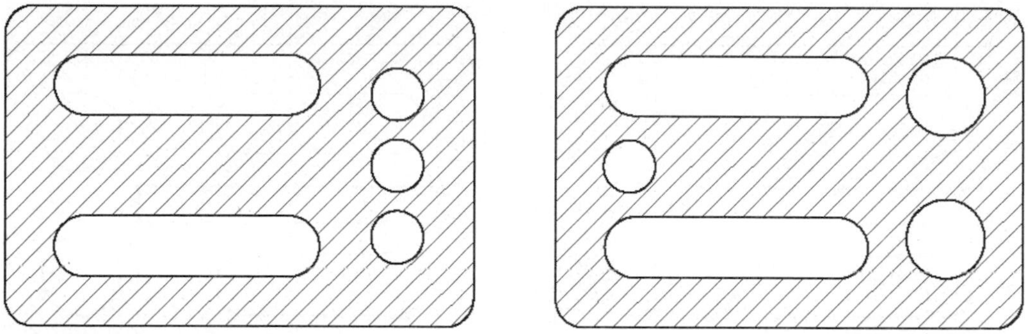

FIGURE 9.17

As illustrated in Figure 9.18 (Left), the length of the slots needs to be shortened. Also, hole "A" needs to be deleted. Use the ERASE command to delete hole "A." Use the STRETCH command to shorten the slots. Use the crossing box (pick right to left) at "B" to select the slots and stretch them one unit to the left.

The result of the editing operations is shown in Figure 9.18 (Right). In this figure, associative hatching has allowed islands to be modified while the hatch pattern still maintains its associativity.

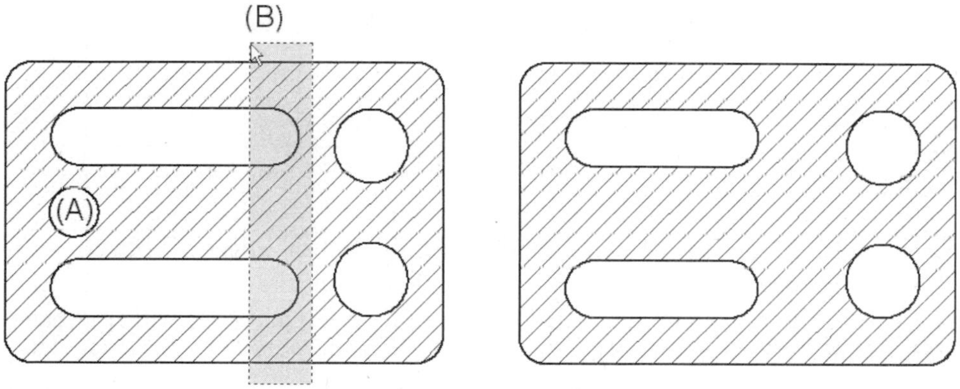

FIGURE 9.18

TRIMMING HATCH AND FILL PATTERNS

Once a hatch pattern is placed in the drawing, it does not update itself to any new additions in the form of closed shapes in the drawing. Illustrated in Figure 9.19 (Left), a rectangle was added at the center of the object. Notice, however, that the hatch pattern cuts directly through the rectangle. The hatch pattern does not have the intelligence to recognize the new boundary.

On the other hand, the TRIM command could be used to select the rectangle as the cutting edge and select the hatch pattern inside the rectangle to remove it, as shown in Figure 9.19 (Right). After this trimming operation is performed, the rectangle is now recognized as a valid member of the associative hatch.

NOTE

If you are unable to trim a new object placed in a hatch pattern, change your Island Detection Style from "Outer" to "Normal" and reapply the hatching. These styles will be discussed later in this chapter—"Advanced Hatching Techniques."

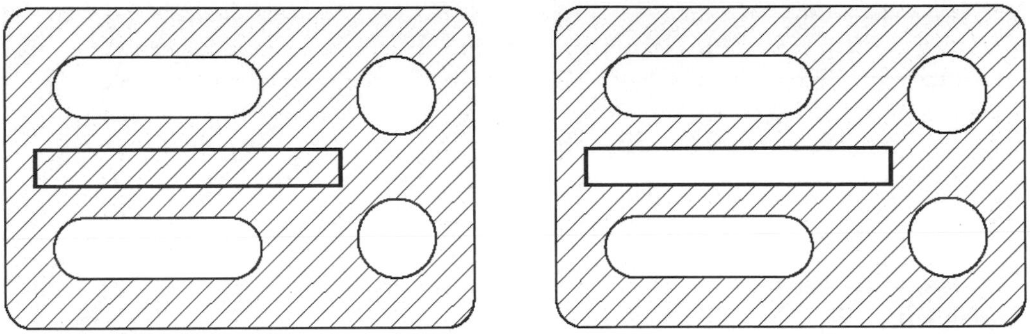

FIGURE 9.19

Hatch patterns can even be trimmed when the objects you add are not fully within the hatch boundary. In Figure 9.20 (Left), the rectangle is constructed and hatched with the ANSI31 pattern. In Figure 9.15 (Right), two circles have been added to the ends of the rectangle. You now want to remove the hatch pattern from both halves of the circles.

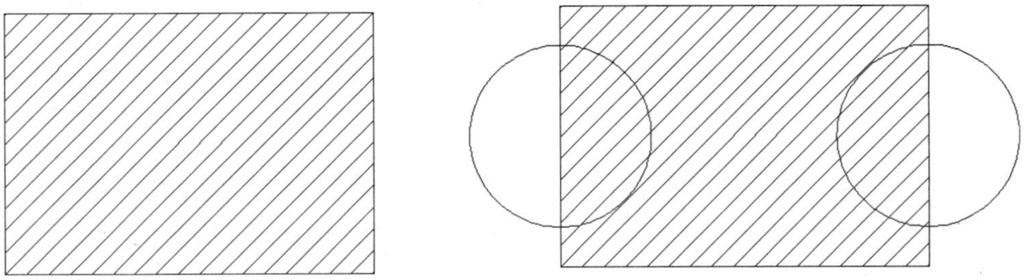

FIGURE 9.20

To accomplish this, enter the TRIM command and select the edges of both circles as cutting edges, as shown in Figure 9.21 (Left). Then pick the hatch patterns inside both circles. The results are illustrated on the right in Figure 9.21, with the hatch pattern trimmed out of both circles.

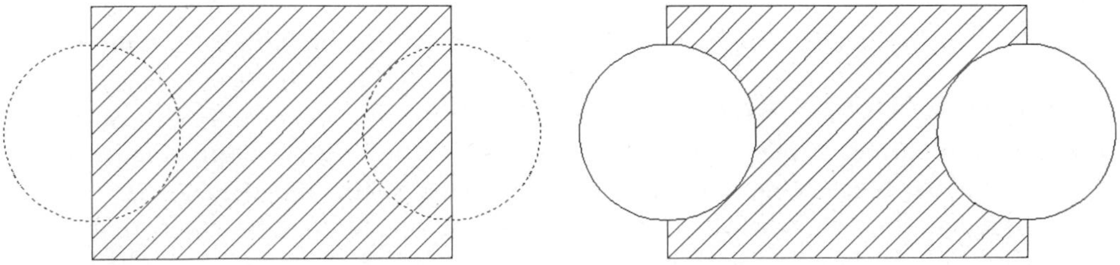

FIGURE 9.21

NOTE

Exploding a hatch pattern will result in the loss of associativity and should be avoided if at all possible. The Modify command operations just discussed and the editing techniques discussed in the next section would be unavailable. It should also be noted, that the `MIRROR` command can be used without loss of associativity (the pattern itself is not mirrored).

EDITING HATCH PATTERNS

TRY IT!

Open the drawing file 09_Hatchedit. The pattern needs to be increased to a new scale factor of 3 units and the angle of the pattern needs to be rotated by 90°. When utilizing the Drafting & Annotation Workspace, simply pick the hatch pattern to display the Ribbon with the Hatch Editor tab activated, as shown in Figure 9.22. The current scale value of the pattern is 1 unit and the angle is 0°. Change the angle and scale setting and the hatch display is automatically updated. Click the Close Hatch Editor button to complete the operation. The edited hatch pattern is shown in Figure 9.22 (Right).

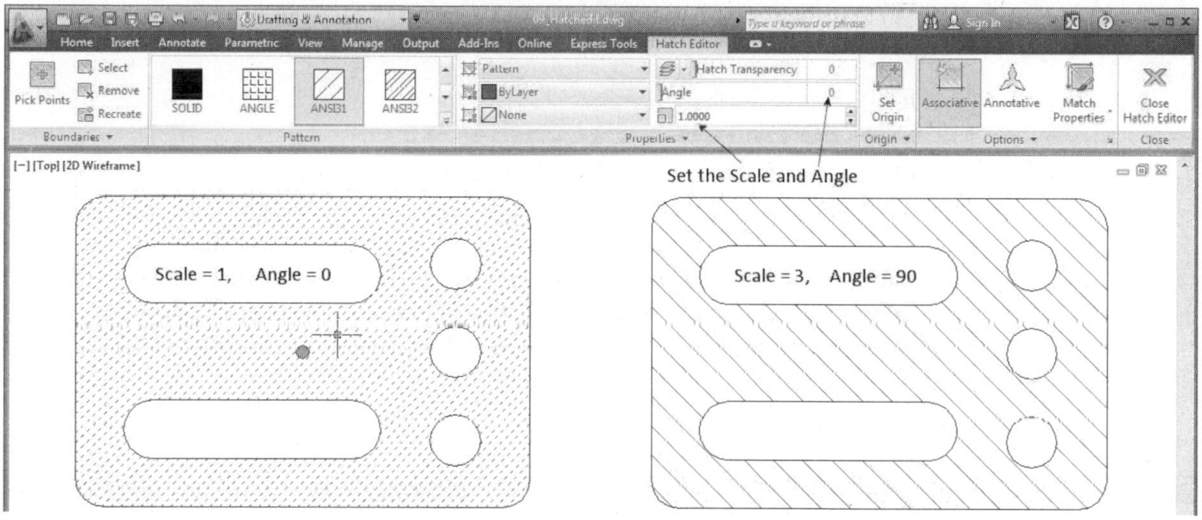

FIGURE 9.22

Hatch patterns can also be edited utilizing the `HATCHEDIT` command. This command can be found in the Ribbon, as shown in Figure 9.23. Issuing this command prompts you to select the hatch pattern to edit. Clicking on the hatch pattern displays the Hatch Edit dialog box, also shown in Figure 9.23. The same settings available in the Hatch Editor tab of the Ribbon can be found in this dialog box. To perform the same editing operation just discussed click in the Scale field to change the scale from the current value of 1 unit to the new value of 3 units. Next, click in the Angle field and change the angle of the hatch pattern from the current value of 0° to the new value of 90°. Clicking the OK button in the Hatch Edit dialog box returns you to the drawing editor and updates the hatch pattern to these changes.

NOTE

The Hatch Edit dialog box can also be activated by pre-selecting a hatch pattern, right-clicking and selecting Hatch Edit... from the provided shortcut menu. Double-clicking a hatch pattern launches a Quick Properties palette where you can also change the angle and scale.

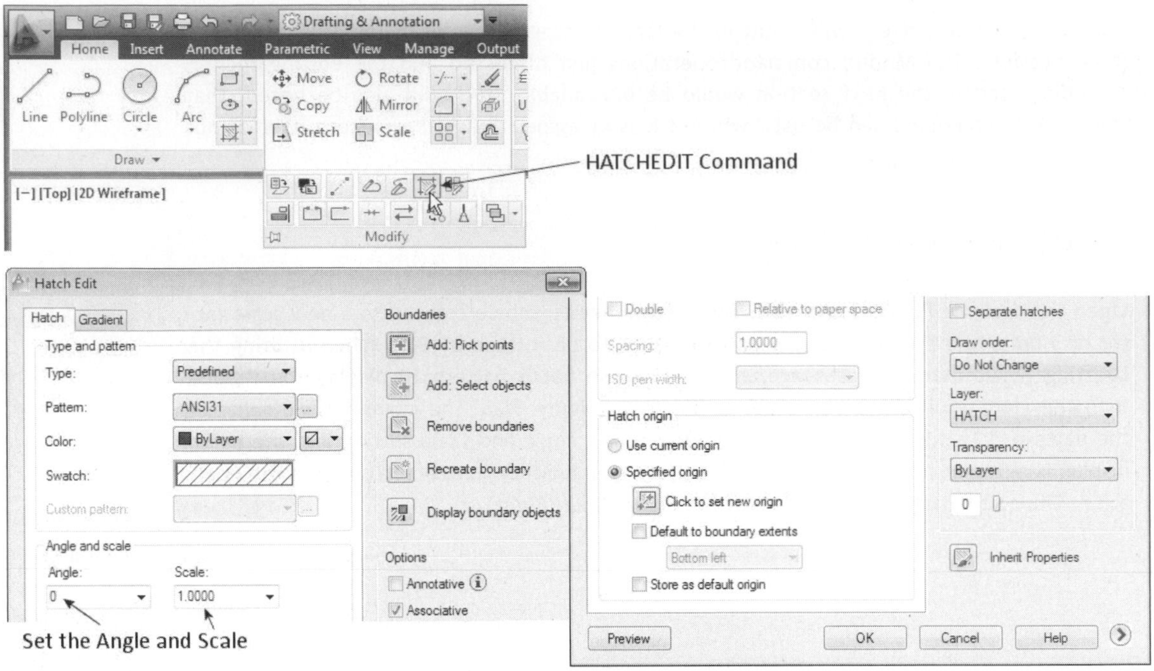

FIGURE 9.23

ADVANCED HATCHING TECHNIQUES

Advanced features such as Island Detection Styles, Boundary Retention, Boundary Sets, Gap Tolerances, and Inherit Options, are available through the Ribbon's Hatch Creation tab. To view these features, expand the appropriate Ribbon panel: Boundaries, Properties, Origin, or Options, as shown in Figure 9.24. If desired, these options are also available through the Hatch and Gradient dialog box, which can be displayed by clicking the arrow in the lower-right corner of the Options panel.

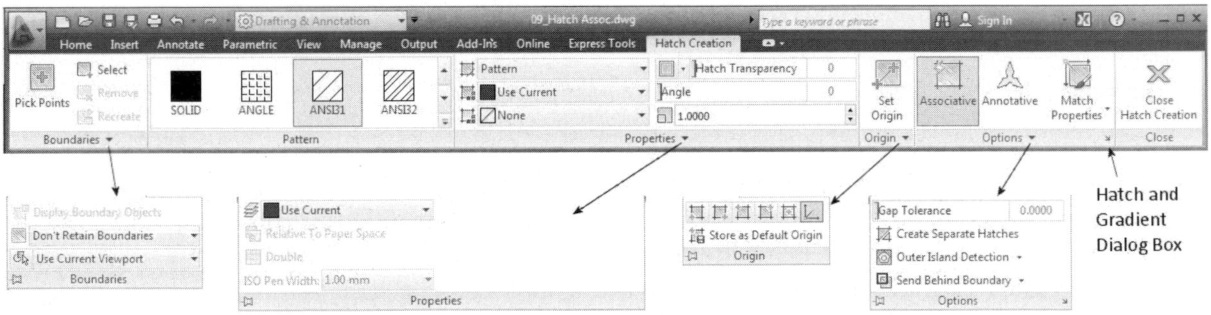

FIGURE 9.24

Island Detection Styles

The illustration in Figure 9.25 shows three detection styles and how they affect the crosshatching of Islands. The Normal style provides hatching in every other island. When you pick within the outermost boundary, the hatching begins, skips the next inside boundary, hatches the next innermost boundary, and so on. The Outer boundary style hatches only the outermost boundary of the object. The Ignore style ignores the islands and hatches the entire object.

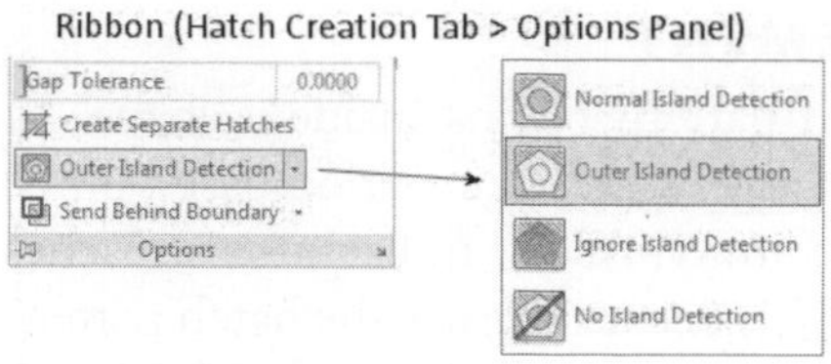

FIGURE 9.25

Gap Tolerance

Your success in hatching an object relies heavily on making sure the object being hatched is completely closed. This requirement can be relaxed with the addition of a Gap Tolerance field, as shown in Figure 9.26. In this example, a gap tolerance of 0.20 has been entered. However, the object in the middle has a gap of 0.40. Since the object's gap is larger than the tolerance value, an alert dialog box appears, as shown in Figure 9.26 (Right). You must set the gap tolerance larger than the gap in the object or completely close the shape.

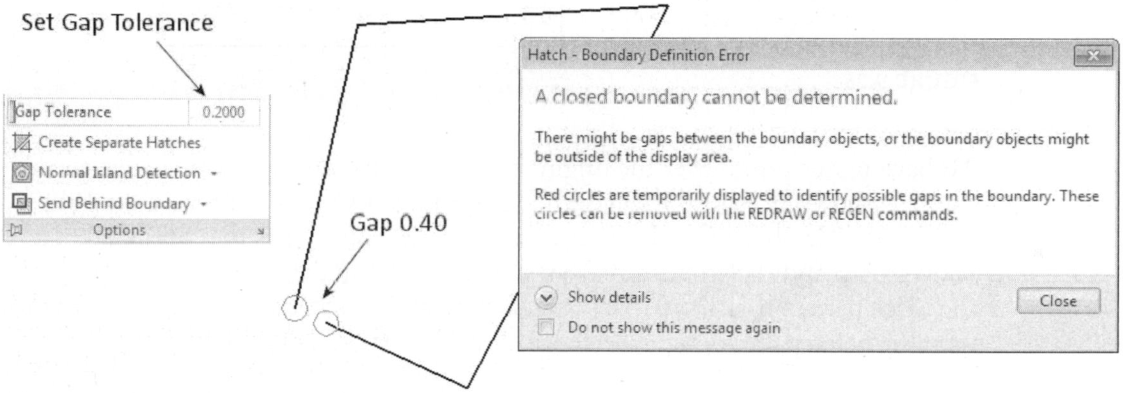

FIGURE 9.26

The illustration in Figure 9.27 (Left) shows a gap of 0.16. With a gap tolerance setting of 0.20, another alert dialog box warns you that the shape is not closed but if the gap setting is larger than the gap in the object, the object will be hatched. The results are shown in Figure 9.27 (Right).

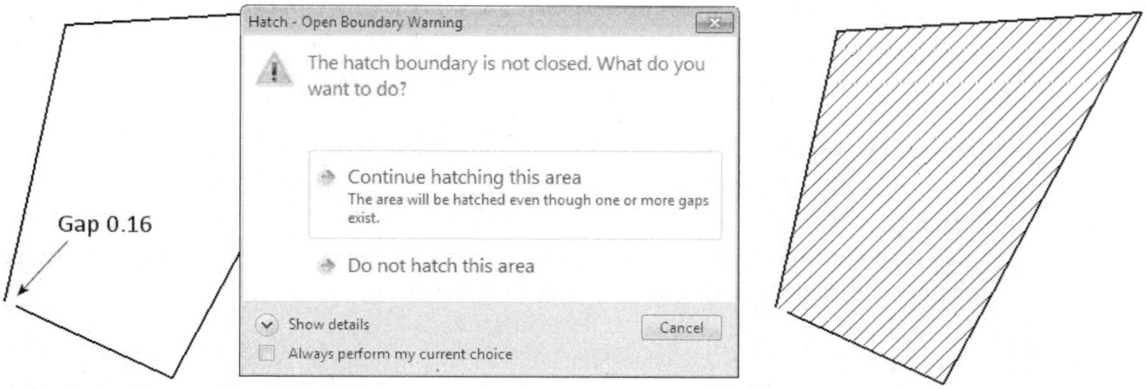

FIGURE 9.27

PRECISION HATCH PATTERN PLACEMENT

At times, you want to control where the hatch pattern begins inside a shape. In Figure 9.28, a brick pattern was applied without any regard to insertion base points. In this image, the pattern's start point appears arbitrary. As the brick pattern reaches the edges of the rectangle, the pattern just ends. This occurs because the hatch pattern uses a current origin point (0,0) to be constructed from. Settings for the hatch pattern's origin can be found in the Ribbon's Hatch Creation tab, as shown in Figure 9.28 (Left).

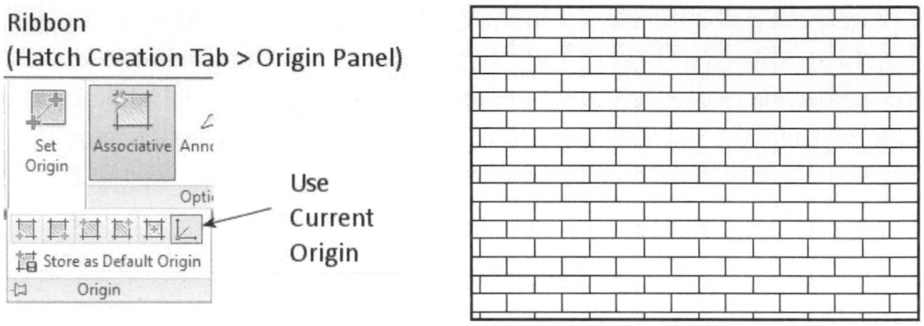

FIGURE 9.28

To have more control over the origin of the hatch pattern, you can change the settings, as shown in Figure 9.29 (Left). Click the "Set Origin" button in the Ribbon. If using the Hatch Edit or Hatch and Gradient dialog box, pick the "Specified origin" radio button and click the button "Set Origin." You can then select the new origin of the hatch pattern back in your drawing. The results are shown in Figure 9.29 (Right), with the hatch pattern origin located at the lower-left corner of the rectangle.

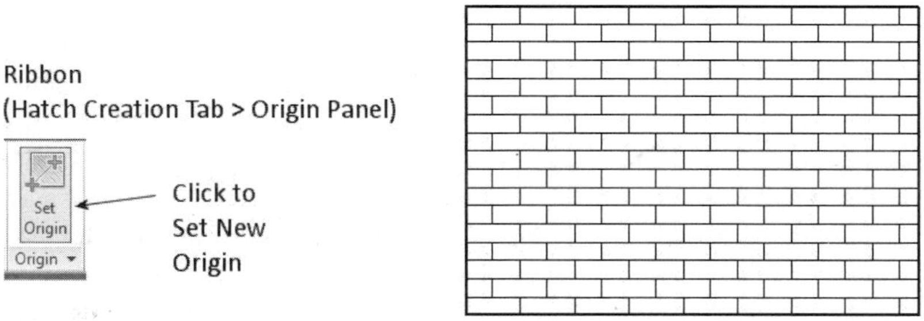

FIGURE 9.29

HATCHING AND MULTI-FUNCTIONAL GRIPS

Multi-functional grips provide a simple and quick method for editing an existing hatch pattern. Selecting the pattern and then hovering your cursor over the round grip displayed, provides a shortcut menu with options to stretch, change the origin, change the angle, and change the scale of the pattern.

TRY IT!

Open the drawing file 09_Hatch Multi-Functional Grip. The drawing contains a rectangle with the default ANSI37 hatch pattern applied, as shown in Figure 9.30 (Left). Pick the hatch pattern and hover the cursor over the grip to display the shortcut menu shown in Figure 9.30 (Center). Select the Hatch Scale option and enter a value of "2" to obtain the results shown in the image on the right.

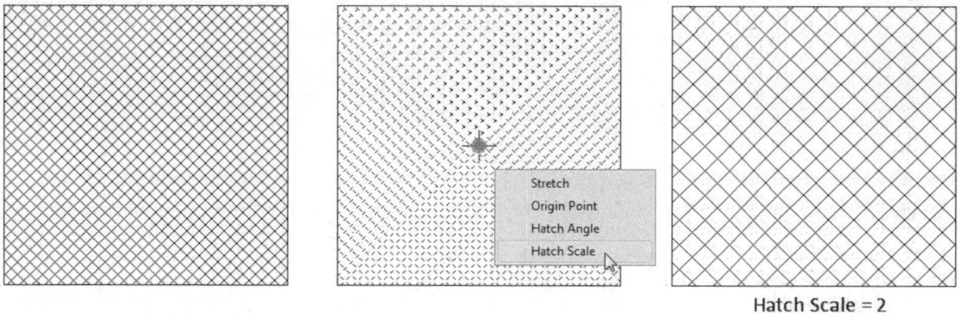

FIGURE 9.30

Use the grip shortcut menu to change the Hatch Angle to "45", as shown in Figure 9.31 (Left). For the final modification, select the Hatch Point option and pick the lower left corner of the rectangle as the new origin point as shown in Figure 9.31 (Center). The results of the multi-functional grip modifications are shown in Figure 9.31 (Right).

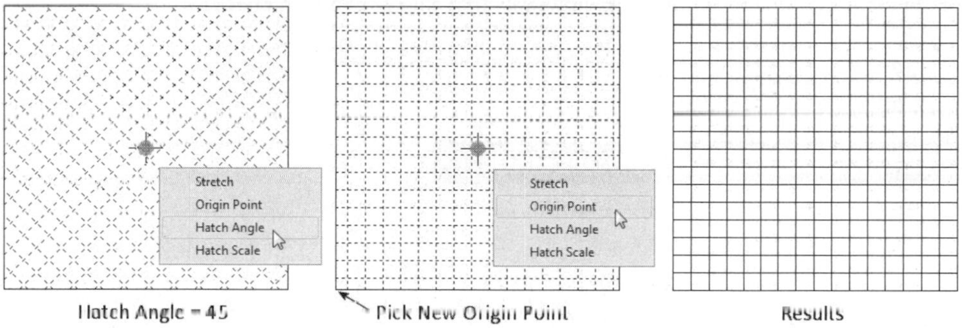

FIGURE 9.31

INHERIT HATCHING PROPERTIES

TRY IT!

Open the drawing file 09_Hatch Inherit. The illustration in Figure 9.32 (Left) consists of a simple assembly drawing. At least three different hatch patterns are displayed to define the different parts of the assembly. Unfortunately, a segment of one of the parts was not hatched, and it is unclear what pattern, scale, and angle were used to place the pattern.

Whenever you are faced with this problem, click the Match Properties button from the Ribbon, as shown in Figure 9.32 (Center). Clicking on the pattern, as shown in the image on the left, sets the pattern, scale, and angle to match the selected pattern.

To complete the hatch operation, click an internal point in the empty area. Right-click and select ENTER from the shortcut menu. The hatch pattern is placed in this area, and it matches that of the other patterns, as shown in Figure 9.32 (Right).

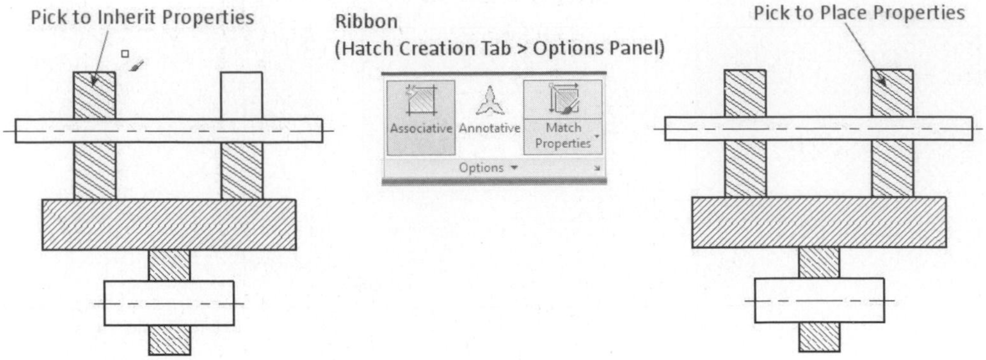

FIGURE 9.32

NOTE

By default, Object Snaps do not work on hatch patterns. This is typically advantageous, for example, while placing dimensions in a drawing view, you will want to snap to object outlines and not to the end of section lines. To change this setting, activate the Options dialog box utilizing the `OPTIONS` command. In the Drafting tab uncheck the box next to "Ignore Hatch Objects." It is recommended, however, that this box remain checked.

TUTORIAL EXERCISE: 09_COUPLER.DWG

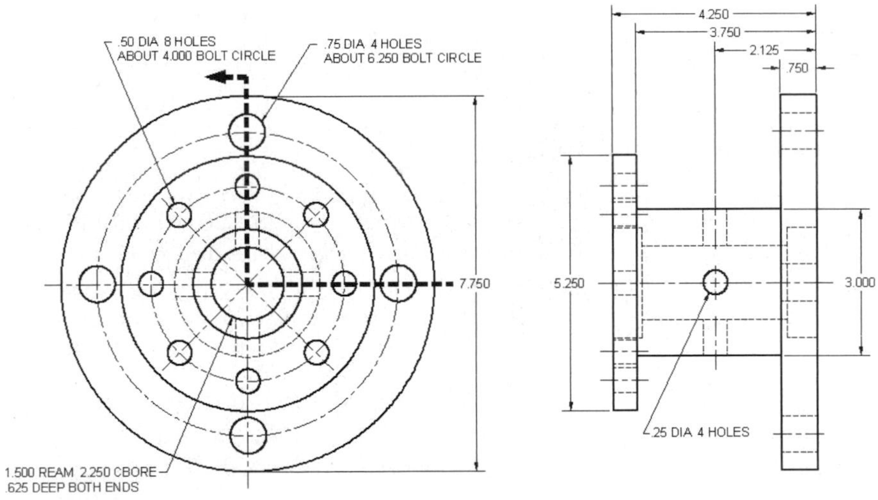

FIGURE 9.33

Purpose

This tutorial is designed to use the MIRROR and HATCH commands to convert 09_Coupler to a half section, as shown in Figure 9.33.

System Settings

Open an existing drawing called 09_Coupler. Follow the steps in this tutorial for converting the upper half of the object to a half section. All Units, Limits, Grid, and Snap values have been previously set.

Layers

The following layers are already created:

Name	Color	Linetype
Object	White	Continuous
Center	Yellow	Center
Hidden	Red	Hidden
Hatch	Magenta	Continuous
Cutting Plane Line	Yellow	Dashed
Dimension	Yellow	Continuous

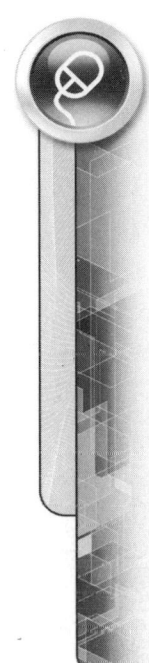

Suggested Commands

You will be converting one-half of the object to a section by erasing unnecessary hidden lines and using the Layer Control box to change the remaining hidden lines to the Object layer. The HATCH command along with the ANSI31 hatch pattern will be used to hatch the upper half of the 09_Coupler on the Hatch layer.

STEP 1

Use the MIRROR command to copy and flip the upper half of the Side view and form the lower half. When in Object Selection mode, use the Remove option to deselect the main centerline, hole centerlines, and hole, as shown in Figure 9.34 (Left). If these objects are included in the mirror operation, a duplicate copy of these objects will be created.

Command: **MI** *(For MIRROR)*

Select objects: *(Pick a point at "A")*

Specify opposite corner: *(Pick a point at "B")*

Select objects: *(Hold down the SHIFT key and pick centerlines "C" and "D" to remove them from the selection set)*

Select objects: *(With the SHIFT key held down, pick a point at "E")*

Specify opposite corner: *(With the SHIFT key held down, pick a point at "F")*

Select objects: *(Press ENTER to continue)*

Specify first point of mirror line: *(Select the endpoint of the centerline near "C")*

Specify second point of mirror line: *(Select the endpoint of the centerline near "D")*

Erase source objects? [Yes/No] <N>: *(Press ENTER to perform the mirror operation)*

Begin converting the upper half of the Side view to a half section by using the ERASE command to remove any unnecessary hidden lines and centerlines from the view, as shown in Figure 9.34 (Right).

Command: E *(For ERASE)*

Select objects: *(Carefully select the hidden lines labeled "G," "H," "J," and "K")*

Select objects: *(Select the centerline labeled "L")*

Select objects: *(Press* ENTER *to execute the erase command)*

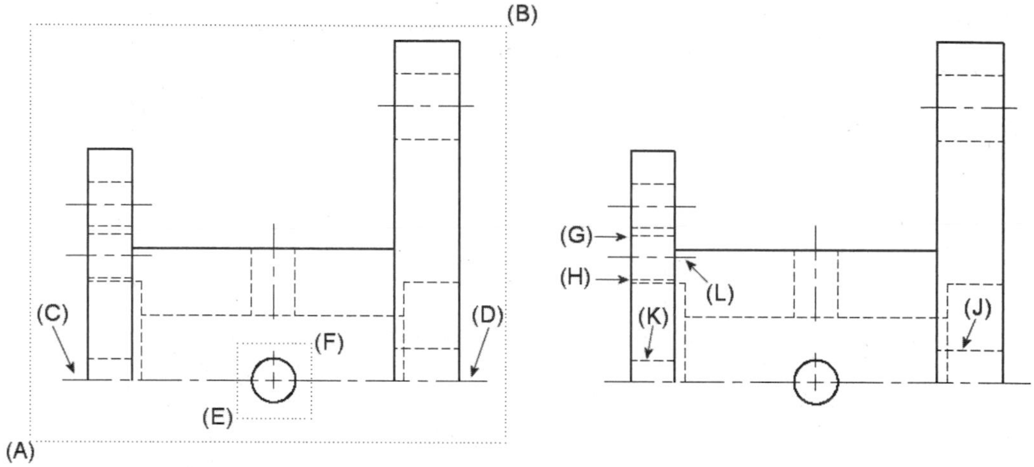

FIGURE 9.34

STEP 2

Since the remaining hidden lines actually represent object lines when shown in a full section, use the Layer Control box in Figure 9.35 to convert all highlighted hidden lines from the Hidden layer to the Object layer.

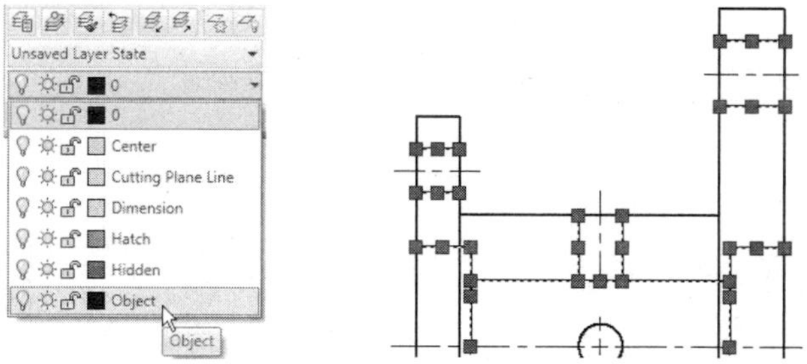

FIGURE 9.35

STEP 3

Remove unnecessary line segments from the upper half of the converted section using the TRIM command. Use the horizontal line at "A" as the cutting edge, and select the two vertical segments at "B" and "C" as the objects to trim, as shown in Figure 9.36 (Left).

Make the Hatch layer the current layer. Then, activate the HATCH command. Use the pattern "ANSI31" and keep all default settings. Select points inside areas "D," "E," "F," and "G," as shown in Figure 9.36 (Right). When finished selecting these internal points, right-click and select ENTER from the shortcut menu.

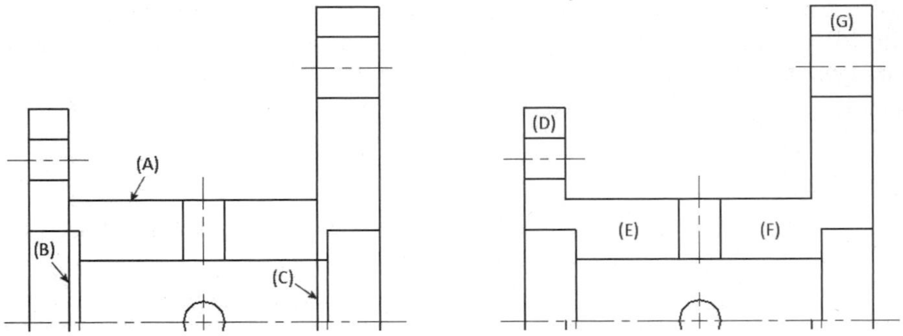

FIGURE 9.36

STEP 4

The complete hatched view is shown in Figure 9.37.

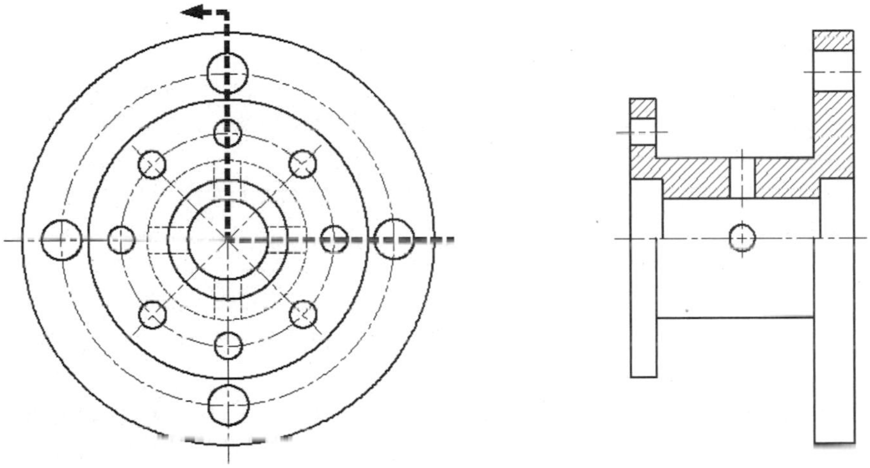

FIGURE 9.37

TUTORIAL EXERCISE: 09_ELEVATION.DWG

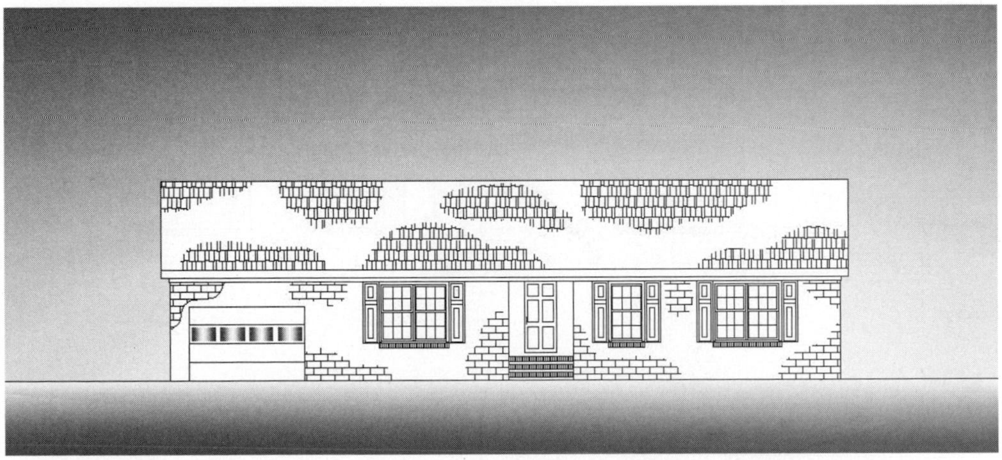

FIGURE 9.38

Purpose

In this tutorial, hatch patterns will be added to the drawing 09_Elevation by utilizing the Match Properties button from the Hatch Creation tab of the Ribbon. Gradient hatch patterns will also be applied to this drawing for presentation purposes.

System Settings

Since this drawing is provided, edit an existing drawing called 09_Elevation. Follow the steps in this tutorial for completing an elevation of the residence.

Layers

The following layers are already created:

Name	Color	Linetype
Elevations	White	Continuous
Exterior Brick	Red	Continuous
Gradient Background	Blue	Continuous
Roof Boundaries	153	Continuous
Roof Hatch	73	Dashed
Sill	30	Continuous
Wall Boundaries	Magenta	Continuous

Suggested Commands

Open up the existing drawing file 09_Elevation and notice the appearance of existing roof and brick patterns. Use the Match Properties feature of the HATCH command to transfer the roof and brick patterns to the irregular-shaped areas of this drawing. Next, a gradient pattern will be added to the upper and lower parts of the drawing for the purpose of creating a background. Another gradient pattern will be added to the garage door windows. In this way, the gradient patterns will be used to enhance the elevation of the house.

STEP 1

Open the drawing 09_Elevation. Notice the appearance of the roof and brick hatch patterns in Figure 9.39. The roof and brick patterns need to be applied to the other irregular areas of this house elevation.

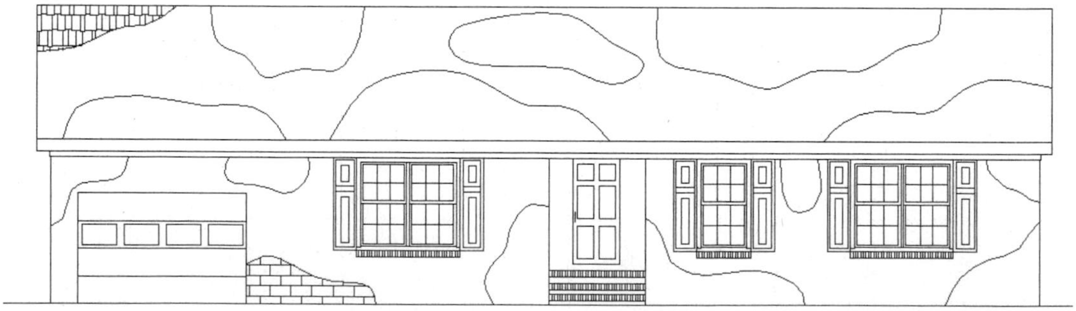

FIGURE 9.39

STEP 2

Activate the HATCH command and click the Match Properties button, as shown in Figure 9.40 (Left). This button allows you to click an existing hatch pattern and have the pattern name, scale, and angle transfer to the proper fields in the Ribbon panels. In other words, you do not have to figure out the name, scale, or angle of the pattern. It should also be noted that the layer is also transferred. The only requirement is that a hatch pattern already exists in the drawing.

Notice the appearance of a glyph that is similar to the Match Properties icon. Click the existing roof pattern, as shown on the right in Figure 9.40.

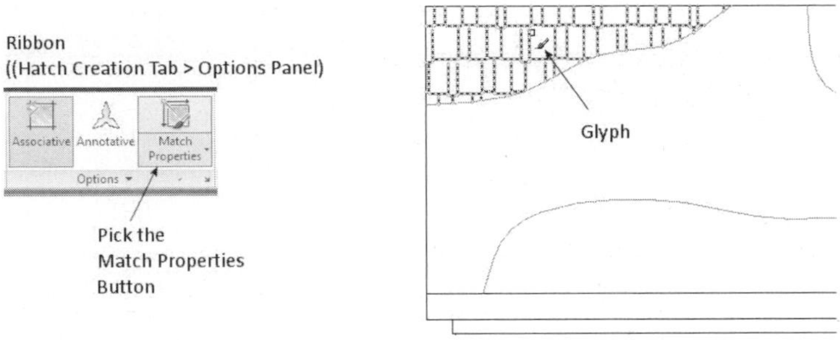

FIGURE 9.40

STEP 3

After you pick the existing hatch pattern shown in the previous image, your cursor changes appearance again. Now pick internal points inside each irregular shape in the roof area as shown in Figure 9.41. Notice that each one previews the hatch pattern. When you're finished, right-click and select ENTER from the shortcut menu.

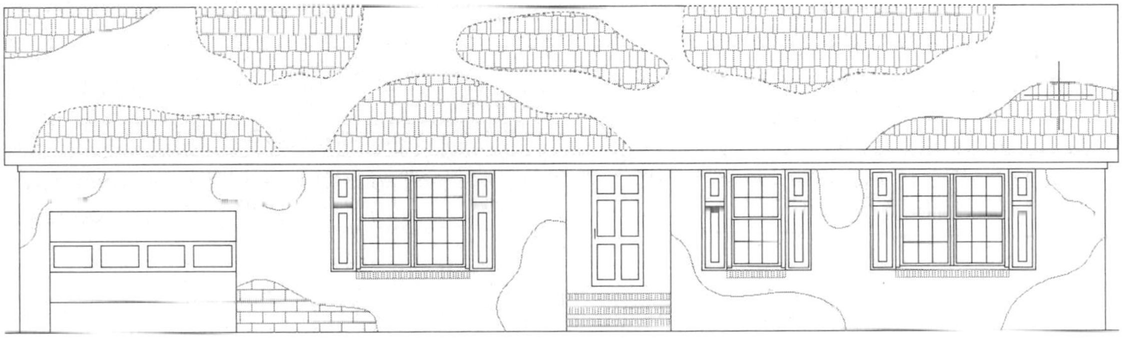

FIGURE 9.41

STEP 4

Again, activate the HATCH command and click the Match Properties button. Pick the existing brick pattern to match the options and click inside the remaining irregular shapes in Figure 9.42 until each one previews the pattern. Right-click and select ENTER if the results are correct.

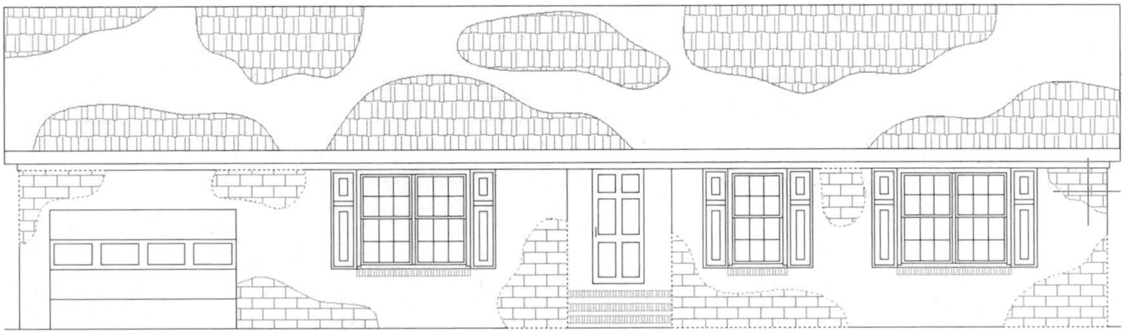

FIGURE 9.42

STEP 5

To give the appearance that the hatch patterns are floating on the roof and wall, use the Layer Control box to turn off the two layers that control the boundaries of these patterns, namely, Roof Boundaries and Wall Boundaries. Your display should appear similar to Figure 9.43.

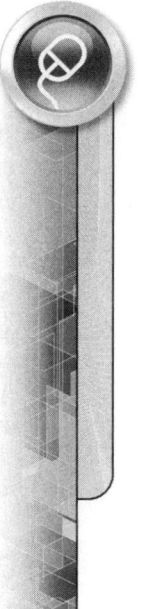

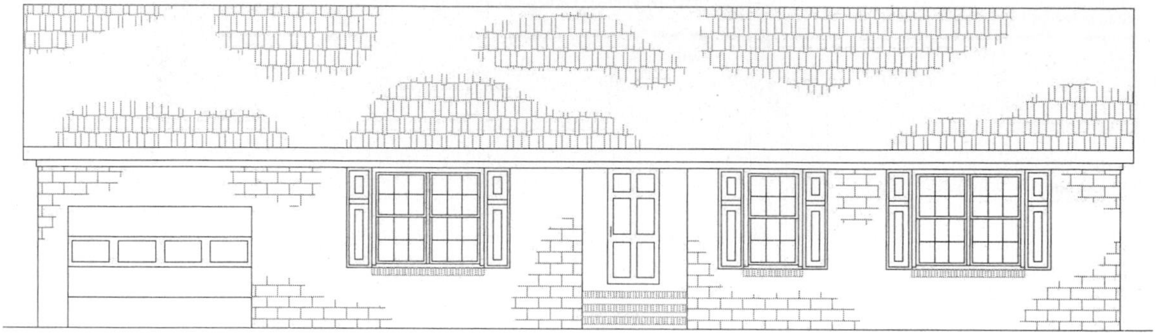

FIGURE 9.43

STEP 6

Several gradient hatch patterns will now be applied to the outer portions of the elevation. First perform a ZOOM-EXTENTS. Then change the current layer to Gradient Background using the Layer Control box. Activate the GRADIENT command from the Ribbon, as shown in Figure 9.44.

```
Command: Z (For ZOOM)
Specify corner of window, enter a scale factor (nX or nXP), or
[All/Center/Dynamic/Extents/Previous/Scale/Window/
Object] <real time>:
E (For Extents)
```

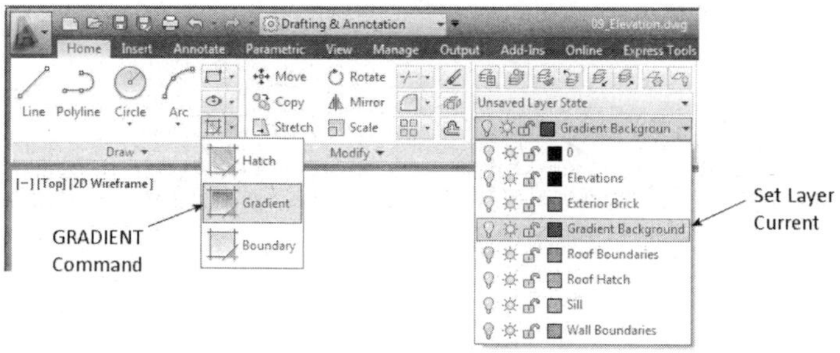

FIGURE 9.44

NOTE

The HATCH command can be used to create gradient hatch patterns. The GRADIENT command is simply more convenient because it pre-selects gradient patterns.

STEP 7

You will now apply a gradient hatch pattern to the upper portion of the elevation plan. Pick the GR_CURVED pattern, as shown in Figure 9.45 (Left), and then pick an internal point in the figure of the elevation. Right-click and select ENTER to accept the results.

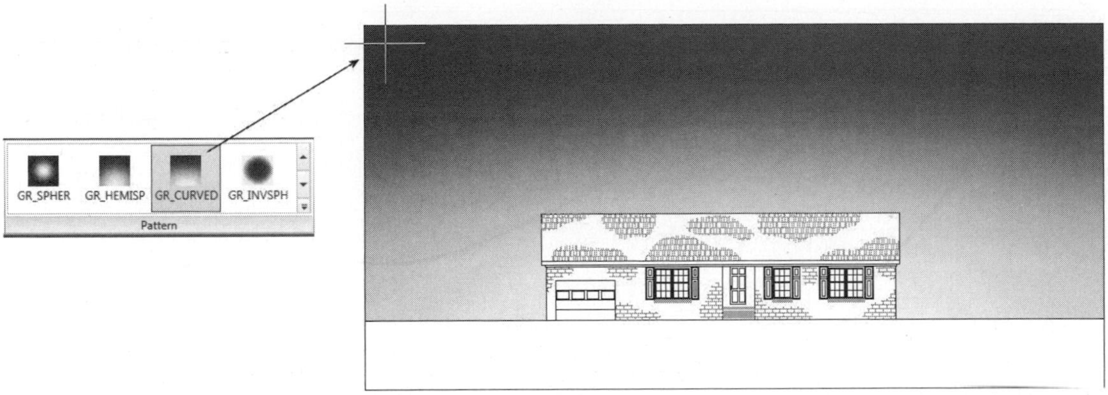

FIGURE 9.45

STEP 8

Apply another gradient hatch pattern to the lower portion of the elevation plan. Activate the GRADIENT command and pick the GR_INVCUR pattern, as shown in Figure 9.46 (Left). Then Pick an internal point in the figure of the elevation. Right-click and select ENTER to place the hatch pattern.

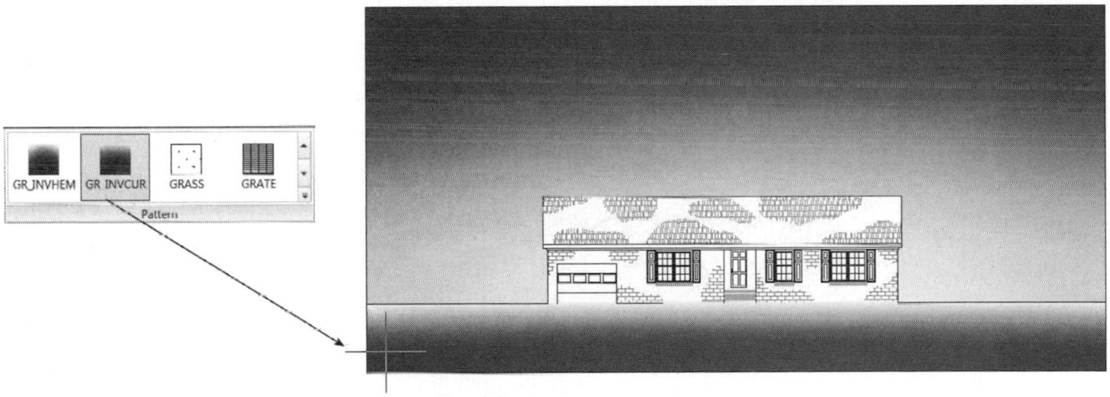

FIGURE 9.46

STEP 9

Apply gradient patterns (GR_CYLIN) to each of the garage windows, as shown in Figure 9.47, by using the GRADIENT command for each of the four windows.

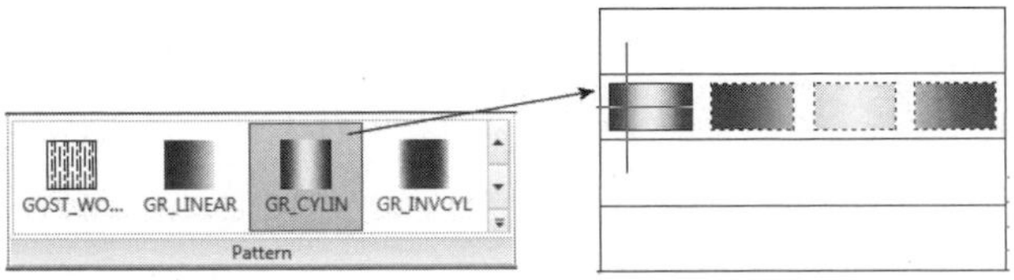

FIGURE 9.47

STEP 10

If the boundary lines that define the elevation disappear, verify that the Send Behind Boundary setting is selected in the Options panel of the Ribbon, as shown in Figure 9.48 here.

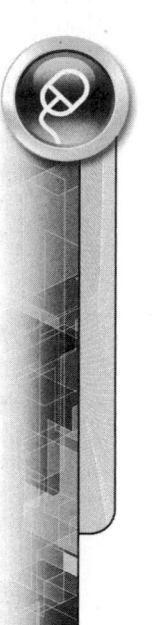

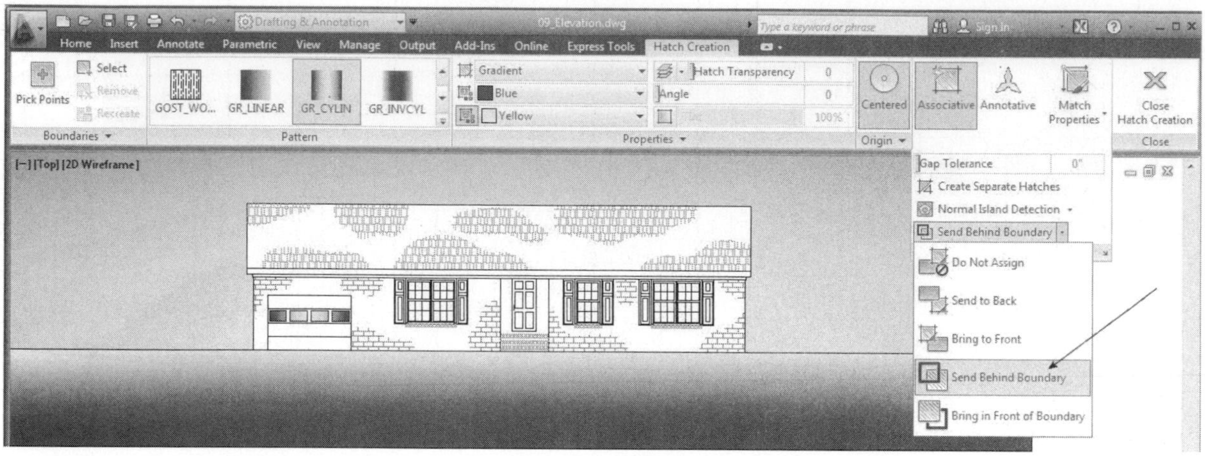

FIGURE 9.48

PROBLEMS FOR CHAPTER 9

PROBLEM 9-1

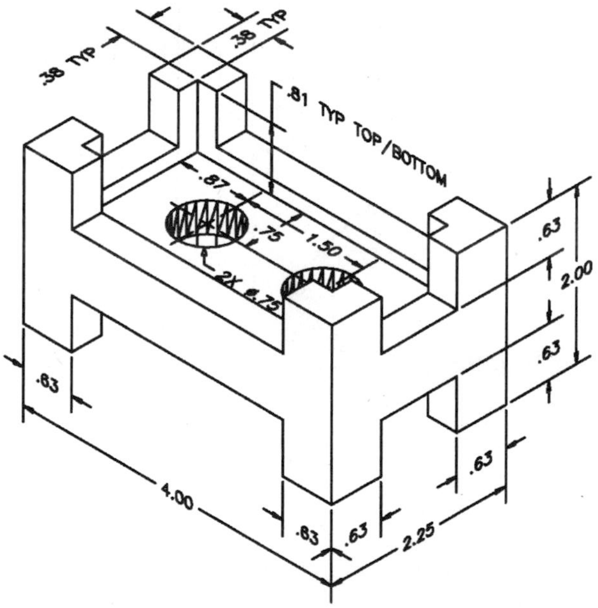

Center a three-view drawing and make the Front view a full section.

PROBLEM 9-2

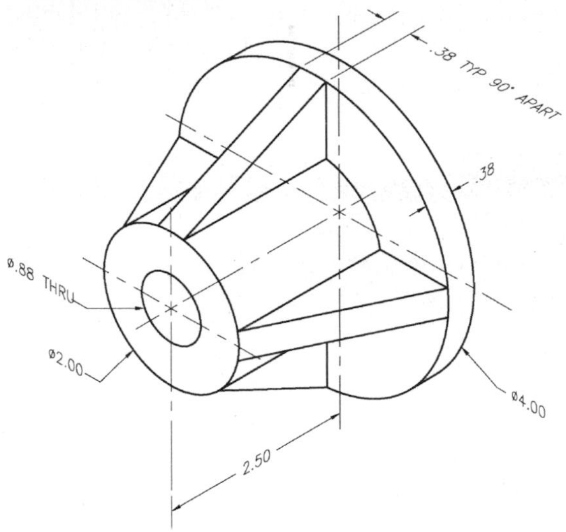

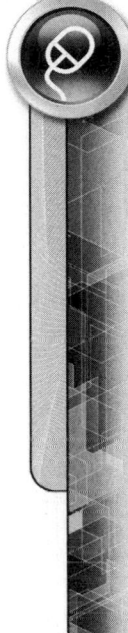

Center two views within the work area and make one view a full section. Use correct drafting practices for the ribs.

PROBLEM 9-3

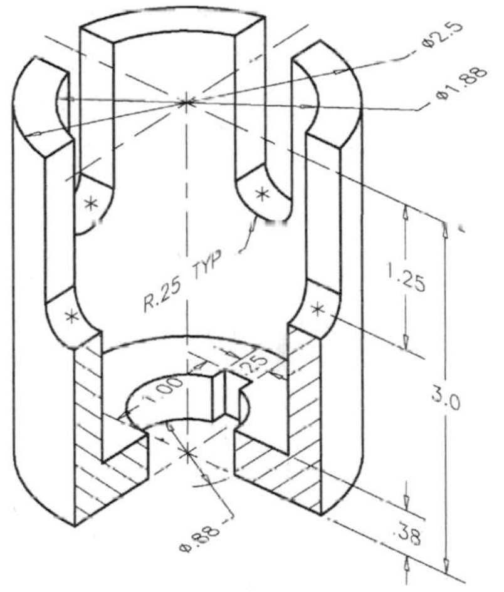

Center two views within the work area and make one view a half section.

PROBLEM 9-4

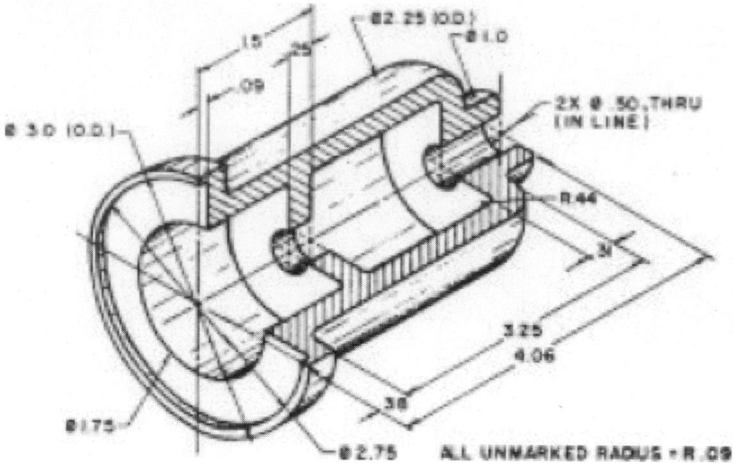

Center two views within the work area and make one view a half section.

PROBLEM 9-5

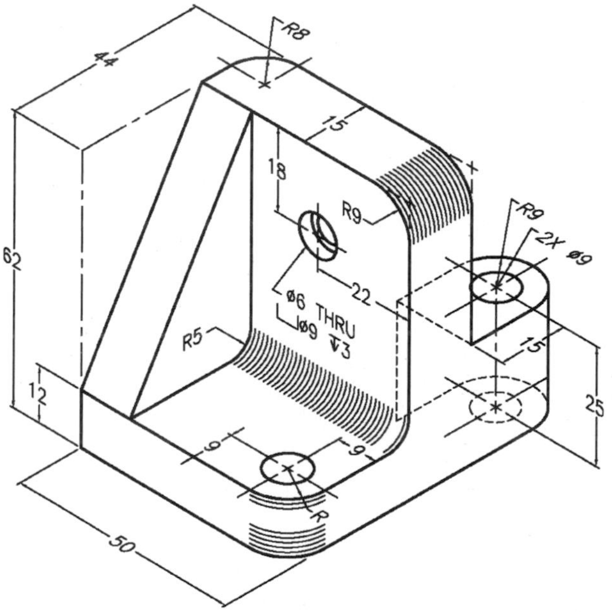

Center three views within the work area and make one view an offset section.

PROBLEM 9-6

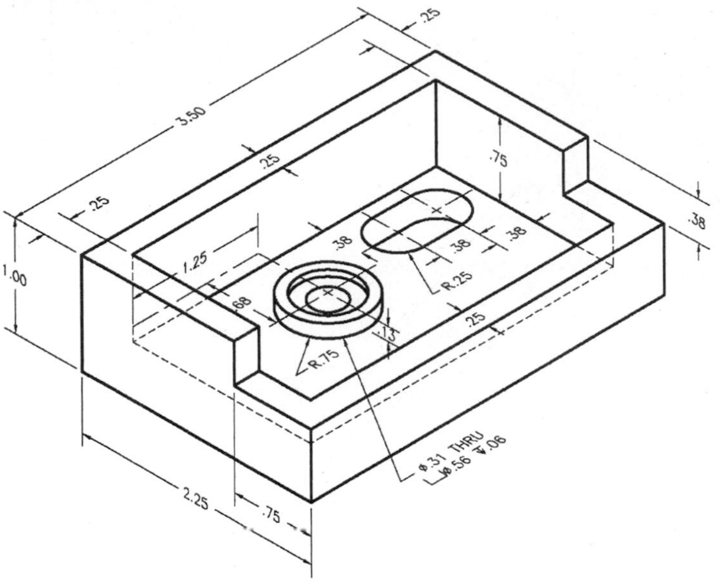

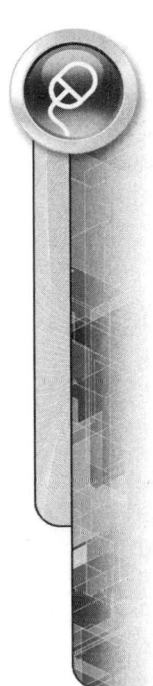

Center three views within the work area and make one view an offset section.

PROBLEM 9-7

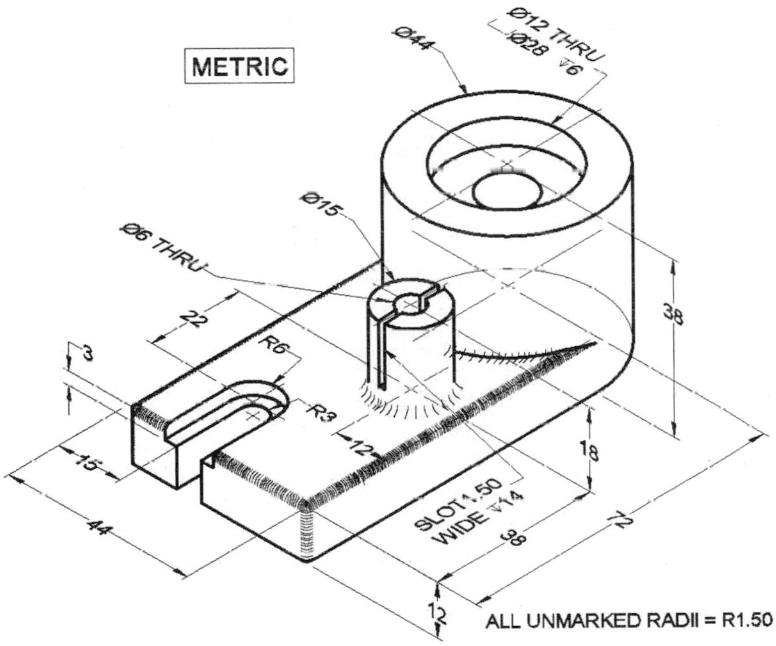

Center three views within the work area and make one view an offset section.

PROBLEM 9-8

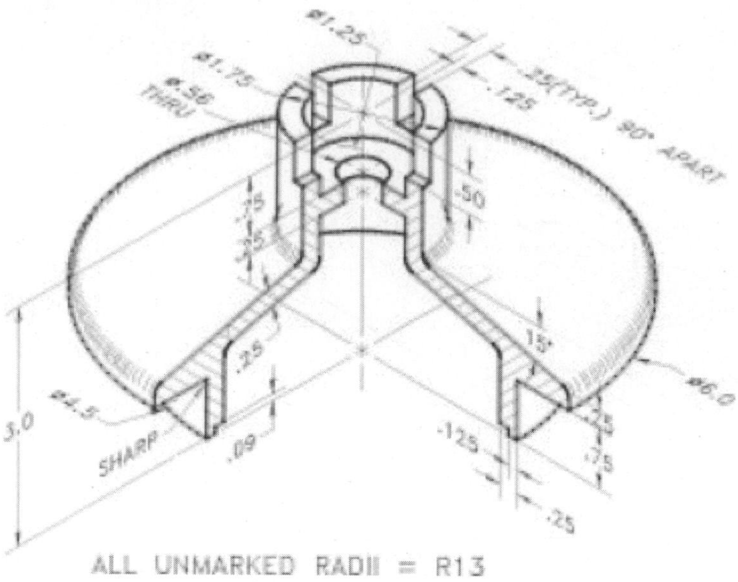

Center two views within the work area and make one view a half section.

PROBLEM 9-9

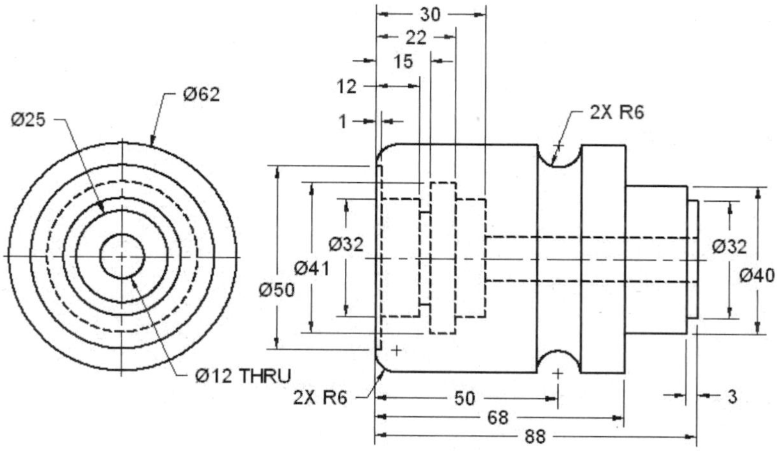

Center the required views within the work area and make one view a broken-out section to illustrate the complicated interior area.

PROBLEM 9-10

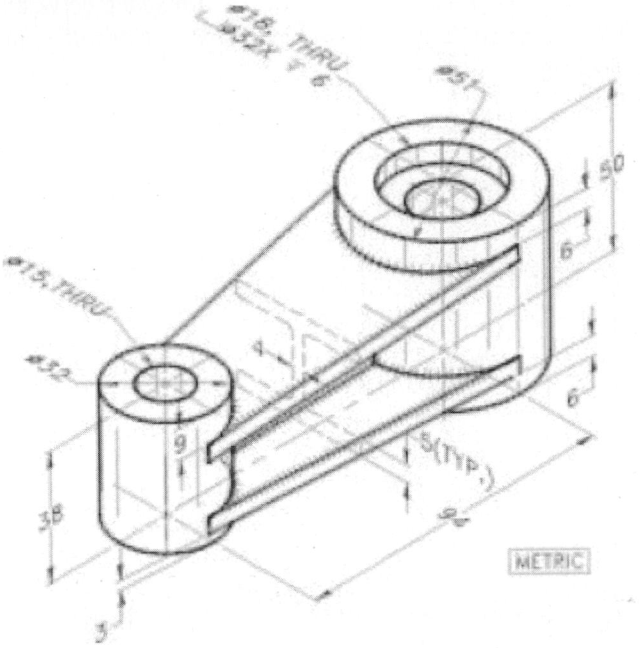

Create the required views within the work area and add a full removed section called A-A.

PROBLEM 9-11

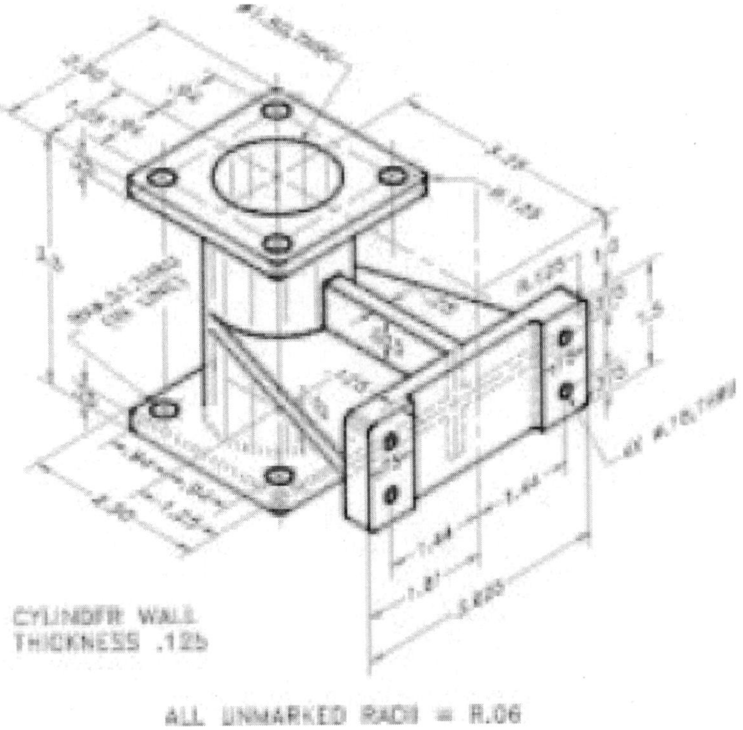

Center the required views within the work area and add a full removed section called A-A.

DIRECTIONS FROM PROBLEMS 9-12 THROUGH 9-14

Center the required views within the work area. Leave a 1" or 25-mm space between the views. Make one view a full section to fully illustrate the object. Consult your instructor if you need to add dimensions.

PROBLEM 9-12

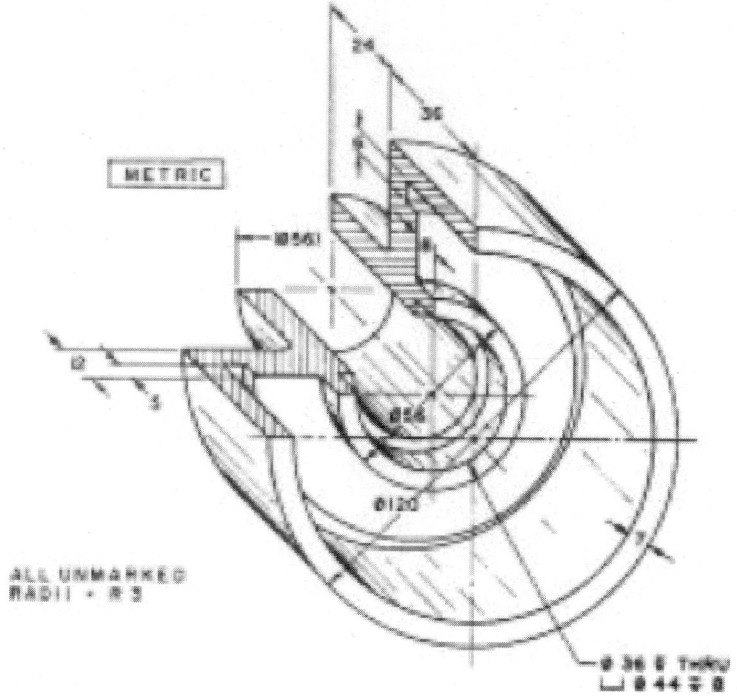

PROBLEM 9-13

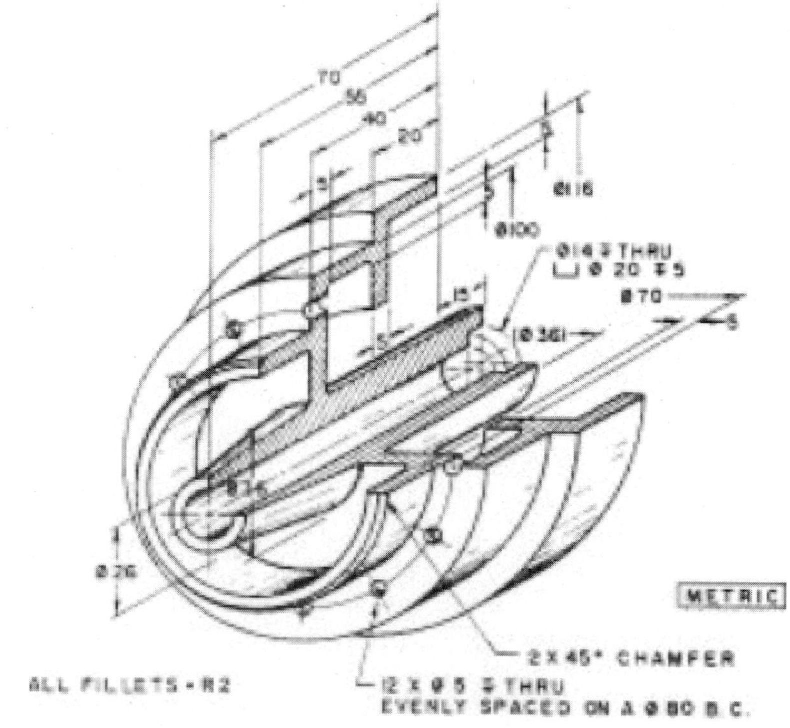

PROBLEM 9-14

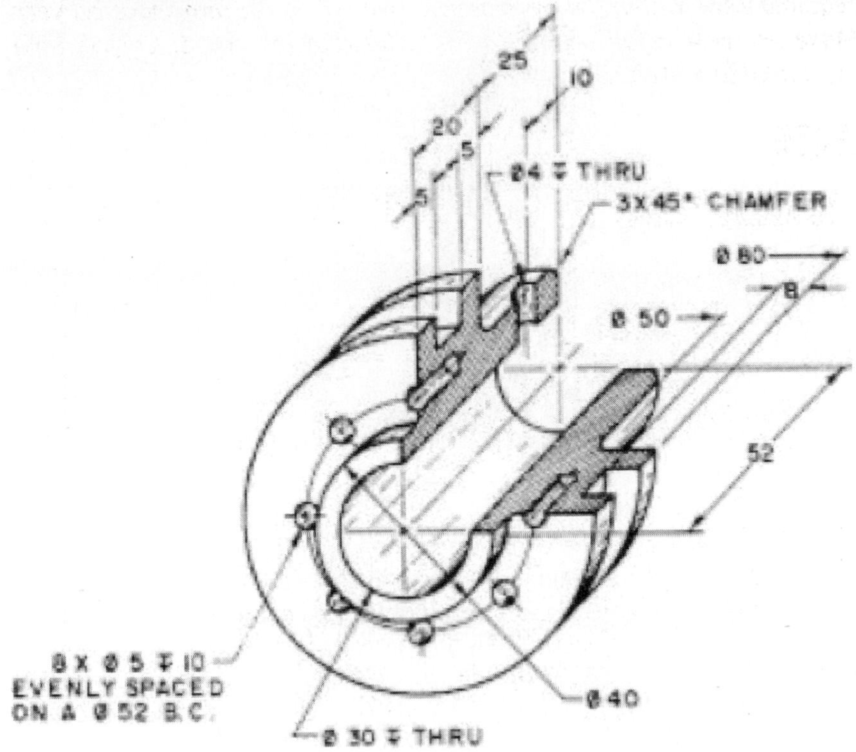
25
10
20
5
5
Ø4 ↧ THRU
3X45° CHAMFER
Ø 80
8
Ø 50
52
8 X Ø 5 ↧ 10
EVENLY SPACED
ON A Ø 52 B.C.
Ø 40
Ø 30 ↧ THRU

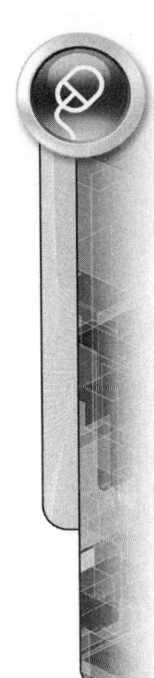

CHAPTER

10

Adding Dimensions to Your Drawing

Once orthographic views have been laid out, a design is not ready for production until dimensions describing the width, height, and depth of the object are added to the drawing. These dimensions must be applied correctly and organized in such a way as to ensure the drawing is interpreted correctly. Confusion caused by missing, extra, incorrect, or poorly placed dimensions may lead to incorrectly manufactured parts. AutoCAD's dimensioning commands will help you to properly create and organize dimensions on your drawings. This chapter begins with a discussion of the basic dimension commands, which include `DIMLINEAR`, `DIMALIGNED`, `DIMCONTINUE`, `DIMBASELINE`, `DIMDIAMETER`, `DIMRADIUS`, `DIMANGULAR`, `QLEADER`, and `MLEADER`. The discussion will then focus on more specialized dimensioning commands. The last part of this chapter will concentrate on dimension editing.

METHODS OF CHOOSING DIMENSION COMMANDS

A number of tools are available for choosing dimension-related commands. The Drafting & Annotation Workspace has two tabs available on the Ribbon for selecting these commands, as shown in Figure 10.1. The Home tab includes the basic dimensioning commands, while the Annotate tab includes more advanced tools in addition to the basic commands. Another way of activating dimension commands is through the keyboard. These commands tend to get long; the `DIMLINEAR` command is one example. To spare you the effort of entering the entire command, most dimension commands have been abbreviated to three letters. For example, `DLI` is the alias for the `DIMLINEAR` command.

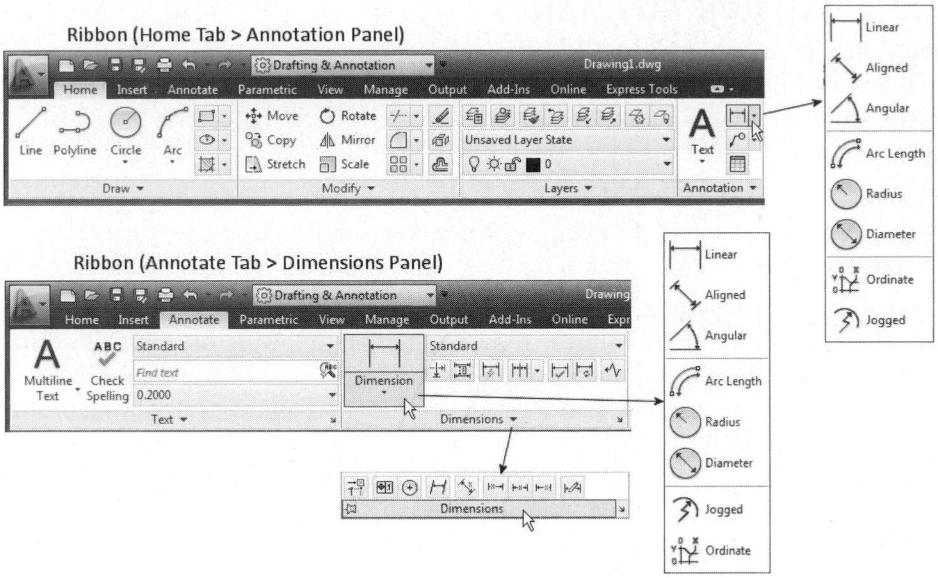

FIGURE 10.1

Study the following table for a brief description of available dimensioning tools.

Button	Tool	Shortcut	Function
	Linear	DLI	Creates a horizontal, vertical, or rotated dimension
	Aligned	DAL	Creates a linear dimension parallel to an object
	Arc Length	DAR	Dimensions the total length of an arc
	Ordinate	DOR	Creates an ordinate dimension based on the current position of the User Coordinate System (UCS)
	Radius	DRA	Creates a radius dimension
	Jogged	DJO	Creates a jogged dimension
	Diameter	DIA	Creates a diameter dimension
	Angular	DAN	Creates an angular dimension
	Quick Dimension	QDIM	Creates a quick dimension
	Continue	DCO	Creates a continued dimension
	Baseline	DBA	Creates a baseline dimension
	Dimension Space	—	Adjusts the spacing of parallel linear and angular dimensions to be equal
	Dimension Break	—	Creates breaks in dimension, extension, and leader lines
	Tolerance	TOL	Activates the Tolerance dialog box
	Center Mark	DCE	Places a center mark inside a circle or an arc
	Inspection	—	Creates an inspection dimension
	Jogged Linear	DJL	Creates a jogged linear dimension
	Edit	DED	Used to edit a dimension
	Text Edit	—	Used to edit the text of a dimension
	Update	—	Used for updating existing dimensions to changes in the current dimension style
	Dimstyle Dialog	D	Activates the Dimension Style Manager dialog box

BASIC DIMENSION COMMANDS

Dimensioning is an essential part of technical drawings and, as seen in the previous table, the list of dimensioning commands is long. We will start our discussion with the basic dimension commands: DIMLINEAR, DIMALIGNED, DIMCONTINUE, DIMBASELINE, DIMDIAMETER, DIMRADIUS, DIMANGULAR, QLEADER, and MLEADER.

LINEAR DIMENSIONS

The Linear Dimensioning mode generates either a horizontal or vertical dimension, depending on the location of the dimension. The following prompts illustrate the generation of a horizontal dimension with the DIMLINEAR command. Notice that identifying the dimension line location at "C" in Figure 10.2 automatically generates a horizontal dimension.

TRY IT!

Open the drawing file 10_Dim Linear1. Verify that OSNAP is on and set to Endpoint. Use the following command sequence and image for performing this dimensioning task.

Command: DLI *(For DIMLINEAR)*

Specify first extension line origin or <select object>: *(Select the endpoint of the line at "A")*

Specify second extension line origin: *(Select the other endpoint of the line at "B")*

Specify dimension line location or [Mtext/Text/Angle/Horizontal/Vertical/Rotated]: *(Pick a point near "C" to locate the dimension)*

Dimension text = 8.00

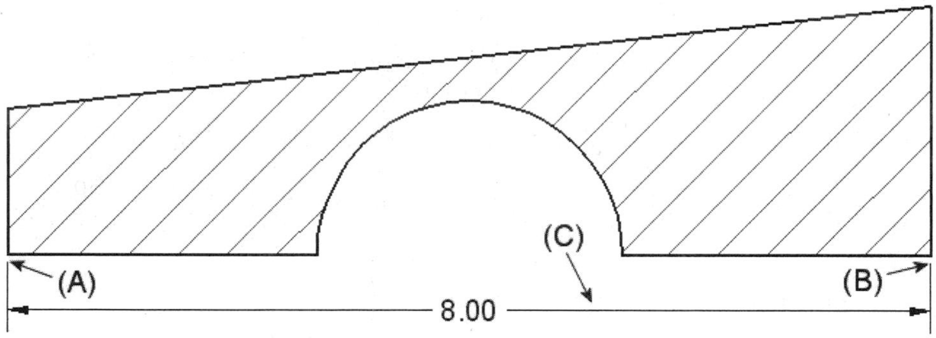

FIGURE 10.2

The linear dimensioning command is also used to generate vertical dimensions. The following prompts illustrate the generation of a vertical dimension with the DIMLINEAR command. Notice that identifying the dimension line location at "C" in Figure 10.3 automatically generates a vertical dimension.

TRY IT!

Open the drawing file 10_Dim Linear2. Verify that OSNAP is on and set to Endpoint. Use the following command sequence and image for performing this dimensioning task.

Command: DLI *(For DIMLINEAR)*

Specify first extension line origin or <select object>: *(Select the endpoint of the line at "A")*

Specify second extension line origin: *(Select the endpoint of the line at "B")*

Specify dimension line location or [Mtext/Text/Angle/Horizontal/Vertical/Rotated]: *(Pick a point near "C" to locate the dimension)*

Dimension text = 1.14

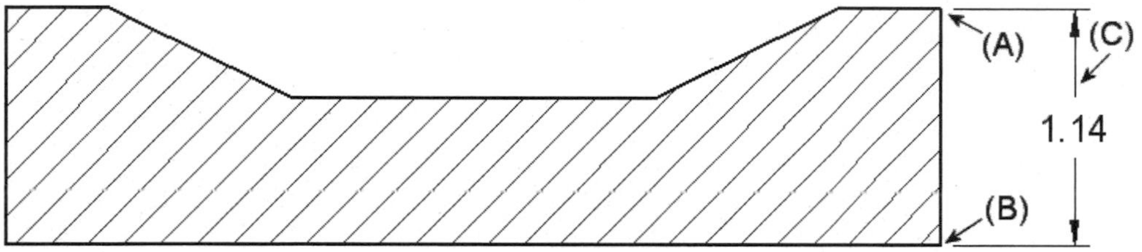

FIGURE 10.3

Rather than select two separate endpoints to create a dimension, certain situations allow you to press ENTER and select the object (in this case, the line). This selects the two endpoints and prompts you for the dimension location. The completed dimension is illustrated in Figure 10.4.

TRY IT!

Open the drawing file 10_Dim Linear3. Use the following command sequence and image for performing this dimensioning task.

Command: DLI *(For DIMLINEAR)*

Specify first extension line origin or <select object>: *(Press* ENTER *to select an object)*

Select object to dimension: *(Select the line at "A")*

Specify dimension line location or [Mtext/Text/Angle/Horizontal/Vertical/Rotated]: *(Pick a point near "B" to locate the dimension)*

Dimension text = 9.17

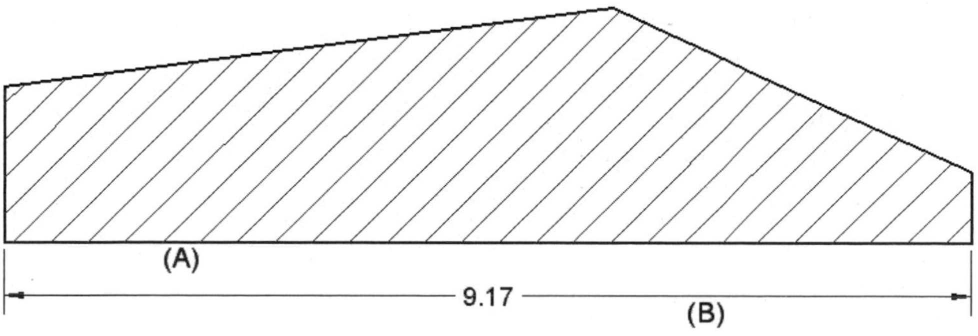

FIGURE 10.4

ALIGNED DIMENSIONS

The Aligned Dimensioning mode generates a dimension line parallel to the distance specified by the location of two extension line origins, as shown in Figure 10.5.

TRY IT!

Open the drawing file 10_Dim Aligned. Verify that OSNAP is on and set to Endpoint. The following prompts and Figure 10.5 illustrate the creation of an aligned dimension with the `DIMALIGNED` command.

```
Command: DAL (For DIMALIGNED)

Specify first extension line origin or <select object>:
(Select the endpoint of the line at "A")

Specify second extension line origin: (Select the endpoint of
the line at "B")

Specify dimension line location or [Mtext/Text/Angle]:
(Pick a point at "C" to locate the dimension)

Dimension text = 12.06
```

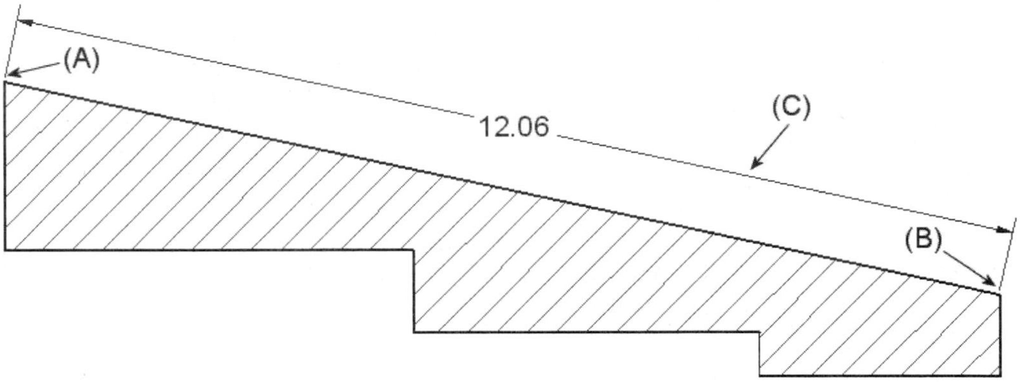

FIGURE 10.5

Rotated Linear Dimensions

This Linear Dimensioning mode can also create a dimension line rotated at a specific angle, as shown in Figure 10.6. The following prompts illustrate the generation of a rotated dimension with the `DIMLINEAR` command and a known angle of 15°. If you do not know the angle, you could easily establish the angle by clicking the endpoints at "A" and "D" in Figure 10.6 when prompted to "Specify angle of dimension line <0>."

TRY IT!

Open the drawing file 10_Dim Rotated. Verify that OSNAP is on and set to Endpoint. Use the following command sequence and image for creating a rotated dimension.

```
Command: DLI (For DIMLINEAR)

Specify first extension line origin or <select object>:
(Select the endpoint of the line at "A")

Specify second extension line origin: (Select the endpoint of
the line at "B")

Specify dimension line location or [Mtext/Text/Angle/Hori-
zontal/Vertical/Rotated]: R (For Rotated)
```

`Specify angle of dimension line <0>:` **15**

`Specify dimension line location or [Mtext/Text/Angle/Horizontal/Vertical/Rotated]:` *(Pick a point at "C" to locate the dimension)*

`Dimension text = 6.00`

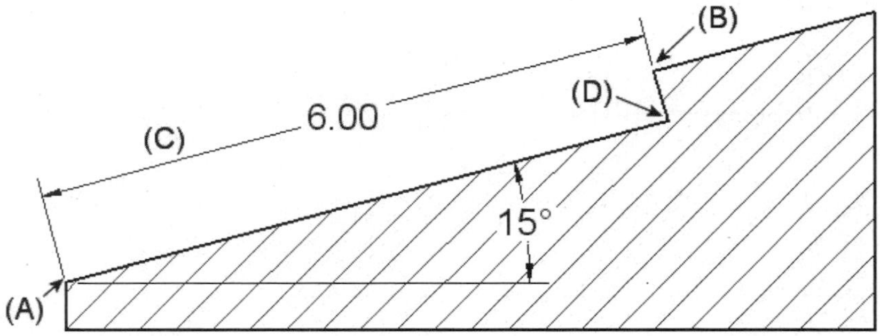

FIGURE 10.6

CONTINUE DIMENSIONS

The importance of grouping dimensions for ease of reading has already been explained. The use of continue dimensions allows you to automatically group dimensions together in a single line. With one dimension already placed with the DIMLINEAR command, you issue the DIMCONTINUE command, which prompts you for the second extension line location. Picking the second extension line location strings the dimensions next to each other or continues the dimension.

TRY IT!

Open the drawing file 10_Dim Continue. Verify that Running OSNAP is on and set to Endpoint. Use the following command sequence and Figure 10.7 for creating continued dimensions.

`Command:` **DLI** *(For DIMLINEAR)*

`Specify first extension line origin or <select object>:` *(Select the endpoint of the line at "A")*

`Specify second extension line origin:` *(Select the endpoint of the line at "B")*

`Specify dimension line location or [Mtext/Text/Angle/Horizontal/Vertical/Rotated]:` *(Pick a point to locate the 1.75 horizontal dimension)*

`Dimension text = 1.75`

`Command:` **DCO** *(For DIMCONTINUE)*

`Specify a second extension line origin or [Undo/Select] <Select>:` *(Select the endpoint of the line at "C")*

`Dimension text = 1.25`

`Specify a second extension line origin or [Undo/Select] <Select>:` *(Select the endpoint of the line at "D")*

`Dimension text = 1.50`

`Specify a second extension line origin or [Undo/Select] <Select>:` *(Select the endpoint of the line at "E")*

Dimension text = 1.00

Specify a second extension line origin or [Undo/Select] <Select>: *(Press* ENTER *when finished)*

Select continued dimension: *(Press* ENTER *to exit this command)*

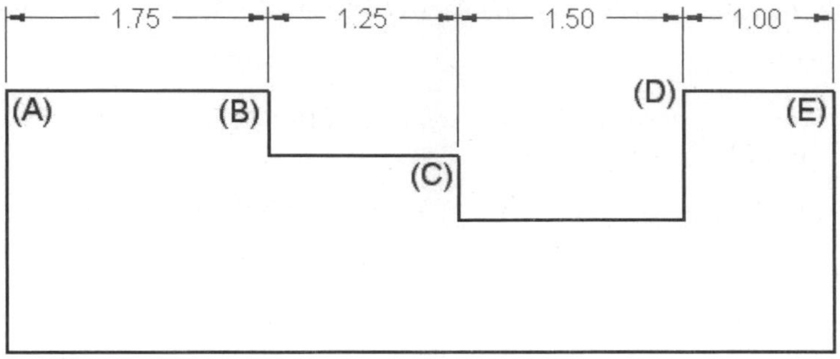

FIGURE 10.7

BASELINE DIMENSIONS

Yet another aid in grouping dimensions is the DIMBASELINE command. Continue dimensions place dimensions next to each other; baseline dimensions establish a base or starting point for the first dimension, as shown in Figure 10.8. Any dimensions that follow in the DIMBASELINE command are calculated from the common base point already established. This is a very popular mode to use when one end of an object acts as a reference edge. When you place dimensions using the DIMBASELINE command, a default baseline spacing setting of 0.38 units controls the spacing of the dimensions from each other. The DIMBASELINE command is initiated after the DIMLINEAR command.

TRY IT!

Open the drawing file 10_Dim Baseline. Verify that Running OSNAP is on and set to Endpoint. Use the following command sequence and image for creating baseline dimensions.

Command: DLI *(For DIMLINEAR)*

Specify first extension line origin or <select object>: *(Select the endpoint of the line at "A")*

Specify second extension line origin: *(Select the endpoint of the line at "B")*

Specify dimension line location or [Mtext/Text/Angle/Horizontal/Vertical/Rotated]: *(Locate the 1.75 horizontal dimension)*

Dimension text = 1.75

TIP

For the DIMBASELINE and DIMCONTINUE commands to work correctly, you must create the DIMLINEAR command in the correct direction. Picking from "B" to "A" would establish the wrong baseline. Instead of picking points with the DIMLINEAR command, you can use the "Select object" prompt but be sure to select the line closer to the "A" end.

Command: DBA *(For DIMBASELINE)*

Specify a second extension line origin or [Undo/Select] <Select>: *(Select the endpoint of the line at "C")*

Dimension text = 3.00

Specify a second extension line origin or [Undo/Select]
<Select>: *(Select the endpoint of the line at "D")*

Dimension text = 4.50

Specify a second extension line origin or [Undo/Select]
<Select>: *(Select the endpoint of the line at "E")*

Dimension text = 5.50

Specify a second extension line origin or [Undo/Select]
<Select>: *(Press* ENTER *when finished)*

Select base dimension: *(Press* ENTER *to exit this command)*

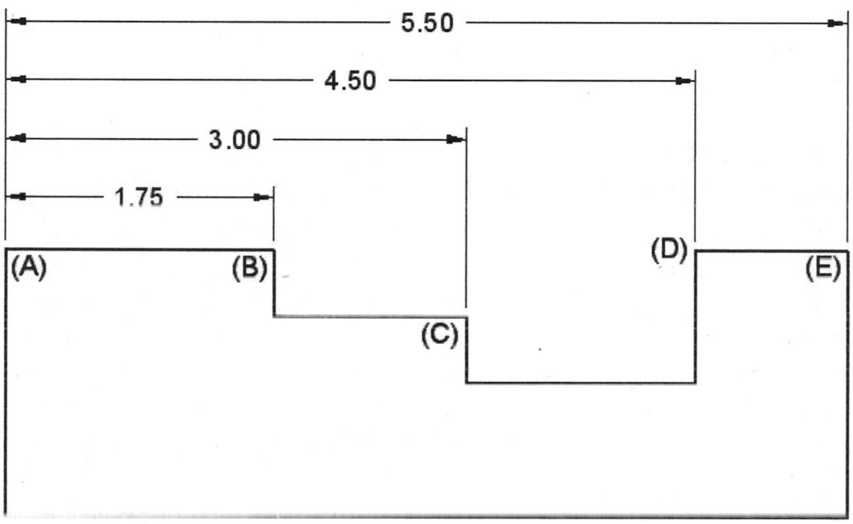

FIGURE 10.8

DIAMETER AND RADIUS DIMENSIONING

Arcs and circles should be dimensioned in the view where their true shape is visible. Arcs are typically dimensioned by giving their radius and circles by their diameter. The mark in the center of the circle or arc indicates its center point, as shown in Figure 10.9. You may place the dimension text either inside or outside the circle; you may also use grips to aid in the dimension text location of a diameter or radius dimension. When dimensioning a small radius, an arc extension line is formed, depending on where you locate the radius dimension text.

Open the drawing file 10_Dim Radial. Use the following command sequence and Figure 10.9 for placing diameter and radius dimensions.

TRY IT!

Command: DDI *(For DIMDIAMETER)*

Select arc or circle: *(Select the edge of the large circle)*

Dimension text = 2.50

Specify dimension line location or [Mtext/Text/Angle]:
(Pick a point to locate the diameter dimension)

Command: DDI *(For DIMDIAMETER)*

Select arc or circle: *(Select the edge of the small circle)*

Dimension text = 1.00

Specify dimension line location or [Mtext/Text/Angle]: *(Pick a point to locate the diameter dimension)*

Command: DRA *(For DIMRADIUS)*

Select arc or circle: *(Select the edge of the large arc)*

Dimension text = 1.00

Specify dimension line location or [Mtext/Text/Angle]: *(Pick a point to locate the radius dimension)*

Command: DRA *(For DIMRADIUS)*

Select arc or circle: *(Select the edge of the small arc)*

Dimension text = .50

Specify dimension line location or [Mtext/Text/Angle]: *(Pick a point to locate the radius dimension)*

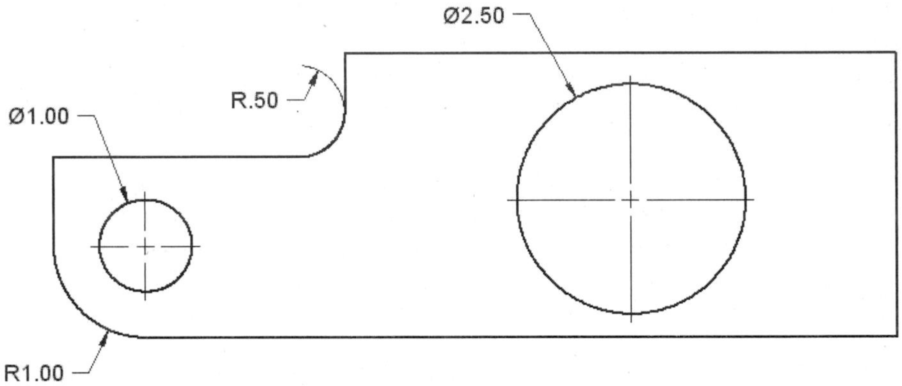

FIGURE 10.9

DIMENSIONING ANGLES

Dimensioning angles requires two lines forming the angle, an arc, or three points, one of which defines the vertex. The dimension line for an angular dimension is actually a curved arc whose center is at the vertex of the two lines, as shown in Figure 10.10.

TRY IT!

Open the drawing file 10_Dim Angle. Use the following command sequence and Figure 10.10 for performing this operation on both angles.

Command: DAN *(For DIMANGULAR)*

Select arc, circle, line, or <specify vertex>: *(Select line "A")*

Select second line: *(Select line "B")*

Specify dimension arc line location or [Mtext/Text/Angle/Quadrant]: *(Pick a point near "C" to locate the dimension)*

Dimension text = 53

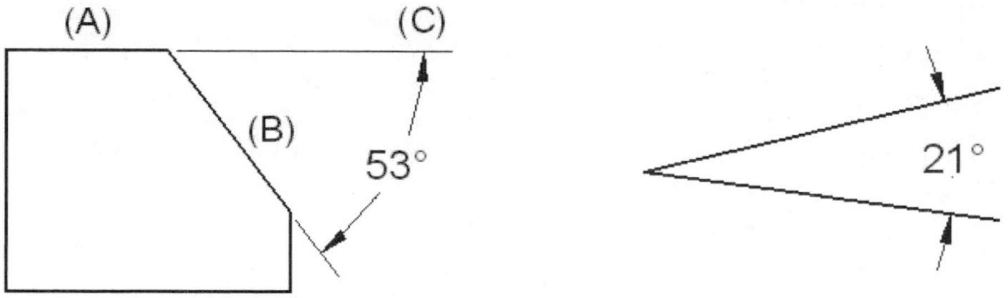

FIGURE 10.10

NOTE

In the previous Try-It exercise, where you locate the dimension, picking "C" will create the dimension at one of four possible angular solutions. Also, depending on where you pick "C," extension lines are automatically created when necessary.

LEADER LINES

A leader line is a thin, solid line leading from a note or dimension and typically ending with an arrowhead, as illustrated at "A" in Figure 10.11. The arrowhead should always terminate at an object line or an arc. A leader to a circle or arc should be radial; this means it is drawn so that if extended, it would pass through the center of the circle, as illustrated at "B." Example "C" demonstrates the incorrect directional placement of an arrow. Leaders should cross as few lines as possible, but in particular, avoid crossing dimension or other leader lines. The short horizontal shoulder (typically 0.125 inch) of a leader should meet the dimension text, as illustrated at "A." Some companies may allow other text placement styles, such as the practice to underline the dimension with the horizontal shoulder, as illustrated at "C." Avoid leader lines that are near horizontal or vertical angles. Some companies require specific angles (30°, 45°, or 60°) for leader lines. You should check your company's standard practices to ensure that your leaders are acceptable.

Illustrated in "D" are two leaders attached to local notes. Notice that the two leaders have different terminators: arrows and dots. Dots are sometimes used as terminators when the note is referring to an entire surface.

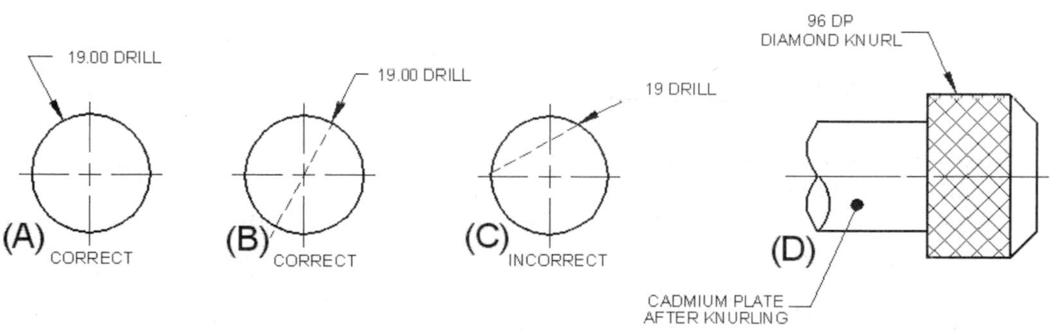

FIGURE 10.11

THE QLEADER COMMAND

The QLEADER command, or Quick Leader, provides numerous controls for placing leaders in your drawing. The QLEADER command, while still popular, is being quickly replaced by the more powerful MLEADER command (discussed in the next section). The command is no longer found in the menus and must be activated from the Command prompt (LE is the alias).

TRY IT!

Open the drawing file 10_Qleader. Study the prompt sequence below and Figure 10.12 for this command.

Command: **LE** *(For QLEADER)*

Specify first leader point, or [Settings]<Settings>: **Mid**

to *(Pick the inclined line near "A")*

Specify next point: *(Pick a point at "B")*

Specify next point: *(Press* ENTER *to continue)*

Specify text width <0.00>: *(Press* ENTER *to accept default)*

Enter first line of annotation text <Mtext>: *(Press* ENTER *to display the Multiline Text Editor.*

Enter "**.125 X 45%%D CHAMFER.**" *Close the Text Editor to place the leader)*

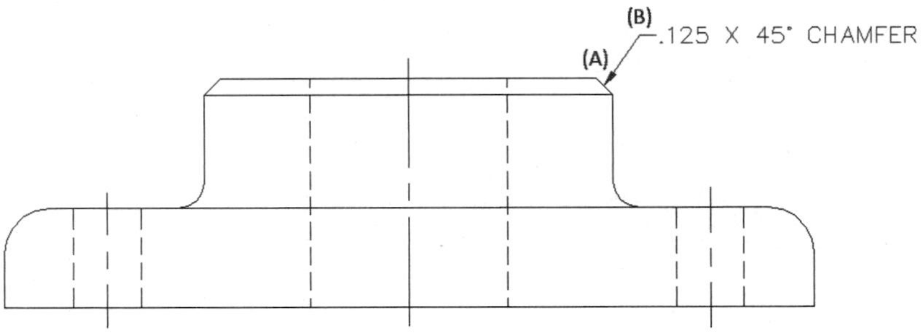

FIGURE 10.12

Pressing ENTER at the first Quick Leader prompt displays the Leader Settings dialog box that consists of three tabs used for controlling leaders.

Command: **LE** *(For QLEADER)*

Specify first leader point, or [Settings] <Settings>: *(Press* ENTER *to accept the default value and display the Leader Settings dialog box in Figure 10.13)*

The Annotation tab, shown in Figure 10.13 (Left), deals with the object placed at the end of the leader. By default, the MText radio button is selected, allowing you to add a note through the Multiline Text Editor. You could also copy an object at the end of the leader, have a geometric tolerancing symbol placed in the leader, have a predefined block placed in the leader, or leave the leader blank.

The Leader Line & Arrow tab in Figure 10.13 (Center) allows you to draw a leader line consisting of straight segments or in the form of a spline object. You can control the number of points used to define the leader; a maximum of three points is more than enough to create a standard leader. You can even change the arrowhead type.

The Attachment tab in Figure 10.13 (Right) allows you to control how text is attached to the end of the leader through various justification modes. The settings in the figure are the default values.

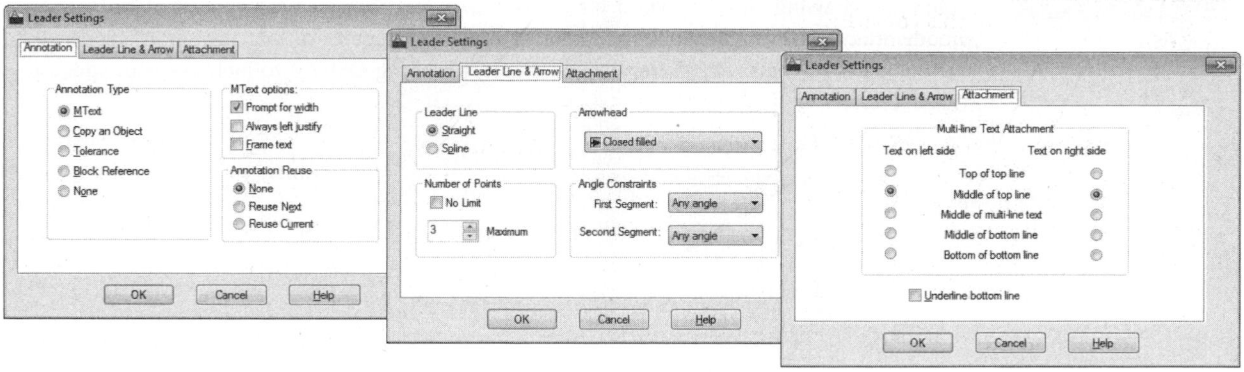

FIGURE 10.13

ANNOTATING WITH MULTILEADERS

The MLEADER command can be used to create simple leaders but additional controls are available to align, collect, add, and remove leader lines. Multileaders are also controlled by their own style, which will be discussed in Chapter 11. Choose multileader tools from either the Home or Annotate tab of the Ribbon, as shown in Figure 10.14.

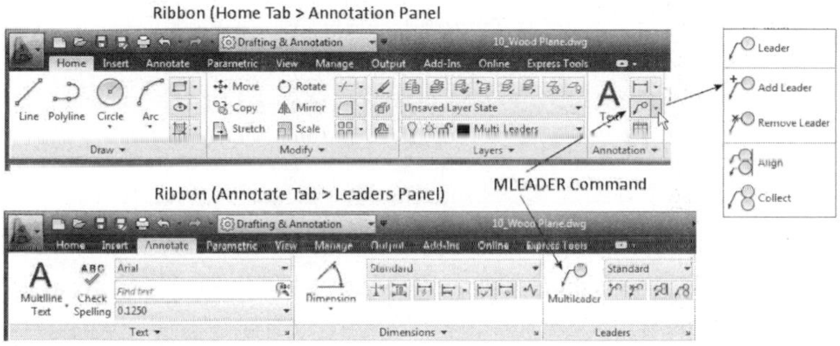

FIGURE 10.14

The following table illustrates each multileader tool along with a brief description.

Button	Tool	Function
	MLEADER	Creates a multileader
	MLEADEREDIT	Adds a multileader
	MLEADEREDIT	Removes a multileader
	MLEADERALIGN	Aligns a number of multileaders
	MLEADERCOLLECT	Collects a number of multileaders
	MLEADERSTYLE	Launches the Multileader Style Properties dialog box

Creating Multileaders

The following Try It! exercise allows you to place a number of multileaders that identify various parts of a wood plane.

TRY IT!

Open the drawing file 10 _Wood_Plane. This drawing consists of a typical plane used in woodworking. Use the MLEADER command to place a number of multileaders that identify various parts of the wood plane. Using a Nearest running OSNAP to increase your speed in this exercise.

```
Command: MLEADER
Specify leader arrowhead location or [leader Landing first/
Content first/Options] <Options>: (Pick a point at "A")
Specify leader landing location: (Pick a point at "B." When
the Multi-Text editing box appears, enter the name of the wood
plane part. When finished, click the Close Text Editor button
(or pick in a blank area of the drawing area) and place the
multileader, as shown in Figure 10.15 on the left)
```

Continue placing additional multileaders to identify additional wood plane parts, as shown in Figure 10.15 (Right).

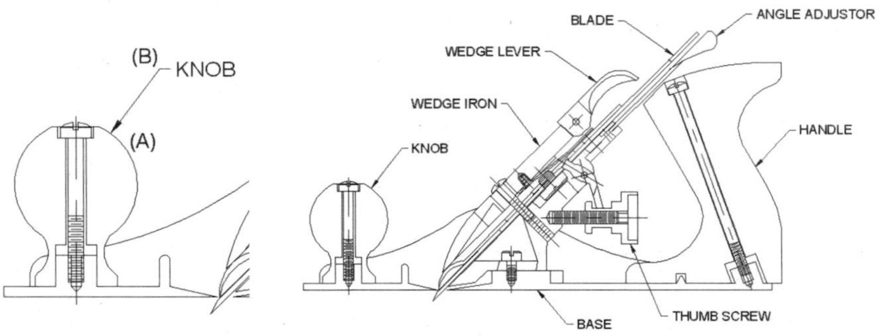

FIGURE 10.15

Aligning and Adding Multileaders

As numerous leaders are placed in a drawing, the time comes when they need to be aligned with each other to promote a neat and organized drawing. The next Try It! exercise will show how to align leaders. Then you will add additional leaders to an existing leader line.

TRY IT!

Open the drawing file 10_Multileaders Align. A number of multileaders are already created; however, notice that each leader is out of alignment with the others. Begin the alignment process by activating the Ribbon and clicking the Align Multileaders button found under the Annotate tab, as shown in Figure 10.16. First select all of the multileaders present in Figure 10.16. At this point you could select one of the existing multileaders to align to. Instead, type `O` **for** `OPTIONS`**; then type** `D` **for** `DISTRIBUTE` **and pick the two points, as shown in Figure 10.16.**

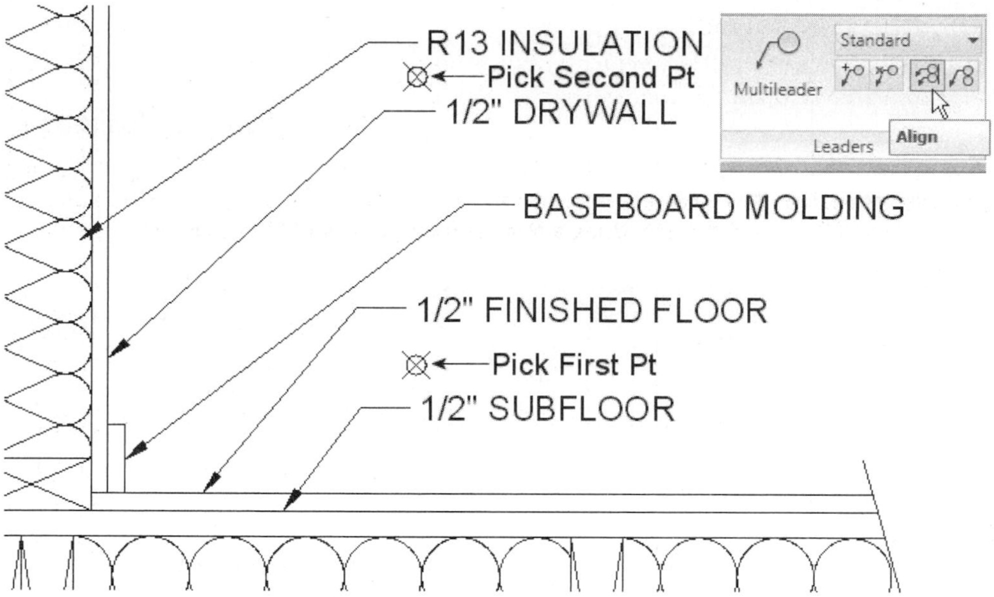

FIGURE 10.16

The results are illustrated in Figure 10.17 with all multileaders aligned with each other and equally spaced due to the distribute option. This makes the leaders more presentable in the drawing.

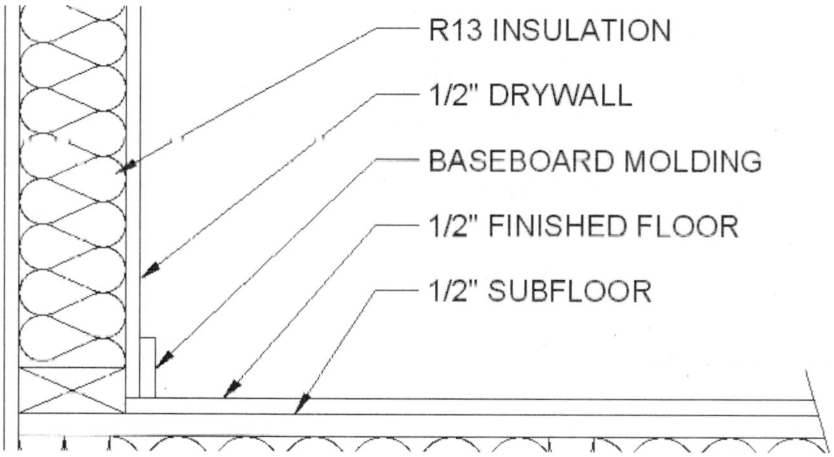

FIGURE 10.17

Continue with this Try It! exercise by panning to the lower portion of the architectural detail until your image appears similar to Figure 10.18. One concrete block is already called out with a multileader. With the Ribbon still present, click the add leader button, as shown in Figure 10.18. When prompted to select a multileader, click the existing leader identified by concrete block. When prompted to specify the leader arrowhead location, use OSNAP Nearest to pick the edges of the other two concrete block symbols, as shown in Figure 10.18. The additional leaders will be added to the existing concrete block leader.

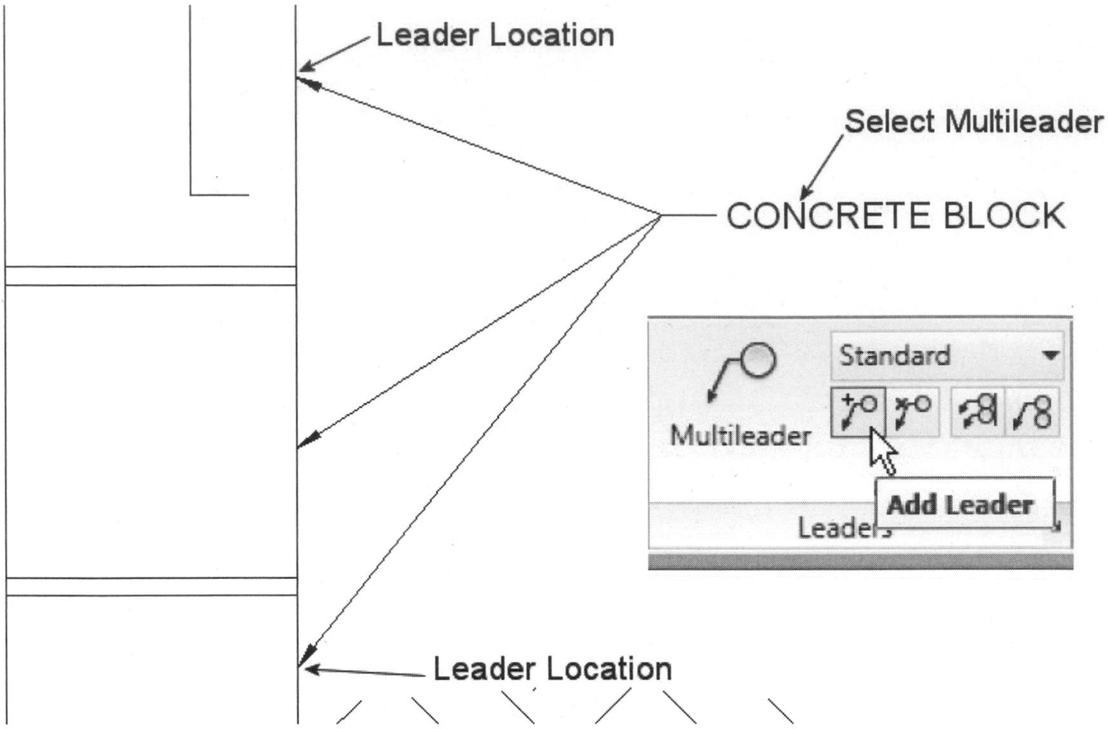

FIGURE 10.18

Collecting Multileaders

In the previous Try It! exercise, typical wall section elements were identified by multileaders with text. Multileaders can also be connected to blocks or symbols. Blocks will be covered in great detail in Chapter 16. Typical examples of multileader blocks include circles, boxes, and triangles, to name a few. Text is typically placed inside the symbols. In Figure 10.19, circles with text, often referred to as balloons, are used to identify items located in a parts list. These multileader blocks can either be displayed individually or can be grouped or collected using a single multileader. The next Try It! exercise illustrates this.

TRY IT!

Open the drawing file 10_Multileaders_Collect. Four multileaders have been placed using circles as blocks. Begin by clicking the Collect Multileaders button found under the Annotate tab of the Ribbon, as shown in Figure 10.19. When prompted to select the multileaders to group, pick items 1 through 4 in order.

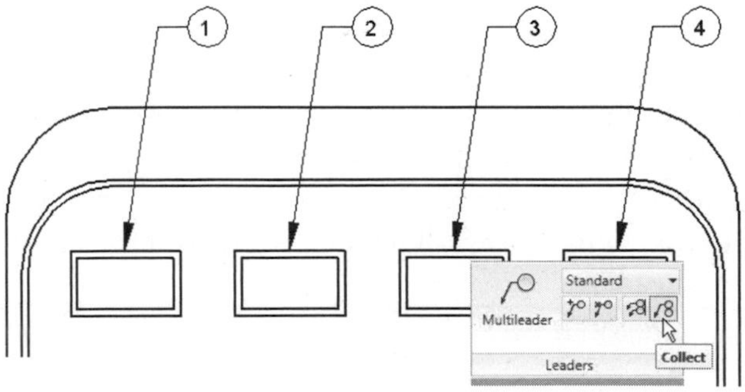

FIGURE 10.19

When prompted to select a new location, pick the location as shown in Figure 10.20. Notice how all four blocks are collected under the common leader.

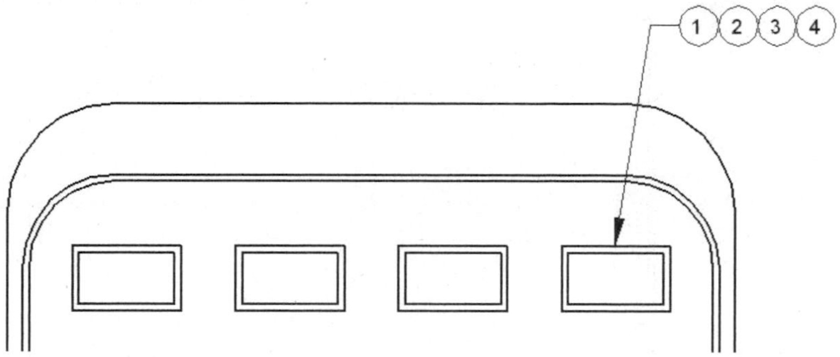

FIGURE 10.20

ADDING ORDINATE DIMENSIONS

The plate in Figure 10.21 (Left) consists of numerous drill holes, a slot, and numerous 90°-angle cuts along the perimeter. This object is not considered difficult to draw or make because it consists mainly of drill holes. However, conventional dimensioning techniques make the plate appear complex because a dimension is required for the location of every feature in both the X and Y directions. Add standard dimension components, such as extension lines, dimension lines, and arrowheads, and it is easy to get lost in the complexity of the dimensions even on this simple object.

A possible solution is the ordinate or datum dimensioning system, which is illustrated in Figure 10.21 (Right). Here, dimension lines or arrowheads are not drawn; instead, one extension line is constructed from the selected feature to a location specified by you. A dimension is added to identify this feature in either the X or Y direction. It is important to understand that all dimension calculations occur in relation to the current User Coordinate System (UCS), or the current 0,0 origin. In Figure 10.21, with the 0,0 origin located in the lower-left corner of the plate, all dimensions in the horizontal and vertical directions are calculated in relation to this 0,0 location. Hole sizes are called out with the `DIMDIAMETER` command. The following illustrates a typical ordinate dimensioning command sequence:

```
Command: DOR (For DIMORDINATE)

Specify feature location: (Select a feature using an OSNAP
option)

Specify leader endpoint or [Xdatum/Ydatum/Mtext/Text/
Angle]: (Locate a point outside of the object)

Dimension text = Calculated value
```

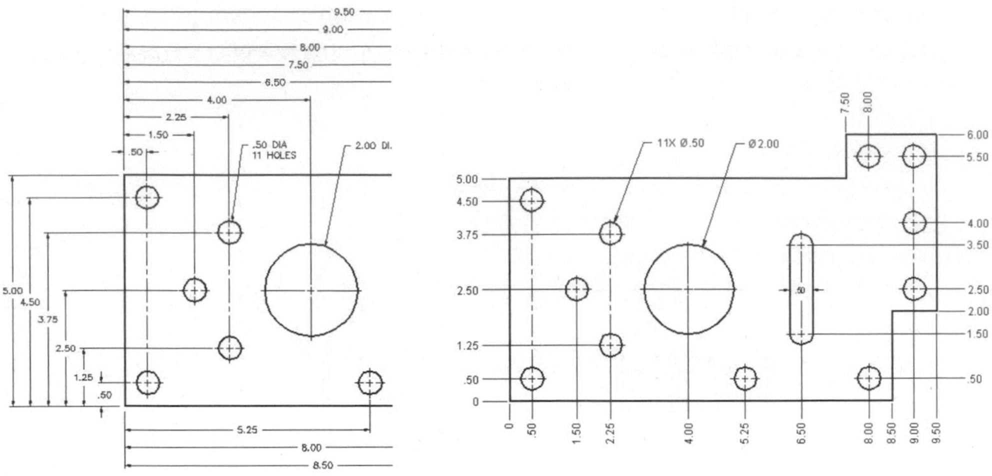

FIGURE 10.21

To understand how to place ordinate dimensions, see the example in Figure 10.22 and the prompt sequence below. Before you place any dimensions, a new User Coordinate System should be established at a convenient location on the object with the UCS command and the Origin option. All ordinate dimensions will reference this new origin. At the Command prompt, enter DOR (for DIMORDINATE) to begin ordinate dimensioning. Select the quadrant of the arc at "A" as the feature. For the leader endpoint, pick a point at "B." For the following exercise, Snap mode has been activated to assist in locating the leader endpoints. This helps in keeping all ordinate dimensions in line with one another.

TRY IT!

Open the drawing file 10_Dim Ordinate. Follow the next series of figures and the following command sequences to place ordinate dimensions.

Command: **DOR** *(For DIMORDINATE)*

Specify feature location: **Qua**

of *(Select the quadrant of the slot at "A," as shown in Figure 10.22 (Left))*

Specify leader endpoint or [Xdatum/Ydatum/Mtext/Text/Angle]: *(Locate a point at "B," as shown in Figure 10.22 (Left))*

Dimension text = 1.50

As in the previous example highlighting horizontal ordinate dimensions, placing vertical ordinate dimensions is identical, as shown in Figure 10.22 (Right). With the UCS still located in the lower-left corner of the object, select the feature at "C," using either the Endpoint or Quadrant mode. Pick a point at "D" to locate the dimension. Instead of using Snap mode, try object snap tracking from the endpoint of an existing dimension to align the leader endpoints.

Command: DOR *(For DIMORDINATE)*

Specify feature location: Qua

of *(Select the quadrant of the slot at "C," as shown in Figure 10.22 (Right))*

Specify leader endpoint or [Xdatum/Ydatum/Mtext/Text/Angle]: *(Locate a point at "D," as shown in Figure 10.22 (Right))*

Dimension text = 3.00

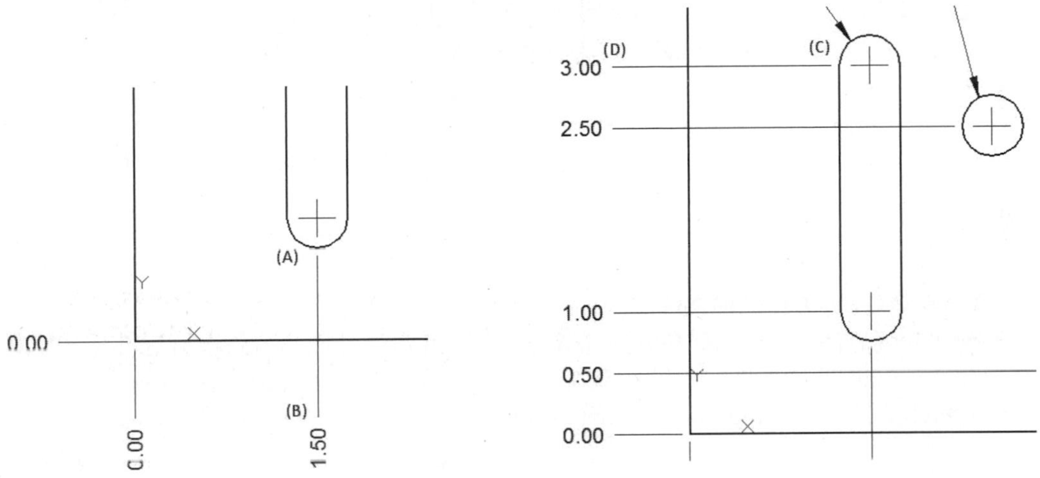

FIGURE 10.22

When spaces are tight to dimension to, two points not parallel to the X or Y axis will result in an "offset" being drawn, as shown in Figure 10.23 (Left). It is still helpful to snap or track when locating the leader endpoint; however, be sure Ortho is turned off.

Command: DOR *(For DIMORDINATE)*

Specify feature location: End

of *(Select the endpoint of the line at "A," as shown in Figure 10.23 (Left))*

Specify leader endpoint or [Xdatum/Ydatum/Mtext/Text/Angle]: *(Locate a point at "B," as shown in Figure 10.23 (Left))*

Dimension text = 2.00

Ordinate dimensioning provides a neat and efficient way of organizing dimensions. Only two points are required to place the dimension that references the current location of the UCS, as shown in Figure 10.23 (Right). Origin-based dimensioning systems are often found on mechanical drawings where numerically controlled machining operations are used.

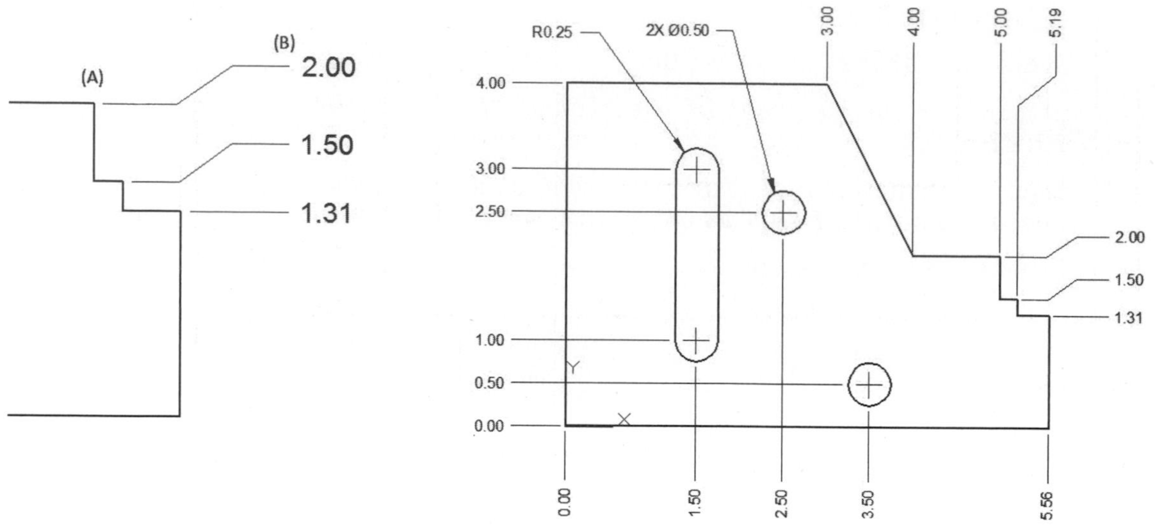

FIGURE 10.23

THE QDIM COMMAND

The QDIM, or Quick Dimension, command provides a way to create a series of dimensions all in one operation. You identify a number of valid corners representing intersections or endpoints of an object and all dimensions are placed according to the mode selected.

Continuous Mode

In Figure 10.24, a crossing box ("A" and "B") is used to identify all corners to dimension to.

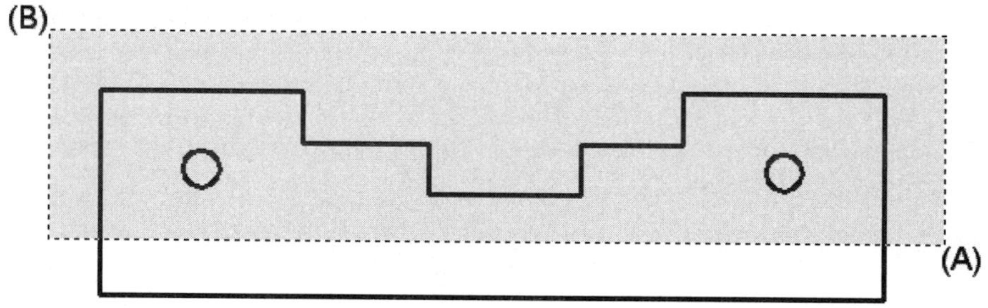

FIGURE 10.24

When you have finished identifying the crossing box and pressed the ENTER key, a preview of the dimensioning mode appears, as in Figure 10.25. During this preview mode, you can right-click and have a shortcut menu appear. This allows you to select other dimension modes.

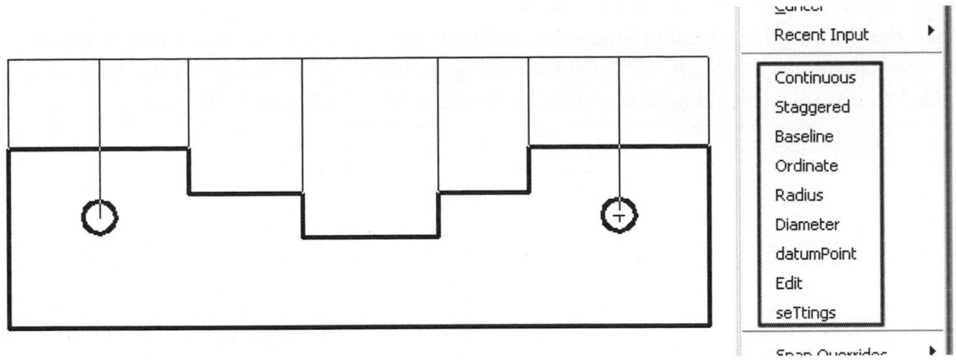

FIGURE 10.25

When the dimension line position "C" is identified, as in Figure 10.26, all continued dimensions are placed.

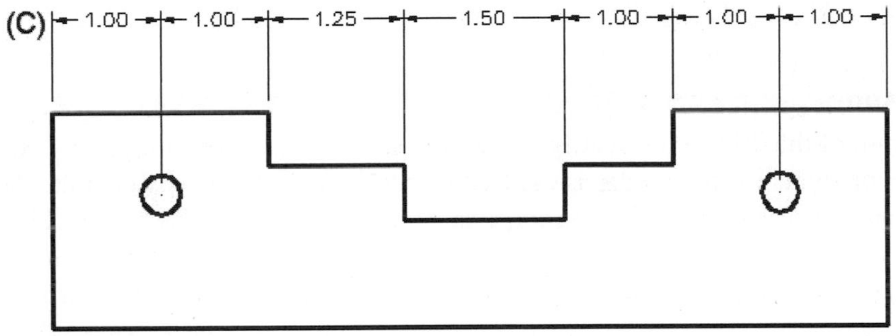

FIGURE 10.26

TRY IT!

Open the drawing file 10_Qdim Continuous. Follow the command sequence below and previous images for performing this dimensioning task.

```
Command: QDIM
Select geometry to dimension: (Pick a point at "A")
Specify opposite corner: (Pick a point at "B")
Select geometry to dimension: (Press ENTER to continue)
Specify dimension line position, or
[Continuous/Staggered/Baseline/Ordinate/Radius/Diameter/
datumPoint/Edit/settings]
<Continuous>: (Change to a different mode or locate the
dimension line at "C")
```

Staggered Mode

By default, the QDIM command places continued dimensions. Before locating the dimensions, use the Staggered option for placing the staggered dimensions, as shown on the right in Figure 10.27. The process with this style begins with adding dimensions to inside details and continuing outward until all features are dimensioned.

TRY IT!

Open the drawing file 10_Qdim Staggered. Activate the QDIM command and pick the vertical lines labeled "A" through "H," as shown in Figure 10.27 (Left). Change to the Staggered mode. Your display should appear similar to the illustration in Figure 10.27 (Right).

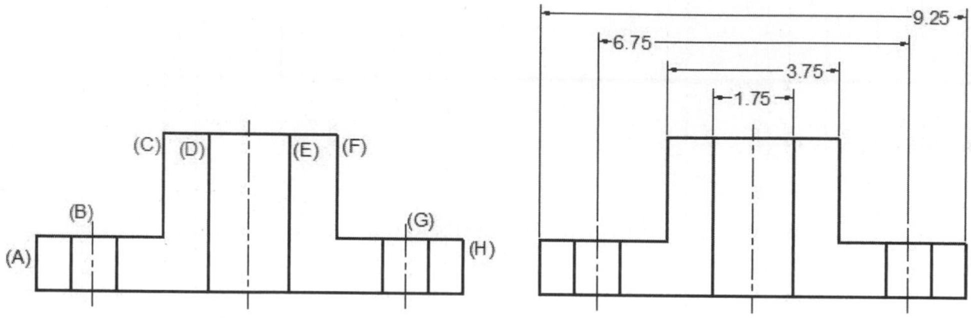

FIGURE 10.27

Baseline Mode

Another option of the QDIM command is the ability to place baseline dimensions. As with all baseline dimensions, an edge is used as the baseline, or datum. By default, all dimensions are calculated from the left edge, as shown in Figure 10.28 (Left). The datumPoint option of the QDIM command allows you to select a different baseline. The right edge of the object in Figure 10.28 (Right) is selected as the new datum or baseline.

TRY IT!

Open the drawing file 10_Qdim Baseline. Activate the QDIM command and identify the same set of objects as in the first Qdim exercise. Change to the Baseline mode. Your display should appear similar to the illustration in Figure 10.28. Try using the datumPoint option to change the baseline.

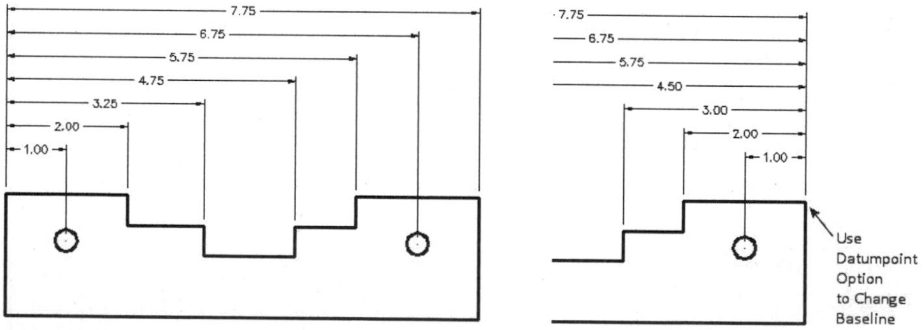

FIGURE 10.28

Ordinate Mode

The Ordinate option of the QDIM command calculates dimensions from a known 0,0 corner. A new User Coordinate System can be established to set the origin (lower-left corner) or you can use the datumPoint option to set the new 0,0 point.

TRY IT!

Open the drawing file 10_Qdim Ordinate. Activate the `QDIM` command and identify the same set of objects as in the first Qdim exercise. Change to the Ordinate mode. Your display should appear similar to the illustration in Figure 10.29.

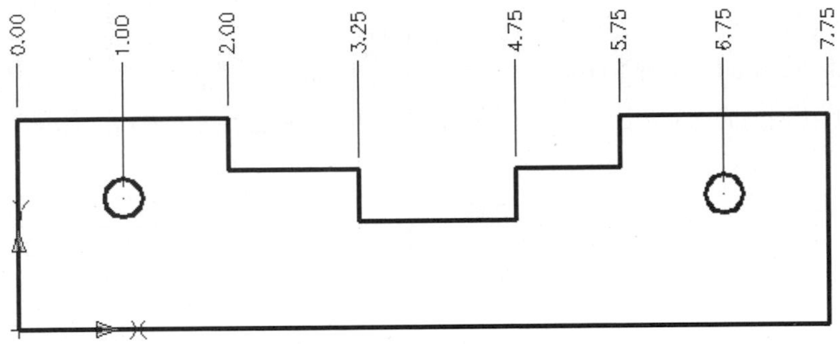

FIGURE 10.29

Radius and Diameter Mode

The `QDIM` command also has options that can be applied to radius and diameter dimensions. In Figure 10.30, all arcs and circles are selected. Activating either the Radius or Diameter option displays the results in the illustrations. When locating the diameter dimensions, you can specify the angle of the leader. A predefined leader length is applied to all dimensions. Grips could be used to relocate the dimensions to better places.

TRY IT!

Open the drawing file 10_Qdim Radius. Use the illustration in Figure 10.30 (Left) and the command sequence below for placing a series of radius dimensions using the QDIM command.

```
Command: QDIM
Select geometry to dimension: (Select the four arcs in Figure 10.30 on the left)
Select geometry to dimension: (Press ENTER to continue)
Specify dimension line position, or
[Continuous/Staggered/Baseline/Ordinate/Radius/Diameter/datumPoint/Edit/seTtings]
<Continuous>: R (For Radius)
```

TRY IT!

Open the drawing file 10_Qdim Diameter. Use the illustration in Figure 10.30 (Right) and the command sequence below for placing a series of diameter dimensions using the `QDIM` command.

```
Command: QDIM
Select geometry to dimension: (Select the four circles in Figure 10.30 on the right)
```

Select geometry to dimension: *(Press* ENTER *to continue)*

Specify dimension line position, or

[Continuous/Staggered/Baseline/Ordinate/Radius/Diameter/datumPoint/Edit/seTtings]

<Continuous>: D *(For Diameter)*

Specify dimension line position, or

[Continuous/Staggered/Baseline/Ordinate/Radius/Diameter/datumPoint/Edit/seTtings]

<Diameter>: *(Pick a point to locate the diameter dimension)*

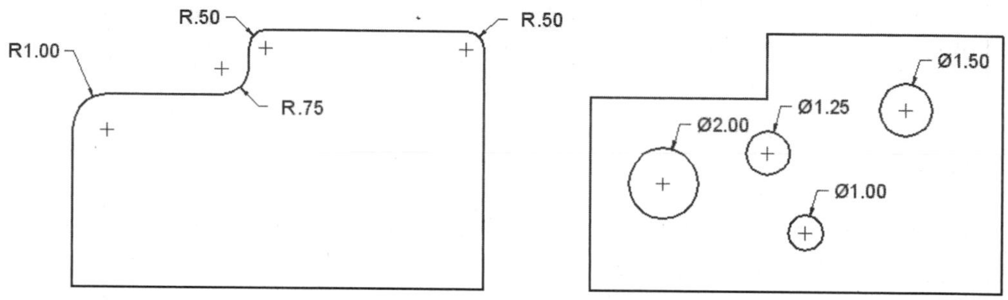

FIGURE 10.30

TIP

The QDIM command can also be used to edit an existing group of dimensions. Activate the command, select all existing dimensions with a crossing box, and enter an option to change the style of all dimensions.

SPACING DIMENSIONS

This tool is used to equalize the spacing between parallel linear or angular dimensions. You can use the Auto option to have this tool calculate the dimension spacing distance based on the dimension text height or you can provide a specific value for the dimensions to be separated by. These parameters work well for baseline dimensions. In the case of continue dimensions, entering a value of 0 lines up all continue dimensions.

TRY IT!

Open the drawing file 10_Dim Spacing. Use the following command sequence and Figure 10.31 for performing this operation on both baseline and continue dimensions.

Command: DIMSPACE

Select base dimension: *(Select dimension "A")*

Select dimensions to space: *(Select dimensions "B," "C," and "D")*

Select dimensions to space: *(Press* ENTER *to continue)*

Enter value or [Auto] <Auto>: A *(For Auto)*

Command: DIMSPACE

Select base dimension: *(Select dimension "E")*

Select dimensions to space: *(Select dimensions "F," "G," and "H")*

Select dimensions to space: *(Press* ENTER *to continue)*

Enter value or [Auto] <Auto>: 0

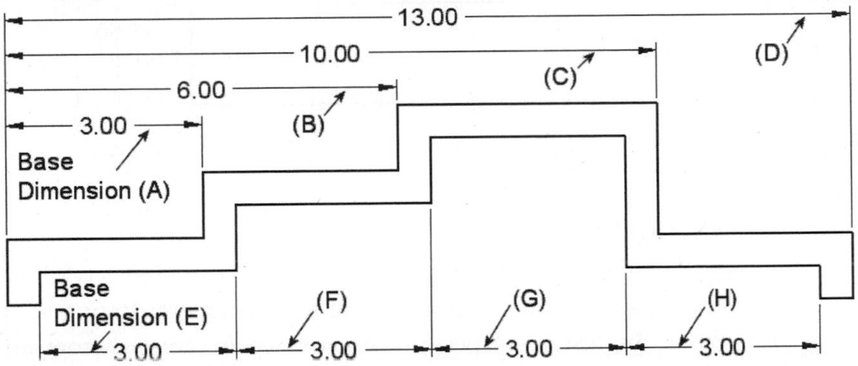

FIGURE 10.31

APPLYING BREAKS IN DIMENSIONS

Use this tool to break dimension, extension, or leader lines. Dimension breaks can be added to the following dimension types; Linear, Angular, Diameter, Radius, Jogged, Ordinate, and Multileaders. The following objects act as cutting edges when producing dimension breaks: Arcs, Circles, Dimensions, Ellipses, Leaders, Lines, Mtext, Polylines, Splines, and Text. Dimension breaks cannot be placed on an arrowhead or dimension text.

The size of the break is controlled through the Symbols and Arrows tab of the Dimension Style Manager dialog box, which will be explained in Chapter 11.

TRY IT!

Open the drawing file 10_Dim Break Mech. Use the following command sequence and Figure 10.32 for creating numerous breaks in a dimensions extension line.

Command: DIMBREAK

Select dimension to add/remove or [Multiple]: *(Pick the 2.00 vertical dimension on the right)*

Select object to break dimension or [Auto/Manual/Remove] <Auto>: *(Press* ENTER *to automatically break this dimension)*

Command: DIMBREAK

Select dimension to add/remove or [Multiple]: *(Pick the dimension at "A")*

Select object to break dimension or [Auto/Restore/Manual] <Auto>: *(Pick the dimension at "B")*

Select object to break dimension: *(Press* ENTER *to exit this command)*

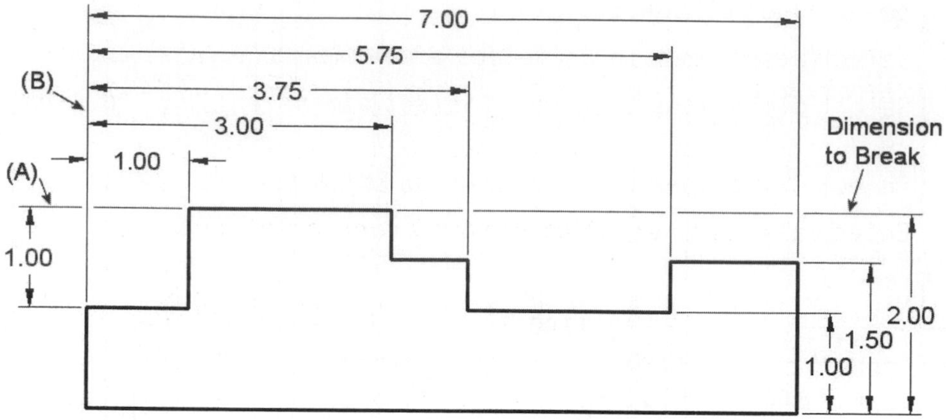

FIGURE 10.32

INSPECTION DIMENSIONS

An inspection dimension identifies a distance that needs to be verified according to labeled criteria. Part of the inspection dimension includes a parameter stating how often the dimension should be tested. Inspection dimensions are created from an already existing dimension.

Before converting an existing dimension to an inspection dimension, a dialog box displays, as shown in Figure 10.33. The dialog box displays the shape, label, and inspection rate. You can choose from a round or angular frame; you can even have no frame applied to the inspection dimension. The inspection label is present at the leftmost portion of the inspection dimension; the actual dimension value is located in the center section of the inspection dimension. The inspection rate located at the rightmost section of the inspection dimension is used to indicate the frequency at which the dimension value is inspected.

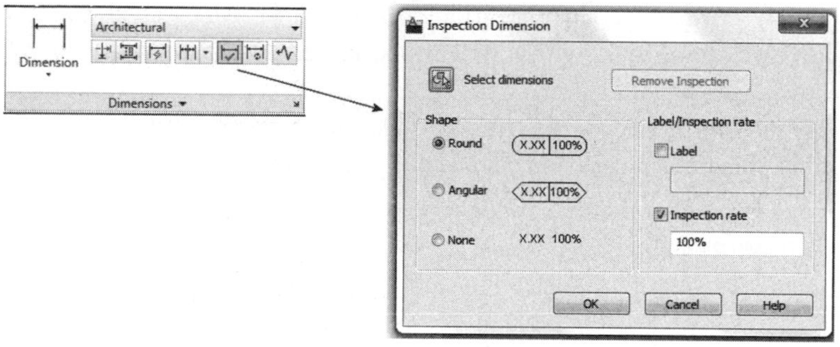

FIGURE 10.33

TRY IT!

Open the drawing file 10_Dim Inspection. Use the following command sequence and Figure 10.34 for creating an inspection dimension.

Command: DIMINSPECT

(When the Inspection Dimension dialog box displays, change the shape to Angular and place checks in the label and Inspection rate boxes, as shown in Figure 10.34 (Right). Enter "A" as the label designation and click the Select dimensions button)

Select dimensions: *(Pick the 8.00 vertical dimension on the right)*

Select dimensions: *(Press ENTER to return to the dialog box; click OK)*

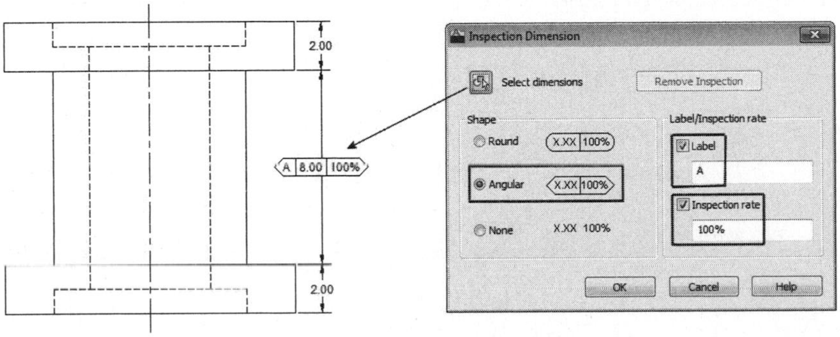

FIGURE 10.34

ADDING JOGGED DIMENSIONS

At times, you may need to add a radius dimension to a large arc. When creating the radius dimension, the center for the dimension is placed outside the boundaries of the drawing. To give better control over these situations, you can create a jog in the radius. This is represented by a zigzag appearance, or jog, in the leader holding the radius dimension.

TRY IT!

Open the drawing file 10_Dim Jogged. Use the following command sequence and Figure 10.35 for creating a jogged radius dimension.

Command: DJO *(For DIMJOGGED)*

Select arc or circle: *(Select the arc in Figure 10.35)*

Specify center location override: *(Locate the center at "A")*

Dimension text = 23.1588

Specify dimension line location or [Mtext/Text/Angle]: *(Locate the dimension line at "B")*

Specify jog location: *(Pick a point at "C")*

ADDING ARC DIMENSIONS

You can also dimension the length of an arc using the Dimension Arc command (DIMARC, or the shortcut DAR). After you select the arc, extension lines are created at the endpoints of the arc and a dimension arc is constructed parallel to the arc being dimensioned. The arc symbol is placed with the dimension text. This symbol can also be located above the dimension text. This technique will be covered in Chapter 11.

```
Command: DAR (For DIMARC)
Select arc or polyline arc segment: (Select the arc)
Specify arc length dimension location, or [Mtext/Text/
Angle/Partial]: (Locate the dimension at "D")
Dimension text = 10.8547
```

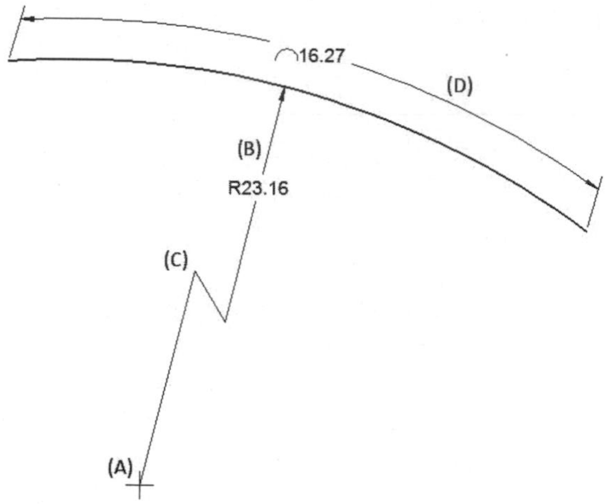

FIGURE 10.35

LINEAR JOG DIMENSIONS

Jog lines are used to represent a dimension value that does not display the actual measurement. Typically, the actual measurement value of the dimension is smaller than the displayed value. The Linear Jog dimension tool is used for creating this type of jogged dimension.

TRY IT!

Open the drawing file 10_Dim Linear Jog. Use the following command sequence and Figure 10.36 for creating a jog along a linear dimension.

```
Command: DJL (For DIMJOGLINE)
Select dimension to add jog or [Remove]: (Select dimension
"A")
Specify jog location (or press Enter): (Pick the jog location
at "B")
```

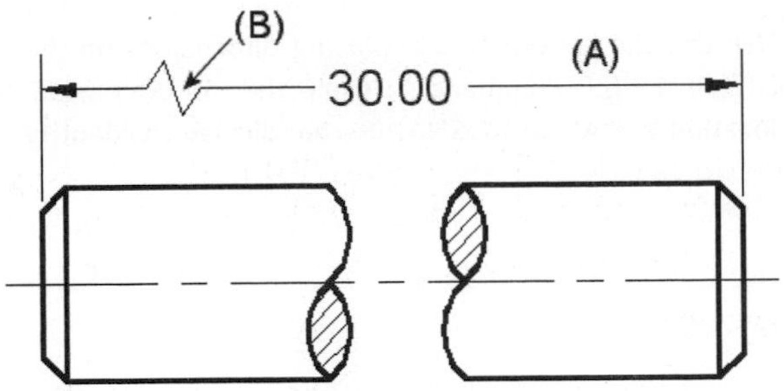

FIGURE 10.36

DIMENSIONING SLOTS

Two methods of dimensioning slot are illustrated in Figure 10.37. The method is usually determined by the machining process used to create the slot. The first method is often used for milled slots. You call out the slot by locating the center-to-center distance between the two arcs, followed by the overall depth. A radius dimension showing only an "R" indicates that the arc is fully rounded (the radius value is not needed; it is half the depth). The second method involves providing the overall width and depth of the slot, this technique is often provided if the slots are created by a stamping or punching operation.

TRY IT!

Open the drawing file 10_Dim Slot1. Use this file for practice in placing dimensions on the slot, as shown in Figure 10.37. Use the Text option of the DIMRADIUS command to replace the actual radius dimension with an "R."

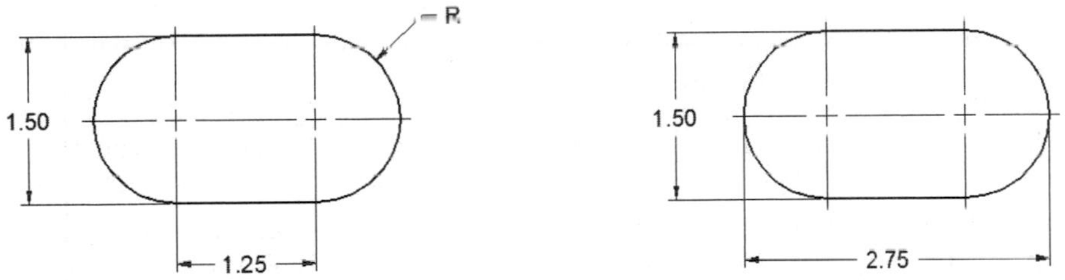

FIGURE 10.37

TIP

Using the Text option in the `DIMRADIUS` command is an efficient way of placing a local note on any arc or circle—it ensures the leader arrow always points toward the center. The Text option is available in several of the basic dimension commands; use it when you want to replace the actual dimension value with a note. Typing in a "<>" will return the actual dimension value. Try using the Mtext option when you want to add a prefix or suffix to a dimension value (a "2X" in front of a dimension or a "TYP" after).

A more complex example in Figure 10.38 involves slots formed by curves and angles. Here, the radius of the circular center arc is called out. Angles reference each other for accuracy. The overall width of the slot is dimensioned, which happens to be the diameter of the arcs at opposite ends of the slot.

TRY IT!

Open the drawing file 10_Dim Slot2. Use this file for practice in placing dimensions on the angular slot in Figure 10.38. Use the DIMALIGNED command to place the .48 dimension. Use OSNAP-Nearest to identify the location at "A" and OSNAP-Perpendicular to identify the location at "B."

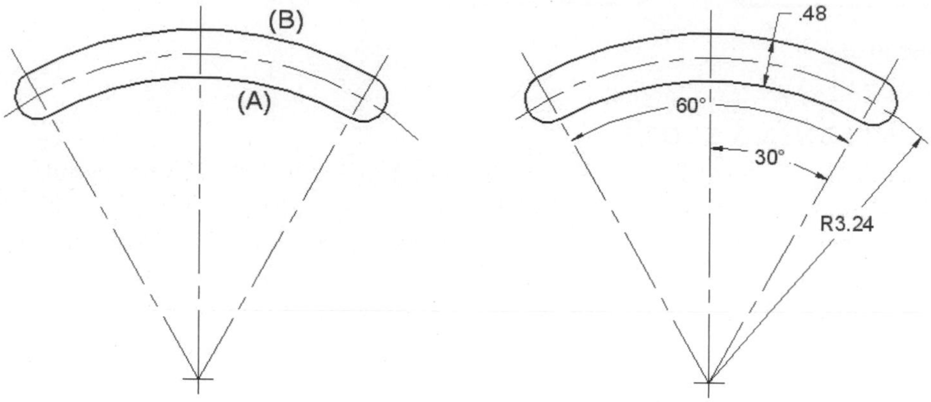

FIGURE 10.38

EDITING DIMENSIONS

Use the DIMEDIT command to add text to the dimension value, rotate the dimension text, rotate the extension lines for an oblique effect, or return the dimension text to its home position. Figure 10.39 shows the effects of adding text to a dimension and rotating the dimension to a user-specified angle.

TRY IT!

Open the drawing file 10_Dim Dimedit. Follow the next series of figures and Command prompt sequences for accomplishing this task.

Command: DED *(For DIMEDIT)*

Enter type of dimension editing [Home/New/Rotate/Oblique] <Home>: N *(For New. This displays the Multiline Text Editor. Add the text* "TYPICAL" *on the right side of the value (use right arrow key). When finished, close the Text Editor.)*

Select objects: *(Select the 5.00 dimension)*

Select objects: *(Press* ENTER *to perform the dimension edit operation)*

Command: DED *(For DIMEDIT)*

Enter type of dimension editing [Home/New/Rotate/Oblique] <Home>: R *(For Rotate)*

Specify angle for dimension text: 10

Select objects: *(Select the 5.00 dimension)*

Select objects: *(Press* ENTER *to perform the dimension edit operation)*

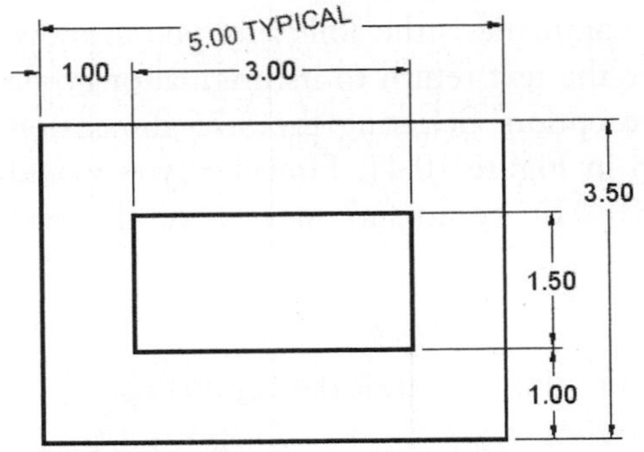

FIGURE 10.39

Regular AutoCAD modify commands can also affect dimensions. When you use the STRETCH command on the object in Figure 10.40 (Left), points "A" and "B" identify a crossing window. Point "C" is the base point of displacement. The results are displayed in Figure 10.40 (Right). Not only did the object lines stretch to the new position, but the dimensions also all updated themselves to new values.

Command: S *(For STRETCH)*

Select objects to stretch by crossing-window or crossing-polygon...

Select objects: *(Pick a point at "A")*

Specify opposite corner: *(Pick a point at "B" to activate the crossing window)*

Select objects: *(Press* ENTER *to continue)*

Specify base point or [Displacement] <Displacement>: *(Pick a point at "C"; it could also be anywhere on the screen)*

Specify second point or <use first point as displacement>: *(Use the Direct Distance mode; with ORTHO or POLAR mode on, move your cursor to the right and type* .75 *to perform the stretch)*

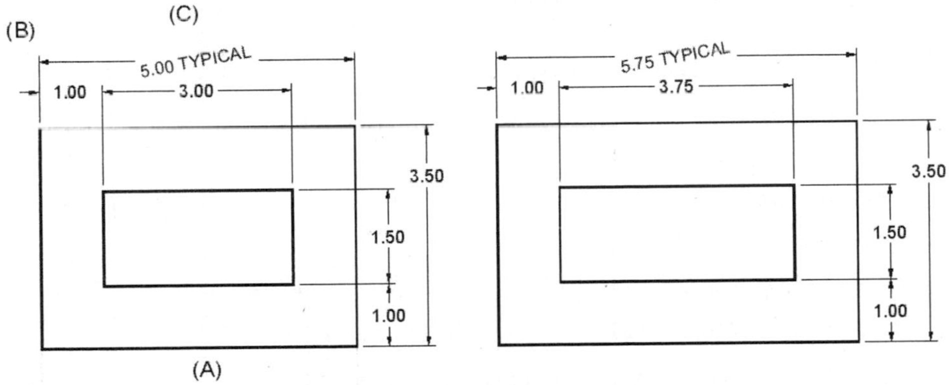

FIGURE 10.40

One of the other options of the DIMEDIT command is the ability for you to move and rotate the dimension text and still have the text return to its original or home location. This is the purpose of the Home option. Selecting the 5.75 dimension returns it to its original position, as shown in Figure 10.41. However, you would have to use the New option of the DIMEDIT command to remove the text "TYPICAL."

```
Command: DED (For DIMEDIT)

Enter type of dimension editing [Home/New/Rotate/Oblique]
<Home>: H (Press ENTER to continue)

Select objects: (Select the 5.75 dimension)

Select objects: (Press ENTER to perform the dimension edit
operation)
```

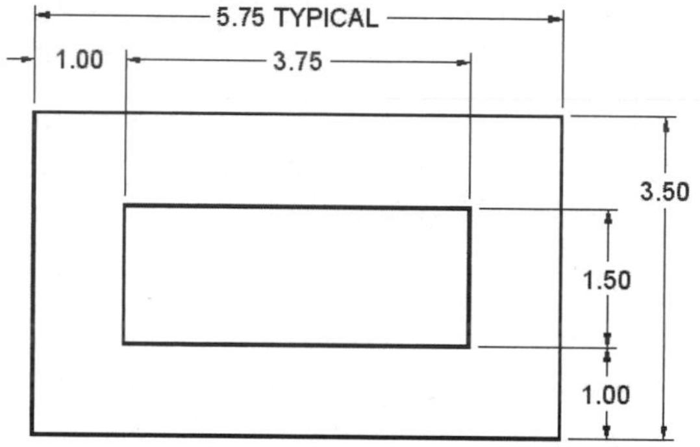

FIGURE 10.41

The DIMEDIT command also has an Oblique option that allows you to enter an obliquing angle, which rotates the extension lines and repositions the dimension line. This option is useful if you are interested in placing dimensions on isometric drawings.

TRY IT!

Open the drawing file 10_Dim Oblique. Follow the next series of images and Command prompt sequence for accomplishing this task.

```
Command: DED (For DIMEDIT)

Enter type of dimension editing [Home/New/Rotate/Oblique]
<Home>: O (For Oblique)

Select objects: (Select the 2.00 and 1.00 dimensions in
Figure 10.42 (Left))

Select objects: (Press ENTER to continue)

Enter obliquing angle (Press ENTER for none): 150
```

The results are illustrated in Figure 10.42 (Right), with both dimensions being repositioned with the Oblique option of the DIMEDIT command. Notice that the extension and dimension lines were affected; however, the dimension text remained the same.

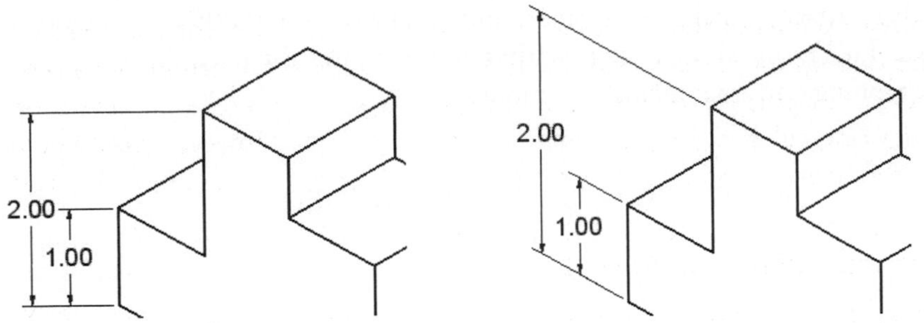

FIGURE 10.42

Use the Oblique option to complete the editing of this drawing by rotating the dimension at "A" at an obliquing angle of 210°. An obliquing angle of −30° was used to rotate the dimensions at "B" and "C." The dimensions at "D" require an obliquing angle of 90°, as shown in Figure 10.43. This represents proper isometric dimensions, except for the orientation of the text.

NOTE

To properly orientate the dimension text in an isometric drawing, you must rotate both the text and the letters to orient them to the oblique dimension. For the dimensions in the previous image, rotate the text −30° and then slant the letters by setting the text style's oblique angle to −30°.

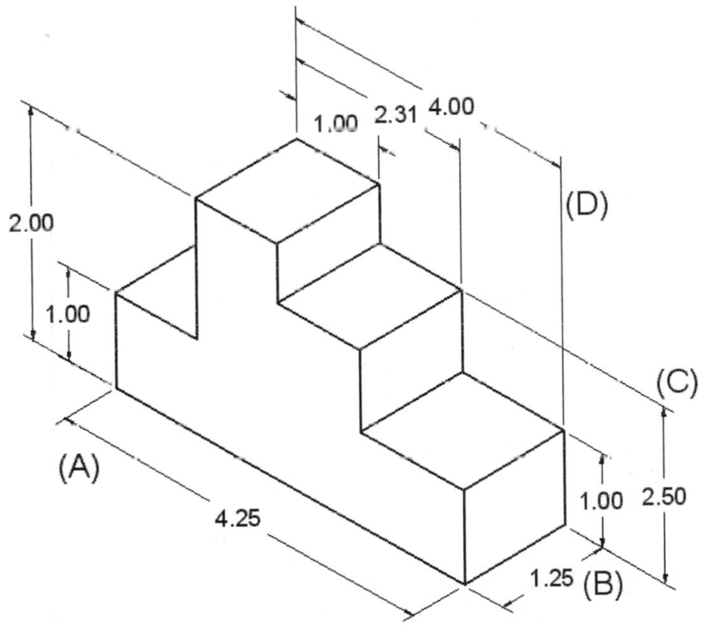

FIGURE 10.43

GEOMETRIC DIMENSIONING AND TOLERANCING (GDT)

In the object in Figure 10.44, a note is used to try and explain a required tolerance for the part. The note is indicating that the leg surface should be 90° to the base surface within a tolerance of 0.005 units. This note could be misinterpreted in several ways: which leg surface, which base surface, and what does the 0.005 unit tolerance zone look like. This is one simple example that demonstrates the need for using geometric dimensioning and tolerancing techniques.

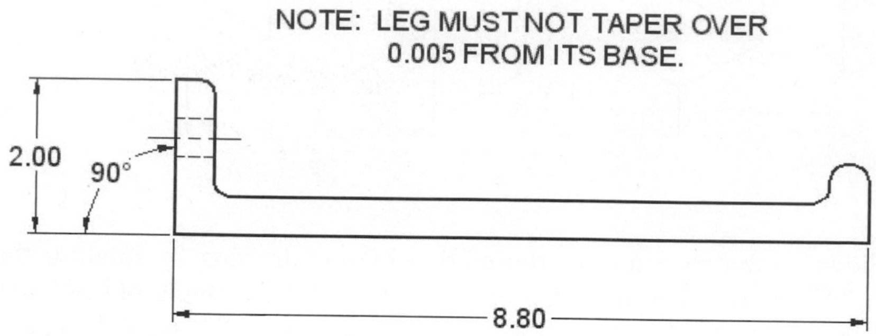

FIGURE 10.44

Figure 10.45 shows the same object complete with dimensioning and tolerancing (GDT) symbols. The letter "A" inside the rectangle identifies the reference surface (base surface), which establishes a datum to measure from. The tolerance symbol (feature control frame) is attached with a leader to the surface that must not taper. The tolerance indicates that every point on this surface must be within two parallel planes that are 0.005 units apart and those planes are to be perpendicular to the datum identified as "A." GDT specifications utilize standard symbols and provide detailed instruction on how tolerance zones are to be interpreted.

Using geometric tolerancing symbols ensures accurate parts with less error in interpreting the dimensions.

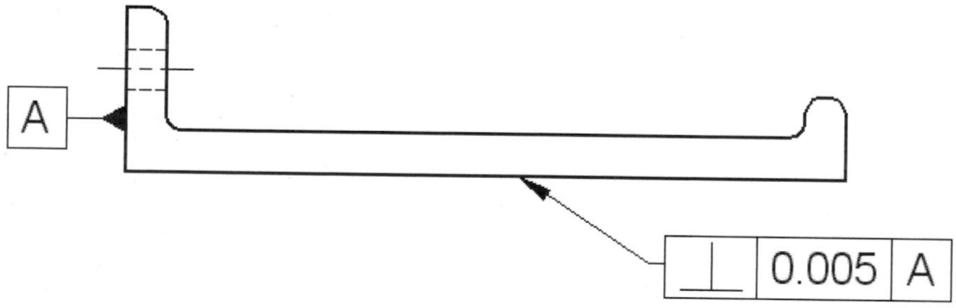

FIGURE 10.45

GDT Symbols

Entering `TOLERANCE` or `TOL` at the Command prompt brings up the main Geometric Tolerance dialog box, as shown in Figure 10.46 (Left). This box contains all tolerance zone boxes and datum identifier areas. Clicking on one of the dark boxes under the Sym area displays the Symbol dialog box illustrated in Figure 10.46 (Right). This dialog box contains all geometric tolerancing symbols. Choose the desired symbol by clicking the specific icon; this returns you to the main Geometric Tolerance dialog box.

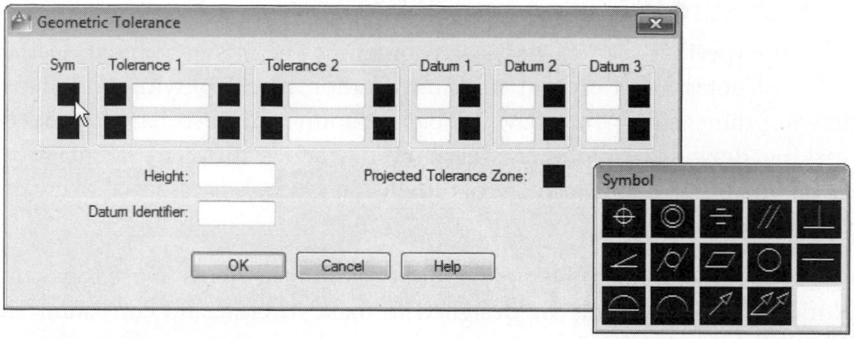

FIGURE 10.46

The illustration in Figure 10.47 shows a chart outlining the geometric tolerancing symbols. Alongside each symbol is the characteristic controlled by the symbol. Tolerances of form such as Flatness and Straightness are applied to surfaces without a datum being referenced. On the other hand, tolerances of orientation such as Angularity and Perpendicularity require datums as reference.

Symbol	Purpose	Symbol	Purpose
⏥	Flatness	⊥	Perpendicularity
—	Straightness	//	Parallelism
○	Roundness	⌖	Position
⌭	Cylindricity	◎	Concentricity
⌒	Profile of a Line	⌯	Symmetry
⌓	Profile of a Surface	↗	Circular Runout
∠	Angularity	⌰	Total Runout

FIGURE 10.47

With the symbol placed inside this dialog box, you now assign such items as tolerance values, material condition modifiers, and datum references. In Figure 10.48, the tolerance of Parallelism is to be applied at a tolerance value of 0.005 units to Datum "A."

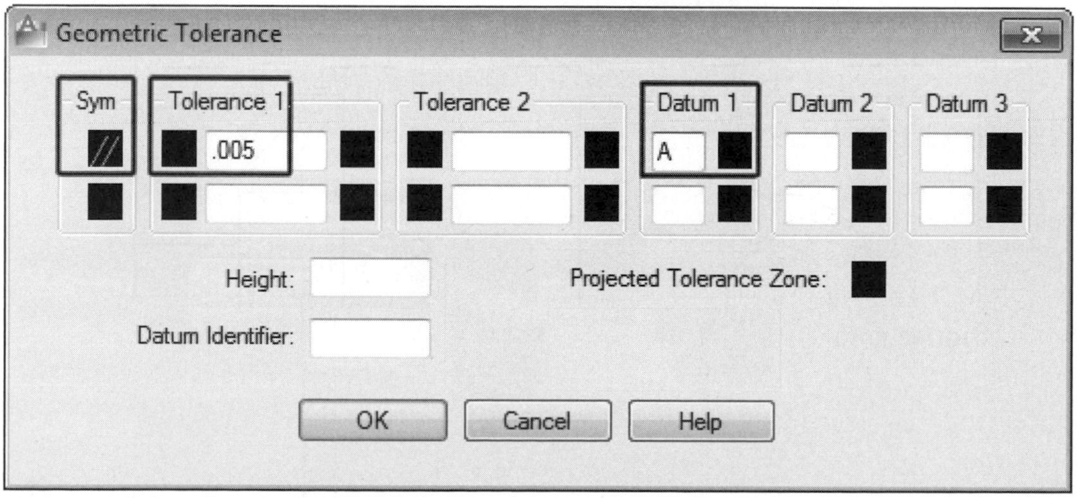

FIGURE 10.48

DIMENSION SYMBOLS

As discussed in the previous section, notes can sometimes be misinterpreted. Using symbols instead of notes helps prevent misinterpretation by simplifying and standardizing drawing dimensions. In today's global economy, this also has the added benefit of making drawing requirements clear even when a different language is being used. The proper use and placement of dimension symbols is essential to creating clear, concise drawings.

Figure 10.49 shows some of the more popular dimensioning symbols in use today on drawings. Notice how the symbols are designed to make as clear and consistent an interpretation of the dimension as possible. As an example, the Deep or Depth symbol displays an arrow pointing down. This symbol is typically used to identify how far into a part a drill hole goes. The Counterbore symbol identifies a large diameter hole used to recess a bolt head and resembles the side view of such a hole, and so on.

Symbol	Description	Symbol	Description
⌒	Arc Length	2X	Number of Times
[X.XX]	Basic Dimension	R	Radius
⌲	Conical Taper	(X.XX)	Reference Dimension
⌴	Counterbore/Spotface	SØ	Spherical Diameter
⌵	Countersink	SR	Spherical Radius
↧	Deep or Depth	⌳	Slope
Ø	Diameter	□	Square
X.XX (underlined)	Not to Scale		

FIGURE 10.49

CHARACTER MAPPING FOR DIMENSION SYMBOLS

Illustrated in Figure 10.50 is a typical dimension callout for a counterbore hole. The first line indicates the diameter of the through hole, while the second line provides information about the counterbore hole—its diameter and depth. The counterbore and depth symbols were generated with the `MLEADER` command along with the Character Mapping dialog box.

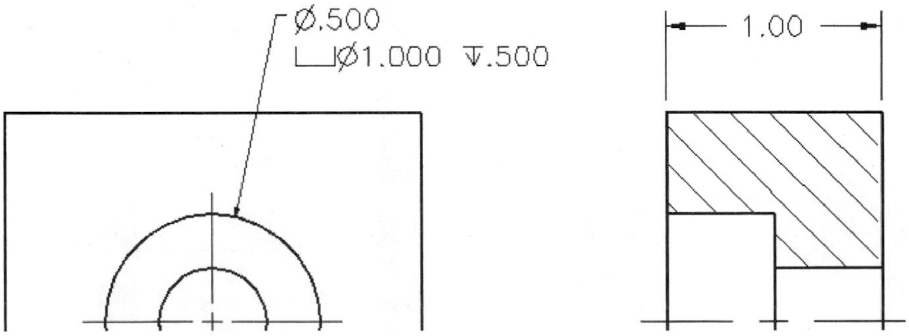

FIGURE 10.50

These extra symbol characters can be found by right-clicking inside the Multileader Text Editor field or from the Ribbon, click Symbol followed by Other..., as shown in Figure 10.51.

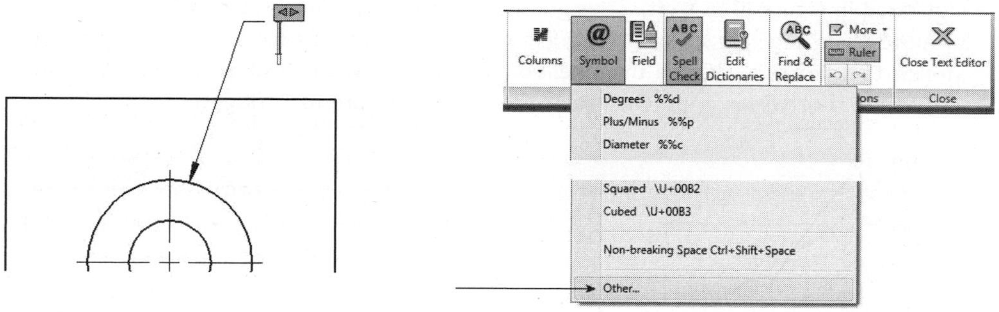

FIGURE 10.51

Clicking Other... displays the Character Map dialog box, as shown in Figure 10.52 (Left). Notice, in the upper-left corner, that the current font is GDT, which holds all geometric tolerancing symbols along with the special dimensioning symbols such as counterbore, deep, and countersink. Be sure your current font is set to GDT. Once you identify a symbol, double-click it. A box appears around the symbol. Also, the symbol appears in the Characters to copy area in the lower-left corner of the dialog box. Click the Copy button to copy this symbol to the Windows Clipboard. Then close the Character Map. Return to the Multileader Text Editor and press `CTRL + V`, which performs a paste operation. You may also paste the symbol by right-clicking in the Text Editor and clicking Paste. Because the counterbore symbol was copied to the Windows Clipboard, it pastes into the Text Editor, as shown in Figure 10.52 (Right).

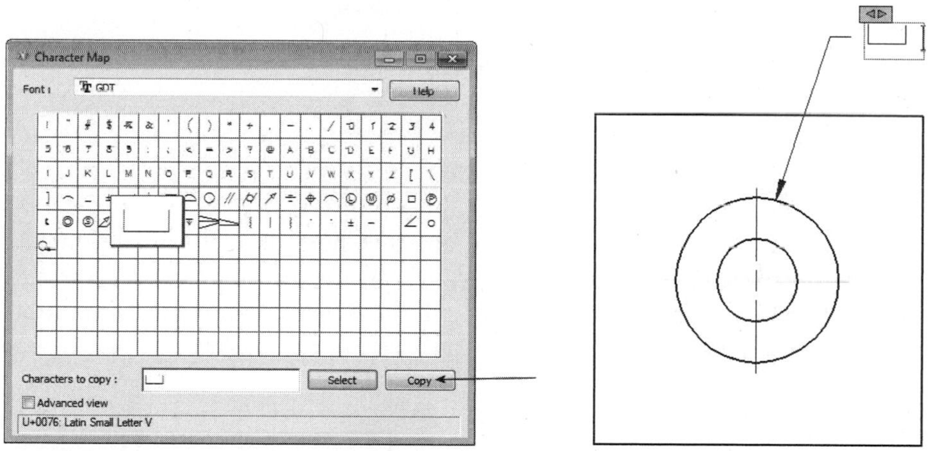

FIGURE 10.52

GRIPS AND DIMENSIONS

Grips have a tremendous amount of influence on dimensions. Grips allow dimensions to be moved to better locations; grips also allow the dimension text to be located at a better position along the dimension line.

TRY IT!

Open the drawing file 10_Dimgrip1. In the example shown in Figure 10.53, notice the various unacceptable dimension placements. The 1.38 and 1.50 horizontal dimensions lie too close to the object line. To relocate these dimensions to a better position, click both the dimensions. Notice that the grips appear and the dimensions highlight. Now click the grip near "A," as illustrated in Figure 10.53 (Left). (This grip is shared by both dimensions and will allow you to move them together.) When the grip turns red, the Stretch mode is currently active. Stretch the dimensions above the extension line but below the 3.50 dimension. Press the `ESC` **key to turn off the grips. The results are shown in Figure 10.53 (Right).**

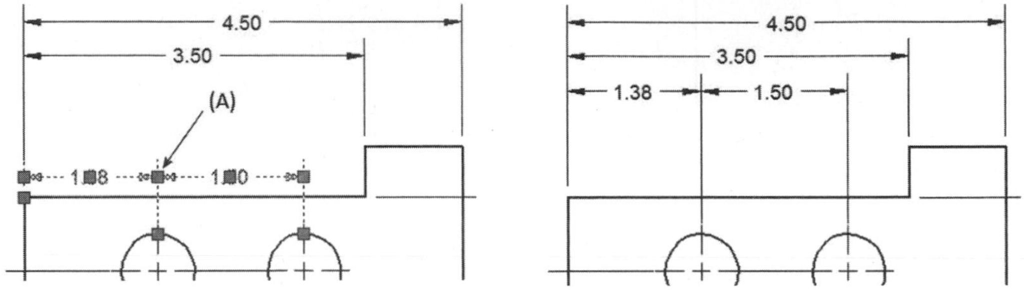

FIGURE 10.53

Notice that the two vertical dimensions shown in Figure 10.54 (Left) do not line up with each other; this would be poor practice. Pick both dimensions and notice the appearance of the grips, in addition to both dimensions being highlighted. Click the upper grip at "A" of the 1.50 dimension. When this grip turns red and places you in Stretch mode, select the grip at "B" of the 0.50 opposite dimension. The result will be that both dimensions now line up with each other, as shown in Figure 10.54 (Right).

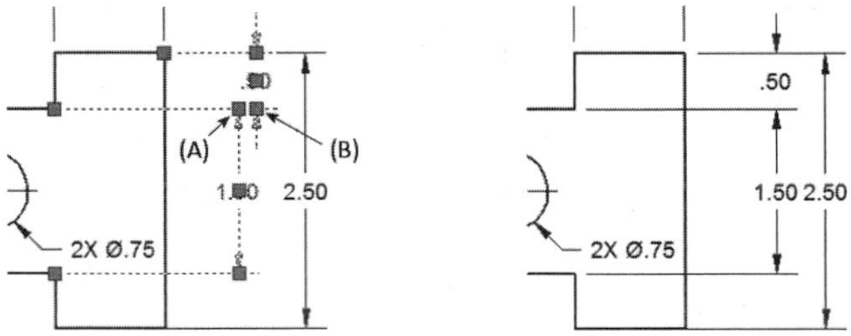

FIGURE 10.54

Illustrated in Figure 10.55 (Left), the 2.50 dimension text is too close to the 1.50 vertical dimension on the right side of the object. Click the 2.50 dimension; the grips appear and the dimension highlights. Click on the grip representing the text location at "A." When this grip turns red, stretch the dimension text to a better location, as shown in Figure 10.55 (Right).

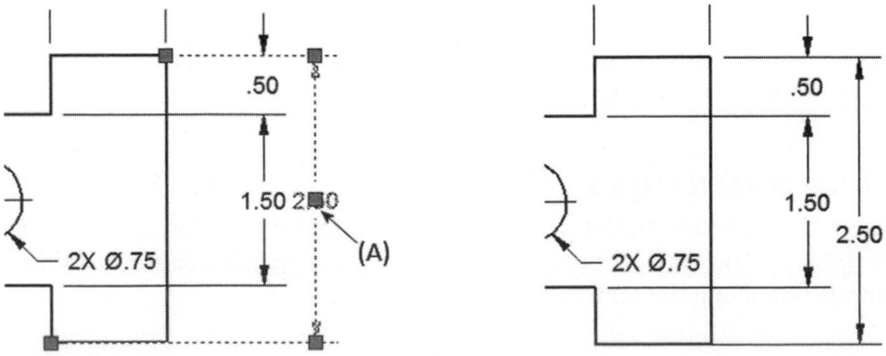

FIGURE 10.55

Grips also provide a simple method to correct the placement of diameter and radius dimensions. As shown in Figure 10.56 (Left), click the diameter dimension; the grips appear in addition to the diameter dimension being highlighted. Click the grip that locates the dimension text at "A." When this grip turns red, relocate the diameter dimension text to a better location using the Stretch mode, as shown in Figure 10.56 (Right).

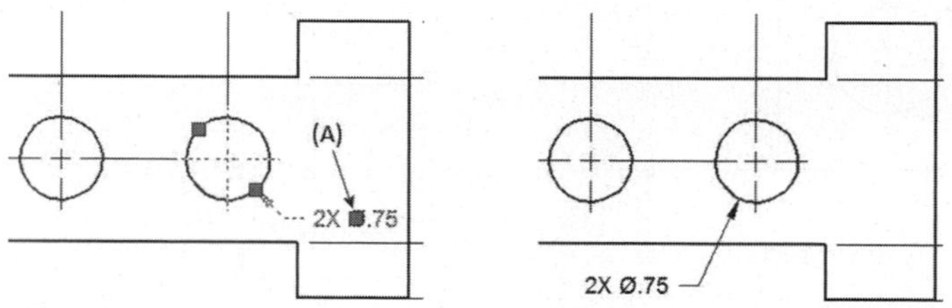

FIGURE 10.56

The final modifications concern incorrectly selected dimension origin points. Select the 0.87 vertical dimension, as shown in Figure 10.57 (Left). Notice that the grip at "A" was mistakenly placed at the end of an extension line. Select this grip, and once it turns red, stretch the grip to the endpoint at "B." The dimension value automatically updates to 0.75, as shown in Figure 10.57 (Right).

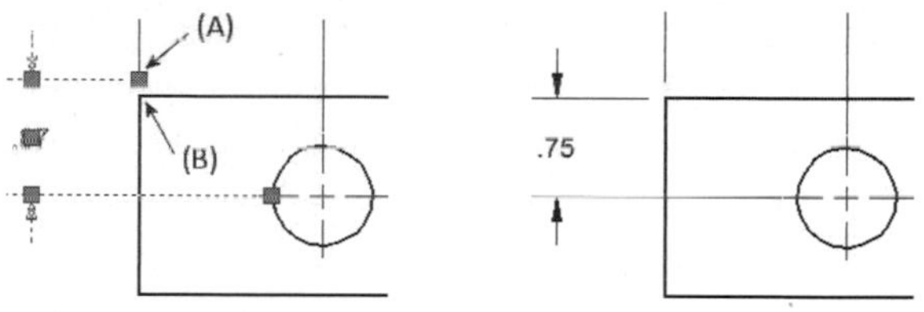

FIGURE 10.57

For the final modification, notice that the 2.50 vertical dimension shown in Figure 10.58 (Left), doesn't display a gap between the object line and extension line. This is due to the dimension origin point being incorrectly placed. Select the dimension to display the grips and pick the grip at "A," as shown in Figure 10.58 (Center). When the grip turns red, relocate the origin to the endpoint at "B." The appropriate extension line gap will be created automatically.

The completed object, with dimensions edited through grips, is displayed in Figure 10.58 (Right).

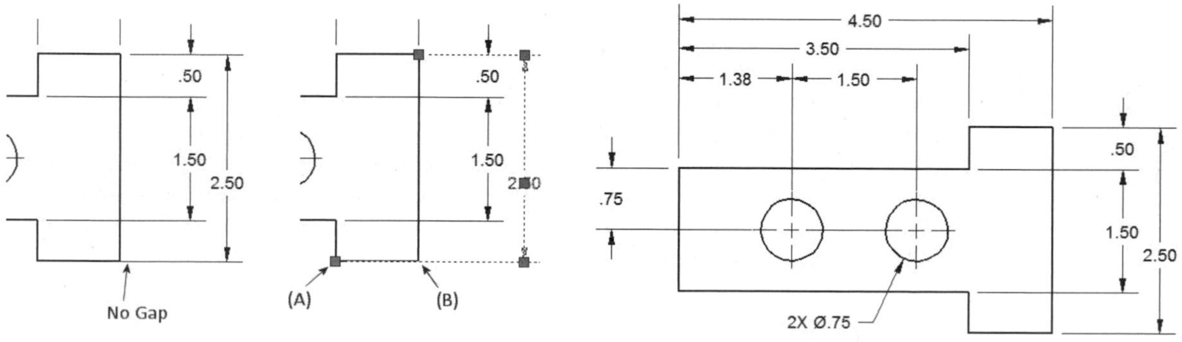

FIGURE 10.58

DIMENSIONS AND MULTI-FUNCTIONAL GRIPS

When you hover your cursor over a dimension grip, a dynamic menu appears with various modification options. Hovering over a dimension text grip, as shown in Figure 10.59 (Left), provides a series of options related to relocating the text. The results of selecting some of these modifications options are displayed on the right. The Stretch option is the default and it is not necessary to select it from the menu.

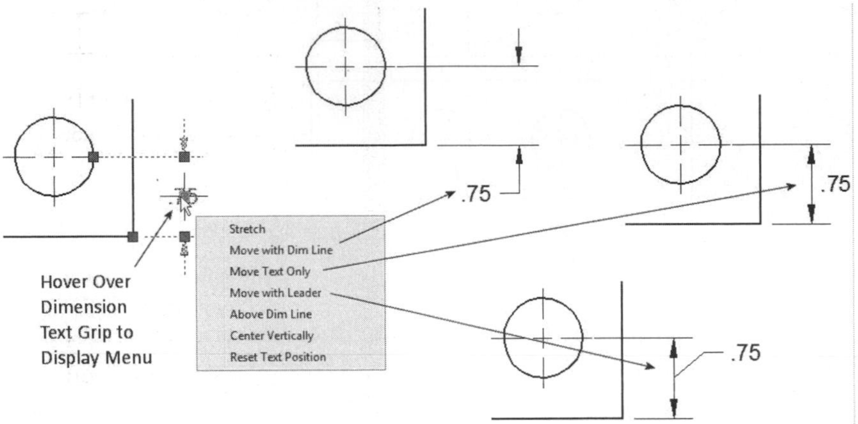

FIGURE 10.59

Hovering your cursor over an extension line grip also displays a menu, as shown in Figure 10.60. These options allow you to perform operations such as adding continue or baseline dimensions. There is also an option that allows you to flip the dimension arrows.

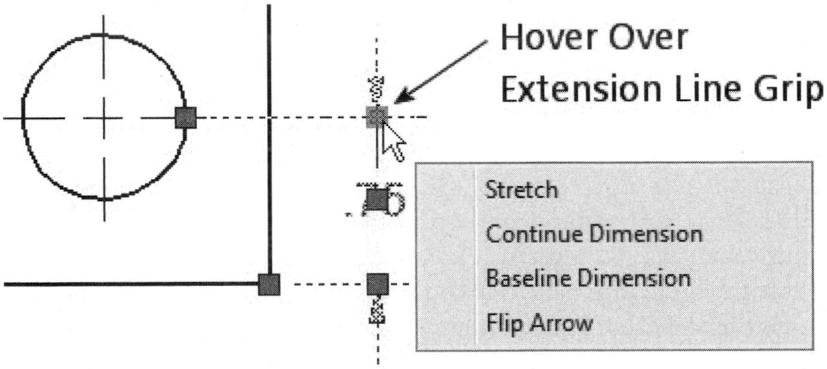

FIGURE 10.60

TRY IT!

Open the drawing file 10_Dimgrip2. In the example shown in Figure 10.61, notice that several dimensions are missing. To correct this, grips will be used to add both continue and baseline dimensions. Hover your cursor over the extension line grip at "A" to display the menu, as shown in Figure 10.61 (Left). Select the Continue option, and once prompted, select the circle quadrants at "B" and "C;" right-click to complete the operation. The results are shown in Figure 10.61 (Right).

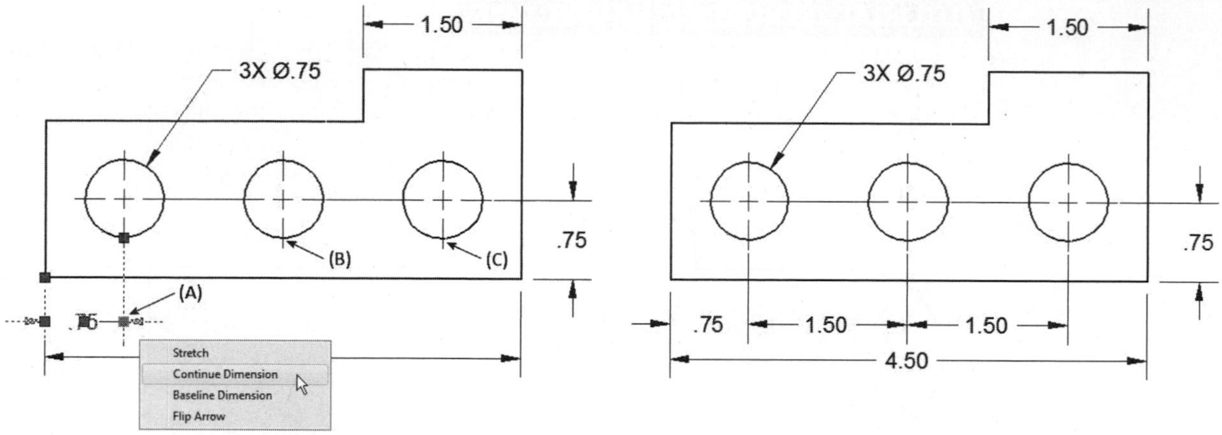

FIGURE 10.61

To add the baseline dimensions, hover your cursor over the extension line grip at "A," as shown in Figure 10.62 (Left). Select the Baseline option from the menu, and once prompted, select the endpoints at "B" and "C;" right-click to complete the operation. The results are shown on the right.

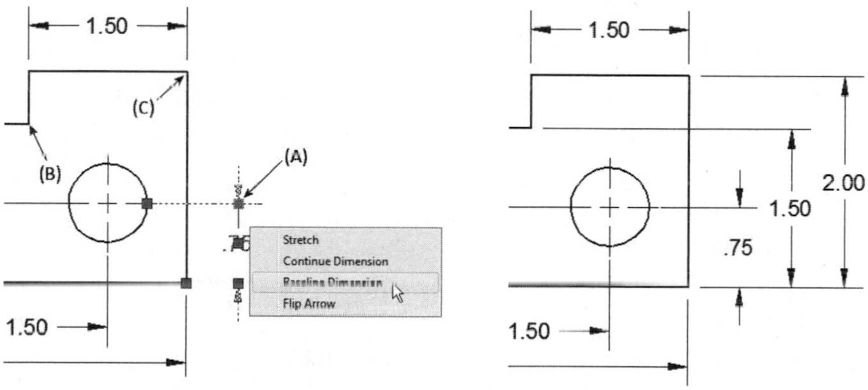

FIGURE 10.62

For a final modification, use the Flip Arrow option to change the arrow direction on the 1.50 horizontal dimension shown in Figure 10.63 (Left). Perform this operation at both extension line grips, identified as "A" and "B" in Figure 10.63. The results are shown on the right.

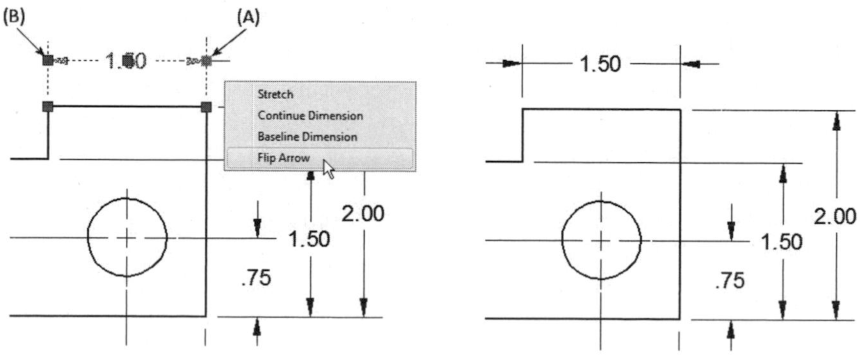

FIGURE 10.63

TUTORIAL EXERCISE: 10_FIXTURE.DWG

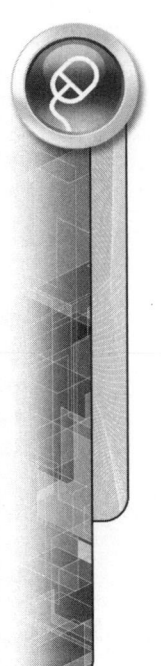

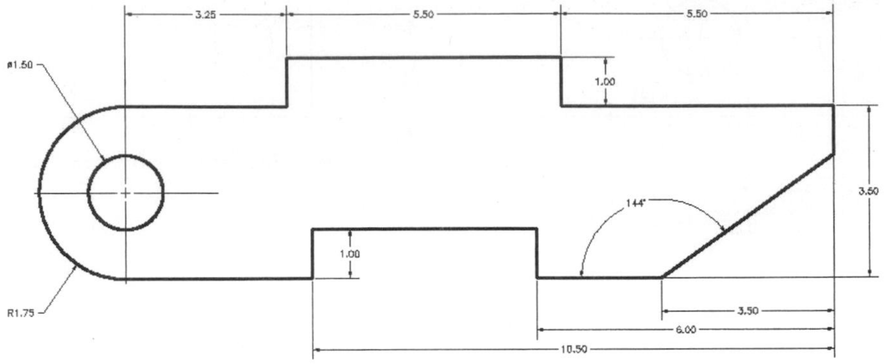

FIGURE 10.64

Purpose

The purpose of this tutorial is to add dimensions to the drawing of 10_Fixture.

System Settings

The drawing in the previous image is already constructed. Follow the steps in this tutorial for adding dimensions. Be sure the Endpoint and Intersection Object Snap modes are set.

Layers

Layers have already been created for this tutorial.

Suggested Commands

Use the DIMCENTER, DIMLINEAR, DIMCONTINUE, DIMBASELINE, DIMDIAMETER, and DIMRADIUS commands for placing dimensions throughout this tutorial.

STEP 1

Open the drawing file 10_Fixture. A series of linear, baseline, continue, radius, and diameter dimensions will be added to this view. Before continuing, verify that running OSNAP is set to Endpoint and Intersection modes and that OSNAP is turned on.

A number of dimension styles have been created to control various dimension properties. You will learn more about dimension styles in Chapter 11. For now, make the Center Mark dimension style current by clicking its name in the Dimension Styles Control box (located under the Annotate tab of the Ribbon), as shown in Figure 10.65.

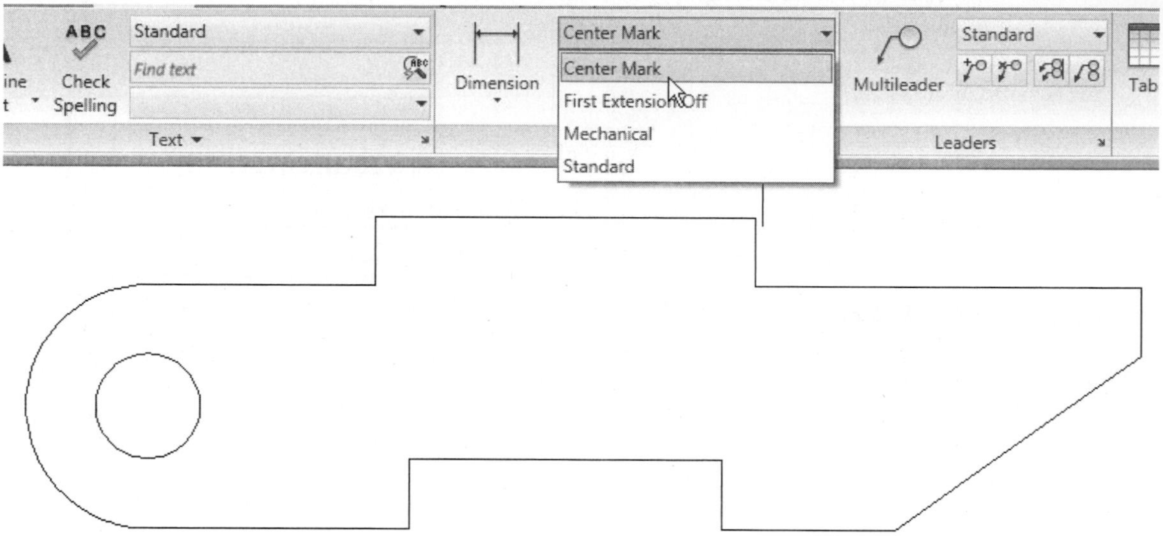

FIGURE 10.65

STEP 2

Using the `DIMCENTER` command (`DCE`), add a center mark to identify the center of the circular features. Touch the edge of the arc to place the center mark. The Center Mark dimension style already allows a small and long dash to be constructed. See Figure 10.66.

`Command: DCE` *(For DIMCENTER)*

`Select arc or circle:` *(Pick the edge of the arc in Figure 10.66)*

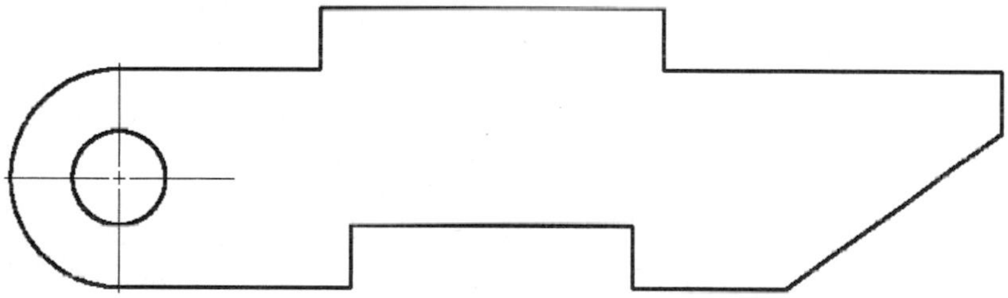

FIGURE 10.66

STEP 3

Now that the center mark is properly placed, it is time to begin adding the horizontal and vertical dimensions. Before performing these tasks, make the Mechanical dimension style current by clicking its name in the Dimension Styles Control box (located under the Annotate tab of the Ribbon), as shown in Figure 10.67.

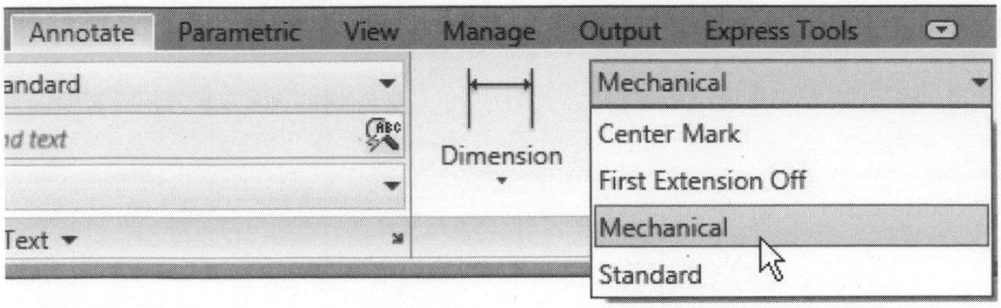

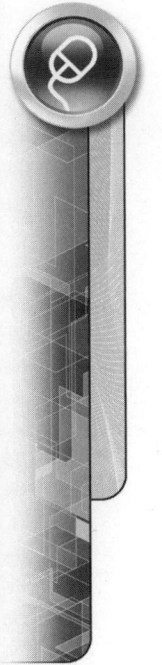

FIGURE 10.67

STEP 4

Verify in the Status bar that OSNAP is turned on. Then place a linear dimension of 5.50 units from the intersection at "A" to the intersection at "B," as shown in Figure 10.68.

Command: DLI *(For DIMLINEAR)*

Specify first extension line origin or <select object>: *(Pick the intersection at "A")*

Specify second extension line origin: *(Pick the intersection at "B")*

Specify dimension line location or [Mtext/Text/Angle/Horizontal/Vertical/Rotated]: *(Locate the dimension in Figure 10.68)*

Dimension text = 5.50

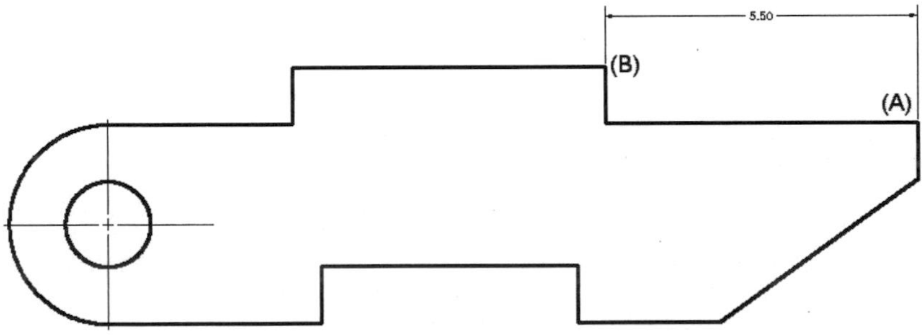

FIGURE 10.68

STEP 5

Place two Continue dimensions at intersection "A" and endpoint "B" in Figure 10.69. These Continue dimensions know to calculate the new dimension from the second extension line of the previous dimension.

Command: DCO *(For DIMCONTINUE)*

Specify a second extension line origin or [Undo/Select] <Select>: *(Pick the intersection at "A")*

Dimension text = 5.50

Specify a second extension line origin or [Undo/Select] <Select>: *(Pick the endpoint at "B")*

Dimension text = 3.25

Specify a second extension line origin or [Undo/Select] <Select>: *(Press* ENTER*)*

Select continued dimension: *(Press* ENTER*)*

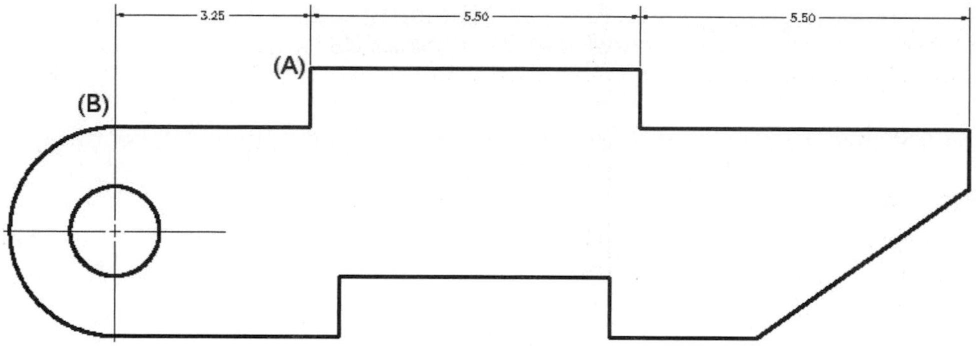

FIGURE 10.69

STEP 6

Place a linear dimension of 3.50 units from the intersection at "A" to the intersection at "B," as shown in Figure 10.70.

Command: DLI *(For DIMLINEAR)*

Specify first extension line origin or <select object>: *(Pick the intersection at "A")*

Specify second extension line origin: *(Pick the intersection at "B")*

Specify dimension line location or [Mtext/Text/Angle/Horizontal/Vertical/Rotated]: *(Locate the dimension in Figure 10.70)*

Dimension text = 3.50

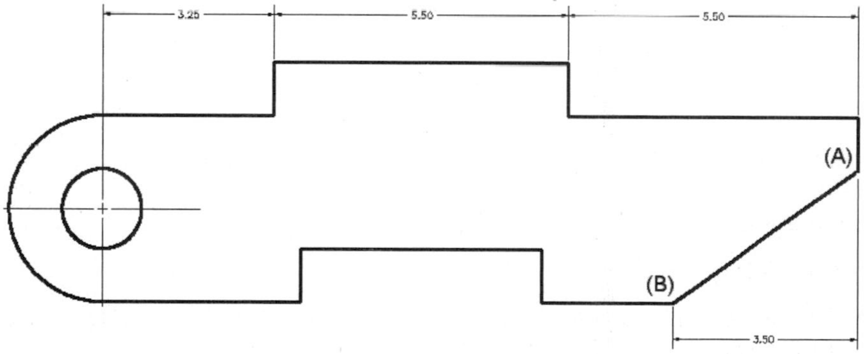

FIGURE 10.70

STEP 7

Add a series of baseline dimensions, as shown in Figure 10.71. All baseline dimensions are calculated from the first extension line of the previous dimension (the linear dimension measuring 3.50 units).

Command: DBA *(For DIMBASELINE)*

Specify a second extension line origin or [Undo/Select] <Select>: *(Pick the intersection at "A")*

Dimension text = 6.00

Specify a second extension line origin or [Undo/Select] <Select>: *(Pick the intersection at "B")*

Dimension text = 10.50

Specify a second extension line origin or [Undo/Select] <Select>: *(Press* ENTER*)*

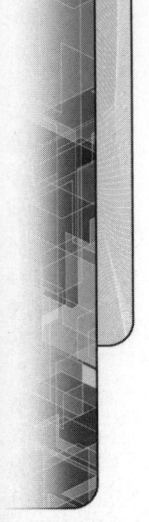

Select base dimension: *(Press* ENTER*)*

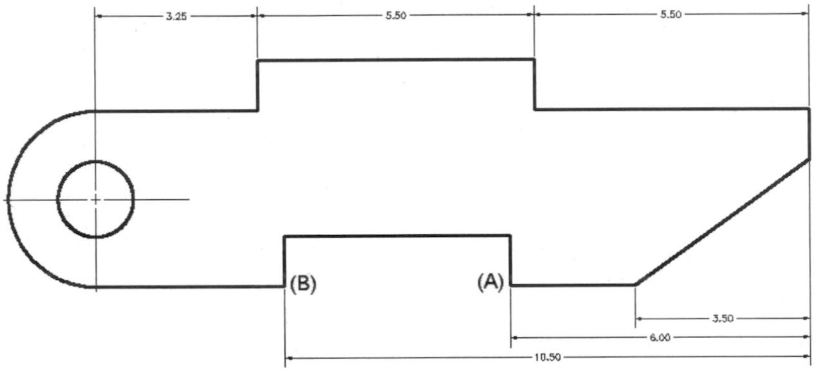

FIGURE 10.71

STEP 8

Add a diameter dimension to the circle using the DIMDIAMETER command (DDI). Then add a radius dimension to the arc using the DIMRADIUS command (DRA), as shown in Figure 10.72.

Command: DDI *(For DIMDIAMETER)*

Select arc or circle: *(Pick the edge of the circle at "A")*

Dimension text = 1.50

Specify dimension line location or [Mtext/Text/Angle]: *(Locate the diameter dimension in Figure 10.72)*

Command: DRA *(For DIMRADIUS)*

Select arc or circle: *(Pick the edge of the arc at "B")*

Dimension text = 1.75

Specify dimension line location or [Mtext/Text/Angle]: *(Locate the radius dimension in Figure 10.72)*

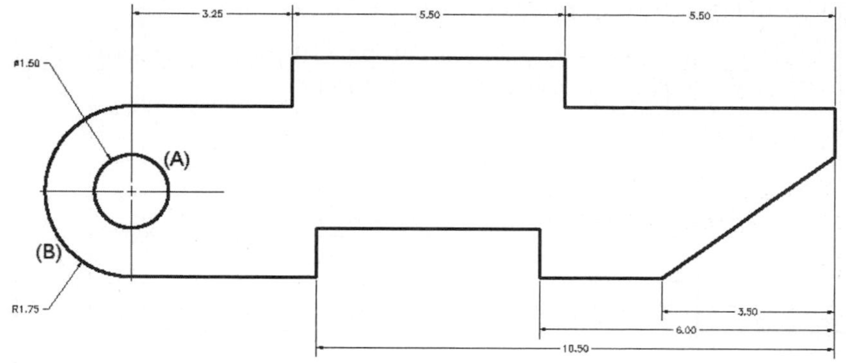

FIGURE 10.72

STEP 9

Place a vertical dimension of 3.50 units using the DIMLINEAR command (DLI). Pick the first extension line origin at the intersection at "A" and the second extension line origin at the intersection at "B," as shown in Figure 10.73.

Command: DLI *(For DIMLINEAR)*

Specify first extension line origin or <select object>:

(Pick the intersection at "A")

Specify second extension line origin: *(Pick the intersection at "B")*

Specify dimension line location or [Mtext/Text/Angle/Horizontal/Vertical/Rotated]: *(Locate the dimension in Figure 10.73)*

Dimension text = 3.50

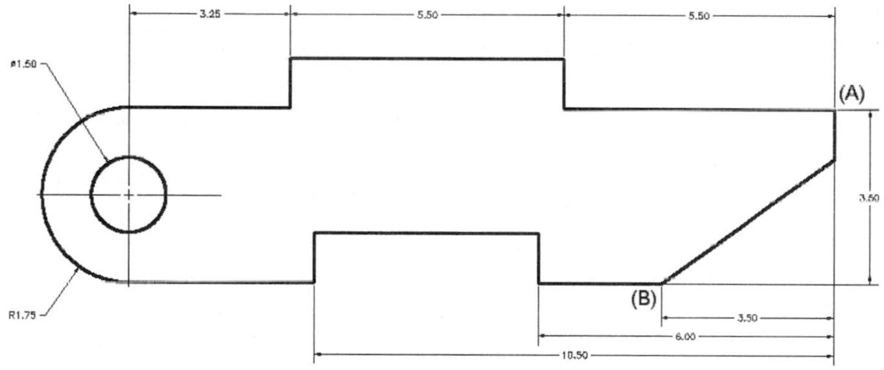

FIGURE 10.73

STEP 10

Add an angular dimension of 144° using the DIMANGULAR command (DAN), as shown in Figure 10.74.

Command: DAN *(For DIMANGULAR)*

Select arc, circle, line, or <specify vertex>: *(Pick the line at "A")*

Select second line: *(Pick the line at "B")*

Specify dimension arc line location or [Mtext/Text/Angle/Quadrant]: *(Locate the dimension in Figure 10.74)*

Dimension text = 144

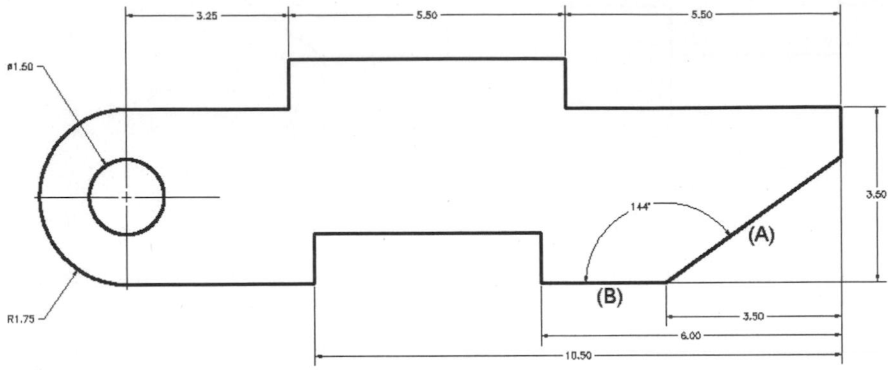

FIGURE 10.74

STEP 11

The next series of dimensions involves placing linear dimensions so that one extension line is constructed by the dimension and the other is actually an object line that already exists. You need to turn off one extension line while leaving the other extension line turned on. This is accomplished by using an existing dimension style. Make the First Extension Off dimension style current by clicking its name in the Dimension Styles Control box (located under the Annotate tab of the Ribbon), as shown in Figure 10.75. This dimension style will suppress or turn off the first extension line while leaving the second extension line visible. This prevents the extension line from being drawn on top of the object line.

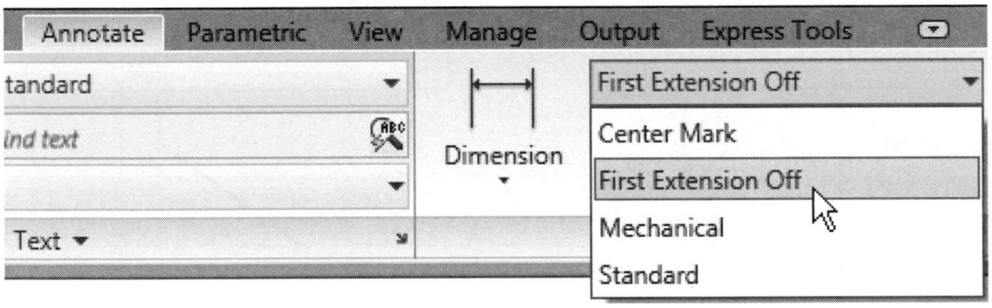

FIGURE 10.75

STEP 12

Add the two linear dimensions to the slots by using the following command sequences and Figure 10.76 as a guide.

Command: DLI *(For DIMLINEAR)*

Specify first extension line origin or <select object>:

(Pick the intersection at "A")

Specify second extension line origin:

(Pick the intersection at "B")

Specify dimension line location or [Mtext/Text/Angle/Horizontal/Vertical/Rotated]: *(Locate the dimension in Figure 10.76)*

Dimension text = 1.00

Command: DLI *(For DIMLINEAR)*

Specify first extension line origin or <select object>:

(Pick the intersection at "C")

Specify second extension line origin:

(Pick the intersection at "D")

Specify dimension line location or [Mtext/Text/Angle/Horizontal/Vertical/Rotated]: *(Locate the dimension in Figure 10.76)*

Dimension text = 1.00

The completed fixture drawing with all dimensions is shown in Figure 10.76.

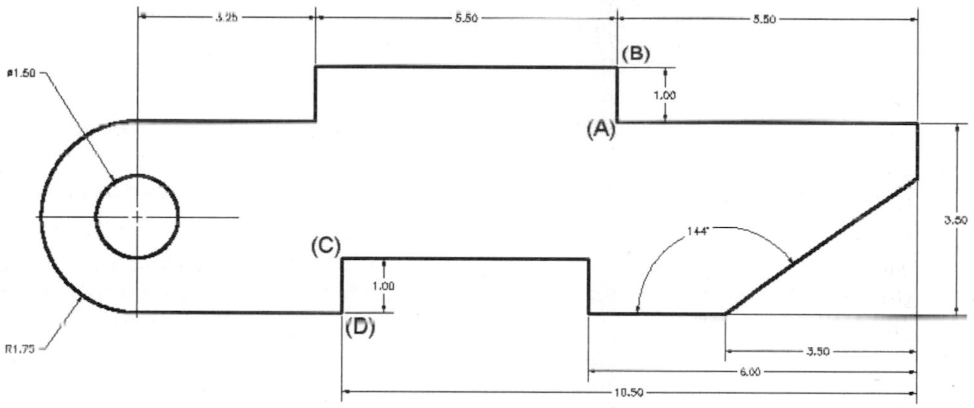

FIGURE 10.76

TUTORIAL EXERCISE: 10_DIMENSION VIEWS.DWG

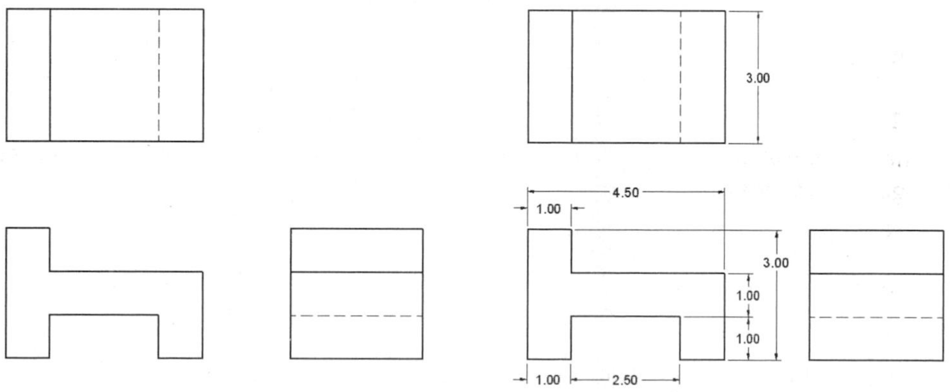

FIGURE 10.77

Purpose

The purpose of this tutorial is to add dimensions to the three-view drawing named 10_Dimension Views.dwg.

System Settings

The drawing in the previous image is already constructed. Dimensions must be added to various views to call out overall distances, in addition to features such as cuts and slots. Be sure the Object Snap modes Endpoint, Center, Intersect, and Extension are set.

Layers

Layers have already been created for this tutorial.

Suggested Commands

Use the DIMLINEAR command for horizontal and vertical dimensions on different views in this drawing.

STEP 1

Open the drawing file 10_Dimension Views.dwg. Linear dimensions are placed identifying the overall length, width, and depth dimensions. The DIMLINEAR command is used to perform this task, as shown in Figure 10.78 (Left).

STEP 2

Detail dimensions that identify cuts and slots are placed. Because these cuts are visible in the Front view, the dimensions are placed there, as shown in Figure 10.78 (Center). The DIMCONTINUE command (DCO) could be used for the second 1.00 dimension.

STEP 3

Once the spaces between the Front, Top, and Right Side views are used up by dimensions, the outer areas are used for placing additional dimensions such as the two horizontal dimensions in Figure 10.78 (Right). Again, use DIMCONTINUE (DCO) for the second dimension. Use grips to adjust dimension text locations if necessary.

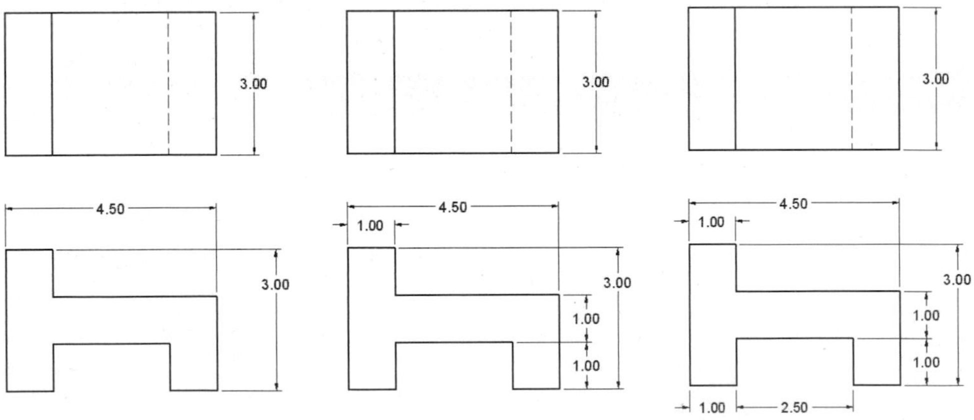

FIGURE 10.78

PROBLEMS FOR CHAPTER 10

DIRECTIONS FOR PROBLEMS 10-1 THROUGH 10-18

1. Open each existing drawing.
2. Use proper techniques to dimension each drawing.
3. Follow the steps involved in completing 10_Dimension Views.dwg.

PROBLEM 10-1

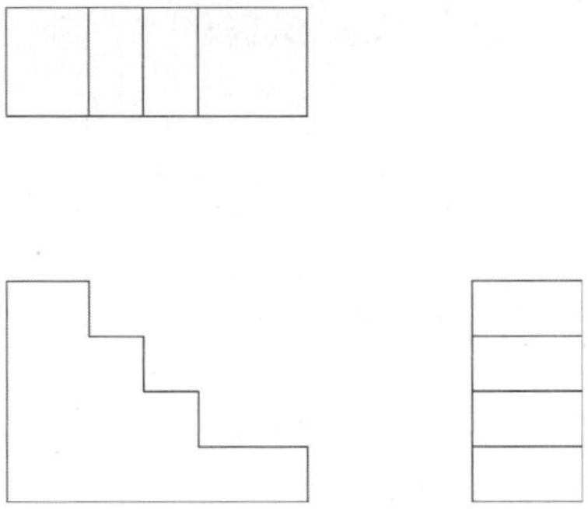

PROBLEM 10-4

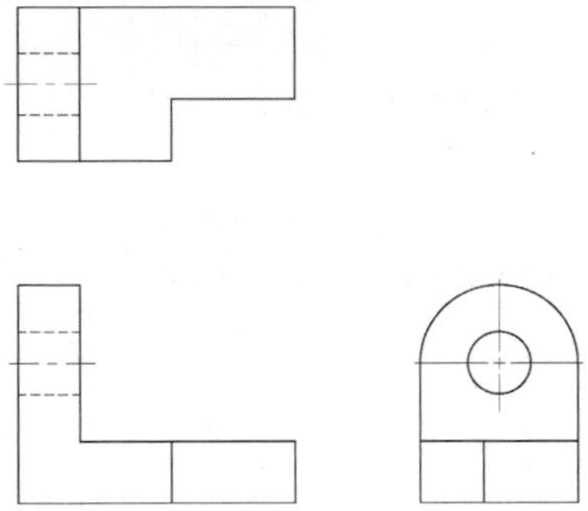

PROBLEM 10-2

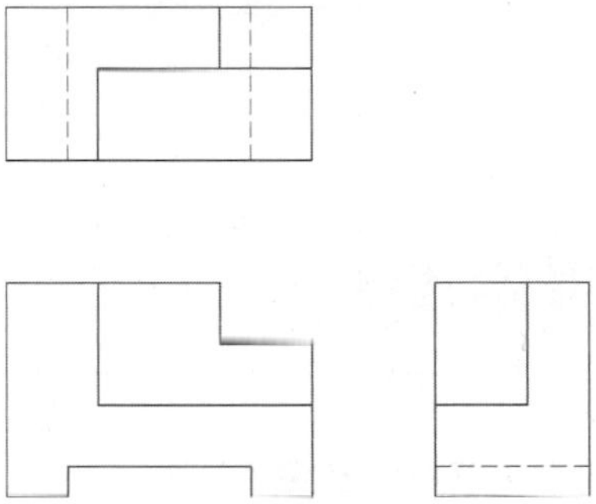

PROBLEM 10-5

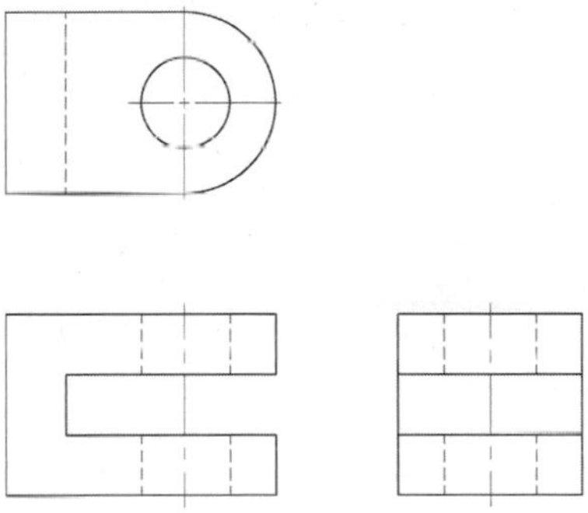

PROBLEM 10-3

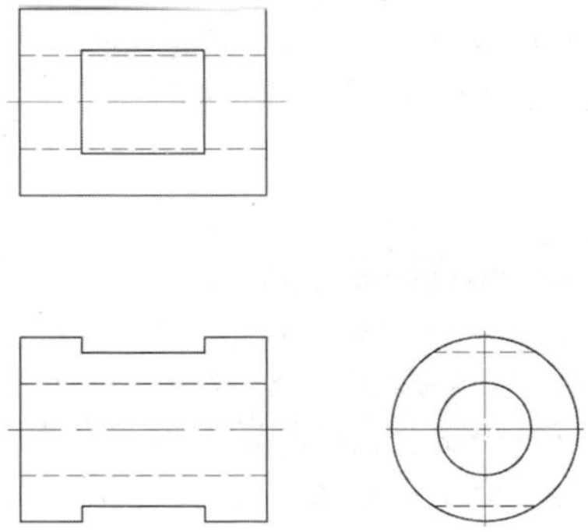

PROBLEM 10-6

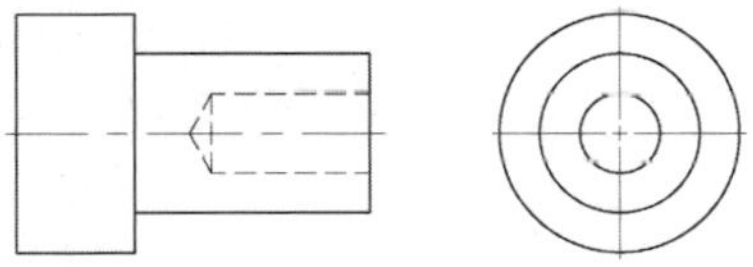

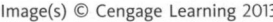

PROBLEM 10-7

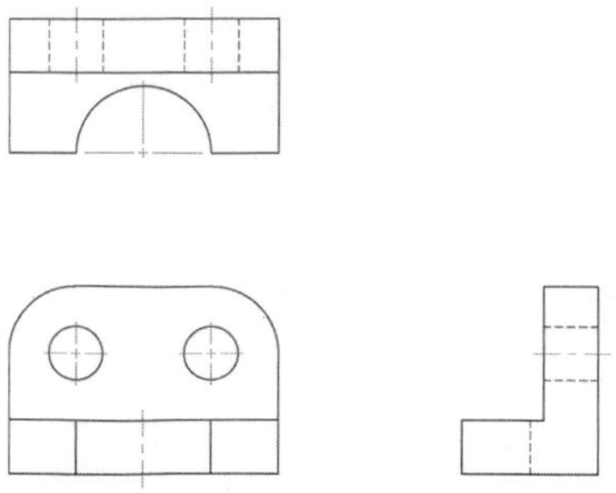

PROBLEM 10-8

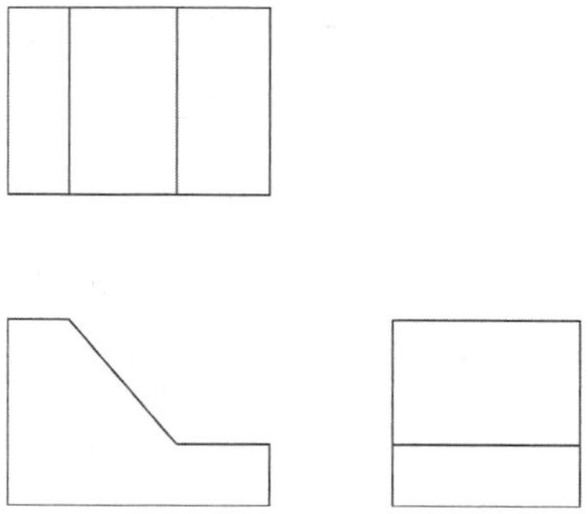

PROBLEM 10-9

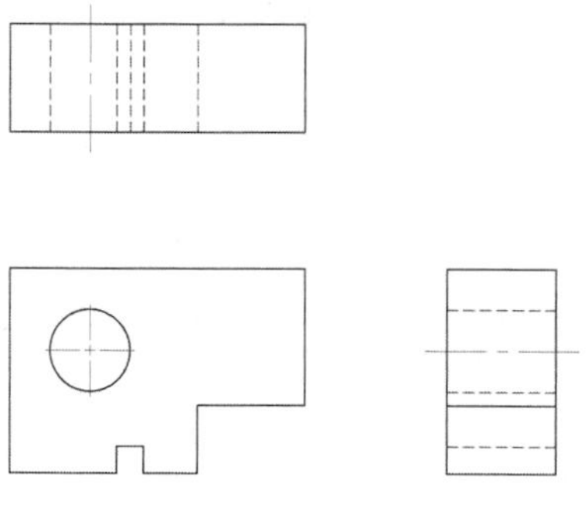

PROBLEM 10-10

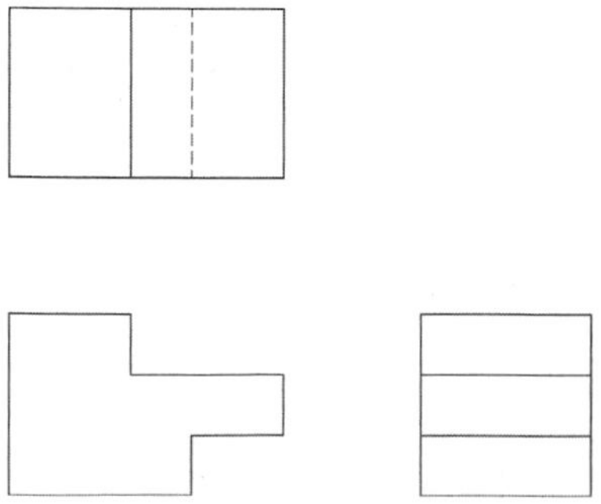

PROBLEM 10-11

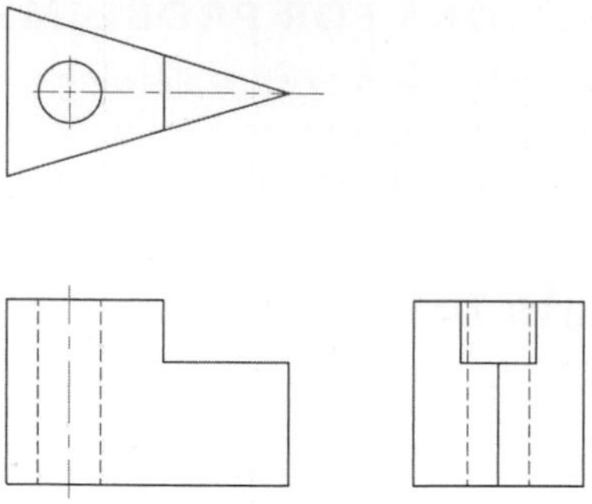

PROBLEM 10-12

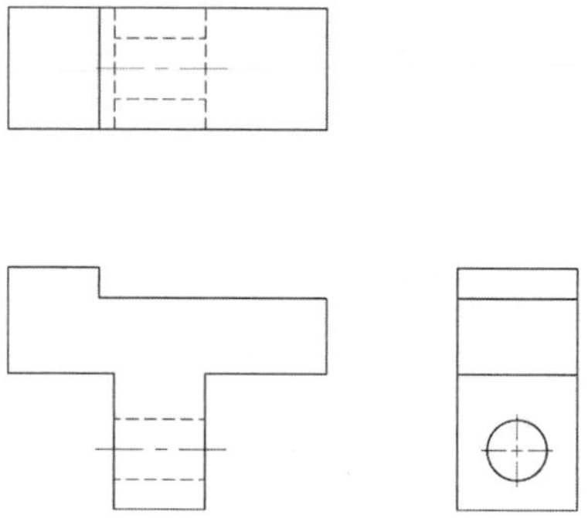

PROBLEM 10-13

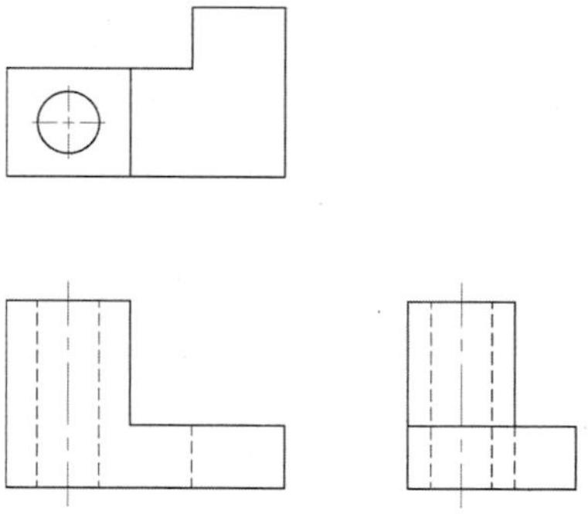

PROBLEM 10-14

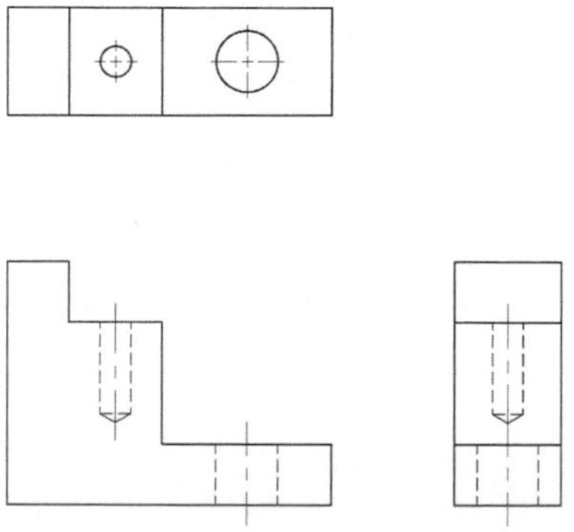

PROBLEM 10-15

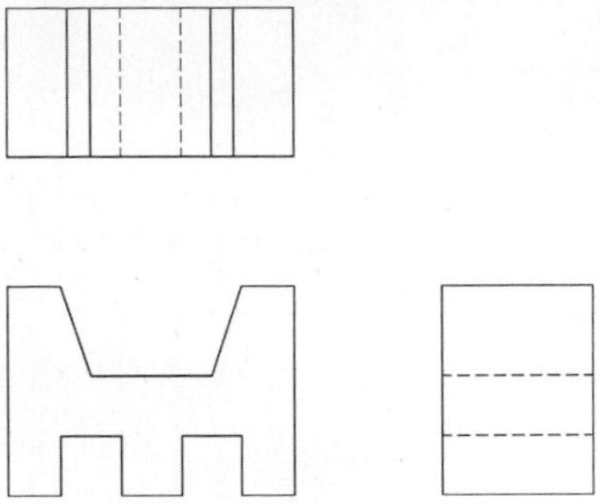

PROBLEM 10-16

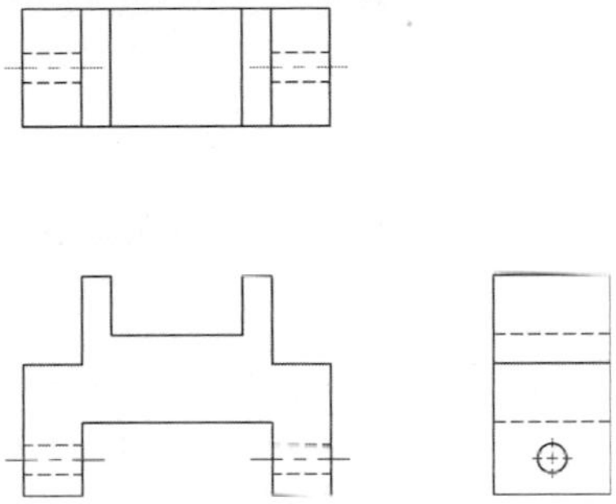

PROBLEM 10-17

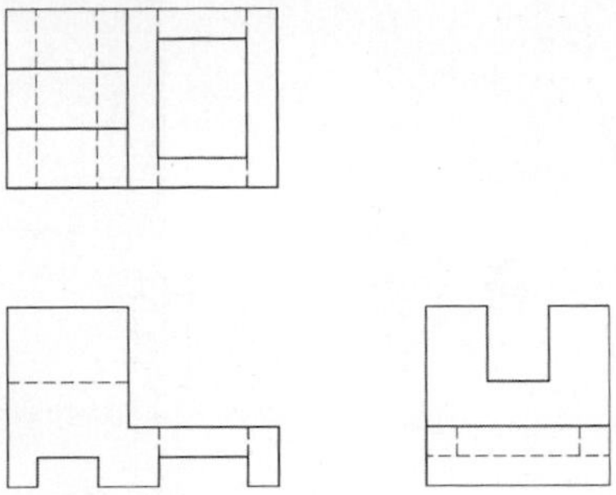

PROBLEM 10-18

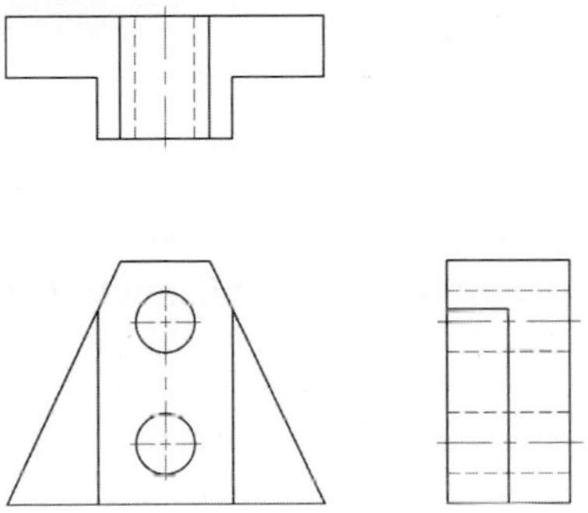

CHAPTER

11

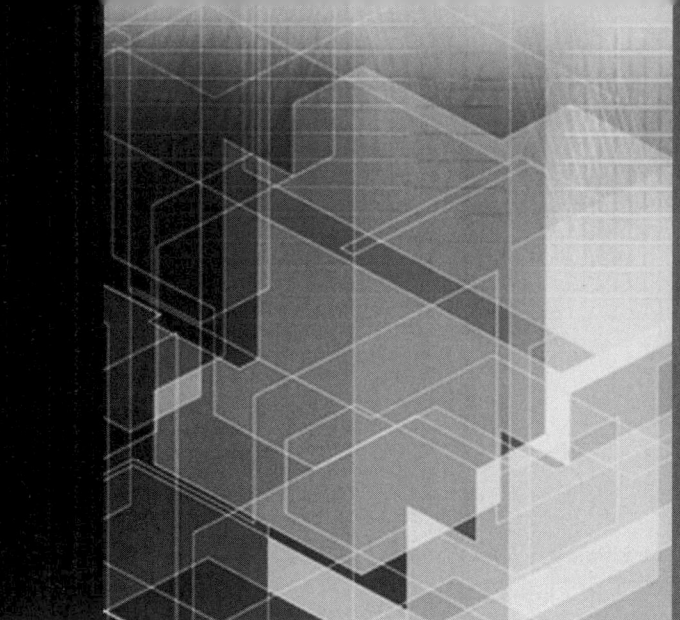

Managing Dimension Styles

Dimensions have different settings that affect how they behave and appear. These settings include the control of the dimension text height, the size and type of arrowhead used, and whether the dimension text is centered in the dimension line or placed above the dimension line. These are but a few of the numerous settings available to you. In fact, some settings are used mainly for architectural applications, while other settings are used only for mechanical applications. As a means of managing these settings, dimension styles are used to group a series of dimension settings under a unique name. These styles can then be assigned to specific dimensions. This chapter will cover in detail the Dimension Style Manager dialog box and the following tabs associated with it: Lines; Symbols and Arrows; Text; Fit; Primary Units; Alternate Units; and Tolerances. Additional topics include dimension style types, overriding a dimension style, modifying the dimension style of an object, and multileader styles.

THE DIMENSION STYLE MANAGER DIALOG BOX

AutoCAD CERTIFIED ASSOCIATE EXAM 2013 — AutoCAD CERTIFIED PROFESSIONAL EXAM 2013

Begin the process of creating a dimension style by choosing a Dimension Style icon from either the Home or Annotate tabs of the Ribbon, as shown in Figure 11.1. Entering the keyboard command `DIMSTYLE` or `D` is another way of accessing a Dimension Style.

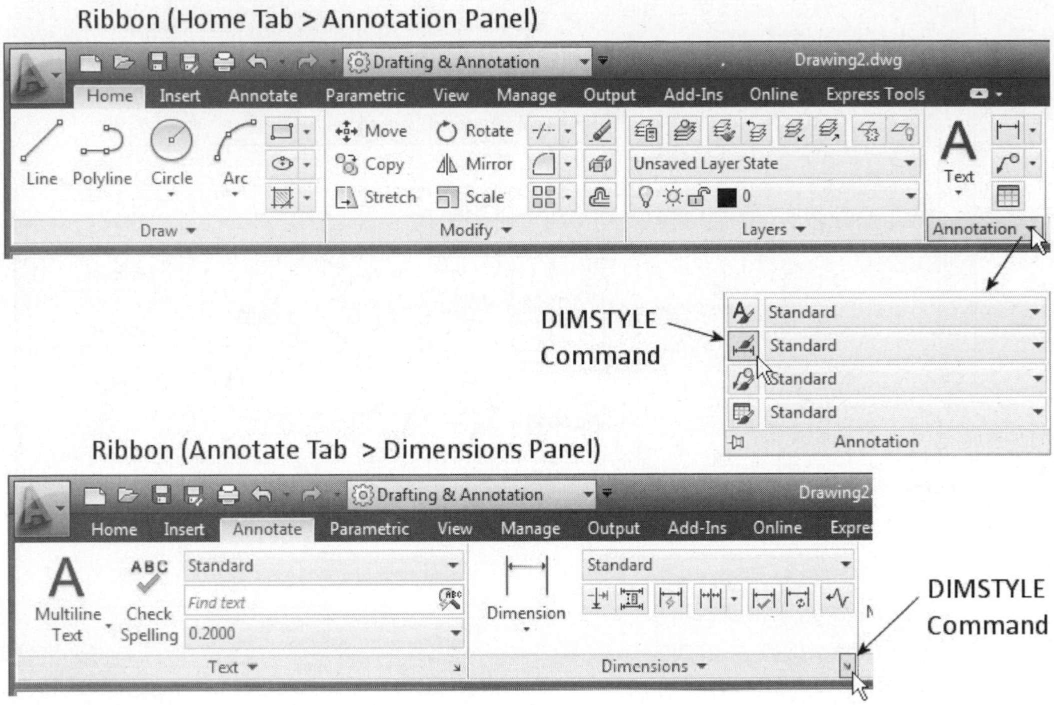

FIGURE 11.1

Activating the DIMSTYLE command utilizing one of the methods shown will launch the Dimension Style Manager dialog box, as shown in Figure 11.2. The available styles are listed on the left side of the dialog box. The current dimension style is listed as Standard. There is also an Annotative dimension style present; this particular style will be discussed in detail in Chapter 19. These dimension styles are automatically available when you create any new drawing. In the middle of the dialog box is a preview area that displays how various dimensions will appear based on the current value of the dimension settings. On the right side of the dialog box, a series of buttons are available to set a dimension style to current, create a new dimension style, make modifications to an existing dimension style, create or modify an override style, and compare the differences and similarities of two dimension styles.

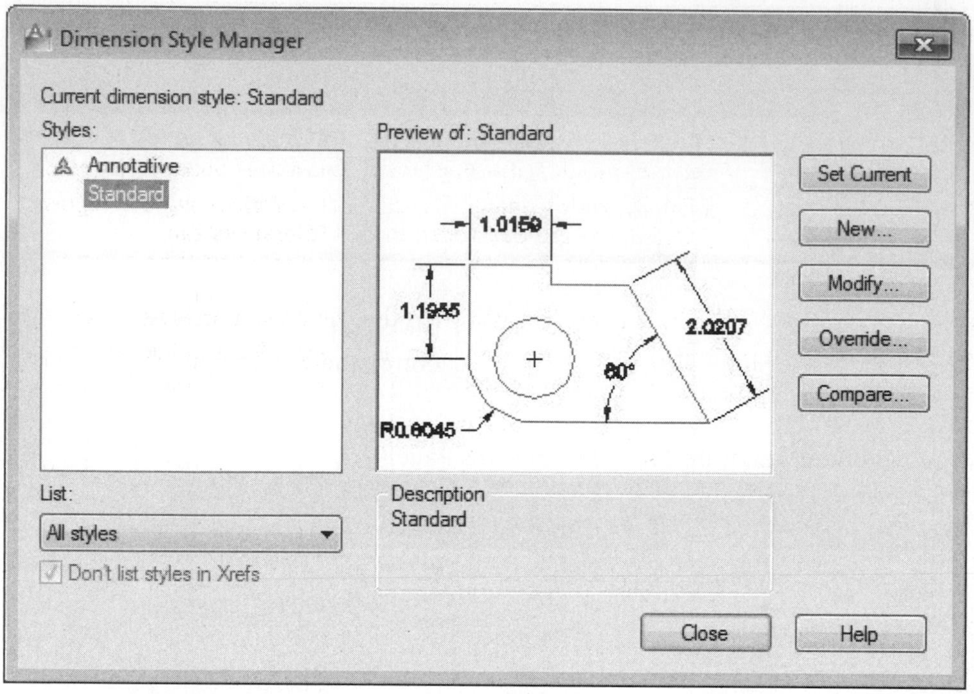

FIGURE 11.2

NOTE

When starting a drawing with a metric template an ISO-25 dimension style will also be listed and will be set current.

To create a new dimension style, click the New... button. This activates the Create New Dimension Style dialog box, as shown at the top of Figure 11.3. Enter a new name such as Mechanical in the New Style Name area. Then click the Continue button. This takes you to the New Dimension Style: Mechanical dialog box, as shown at the bottom of Figure 11.3, where a number of tabs organize all the dimension settings.

Refer to the following table for a brief description of each tab located in the Dimension Style Manager dialog box:

Tab	Description
Lines	The Lines tab deals with settings that control dimension and extension lines.
Symbols and Arrows	The Lines and Arrows tab controls arrowheads, center marks, dimension breaks, arc length, and jog dimension settings.
Text	The Text tab contains the settings that control the appearance, placement, and alignment of dimension text.
Fit	The Fit tab contains various fit options for placing dimension text and arrows when there isn't enough room to fit them between the extension lines. This tab also controls the scale for dimension features.

(Continued)

Tab	Description
Primary Units	You control the linear and angular units used in dimensioning through the Primary Units tab.
Alternate Units	If you need to display primary and secondary units in the same drawing, the Alternate Units tab is used.
Tolerances	Finally, for mechanical applications, various ways to show tolerances are controlled in the Tolerances tab.

When you make changes to any of the settings under the tabs, a preview image provided in each tab updates to show these changes. This provides a quick way of previewing how your dimensions will appear in your drawing.

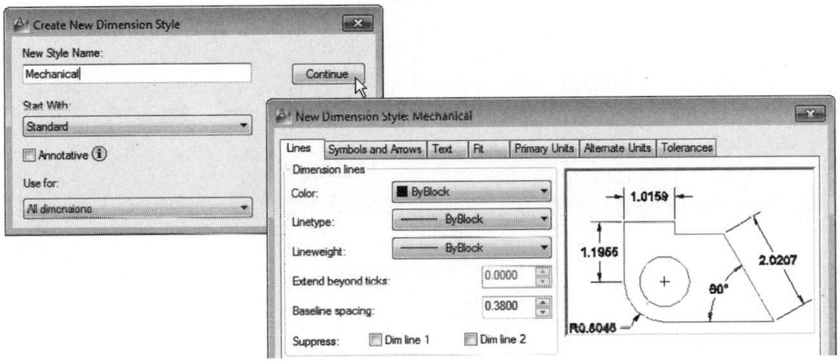

FIGURE 11.3

When you are finished making changes, click the OK button. This returns you to the Dimension Style Manager dialog box shown in Figure 11.4. Notice that the Mechanical dimension style has been added to the list of styles. Any changes made in the tabs are automatically saved to the dimension style. Verify Mechanical is the current dimension style, if not, highlight it in the list and then pick the Set Current button. Clicking the Close button returns you to your drawing.

Let's get a closer look at all the tabs and their settings. Type the letter D to reopen the Dimension Style Manager. Make sure Mechanical is highlighted and click the Modify ... button.

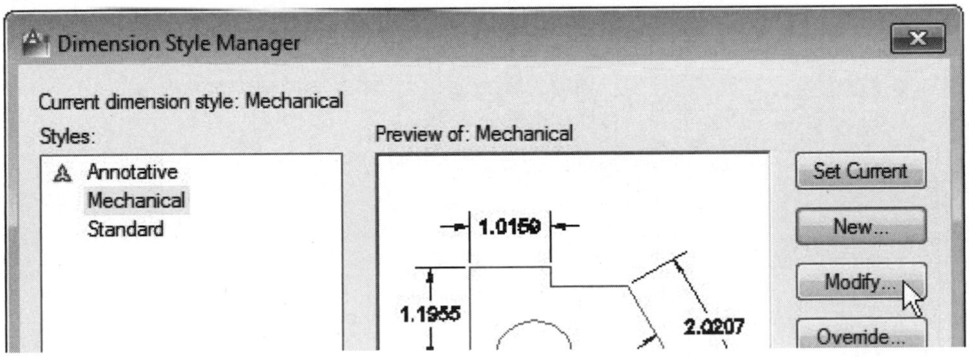

FIGURE 11.4

THE LINES TAB

The Modify Dimension Style: Mechanical dialog box in Figure 11.5 displays the Lines tab, which will now be discussed in greater detail. This tab consists of two main areas dealing with dimension lines and extension lines.

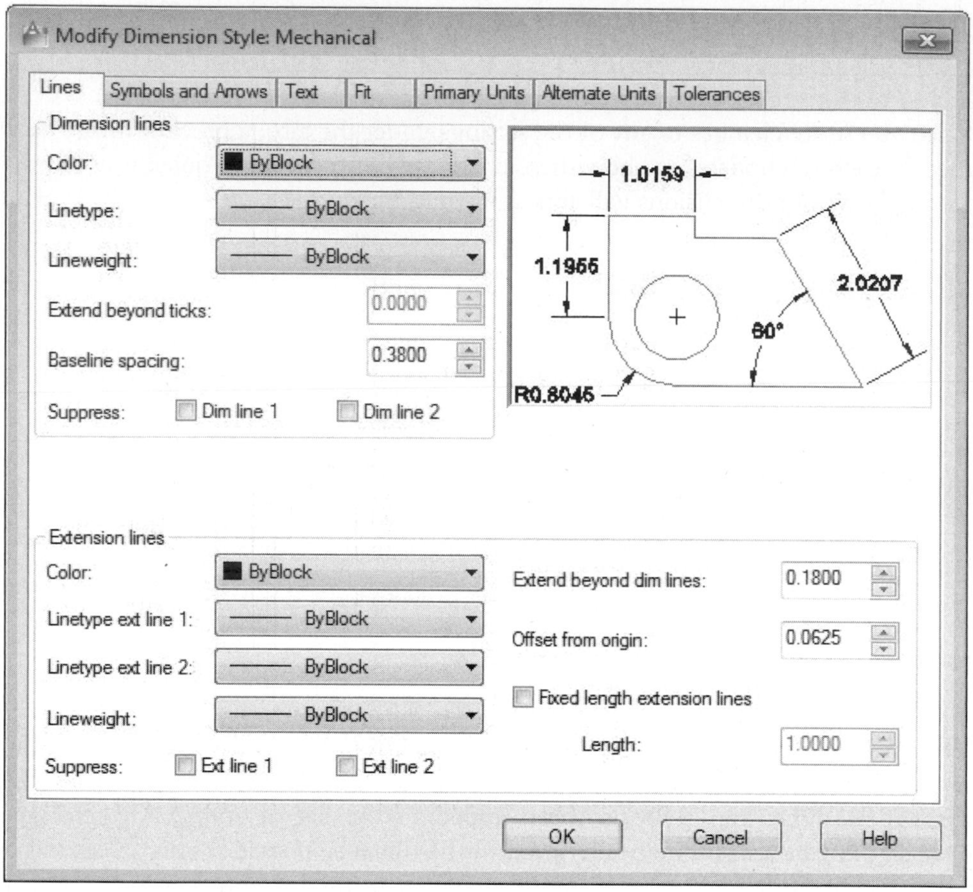

FIGURE 11.5

Dimension Line Settings

Use the Dimension lines area, shown in Figure 11.6, to control the color, lineweight, visibility, and spacing of the dimension line.

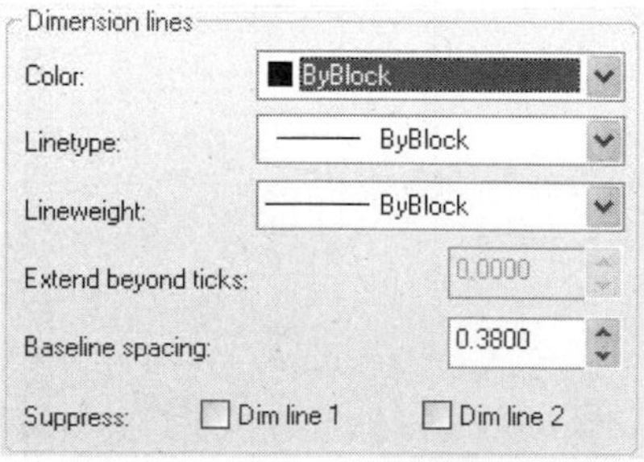

FIGURE 11.6

The Baseline spacing setting in the Dimension lines area controls the spacing of baseline dimensions, as shown in Figure 11.7 (Left). When utilizing the DIMBASELINE command (or DBA for short), the automatic spacing provided is determined by this setting.

By default, dimension line suppression is turned off. To turn on suppression of dimension lines, place a check in the Dim line 1 or Dim line 2 box next to Suppress. This operation turns off the display of dimension lines for all dimensions placed under this dimension style. See the illustration in Figure 11.7 (Right). This may be beneficial where tight spaces require that only the dimension text be placed.

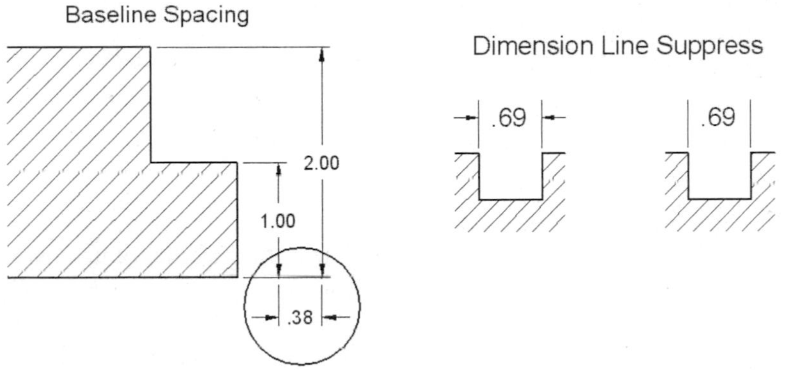

FIGURE 11.7

Extension Line Settings

The Extension lines area controls the color, the distance the extension extends past the arrowhead, the distance from the object to the end of the extension line, and the visibility of extension lines, as shown Figure 11.8. Additional controls allow you to set a different linetype for extension lines and provide the ability to use a fixed length value for all extension lines.

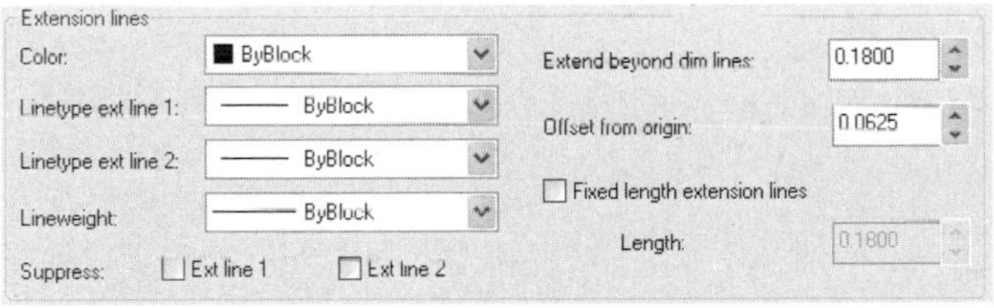

FIGURE 11.8

The Extend beyond dim lines setting controls how far the extension extends past the arrowhead or dimension line, as shown in Figure 11.9 (Left). By default, a value of 0.18 is assigned to this setting.

The Offset from origin setting controls how far away from the object the extension line will start, as shown in Figure 11.9 (Right). By default, a value of .06 is assigned to this setting.

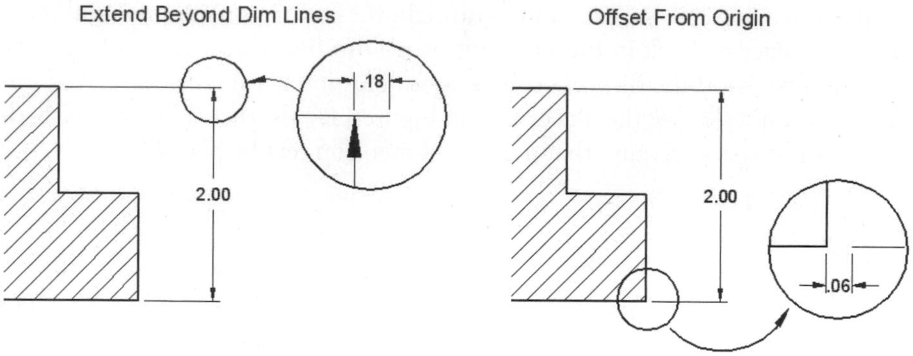

FIGURE 11.9

The Suppress Ext line 1 and Ext line 2 check boxes control the visibility of extension lines. They are useful when you need to locate a dimension inside an object, but want to avoid placing the extension line on top of the object line. Placing a check in the Ext line 1 box of Suppress turns off the first extension line. Similarly, when you place a check in the Ext line 2 box of Suppress, the second extension line is turned off. Suppressing extension lines by checking these boxes suppresses all extension lines for dimensions placed under this dimension style. Study the examples in Figure 11.10 to get a better idea about how suppression of extension lines operates.

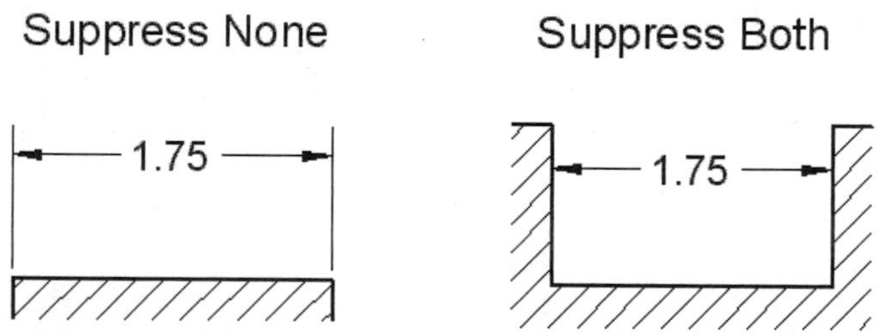

FIGURE 11.10

Extension Linetypes

You can assign linetypes presently loaded in your drawing to extension lines, as illustrated in Figure 11.11. In this example, the Center linetype is assigned to the extension lines. This provides an efficient means of applying custom linetypes to dimensions where appropriate.

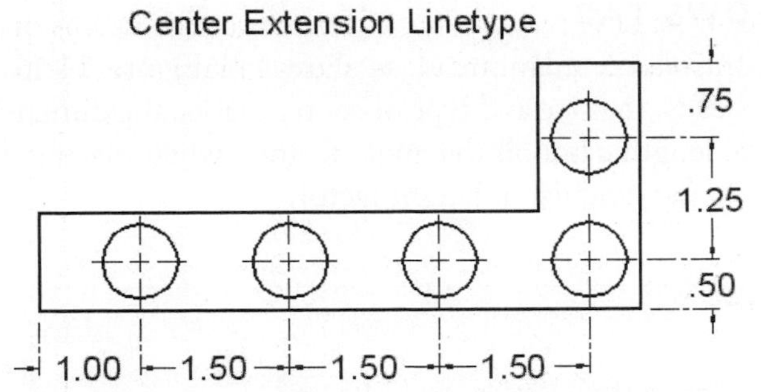

FIGURE 11.11

Fixed Length Extension Lines

Fixed length extension lines is also a setting found under the Extension lines area of the Lines tab. It sets a user-defined length for all extension lines. In the illustration in Figure 11.12 (Left), extension lines of varying lengths are displayed. All extension lines begin by default .0625 units away from the object. Illustrated on the right is an example of how extension lines are displayed at a fixed length of 0.25 units. Notice that all extension lines are the same length in this example. This value is calculated from where the endpoint of the arrow intersects with the extension to the bottom of the extension line. As you can see, this results in very short extension lines that are all the same length.

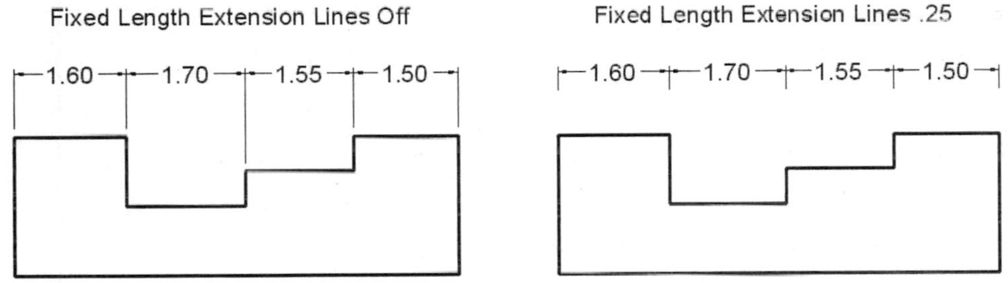

FIGURE 11.12

THE SYMBOLS AND ARROWS TAB

The Symbols and Arrows tab controls six main areas, as shown in Figure 11.13, namely, the size and type of arrowheads, the size and type of center marks, the dimension break size, the style of the arc length symbol, the angle formed when placing a radius jog dimension, and the linear jog dimension height factor.

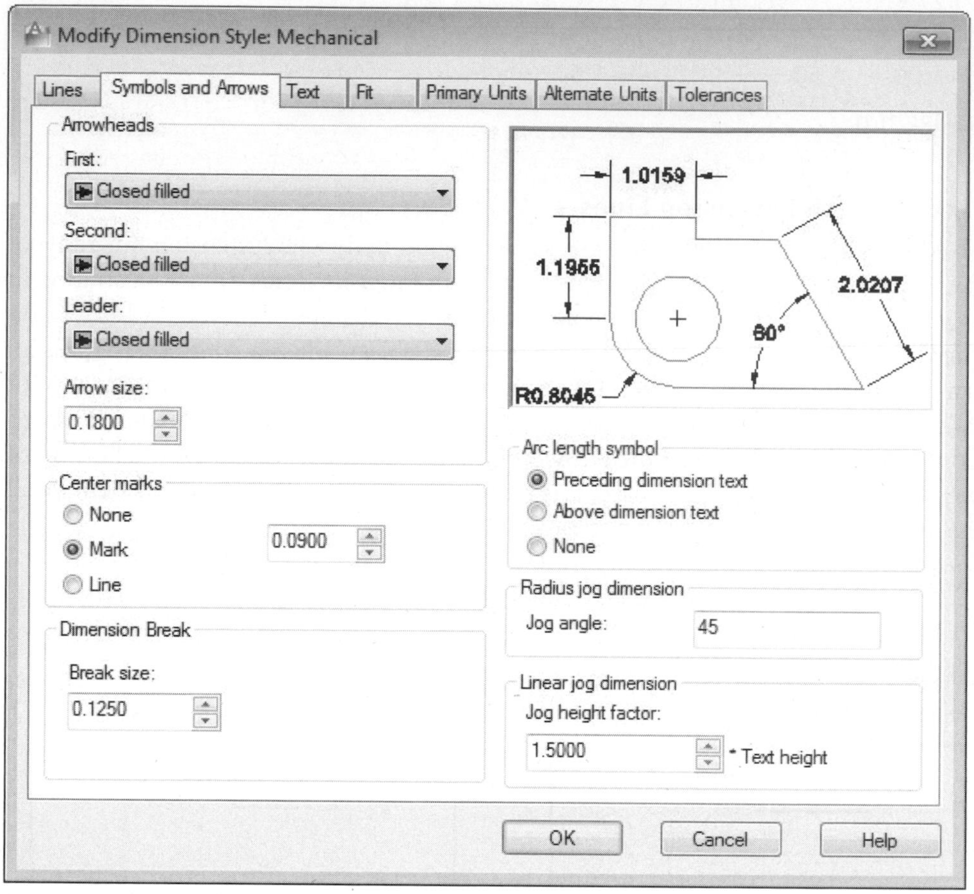

FIGURE 11.13

Arrowhead Settings

Use the Arrowheads area to control the type of arrowhead terminator used for dimension lines and leaders, as shown in Figure 11.14. This dialog box also controls the size of the arrowhead.

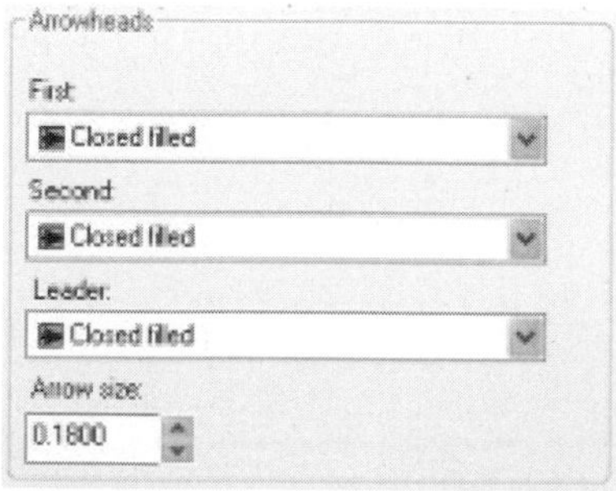

FIGURE 11.14

Clicking on the First box displays a number of arrowhead terminators. Choose the desired terminator from the list, as shown in Figure 11.15 (Left). When you make a selection in the First box, the Second box automatically updates to the same selection made in the First box. If you choose an arrowhead from the Second field, the first and second arrowheads may be different at opposite ends of the dimension line; this is desired in some applications. Choosing a terminator in the Leader box displays the arrowhead that will be used whenever you place a leader.

> **NOTE**
>
> **Leaders placed with the QLEADER command are controlled by the Dimension Style Manager. Leaders placed with the MLEADER command are controlled by the Multileader Style Manager, which is discussed at the end of this chapter.**

Illustrated in Figure 11.15 (Right) is the complete set of arrowheads, along with their names. The last arrow type in Figure 11.15 (Left) is a User Arrow, which allows you to create your own custom arrowhead.

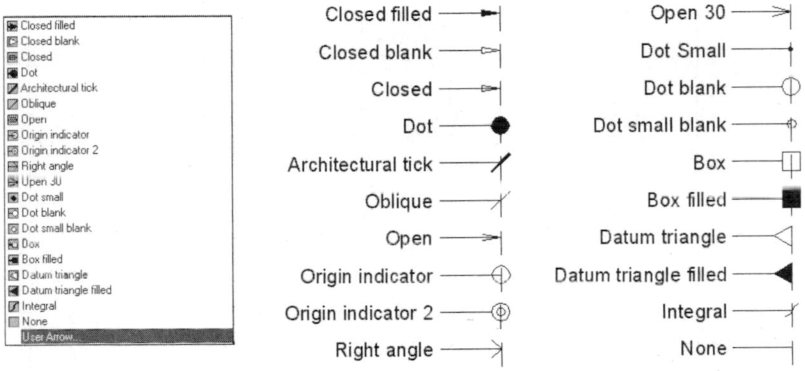

FIGURE 11.15

Use the Arrow size setting to control the size of the arrowhead terminator. By default, the arrow size is set to 0.18 units, as shown in Figure 11.16.

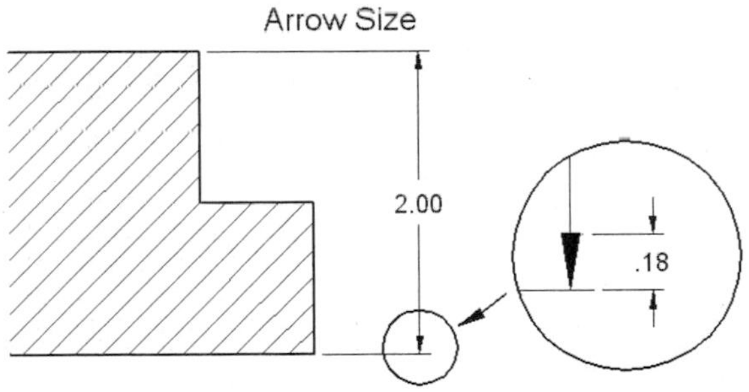

FIGURE 11.16

Center Mark Settings

The Center marks for circles area allows you to control the type of center marker used when identifying the centers of circles and arcs. You can make changes to these settings by clicking the three center mark modes: None, Mark, or Line, as shown in

Figure 11.17 (Left). The Size box controls the size of the small plus mark (+) and the distance the line extends past the circle or arc. If Mark or Line is chosen, you can place center marks in your drawing by using the DIMCENTER, DIMRADIUS, or DIMDIAMETER commands.

The three types of center marks are illustrated in Figure 11.17 (Right). The Mark option places a plus mark in the center of the circle. The Line option places the plus mark and extends a line past the edge of the circle. The None option will not display a center mark.

TIP

To better control center marks, try placing them with the DIMCENTER (DCE) command. Set the center mark mode to None when using the DIMRADIUS and DIMDIAMETER commands. The center mark lines can then be modified (erased or stretched) since they are not part of the radius or diameter dimension. Creating styles for dimension types is a good way of setting this up—discussed later in this chapter.

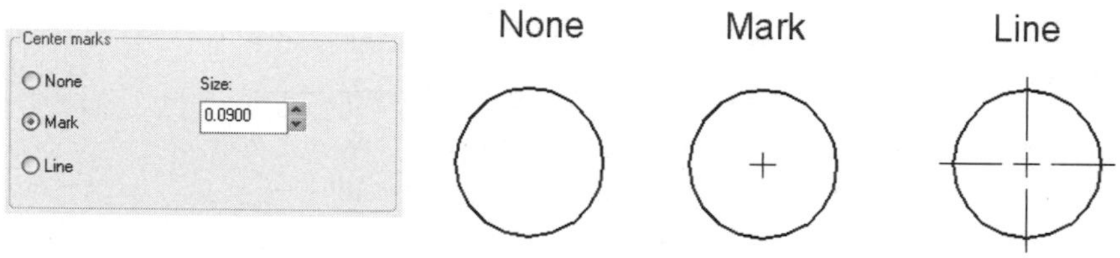

FIGURE 11.17

Dimension Break Settings

Use this area to set the distance when breaking extension or dimension lines with the DIMBREAK command. Figure 11.18 compares the default break distance of 0.125 with a new break distance of 0.25 and how the break is affected.

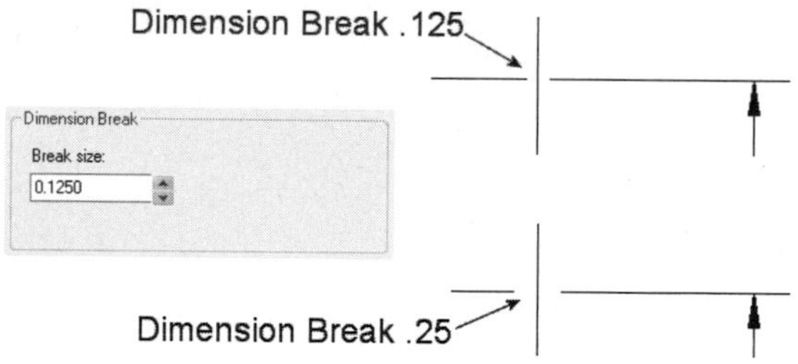

FIGURE 11.18

Controlling Dimension Arc Length Symbol

When placing an arc length dimension (DIMARC command), you can control where the arc symbol is placed. In Figure 11.19, notice the three settings, namely, the arc symbol preceding the dimension text, the arc symbol placed above the dimension text, or no arc symbol displayed.

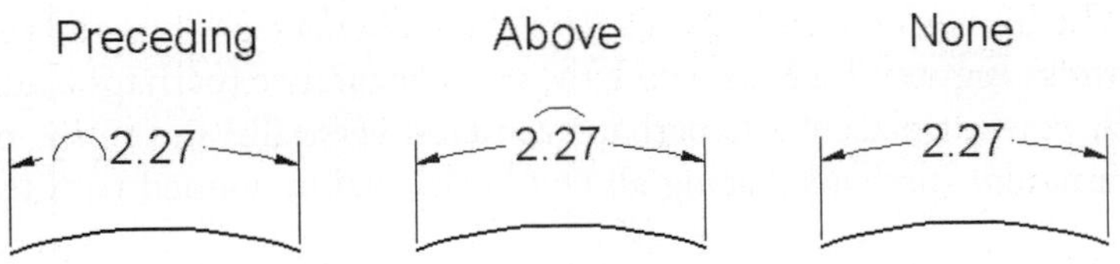

FIGURE 11.19

Controlling the Radius Jog Dimension Angle

You can control the angle of a jog dimension (DIMJOGGED command), as shown in Figure 11.20. By default, a 45° angle defines the jog, as shown on the left in Figure 11.20. Notice the effects of entering a 30° or 70° angle in the other examples in this image.

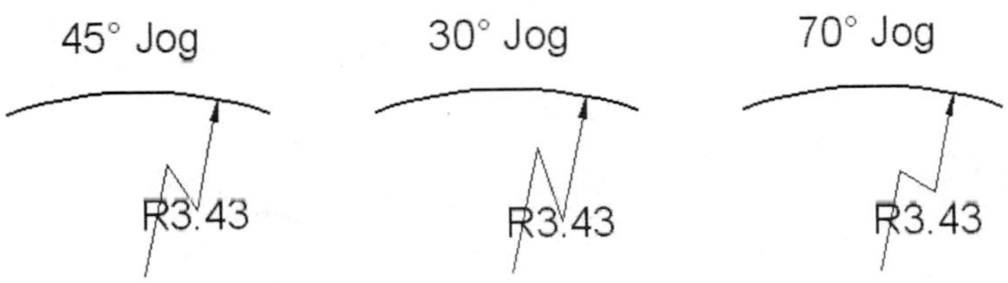

FIGURE 11.20

Controlling the Linear Jog Dimension Height Factor

This area is used to set a multiplication factor for jogs created with the DIMJOGLINE command. Jog height is determined by multiplying this factor by the current dimension text height. Figure 11.21 compares the default 1.50 jog height factor with that of a 3.00 jog height factor. Notice that the lower jog is twice the size of the upper jog.

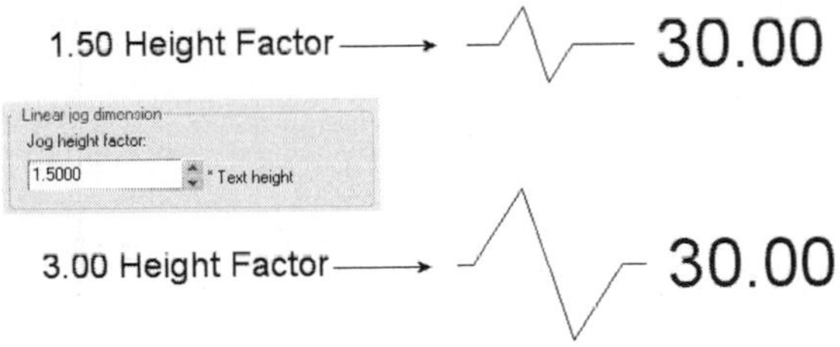

FIGURE 11.21

THE TEXT TAB

Use the Text tab shown in Figure 11.22 to change the text appearance (perhaps a new text style and height), the text placement (perhaps centered vertically and horizontally), and the text alignment (perhaps placing all text horizontal or aligned with the dimension line).

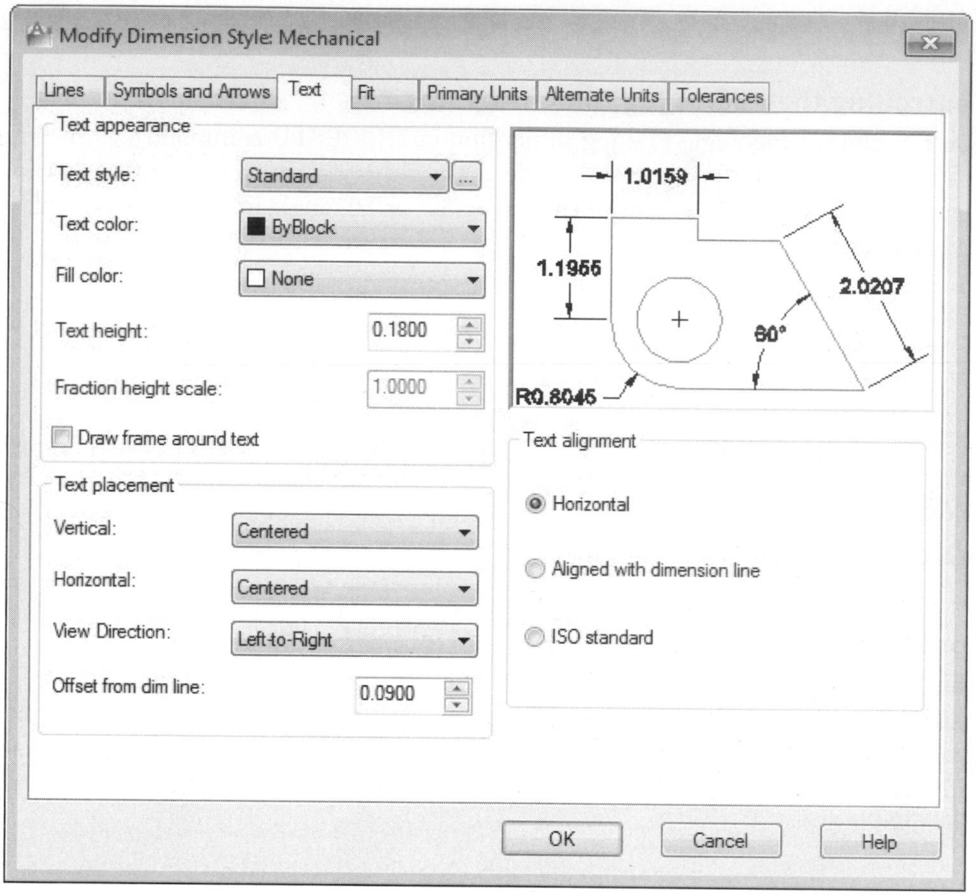

FIGURE 11.22

Text Appearance

The Text height setting controls the size of the dimension text, as shown in Figure 11.23 (Right). By default, a value of 0.18 is assigned to this setting.

TIP

Verify the text style assigned in the text appearance area has a height of .00. If the text style has a size other than zero specified, it will override the setting for the height in the dimension style.

Placing a check in the Draw frame around text box draws a rectangular box around all dimensions, as shown in Figure 11.23 (Left). This is used in geometric dimensioning and tolerancing to identify basic dimensions whose tolerances are associated with feature control frames.

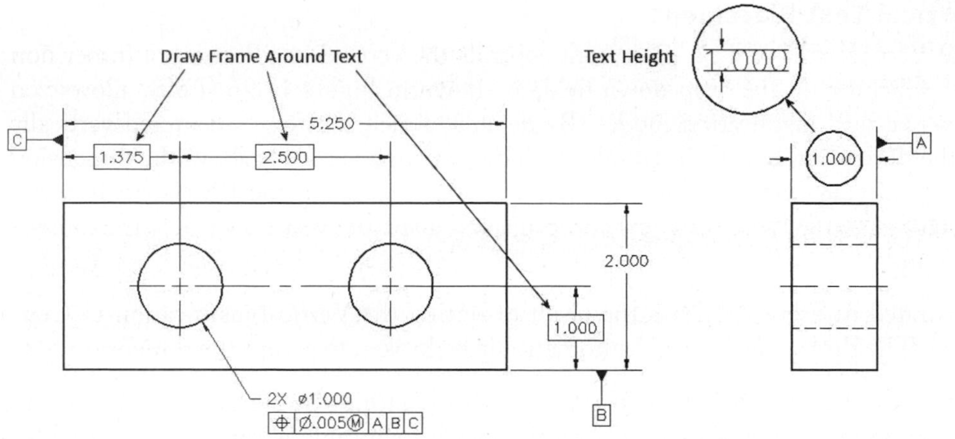

FIGURE 11.23

If the primary dimension units are set to architectural or fractional, the Fraction height scale activates. Changing this value affects the height of fractions that appear in the dimension. In Figure 11.24, a fractional height scale of 0.5000 will make the fractions as tall as the primary dimension number in the preview window. Decreasing the text height of fractions, however, can make them hard to read; therefore, care should be exercised.

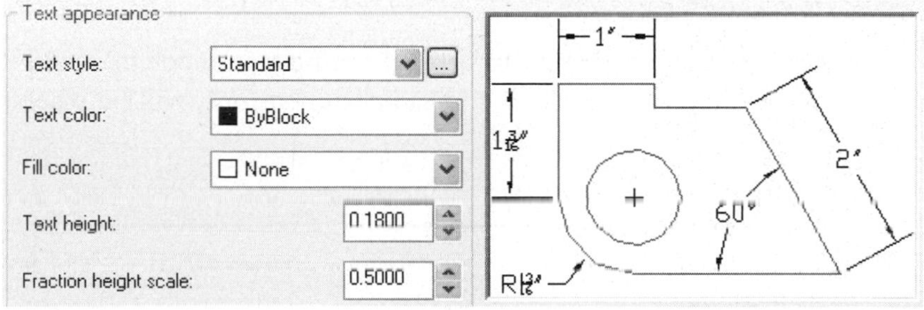

FIGURE 11.24

Text Placement

The Text placement area, shown in Figure 11.25, allows you to control the vertical and horizontal placement of dimension text. You can also set an offset distance from the dimension line for the dimension text.

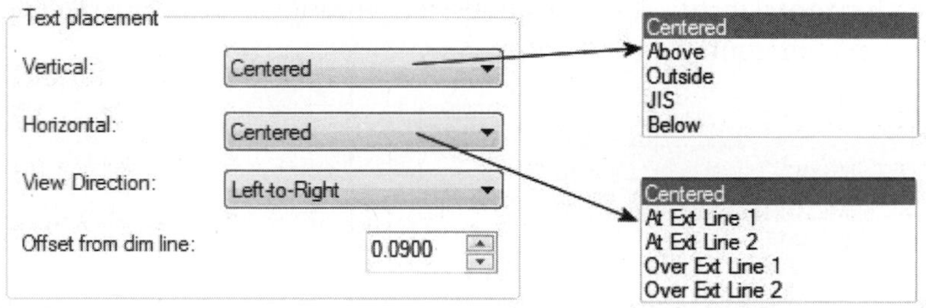

FIGURE 11.25

Vertical Text Placement

The Vertical area of Text placement controls the vertical justification of dimension text. Clicking on the drop-down field, as shown in Figure 11.26 (Left), allows you to set vertical justification modes. By default, dimension text is centered vertically in the dimension line. Other modes include justifying vertically above the dimension line, justifying vertically outside the dimension line, using the JIS (Japan International Standard) for placing text vertically, and justifying vertically below the dimension line.

Illustrated in Figure 11.26 is the result of setting the Vertical justification to Centered. The dimension line will automatically be broken to accept the dimension text.

Also illustrated in Figure 11.26 is the result of setting the Vertical justification to Above. Here the text is placed directly above a continuous dimension line. This mode is popular for architectural applications. Finally, the results for setting the Vertical justification to Below, is illustrated in Figure 11.26.

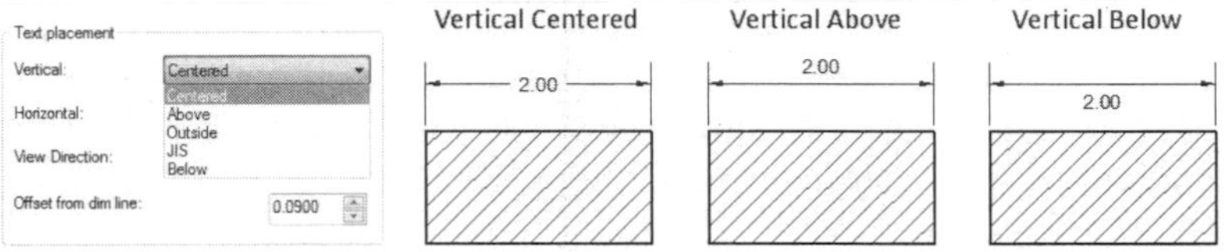

FIGURE 11.26

Figure 11.27 illustrates the result of setting the Vertical justification mode to outside. All text, including that contained in angular and radial dimensions, will be placed outside the dimension lines and leaders.

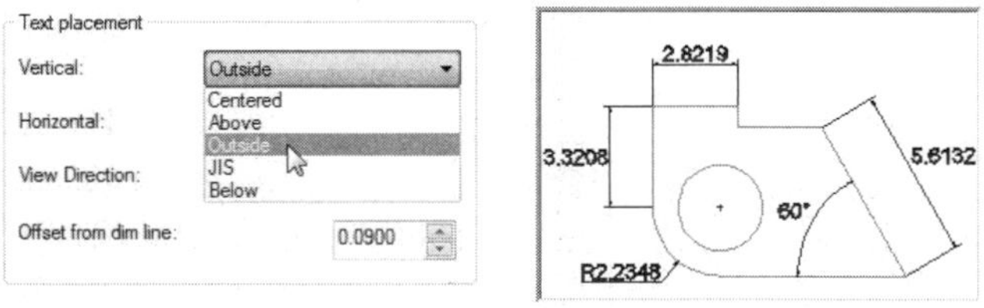

FIGURE 11.27

Horizontal Text Placement

At times, dimension text needs to be better located in the horizontal direction; this is the purpose of the Horizontal justification area. Illustrated in Figure 11.28 are the five modes of justifying text horizontally.

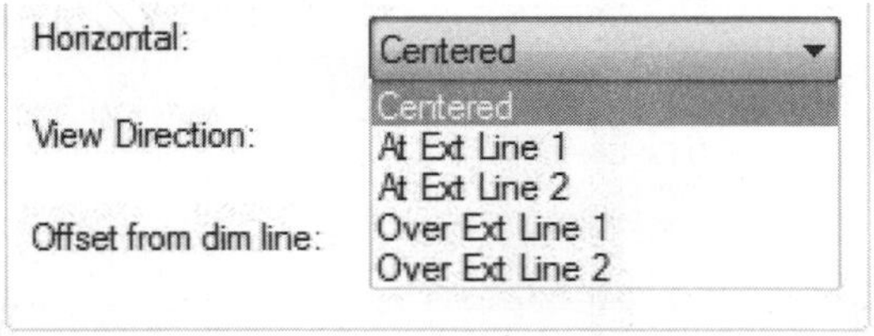

FIGURE 11.28

Selecting the Centered option of the Horizontal justification area displays the dimension text, as shown in Figure 11.29 (Left). This is the default setting because it is the most commonly used text justification mode in dimensioning.

Clicking the At Ext Line 1 option displays the dimension text, as shown in Figure 11.29 (Center), where the dimension text slides close to the first extension line. This option is sometimes used to stagger the position of text in such a way that it will not interfere with other dimension text. Notice the corresponding option to have the text positioned nearer to the second extension line.

Clicking the Over Ext Line 1 option displays the dimension text parallel to and over the first extension line, as shown in Figure 11.29 (Right). Notice the corresponding option to position dimension text over the second extension line.

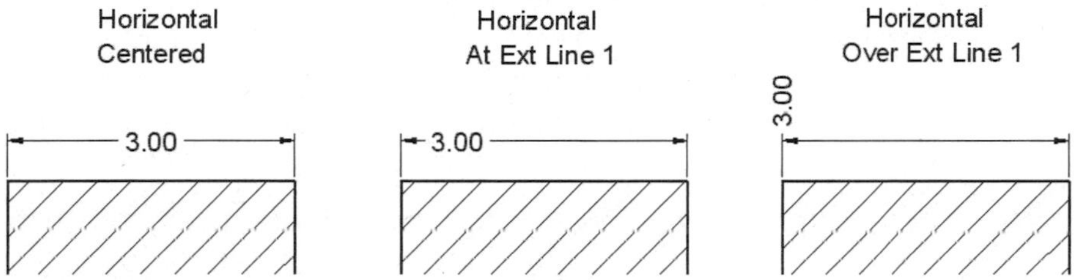

FIGURE 11.29

The next item deals with the View Direction for dimension text. The choices are Left-to-Right or Right-to-Left. It is unlikely that you will need to change this from the default viewing direction which is Left-to-Right. Text should appear such that it can be read from the bottom or right side of a drawing. Text with a Right-to-left setting is displayed such that the text is read from the top (upside down) and left side of the drawing. The last item of the Text placement area deals with setting an offset distance from the dimension line. When you place dimensions, a gap is established between the inside ends of the dimension lines and the dimension text. Entering different values for this setting changes the size of this gap. Study the examples in Figure 11.30 in the middle and on the right, which have different text offset settings. Entering a value of zero (0) forces the dimension lines to touch the edge of the dimension text. Negative values are not supported.

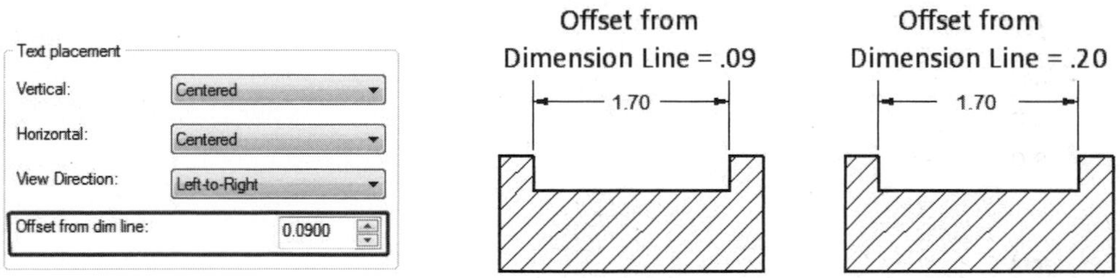

FIGURE 11.30

Text Alignment

Use the Text alignment area shown in Figure 11.31 (Left) to control the orientation of dimension text. Text can be placed either horizontally or parallel (aligned) to the edge of the object being dimensioned. Click on the appropriate radio button to turn the desired text alignment mode on.

If the Horizontal radio button is clicked, all text will be read horizontally, as shown in Figure 11.31 (Center). This includes text located inside and outside the extension lines.

Clicking the Aligned with dimension line radio button displays the alignment results shown in Figure 11.31 (Right). Here all text is read parallel to the edge being dimensioned. Not only will vertical dimensions align the text vertically, but the 2.06 dimension used to dimension the incline is also parallel to the edge. Care should be exercised when using this setting because some dimension text may be improperly orientated; it is considered poor practice to have to view a drawing from the left side or upside down to read text. An ISO (International Standards Organization) standard radio button is also available. This setting is similar to the aligned mode with the exception that radial and diameter dimension text is placed horizontally.

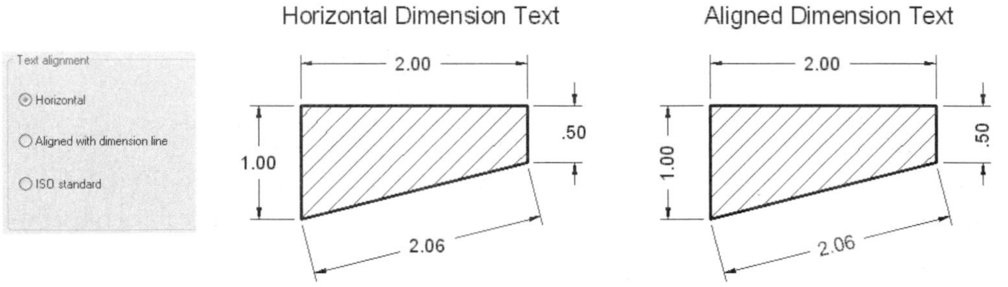

FIGURE 11.31

THE FIT TAB

Use the Fit tab, as shown in Figure 11.32, to control how text and/or arrows are displayed if there isn't enough room to place both inside the extension lines. Settings are also available to place dimension text beside or over the dimension with or without a leader line. The Scale for dimension features area of this tab is very important and is used to size dimensions to the drawing scale or scales being used. You can even fine-tune the placement of text manually and force the dimension line to be drawn between extension lines.

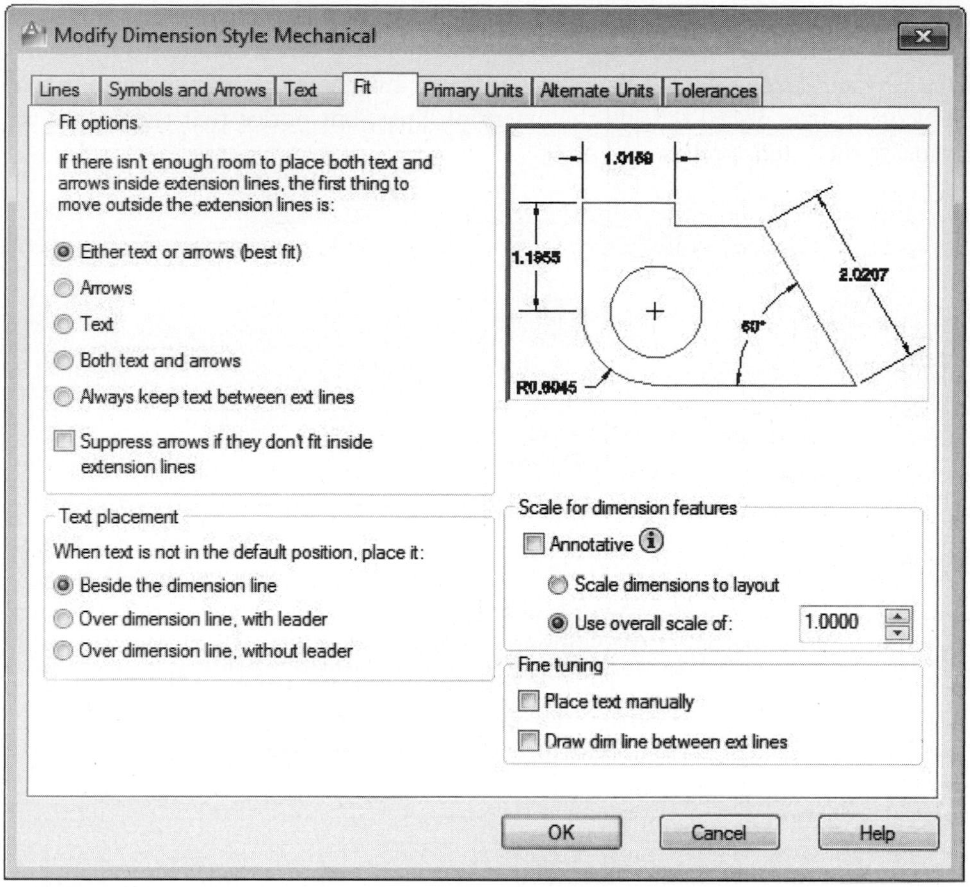

FIGURE 11.32

Fit Options

The Fit options area, shown in Figure 11.33 (Left), has the radio button set for Either text or arrows (best fit). If insufficient room exists between extension lines, AutoCAD will decide to move either text or arrows outside the extension lines. It will determine the item that fits the best. This setting is illustrated in the preview area in Figure 11.33 (Right). Although the preview image appears identical when you click the radio button for Arrows, this setting tells AutoCAD to move the arrowheads outside the extension lines first if there isn't enough room to fit both text and arrows. This leaves the text inside if there is enough room.

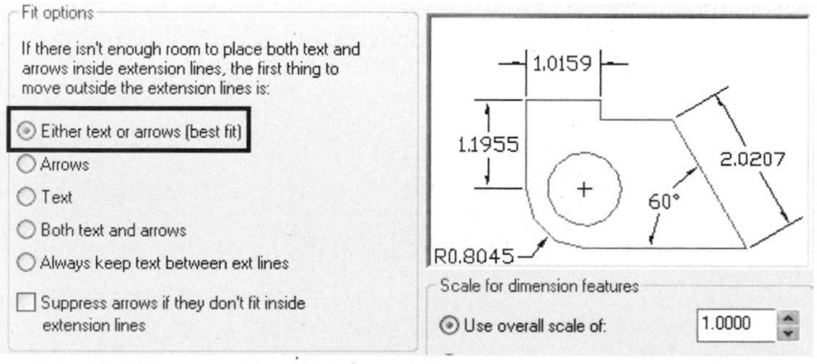

FIGURE 11.33

Clicking the Text radio button of the Fit options area updates the preview image, as shown in Figure 11.34. Here you are electing to move the dimension text outside first if the text and arrows do not fit. Since the value 1.0159 is the only dimension that does not fit, it is placed outside the extension lines, but notice that the arrows are left inside the extension lines.

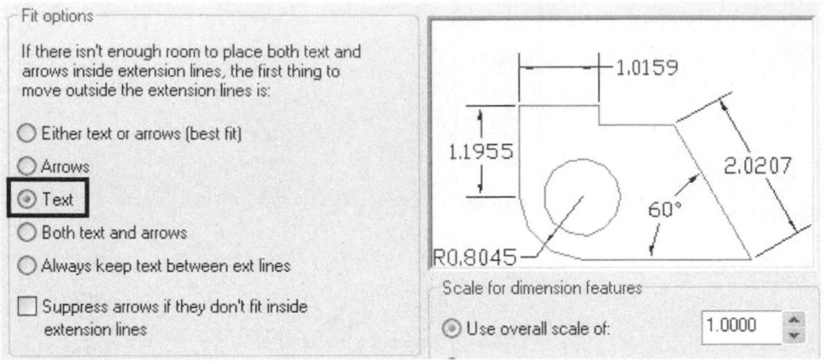

FIGURE 11.34

Clicking the Both text and arrows radio button updates the preview image, as shown in Figure 11.35, where the 1.0159 dimension text and arrows are both placed outside the extension lines.

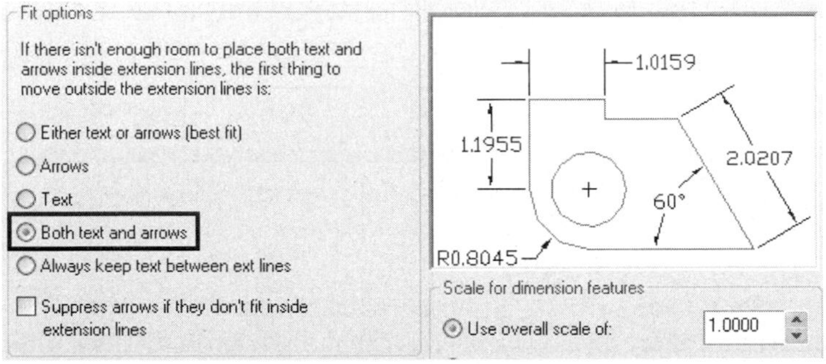

FIGURE 11.35

If you click the radio button for Always keep text between ext lines, the result is illustrated in the preview image in Figure 11.36. Here all dimension text, including the radius dimension, is placed between the extension lines.

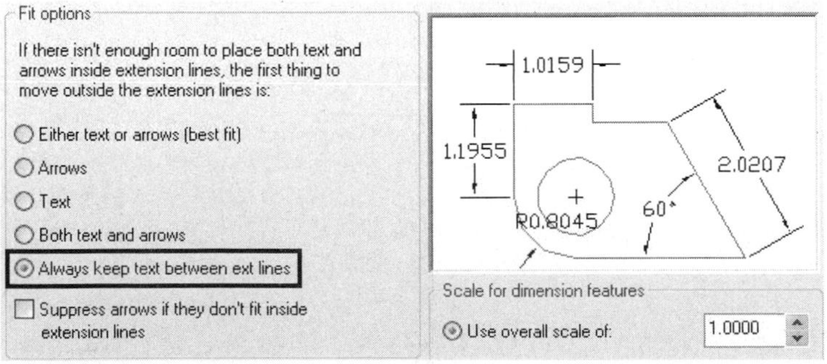

FIGURE 11.36

If you click the radio button for Either the text or the arrows (best fit) and you also place a check in the box for Suppress arrows if they don't fit inside the extension lines, the dimension line is turned off only for dimensions that cannot fit the dimension text and arrows. This is illustrated in Figure 11.37.

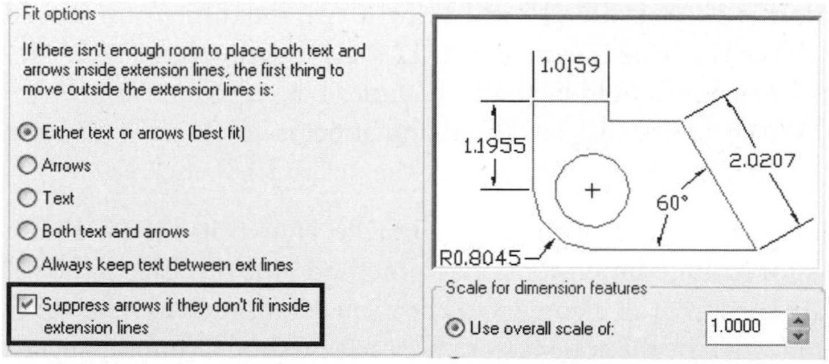

FIGURE 11.37

Text Placement

You control the placement of dimension text if it is not in the default position. Your choices are, shown in Figure 11.38. Beside the dimension line, which is the default, Over dimension line, with leader, or Over dimension line, without leader. If the text is moved from its default position, such as demonstrated earlier when the Text radio button was selected in the Fit options area, it is placed beside the dimension line. This is because of the default setting and illustrated in Figure 11.38 (Left).

If you click the radio button for Text in the Fit options area and you click the radio button for Over dimension line, with leader, you get the result that is illustrated in Figure 11.38 (Center). For the 1.0159 dimension that does not fit, the text is placed outside the dimension line with the text connected to the dimension line with a leader.

If you click the radio button for Text in the Fit options area and you click the radio button for Over dimension line, without leader, the result is illustrated in Figure 11.38 (Right). For the 1.0159 dimension that does not fit, the text is placed outside the dimension. No leader is used.

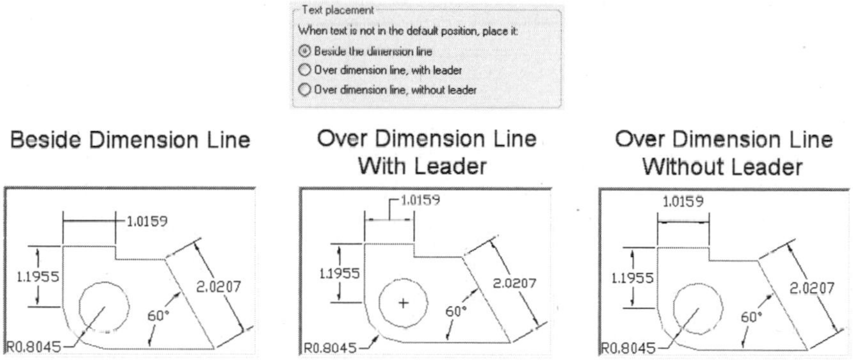

FIGURE 11.38

Scale for Dimension Features

Dimensions placed in CAD drawings must be properly scaled to appear correct on the final plotted sheet. It is interesting to note that when a drawing is created with pencil and paper, the object on the drawing is scaled to fit on the sheet of paper, but the text and dimensions are always created at a fixed size. Whether you are drawing the gear in a wrist watch or a residential floor plan, the dimensions will be created at the same fixed size. In CAD, the opposite is true. Objects are drawn actual size and the dimensions and text are scaled to the objects. If in an architectural drawing, for example, a dimension with 1/8-inch arrows and text were placed on a 24-foot long line, you wouldn't be able to see the arrows or text once you zoom out far enough to see the entire 24-foot line. The dimension needs to be increased in size by a scale factor. If the drawing scale is 1" = 1'-0" (1" = 12" or 1/12), the dimension should be scaled by the reciprocal of the scale fraction—a 1/12 scale drawing has a scale factor of 12/1 or 12. The dimension should be made 12 times bigger to appear proportional in the drawing. When the drawing is plotted and appears 1/12 scale (12 times smaller), the dimension text and arrow will appear at the desired 1/8-inch size.

The Scale for dimension features area of the dialog box, as shown in Figure 11.39, provides three methods for scaling dimensions. Each method globally affects the dimension settings that are specified; all elements of the dimension, such as text, spacing, and arrows, scale uniformly by the scale factor. The first method, Annotative, can either be turned on or off, and automatically sets the scale factor based on the Annotation scale that is specified on the Status bar. The Status bar setting should be made before placing the dimensions to have them properly sized. The second method is activated by picking the radio button next to Scale dimensions to layout. Selecting this option allows you to have dimensions automatically scaled to Paper Space units inside a layout. The scale of the viewport will control the scale of the dimensions, and should be set prior to placing the dimensions. These first two methods allow you to set multiple scales for use in multiple viewports. This concept will be discussed in greater detail in Chapter 19. If multiple scales are not required, the third method provides a simple scaling solution and is activated by clicking the radio button next to Use overall scale of. A text box is provided for you to enter the scale factor value. The illustrations in Figure 11.39 show the effects of changing the overall scale factor from 1.00 (for a 1/1 drawing) to 2.00 (for a 1/2 drawing). The dimension text, arrows, origin offset, and extension beyond the arrow have all doubled in size.

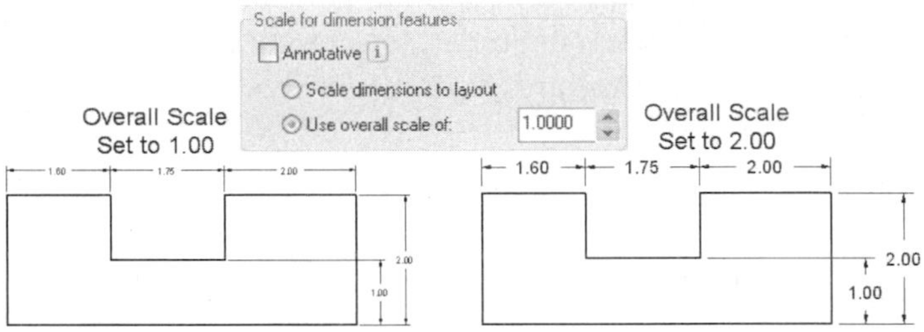

FIGURE 11.39

TRY IT!

Open the drawing file 11_Dimscale. A simple floor plan is displayed in Figure 11.40 (Left). This floor plan is designed to be plotted at a scale of 1/2" = 1'0" (same as 1:24 scale). Also displayed in this floor plan are dimensions in the magenta color. However, the dimensions are too small to be viewed. Set the Scale for dimension features found under the Fit tab of the Dimension Style dialog box to the scale factor—24. The dimension values and tick marks should now be visible, as shown in Figure 11.40 (Right).

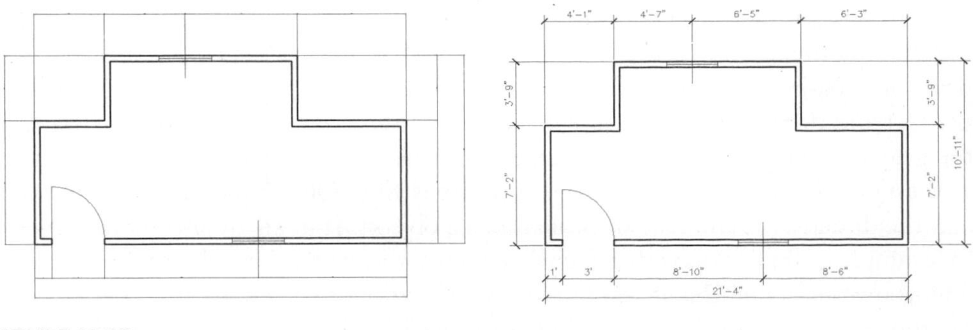

FIGURE 11.40

Fine-Tuning

You have two options to add further control for dimension text placement, in the Fine-tuning area, as shown in Figure 11.41 (Left). You can have total control for horizontally justifying dimension text if you place a check in the box for Place text manually when dimensioning. You can also force the dimension line to be drawn between extension lines by placing a check in this box. The results are displayed in Figure 11.41 (Right).

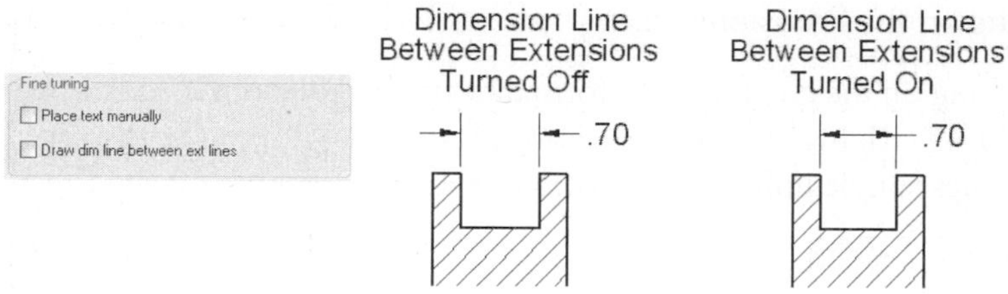

FIGURE 11.41

THE PRIMARY UNITS TAB

Use the Primary Units tab, shown in Figure 11.42, to control settings affecting the dimension text units. This includes the type of units the dimensions will be constructed in (decimal, engineering, architectural, and so on), the units precision, and whether the dimension text requires a prefix or suffix.

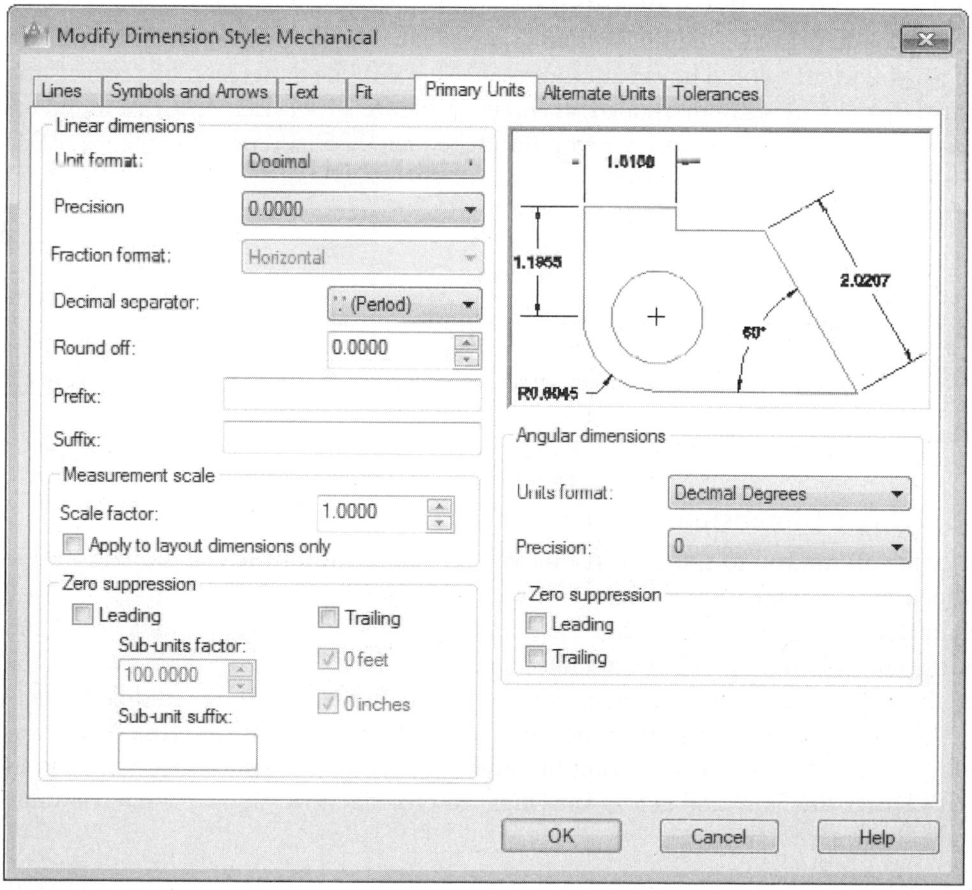

FIGURE 11.42

Linear and Angular Dimension Units

The Linear dimensions area of the Primary Units tab, as shown in Figure 11.42, has various settings that deal with primary dimension units. A few of the settings deal with the format when working with fractions, which activates only if you are working in architectural or fractional units. When utilizing decimal units, the default, you can also designate the decimal separator as a Period, Comma, or Space.

Clicking on the box for Unit format, as shown in Figure 11.43 (Left), displays the types of units you can apply to dimensions. You also control the precision of the primary units by clicking on the Precision box.

In a similar way, clicking on the box for Units format in the Angular dimensions area, as shown in Figure 11.43 (Right), displays the various formats angles can be displayed in. The precision of these angle units can also be controlled by clicking on the Precision box.

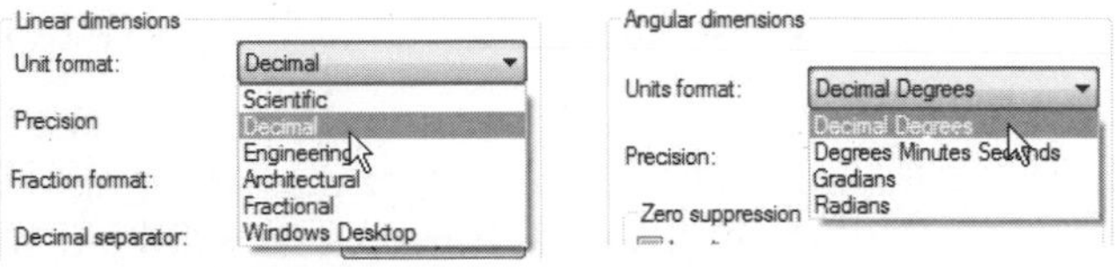

FIGURE 11.43

Rounding Off Dimension Values

Use a Round off value to round off all dimension distances to the nearest unit based on the round off value. The default round off value is 0; the dimension text reflects the actual distance being dimensioned, as shown in Figure 11.44 (Left). With a round off value set to 0.25, the dimension text reflects the next 0.25 increment, namely 2.50, as shown in Figure 11.44 (Right).

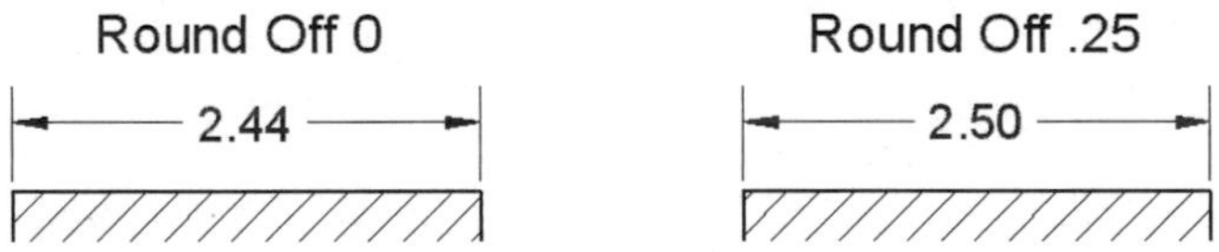

FIGURE 11.44

Applying a Dimension Prefix or Suffix

A prefix allows you to specify text that will be placed in front of the dimension text whenever you place a dimension. Use the Suffix box to control the placement of a character string immediately after the dimension value. Examples of both Prefix and Suffix control boxes are illustrated in Figure 11.45. In the illustration on the left, the prefix “2X” was added to the dimension to indicate two places, in the middle the suffix “'”(apostrophe) was added to indicate feet, and on the right the suffix “mm” was added to the dimension, signifying millimeters. It should be noted that a space was placed after the “2X” and before the “mm” to provide separation between the value and abbreviation.

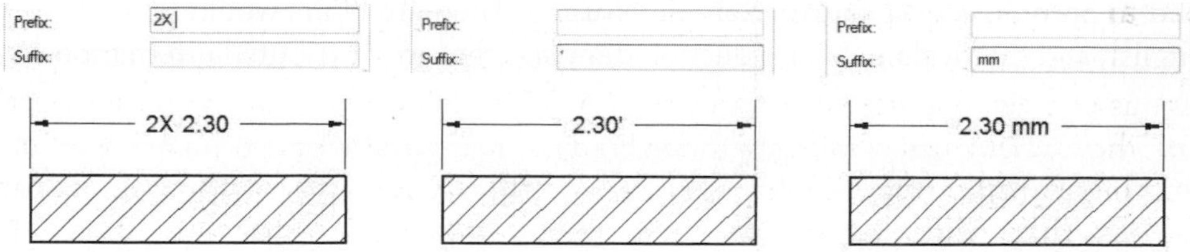

FIGURE 11.45

Measurement Scale

The Measurement scale area, shown in Figure 11.46, allows you to change dimension values by a multiplier. When a dimension distance is calculated, the current value set in the Scale factor field is multiplied by the dimension to arrive at a new dimension value.

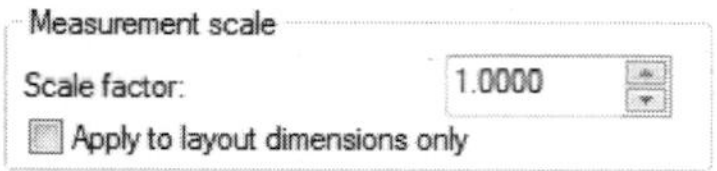

FIGURE 11.46

Illustrated in Figure 11.47 (Left), and with a linear scale value of 1.00, the dimension distances are taken at their default values.

Illustrated in Figure 11.47 (Right), the linear scale value has been changed to 2.00 units. This means 2.00 will multiply every dimension distance; the result is that the previous 3.00 and 2.00 dimensions are changed to 6.00 and 4.00, respectively. In a similar fashion, having a linear scale value set to 0.50 will reduce all dimension values to half their original values. This technique can be used to change dimensions distances that are placed on an object not drawn to actual size. If, for example, you scaled an object up four times in a detail, you could change the scale setting to 0.25 so the dimension values would appear correct.

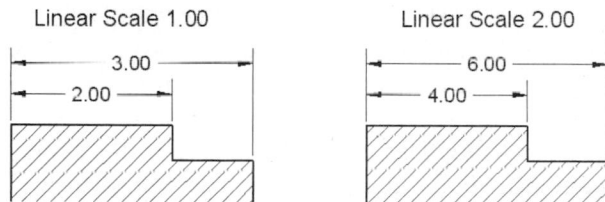

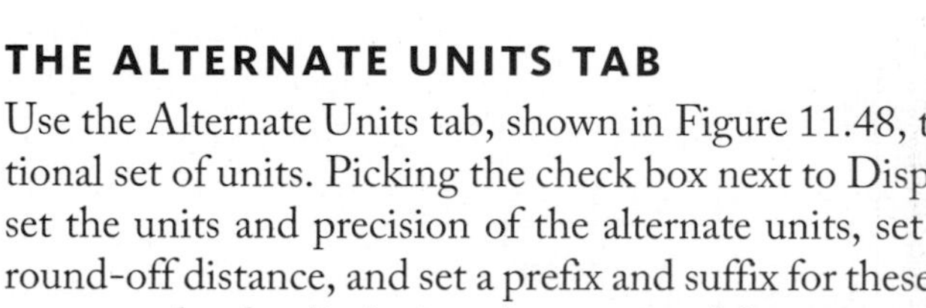

FIGURE 11.47

THE ALTERNATE UNITS TAB

Use the Alternate Units tab, shown in Figure 11.48, to enable the display of an additional set of units. Picking the check box next to Display alternate units allows you to set the units and precision of the alternate units, set a multiplier for alt units, use a round-off distance, and set a prefix and suffix for these units. You also have two placement modes for displaying these units: After primary value or Below primary value. By default, alternate units are placed beside primary units. The alternate units are enclosed in square brackets. With two sets of units being displayed, your drawing could tend to become very busy. An example of using Alternate Units would be to display English and metric dimension values on drawings that are distributed internationally.

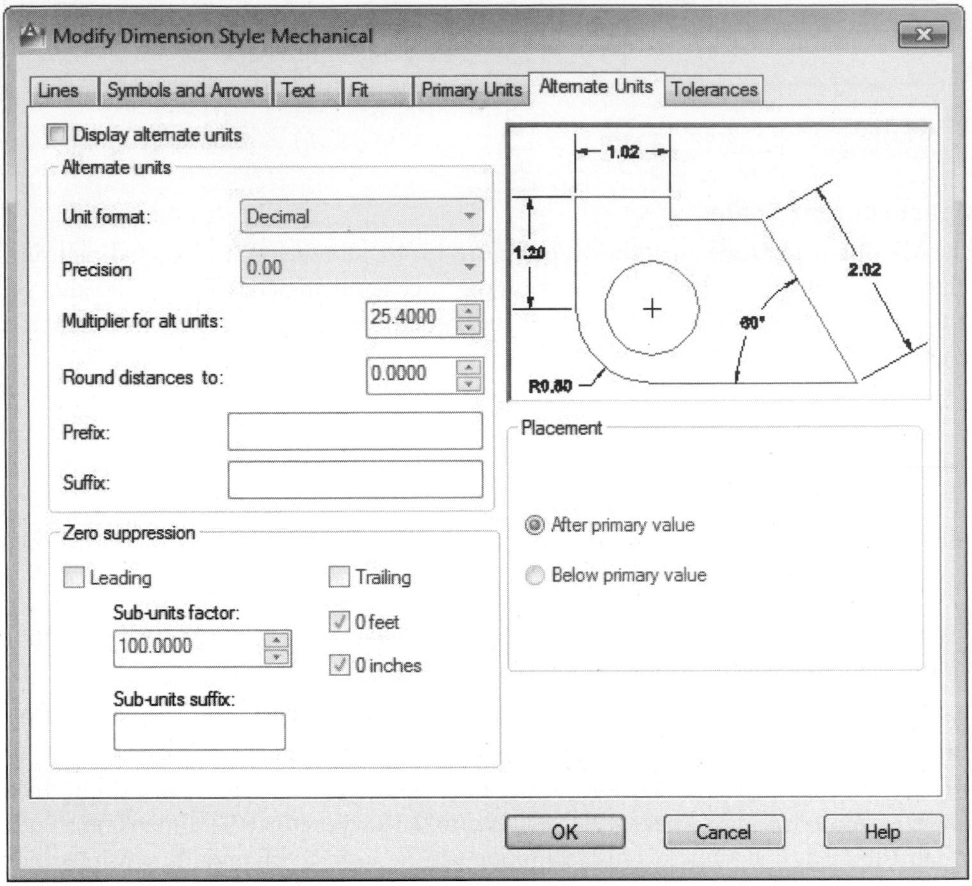

FIGURE 11.48

Alternate Units

Once alternate units are enabled, all items in the Alternate Units area become active, as shown in Figure 11.49 (Left). By default, the alternate unit value is placed in brackets alongside the calculated dimension value, as shown in Figure 11.49 (Right). This value depends on the current setting in the Multiplier for all units field. This factor, set to 25.40 in Figure 11.49, is used as a multiplier for all calculated alternate dimension values. With the primary units in inches, the alternate units would display in millimeters, since 1 inch equals 25.40 millimeters.

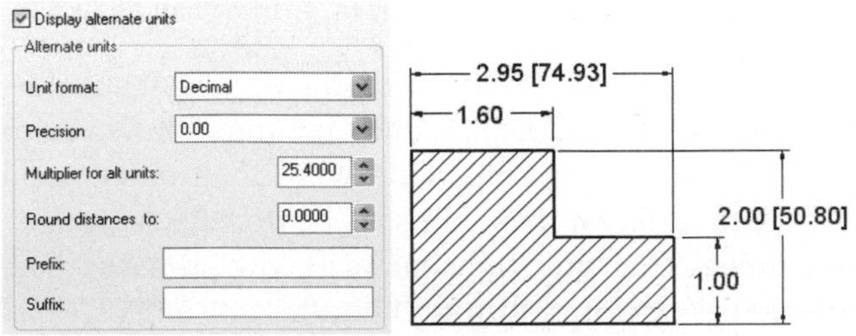

FIGURE 11.49

You could also click the radio button next to Below primary value, as shown in Figure 11.50 (Left). This places the primary dimension above the dimension line and the alternate dimension below it, as shown in Figure 11.50 (Right).

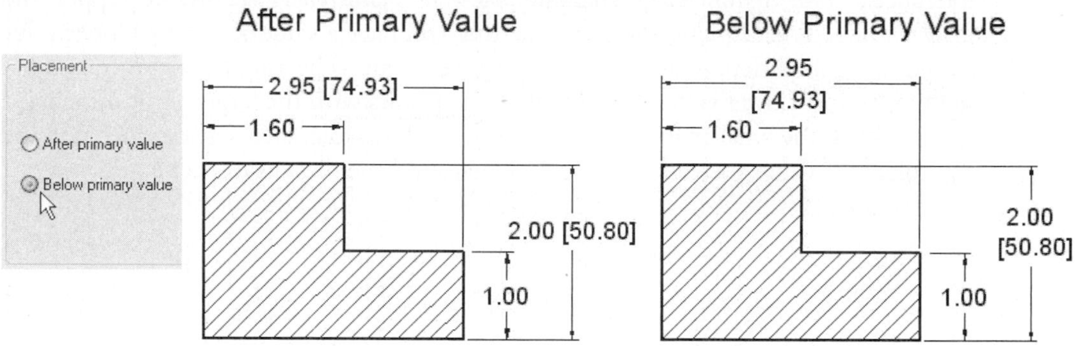

FIGURE 11.50

THE TOLERANCES TAB

The Tolerances tab shown in Figure 11.51 consists of various fields used to control the five types of tolerance settings: None, Symmetrical, Deviation, Limits, and Basic. Depending on the type of tolerance being constructed, an Upper value and Lower value may be set to call out the current tolerance variance. The Vertical position setting allows you to determine where the tolerance will be drawn in relation to the location of the body text. The Scaling for height setting controls the text size of the tolerance.

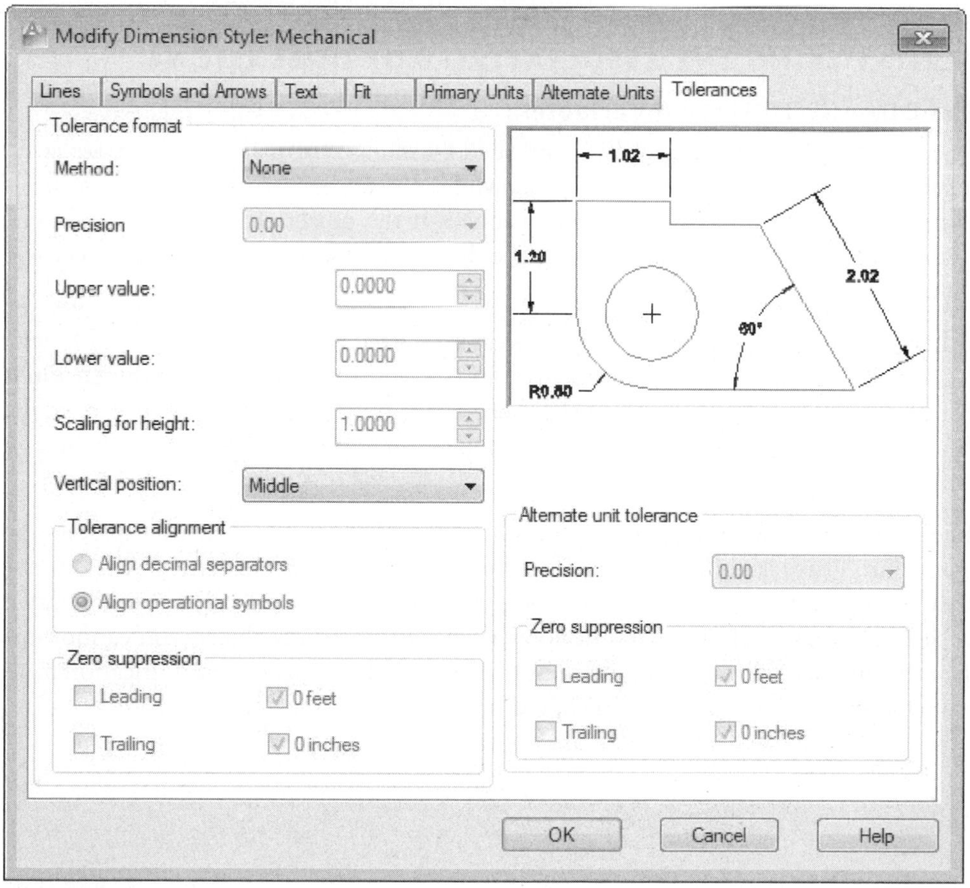

FIGURE 11.51

Tolerance Format

The five tolerance types available in the drop-down list are illustrated in Figure 11.52. A tolerance setting of None uses the calculated dimension value without applying any tolerances. The Symmetrical tolerance uses the same value set in the Upper and Lower value. The Deviation tolerance setting will have a value set in the Upper value and an entirely different value set in the Lower value. The Limits tolerance will use the Upper and Lower values and calculate the results with the larger limit dimension placed above the smaller limit dimension. The Basic tolerance setting does not add any tolerance value; instead, a box is drawn around the dimension value.

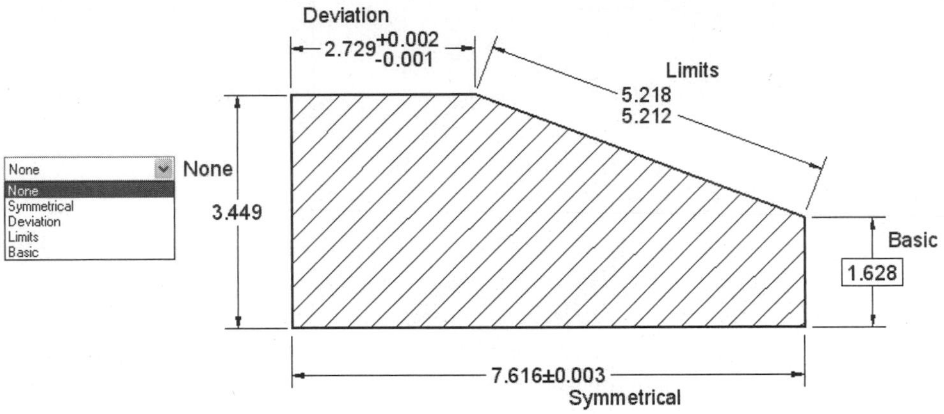

FIGURE 11.52

CONTROLLING THE ASSOCIATIVITY OF DIMENSIONS

The DIMASSOC System Variable

The associativity of dimensions is controlled by the `DIMASSOC` system variable. By default, this value is set to 2 for new drawings. With this setting, the dimension is associated with the object being dimensioned; if the object changes size or if some element of an object changes location, the dimension associated with the object will change as well.

When this variable is set to 1, the dimension is referred to as non-associative. Dimensions with this setting typically exist in older AutoCAD drawings. This dimension is not associated with the object being dimensioned; however, it is possible to have the dimension value automatically updated if the editing operation performed includes a dimension origin point.

The `DIMASSOC` system variable can also be set to 0, which will create exploded dimensions. There is no association, even, between the various elements of the dimension. All dimension lines, extension lines, arrowheads, and dimension text are drawn as separate objects. Because you no longer have a dimension object, you cannot update the dimension through its style or utilize any of the dimension modification tools available. Exploding a dimension should be avoided in almost all cases.

TRY IT!

Open the drawing 11_Dimassoc2. The drawing contains a polyline object, as shown in Figure 11.53 (Left), that has dimensions placed with the DIMASSOC system variable set to 2. Click the polyline to activate its grips. Click the grip at "A" and stretch the polyline vertex up and to the left. The results are shown in the image on the right; notice that all dimension locations, orientations, and values are automatically updated. Click the polyline vertex at "B" and stretch this vertex up and to the right. The same results occur, with the dimension elements being updated based on the new location of the object. Use grips to realign any dimensions that are not placed where you intended.

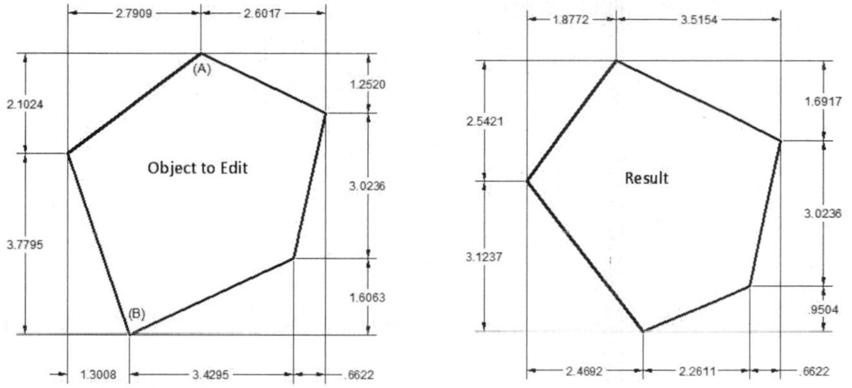

FIGURE 11.53

TIP

When you make changes to the DIMASSOC system variable, these changes are stored in the drawing file.

Reassociating Dimensions

It is possible to change a non-associative dimension into an associative dimension through the process called Reassociation. This command allows you to pick the non-associative dimension and reestablish its endpoints with new endpoints located on the object being dimensioned. The following Try It! will demonstrate this capability.

TRY IT!

Open the drawing 11_Dim Reassoc. The vertical 1.60 dimension is not associated with the polyline object in this example, because it was generated while the DIMASSOC variable was set to 1. It should be noted that it is also possible for disassociation to occur if a dimension's defpoint (the extension line endpoint—definition point) is not connected properly or it became disconnected from an object. Using grips to reconnect the dimension to the object will not restore the associativity. Activate the DIMREASSOCIATE command from the Ribbon's Annotate tab, as shown in Figure 11.54 (Left). Pick the 1.60 dimension as the object to reassociate. When the blue X appears, pick endpoints on the object to perform the reassociation, as shown in the middle of Figure 11.54.

Command: DIMREASSOCIATE

Select dimensions to reassociate...

Select objects or [Disassociated]: *(Pick the vertical 1.60 dimension)*

Select objects or [Disassociated]: *(Press* ENTER *to continue)*

Specify first extension line origin or [Select object] <next>: *(When the blue X appears at the end of the proper extension line, pick the endpoint of the object at "B")*

Specify second extension line origin <next>: *(The blue X will move to the opposite endpoint of the dimension. Pick the endpoint of the object at "C" in order to complete the dimension reassociation)*

You can test to see whether this dimension is associative by selecting the object, picking either grip at "B" or "C," and stretching the object. The dimension should now update to the changes in the object's shape.

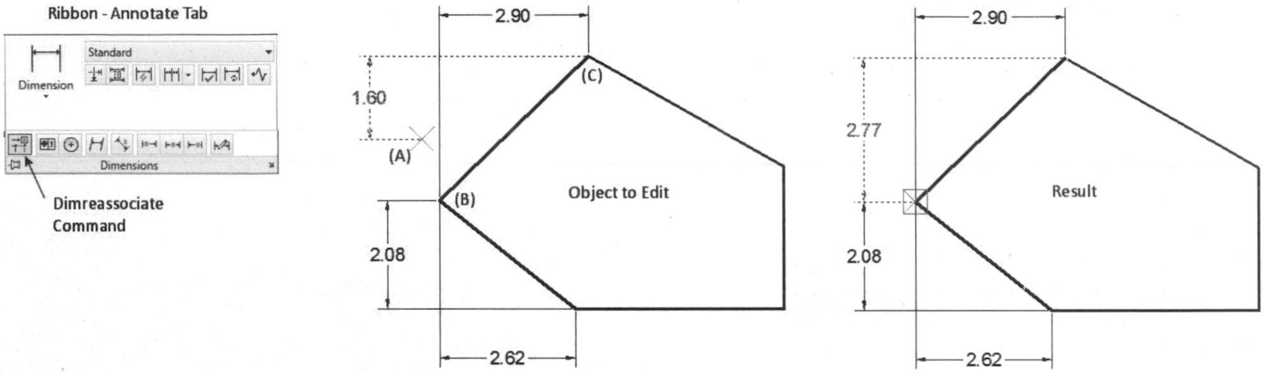

FIGURE 11.54

You will be able to identify associated dimensions by clicking the dimension while in the DIMREASSOCIATE command. The familiar X is surrounded by a blue box, as shown in Figure 11.54 (Right). If you do not see the blue box, the dimension is not associative.

Reassociating works on associative dimensions that become disconnected from an object and on non-associative dimensions that have their dimassoc value set to 1. The process of Reassociation does not have any effect on exploded dimensions (dimassoc set to 0).

While AutoCAD gives you this ability to reassociate dimensions, it would be very time consuming to perform this task on hundreds of dimensions in a drawing. Reassociation is ideal when dealing with only a few dimensions.

Monitoring Associative Annotations

Maintaining associative annotations in a drawing is essential when performing editing operations. Because dimensions remain connected to drawing objects as they are moved, rotated, or scaled, modifications are simple to perform and completed quickly. The Annotation Monitor button on the Status Bar, shown in Figure 11.55, provides a tool that helps you maintain associativity by identifying any annotations that become disassociated. If this happens, an Annotation Monitor icon at the right end of the Status Bar turns bright red, and yellow alert icons are displayed adjacent to any disassociated annotations. Selecting an alert icon, as shown in Figure 11.55 (Right), provides a shortcut menu that allows you to reassociate or delete the annotation.

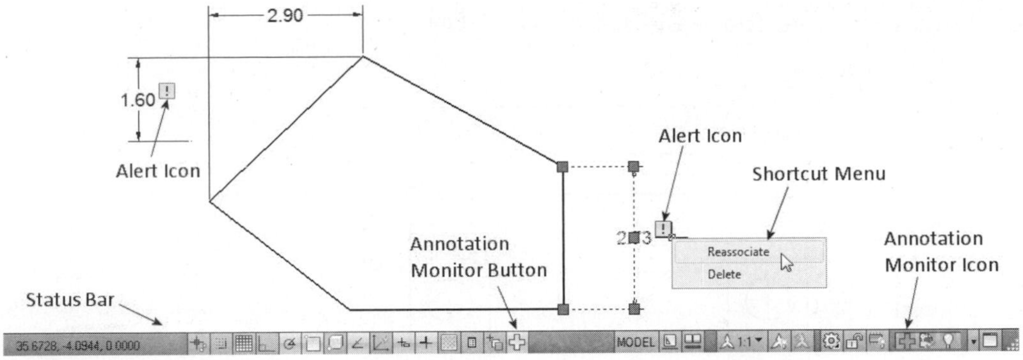

FIGURE 11.55

TRY IT!

Open the drawing 11_Dim Assoc Monitor. Turn on the Annotation Monitor mode by selecting the button in the Status Bar. Notice that an Annotation Monitor icon is also displayed at the right end of the Status Bar. Use the ERASE command to delete the outer rectangle, as shown in Figure 11.56 (Left). The Annotation Monitor icon turns red and yellow alert icons are displayed at the now disassociated annotations. Pick the alert icon for the bottom horizontal dimension and select Delete from the shortcut menu as shown in the middle image. For the remaining annotations (two linear dimensions and a multileader), use the Reassociate option from the shortcut menus to reassociate the annotations to the inner rectangle. The results are shown in Figure 11.56 (Right).

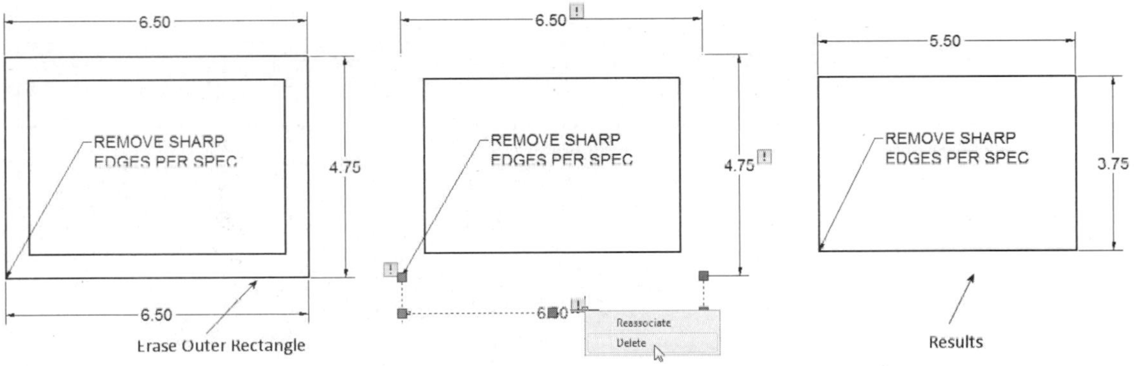

FIGURE 11.56

USING DIMENSION TYPES IN DIMENSION STYLES

In addition to creating dimension styles, you can assign settings to specific dimension types. These dimension types are sub-styles to the main style. The purpose of using dimension types is to reduce the number of dimension styles defined in a drawing. For example, the object in Figure 11.57 consists of linear dimensions with three-decimal-place accuracy, a radius dimension with one-decimal-place accuracy, and an angle dimension with a box surrounding the number. Normally you would have to create three separate dimension styles to create this effect. However, the linear, radius, and angular dimension settings can instead exist as dimension types under a single dimension style.

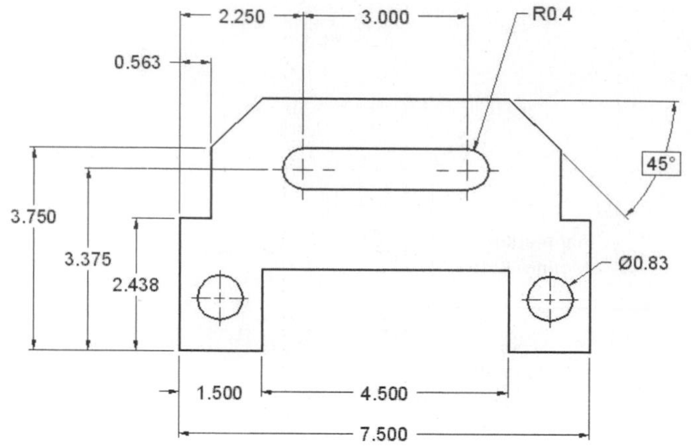

FIGURE 11.57

Four dimension types were created under the Mechanical dimension style, as shown in Figure 11.58. When the Mechanical style is current these types are also active. One of the dimension types shown is Angular. Before creating an angular dimension, you create the dimension type and make changes in various tabs located in the Modify Dimension Style dialog box. These changes will apply only to the angular dimension type.

To expose the dimension types, click the New button in the main Dimension Style Manager dialog box.

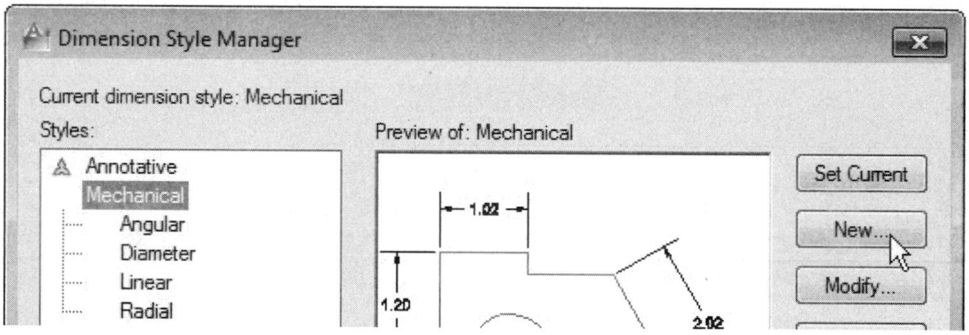

FIGURE 11.58

When the Create New Dimension Style dialog box appears, as shown in Figure 11.59, click on the Use for box. If the default, All dimensions, is selected, you are creating a new dimension style affecting all dimension types. On the other hand, if you select one of the other options (linear, angular, radius, diameter, ordinate, or Leaders and Tolerances), you are setting up a dimension type under the current style. Highlight the Leaders and Tolerances option and then click the Continue button.

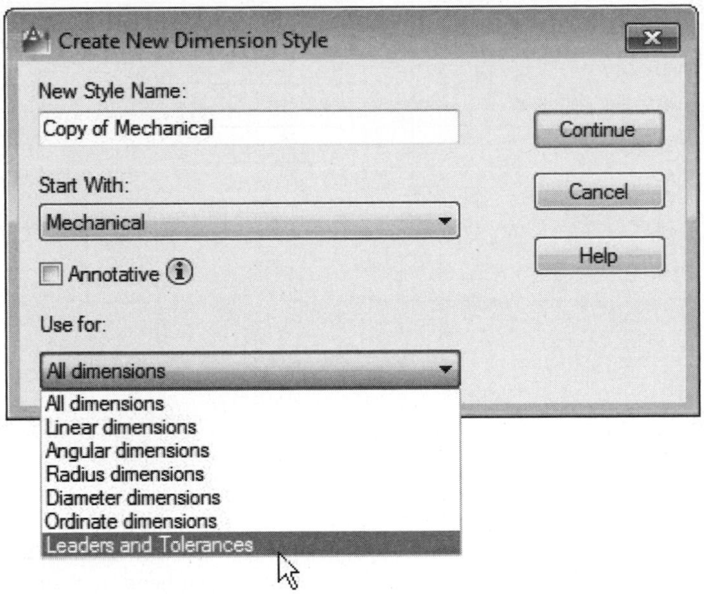

FIGURE 11.59

This takes you to the New Dimension Style: Mechanical: Leader dialog box, as shown in Figure 11.60. Any changes you make in the tabs will apply only to dimensions created with the QLEADER or LE command. The use of dimension types is an efficient means of organizing and simplifying your dimension styles.

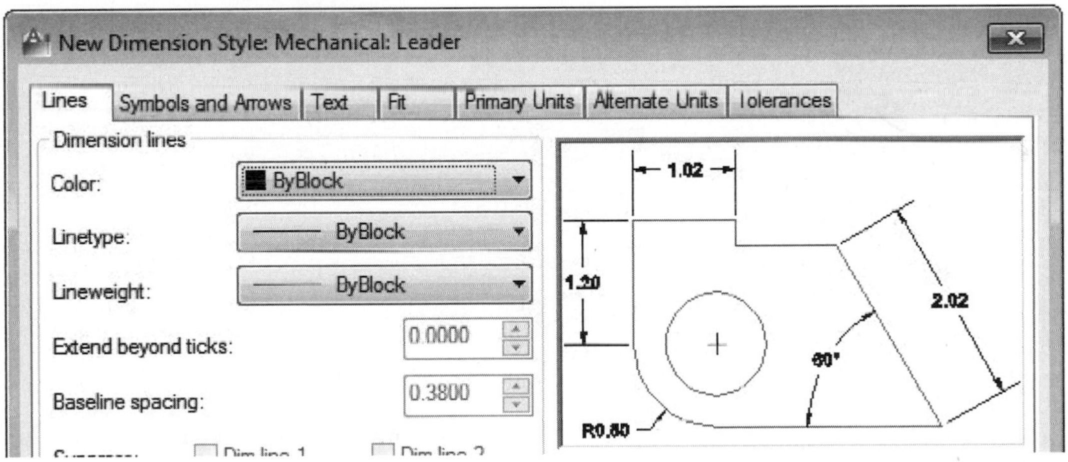

FIGURE 11.60

TRY IT!

Open the drawing file 11_Dimension Types. A series of Dimension Types have already been created. Use the illustration in Figure 11.61 and add all dimensions to this object.

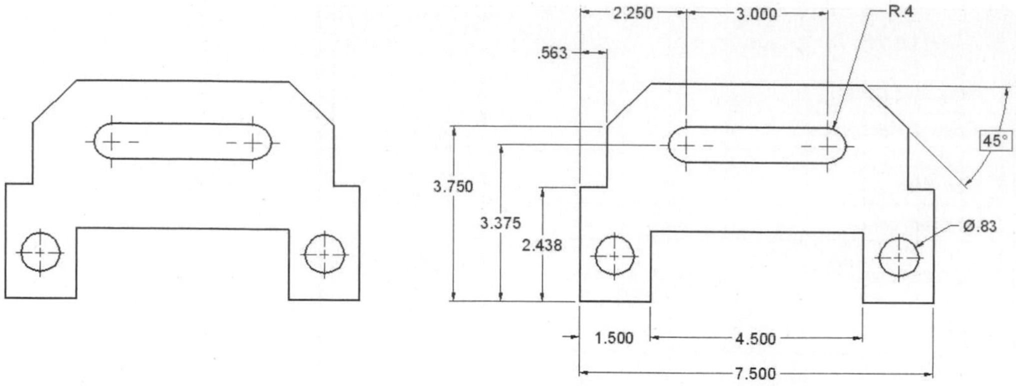

FIGURE 11.61

OVERRIDING A DIMENSION STYLE

TRY IT!

Open the drawing file 11_Dimension Override. For those cases in which you need to change the settings of some dimensions but don't want to save a style, a dimension override would be used. First launch the Dimension Style Manager dialog box. Clicking the Override button, as shown in Figure 11.62 (Left), displays the Override Current Style: Mechanical dialog box. Under the Lines tab, check the Ext line 1 and Ext line 2 boxes in the Extension lines area for suppression, as shown in Figure 11.62 (Right). This will allow you to create a dimension without displaying extension lines.

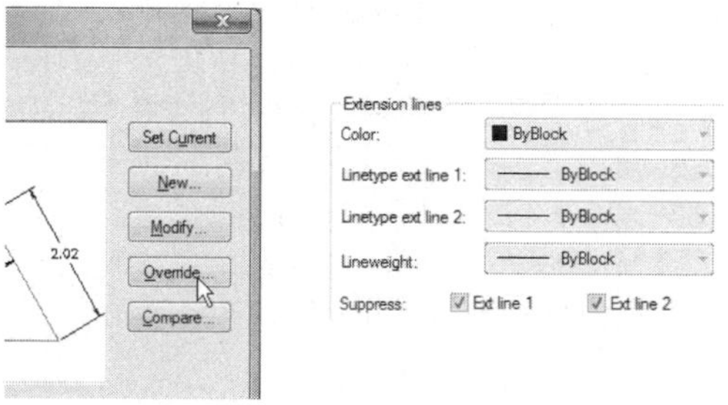

FIGURE 11.62

Clicking the OK button returns you to the main Dimension Style Manager dialog box, as shown in Figure 11.63. Notice that, under the Mechanical style, a new dimension type has been created called <style overrides> (Angular, Diameter, Linear, and Radial were already existing in this example). Also notice that the image in the Preview box shows sample dimensions displayed without extension lines. The <style overrides> dimension style is also the current style. Click the Close button to return to your drawing.

NOTE

If the Preview box does not show the correct sample image when you create the new <style overrides>, close and immediately reopen the Dimension Style Manager dialog box. The Preview box will show the correct sample image.

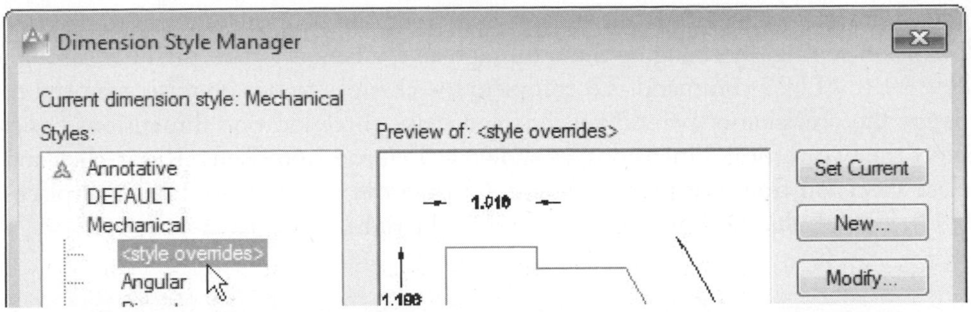

FIGURE 11.63

In the illustration in Figure 11.64 (Right), a linear dimension that identifies the vertical distance of 1.250 is placed. Notice that extension lines are not drawn on top of existing object lines, due to the style override.

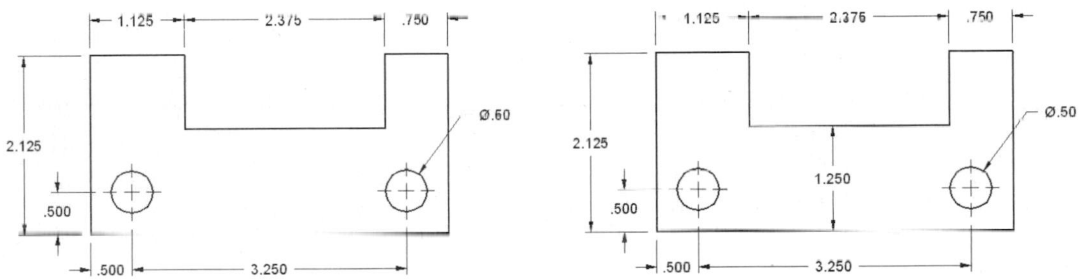

FIGURE 11.64

Unfortunately, if you start placing other dimensions, these will also lack extension lines. The Dimension Style Manager dialog box is once again activated. Clicking an existing dimension style such as Mechanical and then clicking the Set Current button displays the AutoCAD Alert box shown in Figure 11.65. If you click OK, the style override disappears from the listing of dimension types. If you would like to save the overrides under a name, click the Cancel button, right-click the <style overrides> listing, and rename this style to a new name. This preserves the settings under this new name.

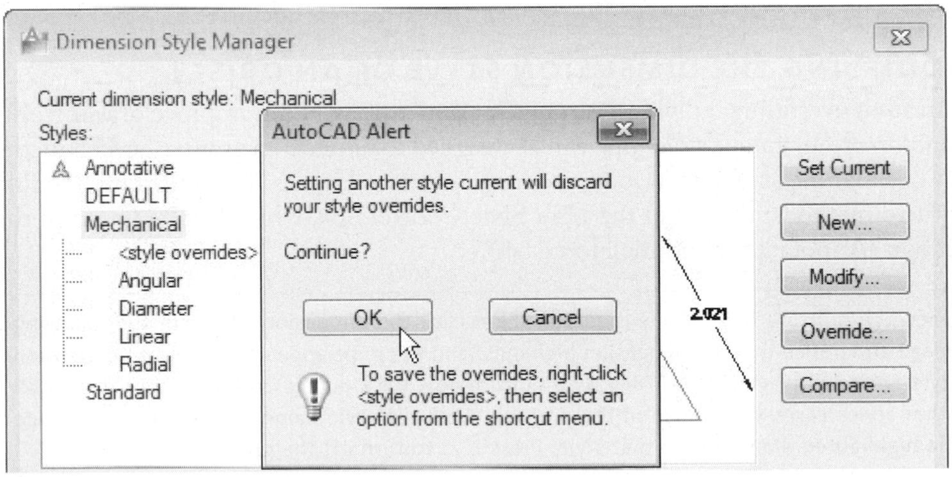

FIGURE 11.65

Instead of setting up an Override style, it can sometimes be more efficient to override dimension settings by changing them through an available shortcut menu or through the PROPERTIES command. To complete the changes to our exercise, we need to change the precision of two dimensions and make three location dimensions basic. Select the two vertical dimensions, as shown in Figure 11.66 (Left). Right-click and select Precision from the shortcut menu. Change the precision to 2 decimal places (0.00). The results are shown in the image on the right.

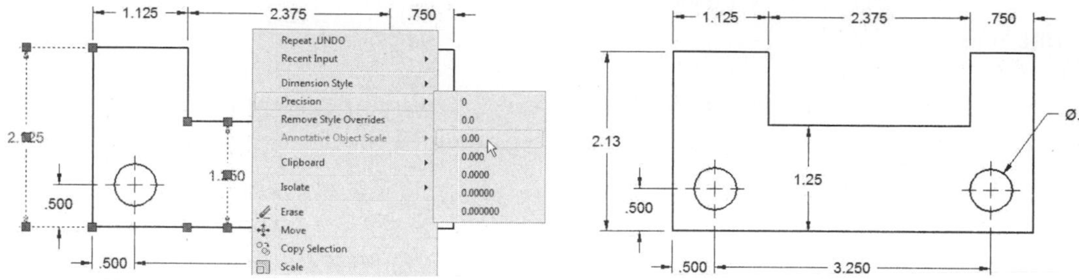

FIGURE 11.66

Select the three dimensions that locate the holes, as shown in Figure 11.67 (Left). Activate the PROPERTIES command (use the shortcut menu). Scroll down until you find the Tolerances heading. Expand the Tolerance Display list box and select Basic. The final results are shown in the image to the right.

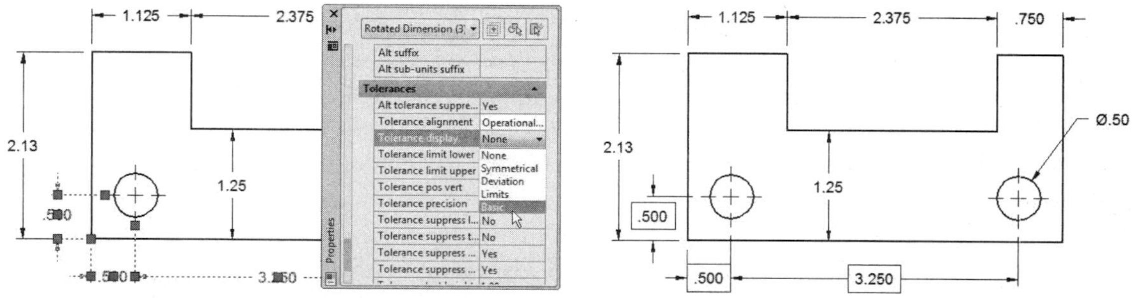

FIGURE 11.67

MODIFYING THE DIMENSION STYLE OF AN OBJECT

Instead of overriding individual settings for dimensions, it can improve drawing efficiency to create unique styles that can be assigned to dimension groups. Several methods of assigning styles will be outlined in this section. Figure 11.68 illustrates the first of these methods: the use of the Dim Style Control box, which can be found in the Ribbon (Annotate tab and the Home tab).

TRY IT!

Open the drawing file 11_Dimension Edit. First click the dimension shown in the following image and notice that the dimension highlights and the grips appear. The current dimension style is also displayed in the Dim Style Control box. Click on the control box to display all other styles currently defined in the drawing. Click the style name TOLERANCE to change the highlighted dimension to that style. Press ESC to turn off the grips.

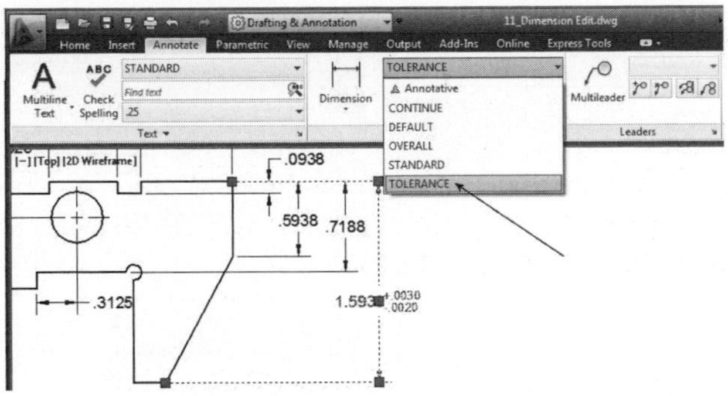

FIGURE 11.68

The second method of changing dimension styles is illustrated in Figure 11.69. In this example, select the dimension shown in this image; it highlights and grips appear. Right-clicking displays the cursor menu. Choosing the Dimension Style heading displays the cascading menu of all dimension styles defined in the drawing. Click the TOLERANCE style to change the highlighted dimension to the new style.

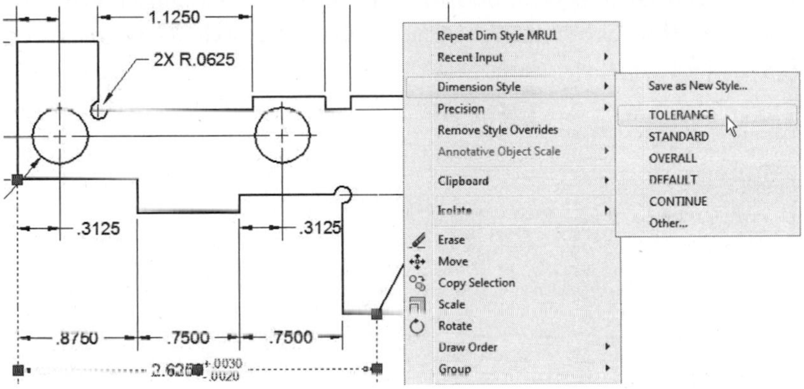

FIGURE 11.69

The third method of changing dimension styles begins with selecting the overall length dimension shown in Figure 11.70 (Left). Again, the dimension highlights and grips appear. Clicking the Properties button (View tab > Palettes panel) displays the Properties Palette, as shown in Figure 11.70 (Right). Scroll to the Misc heading and click on the list box next to Dim style; all dimension styles will appear in the field. Click the OVERALL style to change the highlighted dimension to the new style. Close the Properties Palette and press the ESC key to turn off the grips.

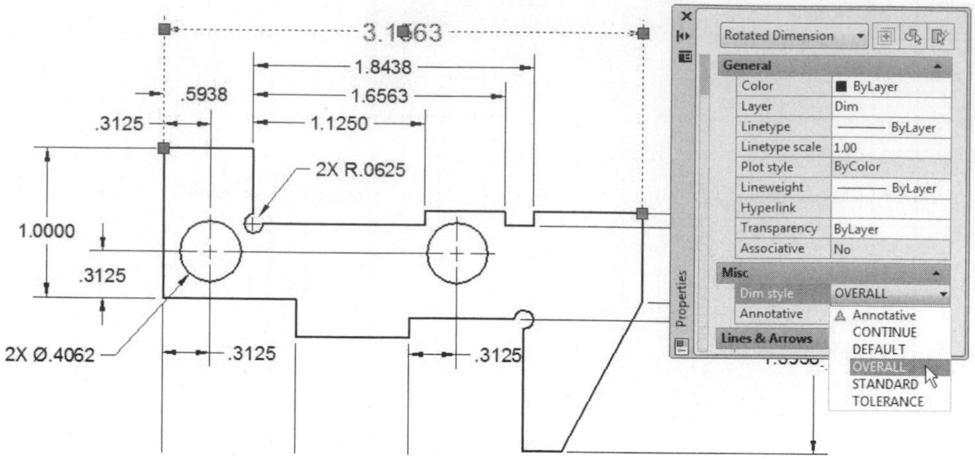

FIGURE 11.70

CREATING MULTILEADER STYLES

As with the Dimension Style Manager, you can create different styles of multileaders using the Multileader Style Manager. Choosing Multileader Style either from the Home or Annotate tab of the Ribbon launches the Multileader Style Manager dialog box shown in Figure 11.71.

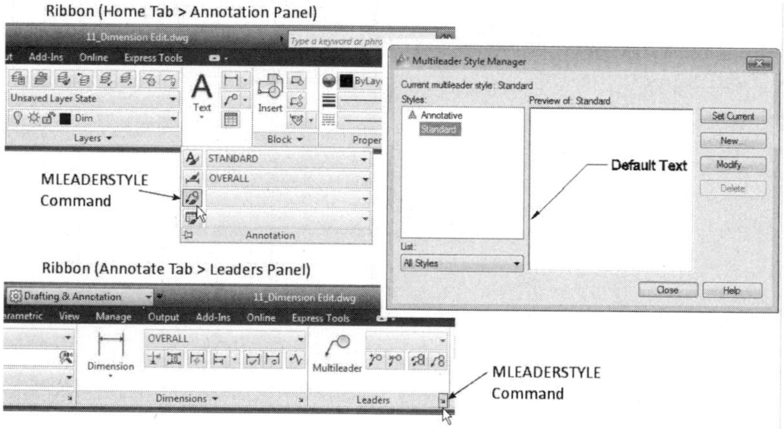

FIGURE 11.71

Clicking the New button launches the Create New Multileader Style dialog box, as shown in Figure 11.72. In the following example, a new style called Circle Balloons will be created. This style will consist of a circle with text inside.

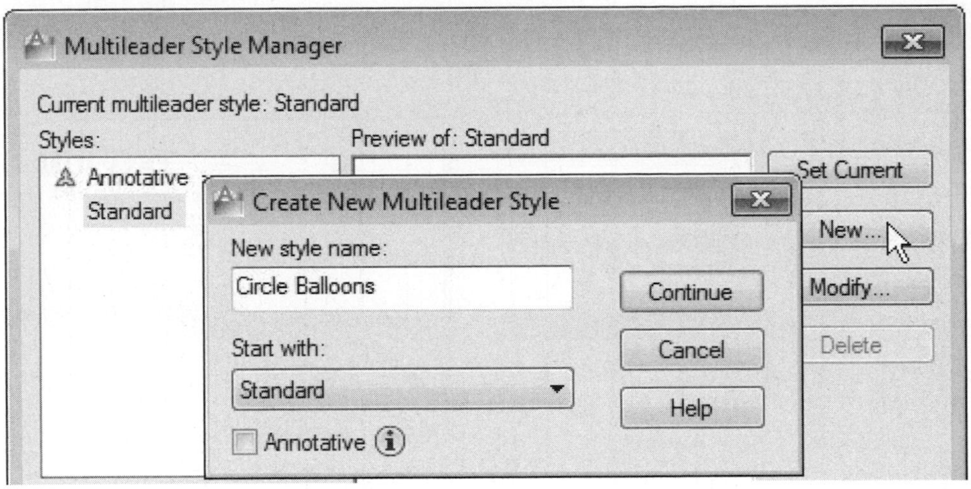

FIGURE 11.72

After providing the new style name, click the Continue button to make available three tabs that control the appearance of the multiline style, as shown in Figure 11.73. The first tab, Leader Format, controls the leader and includes type of leader (Straight, Spline, or None), the size of the arrowhead, and the leader break distance. The second tab, Leader Structure, controls the number of points that make up the leader, the length of the leader landing, and the scale factor of the leader. The last tab, Content, deals with the contents of the leader. By default, the Multileader type is made up of Mtext objects. Other options include Block or None. The other content includes basic text options and the type of leader connection.

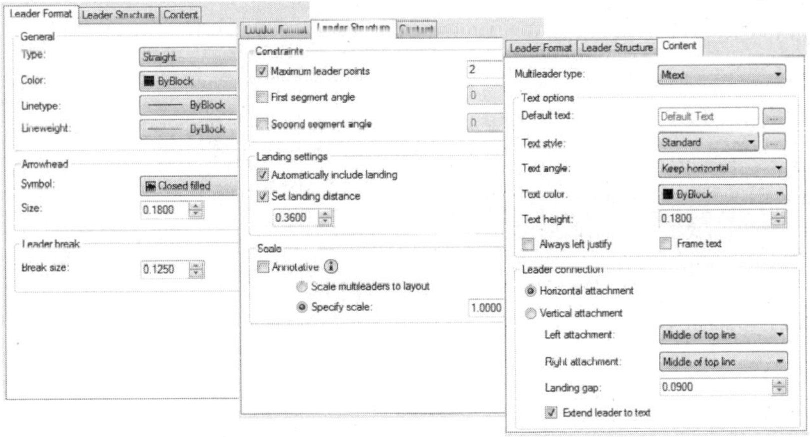

FIGURE 11.73

While in the Content tab, changing Mtext to Block activates block options, as shown in Figure 11.74. Changing the source block to circle updates the preview image of the multileader.

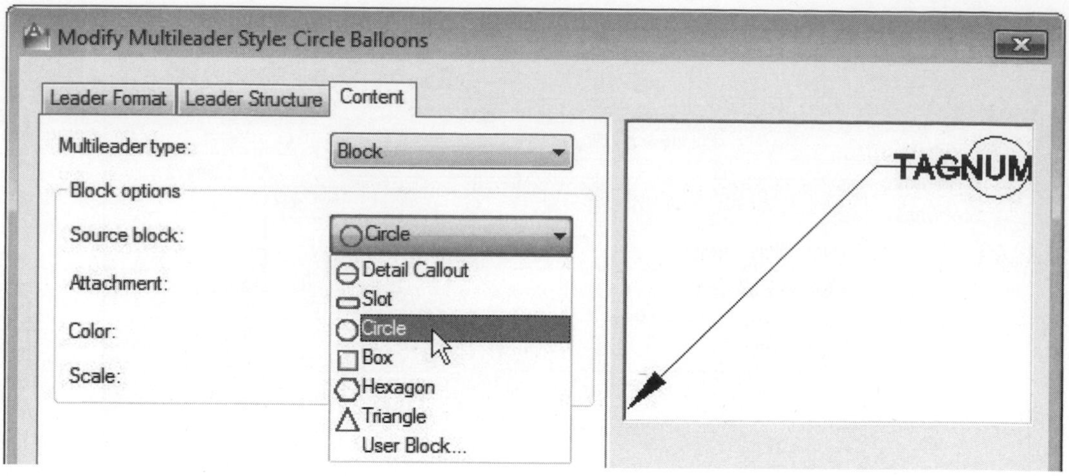

FIGURE 11.74

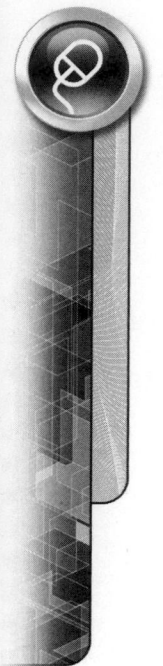

Returning back to the Multileader Style Manager dialog box allows you to make the new style, namely Circle Balloons, the current style, as shown in Figure 11.75 (Left). The results of using this style in a drawing are shown in Figure 11.75 (Right).

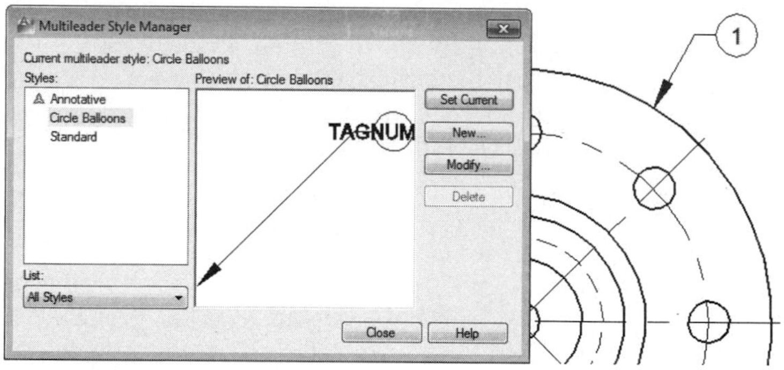

FIGURE 11.75

TUTORIAL EXERCISE: 11_DIMEX.DWG

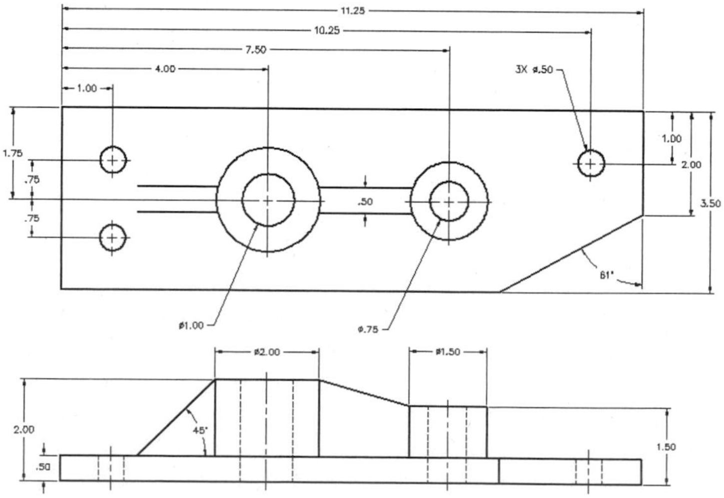

FIGURE 11.76

Purpose

The purpose of this tutorial is to place dimensions on the drawing of the two-view object illustrated in Figure 11.76.

System Settings

No special system settings need to be made for this drawing file.

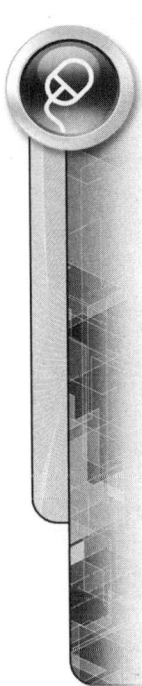

Layers

The drawing file 11_Dimex.Dwg has the following layers already created for this tutorial.

Name	Color	Linetype
Object	Magenta	Continuous
Hidden	Red	Hidden
Center	Yellow	Center
Dim	Yellow	Continuous

Suggested Commands

Open the drawing called 11_Dimex. The following dimension commands will be used: DIMSTYLE, DIMLINEAR, DIMCONTINUE, DIMCENTER, DIMRADIUS, DIMDIAMETER, DIMANGULAR, DIMEDIT, and QDIM. Use the ZOOM command to get a closer look at details and features that are being dimensioned.

STEP 1

To prepare for the dimensioning of the drawing, type D to activate the Dimension Style Manager dialog box. Click the New button, which activates the Create New Dimension Style dialog box shown in Figure 11.77. In the New Style Name area, enter MECHANICAL. Click the Continue button to create the style, as shown in Figure 11.77.

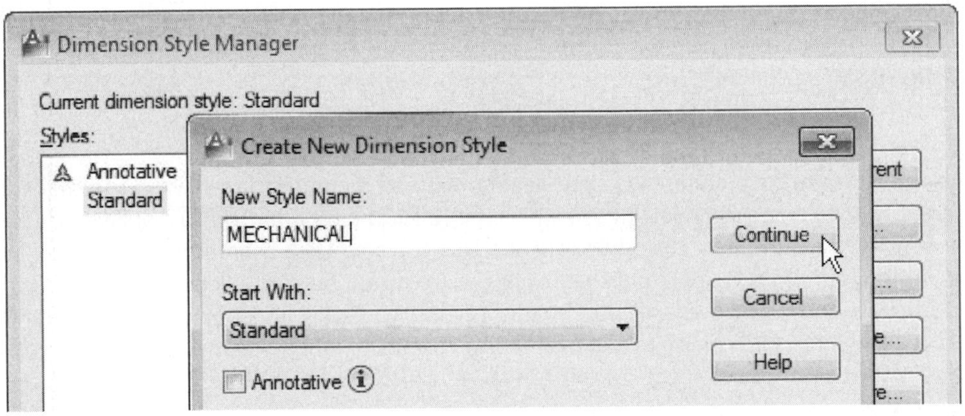

FIGURE 11.77

STEP 2

When the New Dimension Style: MECHANICAL dialog box appears, make the following changes in the Lines tab, as shown in Figure 11.78 (Left): Change the size of the Extend beyond dim lines from a value of 0.18 to a new value of .07. Switch to the Symbols and Arrows tab and change the Arrow size from 0.18 to 0.12. Also change the Center marks to Line, as shown in Figure 11.78 (Right).

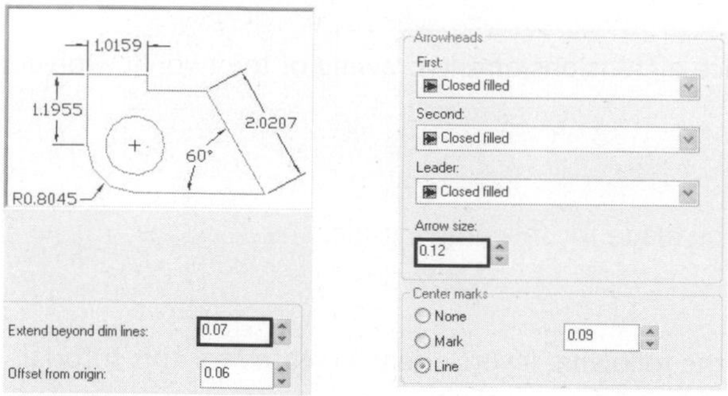

FIGURE 11.78

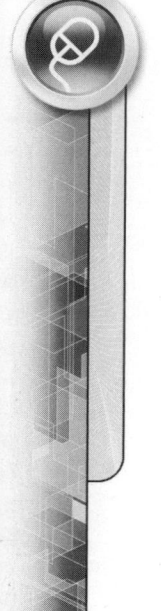

STEP 3

Next, click the Text tab and change the Text height from a value of 0.18 to a new value of 0.12, as shown in Figure 11.79 (Left). Then click the Primary Units tab and change the number of decimal places from 4 to 2. Also place checks in the boxes next to Leading in the Zero suppression areas, as shown in Figure 11.79 (Right). This turns off the leading zero for dimension values under 1 unit. When finished, click the OK button to return to the main Dimension Style Manager dialog box.

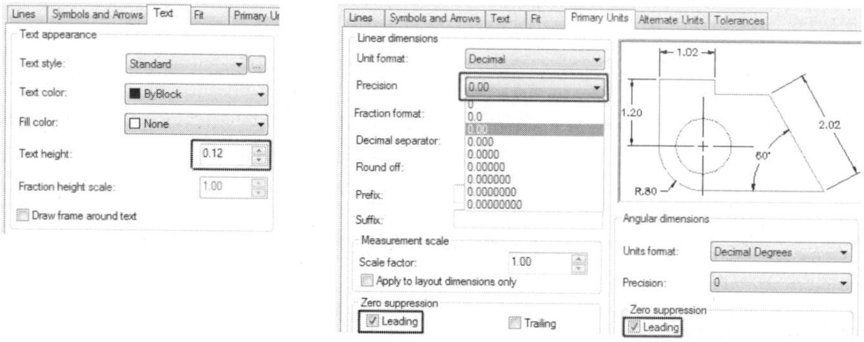

FIGURE 11.79

STEP 4

You will now create a dimension type dealing with all diameter dimensions and make a change to the center mark settings for this type. Click the New button to display the Create New Dimension Style dialog box. In the Use for box, click Diameter dimensions, as shown in Figure 11.80 (Left), and then click the Continue button. Click the Symbols and Arrows tab and change the Center marks setting to None, as shown in Figure 11.80 (Right). This prevents center marks from being displayed when placing diameter dimensions. When finished, click the OK button to return to the main Dimension Styles Manager dialog box.

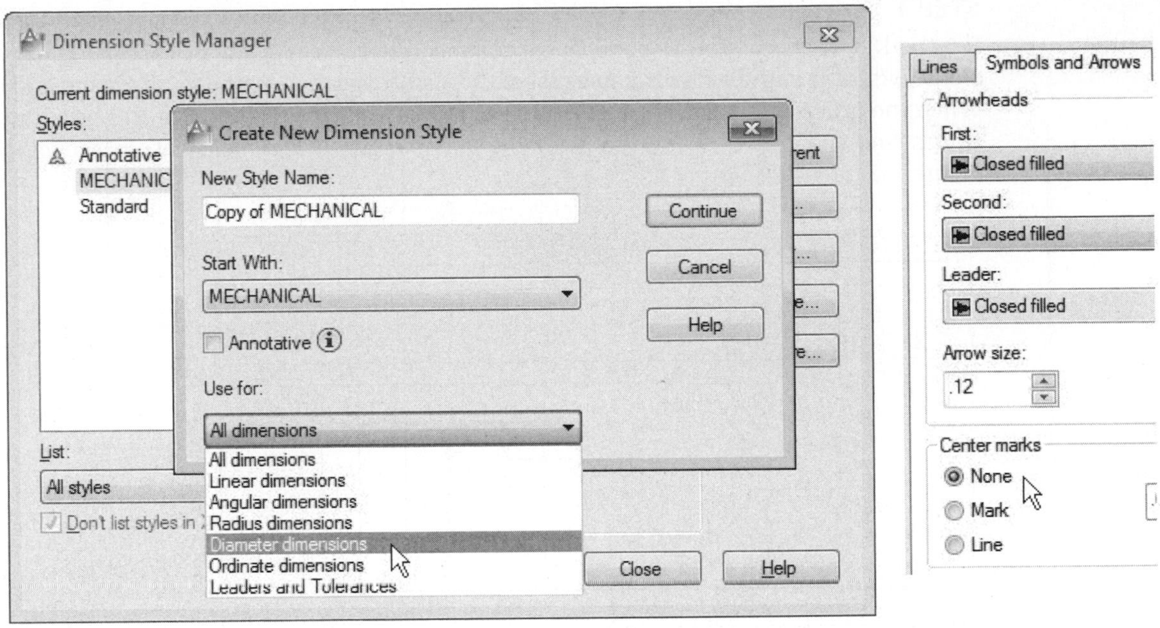

FIGURE 11.80

STEP 5

In the Dimension Style Manager dialog box, verify that MECHANICAL is the current dimension style. If not, select MECHNICAL in the Styles area and then click the Set Current button. Your display should appear similar to Figure 11.81. Click the Close button to save all changes and return to the drawing.

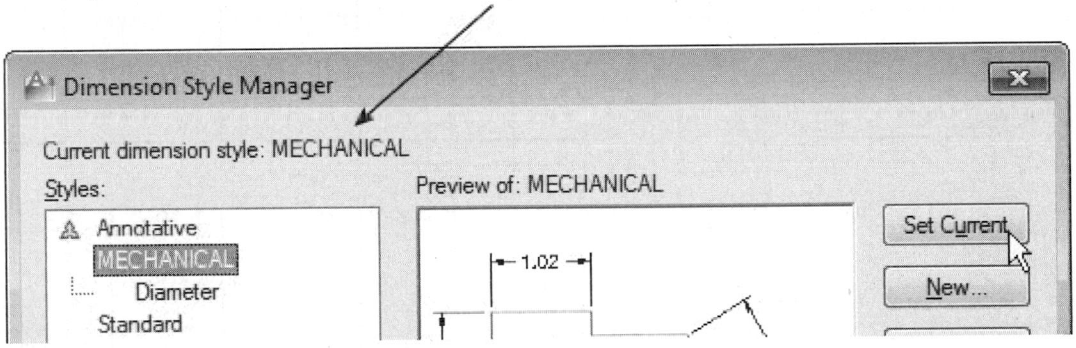

FIGURE 11.81

STEP 6

Make the Center layer current. Begin placing center markers to identify the centers of all circular features in the Top view. Use the `DIMCENTER` command (or the shortcut `DCE`) to perform this operation on circles "A" through "E," as shown in Figure 11.82.

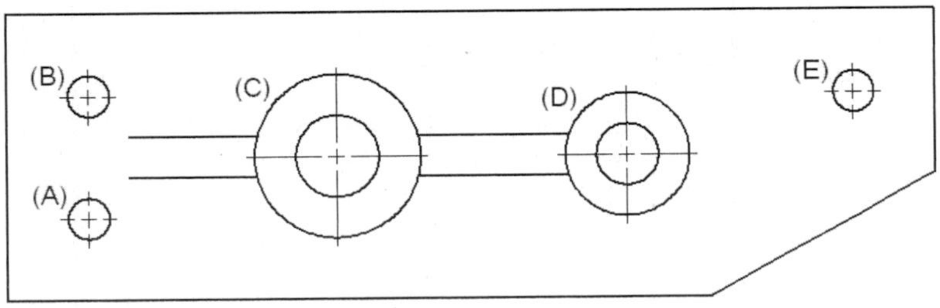

FIGURE 11.82

STEP 7

Set the Dim layer current. Use the QDIM command to place a string of baseline dimensions. First select the individual lines labeled "A" through "F," as shown in Figure 11.83. When the group of dimensions previews as continued dimensions, change this grouping to baseline and click a location to place the baseline group of dimensions, as shown in Figure 11.83.

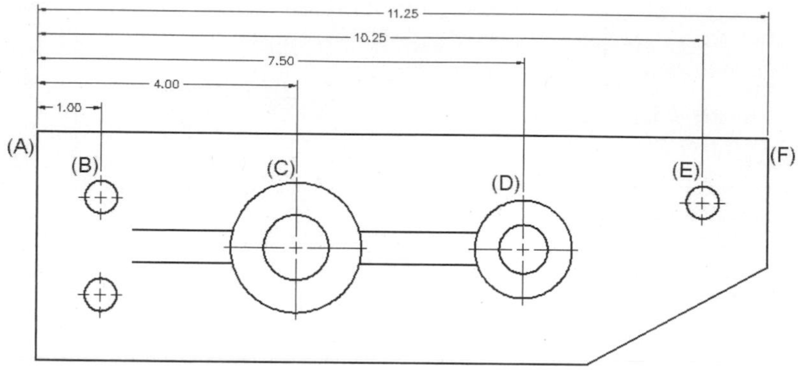

FIGURE 11.83

STEP 8

Magnify the left side of the Top view using the ZOOM command. Then use the DIMLINEAR command (or DLI shortcut) to place the 0.75 vertical dimension, as shown in Figure 11.84 (Left). Next, use the DIMCONTINUE command (or DCO shortcut) to place the next dimension in line with the previous dimension, as shown in Figure 11.84 in the middle. Then use the DIMLINEAR command (or DLI shortcut) to place the 1.75 vertical dimension, as shown in Figure 11.84 (Right). Use grips to place the text as shown.

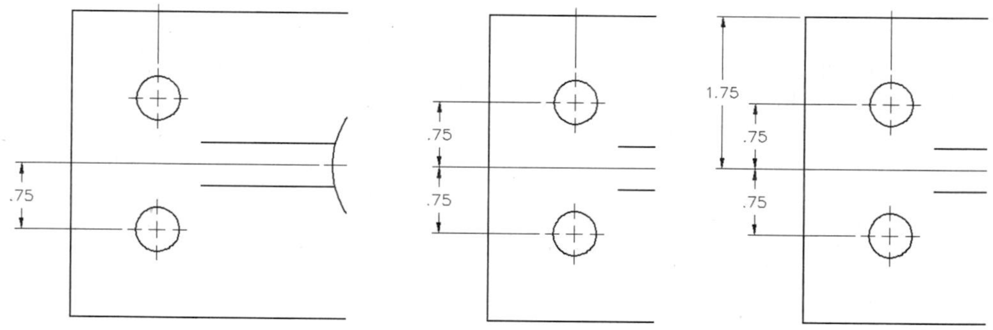

FIGURE 11.84

STEP 9

Use the PAN command to slide over to the right side of the Top view while keeping the same zoom percentage. Then use the QDIM command and select the lines "A" through "C" in Figure 11.85. When the group of dimensions previews, change the mode to Baseline, if necessary, and click a location to place the baseline group of dimensions, as shown in Figure 11.85. You may have to identify a new base point (datumPoint) in order for your dimension to match the illustration. Then use the DIMANGULAR command (or DAN shortcut) to place the 61° dimension in Figure 11.85.

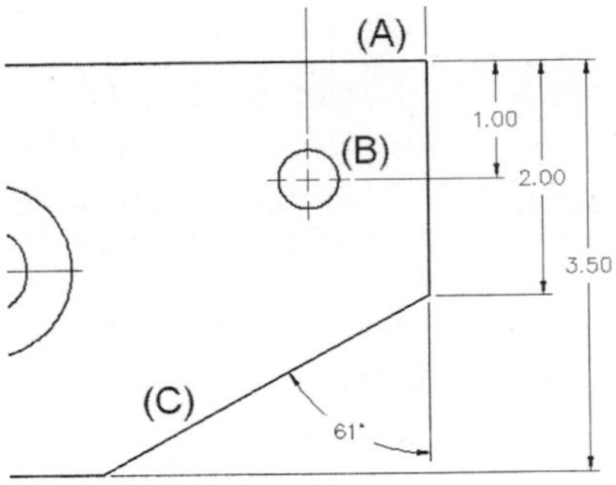

FIGURE 11.85

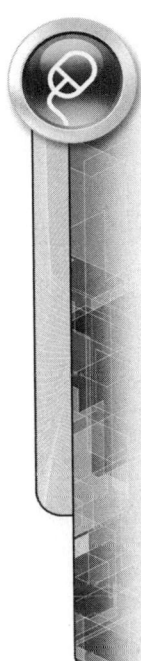

STEP 10

Use the ZOOM command and the Extents option to display both the Front and Top views. Then use the DIMDIAMETER command (or DDI shortcut) to place two diameter dimensions, as shown in Figure 11.86.

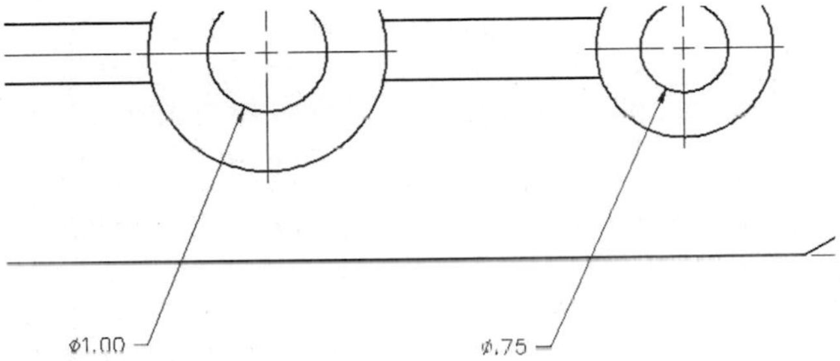

FIGURE 11.86

STEP 11

Place a diameter dimension using the DIMDIAMETER (DDI) command on the circle, as shown in Figure 11.87.

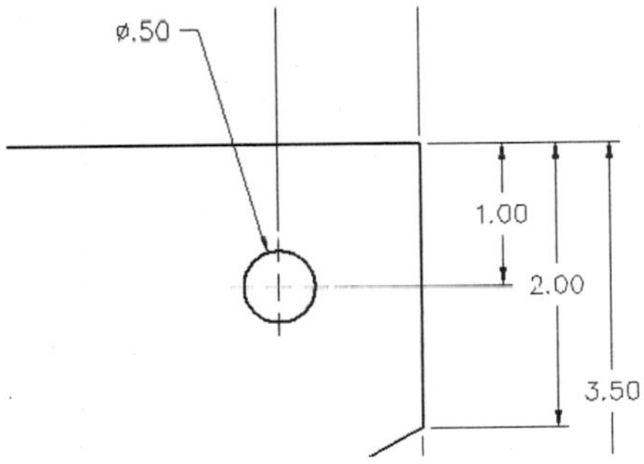

FIGURE 11.87

STEP 12

Since the two other smaller holes share the same diameter value, double-click the dimension or use the DDEDIT command (or ED shortcut) to edit this dimension value. Clicking the diameter value activates the Text Formatting dialog box. Begin by typing 3X to signify three holes of the same diameter, as shown in Figure 11.88 (Left). Click the Close Text Editor button to return to your drawing. The results are illustrated in Figure 11.88 (Right).

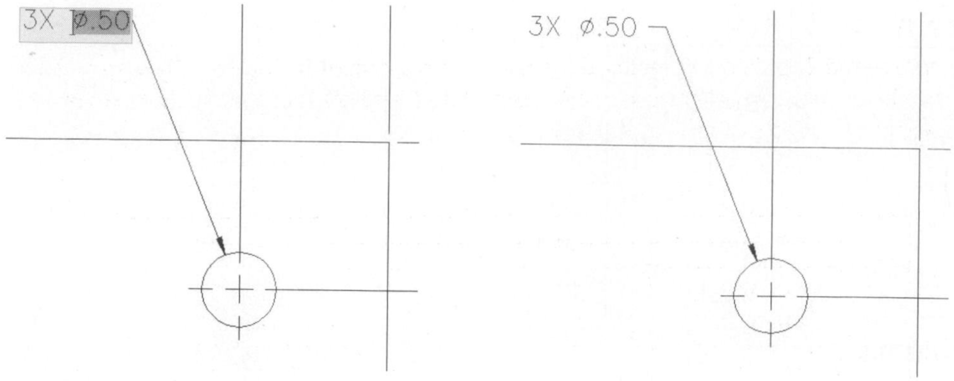

FIGURE 11.88

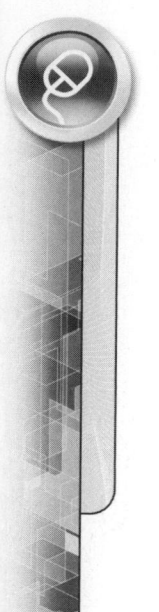

STEP 13

One more dimension needs to be placed in the Top view: the 0.50 width of the rib. Unfortunately, because of the placement of this dimension, extension lines will be drawn on top of the object's lines. This is poor practice. To remedy this, a dimension override will be created. To do this, activate the Dimension Style Manager dialog box and click the Override button, as shown in Figure 11.89 (Left). This displays the Override Current Style: MECHANICAL dialog box. In the Lines tab, place checks in the boxes to Suppress (turn off) Ext line 1 and Ext line 2 in the Extension lines area, as shown in Figure 11.89 (Center).

Click the OK button and notice the new <style overrides> listing under MECHANICAL, as shown in Figure 11.89 (Right). Notice also that the dialog box's preview image lacks extension lines. Click the Close button to return to the drawing.

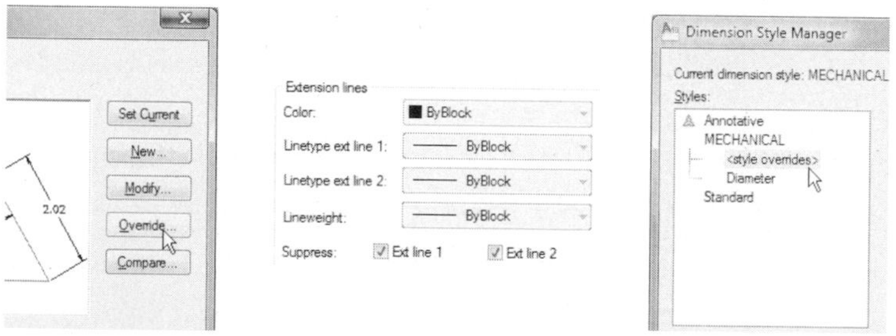

FIGURE 11.89

STEP 14

Pan to the middle area of the Top view and place the 0.50 rib-width dimension while in the dimension style override, as shown in Figure 11.90. This dimension should be placed without having any extension lines visible. Also, use grips to drag the dimension text inside the visible object lines (temporarily turn off OSNAP to better accomplish this task).

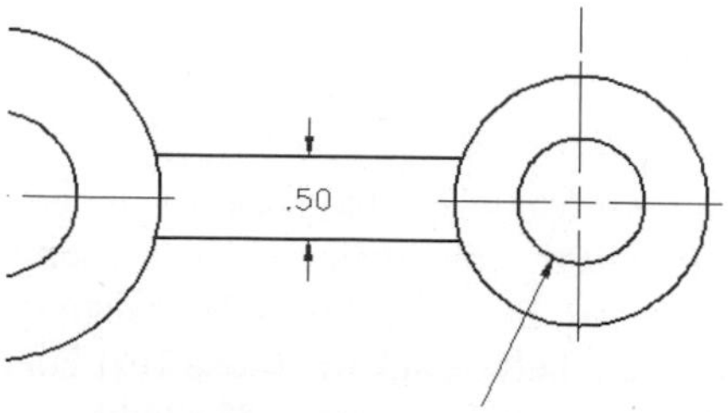

FIGURE 11.90

STEP 15

The completed Top view, including dimensions, is illustrated in Figure 11.91. Use this figure to check that all dimensions have been placed and all features such as holes and chamfers have been properly identified.

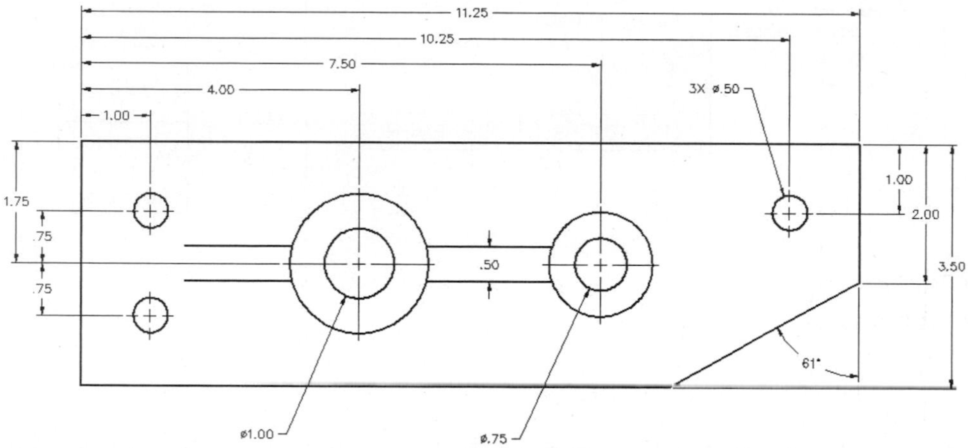

FIGURE 11.91

STEP 16

The Front view in Figure 11.92 will be the focus for the next series of dimensioning steps. Again use the ZOOM and PAN commands whenever you need to magnify or slide to a better drawing view position.

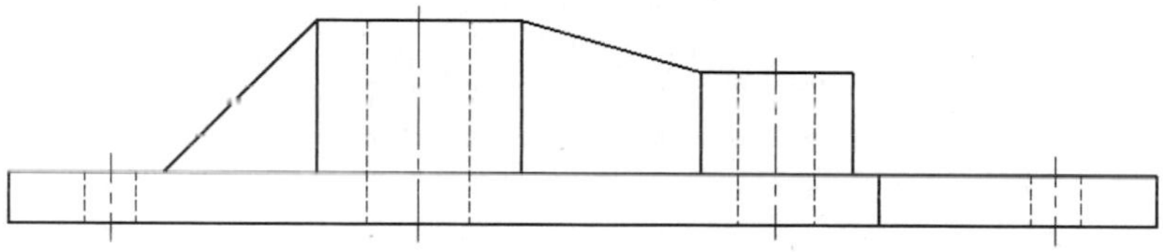

FIGURE 11.92

STEP 17

Before continuing, activate the Dimension Style Manager dialog box, click the MECHANICAL style, and then click the Set Current button. An AutoCAD Alert box displays, as shown in Figure 11.93. Making MECHANICAL current discards any style overrides. Click the OK button to discard the changes because you need to return to having extension lines visible in linear dimensions. Click Close to dismiss the Dimension Style Manager dialog box.

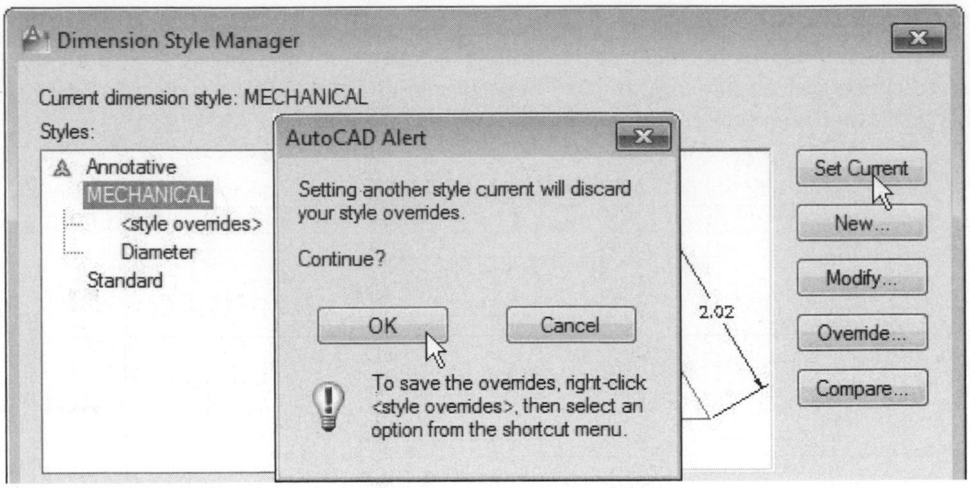

FIGURE 11.93

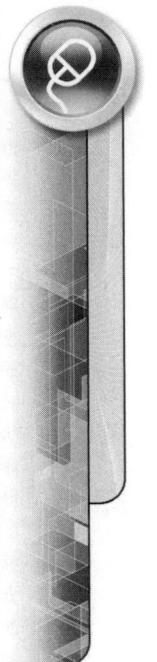

STEP 18

Place two horizontal linear dimensions that signify the diameters of the two cylinders using the `DIMLINEAR` command, as shown in Figure 11.94. Also add the 1.50 vertical dimension to the side of the Front view, as shown in Figure 11.94.

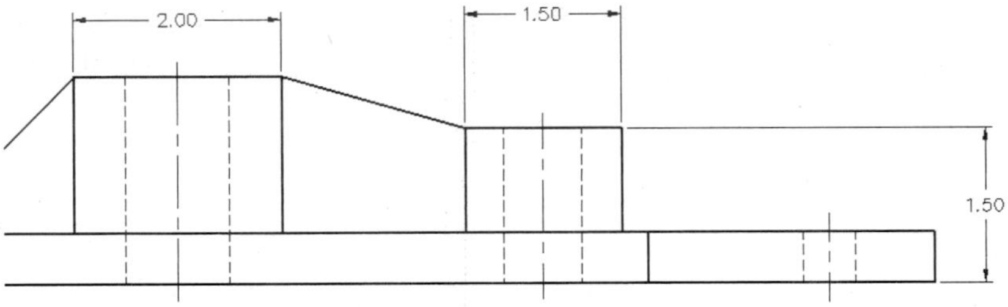

FIGURE 11.94

STEP 19

Use the `QDIM` command to place the vertical baseline dimensions, as shown in Figure 11.95. Use the DIMANGULAR command (or DAN shortcut) to place the 45° angular dimension in Figure 11.95.

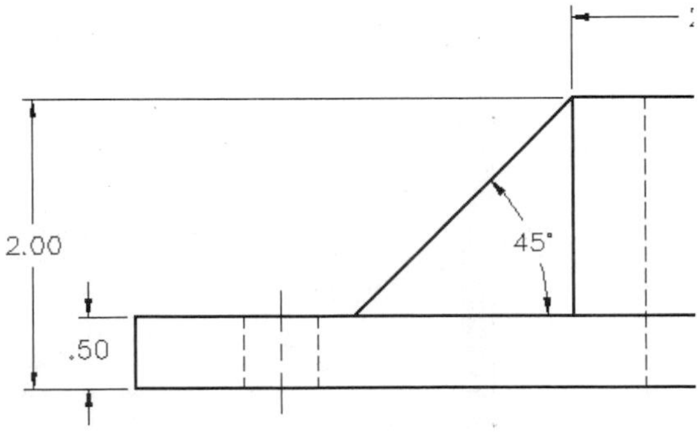

FIGURE 11.95

STEP 20

The remaining step is to add the diameter symbol to the 2.00 and 1.50 dimensions, since the dimension is placed on a cylindrical object. Use the DIMEDIT command (or DED for shortcut) with the New option to accomplish this. When the text edit box appears, enter the diameter symbol by picking Symbol and Diameter from the Ribbon menu, as shown in Figure 11.96. You could instead, create the diameter symbol by entering "%%C". Close the text editor and pick the two dimensions (3.00 and 1.50) to add the diameter symbols.

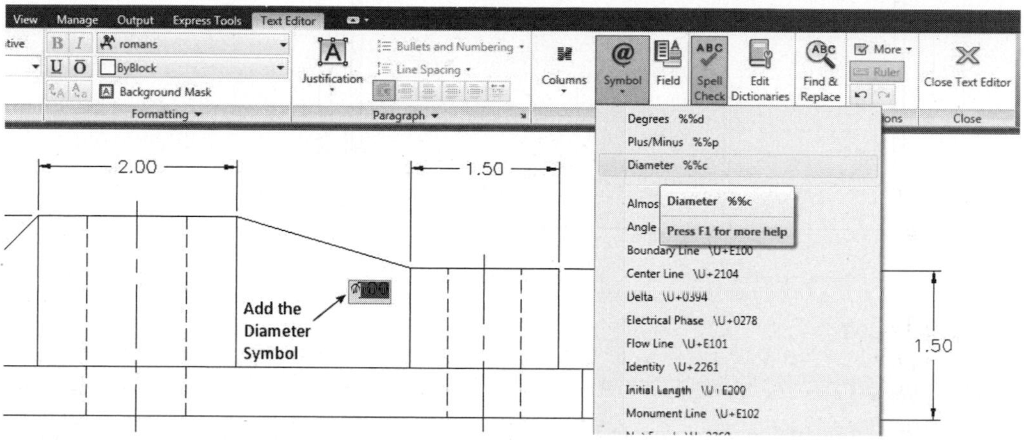

FIGURE 11.96

Once you have completed all dimensioning steps, your drawing should appear similar to Figure 11.97.

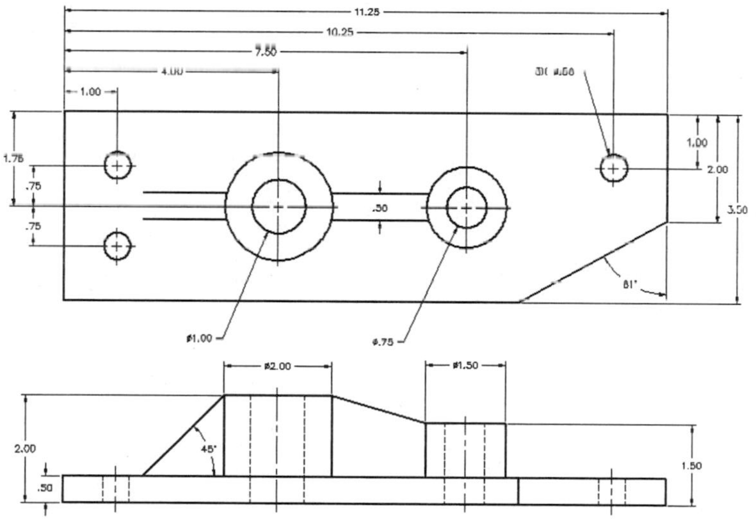

FIGURE 11.97

TUTORIAL EXERCISE: 11_ARCHITECTURAL DIMENSION.DWG

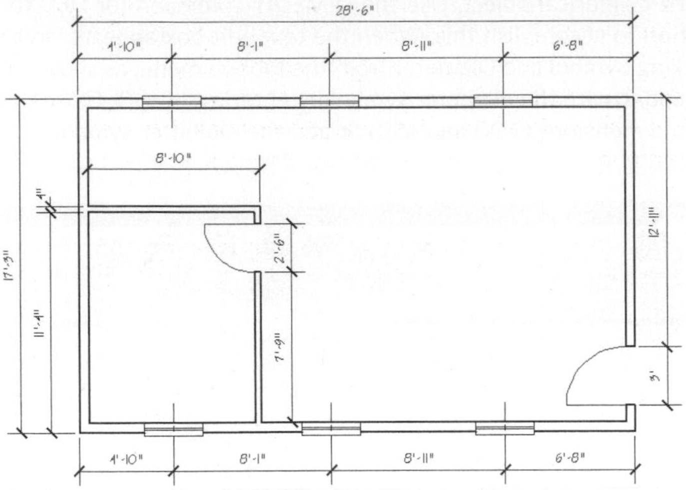

FIGURE 11.98

Purpose

This tutorial is designed to develop and apply dimension styles to an architectural drawing project shown in Figure 11.98. The dimensions will be applied to a Model Space environment.

System Settings

No special settings are needed for this project.

Layers

Layers have already been created for this drawing.

Suggested Commands

An architectural dimension style will be created to reflect the settings required for an architectural drawing. Various horizontal and vertical dimensions will be placed around the floor plan to call out distances between windows, doors, and walls.

STEP 1

Open the drawing file 11_Architectural Dimension, as shown in Figure 11.99. Check to make sure that the dimension layer is the current layer.

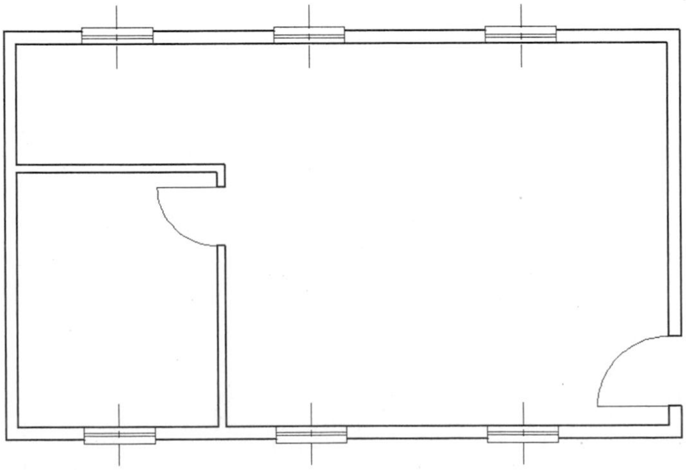

FIGURE 11.99

STEP 2

The Dimension Style Manager will be used to set up characteristics of the dimensions used in architectural applications. Activate the Dimension Style Manager dialog box (DIMSTYLE command or D for shortcut). Clicking the New. . . button launches the Create New Dimension Style dialog box, as shown in Figure 11.100. Replace Copy of Standard in the New Style Name field with Architectural. This dimension style will initially be based on the current standard dimension style and applied to all dimensions that are placed in the drawing. Click the Continue button to make various changes to the dimension settings.

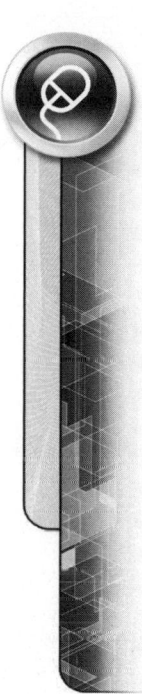

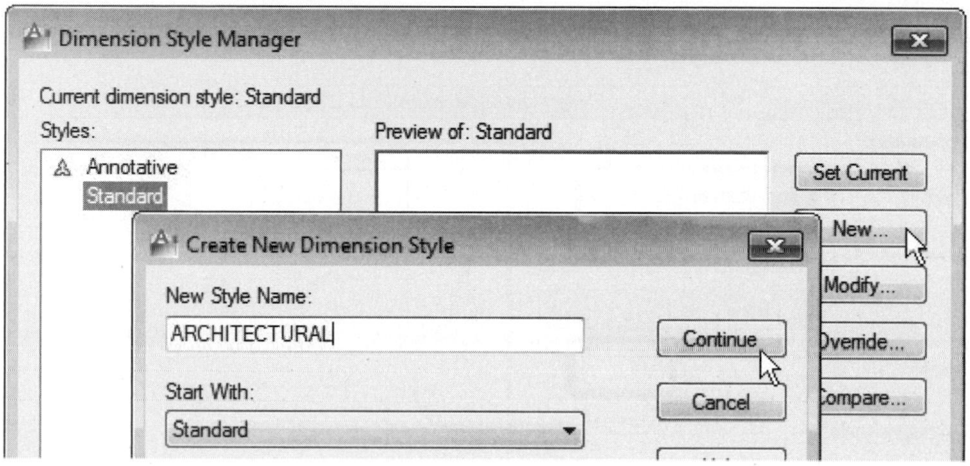

FIGURE 11.100

STEP 3

Note that the name Architectural is now displayed in the title bar of the New Dimension Style dialog box, as shown in Figure 11.101. In the Symbols and Arrows section, change the arrowhead style to Architectural tick. When you change the First arrowhead, it also changes the Second arrowhead. Also change the arrow size to 1/8", as shown in Figure 11.101.

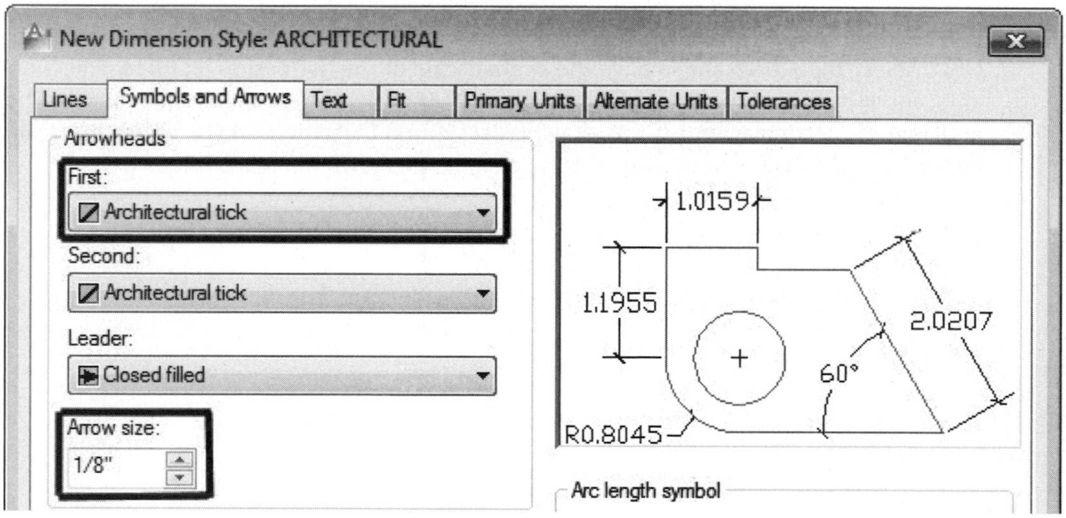

FIGURE 11.101

STEP 4

Select the Text tab and make changes to the following items, as shown in Figure 11.102. Change the Text style from Standard to Architectural (this text style already exists in the drawing). As with the arrowheads, change the Text height to 1/8". Next, change the Vertical Text placement from Centered to Above. This will place dimension text above the dimension line, which is the architectural standard. Finally, in the Text alignment area, click the radio button next to Aligned with dimension line. This forces the dimension text to display parallel to the dimension line, which represents another architectural standard.

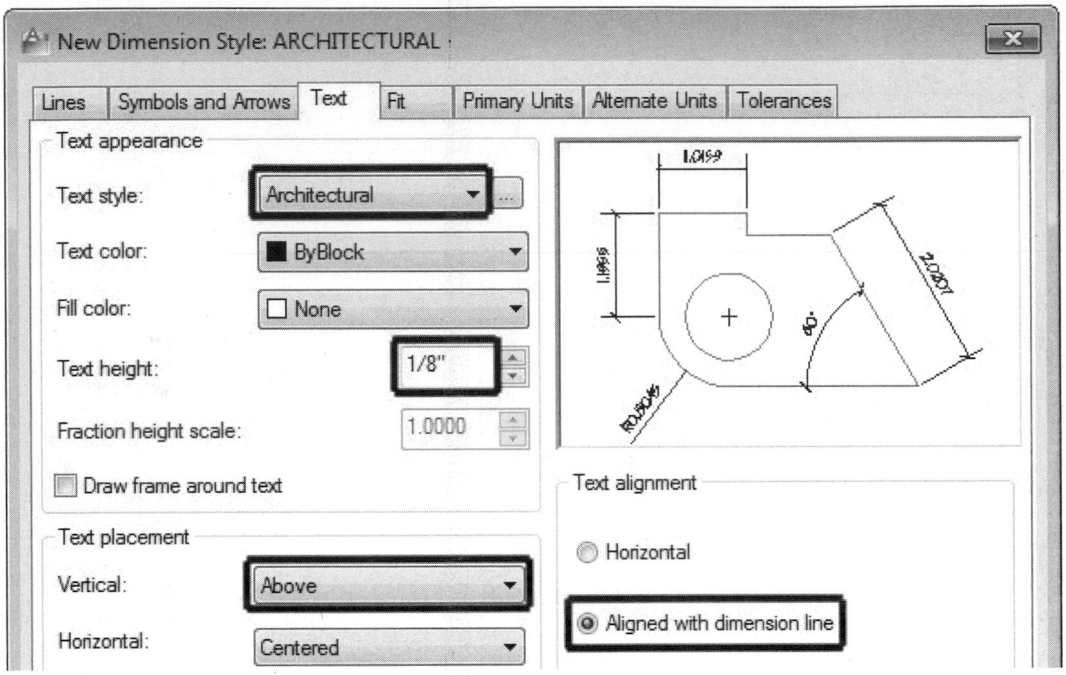

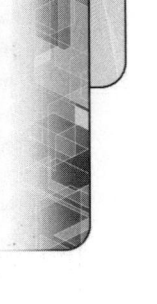

FIGURE 11.102

STEP 5

Next, select the Fit tab and change the overall scale of the dimension features. The overall scale is based on the plotting scale when plotting is done from the model tab. The anticipated plotting scale for this drawing is 1/4" = 1' (same as 1:48). Place the scale factor value of 48 in the Use overall scale of field, as shown in Figure 11.103.

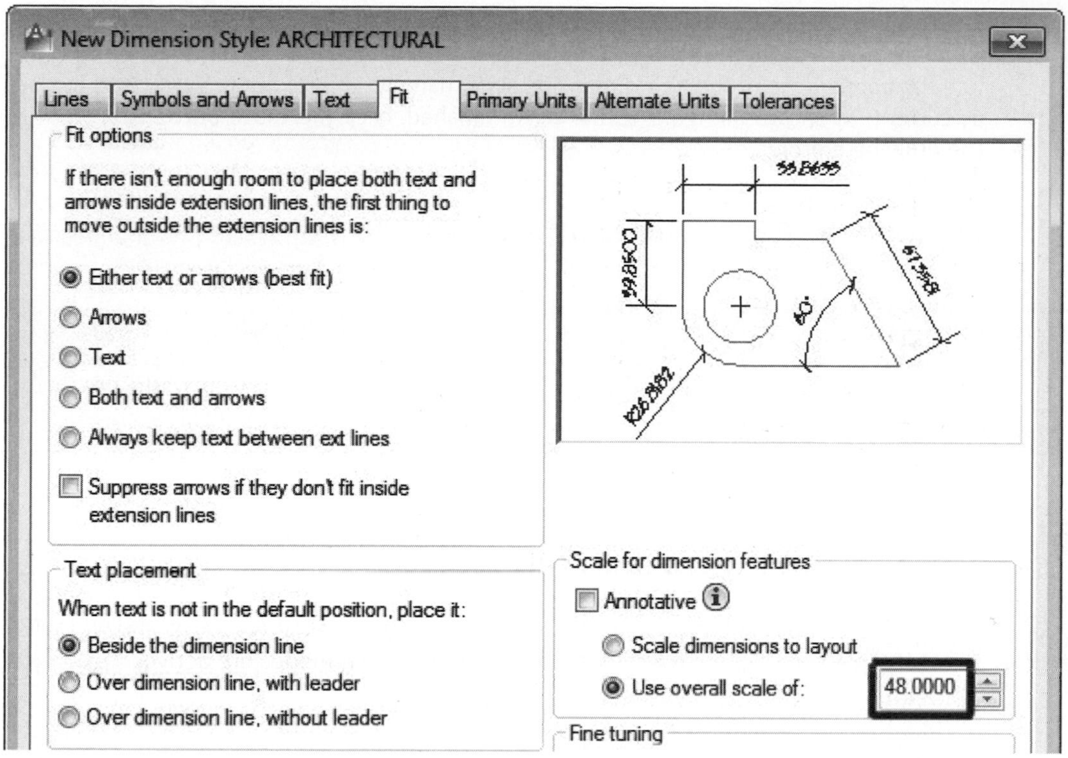

FIGURE 11.103

STEP 6

Next, choose the Primary Units tab. In the Linear dimensions section, change the Unit format option to Architectural. Also, change the precision to 0'-0". Your display should appear similar to Figure 11.104.

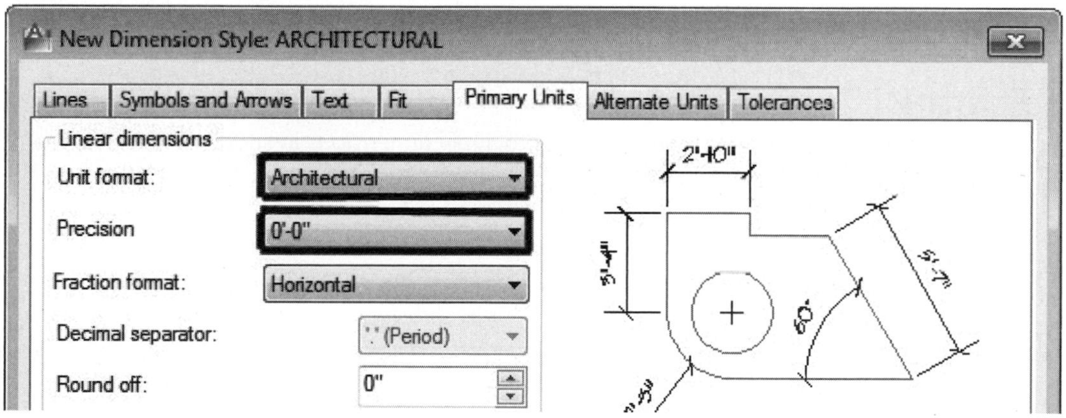

FIGURE 11.104

STEP 7

Clicking the OK button returns you to the main Dimension Style Manager dialog box. Notice the appearance of the new dimension style, namely Architectural. Verify that it is set as current, as shown in Figure 11.105. When finished, click the Close button to return to the drawing editor.

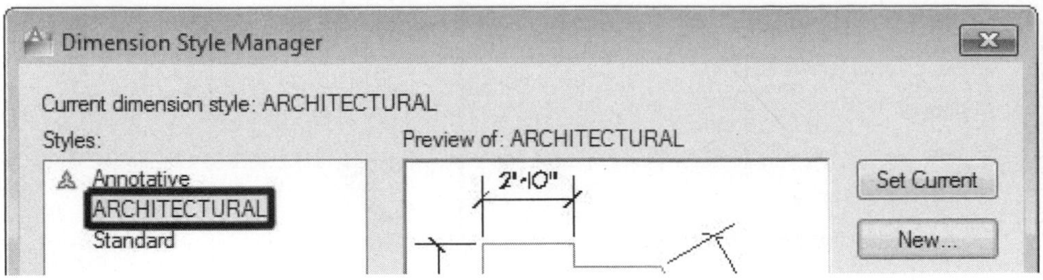

FIGURE 11.105

STEP 8

Place a horizontal dimension using the `DIMLINEAR` command (`DLI`), as shown in Figure 11.106. Be sure that OSNAP Endpoint and Intersection modes are active.

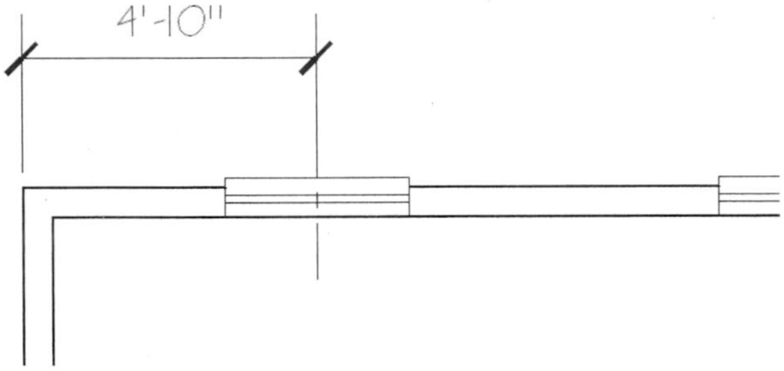

FIGURE 11.106

STEP 9

Use the `DIMCONTINUE` command (`DCO`) to apply dimensions in succession to dimension the window locations, as shown in Figure 11.107.

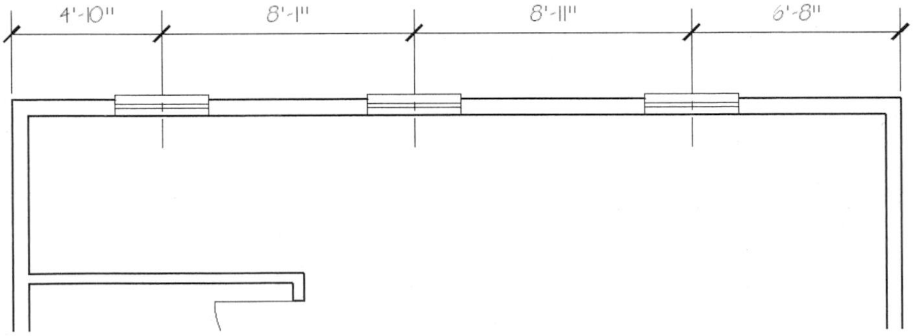

FIGURE 11.107

STEP 10

Next add an overall length dimension using the DIMLINEAR command (DLI), as shown in Figure 11.108.

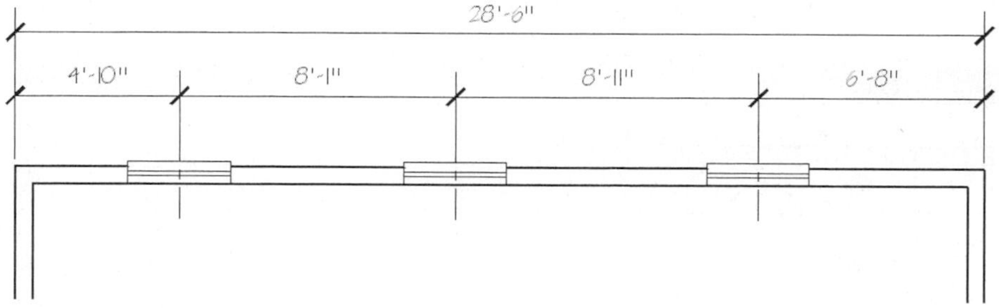

FIGURE 11.108

STEP 11

Continue adding other dimensions to the floor plan, including those defining vertical and interior dimensions of the building, as shown in Figure 11.109.

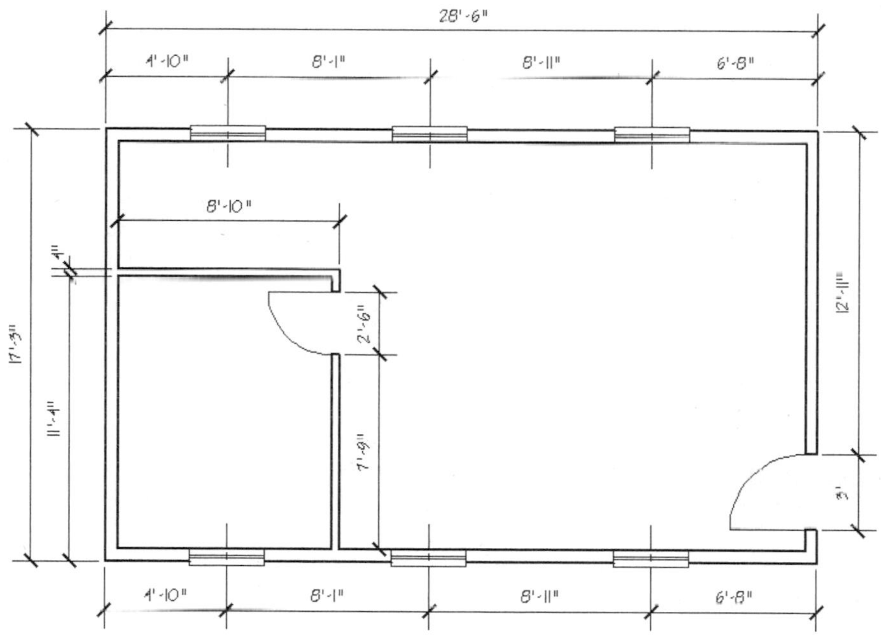

FIGURE 11.109

PROBLEMS FOR CHAPTER 11

DIRECTIONS FOR PROBLEMS 11-1 THROUGH 11-18:

1. Open each drawing.
2. Use proper techniques to dimension each drawing.

PROBLEM 11-1

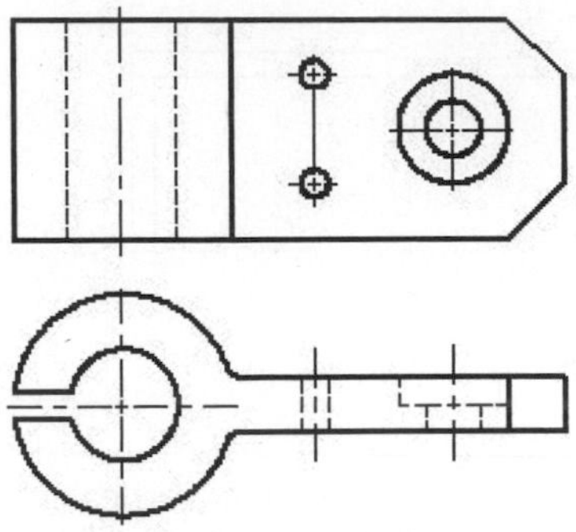

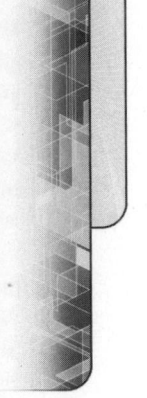

ALL FILLETS = R.125

PROBLEM 11-2

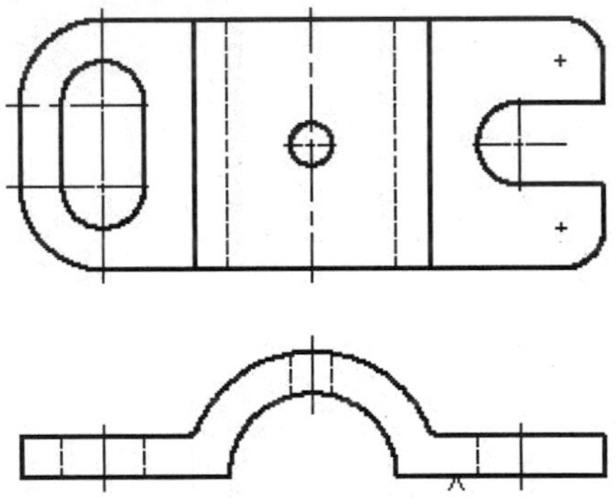

ALL FILLETS = R.12

PROBLEM 11-3

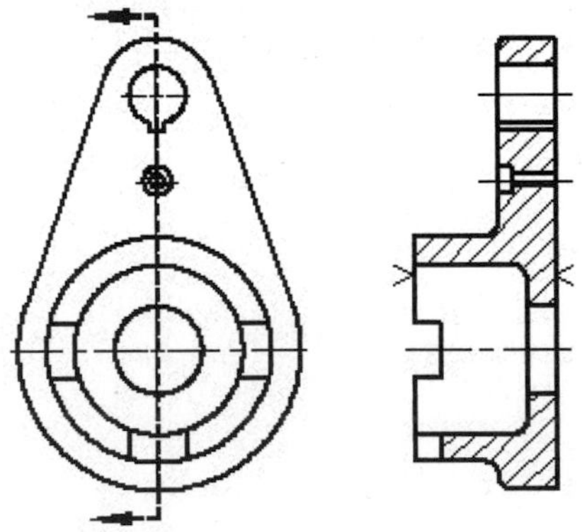

ALL FILLETS = R.12

PROBLEM 11-4

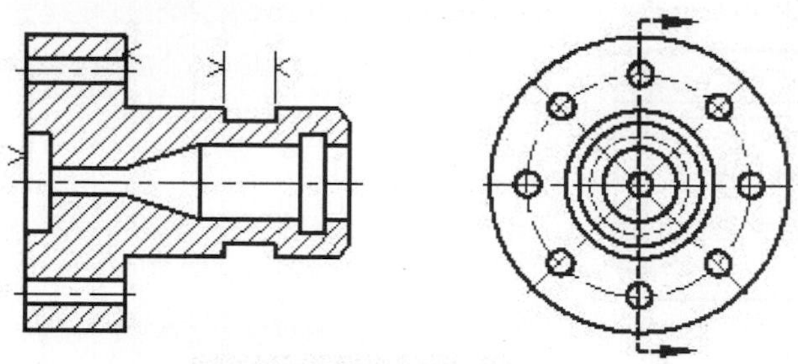

ALL CHAMFERS .125 x 45°

PROBLEM 11-5

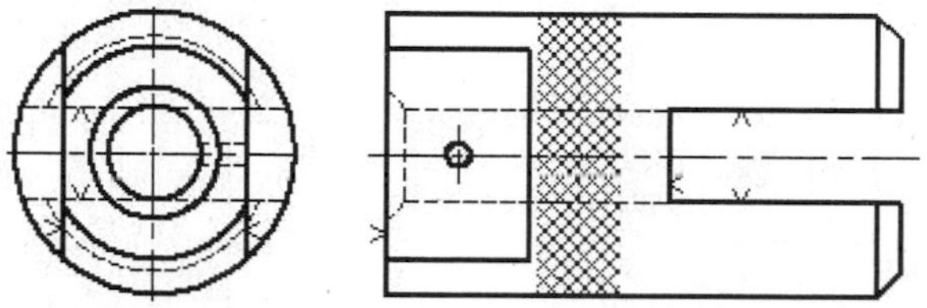

PROBLEM 11-6

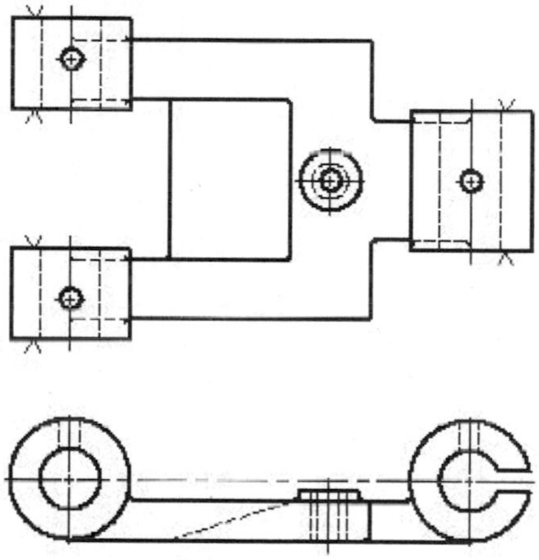

ALL FILLETS = R.12

PROBLEM 11-7

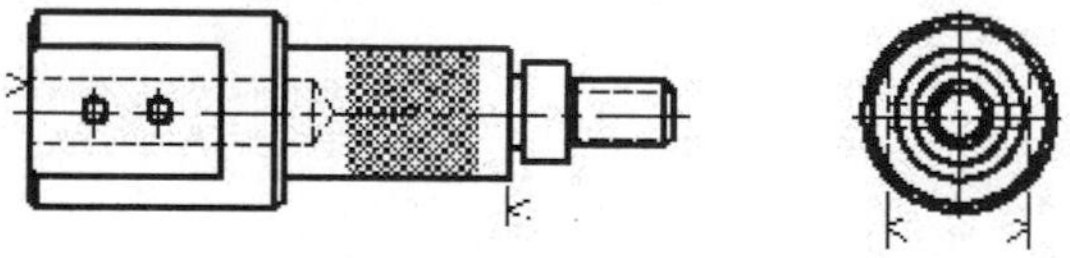

ALL CHAMFERS = .125 x 45°

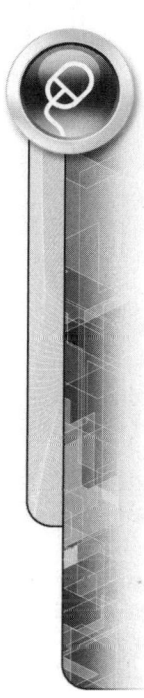

PROBLEM 11-8

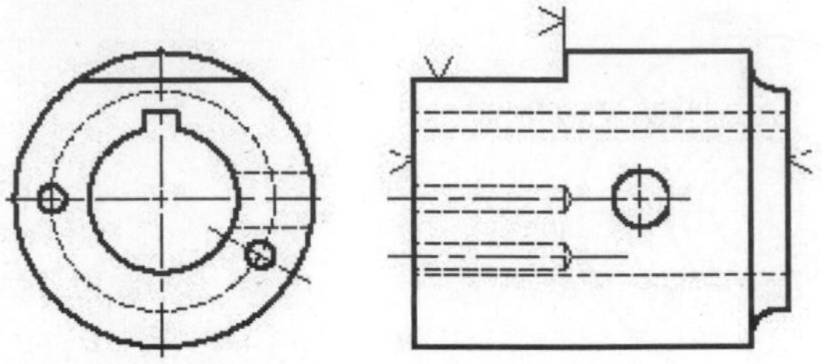

PROBLEM 11-9

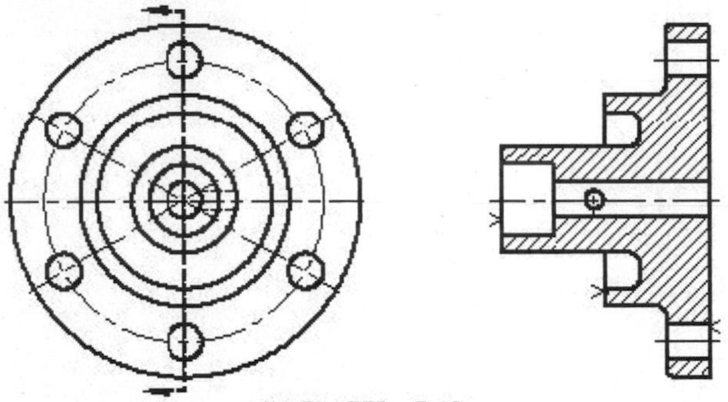

ALL FILLETS = R.12

PROBLEM 11-10

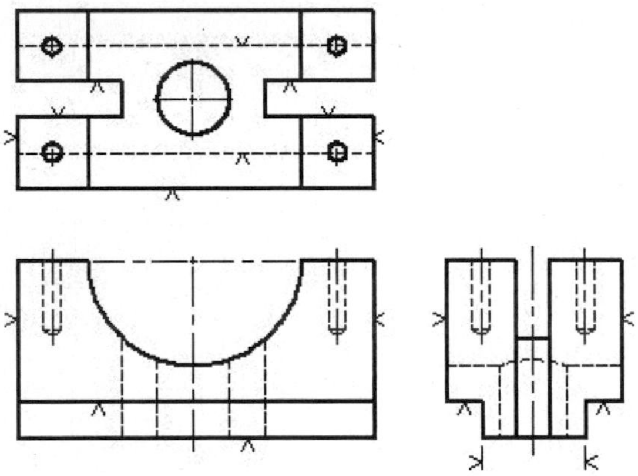

PROBLEM 11-11

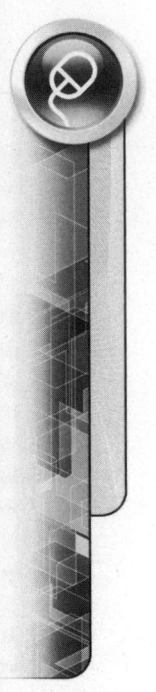

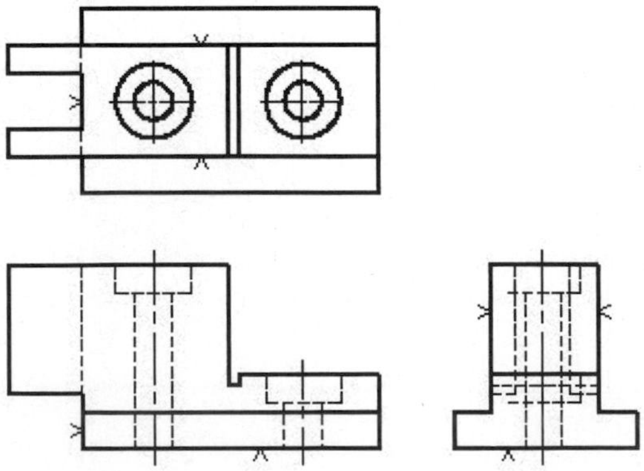

PROBLEM 11-12

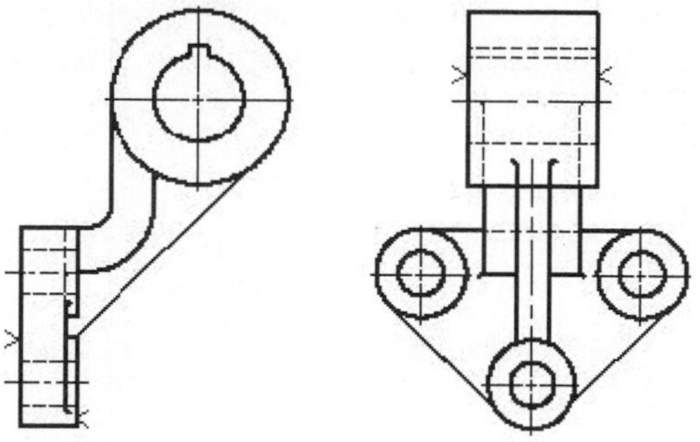

ALL FILLETS = R.0625

PROBLEM 11-13

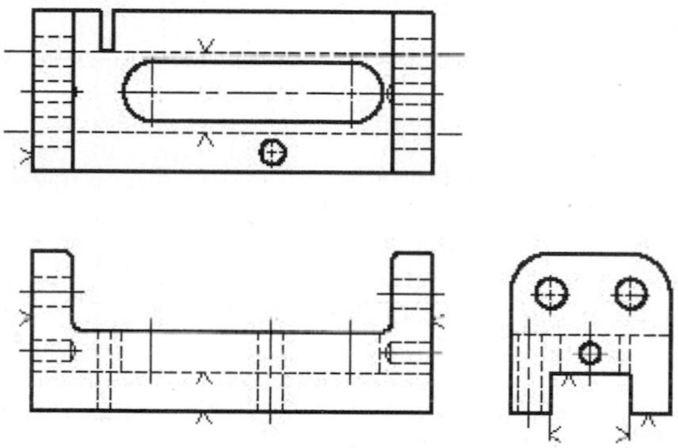

ALL FILLETS = R.12

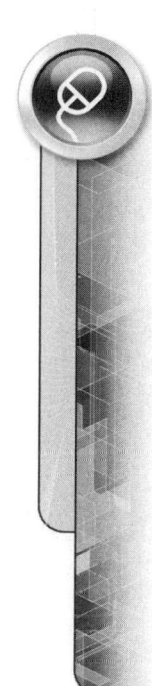

PROBLEM 11-14

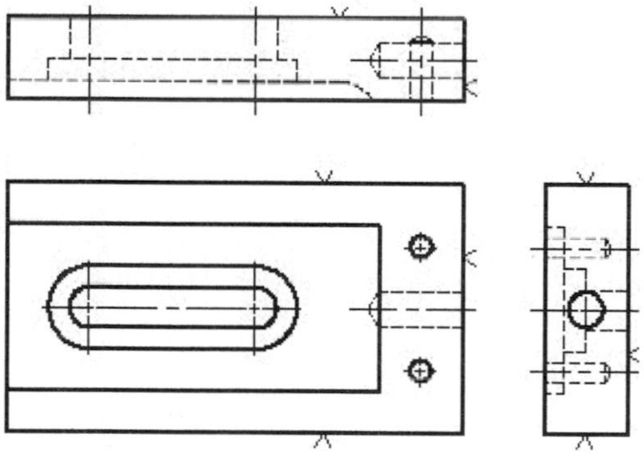

PROBLEM 11-15

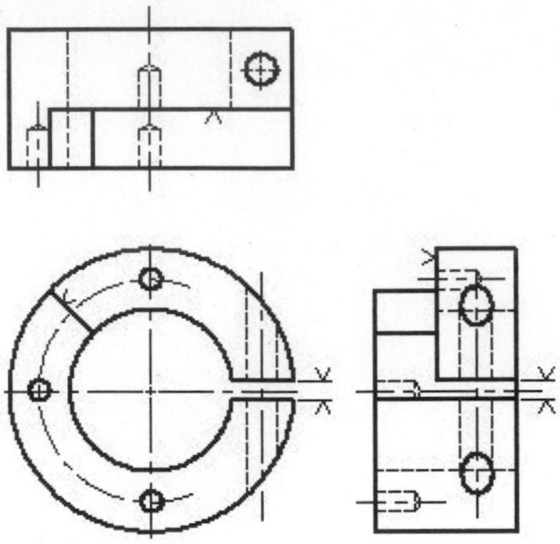

PROBLEM 11-16

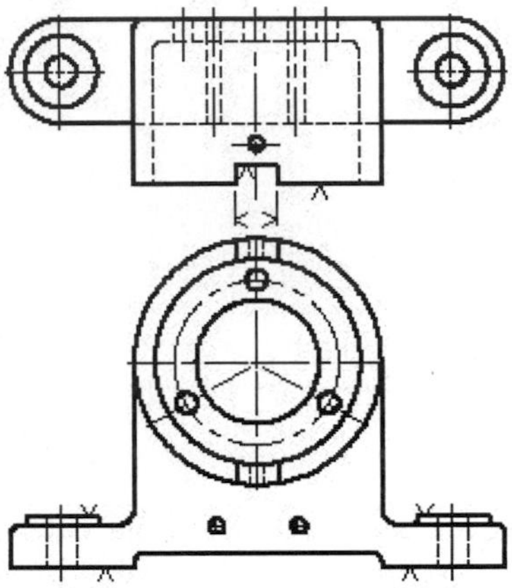

ALL FILLETS = R.12

PROBLEM 11-17

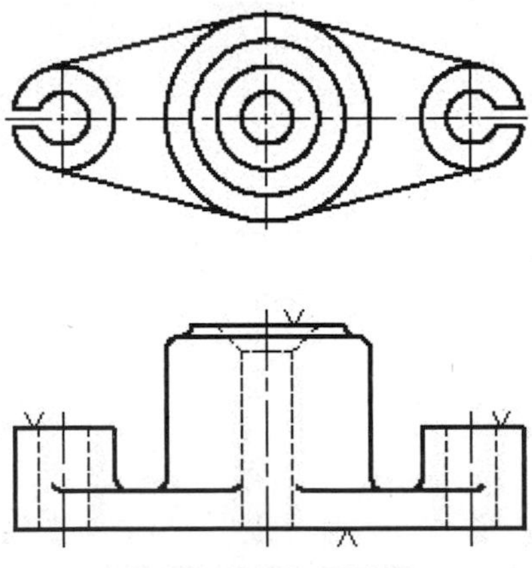

ALL FILLETS = R.12

PROBLEM 11-18

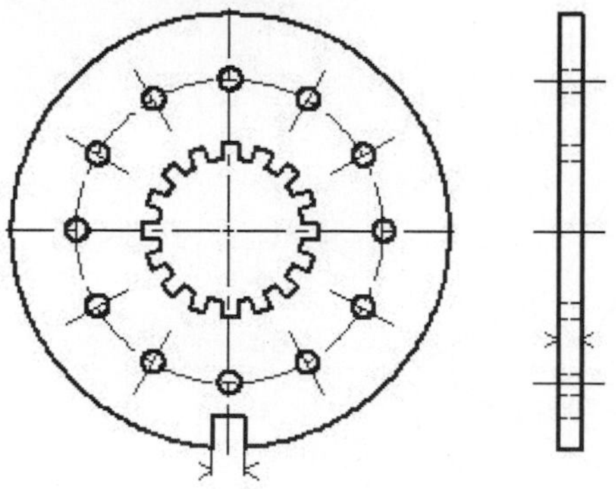

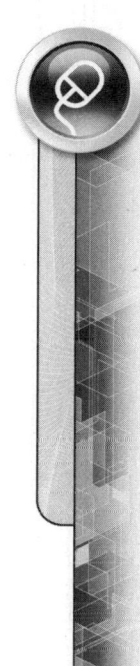

DIRECTIONS FOR PROBLEMS 11-19 THROUGH 11-24:

1. Use the isometric drawings provided to develop orthographic view drawings, showing as many views as necessary to communicate the design.
2. Fully dimension your drawings.

PROBLEM 11-19

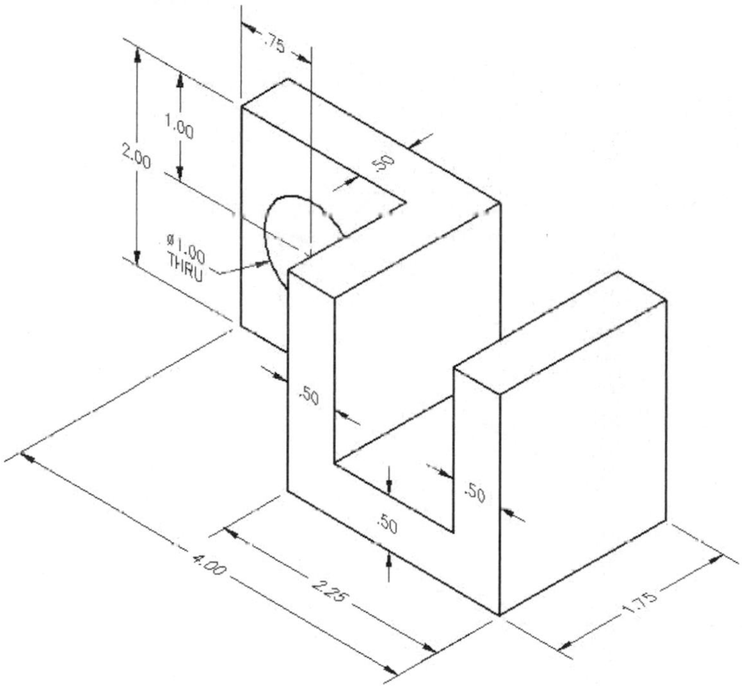

PROBLEM 11-20

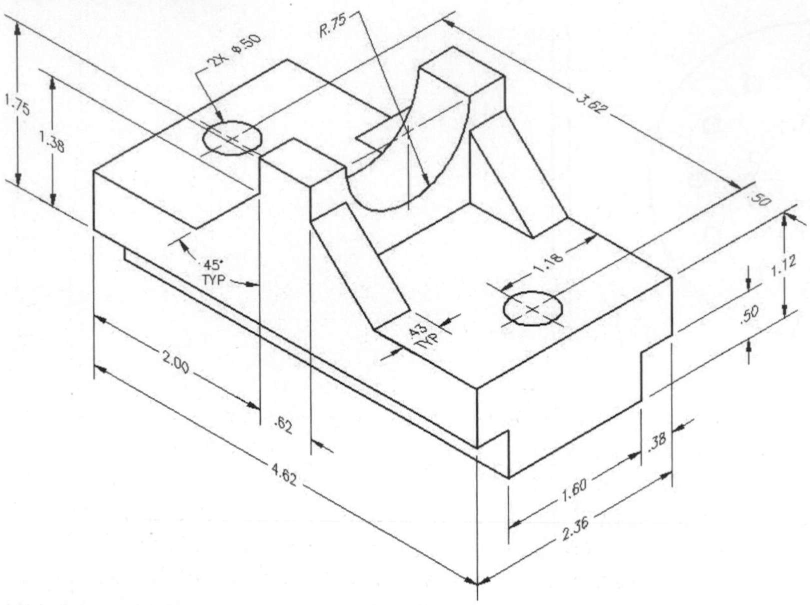

PROBLEM 11-21

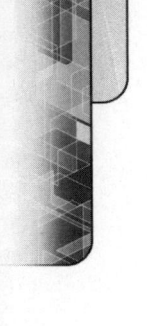

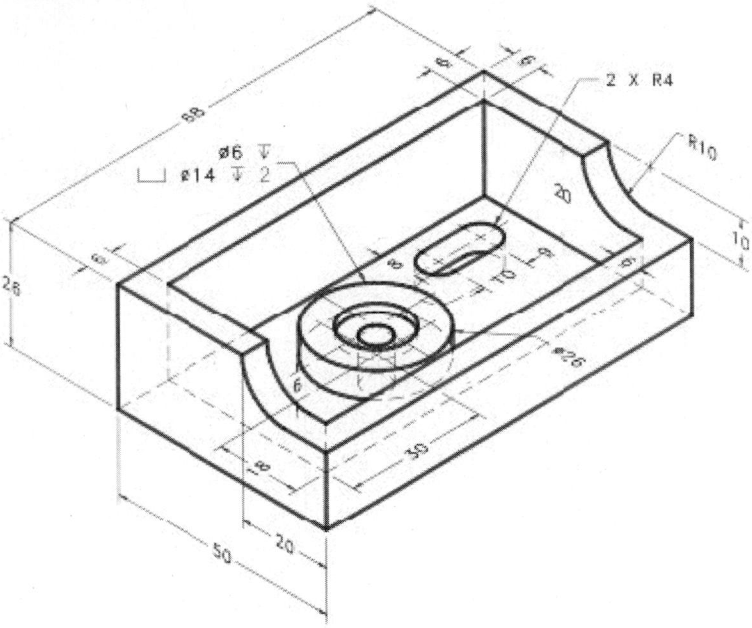

PROBLEM 11-22

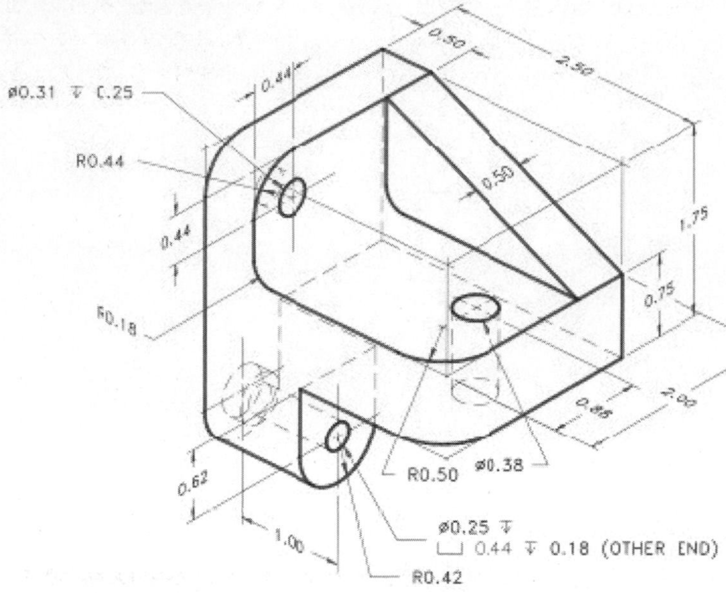

PROBLEM 11-23

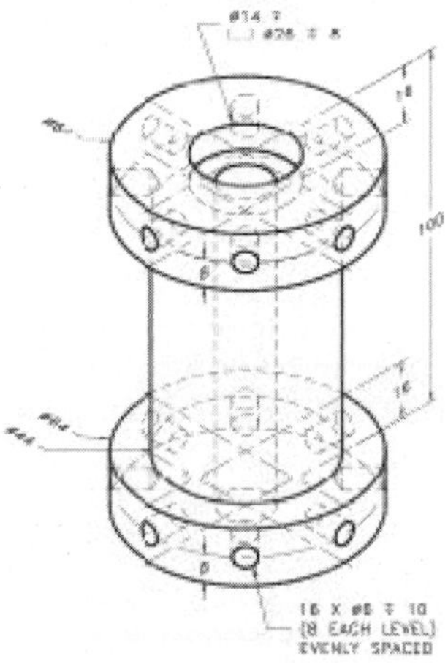

PROBLEM 11-24

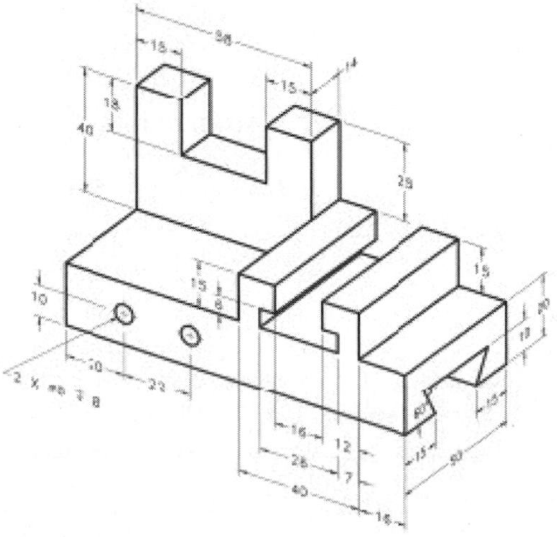

Image(s) © Cengage Learning 2013

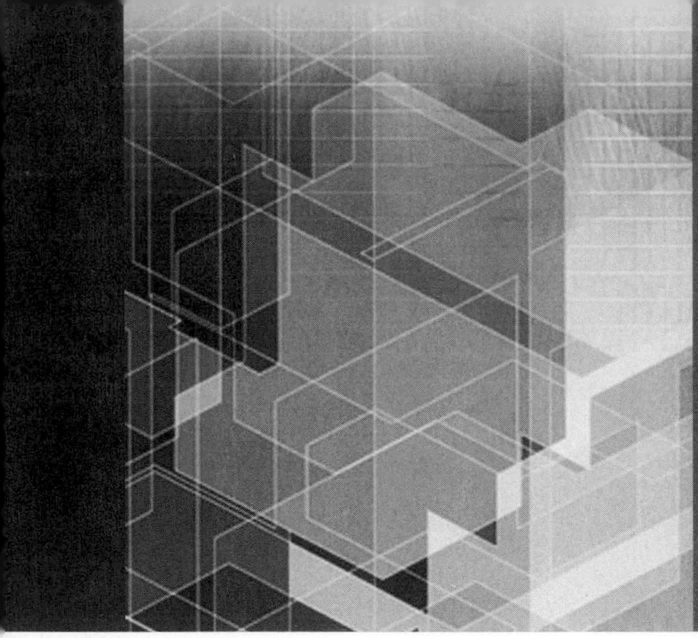

CHAPTER 12

Analyzing 2D Drawings

This chapter will show how a series of commands may be used to calculate distances and angles of selected objects. It will also show how surface areas may be calculated on complex geometric shapes. The following pages highlight AutoCAD's Measure Geometry commands and show how you can use them to display useful information on an object or group of objects.

USING MEASURE GEOMETRY COMMANDS

You can choose AutoCAD's Measure Geometry commands by using the Home tab of the Ribbon as illustrated in Figure 12.1.

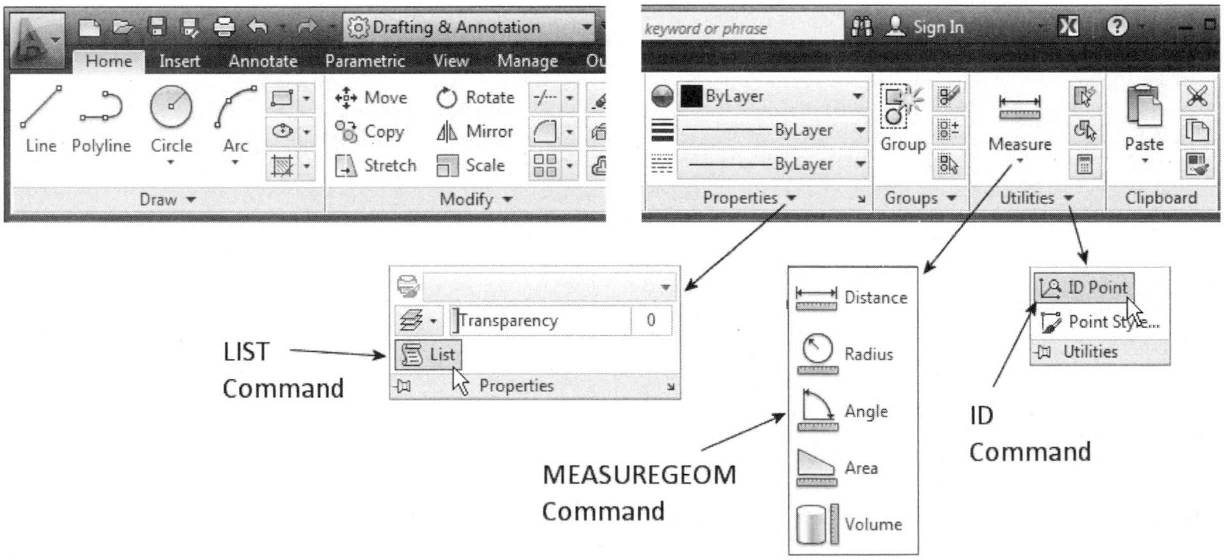

FIGURE 12.1

NOTE

The results provided by the MEASUREGEOM, ID, and LIST commands, discussed in the following pages, are controlled by the Drawing Units dialog box. Use the UNITS command to activate this dialog box and set the unit type and precision desired before using these commands.

FINDING THE AREA OF AN ENCLOSED SHAPE

The ARea option of the MEASUREGEOM command is used to calculate the area of any drawing shape. Several methods are available and one of the simplest is through the selection of a series of points. Select the endpoints of all vertices in the following image with the OSNAP-Endpoint option. Once you have selected the first point along with the remaining points in either a clockwise or counterclockwise pattern, respond to the prompt "Next point:" by pressing ENTER to calculate the area of the shape. As these points are picked, a transparent green mask appears identifying the area being created. Along with the area is a calculation for the perimeter.

TRY IT!

Open the drawing file 12_Area1. Use Figure 12.2 and the prompt sequence below for finding the area by identifying a series of points.

Command: MEA *(For MEASUREGEOM)*

Enter an option [Distance/Radius/Angle/ARea/Volume] <Distance>: AR *(For ARea)*

Specify first corner point or [Object/Add area/Subtract area/eXit] <Object>: *(Select endpoint at "A")*

Specify next point or [Arc/Length/Undo]: *(Select endpoint at "B")*

Specify next point or [Arc/Length/Undo]: *(Select endpoint at "C")*

Specify next point or [Arc/Length/Undo/Total] <Total>: *(Select endpoint at "D")*

Specify next point or [Arc/Length/Undo/Total] <Total>: *(Select endpoint at "E")*

Specify next point or [Arc/Length/Undo/Total] <Total>: *(Press ENTER for the total area)*

Area = 25.25, Perimeter = 20.35

Enter an option [Distance/Radius/Angle/ARea/Volume/eXit] <ARea>: X *(For eXit)*

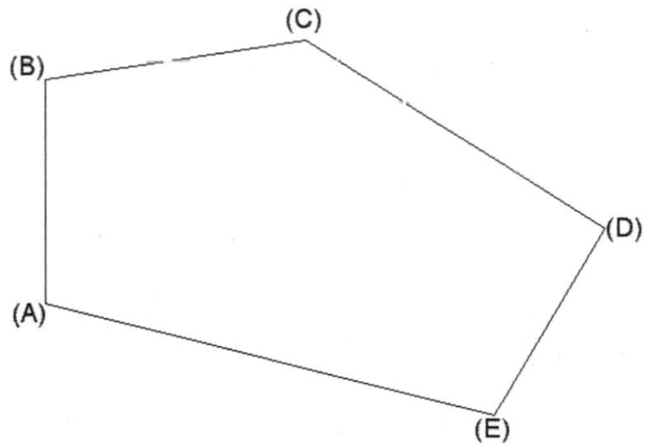

FIGURE 12.2

FINDING THE AREA OF AN ENCLOSED POLYLINE OR CIRCLE

The method of finding the area in the previous example involved identifying the corners and intersections of an enclosed area by a series of points. For a complex area, this could be a very tedious operation. As a result, the `ARea` option has a built-in Object modifier that calculates the area and perimeter of a polyline and the area and circumference of a circle. Finding the area of a polyline can only be accomplished if one of the following conditions is satisfied:

- The shape must have already been constructed through the `PLINE` command.
- The shape must have already been converted to a polyline through the `PEDIT` command if originally constructed from individual objects.

TRY IT!

Open the drawing file 12_Area2. Use Figure 12.3 and the prompt sequence below for finding the area of both shapes.

```
Command: MEA (For MEASUREGEOM)

Enter an option [Distance/Radius/Angle/ARea/Volume]
<Distance>: AR (For ARea)

Specify first corner point or [Object/Add area/Subtract
area/eXit] <Object>: O (For Object)

Select objects: (Select the polyline at "A")

Area = 24.88, Perimeter = 19.51

Enter an option [Distance/Radius/Angle/ARea/Volume/eXit]
<ARea>: (Press ENTER to repeat the ARea option)

Specify first corner point or [Object/Add area/Subtract
area/eXit] <Object>: O (For Object)

Select objects: (Select the circle at "B")

Area = 7.07, Circumference = 9.42

Enter an option [Distance/Radius/Angle/ARea/Volume/eXit]
<ARea>: X(For eXit)
```

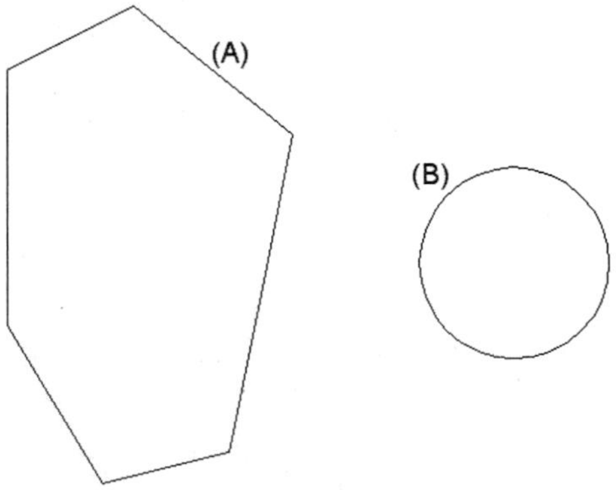

FIGURE 12.3

TRY IT!

Open the drawing file 12_Extrude1. In Figure 12.4, convert all line segments into a single polyline object using the PEDIT command and the Join option (see the following Command prompt sequence). Use the MEASUREGEOM command and Object option to answer Question 1 regarding 12_Extrude1.

Command: PE *(For PEDIT)*

Select polyline or [Multiple]: *(Pick any line segment)*

Object selected is not a polyline

Do you want to turn it into one <Y>: *(Press ENTER for Yes)*

Enter an option [Close/Join/Width/Edit vertex/Fit/Spline/Decurve/Ltype gen/Reverse/Undo]: J *(For Join)*

Select objects: *(Use a Window to select all lines)*

Select objects: *(Press ENTER)*

69 segments added to polyline

Enter an option [Close/Join/Width/Edit vertex/Fit/Spline/Decurve/Ltype gen/Reverse/Undo]: *(Press ENTER to complete the operation)*

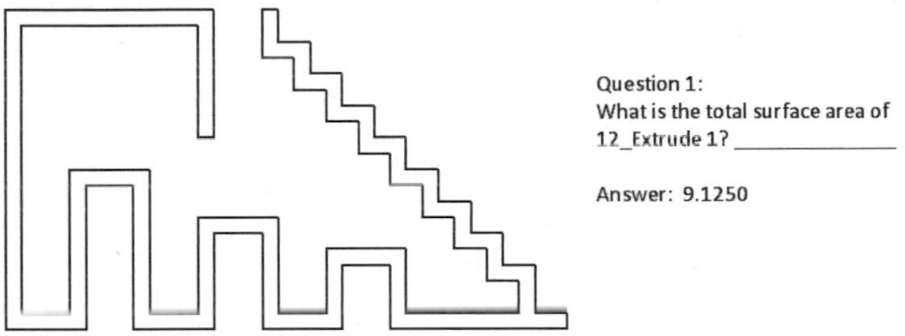

Question 1:
What is the total surface area of 12_Extrude 1? ____________
Answer: 9.1250

FIGURE 12.4

TRY IT!

Open the drawing file 12_Extrude2. In Figure 12.5, use the BOUNDARY command to trace a polyline on the top of all the line segments. To help separate the Boundary polyline from the existing Object lines, make the Boundary layer current prior to executing the Command prompt sequence shown. Use the MEASUREGEOM command and Object option to answer Question 1 regarding 12_Extrude2.

Command: BO *(For BOUNDARY)*

(When the Boundary Creation dialog box appears, click the Pick points button.)

Pick internal point: *(Pick a point inside of the object)*

Selecting everything...

Selecting everything visible...

Analyzing the selected data...

Analyzing internal islands...

Pick internal point: *(Press ENTER to create the boundaries)*

BOUNDARY created 1 polyline

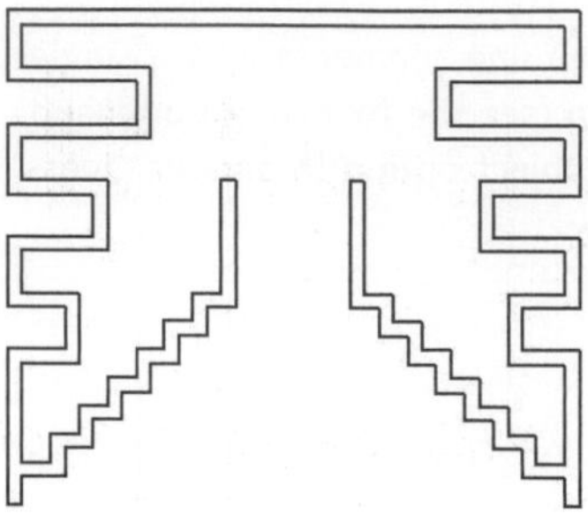
Question 1:
What is the total surface area of 12_Extrude 2? ___________
Answer: 28.9362

FIGURE 12.5

FINDING THE AREA OF A SHAPE BY SUBTRACTION

Some shapes may include holes or cutouts that should not be included in the area calculation. In this case, the steps you use to calculate the total surface area are: (1) calculate the area of the outline and (2) subtract the objects inside the outline. All individual objects, except circles, should first be converted to polylines through the PEDIT command. Next, find the overall area and add it to the database using the Add mode of the ARea option. Exit the Add mode and remove the inner objects using the Subtract mode of the AREA command. Care must be taken when selecting the objects to subtract. If an object is selected twice, it is subtracted twice and may yield an inaccurate area in the final calculation. As you add objects, a transparent green mask appears to identify the area being added; when you subtract objects, a transparent brown mask appears to identify the area being subtracted, as shown in Figure 12.6 (Right).

In Figure 12.6, the total area with the circle and rectangle removed is 30.4314.

TRY IT!

Open the drawing file 12_Area3. Use Figure 12.6 and the prompt sequence to verify this area calculation.

Command: Command: MEA *(For MEASUREGEOM)*

Enter an option [Distance/Radius/Angle/ARea/Volume] <Distance>: AR *(For ARea)*

Specify first corner point or [Object/Add area/Subtract area/eXit] <Object>: A *(For Add area)*

Specify first corner point or [Object/Subtract area]: O

(For Object)

(ADD mode) Select objects: *(Select the polyline at "A")*

Area = 47.5000, Perimeter = 32.0000

Total area = 47.5000

(ADD mode) Select objects: *(Press* ENTER *to exit ADD mode)*

Area = 47.5000, Perimeter = 32.0000

Total area = 47.5000

Specify first corner point or [Object/Subtract area/eXit]: S *(For Subtract area)*

Specify first corner point or [Object/Add area/eXit]: O

(For Object)

(SUBTRACT mode) Select objects: *(Select the rectangle at "B")*

```
Area = 10.0000, Perimeter = 13.0000
Total area = 37.5000
(SUBTRACT mode) Select objects: (Select the circle at "C")
Area = 7.0686, Circumference = 9.4248
Total area = 30.4314
(SUBTRACT mode) Select objects: (Press ENTER to exit SUBTRACT
mode)
Specify first corner point or [Object/Add area eXit]: X (For
eXit)
Enter an option [Distance/Radius/Angle/ARea/Volume/eXit]
<ARea>: X (For eXit)
```

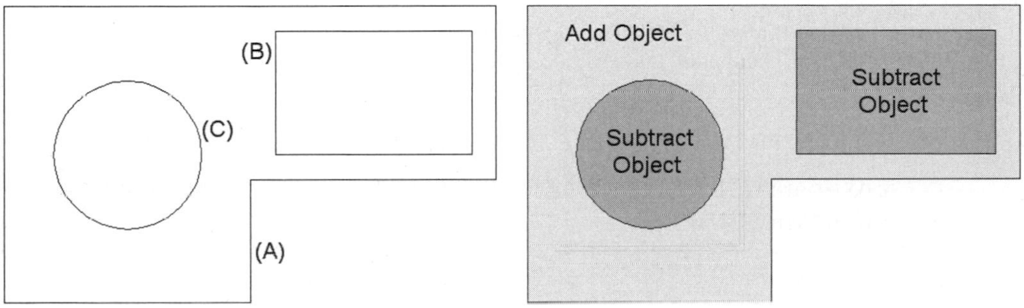

FIGURE 12.6

TRY IT!

Open the drawing file 12_Shield. In Figure 12.7, use the BOUNDARY **command to trace a polyline on the top of all the line segments. Subtract all four-sided shapes from the main shape using the** ARea **option of the MEASUREGEOM command. Answer Question 1 regarding 12_Shield.**

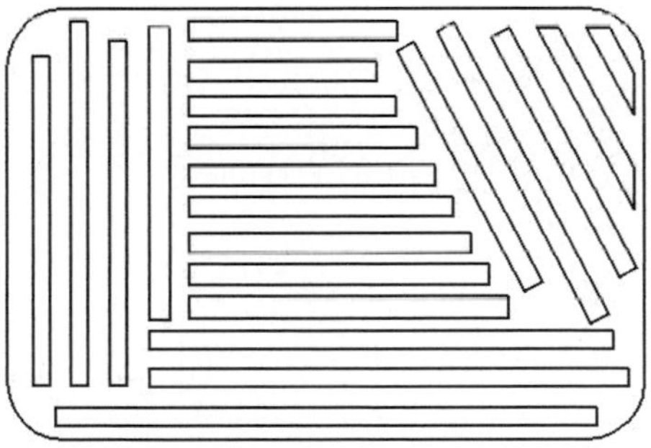

Question 1:
What is the total surface area of 12_Shield with all inner, four-sided shapes removed? ____________

Answer: 44.2246

FIGURE 12.7

MEASURING DISTANCE

The Distance option of the MEASUREGEOM command calculates the linear distance between two selected points, as shown in Figure 12.8. This option can be used to determine the distance of a line, the distance between two unrelated points, or the distance from the center of one circle to the center of another circle. This method of calculating distances also graphically shows the angle in the XY plane, the angle from the XY plane (always zero for a 2D drawing), and the delta XYZ coordinate values. The angle in the XY plane is given in the current angular mode set by the Drawing

Units dialog box. The delta XYZ coordinate is a relative coordinate value taken from the first point identified by the Distance option of the MEASUREGEOM command to the second point.

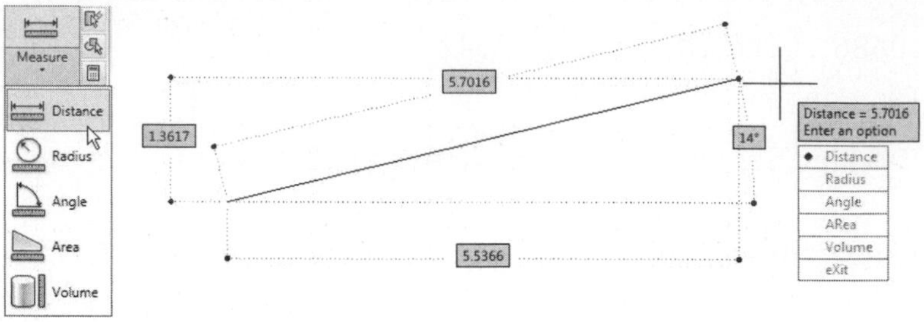

FIGURE 12.8

TRY IT!

Open the drawing file 12_Distance. Use Figure 12.9 and the prompt sequences below for the MEASUREGEOM command and Distance option.

Command: MEA *(For MEASUREGEOM)*

Enter an option [Distance/Radius/Angle/ARea/Volume] <Distance>: D *(For Distance)*

Specify first point: *(Select the endpoint at "A")*

Specify second point or [Multiple points] : *(Select the endpoint at "B")*

Distance = 6.36, Angle in XY Plane = 45.0000, Angle from XY Plane = 0.0000

Delta X = 4.50, Delta Y = 4.50, Delta Z = 0.00

Enter an option [Distance/Radius/Angle/ARea/Volume/eXit] <Distance>: X *(For eXit)*

A Multiple points option is available, which allows you to calculate a running distance between all points selected.

Command: MEA *(For MEASUREGEOM)*

Enter an option [Distance/Radius/Angle/ARea/Volume] <Distance>: D *(For Distance)*

Specify first point: *(Select the endpoint at "A")*

Specify second point or [Multiple points]: M *(For Multiple Points)*

Specify next point or [Arc/Length/Undo/Total] <Total>: *(Select the endpoint at "B")*

Distance - 6.36

Specify next point or [Arc/Close/Length/Undo/Total] <Total>: *(Select the endpoint at "C")*

Distance - 13.37

Specify next point or [Arc/Close/Length/Undo/Total] <Total>: *(Select the endpoint at "D")*

Distance - 22.51

```
Specify next point or [Arc/Close/Length/Undo/Total]
<Total>: (Press ENTER to continue)

Distance - 22.51

Enter an option [Distance/Radius/Angle/ARea/Volume/eXit]
<Distance>: X (For eXit)
```

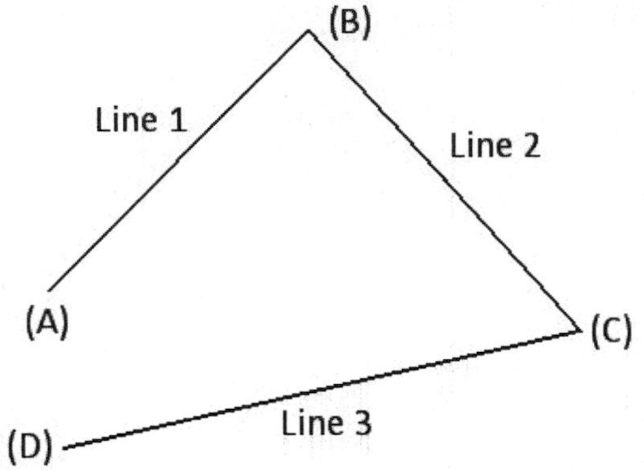

FIGURE 12.9

INTERPRETATION OF ANGLES WHEN MEASURING DISTANCE

Previously, it was noted that the Distance option of the MEASUREGEOM command yields information regarding distance, delta XYZ coordinate values, and angle information. Of particular interest is the angle in the XY plane formed between two points. Picking the endpoint of the line segment at "A" as the first point followed by the endpoint of the line segment at "B" as the second point displays an angle of 42°, as shown in Figure 12.10 (Left). This angle is formed from an imaginary horizontal line drawn from the endpoint of the line segment at "A" in the zero direction.

Take care when using the Distance option to find an angle on an identical line segment, illustrated in Figure 12.10 on the right compared to the example on the left. However, notice that the two points for identifying the angle are selected differently. With the Distance option, you select the endpoint of the line segment at "B" as the first point, followed by the endpoint of the segment at "A" for the second point. A new angle in the XY plane of 222° is formed. In Figure 12.10 (Right), the angle is calculated by the construction of a horizontal line from the endpoint at "B," the new first point of the Distance option. This horizontal line is also drawn in the zero direction. Notice the relationship of the line segment to the horizontal baseline. In other words, be careful when identifying the order of line segment endpoints for extracting angular information.

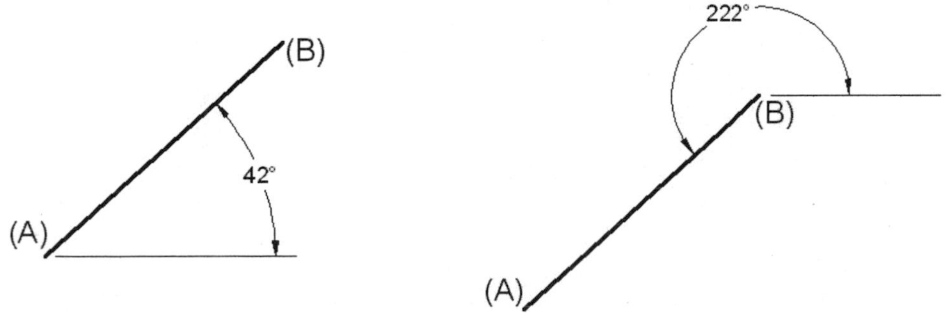

FIGURE 12.10

MEASURING A RADIUS OR DIAMETER

The Radius option of the `MEASUREGEOM` command allows you to quickly measure the radius of any selected arc or circle. Also displayed is the diameter of the arc or circle. The command sequence and example are both illustrated in Figure 12.11.

```
Command: MEA (For MEASUREGEOM)

Enter an option [Distance/Radius/Angle/ARea/Volume] <Distance>: R (For Radius)

Select arc or circle: (Select the edge of an arc)

Radius = 0.5000

Diameter = 1.0000

Enter an option [Distance/Radius/Angle/ARea/Volume/eXit]
<Radius>: X (For eXit)
```

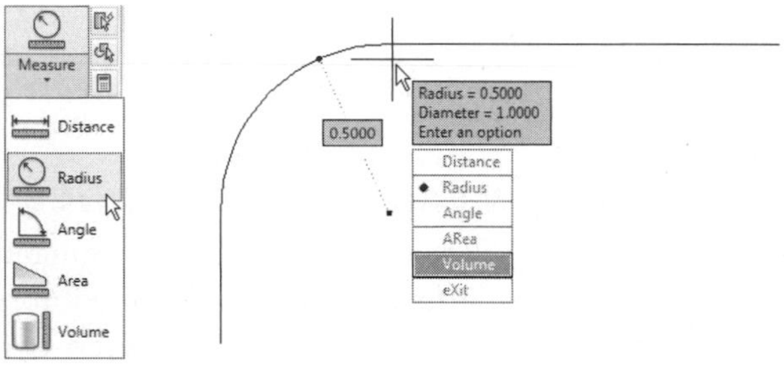

FIGURE 12.11

MEASURING AN ANGLE

The Angle option of the `MEASUREGEOM` command allows you to measure angles based on arcs, circles, lines, or by a specified vertex. Figure 12.12 illustrates an angle calculation based on two line segments.

```
Command: MEA (For MEASUREGEOM)

Enter an option [Distance/Radius/Angle/ARea/Volume]
<Distance>: A (For Angle)

Select arc, circle, line, or <Specify vertex>: (Select
a line)

Select second line: (Select a second line)

Angle = 37°

Enter an option [Distance/Radius/Angle/ARea/Volume/eXit]
<Angle>: X (For eXit)
```

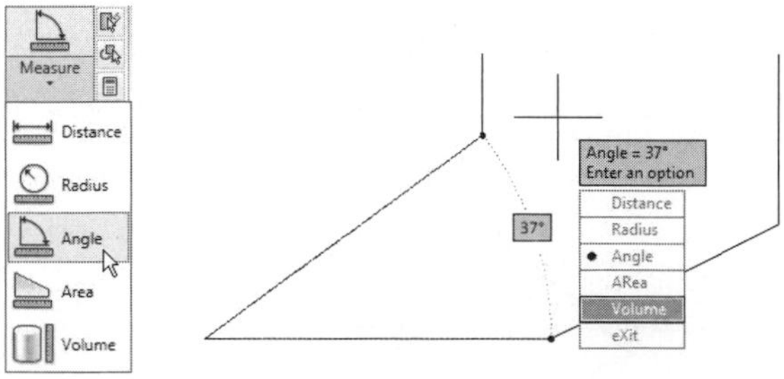

FIGURE 12.12

MEASURING A VOLUME

The Volume option of the MEASUREGEOM command is used to calculate the volume of a drawing shape. The methods of finding the volume are similar to finding the area, with the additional step of specifying a height.

Open the drawing file 12_Volume. Use Figure 12.13 and the prompt sequence below for finding the volume of the shape. Turn Dynamic Input off on the status bar for this exercise.

TRY IT!

```
Command: MEA (For MEASUREGEOM)
Enter an option [Distance/Radius/Angle/ARea/Volume]
<Distance>: V (For Volume)
Specify first corner point or [Object/Add volume/Subtract
volume/eXit]
<Object>: A (For Add volume)
Specify first corner point or [Object/Subtract volume/
eXit]: O (For Object)
(ADD mode) Select objects: (Select the polyline at "A")
Specify height: 0.75
Volume = 61.77
Total Volume = 61.77
(ADD mode) Select objects: (Press ENTER to exit ADD mode)
Volume = 61.77
Total Volume = 61.77
Specify first corner point or [Object/Subtract volume/
eXit]: S (For Subtract volume)
Specify first corner point or [Object/Add volume/eXit]: O
(For Object)
(SUBTRACT mode) Select objects: (Select the circle at "B")
Specify height: 0.75
Volume = 21.21
Total Volume = 40.56
(SUBTRACT mode) Select objects: (Press ENTER to exit SUBTRACT
mode)
Volume = 21.21
Total Volume = 40.56
Specify first corner point or [Object/Add volume/eXit]: X
(For eXit)
Total Volume = 40.56
Enter an option [Distance/Radius/Angle/ARea/Volume/eXit]
<Volume>: X (For eXit)
```

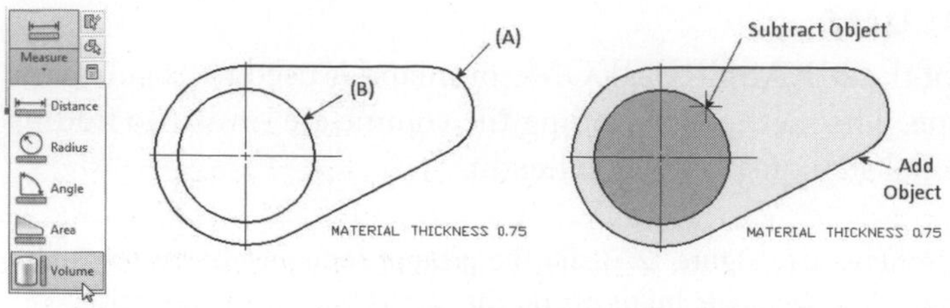

FIGURE 12.13

THE ID (IDENTIFY) COMMAND

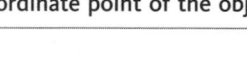

The ID command is probably one of the more straightforward of the Inquiry commands. ID can stand for "Identify" or "ID Point" and allows you to obtain the current absolute coordinate listing of a point.

In Figure 12.14, the coordinate value of the center of the circle at "A" was found through the use of ID and the OSNAP-Center mode. The coordinate value of the starting point of text string "B" was found with ID and the OSNAP-Insert mode. The coordinate value of the midpoint of the line segment at "CD" was found with ID and the OSNAP-Midpoint mode. Finally, the coordinate value of the current position of point "E" was found with ID and the OSNAP-Node mode.

TRY IT!

Open the drawing file 12_ID. Follow the prompt sequences below for calculating the XYZ coordinate point of the objects in Figure 12.15.

```
Command: ID
Specify point: Cen
of (Select circle at "A")
X = 2.00 Y = 7.00 Z = 0.00

Command: ID
Specify point: Ins
of (Select the text at "B")
X = 5.54 Y = 7.67 Z = 0.00

Command: ID
Specify point: Mid
of (Select line "CD")
X = 5.13 Y = 3.08 Z = 0.00

Command: ID
Specify point: Nod
of (Select the point at "E")
X = 9.98 Y = 1.98 Z = 0.00
```

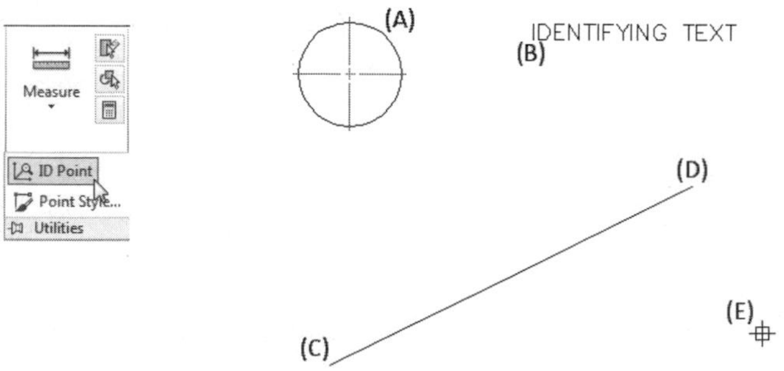

FIGURE 12.14

THE LIST COMMAND

Use the LIST command to obtain information about an object or group of objects. In Figure 12.15, two rectangles are displayed along with a circle, but are the rectangles made up of individual line segments or a polyline object? Using the LIST command on each object informs you that the first rectangle at "A" is a polyline. In addition to the object type, you also can obtain key information such as the layer that the object resides on, area and perimeter information for polylines, and circumference information for circles.

TRY IT!

Open the drawing file 12_List. Study the prompt sequence below for using the list command. Repeat the LIST command for the other two objects identified as "B" and "C."

Command: LI *(For LIST)*

Select objects: *(Select the object at "A")*

Select objects: *(Press* ENTER *to list the information on this object)*

LWPOLYLINE Layer: "Object"

Space: Model space

Handle = 3a

Closed

Constant width 0.0000

area 3.3835

perimeter 7.6235

at point X= 4.7002 Y= 4.1846 Z= 0.0000

at point X= 7.1049 Y= 4.1846 Z= 0.0000

at point X= 7.1049 Y= 5.5916 Z= 0.0000

at point X= 4.7002 Y= 5.5916 Z= 0.0000

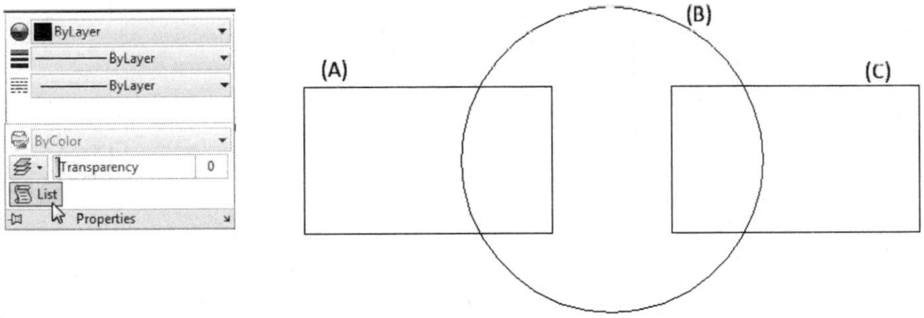

FIGURE 12.15

ADDITIONAL INQUIRY COMMANDS

The following Inquiry commands may not be used all that often. However, they may come in handy when you want a readout of the time spent in the drawing editor or when you want to review the status of a drawing. You can even get a list of all system variables used in any AutoCAD drawing session.

Command	Shortcut	Description
Time	TIME	Displays the time spent in the drawing editor.
Status	STATUS	Displays important information on the current drawing.
Set Variable	SETVAR	Used for making changes to one of the many system variables internal to AutoCAD.

USING FIELDS IN AREA CALCULATIONS

Fields can be of great use when performing area calculations on geometric shapes. In Figure 12.16, the area of the room of the partial office plan will be calculated using fields. When the geometric shape changes, the field can be updated to display the new area calculation.

TRY IT!

Open the drawing file 12_Field. Begin this process by first adding the following multiline text object, (AREA=). Then click the Insert Field button in Text Editor tab of the Ribbon.

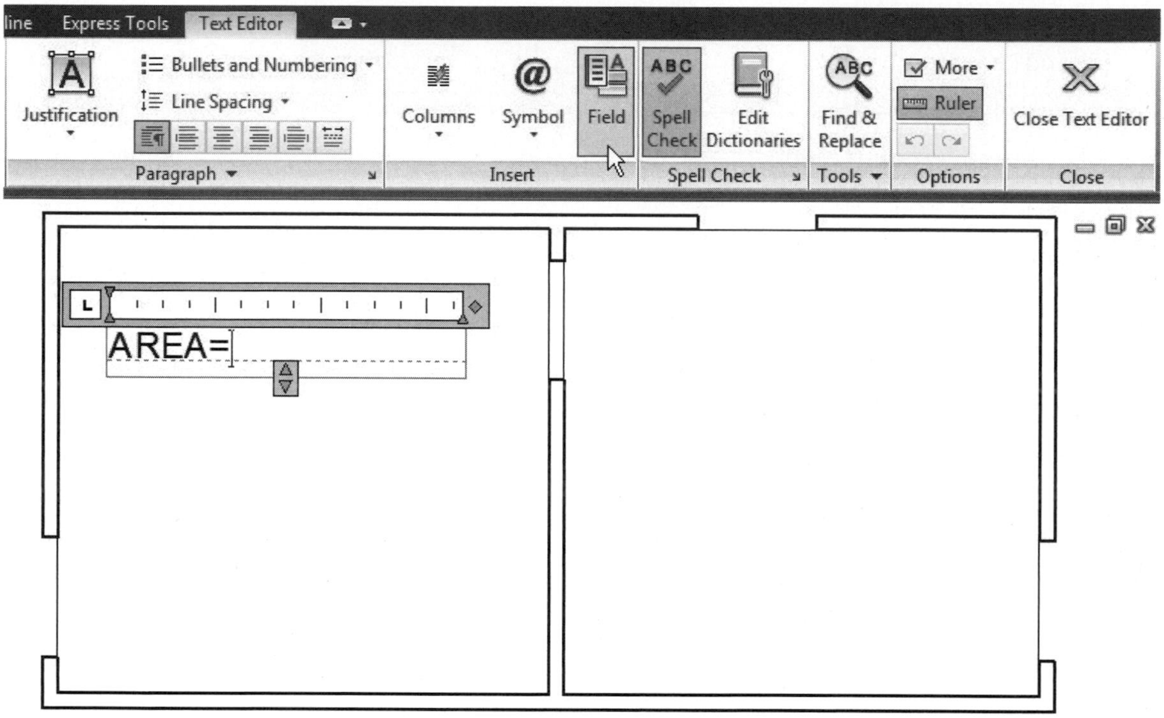

FIGURE 12.16

This action activates the Field dialog box, as shown in Figure 12.17. Locate Object under the Field names area. Under the Object Type heading, click the Select object button.

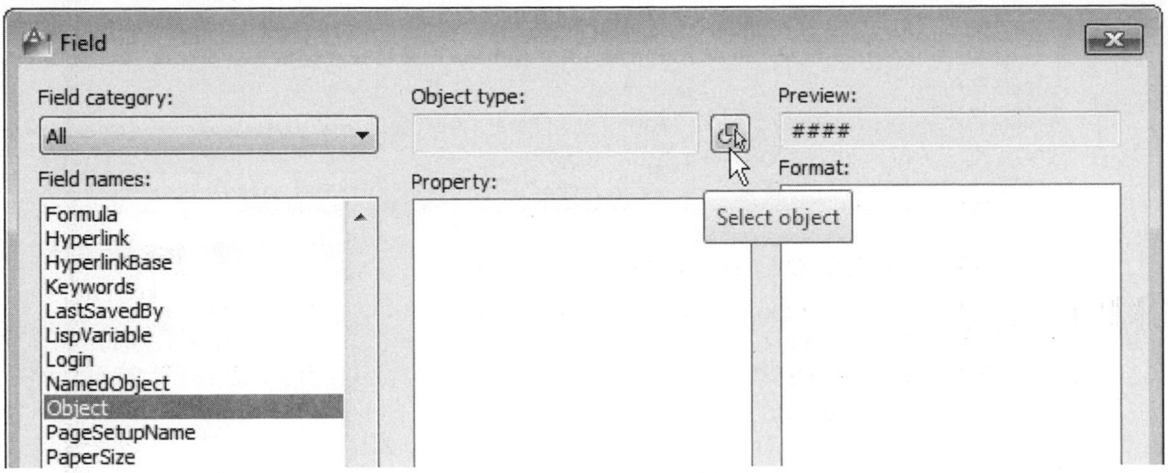

FIGURE 12.17

Clicking the Select object button in the previous image returns you to the drawing editor. Once you are there, select the rectangular polyline shape, as shown in Figure 12.18.

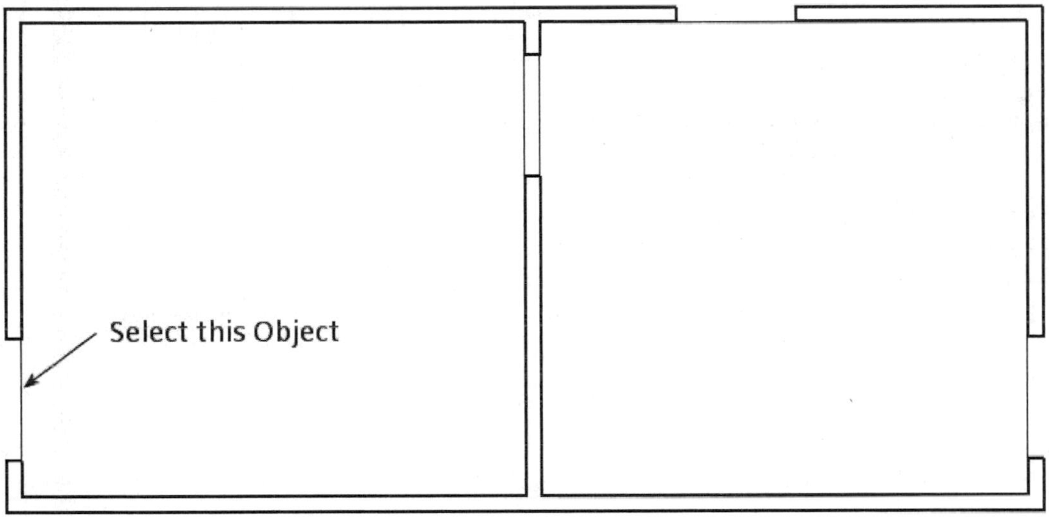

FIGURE 12.18

This returns you back to the Field dialog box, as shown in Figure 12.19. Verify that the property changed to Area; change the Format to Architectural. The Preview area shows the current area calculation. Click the OK button to exit the Field dialog box and then close the Text Editor.

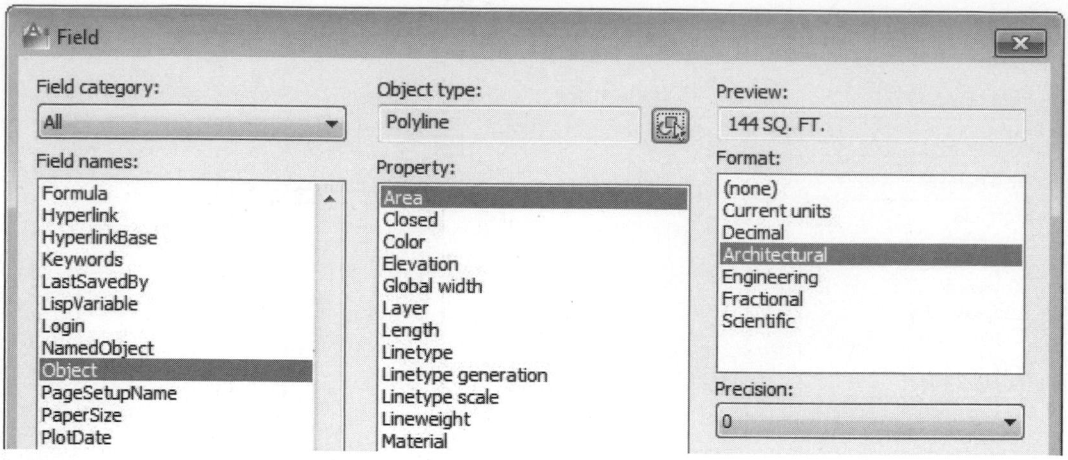

FIGURE 12.19

The results are illustrated in Figure 12.20. A field consisting of the square footage information is added to the end of the multiline text. Notice also that the field is identified by the text with the gray background.

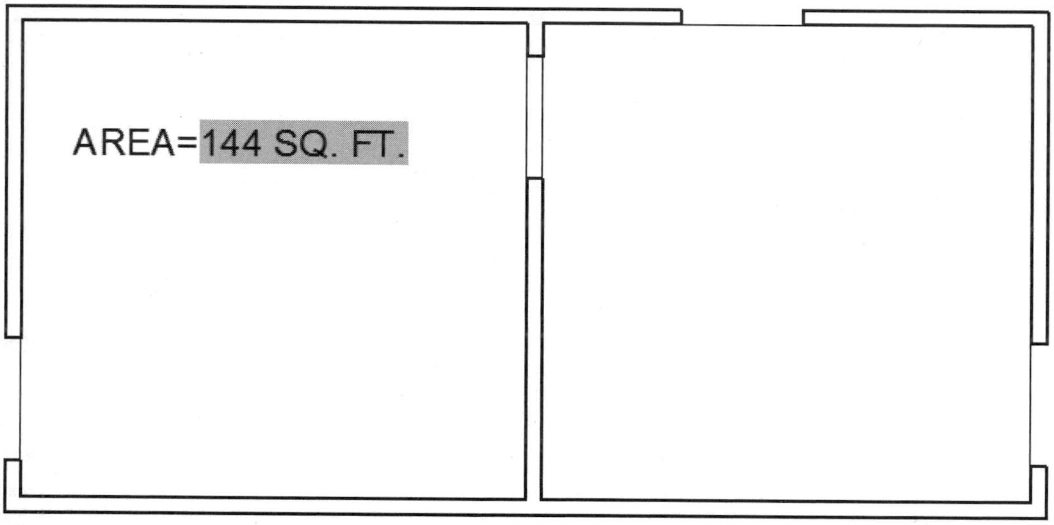

FIGURE 12.20

In Figure 12.21, the room was stretched 2 feet to the right. Regenerating the drawing screen (REGEN command) updates the field to reflect the change in the square footage calculation, as shown in Figure 12.21.

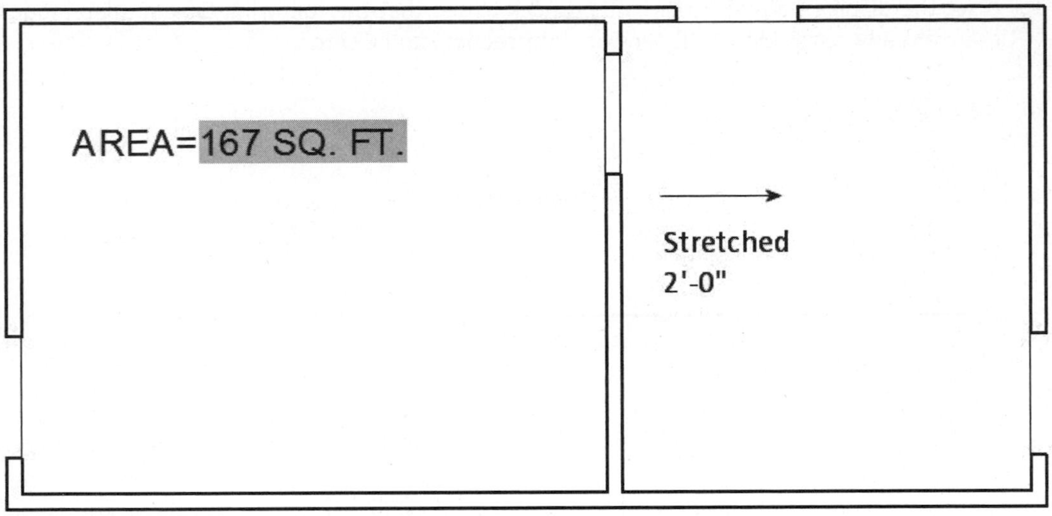

FIGURE 12.21

TUTORIAL EXERCISE: 13_C-LEVER.DWG

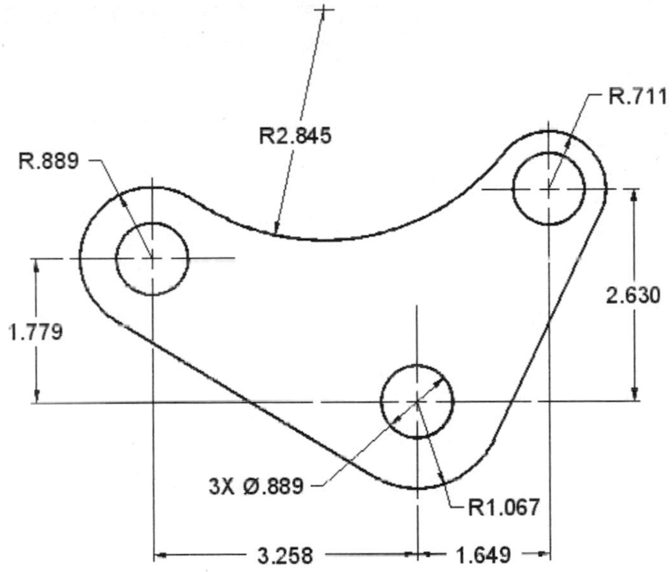

FIGURE 12.22

Purpose

This tutorial will begin by constructing the C-Lever object in Figure 12.22. Numerous questions will be asked about the object, requiring the use of the following options of the MEASUREGEOM command: Area, Distance, Angle, and Radius.

System Settings

Start a new drawing utilizing the Acad.dwt template. Use the Drawing Units dialog box and change the number of decimal places past the zero from four units to three units. Keep the default drawing limits at (0.000,0.000) for the lower-left corner and (12.000,9.000) for the upper-right corner. Use the ZOOM command and All option to set the display screen to the limits. Check to see that the following Object Snap modes are already set: Endpoint, Extension, Intersection, and Center.

Layers

Create the following layers with the format:

Name	Color	Linetype
Boundary	Magenta	Continuous
Object	Yellow	Continuous

Suggested Commands

You will begin drawing the C-Lever at absolute coordinate 7.000,3.375. After laying out all circles, you will draw tangent lines and arcs. Use the `TRIM` command to clean up unnecessary objects. To prepare to answer the `ARea` option of the `MEASUREGEOM` command question, convert the profile of the C-Lever to a polyline using the `BOUNDARY` command. Other questions pertaining to distances, angles, and point identifications follow. Do not dimension this drawing.

STEP 1

Make the Object layer current. Then construct one circle of 0.889 diameter with the center of the circle at absolute coordinate 7.000,3.375, as shown in Figure 12.23. Construct the remaining circles of the same diameter by using the `COPY` command. Be sure Dynamic Input is turned on. Turn on OSNAP Center and pick the center of the first circle drawn at coordinate 7.000,3.375. Then enter coordinates to identify the second point of the remaining circles; namely 1.649,2.630 and -3.258,1.779. The negative value in one of the coordinates will copy to the left from the previous point.

Command: **C** *(For CIRCLE)*

Specify center point for circle or [3P/2P/Ttr (tan tan radius)]: **7.000,3.375**

Specify radius of circle or [Diameter]: **D** *(For Diameter)*

Specify diameter of circle: **0.889**

Command: **CP** *(For COPY)*

Select objects: **L** *(For Last)*

Select objects: *(Press* ENTER *to continue)*

Specify base point or [Displacement] <Displacement>: *(Be sure OSNAP Center is turned on; then select the center of the existing circle)*

Specify second point or <use first point as displacement> **1.649,2.630**

Specify second point or [Exit/Undo] <Exit>: **-3.258,1.779**

Specify second point or [Exit/Undo] <Exit>: *(Press* ENTER *to exit this command)*

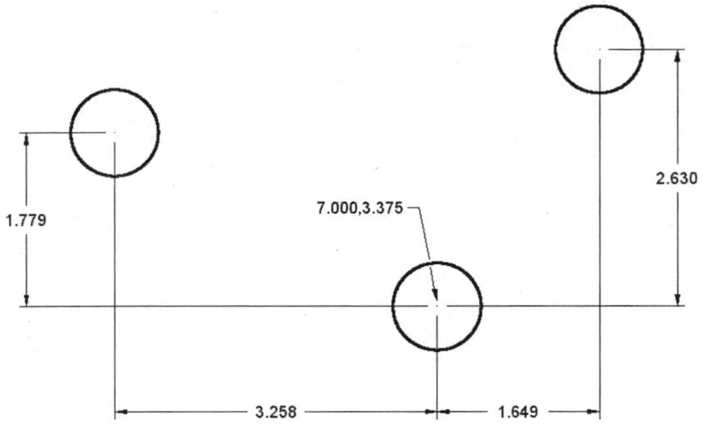

FIGURE 12.23

STEP 2

Construct three more circles, as shown in Figure 12.24. Even though these objects actually represent arcs, circles will be drawn now and trimmed later to form the arcs.

Command: C *(For CIRCLE)*

Specify center point for circle or [3P/2P/Ttr (tan tan radius)]: *(Select the edge of the circle at "A" to snap to its center)*

Specify radius of circle or [Diameter] <0.445>: **1.067**

Command: C *(For CIRCLE)*

Specify center point for circle or [3P/2P/Ttr (tan tan radius)]: *(Select the edge of the circle at "B" to snap to its center)*

Specify radius of circle or [Diameter] <1.067>: **0.889**

Command: C *(For CIRCLE)*

Specify center point for circle or [3P/2P/Ttr (tan tan radius)]: *(Select the edge of the circle at "C" to snap to its center)*

Specify radius of circle or [Diameter] <0.889>: **0.711**

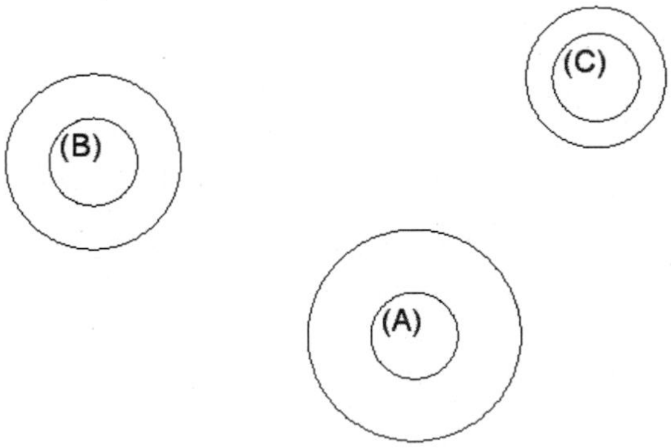

FIGURE 12.24

STEP 3

Construct lines tangent to the three outer circles, as shown in Figure 12.25.

Command: L *(For LINE)*

Specify first point: **Tan**

to *(Select the outer circle near "A")*

Specify next point or [Undo]: **Tan**

to *(Select the outer circle near "B")*

Specify next point or [Undo]: *(Press* ENTER *to exit this command)*

Command: L *(For LINE)*

Specify first point: **Tan**

to *(Select the outer circle near "C")*

Specify next point or [Undo]: **Tan**

to *(Select the outer circle near "D")*

Specify next point or [Undo]: *(Press* ENTER *to exit this command)*

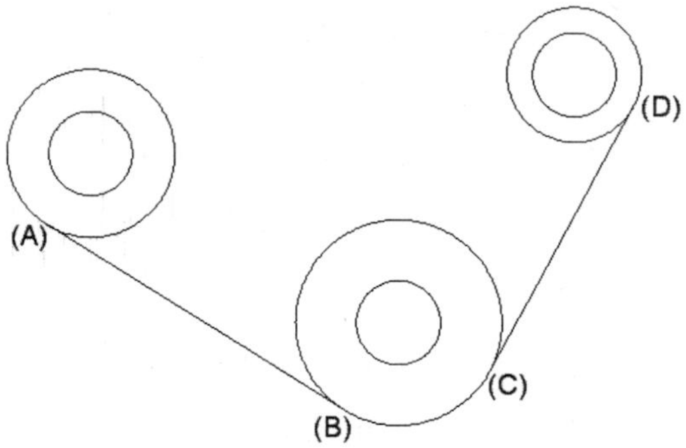

FIGURE 12.25

STEP 4

Construct a circle tangent to the two circles in Figure 12.26, using the CIRCLE command with the Tangent-Tangent-Radius (Ttr) option.

Command: C *(For CIRCLE)*

Specify center point for circle or [3P/2P/Ttr (tan tan radius)]: **T** *(For Ttr)*

Specify point on object for first tangent of circle: *(Select the outer circle near "A")*

Specify point on object for second tangent of circle: *(Select the outer circle near "B")*

Specify radius of circle <0.711>: **2.845**

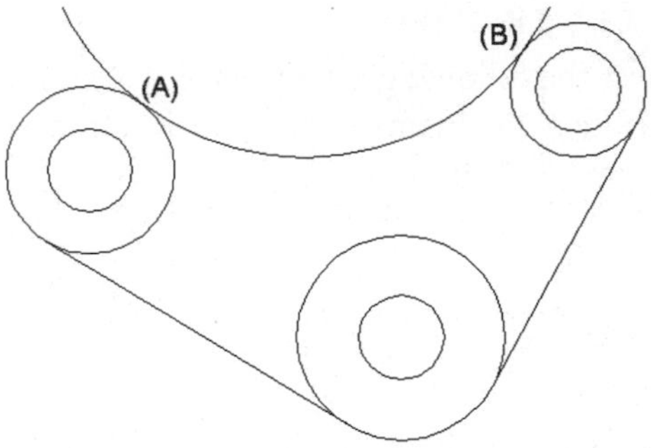

FIGURE 12.26

STEP 5

Use the TRIM command to clean up and form the finished drawing. Pressing ENTER at the "Select objects": prompt selects all objects as cutting edges although the objects do not highlight, as shown in Figure 12.27. Study the following prompts for selecting the objects to trim.

Command: TR *(For TRIM)*

Current settings: Projection=UCS Edge=None

Select cutting edges...

Select objects or <select all>: *(Press* ENTER*)*

Select object to trim or shift-select to extend or

[Fence/Crossing/Project/Edge/eRase/Undo]: *(Select the circle at "A")*

Select object to trim or shift-select to extend or

[Fence/Crossing/Project/Edge/eRase/Undo]: *(Select the circle at "B")*

Select object to trim or shift-select to extend or

[Fence/Crossing/Project/Edge/eRase/Undo]: *(Select the circle at "C")*

Select object to trim or shift-select to extend or

[Fence/Crossing/Project/Edge/eRase/Undo]: *(Select the circle at "D")*

Select object to trim or shift-select to extend or

[Fence/Crossing/Project/Edge/eRase/Undo]: *(Press* ENTER *to exit this command)*

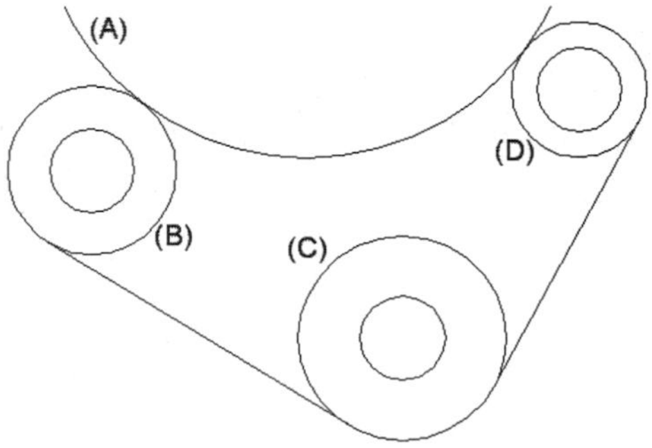

FIGURE 12.27

CHECKING THE ACCURACY OF C-LEVER.DWG

Once the C-Lever has been constructed, answer the following questions to determine the accuracy of this drawing. Use Figure 12.28 to assist in answering the questions.

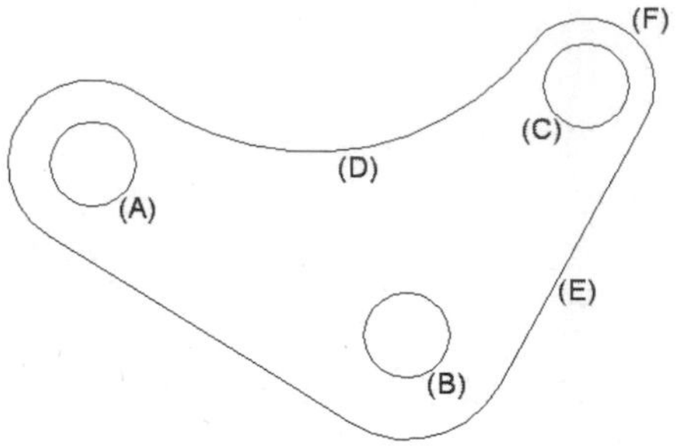

FIGURE 12.28

1. What is the total area of the C-Lever with all three holes removed?
   Answer: _______
2. What is the total distance from the center of circle "A" to the center of circle "B"?
   Answer: _______
3. What is the angle formed in the XY plane from the center of circle "C" to the center of circle "B"?
   Answer: _______
4. What is the delta X,Y distance from the center of circle "C" to the center of circle "A"?
   Answer: _______
5. What is the absolute coordinate value of the center of arc "D"?
   Answer: _______
6. What is the total length of line "E"?
   Answer: _______
7. What is the total length of arc "F"?
   Answer: _______

A solution for each question follows, complete with the method used to arrive at the answer. Apply these methods to any type of drawing that requires the use of MEASUREGEOM commands.

SOLUTIONS TO THE QUESTIONS ON C-LEVER

Question 1

What is the total area of the C-Lever with all three holes removed?

First make the Boundary layer current. Then use the BOUNDARY command and pick a point inside the object at "Y" in Figure 12.29. This traces a polyline around all closed objects on the Boundary layer.

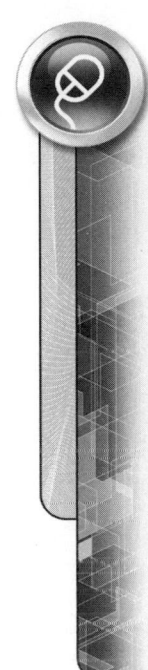

Command: BO *(For BOUNDARY)*

(When the Boundary Creation dialog box appears, click the Pick Points button.)

Pick internal point: *(Pick a point inside of the object at "Y")*

Selecting everything...

Selecting everything visible...

Analyzing the selected data...

Analyzing internal islands...

Pick internal point: *(Press* ENTER *to create the boundaries)*

BOUNDARY created 4 polylines

Next, turn off the Object layer. All objects on the Boundary layer should be visible. Then use the MEASUREGEOM command and the ARea option to add and subtract objects to arrive at the final area of the object.

Command: MEA ***(For MEASUREGEOM)***

Enter an option [Distance/Radius/Angle/ARea/Volume] <Distance>: AR *(For ARea)*

Specify first corner point or [Object/Add area/Subtract area/eXit] <Object>: A *(For Add area)*

Specify first corner point or [Object/Subtract area/eXit]: O *(For Object)*

(ADD mode) Select objects: *(Select the edge of the shape near "X")*

Area = 15.611, Perimeter = 17.771

Total area = 15.611

(ADD mode) Select objects: *(Press ENTER to continue)*

Area = 15.611, Perimeter = 17.771

Total area = 15.611

Specify first corner point or [Object/Subtract area/eXit]: S *(For Subtract area)*

Specify first corner point or [Object/Add area/eXit]: O *(For Object)*

(SUBTRACT mode) Select objects: *(Select circle "A")*

Area = 0.621, Circumference = 2.793

Total area = 14.991

(SUBTRACT mode) Select objects: *(Select circle "B")*

Area = 0.621, Circumference = 2.793

Total area = 14.370

(SUBTRACT mode) Select objects: *(Select circle "C")*

Area = 0.621, Circumference = 2.793

Total area = 13.749

(SUBTRACT mode) Select objects: *(Press ENTER)*

Area = 0.621, Circumference = 2.793

Total area = 13.749

Specify first corner point or [Object/Add area eXit]: X *(For eXit)*

Enter an option [Distance/Radius/Angle/ARea/Volume/eXit] <ARea>: X *(For eXit)*

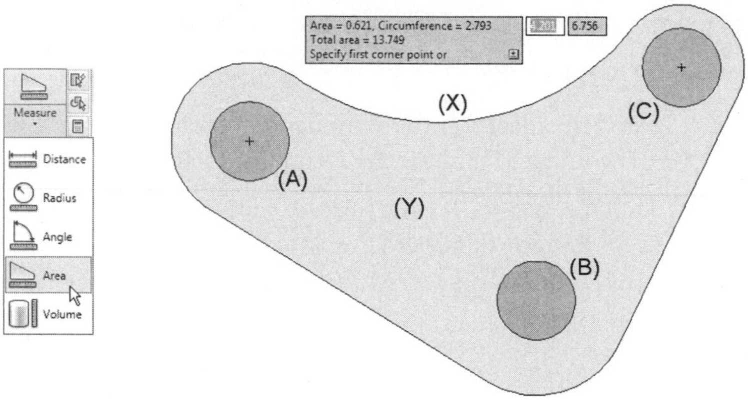

FIGURE 12.29

The total area of the C-Lever with all three holes removed is 13.749.

Question 2

What is the total distance from the center of circle "A" to the center of circle "B"?

Use the MEASUREGEOM command along with the Distance option to calculate the distance from the center of circle "A" to the center of circle "B" in Figure 12.30. Be sure to use the OSNAP-Center mode for locating the centers of all circles.

Command: MEA *(For MEASUREGEOM)*

Enter an option [Distance/Radius/Angle/ARea/Volume] <Distance>: D *(For Distance)*

Specify first point: *(Select the center of the circle at "A")*

Specify second point or [Multiple points]: *(Select the center of the circle at "B")*

Distance = 3.712, Angle in XY Plane = 331, Angle from XY Plane = 0

Delta X = 3.258, Delta Y = −1.779, Delta Z = 0.000

Enter an option [Distance/Radius/Angle/ARea/Volume/eXit] <Distance>: X *(For eXit)*

The total distance from the center of circle "A" to the center of circle "B" is 3.712.

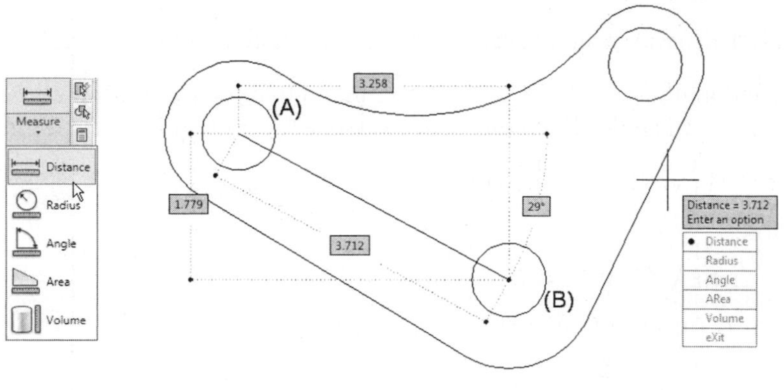

FIGURE 12.30

Question 3

What is the angle formed in the XY plane from the center of circle "C" to the center of circle "B"?

Use the Distance option of the `MEASUREGEOM` command to calculate the angle from the center of circle "C" to the center of circle "B" in Figure 12.31. Be sure to use the OSNAP-Center mode for locating the centers of all circles.

```
Command: MEA (For MEASUREGEOM)
Enter an option [Distance/Radius/Angle/ARea/Volume] <Distance>: D (For Distance)
Specify first point: (Select the center of the circle at "C")
Specify second point or [Multiple points]: (Select the center of the circle at "B")
Distance = 3.104, Angle in XY Plane = 238, Angle from XY Plane = 0
Delta X = -1.649, Delta Y = -2.630, Delta Z = 0.000
Enter an option [Distance/Radius/Angle/ARea/Volume/eXit] <Distance>: X (For eXit)
```

The angle formed in the XY plane from the center of circle "C" to the center of circle "B" is 238°. Notice the two different angle calculations. While an angle of 122° is displayed in Figure 12.31 (Right), the angle of 238° is correct. This is due to the angle in the XY plane which is calculated in the counterclockwise direction. The angle of 122° was calculated in the clockwise direction.

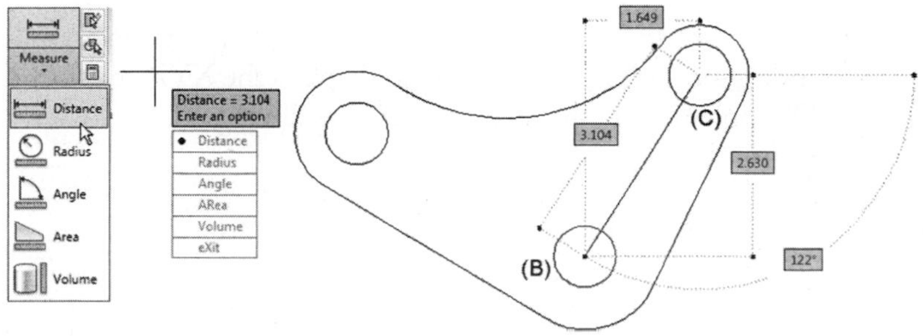

FIGURE 12.31

Question 4

What is the delta X,Y distance from the center of circle "C" to the center of circle "A"?

Use the Distance option of the MEASUREGEOM command to calculate the delta XYZ distance from the center of circle "C" to the center of circle "A" in Figure 12.32. Since this is a 2D problem, only the X and Y values will be used. Be sure to use the OSNAP-Center mode. Notice that additional information is given when you use the Distance option.

```
Command: MEA (For MEASUREGEOM)
Enter an option [Distance/Radius/Angle/ARea/Volume] <Distance>: D (For Distance)
Specify first point: (Select the center of circle "C")
Specify second point or [Multiple points]: (Select the center of circle "A")
Distance = 4.980, Angle in XY Plane = 190, Angle from XY Plane = 0
Delta X = -4.907, Delta Y = -0.851, Delta Z = 0.000
Enter an option [Distance/Radius/Angle/ARea/Volume/eXit] <Distance>: X (For eXit)
```

The delta X,Y distance from the center of circle "C" to the center of circle "A" is -4.907, -0.851.

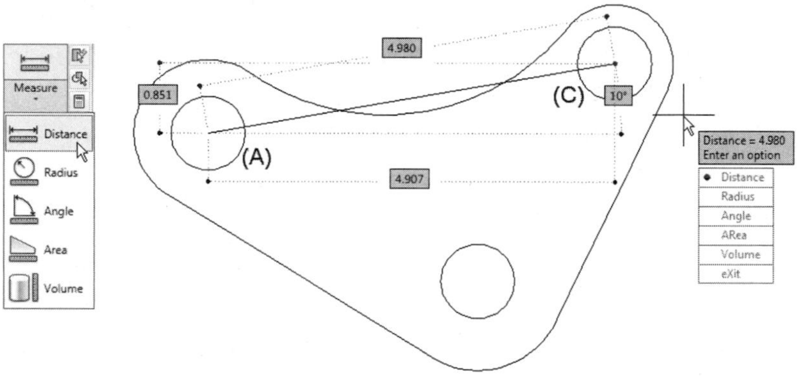

FIGURE 12.32

Question 5

What is the absolute coordinate value of the center of arc "D"?

The ID command is used to get the current absolute coordinate information on a desired point, as shown in Figure 12.33. This command displays the XYZ coordinate values. Since this is a 2D problem, only the X and Y values will be used.

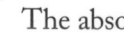

```
Command: ID
Specify point: Cen
of (Select the edge of the arc at "D")
X = 5.869 Y = 8.223 Z = 0.000
```

The absolute coordinate value of the center of arc "D" is 5.869,8.223.

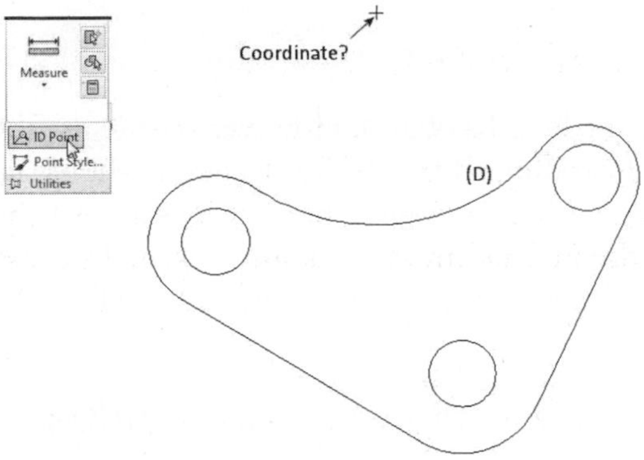

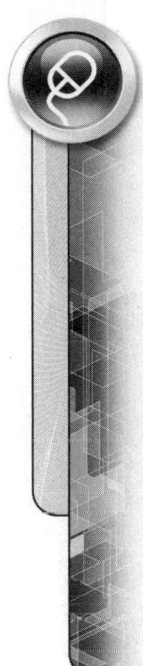

FIGURE 12.33

Question 6

What is the total length of line "E"?

Use the Distance option of the MEASUREGEOM command to find the total length of line "E" in Figure 12.34. Be sure to use the OSNAP-Endpoint mode. Notice that additional information is given when you use the Distance option. For the purpose of this question, we will be looking only for the distance.

Command: MEA *(For MEASUREGEOM)*

Enter an option [Distance/Radius/Angle/ARea/Volume] <Distance>: D *(For Distance)*

Specify first point: *(Select the endpoint of the line at "X")*

Specify second point or [Multiple points]: *(Select the endpoint of the line at "Y")*

Distance = 3.084, Angle in XY Plane = 64, Angle from XY Plane = 0

Delta X = 1.328, Delta Y = 2.783, Delta Z = 0.000

Enter an option [Distance/Radius/Angle/ARea/Volume/eXit] <Distance>: X *(For eXit)*

The total length of line "E" is 3.084.

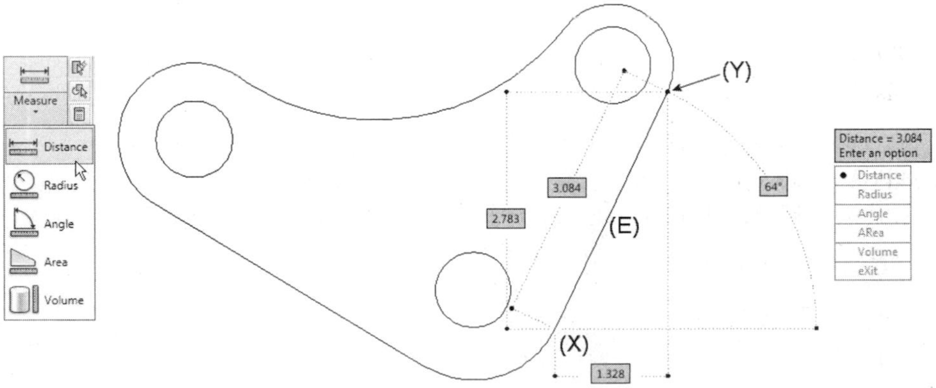

FIGURE 12.34

Question 7

What is the total length of arc "F"?

The LIST command is used to calculate the lengths of arcs. However, a little preparation is needed before you perform this operation. If arc "F" is selected, as shown in Figure 12.35 (Left), notice that the entire outline is selected because it is a polyline. Use the EXPLODE command to break the outline into individual objects. Use the LIST command to get a listing of the arc length, as shown in Figure 12.35 (Right).

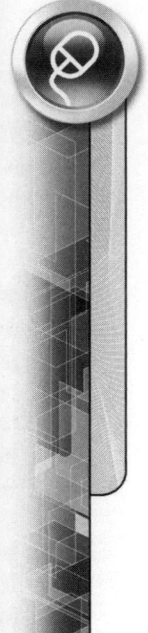

```
Command: X (For EXPLODE)
Select objects: (Select the edge of the dashed polyline in the following image)
Select objects: (Press ENTER to perform the explode operation)
Command: LI (For LIST)
Select objects: (Select the edge of the arc at "F" in the following image)
Select objects: (Press ENTER to continue)
ARC Layer: "Boundary"
Space: Model space
Handle = 1ce
center point, X= 8.649 Y= 6.005 Z= 0.000
radius 0.711
start angle 334
end angle 141
length 2.071
```

The total length of arc "F" is 2.071.

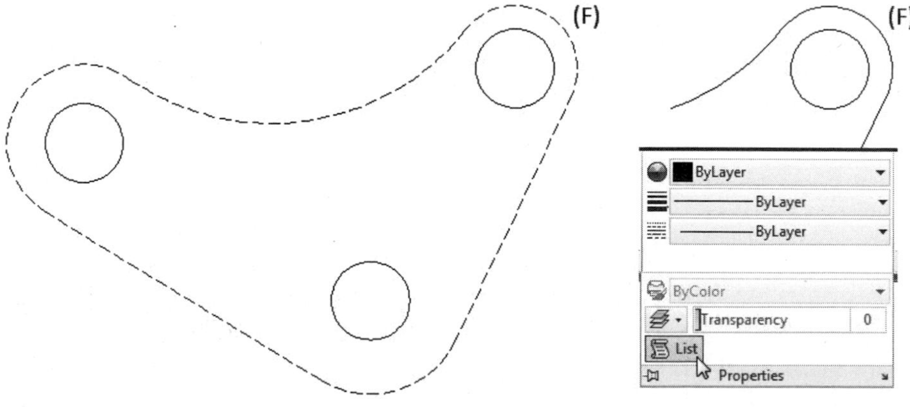

FIGURE 12.35

TUTORIAL EXERCISE: 12_FIELD_CALC.DWG

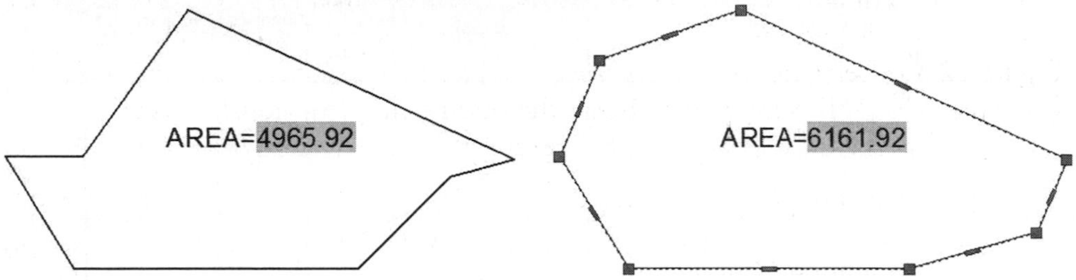

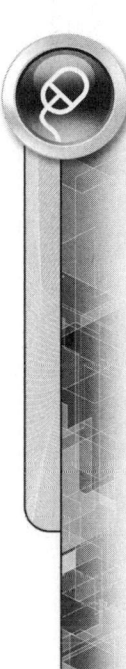

FIGURE 12.36

Purpose

This tutorial is designed to utilize the Field function of AutoCAD to calculate the area of an object, as shown in Figure 12.36.

System Settings

No changes to system settings are required for this drawing.

Layers

Layers already exist in this drawing.

Suggested Commands

In this tutorial, you will open the drawing 12_Field_Calc.Dwg and activate the multiline text command. Set the text height to 5 and add a text header dealing with the Area. Then activate the Field dialog box and select a polyline object in the drawing. This retrieves the area information and adds this text calculation to the Text Formatting dialog box. Once the field is added, stretching the original polyline shape and regenerating the drawing recalculates the area.

STEP 1

Open the drawing 12_Field_Calc.Dwg. Activate the MTEXT command and select two corners to establish the text box. In the Text Editor Ribbon, change the text height to 5, and add the text (AREA=), as shown in Figure 12.37. Continue by clicking on the Insert Field button.

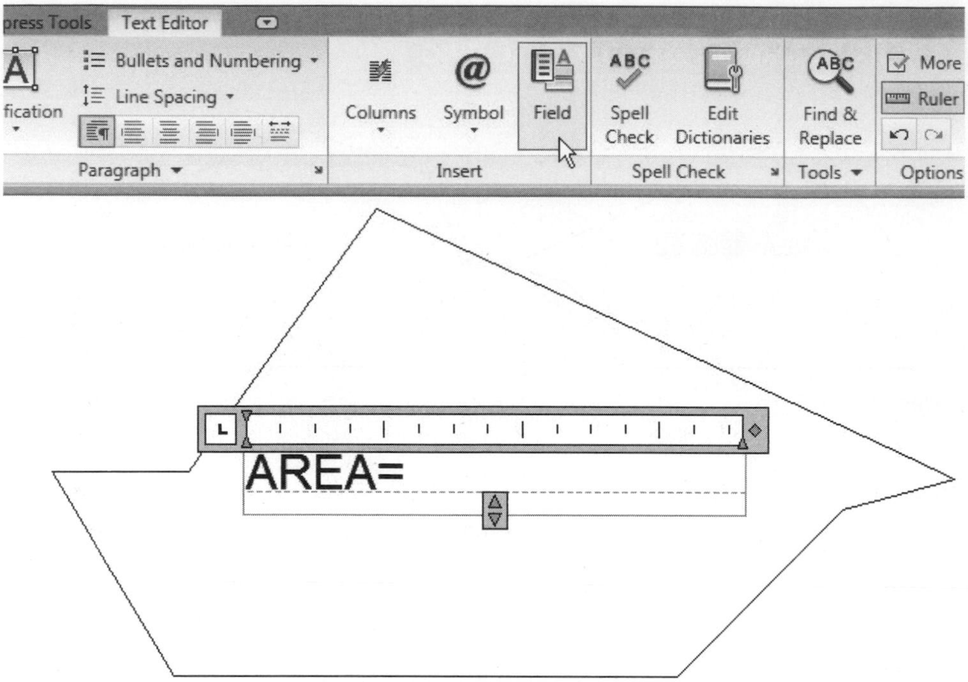

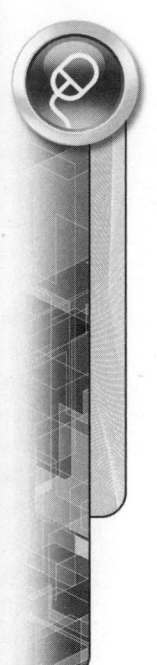

FIGURE 12.37

STEP 2

When the Field dialog box appears, select the Object Field name, as shown in Figure 12.38. This prepares the dialog box to display field information based on the object that you select. Continue by clicking the Select object button, as shown in Figure 12.38.

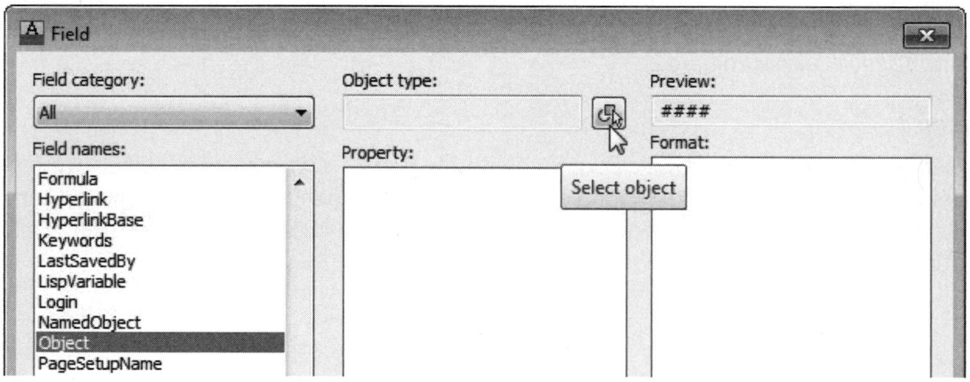

FIGURE 12.38

STEP 3

Clicking the Select object button returns you to the drawing editor, where you can select the polyline shape of the object. This action again returns you to the Field dialog box, as shown in Figure 12.39. Verify or, if necessary, make the following changes in this dialog box: Under the Field names category, pick Object; under the Property category, pick Area; under the Format category, pick Decimal. Click the OK button to exit the Field dialog box and then close the Text Editor.

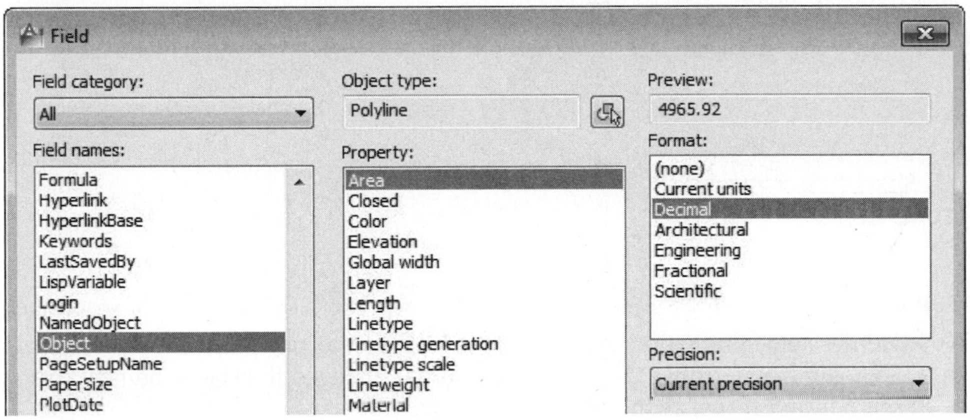

FIGURE 12.39

STEP 4

The results are illustrated in Figure 12.40. A field consisting of the square footage information is added to the end of the multiline text.

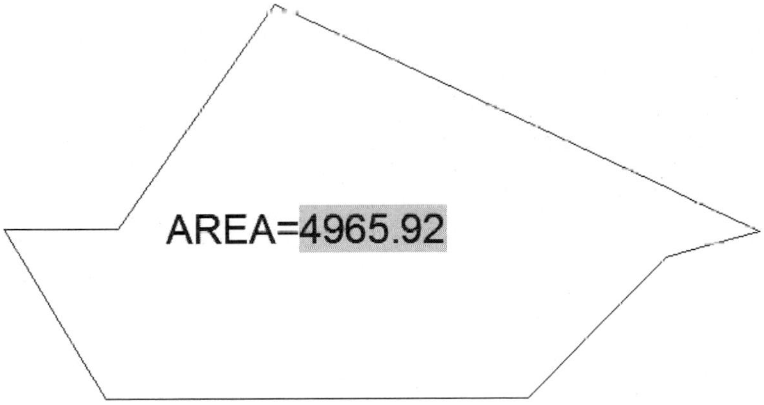

FIGURE 12.40

STEP 5

Now click the polyline shape to display grips, choose a number of grips, and stretch the vertices of the polyline to new locations, as shown in Figure 12.41. Your display may not match Figure 12.41 exactly. You will notice that the square footage does not automatically update itself.

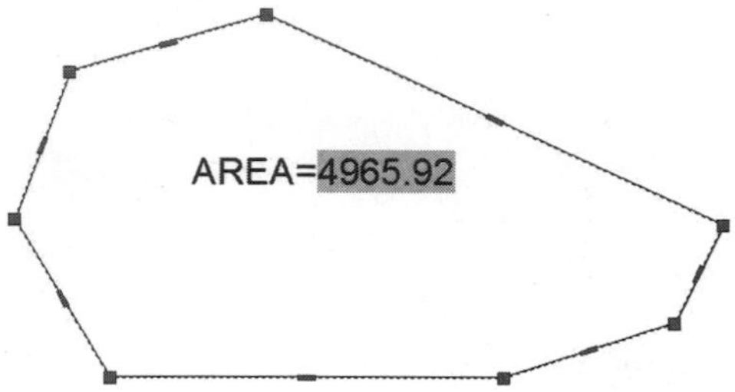

FIGURE 12.41

STEP 6

To update the field information and change the area for the new polyline shape, activate the REGEN command. Notice that the area field has changed to reflect the new size of the closed polyline shape, as shown in Figure 12.42.

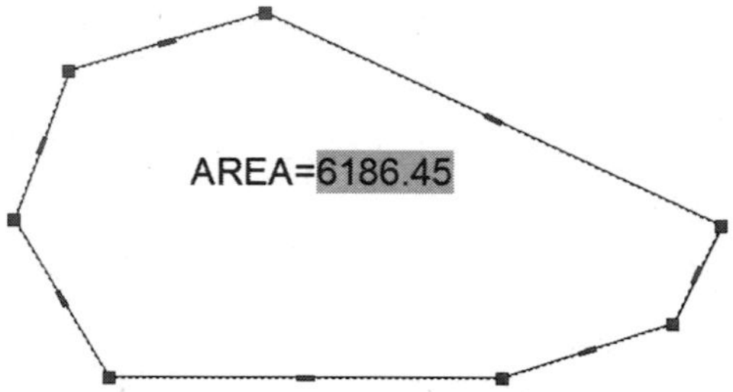

FIGURE 12.42

PROBLEMS FOR CHAPTER 12

PROBLEM 12–1 ANGLEBLK.DWG

Use the Drawing Units dialog box to set the units to decimal. Set the precision to two. Be sure the system of angle measure is set to decimal degrees and the number of decimal places for the display of angles is zero.

Keep all remaining default unit values. Keep the default settings for the drawing limits.

Begin the drawing shown in the following image by locating the lower-left corner of Angleblk, identified by "X" at coordinate (2.35,3.17).

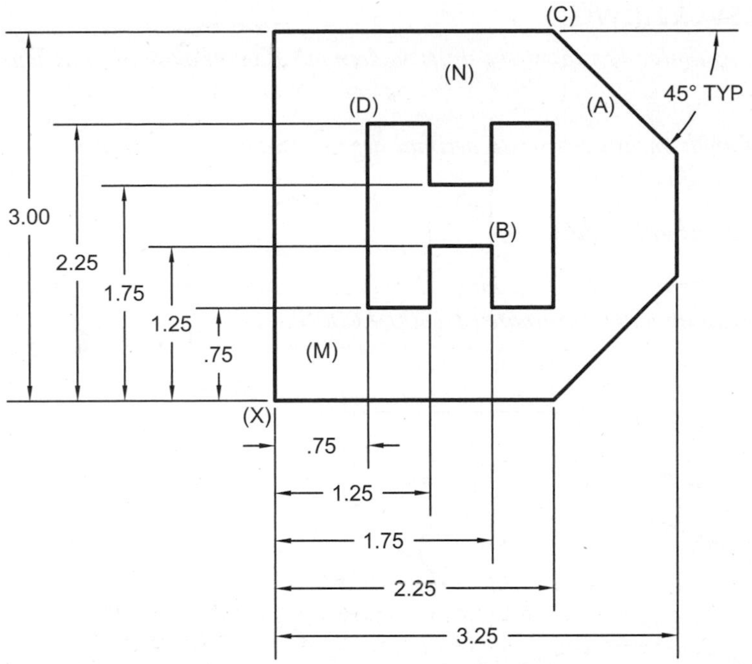

Refer to the drawing of the Angleblk and answer the following questions.

1. What is the total surface area of the Angleblk with the inner "H" shape removed?

   Answer: ____________

2. What is the total area of the inner "H" shape?

   Answer: ____________

3. What is the total length of line "A"?

   Answer: ____________

4. What is the absolute coordinate value of the endpoint of the line at "B"?

   Answer: ____________

5. What is the absolute coordinate value of the endpoint of the line at "C"?

   Answer: ____________

6. Use the STRETCH command and extend the inner "H" shape a distance of 0.37 units in the 180° direction. Use "N" as the first corner of the crossing window and "M" as the other corner. Use the endpoint of "D" as the base point of the stretching operation. What is the new surface area of Angleblk with the inner "H" shape removed?

   Answer: ____________

7. Use the SCALE command with the endpoint of the line at "D" as the base point. Reduce the size of just the inner "H" using a scale factor of 0.77. What is the new surface area of Angleblk with the inner "H" removed?

   Answer: ____________

PROBLEM 12–2 LEVER2.DWG

Use the Drawing Units dialog box to set the units to decimal. Keep the precision at four places.

Be sure the system of angle measure is set to decimal degrees and the number of decimal places for the display of angles is zero.

Keep the remaining default unit values.

Begin the drawing shown in the following image by locating the center of the 1.0000-diameter circle at absolute coordinate (2.2500,4.0000).

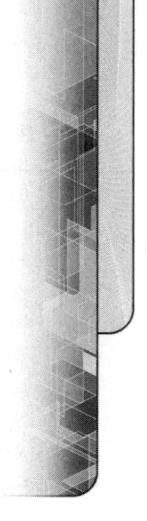

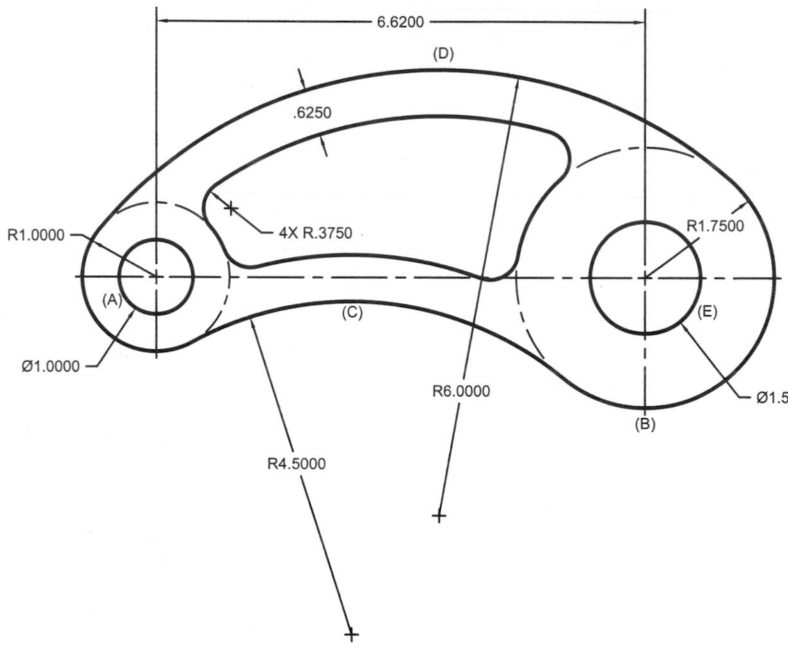

Refer to the drawing of Lever2 and answer the following questions.

1. What is the total area of Lever2 with the inner irregular shape and both holes removed?

   Answer: ____________

2. What is the absolute coordinate value of the center of the 4.5000-radius arc "C"?

   Answer: ____________

3. What is the absolute coordinate value of the center of the 6.0000-radius arc "D"?

   Answer: ____________

4. What is the total length of arc "C"?

   Answer: ____________

5. What is the distance from the center of the 1.0000-diameter circle "A" to the intersection of the circle and centerline at "B"?

   Answer: ____________

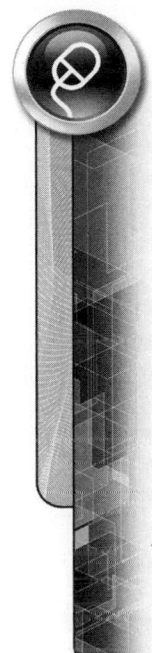

6. What is the angle formed in the XY plane from the upper quadrant of arc "D" to the center of the 1.5000-diameter circle "E"?

   Answer: ____________

7. What is the delta X,Y distance from the upper quadrant of arc "C" to the center of the 1.0000-hole "A"?

   Answer: ____________

INTERMEDIATE-LEVEL DRAWINGS

PROBLEM 12–3 GASKET2.DWG

Use the Drawing Units dialog box to set the units to decimal. Set the precision to two. Be sure the system of angle measure is set to decimal degrees and the number of decimal places for the display of angles is zero.

Keep the remaining default unit values.

Begin the drawing by locating the center of the 6.00 × 3.00 rectangle at absolute coordinate (6.00,4.75).

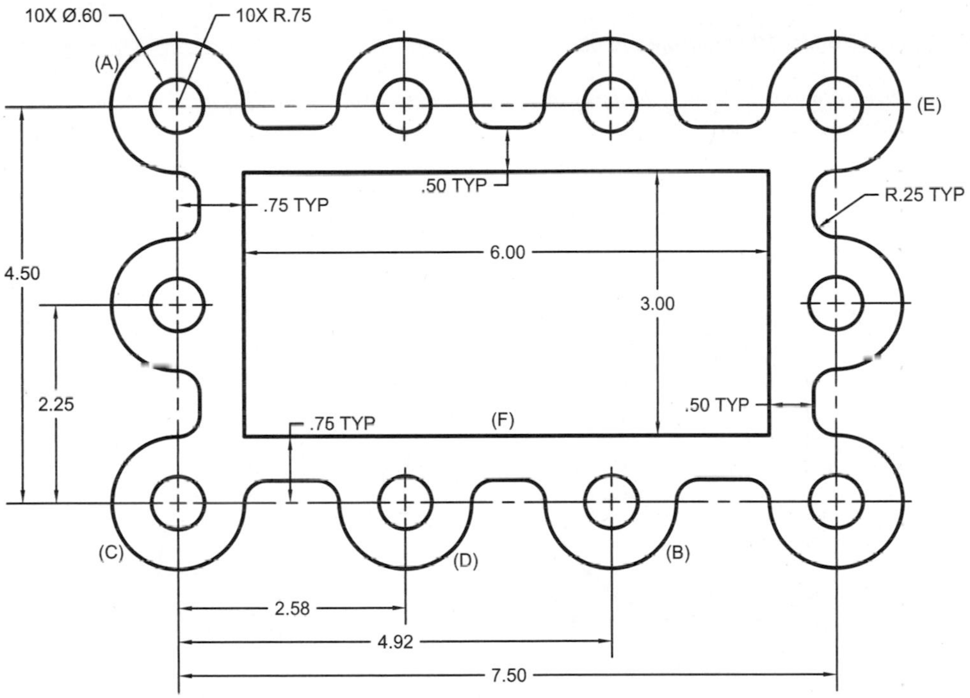

Refer to the drawing of Gasket2 and answer the following questions.

1. What is the total surface area of Gasket2 with the rectangle and all ten holes removed?

   Answer: ____________

2. What is the distance from the center of arc "A" to the center of arc "B"?

   Answer: ____________

3. What is the length of arc segment "C"?

   Answer: ____________

4. What is the absolute coordinate value of the center of the 0.75-radius arc "D"?

   Answer: ___________

5. What is the angle formed in the XY plane from the center of the 0.75-radius arc "D" to the center of the 0.75-radius arc "A"?

   Answer: ___________

6. What is the delta X,Y distance from the intersection at "E" to the midpoint of the line at "F"?

   Answer: ___________

7. Use the SCALE command to reduce the size of the inner rectangle. Use the midpoint of the line at "F" as the base point. Use a scale factor of 0.83 units. What is the new total surface area with the rectangle and all ten holes removed?

   Answer: ___________

PROBLEM 12–4 HANGER.DWG

Use the Drawing Units dialog box to set units to decimal. Set precision to three. Be sure the system of angle measure is set to decimal degrees and the number of decimal places for the display of angles is zero. Keep all remaining default unit values. Use the LIMITS command and set the upper-right corner of the screen area to a value of (250.000,150.000). Perform a Zoom All operation.

Begin drawing the hanger by locating the center of the 40.000-radius arc at absolute coordinate (55.000,85.000).

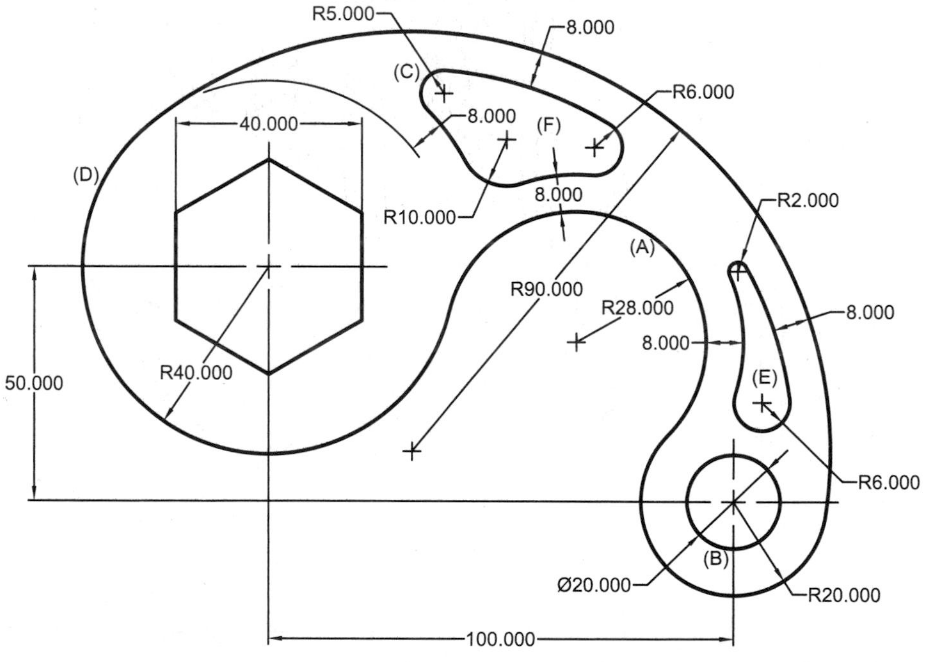

Refer to the drawing of the hanger and answer the following questions.

1. What is the area of the outer profile of the hanger?

   Answer: ___________

2. What is the area of the hanger with the polygon, circle, and irregular shapes removed?

   Answer: ___________

3. What is the absolute coordinate value of the center of the 28.000-radius arc "A"?

   Answer: ____________

4. What is the absolute coordinate value of the center of the 20.000-diameter circle "B"?

   Answer: ____________

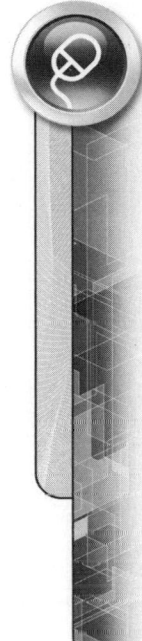

5. What is the angle formed in the XY plane from the center of the 5.000-radius arc "C" to the center of the 40.000-radius arc "D"?

   Answer: ____________

6. What is the total area of irregular shape "E"?

   Answer: ____________

7. What is the total area of irregular shape "F"?

   Answer: ____________

PROBLEM 12–5 HOUSING1.DWG

Begin constructing Housing1 by keeping the default units set to decimal but changing the precision to three. Be sure the system of angle measure is set to decimal degrees and the number of decimal places for the display of angles is zero. Place the center of the 1.500-radius circular centerline at absolute coordinate (6.500,5.250).

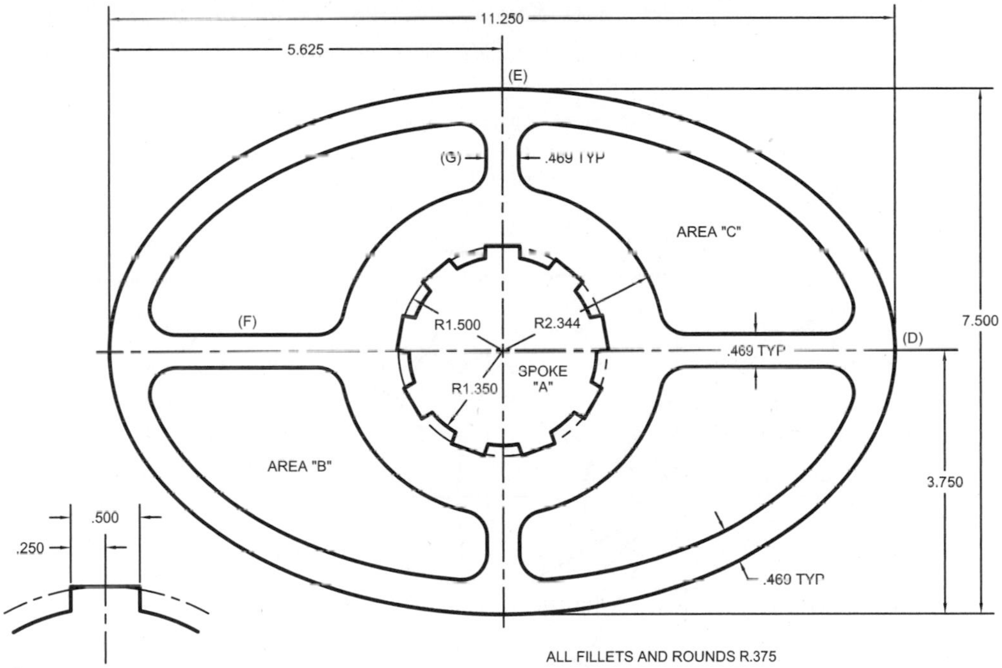

Refer to the drawing of Housing1 and answer the following questions.

1. What is the perimeter of Spoke "A"?

   Answer: ____________

2. What is the perimeter of Area "B"?

   Answer: ____________

3. What is the total area of Area "C"?
   Answer: ____________
4. What is the absolute coordinate value of the intersection of the ellipse and centerline at "D"?
   Answer: ____________
5. What is the total surface area of Housing1 with the spoke and all slots removed?
   Answer: ____________
6. What is the distance from the midpoint of the horizontal line segment at "F" to the midpoint of the vertical line segment at "G"?
   Answer: ____________
7. Increase Spoke "A" in size using the SCALE command. Use the center of the 1.500-radius arc as the base point. Use a scale factor of 1.115 units. What is the new total area of Housing1 with the spoke and all slots removed?
   Answer: ____________

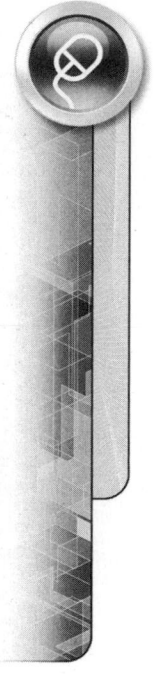

PROBLEM 12–6 ROTOR.DWG

Start a new drawing called Rotor, as shown in the following image. Keep the default setting of decimal units precision to three.

Be sure the system of angle measure is set to decimal degrees and the number of decimal places for the display of angles is zero. Keep all remaining default unit values. Begin the drawing by constructing the center of the 6.250-unit-diameter circle at "A" at absolute coordinate (5.500,5.000).

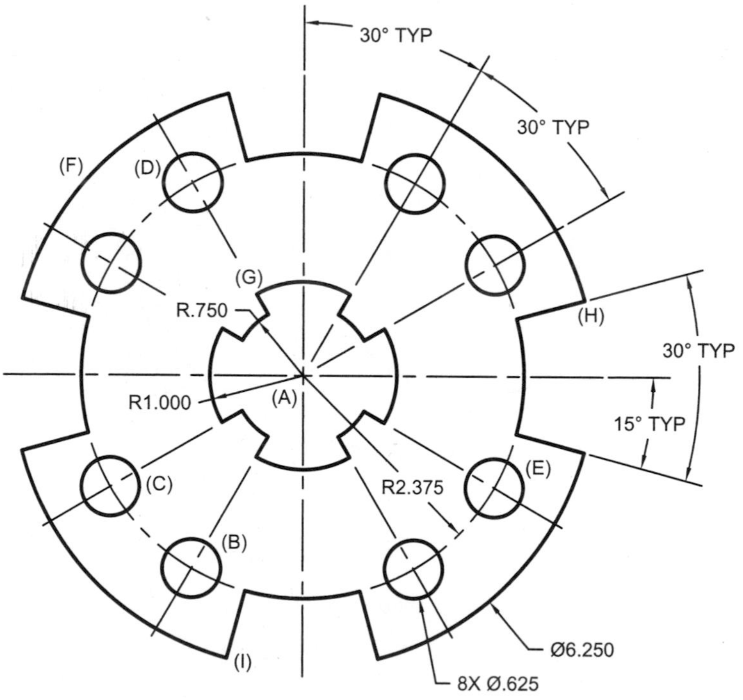

Refer to the drawing of the rotor and answer the following questions.

1. What is the absolute coordinate value of the center of the 0.625-diameter circle "B"?

   Answer: ____________

2. What is the total area of the rotor with all eight holes and the center slot removed?

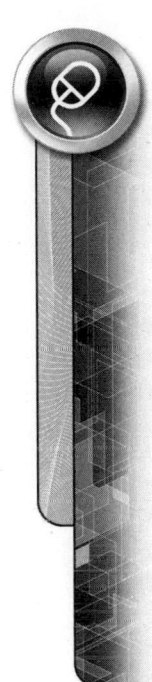

   Answer: ____________

3. What is the total length of arc "F"?

   Answer: ____________

4. What is the distance from the center of the 0.625 circle "C" to the center of the 0.625 circle "D"?

   Answer: ____________

5. What is the angle formed in the XY plane from the center of the 0.625 circle "B" to the center of the 0.625 circle "E"?

   Answer: ____________

6. What is the delta X,Y distance from the intersection at "H" to the intersection at "I"?

   Answer: ____________

7. Use the SCALE command to increase the size of just the center slot "G." Use the center of arc "F" as the base point. Use a scale factor of 1.500 units. What is the new surface area of the rotor with the center slot and all eight holes removed?

   Answer: ____________

PROBLEM 12–7 PATTERN5.DWG

Begin the construction of Pattern5 by keeping the default units set to decimal but changing the precision to zero. Be sure the system of angle measure is set to decimal degrees and the number of decimal places for the display of angles is zero. Begin the drawing by placing Vertex "A" at absolute coordinate (190,30).

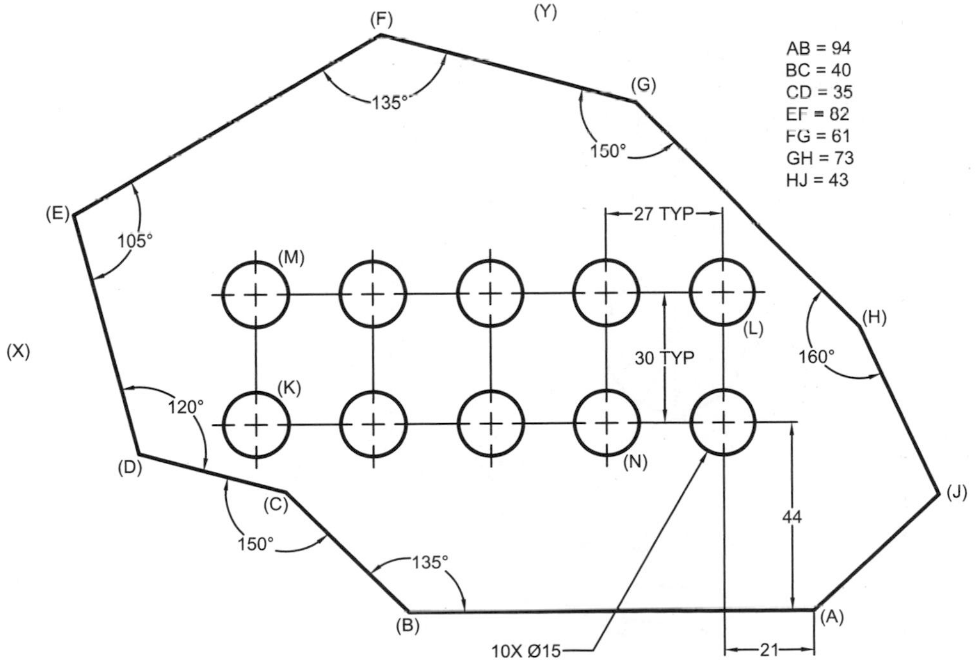

Refer to the drawing of Pattern5 and answer the following questions.

1. What is the distance from the intersection of vertex "J" to the intersection of vertex "A"?

   Answer: ____________

2. What is the perimeter of Pattern5?

   Answer: ____________

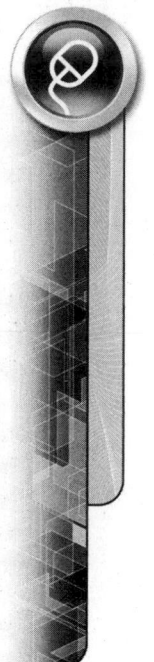

3. What is the total area of Pattern5 with all ten holes removed?

   Answer: ____________

4. What is the distance from the center of the 15-diameter hole "K" to the center of the 15-diameter hole "L"?

   Answer: ____________

5. What is the angle formed in the XY plane from the center of the 15-diameter hole "M" to the center of the 15-diameter hole "N"?

   Answer: ____________

6. What is the absolute coordinate value of the intersection at "F"?

   Answer: ____________

7. Use the STRETCH command to lengthen Pattern5. Use "Y" as the first point of the stretch crossing box. Use "X" as the other corner. Pick the intersection at "F" as the base point and stretch Pattern5 a total of 23 units in the 180° direction. What is the new total area of Pattern5 with all ten holes removed?

   Answer: ____________

PROBLEM 12–8 RATCHET.DWG

Use the Drawing Units dialog box to change the precision to two. Be sure the system of angle measure is set to decimal degrees and the number of decimal places for the display of angles is zero. Keep the remaining default unit values. Begin by drawing the center of the 1.00-radius arc of the ratchet at absolute coordinate (6.00,4.50), as shown in the following image.

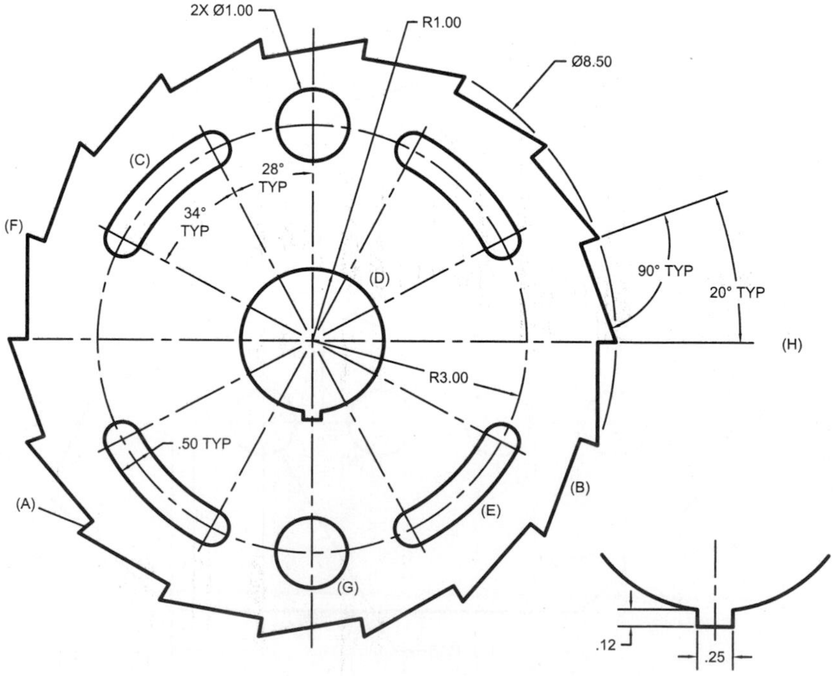

Refer to the drawing of the ratchet and answer the following questions.

1. What is the total length of the short line segment "A"?

   Answer: ____________

2. What is the total length of line "B"?

   Answer: ____________

3. What is the total length of arc "C"?

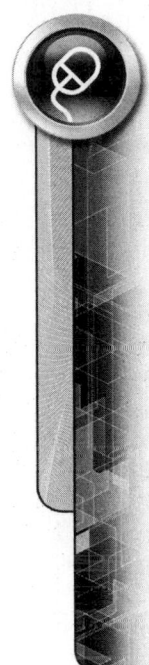

   Answer: ____________

4. What is the perimeter of the 1.00-radius arc "D" with the 0.25 x 0.12 keyway?

   Answer: ____________

5. What is the total surface area of the ratchet with all four slots, the two 1.00-diameter holes, and the 1.00-radius arc with the keyway removed?

   Answer: ____________

6. What is the absolute coordinate value of the endpoint at "F"?

   Answer: ____________

7. What is the angle formed in the XY plane from the endpoint of the line at "F" to the center of the 1.00-diameter hole "G"?

   Answer: ____________

PROBLEM 12–9 GENEVA.DWG

Start a new drawing called Geneva. Keep the default settings of decimal units, and set precision to two. Be sure the system of angle measure is set to decimal degrees and the number of decimal places for the display of angles is zero. Begin the drawing by constructing the 1.50-diameter arc at absolute coordinate (7.50,5.50).

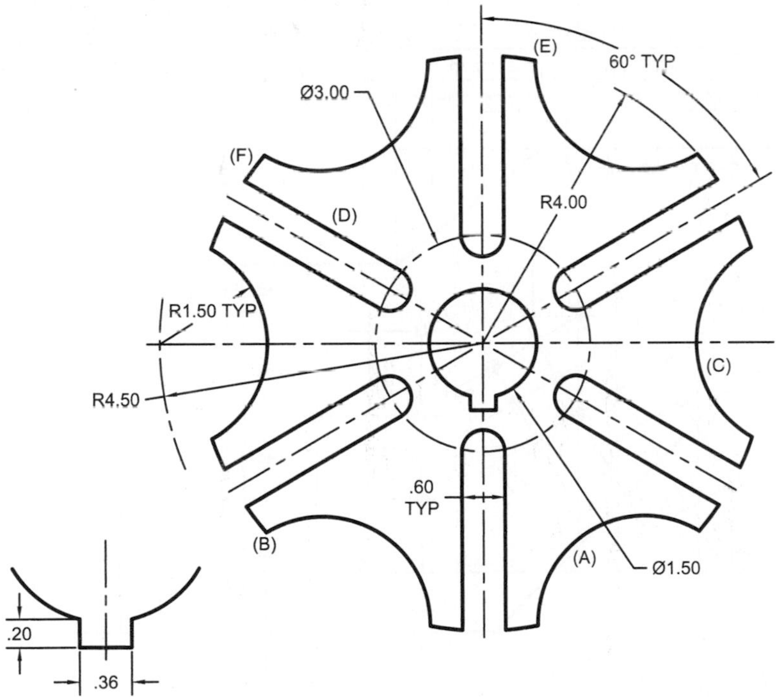

Refer to the drawing of the Geneva and answer the following questions.

1. What is the total length of arc "A"?

   Answer: ____________

2. What is the angle formed in the XY plane from the intersection at "B" to the center of arc "C"?

   Answer: ____________

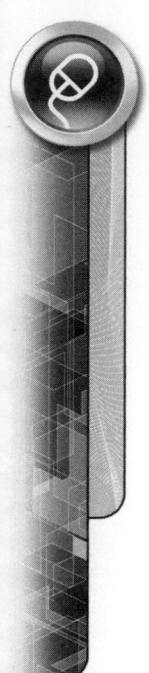

3. What is the absolute coordinate value of the midpoint of line "D"?

   Answer: ____________

4. What is the total area of the Geneva with the 1.50-diameter hole and keyway removed?

   Answer: ____________

5. What is the total distance from the midpoint of arc "F" to the center of arc "A"?

   Answer: ____________

6. What is the delta X,Y distance from the intersection at "E" to the center of arc "C"?

   Answer: ____________

7. Use the SCALE command to reduce the Geneva in size. Use 7.50,5.50 as the base point; use a scale factor of 0.83 units. What is the absolute coordinate value of the intersection at "E"?

   Answer: ____________

PROBLEM 12–10 STRUCTURAL APPLICATION—GUSSET.DWG

Begin the construction of the Gusset by keeping the default units set to decimal but changing the precision to two.

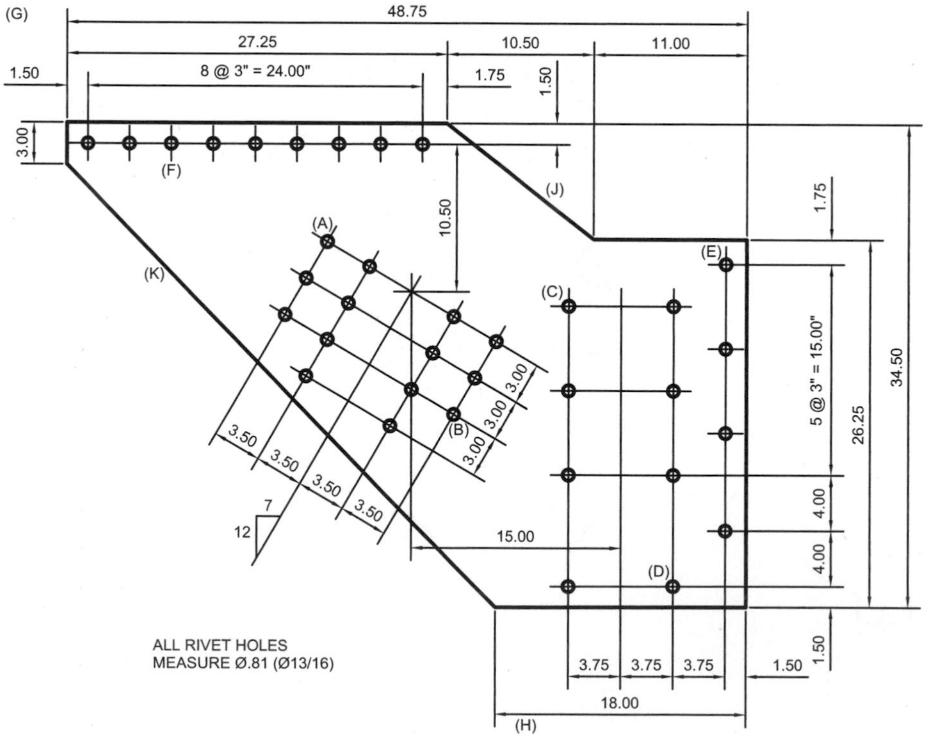

Refer to the drawing of the Gusset and answer the following questions.

1. What is the total area of the gusset plate with all thirty-five rivet holes removed?

   Answer: ____________

2. What is the delta X,Y distance from the center of hole "A" to the center of hole "B"?

   Answer: ____________

3. What is the angle formed in the XY plane from the center of hole "C" to the center of hole "D"?

   Answer: ____________

4. What is the angle formed in the XY plane from the center of hole "E" to the center of hole "F"?

   Answer: ____________

5. What is the total length of line "J"?

   Answer: ____________

6. What is the total length of line "K"?

   Answer: ____________

7. Stretch the gusset plate directly to the left using a crossing box from "H" to "G" and at a distance of 3.00 units. What is the new area of the gusset plate with all thirty-five rivet holes removed?

   Answer: ____________

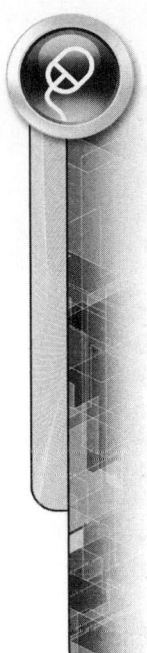

CHAPTER 13

Creating Parametric Drawings

Geometric constraints allow you to create geometric relationships between selected objects. An example would be to apply the Equal constraint to all holes on an object. Then when one of the holes is dimensioned with a dimensional constraint and you change the value of the hole, all holes that share the same Equal constraint change to reflect the currently dimensioned diameter. In this chapter, you will learn the constraint types and how to apply them to drawing objects. You will also be shown the power of controlling the objects in a design through the use of parameters. A number of Try It! exercises are available to practice with the various methods of constraining objects. Two tutorials are also available at the end of this chapter to guide you along with assigning constraints to objects.

DISPLAYING PARAMETRIC MENUS

Commands used for accessing geometric and dimensional constraints can all be accessed from the Parametric tab of the Ribbon, as shown in Figure 13.1.

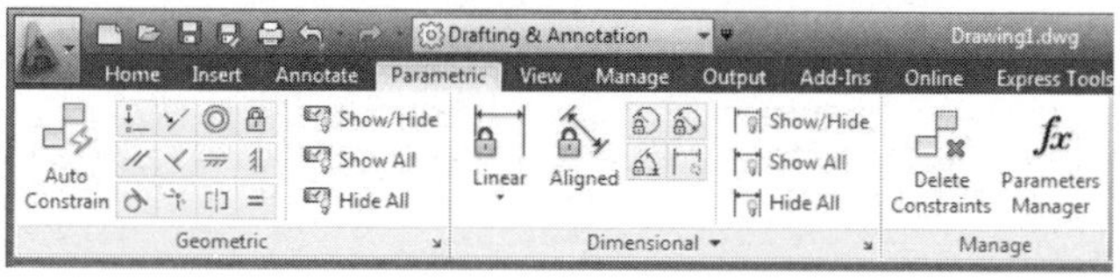

FIGURE 13.1

GEOMETRIC CONSTRAINTS

The following table illustrates the various geometric constraints available complete with constraint name, constraint icon, and a description of the constraint.

Constraint	Icon	Description
Coincident	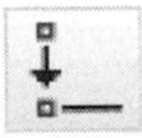	A point is constrained to lie on another point or curve (line, arc, etc.).
Collinear	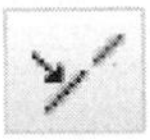	Two selected lines will line up along a single axis. If the first line moves, so will the second. The two line segments do not have to be touching.
Concentric	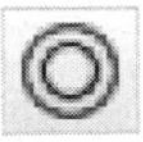	Arcs or circles will share the same center point.
Equal	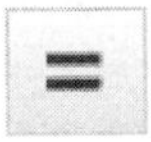	If two arcs or circles are selected, they will have the same radius or diameter. If two lines are selected, they will become the same length. If you select multiple similar objects such as lines, circles, and arcs before using the Equal constraint, the constraint is applied to all of them.
Fix	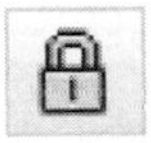	Applying a fixed constraint to a point or points will prevent the selected objects from moving.
Horizontal	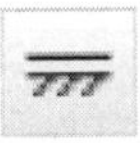	Lines are positioned parallel to the X-axis.
Parallel	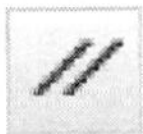	Lines will be repositioned so that they are parallel to one another. The first line sketched will stay in its position and the second will move to become parallel to the first.
Perpendicular	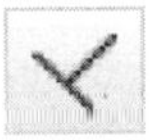	Lines will be repositioned at 90° angles to one another. The first line selected will stay in its position and the second will rotate until the angle between them is 90°.
Smooth	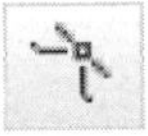	A spline and another spline, line, or arc that connect at an endpoint with a coincident constraint.
Symmetric	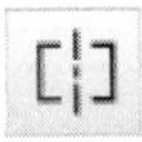	Selected points defining the selected geometry are made symmetric about the selected line.
Tangent	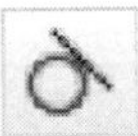	An arc, circle, or line will become tangent to another arc or circle.
Vertical	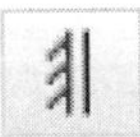	Lines are positioned parallel to the Y-axis.

METHODS OF CHOOSING CONSTRAINTS

Geometric constraints can be created through the Geometric panel of the Ribbon as shown in the following image on the left. Constraints can also be created by toggling on the Infer Constraints button on the Status Bar, as shown on the right in Figure 13.2.

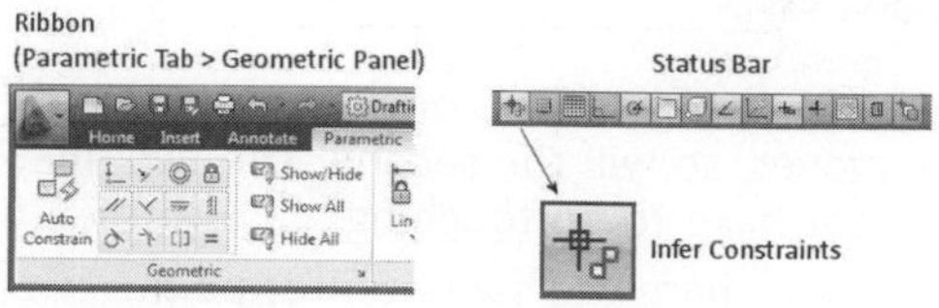

FIGURE 13.2

DISPLAYING CONSTRAINTS

As constraints are placed, they can be viewed through the Show All button located in the Geometric panel of the Ribbon, as shown in Figure 13.3 (Left). When dealing with complicated objects involving numerous constraints, you can use the Hide All button to turn off all constraints, as shown in Figure 13.3 (Center). The Show/Hide button allows you to show or hide constraints for selected objects, as shown in Figure 13.3 (Right).

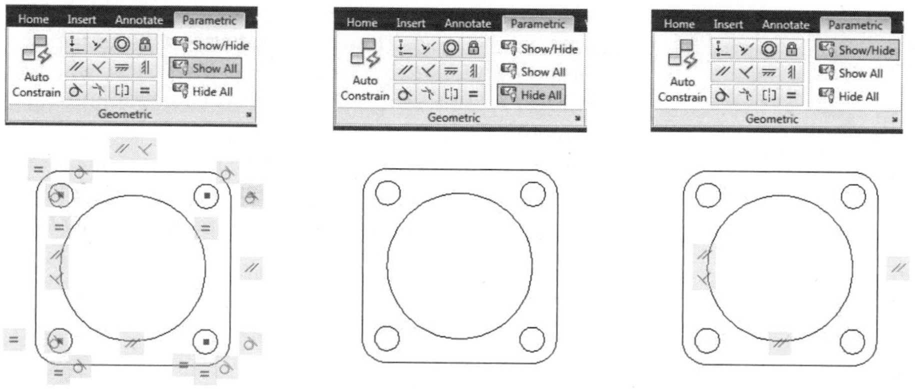

FIGURE 13.3

NOTE

Figure 13.3 shows Geometric constraints. A similar set of buttons is available for showing and hiding Dimensional constraints.

DELETING CONSTRAINTS

As constraints are placed, at times they may need to be deleted. To perform this, right-click on the constraint and pick Delete from the menu, as shown in Figure 13.4. You could also pick the constraint and hit the Delete key in any standard keyboard to remove the constraint from the object.

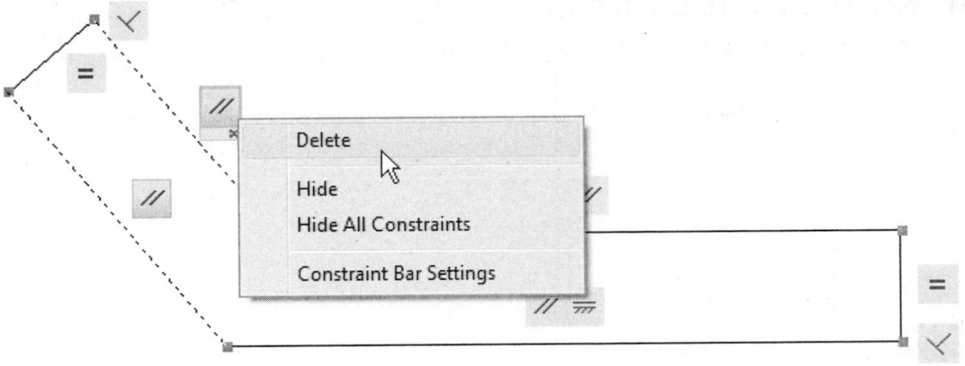

FIGURE 13.4

THE CONSTRAINT SETTINGS DIALOG BOX

You can control the display and behavior of constraints through the Constraint Settings dialog box. Clicking on the small arrow in the lower-right corner of the Geometric panel, as shown in Figure 13.5 in the upper-left, displays the Constraint Settings dialog box, as shown in Figure 13.5 (Right). The dialog can also be displayed by right-clicking the Infer Constraints button on the Status bar and picking Settings. There are three tabs available in the dialog box. Here we will look at the Geometric tab. The Dimensional and AutoConstrain tabs will be discussed later in the chapter.

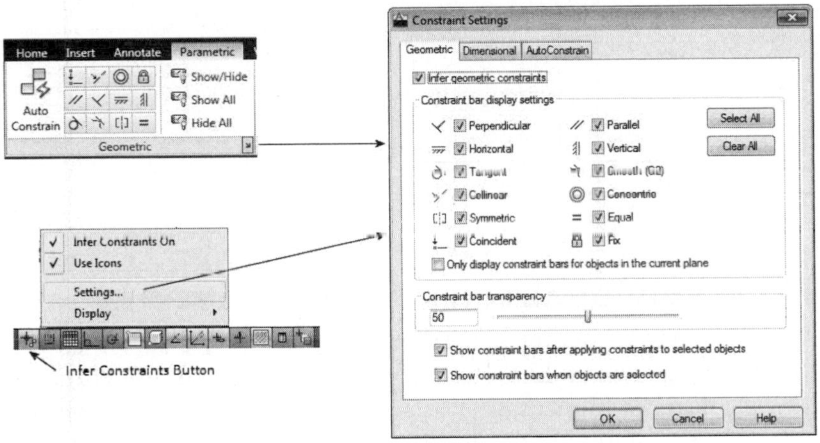

FIGURE 13.5

Figure 13.6 (Left) shows the Geometric tab of the Constraint Settings dialog box. In the "Constraint bar display setting" area of the dialog box, all constraints are checked. Checking a constraint means that each time that constraint is created in your drawing, a glyph (constraint bar) will be displayed to show its existence and location. The drawing shown in Figure 13.6 (Right) has the following geometric constraints applied: Perpendicular, Parallel, Tangent, and Equal. If any of these constraints types had not been checked in the dialog box, the constraints would still exist but the glyph would not appear even after selecting "Show All" from the Ribbon.

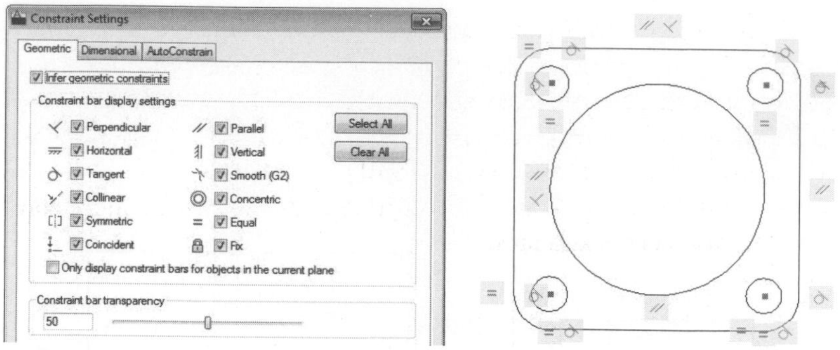

FIGURE 13.6

Changing the Constraint bar display settings can help to isolate geometric constraint types on your drawings. For example, to display only the glyphs for the Equal constraints, pick the Clear All button and place a check in the check box for the Equal constraint, as shown in Figure 13.7 (Left). The results are shown in the following image on the right.

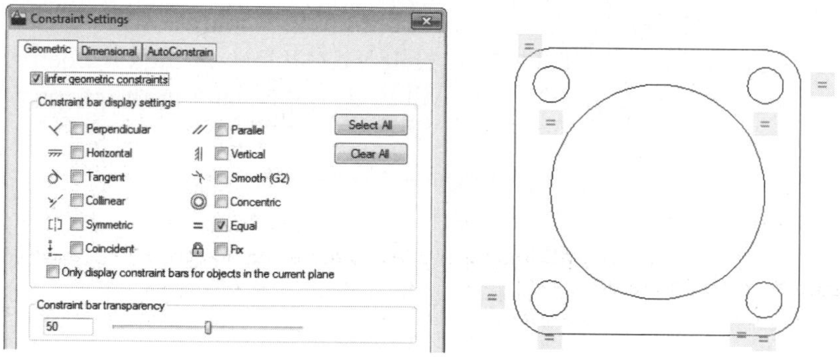

FIGURE 13.7

DRAWING WITH INDIVIDUAL LINES VERSUS POLYLINES

How you construct lines may impact the kind of results you get when applying constraints to line segments. In Figure 13.8, a number of line segments were drawn. When a dimensional constraint (dimension) was applied to the height of the object, only the vertical line segment moved to the dimension. The remaining line segments were unaffected by the dimensional constraint. This means that before applying dimensional constraints to an object consisting of individual line segments, the endpoints of the lines must first be constrained using the Coincident constraint. This will be covered in a later segment of this chapter.

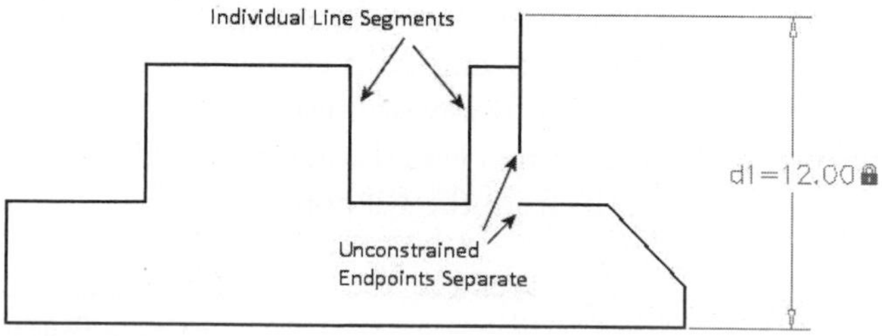

FIGURE 13.8

Figure 13.9 is similar to Figure 13.8 with the exception that all line segments were converted to a single polyline before applying geometric constraints or dimensional constraints. Placing the dimensional constraint for the total height of 12.00 units stretches the polyline shape and does not result in gaps in the object. It is not necessary to add Coincident constraints between segments in the polyline shape. This method of converting objects to a polyline for constraints also works for more complicated shapes that involve curves and arcs.

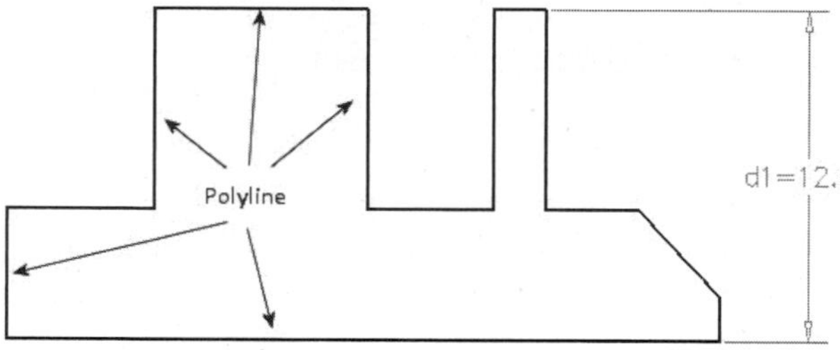

FIGURE 13.9

APPLYING HORIZONTAL AND VERTICAL CONSTRAINTS

Applying a horizontal or vertical constraint to an object is very straightforward. The results are objects that are constrained either horizontally or vertically. If you change your mind and wish to add a dimension consisting of an angle, you must first delete the horizontal or vertical constraint. The following Try It! illustrates the creation of Horizontal and Vertical constraints.

TRY IT!

Open the drawing file 13_Constraints Hor Ver. A number of line segments are drawn inclined, and the Horizontal and Vertical constraints will be applied to force the lines to be horizontal and vertical. Figure 13.10 (Left) shows the object prior to having constraints applied. Moving your cursor over the object will highlight all lines signifying that the object consists of a single polyline shape, as shown in Figure 13.10 (Right).

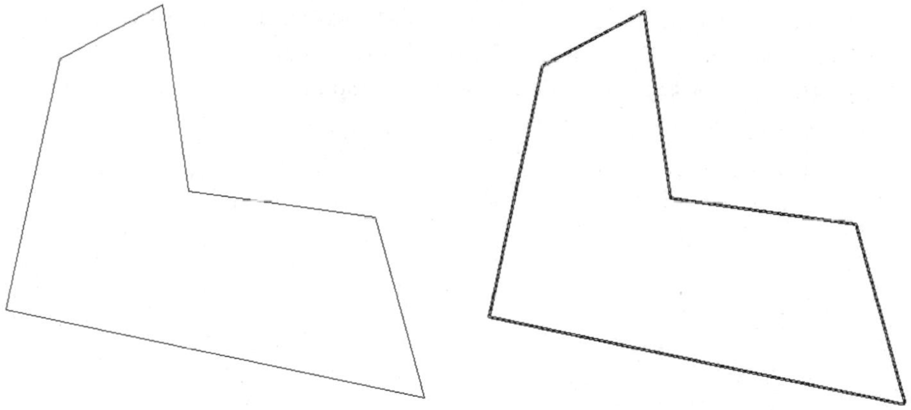

FIGURE 13.10

Begin applying the Horizontal constraints to the three lines, as shown in Figure 13.11 (Left). Notice all three lines snap to horizontal and display the Horizontal constraint glyph. Your object may appear slightly different depending on where you picked the lines. If your constraints do not display, click the Show All button located in the Ribbon. Next, apply the Vertical constraint to the three lines, as shown in Figure 13.11 (Right). All three lines should snap to vertical and display the Vertical constraint glyph.

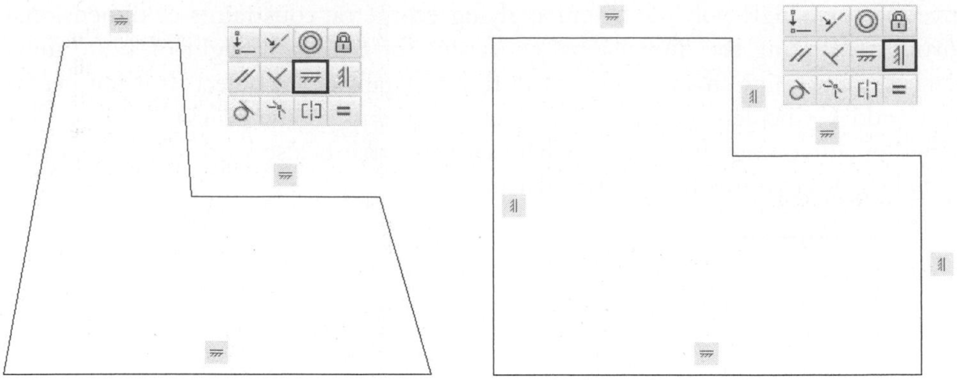

FIGURE 13.11

APPLYING PARALLEL AND PERPENDICULAR CONSTRAINTS

Like the Horizontal and Vertical constraints, the Parallel and Perpendicular constraints are simple to apply. You will, however, be selecting two objects and the order of selection determines the result. The following Try It! exercise will direct you to select a first object. When selecting the second object, the Parallel or Perpendicular constraint is placed in relation to the first object selected. Study the sequence of steps and illustrations to become more familiar with Parallel and Perpendicular constraints.

TRY IT!

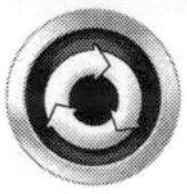

Open the drawing file 13_Constraints Per Par. A number of line segments are drawn at random angles and the Parallel and Perpendicular constraints will be applied to force the lines to be parallel and perpendicular (at a 90° angle). Moving your cursor over the object will highlight all lines signifying that the object consists of a single polyline shape. Also notice the appearance of the Constraints glyph. The glyph appears when you move your cursor over an object that has constraints applied. To make the parallel and perpendicular constraints more predictable, the bottom line of the object has already had the Horizontal and Fix constraints applied. Applying the Fix constraint prevents the line from moving while the other constraints are being placed. Pick the Show All button in the Geometric panel of the Ribbon to display the constraints, if necessary.

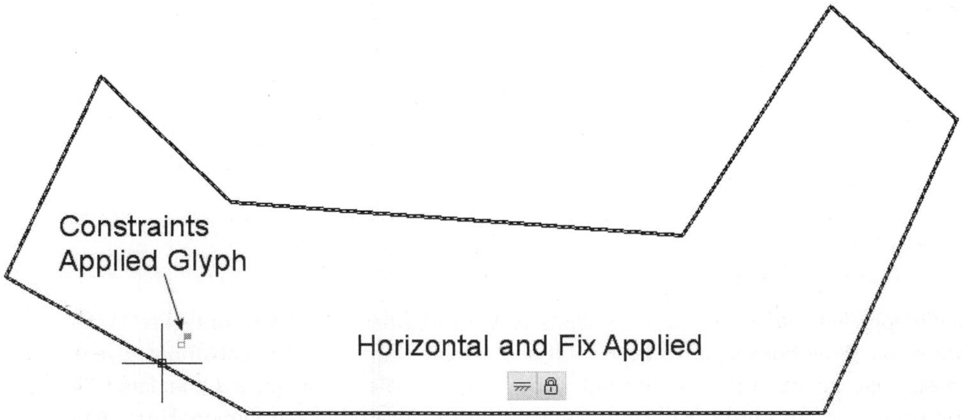

FIGURE 13.12

Begin by clicking on the Parallel constraint button. When prompted to select objects, pick line "A" followed by line "B," and because line "A" was picked first, line "B" will be parallel to line "A." Continue by applying Parallel constraints to the lines on the left side of the object, as shown in Figure 13.13. Pick line "C" followed by line "D" to make line "D" parallel to line "C." Apply the last Parallel constraint using lines "E" and "F" as guides. When finished, your object should appear similar to the one shown in Figure 13.13. The angles will be the same but the spacing of your object may be different depending on which end of line segments were picked.

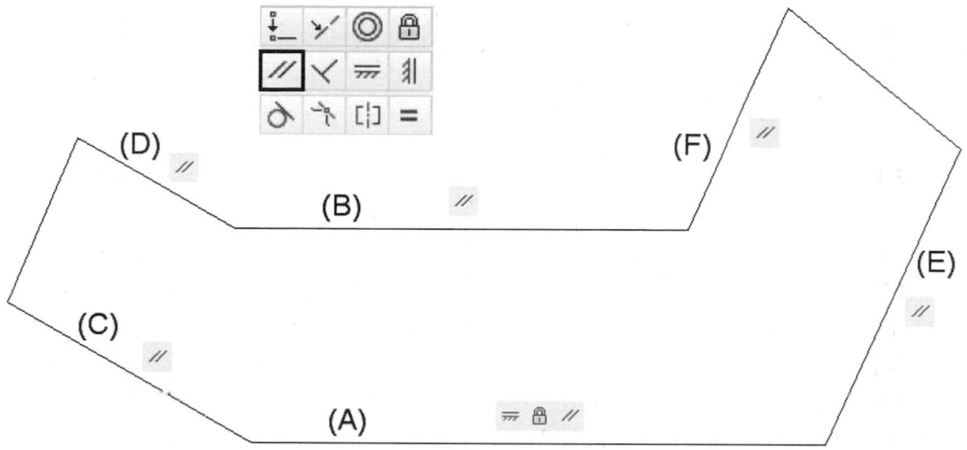

FIGURE 13.13

Next, click on the Perpendicular constraint button. When prompted to select objects, pick line "A" followed by line "B." Because line "A" was picked first, selecting line "B" makes it perpendicular to line "A." Continue by applying a Perpendicular constraint to the lines on the right side of the object, as shown in Figure 13.14. Pick line "C" followed by line "D" to make line "D" perpendicular to line "C."

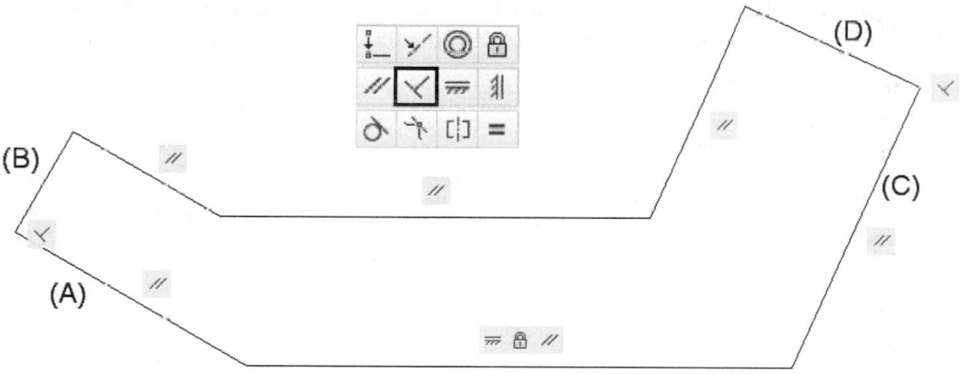

FIGURE 13.14

APPLYING COINCIDENT CONSTRAINTS

We have already discussed the advantages polylines have over regular line segments with placing constraints. This does not mean it is wrong to work with individual line segments. In this case when working with the endpoints of individual line segments, use the Coincident constraint. Follow the prompts that tell you to select the endpoint of the first object and then the second object. Both endpoints will snap together; OSNAP-Endpoint is not enough to automatically apply the Coincident constraint.

TRY IT!

Open the drawing file 13_Constraints Coincident, as shown in Figure 13.15. The line segments are drawn but do not meet; a Coincident constraint can be applied to the ends of the line segments.

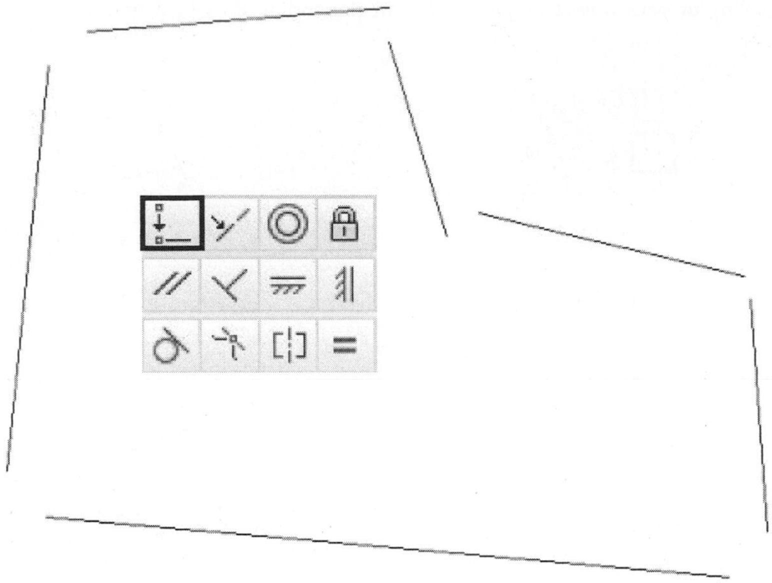

FIGURE 13.15

Click on the Coincident Constraint button and pick the endpoint of the first line, as shown in Figure 13.16 (Left). Continue by selecting the endpoint of the second line, as shown in Figure 13.16 (Center). This action will join the endpoints of both line segments and place a blue dot, as shown in Figure 13.16 (Right). The presence of the blue dot signifies that the Coincident constraint was applied. Also, if you move your cursor over the blue dot, the Coincident constraint glyph will display.

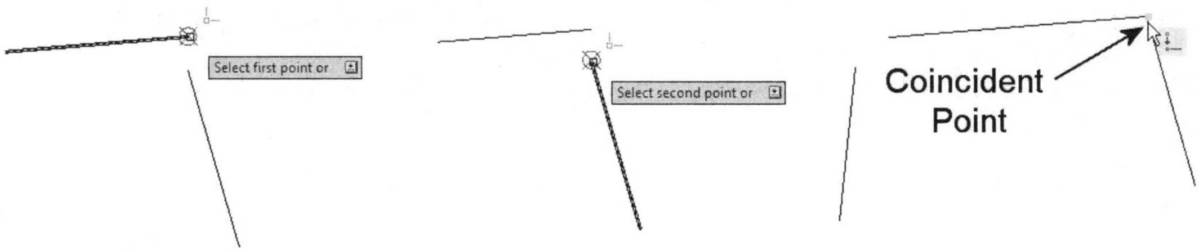

FIGURE 13.16

While this creates the coincident constraint at one corner of the object, this constraint must now be applied to the remaining corners, as shown in Figure 13.17.

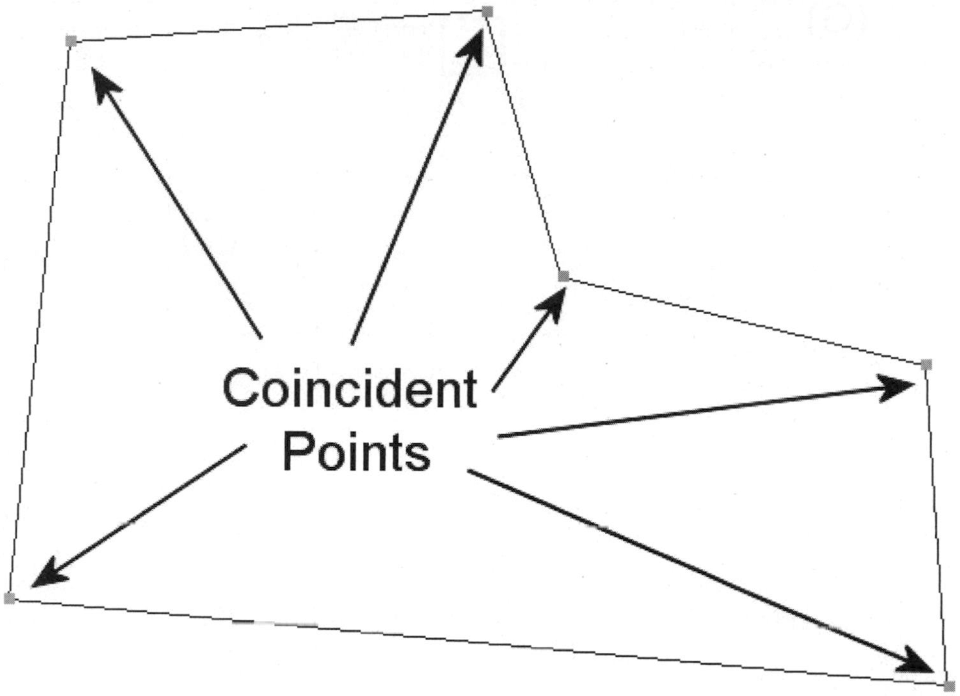

FIGURE 13.17

APPLYING COLLINEAR CONSTRAINTS TO LINES

Applying a Collinear constraint keeps the selected objects in-line with each other. The lines do not have to be touching in order to perform this task. Study the following Try It! and images designed to place Collinear constraints.

TRY IT!

Open the drawing file 13_Constraints Collinear. The object in Figure 13.18 consists of a single polyline object. This will help keep the endpoints intact when applying the Collinear constraints. As with most of the other constraint methods, order is important. Pick the Collinear constraint button and for the first object, pick line "A" and for the second object, pick line "B." This will constrain line "B" to line "A" along the same axis or line of sight. Continue applying Collinear constraints to the lines in the following order; First line "C," Second line "D"; First line "E," Second line "F"; and First line "G," Second line "H."

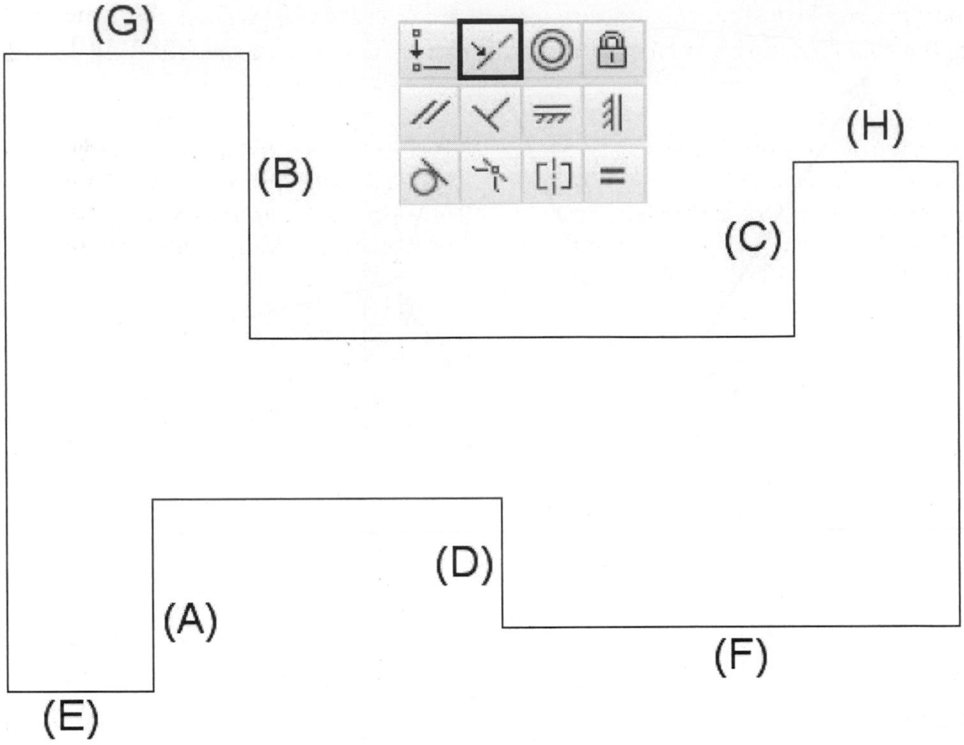

FIGURE 13.18

The results of applying Collinear constraints are shown in Figure 13.19.

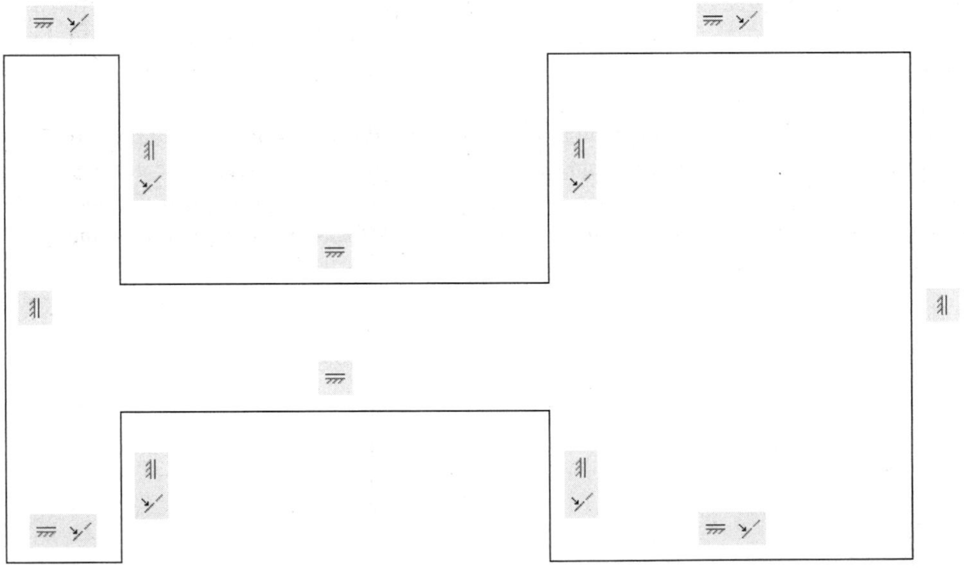

FIGURE 13.19

APPLYING A CONCENTRIC CONSTRAINT

Another useful constraint forces the centers of arcs and circles to share the same centerpoint; this constraint is called Concentric and is illustrated in the following Try It! exercise.

TRY IT!

Open the drawing file 13_Constraints Concentric. A number of constraints have already been to the line segments. Click on the Concentric button and pick the first arc at "A" followed by the second arc at "B," as shown in Figure 13.20 (Left). The results are shown in Figure 13.20 (Right). Notice how both arcs now share the same centerpoint which is another way of saying they are Concentric with each other.

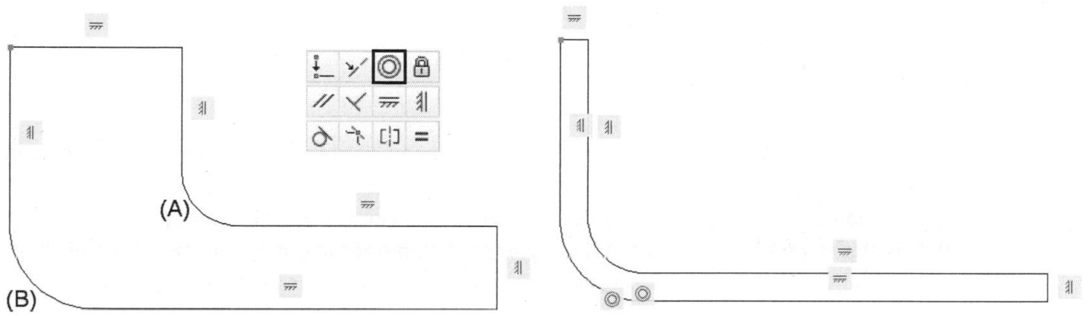

FIGURE 13.20

APPLYING TANGENT CONSTRAINTS

Tangent constraints are essential in making lines tangent to curves or for making curves tangent to other curves. Study the next Try It! exercise for applying Tangent constraints.

TRY IT!

Open the drawing file 13_Constraints Tangent. The object in Figure 13.21 (Left) consists of a polyline perimeter. Included in the perimeter are a number of lines and arc segments. This object looks complete until displaying grips, making the grips active, and stretching the grips to new locations. The results may be similar to Figure 13.21 (Right). While the lines and arcs originally appeared tangent, they were not constrained to be tangent.

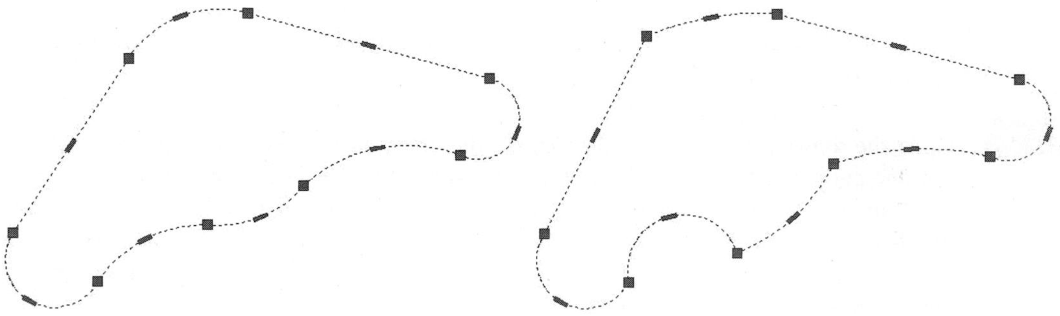

FIGURE 13.21

Begin placing Tangent constraints by first picking the Tangent constraint button and selecting the line segment as the first object, as shown in Figure 13.22 (Left). Next, pick the arc as the second segment, as shown in Figure 13.22 (Center). The results are illustrated in Figure 13.22(Right) with the Tangent constraint being confirmed by the appearance of the Tangent constraint glyph.

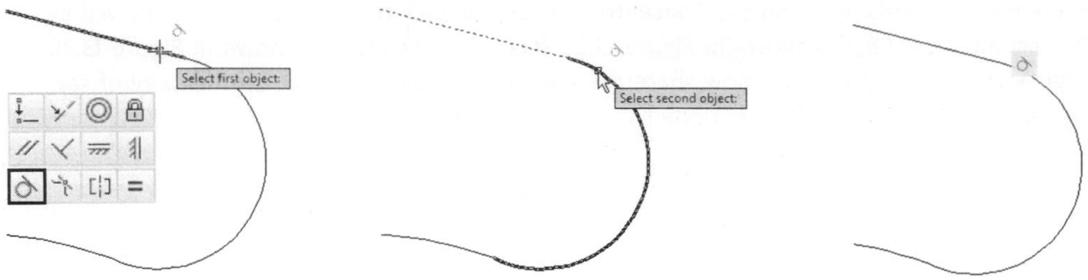

FIGURE 13.22

Continue adding Tangent constraints to all lines and arcs until your object appears as shown in Figure 13.23 (Left). To test if the Tangent constraints will keep their tangencies, click on the object and begin picking grips and stretch them to new locations. Take care when performing this as the stretch operations can display unpredictable results. If it becomes difficult dragging a grip to a different location, undo the operation and try dragging a different grip.

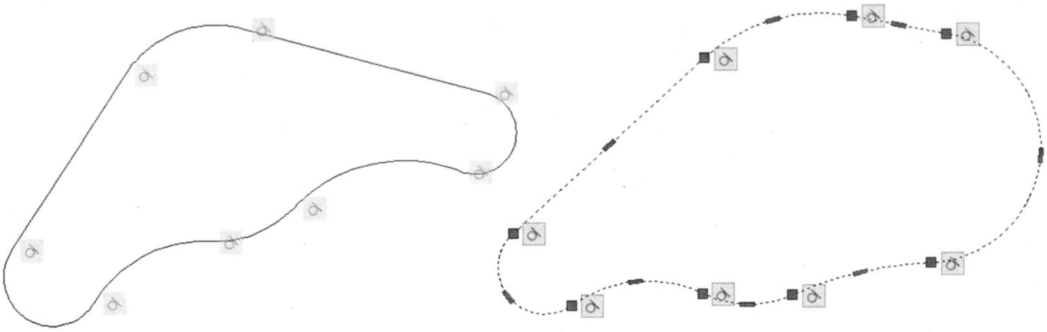

FIGURE 13.23

APPLYING EQUAL CONSTRAINTS

The Equal constraint is used to make "like objects" the same size. This allows you to make such items as circles or arcs equal to the point where adding one Dimensional constraint will affect all circles or arcs. The equal constraint also works with line segments. Study the next Try It! exercise for placing Equal constraints first on the circles and then on the arcs.

TRY IT!

Open the drawing file 13_Constraints Equal. The object in Figure 13.24 (Left) consists of a polyline perimeter that has various fillet radii. Also inside the shape are five circles of various diameters. First click the Equal constraint button, pick the circle at "A" and then select any other circle. Continue making the other circles equal to "A." The results are illustrated in Figure 13.24 (Right) with all five circles having the same diameter. It also means that if one of the circles has a Dimensional constraint placed, the other circles change to the same diameter.

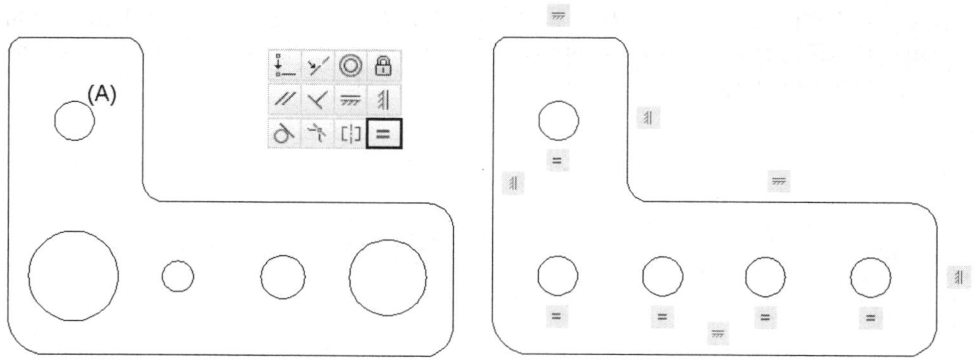

FIGURE 13.24

Next, apply the Equal constraint to make all of the radii the same value. Click on the Equal constraint button and before selecting the arc, enter the Multiple option. Then select the arc in the lower left corner of the object. You will then be prompted to select the objects you want to make equal to the first. Select the remaining arcs and press Enter. The results are illustrated in Figure 13.25 (Right). Unfortunately a few of the arcs appear to have changed their position in the perimeter of the polyline. This is to be expected when applying constraints. A series of Tangent constraints will need to be applied in order to repair the issue with the arcs.

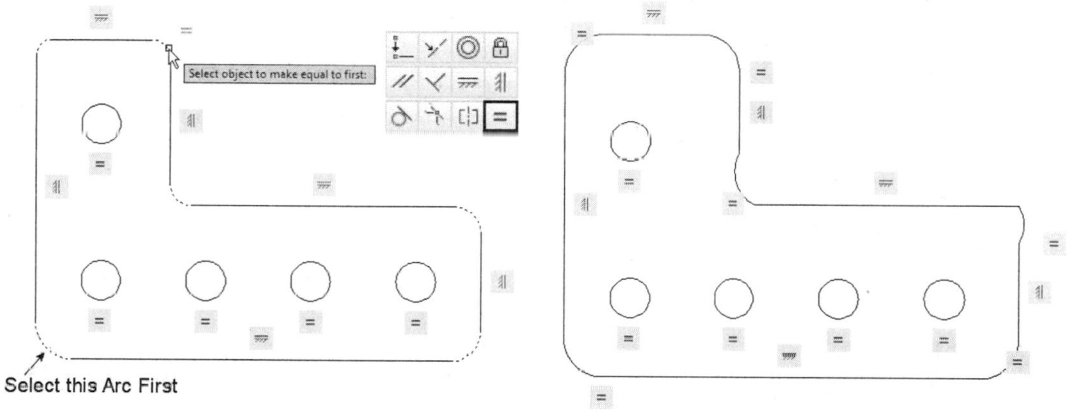

FIGURE 13.25

APPLYING A FIX CONSTRAINT

A Fix constraint fixes the position of an object so it cannot be moved. This constraint helps you to lock down your design. As you apply more geometric and dimensional constraints, the fixed component will remain unchanged in the drawing. Study the next Try It! exercise on how the Fix constraint affects an object.

TRY IT!

Open the drawing file 13_Constraints Fix. Click on the Fix constraint button, as shown in Figure 13.26 (Left), and pick the lower left corner of the object. You will notice the appearance of the padlock glyph signifying that a Fix constraint is about to be placed. The results are shown in the following image on the right. Notice the addition of the Fix constraint to the drawing.

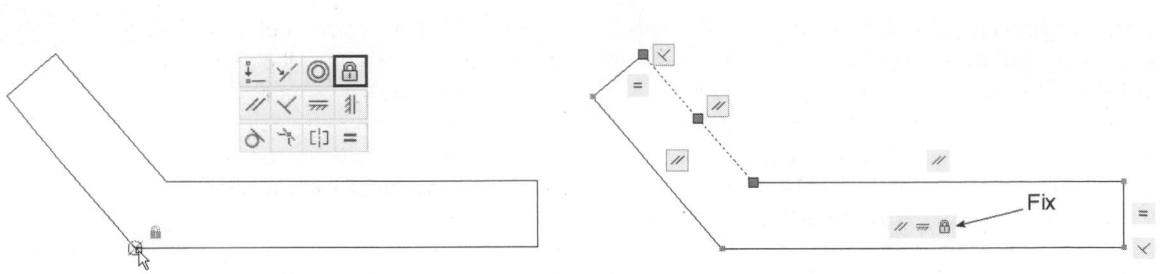

FIGURE 13.26

APPLYING A SYMMETRIC CONSTRAINT

To maintain symmetry in your design a Symmetric constraint can be utilized. You first select two objects or points that are to be symmetric and then you pick the symmetry line. Study the next Try It! exercise on how this constraint affects objects.

TRY IT!

Open the drawing file 13_Constraints Symmetric. Click on the Symmetric constraint button, as shown in Figure 13.27 (Left), and pick the small circle "A" as the first object. Pick the larger circle "B" as the second object and the vertical line "C" as the symmetry line. The results are shown in Figure 13.27 (Center). Place another Symmetric constraint on the end lines. Pick the left end line "D" as the first object, the right end line "E" as the second object, and the vertical line "C" as the symmetry line. The final results are shown in Figure 13.27 (Right).

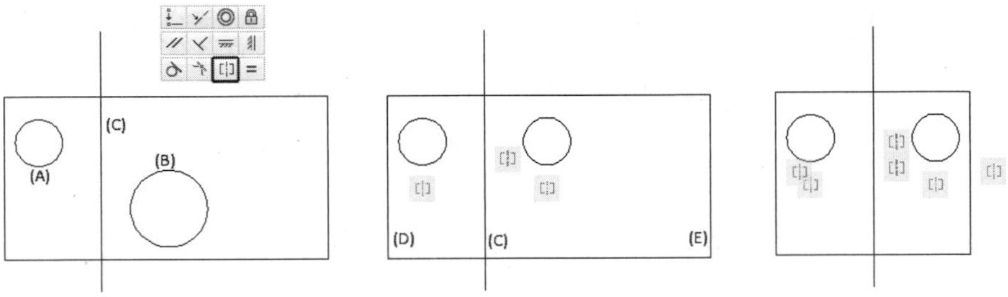

FIGURE 13.27

AUTO CONSTRAINING

The Auto Constrain button allows for constraints to be applied to selected objects in a drawing automatically. Choose this command from the Geometric panel, as shown in Figure 13.28. Also follow the next Try It! exercise to see how the Auto Constrain command applies constraints to objects in a drawing.

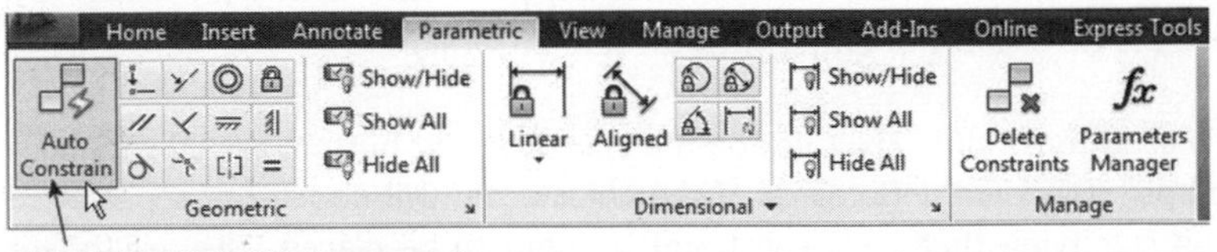

AUTOCONSTRAIN Command

FIGURE 13.28

TRY IT!

Open the drawing file 13_Constraints Auto. The object in Figure 13.29 (Left) consists of individual lines and three circles. Clicking on the Auto Constrain button and selecting all objects including circles and arcs will display the drawing as shown in Figure 13.29 (Right). Notice how various constraints such as Concentric, Tangent, Parallel, Equal, and Perpendicular were applied, as shown in Figure 13.29 (Right). You can even see the individual blue dots at the intersections of all lines and arcs signifying that Coincident constraints were applied through the Auto Constrain operation.

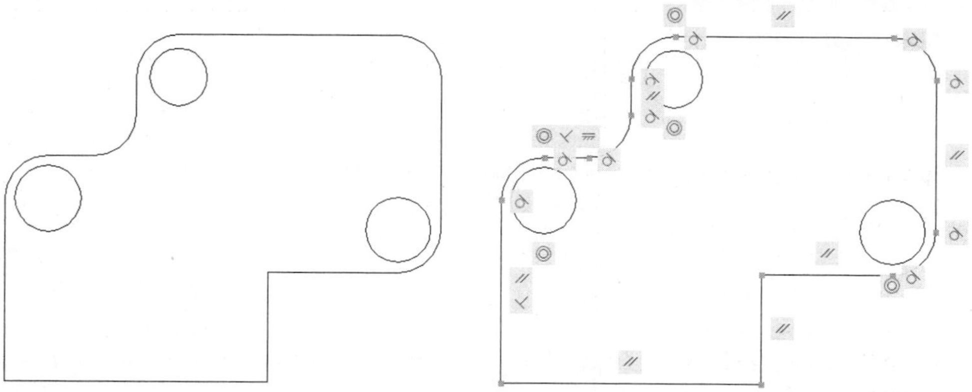

FIGURE 13.29

When placing constraints automatically through the Auto Constraint command, a command prompt appears allowing you to type S for Settings, which takes you to the AutoConstrain tab of the Constraint Settings dialog box. Use this dialog box to control which constraints are available and in what order they are applied, as shown in Figure 13.30.

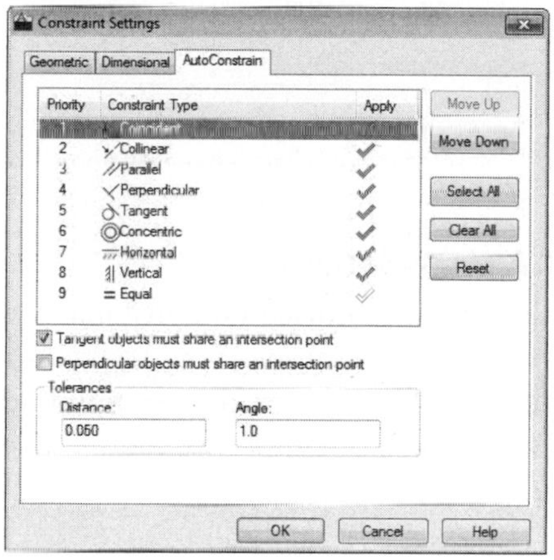

FIGURE 13.30

INFER CONSTRAINTS

Constraints can also be applied automatically by turning on Infer constraints. Toggle Infer constraints on and off by picking the Infer constraints button on the Status bar, as shown in Figure 13.31. Right-clicking the Status bar button displays a shortcut menu, where you can pick Settings to display the Constraint Settings dialog box. Infer constraints can also be activated here by placing a check in the Infer geometric constraints box. With this feature on, constraints are applied automatically as objects are created. Follow the next Try It! exercise to see how Infer constraints are applied to objects in a drawing.

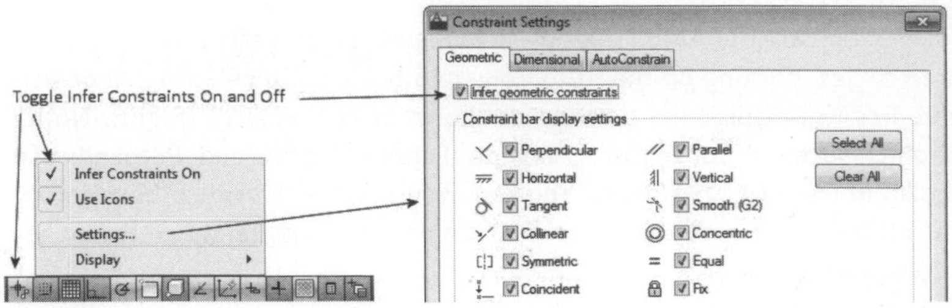

FIGURE 13.31

TRY IT!

Open the drawing file 13_Constraints Infer. In the Status bar ensure Infer Constraints and Polar Tracking are both toggled on. Set Polar Tracking to 30°. Use the LINE command to draw a line in the 0° direction and notice that a horizontal constraint is automatically applied, as shown in Figure 13.32 (Left). Draw a second line in the 60° direction and then one in the 90° direction. Notice the Vertical and Coincident (blue dots) constraints applied, as shown in Figure 13.32 (Center). Draw a line in the 180° direction and then use the Close option to complete the drawing. A Perpendicular constraint, along with additional Coincident constraints, is applied as shown in Figure 13.32 (Right). Toggle off the Infer Constraints button.

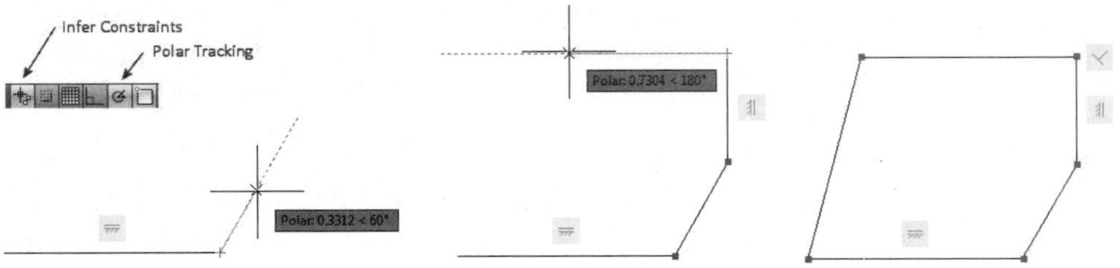

FIGURE 13.32

ESTABLISHING DIMENSIONAL RELATIONSHIPS

Once geometric constraints are applied, the next step is to place Dimensional constraints as a means of putting limits on the geometry. For example, you would add a Dimensional constraint to call out the total length of an object; or you could add a Dimensional constraint to call out the diameter of a circle or the radius of an arc. Dimensional constraints can be selected from the Dimensional panel of the Ribbon, as shown in Figure 13.33.

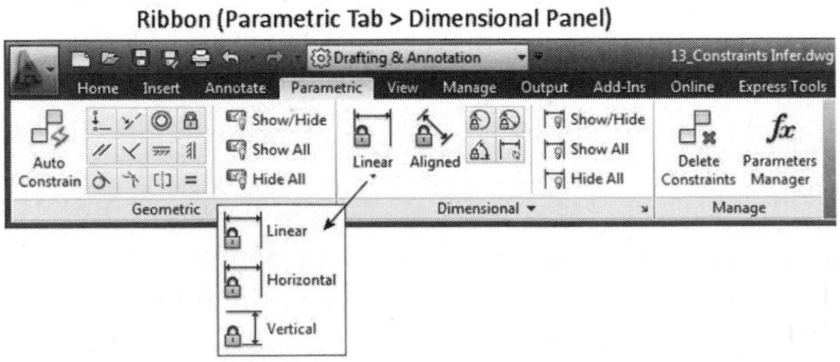

FIGURE 13.33

DIMENSION NAME FORMAT

Dimensional constraints can be displayed in three different formats when placed. These formats are all displayed in Figure 13.34 and include Name, Value, and Name and Expression. To change the formats you utilize the same Constraint Settings dialog box shown earlier; however, the Dimensional tab must be selected.

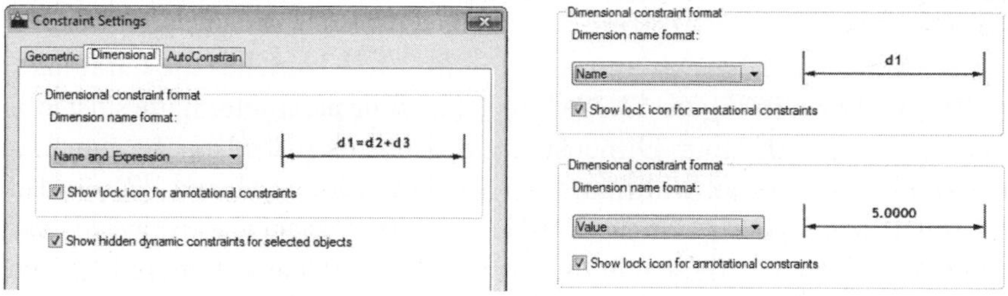

FIGURE 13.34

ADDING DIMENSIONAL CONSTRAINTS

Figure 13.35 illustrates the various types of Dimensional constraints that can be placed in a drawing. These include Linear, Aligned, Radial, Diameter, and Angular.

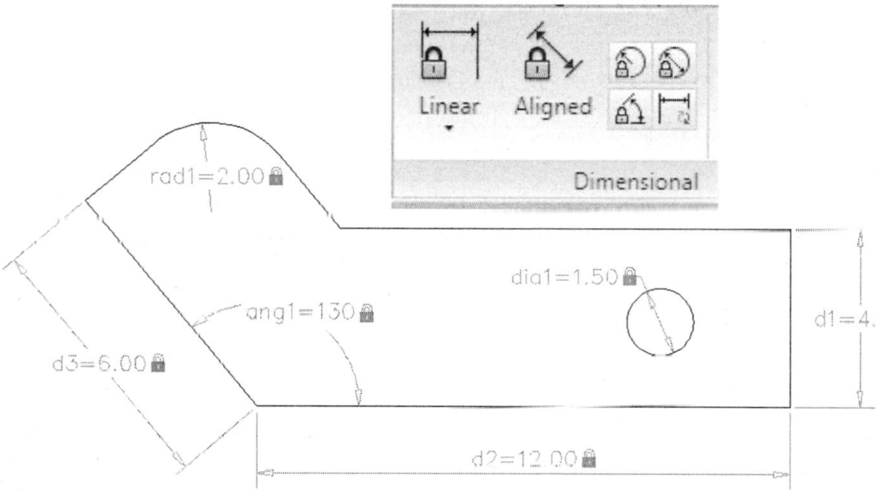

FIGURE 13.35

WORKING WITH PARAMETERS

Another powerful feature of working with Dimensional constraints is assigning parameters to various objects for the purpose of changing a dimension and then having other related dimensions change automatically. This study begins with picking the Parameters Manager button from the Manage panel of the Ribbon, as shown in Figure 13.36.

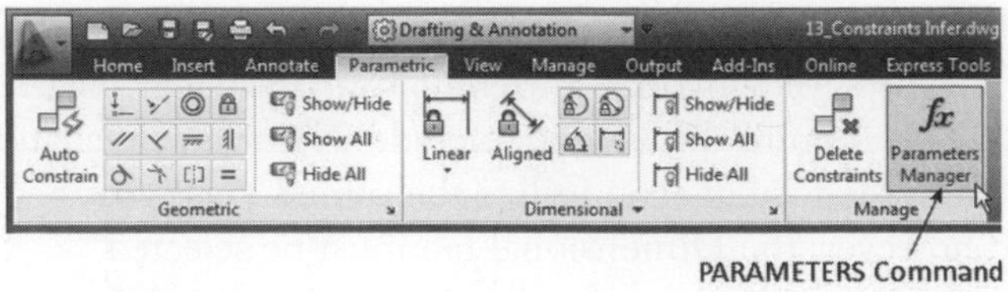

FIGURE 13.36

Clicking on the Parameters Manager button in Figure 13.36 launches the Parameters Manager palette, as shown in Figure 13.37 (Left). The four parameter names listed in the palette correspond to the four Dimensional constraints placed in the drawing. Here is the problem: the overall width and height dimensions (d1 and d2) need to control the dimensions locating the circle (d3 and d4). Whenever the width and height dimensions change, the circle is supposed to remain centered in the rectangle.

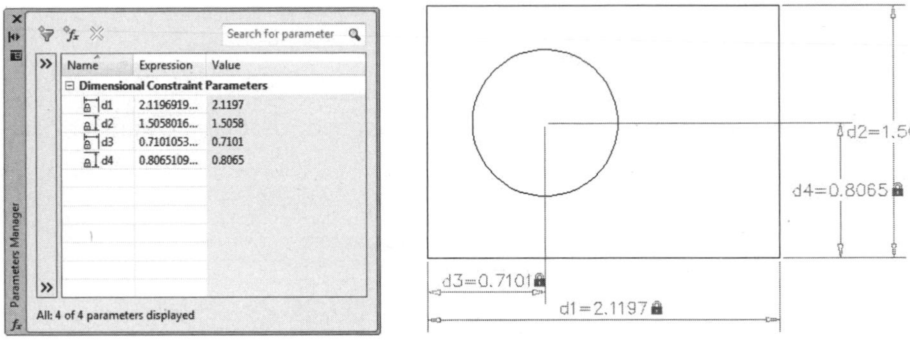

FIGURE 13.37

While inside of the Parameters Manager palette, locate the dimension identified by d3. In the Expression column, change the number to d1/2, as shown in Figure 13.38 (Left). This will divide the value of d1 by 2. The results are displayed in Figure 13.38 (Right). Notice that no matter what the d1 value is, the d3 value will always be one half of the current d1 value.

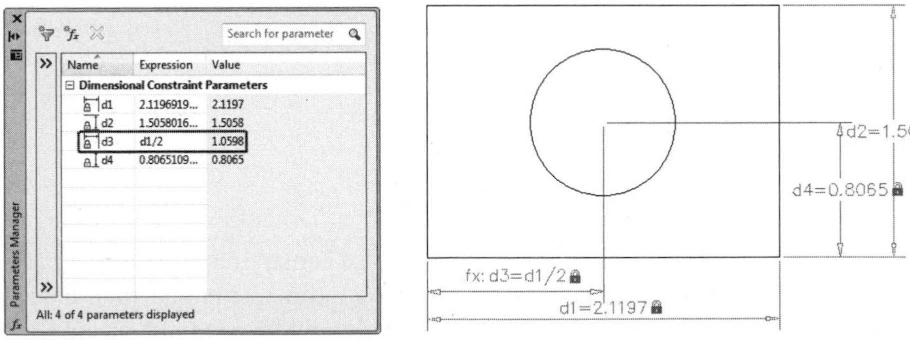

FIGURE 13.38

Next, while inside of the Parameters Manager palette, locate the dimension identified by d4. In the Expression column, change the number to d2/2, as shown in Figure 13.39 (Left). This will divide the value of d2 by 2. The results are displayed in Figure 13.39 (Right). Notice that no matter what the d2 value is, the d4 value will always be one half of the current d2 value.

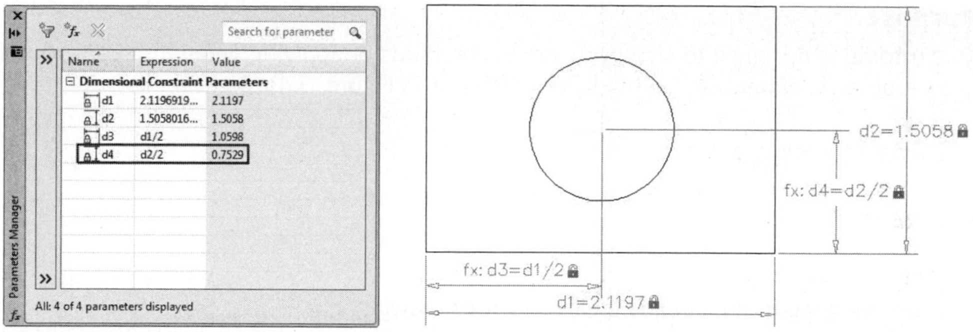

FIGURE 13.39

Now test if changing both the width and height dimensions will keep the circle centered. In Figure 13.40, change the d1 dimension value to 3.00 and the d2 dimension value to 2.00. Notice that in Figure 13.40 the two locator dimensions d3 and d4 adjust to keep the circle centered in the rectangle. Try repeating the steps demonstrated to see if you can keep the circle centered in the box by just creating a rectangle with a circle inside and placing the four dimensional constraints.

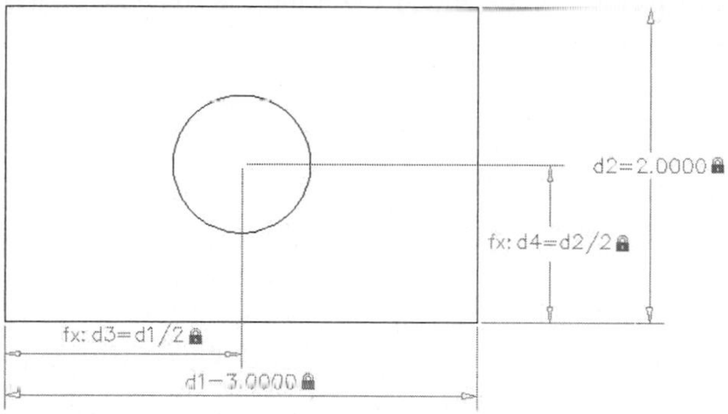

FIGURE 13.40

TUTORIAL EXERCISE: 13_LEVER.DWG

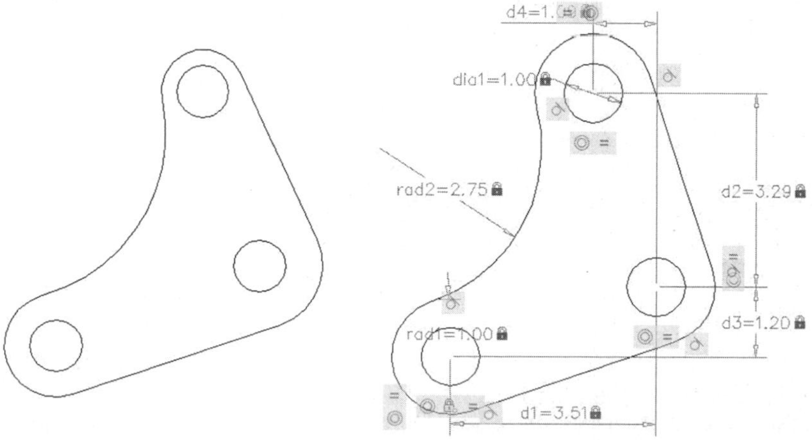

FIGURE 13.41

Purpose

This tutorial is designed to show you various methods of applying geometric constraints and parametric dimensions to the lever, as shown in Figure 13.41 (Right).

System Settings

All system settings are already set for this drawing.

Layers

The following layers are already created for this drawing:

Name	Color	Linetype
Dimensions	Red	Continuous
Object	Black	Continuous

Suggested Commands

You will begin by opening the drawing 13_Lever.Dwg. The objects are not constrained, which can be demonstrated by dragging on the existing objects and observing the results. Undo a stretching of grips before continuing on. Next, a series of constraints will be applied to the objects. Dimensional constraints will be added to define the shape. Once all constraints and dimensions are placed, changes will be made to selected dimensions to display a different lever shape.

STEP 1

Before adding constraints, let's take a closer look at the object. For example, hovering over the outer line of the lever, as shown in Figure 13.42 (Left), highlights the entire perimeter. This is a sign that the perimeter was created as a polyline and no coincident constraints (from one endpoint to another endpoint) need to be applied. Also, experiment with the outer perimeter of the lever by clicking on the perimeter to display the grips and then stretch various endpoints, as shown in Figure 13.42 (Right). What may appear as tangent edges on the left will show a different result on the right. Undo to get back to the original object. Various constraints will now be added to the lever. This will include geometric constraints as well as dimensional constraints.

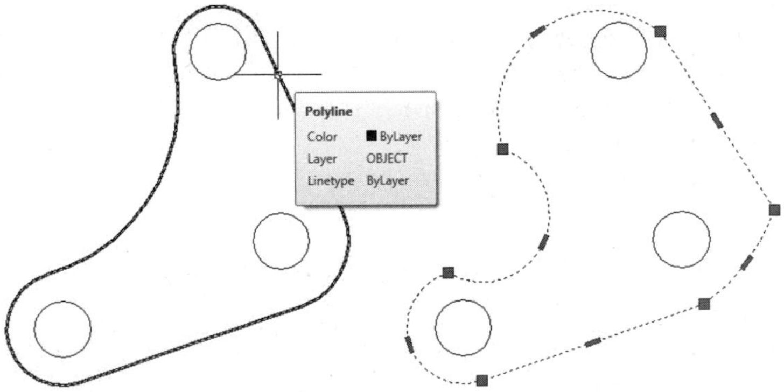

FIGURE 13.42

STEP 2

From the Geometric panel on the Ribbon, identify and click the Tangent constraint to make it active. Then select near the endpoints of the two arcs, as shown in Figure 13.43 (Left). Do the same for edges that consist of lines and arcs. The results are illustrated in

Figure 13.43 (Right). Notice the placement of the Tangent constraint glyphs; there should be six in total to make the lines and arcs tangent with each other. As constraints are added to objects, they may become too numerous to interpret. You can turn off the constraints before placing a new set by clicking on the Hide All button, also shown in Figure 13.43 (Right).

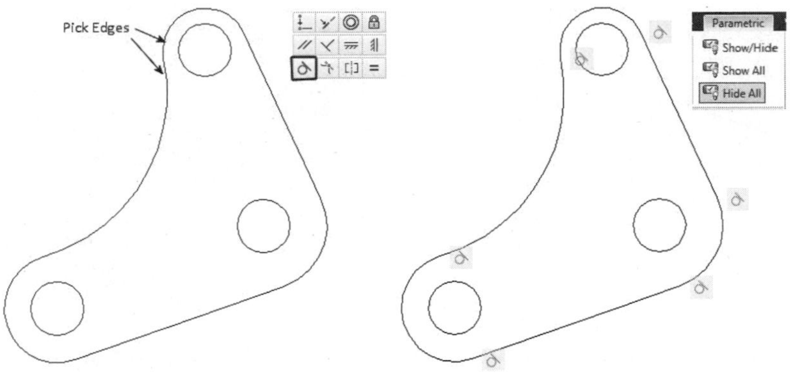

FIGURE 13.43

STEP 3

Next from the Geometric panel, identify and click the Equal constraint to make it active. Use the Multiple option and select the three arcs, as shown in Figure 13.44 (Left). The results are illustrated in Figure 13.44 (Right). Notice the placement of the Equal constraint glyphs; there should be three to identify the three arcs being affected. You may have to turn off the Tangent glyphs to clearly see all three Equal glyphs.

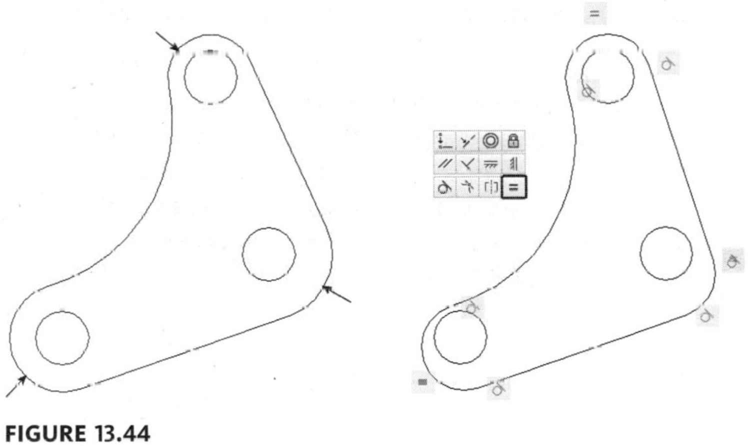

FIGURE 13.44

STEP 4

You may have noticed that after applying the equal constraints to the arcs, the circles inside of the lever may have slightly shifted to a new location. To have the circles share the same center point as the arcs, click the Concentric constraint from the Ribbon to make it active. Then select a circle and an arc, as shown in Figure 13.45 (Left). Apply this constraint to the remaining two sets of circles and arcs. The results are illustrated in the following image on the right. The circles share the same center point as the arc segments.

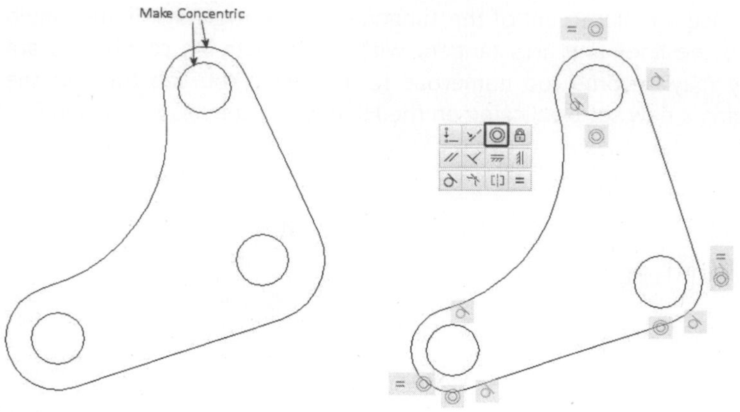

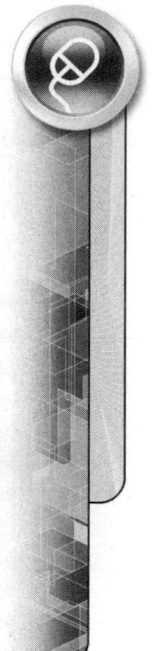

FIGURE 13.45

STEP 5

This step may not be required but will be shown anyway to illustrate the effect of the Fix constraint.

At this point in the design, the lever would move freely while keeping its shape and the circle locations intact. To keep the design from moving, and at the same time to lock down the design, a Fix constraint will be applied. Choose this constraint from the Ribbon, as shown in Figure 13.29 (Left). Then click the edge of the lower left circle, as shown in Figure 13.46 (Right). A "padlock" glyph will appear. This locks the center point of the circle down. It must be pointed out that the other circles and arcs would stretch since no Dimensional constraints are defined yet. However, the center point of the lower left circle is definitely locked.

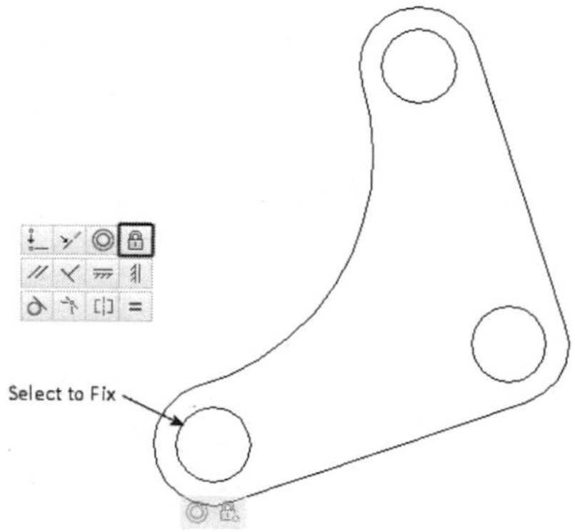

FIGURE 13.46

STEP 6

The last set of geometric constraints to apply deals with the three circles. Since we want the three circles to share the same diameter value, the Equal constraint will be applied. Choose Equal from the list of icons in the Ribbon, as shown in Figure 13.47 (Left). Use the Multiple option and select the three circles to display the Equal constraint, as shown in Figure 13.47 (Right).

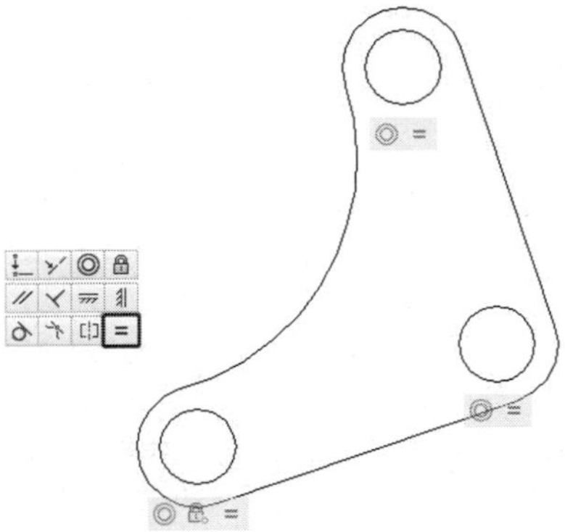

FIGURE 13.47

STEP 7

At any point, you can either Show All or Hide All constraints, as shown in Figure 13.48. You could also click on the Show/Hide button to display the constraints of specific selected objects.

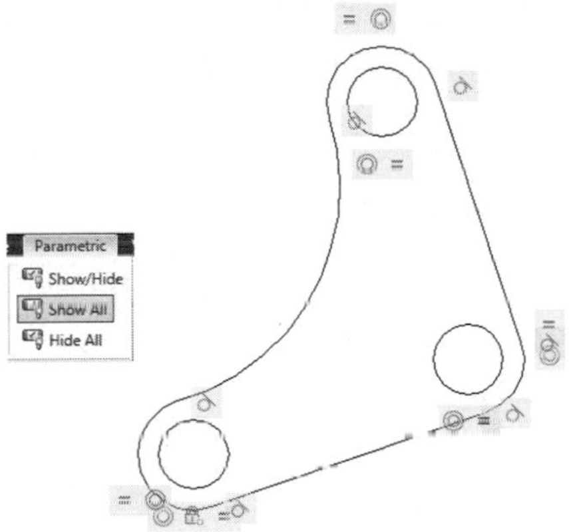

FIGURE 13.48

STEP 8

With the constraints applied to the various objects of the lever, it is time to further lock the design down by adding Dimensional constraints. Controls for applying and displaying Dimensional constraints are provided in the Dimensional panel of the Ribbon, as shown in Figure 13.49 (Left). From the panel, click on the Linear dimensional constraint and then click on the circles, as shown in Figure 13.49 (Right), to place these dimensions (d1, d2, d3, and d4). As a dimension is placed, it also highlights allowing you to change its value. If the dimension does not appear, click the Show All button. Place the dimensions using the values shown in Figure 13.49.

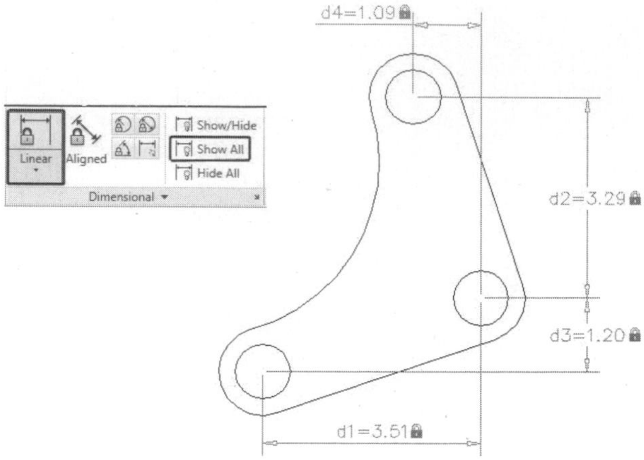

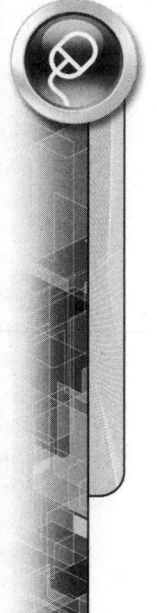

FIGURE 13.49

STEP 9

Using the Figure 13.50 (Right) as a guide, place the radius dimension on one of the three small arcs (rad1 = 1.00), place another radius dimension for the large arc (rad2 = 2.75), and finally place a diameter dimension on one of the three circles (dia1 = 1.00).

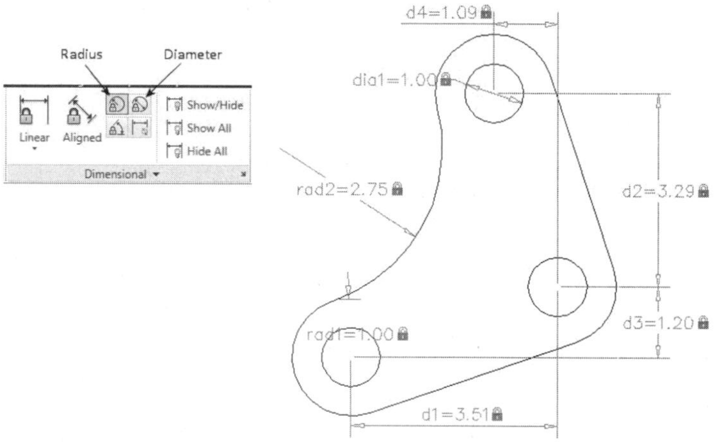

FIGURE 13.50

STEP 10

Based on the previously applied geometric constraints and dimensional constraints, your design is now fully constrained. Test the design to see if it moves. Clicking on the arc displays the radius dimension along with the grips. Clicking on any grip should limit any movement, as shown in Figure 13.51 (Left). In the same way, clicking on the top circle displays the dimensions that relate to this object, namely, the diameter and linear dimensions that locate the hole. Clicking on any grip on the circle will not allow any movement or change in size.

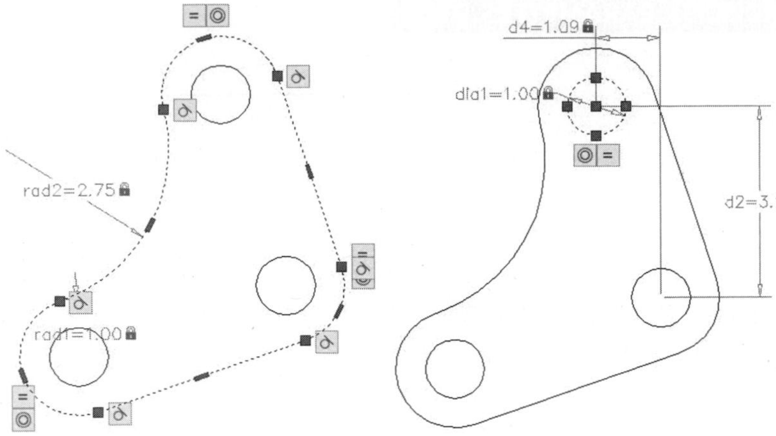

FIGURE 13.51

STEP 11

To test the design further, change the dimension values by double-clicking on them. Make the changes using Figure 13.52 (Left) as a guide: Change d1 from 3.51 to 5.00; change d2 from 3.29 to 2.50; finally change rad2 from 2.75 to 3.50. Notice how the design has changed based on the new dimensional constraints, as shown in Figure 13.52 (Right).

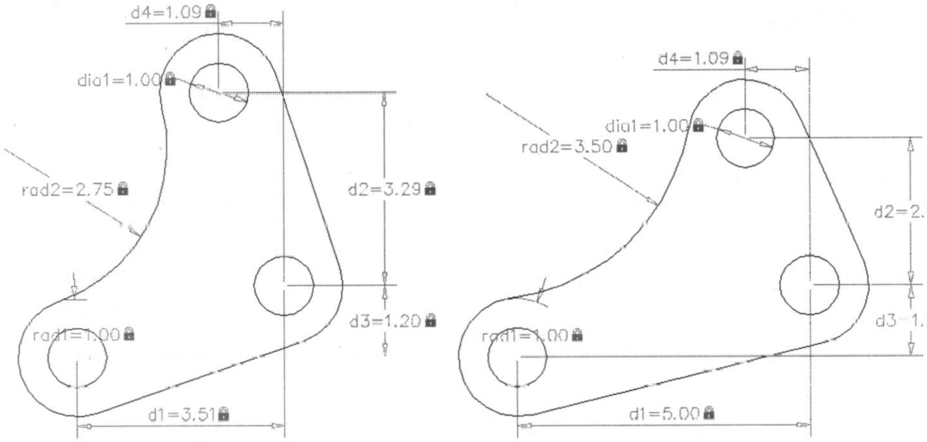

FIGURE 13.52

TUTORIAL EXERCISE: 13_PARAMETERS.DWG

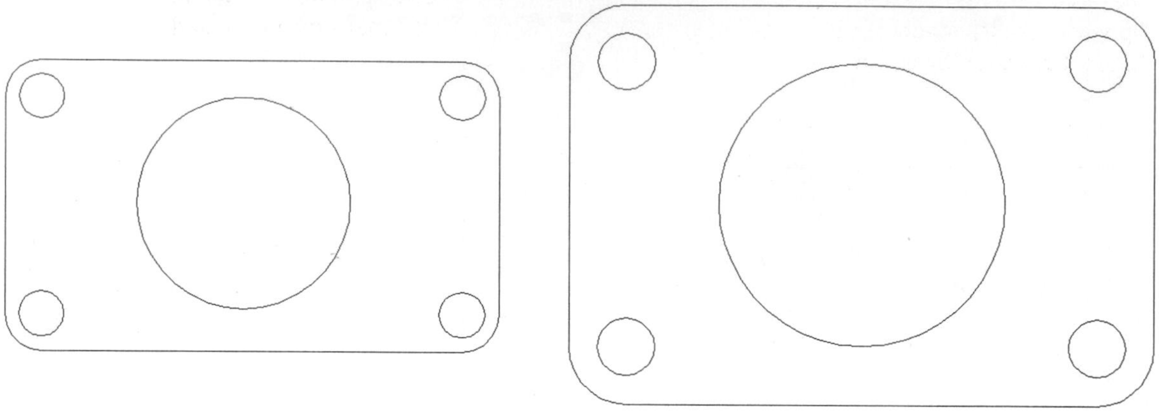

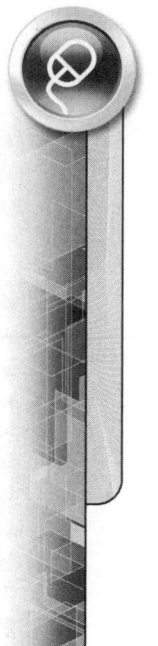

FIGURE 13.53

Purpose

This tutorial is designed to apply geometric constraints and dimensions, and assign parameters to certain objects. You will see how changing a few parameters will affect the final object.

System Settings

All system settings are already set for this drawing.

Layers

The following layers are already created for this drawing:

Name	Color	Linetype
Dimensions	Red	Continuous
Object	Black	Continuous

Suggested Commands

Begin by opening the drawing 13_Parameters.Dwg. You will use the Auto Constrain button to apply constraints to selected objects. Use equal constraints on all arcs and small circles. You will then display the Parameters Manager palette to change equation values based on the parametric dimensions assigned to the object.

STEP 1

Click the Auto Constrain button, as shown in Figure 13.54 (Left), where you will be prompted to select the objects to automatically apply constraints. Select the outer profile and the four small circles, as shown in Figure 13.54 (Right).

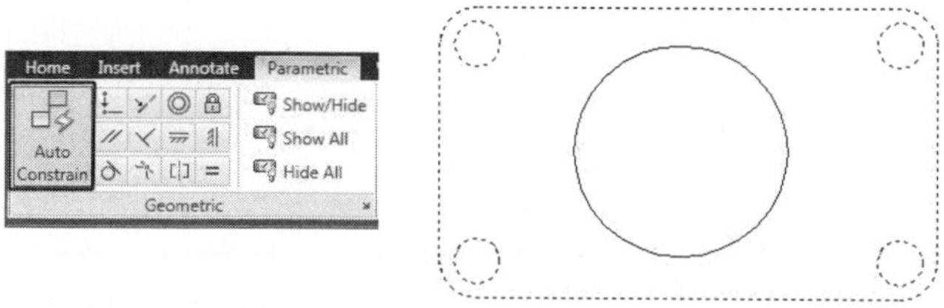

FIGURE 13.54

STEP 2

You should notice all constraints applied to the object, as shown in Figure 13.55 (Left). These constraints include Parallel, Perpendicular, Tangent, Concentric, and Horizontal. Upon closer inspection of the upper left corner of the object, notice that Concentric and Tangent constraints are applied.

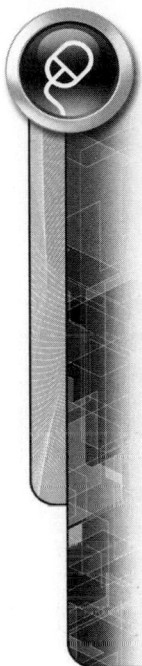

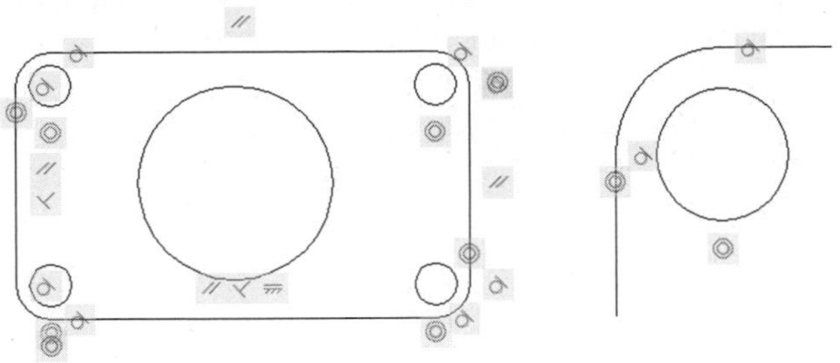

FIGURE 13.55

STEP 3

Even though the Auto Constrain command placed a number of constraints on the object, it did not place all constraints that we will need to better define the object. Missing are Equal constraints for the four corner arcs. Click the Equal constraint button, as shown in Figure 13.56 (Left). Use the Multiple option and pick the four arcs. This will result in all four corner arcs being equal to each other. Also, notice the Equal glyph attached near each arc, as shown in Figure 13.56 (Right).

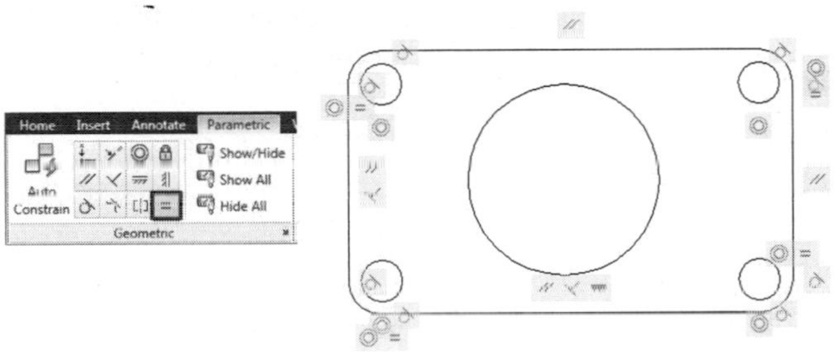

FIGURE 13.56

STEP 4

As with the arcs in the previous step, also missing are Equal constraints for the four small circles. Click the Equal constraint button, as shown in Figure 13.57 (Left). Use the Multiple option and pick the four circles. This will result in all four small circles being equal to each other. Also notice the Equal glyph attached near each circle, as shown in Figure 13.57 (Right). Before continuing on to the next step, hide all constraints as you prepare to add dimensional constraints to the object.

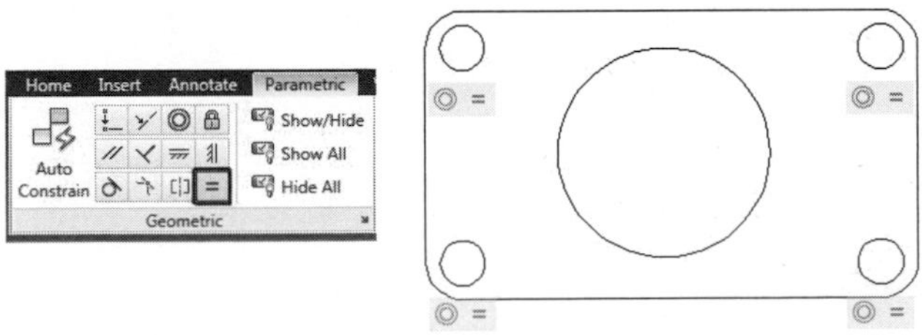

FIGURE 13.57

STEP 5

Begin adding dimensional constraints for the overall length and height of the object. Click the Linear dimensional button, as shown in Figure 13.58 (Left). Then place the horizontal linear dimension and change the dimension value to 8.00. Next place the vertical linear dimension and change its dimension value to 5.00, as shown in Figure 13.58 (Right). These dimensional constraints change the actual size of the object, unlike traditional AutoCAD dimensions such as DLI (DimLinear) and DRA (DimRadius) which can reflect a change in object size (associative) but cannot change the size of an object (parametric).

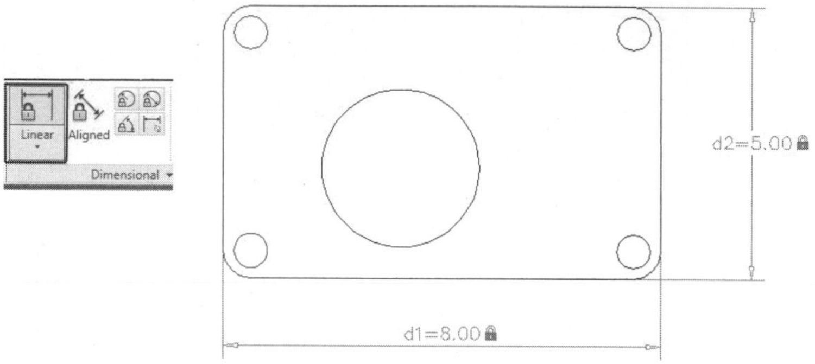

FIGURE 13.58

STEP 6

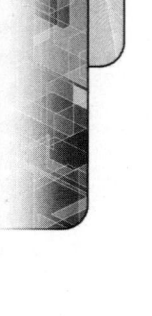

Next click on the Radius dimensional button, as shown in Figure 13.58 (Left). Then place a radius dimension by clicking on the arc in the upper left corner. If needed, change the radius value to 0.50, as shown in Figure 13.59 (Right). Notice all radii change to the value of 0.50 since the Equal constraint was applied to all arcs in a previous step.

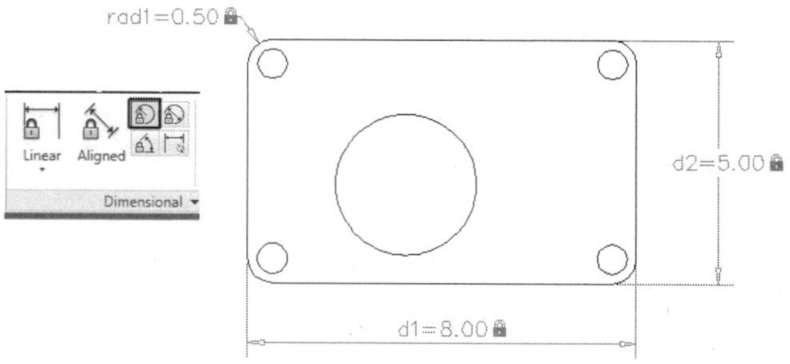

FIGURE 13.59

STEP 7

As with the radius dimension, click on the Diameter dimensional button, as shown in Figure 13.60 (Left). Then place a diameter dimension by clicking on the circle in the upper left corner. If needed, change the diameter value to 0.50, as shown in Figure 13.60 (Right). Notice that all circle diameters change to the value of 0.50 since the Equal constraint was applied to all circles in a previous step.

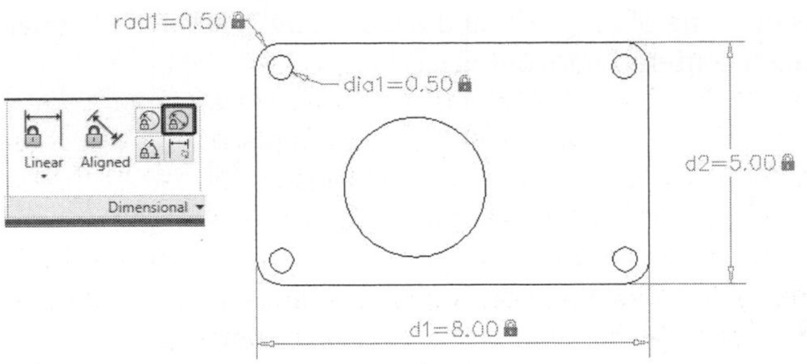

FIGURE 13.60

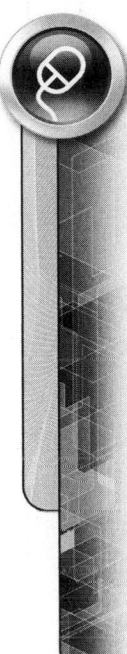

STEP 8

Add another diameter constraint to the large hole near the center of the object. Change the dimension value to 2.50, as shown in Figure 13.61.

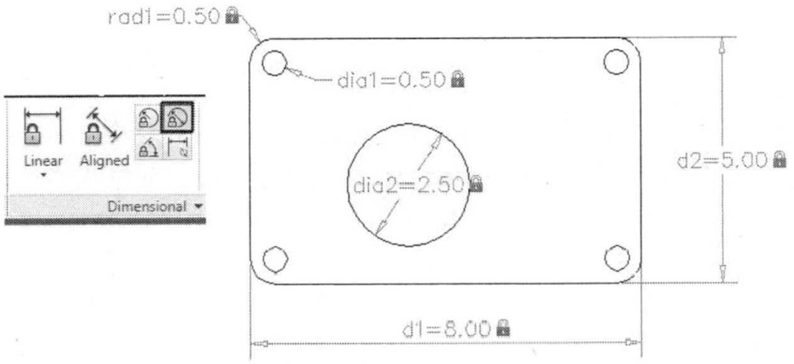

FIGURE 13.61

STEP 9

Add two more Linear dimensional constraints, as shown in Figure 13.62. These dimensions will define the location of the large 2.50-diameter circle. Rather than changing the values, use the default values for both of these linear dimensions. Their values will be changed when parameters are assigned in Steps 10 and 11.

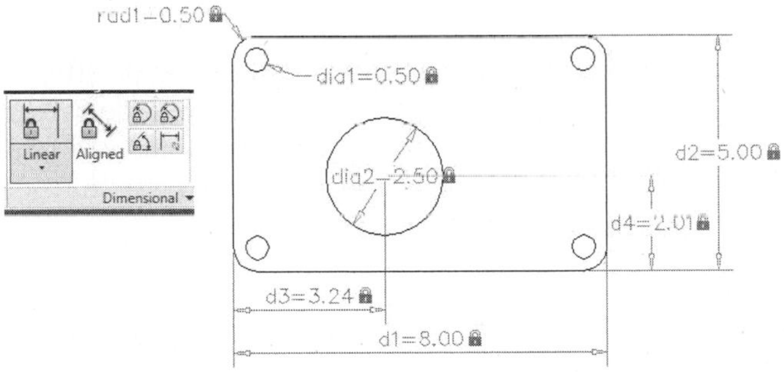

FIGURE 13.62

STEP 10

Parameters will now be added to the existing dimensions. Clicking on the Parameters Manager button will launch the Parameters Manager palette, as shown Figure 13.63 (Left). Notice the dimension labeled d3. This represents the horizontal dimension that locates the large circle. We really want the circle to be horizontally located at the center of the object. To do this, double-click on the Expression next to d3 and change it to d1/2.

This will take the overall dimension value of 8.00 (d1) and divide it by 2. No matter what d1 changes to, the circle will remain centered horizontally.

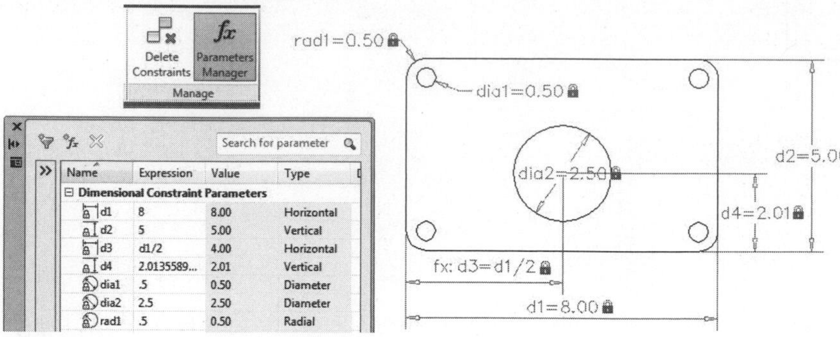

FIGURE 13.63

STEP 11

Now notice the dimension labeled d4. This represents the vertical dimension that locates the large circle. We really want the circle to be located vertically at the center of the object. To do this, double-click on the Expression next to d4 and change it to d2/2. This will take the overall dimension value of 5.00 (d2) and divide it by 2. No matter what d2 changes to, the circle will remain centered vertically, as shown in Figure 13.64 (Right).

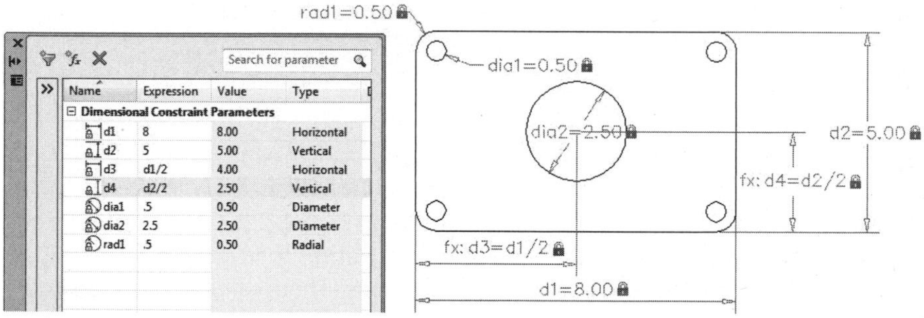

FIGURE 13.64

STEP 12

For the next parameter, identify the dimension dia1 in the Parameter Manager palette, as shown in Figure 13.65 (Left), and change the Expression to rad1. Whenever the radius value changes (which is controlled by the rad1 parameter) the diameter value will also change. Notice in Figure 13.65 (Right) that the radius parameter is currently set to 0.50, which is the same for the diameter values of the four small holes.

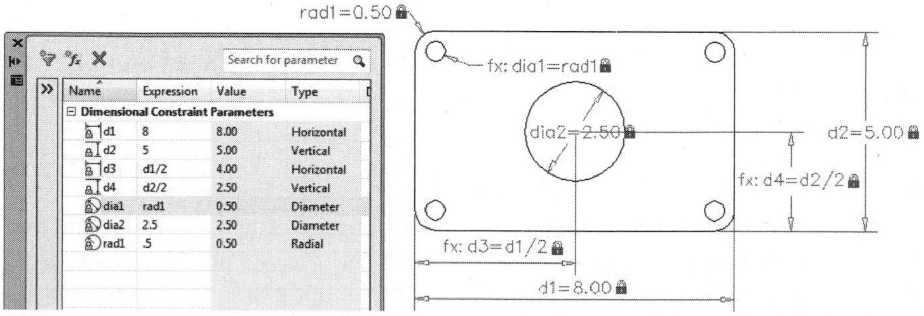

FIGURE 13.65

STEP 13

Finally, identify the parameter dia2 in the Parameter Manager palette and change the Expression to rad1 * 5, as shown Figure 13.66 (Left). Since the * is the same as a multiplication operation, the radius value in rad1 will be multiplied five times. This results in the circle diameter, as shown in Figure 13.66 (Right).

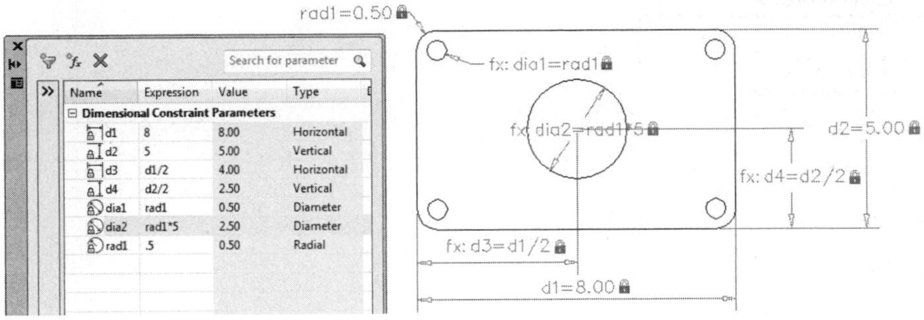

FIGURE 13.66

STEP 14

Test that the parameters work by entering new values in the Parameters Manager palette, as shown in Figure 13.67 (Left). Change the expression in d1 from 8.00 to 11.25. Change the expression in d2 from 5.00 to 7.75. Finally change the rad1 expression from 0.50 to 1.10. The results of making these changes is illustrated in Figure 13.67 (Right).

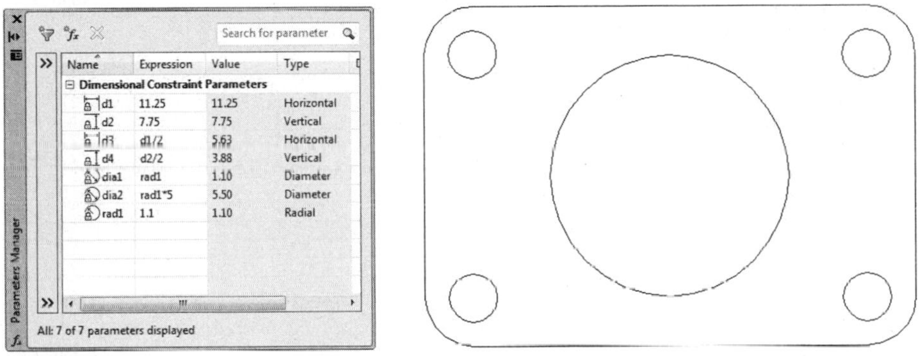

FIGURE 13.67

STEP 15

In this final step we will create a filter, which can be used to help organize your dimensional constraints. Click the "Expands Parameters filter tree" button on the left side of the Parameters Manager palette. Click the "Creates a new parameter group" button and enter Data Fields into the edit box. An empty group named "Data Fields" is created, as shown in Figure 13.68 (Left). Click "All" at the top of the group list, to display all the parameters, as shown in Figure 13.68 (Center). Hold down the CTRL key and select d1, d2, and rad1. Drag the highlighted dimensional constraints to the Data Fields group. This group now displays only those parameters that are just values, as shown in Figure 13.68 (Right).

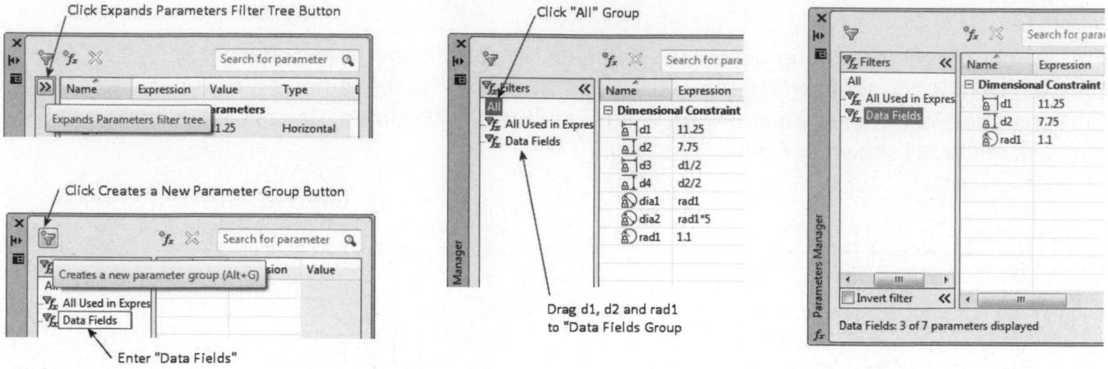
Click Expands Parameters Filter Tree Button
Search for parameter
Name
Expression
Value
Type
Expands Parameters filter tree.
Horizontal
Click Creates a New Parameter Group Button
Search for parameter
Creates a new parameter group (Alt+G)
Value
All Used in Expres
Data Fields
Enter "Data Fields"
Click "All" Group
Filters
All
All Used in Expres
Data Fields
Name
Expression
Dimensional Constraint
d1 11.25
d2 7.75
d3 d1/2
d4 d2/2
dia1 rad1
dia2 rad1*5
rad1 1.1
Manager
Drag d1, d2 and rad1
to "Data Fields Group
Filters
All
All Used in Expres
Data Fields
Name
Expression
Dimensional Constraint
d1 11.25
d2 7.75
rad1 1.1
Invert filter
Parameters Manager
Data Fields: 3 of 7 parameters displayed

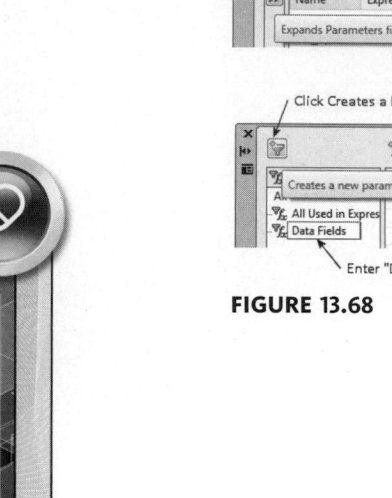

FIGURE 13.68

CHAPTER

14

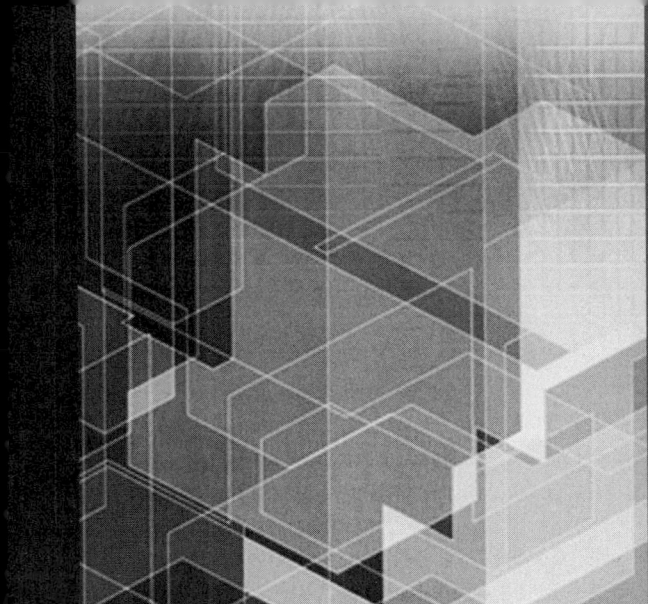

Working with Drawing Layouts

CREATING DRAWING LAYOUTS

Up to this point in the book, all drawings have been constructed in Model Space. This consisted of drawings composed of geometry such as arcs, circles, lines, and even dimensions. You can even plot your drawing from Model Space; however, this gets a little tricky, especially if multiple details need to be arranged on a single sheet at different scales.

The preferred method is to work in two separate spaces on your drawing, namely Model Space and Layout mode, sometimes referred to as Paper Space. Typically, part geometry and dimensions are drawn in Model Space. However, items such as notes, annotations, and title blocks are laid out separately in a drawing layout, which is designed to simulate an actual sheet of paper. To arrange a single view of a drawing or multiple views of different drawings, you arrange a series of viewports in the drawing layout to view the images. These viewports are mainly rectangular in shape but can be created in any shape and size. A viewports tool allows you to scale the image inside the viewport. In this way, a series of images may be scaled differently even though the drawing layout will be plotted out at a scale of 1:1.

This chapter introduces you to the controls used in a drawing layout to manage information contained in a viewport. A tutorial exercise is provided to help you practice creating drawing layouts.

MODEL SPACE

Before starting on the topic of Paper Space, you must first understand the environment in which all drawings are originally constructed, namely Model Space. It is here that the drawing is drawn full size or in real-world units. Model Space is easily identified by a dark background screen color, although the color setting can be changed. It can also be identified by the appearance of the User Coordinate System icon located in the lower-left corner of the active drawing area, as shown in Figure 14.1. This icon is associated with the Model Space environment. Another indicator that you are in Model Space is the presence of the Model tab, located just below the User Coordinate System icon, also shown in Figure 14.1.

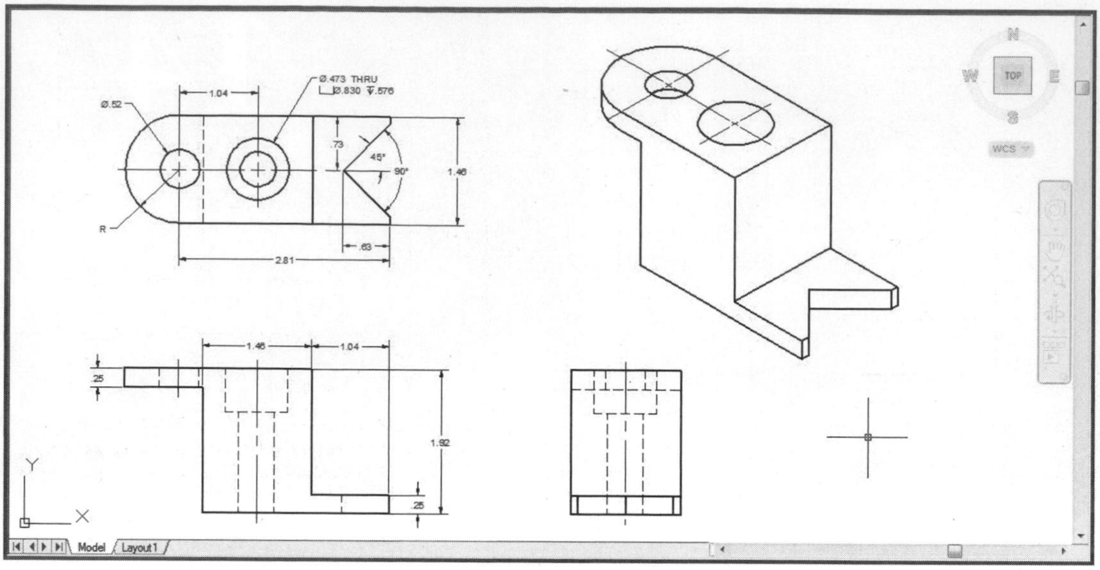

FIGURE 14.1

MODEL SPACE AND LAYOUTS

You have already seen that Model Space is the environment set aside for constructing your drawing in real-world units or full size, as shown in Figure 14.2 (Left).

Clicking the Layout1 tab found in the lower-left corner of the display screen activates the layout environment. Paper Space is considered an area used to lay out your drawing before it is plotted. Layouts are also used to place title block information and notes associated with the drawing. The drawing illustrated in Figure 14.2 (Right) has been laid out in Paper Space and shows a sheet of paper with the drawing surrounded by a viewport. The dashed lines along the outer perimeter of the sheet are referred to as margins. Anything inside the margins will plot; for this reason, this is called the printable area. Notice also at the bottom of the screen that the Layout1 tab is activated. Both tabs at the bottom of the screen can be used to easily display your drawing in either Model Space or Paper Space. Also notice the icon in the lower-left corner of the illustration; this icon is in the form of a triangle and is used to quickly identify the Paper Space environment.

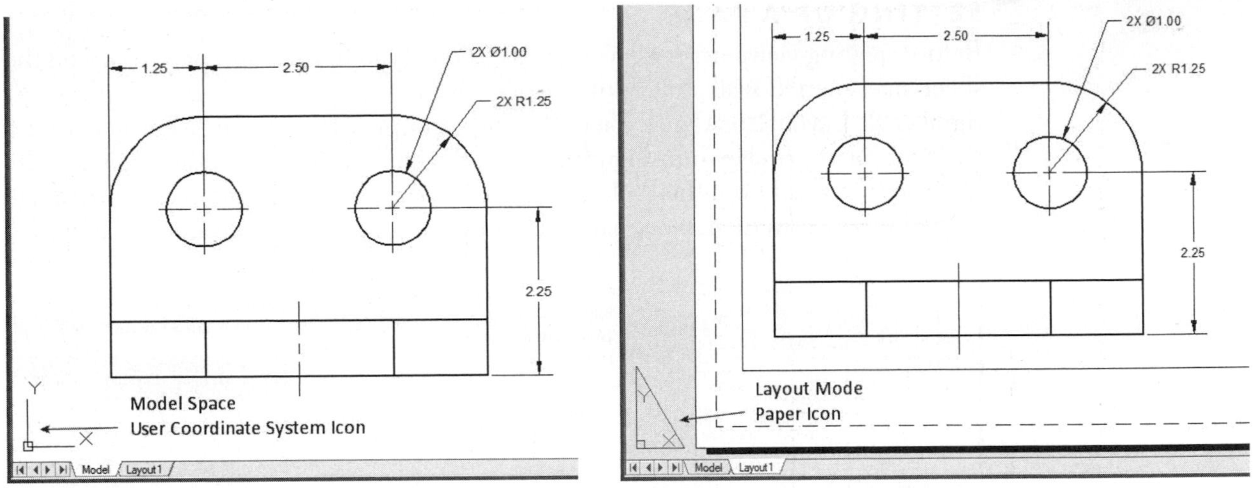

FIGURE 14.2

NOTE

If the Model and Layout1 tab are not displayed (hidden), right-click a blank part of the screen, choose Options to display the Options dialog box. In the Display tab and Layout Elements area, place a check in the box to Display Layout and Model tabs.

LAYOUT FEATURES

When you draw and plot from Model Space, you will have to scale certain objects before they are plotted. For example, if you are using a scale of 1:2 you will need to double the size of your title block and your general notes before you plot. Your real world unit drawing will appear half size, as required to fit on the plotted sheet. The title block and notes will also appear the correct size, but only because they were initially doubled in size before they were plotted half size. When you plot from Model Space, you plot to the selected scale. When you plot from a layout, you always plot 1:1. If the title block and general notes had been placed in Paper Space instead of Model Space, they wouldn't require scaling because they are plotted to size (1:1). Not having to scale items placed in Paper Space is the fundamental reason for using the layout environment. Here are a few other reasons for arranging a drawing in a layout:

- Layouts are based on the actual sheet size. If you are plotting a drawing on a D-size sheet of paper, you use the actual size (36 × 24) in Paper Space.
- Title blocks and annotations do not have to be scaled when placed in a layout.
- Viewports created in a layout are user-defined. Viewports created in Model Space are dependent on a configuration set by AutoCAD. In a layout, the viewports can be of different sizes and shapes, depending on the information contained in the viewport.
- Multiple viewports can be created in a single layout, as in Model space. However, the images assigned to different layout viewports can be assigned different scales. Also, the control of layers in a layout is viewport-dependent. In other words, layers turned on in one viewport can be turned off or frozen in another viewport.
- All drawings, no matter how many viewports are created or details arranged, are plotted out at a scale of 1 = 1.

SETTING UP A PAGE

Before creating viewports in a layout, it is customary to first set up a page based on the sheet size, plotted scale, and plot style. To begin the process of setting up a page, right-click Layout1 or Quick View Layouts, and pick Page Setup Manager from a shortcut menu, as shown in Figure 14.3 on the left and in the middle. This process can also be started from the Ribbon by selecting the Layout tab and picking the Page Setup button from the Layout panel, as shown in the following image on the right.

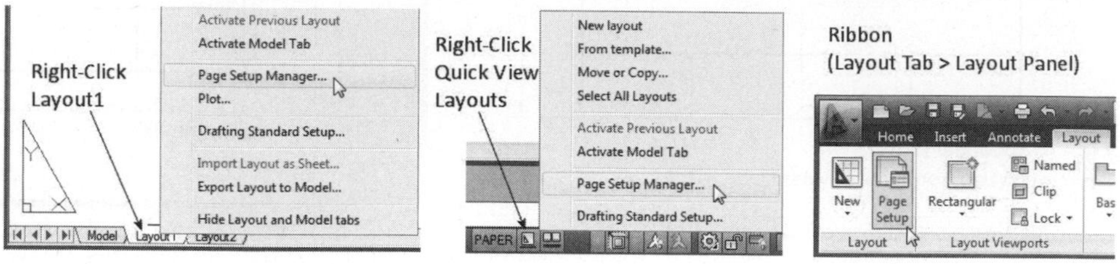

FIGURE 14.3

Performing the previous task displays the Page Setup Manager dialog box, as shown in Figure 14.4 (Left). The purpose of this initial dialog box is to create a series of page setups under unique names. These page setup names can hold information such as plotter name, page sizes, and plot style tables used for pen assignments when plotting. To create a page setup, click the New button to display the New Page Setup dialog box, as shown in Figure 14.4 (Right). When creating the new page setup name, try to give the page a name that will give you a hint for its intended purpose. In this dialog box, a new page setup name was entered: C-Size (DWF6). The "C" refers to the size of the drawing sheet. The DWF6 refers to the type of plot device. DWF stands for Drawing Web Format and is used to plot a drawing out to a file. Once in this DWF form, the file can be sent to others literally around the globe for review.

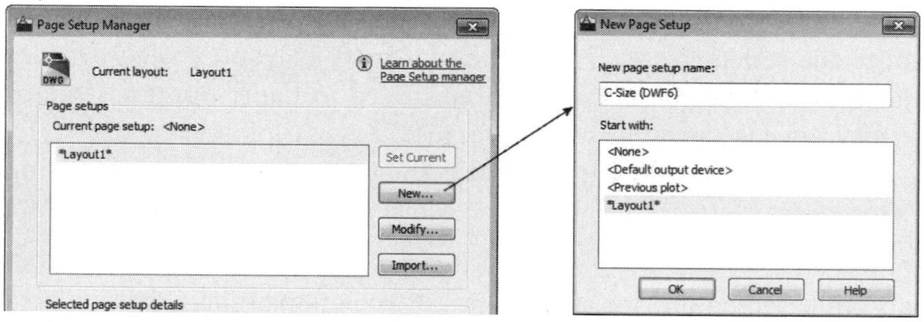

FIGURE 14.4

Clicking the OK button in the New Page Setup dialog box displays a larger, more comprehensive Page Setup dialog box, as shown in Figure 14.5. The following areas will now be explained:

Printer/plotter—Area A: Use this area to choose the plotter you will be outputting to. Even though you may not be ready to plot, you can still designate a plotter. In this image the DWF6 ePlot.pc3 file will be used.

Paper size—Area B: This area holds all paper sizes. It is important to realize that the paper sizes available are based on the plotter you selected back in Area A.

Plot style table (pen assignments)—Area C: This area holds available plot styles, which allow you to set plot properties such as color and lineweight.

Plot scale—Area D: Another important area to take note of is the Plot scale. This is the plotted scale that the layout will be based on. Since a major advantage of creating a layout is to plot at a scale of 1 to 1, the scale reflects this practice.

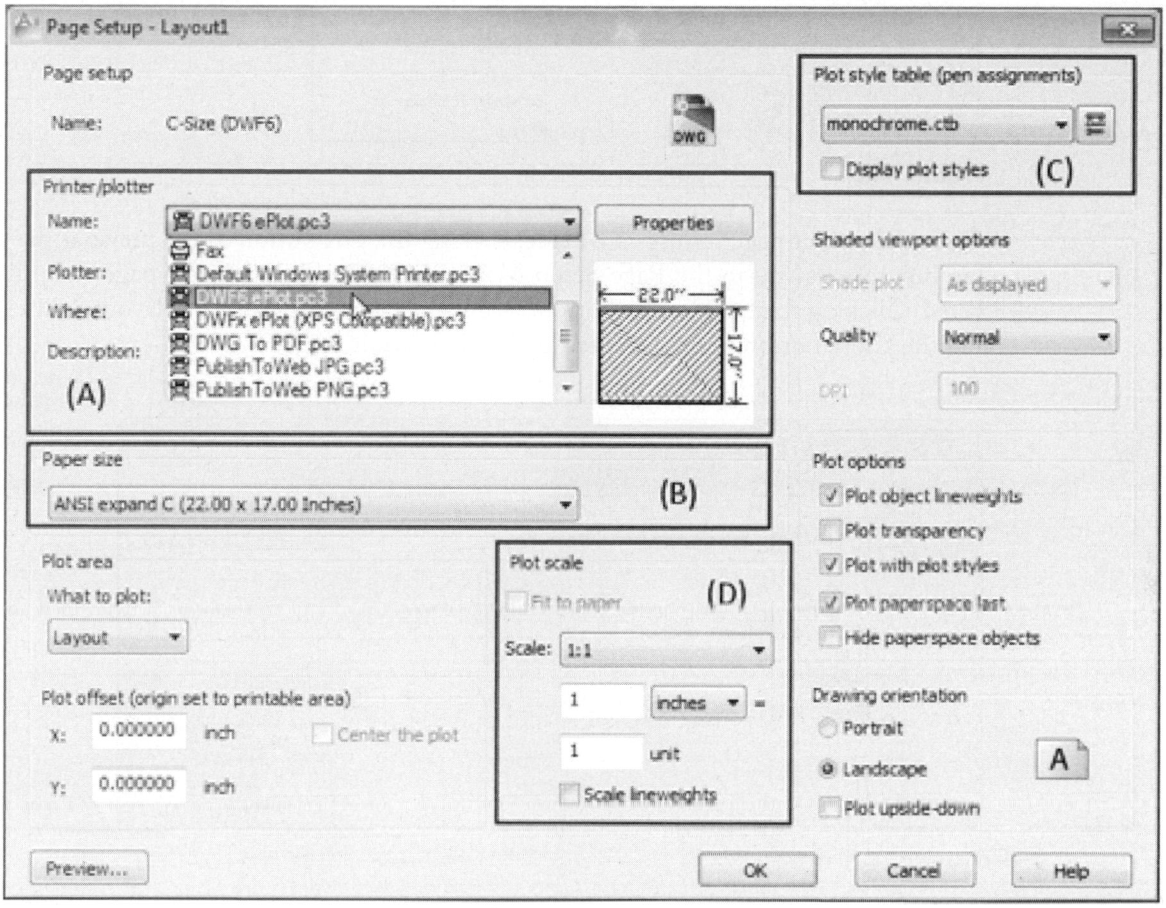

FIGURE 14.5

A more detailed look at the available paper sizes is illustrated in Figure 14.6 (Left). Again, these paper sizes are a direct result of the printer/plotter you are using. Notice in this image a number of paper sizes that have the word "expand" associated with them. These paper sizes have a larger printable area as opposed to the paper sizes that are not identified with "expand."

Illustrated in Figure 14.6 (Right) are the default Plot style tables that are available when AutoCAD is loaded. Plot style tables control the appearance of a printed drawing, which can include color, linetype, and even lineweight. For example, the acad.ctb table is designed for a color plot. Whatever objects are red will plot out in the color red (provided you are using a color printer). The monochrome.ctb plot style takes all colors and plots them out as black objects on a white sheet of paper. Notice other color tables that plot out objects in different shades of gray (Grayscale.ctb) and a number of screening plot styles. These are useful when you want to fade away a group of objects. A screening factor of 25% would have the objects faded more than a screen factor of 50%, and so on.

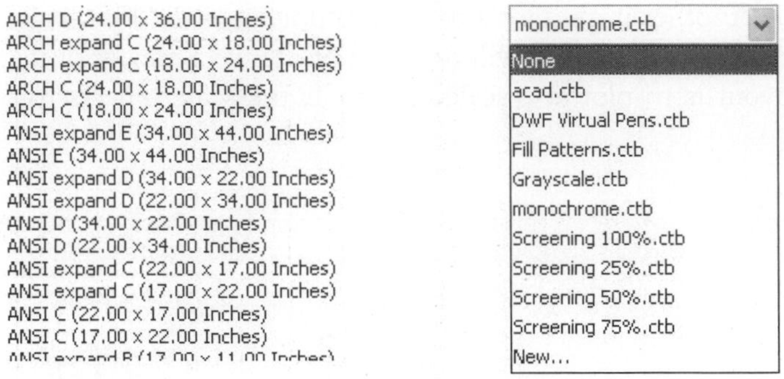

FIGURE 14.6

Once the proper page settings are made, clicking the OK button in the previous dialog box returns you to the Page Setup Manager. Notice that the new page setup is listed. Clicking the Set Current button assigns the new settings to your layout. Clicking the Close button returns you to the drawing layout tab.

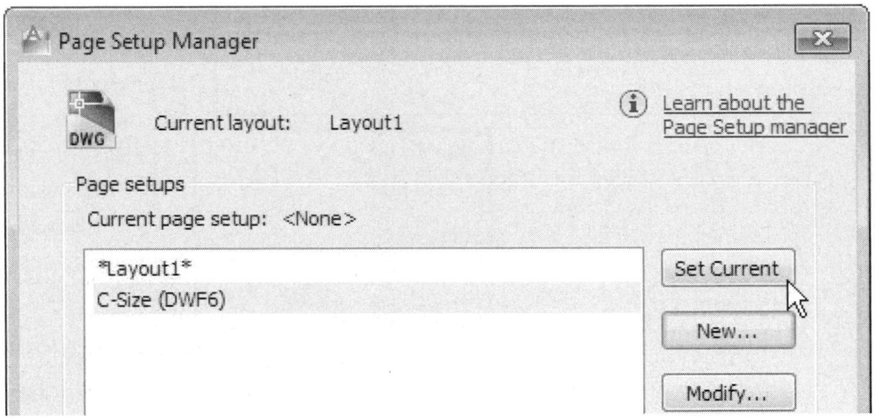

FIGURE 14.7

NOTE

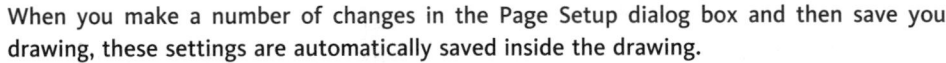

When you make a number of changes in the Page Setup dialog box and then save your drawing, these settings are automatically saved inside the drawing.

FLOATING MODEL SPACE

Notice the appearance of the drafting triangle icon in the lower-left corner of the display screen. This signifies that you are presently in the Paper Space environment. Another indicator is that the Paper button will activate in the status bar located at the bottom of the display screen. While in Paper Space, operations such as adding a border, title block, and general notes are usually performed. Paper Space is also the area in which you plot your drawing.

When a viewport is created in a layout, the drawing image automatically fills up the viewport. Operations can be performed in Model Space without leaving the Layout environment; this is referred to as working in floating Model Space. To activate floating Model Space, double-click inside a viewport. The UCS icon will normally display, as shown in Figure 14.8 (Right). Notice that the icon appears inside the viewport. You can also activate floating model space by clicking the Paper button in the status bar and changing it to Model.

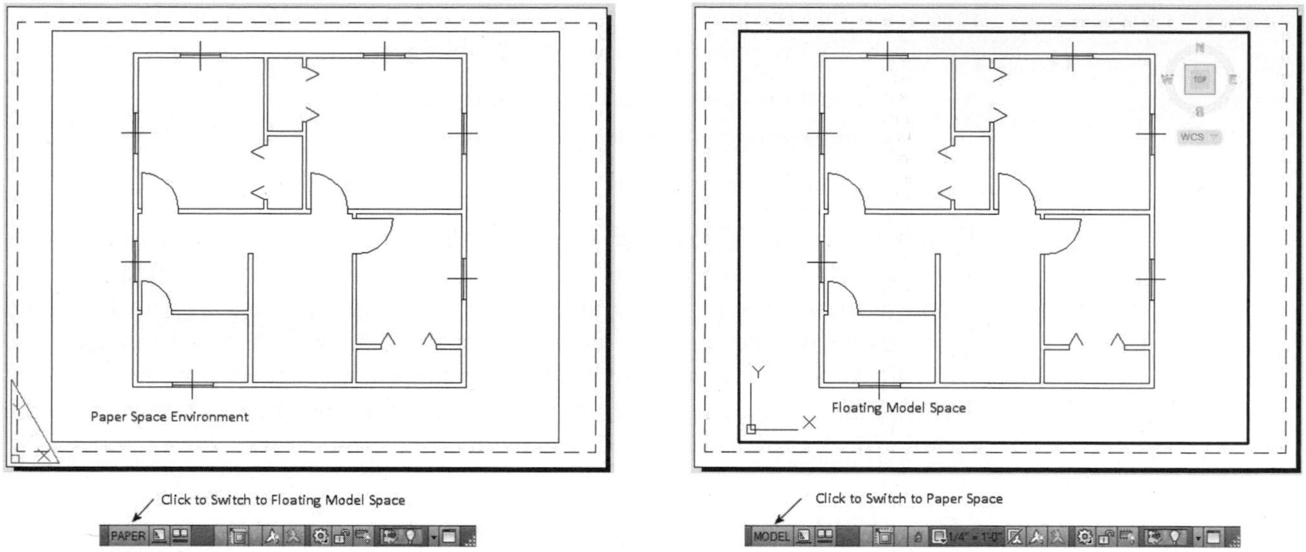

FIGURE 14.8

> **NOTE**
>
> If you need to switch back to Paper Space, double-click outside the viewport. The Paper Space drafting triangle reappears.

SCALING VIEWPORT IMAGES

One of the more important operations to perform on a viewport is the scaling of the image to Paper Space. If plotting from Model Space, you have to plot the drawing to scale, but from a layout you are simply zooming the image so that it appears at scale in the viewport. It was previously mentioned that by default, images that are brought into viewports display in a ZOOM-Extents appearance. To scale an image to a viewport, use the next series of steps and refer to Figure 14.9.

- Click on a Layout tab to enter the layout environment.
- Double-click in a viewport to enter floating Model Space or, from Paper Space, click the edge of the viewport that will be scaled.
- On the Status Bar expand the Viewport Scale box to display the available scales.
- Select the scale that sizes (zooms) the image to fit properly in the viewport.

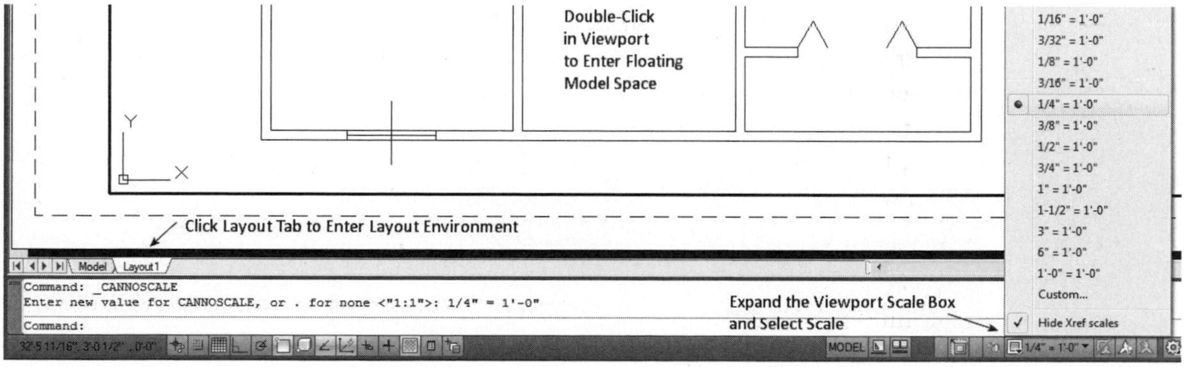

FIGURE 14.9

If you find that the image is very close to fitting but is cut off by one edge of a viewport, there are two options to resolve the problem:

- Stretch the viewport (using grips or the STRETCH command).
- Pan the image into position.

A combination of both methods usually proves to be the most successful.

To pan the image, double-click inside the viewport to start floating Model Space. You will notice the thick borders on the viewport. Using the PAN command or holding down the wheel on a wheel mouse, move the image to fit the viewport. Be careful not to zoom (roll the wheel on the mouse) because this will change the image scale. Once the image is panned into position, double-click outside the viewport on the drawing surface (usually the area surrounding the drawing sheet) to return from the floating Model Space back to the layout space.

CONTROLLING THE LIST OF SCALES

You will notice that when setting the scale of the image inside the Paper Space viewport, the list of scales is very long. You may find yourself using only certain scales and ignoring others. You could also be confronted with assigning a scale that is not present on the list. All these situations can be easily handled through the use of the Edit Drawing Scales dialog box, which can be activated by picking Custom from the Viewport Scale box on the Status Bar, as shown in Figure 14.10.

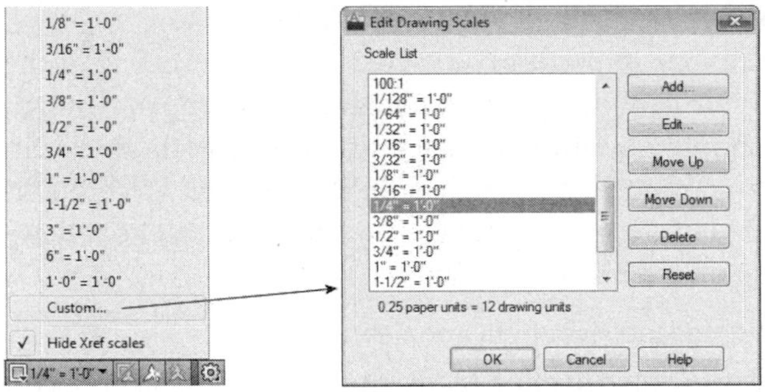

FIGURE 14.10

Various buttons that allow you to add, edit, or delete scales from the list are available. Clicking the Add button displays the Add Scale dialog box, as shown in Figure 14.11 (Left). In this dialog box, a new scale of 1:1000 is being created since it is not available in the default scale list. To edit an existing scale, click the Edit button to display the Edit Scale dialog box, as shown in Figure 14.11 (Right). Here a scale is being edited to reflect the scale of 1:500.

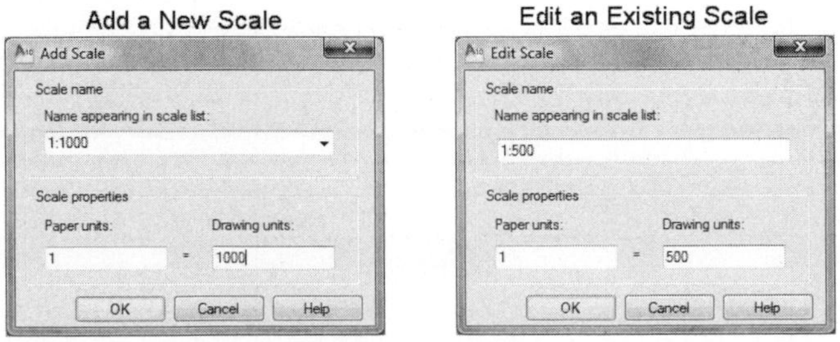

FIGURE 14.11

Clicking the Delete button removes a scale from this list. You can also rearrange scales so that their order is more convenient for you. Simply click the Move Up button to move a selected scale up the list or the Move Down button to arrange the selected scale near the bottom of the list.

LOCKING VIEWPORTS

When you have scaled the contents of a floating Model Space viewport, it is very easy to accidentally change this scale by rolling the wheel of the mouse or performing another zoom operation. To prevent accidental panning and zooming of a viewport image once it has been scaled, it is considered good practice to lock the viewport. To perform this task, use these steps and refer to Figure 14.12:

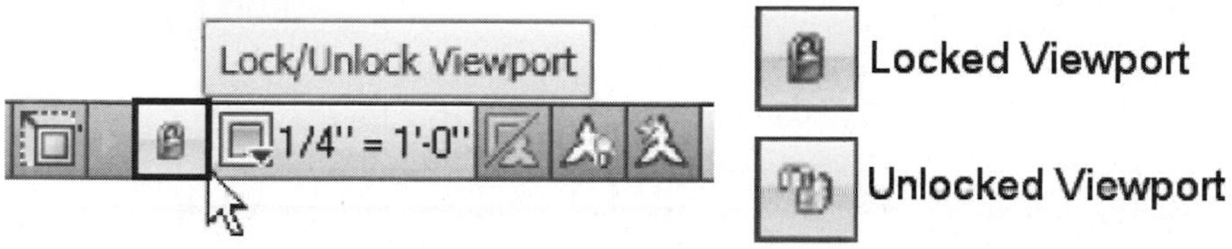

FIGURE 14.12

- Click on a Layout tab to enter the layout environment.
- Double-click in a viewport to enter floating Model Space or, from Paper Space, click the edge of the viewport that will be locked. Click the Lock/Unlock Viewport button on the Status Bar, as shown in Figure 14.12.
- From Paper Space you have the advantage of being able to select more than one viewport. Also, if Quick Properties are activated, as shown in Figure 14.13 (Right), you can lock the display by selecting "Yes" in the palette.
- From Paper Space, you can also use a shortcut menu to lock the viewport. Select the viewport or viewports and right-click to display the menu, as shown in Figure 14.13 (Left). Select Display Locked followed by "Yes."

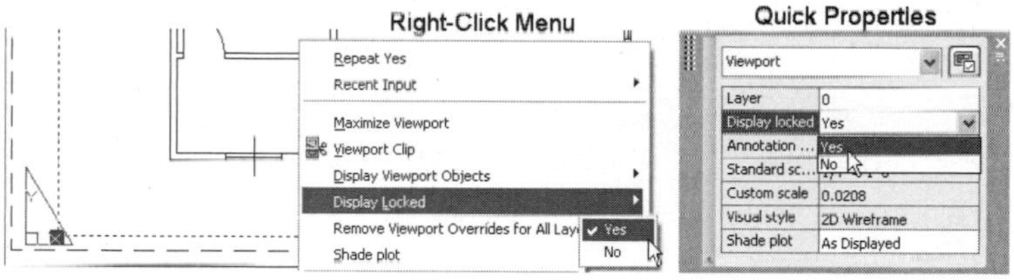

FIGURE 14.13

> **NOTE**
>
> **When you lock a viewport, your scale will be grayed out on the Status Bar. If you need to pan or change the scale of an image inside a viewport, you must first unlock the viewport.**

MAXIMIZING A VIEWPORT

The VPMAX command is designed to maximize the size of a viewport in a layout. This is especially helpful when editing drawings with small viewports. It also eliminates the need to constantly switch between Model Space and a Layout. The VPMAX command can be activated by double-clicking a viewport. You could also click the

Maximize Viewport button located in the status bar at the bottom of the display screen, as shown in Figure 14.14.

NOTE

This button is visible only when you are in a Layout; it is not displayed in Model Space. Also, if numerous viewports are created, arrow buttons appear, allowing you to move between viewports.

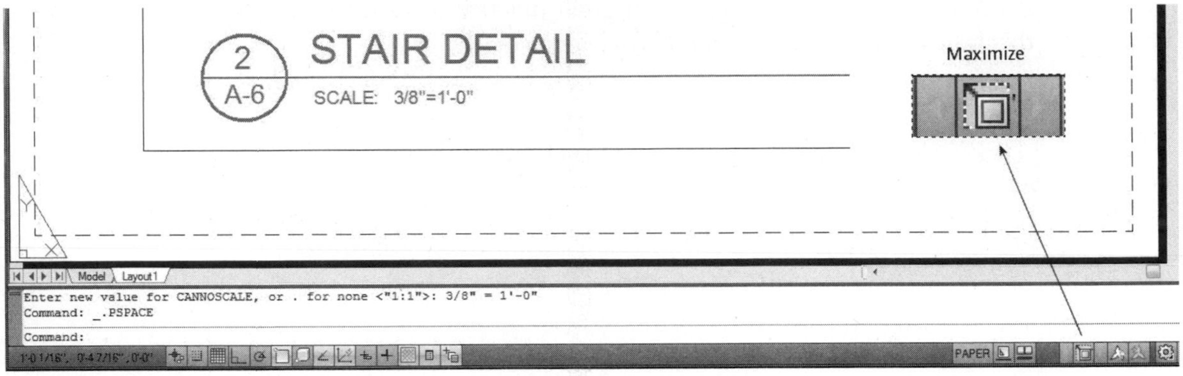

FIGURE 14.14

Figure 14.15 displays a drawing that has its viewport maximized. To return to Paper Space or Layout mode, double-click the red border or click the Minimize Viewport button.

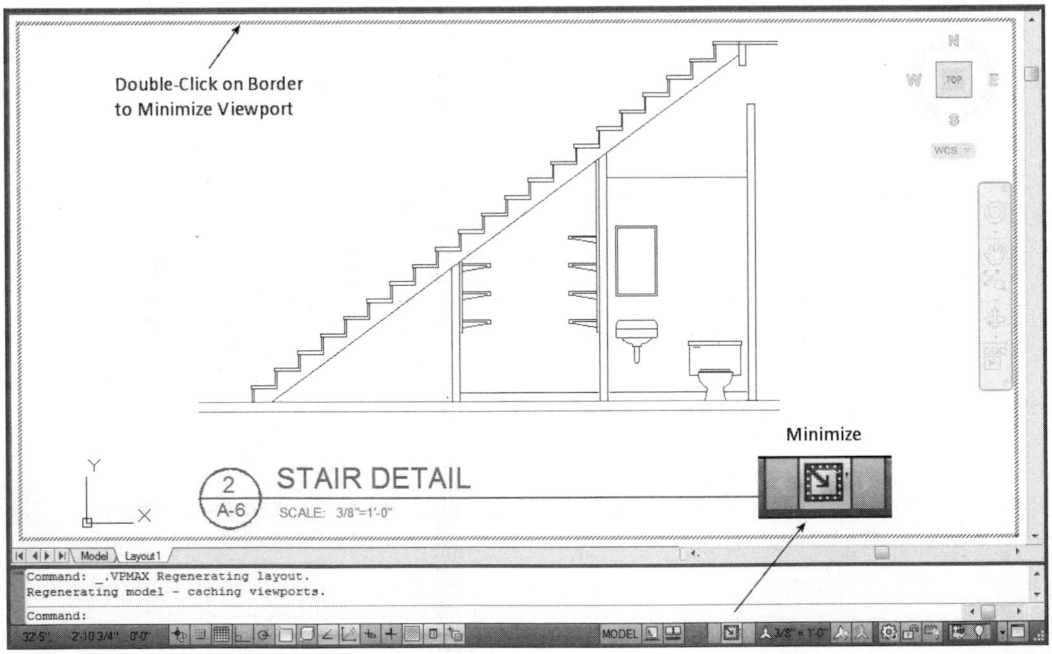

FIGURE 14.15

Hiding Layout and Model Tabs

The Layout and Model tabs will normally be displayed, as shown in Figure 14.16. As a means of regaining screen space, you can hide the layout and model tabs. To do this, right-click one of the tabs and choose Hide Layout and Model tabs from the menu, as shown in Figure 14.16.

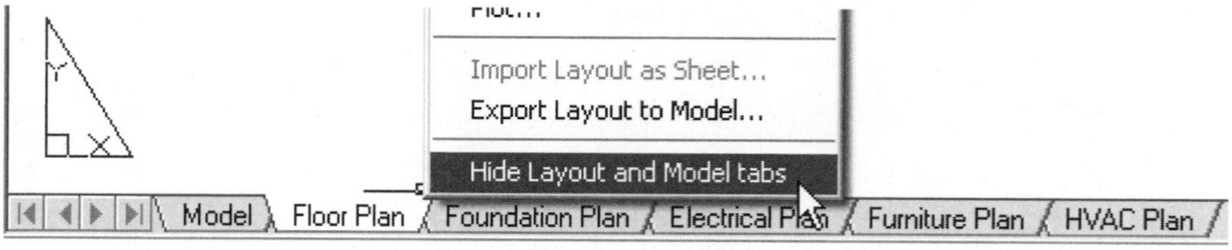

FIGURE 14.16

To redisplay the tabs on the screen, right-click the Model or Layout buttons, right-click, and pick Display Layout and Model Tabs, as shown in Figure 14.17.

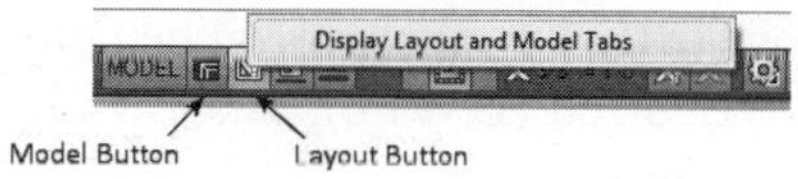

FIGURE 14.17

CREATING A LAYOUT

TRY IT!

Open the drawing file 14_Gasket1. The illustration in Figure 14.18 (Left) shows the drawing originally created in Model Space. This drawing is required to be laid out in Paper Space.

Click the Layout1 tab to activate the Layout environment. A layout is automatically created, as shown in Figure 14.18 (Right).

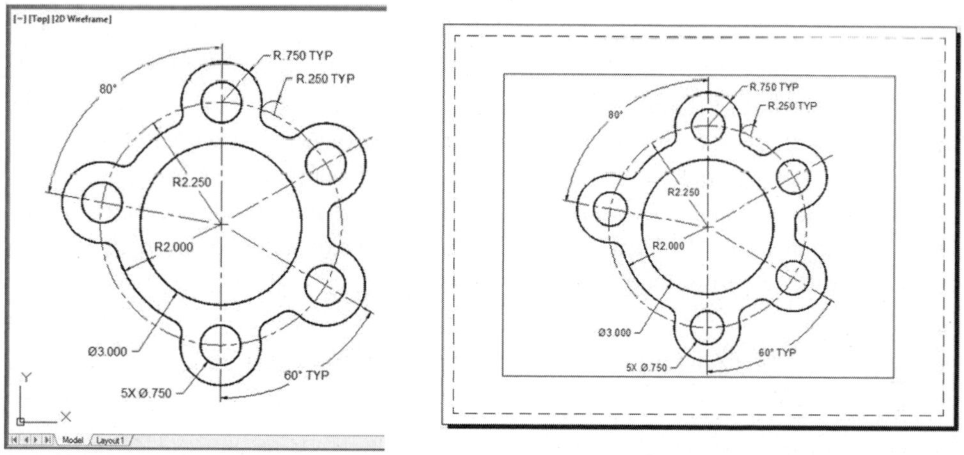

FIGURE 14.18

It is good practice to rename the Layout1 tab located at the bottom of your screen. Do this by double-clicking on the tab or right-clicking and selecting Rename from the shortcut menu. Rename Layout1 to One View Drawing, as shown in Figure 14.19 (Right).

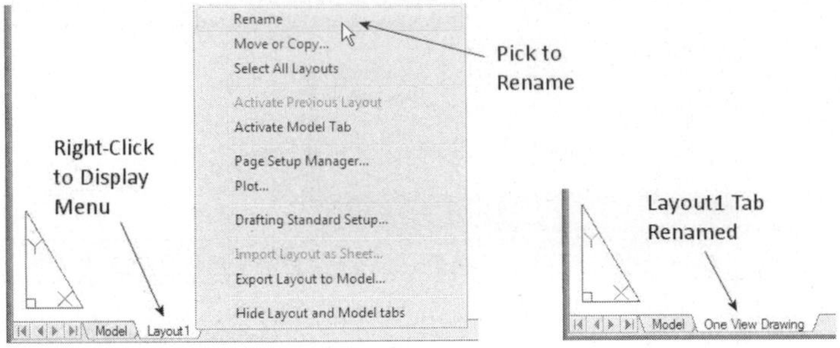

FIGURE 14.19

Next, right-click the One View Drawing icon and pick Page Setup Manager from the menu, as shown in Figure 14.20 (Left). This launches the Page Setup Manager dialog box, illustrated in Figure 14.20 (Center). Click the New button. When the New Page Setup dialog box appears, enter the name B-Sized (DWF6), as shown in Figure 14.20 (Right). When finished, click the OK button.

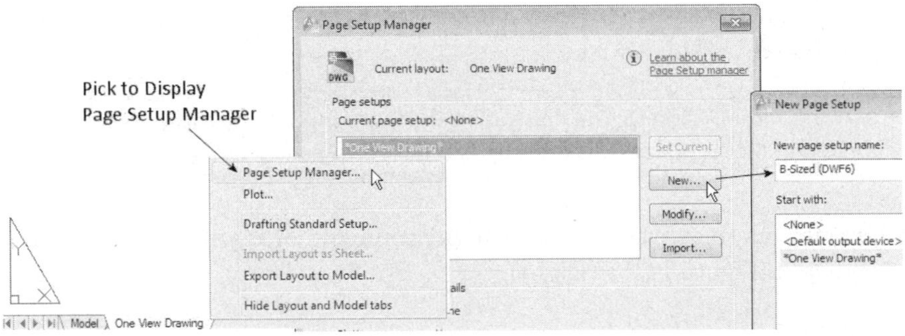

FIGURE 14.20

When the Page Setup dialog box appears, as shown in Figure 14.21, click on the Printer/plotter field and change the name of the plot device to DWF6 ePlot. Next, click in the Paper size field and select ANSI expand B (17.00 × 11.00 inches) from the available list. Finally, click on the field under Plot style table and change the plot style to Monochrome. When you are finished making these changes, click the OK button in the Page Setup dialog box.

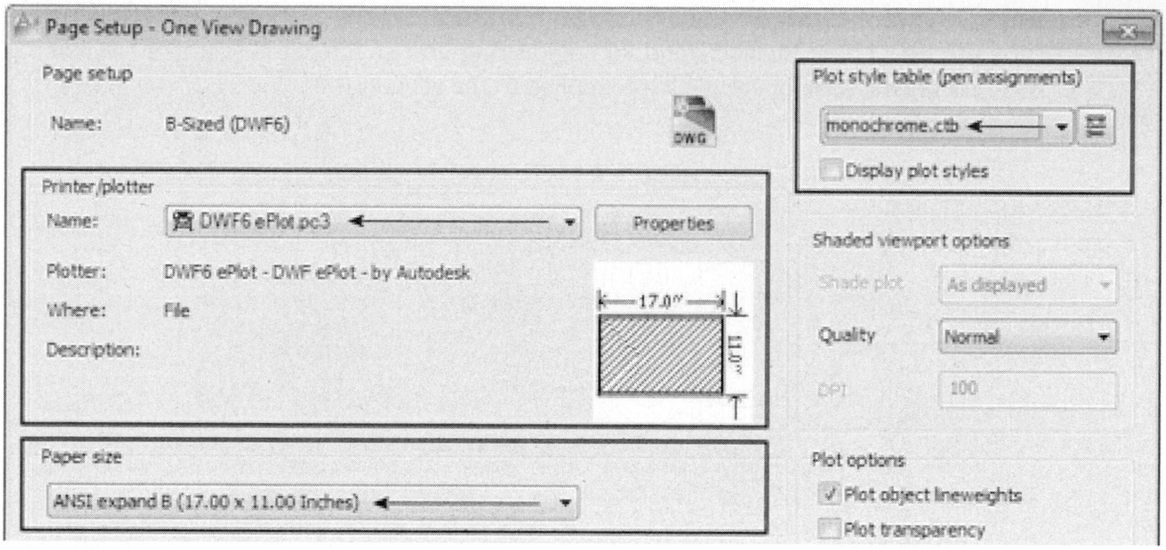

FIGURE 14.21

When you return to the Page Setup Manager dialog box, double-click the B-Size (DWF6) layout to make it current, as shown in Figure 14.22. Click the Close button to continue with laying out this drawing.

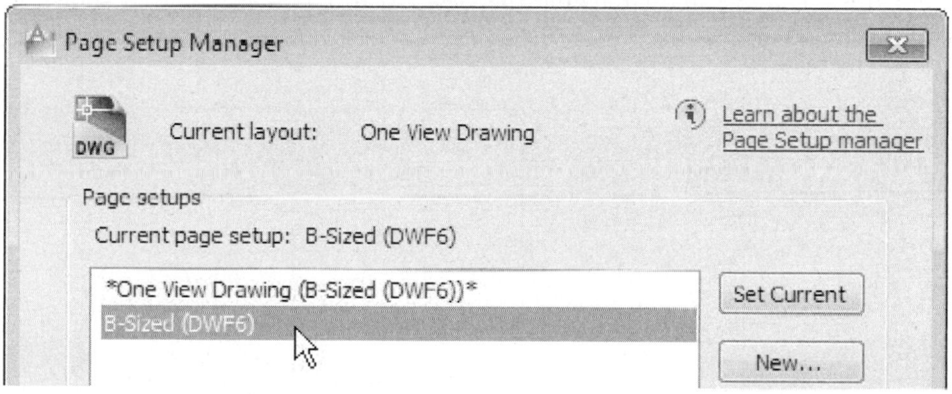

FIGURE 14.22

Although you increased the size of the sheet, the image of the gasket in the viewport did not grow in size. The viewport size and the viewport scale will need to be set. The size of the viewport depends on the scale applied in this next step.

Rather than activating floating Model Space, simply click the edge of the viewport. Expand the Viewport Scale box located on the right side of the status bar, as shown in Figure 14.23. The number inside the field in this image is the current scale of the image inside the viewport. Whenever you create a drawing layout, the image is automatically zoomed to the extents of the viewport. For this reason, the desired scale will never be correctly displayed in this area.

To have the image properly scaled inside the viewport, click one of the scales from the list. These include standard engineering and architectural scales. The image is assigned a scale of 1:1, which is applied to the viewport.

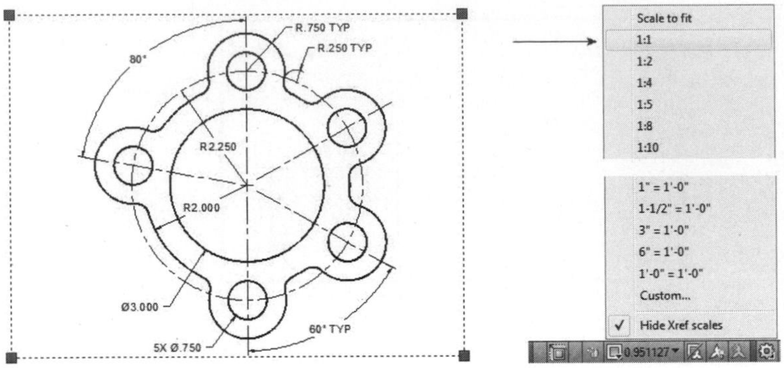

FIGURE 14.23

You can see in Figure 14.23 (Left) that the gasket has grown inside the viewport. The viewport needs to be increased in size to accommodate the scaled image of the gasket. Grips should already be displayed on the viewport, but if they aren't pick the viewport once to display grips at the corners, as shown in Figure 14.23 (Left). Use these grips to size the viewport to the image. Accomplish this by clicking one of the grips in the corner and stretching the viewport to a new location, as shown in Figure 14.24 (Right).

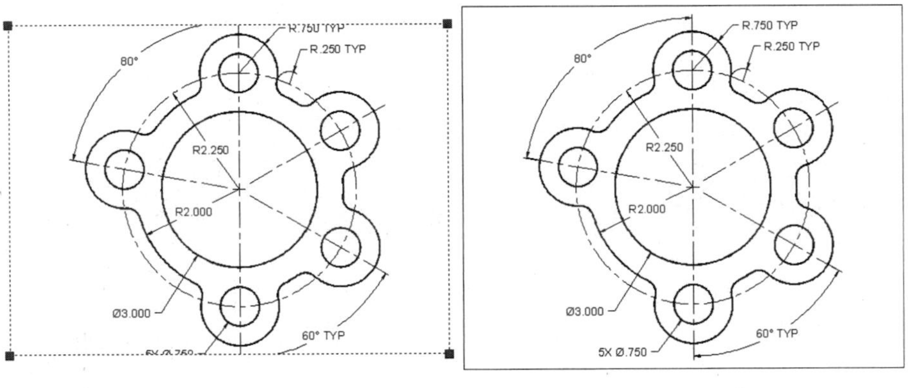

FIGURE 14.24

The results of this operation are displayed in Figure 14.25. At this point, it would be advisable to lock the viewport to prevent accidentally changing the viewport scale. Once locked, press ESC to remove the object highlight and the grips.

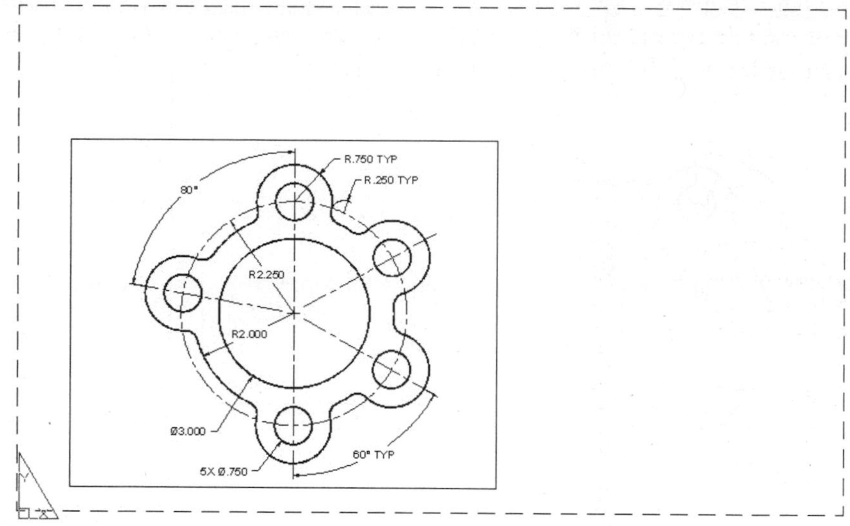

FIGURE 14.25

Before the drawing is plotted, a title block containing information such as drawing scale, date, title, company name, and designer is inserted in the Paper Space environment. After verifying you are in Paper Space, select the Insert button from the Ribbon (Home tab > Block panel), as shown in Figure 14.26 (Left). This activates the Insert dialog box, shown in Figure 14.26 (Right). This feature of AutoCAD will be covered in greater detail in Chapter 16. Be sure the block ANSI B title block is listed in the Name field.

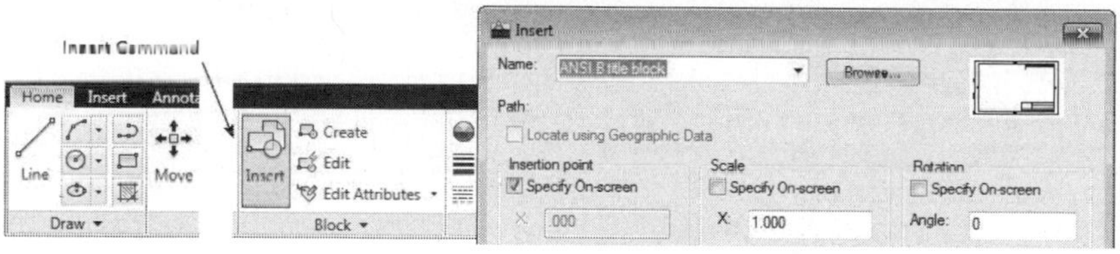

FIGURE 14.26

Click the OK button to place the title block in Figure 14.27. Be sure the title block displays completely inside the paper margins; otherwise, part of it may not plot. At this point, the MOVE command can be used to relocate the viewport containing the drawing image to a better location inside the title block border.

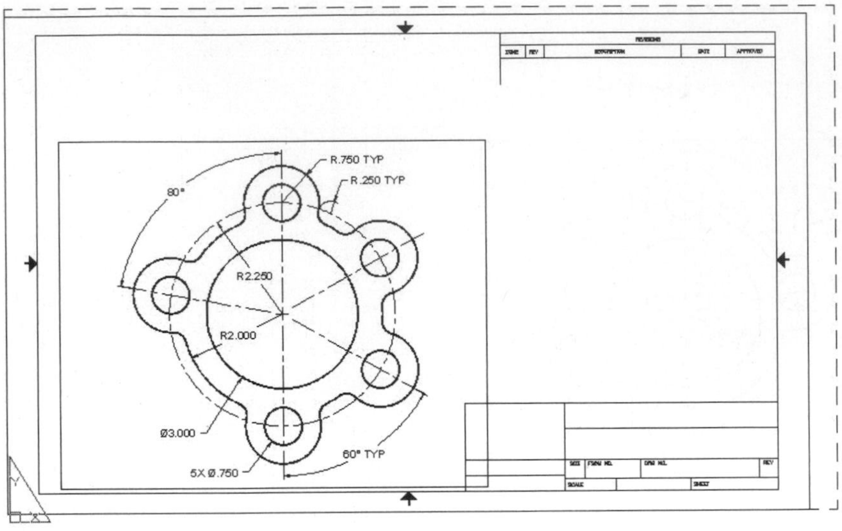

FIGURE 14.27

Unfortunately, if the drawing were to be plotted out at this point, the viewport would also plot. It is considered good practice to assign a layer to the viewport and then turn that layer off before plotting. In Figure 14.28, the viewport is first selected and grips appear. Click on the Layer Control box and click on the Viewports layer. The viewport is now on the Viewports layer. Then click the lightbulb icon in the Viewports layer row to turn it off.

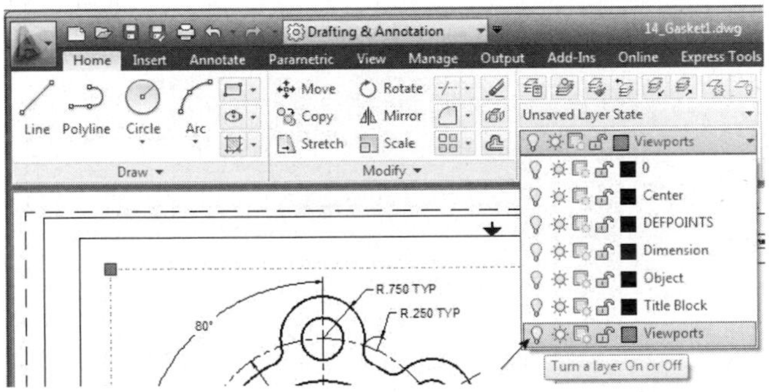

FIGURE 14.28

The drawing display will appear similar to Figure 14.29, with the drawing laid out and properly scaled in Paper Space.

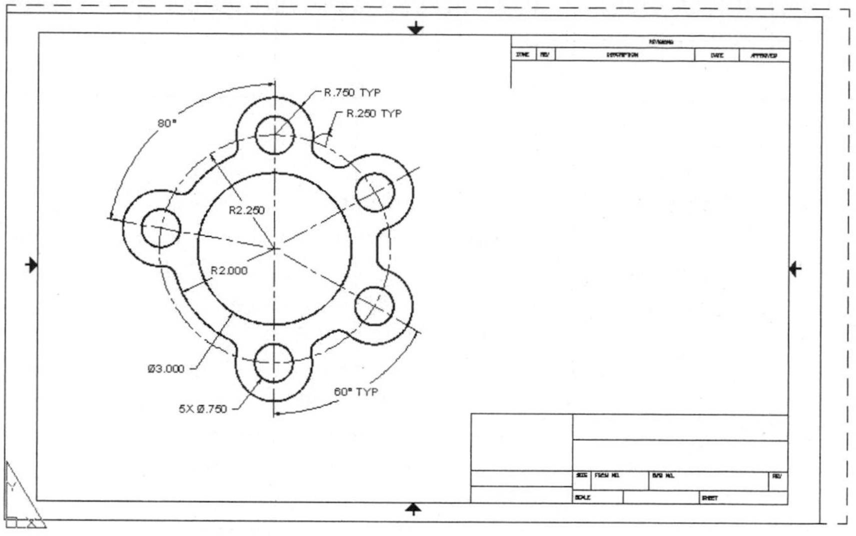

FIGURE 14.29

USING A WIZARD TO CREATE A LAYOUT

The AutoCAD Create Layout wizard can be helpful in properly completing the steps involved in laying out a drawing in the Paper Space environment.

Activating the LAYOUTWIZARD command displays the Create Layout-Begin dialog box, as shown in Figure 14.30. You cycle through the different categories, and when you are finished, the drawing layout is displayed, complete with title block and viewport.

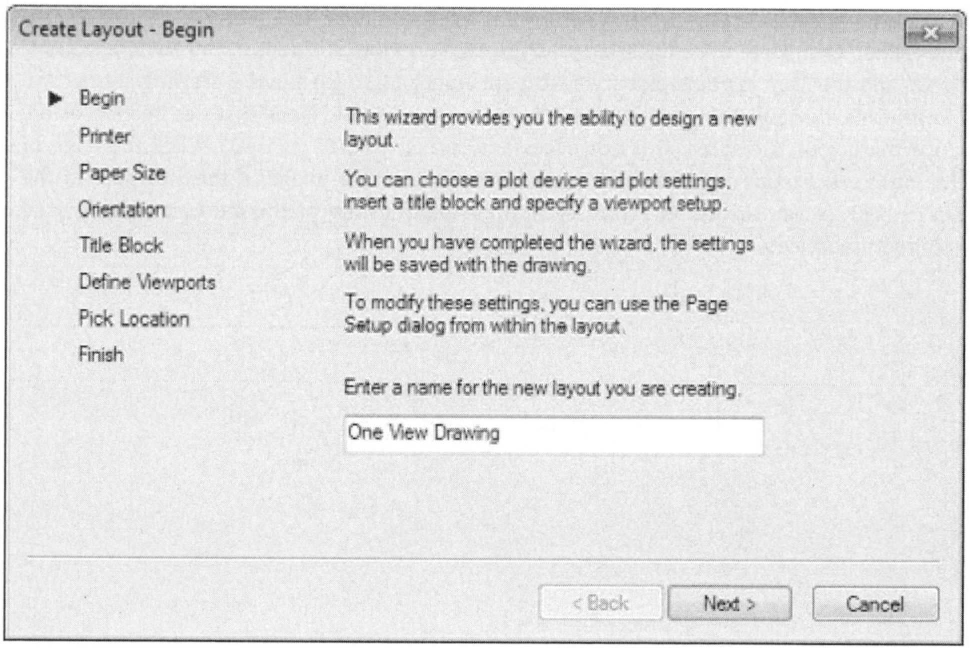

FIGURE 14.30

Functions of the Create Layout wizard dialog box are briefly described in the following table:

Create Layout Wizard

Category	Purpose
Begin	Change the name of the layout here.
Printer	Use the Printer category to select a configured plotter from the list provided.
Paper Size	The Paper Size category displays all available paper sizes supported by the currently configured plotter.
Orientation	Use the Orientation category to designate whether to plot the drawing in Landscape or Portrait mode.
Title Block	Depending on the paper size, choose a corresponding Title Block to be automatically inserted into the layout sheet.
Define Viewports	The Define Viewports category is used to either create a viewport or leave the drawing layout empty of viewports.
Pick Location	The Pick Location category creates the viewport to hold the image in Paper Space. If you click the Next > button, the viewport will be constructed to match the margins of the paper size. If you click the Select location < button, you return to the drawing and pick two diagonal points to define the viewport.
Finish	The Finish category alerts you to a new layout name that will be created. Once it is created, modifications can be made through the Page Setup dialog box.

TRY IT!

Open the drawing file 14_Stair Detail. Use either the Page Setup Manager dialog box (demonstrated in the previous Try It! exercise) or use the Create Layout wizard to arrange the stair detail in Paper Space so that your drawing appears similar to Figure 14.31. Create a layout called "One View Drawing" that uses the DWF6 ePlot plot device, the ANSI expand C sheet, and the Monochrome plot style table. In Figure 14.31, we see the drawing sheet complete with a viewport that holds the image and an ANSI C title block that has been inserted. Once the layout is created, it is not necessarily scaled to 3/8" = 1'-0". Click the edge of the viewport. Expand the Viewport Scale box and scale the image in the viewport to the 3/8 = 1'-0" scale. Notice, in Figure 14.31, a complete listing of the more commonly used architectural scales.

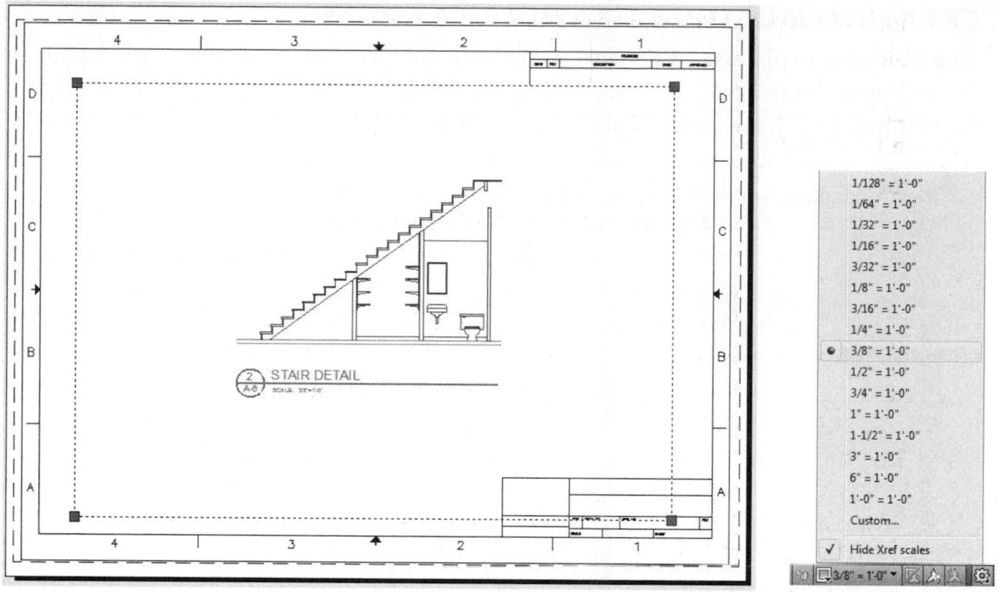

FIGURE 14.31

Changing the viewport to a different layer and then turning that layer off displays the completed layout of the stair detail in Figure 14.32. The final step is to plot the drawing. In a drawing laid out in Paper Space, the image is scaled inside the viewport and the drawing is plotted at a scale of 1:1.

NOTE

Another common practice is to set the layer that contains all viewports to a No Plot state in the Layer Properties Manager palette. The viewport will always remain visible in your layout, yet it will not appear in your plots.

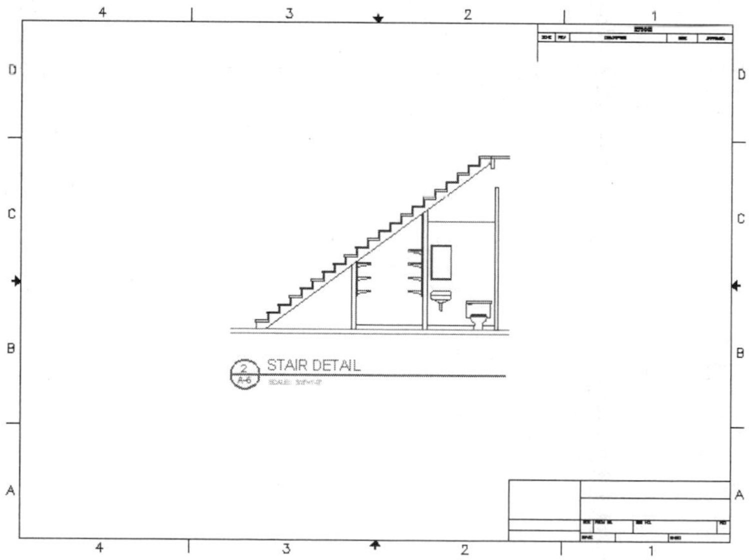

FIGURE 14.32

CREATING MULTIPLE DRAWING LAYOUTS

The methods explained so far have dealt with the arrangement of a single layout in Paper Space. AutoCAD provides for greater flexibility when working in Paper Space by enabling you to create multiple layouts of a drawing. For example, in Figure 14.33, which shows a floor plan complete with electrical plan, individual layouts could be created to display separate images of the floor and electrical plans.

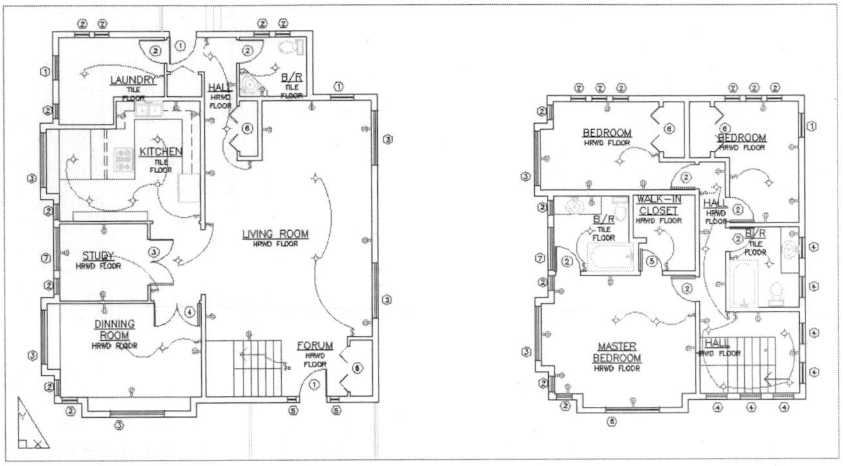

FIGURE 14.33

To create multiple layouts, first verify that the Model and Layout tabs are displayed on the screen. If they are not, right-click the Model or Layout1 button on the Status Bar and select "Display Model and Layout tabs." To create a new layout, right-click the Model or Layout1 tab and select "New layout," as shown in Figure 14.34 (Left).

Instead of creating a new layout, it can often be more efficient to copy a layout. When you copy a layout, all information inside the viewport, such as title block, viewport, and viewport scale/layer state information, is copied. Selecting "Move or Copy…" from the shortcut menu, as shown in Figure 14.34 (Center), will display the Move or Copy dialog box. Moving layouts allows you to change their order. To copy a layout you must check the "Create a copy" box in the Move or Copy dialog box, as shown in Figure 14.34 (Right).

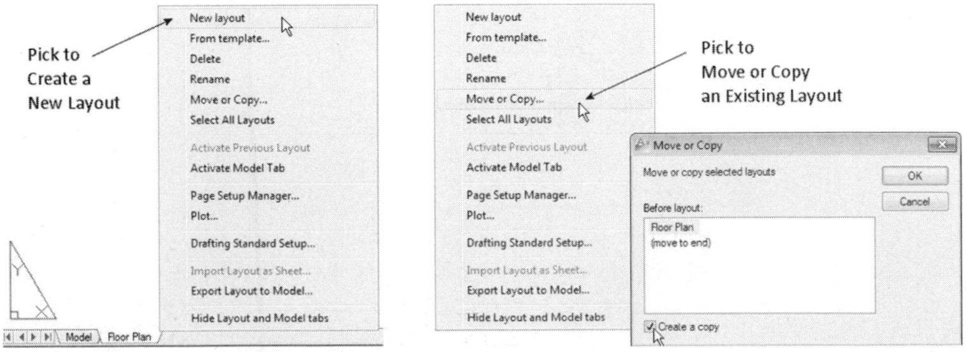

FIGURE 14.34

Once you create a new layout or copy an existing one, you should rename it. In Figure 14.35, the Foundation Plan layout was renamed to Furniture Plan. This was accomplished by double-clicking on the Foundation Plan layout tab. This will highlight all text in the tab. Entering new text renames the layout. You can, of course, also rename a layout by selecting "Rename" from the shortcut menu.

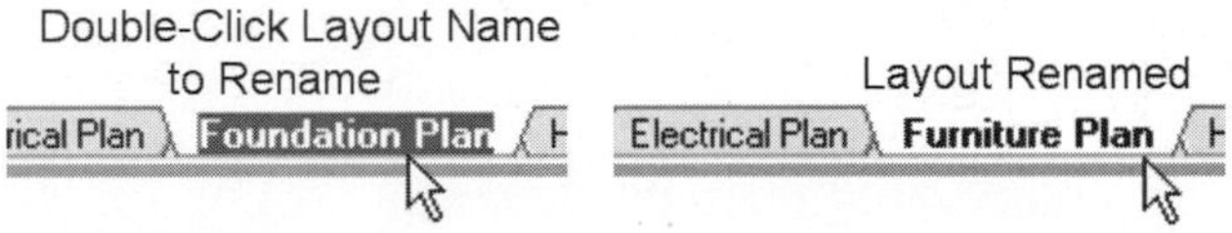

FIGURE 14.35

A drag and drop method is also available to copy and move layouts. To copy a layout, select the layout tab and then press the CTRL key while dragging your cursor. You will notice a small page icon with a "plus" sign indicating the Floor Plan layout will be copied. The results are displayed in Figure 14.36 (Right). Notice the addition of the layout Floor Plan (2) in the list of layouts.

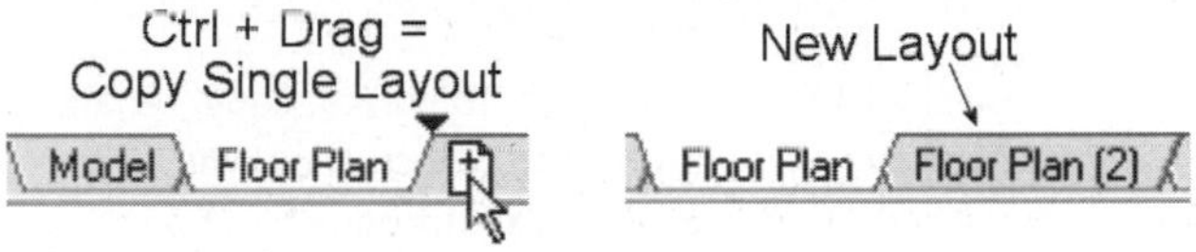

FIGURE 14.36

You can also copy multiple layouts. In this example, first click on the Floor Plan and hold down the `SHIFT` key while you select the Electrical Plan layout tab, as shown in Figure 14.37 (Left). Then press the `CTRL` key while dragging your cursor. You will notice a small multiple page icon with a "plus" sign indicating the Floor Plan and Electrical Plan layouts will be copied. The results are displayed in Figure 14.37 (Right). Notice the addition of the layout Floor Plan (2) and Electrical Plan (2) in the list of layouts.

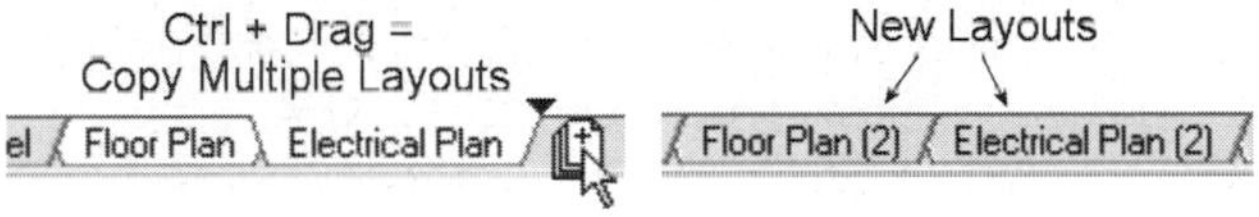

FIGURE 14.37

When you want to move a layout to a new location and, in effect, change the order of the layouts, click on the layout and drag it to its new location, as shown in Figure 14.38. In this example, the `CTRL` key is not utilized since you are moving the layout and not copying it.

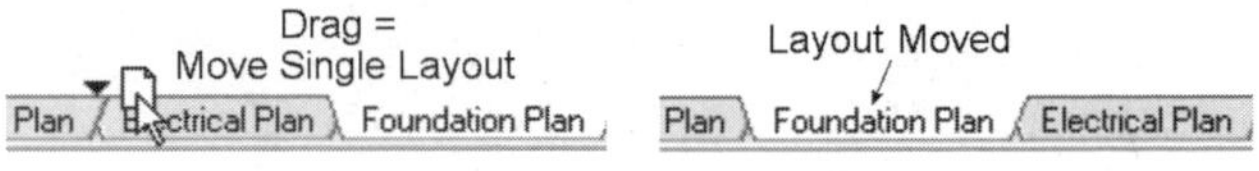

FIGURE 14.38

At times, layouts need to be deleted and removed from the drawing entirely. To delete the HVAC Plan layout in the next example, right-click on the HVAC Plan layout tab and select Delete from the menu, as shown in Figure 14.39 (Left). An AutoCAD alert box will prompt you to delete the layout by clicking on the OK button. Notice from the alert box that the Model tab cannot be deleted.

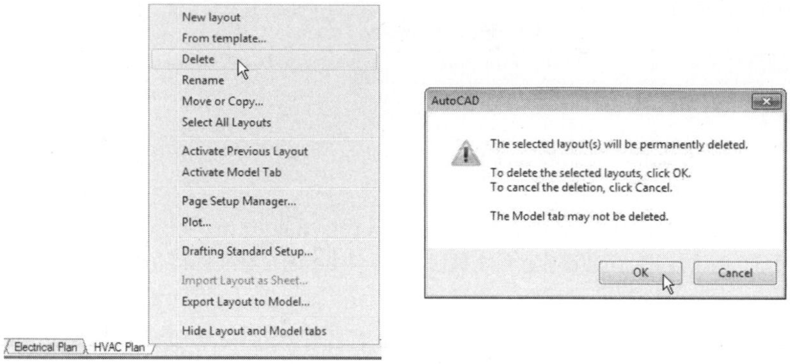

FIGURE 14.39

NOTE

Instead of using the layout tabs, you can use the LAYOUT command to create, copy, rename and delete layouts. The New and Template options of this command, which allow you to create new layouts, can be accessed from the Ribbon (Layout tab > Layout panel).

TRY IT!

Open the drawing 14_Facilities_Plan. Notice that a number of layouts were created that correspond to the various room numbers, as shown in Figure 14.40. However, the room numbers are out of order and need to be rearranged starting with the lowest room number and going to the highest room number. The Overall layout should be reordered directly after the Model tab. Use the drag and drop technique illustrated earlier to rearrange all the layout tabs in order.

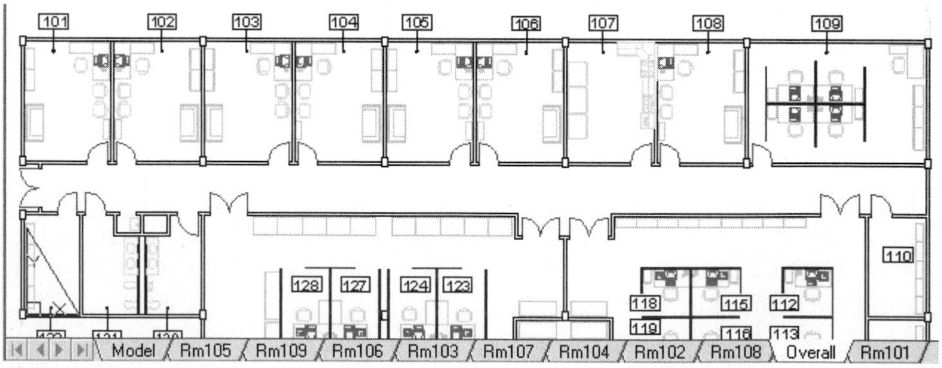

FIGURE 14.40

After performing the reordering operation, your screen should appear similar to Figure 14.41.

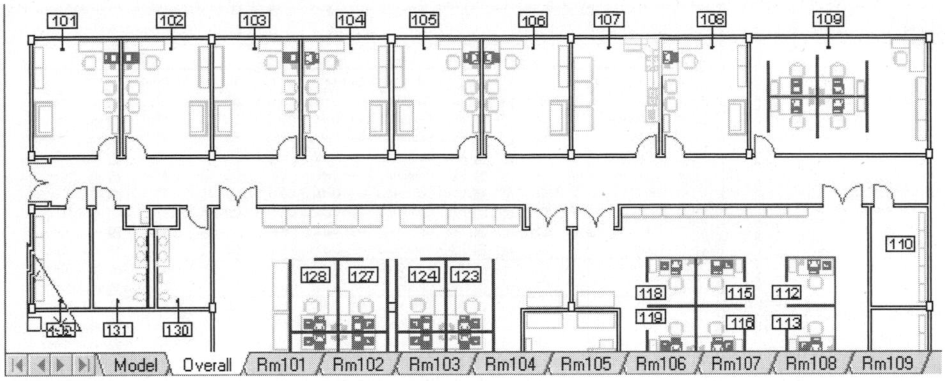

FIGURE 14.41

USING LAYERS TO MANAGE MULTIPLE LAYOUTS

In Figure 14.42, a Floor Plan and Electrical Plan layout are already created. Also, the Floor Plan layout is currently active. You want to see only the floor plan inside this viewport and none of the objects on electrical layers.

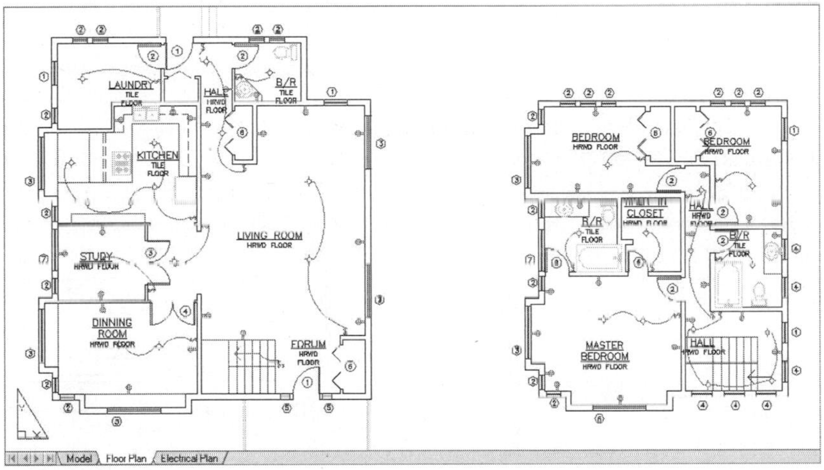

FIGURE 14.42

First, double-click inside the viewport in order to switch to floating Model Space. Activating the Layer Properties Manager palette displays the layer information, as shown in Figure 14.43. When you are in floating Model Space and use this palette, a number of additional layer modes are added. Two of these modes provide for the ability to freeze layers only in the active viewport and the ability to freeze layers in new viewports. Freezing layers in all viewports is not an effective means of controlling layers when you create multiple layouts. You need to be very familiar with the layers created in order to perform this task. To display only the floor plan information, notice that two layers (Lighting and Power) have been frozen under the Viewport Freeze heading, as shown in Figure 14.43.

All other layers remain visible in this viewport.

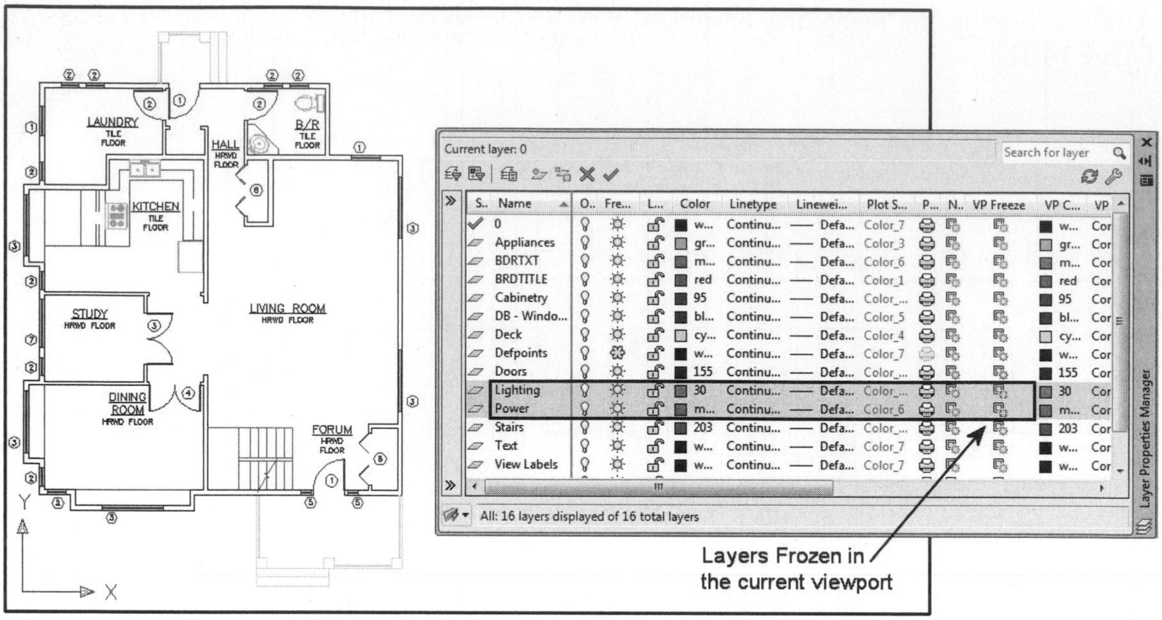

FIGURE 14.43

The result is displayed in Figure 14.44 with only the floor plan information visible in the viewport. The electrical layers have been frozen only in this viewport and do not display.

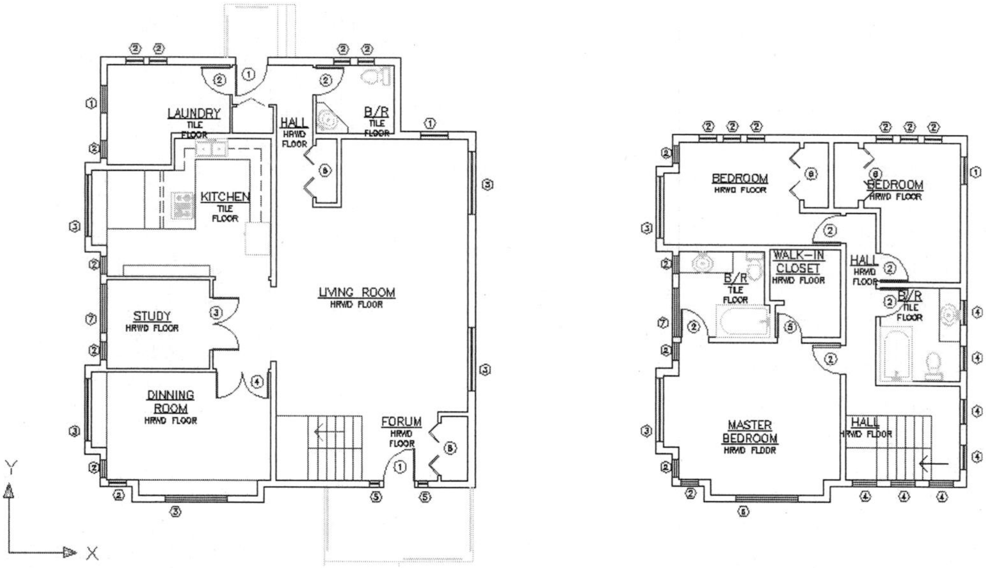

FIGURE 14.44

NOTE

You can also freeze layers in the current viewport through the Layer control box, as shown in Figure 14.45.

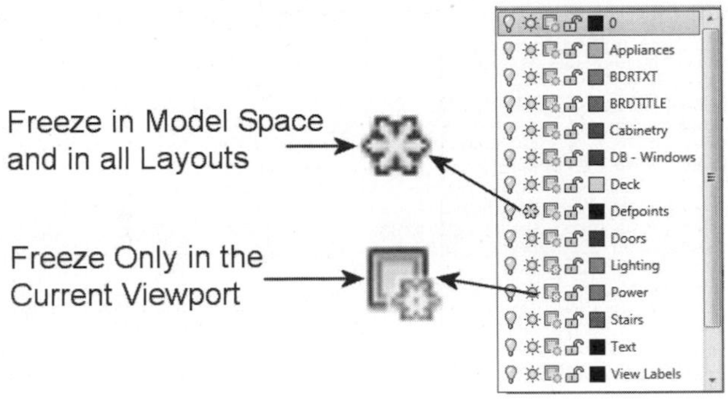

FIGURE 14.45

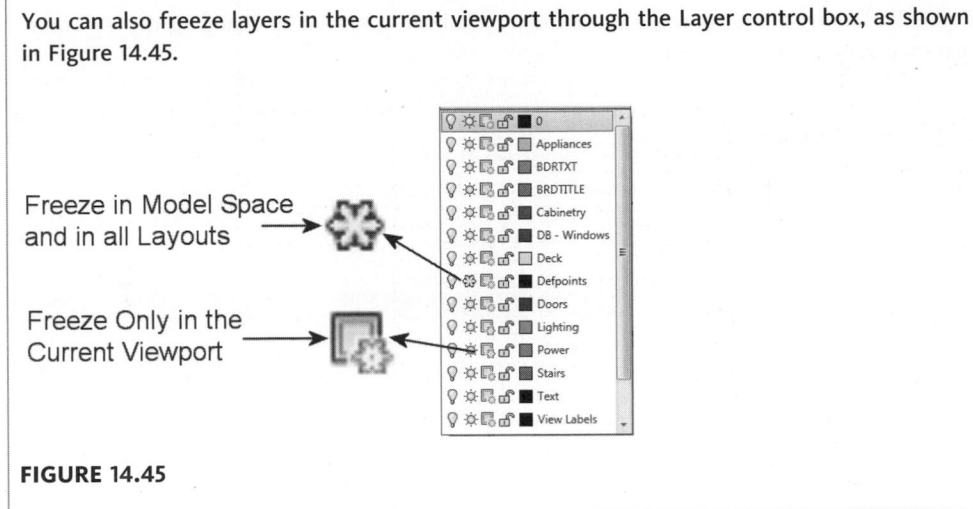

ADDITIONAL LAYER TOOLS THAT AFFECT VIEWPORTS

Additional controls on layers in viewports are available through the Layer Properties Manager palette. Figure 14.46 illustrates the Electrical Plan layout that was created in the previous segment. Sometimes, you want to add special effects to your layouts such as changing the color, linetype, or even the lineweight just in the current viewport.

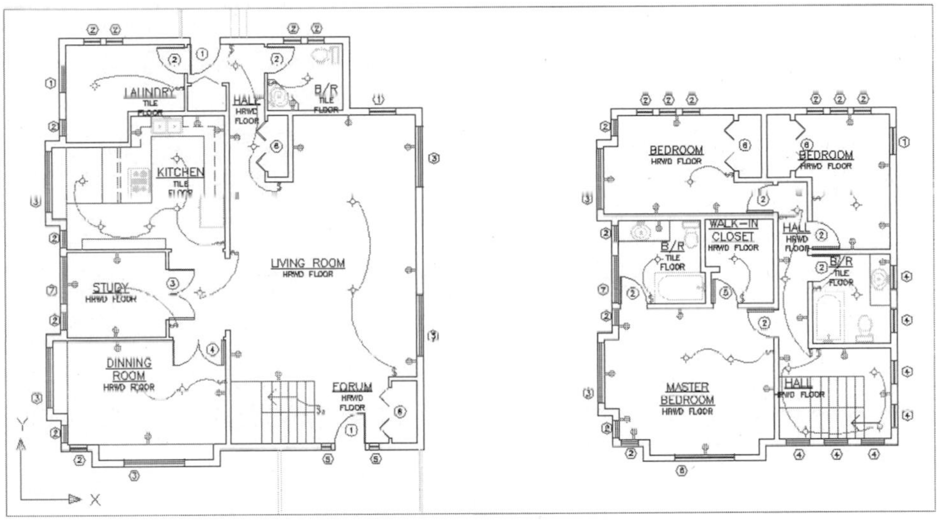

FIGURE 14.46

After double-clicking inside the Electrical Plan layout, launch the Layer Properties Manager palette to display all layers in the drawing. For the purposes of this example, all layers except for those dealing with electrical components will have the color changed to a light gray. The reason is to dim or fade the floor plan out in order to give emphasis on the electrical layers. In Figure 14.47, a number of layers are selected and then their colors changed to 9 or light gray through the VP Color column. Notice also in Figure 14.47 the presence of the VP Linetype, VP Lineweight, VP Transparency, and VP Plot Style columns. These changes are only present in the current viewport.

NOTE

A similar fading effect in the viewport could have been achieved by changing the VP Transparency instead of the VP Color.

S..	Name	VP Freeze	VP Color	VP Linetype	VP Lineweight	VP Transparency	VP Plot Style
	0		white	Continuous	—— Default	0	Color_7
	Appliances		9	Continuous	—— Default	0	Color_9
	BDRTXT		magenta	Continuous	—— Default	0	Color_6
	BRDTITLE		red	Continuous	—— Default	0	Color_1
	Cabinetry		9	Continuous	—— Default	0	Color_9
	DB - Wind...		blue	Continuous	—— Default	0	Color_5
	Deck		9	Continuous	—— Default	0	Color_9
	Defpoints		white	Continuous	—— Default	0	Color_7
	Doors		9	Continuous	—— Default	0	Color_9
	Lighting		30	Continuous	—— Default	0	Color_30
	Power		magenta	Continuous	—— Default	0	Color_6
	Stairs		203	Continuous	—— Default	0	Color_203
	Text		white	Continuous	—— Default	0	Color_7
	View Labels		white	Continuous	—— Default	0	Color_7

FIGURE 14.47

The results of performing this operation are illustrated in Figure 14.48 where the floor plan components display faintly while the electrical components stand out.

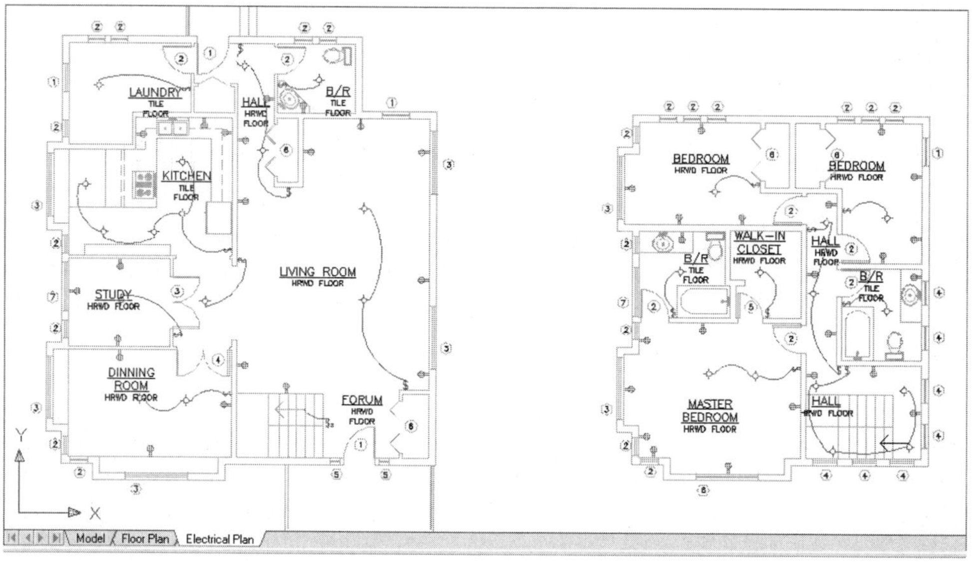

FIGURE 14.48

In the Layer Properties Manager, those layers that have had either the color, linetype, lineweight, transparency, or plot style overridden will be displayed with a light blue background. If you need to return these properties back to their original state, highlight the layers identified with the blue background and right-click. The menu illustrated in Figure 14.49 will appear. Click "Remove Viewport Overrides for" followed by "Selected Layers" and then "In Current Viewport only."

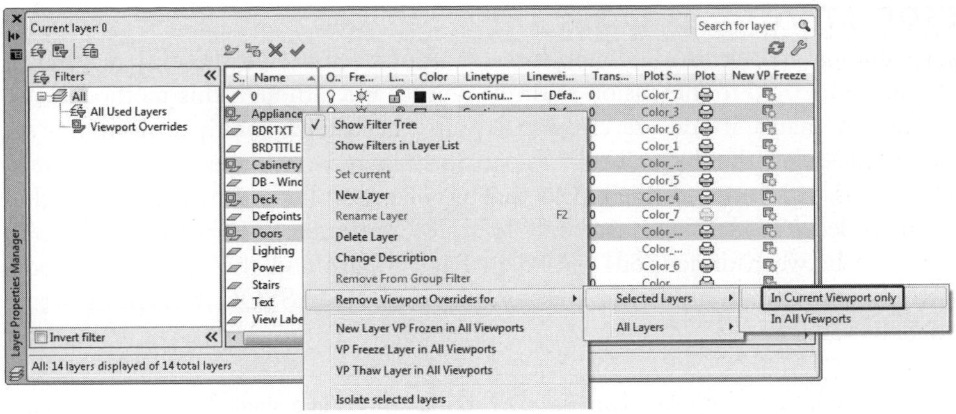

FIGURE 14.49

Whenever overriding a viewport property, a special layer group is automatically created to help you manage these items. In Figure 14.50, the Layer Properties Manager palette, notice the new layer group called Viewport Overrides. Clicking this name displays the layers that had their color overridden in the previous step.

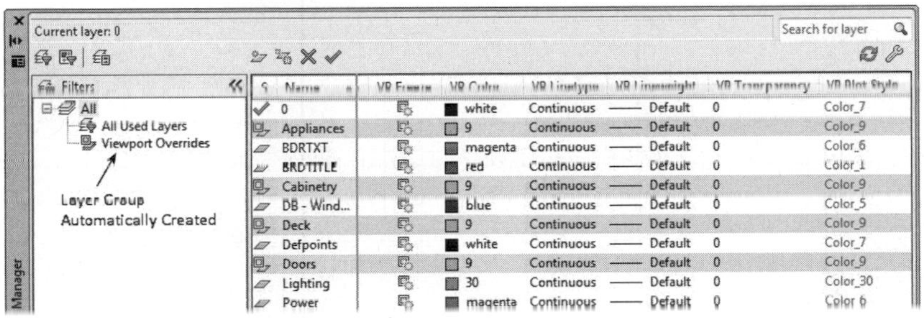

FIGURE 14.50

TIP
The information contained in the Layer Properties Manager palette can become overwhelming. To better manage this information, a special menu exists that allows you to turn off those layer states that you do not use on a regular basis. To activate this menu, move your cursor into one of the layer state headings and right-click to display the menu shown in Figure 14.51. Additional controls allow you to maximize all columns or a single column. You can even reset all columns back to their default widths.

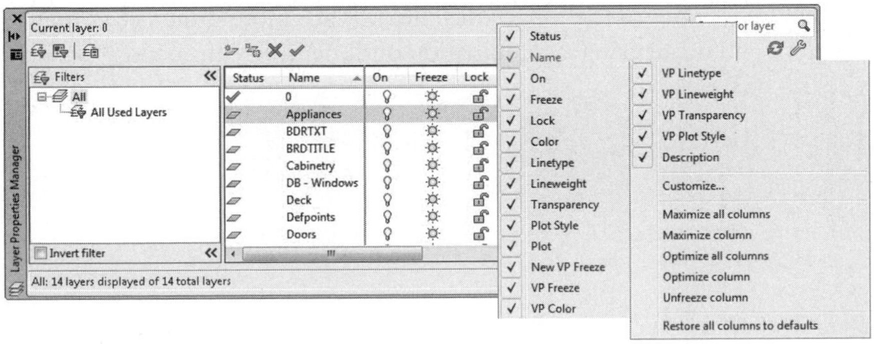

FIGURE 14.51

ASSOCIATIVE DIMENSIONS AND LAYOUTS

Dimensioning all objects in Model Space was once considered the only reasonable and reliable method to dimension multiple objects and although this method is still preferred by many, if you are creating layouts that utilize multiple viewports set at different scales, dimensioning in Paper Space may be a better option. Dimensions, just as with text and title blocks that are placed in Paper Space, are created at their intended size. It is not necessary to scale the dimensions to the objects. An association between dimensions placed in Paper Space and the objects in Model Space is automatically established as long as the DIMASSOC variable is set to its default value of 2.

TRY IT!

Open the drawing file 14_Inlay. Two viewports are arranged in a single layout called Inlay Floor Tile, as shown in Figure 14.52. The images inside these viewports are scaled differently; the main inlay pattern in the left viewport is scaled to 3/4" = 1'-0" and the detail image in the right viewport has been scaled to 3" = 1'-0". The dimension style scale is set for 1:1 and the DIMASSOC variable is currently set to 2.

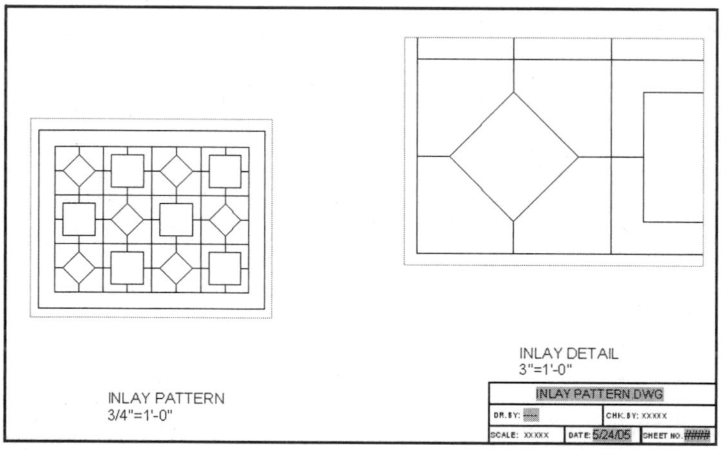

FIGURE 14.52

While in Paper Space, use the `DIMLINEAR` command and place a few linear dimensions on the main inlay plan in the left viewport. Notice how object snaps, for Model Space objects, are available for creating the Paper Space dimensions. Now switch and add a few linear dimensions in the right viewport. Zoom in to a few of the dimensions and see that they reflect the Model Space distances. Place more dimensions on the main inlay pattern and detail using a combination of dimension commands. Observe the correct values being placed.

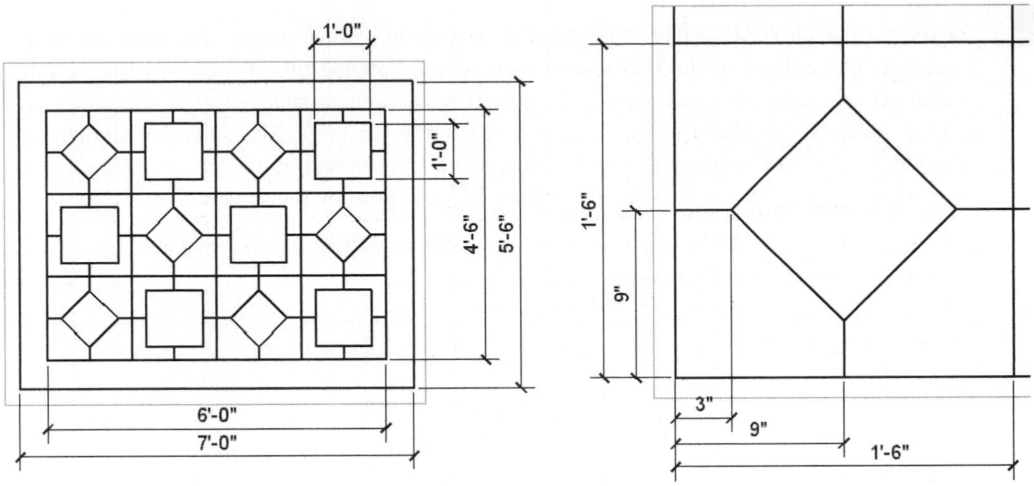

FIGURE 14.53

Next, click and drag each viewport to a more convenient location on the drawing screen. Notice that the dimensions move along with the viewports.

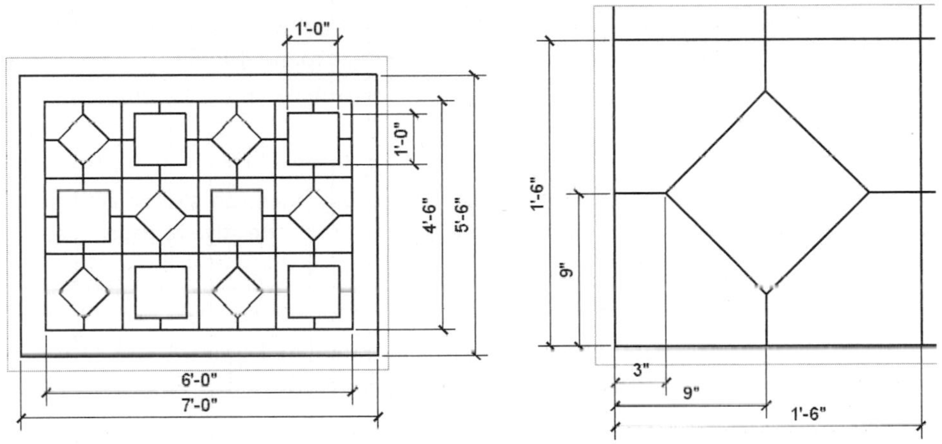

FIGURE 14.54

NOTE

If you drag viewports around and the dimensions do not keep up with the viewports, try entering the `DIMREGEN` command from the keyboard. This should update all dimensions to the new viewport positions. If a dimension is no longer associated with an object, you will need to recreate the dimension or use the DIMREASSOCIATE command to reestablish the link.

HATCH SCALING RELATIVE TO PAPER SPACE

While in a layout, you have the opportunity to scale a hatch pattern relative to the current Paper Space scale in a viewport. In this way, you can easily have AutoCAD calculate the hatch pattern scale, since this is determined by the scale of the image inside a viewport.

TRY IT!

Open the file 14_Valve Gasket. This drawing consists of a small gasket that does not have crosshatching applied to the thin inner border. If you hatched this shape using the default hatch settings while in Model Space, the gasket would appear similar to the illustration in Figure 14.55. Notice that the hatch pattern spacing is too large. Typically, you should set the hatch scale to the drawing scale factor which is determined by taking the reciprocal of the drawing scale (for this drawing the scale is 10:1, so the scale factor would be 1/10 or 0.1).

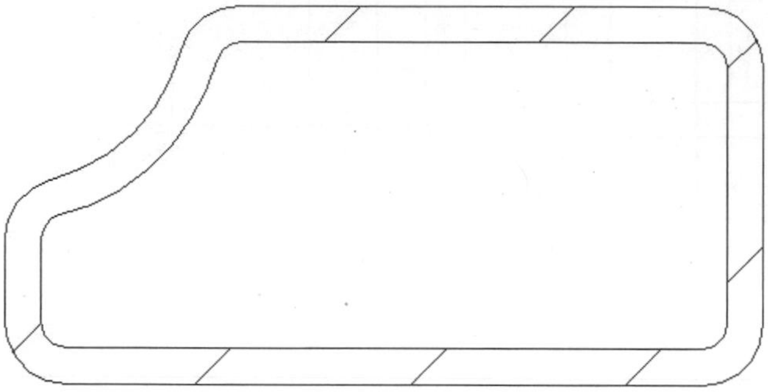

FIGURE 14.55

Instead of calculating the hatch scale, try the following approach. From Paper Space, double-click in the viewport to enter floating Model Space. Activate the Hatch command in the Hatch Creation tab of the Ribbon, expand the Properties panel and select "Relative To Paper Space," as shown in Figure 14.56 (Left). This feature is grayed out if you try to hatch in Model Space.

NOTE

If you are not using the Ribbon, this Paper Space scaling operation can also be activated from the Hatch and Gradient dialog box by placing a check in the "Relative to paper space" check box.

The finished gasket with hatching applied is illustrated in Figure 14.56 (Right). Notice how the hatch scale is based on the viewport scale.

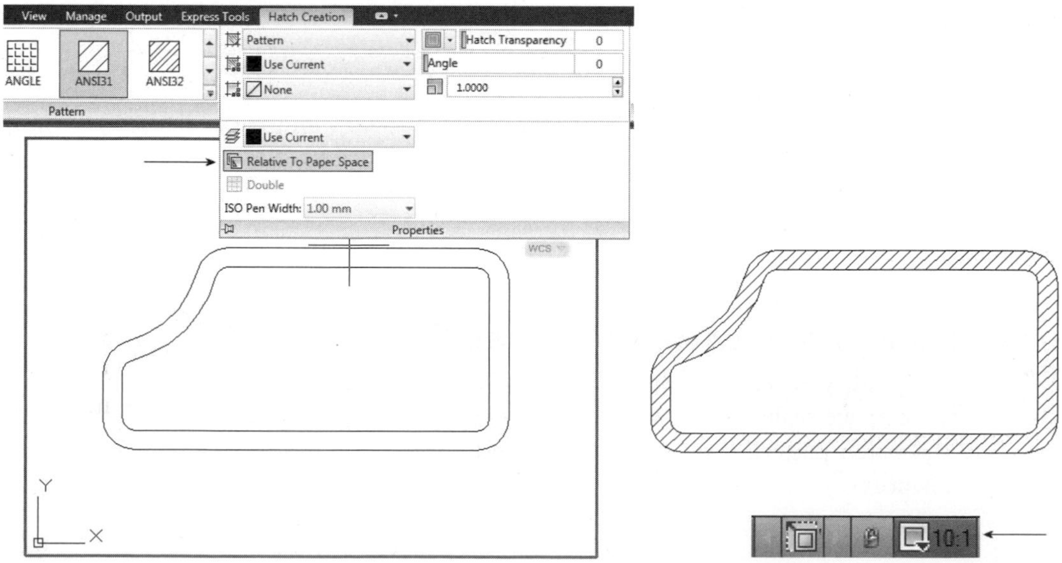

FIGURE 14.56

TRY IT!

Open the drawing file 14_Hatch Partial Plan. The inner walls of the object in Figure 14.57 need to be crosshatched. Apply the technique of making the hatch scale relative to Paper Space. The results are illustrated in Figure 14.57 (Right).

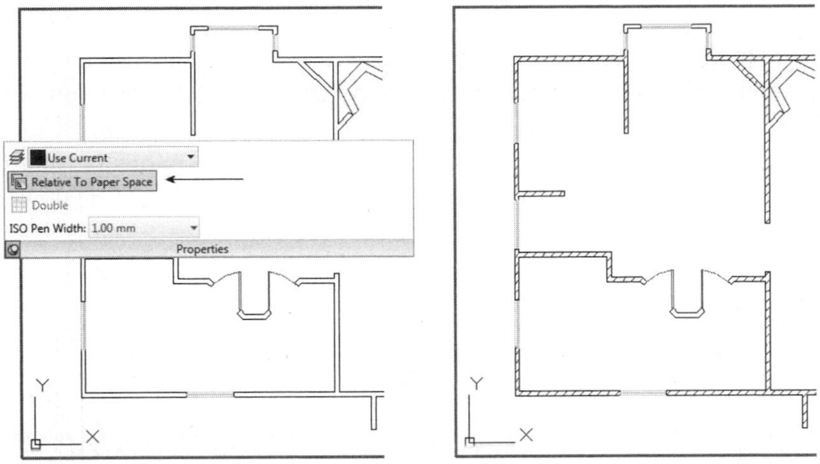

FIGURE 14.57

QUICK VIEW LAYOUTS

The Quick View tool allows you to preview and switch between open drawings and layouts associated with drawings. Two modes are available through the Quick View tool; namely Quick View Layouts and Quick View Drawings. Each tool is activated from the status bar located at the bottom of the display screen, as shown in Figure 14.58.

Clicking on the Quick View Layouts button, as shown in Figure 14.58 (Left), will display the model space and layouts of the current drawing in a row.

Clicking on the Quick View Drawings button, as shown in Figure 14.58 (Right), will display all drawings currently opened.

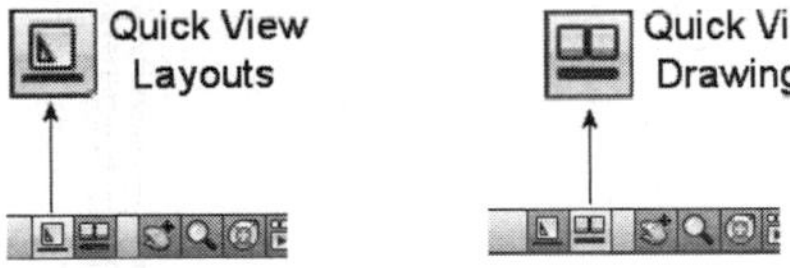

FIGURE 14.58

The following image illustrates various tabs that represent drawing layouts. If you are unsure of the correct layout, moving your cursor over a layout tab, such as HVAC Plan, will preview this layout as, shown in Figure 14.59. If this is the correct layout, pick the tab to launch the layout drawing.

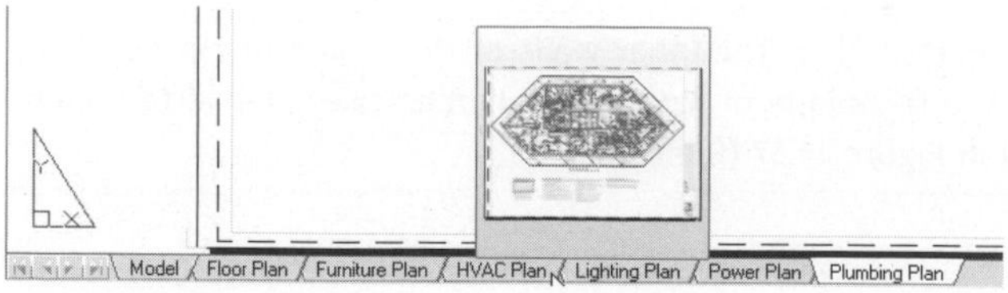

FIGURE 14.59

To preview images of multiple layouts of a drawing, click on the Quick View Layouts button at the bottom of the display screen, as shown in Figure 14.60. Once a number of layouts are displayed, you can click on an image to make this layout current. Additional buttons are displayed on each image that allow you to plot and publish a drawing from this image. These images can even be resized dynamically by holding down the CTRL key as you roll the wheel on a mouse.

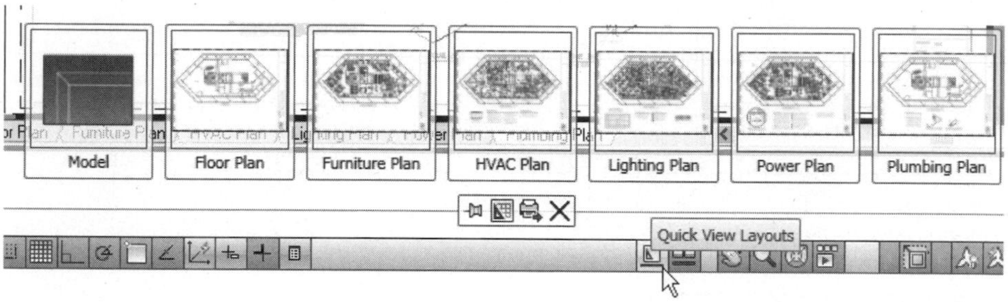

FIGURE 14.60

Figure 14.61 illustrates a special toolbar that is displayed below the Quick View Layout images. These buttons are explained from left to right. The Pin icon allows you to pin images of layouts so they will always be visible even while you are working on a drawing; the New Layout icon creates a layout and displays as a Quick View image at the end of the row; the Publish icon launches the Publish dialog box for the purpose of publishing layouts; and the Close icon closes the Quick View layout images.

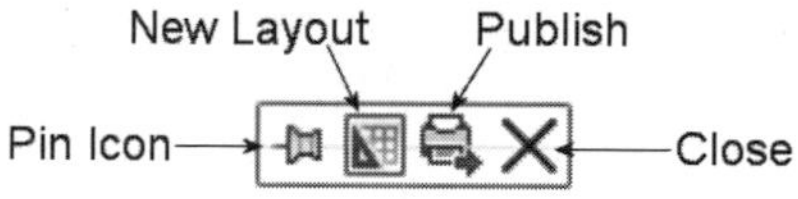

FIGURE 14.61

QUICK VIEW DRAWINGS

As with Quick View layouts, Quick View drawings allow you to display every drawing currently open, as shown in Figure 14.62. The image of the current drawing will appear highlighted.

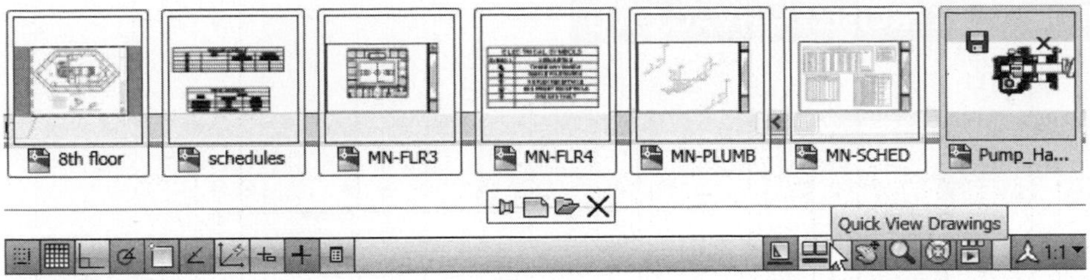

FIGURE 14.62

If you move your cursor over an image of a drawing that contains layouts, all layouts for that drawing are displayed above the Quick View drawing, as shown in Figure 14.63. From there, you can make drawings or layouts current by clicking on the image.

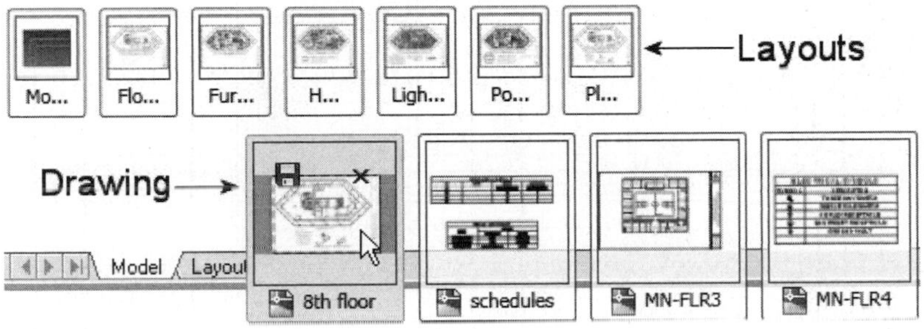

FIGURE 14.63

Figure 14.64 illustrates a special toolbar that is displayed below the Quick View Drawing images. These buttons are explained from left to right. The Pin icon allows you to pin images of drawings so they will always be visible even while you are working on a drawing; the New icon creates a new drawing file and displays the file as a Quick View image at the end of the row; the Open icon launches the Open dialog box for the purpose of opening drawing files; and the Close icon closes the Quick View Drawing images.

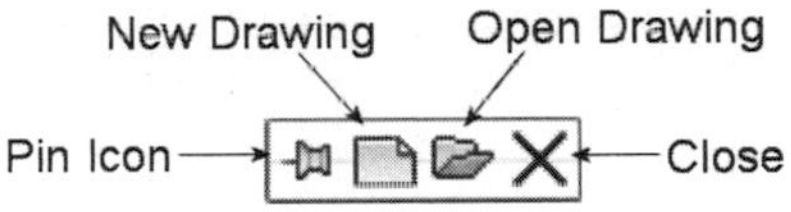

FIGURE 14.64

TUTORIAL EXERCISE: 14_HVAC.DWG

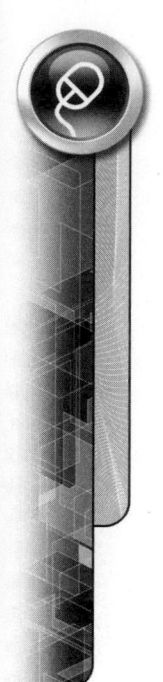

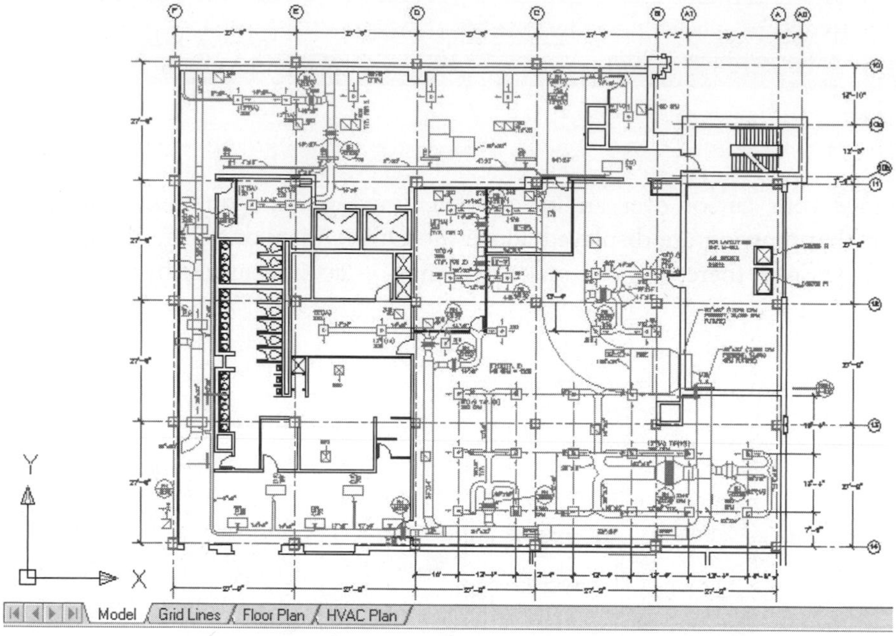

FIGURE 14.65

Purpose

This tutorial is designed to create multiple layouts of the HVAC drawing in Figure 14.65 in Paper Space.

System Settings

All unit, limit, and plotter settings have already been made in this drawing. This exercise utilizes the model and layout tabs. If they are not displayed, right-click on a blank part of the display screen and choose Options from the menu. When the Options dialog box displays, click on the Display tab and under the Layout Elements heading, place a check in the box next to Display Layout and Model tabs.

Layers

Layers have already been created for this exercise.

Suggested Commands

You will begin by opening the drawing file 14_HVAC Dwg. Layout1 has already been created for you, An architectural title block has been inserted into Paper Space and the image has already been scaled to 1/8" = 1'-0". This layout, which will be renamed Grid Lines, will be modified by freezing the layers that pertain to the floor and HVAC. This is accomplished while in floating Model Space such that the layers are only frozen in that viewport. From this layout, create another layout called Floor Plan. While in floating Model Space, freeze the layers that pertain to the grid lines and HVAC plans. From this layout, create another layout called HVAC Plan. While in floating Model Space, freeze the layers that pertain to the grid lines. Turn off all viewports and edit drawing titles for each layout.

STEP 1

Illustrated in Figure 14.66 is a layout of an HVAC plan (heating, venting, and air conditioning). A viewport already exists and the image inside the viewport has been scaled to a value of 1/8" = 1'-0". A border also exists in this layout. The goal is to create two extra layouts that show different aspects of the HVAC plan. One layout will display only the grid pattern used to lay out the plan. Another layout will show just the floor plan information. The third layout will show the HVAC and floor plans together.

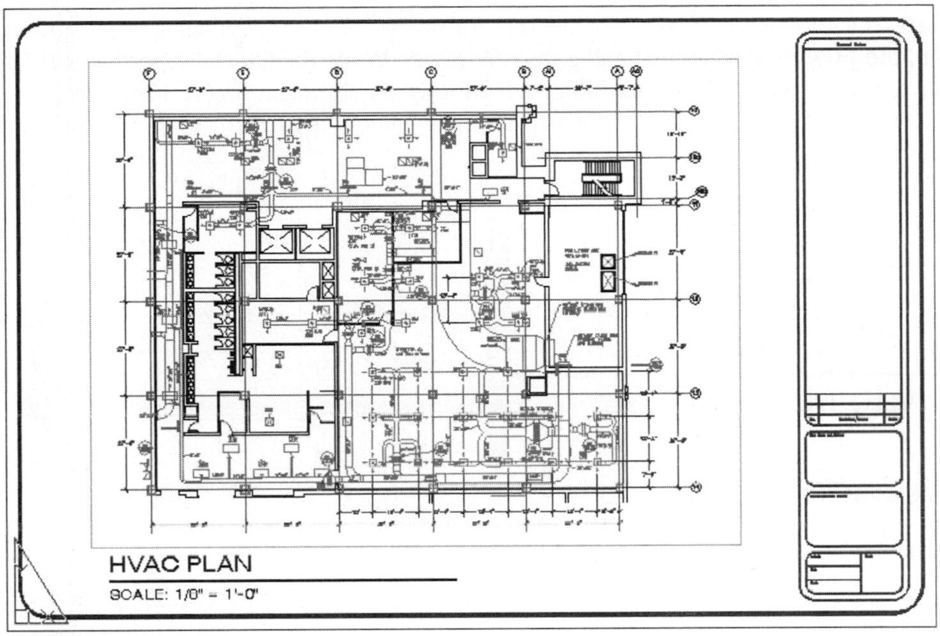

FIGURE 14.66

STEP 2

Before you start to create the new layouts, the viewport present in the existing layout needs to be locked. This will prevent any accidental zooming in and out while inside floating Model Space. To lock a viewport, click the edge of the viewport and pick the Lock/Unlock Viewport icon shown in Figure 14.67. When the viewport is locked, you cannot change the Viewport Scale. Another method used for locking a viewport is to click the edge of the viewport, right-click, and choose Display Locked followed by Yes from the menu, Locking viewports is good practice and should be performed before creating any extra layouts. Now that this viewport is locked, the copied viewports created in other layouts will also be locked.

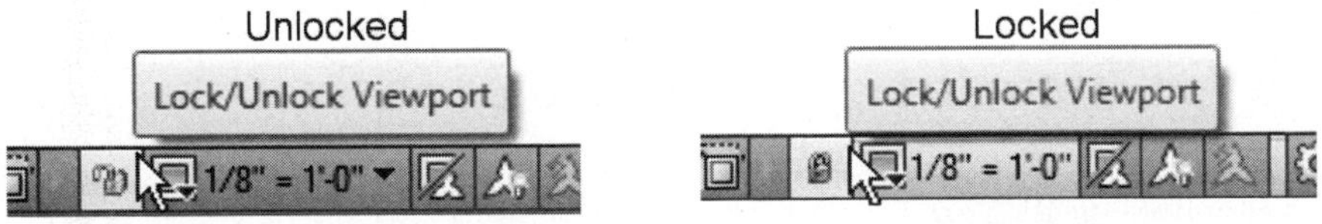

FIGURE 14.67

STEP 3

Next, double-click on the Layout1 tab, located in the lower-left corner of the display screen. Rename this layout to Grid Lines, as shown in Figure 14.68. It is always good practice to give your layouts meaningful names. Press ENTER to accept the change.

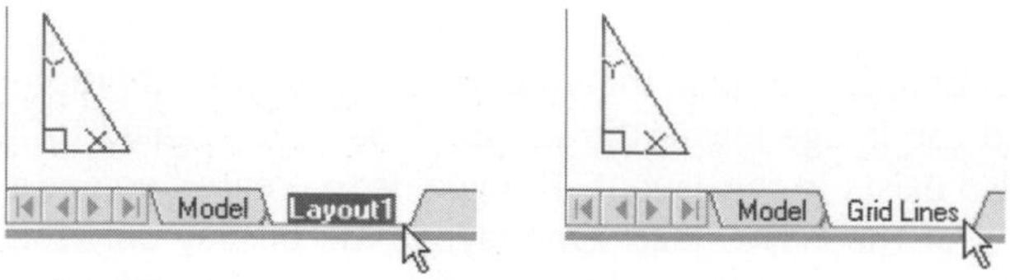

FIGURE 14.68

STEP 4

Activate the Quick Properties tool on the Status Bar. Click the drawing title (HVAC PLAN); the title should highlight and grips will appear. When the Quick Properties palette displays, change HVAC PLAN to GRID LINES PLAN, as shown in Figure 14.69. Pressing ENTER automatically updates the text to the new value. Press ESC to remove the grip and dismiss the Quick Properties palette.

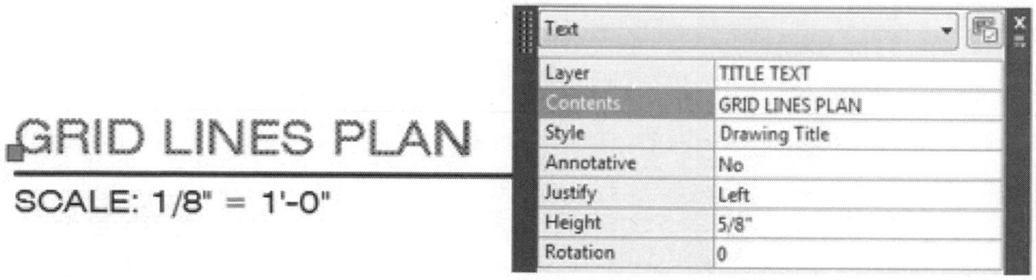

FIGURE 14.69

STEP 5

Prepare to create the second layout by first holding down the CTRL key while pressing and dragging the Grid Lines tab, as shown in Figure 14.70 (Left). Release the mouse button and the new layout, Grid Lines (2), is created, as shown in Figure 14.70 (Right).

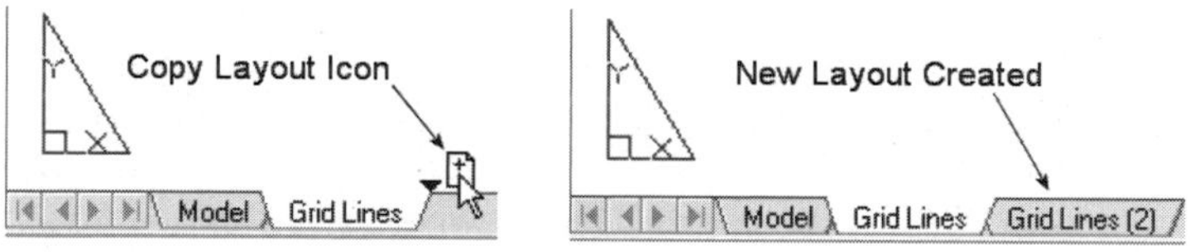

FIGURE 14.70

STEP 6

Before continuing, double-click the new layout Grid Lines (2), as shown in Figure 14.71 (Left). Change this current layout name from Grid Lines (2) to the new name Floor Plan, as shown in Figure 14.71 (Right).

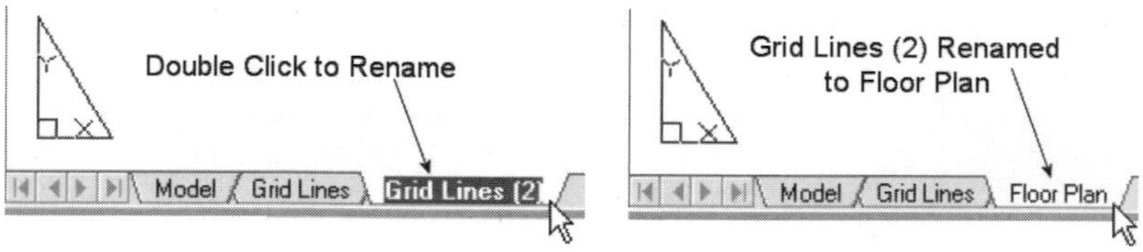

FIGURE 14.71

STEP 7

Use the method described in previous steps to help create one final layout. Create a copy of the Floor Plan layout and change its name to HVAC Plan. When you are finished creating and renaming all the layouts, the lower-left corner of your display should appear similar to Figure 14.72.

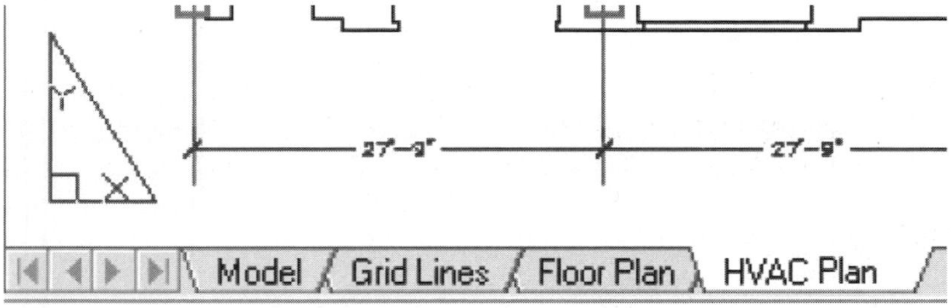

FIGURE 14.72

STEP 8

Click the Grid Lines tab. Double-click anywhere inside the viewport, as shown in Figure 14.73; this places you in floating Model Space. You will need to turn off all layers that deal with the floor plan and HVAC, which leave only the layers with the grid lines visible.

Activate the Layer control box, as shown in Figure 14.73, and freeze the layers DOORS, FLOOR, HVAC DIM, and HVAC SUP in the current viewport by clicking the appropriate icon. The frozen layers pertain to the floor plan and HVAC plan.

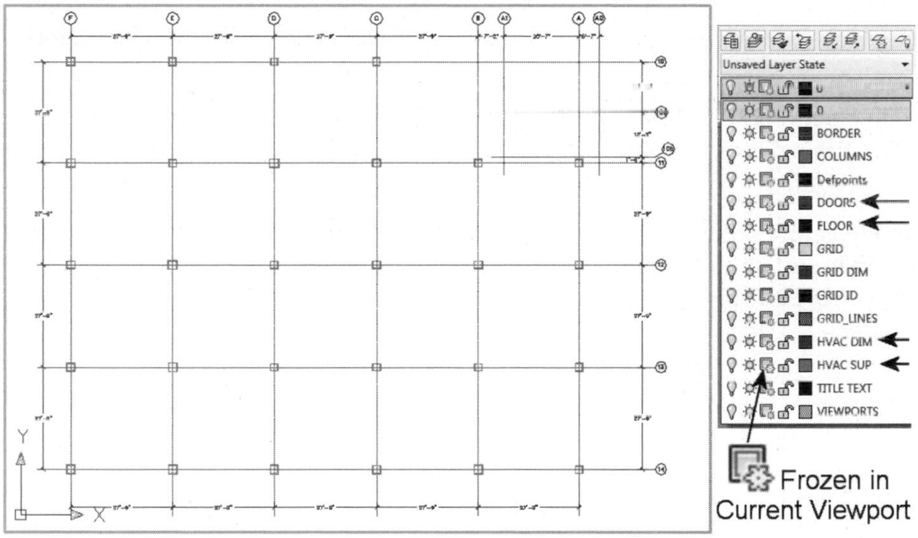

FIGURE 14.73

Double-click anywhere outside the viewport to return to Paper Space. Your drawing should appear similar to Figure 14.74. Notice that only the columns and grid lines appear visible in this layout.

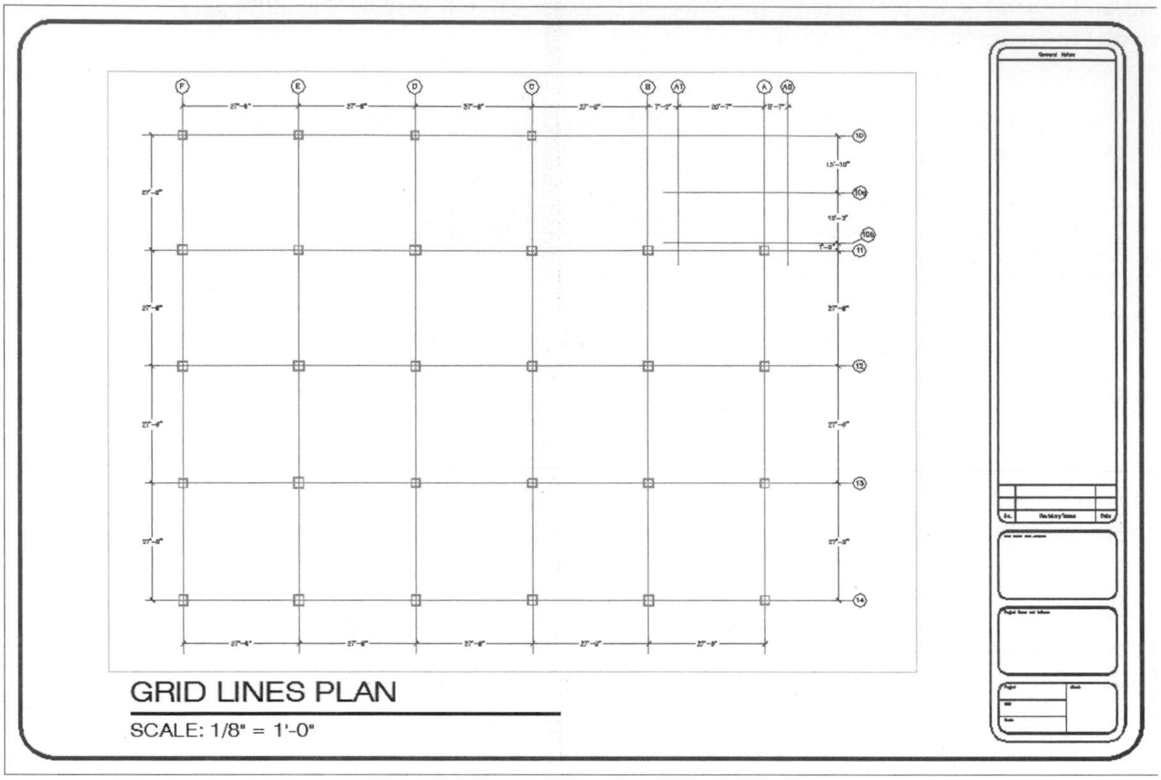

FIGURE 14.74

STEP 9

Click the Floor Plan tab. Double-click anywhere inside the viewport, as shown in Figure 14.75; this places you in floating Model Space. You will need to turn off all layers that deal with the grid lines and HVAC.

Activate the Layer control box, shown in Figure 14.75, and freeze the layers in the current viewport that pertain to the grid lines plan and HVAC plan (DOORS, GRID, GRID DIM, GRID ID, GRID_LINES, HVAC DIM, and HVAC SUP).

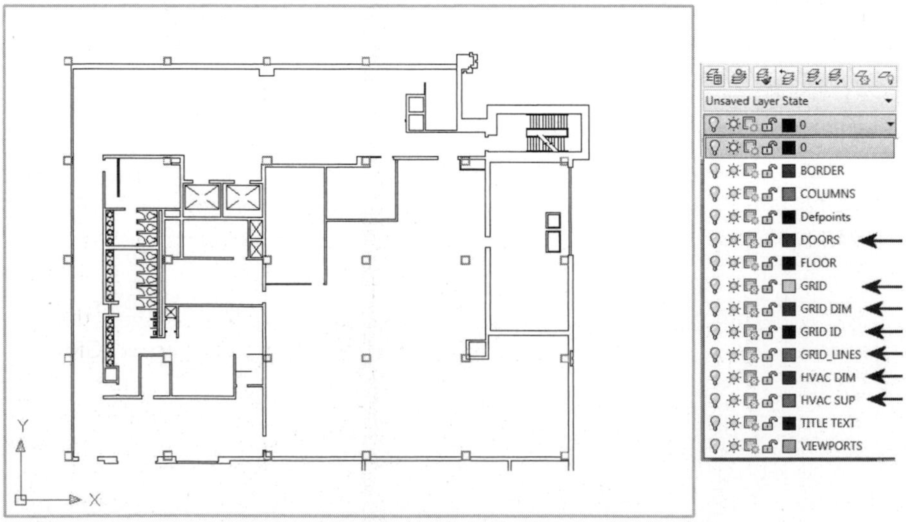

FIGURE 14.75

Double-click anywhere outside the viewport to return to Paper Space. Use the Quick Properties palette to change the drawing title from GRID LINES PLAN to FLOOR PLAN, as shown in Figure 14.76. Do this by first selecting GRID LINES PLAN, which will launch the Quick Properties palette allowing you to change the content of the text. When finished, press ESC to remove the palette and grip.

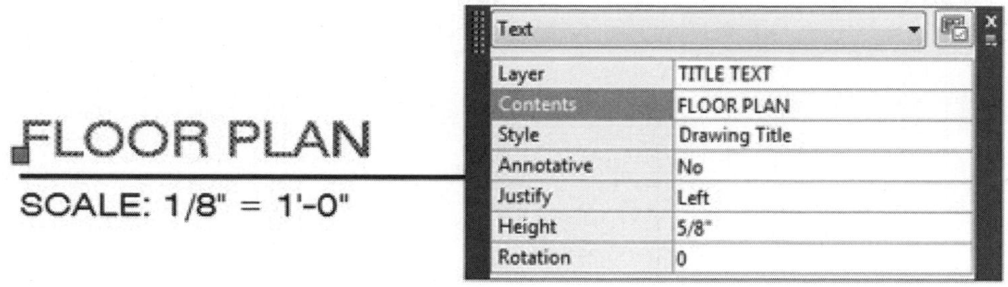

FIGURE 14.76

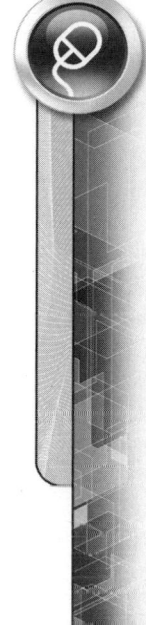

Your drawing now displays only floor plan information, as shown in Figure 14.77, while the HVAC and GRID layers are frozen only in this viewport.

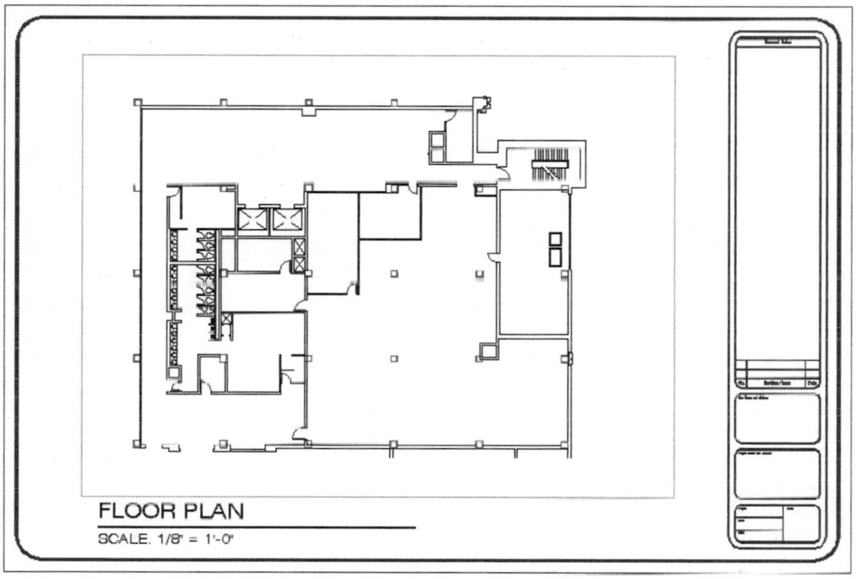

FIGURE 14.77

STEP 10

Click the HVAC Plan tab. Double-click anywhere inside the viewport, as shown in Figure 14.78; this places you in floating Model Space. You will need to turn off all layers that deal with the grid lines plan.

Activate the Layer control box, as shown in Figure 14.78, and freeze the layers in the current viewport that pertain to the grid lines plan (COLUMNS, GRID, GRID DIM, GRID ID, and GRID_LINES).

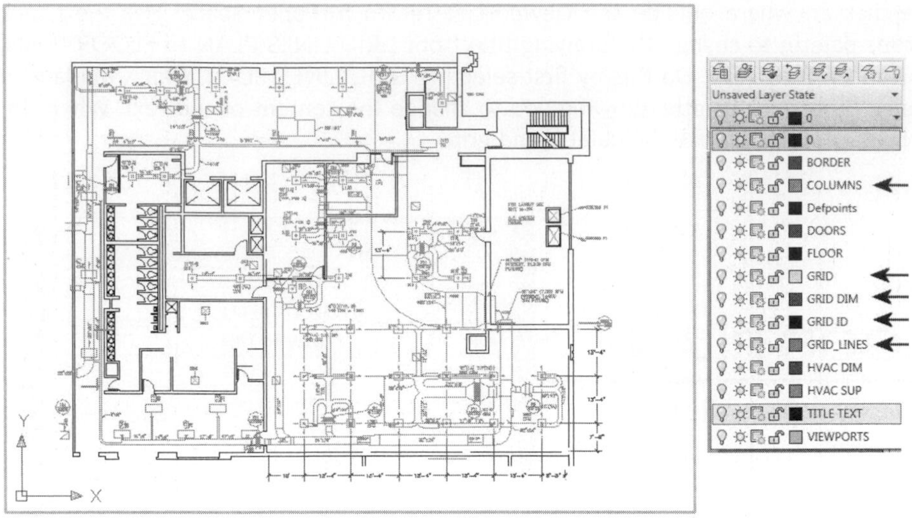

FIGURE 14.78

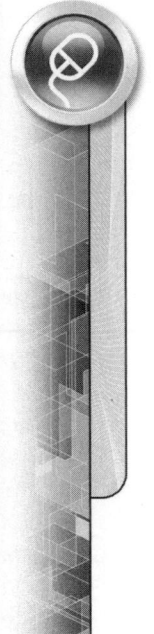

Double-click anywhere outside the viewport to return to Paper Space. Use the Quick Properties palette to change the drawing title from GRID LINES PLAN, as shown in Figure 14.79, to HVAC PLAN. When finished, press ESC to remove the palette and grip.

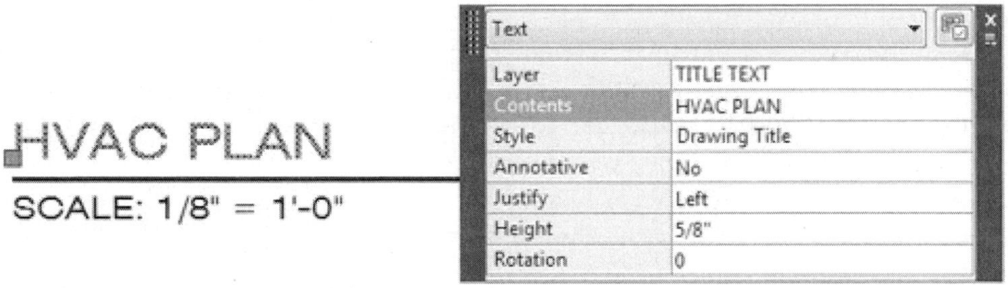

FIGURE 14.79

Your drawing should appear similar to Figure 14.80. In this viewport, you see the floor plan and HVAC ductwork but no grid lines or columns.

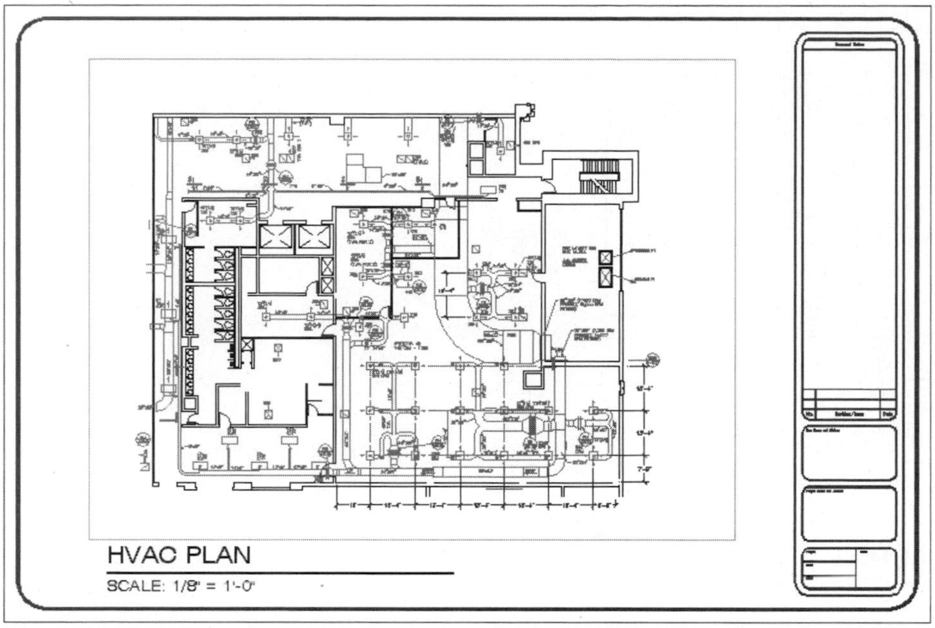

FIGURE 14.80

STEP 11

One additional layer control technique needs to be performed. To better distinguish the objects that represent the HVAC Plan, the floor plan layer needs to be changed to a different color. This change in color must only occur in the current viewport and not affect any other viewports or even objects in Model Space. Double-click inside of this viewport and launch the Layer Properties Manager palette. Click the Floor layer and change the color under the VP Color column to 9, as shown in Figure 14.81. This color represents light gray and affects only the Floor Plan layer. Close the dialog box.

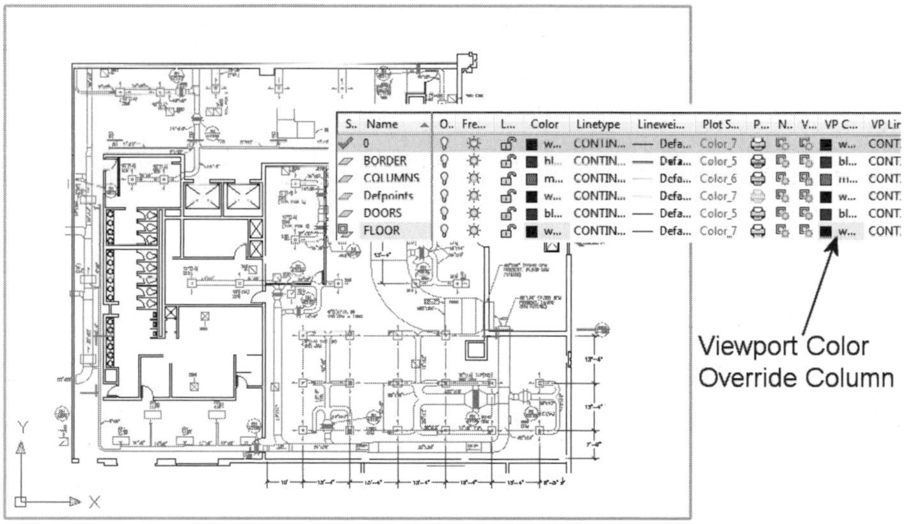

FIGURE 14.81

The results are illustrated in Figure 14.82. Notice how the HVAC objects stand out compared to the floor plan objects.

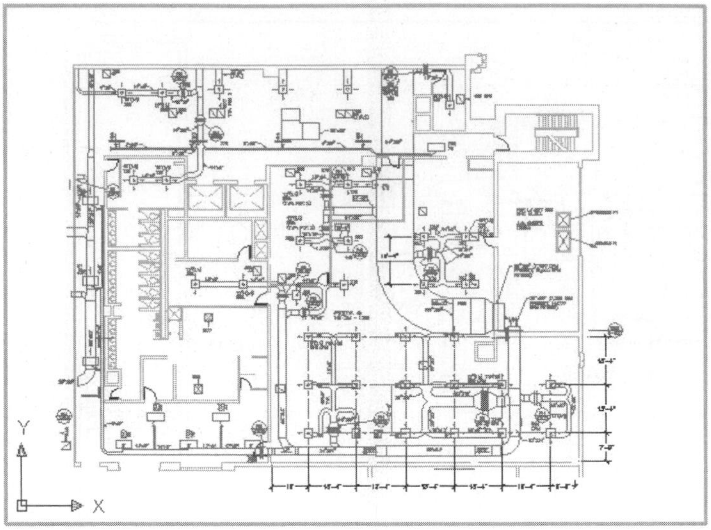

FIGURE 14.82

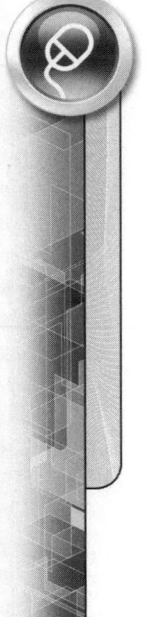

STEP 12

In the event you need to change the color of the Floor layer back to its original color assignment, activate the Layer Properties Manager palette and right-click Color 9 under the VP Color column to display the menu, as shown in Figure 14.83. From this menu, click Remove Viewport Overrides for followed by Color and then In Current Viewport only. This changes the color of the Floor layer back to its original color assignment.

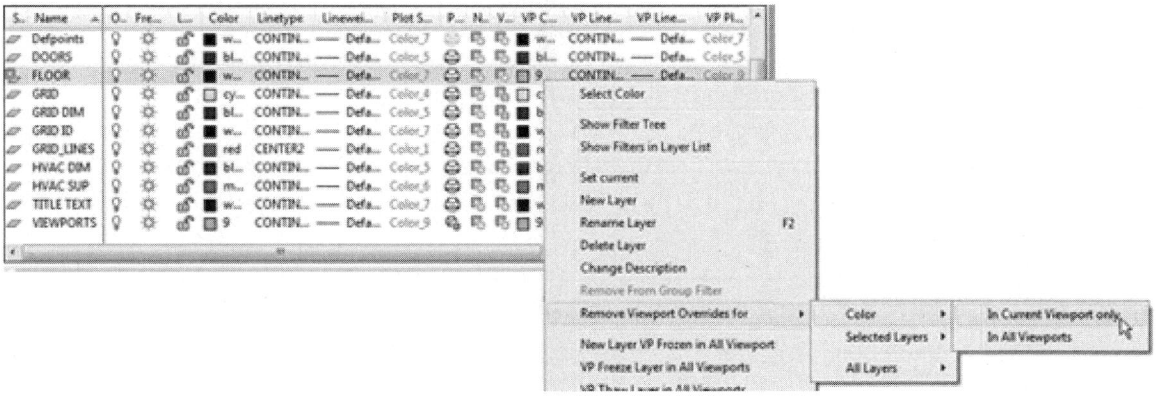

FIGURE 14.83

STEP 13

Turn off the Viewports layer or set it to No Plot. The completed drawing is displayed in Figure 14.84. This exercise illustrated how to create multiple layouts. It also illustrated how to freeze layers in one viewport and have the same layers visible in other viewports. If the Viewports layer is set to No Plot through the Layer Properties Manager palette, performing a plot preview will not display the Viewports layer.

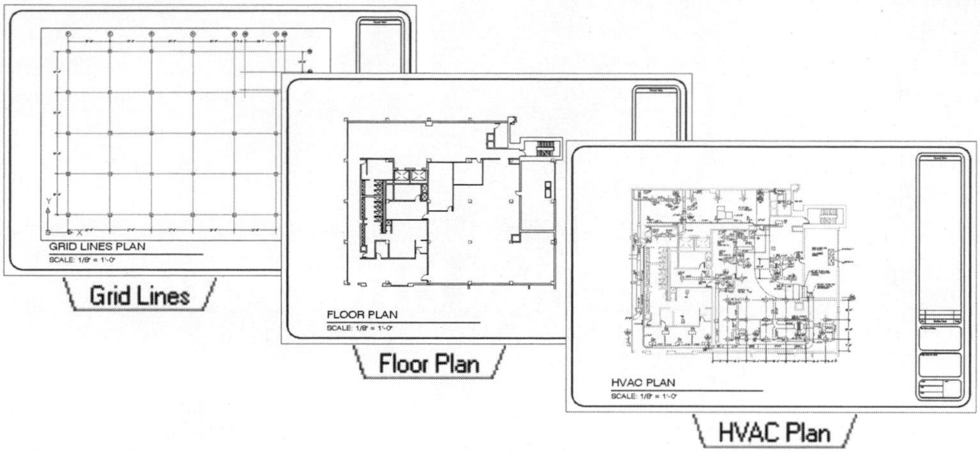
GRID LINES PLAN
SCALE: 1/8" = 1'-0"
Grid Lines
FLOOR PLAN
SCALE: 1/8" = 1'-0"
Floor Plan
HVAC PLAN
SCALE: 1/8" = 1'-0"
HVAC Plan

FIGURE 14.84

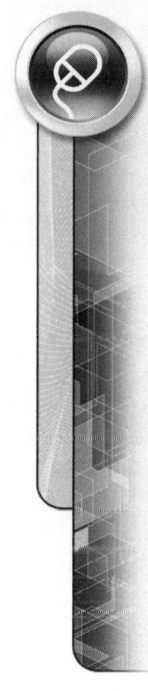

CHAPTER

15

Plotting Your Drawings

This chapter discusses plotting through a series of tutorial exercises designed to perform the following tasks:

- Configure a new plotter
- Plot from a drawing layout (Paper Space)
- Control lineweights
- Create a Color-Dependent plot style table
- Publish multiple drawing sheets

CONFIGURING A PLOTTER

Before plotting, you must first establish communication between AutoCAD and the plotter. This is called configuring. From a list of supported plotting devices, you choose the device that matches the model of plotter you own. This plotter becomes part of the software database, which allows you to choose this plotter at any time. If you have more than one output device, each device must be configured before being used. The following tutorial exercise demonstrates the configuration process used in AutoCAD.

TUTORIAL EXERCISE: 15_CONFIGURING A PLOTTER

Purpose

Use the Add-A-Plotter wizard to configure a plotter.

STEP 1

Begin the plotter configuration process by choosing Manager Plotters from the Print heading of the Application Menu, as shown in Figure 15.1 (Left). This activates the Plotters program group, as shown in Figure 15.1 (Right), which lists all valid plotters that are currently configured. The listing in this image displays the default plotters configured after the software is loaded. Except for the DWF devices, which allow you to publish a drawing for viewing over the Internet, or a popular DWG to PDF device that allows you to create a PDF (Adobe) document directly from an AutoCAD DWG file, a plotter has not yet been configured. Double-click the Add-A-Plotter Wizard icon to continue with the configuration process.

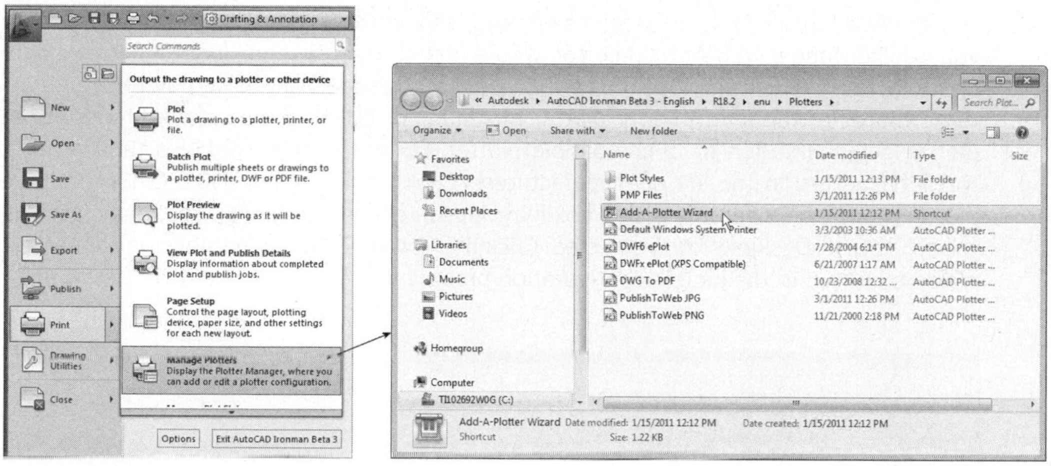

FIGURE 15.1

On your screen, the Window displaying the Plotters program group may appear different from the one shown in Figure 15.1, depending on the operating system you are using.	**NOTE**

STEP 2

Double-clicking the Add-A-Plotter Wizard icon displays the Add Plotter – Introduction Page dialog box, as shown in Figure 15.2. This dialog box states that you are about to configure a Windows or non-Windows system plotter. This configuration information will be saved in a file with the extension .PC3. Click the Next > button to continue on to the next dialog box.

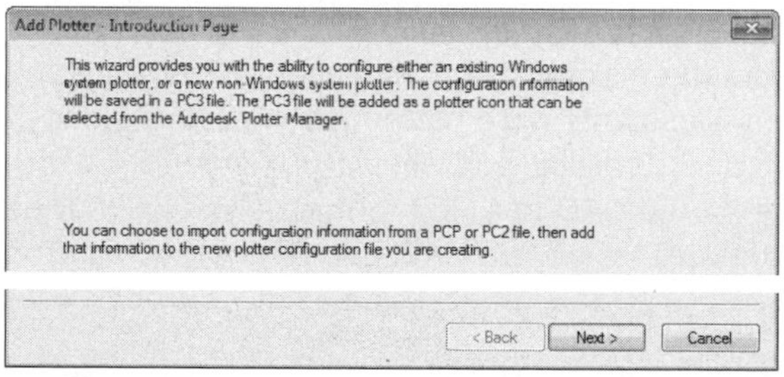

FIGURE 15.2

STEP 3

In the Add Plotter – Begin dialog box shown in Figure 15.3 (Left), decide how the plotter will be controlled by the computer you are currently using, by a network plot server, or by an existing system printer where changes can be made specifically for AutoCAD. Click the radio button next to My Computer. Then click the Next > button to continue on to the next dialog box.

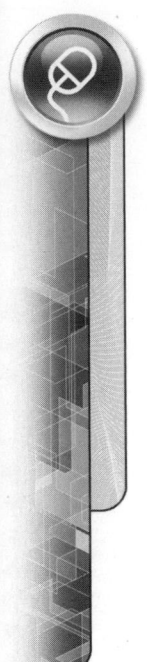

Use the Add Plotter-Plotter Model dialog box, as shown in Figure 15.3 (Right), to associate your plotter model with AutoCAD. You would first choose the appropriate plotter manufacturer from the list provided. Once this is done, all models supported by the manufacturer appear to the right. If your plotter model is not listed, you are told to consult the plotter documentation for a compatible plotter. For the purposes of this tutorial, click Hewlett-Packard in the list of Manufacturers. Click the DesignJet 750C C3196A for the plotter model. A Driver Info dialog box may appear, giving you more directions regarding the type of HP DeskJet plotter selected. Click the Continue button to move on to the next dialog box used in the plotter configuration process.

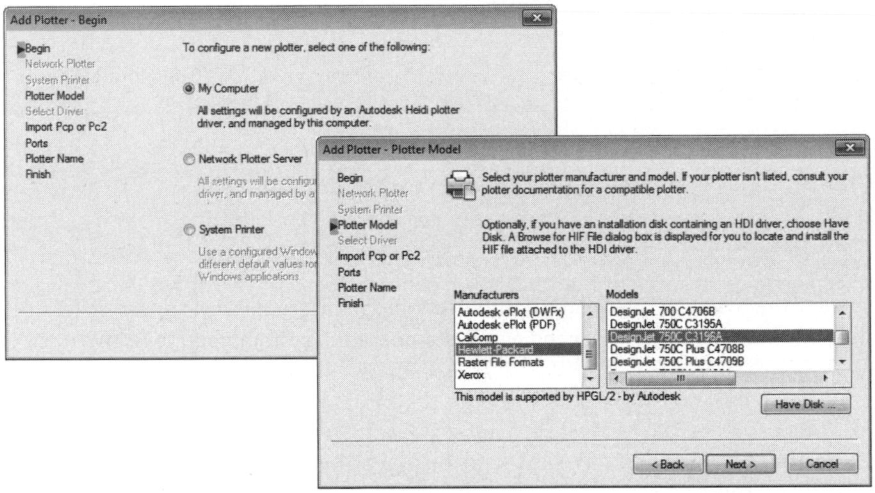

FIGURE 15.3

STEP 4

PCP and PC2 files have been in existence for many years. They were designed to hold plotting information such as pen assignments. In this way, you use the PCP or PC2 files to control pen settings instead of constantly making pen assignments every time you perform a plot; at least this is how pen assignments were performed in past versions of AutoCAD. The Add Plotter-Import PCP or PC2 dialog box, shown in Figure 15.4 (Left), allows you to import those files for use in AutoCAD in a PC3 format. If you will not be using any PCP or PC2 files from previous versions of AutoCAD, click the Next > button to move on to the next dialog box.

In the Add Plotter-Ports dialog box, shown in Figure 15.4 (Right), click the port used for communication between your computer and the plotter. The LPT1 port will be used for the purposes of this tutorial. Select the port from the list and then click the Next > button to continue on to the next dialog box.

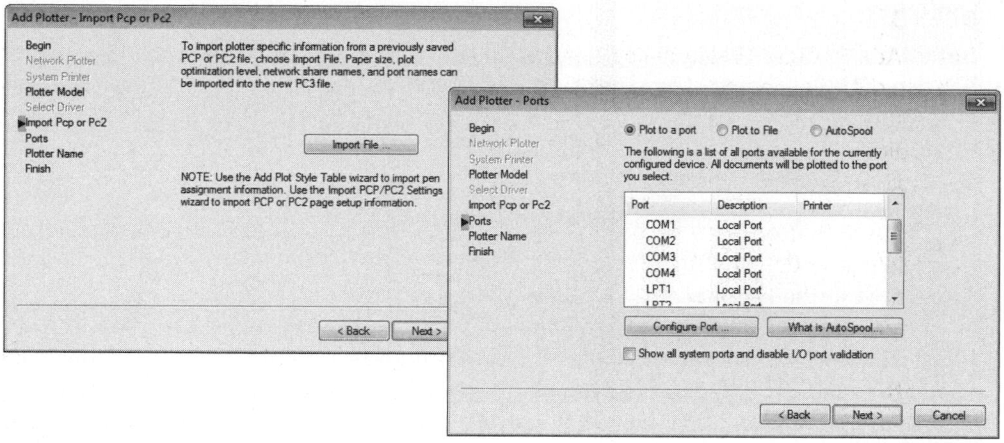

FIGURE 15.4

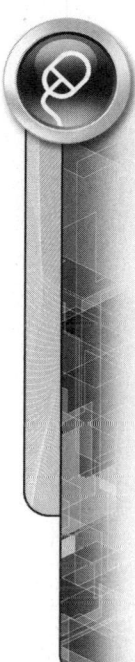

STEP 5

In the Add Plotter – Plotter Name dialog box, shown in Figure 15.5 (Left), you have the option of giving the plotter a name other than the name displayed in the dialog box. For the purpose of this tutorial, accept the name that is given. This name will be displayed whenever you use the Page Setup and Plot dialog boxes.

The last dialog box is displayed in Figure 15.5 (Right). In the Add Plotter – Finish dialog box, you can modify the default settings of the plotter you just configured. You can also test and calibrate the plotter if desired. Click the Finish button to dismiss the Add Plotter – Finish dialog box.

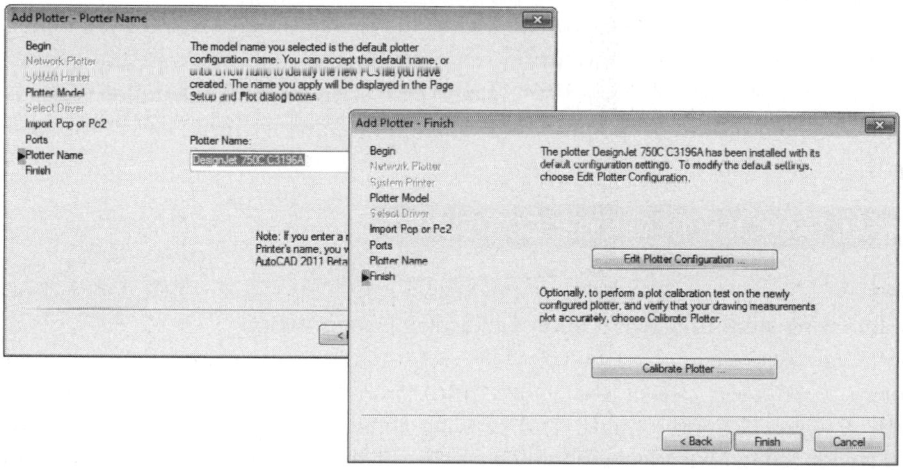

FIGURE 15.5

STEP 6

Exiting the Add Plotter dialog box returns you to the Plotters program group, shown in Figure 15.6. Notice that the icon for the DesignJet 750C plotter has been added to this list. This completes the steps used to configure the DesignJet 750C plotter. Follow these same steps if you need to configure another plotter.

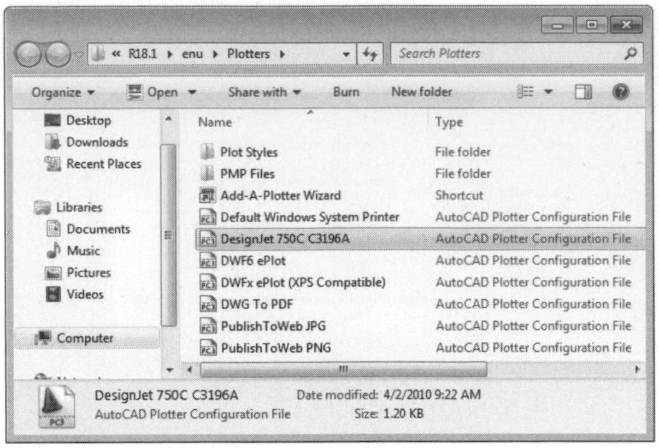

FIGURE 15.6

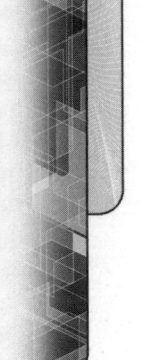

PLOTTING FROM A LAYOUT

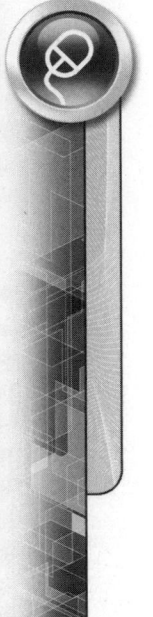

To plot a drawing, select the PLOT command using one of the following methods:

- From the Quick Access toolbar
- From the Application Menu (Print > Plot)
- From the Ribbon > Output Tab > Plot Panel
- From the Menu Bar (File > Plot)
- From the Standard toolbar of the AutoCAD Classic workspace
- From the keyboard (PLOT or CTRL + P)

Prior to issuing the PLOT command, select the Layout tab to enter the Paper Space environment. Verify that the title block information is correct, general notes are complete, and all viewports are scaled properly. Entering the command will show a dialog box, where you will provide information required for the plot, such as: the printer name, paper size, plot area, plot scale (1:1), and plot style table. A tutorial exercise is provided to step you through the process of plotting a drawing from Layout mode.

TUTORIAL EXERCISE: 15_CENTER GUIDE.DWG

Purpose

Use the following steps to plot 15_Center Guide.dwg from a layout.

STEP 1

Open the drawing 15_Center Guide. This drawing should already be laid out in Paper Space. A layout called Four Views should be present at the bottom of the screen next to the Model tab.

STEP 2

Begin the process of plotting this drawing by choosing the PLOT command from the Quick Access toolbar or Application Menu, as shown in Figure 15.7.

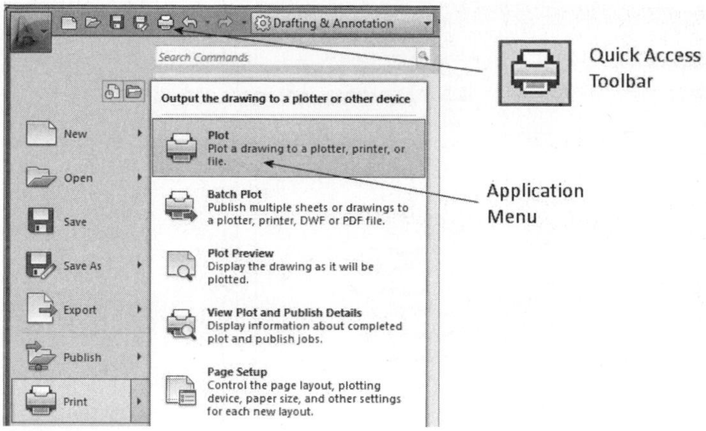

FIGURE 15.7

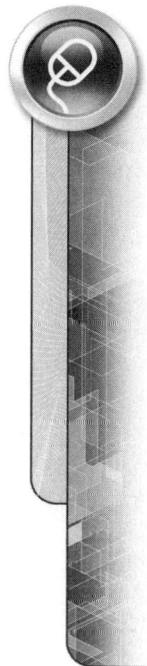

This activates the Plot dialog box, as shown in Figure 15.8. For the purposes of this tutorial, the DWF6 ePlot is being used as the output device. You can opt to create additional copies of your plot; by default this value is set to 1. The DWF6 ePlot can plot only one copy, so this value is grayed out. Next, make sure the paper size is currently set to ANSI expand C (22.00 × 17.00 inches).

In the Plot area, the Layout radio button is selected. Since you created a layout, this is the obvious choice. The Extents mode allows you to plot the drawing based on all objects that make up the drawing. The Extents mode works well as long as you don't draw outside the title block border. Plotting the Display plots your current drawing view, but be careful: If you are currently zoomed in to your drawing, plotting the Display will plot only this view. In this case, it would be more practical to use Layout or Extents to plot. When you plot a layout in Paper Space, the Plot scale will be set to 1:1. Since you pre-scaled the drawing to the Paper Space viewport using the Viewport Scale box on the Status Bar, all drawings in Paper Space are designed to be plotted at this scale. The Plot offset is designed to move or shift the location of your plot on the paper if it appears off center.

Click the More Options button in the lower-right corner of the dialog box to view additional settings. In the Drawing orientation area, make sure that the radio button adjacent to Landscape is selected. You could also plot the drawing out in Portrait mode, where the short edge of the paper is the top of the page. For special plots, you could even plot the drawing upside down. In the Plot options area, you have more control over plots by applying lineweights, plot transparency, using existing plot styles, plotting Paper Space last (Model Space first), or even hiding objects. The Save changes to layout option should be checked, if you want to save the current settings for future plots.

One other area to change is in the Plot style table area, located in the upper-right corner of the dialog box. This area controls the appearance of the plot, for instance, a colored plot versus a monochromatic plot (black lines on white paper). You can even create your own plot style, which will be discussed later in this chapter. For the purpose of this tutorial, monochrome will be used, as shown in Figure 15.8.

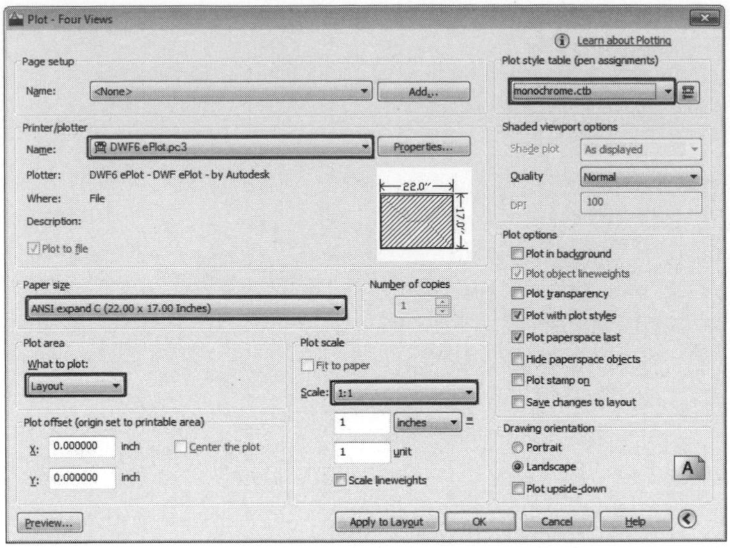

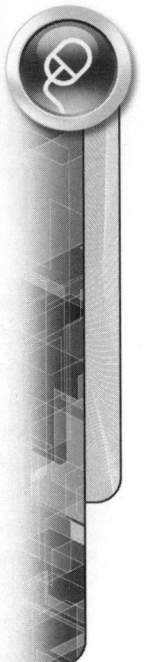

FIGURE 15.8

STEP 3

You should always preview your plot before sending the plot information to the plotter. In this way, you can determine whether the entire drawing will plot based on the sheet size (this includes the border and title block). Clicking the Preview. . . button in the lower-left corner of the dialog box activates the image shown in Figure 15.9. The sheet size is shown along with the border and four-view drawing. Right-clicking anywhere on this preview image displays the cursor menu, allowing you to perform various display functions such as ZOOM and PAN to assist with the verification process. If everything appears satisfactory, click the Plot option to send the drawing information to the plotter. Clicking the Exit option returns you to the Plot dialog box, where you can make any necessary changes.

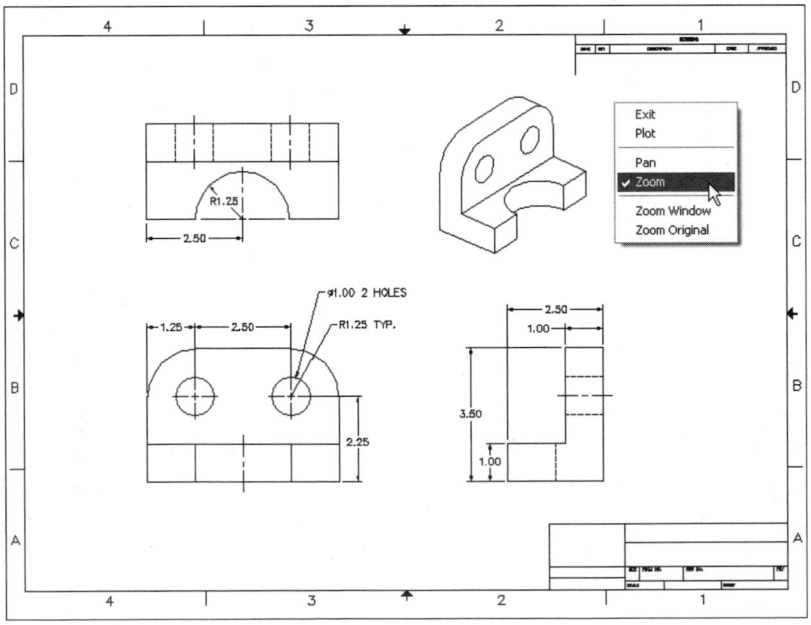

FIGURE 15.9

NOTE

You can also plot from Model Space. However, this is typically more involved because you must bring borders, title blocks, and notes into Model Space utilizing a scale that is determined by a scale factor.

ENHANCING YOUR PLOTS WITH LINEWEIGHTS

This section on plotting describes the process of assigning lineweights to objects and then having the lineweights appear in the finished plot. Follow the next series of steps to assign lineweights to a drawing before it is plotted.

TUTORIAL EXERCISE: 15_V_STEP.DWG

Purpose

This tutorial is designed to show the methods to properly display and plot lineweights on a drawing.

STEP 1

Open the drawing called 15_V_Step.Dwg, shown in Figure 15.10, and notice that you are currently in Model Space (the Model tab is current at the bottom of the screen).

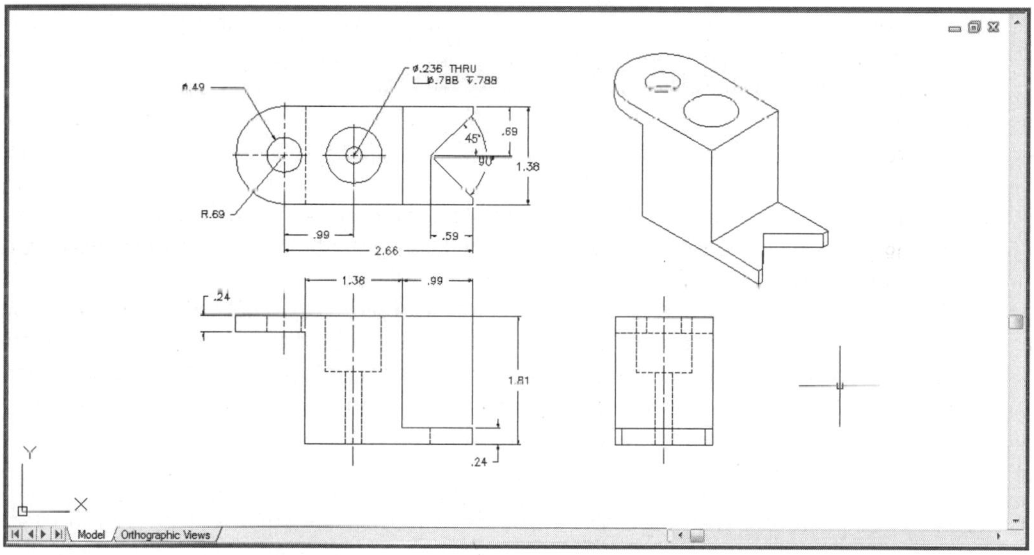

FIGURE 15.10

STEP 2

From the illustration of the drawing in Figure 15.10, all lines on the Object layer need to be assigned a lineweight of 0.50 mm. There is also a title block that will be used with this drawing. The Title Block layer needs a lineweight assignment of 0.80 mm. Click in the Layer Properties Manager palette, shown in Figure 15.11, and make these lineweight assignments. When you are finished, click OK to save the lineweight assignments and return to Model Space.

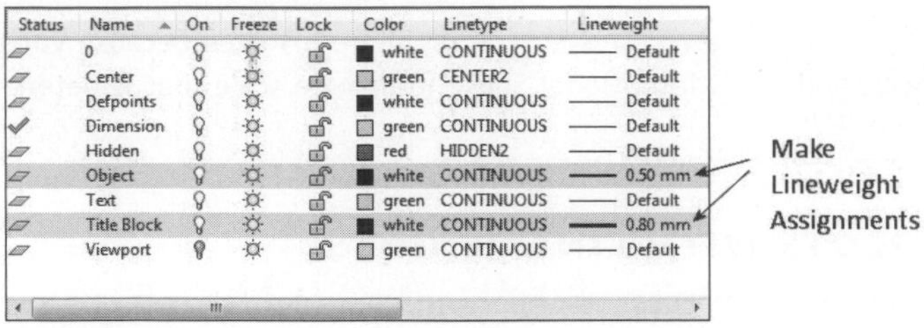

Status	Name	On	Freeze	Lock	Color	Linetype	Lineweight
	0				white	CONTINUOUS	Default
	Center				green	CENTER2	Default
	Defpoints				white	CONTINUOUS	Default
✓	Dimension				green	CONTINUOUS	Default
	Hidden				red	HIDDEN2	Default
	Object				white	CONTINUOUS	0.50 mm
	Text				green	CONTINUOUS	Default
	Title Block				white	CONTINUOUS	0.80 mm
	Viewport				green	CONTINUOUS	Default

FIGURE 15.11

STEP 3

Click the LWT button in the Status bar to display the lineweights, as shown in Figure 15.12 (Right).

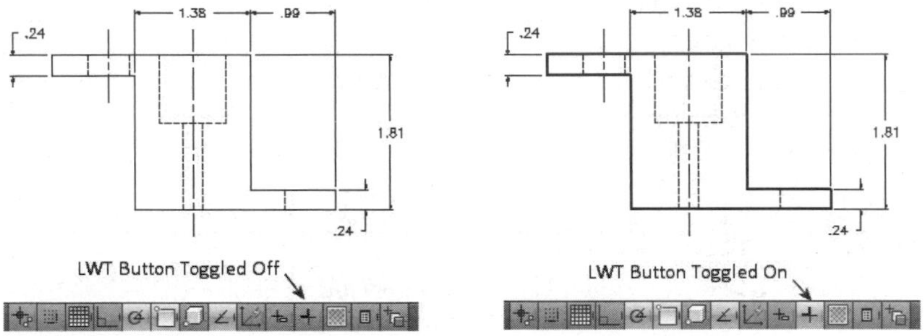

FIGURE 15.12

Right-click the LWT button on the Status bar to display the shortcut menu, as shown in Figure 15.12 (Left). Click Settings... to display the Lineweight Settings dialog box, as shown in Figure 15.12 (Right). Notice the default lineweight is set to 0.25 mm. This value creates a "thin" line and is assigned to the Dimension, Hidden, Text, and Viewport layers. By setting the Object layer's lineweight to 0.50 mm, we doubled the weight and all objects assigned to this layer will plot with a "thick" lineweight. The Title Block layer was assigned a "very thick" (0.80 mm) lineweight. Properly assigned lineweights makes a drawing easier to interpret by making the object stand out on the sheet.

The drawing plot will utilize the assigned lineweights; however, to display weights on the screen you must toggle on the LWT button. If the weights are still not displayed satisfactorily, adjust the slider bar in the area called Adjust Display Scale. This controls only the way lineweights display on the screen—not plot. Click the OK button to return to the drawing and observe the results.

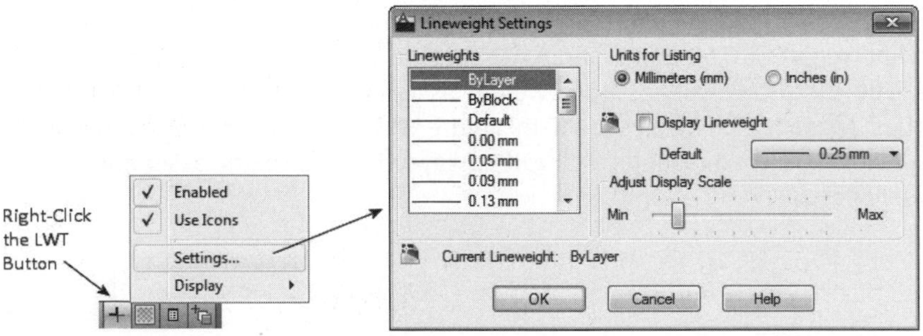

FIGURE 15.13

STEP 4

Clicking the Orthographic Views tab switches you to Paper Space, as shown in Figure 15.14. The lineweights do not appear in Paper Space at first glance. Zooming in to the drawing displays all lineweights at their proper assigned widths.

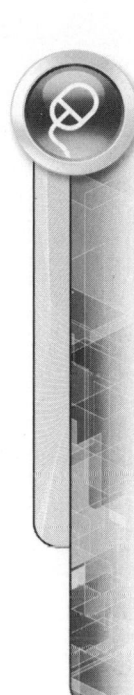

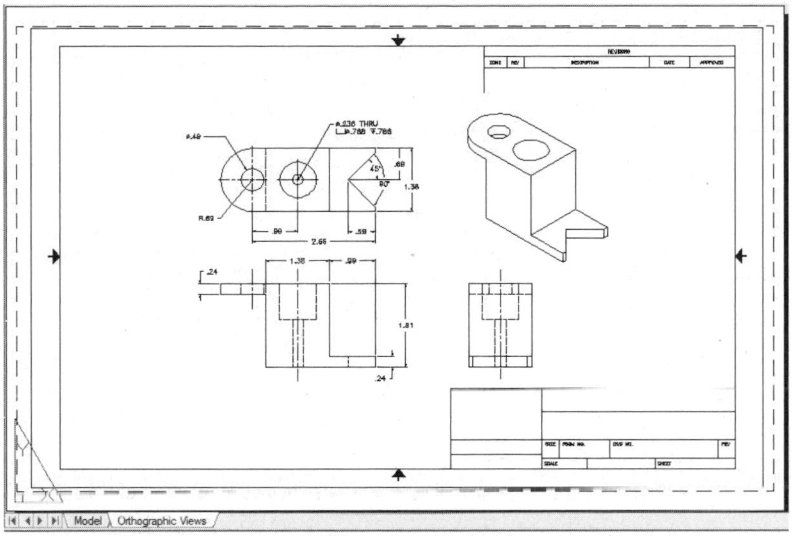

FIGURE 15.14

STEP 5

Activate the Plot dialog box, as shown in Figure 15.15. Verify that the current plot device is the DWF6 ePlot. If this is not the device, click the Name drop-down list to activate this plotter.

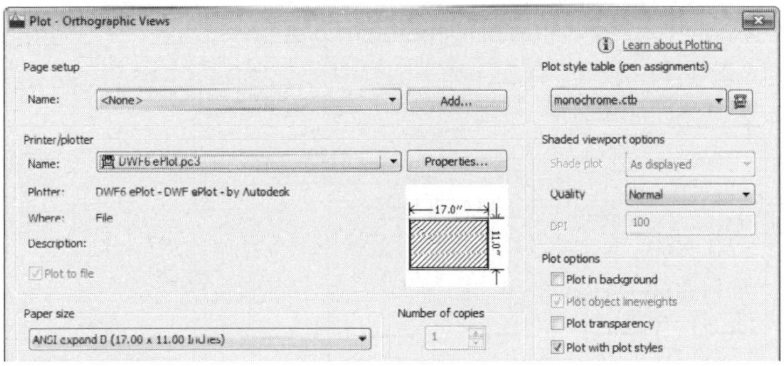

FIGURE 15.15

STEP 6

Click the Preview... button; your display should appear similar to Figure 15.16. Notice that the viewport is not present in the plot preview. The Viewports layer was either turned off or set to a non-plot state inside the Layer Properties Manager palette. To view the lineweights in Preview mode, zoom in to segments of your drawing.

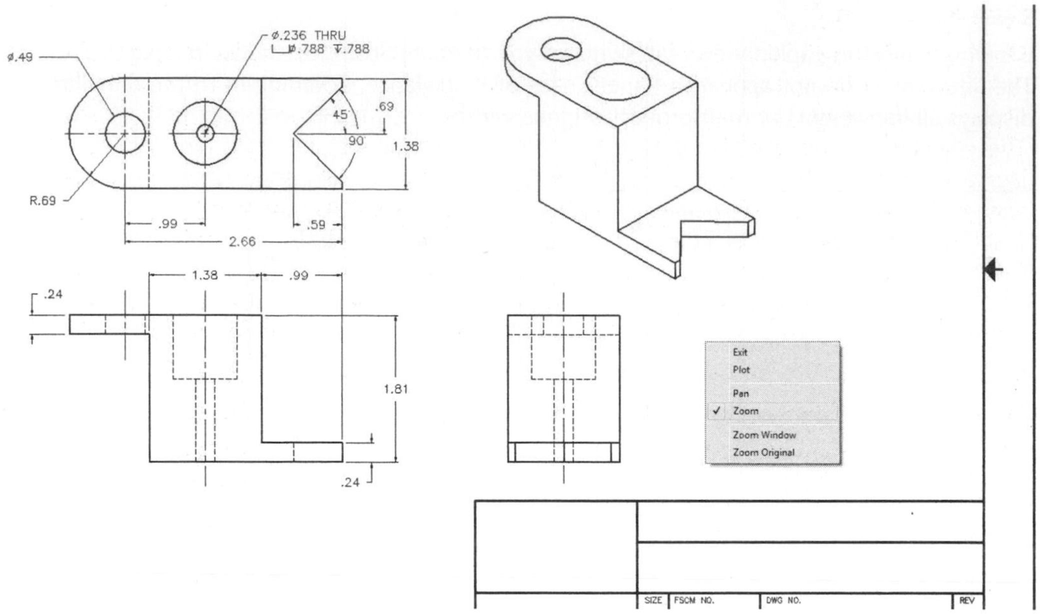

FIGURE 15.16

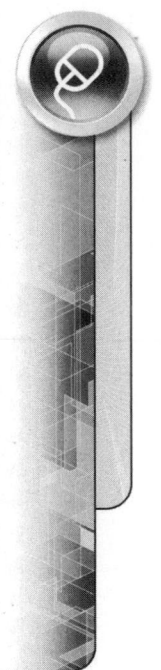

CREATING A COLOR-DEPENDENT PLOT STYLE TABLE

This section of the chapter is devoted to the creation of a Color-Dependent plot style table. Once the table is created, it will be applied to a drawing. From there, the drawing will be previewed to see how this type of plot style table affects the final plot.

TUTORIAL EXERCISE: 15_COLOR_R-GUIDE.DWG

Purpose

This tutorial demonstrates how to create a Color-Dependent plot style.

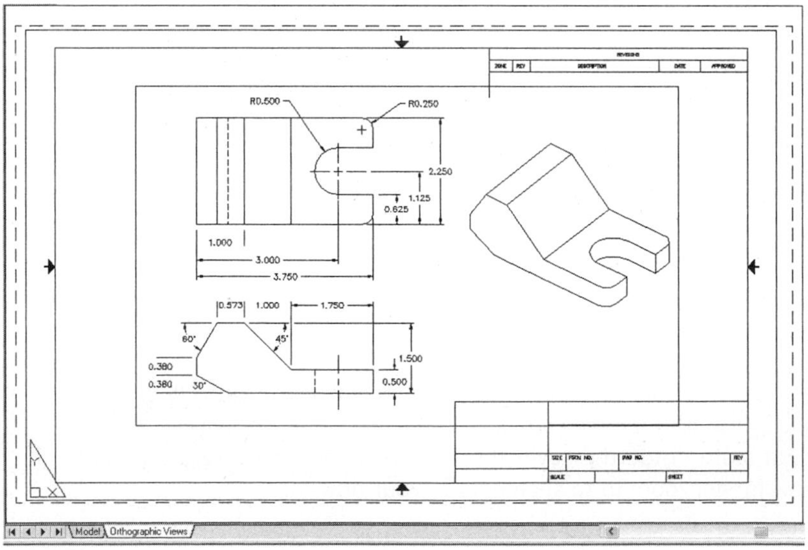

FIGURE 15.17

System Settings

Open the drawing 15_Color_R-Guide.Dwg. Your display should appear similar to Figure 15.17. A two-view drawing together with an isometric view is arranged in a layout called Orthographic Views. The drawing is also organized by layer names and color assignments. The objective of this exercise is to create a Color-Dependent plot style table where all layers will plot out black. Also, through the Color-Dependent plot style table, the hidden lines will be assigned a lineweight of 0.30 mm, object lines 0.70 mm, and the title block 0.80 mm, as shown in the following table. Follow the next series of steps to perform this task.

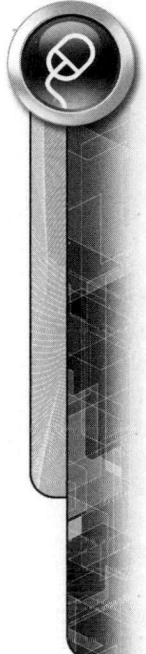

Color	Layer	Lineweight
Red	Hidden	0.30
Yellow	Center	Default
Green	Viewports	Default
Cyan	Text	Default
Blue	Title Block	0.80
Magenta	Dimensions	Default
Black	Object	0.70

STEP 1

Begin the process of creating a Color-Dependent plot style table by choosing Manage Plot Styles from the Print heading of the Application Menu, as shown in Figure 15.18 (Left). This activates the Plot Styles program group, as shown in Figure 15.18 (Right). Various Color-Dependent and Named plot styles already exist in this program group. To create a new plot style, double-click the Add-A-Plot Style Table Wizard.

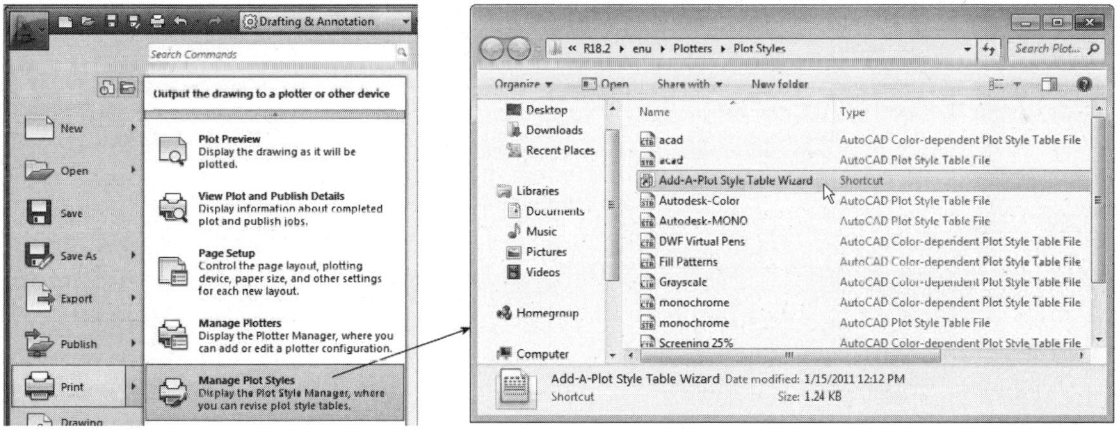

FIGURE 15.18

STEP 2

A partial illustration of the Add Plot Style Table dialog box is shown in Figure 15.19, and introduces you to the process of creating plot style tables. Plot styles contain plot definitions for color, lineweight, linetype, end capping, fill patterns, and screening. You are presented with various choices in creating a plot style from scratch, using the parameters in an existing plot style, or importing pen assignment information from a PCP, PC2, or CFG file. You also have the choice of saving this plot style information in a CTB (Color-Dependent) or STB (Named) plot style. This chapter will discuss only the creation of a Color-Dependent plot style. Click the Next > button to display the next dialog box.

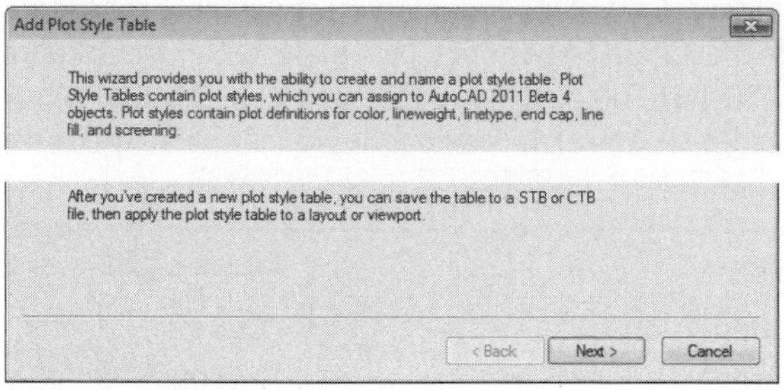

FIGURE 15.19

STEP 3

In the illustration of the Add Plot Style Table – Begin dialog box, as shown in Figure 15.20 (Left), four options are available for you to choose, depending on how you want to create the plot style table. Click the Start from scratch radio button to create this plot style from scratch. If you have made pen assignments from previous releases of AutoCAD, you can import them through this dialog box. Click the Next > button to display the next dialog box. This Plot Style Table will be started from scratch.

In the illustration of the Add Plot Style Table – Pick Plot Style Table dialog box, as shown in Figure 15.20 (Right), click the Color-Dependent Plot Style Table radio button to make this the type of plot style you will create. Click the Next > button to display the next dialog box.

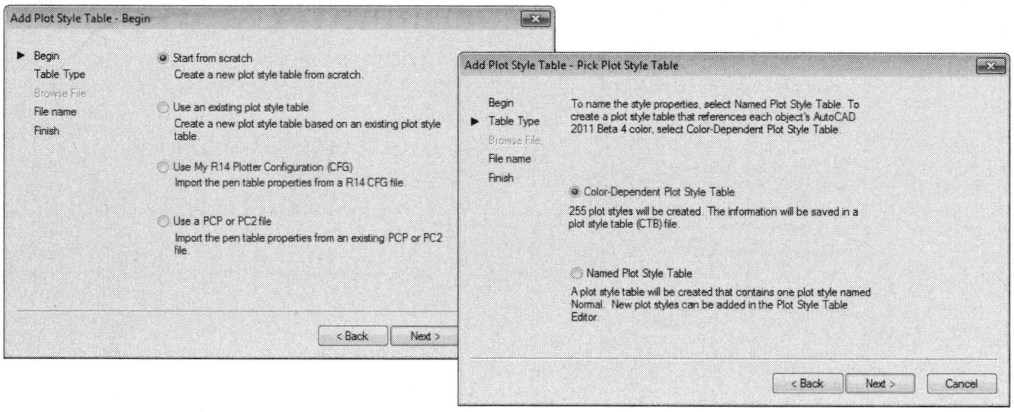

FIGURE 15.20

STEP 4

Use the Add Plot Style Table – File Name dialog box, as shown in Figure 15.21 (Left), to assign a name to the plot style table. Enter the name Ortho_Drawings in the File name area. The extension CTB is automatically added to this file name. Click the Next > button to display the next dialog box.

The Finish dialog box, as shown in Figure 15.21 (Right), alerts you that a plot style called Ortho_Drawings.ctb has been created. However, you want to have all colors plot out black and you need to assign different lineweights to a few of the layers. To accomplish this, click the Plot Style Table Editor... button as shown in Figure 15.21 (Right).

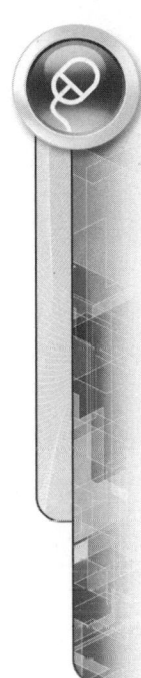

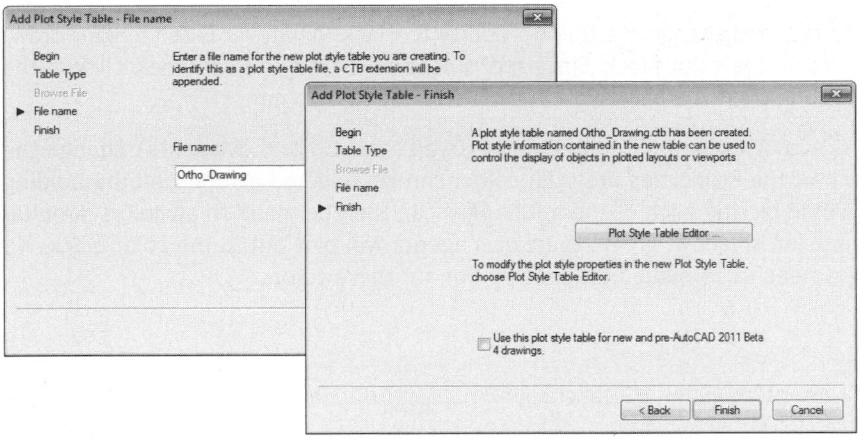

FIGURE 15.21

STEP 5

Clicking the Plot Style Table Editor... button displays the Plot Style Table Editor dialog box, as shown in Figure 15.22. Notice the name of the plot style table present at the top of the dialog box. Also, three tabs are available for making changes to the current plot style table (Ortho-Drawings.ctb). The first tab is General and displays file information about the current plot style table being edited. It is considered good practice to add a description to further document the purpose of this plot style table. It must be pointed out at this time that this plot style table will be used on other drawings besides the current one. Rather, if layers are standard across projects, the same plot style dialog box can be used. This is typical information that can be entered in the Description area.

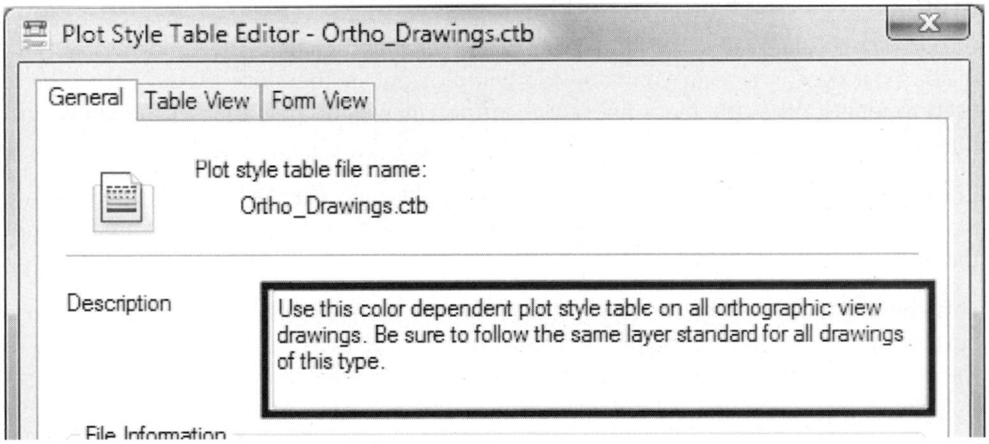

FIGURE 15.22

STEP 6

Clicking the Table View tab activates the dialog box, as shown in Figure 15.23 (Left). You can use the horizontal scroll bar to get a listing of all 255 colors along with special properties that can be changed. This information is presented in a spreadsheet format. You can click in any of the categories under a specific color and make changes, which will be applied to the current plot style table. The color and lineweight changes will be made through the next tab.

Clicking the Form View tab displays the dialog box shown in Figure 15.23 (Right). Here the colors are arranged vertically with the properties displayed on the right.

Click Color 1 (Red) and change the Color property to Black. Whatever is red in your drawing will plot out in the color black. Since red is used to identify hidden lines, click in the Lineweight area and set the lineweight for all red lines to 0.30 mm.

For Color 2 (Yellow), Color 3 (Green), Color 4 (Cyan), and Color 6 (Magenta), change the color to Black in the Properties area. All colors can be selected at one time by holding down CTRL while picking each of them. Changes can then be made to all colors simultaneously. Whatever is yellow, green, cyan, or magenta will plot out in the color black. No other changes need to be made in the dialog box for these colors.

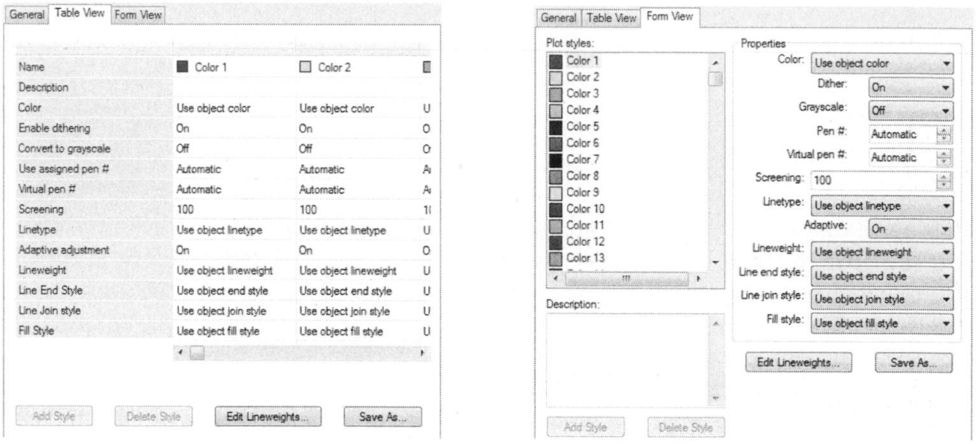

FIGURE 15.23

STEP 7

Click Color 5 (Blue), as shown in Figure 15.24 (Left), and change the Color property to Black. Whatever is blue in your drawing will plot out in the color black. Since blue is used to identify the title block lines, click in the Lineweight area and set the lineweight for all blue lines to 0.80 mm.

Click Color 7 (Black), as shown in Figure 15.24 (Right), and change the Color property to Black. Since black is used to identify the object lines, click in the Lineweight area and set the lineweight for all object lines to 0.70 mm.

This completes the editing process of the current plot style table. Click the Save & Close button; this returns you to the Finish dialog box. Clicking the Finish button returns you to the Plot Styles program group. Close this box to display your drawing.

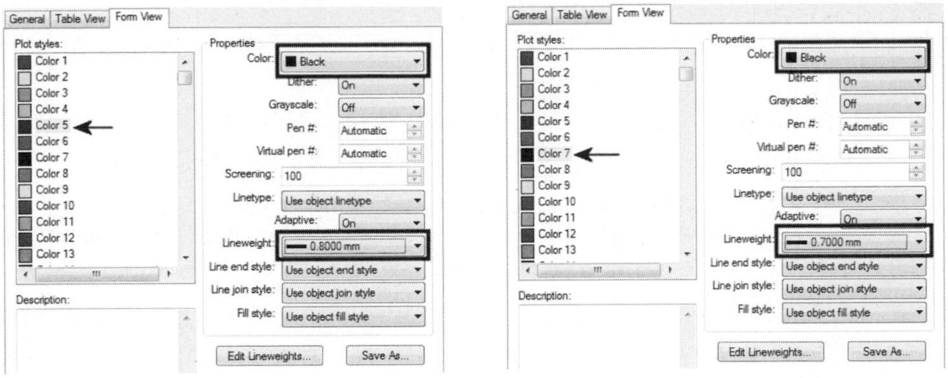

FIGURE 15.24

STEP 8

Activate the Plot dialog box. Verify that the plotter is the DWF6 ePlot plotter (this plotter has been used in all plotting tutorials throughout this chapter). In the Plot style table (pen assignments) area, make the current plot style Ortho_Drawings.ctb, as shown in Figure 15.25 (Left). Notice at the top of the dialog box that the plot style table will be saved to this layout. This means that if you need to plot this drawing again in the future, you will not have to look for the desired plot style table.

Verify in the lower-right corner of the dialog box under Plot options that you will be plotting with plot styles, as shown in Figure 15.25 (Right). Click the Preview button to display the results.

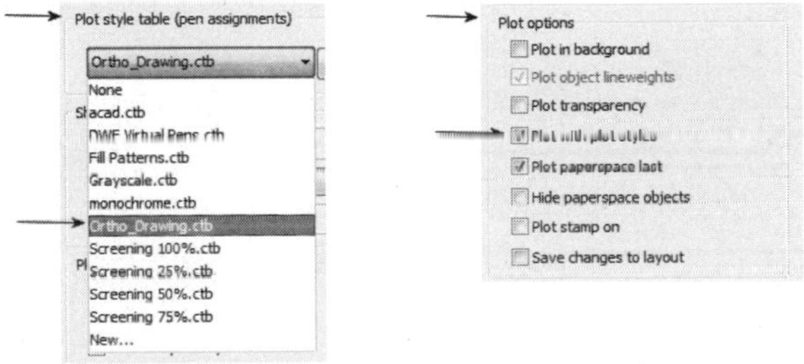

FIGURE 15.25

STEP 9

The results of performing a plot preview are illustrated in Figure 15.26. Notice that all lines are black even though they appear in color in the drawing file.

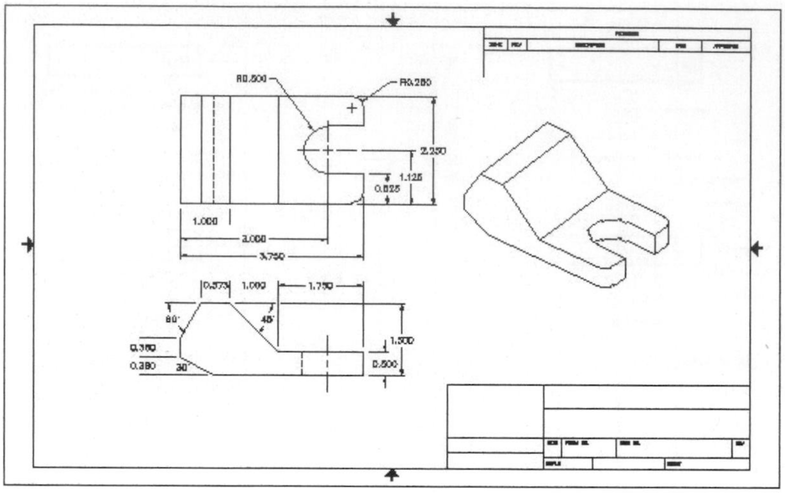

FIGURE 15.26

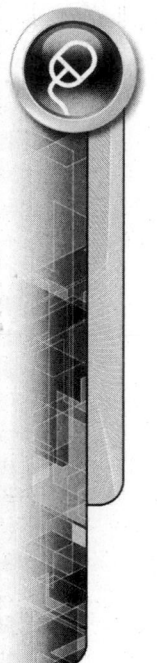

Notice that when you zoom in on a part of the preview, different lineweights appear, as in Figure 15.27, even though they all appear the same in the drawing file. This is the result of using a Color-Dependent plot style table on this drawing. This file can also be attached to other drawings that share the same layer names and colors.

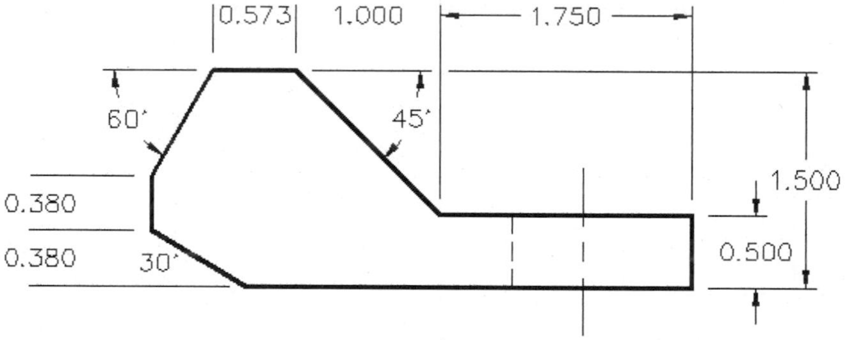

FIGURE 15.27

STEP 10

If changes need to be made to the plot style, select the Edit... button from the Plot dialog box, as shown in Figure 15.28 (Left). This displays the Plot Style Table Editor dialog box, as shown in Figure 15.28 (Right). Make and save any necessary changes. Also shown in Figure 15.28 (Left) is the New... button, which can be used to create new plot styles.

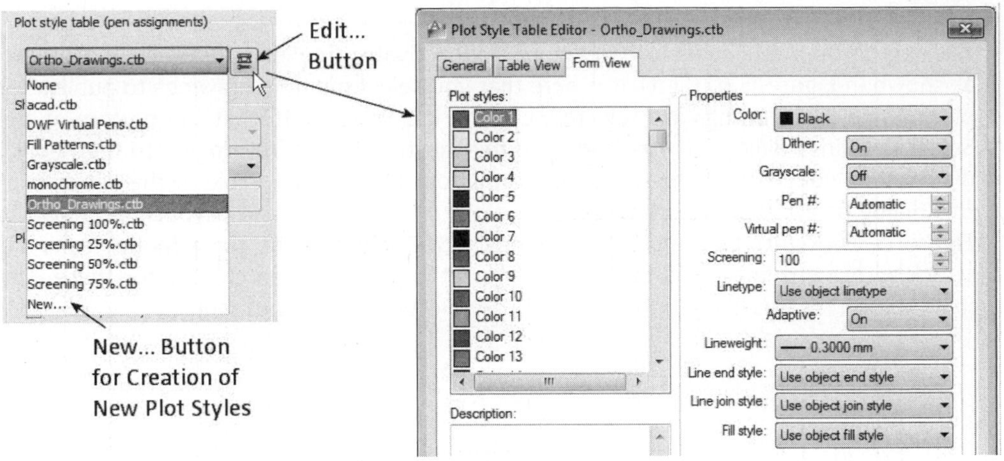

FIGURE 15.28

PUBLISHING MULTIPLE DRAWING SHEETS

You also have the ability to arrange a number of drawing layouts from other drawings under a single dialog box and perform the plot in this manner. This is accomplished through the Publish dialog box.

TUTORIAL EXERCISE: 15_PUBLISHING MULTIPLE SHEETS

Purpose

This tutorial demonstrates how to plot multiple drawing layouts through the PUBLISH command.

STEP 1

Clicking on Publish from the Application Menu, as shown in Figure 15.29 (Left), will activate the Publish dialog box, as shown in Figure 15.29 (Right). By default, the current drawing displays in the list area. You can elect to plot these drawings or add sheets to be published. In Figure 15.29 (Right), the default sheets were removed leaving the list area blank. A number of the controls will now be explained through the various steps that follow. The first step is to populate the list area of the dialog box with drawings that might be located in different folders. Clicking on the Add Sheets icon begins this process.

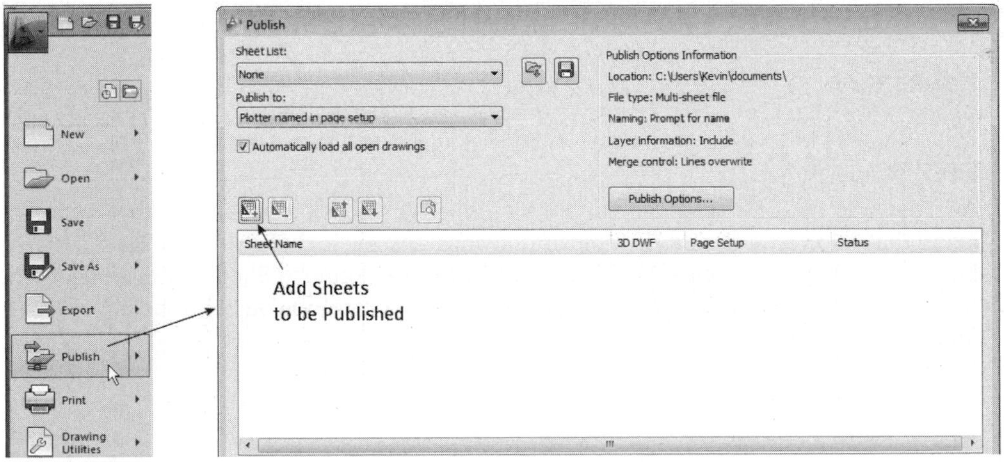

FIGURE 15.29

STEP 2

Clicking on the Add Sheets icon in Figure 15.29 activates the Select drawings dialog box, as shown in Figure 15.30 (Left). It is here that you select multiple drawings to publish (or plot). Once the drawings are selected, clicking the Select button at the bottom of the Select Drawings dialog box (not shown in this illustration) will return you to the Publish dialog box. Notice all of the drawing information that is now visible in the list area of this dialog box. Listed are drawings to be plotted from Model and Layouts, as shown in Figure 15.30 (Right), by the different icons. Notice also that the three Model icons are struck with a red line signifying that a page setup has not been created for these. The main reason for this is that most drawings are plotted from layout mode.

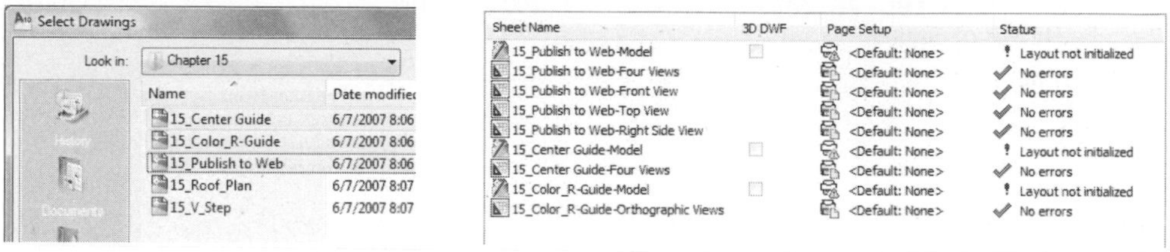

FIGURE 15.30

STEP 3

If you want to plot only layouts, right-click and pick Remove All from the menu. This will make the list area blank again. Before adding sheets, right-click and remove the check next to Include Model When Adding Sheets. Now when you click the Add Sheets button, only the layout sheets are displayed in the list area of the Publish dialog box, as shown in Figure 15.31.

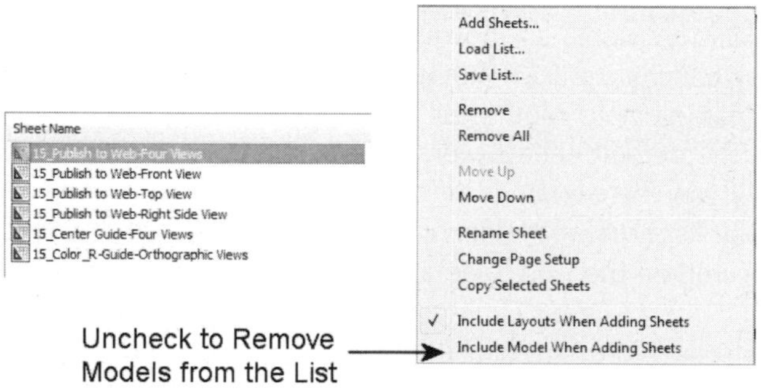

FIGURE 15.31

STEP 4

Another way of controlling the contents of what will be plotted is through a series of icons that allow you to add or remove drawings and even change their order. The five buttons illustrated in Figure 15.32 include Add Sheets, Remove Sheets, Move Sheet Up, Move Sheet Down, and Preview. This provides more control as you build the list of drawings to be published.

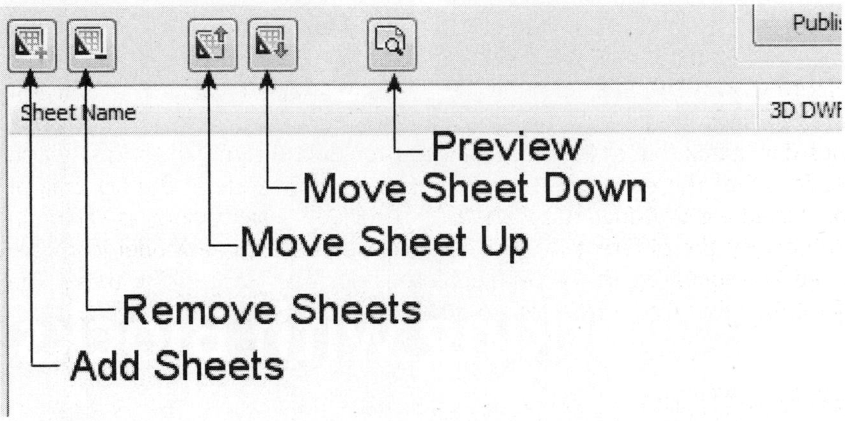

FIGURE 15.32

STEP 5

When it is time to plot the drawings, click on the Publish button located at the bottom of the Publish dialog box, as shown in Figure 15.33 (Left). Before the publish operation is executed, an alert dialog box appears asking if you want to save the current list of sheets under a name, as shown in Figure 15.33 (Right). The purpose of creating a name is to retrieve this information later if you want to perform plots on the same drawings and eliminate the need to build the list of drawings from scratch. Whether you create a name or not, clicking on the Yes or No buttons will begin publishing all drawings in the background while you still have the current drawing present on the screen. There are many more features to the Publish dialog box and this series of steps will get you started in arranging numerous drawings sheets to be plotted.

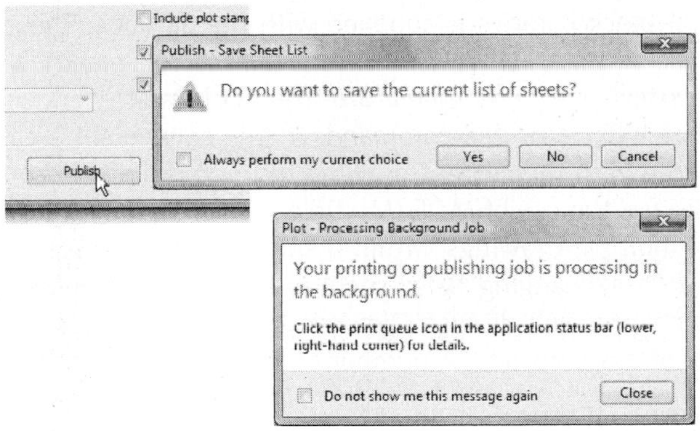

FIGURE 15.33

CHAPTER

16

Working with Blocks

This chapter begins the study of how blocks are created and merged into drawing files. This is a major productivity enhancement and is often compared to templates used to create symbols in manual drawings. Blocks are like electronic templates; once symbols, or any group of objects, are saved as blocks, they can be recalled and inserted as many times as necessary. Blocks are typically inserted in the current drawing but can be inserted in any drawing by utilizing the proper commands and techniques. The chapter continues by discussing other topics such as redefining blocks and the effects blocks have on table objects. This is followed by a discussion about using the Insert dialog box and the DesignCenter to bring blocks into drawings. The DesignCenter is a special feature that allows blocks to be inserted in drawings with drag-and-drop techniques. As an added bonus to AutoCAD users, a series of block libraries is supplied with the package. These block libraries include such application areas as mechanical, architectural, electrical, piping, and welding, to name just a few. Yet another feature, the Tool Palette, allows you to organize blocks, as well as hatch patterns and commands in one convenient area. These object types can then be shared with the current drawing through drag-and-drop techniques. This chapter will also discuss the use of MDE (Multiple Design Environment). This feature allows the opening of multiple drawings within a single AutoCAD session and provides a convenient method of exchanging data, such as blocks, between one drawing and another. Finally, this chapter will discuss Dynamic blocks. These custom blocks provide even greater control of block objects. A number of dynamic block techniques will be explored.

WHAT ARE BLOCKS?

Blocks usually consist of smaller components of a larger drawing. Typical examples include doors and windows for floor plans, nuts and bolts for mechanical assemblies, and resistors and transistors for electrical schematics. In Figure 16.1, which shows an electrical schematic, all resistors, capacitors, tetrodes, and diodes are considered blocks that make up the total drawing of the electrical schematic. The capacitor is highlighted as one of these components.

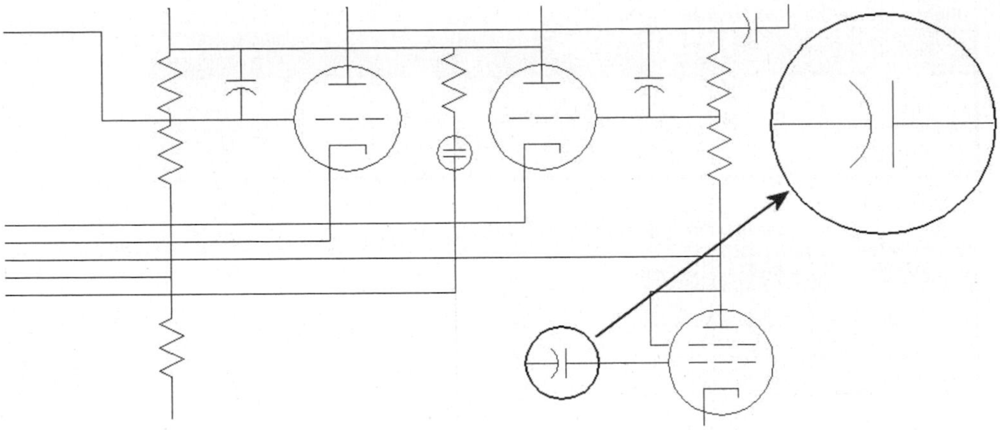

FIGURE 16.1

Blocks are created and then inserted in a drawing. When creating the block, you must first provide a name for the block. The capacitor illustrated in Figure 16.2 (Left) was assigned the name CAPACITOR. Also, when you create a block, an insertion point is required. This acts as a reference point from which the block will be inserted. In the illustration in Figure 16.2 (Left), the insertion point of the block is the left end of the line.

At times, blocks have to be rotated into position. In the illustration in Figure 16.2 (Right), notice that one capacitor is rotated at a 45° angle while the other capacitor is rotated 270°. In this way, the same block can be used numerous times even though it is positioned differently in the drawing.

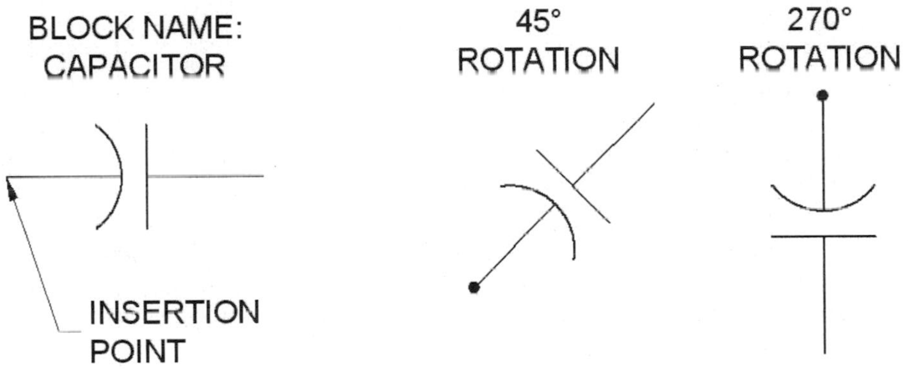

FIGURE 16.2

CREATING A LOCAL BLOCK

Use the BLOCK command to create a new block from selected objects. Choose this command from one of the following:

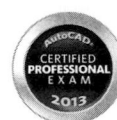

- From the Ribbon > Home Tab > Block Panel > Create Button
- From the Ribbon > Insert Tab > Block Definition Panel > Create Button
- From the Menu Bar (Draw > Block > Make...)
- From the Draw toolbar
- From the keyboard (B or BLOCK)

Access of this command from the Ribbon is illustrated in Figure 16.3.

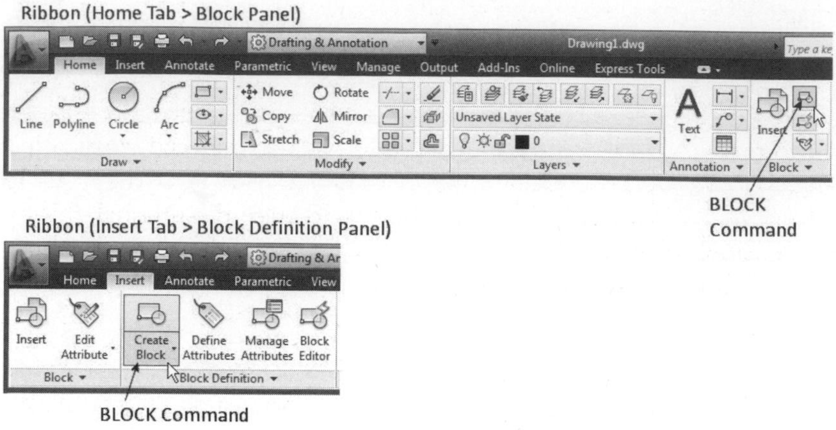

FIGURE 16.3

The illustration in Figure 16.4 (Left) is a drawing of a hex head bolt. This drawing consists of one polygon, representing the hexagon, a circle, indicating that the hexagon is circumscribed about the circle, and two centerlines. Rather than copy these individual objects numerous times throughout the drawing, you can create a block using the `BLOCK` command. Issuing the command activates the Block Definition dialog box, illustrated in Figure 16.4 (Right).

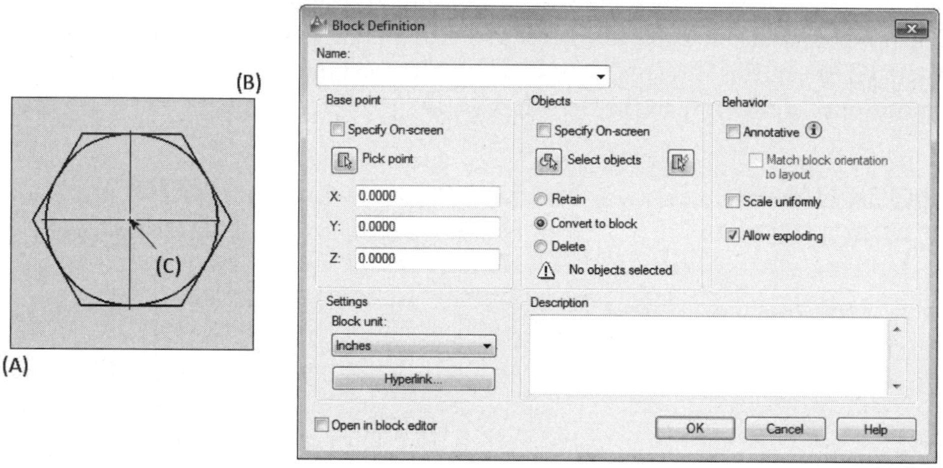

FIGURE 16.4

Once you create a block, the numerous objects that make up the block are combined into a single object. This dialog box allows for the newly created block to be merged into the current drawing file. The block will exist in the drawing database even though you have not yet inserted it into the drawing. This means that a block is available only in the drawing it was created in; it cannot be shared directly with other drawings. (The `WBLOCK` command is used to create global blocks that can be inserted in any drawing file. This command will be discussed later on in this chapter.)

To create a block through the Block Definition dialog box, enter the name of the block, such as "Hexbolt," in the Name field, as shown in Figure 16.5. You can use up to 255 alphanumeric characters, and spaces are allowed when naming a block.

Next, click the Select objects button; this returns you to the drawing editor. Create a window from "A" to "B" around the entire hex bolt to select all objects, as shown in the previous image. When finished, press ENTER. This returns you to the Block Definition dialog box, where a previewed image of the block you are about to create is displayed in the upper-right corner of the dialog box, as shown in Figure 16.5. Whenever you create a block, you can elect to allow the original objects that made up the hex bolt to remain on the screen (Retain button), to be replaced with an instance of the block (Convert to Block button), or to be removed after the block is created (Delete button). Click the Delete radio button to erase the original objects after the block is created. If the original objects that made up the hex bolt are unintentionally removed from the screen during this creation process, the OOPS command can be used to retrieve all original objects to the screen.

The next step is to create a base point or insertion point for the block. This is considered a point of reference and should be identified at a key location along the block. By default, the Base point in the Block Definition dialog box is located at the drawing origin (0,0,0). To enter a more appropriate base-point location, click the Pick point button; this returns you to the drawing editor and allows you to pick the center of the hex bolt at "C," as shown in Figure 16.4. Once this point is selected, you are returned to the Block Definition dialog box; notice how the Base point information now reflects the key location along the block, as shown in Figure 16.5.

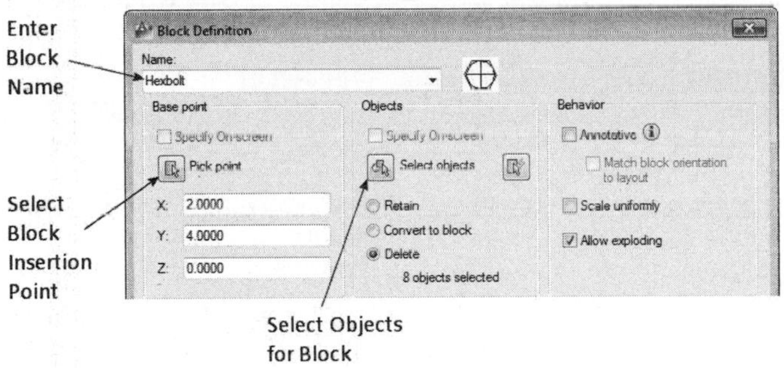

FIGURE 16.5

Yet another feature of the Block Definition dialog box allows you to add a description for the block. Many times the name of the block hides the real meaning of the block, especially if the block name consists of only a few letters. Click on the Description field and add the following statement: "This is a hexagonal head bolt." This allows you to refer to the description in case the block name does not indicate the true intended purpose for the block, as shown in Figure 16.6. Specifying the Block unit determines the type of units utilized for scaling the block if it is inserted into another drawing (perhaps through the DesignCenter).

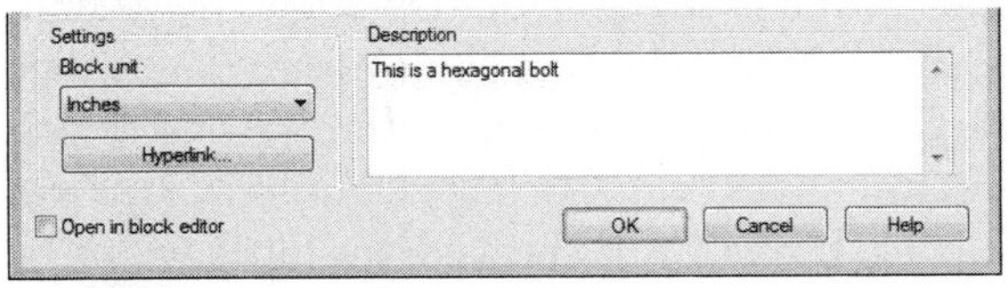

FIGURE 16.6

As the block is written to the database of the current drawing, the individual objects used to create the block automatically disappear from the screen (due to our earlier selection of the Delete radio button). When the block is later inserted in a drawing, it will be placed in relation to the insertion point.

NOTE

In addition to creating blocks by grouping several objects into one, you can also write an entire drawing out to a file using the WBLOCK command. Entering `W` at the command prompt will display the Write Block dialog box. As with the Block Definition dialog box, you will designate a base point and select the objects. You will also be required to enter a file name in order for the objects to be written out to a drawing file.

INSERTING BLOCKS

Once blocks are created, they are typically merged or inserted in the drawing through the INSERT command. Activate the command through one of the following methods:

- From the Ribbon > Home Tab > Block Panel > Insert Button
- From the Ribbon > Insert Tab > Block Panel > Insert Button
- From the Menu Bar (Insert > Block...)
- From the Draw toolbar
- From the keyboard (`I` or `INSERT`)

Access of this command from the Ribbon is illustrated in Figure 16.7.

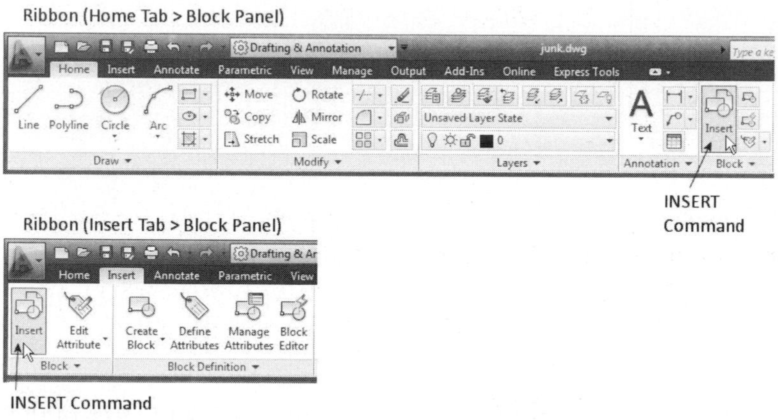

FIGURE 16.7

Issuing the command, utilizing one of the methods just discussed, will activate the Insert dialog box, as shown in Figure 16.8. This dialog box is used to dynamically insert blocks. First, by clicking the Name drop-down list box, select a block from the current drawing (clicking the Browse button locates global blocks or drawing files). After you identify the name of the block to insert, the point where the block will be inserted must be specified, along with its scale and rotation angle. By default, the Insertion point area's Specify On-screen box is checked. This means you will be prompted for the insertion point at the Command prompt area of the drawing editor. The default values for the scale and rotation insert the block at the original size and orientation. An Explode option is available. If this box is checked, the block would be inserted and then exploded back to its individual objects. Generally, you should avoid this option, because it is more efficient to work with blocks than with individual objects. Once the name of the block, such as Hexbolt, is selected, click the OK button.

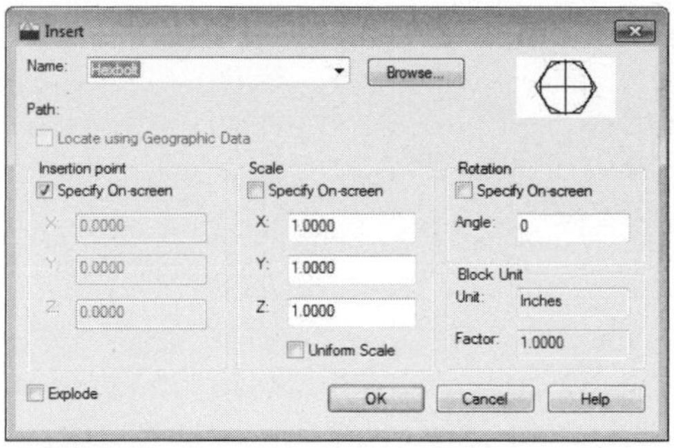

FIGURE 16.8

The following prompts complete the block insertion operation:

> Specify insertion point or [Basepoint/Scale/X/Y/Z/Rotate]: *(Mark a point at "A" in Figure 16.9 to insert the block)*

If the Specify On-screen boxes are also checked for Scale and Rotation, the following prompts will complete the block insertion operation.

> Specify insertion point or [Basepoint/Scale/X/Y/Z/Rotate]: *(Mark a point at "A," as shown in Figure 16.9, to insert the block)*
>
> Enter X scale factor, specify opposite corner, or [Corner/XYZ] <1>: *(Press* ENTER *to accept default X scale factor)*
>
> Enter Y scale factor <use X scale factor>: *(Press* ENTER *to accept default)*
>
> Specify rotation angle <0>: *(Press* ENTER *to accept the default rotation angle and insert the block, as shown in Figure 16.9)*

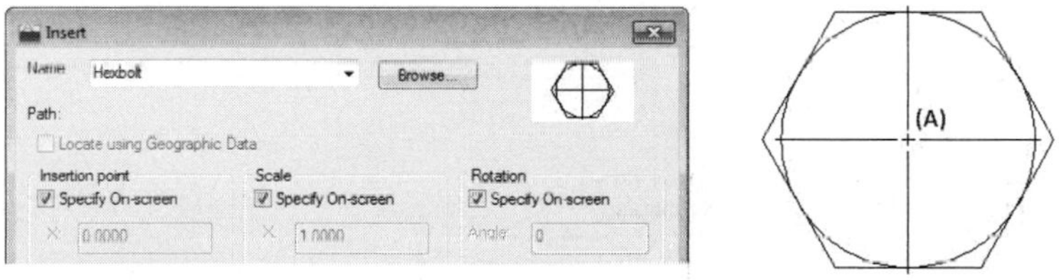

FIGURE 16.9

SCALING AND ROTATING BLOCK INSERTIONS

In the Insert dialog box, typically you will only want to check the Specify On-screen box for the Insertion point, as shown in Figure 16.10 (Left). Completing the Scale and Rotation areas in the dialog box allows you to preview the block as it is inserted. The scale is set to double the size of the block and the rotation is set to turn the block 30° in the counterclockwise direction. The results are shown in Figure 16.10 (Right). Checking the Uniform Scale check box allows you to change the scale in the X and Y direction by entering a single value.

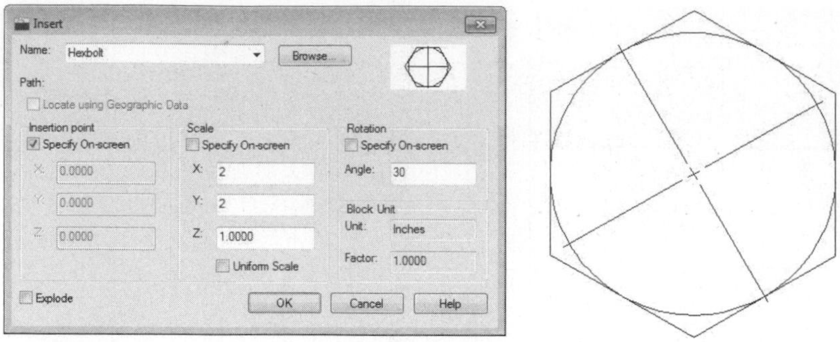

FIGURE 16.10

Figure 16.11 shows the results of entering different scale factors and rotation angles when blocks are inserted in a drawing file. The image at "A" shows the block inserted in a drawing with its default scale and rotation angle values. The image at "B" shows the result of inserting the block with a scale factor of 0.50 and a rotation angle of 0°. The image appears half its normal size. The image at "C" shows the result of inserting the block with a scale factor of 1.75 and a rotation angle of 0°. The image at "D" shows the result of inserting the block with different X and Y scale factors. In this image, the X scale factor is 0.50 while the Y scale factor is 2.00 units. Notice how out of proportion the block appears. There are certain applications where different scale factors are required to produce the desired effect. The image at "E" shows the result of inserting the block with the default scale factor and a rotation angle of 30°. As with all rotations, a positive angle rotates the block in the counterclockwise direction; negative angles rotate in the clockwise direction.

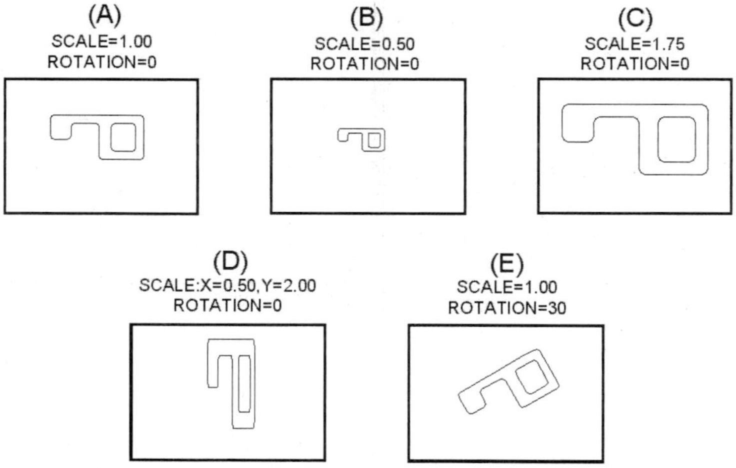

FIGURE 16.11

TIP

Entering a negative value in the scale text box in the Insert dialog box allows you to mirror a block during insertion. A value of -1.00 for the X scale factor mirrors the block over a vertical axis and a value of -1.00 for the Y scale factor mirrors the block over a horizontal axis.

INSERTING GLOBAL BLOCKS (FILES)

If blocks are already defined as part of the database of the current drawing (local blocks), they may be selected from the Name drop-down list box in the Insert dialog box, as shown in Figure 16.12 (Right).

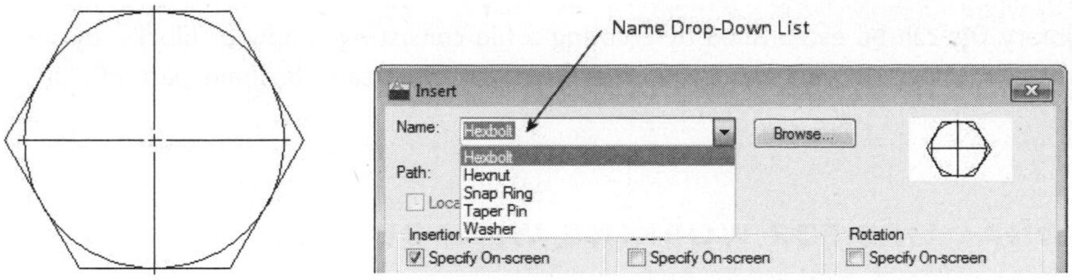

FIGURE 16.12

TIP

Local blocks are often available through template files. When you start a new drawing using one of these templates, the blocks become part of your drawing's database. Images of the blocks may not appear on your screen but you can view a list of the available blocks by activating the Insert dialog box and expanding the Name drop-down list box.

The INSERT command can also be used to insert objects from outside of the current drawing database (global blocks). For inserting global blocks in a drawing, select the Insert dialog box Browse button, which displays the Select Drawing File dialog box, as illustrated in Figure 16.13. This is the same dialog box associated with opening drawing files. In fact, there is no real difference between global blocks and any other AutoCAD drawing file. Global blocks are simply drawing files created for the purpose of being inserted. Select the desired folder where the global block or drawing file is located; then select the name of the drawing. This returns you to the main Insert dialog box with the file now available in the Name drop-down list box and ready for insertion.

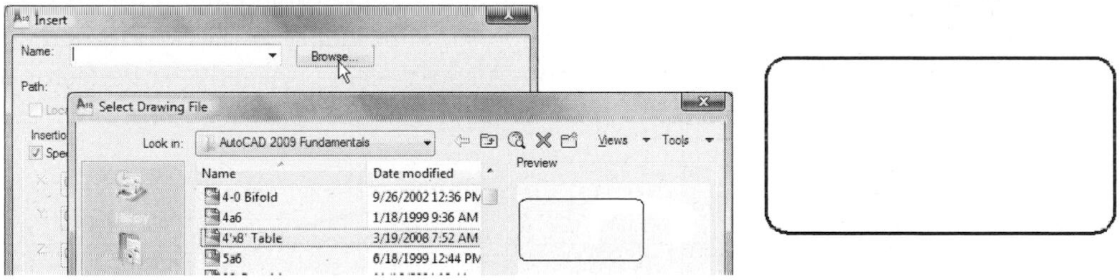

FIGURE 16.13

The WBLOCK command, mentioned earlier, provides a method of creating drawing files specifically designed to be used as a global block. This command allows you to select objects and establish a base point for the drawing that is created. When you insert this file, a logical insertion point is utilized. When you insert a typical drawing file, the insertion point is the origin (0,0,0). However, you can use the BASE command to select a base point, other than the origin, for any drawing file. Another important consideration for a global block is the file content. When you insert a file, you not only insert the drawing entities but also their styles, layers, and blocks. Typically, a good global block is a very simple drawing file consisting of very few layers and styles with a logical base point established.

TIP

A block library file can be established by creating a file consisting simply of blocks. By inserting this file into your current one, those blocks automatically become part of your database.

ADDITIONAL TIPS FOR WORKING WITH BLOCKS

Create Blocks on Layer 0

Blocks are best controlled, when dealing with layer colors, linetypes, and lineweights, by being drawn on Layer 0 because it is considered a neutral layer. By default, Layer 0 is assigned the color White and the Continuous linetype. Objects drawn on Layer 0 and then converted to blocks take on the properties of the current layer when inserted in the drawing. The current layer controls color, linetype, and lineweight.

Create Blocks Full Size If Applicable

The illustration in Figure 16.14 (Left) shows a drawing of a refrigerator, complete with dimensions. In keeping with the concept of drawing in real world units in CAD or at full size, individual blocks must also be drawn at full size in order for them to be inserted in the drawing at the correct proportions. For this block, construct a rectangle 28 units in the X direction and 24 units in the Y direction. Create a block called Refrigerator by picking the rectangle. When inserting this block, the refrigerator appears the correct size no matter what the scale is used on the drawing.

One exception to the full-size rule is illustrated in Figure 16.14 (Right). Rather than create each door block separately to account for different door sizes, create the door so as to fit into a 1-unit-by-1-unit square. The purpose of drawing the door block inside a 1-unit square is to create only one block of the door and insert it at a scale factor matching the required door size. For example, for a 2'-8" door, enter 2'8 or 32 when prompted for the X and Y scale factors. For a 3'-0" door, enter 3' or 36 when prompted for the scale factors. Numerous doors of different types can be inserted in a drawing using only one block of the door. Also try using a negative scale factor to mirror the door as it is inserted.

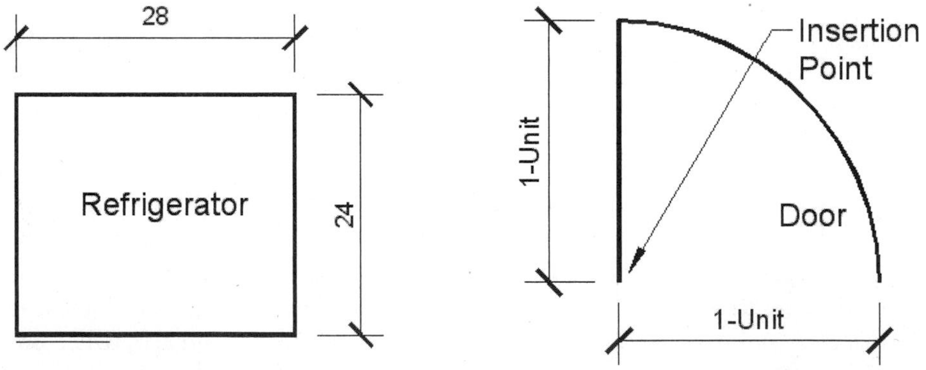

FIGURE 16.14

Use Grid When Proportionality, but Not Scale, Is Important

Sometimes blocks represent drawings in which the scale of each block is not important. In the previous example of the refrigerator, scale was very important in order for the refrigerator to be drawn according to its full-size dimensions. This is not the case, as shown in Figure 16.15 (Left), of the drawing of the resistor block. Electrical schematic blocks are generally not drawn to any specific scale; however, it is important that all blocks are proportional to one another. Setting up a grid is good practice in keeping all blocks at the same proportions. Whatever the size of the grid is, all blocks are designed around the same grid size. The result is shown in Figure 16.15 (Right); with four blocks being drawn with the same grid, their proportions look acceptable.

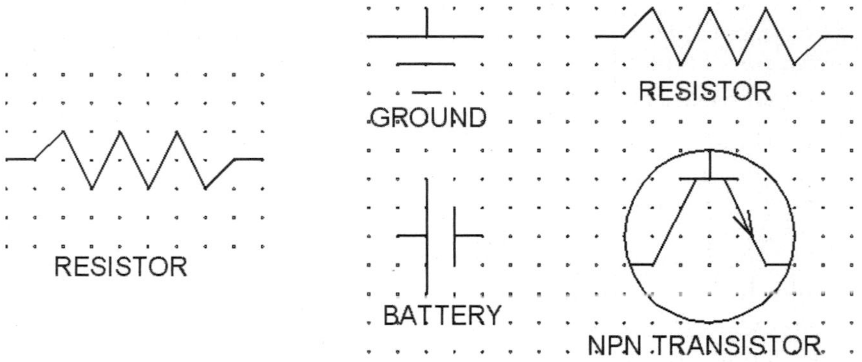

FIGURE 16.15

TRIMMING AND EXTENDING TO BLOCK OBJECTS

Block objects can be used in both the TRIM and EXTEND commands. As you select cutting edges on the block to trim to or boundary edges on the block to extend to, the command isolates the edges from the remainder of the objects that make up the block.

TRY IT!

To test the trim feature, open the drawing 16_Trim Plates and use Figure 16.16 as a guide. The bolt is a block. Activate the TRIM command, pick the edges of the bolt at "A" and "B" as cutting edges, and pick the lines at "C" through "E" as the objects to trim.

The result of using the TRIM command is illustrated in Figure 16.16 (Right).

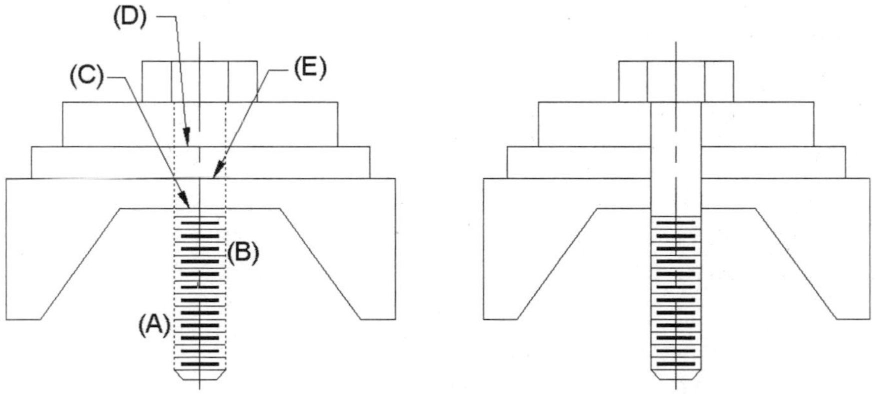

FIGURE 16.16

TRY IT!

To test the extend feature, open the drawing 16_Extend Plates and use Figure 16.17 as a guide. Again the bolt is a block. Activate the EXTEND command, pick the edges of the bolt at "A" and "B" as boundary edges, and pick the lines at "C" through "H" as the objects to extend.

The result of using the EXTEND command on block objects as cutting edges is illustrated in Figure 16.17 (Right).

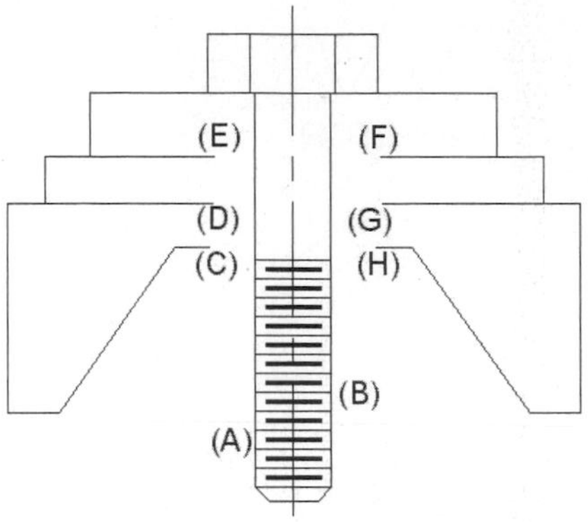

FIGURE 16.17

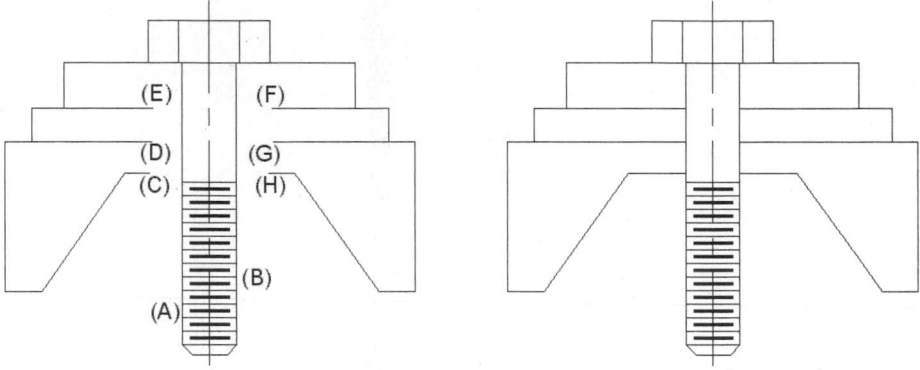

EXPLODING BLOCKS

It has already been mentioned that as blocks are inserted in a drawing file, they are considered one object even though they consist of numerous individual objects. At times it is necessary to break up a block into its individual parts. The EXPLODE command is used for this.

Illustrated in Figure 16.18 (Left) is a block that has been selected. Notice that the entire block highlights and the insertion point of the block is identified by the presence of the grip. Using the EXPLODE command on a block results in Figure 16.18 (Right). Here, when one of the lines is selected, only that object highlights. Exploding a block breaks up the block into its individual objects. As a result, you must determine when it is appropriate to explode a block.

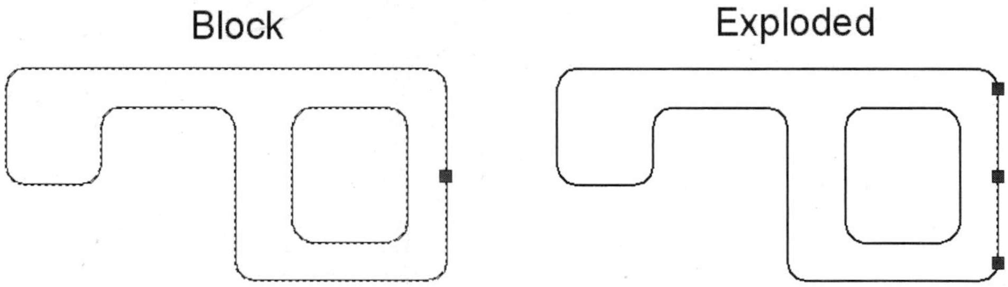

FIGURE 16.18

MANAGING UNUSED BLOCK DATA WITH THE PURGE DIALOG BOX

If you explode a block or if there are blocks in a drawing that are not being utilized, you may want to remove or purge them from the database. AutoCAD stores named objects (blocks, dimension styles, layers, linetypes, multiline styles, plot styles, shapes, table styles, and text styles) with the drawing. When the drawing is opened,

AutoCAD determines whether other objects in the drawing reference each named object. If a named object is unused and not referenced, you can remove the definition of the named object from the drawing by using the `PURGE` (or `PU`) command. This is a very important productivity technique used for compressing or cleaning up the database of the drawing. Picking Drawing Utilities from the Application Menu, followed by Purge, displays the Purge dialog box, as shown in Figure 16.19.

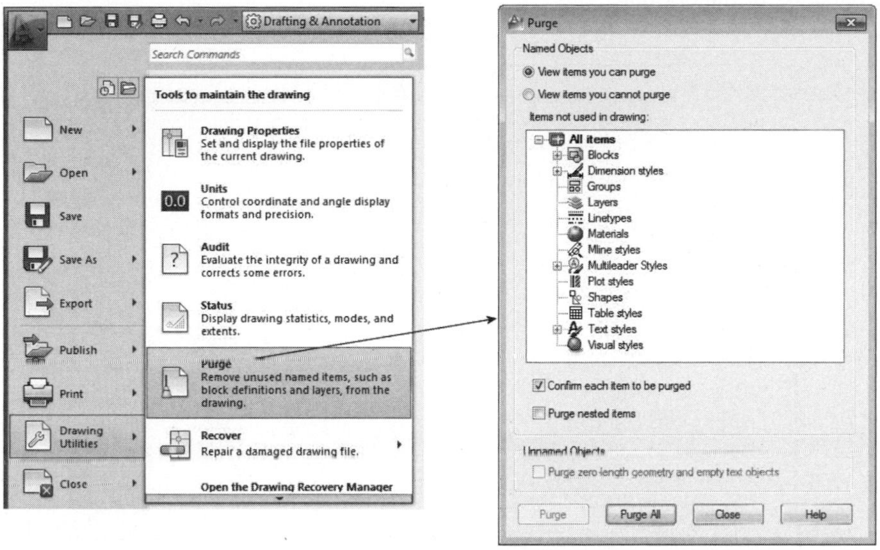

FIGURE 16.19

The Blocks category is expanded by clicking "+." This produces the list of all the items that are currently unused in the drawing that can be removed, as shown in Figure 16.20 (Left). Clicking the item DIGITIZE, right-clicking, and then picking Purge from the shortcut menu (or simply clicking the button at the bottom of the dialog box) displays the Confirm Purge dialog box, as shown in Figure 16.20 (Center). Click the "Purge this item" button and the block is removed from the listed items, as shown in Figure 16.20 (Right).

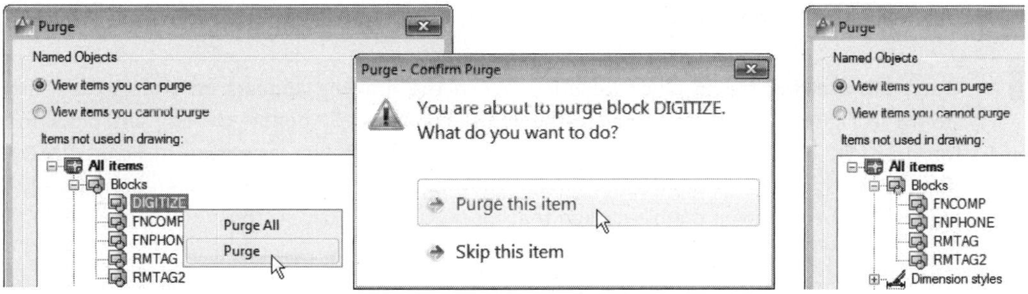

FIGURE 16.20

Other controls are available in the Purge dialog box, such as the ability to view items that cannot be purged and the capability of purging nested items. An example of purging a nested item would be purging a block definition that lies inside another block definition.

TIP

The layer "0," Standard text and dimension styles, and the Continuous linetype cannot be purged from a drawing.

TRY IT!

Open the drawing file 16_Purge and activate the Purge dialog box, as shown in Figure 16.21 (Left). Click the "+" to expand a few of the categories. You can hold down the CTRL key while selecting specific items to purge, but for this exercise, do not select any items before clicking the Purge All button. A Confirm Purge dialog box will be displayed, as shown in Figure 16.21 (Right). Click the "Purge this item" button until all items, including blocks, layers, linetypes, text styles, dimension styles, and multiline styles, have been removed from the database of the drawing.

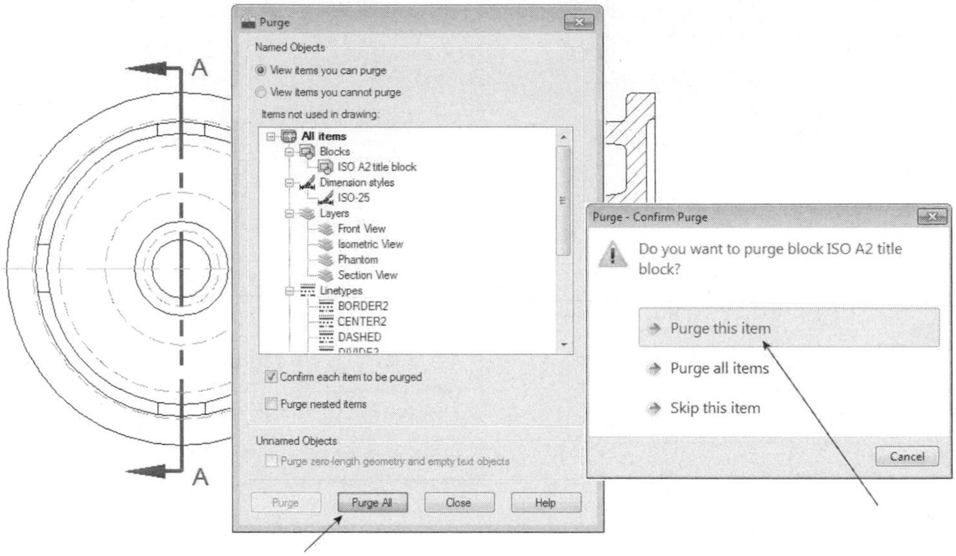

FIGURE 16.21

EDITING AND REDEFINING BLOCKS

At times, a block needs to be edited. Rather than erase all occurrences of the block in a drawing, you can redefine it. This is considered a major productivity technique because all blocks that share the same name as the block being redefined will automatically update to the latest changes. The illustration in Figure 16.22 (Left) shows various blocks inserted in a drawing. The block name is Step Guide and the insertion point is at the lower-left corner of the object.

TRY IT!

Open the drawing file 16_Block Redefine. When the drawing appears on your screen, as shown in Figure 16.22 (Left), activate the BEDIT command by double-clicking any block in the drawing. This launches the Edit Block Definition dialog box, as shown in Figure 16.22 (Right). A list of all blocks defined in the drawing appears. Step Guide is automatically highlighted because you double-clicked that block. Click the OK button.

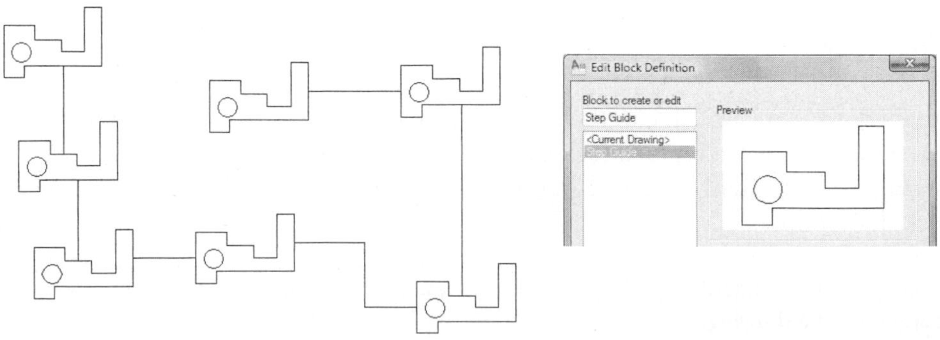

FIGURE 16.22

After you click the OK button, the Block Editor Ribbon appears, as shown in Figure 16.23. A majority of the tools displayed in the palette on the left are designed for creating dynamic blocks, which will be explained in detail later in this chapter. For now, close the palette and use this environment to make changes to the geometry that makes up the Step Guide.

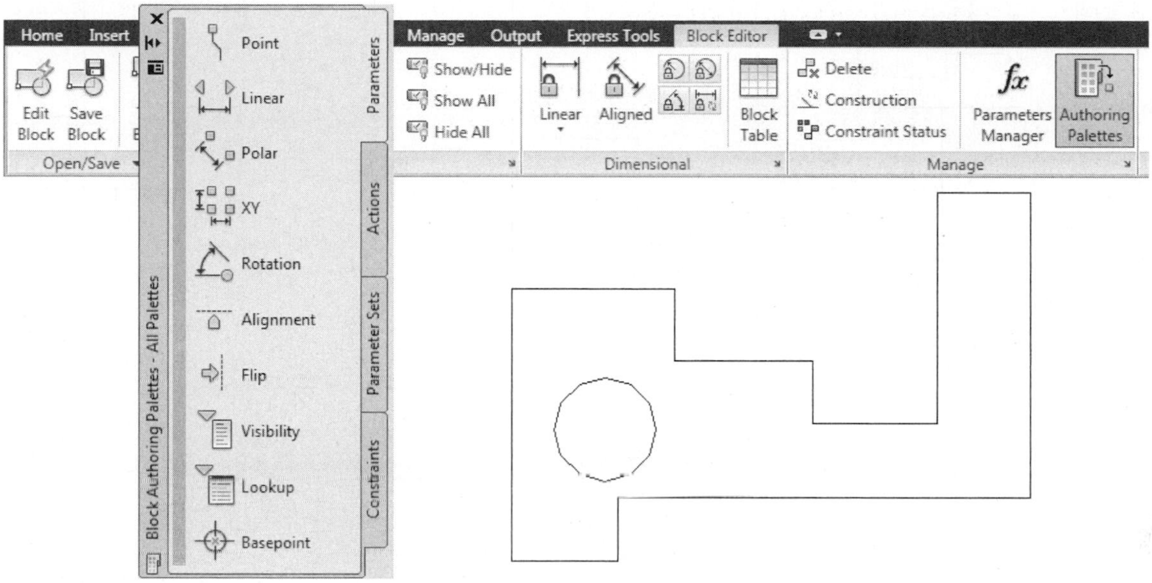

FIGURE 16.23

You can add new objects, change the properties of objects, or make modifications to this object. For this example, erase the circle and stretch the upper-right corner of the object down by 2 units, as shown in Figure 16.24.

To return to your drawing, you must first save the changes made to the object by clicking the Save Block button, also shown in Figure 16.24. To exit the Block Editor area, click the Close Block Editor button. If you do not save the block initially, you will be prompted to save or discard the changes when you close the editor.

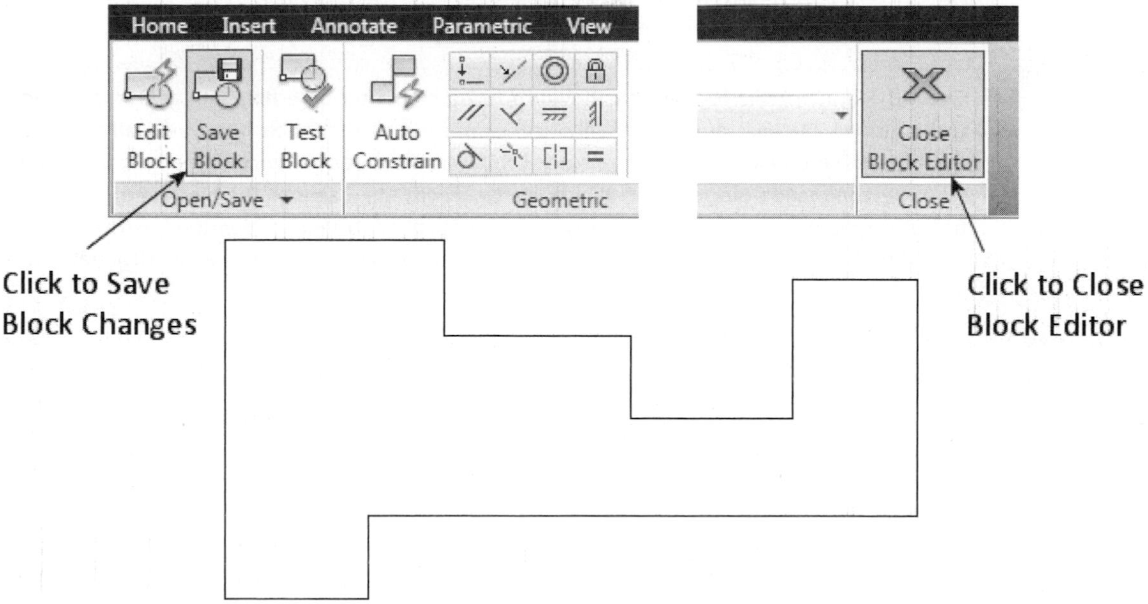

FIGURE 16.24

When you return to the drawing editor, notice that all Step Guide blocks have been updated to the changes made in the Block Editor. This provides a very productive method of making changes to all blocks in a drawing. This method works only if blocks have not been exploded in a drawing. It is for this reason that you must exercise caution when exploding blocks in a drawing.

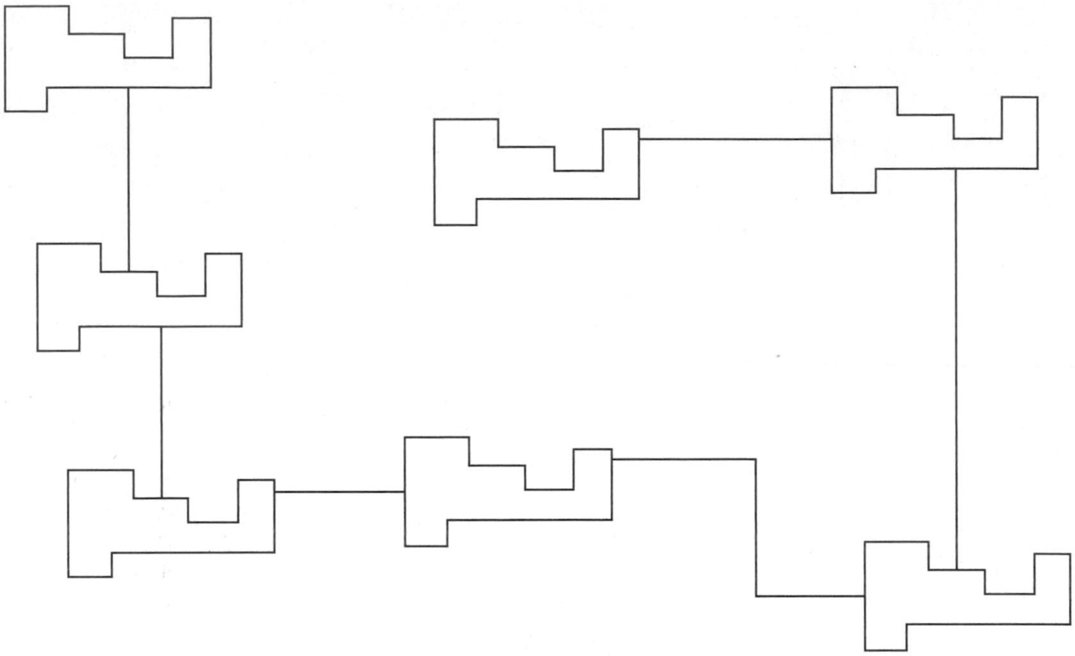

FIGURE 16.25

BLOCKS AND THE DIVIDE COMMAND

The DIVIDE command was already covered in Chapter 5. It allows you to select an object, give the number of segments, and place point objects at equally spaced distances depending on the number of segments. The DIVIDE command has a Block option that allows you to place blocks at equally spaced distances.

TRY IT!

Open the drawing file 16_Speaker. In Figure 16.26, a counter-bore hole and centerline need to be copied 12 times around the elliptical centerline so that each hole is equally spaced from others. (The illustration of the block CBORE is displayed at twice its normal size.) The DIVIDE command's Block option allows you to specify the name of the block and the number of segments. In the following image, the elliptical centerline is identified as the object to divide. Follow the command sequence to place the block CBORE in the elliptical pattern:

Command: DIV *(For DIVIDE)*

Select object to divide: *(Select the elliptical centerline)*

Enter the number of segments or [Block]: B *(For Block)*

Enter name or block to insert: CBORE

Align block with object? [Yes/No] <Y>: *(Press ENTER to accept)*

Enter the number of segments: 12

The results are illustrated in the following image on the right. Notice how the elliptical centerline is divided into 12 equal segments by 12 blocks called CBORE. In this way, any object may be divided through the use of blocks.

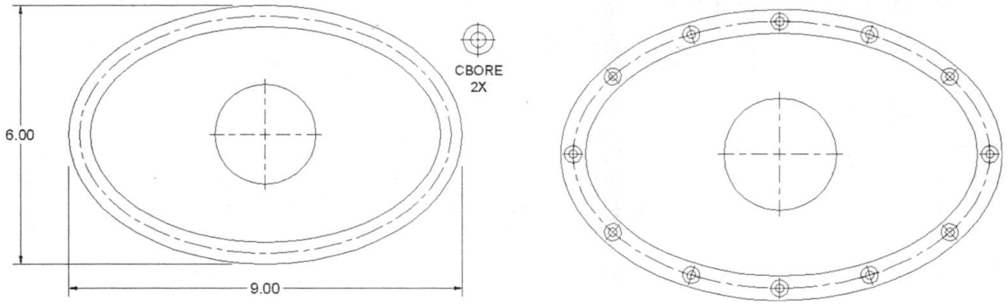

FIGURE 16.26

BLOCKS AND THE MEASURE COMMAND

As with the `DIVIDE` command, the `MEASURE` command offers increased productivity when you measure an object and insert blocks at the same time.

TRY IT!

Open the drawing file 16_Chain. In Figure 16.27, a polyline path will be divided into 0.50-length segments using the block CHAIN2. The perimeter of the polyline was calculated by the `LIST` command to be 22.00 units, which is evenly divisible by 0.50 and will allow the insertion of 44 blocks. Follow the Command prompt sequence below for placing a series of blocks called CHAIN2 around the polyline path to create a linked chain.

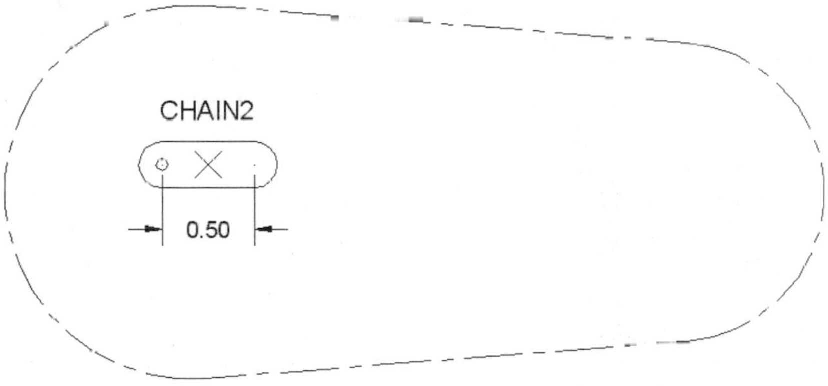

FIGURE 16.27

Command: **ME** (*For MEASURE*)

Select object to measure: (*Pick the polyline path*)

Specify length of segment or [Block]: **B** (*For Block*)

Enter name of block to insert: **CHAIN2**

Align block with object? [Yes/No] <Y>: (*Press ENTER to accept*)

Specify length of segment: **0.50**

The result is illustrated in Figure 16.28 (Left), with all chain links being measured along the polyline path at increments of 0.50 units.

Answering No to the prompt "Align block with object? [Yes/No] <Y>" displays the results, as shown in Figure 16.28 (Right). Here all blocks are inserted horizontally and travel in 0.50 increments. While the polyline path has been successfully measured, the results are not acceptable for creating the chain.

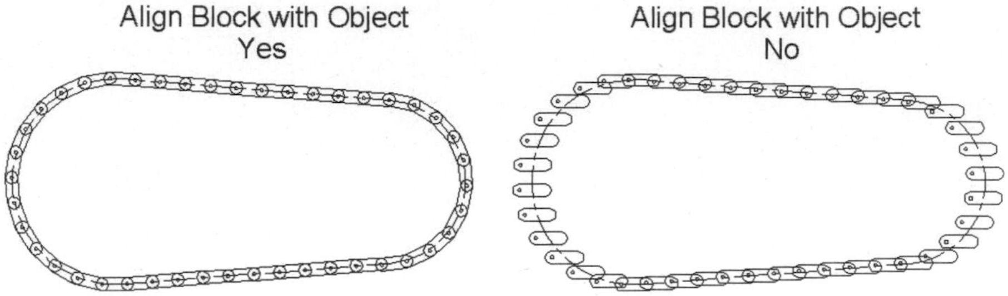

FIGURE 16.28

RENAMING BLOCKS

Blocks can be renamed to make their meanings more clear through the RENAME command. This command is not in the Ribbon but can be found in the Format Menu Bar, as shown in Figure 16.29 (Left), and, when selected, displays a dialog box similar to the one illustrated in Figure 16.29 (Right). Clicking Blocks in the dialog box lists all blocks defined in the current drawing. One block with the name REF1 was abbreviated, and we wish to give it a full name. Clicking the name REF1 pastes it in the Old Name field. Type the desired full name REFRIGERATOR in the Rename To field, and click the Rename To button to rename the block.

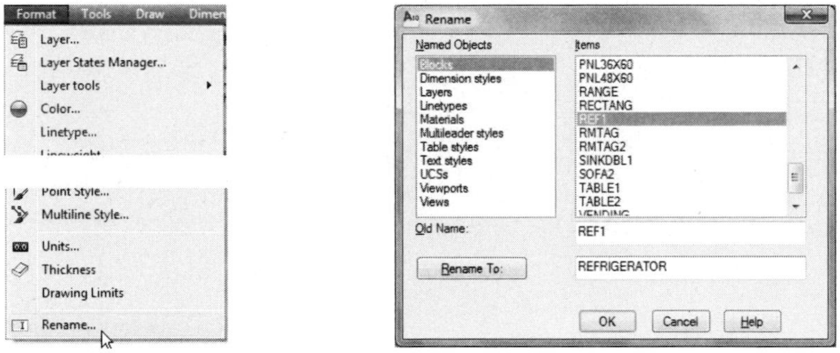

FIGURE 16.29

TABLES AND BLOCKS

In Figure 16.30, electrical symbols and their descriptions are arranged in the legend to call out the symbols in a table. To add blocks to a table, click inside the cell that will hold the block and right-click to display the menu; then click Insert, followed by Block.

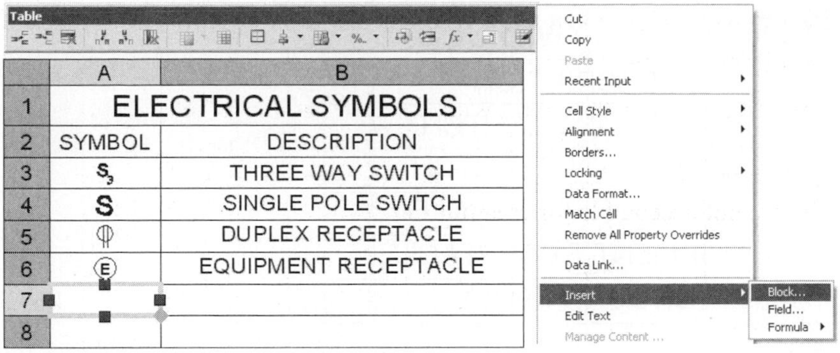

FIGURE 16.30

NOTE

If you are utilizing the Ribbon, picking inside a table will automatically activate the Table Cell tab, and you can pick Block from the Insert panel to place a block in the table. You can also type the `TINSERT` command in the Command prompt to insert a block into a table cell. You will be prompted to select a cell in which to insert the block and then the Insert Block in Table dialog box will appear.

When the Insert a Block in a Table Cell dialog appears, locate the name of the block to insert, as shown in Figure 16.31.

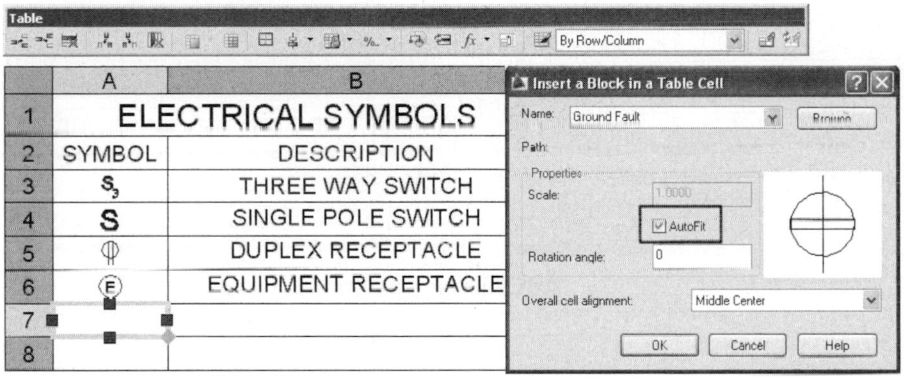

FIGURE 16.31

The results are illustrated in Figure 16.32, where the Ground Fault block is inserted into the highlighted cell of the table. Because the AutoFit function is checked in Figure 16.31, the block will be scaled to fit inside the cell, no matter how large or small the block is. The title of the symbol was also added to the table by double-clicking inside the cell to launch the Text Formatting toolbar.

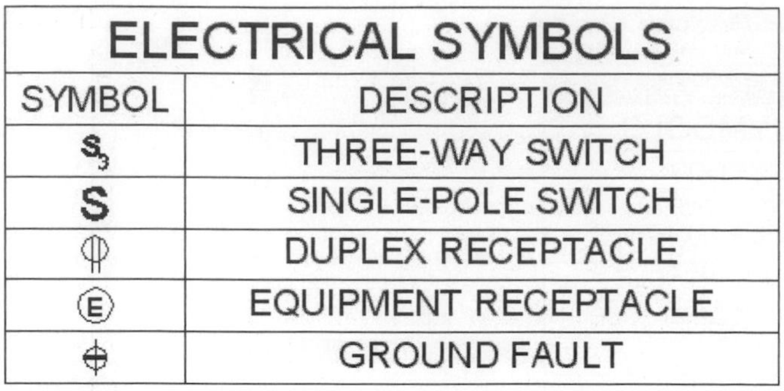

ELECTRICAL SYMBOLS	
SYMBOL	DESCRIPTION
S_3	THREE-WAY SWITCH
S	SINGLE-POLE SWITCH
	DUPLEX RECEPTACLE
(E)	EQUIPMENT RECEPTACLE
	GROUND FAULT

FIGURE 16.32

INTRODUCING THE DESIGNCENTER

DesignCenter provides an additional means of inserting blocks and drawings even more efficiently than through the Insert dialog box. This feature has the distinct advantage of inserting specific blocks internal to one drawing into another drawing. Activate the ADCENTER command through one of the following methods:

- From the Ribbon > Insert Tab > Content Panel > Design Center Button
- From the Menu Bar (Tools > Palettes > DesignCenter)
- From the Standard toolbar
- From the keyboard (`ADC` or `ADCENTER`) (CTRL+2)

Access of this command from the Ribbon is illustrated in the following image.

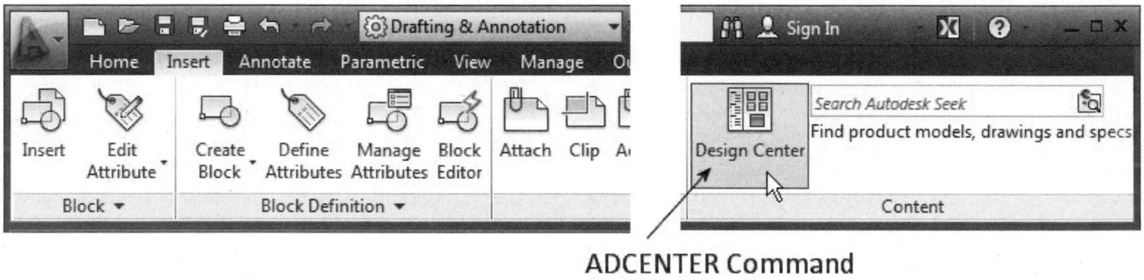

FIGURE 16.33

When used for the first time, the DesignCenter loads in the middle of the screen. To provide additional screen area, you may want to resize the palette, dock it to the side of the screen, or hide it. An Auto-hide feature for the DesignCenter allows it to hide (collapse) once you move the cursor outside the DesignCenter window. This clears the drawing area when the DesignCenter is not being used. To expand it again, simply move the cursor over the DesignCenter title bar. Right-click the title bar to display the shortcut menu shown in Figure 16.34. This menu allows you to turn the Auto-hide and docking features on or off as desired. You can unload the DesignCenter by clicking the X in the title bar, selecting DesignCenter from one of the menus, or entering the `ADCCLOSE` command at the keyboard.

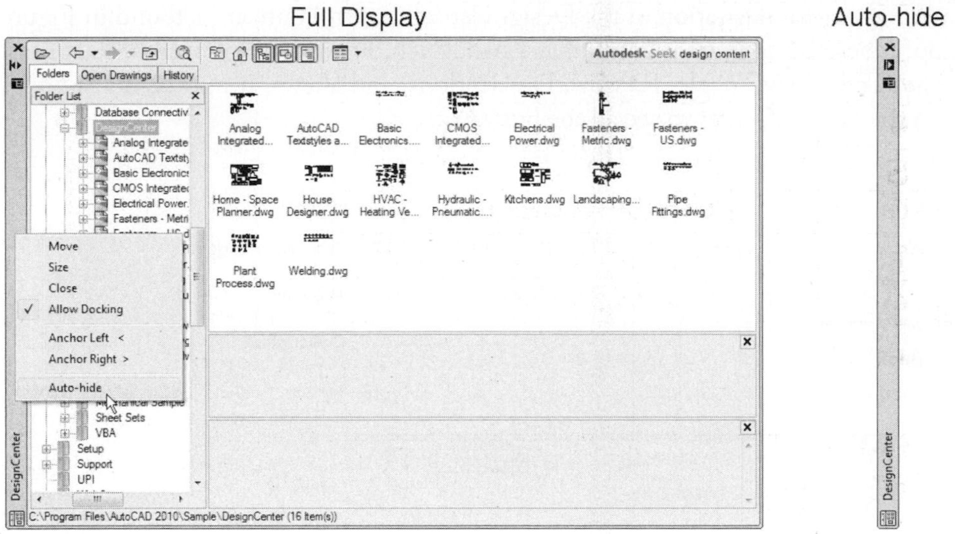

FIGURE 16.34

The main function of the DesignCenter is to transfer blocks between drawings. Block libraries can be prepared in different formats in order for them to be used through the DesignCenter. One method is to place all global blocks in one folder. The DesignCenter identifies this folder and graphically lists all drawing files to be inserted. Another method of organizing blocks is to create one drawing containing all local blocks. When this drawing is identified through the DesignCenter, all blocks internal to this drawing display in the DesignCenter palette area.

DESIGNCENTER COMPONENTS

The DesignCenter is isolated in Figure 16.35. The following components of the palette are identified: Control buttons, Tree View or Navigation Pane, Palette or Content Pane, Preview, Description, and Shortcut Menu.

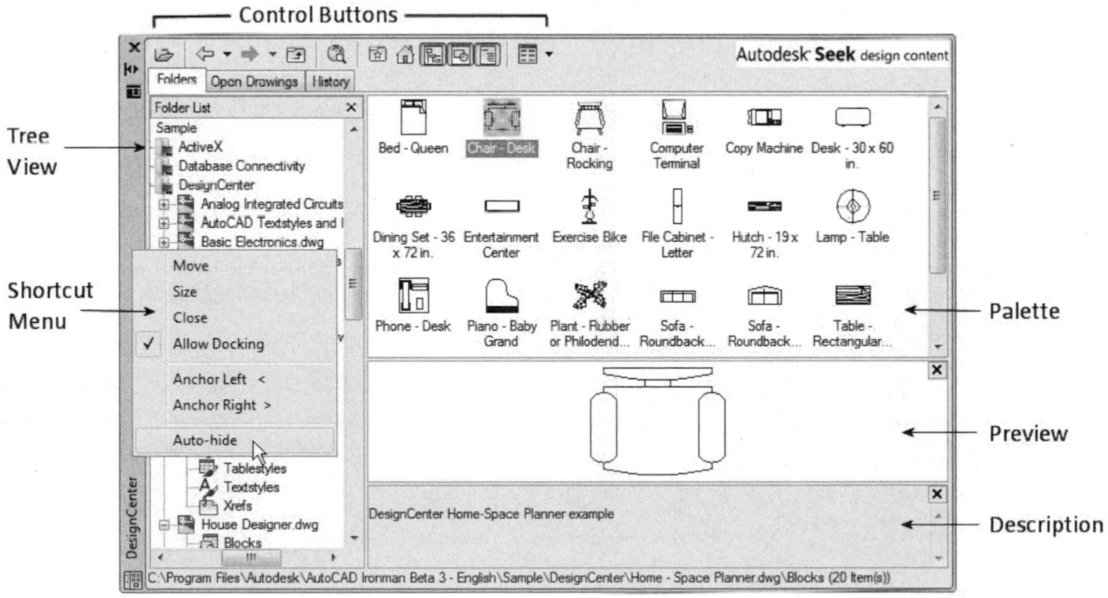

FIGURE 16.35

A more detailed illustration of the DesignCenter Control buttons is found in Figure 16.36. These buttons are identified as Load, Back, Forward, Up, Search, Favorites, Home, Tree View Toggle, Preview, Descriptions, and Views. It may be necessary to resize the DesignCenter to see all the buttons.

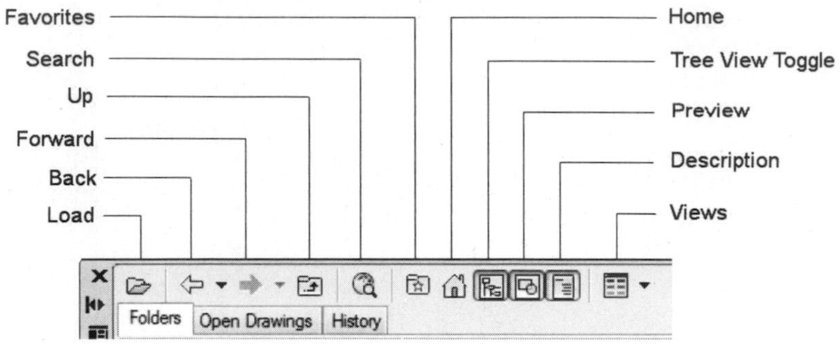

FIGURE 16.36

USING THE TREE VIEW

Clicking the Tree View button expands or contracts the DesignCenter to look similar to the illustrations in Figure 16.37. When Tree View is turned on, the DesignCenter divides into two major areas: the familiar Palette area where the symbols are located, and the Tree View area that shows the folder structure. Turning Tree View off hides the folder structure and expands the palette where the symbols are located.

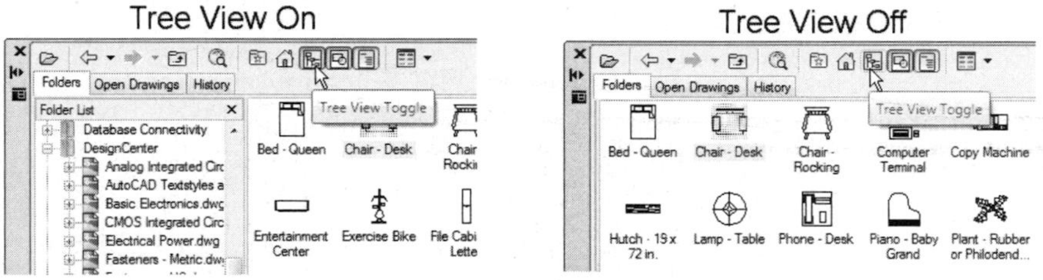

FIGURE 16.37

Clicking a drawing in the Tree View displays the drawing objects, as shown in Figure 16.38, that can be shared through the DesignCenter (Blocks, Styles, Layers, Layouts, Linetypes, and Xrefs).

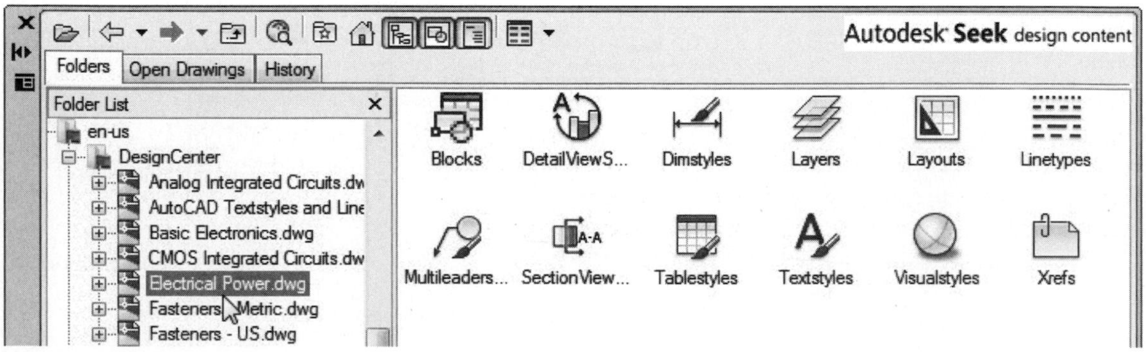

FIGURE 16.38

Double-click the Blocks icon (or click the "+" symbol next to the drawing name and choose Blocks in the Tree View) to display the blocks available in the selected drawing, as shown in Figure 16.39.

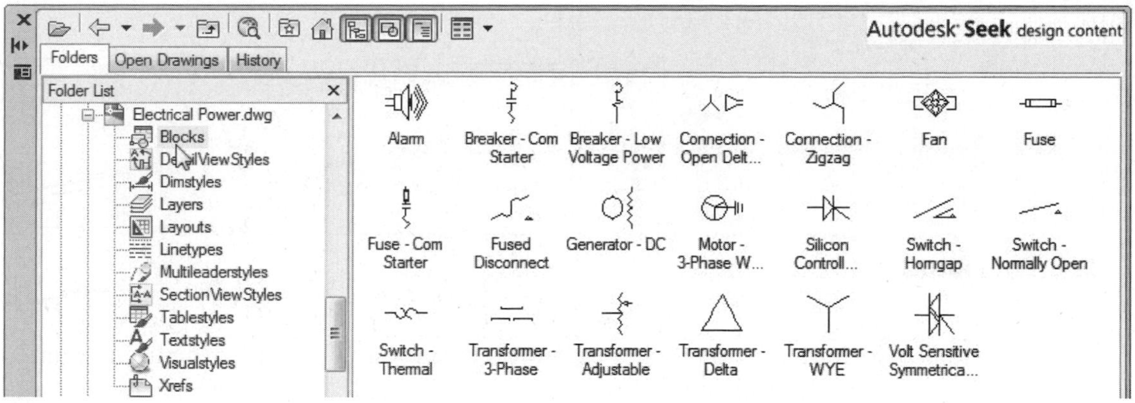

FIGURE 16.39

INSERTING BLOCKS THROUGH THE DESIGNCENTER

Figure 16.40 displays a typical floor plan drawing along with the DesignCenter (docked to the left side of the screen), showing the blocks identified in the current drawing. If no blocks are found internal to the drawing, the Palette area will be empty.

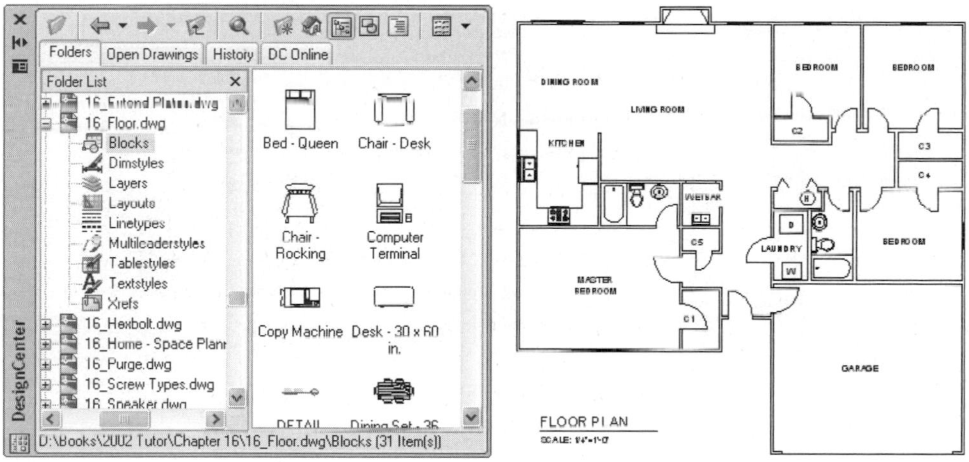

FIGURE 16.40

The DesignCenter operates on the "drag-and-drop" principle. Select the desired block located in the Palette area of the DesignCenter, drag it out, and drop it into the desired location of the drawing. Figure 16.41 shows a queen-size bed dragged and dropped into the bedroom area of the floor plan.

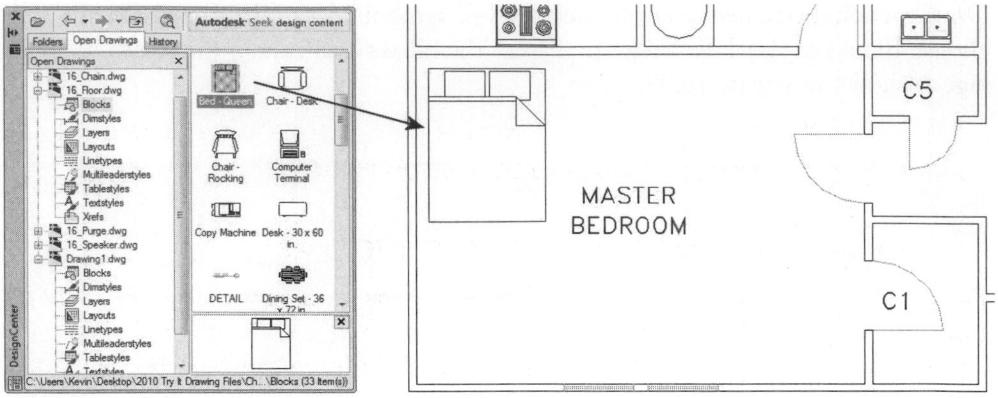

FIGURE 16.41

When performing a basic drag-and-drop operation with the left mouse button, all you have to do is identify where the block is located and drop it into that location (the use of running object snaps can ensure that the blocks are dropped in a specific location). What if the block needs to be scaled or rotated? Right-click the block in the Palette area and a shortcut menu is provided, as shown in the following image. Selecting Insert Block from the shortcut menu displays the Insert dialog box, covered earlier in this chapter, which allows you to specify different scale and rotation values. Figure 16.42 shows a rocking chair inserted and rotated into position in the corner of the room. Instead of right-clicking a block to display the shortcut menu, double-click the block in the Palette area and the Insert dialog box is provided immediately.

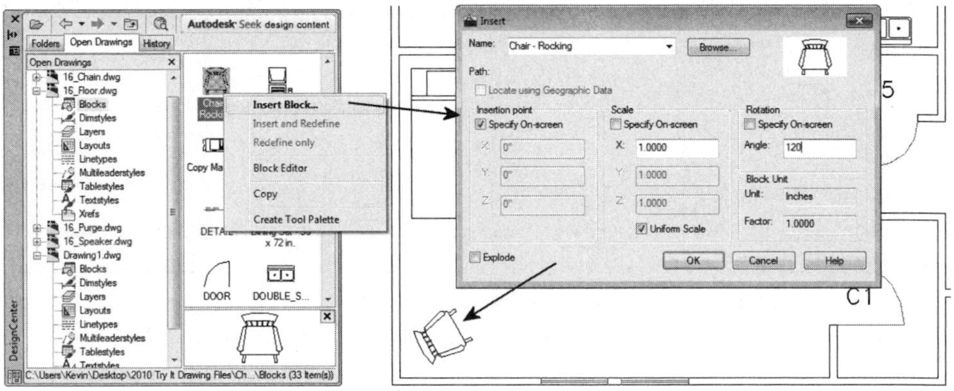

FIGURE 16.42

INSERTING BLOCKS USING THE TOOL PALETTE

Generally, tool palettes allow you to organize blocks and hatch patterns for insertion into your drawing. This feature is somewhat similar to the DesignCenter in its ability to drag and drop blocks, layers, dimension styles, text styles, and hatch patterns into a drawing. The Tool Palette, however, is specific to blocks, commands, and hatch patterns. You can organize and customize the Tool Palette to meet your individual drawing needs.

Depending on the workspace you are utilizing, the Tool Palette may already be displayed on your screen. If the Tool Palette is not visible, it can be activated by using one of the following methods:

- From the Ribbon > View Tab > Palettes Panel > Tool Palettes Button
- From the Menu Bar (Tools > Palettes > Tool Palettes)
- From the Standard toolbar
- From the keyboard (TP or TOOLPALETTES) (CTRL+3)

Access of this command from the Ribbon is illustrated in Figure 16.43. Once displayed, the Tool Palette provides a number of sample tabs for you to experiment with, as shown in Figure 16.43 (Right). Notice in this figure how block icons are displayed with a lightning bolt graphic. This means that the block is considered dynamic. This feature of creating and using dynamic blocks will be covered later in this chapter.

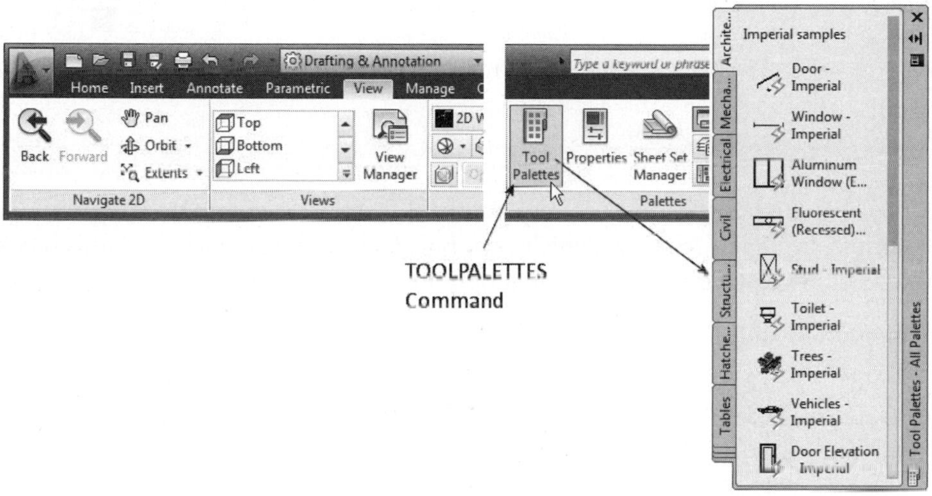

FIGURE 16.43

Additional tool palettes are available to assist in your design capabilities. To access these extra palettes, click the area located in the lower-left corner of any tool palette, as shown in Figure 16.44. A long list activates alongside the existing palette. A number of palettes consist of blocks that can be dragged and dropped into a drawing. Other palettes such as Draw and Modify consist of commands while the Hatch and Fills palette consists of hatch patterns.

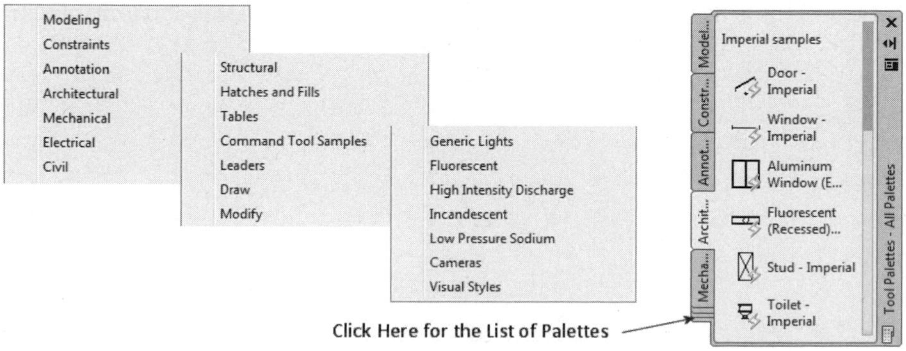

FIGURE 16.44

TRY IT!

Open the drawing file 16_Palette. Display the Tool Palette and click the lower-left corner of the palette. Select Architectural from the displayed list. Drag and drop the Vehicles-Imperial block into the drawing using the Endpoint at (A), as shown in Figure 16.45 (Left). Next, display the Hatch and Fills palette and drag and drop the Ar-Conc hatch pattern into the rectangle near (B), as shown in Figure 16.45 (Right).

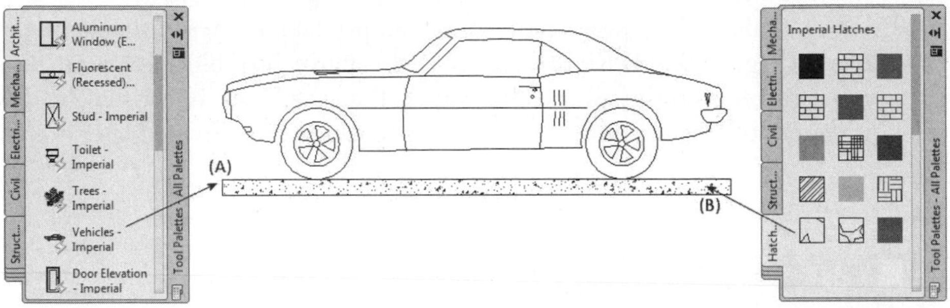

FIGURE 16.45

The Tool Palette has a number of very powerful features to automate its operation. For example, in Figure 16.46, a Vehicles symbol is selected. Right-clicking this block activates a shortcut menu. The following options are available:

Cut—Cutting the block from the Tool Palette to the Windows clipboard

Copy—Copying the block to the Windows clipboard

Delete—Deleting this block from the Tool Palette

Rename—Changing the name of this block in the Tool Palette

Properties—Changing the properties of this block in the Tool Palette

The Properties option allows you to modify the object's properties to suit your specific needs. Select the Properties option, as shown in Figure 16.46 (Left). This activates the Tool Properties dialog box, as shown in Figure 16.46 (Center). All the information in the fields can be changed and applied to this symbol. For example, suppose you need to change the insertion scale of this block for a number of drawings. Changing the scale in the Tool Properties dialog box changes the block's scale as it is inserted into the drawing. This feature is also available for hatch patterns when using the Tool Palette. Scrolling down this dialog box displays a heading for Custom Properties. In the illustration in Figure 16.46 (Right), numerous versions of the block are displayed. This is because the item selected was constructed as a dynamic block.

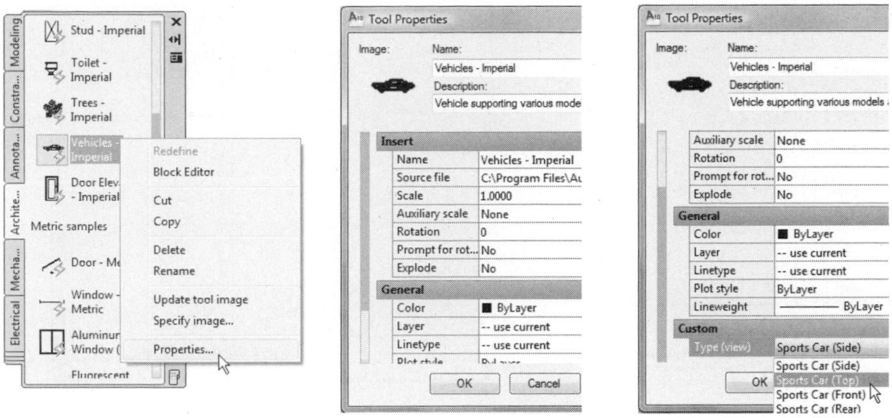

FIGURE 16.46

Creating a New Tool Palette

The real power of the Tool Palette is your ability to group the blocks, hatch patterns and commands you commonly use into one easy to access area. You may, for example, want to create new palettes for specific work tasks or projects. The process for creating new tool palettes is very easy and straightforward. First, right-click anywhere inside the Tool Palette to display the menu illustrated in Figure 16.47 (Left). Options of this Tool Palette menu include:

> **Allow Docking**—Allows the Tool Palette to be docked to the sides of your display screen. Removing the check disables this feature.
>
> **Auto-Hide**—When checked, this feature collapses the Tool Palette so only the thin blue strip is displayed. When you move your cursor over the blue strip, the Tool Palette redisplays.
>
> **Transparency**—Activates a Transparency dialog box, which controls the opaqueness of the Tool Palette.
>
> **View Options**—Activates the View Options dialog box, which controls the size of the hatch and block icons and whether the icon is labeled or not.
>
> **New Palette**—Creates a blank Tool Palette.
>
> **Delete Tool Palette**—Deletes the Tool Palette. A warning dialog box appears asking whether you really want to perform this operation.
>
> **Rename Tool Palette**—Renames the Tool Palette.
>
> **Customize Palettes**—Activates the Customize dialog box, which allows you to create a new Tool Palette.

TIP

Like with the DesignCenter, take advantage of the Dock and Auto-Hide features to free up screen area. The Transparency option can also be utilized on palettes to allow you to see more of your display screen. A slider bar allows you to set the amount of opaqueness of the Tool Palette, allowing you to see the screen through the palette.

Click the New Palette option, as shown in Figure 16.47 (Left). This automatically creates a blank Tool Palette. As illustrated in Figure 16.47 (Right), a new Tool Palette name, Electrical, has been entered in the field.

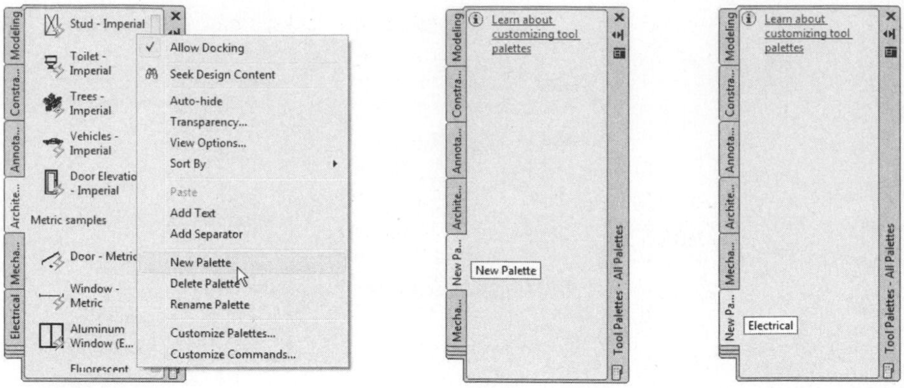

FIGURE 16.47

Once the Tool Palette name is entered, a tab is created for this palette, as shown in Figure 16.48 (Right). One way to add blocks to this new palette is to activate the DesignCenter, search for the folder that contains the symbols you wish to place in the Tool Palette, and drag and drop these blocks from the DesignCenter into the Tool Palette, as shown in Figure 16.48 (Left). Blocks and hatch patterns can also be added to a palette from an open drawing. Select the object in the drawing to highlight it and then drag and drop it onto the palette. Be sure, when dragging blocks, not to pick on a grip (blue box) or you will be simply moving the block.

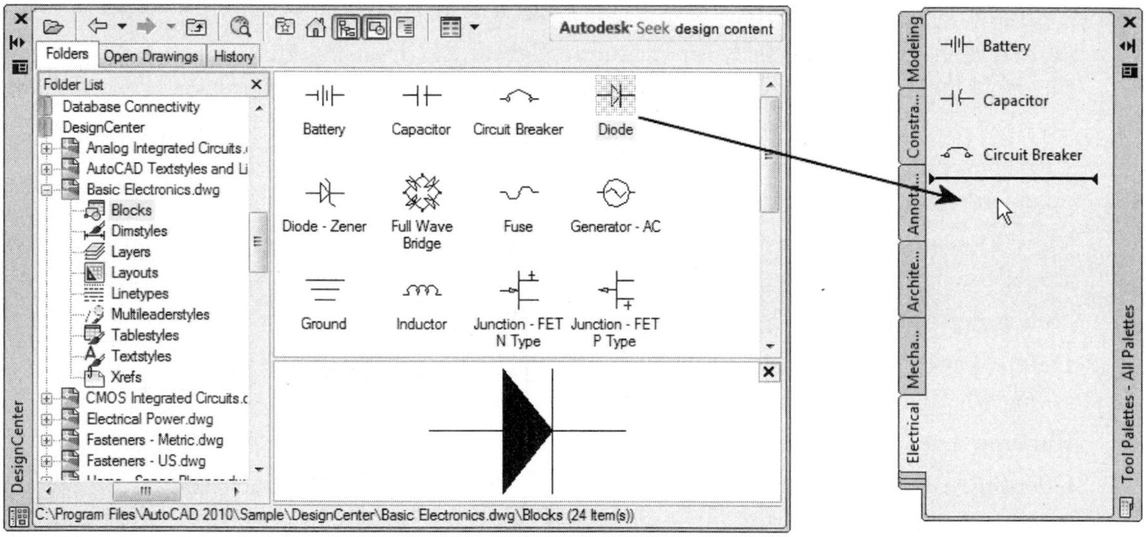

FIGURE 16.48

If you want to create a new Tool Palette from one whole drawing in the DesignCenter, activate the DesignCenter and go to the drawing that contains the blocks. Right-clicking this drawing displays the menu, as shown in Figure 16.49 (Left). Clicking the Create Tool Palette option creates the Tool Palette using the same name as the DesignCenter drawing. The new Tool Palette contains all blocks from this drawing, as shown in Figure 16.49 (Right).

TIP

If you right-click a folder and select Create Tool Palette of Blocks, a new Tool Palette will be created with the name of the folder and will contain the drawings from that folder.

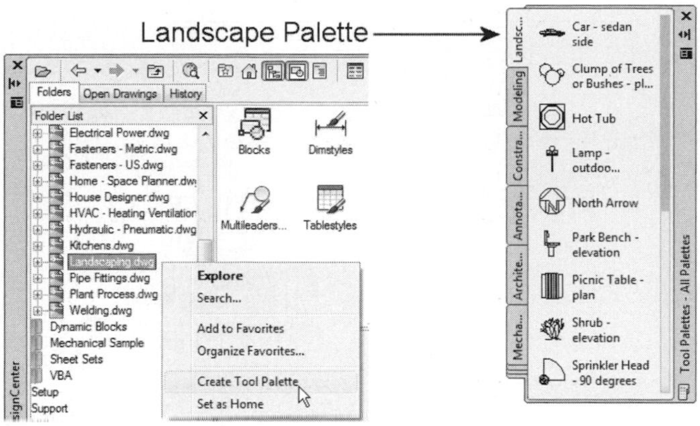

FIGURE 16.49

WORKING WITH MULTIPLE DRAWINGS

The Multiple Design Environment allows users to open multiple drawings within a single session of AutoCAD, as shown in Figure 16.50. This feature, like Design-Center, allows the sharing of data between drawings. You can easily copy and move objects, such as blocks, from one drawing to another.

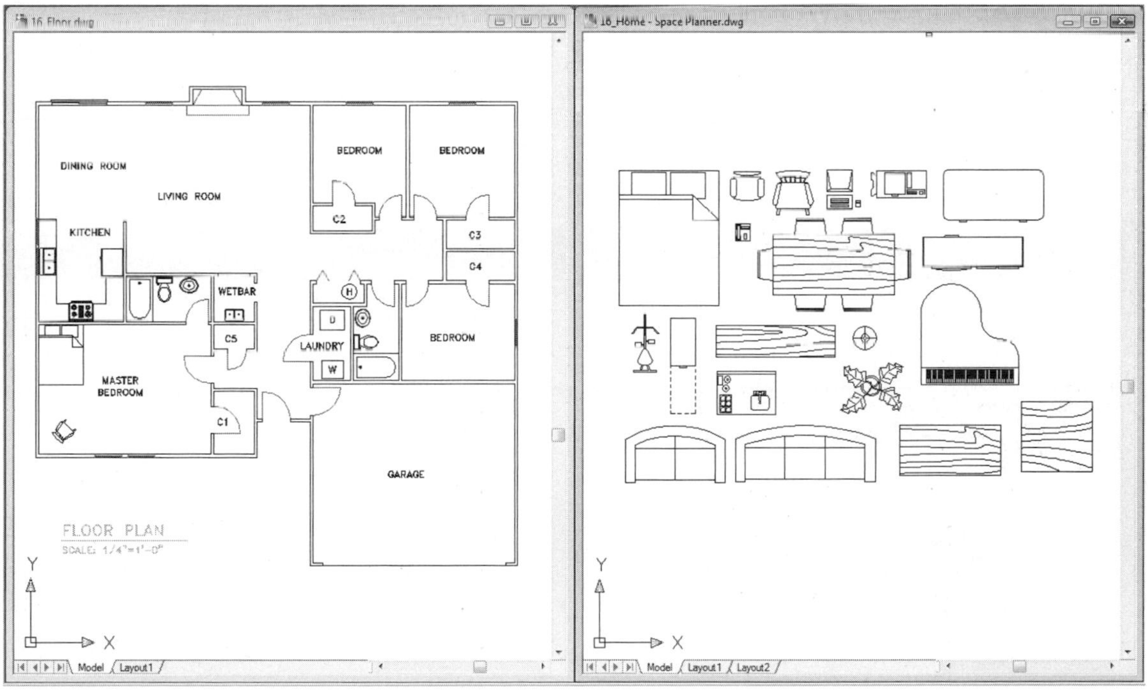

FIGURE 16.50

Opening Multiple Drawings

Repeat the OPEN command as many times as necessary to open all drawings you will need. In fact, you can select multiple drawings in the Select File dialog box by holding down CTRL or SHIFT as you select the files, as shown in Figure 16.51.

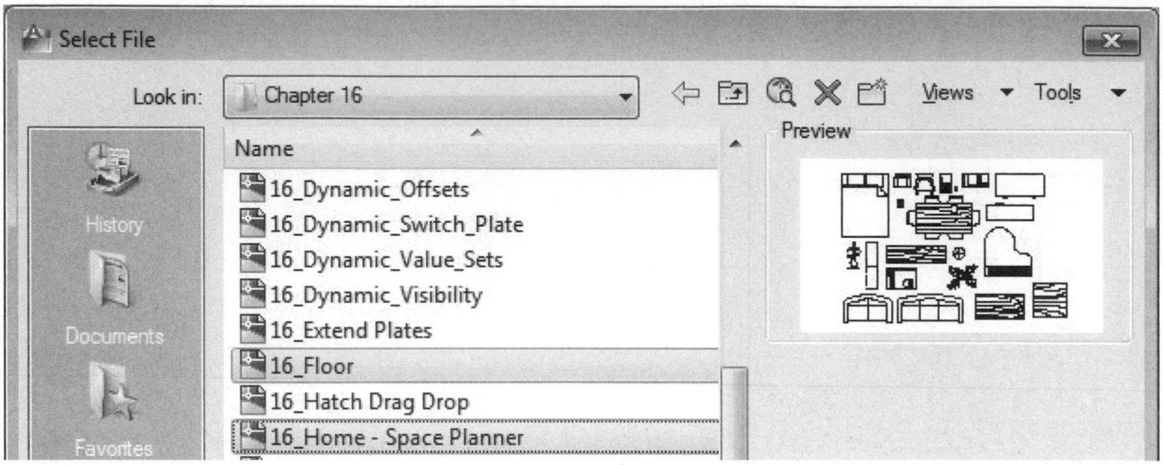

FIGURE 16.51

Once the drawings are open, use CTRL+F6 or CTRL+TAB to switch back and forth between the drawings. To efficiently work between drawings, you may wish to tile or cascade the drawing windows utilizing the User Interface Panel of the View Ribbon, as shown in Figure 16.52. A list of all the open drawings can be displayed by expanding the Switch Windows button. Selecting one of the file names displayed in the list is another convenient way to switch between drawings. Remember to use the CLOSE command (Application Menu > Close > Current Drawing) to individually close any drawings that are not being used. To close all drawings in a single operation, the CLOSEALL command (Application Menu > Close > All drawings) can be used. If changes were made to any of the drawings, you will be prompted to save those changes before the drawing closes.

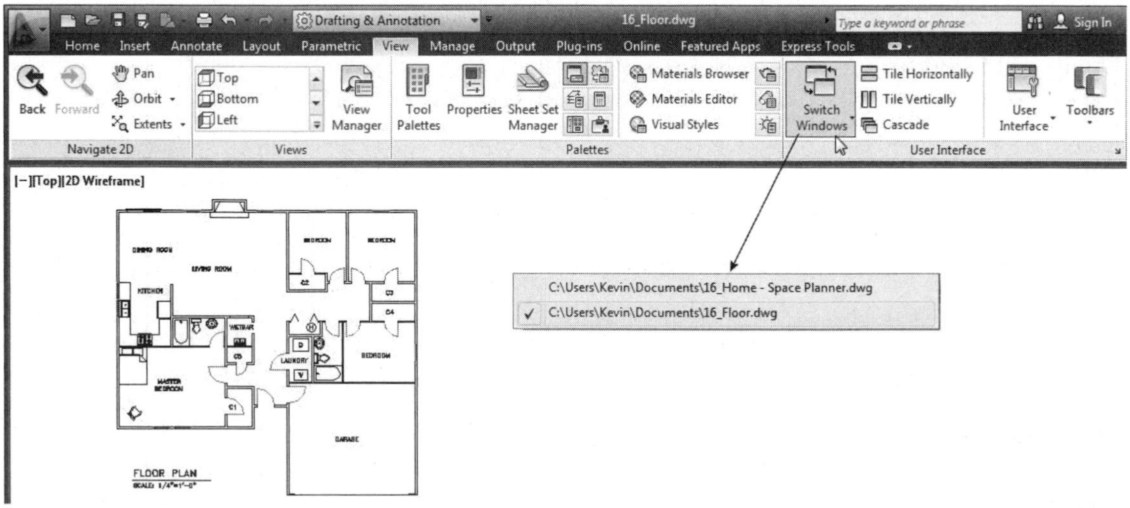

FIGURE 16.52

Working between Drawings

Once your drawings are opened and arranged on the screen, you are ready to cut and paste, copy and paste, or drag and drop objects between drawings. The cut and paste and copy and paste methods utilize the Windows clipboard. The first step is to cut or copy objects from a drawing. The object information is stored on the Windows clipboard until you are ready for the second step, which is to paste the objects into that same drawing or any other open drawing. These operations are not limited to AutoCAD. In fact, you can cut, copy, and paste between different Windows applications.

Use one of the following commands to cut and copy your objects:

CUTCLIP—To remove selected objects from a drawing and store them on the clipboard

COPYCLIP—To copy selected objects from a drawing and store them on the clipboard

COPYBASE—Similar to the `COPYCLIP` command, but allows the selection of a base point for locating your objects when they are pasted

Use one of the following commands to paste your objects:

PASTECLIP—Pastes the objects at the location selected

PASTEBLOCK—Similar to `PASTECLIP` command but objects are inserted as a block and an arbitrary block name is assigned

PASTEORIG—Pastes clipboard information into the current drawing using the same coordinates from the originating drawing

These commands can be typed at the keyboard, selected from the Home Ribbon, as shown in Figure 16.53 (Left), or selected by right-clicking the display screen when a command is not in progress, as shown in Figure 16.53 (Right).

Objects may also be copied between drawings with drag-and-drop operations. After selecting the objects, place the cursor over the objects (without selecting a grip) and then drag and drop the objects in the new location. Dragging with the right mouse button depressed provides a shortcut menu allowing additional control, such as to paste the object as a block or perform a move operation instead of a copy.

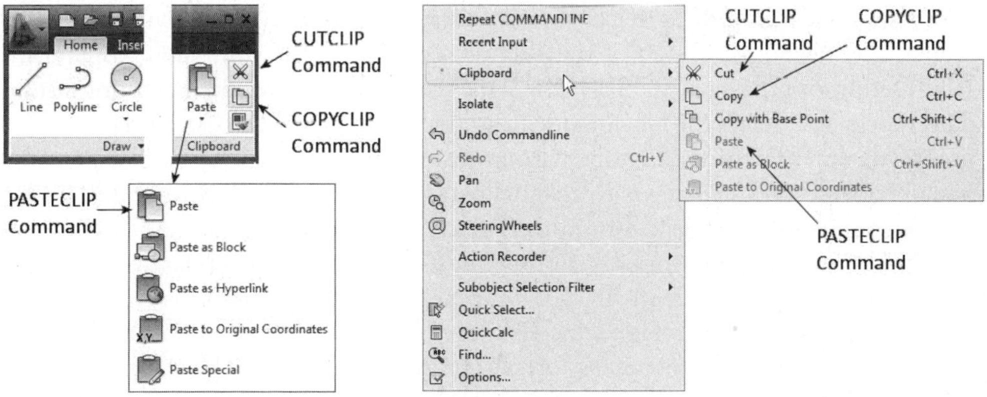

FIGURE 16.53

ADVANCED BLOCK TECHNIQUES—CREATING DYNAMIC BLOCKS

Dynamic blocks are blocks that change in appearance whenever they are edited through grips or through custom tables embedded inside the block. In Figure 16.54, a dynamic block called Drawing Title is available under the Annotation tab of the Tool Palettes, as shown in Figure 16.54 (Right). The block can be identified in the palette as being dynamic by the appearance of a lightning bolt icon. This block is dragged and dropped into the current drawing. Clicking this block displays the normal insertion point grip. However, notice a second grip, which appears as an arrow pointing in the right direction. This grip identifies a stretching action associated with the dynamic block. When you click this arrow and move your cursor, that portion of the dynamic block changes, giving the block a different appearance without redefining or exploding the block. In this example, the arrow grip is designed to stretch the line.

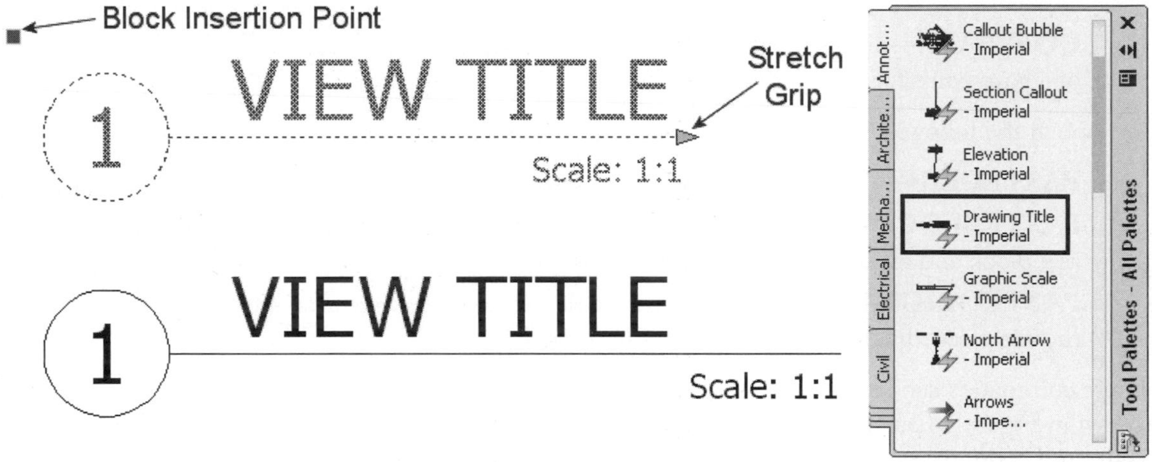

FIGURE 16.54

The creation process begins with the BEDIT command. Generally, the simplest way to activate this command is to double-click the block. Some blocks, such as this example, contain attributes (discussed in the next chapter) and double-clicking activates an attribute editor instead. In this case, activate the command by highlighting the block, right-clicking and picking Block Editor from the provided shortcut menu. The command can also be selected from either the Home (Block panel) or Insert (Block Definition panel) tab of the Ribbon. Activating the command displays the Edit Definitions dialog box. This is the same dialog box used when redefining a standard block. You select the block name from the list and click the OK button. At this point, you enter the Block Editor environment, as shown in Figure 16.55. To create a dynamic block, the first step is to assign a parameter to the block. In Figure 16.55, a Linear Parameter was selected from the Parameters tab of the Block Authoring palette and assigned to the geometry, in this case a line segment. This parameter usually is named Distance by default. In this example, the default name was renamed to a term with more meaning, Title Line Length. The Properties Palette is used for this renaming task. Once a parameter is present in the Block Editor, the second step is to link an action item, such as Stretch, to the parameter. It is the action item that allows the block to change automatically when edited. In this example, two action items are present, namely, Stretch and Move. The Stretch action allows the line to be stretched to different lengths. The Move action is required in order to move the VPSCALE attribute along with the line as it is being stretched.

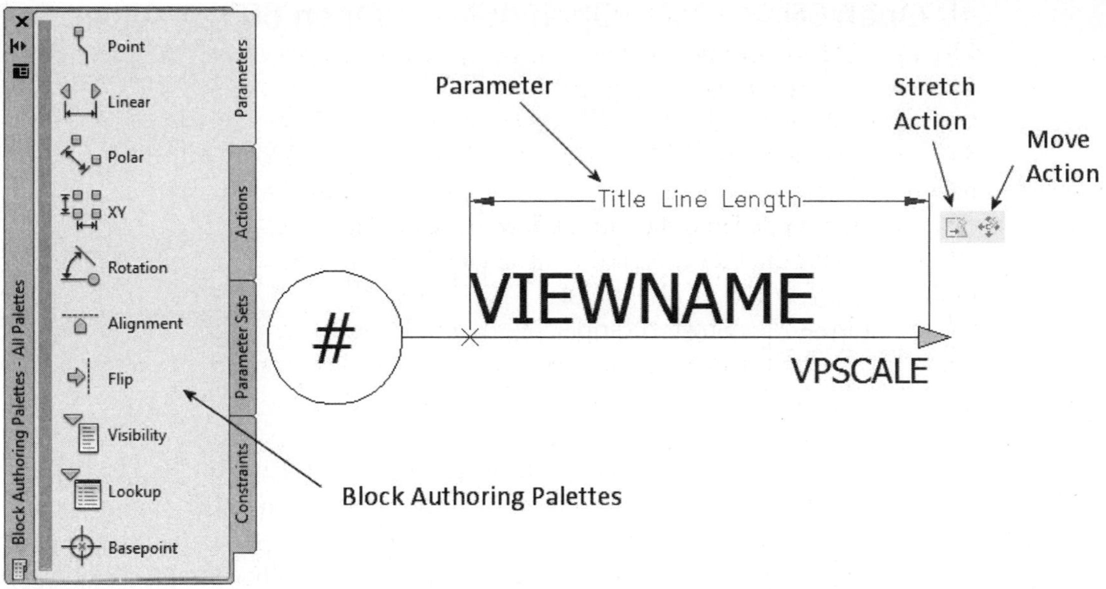

FIGURE 16.55

When you are satisfied with the parameter and action assignments and wish to test the features out on the block, you first click the Save Block button to save the changes to the block name and then click the Close Block Editor button, as shown in Figure 16.56, to return to the drawing editor.

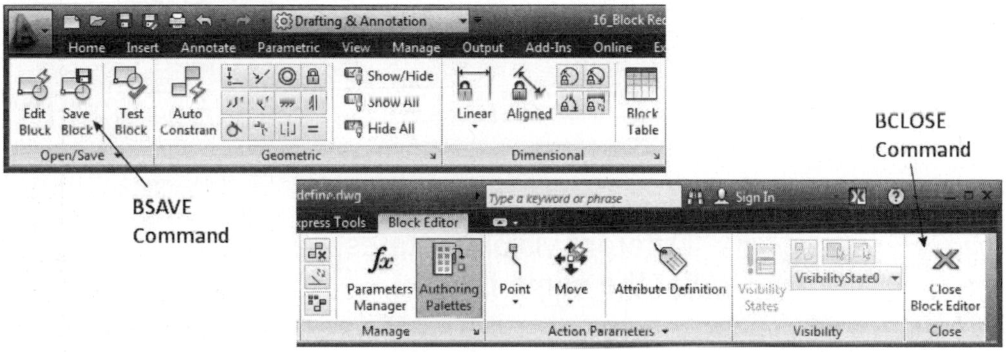

FIGURE 16.56

Figure 16.57 illustrates another, more powerful example of using dynamic blocks. Normally, you would have to create four different blocks in order to show the various door swings. Through the use of Visibility States, all four doors pictured in Figure 16.57 belong to a single block name. You simply pick the desired door opening from a list that displays with the block. Another feature of dynamic blocks illustrated in Figure 16.57 is the ability to flip the door to different locations. Two Flip Grips allow you to flip the door along horizontal or vertical hinges. Notice also a Stretch Grip. This is present to stretch the door based on different wall thicknesses. As you can see, dynamic blocks can easily become a major productivity tool used to reduce the overall number of differently named blocks in your drawing.

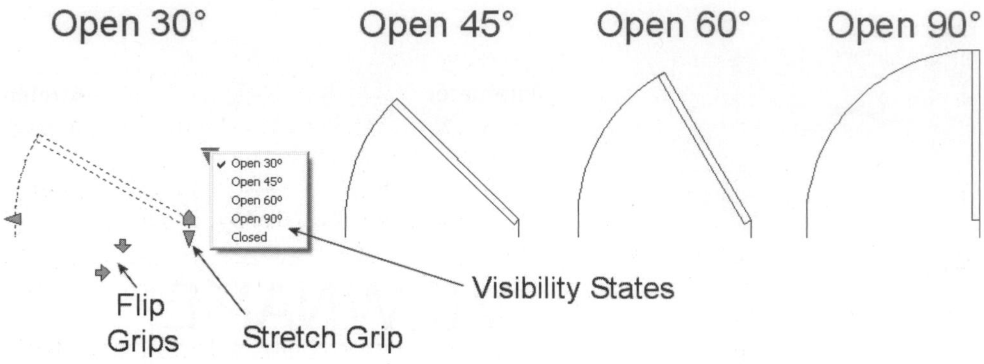

FIGURE 16.57

Yet another example of dynamic blocks is illustrated in Figure 16.58. Through the use of Visibility States, you have the ability to consolidate a number of blocks under a single name. In this image, notice the block name is Trees – Imperial, as shown on the right in the Tool Palette. This single block name actually contains 12 different tree blocks. When you insert one of these dynamic tree blocks, the Visibility States grip appear. Clicking the grip activates the list of trees. You pick the tree from the list and the previous tree block changes based on what you select from the list.

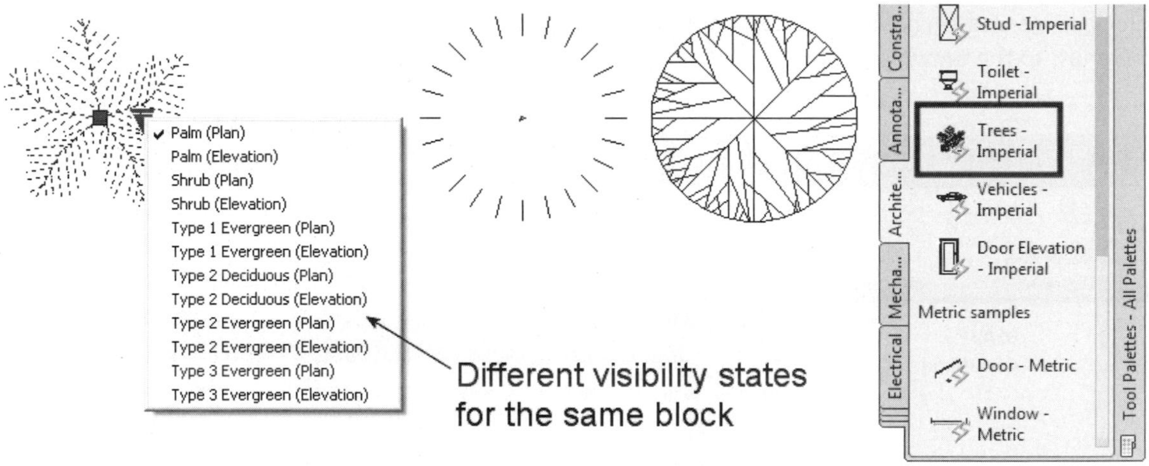

FIGURE 16.58

TRY IT!

Start a new drawing from scratch (Acad.dwt). Display the Tool palette (Ribbon – View tab > Palettes panel > Tool Palette button) and select the Architectural tab. Drag and drop the Door – Imperial and Trees – Imperial blocks into the drawings. Zoom out and pick each block to display the dynamic block grips. Experiment by clicking the various grip types.

The following table gives a brief description of each dynamic block grip.

Dynamic Block Grip	Name	Description
■	Insertion Grip	The original grip that displays at the insertion point of a block; can also be used for moving the block to a new location
► ▼	Lengthen	Allows the block to be lengthened or shortened by stretching, scaling, or arraying
⌂	Alignment	Positions or aligns the block based on an object; the positioning occurs when you move the block near the object
⇨ ⇩	Flip	Allows the entire block or items inside the block to be flipped
●	Rotate	Allows the entire block or items inside the block to be rotated
▼	List	Allows you to choose from a list of items

Before you begin the process of designing dynamic blocks, here are a few items to consider:

- For what intended purpose are you designing this dynamic block?
- How do you want this block to change when it is being edited?
- What parameters are needed in order to create changes in the block?
- What actions need to be assigned to the parameters?
- Do you need the block to contain various size values in order to make incremental changes?
- Do you want to organize various blocks under a single name and control what is displayed through Visibility States?
- Do you want to create a table consisting of different values and change the size of a block through the table?

The next series of exercises will allow you to experiment with various capabilities of dynamic blocks.

Working with Parameters and Actions

This exercise is designed to familiarize you with the basics of assigning parameters and actions to a block and testing its dynamic nature.

TRY IT!

Open the drawing file 16_Dynamic_Basics. An existing block called Table01 is already created. The size of this table is 6' by 4'. Double-click the existing block Table01 to launch the Edit Block Definition dialog box, as shown in Figure 16.59 (Right).

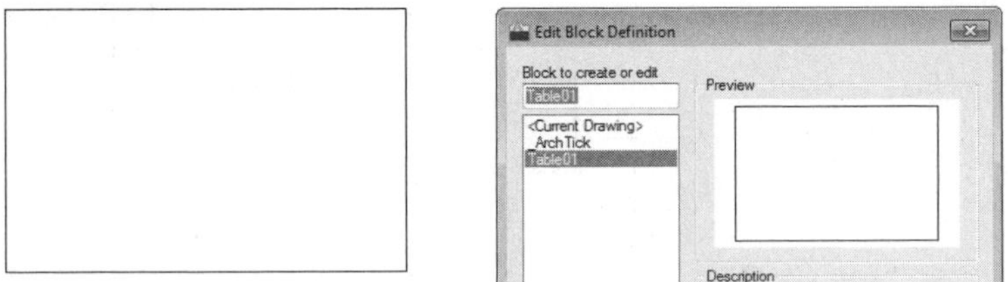

Double-Click Block

FIGURE 16.59

Clicking the OK button in the Edit Block Definition dialog box launches the Block Editor, as shown in Figure 16.60. You can make changes to the geometry of the existing block or you can assign parameters and actions, making the block dynamic. These assignments are made through the Block Authoring Palettes, as shown in Figure 16.60. Additional tools are available on the Block Editor Ribbon displayed at the top of the screen.

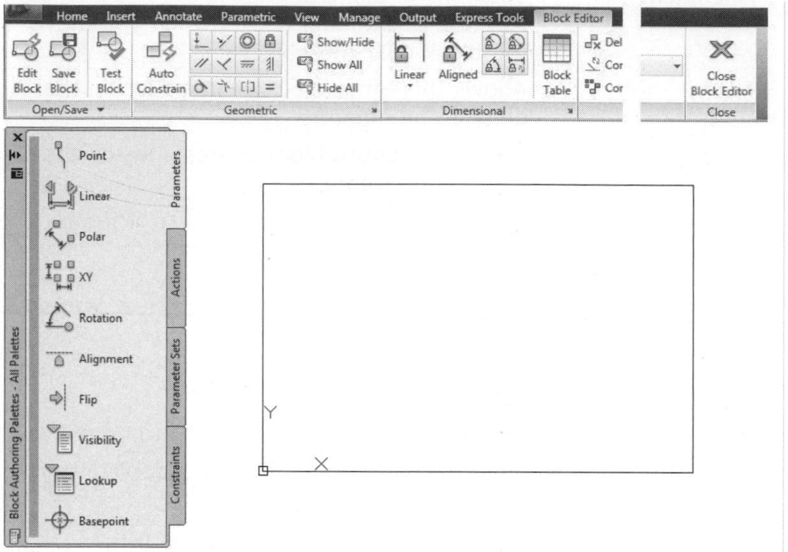

FIGURE 16.60

We want the ability to change the overall width of this table block. To do this, first make sure the Parameters tab is active and click the Linear Parameter tool, as shown in Figure 16.61. You will be directed through a series of prompts, which can be found in the Command prompt area.

Just as in dimensioning, you can pick a starting point and endpoint for the parameter, namely, the endpoints of the bottom line of the rectangle. You can also specify the label location, as shown in Figure 16.61.

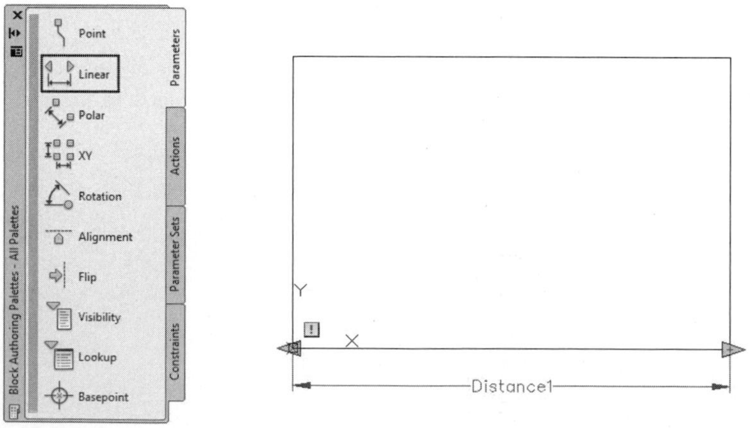

FIGURE 16.61

Adding a parameter is the first step in creating a dynamic block. An action must now be associated with this parameter. Select the Actions tab on the palette. Since you want the ability to change the width of the table, select the Stretch action, which will be used to accomplish this task. As with adding the parameter, the Stretch action comes with a lengthy series of Command prompts.

When initiating the Stretch action, first pick the existing Distance parameter. Since the parameter was created with two endpoints, specify the endpoint to associate with the Stretch action. Select the rightmost parameter endpoint. Next, create a crossing box around the area to Stretch, as shown in Figure 16.62. Then select the rectangle as the object to stretch. Upon completion, a Stretch action icon is displayed.

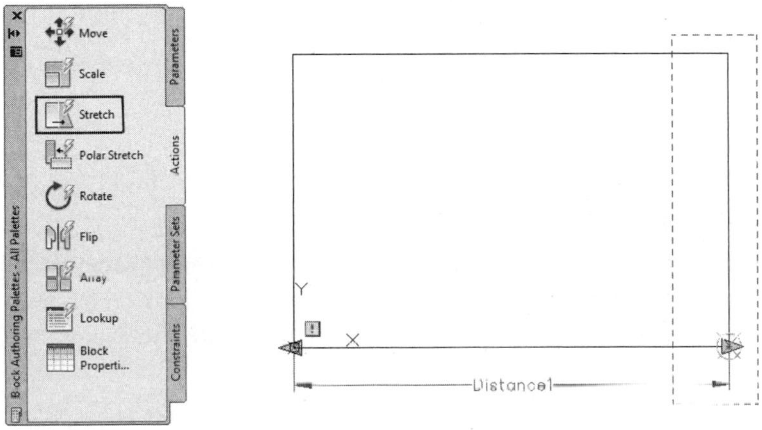

FIGURE 16.62

Figure 16.63 illustrates all the components that make up the dynamic block. The Stretch icon should be positioned near the side of the object the action occurs at. Notice also the Alert icon. One more action needs to be created in order for the Alert icon to disappear. For the purpose of this exercise, we will not need an additional action because we will be stretching this rectangle only to the right. Click the Save Block button to save these changes to the block. To return to the drawing editor, click the Close Block Editor button.

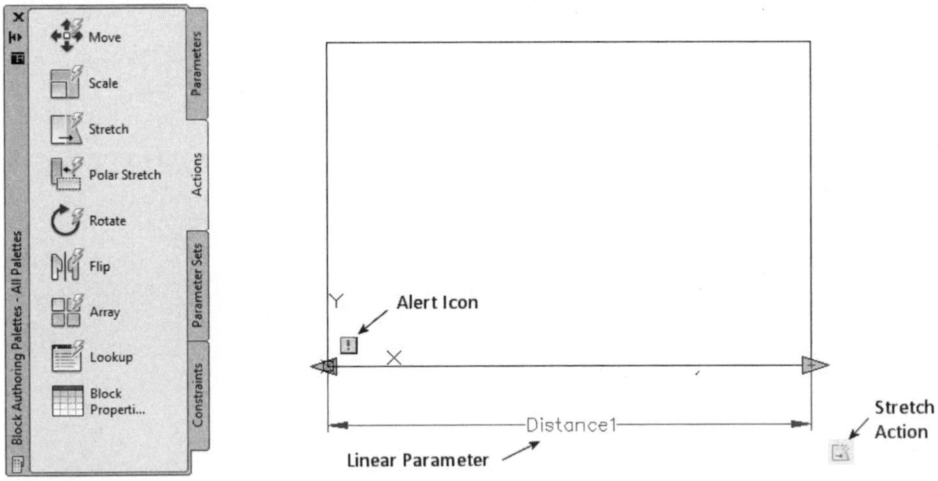

FIGURE 16.63

While back in the drawing, add a dimension to the existing block. Then click the rectangle (block) and notice that an arrow appears. This arrow represents the parameter. Clicking the right arrow stretches the block to the right or left. Notice also that the dimension changes to the new value.

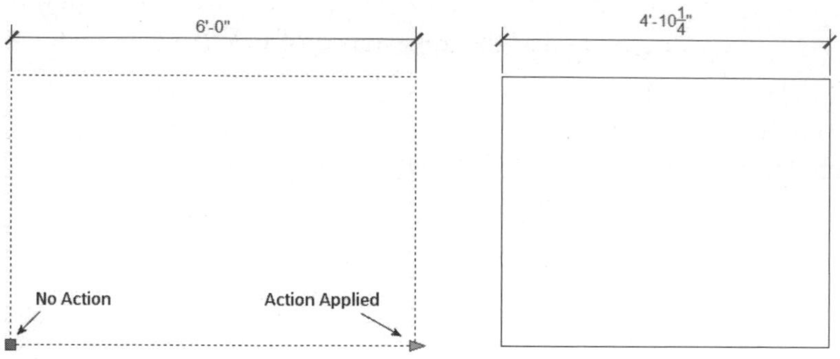

FIGURE 16.64

Working with Value Set Properties for Parameters

Value sets are associated with parameters and are used to define custom property values for block references. Three types of value sets can be defined in a block reference; they are List, Increment, and None. This exercise will demonstrate the Increment value set.

TRY IT!

Open the drawing file 16_Dynamic_Value_Sets. We will be using the same table block from the previous Try it! exercise. However, instead of stretching the table to the right at random lengths, we want to better control the stretching operation. We want to stretch this table in increments of 6 inches. We also do not want the table to get narrower than 4 feet or wider than 8 feet. We will be able to control these items through the Value Set associated with the Distance parameter.

Double-click the block to activate the Edit Block Definition dialog box and click the OK button to enter the Block Authoring environment. Use the PROPERTIES command to activate the Properties Palette, and click the Distance parameter. Change the parameter name from Distance to Table Width, as shown in Figure 16.65. This gives the parameter a more meaningful name. Actions can also be renamed through the Properties Palette.

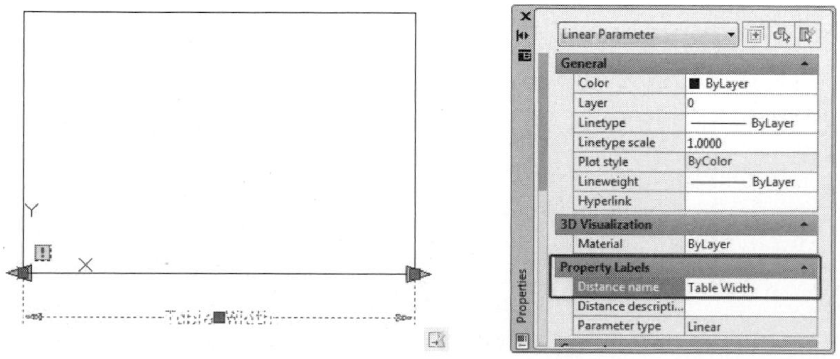

FIGURE 16.65

With the Properties Palette active and the Table Width parameter selected, scroll down to the bottom of the Properties Palette and notice the Value Set category. It is here that you activate one of the three Value Set types. For this example, make the Increment type active. This also displays other information related to this type. Change the Distance Increment field to 6 inches, change the Distance Minimum field to 4 feet, and change the Distance Maximum field to 8 feet, as shown in Figure 16.66. As you make these changes, notice the appearance of tick marks signifying these three values.

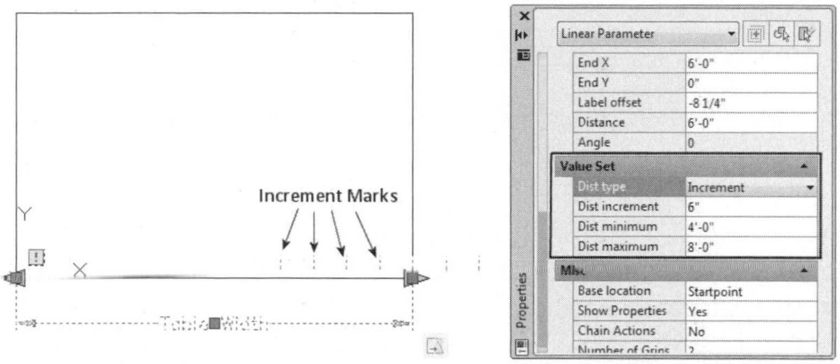

FIGURE 16.66

When you have finished creating these Value Sets for the Table Width parameter, save these changes to the current block name and then close the Block Editor. This returns you to your drawing. Click the block and then the rightmost parameter arrow. Notice the appearance of the tick marks, as shown in Figure 16.67. The tick marks are in increments of 6 inches. The tick mark on the far left begins at the 4 feet distance of the table. The tick mark on the far right ends at the 8 feet distance. Slide your cursor to the right or left and notice your cursor snapping to the increment tick marks. Clicking one of the tick marks adjusts the table to that distance.

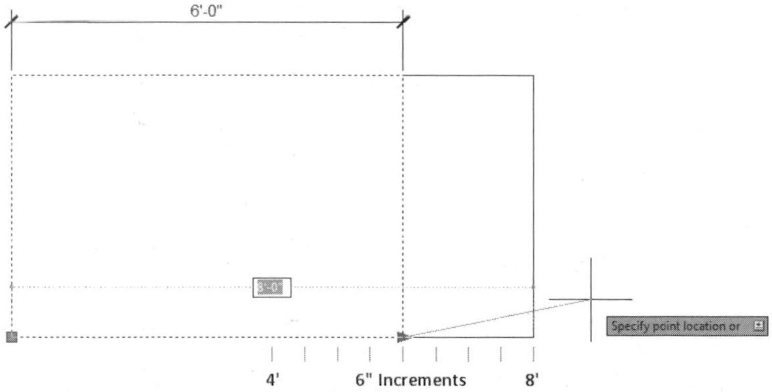

FIGURE 16.67

NOTE

To gain additional experience in the creation of dynamic blocks with different parameters and actions, refer to the 2 tutorial exercises at the end of this chapter and to available additional support files for Chapter 16 found at the Student Companion site from CengageBrain. Refer to the Introduction section of this text, for information on how to access these files.

TUTORIAL EXERCISE: 16_ELECTRICAL SCHEMATIC

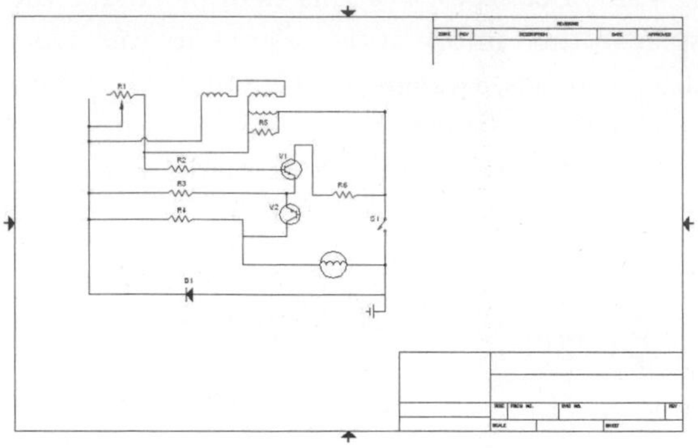

FIGURE 16.68

Purpose

This tutorial is designed to lay out electrical blocks such as resistors, transistors, diodes, and switches to form an electrical schematic, as shown in Figure 16.68.

System Settings

Create a new drawing called ELECT_SYMB to hold the electrical blocks. Keep all default units and limits settings. Set the Grid to 0.0750 and the Snap to 0.0375 units. This will be used to assist in the layout of the blocks. Turn off Object Snap and Polar Tracking modes, which can make it difficult to select some snap points. Once this drawing is finished, create another new drawing called ELECTRICAL_SCHEM1. This drawing will show the layout of an electrical schematic. Keep the default units but use the `LIMITS` command to set the upper-right corner of the display screen to (17.0000, 11.0000). Turn Object Snap and Polar Tracking back on for this drawing.

Layers

Create the following layers for ELECTRICAL_SCHEM1 with the format:

Name	Color	Linetype
Border	White	Continuous
Blocks	Red	Continuous
Wires	Blue	Continuous

Suggested Commands

This tutorial begins by creating a new drawing called ELECT_SYMB, which will hold all electrical blocks. You will use the image below as a guide in drawing all electrical blocks. A grid with spacing 0.0750 would provide further assistance with the drawing of the blocks. Once all blocks are drawn, the `BLOCK` command is used to create blocks out of the individual blocks. You will save this drawing and create a new drawing file called ELECTRICAL_SCHEM1. The DesignCenter will be used to drag and drop the internal blocks from the drawing ELECT_SYMB into the new drawing, ELECTRICAL_SCHEM1. Finally, you will connect all blocks with lines that represent wires and electrical connections, add block identifiers, and save the drawing.

STEP 1

Begin a new drawing and call it ELECT_SYMB. Be sure to save this drawing to a convenient location. It will be used along with the DesignCenter to bring the internal blocks into another drawing file. Activate the Drafting Setting dialog box by right-clicking on the Snap or Grid button on the Status bar and selecting Settings... from the shortcut menu. Make the settings as shown in Figure 16.69.

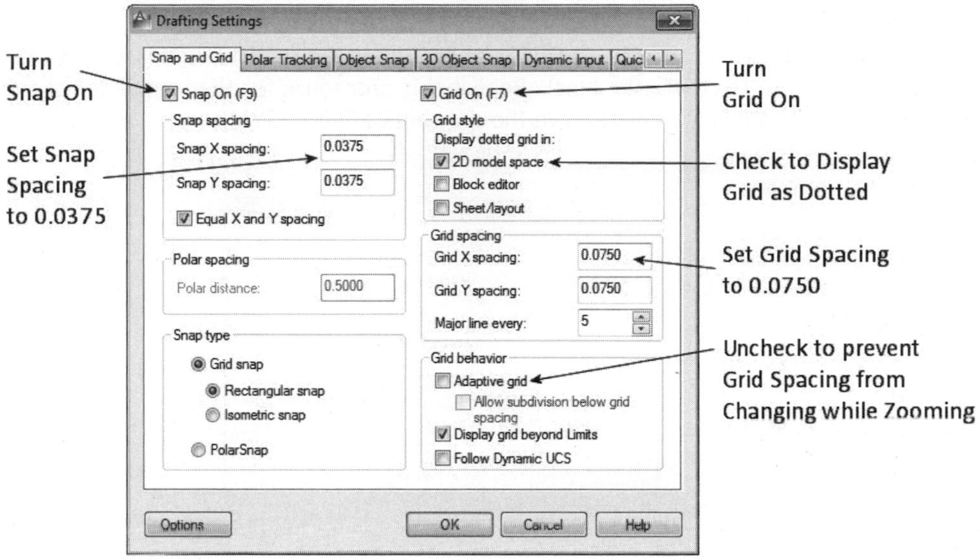

FIGURE 16.69

STEP 2

Using a grid spacing of 0.075 units and Figure 16.70 as a guide, construct each block. The grid/snap will help keep all blocks proportional to each other. Create all the blocks on the neutral layer 0. Also, the X located on each block signifies its insertion point (do not draw the X).

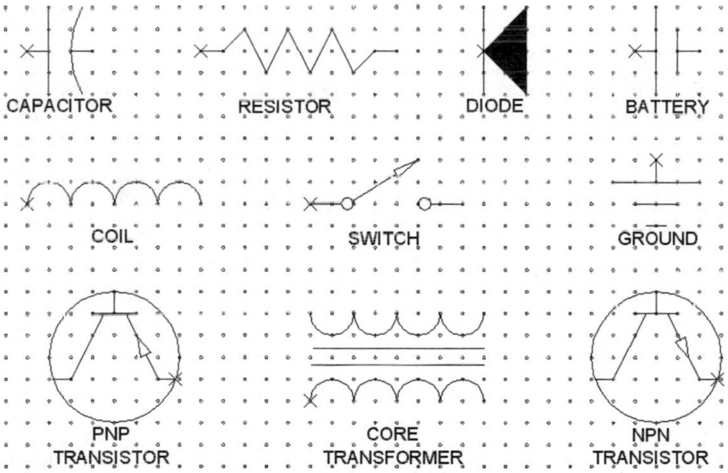

FIGURE 16.70

STEP 3

For a more detailed approach to creating the blocks, the resistor block will be created. Using the illustration in Figure 16.71 (Left) as a guide, first construct the resistor with the `LINE` command while using the grid/snap to connect points until the resistor is created. Be sure the Snap mode is active for this procedure. Then issue the `BLOCK` command, which will display the dialog box illustrated in Figure 16.71 (Right). This will be a local block to the drawing ELECT_SYMB; name the block RESISTOR. In the Select objects area of the dialog box, select all lines that represent the resistor (if you labeled the symbols, do not select the text as part of the block). Select the Retain radio button so that the symbol will remain visible in your drawing. Returning to the dialog box will show the objects selected in a small image icon on the right of the dialog box. Next, identify the base point of the resistor at the (A). It is very important to use Snap or an Object Snap mode when picking the base point. When finished with this dialog box, click the OK button to create the block. Repeat this procedure for the remainder of the electrical symbols. When complete, save the drawing as ELECT_SYMB.

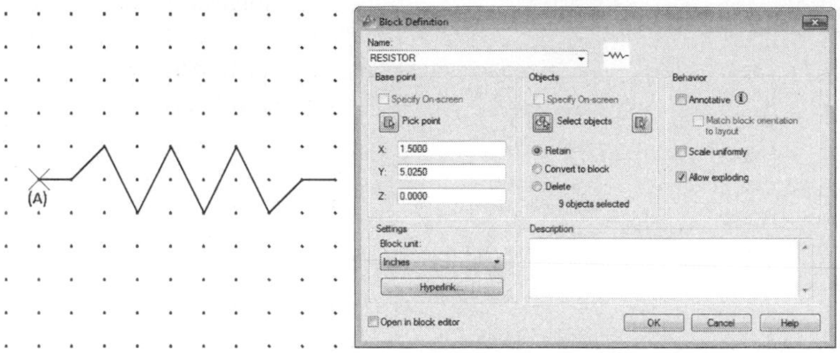

FIGURE 16.71

STEP 4

Begin another new drawing called ELECTRICAL_SCHEM1. Activate the DesignCenter. Click the Tree View icon and expand the DesignCenter to display the folders on your hard drive. Locate the correct folder in which the drawing ELECT_SYMB was saved and click it. Double-click ELECT_SYMB and select Blocks to display the local blocks, as illustrated in Figure 16.72. These local blocks can now be inserted in any AutoCAD drawing file.

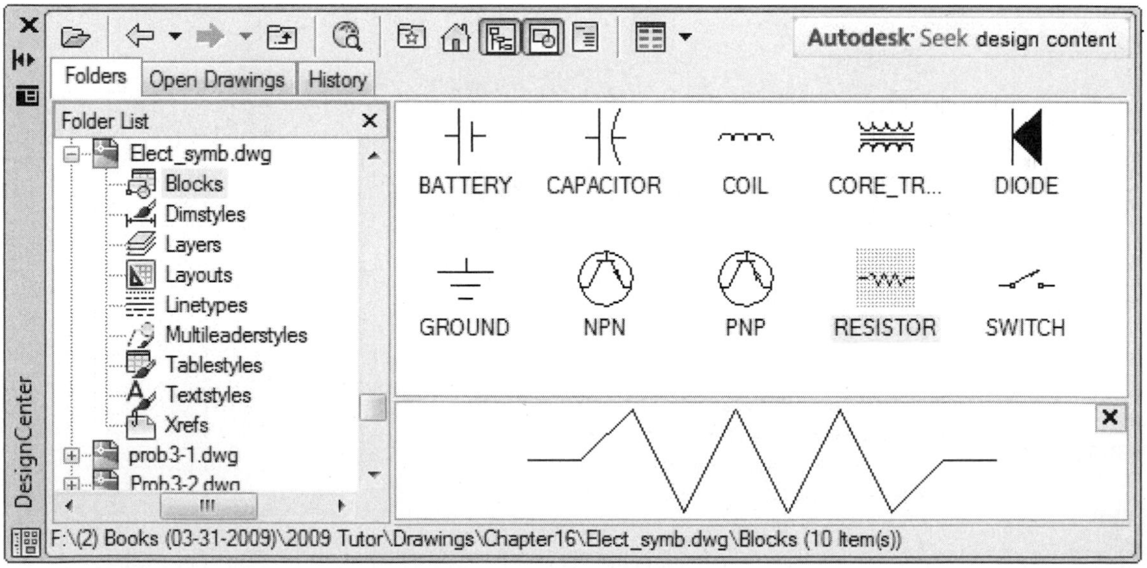

FIGURE 16.72

STEP 5

Make the Blocks layer current and begin dragging and dropping the blocks into the drawing, as shown in Figure 16.73. Double-clicking an image of a block located in the DesignCenter launches the Insert dialog box should you need to change the scale or rotation angle of the block. In the case of the electrical schematic, it is not critical to place the blocks exactly because they are moved to better locations when connected to lines that represent wires.

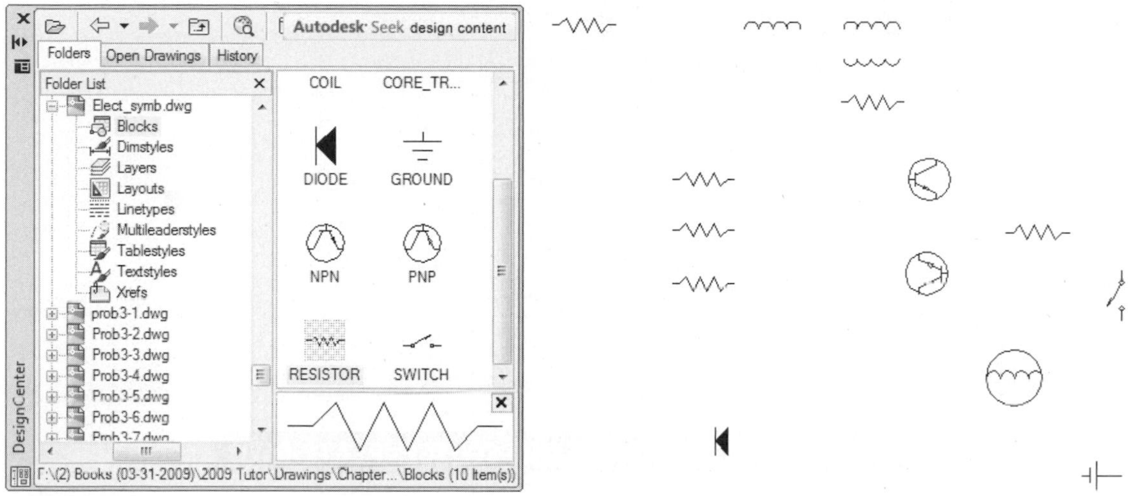

FIGURE 16.73

STEP 6

Make the Wires layer current and begin connecting all the blocks to form the schematic, as shown in Figure 16.74. Some type of Object Snap mode such as Endpoint or Insert must be used for the wire lines to connect exactly with the endpoints of the blocks. If spaces get tight or if the blocks look too crowded, move the block to a better location and continue drawing lines to form the schematic. Hint: The STRETCH command is a fast and easy way to reposition the blocks without having to reconnect the wires.

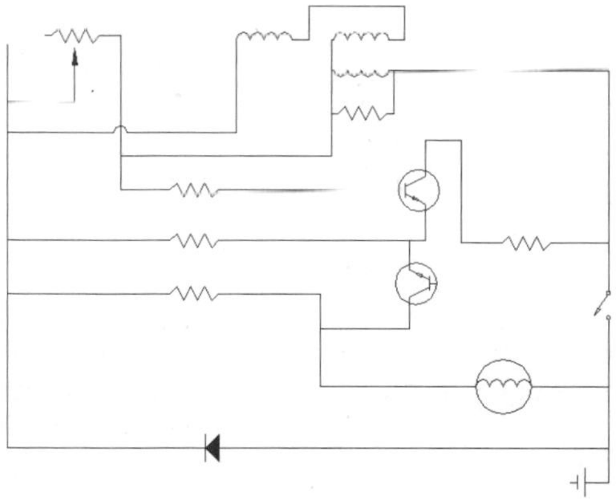

FIGURE 16.74

STEP 7

Add text (place it on the Blocks layer) identifying all resistors, transistors, diodes, and switches by number, as show in Figure 16.75. The text height used was 0.12; however, this could vary depending on the overall size of your schematic. Also, certain areas of the schematic show connections with the presence of a dot at the intersection of the connection. Use the DONUT command with an inside diameter of 0.00 and an outside diameter of 0.05. Place donuts on the Wires layer in all locations, as shown in Figure 16.75.

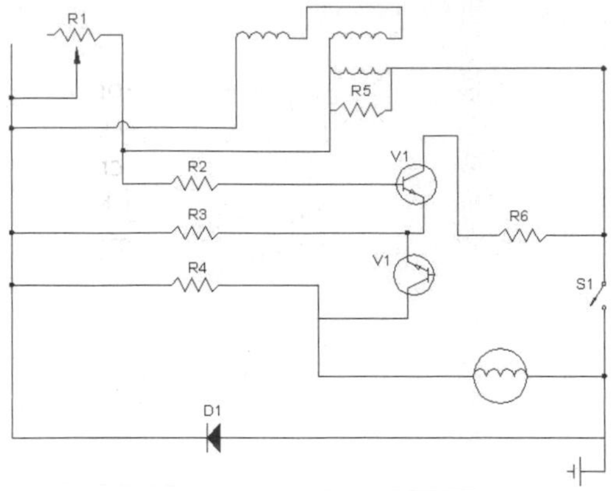

FIGURE 16.75

TUTORIAL EXERCISE: 16_DYNAMIC_SWITCH_PLATE.DWG

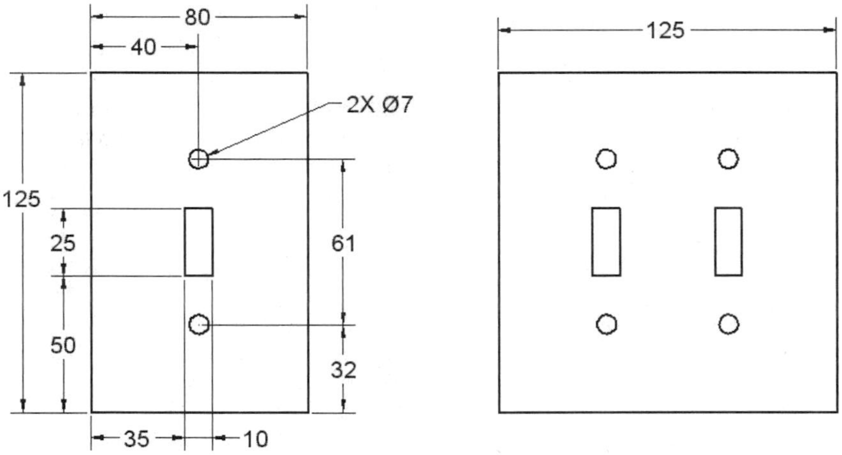

FIGURE 16.76

Purpose

This tutorial is designed to link an Array action to a dynamic block. As the dynamic block increases in width, items associated with the block can be added, depending on the width.

System Settings

No special system settings need to be made for this exercise.

Layers

No special layers need to be created for this exercise.

Suggested Commands

This application will utilize a switch plate consisting of a basic rectangular shape and three openings: two circular openings that represent screw holes and a rectangular opening used to accept the switch. As the switch plate increases in width, extra openings will automatically be added to the block.

STEP 1

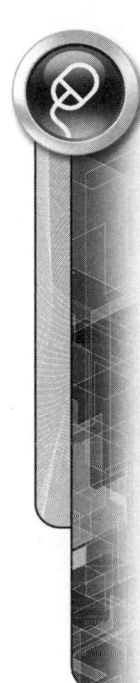

Open the drawing file 16_Dynamic_Switch_Plate. A block consisting of a single-opening switch plate is displayed in the drawing editor. This exercise will illustrate how to stretch the switch plate in 45 mm increments and have additional switch place openings created. Double-click the switch-plate block to activate the Edit Block Definition dialog box, as shown in Figure 16.77. Click the OK button to enter the Block Authoring environment.

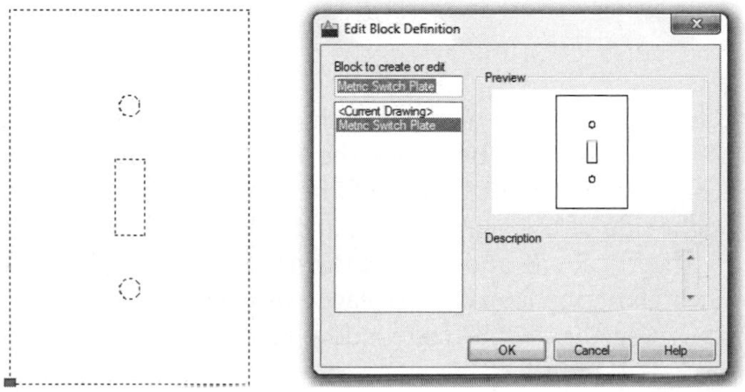

FIGURE 16.77

STEP 2

Add a Linear Parameter to the top of the switch plate. Select the parameter and launch the Properties Palette. Change the name of the Distance parameter to Plate Width, as shown in Figure 16.78.

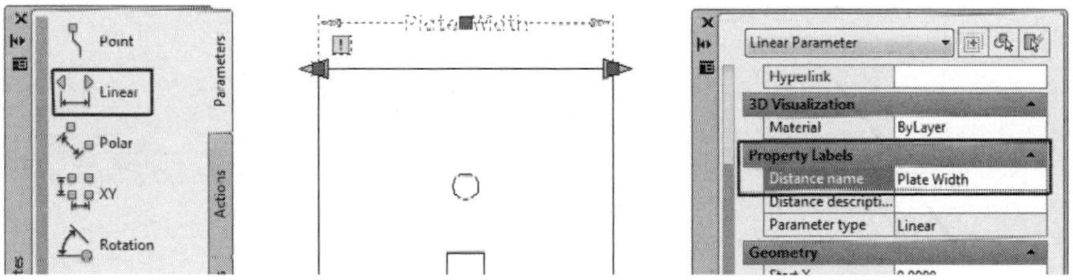

FIGURE 16.78

STEP 3

With the Plate Width parameter still selected, scroll down to the Value Set area and make the changes, as shown in Figure 16.79; set the Distance type to Increment, the Distance increment to 45, the Distance minimum to 80, and the Distance maximum to 305. When the switch plate is at its minimum distance of 80 mm, a single opening is displayed. When the switch plate is at its maximum distance of 305 mm, six openings are displayed.

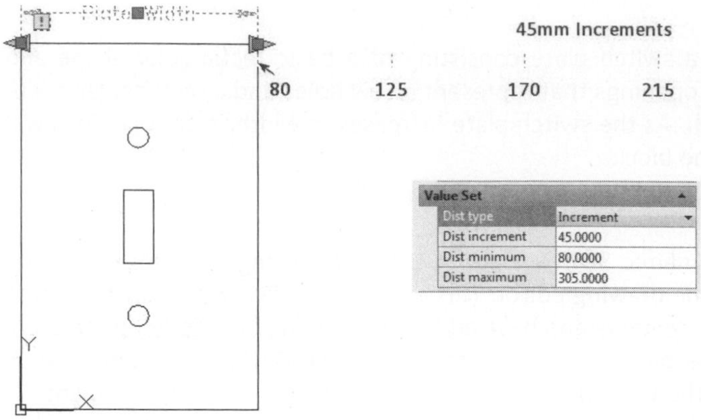

FIGURE 16.79

STEP 4

Next, assign a Stretch Action to the right side of the switch plate, as shown in Figure 16.80 (Left). After activating the Stretch Action, pick the Plate Width parameter, select the base point located in the upper-right corner of the switch plate, create a crossing box to surround the right side of the switch plate, and select the entire right side of the switch plate (the right vertical line and the two connected horizontal lines) as the objects to stretch. The Stretch Action icon will be placed next to the parameter, as shown in Figure 16.80 (Right). Use the Properties window to rename the Stretch action to Stretch Right.

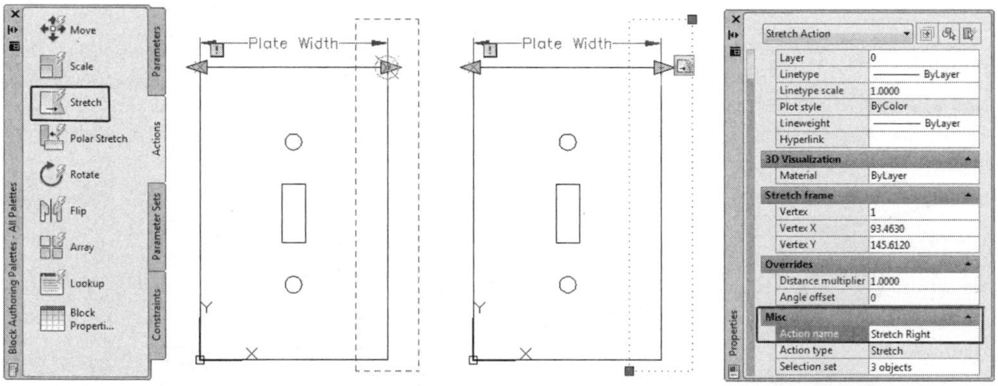

FIGURE 16.80

STEP 5

Now assign a Stretch Action to the left side of the switch plate using the same steps performed in the previous step. After activating the Stretch Action, pick the Plate Width parameter, select the base point located in the upper-left corner of the switch plate, create a crossing box to surround the left side of the switch plate, and select the entire left side of the switch plate (the left vertical line and the two connected horizontal lines) as the objects to stretch. The Stretch Action icon will be placed next to the other Action icon, as shown in Figure 16.81. Use the Properties window to rename the Stretch action to Stretch Left.

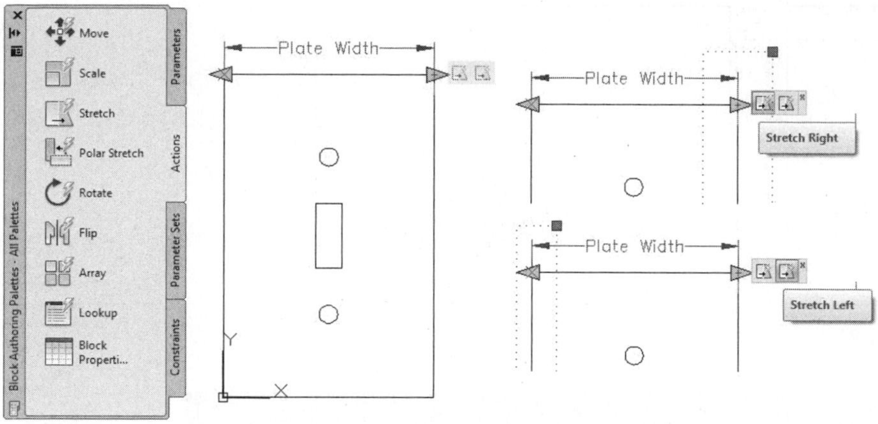

FIGURE 16.81

STEP 6

If you experiment by stretching this new block back in the drawing editor, you will find that the switch plate will stretch; however, the screw holes and switch opening will remain in their original positions. We want these openings to be copied or arrayed at a specified distance, depending on how much the switch plate stretches. To perform this operation, click the Array Action button back in the Block Authoring palette. Select the Plate Width parameter, select all three openings, as shown in Figure 16.82, and enter a Column Width distance of 45 mm as the distance from one set of openings to another. The Array Action icon will be placed next to the other Action icons.

Click the Save Block button to save these changes, and then click the Close Block Editor button to return to the drawing editor.

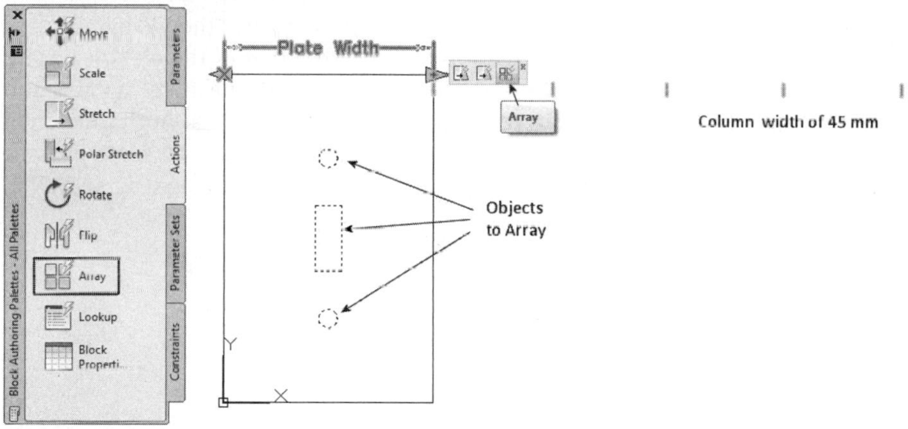

FIGURE 16.82

STEP 7

When you return to the drawing editor, click the block of the switch plate, pick the right arrow grip, and begin stretching the switch plate. When you stretch the switch plate to a new length of 305 mm, which represents the maximum distance, a series of six switch openings are created, as shown in Figure 16.83 (Right). This completes this tutorial exercise on creating a dynamic block of a switch plate.

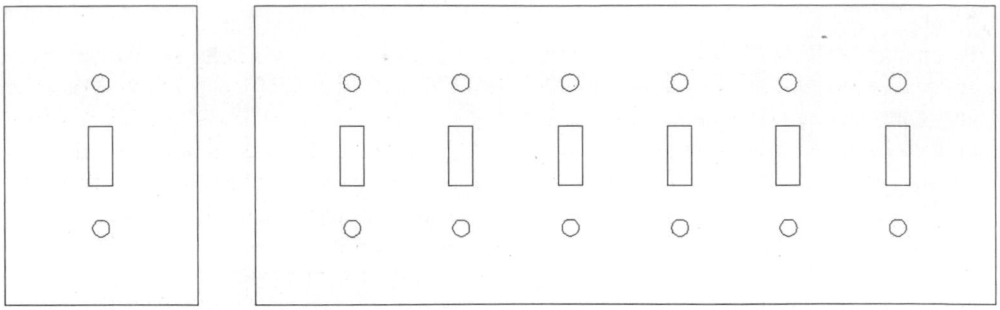

FIGURE 16.83

TUTORIAL EXERCISE: 16 _DYNAMIC_ELECTRICAL LIST.DWG

Purpose

This tutorial is designed to create visibility states to have numerous electrical symbols contained in a single block name.

System Settings

No special system settings need to be made for this tutorial.

Layers

No special layers need to be created for this tutorial.

Suggested Commands

The concept of creating visibility states in dynamic blocks begins with the creation of a block using the Edit Block Definition dialog box. You then create a visibility state while inside the Block Authoring environment. During this process, you insert an electrical symbol and immediately explode it into individual objects. You then enter the Visibility States dialog box and change the name of the current state to match that of the electrical symbol. You then create a new visibility state, name this state the next electrical symbol, and hide all existing objects in this new state. When you return to the Block Authoring environment, you insert the next electrical symbol, explode it, and follow the same process for arranging a number of electrical symbols in the same visibility state, as shown in Figure 16.84.

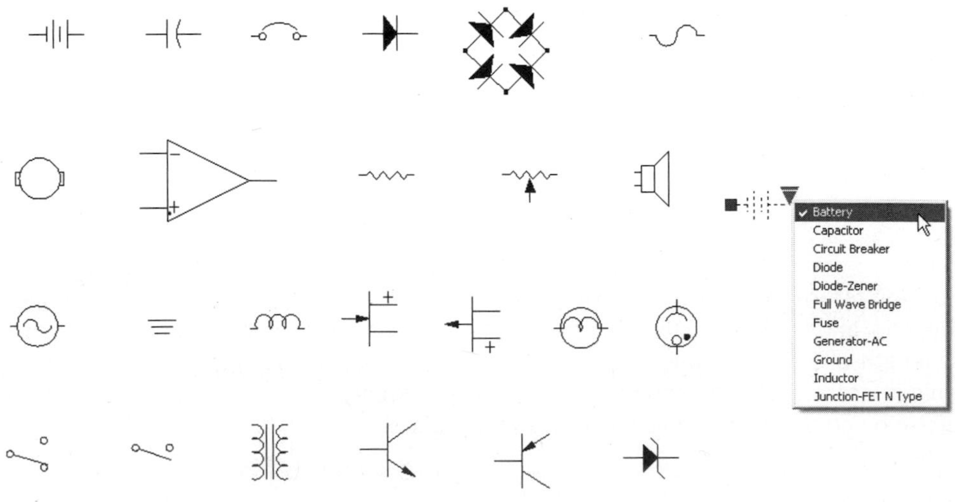

FIGURE 16.84

STEP 1

Open the drawing 16_Dynamic_Electrical_List. Before creating visibility states, you must create a new general block name that will hold all individual electrical symbols. You must also enter the block editor to create the visibility states. First, click the Block Editor button located in the Ribbon, as shown in Figure 16.85 (Left). When the Edit Block Definition dialog box appears, as shown in Figure 16.85 (Right), enter Electrical Symbols as the block name and click the OK button. This action opens the new block name in the Block Editor.

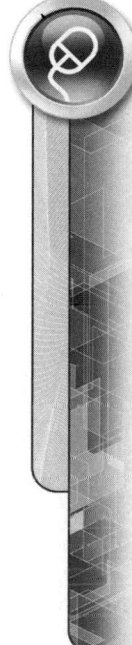

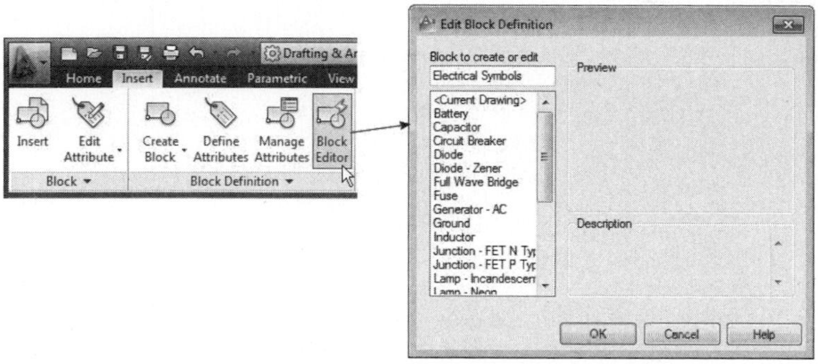

FIGURE 16.85

STEP 2

Insert the first block, called Battery, in the Block Editor, as shown in Figure 16.86. Use an insertion point of 0,0. In fact, all individual electrical blocks will be inserted at 0,0 for consistency. Once the block appears in the Block Editor, use the EXPLODE command to break up the block into individual objects. Then, locate the Visibility Parameter in the Parameters tab of the Block Authoring Palette and place the marker to the right and below the battery symbol, as shown in Figure 16.86.

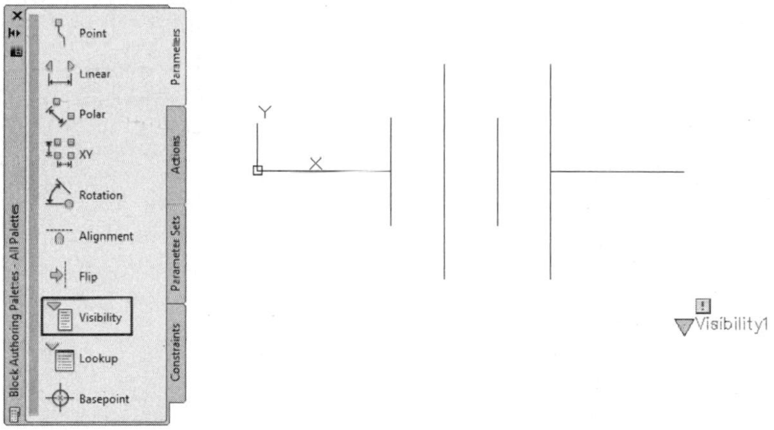

FIGURE 16.86

STEP 3

In the Block Editor Ribbon, click the Visibility States button in the Visibility panel, as shown in Figure 16.87.

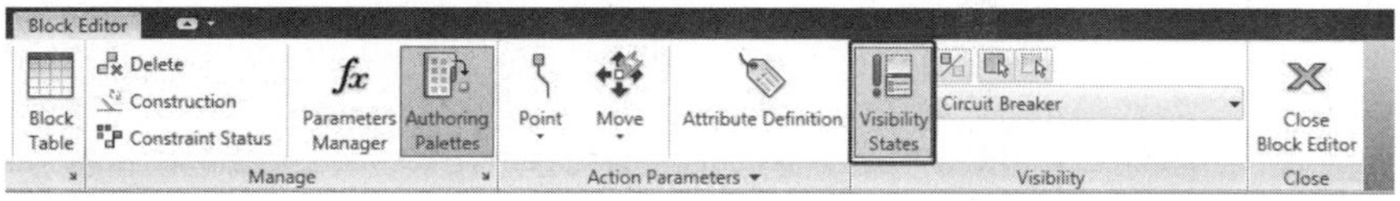

FIGURE 16.87

STEP 4

This launches the Visibility States dialog box, as shown in Figure 16.88. When you use this dialog box for the first time, a default state called VisibilityState0 is already created. Use the Rename button to change the name of this visibility state to Battery. This state coincides with the object currently displayed in the Block Authoring screen. Next, a new visibility state needs to be created by clicking the New button in the Visibility States dialog box.

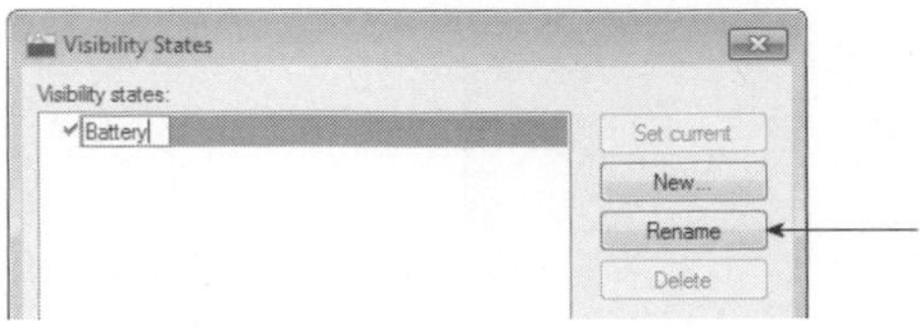

FIGURE 16.88

STEP 5

This launches the New Visibility State dialog box, as shown in Figure 16.89 (Left). In the Visibility state name field, enter the name Capacitor. Also, be sure to check the radio button next to Hide all existing objects in new state. This will hide all objects when you return to the Block Editor. Click the OK button to close the New Visibility State dialog box. Click OK to also close the Visibility States dialog box.

STEP 6

When you return to the Block Editor, insert the block Capacitor at an insertion point of 0,0 and explode the block back into its individual objects. The objects that remain in the Block Editor form the Capacitor visibility state.

Launch the Visibility States dialog box again. Create another new visibility state called Circuit Breaker, and make sure Hide all existing objects in new state is selected. Click OK to close the dialog boxes. Insert the Circuit Breaker block at an insertion point of 0,0 and explode the block back into its individual objects.

Repeat the procedure for creating more visibility states based on the other electrical block symbols. Close the Block Editor and save the changes to the block.

STEP 7

Insert an Electrical Symbols block in the drawing. Pick the block and click the List grip to display the available symbols, as shown in Figure 16.89 (Right). Test the visibility states by selecting them from the list. If you created visibility states of all electrical symbols, your list will be longer.

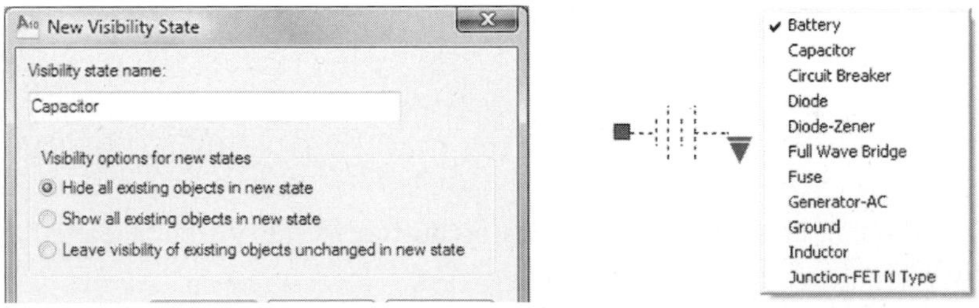

FIGURE 16.89

PROBLEMS FOR CHAPTER 16

PROBLEM 16-1

Construct this battery pack using the block library supplied with AutoCAD: Basic Electronics.dwg. This drawing is not to scale.

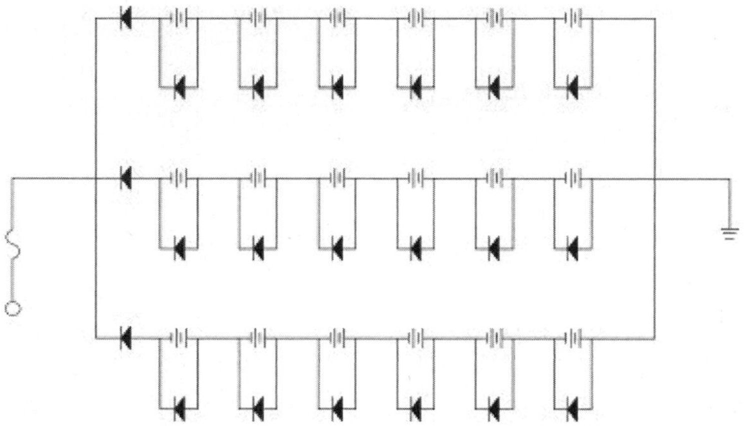

PROBLEM 16-2

Construct this electrical schematic using the block library supplied with AutoCAD: Basic Electronics.dwg. This drawing is not to scale.

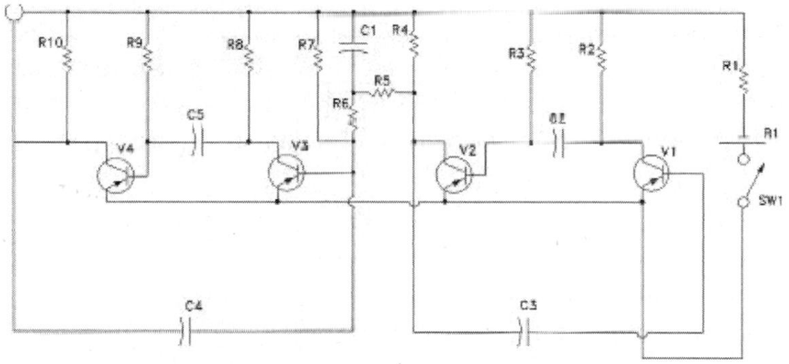

PROBLEM 16-3

Construct this electrical schematic using the block library supplied with AutoCAD: Basic Electronics.dwg. This drawing is not to scale.

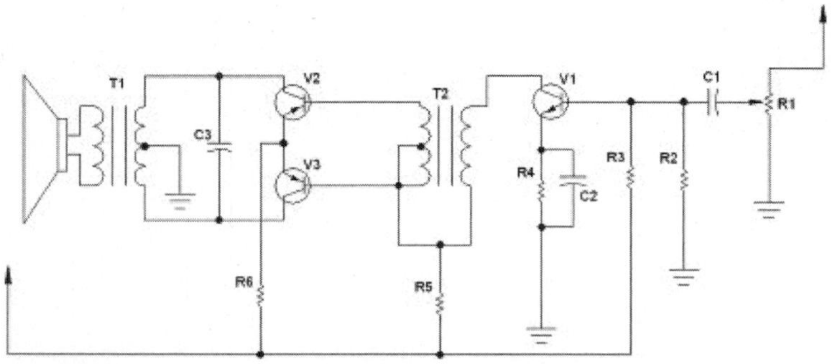

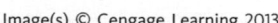

PROBLEM 16-4

Construct this logic gate schematic using the following block libraries supplied with AutoCAD: CMOS Integrated Circuits.dwg and Basic Electronics.dwg. This drawing is not to scale.

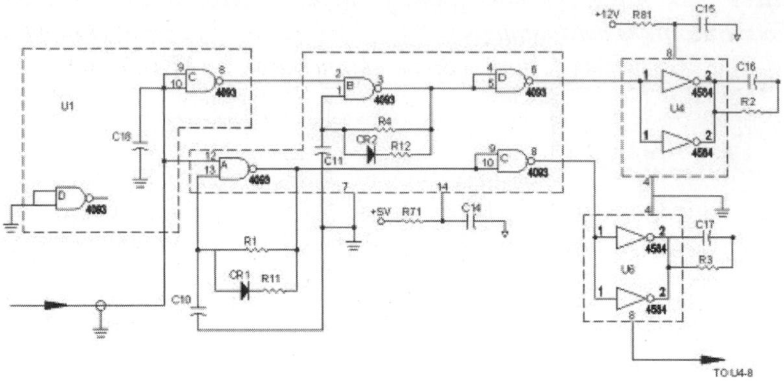

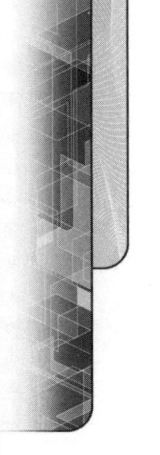

INTERMEDIATE LEVEL DRAWINGS

PROBLEM 16-5

Construct this instrumentation pipe diagram using the following block library supplied with AutoCAD: Pipe Fittings.dwg. This drawing is not to scale.

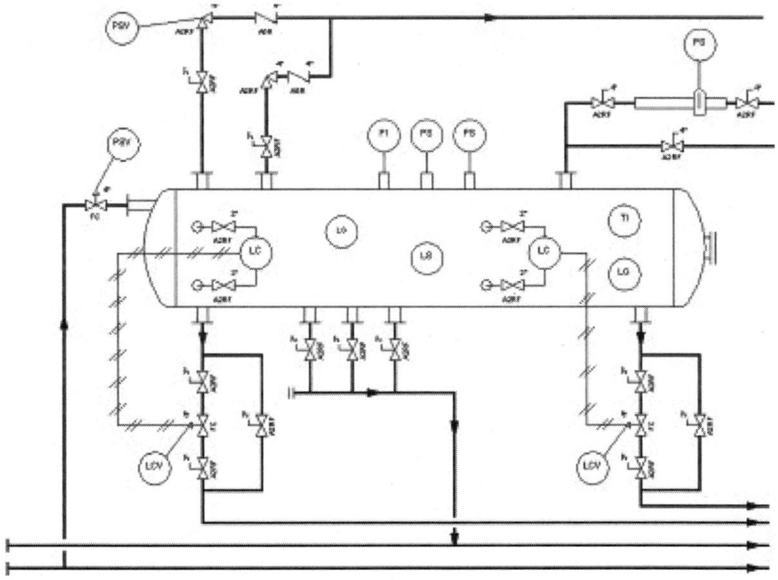

PROBLEM 16-6

Construct the floor plan of the house using the dimensions provided. Create the appropriate layers to separate the walls from the dimensions, and so on. Once the floor plan is completed, create an interior plan consisting of furniture and appliances. Place the blocks in their appropriate spaces using the designated room titles. Use the following block libraries supplied with AutoCAD: House Designer.dwg, Home – Space Planner.dwg, and Kitchens.dwg.

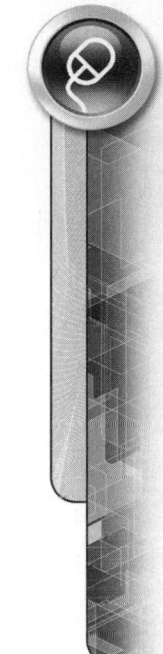

Use the following suggestions for the creation of doors and windows:

All windows measure 2'-8" wide.

All bedroom doors measure 2'-6" wide.

The bathroom door measures 2'-0" wide.

The main entrance door measures 3'-0" wide.

The kitchen and laundry doors each measure 2'-8" wide.

All closet openings measure 4'-0" wide.

All interior walls measure 4".

The exterior walls with brick veneer measures 5".

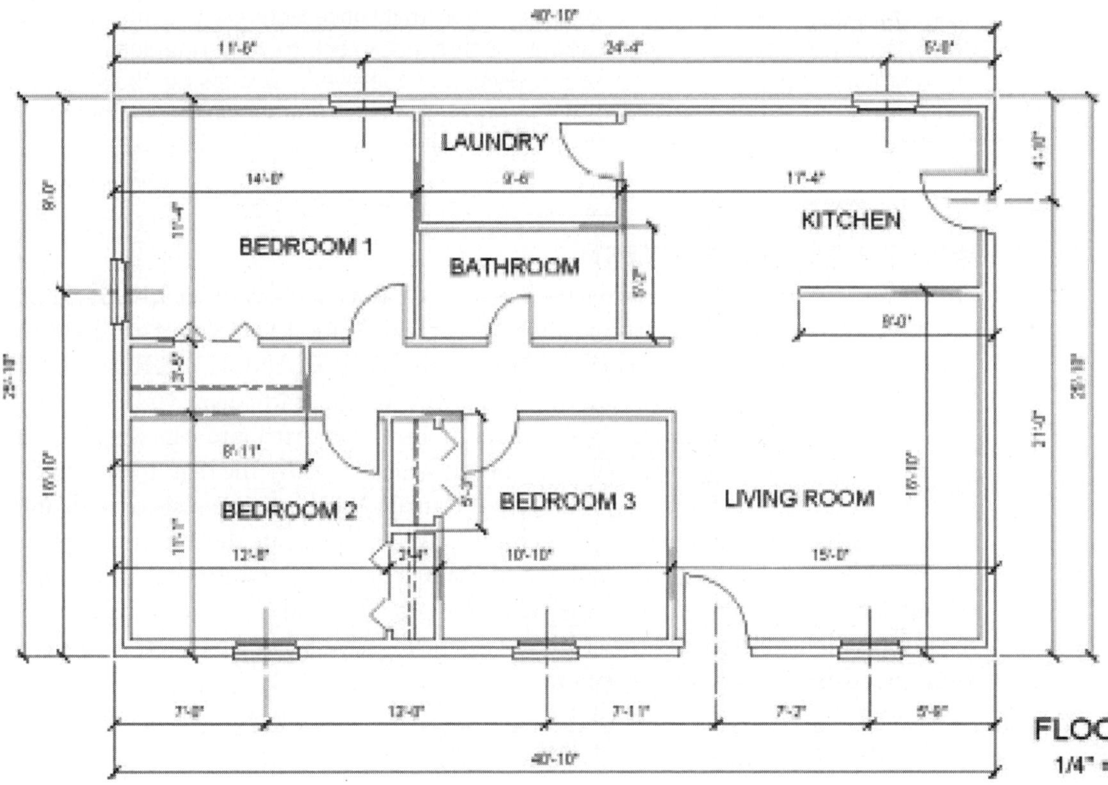

FLOOR PLAN
1/4" = 1'-0"

CHAPTER

17

Working with Attributes

This chapter begins the study of how to create, display, edit, and extract attributes. Attributes consist of intelligent text data that is attached to a block. The data could consist of part description, part number, catalog number, and price. Whenever the block is inserted in a drawing, you are prompted for information that, once entered, becomes attribute data. Attributes could also be associated with a title block for entering such items as drawing name, who created, checked, revised, and approved the drawing, drawing date, and drawing scale. Once included in the drawing, the attributes can be extracted and shared with other programs.

WHAT ARE ATTRIBUTES?

An attribute may be considered a label that is attached to a block. This label is called a tag and can contain any type of information that you desire. Examples of attribute tags are illustrated in Figure 17.1 (Left). The tags are RESISTANCE, PART_NAME, WATTAGE, and TOLERANCE. They relate to the particular symbol they are attached to, in this case an electrical symbol of a resistor. Attribute tags are placed in the drawing with the symbol. When the block, which includes the tags and symbol, is inserted, AutoCAD requests the values for the attributes. The same resistor with attribute values is illustrated in Figure 17.1 (Right). When you create the attribute tags, you determine what information is requested, the actual prompts, and the default values for the information requested. Once the values are provided and inserted in the drawing, the information contained in the attributes can be displayed, extracted, and even shared with spreadsheet or database programs.

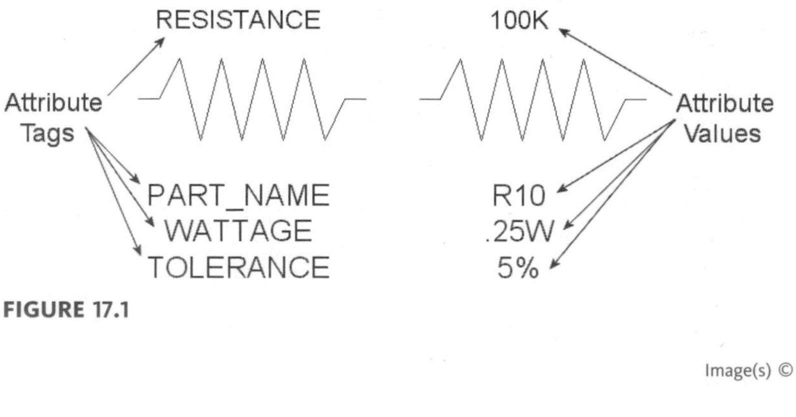

FIGURE 17.1

Once attributes are inserted in the drawing, the following additional tasks can be performed on attributes:

1. Attributes can be turned on, turned off, or displayed normally through the use of the ATTDISP command.
2. The individual attributes can be changed through the Enhanced Attribute Editor dialog box, which is activated through the EATTEDIT command.
3. Attributes can be globally edited using the -ATTEDIT and FIND commands.
4. Once edited, attributes can be extracted into tables or text files through the Data Extraction wizard, which is activated by the DATAEXTRACTION command.
5. The text file created by the attribute extraction process can be exported to a spreadsheet or database program.

The following commands, which will be used throughout this chapter, assist in the creation, editing, and manipulation of attributes:

ATTDEF—Activates the Attribute Definition dialog box used for the creation of attributes

ATTDISP—Used to control the visibility of attributes in a drawing

-ATTEDIT—Used to edit attributes singly or globally; the hyphen (-) in front of the command activates prompts from the command line

FIND—Used to find and replace text including attribute values

BATTMAN—Activates the Block Attribute Manager dialog box, used to edit attribute properties of a block definition

EATTEDIT—Activates the Enhanced Attribute Editor dialog box, used to edit the attributes of a block

DATAEXTRACTION—Activates the Data Extraction wizard, used to extract attributes and other drawing information.

CREATING ATTRIBUTES THROUGH THE ATTRIBUTE DEFINITION DIALOG BOX

Selecting Define Attributes from the Block Definition panel of the Ribbon, shown in Figure 17.2 (Left), activates the Attribute Definition dialog box, illustrated in Figure 17.2 (Right). The following components of this dialog box will be explained in this section: Attribute Tag, Attribute Prompt, Attribute Default, and Attribute Mode.

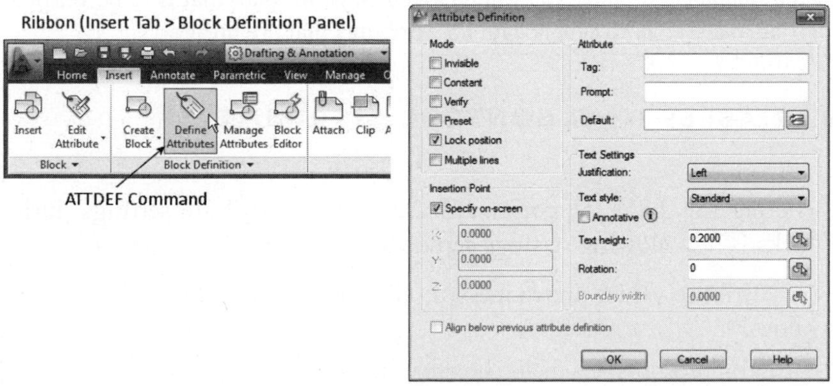

FIGURE 17.2

Attribute Tag

A tag is the name given to an attribute. Typical attribute tags could include PART_NAME, CATALOG_NUMBER, PRICE, DRAWING_NAME, and SCALE. The underscore is used to separate words because spaces are not allowed in tag names.

Attribute Prompt

The attribute prompt is the text that appears on the text line when the block containing the attribute is inserted in the drawing. If you want the prompt to be the same as the tag name, enter a null response by leaving the Prompt field blank. If the Constant mode is specified for the attribute, the prompt area is not available.

Attribute Default

The attribute default is the value displayed when the attribute is inserted in the drawing. This value can be accepted or a new value entered as desired. The attribute value is handled differently if the attribute mode selected is Constant or Preset.

Attribute Modes

Invisible—This mode is used to determine whether the label is displayed when the block containing the attribute is inserted in the drawing. If you later want to make the attribute visible, you can use the `ATTDISP` command.

Constant—Use this mode to give every attribute the same value. This might be very useful when the attribute value is not subject to change. However, if you designate an attribute to contain a constant value, it is difficult to change it later in the design process.

Verify—Use this mode to verify that every value is correct. This is accomplished by prompting the user twice for the attribute value.

Preset—This allows for the presetting of values that can be changed. However, you are not prompted to enter the attribute value when inserting a block. The attribute values are automatically set to their default values.

Lock Position—This mode locks the location of the attribute located inside of the block reference. When unlocked, the attribute can be moved relative to the rest of the block using grips.

Multiple lines—Specifies that the attribute value can contain multiple lines of text.

NOTE

The effects caused by invoking the Verify and Preset modes are apparent only when the `ATTDIA` system variable is off (the Enter Attributes dialog box is not displayed). When entering data at the Command prompt, you will be asked twice for data that is to be verified and you will not be asked at all to supply data for attributes that are preset.

SYSTEM VARIABLES THAT CONTROL ATTRIBUTES

ATTREQ

Determines whether the `INSERT` command uses default attribute settings during insertion of blocks. The following settings can be used:

0 No attribute values are requested; all attributes are set to their default values.

1 Turns on prompts or a dialog box for attribute values, as specified by attdia.

```
Command: ATTREQ
Enter new value for ATTREQ <1>:
```

ATTMODE

Controls the display of attributes. The following settings can be used:

0 Off: Makes all attributes invisible.

1 Normal: Retains current visibility of each attribute; visible attributes are displayed, invisible attributes are not.

2 On: Makes all attributes visible.

```
Command: ATTMODE
Enter new value for ATTMODE <1>:
```

ATTMULTI

Controls the creation of multiline attributes. The following settings can be used:

0 Off: Cannot be created but they can still be viewed and edited.

1 On: Can create multiline attributes.

```
Command: ATTMULTI
Enter new value for ATTMULTI <1>:
```

ATTDIA

Controls whether the `INSERT` command uses a dialog box for attribute value entry. The following settings can be used:

0 Issues prompts on the command line.

1 Initiates a dialog box for attribute value entry.

```
Command: ATTDIA
Enter new value for ATTDIA <1>:
```

TUTORIAL EXERCISE: 17_B TITLE BLOCK.DWG

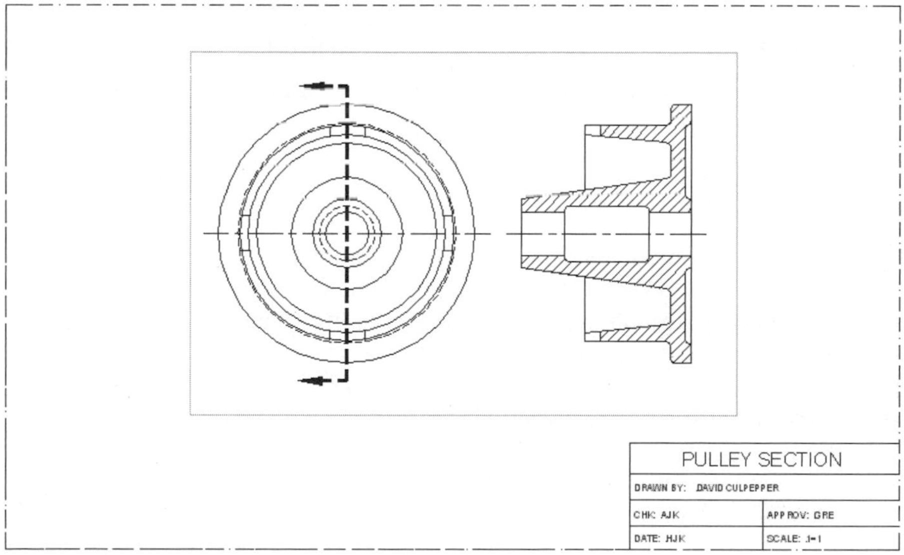

FIGURE 17.3

Purpose

This tutorial is designed to assign attributes to a title block. The completed drawing is illustrated in Figure 17.3.

System Settings

Because the drawings 17_B Title Block and 17_Pulley Section are provided, all units and drawing limits have already been set.

Layers

All layers have already been created for this drawing.

Suggested Commands

You will open the drawing 17_B Title Block. Using the Attribute Definition dialog box (activated through the ATTDEF command), assign attributes consisting of the following tag names: DRAWING_NAME, DRAWN_BY, CHECKED_BY, APPROVED_BY, DATE, and SCALE. Save this drawing file with its default name. This file is considered a global block and will become a block reference object once it is inserted into another drawing. Open the drawing 17_Pulley Section, insert the title block file into this drawing, and answer the attribute prompts designed to complete the title block information.

STEP 1

Open the drawing 17_B Title Block and observe the title block area, shown in Figure 17.4. Various point objects are present to guide you in the placement of the attribute information. All points are located on the Points layer.

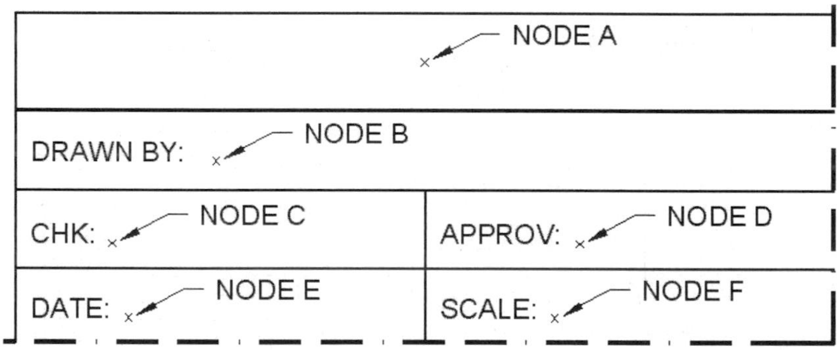

FIGURE 17.4

STEP 2

Activate the Attribute Definition dialog box (ATTDEF command), as shown in Figure 17.5 (Left). Leave all items unchecked in the Mode area of this dialog box with the exception of Lock position. In the Attribute area, make the following changes: Enter DRAWING_NAME in the Tag field. In the Prompt field, enter: What is the name of the drawing? In the Default field, enter UNNAMED. In the Text Settings area, change the Justification to Middle and the Text height to 0.25 units. When finished, click the OK button. This returns you to your drawing. Using OSNAP-Node, pick the point at "A," as shown in Figure 17.5 in the upper right. The DRAWING_NAME tag is added to the title block area, as shown in Figure 17.5 in the lower right.

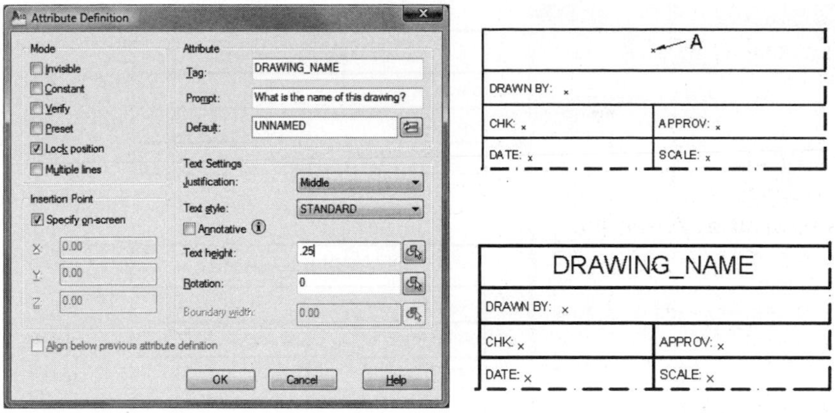

FIGURE 17.5

STEP 3

To define the next attribute, activate the Attribute Definition dialog box, as shown in Figure 17.6 (Left). Leave all items unchecked in the Mode area of this dialog box except for Lock position. In the Attribute area, make the following changes: Enter DRAWN_BY in the Tag field. In the Prompt field, enter: Who created this drawing? In the Default field, enter UNNAMED. In the Text Settings area, verify that Justification is set to Left, and change the Text height to 0.12 units. When finished, click the OK button. This returns you to your drawing. Using OSNAP-Node, pick the point at "B," as shown in Figure 17.6 in the upper right. The DRAWN_BY tag is added to the title block area, as shown in Figure 17.6 in the lower right.

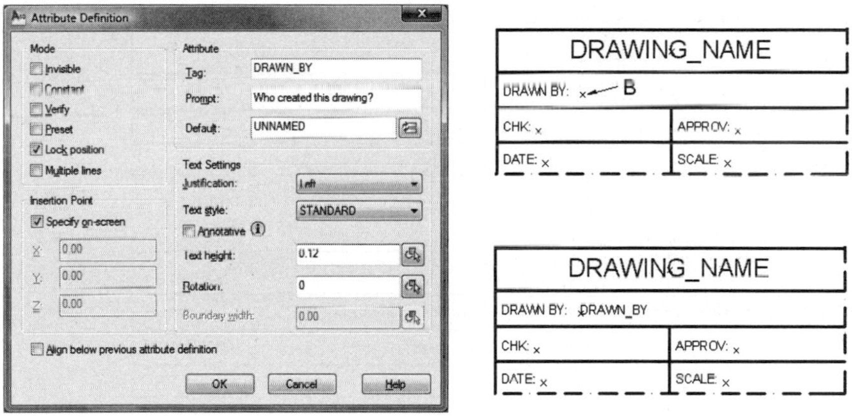

FIGURE 17.6

STEP 4

Activate the Attribute Definition dialog box, as shown in Figure 17.7 (Left). Leave all items unchecked in the Mode area of this dialog box except for Lock position. In the Attribute area, make the following changes: Enter CHECKED_BY in the Tag field. In the Prompt field, enter: Who will be checking this drawing? In the Default field, enter CHIEF DESIGNER. In the Text Settings area, verify that Justification is set to Left and the Text height is 0.12 units. When finished, click the OK button. This returns you to your drawing. Using OSNAP-Node, pick the point at "C," as shown in Figure 17.7 on the upper right. The CHECKED_BY tag is added to the title block area, as shown in Figure 17.7 on the lower right.

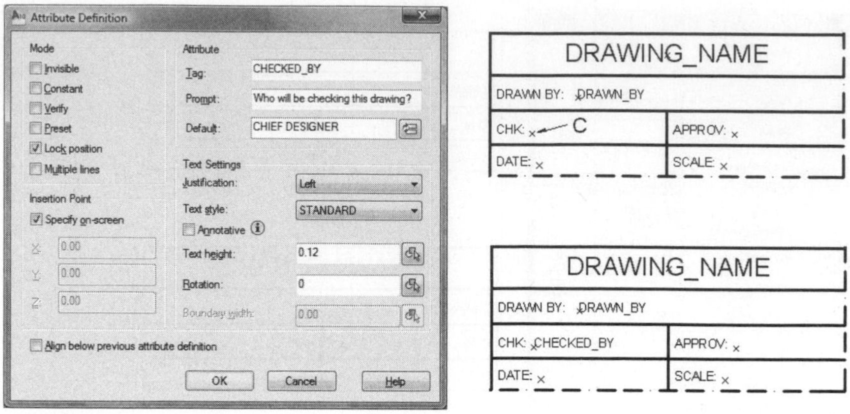

FIGURE 17.7

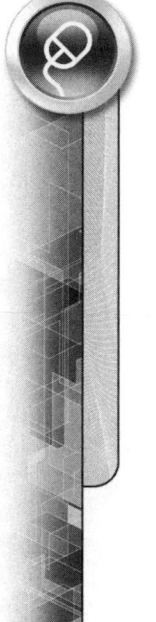

STEP 5

Activate the Attribute Definition dialog box, as shown in Figure 17.8. Leave all items unchecked in the Mode area of this dialog box except for Lock position. In the Attribute area, make the following changes: Enter APPROVED_BY in the Tag field. In the Prompt field, enter: Who will be approving this drawing? In the Default field, enter CHIEF ENGINEER. In the Text Settings area, verify that Justification is set to Left and the Text height is 0.12 units. When finished, click the OK button. This returns you to your drawing. Using OSNAP-Node, pick the point at "D," as shown in Figure 17.8 in the upper right. The APPROVED_BY tag is added to the title block area, as shown in Figure 17.8 in the lower right.

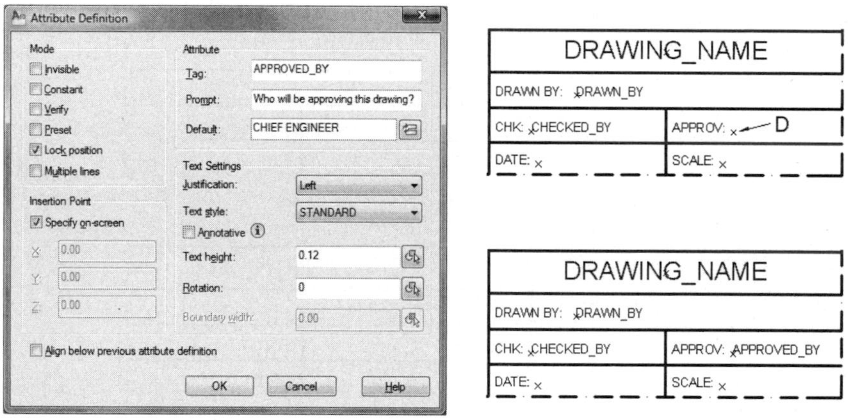

FIGURE 17.8

STEP 6

Activate the Attribute Definition dialog box, as shown in Figure 17.9 (Left). Leave all items unchecked in the Mode area of this dialog box except for Lock position. In the Attribute area, make the following changes: Enter DATE in the Tag field. In the Prompt field, enter: When was this drawing completed? In the Default field, enter UNDATED. In the Text Settings area, verify that Justification is set to Left and the Text height is 0.12 units. When finished, click the OK button. This returns you to your drawing. Using OSNAP-Node, pick the point at "E," as shown in Figure 17.9 in the upper right. The DATE tag is added to the title block area, as shown in Figure 17.9 in the lower right.

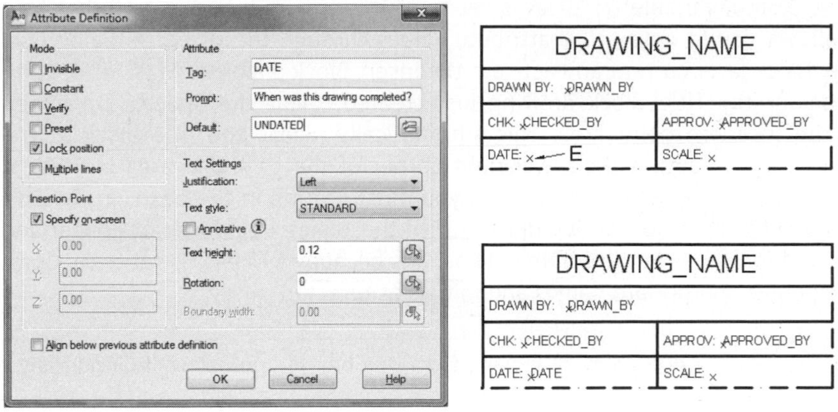

FIGURE 17.9

STEP 7

Activate the Attribute Definition dialog box, as shown in Figure 17.10 (Left). Leave all items unchecked in the Mode area of this dialog box except for Lock position. In the Attribute area, make the following changes: Enter SCALE in the Tag field. In the Prompt field, enter: What is the scale of this drawing? In the Default field, enter 1 = 1. In the Text Settings area, verify that Justification is set to Left and the Text height is 0.12 units. When finished, click the OK button. This returns you to your drawing. Using OSNAP-Node, pick the point at "F," as shown in Figure 17.10 in the upper right. The SCALE tag is added to the title block area, as shown in Figure 17.10 in the lower right.

You have now completed creating the attributes for the title block; turn off the Points layer. Close and save the changes to this drawing with its default name.

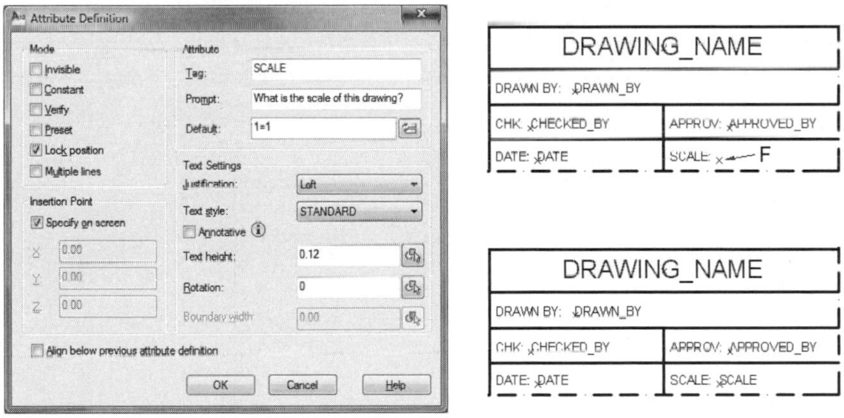

FIGURE 17.10

STEP 8

To test the attributes and see how they function in a drawing, first open the drawing file 17_Pulley Section. Notice that this image is viewed from inside a Layout (or Paper Space). The current page setup is based on the DWF6-ePlot.pc3 file, with the current sheet size as ANSI Expand B (17.00 × 11.00 inches). The viewport that holds the two views of the Pulley has the No Plot state assigned in the Layer Properties Manager dialog box. You could turn the Viewports layer off to hide the viewport if you prefer. You will now insert the title block with attributes in this drawing.

Verify that the system variable ATTDIA is set to 1 (This can be typed in from the keyboard). This allows you to enter your attribute values through the use of a dialog box. Make the Title Block layer current and activate the Insert Block dialog box (INSERT command). Browse for the Title Block drawing just created. Clear the Specify On-screen checkbox in the Insertion point area. This automatically places the title block at the 0,0,0 location of the layout, in the lower-right corner of the printable area indicators. Click OK. Before the title block can be placed, you must first fill in the boxes in the Edit Attributes dialog box, as illustrated in Figure 17.11 (if this dialog box does not appear and you are prompted for values at the command line, it is because ATTDIA is set to 0). Complete all boxes; enter appropriate names and initials as directed.

NOTE

Pressing the TAB key is a quick way of moving from one box to another while inside any dialog box.

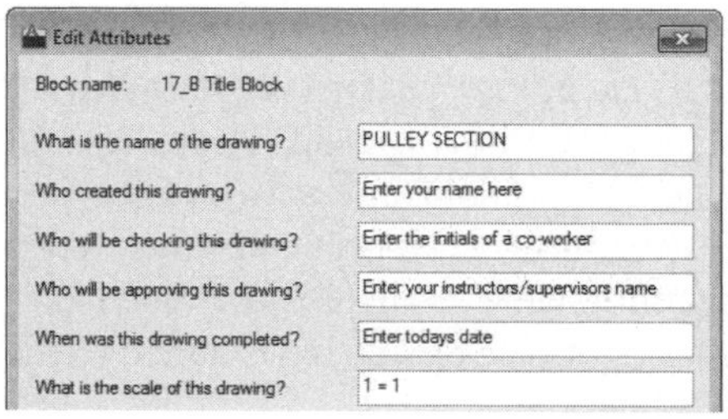

FIGURE 17.11

The completed drawing with title block and attributes inserted is illustrated in Figure 17.12.

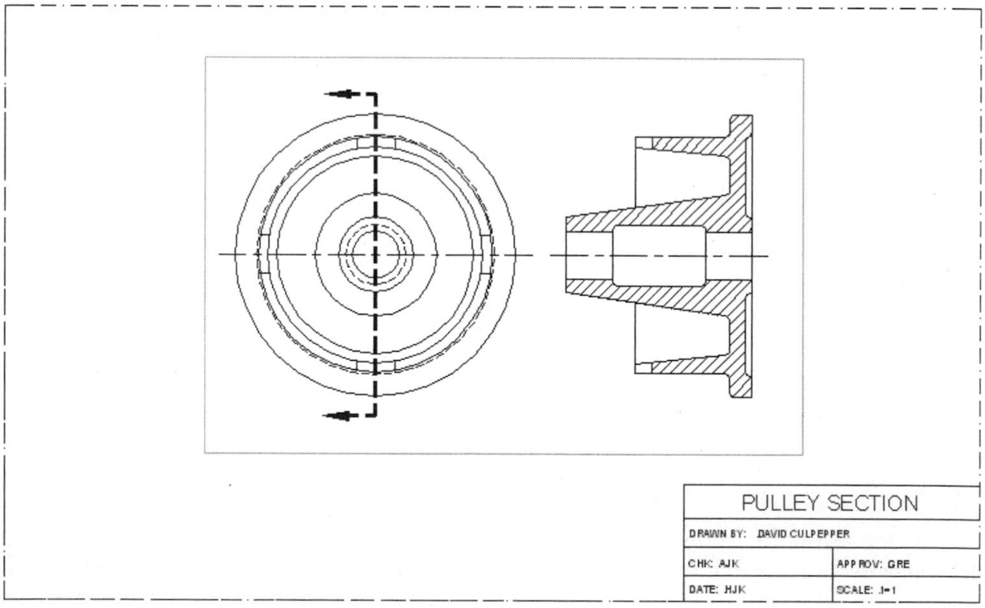

FIGURE 17.12

CREATING MULTIPLE LINES OF ATTRIBUTES

The previous example illustrated the creation of individual attributes for the title block. Attributes can also be created with multiple lines of text. Located in the Attribute Definition dialog box is a Multiple lines mode. When checked, the default attribute value is grayed out and an Open Multiline Editor button is present. After creating the tag CLIENT_INFORMATION, as shown in Figure 17.13, click the Open Multiline Editor button.

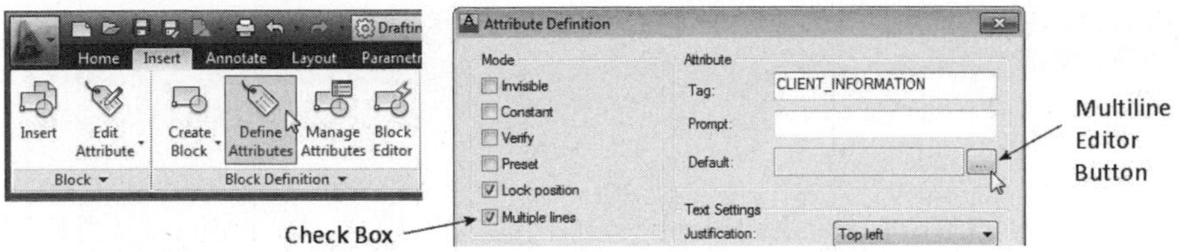

FIGURE 17.13

After picking a point to locate the attribute text, a simplified version of the Text Formatting Toolbar is provided. Utilize the toolbar to create the multiple lines of attributes. Illustrated in the following image on the left are a number of information and address attributes that are created. Clicking the OK button in the toolbar returns you to the Attribute Definition dialog box, as shown in Figure 17.14 (Right). Notice that all of the information created through the Text Formatting is grouped in the Default field.

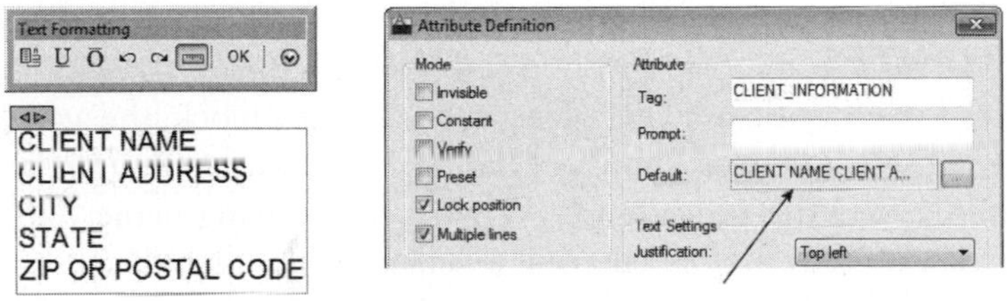

FIGURE 17.14

After creating the multiple line attribute, the attribute tag is grouped with geometry and a new block definition is created with the `BLOCK` command, as shown in Figure 17.15.

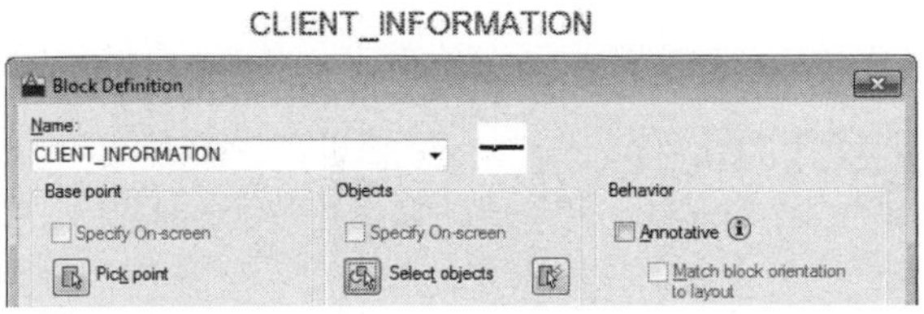

FIGURE 17.15

When inserting the block containing the multiple line attribute, all lines of the attribute display as shown in the following image on the left. If the ATTDIA system variable is turned on (set to 1), the Edit Attributes dialog box displays as shown in Figure 17.16 (Right). To change the various attribute fields, click the Open Multiline Editor button.

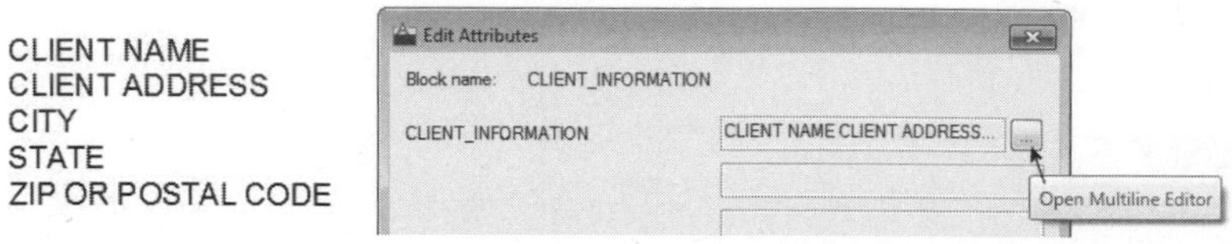

FIGURE 17.16

When the Text Formatting Toolbar displays in Figure 17.17 (Left), replace each field such as CLIENT NAME and CLIENT ADDRESS, with actual names and locations, as shown in Figure 17.17 (Right).

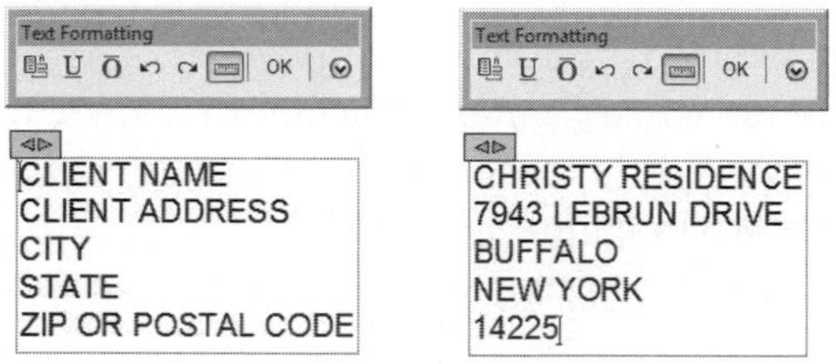

FIGURE 17.17

FIELDS AND ATTRIBUTES

When defining attributes for a specific drawing object such as a title block, the attribute value can be converted into a field, as shown in Figure 17.18. In this figure, the attribute tag (DRAWING_NAME) and prompt (What is the drawing name?) are created using conventional methods. Before entering a default value, click the Insert field button to display the Field dialog box. If this button is not available, turn off "Multiple lines" in the Mode area. In the Field dialog box, the Document field category is being used to assign a field called Filename. Notice some of the available options selected for this field: the "Format" is set to Uppercase, the "Filename only" radio button was selected, and the box for "Display file extension" was unchecked.

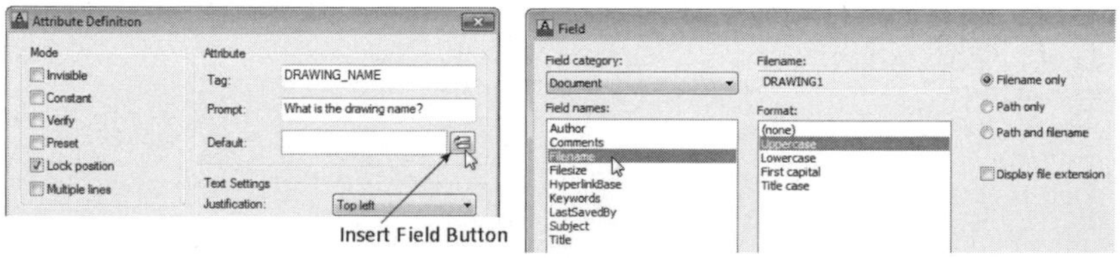

FIGURE 17.18

When you work with attribute values as fields, the actual default entry is displayed with a gray background, as shown in Figure 17.19. In this example, DRAWING1 was inserted into this cell from the Field dialog box in the previous figure. When a title block that contains this attribute is inserted into a drawing, the field will be automatically populated by the current drawing file name in uppercase letters without a file extension.

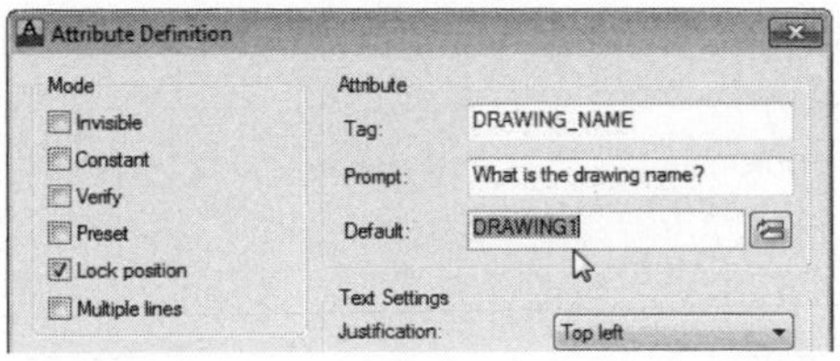

FIGURE 17.19

CONTROLLING THE DISPLAY OF ATTRIBUTES

You don't always want attribute values to be visible in a drawing. The ATTDISP command is used to determine the visibility of the attribute values. This command can be entered from the keyboard or can be selected from the Insert Ribbon, as shown in Figure 17.20.

```
Command: ATTDISP
Enter attribute visibility setting [Normal/ON/OFF]
<Normal>:
(Enter the desired option)
```

The following three modes are used to control the display of attributes:

Visibility Normal (Retain Display)

This setting displays attributes based on the mode set through the Attribute Definition dialog box. If some attributes were created with the Invisible mode turned on, these attributes will not be displayed with this setting, which makes this setting popular for displaying certain attributes and hiding others.

Visibility On (Display All)

Use this setting to force all attribute values to be displayed on the screen. This affects even attribute values with the Invisible mode turned on.

Visibility Off (Hide All)

Use this setting to force all attribute values to be turned off on the screen. This is especially helpful in busy drawings that contain lots of detail and text.

All three settings are illustrated in Figure 17.20. With ATTDISP set to Normal, attribute values defined as Invisible are not displayed. This is the case for the WATTAGE, TOLERANCE, and TYPE tags. The Invisible mode was turned on inside the Attribute Definition dialog box when they were created. As a result, these values are not visible. With ATTDISP set to on, all resistor values are forced to be visible. When set to off, this command makes all resistor attribute values invisible.

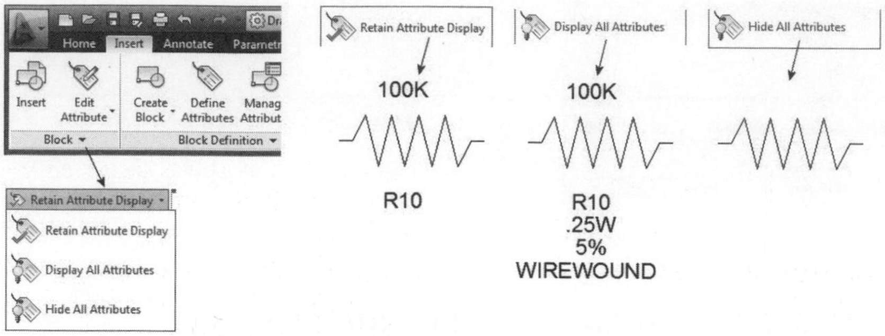

FIGURE 17.20

EDITING ATTRIBUTES

Generally, there are three commands that will be used to edit attributes: EATTEDIT, BATTMAN, and -ATTEDIT. All three commands are available through the Ribbon, as shown in Figure 17.21 (Left). The EATTEDIT command can also be activated by double-clicking a block with attributes. This launches the Enhanced Attribute Editor dialog box which allows you to make changes to that block's attributes. The BATTMAN command activates the Block Attribute Manager dialog box, which allows you to manage all blocks in a drawing. -ATTEDIT is a command line attribute editor whose purpose is to edit multiple attributes values in one operation. The FIND command, which can be accessed from the Annotate tab of the Ribbon, as shown in Figure 17.21 (Right), will also be introduced as an alternative to the -ATTEDIT command. It offers a simpler to use interface with a dialog box.

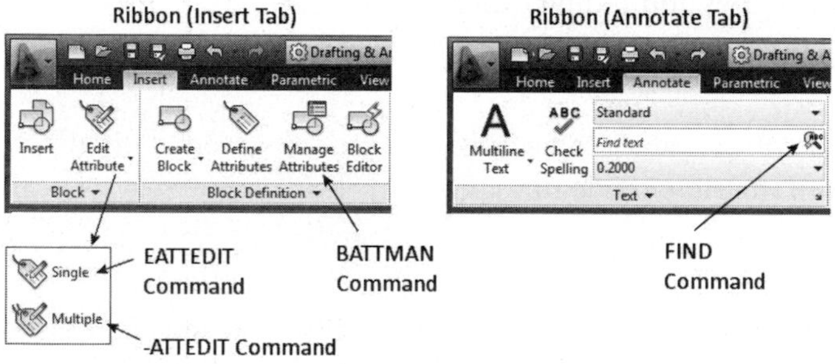

FIGURE 17.21

The following table gives a brief description of the Attribute modification tools available.

Button	Tool	Command	Function
	Edit Attribute	EATTEDIT	Launches a dialog box designed to edit the attributes in a block reference
	Block Attribute Manager	BATTMAN	Launches a dialog box used for managing attributes for blocks in the current drawing
	Synchronize Attributes	ATTSYNC	Updates all instances of the selected block with the currently defined attribute properties
	Data Extraction	DATAEXTRACTION	Exports the attributes found in a block reference out to a table or external file

THE ENHANCED ATTRIBUTE EDITOR DIALOG BOX

Click the Edit Attribute button that is found on the Ribbon to display the Enhanced Attribute Editor dialog box, as shown in Figure 17.22. This dialog box can also be activated by double-clicking the block or entering EATTEDIT at the Command prompt. In the image, the Bathtub block was selected. Notice that the attribute value is based on the selected tag. The value can then be modified in the Value field. To change to a different tag, select it with your cursor. Notice that this dialog box has three tabs. The Attribute tab allows you to select the attribute tag that will be edited.

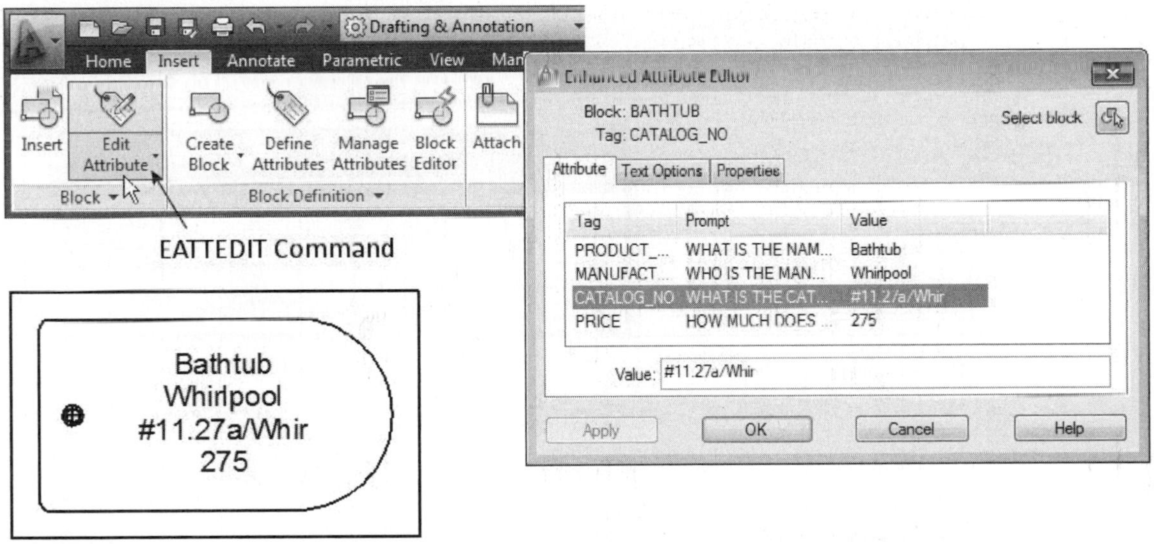

FIGURE 17.22

Clicking the Text Options tab displays the dialog box illustrated in Figure 17.23 (Left). Use this area to change the properties of the text associated with the attribute tag selected, such as text style, justification, height, rotation, and so on. Once the changes are made, click the Apply button.

Clicking the Properties tab displays the dialog box illustrated in Figure 17.23 (Right). Use this area to change properties such as layer, color, lineweight, and so on. When you have completed the changes to the attribute, click the Apply button.

All three tabs of the Enhanced Attribute Editor dialog box have a Select block button visible in the upper-right corner of each dialog box. Once you have edited a block, use this button to select a different block for editing, if desired.

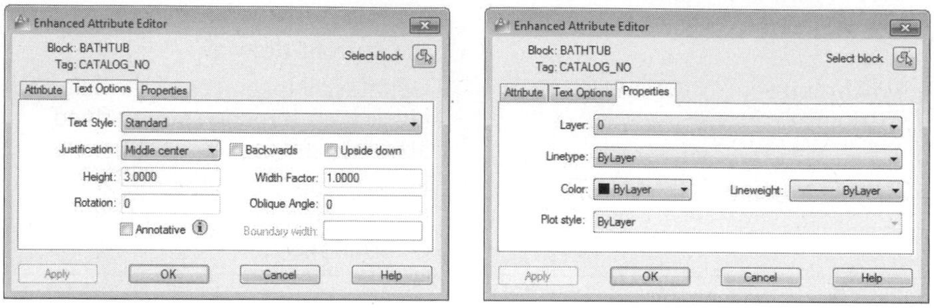

FIGURE 17.23

THE BLOCK ATTRIBUTE MANAGER

Clicking the Manage Attributes button in the Block Definition panel of the Ribbon or entering BATTMAN at the Command prompt displays the Block Attribute Manager dialog box, as shown in Figure 17.24. This dialog box is displayed as long as attributes are defined in your drawing. It does not require you to pick any blocks because it searches the database of the drawing and automatically lists all blocks with attributes. These are illustrated in the drop-down list, also shown in Figure 17.24. As will be demonstrated, the Block Attribute Manager dialog box allows you to edit attributes globally.

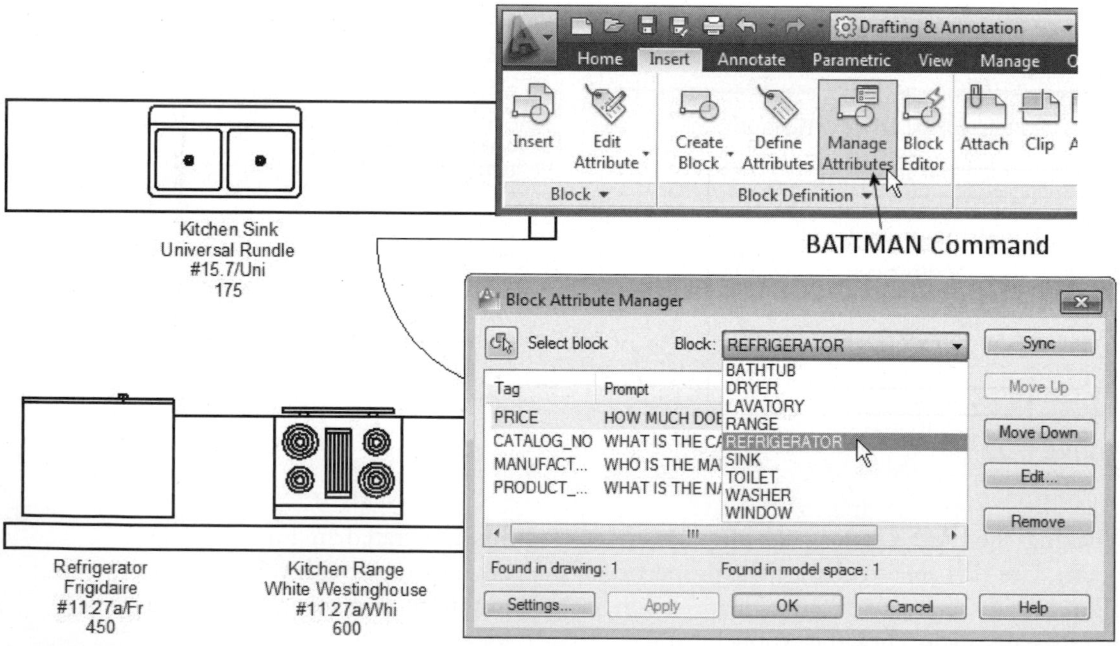

FIGURE 17.24

After selecting a block from the drop-down list, clicking the Edit button displays the Edit Attribute dialog box, as shown in Figure 17.25. Three tabs similar to those in the Enhanced Attribute Editor dialog box are displayed. Use these tabs to edit the attribute characteristics (mode, prompt, etc.), the text options (style, justification, etc.), or the properties (layer, color, etc.) of the attribute. Notice at the bottom of the dialog box that the Auto preview changes checkbox is selected. This setting allows you to preview changes on the display screen as they are made.

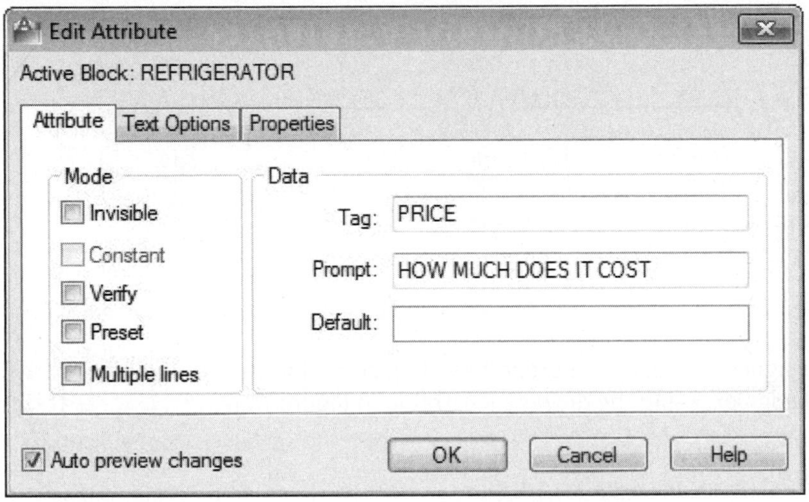

FIGURE 17.25

Clicking the Settings button in the main Block Attribute Manager dialog box displays the Settings dialog box, as shown in Figure 17.26 (Left). If you click one of the properties such as Height and click the OK button, this property will be displayed as a new column in the main Block Attribute Manager dialog box, as shown in Figure 17.26 (Right).

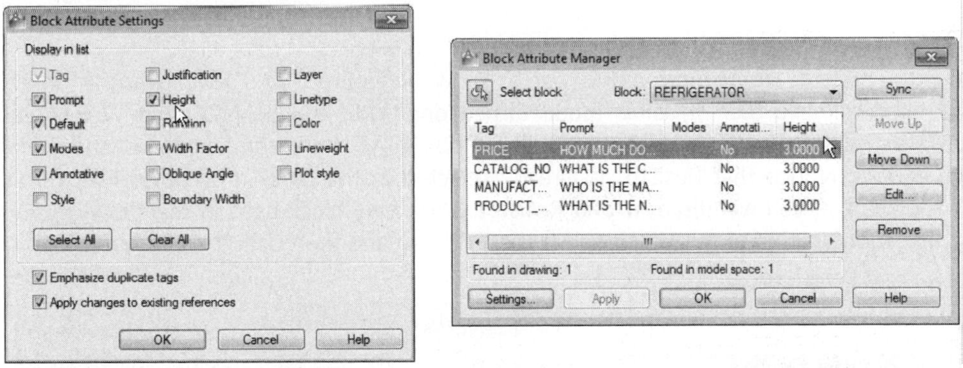

FIGURE 17.26

TUTORIAL EXERCISE: 17_BLOCK ATTRIB MGR.DWG

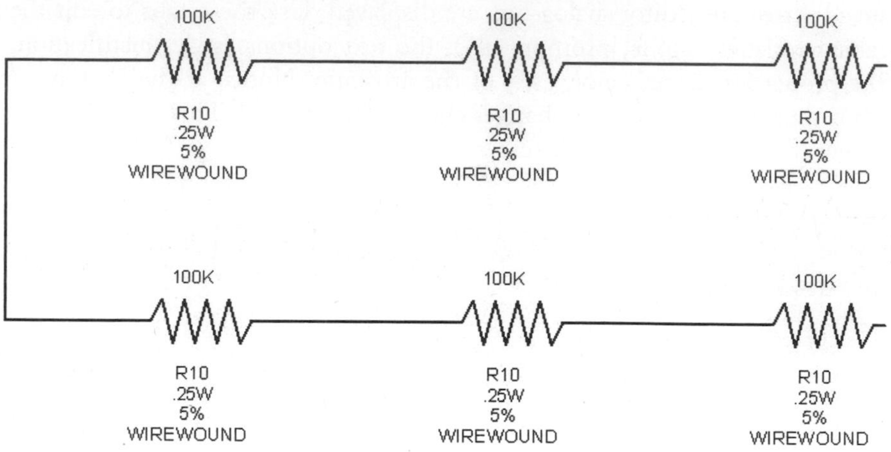

FIGURE 17.27

Purpose

This tutorial is designed to modify the properties of block attributes with the Block Attribute Manager dialog box, using the drawing illustrated in Figure 17.27.

System Settings

All units and drawing limits have already been set. Attributes have also been assigned to the resistor block symbols.

Layers

All layers have already been created for this drawing.

Suggested Commands

The Block Attribute Manager dialog box will be used to turn the Invisible mode on for a number of attribute values. These changes will be automatically seen as they are made.

STEP 1

Open the drawing file 17_Block Attrib Mgr, as shown in Figure 17.27. The following attribute tags need to have their Invisible mode turned on: TYPE, TOLERANCE, and WATTAGE. Activate the Block Attribute Manager dialog box (BATTMAN command), as shown in Figure 17.28. Notice that Resistor is already selected in the Block drop-down list. If you expand the list, you will discover that Resistor is the only block used in this drawing. For the first modification to this block, select the TYPE tag and then click the Edit button.

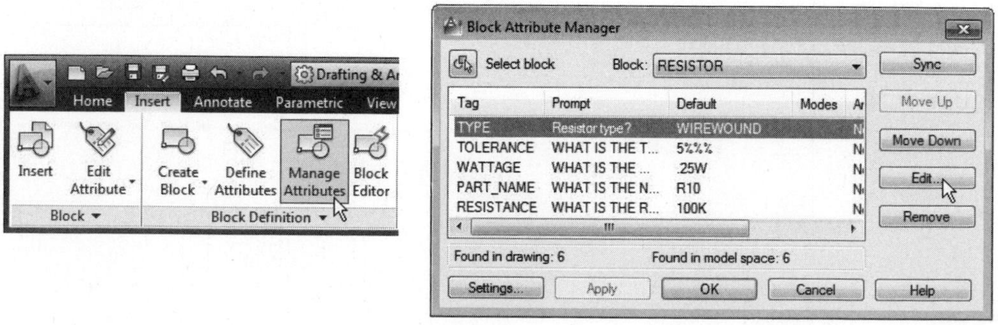

FIGURE 17.28

STEP 2

This takes you to the Edit Attribute dialog box, as shown in Figure 17.29. In the Attribute tab, place a check next to the Invisible mode and click the OK button. Notice that the changes are global and take place automatically. The attribute value WIREWOUND is no longer visible for any of the resistor blocks.

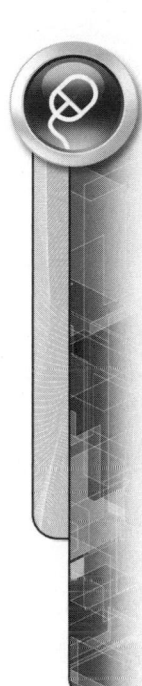

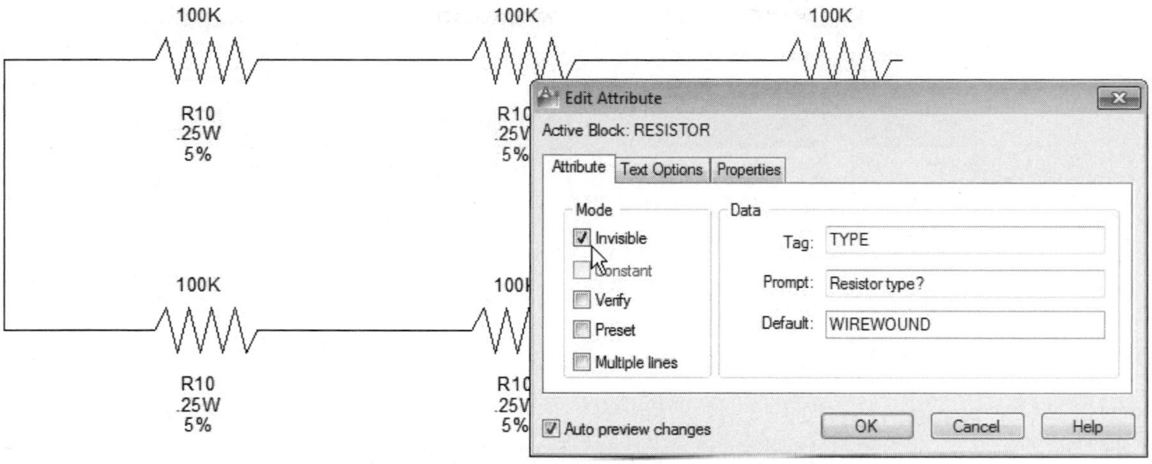

FIGURE 17.29

STEP 3

Turn the Invisible mode on for the TOLERANCE and WATTAGE tags. Your display should appear similar to Figure 17.30, with all attributed tags invisible except for PART_NAME, and RESISTANCE.

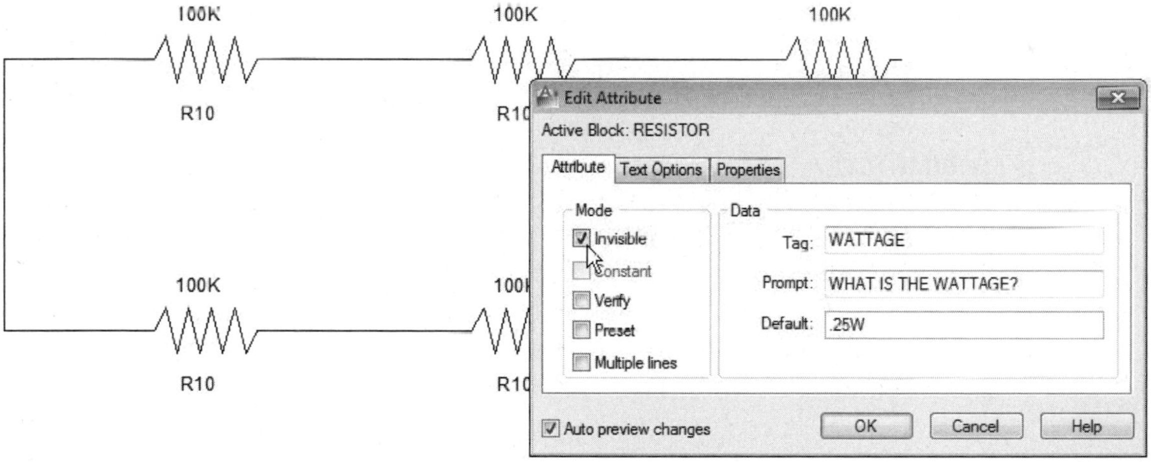

FIGURE 17.30

EDITING ATTRIBUTE VALUES GLOBALLY

Global editing is used to edit multiple attributes at one time. The criteria you specify will limit the set of attributes selected for editing. While the BATTMAN command provided a method to edit attribute properties, the -ATTEDIT and FIND commands can be used to globally edit specific attribute values.

Choose the -ATTEDIT command from the Insert Ribbon by selecting Multiple, as shown in Figure 17.31 (Left). The FIND command can be activated from the Annotate Ribbon, as shown in Figure 17.31 (Right).

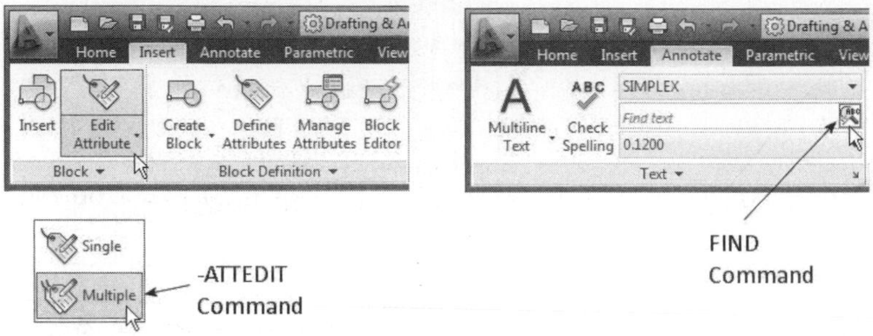

FIGURE 17.31

Let us examine how this process of globally editing attributes works. In Figure 17.32, suppose the resistance value of 100K needs to be changed to 150K on all blocks containing this attribute value. You could accomplish this by editing the attributes one at a time. However, this could be time consuming especially if the edit affects a large number of attributes. A more productive method would be to edit the group of attributes globally. Follow the examples below for accomplishing this task.

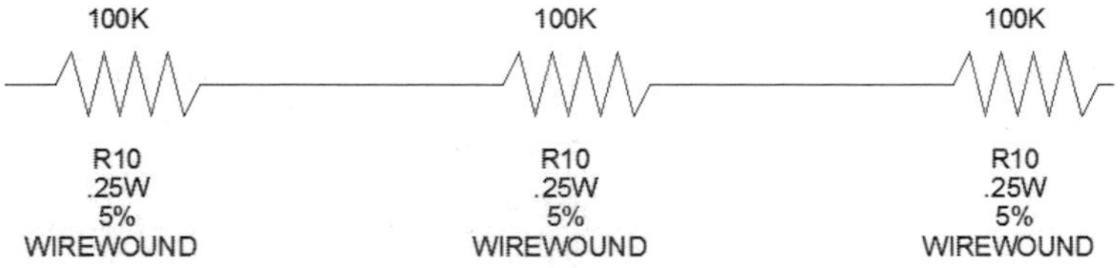

FIGURE 17.32

Issuing the -ATTEDIT command provides a command line version of an attribute editor. Carefully following the prompts provides a powerful global editor that allows you to filter through drawing data and make the necessary value updates. Follow the command sequence below to make the changes discussed:

```
Command: -ATTEDIT
Edit attributes one at a time? [Yes/No] <Y>: N (For No)
Performing global editing of attribute values.
Edit only attributes visible on screen? [Yes/No] <Y>:
(Press ENTER to accept this default)
Enter block name specification <*>: (Press ENTER to select
all blocks)
Enter attribute tag specification <*>: RESISTANCE
Enter attribute value specification <*>: 100K
```

```
Select Attributes: (Select all attributes using a window
selection box)
Select Attributes: (Press ENTER)
6 attributes selected.
Enter string to change: 100
Enter new string: 150
```

The FIND command can perform the same type of global value editing operation. In the next example you will change the tolerance value from 5% to 10%. Enter 5% in the text box of the Annotate Ribbon, as shown in Figure 17.33 (Left). Pick the Find Text button to activate the Find and Replace dialog box. Verify that 5% is entered in the Find What edit box. Then enter 10% in the Replace With edit box as shown in Figure 17.33 (Center). Clicking the Replace All button will display a second dialog box stating how many matches were found and how many objects were changed as shown in Figure 17.33 (Right).

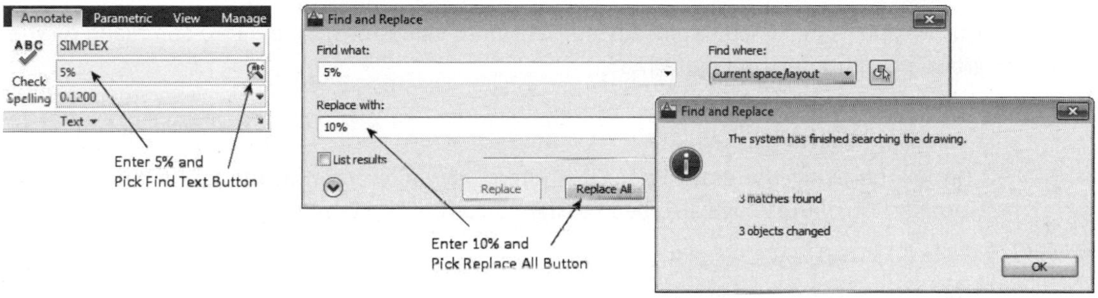

FIGURE 17.33

The result is shown in Figure 17.34, where all attributes values were changed globally.

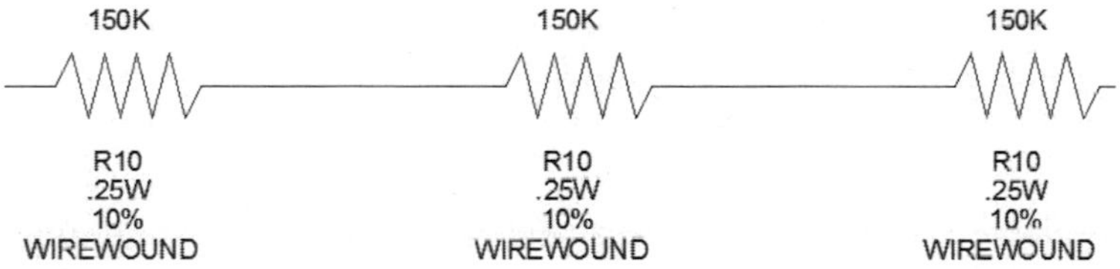

FIGURE 17.34

NOTE

To ensure the FIND command will edit block attribute values, expand the Find and Replace dialog box and verify that a check is placed next to the block attribute value. In order for the FIND command to identify block attribute values, they must not be assigned the Invisible mode. You can use the BATTMAN command to temporarily remove the Invisible mode if needed.

TUTORIAL EXERCISE: 17_COMPUTERS.DWG

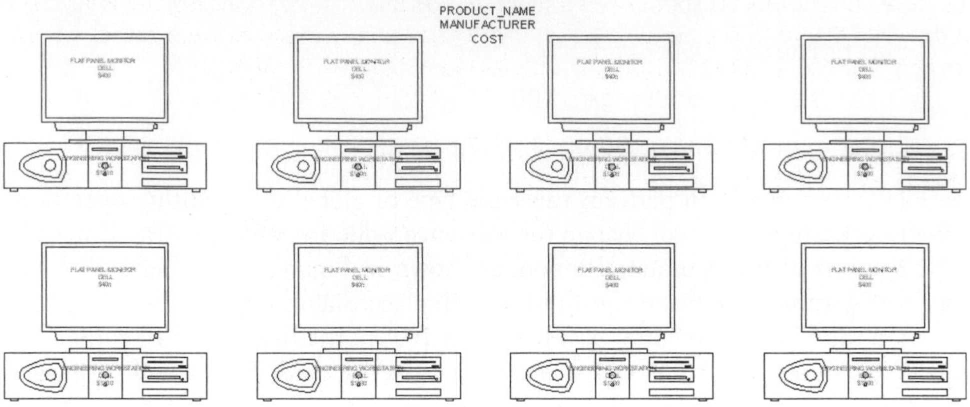

FIGURE 17.35

Purpose

This tutorial is designed to globally edit attribute values using the `FIND` command on the group of computers in Figure 17.35.

System Settings

You will be using the drawing 17_Computers. All units and drawing limits have already been set. Attributes have also been assigned to various computer components.

Layers

All layers have already been created for this drawing.

Suggested Commands

Utilize the `FIND` command to remove all dollar signs from the COST attribute values. This editing operation would be necessary if you needed to change the data in the COST box from character to numeric values for an extraction operation (extracting attribute data will be discussed in more detail later in this chapter).

STEP 1

Open the drawing 17_Computers. Figure 17.36 shows an enlarged view of two of the computer workstations. For each workstation, $400 needs to be changed to 400, and $1300 needs to be changed to 1300. In other words, the dollar sign needs to be removed from all values for each workstation.

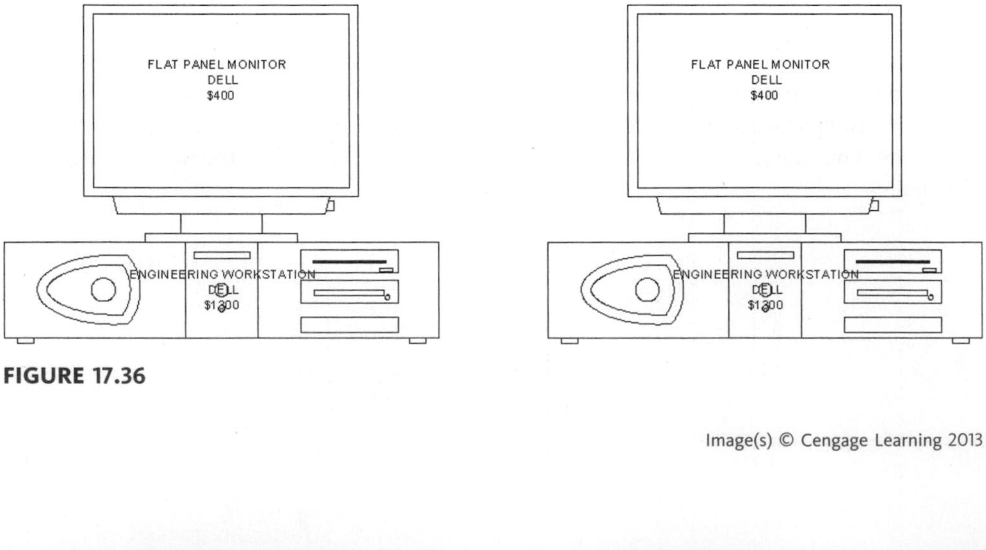

FIGURE 17.36

STEP 2

Launch the Find and Replace dialog box and enter $ in the Find What edit box. Then leave the Replace With edit box empty, as shown in Figure 17.37 (Left). Clicking the Replace All button will display a second dialog box stating how many matches were found (16) and how many objects were changed (16), as shown in Figure 17.37 (Right).

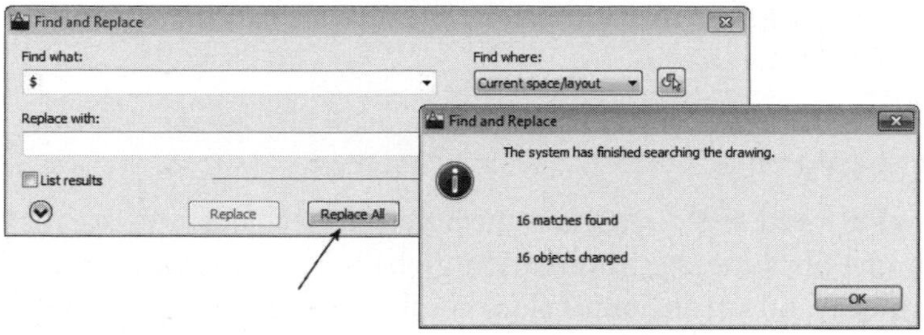

FIGURE 17.37

The results are illustrated in Figure 17.38, with all dollar signs removed from all attribute values through the Find and Replace dialog box.

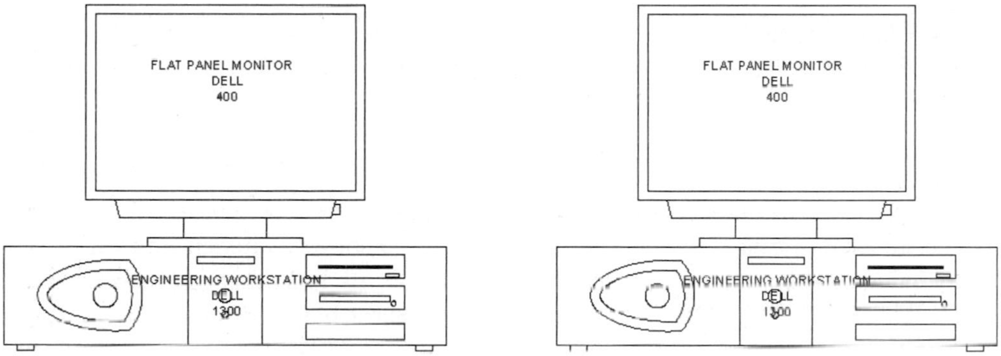

FIGURE 17.38

NOTE

The -ATTEDIT command could also be used to globally edit all attribute values and remove all dollar signs from all blocks. The command sequence is listed as follows:

```
Command: -ATTEDIT
Edit attributes one at a time? [Yes/No] <Y>: N (For No)
Performing global editing of attribute values.
Edit only attributes visible on screen? [Yes/No] <Y>:
(Press ENTER to accept this default)
Enter block name specification <*>: (Press ENTER to accept
this default)
Enter attribute tag specification <*>: COST
Enter attribute value specification <*>: (Press ENTER to
accept this default)
```

Select Attributes: *(Select all attributes using a window selection box)*

Select Attributes: *(Press* ENTER*)*

16 attributes selected.

Enter string to change: **$**

Enter new string: *(Press* ENTER*; this removes the dollar sign from all attribute values)*

REDEFINING ATTRIBUTES

If more sweeping changes need to be made to attributes, a mechanism exists that allows you to redefine the attribute tag information globally. You follow a process similar to redefining a block, and the attribute values are also affected. A few examples of why you would want to redefine attributes might be to change their mode status (Invisible, Constant, etc.), change the name of a tag, reword a prompt, change a value to something completely different, add a new attribute tag, or delete an existing tag entirely.

Any new attributes assigned to existing block references will use their default values. Old attributes in the new block definition retain their old values. If you delete an attribute tag, AutoCAD deletes any old attributes that are not included in the new block definition.

Exploding a Block with Attributes

Before redefining an attribute, you should first copy an existing block with attributes and explode it. This returns the block to its individual objects and return the attribute values to their original tag information. Figure 17.39 shows a kitchen sink with attribute values on the left and, on the right, the same sink but this time with attribute tags. The EXPLODE command was used on the right block to return the attribute values to their tags.

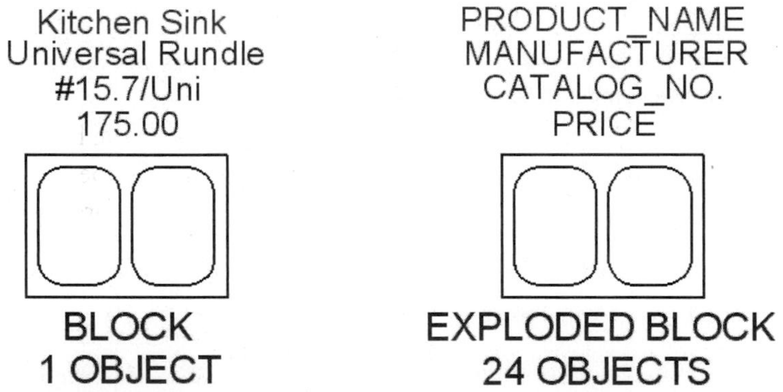

FIGURE 17.39

Using the Properties Palette to Edit Attribute Tags

A useful tool in making changes to attribute information is the Properties Palette. For example, clicking the CATALOG_NO. tag and then activating the Properties Palette, shown in Figure 17.40, allows you to make changes to the attribute prompt and values. You could even replace the attribute tag name with something completely different without having to use the Attribute Definition dialog box, as shown in the middle in Figure 17.40. Scrolling down the Properties Palette, as shown on the right in Figure 17.40, exposes the four attribute modes. If an attribute was originally created to be visible, you can make it invisible by changing the Invisible modifier here from No to Yes.

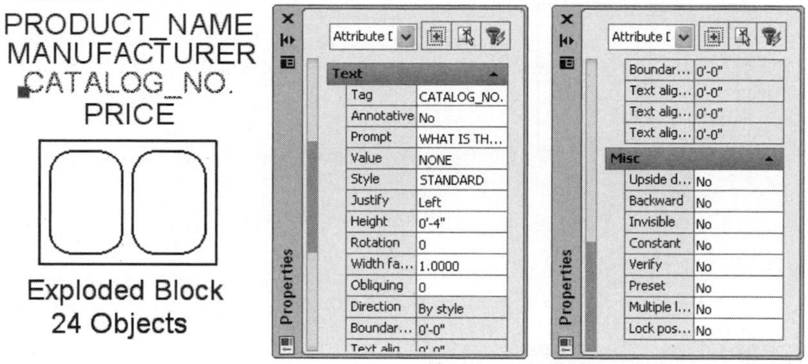

FIGURE 17.40

You can also double-click on an attribute tag to display the Edit Attribute Definition dialog box. This dialog box provides a quick and easy way to make changes to the Tag, Prompt, and Default values, as shown in Figure 17.41. Double-clicking the CATALOG_NO. attribute tag (`DDEDIT` command), as illustrated in Figure 17.41, displays the Edit Attribute Definition dialog box.

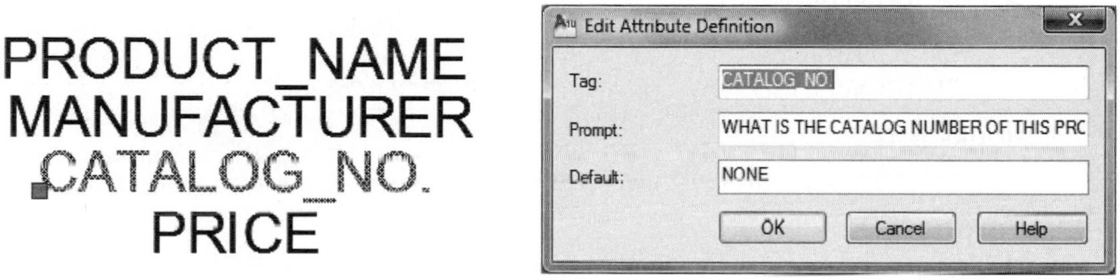

FIGURE 17.41

Redefining the Block with Attributes

Once you have made changes to the attribute tags, you are ready to redefine the block along with the attributes with the ATTREDEF command. This command does not appear in the Ribbon or Menu bar and must be typed in. Use the following prompt sequence and Figure 17.42 for accomplishing this task.

> `Command:` **`ATTREDEF`**
>
> `Enter name of the block you wish to redefine:` **`SINK`**
>
> `Select objects for new Block...`
>
> `Select objects:` *`(Pick a point at "A" in Figure 17.42)`*
>
> `Specify opposite corner:` *`(Pick a point at "B")`*
>
> `Select objects:` *`(Press ENTER to continue)`*
>
> `Specify insertion base point of new Block:` **`MID`**
>
> `of` *`(Pick the midpoint of the sink at "C")`*

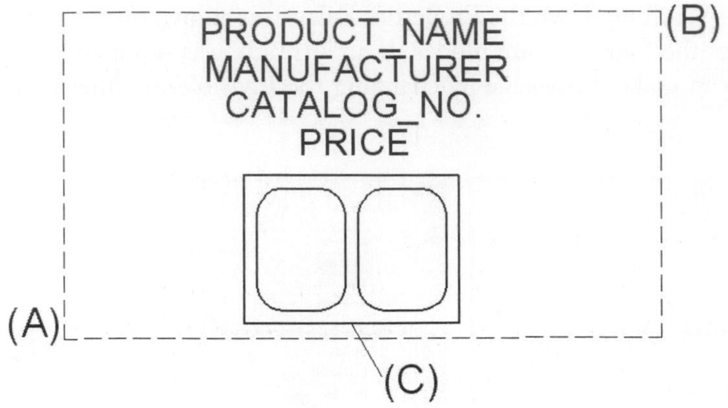

FIGURE 17.42

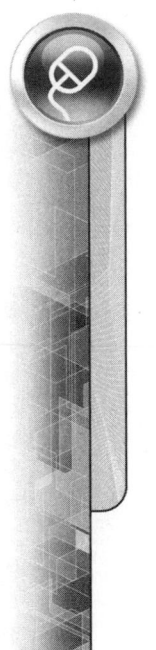

TUTORIAL EXERCISE: 17_RESISTORS.DWG

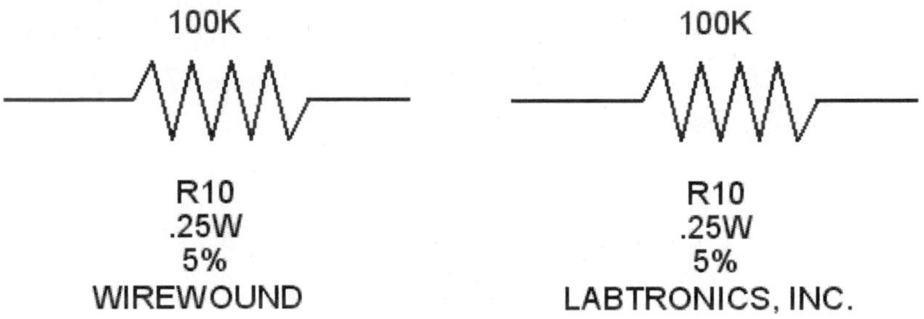

FIGURE 17.43

Purpose

This tutorial is designed to redefine attribute tags and update existing attribute values, as shown in Figure 17.43.

System Settings

All units and drawing limits have already been set. Attributes have also been assigned to computer components.

Layers

All layers have already been created for this drawing.

Suggested Commands

The DDEDIT command will be used to edit various characteristics of attributes.

STEP 1

Open the drawing file 17_Resistors. In Figure 17.44 a series of resistors is arranged in a partial circuit design. A number of changes need to be made to the original attribute definitions. Once the changes are made, the block and attributes will be redefined and all the blocks will automatically be updated.

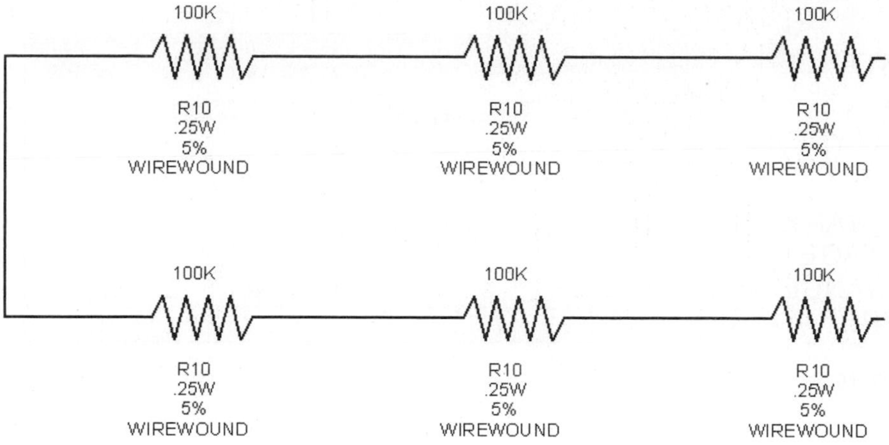

FIGURE 17.44

STEP 2

Copy one of the resistor symbols to a blank part of your screen. Then explode this block. This should break the block down into individual objects and return the attribute values to their tags, as shown in Figure 17.45 (Left).

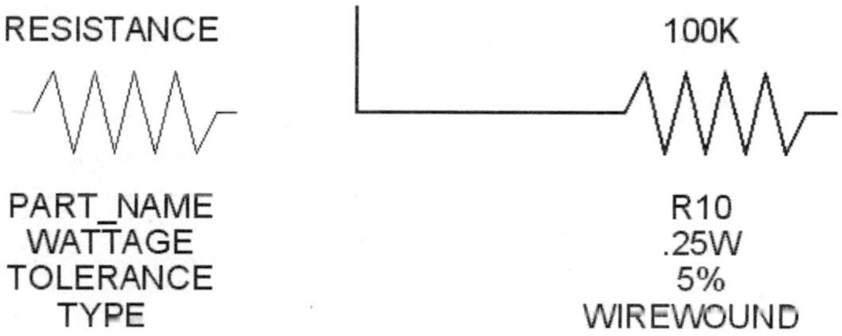

FIGURE 17.45

STEP 3

In this step, the tag TYPE needs to be replaced with a new tag. This also means creating a new prompt and default value. Double-click the TYPE tag to display the Edit Attribute Definition dialog box, as shown in Figure 17.46, and change the Tag from TYPE to SUPPLIER. In the Prompt field, change the existing prompt to Supplier Name? In the Default field, change WIREWOUND to LABTRONICS, INC, as shown in Figure 17.46. When finished, click the OK button to accept the changes and dismiss the dialog box.

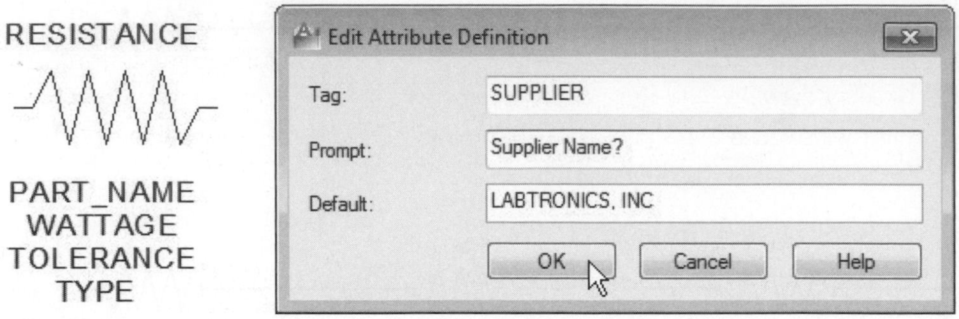

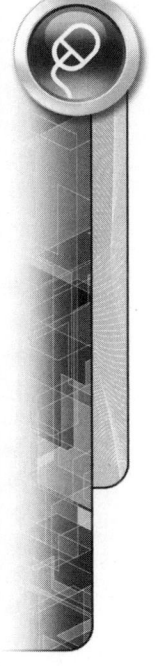

FIGURE 17.46

STEP 4

After these changes are made to the exploded block, the ATTREDEF command will be used to redefine the block and automatically update all resistor blocks to their new values and states. Activate the ATTREDEF command. Enter RESISTOR as the name of the block to redefine. Select the block and attributes, as shown in Figure 17.47 (Left). Pick the new insertion point at "C"; as the drawing regenerates, all blocks are updated to the new attribute values, as shown in Figure 17.47 (Right).

Command: **ATTREDEF**

Enter name of the block you wish to redefine: **RESISTOR**

Select objects for new Block...

Select objects: *(Pick a point at "A" in Figure 17.47 (Right))*

Specify opposite corner: *(Pick a point at "B")*

16 found

Select objects: *(Press* ENTER *to continue)*

Specify insertion base point of new Block: **End** *(For Object Snap Endpoint mode)*

of *(Pick the endpoint of the resistor at "C")*

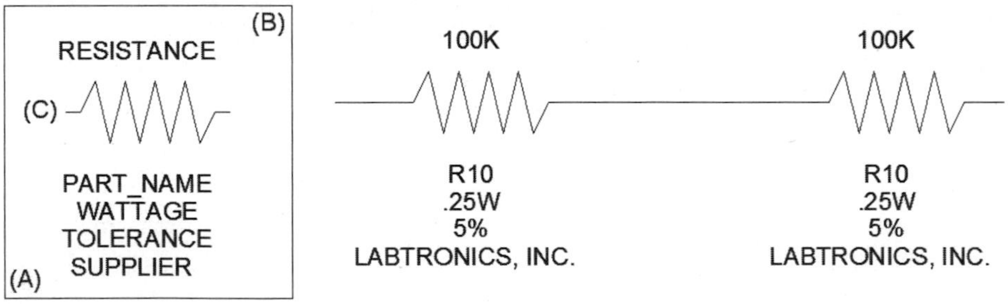

FIGURE 17.47

EXTRACTING ATTRIBUTES

Once attributes have been created, displayed, and edited, one additional step would be to extract attributes out to a file. The attributes can then be imported into Microsoft Excel or even brought into an existing AutoCAD drawing as a table object. This process is handled through the Data Extraction wizard, which can be activated by selecting Extract Data from the Ribbon, as shown in Figure 17.48.

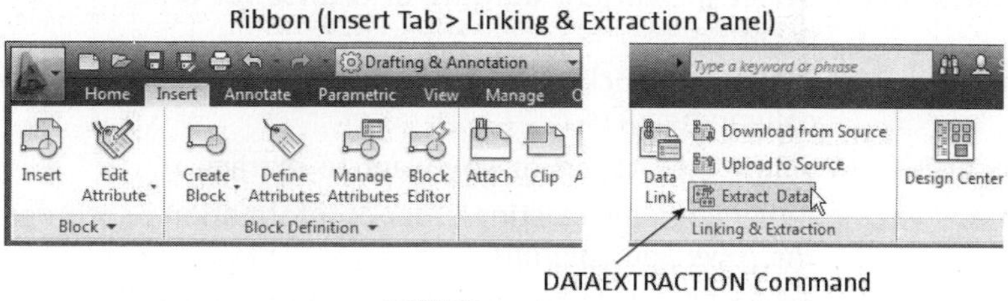

FIGURE 17.48

Activating the DATAEXTRACTION command displays the Data Extraction wizard, shown in Figure 17.49. Use this wizard to step you through the attribute extraction process. You can start with a template or an extraction file, if you have previously created one. The templates are saved with a BLK extension and the extraction files with a DXE extension. Each extraction you create will be saved and can be reused at any time to extract the latest information. You will be able to individually select the attributes as well as specific block information to extract. You can preview the extraction file, save the template, and export the results in either TXT (tab-separated file), CSV (comma separated), XLS (Microsoft Excel), or MDB (Microsoft Access) format or extract the results into an AutoCAD table object.

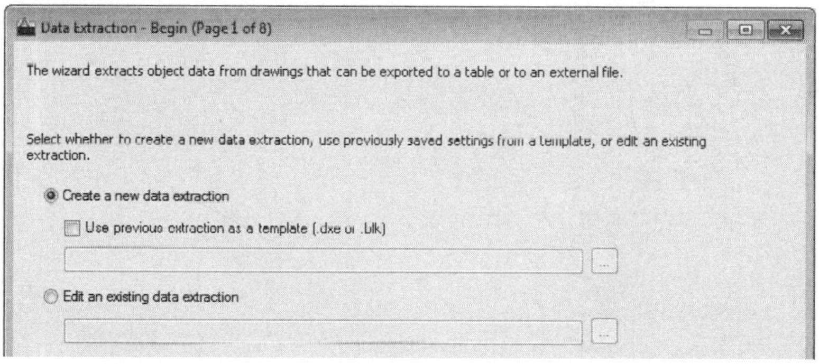

FIGURE 17.49

The following table gives a brief description of each page used for extracting attributes using the Attribute Extraction wizard.

Attribute Extraction Wizard

Page and Title	Description
Page 1—Begin	Create a new data extraction or edit an existing data extraction
Page 2—Define Data Source	Select the current drawing or browse for other drawings
Page 3—Select Objects	Select the blocks and attributes to include as row and column headers
Page 4—Select Properties	Select the properties you want to extract
Page 5—Refine Data	Either keep the existing order of information or reorder the rows and columns
Page 6—Choose Output	Select to output to a table or to an external file
Page 7—Table Style	Select a table style
Page 8—Finish	Click this button to finish the extraction

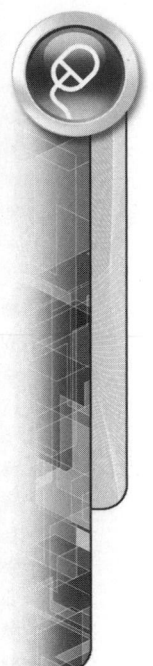

TUTORIAL EXERCISE: 17_CAD DEPT MODULES.DWG

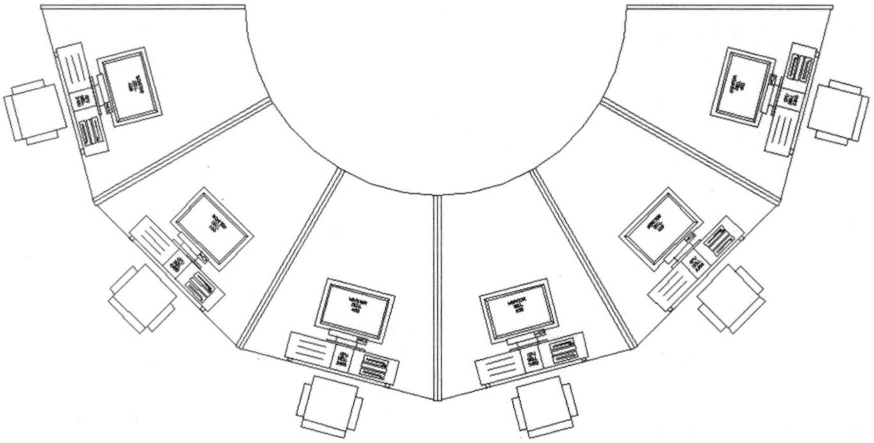

FIGURE 17.50

Purpose

This tutorial is designed to extract attribute data through the use of the Data Extraction wizard for the drawing 17_CAD Dept Modules, illustrated in Figure 17.50.

System Settings

All units and drawing limits have already been set. Attributes have also been assigned to computer components.

Layers

All layers have already been created for this drawing.

Suggested Commands

This drawing appears inside a layout that has already been created. A table consisting of the extracted attributes will be inserted into this layout. Follow the steps provided by the Data Extraction wizard to extract the attributes out to an AutoCAD table.

STEP 1

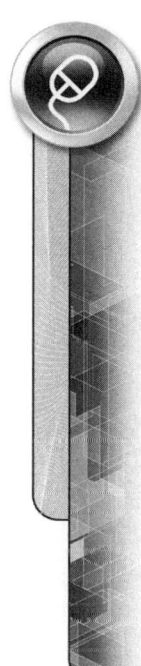

Open the drawing file 17_CAD Dept Modules.dwg. Activate the Data Extraction wizard, as shown in Figure 17.51, by clicking Extract Data from the Insert Ribbon, as shown on the left in the figure. Verify that the "Create a new data extraction" button is selected. Click the Next button to continue to the next step.

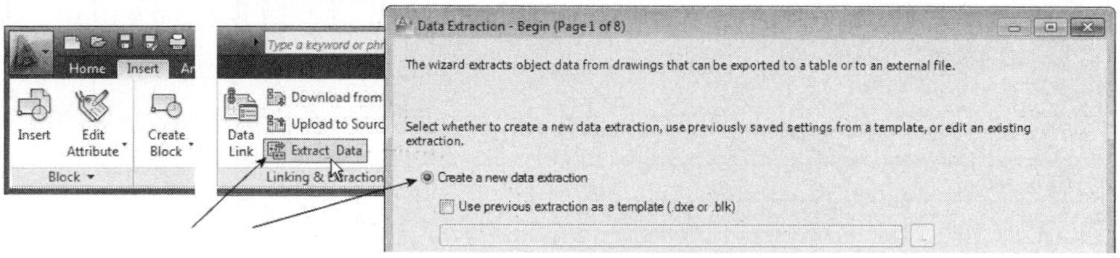

FIGURE 17.51

When the Save Data Extraction As dialog box appears as in Figure 17.52, enter the name of the data extraction file as CAD Dept Modules and select an appropriate folder in the Save in drop-down list. When finished, click the Save button.

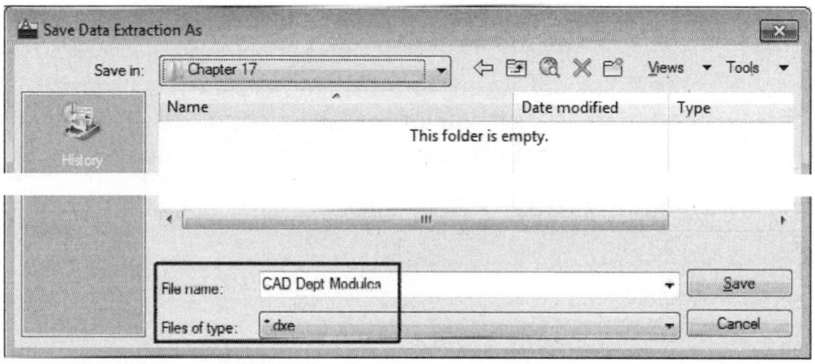

FIGURE 17.52

STEP 2

In this next step, you pick the data source from which the attributes will be extracted. When the Define Data Source dialog box appears, verify that Drawings/Sheet set is picked and that the Include current drawing box is checked, as shown in Figure 17.53. Click the Next > button to continue.

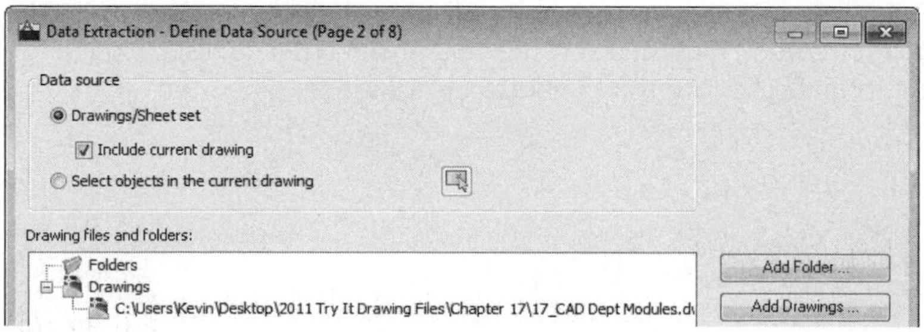

FIGURE 17.53

STEP 3

When the Select Objects dialog box displays, remove the check from the Display all object types box and verify the Display blocks only button is selected. To further filter the available data for extraction, check the Display blocks with attributes only box, as shown in Figure 17.54. When the three items display at the top of this dialog box, remove the check from the box next to Architectural Title. The remaining two items, CPU and Monitor, should remain checked. Click the Next > button to continue.

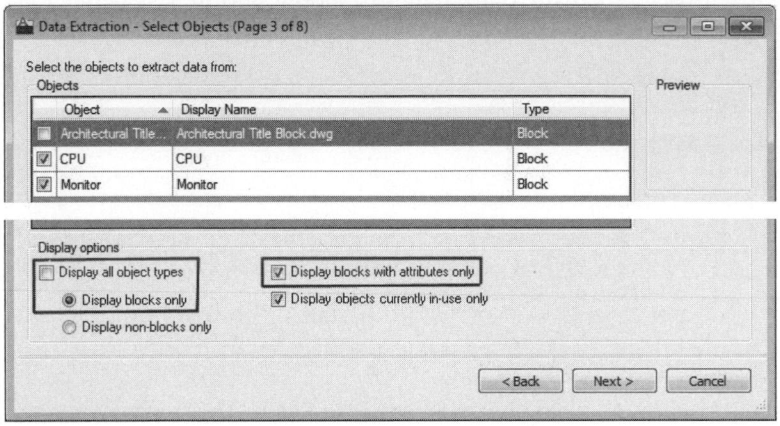

FIGURE 17.54

STEP 4

When the Select Properties dialog box appears, clear all of the checks from the Category filter boxes on the right side of the dialog box except for Attribute, as shown in Figure 17.55. Also, verify that all three properties on the left side of the dialog box, COST, MANUFACTURER, and PRODUCT_NAME, are checked. Click the Next > button to continue.

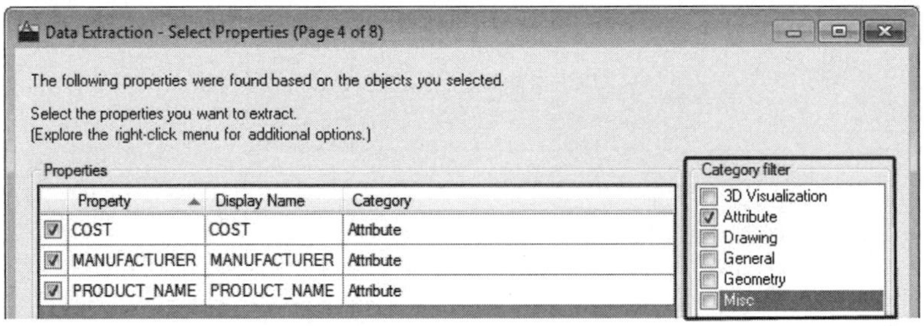

FIGURE 17.55

STEP 5

When the Refine Data dialog box appears, keep the display of all information as is. This dialog box allows you to reorder, hide, and sort the columns in order to display the information in a different format. Click the Next > button to continue.

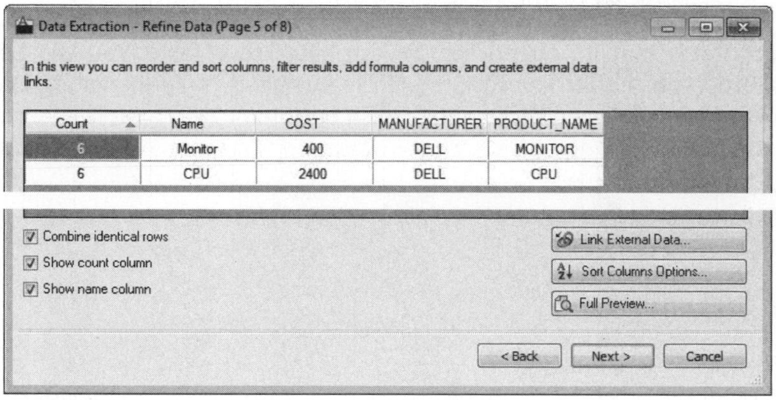

FIGURE 17.56

STEP 6

When the Choose Output dialog box displays, place a check in the box next to Insert data extraction table into drawing, as shown in Figure 17.57. This creates an AutoCAD table that displays in the current drawing. The other check box, if selected, allows you to export extraction data to an external file (spreadsheet, database or text file). Click the Next > button to continue.

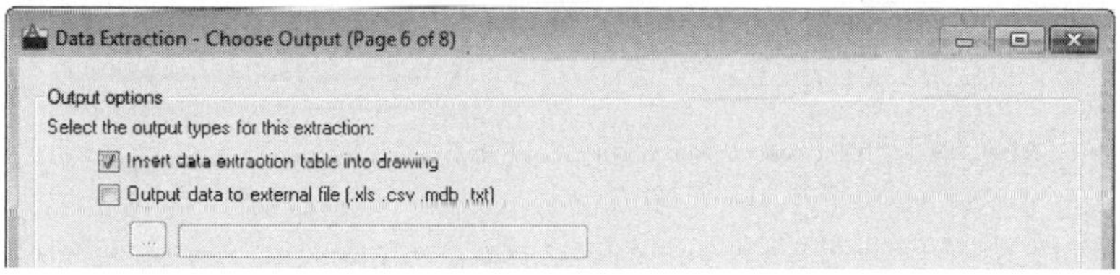

FIGURE 17.57

STEP 7

This next dialog box deals with the components of the table that the attributes will be extracted to. In Figure 17.58, you need to enter a title for the table. In this exercise, the title of this table will be ATTRIBUTE EXTRACTION RESULTS. You can also select a table style. In this exercise, the Standard table style will be utilized and a preview is displayed on the right side of the dialog box. Click the Next button to continue.

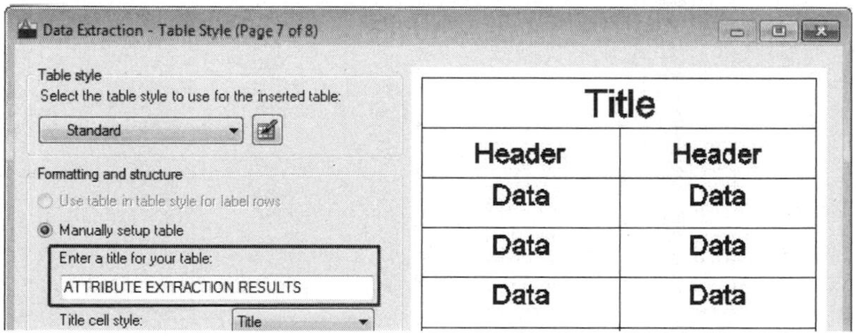

FIGURE 17.58

STEP 8

The final dialog box states that if you extract the attributes to a table, you will be prompted for an insertion point after you click the Finish button. If you are creating external files, these will be created when clicking the Finish button.

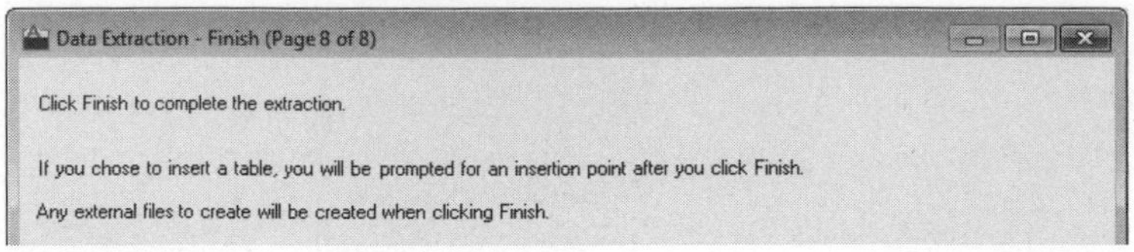

FIGURE 17.59

STEP 9

Clicking the Finish button in the previous step brings you back into the drawing editor. You will be prompted to insert the table; do so in a convenient location of the drawing, as shown in Figure 17.60. In this image, notice that the Count, Block Name, COST, MANUFACTURER, and PRODUCT_NAME were all extracted.

ATTRIBUTE EXTRACTION RESULTS				
Count	Name	COST	MANUFACTURER	PRODUCT_NAME
6	CPU	2400	DELL	CPU
6	Monitor	400	DELL	MONITOR

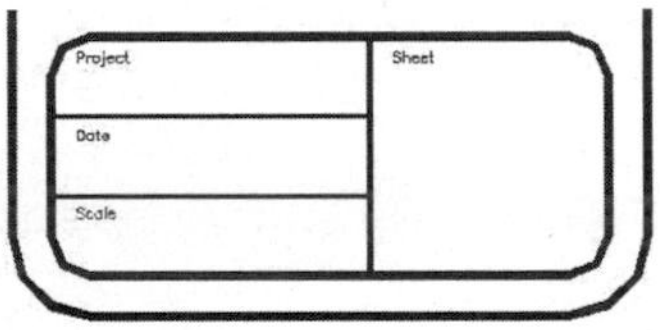

FIGURE 17.60

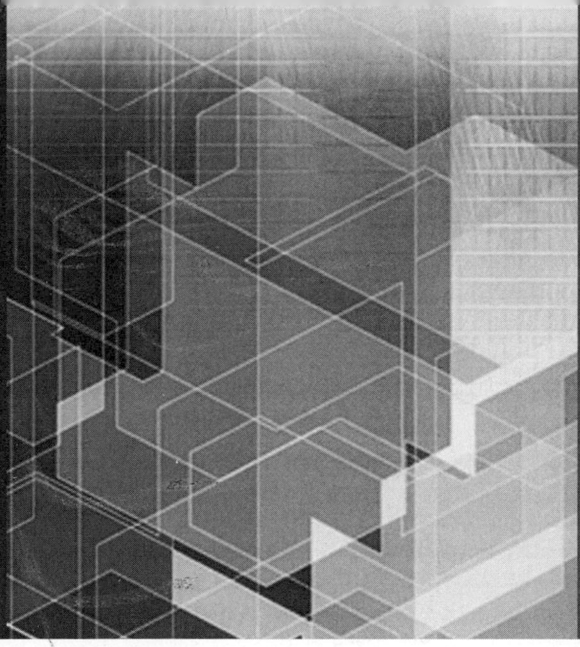

CHAPTER

18

Working with File References

This chapter begins the study of File References and how they are managed in an AutoCAD drawing. External drawings, raster images, and PDF files can all be attached to your drawing. Attaching a file reference is significantly different than the inserting process that was discussed in the chapter on working with blocks. When changes are made to a file that has been externally referenced (attached) into another drawing, these changes are updated automatically. There is no need to redefine the file reference in the same way blocks are redefined. File management is essential when sharing data between files. The `ETRANSMIT` command is used to collect all support files that make up a drawing. This guarantees to the individual reviewing the drawing that all support files come with the drawing. Working with raster images is also discussed in this chapter. Raster images can be attached to a drawing file for presentation purposes. The brightness, contrast, and fade factor of the images can also be adjusted. To help you control the order in which images display in your drawing, the `DRAWORDER` command will be explained.

COMPARING EXTERNAL FILE REFERENCES AND BLOCKS

You have already seen in Chapter 16 how easy and productive it is to insert blocks and drawing files into other drawing files. One advantage in performing these insertions is the grouping of numerous objects into a single entity that can be placed repeatedly in a drawing. Another advantage of blocks is their ability to be redefined in a drawing.

External references are similar to blocks in that they act as a single entity in your drawing. However, external references are attached to the drawing file, whereas blocks are inserted in the drawing. This attachment actually sets up a relationship between the current drawing file and the external referenced file. For example, take a floor plan file and externally reference it into a current drawing file. All objects associated with the floor plan are brought in, including their current layer and linetype qualities. With the floor plan acting as a guide, such items as lighting, power, plumbing, and furniture can be added to the floor plan and saved to the current drawing file. Now a design change needs to be made to the floor plan. Open the floor plan, stretch a few doors and walls to new locations, and save the file. The next time the drawing holding the electrical, plumbing, and furniture arrangements is opened, the changes to the floor plan are made automatically to the current drawing file. This is one of the primary advantages of external file references over blocks. Another important advantage is the control of file size. Inserting files into a drawing can create a large cumbersome file, while attaching them keeps them separate and manageable. One disadvantage of an external reference is that items such as layers and other blocks that belong to the external reference can be viewed but have limited capabilities for manipulation.

CHOOSING EXTERNAL REFERENCE COMMANDS

The EXTERNALREFERENCES (XREF or XR is the alias) command activates a palette that allows you to attach and control external file references in a drawing. This command can be accessed through the Ribbon, as shown in Figure 18.1.

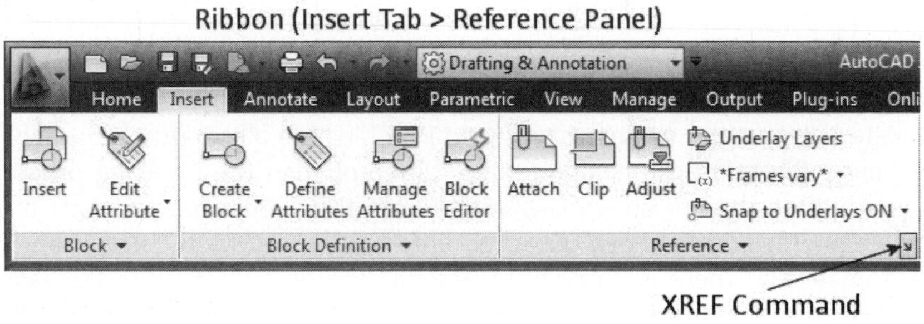

FIGURE 18.1

Once the command is activated, the External References palette displays, as shown in Figure 18.2. In this image, the drawing file 18_Floor Plan is listed at the top of this palette. Underneath this file are a numbers of additional files dealing with the furniture, hvac, lighting, numbers, plumbing, and power; these files represent external references that were attached to the 18_Floor Plan drawing file.

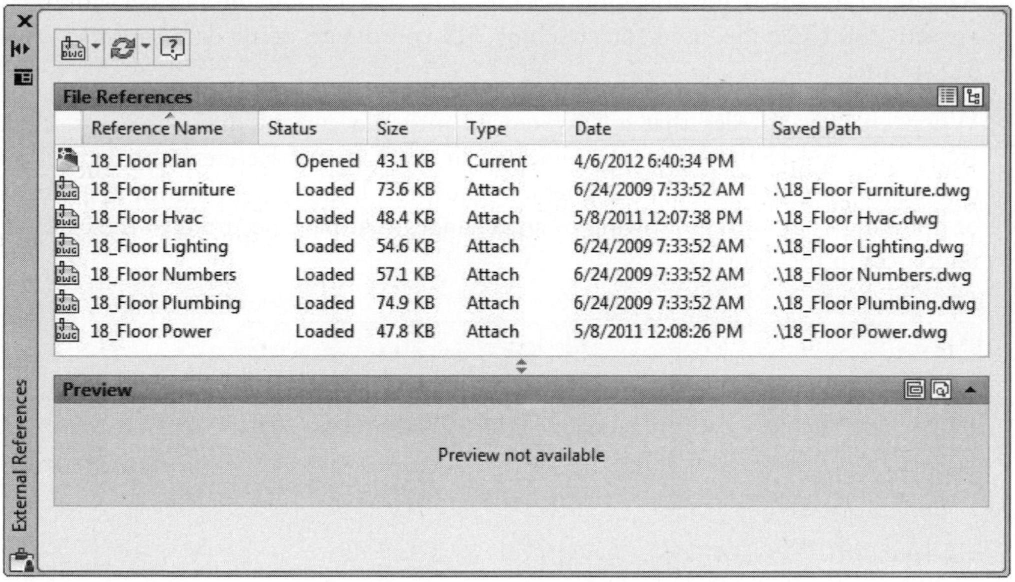

FIGURE 18.2

The lower portion of the External References palette, illustrated in Figure 18.3, has either a Preview or Details heading. Clicking the Details button displays the information about the selected external reference, as shown in Figure 18.3 (Left). In this example, the 18_Floor Furniture external reference is selected and has various items such as status, file size, type, and date, to name a few listed. Clicking the Preview button displays a preview of the selected external reference, as shown in Figure 18.3 (Right).

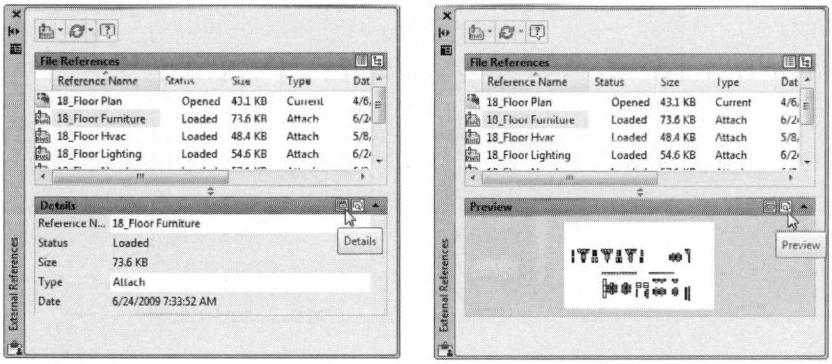

FIGURE 18.3

Located in the upper-left corner of the External References palette are buttons that are used for attaching or reloading external references. Clicking the button located on the left, as shown in Figure 18.4 (Left), displays the file types that can be attached: Attach DWG—for attaching an AutoCAD drawing file, Attach Image—used for attaching a raster image, Attach DWF—used for attaching a DWF (Drawing Web Format) file, Attach DGN—used for attaching a MicroStation® DGN drawing file,

Attach PDF—used for attaching a PDF (Portable Document Format) file, and Attach Point Cloud—used for attaching 3D coordinate point data captured by a 3D scanner.

When changes to files that are referenced into the current drawing occur, click Refresh to update the information located in the External References palette. You can also choose Reload All References, as shown in Figure 18.4 (Right), as a means of updating your current drawing to any changes that have occurred to the external references in the drawing.

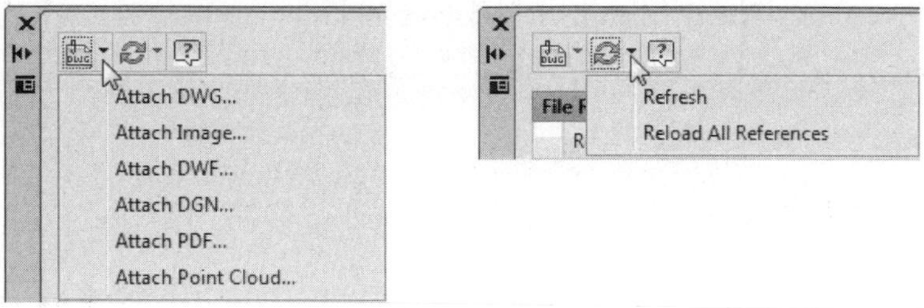

FIGURE 18.4

External references can also be attached and reloaded by right-clicking on the palette area and activating a shortcut menu, as shown in Figure 18.5 (Left). Once the External References palette is populated with various references, select one of the references and right-click to display the shortcut menu shown in Figure 18.5 (Right). With this menu you can elect to open the original drawing file being referenced or click Attach to merge another drawing file. The Unload, Reload, Detach, and Bind options are discussed in detail later in this chapter.

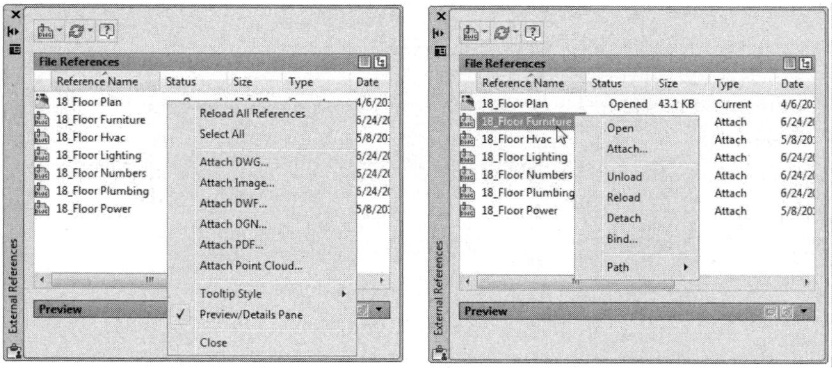

FIGURE 18.5

The following table outlines the main tools that will be utilized when working with file references.

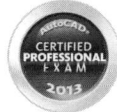

Button	Tool	Function
	External References	Launches the External References Palette, used for managing external references in a drawing
	Attach Xref	Launches the Select Reference File dialog box, used for attaching an external drawing file into a drawing
	Clip Xref	Used to isolate a portion of an external reference by clipping away unnecessary objects
	Xbind	Used to convert named objects, such as blocks and layers that belong to an external reference, to usable items
	Xref Frame	Controls the frame used to clip an external reference

ATTACHING AN EXTERNAL REFERENCE

Use the ATTACH command to attach an external file into the current drawing file. While the XREF command can also be used to attach a file reference, this command takes you directly to the Select Reference file dialog box. In the dialog box you select the file that will be attached to the current drawing.

NOTE

To accomplish the Try It! exercises in this chapter, copy the Chapter 18 files to a convenient location/folder on your hard drive. This way you will be able to modify and save the drawings as required.

TRY IT!

Start a new drawing file from scratch. Activate the ATTACH command or the EXTERNALREFERENCES command with the Attach Dwg button to display the Select Reference File dialog box, as shown in Figure 18.6. This dialog box is very similar to the one used for selecting drawing files to initially load into AutoCAD. Verify that the Files of type list box is set to Drawings and locate the folder with the Chapter 18 files. Click the file 18_Asesmp1 to attach and click the Open button.

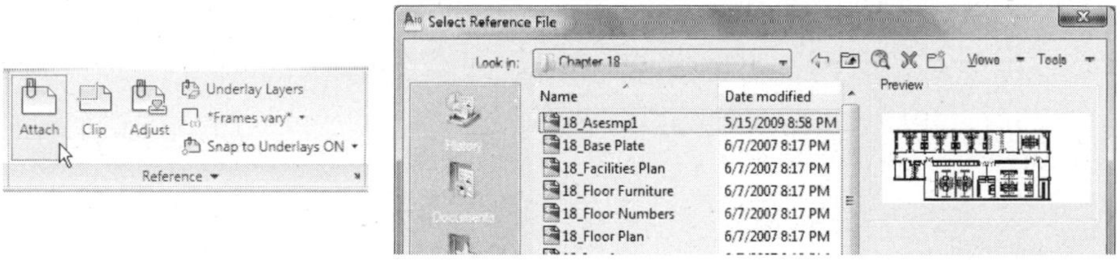

FIGURE 18.6

Once the file 18_Asesmp1 is selected, the Attach External Reference dialog box appears, as shown in Figure 18.7. Some of the information contained in this dialog box is similar to the information in the Insert dialog box, such as the insertion point, scale, and rotation angle of the external reference. Notice in the upper right part of the dialog box is the path information associated with the external reference. If, during file management, an externally referenced file is moved to a new location, the new path of the external reference must be reestablished; otherwise, it does not load into the drawing it was attached to. Remove the check from the Specify On-screen box under the Insertion point. This inserts this external reference at absolute coordinate 0,0,0. Clicking the OK button returns you to the drawing editor and attaches the file to the current drawing.

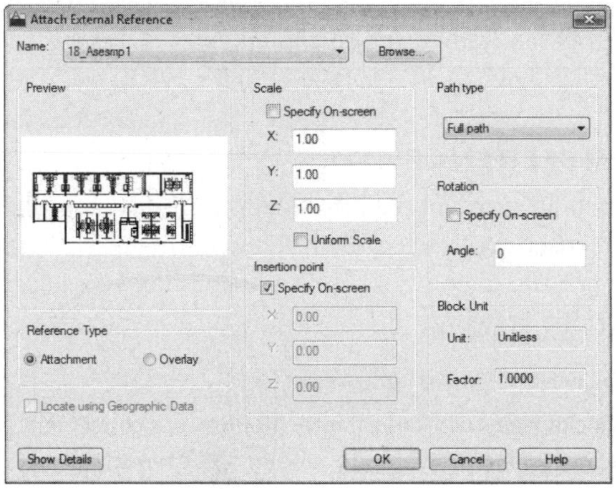

FIGURE 18.7

Due to the difference in scale between the attached drawing and the display screen, you will not see the floor plan until you perform a ZOOM-All. Your results should appear similar to the illustration in Figure 18.8. By default, your external reference will appear faded. Expand the Reference panel on the Insert Ribbon and move the Xref fading slider, shown in Figure 18.8 (Top-Right), to set the amount of fading desired. It should be noted that this setting only affects the screen display and not plots. The fading tool can help you distinguish on a drawing, between those items that belong and those that are referenced.

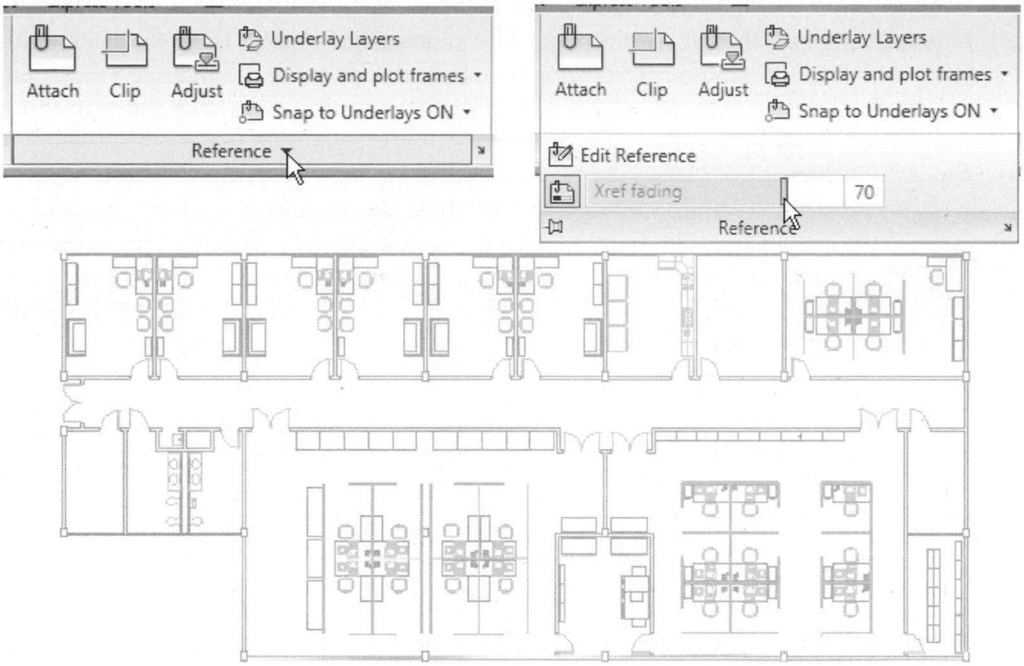

FIGURE 18.8

With the external reference attached to the drawing file, layers and blocks that belong to the external reference are displayed in a unique way. Illustrated in Figure 18.9 are the current layers that are part of the drawing. However, notice how a number of layers begin with the same name (18_Asesmp1); also, what appears to be a vertical bar separates 18_Asesmp1 from the names of the layers. Actually, the vertical bar represents the Pipe symbol on the keyboard, and it designates that the layers belong to the external reference, namely 18_Asesmp1. These layers can be turned off, can be frozen, or can even have the color changed. However, you cannot make these layers current for drawing because they belong to the external reference (xref-dependent layers).

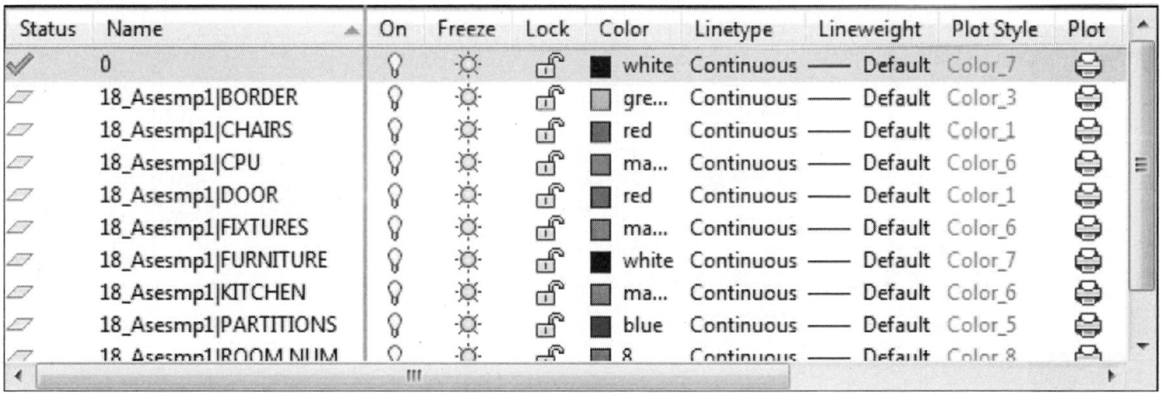

Status	Name	On	Freeze	Lock	Color	Linetype	Lineweight	Plot Style	Plot
✓	0	♀	☼	🔓	■ white	Continuous	—— Default	Color_7	🖨
⬜	18_Asesmp1|BORDER	♀	☼	🔓	□ gre...	Continuous	—— Default	Color_3	🖨
⬜	18_Asesmp1|CHAIRS	♀	☼	🔓	■ red	Continuous	—— Default	Color_1	🖨
⬜	18_Asesmp1|CPU	♀	☼	🔓	■ ma...	Continuous	—— Default	Color_6	🖨
⬜	18_Asesmp1|DOOR	♀	☼	🔓	■ red	Continuous	—— Default	Color_1	🖨
⬜	18_Asesmp1|FIXTURES	♀	☼	🔓	■ ma...	Continuous	—— Default	Color_6	🖨
⬜	18_Asesmp1|FURNITURE	♀	☼	🔓	■ white	Continuous	—— Default	Color_7	🖨
⬜	18_Asesmp1|KITCHEN	♀	☼	🔓	■ ma...	Continuous	—— Default	Color_6	🖨
⬜	18_Asesmp1|PARTITIONS	♀	☼	🔓	■ blue	Continuous	—— Default	Color_5	🖨
⬜	18_Asesmp1|ROOM NUM	♀	☼	🔓	■ 8	Continuous	—— Default	Color_8	🖨

FIGURE 18.9

One of the real advantages of using external references is the way they are affected by drawing changes. To demonstrate this, first save your drawing file under the name 18_Facilities. Close this drawing and open the original floor plan called 18_Asesmp1.

Use the Insert dialog box to insert a block called ROOM NUMBERS, as shown in Figure 18.10, into the 18_Asesmp1 file. Use an insertion point of 0,0,0 for placing this block and click OK in the Insert dialog box. Save this drawing file and open the drawing 18_Facilities.

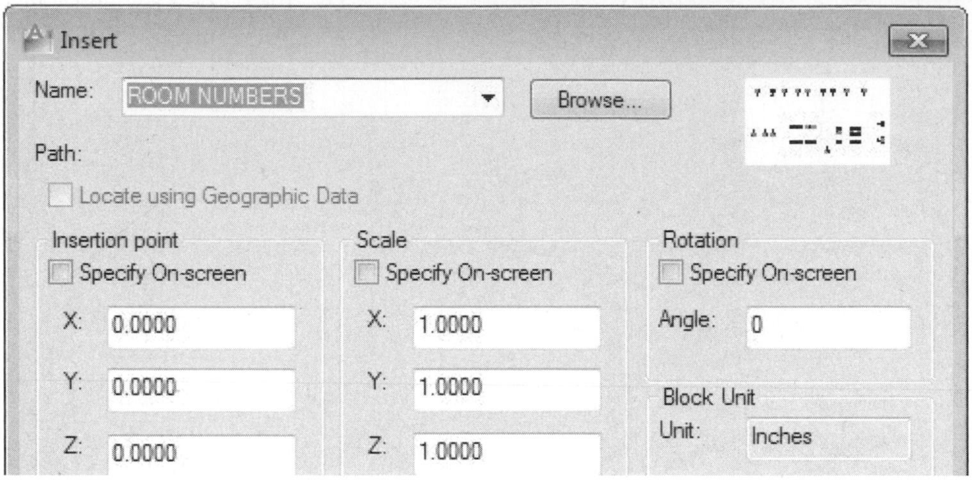

FIGURE 18.10

In Figure 18.11, the latest version of 18_Asesmp1 is loaded and displayed. Notice how room tags have been added to the floor plan. Once the drawing holding the external reference is opened, the attached file is reloaded and the room tags are automatically displayed.

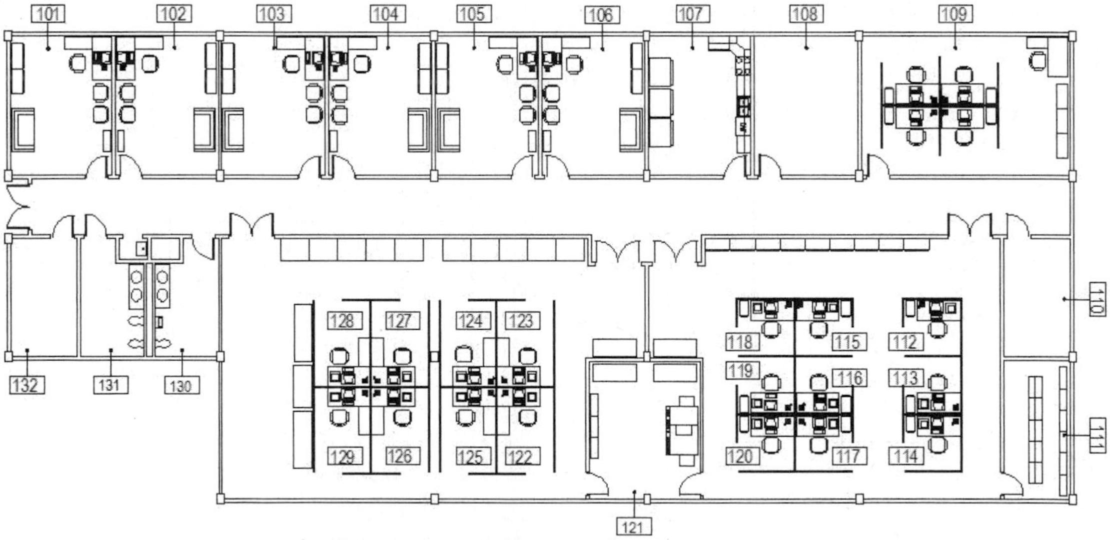

FIGURE 18.11

OVERLAYING AN EXTERNAL REFERENCE

Suppose that, in the last example of the facilities floor plan, some design groups need to see the room number labels while other design groups do not. This can be accomplished by overlaying an external reference instead of attaching it. All design groups will see the information if the external reference is attached. If information is overlaid and the entire drawing is externally referenced, the overlaid information does not display. This option is illustrated in the next Try It! exercise.

TRY IT!

Open the drawing 18_Floor Plan. Then use the Attach External Reference dialog box, shown in Figure 18.12, to attach the file 18_Floor Furniture (the steps for accomplishing this were demonstrated in the previous Try It! exercise). Use an insertion point of 0,0,0 and be sure the reference type is Attachment.

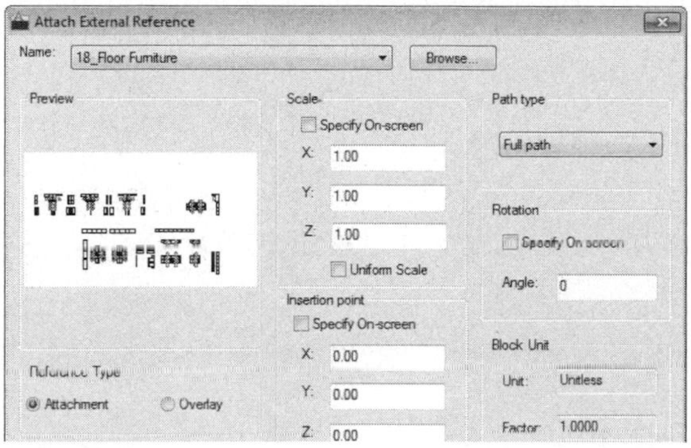

FIGURE 18.12

Your display should be similar to the illustration in Figure 18.13. You want all design groups to view this information.

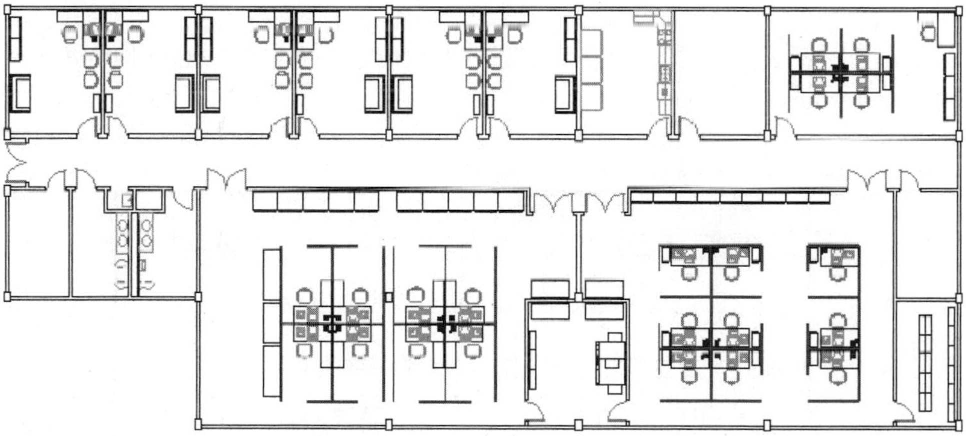

FIGURE 18.13

Now we will overlay an external reference. Activate the ATTACH command to display the Select Reference File dialog box, as shown in Figure 18.14, and pick the file 18_Floor Numbers and click the Open button.

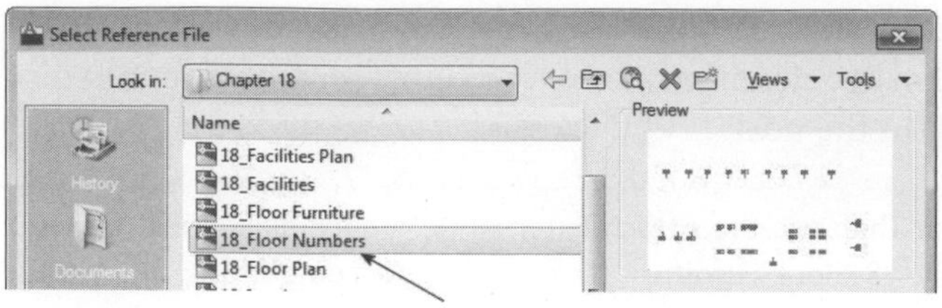

FIGURE 18.14

When the Attach External Reference dialog box appears, as shown in Figure 18.15, change the Reference Type by clicking the radio button next to Overlay. Under the Insertion Point heading, be sure this external reference will be inserted at 0,0,0. When finished, click the OK button.

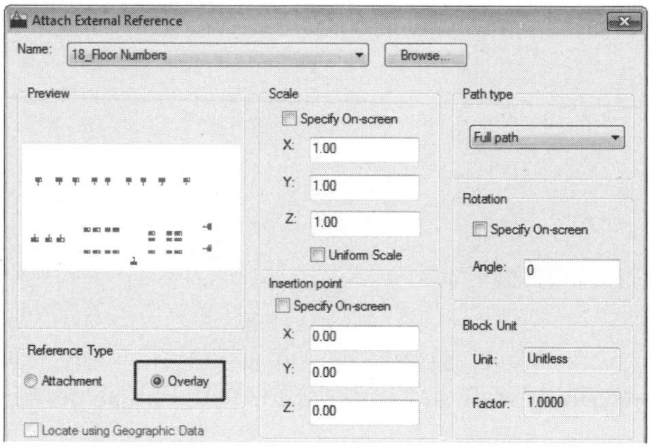

FIGURE 18.15

Your display should appear similar to Figure 18.16. Save this drawing file under its original name of 18_Floor Plan and then close the drawing file.

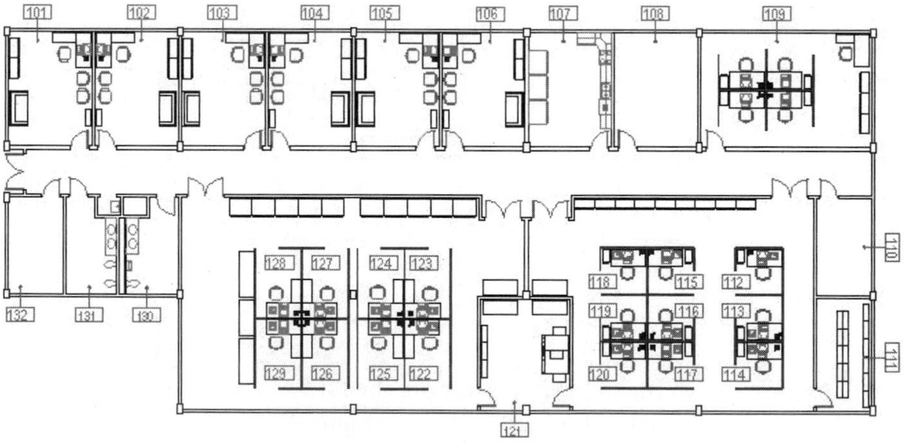

FIGURE 18.16

Now start a new drawing file from scratch. Click the Attach button on the Insert Ribbon and attach the drawing file 18_Floor Plan. When the External Reference dialog box appears, as shown in Figure 18.17, be sure that the Reference Type is reset to Attachment and that the drawing will be inserted at 0,0,0. When finished, click the OK button.

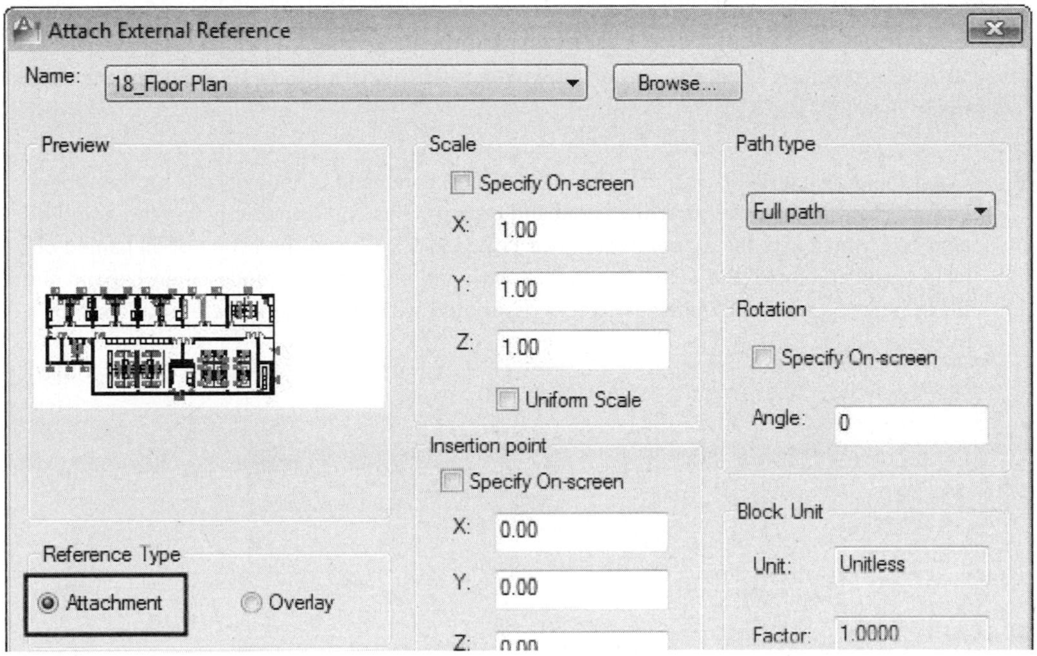

FIGURE 18.17

Perform a ZOOM-All operation and observe the results, as shown in Figure 18.18. Notice that the room numbers do not display because they were originally overlaid in the file 18_Floor Plan. This completes the Try It! exercise.

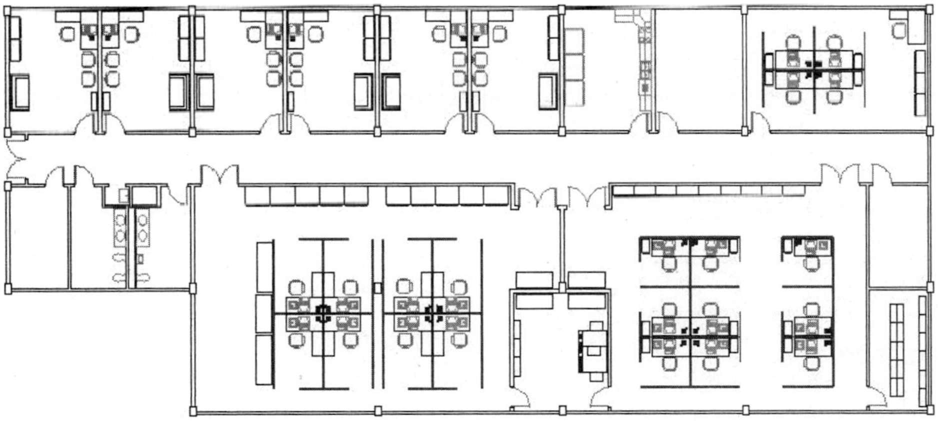

FIGURE 18.18

THE XBIND COMMAND

Earlier, it was mentioned that layers belonging to external references cannot be used in the drawing they were externally referenced into. The same is true with blocks that were referenced into the drawing. When these blocks are listed, the name of the external reference is given first, with the Pipe symbol following, and finally the actual name of the block, as in the example 18_Asesmp1|DESK2 where DESK2 is the block and 10_Asesmp1 is the external reference. As with layers, these blocks cannot be used in the drawing. The Pipe symbol indicating that the object belongs to the external reference. There is a way, however, to convert a block or layer into a usable object through the XBIND command.

TRY IT!

Open the drawing file 18_Facilities Plan. The XBIND command is not available on the Ribbon and will have to be entered at the command prompt. This command activates the Xbind dialog box, which lists the external reference 18_ASESMP1. Expand the listing of all named objects, such as blocks and layers associated with the external reference. Then expand the Block heading to list all individual blocks associated with the external reference, as shown in Figure 18.19.

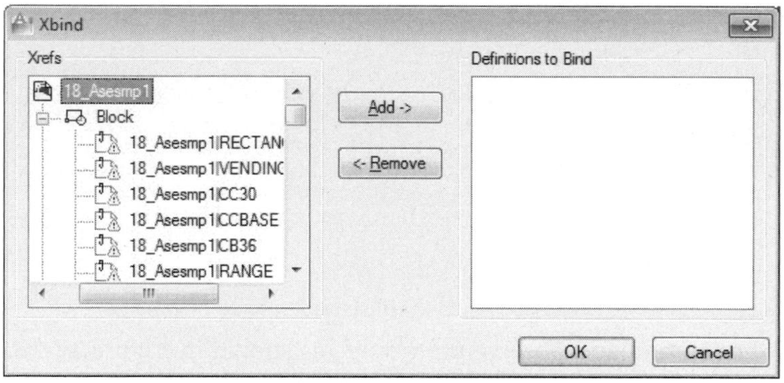

FIGURE 18.19

Click the block 18_Asesmp1|DESK2 in the listing on the left; then click the Add -> button. This moves the block name over to the right under the listing of Definitions to Bind, as shown in Figure 18.20. Do the same for 18_Asesmp1|DESK3 and 18_Asesmp1|DESK4. When finished adding these items, click the OK button to bind the blocks to the current drawing file.

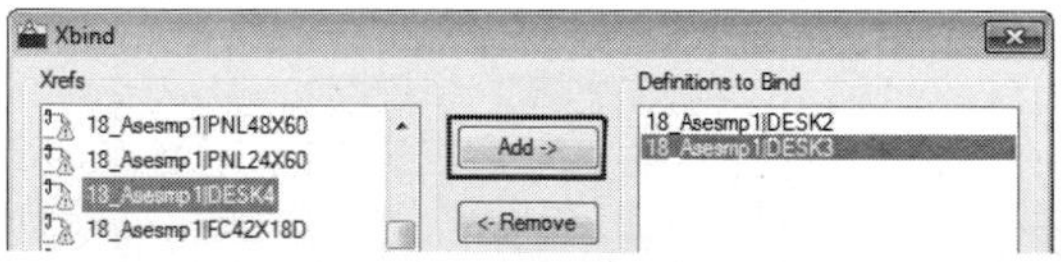

FIGURE 18.20

Test to see that new blocks have in fact been bound to the current drawing file. Activate the Insert dialog box, click the Name drop-down list box, and notice the display of the blocks, as shown in Figure 18.21. The three symbols just bound from the external reference still have the name of the external reference, namely, 18_Asesmp1.

However, instead of the Pipe symbol separating the name of the external reference and block names, the characters 0 are now used. This is what designates that the blocks are valid in the drawing. Now these three blocks can be inserted in the drawing file even though they used to belong only to the external reference 18_Asesmp1.

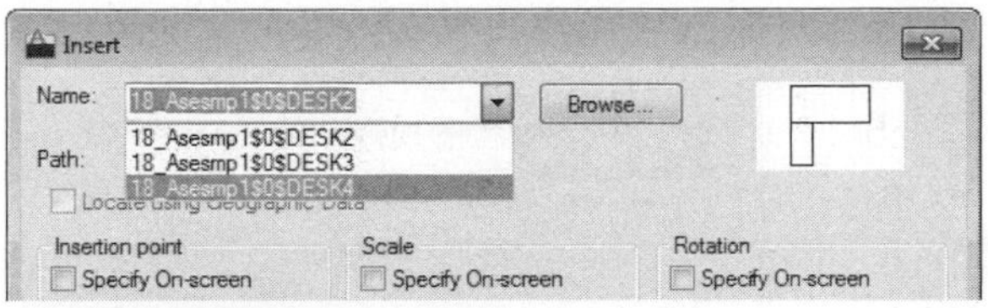

FIGURE 18.21

IN-PLACE REFERENCE EDITING

In-Place Reference Editing allows you to edit a reference drawing from the current drawing, which is externally referencing it. You then save the changes back to the original drawing file. This provides an efficient way of making a change to a drawing file from an externally referenced file.

TRY IT!

Open the drawing file 18_Pulley Assembly shown in Figure 18.22. Both holes located in the Base Plate, Left Support, and Right Support need to be stretched 0.20 units toward the center of the assembly. This will center the holes along the Left and Right Supports. Since all images that make up the Pulley Assembly are external references, the In-Place Reference Editing feature will be illustrated.

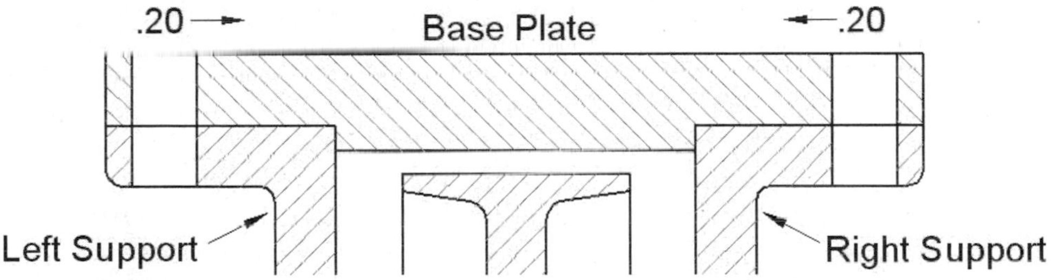

FIGURE 18.22

To begin, access the REFEDIT command by selecting Edit Reference from the Insert Ribbon, as shown in Figure 18.23 (Top-Left).

Once in the REFEDIT command, you are prompted to select the reference you wish to modify. Pick the Base Plate, as shown in Figure 18.23 (Lower-Left). This displays the Reference Edit dialog box, as shown in Figure 18.23 (Right). The reference to be edited, which is 18_Base Plate in our case, should be displayed. Nested references may also be displayed; one of these could be selected for editing instead, if desired. Clicking the OK button returns you to the screen.

NOTE

Double-clicking an external reference also launches the Reference Edit dialog box. Also, by picking an external reference, an External Reference tab is displayed on the Ribbon which can be used to activate In-Place Reference Editing.

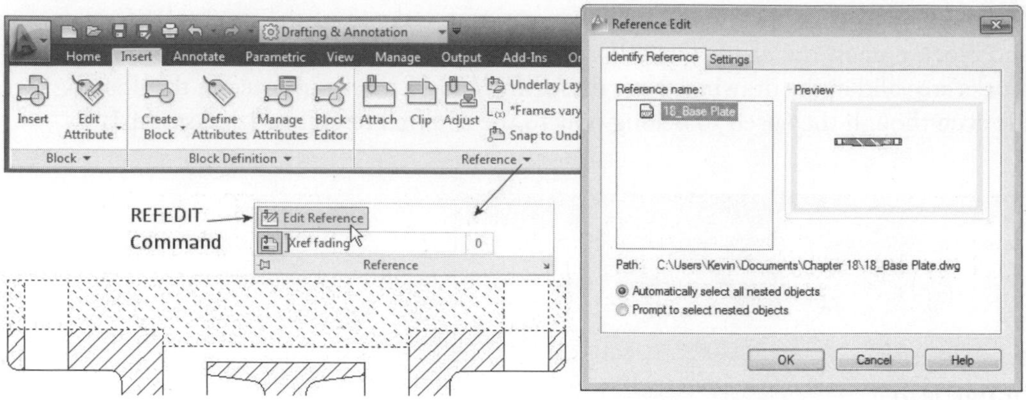

FIGURE 18.23

Notice that an Edit Reference panel is automatically provided on the Ribbon, as shown in Figure 18.24. To better identify the external reference being edited, the other external references in the drawing take on a slightly faded appearance. This faded effect returns to normal when you exit the Refedit process. You can now make modifications to the Base Plate by using the STRETCH command to stretch both holes a distance of 0.20 units to the inside.

NOTE

If you are not utilizing the Ribbon, a Refedit toolbar will automatically be displayed instead of the Edit Reference panel for In-Place Reference Editing.

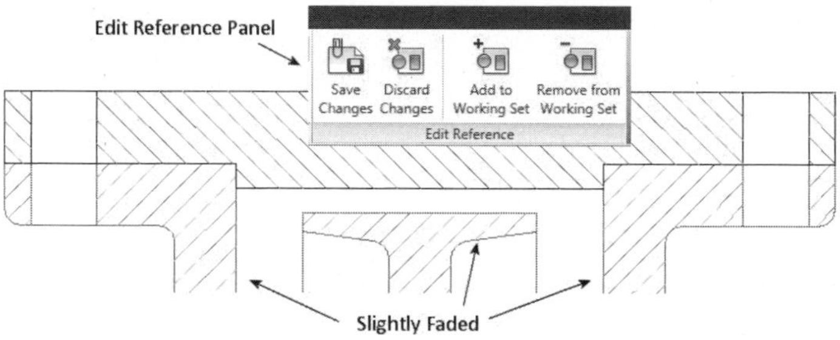

FIGURE 18.24

When performing the STRETCH operation, notice that even if the crossing window were to extend across both parts, only the holes in the Base Plate will be modified, as shown in Figure 18.25. When you are satisfied with the changes, select the Save Changes button on the Edit Reference panel. An AutoCAD alert box will ask you to confirm the saving of reference changes, as shown in Figure 18.25.

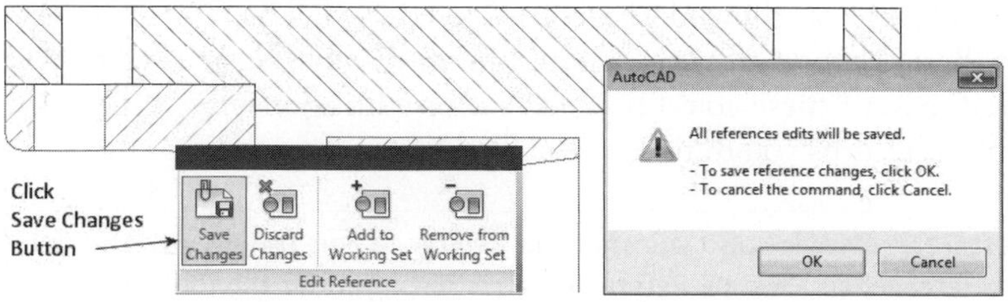

FIGURE 18.25

After you click OK, the results can be seen, as shown in Figure 18.26. Notice that only the objects that belong to the external reference (in this case, the holes that are part of the Base Plate) are affected. Also, the Edit Reference panel on the Ribbon automatically closes.

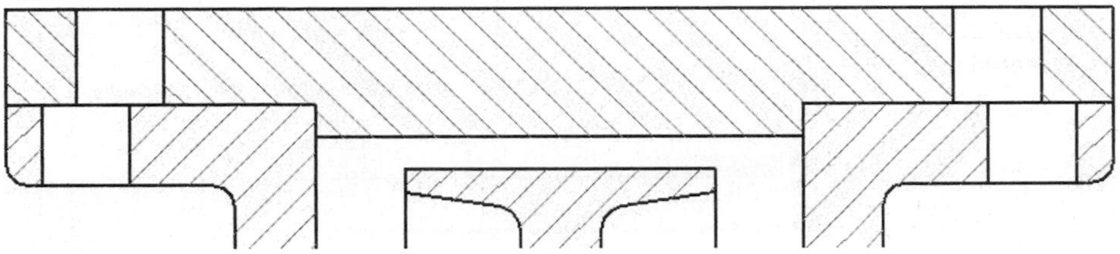

FIGURE 18.26

Perform the same series of steps using In-Place Reference Editing separately for the Left and Right supports. Stretch the hole located on the Left Support a distance of 0.20 units to the right. Stretch the hole located on the Right Support a distance of 0.20 units to the left. The final results of this In-Place Reference Editing Try It! exercise are displayed in Figure 18.27.

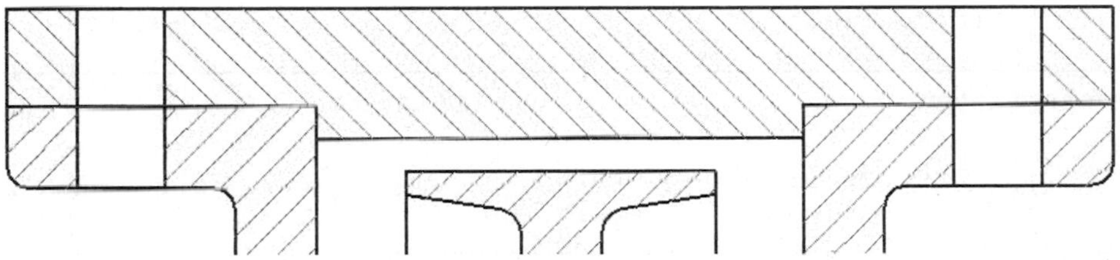

FIGURE 18.27

BINDING AN EXTERNAL REFERENCE

Binding an external reference to a drawing makes the external reference a permanent part of the drawing and no longer an externally referenced file. In actuality, the external reference is converted into a block object. To bind an Xref drawing, including all its xref-dependent named objects, such as blocks, dimension styles, layers, linetypes, and text styles, use the Bind option of the External Reference palette. Since binding an external reference breaks the link with the original drawing file and combines them, this becomes a very popular technique at the end of a project when archiving final drawing files is important.

To bind an external reference, activate the External References palette by clicking the External References arrow located in the lower-right corner of the Reference panel on the Ribbon. Select the external reference that you want to bind to the current drawing, right-click, and pick Bind from the menu, as shown in Figure 18.28 (Center). This option activates the Bind Xrefs dialog box, illustrated in Figure 18.28 (Right). Two options are available inside this dialog box: Bind and Insert.

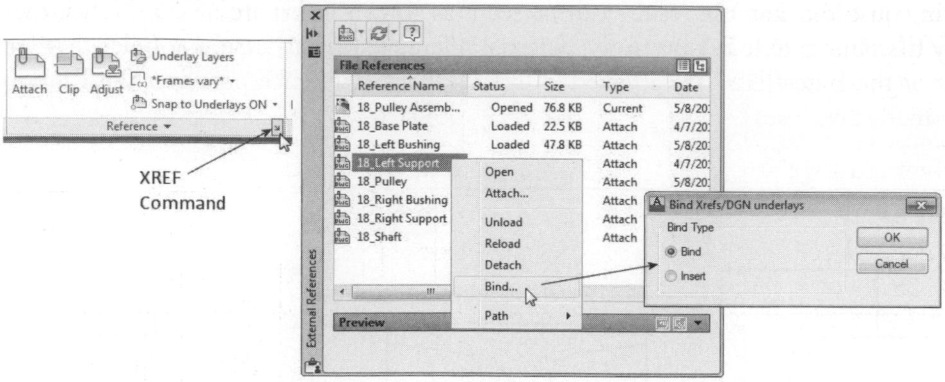

FIGURE 18.28

The Bind option binds to the current drawing file all blocks, layers, dimension styles, and so on that belonged to an external reference. After you perform this operation, layers can be made current and blocks inserted in the drawing. For example, a typical block definition belonging to an external reference is listed in the symbol table as XREFname|-BLOCKname. Once the external reference is bound, all block definitions are converted to XREFname0BLOCKname. In the layer display in Figure 18.29, the referenced layers were converted to usable layers with the Bind option.

Referenced Layers	Converted Layers
18_Pulley|Hatch	18_Pulley0Hatch
18_Pulley|Object	18_Pulley0Object
18_Pulley|Text	18_Pulley0Text

NOTE

The same naming convention is true for blocks, dimension styles, and other named items. The result of binding an external reference is similar to a drawing that was inserted into another drawing.

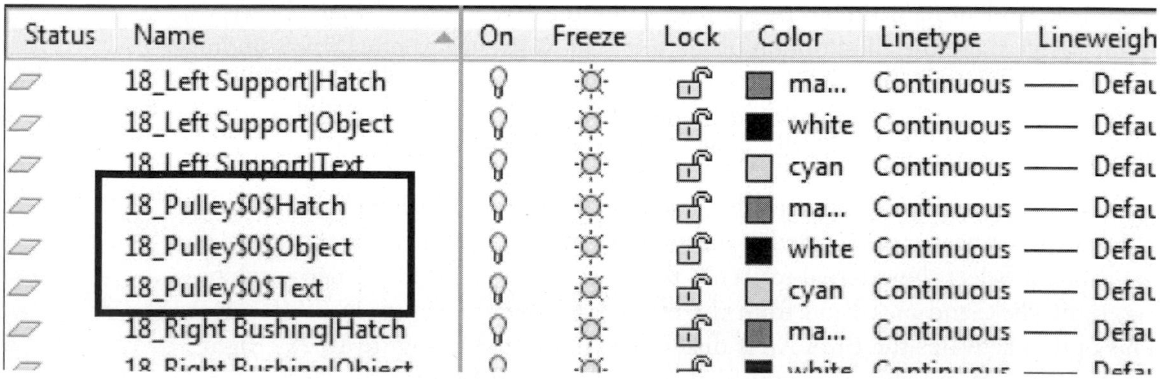

FIGURE 18.29

The Insert option of the Bind Xrefs dialog box is similar to the Bind option. However, instead of named items such as blocks and layers being converted to the format XREFname0-BLOCKname, the name of the external reference is stripped, leaving just the name of the block, layer, or other named item (BLOCKname, LAYERname, etc.). In Figure 18.30, the referenced layers listed in the following table were converted to usable layers with the Insert option:

Referenced Layers	Converted Layers
18_Pulley|Hatch	Hatch
18_Pulley|Object	Object
18_Pulley|Text	Text

NOTE

It can be advantageous to use the Bind option over the Insert option. This way, you can identify the layers that were tied to the previously used external reference and control them individually.

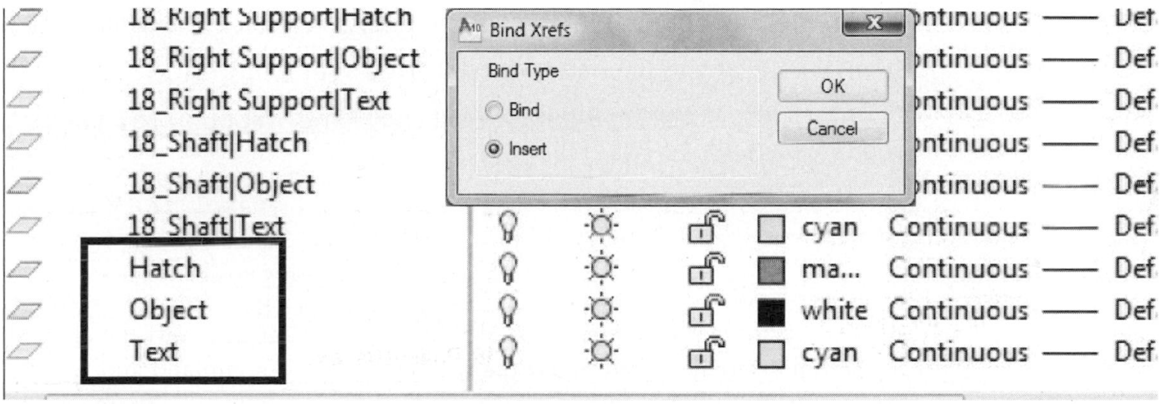

FIGURE 18.30

CLIPPING AN EXTERNAL REFERENCE

Typically, attaching an external reference displays the entire reference file, however, you have the option of displaying a portion of the file. This is accomplished by clipping the external reference with the `CLIP` or `XCLIP` command. To perform a clipping operation, pick the Clip button from the Insert Ribbon, as shown in Figure 18.31.

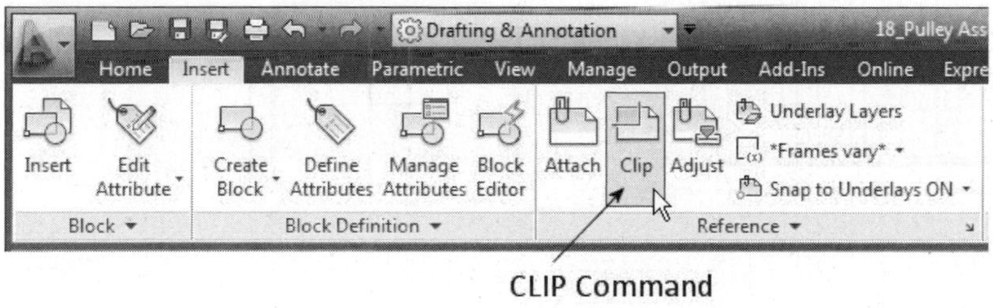

FIGURE 18.31

NOTE

The Clip button on the Ribbon actually activates the `CLIP` command. This more generic version of the `XCLIP` command allows you to clip external references, images, viewports, point clouds, and underlays.

Clipping is useful when you want to emphasize a particular portion of your external reference file. Clipping boundaries include polylines in the form of rectangles, regular polygonal shapes, or even irregular polyline shapes. All polylines must form closed shapes. Also, clipping can take two forms, namely Outside and Inside modes. With the Outside mode, objects outside of the clipping boundary are hidden. With the Inside mode, objects inside of the clipping boundary are hidden.

TRY IT!

Open the drawing file 18_Facilities Plan. Follow the illustration in Figure 18.32 (Left) and the command prompt sequence below for performing an outside clipping operation.

`Command: CLIP`

`Select Object to clip:` *(Pick the external reference)*

`Enter clipping option`

`[ON/OFF/Clipdepth/Delete/generate Polyline/New boundary] <New>: N` *(For New)*

`Outside mode - Objects outside boundary will be hidden.`

`Specify clipping boundary or select invert option:`

`[Select polyline/Polygonal/Rectangular/Invert clip] <Rect-angular>:` *(Press ENTER)*

`Specify first corner:` *(Pick a point at "A," as shown in the following image on the left)*

`Specify opposite corner:` *(Pick a point at "B")*

The results are displayed in Figure 18.32 (Right). If you want to return the clipped image to the full external reference, use the `CLIP` or `XCLIP` command and use the OFF clipping option. This temporarily turns off the clipping frame. To permanently remove the clipping frame, use the Delete clipping option.

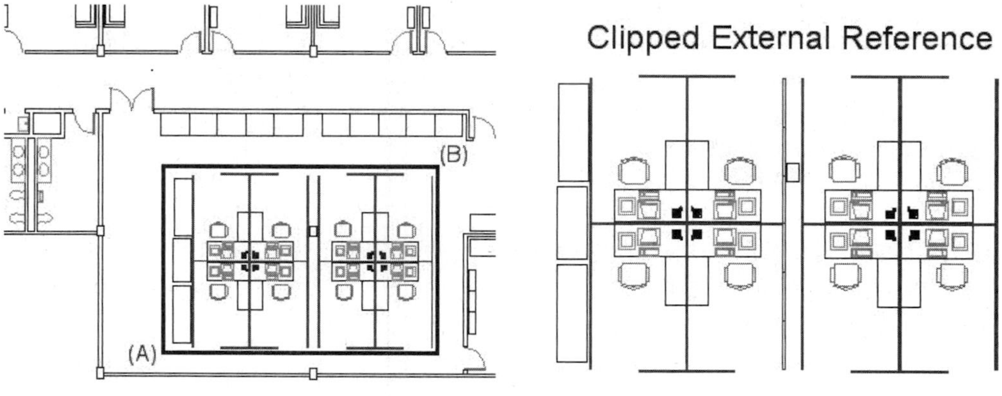

FIGURE 18.32

Figure 18.33 illustrates the results of clipping an external reference based on Inside mode. In this example, the objects defined inside of the clipping boundary are removed.

Command: CLIP

Select Object to clip: *(Pick the external reference, as shown in the previous image)*

Select objects: *(Press ENTER to continue)*

Enter clipping option

[ON/OFF/Clipdepth/Delete/generate Polyline/New boundary] <New>: N *(For New)*

Outside mode - Objects outside boundary will be hidden.

Specify clipping boundary or select invert option:

[Select polyline/Polygonal/Rectangular/Invert clip] <Rectangular>: I *(For Invert clip)*

Inside mode - Objects inside boundary will be hidden.

Specify clipping boundary or select invert option:

[Select polyline/Polygonal/Rectangular/Invert clip] <Rectangular>: *(Press ENTER)*

Specify first corner: *(Pick a point at "A," as shown in the previous image)*

Specify opposite corner: *(Pick a point at "B," as shown in the previous image. The results are shown in the following image)*

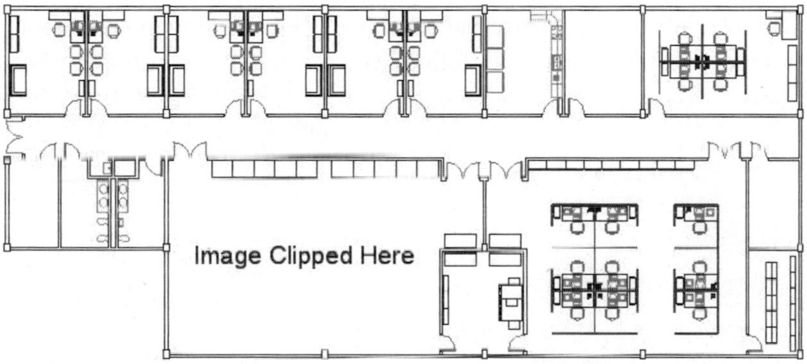

FIGURE 18.33

OTHER OPTIONS OF THE EXTERNAL REFERENCES PALETTE

The following additional options of the External References palette will now be discussed, using Figure 18.34 as a guide.

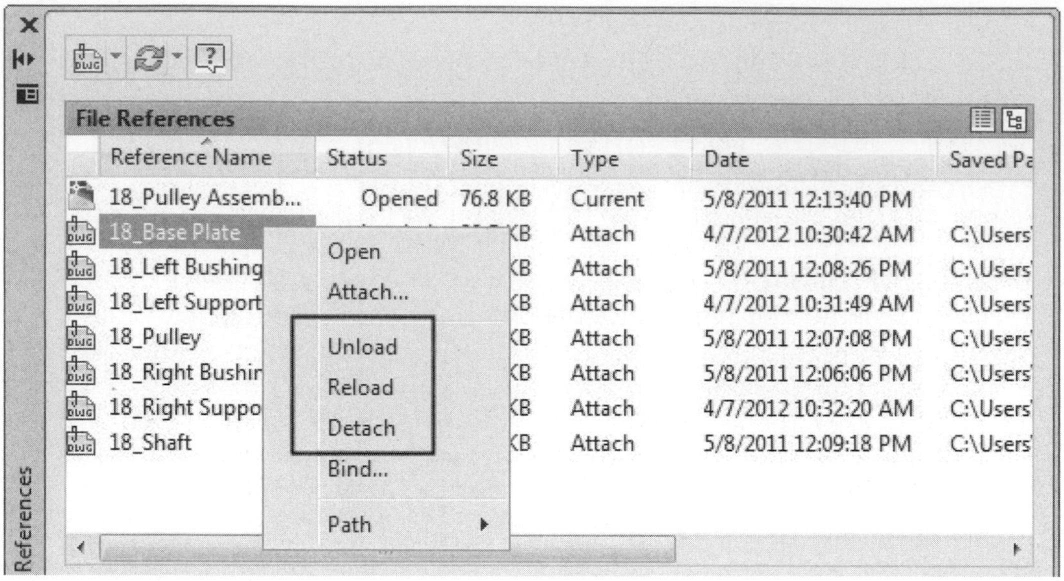

FIGURE 18.34

Unload

Unload is similar to the Detach option, with the exception that the external reference is not permanently removed from the database of the drawing file. When an external reference is unloaded, it is still listed in the External References palette, as shown in Figure 18.35. Notice that an arrow (facing down) is displayed to signify the unloaded status. Since this option suppresses the external reference from any drawing regenerations, it is used as a productivity technique. Reload the external reference when you want it returned to the screen.

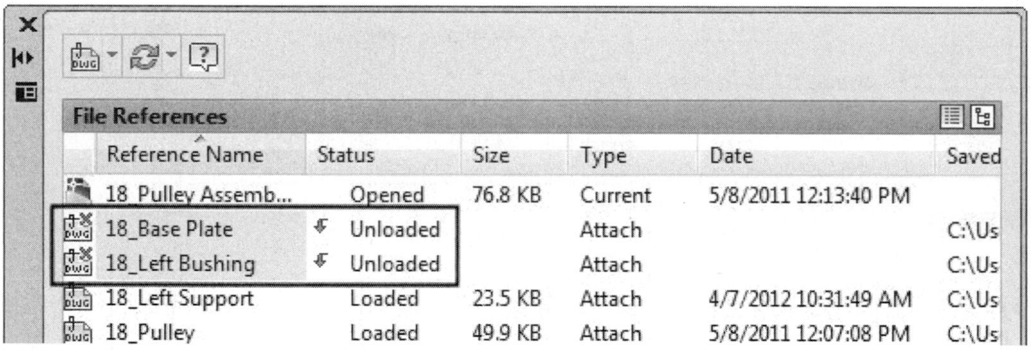

FIGURE 18.35

Reload

This option can be used to reload a file that was previously unloaded. Because it loads the most current version of an external reference, it is also used to update any external references that may have changed while you are working on a drawing that references them. This option works well in a networked environment, where all files reside on a file server.

Detach

Use this option to permanently detach or remove an external reference from the database of a drawing.

EXTERNAL REFERENCE NOTIFICATION TOOLS

A series of notification tools are available to assist with managing external references. When a drawing consisting of external references opens, an icon appears in the lower-right corner of your Status Bar, as shown in Figure 18.36 (Left). Clicking this icon launches the External References palette. In the event that a source file was changed or modified, the next time you return to the external reference drawing, the icon will include a yellow caution triangle. A pop up window is also displayed, as shown in Figure 18.36 (Right), informing you that the external reference was changed. This example refers to the Pulley Assembly, in which the Base Plate was modified and saved. When you return to the Pulley Assembly, the yellow caution icon signifies the change to the file 18_Base Plate. A link "Reload 18_Base Plate" is available in the pop-up window, which can be selected to update the referenced file.

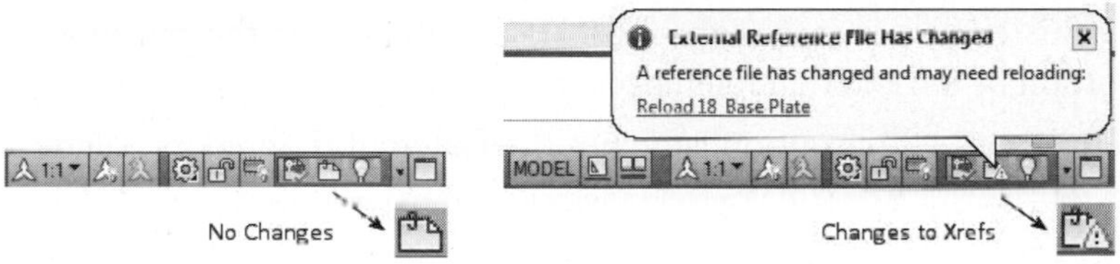

FIGURE 18.36

You can instead, click the yellow caution icon to launch the External References palette. Notice the appearance of a yellow caution triangle, as shown in Figure 18.37. The status of this file alerts you that 18_Base Plate needs to be reloaded in order for the change to be reflected in the Pulley Assembly drawing. A button at the top of the palette is available to reload all the external references.

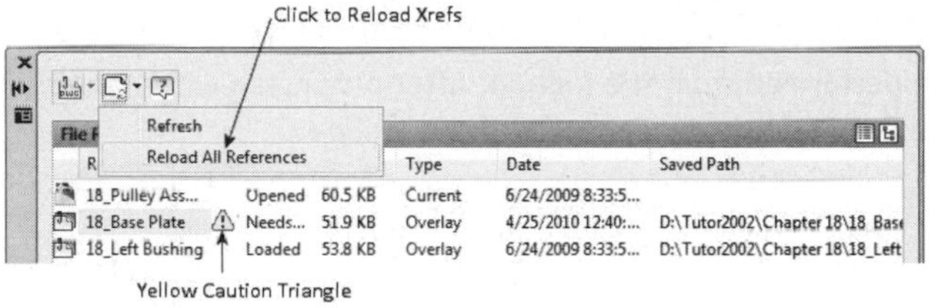

FIGURE 18.37

You can also reload 18_Base Plate by right-clicking the file name and picking Reload from the shortcut menu, as shown in Figure 18.38. Once you reload the file the yellow caution triangle is removed.

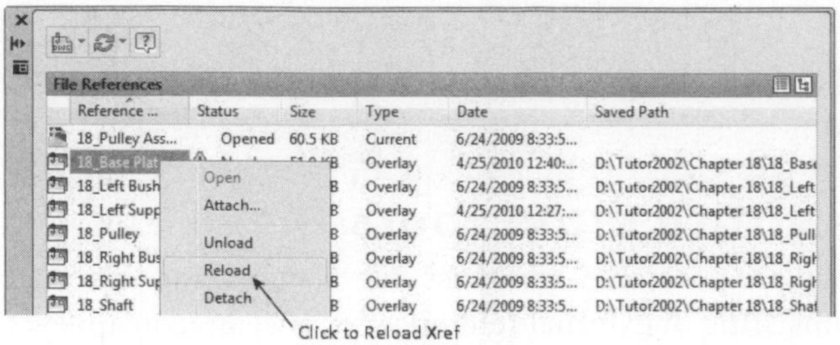

FIGURE 18.38

Illustrated in Figure 18.39 is another feature of external references. The left support of the pulley assembly was selected. When you right-click, the shortcut menu appears, as shown in Figure 18.39 (Right). Use this menu to perform the following tasks:

Edit Xref In-place—This option activates the Reference Edit dialog box for the purpose of editing the external reference in-place.

Open Xref—This option opens the selected external reference in a separate window.

Clip Xref—This option launches the XCLIP command for the purpose of clipping a portion of the external reference.

External References—This option launches the External References palette.

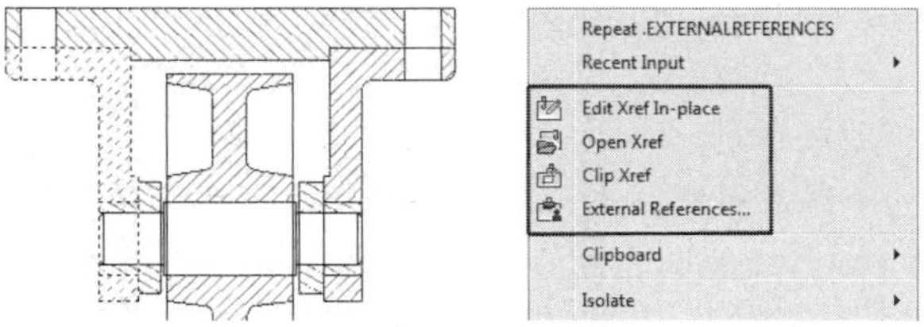

FIGURE 18.39

NOTE

These same four tasks can also be performed from the Ribbon. After picking an external reference, an External Reference tab is automatically displayed on the Ribbon.

USING ETRANSMIT

ETransmit is an AutoCAD utility that is helpful in reading the database of your drawing and listing all support files needed. Once these files are identified, you can have this utility gather all files into one zip file. You can then copy these files to a disk or CD, or transmit the files over the Internet. Clicking eTransmit…, located under the Publish heading of the Application Menu in Figure 18.40 (Left), displays the Create Transmittal dialog box, as shown in Figure 18.40 (Center). Notice in the Files Tree tab a listing of all support files grouped by their specific category. For instance, an External References category exists. Clicking the "+" sign lists all external references set to be transmitted.

Clicking the Files Table tab, as shown in the following image on the right, displays a list of all files that will be included in the transmittal set.

When you are finished examining all the support files, clicking the OK button takes you to the Specify Zip File dialog box. It is here that you enter a file name. All support files listed under the Files Tree tab will be grouped into a single zip file for easy sharing with other individuals or companies.

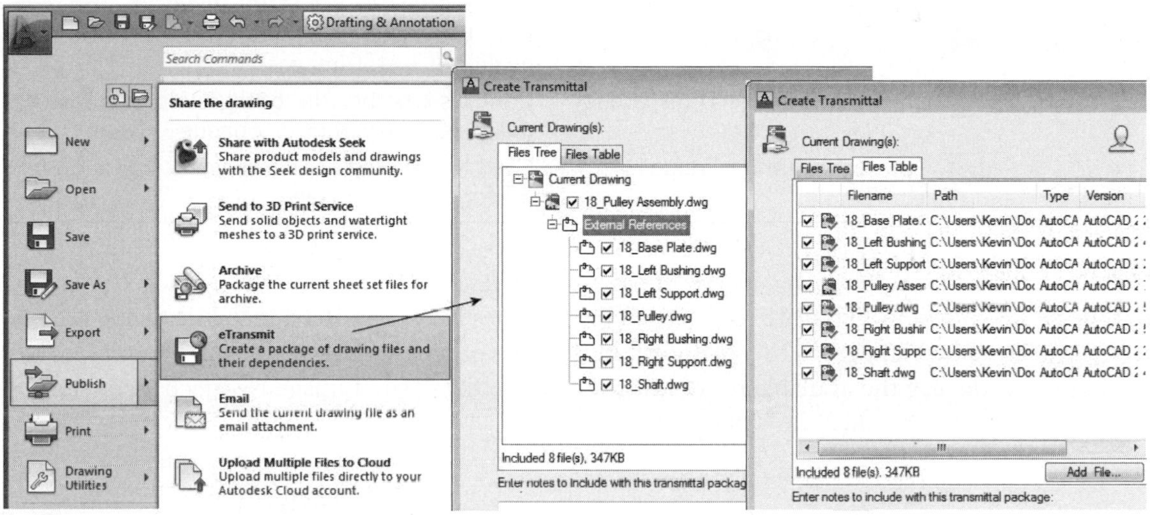

FIGURE 18.40

WORKING WITH RASTER IMAGES

Raster images in the form of JPG, GIF, TIF, and so on can easily be merged with your vector-based AutoCAD drawings. The addition of raster images can give a new dimension to your drawings. Typical examples of raster images are digital photographs of an elevation of a house or an isometric view of a machine part. Whatever the application, working with raster images is very similar to what was just covered with referenced drawings. You attach the raster image to your drawing file. Once it is part of your drawing, additional tools are available to manipulate and fine-tune the image for better results. To attach a raster image to a drawing, you can use the Attach button from the Insert Ribbon or you can use the External References palette (Attach Image... button), as shown in Figure 18.41.

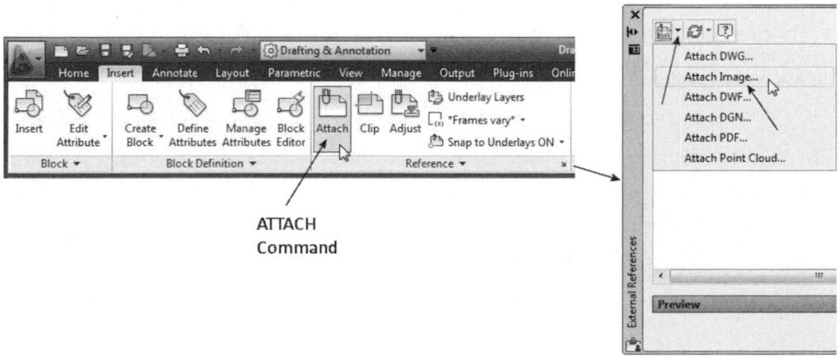

FIGURE 18.41

The following table outlines the main tools that are available when working with raster images.

Button	Tool	Function
	Attach Image	Used for attaching a raster image to a drawing
	Clip Image	Used for cropping or clipping the raster image
	Adjust Image	Used for adjusting the brightness, contrast, and fade factor of a raster image
	Image Quality	Uses two settings, high or draft, to control the quality of raster images
	Image Transparency	Controls whether the background of a raster image is transparent or opaque
	Image Frame	Used for turning the rectangular frame on or off for a raster image

To attach a raster image utilizing the Insert Ribbon, click the Attach button, as shown in Figure 18.42 (Left). This launches the Select Reference File dialog box, as shown in Figure 18.42 (Right). In the Files of type list box select "All image types" to display the available raster images. Choose the desired image to attach from the list and then click the Open button.

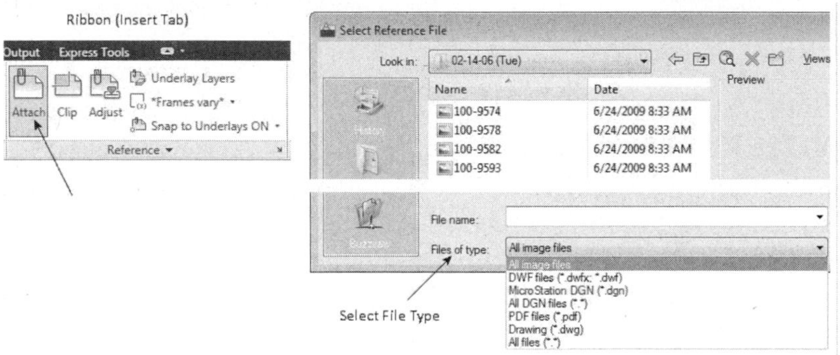

FIGURE 18.42

After choosing the correct raster image, the Attach Image dialog box appears, as shown in Figure 18.43 (Left). From this dialog box, change the insertion point, scale, or rotation parameters if desired. Clicking the OK button returns you to your drawing, where you pick an insertion point and change the scale factor of the image by dragging your cursor until the desired image size is determined. A typical raster image is illustrated in Figure 18.43 (Right).

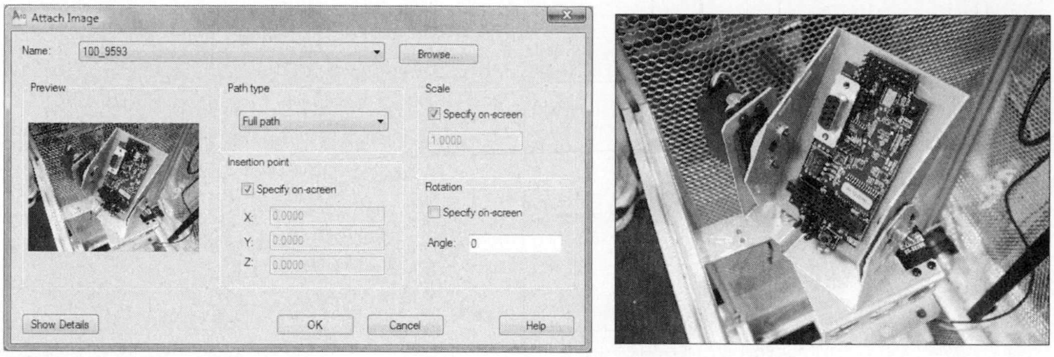

FIGURE 18.43

The same External References palette used for managing external referenced drawing files is used for managing image files, as shown in Figure 18.44.

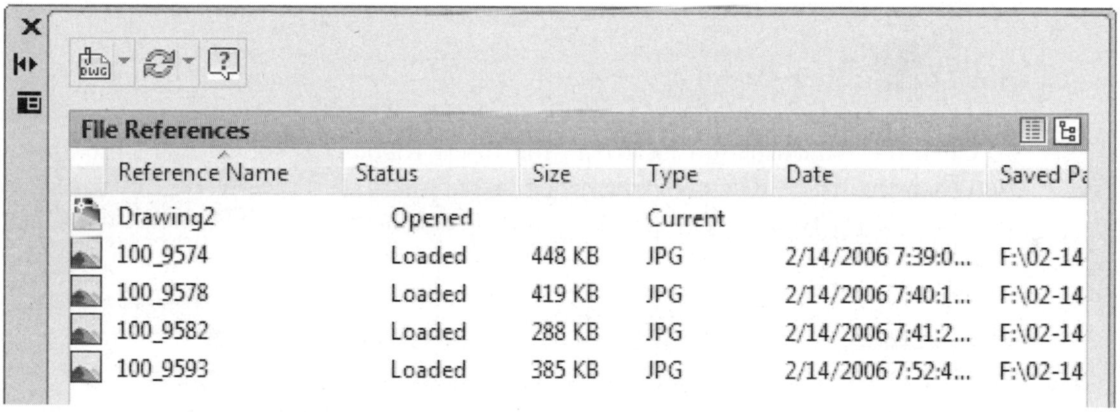

File References

Reference Name	Status	Size	Type	Date	Saved Pa
Drawing2	Opened		Current		
100_9574	Loaded	448 KB	JPG	2/14/2006 7:39:0...	F:\02-14
100_9578	Loaded	419 KB	JPG	2/14/2006 7:40:1...	F:\02-14
100_9582	Loaded	288 KB	JPG	2/14/2006 7:41:2...	F:\02-14
100_9593	Loaded	385 KB	JPG	2/14/2006 7:52:4...	F:\02-14

FIGURE 18.44

TRY IT!

Begin this exercise on working with raster images by opening the drawing file 18_Linkage, as illustrated in Figure 18.45. In this exercise a raster image will be attached and placed in the blank area to the right of the two-view drawing.

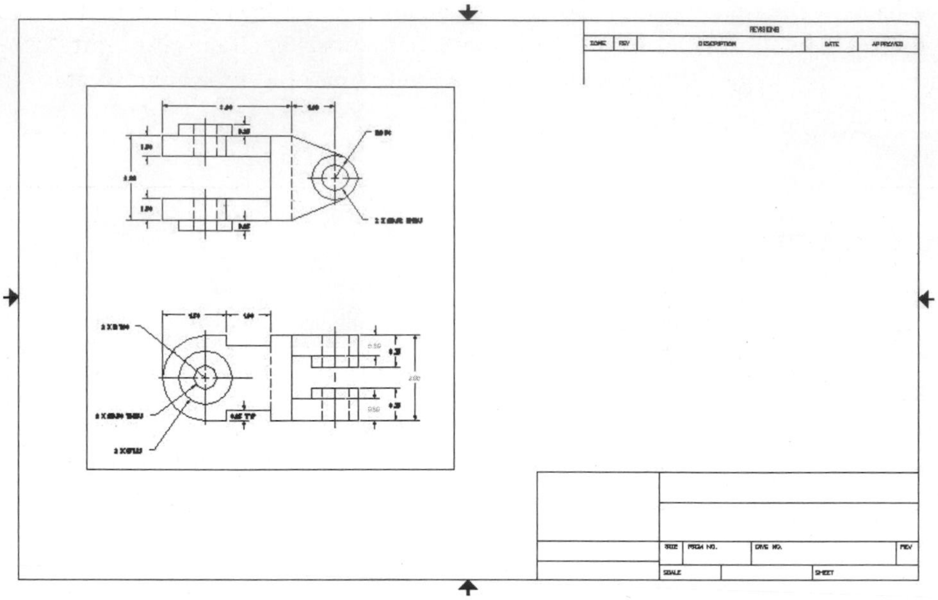

FIGURE 18.45

Click the Attach button located in the Insert Ribbon, as shown in Figure 18.46. When the Select Reference File dialog box appears, click the file 18_Linkage1.jpg. The Files of type list box should be set to “All image files.” Then click the Open button.

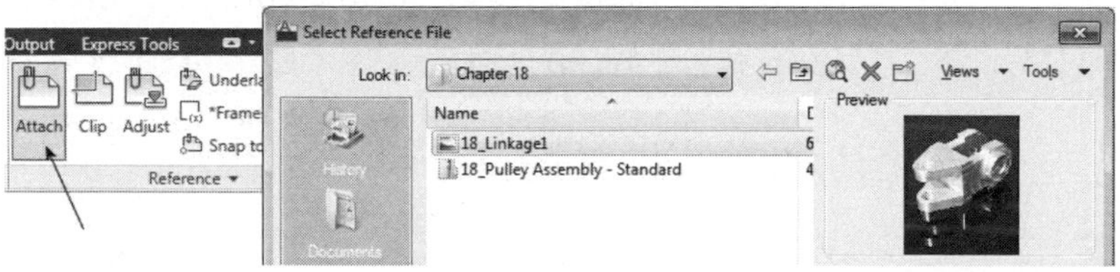

FIGURE 18.46

When the Attach Image dialog box appears, as shown in Figure 18.47, leave all default settings. You will be specifying the insertion point and scale on the drawing screen. Click the OK button to continue.

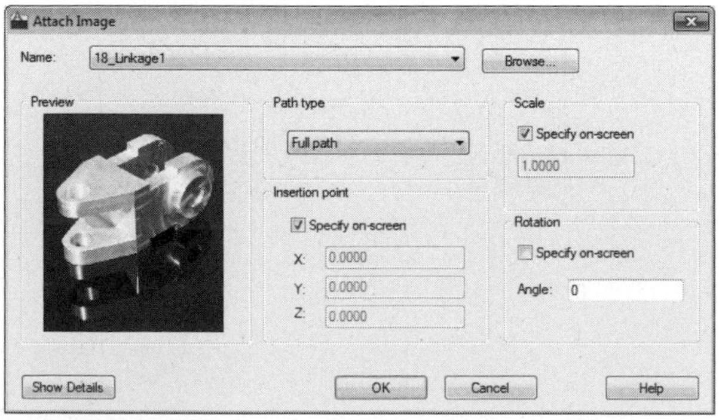

FIGURE 18.47

Once back in the drawing file, pick a point anchoring the lower-left corner of the graphic, as shown in Figure 18.48. Now move your cursor in an upward right direction to scale the image as necessary. Pick a second point at a convenient location to display the image.

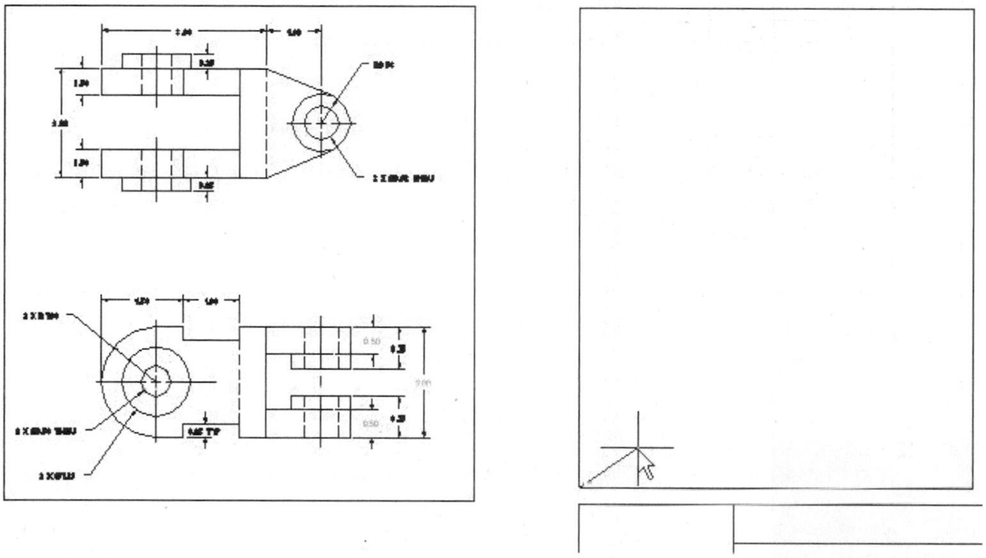

FIGURE 18.48

After the image is attached, you can adjust its size very easily. Click the edge of the image (the image frame) and notice the grips appearing at the four corners of the image. Clicking a grip and then moving your cursor increases or decreases the image's size, as shown in Figure 18.49.

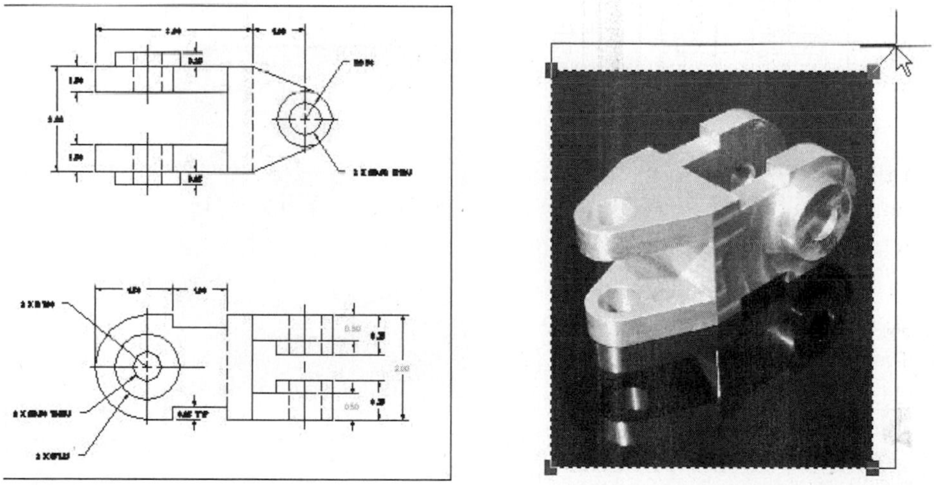

FIGURE 18.49

From the Reference panel of the Insert Ribbon, shown in Figure 18.50, identify the Adjust button and click it. Clicking the edge of the raster image frame displays a command prompt sequence that allows you to adjust the Brightness, Contrast, and Fade settings for the image.

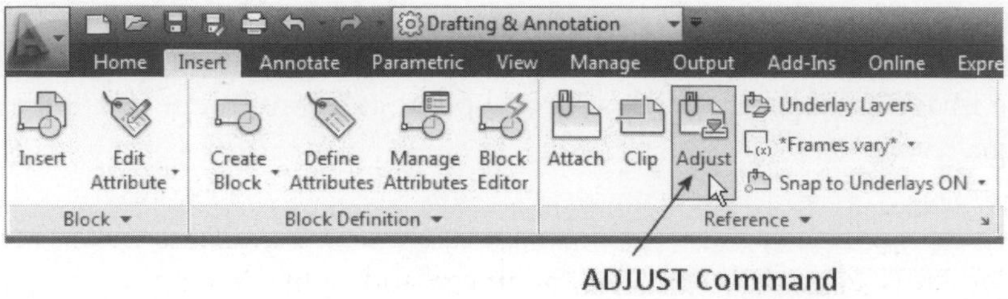

FIGURE 18.50

Instead of picking the Adjust button, select the image and an Image tab will automatically appear on the Ribbon, as shown in Figure 18.51. The Adjust panel provides the same adjustments for brightness, contrast, and fade.

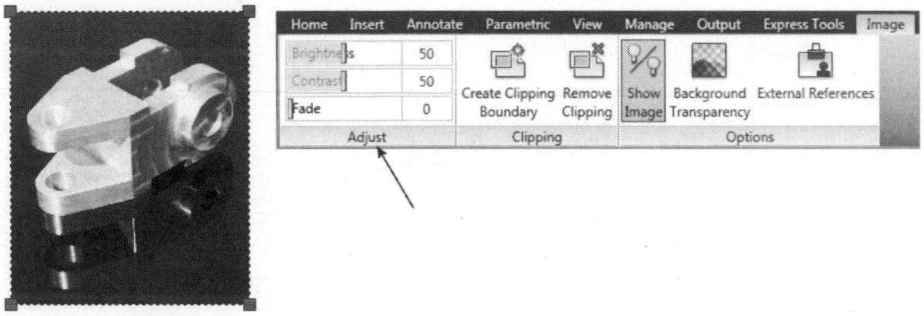

FIGURE 18.51

The final results are illustrated in Figure 18.52, with the vector drawing and raster image sharing the same layout.

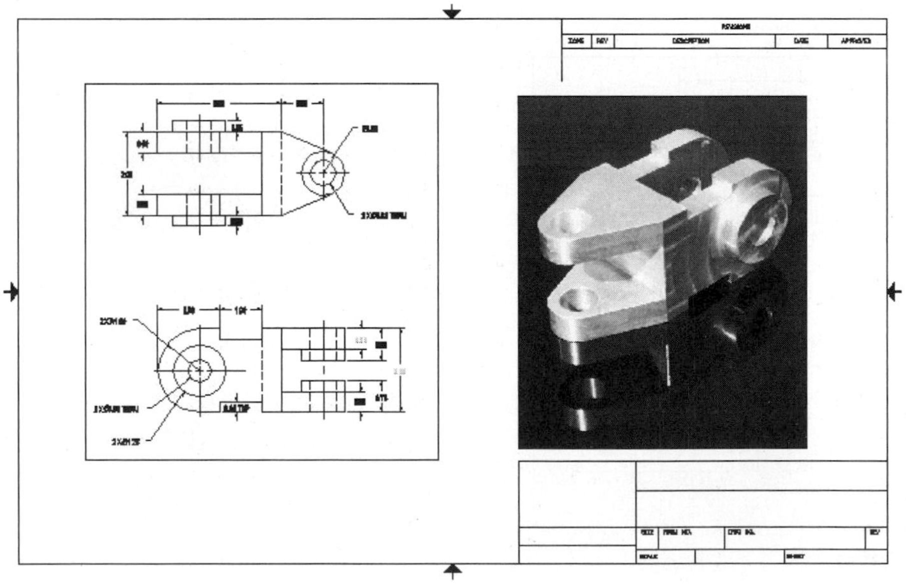

FIGURE 18.52

CONTROLLING IMAGES THROUGH DRAWORDER

With the enhancements made to raster images and the ability to merge raster images with vector graphics, it is important to control the order in which these images are displayed. The DRAWORDER command is used to provide this level of control over raster images. The DRAWORDER command can be selected from the Home Ribbon, as shown in Figure 18.53 (Left). Draw order controls are also available through a shortcut menu which is actuated by picking the image and right-clicking, as shown in Figure 18.53 (Right).

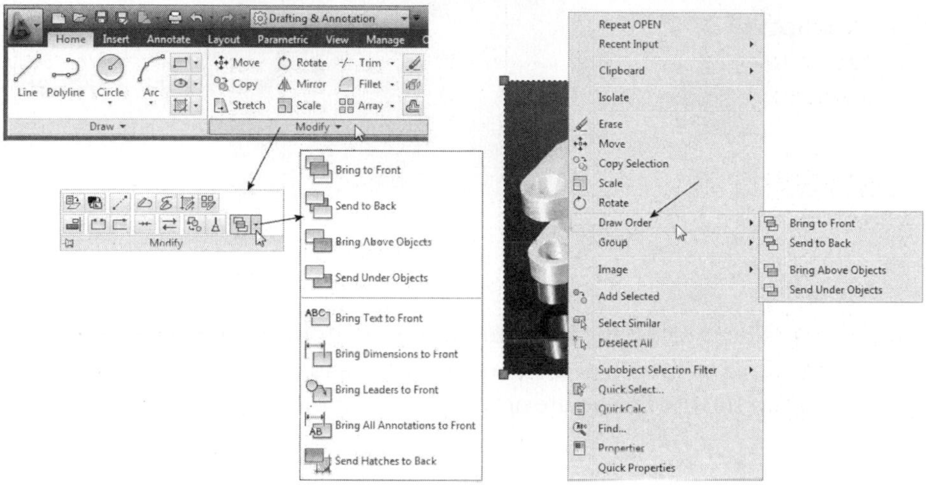

FIGURE 18.53

Four modes of the Draw Order tools are described as follows:

Button	Tool	Function
	Bring to Front	The selected object is brought to the top of the drawing order
	Send to Back	The selected object is sent to the bottom of the drawing order
	Bring Above Objects	The selected object is brought above a specified reference object
	Send Under Objects	The selected object is sent below the specified reference object

Figure 18.54 displays three images, two of which are partially hidden due to the size of the large middle image. We really want both small images to be visible and have the larger middle image sent to the back of the image arrangement.

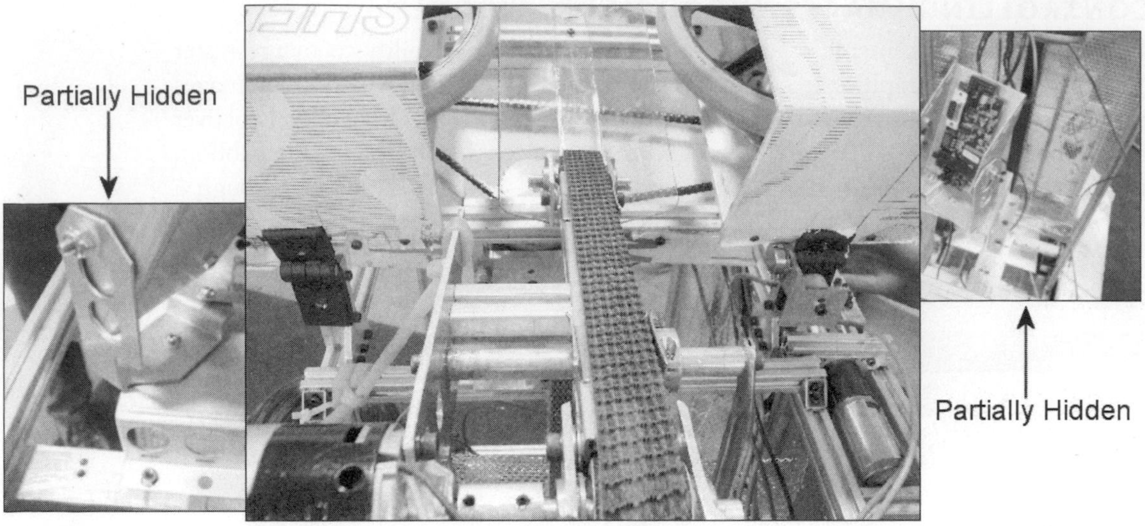

FIGURE 18.54

To correct the problem, select the Bring to Front tool. Both small images are selected as the objects to bring to the top of the drawing order, and the results are illustrated in Figure 18.55. Both smaller images are now at the top of the drawing order and can be viewed in full.

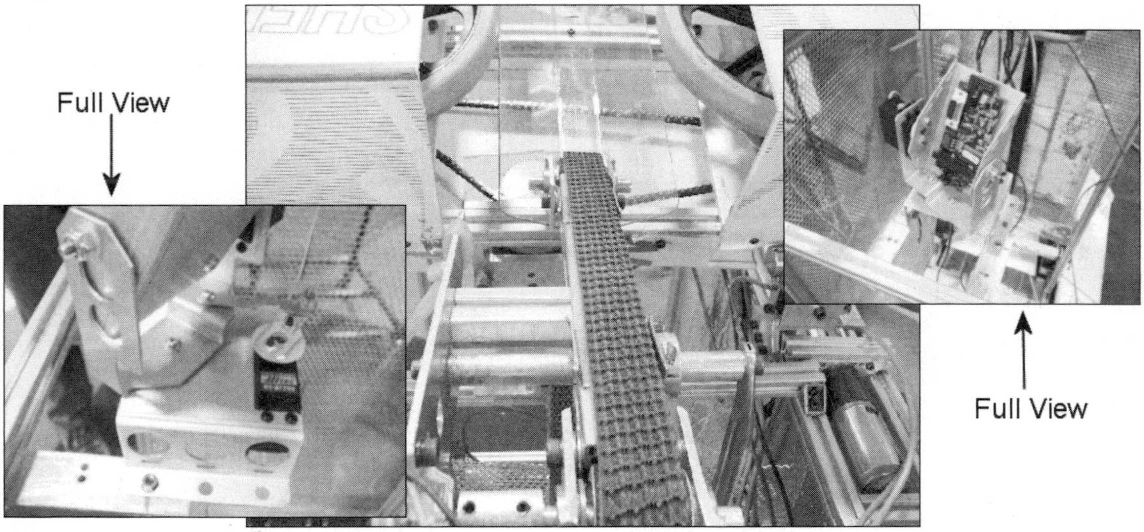

FIGURE 18.55

TUTORIAL EXERCISE: 18_EXTERNAL REFERENCES

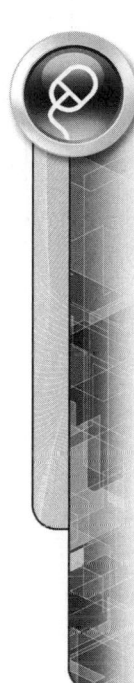

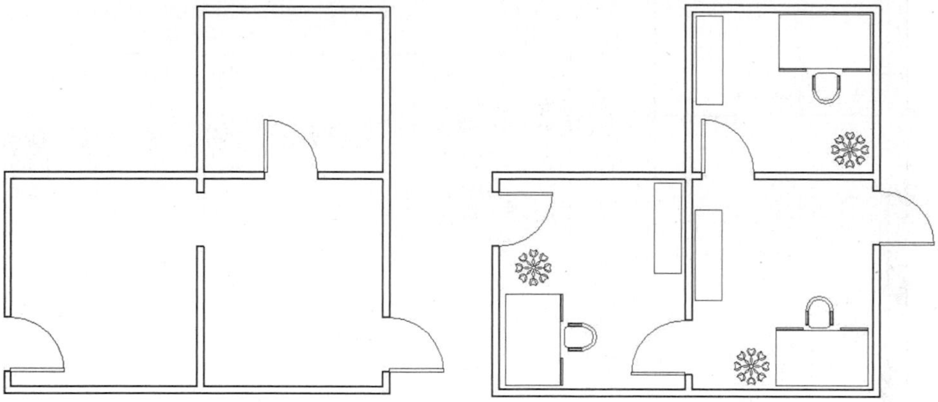

FIGURE 18.56

Purpose

This tutorial is designed to use the office floor plan (shown in Figure 18.56 (Left)) to create an interior plan consisting of various interior symbols such as desks, chairs, shelves, and plants (shown in Figure 18.56 (Right)). The office floor plan will be attached to another drawing file through the XREF command.

System Settings

Since these drawings are provided, all system settings have been made.

Layers

The creation of layers is not necessary because layers already exist for both drawing files you will be working on.

Suggested Commands

Begin this tutorial by opening the drawing 18_Office.Dwg and viewing its layers and internal block definitions. Then open the drawing 18_Interiors.Dwg and view its layers and internal blocks. The file 18_Office.Dwg will then be attached to 18_Interiors.Dwg. Once this is accomplished, chairs, desks, shelves, and plants will be inserted in the office floor plan for laying out the office furniture. Once 18_Interiors.Dwg is saved, a design change needs to be made to the original office plan; open 18_Office.Dwg and stretch a few doors to new locations. Save this file and open 18_Interiors.Dwg; notice how the changes to the doors are automatically made. The Xbind dialog box will also be shown as a means for making a block that had previously belonged to an external reference usable in the file 18_Interiors.Dwg.

STEP 1

Open 18_Office.Dwg and observe a simple floor plan consisting of three rooms, as shown in Figure 18.57 (Left). Furniture will be laid out using the floor plan as a template.

STEP 2

While in 18_Office.Dwg, use the Layer Properties Manager palette and observe the layers that exist in the drawing for such items as doors, walls, and floor, as shown in Figure 18.57 (Right). These layer names will appear differently once the office plan is attached to another drawing through the XREF command.

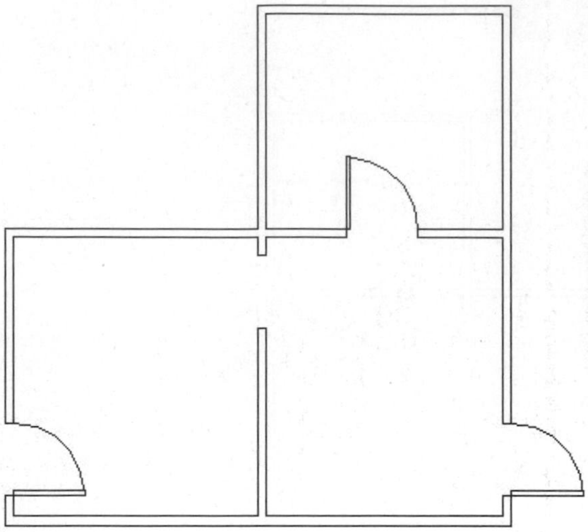

Status	Name	On	Freeze	Lock	Colc
	0				re
	CD				re
	DOORS				cy
	FLOOR				re
✓	WALLS				gı

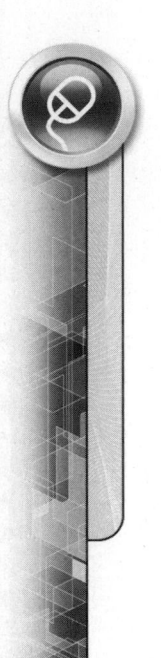

FIGURE 18.57

STEP 3

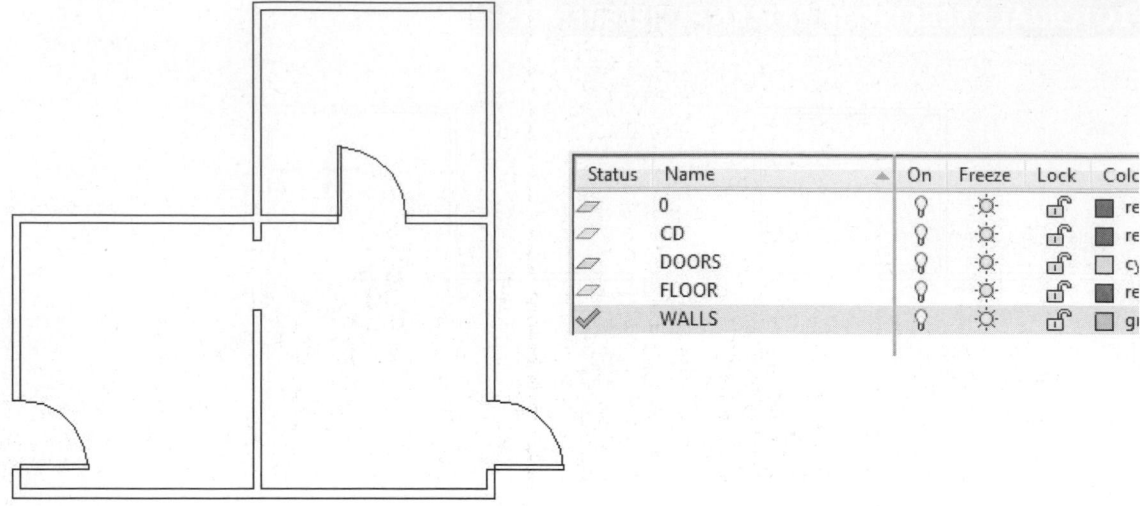

While in the office plan, activate the Insert dialog box through the `INSERT` command. At times, this dialog box is useful for displaying all valid blocks in a drawing. Clicking the Name drop-down list displays the results shown in Figure 18.58. Two blocks are currently defined in this drawing; as with the layers, once the office plan is merged into another drawing through the `XREF` command, these block names will change. When you have finished viewing the defined blocks, close 18_Office.Dwg.

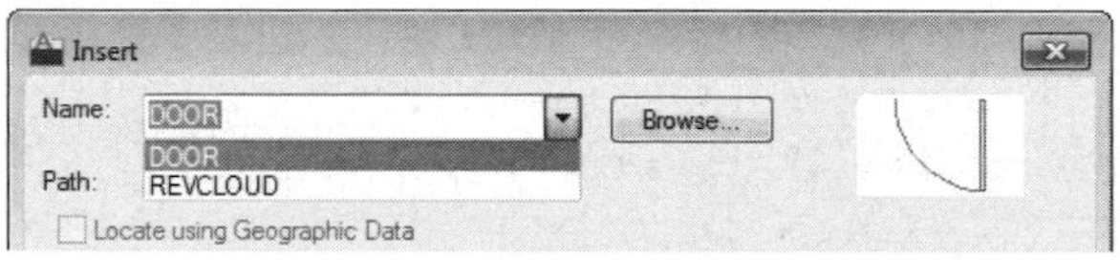

FIGURE 18.58

STEP 4

This next step involves opening 18_Interiors.Dwg and looking at the current layers found in this drawing. Once this drawing is open, use the Layer Properties Manager palette to observe that layers exist in this drawing for such items as floor and furniture, as shown in Figure 18.59.

Status	Name	On	Freeze	Lock	Color	Linetype	Lineweight	Transparency
	0				white	CONTINUOUS	—— Default	0
	FLOOR				white	CONTINUOUS	—— Default	0
✓	FURNITURE				red	CONTINUOUS	—— Default	0

FIGURE 18.59

STEP 5

As with the office plan, activate the Insert dialog box through the `INSERT` command to view the blocks internal to the drawing, as shown in Figure 18.60. The four blocks listed consist of various furniture items and will be used to lay out the interior plan.

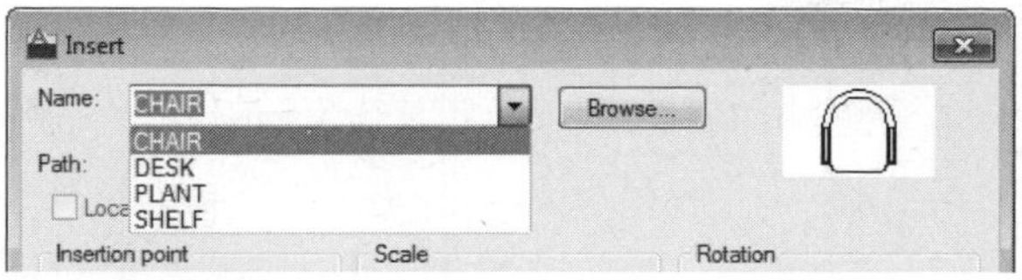

FIGURE 18.60

STEP 6

Verify that 18_Office.Dwg is closed and that 18_Interiors.Dwg is still open and active. The office floor plan will now be attached to the interior plan. Make the Floor layer current. Rather than insert the office plan as a block, use the External References palette to attach the drawing. This palette will activate when you enter the XREF command from the keyboard or choose the External References arrow from the lower-right corner of the Reference panel on the Insert Ribbon, as shown in Figure 18.61 (Left). After the palette displays, click the Attach DWG button (use the drop-down list, if the button is not displayed), as shown in Figure 18.61 (Right).

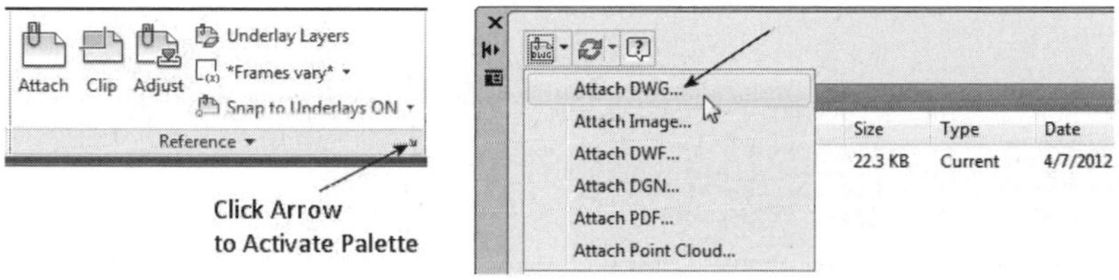

FIGURE 18.61

STEP 7

Clicking Attach DWG, as shown in Figure 18.61, displays the Select Reference File dialog box, as shown in Figure 18.62. Find the appropriate folder and click the drawing file 18_Office.Dwg.

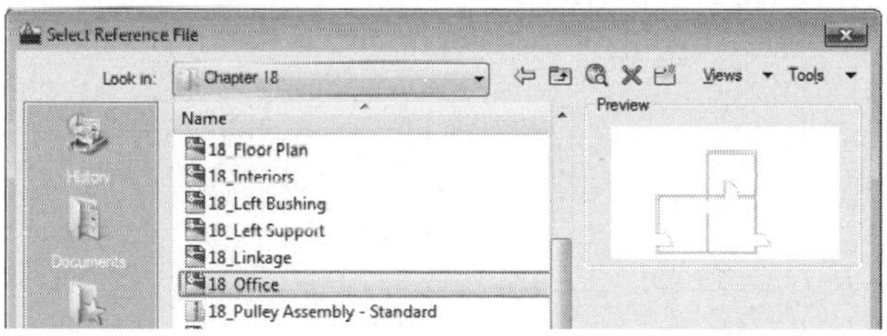

FIGURE 18.62

STEP 8

Selecting the file 18_Office.Dwg displays the Attach External Reference dialog box, as shown in Figure 18.63 (Left). Notice that 18_OFFICE is the name of the external reference file chosen for attachment in the current drawing. Verify that the Reference Type is an attachment. If selected, remove the checkmarks from the Specify On-screen boxes for the Insertion point, Scale, and Rotation. In the Attach External Reference dialog box, click the OK button to attach 18_Office.Dwg to 18_Interiors.Dwg.

The floor plan is now attached to the Interiors drawing at the insertion point 0,0,0, as shown in Figure 18.63 (Right).

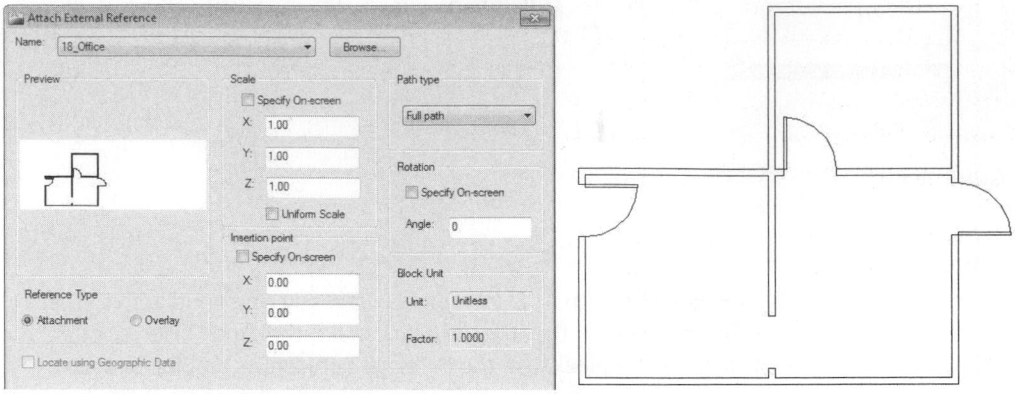

FIGURE 18.63

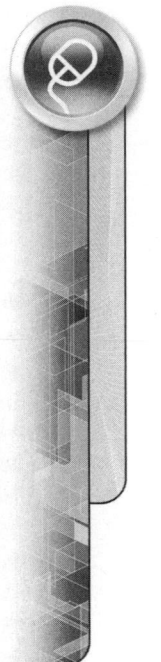

With the floor plan attached to the interiors drawing, notice how both drawings appear in the External References palette, as shown in Figure 18.64. When you have finished studying the information located in this palette, you can dismiss the palette from the screen by clicking the "X" located in the upper-left corner.

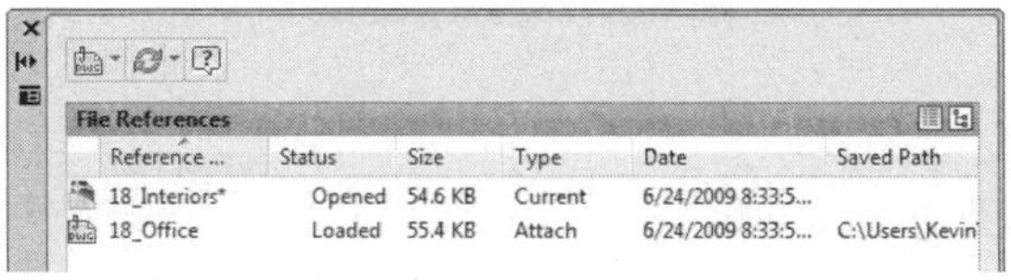

FIGURE 18.64

STEP 9

Once again, activate the Layer Properties Manager palette, paying close attention to the display of the layers. Using Figure 18.65 as a guide, you can see the familiar layers of Floor and Furniture. However, notice the group of layers beginning with 18_Office; the layers actually belonging to the external reference file have the designation of XREF|LAYER. For example, the layer 18_Office|DOORS represents a layer located in the file 18_Office.Dwg that holds all door symbols. The "|," or Pipe symbol, is used to separate the name of the external reference from the layer. The layers belonging to the external reference file may be turned on or off, locked, or even frozen. However, these layers cannot be made current for drawing on.

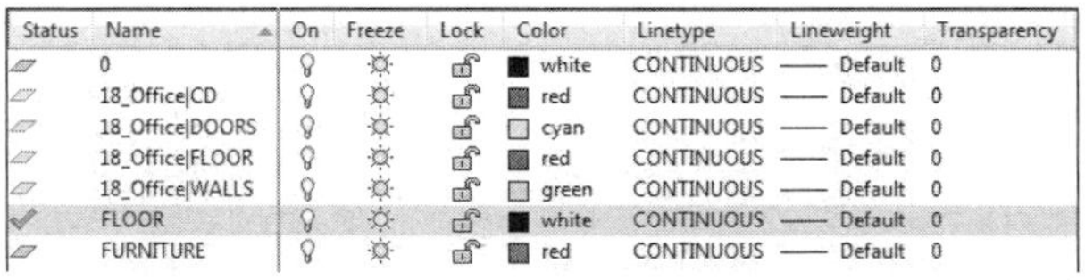

FIGURE 18.65

STEP 10

Make the Furniture layer current and begin inserting the desk, chair, shelf, and plant symbols in the drawing using the INSERT command or through the DesignCenter. The external reference file 18_OFFICE is to be used as a guide throughout this layout. It is not important that your drawing match exactly the image shown in Figure 18.65 (Left). After positioning all symbols in the floor plan, save your drawing under its original name of 18_Interiors.Dwg but do not close the file.

Even with the interior drawing file still open, make the original drawing file, 18_Office.Dwg, current by opening this file, and make the following modifications: stretch all three doors as indicated to new locations and mirror one of the doors so it is positioned closer to the wall; stretch the wall opening over to the other end of the room, as shown in Figure 18.66 (Right). Finally, save these changes under the original name of 18_Office.Dwg and close the file.

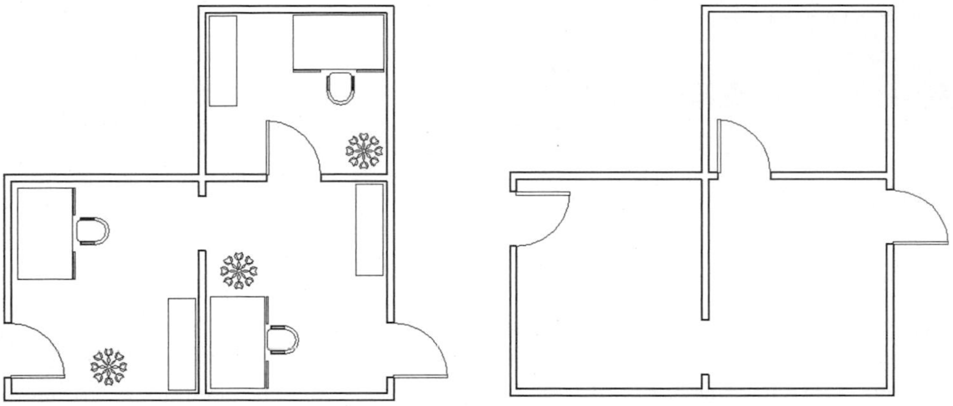

FIGURE 18.66

STEP 11

After you close the office plan, your display returns to the interiors plan. At this point, nothing in the drawing will appear to have changed. However, notice in the lower-right corner of your display screen that a pop-up window indicates that a reference file needs reloading. Clicking Reload 18_Office will update the office plan in the interiors drawing. When an external reference needs to be reloaded a Manage Xrefs button is also displayed on the Status Bar. Right-clicking this button displays two options. One of the options allows you to reload all external reference files. Once you reload the file, notice the effects on the floor plan, illustrated in Figure 18.67 (Left), where changes to the office plan are now reflected in the interiors drawing.

In this case, observe how some of the furniture is now in the way of the doorways. If the office plan had been inserted in the interiors drawing as a block, these changes would not have occurred this easily.

Because of the changes in the door openings, edit the drawing by moving the office furniture to better locations. The illustration in Figure 18.67 (Right) can be used as a guide, although your drawing may appear slightly different.

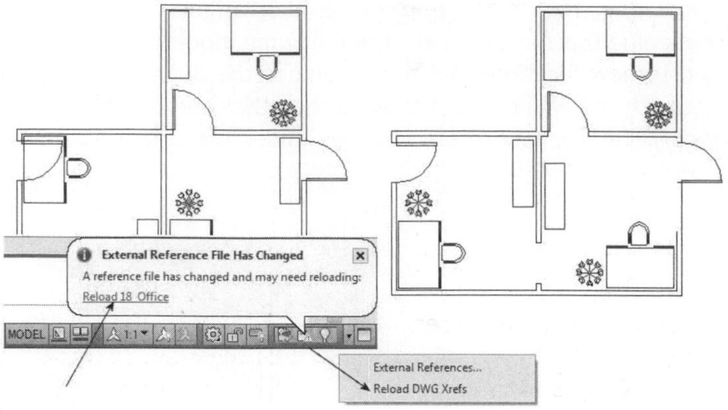

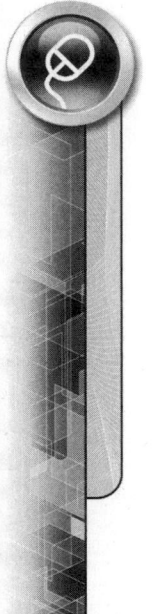

FIGURE 18.67

STEP 12

A door needs to be added to an opening in one of the walls of the office plan. However, the door symbol belongs to the externally referenced file 18_Office.Dwg. The door block is defined as 18_Office|DOOR; the "|" character is not valid in the naming of the block and, therefore, cannot be used in the current drawing. The block must first be bound to the current drawing before it can be used. Use the `XBIND` command to make the external reference's Door block available in the 18_Interiors.Dwg. Activating this command displays the Xbind dialog box, as shown in Figure 18.68. While in this dialog box, click the "+" symbol next to the file 18_Office and then click the "+" symbol next to Block. This displays all blocks that belong to the external reference.

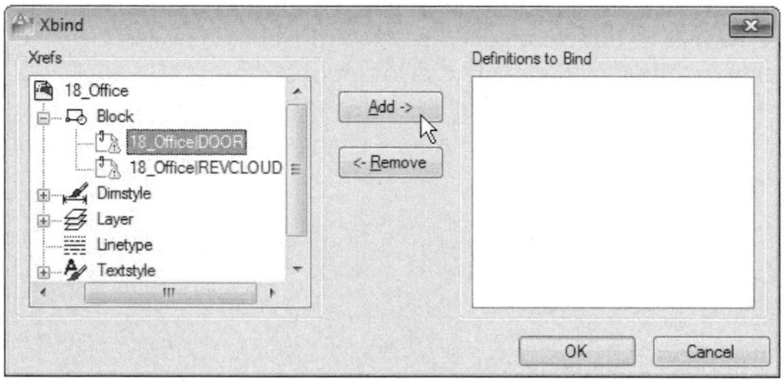

FIGURE 18.68

STEP 13

Clicking 18_OFFICE|DOOR followed by the Add -> button, as shown in Figure 18.68, moves the block of the door to the Definitions to Bind area in Figure 18.69. Click OK to dismiss the dialog box; the door symbol is now a valid block that can be inserted in the drawing.

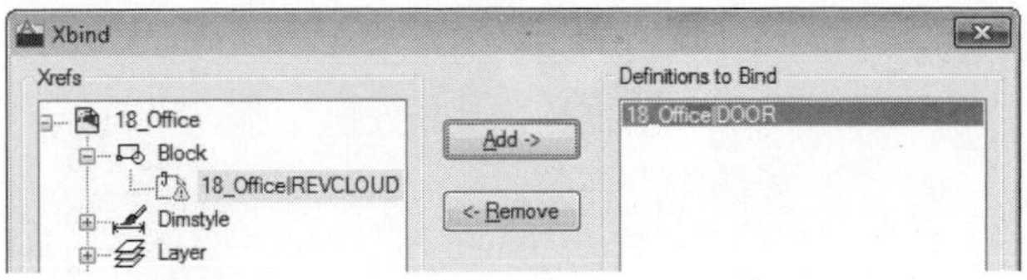

FIGURE 18.69

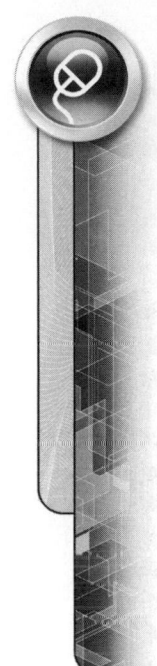

STEP 14

Make the Floor layer current. Activate the Insert dialog box through the INSERT command, click the Name drop-down list, and notice the name of the door, as shown in Figure 18.70 (Left). It is now listed as 18_Office0DOOR; the "|" character was replaced by the "0," making the block valid in the current drawing. This is AutoCAD's standard way of converting blocks that belong to external references to blocks that can be used in the current drawing file. This same procedure works on layers belonging to external references as well.

STEP 15

Insert the door symbol into the open gap, as shown in Figure 18.70 (Right), to complete the drawing.

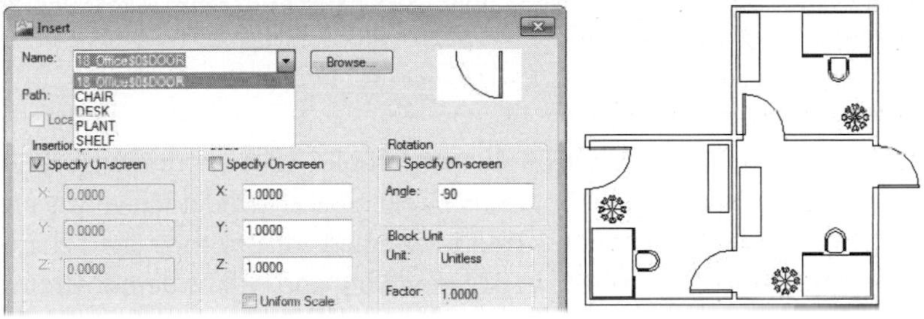

FIGURE 18.70

CHAPTER 19

Advanced Layout Techniques

This chapter is a continuation of Chapter 14, "Working with Drawing Layouts." First, we will discuss viewport creation and then we will demonstrate how multiple images of the same drawing file can be laid out in new viewports at different scales. This demonstration will include a number of layering and dimensioning techniques to achieve success. This chapter continues by demonstrating how to create annotation styles and manage annotation scales.

CREATING NEW VIEWPORTS

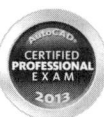

Use the VPORTS command to create a single or multiple viewport drawing. This command can be utilized in either Model Space or Paper Space. Model viewports are created in Model Space and are sometimes referred to as "tiled" due to their rigid shape constraints. These viewports are often utilized in 3D modeling and will be demonstrated in a later chapter. Layout viewports created in Paper Space are sometimes referred to as "floating" and offer a more flexible and powerful variation. In this chapter, our focus is on layout viewports. Choose the VPORTS command from one of the following:

- From the Ribbon > Layout Tab > Layout Viewports Panel > Named Button
- From the Menu Bar (View > Viewports > New Viewports...)
- From the Viewports toolbar
- From the keyboard (VPORTS)

Activate this command from the Ribbon by selecting the Named button, as shown in Figure 19.1. Additional methods of creating viewports, which will be discussed later in this chapter, are available through the use of the -VPORTS command. Placing a dash (-) in front of the command provides additional options on the command line instead of displaying a dialog box.

Ribbon (Layout Tab > Layout Viewports Panel)

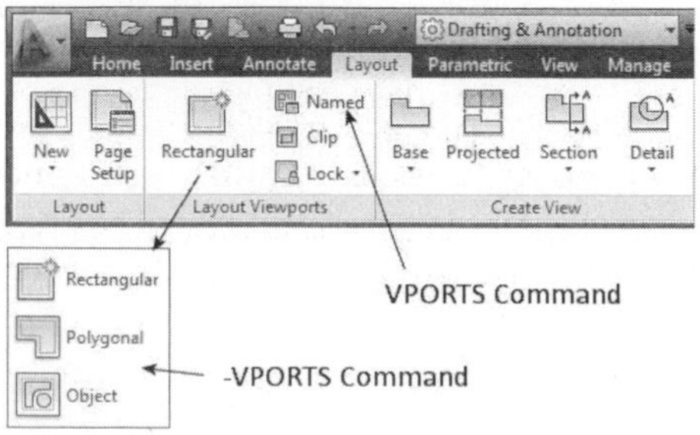

FIGURE 19.1

NOTE

The buttons in the Layout Viewports panel of the Ribbon, as shown in Figure 19.1, will be grayed out in Model Space and are only available after you select a layout tab and enter Paper Space.

By default, when a layout is activated, a viewport is automatically created for you. Often this viewport, after moving and resizing, is sufficient for your drawing layout. If not, the VPORTS command can be used to generate additional viewports. In some cases, you may want to erase any existing viewports and layout one of the standard predefined viewport arrangements available.

Activating the VPORTS command displays the Viewports dialog box shown in Figure 19.2. The Named Viewports tab of the dialog box, shown activated in Figure 19.2 (Left), is used to create and restore saved viewport configurations. The New Viewports tab, shown activated in the image on the right, is used to generate standard viewport arrangements. Select an arrangement from the list provided, such as "Three Left" as shown in Figure 19.2 (Right). After verifying your selection in the Preview window, click the OK button and pick two diagonal corners that will establish the location and size of the new viewports.

NOTE

In the New Viewports tab of the Viewports dialog box, the "Single" configuration can be utilized to create individual rectangular viewports; however, a more efficient method is to utilize the Rectangular button in the Layout Viewports panel of the Layout Ribbon (-VIEW command.)

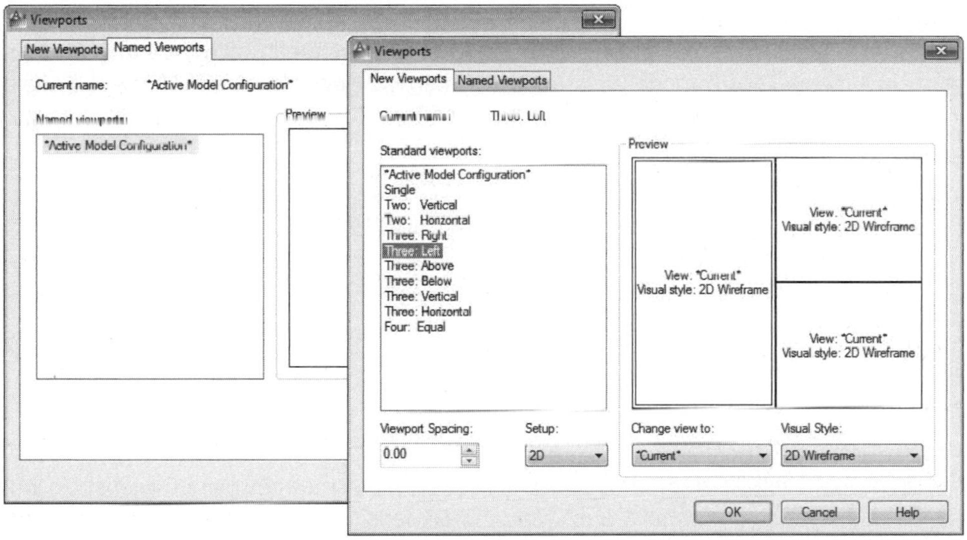

FIGURE 19.2

ARRANGING DIFFERENT VIEWS OF THE SAME DRAWING

This chapter begins immediately with a tutorial exercise designed to lay out two views of the same drawing in the same layout. Both images will be scaled at different values inside their respective viewports. Follow the next series of steps for creating this type of layout.

TUTORIAL EXERCISE: 19_ROOF PLAN.DWG

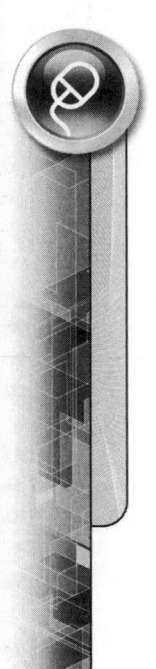

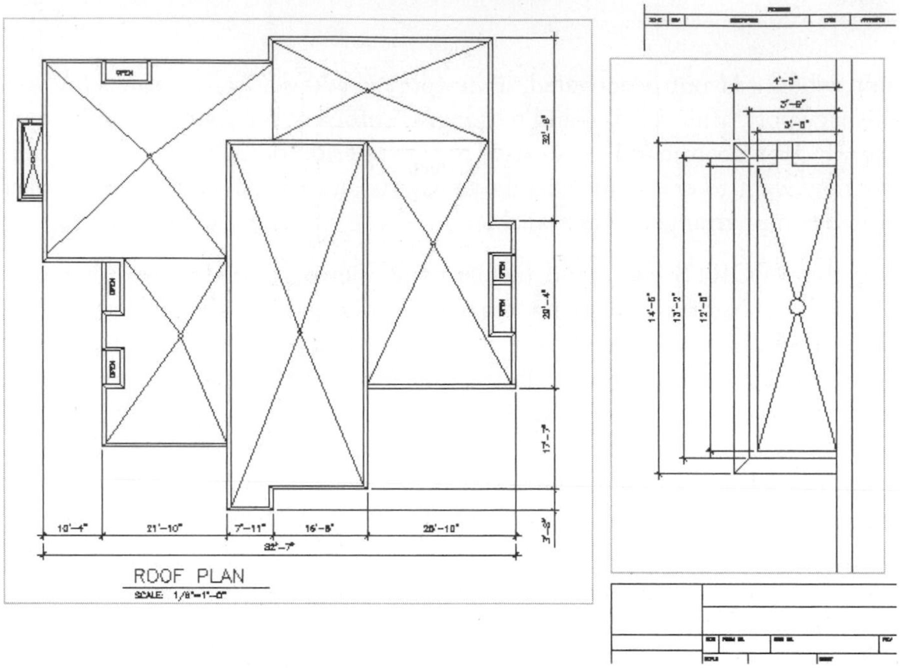

FIGURE 19.3

Purpose

This tutorial exercise is designed to lay out the two architectural views displayed in Figure 19.3 in the Layout mode (Paper Space). The two views consist of an overall and a detailed view of a roof plan.

System Settings

In this tutorial you will open an existing drawing called 19_Roof Plan. Keep all default settings for the units and limits of the drawing.

Layers

Layers have already been created for this drawing.

Suggested Commands

A new viewport will be created to house a detail view of the roof plan. Dimensions will also be added to this viewport. However, the Dimension scale setting of the drawing needs to be set to "Scale dimension to layout." This will enable the scale of the dimensions in the new viewport to match those dimensions in the main roof plan viewport.

STEP 1

Open the drawing file 19_Roof Plan. Use the -VPORTS command (Layout Ribbon > Layout Viewports panel > Rectangular button) to create a new viewport, as shown in Figure 19.4. This viewport should be long and narrow to accommodate the detail. Notice that when you are creating the viewport, the entire image zooms to the extents of the viewport. This is normal and you will arrange the detail in the next series of steps.

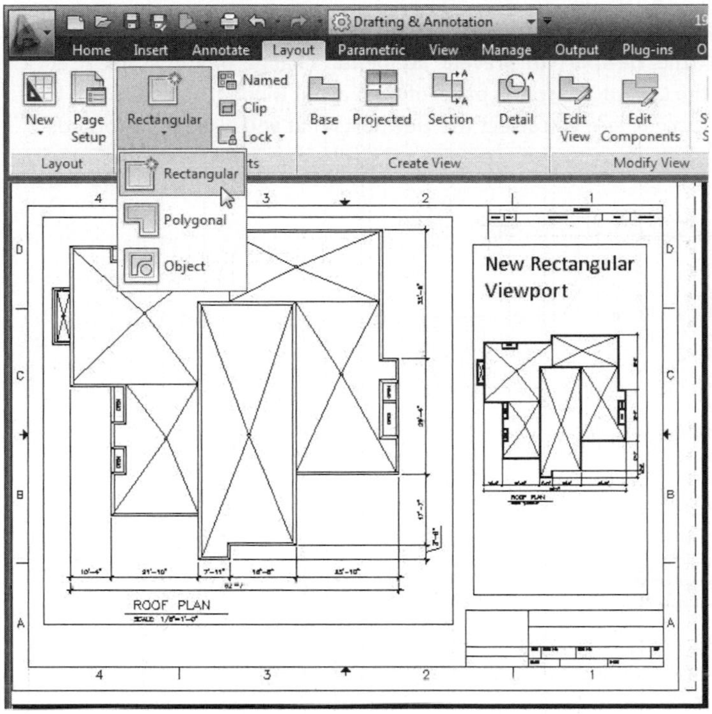

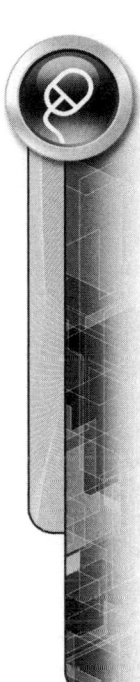

FIGURE 19.4

STEP 2

Double-click inside the new viewport to make it active. Then pan the area of the roof plan so it appears similar to Figure 19.5. When applying a scale to this viewport, it is easier to see the results if the image you are detailing is approximately in the middle of the viewport. If it is not, it may be difficult to locate the area of the detail.

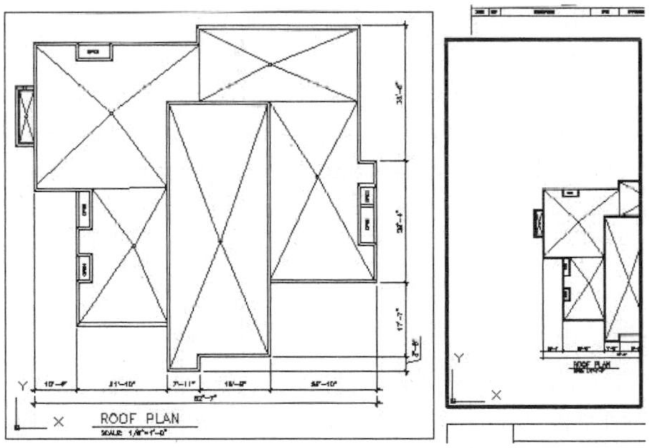

FIGURE 19.5

STEP 3

Expand the Viewport Scale list box on the Status Bar and pick 1/2"= 1"-0". This should increase the size of the detail in this viewport. Use the PAN command to center the long rectangular portion of the roof in the viewport, as shown in Figure 19.6.

NOTE

Once the drawing is properly scaled and panned in the viewport, use the Lock Viewport button on the Status Bar to lock the viewport to prevent accidental change of the scale. A Lock button is also available in the Layout Viewports panel of the Layout Ribbon. Selecting this button returns you to Paper Space where you select the viewports that you want to be locked.

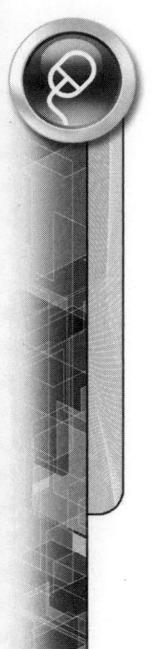

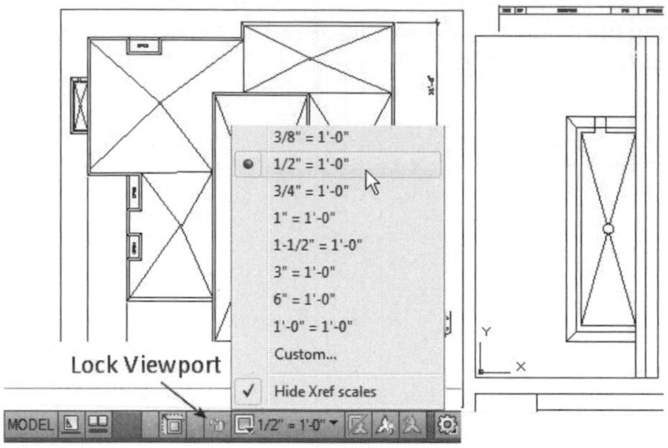

FIGURE 19.6

STEP 4

Before adding dimensions, one more item needs to be taken care of. First activate the main roof plan viewport by clicking inside it to make it active. Then launch the Dimension Styles Manager dialog box, click the Modify button, and pick the Fit tab. Under the Scale for dimension features area, click the radio button next to Scale dimensions to layout, as shown in Figure 19.7. This option scales dimensions to the current viewport scale, which is 1/8" = 1'-0" (1:96) for this viewport. When finished, click the OK button to dismiss this dialog box and close the main Dimension Style Manager dialog box. If the dimension scale does not automatically set for you (the dimensions are unreadable), you will have to update all dimensions in this viewport to reflect the changes to this dimension style. This can be accomplished by picking the Update button from the Dimensions panel of the Annotate Ribbon and selecting all the dimensions in the viewport.

NOTE

You can verify that the scale is correct by using the Properties palette or the `LIST` **command on a dimension and noting that the** `DIMSCALE` **system variable has been overridden and set to 96.**

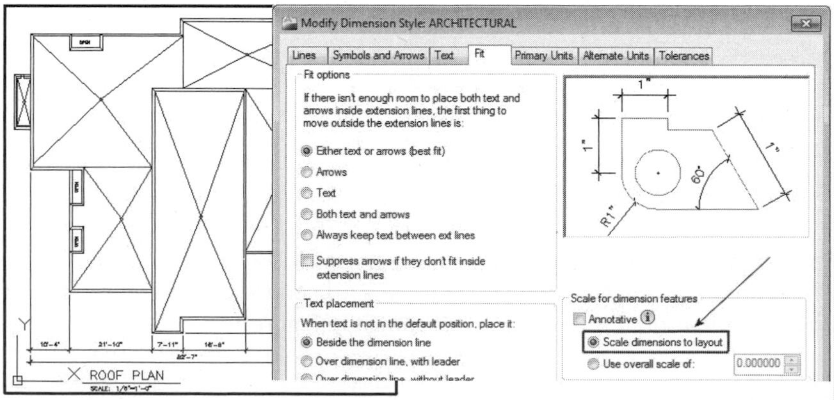

FIGURE 19.7

STEP 5

Make the Detail Dim layer current. This layer will be used to hold all dimensions that will be added to the detail view. Activate the viewport containing the roof detail and place the linear dimensions, as shown in Figure 19.8. Notice that the sizes of these dimensions exactly match those in the main floor plan. The scale of the viewport, which we set earlier to 1/2" = 1'-0", sets the scale of the dimensions. This occurs automatically because the "Scale dimensions to layout" button was selected, as shown in the previous step. You should also notice that as these dimensions are placed in the detail, they also appear in the main roof plan viewport, as shown in Figure 19.8.

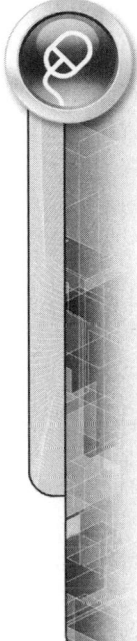

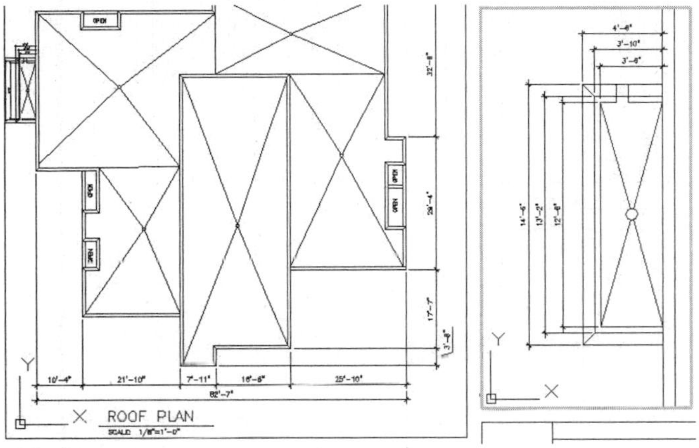

FIGURE 19.8

STEP 6

Activate the main roof plan viewport by clicking inside it. Then launch the Layer Properties Manager palette. Identify the Dim Detail layer and freeze it in the current viewport by clicking the button, as shown in Figure 19.9. This action will freeze the Dim Detail layer in the main roof plan viewport while keeping the layer visible in the detail viewport.

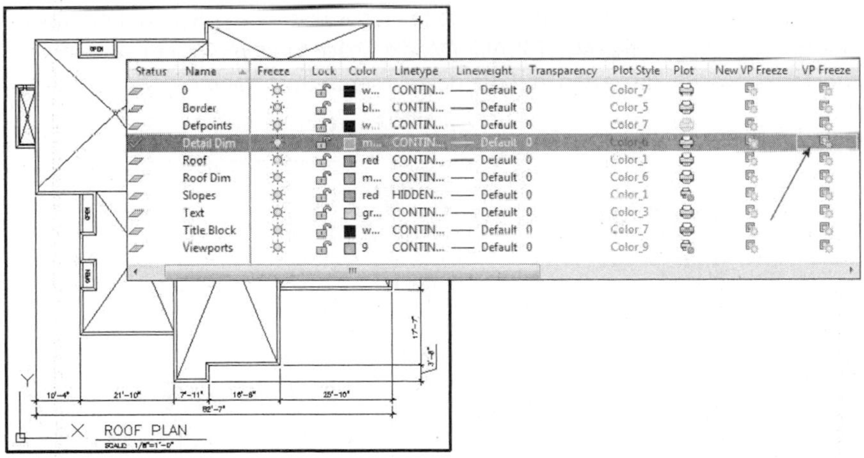

FIGURE 19.9

STEP 7

The completed layout is shown in Figure 19.10.

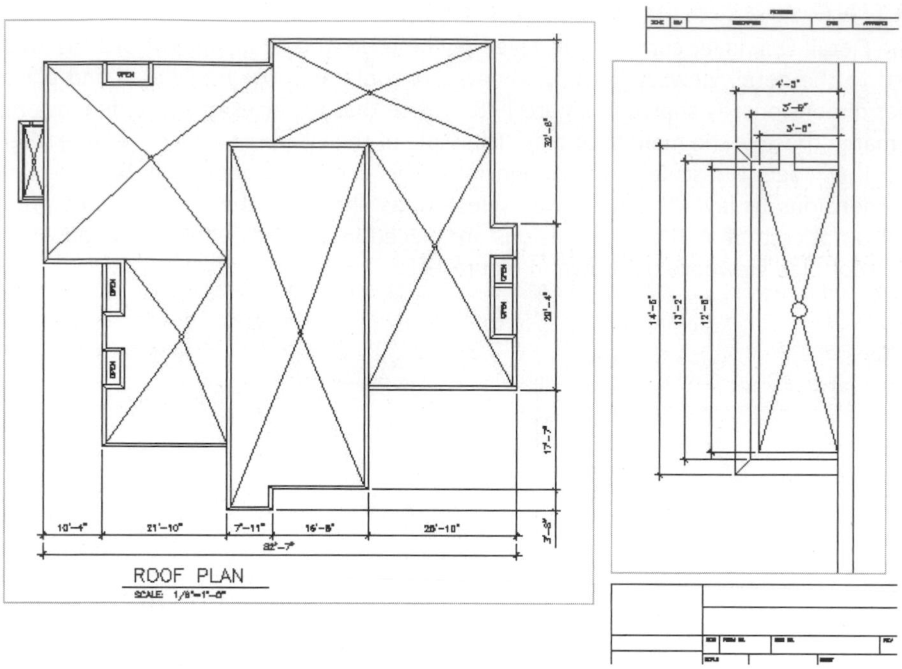

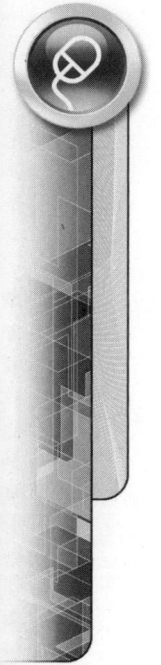

FIGURE 19.10

STEP 8

Switch back to Model Space and notice the appearance of dashed lines representing the slope lines of the roof, as shown in Figure 19.11 (Left). Switch back to Paper Space and notice that the dashed lines disappear. To display the dashed lines in the Paper Space viewports, change the LTSCALE value to 1.00; the dashed lines will appear as shown in Figure 19.11 (Right). The ability to have linetypes display correctly in each viewport in Paper Space is controlled by the PSLTSCALE (Paper Space Linetype Scale) system variable, which is also set to a default value of 1.00. In general, when utilizing layouts, you should leave the LTSCALE set to 1.00 and linetypes will automatically be displayed in the viewport at the designated viewport scale. If you return to Model Space, the dashes are gone again. Instead of changing the LTSCALE, set the Annotation Scale in the Status Bar to 1/8" = 1'-0" and perform a drawing regeneration (REGEN command). The ability to have linetypes display to the annotation scale in Model Space is controlled by the MSLTSCALE (Model Space Linetype Scale) system variable. Annotation scales will be discussed in more detail later in this chapter.

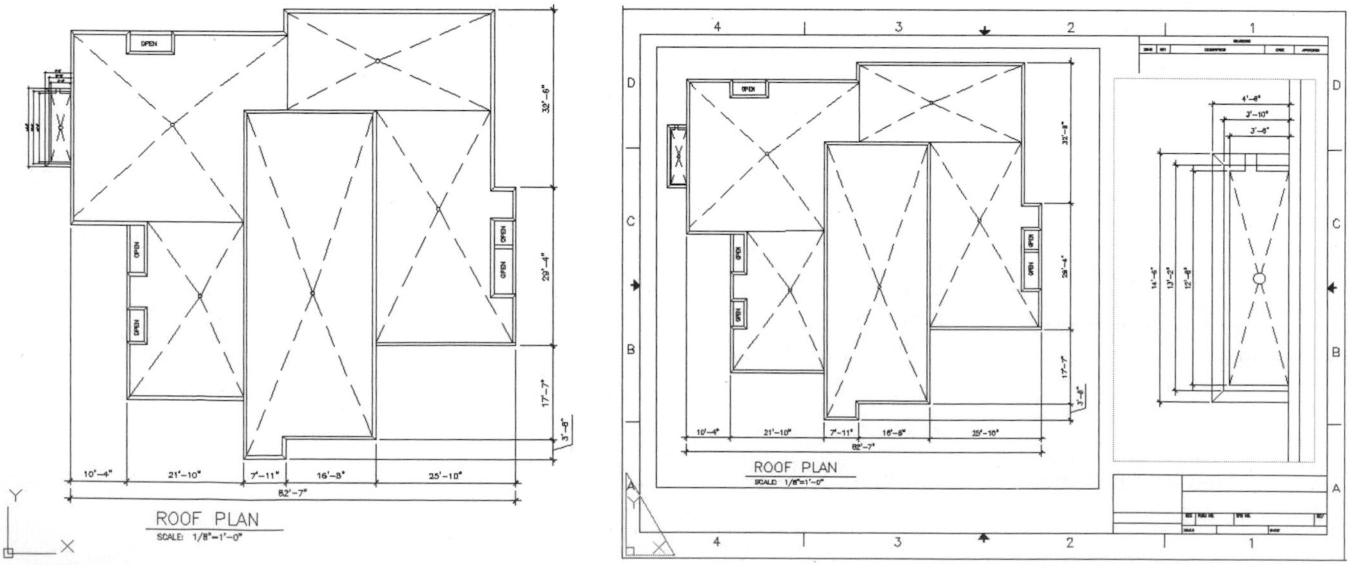

FIGURE 19.11

CREATING A DETAIL PAGE IN LAYOUT MODE

This next discussion focuses on laying out on the same sheet a series of details in multiple viewports, which can be at different scales. As the viewports are laid out in Paper Space and images of the drawings are displayed in floating Model Space, all images will appear in all viewports. This is not a major problem, because layers can be frozen in specified floating viewports.

TRY IT!

Open the drawing file 19_Bearing Details. Figure 19.12 shows three objects: a body, a bushing, and a bearing all arranged in Model Space. The body and bearing will be laid out at a scale of 1:1. The bushing will be laid out at a scale of 2:1 (enlarged to twice its normal size). The dimension scales have been set for all objects in order for all dimension text to be displayed at the same size. Also, layers have been created for each object.

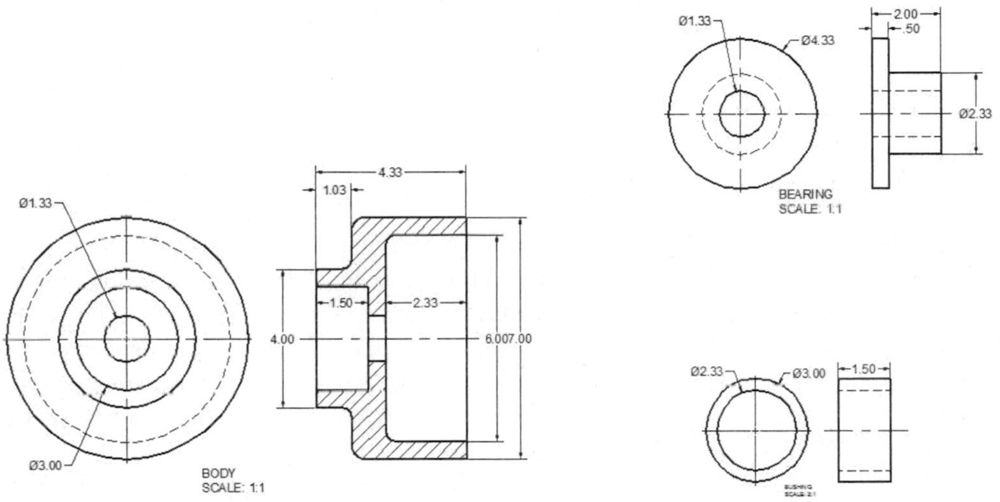

FIGURE 19.12

The following items have already been added to this drawing:

- A VPORTS layer was created. This layer holds all viewport information, and, when the drawing is completed, can be frozen or turned off.
- In Layout mode, an ANSI D-size sheet was selected for Page Setup.
- A D-size title block was inserted onto the Title Block layer.

Click the layout name ANSI-D. Except for the title block, the drawing sheet is empty. Using the Viewports dialog box (`VPORTS` command) select the Three: Left arrangement. Click OK and pick points "A" and "B" to create the three viewports, as shown in Figure 19.13. The exact size of these viewports is not important at this time since you will have to adjust them at the end of the exercise. When creating these viewports, notice that all three objects (body, bearing, and bushing) appear in all three viewports. You will now freeze layers inside these viewports in order to show a different part in each viewport.

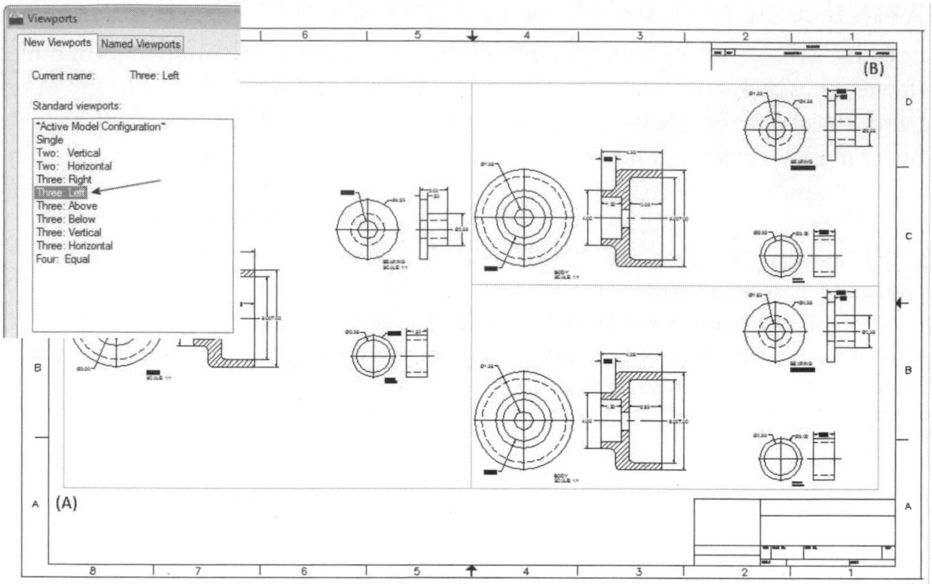

FIGURE 19.13

Begin the process of isolating one part per viewport by making the large viewport on the left active by double-clicking inside it. You will notice the viewport taking on the familiar thick border appearance, and the UCS icon is present in the lower-left corner of the viewport, as shown in Figure 19.14 (Left). Then activate the Layer Properties Manager palette. Only those layers that begin with Body need to be visible in this viewport. Pick all layers that begin with Bearing and Bushing and freeze these layers only in this viewport, as shown in Figure 19.14 (Right).

NOTE

To select multiple layers at one time, hold down the `CTRL` key as you select each layer name.

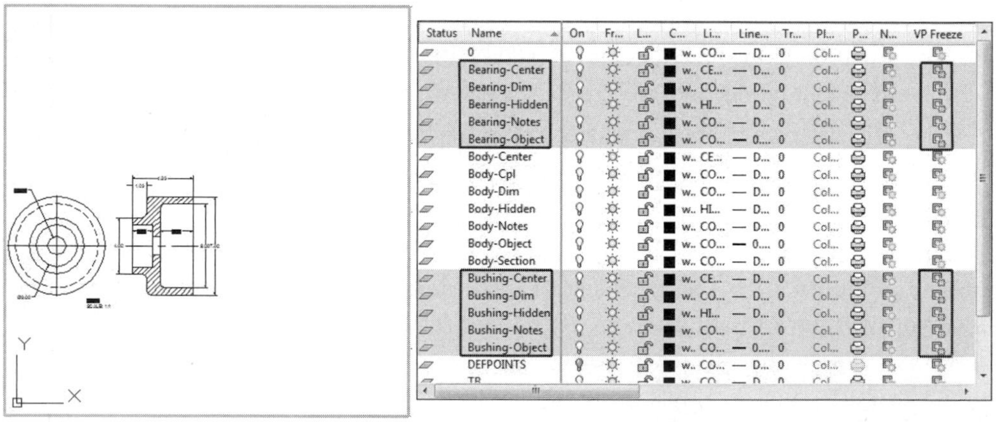

FIGURE 19.14

Next, click inside the viewport located in the upper-right corner to make it active, as shown in Figure 19.15 (Left). Activate the Layer Properties Manager palette. Only those layers that begin with Bearing need to be visible in this viewport. Pick all layers that begin with Body and Bushing and freeze these layers only in this viewport, as shown in Figure 19.15 (Right).

NOTE

To select all the layers at one time, select the first layer (Body-Center) then hold down the `SHIFT` **key as you select the last layer name (Bushing-Object).**

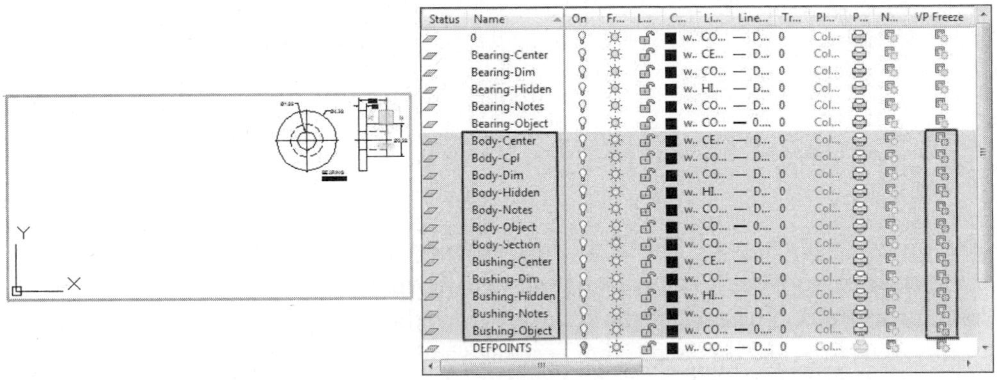

FIGURE 19.15

Finally, click inside the viewport located in the lower-right corner to make it active, as shown in Figure 19.16 (Left). Activate the Layer Properties Manager palette. Only those layers that begin with Bushing need to be visible in this viewport. Pick all layers that begin with Body and Bearing and freeze these layers only in this viewport, as shown in Figure 19.16 (Right).

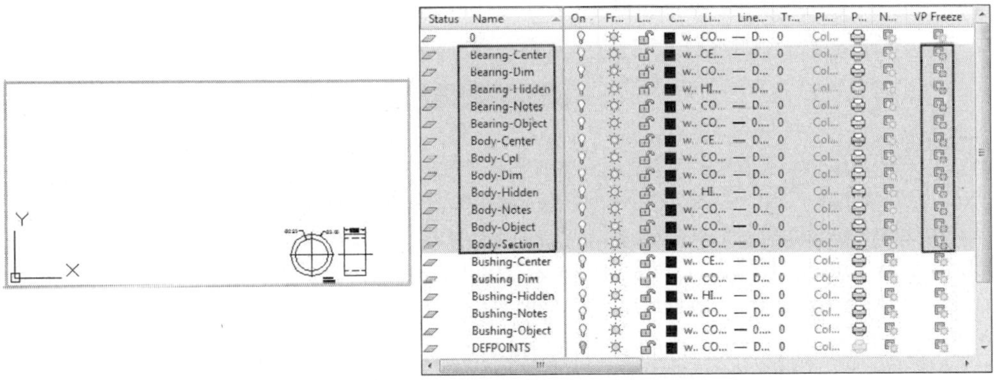

FIGURE 19.16

When finished, each viewport should contain a different part. Items dealing with the part body are visible in the large viewport on the left. The bearing views are visible in the upper-right viewport and the bushing views are visible in the lower-right viewport. One other step is needed to better organize your work. Click in each port and pan each image so it appears centered in each viewport, as shown in Figure 19.17. Return to Paper Space by double-clicking outside the viewports.

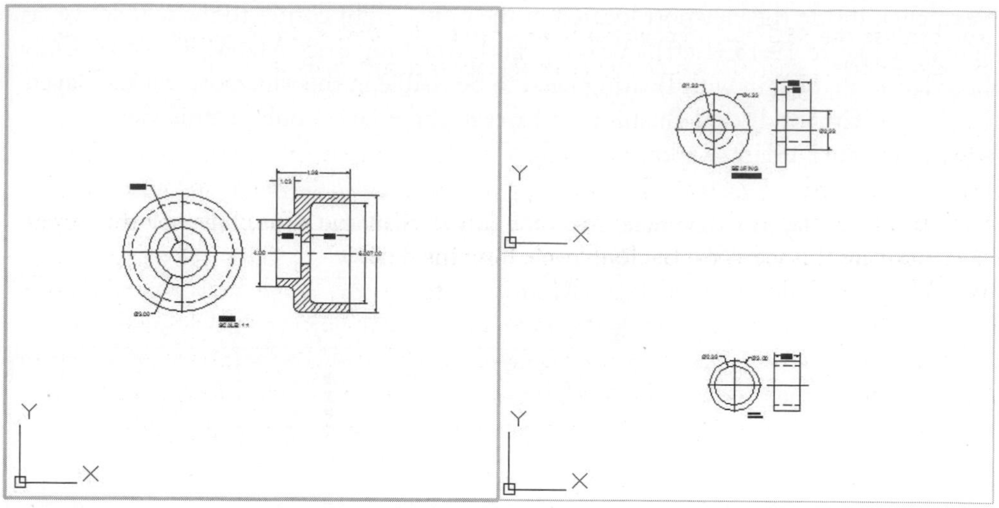

FIGURE 19.17

Each image will now be scaled to each viewport. Two of the viewports share the same scale factor. Click the edge of the two viewports to highlight them as shown in Figure 19.18. Then select the scale 1:1 from the Viewports Scale list box on the Status Bar to scale the body and bearing views to their respective viewports, as shown in Figure 19.18. Press ESC to remove the grips.

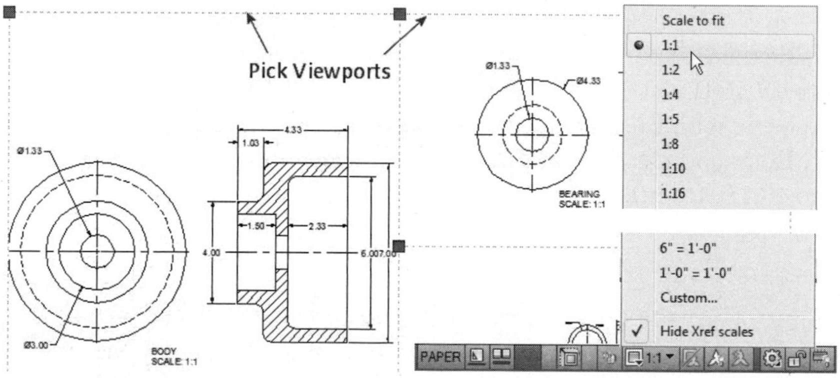

FIGURE 19.18

Next, click the edge of the viewport located in the lower-right corner to highlight it. Then set the viewport scale to 2:1 to scale the bushing views to this viewport, as shown in Figure 19.19. The scale of 2:1 will double the size of this view since it is smaller than the others in Model Space.

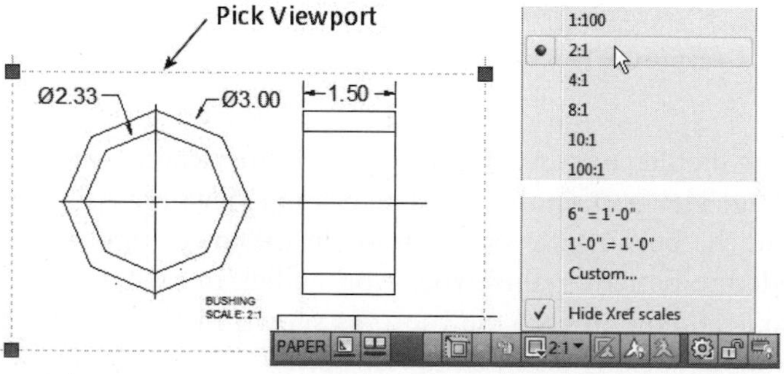

FIGURE 19.19

NOTE

You can use the REGENALL command to regenerate the drawing in all viewports; this will make the circles, such as those in Figure 19.19, appear round on the screen.

Adjust the viewports by clicking the edges, picking corner grips, and stretching the viewports, in order to see all the drawings and dimensions. The dimensions in the lower-right corner of the viewport appear larger than the others due to the larger scale. Activate this viewport by double-clicking inside it. Then launch the Dimension Style Manager dialog box, click the Modify button, select the Fit tab, and click the button next to Scale dimensions to layout; all dimensions inside this viewport will be automatically scaled to the viewport. You will have to update the dimension scales in other viewports by selecting the Update button from the Dimensions panel of the Annotate Ribbon. The final layout, consisting of three different details utilizing two different scales, is shown in Figure 19.20.

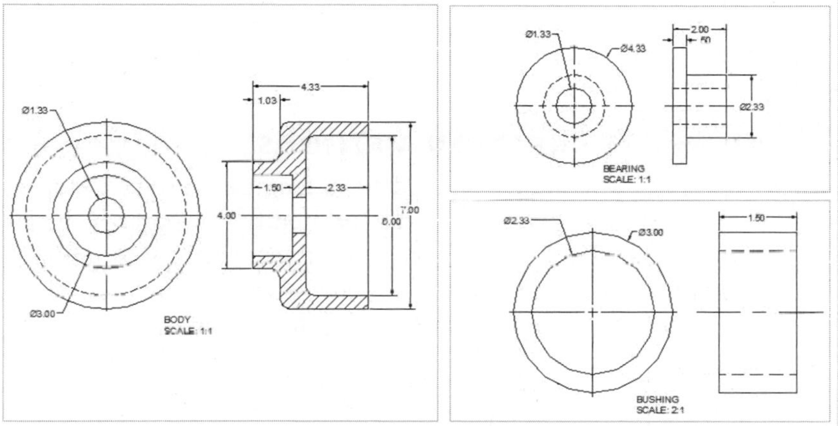

FIGURE 19.20

NOTE

Instead of updating dimension scales in a viewport, it is usually more productive to set the dimension style scale to "Scale dimensions to layout" before placing dimensions in the drawing. This method does, however, require that dimensions be placed in floating model space so that their scale can be set to the current viewport scale as the dimensions are created.

In Figure 19.21 a Plot Preview on the drawing just completed is illustrated. Notice the absence of any viewports. This is due to the No Plot setting being applied to the VPORTS layer.

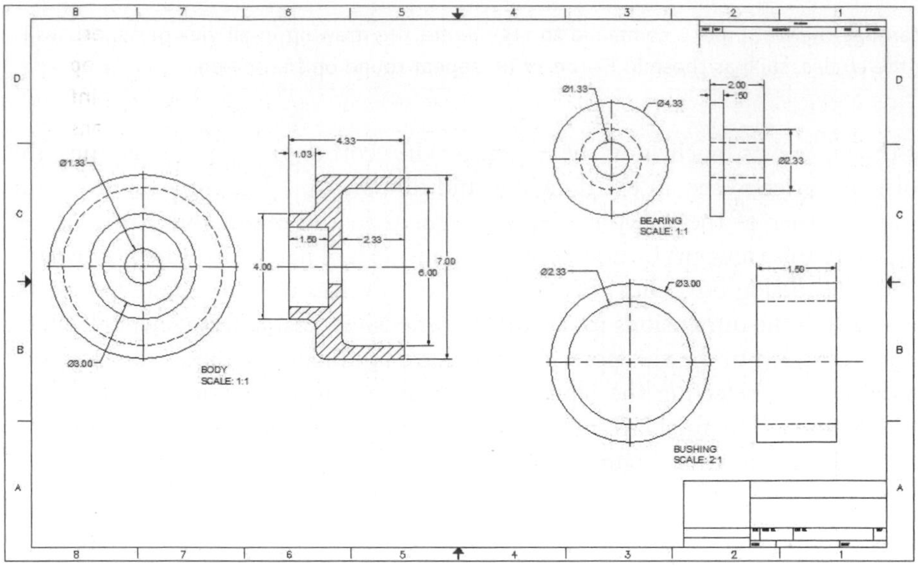

FIGURE 19.21

ADDITIONAL VIEWPORT CREATION METHODS

When constructing viewports, you are not limited to rectangular or square shapes. Other commands and options are available to create or modify viewports. You can clip an existing viewport to reflect a different shape, convert an existing closed object into a viewport, construct a multisided closed or polygonal viewport, or select predefined arrangements through the Viewports dialog box. You can utilize the Layout Viewports Panel of the Layout Ribbon to access these tools, as shown in Figure 19.22.

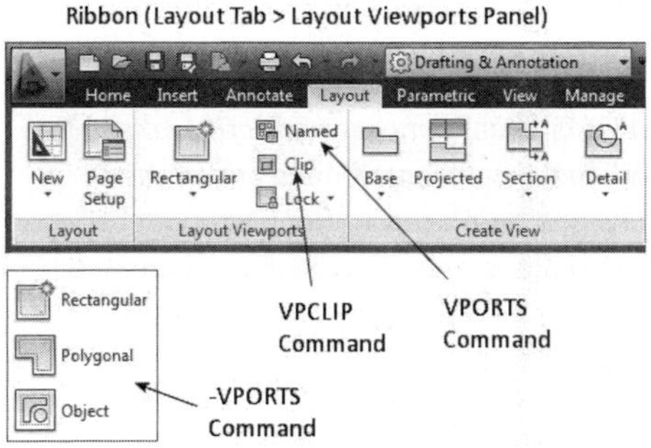

FIGURE 19.22

Refer to the following table for a brief description of these available viewport tools.

Button	Command	Description
	VPORTS	Displays the Viewports dialog box
	-VPORTS Rectangular	Creates a single rectangular viewport
	-VPORTS Polygonal	Is used for creating a polygonal viewport
	-VPORTS Object	Converts existing object into a viewport
	VPCLIP	Is used for clipping an existing viewport

TRY IT!

Open the drawing file 19_Floor Viewports and verify that Vports is the current layer. You will convert the large rectangular viewport into a polygonal viewport by a clipping operation. First click the Clip button. Pick the rectangular viewport and begin picking points to construct a polygonal viewport around the perimeter of the floor plan dimensions, as shown in Figure 19.23. You can turn Polar and Object Snap Tracking on to assist with this operation. The following command sequence will also aid with this operation.

Command: VPCLIP

Select viewport to clip: *(Select the rectangular viewport)*

Select clipping object or [Polygonal] <Polygonal>: *(Press ENTER)*

Specify start point: *(Pick at "A")*

Specify next point or [Arc/Length/Undo]: *(Pick at "B")*

Specify next point or [Arc/Close/Length/Undo]: *(Pick at "C")*

Specify next point or [Arc/Close/Length/Undo]: *(Pick at "D")*

Specify next point or [Arc/Close/Length/Undo]: *(Pick at "E")*

Specify next point or [Arc/Close/Length/Undo]: *(Pick at "F")*

Specify next point or [Arc/Close/Length/Undo]: C *(To close the shape)*

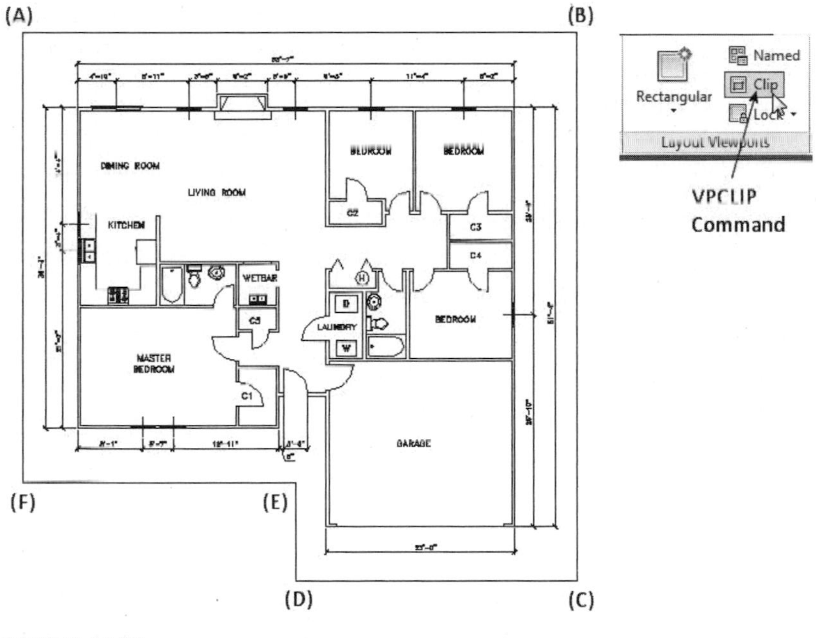

FIGURE 19.23

When you are finished, move the viewport with the image of the floor plan to the right of the screen. Then construct a circle in the upper-left corner of the title block. Pick the Object button and select the circle you just constructed. Notice that the circle converts to a viewport with the entire floor plan displayed inside its border. Double-click inside this new viewport to make it current and change the scale of the image to the 1/2" = 1'-0" scale using the Viewports Scale list box on the Status Bar.

Command: -VPORTS

Specify corner of viewport or[ON/OFF/Fit/Shadeplot/Lock/Object/Polygonal/Restore/LAyer/2/3/4] <Fit>: O *(For Object)*

Select object to clip viewport: *(Pick the circle)*

Pan inside the circular viewport until the laundry and bathroom appear, as shown in Figure 19.24. When finished, double-click outside the edge of the viewport to switch to Paper Space. Adjust the size of the circular viewport with grips if the image is too large or too small.

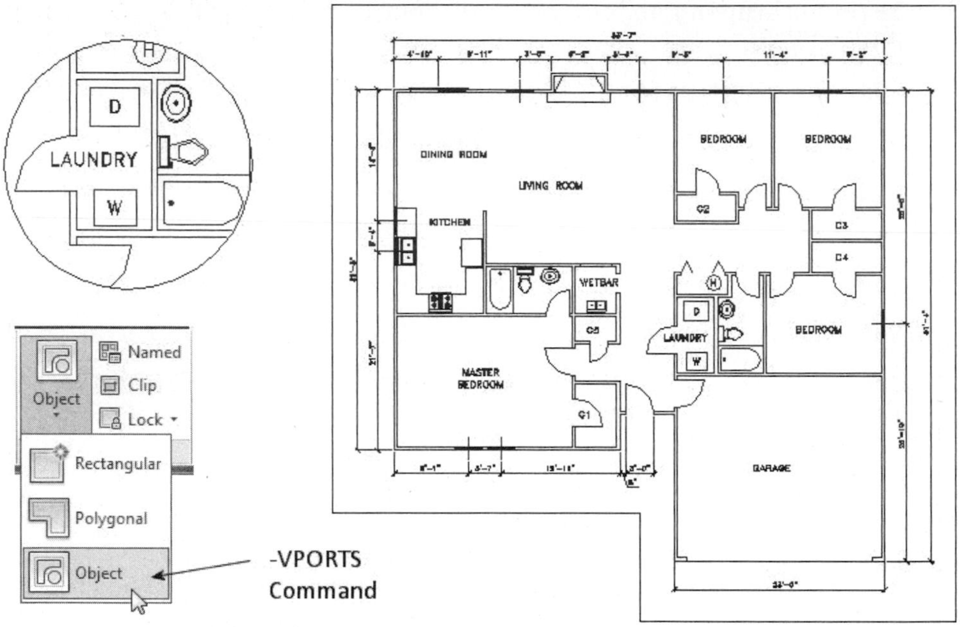

FIGURE 19.24

Click the Polygonal button and construct a multisided viewport similar to the one located in Figure 19.25. You can use Polar and Object Snap Tracking to help create the shape, although it is not necessary for the lines to be orthogonal since the viewport will not be plotted.

Command: -VPORTS

Specify corner of viewport or[ON/OFF/Fit/Shadeplot/Lock/Object/Polygonal/Restore/LAyer/2/3/4] <Fit>: P *(For Polygonal)*

Specify start point: *(Pick at "A")*

Specify next point or [Arc/Length/Undo]: *(Pick at "B")*

Specify next point or [Arc/Close/Length/Undo]: *(Pick at "C")*

Specify next point or [Arc/Close/Length/Undo]: *(Pick at "D")*

Specify next point or [Arc/Close/Length/Undo]: *(Pick at "E")*

Specify next point or [Arc/Close/Length/Undo]: *(Pick at "F")*

Specify next point or [Arc/Close/Length/Undo]: C *(To close the shape)*

Regenerating model.

As the image of the floor plan appears in this new viewport, double-click inside the new viewport to make it current. Scale the image inside the viewport to the scale 1/2"= 1'-0". Pan until the kitchen area and master bathroom are visible. Your display should appear similar to Figure 19.25.

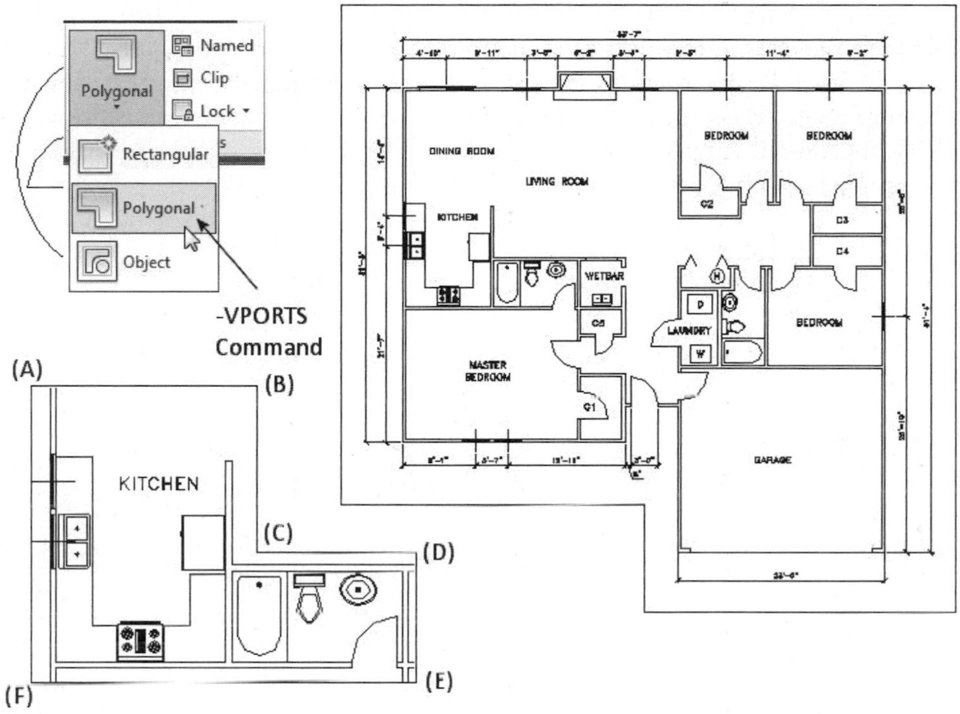

FIGURE 19.25

Double-click outside this viewport to return to Paper Space. Make any final adjustments to viewports using grips. When finished, turn off the VPORTS layer. Your display should appear similar to Figure 19.26.

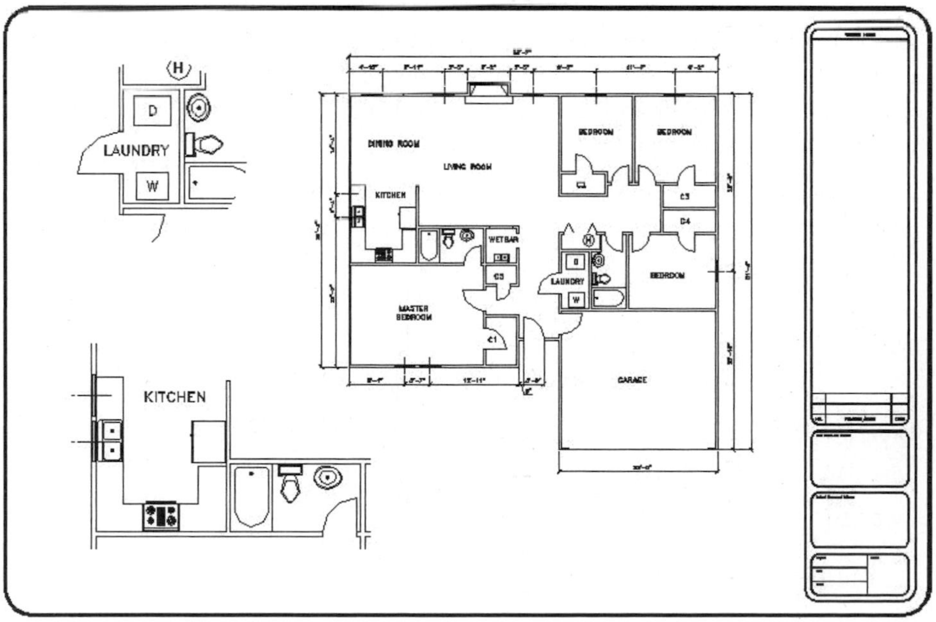

FIGURE 19.26

ROTATING VIEWPORTS

Viewports along with the view inside can easily be rotated using the traditional ROTATE command. First, verify that the VPROTATEASSOC system variable is activated before performing the rotation operation. This system variable must be entered in from the Command prompt and the spelling must be exact. Figure 19.27 (Left) represents a rectangular viewport with a land plat arranged inside. When using the ROTATE command on the viewport, the results are displayed in Figure 19.27. If VPROTATEASSOC is turned off or is set to 0 (zero), only the viewport rotates, as shown in Figure 19.27 (Center). To rotate the viewport and the image inside, verify VPROTATEASSOC is turned on, or set it to 1. The results are displayed in Figure 19.27 (Right) with the viewport and image rotating to 30 degrees.

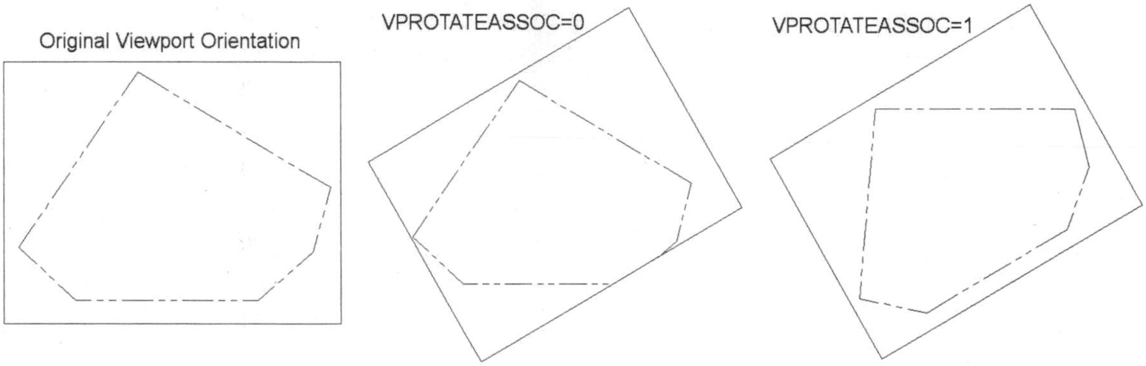

FIGURE 19.27

MATCHING THE PROPERTIES OF VIEWPORTS

In Chapter 7, the MATCHPROP (Match Properties) command was introduced as a means of transferring all or selected properties from a source object to a series of destination objects. In addition to transferring layer information, dimension styles, hatch properties, and text styles from one object to another, you can also transfer viewport information from a source viewport to other viewports. Information such as viewport layer and the viewport scale are a few of the properties to transfer to other viewports. When you enter the MATCHPROP command and select the Settings option, the dialog box in Figure 19.28 will appear. When transferring viewport properties, be sure the Viewport option of this dialog box is checked.

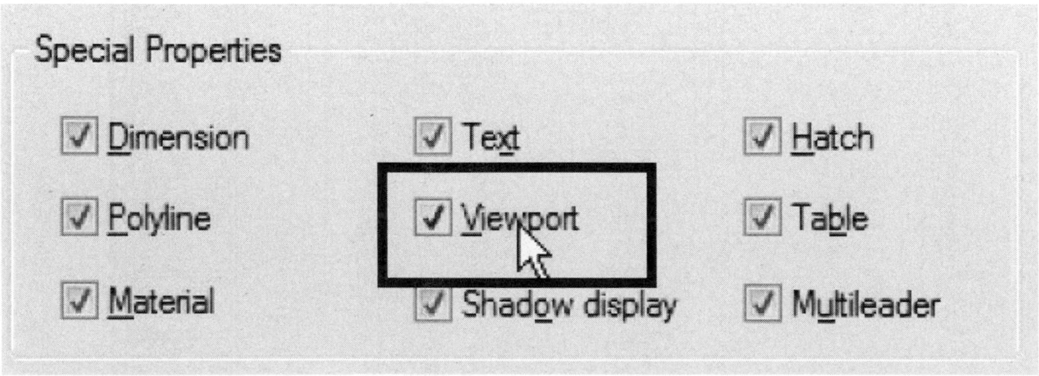

FIGURE 19.28

TRY IT!

Open the drawing file 19_Matchprop Viewports, as shown in Figure 19.29. This drawing consists of four viewports holding different object types. The object in the first viewport (labeled "A" in Figure 19.29) is correctly scaled to 1:1. The first viewport is also placed on the correct layer. All other viewports do not belong to the Viewports layer and are scaled differently. These three viewports need to have the same properties as the first; namely, all viewports need to belong to the Viewports layer and all images inside all viewports need to be scaled to 1:1. Rather than perform these operations on each individual viewport, the `MATCHPROP` command will be used to accomplish this task.

Command: MA *(For MATCHPROP)*

Select source object: *(Pick the edge of Viewport "A")*

Current active settings: Color Layer Ltype Ltscale Lineweight Transparency Thickness PlotStyle Dim Text Hatch Polyline Viewport Table Material Shadow display Multileader
Select destination object(s) or [Settings]: *(Pick the edge of Viewport "B")*

Select destination object(s) or [Settings]: *(Pick the edge of Viewport "C")*

Select destination object(s) or [Settings]: *(Pick the edge of Viewport "D")*

Select destination object(s) or [Settings]: *(Press ENTER to exit this command)*

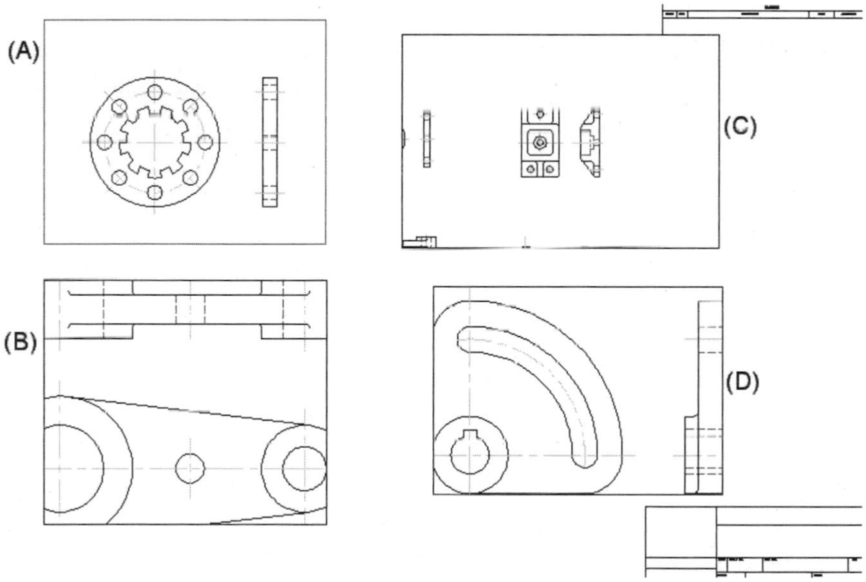

FIGURE 19.29

The results are illustrated in Figure 19.30. All viewports share the same layer and are scaled to 1:1 after using the `MATCHPROP` command.

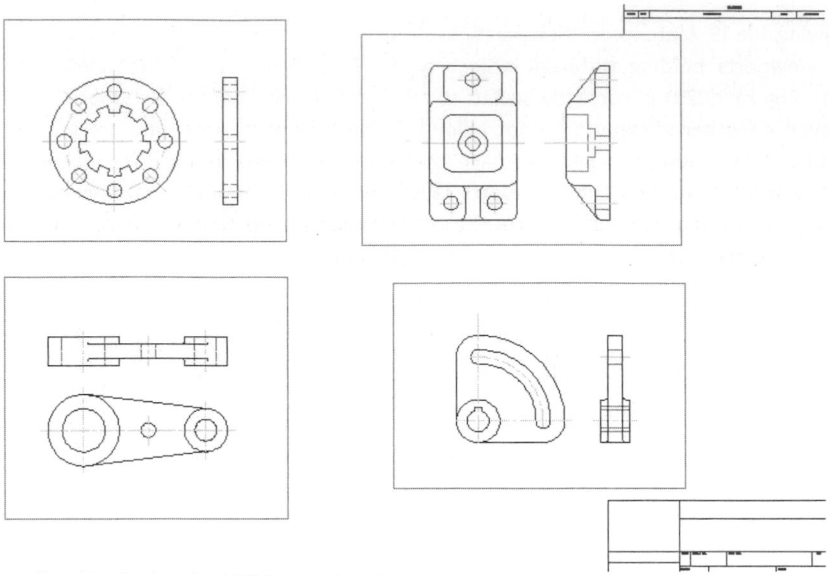

FIGURE 19.30

ANNOTATION SCALE CONCEPTS

When working with multiple viewports set at different scales, it is essential that text, dimensions, and hatching be properly scaled to the objects. In Figure 19.31, the layout on the left is scaled to 1:1 while the layout on the right is scaled to 2:1, or double the original size. Notice how the dimensions, text, and hatching appear smaller on the left than the right. This is the typical problem encountered by individuals utilizing multiple viewports and scales in drawing layouts. One fix for this problem is to create extra layers for the text, dimension, and text objects. Then the objects are created multiple times at multiple scales and then assigned to the extra layers. Depending on the viewport scale, certain objects were frozen while only those scaled properly were kept visible in that viewport. While this process still works, it tends to be very time consuming and confusing. Using Annotative Scales greatly simplifies this process.

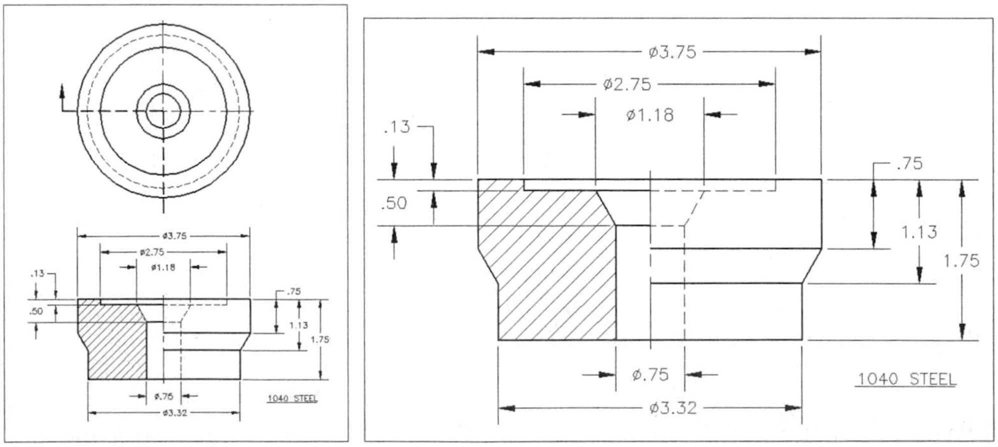

FIGURE 19.31

Figure 19.32 is almost identical to the previous one with the exception that the dimensions, text, and hatching are of the same size even though the scales assigned to the viewports are different. Instead of using the cumbersome layer assignments and object duplication, an Annotative property was assigned. This property performs two basic functions. First, it allows you to automate the process of scaling annotations. This means that when one or more Annotation Scales are assigned to dimensions, text, and hatching in a drawing, they are automatically scaled for you and will appear the correct size in your layouts. The second function of Annotation Scales, is to ensure that only the correctly scaled objects appear in each viewport. If dimensions, text, or hatching is not assigned the Annotation Scale assigned to a viewport; it will not appear in that viewport.

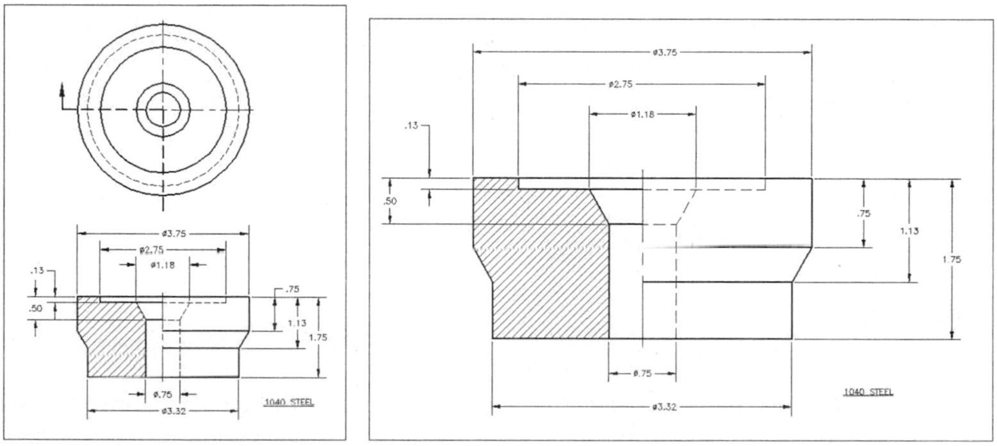

FIGURE 19.32

Figure 19.33 illustrates the methods for making the Annotative property setting for text, dimensions, and hatching. Text and dimensions are assigned this property through their style. Checking the Annotative box in the style dialog box, as shown for text in Figure 19.33 on the left and for dimensions in the middle image, assigns the Annotative property to any objects associated with the style. For assigning the property to hatching, an Annotation button is selected from the Ribbon as the hatching is created, as shown in Figure 19.33 (Right). The Annotative property is also available when creating attributes, blocks, and multileader styles.

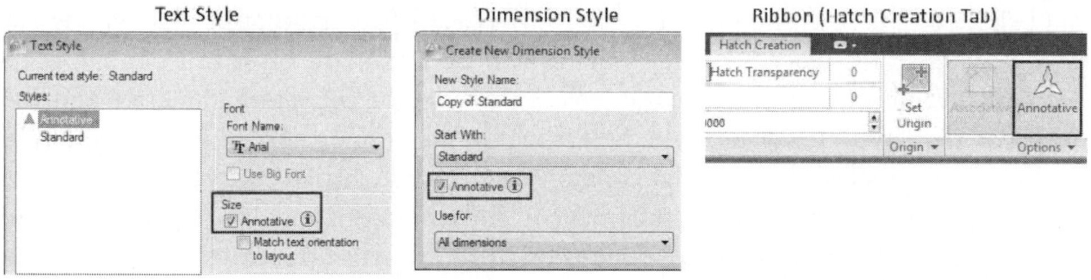

FIGURE 19.33

CREATING AN ANNOTATIVE STYLE

Figure 19.34 illustrates the typical Text Style dialog box with a number of text styles already created. When creating a new text style or modifying an existing text style, place a check in the box next to Annotative, as shown in Figure 19.34.

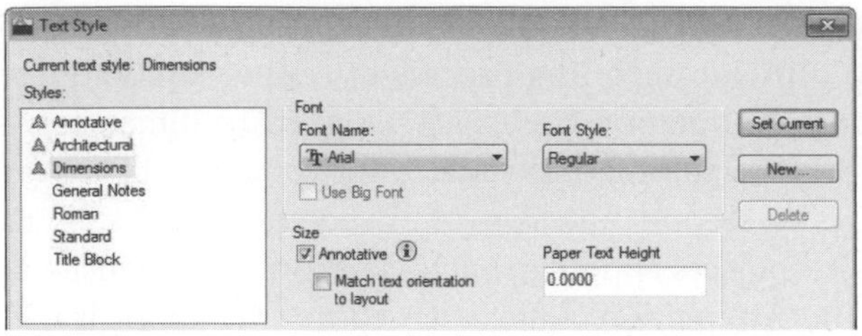

FIGURE 19.34

Annotative styles can be distinguished from traditional styles by a scale icon that appears next to the styles name, as shown in Figure 19.35. In this image, notice the scale icon present in Dimension Style Manager and the Multileader Style Manager.

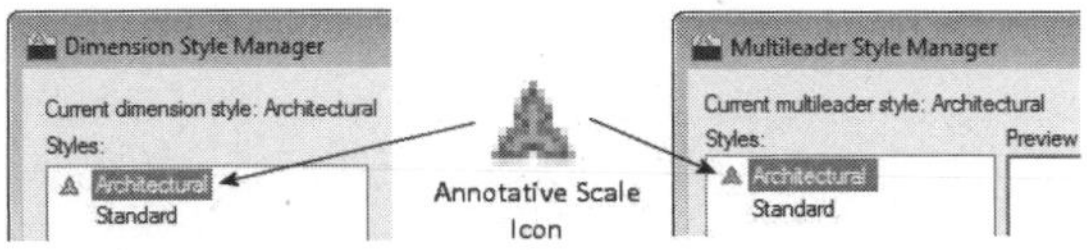

FIGURE 19.35

Figure 19.36 illustrates an mtext object that was created under the control of an Annotative text style. When you hover your cursor over this text object, the Annotative icon appears as shown in Figure 19.36. This provides a quick way of identifying annotative text compared with regular, traditional text.

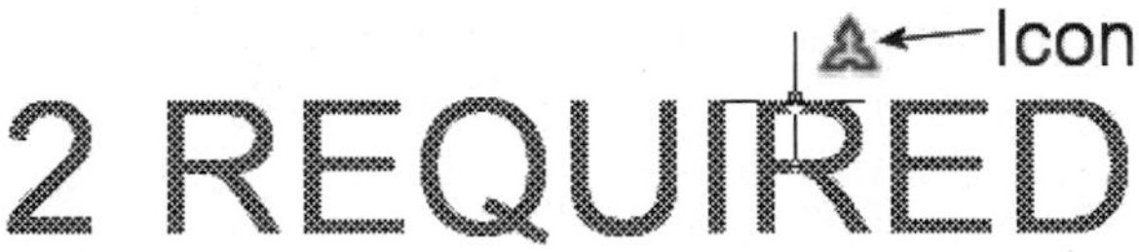

FIGURE 19.36

ANNOTATIVE SCALING TECHNIQUES

While in model space, set the Annotation scale to the appropriate scale for your drawing (1-1/2" = 1'0"), as shown in Figure 19.37 (Left). Any annotative objects, such as text or dimensions, placed in the drawing will be automatically sized per the scale specified. While in a drawing layout, select the edge or activate a viewport and the status bar displays Viewport Scale. Selecting the button, as shown in Figure 19.37 (Right), displays a list of scales that are identical to those displayed for the Annotative scales. Once the scale is selected, the image in the viewport zooms to the size required by the scale and any annotative dimensions or text will appear correctly sized in the layout.

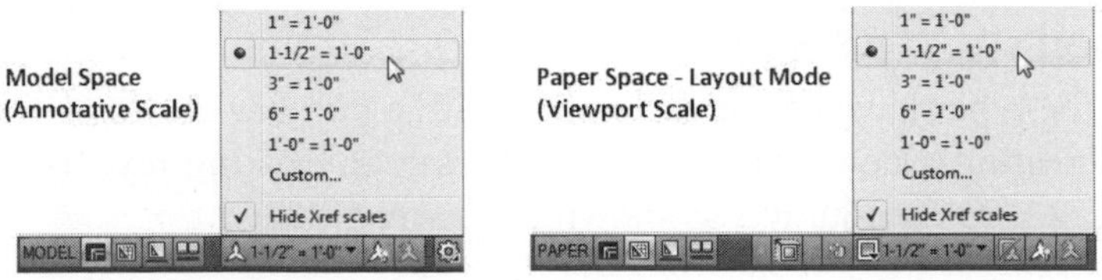

FIGURE 19.37

NOTE

Do not use the Viewports toolbar, as shown in Figure 19.38 (Left), to scale viewports when utilizing annotation scales. Use the status bar Viewport Scale button only, as shown in Figure 19.38 (Right), to ensure proper automatic scaling and viewing of annotative objects.

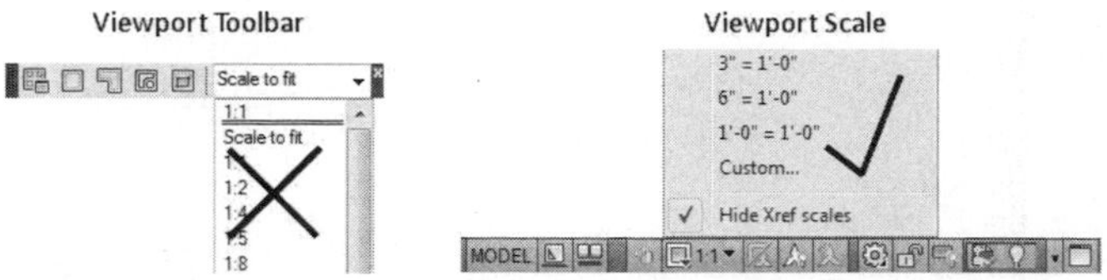

FIGURE 19.38

VIEWING CONTROLS FOR ANNOTATIVE SCALES

Located in the Status bar are two additional buttons used for controlling how annotative scales are viewed in the current viewport. The two buttons are "Annotation Visibility" and "Automatically Add Scales to Annotative Objects when the Annotation Scale Changes."

The Annotation Visibility button, as shown in Figure 19.39, is either on or off. When Annotation Visibility is turned off in a layout, only annotative objects that use or match the current scale display. This means that if you change the Viewport Scale to a different value, the annotation objects will disappear from the screen. This setting prevents you from having to freeze dimensions, text, and hatching in viewports where they would appear at the wrong size. Typically, you will want to leave this setting off to take full advantage of this feature.

When Annotation Visibility is turned on in a layout, all annotative objects that use all scales will display.

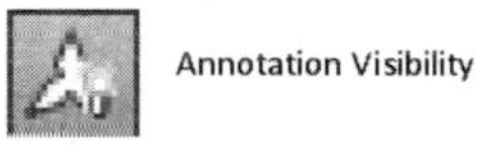

FIGURE 19.39

The second button controls the automatic adding of scales to annotative objects, as shown in Figure 19.40. When this button is turned off, annotative scales are not automatically added to objects inside a viewport. It is interesting to note that objects can have more than one annotative scale assigned to them. This allows the same annotative object to appear at the correct size in multiple viewports. Typically, this feature is best turned off unless numerous scales will be utilized.

When this button is turned on, all annotative objects are automatically updated to include any scales set for a viewport.

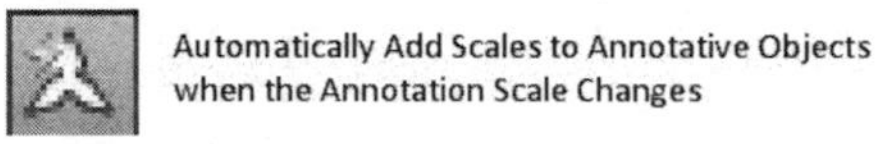

FIGURE 19.40

Figure 19.41 illustrates selected objects that have been assigned multiple scales. A representation of the annotative object is created for each scale. If you hover your cursor over an annotative object, a multiple scales icon is displayed and signifies that the annotative object has multiple scales associated with it.

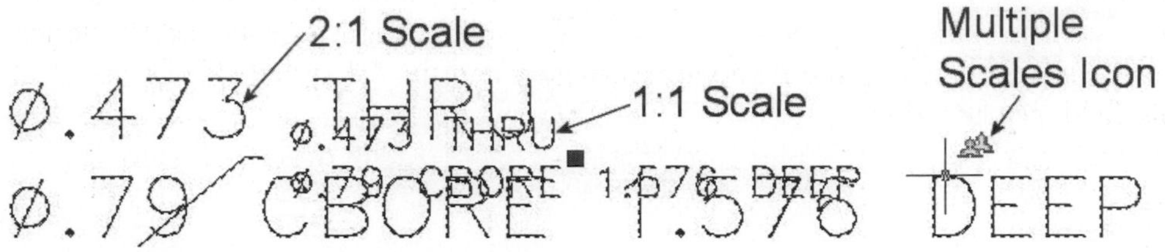

FIGURE 19.41

While there is no limit to the number of scales that can be used to represent annotative objects, too many scales can be difficult to interpret, as shown in Figure 19.42.

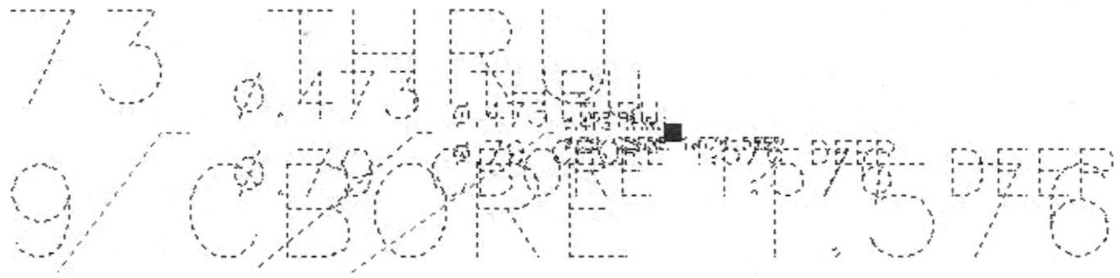

FIGURE 19.42

To add or delete annotation scales from selected objects, first enter Model Space, click the annotation object, and right-click to display the menu, as shown in Figure 19.43 (Left), and then select the Annotation Object Scale item. Clicking Add/Delete Scales… displays the Annotation Object Scale dialog box shown in Figure 19.43 (Right). Click the Add button and you will be provided a list of scales to choose from. To delete a scale, select it from the list shown, such as 2:1, and click the now activated Delete button. Note that this operation adds or deletes the annotation scale for only the selected objects.

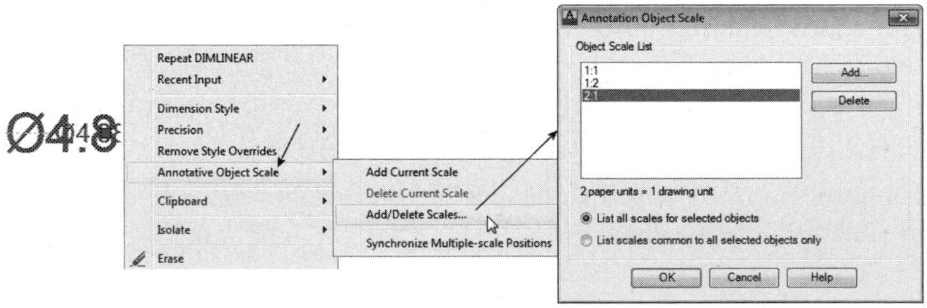

FIGURE 19.43

NOTE

The Annotation Object Scale dialog box can also be accessed by entering `OBJECTSCALE` **at the Command prompt. You will then be prompted to select the annotative objects.**

Yet another method of activating the Annotation Object Scale dialog box is illustrated in Figure 19.44. In this example, an annotative dimension was first selected; then the Properties Palette was displayed, as shown in Figure 19.44 (Left). Notice under the Misc heading the Annotative category and the 2:1 scale field. Clicking on this field displays three dots or ellipsis. Clicking the ellipsis button launches the Annotation Object Scale dialog box. Clicking the Add… button displays the list of scales to add to the selected annotation object.

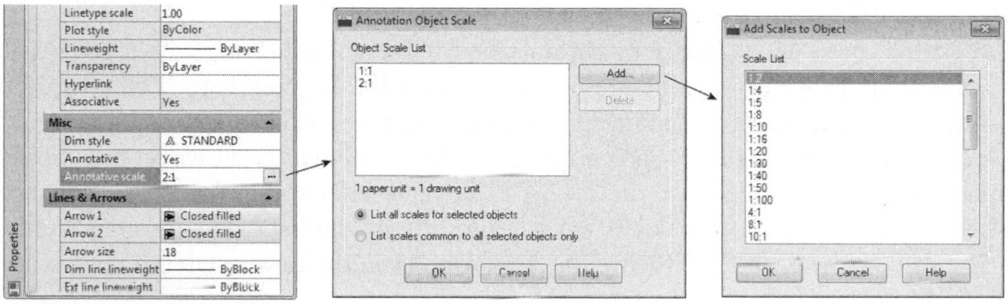

FIGURE 19.44

NOTE

To delete an annotation scale from the list for all annotation objects, enter the `OBJECTSCALE` **command. At the Select Annotation Objects prompt, enter All to select all annotation objects. When the list of scales appears in the Annotation Object Scale dialog box, pick the scale from this list to delete.**

ANNOTATIVE LINETYPE SCALING

The following Try It! exercise demonstrates how Annotative scale controls the scale of linetypes displayed in Model Space. The system variable MSLTSCALE should be set to a value of 1 in order for the linetypes to display properly in Model Space. The default value is 1 but it may be set to 0 in older drawings. The system variable PSLTSCALE should also be set to a value of 1 in order for the linetypes to display properly in Paper Space.

TRY IT!

Open the drawing 19_Land Plat. The drawing opens up in Model Space, as shown in Figure 19.45 (Left). Also, the current value of LTSCALE (Linetype Scale) is 1.00. Since this drawing will be plotted at a scale of 1:30, the land plat outline is too large to view the linetypes. You could zoom in to a segment of the plat; however, even in this magnified view the number of short and long dashes is numerous. This problem, however, is fixed when switching to the Paper Space layout. Here, as shown in Figure 19.45 (Right), the linetypes are properly scaled thanks to the Paper Space linetype scaling function being turn on (PSLTSCALE = 1) by default.

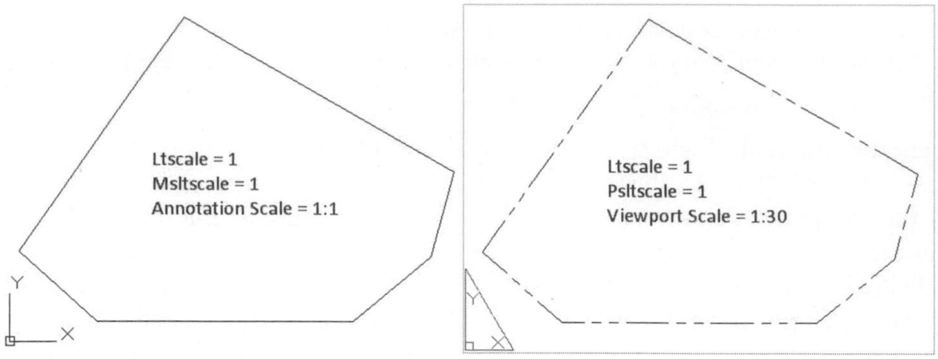

FIGURE 19.45

Switch back to Model Space by clicking the Model tab and change the LTSCALE value from 1.00 to 30.00. The results are displayed in Figure 19.46 (Left) with the linetypes representing the outline of the property being visible. Then click the layout tab and observe the results shown in Figure 19.46 (Right). Even with Paper Space linetype scaling turned on, the LTSCALE value affects the Paper Space image where the linetype scale value is too large to display the linetypes. In other words, you display the linetypes either in the Model tab or in the Layout tab but not both.

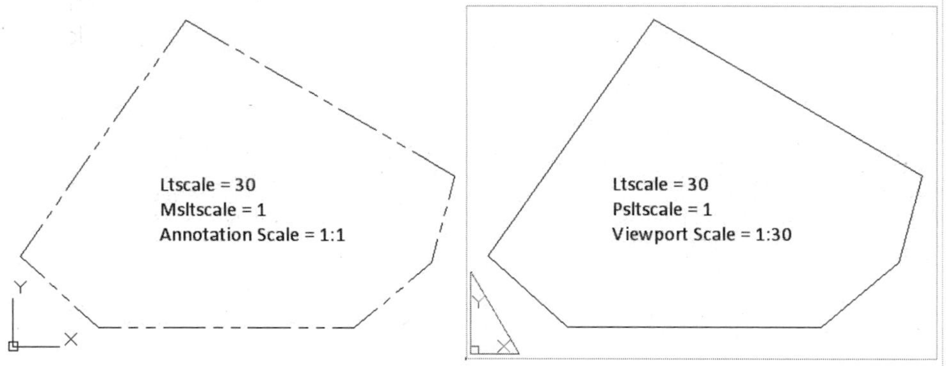

FIGURE 19.46

The solution to this dilemma of the scaling of linetypes is illustrated in Figure 19.47. Return to Model Space by selecting the Model tab and set the LTSCALE back to 1.00. In the Status Bar set the Annotation Scale to 1:30 and perform a REGEN, as illustrated in Figure 19.47 (Left). The linetype scale appears correct thanks to the Model Space linetype scale being turned on (MSLTSCALE = 1). When you are switching to the Layout tab, the linetypes are still visible in this environment thanks to the Paper Space linetype scale being turned on (PSLTSCALE = 1), as illustrated in Figure 19.47 (Right).

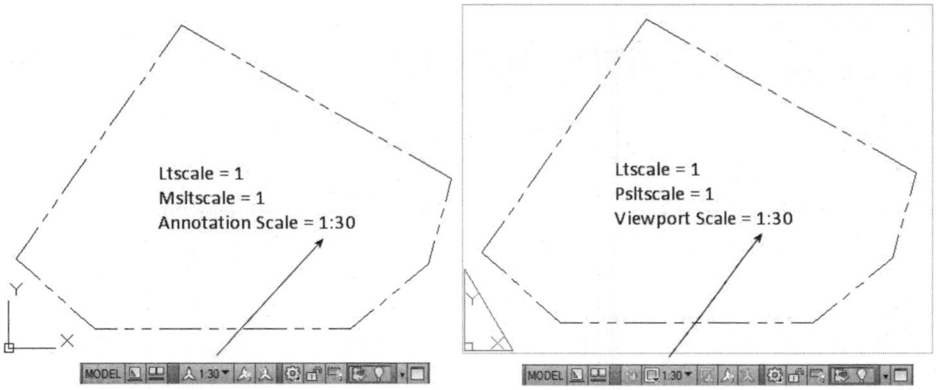

FIGURE 19.47

CREATING AN ANNOTATIVE TEXT STYLE

The following Try It! exercise demonstrates how annotative scales affect text added to a drawing.

TRY IT!

Open the drawing 19_Anno_Duplex. Use the following steps and images for creating an annotative text style and applying it to this drawing.

1. Create a new text style called Room Names, as shown in Figure 19.48. Assign the Arial font and check the box next to Annotative. Apply and close the dialog box.

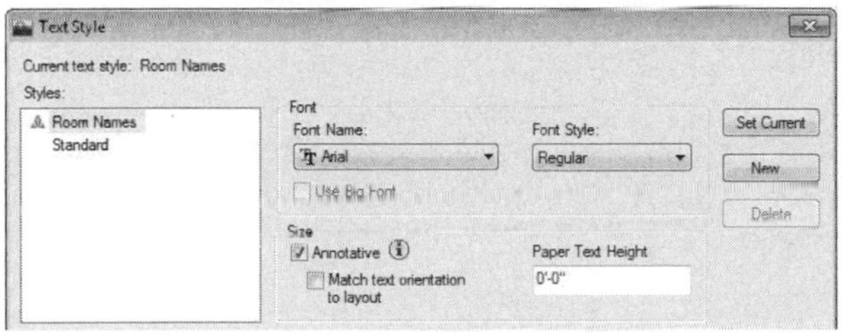

FIGURE 19.48

2. Start the MTEXT command. Notice the mtext height reads 3/16 in the Command prompt, as shown in Figure 19.49. In a full-size drawing, this text will not be readable. Press the ESC key to exit the command.

```
MTEXT Current text style:  "Room Names"  Text height:  3/16"  Annotative:  Yes
Specify first corner:
```

FIGURE 19.49

3. While still in Model Space, change the current Annotation Scale to 1/4"= 1'-0", as shown in Figure 19.50 (Left).

   Reenter the MTEXT command and notice that the height of the text has changed to 9 5/8" based on the current annotation scale. Add the text BEDROOM 1 to the room then close the Text Editor. Hover your cursor over the text and notice the appearance of the scale icon signifying that this text is considered an annotative object, as shown in Figure 19.50 (Right).

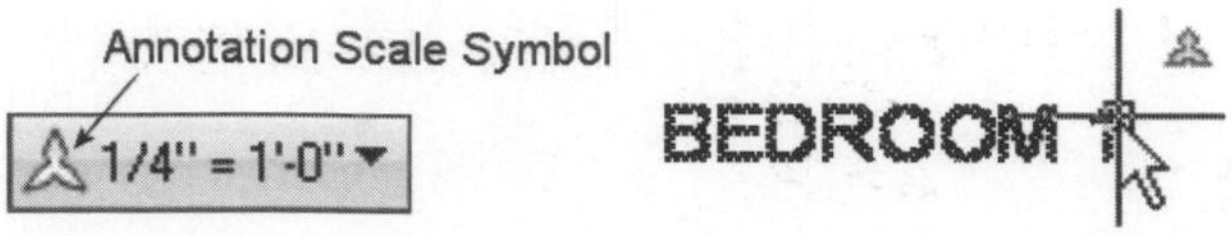

FIGURE 19.50

4. Continue using MTEXT to add names to all of the remaining rooms of the duplex, as shown in Figure 19.51.

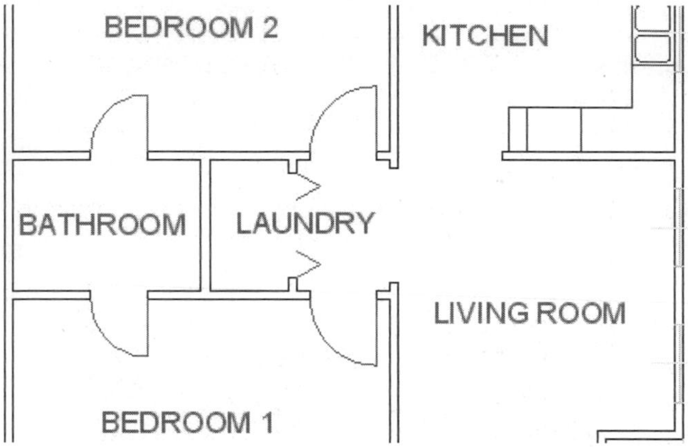

FIGURE 19.51

5. Switch to the B-Size (DWF6) layout. None of the room names display. This is because the correct scale has not been set for this viewport. Click the edge of the viewport and change the Viewport Scale to 1/4"= 1'-0", as shown in Figure 19.52 (Left).

   Notice that the image inside of the viewport changes to reflect the current scale. Also, the text reappears since it matches the original annotative scale of 1/4"= 1'-0", as shown in Figure 19.52 (Right).

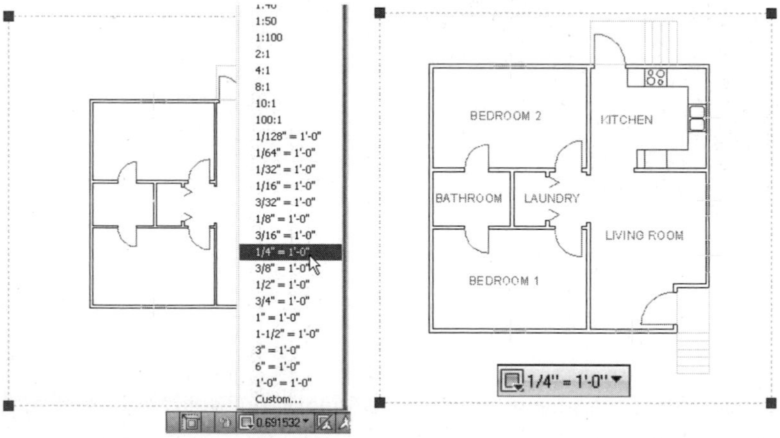

FIGURE 19.52

6. Set the "Automatically Add Scales to Annotative Objects when the Annotation Scale Changes" button to on, as shown in Figure 19.53 (Left).

Click the edge of the viewport and change the scale of the viewport to 3/16"= 1'-0", as shown in Figure 19.53 (Right). Even though the floor plan inside of the viewport is smaller, the size of the text is the same as with the previous scale.

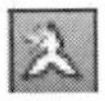

Automatically Add Scales Turned On

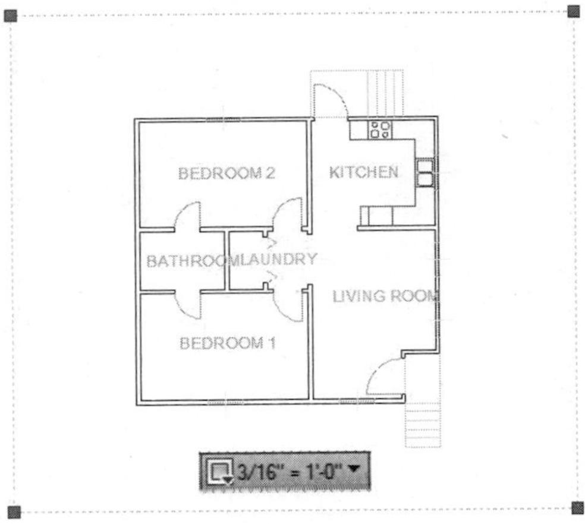

FIGURE 19.53

7. Switch back to Model Space by clicking the Model tab. Hover your cursor over one of the text objects and notice that the icon indicates that multiple scales are applied. Pick the text. Look carefully at the selected text; two text objects actually appear, one smaller than the other, as shown in Figure 19.54. The two text objects reflect the larger and smaller annotative scales that were used on the viewport back in the layout.

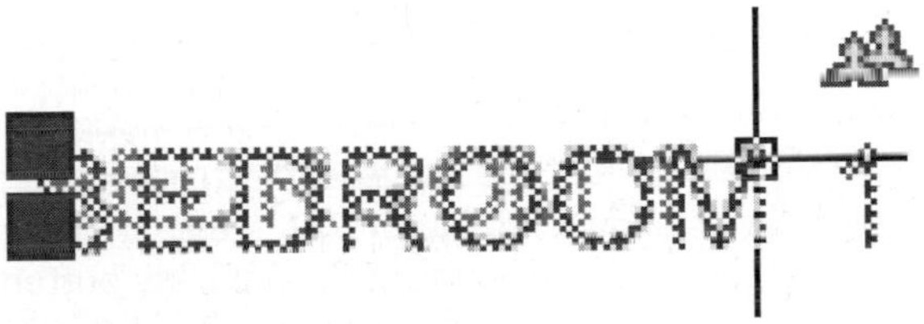

FIGURE 19.54

8. With the text object still selected, activate the Properties Palette and observe the information contained under the Text heading, as shown in Figure 19.55 (Left).

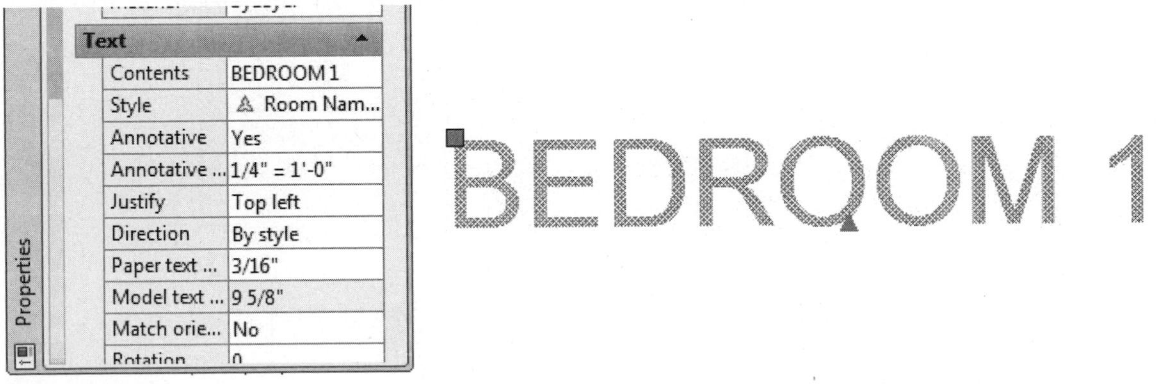

FIGURE 19.55

9. Click inside the Annotation Scale field. When the three dots (ellipsis) appear, click these, as shown in Figure 19.56 (Left). The Annotation Object Scale dialog box is launched, as shown in Figure 19.56 (Right). Notice that the two annotation scales assigned to the text are listed. Click the Cancel button and close the Properties palette when you are finished.

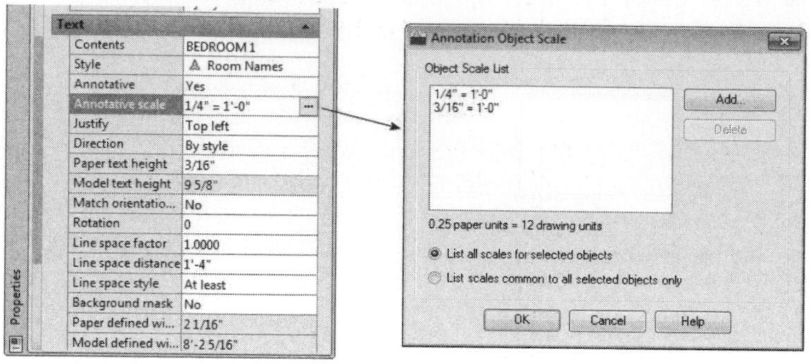

FIGURE 19.56

10. Switch back to the B-Size (DWF6) layout and turn off the "Automatically Add Scales to Annotative Objects when the Annotation Scale Changes" button and verify that the Annotation Visibility button is also turned off, as shown in Figure 19.57. Typically, these buttons should remain off in a layout when working with annotation scales so that annotative objects will appear only in the viewports they are correctly sized for and you will not be continuously creating scales every time you change a viewport scale.

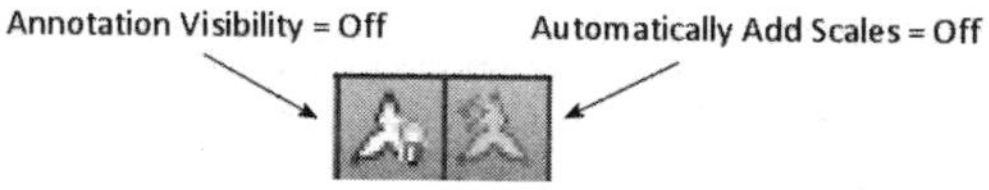

FIGURE 19.57

11. Change the scale of the viewport to 1/2" = 1'-0". The image will increase in size; however, the room names disappear, as shown in Figure 19.58 (Left), because the Annotation Visibility button was turned off. If you turn the Visibility button on, you will see the room names, but they will not appear the correct size because they are not assigned the current viewport scale.

    Change the viewport scale back to 1/4" = 1'-0", as shown in Figure 19.58 (Right).

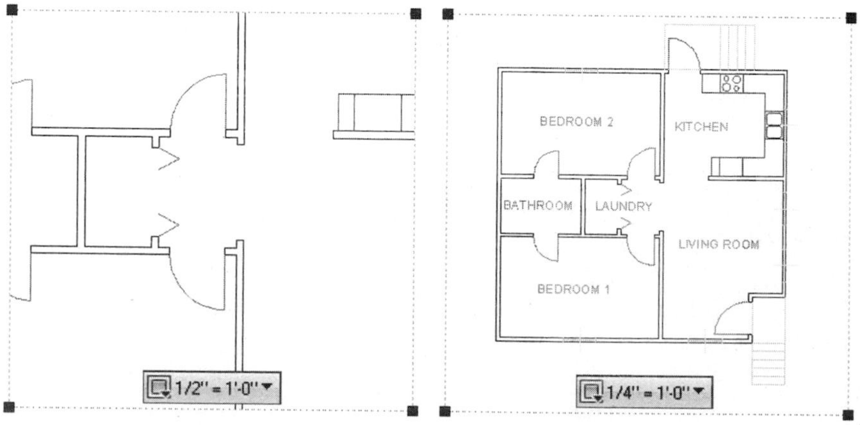

FIGURE 19.58

CREATING AN ANNOTATIVE DIMENSION STYLE

The following Try It! exercise demonstrates the effects that annotative scales have on adding dimensions to a drawing.

TRY IT!

Open the drawing 19_Anno_Dimension. This file picks up from the previous exercise. Two annotative scales are already set for the text objects in this drawing; namely the 1/4"= 1'-0" and 3/16"= 1'-0" scales. Use the following steps and images for creating an annotative dimension style and applying it to this drawing.

1. Create a new dimension style called Arch_Anno, as shown in Figure 19.59. Place a check in the Annotative box, and then click the Continue button.

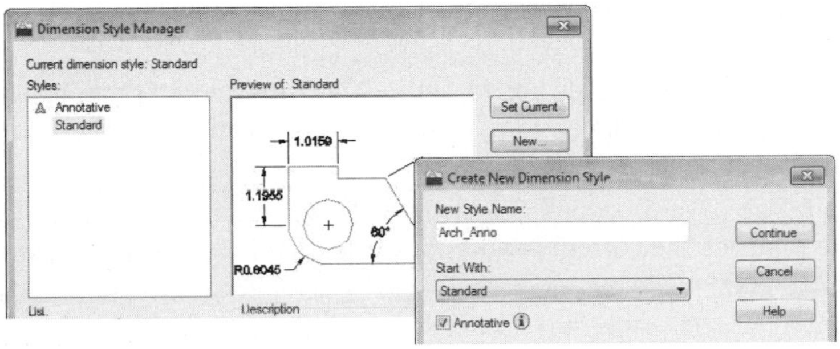

FIGURE 19.59

2. Use the table below for making changes while inside of the Dimension Styles Manager dialog box.

Dimension Styles Dialog Box

Tab	Setting	Change To
Symbols and Arrows	Arrowheads	Architectural Tick
Symbols and Arrows	Arrow Size	1/8"
Text	Text Height	1/8"
Text	Text Placement - Vertical	Above
Text	Text Alignment	Align with dimension line
Primary Units	Unit Format	Architectural
Primary Units	Precision	0'-0"

Verify in the Fit tab that the Annotative box is checked in the Scale for dimension features area, as shown in Figure 19.60. After making changes to the dimension settings, click the OK button.

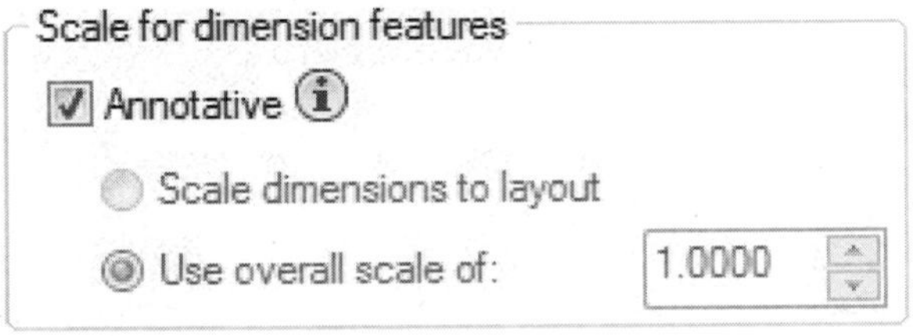

FIGURE 19.60

3. When you return back to the main Dimension Styles Manager dialog box, notice the Arch_Anno dimension style present in the list. Notice also the appearance of the scale icon next to this dimension style name, signifying that all dimensions in this style will be controlled by the annotative scale feature. Click the Close button to exit the Dimension Styles Manager dialog box and continue with this exercise.

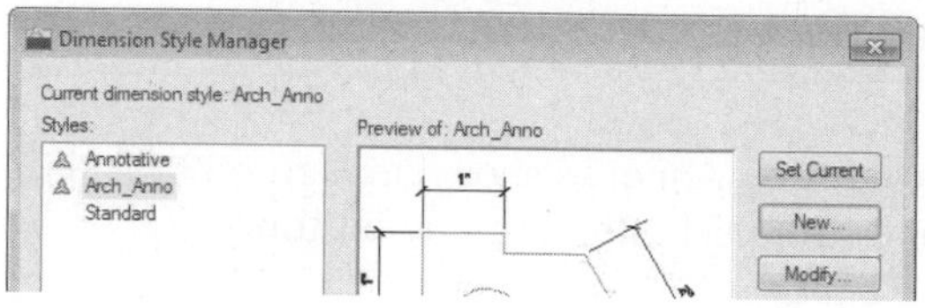

FIGURE 19.61

4. With the Annotation Scale already set to 1/4"= 1'-0" in Model Space, begin placing linear and continue dimensions in the various locations of the floor plan, as shown in Figure 19.62 (Left).

   Switch to the B-Size (DWF6) layout. Select the viewport and observe that the viewport scale matches the current annotation scale of 1/4"= 1'-0", as shown in Figure 19.62 (Right).

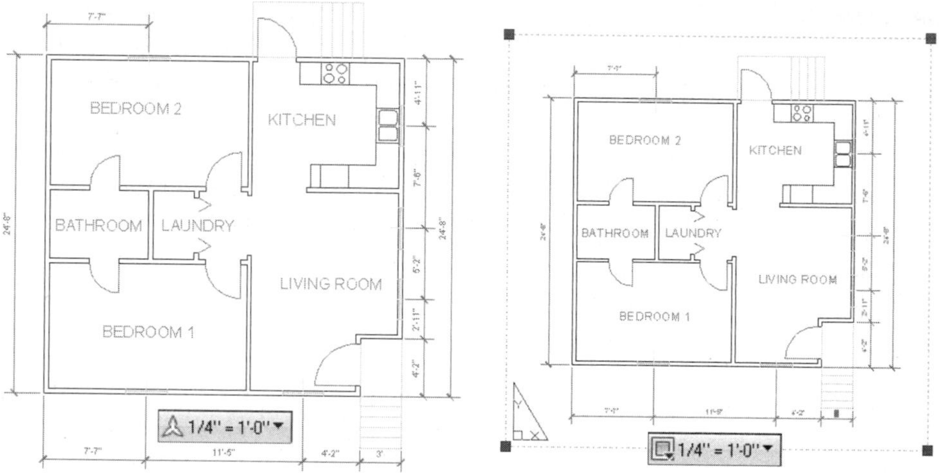

FIGURE 19.62

5. Turn on the Automatically Add Scales button, as shown in Figure 19.63.

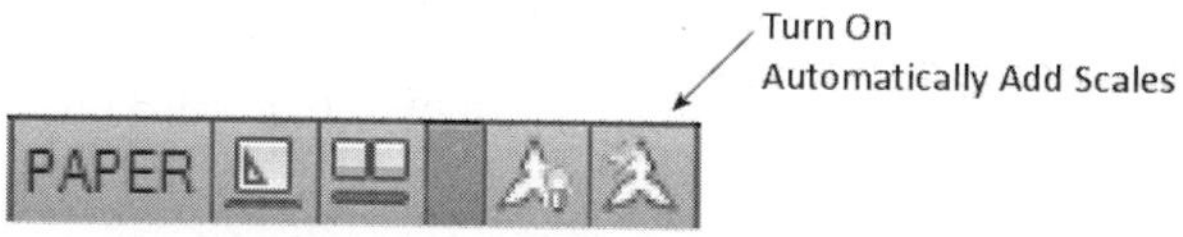

FIGURE 19.63

6. Select the viewport and change the Viewport Scale to 3/16"= 1'-0", as shown in Figure 19.64 (Left). Notice in this image that as the image of the floor plan gets smaller, the dimensions remain their original plotting height of 1/8".

Turn the Automatically Add Scales button back off. Select the viewport and switch the Viewport Scale back to 1/4" = 1'-0", as shown in Figure 19.64 (Right).

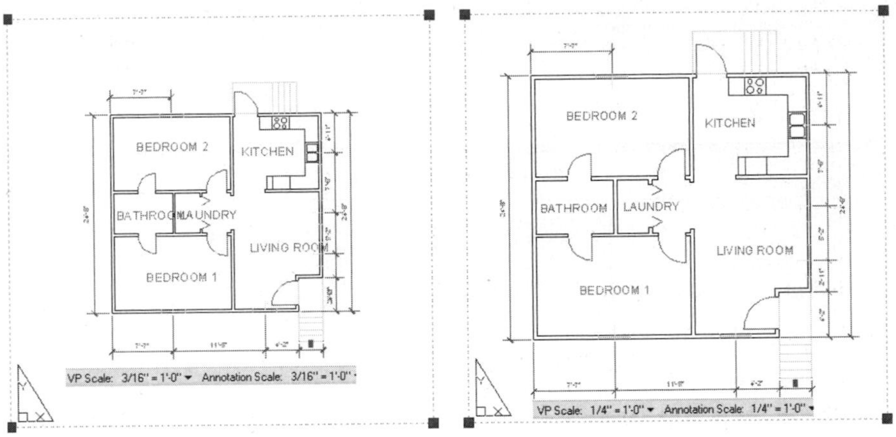

FIGURE 19.64

WORKING WITH ANNOTATIVE HATCHING

The following Try It! exercise demonstrates how annotative scales affect an object that is crosshatched.

TRY IT!

Open the drawing 19_Anno_Hatch. This file picks up from the previous exercise. Two annotative scales are already set for the text and dimensions in this drawing, namely the 1/4" = 1'-0" and 3/16" = 1'-0" scales. Use the following steps and images for working with annotative hatching in this drawing.

1. Activate the HATCH command and select the Annotative button from the Options panel of the Hatch Creation Ribbon, as shown in Figure 19.65.

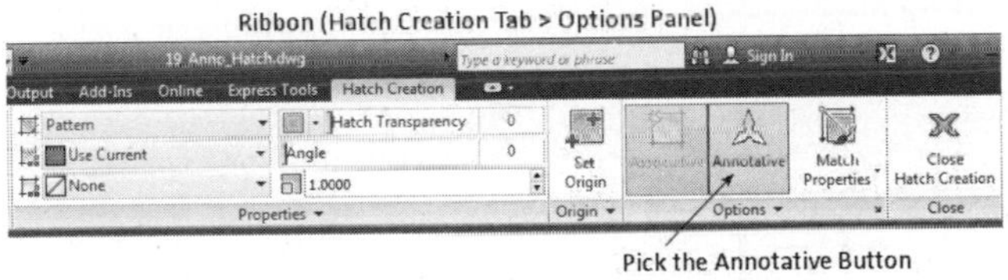

FIGURE 19.65

2. Pick all internal areas that represent the floor plan, as shown in Figure 19.66 (Left). Select the Close Hatch Creation button to place the hatch pattern as shown in Figure 19.66 (Right).

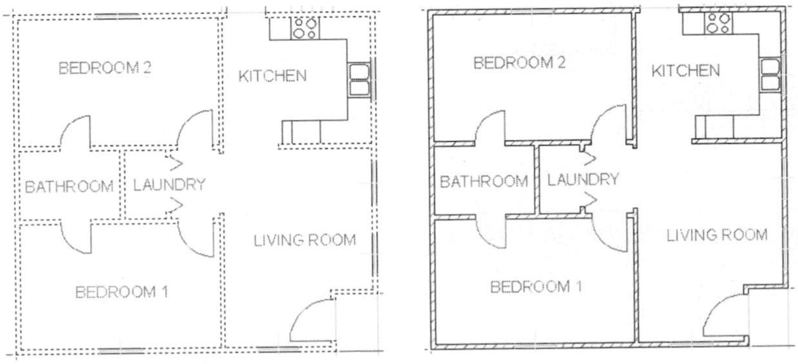

FIGURE 19.66

3. Switch to the B-Size (DWF6) layout and notice that the hatch pattern is visible based on the current Viewport and Annotation scales of 1/4" = 1'-0", as shown in Figure 19.67.

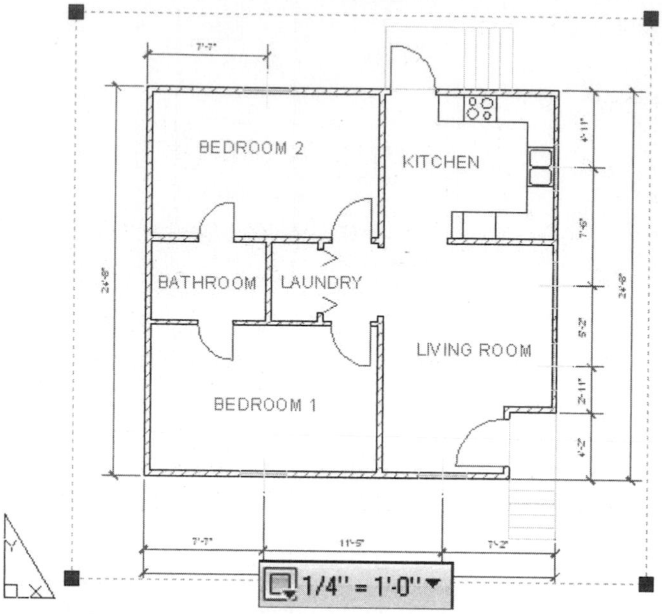

FIGURE 19.67

4. Select the viewport and change the Viewport Scale to 3/16" = 1'-0". Notice that as the image of the floor plan gets smaller, the hatch pattern disappears because the hatch scale does not match the viewport scale. Turn the Annotation Visibility button on to see the hatching but notice that the hatch is incorrectly sized for this viewport scale, as shown in Figure 19.68.

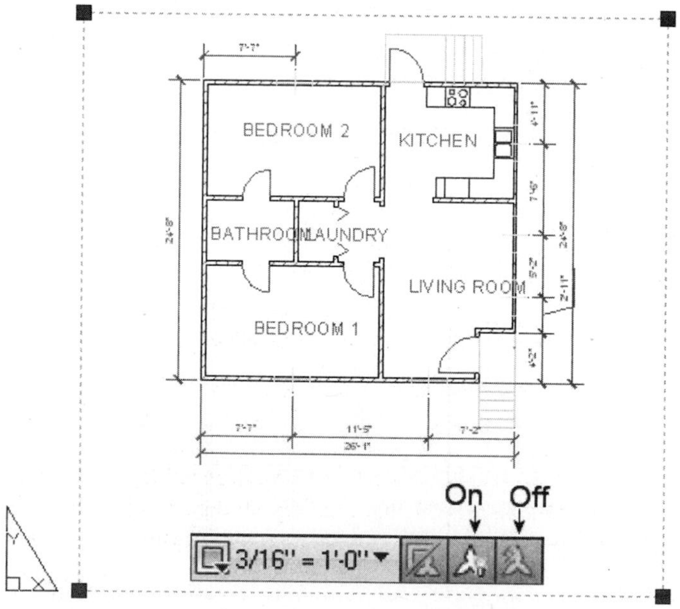

FIGURE 19.68

5. Double-click in the viewport to enter floating model space. Lock the viewport, so the viewport scale doesn't change as you zoom into an area that is hatched. Pick the hatch pattern and then right-click to display the shortcut menu, as shown in Figure 19.69 (Left). Move your cursor over Annotative Object Scale and then pick Add/Delete Scales... to display the Annotative Object Scale dialog box, as shown in Figure 19.69 (Right).

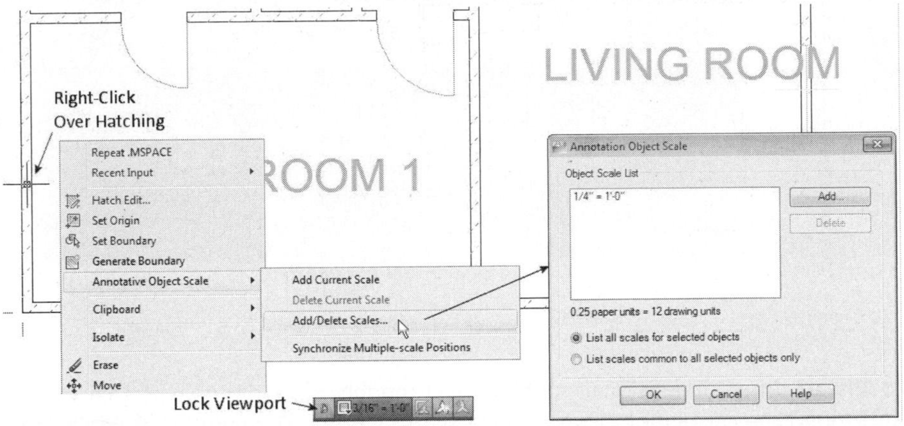

FIGURE 19.69

6. Pick the Add button in the Annotation Object Scale dialog box to display the Add Scales to Object dialog box as shown in Figure 19.70 (Right). Select the 3/16"= 1'-0" scale in the list. As you click OK to dismiss the dialog boxes, notice that the hatching automatically updates to the new added scale. Perform a ZOOM Extents. Turn off viewport locking and Annotation Visibility. Verify that the Automatically Add Scales button is also turned off. Try changing the viewport scales and verify that the hatching, text, and dimensions all display correctly for the scales selected: 1/4"=1'-0" and 3/16"=1'-0".

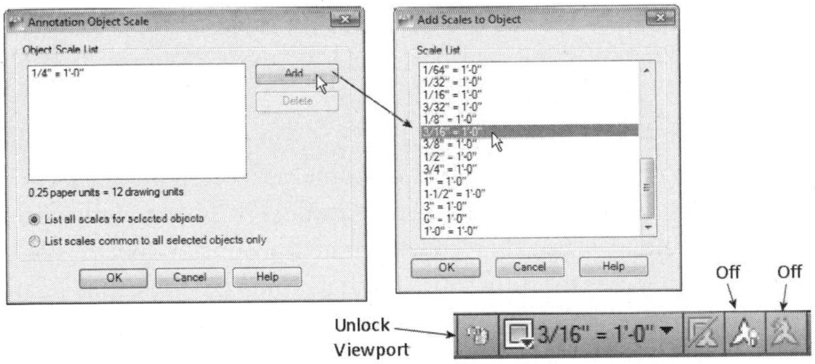

FIGURE 19.70

NOTE

An additional Tutorial Exercise (19_Architectural Details.Dwg) on advanced layout techniques is available in the support files for Chapter 19 found at the Student Companion site from CengageBrain. Refer to the Introduction section of this text for information on how to access these files.

CHAPTER

20

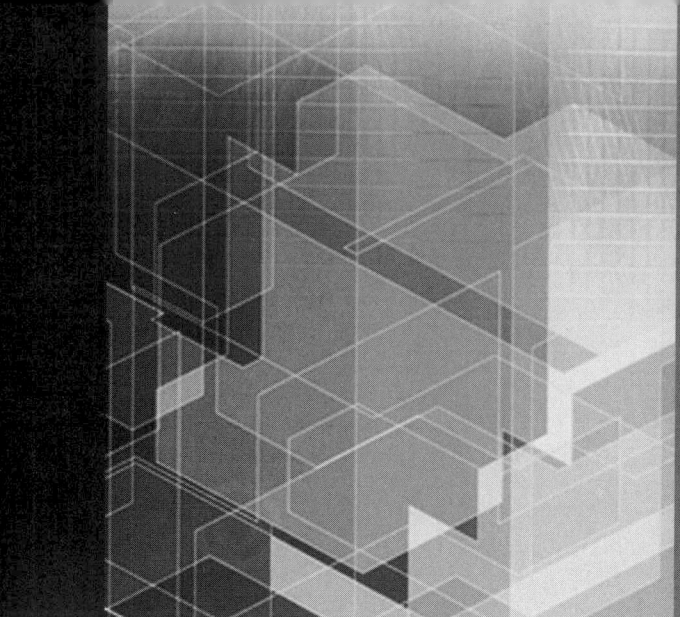

Solid Modeling Fundamentals

Solid models are three-dimensional mathematical models of actual objects that can be analyzed through the calculation of such items as mass properties, center of gravity, surface area, moments of inertia, and much more. Before creating solid models, an understanding of the User Coordinate System (UCS) and 3D visualization tools is helpful. Various options of the `UCS` command will be discussed and 3D viewing tools such as the 3DOrbit, ViewCube, and Steering Wheel will be covered in this chapter. The solid model creation process often starts by defining objects as a series of primitives. Boxes, cones, cylinders, spheres, and wedges are all examples of primitives. These building blocks can then be joined together, subtracted from each other, or combined through an intersection process to create more sophisticated models. Next, we will look at profile-based commands such as `EXTRUDE`, `REVOLVE`, `SWEEP`, and `LOFT`. These commands allow you to create more complicated 3D models from simple 2D shapes or profiles. Finally, we will look at some commands that don't fit neatly into any of the categories just discussed, but are helpful for the creation of solid models: `POLYSOLID`, `PRESSPULL`, `HELIX`, and `MASSPROP`.

WORKSPACES FOR 3D MODELING

As a means of allowing you to work in a dedicated custom, task-orientated environment, predefined workspaces are already created in AutoCAD. These workspaces consist of menus, toolbars, and palettes that are organized around a specific task. When you select a workspace, only those menus, toolbars, and palettes that relate directly to the task are displayed.

The workspaces can be found by picking the Workspace Switching drop-down list next to the Quick Access toolbar, as shown in Figure 20.1 (Left), or from the Status Bar located at the bottom of the display screen, as shown in Figure 20.1 (Right). There are two 3D specific workspaces: 3D Basics and 3D Modeling. The emphasis in this chapter will be on the more robust 3D Modeling workspace. The 3D Basics workspace, while providing common 3D commands, will not display several of the commands and options that we will be discussing.

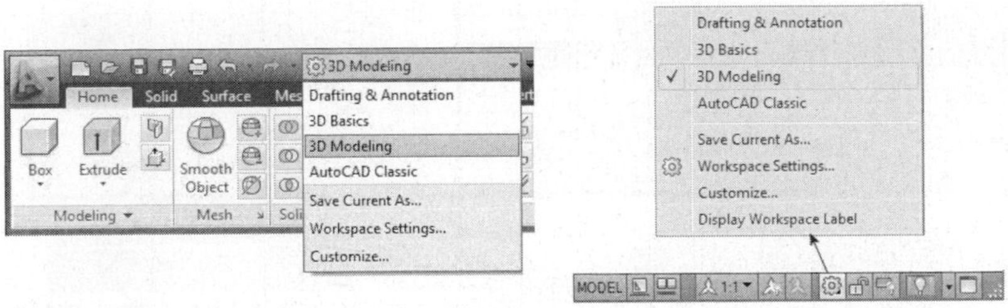

FIGURE 20.1

The 3D Modeling Workspace

Clicking on the 3D Modeling workspace in Figure 20.1 displays a Ribbon that consists of buttons and controls used primarily for 3D modeling, 3D navigation, controlling lights, controlling visual styles, creating and applying materials, and producing renderings, as shown in Figure 20.2. The use of the Ribbon eliminates the need to display numerous toolbars, which tend to clutter up your screen. This enables you to have more screen real estate for constructing your 3D models.

When the Ribbon first displays, the Home tab is active. At this point, the following panels are available based on the 3D Modeling workspace: Modeling, Mesh, Solid Editing, Draw, Modify, Section, Coordinates, View, Selection, Layers, and Groups as shown in Figure 20.2.

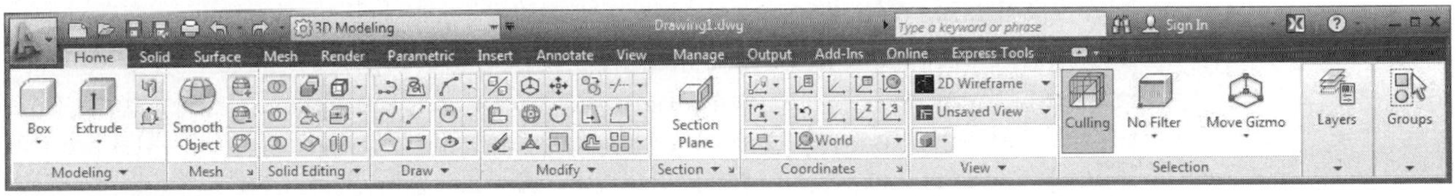

FIGURE 20.2

Additional Tabs of the 3D Modeling Workspace Ribbon

Shown in Figure 20.3 are additional helpful tabs used for working in a 3D environment: the Solid tab for creation of solid models, the Surface and Mesh tabs for creating surface models, and the Render tab for creating realistic images of solid and surface models.

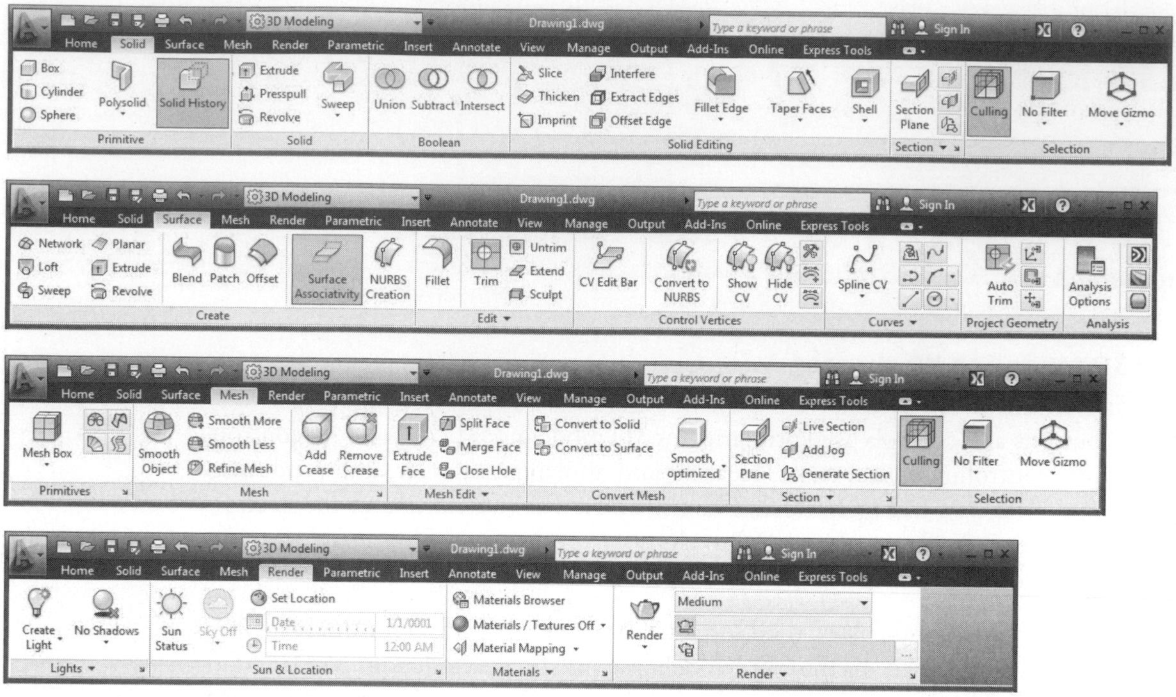

FIGURE 20.3

The 3D Basics Workspace

This workspace, like the 3D Modeling workspace, provides a Ribbon with 3D-related commands. If your modeling projects only require basic 3D operations, this provides an easy-to-use, efficient work environment. Shown in Figure 20.4 are the Ribbon's Home and Render tabs.

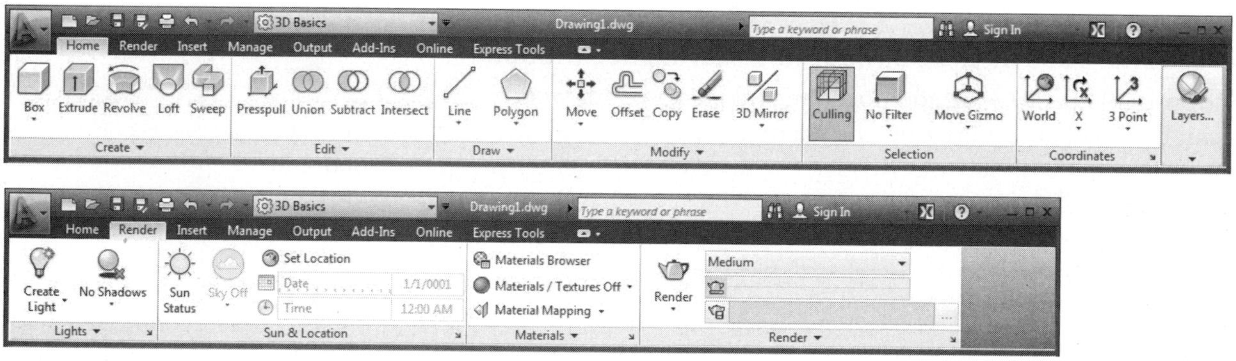

FIGURE 20.4

CREATING A USER COORDINATE SYSTEM (UCS)

Two-dimensional computer-aided design still remains the most popular form of representing drawings for most applications. However, in applications such as architecture and manufacturing, 3D models are becoming increasingly popular for creating rapid prototype models or for creating tool paths from the 3D model. To assist in this creation process, UCS are used to create construction planes where features such as holes and slots are located. In the illustration in Figure 20.5 (Left), a model of a box is displayed along with the UCS icon. The appearance of the UCS icon can change depending on the current visual style the model is displayed in. Visual styles are discussed later in this chapter. For now, the UCS icons illustrated on the right can take on a 2D or 3D appearance. Both UCS icon examples show the positive directions of the three UCS axes.

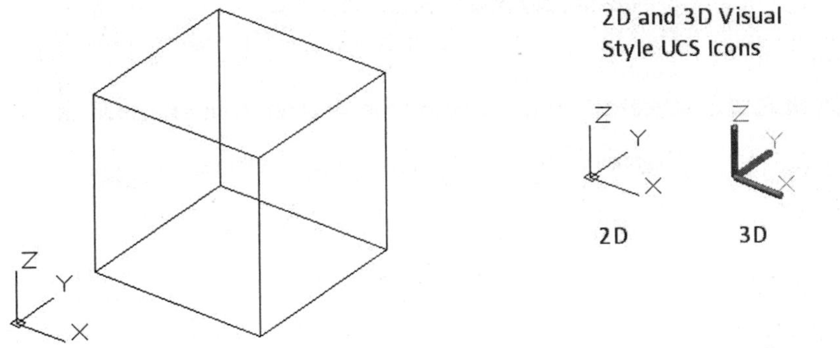

FIGURE 20.5

The process of creating a UCS begins with an understanding of how the UCS command operates. The command line sequence follows, along with available options; the command and options can also be selected from the Coordinates panel of the View Ribbon, as illustrated in Figure 20.6.

```
Command: UCS
Current ucs name: *WORLD*
Specify origin of UCS or [Face/NAmed/OBject/Previous/View/
World/X/Y/Z/ZAxis]
<World>: N (For New)
Specify origin of new UCS or [ZAxis/3point/OBject/Face/
View/X/Y/Z] <0,0,0>:
```

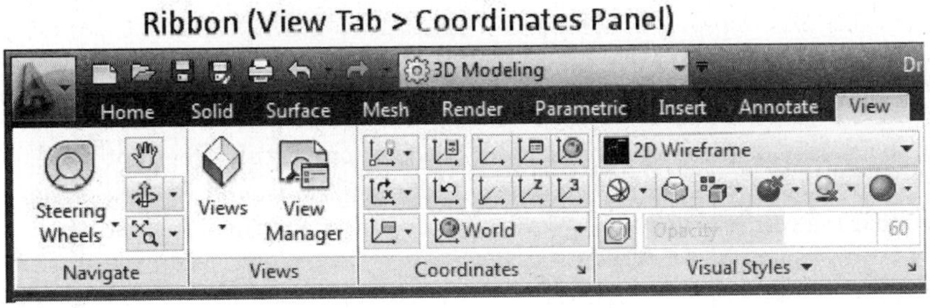

FIGURE 20.6

The following table gives a brief description of each UCS mode.

Button	Tool	Function
	UCS	Activates the UCS command located in the Command prompt area
	World	Switches to the World Coordinate System from any previously defined UCS
	UCS Previous	Sets the UCS icon to the previously defined User Coordinate System
	Face UCS	Creates a UCS based on the selected face of a solid object
	Object	Creates a UCS based on an object selected
	View	Creates a UCS parallel to the current screen display
	Origin	Used to specify a new origin point for the current UCS
	Z Axis Vector	Creates a new UCS based on two points that define the Z axis
	3 Point	Creates a new UCS by picking three points
	X	Used for rotating the current UCS along the X-axis
	Y	Used for rotating the current UCS along the Y-axis
	Z	Used for rotating the current UCS along the Z-axis
	Apply	Sets the current UCS setting to a specific viewport(s)

THE UCS–SPECIFY ORIGIN OF UCS AND ORIGIN OPTION

The default sequence of the `UCS` command defines a new UCS by first defining a new origin (translating—picking a new 0,0,0 position) and second, if needed, changing the direction of the X-axis (rotating the coordinate system).

The illustration in Figure 20.7 (Left) is a sample model with the current coordinate system being the World Coordinate System. Define a new UCS using the default command sequence.

TRY IT!

Open the drawing file 20_Ucs Origin. Activate the UCS command. Identify a new origin point for 0,0,0 at "A," as shown in Figure 20.7 (Center). This should move the UCS icon to the point that you specify. If the icon remains in its previous location and does not move, use the `UCSICON` command with the Origin option to display the icon at its new origin point. To prove that the corner of the box is now 0,0,0, construct a circle at the bottom of the 5″ cube, as shown in Figure 20.7 (Right).

```
Command: UCS
Current ucs name: *WORLD*
Specify origin of UCS or [Face/NAmed/OBject/Previous/View/
World/X/Y/Z/ZAxis]
```

```
<World>: End
of (Select the endpoint of the line at "A")
Specify point on X-axis or <Accept>: (Press ENTER to accept
since you do not want to rotate the coordinate system)
Command: C (For CIRCLE)
Specify center point for circle or [3P/2P/Ttr (tan tan
radius)]: 2.5,2.5,0
Specify radius of circle or [Diameter]: 2
```

Notice a small "box" at the corner of the Coordinate System icon (see the illustration in Figure 20.7 (Left)). This indicates that the World Coordinate System (WCS) is current. Translating or rotating the coordinate system defines a UCS and the small "box" will no longer be displayed. We changed the origin to help locate the circle in the box.

It is also important to note that the circle shown in Figure 20.7 (Right) is a 2D object and can only be drawn in or parallel to the XY plane (for example, the bottom or top of the box). To draw a circle in the side of the box, we will need to not only move (translate) our coordinate system but rotate it as well.

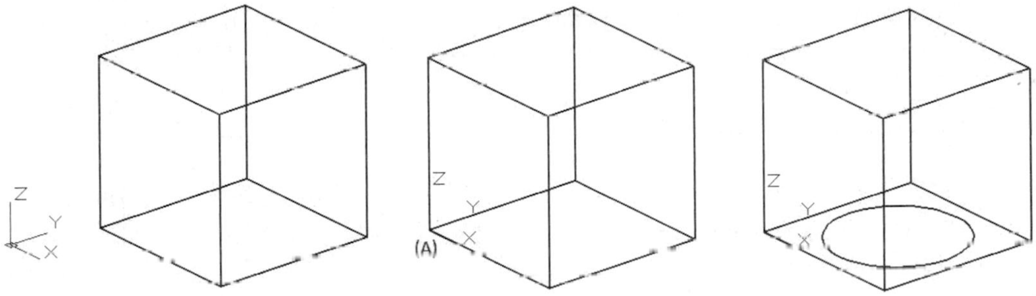

FIGURE 20.7

If you select the UCS command's Origin option through the Ribbon, you are prompted to select a new 0,0,0 position but not to rotate the coordinate system. The Origin option is not listed in the command prompts but can be entered as shown below.

```
Command: UCS
Current ucs name: *WORLD*
Specify origin of UCS or [Face/NAmed/OBject/Previous/View/
World/X/Y/Z/ZAxis]
<World>: O (For Origin)
Specify new origin point <0,0,0>: (Pick desired point)
```

THE UCS–3POINT OPTION

Use the 3point option of the UCS command to specify a new UCS by identifying an origin and new directions of its positive X- and Y-axes (translate and rotate). This option, like the Origin option, is not listed in the command sequence but can be entered anyway. The option is displayed in the Ribbon.

TRY IT!

Open the drawing file 20_Ucs 3p. The illustration in Figure 20.8 (Left) shows a 3D cube in the World Coordinate System. To construct objects on the front panel, first define a new UCS parallel to the front. Use the following command sequence to accomplish this task.

```
Command: UCS
Current ucs name: *WORLD*
Specify origin of UCS or [Face/NAmed/OBject/Previous/View/
World/X/Y/Z/ZAxis]
<World>: 3 (For 3point)
Specify new origin point <0,0,0>: End
of (Select the endpoint of the model at "A" as shown in the
middle of the following image)
Specify point on positive portion of X-axis <>: End
of (Select the endpoint of the model at "B")
Specify point on positive-Y portion of the UCS XY plane <>:
End
of (Select the endpoint of the model at "C")
```

With the Y-axis in the vertical position and the X-axis in the horizontal position, any type of object can be constructed along this plane, such as the polygons shown in Figure 20.8 (Right).

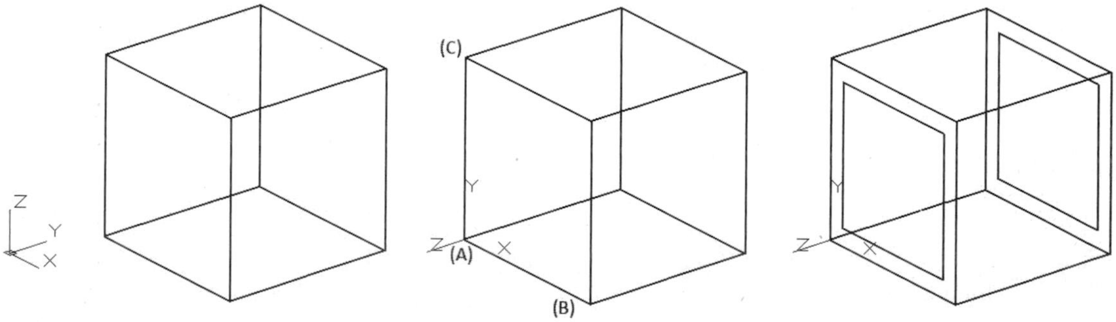

FIGURE 20.8

TRY IT!

Open the drawing file 20_Incline. The 3point method of defining a new UCS is quite useful in the example shown in Figure 20.9, where a UCS needs to be aligned with the inclined plane. Use the intersection at "A" as the origin of the new UCS, the intersection at "B" as the direction of the positive X-axis, and the intersection at "C" as the direction of the positive Y-axis.

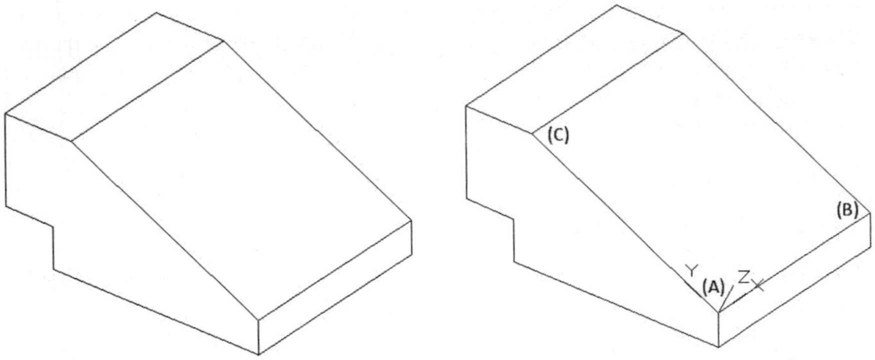

FIGURE 20.9

THE UCS–X/Y/Z ROTATION OPTIONS

Using the X/Y/Z rotation options rotates the current user coordinate around the specific axis. Select the X, Y, or Z option to establish the axis that will act as the pivot; a prompt appears asking for the rotation angle about the pivot axis. The right-hand rule is used to determine the positive direction of rotation around an axis. Think of the right hand gripping the pivot axis with the thumb pointing in the positive X, Y, or Z direction. The curling of the fingers on the right hand determines the positive direction of rotation. Figure 20.10 illustrates positive rotation about each axis. By viewing down the selected axis, the positive rotation is seen to be counterclockwise.

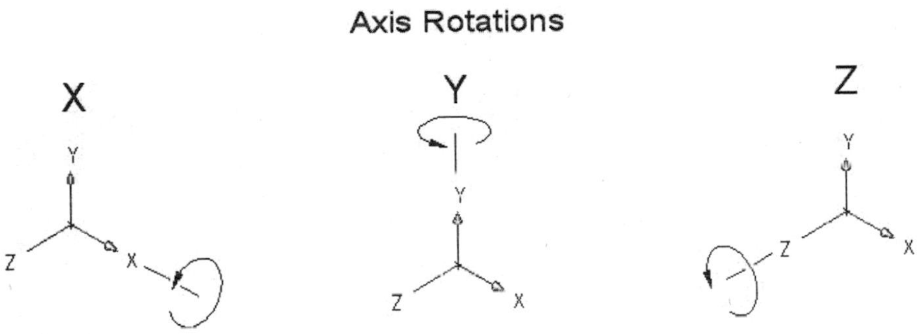

FIGURE 20.10

TRY IT!

Open the drawing file 20_Ucs Rotate. Given the cube shown in Figure 20.11 (Left) in the World Coordinate System, the X option of the UCS **command will be used to stand the icon straight up by entering a 90° rotation value, as in the following prompt sequence.**

```
Command: UCS
Current ucs name: *WORLD*
Specify origin of UCS or [Face/NAmed/OBject/Previous/View/
World/X/Y/Z/ZAxis]
<World>: X
Specify rotation angle about X-axis <90>: (Press ENTER to
accept 90° of rotation)
```

The X-axis is used as the pivot of rotation; entering a value of 90° rotates the icon the desired degrees in the counterclockwise direction, as shown in Figure 20.11 (Right).

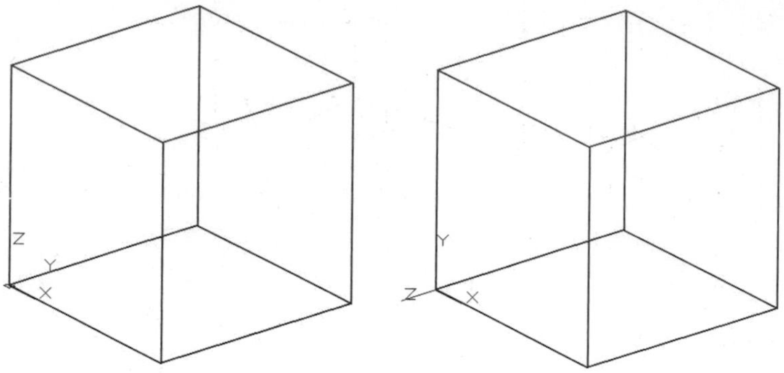

FIGURE 20.11

THE UCS–OBJECT OPTION

Another option for defining a new UCS is to select an object and have the UCS align to that object (translate and rotate).

TRY IT!

Open the drawing file 20_Ucs Object. Given the 3D cube with a circle drawn on the right face shown in Figure 20.12 (Left), use the following command sequence and the illustration on the right to align the coordinate system with the circle on the proper plane.

`Command:` **`UCS`**

`Current ucs name: *WORLD*`

`Specify origin of UCS or [Face/NAmed/OBject/Previous/View/World/X/Y/Z/ZAxis]`

`<World>:` **`OB`** *`(For Object)`*

`Select object to align UCS:` *`(Select the circle shown in the following image on the right)`*

The type of object selected determines the alignment (translation and rotation) of the UCS. In the case of the circle, the center of the circle becomes the origin of the UCS. The point where the circle was selected becomes the point through which the positive X-axis aligns. Other types of objects that can be selected include arcs, dimensions, lines, points, plines, solids, traces, 3DFACE, text, and blocks.

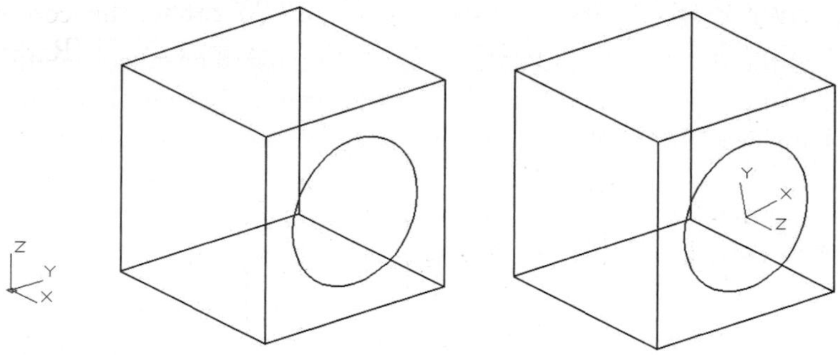

FIGURE 20.12

THE UCS–FACE OPTION

The Face option of the UCS command allows you to establish a new coordinate system aligned to the selected face of a 3D solid, surface, or mesh object.

> **Open the drawing file 20_Ucs Face. Given the current UCS, as shown in Figure 20.13 (Left), and a solid box, follow the command sequence below to align the UCS with the Face option.**

TRY IT!

Command: UCS

Current ucs name: *WORLD*

Specify origin of UCS or [Face/NAmed/OBject/Previous/View/World/X/Y/Z/ZAxis]

<World>: F *(For Face)*

Select face of solid, surface, or mesh: *(move the cursor over the right face of the solid model at point "A," as shown in Figure 20.13 (Center) - the face that will be selected is highlighted. Try moving the cursor over the face from different directions until the UCS appears as shown in the as shown in Figure 20.13 (Right). Pick to place the UCS.)*

Enter an option [Next/Xflip/Yflip] <accept>: *(Press* ENTER *to accept the UCS position)*

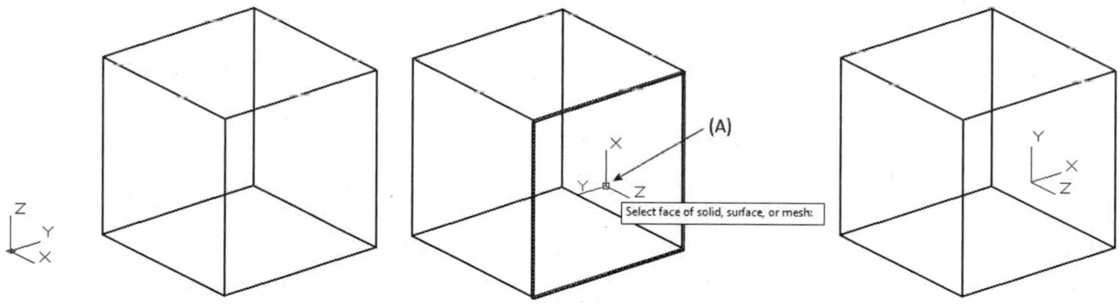

FIGURE 20.13

THE UCS–VIEW OPTION

The View option of the UCS command allows you to establish a new coordinate system where the XY plane is perpendicular to the current screen-viewing direction; in other words, it is parallel to the display screen.

TRY IT!

Open the drawing file 20_Ucs View. Given the current UCS, as shown in Figure 20.14 (Left), follow the prompts below along with the illustration on the right to align the UCS with the View option.

```
Command: UCS
Current ucs name: *WORLD*
Specify origin of UCS or [Face/NAmed/OBject/Previous/View/
World/X/Y/Z/ZAxis]
<World>: V (For View)
```

The results are displayed in Figure 20.14 (Right), with the UCS aligned parallel to the display screen.

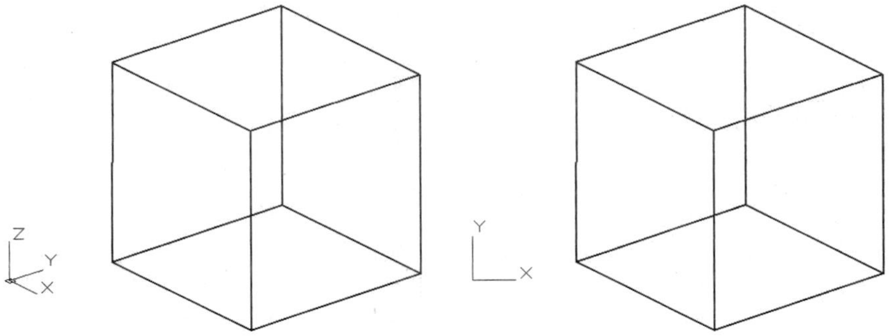

FIGURE 20.14

USING DYNAMIC UCS MODE

While inside a command, such as the UCS command, you can automatically switch the plane of the UCS by simply hovering your cursor over the face of a 3D solid object. This special function is available when the DUCS (Dynamic UCS) button is turned on in the Status Bar, as shown in Figure 20.15. The next Try It! exercise illustrates how this method of manipulating the UCS dynamically is accomplished.

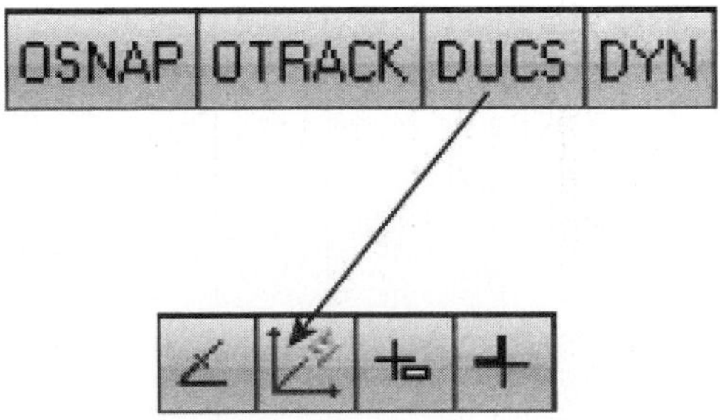

FIGURE 20.15

TRY IT!

Open the drawing file 20_Dynamic_UCS and verify the Dynamic UCS button is activated. Given the current UCS, as shown in Figure 20.16 (Left), follow the prompts below, along with the illustrations, to dynamically align the UCS to a certain face and location.

In this first example of dynamically setting the UCS, hover your cursor along the front face of the object until it highlights, as illustrated in Figure 20.16 (Left), and pick the endpoint at "A" to locate the UCS as shown on the right in Figure 20.16.

Command: **UCS**

Current ucs name: *WORLD*

Specify origin of UCS or [Face/NAmed/OBject/Previous/View/World/X/Y/Z/ZAxis]

<World>: *(Move the Dynamic UCS icon over the front face, as shown in Figure 20.16 (Left). Then pick the endpoint at "A")*

Specify point on X-axis or <Accept>: *(Press* ENTER *to accept)*

With the new UCS defined, it is good practice to save the position of the UCS under a unique name. These named UCS can then be easily retrieved for later use.

Command: **UCS**

Current ucs name: *NO NAME*

Specify origin of UCS or [Face/NAmed/OBject/Previous/View/World/X/Y/Z/ZAxis]

<World>: **NA** *(For NAmed)*

Enter an option [Restore/Save/Delete/?]: **S** *(For Save)*

Enter name to save current UCS or [?]: **Front**

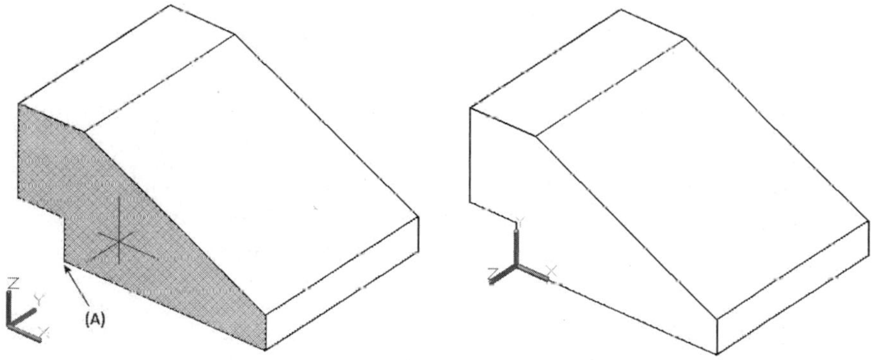

FIGURE 20.16

This next example requires you to pick the endpoint to better define the X-axis while dynamically locating the UCS. Hover your cursor along the top face of the object illustrated in Figure 20.17 (Left) and pick the endpoint at "A" to locate origin of the UCS. Continue by picking the endpoint at "B" as the X-axis. Save this UCS as "Top."

```
Command: UCS
Current ucs name: Front
Specify origin of UCS or [Face/NAmed/OBject/Previous/View/World/X/Y/Z/ZAxis]
<World>:
(Move the Dynamic UCS icon over the top face as shown in Figure 20.17 (Left). Then pick the endpoint at "A")
Specify point on the X-axis or <Accept>: (Pick the endpoint at "B" to align the X-axis)
Specify point on the XY plane or <Accept>: (Press ENTER to accept)
Command: UCS
Current ucs name: *NO NAME*
Specify origin of UCS or [Face/NAmed/OBject/Previous/View/World/X/Y/Z/ZAxis]
<World>: NA (For NAmed)
Enter an option [Restore/Save/Delete/?]: S (For Save)
Enter name to save current UCS or [?]: Top
```

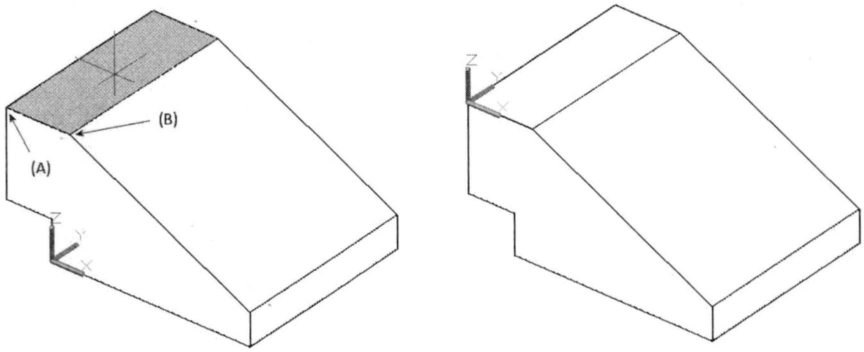

FIGURE 20.17

Next, hover your cursor along the side face of the object illustrated in Figure 20.18 (Left) and pick the endpoint at "A" to locate the origin of the UCS. Continue by picking the endpoint at "B" as the X-axis and the endpoint at "C" to define the XY plane. Save this UCS as "Side."

Command: **UCS**

Current ucs name: Top

Specify origin of UCS or [Face/NAmed/OBject/Previous/View/World/X/Y/Z/ZAxis]

<World>: *(Move the Dynamic UCS icon over the side face as shown in Figure 20.18 (Left). Then pick the endpoint at "A")*

Specify point on X-axis or <Accept>: *(Pick the endpoint at "B" to align the X-axis)*

Specify point on the XY plane or <Accept>: *(Pick the endpoint at "C" to align the XY plane)*

Command: **UCS**

Current ucs name: *NO NAME*

Specify origin of UCS or [Face/NAmed/OBject/Previous/View/World/X/Y/Z/ZAxis]

<World>: **NA** *(For NAmed)*

Enter an option [Restore/Save/Delete/?]: **S** *(For Save)*

Enter name to save current UCS or [?]: **Side**

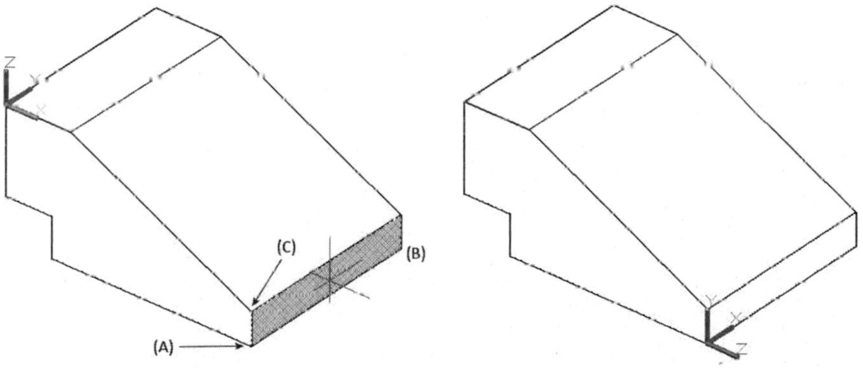

FIGURE 20.18

Finally, hover your cursor along the inclined face of the object illustrated in Figure 20.19 (Left) and pick the endpoint at "A" to locate the origin of the UCS. Continue by picking the endpoint at "B" as the X-axis and the endpoint at "C" to define the XY plane. Save this UCS as "Auxiliary."

Command: **UCS**

Current ucs name: Side

Specify origin of UCS or [Face/NAmed/OBject/Previous/View/World/X/Y/Z/ZAxis]

<World>: *(Move the Dynamic UCS icon over the inclined face as shown in Figure 20.19 (Left). Then pick the endpoint at "A")*

Specify point on X-axis or <Accept>: *(Pick the endpoint at "B" to align the X-axis)*

Specify point on the XY plane or <Accept>: *(Pick the endpoint at "C" to align the XY plane)*

Command: **UCS**

Current ucs name: *NO NAME*

Specify origin of UCS or [Face/NAmed/OBject/Previous/View/World/X/Y/Z/ZAxis]

<World>: **NA** *(For NAmed)*

Enter an option [Restore/Save/Delete/?]: **S** *(For Save)*

Enter name to save current UCS or [?]: **Auxiliary**

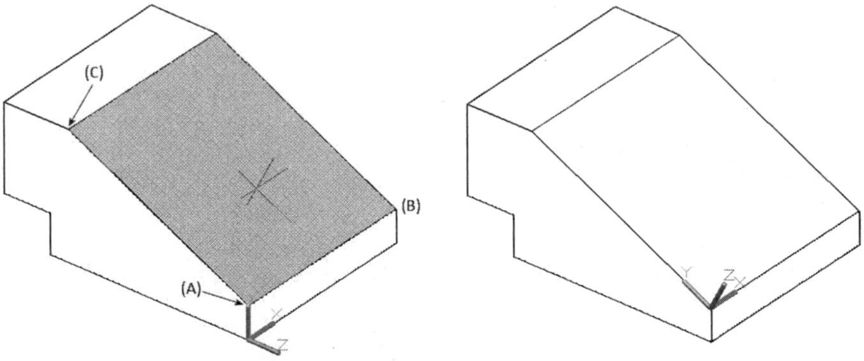

FIGURE 20.19

USING THE UCS DIALOG BOX

As stated earlier, considerable drawing time can be saved by assigning a name to a UCS. Once numerous UCS have been defined in a drawing, using their names instead of re-creating each coordinate system easily restores them. You can accomplish this on the command line by using the Save and Restore options of the UCS command. Another method of retrieving previously saved UCS is to choose Named UCS from the Coordinates panel of the Ribbon illustrated in Figure 20.20 (Left) (the UCSMAN command). This displays the UCS dialog box, as shown in the following image in the middle, with the Named UCSs tab selected. All UCS previously defined in the drawing are listed here. To make one of these coordinate systems current, highlight the desired UCS name and select the Set Current button. A named UCS can also be made current by simply double-clicking it. To define (save) a coordinate system, you must have first translated and rotated the UCS into a desired new position; then use the dialog box to select the "Unnamed" UCS and rename it. The

UCS dialog box provides a quick method of saving and restoring previously defined coordinate systems without entering them at the keyboard.

Clicking on the Orthographic UCSs tab of the UCS dialog box allows you to rotate the UCS so that the XY plane is parallel and oriented to one of the six orthographic views: Front, Top, Back, Right Side, Left Side, and Bottom, as shown in Figure 20.20 (Right).

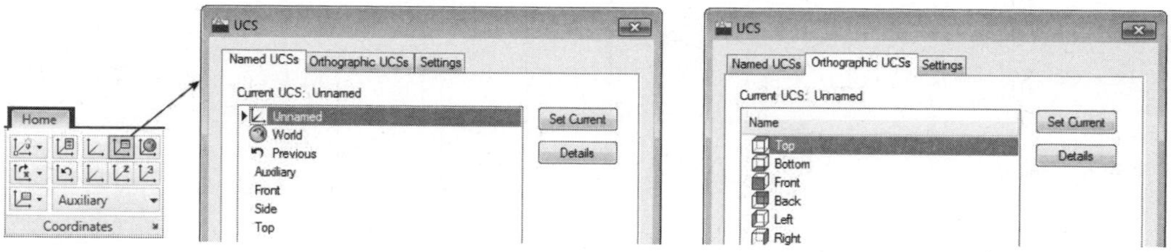

FIGURE 20.20

NOTE

Two additional methods for quickly changing from one named UCS to another are illustrated in Figure 20.21. The first is found by expanding the drop-down list found in the Coordinates panel of the Ribbon, as shown in Figure 20.21 (Left). The second involves expanding a drop-down list located under the ViewCube, as shown in Figure 20.21 (Right). Selecting one of the named UCSs from either of these lists will immediately restore it.

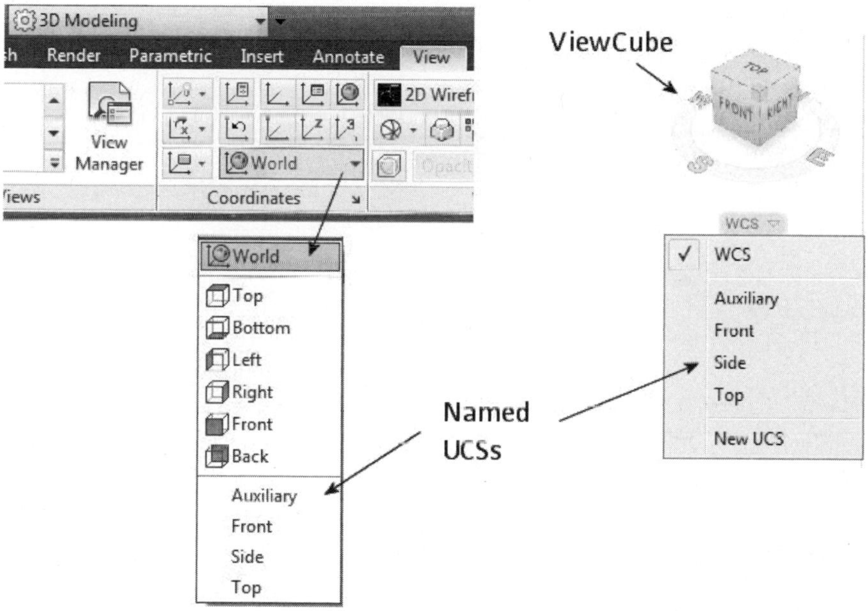

FIGURE 20.21

CONTROLLING THE DISPLAY OF THE UCS ICON

The UCS dialog box (the UCSMAN command) can also be used to control the display of the UCS icon. Selecting the Settings tab, as shown in Figure 20.22, allows you to turn the icon on or off and choose whether it will be displayed at the origin or not. By default, the icon is turned on. If you want to turn it off, remove the check from the

"On" box. Generally, when working in 3D, you will want to have the icon on. If you remove the check from the "Display at UCS origin point" box, the icon will move to the lower-left corner of your screen. This can be useful if the icon is interfering with the view of your model. Return the check to again display the icon at the current origin (0,0,0). It should be noted that the icon will remain in the corner if it can't be fully displayed at the origin because that location is not on the screen.

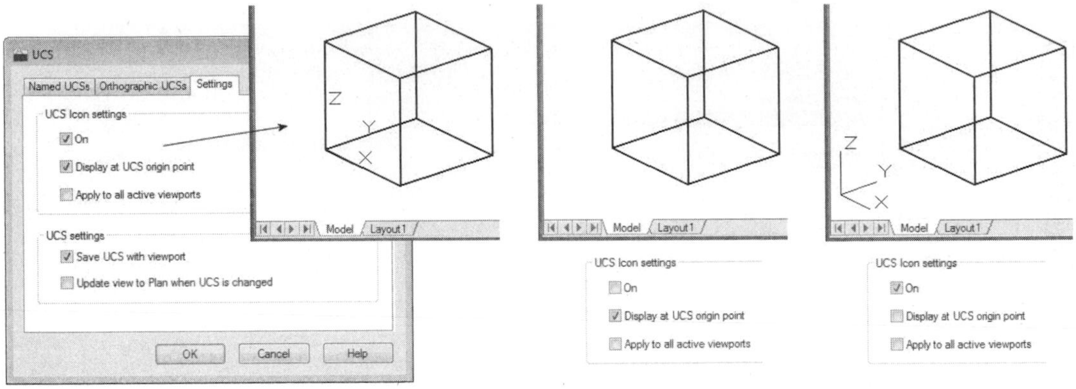

FIGURE 20.22

NOTE

The UCSICON command can also be used to turn the coordinate system icon on or off and determine whether it will be displayed at the origin or always in the lower-left corner (the Noorigin option). This command can be accessed from the Coordinates panel of the Ribbon as shown in Figure 20.23.

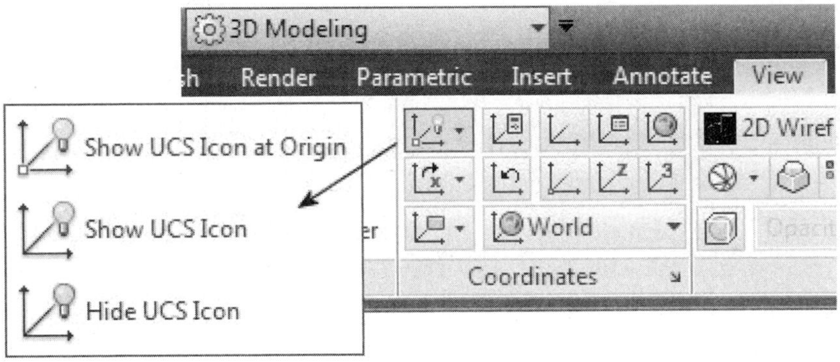

FIGURE 20.23

MODEL SPACE (TILED) VIEWPORTS FOR 3D

When you design in 3D space, it is sometimes helpful to see your model from several different orientations at the same time. The VPORTS command, which was used in Chapter 19 to create floating viewports in layouts, can also be used to create tiled viewports in Model Space. Although tiled viewports are not as powerful as floating viewports, they do allow you to create different views and User Coordinate Systems in each viewport. The Viewports dialog box has a 3D setup option, which makes the creation of tiled viewports simple for 3D applications.

TRY IT!

Open the drawing file 20_Tiled Vports. When the 3D solid displays, activate the VPORTS command by picking the Named button on the Model Viewports panel of the View Ribbon, as shown in Figure 20.24. When the Viewports dialog box displays, select the New Viewports tab, select "Four: Equal" as the viewport configuration, and "3D" as the Setup as shown in the following image on the right. A preview of the configuration is shown in the dialog box.

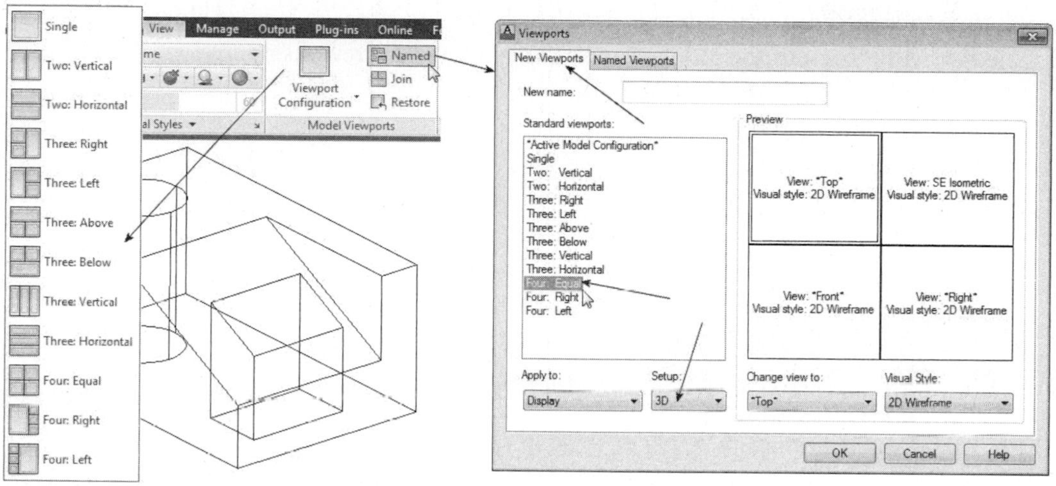

FIGURE 20.24

Clicking OK in the Viewports dialog box returns you to the screen and four tiled viewports are displayed as shown in Figure 20.25. Notice that each viewport has a different view and a different UCS. Click inside a viewport to make it active and you are ready to perform any draw or modification operation desired. Return to a single viewport by returning to the Viewports dialog box and selecting a Single viewport configuration or select Single from the View Configuration drop-down list in the Model Viewports panel of the Ribbon. The active viewport will fill the screen area.

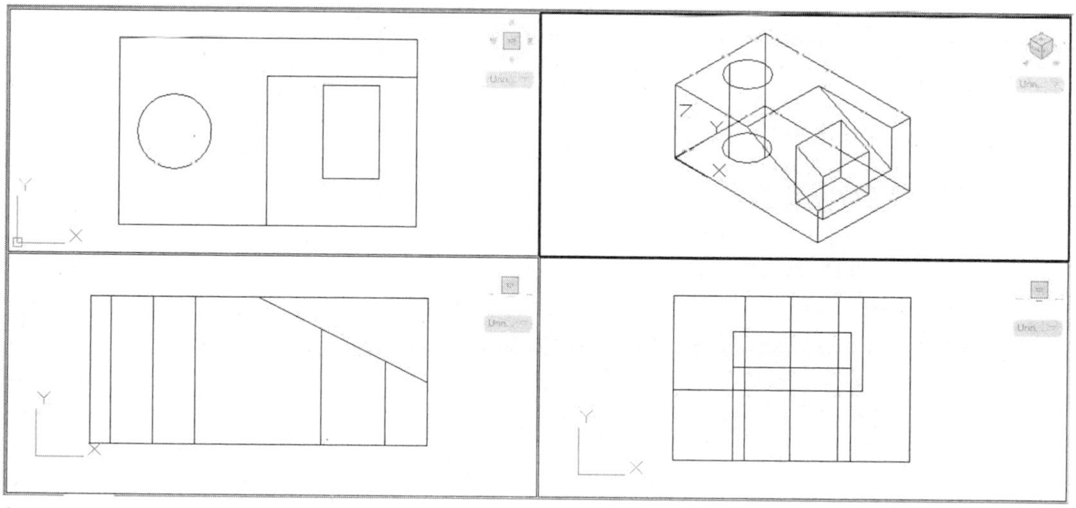

FIGURE 20.25

THE PLAN COMMAND

When working in 2D, you are creating objects on the XY plane, and you are viewing those objects as if looking down the Z-axis. It can sometimes be helpful to see your 3D model from this same 2D viewpoint. This is the purpose of the `PLAN` command. For example, illustrated in Figure 20.26 (Left) is a 3D model in which the X- and Y-axes are positioned along the front face of the object. Activating the `PLAN` command and accepting the default value of <Current> displays the model as shown in Figure 20.26 (Right). This gives you a 2D view of the solid model, which can be used to better see how certain features such as holes and slots are located along a 2D plane. To switch back to the 3D view, perform a ZOOM-Previous operation.

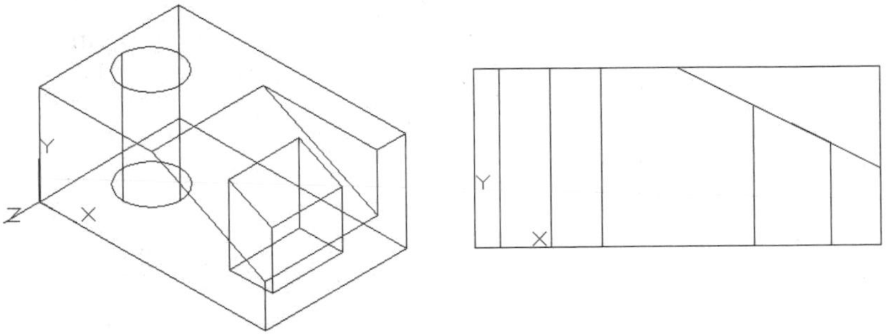

FIGURE 20.26

> **NOTE**
>
> A system variable called UCSFOLLOW is available to automate the `PLAN` command. Here is how it works. Make a viewport active and then set UCSFOLLOW to a value of 1 (or turned on), a plan view is automatically generated in the designated viewport whenever you change to a new UCS. Setting UCSFOLLOW to 0 (zero) turns this mode off.

VIEWING 3D MODELS WITH ORBIT

Various methods are available for viewing a model in 3D. One of the more efficient ways is through the Orbit commands: `3DORBIT` (3D "Constrained" Orbit), `3DFORBIT` (3D "Free" Orbit), and `3DCORBIT` (3D "Continuous" Orbit). These commands can be selected from the Ribbon, as shown in Figure 20.27 (Left), and from the Navigation Bar, as shown in Figure 20.27 (Right).

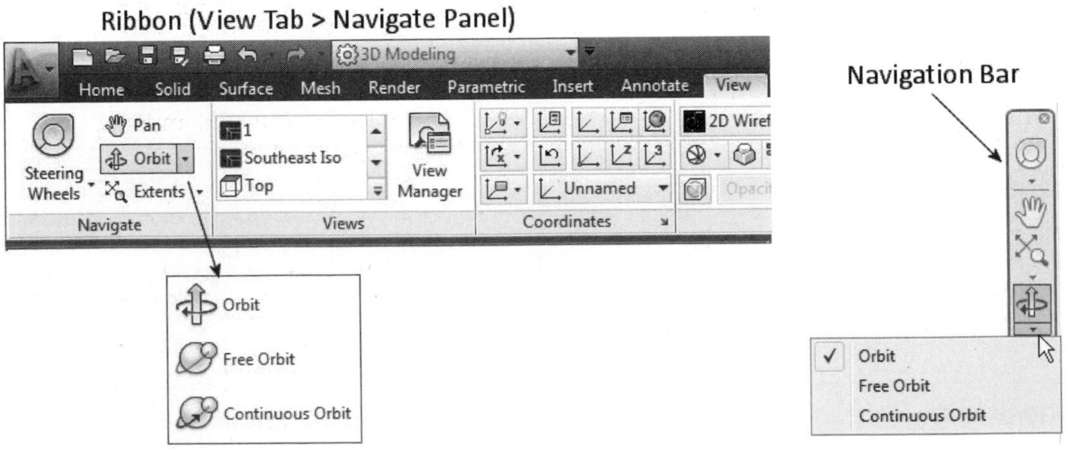

FIGURE 20.27

NOTE

The Navigation Bar and ViewCube are by default displayed on the right side of your screen. If desired, they can be toggled on and off through the Ribbon (View tab > User Interface panel > User Interface button).

VIEWING WITH FREE ORBIT

Choosing the Free Orbit button (3DFORBIT) displays your model, similar to Figure 20.28. Use the large circle (arcball) to guide your model through a series of dynamic rotation maneuvers. For instance, moving your cursor outside the large circle at "A" allows you to dynamically rotate (drag) your model only in a circular direction. Moving your cursor to either of the circle quadrant identifiers at "B" and "C" allows you to rotate your model horizontally. Moving your cursor to either of the circle quadrant identifiers at "D" or "E" allows you to rotate your model vertically. Moving your cursor inside the large circle at "F" allows you to dynamically rotate your model to any viewing position.

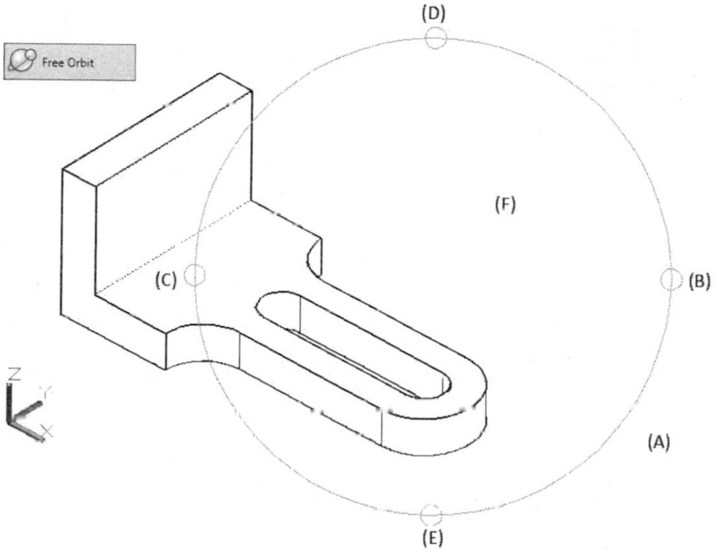

FIGURE 20.28

VIEWING WITH CONSTRAINED ORBIT

Another way to rotate a 3D model is through the Constrained Orbit button (3DORBIT). This command is similar to the free orbit command, however, an arcball will not be displayed and the amount of rotation is limited. A constrained orbit prevents possible disorientation by stopping you from rolling the 3D model completely over. If you attempt to orbit above or below the 3D model, the orbiting stops when you reach the top or bottom. While in this command, the arcball can be displayed by depressing the SHIFT key.

NOTE

A quick and efficient way of activating the Constrained Orbit tool is to press and hold down the SHIFT key while pressing on the middle button or wheel of the mouse.

VIEWING WITH CONTINUOUS ORBIT

Performing a continuous orbit (the 3DCORBIT command) rotates your 3D model continuously. After entering the command, press and drag your cursor in the direction you want the continuous orbit to move. Then, when you release the mouse button, the 3D model continues to rotate in that direction.

VIEWING 3D MODELS WITH THE VIEWCUBE

To further assist in rotating and viewing models in 3D, a ViewCube is available. The ViewCube can be toggled on and off through the Ribbon (View tab > User Interface panel > User Interface button). The basic function of this tool is to view your model in either orthographic or isometric views. By default, the ViewCube is displayed in the upper-right corner of the graphics screen. Also by default, the ViewCube takes on a transparent appearance signifying it is currently inactive, as shown in the following image on the left. Moving your cursor over the ViewCube activates this tool and it takes on an opaque appearance. Right-clicking on the ViewCube will display a menu for controlling display settings, as shown in Figure 20.29 (Center). When viewing a model in Face mode (orthogonally), the ViewCube takes on the appearance as shown in Figure 20.29 (Right). It is there where you can click on the two arrows to rotate the model.

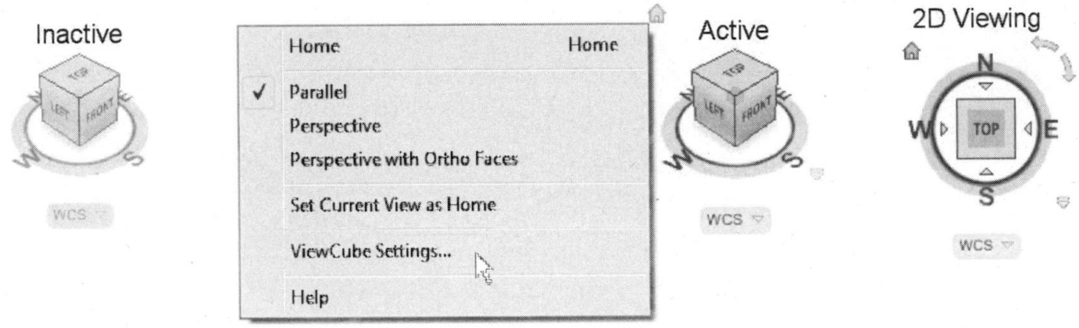

FIGURE 20.29

Clicking on ViewCube Settings, as shown in Figure 20.29, displays the ViewCube Settings dialog box, as shown in Figure 20.30. You can control the opacity level of the ViewCube in addition to its size and on-screen position.

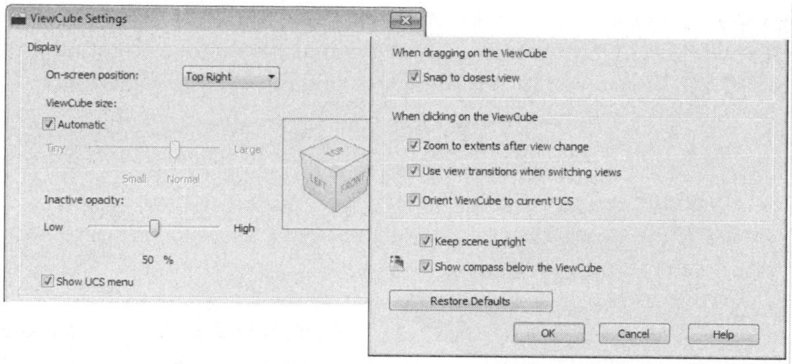

FIGURE 20.30

Various viewing modes are available through the ViewCube depending on what part of the ViewCube is picked. These modes are all displayed in Figure 20.31. For example, clicking on one of the corners of the ViewCube will display a model in an isometric mode. If you want to view a model orthogonally, then one of the six face-viewing modes would work. The edge pivot mode pivots a model along a selected edge. Once in an orthogonal view, such as a Top View, rotation arrows display that allow you to rotate or swivel a model. To return to the default view of a model, click on the house icon signifying the home position.

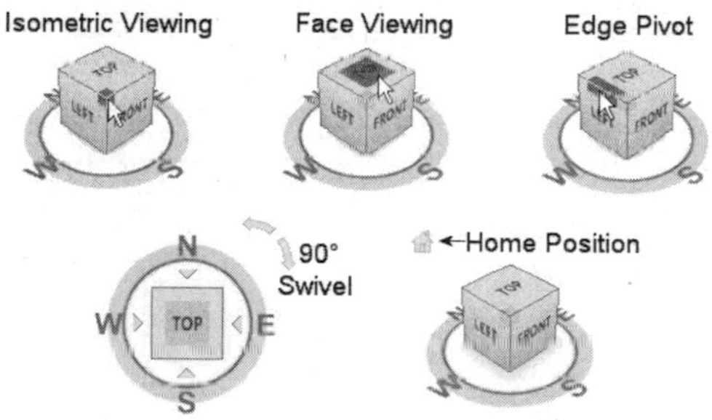

FIGURE 20.31

An example of how the ViewCube operates is illustrated in Figure 20.32. An isometric, edge, or face view can be displayed by picking the box; however, it is also possible to orbit the model by pressing and holding down the left mouse button over the box. In this case, moving the mouse will rotate the model dynamically, as shown in Figure 20.32.

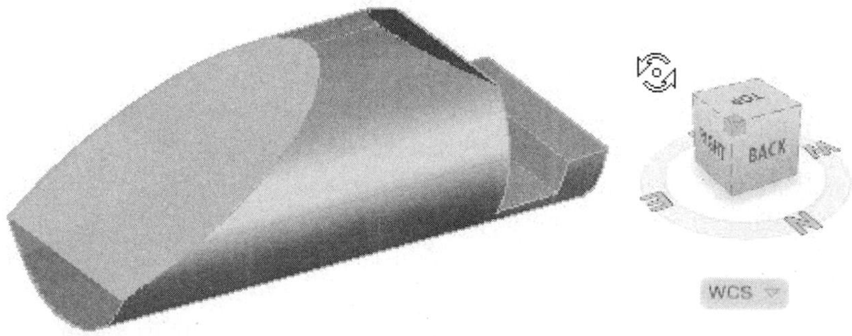

FIGURE 20.32

An example of displaying a model in its default location is illustrated in Figure 20.33. No matter how your model is currently displayed, you can always return the model to its default location by clicking on the Home icon whenever the ViewCube is displayed.

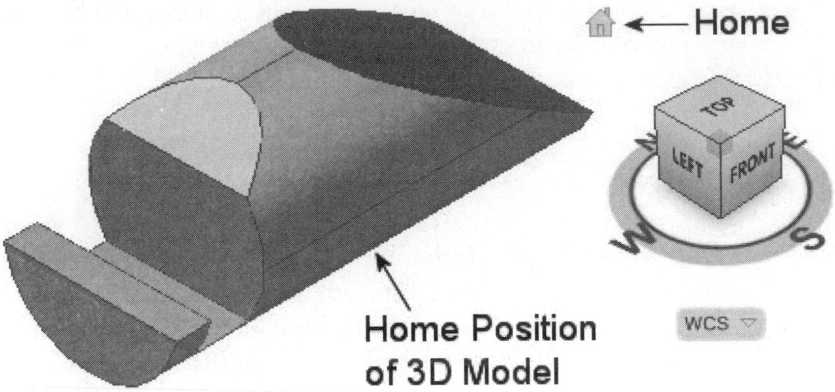

FIGURE 20.33

NOTE

A new home view can be created by orbiting to a new view orientation, right-clicking the Home icon and selecting "Set Current View as Home" from the provided shortcut menu.

Based on how a model is created, it is very easy to view the model orthographically, as shown in the following image. This is accomplished by clicking on any one of the six standard face-viewing modes located in the ViewCube. The following image is displayed from its top viewing mode of the ViewCube.

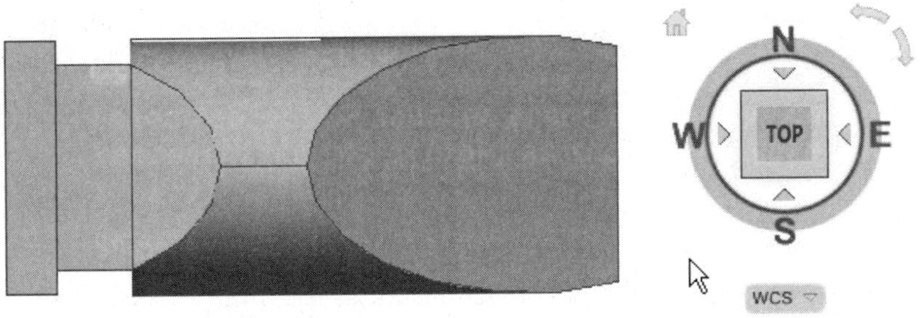

FIGURE 20.34

USING THE STEERING WHEEL

Another tool called the Steering Wheel (or Navigation Wheel) is available to assist in viewing models in 3D. The Steering Wheel can be activated from the Ribbon or Navigation Bar, as shown in Figure 20.35 (Left). When the Steering Wheel displays, eight modes are available that allow you to perform the following operations on a 3D model: Zoom, Rewind, Pan, Orbit, Center, Walk, Look, and Up/Down. Zoom mode allows you to zoom in or out. Rewind mode allows you to use a series of images created to zoom to previous views. Pan mode lets you move the view to a new location. Orbit mode allows you to rotate the view. Center mode centers the view based on the position of your cursor on the model. Walk mode allows you to swivel the viewpoint. Look mode moves the view without rotating the viewpoint. Up/Down mode changes the viewpoint of the model vertically. To assist with the selection of these modes, a tooltip is available to illustrate the purpose of each mode, as shown in Figure 20.35. You can even click on the arrow in the lower-right corner of the Steering Wheel to display the menu, as shown in Figure 20.35 (Right). Use this menu to change settings associated with the Steering Wheel or launch the Steering Wheel Settings dialog box in order to make further changes or restore the Steering Wheel back to its original settings.

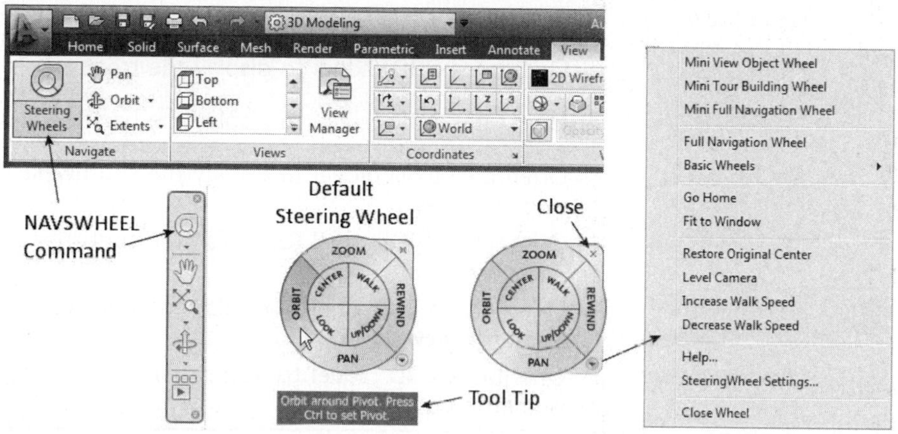

FIGURE 20.35

THE VIEW COMMAND

The VIEW command can be accessed from the Home tab and the View tab of the Ribbon, as shown in Figure 20.36. This command provides a quick and simple way to display your 3D models from preset viewing directions. Select one of the four isometric or six orthographic viewpoints available. If you select one of the orthographic viewing directions, it also automatically changes the UCS to correspond to that view. Selecting the View Manager button displays the View Manager dialog box, which can also be used to select preset viewing directions or it can be used to save and restore named views.

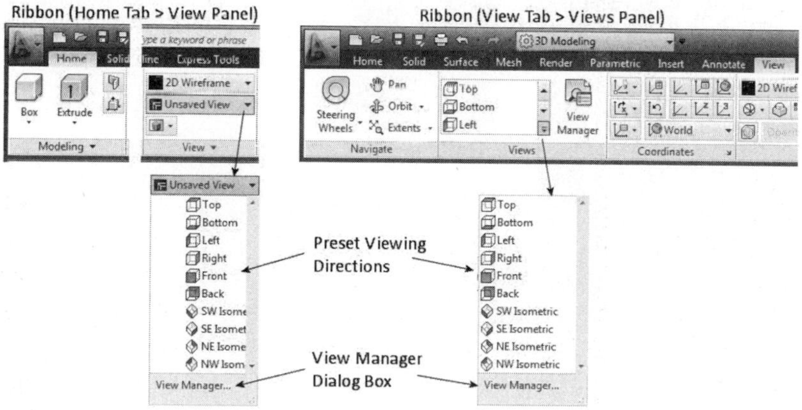

FIGURE 20.36

The following table gives a brief description of the available View modes.

Button	Tool	Function
	View Manager	Launches the View Manager dialog box used for creating named views
	Top View	Orientates a 3D model to display the top view
	Bottom View	Orientates a 3D model to display the bottom view
	Left View	Orientates a 3D model to display the left view
	Right View	Orientates a 3D model to display the right view
	Front View	Orientates a 3D model to display the front view
	Back View	Orientates a 3D model to display the back view
	SW Isometric	Orientates a 3D model to display the southwest isometric view
	SE Isometric	Orientates a 3D model to display the southeast isometric view
	NE Isometric	Orientates a 3D model to display the northeast isometric view
	NW Isometric	Orientates a 3D model to display the northwest isometric view

SHADING SOLID MODELS

Various shading modes are available to help you better visualize the solid model you are constructing. Access the shading modes from the Home tab and View tab of the Ribbon, as shown in Figure 20.37.

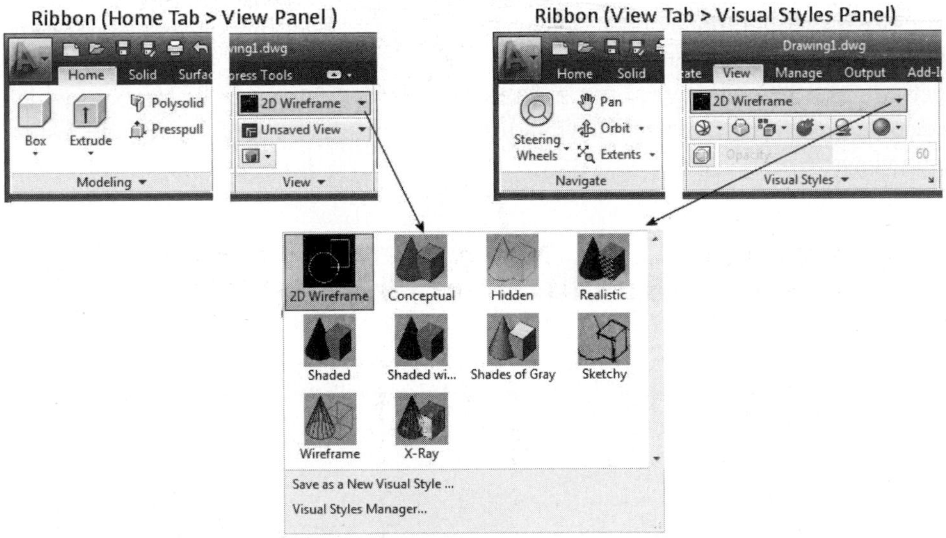

FIGURE 20.37

The following table gives a brief description of the predefined visual style modes available.

Button	Visual Style	Description
	2D Wireframe	Displays the 3D model as a series of lines and arcs that represents boundaries
	Wireframe	Displays the 3D model that is similar in appearance to the 2D option; the UCS icon appears as a color-shaded image
	Hidden	Displays the 3D model with hidden edges removed
	Realistic	Shades the objects and smooths the edges between polygon faces; if materials are attached to the model, they will display when this visual style is chosen
	Conceptual	This mode also shades the objects and smooths the edges between polygon faces; however, the shading transitions between cool and warm colors
	Shaded	Displays model with smooth shading
	Shaded with Edges	Displays model with smooth shading and visible edges
	Shades of Gray	Displays model with monochromatic shades of gray
	Sketchy	Displays model with a hand drawn effect
	X-Ray	Displays model partially transparent

Figure 20.38 illustrates five of the available visual styles.

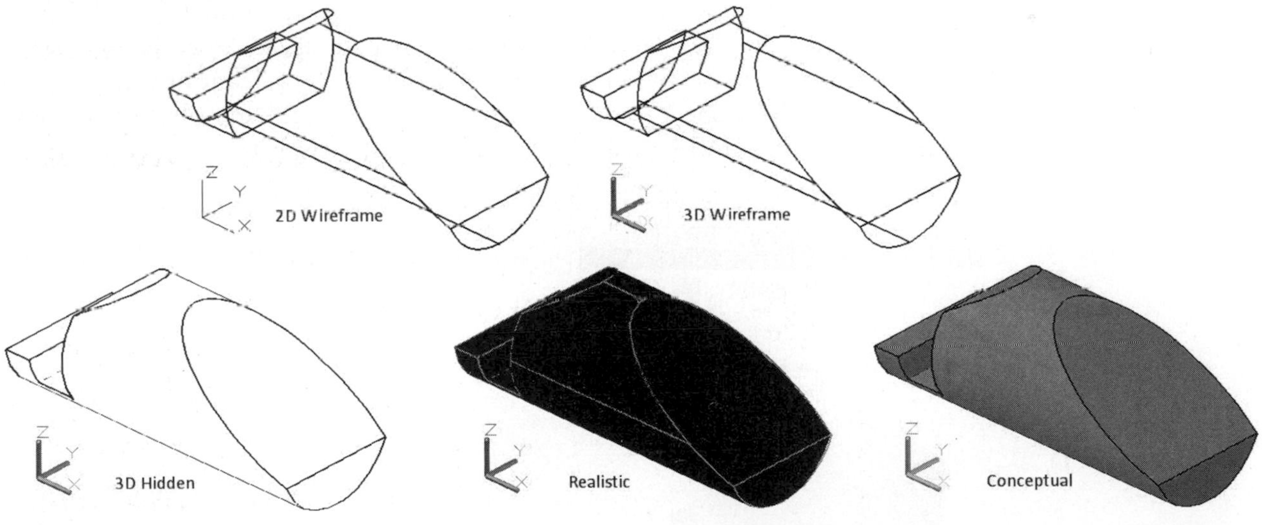

FIGURE 20.38

TRY IT!

Open the drawing file 20_Visual Styles. Experiment with the series of shading modes. Your results should be similar to the images provided in the previous illustrations.

> **NOTE**
>
> **When you perform such operations as Free, Constrained, and Continuous Orbit, the current visual style mode remains persistent. This means that if you are in a Realistic visual style and you rotate your model using one of the previously mentioned operations, the model remains shaded throughout the rotation operation. While performing an orbit operation, a shortcut menu is available which allows you to change the current visual style mode.**

CREATING A VISUAL STYLE

Custom visual styles can be created to better define how a 3D model will appear when shaded. Click on the Visual Styles Manager arrow located on the Visual Styles panel of the View Ribbon, as shown in Figure 20.39 (Left), to launch the Visual Styles Manager palette, as shown on the right of Figure 20.39. All the currently available visual styles are displayed at the top of the palette. Use the palette for creating new visual styles and applying them to the current viewport.

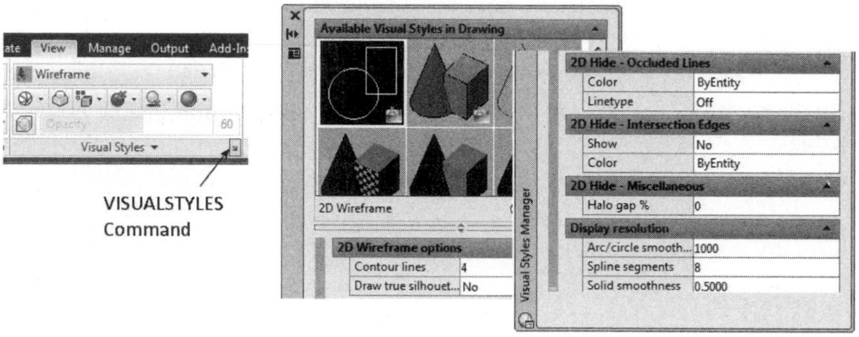

FIGURE 20.39

IN-CANVAS VIEWPORT CONTROL

The In-canvas Viewport Control is displayed in the upper left corner of a viewport, as shown in Figure 20.40. If multiple viewports are being utilized, an in-canvas control is provided for each one. This convenient tool allows you to control the display of viewing tools (ViewCube, Steering Wheels, and Navigation Bar), select preset viewing directions, and set the visual style for each of your viewports.

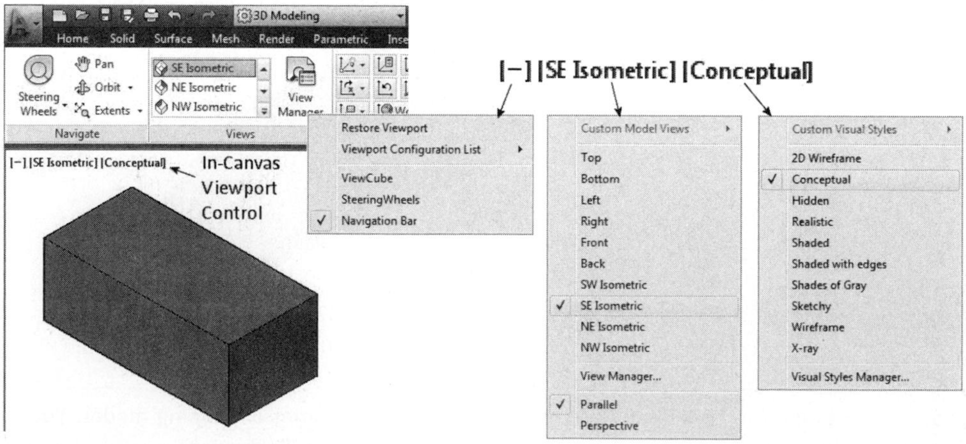

FIGURE 20.40

SOLID MODELING COMMANDS

Solid modeling commands can be accessed from the Home tab or the Solid tab of the Ribbon, as shown in Figure 20.41. Generally, most of the commands utilized for creating 3D solid models can be divided into three groups: primitive commands (BOX, CYLINDER, CONE, SPHERE, WEDGE, PYRAMID, and TORUS), Boolean commands (UNION, SUBTRACT, and INTERSECT), and profile-based commands (EXTRUDE, LOFT, REVOLVE, and SWEEP). Primitive commands are considered the building blocks of the solid model and are used to construct basic "primitives." Boolean commands allow you to combine primitives into a composite solid. Profile-based commands start with a 2D drawing (profile) and allow the creation of more complex models.

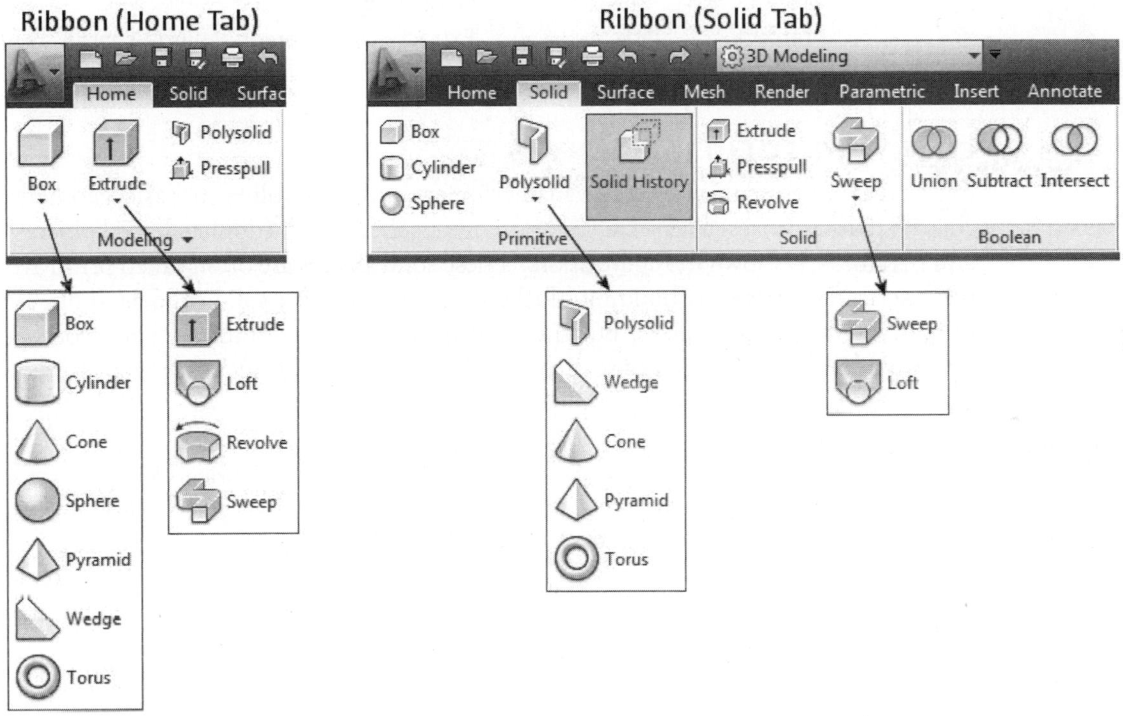

FIGURE 20.41

The following table gives a brief description of solid modeling commands.

Button	Tool	Shortcut	Function
	Polysolid	POLYSOLID	Creates a solid shape based on a direction, width, and height of the solid
	Box	BOX	Creates a solid box
	Wedge	WE	Creates a solid wedge
	Cone	CONE	Creates a solid cone
	Sphere	SPHERE	Creates a solid sphere
	Cylinder	CYL	Creates a solid cylinder
	Torus	TOR	Creates a solid torus
	Pyramid	PYR	Creates a solid pyramid
	Helix	HELIX	Creates a 2D or 3D helix
	Extrude	EXT	Creates a solid by extruding a 2D profile

(continued)

Button	Tool	Shortcut	Function
	Presspull	PRESSPULL	Presses or pulls closed areas resulting in a solid shape or a void in a solid
	Sweep	SWEEP	Creates a solid based on a profile and a path
	Revolve	REV	Creates a solid by revolving a 2D profile about an axis of rotation
	Loft	LOFT	Creates a lofted solid based on a series of cross-section shapes
	Union	UNI	Joins two or more solids together
	Subtraction	SU	Removes one or more solids from a source solid shape
	Intersect	INT	Extracts the common volume shared by two or more solid shapes

CREATING SOLID PRIMITIVES

Seven different commands are available for creating 3D solids in basic geometric shapes; namely boxes, wedges, cylinders, cones, spheres, tori (donut-shaped objects), and pyramids as shown in Figure 20.42. These solid shapes are often called primitives because they are used as building blocks for more complex solid models. They are seldom useful by themselves, but these primitives can be combined and modified into a wide variety of geometric shapes. The following command sequences allow you to practice creating one example of each primitive.

TRY IT!

Open the drawing file 20_Primitive_Examples, which represents a drawing void of any objects. Follow the next series of prompts to create the seven primitive objects illustrated in Figure 20.42. Verify that Polar Tracking is turned on in the Status Bar.

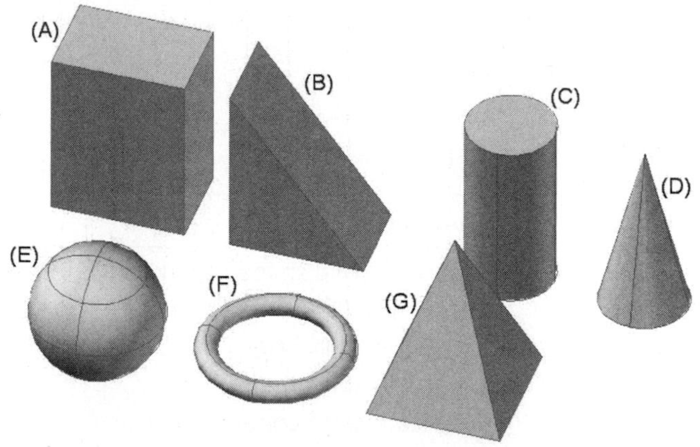

FIGURE 20.42

Creating Box Primitives

Box primitives consist of brick-shaped solid objects. They have six rectangular sides, which are either perpendicular or parallel to one another. Boxes are probably the most often used primitive, as many of the objects that are modeled are made up of rectangles and squares. There are several options available for creating boxes; the following prompt sequence illustrates the length option (see Figure 20.42 at "A").

```
Command: BOX
Specify first corner or [Center]: 4.00,9.00
Specify other corner or [Cube/Length]: L (For Length)
Specify length <5.0000>: 3.00 (point your cursor in the positive X direction)
Specify width <2.0000>: 2.00 (point your cursor in the positive Y direction)
Specify height or [2Point] <1.0000>: 4.00
```

Creating Wedge Primitives

Wedges are like boxes that have been sliced diagonally edge to edge. They have a total of five sides, three of which are rectangular and two are triangular. The top rectangular side slopes down in the X direction. The two sides opposite this sloping side are perpendicular to each other. The bottom rectangular surface is on the XY plane. The following prompt sequence illustrates the length option for creating a wedge (see Figure 20.42 at "B").

```
Command: WE (For WEDGE)
Specify first corner or [Center]: 8.00,9.00
Specify other corner or [Cube/Length]: L (For Length)
Specify length <5.0000>: 3.00
Specify width <2.0000>: 2.00
Specify height or [2Point] <3.0000>: 4.00
```

Creating Cylinder Primitives

Cylinders are probably the second most often used primitive. Cylinders can be created as either circular or elliptical. The following prompt sequence illustrates the diameter option for creating a cylinder (see Figure 20.42 at "C").

```
Command: CYL (For CYLINDER)
Specify center point of base or [3P/2P/Ttr/Elliptical]: 14.00,10.00
Specify base radius or [Diameter] <1.5000>: D (For Diameter)
Specify diameter <3.0000>: 2.00
Specify height or [2Point/Axis endpoint] <4.0000>: 4.00
```

Creating Cone Primitives

Cone primitives are closely related to cylinders. They have the same round or elliptical cross section; but they taper either to a point or a specified height with different radius forming a truncated cone. The following prompt sequence illustrates the radius option for creating a cone (see Figure 20.42 at "D").

```
Command: CONE
Specify center point of base or [3P/2P/Ttr/Elliptical]: 17.00,10.00
```

Specify base radius or [Diameter] <1.0000>: **1.00**

Specify height or [2Point/Axis endpoint/Top radius] <4.0000>: **4.00**

Creating Sphere Primitives

Creating spheres is the most straightforward process in creating primitives. Specify the sphere's center point and then specify either the radius or the diameter of the sphere. The following prompt sequence illustrates the diameter option for creating a sphere (see Figure 20.42 at "E").

Command: **SPHERE**

Specify center point or [3P/2P/Ttr]: **6.00,6.00**

Specify radius or [Diameter] <2.0000>: **D** *(For Diameter)*

Specify diameter <4.0000>: **3.00**

Creating a Torus Primitive

Although the torus is not often needed, it can be used to create some interesting solid primitives. The basic shape of a torus is that of a donut, but it can also take on a football shape, depending on the size values provided. To be made properly, this primitive requires a torus radius value and a tube radius value, as shown in Figure 20.43. The following prompt sequence illustrates creating a torus (see Figure 20.42 at "F").

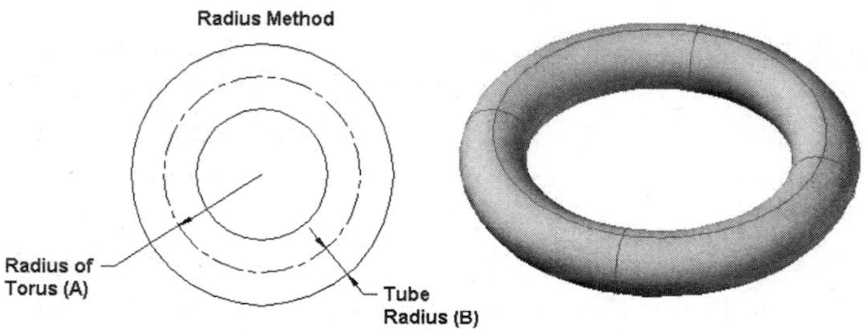

FIGURE 20.43

Command: **TOR** *(For TORUS)*

Specify center point or [3P/2P/Ttr]: **10.00,6.00**

Specify radius or [Diameter] <1.5000>: **1.50**

Specify tube radius or [2Point/Diameter]: **0.25**

Creating Pyramid Primitives

The method for creating solid pyramids is similar to the one used for creating cones with the exception that the base of the pyramid consists of edges and are noncircular in shape. The following prompt sequence illustrates the creation of a pyramid (see Figure 20.42 at "G").

Command: **PYR** *(For PYRAMID)*

4 sides Circumscribed

Specify center point of base or [Edge/Sides]: **14.00,6.00**

Specify base radius or [Inscribed] <1.5000>: **1.50** *(point your cursor in the positive X direction)*

Specify height or [2Point/Axis endpoint/Top radius] <4.0000>: **4.00**

USING BOOLEAN OPERATIONS ON SOLID PRIMITIVES

To combine one or more primitives to form a composite solid, a Boolean operation is performed. Boolean operations must act on at least a pair of primitives, regions, or solids. These operations in the form of commands can be selected from the Solid Editing panel of the Home Ribbon or the Boolean panel of the Solid Ribbon, the Modeling toolbar, or Solid Editing toolbar. Boolean operations allow you to add two or more objects together, subtract a single object or group of objects from another, or find the overlapping volume—in other words, form the solid common to both primitives. Displayed in Figure 20.44 are the UNION, SUBTRACT, and INTERSECT commands that you use to perform these Boolean operations.

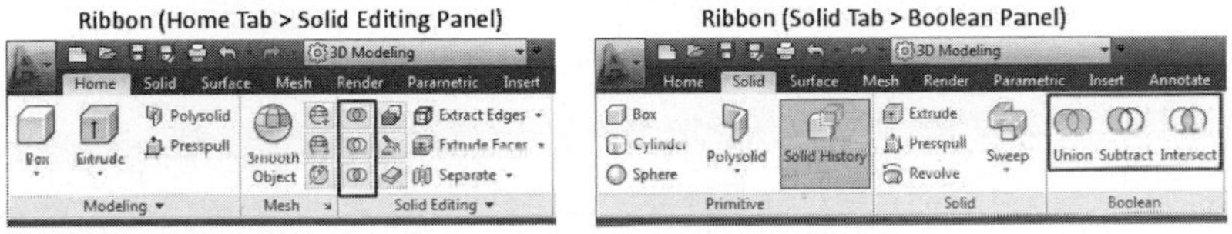

FIGURE 20.44

In Figure 20.45, a cylinder has been constructed along with a box. Depending on which Boolean operation you use, the results could be quite different. In Figure 20.45 at "Union," both the box and cylinder are considered one solid object. This is the purpose of the UNION command: to join or unite two solid primitives into one. The image at "Subtract" goes on to show the result of removing or subtracting the cylinder from the box—a hole is formed inside the box as a result of using the SUBTRACT command. The image at "Intersect" illustrates the intersection of the two solid primitives or the area that both solids have in common. This solid is obtained through the INTERSECT command. All Boolean operation commands can work on numerous solid primitives; that is, if you want to subtract numerous cylinders from a box, you can subtract all of the cylinders at one time.

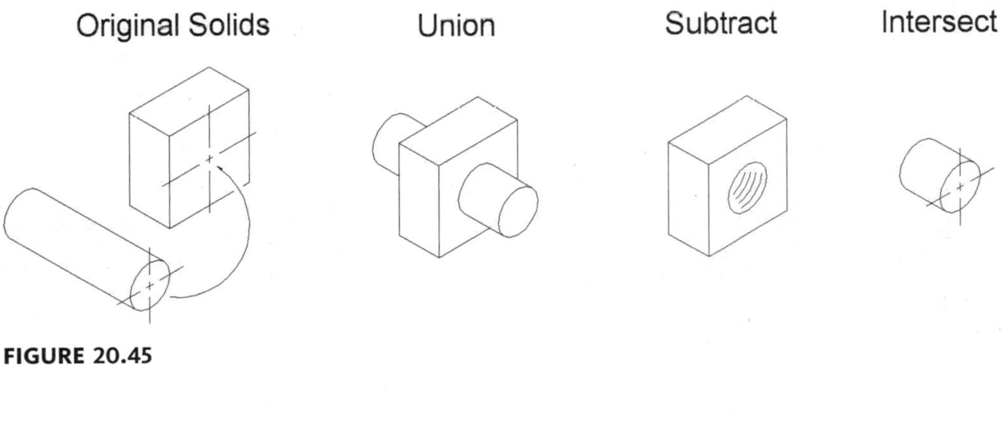

FIGURE 20.45

CREATING SOLID UNIONS

This Boolean operation, the UNION command, joins two or more selected solid objects together into a single solid object.

- From the Ribbon of the 3D Modeling workspace (Home and Solid tabs)
- From the Menu Bar (Modify > Solids Editing > Union)
- From the Solids Editing or Modeling toolbars
- From the keyboard (UNI or UNION)

TRY IT!

Open the drawing file 20_Union. Use the following command sequence and Figure 20.46 for performing this task.

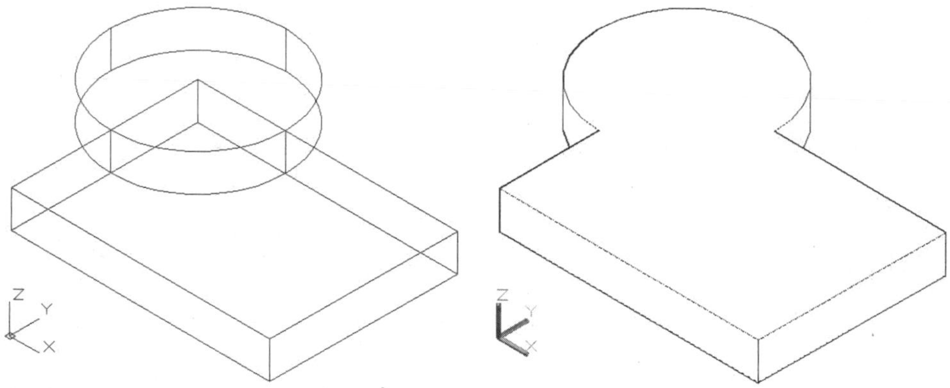

FIGURE 20.46

Command: UNI *(For UNION)*

Select objects: *(Pick the box and cylinder)*

Select objects: *(Press ENTER to perform the union operation)*

SUBTRACTING SOLIDS

Use the SUBTRACT command to subtract one or more solid objects from a source object, as shown in Figure 20.47. Choose this command in one of the following ways:

- From the Ribbon of the 3D Modeling workspace (Home and Solid tabs)
- From the Menu Bar (Modify > Solids Editing > Subtract)
- From the Solids Editing or Modeling toolbars
- From the keyboard (SU or SUBTRACT)

TRY IT!

Open the drawing file 20_Subtract. Use the following command sequence and Figure 20.47 for performing this task.

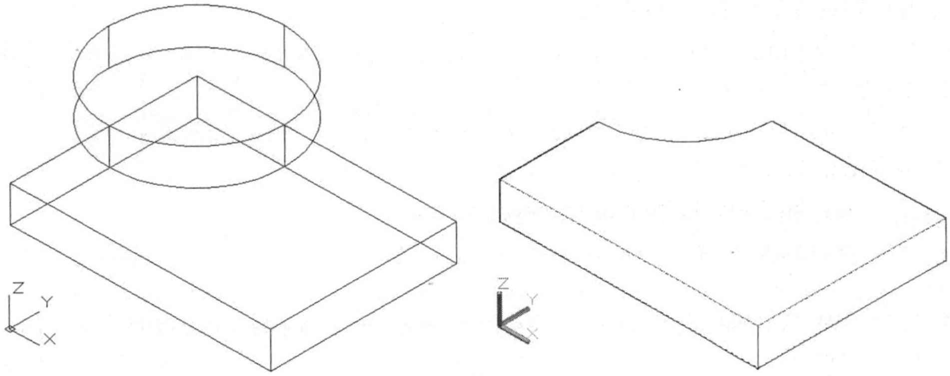

FIGURE 20.47

```
Command: SU (For SUBTRACT)
Select solids, surfaces, and regions to subtract from...
Select objects: (Pick the box)
Select objects: (Press ENTER to continue with this command)
Select solids, surfaces, and regions to subtract...
Select objects: (Pick the cylinder)
Select objects: (Press ENTER to perform the subtraction operation)
```

CREATING INTERSECTIONS

Use the INTERSECT command to find the solid common to a group of selected solid objects, as shown in Figure 20.48. Choose this command in one of the following ways:

- From the Ribbon of the 3D Modeling workspace (Home and Solid tabs)
- From the Menu Bar (Modify > Solids Editing > Intersect)
- From the Solids Editing or Modeling toolbars
- From the Keyboard (IN or INTERSECT)

TRY IT!

Open the drawing file 20_Intersection. Use the following command sequence and Figure 20.48 for performing this task.

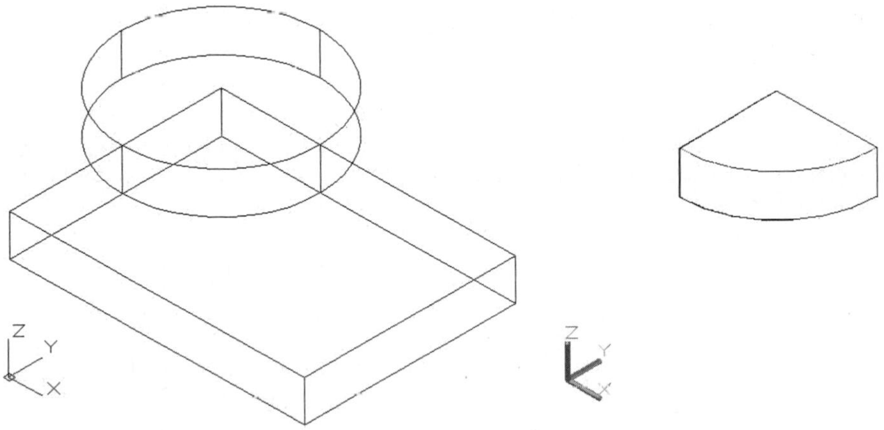

FIGURE 20.48

Command: IN (For INTERSECT)

Select objects: *(Pick the box and cylinder)*

Select objects: *(Press ENTER to perform the intersection operation)*

3D Applications of Unioning Solids

Figure 20.49 shows an object consisting of one horizontal solid box, two vertical solid boxes, and two extruded semicircular shapes. All primitives have been positioned with the MOVE command. To join all solid primitives into one solid object, use the UNION command. The order of selection of these solids for this command is not important.

TRY IT!

Open the drawing file 20_3D App Union. Use the following prompts and Figure 20.49 for performing these tasks.

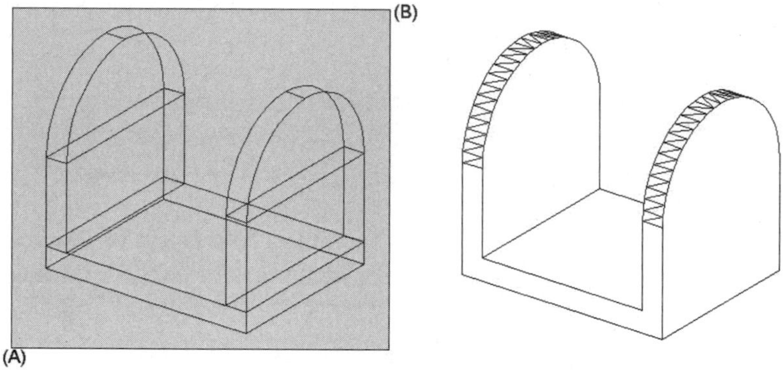

FIGURE 20.49

Command: UNI *(For UNION)*

Select objects: *(Pick your screen at "A")*

Select objects: *(Pick your screen at "B")*

Select objects: *(Press ENTER to perform the union operation)*

3D Applications of Moving Solids

Using the same problem from the previous example, let us now add a hole in the center of the base. The cylinder will be created through the CYLINDER command. It will then be moved to the exact center of the base. You can use the MOVE command along with OSNAPs to accomplish this. See Figure 20.50.

Command: CYL *(For CYLINDER)*

Specify center point of base or [3P/2P/Ttr/Elliptical]: 3.00,3.00

Specify base radius or [Diameter]: 0.75

Specify height or [2Point/Axis endpoint]: 0.25

Command: M *(For MOVE)*

Select objects: *(Select the cylinder at "A")*

Select objects: *(Press ENTER to continue with this command)*

Specify base point or [Displacement] <Displacement>: **Cen**

of *(Select the bottom of the cylinder at "A")*

Specify second point or <use first point as displacement>:
M2P *(To activate Midpoint between two points)*

First point of mid: **End**

of *(Select the endpoint of the bottom of the base at "B")*

Second point of mid: **End**

of *(Select the endpoint of the bottom of the base at "C")*

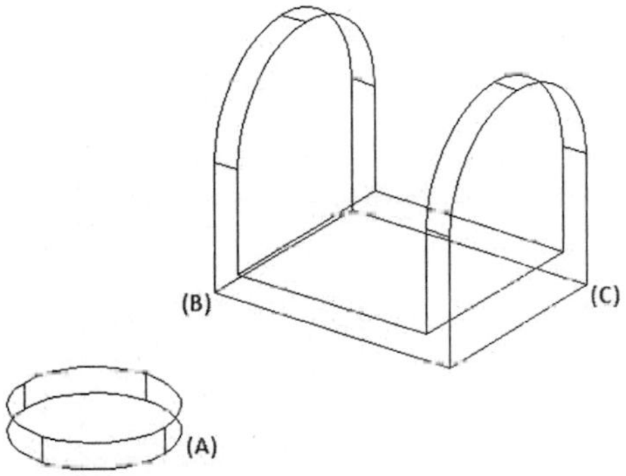

FIGURE 20.50

3D Applications of Subtracting Solids

Now that the solid cylinder is in position, use the SUBTRACT command to remove the cylinder from the base of the main solid and create a hole in the base, as shown in Figure 20.51.

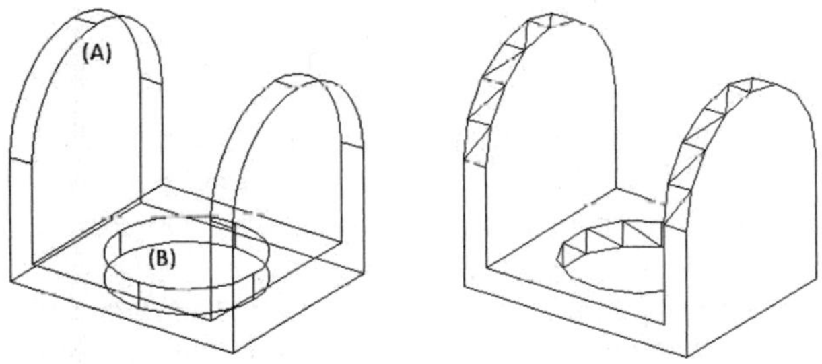

FIGURE 20.51

Command: SU *(For SUBTRACT)*

Select solids, surfaces, and regions to subtract from...

Select objects: *(Select the main solid as source at "A")*

Select objects: *(Press ENTER to continue with this command)*

```
Select solids, surfaces, and regions to subtract...
Select objects: (Select the cylinder at "B")
Select objects: (Press ENTER to perform the subtraction operation)
Command: HIDE (For hidden line removal view)
Command: RE (For REGEN to return to wireframe view)
```

INTERSECTION APPLICATIONS

Intersections remain one of the more difficult concepts to grasp in solid modeling. However, once you become comfortable with this concept, it can be one of the more powerful solid modeling tools available. Creating an intersection involves creating a solid model from the common volumes of two or more overlapping solids. The objects labeled "A" and "B" in Figure 20.52 represent existing solid models. These models are then moved to overlap each other, as shown at "C" in Figure 20.52. After creating the intersection, the nonoverlapping portions of the model are removed leaving the solid model, as shown at "D" in Figure 20.52.

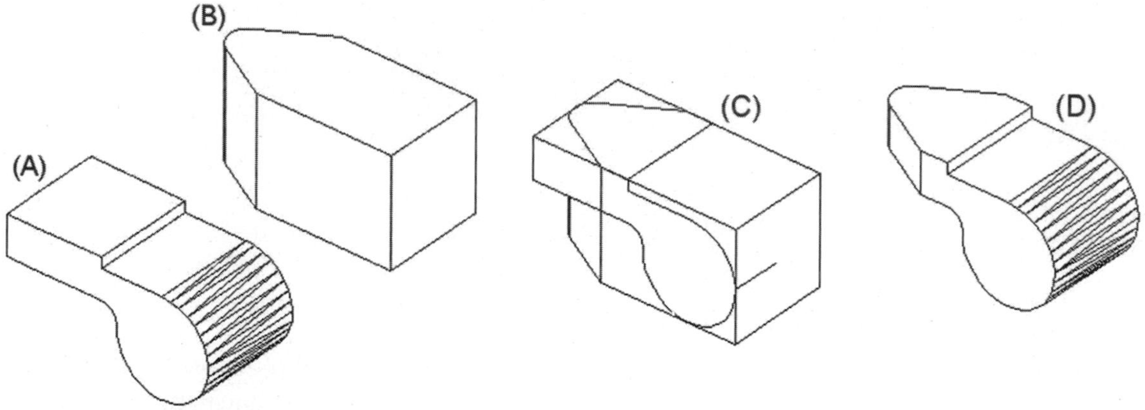

FIGURE 20.52

TRY IT!

Open the drawing file 20_Int1. In Figure 20.53, the object at "C" represents the finished model. Look at the sequence beginning at "A" to see how to prepare the solid primitives for an Intersection operation. Two separate 3D Solid objects are created at "A." One object represents a block that has been filleted along with the placement of a hole drilled through. The other object represents the U-shaped extrusion. With both objects modeled, they are moved on top of each other at "B." The OSNAP-Midpoint was used to accomplish this. Finally, the `INTERSECT` command is used to find the common volume shared by both objects; namely the illustration at "C."

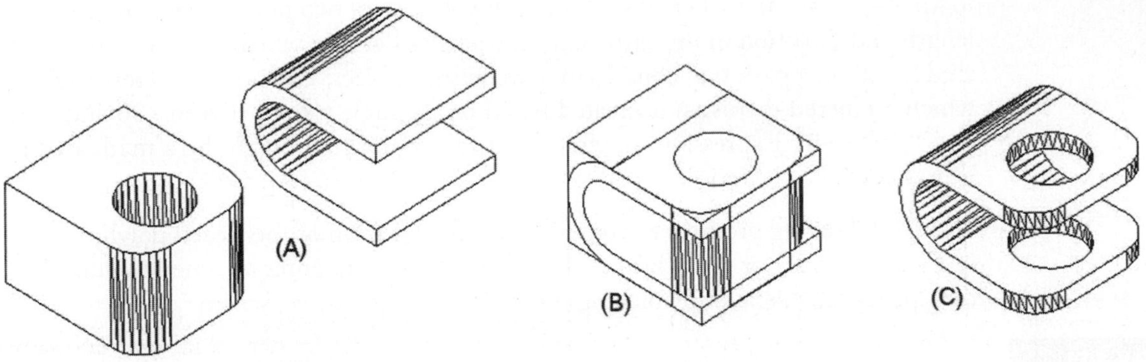

FIGURE 20.53

TRY IT!

Open the drawing file 20_Int2. The object in Figure 20.54 is another example of how the INTERSECT command may be applied to a solid model. For the results at "B" to be obtained from a cylinder that has numerous cuts, the cylinder is first created as a separate model. Then the cuts are made in another model at "A." Again, both models are moved together (use the Quadrant and Midpoint OSNAP modes), and then the INTERSECT command is used to achieve the results at "B." Before undertaking any solid model, first analyze how the model is to be constructed. Using intersections can create dramatic results, which would normally require numerous union and subtraction operations.

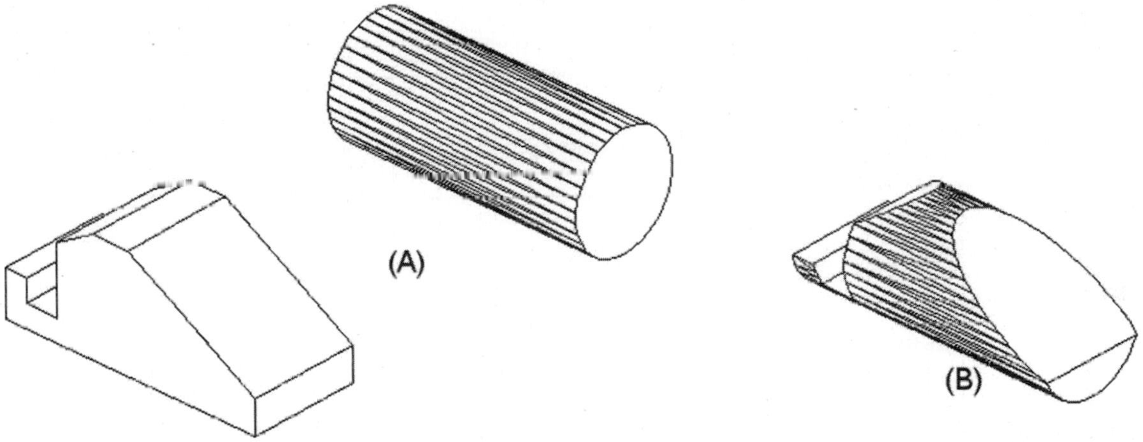

FIGURE 20.54

CREATING SOLID EXTRUSIONS

The EXTRUDE command creates a solid by extrusion. Choose this command in one of the following ways:

- From the Ribbon of the 3D Modeling workspace (Home and Solid tabs)
- From the Menu Bar (Draw > Modeling > Extrude)
- From the Modeling Toolbar
- From the keyboard (EXT or EXTRUDE)

Only regions or closed, single-entity objects such as circles and closed polylines can be extruded into a solid. If the profile being extruded is not closed, a surface instead of a solid model will be created. Other options of the EXTRUDE command include the following: Extrude by Direction, in which you specify two points that determine the length and direction of the extrusion; Extrude by Path, in which the extrusion is created based on a path that consists of a predefined object; Extrude by Taper Angle, in which a tapered extrusion is created based on an angle value between -90 and +90; and Extrude by Expression, in which you can control the height by a mathematical expression.

Use the following prompts to construct a solid extrusion of the closed polyline object in Figure 20.55. For the height of the extrusion, you can enter a numeric value or you can specify the distance by picking two points on the display screen.

TRY IT!

Open the drawing file 20_Extrude1. Use the following command sequence and Figure 20.55 for performing this task.

```
Command: EXT (For EXTRUDE)
Current wireframe density: ISOLINES=4, Closed profiles creation mode = Solid
Select objects to extrude or [MOde]: (Select the polyline object at "A")
Select objects to extrude or [MOde]: (Press ENTER to continue with this command)
Specify height of extrusion or [Direction/Path/Taper angle/Expression]: 1.00
```

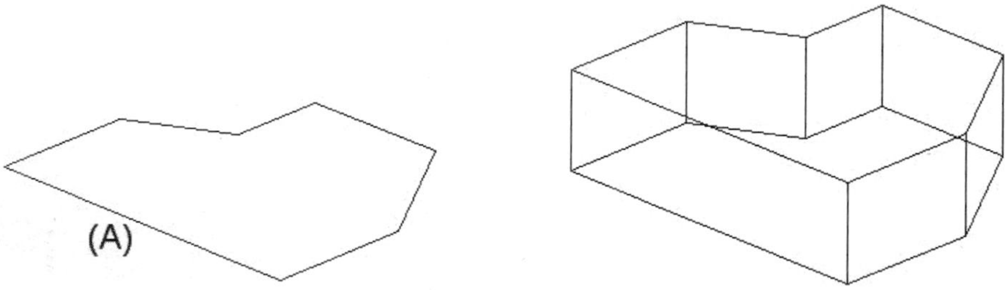

FIGURE 20.55

You can create an optional taper along with the extrusion by utilizing the Taper angle option provided.

TRY IT!

Open the drawing file 20_Extrude2. Use the following command sequence and Figure 20.56 for performing this task.

```
Command: EXT (For EXTRUDE)
Current wireframe density: ISOLINES=4, Closed profiles creation mode = Solid
Select objects to extrude or [MOde]: (Select the polyline object at "B" in the following image)
Select objects to extrude or [MOde]: (Press ENTER to continue with this command)
```

Specify height of extrusion or [Direction/Path/Taper angle/ Expression]: T *(For Taper angle)*

Specify angle of taper for extrusion or [Expression] <0>: 15

Specify height of extrusion or [Direction/Path/Taper angle/ Expression]: 1.00

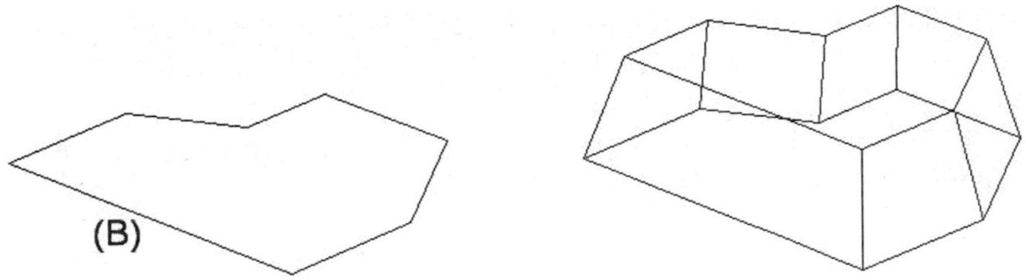

FIGURE 20.56

You can also create a solid extrusion by selecting a path to be followed by the object being extruded. Typical paths include regular and elliptical arcs, 2D and 3D polylines, or splines. The extruded pipe in Figure 20.57 was created using the following steps: First the polyline path was created. Then, a new UCS was established through the UCS command along with the Z-axis option; the new UCS was positioned at the end of the polyline with the Z-axis extending along the polyline. A circle was constructed with its center point at the end of the polyline. Finally, the circle was extruded along the polyline path.

Open the drawing file 20_Extrude Pipe. Use the following command sequence and Figure 20.57 for performing this task.

TRY IT!

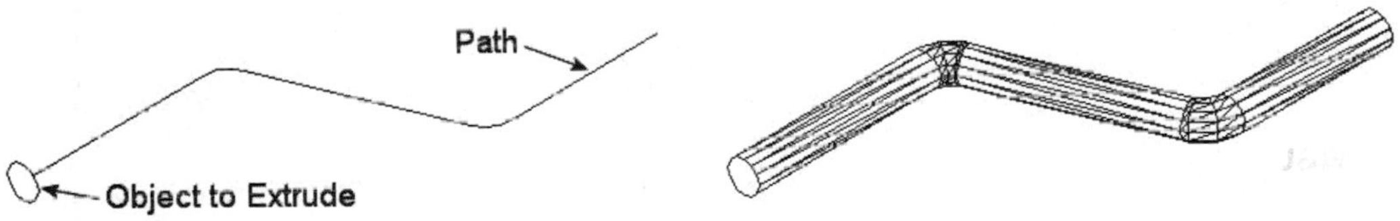

FIGURE 20.57

Command: EXT *(For EXTRUDE)*

Current wireframe density: ISOLINES=4, Closed profiles creation mode = Solid

Select objects to extrude or [MOde]: *(Select the small circle as the object to extrude)*

Select objects to extrude or [MOde]: *(Press* ENTER *to continue)*

Specify height of extrusion or [Direction/Path/Taper angle/ Expression]: P *(For Path)*

Select extrusion path or [Taper angle]: *(Select the polyline object representing the path)*

Extruding an Existing Face

Once a solid object is created, existing faces of the model can be used as profiles to further extrude shapes. Illustrated in Figure 20.58 (Left) is a wedge-shaped 3D model. After entering the EXTRUDE command, press and hold down the CTRL key and select the inclined face. You can either enter a value or drag your cursor to define the height of the extrusion, as shown in Figure 20.58 (Center). The results are shown in Figure 20.58 (Right). The new extruded shape created from the inclined face is considered a separate solid object. If both shapes need to be considered one, use the UNION command and select both extruded boxes to join them as one solid.

TRY IT!

Open the drawing file 20_Extrude Face. Use the following command sequence and Figure 20.58 performing an extrusion on an existing face.

Command: EXT *(For EXTRUDE)*

Current wireframe density: ISOLINES=4, Closed profiles creation mode = Solid

Select objects to extrude or [MOde]: *(Press and hold down the CTRL key and select the inclined face as shown in the following image on the left)*

Select objects to extrude or [MOde]: *(Press ENTER to continue)*

Specify height of extrusion or [Direction/Path/Taper angle/Expression]: 5.00

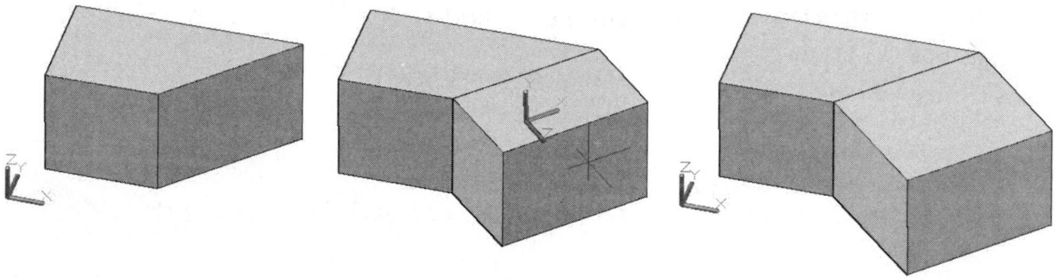

FIGURE 20.58

CREATING REVOLVED SOLIDS

The REVOLVE command creates a solid by revolving an object about an axis of revolution. Choose this command in one of the following ways:

- From the Ribbon of the 3D Modeling workspace (Home and Solid tabs)
- From the Menu Bar (Draw > Modeling > Revolve)
- From the Modeling Toolbar
- From the keyboard (REV or REVOLVE)

Only regions or closed, single-entity objects, such as polylines, polygons, circles, ellipses, and 3D polylines, can be revolved to create a solid. Revolving an open profile will generate a surface instead of a solid model. If a group of objects is not in the form of a single entity, group them together using the PEDIT command or create a closed polyline/region with the BOUNDARY command. Figure 20.59 represents a revolved 3D Solid object.

TRY IT!

Open the drawing file 20_Revolve1. In this exercise, a composite solid will be created from a solid primitive and a profile-based solid. The CYLINDER command was used to construct the cylindrical primitive and the REVOLVE command will be used to create the revolved solid. Use the following command sequence on the objects, as illustrated in Figure 20.59 (Left), for creating a revolved solid.

Command: REV *(For REVOLVE)*

Current wireframe density: ISOLINES=4, Closed profiles creation mode = Solid

Select objects to revolve or [MOde]: *(Select profile "A" as the object to revolve)*

Select objects to revolve or [MOde]: *(Press* ENTER *to continue with this command)*

Specify axis start point or define axis by [Object/X/Y/Z] <Object>: O *(For Object)*

Select an object: *(Select line "B")*

Specify angle of revolution or [STart angle/Reverse/EXpression] <360>: *(Press* ENTER *to accept the default and perform the revolving operation)*

Use the Center option of OSNAP along with the MOVE command to position the revolved solid inside the cylinder.

Command: M *(For MOVE)*

Select objects: *(Select the revolved solid in Figure 20.59)*

Select objects: *(Press* ENTER *to continue with this command)*

Specify base point or [Displacement] <Displacement>: Cen

of *(Select the center of the revolved solid at "C")*

Specify second point or <use first point as displacement>: Cen

of *(Select the center of the cylinder at "D")*

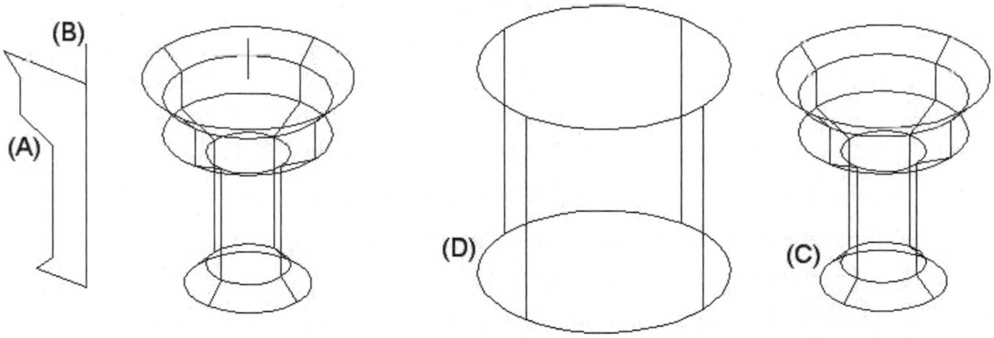

FIGURE 20.59

Once the revolved solid is positioned inside the cylinder, use the SUBTRACT command to subtract the revolved solid from the cylinder, as shown in Figure 20.60.

Use the HIDE command to perform a hidden line removal at "B" to check that the solid is correct (this would be difficult to interpret in wireframe mode).

Command: SU *(For SUBTRACT)*

Select solids, surfaces, and regions to subtract from...

Select objects: *(Select the cylinder as source)*

Select objects: *(Press* ENTER *to continue with this command)*

Select solids, surfaces, and regions to subtract...

Select objects: *(Select the revolved solid)*

Select objects: *(Press* ENTER *to perform the subtraction operation)*

Command: HI *(For HIDE)*

Command: RE *(For REGEN to return to wireframe view)*

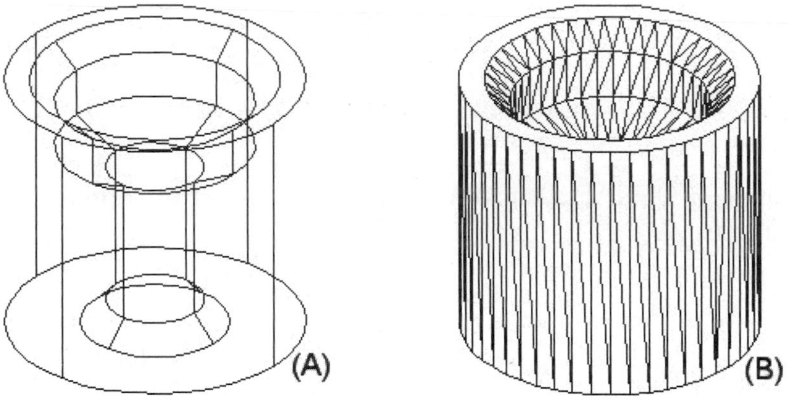

FIGURE 20.60

NOTE

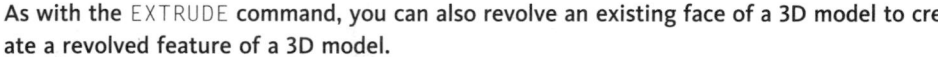

As with the EXTRUDE command, you can also revolve an existing face of a 3D model to create a revolved feature of a 3D model.

CREATING A SOLID BY SWEEPING

The SWEEP command creates a solid by sweeping a profile along an open or closed 2D or 3D path. The result is a solid in the shape of the specified profile along the specified path. If the sweep profile is closed, a solid is created. If the sweep profile is open, a swept surface is created. Choose this command in one of the following ways:

- From the Ribbon of the 3D Modeling workspace (Home and Solid tabs)
- From the Menu Bar (Draw > Modeling > Sweep)
- From the Modeling toolbar
- From the keyboard (SWEEP)

TRY IT!

Open the drawing file 20_Sweep. Use the following command sequence and Figure 20.61 for performing this task.

Illustrated in Figure 20.61 (Left) is an example of the geometry required to create a swept solid. Circles "A" and "B" represent profiles, while arc "C" represents the path of the sweep. Notice in this illustration that the circles do not have to be connected to the path; however, both circles must be constructed in the same plane in order for both to be included in the sweep operation. Use the following command sequence for creating a swept solid. The results of this operation are illustrated in Figure 20.61 (Right).

Command: SWEEP

Current wireframe density: ISOLINES=4, Closed profiles creation mode = Solid

Select objects to sweep or [MOde]: *(Pick circles "A" and "B")*

Select objects to sweep or [MOde]: *(Press* ENTER *to continue)*

Select sweep path or [Alignment/Base point/Scale/Twist]: *(Pick arc "C")*

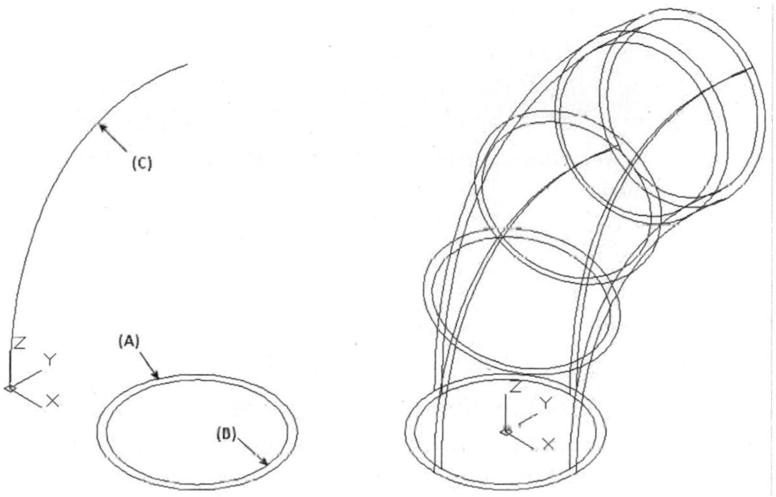

FIGURE 20.61

Illustrated on the left in Figure 20.62 is the shaded solution to sweeping two circles along a path consisting of an arc. Notice, however, that an opening is not created in the shape; instead, the inner swept shape is surrounded by the outer swept shape. Both swept shapes are considered individual objects. To create the opening, subtract the inner shape from the outer shape using the SUBTRACT command. The results are illustrated in Figure 20.62 (Right). The Conceptual visual style was utilized to better display the results.

Command: SU *(For SUBTRACT)*

Select solids, surfaces, and regions to subtract from:

Select objects: *(Select the outer sweep shape)*

Select objects: *(Press* ENTER*)*

Select solids, surfaces, and regions to subtract...

Select objects: *(Select the inner sweep shape)*

Select objects: *(Press* ENTER *to perform the subtraction)*

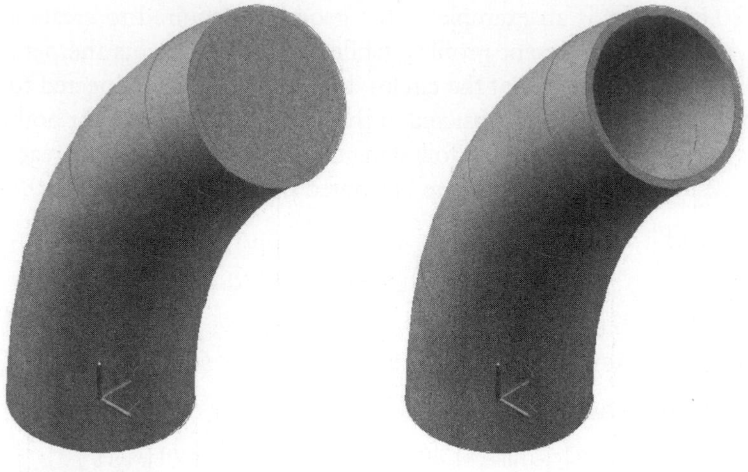

FIGURE 20.62

CREATING A SOLID BY LOFTING

The `LOFT` command creates a solid based on a series of cross sections. These cross sections define the shape of the solid. If the cross sections are open, a surface loft is created. If the cross sections are closed, a solid loft is created. When you are performing lofting operations, at least two cross sections must be created. Choose this command in one of the following ways:

- From the Ribbon of the 3D Modeling workspace (Home and Solid tabs)
- From the Menu Bar (Draw > Modeling > Loft)
- From the Modeling toolbar
- From the keyboard (`LOFT`)

Figure 20.63 illustrates a lofting operation based on open spline-shaped objects. In the illustration on the left in Figure 20.63, the cross sections of the plastic bottle are selected individually and in order starting with the left of the bottle and ending with the profile on the right of the bottle. The results are illustrated in Figure 20.63 (Right), with a surface that is generated from the open profiles.

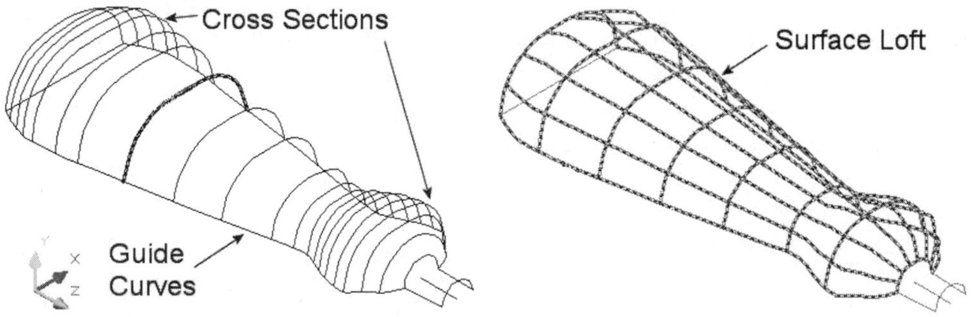

FIGURE 20.63

In Figure 20.64, instead of open profiles, all of the cross sections consist of closed profile shapes. The same rules apply when creating lofts; all profiles must be selected in the proper order. The results of lofting closed profiles are illustrated in Figure 20.64 (Right) with the creation of a solid shape.

```
Command: LOFT

Current wireframe density: ISOLINES=4, Closed profiles
creation mode = Solid

Select cross sections in lofting order or [POint/Join
multiple edges/MOde]: (Select all cross sections from the
rear to the front)

Select cross sections in lofting order or [POint/Join
multiple edges/MOde]: (Press ENTER to continue)

Enter an option [Guides/Path/Cross sections only/Settings]
<Cross sections only>: G (For Guides)

Select guide profiles or [Join multiple edges]: (Select both
guide curves)

Select guide profiles or [Join multiple edges]: (Press ENTER
to create the loft)
```

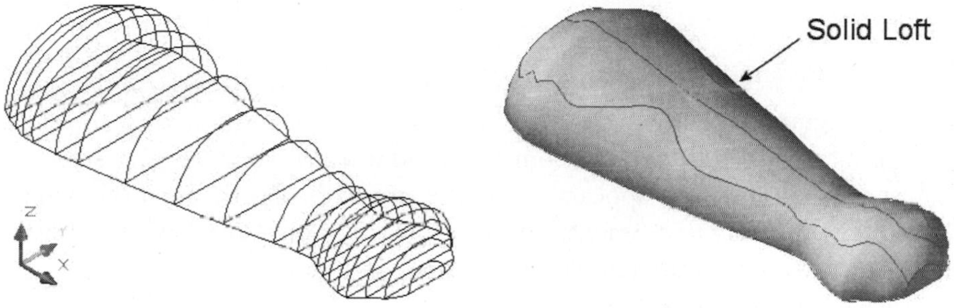

FIGURE 20.64

TRY IT!

Open the drawing file 20_Bowling Pin. You will first create a number of cross-section profiles in the form of circles that represent the diameters at different stations of the bowling pin. The LOFT **command is then used to create the solid shape.**

Since you will be using absolute coordinates in this exercise, begin by turning Dynamic Input (DYN) off. Then, create all the circles that make up the cross sections of the bowling pin. Use the table shown on the left in Figure 20.65 to construct the circles shown on the right in the figure.

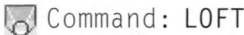

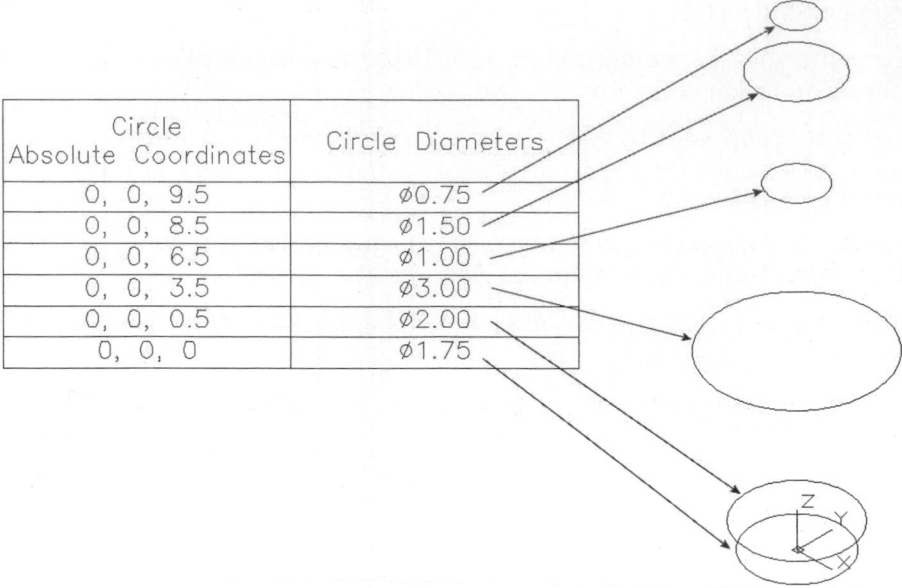

Circle Absolute Coordinates	Circle Diameters
0, 0, 9.5	⌀0.75
0, 0, 8.5	⌀1.50
0, 0, 6.5	⌀1.00
0, 0, 3.5	⌀3.00
0, 0, 0.5	⌀2.00
0, 0, 0	⌀1.75

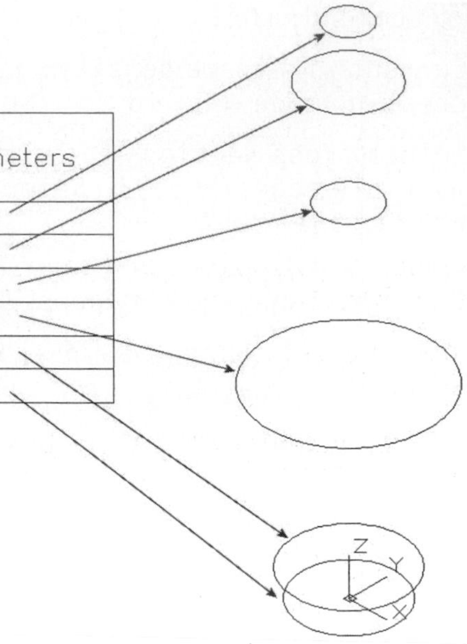

FIGURE 20.65

With all profiles created, activate the `LOFT` command and pick the cross sections of the bowling pin beginning with the bottom circle and working your way up to the top circle. A List grip automatically appears. Select the grip to display the menu shown in Figure 20.66 (Left). Verify that Smooth Fit is selected, and press the ENTER key to produce the loft that is illustrated in wireframe mode in Figure 20.66 (Center).

```
Command: LOFT

Current wireframe density: ISOLINES=4, Closed profiles
creation mode = Solid

Select cross sections in lofting order or [POint/Join
multiple edges/MOde]: (Select the six cross sections of the
bowling pin in order)

Select cross sections in lofting order or [POint/Join
multiple edges/MOde]: (Press ENTER to continue)

Enter an option [Guides/Path/Cross sections only/Settings]
<Cross sections only>: (Press ENTER to accept this default
value and create the loft)
```

To complete the bowling pin, use the `FILLET` command to round off the topmost circle of the bowling pin. Then view the results by clicking on the Realistic or Conceptual visual style. Your display should appear similar to the illustration shown in Figure 20.66 (Right). Select the model, and pick the provided grip. Try modifying the model by changing some of the settings provided in the shortcut menu.

Command: F *(For FILLET)*

Current settings: Mode = TRIM, Radius = 0.00

Select first object or [Undo/Polyline/Radius/Trim/Multiple]: R *(For Radius)*

Specify fillet radius <0.00>: **0.50**

Select first object or [Undo/Polyline/Radius/Trim/Multiple]: *(Pick the edge of the upper circle)*

Enter fillet radius <0.50>: *(Press* ENTER *to accept this value)*

Select an edge or [Chain/Radius]: *(Press* ENTER *to perform the fillet operation)*

1 edge*(s)* selected for fillet.

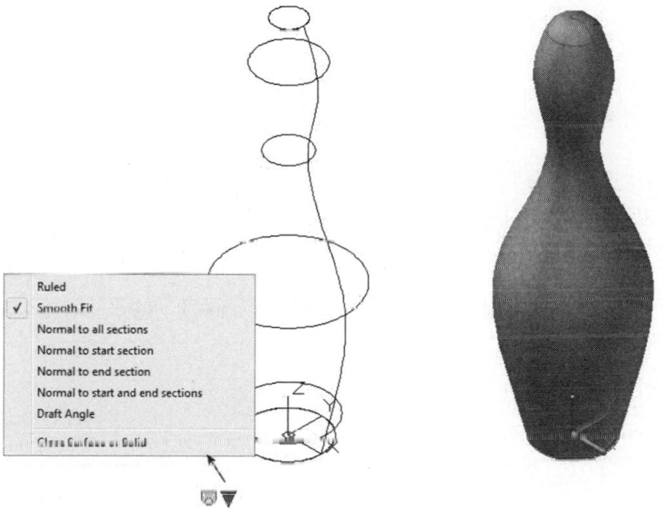

FIGURE 20.66

CREATING POLYSOLIDS

A polysolid is created in a fashion similar to the one used to create a polyline. The POLYSOLID command, however, creates a 3D object. Choose this command in one of the following ways:

- From the Ribbon of the 3D Modeling workspace (Home and Solid tabs)
- From the Menu Bar (Draw > Modeling > Polysolid)
- From the Modeling toolbar
- From the keyboard (POLYSOLID)

The plan view of a polysolid, as shown in Figure 20.67 (Left), appears similar to a polyline where a width has been entered. The main difference between polysolids and polylines is that you can designate the height of a polysolid. In this way, polysolids are ideal for creating wall-shaped models, as illustrated in the isometric view shown in Figure 20.67 (Right).

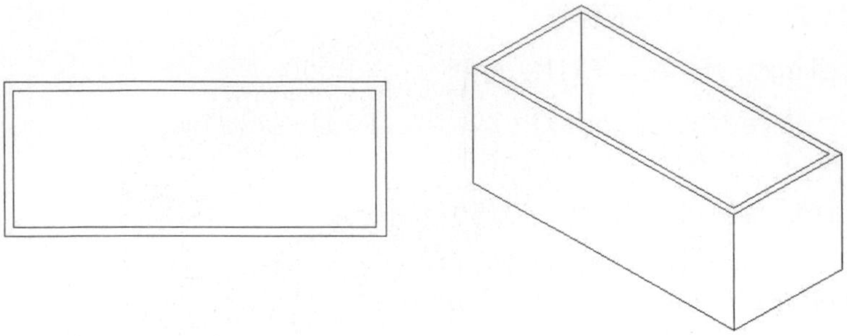

FIGURE 20.67

TRY IT!

Open the drawing file 20_Polysolid Walls. Use the following command sequence and Figure 20.68 to construct the 3D walls using the POLYSOLID **command.**

Command: **POLYSOLID**

Polysolid Height = 0'-4", Width = 0'-0 1/4", Justification = Center

Specify start point or [Object/Height/Width/Justify] <Object>: **H** *(For Height)*

Specify height <0'-4">: **8'**

Specify start point or [Object/Height/Width/Justify] <Object>: **W** *(For Width)*

Specify width <0'-0 1/4">: **4**

Specify start point or [Object/Height/Width/Justify] <Object>: *(Pick a point in the lower-left corner of the display screen)*

Specify next point or [Arc/Undo]: *(Move your cursor to the right and enter* **30'** *)*

Specify next point or [Arc/Undo]: *(Move your cursor up and enter* **10'** *)*

Specify next point or [Arc/Close/Undo]: *(Move your cursor to the left and enter* **5'** *)*

Specify next point or [Arc/Close/Undo]: *(Move your cursor up and enter* **5'** *)*

Specify next point or [Arc/Close/Undo]: *(Move your cursor to the left and enter* **25'** *)*

Specify next point or [Arc/Close/Undo]: **C** *(For Close)*

When finished, use the In-canvas viewport control to select a SE Isometric view and a Conceptual visual style to view the results.

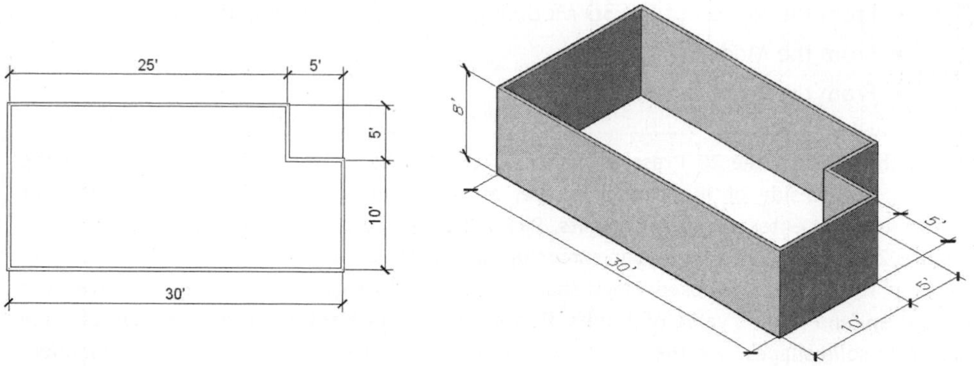

FIGURE 20.68

Polysolids can also be created from existing 2D geometry such as lines, polylines, arcs, and even circles. Illustrated in Figure 20.69 (Left) is a 2D polyline that has had fillets applied to a number of corners. Activating the POLYSOLID command and selecting the polyline changes the appearance of the object to match the illustration, as shown in Figure 20.69 (Center). Here, a width has been automatically applied to the polysolid. When the polysolid is viewed in 3D using a preset viewing direction such as SE Isometric, the polysolid is displayed with a height, as shown in Figure 20.69 (Right).

```
Command: POLYSOLID
Polysolid Height = 0'-4", Width = 0'-0 1/4", Justification =
Center
Specify start point or [Object/Height/Width/Justify]
<Object>: (Press ENTER to accept Object)
Select object: (Pick the polyline object as shown in
Figure 20.69 (Left))
```

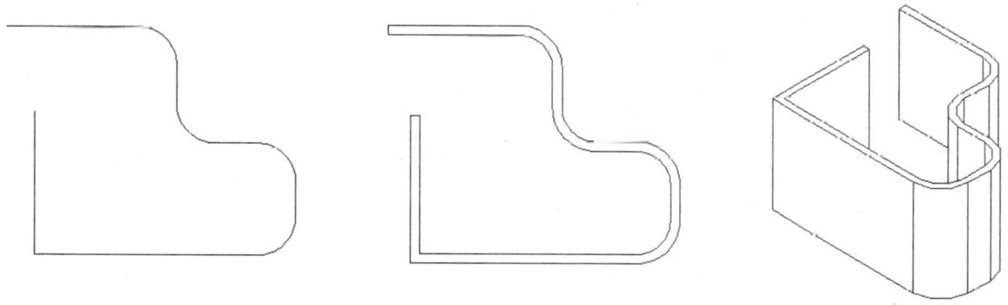

FIGURE 20.69

CREATING A SOLID USING PRESSPULL

An additional technique used for constructing solid models is available to speed up the construction and modification processes. This technique is called pressing and pulling, which is activated with the PRESSPULL command. Any closed area that can be hatched can be manipulated using the PRESSPULL command.

- From the Ribbon of the 3D Modeling workspace (Home and Solid tabs)
- From the Modeling toolbar
- From the keyboard (PRESSPULL)

TRY IT!

Open the drawing file 20_PressPull_Shapes. Activate the PRESSPULL command from the Ribbon, pick inside of the circular shape, as shown in Figure 20.70 (Left), move your cursor up, and enter a value of 6 units. Pick inside of the closed block shape, as shown in Figure 20.70 (Center), move your cursor up, and enter a value of 4 units. Pick one more time in the remaining closed block shape, as shown in Figure 20.70 (Right), move your cursor up, and enter a value of 2 units. Performing this task results in the creation of three separate solid shapes. Use the UNION command to join all shapes into one 3D solid model.

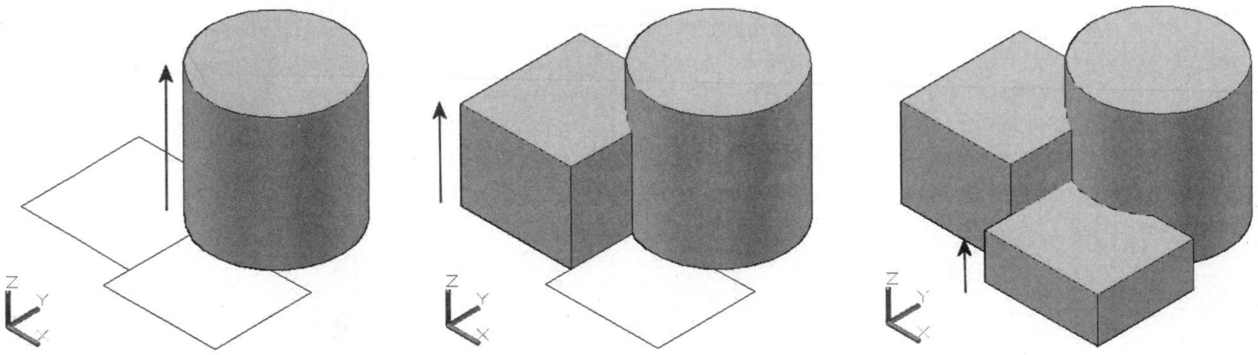

FIGURE 20.70

NOTE

Multiple bounded areas can be included in a single press-and-pull operation by holding down the SHIFT key as the areas are selected. A Multiple option is also available in the command sequence for performing this operation.

Besides creating a solid from a closed shape, this command can also be used to remove material by dragging a bounded area back through the model. The next Try It! exercise demonstrates this technique.

TRY IT!

Open the drawing file 20_PressPull Hole. Activate the PRESSPULL command and pick inside of the circular shape, as shown in Figure 20.71 (Left). Moving your cursor up creates the cylinder, as shown in Figure 20.71 (Right).

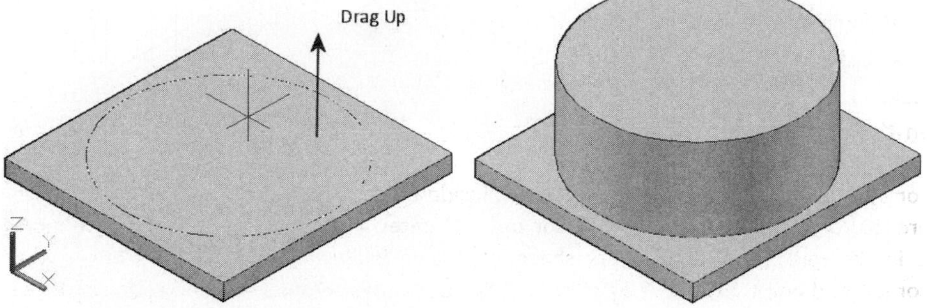

FIGURE 20.71

Undo the previous operation, reactivate the PRESSPULL command and pick inside of the circular shape, as shown in Figure 20.72 (Left). However, instead of moving your cursor up to form a cylinder, drag your cursor down into the thin block to automatically perform a subtraction operation and create a hole in the block.

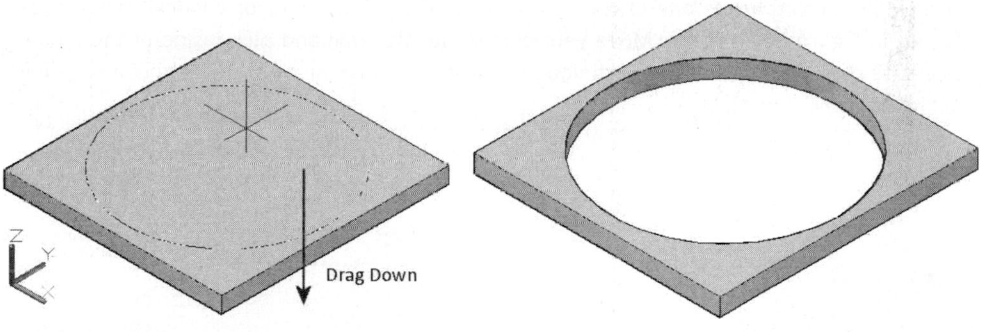

FIGURE 20.72

Another technique to activate the press-and-pull feature is to press and hold the CTRL + SHIFT + E keys, and then pick the area. You can perform the operation the same way as using the PRESSPULL command from the Ribbon. The next Try It! exercises illustrate this technique.

TRY IT!

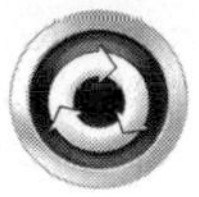

Open the drawing file 20_PressPull Plan. Press and hold down the CTRL + SHIFT + E keys while picking inside of the bounding area created by the inner and outer lines, as shown in Figure 20.73 (Left). Move your cursor up and enter a value of 8' to construct a solid model of the walls, as shown in Figure 20.73 (Right).

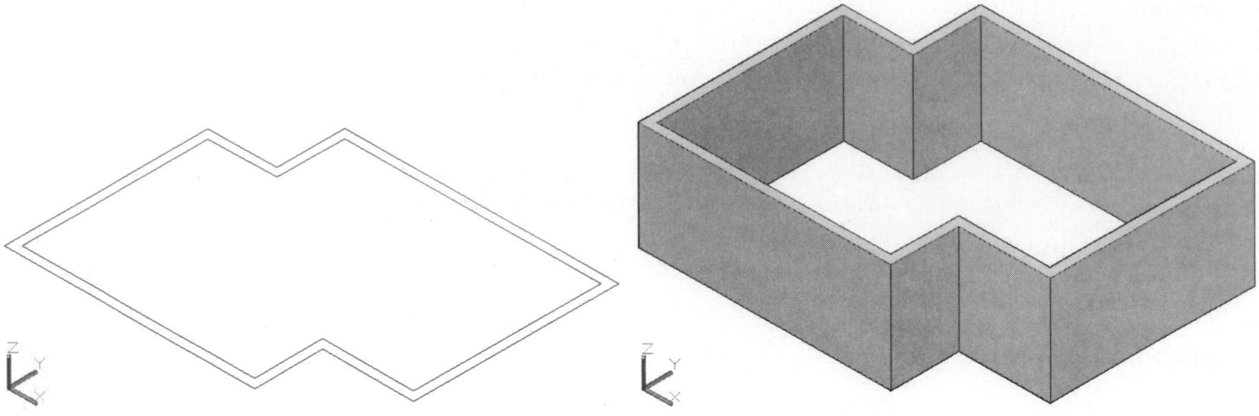

FIGURE 20.73

TRY IT!

Open the drawing file 21_PressPull Openings. Press and hold down the CTRL + SHIFT + E keys while picking inside one of the rectangles that signify a door or window opening, as shown in Figure 20.74 (Left). Move your cursor into the wall and pick inside of the model to create the opening. Use this technique for other openings as well. The results are shown in Figure 20.74 (Right).

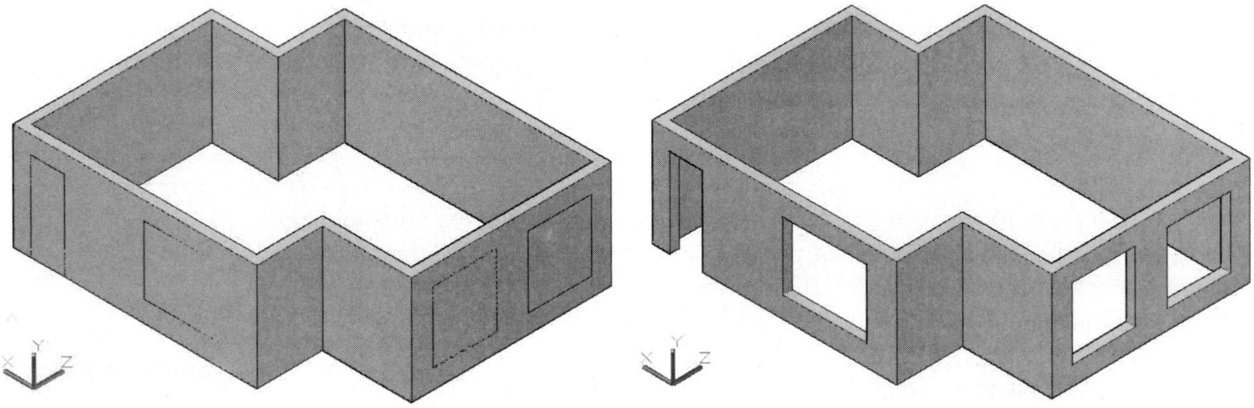

FIGURE 20.74

USING PRESS AND PULL ON BLOCKS

Pressing and pulling to create solid shapes is not just limited to closed shapes such as rectangles, circles, or polylines. The press-and-pull feature can also be used on block objects. In some cases, depending on how you use the press-and-pull feature, it is possible to convert the 2D block into a 3D solid model. The next Try It! exercise illustrates this.

TRY IT!

Open the drawing file 20_PressPull Bed. All objects that describe this 2D bed are made up of a single block. You will use the PRESSPULL command to highlight certain closed boundary areas and pull the area to a new height. In this way, PRESSPULL is used to convert a 2D object into a 3D model.

Activate the PRESSPULL command, move your cursor inside the area, as shown in Figure 20.65 (Left), pick and move your cursor up. Type in a value of 12 to extrude this area a distance of 12 units up, as shown in Figure 20.65 (Right).

Command: PRESSPULL

Select object or bounded area: *(Move your cursor over the area, as shown in Figure 20.65 (Left) and pick.)*

Specify extrusion height or [Multiple]: 12

1 extrusion(s) created

Select object or bounded area:

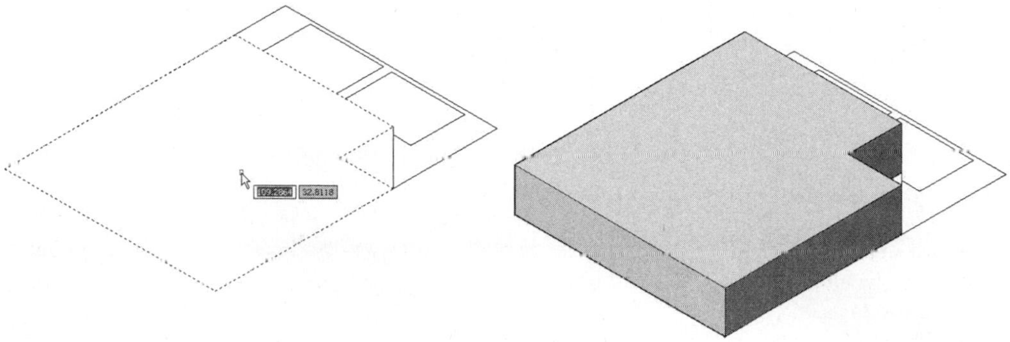

FIGURE 20.75

The PRESSPULL command is still active. Move your cursor inside the small triangular area, as shown in Figure 20.66 (Left), pick, and move your cursor up. Type in a value of 12.5 to extrude this area a distance of 12.5 units up, as shown in Figure 20.66 (Right).

Command: PRESSPULL

Select object or bounded area: *(Move your cursor over the triangular area, as shown in Figure 20.76 (Left) and pick.)*

Specify extrusion height or [Multiple]: 12.5

1 extrusion(s) created

Select object or bounded area:

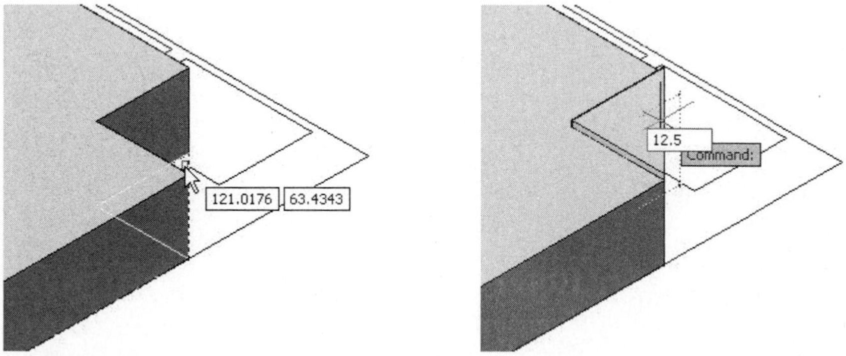

FIGURE 20.76

The PRESSPULL command is still active. Move your cursor inside the area, as shown in Figure 20.77 (Left), pick and move your cursor up. Type in a value of 12 to extrude this area a distance of 12 units up, as shown in Figure 20.77 (Right).

Command: PRESSPULL

Select object or bounded area: *(Move your cursor over the back area of the bed, as shown in Figure 20.77 (Left) and pick.)*

Specify extrusion height or [Multiple]: 12

1 extrusion(s) created

Select object or bounded area:

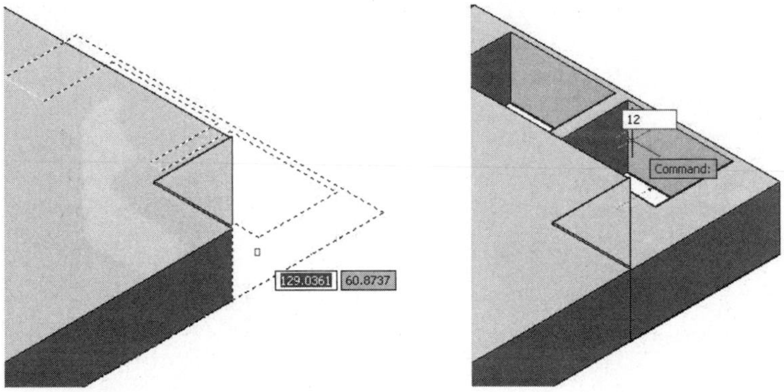

FIGURE 20.77

The PRESSPULL command is still active. Move your cursor inside the rectangular area represented as a pillow, as shown in Figure 20.78 (Left), pick, and move your cursor up. Type in a value of 15 to extrude this area a distance of 15 units up, as shown in Figure 20.78 (Right).

Command:PRESSPULL

Select object or bounded area: *(Move your cursor over the area representing the pillow, as shown in Figure 20.78 (Left) and pick.)*

Specify extrusion height or [Multiple]: 15

1 extrusion(s) created

Select object or bounded area:

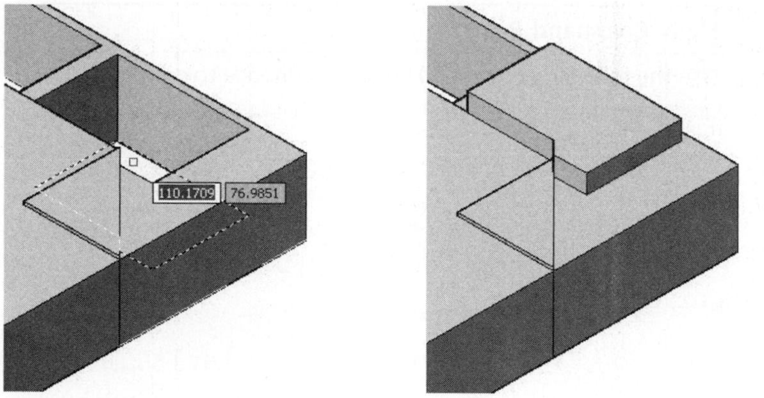

FIGURE 20.78

Perform the same press-and-pull operation on the second pillow. Extrude this shape a value of 15 units, as shown in Figure 20.79 (Left). Since the 3D objects created by the press-and-pull operations are all considered individual primitives, use the UNION command to join all primitives into a single 3D solid model. An alternate step would be to use the FILLET command to round off all corners and edges of the bed and pillows, as shown in Figure 20.79 (Right).

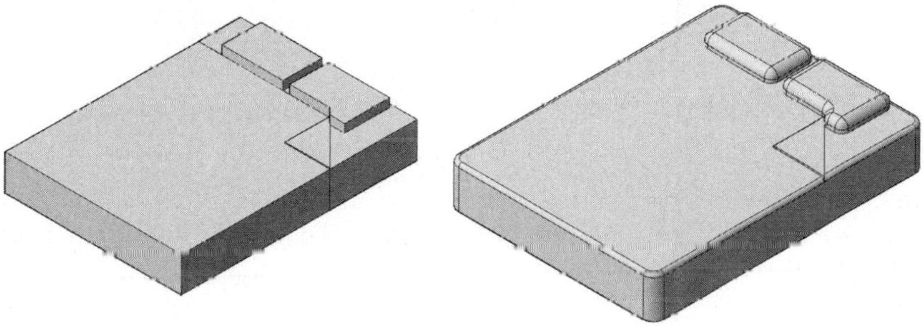

FIGURE 20.79

CREATING A HELIX

The HELIX command creates a 2D or 3D spiral object. Choose this command in one of the following ways:

- From the Ribbon of the 3D Modeling workspace (Home tab—expand the Draw panel)
- From the Menu Bar (Draw > Helix)
- From the Modeling toolbar
- From the keyboard (HELIX)

TRY IT!

Open the drawing file 20_Helix. Use the following command sequences and images for creating a 2D, 3D, and spiral helix.

Using the table below, experiment with the following command prompts for constructing a 2D helix, 3D helix, and 3D spiral, as shown in the illustrations on the left in Figure 20.80.

Helix Type	Helix Command Prompt
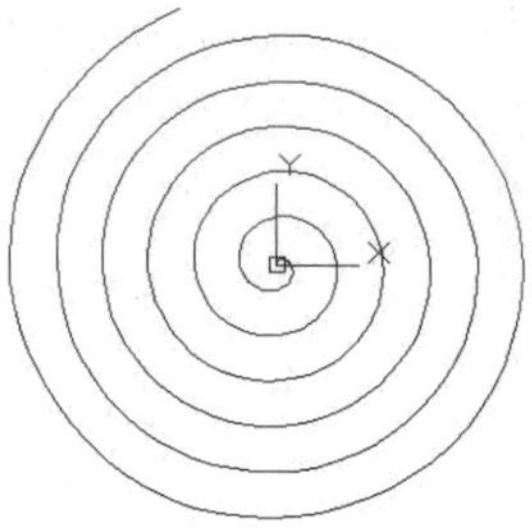	Use the following command prompt sequence to create a 2D Helix: Command: **HELIX** Number of turns = 3.0000 Twist = CCW Specify center point of base: **0,5,0** Specify base radius or [Diameter] <1.0000>: **1.00** Specify top radius or [Diameter] <1.0000>: **0** Specify helix height or [Axis endpoint/ Turns/turn Height/tWist] <1.0000>: **T** Enter number of turns <3.0000>: **6** Specify helix height or [Axis endpoint/ Turns/turn Height/tWist] <1.0000>: **0**
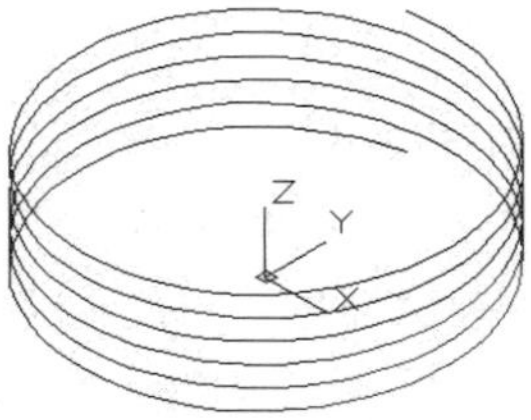	Use the following command prompt sequence to create a 3D Helix: Command: **HELIX** Number of turns = 3.0000 Twist = CCW Specify center point of base: **0,15,0** Specify base radius or [Diameter] <1.0000>: **3** Specify top radius or [Diameter] <3.0000>: **3** Specify helix height or [Axis endpoint/ Turns/turn Height/tWist] <1.0000>: **T** Enter number of turns <3.0000>: **6** Specify helix height or [Axis endpoint/ Turns/turn Height/tWist] <1.0000>: **2**
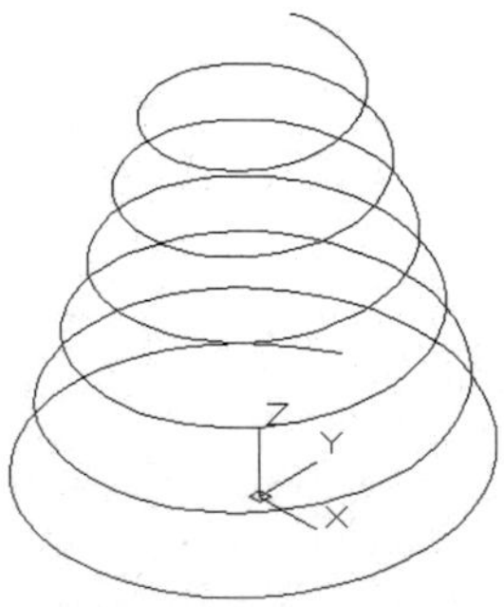	Use the following command prompt sequence to create a 3D Spiral: Command: **HELIX** Number of turns = 3.0000 Twist = CCW Specify center point of base: **0,30,0** Specify base radius or [Diameter] <1.0000>: **5** Specify top radius or [Diameter] <5.0000>: **2** Specify helix height or [Axis endpoint/ Turns/turn Height/tWist] <1.0000>: **T** Enter number of turns <3.0000>: **6** Specify helix height or [Axis endpoint/ Turns/turn Height/tWist] <1.0000>: **10**

FIGURE 20.80

HELIX APPLICATIONS

Typically, a wireframe model of a helix does not fully define how an object like a spring should look. It would be beneficial to show the spring as a thin wire wrapping around a cylinder to form the helical shape. To produce this type of object, use the wireframe of a circle as the object to sweep around the helix to produce the spring.

TRY IT!

Open the drawing file 20_Helix Spring. A helix is already created, along with small circular profile. With the helix as a path and the circle as the object to sweep, use the following command prompt and Figure 20.81 to create a spring.

Command: SWEEP

Current wire frame density: ISOLINES=4, Closed profiles creation mode = Solid

Select objects to sweep or [MOde]: *(Select the small circle)*

Select objects to sweep or [MOde]: *(Press ENTER to continue)*

Select sweep path or [Alignment/Base point/Scale/Twist]: *(Select the helix)*

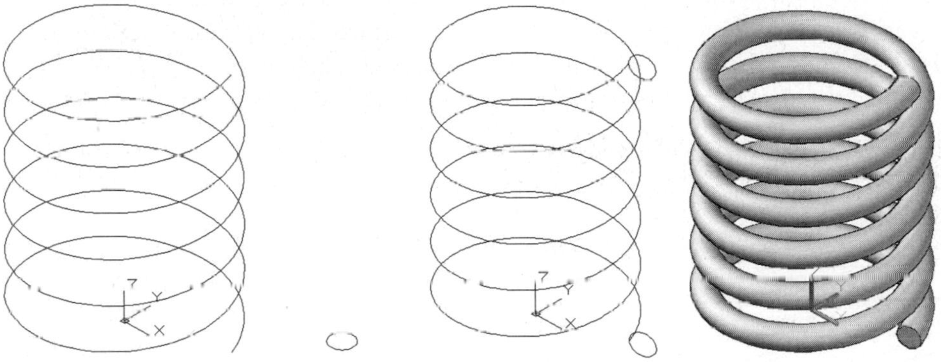

FIGURE 20.81

OBTAINING MASS PROPERTIES OF A SOLID MODEL

The MASSPROP command calculates the mass properties of a solid model. Choose this command in one of the following ways:

- From the Menu Bar (Tools > Inquiry > Region/Mass Properties)
- From the Inquiry toolbar
- From the keyboard (MASSPROP)

TRY IT!

Open the drawing file 20_Tee Massprop. Use the MASSPROP command to calculate the mass properties of a selected solid, as shown in Figure 20.82. All calculations are based on the current position of the UCS. You will be given the option of writing this information to a file if desired.

Command: MASSPROP

Select objects: *(Select the model)*

Select objects: *(Press ENTER to continue with this command)*

```
----------------   SOLIDS   ----------------

Mass:                     50.1410
Volume:                   50.1410
Bounding box:          X: 5.0974  --  10.6984
                       Y: 1.3497  --  4.5516
                       Z: 0.0000  --  5.0000
Centroid:              X: 7.2394
                       Y: 2.9507
                       Z: 2.5000
Moments of inertia:    X: 865.9024
                       Y: 3118.9600
                       Z: 3184.2327
Products of inertia:  XY: 1071.0661
                      YZ: 369.8746
                      ZX: 907.4749
Radii of gyration:     X: 4.1556
                       Y: 7.8869
                       Z: 7.9690
Principal moments and X-Y-Z directions about centroid:
                       I: 115.9690 along [1.0000 0.0000 0.0000]
                       J: 177.7544 along [0.0000 1.0000 0.0000]
                       K: 119.8560 along [0.0000 0.0000 1.0000]
```

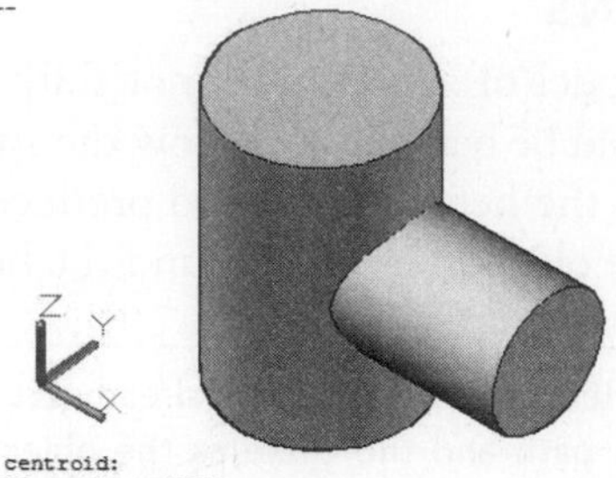

FIGURE 20.82

SYSTEM VARIABLES THAT AFFECT SOLID MODELS

System variables are available to control the appearance of your solid model whenever you perform shading or hidden line removal operations. Three in particular are useful: `ISOLINES`, `FACETRES`, and `DISPSILH`. The following text describes each system variable in detail.

The ISOLINES (Isometric Lines) System Variable

Tessellation refers to the lines that are displayed on any curved surface, such as those shown in Figure 20.83, to help you visualize the surface. Tessellation lines are automatically formed when you construct solid primitives such as cylinders and cones. These lines are also calculated when you perform solid modeling operations such as `SUBTRACT` and `UNION`.

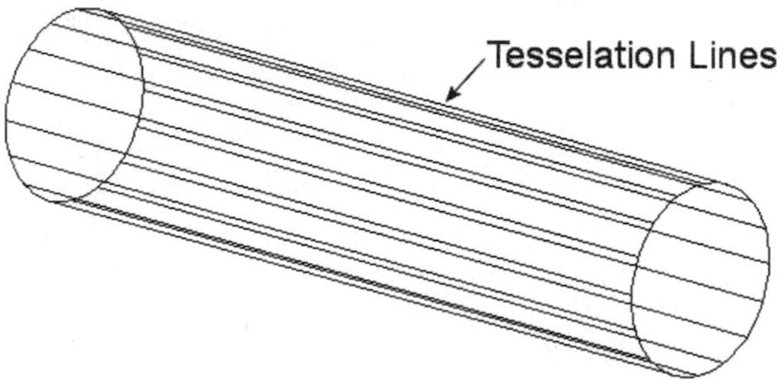

FIGURE 20.83

The number of tessellation lines per curved object is controlled by the system variable called `ISOLINES`. By default, this variable is set to a value of 4. Figure 20.84 shows the results of setting this variable to other values, such as 9 and 20. After the isolines have been changed, regenerate the screen to view the results. The more lines used to describe a curved surface, the more accurate the surface will look in wireframe mode; however, it will take longer to process screen regenerations.

TRY IT!

Open the drawing file 20_Tee Isolines. Experiment by changing the `ISOLINES` system variable to numerous values. Perform a drawing regeneration using the `REGEN` command each time that you change the number of isolines.

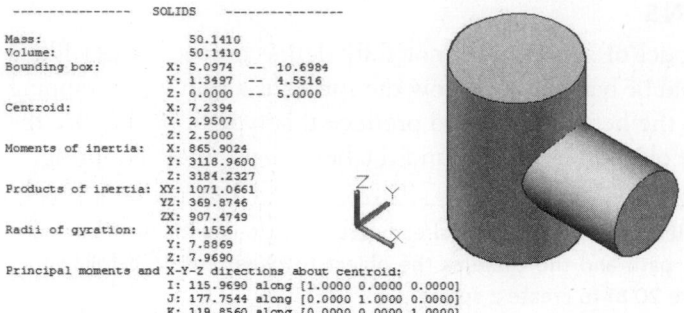

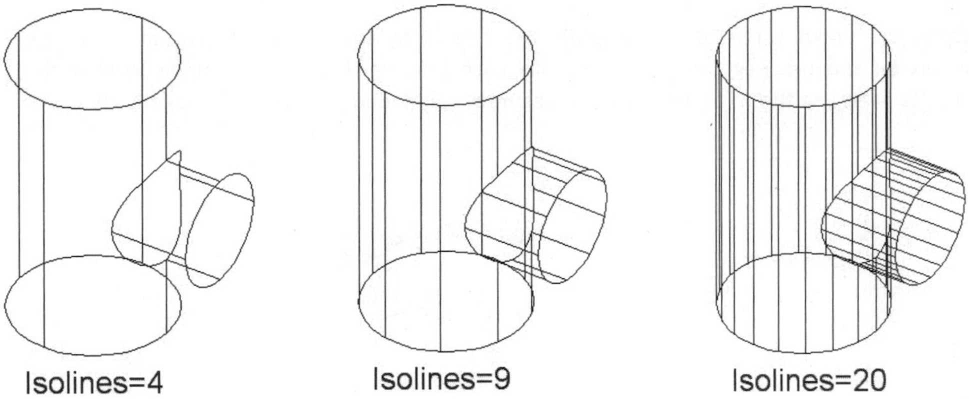

FIGURE 20.84

The FACETRES (Facet Resolution) System Variable

When you perform hidden line removals on solid objects, the results are similar to those displayed in Figure 20.85. The curved surfaces are now displayed as flat triangular surfaces (faces), and the display of these faces is controlled by the FACETRES system variable. The cylinder with FACETRES set to 0.50 processes much more quickly than the cylinder with FACETRES set to 2, because there are fewer surfaces to process in such operations as hidden line removals. However, the image with FACETRES set to a large value (as high as 10) shows a more defined circle. The default value for FACETRES is 0.50, which seems adequate for most applications.

TRY IT!

Open the drawing file 20_Tee Facetres. Experiment by changing the FACETRES system variable to numerous values (valid values are 0.01 to 10). Perform a HIDE each time you change the number of facet lines to view the results.

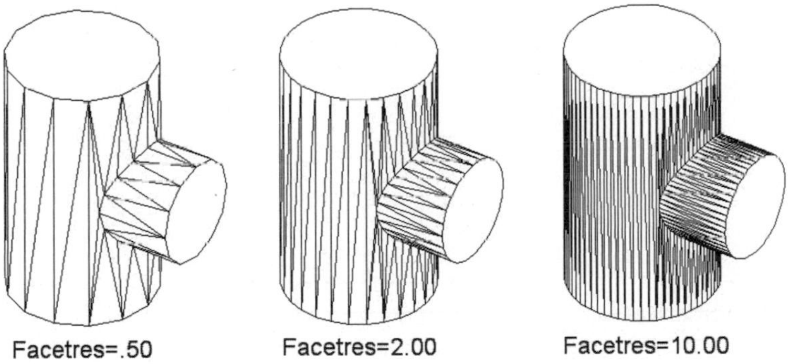

FIGURE 20.85

The DISPSILH (Display Silhouette) System Variable

To have the edges of your solid model take on the appearance of an isometric drawing when displayed as a wireframe, use the DISPSILH system variable. This system variable means "display silhouette" and is used to control the display of silhouette curves of solid objects while in either the wireframe or hide mode. Silhouette edges are turned either on or off, with the results displayed in Figure 20.86; by default, they are turned off. When a hide is performed, this system variable controls whether faces are drawn or suppressed on a solid model.

TRY IT!

Open the drawing file 20_Tee Dispsilh. Experiment by turning the display of silhouette edges on and off. Regenerate your display each time you change this mode to view the results. Also use the HIDE command to see how the hidden line removal image is changed.

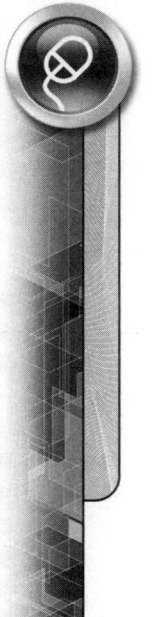

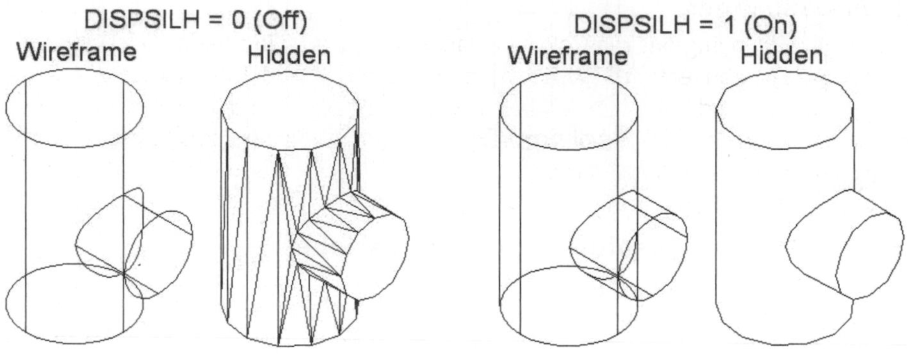

FIGURE 20.86

TUTORIAL EXERCISE: 20_COLLAR.DWG

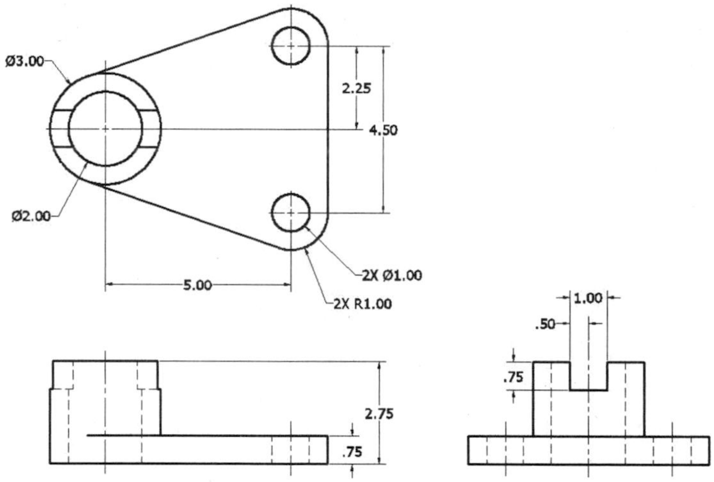

FIGURE 20.87

Purpose

This tutorial is designed to construct a solid model of the Collar using the dimensions in Figure 20.87.

System Settings

Start a new drawing from scratch. Use the current limits set to 0,0 for the lower-left corner and (12,9) for the upper-right corner. Change the number of decimal places from four to two using the Drawing Units dialog box.

Layers

Create the following layer:

Name	Color	Linetype
Model	Cyan	Continuous

Suggested Commands

Begin this tutorial by laying out the Collar in plan view and drawing the basic shape outlined in the top view. Convert the objects to a polyline and extrude the objects to form a solid. Draw a cylinder and combine this object with the base. Add another cylinder and then subtract it to form the large hole through the model. Add two small cylinders and subtract them from the base to form the smaller holes. Construct a solid box and subtract it to form the slot across the large cylinder.

STEP 1

Begin the Collar by setting the Model layer current. Then draw the three circles shown in Figure 20.88 (Left) using the CIRCLE command. Perform a ZOOM-All after all three circles have been constructed.

STEP 2

Draw lines tangent to the three arcs using the LINE command and the OSNAP-Tangent mode, as shown in Figure 20.88 (Right). Notice that the UCS icon has been turned off in Figure 20.88.

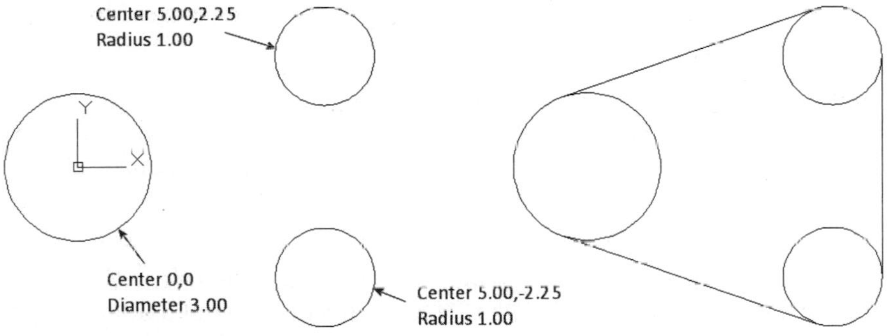

FIGURE 20.88

STEP 3

Use the TRIM command to trim the circles. When prompted to select the cutting edge object, press ENTER; this makes cutting edges out of all objects in the drawing. Perform this trimming operation so your display appears similar to the illustration in Figure 20.89 (Left).

STEP 4

Prepare to construct the base by viewing the object in 3D. Select a viewpoint by choosing SE Isometric from the In-canvas viewport control (located in the upper left corner of the viewport). Your display should appear similar to the illustration in Figure 20.89 (Right).

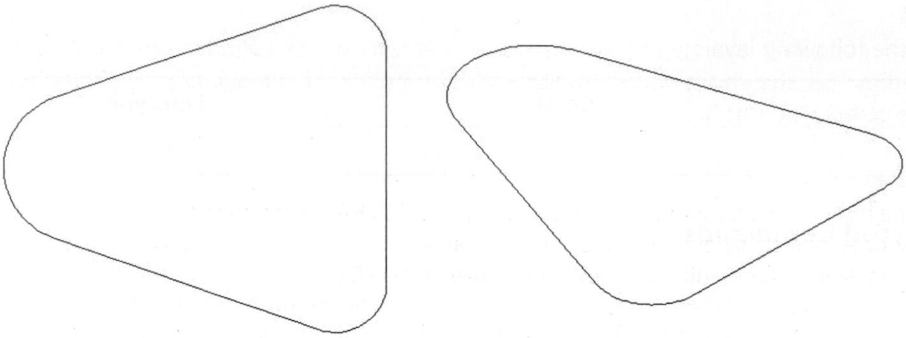

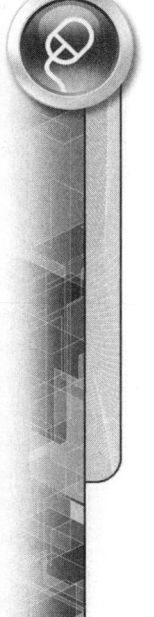

FIGURE 20.89

STEP 5

Convert all objects to a polyline using the Join option of the PEDIT command, as shown in Figure 20.90 (Left).

STEP 6

Use the EXTRUDE command to extrude the base to a thickness of 0.75 units, as shown in Figure 20.90 (Right).

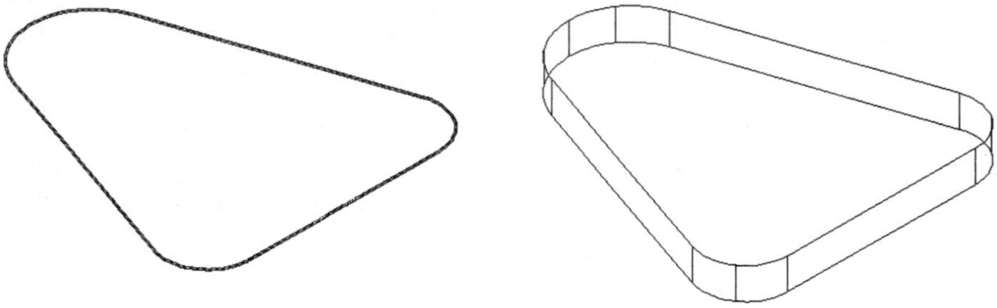

FIGURE 20.90

STEP 7

Turn off Dynamic UCS on the Status Bar. Create a cylinder using the CYLINDER command. Begin the center point of the cylinder at 0,0,0, with a diameter of 3.00 units and a height of 2.75 units. You may have to perform a ZOOM-All to display the entire model, as shown in Figure 20.91 (Left).

STEP 8

Merge the cylinder just created with the extruded base using the UNION command, as shown in Figure 20.91 (Right).

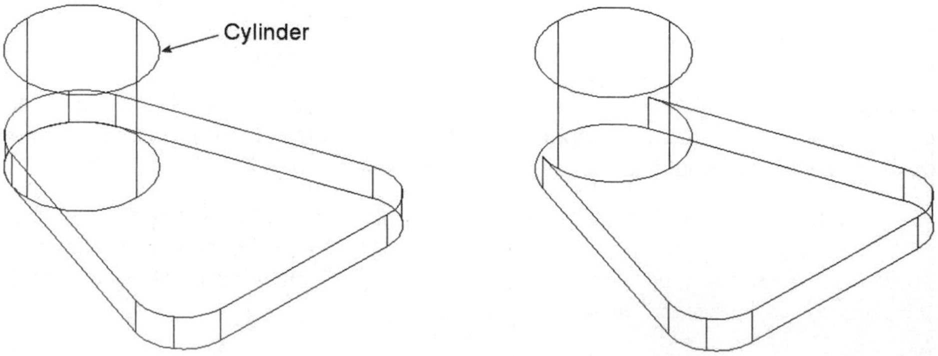

FIGURE 20.91

STEP 9

Use the CYLINDER command to create a 2.00-unit-diameter cylinder representing a through hole, as shown in Figure 20.92 (Left). The height of the cylinder is 2.75 units, with the center point at 0,0,0.

STEP 10

To cut the hole through the outer cylinder, use the SUBTRACT command. Select the base as the source object; select the inner cylinder as the object to subtract. Use the HIDE command to view the results, as shown in Figure 20.92 (Right).

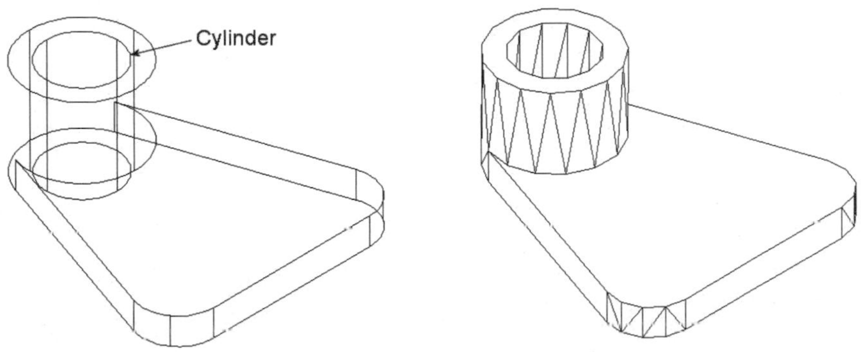

FIGURE 20.92

STEP 11

Use the REGEN command to regenerate your screen and return to Wireframe mode. Begin placing the two small drill holes (1.00 diameter and 0.75 high) in the base using the CYLINDER command. Use the OSNAP-Center mode to place each cylinder at the center of arcs "A" and "B," as shown in Figure 20.93 (Left).

STEP 12

Subtract both 1.00-diameter cylinders from the base of the model using the SUBTRACT command. Use the HIDE command to view the results, as shown in Figure 20.93 (Right).

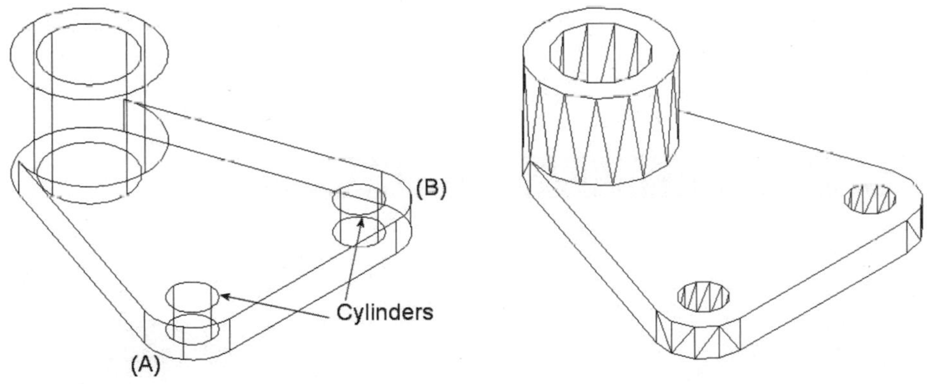

FIGURE 20.93

STEP 13

Begin constructing the rectangular slot that will pass through the two cylinders. Use the BOX command and Center option to accomplish this. Locate the center of the box at 0,0,2.75 and make the box 4 units long (X-direction), 1 unit wide (Y-direction), and 1.50 units high (Z-direction), as shown in Figure 20.94.

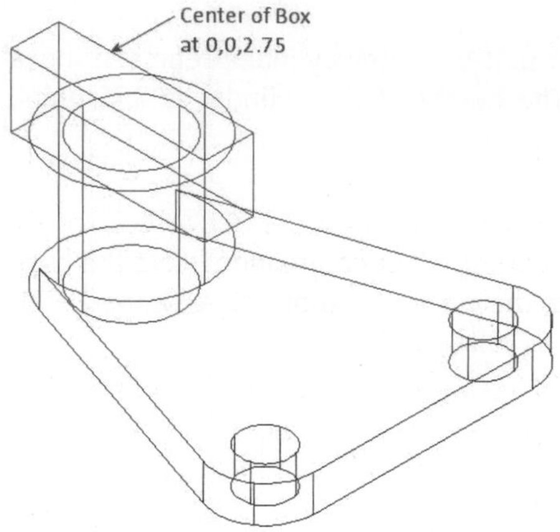

FIGURE 20.94

STEP 14

Use the SUBTRACT command to subtract the rectangular box from the solid model, as shown in Figure 20.95 (Left).

STEP 15

Change the facet resolution to a higher value using the FACETRES system variable and a value of 5. Then perform a hidden line removal to see the appearance of the completed model, as shown in Figure 20.95 (Right).

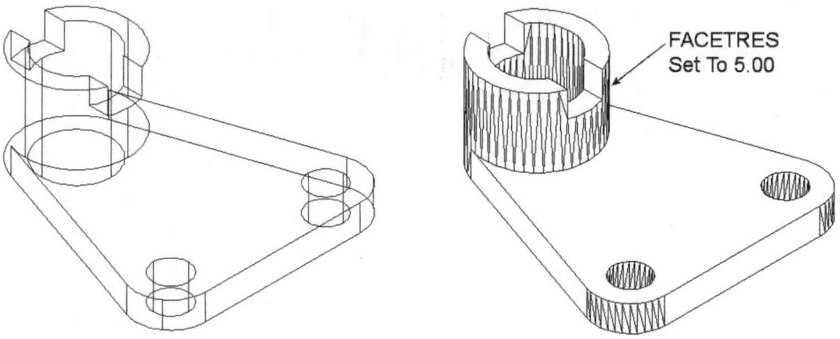

FIGURE 20.95

TUTORIAL EXERCISE: 20_VACUUM ATTACHMENT.DWG

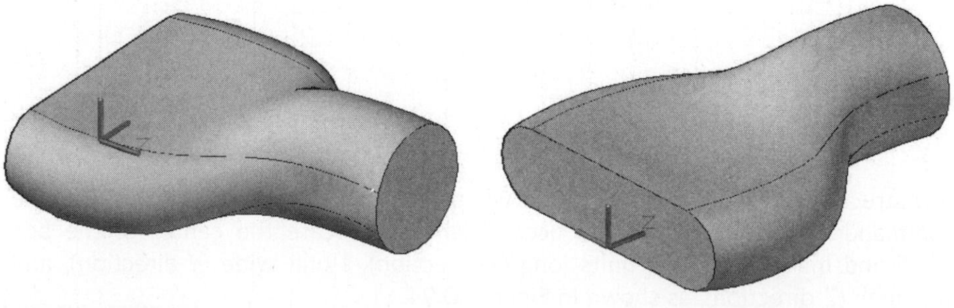

FIGURE 20.96

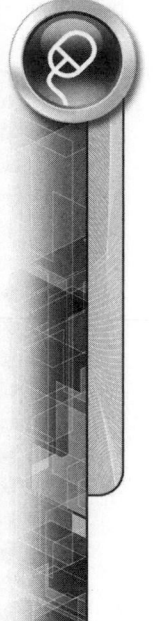

Purpose

This tutorial is designed to construct a solid model of the vacuum cleaner attachment using the Loft tool, as shown in Figure 20.96.

System Settings

Begin a new drawing from scratch. Use the current limits, which are set to 0,0 for the lower-left corner and (12,9) for the upper-right corner. Change the number of decimal places from four to two using the Drawing Units dialog box. Turn off Dynamic Input on the Status Bar.

Layers

Create the following layer:

Name	Color	Linetype
Model	Cyan	Continuous

Suggested Commands

Begin this tutorial by laying out a slot shape and circle. Next move the midpoint of the bottom of the slot to 0,0,0 and then copy it to 0,0,2. Move the bottom quadrant of the circle to 0,0,3.5 and then copy it to 0,0,5. These steps form the four cross sections used for creating the loft.

STEP 1

Begin the construction of this 3D model by constructing 2D geometry that will define the final shape of the vacuum cleaner attachment. Use the PLINE command to construct the slot shape as a closed polyline, using the dimensions shown in Figure 20.97 (Left). Then construct the circle using the diameter dimension shown in Figure 20.97 (Right).

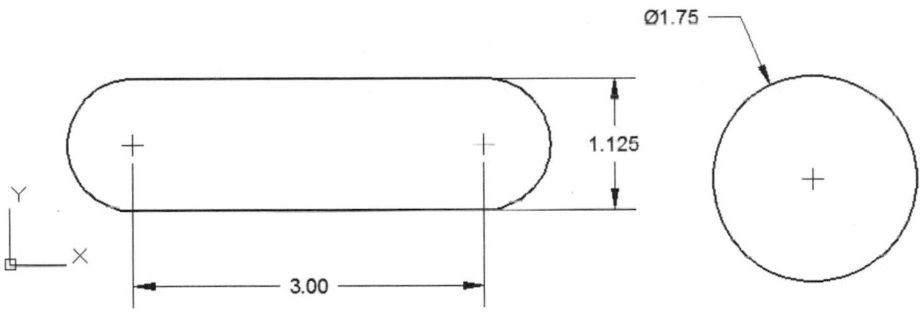

FIGURE 20.97

STEP 2

Switch to SE Isometric viewing, as shown in Figure 20.98. Use the In-canvas viewport control to establish this viewing direction. Before continuing, be sure that Dynamic Input is turned off for this segment. Next, move the slot shape from the midpoint of the bottom line to point 0,0,0.

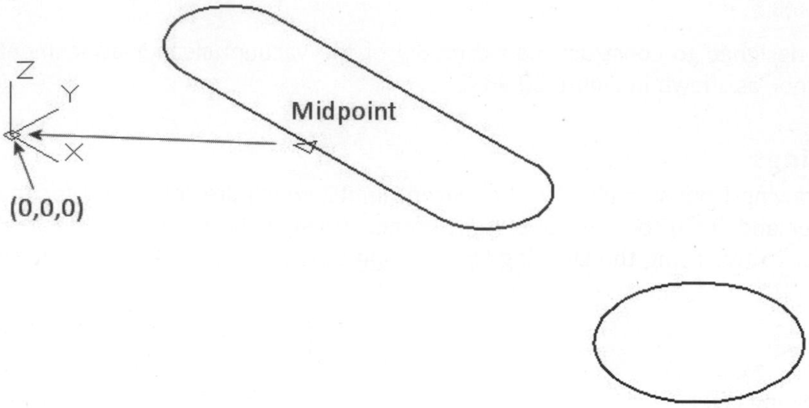

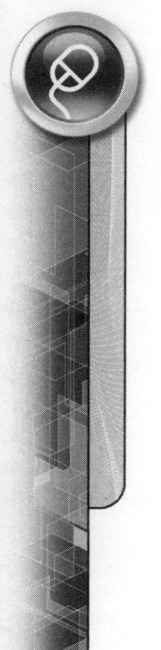

FIGURE 20.98

STEP 3

With the slot moved to the correct position, use the COPY command to create a duplicate shape of the slot using a base point of 0,0,0 and a second point at 0,0,2. When finished, your display should appear similar to Figure 20.99.

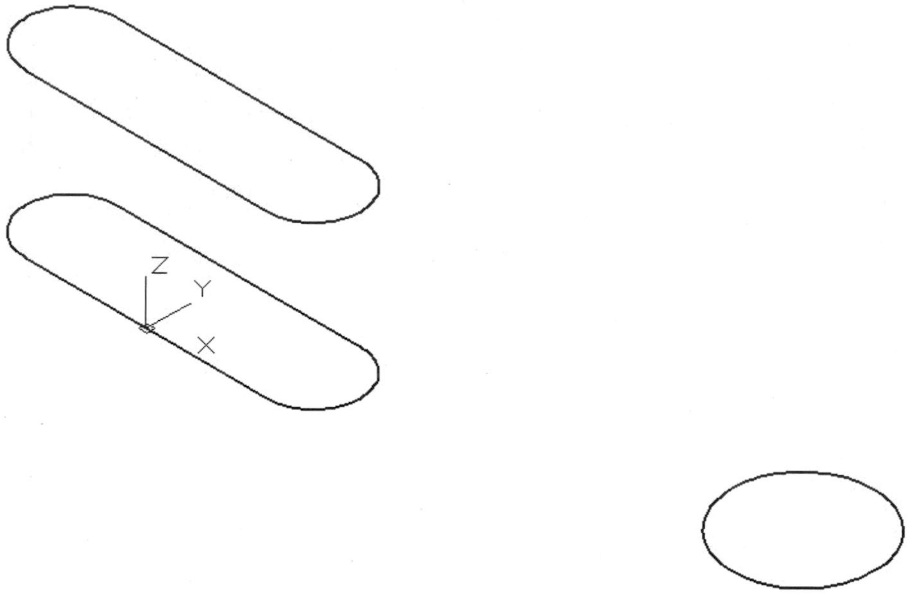

FIGURE 20.99

STEP 4

You will next move the circle. Enter the MOVE command and pick the base point of the move using the bottom quadrant of the circle, as shown in Figure 20.100 (Right). For the second point of displacement, enter the coordinate value of 0,0,3.5 from the keyboard.

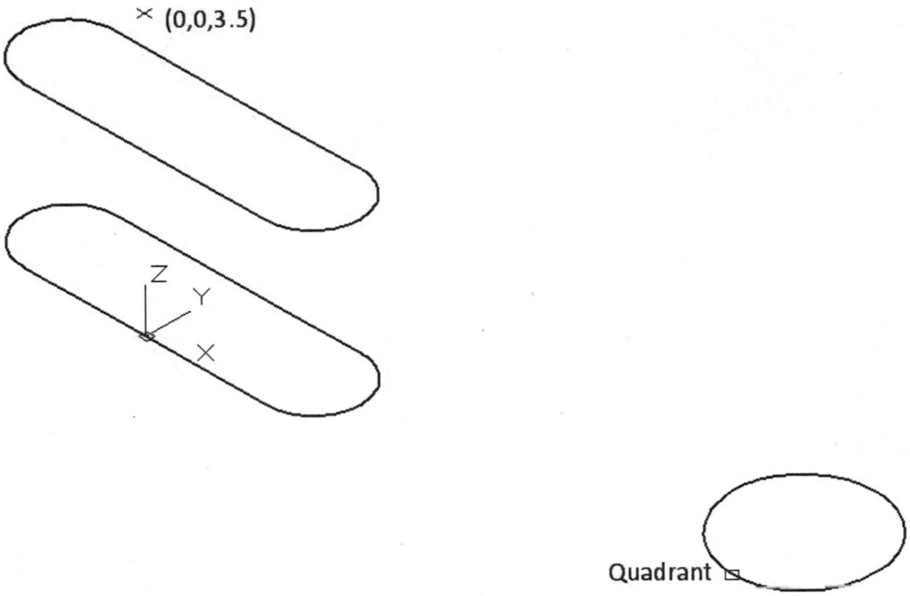

FIGURE 20.100

STEP 5

With the circle moved to the correct location, use the COPY command to duplicate this circle. Copy this circle from the bottom quadrant to a second point located at 0,0,5, as shown in Figure 20.101 (Left). When finished, your display should appear similar to the illustration in Figure 20.101 (Right). These form the four cross sectional shapes that make up the vacuum cleaner attachment.

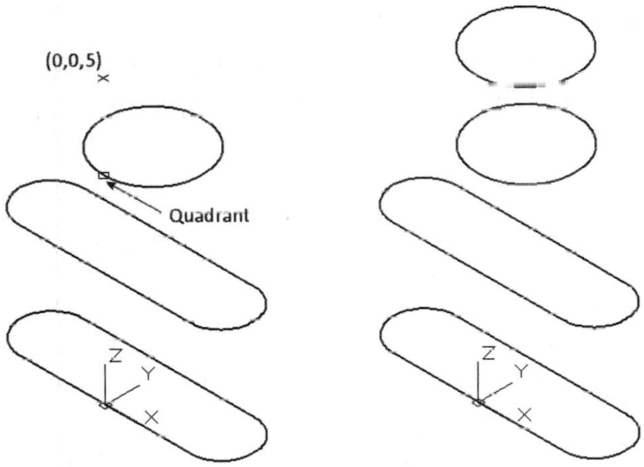

FIGURE 20.101

STEP 6

Issue the PLAN command through the keyboard and observe that the bottom midpoints of both slots and the bottom quadrants of both circles are all aligned to 0,0,0, as shown in Figure 20.102. When finished, perform a ZOOM-Previous operation to return to the SE Isometric view.

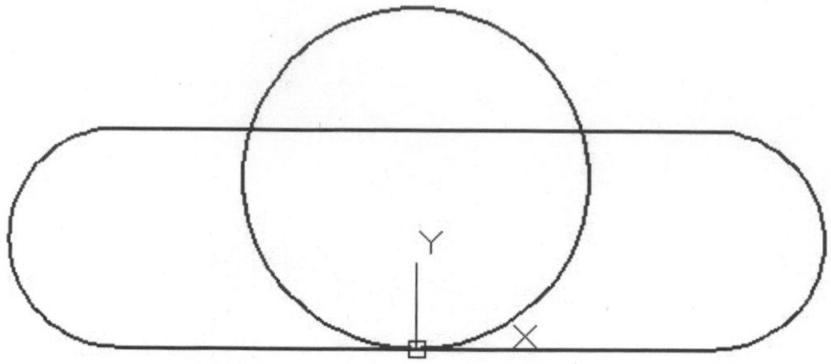

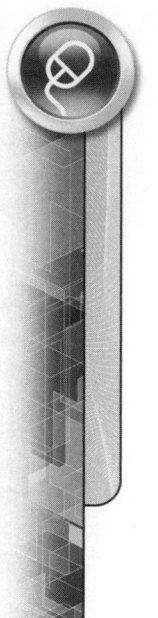

FIGURE 20.102

STEP 7

Use the 3DFORBIT (Free Orbit) command and rotate your model so it appears similar to Figure 20.103.

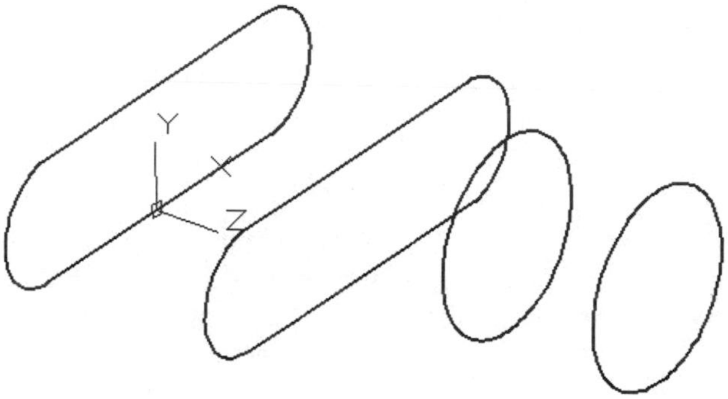

FIGURE 20.103

STEP 8

Activate the LOFT command and pick the four cross sections in the order labeled 1 through 4, as shown in Figure 20.104 (Left). When the List grip appears, keep the default Surface control at cross-sections setting at Smooth Fit and press ENTER, as shown in Figure 20.104 (Right).

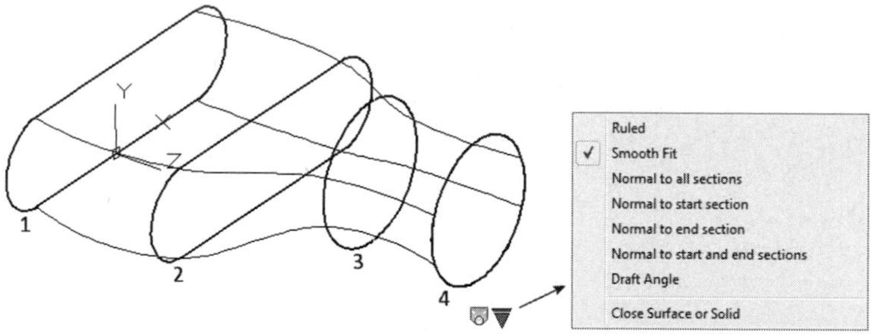

FIGURE 20.104

STEP 9

The results are illustrated in Figure 20.105.

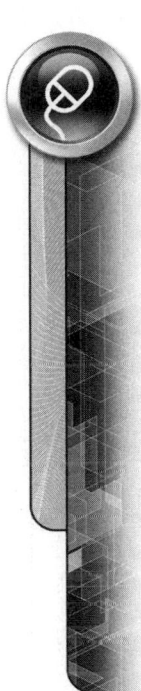

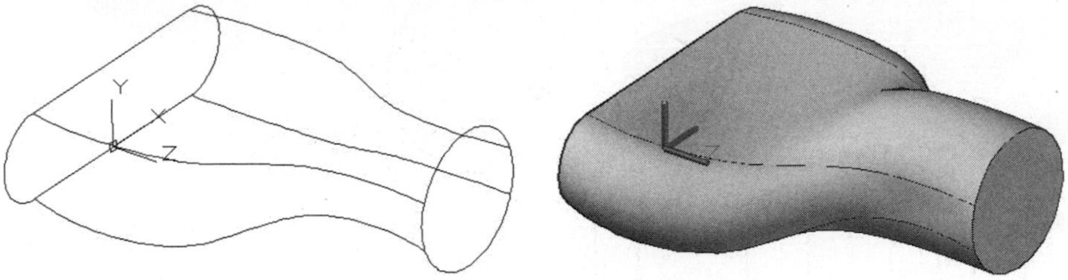

FIGURE 20.105

STEP 10

An alternate step would be to activate the SOLIDEDIT command and use the Shell option to produce a thin wall of 0.10 units on the inside of the vacuum attachment. This command and option will be discussed in greater detail in Chapter 21. Use the following command prompt sequence and Figure 20.106 to perform this operation.

Command: SOLIDEDIT

Solids editing automatic checking: SOLIDCHECK=1

Enter a solids editing option [Face/Edge/Body/Undo/eXit] <eXit>: B *(For Body)*

Enter a body editing option

[Imprint/seParate solids/Shell/cLean/Check/Undo/eXit] <eXit>: S *(For Shell)*

Select a 3D solid: *(Select the vacuum attachment)*

Remove faces or [Undo/Add/ALL]: *(Pick the face at "A")*

Remove faces or [Undo/Add/ALL]: *(Pick the face at "B")*

Remove faces or [Undo/Add/ALL]: *(Press* ENTER *to continue)*

Enter the shell offset distance: 0.10

Solid validation started.

Solid validation completed.

Enter a body editing option

[Imprint/seParate solids/Shell/cLean/Check/Undo/eXit] <eXit>: *(Press* ENTER*)*

Solids editing automatic checking: SOLIDCHECK=1 Enter a solids editing option [Face/Edge/Body/Undo/eXit] <eXit>: *(Press* ENTER*)*

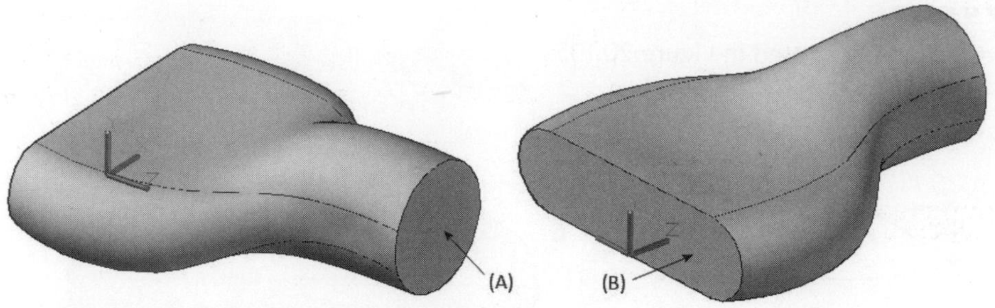

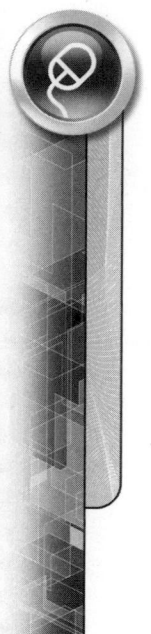

FIGURE 20.106

STEP 11

When finished completing the Shell option of the `SOLIDEDIT` command, your model should appear similar to Figure 20.107.

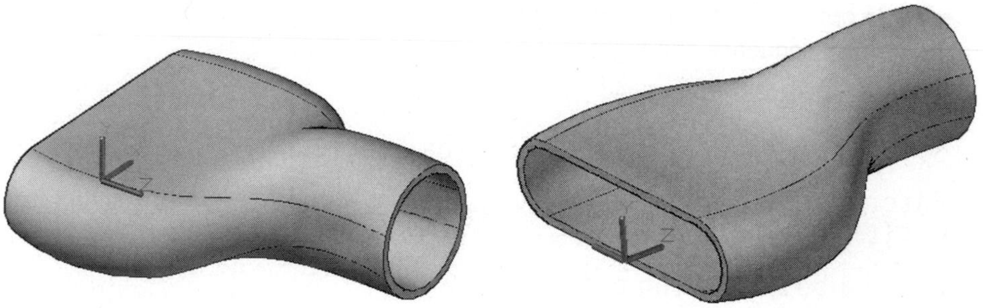

FIGURE 20.107

PROBLEMS FOR CHAPTER 20

1. Create a 3D solid model of each object on Layer "Model."
2. When finished, calculate the volume of the solid model using the MASSPROP command.

PROBLEM 20-1

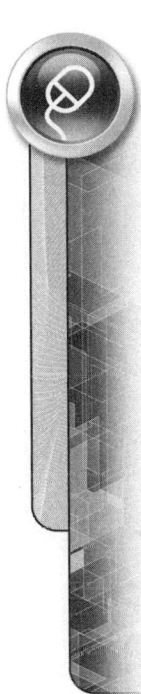

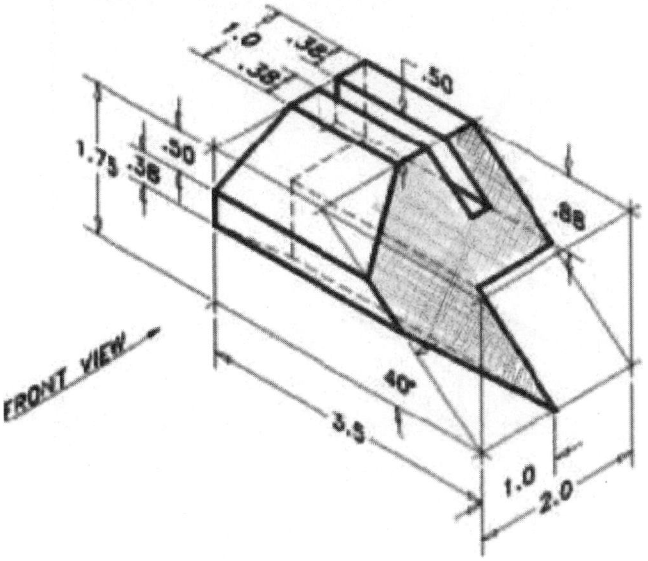

PROBLEM 20-2

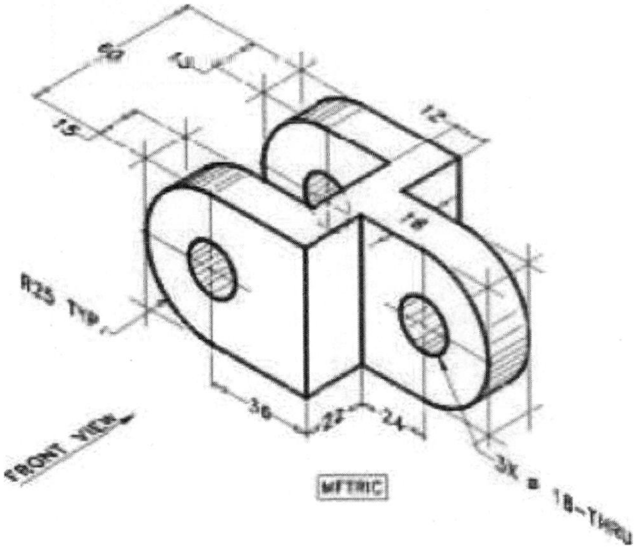

PROBLEM 20-3

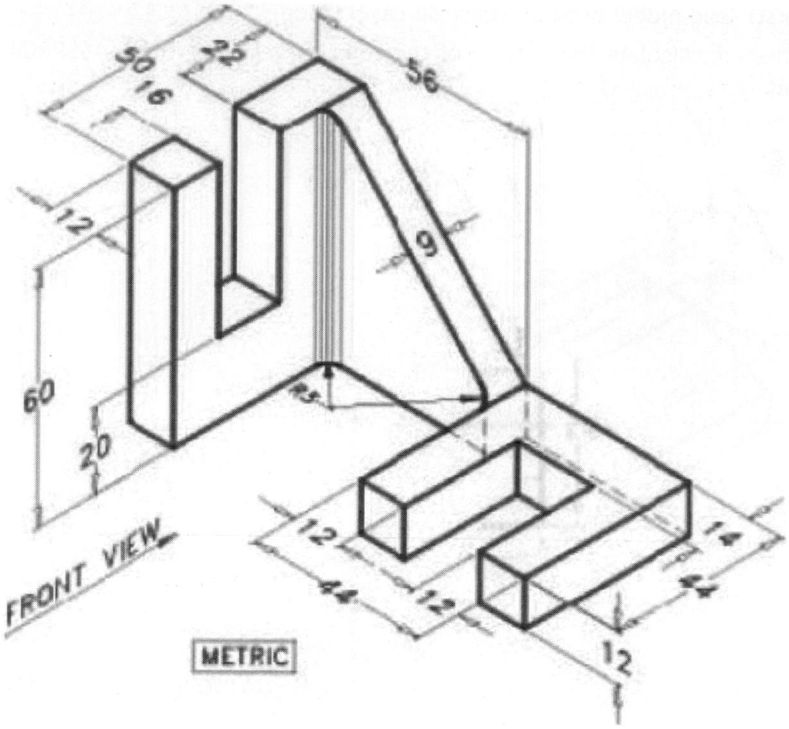

PROBLEM 20-4

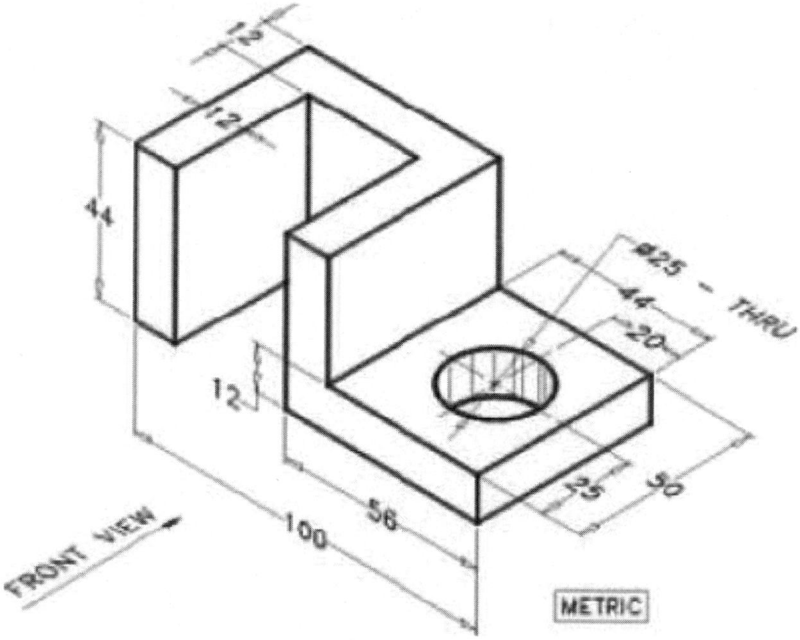

PROBLEM 20-5

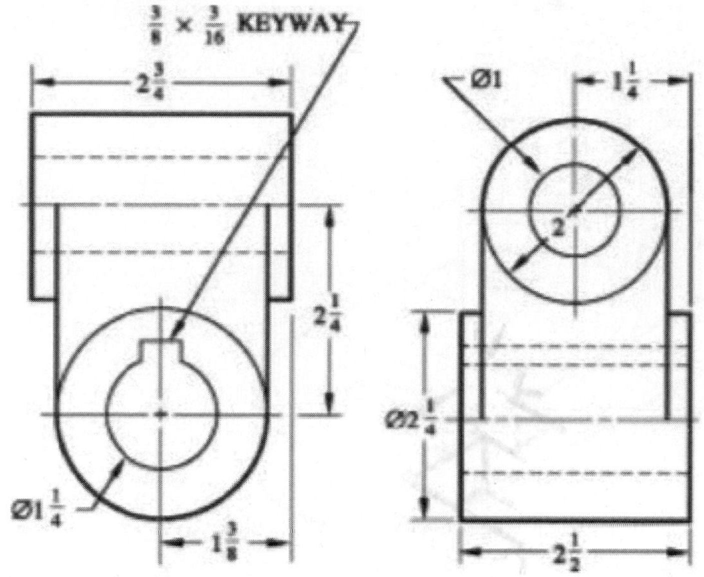

PROBLEM 20-6

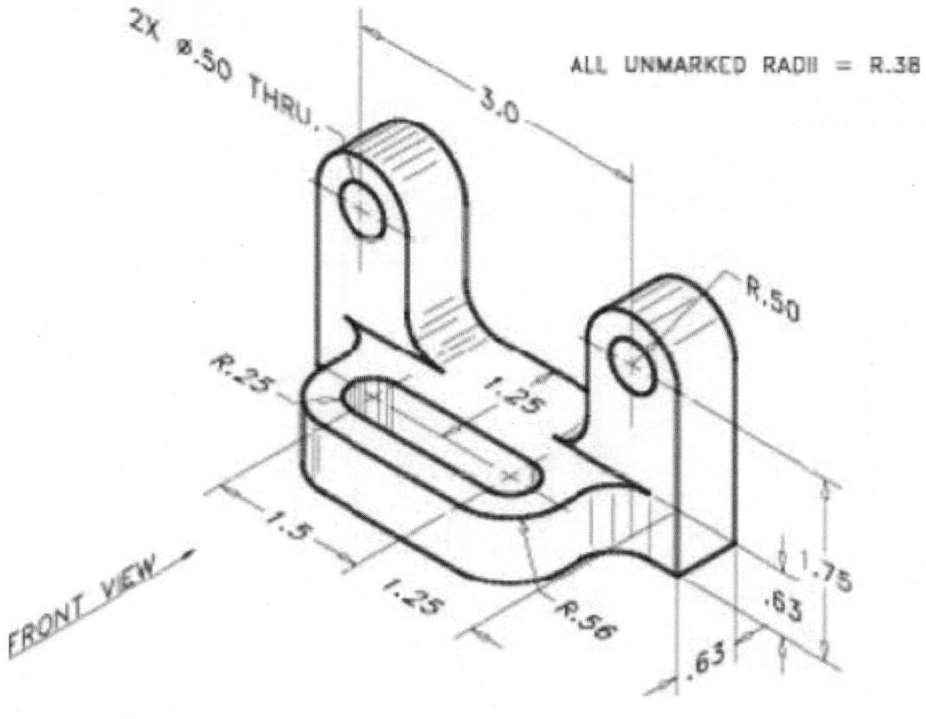

PROBLEM 20-7

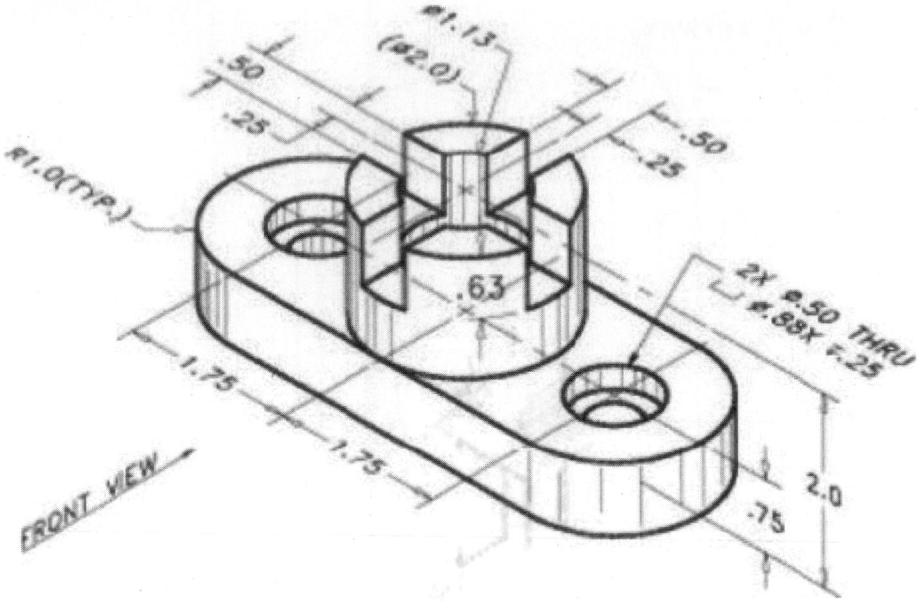

PROBLEM 20-8

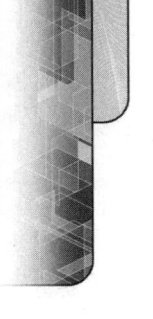

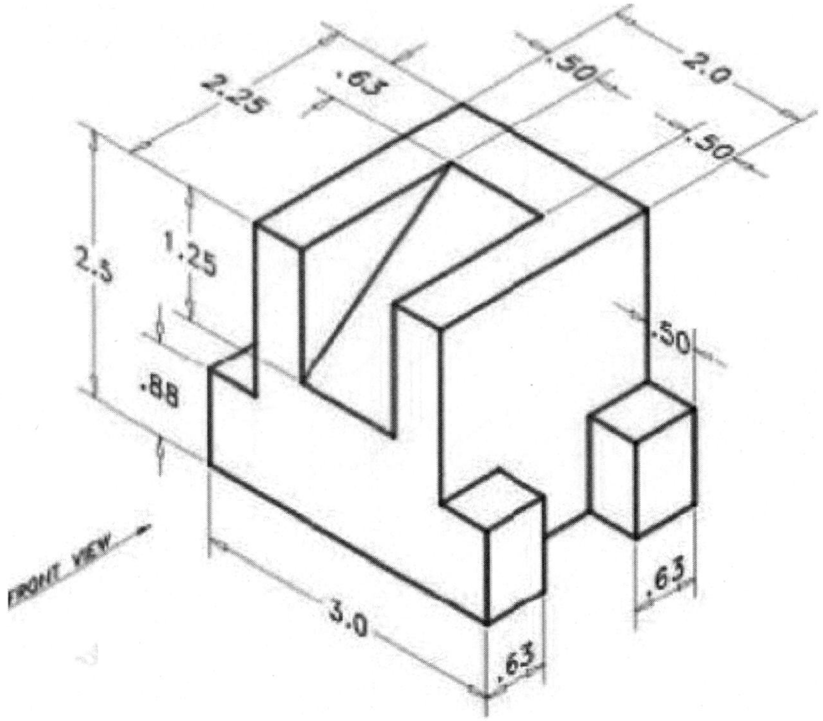

PROBLEM 20-9

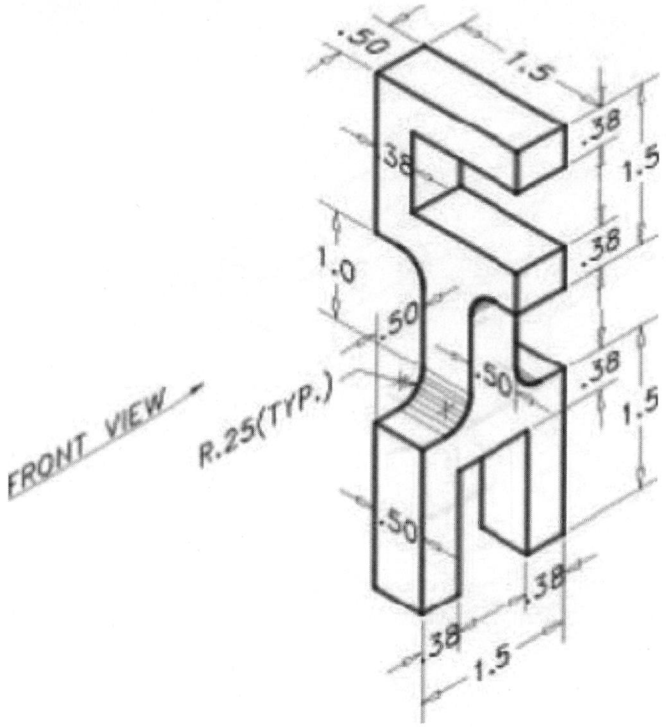

PROBLEM 20-10

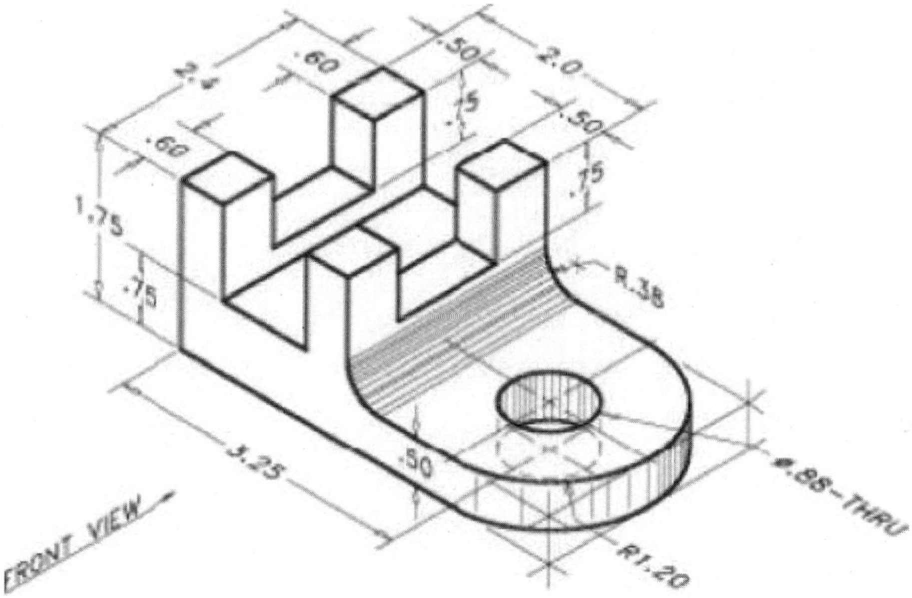

PROBLEM 20-11

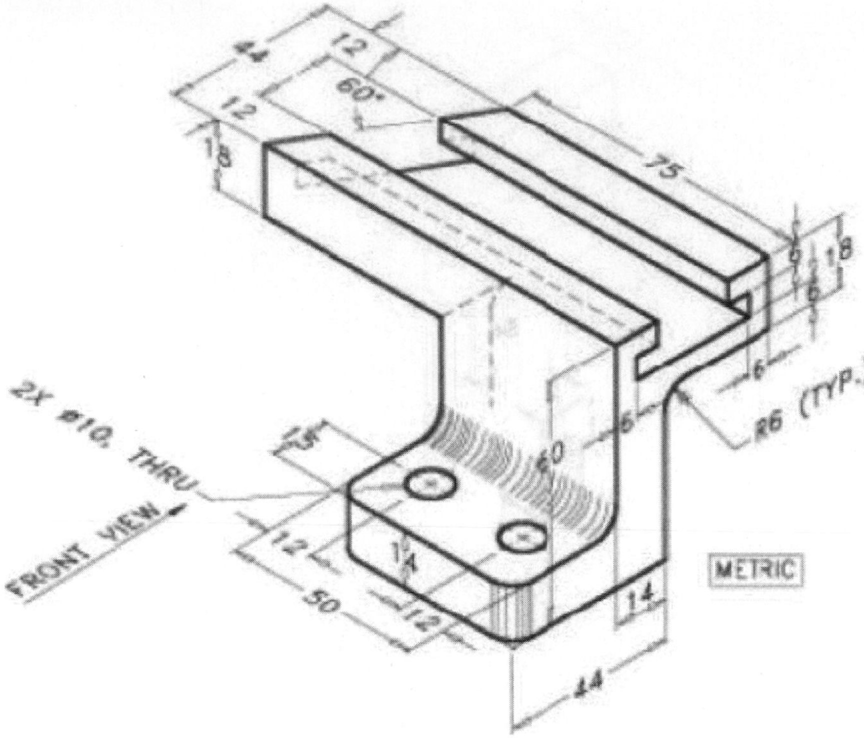

PROBLEM 20-12

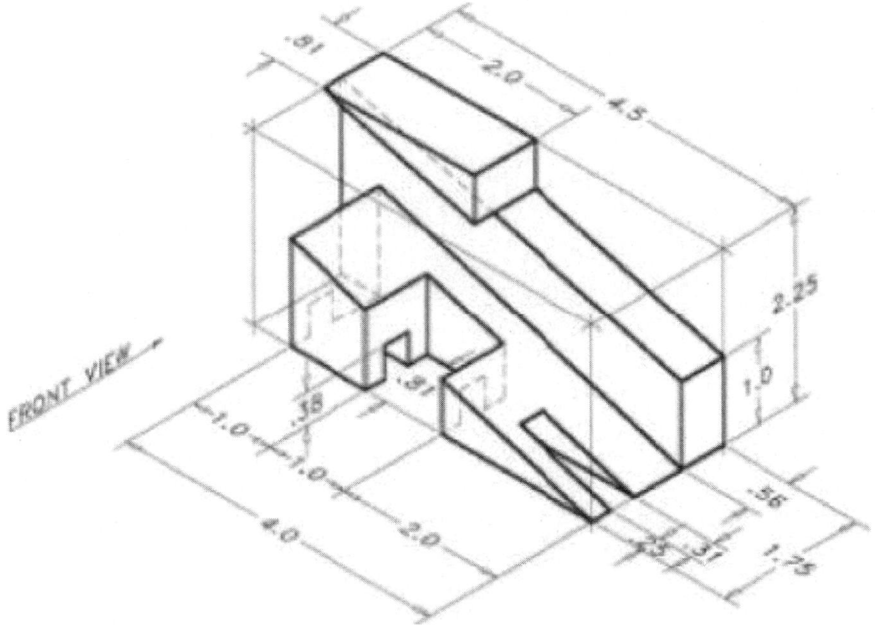

PROBLEM 20-13

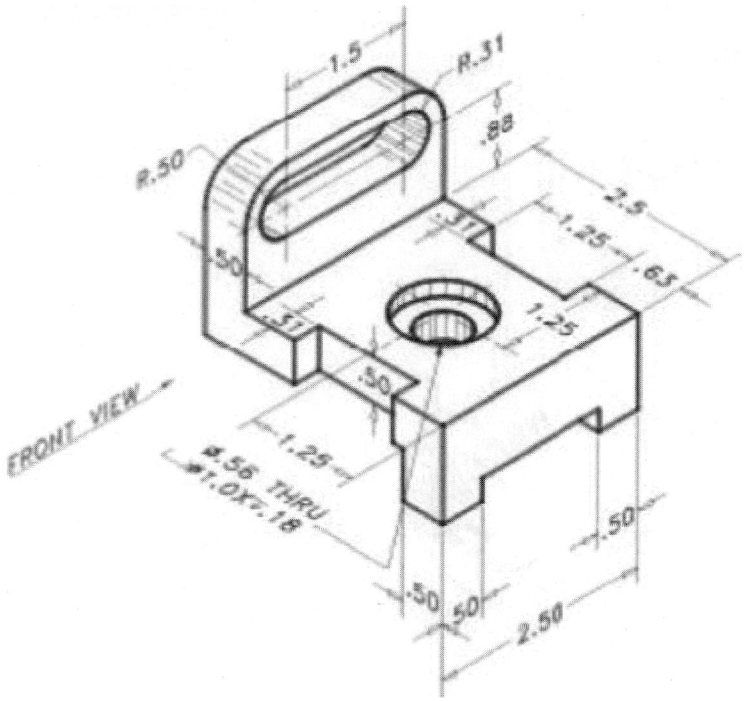

PROBLEM 20-14

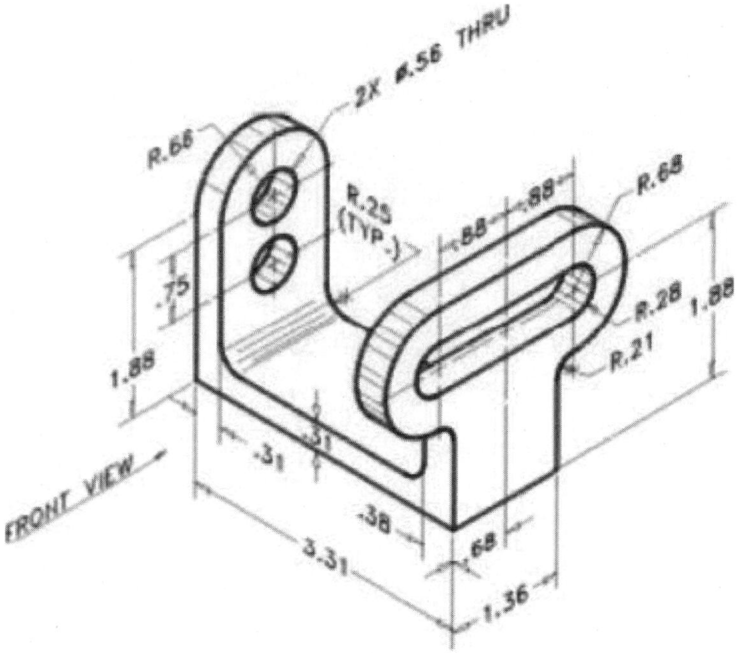

PROBLEM 20-15

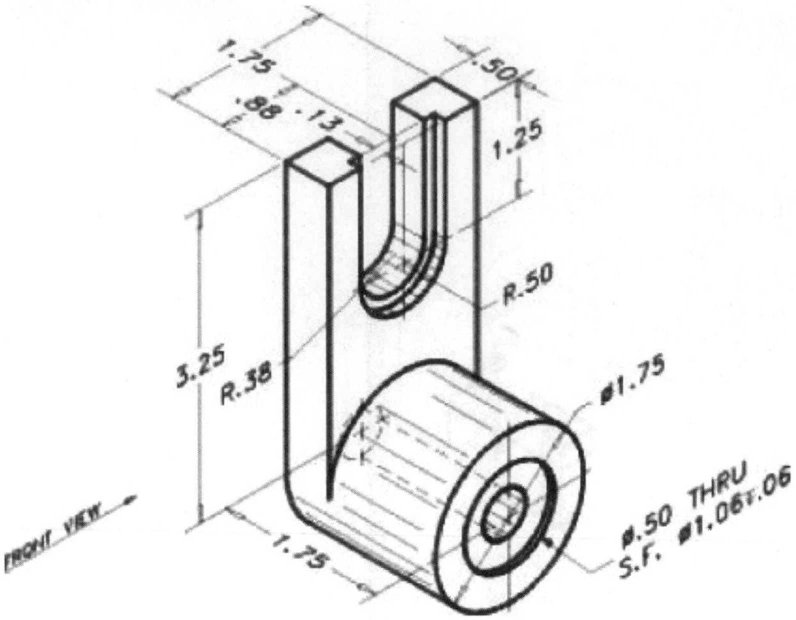

PROBLEM 20-16

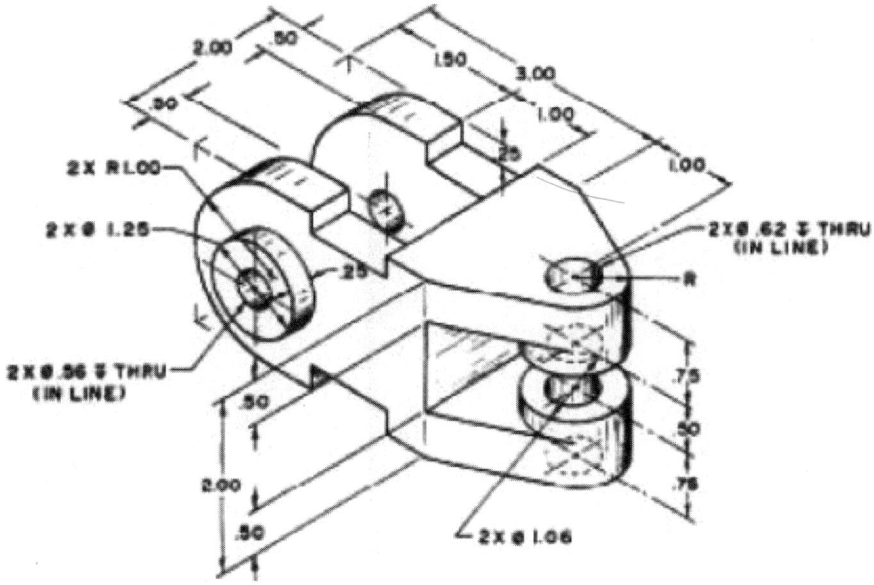

PROBLEM 20-17

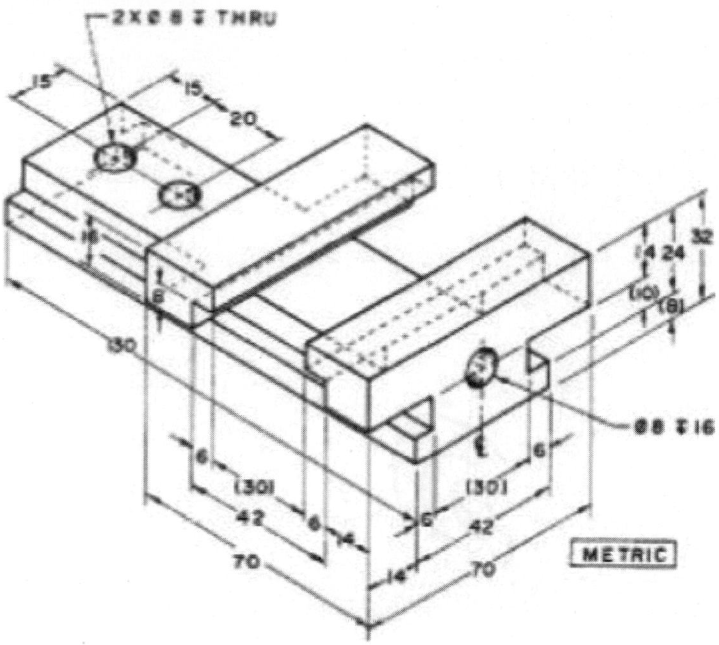

PROBLEM 20-18

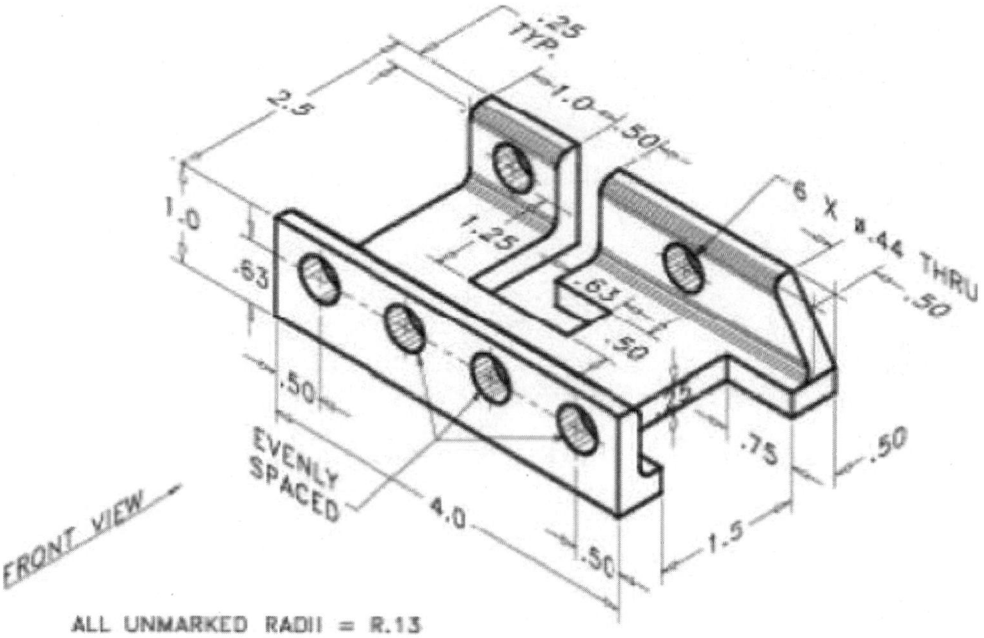

ALL UNMARKED RADII = R.13

PROBLEM 20-19

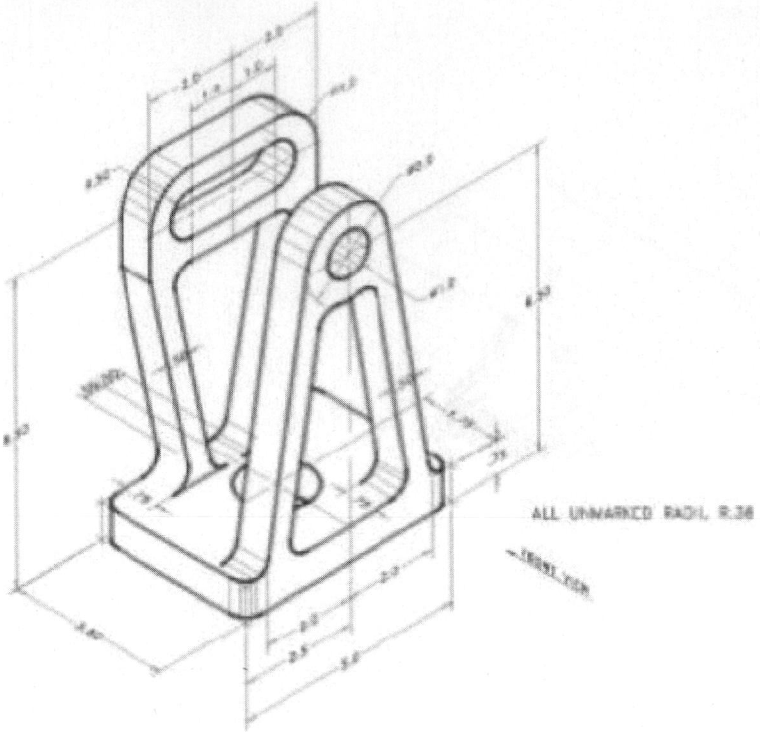

PROBLEM 20-20

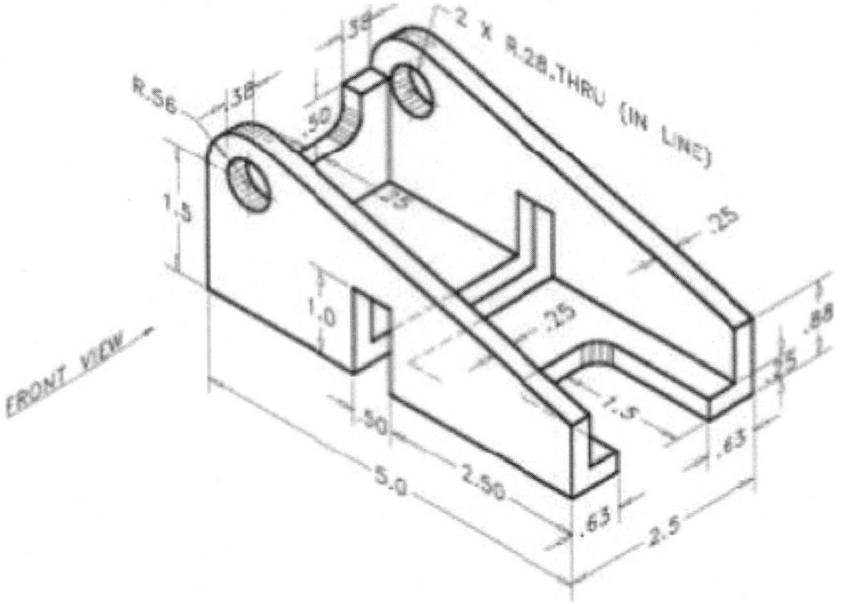

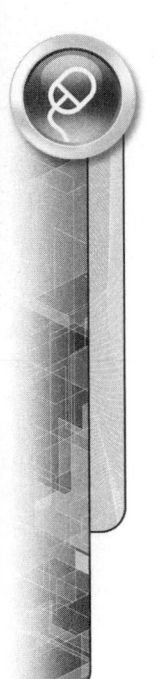

CHAPTER

21

Concept Modeling, Editing Solids, and Surface Modeling

This chapter begins with the study of creating concept models using a pick-and-drag method. The use of subshapes for editing solid models is discussed, as is the use of grips for editing solids. Once a solid model is created, various methods are available to edit the model. The methods discussed in this chapter include the following 3D operation commands: FILLETEDGE, CHAMFEREDGE, 3DMOVE, 3DROTATE, 3DALIGN, MIRROR3D, 3DSCALE, and 3DARRAY. Also included in this chapter is a segment on using the many options of the SOLIDEDIT command. Various 3D models will be opened and used to illustrate these commands and options. A look at surface modeling commands is also included. The emphasis will be on procedural and mesh surface modeling techniques.

CONCEPTUAL MODELING

Chapter 20 dealt with the basics of creating solid primitives and how to join, subtract, or intersect these shapes to form composite solid models. Sometimes you do not need the regimented procedures outlined in Chapter 20 to get your point across about a specific product you have in mind. You can simply create a concept solid model by dragging shapes together to form your idea. Various concept-modeling techniques will be explained in the next segment of this chapter.

Dragging Basic Solid Shapes

Constructing a 3D model with a concept in mind is easier than ever. Where exact distances are not important, Figure 21.1 illustrates the construction of a solid block using the BOX command. First, view your model in one of the many 3D viewing positions, such as SE (Southeast) Isometric. Next enter the BOX command and pick first and second corner points for the box on the screen, as shown in Figure 21.1 (Left). When prompted for the height of the box, move your cursor up and notice the box increasing in height, as shown in Figure 21.1 (Right). Click to locate the height. All solid primitives can be constructed using this technique.

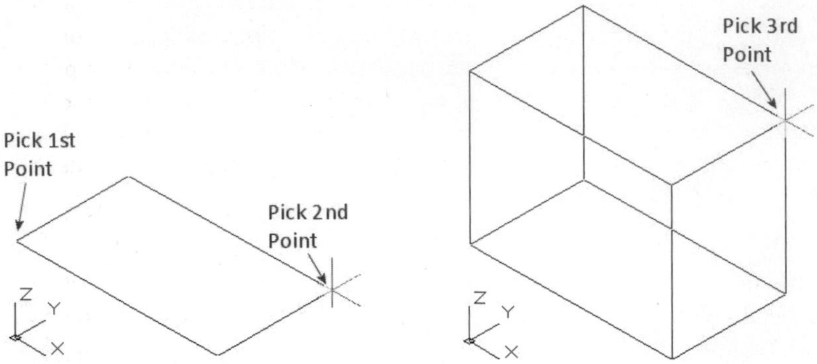

FIGURE 21.1

TRY IT!

Open the drawing file 21_Concept Box. This drawing file does not contain any objects. Also, it is already set up to be viewed in the Southeast Isometric position (SE Iso). Activate the `BOX` command, pick two points to define the first and second corner points of the rectangular base, and move your cursor up and pick to define the height of the box. You could also experiment using this technique for creating cylinders, pyramids, spheres, cones, wedges, and a torus. When finished, exit this drawing without saving any changes.

Using Dynamic UCS to Construct on Faces

Once a basic shape is created, it is very easy to create a second shape on an existing face with Dynamic UCS (DUCS) mode. Here is how it works. In Figure 21.2, a cylinder will be constructed on one of the faces. Turn on DUCS (located in the Status Bar), activate the `CYLINDER` command, and when prompted for the base or center point, hover your cursor over the face, as shown in Figure 21.2 (left). The face highlights (the edge appears dashed) to indicate that it has been acquired. The Dynamic UCS cursor adjusts itself to this face by aligning the XY plane parallel to the face. When you click a point on this face for the start of the cylinder, the UCS icon changes to reflect this change and the base of the cylinder can be seen, as shown in Figure 21.2 (Center). Pick a point to specify the radius. To specify the height of the cylinder, simply drag your cursor away from the face and you will notice the cylinder taking shape, as shown in Figure 21.2 (Right). Clicking a point defines the height and exit the `CYLINDER` command.

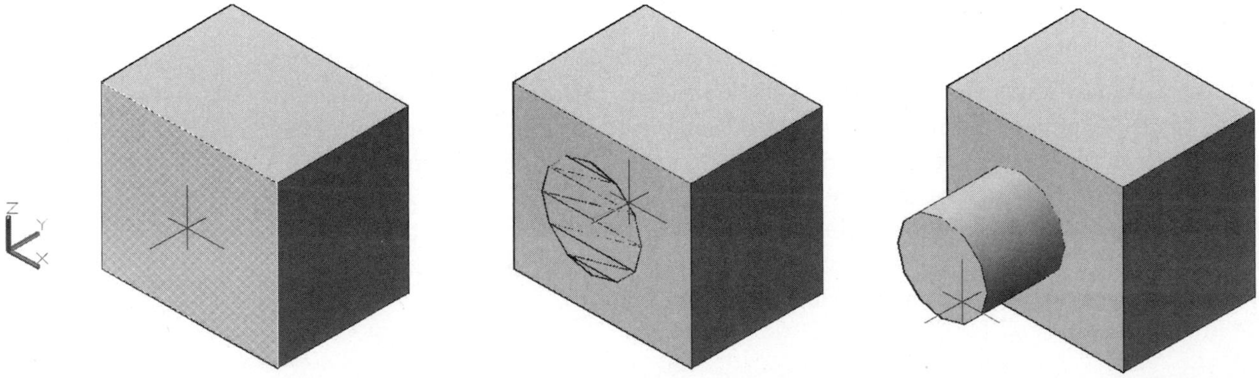

FIGURE 21.2

TRY IT!

Open the drawing file 21_Concept Cylinder. Issue the CYLINDER command and, for the base or center point, hover your cursor over the front face in Figure 21.2 and pick a point. The UCS changes to reflect this new position. Move your cursor until you see the circle forming on the face and pick to define the radius of the cylinder. When prompted for the height, move your cursor forward and notice the cylinder taking shape. Pick to define the cylinder height and to exit the command.

When the cylinder is created, it is considered an individual object separate from the solid block, as shown in Figure 21.3 (Left). After the primitives are constructed, the UNION command is used to join all primitives together as a single solid object, as shown in Figure 21.3 (Right).

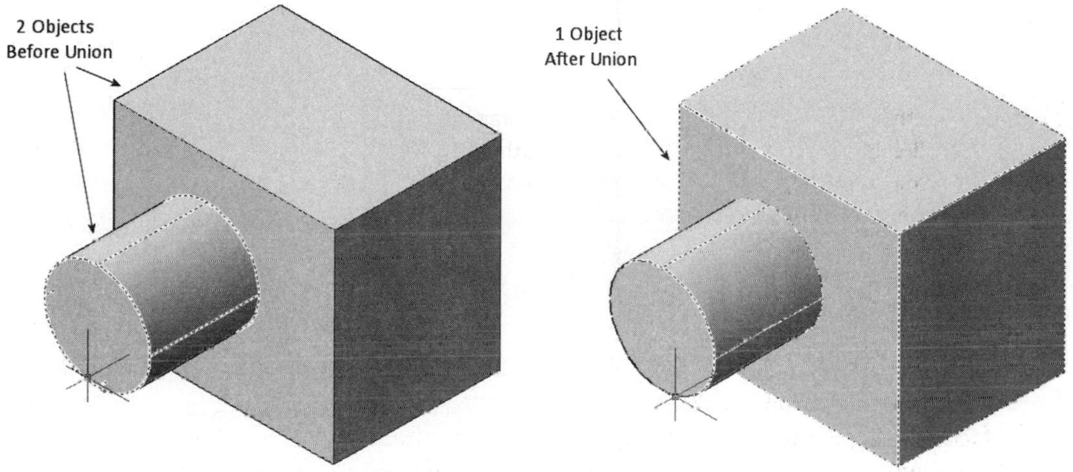

FIGURE 21.3

USING GRIPS TO MODIFY SOLID MODELS

Whenever a solid primitive is selected when no command is active, grips are displayed, as with all types of objects that make up a drawing. The grips that appear on solid primitives, as shown in Figure 21.4, range in shape from squares to arrows. You can perform an edit operation by selecting either the square or arrow shapes. The type of editing that occurs depends on the type of grip selected.

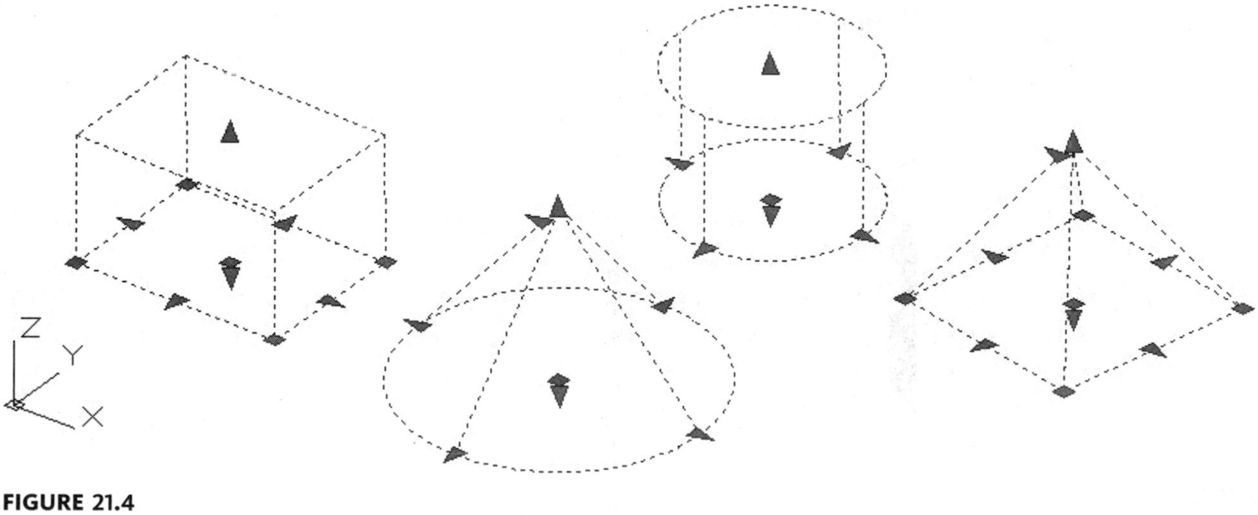

FIGURE 21.4

Key Grip Locations

The difference between the square and arrow grips is illustrated in Figure 21.5 of a solid box primitive. The square grip located in the center of a primitive allows you to change the location of the solid. Square grips displayed at the corner (vertex) locations of a primitive allow you to resize the base shape. The arrows located along the edges of the rectangular base allow each individual side to be modified. Arrow grips that point vertically also appear in the middle of the top and bottom faces of the box primitive. These grips allow you to change the height of the primitive.

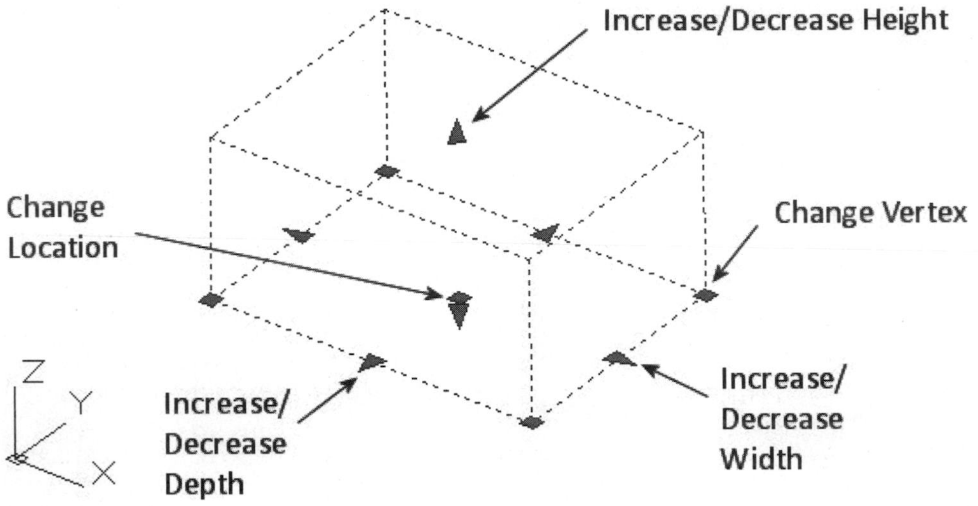

FIGURE 21.5

Grip Editing a Cone

Figure 21.5 outlined the various types of grips that display on a box primitive. The grips that appear on cylinders, pyramids, cones, and spheres have similar editing capabilities. The following Try It! exercise illustrates the effects of editing certain arrow grips on a cone.

TRY IT!

Open the drawing file 21_GripEdit Cone. Click the cone and the grips appear, as shown in Figure 21.6 (Left). Click the arrow grip at "A" and stretch the base of the cone in to match the object illustrated in Figure 21.6 (Right). Next, click the arrow grip at "B" (not the grip that points up) and stretch this grip away from the cone to create a top surface similar to the object illustrated in Figure 21.6 (Right). Experiment further with grips on the cone by clicking the top grip, which points up, and stretch the cone (now referred to as a frustum) up.

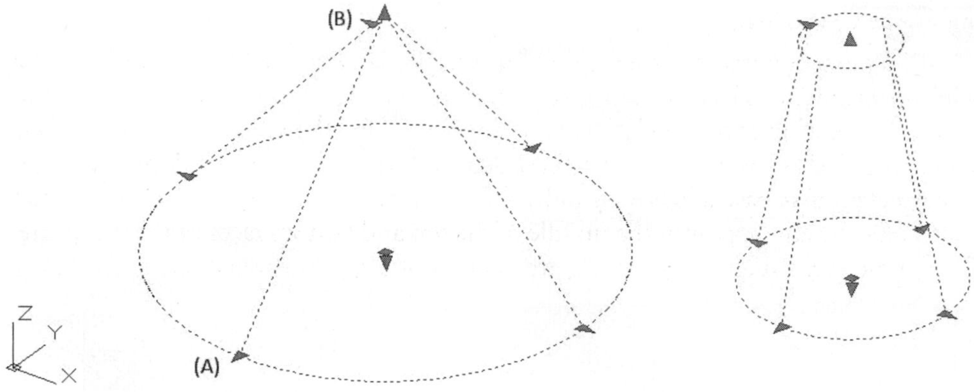

FIGURE 21.6

Editing with Grips and Dynamic Input

You have just seen how easy it is to select grips that belong to a solid primitive and stretch the grips to change the shape of the primitive. This next Try It! exercise deals with grip editing of primitives in an accurate manner. To accomplish this, Dynamic Input must be turned on.

TRY IT!

Open the drawing file 21_GripEdit Shape01. This 3D model consists of three separate primitives, namely, two boxes and one cylinder. Notice that the visual style Conceptual is active. Turn off the CULLINGOBJ system variable to better view hidden objects, including grips, when a 3D visual style is active. This system variable can be deactivated by picking the Culling button in the Selection panel of the Home Ribbon, as shown in Figure 21.7 in the upper-left. To accurately change the size of objects in this exercise, you will need to turn Dynamic Input on in the Status Bar. Also, right-click the Dynamic Input icon and select Settings from the shortcut menu provided. In the Drafting Settings dialog box provided, check the option to "Enable Dimension Input where possible", as shown in Figure 21.7 in the lower-left. The height of the cylinder needs to be lowered. To accomplish this, click the cylinder and observe the positions of the grips. The total height of 6.0000 needs to be changed to 3.50 units. Click the arrow grip at the top of the cylinder that is pointing up, press the TAB key to highlight the overall height value of 6.0000, as shown in Figure 21.7 (Center). Change this value to 3.50. Pressing ENTER to accept this new value lowers the cylinder, as shown in the following image on the right. Feel free to experiment with both boxes by either increasing or decreasing their heights using this grip-editing method.

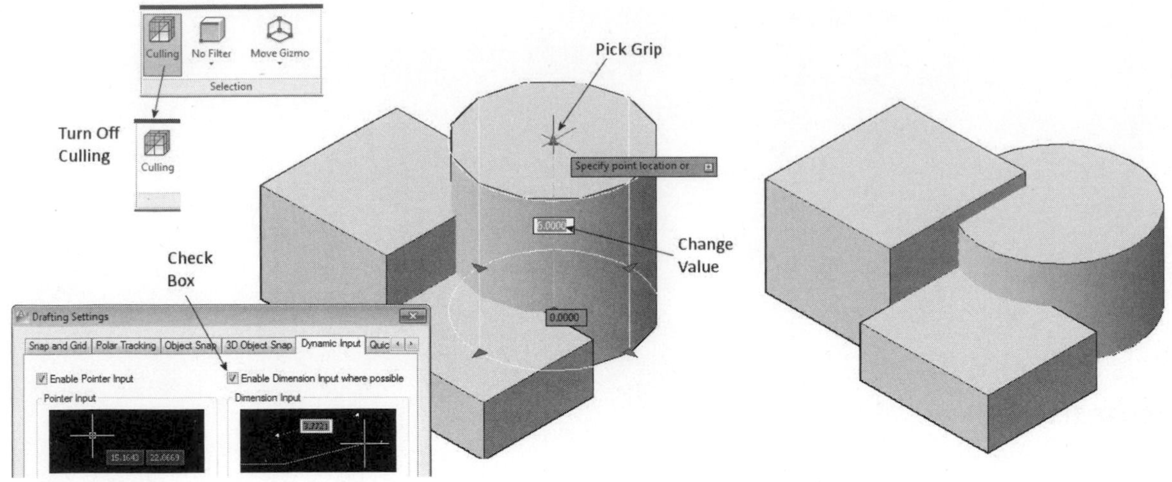

FIGURE 21.7

MANIPULATING SUBOBJECTS

A subobject is the face, edge, or vertex (corner) of a solid. Figure 21.8 illustrates a 3D box and pyramid. Pressing the CTRL key and picking near the center of a face, the middle of an edge, or the intersection of a corner, allow you to select a subobject. Notice in this illustration that each subobject selected has a grip associated with it. If you accidentally select a subobject, press CTRL + SHIFT and pick it again to deactivate it.

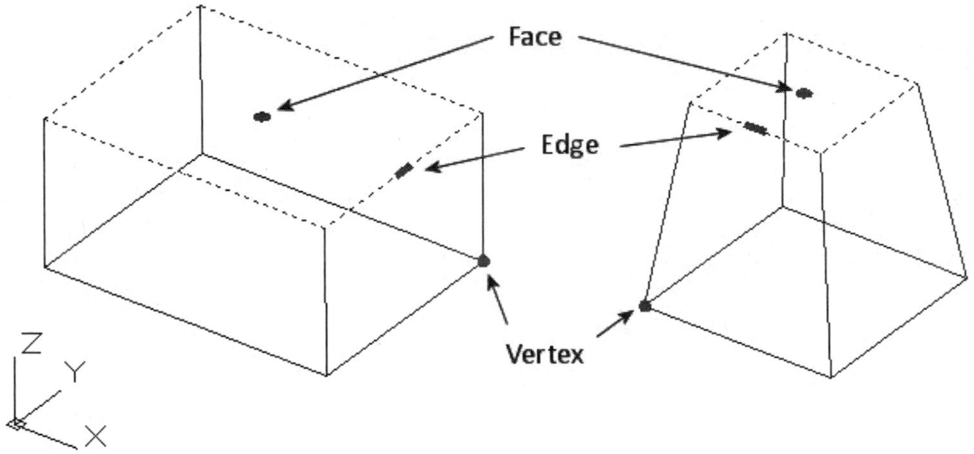

FIGURE 21.8

Once a subobject is selected, click the grip to activate the grip Stretch, Move, Rotate, Scale, or Mirror mode. You can drag your cursor to a new location or enter a direct distance value from the keyboard. In Figure 21.9, the edge of each solid object was selected as the subobject and dragged to a new location.

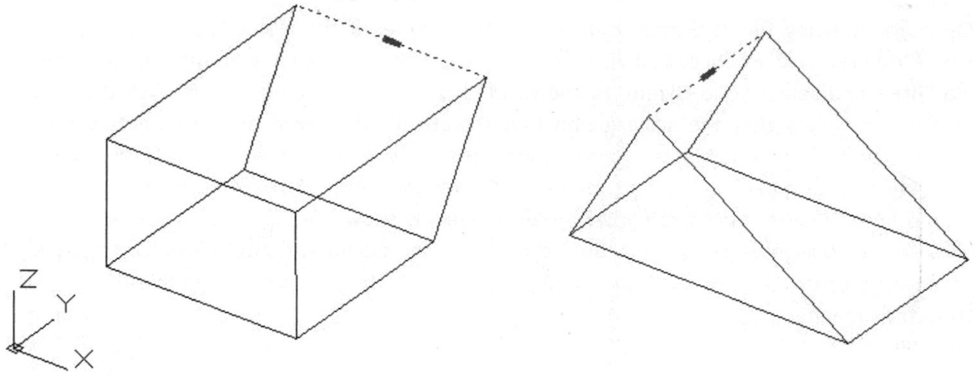

FIGURE 21.9

Filter and Gizmo Tools

To make it easier to pick vertices, faces, or edges, a special Selection panel is available in the Home tab of the Ribbon, as shown in Figure 21.10. Turning Culling (CULLINGOBJ) on and off, as discussed earlier, controls whether or not hidden objects are highlighted as a cursor is moved over them. Clicking the down arrow in No Filter (SUBOBJSELECTIONMODE) will display a drop-down menu that allows you to pick which subobject type will be highlighted when the cursor is moved over them. The final subobject tool (DEFAULTGIZMO) allows you to select the default gizmo type displayed when you select an object utilizing a 3D visual style. Here, you switch from different gizmos to either move, rotate, or scale model vertices, faces, or edges.

NOTE

The Selection panel can also be found in the Solid and Mesh tabs of the Ribbon in the 3D Modeling workspace.

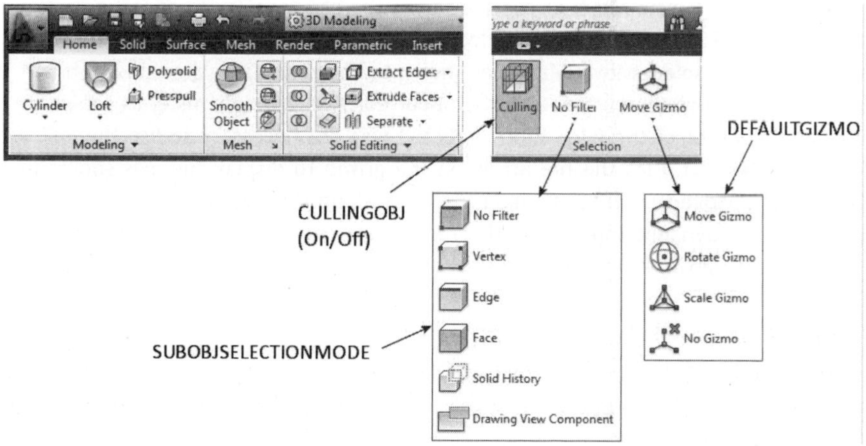

FIGURE 21.10

TRY IT!

Open the drawing file 21_Gizmo Scale. In the View panel of the Home Ribbon, select a SE Isometric view and a Conceptual visual style. In the Selection panel, turn off Culling, select No Filter, and select Scale Gizmo, as shown in Figure 21.11 in the upper-left. Pick the solid model and notice that the scale gizmo is automatically displayed at the objects center. Gizmos can be used on objects as well as subobjects. Move your cursor over the gizmo and right-click to display the shortcut menu, as shown in Figure 21.11 (Left). Select Relocate Gizmo and pick the lower-left corner of the model, as shown in Figure 21.11 (Center). This allows us to scale the model from this corner. Pick the gizmo and enter 0.75 for the scale factor. The result is shown in Figure 21.11 (Right). Press the ESC key to remove the gizmo from the screen.

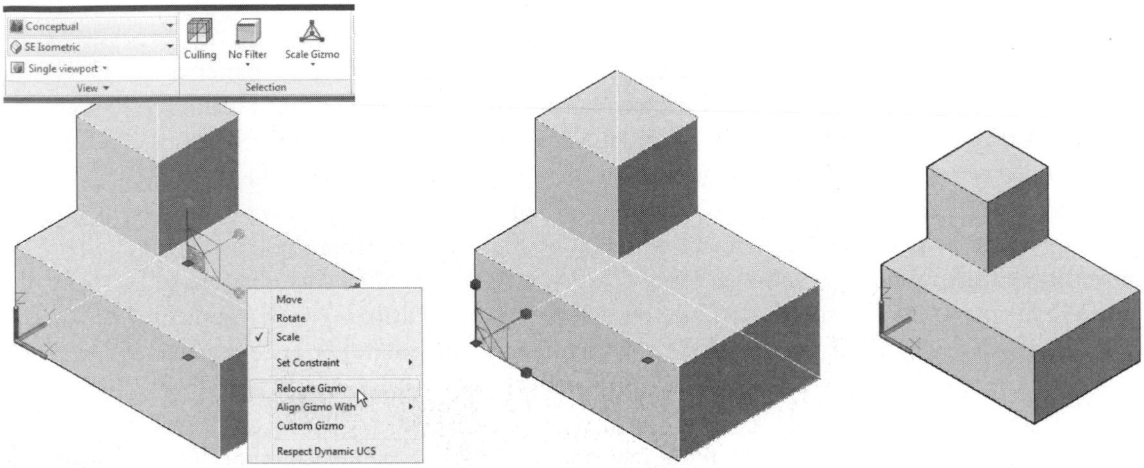

FIGURE 21.11

TRY IT!

Open the drawing file 21_Gizmo Move or continue from the previous Try It! exercise. In the Status Bar turn Dynamic Input on. In the Selection panel of the Home Ribbon, turn off Culling, select Edge, and select Move Gizmo, as shown in Figure 21.12 in the upper-left. An edge filter glyph appears at the cursor. Pick the upper edge of the right face, as shown in Figure 21.12 (Left). Because of the edge filter, notice it was not necessary to use the CTRL key to select the subobject. Pick the red arrow on the gizmo to display the axis shown in Figure 21.12 (Center). Move the cursor to slide the edge along the designated axis. Enter a value of 1.00 into the dynamic input box. The results are shown in Figure 21.12 (Right). Press the ESC key to remove the gizmo from the screen.

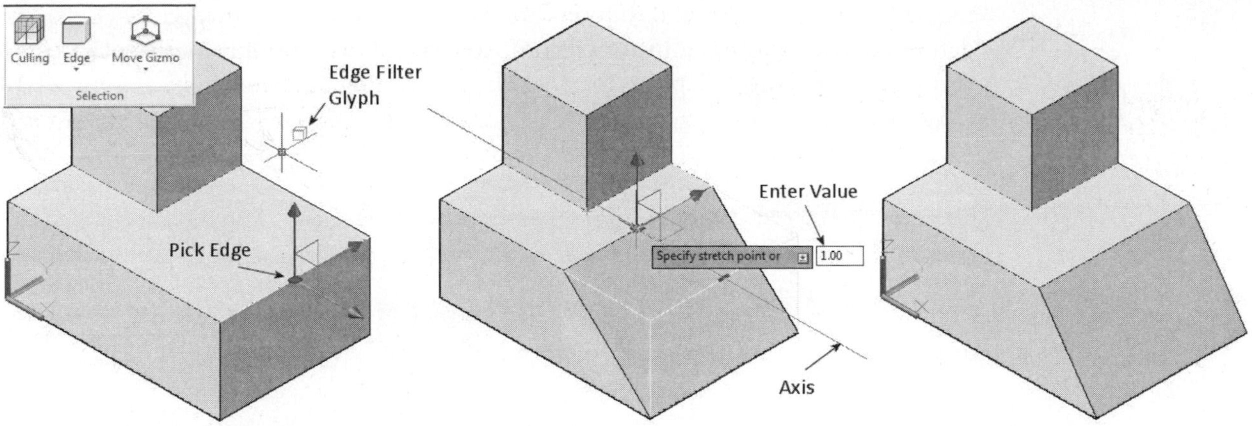

FIGURE 21.12

TRY IT!

Open the drawing file 21_Gizmo Rotate or continue from the previous Try It! exercise. In the Status Bar turn Dynamic Input on. In the Selection panel of the Home Ribbon, turn off Culling, select Face, and select Rotate Gizmo, as shown in Figure 21.13 (Left). A face filter glyph appears at the cursor. Pick in the center of the upper face, as shown in Figure 21.13 (Left). The rotate gizmo displays in the center of the face. Because of the face filter, notice it was not necessary to use the CTRL key to select the subobject. Pick the green circle on the gizmo to display the axis shown in Figure 21.13 (Center). Move the cursor to rotate the face around the designated axis. Enter a value of 30 into the dynamic input box. The results are shown in Figure 21.13 (Right). The Front view was selected on the ViewCube to better show the results of the rotation. Press the ESC key to remove the gizmo from the screen.

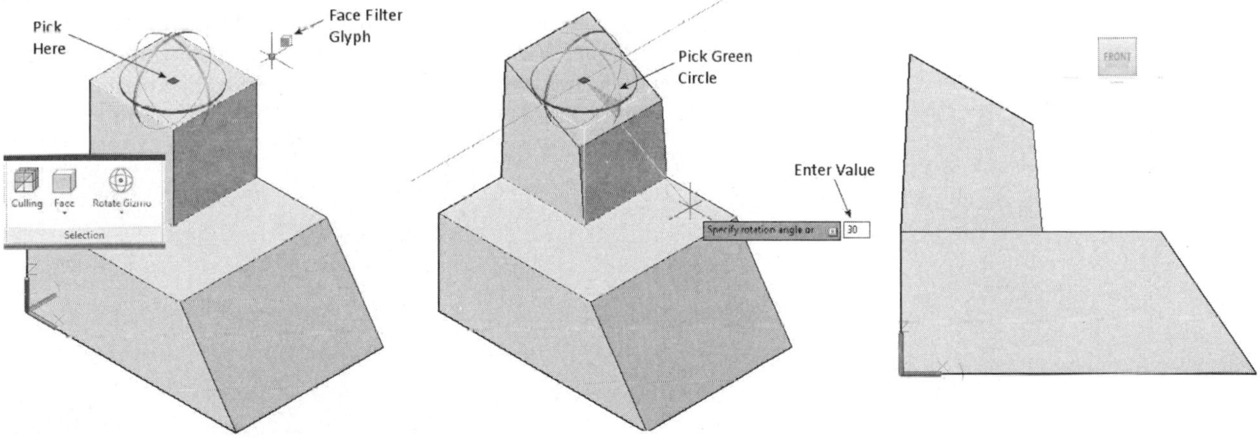

FIGURE 21.13

Editing Subobjects with a Solid History

You have seen how easy it is to isolate a subobject of a solid model by utilizing a filter or pressing the CTRL key while selecting the subobject. The same technique can be used to isolate a primitive that is already consumed or made part of a composite solid model. This is only possible, however, if the history of the solid composite is recorded. By default the SOLIDHIST system variable is turned off and a record is not kept of the original objects that make up the solid. This system variable can be toggled on and off from the Ribbon (Solid tab), as shown in Figure 21.14 (Left). To determine if a solid has a history record, you can activate the Properties palette and verify that "Record" displays for the History setting, as shown in Figure 21.14 (Right).

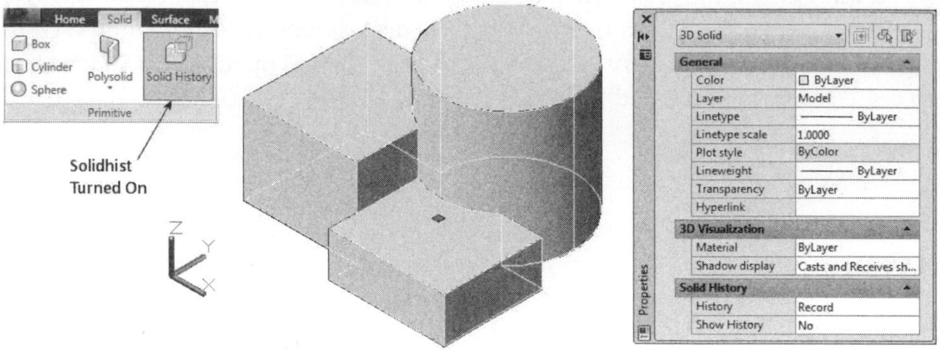

FIGURE 21.14

If the history is recorded for a composite solid, you will be able to edit a specific primitive while leaving other primitives of the solid model unselected. The next Try It! exercise illustrates this technique.

TRY IT!

Open the drawing file 21_GripEdit Shape02. This illustration represents three separate primitives joined into one solid model using the UNION command. The diameter of the cylinder needs to be increased. First, verify that the Culling button is not activated; this will make it easier to view the grips. Then, press and hold down the CTRL key while clicking the cylinder. Notice that only this primitive highlights and displays various grips, as shown in Figure 21.15 (Left). Click the arrow grip at the side of the cylinder, and drag the geometry out approximately 2.00 units, as shown in Figure 21.15 (Center). Pick to increase the cylinder's diameter, as shown in Figure 21.15 (Right). Feel free to experiment with the other primitives by either increasing or decreasing their sizes using this grip-editing method.

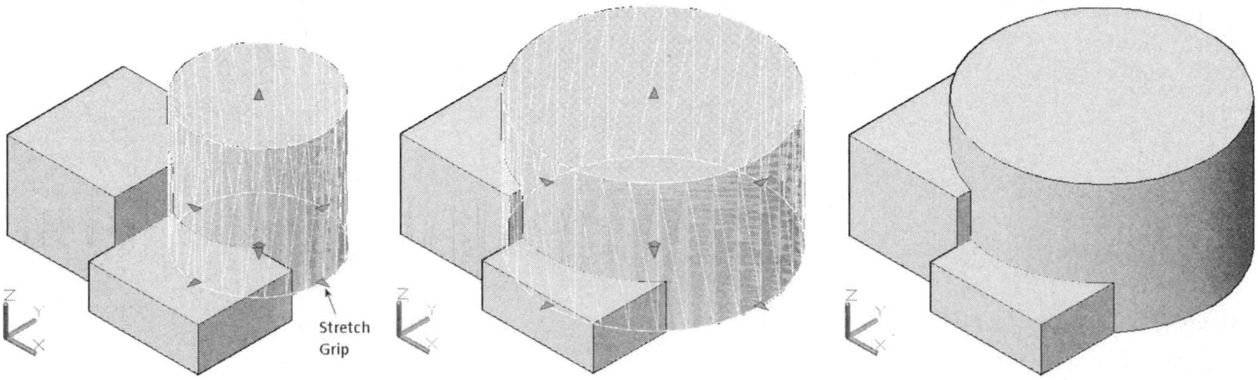

FIGURE 21.15

ADDING EDGES AND FACES TO A SOLID MODEL

By imprinting regular objects such as lines, circles, or polylines onto a solid model you can create additional edges and faces. These can then be used to change the shape of the model using the subobject techniques previously discussed. For example, the line segment shown in Figure 21.16 (Left) is drawn directly onto the face of a solid model. This line, once imprinted, becomes part of the solid and, in our case, creates an edge, which divides the top face into two faces. These new subobjects can now be used to modify the solid model. The command used to perform this operation is `IMPRINT`, which can be found in the Solid tab of the Ribbon, as shown in Figure 21.16 (Right).

`Command: IMPRINT`

`Select a 3D solid or surface:` *(Select the 3D solid model)*

`Select an object to imprint:` *(Pick the line segment constructed across the top surface of the solid model)*

`Delete the source object [Yes/No] <N>: Y` *(For Yes)*

`Select an object to imprint:` *(Press* `ENTER` *to perform this operation and exit the command)*

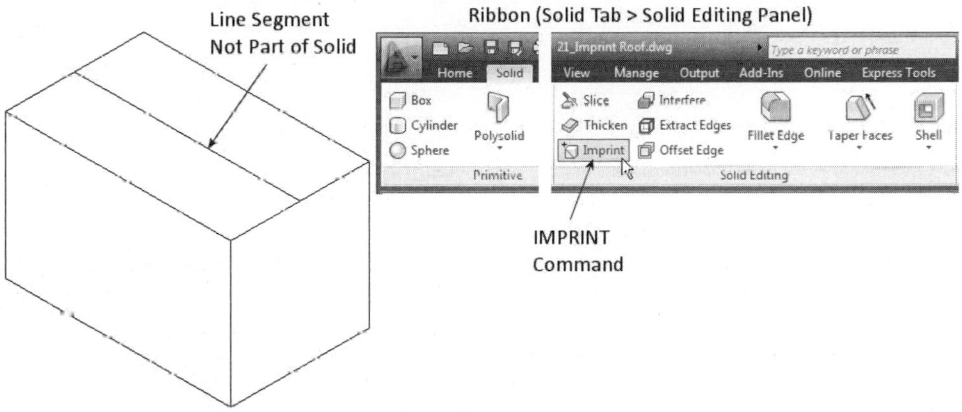

FIGURE 21.16

After the imprint operation is performed on the line segment, this object becomes part of the solid model. When one of the new faces or the edge is selected as a subobject, the grip can be selected and dragged up or down. The results can dramatically change the shape of the solid model.

TRY IT!

Open the drawing file 21_Imprint Roof1. First, verify that Ortho Mode is turned on and that "No Gizmo" is selected in the Selection panel of the Solid Ribbon. Issue the IMPRINT command, select the solid block, as shown in Figure 21.17 (Left), and pick the single line segment as the object to imprint. Answer "Yes" to deleting the source object and complete the command. Next, press and hold down the CTRL key while clicking the imprinted line (or use an Edge filter, if desired). Notice that only this line highlights and displays a grip at its midpoint, as shown in Figure 21.17 (Center). Pick this grip, slowly move your cursor up, and notice the creation of the roof peak. You could also move your cursor down to create a V-shaped object. Enter a value or pick to complete the operation.

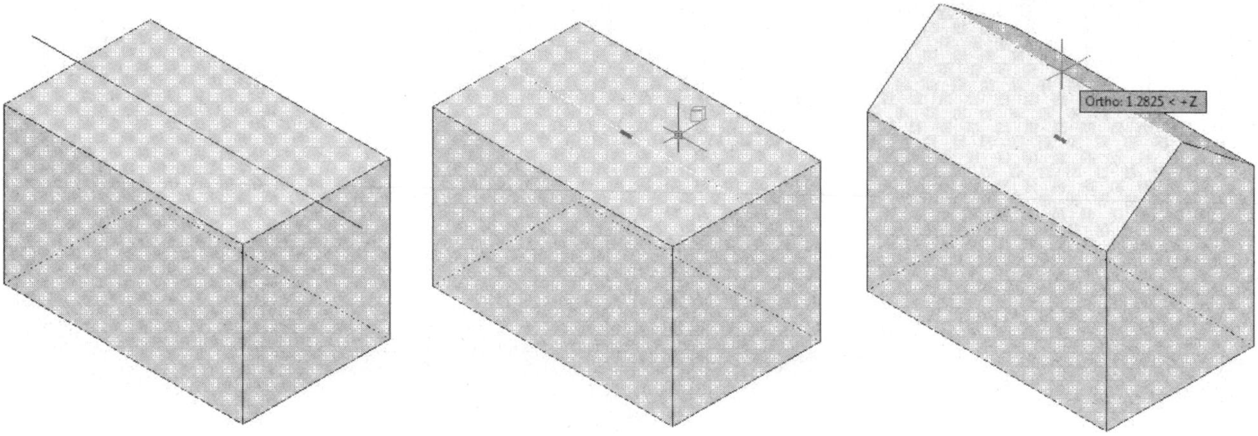

FIGURE 21.17

TRY IT!

Open the drawing file 21_Imprint Roof2. Verify that Ortho Mode is turned on and that "No Gizmo" is selected in the Selection panel of the Solid Ribbon. In this exercise, the line segment has already been imprinted on the solid for you. Next, press and hold down the CTRL key, while clicking in the center of the face, as shown in Figure 21.18 (Left). When the grip displays, hover your cursor over the grip to display the menu, as shown in Figure 21.18 (Center). Select Move Face and move your cursor straight up. Enter a value of 1.00 and then press the ESC key to complete the operation. The results are displayed in Figure 21.18 (Right).

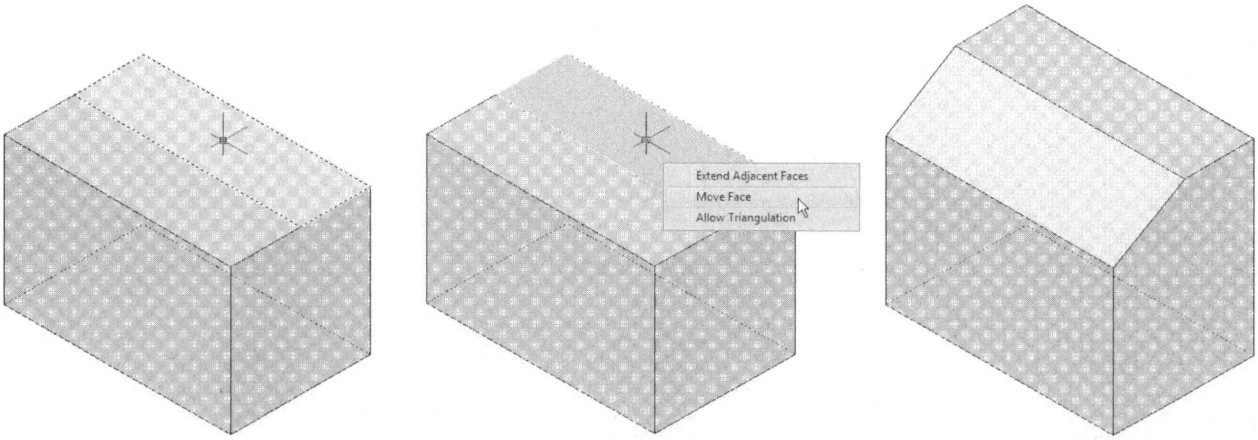

FIGURE 21.18

Continue with the previous exercise. This time, however, turn off Ortho Mode and in the Selection panel of the Solid Ribbon, select "Move Gizmo" and "Face". Since the Face filter is active, it will not be necessary to hold down the control key. Simply move your cursor to the center of the face and pick. Notice the move gizmo appears, as shown in Figure 21.19 (Left). Select the blue vertical axis and move your cursor down as shown in the image in the middle. Enter a value of 2.00 and then press the ESC key to complete the operation. The results are displayed in Figure 21.19 (Right).

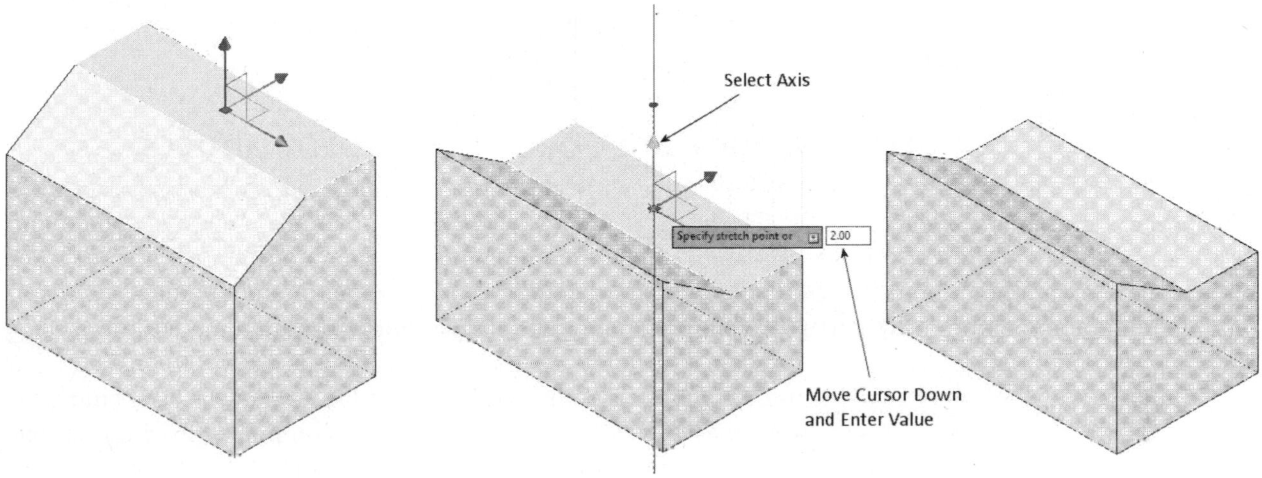

FIGURE 21.19

PRESSPULL A SOLID MODEL FACE

It was demonstrated in Chapter 20 how the PRESSPULL command could be used to create a solid model from a bounded 2D area. This command can also be used to edit an existing solid by picking on the bounded area of a solid (its face) and dragging the cursor to add or subtract material from the solid.

TRY IT!

Open the drawing file 21_Presspull. Use the following prompt sequence and Figure 21.20 for modifying the following solid.

```
Command: PRESSPULL
Select object or bounded area: (Move your cursor over the top
face and pick, as shown in Figure 21.20 (Left).)
Specify extrusion height or [Multiple]: 2.00
1 extrusion(s) created.
Select object or bounded area: (Move your cursor over the side
face and pick, as shown in Figure 21.20 (Center).)
Specify extrusion height or [Multiple]: (Move your cursor to
pull the cylinder through the solid and pick.)
1 extrusion(s) created
Select object or bounded area: (Press ENTER to complete the
command. The results are shown in Figure 21.20 (Right))
```

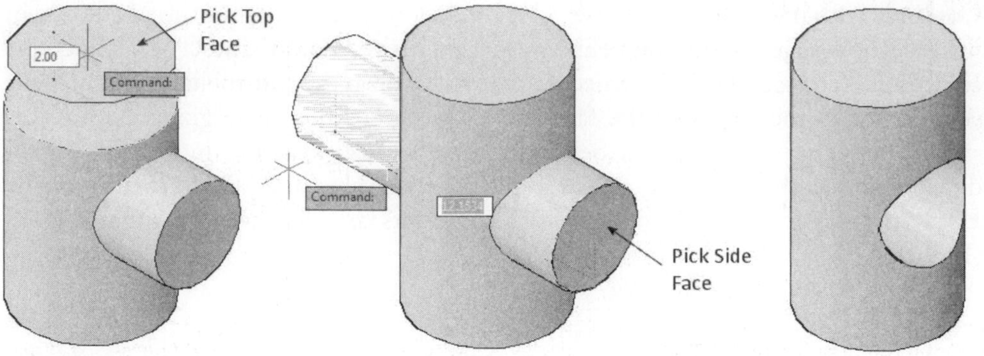

FIGURE 21.20

ADDITIONAL METHODS FOR EDITING SOLID MODELS

Many of the Modify commands that you used in 2D drawings can be used on 3D models, such as `FILLET`, `CHAMFER`, `MOVE`, `ALIGN`, `ROTATE`, `MIRROR`, and `SCALE`. However, each of these commands has a 3D version: `FILLETEDGE`, `CHAMFEREDGE`, `3DMOVE`, `3DALIGN`, `3DROTATE`, `MIRROR3D`, and `3DSCALE`. Typically, these 3D commands are more efficient for working on models. For example, the `ROTATE` command allows you to only rotate objects around the Z-axis. You can change the user coordinate system to change the direction of the axis, but this is not necessary with the `3DROTATE` command. This command provides various options for defining a new axis direction. Arraying objects in 3D can be accomplished with the same commands that were discussed back in Chapter 4 for creating 2D arrays: `ARRAYRECT`, `ARRAYPATH`, and `ARRAYPOLAR`. By utilizing the Level option, objects can also be arrayed in the Z direction. Another important command for editing solid models is `SOLIDEDIT`. This command has numerous options for modifying solid models. The 3D editing commands can be found in the Modify and Solid Editing panels of the Ribbon, as illustrated in Figure 21.21. Various 3D drawings will be opened and used to illustrate these commands.

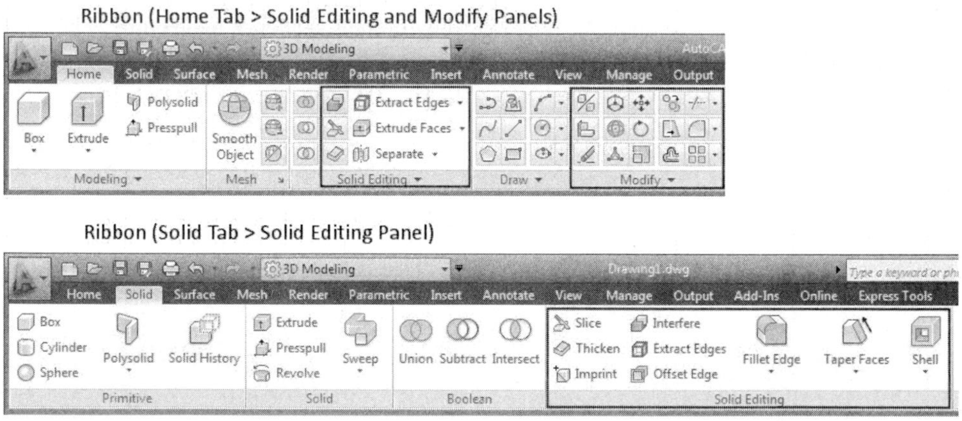

FIGURE 21.21

FILLETING SOLID MODELS

Filleting of simple or complex objects is easily handled with the FILLETEDGE command. The FILLET command can also be used on solid models and has very similar prompts.

- From the Ribbon > Solid Tab > Solid Editing Panel (3D Modeling Workspace)
- From the Menu Bar (Modify > Solid Editing > Fillet edges)
- From the Solid Editing Toolbar
- From the keyboard (FILLETEDGE)

TRY IT!

Open the drawing file 21_Tee Fillet. Use the following prompt sequence and Figure 21.22 for performing this task.

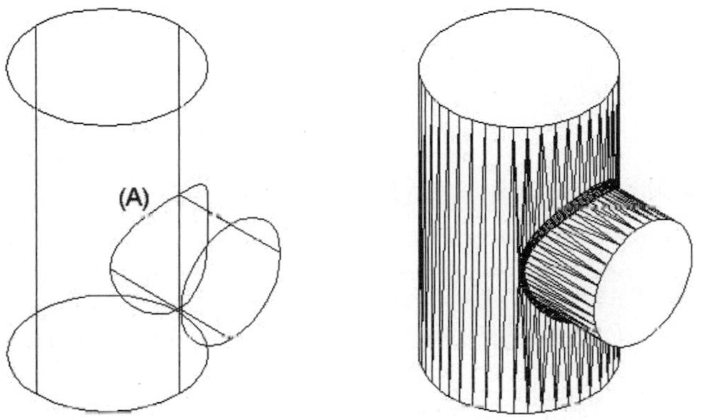

FIGURE 21.22

Command: FILLETEDGE

Radius = 1.0000

Select an edge or [Chain/Loop/Radius]: R *(For Radius)*

Enter fillet radius or [Expression]<1.0000>: 0.25

Select an edge or [Chain/Loop/Radius]: *(Select the edge at "A," which represents the intersection of both cylinders)*

Select an edge or [Chain/Loop/Radius]: *(Press* ENTER*)*

1 edge(s) selected for fillet.

Press Enter to accept the fillet or [Radius]: *(Press* ENTER *to complete the fillet operation)*

TRY IT!

Open the drawing file 21_Slab Fillet. A group of objects with a series of edges can be filleted with the Chain option of the FILLETEDGE command. Use the following prompt sequence and Figure 21.23 for performing this task.

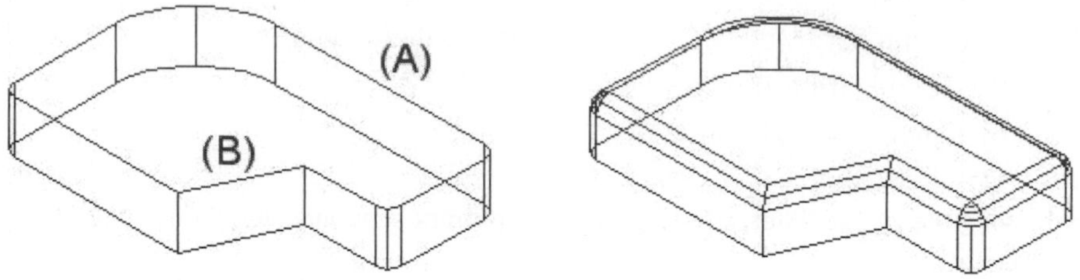

FIGURE 21.23

Command: FILLETEDGE

Radius = 1.0000

Select an edge or [Chain/Loop/Radius]: R *(For Radius)*

Enter fillet radius or [Expression]<1.0000>: 0.50

Select an edge or [Chain/Loop/Radius]: C *(For chain mode)*

Select an edge chain or [Edge/Radius]: *(Select edge "A"; notice how the selection is chained until it reaches an abrupt corner)*

Select an edge chain or [Edge/Radius]: *(Select the edge at "B")*

Select an edge chain or [Edge/Radius]: *(Press* ENTER*)*

10 edge*(s)* selected for fillet.

Press Enter to accept the fillet or [Radius]: *(Press* ENTER *to complete the fillet operation)*

NOTE

In the previous Try It! exercise, the same results could have been achieved by using the Loop option. A loop is similar to a chain operation with the exception that it will include all edges related to a highlighted face. Try repeating the exercise utilizing the Loop option.

CHAMFERING SOLID MODELS

The CHAMFEREDGE command applies a chamfer to any selected edges of a solid model or of a surface, as will be demonstrated later in the chapter. The Loop option is available to automatically select contiguous edges. The CHAMFER command can also be used on solid models and has similar prompts.

- From the Ribbon > Solid Tab > Solid Editing Panel (3D Modeling Workspace)
- From the Menu Bar (Modify > Solid Editing > Chamfer edges)
- From the Solid Editing Toolbar
- From the keyboard (CHAMFEREDGE)

TRY IT!

Open the drawing file 21_Slab Chamfer. Use the following prompt sequence and Figure 21.24 for performing this task.

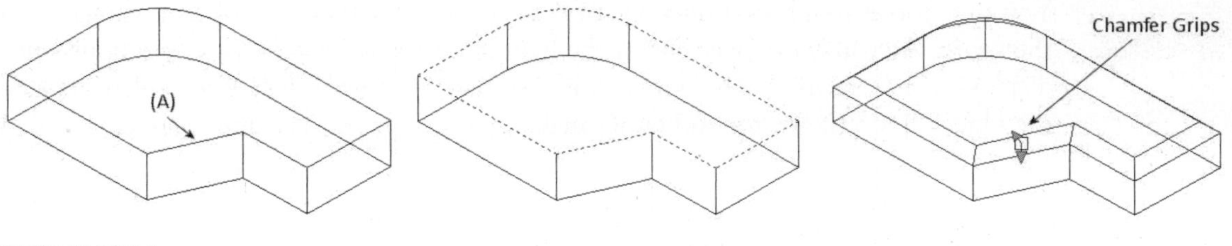

FIGURE 21.24

Command: CHAMFEREDGE

Distance1 = 0.5000, Distance2 = 0.5000

Select an edge or [Loop/Distance]: L *(To loop all top edges together into one)*

Select edge of loop or [Edge/Distance]: *(Pick any top edge such as the one at "A"; notice that an Edge filter is automatically provided)*

Enter an option [Accept/Next] <Accept>: N *(For Next; notice that the loop does not stop at abrupt corners)*

Enter an option [Accept/Next] <Accept>: *(Press* ENTER *to accept)*

Select edge of loop or [Edge/Distance]: *(Press* ENTER*; notice the appearance of chamfer grips, these can be stretched to change the chamfer distances)*

Press Enter to accept the chamfer or [Distance]: *(Press* ENTER *to complete the chamfer operation)*

MOVING OBJECTS IN 3D

To assist in the positioning of objects in a 3D environment, the 3DMOVE tool is available. Choose this command using one of the following methods:

- From the Ribbon > Home Tab > Modify Panel (3D Modeling Workspace)
- From the Menu Bar (Modify > 3D Operations > 3DMove)
- From the Modeling Toolbar
- From the keyboard (3M or 3DMOVE)

Activating the 3DMOVE command displays the 3D move gizmo, as shown in Figure 21.25, which displays axis handles for the purpose of moving objects a specified direction and distance. After you press ENTER to signify that you are done with the selection process, the move gizmo displays at the center of your object. When using this command, you can select either objects or subobjects to move. To select subobjects, press and hold down the CTRL key as you select.

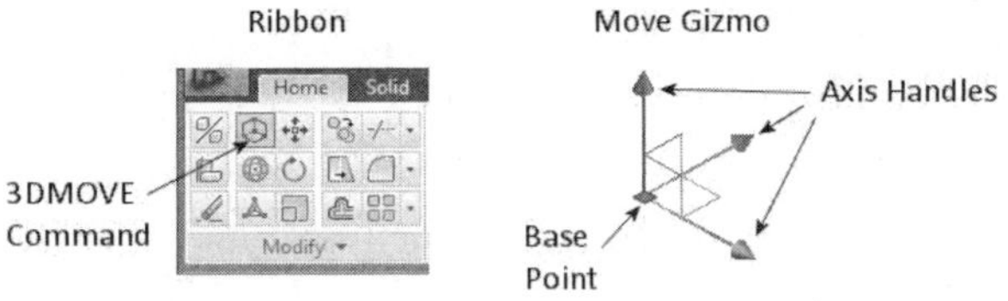

FIGURE 21.25

You then move your cursor over one of the three axis handles to define the direction of the move. As you hover over one of the handles, it turns yellow and a direction vector displays. Click this axis handle to lock in the direction vector. Move the cursor to slide the object along the vector and either pick a point or enter a specific distance value, as shown in Figure 21.26.

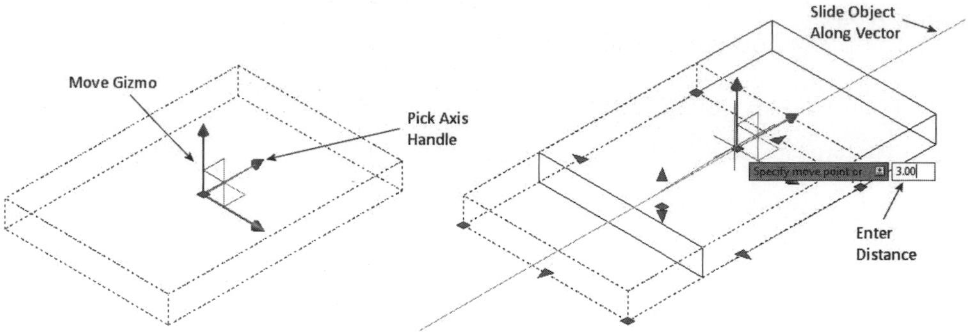

FIGURE 21.26

ALIGNING OBJECTS IN 3D

Use the `3DALIGN` command to specify up to three points to define the source plane of one 3D solid model, followed by up to three points to define the destination plane where the first solid model will be moved and aligned to. Choose this command from one of the following:

- From the Ribbon > Home Tab > Modify Panel (3D Modeling Workspace)
- From the Menu Bar (Modify > 3D Operations > 3DAlign)
- From the Modeling Toolbar
- From the keyboard (3DALIGN)

When you specify points, the first source point is referred to as the base point. This point is always moved to the first destination point. Selecting second and third source or destination points results in the 3D solid model being rotated into position.

TRY IT!

Open the drawing file 21_3DAlign. The objects in Figure 21.27 (Right) need to be positioned or aligned to form the assembled object shown in the small isometric view in this image. At this point, it is unclear at what angle the objects are currently rotated. When you use the 3DALIGN command, it is not necessary to know this information. Rather, you line up source points with destination points. When the three sets of points are identified, the object moves and rotates into position.

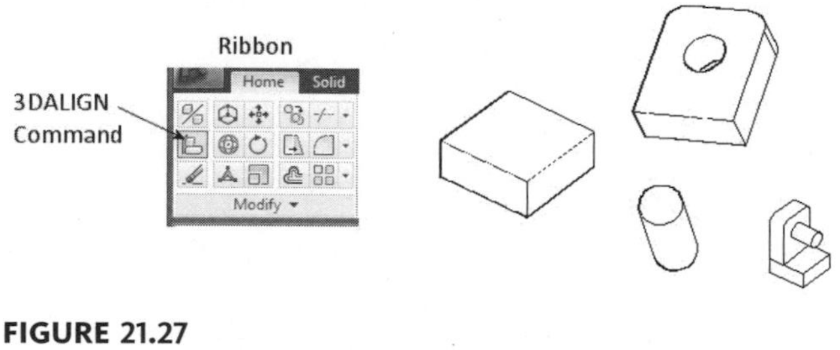

FIGURE 21.27

Follow the provided prompt sequence and the illustration in Figure 21.28 for aligning the hole plate with the bottom base. The first destination point acts as a base point to which the cylinder locates.

Command: 3DALIGN

Select objects: *(Select the object with the hole)*

Select objects: *(Press* ENTER *to continue)*

Specify source plane and orientation...

Specify base point or [Copy]: *(Select the endpoint at "A")*

Specify second point or [Continue] <C>: *(Select the endpoint at "B")*

Specify third point or [Continue] <C>: *(Select the endpoint at "C")*

Specify destination plane and orientation...

Specify first destination point: *(Select the endpoint at "D")*

Specify second destination point or [eXit] <X>: *(Select the endpoint at "E")*

Specify third destination point or [eXit] <X>: *(Select the endpoint at "F")*

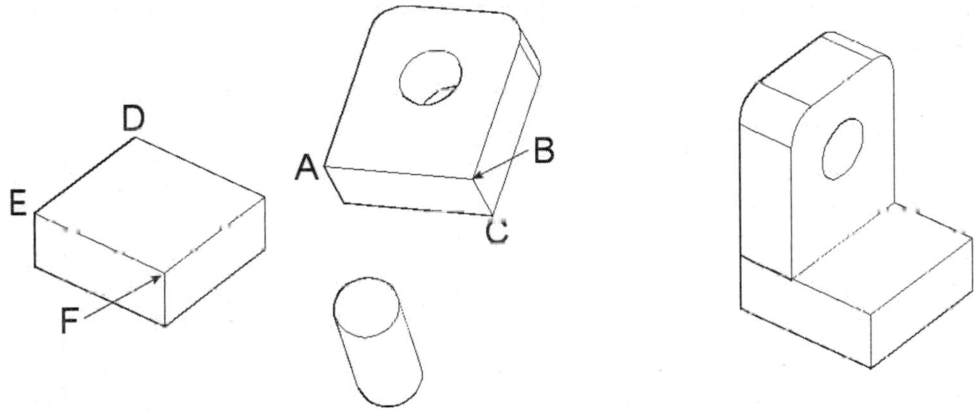

FIGURE 21.28

The results of the previous step are illustrated in Figure 21.29 (Left). Next, align the cylinder with the hole. Circular shapes often need only two sets of source and destination points for the shapes to be properly aligned.

Command: 3DALIGN

Select objects: *(Select the cylinder)*

Select objects: *(Press* ENTER *to continue)*

Specify source plane and orientation...

Specify base point or [Copy]: *(Select the center of circle "A")*

Specify second point or [Continue] <C>: *(Select the center of circle "B")*

Specify third point or [Continue] <C>: *(Press* ENTER *to continue)*

```
Specify destination plane and orientation...
Specify first destination point: (Select the center of circle "C")
Specify second destination point or [eXit] <X>: (Select the center of circle "D")
Specify third destination point or [eXit] <X>: (Press ENTER to exit)
```

The completed 3D model is illustrated in Figure 21.29 (Right). The 3DALIGN command provides an easy means of putting solid objects together to form assembly models.

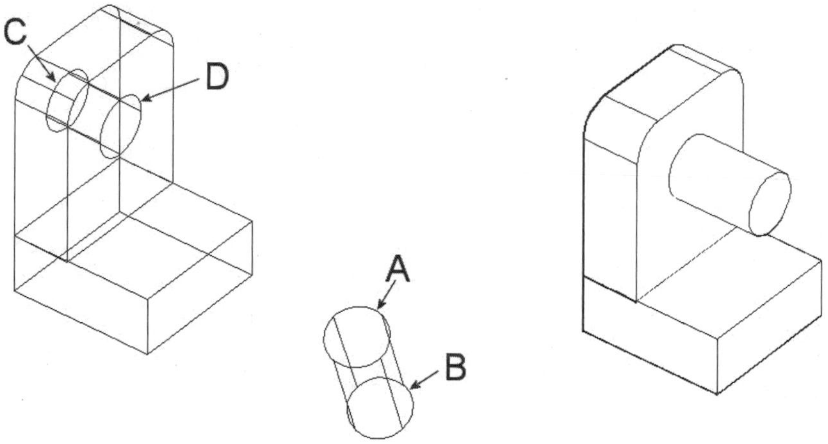

FIGURE 21.29

ROTATING OBJECTS IN 3D

The 3DROTATE command uses a special 3D rotate gizmo to rotate objects around a base point. Choose this command using one of the following methods:

- From the Ribbon > Home Tab > Modify Panel (3D Modeling Workspace)
- From the Menu Bar (Modify > 3D Operations > 3DRotate)
- From the Modeling Toolbar
- From the keyboard (3R or 3DROTATE)

After activating the 3DROTATE command, you are prompted to select the object or objects to rotate. Next, you pick a base point, which will act as the pivot point of the rotation. After you pick this base point, the rotate gizmo appears, as shown in Figure 21.30. You then hover your cursor over an axis handle until it turns yellow and an axis vector appears. If this axis is correct, click it to establish the axis of rotation. Then enter the angle to perform the rotation.

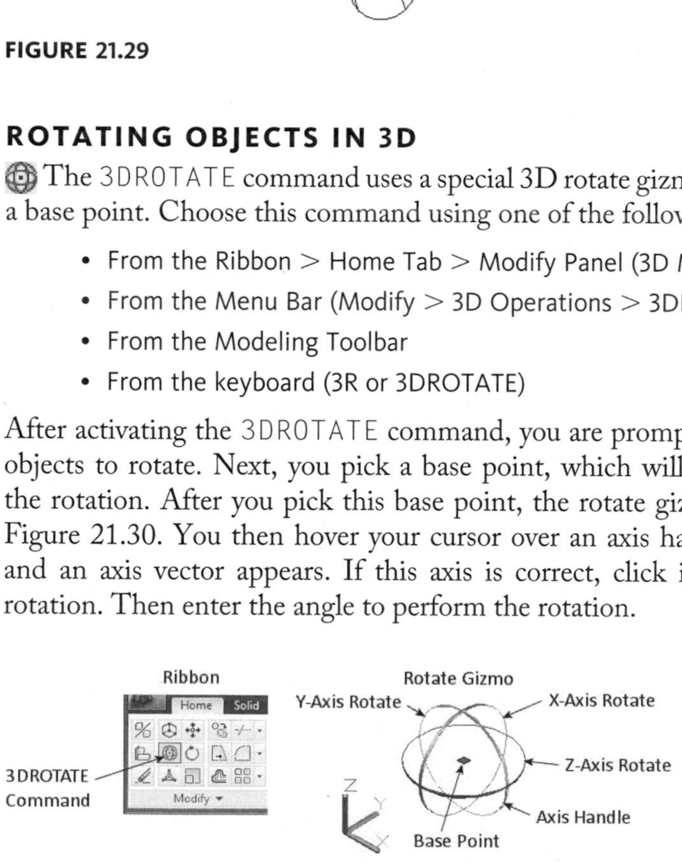

FIGURE 21.30

TRY IT!

Open the drawing 21_3DRot. In Figure 21.31, a base containing a slot needs to be joined with the two rectangular boxes to form a back and side. First select box "A" as the object to rotate in 3D. You will be prompted to define the base point of rotation. Next, you will be prompted to choose a rotation axis by picking the appropriate axis handle on the gizmo. This axis will serve as a pivot point where the rotation occurs. Entering a negative angle of 90° rotates the box in the clockwise direction.

Command: 3R *(For 3DROTATE)*

Current positive angle in UCS: ANGDIR=counterclockwise ANGBASE=0

Select objects: *(Select box "A")*

Select objects: *(Press* ENTER *to continue)*

Specify base point: *(Pick the endpoint at "A")*

Pick a rotation axis: *(When the rotate gizmo appears, hover on the axis handle until the proper axis appears, as shown in Figure 21.31; then pick with the cursor)*

Specify angle start point or type an angle: −90

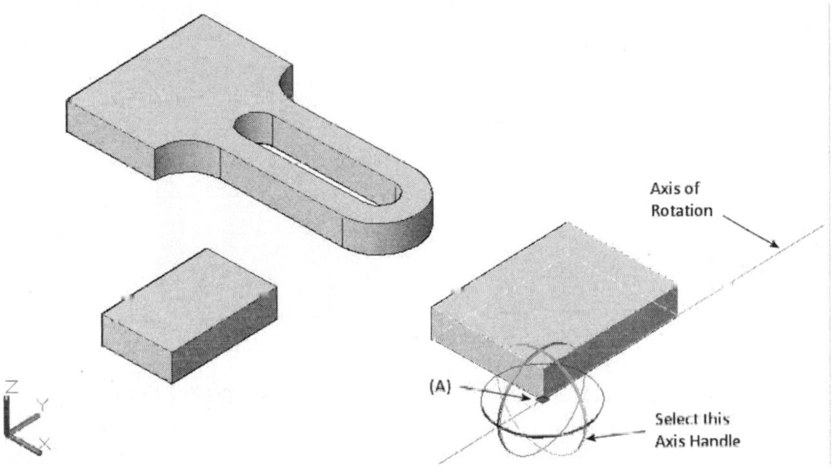

FIGURE 21.31

Next, the second box, as shown in Figure 21.32, is rotated 90° after the proper rotation axis is selected.

Command: 3R *(For 3DROTATE)*

Current positive angle in UCS: ANGDIR=counterclockwise ANGBASE=0

Select objects: *(Select box "A")*

Select objects: *(Press* ENTER *to continue)*

Specify base point: *(Pick the endpoint at "A")*

Pick a rotation axis: *(When the rotate gizmo appears, click the axis handle until the proper axis appears, as shown in Figure 21.32; then pick this axis with the cursor)*

Specify angle start point or type an angle: 90

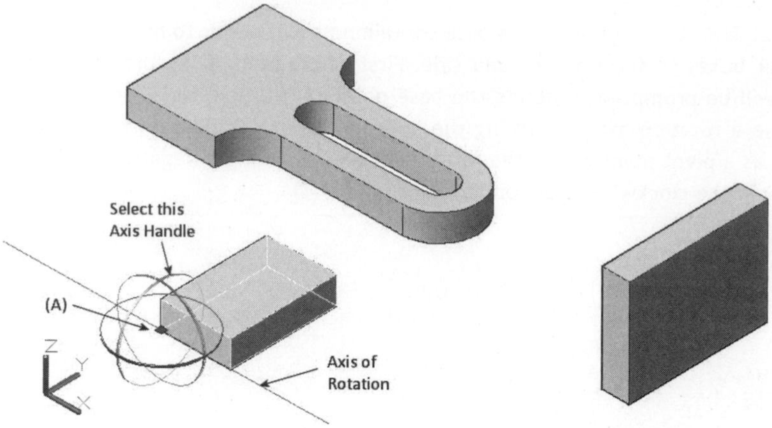

FIGURE 21.32

The results of these operations are illustrated in Figure 21.33 (Left). Once the boxes are rotated to the correct angles, they are moved into position using the `MOVE` command and the appropriate Object Snap modes. Box "A" is moved from the endpoint of the corner at "A" to the endpoint of the corner at "C." Box "B" is moved from the endpoint of the corner at "B" to the endpoint of the corner at "C." Once moved, they are then joined to the model through the `UNION` command, as shown in Figure 21.33 (Right).

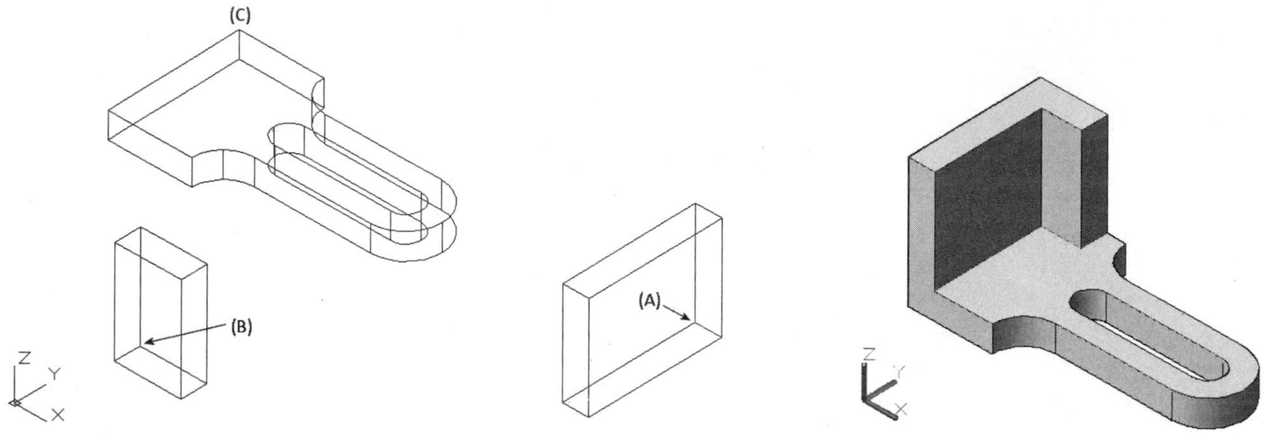

FIGURE 21.33

MIRRORING OBJECTS IN 3D

The `MIRROR3D` command is a 3D version of the `MIRROR` command. In this command, however, instead of flipping over an axis, you mirror over a plane. A thorough understanding of the User Coordinate System is a must in order to properly operate this command. Choose this command from one of the following:

- From the Ribbon > Home Tab > Modify Panel (3D Modeling Workspace)
- From the Menu Bar (Modify > 3D Operations > 3D Mirror)
- From the keyboard (MIRROR3D)

TRY IT!

Open the drawing 21_Mirror3D. As illustrated in Figure 21.34 (Left), only half of the object is created. The symmetrical object in Figure 21.34 (Right) is needed and can easily be created by using the `MIRROR3D` and `UNION` commands.

Command: MIRROR3D

Select objects: *(Select the part)*

Select objects: *(Press* ENTER *to continue)*

Specify first point of mirror plane *(3 points)* or

[Object/Last/Zaxis/View/XY/YZ/ZX/3points] <3points>:

YZ *(For YZ plane)*

Specify point on YZ plane <0,0,0>: *(Select the endpoint at "A")*

Delete source objects? [Yes/No] <N>: N *(Keep both objects)*

Command: UNI *(For UNION)*

Select objects: *(Select both solid objects)*

Select objects: *(Press* ENTER *to perform the union operation)*

Command: HI *(For HIDE; the final solid should appear as illustrated in Figure 21.34 (Right))*

Command: RE *(For REGEN; this will convert the image back to wireframe mode)*

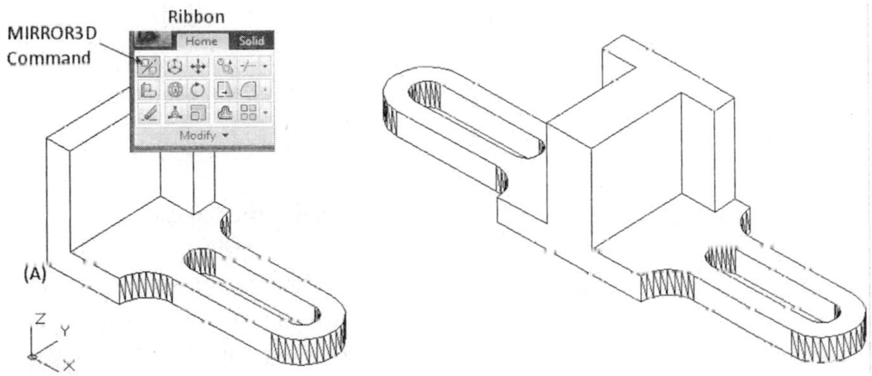

FIGURE 21.34

SCALING OBJECTS IN 3D

To assist in the scaling of objects in a 3D environment, the 3DSCALE tool is available. Choose this command using one of the following methods:

- From the Ribbon > Home Tab > Modify Panel (3D Modeling Workspace)
- From the keyboard (3S or 3DSCALE)

Activating the 3DSCALE command displays the 3D scale gizmo, as shown in Figure 21.35, which displays axis handles for the purpose of scaling objects or subobjects along an axis or plane. Solid models, however, can only be scaled uniformly. After you press ENTER to signify that you are done with the selection process, the scale gizmo displays at the center of your object. Pick a base point for the scaling operation. Select the gizmo and enter a scale factor, as shown in Figure 21.35.

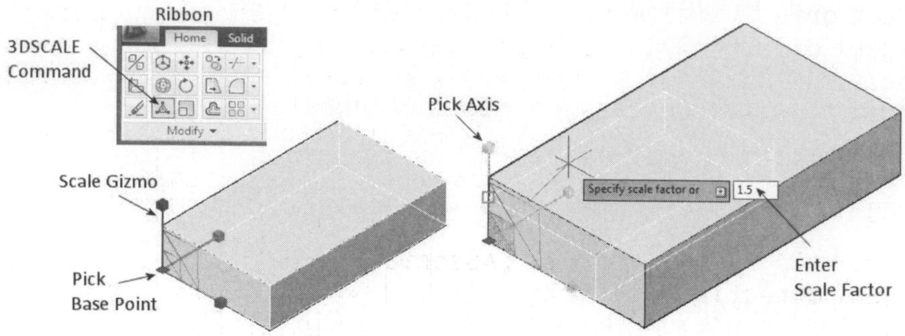

FIGURE 21.35

ARRAYING OBJECTS IN 3D

The ARRAYRECT, ARRAYPOLAR, and ARRAYPATH commands can be used to create either 2D or 3D arrays of objects. An available Level option allows you to duplicate objects in the Z direction as the array is created. It is also possible to provide an incremental increase in elevation when additional rows are specified in the array. It should be noted, however, that you will be unable to perform a boolean operation (union, subtraction, or intersection) on an associative array. You can turn associativity off with the Associative option, which is available in all three array commands, or you can always explode the array after it is created. The later method allows you to take advantage of associative array features until you are ready to generate the composite solid. Choose these command from one of the following:

- From the Ribbon > Home Tab > Modify Panel (3D Modeling Workspace)
- From the Menu Bar (Modify > Array)
- From the Modify Toolbar
- From the keyboard (ARRAYRECT, ARRAYPOLAR, and ARRAYPATH)

TRY IT!

Open the drawing 21_3DArray1. As illustrated in Figure 21.36 (Left), a hexagon shape, cylinder, and box need to be arrayed to create the final model, shown in Figure 21.36 (Right). The Level option will be used to stack the objects on top of each other to increase their height, as shown in Figure 21.36 (Center).

Command: ARRAYPATH

Select objects: *(Select the hexagon shape)*

Select objects: *(Press* ENTER *to continue)*

Type = Path Associative = Yes

Select path curve: *(Select path -centerline)*

Select grip to edit array or [ASsociative/Method/Base point/Tangent direction/Items/Rows/Levels/Align items/Z direction/eXit] <eXit>: M (For Method)

Enter path method [Divide/Measure] <Measure>: D (For Divide)

Select grip to edit array or [ASsociative/Method/Base point/Tangent direction/Items/Rows/Levels/Align items/Z direction/eXit] <eXit>: I (For Items)

Enter number of items along path or [Expression] <32>: 10

Select grip to edit array or [ASsociative/Method/Base point/Tangent direction/Items/Rows/Levels/Align items/Z direction/eXit] <eXit> : L *(For Levels option)*

Enter the number of levels or [Expression] <1>: **2**

Specify the distance between levels or [Total/Expression] <1.5000>: **1** *(This is the height of the polygon)*

Select grip to edit array or [ASsociative/Method/Base point/Tangent direction/Items/Rows/Levels/Align items/Z direction/eXit] <eXit>: *(Press* ENTER *to complete the array operation)*

Command: **ARRAYPOLAR**

Select objects: *(Select the cylinder)*

Select objects: *(Press* ENTER *to continue)*

Type = Polar Associative = Yes

Specify center point of array or [Base point/Axis of rotation]: **Cen**

of *(Select hole in base solid)*

Select grip to edit array or [ASsociative/Base point/Items/Angle between/Fill angle/ROWs/Levels/Rotate items/eXit] <eXit>: I (For Items)

Enter number of items in array or [Expression] <6>: **6**

Select grip to edit array or [ASsociative/Base point/Items/Angle between/Fill angle/ROWs/Levels/Rotate items/eXit] <eXit>: F (For Fill angle)

Specify the angle to fill (+=ccw, −=cw) or [EXpression] <360>: **180** *(half a circle rotated counterclockwise)*

Select grip to edit array or [ASsociative/Base point/Items/Angle between/Fill angle/ROWs/Levels/Rotate items/eXit] <eXit>: L *(For Levels option)*

Enter the number of levels or [Expression] <1>: **3**

Specify the distance between levels or [Total/Expression] <1.5000>: **1** *(This is the height of the cylinder)*

Select grip to edit array or [ASsociative/Base point/Items/Angle between/Fill angle/ROWs/Levels/Rotate items/eXit] <eXit>: *(Press* ENTER *to complete the array operation)*

Command: **ARRAYRECT**

Select objects: *(Select the box)*

Select objects: *(Press* ENTER *to continue)* Type = Rectangular Associative = Yes

Select grip to edit array or [ASsociative/Base point/COUnt/Spacing/COLumns/Rows/Levels/eXit] <eXit>: Cou (For COUnt)

Enter the number of columns or [Expression] <4>: **2**

Enter the number of rows or [Expression] <3>: **2**

Select grip to edit array or [ASsociative/Base point/COUnt/Spacing/COLumns/Rows/Levels/eXit] <eXit>: S (For Spacing)

Specify the distance between columns or [Unit cell] <1.5000>: 4

Specify the distance between rows <1.5000>: 4

Select grip to edit array or [ASsociative/Base point/COUnt/Spacing/COLumns/Rows/Levels/eXit] <eXit>: L *(For Levels option)*

Enter the number of levels or [Expression] <1>: 6

Specify the distance between levels or [Total/Expression] <1.5000>: 1 *(This is the height of the box)*

Select grip to edit array or [ASsociative/Base point/COUnt/Spacing/COLumns/Rows/Levels/eXit] <eXit>: *(Press* ENTER *to complete the array operation)*

Command: X *(For EXPLODE)*

Select objects: All *(You need to explode the associative arrays, if you want to union the solids together, as shown in Figure 21.36 (Right))*

Select objects: *(Hold down the* SHIFT *key and select the base solid to unselect it; you do not want to explode the solid)*

Select objects: *(Press* ENTER *to complete the operation)*

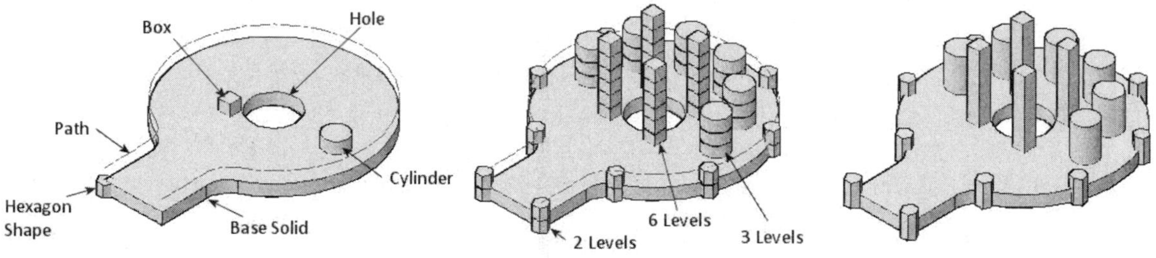

FIGURE 21.36

TRY IT!

Open the drawing 21_3DArray2. As illustrated in Figure 21.37 (Left), six arms need to be arrayed around the hub in the center. A second row of six arms will also be created, 2.50 units to the outside of the first row and 2.00 units in the direction of the rotational axis (this creates a change in elevation). It will be necessary to utilize the Axis of rotation option to establish a different axis of rotation for the polar array. By default polar arrays occur in the XY plane (around the Z-axis). This option is quite handy, by allowing you to rotate around any 3D axis without having to establish a new UCS. It should be noted, however, that it is unnecessary to change the Items and Fill angle options in this exercise, since the default settings create an array consisting of 6 items rotated through 360 degrees.

Command: **ARRAYPOLAR**

Select objects: *(Select the Arm "A")*

Select objects: *(Press* ENTER *to continue)*

Type = Polar Associative = Yes

Specify center point of array or [Base point/Axis of rotation]: **A** *(For Axis of Rotation)*

Specify first point on axis of rotation: **Cen**

of *(Select the center of the back hub circle at "B")*

Specify second point on axis of rotation: **Cen**

of *(Select the center of the front hub circle at "C")*

Select grip to edit array or [ASsociative/Base point/Items/Angle between/Fill angle/ROWs/Levels/Rotate items/eXit] <eXit>: **Row** *(For ROWs)*

Enter the number of rows or [Expression] <1>: **2**

Specify the distance between rows or [Total/Expression] <3.0000>: **2.5**

Specify the incrementing elevation between rows or [Expression] <0.0000>: **2**

Select grip to edit array or [ASsociative/Base point/Items/Angle between/Fill angle/ROWs/Levels/Rotate items/eXit] <eXit>: **AS** *(For ASsociative)*

Create associative array [Yes/No] <Yes>: **N** *(For No)*

Select grip to edit array or [ASsociative/Base point/Items/Angle between/Fill angle/ROWs/Levels/Rotate items/eXit] <eXit>: *(Press* ENTER *to complete the array operation)*

Command: **UNI** *(For UNION)*

Select objects: **All**

Select objects: *(Press* ENTER *to perform the union operation)*

Command: **HI** *(For HIDE, the solid should appear as illustrated in Figure 21.37 (Right))*

Command: **RE** *(For REGEN; this will convert the image back to wireframe mode)*

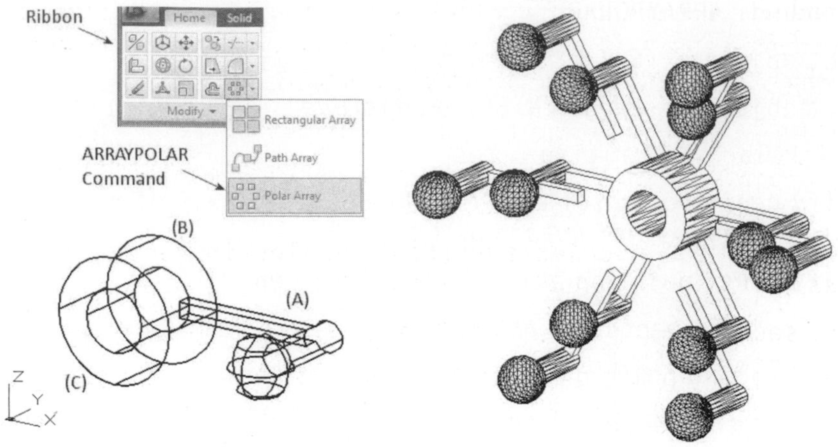

FIGURE 21.37

DETECTING INTERFERENCES OF SOLID MODELS

The INTERFERE command identifies any interference and highlights the solid models that overlap. Choose this command in one of the following ways:

- From the Ribbon > Home Tab > Solid Editing Panel (3D Modeling Workspace)
- From the Menu Bar (Modify > 3D Operations > Interference Checking)
- From the keyboard (INTERFERE)

TRY IT!

Open the drawing file 21_Pipe Interference. Use this command to find any interference shared by a series of solid pipe objects, as shown in Figure 21.38 (Left). Click the Interference checking button, located in the Ribbon, to begin this command.

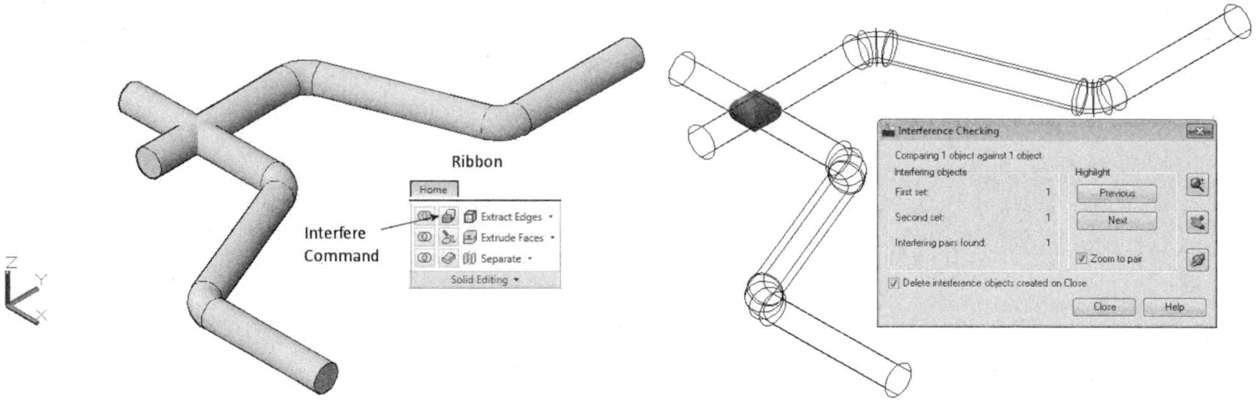

FIGURE 21.38

After launching this command, you will be prompted to select first and second sets of solids. In this example, select one pipe for the first set and the second pipe for the second set. Pressing ENTER at the end of the command sequence changes the solid objects to a wireframe display to expose the areas of the objects considered to be interfering with one another, as shown in Figure 21.38 (Right). An Interference Checking dialog box also appears. The information in the dialog box allows you to verify the first and second sets of interfering objects. If they exist, you can also cycle through additional interferences using the Previous and Next buttons. In addition, a checkbox is provided, which allows you to create the interference solid, if desired.

Command: INTERFERE

Select first set of objects or [Nested selection/Settings]: *(Pick one of the pipe objects)*

Select first set of objects or [Nested selection/Settings]: *(Press* ENTER *to continue)*

Select second set of objects or [Nested selection/checK first set] <checK>: *(Pick the second pipe object)*

Select second set of objects or [Nested selection/checK first set] <checK>: *(Press* ENTER *to perform the interference check)*

SLICING SOLID MODELS

Yet another tool for editing 3D solid models is through the SLICE command. This command creates new solids from the existing ones that are sliced. You can retain one or both halves of the sliced solid. Slicing a solid requires some type of cutting plane. The default method of creating this plane is by picking three points. You can also define the cutting plane by picking a surface, by using another object, or by basing the cutting plane line on the current positions of the XY, YZ, or ZX planes.

Choose the SLICE command in one of the following ways:

- From the Ribbon > Home Tab > Solid Editing Panel (3D Modeling Workspace)
- From the Menu Bar (Modify > 3D Operations > Slice)
- From the keyboard (SL or SLICE)

TRY IT!

Open the drawing file 21_Tee Slice. For this command, the solid model is actually cut or sliced at a plane that you define. In the example in Figure 21.39, this plane is defined by the User Coordinate System. Before the slice is made, you also have the option of keeping either one or both halves of the object. The MOVE command is used to separate both halves, as shown in Figure 21.39 (Right).

Command: SL *(For SLICE)*

Select objects to slice: *(Select the solid object)*

Select objects to slice: *(Press* ENTER *to continue with this command)*

Specify start point of slicing plane or [planar Object/Surface/Zaxis/View/XY/YZ/ZX/3points] <3points>: XY

Specify a point on the XY-plane <0,0,0>: *(Press* ENTER *to accept this default value)*

Specify a point on desired side of the plane or [keep Both sides]: B *(To keep both sides)*

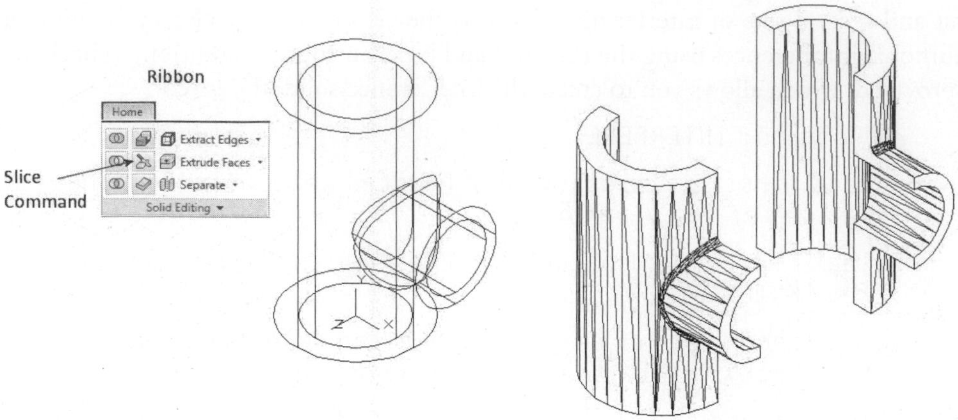

FIGURE 21.39

Slicing a Solid with an Extruded Surface

A solid object can also be sliced by a surface. A surface can be created by performing a 3D operation such as extrusion or revolution on an open object. Once the surface is created, it is positioned inside of the 3D solid model, where a slicing operation is performed. The next Try It! exercise illustrates the use of this technique.

TRY IT!

Open the drawing file 21_Surface Flow. You will first extrude a spline to create a surface. The surface will then be used to slice a solid block. You will keep the bottom portion of the solid.

Before slicing the solid block, first extrude the spline object, as shown in Figure 21.40 (Left), a distance equal to the depth of the block (from "A" to "B"). Since the spline represents an open shape, the result of performing this operation is the creation of a surface instead of a solid, as shown in Figure 21.40 (Right).

Command: EXT *(For EXTRUDE)*

Current wire frame density: ISOLINES=4, Closed profile creation mode = Solid

Select objects to extrude or [MOde]: *(Pick the spline object)*

Select objects to extrude or [MOde]: *(Press* ENTER *to continue)*

Specify height of extrusion or [Direction/Path/Taper angle/Expression]: D *(For Direction)*

Specify start point of direction: *(Pick the endpoint at "A")*

Specify end point of direction: *(Pick the endpoint at "B")*

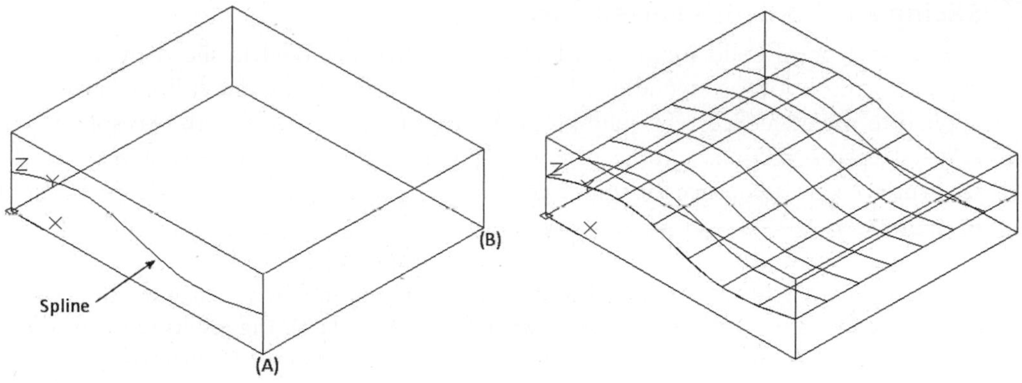

FIGURE 21.40

With the newly created surface positioned inside of the solid block, issue the SLICE command. Pick the solid block as the object to slice and select the surface as the slicing plane, as shown in Figure 21.41 (Left). You will also be prompted to select the portion of the solid to keep. Here is where you pick the bottom of the solid, as shown in Figure 21.41 (Left).

Command: SL *(For SLICE)*

Select objects to slice: *(Pick the solid block)*

Select objects to slice: *(Press ENTER to continue)*

Specify start point of slicing plane or [planar

Object/Surface/Zaxis/View/XY/YZ/ZX/3points] <3points>:
S *(For Surface)*

Select a surface: *(Pick the surface, as shown in Figure 21.41 (Left))*

Select sliced object to keep or [keep Both sides] <Both>:
(Pick the bottom of the solid)

The results are displayed in Figure 21.41 (Right), with the solid block being cut by the surface.

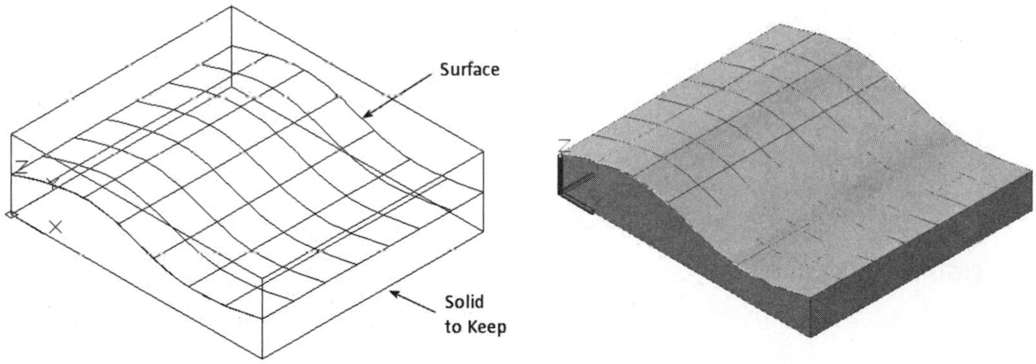

FIGURE 21.41

Slicing a Solid with a Lofted Surface

A unique type of solid model can be created when a model is sliced by a surface created using a lofting operation. In Figure 21.31, four different splines have been applied to the edge faces of a solid block. Using the LOFT command, two splines are selected as cross sections and the other two splines as guides or rails. Once this specialized surface is created, slice the solid block using the surface and keep the lower portion of the model.

TRY IT!

Open the drawing file 21_Terrain. You will create a lofted surface by selecting the two cross sections and the two guide curves, as shown in Figure 21.31 (Left). The results are displayed in Figure 21.31 (Right), with a complex surface being created from the loft operation.

```
Command: LOFT
Current wire frame density: ISOLINES=4, Closed profile
creation mode = Solid
Select cross-sections in lofting order or [POint/Join
multiple edges/MOde]: (Select cross section #1)
Select cross-sections in lofting order or [POint/Join
multiple edges/MOde]: (Select cross section #2)
Select cross-sections in lofting order or [POint/Join
multiple edges/MOde]: (Press ENTER to continue)
Enter an option [Guides/Path/Cross-sections only/Settings]
<Cross-sections only>: G (For Guides)
Select guide profiles or [Join multiple edges]: (Select guide
curve #1)
Select guide profiles or [Join multiple edges]: (Select guide
curve #2)
Select guide profiles or [Join multiple edges]: (Press ENTER
to create the surface)
```

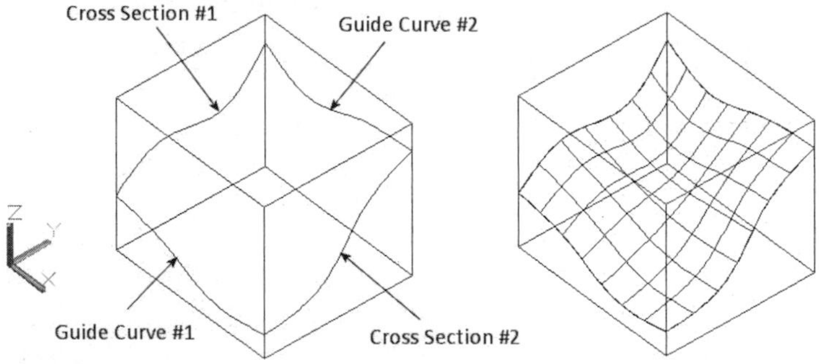

FIGURE 21.42

NOTE

If you have difficulty selecting the splines in the previous exercise, turn Selection Cycling on in the Status Bar.

With the surface created, activate the slice command, pick the solid block as the object to slice, select the surface as the slicing plane, and, finally, pick the lower portion of the solid as the portion to keep, as shown in Figure 21.43 (Left). The results are illustrated in Figure 21.43 (Right).

Command: SL *(For SLICE)*

Select objects to slice: *(Select the solid block)*

Select objects to slice: *(Press* ENTER *to continue)*

Specify start point of slicing plane or [planar Object/Surface/Zaxis/View/XY/YZ/ZX/3points] <3points>:

S *(For Surface)*

Select a surface: *(Select the surface)*

Select sliced object to keep or [keep Both sides] <Both>: *(Pick the lower portion of the solid block)*

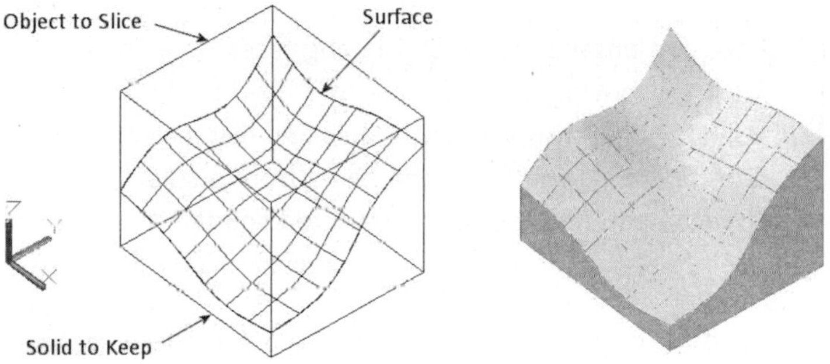

FIGURE 21.43

EDITING SOLID FEATURES

Once features such as holes, slots, and extrusions are constructed in a solid model, the time may come to make changes to these features. This is the function of the SOLIDEDIT command. This command contains numerous options, which can be selected from the Ribbon, as shown in Figure 21.44. The menus are arranged in three groupings, namely, Face, Edge, and Body editing. These groupings are discussed in the pages that follow. Also, it is recommended that instead of entering the SOLIDEDIT command at the Command Prompt, you use the Ribbon to perform editing operations. This eliminates a number of steps and make it easier to locate the appropriate option under the correct grouping.

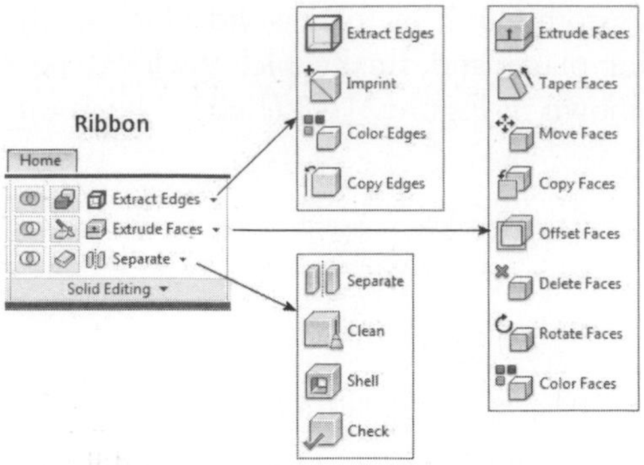

FIGURE 21.44

The following table gives a brief description of each mode for performing solid editing operations.

Button	Tool	Function
	Extrude Faces	Used for lengthening or shortening faces on a solid
	Move Faces	Used for moving a solid shape to a new location
	Offset Faces	Used for offsetting faces at a specified distance on a solid
	Delete Faces	Used for removing faces and fillets from a solid
	Rotate Faces	Used for rotating faces on a selected solid
	Taper Faces	Used for tapering selected faces of a solid at a draft angle along a vector direction
	Copy Faces	Used for copying selected faces of a solid. These copied faces can take the form of regions or bodies.
	Color Faces	Used for assigning unique colors to individual faces
	Copy Edges	Used to copy edges from a solid. These new edges are often used to create new solids.
	Color Edges	Used for assigning unique colors to individual solid edges
	Imprint	Used for adding construction geometry to a solid model
	Clean	Used to remove imprints from a solid
	Separate	Used to separate a solid into multiple parts as long as those parts do not intersect at any point. Solids sometimes act as a single entity even though they appear to be separate solids (unioning solids together that do not touch or removing part of a solid so that the remaining pieces do not touch).
	Shell	Used to create a thin wall in a solid model
	Check	Used to prove a solid is valid

EXTRUDING (FACE EDITING)

Faces may be lengthened or shortened through the Extrude option of the SOLIDEDIT command. A positive distance extrudes the face in the direction of its normal. A negative distance extrudes the face in the opposite direction.

TRY IT!

Open the drawing file 21_Extrude. In Figure 21.45, the highlighted face at "A" needs to be decreased in height.

NOTE

You will achieve better results when selecting a face for the SOLIDEDIT command if you pick on the inside of the face rather than on the edge of the face.

```
Command: SOLIDEDIT
Solids editing automatic checking: SOLIDCHECK=1
Enter a solids editing option [Face/Edge/Body/Undo/eXit]
<eXit>: F (For Face)
Enter a face editing option
[Extrude/Move/Rotate/Offset/Taper/Delete/Copy/coLor/
mAterial/Undo/eXit] <eXit>: E (For Extrude)
Select faces or [Undo/Remove]: (Select the face inside the
area represented by "A")
Select faces or [Undo/Remove/ALL]: (Press ENTER to continue)
Specify height of extrusion or [Path]: -10.00
Specify angle of taper for extrusion <0>: (Press ENTER)
Solid validation started.
Solid validation completed.
Enter a face editing option
[Extrude/Move/Rotate/Offset/Taper/Delete/Copy/coLor/
mAterial/Undo/eXit] <eXit>: (Press ENTER)
Solids editing automatic checking: SOLIDCHECK=1
Enter a solids editing option [Face/Edge/Body/Undo/eXit]
<eXit>: (Press ENTER)
```

The result is illustrated in Figure 21.45 at "B."

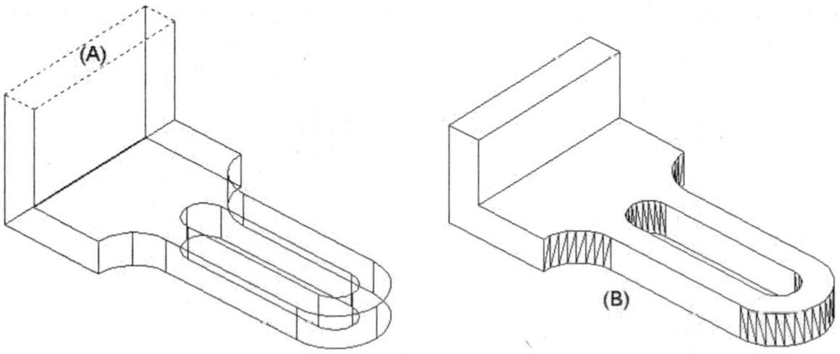

FIGURE 21.45

MOVING (FACE EDITING)

TRY IT!

Open the drawing 21_Move. The object in Figure 21.46 illustrates two intersecting cylinders. The two horizontal cylinders need to be moved 1 unit up from their current location. The cylinders are first selected at "A" and "B" through the SOLIDEDIT **command along with the Move option.**

Command: **SOLIDEDIT**

Solids editing automatic checking: SOLIDCHECK=1

Enter a solids editing option [Face/Edge/Body/Undo/eXit] <eXit>: **F** *(For Face)*

Enter a face editing option

[Extrude/Move/Rotate/Offset/Taper/Delete/Copy/coLor/mAterial/Undo/eXit] <eXit>: **M** *(For Move)*

Select faces or [Undo/Remove]: *(Select both highlighted faces at "A" and "B")*

Select faces or [Undo/Remove/ALL]: *(Press* ENTER *to continue)*

Specify a base point or displacement: *(Pick any point on the screen)*

Specify a second point of displacement: **@0,0,1**

Solid validation started.

Solid validation completed.

Enter a face editing option

[Extrude/Move/Rotate/Offset/Taper/Delete/Copy/coLor/mAterial/Undo/eXit] <eXit>: *(Press* ENTER*)*

Solids editing automatic checking: SOLIDCHECK=1

Enter a solids editing option [Face/Edge/Body/Undo/eXit] <eXit>: *(Press* ENTER*)*

The results are illustrated in Figure 21.46 (Right).

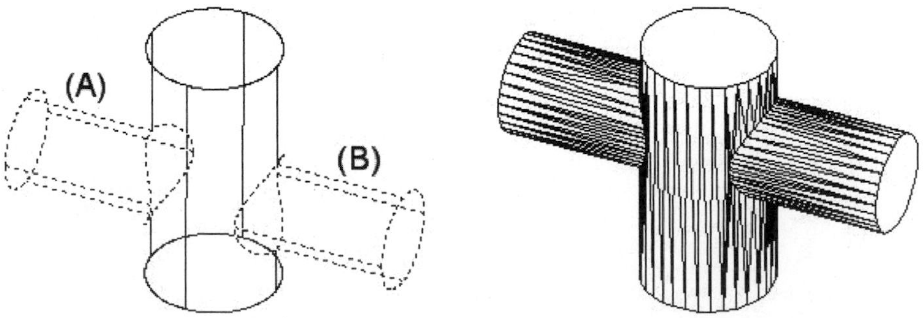

FIGURE 21.46

ROTATING (FACE EDITING)

TRY IT!

Open the drawing file 21_Rotate. In Figure 21.47, the triangular cutout needs to be rotated 45° in the clockwise direction. Use the Rotate Face option of the SOLIDEDIT **command to accomplish this. You must select all faces of the triangular cutout at "A," "B," and "C." During the selection process you will have to remove the face that makes up the top of the rectangular base at "D" before proceeding.**

Command: SOLIDEDIT

Solids editing automatic checking: SOLIDCHECK=1

Enter a solids editing option [Face/Edge/Body/Undo/eXit] <eXit>: F *(For Face)*

Enter a face editing option

[Extrude/Move/Rotate/Offset/Taper/Delete/Copy/coLor/mAterial/Undo/eXit] <eXit>: R *(For Rotate)*

Select faces or [Undo/Remove]: *(Select all faces that make up the triangular extrusion. Pick inside areas at "A," "B," and "C"; select "C" twice to highlight it)*

Select faces or [Undo/Remove/ALL]: R *(For Remove)*

Remove faces or [Undo/Add/ALL]: *(Select the face at "D" to remove)*

Remove faces or [Undo/Add/ALL]: *(Press* ENTER *to continue)*

Specify an axis point or [Axis by object/View/Xaxis/Yaxis/Zaxis] <2points>: Z *(For Zaxis)*

Specify the origin of the rotation <0,0,0>: *(Select the endpoint at "E")*

Specify a rotation angle or [Reference]: -45 *(To rotate the triangular extrusion 45° in the clockwise direction)*

Solid validation started.

Solid validation completed.

Enter a face editing option

[Extrude/Move/Rotate/Offset/Taper/Delete/Copy/coLor/mAterial/Undo/eXit] <eXit>: *(Press* ENTER*)*

Solids editing automatic checking: SOLIDCHECK=1

Enter a solids editing option [Face/Edge/Body/Undo/eXit] <eXit>: *(Press* ENTER*)*

The results are illustrated in Figure 21.47 (Right).

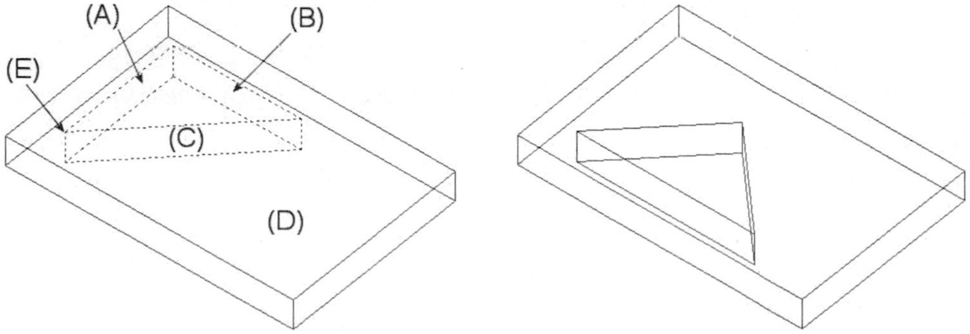

FIGURE 21.47

OFFSETTING (FACE EDITING)

TRY IT!

Open the drawing file 21_Offset. In Figure 21.48, the holes need to be resized. Use the Offset Face option of the SOLIDEDIT command to increase or decrease the size of selected faces. Using positive values increases the volume of the solid. Therefore, the feature being offset gets smaller, similar to the illustration at "B." Entering negative values reduces the volume of the solid; this means that the feature being offset gets larger, as in the figures at "C." Study the prompt sequence and Figure 21.48 for the mechanics of this command option.

Command: SOLIDEDIT

Solids editing automatic checking: SOLIDCHECK=1

Enter a solids editing option [Face/Edge/Body/Undo/eXit] <eXit>: F *(For Face)*

Enter a face editing option

[Extrude/Move/Rotate/Offset/Taper/Delete/Copy/coLor/mAterial/Undo/eXit] <eXit>: O *(For Offset)*

Select faces or [Undo/Remove]: *(Select inside the edges of the two holes at "D" and "E")*

Select faces or [Undo/Remove/ALL]: *(Press* ENTER *to continue)*

Specify the offset distance: .50

Solid validation started.

Solid validation completed.

Enter a face editing option

[Extrude/Move/Rotate/Offset/Taper/Delete/Copy/coLor/mAterial/Undo/eXit] <eXit>: *(Press* ENTER*)*

Solids editing automatic checking: SOLIDCHECK=1

Enter a solids editing option [Face/Edge/Body/Undo/eXit] <eXit>: *(Press* ENTER*)*

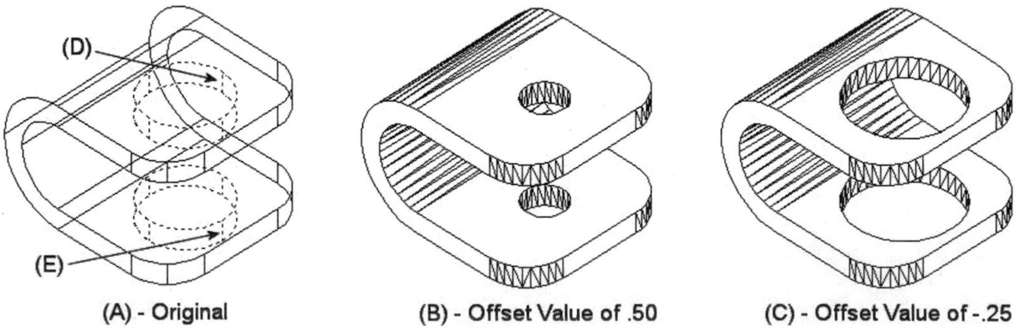

FIGURE 21.48

TAPERING (FACE EDITING)

Open the drawing 21_Taper. The object at "A" in Figure 21.49 represents a solid box that needs to have tapers applied to its sides. Using the Taper Face option of the SOLIDEDIT command allows you to accomplish this task. Entering a positive angle moves the location of the second point into the part, as shown in Figure 21.49 (Center). Entering a negative angle moves the location of the second point away from the part, as shown in Figure 21.49 (Right).

TRY IT!

Command: SOLIDEDIT

Solids editing automatic checking: SOLIDCHECK=1

Enter a solids editing option [Face/Edge/Body/Undo/eXit] <eXit>: F *(For Face)*

Enter a face editing option

[Extrude/Move/Rotate/Offset/Taper/Delete/Copy/coLor/mAterial/Undo/eXit] <eXit>: T *(For Taper)*

Select faces or [Undo/Remove]: *(Select faces "A" through "D")*

Select faces or [Undo/Remove/ALL]: *(Press* ENTER *to continue)*

Specify the base point: *(Select the endpoint at "E")*

Specify another point along the axis of tapering: *(Select the endpoint at "F")*

Specify the taper angle: 10 *(For the angle of the taper)*

Solid validation started.

Solid validation completed.

Enter a face editing option

[Extrude/Move/Rotate/Offset/Taper/Delete/Copy/coLor/mAterial/Undo/eXit] <eXit>: *(Press* ENTER*)*

Solids editing automatic checking: SOLIDCHECK=1

Enter a solids editing option [Face/Edge/Body/Undo/eXit] <eXit>: *(Press* ENTER*)*

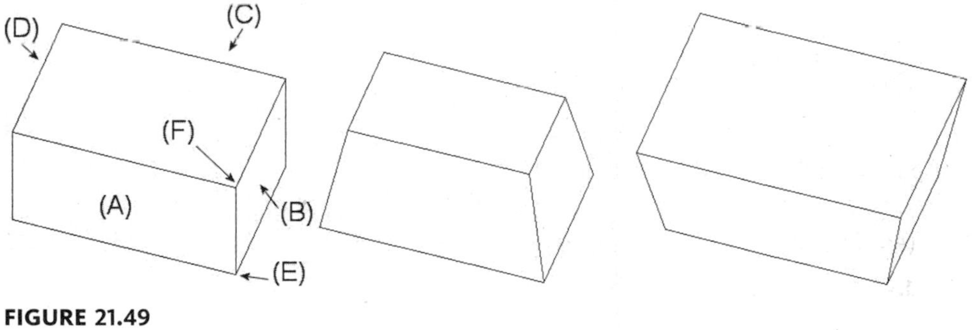

FIGURE 21.49

DELETING (FACE EDITING)

Faces can be erased through the Delete Face option of the SOLIDEDIT command.

TRY IT!

Open the drawing file 21_Delete. In Figure 21.50, select the hole at "A" as the face to erase.

Command: SOLIDEDIT

Solids editing automatic checking: SOLIDCHECK=1

Enter a solids editing option [Face/Edge/Body/Undo/eXit] <eXit>: F *(For Face)*

Enter a face editing option

[Extrude/Move/Rotate/Offset/Taper/Delete/Copy/coLor/mAterial/Undo/eXit] <eXit>: D *(For Delete)*

Select faces or [Undo/Remove]: *(Select inside the hole at "A")*

Select faces or [Undo/Remove/ALL]: *(Press* ENTER *to continue)*

Solid validation started.

Solid validation completed.

Enter a face editing option

[Extrude/Move/Rotate/Offset/Taper/Delete/Copy/coLor/mAterial/Undo/eXit] <eXit>: *(Press* ENTER*)*

Solids editing automatic checking: SOLIDCHECK=1

Enter a solids editing option [Face/Edge/Body/Undo/eXit] <eXit>: *(Press* ENTER*)*

The results are illustrated in Figure 21.50 (Right).

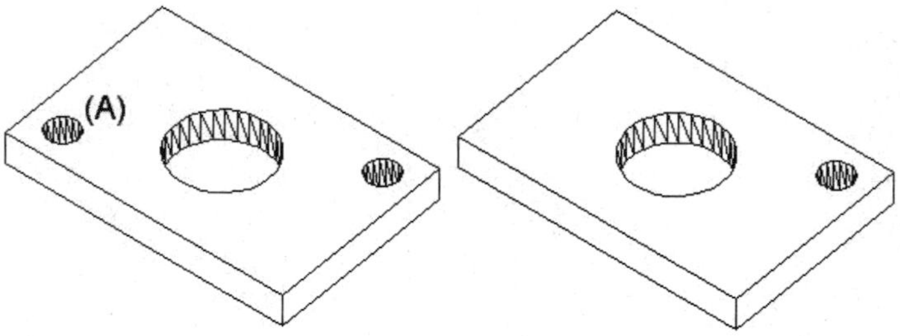

FIGURE 21.50

COPYING (FACE EDITING)

You can copy a face for use in the creation of another solid model using the Copy Face option of the SOLIDEDIT command.

TRY IT!

Open the drawing file 21_Copy. In Figure 21.51, the solid model at "A" will be used to create a region from the face at "B." While in the command, select the face by picking in the area at "B." Notice that all objects making up the face, such as the rectangle and circles, are highlighted. Picking a base point and second point copies the face at "C." The resulting object at "C" is actually a region. The region could be exploded back into individual lines and circles, which could then be used to create a new object. A region can also be extruded by using the `EXTRUDE` **command to create another solid model such as the one illustrated in Figure 21.51 (Right).**

Command: SOLIDEDIT

Solids editing automatic checking: SOLIDCHECK=1

Enter a solids editing option [Face/Edge/Body/Undo/eXit] <eXit>: F *(For Face)*

Enter a face editing option

[Extrude/Move/Rotate/Offset/Taper/Delete/Copy/coLor/mAterial/Undo/eXit] <eXit>: C *(For Copy)*

Select faces or [Undo/Remove]: *(Select the top face in area "B")*

Select faces or [Undo/Remove/ALL]: *(Press* ENTER *to continue)*

Specify a base point or displacement: *(Pick a point to copy from)*

Specify a second point of displacement: *(Pick a point to copy to)*

Enter a face editing option

[Extrude/Move/Rotate/Offset/Taper/Delete/Copy/coLor/mAterial/Undo/eXit] <eXit>: *(Press* ENTER*)*

Solids editing automatic checking: SOLIDCHECK=1

Enter a solids editing option [Face/Edge/Body/Undo/eXit] <eXit>: *(Press* ENTER*)*

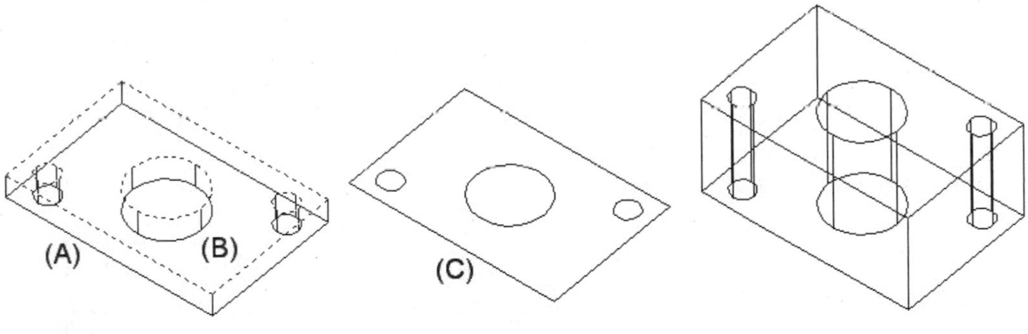

FIGURE 21.51

IMPRINTING (BODY EDITING)

An interesting and powerful method of adding construction geometry to a solid model is though the process of imprinting.

TRY IT!

Open the drawing file 21_Imprint. In Figure 21.52 at "A," a box along with slot is already modeled in 3D. A line was constructed from the midpoints of the top surface of the solid model. The `SOLIDEDIT` command will be used to imprint this line to the model, which results in dividing the top surface into two faces. It should be noted that the `IMPRINT` command could also have been used to perform this imprinting operation.

```
Command: SOLIDEDIT
Solids editing automatic checking: SOLIDCHECK=1
Enter a solids editing option [Face/Edge/Body/Undo/eXit]
<eXit>: B (For Body)
Enter a body editing option
[Imprint/seParate solids/Shell/cLean/Check/Undo/eXit]
<eXit>: I (For Imprint)
Select a 3D solid: (Select the solid model)
Select an object to imprint: (Select line "B")
Delete the source object [Yes/No] <N>: Y (For Yes; this erases
the line)
Select an object to imprint: (Press ENTER to complete the
imprint operation)
Enter a body editing option
[Imprint/seParate solids/Shell/cLean/Check/Undo/eXit]
<eXit>: (Press ENTER)
Solids editing automatic checking: SOLIDCHECK=1
Enter a solids editing option [Face/Edge/Body/Undo/eXit]
<eXit>: (Press ENTER)
```

The segments of the lines that come in contact with the 3D solid remain on the part's surface. However, these are no longer line segments; rather, these lines now belong to the part. The lines actually separate the top surface into two faces.

Use the Extrude Face option of the `SOLIDEDIT` command on one of the newly created faces at "C" and increase the face in height by an extra 1.50 units. The results are illustrated in Figure 21.52 (Right).

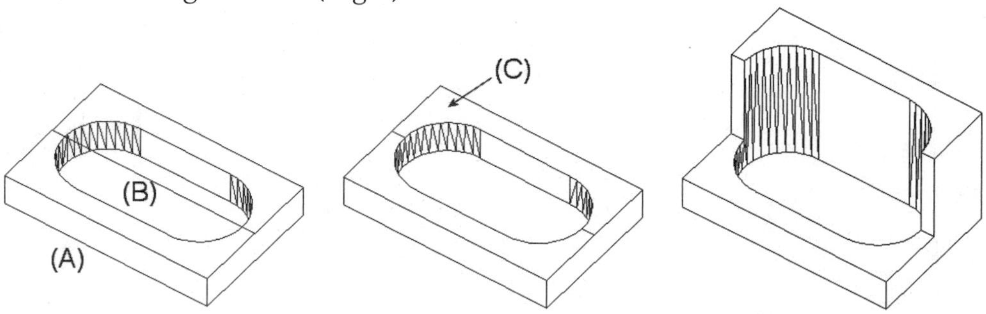

FIGURE 21.52

SEPARATING SOLIDS (BODY EDITING)

Sometimes when performing union and subtraction operations on solid models, you can end up with models that do not actually touch (intersect) but act as a single object. The Separating Solids option of the SOLIDEDIT command is used to correct this condition.

TRY IT!

Open the drawing file 21_Separate. In Figure 21.53, the model at "A" is about to be sliced in half with a thin box created at the center of the circle and spanning the depth of the rectangular shelf.

Use the SUBTRACT command and subtract the rectangular box from the solid object. After you subtract the box, pick the solid at "B" and notice that both halves of the object highlight even though they appear separate. To convert the single solid model into two separate models, use the SOLIDEDIT command followed by the Body and Separate options.

Command: SOLIDEDIT

Solids editing automatic checking: SOLIDCHECK=1

Enter a solids editing option [Face/Edge/Body/Undo/eXit] <eXit>: B *(For Body)*

Enter a body editing option

[Imprint/seParate solids/Shell/cLean/Check/Undo/eXit] <eXit>: P *(For Separate)*

Select a 3D solid: *(Pick the solid model at "B")*

Enter a body editing option

[Imprint/seParate solids/Shell/cLean/Check/Undo/eXit] <eXit>: *(Press* ENTER*)*

Solids editing automatic checking: SOLIDCHECK=1

Enter a solids editing option [Face/Edge/Body/Undo/eXit] <eXit>: *(Press* ENTER*)*

This action separates the single model into two. When you select the model in Figure 21.53 at "C," only one half highlights.

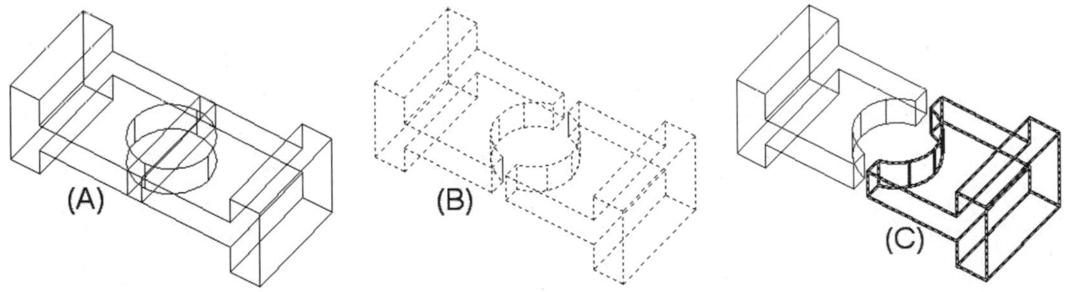

FIGURE 21.53

SHELLING (BODY EDITING)

Shelling is the process of constructing a thin wall inside or outside of a solid model. Positive thickness produces the thin wall inside; negative values for thickness produce the thin wall outside. This wall thickness remains constant throughout the entire model. Faces may be removed during the shelling operation to create an opening.

TRY IT!

Open the drawing file 21_Shell. For simplicity, first rotate your model or your viewpoint such that any faces to be removed are visible. Use the 3DFORBIT (Free Orbit) command and Figure 21.54 to change the model view from the one shown at "A" to the one shown at "B." Use the Hidden Visual Style option selected from a shortcut menu in the 3DFORBIT command to ensure that the bottom surface at "C" is visible. Now the Shell option of the SOLIDEDIT **command can be used to "hollow out" the part and remove the bottom face. An additional Note: only one shell is permitted in a model.**

Command: **SOLIDEDIT**

Solids editing automatic checking: SOLIDCHECK=1

Enter a solids editing option [Face/Edge/Body/Undo/eXit] <eXit>: **B** *(For Body)*

Enter a body editing option

[Imprint/seParate solids/Shell/cLean/Check/Undo/eXit] <eXit>: **S** *(For Shell)*

Select a 3D solid: *(Select the solid model at "B")*

Remove faces or [Undo/Add/ALL]: *(Pick a point at "C"; because the model is already highlighted, it is not obvious that the face is selected but "1 face found, 1 removed" will be indicated)*

Remove faces or [Undo/Add/ALL]: *(Press* ENTER *to continue)*

Enter the shell offset distance: **0.20**

Solid validation started.

Solid validation completed.

Enter a body editing option

[Imprint/seParate solids/Shell/cLean/Check/Undo/eXit] <eXit>: *(Press* ENTER*)*

Solids editing automatic checking: SOLIDCHECK=1

Enter a solids editing option [Face/Edge/Body/Undo/eXit] <eXit>: *(Press* ENTER*)*

The results are illustrated in Figure 21.54 (Right).

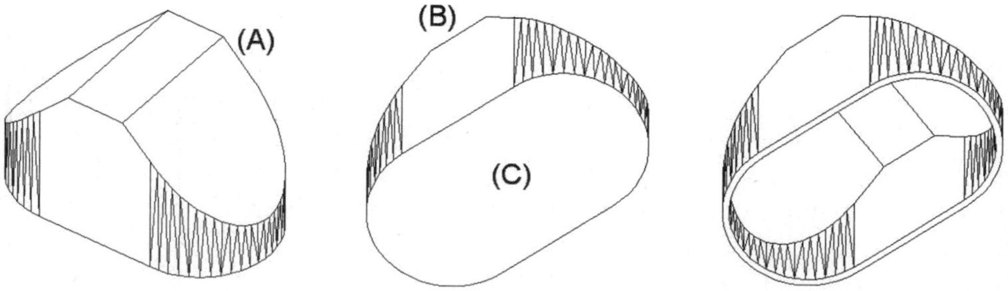

FIGURE 21.54

CLEANING (BODY EDITING)

When imprinted lines that form faces are not used, they can be deleted from a model by the Clean (Body Editing) option of the SOLIDEDIT command.

TRY IT!

Open the drawing file 21_Clean. The object at "A" in Figure 21.55 illustrates lines originally constructed on the top of the solid model. These lines were then imprinted at "B." Since these lines now belong to the model, the Clean option is used to remove them. The results are illustrated in Figure 21.55 (Right).

```
Command: SOLIDEDIT
Solids editing automatic checking: SOLIDCHECK=1
Enter a solids editing option [Face/Edge/Body/Undo/eXit]
<eXit>: B (For Body)
Enter a body editing option
[Imprint/seParate solids/Shell/cLean/Check/Undo/eXit]
<eXit>: L (For Clean)
Select a 3D solid: (Select the solid model at "B")
Enter a body editing option
[Imprint/seParate solids/Shell/cLean/Check/Undo/eXit]
<eXit>: (Press ENTER)
Solids editing automatic checking: SOLIDCHECK=1
Enter a solids editing option [Face/Edge/Body/Undo/eXit]
<eXit>: (Press ENTER)
```

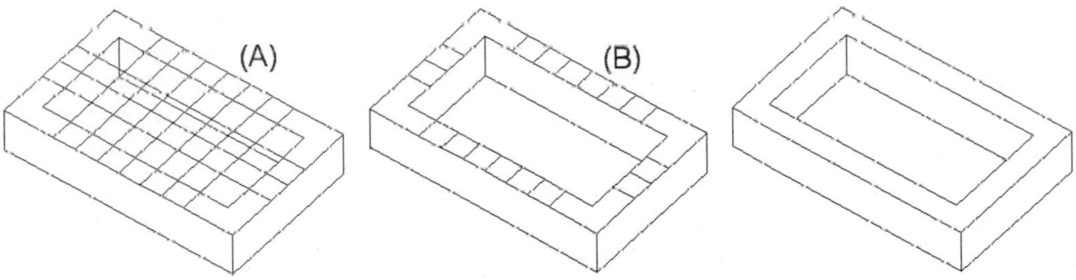

FIGURE 21.55

Additional options of the SOLIDEDIT command include the ability to apply a color or material to a selected face. You can also apply color to an edge or copy edges from a model so that they can be used for construction purposes on other models.

SURFACE MODELING

In addition to creating solid models, AutoCAD contains numerous tools for creating surface models. Surface models are often utilized for models that have complex surface shapes. Generally, there are three types of surfaces: Procedural, NURBS, and Meshes. An example of each is shown in Figure 21.56, although it should be noted that the differences in the surfaces are more in how they are created and edited than in how they appear.

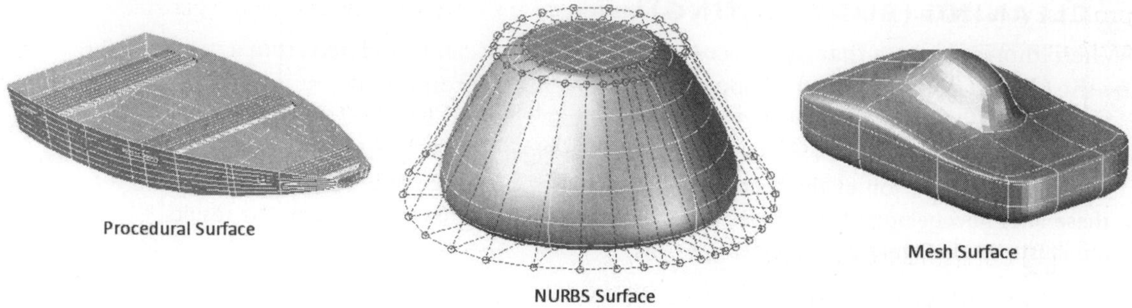

FIGURE 21.56

When in the 3D Modeling workspace, the Ribbon displays a dedicated Surface Modeling tab. Clicking on this tab displays modeling commands for creating procedural and NURBS surfaces, as shown in Figure 21.57. There is also a Mesh tab for creating Mesh surfaces, which will be discussed later in this chapter.

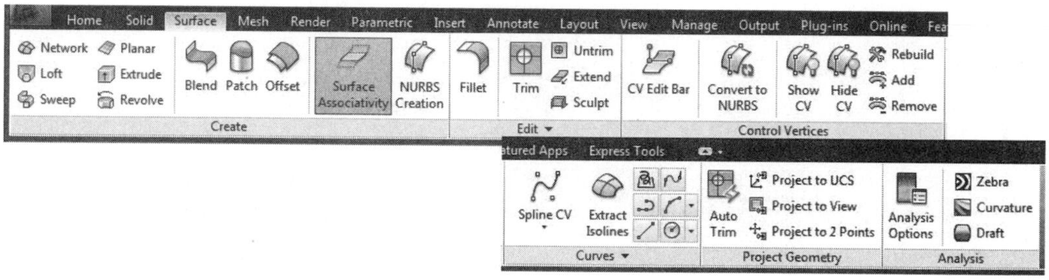

FIGURE 21.57

PROCEDURAL SURFACES

The major advantage of procedural surfaces is that they are associative. When a surface is modified, all the associated surfaces adjust automatically. For the surfaces to be associative, the Surface Associativity button on the Create panel of the Ribbon should be activated. You should also verify that the NURBS Creation button is deactivated. Once these settings are verified, you can utilize the commands in the Create panel, as shown in Figure 21.58, to create your surfaces. These commands create both profile-based surfaces (SURFNETWORK, PLANESURF, LOFT, EXTRUDE, SWEEP, and REVOLVE) and surfaces that are generated from other surfaces (SURFBLEND, SURFPATCH, and SURFOFFSET).

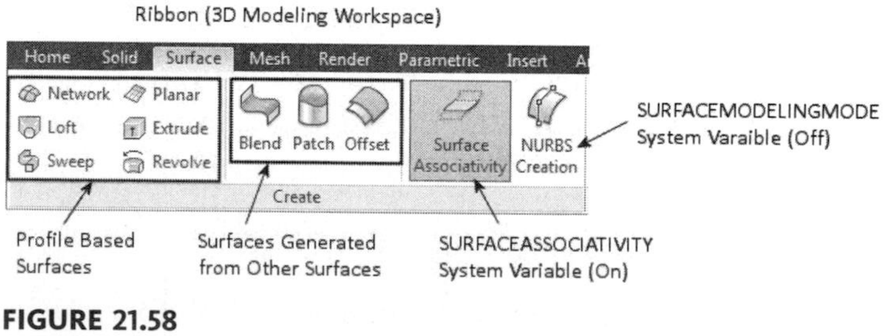

FIGURE 21.58

CREATING PROFILE-BASED SURFACES

The EXTRUDE, REVOLVE, SWEEP, and LOFT commands were used in Chapter 20 to create solid models. Generally, if the profile used during the command operation is closed, a solid is created and if it is open a surface is created. You can, however, set the MOde option to SUrface and create a surface from a closed profile. Besides using profiles, you also now have the capability of using edges from existing models to generate new surfaces.

Open the drawing file 21_Extrude Surface. Use the following command sequence and Figure 21.59 for performing this task. It should be noted that, if the Extrude button is selected from the Surface Ribbon, the Mode will automatically be set to Surface.

TRY IT!

Command: EXT *(For EXTRUDE)*

Current wire frame density: ISOLINES=4, Closed profiles creation mode = Solid

Select objects to extrude or [MOde]: MO *(For MOde)*

Closed profiles creation mode [SOlid/SUrface] <Solid>: SU *(For SUrface)*

Select objects to extrude or [MOde]: *(Select the polyline object at "A")*

Select objects to extrude or [MOde]: *(Press ENTER to continue with this command)*

Specify height of extrusion or [Direction/Path/Taper angle/Expression]: 5.00

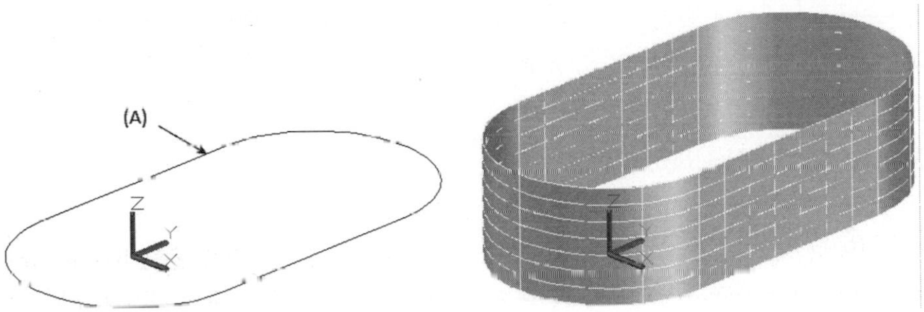

FIGURE 21.59

Next, use the REVOLVE command on the top straight edges of the surface to create an arched top on the part.

Command: REV *(For REVOLVE)*

Current wire frame density: ISOLINES=4, Closed profiles creation mode = Surface

Select objects to revolve or [MOde]: *(Hold down the CTRL key to select the edge at "A" as shown in Figure 21.60)*

Select objects to revolve or [MOde]: *(Press ENTER to continue with this command)*

Specify axis start point or define axis by [Object/X/Y/Z] <Object>: Mid

of (Pick the top of the arc at "B")

Specify axis endpoint: Mid

of (Pick the top of the arc at "C")

Specify angle of revolution or [STart angle/Reverse/EXpression] <360>: 180 *(To complete the revolve operation, as shown in Figure 21.60 (Right))*

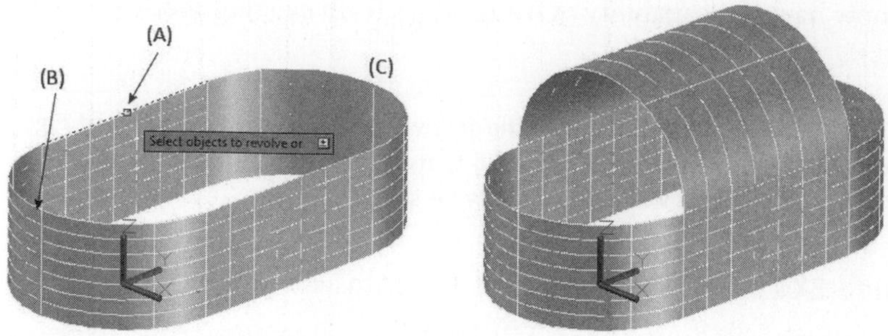

FIGURE 21.60

CREATING A PLANE SURFACE

The PLANESURF command can create simple rectangular surfaces or it can be used to create more complex planar surfaces from existing closed objects.

TRY IT!

Open the drawing file 21_Plane Surface. Use the following command sequence and Figure 21.61 to create two plane surfaces.

```
Command: PLANESURF
Specify first corner or [Object] <Object>: End
of (Pick the corner of the polyline object at "A")
Specify other corner: @6,4
Command: PLANESURF
Specify first corner or [Object] <Object>: (Press ENTER to
issue the Object option)
Select objects: (Select the polyline object)
Select objects: (Press ENTER to complete the operation, as
shown in Figure 21.61 (Right))
```

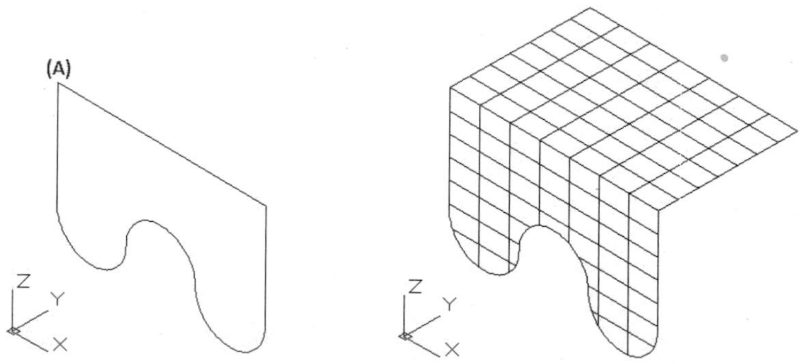

FIGURE 21.61

CREATING A NETWORK SURFACE

The SURFNETWORK command creates a surface from a series of curves. In order to properly define the surface, the curves are selected in two distinct directions. The mesh of the surface will connect the curves in those directions. The edges of existing 3D surface or solid models can be utilized as curves for the new surface.

TRY IT!

Open the drawing file 21_Network Surface. Use the following command sequence and Figure 21.62 to create the network surface.

Command: SURFNETWORK

Select curves or surface edges in first direction: *(Pick the curve at "A")* 1 found

Select curves or surface edges in first direction: *(Pick the curve at "B")* 1 found, 2 total

Select curves or surface edges in first direction: *(Pick the curve at "C")* 1 found, 3 total

Select curves or surface edges in first direction: *(Pick the curve at "D")* 1 found, 4 total

Select curves or surface edges in first direction: *(Pick the curve at "E")* 1 found, 5 total

Select curves or surface edges in first direction: *(Press* ENTER *to continue)*

Select curves or surface edges in second direction: *(Pick the spline at "F")* 1 found

Select curves or surface edges in second direction: *(Pick the spline at "G")* 1 found, 2 total

Select curves or surface edges in second direction: *(Press* ENTER *to complete the operation)*

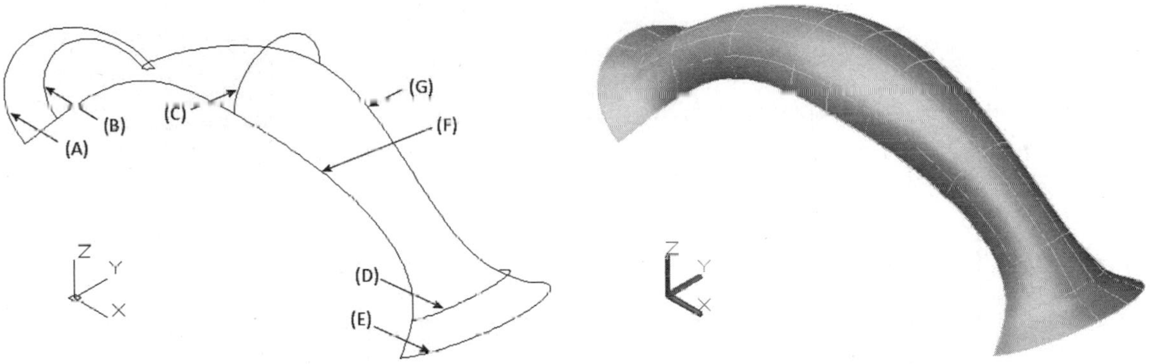

FIGURE 21.62

CREATING SURFACES FROM EXISTING SURFACES

The SURFBLEND, SURFPATCH, and SURFOFFSET commands can all be used to generate a new surface from existing ones. When blending or patching a surface you can change surface continuity and bulge magnitude settings to help control the shape of the new surface. The surface continuity establishes how smoothly the new surface blends with an existing one, while the bulge magnitude helps determine the roundness at the surface intersections.

TRY IT!

Open the drawing file 21_Blend Surface. Two plane surfaces are already created. Use the following command sequence and Figure 21.63 for creating a blend surface between the two existing surfaces.

Command: **SURFBLEND**

Continuity = G1 - tangent, bulge magnitude = 0.5

Select first surface edges to blend or [CHain]: *(Pick the edge at "A")* 1 found

Select first surface edges to blend or [CHain]: *(Pick the edge at "B")* 1 found, 2 total

Select first surface edges to blend or [CHain]: *(Pick the edge at "C")* 1 found, 3 total

Select first surface edges to blend or [CHain]: *(Pick the edge at "D")* 1 found, 4 total

Select first surface edges to blend or [CHain]: *(Press* ENTER *to continue)*

Select second surface edges to blend or [CHain]: *(Pick the edge at "E")* 1 found

Select second surface edges to blend or [CHain]: *(Press* ENTER *to continue)*

Press Enter to accept the blend surface or [CONtinuity/Bulge magnitude]: *(Press* ENTER *to create the surface, as shown in Figure 21.63 (Center))*

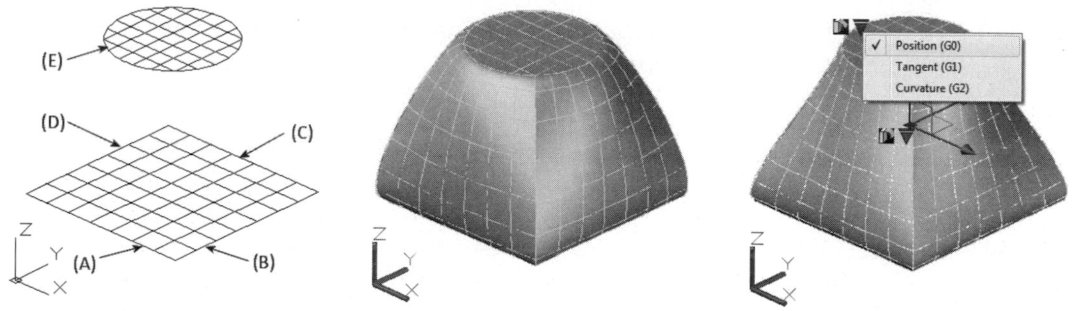

FIGURE 21.63

In Figure 21.63 (Center), the surface continuity was set to Tangent (G1) and the bulge magnitude to 0.5. Try recreating the surface with different settings to see the results. List grips are provided when creating or editing a blend surface and can be used to change the surface continuity. In Figure 21.63 (Right), the top list grip was activated to change the first edge continuity from a Tangent (G1) to a Position (G0) setting.

TRY IT!

Open the drawing file 21_Patch Surface. A lofted surface is already created. Use the following command sequence and Figure 21.64 for creating a patch surface to close the top of the object.

Command: SURFPATCH

Continuity = G0 - position, bulge magnitude = 0.5

Select surface edges to patch or [CHain/CUrves] <CUrves>: *(Pick the edge at "A")* 1 found

Select surface edges to patch or [CHain/CUrves] <CUrves>: *(Press* ENTER *to continue)*

Press Enter to accept the patch surface or [CONtinuity/Bulge magnitude/Guides]: **Con** *(For CONtinuity)*

Patch surface continuity [G0/G1/G2] <G0>: **G1** *(For Tangent)*

Press Enter to accept the patch surface or [CONtinuity/Bulge magnitude/Guides]: *(Press* ENTER *to create the surface, as shown in Figure 21.64 (Center))*

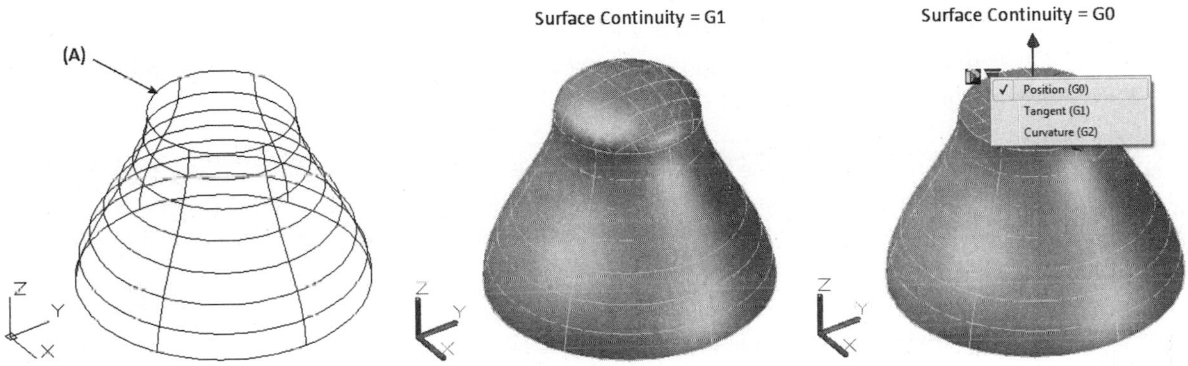

FIGURE 21.64

In Figure 21.64 (Center), the surface continuity was set to Tangent (G1) and the bulge magnitude to 0.5. Try recreating the patch surface with different settings to see the results. A list grip is provided when creating or editing a patch surface and can be used to change the surface continuity. In Figure 21.64 (Right), the list grip was activated to change the edge continuity from Tangent (G1) to the Position (G0) setting.

TRY IT!

Open the drawing file 21_Offset Surface. An extracted surface is already created. Use the following command sequence and Figure 21.65 for creating an offset surface outside the existing surface.

Command: SURFOFFSET

Connect adjacent edges = No

Select surfaces or regions to offset: *(Pick the surface)* 1 found

Select surfaces or regions to offset: *(Press* ENTER *to continue)*

Specify offset distance or [Flip direction/Both sides/Solid/Connect/Expression] <0'-0">: **2'** *(Arrows should be pointing out as shown in Figure 21.65 (Center)—use Flip direction option if pointing inward)*

1 object(s) to offset.

1 offset operation(s) successful.

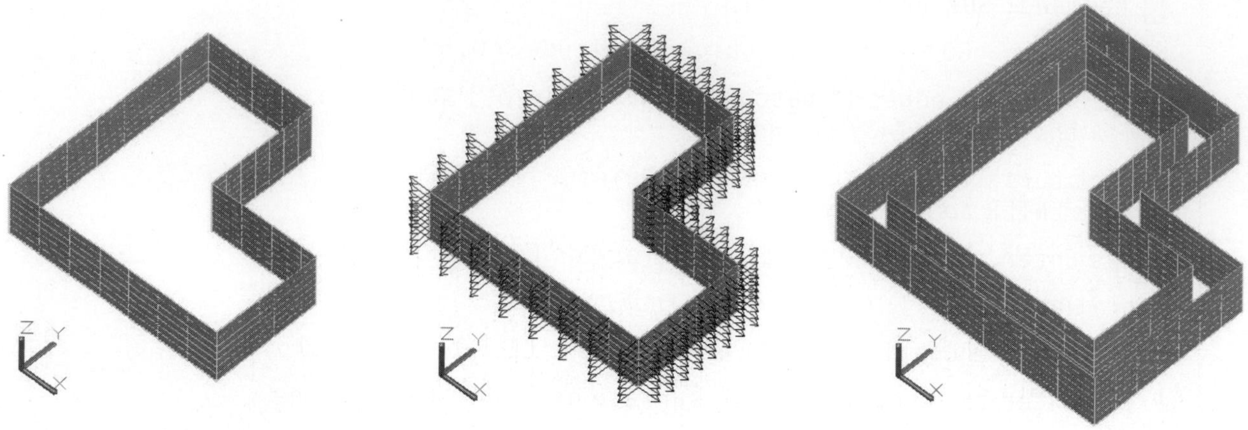

FIGURE 21.65

EDITING SURFACES

The SURFFILLET, SURFTRIM, SURFUNTRIM, and SURFEXTEND commands can all be used to modify existing surfaces. These commands can be activated from the Edit panel of the Surface Ribbon.

TRY IT!

Open the drawing file 21_Fillet Surface. Four plane surfaces are already created. Use the following command sequence and Figure 21.66 for creating fillet surfaces (transition surfaces that are tangent to the existing surfaces). Options for changing the radius and turning off automatic trimming are provided.

```
Command: SURFFILLET
Radius = 1.0000, Trim Surface = yes
Select first surface or region to fillet or [Radius/Trim surface]: R (For Radius)
Specify radius or [Expression] <2.0000>: 2.00
Select first surface or region to fillet or [Radius/Trim surface]: T (For Trim)
Automatically trim surfaces to fillet edge [Yes/No] <No>: N (For No)
Select first surface or region to fillet or [Radius/Trim surface]: (Pick the surface at "A")
Select second surface or region to fillet or [Radius/Trim surface]: (Pick the surface at "B")
Press Enter to accept the fillet surface or [Radius/Trim surfaces]: (Press ENTER to construct the fillet surface)
Command: SURFFILLET
Radius = 2.0000, Trim Surface = no
Select first surface or region to fillet or [Radius/Trim surface]: R (For Radius)
Specify radius or [Expression] <2.0000>: 1.00
Select first surface or region to fillet or [Radius/Trim surface]: T (For Trim)
```

Automatically trim surfaces to fillet edge [Yes/No] <No>: Y *(For Yes)*

Select first surface or region to fillet or [Radius/Trim surface]: *(Pick the surface at "C")*

Select second surface or region to fillet or [Radius/Trim surface]: *(Pick the surface at "D")*

Press Enter to accept the fillet surface or [Radius/Trim surfaces]: *(Press ENTER to construct the fillet surface)*

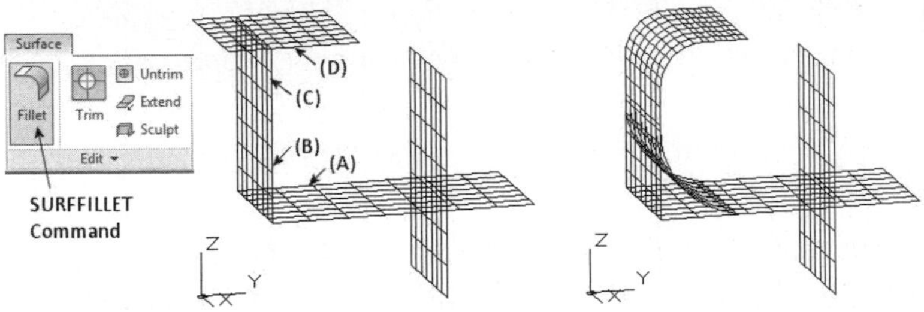

FIGURE 21.66

Open the drawing file 21_Extend Surface or continue from the previous Try It! exercise. Use the following command sequence and Figure 21.67 for creating an extended surface from the edge of an existing surface.

TRY IT!

Command: SURFEXTEND

Modes = Extend, Creation = Append

Select surface edges to extend: *(Pick the edge at "A")*

Select surface edges to extend: *(Press ENTER to continue)*

Specify extend distance [Expression/Modes]: 4 *(The results are shown in Figure 21.67 (Right))*

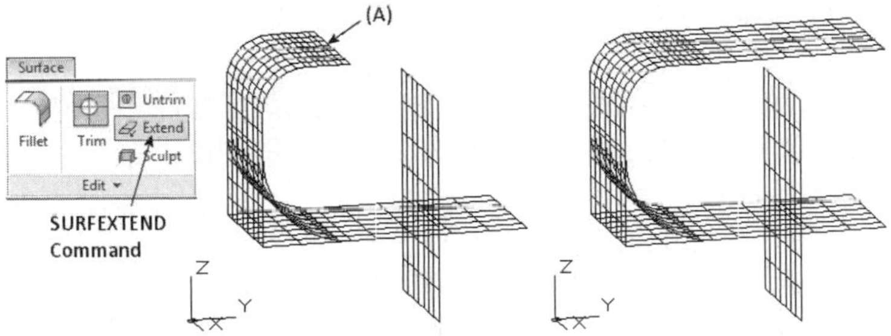

FIGURE 21.67

Open the drawing file 21_Trim Surface or continue from the previous Try It! exercise. Use the following command sequence and Figure 21.68 for trimming and untrimming existing surfaces.

TRY IT!

Command: SURFTRIM

Extend surfaces = Yes, Projection = Automatic

Select surfaces or regions to trim or [Extend/PROjection direction]: *(Pick the two surfaces to trim shown in Figure 21.68 (Left))*

Select surfaces or regions to trim or [Extend/PROjection direction]: *(Press* ENTER *to continue)*

Select cutting curves, surfaces, or regions: *(Pick the cutting surface shown in Figure 21.68 (Center))*

Select cutting curves, surfaces or regions: *(Press* ENTER *to continue)*

Select area to trim [Undo]: *(Pick the two areas to trim shown in Figure 21.68 (Center))*

Select area to trim [Undo]: *(Press* ENTER, the results are shown in Figure 21.68 (Right)*)*

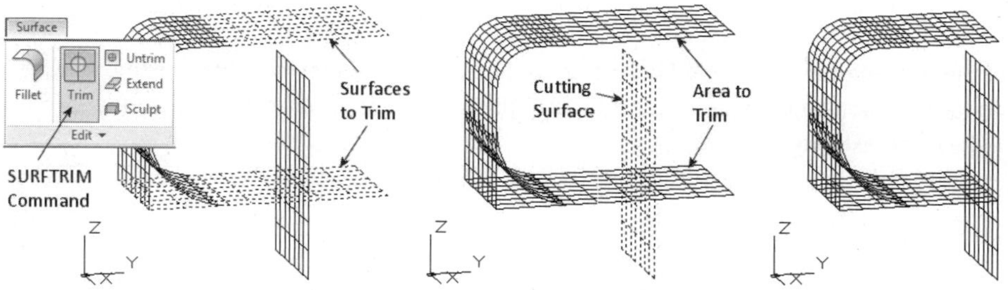

FIGURE 21.68

In Figure 21.68 two areas were trimmed. Use the SURFUNTRIM command to bring back one of the trimmed areas.

Command: SURFUNTRIM

Select edges on surface to un-trim or [SURface]: *(Pick the edge at "A")*

Select edges on surface to un-trim or [SURface]: *(Press* ENTER, the results are shown in Figure 21.69 (Right)*)*

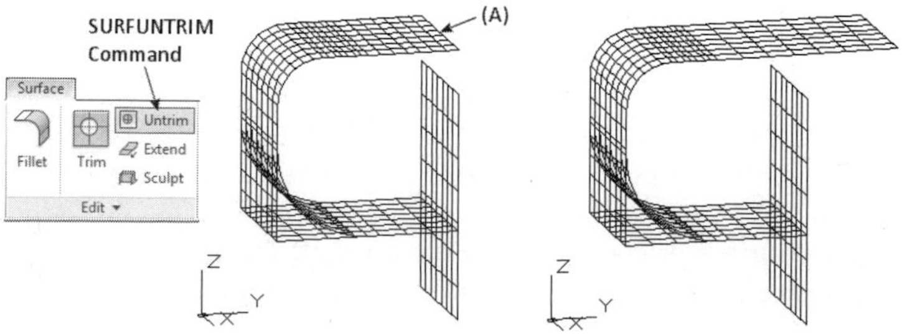

FIGURE 21.69

MESH MODELING

An additional means of producing concept-type surface shapes is through the process of mesh modeling.

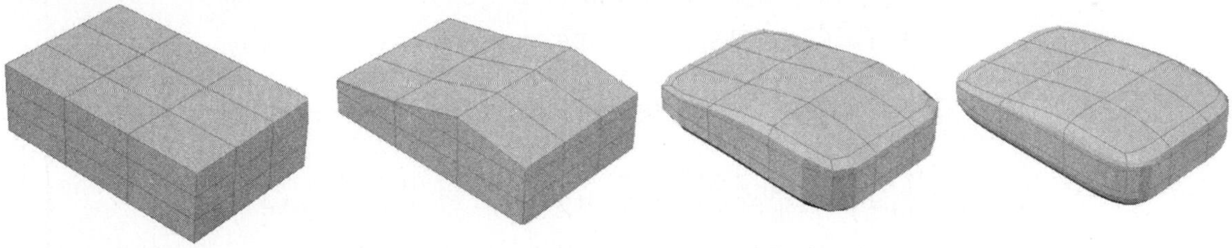

FIGURE 21.70

When in the 3D Modeling workspace, the Ribbon displays a dedicated Mesh Modeling tab. Clicking on this tab displays all mesh modeling commands, as shown in Figure 21.71.

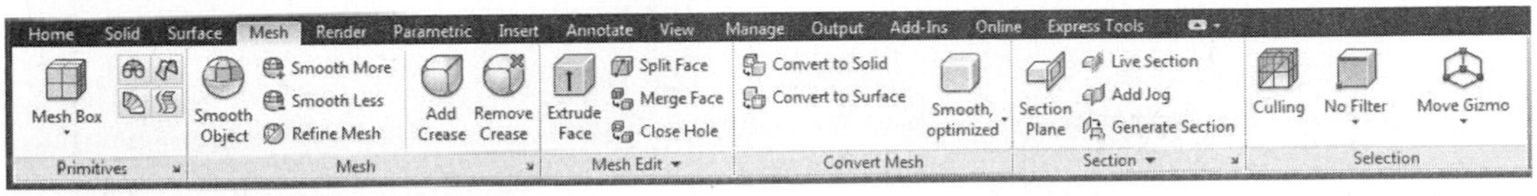

FIGURE 21.71

Some of the most basic of functions when producing mesh models is to begin by constructing a primitive shape, as shown in Figure 21.72. These shapes are similar to the solid model primitives already covered in Chapter 20.

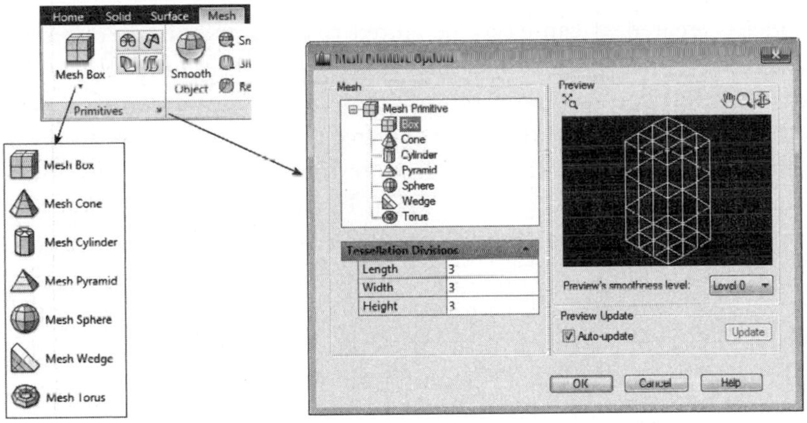

FIGURE 21.72

An example of constructing a mesh model of a mesh box is illustrated in Figure 21.73. You identify 2 points in the X and Y directions, as shown in Figure 21.73 (Left). You then drag to get the 3rd direction, as shown in Figure 21.73 (Right). While in the BOX command, you can switch to Length mode which will allow you to enter the length, width, and height of the model. Notice in Figure 21.72 the number of tesselation lines created in the length, width, and height directions; these values are automatically applied to the mesh model being created. The Mesh Primitive Options dialog box can be activated by picking the arrow in the corner of the Primitives panel on the Ribbon.

```
Command: MESH
Current smoothness level is set to: 0
Enter an option [Box/Cone/Cylinder/Pyramid/Sphere/Wedge/
Torus/SEttings]<Box>: B (For Box)
Specify first corner or [Center]: (Pick a corner point)
Specify other corner or [Cube/Length]: (Pick the other corner
point)
Specify height or [2Point]: (Drag or enter a value for the
height)
```

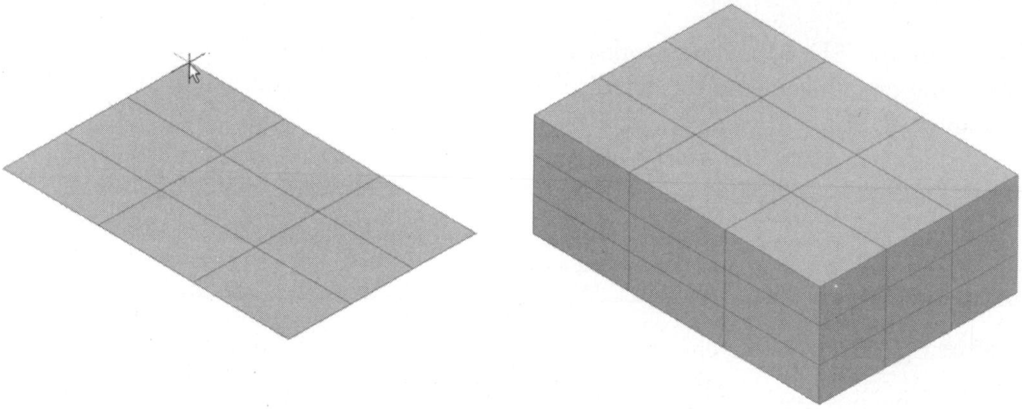

FIGURE 21.73

Working with Subobjects

After a mesh model is created, it can easily be edited to form an entirely different shape. This is due to the number of subobjects that make up the mesh model. Modifying faces, edges, and vertices of a basic primitive mesh can create complex new shapes. To illustrate how this works, the three top front faces need to be dropped down to a predefined level. To select the three faces of the mesh model, hold down the CTRL key and pick all three faces, as shown in Figure 21.74 (Left). You will notice the appearance of the default gizmo, namely Move. Move your cursor to the Z axis and an axis line will appear in Figure 21.74 (Center). At this point, you can drag your cursor up or down. As you perform this you will dynamically see the model changing. In Figure 21.74 (Right), the three faces were moved or dragged down. You can also enter an exact value when performing this operation.

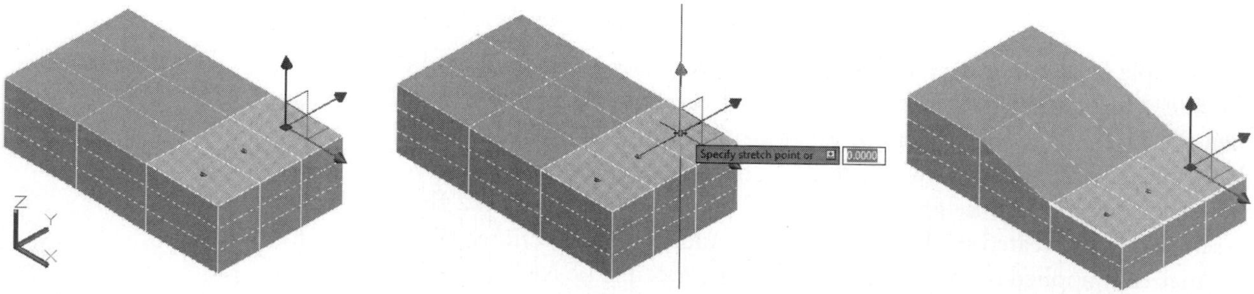

FIGURE 21.74

Figure 21.74 illustrated how to pick faces for editing a mesh model. Figure 21.75 illustrates how you can use an Edge filter, located in the Selection panel, to assist with picking the three back edges of the mesh object, as shown in Figure 21.75 (Center). When performing this operation, the results are shown in Figure 21.75 (Right).

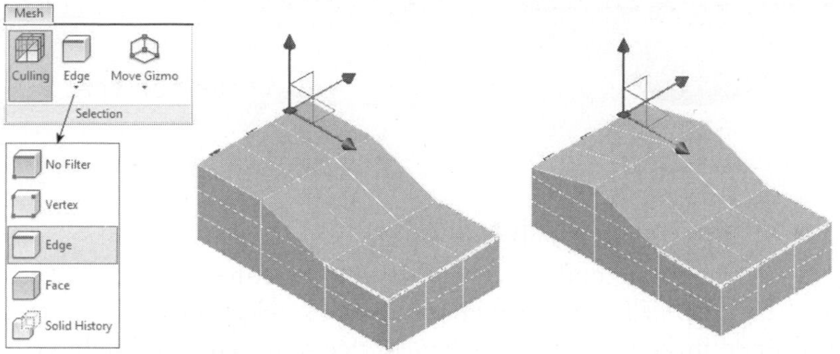

FIGURE 21.75

Smoothing a Mesh Model

Once a mesh model is created as in Figure 21.76 (Left), you can easily smooth the model using a number of techniques. The example in Figure 21.76 (Center) shows the initial affects of making all mesh model edges smooth. In Figure 21.76 (Right), this object displays the highest level of smoothness.

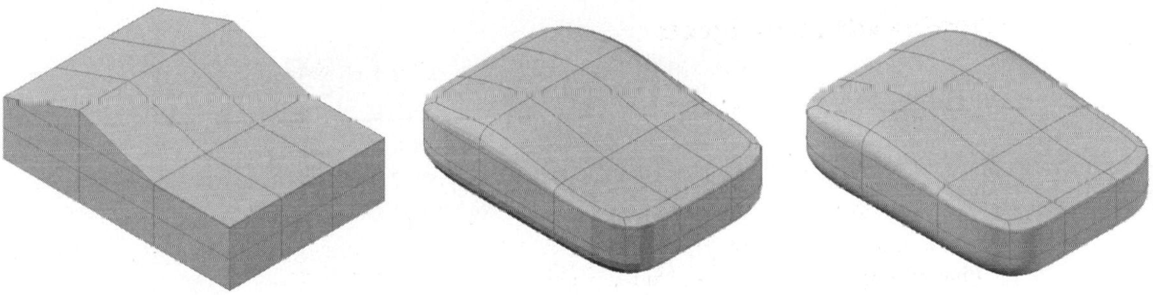

FIGURE 21.76

One simple way of changing the smoothness of a mesh model is through the Quick Properties palette, as shown in Figure 21.77. Verify that Quick Properties is activated on the Status Bar and then pick on the mesh to display this palette. Clicking on Smoothness displays the drop-down menu that allows you to change to four different levels of smoothness. It must be pointed out that for complex mesh models, the Level 4 smoothness may result in a slower than normal processing speed.

Also shown in Figure 21.77 is the Ribbon that displays three additional smoothness modes; namely Smooth More, Smooth Less, and Refine Mesh. If no smoothness is applied to a mesh model and you click on the Smooth More button, a Level I smoothness is applied. Each time the button is clicked the smoothness level is increased. Of course, clicking the Smooth Less button steps you back down through the smoothness levels.

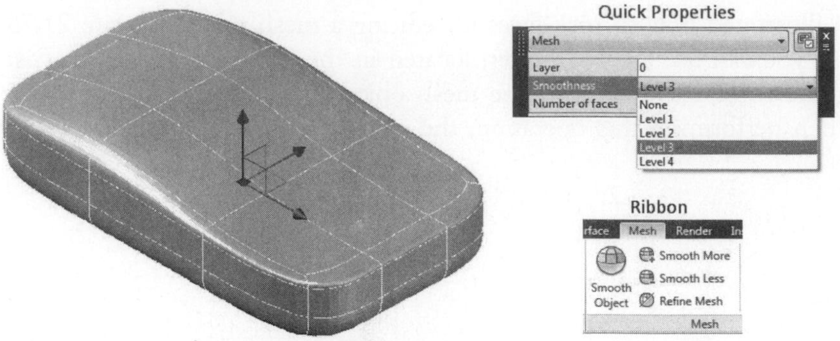

FIGURE 21.77

Refining the Smoothness of a Mesh Model

A more refined display of a mesh model is achieved when selecting the Refine Mesh button, as shown in Figure 21.78 (Left). When prompted to select the mesh pattern, the results are illustrated in Figure 21.78 (Right). Notice how dense and numerous the individual faces and edges are in the mesh model. Use this mode to really fine tune your model. Unfortunately this would involve the selecting of numerous faces and edges. The presence of the increase number of edges also makes the mesh model heavy as far as the file size goes. If you are through reviewing the mesh model in refined mode, undo the operation to return to Figure 21.77.

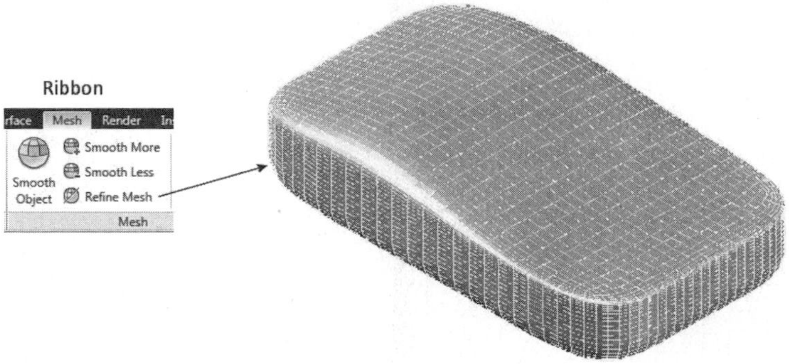

FIGURE 21.78

Scaling a Face in a Mesh Model

Another important function in working with subobjects in a mesh model is through the 3D scale gizmo. In Figure 21.79 of the mesh model, the CTRL key is used to select the top middle face. By default, the 3D move gizmo displays. However you could move your cursor over one of the axis lines and right-click to display the shortcut menu also displayed in the following image. From this menu, the scale gizmo was selected which will be used to scale the highlighted face. You can also scale the face based on a specified plane or axis, which is found in the Set Constraint area of the menu. In this example, the XY plane is selected. Then a scale factor of 2 was entered in at the Command prompt.

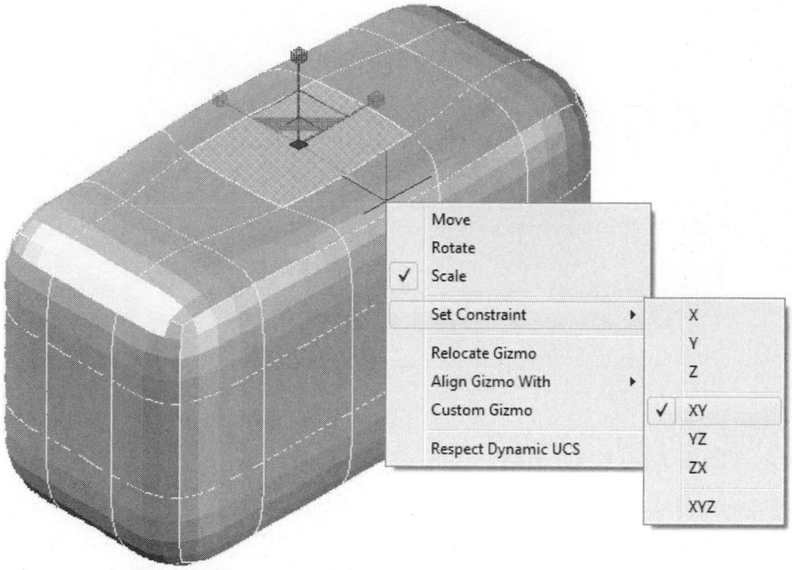

FIGURE 21.79

Figure 21.80 of the mesh model shows how the selected face was scaled by a factor of 2 along the XY plane. This may not be enough to make the face appear as a square. For this reason, another scale operation will be performed based on the Y-axis and a scale factor of 1.25.

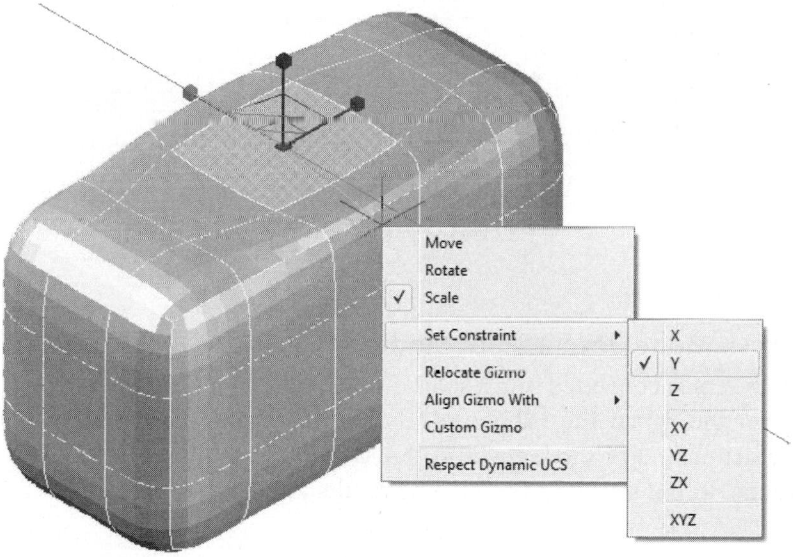

FIGURE 21.80

Extruding a Face in a Mesh Model

Once the surface in Figure 21.80 was scaled to form an approximate square, you can extrude the face up or down to form the shapes displayed in Figure 21.81. In Figure 21.81 (Right), an extrude distance of 1 was applied to the face.

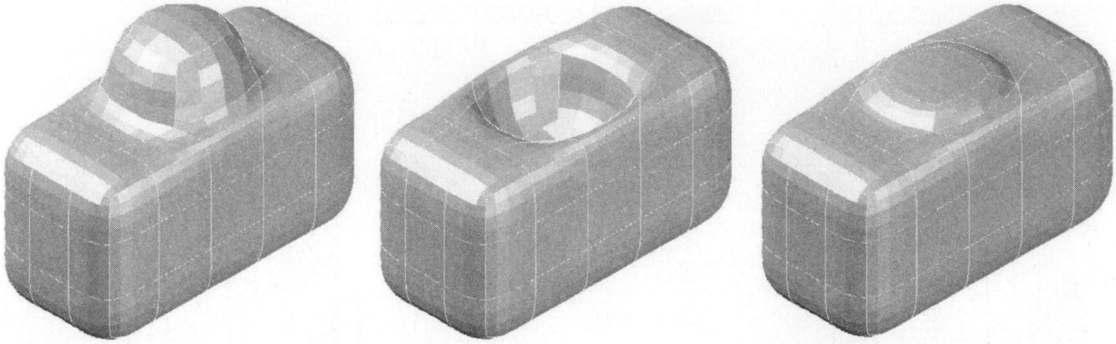

FIGURE 21.81

Creating a Crease in a Mesh Model

A useful tool in mesh modeling is to convert the edges of an extruded face into sharp edges. In the following example, the image on the left has a number of rounded edges highlighted. The results of creating a crease are illustrated on the right where the edges can form a circular cylindrical shape.

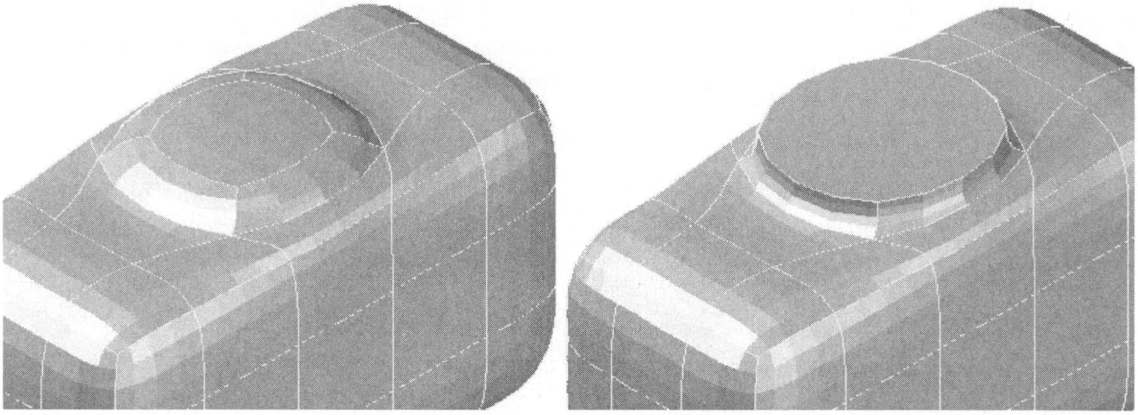

FIGURE 21.82

Converting a Mesh Model into a Solid Model

A mesh model can be easily converted into a solid model. You first decide the smooth method to use from the list in the Ribbon, as shown in Figure 21.83. After the smooth method is defined, click the Convert to Solid command in the Ribbon and select the mesh model in Figure 21.83 (Left). The results are shown in Figure 21.83 (Right).

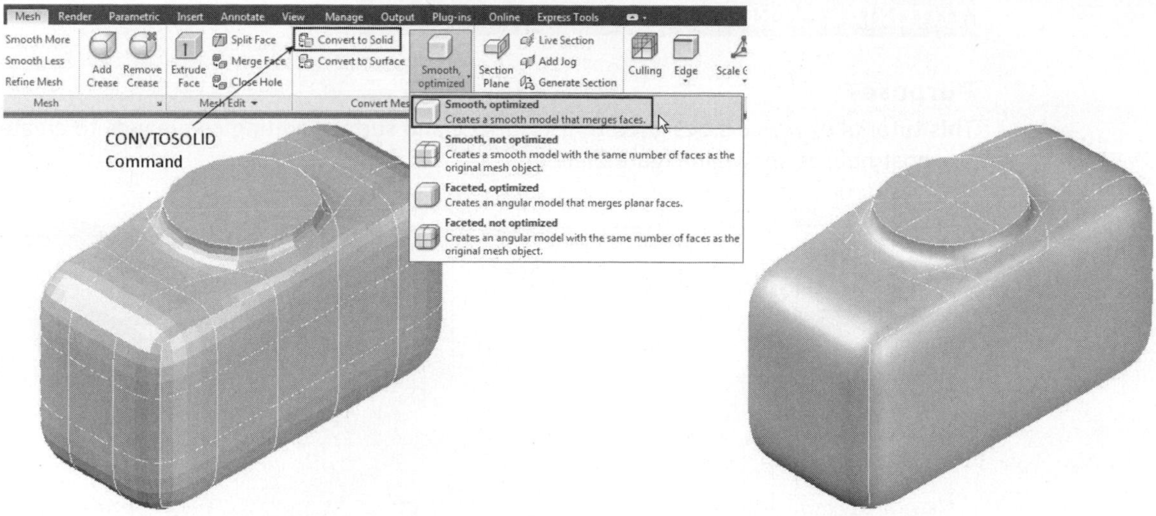

FIGURE 21.83

Now a solid model, a circle was constructed in the top face of the cylinder. The circle was extruded to create a cylinder and then subtracted from the solid to form the hole, as shown in Figure 21.84 (Right).

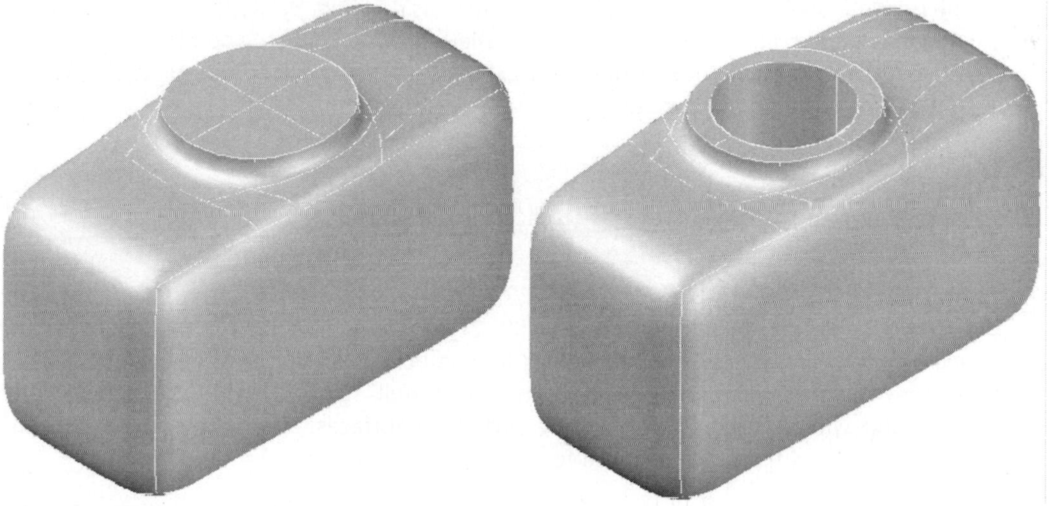

FIGURE 21.84

TUTORIAL EXERCISE: 21_SURFACE BOAT.DWG

Purpose

This tutorial exercise is designed to use surface and surface editing commands to create the boat hull, as shown in Figure 21.85.

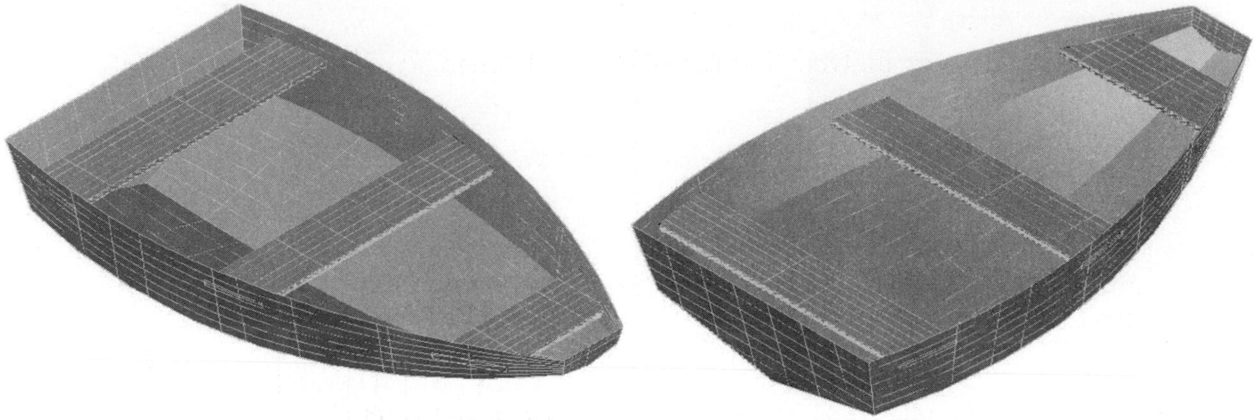

FIGURE 21.85

System Settings

The drawing units and limits are already set for this drawing. Verify that the 3D Modeling workspace is current and check to see that the following Object Snap modes are already set: Endpoint, Center, Intersection, and Extension. Also check to see that OSNAP and POLAR are turned on in the Status Bar.

Layers

Make sure that the current layer is Surface.

Suggested Commands

You will begin by opening the drawing file 21_Surface Boat.dwg. The PLINE command will be used to create a series of frames for the boat hull. These frames will be utilized by the LOFT command to generate the boat hull surface. Next, the PLANESURF command will be used to create bow and stern surfaces. Finally, the EXTRUDE and SURFTRIM commands will be used to create three bench seats in the boat.

STEP 1

Open the drawing 21_Surface Boat. From the 3D Modeling workspaces, activate the Home tab in the Ribbon. Select the 2D Wireframe visual style and a SE Isometric view from the View panel, as shown in Figure 21.86 (Left). From the Coordinates panel select a Right UCS. The User Coordinate System icon should appear, as shown in Figure 21.86 (Right).

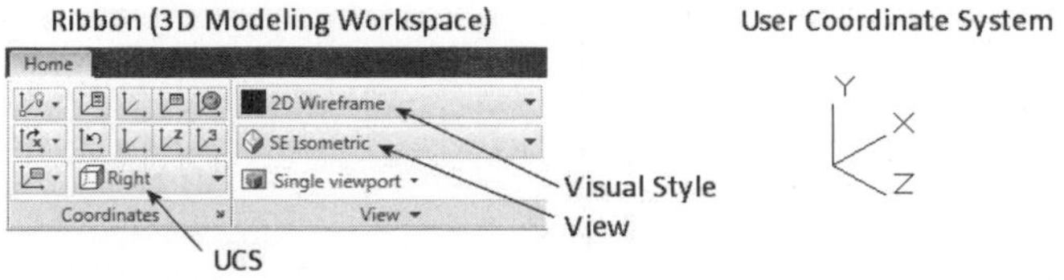

FIGURE 21.86

STEP 2

Activate the `PLINE` command from the Draw panel. Use the command sequence and Figure 21.87 shown to create the first frame.

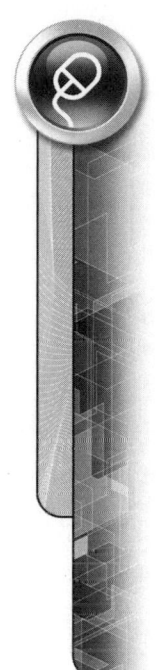

```
Command: PL (For PLINE)
Specify start point: 2',0
Current line-width is 0'-0"
Specify next point or [Arc/Halfwidth/Length/Undo/Width]:
@ -2,-1'
Specify next point or [Arc/Close/Halfwidth/Length/Undo/
Width]:
@ -3'8,0
Specify next point or [Arc/Close/Halfwidth/Length/Undo/
Width]:
@ -2,1'
Specify next point or [Arc/Close/Halfwidth/Length/Undo/
Width]: (Press ENTER to complete the operation)
```

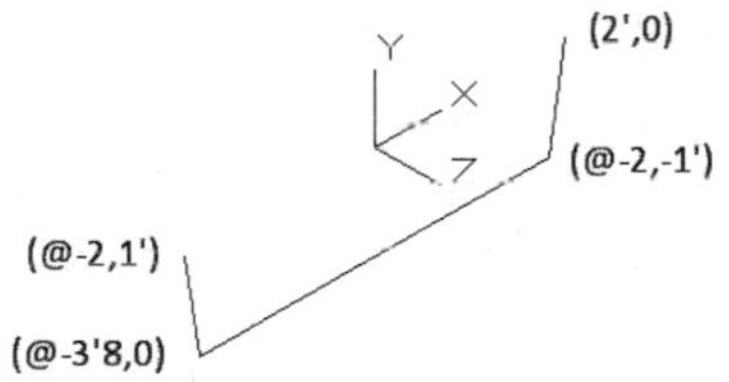

FIGURE 21.87

STEP 3

Modify the first frame utilizing grips. Select the polyline and hover the cursor over the bottom-midpoint grip to display the shortcut menu, as shown on the left in Figure 21.88. Pick "Add Vertex" from the menu and stretch the grip 6.0 inches straight down (270° direction), as shown in Figure 21.88 (Right).

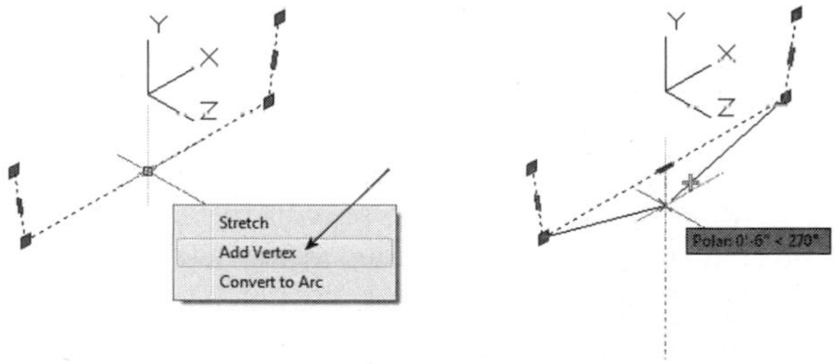

FIGURE 21.88

STEP 4

Change back to the World Coordinate System and use the COPY command (Modify panel) to create four additional frames at 2-foot increments in the "X" direction, as shown in Figure 21.89.

Command: CO *(For COPY)*

Select objects: *(Select the polyline)*

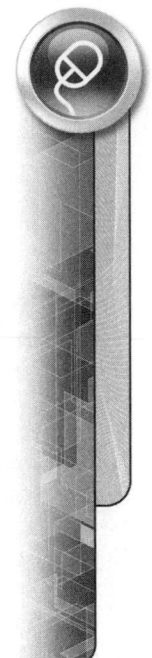

Select objects: *(Press* ENTER *to continue)*

Current settings: Copy mode = Multiple

Specify base point or [Displacement/mOde] <Displacement>: *(Pick any convenient point on the screen)*

Specify second point or <use first point as displacement>: 2' *(Track in "X" direction)*

Specify second point or [Exit/Undo] <Exit>: 4' *(Track in "X" direction)*

Specify second point or [Exit/Undo] <Exit>: 6' *(Track in "X" direction)*

Specify second point or [Exit/Undo] <Exit>: 8' *(Track in "X" direction)*

Specify second point or [Exit/Undo] <Exit>: *(Press* ENTER *to complete the operation)*

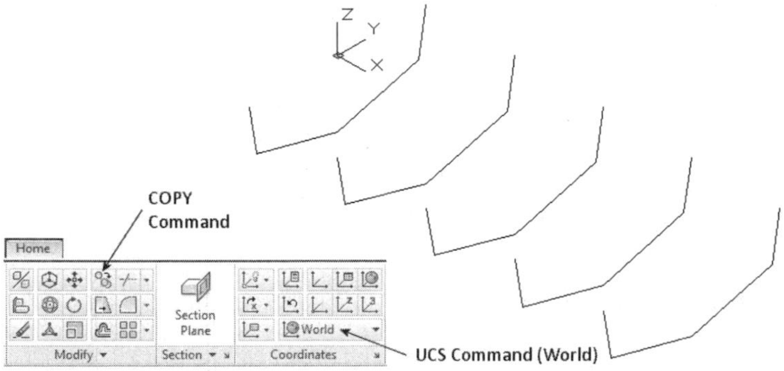

FIGURE 21.89

STEP 5

Modify the second frame utilizing grips. This frame will be widened 6.0 inches—3.0 inches on each side. Select the polyline and pick the left-midpoint grip. The UCS automatically aligns to the polyline. Stretch the polyline 3.0 inches in the 180° direction, as shown in Figure 21.90 (Left). Pick the right-midpoint grip. Stretch the polyline 3.0 inches in the 0° direction, as shown in Figure 21.90 (Right).

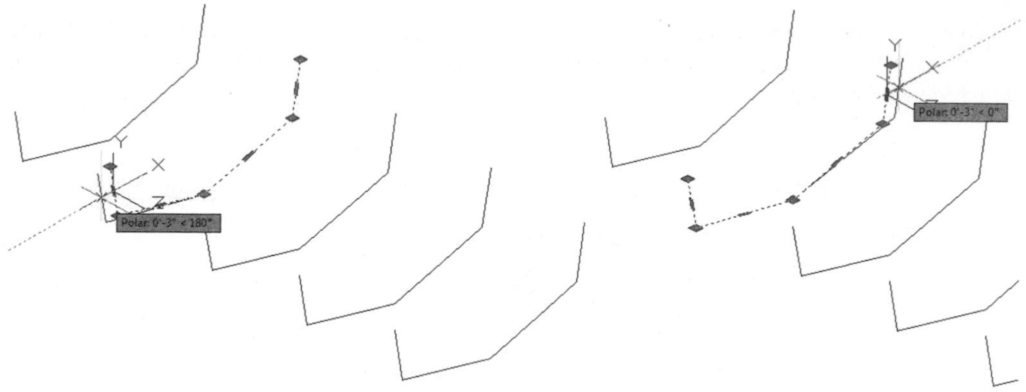

FIGURE 21.90

STEP 6

Modify the third frame utilizing grips. This frame will be widened 2.0 inches—1.0 inch on each side. Select the polyline and pick the left-midpoint grip. The UCS automatically aligns to the polyline. Stretch the polyline 1.0 inch in the 180° direction, as shown in Figure 21.91 (Left). Pick the right-midpoint grip. Stretch the polyline 1.0 inch in the 0° direction, as shown in Figure 21.91 (Right).

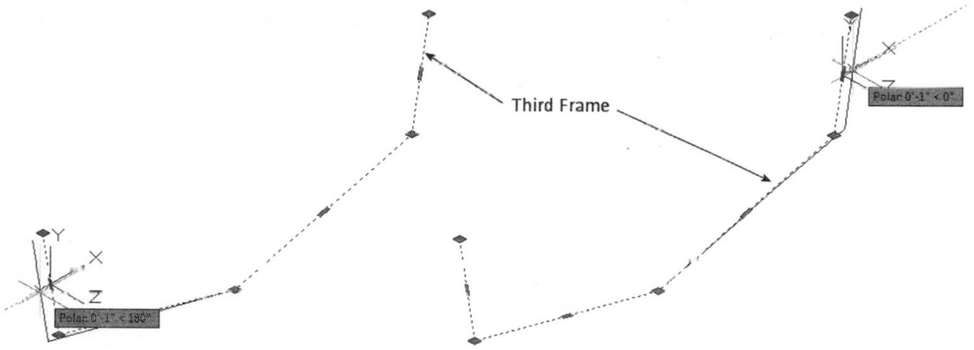

FIGURE 21.91

STEP 7

Use the 3DSCALE command (Modify panel) to resize the fourth frame by a factor of 0.75.

> Command: 3S *(For 3DSCALE)*
>
> Select objects: *(Select the polyline – fourth frame)*
>
> Select objects: *(Press* ENTER *to continue)*
>
> Specify base point: M2P *(This OSNAP can be selected by pressing* CTRL + Right Mouse Button *and picking "Mid Between 2 Points" from the shortcut menu, as shown in Figure 21.92 (Center))*
>
> First point of mid: *(Select the endpoint at "A")*
>
> Second point of mid: *(Select the endpoint at "B" to establish the new base point)*
>
> Pick a scale axis or plane: *(Select the scale gizmo)*
>
> Specify scale factor or [Copy/Reference]: 0.75
>
> Regenerating model.

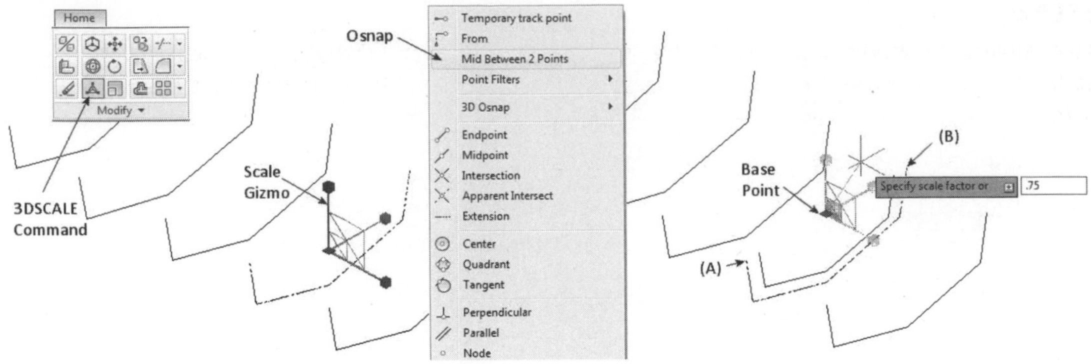

FIGURE 21.92

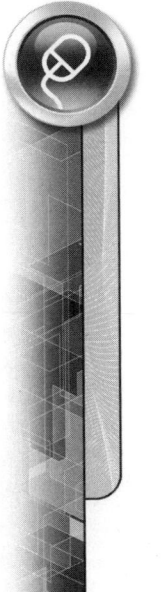

STEP 8

Use the 3DSCALE command (Modify panel) to resize the fifth frame by a factor of 0.25, as shown in Figure 21.93.

Command: **3S** *(For 3DSCALE)*

Select objects: *(Select the polyline – fifth frame)*

Select objects: *(Press* ENTER *to continue)*

Specify base point: **M2P** *(This OSNAP can be selected by pressing* CTRL + Right Mouse Button *and picking "Mid Between 2 Points" from the shortcut menu)*

First point of mid: *(Select the endpoint at "A")*

Second point of mid: *(Select the endpoint at "B" to establish the new base point)*

Pick a scale axis or plane: *(Select the scale gizmo)*

Specify scale factor or [Copy/Reference]: **0.25**

Regenerating model.

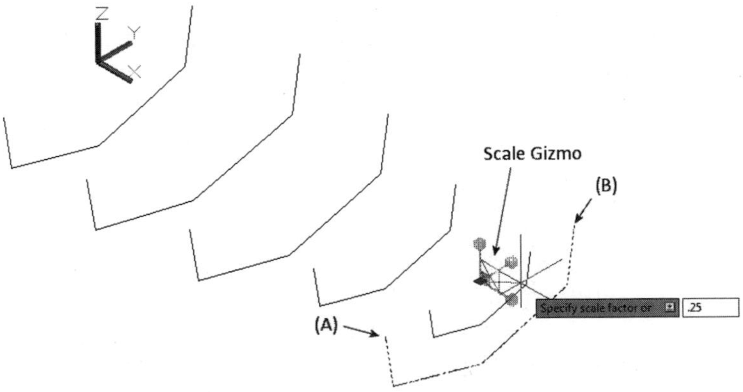

FIGURE 21.93

STEP 9

To complete the modifications to the frames, the first (stern) and fifth (bow) frames need to be rotated. Use the 3DROTATE command and Figure 21.94 to rotate the first frame inward 15°.

Command: 3R *(For 3DROTATE)*

Current positive angle in UCS: ANGDIR=counterclockwise ANGBASE=0

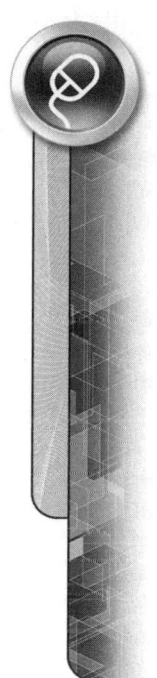

Select objects: *(Select the polyline – first frame)*

Select objects: *(Press ENTER to continue)*

Specify base point: M2P

First point of mid: *(Select the endpoint at "A")*

Second point of mid: *(Select the endpoint at "B" to establish the new base point)*

Pick a rotation axis: *(When the rotate gizmo appears, hover on the axis handle until the proper axis appears, as shown in Figure 21.94; then pick with the cursor)*

Specify angle start point or type an angle: 15

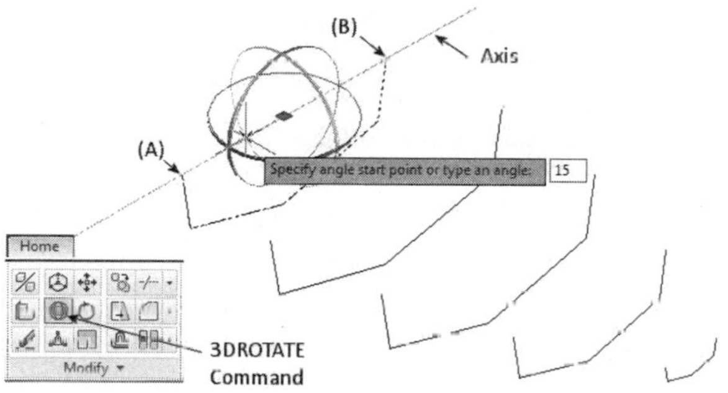

FIGURE 21.94

STEP 10

Use the 3DROTATE command and Figure 21.95 to rotate the fifth frame inward –15°.

Command: 3R *(For 3DROTATE)*

Current positive angle in UCS: ANGDIR=counterclockwise ANGBASE=0

Select objects: *(Select the polyline - fifth frame)*

Select objects: *(Press ENTER to continue)*

Specify base point: M2P

First point of mid: *(Select the endpoint at "A")*

Second point of mid: *(Select the endpoint at "B" to establish the new base point)*

Pick a rotation axis: *(When the rotate gizmo appears, hover on the axis handle until the proper axis appears, as shown in Figure 21.95; then pick with the cursor)*

Specify angle start point or type an angle: –15

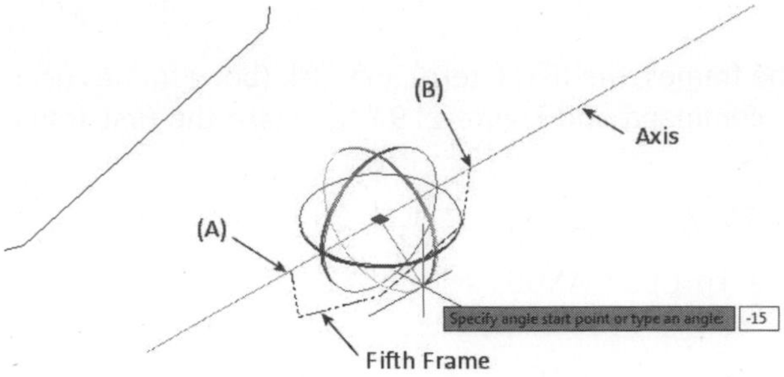

FIGURE 21.95

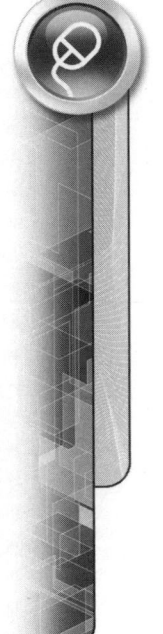

STEP 11

Next, create the first surface using the loft command. Select the Surface tab on the Ribbon and activate the LOFT command, as shown in Figure 21.96 (Left). Select the frames in order to create the lofted surface, as shown Figure 21.96 (Right).

Command: LOFT

Current wire frame density: ISOLINES=4, Closed profiles creation mode = Surface

Select cross sections in lofting order or [POint/Join multiple edges/MOde]: *(Select the frames in order)*

Select cross sections in lofting order or [POint/Join multiple edges/MOde]: *(Press* ENTER *to continue)*

Enter an option [Guides/Path/Cross sections only/Settings] <Cross sections only>: *(Press* ENTER *to create the lofted surface, as shown in Figure 21.96 (Right))*

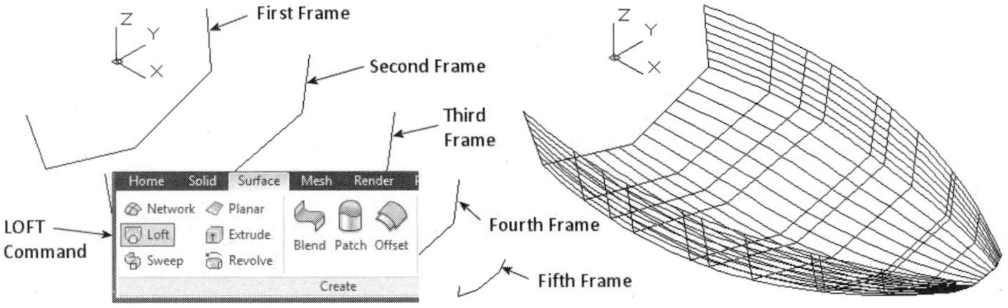

FIGURE 21.96

STEP 12

Next, use the UCS, PLINE, and PLANESURF commands to create a stern surface. Select the Home tab on the Ribbon and activate the 3 Point option of the UCS command in the Coordinates panel, as shown in Figure 21.97 (Left). Use the following command line sequence and Figure 21.97 to locate the user coordinate system in the same plane as the frame.

Command: UCS

Current ucs name: *WORLD*

Specify origin of UCS or [Face/NAmed/OBject/Previous/View/World/X/Y/Z/ZAxis] <World>: 3 *(For 3point)*

Specify new origin point <0,0,0>: *(Pick the endpoint at "A")*

Specify point on positive portion of X-axis <0'-1",2'-0",0'-0">: *(Pick the endpoint at "B")*

Specify point on positive-Y portion of the UCS XY plane <0'-0 1/2",2'-0 3/4",0'-0">: *(Pick the endpoint at "C")*

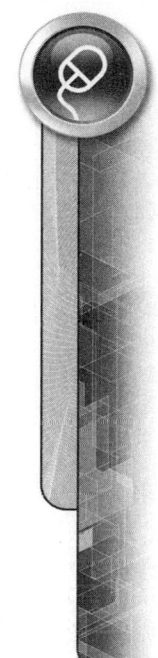

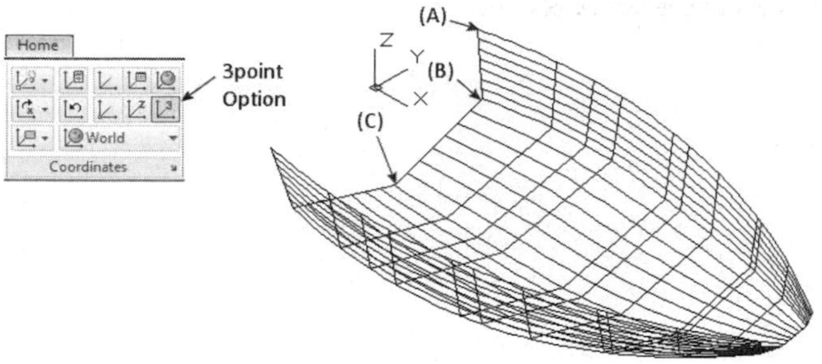

FIGURE 21.97

Use the following command line sequence and Figure 21.98 to create a closed polyline located on the current user coordinate system.

Command: PL *(For PLINE)*

Specify start point: *(Pick the endpoint at "A")*

Current line-width is 0'-0"

Specify next point or [Arc/Halfwidth/Length/Undo/Width]: *(Pick the endpoint at "B")*

Specify next point or [Arc/Close/Halfwidth/Length/Undo/Width]: *(Pick the endpoint at "C")*

Specify next point or [Arc/Close/Halfwidth/Length/Undo/Width]: *(Pick the endpoint at "D")*

Specify next point or [Arc/Close/Halfwidth/Length/Undo/Width]: *(Pick the endpoint at "E")*

Specify next point or [Arc/Close/Halfwidth/Length/Undo/Width]: C *(For CLOSE)*

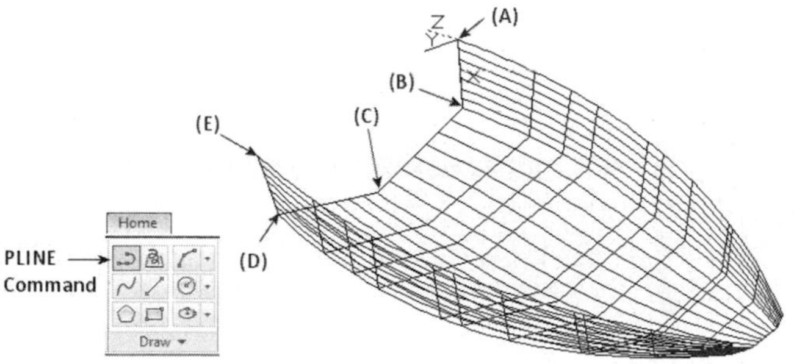

FIGURE 21.98

Use the following command line sequence and Figure 21.99 to create a planar surface for the stern of the boat.

Command: **PLANESURF**

Specify first corner or [Object] <Object>: *(Press* ENTER *to issue the Object option)*

Select objects: *(Select the polyline object)*

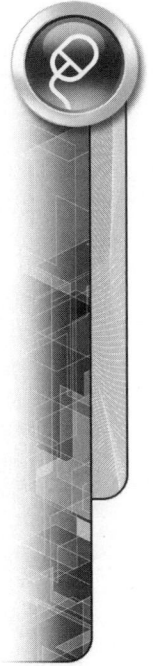

Select objects: *(Press* ENTER *to complete the operation, as shown in Figure 21.99 (Right))*

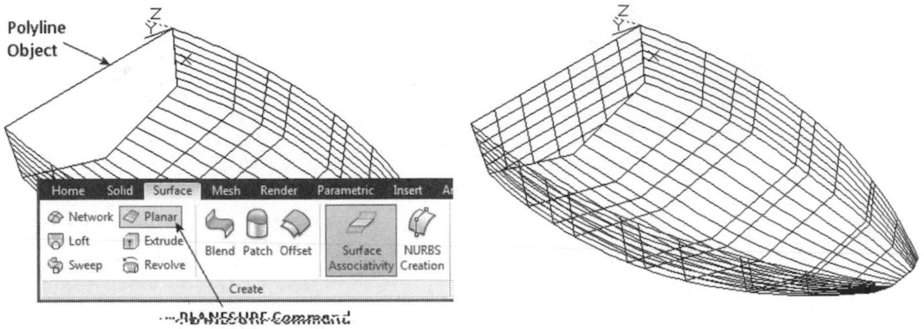

FIGURE 21.99

STEP 13

To complete the boat hull, use the UCS, PLINE, and PLANESURF commands to create a bow surface. Select the Home tab on the Ribbon. Pick SW Isometric from the View panel and activate the 3 Point option of the UCS command in the Coordinates panel, as shown in Figure 21.100 (Right). Use the following command line sequence and Figure 21.100 to locate the user coordinate system in the same plane as the frame.

Command: **UCS**

Current ucs name: * NO NAME *

Specify origin of UCS or [Face/NAmed/OBject/Previous/View/World/X/Y/Z/ZAxis] <World>: **3** *(For 3point)*

Specify new origin point <0,0,0>: *(Pick the endpoint at "A")*

Specify point on positive portion of X-axis <2'-6 1/2",2'–1 1/2",-7'–8 3/4">: *(Pick the endpoint at "B")*

Specify point on positive-Y portion of the UCS XY plane <2'–5 3/4",2'–2 1/2",-7'–8 3/4">: *(Pick the endpoint at "C")*

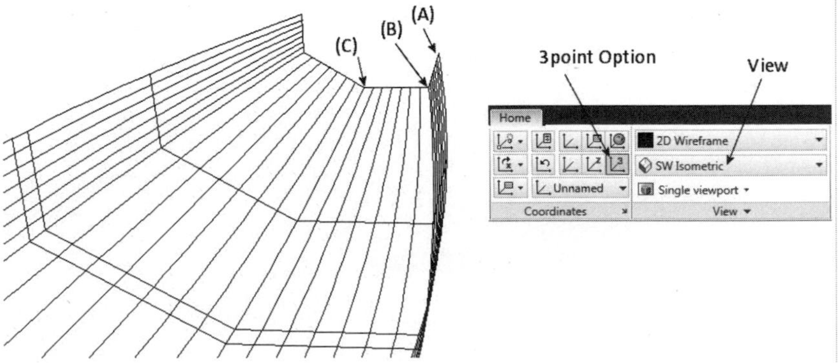

FIGURE 21.100

Use the following command line sequences and Figure 21.101 to create a closed polyline and a planar surface.

Command: PL *(For PLINE)*

Specify start point: *(Pick the endpoint at "A")*

Current line-width is 0'-0"

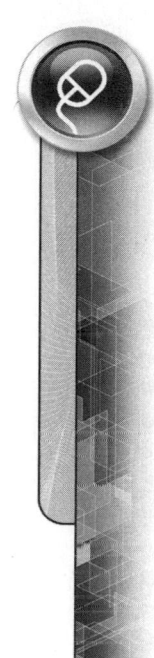

Specify next point or [Arc/Halfwidth/Length/Undo/Width]: *(Pick the endpoint at "B")*

Specify next point or [Arc/Close/Halfwidth/Length/Undo/Width]: *(Pick the endpoint at "C")*

Specify next point or [Arc/Close/Halfwidth/Length/Undo/Width]: *(Pick the endpoint at "D")*

Specify next point or [Arc/Close/Halfwidth/Length/Undo/Width]: *(Pick the endpoint at "E")*

Specify next point or [Arc/Close/Halfwidth/Length/Undo/Width]: C *(For CLOSE)*

Command: PLANESURF

Specify first corner or [Object] <Object>: *(Press ENTER to issue the Object option)*

Select objects: *(Select the polyline object just created)*

Select objects: *(Press ENTER to complete the operation, as shown in Figure 21.101 (Right))*

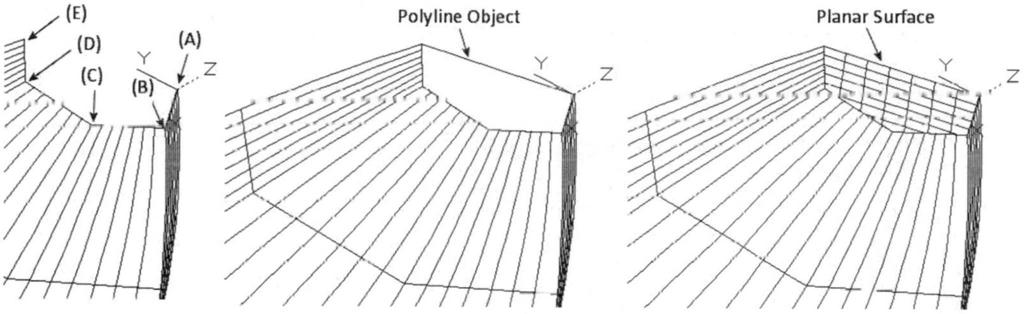

FIGURE 21.101

STEP 14

In the next series of steps, bench seating will be provided for the boat. Select the Home tab on the Ribbon. Pick SE Isometric from the View panel and activate the Front option of the UCS command in the Coordinates panel, as shown in Figure 21.102 (Right). Use the following command line sequences and Figure 21.102 to create polyline objects that will be used to generate the benches.

Command: PL *(For PLINE)*

Specify start point: 6,−4,3'

Current line-width is 0'−0"

Specify next point or [Arc/Halfwidth/Length/Undo/Width]: 10 *(Track in 0° direction)*

Specify next point or [Arc/Close/Halfwidth/Length/Undo/Width]: 1 *(Track in 90° direction)*

Specify next point or [Arc/Close/Halfwidth/Length/Undo/Width]: **10** *(Track in 180° direction)*

Specify next point or [Arc/Close/Halfwidth/Length/Undo/Width]: **C** *(For CLOSE)*

Command: **CO** *(For COPY)*

Select objects: *(Select the polyline)*

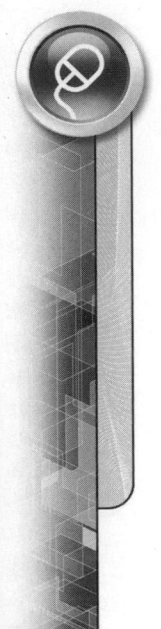

Select objects: *(Press* ENTER *to continue)*

Current settings: Copy mode = Multiple

Specify base point or [Displacement/mOde] <Displacement>: *(Pick any convenient point on the screen)*

Specify second point or <use first point as displacement>: **3'** *(Track in "X" direction)*

Specify second point or [Exit/Undo] <Exit>: **6'** *(Track in "X" direction)*

Specify second point or [Exit/Undo] <Exit>: *(Press* ENTER *to complete the operation)*

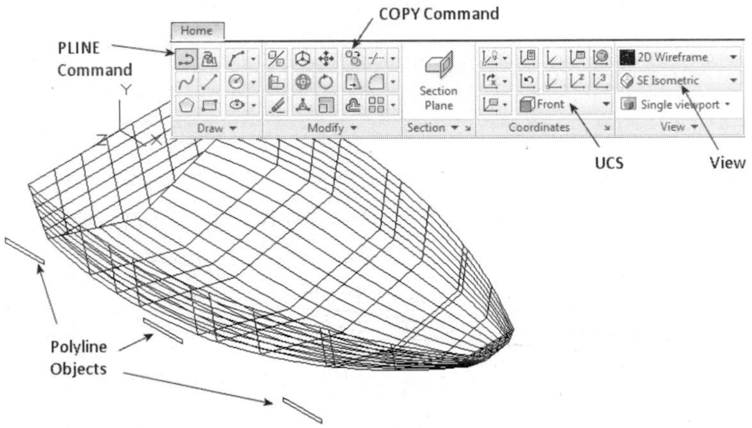

FIGURE 21.102

STEP 15

In this step we will create extruded surfaces from the polyline objects. Select the Surface tab on the Ribbon and activate the EXTRUDE command in the Create panel, as shown in Figure 21.103 (Left). Use the following command line sequence and Figure 21.103 to create the surfaces for the benches.

Command: **EXT** *(For EXTRUDE)*

Current wire frame density: ISOLINES=4, Closed profiles creation mode = Solid

Select objects to extrude or [MOde]: **MO** *(For MOde)*

Closed profiles creation mode [SOlid/SUrface] <Solid>: **SU** *(For SUrface)*

Select objects to extrude or [MOde]: *(Select the polyline objects, as shown in Figure 21.103 (Left))*

Select objects to extrude or [MOde]: *(Press* ENTER *to continue with this command)*

Specify height of extrusion or [Direction/Path/Taper angle/Expression]:

-6' *(Such that the surfaces pass through the boat hull)*

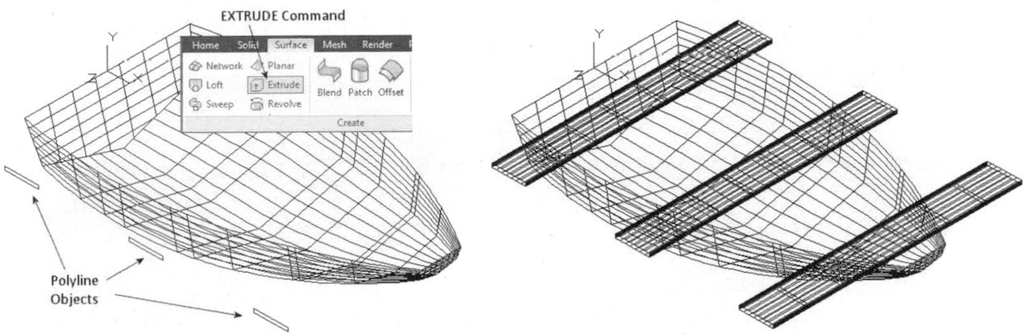

FIGURE 21.103

STEP 16

In this final step we will trim the bench surfaces to the hull. From the Surface tab on the Ribbon, activate the SURFTRIM command in the Edit panel. Use the following command line sequence and Figure 21.104 to trim the bench surfaces and complete the surfaced boat.

Command: SURFTRIM

Extend surfaces = Yes, Projection = Automatic

Select surfaces or regions to trim or [Extend/PROjection direction]: *(Select the three bench surfaces, as shown in Figure 21.104 (Left))*

Select surfaces or regions to trim or [Extend/PROjection direction]: *(Press* ENTER *to continue with this command)*

Select cutting curves, surfaces, or regions: *(Select the hull surface, as shown in Figure 21.104 (Right))*

Select cutting curves, surfaces, or regions: *(Press* ENTER *to continue with this command)*

Select area to trim [Undo]: *(Select the bench ends, as shown in Figure 21.104 (Right))*

Select area to trim [Undo]: *(Press* ENTER *to complete the operation)*

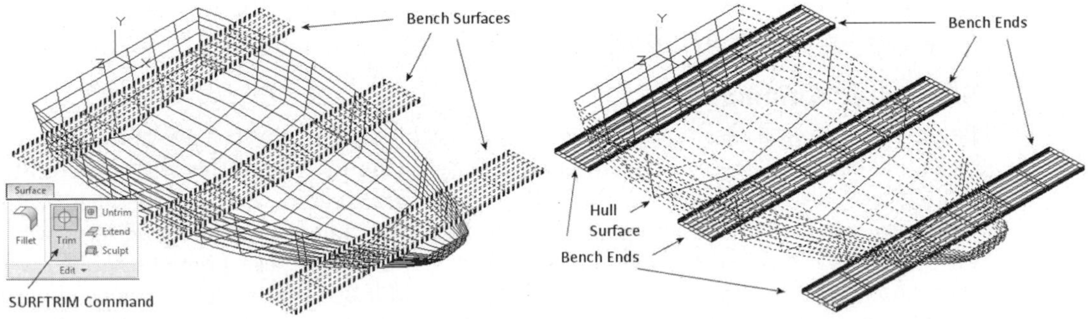

FIGURE 21.104

From the Home tab on the Ribbon, activate the ERASE command in the Modify panel and remove the three polyline objects, as shown in Figure 21.105 (Left). From the View panel, select the Conceptual visual style to better display the surfaces, as shown in Figure 21.105 (Right).

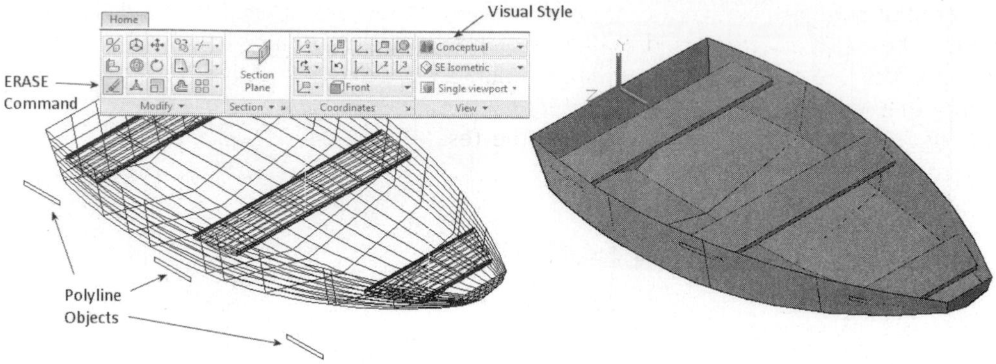

FIGURE 21.105

TUTORIAL EXERCISE: 21_MESH CAMERA.DWG

Purpose

This tutorial exercise is designed to use mesh modeling techniques to design the body of a digital camera, as shown in Figure 21.106.

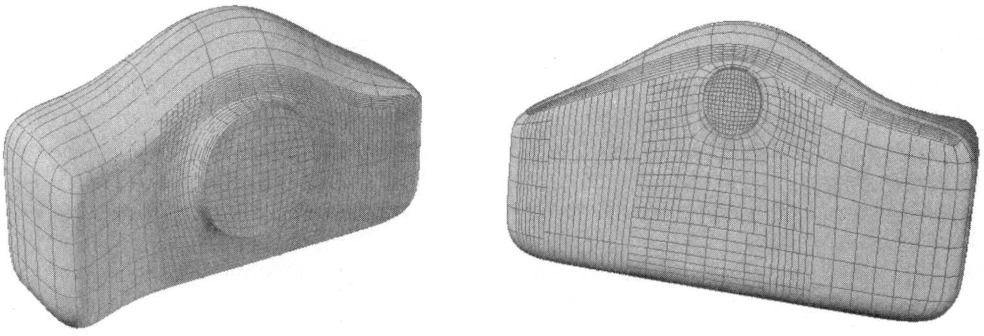

FIGURE 21.106

System Settings

The drawing units, limits, grid, and snap values are already set for this drawing.

Layers

Make sure the current layer is set to Mesh.

Suggested Commands

Begin by creating a mesh box that will serve as the start of the camera body. Various faces and edges will be selected through the use of the CTRL key. This will activate the editing capabilities of subobjects. With the base camera created, you can experiment with the many ways of smoothing the mesh model of the camera body. You will also work on the camera lens and the eye piece of the camera body. These primitive shapes will then be creased to form edges even thought a majority of the model is displayed with smooth edges. Finally, you will convert the mesh model into a solid model where you can add additional features to the camera body.

STEP 1

Open the drawing 21_Mesh Camera. Before creating the mesh model of the camera body, first examine the options available to you regarding the individual primitives used to model. Clicking the arrow located in the Primitives panel of the Mesh Ribbon, as shown in Figure 21.107 (Left), will activate the Mesh Primitive Options dialog box, as shown in Figure 21.107 (Right). Notice the various primitives that are contained in this dialog box such as box, cylinder, cone, and wedge to name a few. In the dialog box, the box object is highlighted. Notice in the dialog box the number of Tesselation Divisions for the box. Verify that the Length is set to 5, the Width set to 3, and the Height is set to 3. You will also notice a preview of the box based on the Tesselation Divisions. Click the OK button to exit this dialog box.

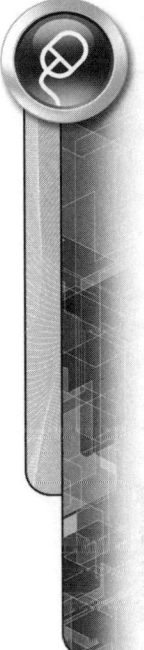

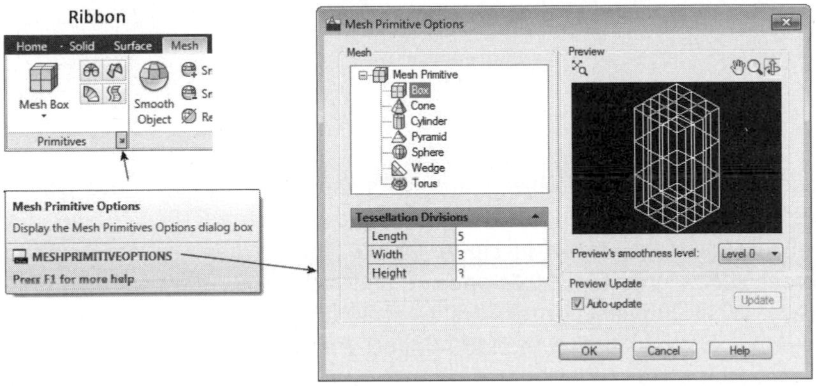

FIGURE 21.107

STEP 2

Click the Mesh Box button located on the Ribbon, as shown in Figure 21.108 (Left). Then create a mesh box by first specifying the corner point of 10,10. For the other corner, type L for Length at the Command prompt. This will allow you to create the mesh box by length, width, and height. Next, move your cursor until the angle readout reads 0° and a value of 180, as shown in Figure 21.108 (Left). Continue moving your cursor, this time in the Y direction and enter a value of 40, which represents the width. Finally move your cursor in the Z direction and enter a value of 80 for the height to complete the creation of the mesh box, as shown in Figure 21.108 (Right).

```
Command: _MESH
Current smoothness level is set to : 0
Enter an option [Box/Cone/CYlinder/Pyramid/Sphere/Wedge/
Torus/SEttings] <Box>: _BOX
Specify first corner or [Center]: 10,10
Specify other corner or [Cube/Length]: L (For Length)
Specify length <180.0000>: 180 (Point cursor in "X"
Direction)
Specify width <40.0000>: 40
Specify height or [2Point] <80.0000>: 80
```

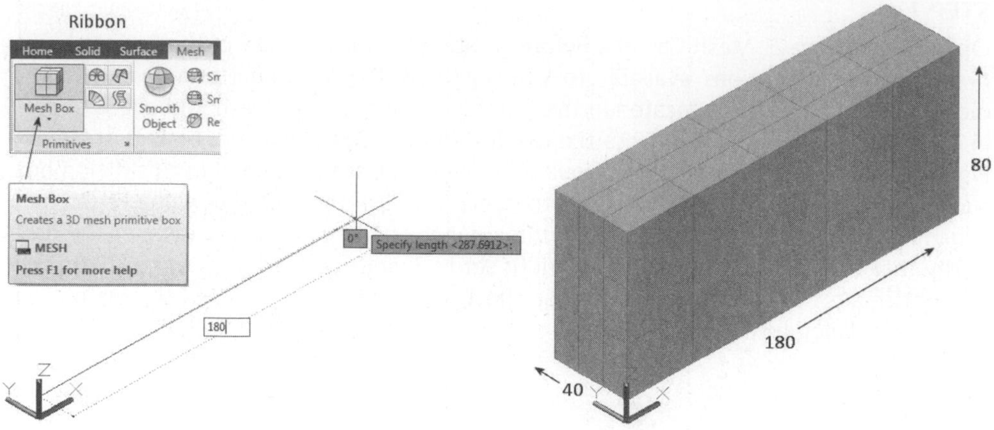

FIGURE 21.108

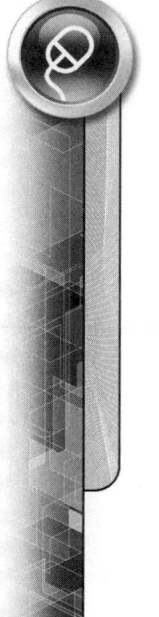

STEP 3

You will now select a number of subobjects and, with the aid of the 3D move gizmo, stretch the faces vertically at a distance of 30. First, hold down the CTRL key and pick the three top middle faces, as shown in Figure 21.109 (Left). When the move gizmo appears, move your cursor until the Z-axis appears and pick. Move the cursor to drag the faces vertically. While dragging, enter a value of 30 and press ENTER to drag the three faces 30 mm up, as shown in Figure 21.109 (Right). Press ESC to remove the grips.

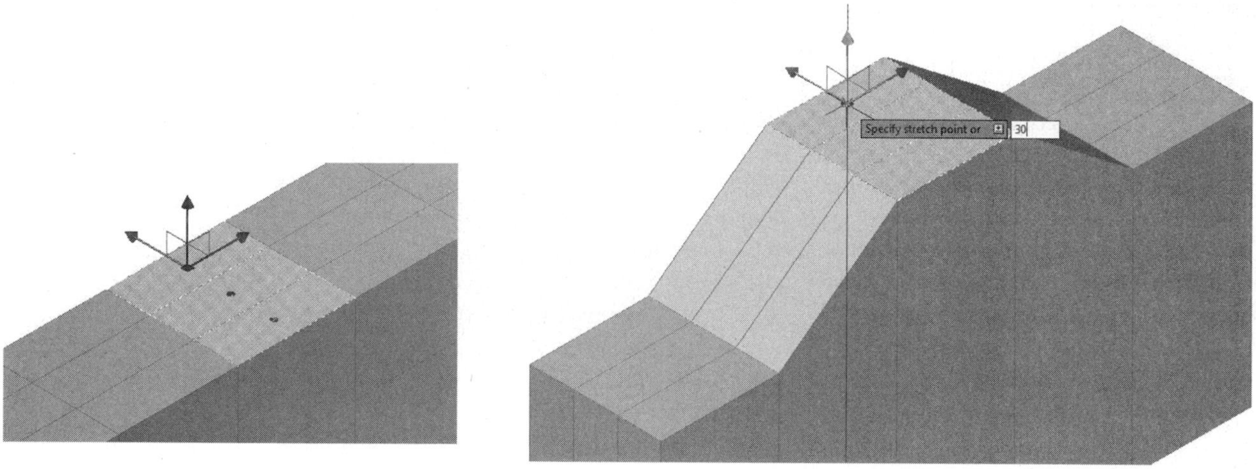

FIGURE 21.109

STEP 4

Next, you will stretch the edges of the mesh box instead of its faces. To better accomplish this, change the selection filter to Edge that is found under the Selection panel of the Ribbon, as shown in Figure 21.110 (Left). Remember... you should be in the Mesh tab of the Ribbon in order to properly access the correct commands. After switching to the Edge selection filter, select the three vertical edges, as shown in Figure 21.110 (Center). When the Move gizmo appears, move your cursor until the Y axis appears and pick. Drag these three edges toward you while entering a value of 20. Pressing ENTER will stretch these edges 20 mm, as shown in Figure 21.110 (Right).

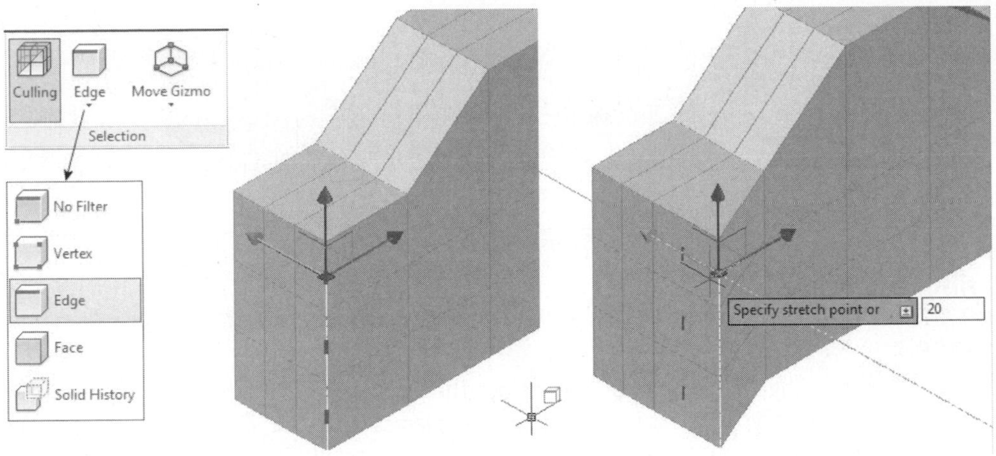

FIGURE 21.110

STEP 5

Change to an SE Isometric view. While still in the Edge selection filter, select the three horizontal edges, as shown in Figure 21.111 (Left). When the 3D move gizmo appears, move your cursor until the Z-axis appears and pick. Drag these three edges down while entering a value of 10. Pressing ENTER will stretch these edges 10 mm, as shown in Figure 21.111 (Right). Turn off the Edge selection filter - No Filter.

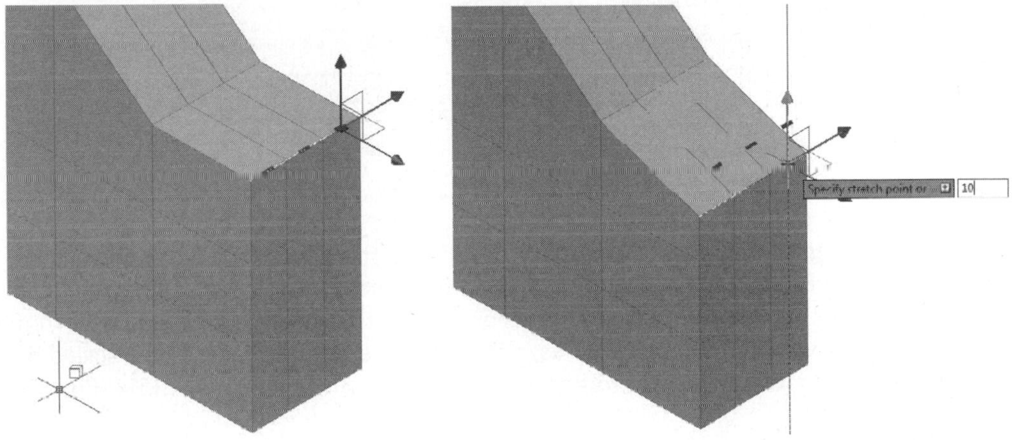

FIGURE 21.111

STEP 6

Change back to a SW Isometric view. At this point your model of the camera body should appear similar to Figure 21.112 (Left). Unfortunately, most digital camera bodies do not consist of sharp edges. To improve the surface model, first activate the Quick Properties button on the status bar; this tool can be used to control the smoothness of the camera body. Picking on any edge of the camera body will display the Quick Properties palette, as shown in Figure 21.112 (Center). The Smoothness setting can range from None to Level 4. Presently, your model is displayed with no smoothness. To resemble a more realistic camera body, change to Level 2. The results are shown in Figure 21.112 (Right). Press ESC to remove the Quick Properties palette.

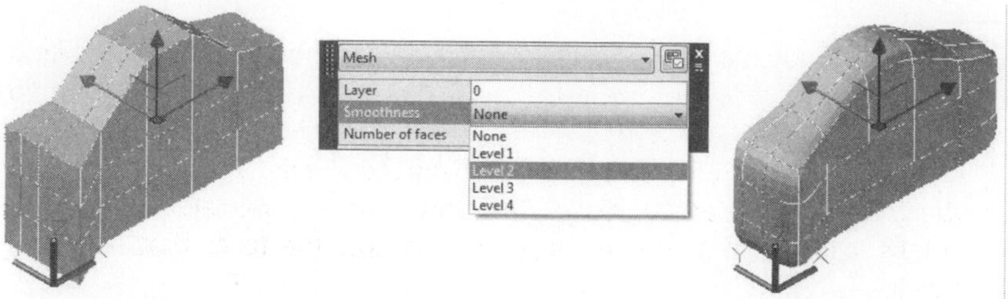

FIGURE 21.112

STEP 7

Next, turn the Edge filter back on and select the three edges, as shown in Figure 21.113 (Left). Then move your cursor over the move gizmo, pick the Z-axis, and stretch the three curving edges down at a distance of 20 mm. The results are shown in Figure 21.113 (Right). Press ESC to remove the grips.

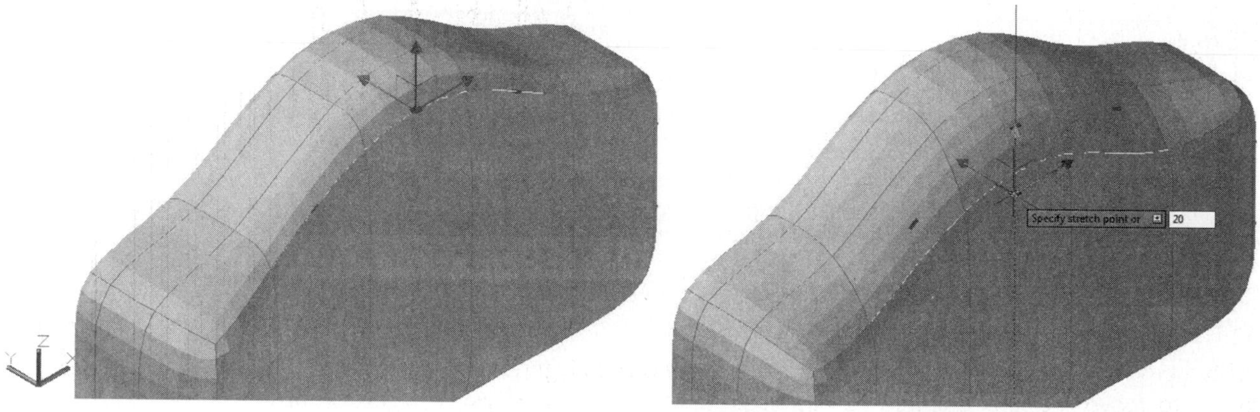

FIGURE 21.113

STEP 8

Next, modify the face where the lens will go. Change the selection type to Face mode. Select the middle face, then right-click anywhere on the move gizmo and choose Scale from the menu. Right-click on the gizmo again and this time, click Set Constraint from the menu followed by the ZX plane. This is the plane the selected face will be scaled by. Entering a scale factor of 1.50 will display your model, as shown in Figure 21.114 (Right).

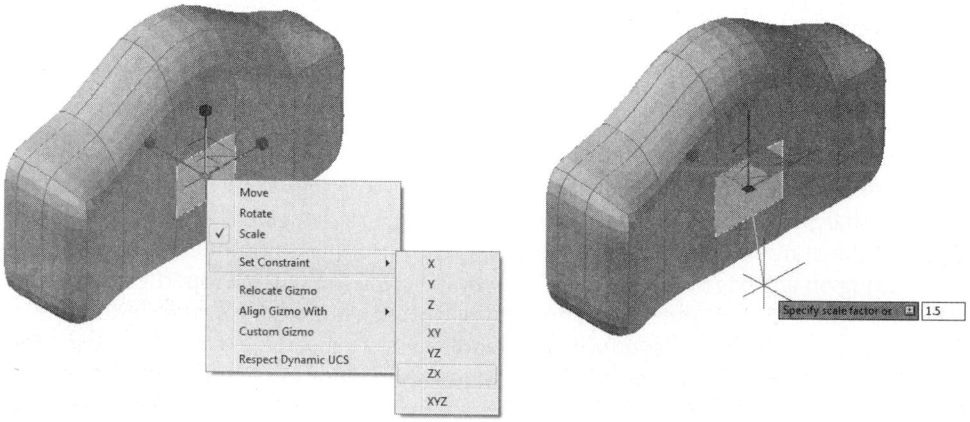

FIGURE 21.114

STEP 9

Unfortunately the four sides of the face do not appear square; they appear more like a rectangle. We really want this face to have equal values in the X and Z directions. To accomplish this, right-click on the scale gizmo to display the menu, as shown in Figure 21.115 (Left), and set the constraint from the menu to the Z-axis. This face will now be scaled only in the Z direction by an additional 1.40 units. The Z scale value of 1.40 will display the camera lens in more of a square shape. Notice how the faces that surround the face update as well.

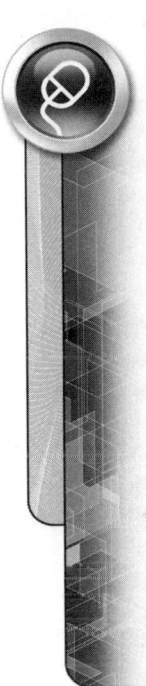

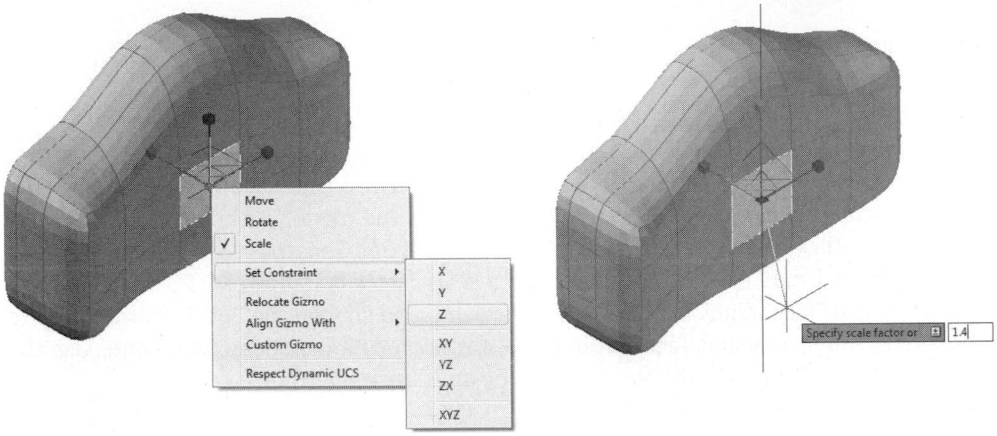

FIGURE 21.115

STEP 10

With the face still active, select the Move Gizmo, as shown in Figure 21.115 (Left). Use the gizmo to move the lens opening up approximately 12 mm, as shown in Figure 21.116 (Right). This distance does not have to be exact.

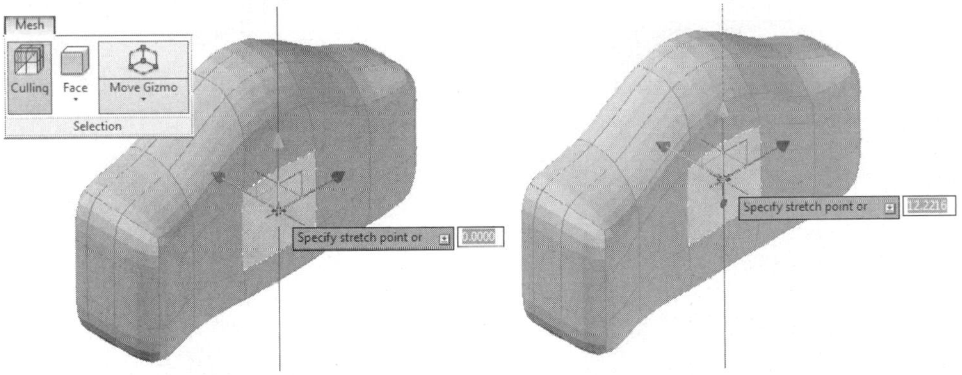

FIGURE 21.116

STEP 11

Click on the Extrude Face tool from Mesh Edit panel of the Ribbon and if necessary, select the square face you have been working on, as shown in Figure 21.117 (Left). The Face selection filter is automatically provided. Drag the lens opening out. You could also have directed the extrusion inside of the part to create a hole or void. Keep dragging the face out and enter a distance of 10 mm. Your mesh model of the camera body should appear similar to Figure 21.117 (Right).

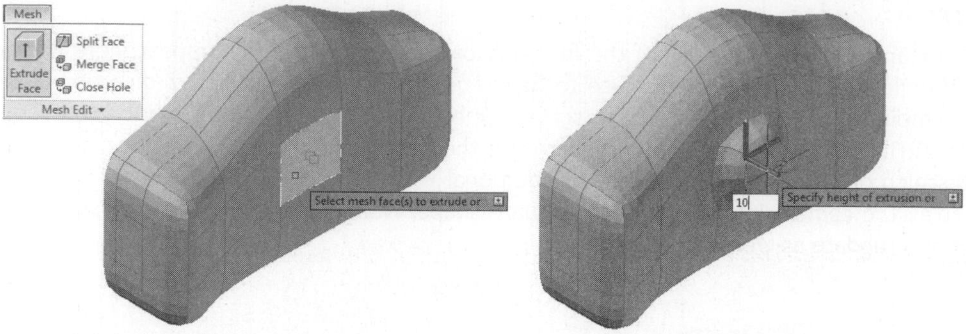

FIGURE 21.117

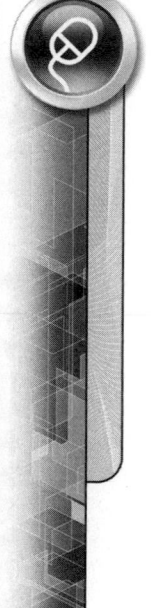

STEP 12

Next, change to a NW Isometric view for the purpose of creating the eye piece. As with the camera lens, select the face, as shown in Figure 21.118 (the Face filter should still be active). Right-click on the gizmo and pick Scale from the menu, as shown in Figure 21.118 (Left). Right-click a second time on the gizmo, select Set Constraint from the menu, and scale this face based on the ZX plane. Enter a scale factor distance of 0.50 mm to reduce the size of this face. Right-click again on the gizmo and change the Set Constraint axis to the X-axis. Then scale this face out in the X direction by a distance of 1.30 mm. Use the Extrude Mesh tool to extrude this face out by a distance of 2 mm.

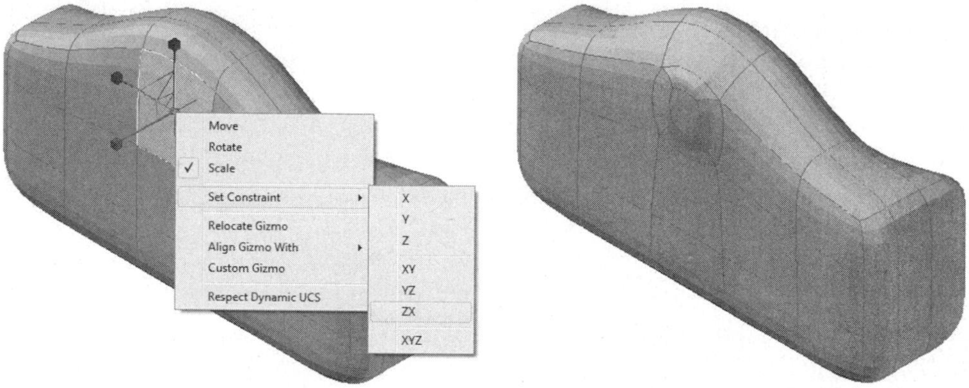

FIGURE 21.118

STEP 13

Use the Extrude Mesh tool to extrude the center face out a second time by 2 mm. Notice how both the eye piece and lens extrusions are smooth.

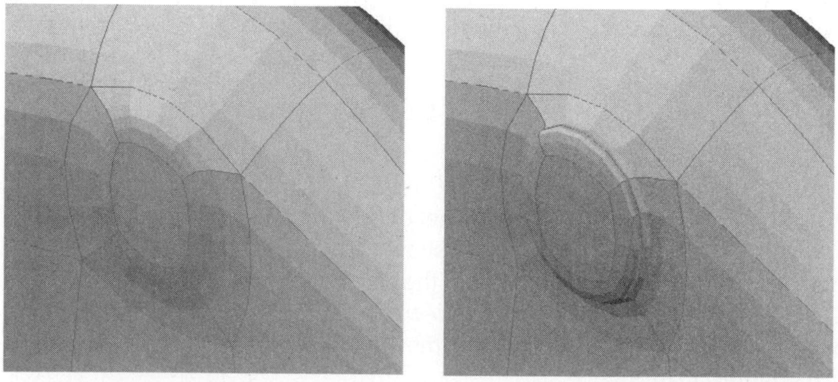

FIGURE 21.119

STEP 14

A crease will now be added to the top face of the eye piece. Activate the Edge filter and then select the Add Crease tool from the Mesh panel of the Ribbon. Select the four edges of the eye piece, as shown in Figure 21.120 (Left). Accept the default crease value as "Always." This creates an edge that is sharp and better defines the eye piece of the camera body, as shown in Figure 21.120 (Right).

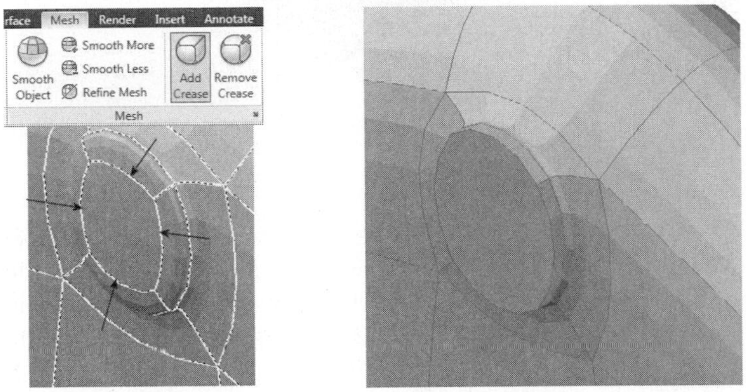

FIGURE 21.120

STEP 15

Change to a SW Isometric view. Next, create a crease on the front of the camera lens using the same techniques used in the previous step. Activate the Add Crease tool and select the front edges of the lens, as shown in Figure 21.121 (Left). Keep the default values to generate the sharp edge, as shown in Figure 21.121 (Right).

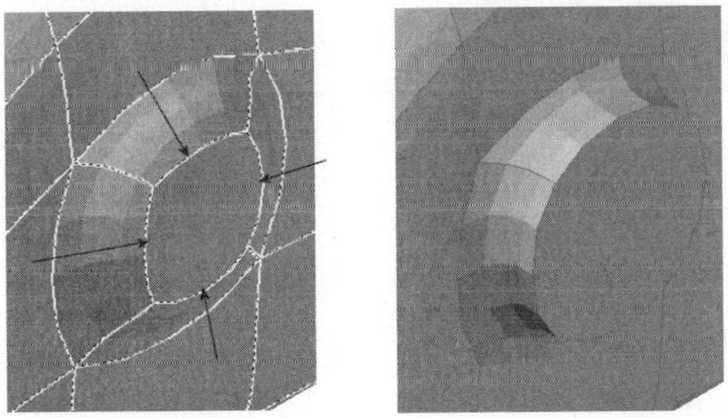

FIGURE 21.121

STEP 16

The results are shown in Figure 21.122. The camera lens displays edges, as shown in Figure 21.122 (Left). The eye piece also has edges that are displayed in Figure 21.122 (Right).

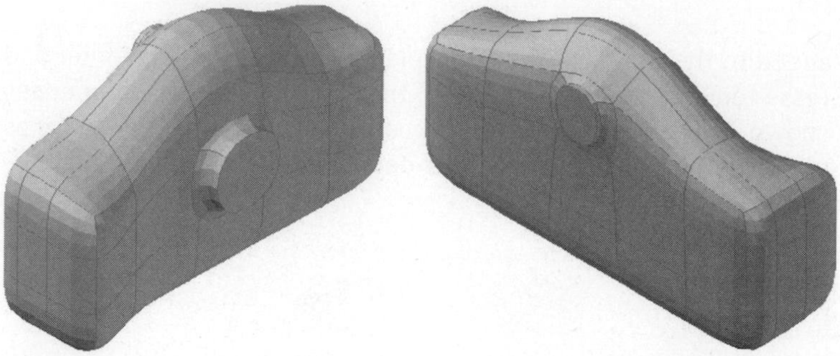

FIGURE 21.122

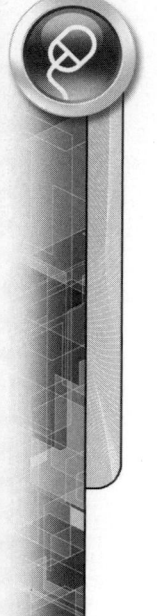

STEP 17

Once the mesh model has been created, it will now be converted into a solid model using the Convert to Solid tool, as shown in Figure 21.123. When prompted to select the model to convert, click anywhere on the mesh model of the camera body, as shown in Figure 21.123 (Left). The results are illustrated in Figure 21.123 (Right). Once the camera body is converted into a solid, you can now use the traditional solid modeling commands to add or subtract features.

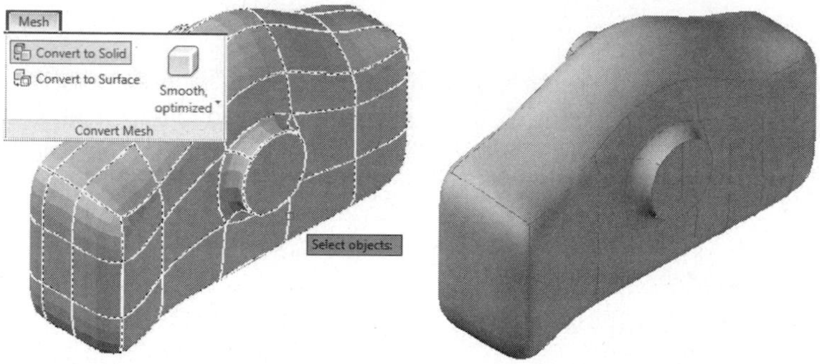

FIGURE 21.123

PROBLEMS FOR CHAPTER 21

PROBLEM 21-1

Utilize the skills listed in this chapter on problems that are found in Chapters 8, 9, and 20. Many of the problems in these chapters apply the editing skills found in this chapter.

One option is to take a drawing that was completed as a 2D drawing and re-create it as a 3D solid or surface.

CHAPTER 22

Creating 2D Drawings from a 3D Solid Model

One of the advantages of creating a 3D image in the form of a solid model is the ability to use the data of the solid model numerous times for other purposes. The purpose of this chapter is to generate 2D multiview drawings from the solid model. This chapter will demonstrate three available methods of generating 2D drawings from 3D objects. The first method utilizes two commands: SOLVIEW and SOLDRAW. The SOLVIEW command is used to create a layout for the 2D views. This command automatically creates layers used to organize visible lines, hidden lines, and dimensions. The SOLDRAW command draws the requested views on the specific layers that were created by the SOLVIEW command (this includes the drawing of hidden lines to show hidden features, and even hatching if a section is requested). The second method utilizes the FLATSHOT command. This command, although limited in its capabilities, efficiently creates 2D views that are saved as blocks. The third method utilizes the model documentation tools: VIEWBASE, VIEWPROJ, VIEWSECTION, and VIEWDETAIL commands. This new process is simple and powerful. Once a base view is created, other section, orthographic, isometric, and detail views are projected with a pick and place technique. It is also important to note that the model is linked to the view, and changes to the model will automatically be updated in the drawing views.

THE SOLVIEW AND SOLDRAW COMMANDS

Once you create the solid model, you can lay out and draw the necessary 2D views using a number of Modeling Setup commands. Expanding the Modeling tab of the Home Ribbon, as shown in Figure 22.1, exposes the Drawing (SOLDRAW), View (SOLVIEW), and Profile (SOLPROF) commands. Only SOLDRAW and SOLVIEW will be discussed in this chapter. The SOLPROF command provides an additional method for creating 2D drawing views, but the process is not as automated.

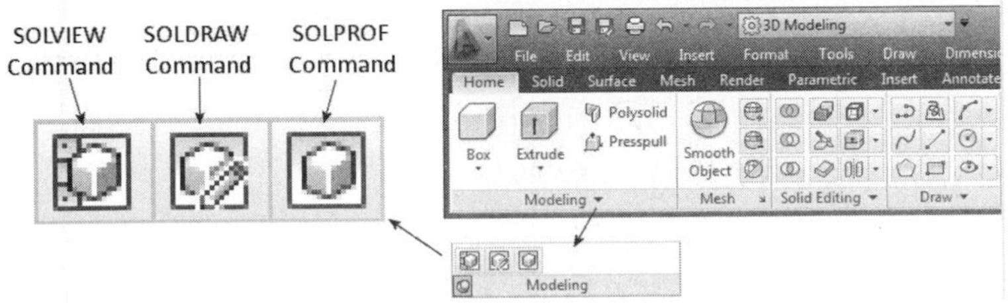

FIGURE 22.1

Once the SOLVIEW command is entered, the display screen automatically switches to the first layout in the Paper Space environment. Using SOLVIEW lays out a view based on responses to a series of prompts, depending on the type of view you want to create. Usually, the initial view that you lay out serves as the starting point for other views and is based on the current User Coordinate System. This needs to be determined before you begin this command. Once an initial view is created, it is easy to generate Orthographic, Auxiliary, and Section views from the original view. If an Isometric view is desired, it is created much like an initial view based on the User Coordinate System (UCS).

As SOLVIEW is used as a layout tool, the images of the views created are still simply plan views of the original solid model. In other words, after you lay out a view, it does not contain any 2D features, such as hidden lines. The SOLDRAW command is used to actually create the 2D profiles once it has been laid out through the SOLVIEW command.

TRY IT!

Open the drawing file 22_Solview. Before using the SOLVIEW command, study the illustration of this solid model, as shown at "A" in Figure 22.2. In particular, pay close attention to the position of the User Coordinate System icon. The current position of the User Coordinate System will start the creation process of the base view of the 2D drawing.

TIP

Before you start using the SOLVIEW command, remember to load the Hidden linetype. This automatically assigns this linetype to any new layer that requires hidden lines for the drawing mode. If the linetype is not loaded at this point, it must be manually assigned later to each layer that contains hidden lines through the Layer Properties Manager palette.

Activating SOLVIEW automatically switches the display to the layout. Since this is the first view to be laid out, the UCS option will be used to create the view based on the current User Coordinate System. The view produced is similar to looking down the Z-axis of the UCS icon. A scale value may be entered for the view. For the View Center, click anywhere on the screen and notice the view being constructed. You can pick numerous times on the screen until the view is in a desired location. The placement of this first view is very important because other views will most likely be positioned relative to this one. When this step is completed, press ENTER to place the view. Next, you are prompted to construct a viewport around the view. Remember to make this viewport large enough for dimensions to fit inside. Once the view is given a name, it is laid out similar to the illustration at "B" in Figure 22.2.

Command: SOLVIEW

Enter an option [Ucs/Ortho/Auxiliary/Section]: U *(For Ucs)*

Enter an option [Named/World/?/Current] <Current>: *(Press ENTER)*

Enter view scale <1.0000>: *(Press ENTER)*

Specify view center: *(Pick a point near the center of the screen to display the view; keep picking until the view is in the desired location)*

Specify view center <specify viewport>: *(Press ENTER to place the view)*

Specify first corner of viewport: *(Pick a point at "D")*

Specify opposite corner of viewport: *(Pick a point at "E")*

Enter view name: FRONT

Enter an option [Ucs/Ortho/Auxiliary/Section]: *(Press ENTER to exit this command)*

Once the view has been laid out through the SOLVIEW command, use SOLDRAW to actually draw the view in two dimensions. The Hidden linetype was loaded for you in this drawing and since it was loaded prior to using the SOLVIEW command, hidden lines will automatically be assigned to layers that contain hidden line information. The result of using the SOLDRAW command is shown at "C" in Figure 22.2. You are no longer looking at a 3D solid model but at a 2D drawing created by this command.

Command: SOLDRAW

(If in Model Space, you are switched to Paper Space)

Select viewports to draw...

Select objects: *(Pick anywhere on the viewport at "C" in Figure 22.2)*

Select objects: *(Press ENTER to perform the Soldraw operation)*

One solid selected.

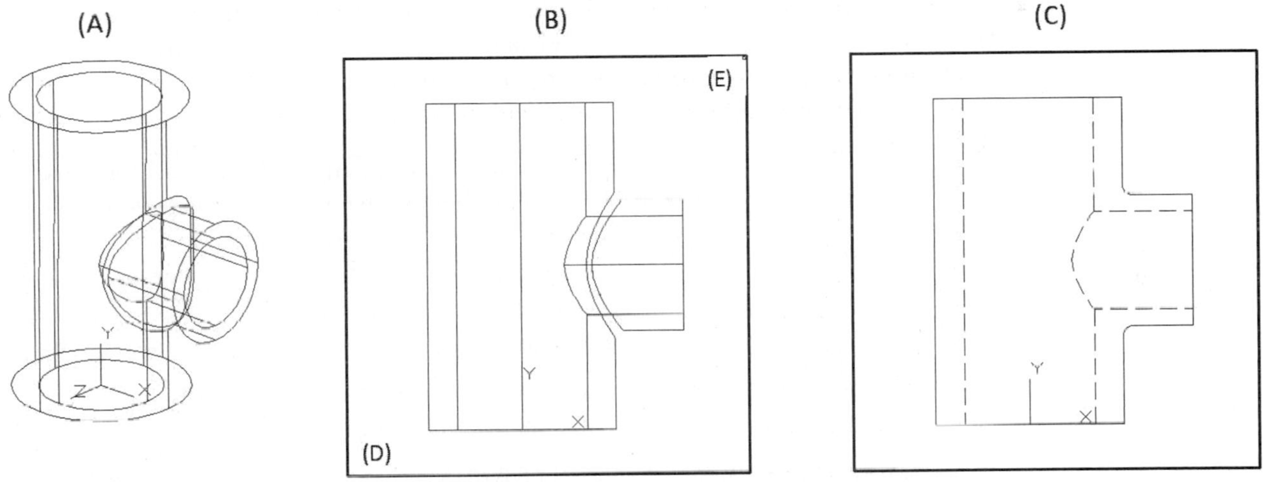

FIGURE 22.2

The use of layers in 2D-view layout is so important that when you run the SOLVIEW command, the layers shown in Figure 22.3 are created automatically. With the exception of Model and 0, the layers that begin with "FRONT" and the VPORTS layer were all created by the SOLVIEW command. The FRONT-DIM layer is designed to hold dimension information for the Front view. FRONT-HID holds all hidden lines information for the Front view; FRONT-VIS holds all visible line information for the Front view. All Paper Space viewports are placed on the VPORTS layer. The Model layer has automatically been frozen in the current viewport to hide the 3D model and show the 2D visible and hidden lines.

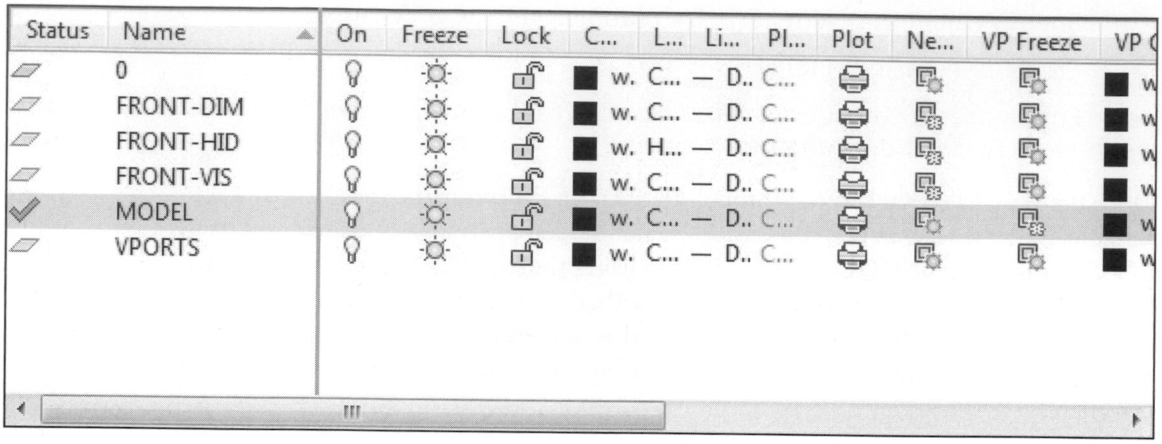
Status
Name
On
Freeze
Lock
C...
L...
Li...
Pl...
Plot
Ne...
VP Freeze
0
FRONT-DIM
FRONT-HID
FRONT-VIS
MODEL
VPORTS
w. C... — D.. C...
w. C... — D.. C...
w. H... — D.. C...
w. C... — D.. C...
w. C... — D.. C...
w. C... — D.. C...

FIGURE 22.3

In order for the view shown in Figure 22.4 (Left) to be dimensioned in model space, three operations must be performed. First, double-click inside the Front view to be sure it is the current floating Model Space viewport. Next, make FRONT-DIM the current layer. Finally, if it is not already positioned correctly, set the User Coordinate System to the current view using the View option of the UCS command. The UCS icon should be similar to the illustration in Figure 22.4 (Left) (you should be looking straight down the Z-axis). Now add all dimensions to the view using conventional dimensioning commands with the aid of Object Snap modes. When you work on adding dimensions to another view, the same three operations must be made in the new view: make the viewport active by double-clicking inside it, make the appropriate dimension layer current, and update the UCS, if necessary, to the current view with the View option.

When you draw the views using the SOLDRAW command and then add the dimensions, switching back to the solid model by clicking the Model tab displays the illustration shown in Figure 22.4 (Right). In addition to the solid model of the object, the constructed 2D view and dimensions are also displayed. All drawn views from Paper Space display with the model. To view just the solid model you would have to use the Layer Properties Manager palette along with the Freeze option and freeze all drawing-related layers.

TIP

The solid model and drawing views are not linked. Any changes made to the solid model will not update the drawing views. If changes are required, you can modify the model and the 2D drawing views as needed. It is recommended that you save your drawing prior to generating the 2D views. It is sometimes more efficient to repair a model and restart the SOLVIEW and SOLDRAW process than it is to edit the model and views.

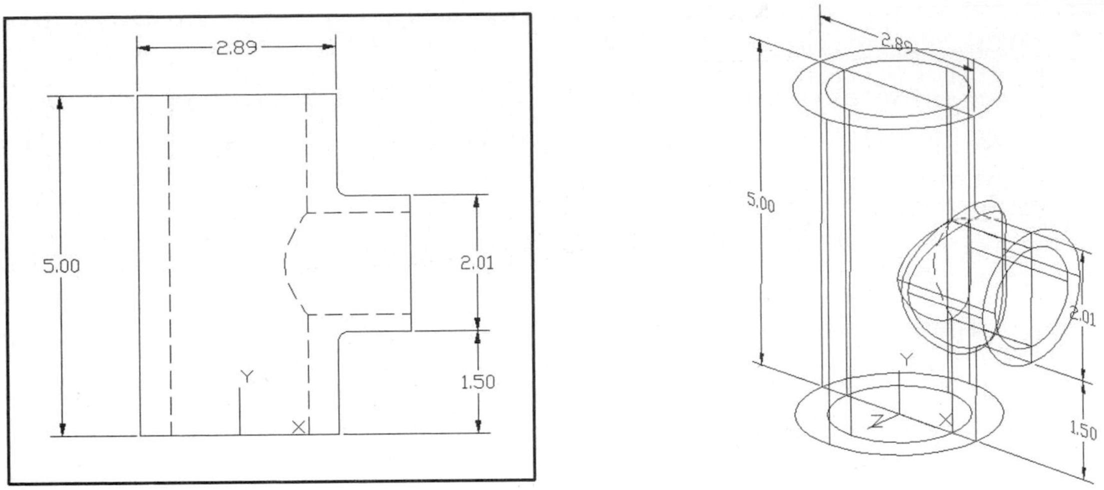

FIGURE 22.4

TIP

The User Interface panel of the Ribbon can be used to turn off the ViewCube and Navigation Bar that display in your viewports (Ribbon > View tab > User Interface panel > User Interface button).

CREATING ORTHOGRAPHIC VIEWS

Once the first view is created, orthographic views can easily be created with the Ortho option of the `SOLVIEW` command.

TRY IT!

Open the drawing file 22_Ortho. Notice that you are in a layout and a Front view is already created. Follow the Command prompt sequence below to create two orthographic views. When finished, your drawing should appear similar to Figure 22.5.

`Command: SOLVIEW`

`Enter an option [Ucs/Ortho/Auxiliary/Section]: O` *(For Ortho)*

`Specify side of viewport to project:` *(Select the top of the viewport at "A"—a midpoint OSNAP will be automatically provided)*

`Specify view center:` *(Pick a point above the front view to locate the top view)*

`Specify view center <specify viewport>:` *(Press ENTER to place the view)*

`Specify first corner of viewport:` *(Pick a point at "B")*

`Specify opposite corner of viewport:` *(Pick a point at "C")*

`Enter view name: TOP`

`Enter an option [Ucs/Ortho/Auxiliary/Section]: O` *(For Ortho)*

`Specify side of viewport to project:` *(Select the right side of the viewport at "D")*

`Specify view center:` *(Pick a point to the right of the front view to locate the right side view)*

Specify view center <specify viewport>: *(Press* ENTER *to place the view)*

Specify first corner of viewport: *(Pick a point at "E")*

Specify opposite corner of viewport: *(Pick a point at "F")*

Enter view name: R_SIDE

Enter an option [Ucs/Ortho/Auxiliary/Section]: *(Press* ENTER *to exit this command)*

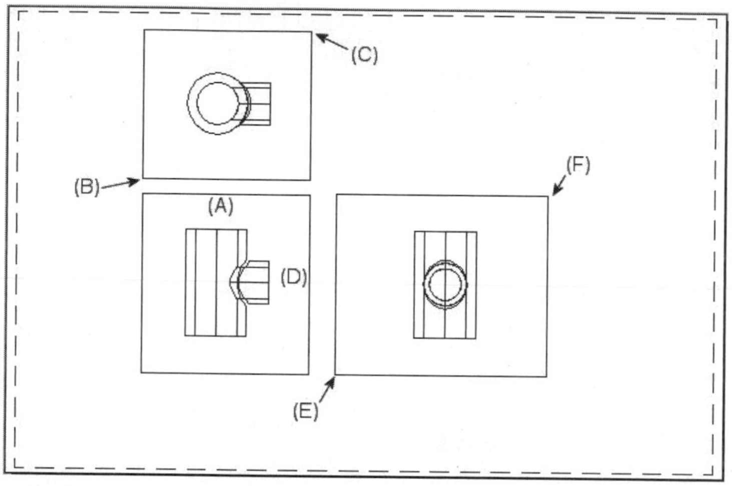

FIGURE 22.5

Running the SOLDRAW command on the three views displays the illustration shown in Figure 22.6. Notice the appearance of the hidden lines in all views. The VPORTS layer is turned off to display only the three views.

Command: SOLDRAW

Select viewports to draw.

Select objects: *(Select the three viewports that contain the Front, Top, and Right Side view information)*

Select objects: *(Press* ENTER *to perform the Soldraw operation)*

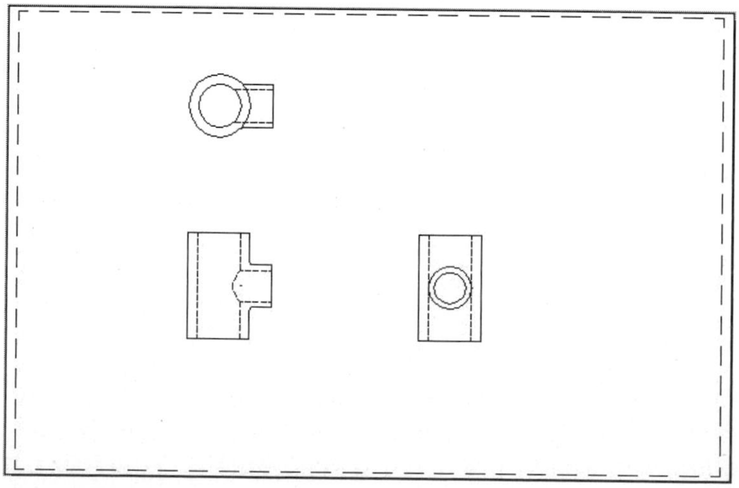

FIGURE 22.6

CREATING AN AUXILIARY VIEW

In the illustration in Figure 22.7 (Left), the true size and shape of the inclined surface containing the large counterbore hole cannot be shown with the standard orthographic view. An auxiliary view must be used to properly show these features.

TRY IT!

Open the drawing file 22_Auxiliary. From the 3D model in Figure 22.7, use the SOLVIEW command to create a Front view based on the current User Coordinate System. The results are shown in Figure 22.7 (Right).

Command: SOLVIEW

Enter an option [Ucs/Ortho/Auxiliary/Section]: U *(For Ucs)*

Enter an option [Named/World/?/Current] <Current>: *(Press* ENTER*)*

Enter view scale <1.0000>: *(Press* ENTER*)*

Specify view center: *(Pick a point to locate the view, as shown in Figure 22.7 (Right))*

Specify view center <specify viewport>: *(Press* ENTER *to place the view)*

Specify first corner of viewport: *(Pick a point at "A")*

Specify opposite corner of viewport: *(Pick a point at "B")*

Enter view name: FRONT

Enter an option [Ucs/Ortho/Auxiliary/Section]: *(Press* ENTER *to exit this command)*

TIP

Normally you do not end the SOLVIEW command after each view is laid out. Once you finish creating a view you simply enter the appropriate option (UCS, Ortho, Auxiliary, or Section) and create the next one. This process can continue until all necessary views are provided.

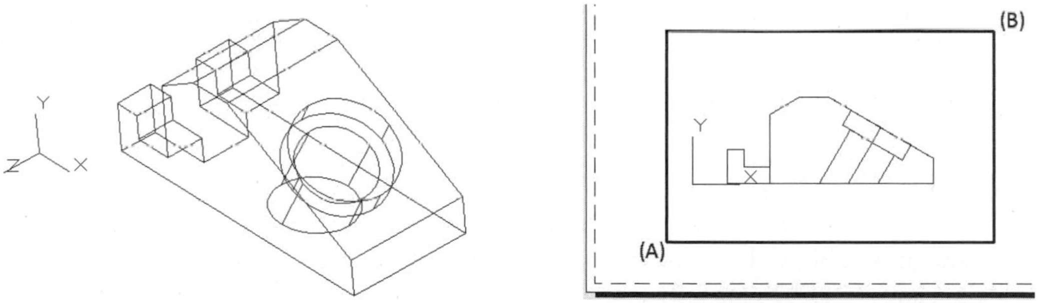

FIGURE 22.7

Now begin the process of constructing an auxiliary view, as shown in Figure 22.8 (Left). After selecting the Auxiliary option of the SOLVIEW command, click the endpoints at "A" and "B" to establish the edge of the surface to view. Pick a point at "C" to indicate the side from which to view the auxiliary view. Notice how the Paper Space icon tilts perpendicular to the edge of the auxiliary view. Pick a location for the auxiliary view and establish a viewport. The result is illustrated in Figure 22.8 (Center).

Command: SOLVIEW

Enter an option [Ucs/Ortho/Auxiliary/Section]: A *(For Auxiliary)*

Specify first point of inclined plane: End

of *(Pick the endpoint at "A")*

Specify second point of inclined plane: End

of *(Pick the endpoint at "B")*

Specify side to view from: *(Pick a point inside of the viewport at "C")*

Specify view center: *(Pick a point to locate the view, as shown in Figure 22.8 (Center))*

Specify view center <specify viewport>: *(Press* ENTER *to place the view)*

Specify first corner of viewport: *(Pick a point at "D")*

Specify opposite corner of viewport: *(Pick a point at "E")*

Enter view name: AUXILIARY

Enter an option [Ucs/Ortho/Auxiliary/Section]: *(Press* ENTER *to exit this command)*

Run the SOLDRAW command and turn off the VPORTS layer. The finished result is illustrated in Figure 22.8 (Right). Hidden lines are displayed only because this linetype was previously loaded.

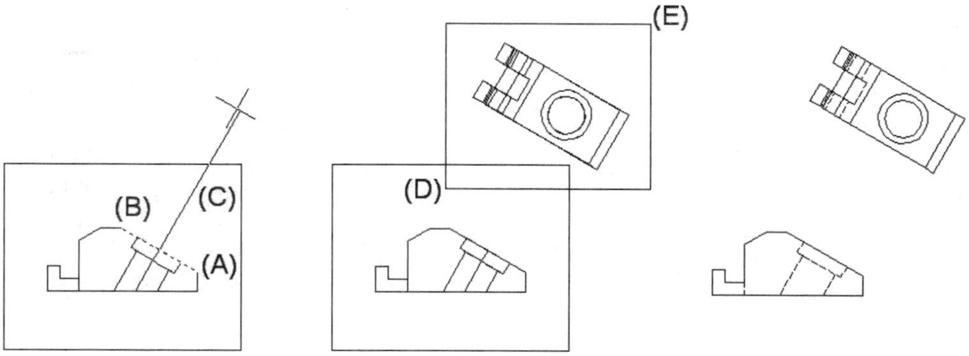

FIGURE 22.8

CREATING A SECTION VIEW

The SOLVIEW and SOLDRAW commands can also be used to create a full section view of an object. This process automatically creates section lines and places them on a layer (*-HAT) for you.

TRY IT!

Open the drawing file 22_Section. From the model illustrated in Figure 22.9 (Left), create a Top view based on the current User Coordinate System, as shown in Figure 22.9 (Right).

Command: SOLVIEW

Enter an option [Ucs/Ortho/Auxiliary/Section]: U *(For Ucs)*

Enter an option [Named/World/?/Current] <Current>: *(Press* ENTER*)*

Enter view scale <1.0000>: *(Press* ENTER*)*

Specify view center: *(Pick a point to locate the view, as shown in Figure 22.9 (Right))*

Specify view center <specify viewport>: *(Press* ENTER *to place the view)*

Specify first corner of viewport: *(Pick a point at "A")*

Specify opposite corner of viewport: *(Pick a point at "B")*

Enter view name: **TOP**

Enter an option [Ucs/Ortho/Auxiliary/Section]: *(Press* ENTER *to exit this command)*

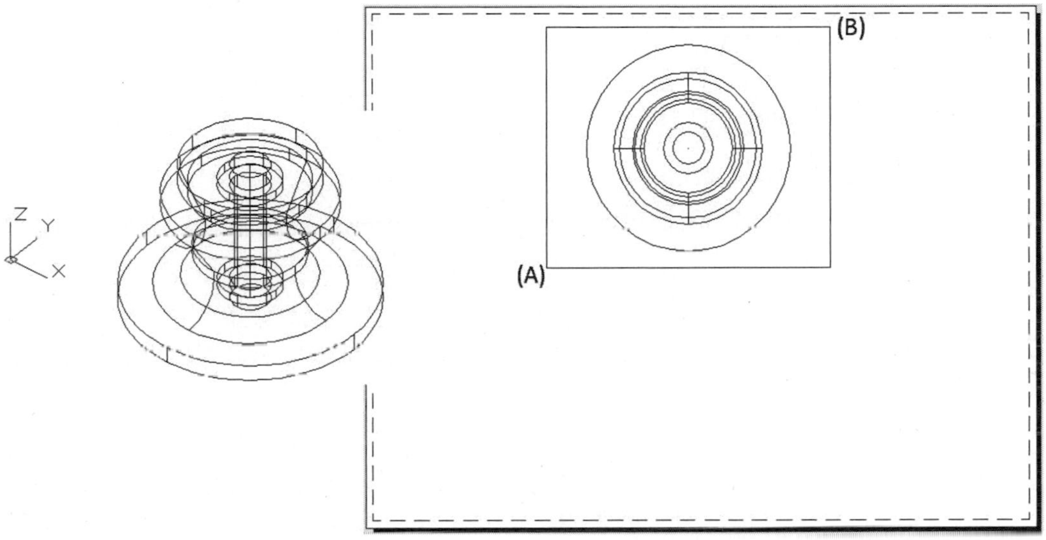

FIGURE 22.9

Begin the process of creating the section. You must first establish the cutting plane line in the Top view, as shown in Figure 22.10 (Left). After the cutting plane line is drawn, select the side from which to view the section. Then locate the section view. This is similar to the process of placing an auxiliary view.

Command: **SOLVIEW**

Enter an option [Ucs/Ortho/Auxiliary/Section]: **S** *(For Section)*

Specify first point of cutting plane: **Qua**

of *(Pick a point at "A")*

Specify second point of cutting plane: *(Turn Ortho on, pick a point at "B")*

Specify side to view from: *(Pick a point inside of the viewport at "C")*

Enter view scale <1.0000>: *(Press* ENTER*)*

Specify view center: *(Pick a point below the top view to locate the view, as shown in Figure 22.10 (Right))*

Specify view center <specify viewport>: *(Press* ENTER *to place the view)*

Specify first corner of viewport: *(Pick a point at "D")*

Specify opposite corner of viewport: *(Pick a point at "E")*

Enter view name: FRONT_SECTION

Enter an option [Ucs/Ortho/Auxiliary/Section]: *(Press* ENTER *to continue)*

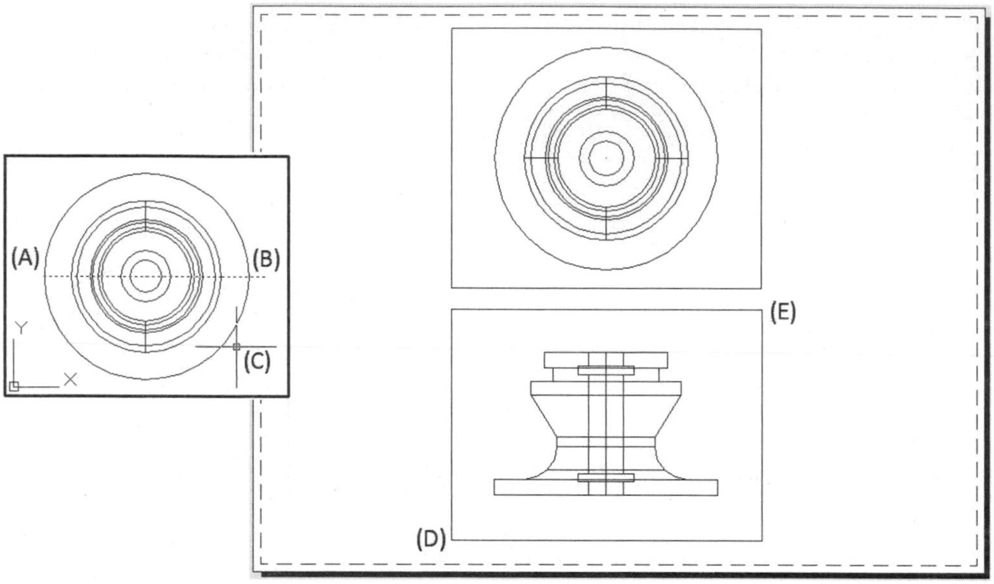

FIGURE 22.10

Running the SOLDRAW command on the viewports results in the illustration in Figure 22.11 (Left). In the illustration in Figure 22.11 (Right), the hatch pattern scale was increased to a value of 2.00 and the viewports were turned off.

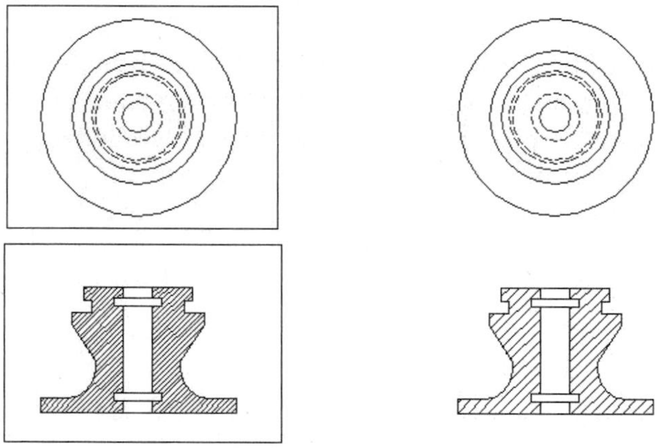

FIGURE 22.11

NOTE

Click inside the viewport containing the section view to enter floating model space. In the Layers panel of the Ribbon, expand the Layers Control box and notice that all the layers dealing with the top view are frozen in the viewport. SOLVIEW automatically freezes the appropriate layers for each view created. Also notice that the layer FRONT_SECTION-HID was frozen in the viewport. This prevents the hidden lines from displaying. Typically, hidden lines are not displayed in sections, auxiliary or isometric views. SOLVIEW does not automatically freeze the hidden line layers in auxiliary and isometric views. You will have to freeze or turn off these layers yourself.

CREATING AN ISOMETRIC VIEW

Once orthographic, section, and auxiliary views are projected, you also have an opportunity to project an isometric view of the 3D model. This type of projection is accomplished using the UCS option of the `SOLVIEW` command and relies entirely on the viewpoint and User Coordinate System setting for your model.

TRY IT!

Open the drawing file 22_Iso. This 3D model should appear similar to the illustration in Figure 22.12 (Left). To prepare this image to be projected as an isometric view, first define a new User Coordinate System based on the current view. See the prompt sequence below to accomplish this task. Your image and UCS icon should appear similar to the illustration in Figure 22.12 (Right).

Command: **UCS**

Current UCS name: *WORLD*

Specify origin of new UCS or [Face/NAmed/OBject/Previous/View/World/X/Y/Z/ZAxis] <World>: **V** *(For View)*

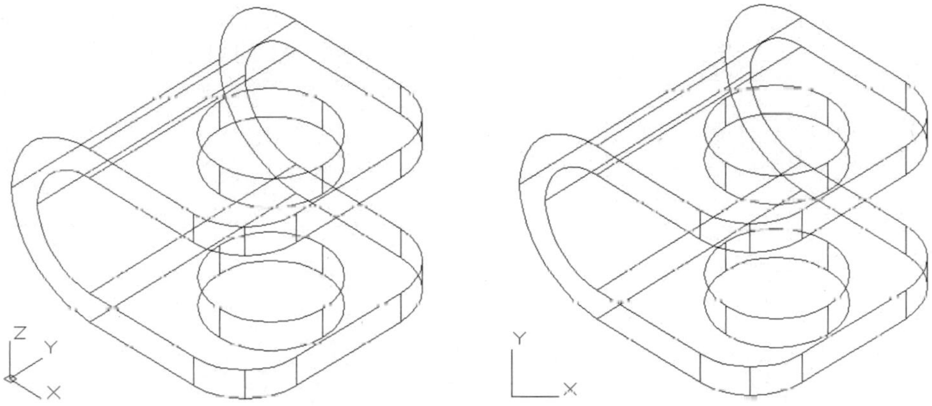

FIGURE 22.12

Next, run the `SOLVIEW` command based on the current UCS. Locate the view and construct a viewport around the isometric, as shown in the sample layout in Figure 22.13. Since dimensions are normally placed in the orthographic view drawings and not on an isometric, you can tighten up on the size of the viewport.

Command: **SOLVIEW**

Enter an option [Ucs/Ortho/Auxiliary/Section]: **U** *(For Ucs)*

Enter an option [Named/World/?/Current] <Current>: *(Press* ENTER*)*

Enter view scale <1.0000>: *(Press* ENTER*)*

Specify view center: *(Pick a point to locate the view in Figure 22.13)*

Specify view center <specify viewport>: *(Press* ENTER *to place the view)*

Specify first corner of viewport: *(Pick a point at "A")*

Specify opposite corner of viewport: *(Pick a point at "B")*

Enter view name: **ISO**

Enter an option [Ucs/Ortho/Auxiliary/Section]: *(Press* ENTER *to exit this command)*

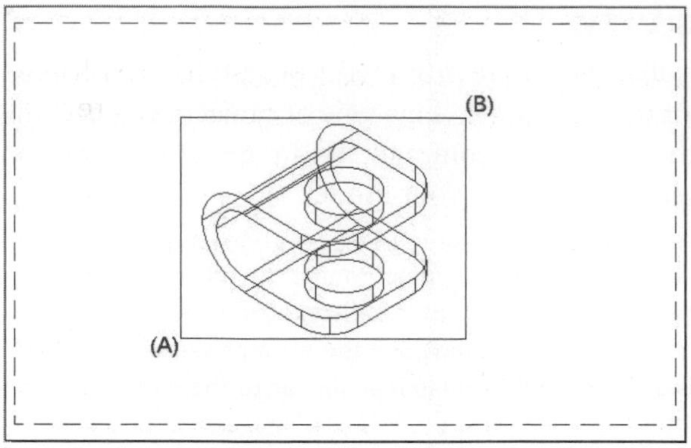

FIGURE 22.13

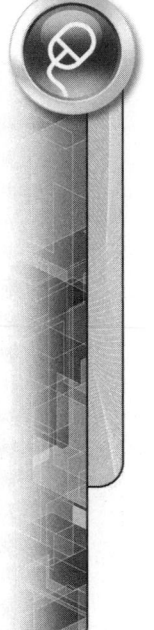

Running the SOLDRAW command on the isometric results in visible lines as well as hidden lines being displayed, as shown in Figure 22.14 (Left). Generally, hidden lines are not displayed in an isometric view. The layer called ISO-HID, which was created by SOLVIEW, contains the hidden lines for the isometric drawing. Use the Layer Control box or the Layer Properties Manager palette to freeze or turn off this layer. The results of this operation are illustrated in Figure 22.14 (Right).

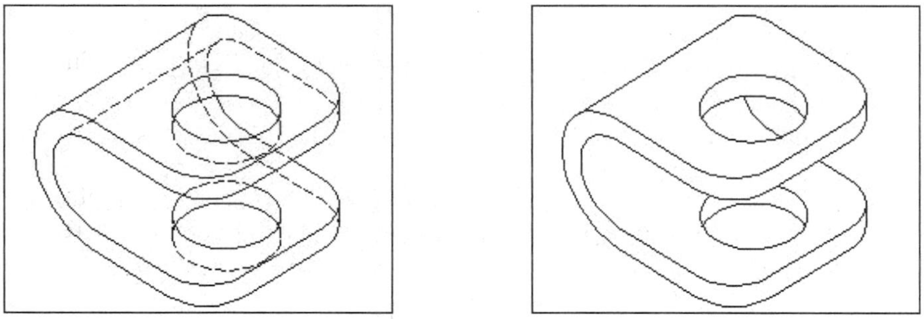

FIGURE 22.14

TUTORIAL EXERCISE: 22_SOLID DIMENSION.DWG

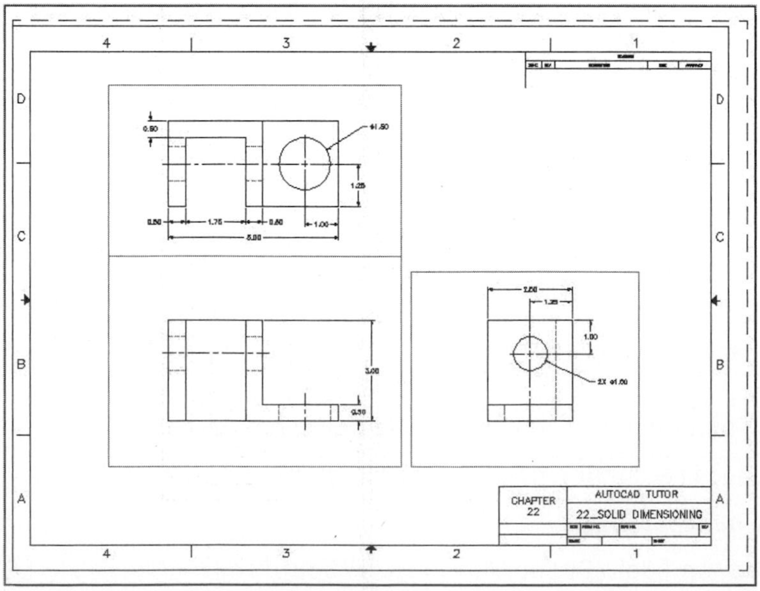

FIGURE 22.15

Purpose

This tutorial exercise is designed to add dimensions to a solid model that has had its views extracted using the SOLVIEW and SOLDRAW commands, as shown in Figure 22.15.

System Settings

Drawing and dimension settings have already been changed for this drawing.

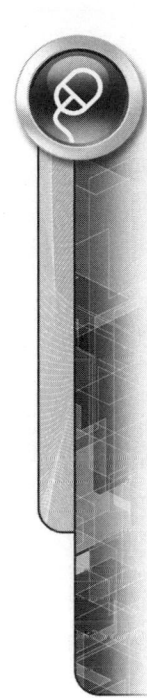

Layers

Layers have already been created for this tutorial exercise.

Suggested Commands

In this tutorial, you will activate the front viewport of the 22_Solid Dimension drawing and make the Front-DIM layer current. Then add dimensions to the Front view. As these dimensions are placed in the Front view, the dimensions do not appear in the other views. This is because the SOLVIEW command automatically creates layers and then freezes those layers in the appropriate viewports. Next, activate the top viewport, make the Top-DIM layer current, and add dimensions to the Top view. Finally, activate the right side viewport, make the Right Side-DIM layer current, and add the remaining dimensions to the Right Side views.

STEP 1

Begin this tutorial by opening the drawing 22_Solid Dimension. Your display should appear similar to Figure 22.16. Viewports have already been created and locked in this drawing. A locked viewport prevents you from changing the image size (scale) by accidentally zooming in a viewport, the image inside of the viewport does not zoom; rather, the entire drawing is affected by the zoom operation. Centerline layers have also been created and correspond to the three viewports. Centerlines have already been placed in their respective views.

Be sure OSNAP is turned on with Endpoint being the active mode.

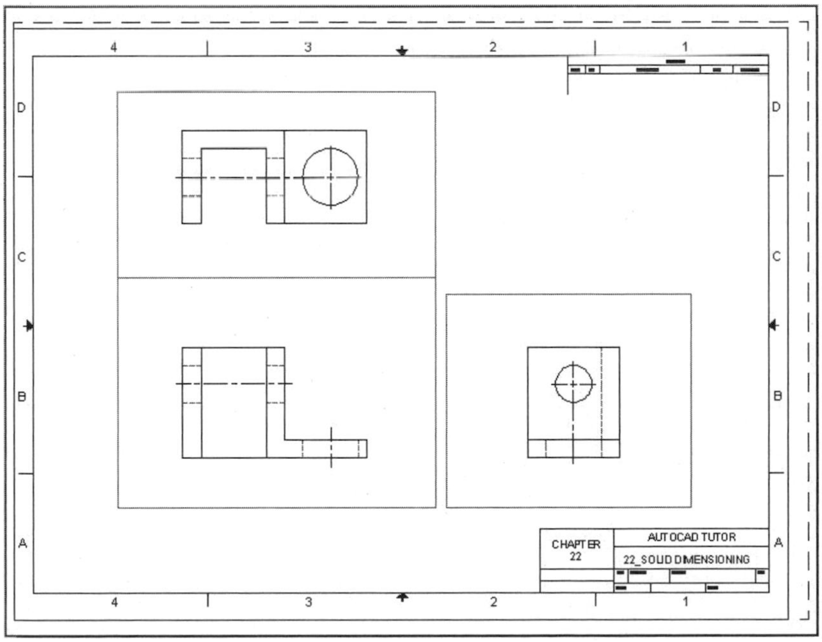

FIGURE 22.16

STEP 2

In the Layer Properties Manager, set up a Dimension property filter to help organize the layers created by the SOLVIEW command. Issue the LAYER command to display the palette, as shown in Figure 22.17 (Left). Notice that Center, Front, Hidden, Right Side, Top, and Visible filters have already been created for you. Pick the New Property Filter button to display the Layer Property Filters dialog box, as shown in Figure 22.17 (Right). Enter "Dimension" as the Filter name and enter "*DIM" in the Filter definition area under Name. Only those layers whose name ends with DIM will be displayed. Filters can assist you when assigning colors, linetypes, and lineweights to the numerous layers created during the Solview process. Click the OK button to create the new filter.

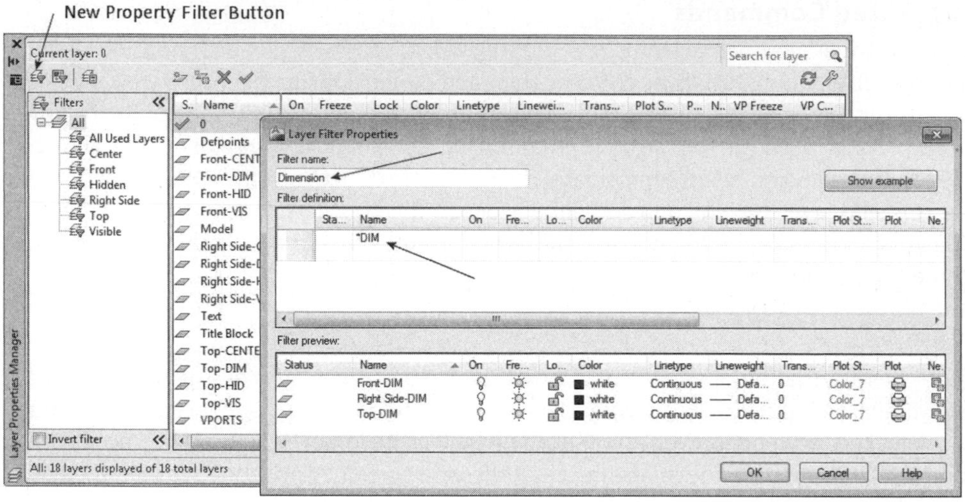

FIGURE 22.17

STEP 3

Activate the viewport that contains the Front view by double-clicking inside of it. You know you have accomplished this if the floating model space icon appears. Then make the Front-DIM layer current, as shown in Figure 22.18.

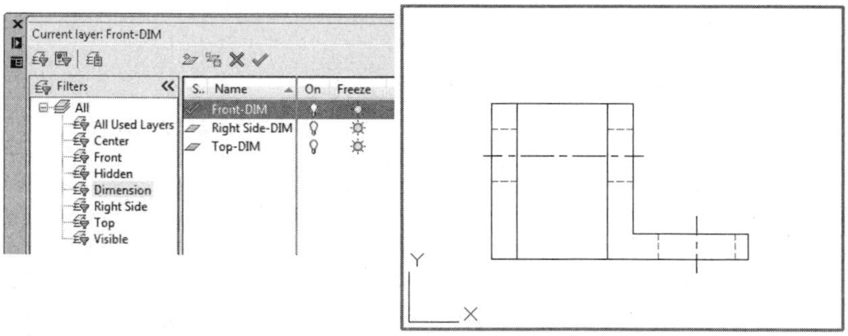

FIGURE 22.18

STEP 4

Activate the Front filter in the Layer Properties Manager. Most of the layers displayed in the Layer Control box were created by the SOLVIEW command. Notice how they correspond to a particular viewport. For example, study the following table regarding the layer names dealing with the front viewport:

Layer Name	Purpose
Front-CENTER	Centerlines
Front-DIM	Dimension lines
Front-HID	Hidden lines
Front-VIS	Visual (Object) lines

Activate the Dimension filter in the Layer Properties Manager and notice the Layer Front-DIM in Figure 22.19. Because the viewport holding the Front view is current, this layer is thawed, meaning the dimensions placed in this viewport will be visible. Notice that the other dimension layers (Right Side-DIM and Top-DIM) are frozen. The SOLVIEW command automatically sets up the dimension layers to be visible in the current viewport and frozen in the other viewports.

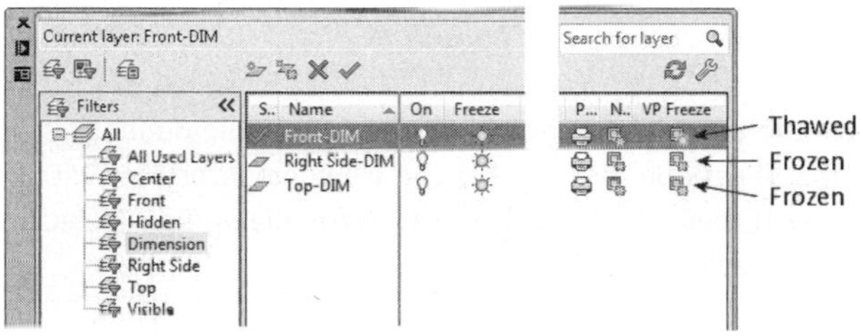

FIGURE 22.19

STEP 5

Add the two dimensions to the Front view, as shown in Figure 22.20. Linear and baseline dimension commands could be used to create these dimensions. Grips can be used to stretch the text of the 0.50 dimension so that it is located in the middle of the extension lines.

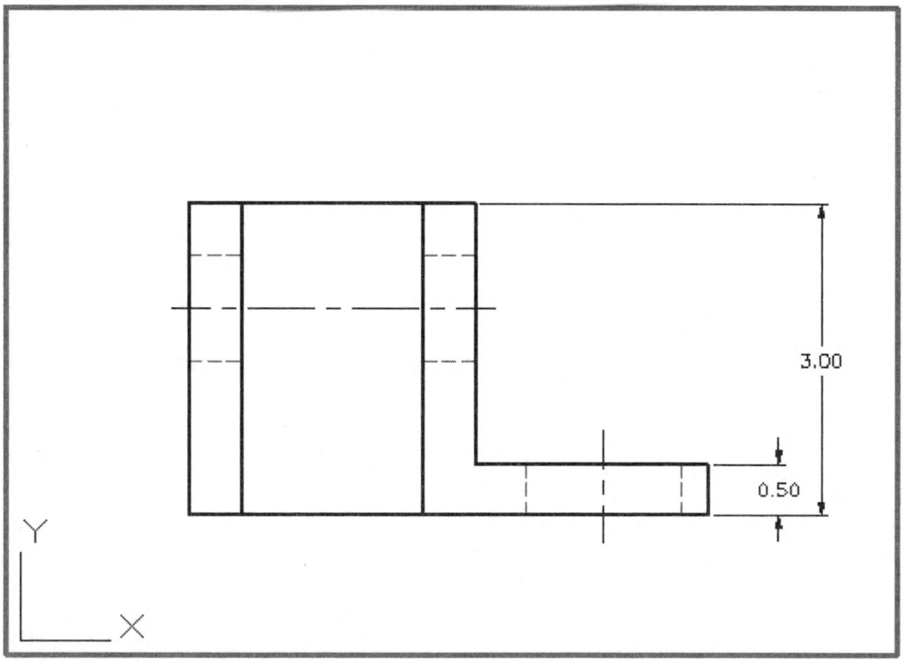

FIGURE 22.20

STEP 6

Activate the viewport that contains the Top view by clicking inside it. Then make the Top-DIM layer current, as shown in Figure 22.21. This layer is designed to show dimensions visible in the top viewport and make dimensions in the front and right side viewports frozen (or invisible).

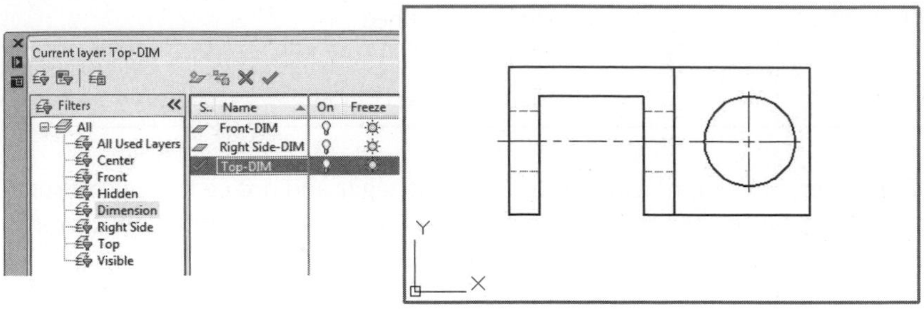

FIGURE 22.21

STEP 7

Add dimensions to the Top view using Figure 22.22 as a guide.

TIP

Because the viewports are locked, use the zoom and pan operations freely while dimensioning in floating model space. Also, if the ViewCube displayed in the viewport interferes with the placement of dimensions, it can be turned off through the Options dialog box (select the 3D Modeling tab > Display Tools in Viewport area).

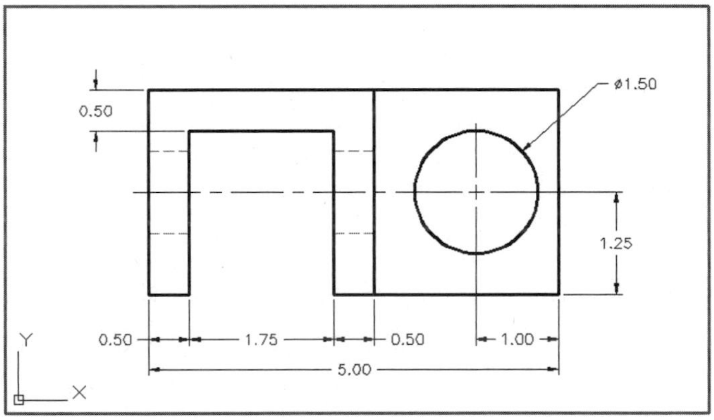

FIGURE 22.22

STEP 8

Activate the viewport that contains the Right Side view by clicking inside the viewport. Then make the Right Side-DIM layer current, as shown in Figure 22.23. This layer is designed to make dimensions visible in the right-side viewport and make dimensions in the front and top viewports frozen (or invisible).

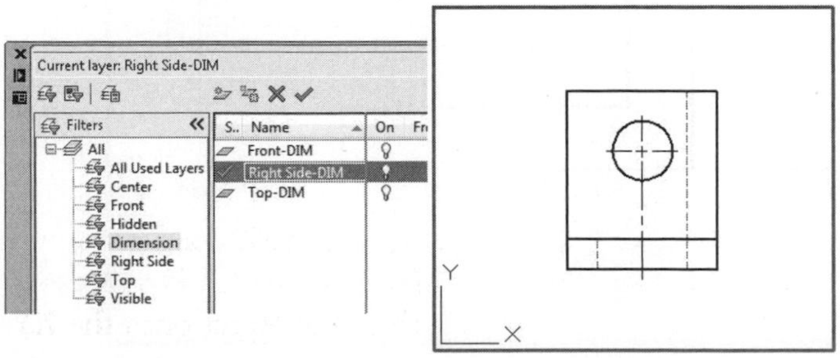

FIGURE 22.23

STEP 9

Add dimensions to the Right Side view using Figure 22.24 as a guide. Edit the diameter dimension text to reflect two holes. Use the DDEDIT command to accomplish this. When the Text Formatting dialog box appears, type "2X" and click the OK button.

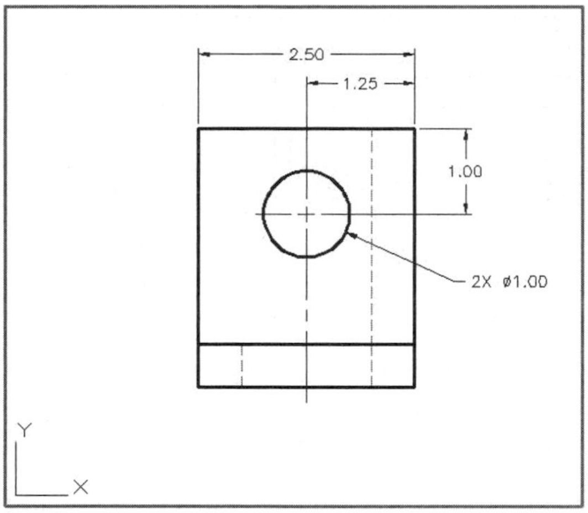

FIGURE 22.24

STEP 10

The completed dimensioned solid model is illustrated in Figure 22.25.

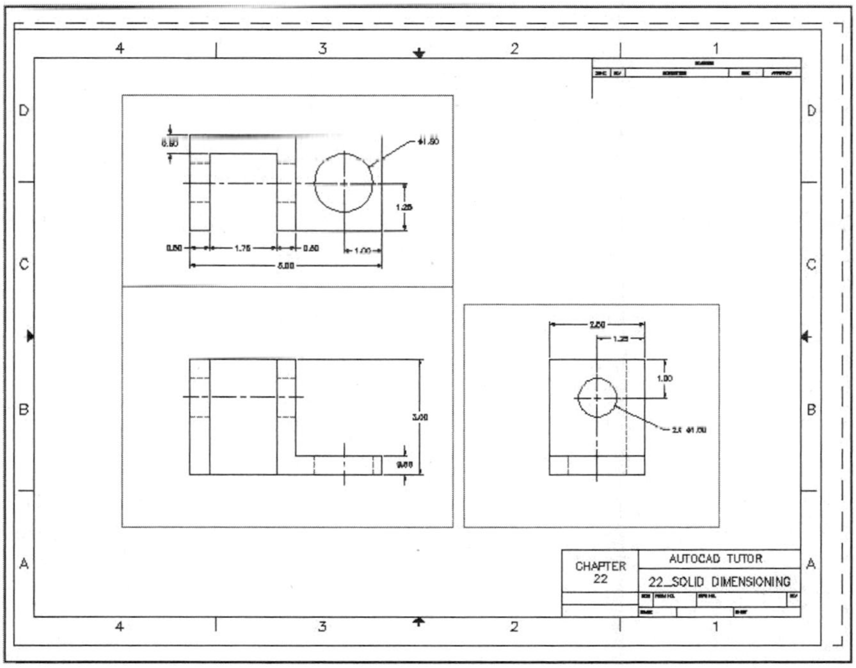

FIGURE 22.25

EXTRACTING 2D VIEWS WITH FLATSHOT

The FLATSHOT command is used to create a quick 2D view of a 3D solid or surface model based on the current view. First align your view of the 3D model. Flatshot creates a 2D object based on that view and then projects this new object onto the XY plane. The object created is in the form of a block and can be inserted and modified, if necessary, since the block consists of 2D geometry. Choose Flatshot from the Section panel of the Ribbon, as shown in Figure 22.26.

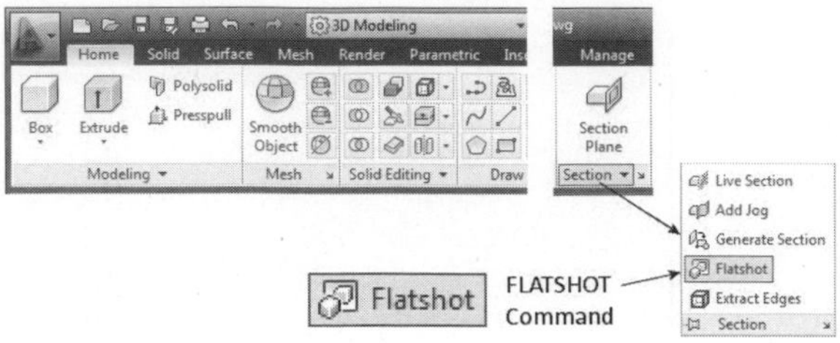

FIGURE 22.26

Begin the process of creating a flattened 2D view from a 3D model by first aligning the screen for the view that you want captured. Illustrated in Figure 22.27 (Left) is a 3D model that is currently being viewed in the Front direction. Notice also the alignment of the XY plane; Flatshot will project the geometry to this plane.

When you activate the `FLATSHOT` command, the dialog box illustrated in Figure 22.27 (Right) displays. The Destination area is used for inserting a new block or replacing an existing block. You can even export the geometry to a file with the familiar DWG extension that can be read directly by AutoCAD.

The Foreground lines area allows you to change the color and linetype of the lines considered visible.

In the Obscured lines area, you have the option of either showing or not showing these lines. Obscured lines are considered invisible to the view and should be assigned the HIDDEN linetype if showing this geometry.

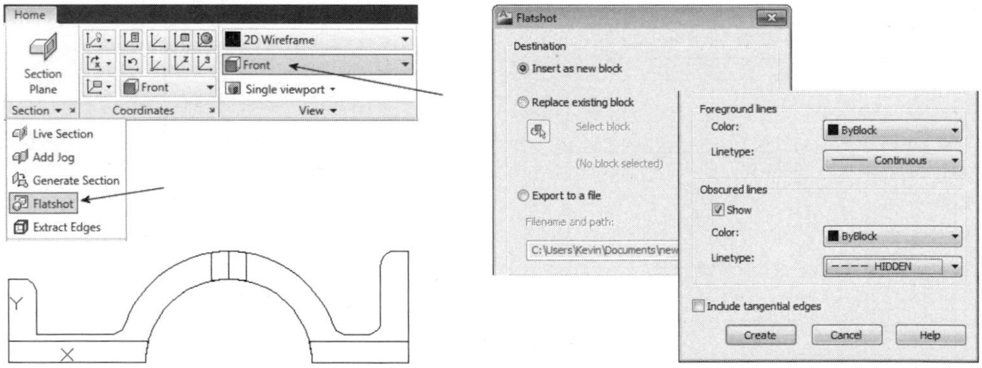

FIGURE 22.27

After clicking the Create button, you will be prompted to insert the block based on the view. In Figure 22.28, the object on the left is the original 3D model, and the object on the right is the 2D block generated by the Flatshot operation. An isometric view is used in Figure 22.28 to illustrate the results performed by Flatshot.

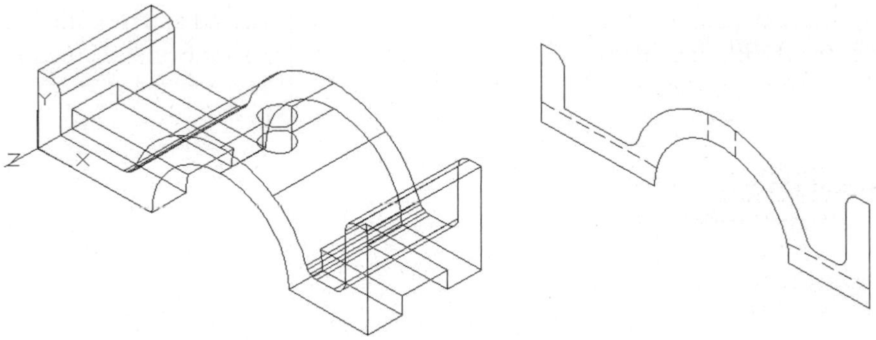

FIGURE 22.28

It was pointed out that Flatshot creates a block. However, during the creation process, you are never asked to input the name of the block. This is because Flatshot creates a block with a randomly generated name (sometimes referred to as anonymous). This name is illustrated in Figure 22.29, where the Rename dialog box (RENAME command) is used to change the name of the block to something more meaningful, such as Front View. In fact, it is considered good practice to rename the block generated by Flatshot to something more recognizable.

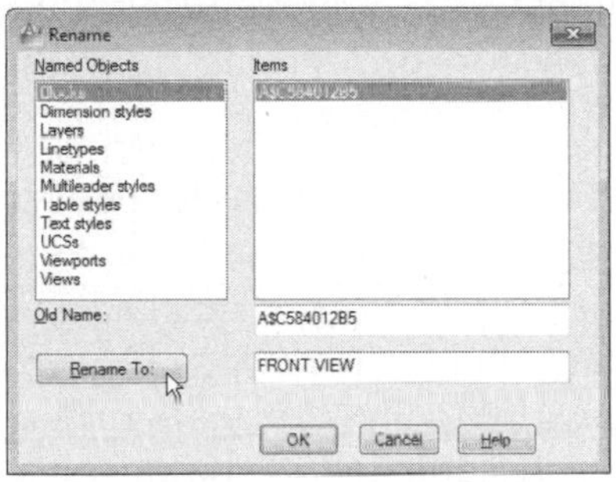

FIGURE 22.29

When performing a Flatshot operation, it is important to remember that the view created is based on the view direction and the placement of the 2D block is based on the current UCS.

TRY IT!

Open the drawing file 22_Flatshot. In this exercise, perform four flatshot operations to create the four views shown in the following image on the right. Notice that the views are all placed on the XY plane of the World Coordinate System. To create the first view, select the Front view from the View panel of the Ribbon and reset the UCS in the Coordinates panel to World, as shown in Figure 22.30. Issue the FLATSHOT command and when the dialog box appears, keep the default settings with the exception of changing the Linetype to Hidden in the Obscured lines area, as shown in Figure 22.30 (Left). Click the Create button and place the view at any convenient spot. The view will not be recognizable from your current viewpoint. Repeat this process changing the view direction to Top, Right, and SE Isometric. Be sure to reset the UCS to World each time. Once all the views are created, change to a top view to view the blocks you created. Align the views properly using the image on the right as a guide. Notice the block generated for the isometric view does not show the obscured lines. This was accomplished by removing the check from the Show box for Obscured lines in the Flatshot dialog box.

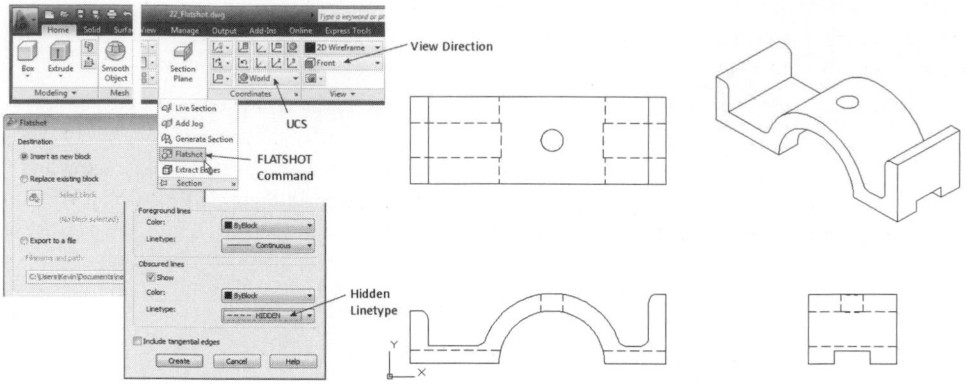

FIGURE 22.30

THE MODEL DOCUMENTATION TOOLS

Utilizing the new model documentation tools to create 2D drawing views from 3D objects provides many advantages over the previous methods discussed. The process is simple and powerful. Although these new tools create a series of layers to contain created entities, unlike the SOLVIEW/SOLDRAW process, there is no need to freeze or thaw layers in a specific viewport to hide entities or annotate the drawings. Annotations are handled in Paper Space, which greatly simplifies the process. Also, unlike the SOLVIEW/SOLDRAW process, the model is linked to the drawing such that changes to the model are automatically updated in the drawing views.

The model documentation tools are found in a new Layout tab on the Ribbon, as shown in Figure 22.31. Located in the Create View panel are the Base View, Projected View, Section View, and Detail View tools that are used to generate 2D drawing views. There are also tools available to manage and modify the views and layouts generated. The appearance of a view can be changed with Modify View tools, such as changing its scale or style (Visible lines, Visible and hidden lines, Shaded with visible lines, and Shaded with visible and hidden lines). Two new style managers are also available to control identifiers and labels for section and detail views.

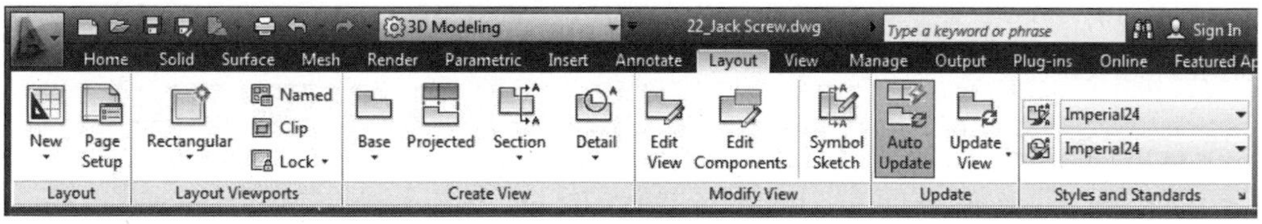

FIGURE 22.31

THE BASE VIEW TOOL

Once a 3D model is created, the model documentation process can begin. The Base View tool, which is found in the Create View panel of the Layout Ribbon, is used to create the first view in a layout. Once an initial view is created, then you can create projected, section, and detail views. When the VIEWBASE command is activated from Model Space, you are prompted to select the object or objects that will generate the base view. Next you are prompted for a layout name. If the layout name exists, that layout will be activated. If the layout name supplied does not exist, a layout with that name will be created for you and activated. Once in the layout, the view is located by picking a point on the screen. If this command is started from Paper Space, it is not necessary to select objects (although a Select option is available) and the layout is already selected.

TRY IT!

Open the drawing file 22_View Base1. Notice that you are in Model Space and a solid model is already created. Follow the Command prompt sequence below to create a base view and three projected views. Locate the views at the points indicated in Figure 22.32 (Left). When finished, your drawing should appear similar to Figure 22.32 (Right).

```
Command: VIEWBASE
Specify model source [Model space/File] <Model space>:
(Press ENTER to except default)
Select objects or [Entire model] <Entire model>: (Select the
model)
Select objects or [Entire model] <Entire model>: (Press ENTER
to continue)
Enter new or existing layout name to make current or [?]
<Layout1>: FOUR VIEWS (Creates a new layout)
Regenerating Layout.
Type = Base and Projected Hidden Lines = Visible and hidden
lines Scale – 1:4
Specify location of base view or [Type/sElect/Orientation/
Hidden Lines/Scale/Visibility] <Type>: (Pick at "A" to place
the front view)
Select option [sElect/Orientation/Hidden
Lines/Scale/Visibility/Move/eXit] <eXit>: (Press ENTER)
Specify location of projected view or <eXit>: (Pick at "B" to
place the top view)
Specify location of projected view or [Undo/eXit] <eXit>:
(Pick at "C" to place the right view)
Specify location of projected view or [Undo/eXit] <eXit>:
(Pick at "D" to place the isometric view)
Specify location of projected view or [Undo/eXit] <eXit>:
(Press ENTER to complete the operation)
Base and 3 projected view(s) created successfully.
```

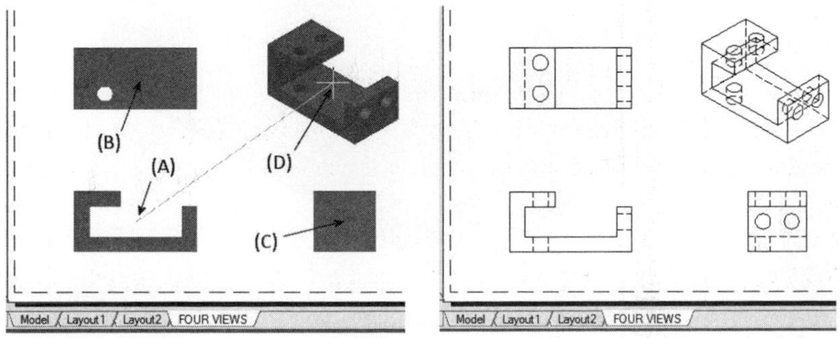

FIGURE 22.32

During the model documentation process, layers are created with a MD_ prefix, as shown in Figure 22.33. The hidden and visible lines that make up the drawing views are placed on the appropriate layers. You can use the Layer Property Manager to change properties such as Linetype, Lineweight, and Color for the drawing objects.

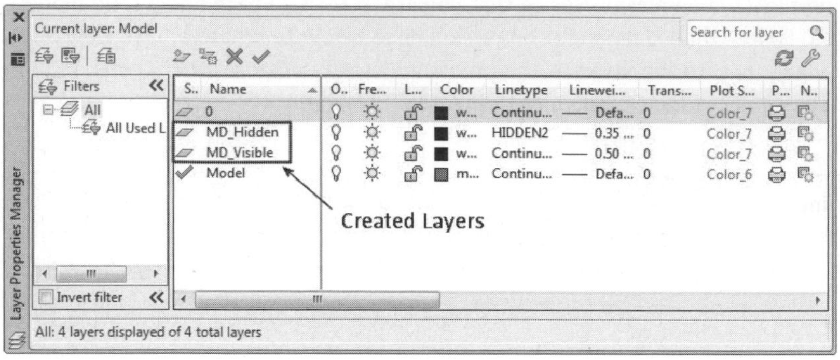

FIGURE 22.33

In the previous Try It! exercise, the base view was placed as a front view, the view style showed visible and hidden lines, and the scale was 1:4. Options are available that will allow you to change these settings during the view creation process.

These options can be accessed through Command prompts, shortcut menus, or through the Ribbon.

TRY IT!

Open the drawing file 22_View Base2 or continue utilizing the previous Try It! Exercise drawing. Select the Model tab to return to Model Space. Select the Base (>From Model Space) button in the Create View panel of the Ribbon, as shown in Figure 22.34 (Left). When prompted to select objects, select the model and press ENTER**. When you are prompted to select a layout, enter "NW ISOMETRIC" and press** ENTER**. Once the layout appears, a front view of the model is displayed at a scale of 1:4. Before placing the view make the following option changes in the Ribbon, as shown in Figure 22.34 (Right): in the Orientation panel, select NW Isometric, and in the Appearance panel, select "Shaded with visible lines and hidden lines," and change the Scale to 1:1. Pick to place the view as shown in Figure 22.34. Select the** OK **button from the Create panel and press** ENTER **to complete the operation.**

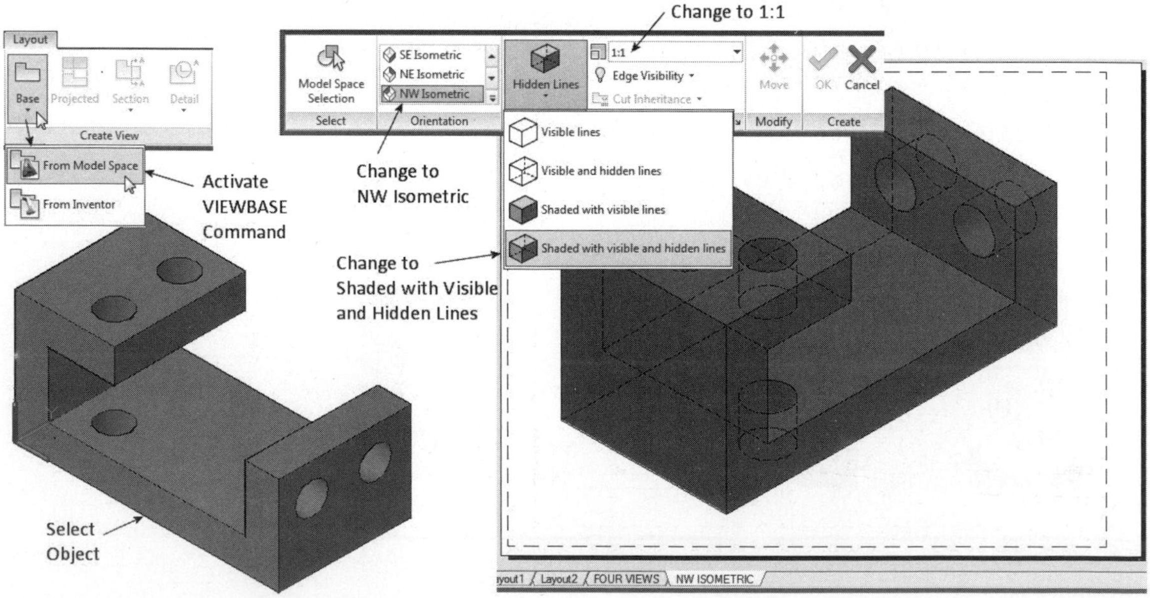

FIGURE 22.34

THE PROJECTED VIEW TOOL

The Projected View tool, which is also found in the Create View panel of the Layout Ribbon, is used to create a new view from an existing view, such as a base view. The existing or parent view can be any type except a detail view. Depending on the direction that the cursor is moved during the commands execution, you can either create orthographic (four possible) or isometric (four possible) views. It is important to note that parent and orthographic projected views will remain aligned during any move operations.

TRY IT!

Open the drawing file 22_View Projected1. Notice that you are in a layout and a base view is already created. Follow the Command prompt sequence below to create four orthographic and four isometric projected views. Locate the views at the points indicated in Figure 22.35 (Left). When finished, your drawing should appear similar to Figure 22.35 (Right).

Command: VIEWPROJ

Select parent view: *(Select the base view)*

Specify location of projected view or <eXit>: *(Pick at "A" to place the right view)*

Specify location of projected view or [Undo/eXit] <eXit>: *(Pick at "B" to place the bottom view)*

Specify location of projected view or [Undo/eXit] <eXit>: *(Pick at "C" to place the left view)*

Specify location of projected view or [Undo/eXit] <eXit>: *(Pick at "D" to place the top view)*

Specify location of projected view or [Undo/eXit] <eXit>: *(Pick at "E" to place the first isometric view)*

Specify location of projected view or [Undo/eXit] <eXit>: *(Pick at "F" to place the second isometric view)*

Specify location of projected view or [Undo/eXit] <eXit>: *(Pick at "G" to place the third isometric view)*

Specify location of projected view or [Undo/eXit] <eXit>: *(Pick at "H" to place the fourth isometric view)*

Specify location of projected view or [Undo/eXit] <eXit>: *(Press* ENTER *to complete the operation)*

8 projected view(s) created successfully.

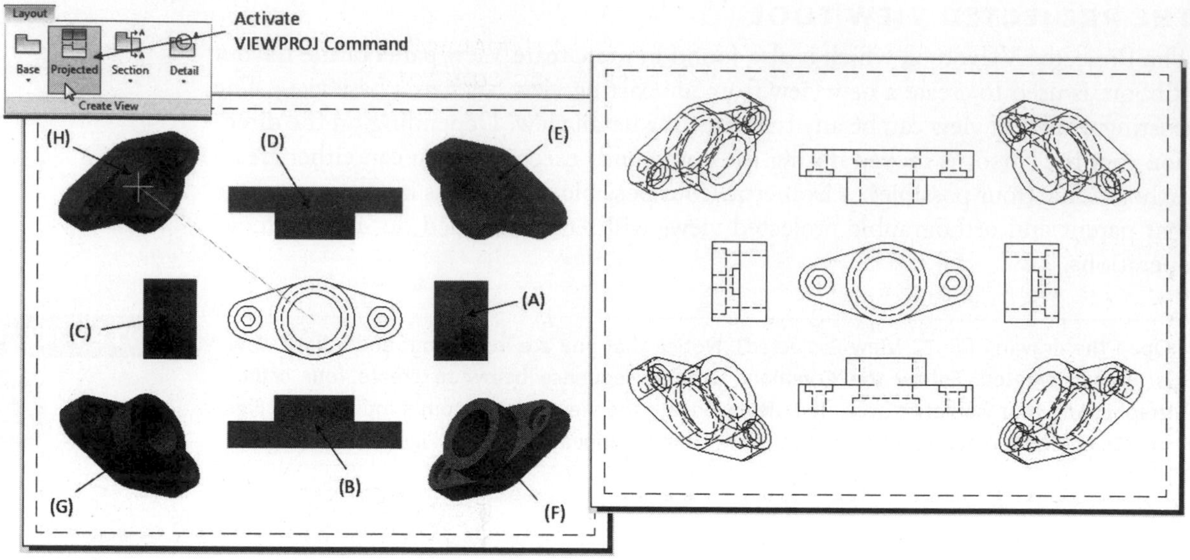

FIGURE 22.35

THE SECTION VIEW TOOL

The Section View tool can be used to create Full, Half, Offset and Aligned sections. Before a section view can be created, however, an existing drawing view must already exist. Sections are created based on a cutting plane line that is laid out on the existing (parent) view. The Polar Tracking and Object Snap Tracking modes are typically used to create accurate cutting plane lines. Identifiers for the cutting plane lines are provided as is a section view label. The section view creation process also automatically creates a MD_Hatching layer for the section lines and a MD_Annotation layer for the cutting plane line, view label and identifiers.

TRY IT!

Open the drawing file 22_View Section1. Notice that the Full & Half layout is active and two base views have already been created for you. A full section will be generated for the model on the left and a half section for the model on the right. Follow the Command prompt sequence below and Figure 22.36 to assist with the creation of the full section view.

Command: VIEWSECTION

Select parent view: *(Select the left base view)*

Hidden Lines = Visible Lines Scale = 1:1 (From parent)

Specify start point or [Type/Hidden Lines/Scale/Visibility/Annotation/hatCh] <Type>: *(Press* ENTER *to except default)*

Select type [Full/Half/OFfset/Aligned/OBject/eXit] <eXit>: F *(For Full)*

Specify start point: *(Move your cursor over the Center Osnap and track to the left. Pick the point at "A")*

Specify end point or [Undo]: *(Track to the right and pick the point at "B")*

Specify location of section view or: *(Pick at "C" to place the full front section view)*

```
Select option [Hidden
lines/Scale/Visibility/Projection/Depth/Annotation/
hatCh/Move/eXit] <eXit>: (Press ENTER to complete the
operation)
Section view created successfully.
```

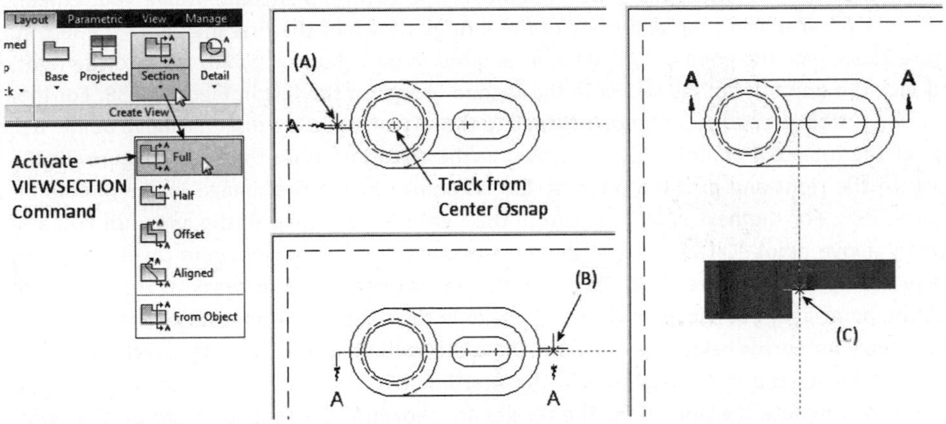

FIGURE 22.36

Begin the process of creating the half section for the model on the right. After selecting the Section (> Half) button in the Create View panel, you will be prompted to select the parent view. Select the view in the upper right, as shown in Figure 22.37 (Left). Next you are prompted to select the starting point for the cutting plane line. Move your cursor over the Center object snap and track to the left. Pick the point at "A," as shown in the Figure 22.37 (Left). When prompted to pick the next point, pick the center of the object at "B." When prompted to pick the end point, pick the point at "C." Next move your cursor below the top view and once the section view is displayed, pick the point at "D" to locate the view, as shown in Figure 22.37 (Center). Press `ENTER` to complete the operation. The results are shown in Figure 22.37 (Right).

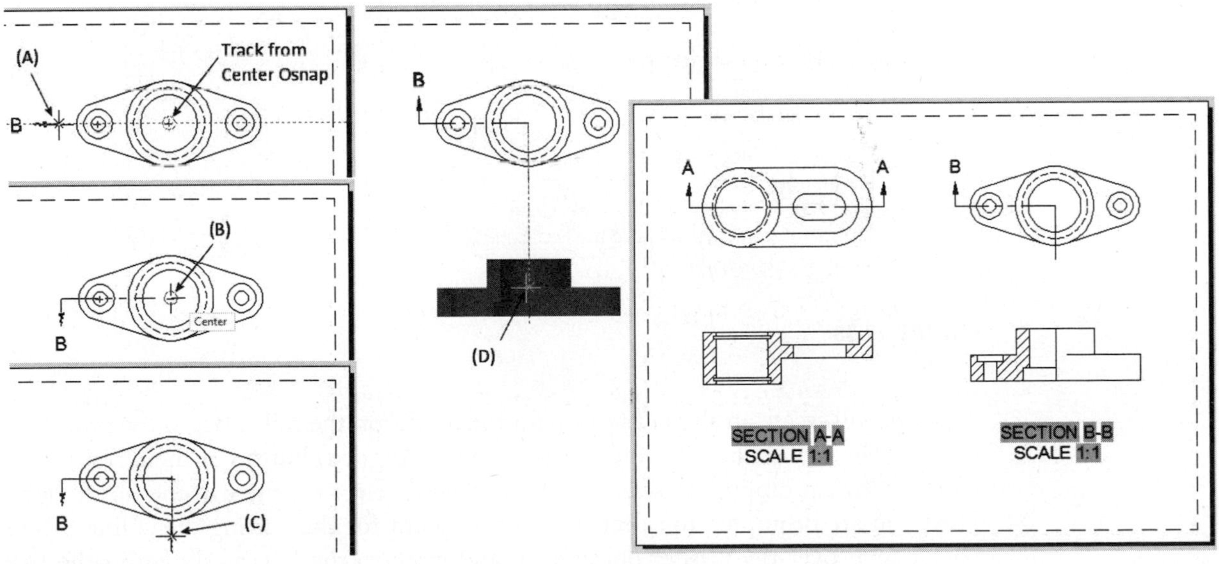

FIGURE 22.37

TRY IT!

Open the drawing file 22_View Section2 or continue using the previous Try It! Exercise drawing. If necessary, select the Offset & Aligned layout to make it active. Notice that two new base views have been created for you. An offset section will be generated for the model on the left and an aligned section for the model on the right. Begin the process of creating the offset section by selecting the Section (> Offset) button in the Create View panel. When prompted to select the parent view, select the view in the upper left. Next you are prompted to select the starting point for the cutting plane line. Move your cursor over the Center object snap and track to the left, as shown in the first image on the left in Figure 22.38. Pick the point at "A." When prompted to pick the next point, track to the right and pick the point at "B," as shown in the second image on the left in Figure 22.38. For the next point, track from the Center object snap to the left until you are directly above point "B". Pick at the intersection point "C," as shown in the third image on the left in Figure 22.38. Track to the right and pick the point at "D," as shown in the first image in the middle in Figure 22.38. For the next point, track from the Center object snap to the left until you are directly above point "D" and pick at the intersection point "E," as shown in the second image in the middle in Figure 22.38. To complete the cutting plane line, track to the right and pick the point at "F" as shown in the third image in the middle in Figure 22.38. Press ENTER and move your cursor below the top view. Once the section view is displayed, pick the point at "G" to locate the view, as shown in the first image on the right in Figure 22.38. Press ENTER to complete the operation. The results are shown in the second image on the right in Figure 22.38.

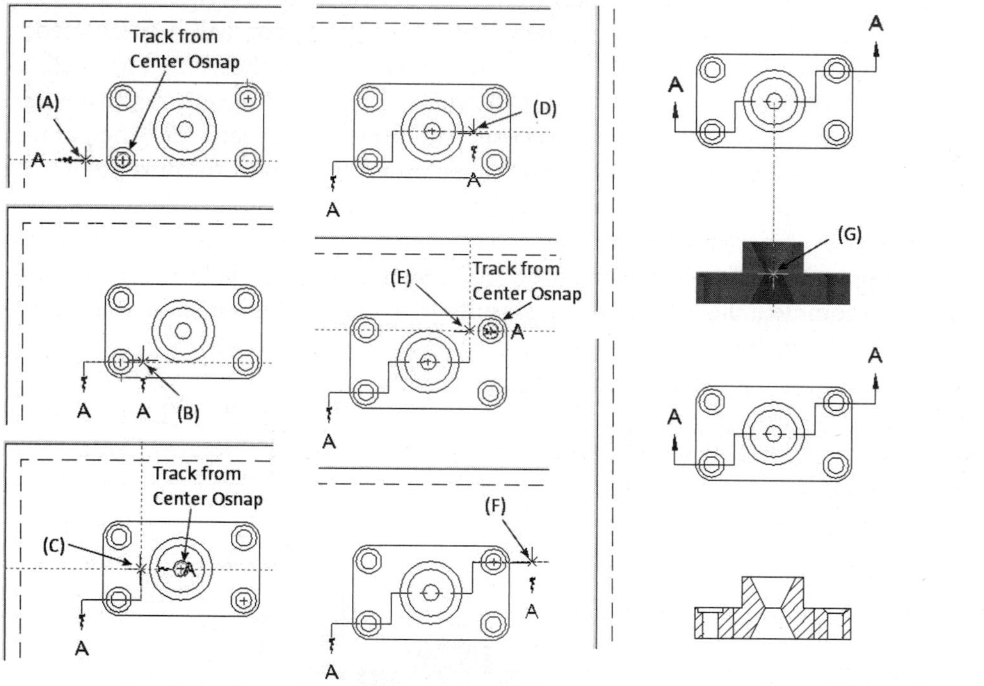

FIGURE 22.38

Next we will create an aligned section for the model on the right. Begin the process of creating the section by selecting the Section (> Aligned) button in the Create View panel. When prompted to select the parent view, select the view in the upper right. Next you are prompted to select the starting point for the cutting plane line. Move your cursor over the Center object snap and track to the left, as shown in the first image on the left in Figure 22.39. Pick the point at "A." When prompted to pick the next point, track to the right and pick the center point at "B," as shown in the second image on the left in Figure 22.39. For the next point, track from the Center object snap down and to the right, as shown in the third image on the left in Figure 22.39. Pick the point "C." Press ENTER and move your cursor below the top view. Once the section view is displayed, pick the point at "D" to locate the view, as shown in the first image on the right in Figure 22.39. Press ENTER to complete the operation. The results are shown in the second image on the right in Figure 22.39.

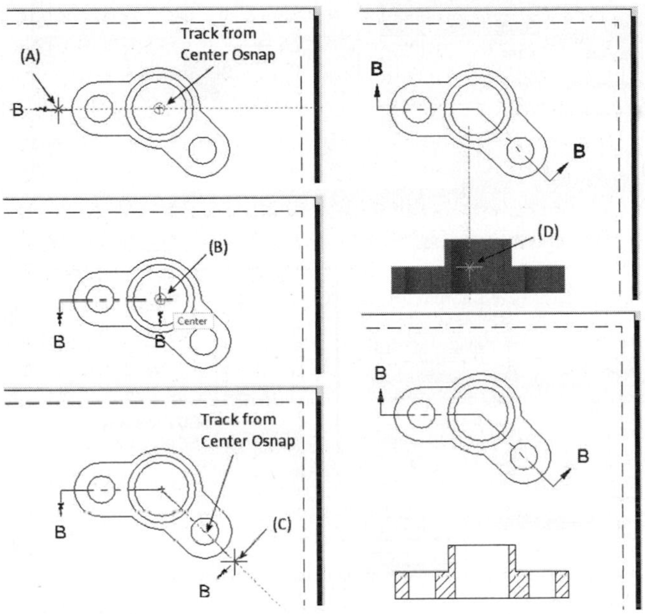

FIGURE 22.39

SECTION VIEW STYLE MANAGER

The Section View Style Manager allows you to modify existing or create new section styles. Various settings organized on four tabs provide controls for the section identifiers, arrows, cutting plane, view label, and hatching. The VIEWSECTIONSTYLE command can be activated by selecting the Section View Style button found in the Styles and Standards panel of the Ribbon, as shown in Figure 22.40 (Left). Activating the command displays the Section View Style Manager, as shown in Figure 22.40 (Right). Pick the New button to create a new style or the Modify button to modify an existing style.

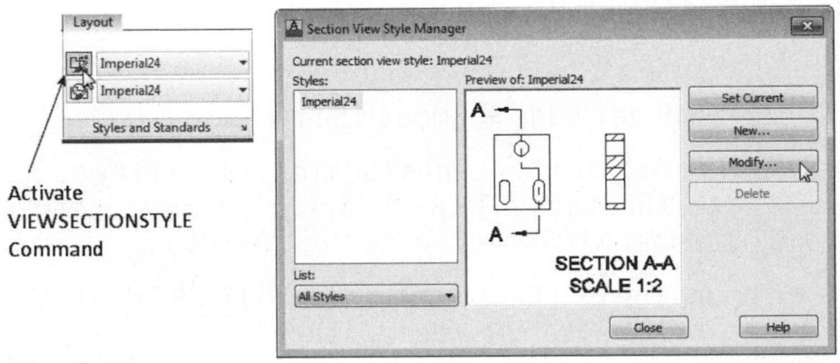

FIGURE 22.40

Figure 22.41 shows the four tabs (Identifier and Arrows, Cutting Plane, View Label, and Hatch) available for modifying section view settings.

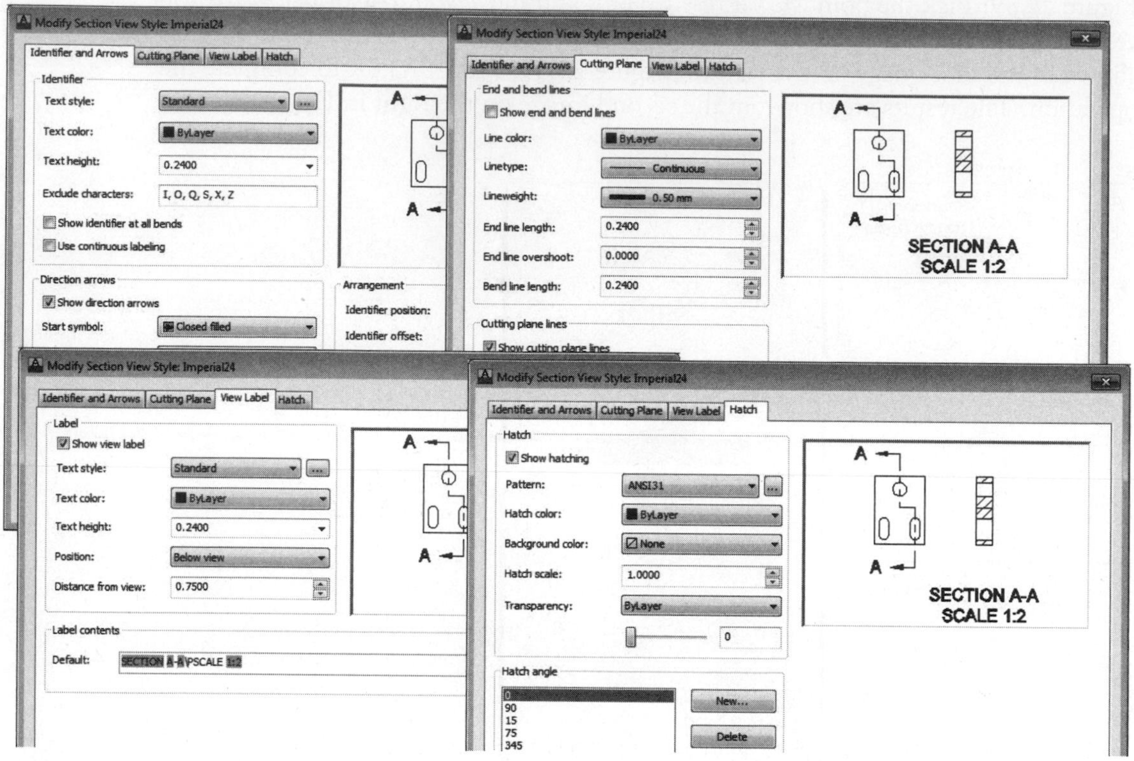

FIGURE 22.41

THE DETAIL VIEW TOOL

The Detail View tool can be used to create Circular or Rectangular boundary details. Like section views, a detail view cannot be created without an existing drawing view. Identifiers are automatically provided as the detail boundary is created. A detail view label is also automatically provided.

TRY IT!

Open the drawing file 22_View Detail1. Notice that the Detail1 layout is active and a base and section view have already been created for you at a scale of 1:2. A detail view with a circular boundary will be created and will be automatically scaled up to 1:1. Follow the Command prompt sequence below and Figure 22.42 to assist with the creation of the detail view.

Command: VIEWDETAIL

Select parent view: *(Select the section view)*

Boundary = Circular Model edge = Smooth Scale = 1:1

Specify center point or [Hidden Lines/Scale/Visibility/Boundary/model Edge/Annotation] <Boundary>: *(Pick at "A" to locate center of boundary)*

Specify size of boundary or [Rectangular/Undo]: *(Pick at "B" to size the circle and locate the identifier)*

Specify location of detail view: *(Pick at "C" to locate the detail)*

Select option [Hidden Lines/Scale/Visibility/Boundary/model Edge/Annotation/Move/eXit] <eXit>: *(Press ENTER to complete operation)*

Detail view created successfully.

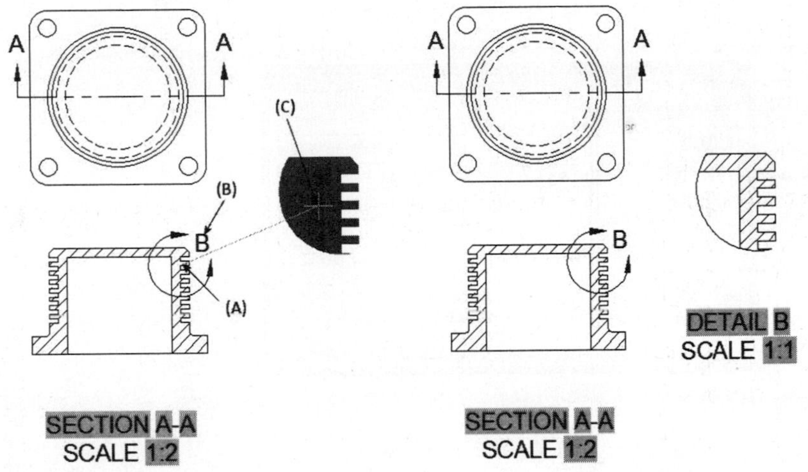

FIGURE 22.42

Undo or erase the detail view that you just created or reopen the 22_View Detail1 drawing. In this part of the exercise you will create a new rectangular boundary detail. Once the old detail is removed, pick the Detail (> Rectangular) button in the Create View panel, as shown in Figure 22.43 (Left). When you are prompted to select the parent view, select the section view. Next you are prompted to select the center of the detail boundary. Pick the point at "A," as shown in Figure 22.43 (Center). Move the cursor up and to the right and pick at "B" to size the boundary and locate the identifier. Finally, move your cursor and pick the point at "C" to locate the detail view. Press ENTER to complete the operation. The results are shown in Figure 22.43 (Right).

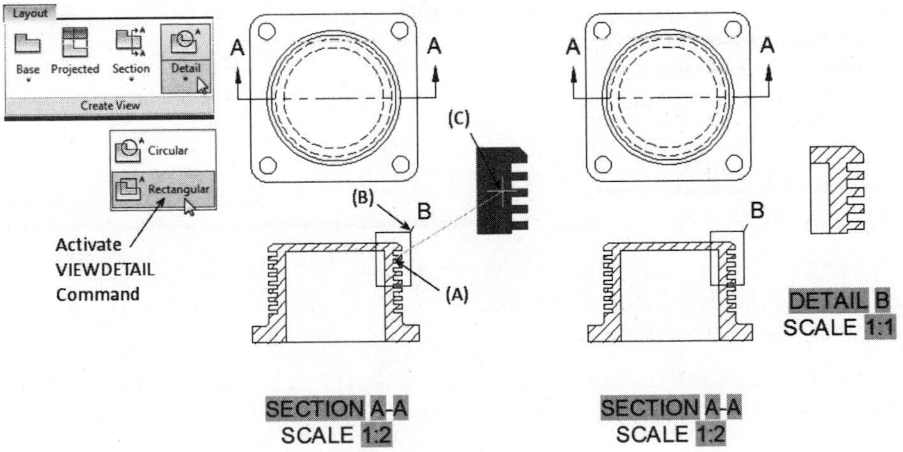

FIGURE 22.43

DETAIL VIEW STYLE MANAGER

The Detail View Style Manager allows you to modify existing or create new detail styles. Various settings organized on three tabs provide controls for the detail identifiers, boundaries, and view labels. The `VIEWDETAILSTYLE` command can be activated by selecting the Detail View Style button found in the Styles and Standards panel of the Ribbon, as shown in Figure 22.44 (Left). Activating the command displays the Detail View Style Manager, as shown in Figure 22.44 (Right). Pick the New button to create a new style or the Modify button to modify an existing style.

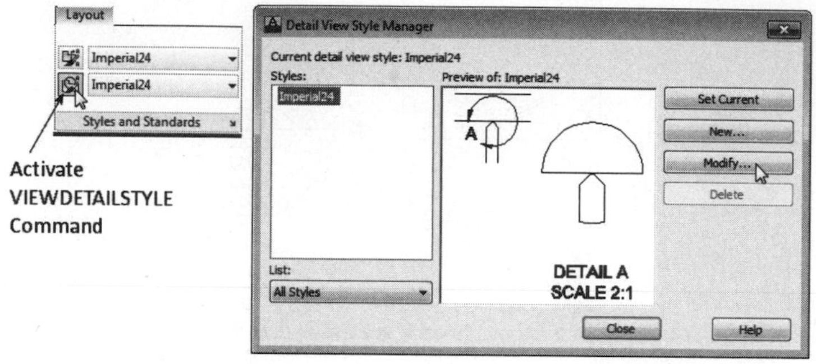

FIGURE 22.44

Figure 22.45 shows the three tabs (Identifier, Detail Boundary, and View Label) available for modifying detail view settings.

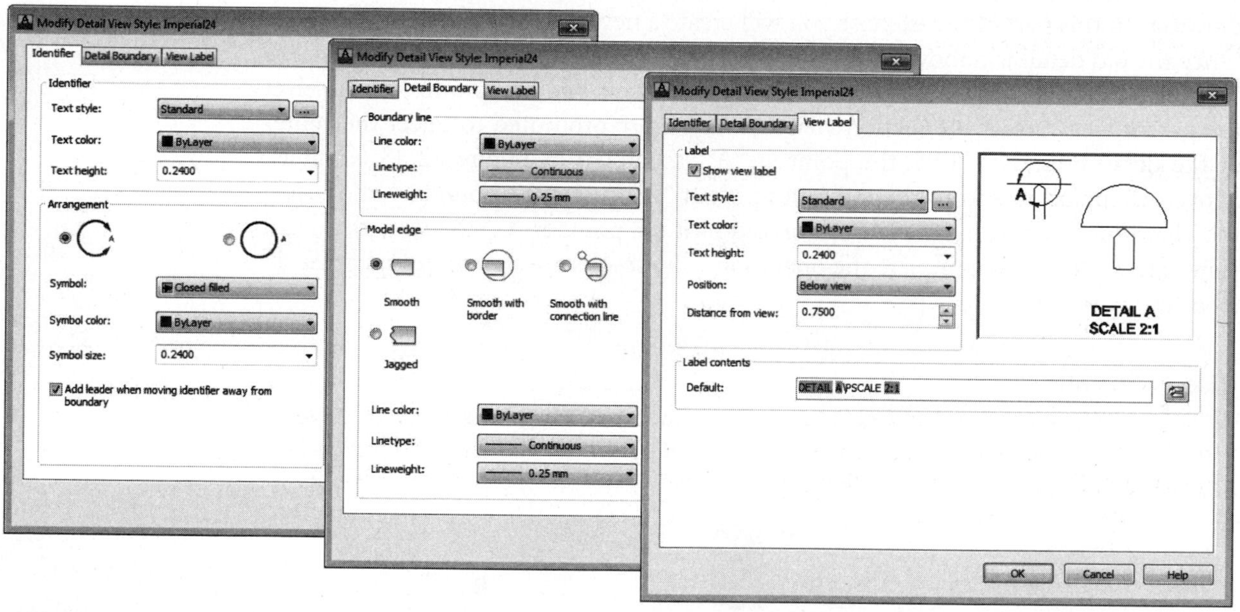

FIGURE 22.45

TUTORIAL EXERCISE: 22_JACK SCREW.DWG

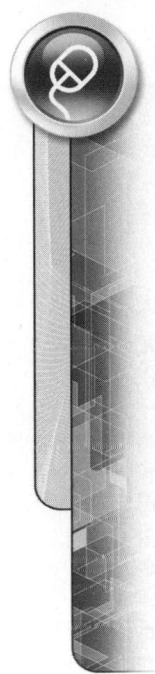

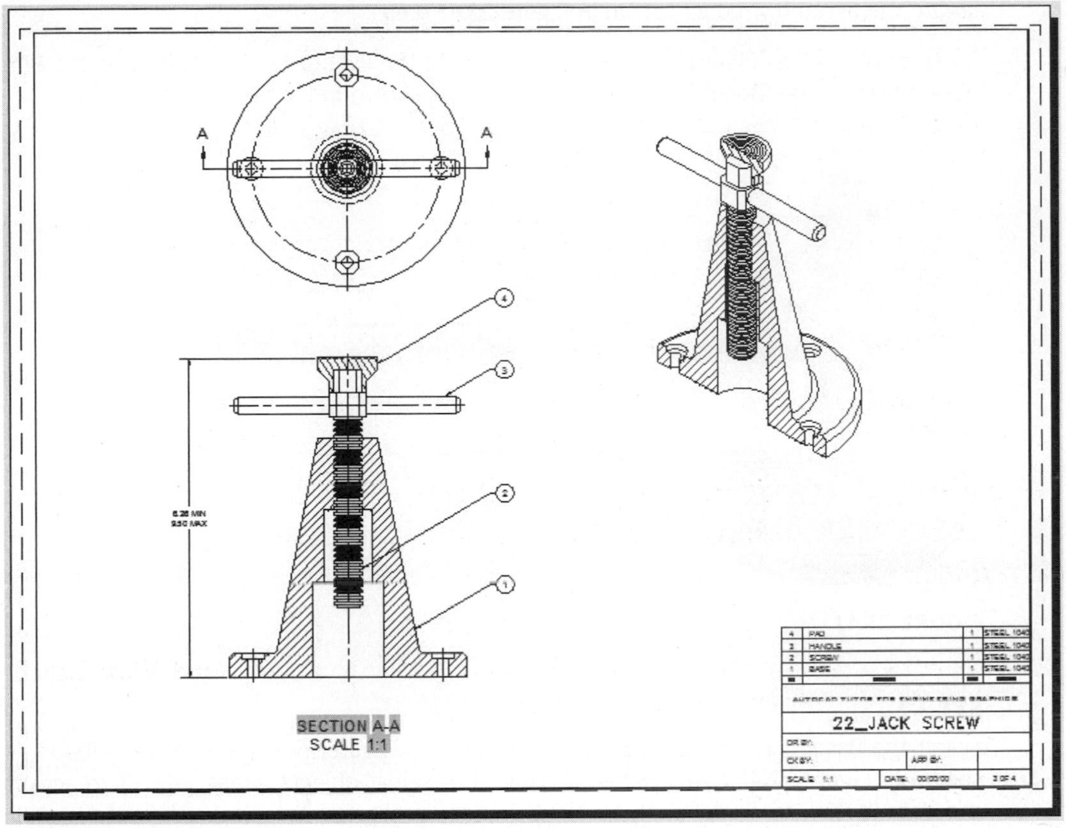

FIGURE 22.46

Purpose

This tutorial exercise is designed to create a set of working drawings for a Jack Screw. Layouts and views will be generated utilizing the new Model Documentation tools found in the Layout Ribbon. Figure 22.46 illustrates an assembly drawing of the Jack Screw that will be created in this tutorial.

System Settings

Drawing and dimension settings have already been set for you in the drawing 22_Jack Screw. A series of layout tabs have been created and consist of "C" size sheets with appropriate title blocks and borders. Some blocks have also been created for you to help with annotation of the layouts.

Layers

Layers have already been created for this tutorial exercise.

Suggested Commands

In this tutorial you will utilize the Model Documentations tools available on the Layout tab shown in Figure 22.47 to generate a set of working drawings for a Jack Screw. The Base View, Section View, and Projected View tools will be demonstrated and modifications will be made to the created views utilizing available editing tools. Associative annotations will also be placed in the drawings to complete the drawing layout process.

STEP 1

Begin this tutorial by opening the drawing 22_Jack Screw. Four solid models that make up the Jack Screw assembly have been created for you. The models were placed together to create the assembly model shown in Figure 22.47 (Left).

Be sure the 3D Modeling workspace is current and the Layout tab is selected on the Ribbon to display the Model Documentation tools, as shown in Figure 22.47 (Right).

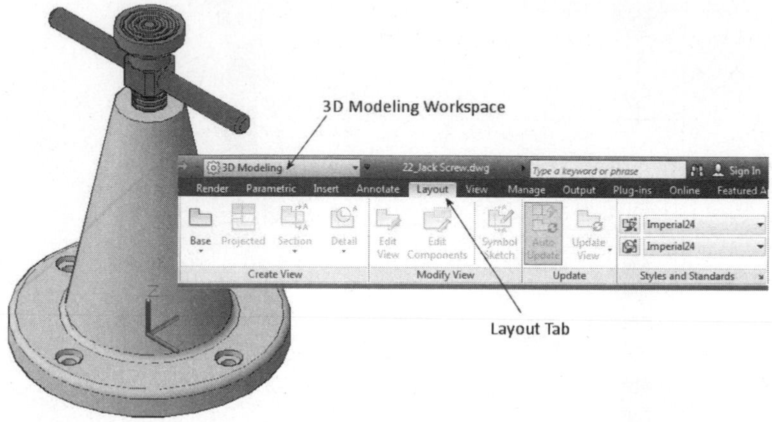

FIGURE 22.47

STEP 2

Select the Base (>From Model Space) button in the Create View panel of the Ribbon as shown in Figure 22.48 (Left). When prompted to select objects, press ENTER to accept the default "Entire model" option. Next you are prompted to select a layout for your new view. Press ENTER to accept "Assembly" as the layout in which you will place your first drawing view. Once the layout appears, a front view of the base is displayed, as shown in Figure 22.48 (Center). We will not need this view, so we will change it by using a shortcut menu. Right-click and select "Orientation" from the menu. Right-click again and select "Top" from the available orientation options. Pick in the upper left area of the drawing, as shown in Figure 22.48 (Right), to place the view. Press ENTER twice to complete the view creation process.

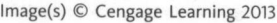

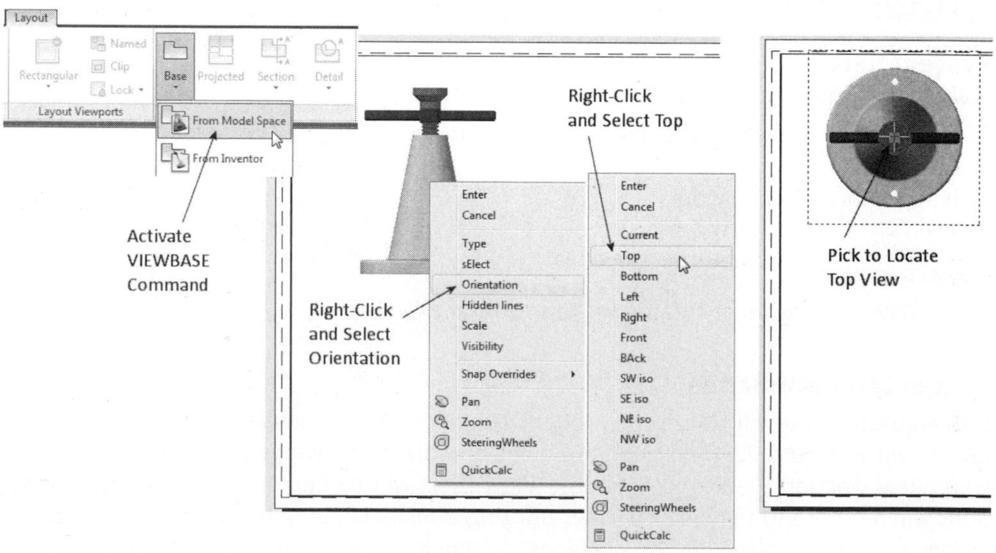

FIGURE 22.48

STEP 3

In this step we will create a full front section from the top view. Begin by selecting the Section (> Full) button in the Create View panel, as shown in Figure 22.49 (Left). When prompted to select the parent view, move your cursor over the top view (a dotted box appears around the view) and pick. Next you are prompted to select the starting point for the cutting plane line. Move your cursor over the center object snap, as shown in Figure 22.49 (Center), and track to the left. Pick the point at "A". When prompted to pick the end point, pick at "B." Next move your cursor below the top view and once the section view is displayed, pick the point at "C" to locate the view, as shown in Figure 22.49 (Right). Press ENTER to complete the operation.

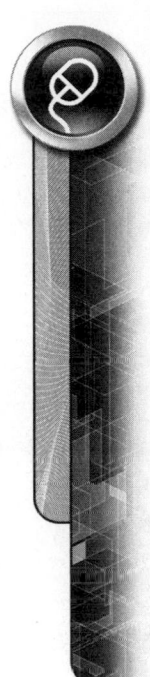

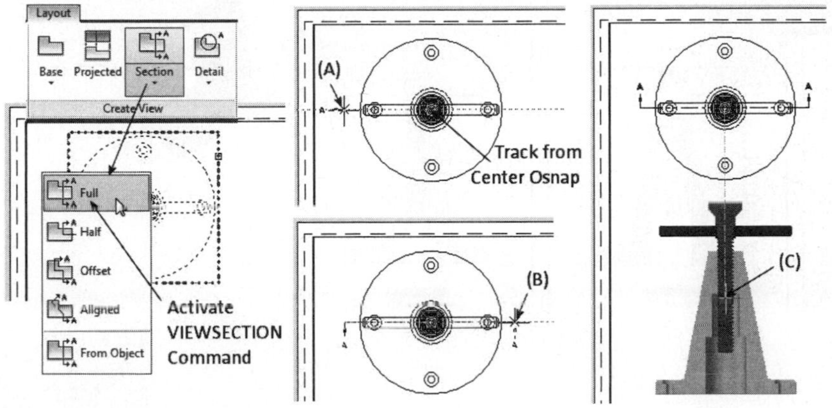

FIGURE 22.49

STEP 4

The results or the full front section created in the previous step is shown in Figure 22.50 (Center). Notice that all four models (Base, Screw, Handle, and Pad) that make up the assembly were sectioned. In assembly sections, components such as shafts and pins are generally not shown sectioned. In our view we will use the Edit Components tool to remove sectioning from the Screw and Handle components. Select the Edit Components button from the Modify View panel of the Ribbon, as shown in Figure 22.50 (Left). When prompted to select components, select the Handle and Screw. After pressing ENTER you are requested to select a Section participation option. Select "None" to remove the sectioning from the Handle and Screw. The results are shown in Figure 22.50 (Right).

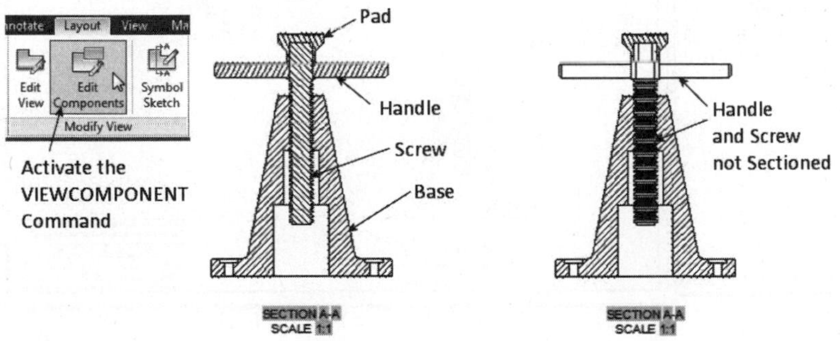

FIGURE 22.50

STEP 5

To complete the layout of the assembly, an isometric view will be generated. Select the Projected button in the Create View panel of the Ribbon, as shown in Figure 22.51 (Left). When prompted to select the parent view, pick the front section. Drag your cursor to the upper right to display an isometric view, as shown in Figure 22.51 (Right). Pick a point on the screen to locate the view. Press ENTER to complete the operation.

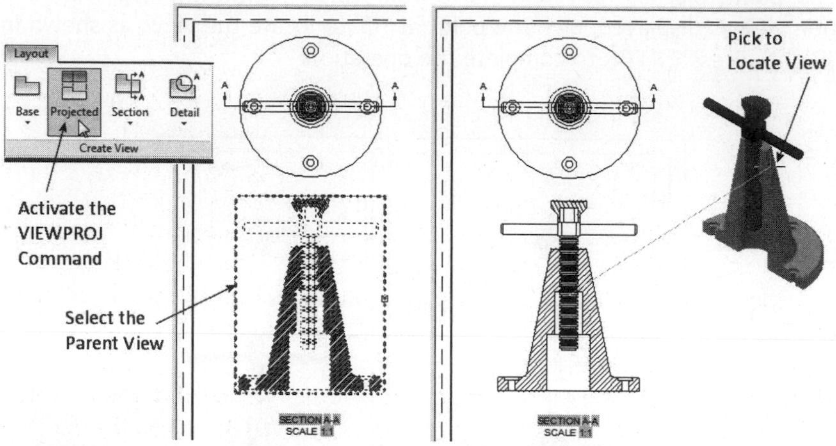

FIGURE 22.51

STEP 6

Before adding annotations to the drawing, it may be necessary to move the views to provide additional room in certain areas of your layout. To move a view, select the view and then activate the provided grip, as shown in Figure 22.52 (Left). The views will remain aligned as they are dragged to a new position. If a parent view is selected, the related views will move with it. The results are shown in Figure 22.52 (Right).

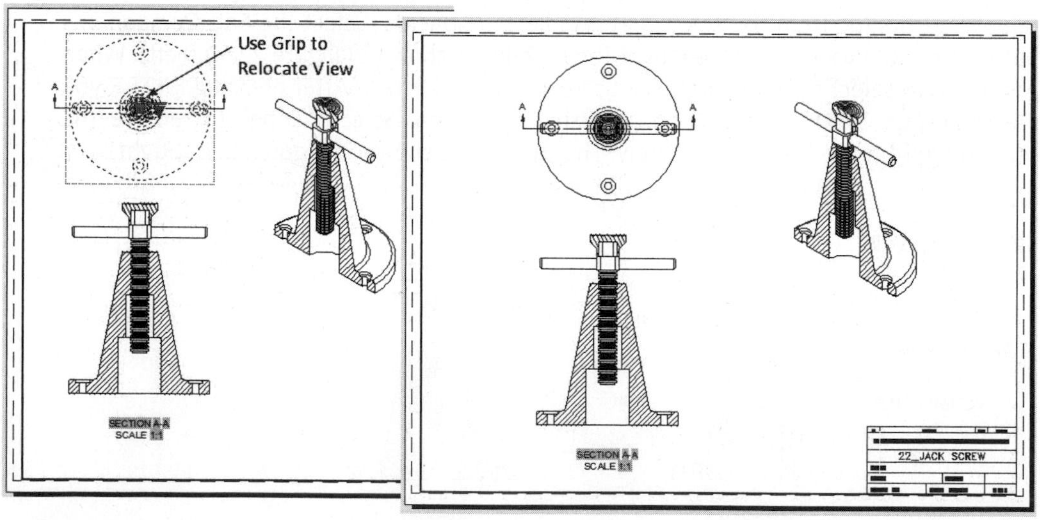

FIGURE 22.52

STEP 7

To complete the layout we need to add centerlines, dimensions, and annotations. All these items will be placed in Paper Space. Make the Center layer current and use the `CIRCLE`, `LINE`, and `DIMCENTER` commands to create the centerlines for this layout. You should erase any center mark lines that are placed over the cutting plane line. See Figure 22.53 (Left) for assistance in placing the drawing's centerlines. Next we will add the dimensions. Set the Dimension layer current and use the `DIMLINEAR` and `MLEADER` commands to create the linear dimension and leaders shown in Figure 22.53 (Center). The dimension shows the operating limits of the Jack Screw from 6.25 to 9.50 inches. The four leaders identify the assembly components. To place balloons on the end of the leaders, as shown in Figure 22.53, you will need to modify the Multileader style (`MLEADERSTYLE` command). In the dialog box, select the Modify button, select the Content tab, set the Multileader type to Block, set the Source block to Circle, and set the Scale to 1.5. For the final annotation, make the Title Block layer current and use the `INSERT` command to create a parts list from blocks. In the Insert dialog box, select the block named "Parts List Line" and place the block just above the title block, as shown in Figure 22.53 (Right). Use the image to provide the attribute data for the block. Insert three more blocks to complete the annotation of this layout.

TIP

Because dimensions are being place in Paper Space, it is recommended that you activate the Annotation Monitor mode in the Status Bar. This will provide an Alert icon if a dimensions becomes disassociated. Picking the icon and selecting Reassociate from the shortcut menu allows you to reassign the dimension definition points.

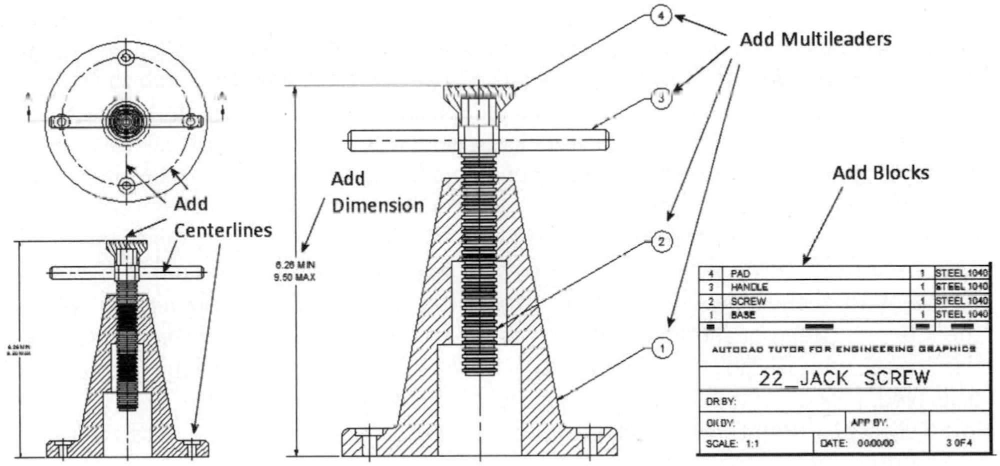

FIGURE 22.53

Figure 22.54 shows the completed drawing for the Assembly layout.

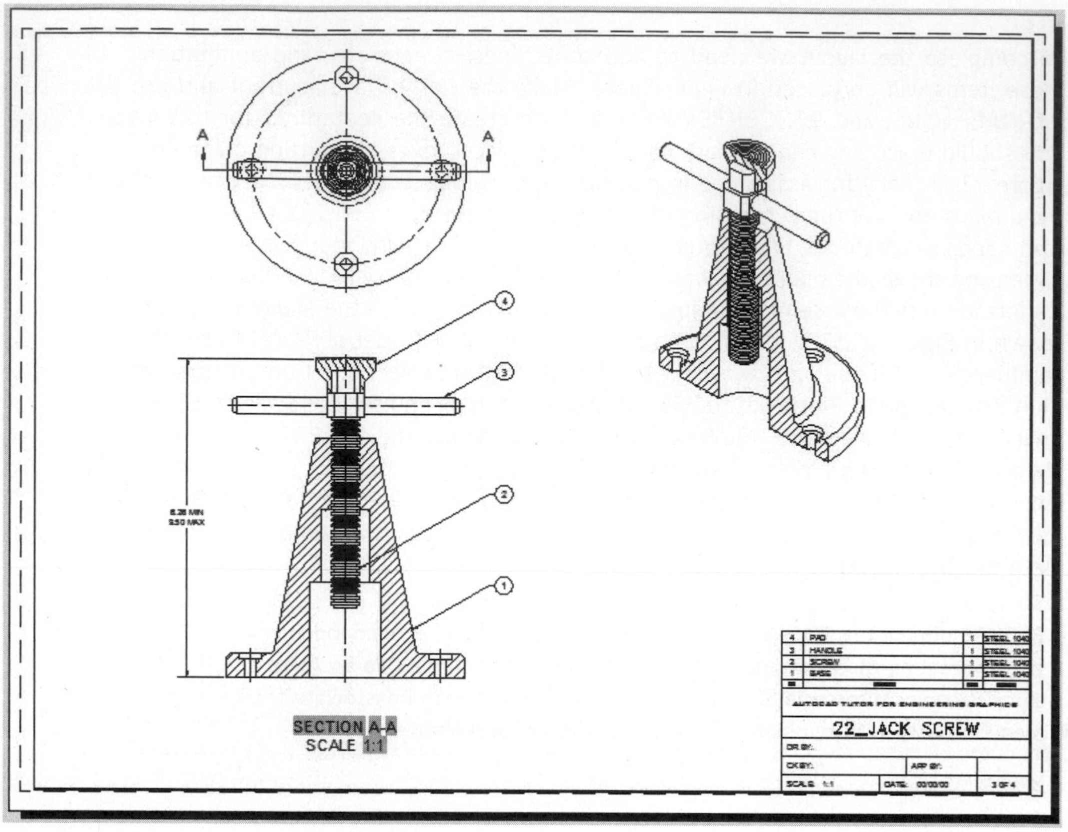

FIGURE 22.54

STEP 8

Now we are ready to create the next layout (Part1&2). Select the Model tab to return to Model Space. Select the Base (>From Model Space) button in the Create View panel of the Ribbon. When prompted to select objects, select the cone-shaped Base, as shown in Figure 22.55 (Left), and press ENTER. When you are prompted to select a layout for your new view, enter "Part1&2" and press ENTER to open the prearranged layout. Be careful, because if the name is mistyped a new blank layout will be created. Once the layout appears a front view of the base is displayed. As in the previous Assembly layout, we will want to change this first view to a top view. In this layout, however, we will use the Ribbon instead of shortcut menus to change options. Select "Top" from the Orientation panel on the Ribbon, as shown in Figure 22.55 (Center). Once the top view is displayed, pick a point in the upper left part of the drawing to place the view, as shown in Figure 22.55 (Right). Pick "OK" from the Create panel of the Ribbon and then press ENTER to complete the operation.

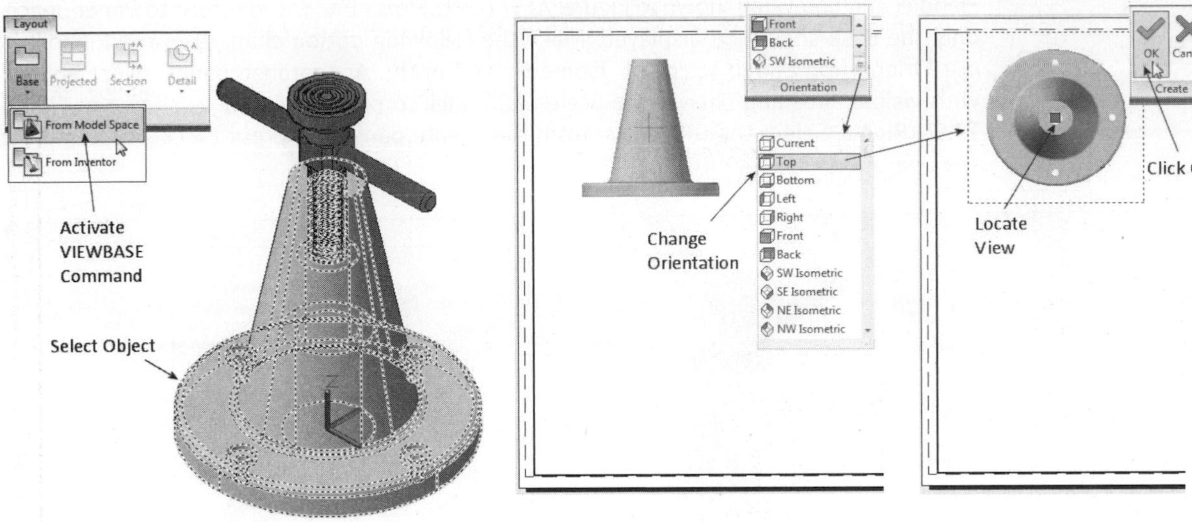

FIGURE 22.55

STEP 9

In this step we will create a full front section for the Base. Begin by selecting the Section (> Full) button in the Create View panel. When prompted to select the parent view, move your cursor over the top view (a dotted box appears around the view) and pick. When prompted to select the starting point for the cutting plane line, move your cursor over the center object snap, as shown in Figure 22.56 (Center), and track to the left. Pick the point at "A". When prompted to pick the endpoint, pick at "B". Next move your cursor below the top view and once the section view is displayed, pick the point at "C" to locate the view, as shown in Figure 22.56 (Right). Press ENTER to complete the operation.

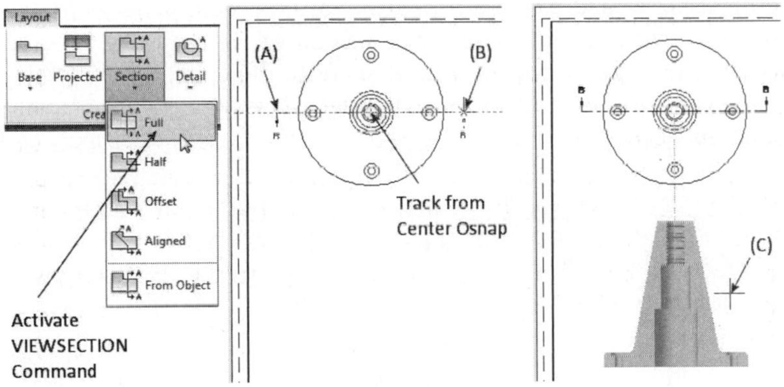

FIGURE 22.56

STEP 10

In this step, an isometric view of the Base will be generated. Instead of creating a projected view from front section as done in the Assembly layout, this time a base view will be created so that the isometric image will not appear sectioned. Select the Base (>From Model Space) button in the Create View panel of the Ribbon. This time you are not prompted to select objects, instead all four models are selected for you. For this view, we only want the base. Pick the Model Space Selection button in the Select panel of the Ribbon. When returned to Model Space, activate the Remove option and select the Pad,

Handle, and Screw, as shown in Figure 22.57 (Left). Press ENTER to return to Paper Space, only the Base should be displayed. Make the following option changes in the Ribbon: in the Orientation panel, select NE Isometric and in the Appearance panel, select Shaded with visible lines and change the Scale to 1:1. Pick to place the view as shown in Figure 22.57 (Right). Select the OK button from the Create panel and press ENTER to complete the operation.

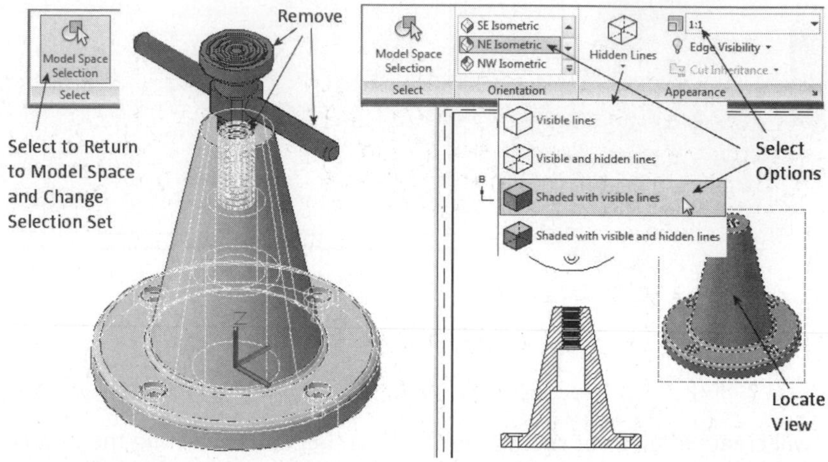

FIGURE 22.57

STEP 11

Next we will add three views for the Screw. Select the Model tab to return to Model Space. Select the Base (>From Model Space) button in the Create View panel of the Ribbon. When prompted to select objects, select the Screw and press ENTER. When you are prompted to select a layout for your new view, press ENTER to accept the "Part1&2" layout. Once the layout appears, a front view of the base is displayed, as shown in Figure 22.58 (Left). Pick the point at "A" and pick the OK button in the Create panel to fix the views location. Drag the cursor straight up and pick a point at "B" to locate the top view, as shown in the middle image. Finally, move the cursor to the upper right and pick at point "C" to locate the isometric view as shown on the right. Press ENTER to complete the generation of the three views.

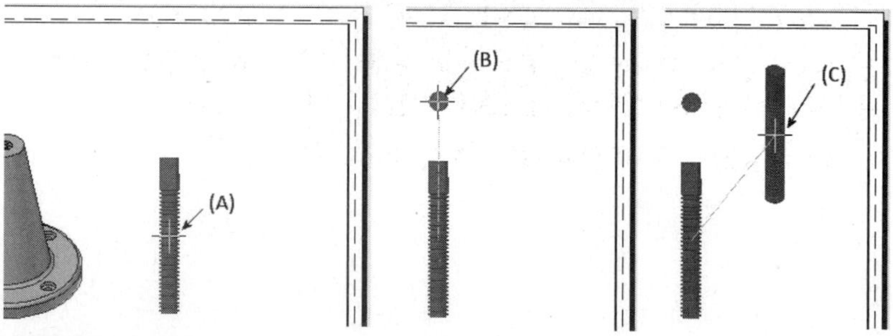

FIGURE 22.58

STEP 12

The three views for the Screw are shown in Figure 22.59 (Left). Notice that the Hidden Lines view style was assigned to the isometric view. To change the style, select the view and pick the Edit View button in the Edit panel of the Ribbon, as shown in Figure 22.59 (Center). When the Appearance panel is displayed, select the Shaded with visible lines style, as shown in Figure 22.59 (Right). Press ENTER to complete the operation. This completes the layout of views for the Base (Part1) and the Screw (Part2).

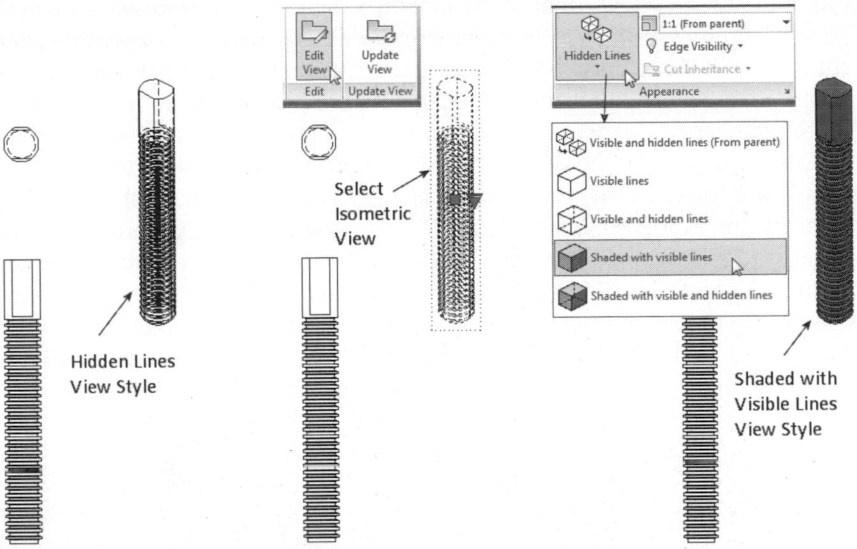

FIGURE 22.59

STEP 13

To complete this layout we need to add centerlines, dimensions, and annotations. This can be accomplished by following the procedures discussed in previous steps (Step 6 and Step 7) and using Figure 22.60 as a guide.

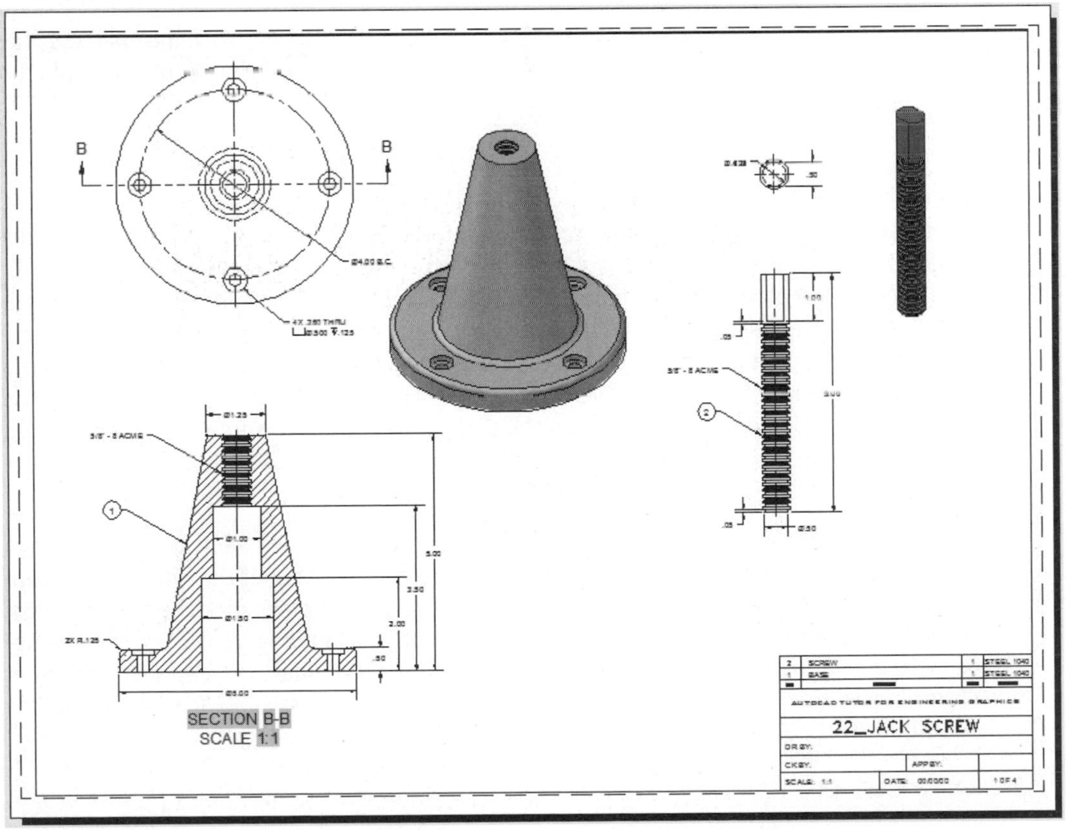

FIGURE 22.60

STEP 14

Now we are ready to create the final layout. We will begin by adding three views for the Handle. Select the Model tab to return to Model Space. Select the Base (>From Model Space) button in the Create View panel of the Ribbon. When prompted to select objects, select the Handle and press ENTER. When you are prompted to select a layout for your new view, enter "Part3&4" and press ENTER. Once the layout appears, a front view of the Handle is displayed, as shown in Figure 22.61 (Left). Before placing the view, change the scale to 2:1 in the Appearance panel of the Ribbon. Pick the point at "A" and select the OK button in the Create panel to fix the views location. Drag the cursor straight up and pick a point at "B" to locate the top view. Next, move the cursor to the upper right and pick at point "C" to locate the isometric view. Press ENTER to complete the generation of the three views. Select the isometric view and use the grip to move the view above the top view, as shown in Figure 22.61 (Right). While the view is still highlighted, pick the Edit View button in the Edit panel of the Ribbon and when the Appearance panel is displayed, select the Shaded with visible lines style, as shown in Figure 22.61 (Right). Press ENTER to complete the operation. This completes the layout of views for the Handle (Part3).

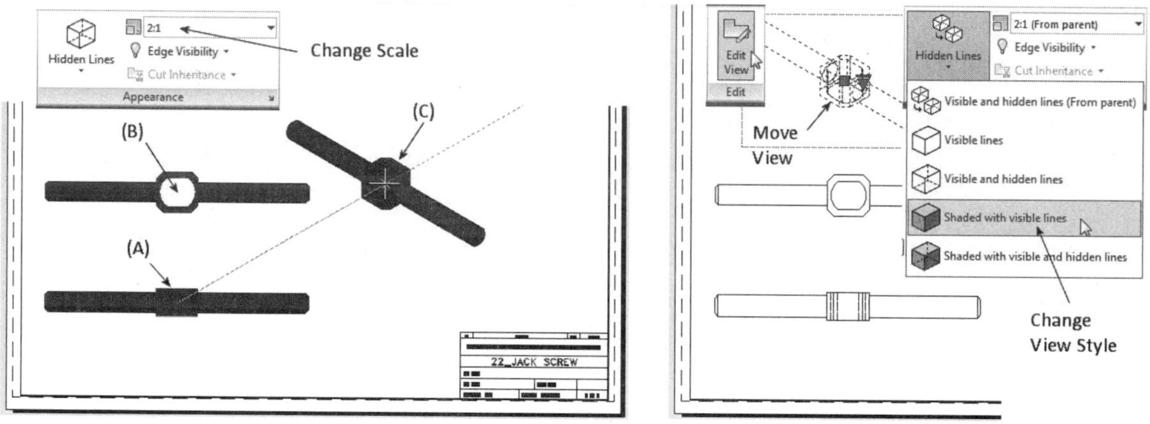

FIGURE 22.61

STEP 15

In this step we will add five views for the Pad. Select the Model tab to return to Model Space. Select the Base (>From Model Space) button in the Create View panel of the Ribbon. When prompted to select objects, select the Pad and press ENTER. When you are prompted to select a layout for your new view, press ENTER to accept the "Part3&4" layout. Once the layout appears, utilize the procedures previously demonstrated to create a top view at a scale of 2:1 with a Visible lines view style, a full front section view from the top view, a bottom view projected from the section view, and an isometric view projected from the section view. Use grips or the MOVE command to reposition the views as necessary. The results are shown in Figure 22.62 (Left).

To complete the layout of the Pad, a rectangular detail view will need to be generated. Select the Detail (>Rectangular) button in the Create View panel of the Ribbon, as shown in Figure 22.62 (Right). When prompted to select the parent view, pick the front section. When prompted for a Center point, pick the point at "A" and drag your cursor to the upper left. Pick at point "B" to define the size of the detail boundary. Finally, pick a point at "C" to locate the detail, as shown in Figure 22.62 (Right). Press ENTER to complete the operation. This completes the layout of views for the Pad (Part4).

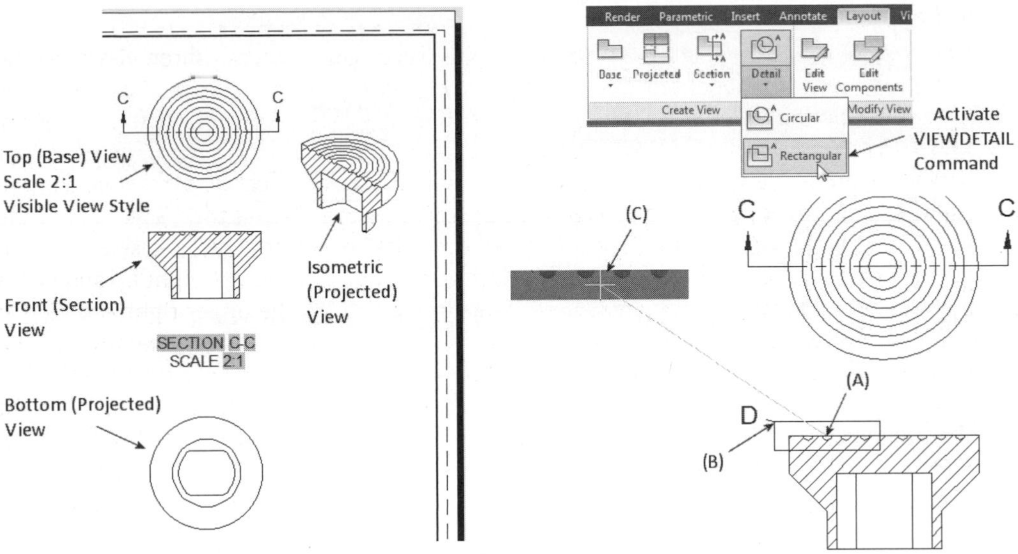

FIGURE 22.62

STEP 16

To complete this layout we need to add centerlines, dimensions, and annotations. This can be accomplished by following the procedures discussed in previous steps (Step 6 and Step 7) and using Figure 22.63 as a guide.

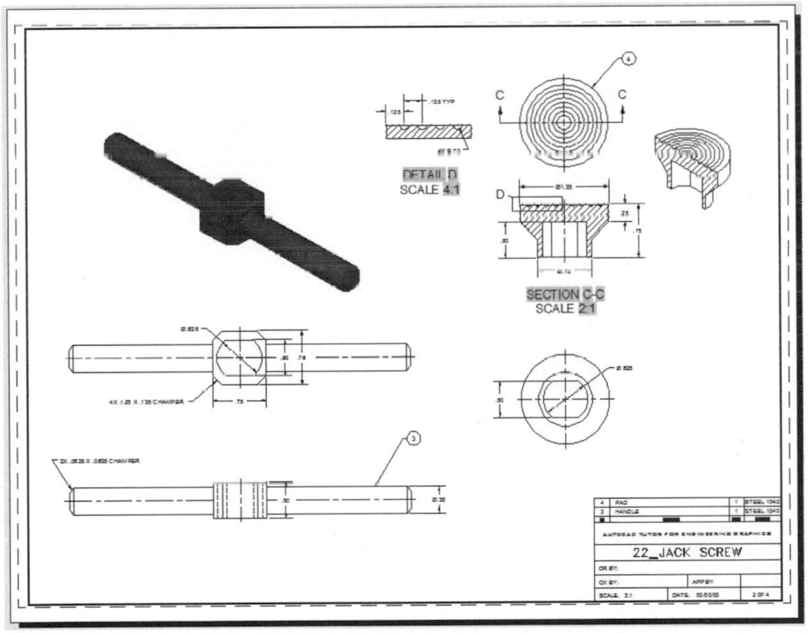

FIGURE 22.63

STEP 17

The labels for the section views and detail view appear a little too large when compared to other annotations in the drawing. To modify the section view label size, select the Section View Style button from the Styles and Standards panel of the Ribbon, as shown in Figure 22.64 (Left). When the Section View Style Manager appears, select the Modify button. In the Modify Section View Style dialog box, select the View Label tab and change the Text height to .18, as shown in Figure 22.64 (Right). Click OK and close the style manager dialog box. The text height will be automatically updated.

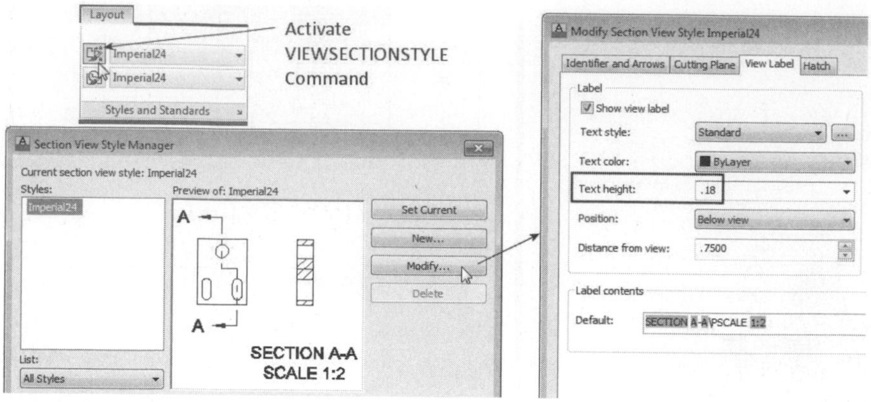

FIGURE 22.64

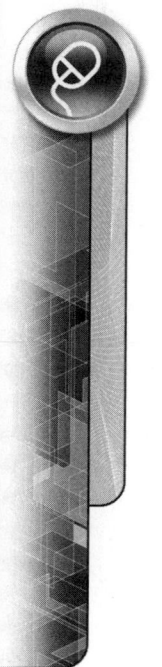

STEP 18

To modify the detail view label size, select the Detail View Style button from the Styles and Standards panel of the Ribbon, as shown in Figure 22.65 (Left). When the Detail View Style Manager appears, select the Modify button. In the Modify Detail View Style dialog box, select the View Label tab and change the Text height to .18, as shown in Figure 22.65 (Right). Click `OK` and close the style manager dialog box. The text height will be automatically updated.

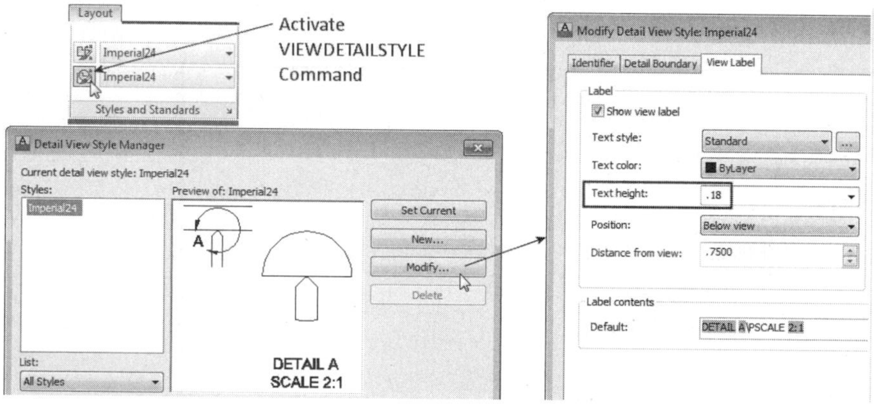

FIGURE 22.65

STEP 19

For this last step, one final modification will be made. The top edge of the Base model is supposed to be filleted. Once the fillet is added to the model, the drawing views will be automatically updated. This is a powerful feature of the model documentation process.

Select the Model tab to return to Model Space. Activate the `FILLETEDGE` command from the Solid Editing panel of the Solid Ribbon. Use the Radius option to set the fillet radius to .125. Pick the top edge of the Base model, as shown in Figure 22.66 (Left). Press `ENTER` twice to complete the operation. Figure 22.66 (Right) illustrates how the drawing views are linked and automatically updated. This completes the Jack Screw tutorial.

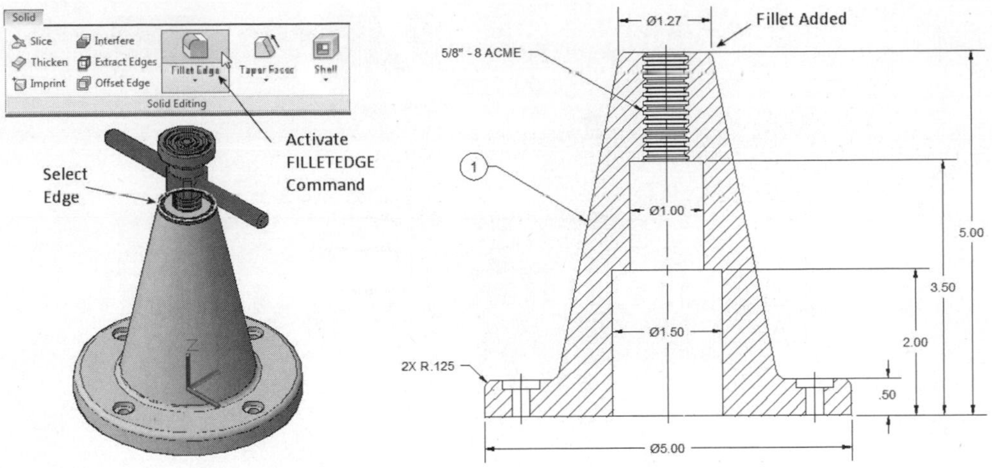

FIGURE 22.66

PROBLEMS FOR CHAPTER 22

1. Create an engineering drawing of Problems 20-1 through 20-20 consisting of Front, Top, Right Side, and Isometric views.
2. Add proper dimensions to each drawing.

CHAPTER

23

Producing Renderings and Motion Studies

This chapter introduces you to renderings in AutoCAD and how to produce realistic renderings of 3D models. Adding lights is one of the tools available to help add realism to your renderings. With this important feature you produce unlimited special effects, depending on the location and intensity of the lights. The Point, Distant, and Spotlights features are discussed, as is the ability to simulate sunlight. You will be able to attach materials to your models to provide additional realism. Materials can be selected from available libraries, or you can even make your own materials. You will be shown how to walk and fly through a 3D model. You will also be given instruction on how to make a motion path animation of a 3D model.

AN INTRODUCTION TO RENDERINGS

Engineering and architectural drawings are able to pack a vast amount of information into a 2D outline drawing supplemented with dimensions, symbols, and notes. However, training, experience, and sometimes imagination are required to interpret them, and many people would rather see a realistic picture of the object. Actually, realistic pictures of a 3D model are more than just a visual aid for the untrained. They can help everyone visualize and appreciate a design, and can sometimes even reveal design flaws and errors.

Shaded, realistic pictures of 3D models are called renderings. Until recently, they were made with colored pencils and pens or with paintbrushes and airbrushes. Now they are often made with computers, and AutoCAD comes with a complete set of rendering tools and is ready for your use. Figure 23.1 shows, for comparison, the solid model of a bracket in its wireframe form, as it looks when the `HIDE` command has been invoked, and when it is rendered.

This chapter is designed to give you an overview on how to create pleasing, photo-realistic renderings of your 3D models.

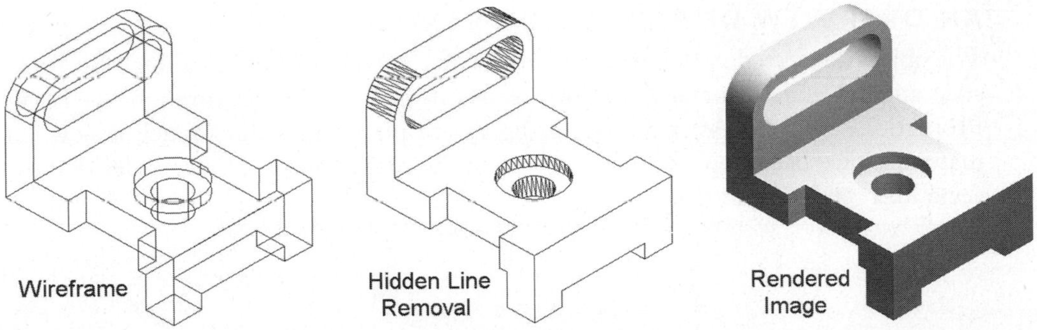

FIGURE 23.1

The tools for creating realistic renderings can be accessed from the Render tab of the Ribbon, as shown in Figure 23.2. Notice the panels available in the Ribbon for accessing light, material, and render commands.

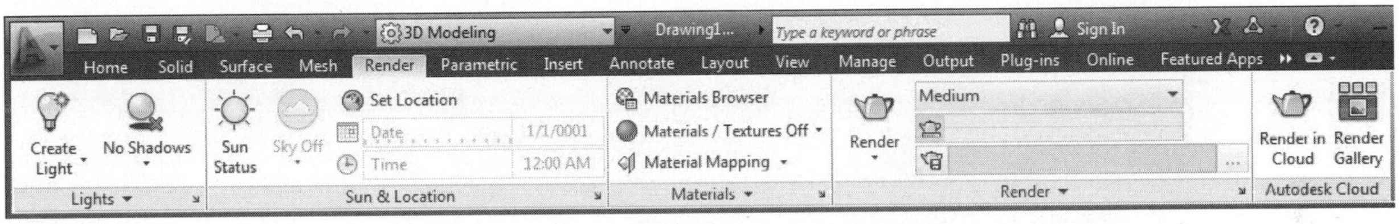

FIGURE 23.2

The following table gives a brief description of available rendering tools.

Button	Tool	Function
	Hide	Performs a hidden line removal on a 3D model
	Render	Switches to the Render window, where a true rendering of the 3D model is performed
	Lights	Contains six additional buttons used for controlling lights in a 3D model
	Light List	Displays the Lights in Model palette used for managing lights that already exist in a 3D model
	Materials Browser	Toggles the Materials Browser palette on and off, used for creating and applying materials to a 3D model
	Materials Editor	Toggles the Materials Editor palette on and off, used for modifying and creating materials
	Planar Mapping	Contains four additional buttons, used for mapping materials to planar, box, cylindrical, and spherical surfaces
	Render Environment	Displays the Rendering Environment dialog box that is used mainly for controlling the amount of fog applied to a 3D model
	Advanced Render Settings	Displays the Advanced Rendering Settings palette, used for making changes to various rendering settings

AN OVERVIEW OF PRODUCING RENDERINGS

The object illustrated in Figure 23.3 consists of a 3D model that has a Realistic visual style applied. The color of the model comes from the color set through the Layer Properties Manager. After a series of lights are placed in a 3D model, and when materials have been applied, the next step in the rendering process is to decide how accurate a rendering to make.

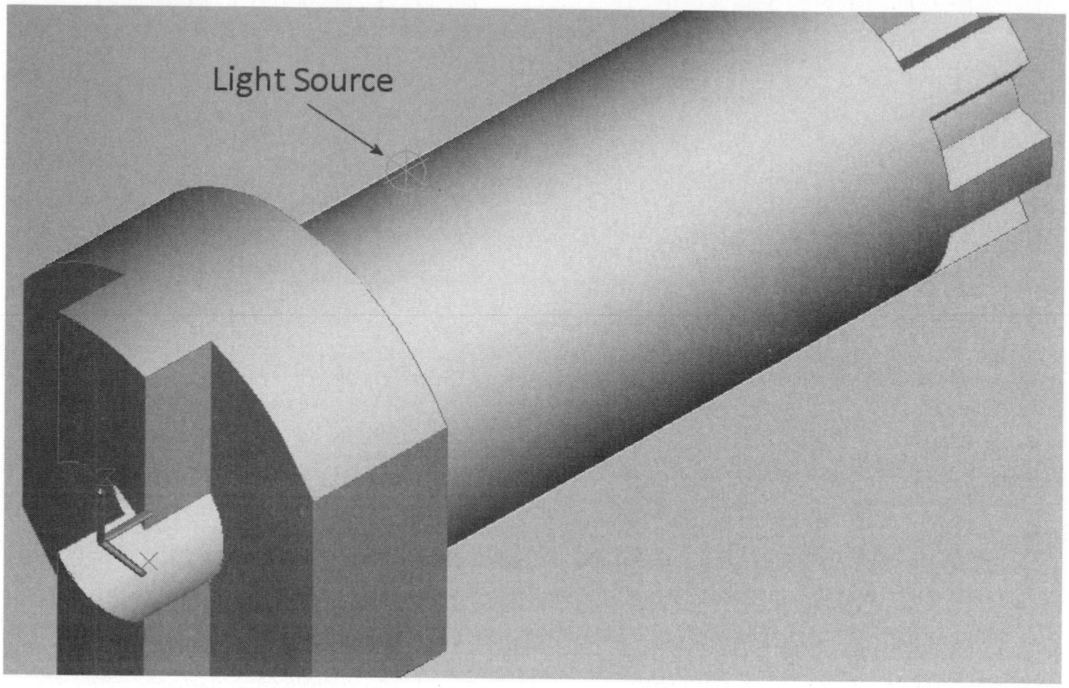

FIGURE 23.3

As shown in Figure 23.4 of the Ribbon, the Render tab displays different rendering modes, or presets, as they are called. The Standard presets range from Draft to Presentation and control the quality of the final rendered image. For example, when you perform a rendering in draft mode, the processing speed of the rendering is very fast; however, the quality of the rendering is very poor. This render preset is used, for example, to perform a quick rendering when you are unsure about the positioning of lights. When you are pleased with the lighting in the 3D model, you can switch to a higher render preset, such as High or Presentation. These modes process the rendering much slower, however, the quality is photo-realistic. To perform a rendering, click the Render button, as shown in Figure 23.4.

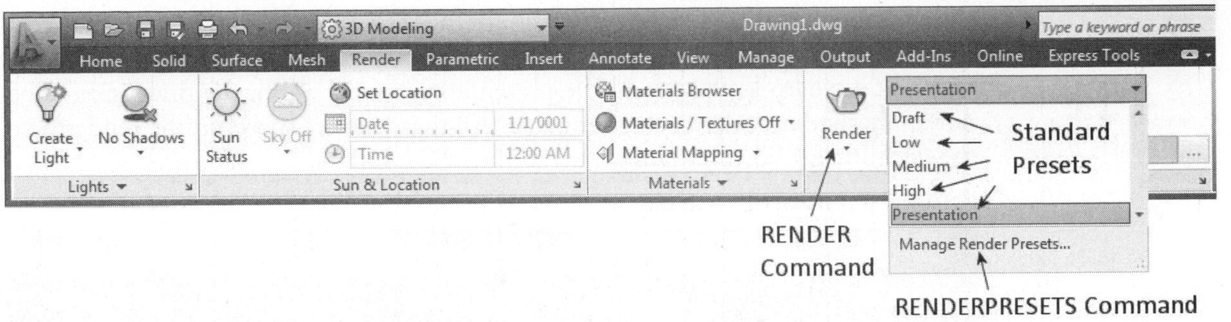

FIGURE 23.4

TIP

Clicking the Manage Render Presets... button, found at the bottom of the list of Standard presets shown in Figure 23.4, activates the Render Presets Manager dialog box. This dialog box allows you to not only manage the Standard presets, but also to create and manage Custom presets.

Clicking the Render button, as shown in Figure 23.4, switches your screen to the render window, as shown in Figure 23.5. Notice the shadows that are cast to the base of the 3D model. Shadow effects are one of many special rendering tools used to make a 3D model appear more realistic. Illustrated on the right of the rendering window is an area used for viewing information regarding the rendered image. Also, when you produce a number of renderings, they are saved in a list at the bottom of the rendering window.

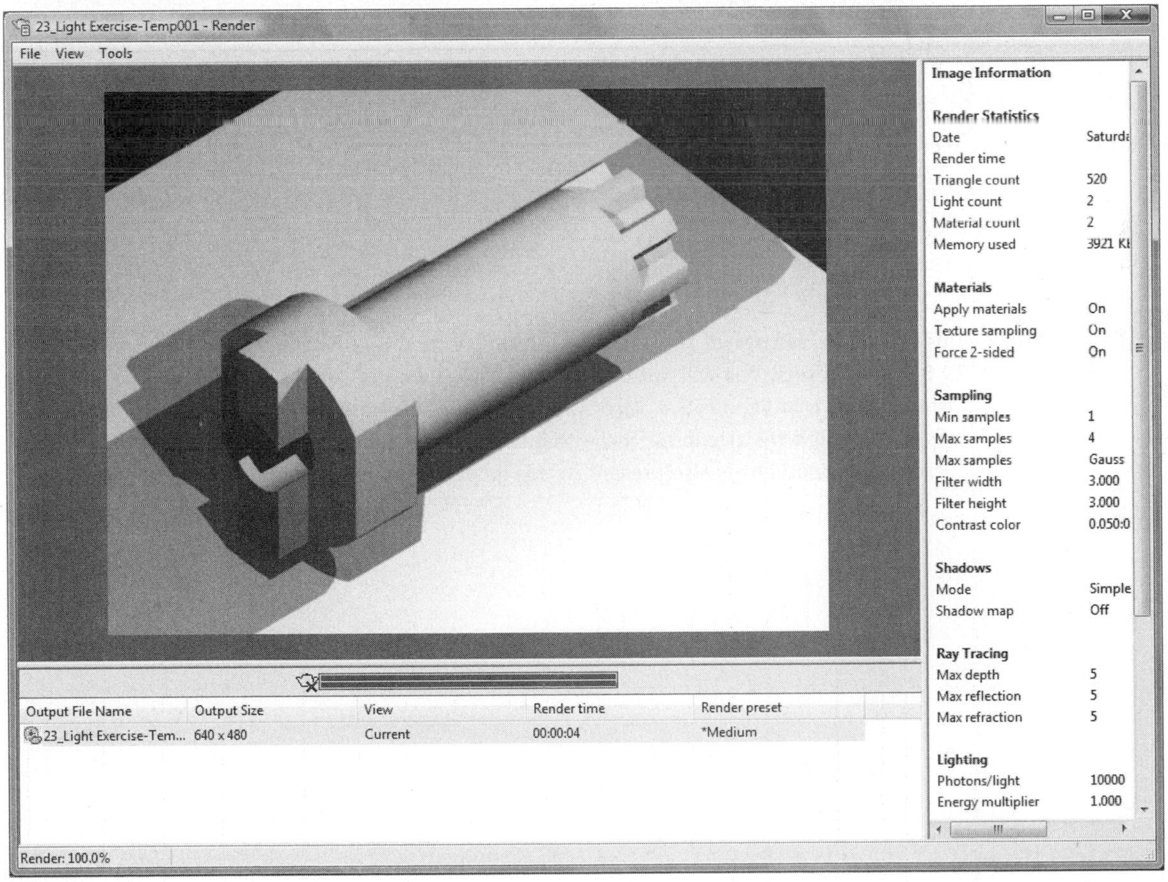

FIGURE 23.5

As mentioned earlier, Figure 23.6 displays the results of performing a Draft versus Presentation rendering. In the draft image, notice that the edges of the model do not look as sharp as they do in the presentation model. Also, the draft image does not apply shadows when being rendered. All these factors speed up the rendering of the draft image; however, the quality of the image suffers.

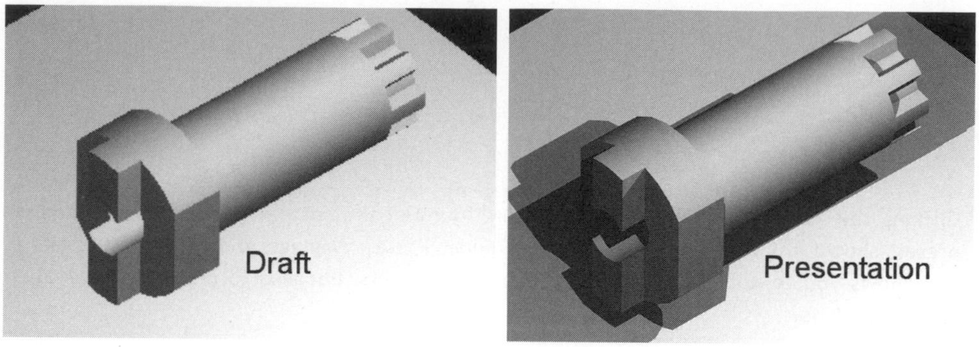

FIGURE 23.6

When you have produced a quality rendered image, you can save this image under one of the many file formats illustrated in Figure 23.7. Supported raster image formats include BMP, PCX, TGA, TIF, JPEG, and PNG.

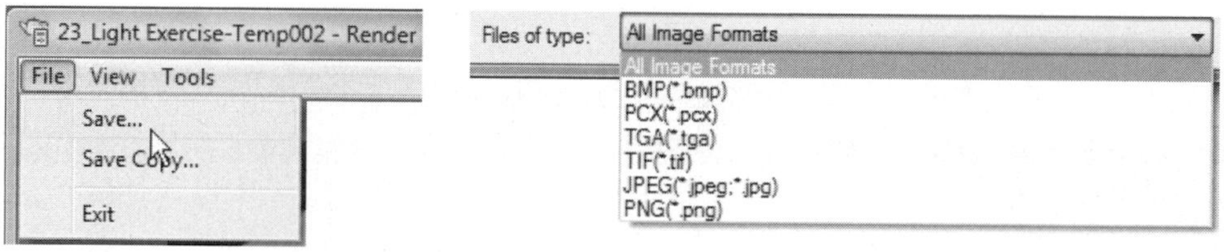

FIGURE 23.7

TRY IT!

This exercise consists of an overview of the rendering process. Open the drawing file 23_Render_Presets. You will notice a container and two glasses resting on top of a flat platform, as shown in Figure 23.8. Also shown in Figure 23.8 are two circular shapes with lines crossing through their centers. These shapes represent lights in the model. A third light representing a spotlight is also present in this model, although it is not visible in Figure 23.8.

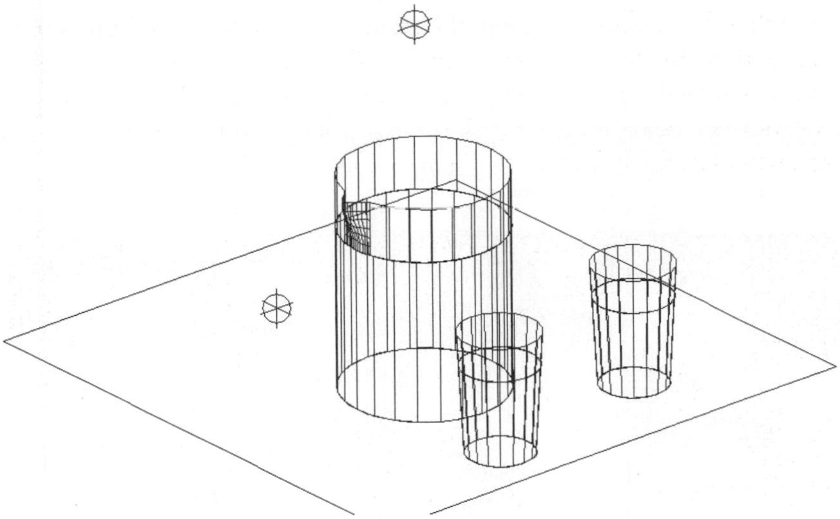

FIGURE 23.8

Activate the 3D Modeling workspace, click on the Render tab, and click the Render button found in the Render panel. The results are illustrated in Figure 23.9, with materials, lights, and a background being part of the rendered scheme. Notice in the Render panel the presence of Medium. This represents one of the many render presets used to control the quality of the rendered image.

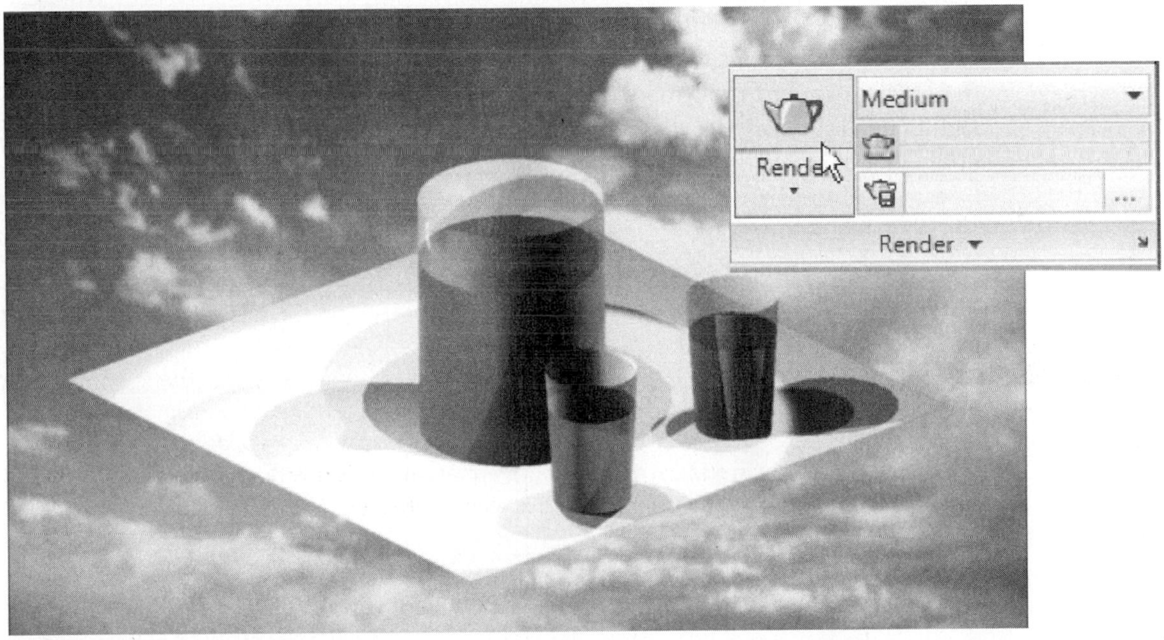

FIGURE 23.9

To see how the Render Preset value affects the rendered image, change the render preset in the panel from Medium to Draft and click on the Render button, as shown in Figure 23.10. Notice that the quality of the rendered image looks choppy; also, shadows are lost. However, the Draft Render Preset is always useful when testing out the lighting of the rendered scene. The processing time of this preset is very fast compared to other render presets, although the quality of the Draft Render Preset does not look very appealing.

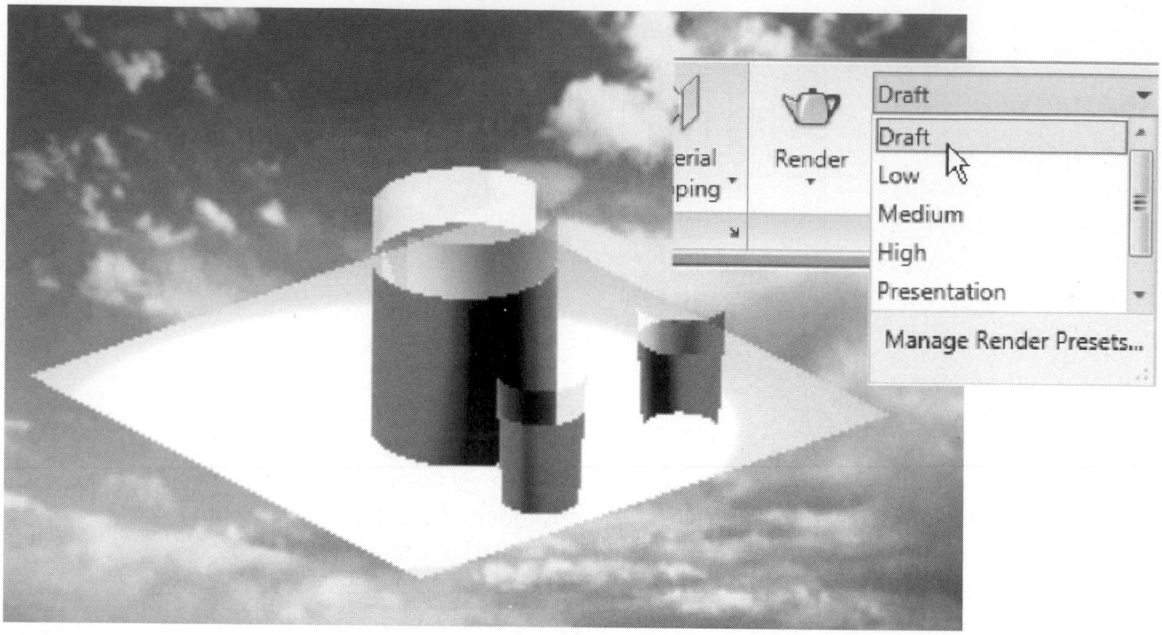

FIGURE 23.10

In Figure 23.10, the lighting looks too bright and overpowering in the rendered image. To edit the intensity of existing lights, click the Lights Arrow button, located in the Lights panel of the Ribbon, as shown in Figure 23.11 (Left). This launches the Lights in Model palette, as shown in Figure 23.11 (Center). Double-clicking Pointlight1 launches the Properties palette for that light, as shown in Figure 23.11 (Right). Locate the Intensity factor under the General category and change the default intensity value of 1.00 to a new value of 0.50. This reduces the intensity of this light by half. Perform this same operation on Pointlight2 and Spotlight1 by changing their intensities from 1.00 to 0.50.

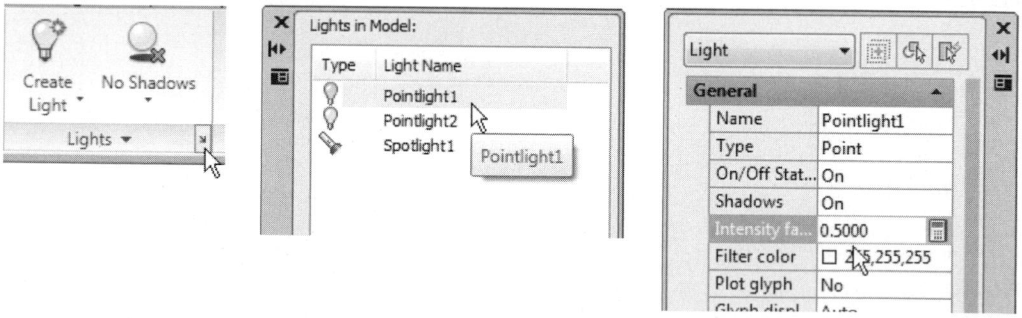

FIGURE 23.11

After changing the intensity of each light in the model, change the render preset from Draft to Presentation in the Ribbon and then click the Render button. The results are displayed in Figure 23.12, with the lights being less intense and the shadows being more pronounced.

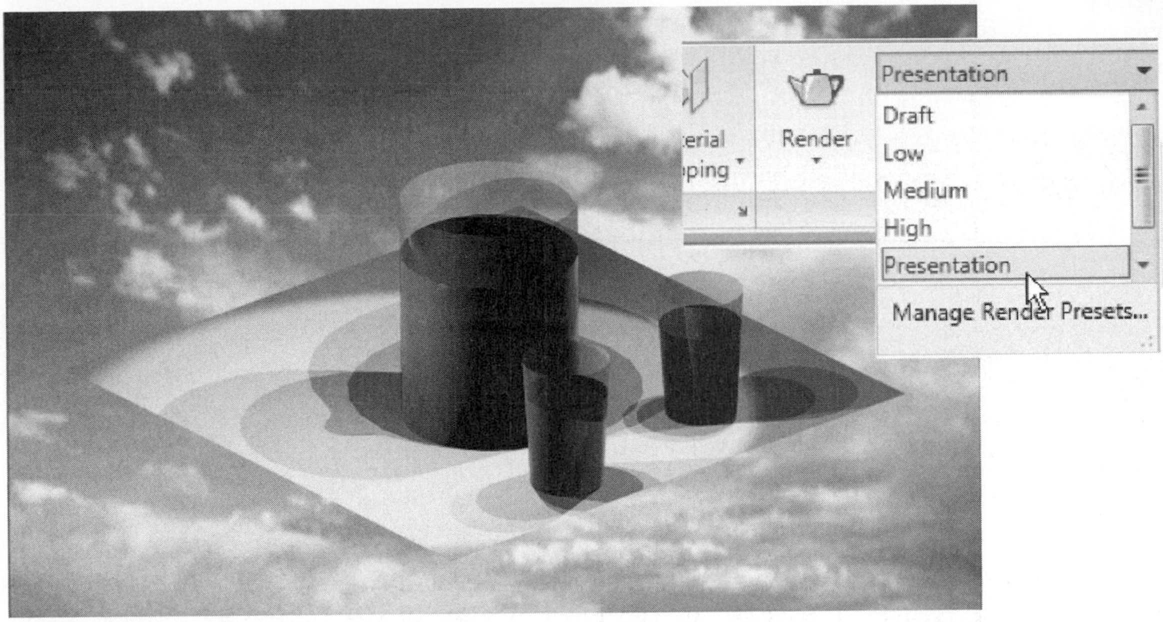

FIGURE 23.12

Next, click the Materials Browser button, located in the Materials panel of the Ribbon, as shown in Figure 23.13 (Left). This button toggles the Materials Browser palette on and off. The top portion of the browser displays all materials created in the model. The bottom portion of the browser displays available materials that can be applied by dragging and dropping them on 3D models. We will experiment more with materials later in this chapter. This concludes the exercise.

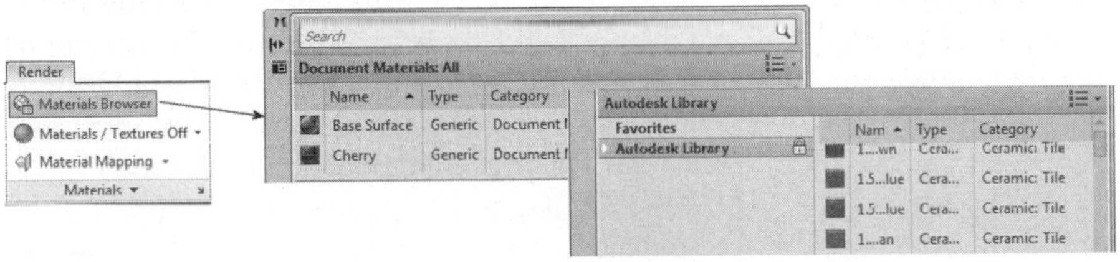

FIGURE 23.13

CREATING AND PLACING LIGHTS FOR RENDERING

The ability to produce realistic renderings is dependent on what type of lighting is used and how these lights are placed in the 3D model. Once the lights are placed, their properties can be edited through the Properties palette. Before placing the light, you need to decide on the type of light you wish to use: Point Light, Spotlight, Distant Light, or Weblight. Figure 23.14 displays the Lights panel of the Ribbon, where access to lighting-related commands is provided.

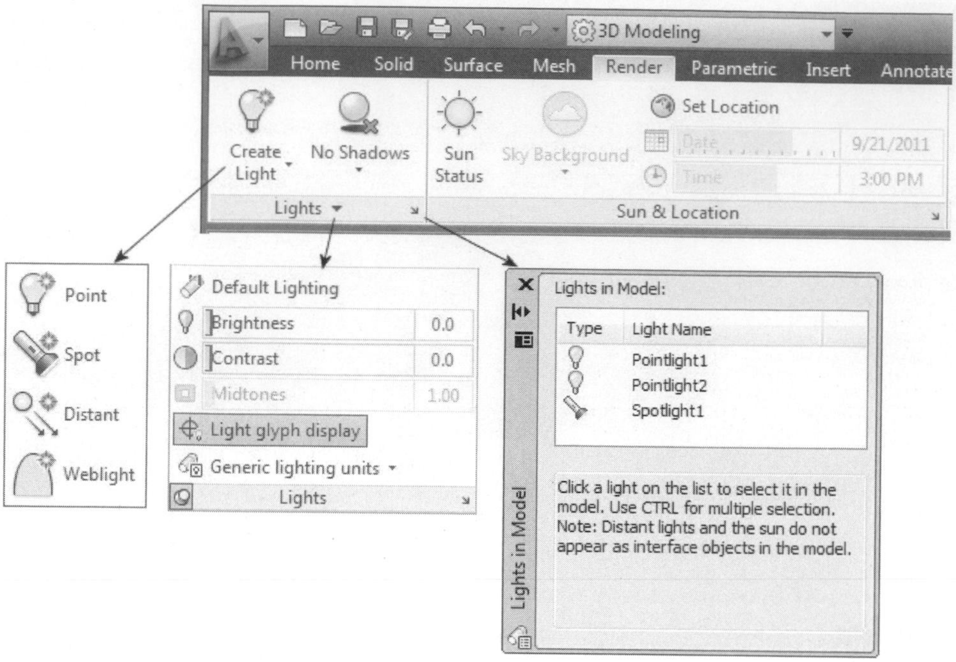

FIGURE 23.14

You can place numerous lights in a single 3D model and then adjust these lights depending on the desired effect. Each of the standard light types also allows for shadows to be cast, giving depth to your 3D model. The following table gives a brief description of lighting-related tools.

Button	Tool	Function
	New Point Light	Creates a new point light, similar to a lightbulb
	New Spotlight	Creates a new spotlight given a source and target
	New Distant Light	Creates a new distant light, similar to the sun
	Light List	Displays a list of all lights defined in a model
	Geographic Location	Displays a dialog box that allows you to select a geographic location from which to calculate the sun angle
	Sun Properties	Displays the Sun Properties palette, which allows you to make changes to the properties of the sun

The Sun and Location panel of the Ribbon, as shown in Figure 23.15, provides access to tools related to a special type of distance light—the sun. Clicking on the arrow in the lower right corner of the panel activates the `SUNPROPERTIES` command, which provides a palette that allows you to change the properties of the sun. Clicking on the Set Location button activates the `GEOGRAPHICLOCATION` command. This command is unique in that it allows you to control the position of the sun depending on a location, as shown in Figure 23.15. Clicking on the button launches the dialog box on the left. It is here where you determine the geographic location from a kml or kmz file, from Google Earth, or entering location values. Clicking on the Location Values area displays the Geographic Location dialog box, as shown in Figure 23.15 (Center). Clicking the Use Map button in this dialog box displays regional maps of the world, as shown in Figure 23.15 (Right), which allow you to produce sun studies based on these locations.

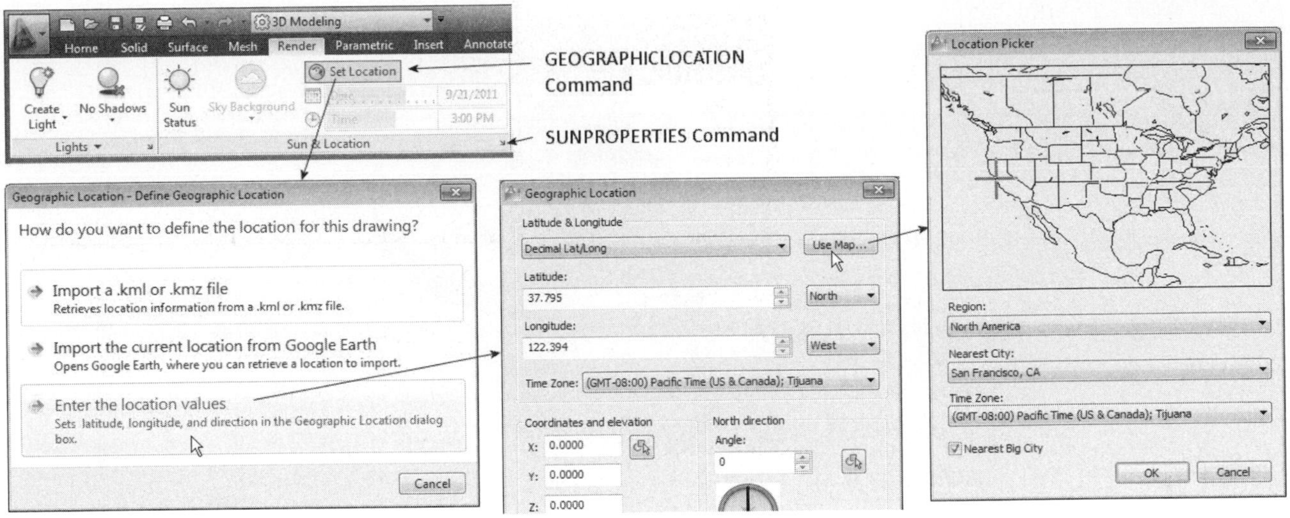

FIGURE 23.15

A number of lights are also available through the Tool Palette, which is activated from the View tab of the Ribbon, as shown in Figure 23.16 (Top). Five light categories are available to give your renderings a more realistic appearance. One of the categories deals with generic lights, as shown in Figure 23.16 (Bottom). Through this category, you can drag and drop point, spot, and distant lights into your 3D model. The remaining categories of lights (Fluorescent, High Intensity Discharge, Incandescent, and Low Pressure Sodium) are classified as photometric.

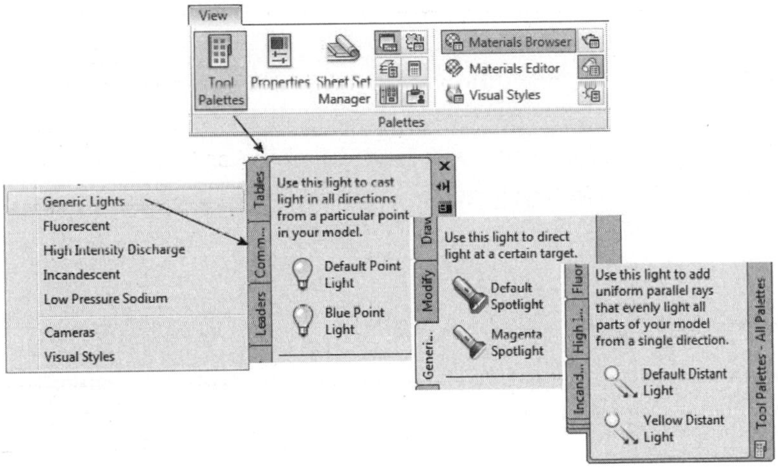

FIGURE 23.16

TRY IT!

Open the file 23_Valve Head Lights. This drawing file, shown in Figure 23.17, contains the solid model of a machine component that we will use for a rendering. You will place a number of point lights and one spotlight to illuminate this model.

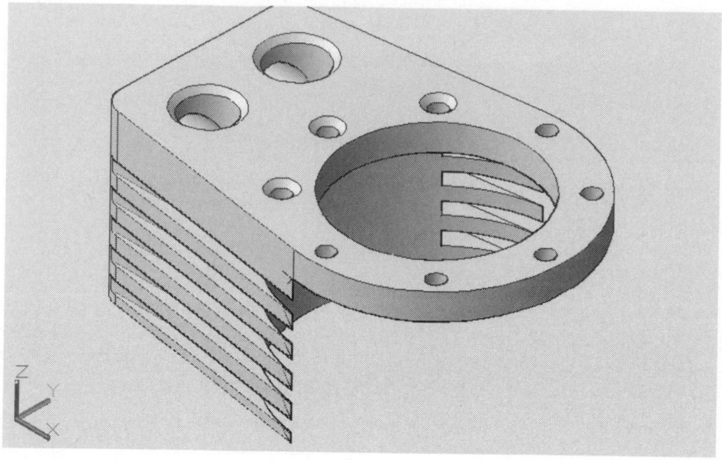

FIGURE 23.17

Switch to a top view, as shown in Figure 23.18. It will be easier to place the lights while viewing the model from this position. Activate the Ribbon, and in the Light panel, click the Point button, as shown in Figure 23.18 (Left).

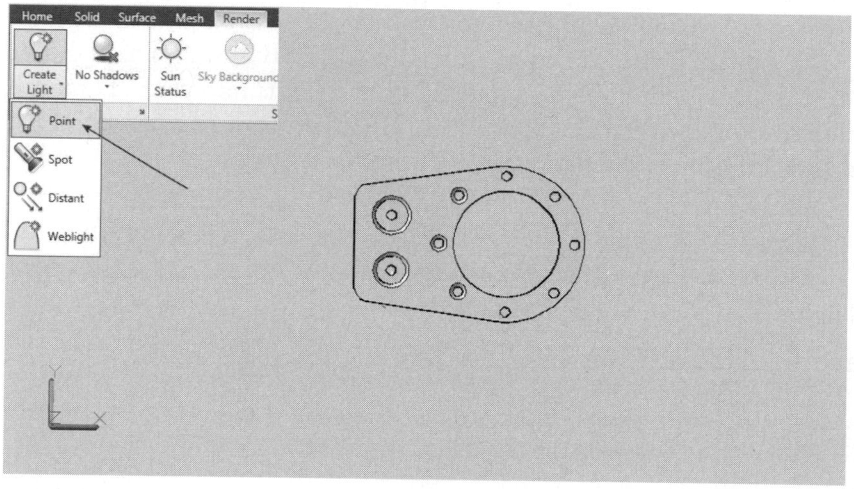

FIGURE 23.18

Place three point lights using the following Command prompts. You will place the lights at the approximate locations indicated in Figure 23.19. You will change the intensity and the default names of each light through the following Command prompts.

For the first light, enter the following information:

```
Command: _POINTLIGHT (Pick from the Lights panel of the
Ribbon)

Specify source location <0,0,0>: (Pick at "A")

Enter an option to change [Name/Intensity/Status/shadoW/
Attenuation/Color/eXit]

<eXit>: I (For Intensity)

Enter intensity (0.00 - max float) <1.0000>: 0.25
```

Enter an option to change [Name/Intensity/Status/shadoW/Attenuation/Color/eXit]

<eXit>: **N** *(For Name)*

Enter light name <Pointlight1>: **Overhead Light**

Enter an option to change [Name/Intensity/Status/shadoW/Attenuation/Color/eXit]

<eXit>: *(Press* ENTER *to create the light)*

For the second light, enter the following information:

Command: **_POINTLIGHT** *(Pick from the Lights panel of the Ribbon)*

Specify source location <0,0,0>: *(Pick at "B")*

Enter an option to change [Name/Intensity/Status/shadoW/Attenuation/Color/eXit]

<eXit>: **I** *(For Intensity)*

Enter intensity *(0.00 - max float)* <1.0000>: **0.25**

Enter an option to change [Name/Intensity/Status/shadoW/Attenuation/Color/eXit]

<eXit>: **N** *(For Name)*

Enter light name <Pointlight2>: **Lower Left Light**

Enter an option to change [Name/Intensity/Status/shadoW/Attenuation/Color/eXit]

<eXit>: *(Press* ENTER *to create the light)*

For the third light, enter the following information:

Command: **_POINTLIGHT** *(Pick from the Lights panel of the Ribbon)*

Specify source location <0,0,0>: *(Pick at "C")*

Enter an option to change [Name/Intensity/Status/shadoW/Attenuation/Color/eXit]

<eXit>: **I** *(For Intensity)*

Enter intensity *(0.00 - max float)* <1.0000>: **0.25**

Enter an option to change [Name/Intensity/Status/shadoW/Attenuation/Color/eXit]

<eXit>: **N** *(For Name)*

Enter light name <Pointlight3>: **Upper Left Light**

Enter an option to change [Name/Intensity/Status/shadoW/Attenuation/Color/eXit]

<eXit>: *(Press* ENTER *to create the light)*

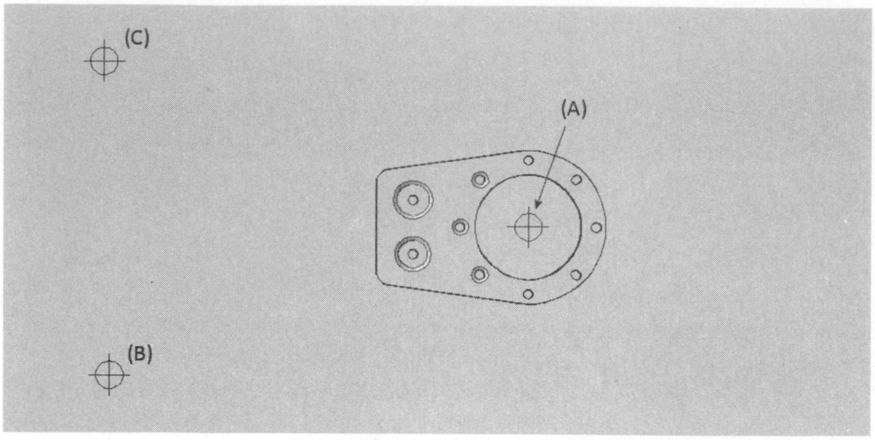

FIGURE 23.19

When you have finished placing all three point lights, activate the View tab on the Ribbon. In the Views panel, select the SE_Zoomed view, as shown in Figure 23.20. This named view was created for you (with the VIEW command) and changes your model to a zoomed-in version of the Southeast Isometric view.

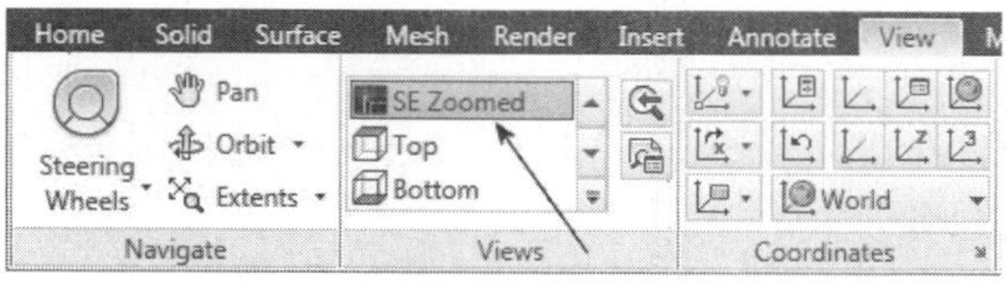

FIGURE 23.20

When all three point lights were placed, unfortunately all of these lights were located on the top of the base plate, as shown in Figure 23.21. The lights need to be assigned an elevation, or Z coordinate. To perform this task, activate the Render tab on the Ribbon and click the arrow at the lower-right corner of the Lights panel to display the Lights in Model palette, as shown in Figure 23.21.

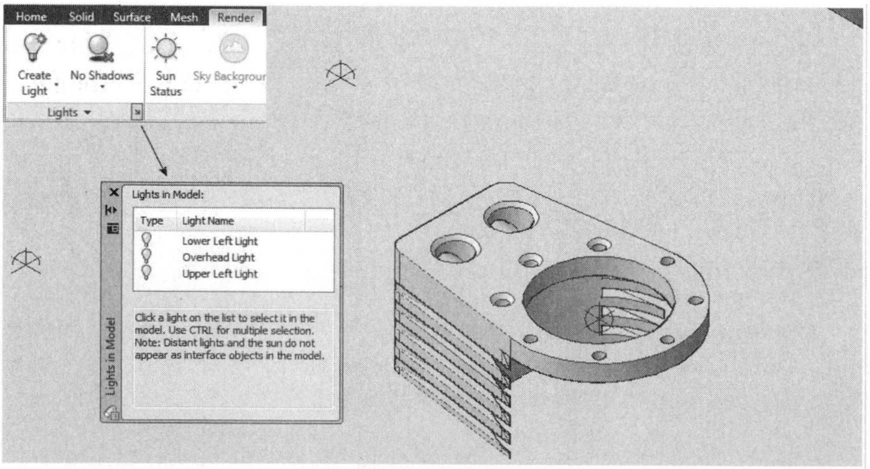

FIGURE 23.21

From the list of lights, double-click the Lower Left Light to display the Properties palette on this light. Locate the Position Z coordinate, located under the Geometry category of this palette, as shown in Figure 23.22, and change the value from 0 to 200. This elevates the Lower Left Light to a distance of 200 mm.

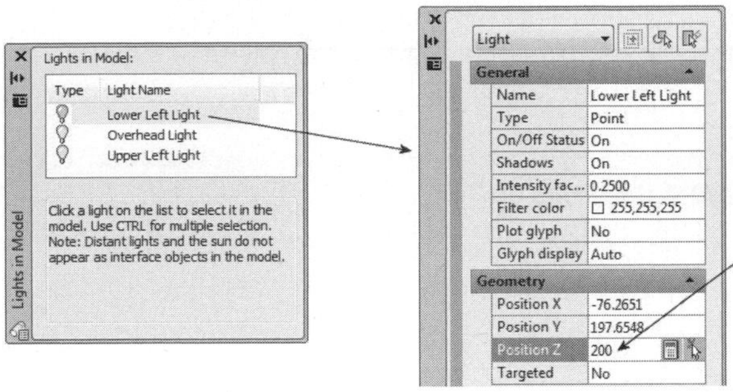

FIGURE 23.22

Continue changing the elevations of the remaining point lights. Change the Position Z coordinate value of the Overhead Light from 0 mm to 300 mm and the Position Z coordinate value of the Upper Left Light from 0 mm to 200 mm. Your display should appear similar to Figure 23.23.

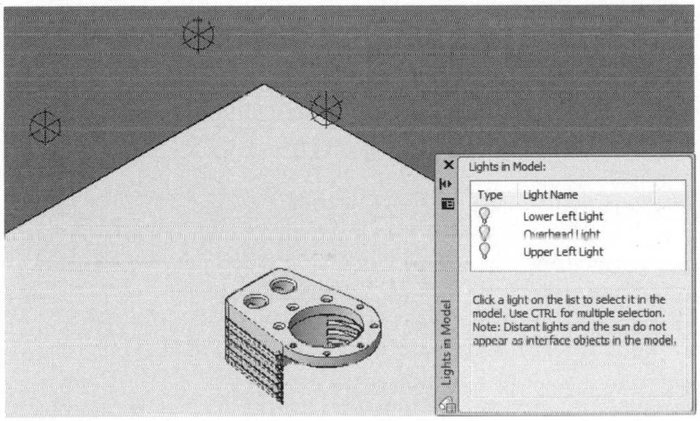

FIGURE 23.23

Next, place a spotlight into the 3D model by clicking the Spot button, as shown in Figure 23.24 (Right). Place the source for the spotlight at the approximate location at "A" and the spotlight target at the center of the bottom of the valve head at "B," as shown in Figure 23.24.

```
Command: SPOTLIGHT (Pick from the Lights panel of the Ribbon)

Specify source location <0,0,0>: (Pick the approximate loca-
tion for the spotlight at "A")

Specify target location <0,0,-10>: (Pick the bottom center of
the valve head at "B")

Enter an option to change

[Name/Intensity/Status/Hotspot/Falloff/shadoW/Attenua-
tion/Color/eXit] <eXit>: I (For Intensity)

<Enter intensity (0.00 - max float) <1.0000>: 0.50
```

Enter an option to change

[Name/Intensity/Status/Hotspot/Falloff/shadoW/Attenuation/Color/eXit] <eXit>: N *(For Name)*

Enter light name <Spotlight5>: **Spotlight**

Enter an option to change

[Name/Intensity/Status/Hotspot/Falloff/shadoW/Attenuation/Color/eXit] <eXit>: *(Press* ENTER *to create the light)*

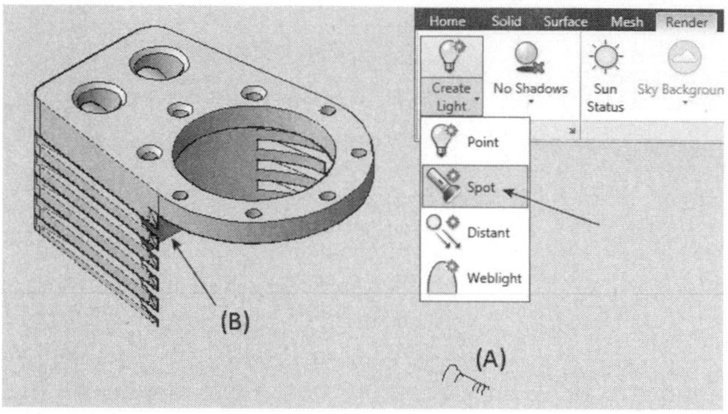

FIGURE 23.24

As with the point lights, the source of the spotlight is located at an elevation of 0 and needs to be changed to a different height. Select the spotlight icon, right-click and select Properties from the shortcut menu to display the Properties palette, and change the Position Z coordinate value to 200 mm, as shown in Figure 23.25 (Right).

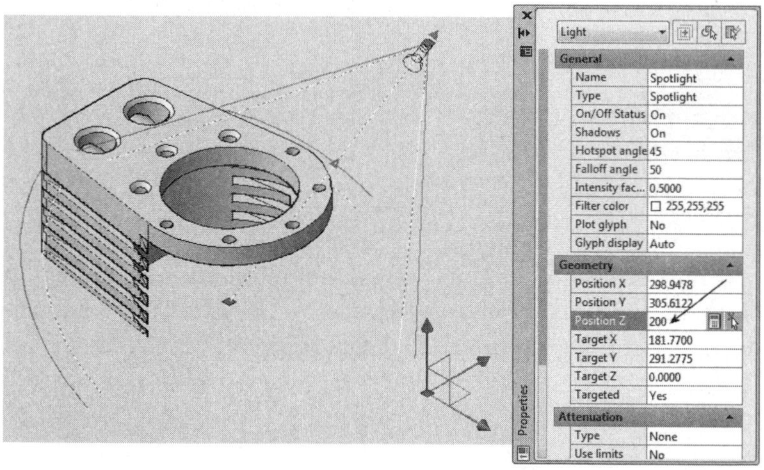

FIGURE 23.25

Finally, check to see that the render preset value is set to Medium and click the Render button to display the model, as shown in Figure 23.26. Shadows are automatically applied to the model from the lights. This concludes the exercise.

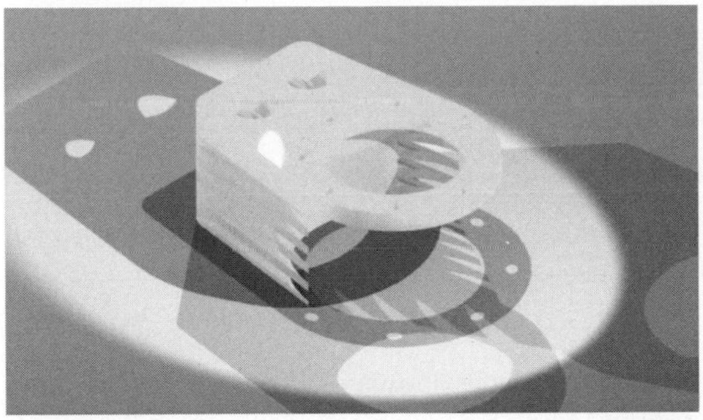

FIGURE 23.26

AN INTRODUCTION TO MATERIALS

Another way to make models more realistic and lifelike is to apply a material to the 3D model. The Material Browser palette, shown in Figure 23.27 (Left), can be utilized to apply, create, and organize materials. The Materials Editor palette, shown in Figure 23.27 (Right), can be used to modify material properties.

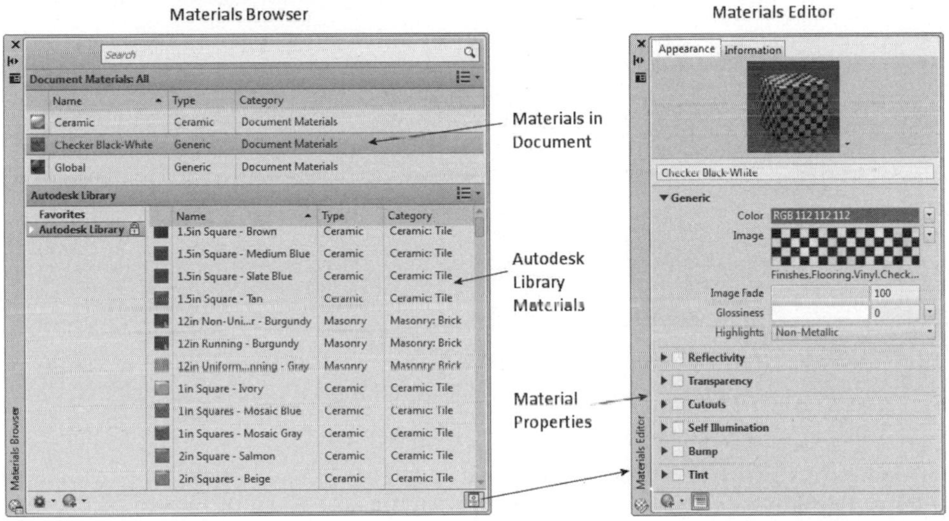

FIGURE 23.27

Both the Materials Browser and Materials Editor palettes can be activated from the Ribbon, as shown in Figure 23.28.

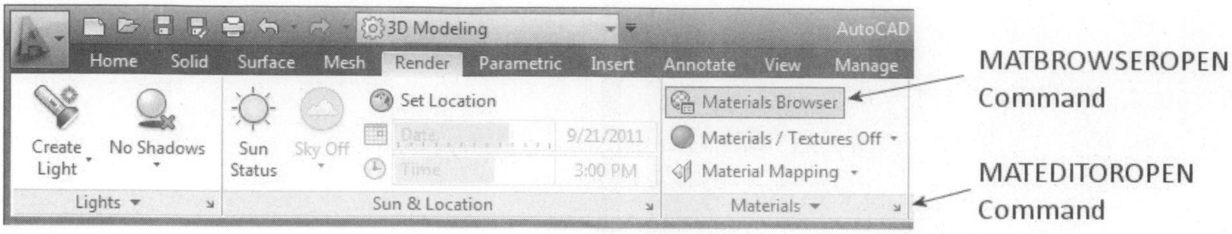

FIGURE 23.28

The top portion of the Materials Browser palette, as shown in Figure 23.29, displays the materials available in the current drawing. Selecting the Document Materials button on the palette provides a list of display control options. Figure 23.29 (Left) shows the default settings, which include displaying all materials in the document, sorted them by name, and showing in a list view format. The image on the right is displaying only those materials applied to the model, sorted by material type, and displayed as thumbnails. To help control the number of materials loaded in a drawing, you may want to use the "Purge All Unused" option to remove listed materials that have not been applied.

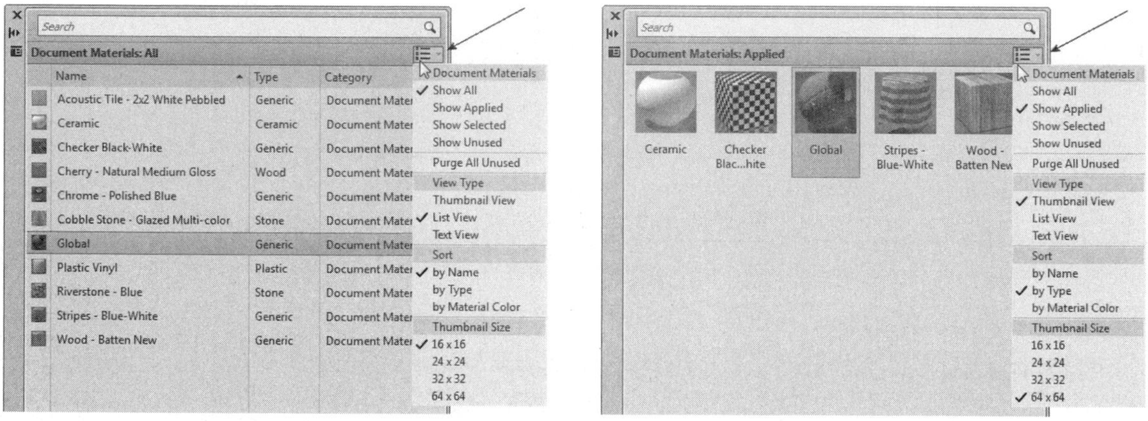

FIGURE 23.29

The bottom portion of the Materials Browser palette, as shown in Figure 23.30, displays material libraries on the left (Library Tree) and materials available in the selected library on the right. More than 700 materials are available through the default Autodesk Library. Additional libraries are available or can be created to store user-defined materials. A Manage button in the lower-left corner of the palette is available to assist with opening, creating, and organizing libraries. A second button in the lower-left corner of the palette is available for the creation of new materials. Picking this button displays a list of templates available to assist with the creation of your own custom material. A button in the lower-right corner of the palette will open the Material Editor palette and will display the settings for any selected material. The final button in this palette, in the upper-right portion of the materials area, is the Library button. This button provides display options allowing you to sort the listed materials (by Name, by Type, by Material Color, or by Category), change the view format (Thumbnail View, List View, or Text View), or change the selected library. A Show/Hide Library Tree option allows you to collapse and expand the Library Tree area. By hiding the tree, more materials can be displayed.

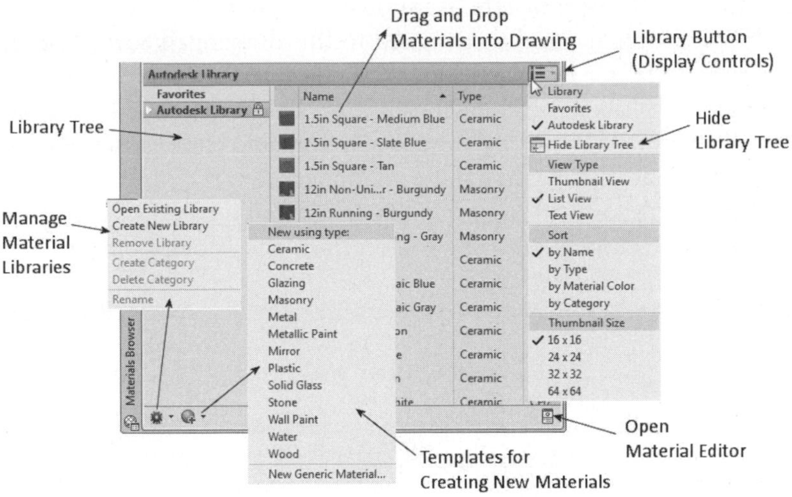

FIGURE 23.30

WORKING WITH MATERIALS

The primary method of applying materials in a drawing is to open the Materials Browser, select a material from the list, and drag and drop the material onto the 3D model. If a material is dropped into a blank area of the drawing, it is loaded but not yet attached to an object. It will appear in the Document Materials area (top portion) of the Browser.

TIP

The drag and drop technique can be used to copy materials between libraries in the Browser or to place your favorite materials on a tool palette.

TRY IT!

Open the drawing file 23_Connecting Rod. A single point light source has already been created and placed in this model. A realistic visual style is currently being applied to the model, as shown in Figure 23.31. You will attach a material from the Materials Browser palette and observe the rendering results. After removing this material, you will create a new material, change a few settings, and observe these rendering results.

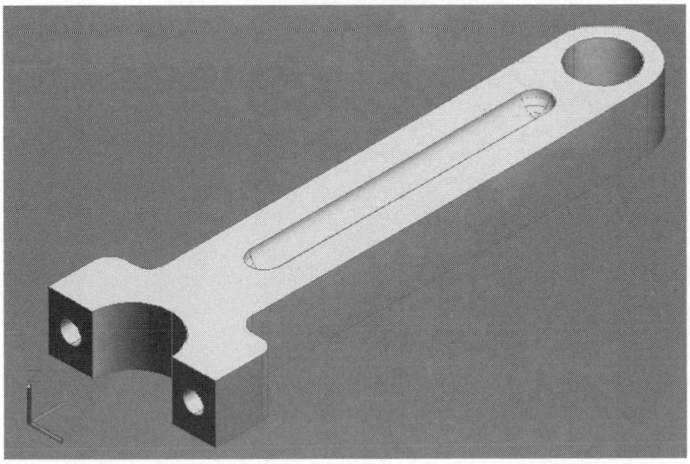

FIGURE 23.31

First, activate the Materials Browser palette. As shown in Figure 23.32 (Left), notice that "Global" is the only material currently assigned to this drawing. To assist in the location of a Copper material for this part, expand the library categories by picking the arrow next to Autodesk Library in the Library Tree. Select the Metals category to display all the available metal materials. The list of available choices is still quite long. To more efficiently find the material, enter "Copper" into the Search text box at the top of the palette, as shown in Figure 23.32 (Right). To better visualize the material choices, select the Thumbnail View option from the Library button. The results are shown in Figure 23.32 (Right).

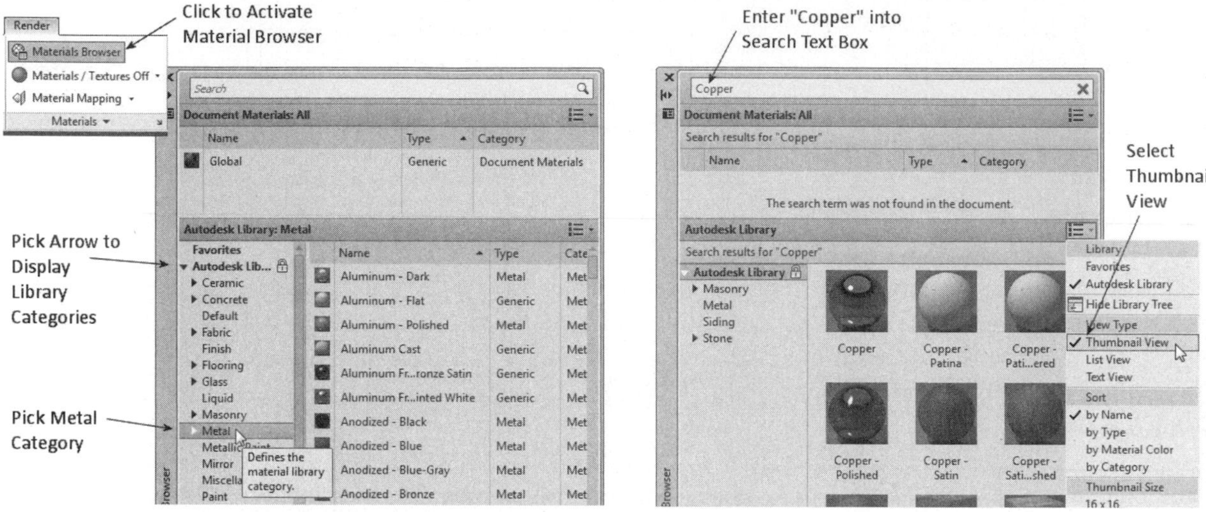

FIGURE 23.32

Verify that materials and textures are turned on in the Materials panel of the Ribbon, as shown in Figure 23.33. Since the Realistic visual style is current for this drawing, this setting is automatically made for you. Drag the Copper material swatch from the Browser and drop it on the Connecting Rod. The results of applying the copper material to the model are shown in Figure 23.33.

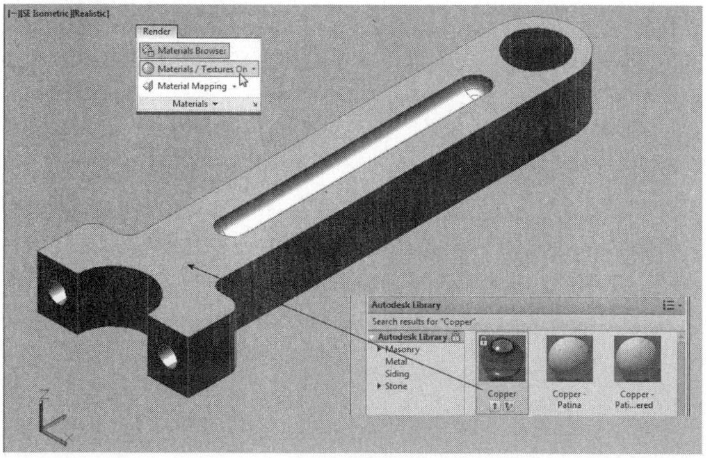

FIGURE 23.33

NOTE

To load a material into a document without attaching it to a model, simply drag the material from the browser and drop it in a blank area of the screen. You can also load a material into a document by moving your cursor over a material and picking one of two buttons that are automatically displayed. These buttons are shown in Figure 23.33. Picking the left button just loads the material, while picking the right loads the material and opens it in the Material Editor palette.

Before producing the rendering of this object, activate the Materials Editor palette by double-clicking the Copper material listed in the Document Materials area of the Browser or right-click on the material and select Edit from the shortcut menu, as shown in Figure 23.34 (Left). Once the Materials Editor palette displays, we will modify the Copper material to brighten the rendering of the object. Place a check next to Self-Illumination to turn this mode on and change its Luminance value to 100 by selecting LED Panel from the drop-down list, as shown in Figure 23.34 (Right).

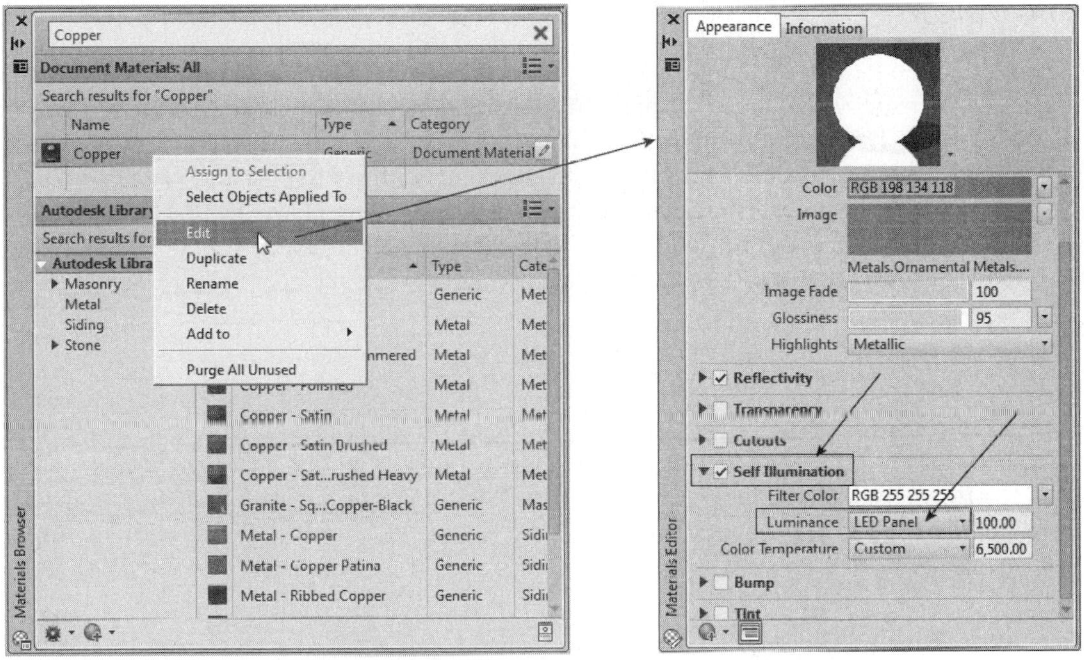

FIGURE 23.34

Perform a rendering with the rendering style set to Medium. The results of the rendering are shown in the following image. In addition to the copper material being applied, shadows are cast along a flat surface from the existing light source.

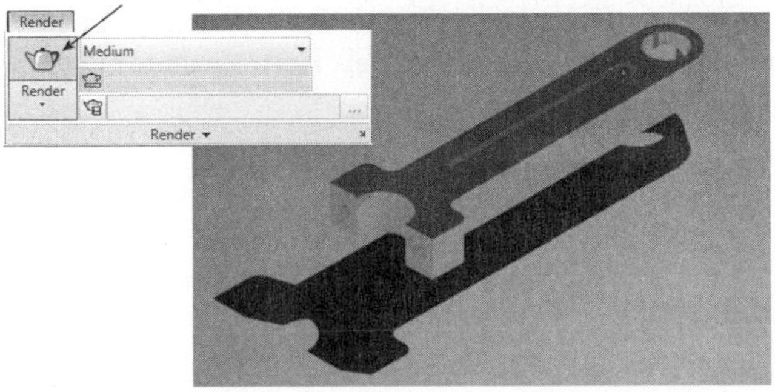

FIGURE 23.35

Before creating a new material for the next part of this exercise, click the "X" in the Search text box to stop the current search. Right-click the Copper material in the Document Materials area of the Browser to display the shortcut menu shown in Figure 23.35 (Center). Selecting Delete from the menu displays the Material in Use dialog box shown in Figure 23.35 (Right). Clicking Yes removes the material from the top portion of the Browser and from the 3D model. Notice how the original realistic color replaces the copper color on your screen.

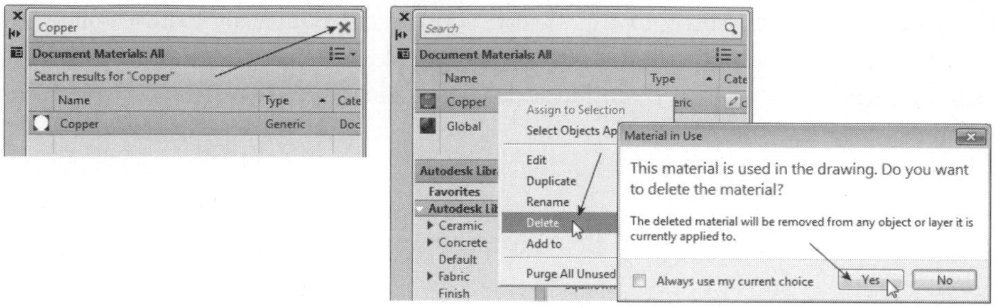

FIGURE 23.36

To create a new material, pick the "Creates a new material" button at the bottom of the Browser. A list of available material templates is displayed, as shown in Figure 23.37 (Left). Select New Generic Material from the list and when the Materials Editor palette is displayed enter Red Metal in the Name text box, as shown in Figure 23.37 (Right).

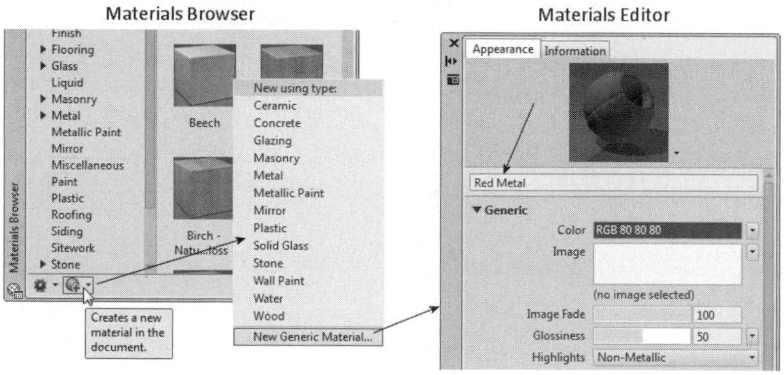

FIGURE 23.37

The first change to our new material will be the color. Click the Color box or select the drop-down arrow next to the box and select Edit Color..., as shown in Figure 23.38 (Left). When the Select Color dialog box appears, change to the Index Color tab and change the color to the one shown in Figure 23.38 (Right). Click the OK button to accept the material and return to the Materials Editor palette.

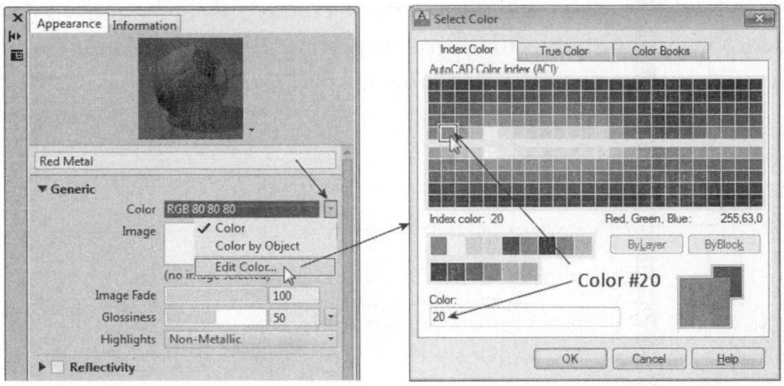

FIGURE 23.38

Now we are ready to apply our material to the Connecting Rod. Notice that Red Metal is now displayed in the Document Materials area of the Materials Browser palette, as shown in Figure 23.39 (Left). Drag and drop this material onto the 3D model, as shown in Figure 23.39 (Right).

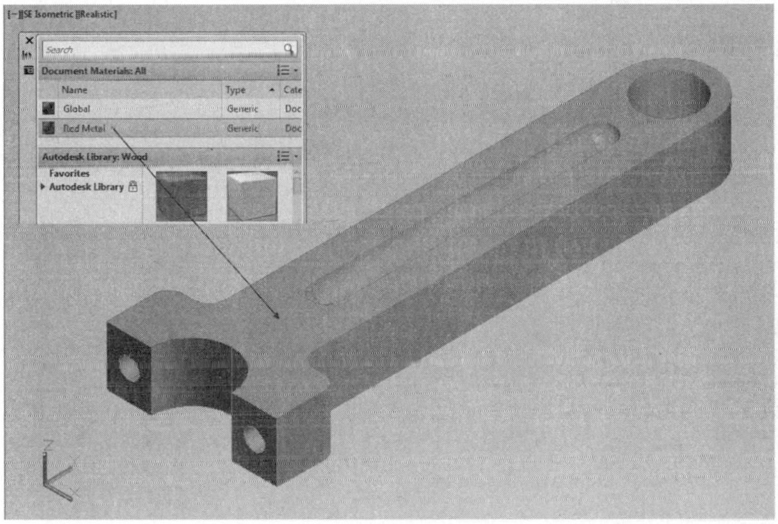

FIGURE 23.39

Producing a rendering of the 3D model with the new material should result in an image similar to Figure 23.40. When finished, exit the rendering mode and return to the 3D model.

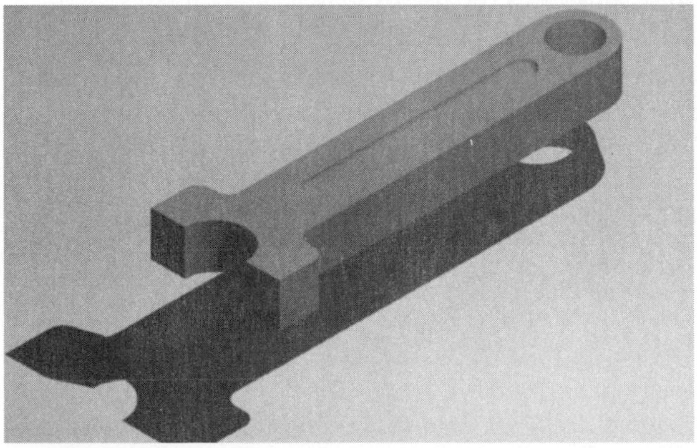

FIGURE 23.40

While the Red Metal material is still active in the Editor palette, place a check in the box next to Transparency to expose the available settings, and change the value to 50 by sliding the bar or entering the value, as shown in Figure 23.41.

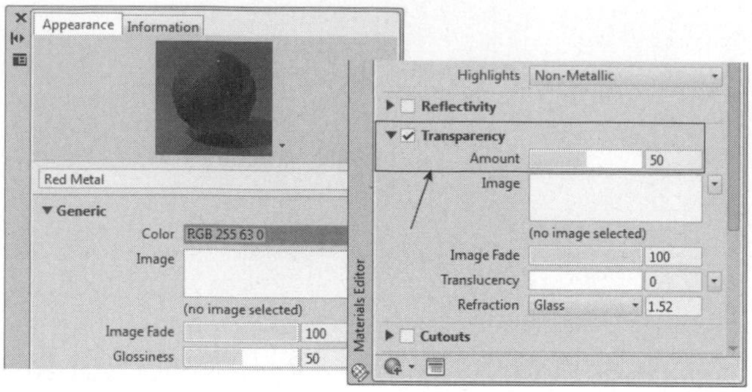

FIGURE 23.41

Producing a new rendering displays the model, as shown in Figure 23.42, complete with transparent material and shadows. This concludes the exercise.

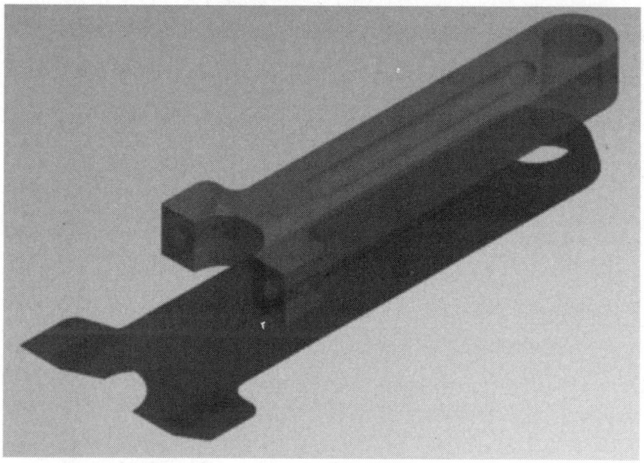

FIGURE 23.42

USING MATERIAL TEMPLATES

Using a materials template is one way to help automate the creation process for new materials. The next Try It! exercise illustrates the use of materials templates.

TRY IT!

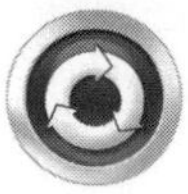

Open the file 23_Piston Mirror, as shown in Figure 23.43. A number of lights have already been placed in this model. Also, the model is being viewed through the Realistic visual style. In this exercise, you will create two new materials. One of the materials will contain mirror properties and be applied to the piston. When you perform a render operation, the reflection of one of the piston rings will be visible in the top of the piston.

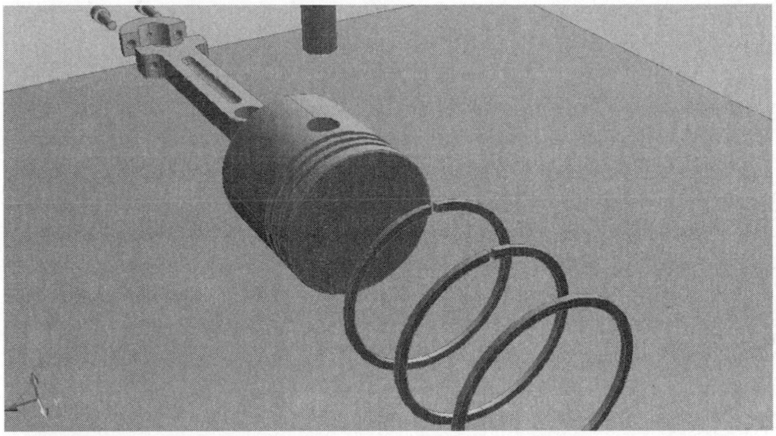

FIGURE 23.43

Begin by launching the Materials Editor palette and expanding the Create Material button, as shown in Figure 23.44 (Center). A number of templates are available to assist with the creation of special material types such as a mirror property. Selecting the Mirror template creates a new material named Default Mirror, as shown in Figure 23.44 (Right).

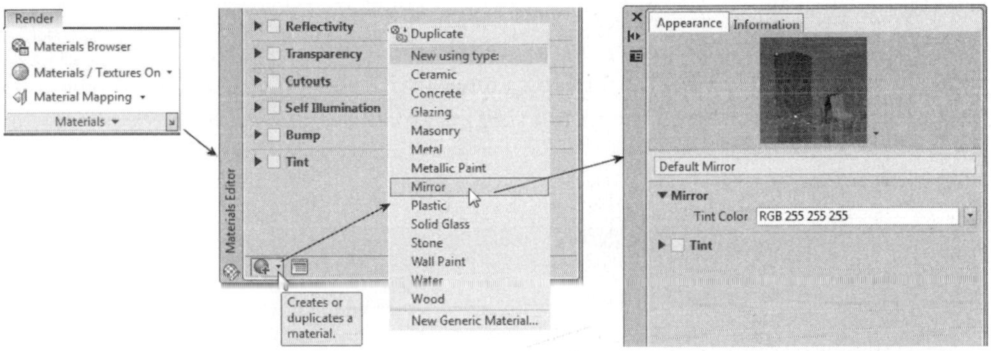

FIGURE 23.44

Change the material name from Default Mirror to Mirror, as shown in Figure 23.45 (Left). To best display the mirror property, a light color will be selected. First, click the Color box or select the drop-down arrow next to the box and select Edit Color…; this launches the Select Color dialog box. Click the Index Color tab and choose one of the shades of yellow, as shown in Figure 23.45 (Right).

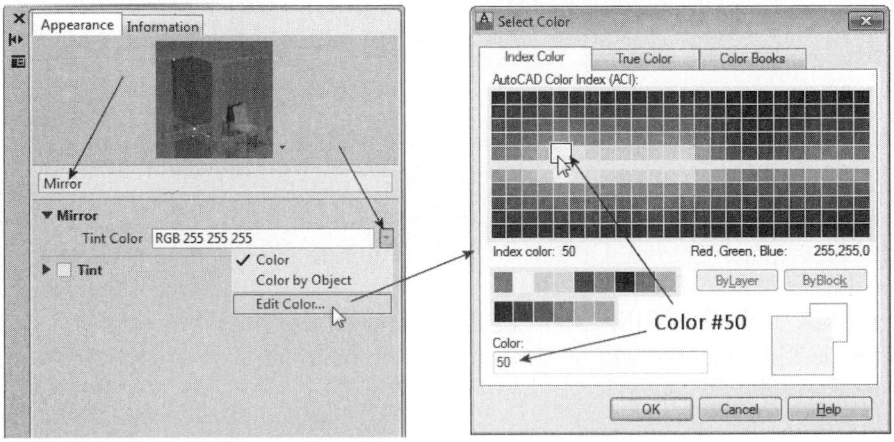

FIGURE 23.45

Next, we will apply this new material to the piston. Open the Materials Browser palette by picking the Browser button on the Editor palette, as shown in Figure 23.46 (Left). From the Browser, drag the Mirror material and drop it on the piston, as shown in Figure 23.46 (Right).

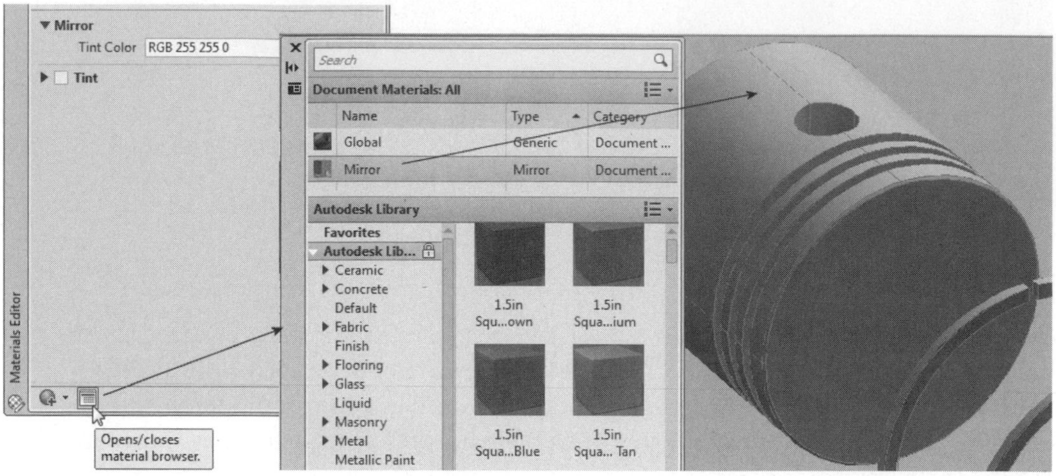

FIGURE 23.46

Test the mirror property by performing a render with the render style set to Medium. Notice in Figure 23.47 that the piston ring and shadows are visible in the piston due to the mirror material.

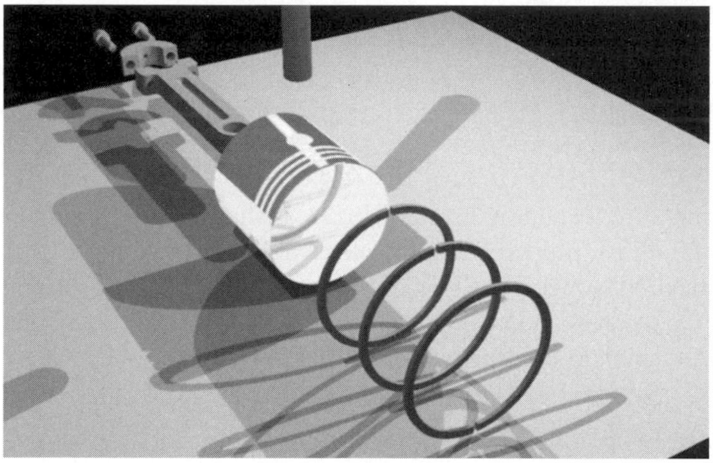

FIGURE 23.47

Create another material called Piston Support Parts. From the Editor palette, expand the list of templates and select Metallic Paint, as shown in Figure 23.48 (Left). Rename the new material from Default Metallic Paint to Piston Support Parts, as shown in Figure 23.48 (Right). Click the Color box or select the drop-down arrow next to the box and select Edit Color…; choose one of the shades of green for this new material. This material will be applied to the remainder of the parts that form the piston assembly.

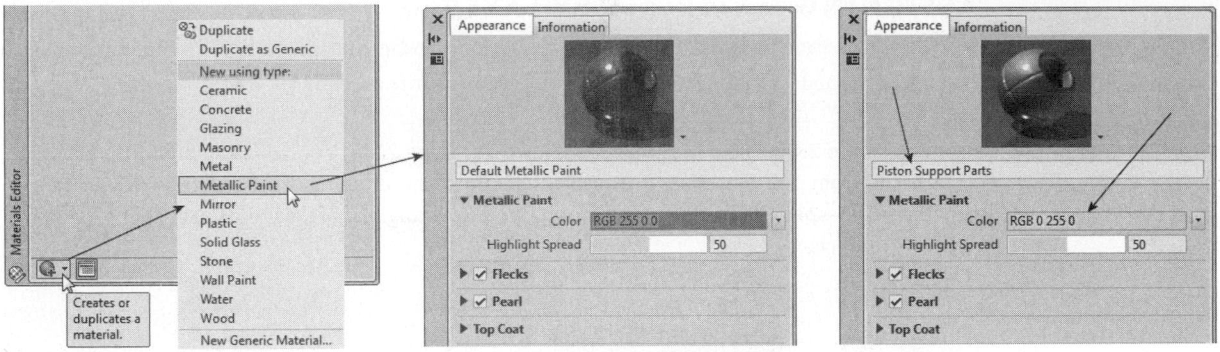

FIGURE 23.48

Preselect the remainder of the piston parts, as shown in Figure 23.49 (leave the main piston set to the mirror material.) In the Materials Browser palette, pick the Piston Support Parts material swatch to attach the material to the preselected 3D models.

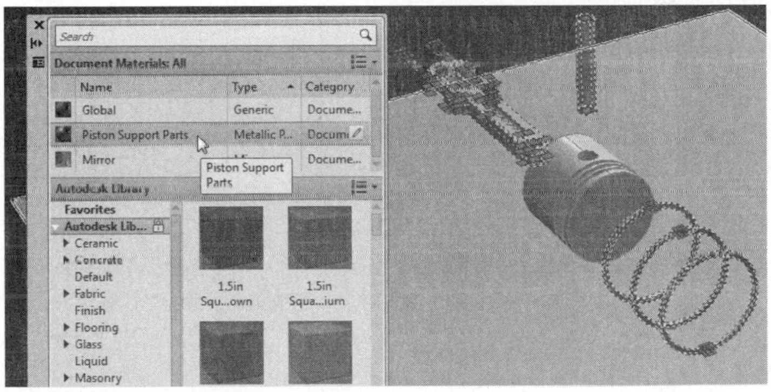

FIGURE 23.49

Perform another rendering test. Your image should appear similar to Figure 23.50. This concludes the exercise.

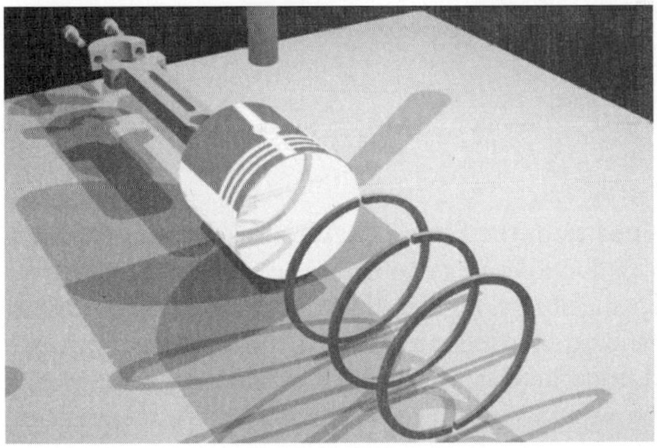

FIGURE 23.50

ASSIGNING MATERIALS BY LAYER

When a number of material assignments need to be made, such as in the case shown in Figure 23.51, it can be advantageous to assign materials by layer.

TRY IT!

Open the file 23_Interior Materials, as shown in Figure 23.51. You will drag and drop existing materials from the Materials Browser palette into the drawing. You will then assign these materials to specific layers and perform the rendering. Lights have already been created for this exercise.

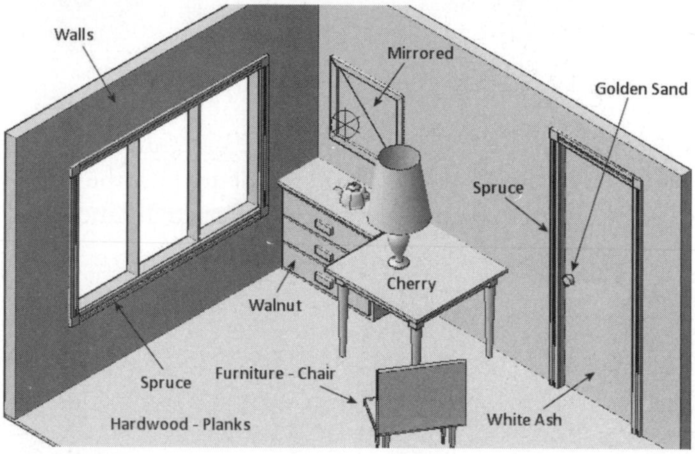

FIGURE 23.51

Activate the Materials Browser palette and notice that two materials have already been created in this drawing. The first material, shown in Figure 23.52 (Left), is a fabric color designed to be applied to the chair. The second material, shown in Figure 23.52 (Right), is a paint color to be applied to the walls of the 3D model.

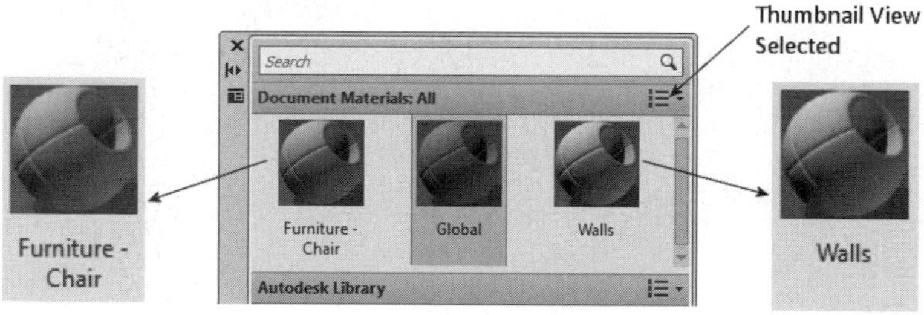

FIGURE 23.52

All other materials will be obtained from the Browser's Autodesk Library. You will load the materials into the drawing for your use. You will be selecting materials from various categories arranged under the library. Picking the arrow next to the Autodesk Library rotates the arrow down and exposes the categories, as shown in Figure 23.53. The first material will be loaded from the Ceramic—Porcelain category, as shown in Figure 23.53. In some cases, you will have to move your cursor over a material and leave it stationary in order for the whole material name to be displayed. Locate the first material, Golden Sand, as shown in Figure 23.53 (Right). Press and hold down your mouse button over this material, drag the material icon into your drawing, and drop it to load it. You can instead load the material by selecting the "Add material to document" button which is displayed as you hover your cursor over the material icon.

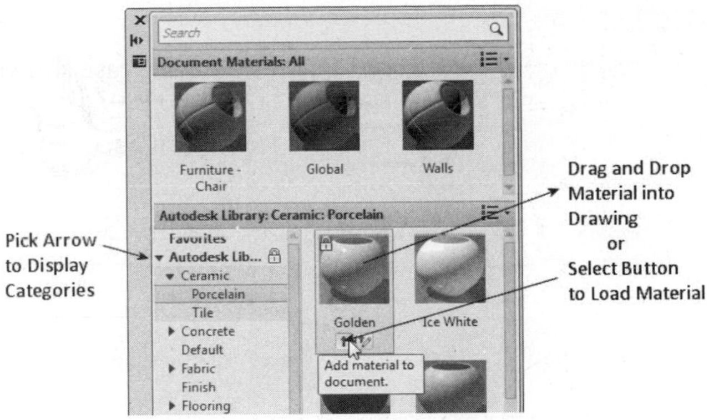

FIGURE 23.53

Continue loading the following materials located in the appropriate categories using the table below as a guide:

Category	Material
Ceramic — Porcelain	Golden Sand
Flooring — Wood	Hardwood — Planks
Glass — Glazing	Mirrored
Wood	Cherry
Wood	Spruce
Wood	Walnut
Wood	White Ash

When you have finished loading all materials, you can check the status of the load by expanding the Materials Browser palette, as shown in Figure 23.54. All materials that can be applied to 3D models in the drawing are displayed in the top portion of the Browser.

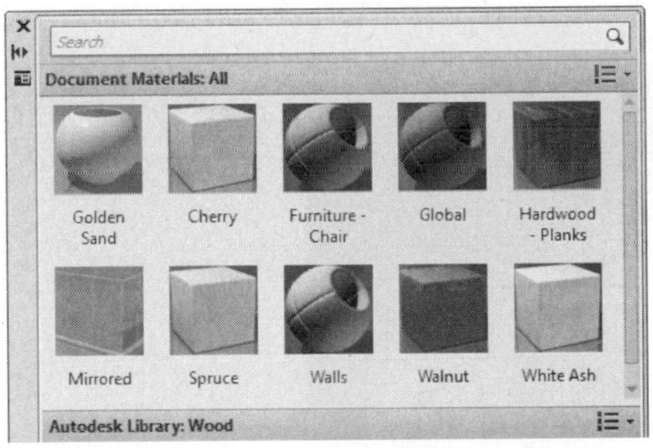

FIGURE 23.54

The next step is to assign a material to a layer. From the Ribbon, click the Attach By Layer button, as shown in Figure 23.55 (Left). This launches the Material Attachment Options dialog box, as shown in Figure 23.55 (Right). Now you will drag a material located in the left column of the dialog box and drop it onto a layer located in the right column of the dialog box. In this example, the Golden Sand material has been dropped onto the Door Knob layer.

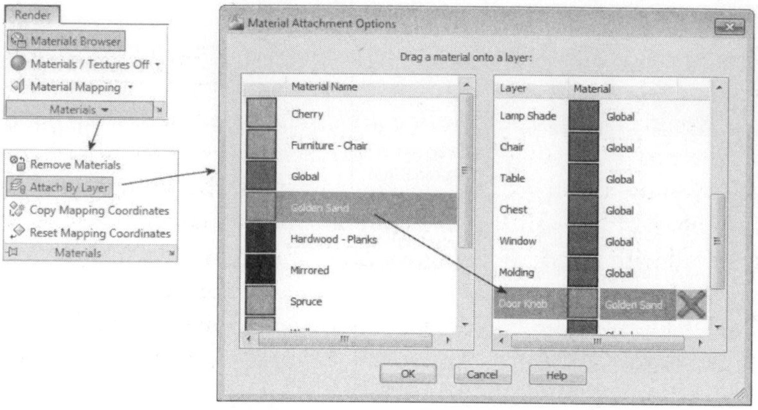

FIGURE 23.55

Using Figure 23.56 as a guide, continue assigning materials to layers by dragging and dropping the materials onto the appropriate layers.

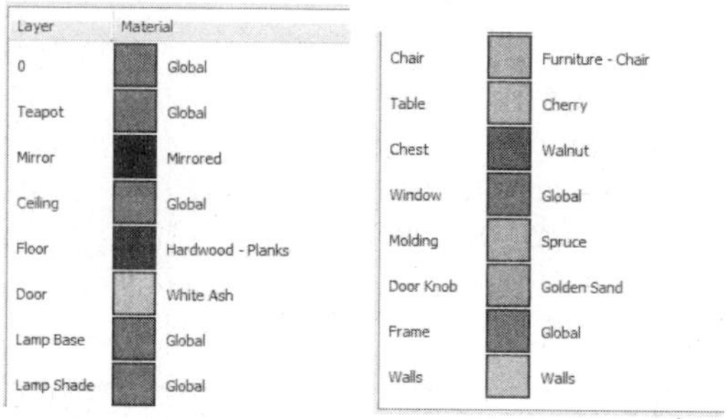

FIGURE 23.56

When all material assignments have been made to the layers, click OK. Render out the design and verify that the materials are properly assigned to the correct 3D objects, as shown in Figure 23.57.

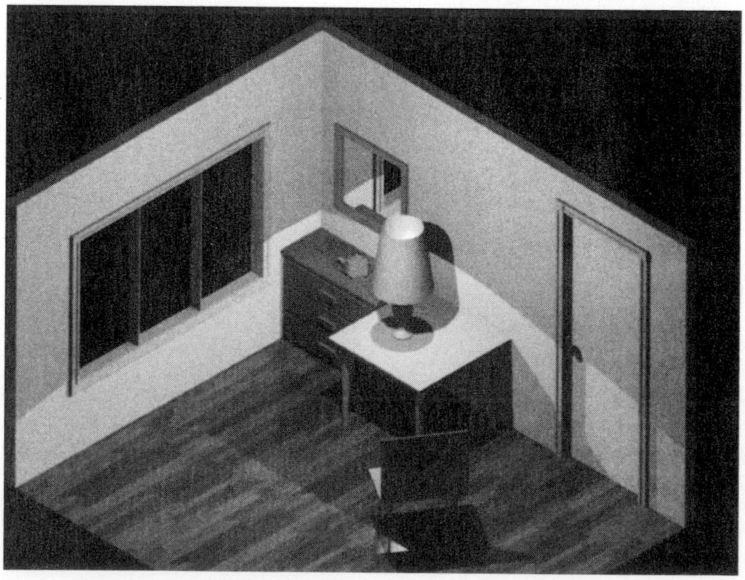

FIGURE 23.57

APPLYING A BACKGROUND

Images can be placed behind a 3D model for the purpose of creating a background effect. Backgrounds can further enhance a rendering. For example, if you have designed a 3D house, you could place a landscape image behind the rendering. To place a background image, the image must be in a raster format such as BMP, TGA, or TIF. The process of assigning a background image begins with the View Manager dialog box (the `VIEW` command), as shown in Figure 23.58. You create a new view and associate a background with the view. In Figure 23.58, the New button is picked, which launches the New View dialog box. Enter a new name for the view and select a background type. Besides Default, three background types are available: Solid, Gradient, and Image. Selecting one of the background types (Image in this example) launches the Background dialog box, where you can specify the background details.

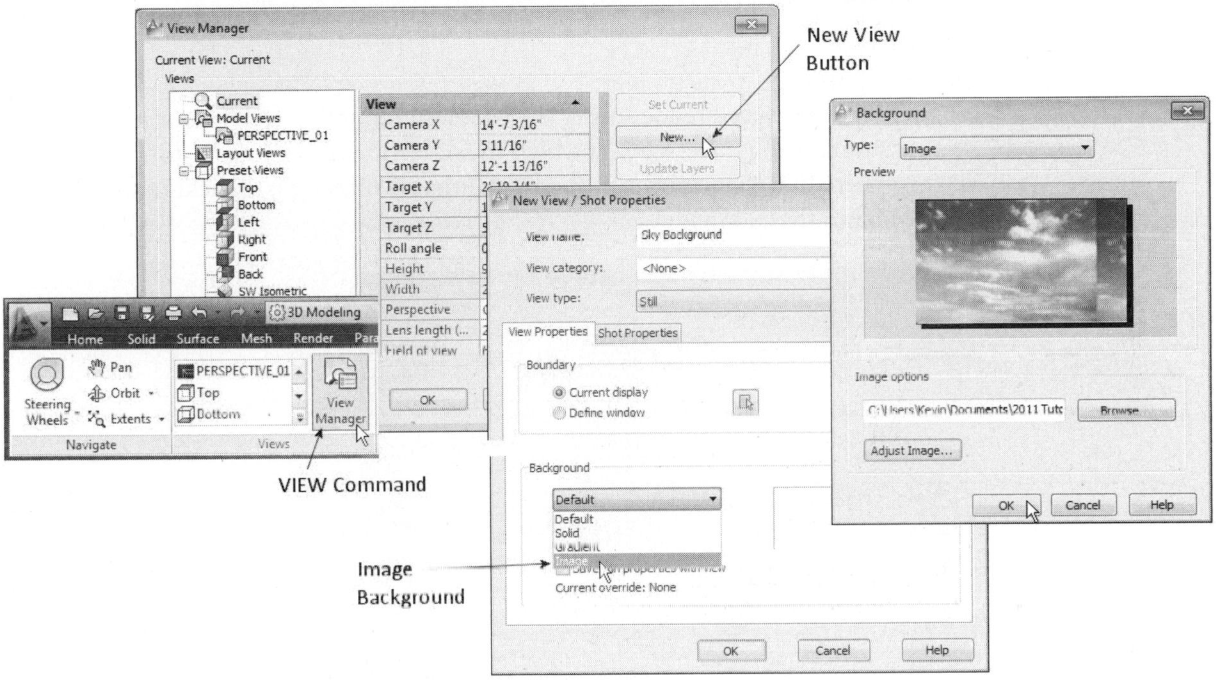

FIGURE 23.58

These background types are explained as follows:

Solid—A solid background means that AutoCAD replaces the default white (or black) background of the drawing screen with another color. You choose the color from the Colors section of the dialog box.

Gradient—A gradient means that the color changes from one end of the screen to the other, such as from red at the bottom to light blue at the top (to simulate a sunset). The option is available to select a two-color or three-color gradient. A Rotation option is also available to rotate the gradient.

Image—You select a raster image for the background. The image can be in BMP (Bitmap), GIF, PNG, TGA (Targa), TIFF (Tagged Image File Format), JFIF (JPEG File Interchange Format), or PCX (PC Paintbrush) format.

TRY IT!

Open the drawing file 23_Piston Background. The Realistic visual style is applied to this model, as shown in Figure 23.59. Also applied are four point lights. A special mirror material is attached to the sphere.

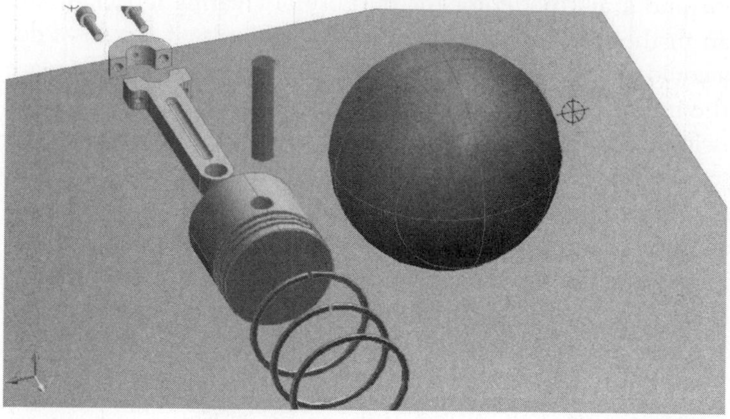

FIGURE 23.59

Begin by performing a render using the Medium rendering style. Your image should appear similar to Figure 23.60, in which the reflections of various piston parts appear in the sphere due to its mirror property. All that is missing is a background that will be applied to the 3D models.

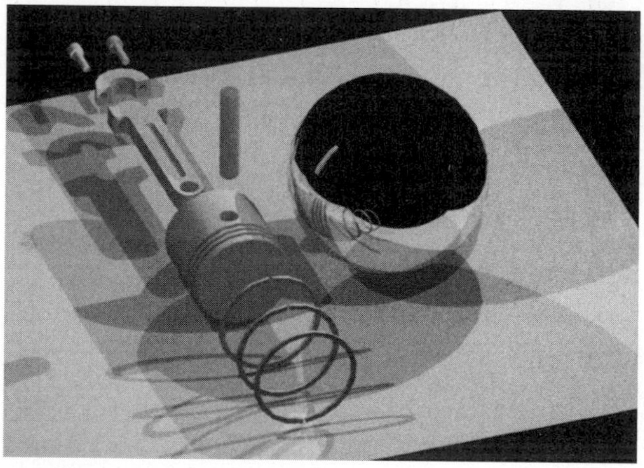

FIGURE 23.60

All visual style and rendering backgrounds are controlled in the View Manager dialog box, as shown in Figure 23.61. Begin the selection of a background by clicking the New button.

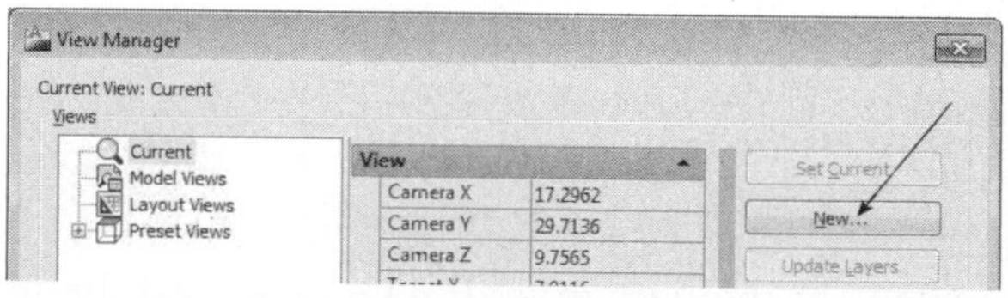

FIGURE 23.61

You must first create a new view and then assign a background to this view. When the New View/Shot Properties dialog box appears, as shown in Figure 23.62 (Left), enter a name for the view, such as Sky Background. In the background area, click on the Default drop-down list and pick Image, as shown in Figure 23.62 (Left). This launches the Background dialog box shown in Figure 23.62 (Right). Click the Browse button to search for valid image files.

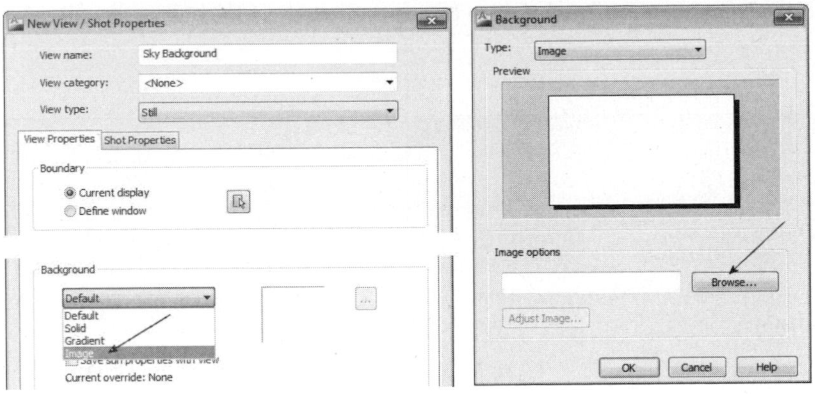

FIGURE 23.62

Once you click the Browse button in the Background dialog box, the Select File dialog box appears, as shown in Figure 23.63. Locate the folder in which the Chapter 23 Try It! exercises are located. From this list, find sky.tga, select it, and click the Open button.

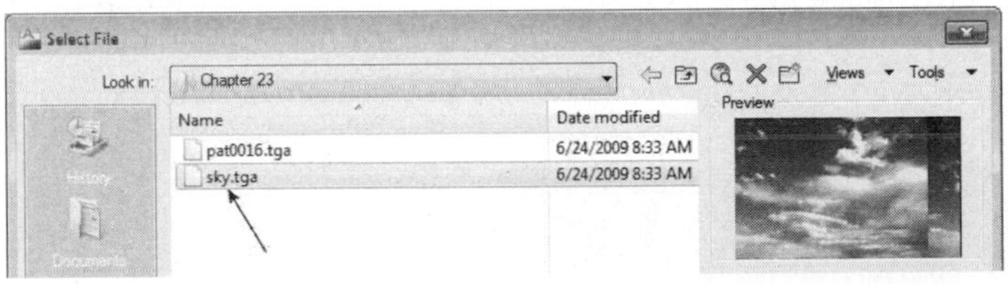

FIGURE 23.63

This takes you to the Background dialog box again. The sky graphic appears small but centered on the sheet, as shown in Figure 23.64 (Left). To control the display of this file in the final rendering, click the Adjust Image button to launch the Adjust Background Image dialog box and change the Image position to Stretch. This should make the sky graphic fill the entire screen, as shown in Figure 23.64 (Right).

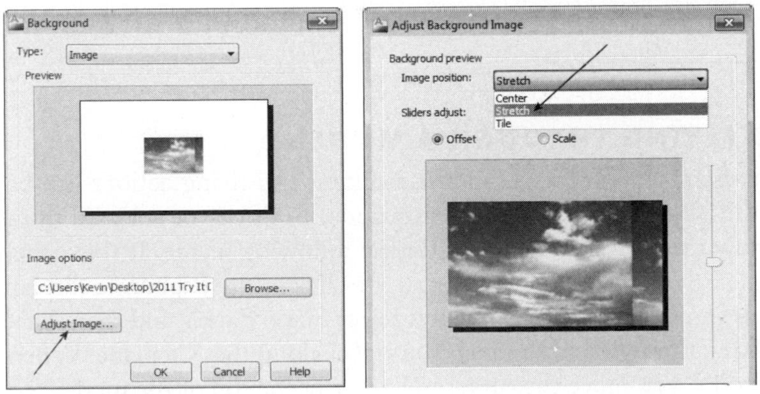

FIGURE 23.64

Click the OK buttons in the Adjust Background Image, Background, and New View/Shot Properties dialog boxes to return to the View Manager dialog box, where the Sky Background is now part of the Views list. Click the Set Current button to make this view current in the drawing and click the OK button to dismiss this dialog box.

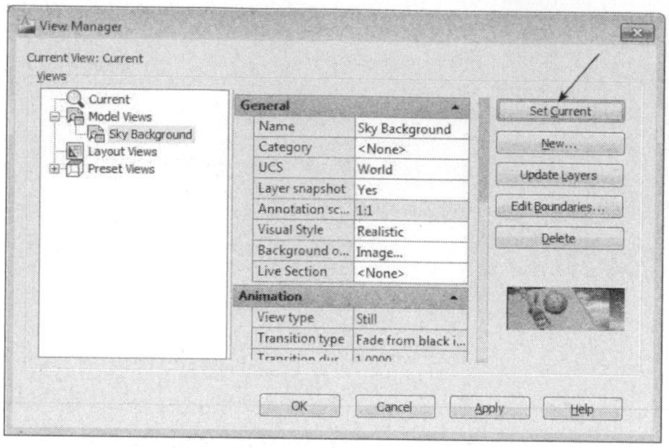

FIGURE 23.65

Perform a render using the Medium rendering style and notice the results, as shown in Figure 23.65. With the image applied, it appears that the flat base sheet is floating in air. Also, since the sphere still has the mirror material property, the sky is reflected here in addition to the piston parts. This concludes the exercise.

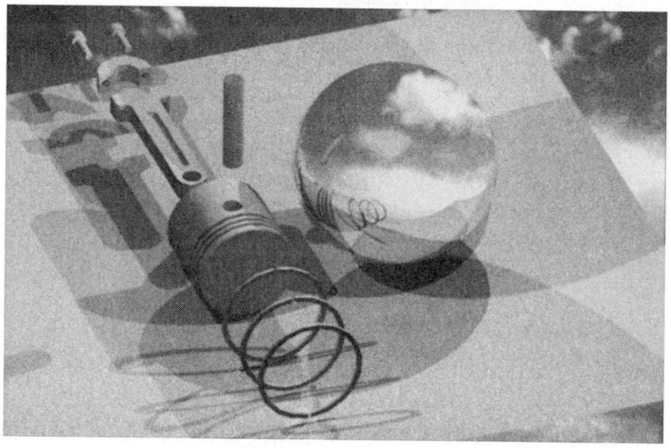

FIGURE 23.66

WALKING AND FLYING THROUGH A MODEL

To further aid with visualization of a 3D model, walking and flying actions can be simulated through the `3DWALK` and `3DFLY` commands. Both can be selected from the Animations panel of the Ribbon, as shown in the following image. If this panel does not appear in the Render tab of the Ribbon, right-click in the Ribbon to display the shortcut menu, as shown in Figure 23.67, and select Show Panels and then Animations. When walking through a 3D model, you travel along the XY plane. When flying through the model, you move the cursor to look over the top of the model.

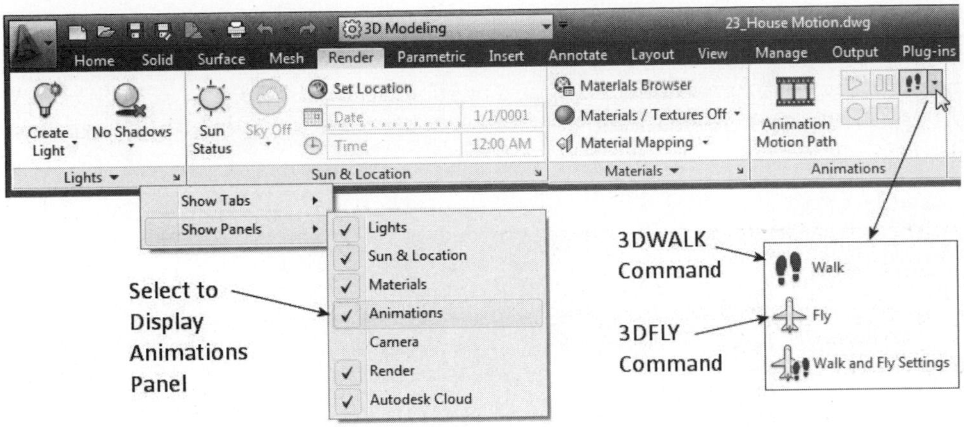

FIGURE 23.67

When you first activate the 3DWALK or 3DFLY command, a warning dialog box appears, as shown in Figure 23.68, stating that you must be in Perspective mode to walk or fly through your model. Click the Change button to enter Perspective mode.

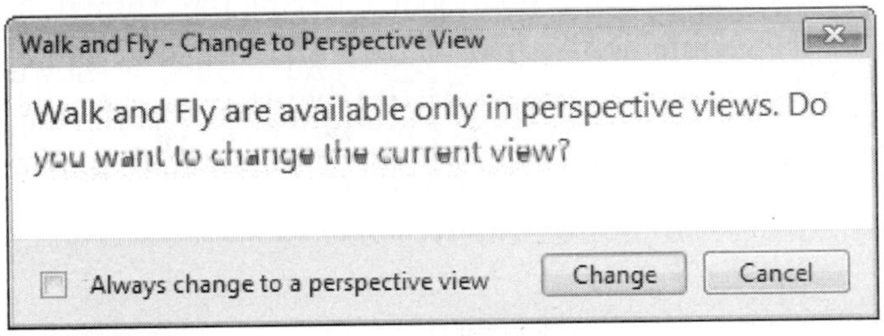

FIGURE 23.68

When you first enter the walk or fly mode, a Position Locator palette appears, as shown in Figure 23.69 (Left). It gives you an overall view of the position of the camera and target in relation to the 3D model. You can drag on the camera location inside the preview pane of the Position Locator to change its position. You can also change the target as you adjust the viewing points of the 3D model. In the drawing, dragging the left mouse button also changes the viewing direction. Use the arrow keys on your keyboard to step in the arrow direction. Click the Walk and Fly Settings button to display the dialog box, as shown in Figure 23.69. Use the Current drawing settings area of the dialog box to change the step size and speed, if desired.

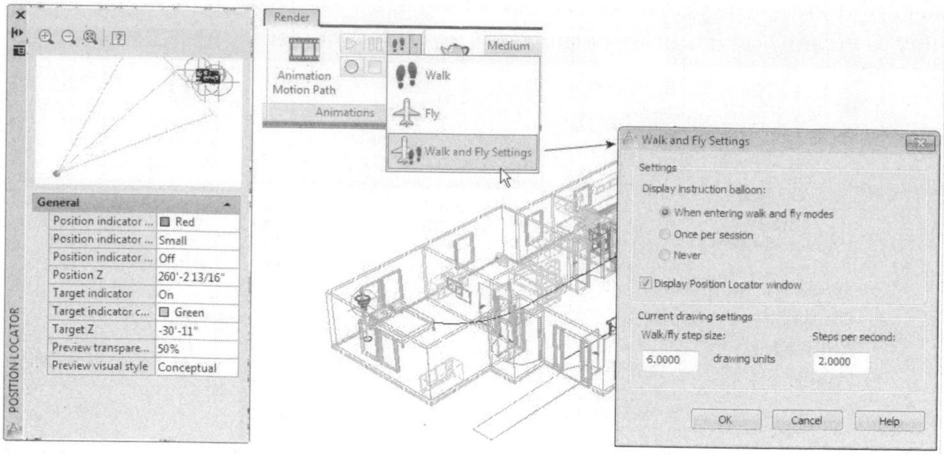

FIGURE 23.69

ANIMATING THE PATH OF A CAMERA

The ability to walk or fly through a model has just been discussed. This last segment will concentrate on creating a motion path animation by which a camera can follow a predefined polyline path to view the contents of a 3D model. Clicking the Animation Motion Path button found in the Animations panel of the Ribbon, launches the Motion Path Animation dialog box shown in Figure 23.70 (Right). You select the path for the camera and target in addition to changing the number of frames per second and the number of frames that will make up the animation.

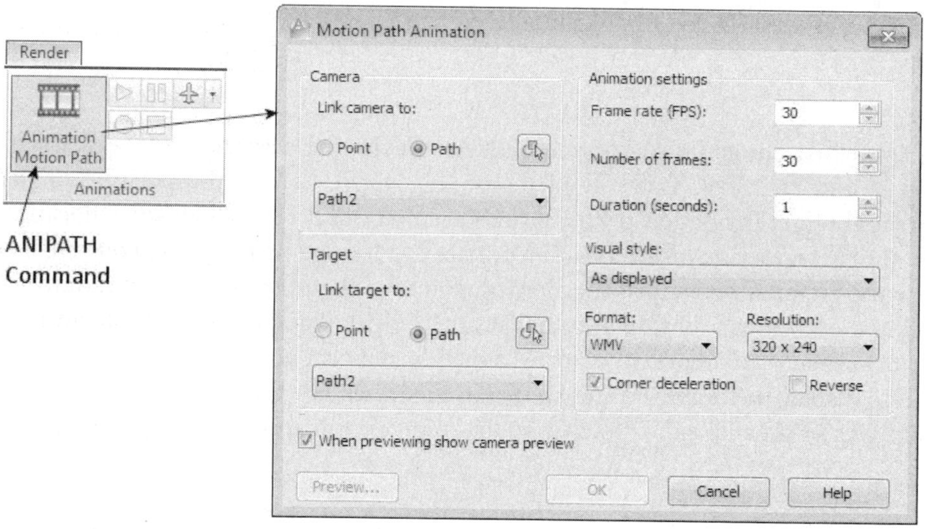

FIGURE 23.70

After making changes to the Motion Path Animation dialog box, you have the opportunity to preview the animation before actually creating it. A sample animation preview is shown in Figure 23.71. You can even see the relative position the camera is in as it passes through the 3D model along the polyline path. After the preview is finished, clicking the OK button in the main Motion Path Animation dialog box creates the animation and writes the results out to a dedicated file format. Supported formats include AVI, MOV, MPG, and WMV. Depending on the resolution and number of frames, this process could take a long time. However, it gives you the capability of creating an animation using any kind of 3D model.

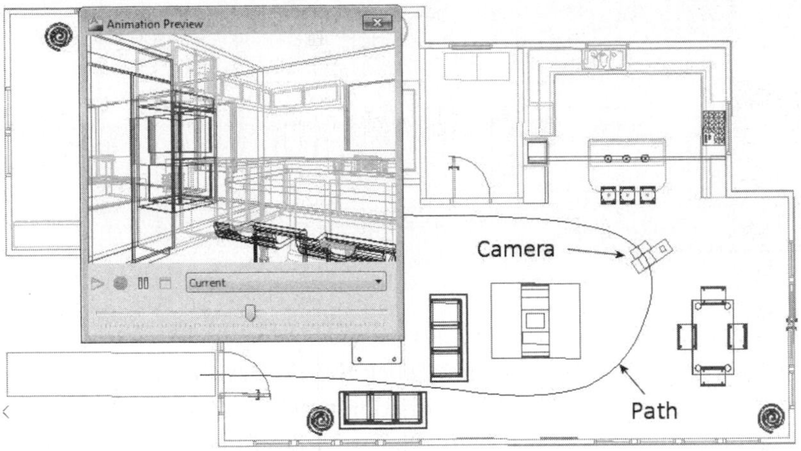

FIGURE 23.71

TRY IT!

Open the drawing file 23_House Motion. A polyline path has already been created at an elevation of 4' to simulate an individual walking through this house. Once you have created the motion path animation, this polyline path will not be visible when the animation is played back.

Begin by clicking Animation Motion Path button, which is found in the Animations panel of the Ribbon. When the Motion Path Animation dialog box appears, as shown in Figure 23.72, make the following changes:

- Click the Select Path button in the Camera area and pick the polyline displayed in the floor plan. If necessary, change the path name to Path1. If an AutoCAD Alert box appears, click the Yes button to override the existing path name.
- Click the Select Path button in the Target area and pick the polyline displayed in the floor plan. If necessary, change the path name to Path2. If an AutoCAD Alert box appears, click the Yes button to override the existing path name. Both the Camera and Target will share the same polyline path.
- Change the Frame rate (frames per second) from 30 to 60.
- Change the Number of frames from 30 to 600. This updates the Duration from 1 to 10 seconds.
- Change the Visual style to Realistic.
- Change the Format to AVI.
- Keep the resolution set to 320 × 240.

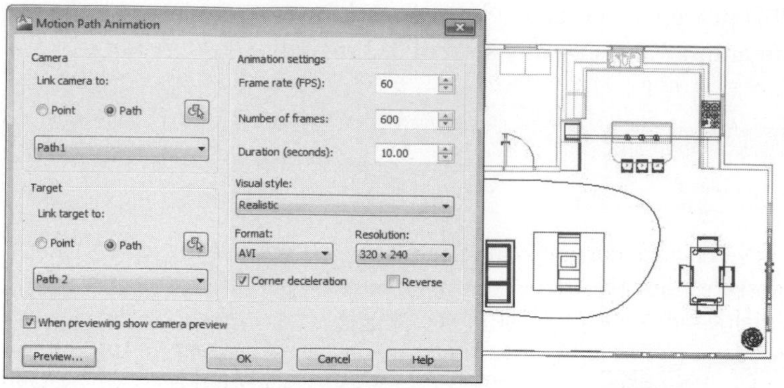

FIGURE 23.72

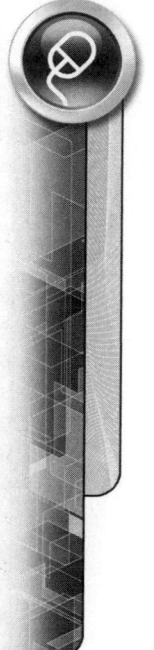

When you have finished making these changes, click the Preview button to preview the results of the motion animation. Play the animation preview as many times as you like. Once you close the preview, you will return to the Motion Path Animation dialog box. Click the OK button and enter the name of the AVI file as House Motion Study. Clicking the OK button in this dialog box begins the processing of the individual frames that will make up the animation. The total processing time to produce the animation should be between 2 and 5 minutes.

When finished, launch one of the many Windows Media Player applications and play the AVI file. This concludes the exercise.

TUTORIAL EXERCISE: SUNLIGHT STUDY

Purpose

This tutorial is designed to simulate sunlight on a specific day and time, and to observe the shadows that are cast by the house.

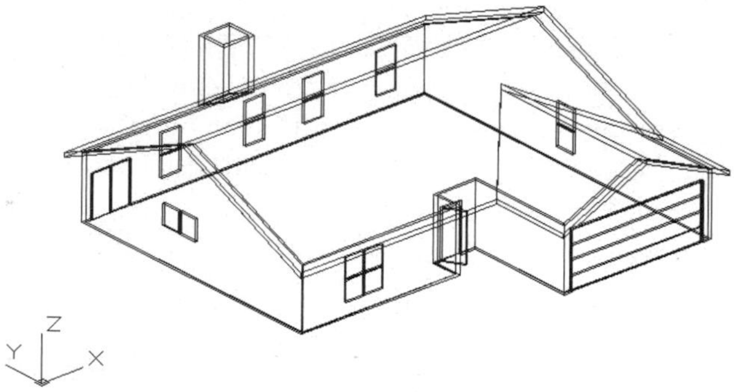

FIGURE 23.73

System Settings

Since this drawing is provided, all system settings have already been made.

Layers

The creation of layers is not necessary.

Suggested Commands

Begin this tutorial by opening up the drawing 23_House Plan Rendering, as shown in Figure 23.73. You will be performing a study based on the current location of the house and the position of the sun on a certain date, time, and geographic location. Shadow casting will be utilized to create a more realistic study.

STEP 1

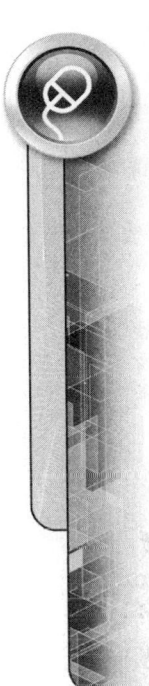

Make the 3D Modeling workspace current. With the Ribbon active, locate the Sun & Location panel and, if necessary, click the Sun Status button, as shown in Figure 23.74 (Left), to turn on the sun. A dialog box appears, informing you that you cannot display sunlight if the default lighting mode is turned on and asking you whether you want to turn off default lighting. Click the area shown in Figure 23.74 (Right) to accept the recommended setting to turn off the default lighting.

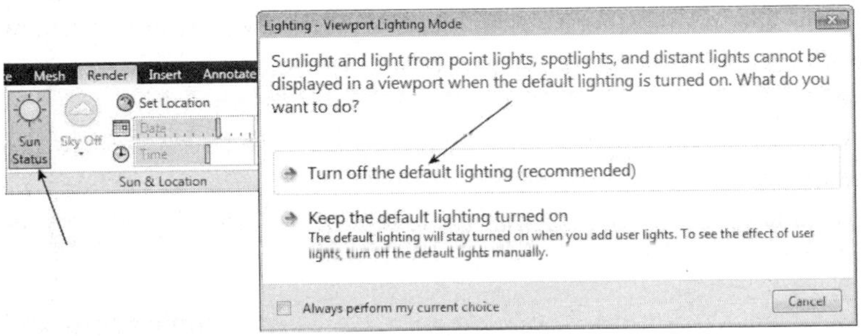

FIGURE 23.74

STEP 2

Next, click the Sun Properties arrow, located in the Ribbon shown in Figure 23.75 (Left), to display the Sun Properties palette shown in Figure 23.75 (Right). In this palette, you can change various properties that deal with the sun, such as shadows, date, time, azimuth, altitude, and source vector of the sun.

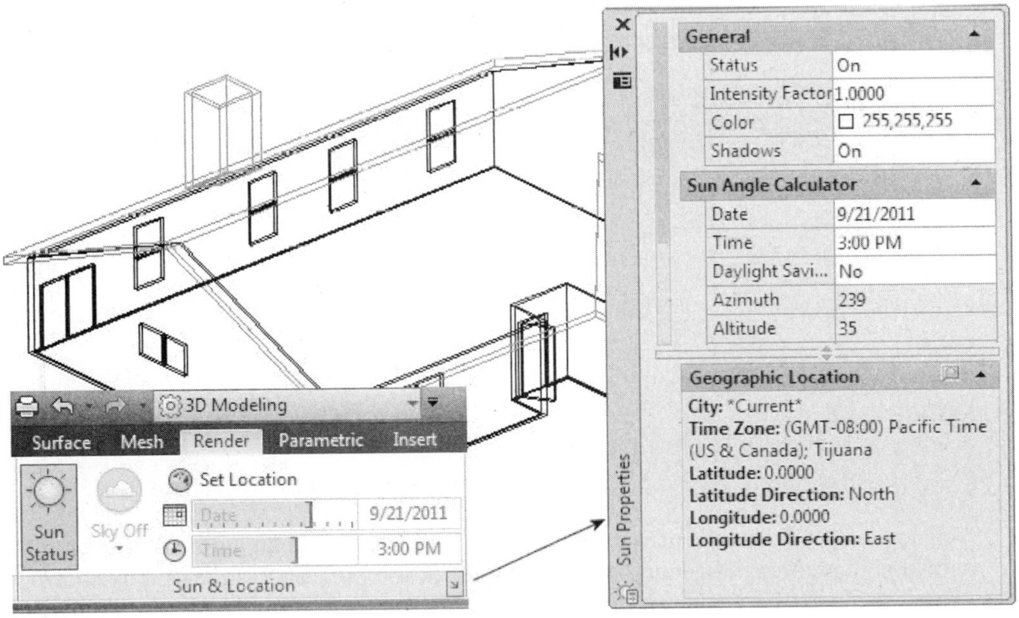

FIGURE 23.75

STEP 3

Click the launch Geographic Location button, as shown in Figure 23.76 (Left), to change the location of the 3D model. When the Define Geographic Location dialog box appears, as shown in Figure 23.76 (Right), click on the area to enter the location values.

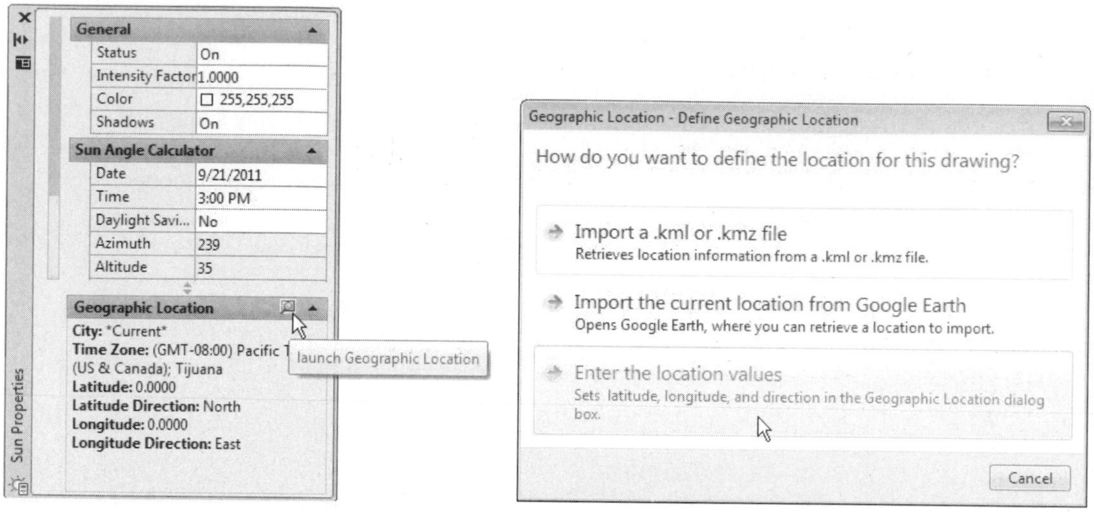

FIGURE 23.76

STEP 4

When the Geographic Location dialog box appears, click on the Use Map button, as shown in Figure 23.77 (Left). This will launch the Location Picker dialog box, as shown in Figure 23.77 (Right). Click the coast of South Carolina and check to see that Charleston, SC, appears. You could also use the Nearest City drop-down list to select the desired location. Other maps from throughout the world also are available. When you are satisfied with the location, click the OK button to leave the Location Picker and Geographic Location dialog boxes.

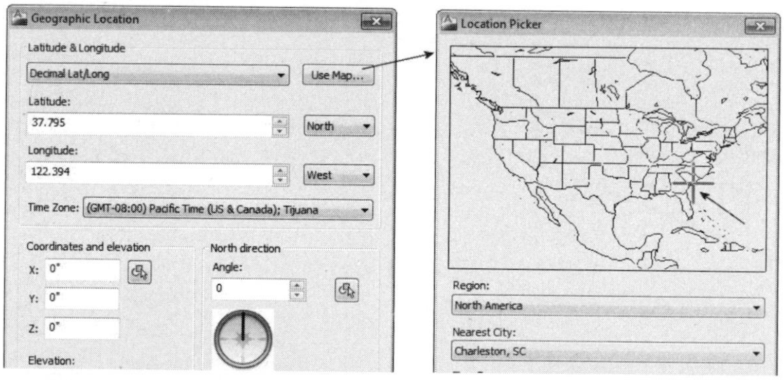

FIGURE 23.77

STEP 5

A dialog box appears, informing you that the time zone has been automatically updated with the change in the geographic location. Click the area of this dialog box shown in Figure 23.78 to accept updated time zone.

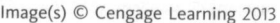

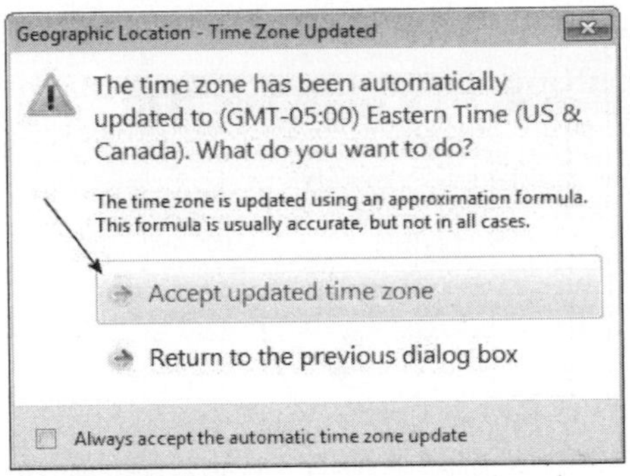

FIGURE 23.78

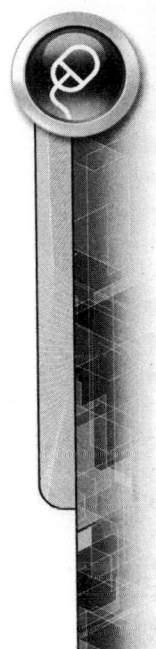

STEP 6

If necessary, click OK to accept the changes to the Geographic Location dialog Box. Next you will change the date and time to perform a sun study when the sun is positioned on a fall day. Click the Date area in the Sun Properties palette, click the three dots (ellipsis), and change the date to October 18, 2012, as shown in Figure 23.79.

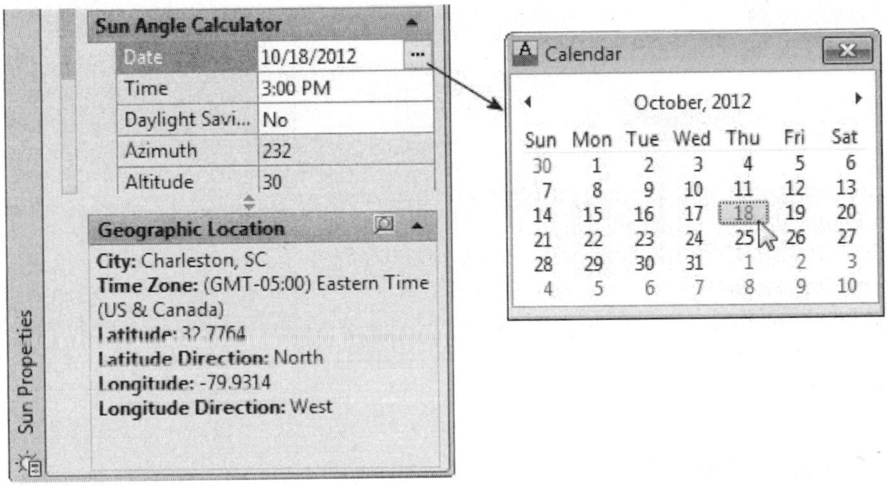

FIGURE 23.79

STEP 7

Then, click on the Time area. When an arrow appears, click it to display a number of times of day in 15-minute increments, and click on 11:00 AM, as shown in Figure 23.80 (Left). When you perform this change, the date and time information is also updated in the Ribbon, as shown in Figure 23.80 (Right).

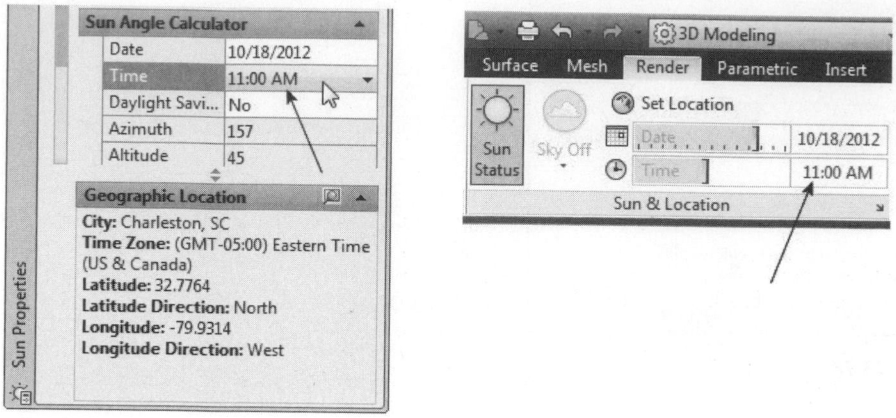

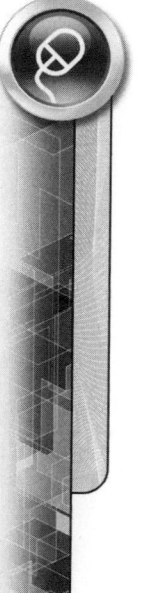

FIGURE 23.80

STEP 8

Clicking the Render button in the Ribbon will render the house, as shown in Figure 23.81. The shadows cast by the house reflect the time of 11:00 AM in mid-October in Charleston, South Carolina, in the United States.

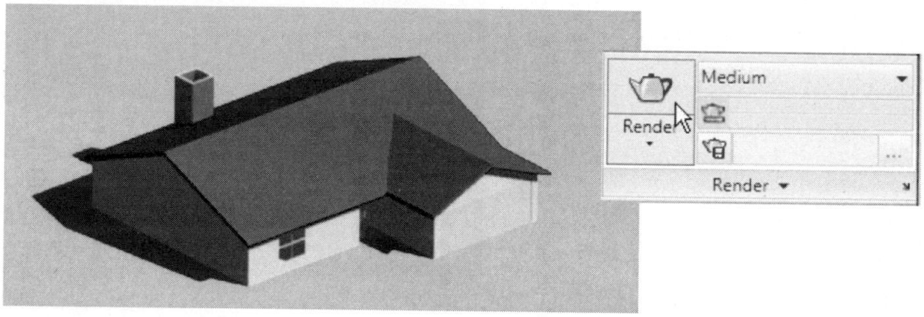

FIGURE 23.81

INDEX

A

D

E

F

G

H

M

N

O

P

Q

R

S

U

V

W

X

Z